Mobile DNA III

Mobile DNA III

Editor in Chief

Nancy L. Craig

Howard Hughes Medical Institute,
Department of Molecular Biology & Genetics,
Johns Hopkins University School of Medicine, Baltimore, MD

Editors

Michael Chandler

Laboratoire de Microbiologie et Génétique Moléculaire,
C.N.R.S., F-31062 Toulouse Cedex, France

Martin Gellert

National Institutes of Health,
Molecular Genetics Section, NIDDK, Bethesda, MD

Alan M. Lambowitz

Institute for Cellular Molecular Biology and
Department of Molecular Biosciences,
University of Texas at Austin, Austin, TX

Phoebe A. Rice

Department of Biochemistry and Molecular Biology,
University of Chicago, Chicago, IL

Suzanne B. Sandmeyer

Departments of Biological Chemistry and
Microbiology and Molecular Genetics,
University of California, Irvine, Irvine, CA

ASM Press, Washington, DC

Library of Congress Cataloging-in-Publication Data

Mobile DNA III / editor in chief, Nancy L. Craig, Johns Hopkins University School of Medicine, Baltimore, MD ; editors, Michael Chandler, Université Paul Sabatier, Toulouse, France, Martin Gellert, National Institutes of Health, Bethesda, MD, Alan M. Lambowitz, University of Texas, Austin, TX, Phoebe A. Rice, University of Chicago, Chicago, IL, Suzanne Sandmeyer, University of California, Irvine, CA.

pages cm

Includes bibliographical references and index.

ISBN 978-1-55581-920-0 (alk. paper)

1. Mobile genetic elements. I. Craig, Nancy Lynn, 1952- editor. II. Chandler, Michael (Molecular microbiologist), editor. III. Gellert, Martin, editor. IV. Lambowitz, Alan, editor. V. Rice, Phoebe A., editor. VI. Sandmeyer, Suzanne, editor. VII. Title: Mobile DNA 3. VIII. Title: Mobile DNA three.

QH452.3.M633 2015

572.8'69--dc23

2015011126

doi:10.1128/9781555819217

Printed in Canada

10 9 8 7 6 5 4 3 2 1

Address editorial correspondence to: ASM Press, 1752 N St., N.W., Washington, DC 20036-2904, USA.

Send orders to: ASM Press, P.O. Box 605, Herndon, VA 20172, USA.
Phone: 800-546-2416; 703-661-1593. Fax: 703-661-1501.
E-mail: books@asmusa.org
Online: http://estore.asm.org

Cover: An artistic representation of the retroviral intasome engaged with target DNA in the nucleus of a host cell. The illustration is based on the crystal structure of the prototype foamy virus strand transfer complex (Protein Databank ID 3OS0; for details see Maertens *et al.*, Nature, 2010, 468, 326-9). Image provided by Dr. Peter Cherepanov, Cancer Research UK, London Research Institute, London, EN6 3LD United Kingdom.

Contents

IV. DNA-ONLY TRANSPOSONS

VI. NON-LTR RETROTRANSPOSONS

Contributors

Frederick W. Alt
Department of Genetics, Harvard Medical School, Boston, MA 02115

Diego Arambula
Department of Microbiology, Immunology and Molecular Genetics, David Geffen School of Medicine, University. of California at Los Angeles, Los Angeles, CA 90095

Peter W. Atkinson
Department of Entomology and Institute for Integrative Genome Biology, University of California, Riverside, CA 92521

Francois-Xavier Barre
Centre de Genetique Moleculaire, C.N.R.S., Gif-sur Yvette, France 91198

Marlene Belfort
Department of Biological Sciences and RNA Institute, University of Albany, State University of New York, Albany, NY 12222

Mireille Betermier
Centre de Génétique Moléculaire, C.N.R.S., Gif-sur Yvette, France 91198

Virginia W. Bilanchone
Departments of Biological Chemistry and Microbiology and Molecular Genetics, University of California, Irvine, Irvine, CA 92697

Vincent Burrus
Departement de biologie, Université de Sherbrooke, Sherbrooke, Québec, J1K 2R1, Canada

Frederic D. Bushman
Department of Microbiology, Perelman School of Medicine, University of Pennsylvania, Philadelphia, PA 19104

Margi I. Butler
Department of Biochemistry, University of Otago, Dunedin, New Zealand 09054

Nicolas Carraro
Departement de biologie, Université de Sherbrooke, Sherbrooke, Québec, J1K 2R1, Canada

George Chaconas
Departments of Biochemistry and Molecular Biology; Microbiology, Immunology & Infectious Diseases, The University of Calgary, Calgary, ALB, Canada AB T2N 4N1

Ronald Chalmers
School of Life Sciences, University of Nottingham, Nottingham, NG7 2UH United Kingdom

Michael Chandler
Laboratoire de Microbiologie et Génétique Moléculaire, C.N.R.S., F-31062 Toulouse Cedex, France

Ju-Lan Chao
Institute of Molecular Biology, Academia Sinica, Taipei, Taiwan 11529

Chao-Yin Cheng
Institute of Molecular Biology, Academia Sinica, Taipei, Taiwan 11529

Peter Cherepanov
Cancer Research UK, London Research Institute, London, EN6 3LD United Kingdom

Corentin Claeys Bouuaert
School of Life Sciences, University of Nottingham, Nottingham, NG7 2UH United Kingdom

Alix Couneloup
Laboratoire de Microbiologie et Génétique Moléculaire, C.N.R.S., F-31062 Toulouse Cedex, France

Robert Craigie
Laboratory of Molecular Biology, NIDDK, National Institutes of Health, Bethesda, MD 20892

M. Joan Curcio
Laboratory of Molecular Genetics, University of Albany, State University of New York, Albany, NY 12208

Damien Dandoy
Institut des Sciences de la Vie, Université Catholique de Louvain, Louvain-la-Neuve, B-1348, Belgium

Kirk W. Deitsch
Department of Microbiology and Immunology, Weill Medical College of Cornell University, New York, NY 10065

AURELIEN J DOUCET
Department of Human Genetics, University of Michigan Medical School, Ann Arbor, MI 48109

SANDRA DUHARCOURT
Institut Jacques Monod, C.N.R.S., Paris, France 75205

GUY DUVAL-VALENTIN
Laboratoire de Microbiologie et Génétique Moléculaire, C.N.R.S., F-31062 Toulouse Cedex, France

FRED DYDA
Laboratory of Molecular Biology, NIDDK, National Institutes of Health, Bethesda, MD 20814

DANNA G. EICKBUSH
Department of Biology, University of Rochester, Rochester, NY 14627

THOMAS H. EICKBUSH
Department of Biology, University of Rochester, Rochester, NY 14627

MICHAEL JOSEPH ELLIS
Department of Biochemistry, University of Western Ontario, London, Ontario, N6A 5C1, Canada

ALAN ENGELMAN
Department of Cancer Immunology and AIDS, Dana-Farber Cancer Institute, Boston, MA 02215

JOSE ANTONIO ESCUDERO
Unité Plasticité du Génome Bactérien, Institut Pasteur, Paris, 75724 France

CAROLINE ESNAULT
Program in Cellular Regulation and Metabolism, NICHD, National Institutes of Health, Bethesda, MD 20892

HSIU-FANG FAN
Department of Life Sciences and Institute of Genome Sciences, National Yang-Ming University, Taipei, Taiwan 00112

OLIVIER FAYET
Laboratoire de Microbiologie et Génétique Moléculaire, C.N.R.S., Toulouse Cedex, France 31062

HARUHIKO FUJIWARA
Department of Integrated Biosciences, Graduate School of Frontier Sciences, University of Tokyo, Kashiwa, Japan 277-8562

CHRISTINE GALLOY
Institut des Sciences de la Vie, Université Catholique de Louvain, B-1348 Louvain-la-Neuve, Belgium

JOSE L. GARCIA-PEREZ
Department of Human DNA Variability, GENYO (Pfizer-University of Granada & Andalusian Regional Government Genomics & Oncology Center), 18016 Granada, Spain

JEFFREY F. GARDNER
Department of Microbiology, University of Illinois at Urbana-Champaign, Urbana, IL 61801

PARTHO GHOSH
Department of Chemistry and Biochemistry, University of California at San Diego, La Jolla, CA 92093

EDITH GOURBEYRE
Laboratoire de Microbiologie et Génétique Moléculaire, C.N.R.S., Toulouse Cedex, France 31062

PIOTR GUGA
Department of Bio-organic Chemistry, Polish Academy of Sciences, Lodz, POLAND

HUATAO GUO
Department of Molecular Microbiology and Immunology, University of Missouri School of Medicine, Columbia, MO 65212

CATHERINE GUYNET
Laboratoire de Microbiologie et Génétique Moléculaire, C.N.R.S., Toulouse Cedex, France 31062

JAMES HABER
Department of Biology and Rosenstiel Basic Medical Sciences Research Center, Brandeis University, Waltham, MA 02454

JAMES HALL
Department of Biology, University of York, York, YO10 5DD United Kingdom

BERNARD HALLET
Institut des Sciences de la Vie, Universté Catholique de Louvain, Louvain-la-Neuve, BELGIUM B-1348

DAVID B. HANIFORD
Department of Biochemistry, The University of Western Ontario, London, Ontario, N6A 5C1 Canada

RASIKA HARSHEY
Department of Molecular Biosciences, Institute of Cellular and Molecular Biology, University of Texas at Austin, Austin, TX 78712

SUSU HE
Laboratoire de Microbiologie et Génétique Moléculaire, C.N.R.S., Toulouse Cedex, France 31062

ALISON BURGESS HICKMAN
Laboratory of Molecular Biology, NIDDK, National Institutes of Health, Bethesda, MD 20814

MATTHEW HIRSCH
Gene Therapy Center, Department of Ophthalmology, University of North Carolina, Chapel Hill, Chapel Hill, NC 27599

STEPHEN H. HUGHES
HIV Drug Resistance Program, National Cancer Institute at Frederick, NIH, Frederick, MD 21702

JOYCE K. HWANG
Program in Cellular and Molecular Medicine; Department of Genetics, Harvard Medical School, Boston, MA 02115

KEN ISHIKAWA
Gene Regulation and Chromosome Biology Laboratory, National Cancer Institute at Frederick, NIH, Frederick, MD 21702

ZOLTAN IVICS
Division of Medical Biotechnology, Paul Ehrlich Institute, Langen, Germany 63225

ZSUZSANNA IZSVAK
Max Delbruck Center for Molecular Medicine, Max Delbruck Center for Molecular Medicine, Berlin, Germany 13092

MAKKUNI JAYARAM
Department of Molecular Biosciences, University of Texas at Austin, Austin, TX 78712

REID C. JOHNSON
Department of Biological Chemistry, University of California, Los Angeles School of Medicine, Los Angeles, CA 90095

AASHIQ H. KACHROO
Department of Molecular Biosciences, University of Texas at Austin, Austin, TX 78712

LAURA KIRKMAN
Department of Internal Medicine, Department of Microbiology and Immunology, Weill Medical College of Cornell University, New York, NY 10065

AMAR J. S. KLAR
Gene Regulation and Chromosome Biology Laboratory, National Cancer Institute at Frederick, NIH, Frederick, MD 21702

KERRI KOBRYN
Department of Microbiology and Immunology, College of Medicine, University of Saskatchewan, Saskatoon, Saskatchewan, S7N 5E5 Canada

HUIRA C. KOPERA
Department of Human Genetics, University of Michigan Medical School, Ann Arbor, MI 48109

MICHAEL LAMBIN
Institut des Sciences de la Vie, Université Catholique de Louvain, B-1348 Louvain-la-Neuve, Belgium

ALAN M. LAMBOWITZ
Institute for Cellular Molecular Biology and Department of Molecular Biosciences, University of Texas at Austin, Austin, TX 78712

LAURA F. LANDWEBER
Department of Ecology and Evolutionary Biology, Princeton University, Princeton, NJ 08544

ARTHUR LANDY
Division of Biology and Medicine, Brown University, Providence, RI 02912

LAURE LAVATINE
Laboratoire de Microbiologie et Génétique Moléculaire, C.N.R.S., F-31062 Toulouse Cedex, France

CHENG-SHENG LEE
Department of Biology and Rosenstiel Basic Medical Sciences Research Center, Brandeis University, Waltham, MA 02454

PASCALE LESAGE
Institut Universitaire d'hématologie, C.N.R.S., F-75475 Paris Cedex 10, France

HENRY L. LEVIN
Program in Cellular Regulation and Metabolism, NICHD, National Institutes of Health, Bethesda, MD 20892

DAMON LISCH
Department of Botany and Plant Pathology, University of California, Berkeley, Berkeley, CA 94720

CELINE LOOT
Unité Plasticité du Génome Bactérien, Institut Pasteur, Paris, 75724 France

SHEILA LUTZ
Laboratory of Molecular Genetics, University of Albany, State University of New York, Albany, NY 12208

CHIEN-HUI MA
Department of Molecular Biosciences, University of Texas at Austin, Austin, TX 78712

DIXIE L. MAGER
Terry Fox Laboratory, British Columbia Cancer Agency and Dept. of Medical Genetics, University of British Columbia, Vancouver, British Columbia, V5Z 1L3, Canada

SHARMISTHA MAJUMDAR
Department of Molecular and Cell Biology, University of California, Berkeley, Berkeley, CA 94720

BRIGITTE MARTY
Laboratoire de Microbiologie et Génétique Moléculaire, C.N.R.S., F-31062 Toulouse Cedex, France

DIDIER MAZEL
Unité Plasticité du Génome Bactérien, Institut Pasteur, Paris, 75724 France

RICHARD MCCULLOCH
Wellcome Trust Centre for Molecular Parasitology, University of Glasgow, Glasgow, G12 8TA, United Kingdom

CAROLINE MIDONET
Centre de Génétique Moléculaire, C.N.R.S., Gif-sur Yvette, 91198 France

JEFF F. MILLER
Department of Microbiology, Immunology and Molecular Genetics; The California Nano Systems Institute, University of California at Los Angeles, Los Angeles, CA 90095

JOHN B. MOLDOVAN
Cellular and Molecular Biology Graduate Program, University of Michigan Medical School, Ann Arbor, MI 48109

SHARON MOORE
Gene Regulation and Chromosome Biology Laboratory, National Cancer Institute at Frederick, NIH, Frederick, MD 21702

JOHN V. MORAN
Departments of Human Genetics; Cellular and Molecular Biology, University of Michigan Medical School, Ann Arbor, Michigan 48109

LIAM MORRISON
Roslin Institute, University of Edinburgh, Midlothian, EH25 9RG United Kingdom

NATHAN NGUYEN
Institut des Sciences de la Vie, Université Catholique de Louvain, B-1348 Louvain-la-Neuve, Belgium

EMILIEN NICOLAS
Institut des Sciences de la Vie, Université Catholique de Louvain, B-1348 Louvain-la-Neuve, Belgium

ALEKSANDRA NIVINA
Unité Plasticité du Génome Bactérien, Institut Pasteur, Paris, 75724 France

STEVEN J NORRIS
Department of Pathology and Laboratory Medicine, University of Texas Health Medical School, Houston, TX 77225

KYLE P. OBERGFELL
Department of Microbiology and Immunology, Northwestern University Feinberg School of Medicine, Chicago, IL 60611

CEDRIC A. OGER
Institut des Sciences de la Vie, Université Catholique de Louvain, B-1348 Louvain-la-Neuve, Belgium

KURT PATTERSON
Departments of Biological Chemistry and Microbiology and Molecular Genetics, University of California, Irvine, Irvine, CA 92697

JOSEPH E. PETERS
Department of Microbiology, Cornell University, Ithaca, NY 14853

RUSSELL POULTER
Department of Biochemistry, University of Otago, Dunedin, 09054 New Zealand

ELLEN J. PRITHAM
Department of Human Genetics, University of Utah, Salt Lake City, UT 84112

PHOEBE A. RICE
Department of Biochemistry and Molecular Biophysics, University of Chicago, Chicago, IL 60637

SANDRA R RICHARDSON
Department of Human Genetics, University of Michigan Medical School, Ann Arbor, MI 48109

DONALD RIO
Department of Molecular and Cell Biology, University of California, Berkeley, Berkeley, CA 94720

DAVID B. ROTH
Department Pathology and Laboratory Medicine and the Center for Personalized Diagnostics, Perelman School of Medicine, University of Pennsylvania, Philadelphia, PA 19104

PHILIPPE ROUSSEAU
Laboratoire de Microbiologie et Génétique Moléculaires, C.N.R.S., F-31062 Toulouse Cedex, France

PAUL A. ROWLEY
Department of Molecular Biosciences, University of Texas at Austin, Austin, TX 78712

MAXIM SALGANIK
Gene Therapy Center/Lineberger Comprehensive Cancer Center, University of North Carolina, Chapel Hill, Chapel Hill, NC 27599

RICHARD JUDE SAMULSKI
Gene Therapy Center, University of North Carolina, Chapel Hill, Chapel Hill, NC 27599

SUZANNE B. SANDMEYER
Departments of Biological Chemistry and Microbiology and Molecular Genetics, University of California, Irvine, Irvine, CA 92697

ANNE SARCOS
Laboratoire de Microbiologie et Génétique Moléculaire, C.N.R.S., F-31062 Toulouse Cedex, France

H. STEVEN SEIFERT
Department of Microbiology and Immunology, Northwestern University Feinberg School of Medicine, Chicago, IL 60611

PATRICIA SIGUIER
Laboratoire de Microbiologie et Génétique Moléculaire, C.N.R.S., F-31062 Toulouse Cedex, France

ANNA MARIE SKALKA
Department of Microbiology, Fox Chase Cancer Center, Philadelphia, PA 19111

MARGARET C.M. SMITH
Department of Biology, University of York, York, YO10 5DD United Kingdom

W. MARSHALL STARK
Institute of Molecular Cell and Systems Biology, University of Glasgow, Glasgow, G12 8WW, United Kingdom

JONATHAN P. STOYE
Division of Virology, MRC National Institute for Medical Research, London, NW7 1AA, United Kingdom

MICHAEL TELLIER
School of Life Sciences, University of Nottingham, Nottingham, NG7 2UH, United Kingdom

JAINY THOMAS
Department of Human Genetic, University of Utah, Salt Lake City, UT 84112

BAO TON HOANG
Laboratoire de Microbiologie et Génétique Moléculaire, C.N.R.S., F-31062 Toulouse Cedex, France

GREGORY D. VAN DUYNE
Department of Biochemistry and Biophysics, Perelman School of Medicine, University of Pennsylvania, Philadelphia, PA 19104

ALESSANDRO VARANI
Departamento do Tecnología, UNESP - Univ. Estadual Paulista, Jaboticaba. SP, BRAZIL 14870

YURI VOZIYANOV
School of Biosciences, Louisiana Tech University, Ruston, LA 71272

MARGARET M. WOOD
Academic Advising Center, Simmons College, Boston, MA 02115

LI WU
Department of Biological Sciences, University of Calgary, Calgary, Alberta, T2N 1N4, Canada

MENG-CHAO YAO
Institute of Molecular Biology, Academia Sinica, Taipei, Taiwan 11529

LENG SIEW YEAP
Program in Cellular and Molecular Medicine; Department of Genetics, Harvard Medical School, Boston, MA 02115

V. TALYA YERLICI
Department of Molecular Biology, Princeton University, Princeton, NJ 8544

KOSUKE YUSA
Stem Cell Genetics, Wellcome Trust Sanger Institute, Cambridge, CB10 1SA, United Kingdom

STEVEN ZIMMERLY
Department of Biological Sciences, University of Calgary, Calgary, Alberta, T2N 1N4, Canada

Preface

Nowhere do the cooperative powers of DNA sequencing, high-resolution protein structure, biochemistry and molecular genetics shine more intensely than on Mobile DNAs. In Mobile DNA II, we knew that almost half of the human genome is comprised of retroelements. What discoveries since Mobile DNA II could surpass that claim? Very simply: everywhere DNA is dynamic, and we now meet the elegant protein machines, co-evolved DNA partners, and diverse RNA choreographers. These pages hold something for every reader, beginning with the introductory overview of mechanisms. Novices will find some of the most lucid reviews of these complex topics available anywhere. Specialists will be able to pick and choose advanced reviews of specific elements, but will be drawn in by unexpected parallels and contrasts among the elements in diverse organisms. Biomedical researchers will find documentation of recent advances in understanding immune-antigen conflict between host and pathogen. Biotechnicians will be introduced to amazing tools for *in vivo* control of designer DNAs. And long-time aficionados will simply fall in love all over again.

Questions still abound about the Transposable Elements (TE) described in this volume. Perhaps none is more profound than the basis and consequences of TE diversity even among related genomes. Active DNA TE show perhaps the most disparate distribution among organisms, being dominant in prokaryotes, and in some animals, including some insects and fish, but with the exception of certain bats, virtually absent in mammals. Plants illustrate expansion of genomes, mediated not only by increasing ploidy, but also by expansion of DNA-based TE and Long Terminal Repeat (LTR) retrotransposons. Although reverse transcriptases are found throughout all kingdoms, autonomous retroelements simply explode together with their non-autonomous partners in mammals with remarkably species-specific types. These differences in mobile DNAs define self and mate, sister species, host and pathogen.

The most striking impression from these pages must be the raw power of genetic material to refashion itself to any purpose. DNA exchange between bacteria and their environment blurs the boundaries between host, transposon, and phage, as organisms secrete and take up DNA, stash genetic material in integrons for future use, conjugate, are attacked by phage and fight back. Delving into mechanisms, we see single-stranded hairpin structures and G quartets that anchor rearrangements in multiple ways; chemically diverse nucleophilic centers–hydroxyls couched in pentose, tyrosine or serine moieties–that covalently bond or attack directly in strand-transfer reactions. Proteins act as clamps to topologically constrain DNA or act as mechanical swivels, linking and unlinking mobilizing strands. Resolution of transposition intermediates might also involve host replication or recombination machinery. More recently discovered helitrons offer unexpected opportunities for expansion of DNA-based elements by rolling-circle replication.

RNA, the primal, catalytic nucleic acid, is in evidence everywhere. In retroelements, RNA partners with reverse transcriptase to deliver on transcriptional expansion of autonomous and non-autonomous TE sequences. Group II introns in bacteria likely gave rise to eukaryotic organelle group II self-splicing, retro-homing introns, Long INterspersed Elements (LINEs), telomerase reverse transcriptases and in addition, spliceosomal introns. Phylogenetic analysis of bacterial genomes previously revealed group II introns, diversity-generating retroelements Diversity-Generating Retroelements (DGR), and retrons, but next generation sequencing now identifies a multitude of novel reverse transcriptases of unknown function.

In ciliates, *Paramecium, Tetrahymena* and *Oxytrichia,* RNA directs massive genome reduction between germ-line and somatic nuclei, mediated by ancient transposase-like enzymes. LINEs containing restriction-enzyme like- or AP-endonucleases dominate in some eukaryotic cells. Others are dominated by LTR retrotransposons and their offspring, the retroviruses; stripped down Penelope-like elements with GIY-YIG endonucleases; DIRS elements with tyrosine recombinases: and attendant non-autonomous elements.

Exceptional elements provide evidence for the interaction of domains over evolutionary time, including LTR retrotransposons encoding envelope proteins, retroviruses replicating intracellularly, and DIRS elements in which retroelement RT/RNaseH is associated with a Crypton-type DNA element tyrosine recombinase.

Nowhere is the sharp focus of structural biology and biochemistry more apparent than in studies of key retroelement enzymes reverse transcriptase and integrase motivated by the quest for inhibitors of human immunodeficiency virus (HIV) replication. Reverse transcriptase structures for multiple retroviruses, as well as now one retrotransposon, demonstrate the robustness of the palm, thumb, fingers model. However, as a caution against too much generalization, subdomains are re-arranged in monomeric, homodimeric, and heterodimeric forms in different enzymes, and catalytic activities operate in *cis* or *trans* within the complex, depending on the enzyme. The structure of full length retrovirus integrase notoriously resisted high resolution structural analysis, but now has rewarded efforts of many labs with key insights (cover of this volume). These include a surprising dimer-dimer interface where active sites are juxtaposed to a trapped, and dramatically bent and widened, major groove target. Whereas LTR retrotransposons target integration to transcriptionally-repressed regions through interactions with heterochromatic protein domains or Pol III-transcribed genes thought to repress Pol II transcription, next generation sequencing has surfaced less dramatic, but significant, retrovirus integration bias, favoring transcriptionally-active regions.

This distribution has been shown now in two cases to be mediated by interactions between integrase and epigenetic mark-associated proteins.

While it has been argued that mobile elements are "selfish DNAs", these pages are replete with examples of the positive contributions of mobile elements to host genome function. Bacterial transposons encode and mobilize selectable markers including antibiotic resistance, detoxifying enzymes, and conjugation and virulence functions. In eukaryotic cells, mobile elements contribute to chromosome structure: constituting centromeres or telomeres in some organisms and seeding heterochromatin in others. TE constitute a large fraction of transcription factor binding sites and provide an ongoing source of novel combinations which are responsive to stress signaling, MAP kinase activation and other developmental signals. Insertions of LINEs and Alu elements affect RNA processing because they encode cryptic splice sites, termination signals, and can target RNA editing.

Exapted mobile DNA coding sequences appear in novel contexts: transposases have evolved into the RAG endonuclease for V(D)J immunoglobulin gene diversification and into heterochromatic factor CENP-B; a reverse transcriptase evolved into telomerase; retrovirus envelope proteins became the trophoblast fusion protein syncytin. There are other examples of TE Open Reading Frames (ORFs) under selection, but with, as yet, unknown functions. Endogenous retroviruses forego prior allegiances and join strategies to resist new infections. For example, Fv–1, a retroviral Gag relic, thwarts incoming retroviruses of similar type. Repeated TE sequences are susceptible to DNA rearrangements via non-allelic recombination, aborted transposition, and generation of pseudo-genes–all of which might ultimately contribute to the resiliency of host genomes.

TE exploit their hosts as well. The bacterial XerCD tyrosine recombinases which function in bacteria to unlink multimeric chromosomes are exploited to integrate phage genomes or mediate invasion of the host chromosome on behalf of certain plasmid-borne mobile elements. Transposases, resolvases and integrases *in vivo* likely associate with host factors as they are joined with host genomes. TE are generally tightly controlled by host regulation so that some display opportunistic bursts of activity during specific windows of development. This is exemplified by yeast Ty transcription in response to MAP kinase signaling and activation of reverse transcription by DNA checkpoints. A common theme more generally is TE activation during stress. Diverse retroelements are derepressed during periods of germ cell development ensuring their vertical spreading in populations.

Despite these examples of cooperation, mobile DNAs are also in conflict with their hosts. RNA interference likely arose in part to combat mobilization of retroelements. Invaders of one sort or another engage in a dizzying unscored dance with their hosts. One result of this conflict is rapid evolution of genes encoding host innate immunity restriction factors, which for retroviruses include ones that prematurely uncoat incoming viruses, starve reverse transcriptases for nucleotides, and deaminate cytidines in replicating cDNA. Some of these same factors also suppress movement of endogenous retroelements.

Programmed variation is used by invaders and hosts alike for purposes of immune evasion or resistance, respectively. Examples include *Salmonella* Hin invertase flipping a promoter sequence to switch between expression of different antigenic flagellar proteins and DGR directing mutagenic reverse transcription of a template transcript coupled with directed conversion of a target expression site. *Neisseria gonorrhea, Borrelia burgdorferi, Trypanosoma brucei*; and *Plasmodium falciparum*, agents of gonorrhea, Lyme disease, sleeping sickness, and malaria, respectively, use amazingly diverse mechanisms to program variation of their antigenic surfaces for immune evasion. To counter this assault, there is programmed variation of host immune proteins. In human immunoglobulin production, a DDE TE-derived RAG site-specific endonuclease initiates V(D)J switching, followed by transcription-activated somatic hyper-mutation (activation-induced cytidine deaminase), nuclease introduction of DSB, and final joint formation by redundant NHEJ pathways.

Next generation sequencing and development of methods for rapid TE mapping have greatly improved understanding of the distribution of TE as well as the utility of transposons for functional genomics. The bacterial Tn5 system has been exploited in particular for *in vitro* mutagenesis and next generation sequencing libraries by collapsing fragmentation and adapter ligation into a single step. P, Hermes, piggyBac, and Sleeping Beauty transposons have wide activity in eukaryotic systems and have been harnessed for genome-wide profiling, gene disruption and tagging, and genome modification. Retroviruses are additionally used in lineage tracing. The controlled, high-frequency mobilization of Mutator has made it indispensable for gene discovery in maize.

In medical research, understanding the impact of DNA mobilization is critical. In addition to individual TE, other mobile DNAs such as plasmids, Integrative Conjugative Elements (ICE) and both transposon-borne and chromosomal integrons are bacterial reservoirs of mobilizable antibiotic resistance. HIV, malaria, and sleeping sickness, and other pathogens, too numerous to mention here, remain threats to global health. Mobile element vectors transposons piggyBac, Sleeping Beauty, lenti-retroviruses and adenoviruses are being used as vectors to introduce exogenous DNAs in research, and in clinical trials. They differ with respect to targeting, excision footprints, payload size, and host activity profiles. Their mechanisms of DNA breakage and joining were among the systems first analyzed, now enabling them to be harnessed and used extensively for genome engineering including with developmentally-regulated expression, inactivation, and self-deletion strategies to enable probing essential or tissue- specific functions.

What challenges remain? One goal is to connect key findings from basic research, to clinical developments in drug resistance and genetic engineering. This volume is based on the considerable increase in understanding of molecular mechanisms of mobilization in the last decade. However, we have likely seen only the tip of the iceberg of how mobile DNAs affect the day to day biology of their hosts.

In the human genome alone, retroelements provide promoters for long non-coding and other RNAs of completely unknown function; Alu elements redirect RNA processing and delivery, and mobilization is occurring during neuronal development and in cancer with unknown consequences, just to mention a few. Finally, endogenization of a gamma retrovirus in Australian koalas is ongoing and those studies should provide insights into retroelement-host interaction. How have transposition events after separation from other great apes contributed to traits that make us human? What transposition events will provide key substrates for future evolution? And of course, perhaps the ultimate question, could we survive as a species were there no transposition?

We give our heartfelt thanks to all the authors who contributed diverse and fascinating chapters to Mobile DNA III. We express special thanks to Patti Kodeck, Administrative Assistant to Editor in Chief N. Craig, who mediated recruitment of and communications with authors and interactions between them and the publisher. Finally, our most sincere thanks to all of our supporters at ASM Press for their dedication in producing this volume, but especially to: Gregory Payne, Senior Editor; Larry Klein, Production Editor; Christine Charlip, Director of Administration; and Cathy Balogh, Administrative Assistant for Production.

Introduction

I

Mobile DNA, 3rd Edition
Nancy L. Craig, Michael Chandler, Martin Gellert, Alan M. Lambowitz, Phoebe A. Rice and Suzanne Sandmeyer

doi:10.1128/microbiolspec.MDNA3-0062-2014

Nancy L. Craig[1]

1 A Moveable Feast: An Introduction to Mobile DNA

INTRODUCTION

DNA has two critical functions: to provide the cell with the information necessary for macromolecular synthesis and to transmit that information to progeny cells. Genome sequence stability is important for both these functions. Indeed, cells devote significant resources to various DNA repair processes that maintain genome structure and repair alterations that can arise from DNA synthesis errors and assaults from both endogenous and exogenous sources. DNA sequence variation, however, provides the substrate for adaptation, selection, and evolution.

Genomes are, however, highly dynamic. Notably, they vary not only at the single or several base pair level (although such changes can be transformative and even deadly), but they also change by DNA rearrangements, that is, the movement of DNA segments that may be many kb (or even longer) in length. Such rearrangements can have enormous impacts on genome structure, function, and evolution.

The DNA rearrangements discussed here generally involve specific DNA sequences, or in some cases RNA sequences, that are recognized and acted on by specialized recombination proteins or recombinases that promote DNA breakage followed by joining of the broken DNAs to new sites. The involvement of a sequence-specific recombinase is what distinguishes site-specific recombination from homologous recombination, which can occur between any two DNA segments as long as they are homologous to each other, as in RecA- and Rad51-dependent recombination. In some cases, the specialized recombinase is a sequence-specific nuclease that targets homologous recombination to a specific DNA site.

In some rearrangements, the recombinase alone breaks, exchanges, and joins DNA by using covalent protein-DNA intermediates. In other cases, DNA synthesis is also essential in these rearrangements. Notably, this DNA synthesis can involve not only conventional DNA synthesis in which a DNA polymerase uses DNA as a template, but also reverse transcription in which a novel DNA polymerase, reverse transcriptase, uses an RNA template to generate new DNA. A very wide number of other cellular processes can influence or be required for DNA rearrangements, including transcriptional activation of particular sites, DNA bending by bending proteins, DNA supercoiling, and many variations in chromatin structure, as well as DNA repair reactions including DNA end joining. Although a purified recombinase may execute DNA breakage and joining *in vitro*, it is critical to remember that this reaction and its consequences will be enormously influenced by its cellular environment.

[1]Department of Molecular Biology and Genetics, Howard Hughes Medical Institute, Johns Hopkins University School of Medicine, Baltimore, MD 21205.

Although DNA rearrangements can provide very useful rapid and focused changes in genetic information, they are also very hazardous. Unrepaired DNA breaks can result in DNA mis-segregation and are often lethal. Not surprisingly then, DNA rearrangements often occur in elaborate nucleoprotein complexes that organize and juxtapose the participating DNAs and promote breakage and joining in carefully choreographed steps. The frequency of DNA rearrangements is usually highly regulated, often by restricting to low levels the recombinase that initiates or mediates the rearrangements.

Mobile DNAs also include a diverse variety of discrete mobile genetic elements, such as transposable elements, that move themselves or copies of themselves from place to place within and between genomes. Thus, in some cases, a copy of the element remains at its original site and there is a new copy at the new insertion site. This type of replicative mechanism leads to an enormously high element copy number, especially in some eukaryotic genomes. The majority of the maize genome, for example, is derived from a particular kind of transposon. High copy numbers of repetitive sequences result in increased susceptibility to nonallelic homologous recombination events that can lead to deletions, inversions, translocations, and other chromosomal rearrangements.

The movement of a transposable element within a single genome can have substantial genotypic and phenotypic consequences. The insertion of a transposable element into a gene can lead to gene disruption but even nearby insertions can effect gene expression as many elements carry regulatory signals, such as enhancers and promoters, as well as splice sites and transcription termination signals. Excision of an element also changes the donor site. Thus, the intracellular translocation of a mobile element results in genetic variation. The range of target sites used by the elements ranges from insertion into specific sites or regions that provide a "safe harbor" for the element with reduced negative consequence on the host, to preferences for actively transcribed regions to facilitate element expression to virtually random insertion, which can thus result in genetic variation anywhere within the host genome.

DNA rearrangements also play a crucial role in the interactions between viral chromosomes and their hosts, as well as the proper replication and segregation of host chromosomes. Many viruses integrate into and excise from host genomes, although in some cases integration is irreversible, such as with HIV-1. All of these reactions involve at least specific sites on the viral genome that are acted upon by site-specific recombinases and which sometimes involve specific sites on the host genome. Recombination between specific sites to promote chromosome monomerization plays a key role in chromosome transmission in bacteria.

The translocation of mobile elements encoding a wide variety of determinants including genes encoding antibiotic resistances, virulence determinants, and diverse metabolic pathways from plasmids to chromosomes and from viruses and DNA fragments that are transduced or transformed into a cell, can also result in permanent chromosomal acquisition of these determinants. This sort of horizontal gene transfer involving mobile elements is rampant in bacteria and contributes greatly to genetic variation. There are also an increasing number of examples of horizontal gene transfer involving mobile elements in eukaryotes.

Perhaps the most profound example of the effect of mobile elements on eukaryotic genome evolution is the nuclear invasion of mobile group II introns into the nuclear genome from bacterial symbionts to form spliceosomal introns.

Cell type can also have substantial impact on DNA rearrangements. The elaborate DNA breakage and joining reactions that underlie immunoglobulin diversification are actually terminal differentiation events restricted to particular somatic cells. There is increasing interest in the somatic movement of transposable elements, which can also have profound organismal impact. The movement of human transposable elements in somatic tissue is associated with a variety of cancers, although it remains to be determined if such events can cause oncogenic transformation or are rather a consequence of transformation. The movement of transposable elements in neuronal tissue in several organisms raises the interesting possibility that such rearrangements are a deliberate strategy for neural plasticity.

Such terminal differentiation events involving DNA rearrangements are incompatible with the bacterial lifestyle, except in a few known cases such as spore formation by a subset of cells. By contrast, reversible DNA inversions that vary promoter or ORF orientation are well known in bacteria.

Thus, DNA rearrangements can contribute substantially to genetic variation. The frequency and potential advantage of the resulting variation must be carefully balanced with genome stability to avoid its potential for population-wide genomic catastrophe.

Although not exclusively so, the focus of this work is on the mechanism and regulation of DNA rearrangements. How do specific DNA (and sometimes RNA) sequences recognize each other and how do they assemble to form the machines in which DNA rearrangements occur? What are the mechanisms for DNA strand breakage and joining? What processes determine when and where these reactions occur? How are actions at multiple DNA sites, for example, the two ends of a transposable element and its target DNA, coordinated? Importantly, how are nonproductive breakage and joining events avoided and how is intact duplex DNA regenerated?

Mobile DNAs are "natural" genome engineers. Although not a focus of this work, many of the mobile elements discussed here have been harnessed to facilitate researcher-directed rearrangements both *in vitro* and *in vivo*. Mobile elements are used for "random" insertional genome mutagenesis both *in vivo* and *in vitro*, as well as for "targeted mutagenesis." Many mobile elements are used as vectors in both homologous and heterologous systems.

TARGETED DNA BREAKS LEAD TO GENE REPLACEMENT

DNA Double Strand Breaks Stimulate Homology-Dependent DNA Repair

Homologous recombination occurs without requirement for any particular sequence, depending only on base pairing between the participating DNA strands. However, the frequency of homologous recombination is stimulated by the presence of broken DNA, in particular double strand breaks. These breaks stimulate recombination because the action of nucleases and helicases at these breaks leads to the generation of DNA with single stranded 3′ trails that are the preferred substrate for DNA pairing mediated by RecA- and Rad51-like proteins. By interacting with a donor site, this pairing of 3′-OH ends can initiate homology-dependent DNA repair, which copies DNA sequence information from the donor site into the broken DNA target site. This repair leads to the replacement, or modification, of an existing gene or insertion of a new gene. The insertion of many mobile DNAs into a new site is targeted by double strand breaks by highly site-specific endonucleases.

There's No Place Like Home: Homing Endonucleases

Homing endonucleases (HENs) are highly site-specific endonucleases (1). Although often associated with other genetic elements (see below), freestanding HEN genes can themselves be mobile DNA elements. If a HEN cleavage site lies within an "empty allele" of DNA that flanks the HEN ORF, cleavage of that target site can initiate homology-dependent DNA synthesis that will transfer a copy of the HEN gene to that double strand break at the nuclease target site (Fig. 1).

HEN genes are also often found in other genetic elements such as self-splicing RNA introns, that is, group I introns, and self-splicing proteins, that is, inteins. Thus, if the HEN introduces a double strand break into the "empty" allele of a site occupied by the intron or intein, targeted DNA repair introduces a copy of the DNA encoding the intron or intein into that target site. Because the RNA intron can splice out of the RNA containing it and the protein intein can splice out of the protein containing it, the insertion of these elements is generally phenotypically silent.

Alternative Life Styles: Switching Mating Type in *Saccharomyces cerevisiae*

These yeasts have two different haploid cell types, mating type **a** and mating type α, which can mate to form diploids. During sporulation, meiotic recombination shuffles the two parental genomes, generating diverse haploid progeny. To facilitate diploid formation, haploids can switch mating type from mating type **a**

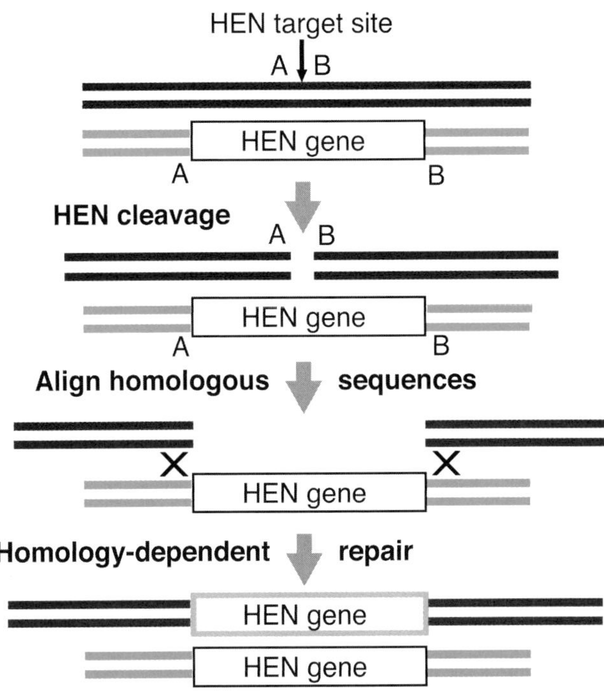

Figure 1 A targeted DNA double strand break can lead to gene insertion. Introduction of a site-specific double strand break by a homing endonuclease (HEN) in a homologous DNA duplex lacking the HEN gene targets homology-dependent DNA synthesis (green) that introduces a copy of the HEN gene to the broken DNA.
doi:10.1128/microbiolspec.MDNA3-0062-2014.f1

to mating type α and from mating type α to mating type **a** (see Chapter 23).

MAT is the mating type expression site, which can express either of two different mating type regulators. One regulator set controls mating type **a** gene expression and the other regulator set controls mating type α gene expression. Mating type switching occurs when the highly site-specific HO nuclease, which is a member of the HEN family, introduces a double strand break into the *MAT* expression site (Fig. 2). This double strand break initiates homology-dependent DNA repair using either one of the two nonexpressed, silent storage copies of mating type information called *HML*α and *HMR***a**, as a template to replace the mating type information at the *MAT* expression site. *HML*α has a silent copy of mating type α information and *HMR***a** has a silent copy of the mating type **a** information. HO expression and the choice of *HML*α or *HMR***a** as a donor site are highly regulated. When the *MAT* expression

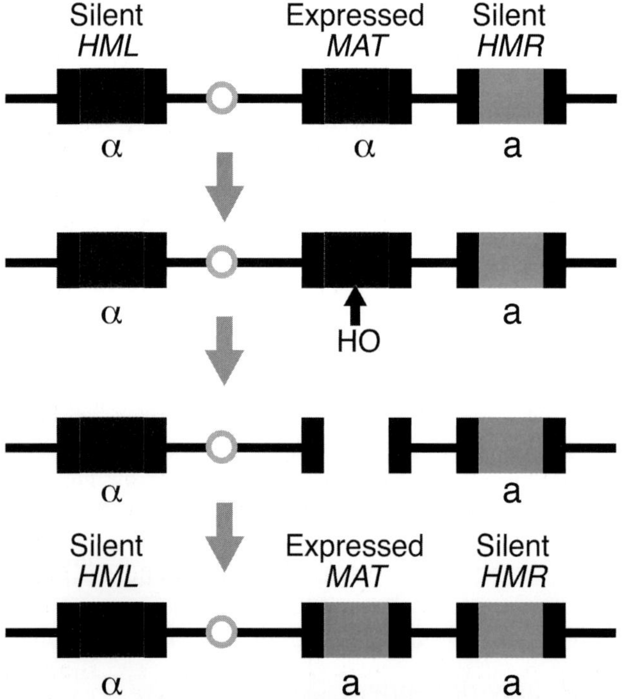

Figure 2 A targeted DNA double strand break causes mating type switching in *Saccharomyces cerevisiae*. Information that specifies either mating type **a** or mating type α is expressed at *MAT* and is also present at silent storage sites *HML* and *HMR*. Introduction of a DNA double strand break in *MAT* by the endonuclease HO targets homology-dependent DNA synthesis at *MAT* using either the silent *HML* and *HMR* storage sites as templates. DNA breaks in *MAT*a cells use silent *HML* α as the donor, thus switching *MAT*a to *MAT* α, resulting in a switch in mating type from **a** to α. Conversely, *MAT*a cleavage results in a switch to *MAT* α.
doi:10.1128/microbiolspec.MDNA3-0062-2014.f2

locus contains mating type information of one type, switching information from the silent locus of the opposite mating type is copied into the *MAT* expression site, thus switching a mating type α cell into a mating type **a** cell or a mating type **a** cell into a mating type α cell.

A different strategy is used for mating and cell type switching in several fission yeasts including *Schizosaccharomyces pombe*. Although these yeasts also have a *MAT* expression locus and two silent storage sites for mating type information, switching occurs, not by targeted DNA repair, but rather by another mechanism involving strand- and site-specific imprinting of one of the DNA strands at the *MAT* expression locus (see Chapter 24).

Taking Evasive Action: Changing Cell Surface Proteins to Elude the Host Immune Defense

A key step in successful pathogen invasion of a new host is evasion of the host's immune response directed against pathogen cell surface antigens. Many pathogens evade the immune system by changing their cell surface proteins, often by DNA rearrangements. The simplest variation system is to alternate expression between two different cell surface proteins. Switching between expression of two different surface protein types occurs in *Salmonella* via a DNA inversion. This DNA inversion flips an invertible segment containing a promoter, which in one orientation, drives expression of one surface flagellar protein, and in the other orientation, drives expression of the alternative flagellar protein (see below; see Chapter 9).

In other systems, homologous recombination between multiple silent gene variants and a single gene expression site underlies the alternate expression of surface protein variants (Fig. 3). The bacterial pathogens *Neisseria gonorrhoeae* and *N. meningitidis* contain about 20 silent variant copies of a surface pilin gene, *pilS*, and a single pilin gene expression site, *pilE* (see Chapter 21). *pilS* gene segments are transferred to *pilE* by RecA-dependent homologous recombination that appears to be stimulated by nicking of a DNA G4 quadraplex upstream of *pilE*. The surface lipoprotein VlsE of the bacterial spirochete *Borrelia burgdorferi*, which causes Lyme disease, also undergoes antigenic variation (see Chapter 22). As in *Neisseria*, there is a single *vlsE* expression site, which contains a required G4 quadraplex, and multiple variant silent *vlsS* genes whose sequences are transferred to *vlsE*. Interestingly, this reaction does not require RecA.

The *Borrelia* genome includes a linear chromosome and multiple linear plasmids with hairpin ends.

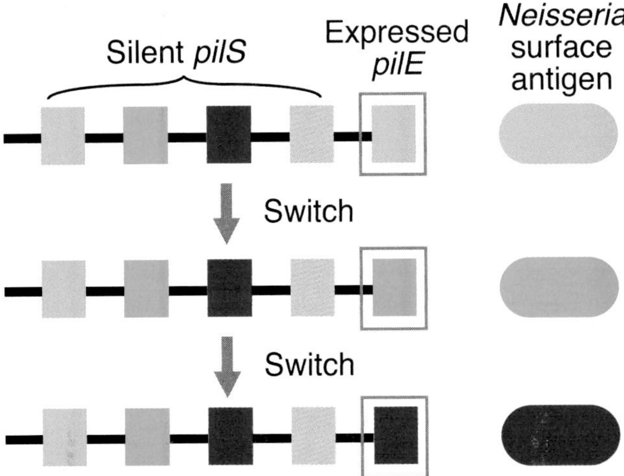

Figure 3 Gene replacement underlies antigenic variation in *Neisseria*. A pilin surface antigen is expressed from the *pilE* site and multiple variant pilin genes are stored in silent *pilS* sites. Homology-dependent repair using template information from a *pilS* gene changes the information in *pilE*, varying the surface antigen.
doi:10.1128/microbiolspec.MDNA3-0062-2014.f3

To complete replication, these ends are processed by the proto-telomerase ResT, a phospho-Tyrosine recombinase (see below; see Chapter 12).

Antigenic variation of the variant surface glycoprotein (VSG) of *Trypanosoma brucei*, the protist that causes sleeping sickness, involves some 2,500 silent *vsg* gene variants and multiple *vsg* expression sites (see Chapter 19). The introduction of double strand breaks into repeats that flank the *vsg* expression sites results in gene switching but how such breaks might be generated remains to be determined.

Antigenic variation of several large multi-gene families of surface proteins also occurs in the malaria parasite *Plasmodium falciparum* (see Chapter 20). The best-studied system is the *var* family with about 60 members clustered in several different arrays. In this case differential gene expression regulated *in situ* mediates antigenic variation, but gene copies are also diversified by recombination.

LESS IS MORE: ACTIVE GENE ASSEMBLY BY DELETION

The Immune System Strikes Back: Immunoglobulin Gene Diversification Allows Detection of Millions of Antigens

In the vertebrate adaptive immune system, the B cells that make antibodies and T cells that make antigen receptors can produce millions of diverse immunoglobulins that can recognize different antigens. Multiple processes underlie these gene diversifications. In both B and T cells, the combinatorial assembly of many different V, D, and J coding gene segments, which encode different protein segments to form the variable regions of immunoglobulin genes and proteins, is mediated by V(D)J recombination (see Chapter 14). The loci encoding these gene segments contain many, in some cases hundreds, of these coding segments separated by nonimmunoglobulin spacer DNA. In B cells, somatic hypermutation (SHM) then targets the assembled V(D)J gene segments to further diversify the variable regions (see Chapter 15). Class switch recombination (CSR) then adds one of several different classes of constant regions to the assembled, mutated V(D)J variable regions to yield antibodies for several different cellular activities (see Chapter 15).

V(D)J Recombination

An immunoglobulin locus contains arrays of multiple V (variable), D (diversity), and J (joining) gene coding segments upstream of a constant gene-coding segment. These V, D, and J gene-coding segments are assembled by the combinatorial joining of different V, D, and J coding gene segments, yielding many different $V_xD_yJ_z$ coding regions by V(D)J recombination (see Chapter 14). This combinatorial gene assembly occurs by the introduction of targeted DNA double strand breaks at the edges of the V, D, and J segments to be joined, resulting in excision of the DNA between the targeted V, D, and J segments. The coding segments are joined by nonhomologous end joining.

A highly specialized site-specific nuclease called RAG makes these initiating double strand breaks, acting at recombination signal sequences (RSSs), at the edges of the V, D, and J gene coding segments (Fig. 4). These RSSs may be either of two slightly different forms, RSS12 or RSS23, and RAG binding to and pairing of one RSS12 type and one RSS23 type site is required for RAG-dependent cleavage. RAG activity is also elaborately regulated *in vivo*. The RSSs are accessible to RAG only when activated by regulated transcription through them and by particular chromatin modifications. The interaction of two RSSs that may be separated by many tens of kb is facilitated by the highly structured 3D organization of the genome in each immunoglobulin locus (2).

The different positions and orientations of the RSS signals at the edges of the V, D, and J coding segments organize the breakage by RAG at particular positions, followed by the joining of V segments to D segments, and the joining of D segments to J segments. The joining

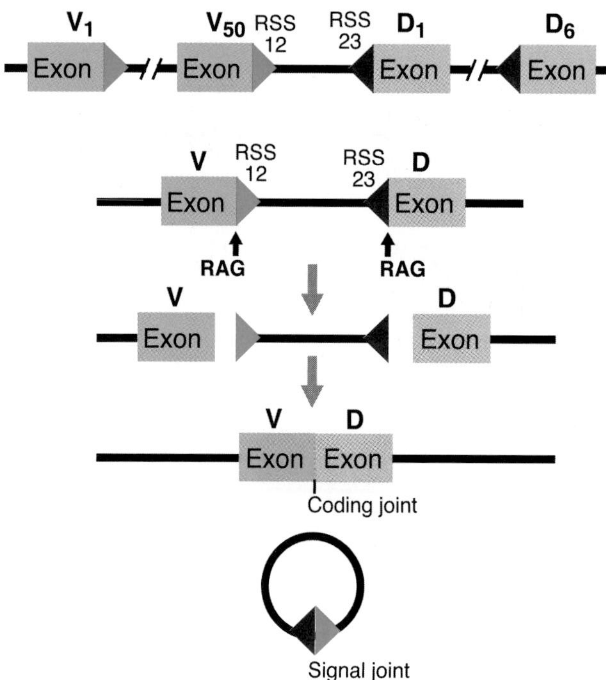

Figure 4 Targeted DNA double strand breaks by RAG mediate immunoglobulin gene assembly during V(D)J recombination. Diverse immunoglobulins that recognize many different antigens result from combinatorial assembly of different V, D, and J coding segments. Site-specific cleavage by RAG, at RSS12 and RSS23 sites that bound the multiple V, D and J segments, results in excision of intervening DNA between the gene segments to be joined and formation of a coding joint by NHEJ. The intervening DNA circularizes, forming a signal joint and is lost from the cell.
doi:10.1128/microbiolspec.MDNA3-0062-2014.f4

of V to D segments and D to J segments creates coding joints that link these protein-coding segments, which will then be transcribed into mRNA. The joining of the RSS12 to RSS23 sites creates the signal joint on the excised intervening DNA segment, which is lost from the cell.

RAG, whose structure has been recently determined (3), is closely related to DDE transposases (see below; see Chapter 25), that bind to and break at terminal inverted repeats (TIRs) at the ends of DDE transposons. Indeed, the DNAs between the coding segments, which are to be joined, are bounded by inverted RSSs just as the ends of DDE transposons are bounded by TIRs. Moreover, the chemical steps by which double strand breaks are made at RSSs by RAG and by which double strand breaks are made at TIRs at the ends of *hAT* transposons such as *Hermes* (see Chapter 35), occur by the same chemical mechanism (4, 5). This mechanistic similarity and the marked similarities between the structures of RAG and *Hermes* (3,6) support the hypothesis that the RAG recombination system evolved from an ancient *Transib* transposon (7, 8).

Somatic Hypermutation

The V(D)J segment in an active immunoglobulin gene in a B cell is then subjected to local, targeted, high-frequency mutagenesis called somatic hypermutation (SHM) to further diversify the antigen binding region (see Chapter 15). This mutagenesis is targeted to particular regions by regulated transcription that recruits the activation-induced cytidine deaminase (AID). Deamination converts cytidine to uracil resulting in a U::G mismatch. Mutagenic repair, likely by error-prone polymerases, then occurs by base excision repair or mismatch repair, resulting in localized, high frequency mutagenesis that further diversifies the antigen-binding region of the antibody.

Class-Switch Recombination

Downstream of the V(D)J coding exon segments are multiple, different "C" (constant) exons that encode various immune system effectors. The V(D)J coding exon and a C exon are joined in the mRNA by RNA splicing. Class-switch recombination (CSR) positions particular, different isotype C exons immediately downstream of the V(D)J exon by DNA excision of the intervening C exons. This excision occurs by targeted double strand breaks at a switch site downstream of the V(D)J exon and the switch site upstream of a particular C exon (Fig. 5). As in SHM, transcription-directed AID changes cytidines to uracil in the switch sites, which are then further processed to introduce double strand breaks. The cleaved switch site downstream of the V(D) J coding segment then joins to the cleaved switch site upstream of a C exon by nonhomologous end joining. The intervening DNA is lost from the cell.

Massive Chromosome Destruction Leads to Active Ciliate Genes

The most spectacular examples of programmed DNA breakage and joining events to restructure chromosomes and genes (9) are those in the ciliates *Paramecium* (see Chapter 17), *Tetrahymena* (see Chapter 16), and *Oxytricha* (see Chapter 18). It should be noted, however, that large-scale chromosome diminution also occurs in some nematodes, insects, and vertebrates (see Chapter 18) (10).

Ciliates are single cell eukaryotes that have two nuclei, a germline micronucleus of ~50 to 100 Mbp, which is transmitted to progeny, and a somatic macronucleus

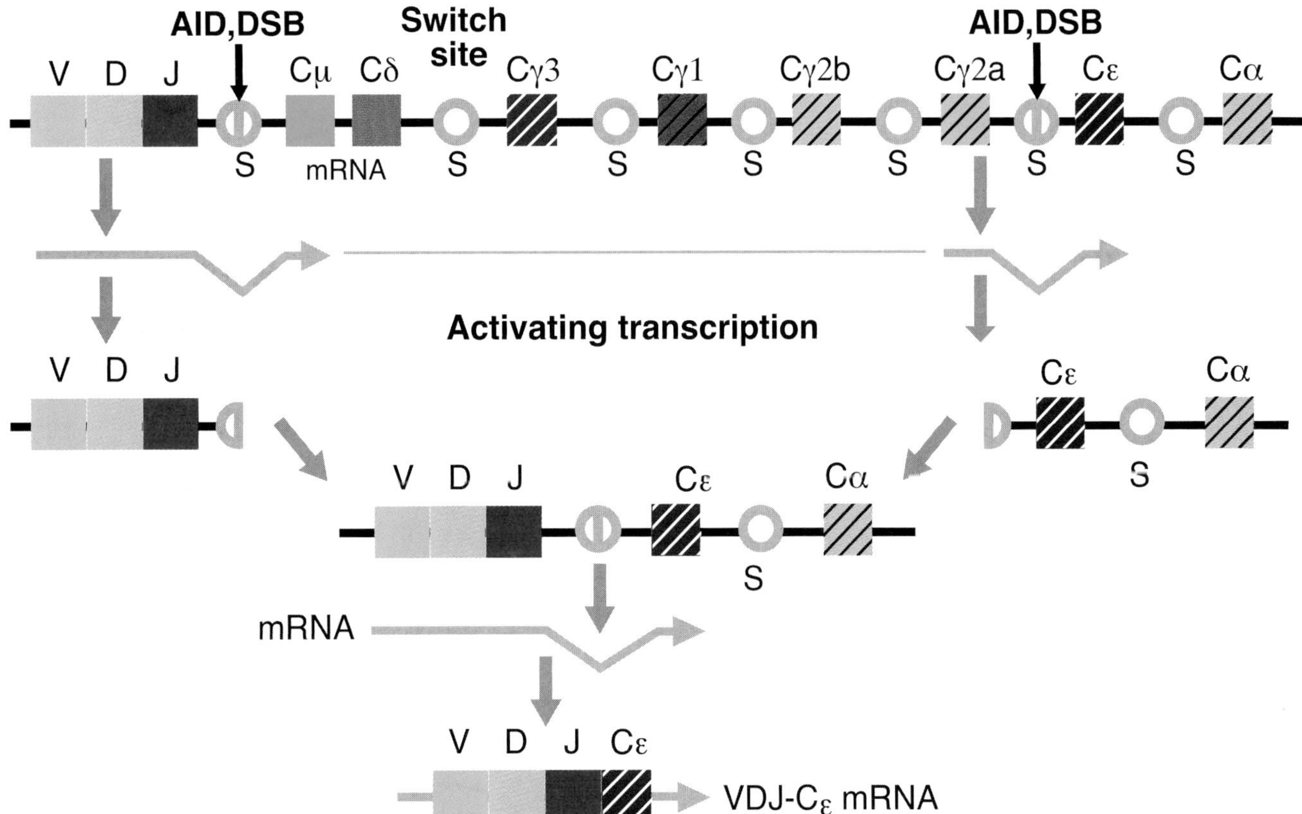

Figure 5 Targeted DNA double strand breaks at switch sites join V(D)J coding segments to different antibody class segments. DNA double strand breaks (DSB) are targeted by transcription-induced activation-induced cytidine deaminase (AID) modification of switch sites downstream of an assembled V(D)J coding region and upstream of different antibody class coding regions. The intervening DNA is excised and combinatorial joining of the V(D)J and antibody class segments by NHEJ results in different classes of antibodies. doi:10.1128/microbiolspec.MDNA3-0062-2014.f5

from which all genes are expressed. In the macronucleus, the much larger micronuclear chromosome has been shattered by double strand breaks into smaller DNA fragments. Macronuclear development also involves excision and loss of a substantial fraction of the micronuclear genome, ranging from ~30% in *Tetrahymena* to almost 95% in *Oxytricha*. These internal eliminated sequences (IESs) lie in and between genes in the micronucleus, and thus, must be carefully excised and the genic material rejoined to maintain gene function. In *Paramecium* and *Tetrahymena*, many of these IESs are excised by a domesticated *piggyBac* transposase (see Chapter 39), which introduces targeted double strand breaks at the edges of the elements, following which the gene segments are rejoined by nonhomologous end joining. A smaller number of IESs are excised by other transposase-like proteins that make targeted DNA double strand breaks. An attractive view is that the IESs began as transposon insertions and that their elimination is an extreme form of gene silencing.

RNA plays a prominent role in these programmed DNA rearrangements. Formation of the macronucleus is completed by the addition of new telomers to the ends of the new "minichromosomes." Experiments in *Tetrahymena* lead to the proposal (11) that a new class of RNA, called scan RNA (scRNA), actually directs the elimination of IES sequences. The scRNAs specific for the sequences to be maintained in the next round of macronuclear development derive from pervasive transcription of the entire genome, which is then compared to the edited macronuclear DNA, thus, identifying the scRNAs.

An even more remarkable role for RNA has been revealed in *Oxytricha* macronuclear development (see Chapter 18). Here, the micronuclear genes are not only interrupted by transposon-like sequences that are removed by *Tc1/mariner* transposases (see below), but gene segment order is highly scrambled. For example, if the linear order of the segments that make up a gene in the macronuclear DNA is ABCDE, their order in the germline micronucleus may be BADCE! How can a sensible gene be assembled from this scrambled genome? The answer lies in gene-length RNAs with the correct gene segment order that are transcribed from the macronucleus and are then transported to the micronucleus where they provide a template for the correctly ordered assembly of genes in the next round of macronuclear development. The mechanism of this assembly process remains to be determined.

TRANSPOSABLE ELEMENTS

Transposable elements, which are present in virtually all genomes, are discrete DNA segments that can move themselves or a copy of themselves within and between genomes. Transposable elements underlie a wide variety of processes, such as the interaction of viral and host chromosomes, reactions that underlie the replication and accurate segregation of chromosomes and the regulation of gene expression. Transposable elements may also encode a wide variety of accessory determinants including antibiotic resistance genes, virulence determinants, and a wide variety of metabolic genes. Moreover, the movement of a transposable element can have a profound effect on host gene expression. Element insertion into a gene can result in gene inactivation. Element insertion near a gene can also alter gene expression as transposable elements can also encode regulatory signals such as promoters, enhancers, splice sites, polyadenylation signals, and transcription termination signals. These variations in genetic information are substrates for adaptation, selection, and evolution.

Other than viruses, it has long been thought that only mobile elements in bacteria carried accessory genes such as antibiotic resistance genes. Multiple examples of mobile elements containing genes or gene fragments have now been observed in eukaryotes, however, most notably with the DNA-only *Helitrons* (see Chapter 40) (12) and versions of *Mutator* transposons called *Pack-MULEs* (see Chapter 36) (13).

Elements that encode their own mobility functions, for example, a transposase, and its cognate recognition sequences at the transposon ends are called "autonomous" elements. Some elements, however, encode only the necessary *cis*-acting sequences at the transposon ends and are mobilized in *trans* by transposase from another autonomous version of the element elsewhere in the genome. These elements are called "nonautonomous" elements.

Some transposable elements move only via DNA intermediates. The DNA intermediates of some other transposable elements are generated by reverse transcription of RNA into double stranded DNA, which then interacts with and inserts into a target site. Transposition of another major class of elements occurs by the interaction of an RNA form of the element directly with a target DNA, followed by reverse transcription of the RNA *in situ* at the target site to generate a new DNA copy of the element. This ability to convert RNA copies of an element into DNA and the resulting amplification likely leads to the very high copy number of retroelements in some organisms.

There are many types of transposable elements that have different structures and move by different mechanisms. Confusingly, but not surprisingly, because many elements were first identified in different organisms and were named in the absence of molecular understanding of how they moved, what turn out to be mechanistically very related elements can have quite different types of names. Conversely, other elements that move by very different mechanisms can have similar names. For example, many bacterial insertion sequences move by the same mechanism, as do many eukaryotic transposable elements. The transposition mechanism of bacterial *IS4* family members, including *IS10* and *IS50* that form *Tn10* and *Tn5* (see Chapter 29), is related to that of the eukaryotic transposable element *piggyBac* (see Chapter 39). The transposable bacterial insertion sequence, *ISY100* (14), uses the same breakage and joining steps as do transposable *ITm* (*Tc1/mariner*) elements found in many eukaryotes (see Chapter 34). Conversely, the integrases of many bacterial viruses use a very different mechanism for DNA breakage and joining than do the integrases that mediate the integration of the DNA forms of retroviruses.

DNA-ONLY TRANSPOSONS

Transposases are sequence-specific DNA binding proteins that also contain a catalytic domain that mediates DNA breakage and joining. Some transposable elements move by breakage and joining mediated only by the transposase, whereas others also involve DNA synthesis and ligation by host proteins to regenerate intact duplex DNA. There are four major classes of DNA-only transposases: DDE transposases, tyrosine-histidine-hydrophobic-histidine (HUH) transposases, tyrosine-transposases, and serine-transposases. DDE transposases break and join DNA by direct transesterification. The other classes of transposases act via covalent-protein DNA intermediates. Eubacteria, archaea, and eukaryotes all contain mobile elements with these four major classes of transposases.

A new class of transposons, casposons, and their novel transposases, was recently proposed based on bioinformatic analysis of bacterial mobile elements that also encode a DNA polymerase (15). These bacterial elements have been called casposons because the protein proposed to be their transposase (integrase) derives from Cas1, a protein component of bacterial adaptive immunity CRISPR-Cas systems (16). These immunity systems take DNA sequences from infecting nucleic acids, such as viruses, and incorporate them into "clustered regularly interspaced short palindromic repeats" (CRISPR) arrays. Transcription of these arrays generates RNA copies of these incorporated sequences, which are used as guide sequences for nucleolytic attack and destruction of re-invading foreign DNAs and their transcripts. Notably, it has recently been shown that a complex of Cas1-Cas2 proteins can integrate a new DNA segment into a target DNA *in vitro* by direct transesterification using a 3′-OH end as a nucleophile (17). This is the same chemical mechanism used for DNA joining as that used by DDE transposases (see below; see Chapter 25) and the closely related retroviral integrases (see Chapter 44). Notably, however, the structure of the Cas1-Cas2 complex (18), which has been determined by X-ray crystallography, has no structural homology with DDE transposases, thus, identifying *Cas1* as a new class of transposase.

Such apparently DNA self-synthesizing elements in eukaryotes called Polintons (Mavericks) that also encode a DDE transposase have also been described bioinformatically (19).

Double Strand DNA Transposons Move via DDE Transposases

DDE transposons are discrete DNA segments bounded by terminal inverted repeats (TIRs) that are the specific binding sites of the cognate DDE transposase (see Chapter 25). The TIRs position the transposase at each end of the element to carry out the DNA breakage and joining reactions. DDE transposases carry out two closely related reactions: (i) the cleavage of a DNA phosphodiester bond using water as the nucleophile to yield 3′-OH and 5′-P ends; and (ii) the joining or "strand transfer" of the 3′-OH transposon end to a target DNA in which the 3′-OH is the nucleophile in a direct transesterification reaction. The active site of a DDE transposase is formed by an RNase H-like fold that closely juxtaposes three conserved acidic amino acids - D, D, and E - to position the essential Mg^{2+} ion cofactors. Thus, DDE transposases are sometime called RNase H-like transposases. As described below, retroviral integrases (see Chapter 44) are also DDE transposases.

There are many superfamiles of DDE transposable elements (20), which are defined by similarities in the transposase sequence, the sequence of the transposon ends, and in some cases their target sequence (Fig. 6). Members of some superfamilies are present in both bacteria and eukaryotes.

Different DDE transposases use different combinations of breakage and joining steps to disconnect at least their 3′ ends from the donor site (Fig. 6) (see Chapter 25 for an overview and element-specific chapters). The key events in transposition of all DDE elements are the release of the 3′-OH transposon ends from the donor site, which then attack and join to the target DNA at staggered positions on the top and bottom target strands. These joining reactions result in the covalent linkage of the 3′ transposon ends to the target DNA. Because of the staggered joining positions on the top and bottom strands, single strand gaps extend from 3′-OHs on the flanking target strands to the 5′ transposon ends. These 3′-OH target ends provide the primers for the DNA synthesis that will repair these gaps or, in some cases, copy the entire element. Repair of these gaps by host proteins can occur by several different mechanisms (see Chapter 31) (21), resulting in target site duplications, which are a hallmark of transposition.

Insertion of the eukaryotic DDE *Spy* element, however, occurs without target site duplication (22). Likely, the excised 3′-OH transposon ends join the target DNA at nonstaggered positions.

Transposition reactions occur within elaborate nucleoprotein complexes called transpososomes, whose assembly is a key control point in transposition, which bring the transposon ends and the target DNA together, such that uncoordinated unproductive events do not occur. In some systems, host DNA bending proteins are important for transpososome assembly. Known transpososomes contain at least a dimer of transposase, the active site of each transposase protomer mediating breakage and joining at one transposon end. Thus, protein–protein interactions for oligomerization are also important features of transposases.

In some systems, multiple transposon-encoded proteins are required for transposition. For example, the *Tn7* transposase contains two different polypeptides, each one cleaving a particular strand (see Chapter 30) (Fig. 7). Some systems, for example, *Mu* (see Chapter 31) and *Tn7* (see Chapter 30) encode an ATP-dependent DNA binding protein involved in target site selection. *Tn7* also encodes additional target site selection proteins.

DDE Transposons

Family	Element	Cut & Paste: Hairpins on transposon ends	Cut & Paste: Hairpins on donor flank	Cut & Paste: 2-strand cleavage	Nick Copy-out Paste	Nick Paste Copy	Reference/ Chapter
IS4	*Tn5, Tn10*	■					29
piggyBac	*piggyBac*	■					39
hAT	*Hermes*		■				35
Transib	RAG (RAG1)		■				14
ITm (IS630-Tc1-mariner) or *Tc1/mariner*	*Mariner*			■			34
	Sleeping Beauty			■			38
	Mos1			■			Richardson et al 2009
	IS100			■			Feng & Colloms 2007
P	*P* element			■			33
Mutator	*MuDR*						36
	MULE Os3378						Zhao et al 2015
	IS256						Hennig & Zeibuhr 2010
IS3	*IS911*				■		27
Tn3	*Tn3* *Tn4430*					■	32

Figure 6 Different families of DDE transposases mediate the transposition of different elements. Different superfamilies of DDE transposases use different combinations of DNA breakage, replication, and joining reactions to move different DNA transposons (see Chapter 25 for details). Some elements move by excision and integration (cut and paste), whereas the movement of other elements involves copying of the element by DNA replication (nick-copy out-paste and nick-paste-copy). Transposases from different families can use related mechanisms. **Richardson JM, Colloms DS, Finnegan DJ, Walkinshaw MD.** 2009. Molecular architecture of the *Mos1* paired-end complex: the structural basis of DNA transposition in a eukaryote. *Cell* **138**:1096–1108; **Feng X, Colloms SD.** 2007. In vitro transposition of *ISY100*, a bacterial insertion sequence belonging to the *Tc1/mariner* family. *Mol Microbiol* **65**:1432–1443; **Zhao D, Ferguson A, Jiang N.** 2015. Transposition of a Rice Mutator-Like Element in the Yeast *Saccharomyces cerevisiae*. *Plant Cell* **27**:132–148; **Hennig S, Ziebuhr W.** 2010. Characterization of the transposase encoded by *IS256*, the prototype of a major family of bacterial insertion sequence elements. *J. Bacteriol* **192**:4153–4163. doi:10.1128/microbiolspec.MDNA3-0062-2014.f6

Cut and Paste Transposons Move by Excision and Integration

Many elements move by a "cut and paste" mechanism in which the transposon is excised from the donor site by double strand breaks (Fig. 8), which can occur by several different pathways (see Chapter 25), all of which expose the 3′-OH transposon ends. Target sites range from nearly random to highly site specific. Note that the gapped donor backbone from which the element excised must also be repaired.

Nick, Copy-Out and Paste: Replicative Transposition of *IS911* and other *IS3* Family Elements

In DNA cut and paste reactions, DNA synthesis is limited to repair reactions at the ends of the newly inserted transposon. The movement of some other DDE transposons, however, involves replication of the entire element. DNA replication is essential to transposition of *IS911* and other members of the widespread *IS3* family (see Chapter 28). In this case, the transposase introduces a nick at only one 3′ end of the transposon, that is, donor cleavage is asymmetric (Fig. 9). The released 3′-OH transposon end then joins intramolecularly to just outside its own 5′ end, circularizing one strand of the transposon. DNA replication then initiates at the flanking target 3′-OH generated upon intramolecular transposon strand joining. Replication of the transposon results in a free, double strand transposon circle in which the transposon ends are closely abutted. The transposase then breaks this junction and inserts the transposon by attack of its 3′-OH ends into a target site. Note that the noncircularized transposon strand at the donor site is also copied by DNA replication. Transposition is thus replicative, resulting in a transposon copy remaining at the donor site and a transposon copy at the new insertion site.

Nick, Paste and Copy: Replicative Transposition of *Tn3* Family Transposons

Members of the DDE *Tn3* transposon family, including the closely related transposon $\gamma\delta$ and *Tn4430* (see Chapter 32), are found in many types of bacterial plasmids, transposing from one plasmid to another. Upon transposition from a donor plasmid to a target plasmid, the primary product of their transposition is a cointegrate in which two copies of the transposon link a copy of the donor plasmid and a copy of the target plasmid (Fig. 10). Thus, transposition of these elements involves generating a copy of the transposon, that is, transposition is replicative. The *Tn3* and *Tn4430* transposases can make single-strand nicks at the 3′ transposon end, exposing a 3′-OH. Assuming that *Tn3* replicative transposition proceeds as with *Mu* (see below and Chapter

Mu-like transposons

Transposon	Transposase 5'	Transposase 3'	ATP-dependent targeting	Alternative target selectors		Cut & paste	Nick, paste and copy	Reference/ Chapter
Mu	?	MuA	MuB			■	■	31
Tn7	TnsA	TnsB	TnsC	TnsD/ *attTn7*	TnsE/replicating DNA	■		30
Tn522	—	p480	p271	—			■	Rowland & Dyke 1990
Tn5090/ Tn5053	—	TniA	TniB	TniQ/*res*			■	Minakhina et al 1999

Figure 7 Some transposons encode DDE Transposases and ATP-utilizing target choice regulators. Bacteriophage *Mu* uses cut and paste transposition to insert into the bacterial genome and replicative transposition to replicate its DNA during lytic growth. MuA is a DDE transposase that breaks and joins DNA and the ATP-dependent regulator MuB controls target DNA selection. Related transposons also encode a transposase, an ATP-dependent target regulator and, in some cases, an additional target type specification protein. **Minakhina S, Kholodii G, Mindlin S, Yurieva O, Nikiforov V.** 1999. *Tn5053* family transposons are *res* site hunters sensing plasmidal *res* sites occupied by cognate resolvases. *Mol Microbiol* **33**:1059–1068; **Rowland S-J, Dyke K.** 1990. *Tn552*, a novel transposable element from *Staphylococcus aureus*. *Mol Microbiol* **4**:961–975.
doi:10.1128/microbiolspec.MDNA3-0062-2014.f7

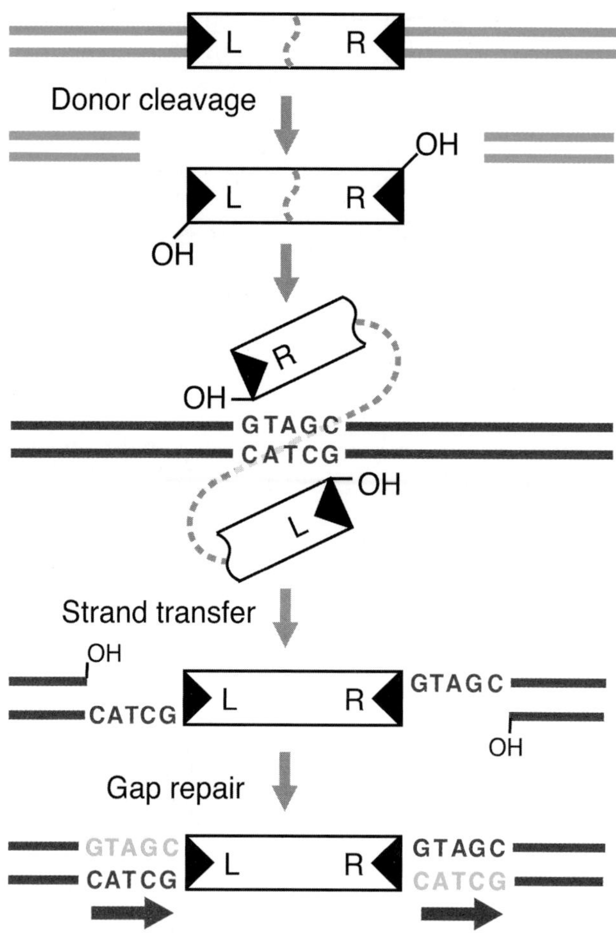

Figure 8 Mechanism of DNA cut and paste transposition by a DDE transposase. The transposase makes DNA double strand breaks at the transposon ends that excise the element from the donor site, exposing the 3′-OH transposon ends. These 3′-OH ends then attack the two target DNA strands at staggered positions by direct transesterification reactions that covalently link the 3′ transposon ends to the target DNA. The staggered end joining positions result in single strand gaps at the 5′ transposon ends that are repaired by host DNA synthesis (green) to generate flanking target site duplications. doi:10.1128/microbiolspec.MDNA3-0062-2014.f8

31), transposition begins with transposase nicking at both 3′ ends of the element. The two released 3′-OH transposon ends then attack a target site on another plasmid, generating a strand transfer product in which the 3′ transposon ends are covalently linked to the target plasmid and the 5′ transposon ends remain covalently linked to the donor site. Target joining of the 3′ transposon ends releases two 3′-OH target ends that flank the newly inserted transposon. DNA replication initiating from these 3′-OHs then copies both strands of the transposon, generating the two transposons that link the donor and target plasmids.

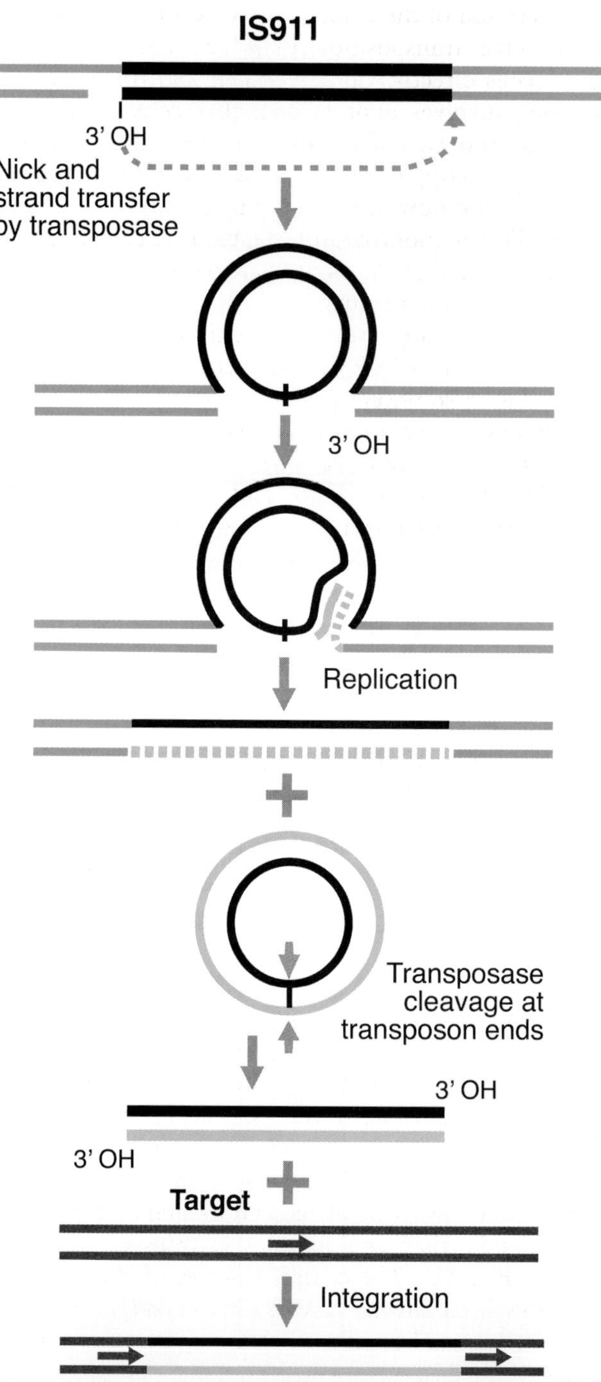

Figure 9 Mechanism of transposition of the nick-copy out and paste transposon *IS911* by a DDE transposase. Transposition begins by transposase nicking at one transposon end. The resulting 3′-OH then attacks its own 5′ end, circularizing the transposon. DNA replication (green) initiated at a flanking target 3′-OH copies both transposon strands, releasing a circularized transposon and repairing the donor site. Transposase then cleaves the transposon ends in the transposon circle, releasing 3′-OH ends that attack the target DNA. doi:10.1128/microbiolspec.MDNA3-0062-2014.f9

Generation of the cointegrate is the only step of *Tn3* element-like transposition in which the transposase participates directly. Completion of *Tn3* transposition, however, involves another step that converts the cointegrate into two plasmids, one, the donor plasmid containing a copy of the transposon, and the other, the target plasmid now also containing a copy of the transposon. This monomerization reaction is called resolution, but note that the same reaction is called excision or deletion in other systems.

This cointegrate resolution requires another recombination system, which is also encoded in the transposon. This resolution system consists of a Serine recombinase called a resolvase that acts at the transposon-encoded *res* site to convert the dimeric cointegrate plasmid into separate donor and target plasmids. We consider the mechanism of such resolution reactions by resolvases below.

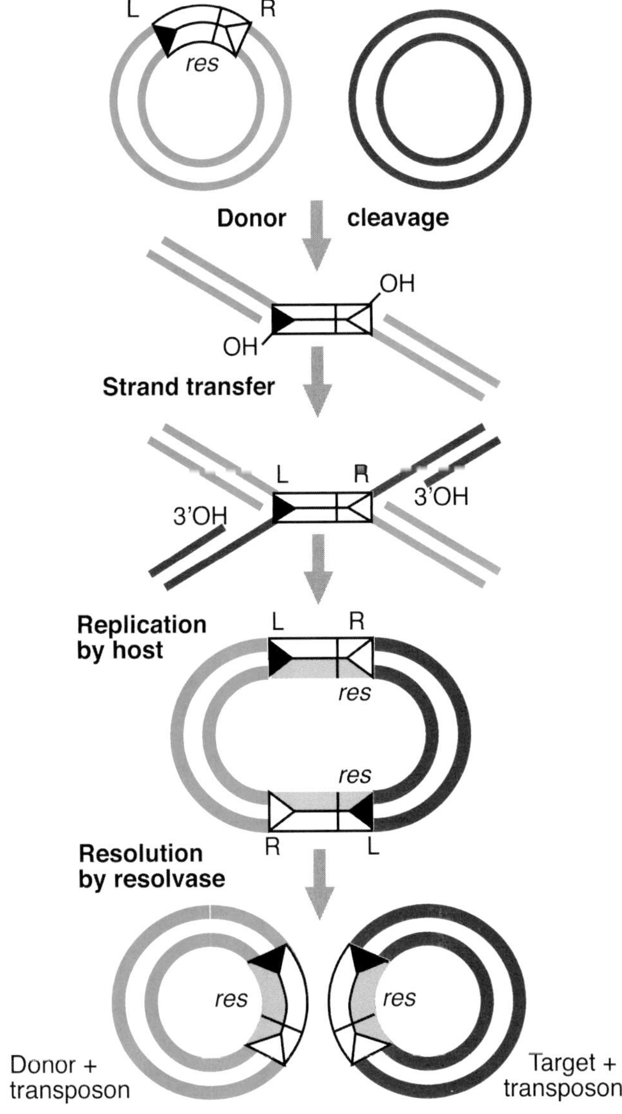

Figure 10 Mechanism of replicative transposition of *Tn3* by a DDE transposase. Transposition begins by transposase nicking to expose both 3′-OH transposon ends that attack and link to the target DNA while the 5′ transposon ends remain linked to the donor plasmid. Replication (green) initiated at the flanking target 3′-OHs copies both strands of the transposon, generating a cointegrate in which two transposon copies link the donor and target plasmids. DNA breakage, strand exchange and rejoining at the directly oriented *res* sites by the Ser_{CSSR} recombinase resolvase that is also encoded by the transposon separates the donor and target plasmids, each containing a transposon copy.
doi:10.1128/microbiolspec.MDNA3-0062-2014.f10

"Bacteriophage *Mu*: A Transposon" and "Transposon *Mu*: A Bacteriophage"

Bacteriophage *Mu* uses DDE transposition in two different steps of its life cycle (see Chapter 31) (23). *Mu* uses cut and paste transposition to integrate randomly into the bacterial chromosome upon lysogenization. *Mu* then uses replicative transposition that uses multiple chromosomal target sites to replicate its DNA during lytic growth. The *Mu* transposition machinery is elaborate: the ends of *Mu* contain multiple transposase binding sites, as well as internal binding sites that enhance transposition by facilitating correct assembly of the nucleoprotein machine that executes transposition. *Mu* also encodes two transposition proteins, the DDE transposase *Mu*A and *Mu*B, which facilitates the interaction of *Mu*A bound to the transposon ends with target DNA. Perhaps the key event in *Mu* transposition is the formation of a *Mu*A tetramer in which two subunits mediate *Mu*A breakage and joining and the other two subunits play critical roles in transpososome assembly (Fig. 11). The central features of this machine have now been directly revealed in a crystal structure of a *Mu*A tetramer bound to both transposon ends and to target DNA (24).

Single Strand DNA Transposons Move via Tyrosine-HUH Transposases

Hallmarks of DDE transposons are their termini inverted repeats and the absence of covalent transposase-DNA bonds during transposition. By contrast, the substrate of a tyrosine-HUH transposase is a single-stranded version of the element as would be transiently present in replication forks. The element lacks terminal inverted repeats but rather contains two internal sets of palindromes that form hairpin structures that are specifically recognized by the transposase. The active sites of tyrosine-HUH transposases contain conserved tyrosine (Tyr) and HUH motifs. To break DNA, the Tyr-OH acts as a nucleophile on a single strand DNA, releasing a free DNA 3′-OH and forming a 5′ phosphotyrosine bond, conserving the high energy of the DNA phosphodiester bond. Attack of a DNA 3′-OH on a 5′-P-Tyr rejoins the DNA and releases the transposase.

Out of and Into Single Strand DNA at Replication Forks

The bacterial IS elements, *IS200*, *IS608*, and *ISDra2*, move via Tyr-HUH transposition (see Chapter 28). The first step in DNA single strand transposition is mediated by a Tyr-HUH transposase dimer and it occurs by two, concerted breakage, exchange, and joining events that excise the element and rejoin the donor site (Fig. 12). Guided by base-pairing interactions between particular transposon sequences, one protomer acts at the upstream end of the element to generate "Donor flank-3′-OH" and "5′-P-Tyr-Transposon end." The other protomer acts at the downstream transposon end to give "Transposon end-3′-OH" and "5′-P-Tyr-Donor flank." The 3′-OH from one protomer then attacks the 5′ phosphotyrosine link on the other protomer. Attack of the "Transposon end-3′-OH" on the "5′-P-Tyr-Transposon end" excises the element as a single strand circle and attack of the "Donor flank-3′-OH" on the "5′-P-Tyr-Donor flank" rejoins the single strand donor site.

Transposon integration into a single strand DNA target site again occurs within a transposase dimer. One protomer cuts the transposon circle, giving "Transposon end-3′-OH" and "5′-P-Tyr- Transposon end." The other protomer cuts the target site strand, giving "Target end-3′-OH" and "5′-P-Tyr-Target end." Rejoining, that is, integration, occurs by the attack of "Transposon end-3′-OH" on the "5′-P-Tyr-Target end" and attack of the "Target-3′-OH" on the "5′-P-Tyr-Transposon end."

These breakage and joining reactions occur at DNA single strands in replication forks. DNA replication is also needed, however, to regenerate intact duplex DNA at both the donor and target sites. Thus, when a replication fork passes through a donor duplex from which the transposon has excised, one daughter duplex lacks the element and replication of the other daughter duplex yields a daughter duplex containing the element. Similarly, when a replication passes through a target site into which the element has inserted, replication of the insertion strand gives rise to a daughter duplex containing the element and replication of the other strand yields a daughter duplex without the element. Whereas there was one copy of the element at the donor site before transposition, following replication, one copy

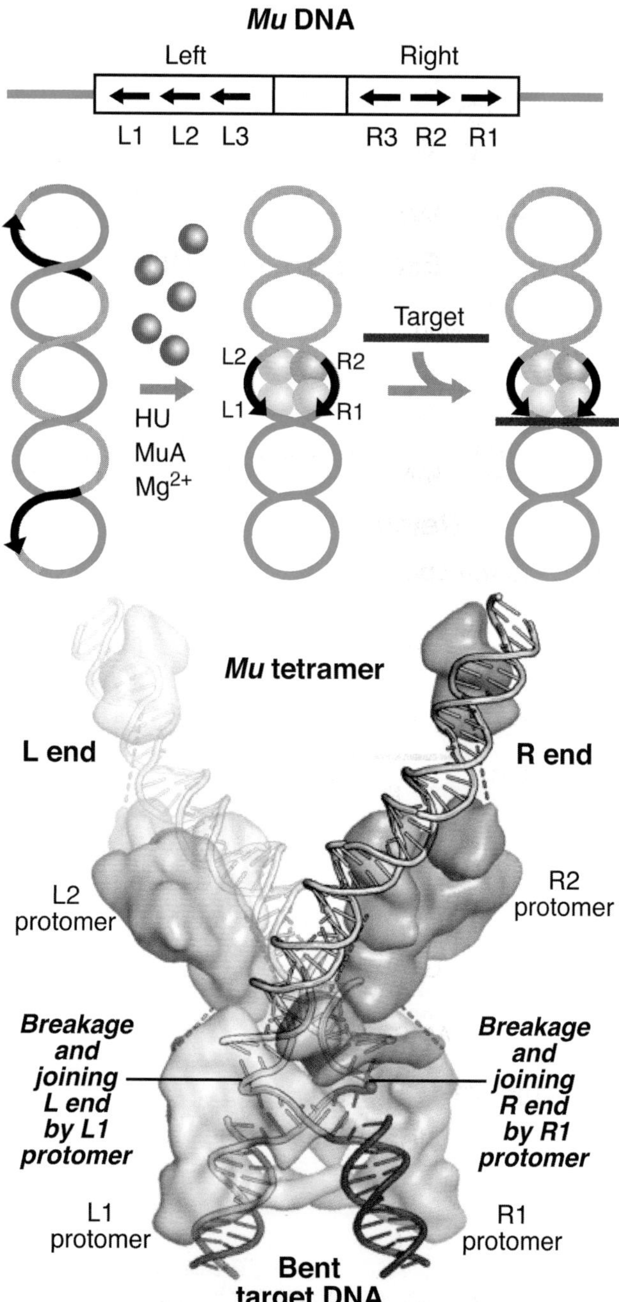

Figure 11 Assembly of an active tetramer of the MuA DDE transposase. The ends of *Mu* contain multiple bindings sites for the MuA transposase. MuA interaction with the ends results in formation of a MuA tetramer that synapses the two ends and activates DNA breakage and joining by the two transposase protomers bound to the outermost *L1* and *R1* sites. Internal MuA binding sites in an enhancer (not shown) facilitate tetramer assembly. Usually recruited by MuB (not shown here), a sharply bent target DNA binds to the *L1* and *R1* promoters, which is attacked by the MuA 3′-OH ends exposed by MuA nicking.
doi:10.1128/microbiolspec.MDNA3-0062-2014.f11

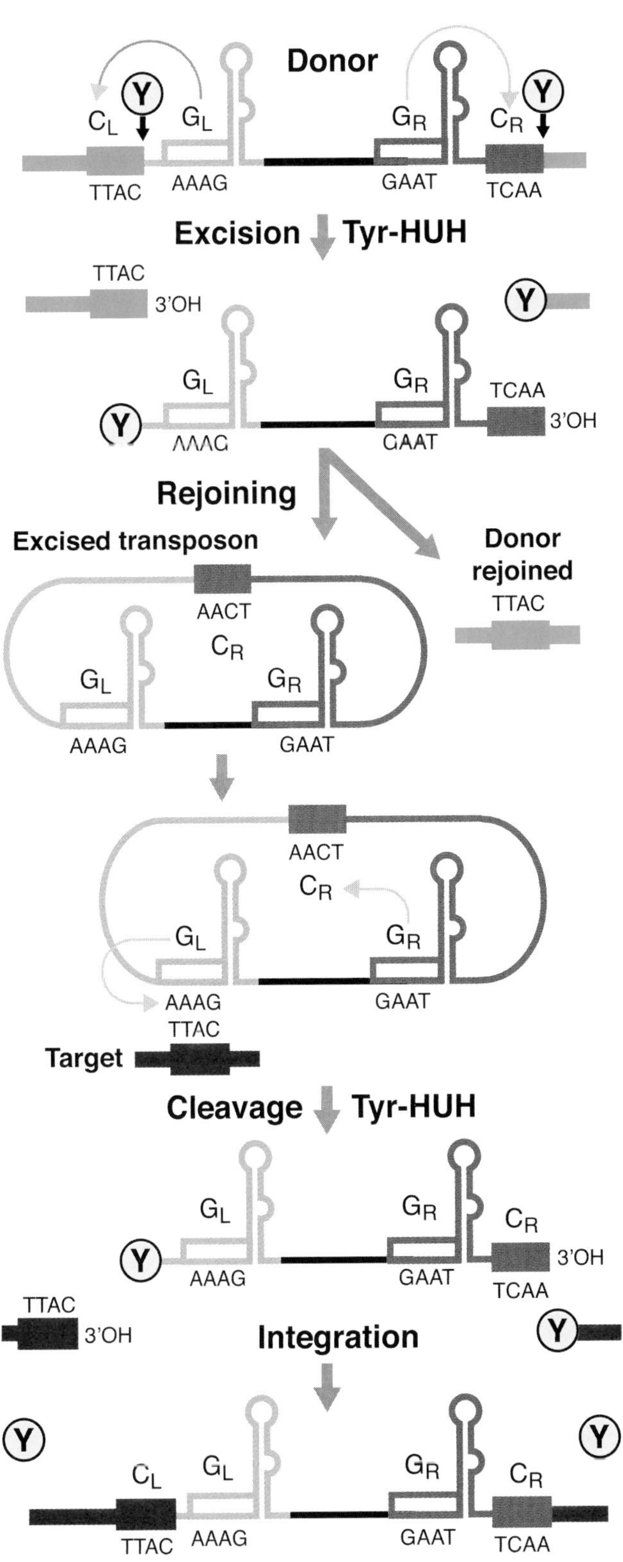

remains at the donor site and there is a new copy of the element at the target site.

Some Tyr-HUH Transposases are Rolling Circle Transposases

Tyr-HUH proteins also mediate rolling-circle replication reactions involved in bacterial plasmid replication and conjugation. In these reactions, after recognition of a DNA signal, such as a particular sequence or hairpin, the Tyr-HUH protein nicks one strand of a plasmid circle forming a DNA 3′-OH end, which serves as a primer for DNA replication that displaces the 5′-P-Tyr containing strand. Replication continues around the circle and then the displaced strand can be circularized by the attack of a 3′-OH on the 5′-P-Tyr containing strand. In transposition of bacterial *IS91*-like elements (25), it is thought that the displaced 5′-P-Tyr strand can join to a target DNA via attack of a target 3′-OH generated by the transposase on the incoming 5′-P-Tyr end. DNA replication at the new insertion site and donor site is required to regenerate intact duplex DNA at both sites.

Eukaryotic DNA transposons called *Helitrons* are thought to use a rolling circle mechanism (see Chapter 40). They encode an ORF thought to be the transposase that contains Tyr-HUH motifs and a helicase motif, which could provide a strand displacement activity (Fig. 13). Some elements also encode another nuclease motif. *Helitrons* are found in diverse eukaryotes but have been analyzed most intensively in the mammalian little brown bat and in maize where they make up several percent of the genome. Their presence in bats is

Figure 12 Mechanism of transposition of *IS200*-like transposable element by a tyrosine (Tyr)-histidine-hydrophobic-histidine (HUH) transposase acting on a single strand DNA substrate. A Tyr-HUH transposase breaks DNA by the attack of the hydroxyl of a higher conserved tyrosine, resulting in a free 3′-OH and a 5′-P-Tyr link. DNA rejoining occurs by attack of the 3′-OH on the 5′-P-Tyr link. Guided by base-pairing interactions between the guide and cleavage sequences at both transposon ends, one protomer of a transposase dimer acts at the upstream transposon end, generating “Donor flank-3′-OH” and “5′-P-Tyr-Transposon end.” The other protomer acts at the downstream transposon end, generating “Transposon end-3′-OH” and “5′-P-Tyr-Donor flank.” The 3′-OHs from one protomer then attack the 5′ phosphotyrosine links on the other protomer. This excises the element as a single strand circle and rejoins the single strand donor site. A transposase dimer then integrates the transposon integration into a single strand DNA target site, guided by the guide and cleavage sequences, by another set of cleavage, strand exchange and rejoining reactions.
doi:10.1128/microbiolspec.MDNA3-0062-2014.f12

particularly notable, as they appear to have been very recently active. Indeed, bats are the only known mammalian source of active endogenous DNA transposons, being host to *Helitrons* (12) and active *piggyBac* elements (26). The impact of *Helitrons* on a genome can be profound as they can acquire and shuffle a wide variety of host sequences when element replication extends into flanking DNA past the end of the element.

A Target Site-Specific Eukaryotic Tyr-HUH Transposon: Adeno-Associated Virus

Adeno-associated virus is a single strand DNA virus. A critical step in its replication requires site specific nicking of a folded region of its termini by a Tyr-HUH endonuclease (see Chapter 37). Although AAV can replicate extrachromosomally, regional specific integration does occur in a region of human chromosome 19.

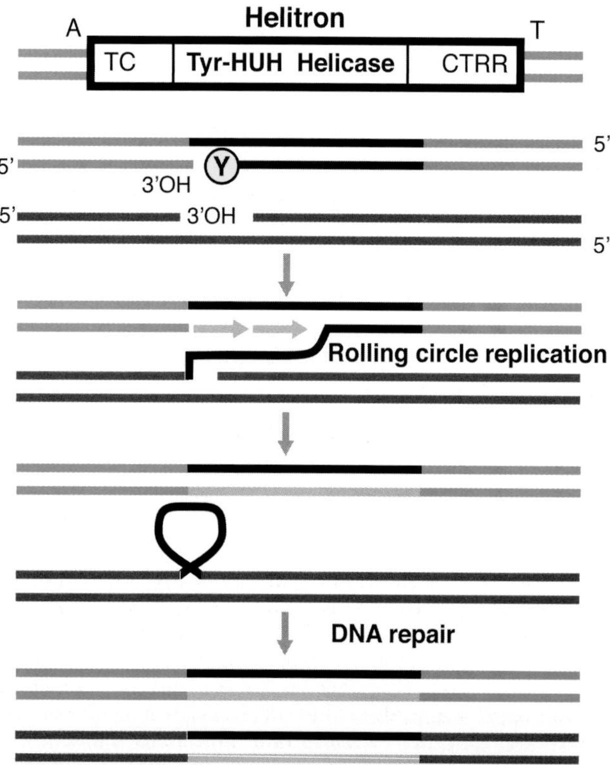

Figure 13 Mechanism of transposition of a *Helitron* by a a tyrosine (Tyr)-histidine-hydrophobic-histidine (HUH)/helicase transposase. The Tyr-HUH transposase acts at the upstream end of the transposon, releasing a 3′-OH donor end and a 5′-P-Tyr end. The broken transposon strand is displaced from the donor DNA by rolling circle replication (green) likely assisted by the helicase and it is covalently linked to the target DNA by attack of a target 3′-OH on the 5′-P-Tyr transposon end. DNA synthesis (green) copies the single transposon strand in the target to generate intact duplex DNA.
doi:10.1128/microbiolspec.MDNA3-0062-2014.f13

Tyrsosine- and Serine-transposases: From Conservative to Libertarian

As with Tyr-HUH transposases, unrelated Tyr- and Ser-transposases also break DNA using a protein nucleophile to release an OH DNA end and a covalent protein-DNA link. Tyr_{TRANSP} use a conserved tyrosine as the nucleophile, yielding DNA 3′-P-Tyrosine and 5′-OH DNA products. Ser_{TRANSP} use a conserved Serine as the nucleophile, yielding DNA 3′-OH and 5′-P-Serine products. DNA rejoining occurs by attack of the DNA-OH on the protein-DNA link. Each transposase class has distinct DNA binding and catalytic domains, but these are not structurally related, thus they represent independent strategies for DNA breakage and joining.

These transposases can promote strand exchange between two recombination sites on two parental DNA duplexes, each site containing two specific transposase binding sites in inverted orientation flanking a short (2 to 8 bp) region of homology. Recombination occurs by DNA cleavage at the outside edges of the regions of homology, also called the crossover region, followed by strand exchange between the duplexes and DNA rejoining (Fig. 14). In these reactions, recombination is reciprocal, that is, no DNA is lost or synthesized, and conservative, that is, no high-energy cofactor is required to rejoin broken DNA because the phosphotyrosine and phosphoserine intermediates preserve the high energy of the phosphodiester bond. This type of recombination is called conservative site-specific recombination (CSSR) and the Tyr- and Ser-transposases that execute such homology-dependent exchange are called Tyr_{CSSR} and Ser_{CSSR} recombinases.

Conservative Site-Specific Recombination

In CSSR recombination sites, the inverted CSSR recombinase binding sites that flank the homology region position the recombinase to cleave at the edges of the homology such that each DNA duplex can bind two recombinase protomers. Recombinase dimers bound to each parental duplex interact with the bound dimers on the other duplex, thereby juxtaposing the substrate DNAs on a recombinase tetramer. The tetramers of Ser_{CSSR} recombinases break and exchange all four DNA strands simultaneously by two pairs of double strand breaks. By contrast, Tyr_{CSSR} recombinases first break, exchange, and rejoin one pair of strands from each duplex and then subsequently break, exchange, and rejoin the other pair of strands. Because the positions of strand exchange lie at the outer edges of the crossover regions, the DNA products are heteroduplex, that is, there is one strand from each parent in the recombination homology regions (Fig. 14). Note also

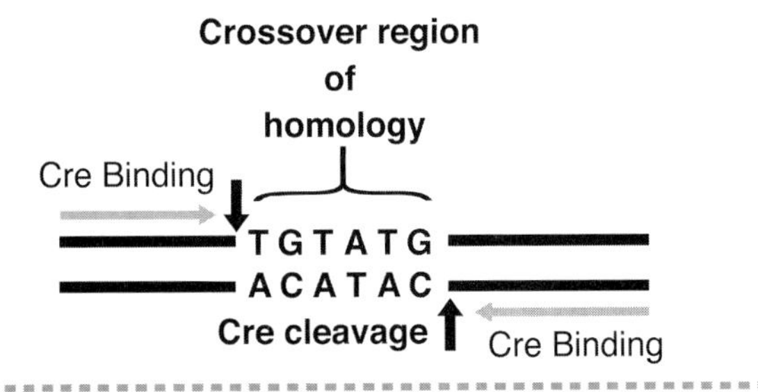

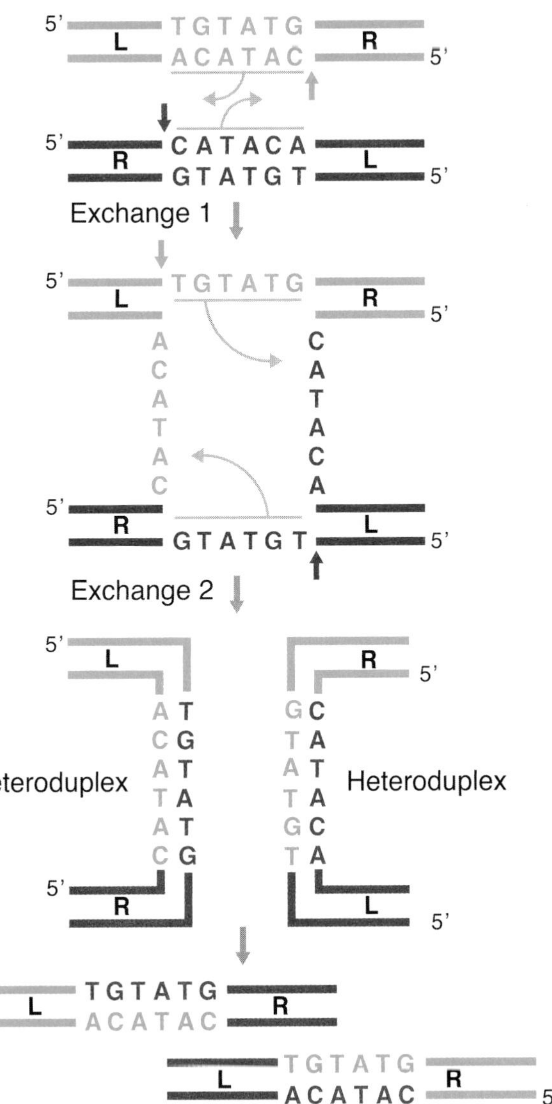

Figure 14 Mechanism of DNA breakage, strand exchange and joining during conservative site-specific recombination by the Tyrosine$_{CSSR}$ recombinase Cre. Cre is a Tyr$_{CSSR}$ recombinase that acts at *lox* recombination sites consisting of two inverted Cre binding sites flanking a conserved central crossover region. Strand breakage, strand exchange and rejoining by Cre occur at the edges of the crossover region. Cre dimers bind to each *lox* site and pair to form the active Cre tetramer that pairs the *lox* sites in antiparallel alignment. Recombination begins by cleavage of the two *lox* sites on strands of the same polarity, making 3′-P-Tyr and 5′-OH ends. Strand exchange and rejoining occurs by the attack of each 5′-OH on the 3′-P-Tyr of the other strand. The second round of strand exchange occurs by Cre cleavage of the other pair of strands, making 3′-P-Tyr and 5′-OH ends. Strand exchange and rejoining occurs by attack of each 5′-OH on the 3′-P-Tyr of the partner *lox* site. Note that because of staggered positions of strand exchange, the recombinant duplexes are heteroduplex in the crossover region, which is bounded by the sites of strand exchange. doi:10.1128/microbiolspec.MDNA3-0062-2014.f14

that the crossover region contains no internal repeats and is thus directional.

Some CSSR sites are far more elaborate than just a pair of inverted CSSR recombinase binding sites flanking a crossover region. There may be multiple additional recombinase binding sites flanking the binding sites at the crossover region that execute strand exchange, as well as binding sites for accessory proteins, which are often architectural DNA bending proteins that facilitate assembly of the elaborate nucleoprotein complexes in which many of these reactions occur. As we will see, assembly of these complexes is a key control point of recombination.

Different products for recombination sites in different orientations

CSSR can mediate several types of DNA rearrangements, depending on the relative orientation of the recombination sites as defined by the direction of the sequence of the crossover region (Fig. 15). When the two recombination sites flank a DNA segment in direct orientation, recombination results in excision of the DNA segment. This reaction can also be called deletion or resolution, depending on the biological context. Recombination between inversely oriented sites results in inversion of the DNA segment between them. Recombination between sites on two different DNAs results in joining of the DNAs, that is, in integration.

Tyr$_{CSSR}$ recombination: DNA nicking, strand swapping and joining

Anti-parallel is the way they go. Although it is often visually convenient to align the 6 to 8 bp cross-

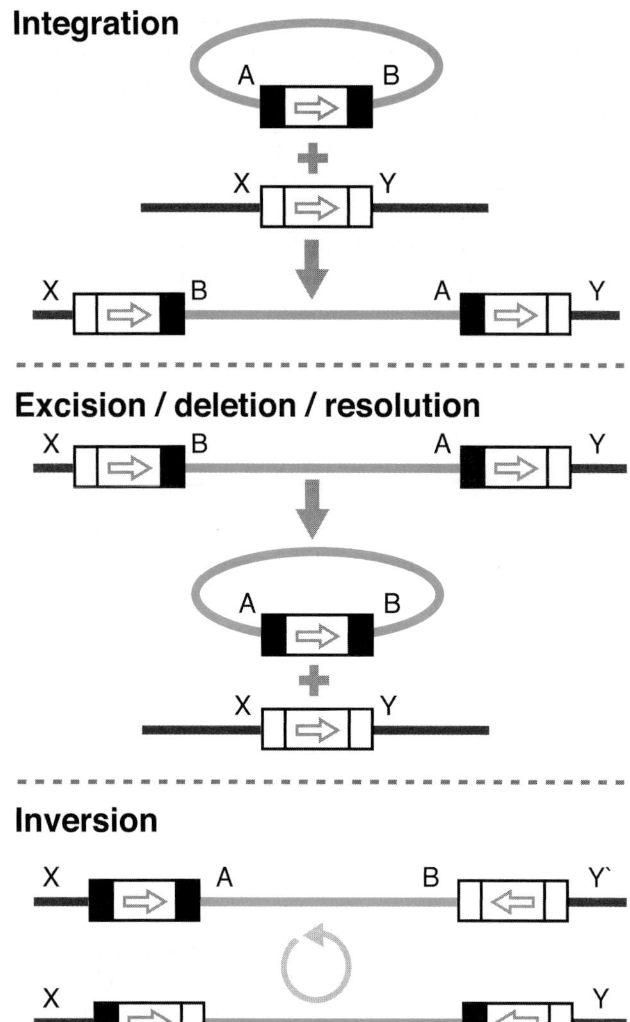

Figure 15 The relative orientation of the substrate recombination sites determines the structure of the recombination products. Sites for CSSR consist of two recombinase binding sites flanking a short crossover region of sequence homology sequence, which lacks repeats and is thus asymmetric. Although the local DNA strand breakage and joining reactions are the same in all cases, the overall structure of the recombination products is determined by the relative orientations of the substrate sites.
doi:10.1128/microbiolspec.MDNA3-0062-2014.f15

over regions of CSSR sites in parallel, strand exchange actually occurs between DNA molecules that are aligned in antiparallel fashion in an active tetramer of the recombinase (Fig. 16). DNA strands are exchanged between duplexes by displacement of short segments of the crossover region such that strand swapping occurs between the duplexes (27). If the sites are nonhomologous, the exchange reactions abort and the donor duplexes rejoin, leaving the duplexes in their parental configuration.

Going it alone: Cre and Flp can do it all. CSSR reactions promoted by Cre (see Chapter 5) and Flp (see Chapter 1) are the simplest of the CSSR reactions. Each recombination site on each parental duplex contains only the crossover homology region with its two flanking Tyr_{CSSR} recombinase binding sites. These systems can recombine sites in any orientation, making them powerful tools for genome engineering (28). Cre dimers bind to each parental duplex and synapse in antiparallel fashion to form the active tetramer.

Danger, dimers ahead: XerCD to the rescue. A hazard of having a circular genome is that if a circular chromosome dimerizes because of homologous recombination, then when segregation occurs one daughter does not receive a chromosome monomer. Thus, many bacteria encode a CSSR system that promotes chromosome monomerization in a reaction also called resolution. The chromosomally encoded Tyr_{CSSR} recombinases XerC and XerD collaborate to promote resolution at a specific chromosomal site called *dif* (see Chapter 7). This system is distinguished by its tight control of when and where recombination occurs by requiring the host protein FtsK, which acts at the septum, to pump DNA into daughter cells.

My way or no way: control, control, and more control. Some CSSR reactions are very tightly regulated, both in time, that is, only when particular proteins are synthesized, and in directionality, that is, once integration occurs excision cannot immediately occur without additional excision proteins because the proteins that mediate integration are not sufficient to mediate excision. The integration/excision cycle of bacteriophage *lambda* is elaborately regulated (see Chapter 4). Integration, that is, recombination between *attachment site Phage* (*attP)* on the phage chromosome and *attachment site Bacterial* (*attB)* on the bacterial chromosome, generates the hybrid sites *attL* and *attR* that flank the newly inserted element. Excision, that is, recombination between the *att*L and *att*R sequences, regenerates *attP* and *attB*. Integration and excision require both the phage-encoded Tyr_{CSSR} integrase and host-encoded integration host factor, a sequence-specific DNA bending protein. By contrast, excision also requires the phage-encoded protein excisionase (Xis), which

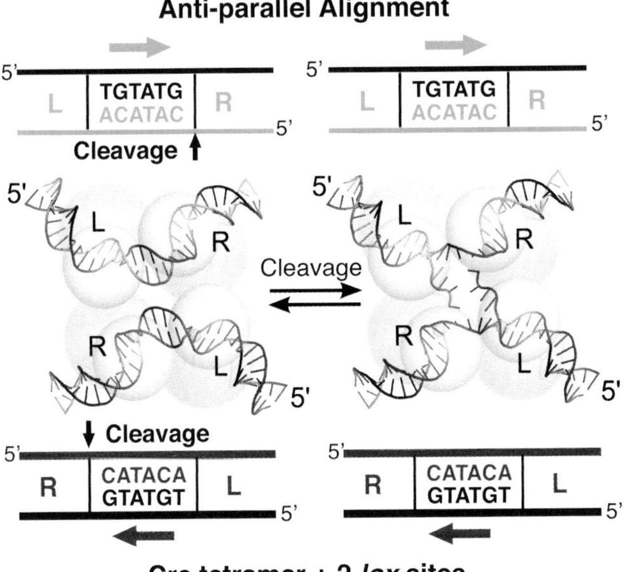

Figure 16 The *lox* sites of Tyrosine$_{CSSR}$ recombinase Cre align in antiparallel orientation in the active tetramer. Dimers of Cre bind to each parental *lox* site and pair to form the active tetramer in which the *lox* sites are aligned in antiparallel fashion. Thus, once cleavage has occurred, strand exchange can occur by local melting and swapping of short, closely juxtaposed DNA segments. Structure graciously provided by Greg Van Duyne.
doi:10.1128/microbiolspec.MDNA3-0062-2014.f16

is expressed only under excision conditions, and the host-encoded FIS protein.

An additional control element is that Int also has two different DNA binding activities (Fig. 17). One binding activity is encoded in the C-terminal region of Int along with the active site catalytic region and it specifically recognizes the sequences of Int binding sites that immediately flank the crossover region. The other Int binding activity is encoded in the Int N-terminal domain and it recognizes binding sites of a different sequence specificity in the arms of *attP* far outside the region of strand exchange (Fig. 17). IHF, Xis, and FIS are sequence-specific DNA binding proteins that bind to the arms of *attP* and promote the DNA bending reactions necessary to form active recombination complexes.

As with Cre and Flp, as well as other Tyr$_{CSSR}$ systems, the active form of *lambda* integrase is a tetramer of Int, each protomer of which acts as one of four sites of strand breakage and joining at the crossover regions in *attP* and *attB*. However, the affinity between the Int C-terminal domains is much lower than for Cre and Flp. Thus, the active Int tetramer is only assembled when Int binds to both its crossover and arm sites and when IHF binds to facilitate the DNA bending necessary for formation of the "integrative intasome." The active Int tetramer likely assembles on *attP* and then captures the much less complex *attB* site, followed by strand exchange.

Once integration occurs generating the new hybrid *attL* and *attR* sites, the configurations of the Int and IHF protein-DNA binding sites in the arms of *attL* and *attR* are very different from those required to assemble the "integrative intasome." Thus, a different "excisive intasome" must be formed to assemble an active Int tetramer at *attL* and *attR* (Fig. 17). Assembly of this "excisive intasome" additionally requires the DNA bending proteins Xis and FIS.

Therefore, although Int and IHF alone can assemble the "integrative intasome" with *attP* and *attB*, Int and IHF alone cannot assemble the "excisive intasome" with *attL* and *attR* so that integration cannot be reversed until Xis and FIS are expressed, which happens only when a particular excision developmental program occurs.

Such directionality control of recombination reactions by changes in the arrangements of flanking accessory sites is a common strategy, especially in bacteriophage integration and excision reactions, which need to be unidirectional. The fundamental difference between these regulated reactions with *lambda* Int and their flanking accessory sites that are necessary to promote formation of the catalytic recombinase tetramer, and the more permissive Cre and Flp recombinases, is that protein–protein interactions between the protomers of Cre and of Flp are sufficiently high that they can bind to the sites of strand exchange and make a reactive tetramer in the absence of accessory proteins or binding sites.

Ser$_{CSSR}$ recombinases exchange broken DNAs by rotation on greasy protein swivels

As with Tyr$_{CSSR}$ recombinases, Ser$_{CSSR}$ recombinases act as tetramers that bind specifically to two sites in inverted orientation that flank a region of homology on each parental duplex, exchange DNA strands between the duplex, followed by rejoining (see Chapter 3). However, in contrast to Tyr$_{CSSR}$ recombinases, which use two consecutive cycles of single strand exchange between two duplexes, Ser$_{CSSR}$ recombinases make two concerted double strand breaks, one on each parental duplex. Once the DNA strands are broken, rotation of two of the recombinase subunits with their bound DNAs occurs which positions the broken ends of one duplex adjacent to the broken ends of the other duplex, thereby promoting strand exchange. Joining of the juxtaposed ends then occurs. This subunit rotation can occur because there is an interface between

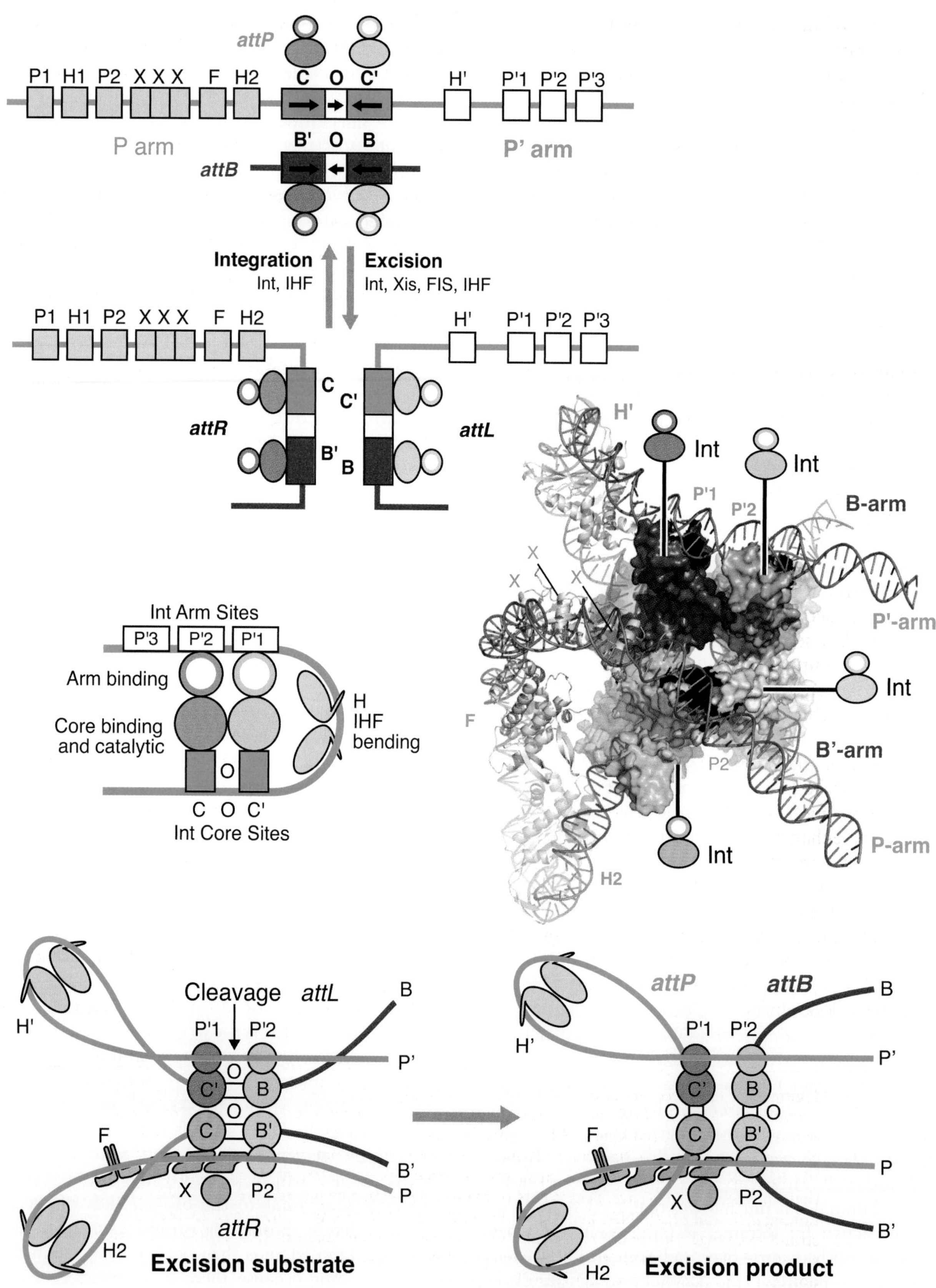
attP
P1 H1 P2 X X X F H2
C O C'
H' P'1 P'2 P'3
P arm
P' arm
B' O B
attB
Integration
Int, IHF
Excision
Int, Xis, FIS, IHF
P1 H1 P2 X X X F H2
H' P'1 P'2 P'3
C
C'
attR
B'
B
attL
H'
Int
Int
P'1
P'2
B-arm
X
X
X
P'-arm
Int Arm Sites
P'3 P'2 P'1
Arm binding
Core binding
and catalytic
H
IHF
bending
O
C O C'
Int Core Sites
Int
F
B'-arm
P2
Int
P-arm
H2
Cleavage
attL
B
H'
P'1
P'2
P'
O
C'
B
O
C
B'
F
B'
X
P2
P
attR
H2
Excision substrate
attP
attB
B
H'
P'1
P'2
P'
C'
B
O
O
C
B'
F
P
X
P2
B'
H2
Excision product

the two sets of dimers that is flat and hydrophobic allowing rotation of the dimers. The flat surface of the interface within the dimers has been visualized by X-ray crystallography.

The Ser_{CSSR} recombinases fall into two categories. Both types contain related catalytic domains with the conserved serine but differ in their DNA binding domains and the complexity of their substrates. The small Ser_{CSSR} recombinases, which include the resolvases and invertases described below, have simple H-T-H DNA binding domains, whereas the large Ser_{CSSR} recombinases, which include phage integrases, have a much more elaborate DNA binding domain.

A small Ser_{CSSR} recombinase mediates inversion. The inversion of a promoter-bearing DNA segment in *Salmonella* directs alternative expression of two surface antigens (see Chapter 9). Inversion is carried out by the Ser_{CSSR} recombinase Hin, which mediates DNA breakage, strand exchange by subunit rotation, and rejoining at *hix* sites that lie in inverted orientation that bound the invertible segment (Fig. 18). Essential to inversion are two host proteins that bind and bend DNA, factor for inversion stimulation (FIS), a sequence-specific DNA binding protein, and HU, a sequence nonspecific DNA bending protein. Also important to inversion is a recombination enhancer sequence usually found between the *hix* sites, which contains multiple binding sites for FIS.

Activation of the DNA breakage and joining reactions that underlie Hin inversion requires the assembly of an Invertasome (Fig. 18). In the Invertasome, FIS bound to the enhancer provides an assembly platform for the two dimers of Hin bound on each *hix* site to form the Hin active tetramer. The conserved active site serines of Hin introduce DNA double strand breaks at each *hix* site and strand exchange occurs by the 180° rotation of one dimer with respect to the other. The broken DNAs rejoin by the attack of the DNA 3′-OH ends on the 5′-P-Ser links.

Notably, FIS-independent Hin mutants have been isolated that can assemble the active Hin tetramer without the aid of FIS and the Enhancer. These mutants, now without directionality control, can recombine *hix* sites in any orientation, as do Cre and Flp with their cognate recombination sites.

A small Ser_{CSSR} recombinase promotes resolution of plasmid dimers to monomers. As described above, a DDE *Tn3* transposase reaction generates *Tn3* cointegrates, which contain two transposon copies linking the donor and target plasmids. The cointegrate can undergo further recombination to separate donor and target plasmids which now each contain a copy of the transposon (Fig. 10). Ser_{CSSR} recombinases called resolvases promote CSSR between directly repeated copies of *res* recombination sites on a cointegrate plasmid to generate plasmid monomers.

In each parental *res* site, resolvase binding sites in inverted orientation flank the region of homology where strand exchange will occur. Resolvase dimers bind to each *res* site and the dimers pair to form the active tetamer that juxtaposes the parental *res* sites. As with the Ser_{cssr} invertases, strand exchange occurs by subunit rotation after the *res* sites are broken.

Res sites also contain additional resolvase binding sites that regulate recombination so that resolution, not inversion, occurs (see Chapter 10). These additional resolvase binding sites assemble with the crossover site dimers to form a synaptosome that contains the active tetramer, facilitated by the interwrapping of DNAs at the other resolvase binding sites. Again, nucleoprotein complex assembly activates the resolvase and assures that only resolution occurs because the activating synaptic structure cannot be formed with *res* sites in inverted orientation.

Plasmid multimers can also result from several other reactions, for example, some plasmid multimers arise from homologous recombination between monomers. Instead of using the chromosomal encoded XerCD

Figure 17 The integration/ excision cycle of bacteriophage *lambda* is elaborately regulated. *Lambda attachment* (*att*) sites contain two Int core binding sites, COC′ and BOB′, in inverted orientation flanking a 7 bp crossover region of homology, "O," where strand exchange occurs. COC′ is flanked by *P* and *P′* arms containing multiple protein binding sites: P1, P2, P′1, P′2, and P′3 sites = Int arm sites, which have a different recognition sequence than Int core C and B sites. Schematics of the excisive intasome paired substrate DNAs containing *attL* and *attR* DNAs and the product *attP* and *attB* DNAs following DNA breakage, strand exchange and rejoining at the "O" regions between the core Int binding sites are shown. H, IHF binding sites; X, Xis binding; F, FIS binding.
doi:10.1128/microbiolspec.MDNA3-0062-2014.f17

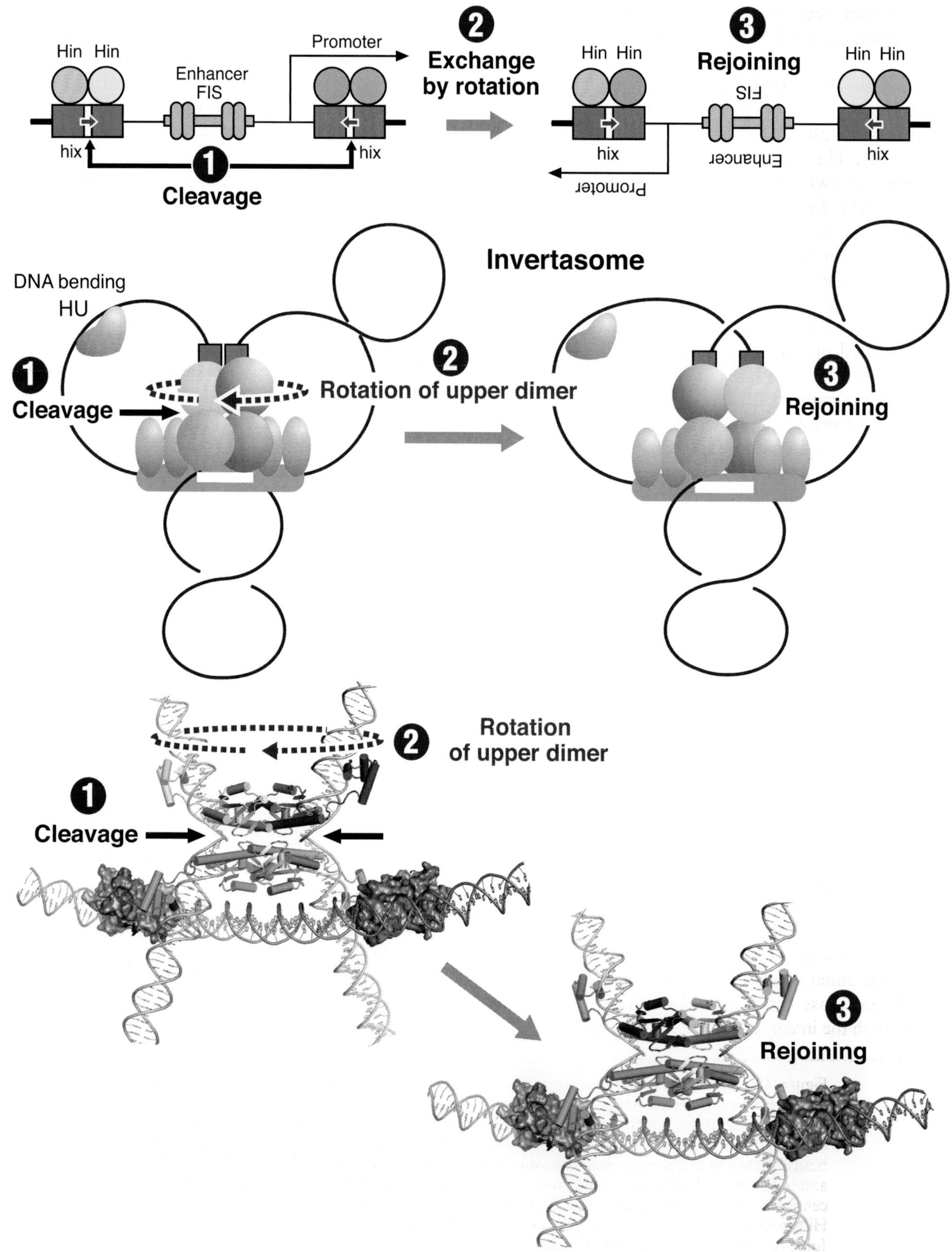

Hin Hin
Enhancer
FIS
Promoter
hix
hix
1
Cleavage
2
Exchange
by rotation
Hin Hin
3
Rejoining
FIS
Hin Hin
hix
Enhancer
hix
Promoter
Invertasome
DNA bending
HU
1
Cleavage
2
Rotation of upper dimer
3
Rejoining
2
Rotation
of upper dimer
1
Cleavage
3
Rejoining

Tyr_{CSSR} system (see Chapter 7), some plasmids encode their own Ser_{CSSR} resolvase and *res* sites.

Large Ser_{CSSR} recombinases mediate phage integration and excision. The other class of Ser_{CSSR} recombinases is called large serine recombinases (see Chapter 11). The Large Ser_{CSSR} recombinases were discovered in two phage integration and excision systems, *ϕC31* from *Streptomyces* and *Bxb1* from *Mycobacteria*. In both systems the integration reaction, *attP* x *attB*, can occur with only the large Ser_{CSSR} integrase in the absence of other phage or host proteins and is highly directional. Excisive recombination, *attL* × *attR*, requires an additional phage-encoded protein, recombination directionality factor (RDF) (Fig. 19). These large Ser_{CSSR} integrases have the same catalytic domain as the small Ser_{CSSR} invertases and resolvases but have much more elaborate DNA binding domains (29, 30). These DNA binding domains, RZ and ZD, are contained in a several hundred amino acid domain at their C-terminal ends. These DNA binding domains bind to their cognate binding sites, *RZ* and *ZD*, which flank the *att* crossover regions in inverted orientation. Notably, the spacing between the RZ and ZD sites is different in *attP* and *attB* such that the conformations of Int bound to these sites, and hence to *attL* and *attR*, are distinct (see below). Each *att* site binds a dimer of integrase.

The integrase C-terminal domain also contains a protein-protein interaction domain, CC, which can interact with the CC domain from another Integrase. Notably, however, although the integrase can bind to all four *att* sites - *attP*, *attB*, *attL*, and *attR* - the conformation of the CC domain is different when bound to the different *att* sites because of spacing differences between multiple DNA binding motifs in *attP* and *attB*. Thus, *attP* CC interacts only with *attB* CC and *attL* CC interacts only with *attR* CC. Therefore, the integrase dimer bound to the *attP* site can pair only with the Integrase dimer bound to the *attB* site (Fig. 19). Similarly, the integrase dimer bound to the *attL* site can pair only with the integrase dimer bound to the *attR* site.

This defined synapsis pathway controls integration and excision. As with small Ser_{CSSR} invertases and resolvases, the active form of integrase is a tetramer assembled from two dimers and breakage and joining occurs only when dimers synapse to form a tetramer. Once the tetramer is assembled, double strand breaks occur at each *att* site, strands are exchanged between the parental duplexes by rotation of integrase dimers as with the resolvases and invertases, and then the exchanged ends are rejoined.

In the presence of the phage-encoded RDF that is expressed only as part of the developmental program for excision, synapsis between integrase dimers bound to *attL* and integrase dimers bound to *attR* can occur, leading to *attL* × *attR* recombination, that is, excision. The mechanism by which RDF acts remains to be determined in molecular detail but an attractive view is that RDF interacts with the Integrase to change the conformation of the CC domain to allow Int-Int interactions when bound to *attL* and *attR*. This RDF strategy of changing integrase conformation is distinct from that in Tyr_{CSSR} systems such as bacteriophage *lambda* in which the proteins uniquely required for excision, Xis and FIS, are both DNA binding and bending proteins.

Tyr- and Ser-Transposases Mediate Transposition: DNA Breaking, Strand Exchange, and Joining in the Absence of Homology at the Region of Strand Exchange

The ICE-element cometh (and goeth)

Integrative conjugative elements (ICEs) are bacterial transposable elements that can excise from the chromosome in a donor cell, transfer between cells by conjugation, and then integrate into the chromosome in the recipient cell (see Chapters 8 and 13). ICEs encode conjugation functions but do not encode their own replication functions. In addition to encoding antibiotic resistance genes and other accessory determinants, ICEs also encode proteins that mediate their integration and excision. Different ICEs vary in their target site selectivity, which varies from integration into only a few sites to integration into many sites. ICE systems use

Figure 18 Inversion of a promoter-containing DNA segment by the Ser_{CSSR} recombinase Hin within an invertasome controls gene expression. Inversion of a DNA segment containing a promoter changes which surface antigen gene is expressed. *hix* recombination sites bound the invertible segment in inverted orientation. Hin dimers bind to each *hix* site and FIS binds to the enhancer segment. The Hin dimers interact with FIS on the Enhancer platform and with each other to form the active Hin tetramer. Cleavage of both *hix* sites occurs in the central region of homology and strands are exchanged by rotation of the upper dimer of Hin, followed by DNA rejoining. Drawing and structures adapted from material from Reid Johnson. doi:10.1128/microbiolspec.MDNA3-0062-2014.f18

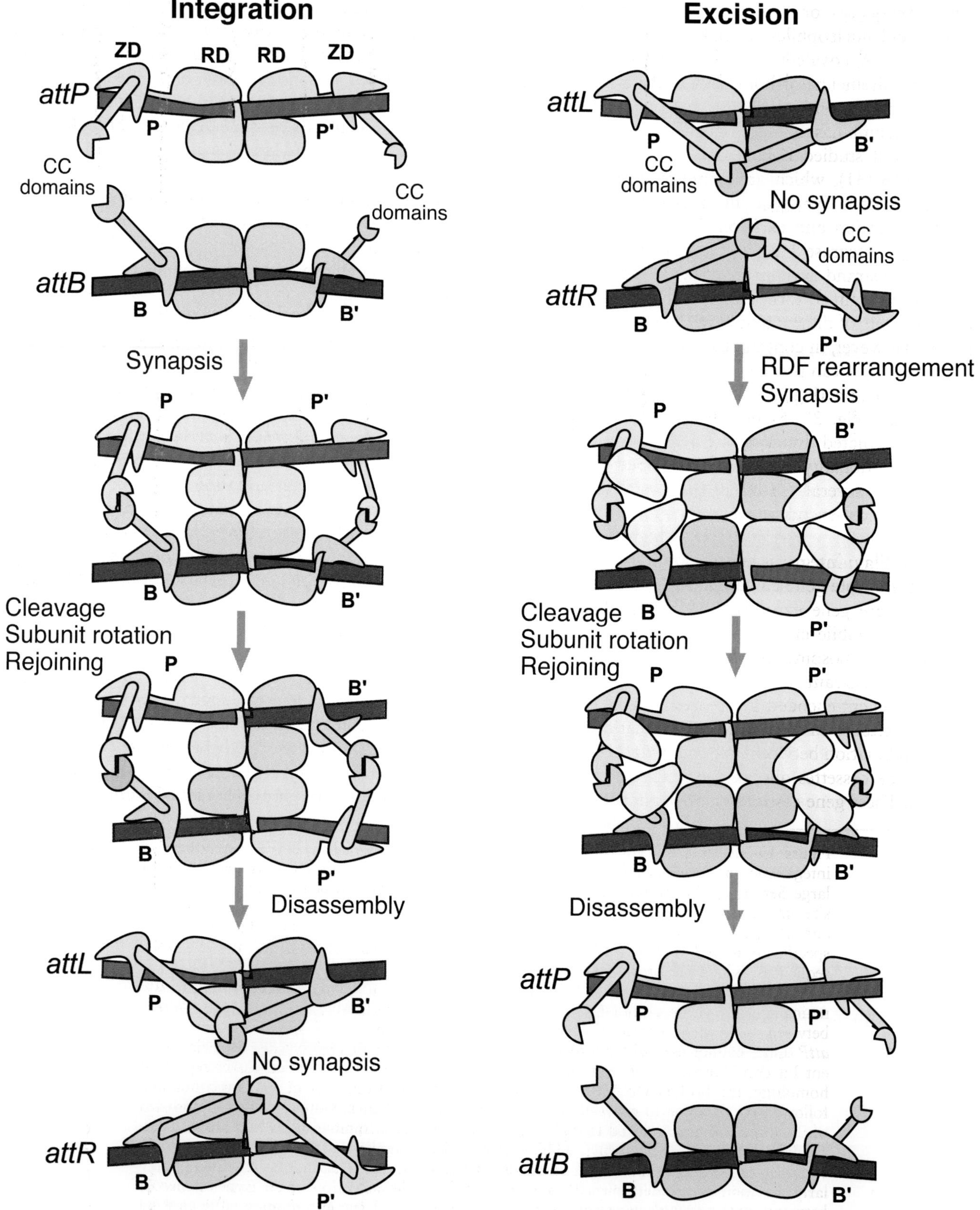

Integration
ZD
RD
RD
ZD
attP
P
P'
CC domains
CC domains
attB
B
B'
Synapsis
P
P'
B
B'
Cleavage
Subunit rotation
Rejoining
P
B'
B
P'
Disassembly
attL
P
B'
No synapsis
attR
B
P'
Excision
attL
P
B'
CC domains
No synapsis
CC domains
attR
B
P'
RDF rearrangement
Synapsis
P
B'
Cleavage
Subunit rotation
Rejoining
B
P'
P
P'
B
B'
Disassembly
attP
P
P'
attB
B
B'

either Tyr_{TRANSP} or Ser_{TRANSP}-transposases that use amino acid nucleophiles to make reversible Tyr-DNA and Ser-DNA covalent intermediates but have much looser requirements for homology at the crossover sites between the recombining sites than do Tyr_{CSSR} and Ser_{CSSR} recombinases.

Two well studied ICEs, *CTnDOT* (see Chapter 8) and *Tn916* (31), which are found in a wide range of bacteria, encode Tyr_{TRANSP} integrases that mediate element integration and excision. In *CTnDOT*, multiple Tyr_{TRANSP} integrase binding sites flank the crossover sites where strand exchange will occur. The bound proteins are thought to form an active tetramer assembled from two sets of dimers bound to the parental substrate DNAs. However, in contrast to Tyr_{CSSR}, recombination with Tyr_{TRANSP} can proceed with nonhomologies in the crossover region (Fig. 20; see Chapter 8, Fig. 2; and Chapter 25, Fig. 3). As occurs in integration/excision systems mediated by Tyr_{CSSR} and Ser_{CSSR} recombinases, directionality is tightly controlled as excision alsorequires several *CTnDOT*-encoded excision proteins and a host-encoded factor.

Plug and Play: integrons capture and express multiple gene cassettes

Integrons are gene expression platforms that may be present on mobile elements such as transposons or in bacterial chromosomes (see Chapter 6). They encode a promoter upstream of an *attI* recombination site that uses an element-encoded Tyr_{TRANSP} integrase to capture gene cassettes, which carry a recombination site *attC*, by recombination between *attI* and *attC* (Fig. 21). Excision of the cassettes allows their capture by other integrons. These gene cassettes can encode a wide variety of determinants ranging from antibiotic resistance to metabolic functions. After recombination, the gene cassettes lie downstream of the *attI* promoter and they are expressed.

Recombination between *attC* × *attI* does not occur by CSSR because there is no region of homology between these sites.

Phage exploitation of XerCD and *dif* sites

As described above, in the highly conserved Tyr_{CSSR} XerCD system (see Chapter 7), the XerCD recombinases act on chromosomal *dif* sites to convert hazardous chromosomal dimers generated by homologous recombination to monomers to facilitate chromosomal segregation. Many phages that do not encode their own recombinase hijack the XerCD system to promote their integration into bacterial chromosomes (32). For example, the integration of *CTXψ*, a filamentous phage, which encodes the diptheria toxin, into the *Vibrio cholerae* genome (33), converts nonpathogenic *V. cholerae* into a pathogen. Despite the fact that XerCD mediates CSSR between *dif* sites that have a region of homology, *CTXψ* integration does not occur by CSSR. As in *attC* containing cassettes in integrons (see Chapter 6), the *CTXψ*, genome is single strand DNA that is folded by several palindromes into a double strand form to be able to recombine with *dif*.

Tyr_{TRANSP} in eukaryotes: Cryptons and Tecs

Eukaryotic DNA elements, called *Cryptons* that contain Tyr_{TRANSP} integrases have been identified in fungi (see Chapter 53) (34, 35). *Tec* elements containing Tyr_{TRANSP} that undergo excision during macronuclear development in the ciliate *Euplotes* have also

Figure 19 A large Ser_{CSSR} integrase can mediate highly regulated cycles of bacteriophage integration and excision. The integration and excision cycle of phage *ϕC31* is mediated by a large $Serine_{CSSR}$ integrase. Integration between *attP* and *attB*, which generates the hybrid sites *attL* and *attR*, requires integrase. Excision between *attL* and *attR*, which generates *attP* and *attB*, requires integrase + recombination directionality factor (RDF). Integrase binds specifically to *attP* and *attB* using its *RZ* and *ZD* domains that recognize *RZ* and *ZD* DNA sequences, which are present on all *att* sites. Note the difference in *RZ* and *ZD* spacing in *attP* vs *attB* such that integrase binds to each in a slightly different conformation. CC domains interact with each other. A dimer of Integrase can bind to both *attP* and *attB*. Pairing between the Int dimers forms the active tetramer, which synapses *attP* and *attB*. However, an *attP* dimer cannot pair with another *attP* nor can *attB* pair with *attB* because of their different Int configurations. Once synapsis occurs, the integrase cleaves the crossover region of homology; the broken DNA ends exchange by rotation of one pair of Integrase subunits, followed by rejoining to generate the hybrid *attL* and *attR* sites. Note that the CC domains of the dimer integrase bound to *attL* and to *attR* interact intramolecularly and are thus unavailable for interdimer pairing. RDF is proposed to interact with integrase to change its conformation when bound to *attL* and *attR*, such that CC domains can interact intermolecularly and pair the dimers on *attL* and *attR* to form the active tetramer. Excision occurs between *attL* and *attR* by cleavage of *attL* and *attR* and subunit rotation, followed by rejoining to generate *attP* and *attB*. doi:10.1128/microbiolspec.MDNA3-0062-2014.f19

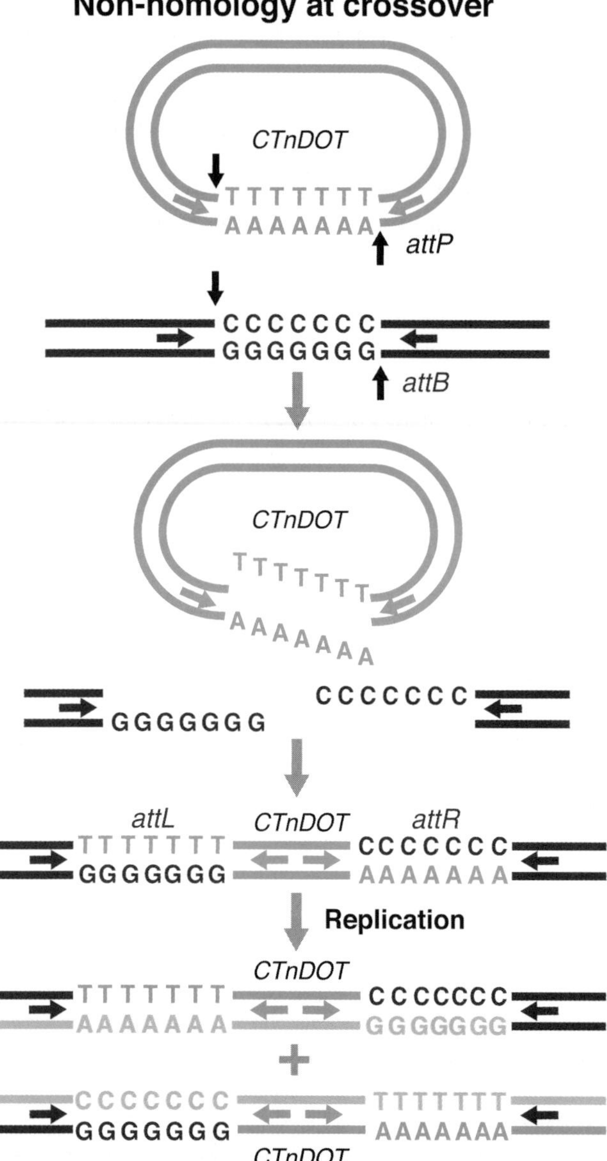

Figure 20 A Tyr_{TRANSP} integrase mediates integration of the ICE *CTnDOT* in the absence of crossover homology between the recombination sites. A hallmark of CSSR crossover regions, which are flanked by inverted recombinase binding sites that promote breakage, exchange and joining at the outer edges of the crossover region, is that the crossover regions are identical. Some Ser and Tyr recombinases can, however, promote recombination between nonhomologous crossover regions. The crossover regions of *attP* and *attB* of *CTnDOT* are nonhomologous such that when strand exchange occurs, the heteroduplex regions contain base pair mismatches as shown in this extreme example. Replication of the recombination product yields two daughter chromosomes with different sequences in *attL* and *attR*. Recombinases that do not require absolute homology can be considered transposases. doi:10.1128/microbiolspec.MDNA3-0062-2014.f20

been identified bioinformatically (36,37). Tyr_{TRANSP} have also been found in eukaryotic retroelements (see below; see Chapter 53).

RETROTRANSPOSONS: DNA→RNA→DNA

In contrast to the mobile elements described above that have DNA substrates, intermediates, and products, mobile elements called retrotransposons move via an RNA intermediate. The movement of a retrotransposon from a donor site to a new insertion site begins with the synthesis of an RNA copy of the element by the host pol II polymerase. This RNA copy is then converted into DNA by an element-encoded RNA-dependent DNA polymerase, reverse transcriptase (see Chapter 46). Notably, in contrast to DNA elements that excise from their donor sites, the retrotransposon donor site is unchanged, and thus, can generate more RNA copies, which are converted into DNA. This replicative capacity likely contributes to the high copy number of retrotransposons in some organisms.

Different types of retrotransposons use different mechanisms to convert the RNA copy of the element into a DNA copy at a new insertion site. The DNA form of retroviruses and retroviral-like retrotransposons is generated by reverse transcription in the cytoplasm far from the nuclear DNA where it is integrated by a DDE transposase, in this case called a retroviral integrase. The DNA cleavage reactions, which for some retrotransposon elements include several nucleotides from their 3′ ends to expose their reactive 3′-OH termini, and the staggered attacks of these 3′OH ends on both strands of the target DNA, occur by the same mechanism as with DDE DNA-only transposons (see below).

By contrast, other retrotransposons, including non-LTR retrotransposons and mobile group II introns, use a quite different strategy in which their RNA copies interact directly with their new DNA insertion site prior to reverse transcription. With these elements, a target DNA 3′-OH is used as the primer for reverse transcription of the template element RNA that generates the DNA form of the element *in situ* in a reaction called target primed reverse transcription (TPRT).

Retroviruses and Retroviral-like Retrotransposons

When integrated into the genome, the central gene-encoding region of retroviruses (see Chapter 48) and retroviral-like retrotranposons is flanked by directly repeated long (hundreds of bp) terminal repeats (LTRs), and thus, these elements are called LTR retrotransposons

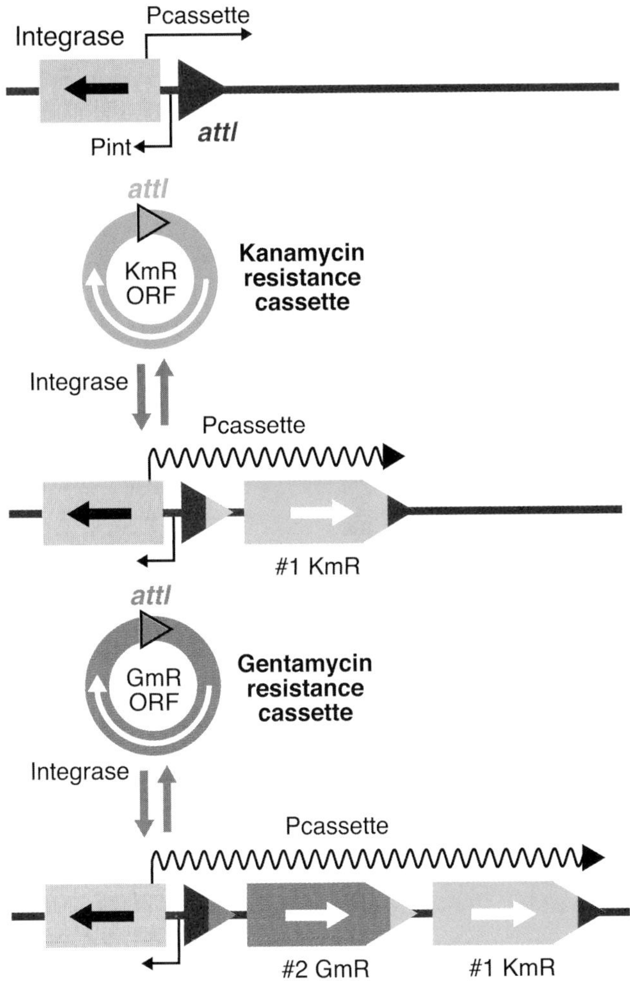

Figure 21 A Tyr_{TRANSP} integrase mediates capture and expression of multiple different gene cassettes. Integrons are gene-expression platforms containing an attachment site (*attI*) and cognate integrase that are found on other mobile DNAs such as transposons and in bacterial chromosomes. The integron can capture, integrate, and express gene cassettes that contain related but often not identical *att* sites, *attC*, by site-specific recombination. Once integration occurs, the $P_{cassette}$ promoter drives expression of the promoter-less cassettes. Some chromosomal integrons contain hundreds of gene cassettes.
doi:10.1128/microbiolspec.MDNA3-0062-2014.f21

(Fig. 22). LTR retrotransposons are widespread in fungi, plants, and animals but retroviruses are found mostly in vertebrates. Although HIV and HTLV are the only known active human retroviruses, many copies of autonomous and nonautonomous retroviral elements are present in the human genome, reflecting past retroviral infections (Fig. 23). These endogenous retroviruses are called ERVs and are thought to be inactive for retrotransposition but provide important regulatory elements to the cell (see Chapter 47).

Transcription of the integrated virus, also called the provirus, by the host polymerase begins within the upstream LTR and terminates in the downstream LTR. The central region of all LTR elements encodes the polyprotein Gag, which includes the RNA binding proteins capsid and nucleocapsid, and Pol (Fig. 22). The Pol domain is typically expressed at lower levels than Gag as a Gag-Pol fusion precursor polyprotein. It can include Protease and does contain reverse transcriptase-RNase H (see Chapter 46) and integrase domains (see Chapters 44 and 45). Integrases of retrovirus and LTR retrotransposons are closely related to the DDE class of DNA transposases.

Retroviruses, which are distinct from retrotransposons, bud from host cells and infect new cells. This process is enabled by an Env, a membrane protein that mediates cellular exit and entry. Following expression of Gag and Gag-Pol, these proteins condense together with some host proteins around two copies of genomic retroviral RNA to form virus-like particles (VLPs). In the case of retroviruses, VLPs can form intracellularly or at the plasma membrane, but in any case, bud from the cell and in the process acquire a membrane containing Env. As budding occurs activation of protease results in processing of the precursor polyproteins into their mature forms. Maturation occurs intracellularly in the case of retrotransposons.

In multiple steps involving both copies of the viral RNA, the element-encoded reverse transcriptase makes a double-strand DNA copy, sometimes called a cDNA, containing the LTRs in the cytoplasm in a large complex containing viral and host proteins. In the case of fungi where LTR retrotransposition has been most extensively studied, the nuclear envelope does not break down and VLPs are likely to be significantly remodeled as the DNA is translocated together with integrase and other proteins into the nucleus.

Integrase then inserts the viral DNA into its new target site. Using the same steps as DNA cut and paste DNA-only transposons, integrase mediates attack by the viral DNA 3′-OH ends on staggered positions on the top and bottom strands of the target DNA (Fig. 24). Therefore, integrated retroviral elements are flanked by target site duplications and LTR elements from yeast to humans follow this same fundamental pathway. Retroviruses can also encode auxiliary determinants, for example, several proteins encoded by HIV limit the efficacy of host restriction factors.

The retroviral family is large and diverse (see Chapter 48) and the study of many different elements has contributed to our current understanding. Not surprisingly, however, much recent work has focused on

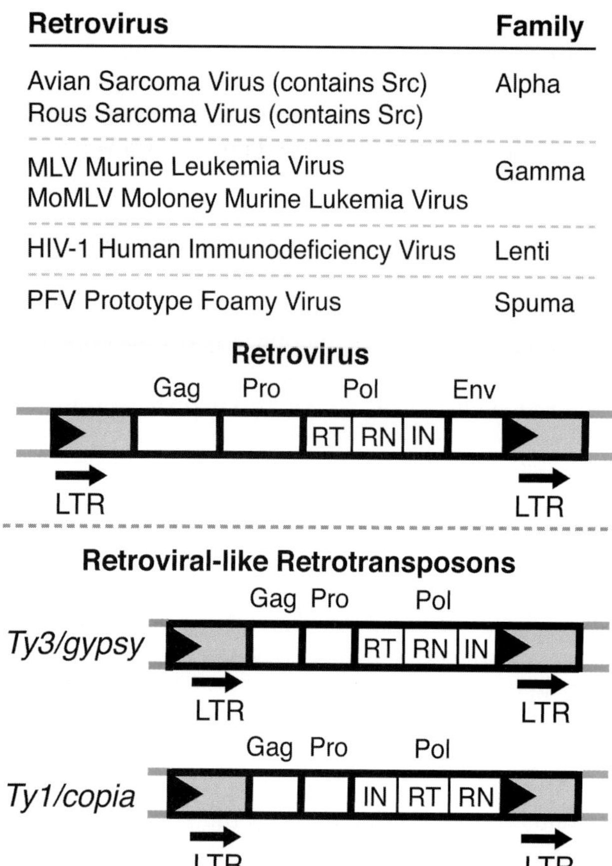

Figure 22 Structures of some well-studied retroviruses and retroviral-like retrotransposons. Retroviruses and the closely related retroviral-like retrotransposons contain related central protein-coding regions flanked by direct long terminal repeats. The central regions of both retroviruses and retroviral-like transposons encode Gag, which includes several nucleic acid domains, a protease that cleaves polyprotein precursors, and Pol, which encodes reverse transcriptase, RNase-H and integrase. Retroviruses also encode Env, a membrane protein that facilitates viral particle exit from host cells and entry into new host cells. In different families of retroviruses, different combinations of protein domains are expressed as fusion polyproteins. Some retroviruses also encode other genes, for example, the avian transforming viruses, ASV and RSV, encode the oncogene Src. Reverse transcriptase and RNase-H convert the two viral RNA copies in the viral particle into the DNA form of the virus, which terminates in direct long terminal repeats. Integrase, which is a DDE transposase, integrates the viral DNA into the host genome. In retroviral-like elements of the yeast *Ty3* and *Drosophila gypsy* family, the order of Gag, Pro and RT, RN and in Pol are the same order as in retroviruses, whereas in the yeast *Ty1* and *Drosophila copia* family, Integrase proceeds RT-RNase H.
doi:10.1128/microbiolspec.MDNA3-0062-2014.f22

HIV-1, contributing greatly to human AIDS treatment including development of drugs that inhibit integration, replication, maturation, and cell fusion.

Much recent work on LTR elements has focused on dissecting the roles of host factors and on understanding target site selection, which has a major impact on the effect of insertions on host cells. Early-on genome-wide screens for budding yeast *Ty1* of the *Ty1/Copia* class (see Chapter 41), *Ty3* of the *Ty3/Gypsy* class (see Chapter 42), and fission yeast *Tf1* of the *Ty3/Gypsy* class (see Chapter 43) elements identified multiple host cofactors and restriction factors for retrotransposition. More recently, siRNA and dominant-negative screens have identified numerous host factors for retroviruses, particularly for HIV-1 (see Chapter 45). Perhaps not surprisingly, host factors for both LTR retrotransposons and retroviruses include RNA helicases, translation factors, replication factors, and nuclear porins.

Both retrotransposons and retroviruses have distinct targeting biases that influence where they insert into the host genome. These preferences are now explained, at least in part, by interactions with chromatin proteins including host transcription factors and proteins that mediate interactions with histone modifications.

HIV-1 inserts preferentially into transcribed regions, perhaps to facilitate transcription of the virus. An intriguing possibility is that one factor in this preference is a preferential association between HIV-1 and components of the nuclear pore complex that are, in turn, preferentially associated with active chromatin. Integrase interacts directly with the host factor LEDGF, which also interacts with histones associated with active genes (see Chapters 44 and 45).

The *S. cerevisiae Ty1, Ty3, and Ty5* elements and *S. pombe Tf1* element (see Chapter 43) display strong target site specificity (Fig. 25). *Ty1* (see Chapter 41) and *Ty3* (see Chapter 42) elements target tRNA genes. *Ty3* is the most selective, inserting site-specifically at *pol III* initiation sites by direct interaction with a *pol III* transcription factor. *Ty1* inserts preferentially within about 1 kb upstream of *pol III* transcription initiation sites, being guided by direct interaction with polymerase with a bias toward specific nucleosome surfaces. This preferential targeting to small genes discourages potentially harmful insertions at other sites. *Ty5* inserts preferentially into heterochromatic regions such as telomers and the silent *HMLα* and *HMRa* storage cassettes. *Tf1* insertions are concentrated in *pol II* promoters and correlate with Sap1 binding sites with preference for stress response genes. This preference could be important for increased genetic diversity in response to stress.

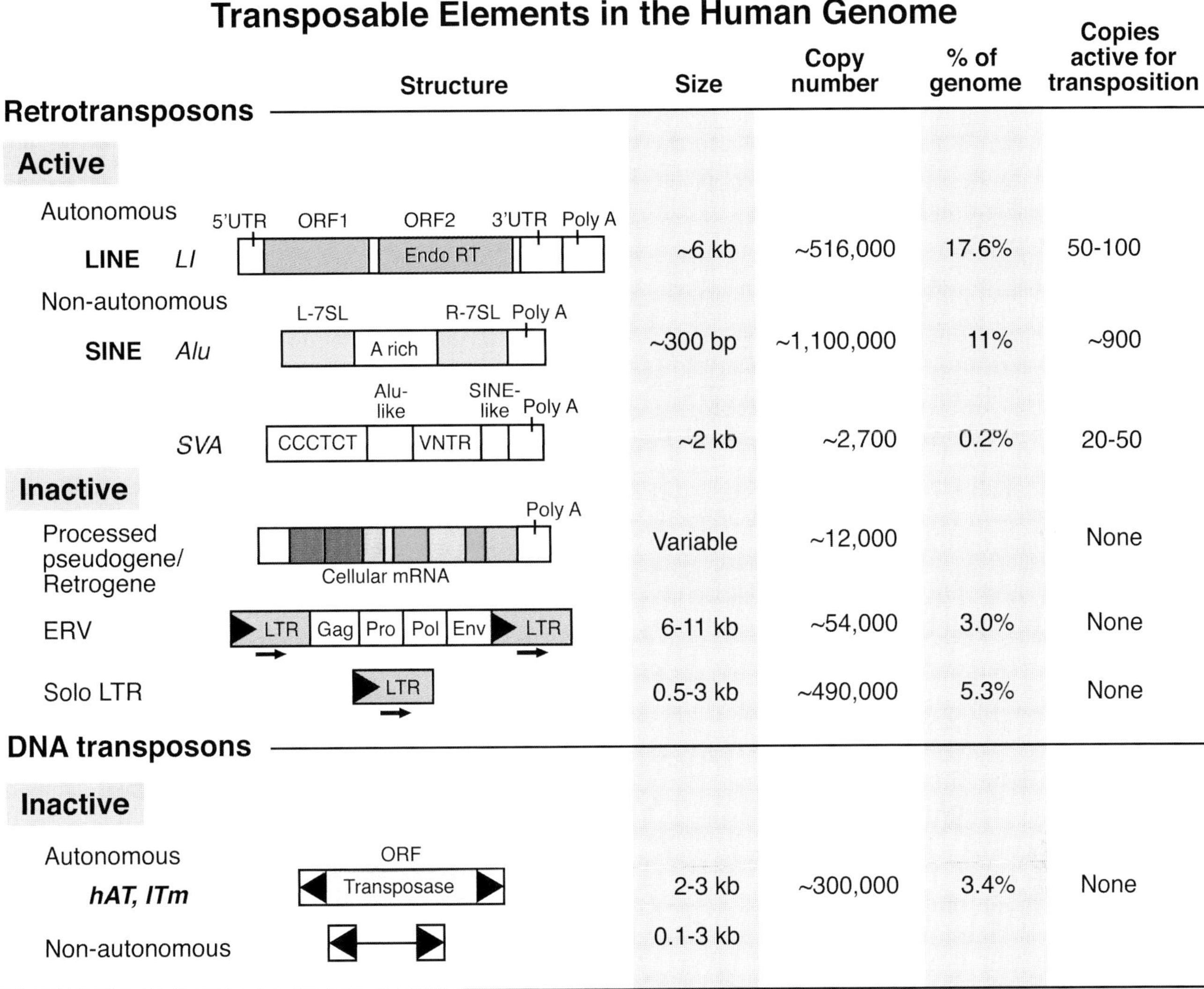

Transposable Elements in the Human Genome
Structure
Size
Copy number
% of genome
Copies active for transposition
Retrotransposons
Active
Autonomous
5'UTR ORF1 ORF2 3'UTR Poly A
Endo RT
LINE L1
~6 kb
~516,000
17.6%
50-100
Non-autonomous
L-7SL R-7SL Poly A
A rich
SINE Alu
~300 bp
~1,100,000
11%
~900
Alu-like SINE-like Poly A
CCCTCT VNTR
SVA
~2 kb
~2,700
0.2%
20-50
Inactive
Poly A
Processed pseudogene/ Retrogene
Cellular mRNA
Variable
~12,000
None
ERV
LTR Gag Pro Pol Env LTR
6-11 kb
~54,000
3.0%
None
Solo LTR
LTR
0.5-3 kb
~490,000
5.3%
None
DNA transposons
Inactive
Autonomous
ORF
Transposase
hAT, ITm
2-3 kb
~300,000
3.4%
None
Non-autonomous
0.1-3 kb

Figure 23 A large fraction of the human genome is comprised of transposable elements. The structures of examples of the major classes of transposable elements found in the human genome are shown. Their total copy numbers and the number of estimated active elements are also shown. The active elements, LINE element *L1* and the SINES *Alu* and SVA, whose movement depends on *L1* proteins, contribute to human genetic variation. Although processed pseudogenes do not retrotranspose, formation of new pseudogenes has occurred in humans. Note the amount of human genome derived from mobile elements, in some cases significantly (DNA elements=3.4%), or in other cases very dramatically (L1=17.6% and Alu =11%), exceeds the fraction of the human genome that encodes ORFs, about 1.5%. doi:10.1128/microbiolspec.MDNA3-0062-2014.f23

Important insights into the mechanism of retrotransposition at the molecular level have come from structural characterization of individual steps of reverse transcription and with the first crystal structure of a retrotransposon reverse transcriptase, as well as RNase H complexed with the oligonucleotide substrates (see Chapter 46) and with achievement of a crystal structure of the prototypic foamy virus integrase tetramer complexed with virus and host target DNA (Chapter 44). The latter structure has allowed modeling of the HIV-1 integrase and a deeper understanding of the mode of action of integrase inhibitors.

Retroviral integrases that are DDE transposases are not the only type of integrase used for the integration of a mobile element with a genome generated by reverse transcriptase. *DIRS*, *Ngaro* and *Viper* are retro-

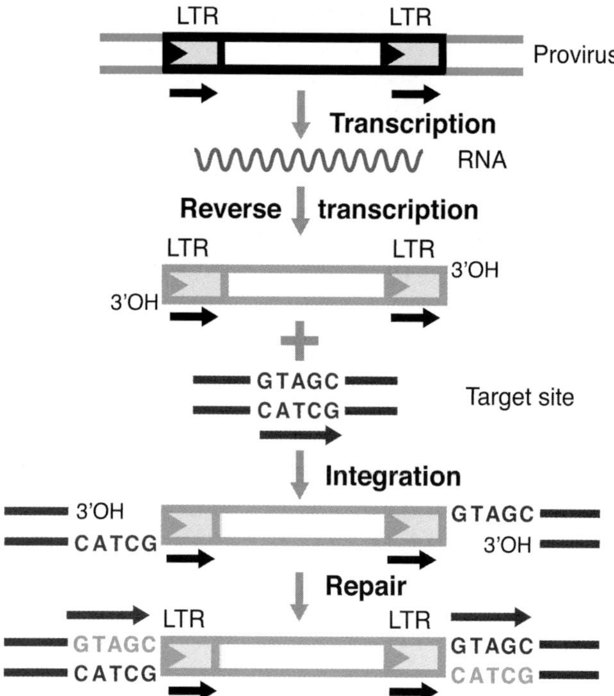

Figure 24 The mechanism of integration of the DNA forms of retroviruses and retroviral-like retrotransposons. The first step in retroviral replication and integration is transcription of the provirus by host RNA polymerase. Two RNA copies are packaged into each virus particle along with the proteins Reverse transcriptase-RNaseH and IN. Reverse transcriptase uses the two viral RNA copies to synthesize a cDNA (green) extending from the 5′ end of one LTR to the 3′ of the other with exposed 3-OHs. The retroviral integrase, a DDE transposase, inserts the cDNA into the target DNA by direct nucleophillic attack of the 3′-OH cDNA ends at staggered positions on the target. The 5′ gaps are repaired (green) by host functions to give target site duplications.
doi:10.1128/microbiolspec.MDNA3-0062-2014.f24

elements that appear to use a $\text{Tyr}_{\text{TRANSP}}$ for integration (see Chapter 55).

Non-LTR Retrotransposons

As with the LTR-containing retroviruses and retroviral-like retrotransposons, the RNA copy of a non-LTR element is generated by a host polymerase and moves from the nucleus to the cytoplasm where it is translated. Non-LTR-encoded proteins always include reverse transcriptase but may also include other mobility functions including nucleases. Although the RNA and reverse transcriptase assemble to form a non-LTR element RNP in the cytoplasm, no reverse transcription occurs until the RNP interacts with the target nuclear DNA where the element will insert. A DNA 3′-OH at the target site serves as the primer for reverse transcription using an RNA copy of the element as a template that generates the DNA copy of the element at the insertion site. This reaction is called target primed reverse transcription.

R2 is a Target Site-Specific Non-LTR Element

One of the best-understood non-LTR elements is *R2* (see Chapter 49), originally found in *Drosophila melanogaster* but now known to be widespread in animals. The signature of this element is that it is highly target site-specific, inserting into a particular site in rDNA (Fig. 26). Other target site-specific elements have also been identified and are often associated with specific sites in repeated DNAs (see Chapter 50).

The *R2* RNA is derived from the rRNA transcript and is excised by a self-cleaving ribozyme that generates the precise 5′ end of the RNA. *R2* encodes a single ORF that has both reverse transcriptase and target site-specific endonuclease, as well as binding sites for the 5′ and 3′ ends of the element (Fig. 26). *R2* insertion involves two *R2* RNPs. One subunit nicks the template DNA at the specific insertion site and then uses the released

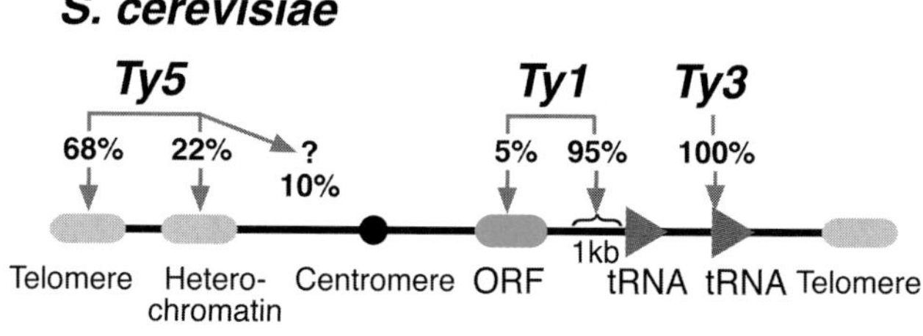

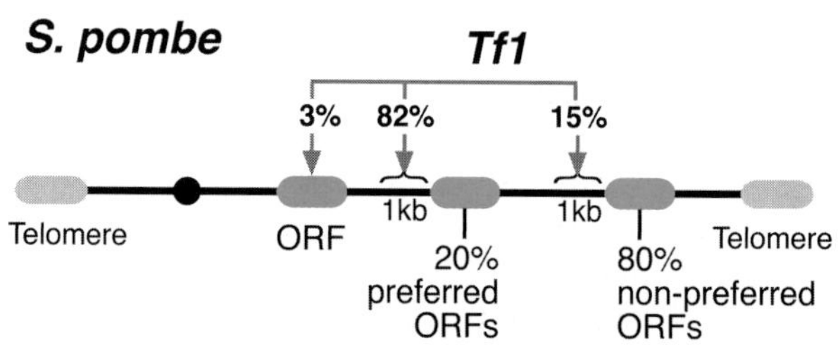

Figure 25 Retroviral-like retrotransposons in yeast are highly target site-selective. Next gene sequencing has been used to map hundreds of thousands of retroelement insertions in yeast. In *Saccharomyces cerevisiae*, *Ty1* and *Ty3* insert at tRNA genes (and other *poll III* genes). *Ty3* is highly site-specific, inserting within a nucleotide or two of the transcript start site and is positioned by interaction with tRNA transcription factors. *Ty1* inserts preferentially on nucleosomal DNA within about 1 kb upstream of transcription start sites. *Ty5* inserts preferentially into heterchromatic regions, including telomers. In *S. pombe*, *Tf1* inserts preferentially within about 1 kb upstream upstream of *pol II* ORF promoters. The regions upstream of some genes are far more attractive to *Tf1* than others.
doi:10.1128/microbiolspec.MDNA3-0062-2014.f25

DNA 3′-OH on the target DNA as the primer for reverse transcription using the *R2* RNA as a template. As polymerization proceeds, this will remove the 5′ end from the subunit. The second RNP then cleaves the top strand, using the released 3′ end of the target DNA as the primer. TPRT-dependent *R2* insertion, like all TPRT events, results in variously sized target site duplications.

LINEs are Another Class of Non-LTR Retrotransposon

Long interspersed elements (LINE) elements are major components of mammalian genomes and have been studied intensively in humans and mice (see Chapter 51). A full-length of the *L1* LINE element is about 6 kb long, encodes two ORFs, ORF1 and ORF2, and ends with a polyA tail (Fig. 23). ORF1 encodes an RNA chaperone protein and ORF2 contains reverse transcriptase and APE-like endonuclease domains. Following translation, the *L1* RNA and the ORF1 and ORF2 proteins assemble into an RNP, which then returns to the nucleus. Formation of this RNP is thought to underlie the preferential *cis*-action of the LINE transposition proteins. Although they have yet to be defined in detail, the RNP also contains multiple host proteins.

The binding site preference of the APE-endonuclease for AT-rich DNA sequences mediates *L1* target site selection. Cleavage of the target DNA by the endonuclease generates a free 3′-OH DNA end that will be used as the primer for reverse transcriptase during TPRT (Fig. 27). Pairing of the *L1* polyA tail with the T-rich DNA strand released by cleavage at the target sites guides APE pairing of the template RNA to the target DNA (Fig. 25). The details of synthesis of the second DNA strand synthesis and the involvement of host proteins remain to be determined. *L1* insertion results in variable length target site duplications. Not surprisingly, multiple regulatory systems modulate the frequency of *L1* transposition.

Not all *L1* transposition events yield straightforward insertions of the full-length element. Many are 5′ truncated or contain internal rearrangements. Notably *L1* transposition can also result in rearrangements of sequences outside the *L1* (see Chapter 51). Transduction of host sequences, both upstream and downstream of *L1*, have been observed as have alterations around the target site such as large deletions.

As with the movement of other mobile genetic elements, *L1* insertions can lead to gene inactivation by gene disruption and there are multiple examples of human disease resulting from *de novo L1* insertions. As with other mobile elements, regulatory signals within *L1* can also affect expression of adjacent genes.

L1 has very successfully colonized the human genome and many other genomes (Fig. 23). Sequences derived from *L1* make up about 17% of the human genome. Although most of these sequences are inactive *L1* fossils, the human genome does contain about 80 to 100 *L1* active elements. Although the great majority of human *L1*s are fixed, that is, occurred before the emergence of modern humans, the ongoing activity of some *L1*s means that *L1* retrotransposition does contribute to human variation. Indeed, there may be millions of "private" *L1* elements that have recently transposed in different human lineages. Comparison of multiple human genomes suggests that a new germline *L1* insertion occurs about once every 100 births.

L1 transposition in the human germline (or in very early embryos) is required to pass new *L1* insertions to progeny. *L1* transposition also occurs, however, in somatic cells. Notably, *L1* transposition has been observed in the human brain and in multiple tumor types, raising interesting questions about the possible contributions of *L1* transposition to somatic phenotypic changes.

SINEs are Nonautonomous Non-LTR Elements

L1 is an autonomous transposable element, that is, it encodes its own mobility proteins as well as the mobile element itself. The *L1* retrotransposition proteins also mobilize other nonautonomous, non-LINE RNAs, a prominent class of which are short interspered elements (SINEs) (as well as mRNAs, see below) (Fig. 23). RNA copies of SINE elements transpose via TPRT that requires only the *L1* ORF2 and, like *L1* insertions, are flanked by variable target site duplications. SINEs do not encode proteins and are derived from short (<400 bp) housekeeping RNAs, such as tRNAs, rRNAs, and 7SL RNA, the RNA component of the signal recognition particle, that transpose repetitively (Fig. 23). A very successful class of human SINE elements is derived from 7SL RNA, which is called *Alu* because their DNA copies contain an AluI restriction site and also have an internal *pol III* promoter. The human genome contains more than a million *Alu* elements, forming about 11% of the genome.

There are multiple families of *Alu* elements but only some are currently active. The frequency of *Alu* transposition is, however, higher than the frequency of *L1* transposition. *Alu*s continue to contribute to variation within the human population again with likely millions of "private" insertions. As with *L1*s, there are multiple examples of *de novo Alu* insertions that have resulted in human disease. In addition to insertional mutagenesis, *Alu* elements also contribute to human variation

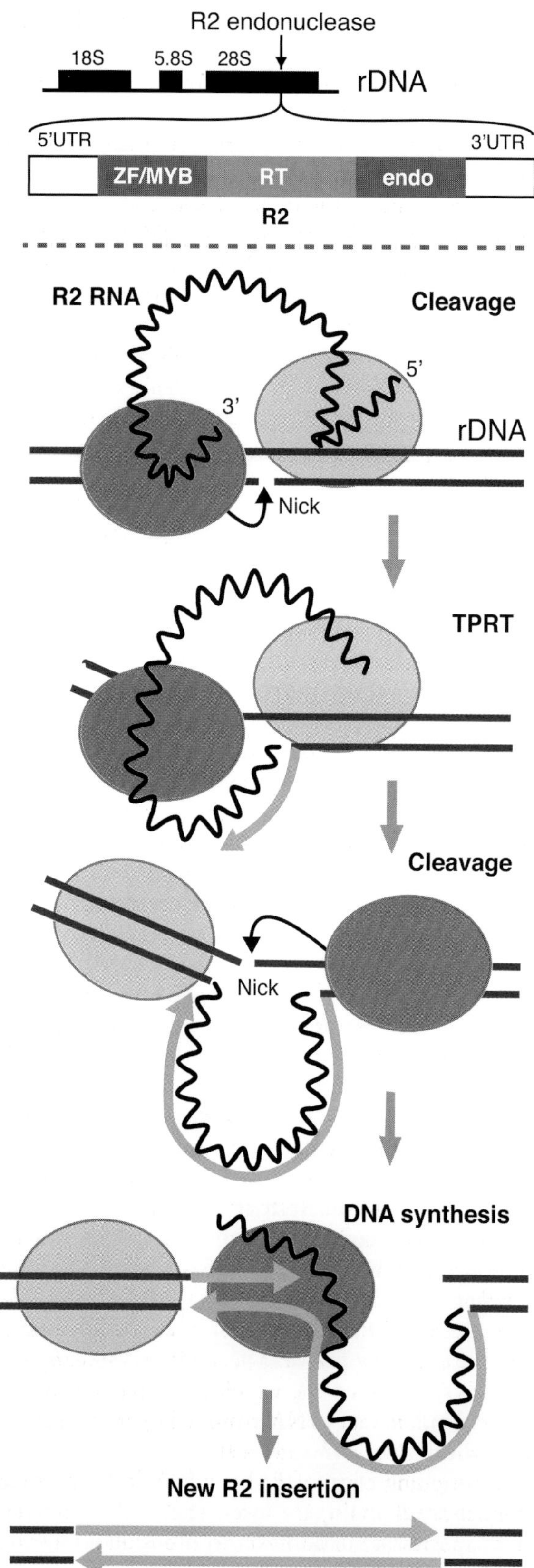

and cause disease by genome rearrangements that result from nonallelic homologous recombination between *Alu* elements that generate in chromosomal deletions, inversions, and translocations.

Another active SINE element in the human genome is SVA (Fig. 23), which is about 2 kbp long and is a composite of other SINE elements. These are very young elements and are present in only a few thousand copies.

Processed Pseudogenes

Pseudogenes are nonfunctional copies of active genes, which accumulate mutations as they are no longer under selection. One class of pseudogenes called processed pseudogenes (38), also called retrogenes (39), have arisen by *L1*-mediated TPRT of mRNAs that have had their introns removed by splicing (Fig. 23). Pseudogenes do have the *L1* TPRT hallmark of target site duplications. Pseudogenes generally lack active promoters, however, because the necessary regulatory sequences are located upstream of transcription start sites, and thus, are not present in mRNAs. Pseudogenes are common in mammalian genomes. For example, there are about 12,000 processed pseudogenes in the human genome compared to about 20,000 protein coding human genes. These elements are polymorphic in the human genome, indicating ongoing insertion.

Mobile Group II Introns: Back to the Future

Group II introns are both introns, which splice from mRNA, and mobile genetic elements, which reverse-splice into target DNA sites where they are converted to DNA via reverse transcription (see Chapter 52). They are widespread in bacteria and are also found in mitochondrial and chloroplast genomes. Notably, they are also the likely progenitors of eukaryotic spliceosomal introns, having invaded the nuclear genome from organelle genomes.

Figure 26 The non-LTR element R2 inserts site-specifically into rDNA by target-primed reverse transcription. R2 inserts into a specific site in the rDNA genes, which is determined by the target-site selectivity of its own endonuclease. Retrotransposition begins with the formation of a RNP containing R2 RNA and two R2 proteins. The R2 subunit bound to the 3′ end of the RNA nicks the target DNA at a specific sequence. The resulting target 3′-OH is used as a primer for reverse transcription (green) by the element-encoded RT that uses R2 RNA as a template. Reverse transcription extends to the end of the RNA template, followed by cleavage of the top strand by the other subunit. DNA synthesis (green) of the other R2 DNA strand completes insertion of the element.
doi:10.1128/microbiolspec.MDNA3-0062-2014.f26

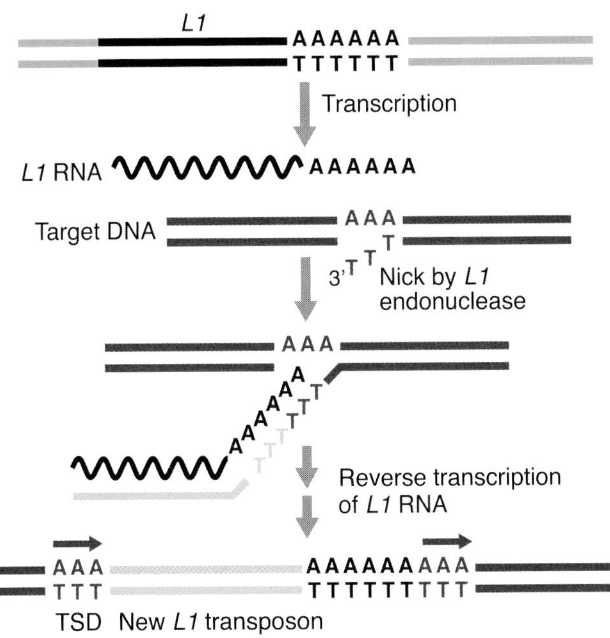

Figure 27 Long interspersed elements (LINE) elements insert by target-primed reverse transcription into target sites cleaved by ORF2 endonuclease. Transcription of the *L1* element initiates retrotransposition. Following synthesis of ORF1 and ORF2 proteins, which associate preferentially with *L1* RNA, the RNP enters the nucleus where the ORF2 APE endonuclease makes a nick in the AT-rich target site. Nicking releases a T-rich strand that pairs with the polyA tail of the *L1* RNA and the 3′-OH of the target DNA is used as the primer for reverse transcription (green) of the template *L1* RNA. Following multiple other processing steps, a new copy of *L1* occupies the target DNA.
doi:10.1128/microbiolspec.MDNA3-0062-2014.f27

Introns encode an RNA that can auto-catalytically excise (Fig. 28). They also encode a multifunctional ORF called an intron-encoded protein (IEP), or sometimes called a maturase, that assists with both RNA-splicing from mRNA and reverse splicing of the RNA into DNA. The IEP also encodes a reverse transcriptase and sometimes an endonuclease that facilitates TPRT by cleaving the target DNA to release a 3′-OH, which will be the primer for reverse transcription of the element RNA.

Following transcription of a parental mRNA, the intron folds into an elaborate structure, which is a self-splicing ribozyme in which intron sequences pair specifically with the exon-intron boundaries. The IEP interacts with the RNA to form a RNP. This ribozyme is highly similar in structure and enzymatic activity to the spliceosomal machinery. Facilitated by the IEP, RNA splicing joins the exonic RNA and generates the spliced lariat RNA.

The mobile group II intron inserts into an intron-less allele of the donor gene.

Bound tightly to the IEP, intron boundary sequences in the RNA base pair specifically with exonic sequences on a single strand of the target DNA, leading to reverse splicing of the RNA into the target DNA. Conversion of this RNA strand into a DNA strand occurs by IEP-mediated TPRT. The 3′-OH DNA required as the primer for reverse transcription is generated by the endonuclease, which cuts the unpaired target DNA strand, or, in a clever variation, the 3′-OH of an Okazaki fragment if reverse splicing occurs into a target single strand adjacent to a replication fork. A wide variety of bacterial proteins, including RNases and DNases, are needed to generate intact duplex DNA at the target site (40).

Because of base pairing between the RNA and the DNA target during reverse transcription, group II retromobility is usually highly target site-specific, the intron inserting into the intron-less allele of its parent gene in a process called "retrohoming." At lower frequency, group II introns insert into targets with relaxed sequence specificity in a process called "retrotransposition." Deliberate manipulation of the intron sequence can result in targeted insertion into a chosen site.

Although the mechanism of group II intron mobility is well understood, the cellular regulation of mobility is relatively unexplored. Intriguing recent work suggests that they may be activated to move under conditions of stress (41).

PLEs are Another Class of Retroelement

Penelope-like elements (*PLE*s) are a widespread class of retroelement named after *Penelope*, an element isolated from *Drosophila virilis* (42), that have, to date, been observed only in eukaryotes. *Penelope* encodes a single ORF that has a reverse transcriptase domain and usually an endonuclease domain that is related to a homing endonuclease. Both activities have been shown to be functional in *Penelope* (43). *PLE*s also often encode spliceosomal introns, which is unusual for an element that presumably moves by a RNA intermediate. The ends of *PLE*s are usually direct repeats several hundred base pairs long that encode self-splicing hammerhead ribozymes (44), which may mediate the excision of the *Penelope* RNA from a larger transcript as occurs with *R2* (see Chapter 49).

An intriguing class of *PLE*s, which lack an obvious endonuclease domain, are associated with telomeres in diverse species (45) and have been postulated to move using the 3′ chromosome ends as a priming site. The

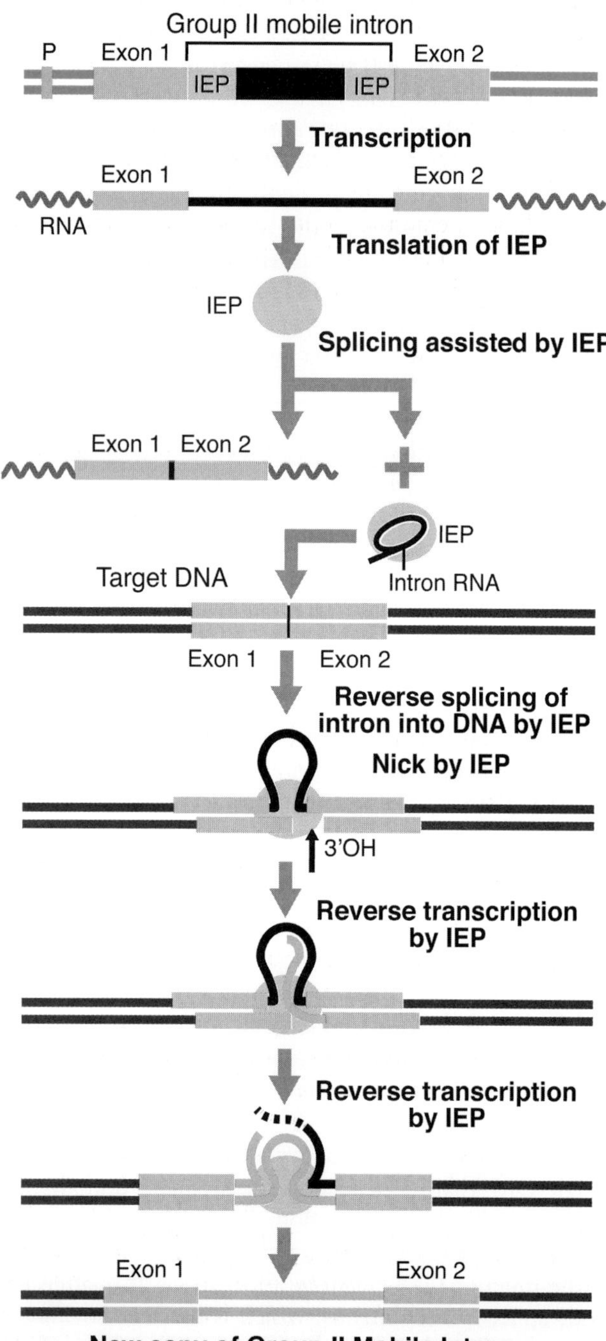

Figure 28 Mechanism of group II intron mobility. Transposition begins with transcription of the gene containing the intron, followed by synthesis of the multifunctional intron-encoded protein (IEP) protein that assists in RNA splicing and reverse splicing into the target DNA and has reverse transcriptase and endonuclease activity. After RNA splicing, the excised lariat RNA remains bound to IEP and then reverse splices into an allele of the gene that lacks the intron. This is highly site-specific, being mediated by base pairing between the intronic sequences and exonic sequences in the target DNA. A 3′-OH on the target DNA provides the primer for reverse transcription (green) of the intron, which generates a target gene containing the intron.
doi:10.1128/microbiolspec.MDNA3-0062-2014.f28

PLE reverse transcriptase is closely related to the reverse transcriptase of telomerase and it has been suggested that *PLE*s may be a missing link between telomerase and modern retroelements (46). Interestingly, the non-LTR elements *Het-A* and *TART* form the telomers of chromosomes in *Drosophila*.

DOMESTICATED REVERSE TRANSCRIPTASES

Bacterial Reverse Transcriptases

It is generally thought that reverse transcriptase originated in bacteria. group II introns are the only known mobile retroelements in bacteria, but they are not the only source of bacterial reverse transcriptase (see Chapter 54).

Retrons: A Retroelement Without a Function

The first reverse transcriptase to be identified in bacteria is encoded by an element called a "retron" (47). A retron is an unusual DNA-RNA molecule in which single strand DNA is covalently linked at its 5′ end to single strand RNA in what is called multicopy single strand DNA (msDNA). msDNA has been found in a number of bacteria including *Escherichia coli* B and can be present in hundreds of copies per cell but its function remains underdetermined.

Diversity-Generating Retroelements

While group II introns and retrons account for about 90% of bacterial reverse transcriptases, there are other reverse transcriptases. Some of these are found in diversity generating retroelements (DGRs) (see Chapter 53), which generate high frequency, targeted mutagenesis into particular gene segments to generate very high levels of sequence diversity. This mutagenesis results from error-prone reverse transcription that changes an A in the mutagenized region to any other nucleotide. Similar to TPRT in the movement of non-LTR elements, a region of the template RNA is used as the template for error-prone reverse transcription. It is not yet known, however, how the mutagenized DNA replaces the original coding region. DGRs have been characterized that diversify phage tail fibers, which mediate phage binding to variable surface proteins in *Bordatella*, and which diversify surface lipoproteins in

Legionella. Genome sequencing has identified hundreds of DGRs in phage and bacterial genomes whose functions remain to be determined.

Domesticated Reverse Transcriptases in Eukaryotes

The most widespread domesticated reverse transcriptase in eukaryotes is telomerase, the ribonucleoprotein complex containing reverse transcriptase that uses an RNA template to add DNA sequences called telomers to the tips of linear chromosomes to complete their end-replication. As noted above, the reverse transcriptase of *Penelope* elements may be a link between ancient retroelements and teleomerases (48). Other eukaryotic genes called RVT genes appear to be domesticated reverse transcriptases (45) but their functions are not yet known.

PERSPECTIVE: "ALL THAT WE KNOW IS STILL INFINITELY LESS THAN ALL THAT REMAINS"

William Harvey, 17th century scientist

Although *Mobile DNA III* includes much about what is known about many interesting aspects of mobile DNA, what is unknown, of course, remains most intriguing and exciting.

Although multiple types of DNA breakage and joining reactions have been analyzed *in vitro* at the biochemical and structural level, we have only a few snapshot views of a limited number of steps in what are active and dynamic processes. Structural analysis of multiple steps in these reactions, as well as single molecule approaches, will provide deeper insights. Biochemical and structural analysis, even of relatives of already known systems, will also be valuable. It is notable that an entirely new class of transposases has been discovered in the past year. Other systems, such as *Helitrons*, seem poised for *in vitro* analysis.

The conditions of *in vitro* reactions are, however, quite different from the cellular environment. Much remains to be learned about mobile DNA in the context of chromatin and three- and four-dimensional genomes. Other than retroviruses and retroviral-like transposons, few genome-wide screens for host factors have been done even in well-developed model organisms. New genetic, biochemical, and proteomic methods will be very useful for such investigations. Advances in microscopy and *in vivo* imaging should also allow more "real-time" analysis of mobile DNA rather than assays that depend on many generations of cell growth.

No doubt, genome sequencing, especially of single cells from different environments in multicellular organisms, as well as genomes of newly discovered organisms, in particular eubacteria and archaea from novel environments, will reveal more examples of mobile DNA. As viruses are the most numerous entities in the biosphere and they are intimately dependent on their hosts, these interactions, both in facilitating and blocking viral growth, are likely to be rich sources of information about DNA rearrangements. It will be especially important to look *de novo* for novel types of mobile DNA rather than just identifying new relatives of already known elements.

One already very well known mechanism for DNA rearrangement is DNA breakage and joining via covalent protein-DNA intermediates of the tyrosine and serine recombinases. These systems are wide spread in eubacteria and archaea. The very successful use of these recombinases for genome engineering in a variety of eukaryotes raises the question of why endogenous systems of this type are not more widespread in eukaryotes.

Another interesting issue concerns the role of RNA in DNA mobility. Certainly, RNA is a very useful template for reverse transcription and various types of retroelements have been extremely successful mobile DNAs. Reverse splicing of group II intron RNA into DNA reveals the awesome power of RNA-DNA chemistry and the accuracy and efficiency of RNA-templated CRISPR-Cas DNA cleavage is also notable. The apparent involvement of RNAs as guides for some aspect(s) of gene assembly in ciliates is also intriguing. Perhaps RNA can play an, as yet unknown, intimate role in templating and/or executing the DNA cleavage and even DNA joining reactions that underlie some DNA rearrangements.

Many adventures in mobile DNA await!

Acknowledgments. *I thank Mick Chandler, Marty Gellert, Alan Lambowitz, Phoebe Rice, and Suzanne Sandmeyer for being such excellent editors. I also thank Patti Kodeck for so successfully managing all the communications between the editors, authors, and ASM, and for keeping us all on track. I am grateful to Mick Chandler, Fred Dyda, Tom Eickbush, Rasika Harshey, Reid Johnson, Haig Kazazian, Dixie Mager, Didier Mazel, John Moran, Phoebe Rice, Suzanne Sandmeyer, and Greg Van Duyne for their help with the figures and answering my questions. Special thanks to Reid, Greg, and Phoebe for providing files of their lovely structures, to Suzanne for her useful comments and to E. Hemingway ("A Moveable Feast") and W. Harvey ("Everything we know...") for being so quotable. I am also very grateful to Patti Kodeck for her skill and patience in editing the too many versions of this manuscript. It was a very special pleasure to work with Helen McComas on the figures – they are very much the better for her skill, suggestions, and willingness to keep revising them. I am an Investigator of the Howard Hughes Medical Institute.*

References

1. **Stoddard B.** 2014. Homing endonucleases from mobile group I introns: discovery to genome engineering. *Mobile DNA* **5:**7.
2. **Chaumeil J, Skok J.** 2013. A new take on v(d)j recombination: transcription driven nuclear and chromatin reorganization in RAG-mediated cleavage. *Front Immunol* **4:**423.
3. **Kim M, Lapkouski M, Yang W, Gellert M.** 2015. Crystal structure of the V(D)J recombinase RAG1-RAG2. *Nature* **518:**507–511.
4. **Jones J, Gellert M.** 2004. The taming of a transposon: V(D)J recombination and the immune system. *Immunol Rev* **200:**233–248.
5. **Zhou L, Mitra R, Atkinson P, Hickman A, Dyda F, Craig NL.** 2004. Transposition of hAT elements links transposable elements and V(D)J recombination. *Nature* **432:**995–1001.
6. **Hickman AB, Ewis H, Li X, Knapp J, Laver T, Doss A-L, Atkinson P, Craig NL, Dyda F.** 2014. A hAT DNA transposase with an unusual subunit organization to bind long asymmetric ends. *Cell* **158:**353–367.
7. **Hencken C, Li X, Craig NL.** 2012. Functional characterization a transposon with a RAG connection. *Nat Struct Mol Biol* **19:**834–836.
8. **Kapitonov V, Jurka J.** 2005. RAG1 core and V(D)J recombination signal sequnces were derived from Transib transposons. *PLoS Biol* **3:**e181.
9. **Bracht J, Fang W, Goldman A, Dolzhenko E, Stein E, Landweber L.** 2013. Genomes on the edge: programmed genome instability in ciliates. *Cell* **152:**406–416.
10. **Streit A.** 2012. Silencing by throwing away: a role for chromatin diminution. *Dev Cell* **23:**918–919.
11. **Madireddi M, Coyne R, Smothers J, Mickey K, Yao M, Allis C.** 1996. Pdd1p, a novel chromodomain-containing protein, links heterochromatin assembly and DNA elimination in Tetrahymena. *Cell* **87:**75–84.
12. **Thomas J, Phillips C, Baker R, Pritham E.** 2014. Rolling-circle transposons catalyze genomic innovation in a mammalian lineage. *Genome Biol Evol* **6:**2595–2610.
13. **Ferguson A, Jiang N.** 2011. Pack-MULEs: recycling and reshaping genes through GC-biased acquisition. *Mob Genet Elements* **1:**135–138.
14. **Feng X, Colloms S.** 2007. *In vitro* transposition of ISY100, a bacterial insertion sequence belonging to the Tc1/mariner family. *Mol Microbiol* **65:**1432–1443.
15. **Krupovic M, Makarova K, Forterre P, Prangishvili D, Koonin E.** 2014. Casposons: a new superfamily of self-synthesizing DNA transposons at the origin of prokaryotic CRISPR-Cas immunity. *BMC Biol* **12:**36.
16. **van der Oost J, Westra E, Jackson R, Wiedenheft B.** 2014. Unravelling the structural and mechanistic basis of CRISPR-Cas systems. *Nat Rev Microbiol* **12:**479–492.
17. **Nunez J, Lee A, Engelman A, Doudna J.** 2015. Integrase-mediated spacer acquisition during CRISPR-Cas adaptive immunity. *Nature* **159:**193–198.
18. **Nunez JK, Kranzusch P, Noeske J, Wright A, Davies CW, Doudna J.** 2014. Cas1-Cas2 complex formation mediates spacer acquisition during CRISPR-Cas adaptive immunity. *Nat Struct Mol Biol* **21:**528–534.
19. **Krupovic M, Koonin EV.** 2015. Polintons: a hotbed of eukaryotic virus, transposon and plasmid evolution. *Nat Rev Microbiol* **13:**105–115.
20. **Yuan Y, Wessler S.** 2011. The catalytic domain of all eukaryotic cut-and-paste transposase superfamilies. *Proc Natl Acad Sci U S A* **108:**7884–7889.
21. **Syvanen M, Hopkins JD, Clements M.** 1982. A new class of mutants in DNA polymerase I that affects gene transposition. *J Mol Biol* **158:**203–212.
22. **Han M, Xu H, Zhang H, Feschotte C, Zhang Z.** 2014. Spy: a new group of eukaryotic DNA transposons without target site duplications. *Genome Biol Evol* **6:**1748–1757.
23. **Harshey R.** 2012. The Mu story: how a maverick phage moved the field forward. *Mobile DNA* **3:**21.
24. **Montano S, Pigli Y, Rice P.** 2012. The mu transpososome structure sheds light on DDE recombinase evolution. *Nature* **491:**413–417.
25. **Garcillan-Barcia M, Bernales I, Mendiola M, de la Cruz F.** 2002. IS91 Rolling-Circle Transposition, p 891–904. *In* Craig NL, Craigie R, Gellert M, Lambowitz A (ed), *Mobile DNA II*. ASM Press, Washington DC.
26. **Mitra R, Li X, Kapusta A, Mayhew D, Mitra R, Feschotte C, Craig NL.** 2013. Functional characterization of piggyBat from the bat Myotis lucifugus unveils the first active mammalian DNA transposon. *Proc Natl Acad Sci U S A* **110:**234–239.
27. **Nunes-Duby S, Azaro M, Landy A.** 1995. Swapping DNA strands and sensing homology without branch migration in lambda site-specific recombination. *Curr Biol* **5:**139–148.
28. **Birling MC, Gofflot F, Warot X.** 2009. Site-specific recombinases for manipulation of the mouse genome. *Methods Mol Biol* **561:**245–263.
29. **Rutherford K, Van Duyne G.** 2014. The ins and outs of serine integrase site-specific recombination. *Curr Opin Struct Biol* **24:**125–131.
30. **Rutherford K, Yuan P, Perry K, Sharp R, Van Duyne G.** 2013. Attachment site recognition and regulation of directionality by the serine integrases. *Nucleic Acids Res* **41:**8341–8356.
31. **Roberts A, Mullany P.** 2011. Tn916-like genetic elements: a diverse group of modular mobile elements conferring antibiotic resistance. *FEMS Microbiol Rev* **35:**856–871.
32. **Das B, Martinez E, Midonet C, Barre F.** 2013. Integrative mobile elements exploiting Xer recombination. *Trends Microbiol* **21:**23–30.
33. **Val M, Bouvier M, Campos J, Sherratt D, Cornet F, Mazel D, Barre F.** 2005. The single-stranded genome of phage CTX is the form used for integration into the genome of *Vibrio cholerae*. *Mol Cell* **19:**559–566.
34. **Goodwin T, Butler M, Poulter R.** 2003. Cryptons: a group of tyrosine-recombinase-encoding DNA transposons from pathogenic fungi. *Microbiology* **149:**3099–3109.
35. **Kojima K, Jurka J.** 2011. Crypton transposons: identification of new diverse families and ancient domestication events. *Mobile DNA* **2:**12.

36. **Doak T, Witherspoon D, Jahn C, Herrick G.** 2003. Selection on the genes of Euplotes crassus Tec1 and Tec2 transposons: evolutionary appearance of a programmed frameshift in a Tec2 gene encoding a tyrosine family site-specific recombinase. *Eukaryot Cell* **2:**95–102.

37. **Jacobs M, Sanchez-Blanco A, Katz L, Klobutcher L.** 2003. Tec3, a new developmentally eliminated DNA element in Euplotes crassus. *Eukaryot Cell* **2:**103–114.

38. **Kazazian H Jr.** 2014. Processed pseudogene insertions in somatic cells. *Mobile DNA* **5:**20.

39. **Richardson S, Salvador-Palomeque C, Faulkner G.** 2014. Diversity through duplication: whole-genome sequencing reveals novel gene retrocopies in the human population. *Bioessays* **36:**475–481.

40. **Coros C, Piazza C, Chalamcharla V, Belfort M.** 2008. A mutant screen reveals RNase E as a silencer of group II intron retromobility in *Escherichia coli*. *RNA* **14:**2634–2644.

41. **Coros C, Piazza C, Chalamcharla V, Smith D, Belfort M.** 2009. Global regulators orchestrate group II intron retromobility. *Mol Cell* **34:**250–256.

42. **Evgen'ev MB, Zelentsova H, Shostak N, Kozitsina M, Barskyi V, Lankenau DH, Corces VG.** 1997. Penelope, a new family of transposable elements and its possible role in hybrid dysgenesis in Drosophila virilis. *Proc Natl Acad Sci U S A* **94:**196–201.

43. **Pyatkov K, Arkhipova I, Malkova N, Finnegan D, Evgen'ev M.** 2004. Reverse transcriptase and endonuclease activities encoded by Penelope-like retroelements. *Proc Natl Acad Sci U S A* **101:**14719–14724.

44. **Cervera A, De la Pena M.** 2014. Eukaryotic penelope-like retroelements encode hammerhead ribozyme motifs. *Mol Biol Evol* **31:**2941–2947.

45. **Gladyshev E, Arkhipova I.** 2011. A widespread class of reverse transcriptase-related cellular genes. *Proc Natl Acad Sci U S A* **108:**20311–20316.

46. **Curcio M, Belfort M.** 1996. Retrohoming: cDNA-mediated mobility of group II introns requires a catalytic RNA. *Cell* **84:**9–12.

47. **Lampson B, Inouye M, Inouye S.** 2005. Retrons, msDNA, and the bacterial genome. *Cytogenet Genome Res* **110:**491–499.

48. **Curcio MJ, Belfort M.** 2007. The beginning of the end: links between ancient retroelements and modern telomerases. *Proc Natl Acad U S A* **104:**9107–9108.

Conservative Site-Specific Recombination

II

Mobile DNA, 3rd Edition
Nancy L. Craig, Michael Chandler, Martin Gellert, Alan M. Lambowitz, Phoebe A. Rice and Suzanne Sandmeyer

doi:10.1128/microbiolspec.MDNA3-0021-2014

Makkuni Jayaram,[1] Chien-Hui Ma,[1] Aashiq H Kachroo,[1] Paul A Rowley,[1] Piotr Guga,[2] Hsui-Fang Fan,[3] and Yuri Voziyanov[4]

2 An Overview of Tyrosine Site-specific Recombination: From an Flp Perspective

INTRODUCTION

Tyrosine family site-specific recombinases (YRs), named after the active site tyrosine nucleophile they utilize for DNA strand breakage, are widely distributed among prokaryotes. They were thought to be nearly absent among eukaryotes, the budding yeast lineage (*Saccharomycetaceae*) being an exception in that a subset of its members houses nuclear plasmids that code for YRs (1, 2). However, YR-harboring DIRS and PAT families of retrotransposons and presumed DNA transposons classified as Cryptons have now been identified in a large number of eukaryotes (3, 4). The presence of functional YRs encoded in Archaeal genomes has been established by a combination of comparative genomics and modeling complemented by biochemical and structural analyses (5, 6). Over 1300 YR sequences mined from bacterial genome databases have been organized into families and subfamilies, providing a better understanding of the evolutionary relationships among them (7). These classifications also encourage investigations into the potential functional significance of YRs whose genes are present as pairs or trios in bacterial and plasmid genomes.

YRs are remarkable enzymes that utilize a common chemical mechanism to bring about a wide array of biological consequences. They range from the choice of lysogenic or lytic developmental pathways in phage λ and related phage, equal segregation of phage, plasmid and bacterial chromosomes by resolving genome dimers or multimers formed by homologous recombination into monomers, to the resolution of hairpin telomeres that mark the termini of certain bacterial and phage genomes (8, 9, 10, 11, 12). In addition, YRs promote the transposition of conjugative mobile elements, the resolution of cointegrate intermediates formed by the Tn3-related toluene catabolic transposon Tn4651, the unidirectional insertion of the *Vibrio cholerae* phage CTXϕ into the host chromosomes and the copy number control of budding yeast plasmids (13, 14, 15, 16, 17). A subset of YRs has been utilized as tools for directed genome manipulations with potentially important biotechnological and medical applications (18).

[1]Department of Molecular Biosciences, UT Austin, Austin, TX 78712; [2]Centre of Molecular and Macromolecular Studies, Polish Academy of Sciences, Department of Bio-organic Chemistry, Lodz, Poland; [3]Department of Life Sciences and Institute of Genome Sciences, National Yang-Ming University, Taipei 112, Taiwan; [4]School of Biosciences, Louisiana Tech University, Ruston, LA 71272.

In this overview of tyrosine site-specific recombination, we present our current understanding of the mechanism of the reaction from biochemical, chemical, structural and topological perspectives, and highlight the utility of this knowledge in addressing problems of fundamental importance in biology and in developing new technologies for biomedical engineering (see also chapters by M. Boocock, A. Landy, A. Segall, G. van Duyne, D. Mazel, F-X Barre, J. Gardner, and G. Chaconas).

THE RECOMBINATION REACTION: SYNAPTIC ORGANIZATION OF DNA PARTNERS AND STRAND EXCHANGE MECHANISM

The biochemically and structurally most well characterized YRs are phage λ integrase (λ Int), phage P1 coded Cre, Flp coded for by the *Saccharomyces cerevisiae* plasmid 2 micron circle and XerCD of *Escherichia coli* (10, 19, 20, 21, 22, 23). They have provided the templates for the chemical and conformational attributes of the strand cleavage and strand exchange steps during tyrosine recombination. The reaction is executed in the context of two core DNA target sites, each bound by two recombinase monomers, brought together in a synaptic complex by protein–protein interactions [Fig. 1(A)].

The association of a recombinase, a monomer in solution, with its binding element (a little over one turn of DNA) activates the scissile phosphate adjacent to it. The amino- and carboxyl-terminal domains of the recombinase cradle the DNA between them through a small number of base-specific and many more phosphate contacts. Recognition specificity is imparted to a significant degree through indirect readout. The 13 bp Flp binding element, for example, contains an A/T-rich segment with a characteristically narrow minor groove, and A to T changes within it are well tolerated with respect to binding (24). At the same time, C to G changes are detrimental to binding, and replacement of guanosine with inosine alleviates this negative effect by eliminating the obstructive 2-amino group from the minor grove. Within a DNA substrate, the scissile phosphates are positioned 6 to 8 bp apart (depending on individual systems) on opposite strands, specifying the extent of the strand exchange region. In general, two identical monomers of a recombinase occupy the two binding elements flanking the strand exchange region in a head-to-head (inverted) fashion. In rare instances, as with XerCD, a target site is bound by one monomer each of XerC and XerD.

Strand cleavage by the active site tyrosine nucleophile utilizes the type IB topoisomerase mechanism, yielding a 3′-phosphotyrosyl intermediate and an adjacent 5′-hydroxyl group (20). Strand exchange involves the nucleophilic attack by the 5′-hydroxyl group on the phosphotyrosyl bond across DNA partners to reseal the strand breaks in the recombinant configuration. The reaction is completed in two temporally separated cleavage-exchange steps, the first yielding a Holliday junction intermediate and the second resolving it into reciprocal recombinants.

INHIBITION OF TYROSINE RECOMBINATION BY AGENTS THAT TARGET HOLLIDAY JUNCTIONS

Short synthetic hexapeptides rich in aromatic amino acids inhibit tyrosine recombination by trapping the Holliday junction intermediate (25, 26, 27) (also chapter by A. Segall). The current model for peptide action, based on gel mobility shift and fluorescence quenching results, together with crystal structure data for a Cre recombinase-Holliday junction-peptide ternary complex, posits that the binding of a peptide dimer across the junction core stabilizes the junction in a nearly square-planar (but nonfunctional) conformation (27, 28, 29). More recent analyses of peptide–junction interactions by a combination of single molecule FRET (fluorescence resonance energy transfer), SAXS (small angle X-ray scattering) and gel mobility shifts suggest that peptide binding yields an ensemble of highly dynamic junction conformations that do not fit the canonical square-planar and stacked X-conformations (unpublished observation). The induced conformational heterogeneity likely results from multiple stacking arrangements of aromatic amino acids with the bases surrounding the junction core, perhaps reflecting an intrinsic property of positively charged hydrophobic peptides. Peptide association with a protein–Holliday junction complex may inhibit subsequent reaction steps by inducing global changes in the junction conformation or local changes in the active site environment. Alternatively, peptide binding could accelerate protein dissociation from the junction, and then inhibit further reaction by inducing unfavorable junction conformations. The concept of inhibiting biologically important nucleic acid transactions by enhancing, rather than constraining, conformational freedom may be broadly applicable to peptide and nonpeptide ligands that recognize specific nucleic acid structures. In addition to inhibiting tyrosine recombination, hexapeptides also impede the unwinding of branched DNA structures by

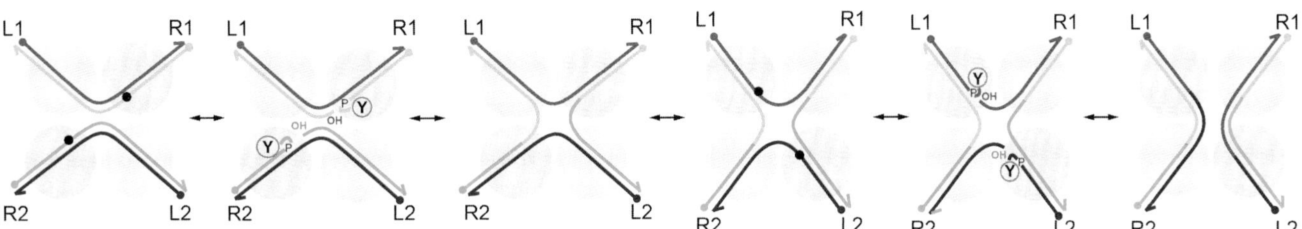

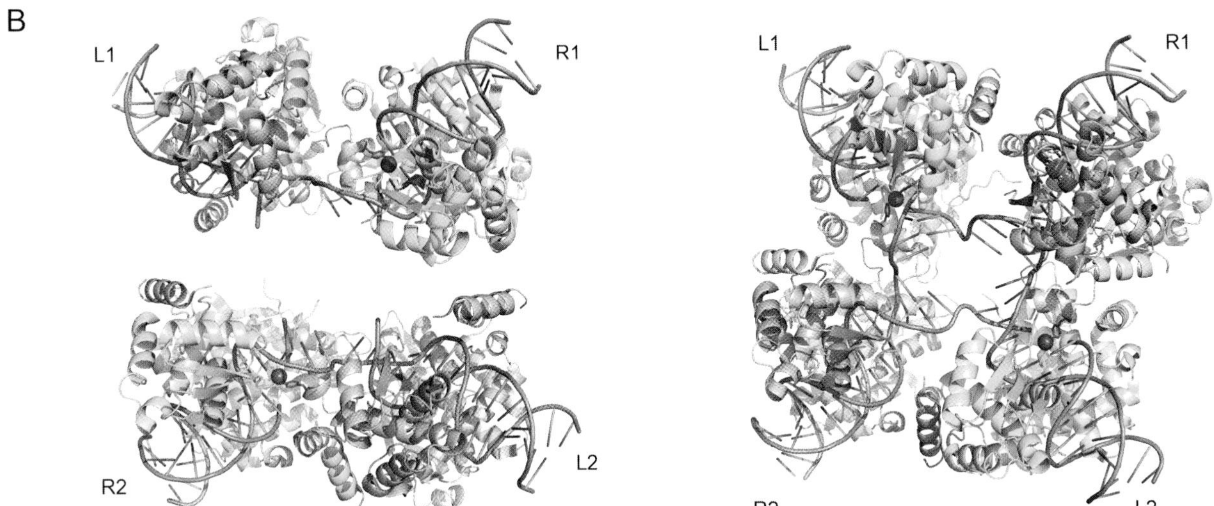

Figure 1 Tyrosine family site-specific recombination. (A) The two target sites, each bound by two recombinase monomers across the strand exchange region, are arranged within the recombination synapse in an almost perfectly planar, antiparallel fashion. The left and right arms of the sites are marked as L1, L2 and R1, R2, respectively. The reaction proceeds by the cleavage/exchange of one pair of strands to form a Holliday junction intermediate, isomerization of the junction, and exchange of the second pair of strands to give the recombinant products (L1R1 + L2R2 → L1R2 + L2R1). The scissile phosphates engaged by the "active" active sites at distinct stages of the reaction are indicated by the filled circles. (B) The "half-of-the-sites" activity, responsible for the two-step strand exchange mechanism, is revealed by the crystal structure of the Flp-DNA complex (34, 36). Within each recombination partner (left), the green Flp monomer (bound at R1 or R2) is poised to promote the cleavage of the scissile phosphate adjacent to it (red circle). The tyrosine nucleophile for cleavage is donated in *trans* by the neighboring Flp monomer (bound at L1 or L2; magenta). Following isomerization of the Holliday junction intermediate (right), there is a switch between the active and inactive Flp pairs, signifying the imminent cleavage of the scissile phosphates adjacent to Flp monomers bound at L1 and L2. The tyrosine nucleophiles are donated across DNA partners, in the R1 to L2 and R2 to L1 configuration.
doi:10.1128/microbiolspec.MDNA3-0021-2014.f1

the RuvG helicase of *E. coli*, and interfere with Holliday junction resolution by the RuvABC complex (28). Consistent with these properties, the inhibitory peptides appear to hold promise as potential antimicrobial agents (30).

SIMPLE AND COMPLEX YRs: CONTROLLING THE DIRECTIONALITY OF RECOMBINATION

Simple YRs such as Cre and Flp are not particular about DNA topology or target site orientation. They can act on supercoiled or nicked circles as well as linear molecules, and promote intra- and intermolecular reactions. They bring about DNA inversion between a pair of sites in head-to-head (inverted) orientation and DNA deletion between sites in head-to-tail (direct) orientation. More complex YRs (λ Int and XerCD, for example), depending on the reaction context, may require DNA supercoiling, and may utilize the interaction between accessory factors and their cognate sites to regulate the reaction and/or impart directionality to it. The crystal structures of λ Int tetramers bound to synapsed DNA partners and the Holliday junction intermediate, together with biochemical data, suggest how interactions of the amino-terminal domains of Int with the 'arm-type' sequences (which also include multiple binding sites for the accessory proteins: IHF, Xis and Fis) can stabilize 2-fold symmetric configurations of the recombination complex (21, 31). The cumulative DNA–protein and protein–protein interactions thus coordinate Int activity at the core recombination sites as well as bias strand exchange towards a particular outcome. Consistent with this model, when the amino-terminal domain of λ Int is fused to the normally unregulated and bias-free Cre, the latter acquires the regulatory features and directionality of Int (32, 33). The action of the Int-Cre chimera on attenuated core target sites containing appropriate embellishments with the arm-type sequences is controlled by the accessory factors as if recombination were being performed by native Int. The conversion of a simple recombinase, whose origin likely traces back to an ancestral type IB topoisomerase, into a complex one by just the addition of a peptide domain suggests a possible "self-promoting" evolutionary scheme for the emergence of the latter class of recombinases. The relevant gene fusion may be performed by the recombinase itself via low frequency crossover events between suitably positioned secondary target sites, which are fortuitously scattered within a genome and may be harbored by the recombinase gene. Alternatively, the "complexity" domain may be acquired via the action of the host's recombination machinery. λ Int may thus be a representative of the evolutionary trajectory from topoisomerase to simple and then complex recombinases (see chapter by A. Landy).

HALF-OF-THE-SITES ACTIVITY OF THE YR ACTIVE SITE

Within the recombination synapse, which has a 2-fold symmetry, only two of the four potential active sites are active at any one time (34, 35) [Fig. 1(B)]. The two active sites responsible for the first cleavage-exchange step become inactive following the isomerization of the Holliday junction intermediate. The other two active sites, which now become activated, resolve the junction into recombinant products. This "half-of-the-sites" activity accounts for the two-step, single-strand exchange mechanism of recombination. Consistent with this mechanism, three Cre or Flp recombinase monomers bound to a three-armed DNA substrate (Y-junction) can yield two functional active sites capable of resolving the junction into a linear and a hairpin product (36, 37, 38).

THE LOCAL GEOMETRY OF PARTNER SITES WITHIN THE RECOMBINATION SYNAPSE

The DNA target sites are almost entirely planar in their paired state, and are arranged in an antiparallel fashion [Fig. 1(A)]. Topological analyses and crystal structure data support this synapse geometry, with strand cleavage and exchange occurring in a diagonal fashion (31, 34, 39, 40, 41, 42, 43, 44). However, a few experiments based on electron microscopy, proximity of DNA ends reported by ligation, and a combination of atomic force microscopy, tangle analysis (see below under "Topological and chiral features of tyrosine recombination") and modeling suggest parallel arrangement of sites or their nonplanar configuration with a potential tetrahedral geometry for the Holliday junction intermediate (45, 46, 47). These could represent transient or intermediate states that precede the functional synapse or isomerization of the Holliday junction. Or, they could be comprised of aberrant complexes. The reactive orientation of key catalytic residues with respect to the scissile phosphates in the crystal structures strongly imply that the antiparallel disposition of the partner sites revealed by them represents the functional geometry of the recombination synapse (19, 31, 39).

THE YR ACTIVE SITE: KEY CATALYTIC RESIDUES

The signature active site motif of YRs consists of a tyrosine nucleophile assisted by an invariant or highly conserved catalytic pentad: Arg, Lys, His, Arg, and His/Trp (23). In addition, a sixth conserved residue, Asp/Glu, appears to contribute indirectly to transition state stabilization by hydrogen bonding to catalytic residues and promoting the integrity of the active site (39). In Flp, the catalytic hexad is comprised of Arg-191, Asp-194, Lys-223, His-305, Arg-308, and Trp-330 with Tyr-343 providing the cleavage nucleophile (Fig. 2). In Cre (see chapter by G. van Duyne), the corresponding

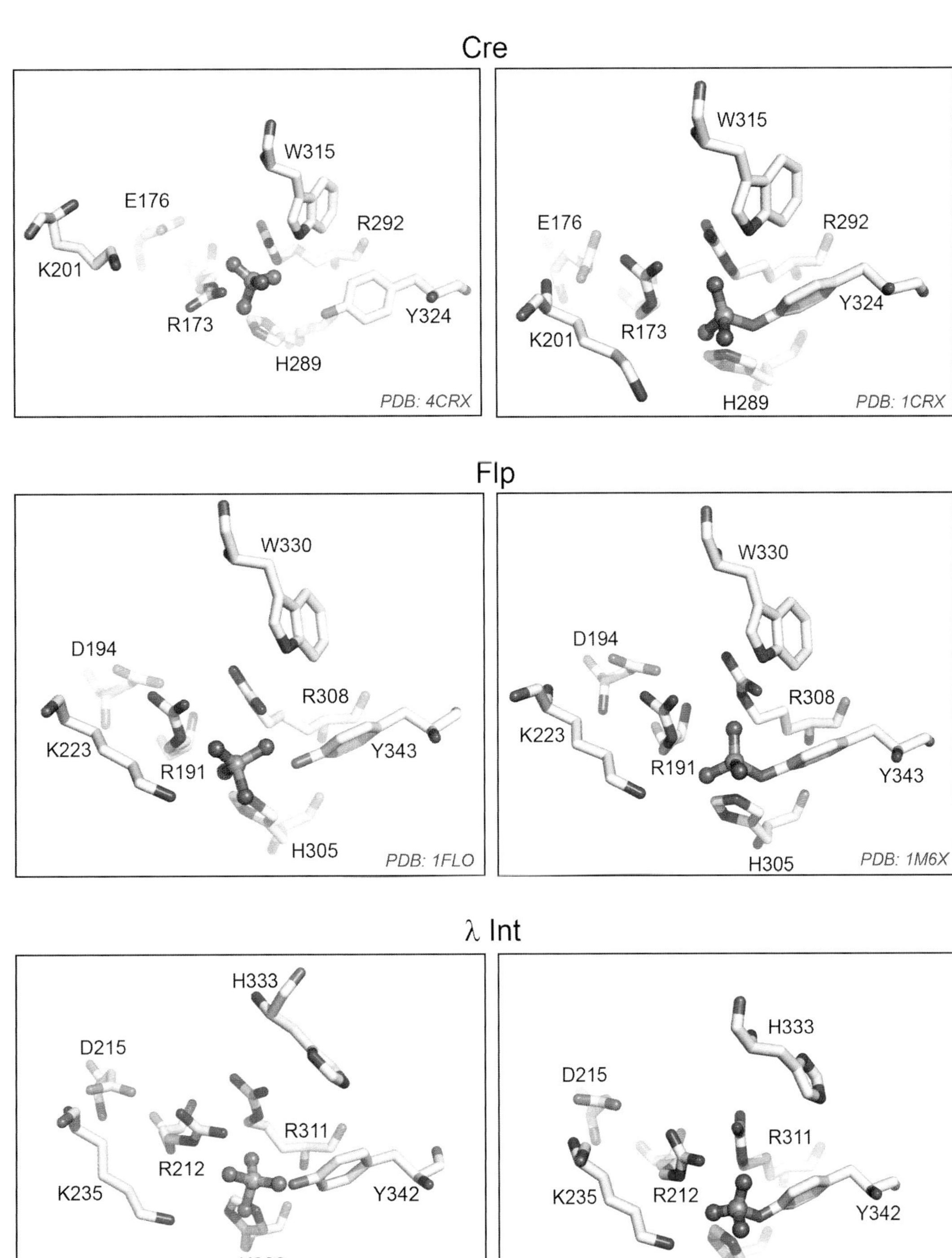

Figure 2 Organization of conserved catalytic residues within the recombinase active site. The arrangements of the catalytic hexad (Arg-Asp/Glu-Lys-His-Arg-His/Trp) and the tyrosine nucleophile in Cre, Flp and λ Int active sites are shown (31, 34, 36, 52, 94, 147). The states of the active site with the scissile phosphate uncleaved and cleaved are shown at the left and right, respectively. The role of the conserved Asp/Glu of the hexad in transition state stabilization is likely indirect, by promoting the structural integrity of the active site. doi:10.1128/microbiolspec.MDNA3-0021-2014.f2

residues are Arg-173, Glu-176, Lys-201, His-289, Arg-292, Trp-315, and Tyr-324. In λ Int, the second and sixth conserved positions are aspartic acid (Asp-215) and histidine (His-333), respectively. Mutational analyses combined with structural data have provided insights into the mechanistic roles of several of these residues.

As would be expected from chemical principles, the two invariant arginine residues balance the negative charges on the nonbridging oxygen atoms of the scissile phosphate group (31, 34, 39). In Cre, Lys-201 appears to function as the general acid that stabilizes the leaving 5′-hydroxyl group during strand cleavage. The absence of this residue can be rescued when the leaving group pK_a is decreased by substituting the 5′-oxygen by sulfur (5′-thiolate) (48). The potential general acid function of Lys-223 of Flp has not been similarly tested. In the type IB vaccinia topoisomerase, Lys-167 (which corresponds to Lys-201 of Cre) collaborates with Arg-130 (corresponding to Arg-173 of Cre and Arg-191 of Flp) to facilitate leaving group departure, perhaps by a proton relay mechanism (49, 50). His-305 of Flp may serve as the general base that abstracts the proton from Tyr-343 to activate it as a nucleophile (51). A tyrosine mimic with reduced pK_a (3-fluorotyrosine, $pK_a = 8.2$ as opposed to tyrosine, $pK_a = 10.0$), when supplied in the context of a short native peptide, can restore the cleavage potential of Flp (H305Q) to a large extent. The predominant majority of YRs contain a histidine rather than tryptophan at the final hexad position. The hydrogen bonding between the indole-nitrogen of Trp-315 and the scissile phosphate observed in the Cre crystal structure (52) seemed to suggest that either histidine or tryptophan at this position helps catalysis through their hydrogen bonding potential. However, this is not the case for Cre or Flp, as replacement of Trp-315 or Trp-330, respectively, by histidine results in lower recombination activity compared to replacement by phenylalanine or tyrosine in Cre and phenylalanine in Flp (39, 53). Consistent with these biochemical results, structural data for Flp suggest that hydrophobic/van der Waals contacts made by Trp-330, over a hydrophobic surface of ~380 Å^2, help position Tyr-343 in its active orientation (53) (Fig. 3). In addition, Trp-330 appears to play a secondary role in stabilizing the 5′-hydroxyl leaving group, as suggested by the stimulation in the cleavage activities of Flp mutants lacking this residue on 5′-thiolate substrates (54). The vanadate transition state mimic structure of Cre-DNA reveals Trp-315 to be located on a turn between two helices (αL and αM), providing a sizable hydrophobic surface on to which the αM helix carrying the Tyr-324 nucleophile docks through van der Waals contacts (39) (see chapter by G. van Duyne).

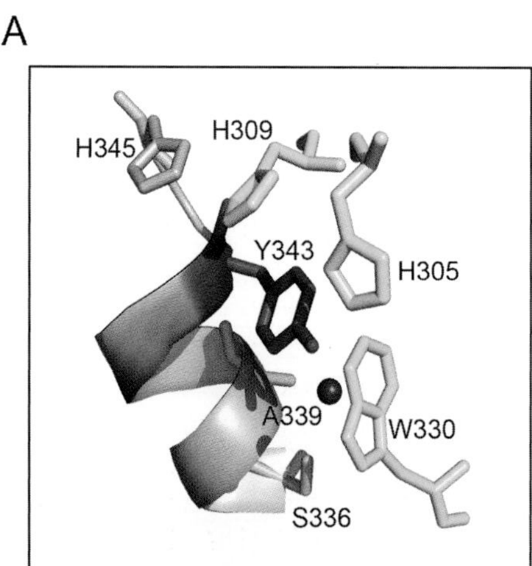

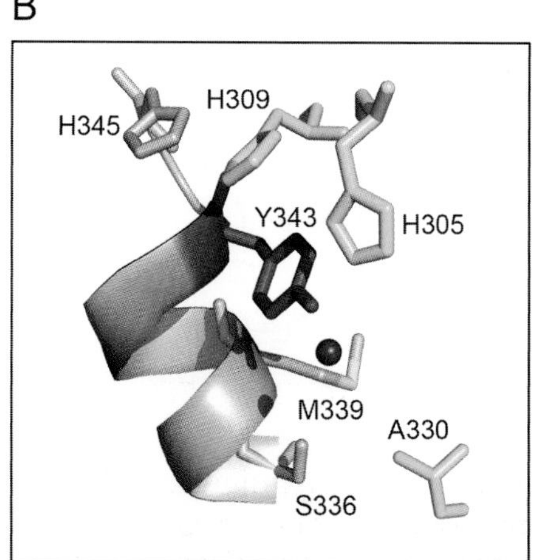

Figure 3 The assembly of the Flp active site in *trans*. (A) In the shared active site of Flp, the tyrosine nucleophile (Tyr-343) is delivered by the donor Flp as part of the M helix (shown in magenta) to the proactive site, whose residues are shown in green, of the recipient Flp. The van der Waals' contacts made by Trp-330 (recipient Flp) with Ser-336 and Ala-339 (donor Flp) are important for the positioning of Tyr-343 (donor) (53). The stacking of His-309 (recipient) over Tyr-343 is stabilized by His-305 (recipient) and His-345 (donor). (B) Consistent with the importance of Trp-330–Ala-339 interaction, the loss of active site function resulting from the W330A substitution can be rescued by the second site A339M mutation, which increases the side chain length at this position (54). The red circle in A and B denotes the scissile phosphate. doi:10.1128/microbiolspec.MDNA3-0021-2014.f3

HIDDEN RNA CLEAVAGE ACTIVITIES OF THE YR ACTIVE SITE

When a ribonucleotide is substituted at the cleavage position or immediately 3´ to it within the Flp target (*FRT*) site, two latent RNase activities of Flp can be unveiled (55, 56, 57). The 2´-hydroxyl, when present as part of the cleavage position nucleotide, can attack the phosphotyrosyl intermediate formed by strand cleavage to give a 2´,3´ cyclic phosphate. Subsequent attack of the cyclic bond by a water nucleophile gives the 3´-phosphate as the end product. This activity, which closely follows the recombination mechanism, has been termed type I RNase. When the 2´-OH is placed on the nucleotide adjacent to the normal cleavage position, it directly attacks the correspondingly shifted scissile phosphate to yield the cyclic phosphate intermediate, which is then hydrolyzed to the 3´-phosphate. This activity, which resembles the classical pancreatic RNase mechanism, has been termed type II RNase. Perhaps the latent RNase activities are the relics of the evolutionary progression of Flp from an elementary nuclease to a recombinase, likely via a topoisomerase. When the interaction between two Flp monomers bound to an *FRT* site is weakened by increasing the length of the strand exchange region, a weak topoisomerase activity of Flp can be unmasked. The type I, but not the type II, RNase activity has been detected in Cre as well (58). The ability of 2´-hydroxyl groups to compete effectively with the tyrosine nucleophile (in the case of Flp) and with the 5´-hydroxyl group (in the case of Cre and Flp), as suggested by their latent RNase activities, speaks to the considerable catalytic flexibility of the tyrosine recombinase active site. These activities also expose the potential threat to recombination by errant nucleophiles that might gain entry into the active site.

ASSEMBLY OF THE YR ACTIVE SITE IN *CIS* OR IN *TRANS*

In general, the active site of a tyrosine recombinase is assembled entirely within a monomer, although its strand cleavage activity may be stimulated by allosteric contact with an adjacent monomer. Flp and the related subfamily of YRs coded for by 2 micron-like plasmids of budding yeasts are unusual in that they assemble an active site at the interface of two neighboring monomers (34, 59, 60) [Fig. 1(B); Fig. 3]. Biochemical and structural evidence suggests that the integrase of SSV1, a virus that infects the extremely thermophilic archaeon *Sulfolobus shibate*, may also harbor a shared active site (61, 62). A *cis* active site (the Cre active site, for example) is responsible for the activation and cleavage of the scissile phosphate engaged by it. A *trans* active site, exemplified by that of Flp, activates the scissile phosphate but relies on the tyrosine nucleophile donated to it for strand cleavage. There are two *trans* modes of DNA cleavage. For the first strand exchange step and formation of the Holliday junction, the tyrosine nucleophile performs cleavage across the strand exchange region within a DNA substrate [Fig. 1(B), left]. For the resolution of the Holliday junction and formation of the recombinant products, the cleavage by tyrosine occurs across partner substrates [Fig. 1(B), right].

Comparison of the crystal structures of Cre and Flp synaptic structures suggests how a simple switch in the connectivity of two helices can switch a *cis* active site into a *trans* active site or vice versa. (23, 34). In the shared active site, the Tyr-343 from an Flp monomer gains entry into the proactive site of the second monomer as part of the protruding "M" helix. Trp-330 of the hexad in the proactive site of Flp may play a particularly important role in helping to dock the M helix by packing against it through contacts with Ser-336, Ala-339, and Tyr-343 (53) [Fig. 3(A)]. These interactions are further augmented by the nearly perfect stacking of His-309 from the recipient Flp over Tyr-343, with likely assistance from His-345 (donor) and His305 (recipient) (34, 63). The extremely weak recombination activity of Flp(W330A) can be restored to nearly wild type level by a second mutation that changes Ala-339 to methionine (54). The longer side-chain of Met-339 located in the M helix can compensate for the lack of Trp-330 in the proactive site, further highlighting the importance of interprotomer hydrophobic interactions in the assembly of the *trans* active site [Fig. 3(B)].

From a purely mechanistic point of view, there is apparently little advantage of a *cis* active site over a *trans* active site or vice versa. By the "Cheshire cat" paradigm, if one were to erase all the amino acid residues from Cre and Flp structures, except for the tyrosine nucleophile and its principal catalytic cohort, the 'catalytic grins' of the two recombinases would look almost identical (64, 65) (Fig. 2). However, the *trans* active site offers considerable advantages in the analysis of recombination mechanism. For example, strand cleavage can be performed by providing exogenous nucleophiles such as hydrogen peroxide or tyramine or by supplying chemically modified tyrosines embedded in a short native Flp peptide (51, 66, 67). As already noted, the potential general base/acid role of His305 of Flp has been inferred from the ability of tyrosine analogs with lowered pK_a to confer cleavage competence on Flp(H305Q).

From a functional perspective, each type of active site may have its strengths and weaknesses. The *trans* active site will delay DNA cleavage until a target site has been occupied by two monomers of the recombinase. However, the time delay between the binding of the two monomers may allow rogue nucleophiles, the abundant water nucleophile for example, to attack the activated phosphodiester bond (68). Since the *cis* active site positions all the catalytic residues, including the tyrosine nucleophile, in concert around the scissile phosphate, the chances for aberrant strand cleavage are minimized. However, cleavage could potentially occur even before the full occupancy of a DNA site by the recombinase. Premature strand breaks may be minimized if the cleavage potential of the *cis* active site is activated by allosteric interactions between recombinase monomers within a DNA site or between partner sites within a recombination synapse. For Flp, whose physiological function is to trigger plasmid DNA amplification by a replication-coupled recombination event (14, 17), DNA cleavage in *trans* may hold special significance. Should an Flp monomer covalently linked to the cleaved phosphate be dislodged from its binding element by the replication machinery, and be unfolded or partially degraded by a protease as a consequence, the Flp monomer bound to the other binding element will be able to promote healing of the DNA break via ligation.

REQUIREMENT OF HOMOLOGY BETWEEN THE STRAND EXCHANGE REGIONS OF RECOMBINATION PARTNERS

According to the generally accepted paradigm, based on evidence from the λ Int, Cre and Flp systems, successful recombination requires perfect homology in the strand exchange regions of the DNA partners, even though the sequence per se of the exchanged segment can be altered in a number of ways without affecting reaction efficiency. The original notion that homology promotes end-to-end branch migration of the Holliday junction through the strand exchange region (69, 70) has been discounted by biochemical and structural evidence (31, 34, 52, 71, 72, 73, 74). The cleaved strands are swapped in a segmental fashion, perhaps as triplets during the first and strand exchange steps (74). Nonhomology would disfavor stable strand exchange, as mismatches in DNA hinder the ligation reaction (73). In this model, recombination is blocked by nonhomology either because the Holliday junction is not formed or because the junction with mismatched DNA is quickly resolved in the parental mode. For Flp, which mediates the exchange of 8 bp rather than 6 bp (the extent of exchange in the Cre system), the strand swaps at the initiation and termination steps of recombination may be separated by an intermediate step of limited branch migration through the central base pairs (75).

The strict requirement of homology in strand exchange has been called into question as the mechanisms of more YRs have been brought to light, in particular, the integrases of integrative conjugative elements (ICEs), also referred to as conjugative transposons (16) (see chapter by J. Gardner). These enzymes mediate strand exchange across overlap regions that include 5 to 6 bp nonhomologies. IntDOT, the integrase of the *Bacteroides* conjugative transposon CTnDOT, utilizes a 2 bp homology within a 7 bp overlap region for the first exchange step, and carries out the second exchange in a homology-independent manner [Fig. 4 (A)]. The resulting heteroduplex DNA is resolved by replication.

The CTXϕ phage of *Vibrio cholera* manipulates the XerCD recombinase of its host for its integration in a rather unusual reaction that also challenges the homology rule in its conventional sense (13) [Fig. 4(B)] (see chapter by F-X Barre). The + strand of the phage DNA folds itself into a forked hairpin structure to generate a pseudo XerCD target site within it. The first strand exchange between the phage and chromosome target sites is mediated by XerC, and utilizes 3 bp of homology adjacent to its cleavage site. The nonhomology adjacent to XerD stops the reaction at this pseudo-Holliday junction stage. Resolution of this structure by replication generates a chromosome with an integrated copy of the phage. As the lysogen is not flanked by functional recombination sites, the integration reaction is irreversible, proscribing the excision of the replicative form of the phage. However, tandem copies of the lysogen generated by successive integration events permits the production of + single-stranded phage genome by a rolling circle type of replication. An analogous strategy of integration via single-strand exchange by a tyrosine recombinase, followed by replication-mediated resolution of the resulting Holliday junction, is also devised during the capture of exogenous gene cassettes by integrons (76), which are important not only for their role in the spread of antibiotic resistance but also for their potential relevance to bacterial genome evolution. As in the case of the *V. cholorae* phage integration, the recombination target site on the cassette is assembled by the folding of a single-stranded DNA region. The phage and integron systems, rather than breaching the homology rule outright, seem to bend it by cleverly manipulating the strand exchange reaction in their favor.

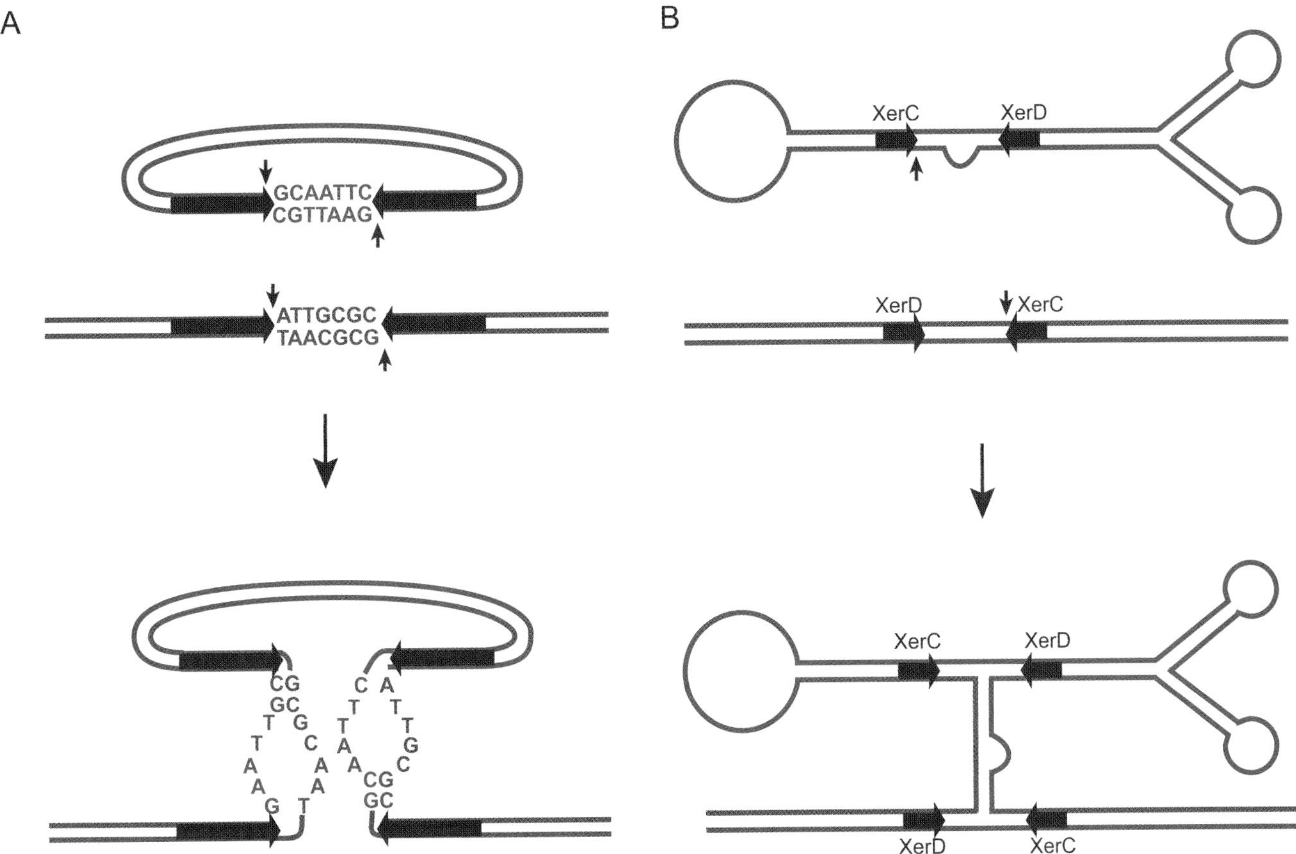

Figure 4 Challenges to the homology rule during tyrosine site-specific recombination. (A) The integrase (IntDOT) of the conjugative transposon CTnDOT catalyzes exchange of both strands between target sites that contain five consecutive nonhomologous positions within the 7 bp segments swapped between them (16). (B) The folded form of the "+" strand of the CTXϕ phage contains an imposter target site for XerCD recombinase of its host bacterium, *V. cholera*. Single-strand exchange mediated by the XerC active site between the phage DNA and the bacterial chromosome results in phage integration (13). The heteroduplex integrant in (A) and the pseudo-Holliday junction in (B) are likely resolved via replication. The flat horizontal arrows indicate recombinase binding sites. The short vertical arrows denote points of strand cleavage. doi:10.1128/microbiolspec.MDNA3-0021-2014.f4

Even the archetypal YRs may violate the homology rule in its strictest sense. Analysis of the Cre reaction between a wild type LoxP (the Cre target site) and mutant LoxPs containing single bp substitutions of the 6 bp strand exchange region reveals several types of outcomes: significant amounts of the Holliday junction intermediate without detectable recombinant product, small amounts of the product with higher amounts of the Holliday junction, small amounts of the Holliday junction with higher amounts of the product and small amounts of the product with no detectable Holliday junction (77). The overall strand exchange, the sum of the Holliday junction and recombinant yields, is most diminished for substitutions adjacent to the scissile phosphate, whose cleavage initiates the first strand exchange step, with one exception. The T to A transversion at position 2 from the initiation end gives a modest amount of the product with much smaller amounts of the Holliday junction. The accumulation of the Holliday junction as nonhomology shifts from the initiation-proximal positions to the central and termination-proximal positions of the strand exchange region is consistent with the normal execution of the first strand exchange step, while the second exchange step is blocked or severely impeded by nonhomology. Stable single-strand exchange by Cre, dictated by the location of nonhomology, is thus somewhat analogous to the formation of the pseudo-Holliday junction by XerCD during CTXϕ integration.

NONHOMOLOGY INDUCED KNOTTING OF SUPERCOILED PLASMIDS BY A YR

An even more flagrant violation of the homology rule is brought to light by Flp reactions between two *FRT* sites, nonhomologous at the central two positions of their 8 bp strand exchange regions and located within negatively supercoiled plasmids (44, 78). While no stable recombinant products result, the reaction ties the plasmids into knots of wide ranging complexity, but all in their parental configuration. When the two sites are in head-to-head orientation (with respect to the six homologous bp), the knots are even numbered; when the sites are in head-to-tail orientation, the knots are odd numbered. The knotting may be explained by two or repeated even rounds of recombination, giving rise initially to unstable recombinants containing mismatched base pairs that then rapidly recombine back to the parental form.

An obvious question is whether the observed knot complexity is due to DNA crossings added during an iterated series of recombination events occurring within a given synapse. A similar increase in topological complexity has been described for the serine recombinases when they attempt to recombine sites that harbor nonhomology in the overlap region (79, 80). However, members of this recombinase family follow a completely different mechanism (see chapter by W. M. Stark). They make concerted double-strand cuts within target sites arranged in a parallel fashion, promote right-handed rotation of the broken DNA through 180 degrees, and reseal the strands in the recombinant configuration. Mismatches between partner strands impede or reverse joining and encourage a second 180 degree rotation to restore complementarity, which favors joining. A repetition of these dual half-rotation steps will progressively increase the complexity of products. Can nonhomology alter the normal synapse geometry of tyrosine recombination and processively generate products of increasing complexity?

An alternative explanation for the knotting reaction by Flp is that the complexity of the knots reflects the topological complexity of the plasmid substrate from which they are generated. Since the pairing of *FRT* sites occurs by random collision, a range of supercoils (crossings between the two DNA domains bordered by the sites; blue × red in Fig. 5) can be trapped by the synapse. The antiparallel geometry of the *FRT* sites requires an odd number of such crossings between head-to-head sites and even number of crossings between head-to-tail sites (Fig. 5). Depending on the number of trapped interdomainal crossings, the first recombination event will generate an unknotted inversion circle plus 3-, 5-, 7- etc. crossing knots from the head-to-head sites [Fig. 5(A)]. Similarly, the products from the head-to-tail sites will be a pair of unlinked deletion circles plus 2-, 4-, 6- etc. crossing catenanes (linked circles) [Fig. 5(B)]. When the parental *FRT* sites are nonhomologous, these products contain mismatches, and are prone to a second round of recombination after dissociation of the original synapse. The addition of one more crossing during this step will convert the knots with odd number crossings from the inversion reaction to knots with even number crossings (4, 6, 8 etc.) [Fig. 5 (A)]. Similarly, the catenanes from the deletion reaction will be converted to fusion knots with odd number crossings (3, 5, 7 etc.) [Fig. 5(B)]. The prediction then is that when the synapse topology is simplified and made unique, the product topology must be correspondingly simple and unique. This indeed is the case (44). When Flp reaction is carried out after assembling the Tn3 resolvase synapse [which traps precisely three interdomainal negative supercoils, as in Fig. 5(A)], and taking care to minimize random entrapment of supercoils, the product yielded by the head-to-head sites is predominantly a 4-noded knot; that yielded by the head-to-tail sites is predominantly a 5-noded knot (44). The topologies of the products from corresponding reactions between two native (homologous) *FRT* sites are a 3-noded knot and a 4-noded catenane. The difference of one in the crossing numbers between the knot and the catenane is consistent with the need to arrange the sites in the same functional geometry, antiparallel, for them to recombine [Fig. 5(B)]. In addition to the three crossings anchored by resolvase, a fourth crossing must be trapped from the supercoiled plasmid substrate to keep the head-to-tail sites antiparallel (see also the section on "Difference topology"). Thus, nonhomology does not block recombination by Flp; nor does it induce processive recombination by altering the normal reaction mechanism. The unstable (mismatched) recombinants resulting from the first recombination event are restored to the more stable parental state by a second dissociative recombination event.

PROBING ACTIVE SITE MECHANISM USING CHEMICAL MODIFICATIONS OF THE SCISSILE PHOSPHATE GROUP

Mechanistic analysis of strand breakage and joining reactions in nucleic acids has greatly benefited from chemical modifications of the phosphate group in the nonbridging oxygen atoms to alter its electronegativity and/or stereochemistry, and in the bridging oxygen atoms to manipulate leaving group properties. The po-

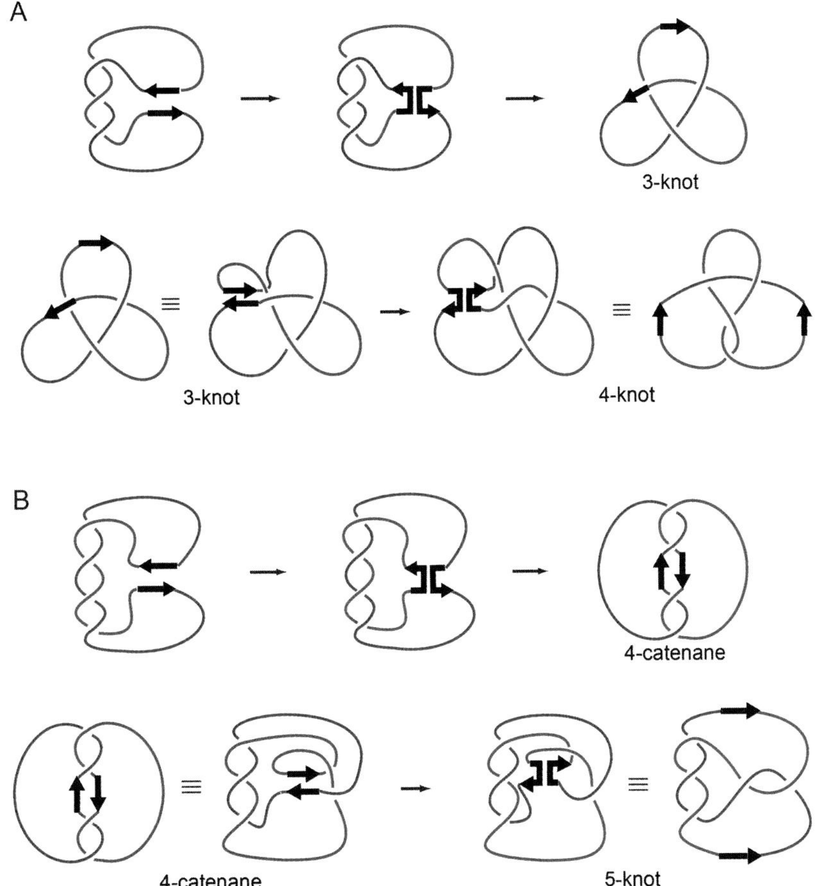

Figure 5 Flp-mediated knotting of supercoiled plasmids by recombination between two *FRT* sites harboring nonhomology within the strand exchange region. (A) The first recombination event between two head-to-head (inverted) *FRT* sites from a synapse containing an odd number of interdomainal (blue × red) supercoil crossings will generate a torus knot with the same number of crossings. The product from a synapse with one blue × red crossing will be an unknotted inversion circle, as it takes a minimum of three crossings to form the simplest knot. In the example shown, a 3-noded knot is formed from a 3-crossing synapse. A second recombination event after dissociation of the first synapse, and the assembly of a *de novo* synapse, can give rise to a twist knot with four crossings. (B) For *FRT* sites in head-to-tail (direct) orientation, the first recombination event from a synapse with an even number of interdomainal crossings yields a catenane with the same number of crossings. The product from a synapse with no crossings will be two unlinked deletion circles. The diagram illustrates the formation of a 4-noded catenane from a 4-crossing synapse. A second round of dissociative recombination can convert the 4-noded catenane into a 5-noded knot. In the reactions shown in (A) and (B), intradomainal supercoils (blue × blue or red × red crossings) are omitted for clarity, as they do not contribute to knot or catenane crossings. The products from the second rounds of recombination revert to the parental configuration. The noncomplementarity in the product formed by recombination between *FRT* sites containing nonhomology in their strand exchange regions encourages a second recombination event that restores base pairing and parental DNA configuration (44).
doi:10.1128/microbiolspec.MDNA3-0021-2014.f5

tential general acid role for Lys-201 and a subsidiary role for Trp-330 in leaving group stabilization during strand cleavage by Cre and Flp, respectively, have been revealed with the help of 5′-thiolate substrates (48, 54). Shuman and colleagues have successfully exploited phosphorothioate (replacement of a nonbridging oxygen by sulfur), methyl phosphonate (MeP; replacement of a nonbridging oxygen by the methyl group) and 5′-

thiolate substrates to investigate the active site mechanisms of vaccinia topoisomerase (81, 82, 83, 84).

Recent studies employing MeP-substrates [Fig. 6(A)] have revealed active site attributes of Cre and Flp that could not have been deduced from reactions of native phosphate containing substrates. These analyses have been performed predominantly using half-site substrates containing a single scissile phosphate or a modified scissile phosphate [Fig. 6(B)], together with recombinase variants harboring specific active site mutations. The chemical synthesis of MeP-half-sites is considerably easier than that of full-sites. Although the half-site reaction involves the breakage of one scissile phosphate within a substrate molecule, it faithfully preserves the chemical mechanism of the normal reaction. Associations of a recombinase-bound half-site can give rise to dimers, trimers and tetramers (85), so that the shared active site assembly and the *trans* mode of DNA cleavage are obeyed during Flp half-site reactions.

The reactivity of Flp variants on MeP-substrates demonstrates that neutralization of the phosphate negative charge in its ground state permits transition state stabilization in the absence of one of the two conserved arginines (either Arg-191 or Arg-308) (68, 86). Flp(R191A) and Flp(R308A) are active in the MeP reaction [Fig. 7(A)], while both these variants are almost completely inactive on phosphate containing DNA substrates. The electrostatic suppression of the lack of a positive charge in the recombinase active site by a compensatory charge substitution in the scissile phosphate of the DNA substrate has been demonstrated for the Cre recombinase as well (87, 88). Not only do Cre(R173A) and Cre(R292A) yield strand cleavage in an MeP-half-site, the double mutant Cre(R173A, R292A) also mediates this reaction. Presumably, the overall electrophilic character of the Cre active site is sufficient to neutralize the diminished negative charge present in the MeP, compared to the phosphate, transition state.

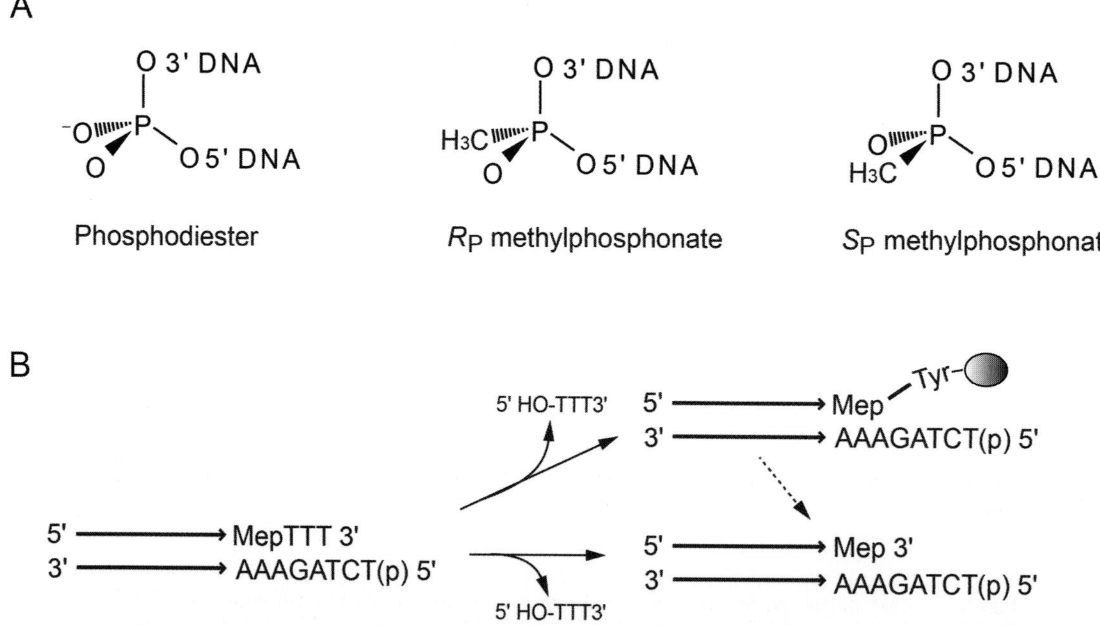

Figure 6 Reactions of half-sites containing methylphosphonate substitution at the scissile phosphate position. (A) The structures of methylphosphonate (MeP) are compared to that of the native phosphate in DNA. There are two possible stereoisomers of MeP (R_P or S_P). (B) The possible reactions of a half-site containing MeP at the scissile phosphate position are illustrated. The 5′-hydroxyl group on the bottom strand of the half-site is blocked by phosphorylation to prevent it from taking part in a pseudo-joining reaction. Attack of the MeP bond by the active site tyrosine will give the MeP-tyrosyl intermediate, which may undergo slow hydrolysis. The hydrolysis product may also be formed by direct water attack on the MeP bond. The two-step (type I) and single-step (type II) reaction pathways are mechanistically analogous to the type I and type II RNA cleavage activities of Flp (see text).
doi:10.1128/microbiolspec.MDNA3-0021-2014.f6

As the methyl substitution of one of the nonbridging oxygen atoms turns the normally symmetric phosphate group into an asymmetric center [Fig. 6(A)], an additional utility of the MeP substrates is in unveiling the stereochemical course of the recombination reaction. Stereochemically pure R_P and S_P forms of the MeP substrates are currently being used to dissect the individual stereochemical contributions of Arg-191 and Arg-308 in Flp, and to probe how other members of the catalytic hexad might influence these contributions.

DISTINCT ACTIVITIES OF Flp(R191A) AND Flp(R308A) IN THE MeP REACTION

The absence of Arg-191 or Arg-308 has strikingly different effects on the activity of Flp on an MeP-half-site [Fig. 7(A)] (68, 86). Flp(R308A) does not utilize the Tyr-343 nucleophile, but promotes direct hydrolysis of the MeP bond. Consistent with this mechanism, the double mutant Flp(R308A,Y343F) also yields the hydrolysis product with similar kinetics and V_{max} (maximal velocity) as Flp(R308A). Apparently, the lack of

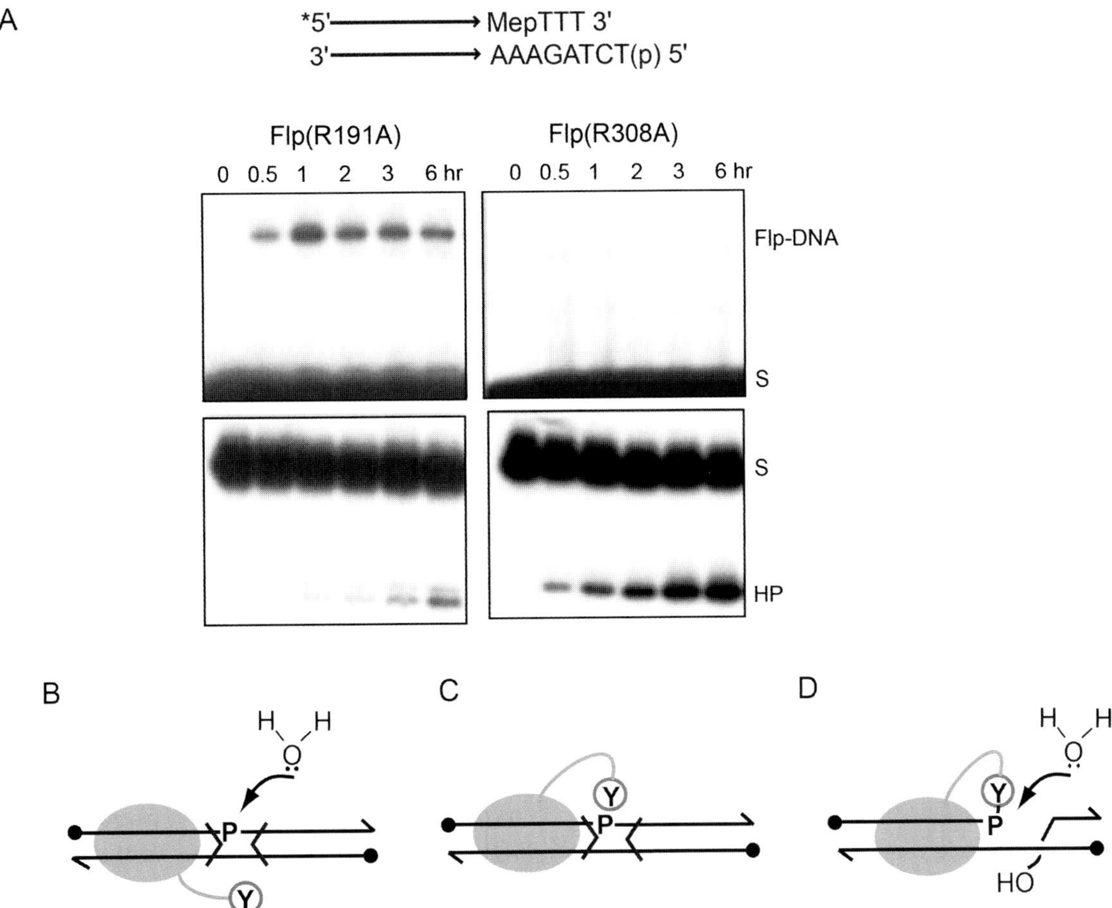

Figure 7 Distinct activities of Flp(R191A) and Flp(R308A) on an MeP-half-site. (A) Flp (R191A) cleaves the MeP-half-site (S) using Tyr-343 to form the protein–DNA adduct (revealed by SDS-PAGE; top) (86). This intermediate is converted to the hydrolysis product (HP) (revealed by denaturing PAGE; bottom) in a subsequent slow reaction. Flp(R308A), by contrast, yields the hydrolysis product directly, without going through the MeP-tyrosyl intermediate (68). (B) The binding of an Flp monomer to *FRT* activates the scissile phosphate, leaving it exposed until the binding of a second Flp monomer delivers Tyr 343 to the active site in *trans*. (C) Concomitant with the binding of a Cre monomer to the LoxP site, Tyr-324 engages the scissile phosphate in *cis*, thus protecting it against direct water attack. (D) As vaccinia topoisomerase, like Cre, assembles its active site in *cis*, the scissile phosphate is protected at the strand cleavage step during DNA relaxation. However, the protein's grip on DNA is loosened during the strand rotation step, leaving the phosphotyrosyl bond vulnerable to attack by water. doi:10.1128/microbiolspec.MDNA3-0021-2014.f7

Arg-308 permits the abundant water nucleophile to access the reaction center, where it out competes Tyr-343 to give a dead-end product. However, the possibility that Arg-308 is required for the positioning of Tyr-343 or its nucleophilic activation cannot be ruled out. The corresponding arginines, Arg-292 of Cre and Arg-410 of *Leishmania* topoisomerase I, are hydrogen bonded to the catalytic tyrosine in their respective vanadate-transition state structures (39, 89). Flp(R191A), by contrast, utilizes Tyr-343 as the cleavage nucleophile to yield the tyrosyl intermediate. Direct hydrolysis in the Flp(R191A) reaction is only a minor component. Cre (R173A) and Cre(R292A) are mechanistically similar to Flp(R191A) in that they promote Tyr-324-mediated cleavage of MeP (87, 88).

POTENTIAL ROLES FOR ACTIVE SITE AND PHOSPHATE ELECTROSTATICS IN PREVENTING FUTILE PHOSPHORYL TRANSFER

As suggested by the MeP reactions, in addition to balancing the phosphate negative charge, Arg-308 of Flp appears to protect the normal reaction course from abortive hydrolysis, perhaps by electrostatically misorienting water nucleophile (which is a dipole) from the activated phosphate. The phosphotyrosyl bond formed by vaccinia topoisomerase during DNA relaxation is apparently protected from hydrolysis by an analogous mechanism, utilizing the negative charge on the scissile phosphate (84). Furthermore, the potential role of the Arg-308 side-chain in orienting or activating Tyr-343 (see above under "Distinct activities of Flp(R191A) and Flp(R308A) in the MeP reaction") suggests an alternative or collaborative mechanism for preventing futile breakage of the DNA backbone by increasing the local concentration of the tyrosine nucleophile. As noted earlier, the need to shield the scissile phosphate from extraneous nucleophiles would be more critical for Flp because of its *trans* active site. The scissile phosphate, activated by the proactive site of a bound Flp monomer, stays exposed until Tyr-343 is provided in *trans* [Fig. 7(B)]. Binding by a Cre or topoisomerase monomer to DNA and the alignment of the tyrosine nucleophile with respect to the scissile phosphate would be nearly concomitant events because of their *cis* active sites [Fig. 7(C)]. In the case of the topoisomerase, which acts as a monomer, the strand rotation step may open the phosphotyrosyl bond to attack by water [Fig. 7(D)], which is prevented by phosphate electrostatics. Such a protective mechanism is likely unnecessary for the recombinases, as the tight organization of the recombinase tetramer-DNA complex (Fig. 1) and the dynamics of strand exchange within it preclude water from accessing the phosphotyrosyl bond. The extrusion of the cleaved strand into the center of the "strand exchange cavity" seen in the structure of the Cre-recombination synapse (52) would be consistent with strand swap being nearly concomitant with strand cleavage.

TYROSINE RECOMBINATION STEP-BY-STEP FROM START TO FINISH: SINGLE MOLECULE ANALYSIS

Single molecule analysis of tyrosine recombination using real-time tethered particle motion (TPM), tethered fluorophore motion (TFM) and fluorescence energy transfer (FRET) have provided deeper insights into the prechemical, chemical, and conformational steps of the reaction pathway by revealing transient states as well as long- and short-range movements of DNA (90, 91, 92, 93). The results of these studies reveal interesting similarities and contrasts among Cre, Flp and λ Int. The kinetics of recombinase binding to target sites and the pairing of bound sites are quite fast in all three cases, ruling out intrinsic barriers to synapsis, at least *in vitro*. There is a strong commitment to recombination following the association of Flp with the *FRT* sites. The formation of nonproductive complexes (those that do not synapse) and wayward complexes (those that do not form the Holliday junction intermediate or complete recombination after synapsis) constitute only minor detractions from the productive pathway (91). The stability of the synapse is enhanced by strand cleavage in the case of Flp and λ Int. However, Cre forms stable synapse even in the absence of strand cleavage (90). Recombination by Flp is efficient, and the frequency of occurrence of the Holliday junction intermediate is quite low (91). λ Int exhibits a strong and early commitment to a directed reaction path, likely assisted by its accessory factors bound to their cognate sites (92). Unidirectionality of an initiated recombination event would be a desirable attribute *in vivo* in bringing about the desired DNA rearrangement, without reversing course midway through a reaction. The Holliday junction formed during Cre recombination, however, is long lived, thanks to a rate-limiting step that follows its isomerization (93). This kinetic barrier affords the opportunity for the reaction to be interrupted and to go backwards, at least *in vitro*. It is possible that the *in vitro* Cre reaction fails to recapitulate the native regulatory features of recombination occurring within the P1 phage genome organized into a nucleoprotein complex.

SINGLE MOLECULE TPM AND FRET ANALYSES AS PROBES FOR THE GEOMETRY OF SITE-PAIRING AND ORDER OF STRAND EXCHANGE

TPM analysis is based on the rationale that the Brownian motion (BM) amplitude of a small polystyrene bead (~200 nm in diameter) attached to one end of a DNA molecule, whose other end is held in place, will be determined by the length of the DNA (Fig. 8). Since individual steps of recombination (binding of recombinase to target sites and their bending, synapsis of two bound sites by DNA looping, formation of the Holliday junction, and excision of DNA between two head-to-tail sites) are accompanied by characteristic changes in DNA length, TPM is well suited for the stepwise analysis of recombination (90, 91, 92). The pre- and post-chemical changes in DNA length can be distinguished by challenging the reaction with SDS. Upon protein dissociation, the prechemical states and completed inversion reactions (from head-to-head sites) will return to the length of the substrate DNA molecules. The Holliday intermediates from the inversion or the excision reaction and the linear product from the latter will retain their reduced "tether" lengths. By performing the excision and inversion reactions in parallel using DNA substrates identical in length and in the location and spacing of the recombination sites, a complete analysis of the reaction path is possible. TPM is also useful for verifying the geometry of a pair of target sites within the recombination synapse, as described below.

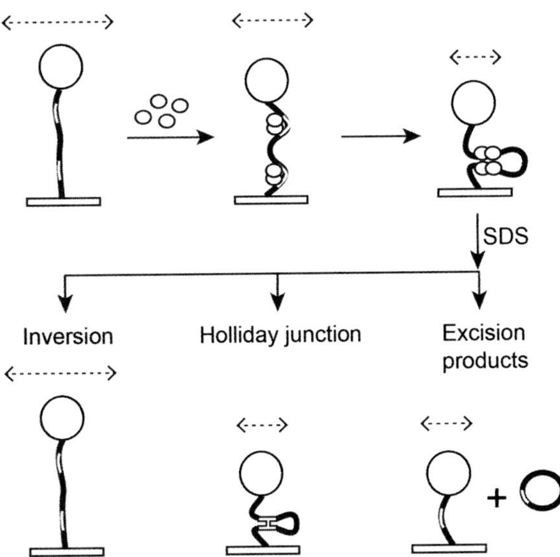

Figure 8 Stepwise analysis of recombination by TPM. The DNA molecule containing two recombination target sites (open boxes) in head-to-head or head-to-tail orientation is attached to a glass slide at one end and tethered to a polystyrene bead at the other. The change in DNA length occurring at individual steps of recombination is reported by the corresponding changes in the BM amplitude of the bead (schematically indicated by the dashed lines with arrowheads at either end). The bending of the sites bound by the recombinase (shown as globules) causes a shortening of DNA, which is magnified upon synapsis. Chemical steps of recombination within the synapse can result in Holliday junction formation or completion of recombination (DNA excision in the case of head-to-tail sites and DNA inversion in the case of head-to-head sites). Upon dissociation of the recombinase from DNA by SDS treatment, the Holliday junction intermediate and the linear excision product will retain their low BM amplitude. The inversion product has the same length, and hence the same BM amplitude, as the starting DNA molecule.
doi:10.1128/microbiolspec.MDNA3-0021-2014.f8

For sites in head-to-head orientation, the entry and exit points of the DNA will be at the same end of the synapse if the sites are aligned in a parallel fashion [Fig. 9(A)]. If the sites are in antiparallel alignment, the DNA will enter and exit the synapse from opposite ends. These situations will be reversed for a pair of sites in the head-to-tail orientation [Fig. 9(B)]. The proximal disposition of the entry and exit points imposes a stronger constraint on the DNA than their distal configuration, and makes it effectively shorter by a small amount. A significant difference in the BM amplitudes of two DNA molecules of identical length harboring a pair of equally spaced recombination sites, head-to-head in one case and head-to-tail in the other, would indicate preferential synapsis in one geometry. If the BM amplitude of the synapsed state is larger for the head-to-head sites, the preferred geometry is antiparallel. This is indeed the observed result for *FRT* sites synapsed by Flp (44). This conclusion is supported by single molecule FRET measurements in two *FRT* sites whose synapse geometry is restricted to being either parallel or antiparallel by a short single-stranded tether joining them (44) (Fig. 10). A change in the FRET state, in the expected direction, upon binding of Flp(Y343F) is observed only for the pair of *FRT* sites constrained in the antiparallel sense.

Given the approximate 2-fold symmetry of the core recombination sites, one might have imagined that they would synapse in a parallel or antiparallel fashion, even if only one of the two arrangements was productive for recombination. Topological and FRET results argue for preferred antiparallel synapsis of *FRT* sites even in the absence of the chemical steps of recombination (41, 44). Perhaps an asymmetric DNA bend within the strand exchange region of an Flp bound *FRT* site may preclude two similarly bent sites from occupying the synapse in a parallel fashion. A sharp bend located at a single bp step at one end of the strand exchange region

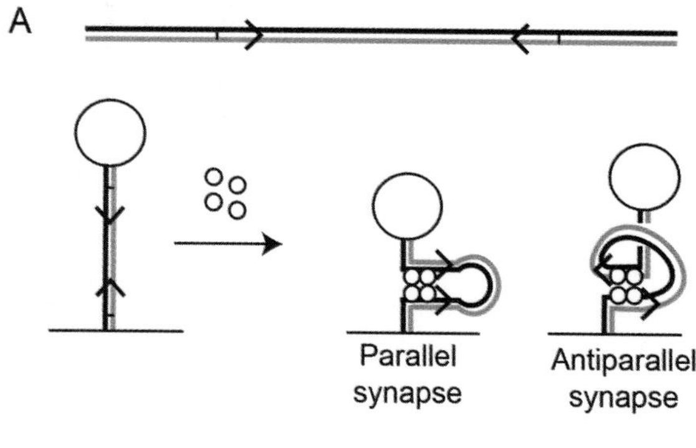

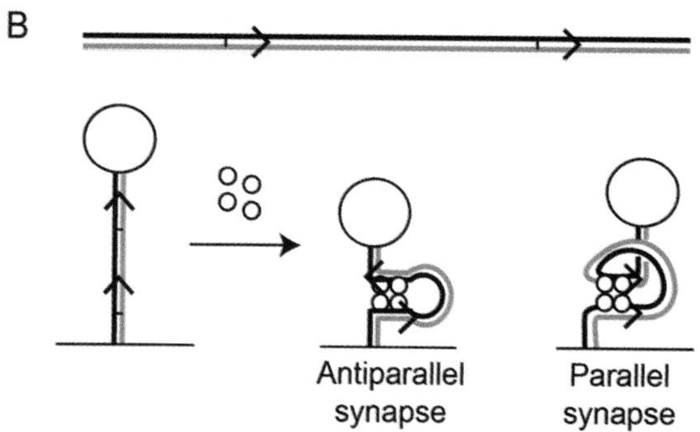

Figure 9 Effect of synapse geometry on the BM amplitude of DNA. (A) The DNA contours for a pair of synapsed head-to-head sites are outlined for their alignment in parallel (left) or antiparallel (right) geometry. (B) Similar diagrams as in (A) represent the antiparallel (left) and parallel (right) synaptic configurations for head-to-tail sites. The effective length of DNA is slightly larger when its entry and exit points are at opposite ends of the synapse than when they are at the same end. For two DNA substrates that differ only in the relative orientation of the recombination sites, a difference in the BM amplitudes of synapsed head-to-head versus head-to-tail sites signifies a preferred geometry of the synapse.
doi:10.1128/microbiolspec.MDNA3-0021-2014.f9

has been observed in the structures of LoxP complexed with cleavage-incompetent mutants of Cre (94).

The synapse geometry raises the question of "order of strand exchange" during recombination. Depending on the location of the asymmetric bend with respect to the scissile phosphate, there are two geometrically equivalent and chemically competent configurations of the antiparallel synapse. One would correspond to "top strand" cleavage, and the other to "bottom strand" cleavage [Fig. 11(A)]. If one of the two synaptic configurations is preferred, the order of strand cleavage/exchange will reflect this preference. Current evidence suggests that Flp performs strand exchange without obvious bias (95, 96), indicating that the two modes of antiparallel synapsis are equally likely. Cre, by contrast, performs ordered strand exchange. FRET analysis with donor–acceptor dye pairs suitably positioned with respect to the strand exchange region demonstrates a preferred synapse, the DNA bend within which is consistent with the biochemically mapped preference in strand cleavage [Fig. 11(B), (C)] (48).

Ordered strand exchange is the norm in the λ Int and XerCD systems as well. The constraints imposed by high-order protein assemblies and DNA topology on the synapsis of the Int-bound core sites can dictate which pair of scissile phosphates is primed for initial cleavage (31, 70, 97). In the case of XerCD, depending on the reaction context, cleavage susceptibility may be determined by the synapse topology organized by accessory factors or may be altered by the presence or absence of an interacting regulatory protein (98, 99).

TOPOLOGICAL AND CHIRAL FEATURES OF TYROSINE RECOMBINATION

Tyrosine recombinases (Cre, Flp and λ Int) in general assemble the synapse by random collision of their target sites (see chapter by M. Boocock). In the case of Int, this randomness appears to be superposed over an intrinsic topological specificity (see below). As noted in discussing the role of homology in Flp recombination, it is the interdomainal crossings trapped during synapsis (blue × red in Fig. 5) that appear as knot or catenane crossings in the recombination products. As pointed out earlier, the inversion reaction results in an unknotted circle (with the blue domain inverted with respect to the red) together with a range of increasingly complex knots; the deletion reaction produces unlinked circles as well as a range of increasingly complex catenanes.

The topology of Cre recombination is sensitive to reaction conditions. Relatively high pH tends to increase the complexity of the products, while lower pH has the opposite effect (43). Computer simulations, combined with DNA cyclization assays, suggest that the topological difference between Cre and Flp can be accounted for by the difference in the presynaptic bends that they induce in their target sites (~35° for LoxP and ~78° for *FRT*) (100). The larger bend tends to localize two presynaptic *FRT* sites within separate branches of a plectonemically supercoiled circle [Fig. 12(A)], while the smaller bend tends to place two presynaptic LoxP sites in the same branch [Fig. 12(B)]. Interbranch recombination results in topologically complex products; intrabranch recombination gives simple products. Thus, protein-induced local changes in the statistical properties of large DNA molecules can strongly influence

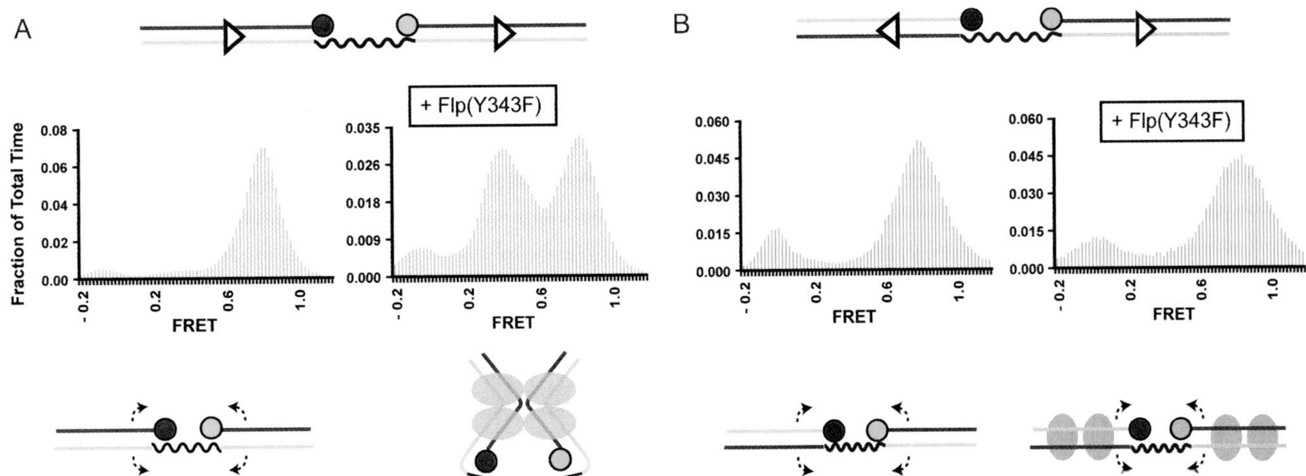

Figure 10 Preferred antiparallel synapsis of a pair of tethered *FRT* sites. (A) The two *FRT* sites, whose orientation is indicated by the arrowheads, are constrained by a single-stranded tether (wavy line) to align only in the antiparallel geometry. The positions of the donor (Cy3) and acceptor (Cy5) fluorophores are indicated by the green and red circles, respectively. The shift towards lower FRET upon Flp(Y343F) binding is consistent with the synapsis of the *FRT* sites as schematically diagrammed (44). (B) In the tethered DNA substrate, analogous to that diagrammed in (A), the *FRT* sites are constrained to pair only in the parallel geometry. Flp(Y343F) binding produces no change in FRET, suggesting the absence of parallel synapsis. doi:10.1128/microbiolspec.MDNA3-0021-2014.f10

their global topology, and dictate the outcomes of the reactions they partake in.

The topological outcomes of XerCD recombination are dictated by the contexts in which the reaction occurs. Recombination between *dif* sites (utilized for the resolution of *E. coli* chromosome dimers) requires the ATP driven DNA translocase FtsK (99). The reaction yields topologically simple products from negatively supercoiled substrates. When *dif*-recombination is activated by the carboxyl-terminal γ-domain of FtsK, which lacks the ATPase function, the products are topologically complex. The topology of recombination between two *cer* sites or two *psi* sites (utilized for the resolution of plasmid dimers) is dictated by accessory protein factors (101). The reaction normally occurs between sites in head-to-tail orientation, and requires negative supercoiling. The unique right-handed 4-crossing catenane produced from *psi* × *psi* recombination, as well as the structure of the Holliday junction formed by *cer* × *cer* strand exchange, conforms to a three-crossing synapse topology. According to tangle analysis, the recombination synapse fits a unique three-dimensional model, with three solutions that correspond to three distinct views obtained by rigid body movements of the synapse and projection on to a planar surface (102).

The FtsK-dependent topology simplification by XerCD recombination is also manifested in the unlinking of catenanes harboring *dif* sites, either in the parallel or antiparallel sense (103). Catenanes with parallel *dif* sites are topologically analogous to catenanes resulting from the replication of circular plasmids and chromosomes, which are unlinked by the type II topoisomerase Topo IV (104). Rather surprisingly, XerCD-FtsK can support the resolution of chromosome catenanes *in vivo* when Topo IV activity is compromised (105). The recombination mechanism would suggest that unlinking by XerCD-FtsK proceeds by removing one crossing at a time (Fig. 13). This intuitive model, based on product distributions observed in *in vitro* reactions with plasmid substrates, has been validated mathematically by a combination of tangle analysis and knot theory under the assumption that each recombination event reduces the topological complexity of the substrate (106, 107).

An intriguing aspect of tyrosine recombination, brought to light primarily from the analysis of λ Int reactions is the apparent chirality of the reaction (108). The chirality of knots and catenanes formed from inversion and deletion reactions, respectively, in negatively supercoiled substrates follows from the right handed chirality of plectonemic negative supercoils. However, quite unexpectedly, even reactions of nicked substrates turn out to be chiral. In the reactions between attP and attB sites in nicked substrates, two inter-domainal right-handed crossings are trapped by the DNA Inter-

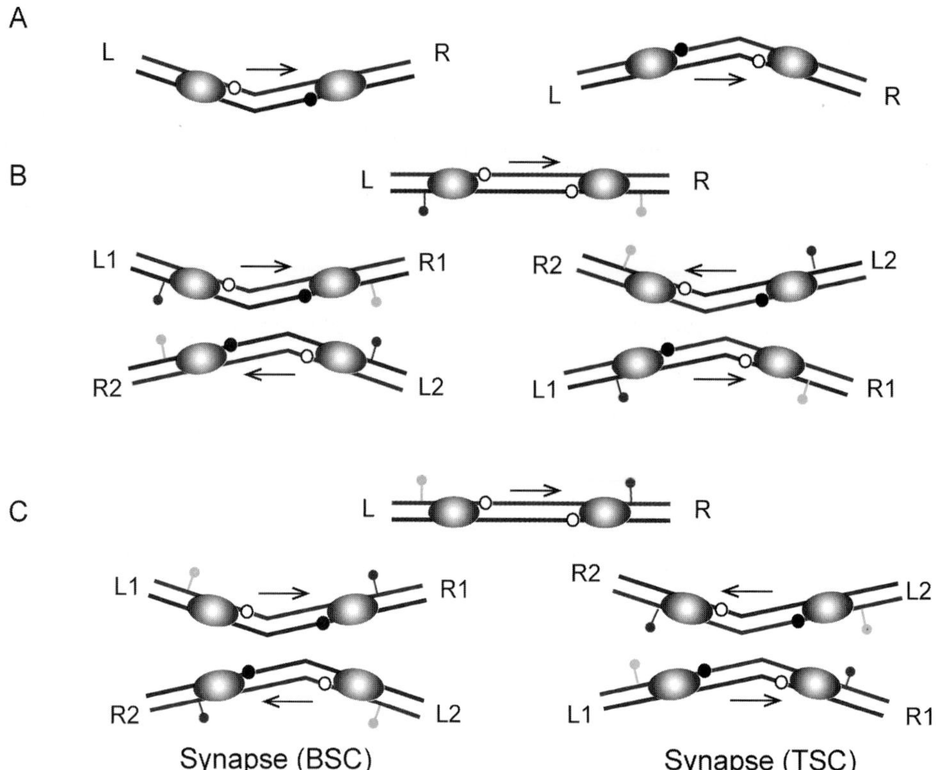

Figure 11 The preferred assembly of one of two possible types of antiparallel synapse can specify the order in which strands are cleaved and exchanged during recombination. (A) A LoxP site bound by Cre is bent asymmetrically, the bend center being located close to one end of the strand exchange region. The two possible asymmetric bends would specify the cleavage of the bottom (blue) or the top strand (red). The scissile phosphates primed for cleavage are indicated by the filled circles; the quiescent ones are shown as open circles. For convenience of orienting the sites, the DNA arms are labeled as L (left) and R (right) as in Fig. 1. (B) Based on the structure of the Cre-LoxP complex, fluorophores can be so positioned as to minimize donor (green)–acceptor (red) distance, and induce efficient FRET when the synapse favoring bottom strand cleavage (shown at the left) is assembled by Cre. In this fluorophore configuration, the FRET efficiency will be low for the synapse favoring top strand cleavage (shown at the right). (C) By reversing the left–right orientation of the fluorophores with respect to the strand exchange region, while maintaining their relative positioning, the synapsis favoring top strand cleavage (right) can be made to acquire the high FRET state. Experimental results indicate a clear preference for the synapse shown at the left in (B) suggesting that recombination is initiated by bottom strand cleavage and exchange (48). doi:10.1128/microbiolspec.MDNA3-0021-2014.f11

actions of Int and its accessory factors; in the reactions between attL and attR sites, the corresponding number is one. It would take only a single additional right handed DNA crossing randomly trapped in the substrate to generate a chiral three-noded knot as the product of an attP-attB inversion reaction. During attP-attB reactions in negatively supercoiled molecules, Int (in conjunction with accessory factors) traps three right handed DNA crossings, one more than that deduced for the same reactions in nicked circular molecules. However, for attL-att R reactions, this number is still one, unchanged between nicked and negatively supercoiled substrates.

The topological and chiral features of a recombination reaction are conveniently and succinctly summarized by tangle diagrams such as those illustrated in Fig. 14. A tangle may be perceived as a three-dimensional ball within which strings representing double stranded DNA may cross in a variety of ways. In the two dimensional projection of a tangle, the entry and exit points of DNA are placed at the NE, NW, SE and SW corners (in a geographical sense). The O_b and

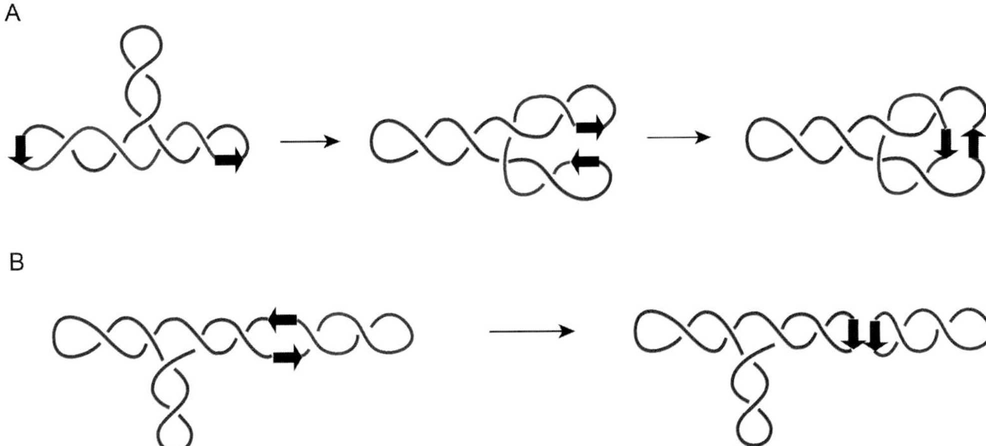

Figure 12 The magnitude of the DNA bend at the recombination target sites influences their localization within the branches of plectonemically supercoiled DNA. (A) The large DNA bend induced by Flp tends to localize presynaptic *FRT* sites in separate plectonemic branches. Recombination between such sites yields topologically complex products. In the example shown, the excision reaction yields a 4-noded catenane. (B) The relatively small DNA bend induced by Cre tends to place presynaptic LoxP sites within the same plectonemic branch, thus simplifying the topology of recombination products. The excision reaction shown here yields two unlinked deletion circles.
doi:10.1128/microbiolspec.MDNA3-0021-2014.f12

O_f tangles harbor DNA constrained by protein binding and 'free' DNA, respectively. The P and R tangles contain the DNA segments that engage in crossover in their parental and recombined states, respectively. O_b for attL-attR reactions, containing one right-handed crossing, is a +1 tangle for nicked as well as negatively supercoiled substrates (as shown In Fig. 14 A, B). For attP-attB reactions in nicked substrates, the O_b tangle is +2; for the same reactions in negatively supercoiled substrates, the Ob tangle is +3.

Most surprisingly, recombination reactions from nicked substrates by Cre and Flp also appear to be chiral, trapping one right-handed inter-domainal crossing in the synapse. This crossing is proposed to predispose the reaction towards a chiral product via a right-handed Holliday junction intermediate. The near perfect planarity of the DNA arms in the crystal structures of Cre and Flp (34, 52) challenges this postulate. Nevertheless, the slight out-of-plane disposition of the Holliday junction arms in the Flp crystal structure is consistent with the proposed right-handed chirality.

An irksome aspect of chirality is the difficulty in accommodating the experimental observation that the linking number change (ΔLk) associated with Flp- or λ Int-mediated inversion reactions between *FRT* sites and attL-attR sites, respectively, is either +2 or −2 (41, 109), and the two outcomes are equally likely for a nearly perfectly relaxed substrate. The right-handed chirality would predict a ΔLk of exclusively +2. For example, the right-handed crossing (a − node) trapped by Int would change its sign (a + node) as a result of DNA inversion (ΔLk = +1 − (−1) = +2). The tangle diagram depicting this change in the node sign in O_b is shown in Fig. 14A. The two suggested tangle solutions to resolve this paradox are shown in Fig. 14B, C. In Fig. 14B, the substrate DNA enclosed by the O_f tangle harbors two + crossings, one to compensate for the − crossing trapped by Int (in the O_b tangle) and an additional one to arrange the recombination sites with the antiparallel geometry in the P (parental) tangle. The inversion of each of these crossings would give ΔLk = −2, [(+1 − (−1)] + [(−2) − (+2)]. The problem, though, is that the energetic cost of introducing additional O_f crossings should make ΔLk = −2 less likely than ΔLk = +2, in violation of the experimental result. In Fig. 14C, the P tangle is switched from an ∞ tangle to a 0 tangle, so as to preserve the tangle notation in O_b, and still produce a ΔLk = −2, [−1 − (+1)]. This is also unsatisfactory, as it accommodates ΔLk = −2 by a slcight of hand(edness). If one follows the contour of the DNA circle, it is obvious that the crossing in O_b is left-handed, not right-handed. Chirality of tyrosine recombination and the ΔLk paradox arising from it remain an enigmatic curiosity that calls for further exploration.

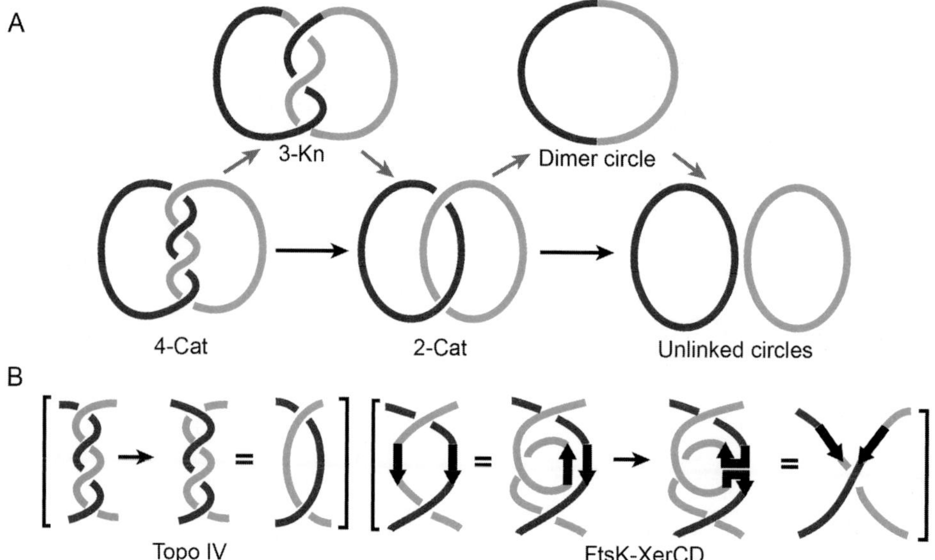

Figure 13 Unlinking of replication catenanes by XerCD-FtsK. (A) The unlinking of replication catenanes in *E. coli* is normally carried out by the type II topoisomerase Topo IV. For a 4-noded replication catenane containing parallel *dif* sites, unlinking by Topo IV will be completed in two steps (the straight path), removing two crossings at each step. Unlinking of the same catenane by FtsK-XerCD-mediated recombination at the *dif* sites requires four steps (the zigzag path), by removal of one crossing at a time. (B) The mechanisms for topology simplification by Topo IV and FtsK-XerCD are illustrated.
doi:10.1128/microbiolspec.MDNA3-0021-2014.f13

CONTRIBUTIONS OF YRs TO BASIC BIOLOGY AND BIOENGINEERING APPLICATIONS

Biochemical, biophysical, and structural studies of YRs have been seminal to unveiling the mechanisms of an important class of phosphoryl transfer reactions in nucleic acids and to understanding conformational dynamics associated with strand exchange between two DNA partners (20, 21, 23, 90, 91, 92, 93). The simple requirements of Cre and Flp have been exploited to carry out specific genetic rearrangements in bacteria, fungi, plants, and animals. By combining the DNA delivery properties of mobile group II introns in bacteria and the DNA exchange potential of tyrosine recombination, a new platform for genome editing via targetrons and recombinases (GETR) has been developed (110). In general, prokaryotic and eukaryotic cells engineered to express a recombinase and harboring its target sites in the genome or housed in an extrachromosomal vector carry out the expected reaction with high efficiency. Directed insertion of a desired foreign DNA into a genome as well as inversions, deletions, or translocations of selected genomic segments can thus be accomplished with reasonable ease. These manipulations have been particularly helpful in addressing fundamental problems in cell and developmental biology. The utilization of controlled and efficient site-specific recombination between homologous chromosomes to generate mosaic flies has provided a technical breakthrough for tracking cell lineages in *Drosophila* (111). Analogous strategies, in conjunction with multicolored reporter genes and live-cell imaging, have expanded the power and range of lineage tracking to higher organisms and facilitated its integration with the monitoring of intracellular signaling pathways (112). Methodologies for tissue-, cell type- and stage-specific induction of recombination activity make it possible to analyze spatial and temporal controls of developmental programs in intricate detail (113, 114). Another, perhaps less widely publicized, utility of Cre and Flp in basic biology is exemplified by "difference topology," an analytical method for tracing the topological path of DNA within high-order DNA–protein complexes (115, 116). Finally, Cre, Flp, and to a limited extent, λ Int have been put to practical use in a number of biotechnology-related applications. A brief description of the principles and practice of difference topology and of the potential impact of site-specific recombination on biotechnology is given below.

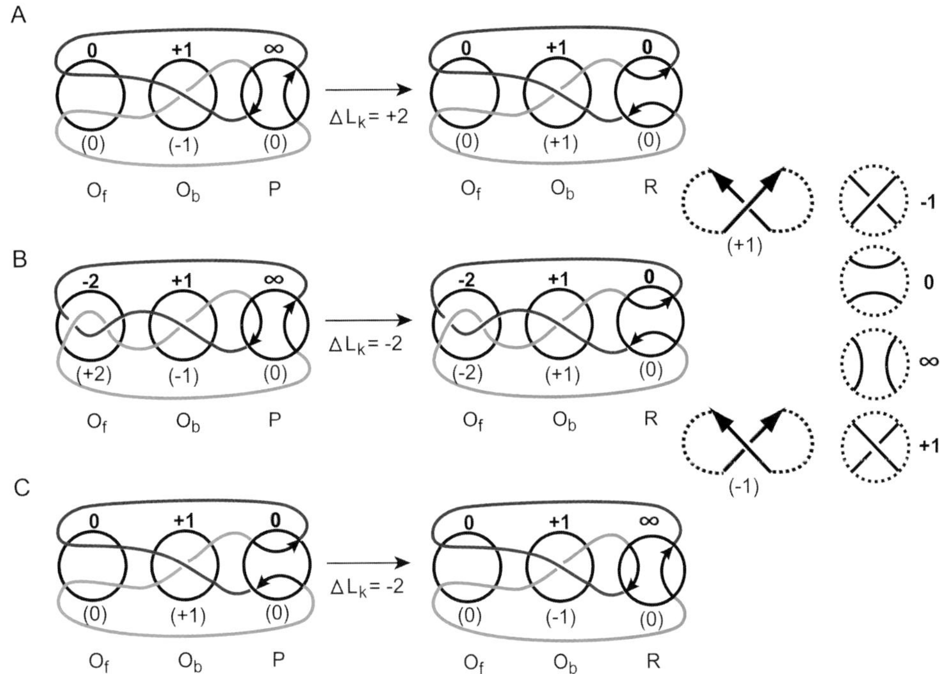

Figure 14 Tangle diagrams of attL- attR recombination performed by λ Int; ΔLk associated with DNA inversion. The λ Int mediated inversion reaction between attL and attR sites in three relaxed circular substrate molecules is represented by tangle diagrams (A-C). The O_b tangle contains inter-domainal DNA crossings trapped by Int (likely assisted by the accessory factors). O_f contains randomly trapped crossings in the 'free' DNA. The core recombination sites reside in the P tangle in anti-parallel geometry. The R tangle represents the post-recombination state of the sites. The tangle notations are shown at the top in bold; the corresponding DNA crossing (node) signs are given at the bottom in parentheses. The convention for the crossing signs (+1 or −1) is illustrated at the right, with the arrow heads denoting the direction (arbitrarily assigned) for the circular DNA axis. The simplest tangles (**0**, **+1**, **−1**, **∞**) are diagrammed at the far right. **A.** In the DNA molecule shown here, one right-handed crossing is trapped in O_b, and none are contained in O_f. In tangle parlance, recombination changes P(**∞**) tangle to the R(**0**) tangle, yielding an unknotted inversion circle. Note that a right-handed crossing in O_b in the substrate is **+1** by the tangle convention, but −1 by the sign convention. In the recombinant product, the crossing sign in O_b becomes +1 because of DNA inversion. The linking number change (ΔLk) accompanying the attL-attR reaction is +2. **B.** The same reaction as in **B** is shown for a molecule with two left-handed crossings present in O_f. The ΔLk for the reaction is −2. **C.** A molecule performing the same reaction as in **A** and **B** is represented with P(**∞**) and R(**0**) switched to P(**0**) and R(**∞**), respectively. The ΔLk associated with recombination is −2 in this case as well. The ΔLk changes are explained in more detail in the text. doi:10.1128/microbiolspec.MDNA3-0021-2014.f14

DIFFERENCE TOPOLOGY: DECIPHERING DNA TOPOLOGY WITHIN DNA–PROTEIN MACHINES

The elimination of topological randomness from Cre and Flp reactions by assembling a unique synapse with the assistance of protein factors (40, 41, 42, 43) is the basis for the analytical tool called difference topology. The method is useful for determining the number of supercoils sequestered by two DNA sites when they functionally interact with each other. As we have seen already, when three negative supercoils are trapped adjacent to Cre or Flp synapse, say by utilizing the resolvase synapse, the inversion and deletion reactions yield a 3-noded knot and a 4-noded catenane, respectively (Fig. 5). The inversion knot faithfully preserves the number of DNA crossings in the external synapse, as three (or an odd number) crossings would bring head-to-head LoxP or *FRT* sites in the antiparallel geometry that promotes recombination. A fourth crossing, easily provided by the negatively supercoiled substrate, is necessary for the deletion reaction, as it takes an even number of crossings to confer antiparallel geometry on head-to-tail sites. By similar arguments, two external negative supercoil crossings would be revealed in the difference topology analysis as a 2-noded catenane for the deletion reaction and a 3-noded knot for the inversion reaction. Random entrapment of supercoils in the hybrid synapse can be avoided by suitable placement of the recombination sites with respect to the external synapse. The crossings in the inversion knot and the deletion catenane, analyzed by gel electrophoresis and electron microscopy, would thus accurately report on the DNA topology of the external synapse. A simplified description of the concepts and experimental applications of difference topology can be found in a recent review (2). Using this analysis, the topology of the interactions among the left and right ends of phage Mu and its transposition enhancer element within the transposition complex organized by the MuA protein has been mapped as a three-branched, five-crossing plectoneme (117) (see chapter by R. M. Harshey).

ENGINEERING OF EUKARYOTIC GENOMES USING YRs

Before the practical utilities of site-specific recombination came into prominence, genome manipulations in higher eukaryotes relied either on nonhomologous or homologous recombination. Nonhomologous recombination promotes the efficient integration of linear DNA molecules into most genomes, but does so randomly. Homologous recombination permits modification of genomic loci with high specificity, but suffers from very low efficiency. Site-specific recombinases circumvent the drawbacks of both inefficiency and promiscuity, but require the prior integration of their cognate target sites into the genome to be modified. Furthermore, targeting multiple loci within a genome is limited by the number of available recombinases with the desired properties. This problem can at least be partly circumvented by taking advantage of the homology rule that dictates successful recombination. Mutually incompatible, but individually functional, target sites may be designed by introducing nonhomologies within their strand exchange regions. An even better solution, at least in principle, is the directed evolution of recombinases with altered target specificities (118, 119, 120, 121). Among YRs, Cre and Flp have been, by far, the enzymes of choice for applications in biotechnology. Variants of λ Int that are not functionally limited by cofactor requirements (122) have so far been running a rather distant third. The integrase of the *Streptomyces* phage ϕC31 and chimeras derived from an activated form of the Tn3 resolvase represent serine recombinases that have shown promise as tools in applied genetics (123, 124). Zinc finger nucleases (ZFNs), transcription activator effector-like nucleases (TALENs) and clustered regulatory interspersed short repeats (CRISPR)-Cas based RNA guided nucleases have complemented and augmented site-specific recombinases in the bioengineer's arsenal for analyzing and reshaping genomes (125, 126, 127).

The optimal performance of a recombinase and the tight regulation of its activity in a given biological context often require amino acid substitutions in the native protein sequence and/or the introduction of allosteric control regions. For example, the preferred growth temperature of budding yeast (~30 °C), in which Flp normally functions, differs from that of mammalian cells (37 °C). A thermo-tolerant variant of Flp (Flpe), with higher activity at 37 °C than Flp, was obtained by mutagenesis coupled to selection (128). Flpe harbors four amino acid changes from Flp: P2S, L33S, Y108N, and S294P. Flpe, which still underperforms Cre in mammalian cells, has been further improved by adding a nuclear localization signal and by 'humanizing' its codons (129). The new variant Flpo is comparable to Cre in its activity in mammalian cells. Continuous expression of a recombinase from a constitutively active promoter could be counterproductive because of the likelihood of the intended reaction being reversed and potential toxic effects arising from rare off-target recombination events. These impediments can be overcome or minimized by conditional recombinase expression from a regulatable promoter (Tet-on or Tet-off, for example) (130). Small molecule effectors are also useful for controlling recombination activity. The recombination potential of Flp or Cre fused to the ligand binding domain of the steroid hormone receptor is activated only in the presence of natural estrogens or synthetic estrogen receptor antagonists (131, 132, 133, 134).

Attempts to increase the target repertoire of a recombinase by generating variants with nonoverlapping specificities have been moderately successful. The basic strategy involves the screening of a large pool of the mutagenized recombinase gene for those that code for "shifted" or "switched" specificities. Simple bacterial genetic assays or more rapid high-throughput cell sorting screens, based on chromogenic reporter genes, have been effectively employed for identifying recombinases with the desired recombination potential (119, 121). Substrate designs that place the target sites and the recombinase genes in *cis* so as to link them by the act of recombination (substrate linked protein evolution; SLiPE) can accelerate screening by simple PCR-based protocols (135). In a distinct cell-free approach, *in vitro* compartmentalization (IVC) has been used to obtain altered specificity variants of λ Int (120). The IVC method relies on compartments of an oil-in-water emulsion in which *in vitro* expressed Int variants and the target sites are encapsulated.

In the Flp system, the strategy of mutagenesis followed by a bacterial dual reporter screen yielded recombination capability initially towards *FRT* sites containing single point mutations in the Flp binding element and subsequently towards sites containing combinations of these mutations (121, 136). It was further demonstrated that hybrid *FRT* sites, harboring distinct specificities in their two binding elements, can be recombined by a binary combination of Flp variants, each with the appropriate monospecific recognition potential (118). A step-wise directed evolution scheme with intermediate DNA shuffling steps is necessary to progressively coax Flp into accepting multiple changes within the *FRT* site. Consistent with the array of DNA contacts made by Flp, and the rather complex mode of substrate recognition, mutations involved in the acquisition of new specificities are distributed among amino acids that directly contact DNA as well as those that are located at monomer–monomer interfaces or in the proximity of catalytic residues.

Altered specificity variants of Cre have been evolved by structure-based substitution of base pairs recognized by Cre and randomization of selected amino acid positions in close proximity to them (119). Structural analysis of a subset of these variants suggests that two target sequences can be functionally recognized by a Cre variant through similar backbone contacts in conjunction with distinct base-specific contacts (137). These alternative modes of recognition are facilitated by a network of water-mediated contacts and an unexpected shift in the DNA backbone configuration. The contributions of water networks and macromolecular plasticity to DNA–protein interactions may thus complicate efforts to evolve new target specificities based on predictive schemes.

Directed evolution of recombinases that can act on naturally occurring sequences in their native biological context would signify a giant step forward in site-specific genome remodeling. Search algorithms such as Target Finder and TargetSiteAnalyzer have been developed to identify genomic sequences that match the size of a given recombination target site, and rank them according to the degree of their resemblance in organization and sequence to the chosen site (138, 139). There are >600,000 potential *FRT*-like sequences in the human genome (roughly one such sequence per 5,000 bp). Their distributions are inversely correlated to the average G/C content of individual chromosomes, in agreement with the A/T richness of the *FRT* site. The highest density (one *FRT*-like sequence per ~4 kb) is in chromosomes 4 and 13 with an average G/C content of 38%, and the lowest (one *FRT*-like sequence per ~8 kb) is in chromosomes 19 and 22 with an average G/C content of 48%. The majority of duplicate *FRT*-like sequences are located in the copies of LINE1, while others form part of the LTRs (long terminal repeats) of endogenous retroviruses, Alu repeats and other repetitive DNA sequences. The potential genomic target sites located by search algorithms not only facilitate the manipulation of genetic loci of interest but also promote stringent specificity by providing sequences for counterselection during the steps of directed evolution of novel specificity recombinases.

Once an appropriately placed "high-ranking" site has been identified, the procedures of progressive directed evolution of the recombinase, aided by structural information in at least some of the cases, can be employed to turn it into an authentic recombination target. This strategy has produced Flp variants that utilize an *FRT*-like sequence located upstream of the human IL-10 gene and analogous sequences found in the human β globin locus (138, 139). Members of the latter set of sites perform well in mammalian cells when they are present on episomal vectors (139). However, further optimization of specificity, recombinase expression, and activity will be required before the system operates efficiently in the native chromosomal context. The step-wise evolutionary approach has proven to be powerful enough to yield a Cre variant capable of deleting a proviral DNA of HIV-1 pseudotype by recombination between LoxP-like sequences located in the LTRs (140).

An important and frequently employed genome engineering reaction is recombinase-mediated cassette exchange (RMCE), a replacement reaction that exchanges two DNA fragments by a double recombination event between sites flanking them at either end (141, 142) (Fig. 15). The two sites harbored by each DNA partner are designed to be incompatible (heterotypic) for intramolecular recombination but compatible (homotypic) for intermolecular recombination in one configuration of the partners. In its early formats, RMCE reactions utilized a single recombinase such as Cre or Flp to perform exchange at both DNA ends (141, 143) [Fig. 15 (A)]. In more recent versions of RMCE, referred to as dual RMCE (144), recombination at each end is mediated by a separate recombinase, Flp and Cre or Flp and the integrase of the λ related HK022 phage [Fig. 15(B)]. To obtain the best replacement results by dual RMCE, the relative expressions of the two recombinases have

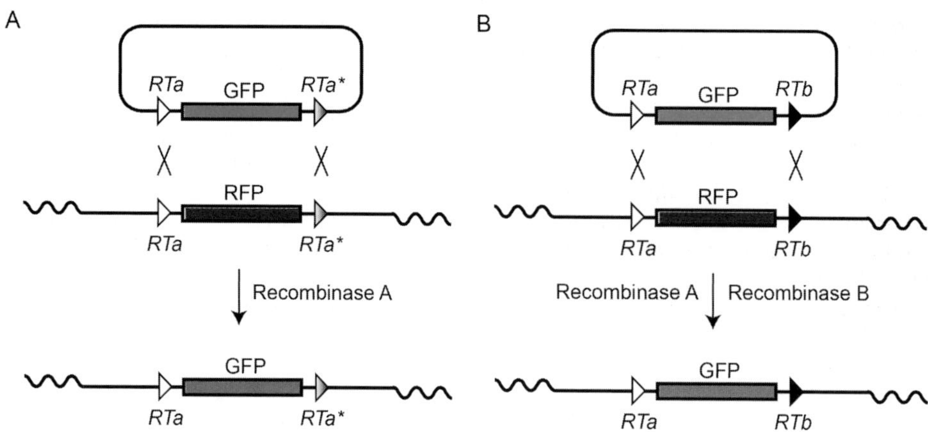

Figure 15 Recombination-mediated cassette exchange. (A) In the classical RMCE, the replacement of a native locus by a donor DNA fragment is mediated by the same recombinase acting on two pairs of target sites (*RTa* and *RTa**) that are compatible only in one configuration of the DNA partners. The reaction may be followed by the replacement of one fluorescent reporter (RFP; red fluorescent protein) by another (GFP; green fluorescent protein). (B) In dual RMCE, the same reaction as in (A) is mediated by two separate recombinases acting on their respective cognate sites (*RTa* and *RTb*).
doi:10.1128/microbiolspec.MDNA3-0021-2014.f15

to be carefully optimized (145, 146). The Cre-Flp pair can yield RMCE in up to 35 to 45% of the transfected cells, while the corresponding yield for the Flp-HK022 Int pair is ~12%.

The Cre-Flp based dual RMCE system is being successfully employed by several mouse genome engineering programs for systematically knocking out protein coding regions, expressing reporter cassettes from cellular promoters, and for amino-terminal protein tagging of gene trap clones *in situ*. The organizations leading these efforts are the International Mouse Phenotyping Consortium (IMPC; www.mousephenotype.org), the European Conditional Mouse Mutagenesis Program (EUCOMM; www.eucomm.org), the KnockOut Mouse Project (KOMP; www.knockoutmouse.org) and the German Gene Trap Consortium (www.genetrap.de). The beneficiaries from these endeavors will be high-throughput genomics and proteomics related to molecular medicine.

EPILOGUE

The intellectual seed for the advances in our understanding of site-specific recombination was sown more than fifty years ago by a simple and elegant model proposed by Allan Campbell for the integration of the phage λ genome into the *E. coli* chromosome. Over these five decades, the study of recombination has been transformed from a geneticists' sanctuary to the playing fields of biochemists and to the roaming grounds of crystallographers and biophysicists. Their collective contributions have unveiled the chemical simplicity, mechanistic elegance, and structural sophistication of the reaction. Genome engineers, biotechnologists, and system biologists have now almost completely taken over the field and seem poised to lead it in new directions.

Acknowledgments. *I thank Phoebe Rice, Ian Grainge, F.-X. Barre and Jeff Gardner for helpful criticisms and suggestions. The work in the Jayaram laboratory on the mechanism of site-specific recombination has been supported by the National Institutes of Health, the National Science Foundation, the Human Frontiers in Science Program, and the Texas Higher Education Coordinating Board. Funding for our recent work on the DNA topology and single molecule analysis of recombination has been provided by the National Science Foundation (MCB-1049925) and the Robert F Welch Foundation (F-1274).*

Citation. Jayaram M, MA C-H, Kachroo AH, Rowley PA, Guga P, Fan H-F, Voziyanov Y. 2014. An overview of tyrosine site-specific recombination: from an flp perspective. Microbiol Spectrum 3(1):MDNA3-0021-2014.

References

1. **Blaisonneau J, Sor F, Cheret G, Yarrow D, Fukuhara H.** 1997. A circular plasmid from the yeast Torulaspora delbrueckii. *Plasmid* **38:**202–209.
2. **Chang KM, Liu YT, Ma CH, Jayaram M, Sau S.** 2013. The 2 micron plasmid of Saccharomyces cerevisiae: a miniaturized selfish genome with optimized functional competence. *Plasmid* **70:**2–17.

3. **Piednoel M, Goncalves IR, Higuet D, Bonnivard E.** 2011. Eukaryote DIRS1-like retrotransposons: an overview. *BMC Genomics* **12:**621.
4. **Poulter RT, Goodwin TJ.** 2005. DIRS-1 and the other tyrosine recombinase retrotransposons. *Cytogenet Genome Res* **110:**575–588.
5. **Cortez D, Quevillon-Cheruel S, Gribaldo S, Desnoues N, Sezonov G, Forterre P, Serre MC.** 2010. Evidence for a Xer/dif system for chromosome resolution in archaea. *PLoS Genet* **6:**e1001166.
6. **Serre MC, El Arnaout T, Brooks MA, Durand D, Lisboa J, Lazar N, Raynal B, van Tilbeurgh H, Quevillon-Cheruel S.** 2013. The carboxy-terminal αN helix of the archaeal XerA tyrosine recombinase is a molecular switch to control site-specific recombination. *PLoS One* **8:**e63010.
7. **Van Houdt R, Leplae R, Mergeay M, Toussaint A, Lima-Mendez G.** 2012. Towards a more accurate annotation of tyrosine-based site-specific recombinases in bacterial genomes. *Mobile DNA* **3:**6.
8. **Austin S, Ziese M, Sternberg N.** 1981. A novel role for site-specific recombination in maintenance of bacterial replicons. *Cell* **25:**729–736.
9. **Azaro MA, Landy A.** 2002. λ integrase and the λ int family, p 118–148. *In* Craig NL, Craigie R, Gellert M, Lambowitz AM (ed), *Mobile DNA II*, ASM Press, Washington, DC.
10. **Barre F-X, Sherratt DJ.** 2002. Xer site-specific recombination: promoting chromosome segregation, p 149–161. *In* Craig NL, Craigie R, Gellert M, Lambowitz AM (ed), *Mobile DNA II*, ASM Press, Washington DC.
11. **Chaconas G, Kobryn K.** 2010. Structure, function, and evolution of linear replicons in Borrelia. *Annu Rev Microbiol* **64:**185–202.
12. **Ravin NV.** 2011. N15: the linear phage-plasmid. *Plasmid* **65:**102–109.
13. **Das B, Martinez E, Midonet C, Barre FX.** 2013. Integrative mobile elements exploiting Xer recombination. *Trends Microbiol* **21:**23–30.
14. **Futcher AB.** 1986. Copy number amplification of the 2 micron circle plasmid of Saccharomyces cerevisiae. *J Theor Biol* **119:**197–204.
15. **Genka H, Nagata Y, Tsuda M.** 2002. Site-specific recombination system encoded by toluene catabolic transposon Tn4651. *J Bacteriol* **184:**4757–4766.
16. **Rajeev L, Malanowska K, Gardner JF.** 2009. Challenging a paradigm: the role of DNA homology in tyrosine recombinase reactions. *Microbiol Mol Biol Rev* **73:** 300–309.
17. **Volkert FC, Broach JR.** 1986. Site-specific recombination promotes plasmid amplification in yeast. *Cell* **46:** 541–550.
18. **Turan S, Zehe C, Kuehle J, Qiao J, Bode J.** 2013. Recombinase-mediated cassette exchange (RMCE)- a rapidly-expanding toolbox for targeted genomic modifications. *Gene* **515:**1–27.
19. **Chen Y, Rice PA.** 2003. New insight into site-specific recombination from Flp recombinase-DNA structures. *Annu Rev Biophys Biomol Struct* **32:**135–159.
20. **Grindley ND, Whiteson KL, Rice PA.** 2006. Mechanisms of site-specific recombination. *Annu Rev Biochem* **75:**567–605.
21. **Radman-Livaja M, Biswas T, Ellenberger T, Landy A, Aihara H.** 2006. DNA arms do the legwork to ensure the directionality of λ site-specific recombination. *Curr Opin Struct Biol* **16:**42–50.
22. **Van Duyne GD.** 2005. λ integrase: armed for recombination. *Curr Biol* **15:**R658–660.
23. **Van Duyne GD.** 2002. A structural view of tyrosine recombinase site-specific recombination, p 93–117. *In* Craig NL, Craigie R, Gellert M, Lambowitz AM (ed), *Mobile DNA II*, ASM Press, Washington, DC.
24. **Whiteson KL, Rice PA.** 2008. Binding and catalytic contributions to site recognition by Flp recombinase. *J Biol Chem* **283:**11414–11423.
25. **Boldt JL, Pinilla C, Segall AM.** 2004. Reversible inhibitors of λ integrase-mediated recombination efficiently trap Holliday junction intermediates and form the basis of a novel assay for junction resolution. *J Biol Chem* **279:**3472–3483.
26. **Cassell G, Klemm M, Pinilla C, Segall A.** 2000. Dissection of bacteriophage λ site-specific recombination using synthetic peptide combinatorial libraries. *J Mol Biol* **299:**1193–1202.
27. **Ghosh K, Lau CK, Guo F, Segall AM, Van Duyne GD.** 2005. Peptide trapping of the Holliday junction intermediate in Cre-loxP site-specific recombination. *J Biol Chem* **280:**8290–8299.
28. **Kepple KV, Boldt JL, Segall AM.** 2005. Holliday junction-binding peptides inhibit distinct junction-processing enzymes. *Proc Natl Acad Sci U S A* **102:**6867–6872.
29. **Kepple KV, Patel N, Salamon P, Segall AM.** 2008. Interactions between branched DNAs and peptide inhibitors of DNA repair. *Nucleic Acids Res* **36:**5319–5334.
30. **Su LY, Willner DL, Segall AM.** 2010. An antimicrobial peptide that targets DNA repair intermediates in vitro inhibits Salmonella growth within murine macrophages. *Antimicrob Agents Chemother* **54:**1888–1899.
31. **Biswas T, Aihara H, Radman-Livaja M, Filman D, Landy A, Ellenberger T.** 2005. A structural basis for allosteric control of DNA recombination by lambda integrase. *Nature* **435:**1059–1066.
32. **Van Duyne GD.** 2009. Teaching Cre to follow directions. *Proc Natl Acad Sci U S A* **106:**4–5.
33. **Warren D, Laxmikanthan G, Landy A.** 2008. A chimeric Cre recombinase with regulated directionality. *Proc Natl Acad Sci U S A* **105:**18278–18283.
34. **Chen Y, Narendra U, Iype LE, Cox MM, Rice PA.** 2000. Crystal structure of a Flp recombinase-Holliday junction complex. Assembly of an active oligomer by helix swapping. *Mol Cell* **6:**885–897.
35. **Lee J, Tonozuka T, Jayaram M.** 1997. Mechanism of active site exclusion in a site-specific recombinase: role of the DNA substrate in conferring half-of-the-sites activity. *Genes Dev* **11:**3061–3071.
36. **Conway AB, Chen Y, Rice PA.** 2003. Structural plasticity of the Flp-Holliday junction complex. *J Mol Biol* **326:**425–434.

37. **Lee J, Whang I, Jayaram M.** 1996. Assembly and orientation of Flp recombinase active sites on two-, three- and four-armed DNA substrates: implications for a recombination mechanism. *J Mol Biol* **257:**532–549.
38. **Woods KC, Martin SS, Chu VC, Baldwin EP.** 2001. Quasi-equivalence in site-specific recombinase structure and function: crystal structure and activity of trimeric Cre recombinase bound to a three-way Lox DNA junction. *J Mol Biol* **313:**49–69.
39. **Gibb B, Gupta K, Ghosh K, Sharp R, Chen J, Van Duyne GD.** 2010. Requirements for catalysis in the Cre recombinase active site. *Nucleic Acids Res* **38:**5817–5832.
40. **Gourlay SC, Colloms SD.** 2004. Control of Cre recombination by regulatory elements from Xer recombination systems. *Mol Microbiol* **52:**53–65.
41. **Grainge I, Buck D, Jayaram M.** 2000. Geometry of site alignment during Int family recombination: antiparallel synapsis by the Flp recombinase. *J Mol Biol* **298:** 749–764.
42. **Grainge I, Pathania S, Vologodskii A, Harshey RM, Jayaram M.** 2002. Symmetric DNA sites are functionally asymmetric within Flp and Cre site-specific DNA recombination synapses. *J Mol Biol* **320:**515–527.
43. **Kilbride E, Boocock MR, Stark WM.** 1999. Topological selectivity of a hybrid site-specific recombination system with elements from Tn3 res/resolvase and bacteriophage P1 loxP/Cre. *J Mol Biol* **289:**1219–1230.
44. **Ma CH, Liu YT, Savva CG, Rowley PA, Cannon B, Fan HF, Russell R, Holzenburg A, Jayaram M.** 2014. Organization of DNA partners and strand exchange mechanisms during Flp site-specific recombination analyzed by difference topology, single molecule FRET and single molecule TPM. *J Mol Biol* **426:**793–815.
45. **Cassell G, Moision R, Rabani E, Segall A.** 1999. The geometry of a synaptic intermediate in a pathway of bacteriophage λ site-specific recombination. *Nucleic Acids Res* **27:**1145–1151.
46. **Huffman KE, Levene SD.** 1999. DNA-sequence asymmetry directs the alignment of recombination sites in the FLP synaptic complex. *J Mol Biol* **286:**1–13.
47. **Vetcher AA, Lushnikov AY, Navarra-Madsen J, Scharein RG, Lyubchenko YL, Darcy IK, Levene SD.** 2006. DNA topology and geometry in Flp and Cre recombination. *J Mol Biol* **357:**1089–1104.
48. **Ghosh K, Lau CK, Gupta K, Van Duyne GD.** 2005. Preferential synapsis of loxP sites drives ordered strand exchange in Cre-loxP site-specific recombination. *Nat Chem Biol* **1:**275–282.
49. **Krogh BO, Shuman S.** 2002. Proton relay mechanism of general acid catalysis by DNA topoisomerase IB. *J Biol Chem* **277:**5711–5714.
50. **Nagarajan R, Kwon K, Nawrot B, Stec WJ, Stivers JT.** 2005. Catalytic phosphoryl interactions of topoisomerase IB. *Biochemistry* **44:**11476–11485.
51. **Whiteson KL, Chen Y, Chopra N, Raymond AC, Rice PA.** 2007. Identification of a potential general acid/base in the reversible phosphoryl transfer reactions catalyzed by tyrosine recombinases: Flp H305. *Chem Biol* **14:** 121–129.
52. **Guo F, Gopaul DN, Van Duyne GD.** 1997. Structure of Cre recombinase complexed with DNA in a site-specific recombinase synapse. *Nature* **389:**40–46.
53. **Chen Y, Rice PA.** 2003. The role of the conserved Trp330 in Flp-mediated recombination. Functional and structural analysis. *J Biol Chem* **278:**24800–24807.
54. **Ma CH, Kwiatek A, Bolusani S, Voziyanov Y, Jayaram M.** 2007. Unveiling hidden catalytic contributions of the conserved His/Trp-III in tyrosine recombinases: assembly of a novel active site in Flp recombinase harboring alanine at this position. *J Mol Biol* **368:**183–196.
55. **Jayaram M, Grainge I, Tribble G.** 2002. Site-specific DNA recombination mediated by the Flp protein of *Saccharomyces cerevisiae*, p 192–218. *In* Craig NL, Craigie R, Gellert M, Lambowitz AM (ed), *Mobile DNA II*, ASM Press, Washington, DC.
56. **Xu CJ, Ahn YT, Pathania S, Jayaram M.** 1998. Flp ribonuclease activities. Mechanistic similarities and contrasts to site-specific DNA recombination. *J Biol Chem* **273:**30591–30598.
57. **Xu CJ, Grainge I, Lee J, Harshey RM, Jayaram M.** 1998. Unveiling two distinct ribonuclease activities and a topoisomerase activity in a site-specific DNA recombinase. *Mol Cell* **1:**729–739.
58. **Sau AK, Tribble G, Grainge I, Frohlich RF, Knudsen BR, Jayaram M.** 2001. Biochemical and kinetic analysis of the RNase active sites of the integrase/tyrosine family site-specific DNA recombinases. *J Biol Chem* **276:** 46612–46623.
59. **Chen JW, Lee J, Jayaram M.** 1992. DNA cleavage in trans by the active site tyrosine during Flp recombination: switching protein partners before exchanging strands. *Cell* **69:**647–658.
60. **Yang SH, Jayaram M.** 1994. Generality of the shared active site among yeast family site-specific recombinases. The R site-specific recombinase follows the Flp paradigm. *J Biol Chem* **269:**12789–12796.
61. **Letzelter C, Duguet M, Serre MC.** 2004. Mutational analysis of the archaeal tyrosine recombinase SSV1 integrase suggests a mechanism of DNA cleavage in trans. *J Biol Chem* **279:**28936–28944.
62. **Zhan Z, Ouyang S, Liang W, Zhang Z, Liu ZJ, Huang L.** 2012. Structural and functional characterization of the C-terminal catalytic domain of SSV1 integrase. *Acta Crystallogr, Sect D: Biol Crystallogr* **68:**659–670.
63. **Grainge I, Lee J, Xu CJ, Jayaram M.** 2001. DNA recombination and RNA cleavage activities of the Flp protein: roles of two histidine residues in the orientation and activation of the nucleophile for strand cleavage. *J Mol Biol* **314:**717–733.
64. **Jayaram M.** 1997. The *cis-trans* paradox of integrase. *Science* **276:**49–51.
65. **Yarus M.** 1993. How many catalytic RNAs? Ions and the Cheshire cat conjecture. *FASEB J* **7:**31–39.
66. **Kimball AS, Lee J, Jayaram M, Tullius TD.** 1993. Sequence-specific cleavage of DNA via nucleophilic attack of hydrogen peroxide, assisted by Flp recombinase. *Biochemistry* **32:**4698–4701.

67. **Lee J, Jayaram M.** 1995. Functional roles of individual recombinase monomers in strand breakage and strand union during site-specific DNA recombination. *J Biol Chem* **270:**23203–23211.

68. **Ma CH, Rowley PA, Maciaszek A, Guga P, Jayaram M.** 2009. Active site electrostatics protect genome integrity by blocking abortive hydrolysis during DNA recombination. *EMBO J* **28:**1745–1756.

69. **Kitts PA, Nash HA.** 1987. Homology-dependent interactions in phage lambda site-specific recombination. *Nature* **329:**346–348.

70. **Nunes-Duby SE, Matsumoto L, Landy A.** 1987. Site-specific recombination intermediates trapped with suicide substrates. *Cell* **50:**779–788.

71. **Arciszewska L, Grainge I, Sherratt D.** 1995. Effects of Holliday junction position on Xer-mediated recombination in vitro. *EMBO J* **14:**2651–2660.

72. **Dixon JE, Sadowski PD.** 1994. Resolution of immobile chi structures by the FLP recombinase of 2 micron plasmid. *J Mol Biol* **243:**199–207.

73. **Lee J, Jayaram M.** 1995. Role of partner homology in DNA recombination. Complementary base pairing orients the 5′-hydroxyl for strand joining during Flp site-specific recombination. *J Biol Chem* **270:**4042–4052.

74. **Nunes-Duby SE, Azaro MA, Landy A.** 1995. Swapping DNA strands and sensing homology without branch migration in λ site-specific recombination. *Curr Biol* **5:**139–148.

75. **Voziyanov Y, Pathania S, Jayaram M.** 1999. A general model for site-specific recombination by the integrase family recombinases. *Nucleic Acids Res* **27:**930–941.

76. **Cambray G, Guerout AM, Mazel D.** 2010. Integrons. *Annu Rev Genet* **44:**141–166.

77. **Lee G, Saito I.** 1998. Role of nucleotide sequences of loxP spacer region in Cre-mediated recombination. *Gene* **216:**55–65.

78. **Azam N, Dixon JE, Sadowski PD.** 1997. Topological analysis of the role of homology in Flp-mediated recombination. *J Biol Chem* **272:**8731–8738.

79. **Kanaar R, Klippel A, Shekhtman E, Dungan JM, Kahmann R, Cozzarelli NR.** 1990. Processive recombination by the phage Mu Gin system: implications for the mechanisms of DNA strand exchange, DNA site alignment, and enhancer action. *Cell* **62:**353–366.

80. **Stark WM, Grindley ND, Hatfull GF, Boocock MR.** 1991. Resolvase-catalysed reactions between res sites differing in the central dinucleotide of subsite I. *EMBO J* **10:**3541–3548.

81. **Krogh BO, Shuman S.** 2000. Catalytic mechanism of DNA topoisomerase IB. *Mol Cell* **5:**1035–1041.

82. **Stivers JT, Jagadeesh GJ, Nawrot B, Stec WJ, Shuman S.** 2000. Stereochemical outcome and kinetic effects of Rp- and Sp phosphorothioate substitutions at the cleavage site of vaccinia type I DNA topoisomerase. *Biochemistry* **39:**5561–5572.

83. **Tian L, Claeboe CD, Hecht SM, Shuman S.** 2005. Mechanistic plasticity of DNA topoisomerase IB: phosphate electrostatics dictate the need for a catalytic arginine. *Structure* **13:**513–520.

84. **Tian L, Claeboe CD, Hecht SM, Shuman S.** 2003. Guarding the genome: electrostatic repulsion of water by DNA suppresses a potent nuclease activity of topoisomerase IB. *Mol Cell* **12:**199–208.

85. **Qian XH, Inman RB, Cox MM.** 1990. Protein-based asymmetry and protein-protein interactions in FLP recombinase-mediated site-specific recombination. *J Biol Chem* **265:**21779–21788.

86. **Rowley PA, Kachroo AH, Ma CH, Maciaszek AD, Guga P, Jayaram M.** 2010. Electrostatic suppression allows tyrosine site-specific recombination in the absence of a conserved catalytic arginine. *J Biol Chem* **285:**22976–22985.

87. **Kachroo AH, Ma CH, Rowley PA, Maciaszek AD, Guga P, Jayaram M.** 2010. Restoration of catalytic functions in Cre recombinase mutants by electrostatic compensation between active site and DNA substrate. *Nucleic Acids Res* **38:**6589–6601.

88. **Ma CH, Kachroo AH, Maciaszek A, Chen TY, Guga P, Jayaram M.** 2009. Reactions of Cre with methylphosphonate DNA: similarities and contrasts with Flp and vaccinia topoisomerase. *PLoS One* **4:**e7248.

89. **Davies DR, Mushtaq A, Interthal H, Champoux JJ, Hol WG.** 2006. The structure of the transition state of the heterodimeric topoisomerase I of Leishmania donovani as a vanadate complex with nicked DNA. *J Mol Biol* **357:**1202–1210.

90. **Fan HF.** 2012. Real-time single-molecule tethered particle motion experiments reveal the kinetics and mechanisms of Cre-mediated site-specific recombination. *Nucleic Acids Res* **40:**6208–6222.

91. **Fan HF, Ma CH, Jayaram M.** 2013. Real-time single-molecule tethered particle motion analysis reveals mechanistic similarities and contrasts of Flp site-specific recombinase with Cre and lambda Int. *Nucleic Acids Res* **41:**7031–7047.

92. **Mumm JP, Landy A, Gelles J.** 2006. Viewing single λ site-specific recombination events from start to finish. *EMBO J* **25:**4586–4595.

93. **Uphoff S, Reyes-Lamothe R, Garza de Leon F, Sherratt DJ, Kapanidis AN.** 2013. Single-molecule DNA repair in live bacteria. *Proc Natl Acad Sci U S A* **110:** 8063–8068.

94. **Guo F, Gopaul DN, Van Duyne GD.** 1999. Asymmetric DNA bending in the Cre-loxP site-specific recombination synapse. *Proc Natl Acad Sci U S A* **96:**7143–7148.

95. **Dixon JE, Sadowski PD.** 1993. Resolution of synthetic chi structures by the FLP site-specific recombinase. *J Mol Biol* **234:**522–533.

96. **Lee J, Tribble G, Jayaram M.** 2000. Resolution of tethered antiparallel and parallel Holliday junctions by the Flp site-specific recombinase. *J Mol Biol* **296:**403–419.

97. **Kitts PA, Nash HA.** 1988. Bacteriophage λ site-specific recombination proceeds with a defined order of strand exchanges. *J Mol Biol* **204:**95–107.

98. **Bregu M, Sherratt DJ, Colloms SD.** 2002. Accessory factors determine the order of strand exchange in Xer recombination at psi. *EMBO J* **21:**3888–3897.

99. **Grainge I, Lesterlin C, Sherratt DJ.** 2011. Activation of XerCD-dif recombination by the FtsK DNA translocase. *Nucleic Acids Res* **39:**5140–5148.

100. **Du Q, Livshits A, Kwiatek A, Jayaram M, Vologodskii A.** 2007. Protein-induced local DNA bends regulate global topology of recombination products. *J Mol Biol* **368:**170–182.

101. **Colloms SD.** 2013. The topology of plasmid-monomerizing Xer site-specific recombination. *Biochem Soc Trans* **41:**589–594.

102. **Vazquez M, Colloms SD, Sumners D.** 2005. Tangle analysis of Xer recombination reveals only three solutions, all consistent with a single three-dimensional topological pathway. *J Mol Biol* **346:**493–504.

103. **Ip SC, Bregu M, Barre FX, Sherratt DJ.** 2003. Decatenation of DNA circles by FtsK-dependent Xer site-specific recombination. *EMBO J* **22:**6399–6407.

104. **Zechiedrich EL, Khodursky AB, Cozzarelli NR.** 1997. Topoisomerase IV, not gyrase, decatenates products of site-specific recombination in Escherichia coli. *Genes Dev* **11:**2580–2592.

105. **Grainge I, Bregu M, Vazquez M, Sivanathan V, Ip SC, Sherratt DJ.** 2007. Unlinking chromosome catenanes in vivo by site-specific recombination. *EMBO J* **26:** 4228–4238.

106. **Jayaram M.** 2013. Mathematical validation of a biological model for unlinking replication catenanes by recombination. *Proc Natl Acad Sci U S A* **110:**20854–20855.

107. **Shimokawa K, Ishihara K, Grainge I, Sherratt DJ, Vazquez M.** 2013. FtsK-dependent XerCD-dif recombination unlinks replication catenanes in a stepwise manner. *Proc Natl Acad Sci U S A* **110:**20906–20911.

108. **Crisona NJ, Weinberg RL, Peter BJ, Sumners DW, Cozzarelli NR.** 1999. The topological mechanism of phage lambda integrase. *J Mol Biol* **289:**747–775.

109. **Nash HA, Pollock TJ.** 1983. Site-specific recombination of bacteriophage λ. The change in topological linking number associated with exchange of DNA strands. *J Mol Biol* **170:**19–38.

110. **Enyeart PJ, Chirieleison SM, Dao MN, Perutka J, Quandt EM, Yao J, Whitt JT, Keatinge-Clay AT, Lambowitz AM, Ellington AD.** 2013. Generalized bacterial genome editing using mobile group II introns and Cre-lox. *Mol Syst Biol* **9:**685–700.

111. **Lee T.** 2013. Generating mosaics for lineage analysis in flies. *Wiley Interdiscip Rev: Dev Biol* **3:**69–81.

112. **Kretzschmar K, Watt FM.** 2012. Lineage tracing. *Cell* **148:**33–45.

113. **Smedley D, Salimova E, Rosenthal N.** 2011. Cre recombinase resources for conditional mouse mutagenesis. *Methods* **53:**411–416.

114. **Zhang DJ, Wang Q, Wei J, Baimukanova G, Buchholz F, Stewart AF, Mao X, Killeen N.** 2005. Selective expression of the Cre recombinase in late-stage thymocytes using the distal promoter of the Lck gene. *J Immunol* **174:**6725–6731.

115. **Harshey RM, Jayaram M.** 2006. The Mu transpososome through a topological lens. *Crit Rev Biochem Mol Biol* **41:**387–405.

116. **Jayaram M, Harshey RM.** 2009. Difference topology: analysis of high-order DNA-protein assemblies, p 139–158. *In* Santosa F, Keel M (ed), *The IMA volumes in mathematics and its applications, Mathematics of DNA structure, function and interactions*. Springer.

117. **Pathania S, Jayaram M, Harshey RM.** 2002. Path of DNA within the Mu transpososome. Transposase interactions bridging two Mu ends and the enhancer trap five DNA supercoils. *Cell* **109:**425–436.

118. **Konieczka JH, Paek A, Jayaram M, Voziyanov Y.** 2004. Recombination of hybrid target sites by binary combinations of Flp variants: mutations that foster interprotomer collaboration and enlarge substrate tolerance. *J Mol Biol* **339:**365–378.

119. **Santoro SW, Schultz PG.** 2002. Directed evolution of the site specificity of Cre recombinase. *Proc Natl Acad Sci U S A* **99:**4185–4190.

120. **Tay Y, Ho C, Droge P, Ghadessy FJ.** 2010. Selection of bacteriophage λ integrases with altered recombination specificity by in vitro compartmentalization. *Nucleic Acids Res* **38:**e25.

121. **Voziyanov Y, Konieczka JH, Stewart AF, Jayaram M.** 2003. Stepwise manipulation of DNA specificity in Flp recombinase: progressively adapting Flp to individual and combinatorial mutations in its target site. *J Mol Biol* **326:**65–76.

122. **Christ N, Droge P.** 2002. Genetic manipulation of mouse embryonic stem cells by mutant λ integrase. *Genesis* **32:**203–208.

123. **Akopian A, He J, Boocock MR, Stark WM.** 2003. Chimeric recombinases with designed DNA sequence recognition. *Proc Natl Acad Sci U S A* **100:**8688–8691.

124. **Brown WR, Lee NC, Xu Z, Smith MC.** 2011. Serine recombinases as tools for genome engineering. *Methods* **53:**372–379.

125. **Carroll D.** 2011. Genome engineering with zinc-finger nucleases. *Genetics* **188:**773–782.

126. **Gaj T, Gersbach CA, Barbas CF 3rd.** 2013. ZFN, TALEN, and CRISPR/Cas-based methods for genome engineering. *Trends Biotechnol* **31:**397–405.

127. **Proudfoot C, McPherson AL, Kolb AF, Stark WM.** 2011. Zinc finger recombinases with adaptable DNA sequence specificity. *PLoS One* **6:**e19537.

128. **Buchholz F, Angrand PO, Stewart AF.** 1998. Improved properties of FLP recombinase evolved by cycling mutagenesis. *Nat Biotechnol* **16:**657–662.

129. **Raymond CS, Soriano P.** 2007. High-efficiency FLP and ϕC31 site-specific recombination in mammalian cells. *PLoS One* **2:**e162.

130. **Garcia-Otin AL, Guillou F.** 2006. Mammalian genome targeting using site-specific recombinases. *Frontiers Biosci* **11:**1108–1136.

131. **Feil R, Brocard J, Mascrez B, LeMeur M, Metzger D, Chambon P.** 1996. Ligand-activated site-specific recombination in mice. *Proc Natl Acad Sci U S A* **93:**10887–10890.

132. **Feil R, Wagner J, Metzger D, Chambon P.** 1997. Regulation of Cre recombinase activity by mutated estrogen receptor ligand-binding domains. *Biochem Biophys Res Commun* **237:**752–757.

133. **Logie C, Stewart AF.** 1995. Ligand-regulated site-specific recombination. *Proc Natl Acad Sci U S A* **92:**5940–5944.

134. **Metzger D, Clifford J, Chiba H, Chambon P.** 1995. Conditional site-specific recombination in mammalian cells using a ligand-dependent chimeric Cre recombinase. *Proc Natl Acad Sci U S A* **92:**6991–6995.

135. **Buchholz F, Stewart AF.** 2001. Alteration of Cre recombinase site specificity by substrate-linked protein evolution. *Nat Biotechnol* **19:**1047–1052.

136. **Voziyanov Y, Stewart AF, Jayaram M.** 2002. A dual reporter screening system identifies the amino acid at position 82 in Flp site-specific recombinase as a determinant for target specificity. *Nucleic Acids Res* **30:** 1656–1663.

137. **Baldwin EP, Martin SS, Abel J, Gelato KA, Kim H, Schultz PG, Santoro SW.** 2003. A specificity switch in selected Cre recombinase variants is mediated by macromolecular plasticity and water. *Chem Biol* **10:**1085–1094.

138. **Bolusani S, Ma CH, Paek A, Konieczka JH, Jayaram M, Voziyanov Y.** 2006. Evolution of variants of yeast site-specific recombinase Flp that utilize native genomic sequences as recombination target sites. *Nucleic Acids Res* **34:**5259–5269.

139. **Shultz JL, Voziyanova E, Konieczka JH, Voziyanov Y.** 2011. A genome-wide analysis of FRT-like sequences in the human genome. *PLoS One* **6:**e18077.

140. **Sarkar I, Hauber I, Hauber J, Buchholz F.** 2007. HIV-1 proviral DNA excision using an evolved recombinase. *Science* **316:**1912–1915.

141. **Schlake T, Bode J.** 1994. Use of mutated FLP recognition target (FRT) sites for the exchange of expression cassettes at defined chromosomal loci. *Biochemistry* **33:** 12746–12751.

142. **Turan S, Galla M, Ernst E, Qiao J, Voelkel C, Schiedlmeier B, Zehe C, Bode J.** 2011. Recombinase mediated cassette exchange (RMCE): traditional concepts and current challenges. *J Mol Biol* **407:**193–221.

143. **Bethke B, Sauer B.** 1997. Segmental genomic replacement by Cre-mediated recombination: genotoxic stress activation of the p53 promoter in single-copy transformants. *Nucleic Acids Res* **25:**2828–2834.

144. **Osterwalder M, Galli A, Rosen B, Skarnes WC, Zeller R, Lopez-Rios J.** 2010. Dual RMCE for efficient re-engineering of mouse mutant alleles. *Nat Methods* **7:** 893–895.

145. **Anderson RP, Voziyanova E, Voziyanov Y.** 2012. Flp and Cre expressed from Flp-2A-Cre and Flp-IRES-Cre transcription units mediate the highest level of dual recombinase-mediated cassette exchange. *Nucleic Acids Res* **40:**e62.

146. **Voziyanova E, Malchin N, Anderson RP, Yagil E, Kolot M, Voziyanov Y.** 2013. Efficient Flp-Int HK022 dual RMCE in mammalian cells. *Nucleic Acids Res* **41:**e125.

147. **Aihara H, Kwon HJ, Nunes-Duby SE, Landy A, Ellenberger T.** 2003. A conformational switch controls the DNA cleavage activity of λ integrase. *Mol Cell* **12:** 187–198.

Mobile DNA, 3rd Edition
Nancy L. Craig, Michael Chandler, Martin Gellert, Alan M. Lambowitz, Phoebe A. Rice and Suzanne Sandmeyer

doi:10.1128/microbiolspec.MDNA3-0046-2014

W. Marshall Stark[1]

The Serine Recombinases

3

INTRODUCTION

Site-Specific Recombination: A Brief Primer

The term site-specific recombination encompasses a group of biological processes that, unlike homologous recombination, promote rearrangements of DNA by breaking and rejoining strands at precisely defined sequence positions. In a canonical site-specific recombination event, two discrete sites (sequences of DNA, typically a few tens of base pairs long) are broken, and the ends are reciprocally exchanged and rejoined, resulting in recombinant products (Fig. 1). Site-specific recombination does not require extensive sequence homology; the sites are identified and brought together by protein–DNA and protein–protein interactions involving specialized recombinase proteins, unlike homologous recombination where DNA–DNA interactions define the loci of strand exchange. "Conservative" site-specific recombination systems form recombinants without any requirement for DNA synthesis or high-energy cofactors. Some other biological processes such as transposition are sometimes categorized with site-specific recombination because of common features including cleavage and rejoining of DNA strands at precise positions defined by protein–DNA interactions, but these processes may require DNA synthesis and/or ligase-mediated rejoining of DNA strands. The systems discussed in this chapter conform to the strict "conservative" definition. General aspects of site-specific recombination have been reviewed elsewhere (1, 2, 3).

Site-specific recombination can have different outcomes depending on the nature of the DNA substrate(s) (Fig. 2). Recombination between two sites, each on a separate linear DNA molecule, results in linear recombinants. Two outcomes are possible, depending on which "half-site" is joined to which, as shown in Fig. 2a. However, typical sites have a polarity, such that the "left half" of one site is joined to the "right half" of the other, and vice versa; thus, only one of these possibilities normally occurs. The origin of the site polarity is discussed below. If the two sites are on separate molecules but one or both molecules are circular (Fig. 2b), recombination will join the two molecules together (this is called integration or fusion). The product molecule contains two sites, oriented in a direct repeat (head-to-tail) relationship. Conversely, recombination of this two-site molecule splits it into two products (this is called excision or resolution). If two sites within a single DNA molecule are in an inverted relationship (Fig. 2c), recombination inverts the orientation of one DNA segment bounded by the sites, relative to the other. In most real site-specific recombination systems, restrictions imposed by the mechanism of recombinase-mediated catalysis allow only some of these possibilities (see below). Site-specific recombination is seemingly isoenergetic; the products, like the substrates, are normal double-stranded DNA molecules. Reactions might

[1]Institute of Molecular, Cell and Systems Biology, University of Glasgow, Bower Building, Glasgow G12 8QQ, Scotland, United Kingdom.

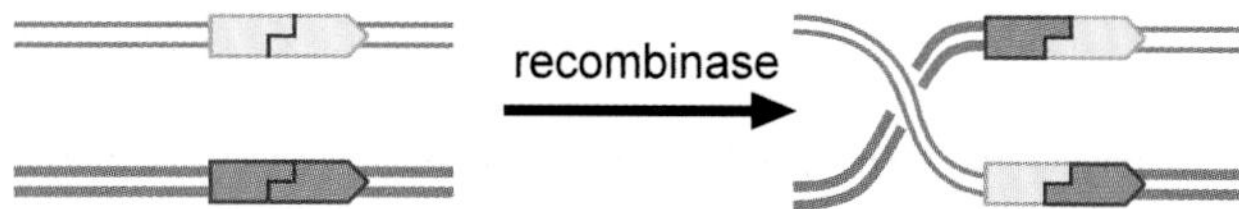

Figure 1 Site-specific recombination. Two sites (pointed boxes) in double-helical DNA (shown as double lines) are recognized by a recombinase protein (not shown), and then cut and rejoined to form recombinants.
doi:10.1128/microbiolspec.MDNA3-0046-2014.f1

therefore be expected to reach a 1:1 equilibrium of substrates and recombinants. However, natural systems have evolved strategies to bias the reaction toward the desired products; some examples are described in the sections that follow.

Conservative site-specific recombination has been adopted widely for diverse programmed DNA rearrangements essential to the biology of bacteria, archaea and the mobile DNA elements that infest them (bacteriophages, plasmids and transposons) (2, 4, 5, 6). Curiously, however, there are only a few known conservative site-specific recombination systems in eukaryotes, and some of these may be associated with bacterial symbionts or bacterial-derived organelles, or may be recent acquisitions from horizontal transfer of mobile DNA (1, 5, 6, 7, 8). Roles of site-specific recombination systems include temperate bacteriophage DNA integration and excision from the host bacterial genomic DNA, transposon cointegrate resolution, monomerization of plasmid multimers, switching of gene expression by inversion of regulatory sequences relative to coding sequences and developmentally programmed excision of intervening genomic sequences. There is no clear distinction of the biological functions of systems based on serine recombinases, the subject of this chapter, from those based on the other large family, the tyrosine recombinases (see Chapter 2, this volume). It seems that Nature has evolved two quite different ways of doing site-specific recombination, both of which are sufficiently "fit for purpose" to survive and prosper in present-day organisms.

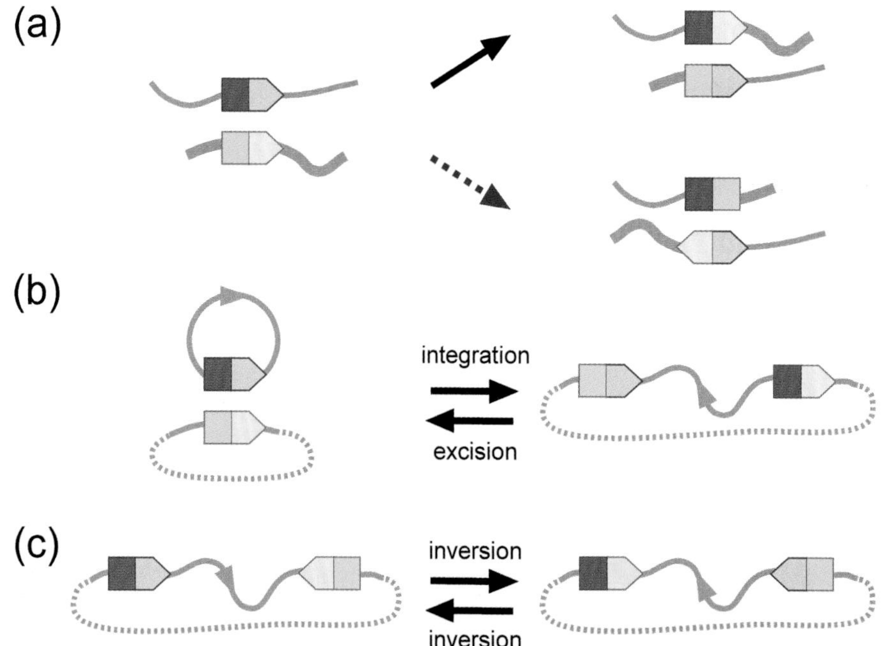

Figure 2 Site-specific recombination outcomes. (a) Recombination between two sites in separate linear DNA molecules results in two linear recombinant products. Usually, the sites have a polarity (indicated by the pointed boxes) such that the lower pathway (red arrow) is forbidden. (b) Recombination between two sites in separate DNA molecules, when one or both of the molecules is circular, results in a single product molecule containing two sites in direct repeat. This is called integration or fusion. The "reverse" reaction splits a molecule containing two sites into two product molecules, one or both of which are circular. This is called resolution, excision, or deletion (depending on the biological context). (c) Recombination between two sites in inverted repeat in a DNA molecule inverts the orientation of one segment of DNA relative to the other. This is called inversion.
doi:10.1128/microbiolspec.MDNA3-0046-2014.f2

Mechanistic Nuts and Bolts

In this section, I will give a brief overview of the molecular mechanisms of conservative site-specific recombination.

In many systems, recombination takes place between two identical sites, and two identical sites are reformed in the recombinants. However, there are examples (notably bacteriophage integrase systems; see "Regulation of recombination by large serine recombinases" below) where recombination is between two different sites. Sites range in length from about 25 up to several hundreds of base pairs. The shortest sites typically have imperfect 2-fold (dyad/palindrome) DNA symmetry, consistent with their observed or inferred property of binding a symmetric dimer of recombinase. The specific phosphodiester linkages that are cut and rejoined during recombination are located close to the center of the site (Fig. 3). Longer sites comprise a "crossover site" conforming to the above description, which binds a recombinase dimer and within which are the points of strand exchange, as well as adjacent "accessory sequences" on one or both sides of the crossover site, which may include binding sites for additional recombinase subunits or other "accessory proteins," or for looping interactions with recombinase subunits bound at the crossover site (Fig. 3a). The roles of the accessory sequences are in regulation of recombinase activity; initiation of catalysis typically depends on their presence and their correct interactions with other components of the system (1, 2, 3).

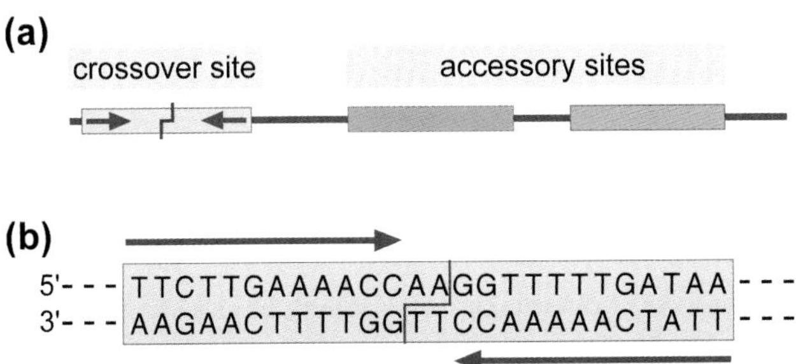

Figure 3 Recombination sites. (a) A typical recombination site. The crossover site, where strand exchange takes place (at the position marked by the staggered red line), binds a recombinase dimer and typically has partial dyad symmetry (indicated by the blue arrows). "Accessory sites," which may be adjacent on one side of the crossover site (as shown), on both sides or more distant, may bind additional recombinase subunits or other proteins, or may make looping interactions with recombinase bound at the crossover site. (b) Example of a real crossover site (*hixL*, a site acted upon by Hin recombinase). The colors and symbols are as in part (a). Hin, like all serine recombinases characterized to date, cuts the DNA at the center of the crossover site with a 2 bp "stagger" as shown. doi:10.1128/microbiolspec.MDNA3-0046-2014.f3

Recombinases do not cut the two DNA strands at the precise center of the site. Instead, the break points are symmetrically positioned off-center, so that there are a few base pairs between the top strand and bottom strand break points. These base pairs are often referred to as the "overlap sequence" because the top and bottom strands of this sequence in the recombinant sites originate from different parent sites. All serine recombinase systems examined in this respect have 2 bp overlap sequences with the strand breaks staggered as shown in Fig. 3b; in contrast, the overlap sequences for tyrosine recombinases vary in length (typically 6 to 8 bp), and the stagger is in the opposite direction. If the "half-sites" that are to be joined to form recombinants do not have complementary overlap sequences, the products would have mismatched base pairs. This scenario can arise if two identical crossover sites are misaligned in the catalytic intermediate such that strand exchange pairs two identical, noncomplementary ends. Serine recombinases do not normally form mismatched recombinants; this is one origin of the site polarity discussed above. However, reactions of "mismatched" sites can have other consequences (see "Subunit rotation" below).

Each crossover site binds a recombinase dimer. A critical subsequent step is when two crossover sites come together; this is called synapsis. The "synapse" or "synaptic complex" that is thus formed comprises the two crossover sites bridged by a recombinase tetramer, and it is in this intermediate that the chemical steps of strand cleavage, exchange and ligation will take place. In regulated systems, crossover-site synapsis is typically a control point that depends on interactions with accessory factors.

In any conservative site-specific recombination event, there are eight chemical steps: four strand cleavages and four ligations. Cleavage occurs when a nucleophilic amino acid functional group at the recombinase active site attacks the scissile phosphodiester bond of a DNA strand; for the serine recombinases, this is the hydroxyl group of a serine residue. The immediate product of cleavage has a broken DNA strand, with a covalent phosphodiester linkage between one DNA end and the recombinase at the break point. Serine recombinases become linked to the 5′ end of the DNA, leaving a 3′-hydroxyl group on the other end at the break. Serine recombinases cleave all four DNA strands in the synaptic complex, creating double-strand breaks at the center of each crossover site. Each half-site thus formed has a recombinase subunit covalently attached to its 5′ end, and 2-nt single-stranded protrusions terminated by a 3′-OH group (Fig. 4). The half-sites are then exchanged and re-ligated, creating recombinants. This mechanism contrasts with that of the tyrosine recombinases, which become linked to the 3′ end of the DNA at the strand break and do not make double-strand break intermediates. Instead, they cleave, exchange and re-ligate pairs of single DNA strands; thus, strand exchange proceeds via a "four-way junction" intermediate with two recombinant and two non-recombinant strands (see Chapter 2 of this volume).

SERINE RECOMBINASES

Some History

Following the discovery of the first site-specific recombinase, λ Int, in the 1970s, it was realized that the product of the *tnpR* gene encoded by the bacterial penicillin resistance transposon Tn*3* has a similar function (9, 10). Detailed characterization of the *tnpR* gene product (resolvase) and the recombination site (*res*) soon followed (11, 12, 13). It quickly became apparent that there was a group of enzymes related to Tn*3* resolvase, encoded by other bacterial transposons and DNA inversion systems. The group came to be known as the resolvase or resolvase/invertase family (14, 15). Pioneering *in vitro* studies of the γδ transposon resolvase (closely related to Tn*3* resolvase) by Reed and Grindley revealed basic mechanistic differences from λ Int and its relatives (16, 17, 18). It was later shown that the resolvase–DNA linkage is via a serine residue, unlike the tyrosine that is used by λ Int and its brethren (19, 20, 21, 22). In the 1990s, the two families came to be referred to as the "serine" and "tyrosine" recombinases (23, 24).

Serine Recombinase Proteins

All serine recombinases possess a characteristic catalytic domain, which implements the chemical steps of strand exchange. I will call it the "SR" (serine recombinase) domain throughout this review. The size of the SR domain is remarkably constant (usually about 150 amino acid residues). Several of its amino acid residues are highly conserved and are now known to contribute to the structure of the active site (3). All known serine recombinases have "attachments" to the SR domain, usually at the C terminus; these vary substantially and their specific properties have roles in definition of the recombinase function (25, 26) (Fig. 5). The recombinases studied in the early days (transposon resolvases and invertases) have a simple configuration with the SR domain at the N-terminus linked to a small C-terminal helix–turn–helix (HTH) DNA-binding domain, giving a total length of ~180–200 residues. These have come to be known as the "small serine recombinases". However, as identification of putative serine recombinases by sequence analysis gathered pace, the diversity of the family became apparent (25, 26) (Fig. 5). Many sequences could be aligned with the entire length of the small serine recombinases but have extensions at the C-terminal end, such as the ISXc5 resolvase. Others have large C-terminal extensions immediately after the SR domain, in place of the HTH domain. An important subgroup of these "large serine recombinases" includes the bacteriophage serine integrases, the first to be identified being that of the *Streptomyces* phage ϕC31 (28). These proteins (~400 to 700 amino acids) have an N-terminal SR domain followed by a complex, variable multidomain region with DNA-binding and regulatory functions, which are still only partially characterized (27). At first, it seemed that a "rule" was that the SR domain should be at the extreme N terminus of the protein, but proteins with a HTH domain preceding the SR domain were then identified and are now known to be transposases (29, 30).

Biological Roles of Serine Recombinases

The role of the patriarchs of the serine recombinase family, the transposon resolvases, is to divide ("resolve") a large circular DNA molecule into two smaller

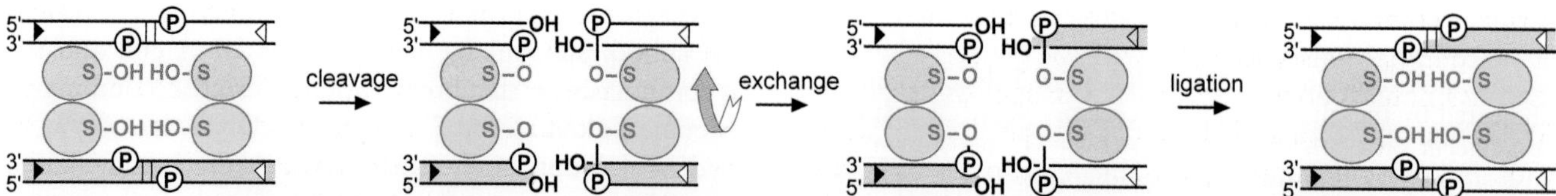

Figure 4 The serine recombinase strand-exchange mechanism. A synaptic complex of two crossover sites bridged by a recombinase tetramer (yellow ovals) is shown. The four subunits are spaced out, so that the catalytic steps can be seen clearly. The catalytic serine residues are indicated by S-OH. The scissile phosphodiesters are represented as circled Ps, and in the first and last panels the 2-bp overlap sequence is indicated by vertical lines. For further details, see text. doi:10.1128/microbiolspec.MDNA3-0046-2014.f4

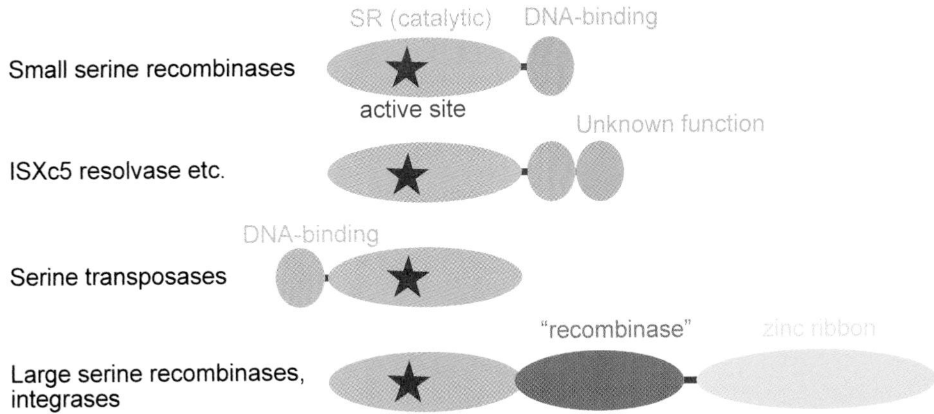

Figure 5 Domain structures of serine recombinases (26). The SR (catalytic) domain (shown in pink; typically ~150 amino acids) is common to all serine recombinases and contains the active site (red star). Small serine recombinases (including resolvases and invertases) have a helix–turn–helix DNA-binding domain (blue; ~40 amino acids) at the C terminus. Some related recombinases such as ISXc5 resolvase have additional C-terminal domains (orange) of unknown function. Serine transposases have a similar helix–turn–helix domain at the N terminus. Large serine recombinases have multiple domains at the C terminus of the SR domain (27). doi:10.1128/microbiolspec.MDNA3-0046-2014.f5

circles. The natural substrate is a "cointegrate" molecule formed by replicative transposition, which contains two transposons, each with a *res* recombination site (31, 32). Closely related resolution systems are encoded by some bacterial plasmids and act to reduce plasmid multimers to monomers. DNA invertases promote flipping of the DNA sequence between two sites, thereby switching between different modes of gene expression (often to evade host defenses against infection). The activity of one invertase, Hin, from *Salmonella typhimurium*, is responsible for the phenomenon of flagellar phase variation studied since the 1920s. Other invertases are encoded in bacterial, bacteriophage and plasmid genomes (33). Bacteriophage serine integrases integrate and excise the DNA genomes of "temperate" or "lysogenic" phages to/from the bacterial host chromosomal DNA, like the famous tyrosine recombinase-based phage λ system (34, 35). Recently, some small serine recombinases with similarity to the DNA invertase group have also been shown to be phage integrases (36). As noted above, a group of transposases have an SR domain with a HTH domain at the N terminus (30). In addition, substantial numbers of proteins in the databases have homology to the SR domain but have unknown functions (M. R. Boocock, personal communication); there may still be surprises in store. More details on the functions of particular groups of serine recombinases are given in Chapters 9, 10 and 11, this volume.

Serine Recombinase Structures

There is now a substantial bank of structural data at atomic resolution on members of the serine recombinase family. In particular, a series of groundbreaking crystal structures of γδ resolvase obtained by the Steitz laboratory in Yale has gone hand in hand with our developing understanding of serine recombinase mechanisms (37, 38, 39, 40, 41, 42). Recently, crystallographic studies of a distantly related serine resolvase, Sin, have transformed our understanding of the regulatory mechanisms and catalytic active site (43, 44). Chapter 10 (this volume) gives an in-depth review of these data. Further insights have come from the structures of the "attachments" to the SR domain. Table 1 summarizes the crystallographic and nuclear magnetic resonance (NMR) structural data on serine recombinases available at the time of submission of this review (45, 46, 47, 48, 49, 50, 51, 52). Small-angle scattering-based methods have also provided important structural information (53).

As an example, Fig. 6a shows the structure of a γδ resolvase dimer bound to the crossover site of the γδ *res* recombination site (40). Each subunit (183 amino acids) comprises an SR domain connected via a short linker peptide to a C-terminal HTH domain. The HTH domains recognize sequence motifs at the ends of the crossover site. Each SR domain comprises a core β-sheet decorated with α-helical and irregular regions, and ends in a long α-helix whose C-terminal region

Table 1 Structural data for serine recombinases

PDB accession no.	Description	Reference(s)
2RSL	γδ resolvase; supersedes 1RSL	37, 39
1GDR	γδ resolvase	38
1GDT	γδ resolvase dimer bound to *res* site I DNA	40
1GHT, 1HX7	γδ resolvase catalytic domain (NMR structures)	45
1RES, 1RET	γδ resolvase DNA-binding domain (NMR structures)	46
1ZR2, 1ZR4	γδ resolvase activated mutant tetramer in cleaved-DNA synaptic intermediate	41
2GM4	γδ resolvase activated mutant tetramer in cleaved-DNA synaptic intermediate	42
2GM5	γδ resolvase mutant tetramer	42
1HCR	Hin invertase C-terminal domain bound to DNA motif	47
1IJ6+	Hin C-terminal domain bound to wild-type and mutant DNA motifs (also 1IJW, 1JJ8, 1JKO, 1JKP, 1JKQ, and 1JKR)	48
3UJ3	Gin activated mutant tetramer. Supersedes 3PLO	49
4M6F	Gin dimer bound to *gix* site DNA	50
2R0Q	Sin tetramer in synaptic complex with res site II DNA	43
3PKZ	Sin activated mutant tetramer	44
4KIS	Bacteriophage A118 integrase (C-terminal part bound to *att* site DNA)	51
4BQQ	Bacteriophage ΦC31 integrase (N-terminal part)	McMahon SA, McEwan AR, Smith MCM, and Naismith, JH, unpublished data
3GUV	Large serine recombinase from *Streptococcus pneumoniae*	Bonanno JB, Freeman J, Bain KT, Do J, Sampathkumar P, Wasserman S, Sauder JM, Burley SK, and Almo SC, unpublished data
3BVP	Bacteriophage TP901-1 integrase catalytic domain	52
3G13	Transposase from CTn7 of *Clostridium difficile*	Bagaria A, Burley SK, and Swaminathan S, unpublished data
3ILX	TnpA transposase from *Sulfolobus solfataricus* ISC1904	Chang C, Bigelow L, Bearden J, and Joachimiak A, unpublished data
3LHF	TnpA transposase from *Sulfolobus solfataricus* IS1921	Stein AJ, Osipiuk J, Marshall N, Bearden J, Davidoff J, and Joachimiak A, unpublished data
3LHK	TnpA transposase from IS607-family element in *Methanocaldococcus jannaschii*	Chang C, Chhor G, Cobb G, and Joachimiak A, unpublished data

contacts the DNA minor groove near the center of the crossover site. The linker between this helix and the HTH domain lies in the minor groove and bears a structural resemblance to the "AT-hook" motif found in other DNA-binding proteins (54). The two SR domains make a complex network of interactions to form a dimer with imperfect 2-fold symmetry. The crossover site DNA is significantly bent but essentially "B-form." The positions of the catalytic serine residues are shown in Fig. 6a; like other putative active-site residues, they are not in contact with the DNA in this structure. The resolvase is therefore considered to be bound to the DNA in a precatalytic conformation. However, subsequent structures, also solved by the Steitz group using "activated" γδ resolvase variants (see below), revealed a catalytic intermediate containing a resolvase tetramer bridging two crossover sites that have each been cleaved in both strands (Fig. 6b) (41, 42).

Taken together, the structural data (Table 1) show that the SR-domain fold is very well conserved throughout the serine recombinase family, despite substantial amino acid sequence divergence.

Serine Recombinase Mechanism

The early studies of Reed and Grindley demonstrated that the resolvase catalytic mechanism was significantly different from that of λ Int, Cre and other members of the "integrase family" (now tyrosine recombinases) (17, 18). DNA cleavage and rejoining by γδ resolvase occur at precise positions within the 28-bp crossover site of *res*. Alteration of the reaction conditions allowed isolation of products with double-strand breaks at the

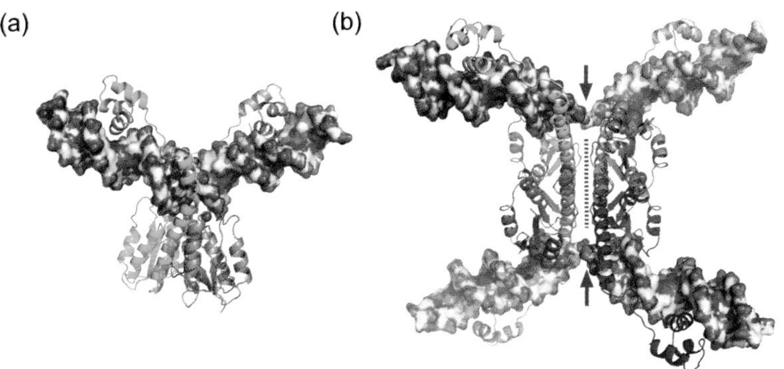

Figure 6 Crystal structures of γδ resolvase–NA complexes. (a) Wild-type γδ resolvase dimer bound to crossover-site DNA (PDB 1GDT; 40). The subunits are in cartoon representation (green and orange). The active site serine residues (α carbons) are indicated by magenta spheres. (b) Activated mutant γδ resolvase tetramer in a cleaved-DNA synaptic intermediate (PDB 1ZR4; 41). The resolvase is rendered as in (a). The active-site serines are covalently linked to DNA ends (see Fig. 4); only two are visible. This view emphasizes the flat interface (marked by a dashed red line) between "rotating pairs" of resolvase subunits. The red arrows indicate positions of double-strand breaks in the DNA. The structure corresponds to the intermediates cartooned in the two central panels of Fig. 4.
doi:10.1128/microbiolspec.MDNA3-0046-2014.f6

center of the crossover site. A resolvase subunit is covalently linked to each 5′ end of the linearized DNA (18) (Fig. 4). The protein–DNA linkage was shown later to be a phosphodiester bond with the resolvase Ser10 (19, 21). The *in vitro* reaction is very efficient; under standard conditions, nearly all the substrate is converted into recombinant products within a few minutes. No cofactors or metal ions such as Mg^{2+} are required for activity. The analysis of Reed and Grindley also revealed that the product of resolution of a supercoiled plasmid substrate *in vitro* was a specific simple catenane in which the two product circles are linked as in a chain, an intriguing observation that led to many further studies and insights (see "Topological studies" below) (16).

Studies on related systems, including the resolution systems of Tn*3* and Tn*21*, and the DNA invertases Gin, Hin and Cin, confirmed the generality of the mechanistic insights from the γδ resolvase system (32, 33). However, the products of the inversion systems, their site structures and their regulation are substantially different, as will be discussed below.

The products with DNA double-strand breaks were presumed to be derived from a recombination intermediate, and suggested a simple "cut-and-paste" mechanism of strand exchange (Fig. 4). Together with the specific, simple catenane or unknotted circle product topologies of resolvases and invertases, respectively, the data suggested that exchange of DNA ends by serine recombinases is a well-ordered process, taking place within a synaptic complex of two crossover sites and a recombinase tetramer, after double-strand cleavage of both sites (55, 56).

Topological Studies

In the absence of protein structural information, most early analysis of the mechanism focused on the DNA reaction products. Analysis of the product topologies from supercoiled circular (plasmid) substrates was especially significant (57). Studies with the tyrosine recombinase λ Int (and later FLP and Cre) had revealed that a supercoiled two-site substrate could give products with a wide range of knot/catenane topologies. These results were interpreted by a "random collision" mechanism of synapsis; that is, the sites collide due to natural random motions of the supercoiled substrate molecule. Various numbers of coils/tangles are trapped as the two sites synapse. A subsequent simple strand exchange mechanism results in products with a range of topologies (Fig. 7). Consistent with a random collision synapsis mechanism, these tyrosine recombinase systems did not distinguish between substrates with sites in different relative orientations: both "head-to-tail" (direct repeat) and "head-to-head" (inverted repeat) arrangements of sites were recombined equally well (57). The serine resolvases and invertases were clearly different. Resolvases yield almost exclusively simple catenane recombination products (Fig. 7a), and invertase recombination products are almost exclusively unknotted circles. Furthermore, resolvases only recombine substrates with sites in direct repeat, and invertases only recombine sites in inverted repeat. These selectivities are very strong (for example, a $>10^4$-fold rate difference for Tn*3* resolvase), and persist even when the sites are separated by several kilobase pairs of DNA. Neither resolvases nor invertases recombine sites on separate supercoiled plasmids. The question therefore arose: how and why do these systems avoid the formation of random collision products? The phenomenon, which became known as "topological selectivity," has been reviewed (57), this volume. To summarize very briefly, the catalytic activity of these serine recombinases is strictly regulated so as to take place only when an elaborate synaptic complex is properly formed. This structure includes the accessory DNA sequences and protein subunits, and involves intertwining of the sites (as shown for resolvase in Fig. 7). The twisting/writhing of the DNA involved in synaptic complex formation is energetically favorable only when the sites come together in a specific way, in a substrate with the correct relationship between the two sites. The regulatory properties of synaptic complexes are discussed further below.

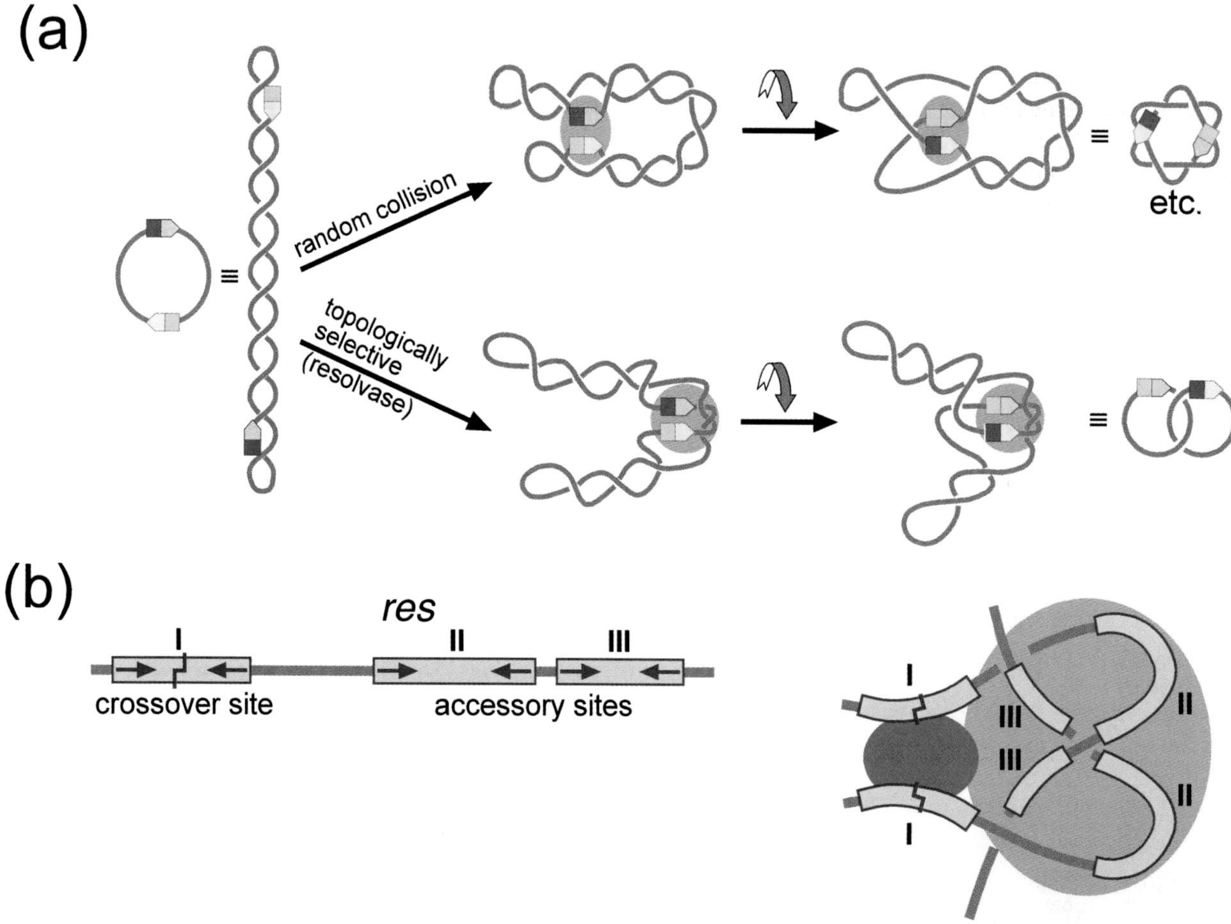

Figure 7 Topologically selective recombination by Tn*3*/γδ resolvase. (a) The reaction pathway of resolvase (lower row) is contrasted with that of a non-selective recombinase (upper row). Random collision of sites results in products with a variety of topologies (a 6-noded catenane is shown as an example here). Selective synapsis by resolvase results in a product with a specific topology (2-noded catenane). (b) Architecture of the synapse. The Tn*3*/γδ *res* site is diagrammed on the left. On the right, the arrangement of DNA in the synapse is shown. The catalytic tetramer bound to the crossover sites (the "catalytic module") is represented as an orange oval, and the eight resolvase subunits bound at the accessory sites (the "regulatory module") are collectively represented by the pink oval. Chapter 10, this volume, gives more details on the structures of this and other synaptic complexes.
doi:10.1128/microbiolspec.MDNA3-0046-2014.f7

Subunit rotation

Pioneering electron microscopy studies by Cozzarelli's group revealed the precise topologies of a series of minor resolvase reaction products (Fig. 8a). These were proposed to be made by repeated rounds of strand exchange equivalent to half-turns of one pair of DNA ends relative to the other, in an intermediate with double-strand breaks in both crossover sites (58, 59, 60). The changes in DNA linkage that accompany the first round of the series (the standard resolution reaction) and its reverse reaction (catenane fusion) were determined and are consistent with this "simple rotation" mechanism (61). However, the simple hypothesis that the recombinase subunits attached to the half-sites rotate along with the DNA ends ("subunit rotation"; Fig. 8b) was difficult for many to accept, because of its radical biochemical implication; one half of the recombinase tetramer must rotate through 180° relative to the other half, but somehow disastrous dissociation of the two halves must be avoided. There is no biochemical precedent for this model; it was a "unicorn in the garden," which would require extraordinarily rigorous testing.

A synapse with the recombining crossover sites on the outside of a recombinase tetramer core ("DNA-out") was argued to be most consistent with the subunit rotation model (61). Later, alternative models that retained a fixed tetramer structure (and thus avoid the dissociation issue) were proposed. Some models placed the two recombining DNA double helices close to each other near the center of the tetramer ("DNA-in"). However, it was very difficult then to account for the observed topological changes after DNA strand exchange. Another model proposed that part of the DNA-out tetramer remains fixed, while the N-terminal parts of two subunits rotate with their attached DNA ends (38, 31).

A strange property of serine recombinase-mediated recombination, first discovered in the Gin DNA invertase system, led to strong experimental support for subunit rotation. If two recombining sites have different 2-bp sequences at their central "overlap," the recombinants that would have mismatched base pairs are not

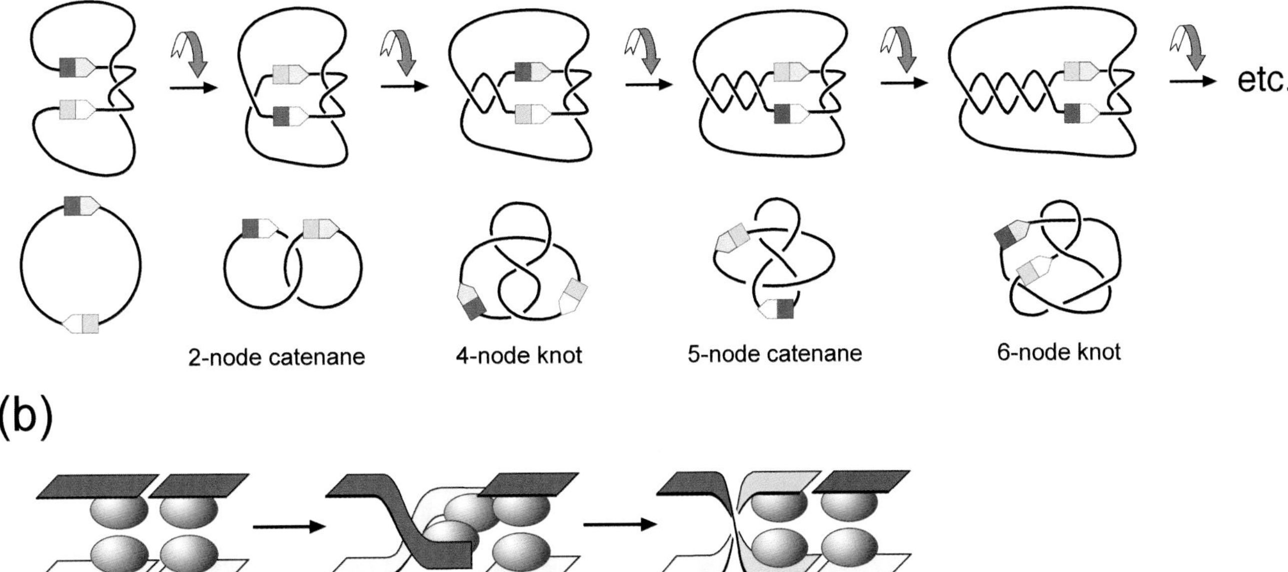

Figure 8 Subunit rotation mechanism of resolvase. (a) Topologies of first round and "iteration products" observed by Cozzarelli's group (58, 59, 60). The upper part shows the products predicted by a rotation mechanism in a resolvase synapse with topology as shown in Fig. 7. The lower panels show the simplified topologies of these products. "Mismatched" substrates (see text) form only the nonrecombinant knot products, starting with the 4-noded knot. (b). Cartoon illustrating the proposed subunit rotation mechanism. DNA is represented as ribbons and recombinase subunits as ovals. The crystal structure of a proposed intermediate in subunit rotation is shown in Fig. 6b.
doi:10.1128/microbiolspec.MDNA3-0046-2014.f8

formed (see above). Instead, a second round of strand exchange ensues, restoring the ends to a nonrecombinant configuration but leaving a record of the transaction as a change in the DNA topology (knotting) (62, 63, 64, 65). This behavior has since been shown to be general to many if not all serine recombinases, and *per se* suggests subunit rotation. In the case of resolvase, reaction of a "mismatched" substrate leads to a 4-noded knot product, consistent with a 360° rotation, and further double rounds of strand exchange give more complex knotted products (65; Fig. 8a). Further analysis of the products supported subunit rotation, but not alternative mechanisms (66, 67, 68, 69, 70). It was shown that knotting of a mismatched substrate proceeds with the DNA linkage changes predicted for subunit rotation, that the recombinase subunits move in concert with the DNA ends to which they are bound and that the knotting reaction of a mismatched substrate proceeds without any intermediate rejoining of the DNA ends that would allow a "reset" of the protein subunits, as would be necessary for all nonrotary mechanisms.

The dimer interface seen in the early γδ resolvase crystal structures (37, 40) is quite rugged and apparently incompatible with subunit rotation. The structural breakthrough came from further crystallographic studies using "activated" γδ resolvase variants. Activated serine recombinase mutants (first identified in the Gin and Cin invertase systems) have lost their dependence on regulation by accessory factors (71, 72, 73, 74). Activated resolvase variants were shown to form synapses *in vitro*, comprising two crossover sites bridged by a resolvase tetramer (75, 76), and low-resolution structural studies confirmed that the sites were bound on the outside of the tetramer (53). The resolvase–DNA cocrystal structures of Li *et al.* and Kamtekar *et al.* (41, 42) revealed a synaptic intermediate with both crossover sites cleaved at their centers and resolvase subunits covalently attached to each 5′ end via the active-site serines, in line with the earlier biochemical experiments (Figs 4 and 6b). The conformation of the resolvase SR domains is dramatically different from that in the dimer structures; the tetramer has a remarkably flat, hydrophobic surface between the two "halves" that are predicted to rotate with respect to each other (Fig. 6b). It is proposed that a flat, greasy interface is maintained throughout rotation, and structure-based modelling has demonstrated that there would be no major energy barriers to this process (41).

Biochemical studies on invertases (69, 70, 77) and serine integrases (78, 79) have provided further support for a subunit rotation mechanism, and recent crystal structures of activated Sin resolvase and Gin invertase variants show recombinase tetramers with flat hydrophobic interfaces but different rotational relationships of the "rotating dimers" compared with the γδ resolvase structures (44, 49). It looks like all serine recombinases work this way.

The Active Site

Each of the four active sites in a serine recombinase tetramer has to perform two chemical steps during a round of recombination (a strand cleavage and then a ligation). We would like to understand the mechanism of catalysis, the contributions of individual subunits, and the choreography and reversibility of the reaction steps.

Alignments of serine recombinases reveal about a dozen well-conserved polar or charged residues that might contribute to catalysis at the active site, and studies involving mutagenesis and *in vitro* biochemical analysis have identified six key residues including the nucleophilic serine (3, 80, 81). Proposed roles for these residues include those typical of phosphoryl transfer enzymes: generation of a strong base to increase the nucleophilicity of the very weakly acidic serine hydroxyl (cleavage reaction) and deoxyribose 3′-hydroxyl (religation reaction) groups, stabilization of the transition state geometry and/or charge, and provision of an acid for protonation of the leaving group during cleavage and re-ligation (82). Other active-site features that must be present include interactions to guide the incoming nucleotide bearing the 3′-OH to the active site for ligation and contacts that detect base pairing (or the lack of it) in the product overlap sequence.

REGULATION OF RECOMBINATION ACTIVITY

Introduction

The "programmed" DNA rearrangements promoted by natural site-specific recombinases typically involve sophisticated regulation to ensure that strand cleavages and subsequent events happen only at the right times and places. Serine recombinase-based systems adapted for resolution, inversion and integration have evolved distinct regulatory strategies, as will be discussed in the following sections.

All site-specific recombinases must have high fidelity for their target sites; off-target reactions are very likely to be deleterious. The C-terminal HTH domains of small serine recombinases recognize sequence motifs at the ends of the crossover site, but their sequence specificity is limited (83) and DNA contacts by the SR (catalytic) domains make a substantial additional contribution to specificity (84). Even so, some variation of the crossover site sequence is tolerated (85, 86). The observed high site specificity of the complete systems is presumably due to tight dependence of catalytic activity on cooperative assembly of all the components including accessory factors.

Formation of the "catalytic module" by bringing together two recombinase dimer-bound crossover sites may be a key regulatory step for most systems. For example, the crossover-site DNA-bound γδ resolvase dimers in the crystals studied by Yang and Steitz (40) do not make a synaptic interaction despite their extremely high concentration, whereas "activated" mutants that are defective in regulation readily form tetramer-containing synaptic complexes (75, 87). This checkpoint apparently prevents wild-type resolvase catalysis until the full synaptic complex including the accessory sites and their bound subunits is correctly assembled (88) (see below).

Regulation of Resolvase Recombination

The serine resolvase systems are described in detail in Chapter 10, this volume. Recombination by resolvases takes place following formation of a specific synaptic complex involving intertwining of the *res* recombination site accessory sequences. As noted above, this complex forms only when the two *res* sites are in direct repeat in a negatively supercoiled DNA molecule. The synaptic complex also plays an important but as yet mysterious role in restricting strand exchange to a single half-turn, so that the first-round simple catenane resolution product is released and inert to further reaction (see Figs 7 and 8). The resolvase-bound accessory sites of Tn*3*/γδ *res* can pair and intertwine to form a "regulatory module" even in the absence of the crossover sites ("site I") (Fig. 7b); this property has been used to impose topological selectivity on normally nonselective recombinases (such as Cre) by putting their crossover sites in place of *res* site I (89, 90). The detailed molecular architecture of the Tn3/γδ regulatory module is still unclear. However, it has been shown that a specific protein interface between resolvase subunits plays a key role in coupling accessory site synapsis to synapsis of the crossover sites and activation of catalysis (43, 88, 91, 92). One hypothesis is that crossover-site synapsis and the dramatic concomitant protein conformational changes (see "Subunit rotation" above; Fig. 6b) are brought about simply by forcing the recombinase dimers into close proximity by their interactions with accessory subunits ("extreme mass action"). Alternatively, the interactions with accessory protein subunits might play an essential role in promoting the required conformational changes.

The arrangements of protein-binding sites within *res*-type recombination sites are quite diverse (93), as exemplified by Sin *res*, which contains just a single accessory binding site for Sin resolvase and a site for an "architectural" DNA-bending protein (HU/IHF). A crystal structure of Sin in a synaptic interaction with accessory DNA has led to a model for the complete synaptic complex formed by that system (43). The intertwining of the DNA and the contacts between Sin subunits bound at the crossover and accessory sites are strikingly similar to the corresponding features in current models of the Tn3/γδ resolvase complex. It seems plausible that many resolution systems adopt a similar strategy for activation of catalysis, despite significantly different regulatory module architectures.

Regulation of Invertase Recombination

The serine invertase systems are described in detail in Chapter 9, this volume. Most research on invertase mechanism has been on the Hin, Gin and Cin systems (33). Like resolvases, the invertases recombine at precise positions within dimer-binding crossover sites, but unlike the resolvases there are no adjacent accessory sequences. However, it was discovered that a sequence quite far from the crossover sites (called the enhancer, or *sis*) which binds an *Escherichia coli* protein FIS (factor for inversion stimulation) was essential for efficient recombination in each system (94, 95, 96, 97, 98). The lengths of DNA between the crossover sites and the enhancer could be varied without loss of activity (99). It is proposed that the invertase-bound crossover sites and FIS-bound enhancer come together to form a synaptic complex, as shown diagrammatically in Fig. 9. The molecular architecture of this complex is still incompletely understood, but structure-based models have been built following characterization of specific invertase–FIS interfaces, and it has been shown recently that Hin subunits make direct contacts with the enhancer DNA (100).

Like resolvases, invertases selectively recombine sites within the same supercoiled molecule, in a specific relative orientation, in this case inverted repeat. However, there is a notable difference. Resolvase has no activity at all on substrates with two *res* sites in an inverted repeat, whereas Hin (or Gin) invertase substrates with *hix* (or *gix*) sites in a direct repeat do not give recombinants but do undergo efficient double rounds of rotational strand exchange, giving knotted nonrecombinant products (62, 63, 64). It was concluded that the synaptic complex (Fig. 9) is formed regardless of relative site orientation, but sites in a direct repeat are thereby misaligned in "antiparallel" such that strand exchange would give recombinants with mismatched base pairs; instead, double rounds of strand exchange give knotted nonrecombinant products (62) (see "Subunit rotation" above).

Regulation of Recombination by Large Serine Recombinases

Serine integrases and other large serine recombinases are the subject of Chapter 11 of this book. The integrases do not display topological selectivity; they adopt a quite different strategy to ensure correct product formation. The phage (*attP*) and bacterial genome (*attB*) crossover sites are not identical, and recombination results in two further nonidentical sites *attL* and *attR*, flanking the integrated prophage DNA (Fig. 10). Unlike phage tyrosine integrases related to λ Int, the large serine integrases (and other large serine recombinases such as the *Clostridium* transposase TnpX) apparently do not have accessory DNA sequences (35, 101, 102). The details that follow derive from *in vitro* studies on the two best-characterized serine integrases, ϕC31 Int and Bxb1 Int (35). The *att* sites (~40 bp) each bind an integrase dimer and have an asymmetric central 2-bp overlap sequence, which has been shown to be the sole determinant of site polarity in *attP*×*attB* recombination (103, 104). An *attP* site only recombines with an *attB* site, not with another *attP*, *attL* or *attR* site; likewise, *attB* only recombines with *attP*. Synapsis is a key selective step; only the "correct" pair of sites (*attP* and *attB*) forms a stable complex (105, 106). The integrase alone does not recombine the "lysogen" sites *attL* and *attR* at all. However, a phage-encoded recombination directionality factor (RDF) protein binds to and transforms integrase so that it efficiently and specifically recombines *attL*×*attR*, whereas *attP*×*attB* recombination is inhibited (107, 108, 109) (Fig. 10). Recent crystallography of the C-terminal part of A118 integrase bound to DNA (51) has led to a structure-based hypothesis for integrase *att* site selectivity (27). Unlike the small serine recombinases, integrases do not require specific connectivities between sites or DNA supercoiling, making them attractive for applications in biotechnology and synthetic biology (see "Serine recombinases in biotechnology and synthetic biology").

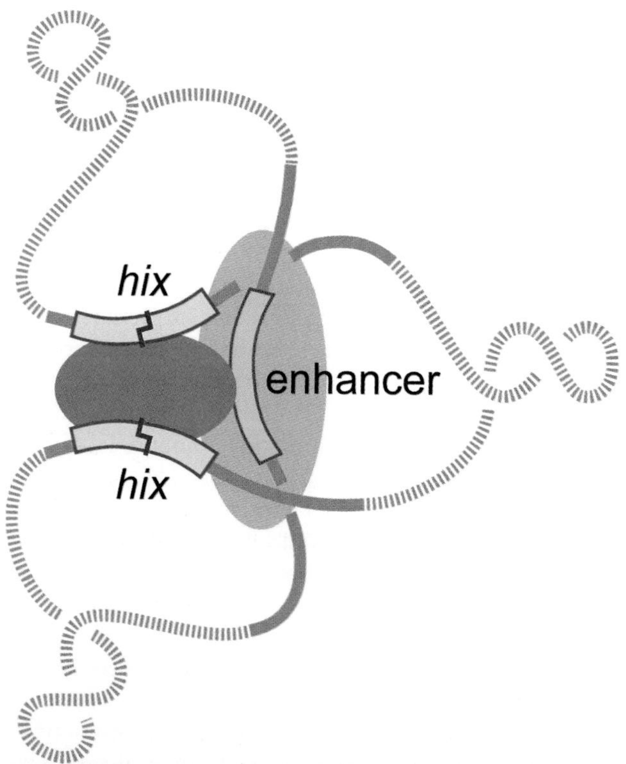

Figure 9 Cartoon of proposed inversion synaptic complex. An invertase tetramer bridging the two crossover sites is shown as an orange oval, and contacts the enhancer DNA (brown) and a FIS dimer bound there (pink). The DNA in the complex is intertwined as shown. Supercoiled loops of DNA outside the complex are shown as dashed lines.
doi:10.1128/microbiolspec.MDNA3-0046-2014.f9

PROTEINS RELATED TO SERINE RECOMBINASES

There are no other families of proteins with known functions that can be unambiguously shown to be related to the serine recombinases. The SR fold has similarity to the structures of a group of 5′-to-3′ exonucleases and to the TOPRIM domain of type IA and type II topoisomerases (110, 111), but the active site residues (and thus presumably the catalytic mechanisms) are quite different, so it is not clear that there is any relationship by descent.

SERINE RECOMBINASES IN BIOTECHNOLOGY AND SYNTHETIC BIOLOGY

Many recombinase systems have been investigated as possible tools for mediation of precise, inducible DNA rearrangements in the fields of experimental genetics, biotechnology and gene therapy (112). However, with one notable exception (ϕC31 integrase), the utilization of serine recombinases has been relatively limited. ϕC31 integrase has been adopted for targeted transgene integration in a number of organisms including

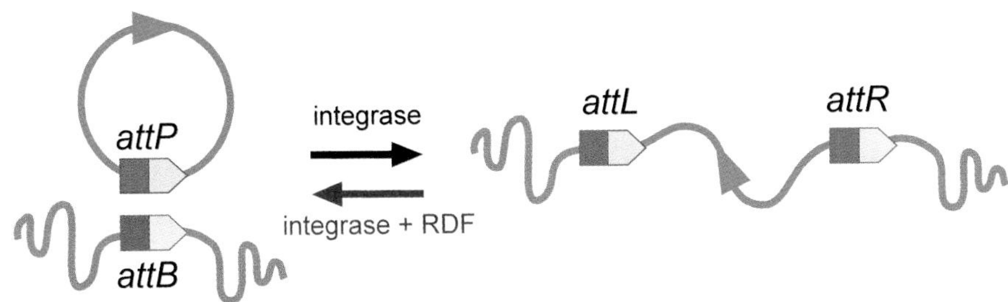

Figure 10 Recombination by serine integrases. In nature, integrase promotes recombination between a crossover site, *attP*, in the circular bacteriophage DNA and a different crossover site, *attB*, in the host bacterial genome (indicated by a squiggly line). Integrase alone does not promote any reaction between the product sites *attL* and *attR*. However, the presence of a bacteriophage-encoded recombination directionality factor (RDF) protein alters the properties of integrase so that it preferentially promotes *attL*×*attR* recombination (red arrow). See text for further details. doi:10.1128/microbiolspec.MDNA3-0046-2014.f10

humans, and it is now in widespread use in experimental research, notably in the *Drosophila* field (112, 113).

Serine recombinases are currently being exploited in the field of synthetic biology, for construction of artificial genetic switches and circuits. Recent studies have shown how all the standard Boolean logic operations can be implemented on gene expression in *E. coli* by the combined action of two orthogonal serine integrases (ϕC31 Int and Bxb1 Int) (114). Other applications are in the assembly and manipulation of metabolic pathway genetic components (113, 115, 116).

ENGINEERING SERINE RECOMBINASES

Site-specific recombination has obvious potential as a tool for "genomic surgery" in organisms of interest to humans, but to realize this potential it will be necessary to engineer recombinases so that they recognize and act on sequences occurring in these organisms. Natural recombinases often require long complex sites, accessory factors, and DNA supercoiling, making this task seem quite daunting. However, the characterization of activated variants of small serine recombinases, which have simplified substrate requirements, has opened up engineering possibilities (71, 74, 117). The small serine recombinases are modular proteins with spatially distinct SR and HTH (DNA-binding) domains (see Fig. 6). Some sequence specificity changes were made by mutating the HTH domain or replacing it with a domain from a related serine recombinase (118, 119, 120). However, much more dramatic retargeting was achieved by linking the SR domain to a zinc-finger DNA-binding domain (121). These "zinc finger recombinases" can be adapted to use a wide range of new "crossover sites" including natural genomic sequences, by engineering zinc-finger-domain specificity, reducing or altering the residual sequence specificity of the SR domain, or using SR domains from different recombinases (86, 122, 123, 124, 125). Recently, transcription activator-like effector (TALE) DNA-binding domains have been used instead of zinc-finger domains to retarget SR domain activity; the modularity of TALE domains and thus the ease of creating new specificities may greatly enhance the applicability of these "designer recombinases" (126).

Acknowledgments. *I apologize to readers that, in this overview chapter on the serine recombinases, it has been painfully necessary for me to cover important aspects of the subject only sketchily or even not at all, and to curtail the reference list. However, much more detail on serine recombinases can be found in the three chapters in this book by Phoebe Rice, Maggie Smith and Reid Johnson. I also take this opportunity to acknowledge the many contributions of colleagues past and present to the advancement of this field, and to look forward to many more exciting developments and insights in the future.*

Citation. Stark WM. 2014. The serine recombinases. Microbiol Spectrum **2**(6):MDNA3-0046-2014.

References

1. **Jayaram M, Grainge I.** 2005. Introduction to site-specific recombination, p 33–81. *In* Mullany P (ed), *The dynamic bacterial genome*. Cambridge University Press, New York, NY.
2. **Nash HA.** 1996. Site-specific recombination: integration, excision, resolution and inversion of defined DNA segments, p 2363–2376. *In* Neidhardt FC (ed), *Escherichia coli and Salmonella typhimurium*. ASM Press, Washington, DC.

3. **Grindley ND, Whiteson KL, Rice PA.** 2006. Mechanism of site-specific recombination. *Ann Rev Biochem* **75:** 567–605.
4. **Hallet B, Sherratt DJ.** 1997. Transposition and site-specific recombination: adapting DNA cut-and-paste mechanisms to a variety of genetic rearrangements. *FEMS Microbiol Rev* **21:**157–178.
5. **Berg DE, Howe MM (ed).** 1989. *Mobile DNA*. ASM Press, Washington DC [see Chapters 1, 5, 28, 29, and 33].
6. **Craig N, Craigie R, Gellert M, Lambowitz A (ed).** 2004. *Mobile DNA II*. ASM Press, Washington DC [see Chapters 6 to 14].
7. **Brembu T, Winge P, Tooming-Klunderud A, Nederbragt AJ, Jakobsen KS, Bones AM.** 2013. The chloroplast genome of the diatom *Seminavis robusta*: new features introduced through multiple mechanisms of horizontal gene transfer. *Mar Genomics* **16:**17–7.
8. **Piednoël M, Gonçalves IR, Higuet D, Bonnivard E.** 2011. Eukaryote DIRS1-like retrotransposons: an overview. *BMC Genomics* **12:**621.
9. **Heffron F, Mccarthy BJ.** 1979. DNA sequence analysis of the transposon Tn3: three genes and three sites involved in transposition of Tn3. *Cell* **18:**1153–1163.
10. **Arthur A, Sherratt DJ.** 1979. Dissection of the transposition process: a transposon-encoded site-specific recombination system. *Mol Gen Genet* **175:**267–274.
11. **Kostriken R, Morita C, Heffron F.** 1981. Transposon Tn3 encodes a site-specific recombination system: identification of essential sequences, genes and actual site of recombination. *Proc Natl Acad Sci U S A* **78:** 4041–4045.
12. **Grindley NDF, Lauth MR, Wells RG, Wityk RJ, Salvo JJ, Reed RR.** 1982. Transposon-mediated site-specific recombination: identification of three binding sites for resolvase at the *res* site of γδ and Tn*3*. *Cell* **30:**19–27.
13. **Kitts PA, Symington LS, Dyson P, Sherratt DJ.** 1983. Transposon-encoded site-specific recombination: nature of the Tn*3* DNA sequences which constitute the recombination site *res*. *EMBO J* **2:**1055–1060.
14. **Plasterk RHA, Brinkman A,van de Putte P.** 1983. DNA inversions in the chromosome of *Escherichia coli* and in bacteriophage Mu: relationship to other site-specific recombination systems. *Proc Natl Acad Sci U S A* **80:** 5355–5358.
15. **Hatfull GF, Grindley NDF.** 1988. Resolvases and DNA-invertases: a family of enzymes active in site-specific recombination, p 357–396. *In* Kucherlapati R, Smith GR (ed), *Genetic recombination*. ASM Press, Washington DC.
16. **Reed RR.** 1981. Transposon-mediated site-specific recombination: a defined *in vitro* system. *Cell* **25:**713–719.
17. **Reed RR.** 1981. Resolution of cointegrates between transposons γδ and Tn*3* defines the recombination site. *Proc Natl Acad Sci U S A* **78:**3428–3432.
18. **Reed RR, Grindley NDF.** 1981. Transposon-mediated site-specific recombination *in vitro*: DNA cleavage and protein-DNA linkage at the recombination site. *Cell* **25:** 721–728.
19. **Reed RR, Moser CD.** 1984. Resolvase-mediated recombination intermediates contain a serine residue covalently linked to DNA. *Cold Spring Harbor Symp Quant Biol* **49:**245–249.
20. **Klippel A, Mertens G, Patschinsky T, Kahmann R.** 1988. The DNA invertase Gin of phage Mu: formation of a covalent complex with DNA via a phosphoserine at amino acid position 9. *EMBO J* **7:**1229–1237.
21. **Hatfull GF, Grindley NDF.** 1986. Analysis of γδ resolvase mutants *in vitro*: evidence for an interaction between serine-10 of resolvase and site I of *res*. *Proc Natl Acad Sci U S A* **83:**5429–5433.
22. **Leschziner A, Boocock MR, Grindley NDF.** 1995. The tyrosine-6 hydroxyl of γδ resolvase is not required for the DNA cleavage and rejoining reactions. *Mol Microbiol* **15:**865–870.
23. **Esposito D, Scocca JJ.** 1997. The integrase family of tyrosine recombinases: evolution of a conserved active site domain. *Nucleic Acids Res* **25:**3605–3614.
24. **Kamali-Moghaddam M, Sundstrom L.** 2000. Transposon targeting determined by resolvase. *FEMS Microbiol Lett* **186:**55–59.
25. **Smith MCM, Thorpe HM.** 2002. Diversity in the serine recombinases. *Mol Microbiol* **44:**299–307.
26. **Rowland SJ, Stark WM.** 2005. Site-specific recombination by the serine recombinases, p 121–150. *In* Mullany P (ed), *The dynamic bacterial genome*. Cambridge University Press, New York, NY.
27. **Van Duyne GD, Rutherford K.** 2013. Large serine recombinase domain structure and attachment site binding. *Crit Rev Biochem Mol Biol* **48:**476–491.
28. **Thorpe HM, Smith MCM.** 1998. *In vitro* site-specific integration of bacteriophage DNA catalysed by a recombinase of the resolvase/invertase family. *Proc Natl Acad Sci U S A*, **95:**5505–5510.
29. **Kersulyte D, Mukhopadhyay AK, Shirai M, Nakazawa T, Berg DE.** 2000. Functional organization and insertion specificity of IS*607*: a chimeric element of *Helicobacter pylori*. *J Bacteriol* **182:**5300–5308.
30. **Boocock MR, Rice PA.** 2013. A proposed mechanism for IS607-family transposases. *Mobile DNA* **4:**24.
31. **Grindley NDF.** 2002. The movement of Tn*3*-like elements: transposition and cointegrate resolution, p 272–302. *In* Craig N, Craigie R, Gellert M, Lambowitz A (ed), *Mobile DNA II*. ASM Press, Washington DC.
32. **Sherratt D.** 1989. Tn*3* and related transposable elements: site-specific recombination and transposition, p 163–184. *In* Berg DE, Howe MM (ed), *Mobile DNA*. ASM Press, Washington DC.
33. **Johnson RC.** 2002. Bacterial site-specific DNA inversion systems, p 230–271. *In* Craig N, Craigie R, Gellert M, Lambowitz A (ed) *Mobile DNA II*. ASM Press, Washington DC.
34. **Groth AC, Calos MP.** 2004. Phage integrases: biology and applications. *J Mol Biol* **335:**667–678.
35. **Smith MCM, Brown WRA, McEwan AR, Rowley PA.** 2010. Site-specific recombination by ϕC31 integrase and other large serine recombinases. *Biochem Soc Trans* **38:**388–394.

36. **Askora A, Kawasaki T, Fujie M, Yamada T.** 2011. Resolvase-like serine recombinase mediates integration/excision in the bacteriophage ϕRSM. *J Biosci Bioeng* **111:**109–116.

37. **Sanderson MR, Freemont PS, Rice PA, Goldman A, Hatfull GF, Grindley NDF, Steitz TA.** 1990. The crystal structure of the catalytic domain of the site-specific recombination enzyme γδ resolvase at 2.7 Å resolution. *Cell* **63:**1323–1329.

38. **Rice PA, Steitz TA.** 1994. Model for a DNA-mediated synaptic complex suggested by crystal packing of γδ resolvase subunits. *EMBO J* **13:**1514–1524.

39. **Rice PA, Steitz TA.** 1994. Refinement of γδ resolvase reveals a strikingly flexible molecule. *Structure* **2:** 371–384.

40. **Yang W, Steitz TA.** 1995. Crystal structure of the site-specific recombinase γδ resolvase complexed with a 34 bp cleavage site. *Cell* **82:**193–207.

41. **Li W, Kamtekar S, Xiong Y, Sarkis GJ, Grindley NDF, Steitz TA.** 2005. Structure of a synaptic γδ resolvase tetramer covalently linked to two cleaved DNAs. *Science* **309:**1210–125.

42. **Kamtekar S, Ho RS, Cocco MJ, Li W, Wenwieser SVCT, Boocock MR, Grindley NDF, Steitz TA.** 2006. Implications of structures of synaptic tetramers of γδ resolvase for the mechanism of recombination. *Proc Natl Acad Sci U S A* **103:**10642–10647.

43. **Mouw KW, Rowland SJ, Gajjar MM, Boocock MR, Stark WM, Rice PA.** 2008. Architecture of a serine recombinase-DNA regulatory complex. *Mol Cell* **30:** 145–155.

44. **Keenholtz RA, Rowland SJ, Boocock MR, Stark WM, Rice PA.** 2011. Structural basis for catalytic activation of a serine recombinase. *Structure* **19:**799–809.

45. **Pan B, Maciejewski MW, Marintchev A, Mullen GP.** 2001. Solution structure of the catalytic domain of γδ resolvase. Implications for the mechanism of catalysis. *J Mol Biol* **310:**1089–1107.

46. **Liu T, DeRose EF, Mullen GP.** 1994. Determination of the structure of the DNA binding domain of γδ resolvase in solution. *Protein Sci* **3:**1286–1295.

47. **Feng JA, Johnson RC, Dickerson RE.** 1994. Hin recombinase bound to DNA: the origin of specificity in major and minor groove interactions. *Science* **263:**348–355.

48. **Chiu TK, Sohn C, Dickerson RE, Johnson RC.** 2002. Testing water-mediated DNA recognition by the Hin recombinase. *EMBO J* **21:**801–814.

49. **Ritacco CJ, Kamtekar S, Wang J, Steitz TA.** 2012. Crystal structure of an intermediate of rotating dimers within the synaptic tetramer of the G-segment invertase. *Nucleic Acids Res* **41:**2673–2682.

50. **Ritacco CJ, Steitz TA, Wang J.** 2014. Exploiting large non-isomorphous differences for phase determination of a G-segment invertase–DNA complex. *Acta Cryst* **D70:** 685–693.

51. **Rutherford K, Yuan P, Perry K, Sharp R, Van Duyne GD.** 2013. Attachment site recognition and regulation of directionality by the serine integrases. *Nucleic Acids Res* **41:**8341–8356.

52. **Yuan P, Gupta K, Van Duyne GD.** 2008. Tetrameric structure of a serine integrase catalytic domain. *Structure* **16:**1275–1286.

53. **Nöllmann M, He J, Byron O, Stark WM.** 2004. Solution structure of the Tn*3* resolvase-crossover site synaptic complex. *Mol Cell* **16:**127–137.

54. **Aravind L, Landsman D.** 1998. AT-hook motifs identified in a wide variety of DNA-binding proteins. *Nucleic Acids Res* **26:**4413–4421.

55. **Krasnow MA, Cozzarelli NR.** 1983. Site-specific relaxation and recombination by the Tn*3* resolvase: recognition of the DNA path between oriented *res* sites. *Cell* **32:**1313–1324.

56. **Stark WM, Boocock MR, Sherratt DJ.** 1989. Site-specific recombination by Tn*3* resolvase. *Trends Genet* **5:** 304–309.

57. **Stark WM, Boocock MR.** 1995. Topological selectivity in site-specific recombination, p 101–129. *In* Sherratt DJ (ed), *Mobile genetic elements.* Frontiers in Molecular Biology series. Oxford University Press, Oxford, UK.

58. **Wasserman SA, Cozzarelli NR.** 1985. Determination of the stereostructure of the product of Tn*3* resolvase by a general method. *Proc Natl Acad Sci U S A* **82:** 1079–1083.

59. **Wasserman SA, Dungan JM, Cozzarelli NR.** 1985. Discovery of a predicted DNA knot substantiates a model for site-specific recombination. *Science* **229:**171–174.

60. **Wasserman SA, Cozzarelli NR.** 1986. Biochemical topology: applications to DNA recombination and replication. *Science* **232:**951–960.

61. **Stark WM, Sherratt DJ, Boocock MR.** 1989. Site-specific recombination by Tn*3* resolvase: topological changes in the forward and reverse reactions. *Cell* **58:**779–790.

62. **Kanaar R, Klippel A, Shekhtman E, Dungan JM, Kahmann R, Cozzarelli NR.** 1990. Processive recombination by the phage Mu Gin system: implications for the mechanism of DNA exchange DNA site alignment, and enhancer action. *Cell* **62:**353–366.

63. **Heichman KA, Moskowitz IPG, Johnson RC.** 1991. Configuration of DNA strands and mechanism of strand exchange in the Hin invertasome as revealed by analysis of recombinant knots. *Genes Dev* **5:**1622–1634.

64. **Moskowitz IP, Heichman KA, Johnson RC.** 1991. Alignment of recombination sites in Hin-mediated site-specific recombination. *Genes Dev* **5:**1635–1645.

65. **Stark WM, Grindley NDF, Hatfull GF, Boocock MR.** 1991. Resolvase-catalysed reactions between *res* sites differing in the central dinucleotide of subsite I. *EMBO J* **10:**3541–3548.

66. **Stark WM, Boocock MR.** 1994. The linkage change of a knotting reaction catalysed by Tn*3* resolvase. *J Mol Biol* **239:**25–36.

67. **McIlwraith MJ, Boocock MR, Stark WM.** 1996. Site-specific recombination by Tn*3* resolvase, photocrosslinked to its supercoiled DNA substrate. *J Mol Biol* **260:**299–303.

68. **McIlwraith MJ, Boocock MR, Stark WM.** 1997. Tn*3* resolvase catalyses multiple recombination events

without intermediate rejoining of DNA ends. *J Mol Biol* **266:**108–121.

69. **Dhar G, Sanders ER, Johnson RC.** 2004. Architecture of the Hin synaptic complex during recombination: the recombinase subunits translocate with the DNA strands. *Cell* **119:**33–45.
70. **Dhar G, McLean MM, Heiss JK, Johnson RC.** 2009. The Hin recombinase assembles a tetrameric protein swivel that exchanges DNA strands. *Nucleic Acids Res* **37:**4743–4756.
71. **Klippel A, Cloppenborg K, Kahmann R.** 1988. Isolation and characterization of unusual *gin* mutants. *EMBO J* **7:**3983–3989.
72. **Haffter P, Bickle TA.** 1988. Enhancer-independent mutants of the Cin recombinase have a relaxed topological specificity. *EMBO J* **7:**3991–3996.
73. **Klippel A, Kanaar R, Kahmann R, Cozzarelli NR.** 1993. Analysis of strand exchange and DNA binding of enhancer-independent Gin recombinase mutants. *EMBO J* **12:**1047–1057.
74. **Arnold PH, Blake DG, Grindley NDF, Boocock MR, Stark WM.** 1999. Mutants of Tn*3* resolvase which do not require accessory binding sites for recombination activity. *EMBO J* **18:**1407–1414.
75. **Sarkis GJ, Murley LL, Leschziner AE, Boocock MR, Stark WM, Grindley NDF.** 2001. A model for the γδ resolvase synaptic complex. *Mol Cell* **8:**623–631.
76. **Leschziner AE, Grindley NDF.** 2003. The architecture of the γδ resolvase crossover site synaptic complex revealed by using constrained DNA substrates. *Mol Cell* **12:**775–781.
77. **Dhar G, Heiss JK, Johnson RC.** 2009. Mechanical constraints on Hin subunit rotation imposed by the Fis/enhancer system and DNA supercoiling during site-specific recombination. *Mol Cell* **34:**746–759.
78. **Bai H, Sun M, Ghosh P, Hatfull GF, Grindley NDF, Marko JF.** 2011. Single-molecule analysis reveals the molecular bearing mechanism of DNA strand exchange by a serine recombinase. *Proc Natl Acad Sci U S A* **108:** 7419–7424.
79. **Olorunniji FJ, Buck DE, Colloms SD, McEwan AR, Smith MCM, Stark WM, Rosser SJ.** 2012. Gated rotation mechanism of site-specific recombination by ϕC31 integrase. *Proc Natl Acad Sci U S A* **109:**19661–19666.
80. **Olorunniji FJ, Stark WM.** 2009. The catalytic residues of Tn*3* resolvase. *Nucleic Acids Res* **37:**7590–7602.
81. **Keenholtz RA, Mouw KW, Boocock MR, Li NS, Piccirilli JA, Rice PA.** 2013. Arginine as a general acid catalyst in serine recombinase-mediated DNA cleavage. *J Biol Chem* **288:**29206–29214.
82. **Mizuuchi K, Baker TA.** 2002. Chemical mechanisms for mobilizing DNA, 12–23. *In* Craig N, Craigie R, Gellert M, Lambowitz A (ed) *Mobile DNA II*. ASM Press, Washington DC.
83. **Abdel-Meguid SS, Grindley NDF, Templeton NS, Steitz TA.** 1984. Cleavage of the site-specific recombination protein γδ resolvase: the smaller of the two fragments binds DNA specifically. *Proc Natl Acad Sci U S A* **81:** 2001–2005.
84. **Rimphanitchayakit V, Grindley NDF.** 1990. Saturation mutagenesis of the DNA site bound by the small carboxy-terminal domain of γδ resolvase. *EMBO J* **9:** 719–725.
85. **Wells RG, Grindley NDF.** 1984. Analysis of the γδ *res* site: sites required for site-specific recombination and gene expression. *J Mol Biol* **179:**667–687.
86. **Proudfoot CM, McPherson AL, Kolb AF, Stark WM.** 2011. Zinc finger recombinases with adaptable DNA sequence specificity. *PLoS ONE* **6:**e19537.
87. **Olorunniji FJ, He J, Wenwieser SVCT, Boocock MR, Stark WM.** 2008. Synapsis and catalysis by activated Tn*3* resolvase mutants. *Nucleic Acids Res* **36:**7181–7191.
88. **Rowland SJ, Boocock MR, McPherson AL, Mouw KW, Rice PA, Stark WM.** 2009. Regulatory mutations in Sin recombinase support a structure-based model of the synaptosome. *Mol Microbiol* **74:**282–298.
89. **Watson MA, Boocock MR, Stark WM.** 1996. Rate and selectivity of synapsis of *res* recombination sites by Tn*3* resolvase. *J Mol Biol* **257:**317–329.
90. **Kilbride E, Boocock MR, Stark WM.** 1999. Topological selectivity of a hybrid site-specific recombination system with elements from Tn*3* *res*/resolvase and bacteriophage P1 *loxP*/Cre. *J Mol Biol* **289:**1219–1230.
91. **Hughes RE, Hatfull GF, Rice PA, Steitz TA, Grindley NDF.** 1990. Cooperativity mutants of the γδ resolvase identify an essential interdimer interaction. *Cell* **63:** 1331–1338.
92. **Murley LL, Grindley NDF.** 1998. Architecture of the γδ resolvase synaptosome: oriented heterodimers identify interactions essential for synapsis and recombination. *Cell* **95:**553–562.
93. **Rowland SJ, Stark WM, Boocock MR.** 2002. Sin recombinase from *Staphylococcus aureus*: synaptic complex architecture and transposon targeting. *Mol Microbiol* **44:**607–619.
94. **Huber HE, Iida S, Arber W, Bickle TA.** 1985. Site-specific DNA inversion is enhanced by a DNA sequence element in *cis*. *Proc Natl Acad Sci U S A* **82:**3776–3780.
95. **Hubner P, Arber W.** 1989. Mutational analysis of a prokaryotic recombinational enhancer with two functions. *EMBO J* **8:**577–585.
96. **Johnson RC, Simon MI.** 1985. Hin-mediated site-specific recombination requires two 26 bp recombination sites and a 60 bp enhancer. *Cell* **41:**781–791.
97. **Kahmann R, Rudt F, Koch C, Mertens G.** 1985. G inversion in bacteriophage Mu DNA is stimulated by a site within the invertase gene and a host factor. *Cell* **41:** 771–780.
98. **Bruist MF, Glasgow AC, Johnson RC, Simon MI.** 1987. Fis binding to the recombinational enhancer of the Hin DNA inversion system. *Genes Dev* **1:**762–772.
99. **Haykinson MJ, Johnson RC.** 1993. DNA looping and the helical repeat *in vitro* and *in vivo*: effect of HU protein and enhancer location on Hin invertasome assembly. *EMBO J* **12:**2503–2512.
100. **McLean MM, Chang Y, Dhar G, Heiss JK, Johnson RC.** 2013. Multiple interfaces between a serine

recombinase and an enhancer control site-specific DNA inversion. *eLife* **2:**e01211.

101. **Adams V, Lucet IS, Lyras D, Rood JI.** 2004. DNA binding properties of TnpX indicate that different synapses are formed in the excision and integration of the Tn*4451* family. *Mol Microbiol* **53:**1195–1207.

102. **Misiura A, Pigli YZ, Boyle-Vavra S, Daum RS, Boocock MR, Rice PA.** 2013. Roles of two large serine recombinases in mobilizing the methicillin-resistance cassette SCC*mec*. *Mol Microbiol* **88:**1218–1229.

103. **Ghosh P, Kim AI, Hatfull GF.** 2003. The orientation of mycobacteriophage Bxb1 integration is solely dependent on the central dinucleotide of *attP* and *attB*. *Mol Cell* **12:**1101–1111.

104. **Smith MCA, Till R, Smith MCM.** 2004. Switching the polarity of a bacteriophage integration system. *Mol Microbiol* **51:**1719–1728.

105. **Ghosh P, Pannunzio NR, Hatfull GF.** 2005. Synapsis in phage Bxb1 integration: selection mechanism for the correct pair of recombination sites. *J Mol Biol* **349:** 331–348.

106. **Smith MCA, Till R, Brady K, Soultanas P, Thorpe H, Smith MCM.** 2004. Synapsis and DNA cleavage in ϕC31 integrase-mediated site-specific recombination. *Nucleic Acids Res* **32:**2607–2617.

107. **Stark WM.** 2011. Cutting out the ϕC31 prophage. *Mol Microbiol* **80:**1417–1419.

108. **Ghosh P, Wasil LR, Hatfull GF.** 2006. Control of phage Bxb1 excision by a novel recombination directionality factor. *PLoS Biol* **4:**e186.

109. **Khaleel T, Younger E, McEwan AR, Varghese AS, Smith MCM.** 2011. A phage protein that binds ϕC31 integrase to switch its directionality. *Mol Microbiol* **80:** 1450–1463.

110. **Artymiuk P, Ceska TA, Suck D, Sayers JR.** 1997. Prokaryotic 5′-3′ exonucleases share a common core structure with gamma-delta resolvase. *Nucleic Acids Res* **25:** 4224–4229.

111. **Yang W.** 2010. Topoisomerases and site-specific recombinases: similarities in structure and mechanism. *Crit Rev Biochem Mol Biol* **45:**520–534.

112. **Bischof J, Basler K.** 2008. Recombinases and their uses in gene activation, gene inactivation, and transgenesis. *Methods Mol Biol* **420:**175–195.

113. **Fogg PCM, Colloms SD, Rosser SJ, Stark WM, Smith MCM.** 2014. New applications for phage integrases. *J Mol Biol* **426:**2703–2716.

114. **Maranhao AC, Ellington AD.** 2013. Endowing cells with logic and memory. *Nat Biotechnol* **31:**413–415.

115. **Zhang L, Zhao G, Ding X.** 2011. Tandem assembly of the epothilone biosynthetic gene cluster by *in vitro* site-specific recombination. *Sci Rep* **1:**141.

116. **Colloms SD, Merrick CA, Olorunniji FJ, Stark WM, Smith MCM, Osbourn A, Keasling JD, Rosser SJ.** 2014. Rapid metabolic pathway assembly and modification using serine integrase site-specific recombination. *Nucleic Acids Res* **42:**e23.

117. **Burke ME, Arnold PH, He J, Wenwieser SVCT, Rowland SJ, Boocock MR, Stark WM.** 2004. Activating mutations of Tn*3* resolvase marking interfaces important in recombination catalysis and its regulation. *Mol Microbiol* **51:**937–948.

118. **Grindley NDF.** 1993. Analysis of a nucleoprotein complex: the synaptosome of γδ resolvase. *Science* **262:** 738–740.

119. **Ackroyd AJ, Avila P, Parker CN, Halford SE.** 1990. Site-specific recombination by mutants of Tn*21* resolvase with DNA recognition functions from Tn*3* resolvase. *J Mol Biol* **216:**633–643.

120. **Schneider F, Schwikardi M, Muskhelishvili G, Dröge P.** 2000. A DNA-binding domain swap converts the invertase Gin into a resolvase. *J Mol Biol* **295:**767–775.

121. **Akopian A, He J, Boocock MR, Stark WM.** 2003. Chimeric site-specific recombinases with designed DNA sequence recognition. *Proc Natl Acad Sci U S A* **100:** 8688–8691.

122. **Gordley RM, Gersbach CA, Barbas CF III.** 2009. Synthesis of programmable integrases. *Proc Natl Acad Sci U S A* **106:**5053–5058.

123. **Sirk SJ, Gaj T, Jonsson A, Mercer AC, Barbas CF III.** 2014. Expanding the zinc-finger recombinase repertoire: directed evolution and mutational analysis of serine recombinase specificity determinants. *Nucleic Acids Res* **42:**4755–4766.

124. **Akopian A, Stark WM.** 2005. Site-specific recombinases as instruments for genomic surgery. *Adv Genet* **55:** 1–23.

125. **Gaj T, Sirk SJ, Barbas CF III.** 2014. Expanding the scope of site-specific recombinases for genetic and metabolic engineering. *Biotechnology and Bioengineering* **111:**1–15.

126. **Mercer AC, Gaj T, Fuller RP, Barbas CF III.** 2012. Chimeric TALE recombinases with programmable DNA sequence specificity. *Nucleic Acids Res* **40:**11163–11172.

Mobile DNA, 3rd Edition
Nancy L. Craig, Michael Chandler, Martin Gellert, Alan M. Lambowitz, Phoebe A. Rice and Suzanne Sandmeyer

doi:10.1128/microbiolspec.MDNA3-0051-2014

Arthur Landy[1]

4

The λ Integrase Site-specific Recombination Pathway

INTRODUCTION

The λ site-specific recombination pathway has enjoyed the sequential attentions of geneticists, biochemists, and structural biologists for more than 50 years. It has proven to be a rewarding model system of sufficient simplicity to yield a gratifying level of understanding within a single (fortuitously timed) professional career, and of sufficient complexity to engage a small cadre of scientists motivated to peal this onion. The initiating highlight of the genetics phase was the insightful proposal by Allan Campbell for the pathway by which the λ chromosome integrates into, and excises from, the *Escherichia coli* host chromosome (1). The breakthrough for the biochemical phase was the purification of λ integrase (Int) and the integration host factor (IHF) by Howard Nash (2, 3). The first major step in the structural phase was the cocrystal structure of IHF bound to its DNA target site by Phoebe Rice and Howard Nash (4). Although the crystal structure of naked Fis protein had been determined earlier (5, 6), the full impact of Fis on understanding the fundamentals of the Int reaction did not come until much later (7, 8).

λ Integrase is generally regarded as the founding member of what is now called the tyrosine recombinase family, even though many family members are not strictly recombinases. Family membership is defined by the creation of novel DNA junctions via an active site tyrosine that cleaves and reseals DNA through the formation of a covalent 3′-phospho-tyrosine high-energy intermediate without the requirement for any high-energy cofactors. Other important, well studied, and highly exploited family members each have their own chapter in this volume of *Mobile DNA III*. Limitations on space prevent the inclusion in this chapter of the many other interesting family members, which comprise a wide range of biological functions and interesting variations on the themes discussed here, including other well-studied members of the heterobivalent subfamily, such as Tn916 (9), HP1 (10), and L5 (11). For previous reviews that include sections on the tyrosine recombinase family and λ Int see references 12, 13, 14, 15, 16, 17, 18, 19, 20, 21, 22, 23, 24, 25, 26, and 27. In this review, I will try to emphasize as much as possible those features of the λ Int pathway that have been reported since, or were not the focus of, earlier reviews, an intention that will consequently highlight recent advances in structural aspects of the pathway.

OVERVIEW OF THE REACTION

The λ Int recombination pathway has evolved to provide a conditional, effectively irreversible, DNA switch

[1]Division of Biology and Medicine, Brown University, Providence, RI.

in the life cycle of the virus. The "cost" (complexity) associated with regulated directionality in the λ Int pathway is what distinguishes it from its Cre and Flp siblings (Fig. 1). As in most of the family members, each recombining partner DNA contains a pair of inverted repeat recombinase binding sites (called core-type sites) that flank a 7 bp over lap region (O) (6 to 8 bp in other systems) that is identical in both DNAs. (Evolution of new core-type and overlap DNA sequences has been proposed to proceed by low frequency λ phage insertions at sites other than the canonical *att*B [28].) DNA cleavage and exchange of the top strands on one side of the overlap region by two active Ints creates a four-way DNA junction

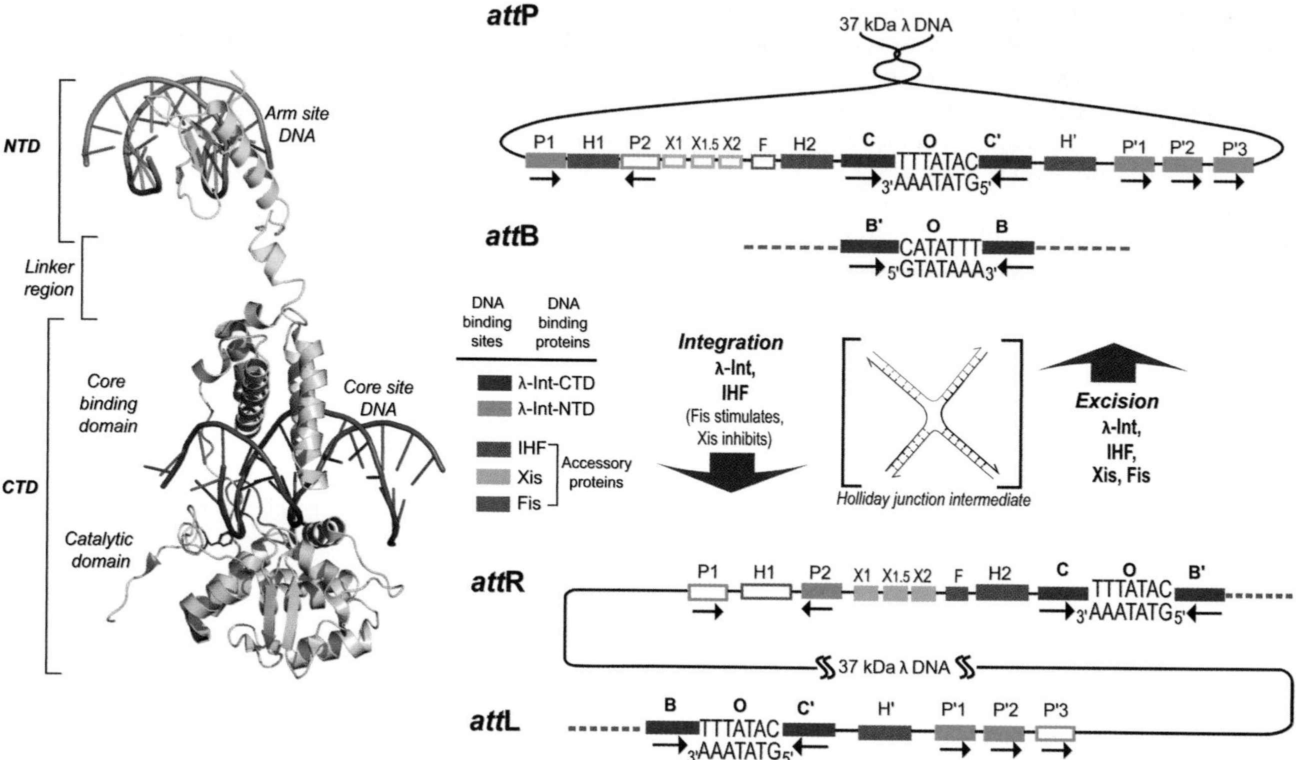

Figure 1 λ Integrase and the overlapping ensembles of protein binding sites that comprise *att* site DNA. The left panel shows the structure of a single λ Int protomer bound via its NTD to an arm site DNA and via its CTD to a core site DNA (adapted from the Int tetrameric structure determined by Biswas *et al.* [44], PDB code 1Z1G). The right panel shows the recombination reactions. Integrative recombination between supercoiled *att*P and linear *att*B requires the virally encoded integrase (Int) (2) and the host-encoded accessory DNA bending protein integration host factor (IHF) (4, 177) and gives rise to an integrated phage chromosome bounded by *att*L and *att*R. Excisive recombination between *att*L and *att*R to regenerate *att*P and *att*B additionally requires the phage-encoded Xis protein (which inhibits integrative recombination) (140) and is stimulated by the host-encoded Fis protein (8). Both reactions proceed through a Holliday junction intermediate that is first generated and then resolved by single strand exchanges on the left and right side of the 7 bp overlap region, respectively. The two reactions proceed with the same order of sequential strand exchanges (not the reverse order) and use different subsets of protein binding sites in the P and P′ arms, as indicated by the filled boxes: Int arm-type P1, P2, P′1, P′2, and P′3 (green); integration host factor (IHF), H1, H2, and H′ (gray); Xis, X1, X1.5, and X2 (gold); and Fis (pink). The four core-type Int binding sites, C, C′, B, and B′ (blue boxes) are each bound in a C-clamp fashion by the CB and CAT domains, referred to here as the CTD. This is where Int executes isoenergetic DNA strand cleavages and ligations via a high-energy covalent 3′-phosphotyrosine intermediate. The CTD of Int and the tetrameric Int complex surrounding the two overlap regions are functionally and structurally similar to the Cre, Flp, and XerC/D proteins. Reprinted with permission from reference 36.
doi:10.1128/microbiolspec.MDNA3-0051-2014.f1

[Holliday junction (HJ)] that is then resolved to recombinant products by the remaining pair of Ints cleaving and exchanging the bottom strands on the other side of the overlap region (Fig. 2). Appended to two of the four core-type sites are additional DNA sequences that encode binding sites for the second (NTD) DNA binding domain of Int and the accessory DNA bending proteins, IHF, Xis, and Fis. As indicated by color coding in Fig. 1, some sites are required only for integrative recombination between *att*P (on the phage chromosome) and *att*B (on the bacterial chromosome), some are required only for excisive recombination between the *att*L and *att*R sites (flanking the integrated prophage), and some sites are required for both reactions. For more detail, see reference 27. It has been suggested that the additional complexity of the λ pathway evolved to regulate the directionality of recombination in response to the physiological state of the host cell (29), a notion that is now well documented in latent human viruses, such as the ubiquitous herpes virus, cf., "... the [herpes] viral genome evolved to sense the infection status of the host... through highly evolved pathogen genomes with the capacity to sense host cytokines..." (30).

HOLLIDAY JUNCTION INTERMEDIATES

A hallmark of the tyrosine recombinase family, discussed here in terms of the λ pathway, is the formation of a four-way DNA junction (HJ) intermediate. For a long time, it was thought that this was a very unstable intermediate because it was difficult to identify without designing elaborate substrates (31, 32, 33). Only many

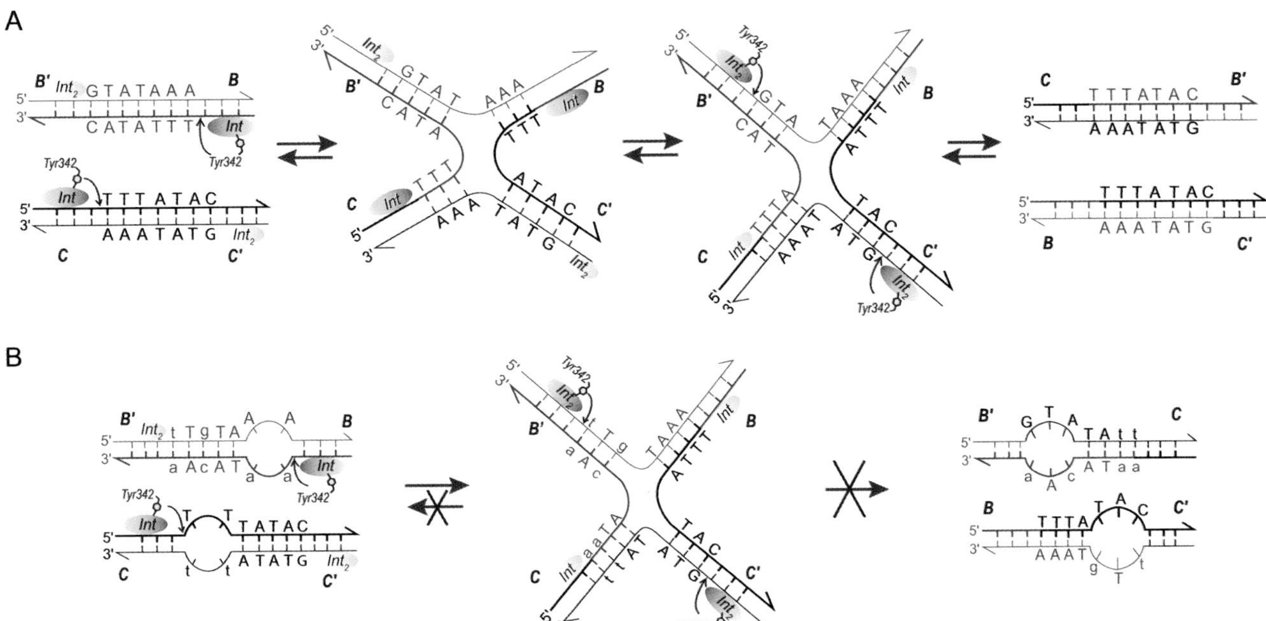

Figure 2 Formation, resolution, and trapping of Holliday junctions (HJ). (A) The top strand of each att site is cleaved via formation of a high-energy phosphotyrosine intermediate and the strands are exchanged (three bases are "swapped") to form the HJ, thus, creating a branch point close to the center of the overlap regions. A conformational change of the complex that slightly repositions the branch point and more extremely repositions the Int protomers leads to the second swap of DNA strands and resolution of the HJ to helical products (44, 178). These features of the reaction suggested the mechanism-based method of trapping HJ complexes shown in (B). The left panel shows the DNA sequence changes made in the 7 bp overlap regions to trap HJ intermediates (lower case letters). Following the first pair of Int cleavages (via the active site Tyr) on one side of the overlap regions (arranged here in antiparallel orientation), the "top" strands are swapped to form the HJ; this simultaneously converts the unpaired (bubble) bases to duplex DNA. On the other side, the sequence differences between the two overlap regions strongly disfavor the second ("bottom") strand swap that would resolve the HJ, because this would generate unpaired bubbles in the product complex (36, 37, 38, 39). This diagram applies to both integrative and excisive recombination (even though the labels refer to integrative recombination). Adapted in part, with permission, from reference 36. doi:10.1128/microbiolspec.MDNA3-0051-2014.f2

years later was it discovered that the standard sodium dodecyl sulfate (SDS) treatment employed for visualizing naked HJ DNA gave misleading results because SDS fails to quench Int ligation activity fast enough to prevent reformation of the initial phosphodiester bonds (see also below) (34).

A striking feature of HJ formation in the Int reaction is that it is always initiated by cleavage and exchange of the same ("top") strands in both integrative and excisive recombination (one of the facts indicating that the two reactions are not the reverse of one another) (Fig. 2) (31, 32, 33, 35). These features of the reaction suggested a mechanism-based method of trapping HJ complexes (outlined in Fig. 2) that would prove useful in studies of the complete higher-order complexes (described below) (36, 37, 38, 39).

Based upon structural snapshots from X-ray crystallographic models, patterns of amino acid residues in the active sites, mutational studies, and biochemical analyses, it is supposed there are only small differences between the pathways of λ Int, Cre, and Flp at the level of HJ formation, structure, and resolution, and likely only minor differences in their respective chemistries of DNA cleavage and ligation. One exception to this generalization is the manner in which the active site tyrosine nucleophile is delivered to the active site. In the case of Flp it is delivered in *trans*, that is, the tyrosine of one protomer in the tetramer is activated as a nucleophile in the active site of its adjacent neighbor (40, 41). While in both Int and Cre the tyrosine nucleophile is in *cis*, its proper positioning within the active site depends upon the nature of an interprotomer interaction between adjacent protomers within the tetrameric complex (42, 43, 44, 45), as discussed further below. In the absence of the NTD DNA binding domain Int can efficiently resolve HJs but it cannot carry out a recombination reaction (discussed further below) (46, 47, 48, 49). It is also clear from mutational analyses that there are Int residues that are specifically critical for HJ resolution but not DNA cleavage (50).

HEXAPEPTIDE INHIBITORS

In a bold and formidable effort to find recombination inhibitors that would trap the HJ intermediate, Anca Segall and her collaborators used deconvolution of synthetic hexapeptide libraries to search for hexapeptides that would block recombination subsequent to the first HJ-forming strand exchange(51, 52). Their most potent peptide inhibitor, WRWYCR, whose active form is a dimer assembled via a disulfide bridge between two peptide monomers, stably traps HJ complexes in all pathways mediated by Int as well as Cre (53, 54). Using this inhibitor, they were able to study the kinetics of HJ resolution under several different conditions and in several different Int-mediated pathways (55, 56). One of their conclusions from these studies was that spermidine stabilizes the "second" HJ isomeric form (the precursor to product formation) (57). Application of a hexapeptide inhibitor to studying the *Bacteriodes* NBU1 recombination pathway revealed that IntN1 recombinase is surprisingly more efficient when it forms HJs in the presence of mismatches, although their resolution to products does require homology (58).

In vitro, the hexapeptides inhibit a range of enzymes involving tyrosine-mediated transesterification, such as vaccinia virus topoisomerase and *E. coli* topoisomerase I (59). Subsequently, they were shown to be bacteriocidal to both Gram positive and Gram negative bacteria, presumably because they can interfere with DNA repair and chromosome dimer resolution by XerC/D. They were also shown to inhibit the excision of several different prophages *in vivo* (60). The *in vivo* successes of the hexapeptide inhibitors motivated the Segall group to search for therapeutically more useful small molecules with similar activities. Indeed, a search of over nine million compounds yielded one potentially interesting compound with properties that suggested the possible value of further searches for functional analogs of the hexapeptide inhibitors (61).

KINETICS

To overcome the difficulty of distinguishing kinetically relevant intermediates from off-pathway species, single molecule experiments were used to determine how binding energy from the multiple protein-DNA interactions is used to achieve efficiency and directionality in the overall Int recombination pathway (34). Protein binding (i.e., associated DNA bending), synapsis between *att*L and *att*R, HJ formation, and recombination were all monitored by changes in the length of a 1353 bp DNA that served as a diffusion-limiting tether of a microscopic bead to the flow chamber bed of a video-enhanced light microscope. In these experiments it was found that stable bent-DNA complexes containing Int, IHF, and Xis form rapidly (<20 s) and independently on *att*L and *att*R, and synapsis under these conditions is extremely rapid (1.0 min^{-1}). These single molecule experiments strongly suggest there are no intrinsic mechanistic features of the pathway that make synapsis slow. While Int-mediated DNA cleavage, before or immediately after synapsis, is required to stabilize the synaptic complexes, those complexes that synapsed (~50% of the total) yield recombinant with an impressive ~100% efficiency. The rate-limiting step of excision occurs after synapsis, but closely precedes or is concomitant with the appearance of a stable HJ. This single molecule result is consistent with the observation that in solution rates of stable HJ formation are similar to the rates of excisive recombination (62).

Given the reversibility of the underlying chemistry of recombination, the apparent irreversibility observed in these experiments of each step of the reaction (except for synapsis) is notable. This result indicates that the overall directionality of excisive recombination is a direct consequence of the sequence of protein–protein and protein-DNA interactions that efficiently drive the reaction forward through nearly every step. It was proposed that the slow step in the reaction is some conformational change that stabilizes the HJ (34). Candidates for this rate-limiting step, such as the scissoring movement of the HJ arms, the shift in the localized bend of the HJ, or the reorientation of the active and inactive pairs of Int protomers, are suggested by comparison of the different X-ray crystal structures of tyrosine family recombinases complexed with their respective four-armed DNAs (15, 22, 43, 44, 63).

A totally different aspect of the kinetics of recombination concerns the process by which λ DNA, once inside the cell, finds its cognate *att*B site. Surprisingly, λ DNA does not carry out an active search but rather remains confined to the point where it entered the cell; it is the directed motion of the bacterial DNA during chromosome replication that delivers *att*B to a waiting, relatively stationary, *att*P (64).

STRUCTURE OF THE Int CTD

Among the most significant recent advances in our understanding of λ Int recombination were those emanating from the X-ray crystallographic studies by the Ellenberger laboratory (44, 45, 65). The second Int fragment to be used by the Ellenberger laboratory for X-ray crystallography, lacked the NTD (arm binding domain) and consisted of residues 75 to 356 (45). Referred to as C75 in the literature and here called the CTD, it corresponds to the two domains comprising the well-studied monovalent family members such as Cre, Flp, and XerC/D. The λ Int CTD is not competent for recombination but it is an efficient topoisomerase, binds weakly to single core-type DNA sites, and resolves λ *att* site HJs (48, 66, 67). The weak binding of the λ Int CTD to single core-type sites was circumvented by trapping covalent Int-*att* site complexes with a "flapped" suicide substrate containing a nick within the overlap region, three bases from the scissile phosphate (Fig. 3A).

As shown in Fig. 3, the λ Int CTD consists of a catalytic domain that is joined to the central binding (CB) domain by a flexible, interdomain linker, residues I160-R176, that is extremely sensitive to proteolytic degradation (45, 48). The CB and catalytic domains of Int both contribute to recognition of the core site, although the former, whose structure has also been determined (68), confers most of the sequence specificity (69, 70). Only two residues from each domain (K95 and N99 in the CB domain and K235 and R287 in the catalytic domain) directly form hydrogen bonds with DNA bases. Interestingly, one of them, K95, interacts with a base, Gua30, that is absent in the B′ site, the weakest of the four core sites (71). The base-specific interactions are consistent with the effects of mutations of these and nearby residues that affect DNA binding specificity (72, 73, 74).

In comparison to the monovalent family members, Cre and Flp (41, 42, 43), the λ Int CTD displays fewer hydrogen bonds and total direct contacts to DNA bases in both its amino- and carboxy-terminal domains. Additionally, the extended unstructured interdomain linker of λ Int appears to be more flexible than the Cre linker (43), suggesting an increase in entropic cost of binding to DNA. Indeed, the helpful and informative *int*-h mutant (E174K), which substitutes a lysine in the middle of the interdomain linker adjacent to the site of DNA cleavage, increases the DNA binding affinity of λ Int and relaxes or eliminates the requirement for IHF during recombination (75, 76). It was proposed that the substituted lysine might enhance DNA binding affinity by contributing a stabilizing interaction with DNA, and/or by constraining the movement of the interdomain linker (45).

A comparison of the structure of the CTD Int covalently bound to DNA (45) with that of the unliganded catalytic domain (65), revealed that the tyrosine342 nucleophile had moved approximately 20 Å into the active site where it forms a 3′-phosphotyrosine linkage with the cleaved DNA (Fig. 4). Additionally, in the tetrameric complex, the eight carboxy-terminal residues (349 to 356) of a protomer extend away from the protein and pack against a neighboring protomer, contributing in *trans* an additional strand (β7) to the sheet formed by strands β1, β2, and β3 of the catalytic domain. This *trans* packing arrangement of β7 is required for appropriate placement of the Tyr342 nucleophile into the active site. This fact, in conjunction with the phenotypes of a number of Int mutants, suggests a dual role for the alternative stacking arrangements of β7.

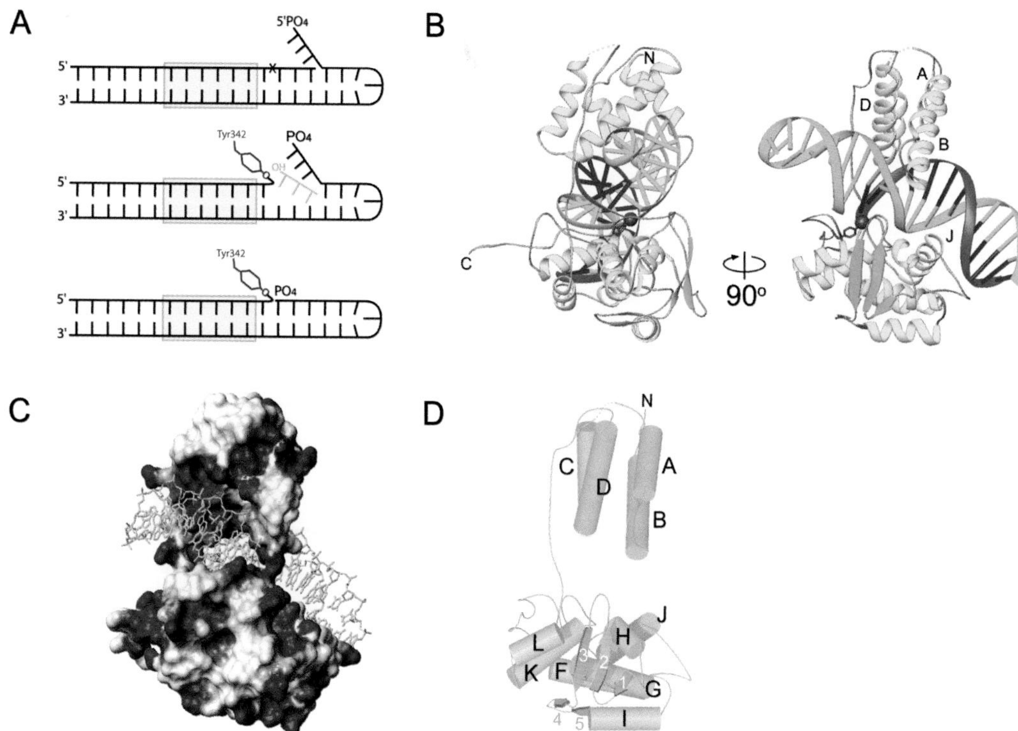

Figure 3 X-ray crystal structure of the Int CTD. (A) With this modified version of previously designed suicide recombination substrates (35, 47) covalently trapped CTD-DNA complexes were stable for weeks. Formation of the phosphotyrosine bond and diffusion of the three base oligonucleotide is followed by annealing of the three base flap to the three nucleotide gap, thus, positioning the 5′-phosphate such that it repels water and shields the phosphotyrosine linkage from hydrolysis. (B) Ribbon diagrams showing the central domain (residues 75 to 160; above the DNA) and the catalytic domain (residues 170 to 356; below the DNA) of λ Int, and their interactions with the major and minor grooves on the opposite sides of the DNA. A long, extended linker (residues I160 to R176) connects these domains. The scissile phosphate that is covalently linked to Y342 is shown as a red sphere. The central domain inserts into the major groove adjacent to the site of DNA cleavage. The catalytic domain makes interactions with the major and minor groove on the opposite side of the DNA, straddling the site of DNA cleavage. (C) The solvent accessible surface of the Int protein is shown, colored according to electrostatic potential. The DNA binding surface is highly positive (blue) and makes numerous interactions with the phosphates of the DNA (cf. Figure 3B). The polypeptide linker between domains joins the central and catalytic domains on one side of the DNA. A salt bridge between the Nζ of K93 and the carbonyl oxygen of S234 bridges between domains on the other side of the DNA, completing the ring-shaped structure that encircles the DNA. (D) The architecture of the λ Int C-75 protein is shown with cylinders and arrows representing helices and β strands, respectively. This view is oriented similarly to that in (A) (right side). The central domain of λ Int lacks helix E, corresponding to the fifth helix of Cre's N-terminal domain, which is involved in subunit interactions. Reprinted with permission from reference 45.
doi:10.1128/microbiolspec.MDNA3-0051-2014.f3

This also suggested an attractive explanation for the findings that a carboxy-terminal deletion of seven residues (commencing with Trp350), and mutations involving residues in or around β7 (all of which were expected to untether the Tyr342) abolished recombination but enhanced the topoisomerase activity of monomers (77, 78, 79). Because these same mutations decrease recombinase activity, the C-terminal tail could also be important in coordinating the catalytic activities of adjacent protomers, as seen in the X-ray crystal structure of the tetrameric higher order recombination complex (Fig. 4C) (44, 80).

In contrast to the large movement of the Tyr342 nucleophile in transitioning from the unliganded to the

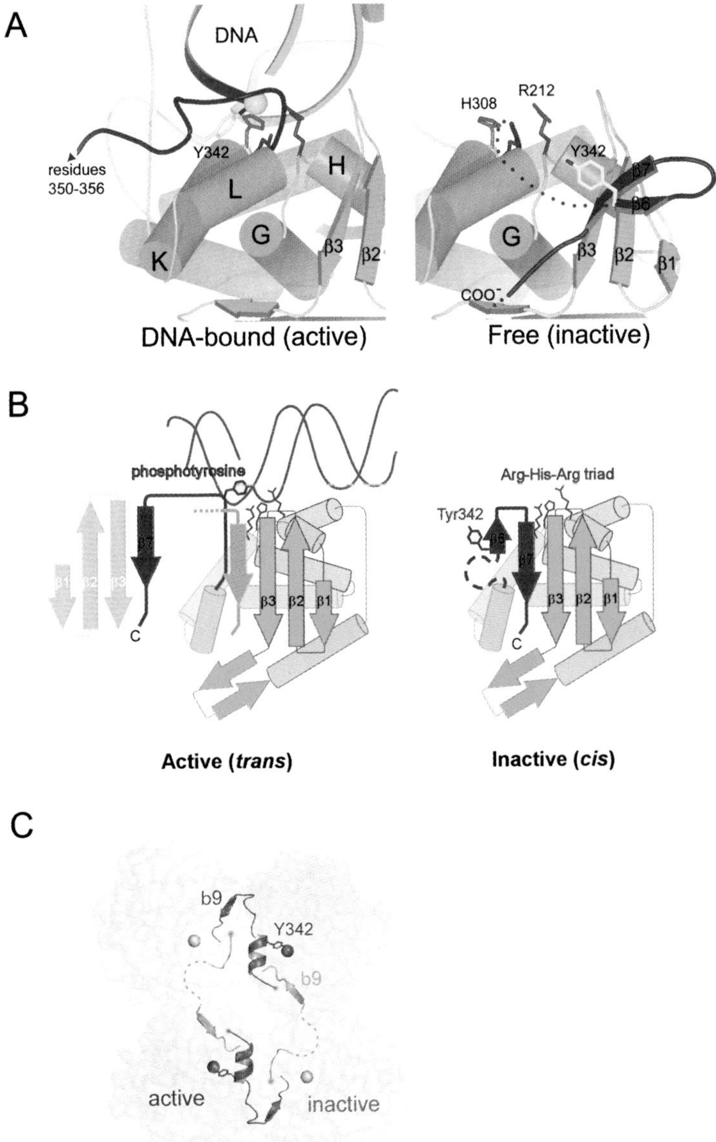

Figure 4 A Remodeling of Int's active site switches DNA cleavage activity on and off. (A) A comparison of the DNA-bound (left), and unbound (right), structures of λ Int shows a dramatic reorganization of the C-terminal region spanning residues 331 to 356 (red). In the absence of DNA, Y342 (yellow) is far from the catalytic triad of R212, H308, and R311 (magenta side chains). In the DNA complex (left panel), Y342 has moved into the active site. Another consequence of the DNA-bound conformation is that the extreme C-terminal residues 349 to 356 extend away from the parent Int molecule and pack against another molecule in *trans*. (B) A cartoon illustrating how the DNA-bound conformation of Int positions the Y342 for cleavage of DNA. The isomerization from the inactive form, in which Y342 is held some distance from the catalytically important Arg212-His308-Arg311 triad (65), to the active conformation seen in complex with DNA, is accompanied by the release of strand β7 and its repacking in *trans* against a neighboring molecule. (C) The assembly of active (orange) and inactive (gray) catalytic sites results from a skewed packing arrangement of λ Int subunits (residues 75 to 356) in the tetramer. The scissile phosphates bound by active and inactive subunits are shown as red and gray spheres, respectively. Reprinted with permission from references 44 and 45. doi:10.1128/microbiolspec.MDNA3-0051-2014.f4

liganded Int, the other four catalytically important residues, R212, K235, H308, and R311, show less than 1 Å movement on average between the two structures, as is also true for most of the other residues in the catalytic domain. The role of these residues in catalysis was established by mutational analyses of several tyrosine recombinase family members, biochemical analyses (especially of topoisomerase I), sequence comparisons of other family members, and shortly thereafter, comparisons with the X-ray crystal structures of other DNA-bound family members (41, 43, 81, 82, 83, 84, 85, 86, 87, 88).

ROLE OF THE Int NTD

The following experiments were carried out to prove it was possible to "de-tune" a monovalent recombinase, for example, Cre, and convert it to a regulated unidirectional recombinase by appending an NTD (89). Cre recombinase is bidirectional, unregulated, does not require accessory proteins, and has a minimal symmetric DNA target. Rather than de-tuning the Cre recombinase its DNA target was attenuated: a single base pair change, previously shown to weaken the interaction between Cre and its DNA binding site (90) was introduced into each of the inverted repeat Cre binding sites and the DNA sequence and spacing between the DNA cleavage sites (the "overlap" region) was changed to the canonical seven base pair sequence of the λ *att* sites. λ P and P′ arms were appended to the modified Cre target sites to generate analogs of the four λ *att* sites.

To complete the recombination pathway, a gene fusion encoding the first 74 residues of λ Int was fused to Cre. The resulting chimeric Cre protein product carried out recombinations between the analogs of the four λ *att* sites with all of the properties of canonical λ Int-dependent pathways: reactions were dependent upon IHF, Xis was required for the excision reaction but inhibited the integration reaction, integrative recombination required the P1 but not the P2 sites, and the excisive reaction required P2 but not P1 (cf. Fig. 1).

It appears from these experiments that the regulated directionality of the λ Int pathway has been conferred on Cre by the appended 74 N-terminal residues of λ Int coupled with the reduction in DNA binding efficiency between Cre and its DNA target sites. These experiments suggest that two simple steps, in no specified order, are all that is required for the evolution of the heterobivalent recombinases from their monovalent siblings. However, they do not rule out an alternative evolutionary trajectory in which the monovalent and heterobivalent site-specific recombinases evolved in parallel from a common, less efficient, precursor.

While the NTD of λ Int was able to confer regulated directionality on the Cre recombinase, it is possible, and even likely, that not all of the λ NTD functions were revealed in these experiments. For example, effects resulting from any interactions between the NTD and the CTD were not studied in those experiments and they would not likely even be manifest in the hybrid protein. One example of such interactions came from studies on the context-dependent effects of the NTD. These studies were prompted by the unexpected finding that the Int CTD (residues 65 to 356, called C65) is more active as a topoisomerase, in binding to core-type sites, cleaving DNA, and resolving synthetic Holliday junctions, than the full length Int. In other words, the NTD is an inhibitor of the primary Int functions (49). Equally surprising was the fact that when the cloned and purified NTD (residues 1 to 65) was added to the cloned and purified CTD, it stimulated all of the primary Int functions, well beyond the levels observed for either CTD or full length Int. In other words, when present in *cis* (i.e., in full length Int), the NTD is an inhibitor of Int functions, but when present in *trans*, it is a stimulator. Resolution of the apparent paradox came with the finding that addition of an oligonucleotide encoding the arm-type DNA sites (P′1–P′2) to full length Int abolished the *cis* NTD inhibition and resulted in the formation of a ternary complex between Int and core and arm-type DNAs.

These results led to the hypothesis of an enhanced dual role for the DNA bending accessory proteins. In addition to their structural function in facilitating the Int-mediated arm-core bridges that comprise the higher-order structure of recombinogenic complexes, they should also be viewed as a requirement to overcome the N-domain inhibition of recombinase functions (49). These data and the resultant hypothesis are consistent with the finding of mutants in one domain that effect the activity of the other (72, 77), and the important observation of Richet *et al.* that Int does not bind well to *att*B unless it part of a higher-order *att*P complex (91).

Residues Met1 to Leu64 comprise the minimal Int fragment that binds to arm-type sites and it does so with almost the same efficiency as full length Int (66, 92). However, an additional six residues are required (Met1 to Ser70) for cooperative binding to the adjacent arm-type sites P′1, P′2, and P′3. The greatest cooperativity in binding, which is between sites P′2 and P′3, depends upon the single bp between them and is resistant to an unopposed three base bulge in the top strand but not in the bottom strand. The asymmetric effect of the unopposed bulge is consistent with DNA bending upon Int binding to the P′ arm sites. Int's affinity for the single sites P′1 or P1 exceeds its net affinity for P′2–P′3 (44, 93). It is interesting that the two lowest affinity arm-type sites, P2 and P′3 are each required for only one of the two recombination reactions, excision and integration, respectively, and are also the outermost sites in their respective pathways. Int binding at P2 is greatly enhanced by its cooperativity with Xis binding at X1, and Int binding at P′3 is enhanced (to a lesser extent) by its cooperativity with Int binding at P′2, thus, rendering the excisive reaction very sensitive to Xis concentration and the integrative reaction more sensitive to Int concentration (66). The latter fit nicely with a very early observation by Enquist *et al.* that integrative recombination is more sensitive than excisive recombination to decreased intracellular levels of Int (94). The Int 1-70 NTD is also equally as competent as full length Int for cooperative interactions with Xis when the two are bound at P2 and X1, respectively (66).

STRUCTURE OF THE NTD

The first view of the NTD structure came from a nuclear magnetic resonance (NMR) analysis of the Met1-Leu64 peptide, which revealed a fold structurally related to the three-stranded β-sheet family of DNA-binding domains. However, it was supplemented with a disordered 10 residue amino-terminal basic tail, that was shown to be important for arm binding by its loss of function upon removing a single positive charge (G2KΔ2R) (92). The importance and role of the amino-terminal basic tail was clearly shown in the subsequent NMR structure of the NTD in complex with its DNA target site (95). Only two other proteins containing this fold have been visualized in complex with their DNA targets: the N-terminal domains from the Tn916 Int protein (96) and from the ethylene responsive factor from *Arabidopsis thaliana* (AtERF1) (97). All three proteins recognize DNA via their unique three-stranded antiparallel β-sheet that is inserted into the major groove of their respective DNA targets. The smaller size of the β-sheet-DNA interface in the λ NTD, relative to the other two proteins, is presumably compensated by the additional contacts of the 11 residue amino-terminal tail that projects deep into the minor groove (95).

STRUCTURE OF A FULL Int TETRAMER COMPLEX

A structural view of the full λ Int did not come until it was cocrystallized with DNA bound at the NTD and CTD, recognition domains for the arm- and core-type DNA sites, respectively. These studies by the Ellenberger laboratory were particularly informative because they represented Int-DNA complexes at three different steps along the recombination pathway (44). One of the structures, a synaptic complex between two COC' core-type sites bound by four CTDs (residues 75 to 356), represented an early step after the first DNA cleavage but before strand exchange. A second structure, with full length Ints in which the cleaved strands had exchanged but ligation was prevented by a modified DNA substrate, represented a post strand-exchange complex. And the third structure was a synthetic Holliday junction intermediate bound by four full length Ints, carrying the Tyr342Phe mutation, that were thus unable to cleave the DNA into products. In the last two structures, the NTDs of the full length Ints were bound to short oligonucleotides containing tandem P′1–P′2 arm-type DNA binding sites. It is likely that the presence of this arm-type DNA occupying the NTD domains was a critical factor in the successful crystallization of the full length Int, and additionally imposed a facilitating (albeit unnatural) 2-fold symmetry. The other factor critical for crystallization was the stable tetrameric arrangement of protomers within each complex.

The tetrameric complexes with full length Int assemble into three distinct layers. The NTD (residues 1 to 63) that binds to arm-type sites is joined to the core-binding domain (CB domain; residues 75 to 175) by a short α-helical segment (residues 64 to 74), and this, in turn, is connected to the C-terminal catalytic domain (residues 176 to 356) through another linker (residues 160 to 176). Together, the three domains of each Int form an ensemble that engages the core and arm DNA targets to form a tightly knit but flexible tetrameric complex (Fig. 5) (44).

The four NTDs are bound by two antiparallel arm DNAs that slightly bend towards each other, with each pair binding the adjacent P′1–P′2 binding sites. The basic N-terminal segment (residues 2 to 10), that was disordered in the NMR structure but shown to be required for recombination activity (92), tucks into the minor groove adjacent to the 3′ side of the arm-type consensus sequence (44).

As noted above, the CB and catalytic domains (which are referred to together as the CTD in this review) are structurally analogous to the full-length monovalent tyrosine recombinases, Cre (42, 43), Flp (41), and XerC/D (98). Thus, it is not surprising that λ Int has a catalytic pocket that resembles the other family members with nearly identical conserved residues (Arg212, Lys235, His308, Arg311, and His333) that

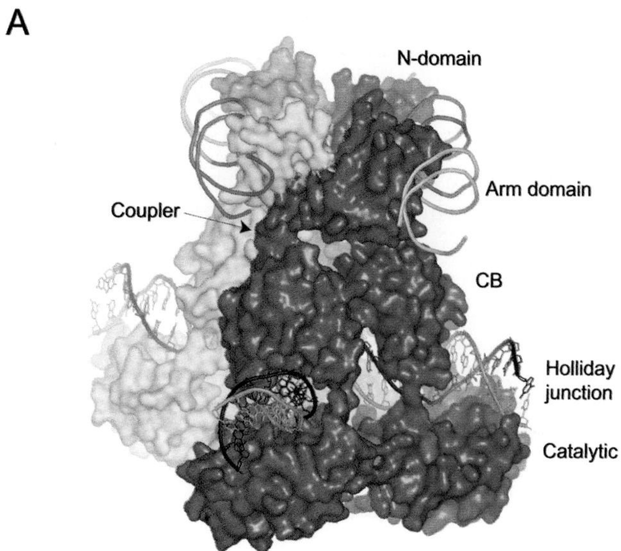

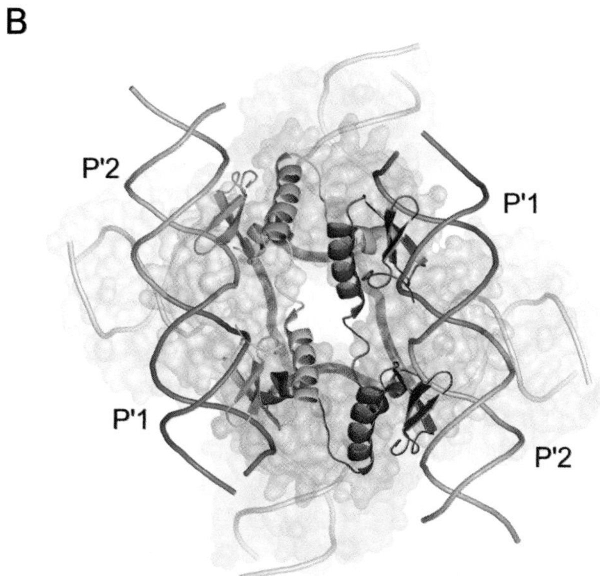

Figure 5 Structure of the λ Int tetramer bound to a Holliday junction and arm DNAs. (A) The domains of Int pack together as three stacked layers, with the NTDs cyclically swapped onto neighboring subunits. The NTD layer embraced by two antiparallel arm DNAs is linked through short α-helical couplers to the CTD, which encircles the branches of the Holliday junction. The active subunits are colored red/green and the inactive subunits are blue/yellow. (B) The 2-fold symmetry of the NTD layer is reflected in the skewed arrangement of the CTDs and the shape of the four-way junction (thick dark gray lines) in the bottom strands reactive isomer. Reprinted with permission from reference 44.
doi:10.1128/microbiolspec.MDNA3-0051-2014.f5

engage the scissile phosphate and Tyr342 nucleophile (44, 45, 65).

Among the factors likely to contribute to λ CTD's lack of recombination function, is the linker (residues 160 to 176) between the CB and catalytic domains. In contrast to the analogous linker in Cre, it lacks the αE helix that contributes many intersubunit interactions that stabilize the Cre tetramer (42, 43). Consequently, the loosely packed CB domains of the λ Int tetramer are able to rotate against each other by as much as 30° in the different isomers that were crystallized (44).

It was particularly interesting that each of the three independent crystal structures determined by Biswas *et al.* (44) illustrates a different conformation of the core DNAs and different subunit packing interactions (Fig. 6). The skewed packing of protomers generates two very different subunit interfaces comprising active versus inactive catalytic sites. In the former, the Tyr342 helix is well ordered and stabilized by electrostatic interactions with two catalytically essential residues. In the latter, the β9 is incompatible with these stabilizing interactions and the region around Tyr342 is disordered (see also Fig. 4C). It should be noted that an α-helical conformation around Tyr342, that was not seen in the active conformation of the earlier crystal structure of the λ CTD (residues 75 to 356) (45), was confirmed by additional crystal structures (in the presence of orthovanadate) to likely be the true active conformation (44).

In the crystal structure of the synaptic, prestrand-exchange, complex, the tetramer deviates strongly from 4-fold symmetry: the scissile phosphates (which can be visualized as the corners of a parallelogram) of the cleaved DNA strands are 39 Å apart while those of the uncleaved strands are 50 Å apart (Fig. 6A, D). This translational offset brings the cleaved 5′ ends closer to the phosphotyrosine of the synapsing partner, thus, facilitating strand exchange and ligation. In the post strand-exchange complex, the core DNAs resemble a HJ intermediate with approximate four-fold symmetry. Here the kink has moved to a more central position, 4 bp away from the cleaved site, bringing the cleavage sites of the bottom strands closer together, and possibly disfavoring reversal of the top strand cleavage (Fig. 6B, E). In the complex with a synthetic preformed HJ, the crossover point was fixed three nucleotides from one pair of cleavage sites, and consequently, these sites are used preferentially for resolution (67, 99). This complex is also highly skewed such that the scissile phosphates, bound by the active protomers, are brought close together (Fig. 6C, F). Although not apparent in the crystal structures, mutational analyses also reveal

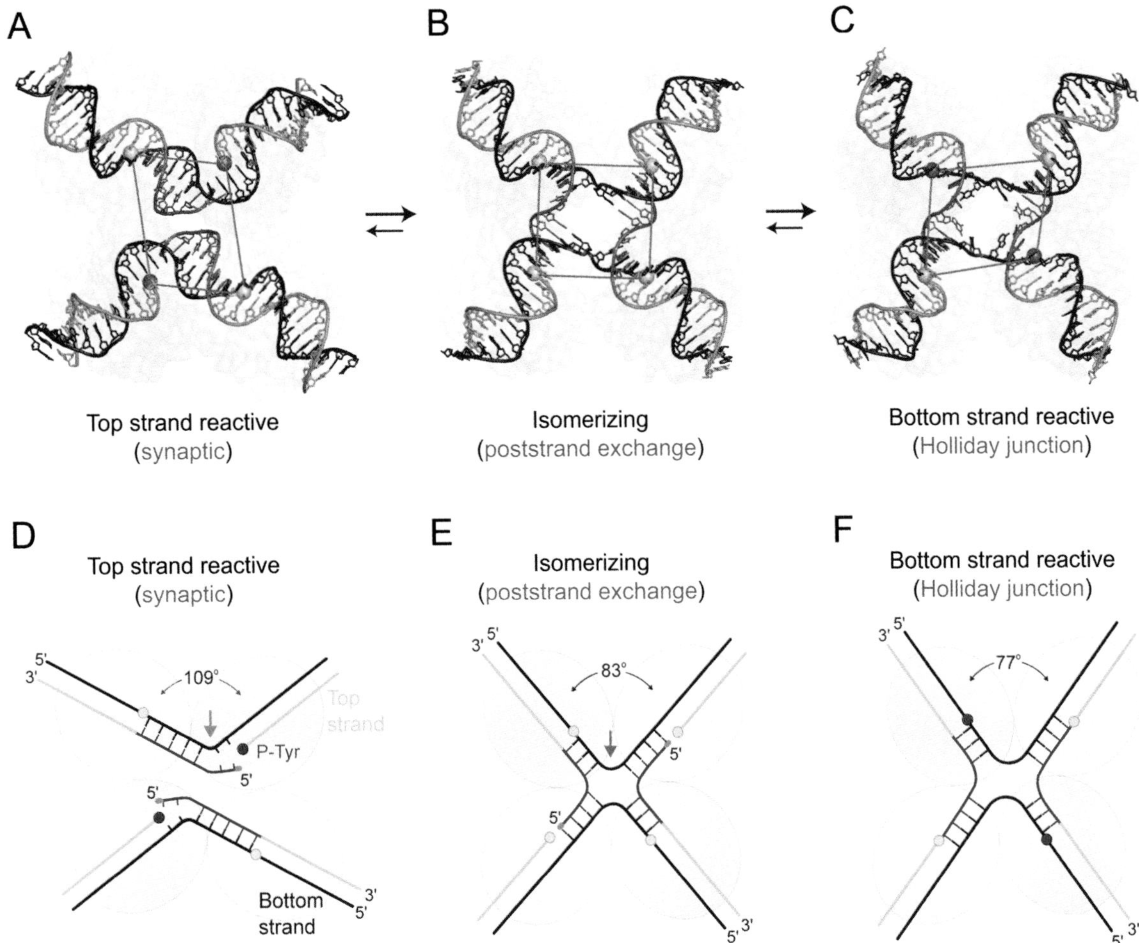

Figure 6 Three different conformations of λ Int tetramers representing distinct steps of the recombination reaction. The core DNAs within the λ-Int$^{(75\text{-}356)}$ synaptic complex (A, D), the λ-Int post- strand exchange complex (B, E), and the λ-Int Holliday junction complex (C, F) are shown along with schematic diagrams illustrating the interbranch angles and position of branch points. The pair of Int subunits in the active conformation (orange/red) is positioned closer to the center of each complex, whereas the inactive pair of subunits (gray) is further apart. Scissile phosphates (spheres) activated for cleavage are colored in red. Reprinted with permission from reference 44. doi:10.1128/microbiolspec.MDNA3-0051-2014.f6

nonequivalent interactions between the NTDs of neighboring Int protomers during HJ resolution (100).

One of the features of the tetrameric Int crystal structures, which was also inferred from solution studies of these small complexes (101), is a cyclically permuted topology, in which each NTD packs on top of the neighboring CB domain. It is now thought that this 2-fold symmetric NTD arrangement does not reflect that of a bona fide (integrative or excisive) recombinogenic complex, but rather is a consequence of the symmetric arm-type sites that are not connected by DNA and bending proteins to the core region, as discussed further below.

INTEGRATION HOST FACTOR

Integration host factor (IHF) was discovered in the very early studies of λ site-specific recombination by virtue of its role as a host-encoded protein that was essential both *in vitro* and *in vivo* for integrative and excisive recombination (102, 103). Its specific architectural role was demonstrated by the observation that IHF bending at the H′ site could stimulate Int binding and cleavage at the low- affinity C′ core site (104). As has not infrequently been the case, an *E. coli* protein discovered for its role in a phage life cycle, turned out to be an important player in the physiology of the cell. IHF is involved in regulation of gene transcription, especially

σ^{54} promoters (105), initiation of DNA replication (106), transposition (107), and phage packaging (108) (for reviews see references 109, 110, 111, 112, 113, 114, 115, 116).

To a large extent, the role and mechanisms of IHF in λ site-specific recombination, the crystal structure of IHF in complex with its DNA target, and the ways in which the physiology and biology of IHF in the host cell can impact recombination *in vivo*, have been reported and discussed prior to, and within, a previous review of λ site-specific recombination (27). More recent studies of IHF have centered on details of its interaction with DNA (117, 118, 119) and the mechanics and features of DNA bending by IHF (120, 121, 122, 123).

IHF is a hetero-dimeric protein consisting of two highly basic polypeptides, α and β, with molecular masses of 11,200 and 10,580 Da, respectively. These two subunits share approximately 30% homology to each other and also to the family of type II DNA-binding proteins that includes major histone-like proteins of *E. coli* such as HU.

The IHF structure, which is very similar to that of HU, is a compact, globular domain, consisting of symmetrically intertwined α and β subunits, from which two long β ribbon arms extend. The arms curl around the DNA and interact exclusively with the minor groove; most of the DNA bending (>160°) occurs at two large kinks, 9 bp apart (Fig. 7) (4). It has also been possible to construct a functional IHF in which the two chains have been fused into one (124).

The sequence preference displayed by IHF does not come from specific side chain contacts: it makes no contacts at all within the major groove and only a few hydrogen bonds to positions in the minor groove. Rather its specificity comes from "indirect readout," based on the sequence dependent structural parameters of its target DNA. Indeed, biochemical and structural studies of a relaxed-specificity mutant of IHF revealed how specificity is determined within the TTR portion of the consensus sequence (117). Within certain constraints, the structure of the DNA was driven by its own sequence and the protein side chains had readjusted to accommodate the different DNA structures.

Evidence that formation of a distinct DNA path was indeed the primary role of IHF came from "bend swap" experiments, where one or more IHF sites were replaced by unrelated DNA bending modules, either intrinsically bent DNA or different DNA bending proteins, such as HU (56, 125, 126). Although able to complement the lack of IHF-induced bending, none of the chimeric constructions performed as efficiently as the wild-type arrangement. The inability of the bend swap chimeras to achieve wild-type efficiency was evidently due to a requirement for considerable precision, as evidenced by the observation that an IHF bend wrongly positioned by just 1 bp, in a loop of constant length between C′ and P′1 of *att*L, could severely reduce excisive recombination efficiency (127).

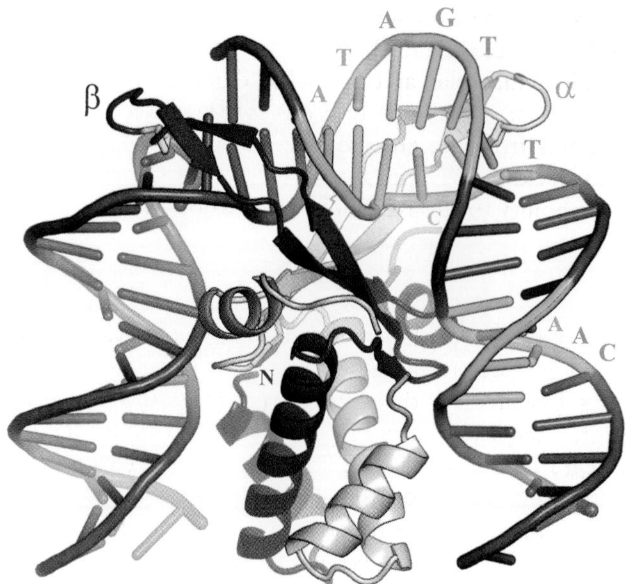

Figure 7 Complex of integration host factor with H′1N. The α and β subunits are shown in white and pink, respectively. The consensus sequence is highlighted in green and interacts mainly with the arm of α and the body of β. The yellow proline at the tip of each arm (P65 α/P64 β) is intercalated between bp 28 and 29 on the left side and 37 and 38 on the right. Reprinted with permission from reference 4.
doi:10.1128/microbiolspec.MDNA3-0051-2014.f7

It is attractive to propose that the evolution of Int's dependence on host-encoded accessory proteins derives, at least in part, from the benefits of linking the regulation and direction of recombination to the physiology of the host cell (see also the discussion of Fis protein below). In this regard, the changing levels of intracellular IHF are potentially interesting. The relative abundance of IHF increases approximately 5- to 7-fold over a 6 h span after entry into stationary phase (128, 129), and decreases when stationary-phase cells are diluted into fresh medium and cell mass begins to double (130). It is interesting that *in vitro* high IHF concentrations tend to inhibit the excisive reaction (131). The *in vivo* downshift in IHF concentration is probably not due to increased protein degradation, as IHF is not unstable in its dimeric form (132, 133), but instead appears to be a consequence of arrested transcription upon entry into exponential phase and increased transcription of the individual subunits upon entry into stationary phase (134). Additionally, there is evidence that IHF may play an essential role in survival from cell starvation: not only is IHF critical for induction of 14 proteins from the glucose starvation stimulon but mutants lacking IHF appear to be severely compromised in their ability to survive glucose starvation (135).

Xis

As noted above, the small phage-encoded Xis protein is the key determinant of directionality in the λ pathway. Essential for the excisive reaction and stimulating more than 10^6-fold *in vivo*, it is also inhibitory for the integrative reaction (136, 137). The NMR structure of $^{1\text{-}55}$Xis^{C28S} revealed an unusual "winged"-helix structure formed by two α-helices that are packed against two extended strands. While this structure itself did not afford critical insights into how Xis plays such a critical role in the λ pathway, it did herald the start of a steady progression towards this goal by the Johnson and Clubb laboratories (138).

A 1.7 Å resolution cocrystal structure of $^{1\text{-}55}$Xis^{C28S} complexed with a 15 bp DNA fragment containing its cognate X2 binding site comprised the second step of the Johnson/Clubb progression and provided a detailed view of the complex, which was largely in accord with their proposals based on the NMR structure of the free protein (Fig. 8A) (139). Although, the Xis-X2 complex is bent only modestly (approximately 25°), and hardly enough to account for the strikingly large curvature observed for a larger Xis-*att*R complex (93), it did suggest a molecular model for the Xis stimulation of Int binding to the adjacent P2 arm-type site (Fig. 8B).

A precursor to a larger and more informative cocrystal structure was the finding that, counter to the previous long-standing notion of two Xis binding sites (X1 and X2), a third Xis was bound at a site between them, called X1.5 (140, 141). The initial EMSA data were supplemented with protein–protein crosslinking experiments to further confirm the trimeric nature of the complex (140). More useful insights for understanding the role of Xis in directing recombination came from the 2.6 Å cocrystal structure of Xis bound to a larger DNA target comprising the entire 33 bp Xis binding region (Fig. 8C) (140). The three Xis proteins bind to this DNA in a head-to-tail orientation that generates a micronucleoprotein filament having approximately 72° of curvature and a slight positive writhe (Fig. 8D).

The differences in the specific interactions at X1 versus X2, combined with the observed nonspecific binding of Xis at the X1.5 site and the range of different interactions at ostensibly similar protein–protein interfaces, foreshadowed experiments showing that the flexible recognition surfaces of Xis result in a relatively promiscuous binder of DNA. The propensity for nonspecific DNA binding was further characterized in an Xis-DNA cocrystal with an 18 bp fragment of DNA (8). While this flexibility of DNA recognition is important for binding at the X1.5 site, where protein–protein interactions with the X1- and X2-bound Xis protomers provide additional stability, it also means that Xis is easily distracted from its *att*R target *in vivo*, where there is a huge excess of nonspecific DNA. Indeed, this latter point explains why excisive recombination *in vivo* is 50 to 200-fold lower in the absence of Fis (see also below) (8). Correspondingly, the Fis dependence of excisive recombination *in vitro* is only seen at limiting concentrations of Xis (142).

Fis

It is ironic, but understandable with hindsight, that although Fis protein was the first component of the λ recombination pathway to be crystallized (5, 6), it was the last component whose biological and molecular role was elucidated (8). Throughout this 16-year period (and beyond), the Johnson lab has played the leading role in studying the many roles and mechanisms of the Fis protein (112, 143, 144, 145). Fis was initially identified as a factor in promoting site-specific recombination by DNA invertases (146, 147) and was shortly thereafter shown to bind cooperatively with Xis at *att*R and to stimulate excisive recombination up to 20-fold when Xis is limiting (142). *In vivo*, the absence of Fis reduced *att*P formation from an induced lysogen by 100 to 1,000-fold (138, 148); it was also shown to be required along with Xis for binding to the *att*R region in the P22 challenge phage system (149).

Fis, like the other host-encoded accessory protein in the λ Int pathway, IHF, is a nucleoid associated protein of global, structural, and regulatory importance. Its role in determining overall chromosome structure is exerted by contributing to the looped-domain architecture of the nucleoid, and by influencing the regulation of genes encoding topoisomerases (150, 151, 152). Fis plays a role in the initiation of DNA replication, in several transposition reactions, and in the regulation of transcription at many different genes by several different mechanisms (for reviews, see references 112, 143, 151, and 153). The large number of critical sites of action for Fis becomes even more significant when considering how dramatically its intracellular levels vary as a function of cellular physiology.

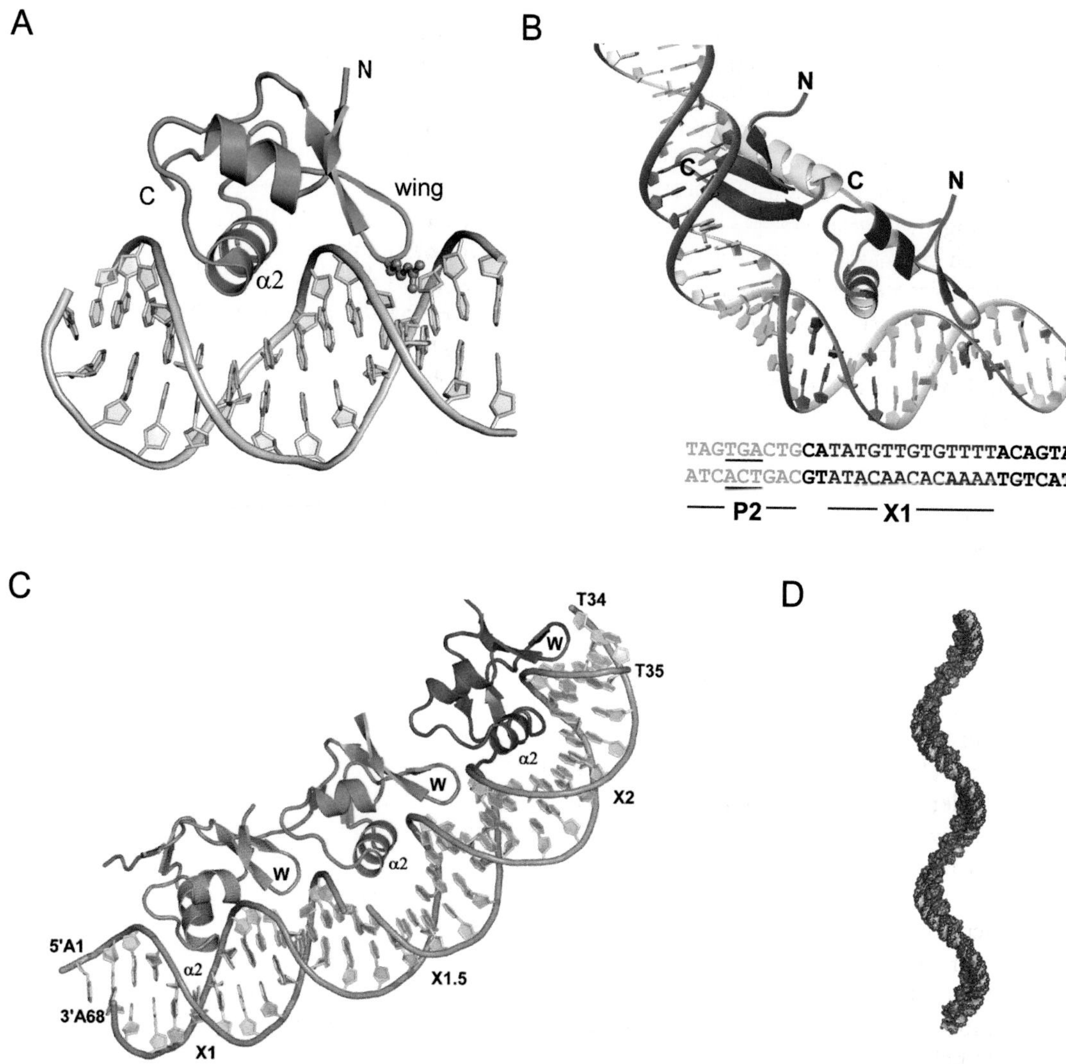

Figure 8 Complex of Xis with DNA. (A) The structure of $^{1\text{-}55}$Xis^{C28S} specifically bound to X2 DNA penetrates adjacent grooves of the duplex by fastening on the phosphodiester backbone. The major groove is filled primarily with helix α2 with the side chains of Glu19, Arg23, and Arg26 playing a major role in specific DNA recognition. The adjacent minor groove is contacted by the "wing" which does not contribute significantly to the specificity of complex formation but does contribute to binding affinity, although to a smaller extent than helix α2. The side-chain of Arg39 (brown) extends along the floor of the minor groove where it makes direct and water-mediated hydrogen bonds. (B) A model for the Int (NTD)-Xis-DNA ternary complex. The Int (NTD) is modeled to interact with the TGA trinucleotide (underlined) of the P2 site (blue) in the DNA major groove. Xis is modeled on the X1 site (magenta) in the same manner as observed in the complex with the X2 site. The C-terminal tail of Xis, which is disordered in solution (not shown), is located adjacent to the C-terminal helix of the NTD of Int to make a protein–protein interaction as shown by mutagenesis and NMR titration data (179). (C) X-ray crystal structure of Xis bound to the Xis binding region reveals the structural basis of cooperative binding. Xis monomers bound to the X1, X1.5, and X2 sites are colored dark salmon, green, and blue, respectively. (D) Structure-based model of an extended Xis-DNA filament. Units of the Xis-DNA$^{X1\text{-}X2}$ crystal structure were stacked end-to-end by superimposing site X1 over X1.5 to assemble a pseudocontinuous helix with a pitch of ~22 nm. Proteins are blue; DNA is orange. Reprinted with permission from reference 139 (A and B) and reference 140 (C and D).
doi:10.1128/microbiolspec.MDNA3-0051-2014.f8

Thompson *et al.* (142) showed that Fis levels drop dramatically when cells entered stationary phase and, more significantly, that occupation of the Fis binding site on *att* site DNA also drops in stationary phase cells. More detailed studies revealed that from these extremely low levels in stationary phase, Fis levels increase 500-fold during the initial lag phase when cells are diluted into fresh medium, and reach a peak of 50,000 to 100,000 copies per cell as the culture enters exponential phase. The control of Fis protein synthesis is at the level of mRNA where it is repressed by Fis protein (154, 155, 156) and stimulated by IHF (157). Transcription from the *fis* promoter, P*fis*, is critically influenced by DksA, a component of the transcription initiation complex that is also required for negative regulation of rRNA promoters (158, 159). DksA, which acts in part by reducing the half-life of (unstable) RNA polymerase-P*fis* promoter complexes, elevates the required concentration of the initiating NTP (CTP) and amplifies the inhibitory effect of ppGpp on P*fis* (154, 155, 158, 160). In so doing, it constrains *fis* expression primarily to early log phase at high growth rates, and it inhibits expression at low growth rates or following amino acid starvation (154, 155). However, as normal growth phase-dependent regulation of *fis* is observed in a Δ*relA* Δ*spoT* strain, other mechanisms can evidently compensate for the role of ppGpp in the pathway (154).

The crystal structures of Fis revealed a globular dimer composed of four tightly intertwined α-helices with two helix-turn-helix (HTH) motifs in each monomer (5, 6). One of the most striking features of the Fis structure was that the D helices, which were proposed to fit into adjacent major grooves of the DNA helix, are only 25 Å apart, approximately 10 Å shorter than the pitch of normal B DNA.

The long-standing hurdle to obtaining Fis-DNA cocrystals was the weak similarity among the many 15 bp sequences capable of forming stable complexes with Fis, thus, making it extremely difficult to derive an optimal consensus sequence for Fis binding. This obstacle was finally overcome by Stella *et al.* (7), who compiled the results of many analyses of Fis binding affinities into an informative hierarchy of DNA sequences, culminating in two 27 bp oligonucleotides whose Fis binding affinities were sufficiently optimized for crystallography (see Fig. 9A). Having established that compression of the central AT-rich minor groove is a critical feature of Fis binding, the authors went on to show that intrinsic DNA bends are unlikely to contribute significantly to Fis binding. Rather, they proposed that Fis initially searches for DNA with an intrinsically narrow minor groove, where AT composition, not sequence, is the critical determinant. Most recently, the Johnson lab has shown that the primary molecular determinant modulating minor groove widths is the 2-amino group on guanine (145).

While intrinsic DNA bends are not very important for targeting Fis binding, the bends induced by bound Fis are critical for its many functions, including DNA compaction, assembly of invertasomes, regulating transcription, and, most importantly for this article, directing the

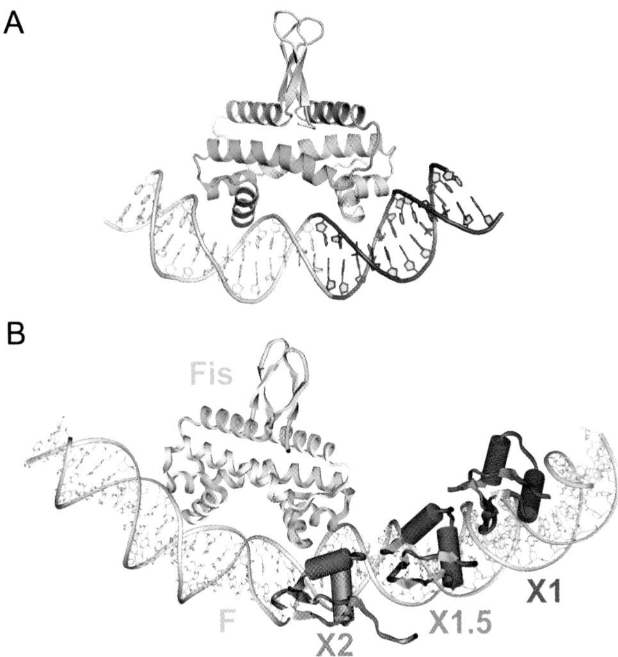

Figure 9 X-ray crystal structure of a Fis dimer complexed with DNA (A) and its relation to Xis binding (B). (A) The C-terminal helix representing the recognition helix of the HTH unit of each subunit is inserted into adjacent major grooves on the concave side of the 21 bp curved DNA. Only base contacts with a single residue, Arg85, are important for binding. The DNA undergoes substantial conformational adjustments, including adoption of ~65° overall curvature, to fit onto the Fis binding surface. The central 5 bp of the DNA interface are not contacted by Fis, but compression of the central minor groove to almost half the width of canonical DNA at the center enables the α-helices to insert into the adjacent major grooves, which do not show any appreciable change in width. (B) Model of the Fis-Xis cooperative complex. The X-ray crystal structure of three $^{\Delta 55}$Xis monomers bound to the X1 (magenta), X1.5 (blue), and X2 (gold) binding sites was superimposed onto the model of the Fis K36E X-ray structure docked to DNA representing the F site. Fis subunits are cyan and yellow. The DNA recognition helices of Xis bound at X2 and the proximal Fis subunit nearly form a continuous protein surface within the major groove. Reprinted with permission from reference 7 (A) and reference 8 (B).
doi:10.1128/microbiolspec.MDNA3-0051-2014.f9

curvature of the *att*R complex. The Johnson lab proposed that cooperative DNA binding between Fis and its partners, which bind immediately adjacent to Fis but generate only a small number of interfacial amino acid residues, is likely facilitated by mutually compatible changes in DNA shape.

Early experiments indicating that the F and X2 sites on *att*R overlap one another had been interpreted to suggest that both sites could not be bound simultaneously by their cognate proteins. However, subsequent experiments using quantitative gel shifts, stoichiometry determinations, nuclease footprinting and protein–protein crosslinking clearly established that Fis and Xis bind to the F and X2 sites simultaneously and cooperatively (8, 141). Most interestingly, Papagiannis *et al.* showed that Fis binds to the *att*R site *in vitro* with approximately 100-fold greater affinity than Xis alone, and *in vivo*, the rate of excision is reduced approximately 100-fold when Fis is absent (8). They proposed that *in vivo* Fis targets the otherwise peripatetic Xis to the X2 site, which then recruits Xis to X1.5 and X1. Based on their Xis-DNA microfilament cocrystal structures and their Fis crystal structures they built a model of the Fis-Xis complex on *att*R DNA, in which their observed protein induced DNA distortions are proposed to favor the cooperative binding of Fis and Xis (Fig. 9B).

PATTERNS OF λ NTD BINDING AND BRIDGING

Prior to considering the patterns of λ NTD binding and bridging in recombination reactions between canonical pairs of *att* sites, it should be noted the λ Int is also capable of efficiently carrying out an IHF-dependent recombination between two identical *att*L sites lacking a P′1 arm site (55). The existence of such a bidirectional pathway lacking the usual complement of components raises interesting questions about the kinds of recombinogenic complexes Int is capable of forming (161, 162, 163) and also underscores the caveat of off-pathway reactions.

The caveat of off-pathway reactions was echoed by a caveat about the artificially imposed symmetry of the NTD domains of the complexes used for X-ray crystallography of Int tetramers bound to Holliday junctions (discussed above). Earlier genetic and nuclease protection experiments had suggested that the patterns of NTD binding to arm-type sites were asymmetric (see Fig. 1) (reviewed in reference 27) and these results were subsequently reinforced by nuclease protection studies on Holliday junction intermediates (trapped with a hexapeptide inhibitor [51]) and biotin interference assays (BIA) (164). The latter, which probe the requirements for protein binding at a particular DNA locus by obstructing the major groove with a biotin bound to the C5 position of designated thymines, was particularly compelling because it monitored a complete integrative or excisive recombination reaction. From these experiments, it became clear that any attempt to understand the architecture and function of canonical recombinogenic complexes would require an analysis of the full ensemble of proteins and DNAs.

A requisite step in moving towards a panoptic investigation of the recombinogenic complexes was the deciphering of which “core-type” and “arm-type” binding sites are joined to one another by Int-mediated bridges. Towards this end, a disulfide trapping technology (165, 166) was used, in conjunction with trapped Holliday junction complexes (see Fig. 2), to introduce disulfide crosslinks at the protein-DNA interfaces between an Int NTD and its cognate arm-type site, and between an Int CTD and its cognate core-type site (36). Trapped nucleo-protein HJ complexes doubly crosslinked to Int were only observed with those *att* sites in which cystamine-labeled arm site and the cystamine-labeled core site are “bridged” by the same Int molecule.

From such analyses, it was concluded that the Int bridges between arm- and core-type sites in the integrative HJ recombination intermediate are: P′1–C′; P′2–C; P′3–B′; and P1–B. The Int bridges in the excisive HJ intermediate are: P′1–C′; P′2–B; and P2–B′. This leaves the C core site as the one that does not form an Int bridge with one of the three arm-type sites required for excisive recombination.

The Int bridges determined by site-directed crosslinking in HJ complexes were confirmed and complemented in full recombination reactions by a genetic approach using two chimeric recombinases. The first, called Crn1, consists of a Cre recombinase fused to the NTD of λ-integrase; it has all the properties of λ Int (described above) (89). This was complemented by construction of a second chimeric recombinase, Crn2, in which the NTD and CTD domains recognized different arm- and core-type DNA target sequences (36). A collection of hybrid *att* sites was constructed in which one of the bridged arm-core pairs (identified by the chemical crosslinking experiments) had the arm and core sequences recognized by Crn2, while the remaining arm-core bridges had the arm and core sequences recognized by Crn1. Using these substrates, it was shown that Crn1 could not carry out recombination unless Crn2 was also present (and vice versa). The results of the chimeric recombination reactions confirmed, and also provided information complementary to, the results from chemical crosslinking (as discussed below).

These results argue strongly against models in which regulated directionality of λ Int recombination depends upon some degree of Int bridge remodeling during the course of the reaction. Furthermore, the monogamous relationship of each arm-core bridged pair throughout the course of the recombination reaction makes it possible to extrapolate from the patterns observed in the HJ recombination intermediate to those predicted for the presynaptic recombination partners and the post HJ recombination products. Inspection of Fig. 10 reveals that for excisive recombination, the presynaptic partners have only intramolecular bridges, suggesting that Int bridging is not a driving force in synapsis of *att*L and *att*R. This is also likely to be the case for integrative recombination, even though the capture of a naked *att*B by a fully assembled (supercoiled) *att*P complex requires two intermolecular bridges (91). It was postulated that the reason *att*B cannot bind Int protomers unless they are part of a higher-order complex stems from the need to overcome the NTD inhibition of CTD function, described above (49), and not from any driving force by intermolecular bridges (see also discussion below).

ARCHITECTURES OF RECOMBINOGENIC COMPLEXES

In an attempt to derive architectural models for the HJ recombination intermediates, the Int bridging results were augmented with in-gel fluorescence resonance energy transfer (FRET) experiments (101, 167, 168, 169) to determine the apparent distances between selected positions within the excisive recombination HJ intermediate (37).

Using the Int bridging data, the apparent FRET distances for the HJ recombination complex, and the 3D structures for all of the protein components in their DNA-bound forms (4, 7, 8, 44, 45, 139, 140), it was possible to computationally build a model for the architecture of the λ excision complex (Fig. 11A, B, C). Insights gained from the excisive complex along with the integrative Int bridging data and 3D structures were used to generate a corresponding model of the integrative complex (Fig. 11D, E, F). Considered individually and together, the two architectures afford a number of interesting insights, as discussed below and in the figure legends (37).

In the excision complex, the P′ and B arms form a left-handed crossing, while the overall path of *att*R DNA indicates a left-handed, nucleosome-like, wrapping by

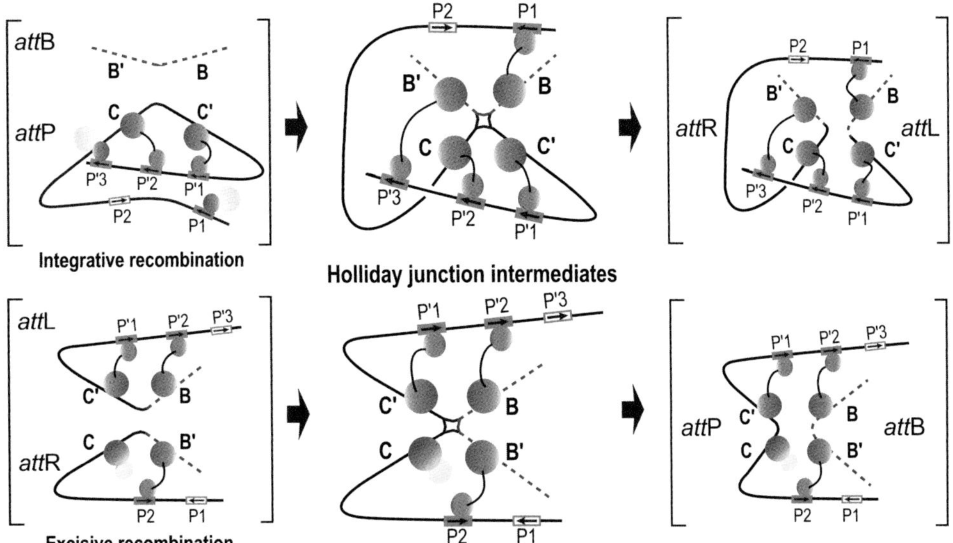

Figure 10 Schematic summary of the Int bridges in integrative and excisive recombination. The middle panel diagrams the Int bridges of the Holliday junction (HJ) recombination intermediates determined by Tong *et al.* (36). In the integrative complex, all four core sites and four of the five arm sites enjoy an Int bridge while the excisive complex engages three of the four core sites and three of the five arm sites. The flanking panels (brackets) depict extrapolations from the HJ complexes to the respective *att* site recombination partners (substrates) and recombinants (products) based on the deduction that Int bridges are not broken and reformed during recombination. doi:10.1128/microbiolspec.MDNA3-0051-2014.f10

IHF, Fis, Xis, and Int. The model thus predicts a negative DNA crossing node in *attL* and left-handed solenoidal wrapping in *attR*, both of which are consistent with negative supercoiling in the normal substrates. In the integrative complex the asymmetric mode of binding to C′, C, and B′ by the P′ arm requires considerable flexibility in the linker segments between the CTDs and NTDs, and this model explicitly predicts the formation of a negative DNA-crossing node in the recombination complex, where the P-arm crosses over P′. The model also features an unusual and flexible P-arm tether that positions Int-B for *att*B binding.

ASYMMETRY AND FLEXIBILITY

The architectures proposed for the recombinogenic complexes differ in several ways from the crystal structures of the HJ-bound Int tetramers bound to arm site DNA duplexes (44). While this would not necessarily have been predicted, it is not surprising, as the crystal structures did not include accessory DNA bending proteins or their cognate DNA sites, which join the core- and arm-type sites. An additional compromise required to form crystals was the substitution of a pair of P′1–P′2-containing oligonucleotides for the canonical asymmetric arrangement of arm-type binding sites. Indeed, subsequent experiments involving biotin-interference mapping of complete recombination reactions (described above) are more consistent with the asymmetric architectures than the symmetric arrangement in the smaller complexes designed for crystallization (164). In contrast to the symmetric and tightly packed NTD organization observed in crystal structures, the models for the architectures of the complexes feature highly asymmetric arrangements of the NTDs. In the former, the domains are swapped, with the NTD of one Int subunit located above the CB domain of an adjacent Int. The latter is incompatible with domain-swapped NTDs and implies considerable flexibility in the CB-NTD linkers.

TOPOLOGY

Excisive recombination between directly repeated *att*L and *att*R sites results in a large fraction of free circles when supercoiling levels are low, similar to that observed for the Cre and Flp recombinases (170). Integrative recombination between directly repeated *att*P and *att*B sites results in catenated circles for supercoiled substrates, implying that the recombination process itself imposes a strand crossing (170, 171, 172). Seah *et al.* (37) argue that the proposed architectures are consistent with these results and explain many of the other topological findings of Crisona *et al.* (170). Additionally, the tightly wrapped nature of the integrative complex model and the inclusion of a negative DNA crossing node are consistent with, and may explain, in part, the requirement for negative supercoiling for efficient integration (91).

CAPTURING THE HOST *att*B SITE

From the time Richet and Nash (91) first showed that *att*B comes naked to a recombination with its fully decorated *att*P partner there has been considerable speculation about the details of this synaptic event. Because of the pseudodyad symmetry of the core-type sites the openings of the bound integrase C-clamps must face in opposite directions (45). While this is not a problem for the monovalent family members it implies that for the fully assembled *att*P complex one of the Int subunits (the one destined to bind the B core site of *att*B) must have the flexibility to wrap around the host chromosome from the opposite face.

Indeed, the architecture proposed for the integrative complex does contain an inherently flexible P-arm that tethers the Int-B subunit and allows for the dynamic binding required to engage the bacterial chromosome and ultimately lock onto the *att*B sequence. The model is also consistent with, and explains, a difference between the two kinds of Int bridging experiments reported by Tong *et al.* (36). Whereas chemical crosslinking of the P1–B Int bridge was the most robust of all the Int bridges, in the genetic analyses, the P1–B Int bridge was the weakest, precisely the difference expected for a flexible arm.

ARCHITECTURAL BASIS FOR DIRECTIONALITY

The source of the strong bias towards the top strands being exchanged first in formation of the HJ (33, 35) is evident from the models in Fig. 11 and Fig. 12. During excisive recombination, both *att*L and *att*R are bent at their core sites in order to promote the bridging interactions that form between core and arm binding sites. The core site bend directions that lead to stable complexes are coupled to IHF-induced bends and commit both *att*L and *att*R to top strand cleavage upon synapsis of the sites. Similarly, only one bend direction of the *att*P core site will lead to stable bridging interactions between C/C′ core sites and P′1/P′2 arm sites. This direction commits *att*P to top strand cleavage in the synaptic complex with *att*B. Thus, the order of strand

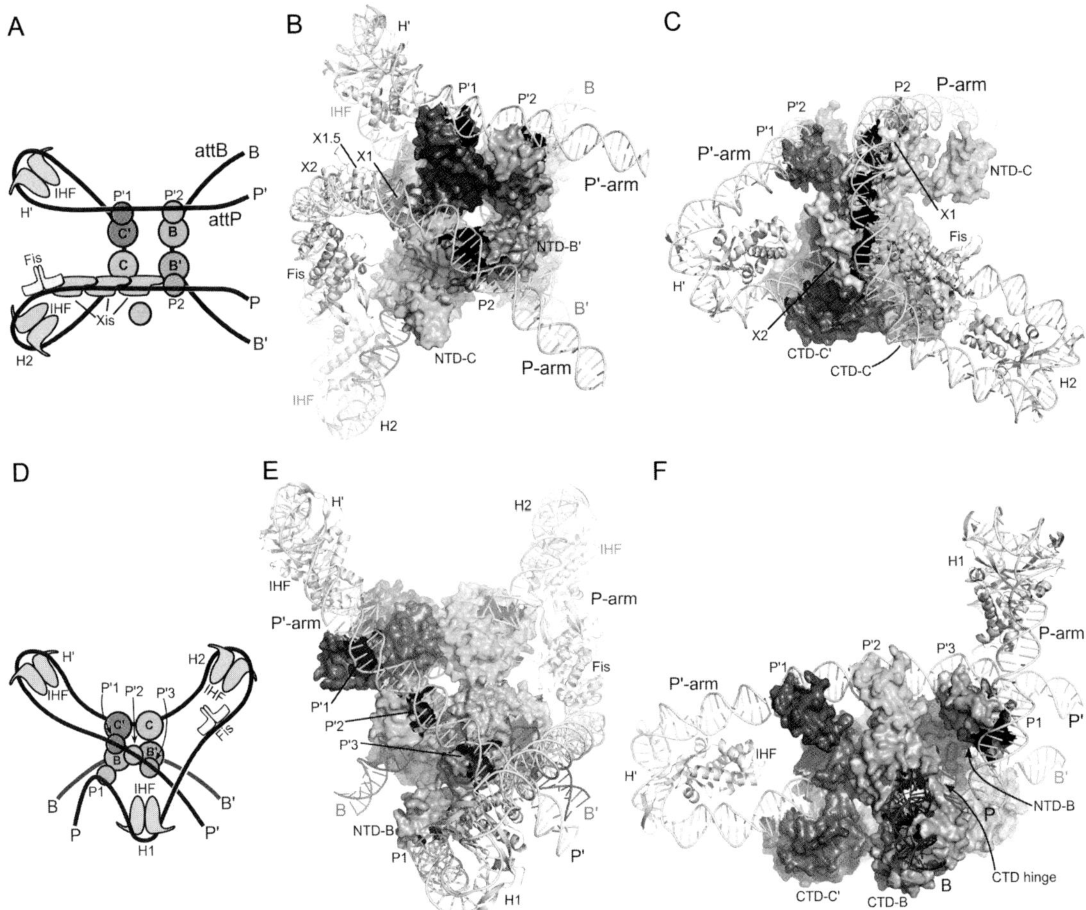

Figure 11 Models of the λ excisive and integrative recombination complexes. (A) Schematic representation of the excisive complex architecture. The excision reaction product resulting from Holliday junction (HJ) resolution is shown. Int subunits (blue, green, magenta, brown) are represented by a small circle (NTD) and a large circle (CTD). Integration host factor (IHF) heterodimers (gray) are shown bound to the H′ and H2 sites. Fis dimer (pink) and Xis (tan) subunits are indicated. (B) Model of the excisive complex in the same "top view" orientation as the schematic drawing in panel A. The NTD of the Int subunit bound at the C core site (NTD-C) is shown separated from the rest of the complex to improve clarity of the P-arm trajectory. (C) Side view of the excisive complex, highlighting the trajectory of the P-arm. IHF bending of the P′ arm at H′ directs the DNA over the CTD domains of the Int tetramer, facilitating engagement of the P′1 and P′2 arm sites by the Int subunits bound at the C′ and B core sites, respectively. In the P-arm of *att*R the phasing of the IHF-induced bend at H2 is different from that at H′; at H2, the P-arm is directed along the plane of the catalytic domain tetramer. An A-tract sequence that is stabilized by Fis binding (7, 8) directs the P-arm upwards, towards the Int CB domains. The cooperative Xis filament (8, 140) then redirects the P-arm across the top of the Int CTD domains, where the P2 site is bound by the Int subunit bound at the B′ core site. The Xis subunit bound at X1 resides close to the position where the NTD of the Int subunit bound at the C core site (Int-C) would be expected. The NTD of Int-C was not docked in a specific location of the excisive complex model, but it seems plausible, even attractive, that this domain could bind nonspecifically to the P-arm near the X1 site, perhaps interacting with Xis. (D) Schematic of the integrative complex architecture. The arm-type binding sites engaged by the four Int subunits are indicated. (E) Model of the integrative complex in the same "top view" as illustrated in panel B. In this orientation, the P-arm rises towards the viewer, crosses over the P′ arm, and is directed back towards the Int tetramer by the IHF bend at the H1 site. (F) Side view of the integrative model, looking approximately down the B core site. The NTD of the Int subunit bound at the B core site (NTD-B) is shown bound at the P1 site, on the flexible P-arm. The CB and catalytic domains of the Int subunit bound at the B site can be seen wrapped around the opposing face of *att*B, with the interdomain hinge indicated. The CTD-NTD linkers were not modeled and are not shown. IHF bending at H′ directs the P′ arm over the CTD domains of the Int tetramer, but in this case the P′1, P′2, and P′3 binding sites are engaged by the Int subunits bound to C′, C, and B′, respectively. As Xis is not present in the integrative complex, the P-arm is directed upwards, parallel to the Int tetramer, and as Fis stimulation of integration has been reported (180, 181), it was included in the model. IHF bound to the H1 site redirects the P-arm back towards the Int tetramer, crossing over the P′ arm in the process. The P1 arm-type site is thereby brought to a position where it can bind the NTD of the Int subunit poised for capture of the B core half-site (Int-B). Reprinted with permission from reference 37. doi:10.1128/microbiolspec.MDNA3-0051-2014.f11

exchange in both pathways is determined prior to synapsis by formation of specific *att*L, *att*R, and *att*P complexes (34).

The architecture of the excisive complex provides a bird's eye view of how Xis mediates its critical role as the regulator of directionality (137, 173) (Fig. 13A). In the absence of Xis, the P-arm would not be directed across the top of the Int CTDs to make the required P2-B′ bridge and the P-arm would not be properly positioned to stabilize a functional *att*R. An additional critical role for Xis is to promote the cooperative binding of the Int NTD at P2 (66, 130, 174).

The architecture of the excisive complex also explains the long-standing question of why the excision reaction does not run efficiently in reverse once *att*B is released (Fig. 13B). After dissociation of attB, the *att*P complex is expected to be less stable because it now only contains a single intramolecular bridge (P′1–C′). Furthermore, this complex has the potential to rearrange, such that the *att*P core bends in the opposite direction and facilitates the formation two intramolecular bridges (P′1–C′ and P′2–C). While this complex resembles a portion of the *att*P substrate complex, it is prevented from proceeding to a competent complex by

A

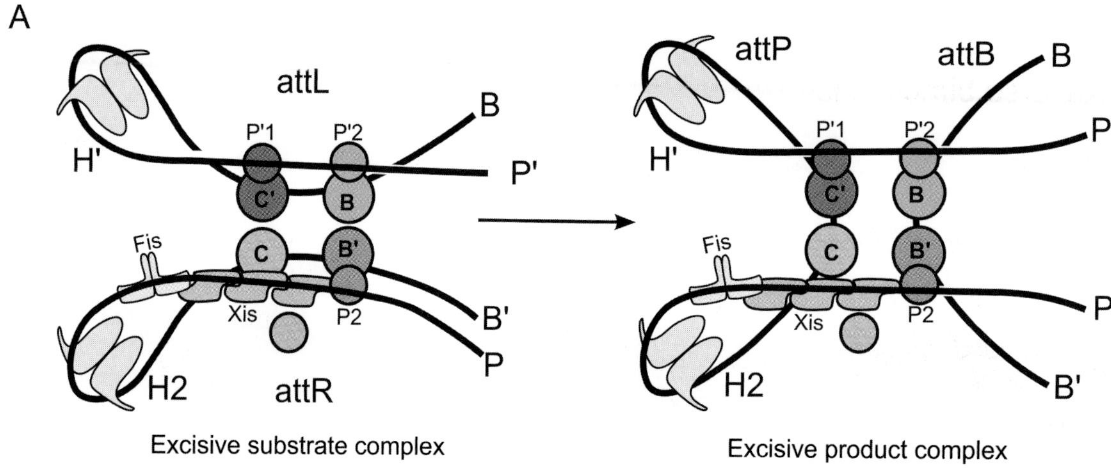

B

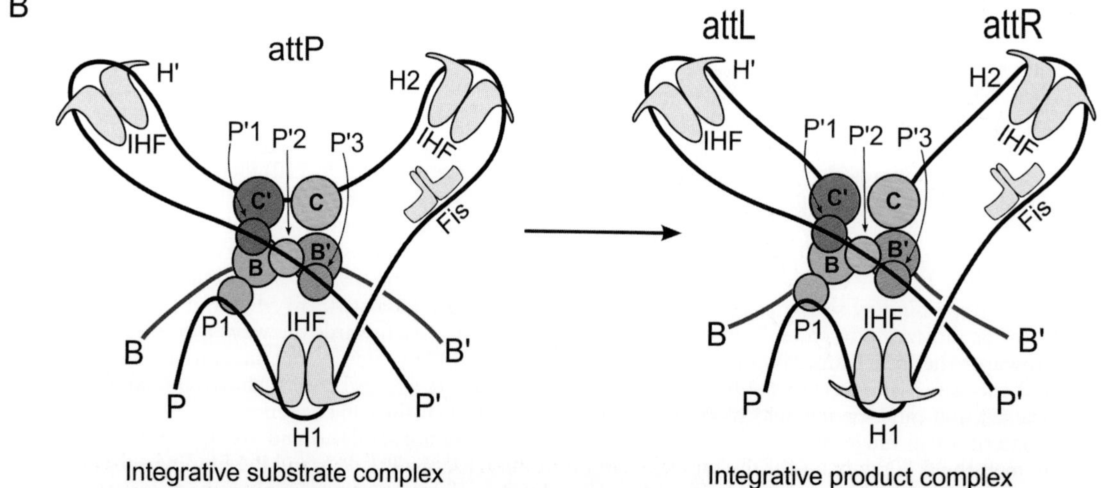

Figure 12 Schematic representation of the excisive and integrative reactions, based on the structural models shown in Fig. 13. Coloring of the protein subunits matches that shown in Fig. 11. Reprinted with permission from reference 37.
doi:10.1128/microbiolspec.MDNA3-0051-2014.f12

A. Excision requires Xis

attL
H'
P'1
P'2
C'
B
B
P'
+
P
attR
P2
(weak binding)
Fis
C
B'
(very weak binding)
H2
B'
→ N.R.

Without Xis to direct P-arm to core sites, attR cannot properly assemble

B. Reverse bimolecular excision reaction does not occur when Xis present

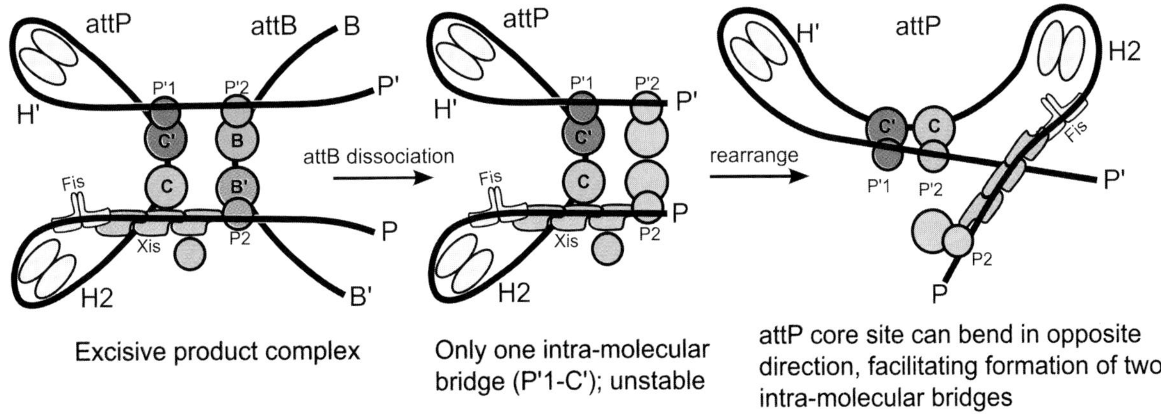

C. Normal integration pathway inhibited by Xis

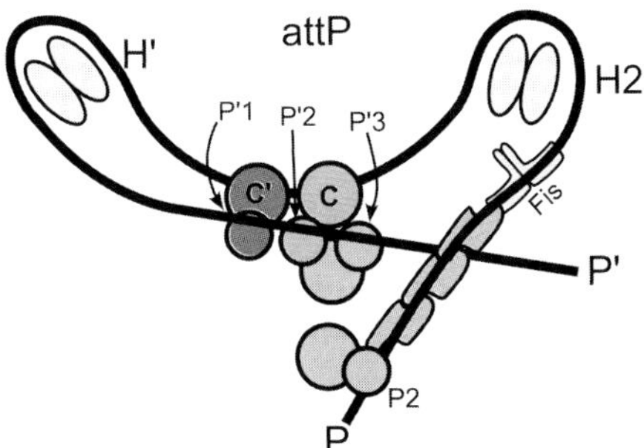

Xis competes for H1 site (see legend);
Int subunit at P1 not positioned by H1 bend

Figure 13 The basis for directionality in λ recombination. (A) An explanation for why Xis is required for excision. (B) Explanation for why the excision pathway is not run efficiently in reverse to perform integration. (C) Explanation for why Xis inhibits the normal integration reaction. The Xis, P2, and H1 sites cannot be occupied simultaneously (22). Schematics follow the same coloring scheme used in Fig. 11 and Fig. 12. Int subunits not bound to a core site are colored gray. doi:10.1128/microbiolspec.MDNA3-0051-2014.f13

the presence of Xis, which prevents the binding of IHF at H1 and Int at P1 (131), leaving the P-arm improperly positioned for the synaptic *att*P complex (cf. Fig. 12B). This same Xis-induced mispositioning of the P-arm is also responsible for the Xis inhibition of integrative recombination (Fig. 13C) (136). The architectures of the two complexes thus explain how the dramatically different roles of the P-arm in integrative versus excisive recombination provides the basis for an effective directionality switch that depends upon the presence or absence of Xis.

SUMMARY AND FUTURE DIRECTIONS

In the 12 years since publication of the last review that focused solely on the λ site-specific recombination pathway (in *Mobile DNA II* [27, 175]), the most dramatic advances in our understanding of the reaction have been in the area of X-ray crystallography. Protein-DNA structures have been determined for the CTD of Int (45), full Int in a tetrameric complex with HJ and arm site DNAs (44), Xis complexed with its DNA target (139), and Fis complexed with its DNA target (7). Including the earlier IHF-DNA structure (4), these results comprise a complete portfolio of all of the DNA-protein interfaces driving the Int pathway. Building on this foundation of structures, it has been possible to derive models for the assembly of components that determine the regulatory apparatus in the P-arm (8, 140), and for the overall architectures that define excisive and integrative recombinogenic complexes (36, 37). The most fundamental additional mechanistic insights derived from the application of hexapeptide inhibitors (53) and single molecule kinetics (34).

On the list of experiments needed to fill out and/or sharpen our understanding of the λ Int reactions, the answers to several structural questions are of high priority. Some of these are probably close at hand, such as structures for the interfaces between Xis and Fis on one side and Xis and the Int NTD on the other side. A more problematic, but very important structural question, concerns the conformations of the linker regions between the Int NTDs and CTDs within a functional recombinogenic complex. Architectures, or structures and stabilities, of the substrate and product complexes are important but will only be useful if accompanied by convincing demonstration they comprise part of the canonical pathway. If it would be too greedy to hope for X-ray crystal structures of the recombinogenic complexes, perhaps high-resolution cryo-electron microscopy might fill the bill. Additional single molecule analyses of the reactions, of HJ formation and resolution, and formation and stability of the products, should provide important mechanistic insights. Finally, without attempting to frame the specifics, it seems reasonable to predict the design of new applications of the λ Int pathway to study questions, and/or solve problems, unimagined at this time. In this vein one might ponder the question raised by J.K. King (176): "Is it possible or not, that in the random process of life, human creativity results in biological behavior related to Lambda genetics?"

Acknowledgments. *This chapter is dedicated to the memory of Howard Nash and Bob Weisberg, who laid the foundations for studying the biochemistry and genetics, respectively, of the Int pathway. I am grateful to Greg Van Duyne, past and present members of my lab, and especially Reid Johnson, for many helpful discussions. I thank Reid Johnson, Tapan Biswas, Hideki Aihara, and Phoebe Rice for helpful comments and for making available figures depicting their work, Joan Boyles for administrative assistance, and especially Nicole Seah for help in preparing the manuscript. The work from my laboratory was supported by NIH grants GM062723 and GM033928.*

Citation. Landy A. 2014. The λ integrase site-specific recombination pathway. Microbiol Spectrum 3(2):MDNA3-0051-2014.

References

1. **Campbell AM.** 1962. Episomes, p 101–145. *In* Caspari EW, Thoday JM (ed), *Advances in Genetics*. Academic Press, New York, NY.
2. **Nash HA.** 1974. Purification of bactriophage λ Int protein. *Nature* **247:**543–545.
3. **Nash HA, Robertson CA.** 1981. Purification and properties of the Escherichia coli protein factor required for λ integrative recombination. *J Biol Chem* **256:**9246–9253.
4. **Rice PA, Yang SW, Mizuuchi K, Nash HA.** 1996. Crystal structure of an IHF-DNA complex: a protein-induced DNA U-turn. *Cell* **87:**1295–1306.
5. **Kostrewa D, Granzin J, Koch C, Choe HW, Labahn J, Kahmann R, Saenger W.** 1991. Three-dimensional structure of the E. coli DNA-binding protein FIS. *Nature* **349:**178–180.
6. **Yuan HS, Finkel SE, Feng JA, Kaczor-Grzeskowiak M, Johnson RC, Dickerson RE.** 1991. The molecular structure of wild-type and a mutant Fis protein: Relationship between mutational changes and recombinational enhancer function or DNA binding. *Proc Natl Acad Sci USA* **88:**9558–9562.
7. **Stella S, Cascio D, Johnson RC.** 2010. The shape of the DNA minor groove directs binding by the DNA-bending protein Fis. *Genes Dev* **24:**814–826.
8. **Papagiannis CV, Sam MD, Abbani MA, Yoo D, Cascio D, Clubb RT, Johnson RC.** 2007. Fis Targets Assembly of the Xis Nucleoprotein Filament to Promote Excisive Recombination by Phage Lambda. *J Mol Biol* **367:**328–343.

9. **Rudy C, Taylor KL, Hinerfeld D, Scott JR, Churchward G.** 1997. Excision of a conjugative transposon *in vitro* by the Int and Xis proteins of Tn 916. *Nucleic Acids Res* **25:**4061–4066.

10. **Hickman AB, Waninger S, Scocca JJ, Dyda F.** 1997. Molecular organization in site-specific recombination: The catalytic domain of bacteriophage HP1 integrase at 2.7 Å resolution. *Cell* **89:**227–237.

11. **Lewis JA, Hatfull GF.** 2003. Control of directionality in L5 integrase-mediated site-specific recombination. *J Mol Biol* **326:**805–821.

12. **Garcia-Russell N, Orchard SS, Segall AM.** 2007. Probing nucleoid structure in bacteria using phage lambda integrase-mediated chromosome rearrangements, p 209–226. *In* Kelly TH, Stanley RM (ed), *Methods Enzymol*, **vol 421**. Academic Press, New York, NY.

13. **Van Houdt R, Leplae R, Lima-Mendez G, Mergeay M, Toussiant A.** 2012. Towards a more accurate annotation of tyrosine-based site-specific recombinases in bacterial genomes. *Mobile DNA* **3:**6.

14. **Warren D, Laxmikanthan G, Landy A.** 2013. Integrase family of site-specific recombinases. *Brenner's Encyclopedia of Genetics* **4:**100–105.

15. **Van Duyne GD.** 2008. Site-specific recombinases, p 303–332. *In Protein-Nucleic Acid Interactions: Structural Biology*. The Royal Society of Chemistry, Cambridge.

16. **Rajeev L, Malanowska K, Gardner JF.** 2009. Challenging a paradigm: the role of DNA homology in tyrosine recombinase reactions. *Microbiol Mol Biol Rev* **73:** 300–309.

17. **Segall AM, Craig NL.** 2006. New wrinkles and folds in site-specific recombination. *Mol Cell* **19:**433–435.

18. **Grindley ND, Whiteson KL, Rice PA.** 2006. Mechanisms of site-specific recombination. *Annu Rev Biochem* **75:**567–605.

19. **Rice PA.** 2005. Resolving integral questions in site-specific recombination. *Nat Struct Mol Biol* **12:** 641–643.

20. **Lee L, Sadowski PD.** 2005. Strand selection by the tyrosine recombinases. *Prog Nucl Acid Res Mol Biol* **80:** 1–42.

21. **Williams KP.** 2002. Integration sites for genetic elements in prokaryotic tRNA and tmRNA genes: sublocation preference of integrase subfamilies. *Nucleic Acids Res* **30:**866–875.

22. **Van Duyne GD.** 2002. A structural view of tyrosine recombinase site-specific recombination, p 93–117. *In* Craig NL, Craigie R, Gellert M, Lambowitz AM (ed), *Mobile DNA II*. ASM Press, Washington DC.

23. **She Q, Brügger K, Chen L.** 2002. Archaeal integrative genetic elements and their impact on genome evolution. *Res Microbiol* **153:**325–332.

24. **Dutton G.** 2002. Site-specific recombinases. *The Scientist* **16:**29–31.

25. **Lewis JA, Hatfull GF.** 2001. Control of directionality in integrase-mediated recombination: examination of recombination directionality factors (RDFs) including Xis and Cox proteins. *Nucleic Acids Res* **29:**2205–2216.

26. **Kobryn K, Chaconas G.** 2001. The circle is broken: telomere resolution in linear replicons. *Curr Opin Microbiol* **4:**558–564.

27. **Azaro MA, Landy A.** 2002. λ Int and the λ Int family, p 118–148. *In* Craig NL, Craigie R, Gellert M, Lambowitz AM (ed), *Mobile DNA II*. ASM Press, Washington DC.

28. **Rutkai E, Gyorgy A, Dorgai L, Weisberg RA.** 2006. Role of secondary attachment sites in changing the specificity of site-specific recombination. *J Bact* **188:** 3409–3411.

29. **Landy A.** 1989. Dynamic, structural and regulatory aspects of lambda site-specific recombination. *Annu Rev Biochem* **58:**913–949.

30. **Reese TA, Wakeman BS, Choi HS, Hufford MM, Huang SC, Zhang X, Buck MD, Jezewski A, Kambal A, Liu CY, Goel G, Murray PJ, Xavier RJ, Kaplan MH, Renne R, Speck SH, Artyomov MN, Pearce EJ, Virgin HW.** 2014. Helminth infection reactivates latent γ-herpesvirus via cytokine competition at a viral promoter. *Science* **345:**573–577.

31. **Hsu PL, Landy A.** 1984. Resolution of synthetic *att*-site Holliday structures by the integrase protein of bacteriophage λ. *Nature* **311:**721–726.

32. **Kitts PA, Nash HA.** 1988. An intermediate in the phage λ site-specific recombination reaction is revealed by phosphorothioate substitution in DNA. *Nucleic Acids Res* **16:**6839–6856.

33. **Kitts PA, Nash HA.** 1988. Bacteriophage λ site-specific recombination proceeds with a defined order of strand-exchanges. *J Mol Biol* **204:**95–107.

34. **Mumm JP, Landy A, Gelles J.** 2006. Viewing single lambda site-specific recombination events from start to finish. *EMBO J* **25:**4586–4595.

35. **Nunes-Düby S, Matsumoto L, Landy A.** 1987. Site-specific recombination intermediates trapped with suicide substrates. *Cell* **50:**779–788.

36. **Tong W, Warren D, Seah N, Laxmikanthan G, Van Duyne GD, Landy A.** 2014. Mapping the Int bridges in the nucleoprotein Holliday junction intermediates of viral integrative and excisive recombination. *Proc Natl Acad Sci USA* **111:**12366–12371.

37. **Seah N, Tong W, Warren D, Van Duyne GD, Landy A.** 2014. Nucleoprotein architectures regulating the directionality of viral integration and excision. *Proc Natl Acad Sci USA* **111:**12372–12377.

38. **Matovina M, Seah N, Hamilton T, Warren D, Landy A.** 2010. Stoichiometric incorporation of base substitutions at specific sites in supercoiled DNA and supercoiled recombination intermediates. *Nucleic Acids Res* **38:**e175.

39. **Sun KQ.** 1990. *A study of DNA-DNA interactions during bacteriophage lambda integrative recombination. Ph.D. Doctoral Thesis in Biochemistry*. University of Illinois at Urbana-Champaign, Champaign, IL.

40. **Conway AB, Chen Y, Rice PA.** 2003. Structural plasticity of the Flp-Holliday junction complex. *J Mol Biol* **326:**425–434.

41. **Chen Y, Narendra U, Iype LE, Cox MM, Rice PA.** 2000. Crystal structure of a Flp recombinase-Holliday

junction complex: assembly of an active oligomer by helix swapping. *Mol Cell* **6:**885–897.

42. **Gopaul DN, Guo F, Van Duyne GD.** 1998. Structure of the Holliday junction intermediate in Cre- loxP site-specific recombination. *EMBO J* **17:**4175–4187.
43. **Guo F, Gopaul DN, Van Duyne GD.** 1997. Structure of Cre recombinase complexed with DNA in a site-specific recombination synapse. *Nature* **389:**40–46.
44. **Biswas T, Aihara H, Radman-Livaja M, Filman D, Landy A, Ellenberger T.** 2005. A structural basis for allosteric control of DNA recombination by λ integrase. *Nature* **435:**1059–1066.
45. **Aihara H, Kwon HJ, Nunes-Düby SE, Landy A, Ellenberger T.** 2003. A conformational switch controls the DNA cleavage activity of Lambda integrase. *Mol Cell* **12:**187–198.
46. **Tirumalai RS, Pargellis CA, Landy A.** 1996. Identification and characterization of the NEM-sensitive site in lambda integrase. *J Biol Chem* **271:**29599–29604.
47. **Pargellis CA, Nunes-Düby SE, Moitoso de Vargas L, Landy A.** 1988. Suicide recombination substrates yield covalent λ integrase- DNA complexes and lead to identification of the active site tyrosine. *J Biol Chem* **263:** 7678–7685.
48. **Tirumalai RS, Healey E, Landy A.** 1997. The catalytic domain of λ site-specific recombinase. *Proc Natl Acad Sci USA* **94:**6104–6109.
49. **Sarkar D, Radman-Livaja M, Landy A.** 2001. The small DNA binding domain of λ Int is a context-sensitive modulator of recombinase functions. *EMBO J* **20:** 1203–1212.
50. **Lee SY, Aihara H, Ellenberger T, Landy A.** 2004. Two structural features of integrase that are critical for DNA cleavage by multimers but not by monomers. *Proc Natl Acad Sci USA* **101:**2770–2775.
51. **Cassell GD, Segall AM.** 2003. Mechanism of inhibition of site-specific recombination by the Holliday junction-trapping peptide WKHYNY: Insights into phage λ integrase-mediated strand exchnage. *J Mol Biol* **327:** 413–429.
52. **Cassell G, Klemm M, Pinilla C, Segall A.** 2000. Dissection of bacteriophage λ site-specific recombination using synthetic peptide combinatorial libraries. *J Mol Biol* **299:**1193–1202.
53. **Boldt JL, Pinilla C, Segall AM.** 2004. Reversible inhibitors of λ Int-mediated recombination efficiently trap Holliday junction intermediates and form the basis of a novel assay for junction resolution. *J Biol Chem* **279:** 3472–3483.
54. **Ghosh K, Lau CK, Guo F, Segall AM, Van Duyne GD.** 2005. Peptide Trapping of the Holliday Junction Intermediate in Cre-loxP Site-specific Recombination. *J Biol Chem* **280:**8290–8299.
55. **Segall AM, Nash HA.** 1996. Architectural flexibility in lambda site-specific recombination: three alternate conformations channel the att L site into three distinct pathways. *Genes Cells* **1:**453–463.
56. **Segall AM, Goodman SD, Nash HA.** 1994. Architectural elements in nucleoprotein complexes:interchangeability of specific and non-specific DNA binding proteins. *EMBO J* **13:**4536–4548.
57. **Boldt JL, Kepple KV, Cassell GD, Segall AM.** 2006. Spermidine biases the resolution of Holliday junctions by phage λ integrase. *Nucl Acids Res* **35:**716–727.
58. **Rajeev L, Segall A, Gardner J.** 2007. The bacteroides NBU1 integrase performs a homology-independent strand exchange to form a Holliday junction intermediate. *J Biol Chem* **282:**31228–31237.
59. **Klemm M, Cheng C, Cassell G, Shuman S, Segall AM.** 2000. Peptide inhibitors of DNA cleavage by tyrosine recombinases and topoisomerases. *J Mol Biol* **299:** 1203–1216.
60. **Gunderson CW, Boldt JL, Authement RN, Segall AM.** 2009. Peptide wrwycr Inhibits the Excision of Several Prophages and Traps Holliday Junctions inside Bacteria. *J Bacteriol* **191:**2169–2176.
61. **Ranjit DK, Rideout MC, Nefzi A, Ostresh JM, Pinilla C, Segall AM.** 2010. Small molecule functional analogs of peptides that inhibit λ site-specific recombination and bind Holliday junctions. *Bioorg Med Chem Lett* **20:** 4531–4534.
62. **Bankhead TM, Etzel BJ, Wolven F, Bordenave S, Boldt JL, Larsen TA, Segall AM.** 2003. Mutations at residues 282, 286, and 293 of phage λ integrase exert pathway-specific effects on synapsis and catalysis in recombination. *J Bacteriol* **185:**2653–2666.
63. **Chen Y, Rice PA.** 2003. New insight into site-specific recombination from Flp recombinase-DNA structures. *Annu Rev Biophys Biomol Struct* **32:**135–139.
64. **Tal A, Arbel-Goren R, Costantino N, Court DL, Stavans J.** 2014. Location of the unique integration site on an Escherichia coli chromosome by bacteriophage lambda DNA *in vivo*. *Proc Natl Acad Sci USA* **111:** 7308–7312.
65. **Kwon HJ, Tirumalai RS, Landy A, Ellenberger T.** 1997. Flexibility in DNA recombination: structure of the λ integrase catalytic core. *Science* **276:**126–131.
66. **Sarkar D, Azaro MA, Aihara H, Papagiannis C, Tirumalai RS, Nunes-Düby SE, Johnson RC, Ellenberger T, Landy A.** 2002. Differential affinity and cooperativity functions of the amino-terminal 70 residues of λ integrase. *J Mol Biol* **324:**775–789.
67. **Azaro MA, Landy A.** 1997. The isomeric preference of Holliday junctions influences resolution bias by λ integrase. *EMBO J* **16:**3744–3755.
68. **Kamadurai HB, Jain R, Foster MP.** 2008. Crystallization and structure determination of the core-binding domain of bacteriophage lambda integrase. *Acta Crystallogr Sect F Struct Biol Cryst Commun* **64:** 470–473.
69. **Kovach MJ, Tirumalai RS, Landy A.** 2002. Site-specific photo-crosslinking of lambda Int. *J Biol Chem* **277:** 14530–14538.
70. **Tirumalai RS, Kwon H, Cardente E, Ellenberger T, Landy A.** 1998. The recognition of core-type DNA sites by λ Integrase. *J Mol Biol* **279:**513–527.
71. **Ross W, Landy A.** 1983. Patterns of λ Int recognition in the regions of strand exchange. *Cell* **33:**261–272.

72. **Cheng Q, Swalla BM, Beck M, Alcaraz R Jr, Gumport RI, Gardner JF.** 2000. Specificity determinants for bacteriophage Hong Kong 022 integrase: analysis of mutants with relaxed core-binding specificities. *Mol Microbiol* **36:**424–436.

73. **Dorgai L, Yagil E, Weisberg RA.** 1995. Identifying determinants of recombination specificity: Construction and characterization of mutant bacteriophage integrases. *J Mol Biol* **252:**178–188.

74. **Yagil E, Dorgai L, Weisberg R.** 1995. Identifying determinants of recombination specificity: Construction and characterization of chimeric bacteriophage integrases. *J Mol Biol* **252:**163–177.

75. **Patsey RL, Bruist MF.** 1995. Characterization of the interaction between the lambda intasome and att B. *J Mol Biol* **252:**47–58.

76. **Lange-Gustafson BJ, Nash HA.** 1984. Purification and properties of Int-H, a variant protein involved in site-specific recombination of bacteriophage λ. *J Biol Chem* **259:**12724–12732.

77. **Han YW, Gumport RI, Gardner JF.** 1994. Mapping the functional domains of bacteriophage lambda integrase protein. *J Mol Biol* **235:**908–925.

78. **Tekle M, Warren DJ, Biswas T, Ellenberger T, Landy A, Nunes-Düby SE.** 2002. Attenuating functions of the C-terminus of λ Integrase. *J Mol Biol* **324:**649–665.

79. **Kazmierczak RA, Swalla BM, Burgin AB, Gumport RI, Gardner JF.** 2002. Regulation of site-specific recombination by the carboxyl terminus of λ integrase. *Nucleic Acids Res* **30:**5193–5204.

80. **Hazelbaker D, Radman-Livaja M, Landy A.** 2005. Receipt of the C-terminal tail from a neighboring Int protomer allosterically stimulates Holliday junction resolution. *J Mol Biol* **351:**948–955.

81. **Rowley PA, Kachroo AH, Ma C-H, Maciaszek AD, Guga P, Jayaram M.** 2010. Electrostatic suppression allows tyrosine site-specific recombination in the absence of a conserved catalytic arginine. *J Biol Chem* **285:**22976–22985.

82. **Grainge I, Jayaram M.** 1999. The integrase family of recombinases: organization and function of the active site. *Mol Microbiol* **33:**449–456.

83. **Krogh BO, Shuman S.** 2000. Catalytic mechanism of DNA topoisomerase IB. *Mol Cell* **5:**1035–1041.

84. **Krogh BO, Shuman S.** 2002. Proton relay mechanism of general acid catalysis by DNA topoisomerase IB. *J Biol Chem* **277:**5711–5714.

85. **Yakovleva L, Lai J, Kool ET, Shuman S.** 2006. Nonpolar nucleobase analogs illuminate requirements for site-specific DNA cleavage by vaccinia topoisomerase. *J Biol Chem* **281:**35914–35921.

86. **Van Duyne GD.** 2001. A structural view of Cre-*lox*P site-specific recombination. *Annu Rev Biophys Biomol Struct* **30:**87–104.

87. **Chen Y, Rice PA.** 2003. The role of the conserved Trp 330 in Flp-mediated recombination. Functional and structural analysis. *J Biol Chem* **278:**24800–24807.

88. **Whiteson KL, Rice PA.** 2008. Binding and catalytic contributions to site recognition by flp recombinase. *J Biol Chem* **283:**11414–11423.

89. **Warren D, Laxmikanthan G, Landy A.** 2008. A chimeric Cre recombinase with regulated directionality. *Proc Natl Acad Sci USA* **47:**18278–18283.

90. **Hartung M, Kisters-Woike B.** 1998. Cre mutants with altered DNA binding properties. *J Biol Chem* **273:**22884–22891.

91. **Richet E, Abcarian P, Nash HA.** 1988. Synapsis of attachment sites during lambda integrative recombination involves capture of a naked DNA by a protein-DNA complex. *Cell* **52:**9–17.

92. **Wojciak JM, Sarkar D, Landy A, Clubb RT.** 2002. Arm-site binding by the lambda integrase protein: solution structure and functional characterization of its amino-terminal domain. *Proc Natl Acad Sci USA* **99:**3434–3439.

93. **Thompson JF, Landy A.** 1988. Empirical estimation of protein-induced DNA bending angles: Applications to λ site-specific recombination complexes. *Nucleic Acids Res* **16:**9687–9705.

94. **Enquist LW, Kikuchi A, Weisberg RA.** 1979. The role of λ integrase in integration and excision. *Cold Spring Harb Symp Quant Biol* **43:**1115–1120.

95. **Fadeev EA, Sam MD, Clubb RT.** 2009. NMR structure of the amino-terminal domain of the lambda integrase protein in complex with DNA: immobilization of a flexible tail facilitates beta-sheet recognition of the major groove. *J Mol Biol* **388:**682–690.

96. **Wojciak JM, Connolly KM, Clubb RT.** 1999. NMR structure of the Tn 916 integrase-DNA complex. *Nat Struct Biol* **6:**366–373.

97. **Allen MD, Yamasaki K, Ohme-Takagi M, Tateno M, Suzuki M.** 1998. A novel mode of DNA recognition by a beta-sheet revealed by the solution structure of the GCC-box binding domain in complex with DNA. *EMBO J* **17:**5484–5496.

98. **Subramanya HS, Arciszewska LK, Baker RA, Bird LE, Sherratt DJ, Wigley DB.** 1997. Crystal structure of the site-specific recombinase, XerD. *EMBO J* **16:**5178–5187.

99. **Nunes-Düby SE, Yu D, Landy A.** 1997. Sensing homology at the strand swapping step in λ excisive recombination. *J Mol Biol* **272:**493–508.

100. **Lee SY, Radman-Livaja M, Warren D, Aihara H, Ellenberger T, Landy A.** 2005. Non-equivalent interactions between amino-terminal domains of neighboring λ integrase protomers direct Holliday junction resolution. *J Mol Biol* **345:**475–485.

101. **Radman-Livaja M, Biswas T, Mierke D, Landy A.** 2005. Architecture of recombination intermediates visualized by In-gel FRET of λ integrase-Holliday junction-arm-DNA complexes. *Proc Natl Acad Sci USA* **102:**3913–3920.

102. **Miller HI, Kikuchi A, Nash HA, Weisberg RA, Friedman DI.** 1979. Site-specific recombination of bacteriophage λ : the role of host gene products. *Cold Spring Harb Symp Quant Biol* **43:**1121–1126.

103. **Miller HI, Kirk M, Echols H.** 1981. SOS induction and autoregulation of the him A gene for site-specific recombination in Escherichia coli. *Proc Natl Acad Sci USA* **78:**6754–6758.

104. **Moitoso de Vargas L, Kim S, Landy A.** 1989. DNA looping generated by the DNA-bending protein IHF and the two domains of lambda integrase. *Science* **244:** 1457–1461.

105. **Sze CC, Laurie AD, Shingler V.** 2001. *In vivo* and *in vitro* effects of integration host factor at the DmpR-regulated σ 54 -dependent Po promoter. *J Bacteriol* **183:** 2842–2851.

106. **Ryan VT, Grimwade JE, Camara JE, Crooke E, Leonard AC.** 2004. Escherichia coli prereplication complex assembly is regulated by dynamic interplay among Fis, IHF and DnaA. *Mol Microbiol* **51:**1347–1359.

107. **Crellin P, Sewitz S, Chalmers R.** 2004. DNA looping and catalysis: The IHF-folded Arm of Tn 10 promotes conformational changes and hairpin resolution. *Mol Cell* **13:**537–547.

108. **Morse BK, Michalczyk R, Kosturko LD.** 1994. Multiple molecules of integration host factor (IHF) at a single DNA binding site, the bacteriophage lambda cos 11 site. *Biochimie* **76:**1005–1017.

109. **Friedman DI.** 1988. Integration host factor: a protein for all reasons. *Cell* **55:**545–554.

110. **Goosen N, van de Putte P.** 1995. The regulation of transcription initiation by integration host factor. *Mol Microbiol* **16:**1–7.

111. **Nash HA, Lin ECC, Lynch AS.** 1996. The HU and IHF proteins: accessory factors for complex protein-DNA assemblies. *In* Lin ECC, Lynch AS (ed), *Regulation of Gene Expression in Escherichia coli*, p. 150–179. R.G. Landes Company, Austin, TX.

112. **Johnson RC, Johnson LM, Schmidt JW, Gardner JF, Higgins NP.** 2005. Major nucleoid proteins in the structure and function of the Escherichia coli chromosome, p 65–132. *In The Bacterial Chromosome*. ASM Press, Washington DC.

113. **Dame RT.** 2005. The role of nucleoid-associated proteins in the organization and compaction of bacterial chromatin. *Mol Microbiol* **56:**858–870.

114. **Swinger KK, Rice PA.** 2004. IHF and HU: flexible architects of bent DNA. *Curr Opin Struct Biol* **14:** 28–35.

115. **Travers A.** 1997. DNA-protein interactions: IHF-the master blender. *Curr Biol* **7:**R252–R254.

116. **Rice PA.** 1997. Making DNA do a U-turn: IHF and related proteins. *Curr Opin Struct Biol* **7:**86–93.

117. **Lynch TW, Read EK, Mattis AN, Gardner JF, Rice PA.** 2003. Integration host factor: putting a twist on protein-DNA recognition. *J Mol Biol* **330:**493–502.

118. **Swinger KK, Lemberg KM, Zhang Y, Rice PA.** 2003. Flexible DNA bending in HU-DNA cocrystal structures. *EMBO J* **22:**3749–3760.

119. **Vander Meulen KA, Saecker RM, Record MT Jr.** 2008. Formation of a wrapped DNA-protein interface: experimental characterization and analysis of the large contributions of ions and water to the thermodynamics of binding IHF to H′ DNA. *J Mol Biol* **377:**9–27.

120. **Dixit S, Singh-Zocchi M, Hanne J, Zocchi G.** 2005. Mechanics of binding of a single integration-host-factor protein to DNA. *Phy Rev Ltrs* **94:**118101–118104.

121. **Sugimura S, Crothers DM.** 2006. Stepwise binding and bending of DNA by *Escherichia coli* integration host factor. *Proc Natl Acad Sci USA* **103:**18510–18514.

122. **Kuznetsov SV, Sugimura S, Vivas P, Crothers DM, Ansarai A.** 2006. Direct observation of DNA bending/unbending kinetics in complex with DNA-bending protein IHF. *Proc Natl Acad Sci USA* **103:**18515–18520.

123. **Sugimura S.** 2005. *Kinetic and steady-state studies of binding and bending of Lambda phage DNA by integration host factor. Ph.D.* Yale University, New Haven, CT.

124. **Corona T, Bao Q, Christ N, Schwartz T, Li J, Droge P.** 2003. Activation of site-specific DNA integration in human cells by a single chain integration host factor. *Nucleic Acids Res* **31:**5140–5148.

125. **Goodman SD, Nicholson SC, Nash HA.** 1992. Deformation of DNA during site-specific recombination of bacteriophage λ: replacement of IHF protein by HU protein or sequence-directed bends. *Proc Natl Acad Sci USA* **89:**11910–11914.

126. **Goodman SD, Nash HA.** 1989. Functional replacement of a protein-induced bend in a DNA recombination site. *Nature* **341:**251–254.

127. **Nunes-Düby SE, Smith-Mungo LI, Landy A.** 1995. Single base-pair precision and structural rigidity in a small IHF-induced DNA loop. *J Mol Biol* **253:**228–242.

128. **Ditto MD, Roberts D, Weisberg RA.** 1994. Growth phase variation of integration host factor level in Escherichia coli. *J Bacteriol* **176:**3738–3748.

129. **Murtin C, Engelhorn M, Geiselmann J, Boccard F.** 1998. A quantitative UV laser footprinting analysis of the interaction of IHF with specific binding sites: Re-evaluation of the effective concentration of IHF in the Cell. *J Mol Biol* **284:**949–961.

130. **Bushman W, Thompson JF, Vargas L, Landy A.** 1985. Control of directionality in lambda site-specific recombination. *Science* **230:**906–911.

131. **Thompson JF, Moitoso de Vargas L, Skinner SE, Landy A.** 1987. Protein–protein interactions in a higher-order structure direct lambda site-specific recombination. *J Mol Biol* **195:**481–493.

132. **Granston AE, Nash HA.** 1993. Characterization of a set of integration host factor mutants deficient for DNA binding. *J Mol Biol* **234:**45–59.

133. **Nash HA, Robertson CA, Flamm E, Weisberg RA, Miller HI.** 1987. Overproduction of Escherichia coli integration host factor, a protein with nonidentical subunits. *J Bacteriol* **169:**4124–4127.

134. **Aviv M, Giladi H, Schreiber G, Oppenheim AB, Glaser G.** 1994. Expression of the genes coding for the Escherichia coli integration host factor are controlled by growth phase, rpoS, ppGpp and by autoregulation. *Mol Microbiol* **14:**1021–1031.

135. **Nystrom T.** 1995. Glucose starvation stimulon of Escherichia coli: Role of integration host factor in starvation survival and growth phase-dependent protein synthesis. *J Bacteriol* **177:**5707–5710.

136. **Abremski K, Gottesman S.** 1982. Purification of the bacteriophage λ xis gene product required for λ excisive recombination. *J Biol Chem* **257:**9658–9662.

137. **Nash HA.** 1975. Integrative recombination of bacteriophage lambda DNA *in vitro*. *Proc Natl Acad Sci USA* **72:**1072–1076.

138. **Sam M, Papagiannis C, Connolly KM, Corselli L, Iwahara J, Lee J, Phillips M, Wojciak JM, Johnson RC, Clubb RT.** 2002. Regulation of directionality in bacteriophage lambda site-specific recombination: structure of the Xis protein. *J Mol Biol* **324:**791–805.

139. **Sam MD, Cascio D, Johnson RC, Clubb RT.** 2004. Crystal structure of the excisionase-DNA complex from bacteriophage lambda. *J Mol Biol* **338:**229–240.

140. **Abbani MA, Papagiannis CV, Sam MD, Cascio D, Johnson RC, Clubb RT.** 2007. Structure of the cooperative excisionase (Xis)-DNA complex reveals a micronucleoprotein filament that regulates phage lambda intasome assembly. *Proc Natl Acad Sci USA* **104:**2109–2114.

141. **Sun X, Mierke DF, Biswas T, Lee SY, Landy A, Radman-Livaja M.** 2006. Architecture of the 99 bp DNA-Six-Protein regulatory complex of the λ *att* Site. *Molecular Cell* **24:**569–580.

142. **Thompson JF, Moitoso de Vargas L, Koch C, Kahmann R, Landy A.** 1987. Cellular factors couple recombination with growth phase: characterization of a new component in the λ site-specific recombination pathway. *Cell* **50:**901–908.

143. **Finkel SE, Johnson RC.** 1992. The FIS protein: it's not just for DNA inversion anymore. *Mol Microbiol* **6:**3257–3265.

144. **Aiyar SE, McLeod SM, Ross W, Hirvonen CA, Thomas MS, Johnson RC, Gourse RL.** 2002. Architecture of Fis-activated transcription complexes at the Escherichia coli rrnB P1 and rrnE P1 promoters. *J Mol Biol* **316:**501–516.

145. **Hancock SP, Ghane T, Cascio D, Rohs R, Di Felice R, Johnson RC.** 2013. Control of DNA minor groove width and Fis protein binding by the purine 2-amino group. *Nucleic Acids Res* **41:**6750–6760.

146. **Johnson RC, Bruist MF, Simon MI.** 1986. Host protein requirements for *in vitro* site-specific DNA inversion. *Cell* **46:**531–539.

147. **Koch C, Kahmann R.** 1986. Purification and properties of the Escherichia coli host factor required for inversion of the G segment in bacteriophage Mu. *J Biol Chem* **261:**15673–15678.

148. **Ball CA, Johnson RC.** 1991. Efficient excision of phage λ from the *Escherichia coli* chromosome requires the Fis protein. *J Bacteriol* **173:**4027–4031.

149. **Numrych TE, Gumport RI, Gardner JF.** 1991. A genetic analysis of Xis and FIS interactions with their binding sites in bacteriophage lambda. *J Bacteriol* **173:**5954–5963.

150. **Skoko D, Yoo D, Bai H, Schnurr B, Yan J, McLeod SM, Marko JF, Johnson RC.** 2006. Mechanism of chromosome compaction and looping by the *Escherichia coli* nucleoid protein Fis. *J Mol Biol* **364:**777–798.

151. **Schneider R, Travers A, Kutateladze T, Muskhelishvili G.** 1999. A DNA architectural protein couples cellular physiology and DNA topology in *Escherichia coli*. *Mol Microbiol* **34:**953–964.

152. **Dorman CJ.** 2009. Nucleoid-associated proteins and bacterial physiology, p 47–64. *In* Laskin A, Sariaslani S, Geoffrey G (ed), *Advances in Applied Microbiology*, Vol 67. Academic Press, New York, NY.

153. **Browning DF, Grainger DC, Busby SJW.** 2010. Effects of nucleoid-associated proteins on bacterial chromosome structure and gene expression. *Curr Opin Microbiol* **13:**773–780.

154. **Ball CA, Osuna R, Ferguson KC, Johnson RC.** 1992. Dramatic changes in Fis levels upon nutrient upshift in *Escherichia coli*. *J Bacteriol* **174:**8043–8056.

155. **Ninnemann O, Koch C, Kahmann R.** 1992. The E. coli fis promoter is subject to stringent control and autoregulation. *EMBO J* **11:**1075–1083.

156. **Nilsson L, Verbeek H, Vijgenboom E, van Drunen C, Vanet A, Bosch L.** 1992. FIS-dependent trans activation of stable RNA operons of Escherichia coli under various growth conditions. *J Bacteriol* **174:**921–929.

157. **Pratt TS, Steiner T, Feldman LS, Walker KA, Osuna R.** 1997. Deletion analysis of the fis promoter region in Escherichia coli: Antagonistic effects of integration host factor and Fis. *J Bacteriol* **179:**6367–6377.

158. **Mallik P, Paul BJ, Rutherford ST, Gourse RL, Osuna R.** 2006. DksA is required for growth phase-dependent regulation, growth rate-dependent control, and stringent control of fis expression in *Escherichia coli*. *J Bacteriol* **188:**5775–5782.

159. **Paul BJ, Ross W, Gaal T, Gourse RL.** 2004. rRNA transcription in *Escherichia coli*. *Annu Rev Genet* **38:**749–770.

160. **Walker KA, Mallik P, Pratt TS, Osuna R.** 2004. The *Escherichia coli* fis promoter is regulated by changes in the levels of its transcription initiation nucleotide CTP. *J Biol Chem* **279:**50818–50828.

161. **Cassell G, Moision R, Rabani E, Segall A.** 1999. The geometry of a synaptic intermediate in a pathway of bacteriophage λ site-specific recombination. *Nucleic Acids Res* **27:**1145–1151.

162. **Segall AM.** 1998. Analysis of higher order intermediates and synapsis in the bent-L pathway of bacteriophage λ site-specific recombination. *J Biol Chem* **273:**24258–24265.

163. **Segall AM, Nash HA.** 1993. Synaptic intermediates in bacteriophage lambda site-specific recombination: Integrase can align pairs of attachment sites. *EMBO J* **12:**4567–4576.

164. **Hazelbaker D, Azaro MA, Landy A.** 2008. A biotin interference assay highlights two different asymmetric interaction profiles for lambda integrase arm-type binding sites in integrative versus excisive recombination. *J Biol Chem* **283:**12402–12414.

165. **Huang H, Harrison SC, Verdine GL.** 2000. Trapping of a catalytic HIV reverse transcriptase template:primer complex through a disulfide bond. *Chem Biol* **7:**355–364.

166. **Verdine GL, Norman DP.** 2003. Covalent trapping of protein-DNA complexes. *Annu Rev Biochem* **72:**337–366.

167. **Lorenz M, Diekmann S.** 2001. Quantitative distance information on protein-DNA complexes determined in polyacrylamide gels by fluorescence resonance energy transfer. *Electrophoresis* **22:**990–998.

168. **Ramirez-Carrozzi VR, Kerppola TK.** 2001. Long-range electrostatic interactions influence the orientation of Fos-Jun Binding at AP-1 sites. *J Mol Biol* **305:**411–427.

169. **Lilley DMJ, Wilson TJ.** 2000. Fluorescence resonance energy transfer as a structural tool for nucleic acids. *Curr Opin Chem Biol* **4:**507–517.

170. **Crisona NJ, Weinberg RL, Peter BJ, Sumners DW, Cozzarelli NR.** 1999. The topological mechanism of phage λ integrase. *J Mol Biol* **289:**747–775.

171. **Pollock TJ, Nash HA.** 1983. Knotting of DNA caused by a genetic rearrangement: evidence for a nucleosome-like structure in site-specific recombination for bacteriophage lambda. *J Mol Biol* **170:**1–18.

172. **Nash HA, Pollock TJ.** 1983. Site-specific recombination of bacteriophage lambda: The change in topological linking number associated with exchange of DNA strands. *J Mol Biol* **170:**19–38.

173. **Gottesman S, Gottesman M.** 1975. Excision of prophage λ in a cell-free system. *Proc Natl Acad Sci USA* **72:**2188–2192.

174. **Numrych TE, Gumport RI, Gardner JF.** 1992. Characterization of the bacteriophage lambda excisionase (Xis) protein: the C-terminus is required for Xis-integrase cooperativity but not for DNA binding. *EMBO J* **11:**3797–3806.

175. **Craig NL, Craigie R, Gellert M, Lambowitz AM.** 2002. *Mobile DNA II*. ASM Press, Washington DC.

176. **King JK.** 2007. Man the misinterpretant: will he discover the universal secret of sexuality encoded within him? *Int J Humanit* **4:**1–19.

177. **Craig NL, Nash HA.** 1984. *E. coli* integration host factor binds to specific sites in DNA. *Cell* **39:**707–716.

178. **Nunes-Düby S, Azaro M, Landy A.** 1995. Swapping DNA strands and sensing homology without branch migration in λ site-specific recombination. *Curr Biol* **5:**139–148.

179. **Warren D, Sam M, Manley K, Sarkar D, Lee SY, Abbani M, Clubb RT, Landy A.** 2003. Identification of the λ integrase surface that interacts with the Xis accessory protein reveals a residue that is also critical for homomeric dimer formation. *Proc Natl Acad Sci USA* **100:**8176–8181.

180. **Esposito D, Gerard GF.** 2003. The *Escherichia coli* Fis protein stimulates bacteriophage λ integrative recombination *in vitro*. *J. Bacteriol* **185:**3076–3080.

181. **Ball CA, Johnson RC.** 1991. Multiple effects of Fis on integration and the control of lysogeny in phage λ. *J Bacteriol* **173:**4032–4038.

Mobile DNA, 3rd Edition
Nancy L. Craig, Michael Chandler, Martin Gellert, Alan M. Lambowitz, Phoebe A. Rice and Suzanne Sandmeyer

doi:10.1128/microbiolspec.MDNA3-0014-2014

Gregory D. van Duyne[1]

5 Cre Recombinase

INTRODUCTION

The use of Cre recombinase to carry out conditional mutagenesis of transgenes and insert DNA cassettes into eukaryotic chromosomes is widespread (1–9). Indeed, a PubMed search for "cre recombinase" in early 2014 returned over 4000 articles. In addition to the numerous *in vivo* and *in vitro* applications that have been reported since Cre was first shown to function in yeast and mammalian cells nearly 30 years ago (10, 11), the Cre–*loxP* system has also played an important role in understanding the mechanism of recombination by the tyrosine recombinase family of site-specific recombinases (12–14). The simplicity of this system, requiring only a single recombinase enzyme and short recombination sequences for robust activity in a variety of contexts (15), has been an important factor in both cases. Cre has also been used in experiments designed to understand the functions of other recombination systems (16–18).

In its physiological role for bacteriophage P1, Cre functions as an intramolecular resolvase. The lysogenic state of P1 normally does not involve integration into the host chromosome (19). Instead, the phage exists as a unit copy episome where Cre ensures faithful segregation by converting P1 chromosome dimers to monomers before cell division (20, 21). The reaction is carried out between 34-bp recombination sites called *loxP*, which are in a directly repeated orientation in the physiological substrates. Hence, Cre functions as a simple version of the Xer system, which performs a similar role for the bacterial chromosome (22).

The site-specific recombination reaction catalyzed by Cre is shown schematically in Fig. 1. Cre exists as a monomer in solution, even at high concentrations (23). However, the enzyme binds cooperatively and with high affinity to the *loxP* DNA sequence, resulting in a dimer of Cre subunits bound to each recombining site (24–27). Two Cre-bound *loxP* sites associate to form a synaptic complex (23, 28, 29), within which strand exchange is catalyzed using a mechanism thought to be shared by all of the tyrosine recombinases (13, 14).

The basic elements of the mechanism shown in Fig. 1 were generally accepted when *Mobile DNA II* was published in 2002 (30). However, several important questions remained unanswered, some aspects of the mechanism had been challenged, and new questions would soon be raised. In particular, it was not clear if a specific synaptic complex was favored to initiate recombination and what the basis for this bias might be. The energetics of synapsis were largely unexplored. The roles and importance of many residues in the active site of Cre and of other tyrosine recombinases were also poorly understood.

In addition to these mechanistic questions, the field of researchers seeking to improve the use of Cre in transgenic mice and other organisms had their own sets of questions and goals. Among these were: Can the

[1]242 Anatomy-Chemistry Building, Perelman School of Medicine, University of Pennsylvania, Philadelphia, PA 19104-6059.

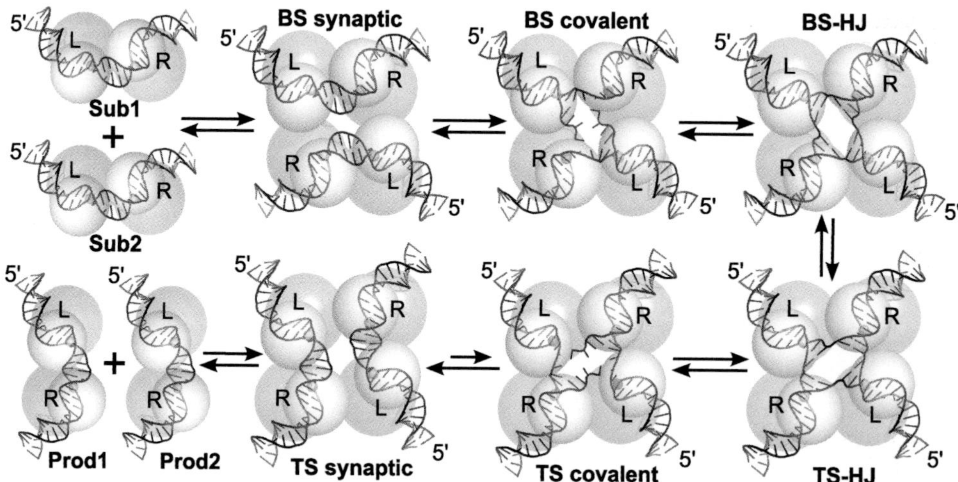

Figure 1 The Cre–*loxP* site-specific recombination reaction. Cre subunits are represented by semi-transparent spheres, with the N-terminal domains (NTDs) in the foreground and the C-terminal domains (CTDs) below. The *loxP* sites are drawn as observed in crystal structures of intermediates in the Cre–*loxP* recombination pathway, with the scissile phosphates drawn as yellow spheres and covalently linked phosphotyrosines as cyan spheres. Two Cre subunits bind cooperatively to the substrate *loxP* sites (Sub1 and Sub2). The structural details of the bending that occurs in the sites at this stage of the reaction are not yet known. The Cre-bound *loxP* sites associate to form an antiparallel synaptic complex where the bottom (red) strands of *loxP* are positioned for cleavage (BS-synapse). Cleavage by Tyr324 in the Cre subunits bound to the right half-sites (R) results in formation of covalent 3′-phosphotyrosine linkages to the bottom strands and release of 5′-hydroxyl ends (BS covalent). Exchange of strands between the two *loxP* sites positions the 5′-hydroxyl groups for attack of the phosphotyrosine linkages on the partner sites, resulting in formation of a four-way Holliday junction intermediate (BS-HJ). Isomerization of BS-HJ to TS-HJ results in activation of the Cre subunits bound to the left half-sites (L), where the top strands (black) are now positioned for cleavage. Cleavage of TS-HJ to form the top-strand covalent intermediate (TS-covalent) and strand exchange to form the top-strand active synapse (TS-synapse) completes the reaction. Dissociation of TS-synapse gives the product *loxP* sites (Prod1 and Prod2). The structure of TS-synaptic is not currently known; a likely candidate is drawn here (discussed in text). doi:10.1128/microbiolspec.MDNA3-0014-2014.f1

loxP site be modified to affect the directionality of the reaction and favor integration? Can Cre variants be engineered to recognize and recombine very different *loxP* sequences? Can more highly regulated versions of Cre be engineered?

In this review, I discuss advances in the Cre recombinase field that have occurred over the past 12 years, since the publication of *Mobile DNA II*. Readers are referred to earlier reviews for discussions relating to fundamental structural and mechanistic aspects of the Cre–*loxP* system (12, 13). Here, the focus will be on those recent contributions that have provided new mechanistic insights into the reaction, including modifications of Cre (Table 1) and/or *loxP* (Table 2) that have led or may lead to useful improvements in genome engineering applications. Progress in strategies of regulating Cre expression, where the focus has been on promoter selection and viral mechanisms of gene delivery, are not discussed because they are less related to the mechanistic aspects of Cre–*loxP* recombination.

WORKING WITH Cre

Part of the reason that Cre has proven to be a useful model system for biochemical and structural studies is the relative ease of expression and purification of the enzyme. Two useful expression and purification strategies have been described. One involves purification of an intein–Cre fusion on chitin beads followed by cleavage and release of Cre from the intein-bound resin (31). The second involves purification of native Cre using cation exchange and hydroxyapatite chromatography, resulting in a single band on silver-stained sodium dodecyl sulfate–polyacrylamide gel electrophoresis (32). Both are rapid, with milligram quantities of pure enzyme per liter of bacterial culture. Purified Cre can be

Table 1 Selected Cre variants

Cre	Properties	References
Wild-type	341 amino acids; 38 kDa	(15)
Y324F	inactive catalytic mutant; partially defective in synapsis	(23, 33, 46)
R173K	poorly active catalytic mutant; partially defective in synapsis	(23, 33, 46)
K201A	inactive catalytic mutant; proficient in synapsis	(23, 58)
A36V	defective in synapsis/cooperativity	(23, 29, 59)
R32V, R32M	defective in cooperativity; improved specificity	(88)
C2(±) #4	preferentially recombines *loxP*-M7 site (Table 2)	(83, 84)
Fre	recombines *loxH* site (Table 2)	(82)
Tre	recombines *loxLTR* sites in HIV-1 prophage (Table 2)	(85)
CreAAF	remodeled Cre–Cre interface in C-terminal domain; basis for heterospecific sites	(87)
His_6-NLS-TAT-Cre, NLS-Cre, TAT-Cre	efficient transduction of mammalian cells	(120–122)
ONBY-Cre	activated by UVA light	(125)
Cre 1-59, Cre 60-341	when fused to heterodimerizing domains, forms active enzyme	(109, 111, 112)
Cre 1-196, Cre 181-341	spontaneously assembles to form active enzyme when coexpressed	(113–115)
DD-Cre	targeted for proteosomal degradation; stabilized by trimethoprim	(116)

highly concentrated under appropriate conditions and can be stored at –80°C for extended periods (32). A variety of affinity tags have also been employed to facilitate purification of Cre fusions.

Assays of Cre activity are straightforward. Cre binds with high affinity to *loxP* sites, with apparent $K_d <$ 1 nM in standard electrophoretic mobility shift assays with *loxP*-containing restriction fragments or PCR products (26). *In vitro* recombination assays can be carried out quantitatively in a number of reaction formats, with product formation occurring within minutes. Rapid quantitative evaluation of the activity of Cre mutants in *Escherichia coli* has also been described, using a single-copy reporter (33).

For *in vivo* experiments making use of Cre to perform deletions, inversions, and exchanges of *loxP*-flanked DNA elements, many resources are available in the form of plasmids, strains, and mice expressing Cre in various tissue-specific and inducible forms. The reader is referred to several recent reviews for application-specific information (34–38).

BINDING OF Cre TO *loxP*

Cre is a two-domain protein, with a 130-residue N-terminal domain (NTD) closely linked to a 211-residue C-terminal domain (CTD; Fig. 2A). The 34-bp *loxP* site contains two 14-bp recombinase-binding elements (RBEs) that are arranged as nearly perfect inverted repeats on opposite sides of an asymmetric 6-bp crossover region (Fig. 2B). When Cre binds to an RBE in the *loxP* site, the two protein domains encircle the DNA duplex, forming a "C-shaped clamp" that forms extensive minor and major groove contacts. Two key binding elements are the αJ helix from the CTD that binds in the major groove near the center of the *loxP* half-site and the αB/αD helices from the NTD that straddle the major groove near the start of the crossover region (Fig. 2B). Cre binds to the first RBE with nanomolar affinity and the second Cre subunit then binds with subnanomolar affinity (26, 27). The basis for cooperative binding was evident from the first Cre–DNA crystal structure, where extensive contacts between the two Cre subunits bound to *loxP* were observed (39) (Fig. 2B).

The Cre–DNA interface is now well-documented in structural terms from several high resolution Cre–DNA crystals structures (23, 33, 39–43) and there is good general agreement between the structural models and both *loxP* mutagenesis studies and the DNA-binding properties of Cre mutants (44–46). The Cre–*loxP* interface is complex, with numerous water-mediated interactions but relatively few direct contacts between protein side chains and bases in the major groove. Developing a simple scheme for understanding Cre–*loxP* binding specificity has therefore been challenging. The most conspicuous polar interactions are made by Arg259, which forms two hydrogen bonds to the C·G base-pairs at positions ±10 (Fig. 3), and Gln90, which hydrogen-bonds to the A·T base-pairs at positions ±5. Of the two, Arg259 plays crucial roles in affinity and specificity, whereas Gln90 plays more minor roles (44).

An example of the complex nature of the Cre–DNA interface was illustrated in a study by Rüfer & Sauer,

Table 2 Selected *loxP* variants

loxP	Sequence[1]	Properties	References
loxP	ATAACTTCGTATAA-TGTATG-CTATACGAAGTTAT	wild-type	(24)
loxS1	ATAACTTCGTATAG-CATATG-CTATACGAAGTTAT	symmetrized using right half-site; biochemical/structural studies	(23, 40, 50)
loxS2	ATAACTTCGTATAA-TGTACA-TTATACGAAGTTAT	symmetrized using left half-site; biochemical studies	(23, 50, 67)
lox511	ATAACTTCGTATAA-TGTATa-CTATACGAAGTTAT	crossover mutant; recombines with *loxP* at very low frequency	(67, 68)
loxFAS	AcAACTTCGTATAt-accttt-CTATACGAAGTTgT	pseudo-*loxP* site in *Saccharomyces cerevisiae* genome	(73)
lox2272	ATAACTTCGTATAA-aGTATc-CTATACGAAGTTAT	alternative crossover; does not recombine with *loxP*	(68)
lox5171	ATAACTTCGTATAA-TGTgTa-CTATACGAAGTTAT	alternative crossover; does not recombine with *loxP*	(68)
lox66	taccgTTCGTATAA-TGTATG-CTATACGAAGTTAT	left RBE mutant	(89)
lox71	ATAACTTCGTATAA-TGTATG-CTATACGAAcggta	right RBE mutant	(89)
lox72	taccgTTCGTATAA-TGTATG-CTATACGAAcggta	produced by lox66 × lox71 recombination; very weak activity	(89)
loxH	ATAtaTaCGTATAt-aGacat-aTATACGtAtaTAT	directed evolution target from human chromosome 22	(82)
loxM7	ATAACTctaTATAA-TGTATG-CTATAtagAGTTAT	directed evolution target	(83)
loxLTR	AcAACaTCcTATtA-caccct-aTATgCcAAcaTgg	*loxP*-like sequence from HIV-1 strain; directed evolution target	(85)

who changed the sequence of the T·A base-pairs at positions ±11 and ±12 near the centers of the RBEs (47). Cre makes van der Waals contacts to the 5-methyl groups of these base-pairs through residues near the amino-terminus of helix αJ (Fig. 3). Recombination decreased by a factor of 10^5 as a result of the substitutions, indicating that substantial disruption of the network of Cre–DNA interactions in this region of the major groove had occurred.

A selection scheme was used to identify changes in Cre that would restore activity, leading to the identification of Glu262 as a key residue that must be altered to permit recognition of the mutant *loxP* site. Curiously, Glu262 is located on the αJ helix, but some distance away from *loxP* residues 11 and 12 (Fig. 3). This residue makes a water-mediated interaction to the C9 base and hydrogen-bonds directly to the phosphate backbone, requiring that a proton be shared between the two acidic groups. Substitution of Glu262 by glutamine or several alternative residues allows Cre to bind and recombine the altered *loxP* sites, presumably by allowing more flexibility in the positioning of αJ in the major groove. The restored function comes at the expense of a significant relaxation in binding specificity at the 11 and 12 positions, as the Cre mutants also recombine the wild-type *loxP* site with high efficiency (47).

A related set of experiments was carried out by Baldwin and colleagues, who investigated *in vitro* recombination of a *loxP* site containing symmetric C→A substitutions at positions ±10 (48). These substitutions abolish the key hydrogen-bonding interaction formed by Arg259, leading to loss of *in vivo* recombination (45). Cre was shown to bind and recombine this substrate weakly *in vitro* and a Cre–*loxP* mutant crystal structure indicated that Glu262 and Glu266 were tethering the displaced Arg259 side chain via salt bridges. Interestingly, a Cre mutant where both glutamates were replaced with glutamine not only bound more efficiently to the *loxP* mutant, but showed an increase rate of cleavage (48). Thus, Cre–DNA interactions far from the *loxP* cleavage sites can have significant effects on catalysis. A subsequent study on alternative substitutions at *loxP* position 10 resulted in even more complex behavior, indicating that the effects of Cre–*loxP* recognition can be manifested at multiple stages in the reaction pathway (49).

Most Cre–DNA binding studies have relied on electrophoretic mobility shift assay (EMSA) -based experiments (25, 26). While these assays perform well in this system, it is difficult to draw conclusions about the effects of divalent ions, polyamines, etc. on binding affinity and cooperativity. Sauer and colleagues have studied the Cre–*loxP* interaction using surface plasmon resonance (SPR) and found a substantial increase in binding cooperativity when 10 mM $MgCl_2$ or 5 mM spermidine is present (27). A much smaller effect had been previously observed from EMSAs (26). The increase in binding cooperativity was suggested to result from a change in DNA conformation and/or bending, since Mg^{2+} and spermidine are known to facilitate such changes. Bending of the *loxP* site is expected to promote interactions between Cre subunits bound to the RBEs (Fig. 2B) (40, 50).

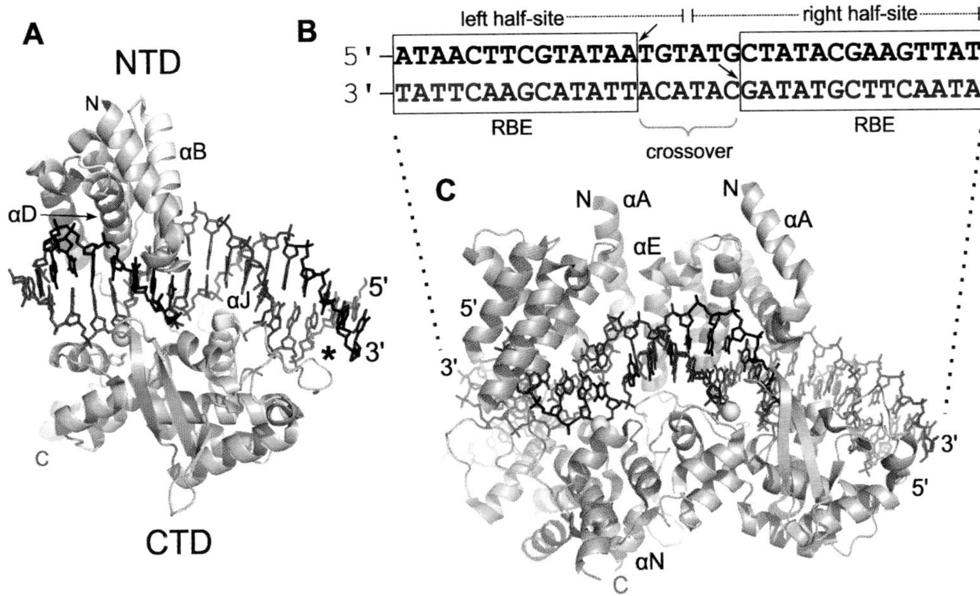

Figure 2 Cre binding to *loxP*. (A) Cre bound to a *loxP* half-site. The two domains of Cre form a "C-shaped clamp" that wraps around the DNA duplex. Helices B and D from the N-terminal domain (NTD) straddle the major groove from one face and helix J sits in the major groove from the opposite face. An asterisk marks the point of insertion of several basic residue side chains (not shown) into the minor groove at the end of the site. (B) Cre bound to the *loxP* site. The *loxP* site is composed of two 14-bp recombinase-binding elements (RBEs) arranged as inverted repeats around an asymmetric 6-bp crossover region. The RBEs differ only in the base pairs adjacent to the crossover region. Arrows indicate the cleavage sites. Two primary sources of Cre–Cre interactions on the *loxP* site are evident: the two NTDs interact via helices A and E and the two C-terminal domains interact where helix-N from one subunit is buried in a hydrophobic pocket of the adjacent subunit. The scissile phosphates are drawn as yellow spheres in (A) and (B) and Tyr324 as yellow sticks in (A). doi:10.1128/microbiolspec.MDNA3-0014-2014.f2

The SPR study raises an important issue regarding the available structural models for understanding Cre–*loxP* binding cooperativity. Most binding experiments monitor association of the first, then the second Cre subunits binding to isolated *loxP* sites. The sites are either maintained at very low concentration (EMSA) or are tethered to the surface of a chip (SPR) under conditions where synaptic complex formation is disfavored. On the other hand, all published structural models of Cre–DNA complexes are of tetrameric assemblies in precleavage, cleaved, or Holliday junction states of the reaction. The interface formed between adjacent Cre subunits is facilitated by the bending that occurs in the *loxP* sites in these complexes and the extensive protein–protein interaction surface formed upon synapsis is therefore likely to play a role in stabilizing the observed DNA deformations. This raises the possibility that Cre-bound *loxP* sites may have somewhat different properties when initially formed compared with the conformations observed in tetrameric complexes. The weaker than expected bending observed in Cre–*loxP* cyclic permutation EMSA experiments would be consistent with this interpretation (50).

SYNAPSIS OF *loxP* SITES AND THE ORDER OF STRAND EXCHANGE

The recombination mechanism illustrated in Fig. 1 requires that the Cre-bound *loxP* sites associate in an antiparallel fashion, generating a two-fold symmetric complex within which sequential strand exchange takes place. The inherent symmetry, or "antiparallel nature" of the synaptic complex is evident in the topological outcome of experiments even when symmetrized sites are used (51). An important feature of a symmetric, antiparallel synaptic complex is that the bending direction of *loxP* and the location of the bend within the crossover region are identical in the two associating sites. An asymmetric, parallel arrangement of sites can also be rationalized, although such synaptic complexes would

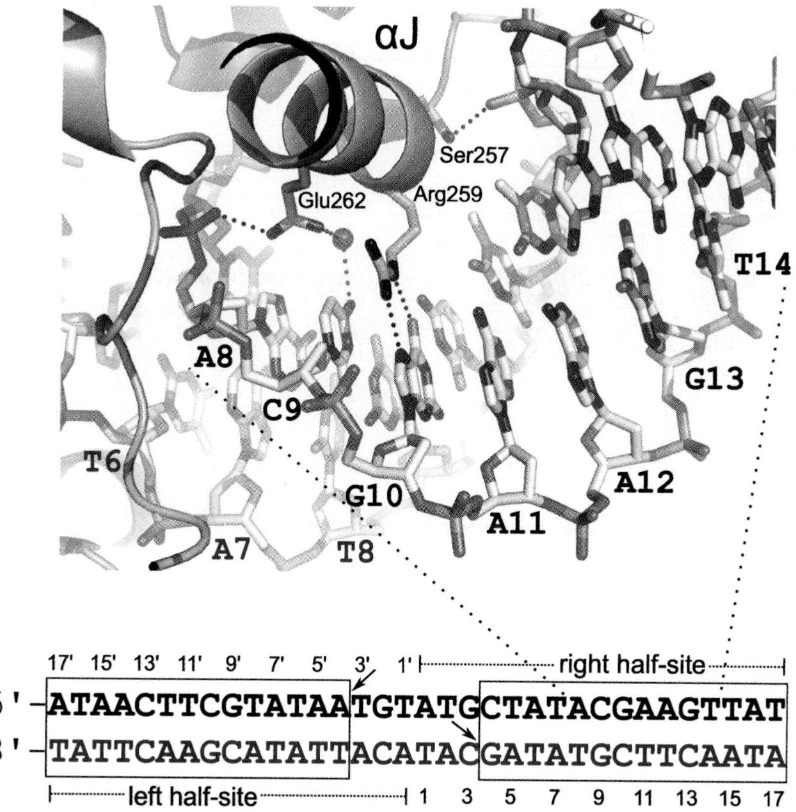

Figure 3 Cre-*loxP* specificity. Polar interactions between helix J and the *loxP* site are shown. The key interaction is between Arg259 and N7/O6 of G10 in the *loxP* site. The Glu262 side chain carboxyl makes a water-mediated interaction to N4 of C9 as well as an unusual hydrogen bond to a backbone phosphate (implying a neutral carboxyl side chain). Ser257 also makes a backbone hydrogen bond. The *loxP* sequence is shown for reference, with the conventional site numbering used by most researchers.
doi:10.1128/microbiolspec.MDNA3-0014-2014.f3

not be expected to undergo efficient strand exchange. For a parallel synaptic complex, the directions (and most likely the locations) of the DNA bends would differ in the two interacting *loxP* sites. If the energies associated with forming and associating *loxP* sites with opposite bend directions were equal, a 1:2:1 distribution of anti-TS:parallel:anti-BS synaptic complexes would be expected, where anti-TS and anti-BS refer to configurations poised to exchange the top strands or bottom strands, respectively (i.e. TS-Synaptic and BS-Synaptic in Fig. 1).

The question of whether a particular antiparallel synaptic complex forms to begin recombination is closely related to the question of whether the top or bottom strands of *loxP* are preferentially exchanged first during recombination. The order of strand exchange question has been particularly confusing in the Cre-*loxP* system because of conflicting reports in the literature. Early work by Hoess and colleagues demonstrated a clear preference for bottom-strand exchange based on isolation and analysis of the HJ intermediate (52). Later reports concluded that the top strands were exchanged first (42, 53), but one of these claims was reversed when it was discovered that the use of a Cre H289A active site mutant was responsible for a change in directionality of the enzyme (54). Biochemical studies by the Sadowski laboratory provided evidence in support of the original findings of Hoess and colleagues and identified those sequence elements in the *loxP* site that are primarily responsible for bottom strand exchange preference (55, 56).

The question of why Cre displays this bias in the reaction pathway was further addressed using 5′-bridging phosphorothioate-containing DNA substrates, a tool that had previously been used to study the role of homology in the λ-integrase system (57). Cre will cleave substrates where the 5′-bridging oxygen of the scissile phosphate has been replaced by sulfur, but the subsequent ligation reaction is sufficiently slow that the reaction can be considered irreversible. Combined with fluorescence resonance energy transfer (FRET) experiments with labeled *loxP* sites and inactive Cre mutants, it was shown that preferential formation of the bottom strand synaptic complex is responsible for establishing the order of strand exchange in the Cre system (58). As discussed below, single molecule FRET (smFRET) experiments later provided additional evidence in support of this idea (59).

An explanation for the bias in Cre-*loxP* synapsis was provided by a synaptic complex crystal structure containing wild-type *loxP* sites and the inactive Cre K201A mutant (23). In that structure, the bend in *loxP* compresses the minor groove at the start of the crossover region in the left half-site (Fig. 1, BS-Synapse). This deformation is more readily accommodated at the T-G step in the left half-site of *loxP* than it is at the C-A step in the right half-site, resulting in a synaptic complex where the left half-sites are preferentially kinked. Consequently, the Cre subunits bound to the right half-sites are activated for cleavage of the *loxP* bottom strands, which are geometrically positioned for exchange to generate the HJ intermediate (12). These reaction steps are illustrated in the top half of Fig. 1.

ENERGETICS OF SYNAPSIS

The ability of Cre to carry out efficient recombination in a variety of cellular and topological contexts is likely related to its ability to form very stable synaptic complexes. The energetics of this process have been studied using sedimentation equilibrium ultracentrifugation (23). Using either wild-type Cre or the inactive K201A mutant, Cre-bound *loxP* sites associate with $K_d = 10$ nM, with only minor effects from divalent ions. An interesting observation from these studies was that synapsis is pH-dependent, with affinity dropping off sharply as the pH is increased above 8.5. This finding explained why Cre–*loxP* synaptic complexes could be identified on native gels, but quantitative experiments using electrophoresis titrations gave inconsistent results. Use of a modified electrophoresis buffer at pH 7.5 led to the development of a native gel electrophoresis assay that reproduced the findings from analytical ultracentrifugation experiments.

Similar synapsis experiments using symmetrized *loxP* sites provided additional insight into the preference for formation of a BS synaptic complex. The loxS1 site contains two right half-sites of *loxP* (Table 2) and a Cre-loxS1 crystal structure showed that bending occurs by negative roll at the C-A step, with resulting compression of the minor groove at GC-rich positions ±4 and ±3 (40). This Cre-bound site associates with K_d ~200 nM, about 20-fold weaker than wild-type *loxP*. The loxS2 site contains two left half-sites (Table 2) and most likely bends with negative roll at the T-G step as observed in the synaptic complex structure containing wild-type *loxP*. Compression of the minor groove involves A·T base pairs in this case. The Cre-bound loxS2 site associates with K_d ~5 nM, an affinity that is two-fold higher than for *loxP*. Hence, the bias towards the BS synaptic complex can be viewed most simply as an energetic preference for bending in the left, AT-rich *loxP* half-site.

The Cre K201A mutant allowed for direct analysis of synapsis in the absence of cleavage and strand exchange. The tight association measured for Cre K201A-*loxP* synapsis is close to that predicted from kinetic modeling studies using wild-type Cre (26), suggesting that loss of Lys201 does not strongly bias the results. As discussed below, Lys201 plays an essential role in catalysis by activating the 5′-hydroxyl leaving group during phosphoryl transfer (58, 60). In general, however, synapsis of *loxP* sites is coupled to other functional elements in the Cre–DNA tetramer. For example, The Cre Y324F and R173K active site mutants are both inactive for strand cleavage (46, 61) but they are also both partially defective in synapsis (23). Arg173 is involved in bending the *loxP* site and Tyr324 plays a role in structural organization of an important region of the CTD, perhaps explaining in part why these residues are required for efficient synapsis. An important consequence of these findings for Cre applications is that use of the Cre Y324F mutant as a negative control will not only eliminate the cleavage and recombination functions of Cre, but will also diminish any structural effects imposed by Cre-mediated synapsis of *loxP* sites in the target substrates.

Crystal structures of synaptic complexes have provided a detailed molecular understanding of the interactions responsible for stabilizing this intermediate (23, 40, 53). An extensive network of contacts between Cre subunits is formed, burying a substantial amount of solvent-accessible surface (Fig. 4). A number of synapsis-deficient Cre mutants have been identified (23, 29, 62), with the same complexity in interpretation as observed for the Cre–DNA interface. Some synapsis-deficient mutants are readily interpretable (e.g., Cre A36V), whereas others are involved more indirectly (e.g., Y324F).

Levine and coworkers have devised an experimental scheme to monitor synapsis and recombination of Cre–*loxP* complexes in real time using FRET (63). Using both intermolecular and intramolecular substrates containing donor/acceptor fluorophores, kinetic constants for a simplified model of the recombination reaction could be extracted. In general, the results are consistent with those obtained using gel-based analysis of reaction kinetics (26). An additional component of this study was the use of Cre-mediated synapsis/recombination kinetics as a tool to monitor the probability of DNA loop closure.

ALTERNATIVE MODELS OF THE SYNAPTIC COMPLEX

Most experiments and structural models involving Cre–*loxP* synaptic complexes have been interpreted in the context of the initial tetrameric assembly that forms immediately before cleavage and strand exchange to form the HJ intermediate. However, the product of top strand cleavage and strand exchange that is formed during resolution of the HJ intermediate is also a synaptic complex (i.e., TS-Synapse in Fig. 1). Kinetic models of the Cre reaction and single molecule experiments have both indicated that the product complex is exceptionally stable (26, 59), implying that synaptic complexes formed by wild-type Cre are likely to be dominated by product, rather than substrate *loxP* sites. The similar synapsis K_d values obtained for wild-type Cre (presumed stable product synapse) and the K201A mutant (stable substrate synapse) suggest that the two synaptic complexes have similar dissociation rates, an idea that was supported by single-molecule studies (64).

There is currently no structural model that has been identified as a product synaptic complex and consequently it is not yet clear what type of DNA-bending is accommodated by the *loxP* sites at the end of the reaction. One formal possibility is that *loxP* is bent in the right half-site, in a manner similar to that observed in the Cre–*loxS1* synaptic complex structure. As it is clear from centrifugation and electrophoresis experiments that synaptic complexes containing a sharp bend in the right half-site are not particularly stable, this model would be difficult to reconcile with the established stability of the product complex.

An alternative explanation may be found in a unique synaptic complex structure containing nonbridging phosphorothioate substitutions in the *loxP* sites (53). In this structure, the *loxP* sites are in an antiparallel top-strand cleavage configuration, but bending is not localized to the right half-site. Instead, a positive roll is distributed across the left half-site where the minor groove is opened towards the synapsed partner (shown as TS-synapse in Fig. 1). This configuration is a strong candidate for the product synaptic complex, where an energetically disfavored right half-site bend is replaced by an alternative left half-site bend that uses positive, rather than negative, roll to bend the site.

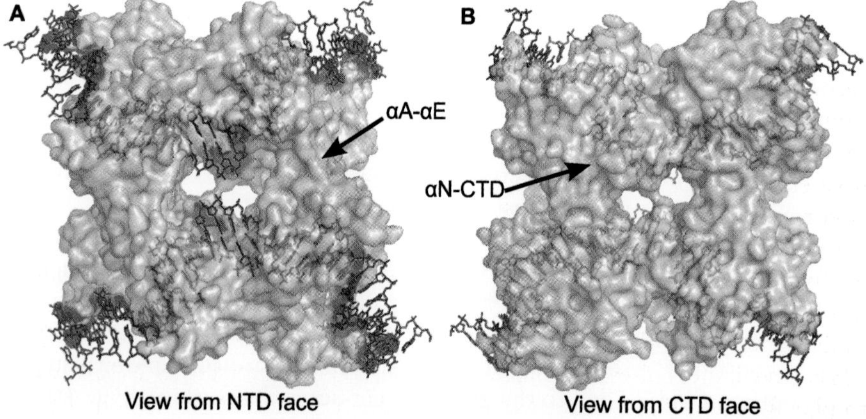

Figure 4 The Cre–*loxP* synaptic complex. The Cre K201A–*loxP* synaptic complex is shown, with the same DNA coloring as in Fig. 1, and Cre subunits drawn in semi-transparent space-filling representations. The *loxP* site is sharply bent in the left half-sites (purple subunits bound) and is undistorted in the right half-sites (green subunits bound). (A) View is from the N-terminal domain face of the complex, with the same orientation as BS-synaptic in Fig. 1. Substantial interactions between helices A and E are indicated. (B) View is from the opposite face of the complex, where the interlocking interactions formed between helix-N of one subunit and the C-terminal domain of the adjacent subunit are emphasized. Note that the same interactions responsible for cooperative binding in Fig. 2B are responsible for stabilizing the synaptic complex shown here. doi:10.1128/microbiolspec.MDNA3-0014-2014.f4

Further investigations will be required to verify this assignment, which implies a strong structural asymmetry in the overall Cre–*loxP* reaction pathway.

These findings raise some intriguing questions about the Cre reaction. For example, if the product synaptic complex is so stable, why does it not form preferentially from Cre-bound *loxP* sites and either run the reaction in reverse or hold the sites together for an extended period in an unproductive state? A kinetic barrier to forming the product complex directly (i.e., without strand exchange) could be part of the answer, but the molecular basis for such a barrier is currently not clear.

An additional question relating to the synaptic complex involves the degree to which the Cre-bound *loxP* sites are co-planar. Topological studies of the integration and excision reactions catalyzed by λ-integrase suggested that the active recombination complex has an intrinsic chirality (65). Related experiments with Flp and Cre recombinase were interpreted to imply that this chirality may be shared by all λ-integrase family members. The simplest explanation for a chirality consistent with this proposal would be an out of plane rotation of the two *loxP* sites during recombination, resulting in a right-handed crossing of the sites (65, 66). There are currently no structural models to support this idea because all tyrosine recombinase crystal structures of synaptic complexes and HJ intermediates feature core half-sites that are nearly co-planar.

REQUIREMENTS FOR *loxP* HOMOLOGY

Efficient Cre–*loxP* recombination requires that the central six base-pairs comprising the crossover region of *loxP* be identical in the two recombining sites (67). Even single base-pair heterologies can severely reduce recombination, as demonstrated in studies where the *loxP* spacer region has been systematically modified (68, 69). The requirement for identical crossover sequences was originally interpreted in the context of branch migration of the Holliday intermediate, but structural models of Cre–*loxP* HJs have indicated that branch migration is unlikely to play a role in this system (42, 43, 70). Instead, sequence identity is required for efficient strand swapping and ligation in both the HJ-forming and HJ-resolving steps in the pathway (71). In molecular terms, formation of canonical base-pairs during strand exchange is thought to be required to establish a geometry that promotes efficient ligation and formation of stable intermediates or products. It is worth noting, however, that crossover sequence identity is not required for all tyrosine recombinases (72).

An important implication of *loxP* homology requirements is that sites with distinct crossover sequences can coexist in DNA substrates without efficient deletion or inversion of the intervening sequences by Cre. This is the basis of the recombinase-mediated cassette exchange reaction, which in its simplest form uses pairs of sites containing incompatible crossover sequences (discussed below). An example of an incompatible site is loxFAS (Table 2), which performs nearly as well as the *loxP* site (73). Systematic studies of the *loxP* crossover region and the flanking base pairs at the RBE borders indicate that changes to this central region of *loxP* all result in some loss of recombination efficency (68, 69).

It is also now clear that the requirement for sequence identity between the crossovers of recombining sites is not absolute. Under some *in vivo* conditions, Cre will carry out recombination between incompatible sites and will carry out deletions for sites arranged as inverted repeats at a measurable frequency (74). In some cases, the sequences of the recombinant products cannot be explained by Cre-mediated strand exchange of heterologous crossovers, suggesting that cellular factors may play a role in a subset of these noncanonical rearrangements (75).

CATALYSIS OF CLEAVAGE AND STRAND EXCHANGE

The Cre active site contains seven residues that are conserved to varying degrees among the tyrosine recombinases (Fig. 5). The structure of a vanadate transition state mimic of the phosphoryl transfer reaction, combined with saturation mutagenesis of each active site residue, has provided insight into the catalytic steps involved in Cre–*loxP* recombination (33, 76). When combined with high-resolution structures of the precleavage (23, 40) and covalent intermediates (39) of the reaction, the positions and hydrogen-bonding patterns of the catalytic residues can be followed through the catalytic steps. These interaction networks are illustrated in Fig. 5.

The Cre active site residues can be divided into three sets. In the first set, Tyr324 and Lys201 are highly conserved, play unique catalytic roles, and no substitutions for either residue results in significant activity. Tyr324 is the nucleophile in the cleavage reaction and Lys201 is the general acid primarily responsible for activating the O5′ leaving group.

In the second set, Arg173 and Arg292 are also highly conserved and play important roles in stabilizing the negative charge of the phosphate in the transition states of the cleavage and ligation reactions. Both residues make double hydrogen bonds to oxygen atoms using their side-chain guanidino groups and likely play dual roles in catalysis. Arg173 hydrogen bonds to O1P and to O5′ of the scissile phosphate and biochemical evidence suggest that the O5′ contact contributes to leaving group activation during cleavage (77). The R173K mutant is inactive *in vitro* and only weakly active *in vivo*, but the R173H mutant shows strong activity both *in vitro* and *in vivo*, indicating that transition state stabilization at O1P is likely to be the primary catalytic role. Arg292 hydrogen bonds to O2P and to the Tyr324 hydroxyl oxygen in the transition state. This suggests a that a secondary role of Arg292 may involve stabilizing the phenolate oxygen that is the leaving group during ligation.

In the third group of residues, His289 and Glu276 are moderately conserved, but Trp315 is more often histidine in the tyrosine recombinase family. Like Cre, the Flp recombinase uses a tryptophan in this position of the active site and biochemical studies have indicated that the primary role of this residue is structural, rather than catalytic (78, 79). Indeed, the Cre W315F mutant has higher activity than W315H, indicating a similar functional role in the Cre system (33).

Neither His289 nor Glu176 are required for catalysis, with several alternative residues showing moderate levels of recombination activity in each case. His289 can hydrogen-bond to O1P and to the Tyr324 hydroxyl oxygen, suggesting a dual stabilization role similar to, but less important than, that seen for Arg292. Glu176

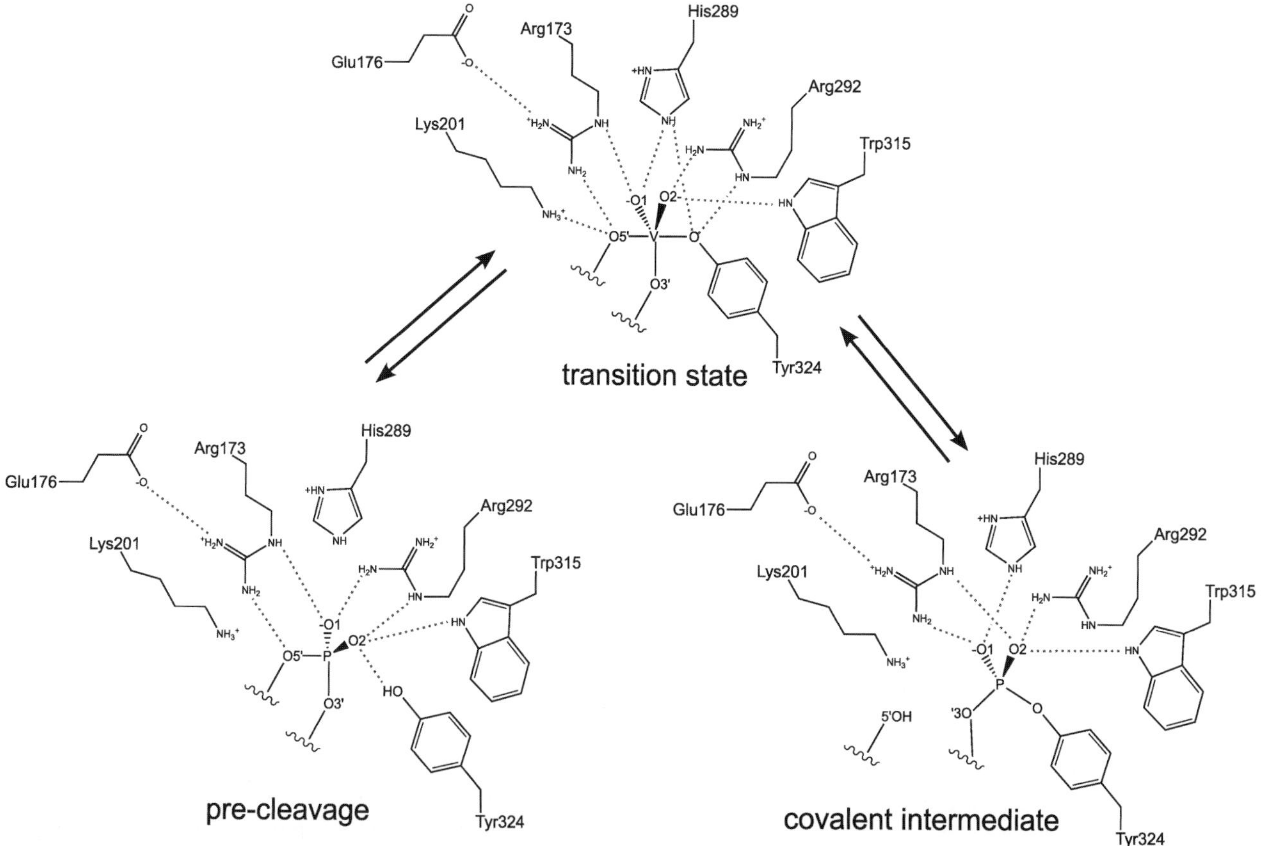

Figure 5 The Cre active site. Hydrogen-bonding interactions between the seven active site residues and the scissile phosphate are shown for the precleavage intermediate, transition state mimic, and covalent intermediate, based on crystal structures of the corresponding complexes. Glu176 forms additional hydrogen bonds to the backbone amides of both Arg173 and Lys201 in each case, which are not shown. The 5′-hydroxyl group is not present in the covalent intermediate structure, but hydrogen bonding to Lys201/Arg173 is likely (not shown). The cleavage reaction proceeds from left to right, whereas the ligation reaction proceeds right to left in this scheme. Note that O1 and O2 of the transition state are labeled in the opposite sense to that used in Gibb et al. (33).
doi:10.1128/microbiolspec.MDNA3-0014-2014.f5

plays a structural role, buttressing the positions of the Arg173 side chain and the main chains of both Arg173 and Lys201.

The role of electrostatic stabilization in Cre catalysis has been probed using methylphosphonate (MeP)-containing *loxP* substrates, where one of the nonbridging oxygen atoms is replaced by a methyl group (80). Both wild-type Cre and Cre R292A will efficiently cleave *loxP* sites containing a scissile MeP, in agreement with a role of electrostatic stabilization by this residue. The Cre active site also protects the MeP-containing active site from both phosphodiester hydrolysis and from hydrolysis of the 3′-phosphotyrosyl linkage, a property that distinguishes Cre from the related Flp recombinase and TopIB systems. Subsequent studies revealed that Arg173 is also not required for cleavage of MeP-containing sites and even the R173A, R292A double mutant is competent for cleavage (81). In the ligation reaction, R292A is competent, R173A is weakly active, and the double mutant is inactive, providing additional support for the role of Arg173 in the activation of the 5′-hydroxyl oxygen.

A number of Cre active site substitutions lead to accumulation of the covalent phosphotyrosyl intermediate (e.g., R292K and H289Q) or the HJ intermediate (e.g., W315F and E176A) of the reaction (33). In many cases, one can rationalize these properties in terms of the transition state structure. For example, the observation that Cre R292K generates large amounts of covalent intermediate can be explained by the loss of hydrogen-bonding to the Tyr324 hydroxyl oxygen by this mutant and a consequent lowering of leaving group activation during the ligation step. This implies that Arg292 has a more important role in leaving group activation during ligation than in activation of the nucleophile during cleavage. The redundant roles and in some cases subtle mutagenesis effects of the less conserved Cre active site residues may reflect the evolutionary fine-tuning that has taken place to strike the correct balance between cleavage and ligation rates. Such a balance is presumably required to efficiently catalyze the sequential strand exchange reactions involved in site-specific recombination.

INSIGHTS FROM SINGLE MOLECULE EXPERIMENTS

Experimental approaches that allow monitoring of individual Cre–*loxP* synaptic complexes undergoing recombination are relatively new to the field, but have already led to several interesting findings. In the first such study, Fan employed tethered particle motion (TPM) to monitor Cre-mediated synapsis and recombination of *loxP* sites (64). In this technique, a DNA substrate containing *loxP* sites is fixed on one end to a surface and on the other end to a bead. The amplitude of brownian motion of the bead is monitored as a function of time and related to the extension of the DNA. Synapsis can be inferred from a shortening of the DNA length and recombination from persistence of the shortened DNA upon treatment with sodium dodecyl sulfate, when the *loxP* sites are arranged as direct repeats. Similar experiments using *loxP* sites arranged as inverted repeats can allow identification of HJ intermediates.

The TPM experiments have indicated the presence of two different rates of decay of Cre–*loxP* synaptic complexes: a faster rate that is associated with dissociation of the complex without recombination ($\sim 3\times10^{-2}$/s) and a slower rate that is associated with dissociation of product *loxP* sites ($\sim 2\times10^{-3}$/s). The slower rate reflects both the recombination process and the stability of the product complex. Interestingly, the K201A mutant also displayed this biphasic dissociation behavior with similar rate constants, despite its lack of ability to carry out recombination.

More recently, Pinkney et al. have combined tethered fluorophore motion (TFM) with single-molecule FRET (smFRET) to provide both a global view of the state of the DNA substrate and a local readout of the recombining *loxP* sites from donor and acceptor groups attached at strategic positions (59). The experiments supported previous work showing that Cre initiates recombination by exchanging the bottom strands and that the strand bias is caused by preferential formation of the anti-BS synaptic complex. These experiments also confirmed findings from the TPM study (64) that a significant fraction of synapsed *loxP* sites dissociate without undergoing recombination.

The additional information provided by smFRET measurements has led to a more detailed picture of the events unfolding during recombination compared with that obtained from TPM or TFM data alone. Two distinct FRET efficiency distributions were observed for synapsis of inactive *loxP* sites by wild-type Cre (those that dissociate without undergoing recombination), indicating that two distinct synaptic structures are formed. Interestingly, only the lower (major population) efficiency distribution is observed for the inactive K201A mutant, and only the higher (minor population) efficiency is observed for the synapsis-deficient A36V mutant. Modeling of the loosely tethered fluorophores onto templates based on Cre–DNA crystal structures suggests that the higher efficiency value corresponds to the distance expected from structural models of the synaptic complex, whereas the lower efficiency results in a distance that is ~10 Å longer than expected (59). The actual relationships between the observed FRET efficiencies and the underlying synaptic structures is not yet clear, but it does appear that there is still much to be learned about the structural and biochemical natures of the complexes that form on the recombination pathway.

The smFRET approach has also provided insight into synaptic complexes that are actively undergoing recombination, showing that the lifetime of the recombination event is about one second and the rate-limiting step is associated with HJ isomerization or the second strand exchange to form products (59). The FRET efficiency distribution for Cre–*loxP* synaptic complexes undergoing recombination was found to be broad, spanning the two distributions obtained for inactive complexes. The breadth of this distribution was interpreted to indicate facile isomerization of the HJ intermediate of the reaction. Finally, modeling of the kinetic data indicated that the product synaptic complex is extremely stable, in agreement with previous reports.

DIRECTED EVOLUTION OF Cre TO RECOGNIZE ALTERNATIVE SITES

There has been considerable interest in the question of whether Cre can be modified to recognize and recombine alternative sequences, including naturally occurring sequences in mammalian and viral genomes. Buchholz and Stewart addressed this question by designing experiments to efficiently evolve Cre to recombine "*loxH*" sites through multiple cycles of PCR-based mutagenesis and selection in *E. coli* (82). The *loxH* site is from human chromosome 22 and differs from *loxP* by four changes in the half-site RBEs and a different crossover sequence (Table 2). The resulting evolved "Fre" recombinases were able to delete *loxH*-flanked gene segments in *E. coli* and in mammalian cells, with diminished activity on *loxP*-containing substrates.

Santoro and Schulz also used directed evolution to generate Cre variants with altered specificity (83). In this case, *E. coli* cells were sorted by fluorescence-activated cell sorting according to the recombination activities of their Cre-expressing plasmids, allowing both positive selection for recombination of the altered *loxP* sequence and negative selection against recombination of wild-type *loxP* sites. A series of modified *loxP* sites was chosen for these experiments based on disruption of recombination by wild-type Cre. This approach led to a Cre variant (called C2(±)#4) that efficiently recombines loxM7 sites containing three changes in each RBE (Table 2), but does not efficiently recombine *loxP* sites. For some altered *loxP* targets, active Cre variants could be identified by positive selection, but not by both positive and negative selection. This suggests that it may be difficult to change the specificity of Cre for some sequences without retention of significant activity on *loxP* (83).

Crystal structures of two Cre–*loxM7* complexes have provided insight into how altered specificity has been achieved in these experiments (84). The first structure contains the Cre variant that prefers *loxM7* over *loxP* and the second contains a relaxed specificity mutant that recombines both substrate sites. The protein–DNA backbone contacts are similar in the two structures, but the contacts made to the DNA bases differ. The structures underscore the crucial role that water molecules play in mediating interactions between Cre and the *loxP* site and in remodeling the specificity of the interfaces.

In a more recent set of experiments, Buchholz and colleagues applied directed evolution strategies to generate a modified Cre (termed Tre) that recognizes a *loxP*-like sequence in the long-terminal repeat (LTR) of an HIV-1 strain (85). This loxLTR sequence contains four changes in the left half-site, seven in the right half-site, and an entirely different crossover sequence (Table 2). By dividing the loxLTR changes into four separate, but manageable target sequences, an active Cre could be evolved following pooling and shuffling of the intermediate coding sequences. The resulting Tre enzyme recombines loxLTR-containing substrates in mammalian cells and can excise the HIV-1 provirus from the chromosomes of human cells. In addition to the important potential application for this Cre variant, the study underscored the feasibility of generating custom recombinases for sequences with considerable differences from *loxP*, including those with different left versus right RBEs.

INCREASING SPECIFICITY BY ENGINEERING Cre–Cre INTERFACES

The success in generating modified recombinases that recombine altered *loxP* sites led to the question of whether different Cre variants could function together on asymmetric sites where the RBEs are different. This idea was tested with sites containing RBE pairs from *loxP* and M7 and mixtures of wild-type Cre and the C2 variant (Tables 1 and 2) (86). Whereas neither recombinase alone will efficiently recombine hybrid *loxP*-M7 sites, the recombinase mixture is effective. These results indicate that dual specificity recombinase pairs such as are found in the Xer system could be useful in Cre applications. The ability to independently manipulate the binding properties of the two recombinases should be an attractive alternative to evolving a single enzyme that simultaneously recognizes both half-sites, as was done in the Tre example discussed in the previous section.

A complication with the two-recombinase approach is introduced by the high level of cooperative binding in the Cre–*loxP* system. Cre binding to a "good" half-site can facilitate binding to a poor half-site through the intersubunit contacts responsible for cooperativity. Indeed, wild-type Cre will recombine the hybrid *loxP*-M7 sites when present at elevated concentrations (86). When two high-affinity Cre variants are present, four distinct types of off-target sequences could in principle be recognized.

To address this issue, Baldwin and colleagues have engineered the interface formed between adjacent Cre subunits (87). They focused on the contacts formed between the C-terminal helix-N (CTH) of Cre and the corresponding binding pocket in the CTD and identified an alternative set of interacting residues that function efficiently with one another, but not with a wild-type Cre partner. Using the hybrid *loxP*-M7 sites as a test system, Cre and the C2 variant were then modified so that a favorable CTH–CTD interaction would form only between Cre and C2 when bound to their corresponding RBEs. The result was an improvement in specificity towards the hybrid sites when both modified recombinases were present, illustrating that this approach may provide a useful contribution to the engineering of custom recombination systems.

Eroshenko and Church have recently identified Cre mutants that have improved specificity for *loxP* sites by modifying intersubunit interactions (88). The authors provided a strong argument that a reduction in binding cooperativity should result in improved accuracy for binding to *loxP* sites relative to off-target sites. The R32V and R32M substitutions were predicted to disrupt an intersubunit salt-bridge between helices (αA and αC) in the NTDs of Cre subunits bound to *loxP*, thereby reducing cooperativity of binding. Although both mutants recombine less efficiently than wild-type Cre on *loxP* sites, the difference in efficiency is even greater for off-target sites and for pseudo-*loxP* sites in the *E. coli* genome. Hence, an increase in recombination fidelity can be achieved at the expense of some recombination efficiency.

Arg32 is involved in a complex network of polar contacts in the interface between Cre subunits, involving Arg-Arg stacking, contributions from three acidic side chains and several tightly associated water molecules. The same network of interactions exists between Cre subunits that stabilize the synaptic complex, suggesting that the R32V and R32M mutants are likely to be deficient in this step of the pathway. The close structural relationship between binding cooperativity and synapsis in the Cre-*loxP* system raises the intriguing question of whether mutants might be identified that primarily affect the former, with minimal effects on the latter. Such mutants might be valuable tools, following on the Eroshenko and Church results.

TURNING Cre INTO AN INTEGRASE

Cre readily catalyzes intermolecular recombination between *loxP* sites, but the reverse intramolecular excision reaction is so efficient that it has been challenging to implement Cre as an effective integrase. This is not surprising, given that Cre normally functions as a resolvase in bacteriophage P1 biology. Two approaches have been successful in carrying out Cre-mediated integrations of DNA segments into chromosomes. The first is referred to as the left element (LE)/right element (RE) strategy, where the recombining sites each have one wild-type half-site and one mutant half-site (Fig. 6A) (89). The most widely used example is the *lox66*/*lox71* combination, where the sites contain 5-bp changes in the left and right RBEs of *loxP*, respectively (Table 2). The products of *lox66* × *lox71* recombination are *loxP* and *lox72* sites, where both half-sites of *lox72* contain the RBE mutations. The diminished ability of *lox72* to undergo recombination provides stability to the product of integration, and so some directionality to the reaction (89). Alternative LE/RE *loxP* sites have been reported more recently, with much higher integration efficiencies (90). The LE/RE approach has also been used to bias Cre-mediated inversion reactions *in vivo* (91, 92).

The second strategy is the recombinase-mediated cassette exchange (RMCE), which employs two incompatible *loxP* variants flanking DNA cassettes on the donor and acceptor molecules (93–96). The sites differ in their crossover sequences and therefore will not recombine with one another with high efficiency. An initial recombination event between compatible sites results in integration of the donor into the acceptor and subsequent recombination between the second set of compatible sites results in a net exchange of DNA cassettes between the two substrates (Fig. 6B). A recent review of the RMCE reaction discusses the issues involved in optimizing the process, including a comparison of Cre versus Flp recombinases as RMCE enzymes (97).

Considerable effort has been devoted to testing which alternative crossover sequence to use for RMCE, since all crossover variants underperform the native *loxP* sequence. The use of lox66/lox71 for one pair and lox2272 sites for the second pair has been reported to result in particularly stable products, even allowing the integration of a normally unstable Cre expression cassette in mouse embryonic stem cells (98). In this case, the sites in the product cassette have incompatible crossovers, plus one of the sites (lox72) is poorly functional. On the other hand, RMCE using *loxP* sites as one pair and loxFAS sites as the second pair has a very high efficiency of integration (9). The loxFAS site contains a relatively efficient *loxP* crossover alternative (Table 2).

A number of variations of Cre-mediated RMCE have also been described. Dual RMCE refers to the use of two different recombinases, each with its own pair of sites (99, 100). Cre, Flp, and the C31 serine integrase have been used for this purpose, eliminating the need to identify incompatible crossover sites for a single recombinase that will resist cassette excision over extended time periods. Schemes for executing successive rounds of RMCE-based integration have also been reported, where each cassette exchange results in inactivation of one product site, but introduction of a new site for use in subsequent reactions (101, 102).

RECONSTITUTION OF ACTIVE Cre FROM FRAGMENTS

The regulation of timing and location of Cre activity in transgenic animals has played a crucial role in the design and execution of experiments where Cre is used as a genetic switch. The use of tissue-specific and ligand-activated promoters to regulate Cre expression is widespread, with a variety of Cre-expressing mice now available (34, 35, 103). However, the basal rates and stochastic nature of expression from many promoters have been associated with cell toxicity and lack of optimal regulation (104–108). An additional level of regulation has been achieved in recent years by requiring that two Cre fragments assemble to form an active enzyme.

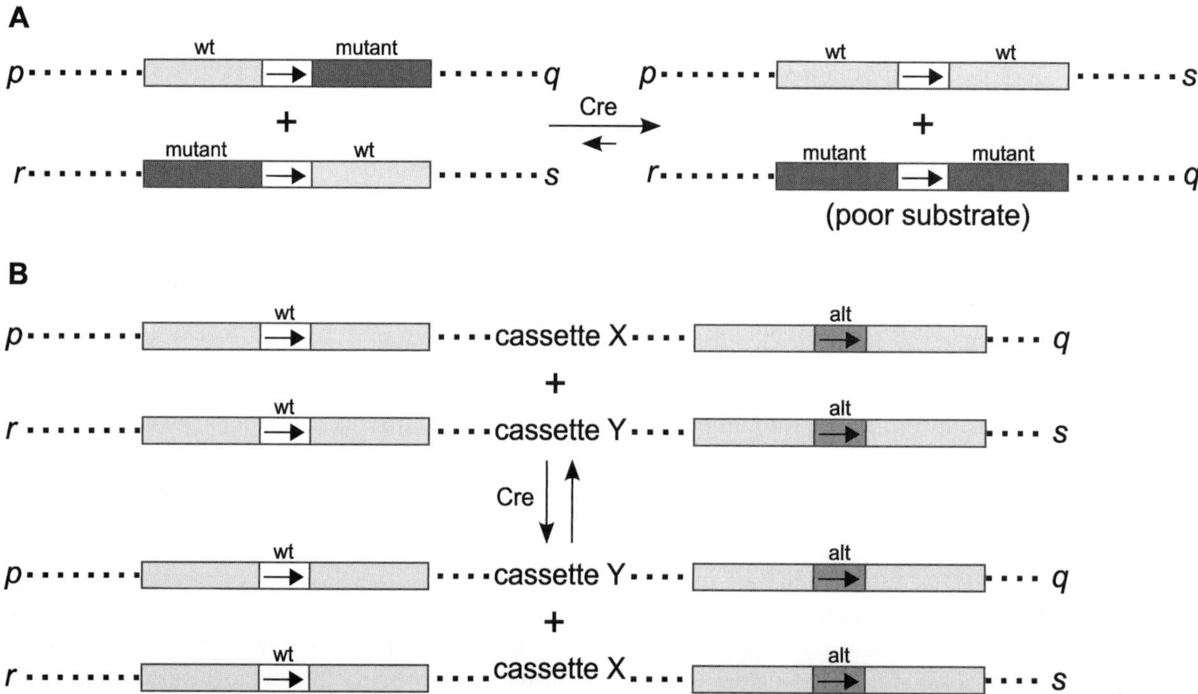

Figure 6 Schemes to achieve directionality in the Cre reaction. (A) The LE/RE strategy. Recombination between sites with mutated (weakened) right and left half-sites results in a wild-type site and a doubly mutated site, which is a poor substrate. The reverse reaction has reduced efficiency. (B) The recombinase-mediated cassette exchange reaction. The DNA segment to be integrated or exchanged is flanked by incompatible *loxP* sites in the donor (p-q) and acceptor (r-s) molecules. The sites are incompatible because their crossover sequences differ. Cre-mediated recombination between one site pair results in the integration of the donor (if circular) into the acceptor (the intermediate is not shown). Subsequent recombination between the second pair of sites excises the intervening DNA, resulting in exchange of cassettes between the two molecules. doi:10.1128/microbiolspec.MDNA3-0014-2014.f6

One approach to the "split Cre" method involves generating fusions of Cre fragments to domains that can be induced by ligands to associate and promote Cre activity (Fig. 7A). The first example of this type used FKBP fused to residues 1 to 59 of Cre and FRB fused to residues 60 to 341 (109). The small molecule rapamycin induces heterodimerization of FKBP and FRB (110), leading to reconstitution of an active Cre subunit. This method of regulation has been demonstrated to function *in vitro* (109) as well as in mice (111).

Residues 1 to 59 of Cre include the first two α-helices of the NTD. This region is important for both cooperative binding to *loxP* (Fig. 2B) and for forming a stable synaptic complex (Fig. 4). Cre A36V disrupts these interactions and as a consequence this mutant is poorly active (23, 29, 59). The Cre 60–341 fragment may resemble Cre A36V in some respects, but with an even stronger phenotype. It is therefore tempting to speculate that synapsis of *loxP* sites is what is primarily being regulated in this split Cre system.

A subsequent study implemented the same Cre fragments, but fused them to the self-associating coiled-coil region from yeast GCN4 and expressed the two fragments under different promoters (112). The fragments spontaneously assemble *in vivo* to form an active enzyme, but with a more complex level of regulation than is possible using Cre expressed from a single promoter.

In an alternative approach, overlapping fragments of Cre were identified that could assemble without the aid of dimerization domains and could be independently regulated with different promoters. The first such implementation used Cre fragments 1 to 196 and 181 to 341, where the overlap is located within the Cre CTD (Fig. 7B) (113). Subsequent studies modified the system to provide different modes of regulation. In the first case, one fragment was expressed in the target cells and the other was added to the cell culture as a purified protein (114). In the second study, the fragments were fused to heterodimeric leucine zipper domains, increasing the efficiency of the original system (115).

DESTABILIZED Cre

The most recent innovation in the quest to regulate Cre activity involves expression of destabilized forms of Cre (116). In this study, an *E. coli* dihydrofolate reductase mutant was fused to Cre (DD-Cre), causing it to be rapidly degraded by the proteosome (117). Addition of trimethoprim stabilizes the dihydrofolate reductase mutant, allowing Cre to escape degradation and carry out recombination of *loxP*-flanked alleles. As trimethoprim can cross the blood–brain barrier, this method of regulation allows both conditional mutagenesis in the central nervous system and a much reduced basal level of Cre activity.

Cre TRANSDUCTION

It was demonstrated in 2001 that recombination can be achieved in mammalian cells by transduction of a purified His_6-NLS-Cre-F fusion protein (118). In this construct, NLS refers to the simian virus 40 nuclear localization sequence and F refers to the 12-residue membrane translocation sequence from the Kaposi fibroblast growth factor FGF-4 (119). This finding implied that one could deliver a one-time pulse of Cre to target cells and avoid the undesirable side-effects resulting from transient transfections or low levels of constitutive expression.

In studies appearing several months later, it was further reported that Cre can be effectively transduced with a variety of membrane translocation sequence-like tags, including NLS-Cre (120–122). Indeed, even unmodified Cre can be used to transduce mammalian cells with moderate efficiency when present at slightly elevated concentrations (120). Optimization of various combinations of tags resulted in identification of His_6-TAT-NLS-Cre as a particularly effective construct, resulting in rapid recombination at micromolar Cre concentrations (121). In this case, TAT is an 11-residue peptide from HIV-TAT that is known to promote the transduction of heterologous proteins across mammalian cell membranes (123). This idea was later applied to human embryonic stem cells and human embryonic stem progeny cells. Addition of His_6-TAT-NLS-Cre fusion to the culture medium at micromolar concentrations resulted in recombination in ~100% of cells containing a *loxP*-flanked reporter, underscoring the usefulness of this approach (124).

The demonstration that purified Cre can be delivered to growing cells in a manner that results in nearly quantitative recombination of genomic *loxP* sites raised a new question of whether chemically modified versions of Cre with even more complex modes of regulation could be used. Deiters and coworkers have answered this question with the development of a light-activated version of Cre (125). Using unnatural amino acid incorporation methodologies in *E. coli*, Tyr324 was replaced by *o*-nitrobenzyl tyrosine (ONBY), effectively blocking the active site nucleophile and rendering Cre inactive. Upon irradiation with 365 nm UVA light, the nitrobenzyl group was efficiently removed, resulting in active Cre and recombination of *loxP* sites.

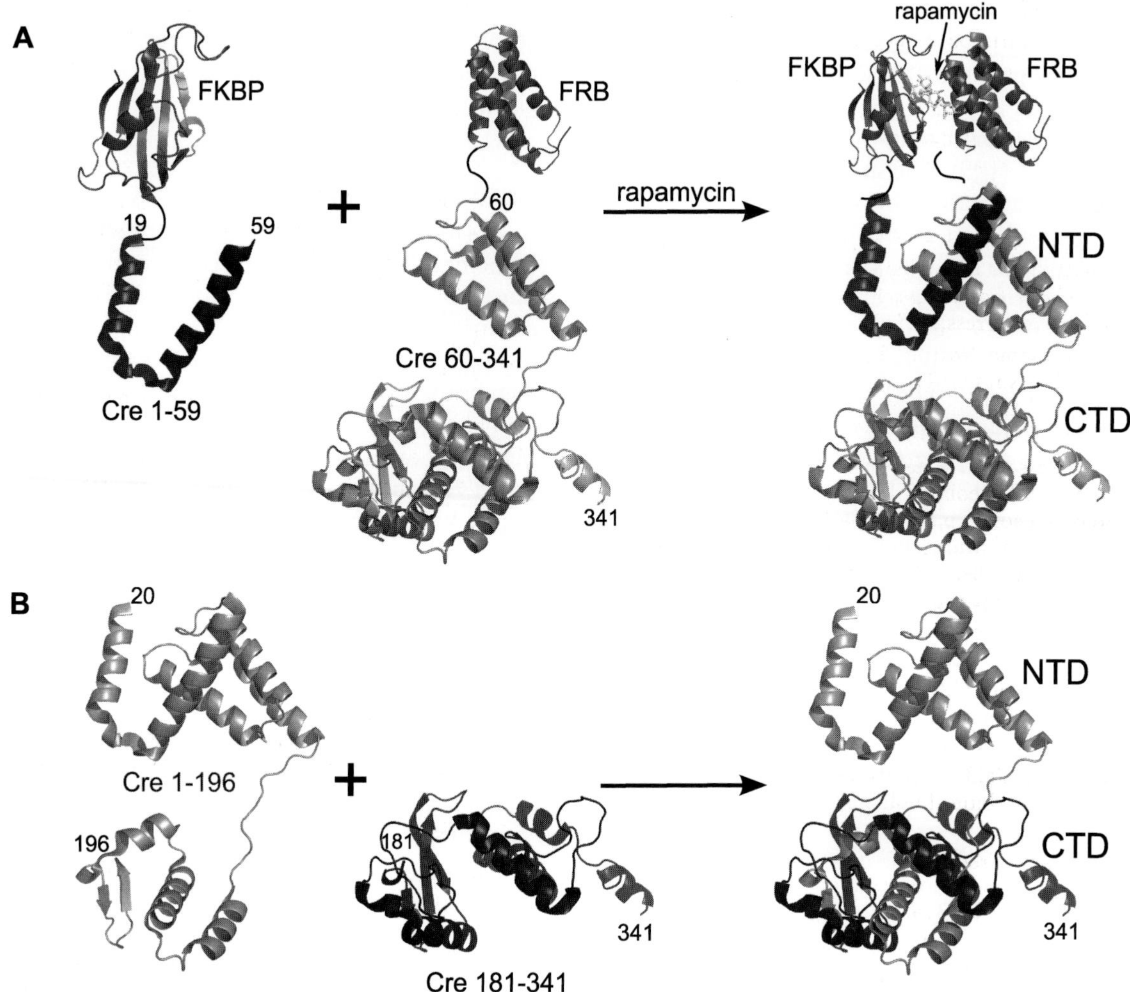

Figure 7 Reconstitution of active Cre from fragments. (A) Cre fragments 1 to 59 and 60 to 341 are inactive individually, but can be re-associated to form an active enzyme if fused to heterodimerizing partners. In the example shown, rapamycin induces dimerization of the FKBP and FRB domains. (B) Overlapping Cre fragments 1 to 196 and 181 to 341 are also inactive individually, but will associate spontaneously to form an active enzyme. doi:10.1128/microbiolspec.MDNA3-0014-2014.f7

When ONBY-Cre was transduced into mammalian cells, Cre could not only be activated by UVA light irradiation, but spatial control of Cre activation could be demonstrated by limiting the activating light to a specific region of a monolayer of cells.

FUTURE PROSPECTS

It would be reasonable to conclude that we now have a good mechanistic understanding of the Cre-*loxP* recombination reaction and that the application of Cre-*loxP* technologies is well-established in many areas of the biological sciences. However, it is also clear from investigations over the past decade that we do not yet know the answers to all the questions posed at the start of this chapter. Understanding and changing the sequence specificity of Cre remains extremely challenging and a complete structural and biochemical understanding of all the steps in the recombination pathway is still lacking. The available experimental data indicate that Cre and the *loxP* sequence have been fine-tuned by evolution, where even single changes to nucleotides in the crossover region can profoundly affect recombination efficiency. In many cases, we still do not understand the origins of these effects. At the time that *Mobile DNA II* was being written, it would have been difficult to predict all of the advances in mechanistic understanding and improvements in regulating Cre activity that would be made in the coming years. We should therefore anticipate at least a few more surprises in the years to come.

Citation. **Van Duyne GD.** 2014. Cre recombinase. Microbiol Spectrum 3(1):MDNA3-0014-2014.

References

1. **Birling M-C, Gofflot F, Warot X.** 2009. Site-Specific Recombinases for Manipulation of the Mouse Genome, p 245–263. *In* Cartwright EJ (ed), *Transgenesis Techniques*. Humana Press, New York.
2. **Wirth D, Gama-Norton L, Riemer P, Sandhu U, Schucht R, Hauser H.** 2007. Road to precision: recombinase-based targeting technologies for genome engineering. *Curr Opin Biotechnol* **18:**411–419.
3. **Brault V, Besson V, Magnol L, Duchon A, Hérault Y.** 2007. Cre/*loxP*-Mediated Chromosome Engineering of the Mouse Genome, p 29–48. *In* Feil PDR, Metzger DD (ed), *Conditional Mutagenesis: An Approach to Disease Models*. Springer, Berlin Heidelberg.
4. **Kos CH.** 2004. Methods in Nutrition Science: Cre/*loxP* System for Generating Tissue-specific Knockout Mouse Models. *Nutr Rev* **62:**243–246.
5. **Branda CS, Dymecki SM.** 2004. Talking about a Revolution: The Impact of Site-Specific Recombinases on Genetic Analyses in Mice. *Dev Cell* **6:**7–28.
6. **Gilbertson L.** 2003. Cre–lox recombination: Cre-ative tools for plant biotechnology. *Trends Biotechnol* **21:** 550–555.
7. **Sauer B.** 2002. Cre/lox: one more step in the taming of the genome. *Endocrine* **19:**221–228.
8. **Wang Y, Yau Y-Y, Perkins-Balding D, Thomson JG.** 2011. Recombinase technology: applications and possibilities. *Plant Cell Rep* **30:**267–285.
9. **Lanza AM, Dyess TJ, Alper HS.** 2012. Using the Cre/lox system for targeted integration into the human genome: loxFAS-*loxP* pairing and delayed introduction of Cre DNA improve gene swapping efficiency. *Biotechnol J* **7:**898–908.
10. **Sauer B, Henderson N.** 1988. Site-specific DNA recombination in mammalian cells by the Cre recombinase of bacteriophage P1. *Proc Natl Acad Sci USA* **85:**5166–70.
11. **Sauer B.** 1987. Functional expression of the cre-lox site-specific recombination system in the yeast *Saccharomyces cerevisiae*. *Mol Cell Biol* **7:**2087–2096.
12. **Van Duyne G.** 2001. A structural view of Cre-*loxP* site-specific recombination. *Annu Rev Biophys Biomol Struct* **30:**87–104.
13. **Van Duyne G.** 2002. A structural view of tyrosine recombinase site-specific recombination, p 93–117. *In Mobile DNA II*. ASM Press, Washington DC.
14. **Grindley N, Whiteson K, Rice P.** 2006. Mechanisms of site-specific recombination. *Annu Rev Biochem* **75:** 567–605.
15. **Abremski K, Hoess R.** 1984. Bacteriophage P1 site-specific recombination. Purification and properties of the Cre recombinase protein. *J Biol Chem* **259:**1509–1514.
16. **Kilbride EA, Burke ME, Boocock MR, Stark WM.** 2006. Determinants of product topology in a hybrid Cre-Tn3 resolvase site-specific recombination system. *J Mol Biol* **355:**185–195.
17. **Warren D, Laxmikanthan G, Landy A.** 2008. A chimeric Cre recombinase with regulated directionality. *Proc Natl Acad Sci USA* **105:**18278–18283.
18. **Gourlay SC, Colloms SD.** 2004. Control of Cre recombination by regulatory elements from Xer recombination systems. *Mol Microbiol* **52:**53–65.
19. **Ikeda H, Tomizawa J.** 1968. Prophage P1, an extrachromosomal element. *Cold Spring Harb Symp Quant Biol* **33:**791–798.
20. **Austin S, Ziese M, Sternberg N.** 1981. A novel role for site-specific recombination in maintenance of bacterial replicons. *Cell* **25:**729–36.
21. **Sternberg N, Hamilton D, Austin S, Yarmolinsky M, Hoess R.** 1981. Site-specific recombination and its role in the life cycle of bacteriophage P1. *Cold Spring Harb Symp Quant Biol* **1:**297–309.
22. **Sherratt D, Soballe B, Barre F, Filipe S, Lau I, Massey T, Yates J.** 2004. Recombination and chromosome segregation. *Philos Trans R Soc Lond B Biol Sci* **359:**61–69.
23. **Ghosh K, Guo F, Van Duyne GD.** 2007. Synapsis of *loxP* sites by Cre recombinase. *J Biol Chem* **282:** 24004–24016.
24. **Hoess R, Abremski K.** 1984. Interaction of the bacteriophage P1 recombinase Cre with the recombining site *loxP*. *Proc Natl Acad Sci USA* **81:**1026–1029.
25. **Mack A, Sauer B, Abremski K, Hoess R.** 1992. Stoichiometry of the Cre recombinase bound to the lox recombining site. *Nucleic Acids Res* **20:**4451–4455.
26. **Ringrose L, Lounnas V, Ehrlich L, Buchholz F, Wade R, Stewart A.** 1998. Comparative kinetic analysis of FLP and cre recombinases: mathematical models for DNA binding and recombination. *J Mol Biol* **284:**363–384.
27. **Rüfer A, Neuenschwander P, Sauer B.** 2002. Analysis of Cre-*loxP* interaction by surface plasmon resonance: influence of spermidine on cooperativity. *Anal Biochem* **308:**90–99.
28. **Hamilton D, Abremski K.** 1984. Site-specific recombination by the bacteriophage P1 lox-Cre system. Cre-mediated synapsis of two lox sites. *J Mol Biol* **178:** 481–486.
29. **Hoess R, Wierzbicki A, Abremski K.** 1990. Synapsis in the Cre-lox site-specific recombination system, p 203–213. *In Structure & Methods*. Adenine Press, New York.
30. **Craig NL.** 2002. *Mobile DNA II*. ASM Press, Washington D.C.
31. **Cantor E, Chong S.** 2001. Intein-mediated rapid purification of Cre recombinase. *Protein Expr Purif* **22:** 135–140.
32. **Ghosh K, Van Duyne G.** 2002. Cre-*loxP* biochemistry. *Methods* **28:**374–383.
33. **Gibb B, Gupta K, Ghosh K, Sharp R, Chen J, Van Duyne GD.** 2010. Requirements for catalysis in the Cre recombinase active site. *Nucleic Acids Res* **38:** 5817–5832.

34. **Murray SA, Eppig JT, Smedley D, Simpson EM, Rosenthal N.** 2012. Beyond knockouts: cre resources for conditional mutagenesis. *Mamm Genome* **23:**587–599.

35. **Smedley D, Salimova E, Rosenthal N.** 2011. Cre recombinase resources for conditional mouse mutagenesis. *Methods* **53:**411–416.

36. **Jones J, Shelton K, Magnuson M.** 2004. Strategies for the use of site-specific recombinases in genome engineering. *Methods Mol Med* **103:**245–258.

37. **Thomson JG, Ow DW.** 2006. Site-specific recombination systems for the genetic manipulation of eukaryotic genomes. *genesis* **44:**465–476.

38. **Sauer B.** 2002. Chromosome manipulation by Cre-lox recombination, p 38–58. *In Mobile DNA II.* ASM Press, Washington D.C.

39. **Guo F, Gopaul D, Van Duyne G.** 1997. Structure of Cre recombinase complexed with DNA in a site-specific recombination synapse. *Nature* **389:**40–46.

40. **Guo F, Gopaul D, Van Duyne G.** 1999. Asymmetric DNA bending in the Cre-*loxP* site-specific recombination synapse. *Proc Natl Acad Sci USA* **96:**7143–7148.

41. **Woods K, Martin S, Chu V, Baldwin E.** 2001. Quasi-equivalence in site-specific recombinase structure and function: crystal structure and activity of trimeric Cre recombinase bound to a three-way Lox DNA junction. *J Mol Biol* **313:**49–69.

42. **Martin S, Pulido E, Chu V, Lechner T, Baldwin E.** 2002. The order of strand exchanges in Cre-*loxP* recombination and its basis suggested by the crystal structure of a Cre-*loxP* Holliday junction complex. *J Mol Biol* **319:**107–127.

43. **Ghosh K, Lau C, Guo F, Segall A, Van Duyne G.** 2005. Peptide trapping of the Holliday junction intermediate in Cre-*loxP* site-specific recombination. *J Biol Chem* **280:**8290–8299.

44. **Kim S, Kim G, Lee Y, Park J.** 2001. Characterization of Cre–*loxP* interaction in the major groove: Hint for structural distortion of mutant Cre and possible strategy for HIV-1 therapy. *J Cell Biochem* **80:**321–327.

45. **Hartung M, Kisters-Woike B.** 1998. Cre mutants with altered DNA binding properties. *J Biol Chem* **273:** 22884–22891.

46. **Wierzbicki A, Kendall M, Abremski K, Hoess R.** 1987. A mutational analysis of the bacteriophage P1 recombinase Cre. *J Mol Biol* **195:**785–794.

47. **Rüfer A, Sauer B.** 2002. Non-contact positions impose site selectivity on Cre recombinase. *Nucleic Acids Res* **30:**2764–2771.

48. **Martin S, Chu V, Baldwin E.** 2003. Modulation of the active complex assembly and turnover rate by protein-DNA interactions in Cre-*loxP* recombination. *Biochemistry (Mosc)* **42:**6814–6826.

49. **Gelato K, Martin S, Wong S, Baldwin E.** 2006. Multiple levels of affinity-dependent DNA discrimination in Cre-*loxP* recombination. *Biochemistry (Mosc.)* **45:** 12216–12226.

50. **Lee L, Chu L, Sadowski P.** 2003. Cre induces an asymmetric DNA bend in its target *loxP* site. *J Biol Chem* **278:**23118–23129.

51. **Grainge I, Pathania S, Vologodskii A, Harshey RM, Jayaram M.** 2002. Symmetric DNA Sites are Functionally Asymmetric Within Flp and Cre Site-specific DNA Recombination Synapses. *J Mol Biol* **320:**515–527.

52. **Hoess R, Wierzbicki A, Abremski K.** 1987. Isolation and characterization of intermediates in site-specific recombination. *Proc Natl Acad Sci USA* **84:**6840–6844.

53. **Ennifar E, Meyer J, Buchholz F, Stewart A, Suck D.** 2003. Crystal structure of a wild-type Cre recombinase-*loxP* synapse reveals a novel spacer conformation suggesting an alternative mechanism for DNA cleavage activation. *Nucleic Acids Res* **31:**5449–5460.

54. **Gelato K, Martin S, Baldwin E.** 2005. Reversed DNA strand cleavage specificity in initiation of Cre-*loxP* recombination induced by the His289Ala active-site substitution. *J Mol Biol* **354:**233–245.

55. **Lee L, Sadowski P.** 2003. Sequence of the *loxP* site determines the order of strand exchange by the Cre recombinase. *J Mol Biol* **326:**397–412.

56. **Lee L, Sadowski P.** 2005. Strand selection by the tyrosine recombinases. *Prog Nucleic Acid Res Mol Biol* **80:** 1–42.

57. **Burgin A, Nash H.** 1995. Suicide substrates reveal properties of the homology-dependent steps during integrative recombination of bacteriophage lambda. *Curr Biol* **5:**1312–1321.

58. **Ghosh K, Lau C, Gupta K, Van Duyne G.** 2005. Preferential synapsis of *loxP* sites drives ordered strand exchange in Cre-*loxP* site-specific recombination. *Nat Chem Biol* **1:**275–282.

59. **Pinkney JNM, Zawadzki P, Mazuryk J, Arciszewska LK, Sherratt DJ, Kapanidis AN.** 2012. Capturing reaction paths and intermediates in Cre-*loxP* recombination using single-molecule fluorescence. *Proc Natl Acad Sci USA* **109:**20871–20876.

60. **Krogh B, Shuman S.** 2000. Catalytic mechanism of DNA topoisomerase IB. *Mol Cell* **5:**1035–1041.

61. **Abremski K, Hoess R.** 1992. Evidence for a second conserved arginine residue in the integrase family of recombination proteins. *Protein Eng* **5:**87–91.

62. **Lee L, Sadowski P.** 2003. Identification of Cre residues involved in synapsis, isomerization, and catalysis. *J Biol Chem* **278:**36905–36915.

63. **Shoura MJ, Vetcher AA, Giovan SM, Bardai F, Bharadwaj A, Kesinger MR, Levene SD.** 2012. Measurements of DNA-loop formation via Cre-mediated recombination. *Nucleic Acids Res* **40:**7452–7464.

64. **Fan H-F.** 2012. Real-time single-molecule tethered particle motion experiments reveal the kinetics and mechanisms of Cre-mediated site-specific recombination. *Nucleic Acids Res* **40:**6208–6222.

65. **Crisona NJ, Weinberg RL, Peter BJ, Sumners DW, Cozzarelli NR.** 1999. The Topological Mechanism of Phage λ Integrase. *J Mol Biol* **289:**747–775.

66. **Vetcher AA, Lushnikov AY, Navarra-Madsen J, Scharein RG, Lyubchenko YL, Darcy IK, Levene SD.** 2006. DNA topology and geometry in Flp and Cre recombination. *J Mol Biol* **357:**1089–1104.

67. **Hoess R, Wierzbicki A, Abremski K.** 1986. The role of the *loxP* spacer region in P1 site-specific recombination. *Nucleic Acids Res* **14:**2287–2300.

68. **Lee G, Saito I.** 1998. Role of nucleotide sequences of *loxP* spacer region in Cre-mediated recombination. *Gene* **216:**55–65.

69. **Sheren J, Langer S, Leinwand L.** 2007. A randomized library approach to identifying functional lox site domains for the Cre recombinase. *Nucleic Acids Res* **35:** 5464–5473.

70. **Gopaul D, Guo F, Van Duyne G.** 1998. Structure of the Holliday junction intermediate in Cre-*loxP* site-specific recombination. *EMBO J* **17:**4175–4187.

71. **Nunes-Duby S, Azaro M, Landy A.** 1995. Swapping DNA strands and sensing homology without branch migration in lambda site-specific recombination. *Curr Biol* **5:**139–148.

72. **Rajeev L, Malanowska K, Gardner JF.** 2009. Challenging a Paradigm: the Role of DNA Homology in Tyrosine Recombinase Reactions. *Microbiol Mol Biol Rev* **73:**300–309.

73. **Sauer B.** 1996. Multiplex Cre/lox recombination permits selective site-specific DNA targeting to both a natural and an engineered site in the yeast genome. *Nucleic Acids Res* **24:**4608–4613.

74. **Aranda M, Kanellopoulou C, Christ N, Peitz M, Rajewsky K, Dröge P.** 2001. Altered directionality in the cre-*loxP* site-specific recombination pathway. *J Mol Biol* **311:**453–459.

75. **Jung U-J, Park S, Lee G, Shin H-J, Kwon M-H.** 2007. Analysis of spacer regions derived from intramolecular recombination between heterologous *loxP* sites. *Biochem Biophys Res Commun* **363:**183–189.

76. **Martin S, Wachi S, Baldwin E.** 2003. Vanadate-based transition-state analog inhibitors of Cre-*loxP* recombination. *Biochem Biophys Res Commun* **308:**529–534.

77. **Krogh B, Shuman S.** 2002. Proton relay mechanism of general acid catalysis by DNA topoisomerase IB. *J Biol Chem* **277:**5711–5714.

78. **Chen Y, Rice PA.** 2003. The Role of the Conserved Trp330 in Flp-mediated Recombination functional and structural analysis. *J Biol Chem* **278:**24800–24807.

79. **Ma C-H, Kwiatek A, Bolusani S, Voziyanov Y, Jayaram M.** 2007. Unveiling hidden catalytic contributions of the conserved His/Trp-III in tyrosine recombinases: assembly of a novel active site in Flp recombinase harboring alanine at this position. *J Mol Biol* **368:**183–196.

80. **Ma C-H, Kachroo AH, Macieszak A, Chen T-Y, Guga P, Jayaram M.** 2009. Reactions of Cre with methylphosphonate DNA: similarities and contrasts with Flp and vaccinia topoisomerase. *PLoS ONE* **4:**e7248.

81. **Kachroo AH, Ma C-H, Rowley PA, Maciaszek AD, Guga P, Jayaram M.** 2010. Restoration of catalytic functions in Cre recombinase mutants by electrostatic compensation between active site and DNA substrate. *Nucleic Acids Res* **38:**6589–6601.

82. **Buchholz F, Stewart A.** 2001. Alteration of Cre recombinase site specificity by substrate-linked protein evolution. *Nat Biotechnol* **19:**1047–1052.

83. **Santoro S, Schultz P.** 2002. Directed evolution of the site specificity of Cre recombinase. *Proc Natl Acad Sci USA* **99:**4185–4190.

84. **Baldwin E, Martin S, Abel J, Gelato K, Kim H, Schultz P, Santoro S.** 2003. A specificity switch in selected cre recombinase variants is mediated by macromolecular plasticity and water. *Chem Biol* **10:**1085–1094.

85. **Sarkar I, Hauber I, Hauber J, Buchholz F.** 2007. HIV-1 proviral DNA excision using an evolved recombinase. *Science* **316:**1912–1915.

86. **Saraf-Levy T, Santoro SW, Volpin H, Kushnirsky T, Eyal Y, Schultz PG, Gidoni D, Carmi N.** 2006. Site-specific recombination of asymmetric lox sites mediated by a heterotetrameric Cre recombinase complex. *Bioorg Med Chem* **14:**3081–3089.

87. **Gelato K, Martin S, Liu P, Saunders A, Baldwin E.** 2008. Spatially directed assembly of a heterotetrameric Cre-Lox synapse restricts recombination specificity. *J Mol Biol* **378:**653–665.

88. **Eroshenko N, Church GM.** 2013. Mutants of Cre recombinase with improved accuracy. *Nat Commun* **4:**2509.

89. **Albert H, Dale EC, Lee E, Ow DW.** 1995. Site-specific integration of DNA into wild-type and mutant lox sites placed in the plant genome. *Plant J* **7:**649–659.

90. **Thomson JG, Rucker EB, Piedrahita JA.** 2003. Mutational analysis of *loxP* sites for efficient Cre-mediated insertion into genomic DNA. *genesis* **36:**162–167.

91. **Oberdoerffer P, Otipoby KL, Maruyama M, Rajewsky K.** 2003. Unidirectional Cre-mediated genetic inversion in mice using the mutant *loxP* pair lox66/lox71. *Nucleic Acids Res* **31:**e140–e140.

92. **Zhang Z, Lutz B.** 2002. Cre recombinase-mediated inversion using lox66 and lox71: method to introduce conditional point mutations into the CREB-binding protein. *Nucleic Acids Res* **30:**e90–e90.

93. **Bouhassira EE, Westerman K, Leboulch P.** 1997. Transcriptional Behavior of LCR Enhancer Elements Integrated at the Same Chromosomal Locus by Recombinase-Mediated Cassette Exchange. *Blood* **90:**3332–3344.

94. **Soukharev S, Miller J, Sauer B.** 1999. Segmental genomic replacement in embryonic stem cells by double lox targeting. *Nucleic Acids Res* **27:**e21.

95. **Feng Y-Q, Seibler J, Alami R, Eisen A, Westerman KA, Leboulch P, Fiering S, Bouhassira EE.** 1999. Site-specific chromosomal integration in mammalian cells: highly efficient CRE recombinase-mediated cassette exchange. *J Mol Biol* **292:**779–785.

96. **Schlake T, Bode J.** 1994. Use of mutated FLP recognition target (FRT) sites for the exchange of expression cassettes at defined chromosomal loci. *Biochemistry (Mosc)* **33:**12746–12751.

97. **Turan S, Galla M, Ernst E, Qiao J, Voelkel C, Schiedlmeier B, Zehe C, Bode J.** 2011. Recombinase-Mediated Cassette Exchange (RMCE): Traditional Concepts and Current Challenges. *J Mol Biol* **407:** 193–221.

98. **Araki K, Araki M, Yamamura K.** 2002. Site-directed integration of the cre gene mediated by Cre recombinase

using a combination of mutant lox sites. *Nucleic Acids Res* **30:**e103–e103.

99. **Lauth M, Spreafico F, Dethleffsen K, Meyer M.** 2002. Stable and efficient cassette exchange under non-selectable conditions by combined use of two site-specific recombinases. *Nucleic Acids Res* **30:**e115–e115.

100. **Osterwalder M, Galli A, Rosen B, Skarnes WC, Zeller R, Lopez-Rios J.** 2010. Dual RMCE for efficient re-engineering of mouse mutant alleles. *Nat Methods* **7:** 893–895.

101. **Kameyama Y, Kawabe Y, Ito A, Kamihira M.** 2010. An accumulative site-specific gene integration system using cre recombinase-mediated cassette exchange. *Biotechnol Bioeng* **105:**1106–1114.

102. **Obayashi H, Kawabe Y, Makitsubo H, Watanabe R, Kameyama Y, Huang S, Takenouchi Y, Ito A, Kamihira M.** 2012. Accumulative gene integration into a pre-determined site using Cre/*loxP*. *J Biosci Bioeng* **113:** 381–388.

103. **Heffner CS, Herbert Pratt C, Babiuk RP, Sharma Y, Rockwood SF, Donahue LR, Eppig JT, Murray SA.** 2012. Supporting conditional mouse mutagenesis with a comprehensive cre characterization resource. *Nat Commun* **3:**1218.

104. **Naiche LA, Papaioannou VE.** 2007. Cre activity causes widespread apoptosis and lethal anemia during embryonic development. *genesis* **45:**768–775.

105. **Jeannotte L, Aubin J, Bourque S, Lemieux M, Montaron S, Provencher St-Pierre A.** 2011. Unsuspected effects of a lung-specific cre deleter mouse line. *genesis* **49:**152–159.

106. **Lewis AE, Vasudevan HN, O'Neill AK, Soriano P, Bush JO.** 2013. The widely used Wnt1-Cre transgene causes developmental phenotypes by ectopic activation of Wnt signaling. *Dev Biol* **379:**229–234.

107. **Harno E, Cottrell EC, White A.** 2013. Metabolic Pitfalls of CNS Cre-Based Technology. *Cell Metab* **18:** 21–28.

108. **Janbandhu VC, Moik D, Fässler R.** 2013. Cre recombinase induces DNA damage and tetraploidy in the absence of *loxP* sites. *Cell Cycle* **13:**462–470.

109. **Jullien N, Sampieri F, Enjalbert A, Herman J-P.** 2003. Regulation of Cre recombinase by ligand-induced complementation of inactive fragments. *Nucleic Acids Res* **31:**e131–e131.

110. **Chen J, Zheng XF, Brown EJ, Schreiber SL.** 1995. Identification of an 11-kDa FKBP12-rapamycin-binding domain within the 289-kDa FKBP12-rapamycin-associated protein and characterization of a critical serine residue. *Proc Natl Acad Sci USA* **92:**4947–4951.

111. **Jullien N, Goddard I, Selmi-Ruby S, Fina J-L, Cremer H, Herman J-P.** 2007. Conditional transgenesis using Dimerizable Cre (DiCre). *PloS One* **2:**e1355.

112. **Hirrlinger J, Scheller A, Hirrlinger PG, Kellert B, Tang W, Wehr MC, Goebbels S, Reichenbach A, Sprengel R, Rossner MJ, Kirchhoff F.** 2009. Split-Cre Complementation Indicates Coincident Activity of Different Genes *in vivo*. *PLoS ONE* **4:**e4286.

113. **Casanova E, Lemberger T, Fehsenfeld S, Mantamadiotis T, Schütz G.** 2003. α Complementation in the Cre recombinase enzyme. *genesis* **37:**25–29.

114. **Seidi A, Mie M, Kobatake E.** 2007. Novel recombination system using Cre recombinase alpha complementation. *Biotechnol Lett* **29:**1315–1322.

115. **Seidi A, Mie M, Kobatake E.** 2009. Recombination system based on cre alpha complementation and leucine zipper fusions. *Appl Biochem Biotechnol* **158:**334–342.

116. **Sando R III, Baumgaertel K, Pieraut S, Torabi-Rander N, Wandless TJ, Mayford M, Maximov A.** 2013. Inducible control of gene expression with destabilized Cre. *Nat Methods* **10:**1085–1088.

117. **Iwamoto M, Björklund T, Lundberg C, Kirik D, Wandless TJ.** 2010. A General Chemical Method to Regulate Protein Stability in the Mammalian Central Nervous System. *Chem Biol* **17:**981–988.

118. **Jo D, Nashabi A, Doxsee C, Lin Q, Unutmaz D, Chen J, Ruley HE.** 2001. Epigenetic regulation of gene structure and function with a cell-permeable Cre recombinase. *Nat Biotechnol* **19:**929–933.

119. **Lin Y-Z, Yao S, Veach RA, Torgerson TR, Hawiger J.** 1995. Inhibition of Nuclear Translocation of Transcription Factor NF-κB by a Synthetic Peptide Containing a Cell Membrane-permeable Motif and Nuclear Localization Sequence. *J Biol Chem* **270:**14255–14258.

120. **Will E, Klump H, Heffner N, Schwieger M, Schiedlmeier B, Ostertag W, Baum C, Stocking C.** 2002. Unmodified Cre recombinase crosses the membrane. *Nucleic Acids Res* **30:**e59–e59.

121. **Peitz M, Pfannkuche K, Rajewsky K, Edenhofer F.** 2002. Ability of the hydrophobic FGF and basic TAT peptides to promote cellular uptake of recombinant Cre recombinase: A tool for efficient genetic engineering of mammalian genomes. *Proc Natl Acad Sci USA* **99:** 4489–4494.

122. **Joshi SK, Hashimoto K, Koni PA.** 2002. Induced DNA recombination by Cre recombinase protein transduction. *genesis* **33:**48–54.

123. **Nagahara H, Vocero-Akbani AM, Snyder EL, Ho A, Latham DG, Lissy NA, Becker-Hapak M, Ezhevsky SA, Dowdy SF.** 1998. Transduction of full-length TAT fusion proteins into mammalian cells: TAT-p27Kip1 induces cell migration. *Nat Med* **4:**1449–1452.

124. **Nolden L, Edenhofer F, Haupt S, Koch P, Wunderlich FT, Siemen H, Brüstle O.** 2006. Site-specific recombination in human embryonic stem cells induced by cell-permeant Cre recombinase. *Nat Methods* **3:**461–467.

125. **Edwards WF, Young DD, Deiters A.** 2009. Light-Activated Cre Recombinase as a Tool for the Spatial and Temporal Control of Gene Function in Mammalian Cells. *ACS Chem Biol* **4:**441–445.

Mobile DNA, 3rd Edition
Nancy L. Craig, Michael Chandler, Martin Gellert, Alan M. Lambowitz, Phoebe A. Rice and Suzanne Sandmeyer

doi:10.1128/microbiolspec.MDNA3-0019-2014

José Antonio Escudero,[1,2,*] Céline Loot,[1,2,*]
Aleksandra Nivina,[1,2,3] and Didier Mazel[1,2]

6 The Integron: Adaptation On Demand

INTRODUCTION

Integrons are genetic platforms that allow bacteria to evolve rapidly through the acquisition, stockpiling, excision, and reordering of open reading frames found in mobile elements named cassettes. The evolutionary potency that integrons provide for bacteria is based on the variety of functions encoded in the cassettes, as well as on the intricate coupling of integron activity to bacterial stress (1).

The structure of any integron includes a stable platform and a variable cassette array. The platform contains: (i) the gene encoding the integrase (IntI), a type of tyrosine recombinase (Y-recombinase) that has evolved structural features to specifically perform the integration and excision of cassettes; (ii) a recombination site for integration of cassettes, the *attI* site, that is found upstream of the *intI* gene (except in integrons from the *Treponema* genus, where it is located downstream); and (iii) a promoter, P_C, within the *intI* gene or between *intI* and the *attI* site, that is oriented towards the integration point and drives the expression of cassettes (2–4) (Fig. 1A). Cassettes are circular nonreplicative elements (5) that generally contain a promoterless gene and a second recombination site: the *attC* site. These genes are rendered functional through their integration into the platform and expression from the P_C promoter. Successive integration events result in the assembly of an array of cassettes, of which only the subset that is closest to the integration site is expressed (Fig. 1B). Cassettes can be randomly excised and further integrated into the first position of the array where their expression is maximal. Hence, the cassette content of an integron is variable, reflecting a history of adaptive events, and represents a low-cost memory of valuable functions for the cell (Fig. 1B).

Origin of Integrons

Integrons are ancient structures that have shaped the evolution of bacteria for hundreds of millions of years (6). They are present in the chromosomes of approximately 17% of the bacterial species for which the genome sequence is available (7), spanning several bacterial phyla. These integrons are commonly referred to as chromosomal integrons (CIs), to distinguish them from plasmid-borne integrons (see below). Phylogenetic studies show that the branching of integrases is coherent with the organismal phylogeny, proving the ancestry and stability of integrons (6, 8, 9). Three large

[1]Institut Pasteur, Unité Plasticité du Génome Bactérien, 75724 Paris, France; [2]CNRS UMR3525, 75724 Paris, France; [3]Paris Descartes, Sorbonne Paris Cité, Paris, France. [*]These authors contributed equally to this work.

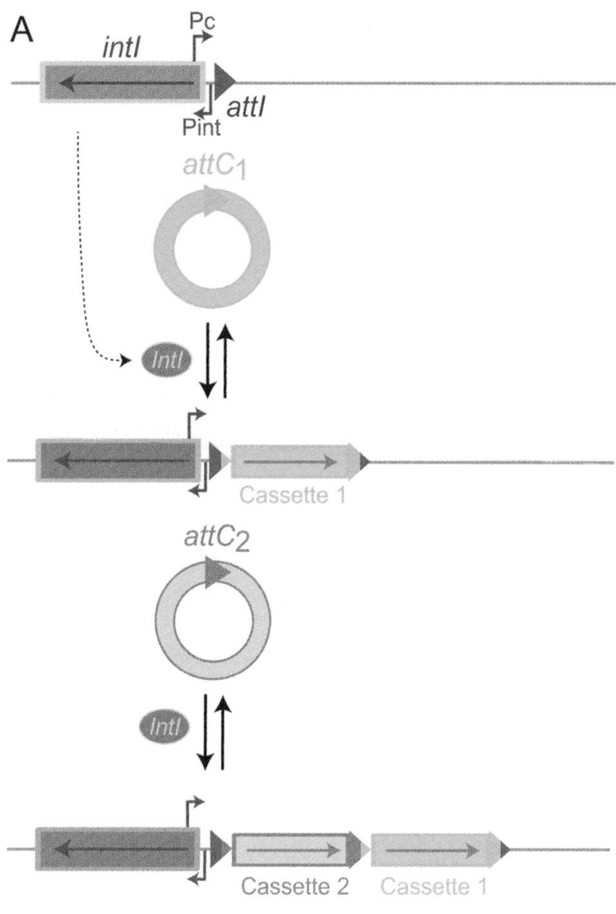

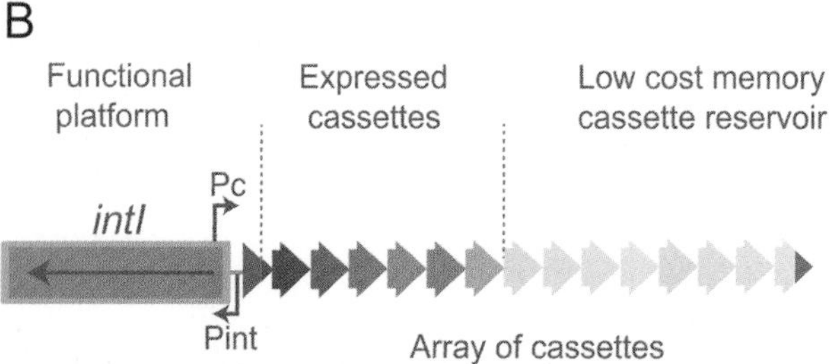

Figure 1 Organization of integrons. (A) Insertion and excision of cassettes: the functional platform, composed of the integrase-encoding *intI1* gene, the cassette (P_C) and integrase promoters (Pint), and the primary *attI* recombination site (red triangle), is shown. Cassette insertion (*attC* × *attI*) and excision (*attC* × *attC*) catalyzed by the IntI integrase are represented. Hybrid *attI* and *attC* sites are indicated. Arrows inside the cassettes indicate the direction of the open reading frame. (B) Expression of cassettes: cassettes of the array are represented by small arrows. Their expression level is reflected by the color intensity of each arrow. Only the first cassettes of the array are expressed, and the subsequent ones can be seen as a low-cost cassette reservoir.
doi:10.1128/microbiolspec.MDNA3-0019-2014.f1

groups can be distinguished in the phylogeny of integrases: (i) the soil–freshwater proteobacteria, (ii) the marine γ-proteobacteria, and (iii) the inverted integrase group (7, 10). The first two groups form ecologically relevant taxons (11), suggesting, together with some phylogenetic incongruences in the branching of integrases, that horizontal transfer of CIs occurs over long evolutionary periods, among bacteria sharing the same environment.

Chromosomal integrons can be very large, embodying a significant percentage of a species' genome. Such is the case of the so-called superintegrons (SI) found in the *Vibrio* genus that can stockpile more than 200 cassettes. The best-studied SI is that of *Vibrio cholerae* (12), which harbors around 175 cassettes, representing 3% of the genome. Although some cassettes of the SI have been characterized (13, 14), mainly through their role in pathogenicity, the function of the majority is yet to be elucidated (a lack of knowledge that holds true for almost all CI cassettes). For instance, of the genes found in *Vibrio* species cassettes, the majority have no known function (10), and the remainder are homologous to proteins with a wide variety of functions, underscoring the genomic plasticity that integrons confer to their hosts. Another interesting feature of many CIs is that the array of *attC* sites shows a high sequence identity, suggesting a relationship between the sequence of the recombination site and the host (6, 15). Although this sequence conservation is not true for all CIs (10), its existence is relevant to our understanding of the hitherto unexplored phenomenon of *de novo* generation of cassettes, as discussed below (see Genesis of integron cassettes).

Mobilization of Integrons: From the Environment to the Hospitals

It is now broadly accepted that sedentary CIs have found their way into clinically relevant bacteria through their association with transposable elements and conjugative plasmids (7, 10, 11). This has allowed for the efficient dispersal of integrons among members of the same and different species, and so their classification as mobile integrons (MIs). It is not surprising that such an evolvable equipment as the integron would be useful for rapid bacterial adaptation in the changing environment shaped by humans. Indeed, MIs played a major role in the early rise of multidrug resistance among clinically relevant bacteria in the 1960s (16), triggering the research that led to the discovery of integrons in the late 1980s (1, 17). Five different MIs (classes 1 to 5) have been described to date, classified according to the sequence of their integrase. Integron classes 1 and 3 (18) are found associated with Tn*402* (19, 20), whereas class 2 integrons are almost exclusively linked to Tn7 derivatives (21). These three classes are the ones that are historically involved in the spread of multiresistance, and are definitively the most important from a medical point of view. Classes 4 and 5 were identified for their role in the rise of trimethoprim resistance in *Vibrio* spp. and were respectively located in a subset of the SXT elements in *V. cholerae* (22) and on the pRSV1 plasmid of *Aliivibrio salmonicida* (GenBank AK277063) (23).

The cargo of MIs is small, with arrays of up to eight cassettes (24). The spread of MIs grants them access to a large pool of cassettes from a variety of genetic backgrounds, as is evidenced by their GC content, codon usage and *attC* sequences. Functions of MI cassettes are strikingly homogeneous, since they encode, almost exclusively, antibiotic resistance genes against almost all families of clinically relevant antibiotics and antiseptics of the quaternary ammonium-compound family (11, 25, 26). In the clinical setting, where antibiotic pressure is high, MIs confer such an adaptive advantage that their presence is nowadays commonplace among Gram-negative pathogens. Class 1 integrons are the most widespread and clinically relevant, since they are detected in 22% to 59% of Gram-negative clinical isolates (27) and have occasionally been identified in some Gram-positive bacteria (28–31). It is also the only class for which evidence of activity *in persona* is available (32). Hence, they represent the major experimental model of the integron.

A UNIQUE RECOMBINATION SYSTEM

The *attI* Site

The *attI* site of the integron, in which cassettes are integrated, is minimally composed of two integrase binding sites termed L and R, forming the *core* site. The recombination point is in the 5′-GTT-3′ triplet located in the R box, with the cleavage taking place between the A and the C on the complementary strand, the bottom strand (bs) (Fig. 2A). The *attI* sites are hardly recognizable, because L binding domains are always degenerate with respect to R and the central regions differ greatly between *attI* sites. *In vitro* experiments have proven the specific binding of integrases to their cognate *attI* sites. In the case of IntI1/*attI1*, four monomers bind to the site, two in the R and L boxes of the core site, as expected, and two within imperfect direct repeats (DR) dubbed DR1 and DR2, located upstream of the core site (33) (Fig. 2A). These accessory sequences enhance, but are dispensable for, recombination in *attI1*. It has been suggested that they could serve as a trap to keep the integrase monomers in the vicinity of the core site (33), but the fact that they have a higher impact in the reaction with an *attC* site compared with that with another *attI* site (34), suggests a role as a topological filter. These accessory sequences do not seem to be a common feature of *attI* sites. In a recent study on the diversity of these sites among environmental samples from Suez and Tokyo Bay only three putative *attI* sites had accessory sequences, summing to a total of only five out of more than 40 sites (taking into account *attI2*, *attI3* and the sites in ref. (35)). Nevertheless, these numbers are merely speculative, as experimental data on the binding of the cognate integrases to these sites is lacking, and some large direct repeats have been overlooked in this study, probably for their lack of a GTT triplet, a feature of unknown importance in these sequences (36). *attI* sites from different integrons diverge significantly, paralleling the pattern observed for integrases (6). Cross recombination assays between noncognate *attI*/IntI partners have proven that integrases recognize preferentially their related site, suggesting the coevolution of both partners. Nevertheless, in some cases cross-talk between two systems can occur. IntI1 from class 1 MIs can, for instance, integrate cassettes into *attI2* and *attI3* sites, albeit 100 times less efficiently than into *attI1*, but not at all into the *attI* site of the *V. cholerae* SI (37).

The *attC* Site

attC sites are an integral part of integron cassettes, terminating each of them, and are necessary for their mobility. These sites differ significantly from canonical Y-recombinase sites. As we will see through the chapter, many of the features that make the integron a unique recombination system are made possible through the peculiarities of *attC* sites.

Two regions of inverted homology, R”-L” and L’-R’, are found in all *attC* sites, separated by a central region that is highly variable in size and sequence, leading to *attC* sites of broadly variable lengths, from 57 to 141 bp (38) (Fig. 2B). *attC* sites have almost no sequence conservation, but instead they display a strikingly conserved palindromic organization that can form secondary structures through DNA intra-strand pairing (Fig. 2C) (39). Upon folding, *attC* sites show a hairpin structure resembling a canonical core site consisting of R and L boxes. This specific single-stranded (ss) structure is the substrate recognized (40, 41) and recombined by the integrase (42). Of both strands in any *attC* site, the bs

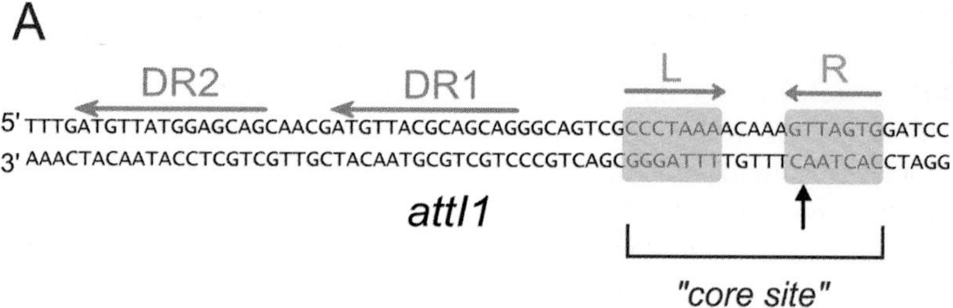
A
DR2
DR1
L
R
5' TTTGATGTTATGGAGCAGCAACGATGTTACGCAGCAGGGCAGTCGCCCTAAAACAAAGTTAGTGGATCC
3' AAACTACAATACCTCGTCGTTGCTACAATGCGTCGTCCCGTCAGCGGGATTTTGTTTCAATCACCTAGG
attI1
"core site"

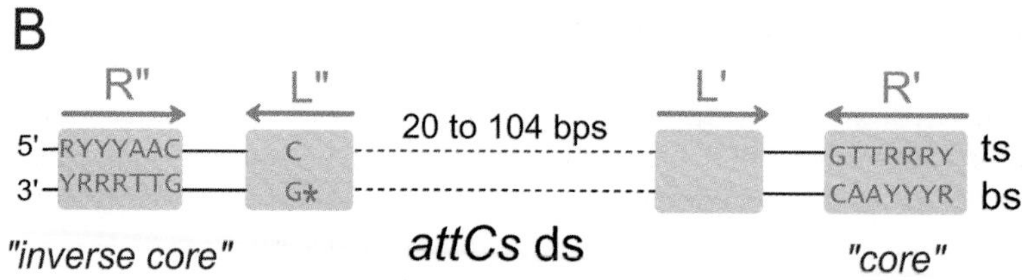
B
R"
L"
20 to 104 bps
L'
R'
5'-RYYYAAC
3'-YRRRTTG
C
G*
GTTRRRY
CAAYYYR
ts
bs
"inverse core"
attCs ds
"core"

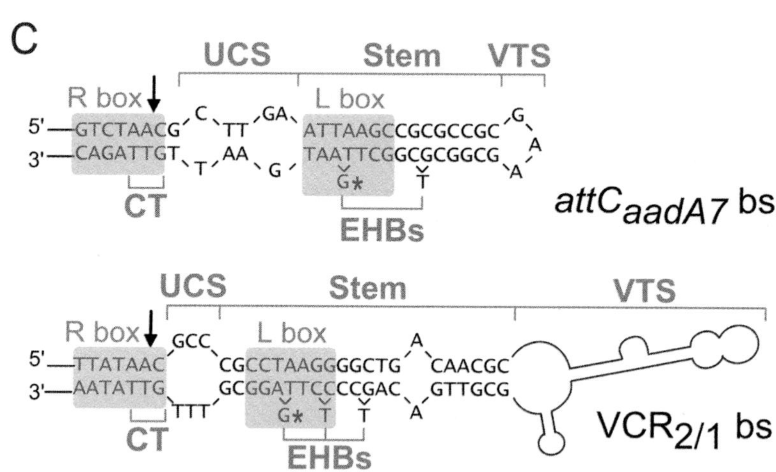
C
UCS
Stem
VTS
R box
L box
5'-GTCTAACG
3'-CAGATTGT
ATTAAGCCGCGCCGC
TAATTCGGCGCGGCG
G*
CT
EHBs
attCaadA7 bs
UCS
Stem
VTS
R box
L box
5'-TTATAAC
3'-AATATTG
GCC
TTT
CGCCTAAGGGGCTG
GCGGATTCCCCGAC
CAACGC
GTTGCG
CT
EHBs
VCR2/1 bs

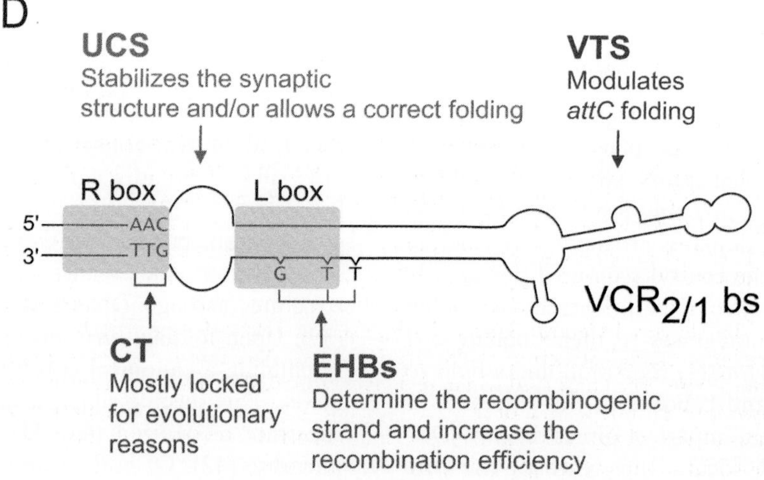
D
UCS
Stabilizes the synaptic structure and/or allows a correct folding
VTS
Modulates attC folding
R box
L box
5'
3'
AAC
TTG
CT
Mostly locked for evolutionary reasons
EHBs
Determine the recombinogenic strand and increase the recombination efficiency
VCR2/1 bs

is approximately 10^3 times more recombinogenic than the top strand (ts) (42).

A comparison of *attC* sites shows that sequence conservation is restricted to two inverted triplets, 5′-**AAC**-3′ and 5′-**GTT**-3′ located in the R" and R' boxes, respectively (Fig. 2B, see ts). Nevertheless, these conserved sequences can be extended to inverted repeat sequences of 7 bp designated as *inverse core* and *core* (Fig. 2B). The *inverse core* consists of an RYYY**AAC** (R: purine; Y: pyrimidine) sequence and is located at the 5′ end of the ds *attC* site and the complementary *core* consisting of **GTT**RRRY is located at the 3′ end (Fig. 2B). Once folded, a conserved CAA triplet is reconstituted within the R box and, consistent with the *attI* sites, the recombination point is located between the C and A nucleotides.

In *attC* sites, the genetic information required for proper recombination is not entirely contained in the primary sequence, but rather determined by specific features of their secondary structures (43, 44). The folding permits the inclusion of a new layer of information and regulation in the site. Three structural features that are common to all known ss *attC* sites emerge from the folding of the bs of *attC* sites (defined in ref. (43)). First, the presence of a set of single bases located on the R"-L" arm of the symmetrical *attC* sequence that have no complementary nucleotides on the R'-L' arm and are present as extrahelical bases (EHBs) in the structured site (Fig. 2C). Depending on the *attC* sites, two or three EHBs can be found (43). These bases have three major roles in the recombination reaction: (i) they determine the recombinogenic strand in the site (the bottom strand), (ii) they serve as stabilizers of the synaptic complex, establishing contacts with protein monomers across the synapse, and (iii) they avoid a second cut in the *attC* site by pulling apart the tyrosine residue of the integrase monomer bound to the L"-L' box (see *attI* × *attC* reaction below) (43, 44). The second structural feature arises from the annealing of the R"-L" and L'-R' arm sequences, which contain spacer regions that are not complementary, leading to the formation of the unpaired central spacer (UCS) between the R"-R' and L"-L' boxes. The structure of the UCS is essential to achieve high-level recombination of the bottom strand, suggesting a dual role for this structure in active site exclusion and in hindering the reverse reaction after the first-strand exchange. The last structural feature is defined as the variable terminal structure (VTS) and corresponds to the sequence located at the end of the stem (Fig. 2C) (43). VTSs vary in length among the various *attC* sites, going from three predicted unpaired nucleotides as in $attC_{aadA7}$, to a complex branched secondary structure in the larger sites such as the *Vibrio cholerae* repeats (VCRs; the *attC* sites of the SI (12)). VTSs have a regulatory role through the modulation of *attC* folding (45), because the length and sequence of VTSs have an impact on the tendency of intrastrand pairing of the *attC* site and hence the propensity to form the recombinogenic structure. This influence is critical in cruciform formation (the extrusion of a hairpin on each strand of a symmetric and paired dsDNA molecule), where *attC* sites containing very large VTSs have a highly unfavorable energy to extrude and form a cruciform even in highly supercoiled DNA. It is noteworthy that during conjugation, where *attC* sites are delivered as ssDNA and folding is favored, the VTS size does not impact the recombination efficiency of the *attC* site (45).

This atypical sequence-independent recognition of ss *attC* sites readily explains how cassettes containing different *attC* sites can be efficiently recombined by the same integrase as well as how the proper orientation

Figure 2 Integron recombination sites. The putative IntI1 binding domains are marked with green boxes. The black arrows show the cleavage points. (A) Sequence of the double-stranded *attI1* site: inverted repeats (R and L) and direct repeats (DR1 and DR2) are indicated with gray arrows. (B) Schematic representation of double-stranded (ds) *attC* sites; inverted repeats (R", L" L', and R') are indicated with gray arrows. The dotted lines represent the variable central part. The conserved nucleotides are indicated. Asterisks (*) show the conserved G nucleotides, which generate extrahelical bases (EHB) in the folded *attC* site bottom strand (bs). The top strand (ts) and bottom strand (bs) are marked. (C) Proposed secondary structures of the $attC_{aadA7}$ and $VCR_{2/1}$ bottom strands: structures were determined by the UNAFOLD online interface at the Institut Pasteur. Structural features of *attC* sites, namely, the Unpaired Central Spacer (UCS), the ExtraHelical Bases (EHBs), the stem and the Variable Terminal Structure (VTS) are indicated. Asterisks (*) show the conserved G extrahelical base. The conserved triplet (CT) is indicated. Primary sequences of the *attC* sites are shown (except for the VTS of the $VCR_{2/1}$ site). (D) Schematic representation of structural features of the $VCR_{2/1}$ site and their roles: the structural features and their roles are indicated. doi:10.1128/microbiolspec.MDNA3-0019-2014.f2

with respect to the P_C promoter is ensured through the recombination of the bs, since the ts recombination would put promoterless genes in antisense orientation relative to P_C.

The Integrase

Integrases belong to the family of Y-recombinases (46), but form a specific subfamily within it, the closest relatives being the chromosome dimer resolution proteins, Xer (47). Integrases possess structural features typical of Y-recombinases, such as patches I to III, boxes I and II and the active site residues RKHRHY (48). Interestingly, compared with the rest of the members of the family, an extra domain of about 19 amino acids is present between patches II and III, containing an α-helix dubbed I2, that is essential for the activity of the protein (49). The crystal structure of the synaptic complex formed by the SI integrase of *V. cholerae* (VchIntIA) with two VCR_{bs}-derivative substrates, has shown that this extra domain has a specific role in the assembly of the complex, in particular in the folding of the hydrophobic pockets, that allows it to accommodate and stabilize the extrahelical bases of the *attC* sites bs (44). The catalytic domain of Y-recombinases is very conserved, and is shared with type IB topoisomerases and telomere resolvases. In all cases, one monomer can only cleave one strand, hence necessitating four monomers to perform the complete recombination reaction. The study of integrases has involved more than 50 mutants of IntI1 and VchIntIA recombinases that have been obtained through directed mutagenesis or directed evolution experiments (49–52). The study of the recombination efficiency and DNA binding affinity of the mutants has confirmed the role of catalytic residues (44), as well as the integrase-specific αI2 helix, that serves to accommodate EHBs. It is noteworthy that *attI* sites do not have EHBs, highlighting the fact that integrases recognize substrates that are structurally different. Little is known about the structural basis of this dual site-recognition, but some results suggest that it is dependent on the strength and flexibility of the interactions among integrase monomers in the synaptic complex, and that this parameter is finely balanced, because a more efficient recognition of the ds *attI* site brings a decrease in the recognition of *attC* sites as a trade-off (50).

RECOMBINATION REACTIONS

Y-recombinases process the recombination between two DNA molecules in a set of archetypal steps. The process starts with the assembly of the synaptic complex, formed by the two DNA substrates and four monomers of the protein. It proceeds through the exchange of a set of strands from each partner molecule, forming a transient Holliday junction (HJ), and ends with the resolution of the junction through a second strand exchange (48). In this section we will see that recombination in the integron is an exception to this model. This is mainly due, as we mentioned before, to the peculiarities of *attC* sites, that impose changes in the process and the mechanism of the recombination reactions. Several of these changes are common to the recombination of the CTX phage of *V. cholerae*, because the ss genome of the phage adopts an *attC*-like structure. Nevertheless, this is not a unified recombination system, since the phage hijacks the host recombinases XerCD (53).

In the next few paragraphs we will try to give a comprehensive view of the integron by providing a detailed explanation of the recombination process followed by the biological meaning of each reaction. As an overview, three different recombination reactions are possible between the *attC* and *attI* sites: (i) the *attI* × *attC* reaction, that integrates cassettes into the integron (Fig. 3), (ii) the *attC* × *attC* reaction leading to the excision of cassettes from the array (Fig. 4); and (iii) the *attI* × *attI* reaction (Fig. 5), a rather cryptic one, that can have biological consequences in the case of multicopy MIs. The combination of excision and integration reactions leads to the shuffling of integron cassettes in the array.

attI × *attC* Recombination

Mechanistic View

The *attI* × *attC* reaction is the most efficient of the reactions catalyzed by the integrase. From a structural point of view it is an atypical one because it involves double-stranded (*attI*) and single-stranded (*attC*) substrates. This poses a mechanistic problem, since the first strand exchange generates an asymmetric and therefore atypical Holliday junction (aHJ) that cannot be resolved through the classical second strand exchange, because cutting the *attC* site twice would lead to linear and thus abortive products (Fig. 3). Hence, a second strand exchange in the bs *attC* must somehow be avoided and the aHJ must be resolved differently. The crystal structure of the synaptic complex of VchIntIA/VCRs bs revealed that the extra-helical "T" in *attC* sites acts to pull the catalytic tyrosine of one integrase monomer away from the phosphate link, avoiding the nucleophilic attack on the L box of the *attC* site (44). Resolution of the junction is then carried

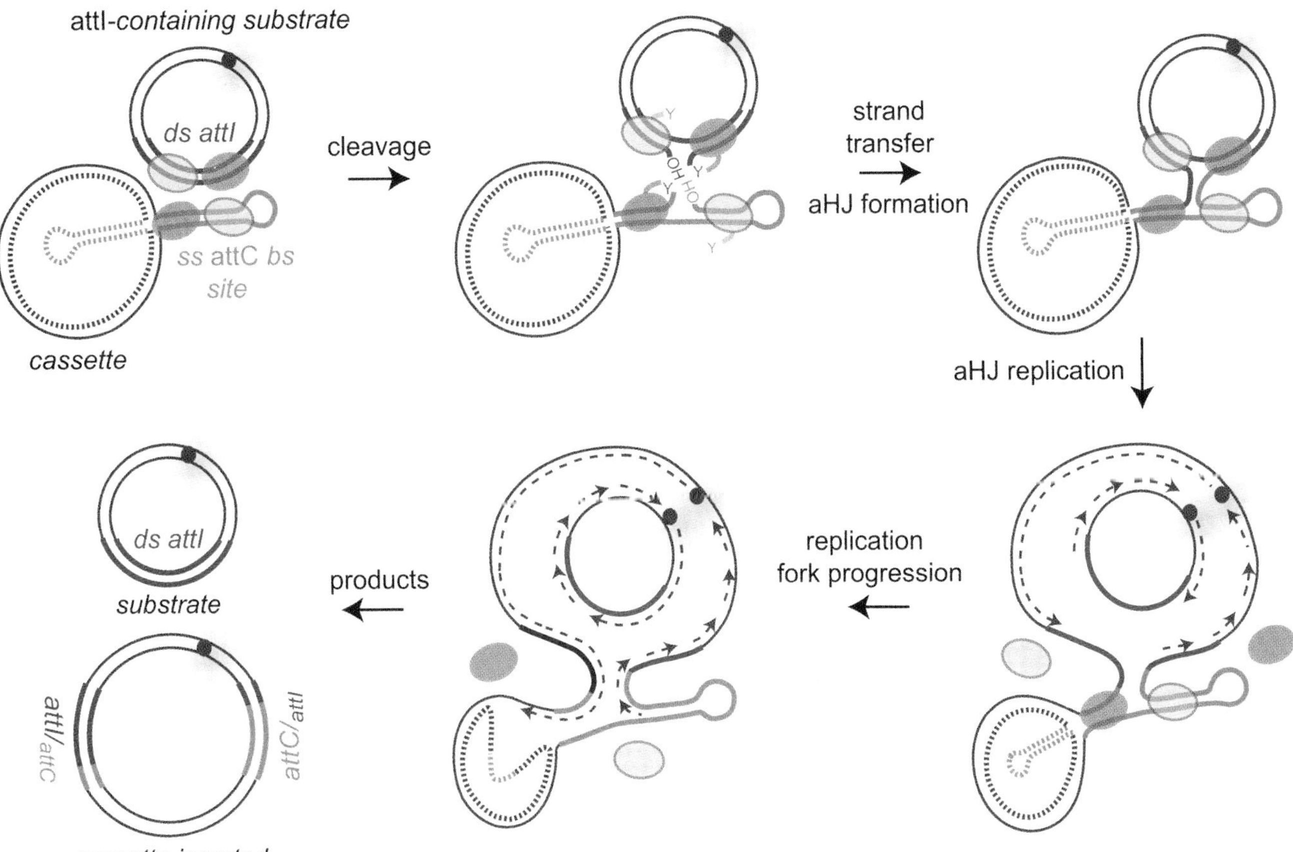

Figure 3 Replicative resolution of integron cassette insertion. Recombination between a double-stranded *attI* site (bold red lines) and a single-stranded bottom *attC* site (bold green lines) terminating a cassette is shown. The top strand of the *attC* site is represented as a dotted line because we do not exactly know the nature of the cassettes (ss or ds). The synaptic complex comprises two DNA duplexes bound by four integrase protomers. The two activated protomers are represented by dark gray ovals. One strand from each duplex is cleaved and transferred to form an atypical Holliday junction (aHJ). Classical resolution gives rise to covalently closed abortive molecules. The non-abortive resolution implies a replication step. The origin of replication is represented by a purple circle and the newly synthesized leading and lagging strands by dashed purple lines. Both products are represented: the initial substrate resulting from the top strand replication, and the molecule containing the inserted cassette resulting from the bottom strand replication. Hybrid *attC* and *attI* sites are indicated.
doi:10.1128/microbiolspec.MDNA3-0019-2014.f3

out by a replicative process without the involvement of a second strand exchange of any kind (Fig. 3) (54). Therefore, cassette insertion is a semiconservative process and implies the release of the initial *attI*-containing substrate.

At the nucleotide level, the DNA cleavage is localized in both substrates between the C and AA (i) on the bs within the R box for the ds *attI* site, and (ii) on the bs within the R box for the ss *attC* site. The reaction results in the formation of a "chimeric" *attI* site in which the last six base-pairs of the R box belong to the last six base-pairs of the *attC* site's *core* of the newly inserted cassette. For the cassettes in the array, their *attC* sites will be ended by the six base-pairs of the *attC* site's *core* of the cassette previously inserted (Fig. 1A). Note that the *attC* site of the last cassette in the array will always be terminated by the last six base-pairs of the *attI* site (Fig. 1A, B).

The imperfect complementarity between *cores* and *inverse cores* contained in "chimeric" *attC* sites does not seem to impede either the propensity of *attC* sites to efficiently fold, or their stability once folded, especially

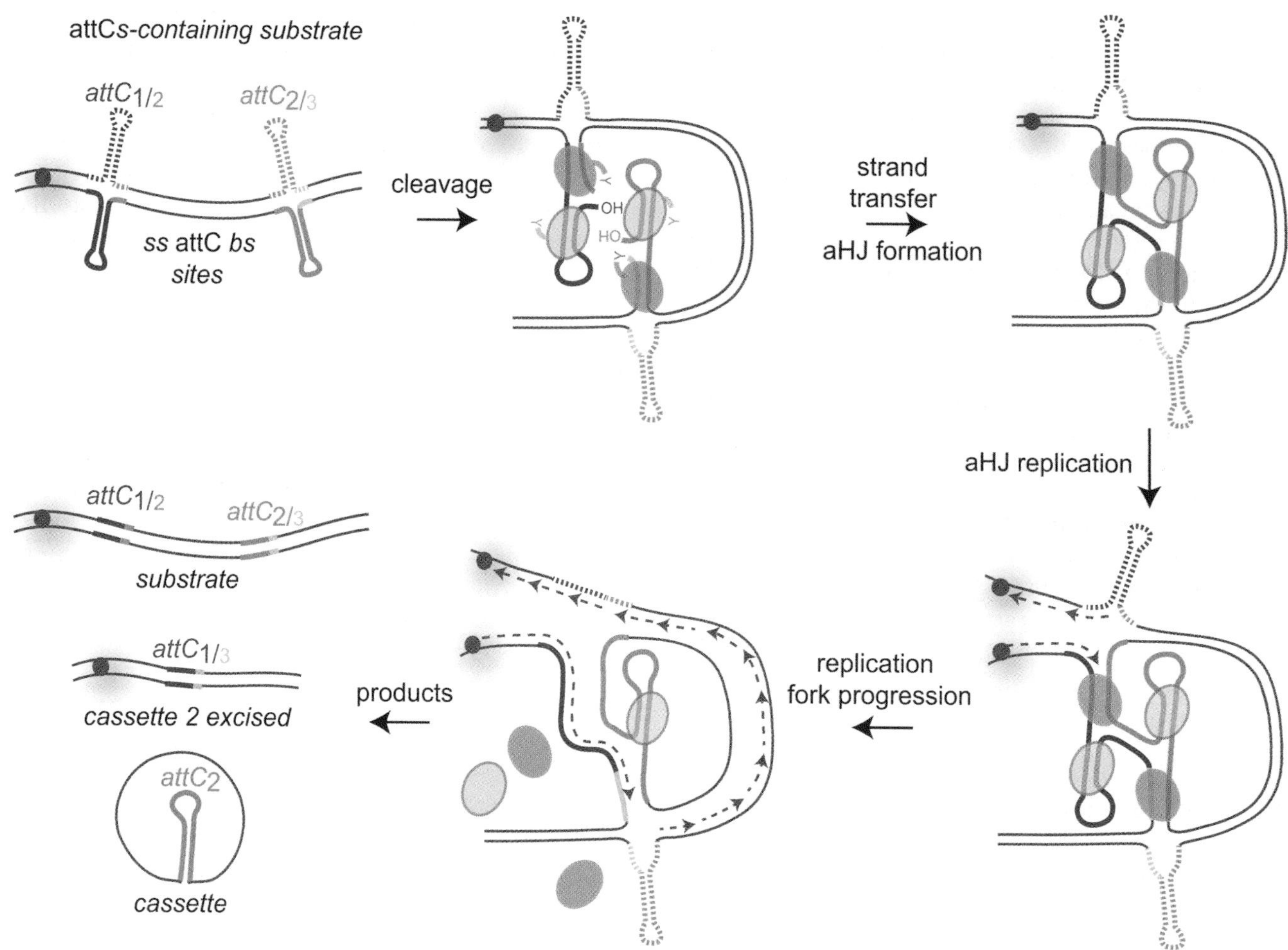

Figure 4 Replicative resolution model of integron cassette excision. Recombination between two single-stranded bottom *attC* sites (bold green and pink lines) is shown. Top strands of *attC* sites are represented as dotted lines. The synaptic complex comprises two DNA duplexes bound by four integrase protomers. The two activated protomers are represented by dark gray ovals. One strand from each duplex is cleaved and transferred to form an atypical Holliday junction (aHJ). The proposed aHJ resolution model implying a replication step is based on the *attC*×*attI* recombination. The origin of replication is represented by a purple circle and the newly synthesized leading and lagging strands by dashed purple lines. Products are represented: on one hand, the initial substrate resulting from the top strand replication, and on the other, the excised cassette (cassette) and the molecule devoid of the excised cassette (cassette 2 excised) both resulting from the bottom strand replication. Hybrid *attC* sites are indicated.
doi:10.1128/microbiolspec.MDNA3-0019-2014.f4

since the integrase is capable of capturing *attC* sites at the beginning of the folding process, stabilizing them and facilitating their complete extrusion (55).

Biological Meaning: Cassette Insertion

Insertion of a circular cassette preferentially occurs at the *attI* site of the integron, rather than at an arbitrary *attC* site within the array (56, 57). This feature is essential for the adaptive role of integrons: it ensures the expression of newly acquired promoterless cassettes from the promoter P_C within the integron, allowing the immediate testing of their adaptive effect. Successive insertions of integron cassettes at the *attI* site lead to the assembly of a cassette array encoding a set of functions that have been useful for adaptation in previous conditions. Thus, it represents a *memory* of adaptive functions that have been valuable for the cell in the past and which can be recalled on demand.

Considering that the *attI* × *attC* reaction is the most efficient one, one could imagine that integrons have a tendency to empty their content by the sequential loss of cassettes in the first position, especially considering that intramolecular reactions are more efficient than intermolecular ones. Here, the different structural nature of the sites allows for an additional layer of regulation of the system. Indeed, it is extremely unlikely that two adjacent *attI* and *attC* sites will be found simultaneously in their reactive form (ds and ss, respectively). Therefore, the equilibrium in the *attI* × *attC* reaction is shifted towards intermolecular rather than intramolecular reactions; i.e., towards integration rather than excision.

Integron cassette insertion can also occur at *attC* sites. As an example, the *catB* cassette of the *V. cholerae* SI moved under experimental conditions to the second position of the array (58). Nevertheless, this type of event is probably disfavored relative to the insertion of integron cassettes at the *attI* site, especially because it requires the *attC* sites carried in the array to be ss at the same time as the incoming cassette. This likely decreases the efficiency of such intermolecular events.

attC × *attC* Recombination

Mechanistic View

Recombination events between two *attC* sites located in the same cassette array lead to the excision of covalently closed integron cassettes (5, 59). Two ss folded *attC* sites must be recruited at the same time by the integrase to generate, after the first strand exchange, the HJ. Once again, the generated HJ is atypical because a second strand exchange of the bottom *attC* site strand would not lead to cassette excision but only to the exchange of the bs of *attC* sites between cassettes. Therefore, we propose, as for the *attI* × *attC* reaction, a resolution of the aHJ by a replicative, and hence semiconservative, mechanism. In this model, the replication of the recombined bs would release a covalently closed single-stranded cassette and the original DNA molecule lacking the excised cassette, while the replication of the nonrecombined top strand would reconstitute the initial substrate (Fig. 4).

If we consider the reaction at the nucleotide level, recombination between two adjacent "chimeric" *attC* sites releases a cassette containing an *attC* site with perfect complementarity between the *core* and *inverse core* (Fig. 4). Therefore, folded *attC* sites contained in circular excised cassettes have a longer stem and probably a more stable secondary structure. This feature could increase the rate of reinsertion of the excised cassettes (see Cassette shuffling below).

Biological Meaning: Integron Cassette Excision

Integron cassette excision requires the simultaneous folding of two ds *attC* sites into their recombinogenic ss form. This event is dependent on the tendency of the site to extrude from ds DNA (see the role of the VTS, in the *attC* section), but can be favored if the fragment containing the *attC* sites is found in ss form. Once the bottom strands of both *attC* sites have recombined, the cassette is presumably separated from the ts and released from the integron via replication (Fig. 4).

Cassette Shuffling

After their excision from the array of an integron, cassettes can be reintegrated into the *attI* site. This relocation is favored, as the *attI* × *attC* reaction is the most efficient of the reactions catalyzed by integrases. Coupling the two reactions allows bacteria to render functional genes with adaptive functions that were kept silent as part of the *low-cost memory* of integrons (see the Expression section in Integron cassettes).

Cassettes excised under an ss form could theoretically be rapidly degraded by host factors and, even if the ss cassette is converted into ds form by DNA synthesis, the resulting molecule is devoid of a replication origin and so is lost after cell division. Therefore, if not rapidly reinserted, an excised cassette is not maintained in the cell. We cannot exclude that both cassette excision and insertion processes are coincidental or almost simultaneous events.

The proposed atypical mechanism of integron cassette excision can even account for cassette duplications if the cassette, after excision, is reintegrated at the *attI* site of the conserved integron (deriving from the replication of the ts). Notably, large integrons such as *V. cholerae* SI, contain duplicated cassettes. Moreover, cassette duplication may lead to generation of cassette diversity, e.g. *aadA1* and *aadA2*, which share 89.3% identity and likely arose via duplication (60).

attI × *attI* Recombination

Mechanistic view

The recombination between two *attI* sites has been described (57, 61), but is 10^3 times less efficient than the *attI* × *attC* reaction. It is, however, considered to follow the same pathway with only one strand exchange and resolution through replication. The cleavage point remains in the CAA of the bs, within the R box of both *attI* sites. Nevertheless, the *attI* × *attI* reaction has structural differences with important implications for the reaction. In this case the reaction is symmetrical, involving two ds partners, for which resolution through

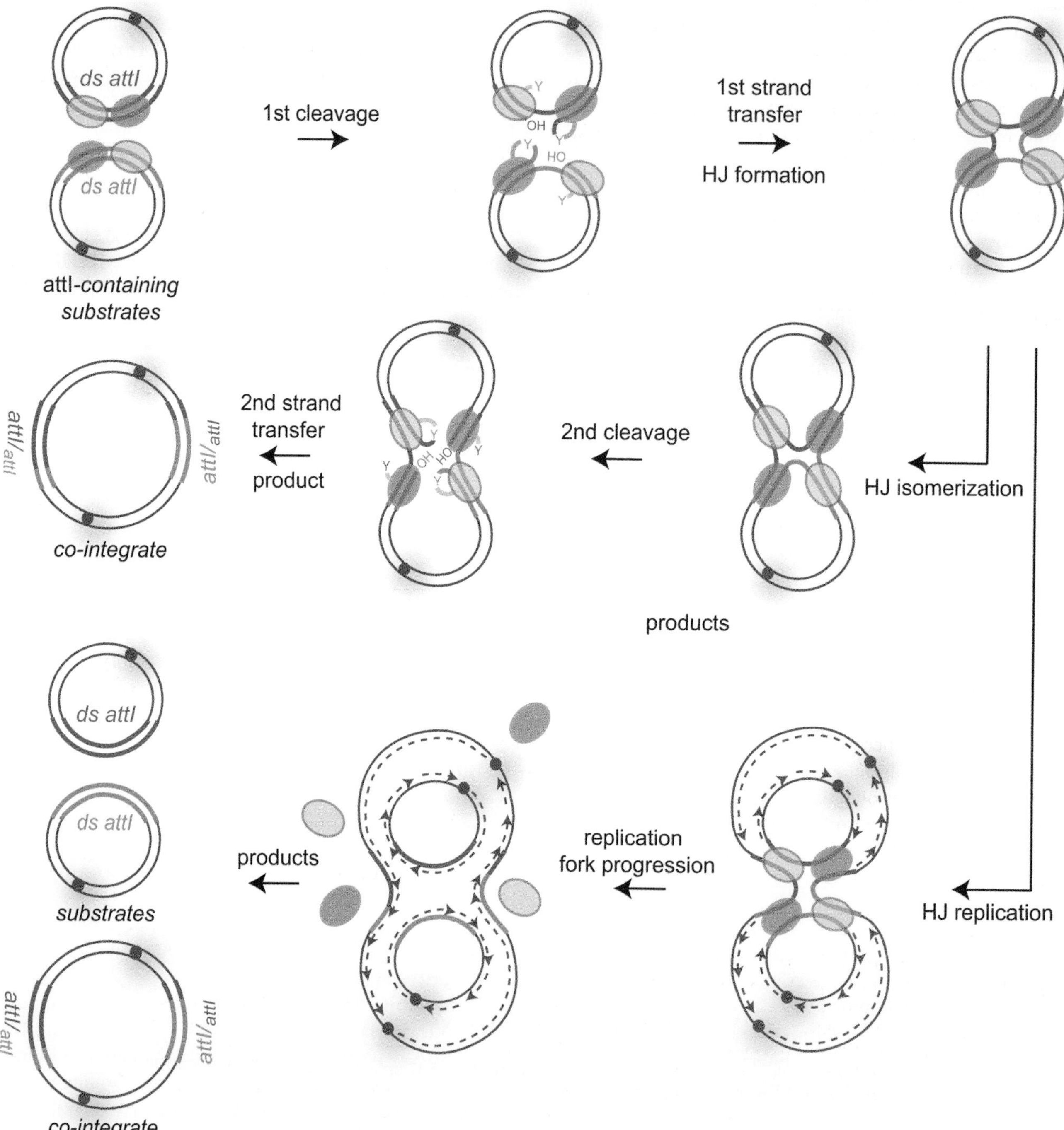
ds attI
ds attI
attI-containing substrates
1st cleavage
OH
HO
Y
1st strand transfer
HJ formation
attI/attI
attI/attI
co-integrate
2nd strand transfer
product
2nd cleavage
HJ isomerization
products
ds attI
ds attI
substrates
attI/attI
attI/attI
co-integrate
products
replication fork progression
HJ replication

a second strand exchange is theoretically possible, since it is neither abortive nor impeded by EHBs. We have recently studied this reaction in depth to elucidate the recombination pathway followed by the integrase when processing two symmetrical *attI1* sites (Fig. 5, HJ replication). We have observed that, in contrast to what happens in the two other reactions, both bottom and top strands of the *attI1* site are reactive. Indeed they can be transferred independently if the transfer is followed by the replicative resolution of the HJ. Interestingly, it is also possible to resolve the *attI1* × *attI1* HJ through the classical second strand exchange (Fig. 5, HJ isomerization) (Escudero et al., in preparation), as for canonical Y-recombinases. The cleavage on the ts occurs within the AAC triplet overlapping the spacer region and the L box (Fig. 2) (Escudero et al., in preparation). We have also studied the influence of each base in the L box on the recombination process and found a higher tolerance for mutation in the central CT base pairs. These results are in accordance with the work from Gravel et al. (33) in which they observed a weak contact between the integrase and the T base in the L box.

Biological Meaning: Inter-Integron Content Rearrangement

The processing of the reaction between two *attI* sites has unveiled an unexpected flexibility of integron integrases, proving that they can recombine structurally distant substrates and switch recombination pathways accordingly. Our data challenge the rather rigid view of site-specific recombination mediated by Y-recombinases. From a biological perspective, it could be argued that this reaction has little evolutionary meaning in CIs, where there is only one copy of the platform and the *attI* × *attI* reaction would most likely produce an undesirable dimer between replicating chromosomes. Interestingly, in the clinical environment, where integrons are plasmid-borne and have become prevalent to the point of redundancy (62, 63), *attI* sites can be found in multiple copies within the cell. In this setting, the reaction between *attI* sites can have biological consequences through the rearrangement of cassette content between different integrons, and is a reaction of unknown relative importance. It is noteworthy that in the case of resolution through second strand exchange, the reaction is not semiconservative.

Recombination at Secondary Sites

The *attI* and *attC* sites can recombine outside the integron at non-specific secondary sites containing GNT sequences. These events have occasionally been observed to occur naturally with *attC* sites, and have been reproduced experimentally with *attI* sites (61). The complete *aadB* integron cassette has been found inserted at a secondary site in the IncQ plasmid RSF1010 just downstream of a known promoter (64). Subsequent excision of these cassettes is unlikely to happen, leading to the stable acquisition of the gene, unless the secondary site contains the canonical GTTRRRY sequence, thereby maintaining the integrity of the newly formed *attC* site (65). In these aberrant integration events, the expression of integron cassettes is conditional upon the presence of a promoter at the insertion point (64).

Regarding *attI* sites, recombination at secondary sites was experimentally observed at very low frequency (61). This reaction disrupts the integrity of the structure, separating the cassettes from the P_C, possibly explaining why these reactions have not been observed in nature.

INTEGRON CASSETTES

Structure

Integron cassettes constitute the variable and mobile part of the integron. They usually contain a single open

Figure 5 Two resolution pathways proposed for *attI* × *attI* recombination. Recombination between two double-stranded *attI* sites (bold green and pink lines) is shown. The first proposed pathway is similar to the classical site-specific recombination catalyzed by Y-recombinases. The synaptic complex comprises two DNA duplexes bound by four recombinase protomers. The first two activated protomers are represented by dark gray ovals. One strand from each duplex is cleaved and transferred to form a HJ. Isomerization of this junction alternates the catalytic activity between the two pairs of protomers (dark and light-gray ovals) ensuring the second strand exchange and recombination product formation (co-integrate). The second pathway proposes a resolution of the HJ by replication. The origin of replication is represented by a purple circle and the newly synthesized leading and lagging strands by dashed purple lines. Products are represented: two initial substrates resulting from the top strand replication and co-integrate resulting from the bottom strand replication. Hybrid *attI* sites are indicated.
doi:10.1128/microbiolspec.MDNA3-0019-2014.f5

reading frame (ORF) devoid of a promoter, and are terminated by the *attC* recombination site recognized by the integrase. Carrying a single ORF, the size of integron cassettes is relatively small, generally lying between 500 and 1000 bp. Integron cassettes can be found in a linear ds form as part of an array within an integron, or, when excised, as a free, nonreplicative circular element that can move between integrons.

The Cassette Array

The ensemble of cassettes found in CIs, their *cargo*, can vary widely from being empty to containing up to 217 cassettes (in the *Vibrio vulnificus* SI), constituting a prodigious reservoir of readily interchangeable genes. MIs typically bear a small cargo of fewer than six cassettes, with the longest array reported being of eight (24).

attC sites found in the integron cassette array of CIs generally show a high degree of identity (> 80% for the VCRs in *V. cholerae*) suggesting a link between the sequence of the site and the bacterial species containing the integron—see also Cassette genesis (6, 7, 11). In contrast, cassettes in MIs show diversity in the length and sequence of their *attC* sites, as well as an inconsistent codon usage in the ORFs encoded. Hence, it is plausible that MIs have access to the vast pool of cassettes found in CIs and that they gather cassettes from different genomic backgrounds. Integron cassettes contained in MIs could therefore be seen as representatives of a specific CI in the environment (6).

Expression

As genes in integron cassettes are generally promoterless, their expression is ensured only when inserted into an integron at the *attI* site, by the proximity of the external P_C promoter carried within the nonmobile functional platform of the integron (Fig. 1). The P_C promoter is located either in the *intI* gene or in the *attI* site. *attC* sites, through their imperfect symmetry, ensure correct orientation of cassette-borne genes relative to the P_C promoter when inserted into *attI* sites. Most studies on cassette expression have been performed using the class I integron system, in which it has been shown that the P_C promoter is located within the *intI1* coding sequence. Occasionally, it may be combined with a second promoter, termed P_C2, and located in *attI1*. P_C variants of different strengths have been identified for both P_C and P_C2 promoters: thirteen for P_C and three for P_C2 (2, 4). There are five main P_C–P_C2 combinations defining five levels of promoter strength. The diversity of strength of these promoters could mediate a differential expression for an identical array of cassettes. The distance between the P_C promoter and integron cassettes affects their expression level. Indeed, expression levels are maximal for the first gene in the array, and gradually decrease for those following, a phenomenon dependent on the distance to the promoter and on the nature of the inserted cassettes (4). For a long time, it was thought that this feature was due to the ability of the folded *attC* sites to act as Rho-independent transcription terminators, impeding the transcription of the full cassette array (4). Finally, contrary to expectations, the transcription of integron cassettes was found not to be affected by the folded *attC* sites. Indeed, mutations of *attC* sites revealed that destabilization of the *attC* secondary structures in the transcript could enhance the expression of the 3′ gene at the translational level but not affect its expression at the transcriptional level. In particular, the presence of a translated ORF was shown to increase translation of the 3′-located gene (66). These results might reflect the capacity of the folded *attC* sites to impede ribosome progression (7, 66), and therefore explain, in part, the expression gradient observed in integron cassette arrays. Altogether, the dependency on P_C for expression of cassettes allows the integron to provide the cell with an adaptive array of functions at low fitness cost. With a large portion of the cassette cargo found too far from the P_C promoter to be expressed, these cassettes are carried at the minimum cost possible, the cost of replication. Although silent, these cassettes remain available for the cell if needed (see cassette reshuffling).

The expression of integron cassettes is also governed by the presence of a binding motif initiating the assembly of ribosomes (67). Some cassette-borne genes are preceded by their own ribosome-binding site (RBS), while others are devoid of this motif. For the latter, translation can be initiated at an upstream RBS site. In class I integrons, a small ORF (*orf11*) preceded by a functional RBS has been found in the *attI1* site. This RBS is present in all transcripts generated from the P_C promoters, and accounts for a significant part of the expression of cassettes devoid of an RBS (68).

Note also that, although rare, integron cassettes sometimes harbor their own promoter, e.g. the *cmlA1* chloramphenicol resistance gene of the In4 class I integron (69, 70), the *ere*(A) erythromycin resistance gene of the pIP1100 plasmid-borne class II integron (71), the *qnrVC1* quinolone resistance genes found in the class I integron (72) and the toxin–antitoxin (TA) gene pairs found in the *V. cholerae* SI—see below for more details (15, 73, 74). The expression of these integron cassettes is assured, regardless of their position in the array.

Diversity and Functions

Integron cassettes seem ubiquitous—they have been recovered from every environment investigated, including soil, riverine sediment, seawater, biofilms, plant surfaces and even eukaryotes' symbionts (36, 75–81).

As mentioned in the Introduction, distinct sets of functions are encoded by cassettes found in CIs and in MIs. On the one hand, MIs carry, almost exclusively, antimicrobial resistance genes. For instance, class I integrons have been associated with more than 130 integron cassettes comprising resistance determinants against almost all antibiotic families, including β-lactams, all aminoglycosides, trimethoprim, chloramphenicol, streptothricin, fosfomycin, macrolides, rifampin, quinolones and antiseptics of the quaternary ammonium-compound family (25, 44). On the other hand, CIs mainly contain cassettes of unknown function. Combined analyses of metagenomic and CI cassettes from *Vibrio* species revealed that up to 65% of cassette-encoded proteins had no known homologs and that 13% had homologs of unknown function (10). The remainder showed a wide range of non-specific functions in metabolism, cellular processes, and information storage. Moreover, cassettes in *Vibrio rotiferianus* DAT722 have been found to be involved in host surface polysaccharide modifications suggesting that integron cassettes may be important in processes such as bacteriophage resistance, adhesion/biofilm formation, protection from grazers and bacterial aggregation (82). Other data also reveal functions mediating interactions with the external environment (i.e., the presence of a signal peptide region, or signatures of multiple transmembrane domains (15, 77)).

Toxin–Antitoxin Cassettes

Among the cassettes of CIs, a distinct type is notable, encoding members of TA families (15, 83, 84). TAs are addiction systems encoding a stable toxin and its labile neutralizing antitoxin. The perpetuity of the system in the genome is assured by the difference in half-lives of toxin and antitoxin. If the TA system is lost, the antitoxin degrades first and the cell suffers toxin-mediated postsegregational killing. TAs have been classified, depending on the nature and mode of action of the antitoxin, into three types: types I and III in which antitoxins are small RNAs that impede the translation of the toxin; and type II systems, in which the antitoxin is a protein that inhibits the toxin through protein complex formation (85). Seventeen cassettes carrying type II TAs have been found in the SI of *V. cholerae* N16961 ((2), Iqbal et al., in preparation). A peculiarity of nine of them is that they are integrated in the opposite orientation relative to the array, and transcription is ensured by their own promoter. Therefore, their expression is independent of the promoter(s) contained in the integron platform. Our laboratory has studied the TA array in *V. cholerae*, and found a remarkable specificity of every toxin for its cognate antitoxin, with no cross talk between toxins and antitoxins of different systems, even between those belonging to the same family (Iqbal et al., in preparation).

From these observations it is tempting to speculate that these TA systems play a role in the stabilization of SIs, preventing the loss of silent cassettes and allowing for the formation of large arrays (15, 74). However, TAs have also been shown to mediate phage resistance (86), and a dual role for these elements cannot be ruled out.

Conventional Annotation

Several web databases have been developed to provide easy access to integron integrase and cassette DNA sequences. Among them, RAC is specialized in the annotations of integron cassettes mainly encoding antibiotic resistance (87), Integrall database has listed more than 8,500 cassettes ((88), http://integrall.bio.ua.pt/), and ACID has collected and stored in its first version 5,622 integron cassettes (89). Annotation systems allowing cassette identification have also been developed, such as XXR ((http://mobyle.pasteur.fr) (15)).

A SYSTEM INTIMATELY CONNECTED TO CELL PHYSIOLOGY

Integrons allow bacteria to adapt rapidly by granting access to an almost infinite array of functions encoded in cassettes. But the elegance of this genetic platform is more easily appreciated in the context of the complex and subtle coupling of its activity to the physiology and needs of its host. In this section we will try to give an overview of the integron–host connection, of which Fig. 6 provides a snapshot scheme that should help the reader throughout the following paragraphs.

Expression of the Integron Elements

The Integrase

Like all elements promoting genetic variation in a cell, integrons must be well controlled (domesticated) to avoid the deleterious effects of an overwhelming recombination activity. The regulation of the integrase has remained elusive until recently, but it had been noticed that integrons and their cassette content are stable in laboratory conditions, while SIs are the most variable loci in the genome of natural isolates of *V. cholerae*

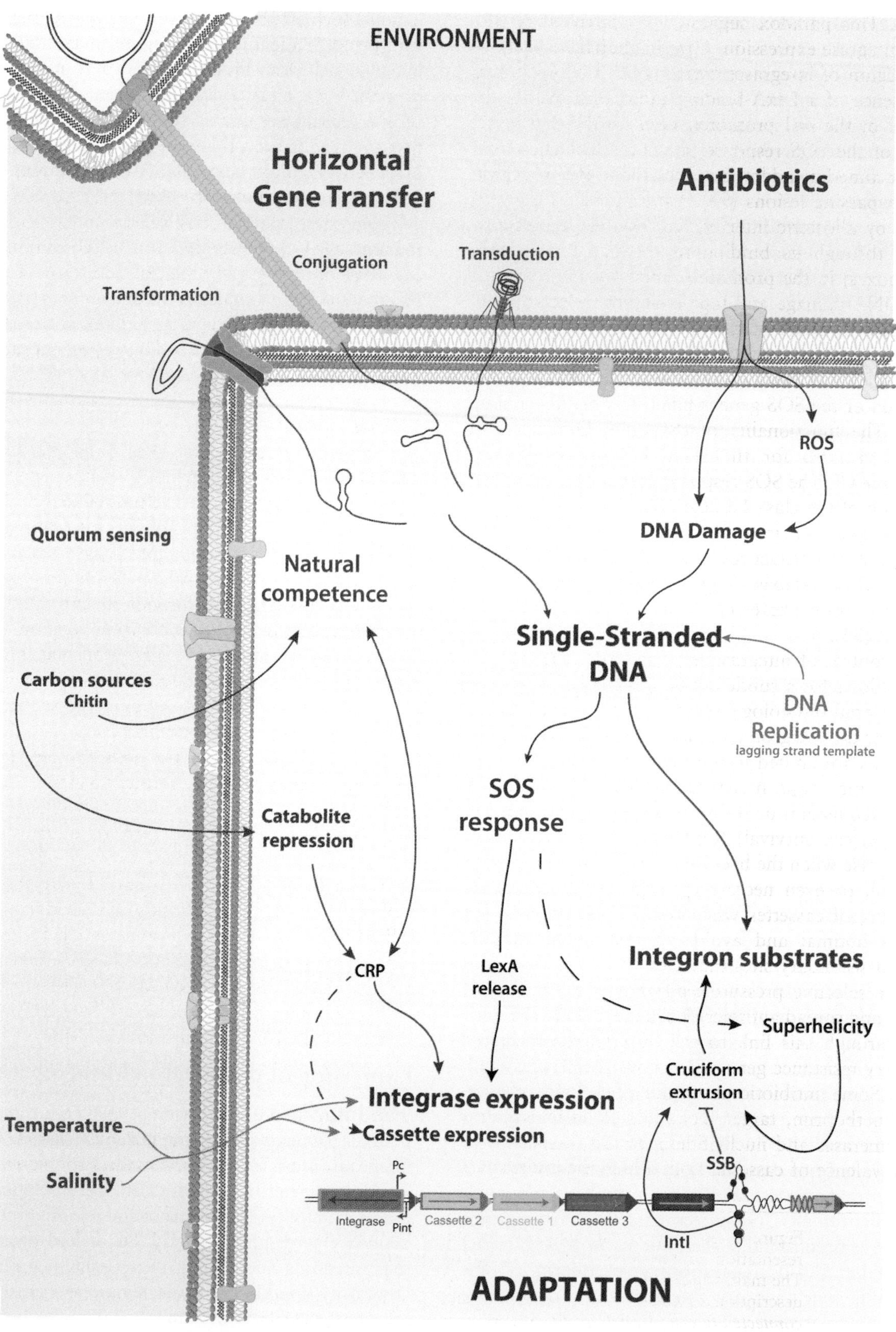
ENVIRONMENT
Horizontal Gene Transfer
Antibiotics
Conjugation
Transduction
Transformation
ROS
DNA Damage
Quorum sensing
Natural competence
Single-Stranded DNA
Carbon sources
Chitin
DNA Replication
lagging strand template
SOS response
Catabolite repression
CRP
LexA release
Integron substrates
Superhelicity
Cruciform extrusion
Integrase expression
Cassette expression
Temperature
Salinity
SSB
Pc
Pint
Integrase
Cassette 2
Cassette 1
Cassette 3
IntI
ADAPTATION

(90–93). This paradox suggested a controlled regulation of integrase expression. Careful analysis of the promoter region of integrases from CIs and MIs revealed the presence of a LexA-binding motif overlapping the –10 box of the *intI* promoter. LexA is the master repressor of the SOS response, a widespread regulatory network aimed at addressing DNA damage by repairing or bypassing lesions (94). LexA represses the SOS regulon by allosteric interference with the RNA polymerase, through its binding to specific DNA motifs (LexA boxes) in the promoter regions of the regulated genes. DNA damage and repair ultimately produce an increase of ss DNA in the cell that is recognized by RecA and promotes its polymerization. RecA nucleofilaments induce autocatalysis of LexA, releasing the repression of the SOS genes and triggering the SOS response. The functionality of the LexA box has been assessed *in vitro* for the class 1 integron and the *V. cholerae* CI. The SOS response increases 4.5-fold the expression of the class 1 integrase, and 37-fold that of the *V. cholerae* integrase (95). The LexA box has been identified in the sequences of a large proportion of integrases in the databases, suggesting that the SOS control of integrase expression is a conserved feature in integrons (96).

The control of integrase expression by the SOS response allows for a subtle coupling of integron activity with bacterial physiology. On one hand, by linking the activity of the integron to an alarm signal, recombination events are limited to the times when bacteria need to evolve and adapt. It is at these precise moments that cassette acquisition or reshuffling can have a dramatic impact on cell survival. Repressing the expression of the integrase when the bacterium is well adapted is also beneficial, or even necessary, because it prevents the reshuffling of cassettes when the configuration of the array is optimal and avoids random recombination events at secondary sites that could be deleterious (97, 98). The selective pressure leading MIs to exclusively capture and spread antimicrobial genes is clearly understood through this link to the SOS response. Indeed MIs carry resistance genes against most classes of antibiotics. Some antibiotic molecules, such as quinolones and trimethoprim, target DNA-related systems (DNA topoisomerases and nucleotide synthesis, respectively). The prevalence of cassettes containing resistance genes against such antibiotics can be explained through the link between the DNA damage produced by the antibiotic and the SOS-induced integron. A matter that had remained elusive was the presence of cassettes containing genes that confer resistance against antibiotics that do not damage DNA and do not induce the SOS response in *Escherichia coli*. This intriguing question led us to the discovery that aminoglycosides and β-lactams, antibiotics targeting the ribosome and the cell wall respectively, do induce the SOS response (even at subinhibitory concentrations) in species other than *E. coli*, such as *V. cholerae*, *Klebsiella pneumoniae* and *Photorhabdus luminescens* (99–101). The underlying mechanism for this seems to be the generation of reactive oxygen species intermediates that oxidize nucleotides and cause DNA damage, ultimately triggering the SOS response.

On the other hand, there is another aspect of the coupling to the SOS response that underscores the subtle adaptation of the integron to the host's needs: ssDNA is simultaneously the triggering signal for integron activity (through the SOS response) and its substrate. As we will see in depth in this chapter, one of the main sources of ssDNA and therefore drivers of integron activation is the entrance of new DNA during conjugation and natural transformation (58, 102), enabling screening of the incoming DNA for new cassettes to capture.

In the case of the *V. cholerae* integrase coding gene (*intIA*), a second layer of control, through the cAMP receptor protein (CRP), has been identified. CRP is the master regulator of the carbon catabolite repression response, adapting cellular metabolism to the type of carbon sources available in the environment. A CRP binding box is present in the promoter region of *intIA* (between P_{int} and P_C), controlling its expression independently of the SOS response and connecting the integron to environmental conditions. It is noteworthy that CRP is also directly and indirectly linked to the uptake of ssDNA, via TfoX and HapR (regulators of natural competence), respectively (102). Finally, low-level induction of the expression of *intIA* has also been observed at high temperature (42°C) (103).

The mechanisms of integrase expression regulation reveal that the integron is intimately connected to bacterial physiology and the environment, allowing it to be active when evolving is either mandatory or possible.

Figure 6 Intimate connection between the integron and cell physiology. A snapshot representation of the links between integrons' activity and bacterial physiology is shown. The main triggering signal for integrase expression is the bacterial SOS response. A detailed description of these connections is depicted in the section entitled: *A system intimately connected to cell physiology.* doi:10.1128/microbiolspec.MDNA3-0019-2014.f6

The Cassettes

As seen previously, cassettes are expressed from the P_C promoter in the integron. The extensive work of Ploy's laboratory on the expression of cassettes in the class I integron has led to a more complex view of the system, with several strength variants of the P_C promoter and an indirect influence of this promoter on the expression of the integrase (2). Indeed, by facing each other, P_{int} and P_C are subjected to transcriptional interference. Strong P_C variants will have a negative effect on P_{int} transcription when the latter is de-repressed during SOS induction. It has been suggested that this can serve as a repression system for the integrase in bacterial species lacking LexA, such as *Acinetobacter baumannii* (Couvé-Deacon, personal communication), where the unregulated expression of the integrase is harmful (98). Also, when present, the P_C2 promoter alters the LexA box, disrupting the SOS regulation of the integrase, but does not interfere transcriptionally with P_{int}, probably because they are close enough for their transcription starts (+1) not to face each other (104). The transcriptional interference between P_C and P_{int} establishes a trade-off between the expression of cassettes and of the integrase. The higher prevalence of weak P_C variants among clinical and nonclinical *E. coli* isolates, and bacteria from wastewater environments, suggests that a flexible cassette array is more important than enhanced expression (105, 106). Since P_C is encoded within the class 1 integrase, P_C variants also have an impact on the sequence of *intI1*. Interestingly, regardless of the transcriptional interference between promoters, the amino acid substitutions associated with the presence of strong promoters entail a decrease in the excision (but not in the integration) activity of the integrase (2).

The analysis of the P_C in the *V. cholerae* SI has recently been carried out in our group (103). As stated before, a CRP binding box is present between P_{int} and P_C, from where the cAMP–CRP complex activates not only the integrase, but also the expression of the cassette array. We have tested several conditions mimicking the different environments in which *V. cholerae* thrives, including different temperatures and salinities. We have observed a CRP-dependent increase in the expression levels of the cassette array during entry into stationary phase, as well as a correlation with temperature.

attC Site Folding

We previously described that *attC* sites are recognized and recombined as ss folded structures, a peculiarity conferring certain advantages to the cell. Indeed, secondary structures represent a way to expand information storage in DNA, in addition to the primary base sequence (107), and more importantly they allow the cell to control integron recombination through a variety of physiological processes that normally regulate the formation of secondary structures. This makes the integron an integrated system. Regulation of *attC* sites is controlled differently depending on whether the site folds as a hairpin from an ss molecule, or as a cruciform from dsDNA.

Single-Strand Pathway

Large fragments of ssDNA, from which *attC* sites can easily fold, can be found in the cell during some of its physiological processes or during the entry of exogenous DNA. Indeed, all horizontal gene transfer mechanisms, namely natural transformation, conjugation and transduction, involve the entry of only one strand of DNA into the recipient cell. Conjugative transfer of DNA containing *attC* sites favors the folding of *attC* sites of various lengths, and hence their recombination (45). At the same time, conjugation induces the expression of the integrase through the SOS response (58), a phenomenon also observed during transformation (102). Altogether, this allows the integron to recruit incoming cassettes.

Three essential cellular processes are a source of ssDNA: transcription, DNA repair, and replication. During replication, the lagging strand is synthesized discontinuously through the assembly of Okazaki fragments, a process in which short stretches of up to 2 kb of the template strand are found transiently in ss form, possibly favoring the folding of *attC* sites. Since the leading strand is synthesized continuously, it is possible that the effect of replication on *attC* site folding depends on the orientation of the integron relative to the replication fork. Indeed, published observations suggest that secondary structures are more easily formed when carried on the template of the lagging strand compared with the template of the leading strand (108). Accordingly, we have demonstrated that the *attC* bs is recombined with an *attI* site at a higher rate when carried on the template of the lagging strand. This suggests that the orientation of the cassette array in the chromosome influences the recombination frequency of *attC* sites, and more generally, that the cellular availability of ssDNA directly impacts the efficiency of integron recombination (107). It is noteworthy that CIs carry the *attC* site bs always on the leading strand template, limiting the excision of cassettes (45). The limited length of ssDNA stretches on the lagging strand template should favor cassette excision events between *attC* sites at less than 2 kb distance from each other, i.e. those resulting in the excision of a limited number of cassettes (1 or 2).

Interestingly, ssDNA production can be triggered in response to stress or specific environmental determinants. For example, in the Gram-positive bacterium *Streptococcus pneumoniae*, competence (and therefore single-strand availability through DNA uptake) is induced by an antibiotic stress response (109). In the Gram-negative human pathogen, *Helicobacter pylori*, competence is increased in conditions of low CO_2 and high pH (110).

Double-Strand Pathway

Another pathway for *attC* site structuring is its extrusion as a cruciform from a dsDNA molecule, a process largely dependent on superhelicity. Formation of cruciforms by annealing of inverted repeats involves deep structural disruption and reorganization of base pairing. Until recently, only extrusions from perfect palindromes have been observed *in vivo*, because imperfections have major negative effects on the overall dynamics of cruciform extrusion (111). We have proven the cruciform extrusion of *attC* sites from dsDNA, including sites with very large VTSs such as VCRs. The positive impact of superhelicity on *attC* folding has been established using topoisomerase I-deficient and gyrase-deficient (the enzymes maintaining supercoiling levels in *E. coli*) strains (112, 113).

Cruciforms are seldom extruded at significant rates under average *in vivo* supercoiling conditions. However, many factors may transiently increase local superhelical density to a level sufficient for cruciform extrusion (107), and hence can impact integron recombination by favoring *attC* site extrusion. Among these factors, we can find biological processes such as transcription and replication (114); growth phase; environmental stimuli, such as antibiotics and growth conditions (115–117); and/or internal stimuli such as the induction of the SOS response that entails higher levels of negative supercoiling (118). Finally, bacterial species can present different levels of superhelicity, resulting in different levels of integron recombination efficiencies according to the host strain (119).

SSB and Integrase Duality

Bacteria have to find a subtle balance between the benefit provided by encoding biological functions in secondary structures and the detrimental effects of possessing an excess of them. Indeed, overly long and stable palindromes could not be maintained *in vivo*, either because they are inviable (i.e. intrinsically toxic to the cell) or because they are genetically unstable (i.e. partially mutated or deleted) (120). It is assumed that *inviability* is caused by an arrest of the replication fork, as it is unable to process these secondary structures, and *instability* is caused by the presence of proteins such as SbcCD, which destroys these structures. This leads to constraints on the size and perfection of the inverted repeats and/or the need of host factor regulation of secondary structure folding.

Among these host factors, the single-stranded DNA-binding protein (SSB) is able to bind cellular ssDNA without sequence specificity. SSB plays extensive cellular roles in DNA replication, repair, and homologous recombination, where it prevents premature annealing, stabilizes and protects the single-stranded DNA, and removes secondary structures (121) by migrating along ssDNA and melting unstable hairpins while stimulating RecA filament elongation (122). SSB assures the stability and viability of *attC* sites by flattening the hairpin. We also demonstrated that the integrase could capture *attC* sites at the moment of their extrusion, efficiently stabilizing and recombining them. Therefore, when expressed, the integrase is able to counteract the effect of SSB (55, 123, 124).

Holliday Junction Resolution

As we have seen, the *attC* × *attI* reaction, which leads to the integration of cassettes, forms an atypical HJ that is resolved through replication, and this is likely to also be the case for the *attC* × *attC* reaction. However, it is unknown whether the replicative resolution is a passive process, involving a replisome that was assembled at the replication origin and implying that the aHJ remains stable until its arrival; or an active one, for which a new replisome is assembled *ad hoc*. The set of proteins involved would be different for both cases: in the first scenario, in *E. coli* it would include DnaA (the replication initiation factor that promotes the unwinding or denaturation of DNA), DnaE (the catalytic subunit of the *pol*III polymerase essential for processive replication) and DnaN (the clamp, presumably essential for all kinds of replication). The second scenario considers that the aHJ could mimic an arrested replication fork, thus involving the local recruitment of replication complexes able to restart halted forks. In bacteria, this essential activity is orchestrated by the PriA DNA helicase, which identifies stalled replication forks via structure-specific DNA binding. We propose that DNA replication start could be initiated by PriA at the aHJ, permitting its resolution (125, 126).

Accessory Host Factors

It has been observed that, contrary to the class I integron system, which recombines equally in *E. coli* and *V. cholerae*, the VchIntIA integrase of the *V. cholerae*

CI recombines at a 2,600-fold higher rate the VCR sites in its original genomic background than in *E. coli* strains (37). Such results suggest either the involvement of one or more host factors in *V. cholerae* that would be absent or too divergent in *E. coli* strains, or the presence of an inhibitory factor in *E. coli*. The nature of CIs and MIs suggests that the first case is more likely, and that the lack of dependence on host factors is what has made class 1 integrons so successful upon mobilization.

A BIOTECHNOLOGICAL TOOL

Several applications of the integron as a bioengineering tool have been developed. For instance, it has been suggested to harness the capacity of the integron to incorporate new cassettes to perform sequence-independent recovery of integron cassettes from genomic libraries (127). The particular utility of this tool lies in its ability to access integron cassettes from unculturable organisms, and the interest of their retrieval is supported by a recent analysis suggesting that the integron cassette metagenome contains a repertoire of genes belonging to new, currently uncharacterized protein families with possible novel functions (128).

An integron-based cloning technique has also been developed, based on the delivery of synthetic cassettes into the genetic backbone of both MIs and CIs via natural transformation (129). This technique can be used with wild-type environmental bacteria, does not require a vector, can yield high frequencies of recombinants in favorable conditions, and recombinants are stable in the absence of selection. The lack of a cloning vector and associated antibiotic resistance genes makes the technique particularly appealing from the biosafety perspective.

The capacity of the integron system to rearrange cassettes has led to its application as an *in vivo* genetic shuffling device (130). In fact, the synthetic integron allows the generation of a large number of genetic combinations and arrangements using site-specific recombination, which is of particular interest for metabolic pathway optimization, where the selection of optimal arrangements of genetic elements can lead to higher production yields. Moreover, the flexibility of *attC* sites suggests the possibility of designing synthetic *attC* sites "à la carte" and embedding them into elements having a distinct function on the sequence level, for instance coding for a protein or a promoter. This avenue of synthetic integron development is being pursued, and might lead to new applications of the system for bioengineering purposes.

EVOLUTIONARY IMPLICATIONS

Integrons are powerful agents of bacterial evolution, granting access to a vast variety of functions encoded in cassettes. As we have seen, integrons are seamlessly coordinated with the physiology of bacteria, so that evolution becomes a somewhat domesticated function rather than a stochastic inevitability. The power of integrons has changed our view on the adaptive capabilities of bacteria as evidenced by their role in the unforeseen rise of multiresistance during the 1960s.

On the Success of MIs

Antibiotic pressure of anthropogenic origin has been extraordinarily high during the last 70 years, leading to the selection of integron mobilization events. Indeed, integrons were not originally present in the genomes of pathogenic enterobacteria where they are today in some cases redundant. It is through the association of integrons with transposable elements and conjugative plasmids that integrons have entered circulation among clinically relevant strains. Yet, it remains unclear why not all MIs have had the same spread or impact on antimicrobial resistance. The success of horizontal gene transfer events has been suggested to depend on the *penetration*, *promiscuity*, *plasticity* and *persistence* capacities (the four Ps) of any given determinant (131). It is not known whether the success of the different MI classes is due to the platform to which each class is associated, and/or to intrinsic characteristics of the integron. Mobilizing platforms can affect the penetrance and promiscuity of MIs through differential transposition/conjugation rates or host range; and their persistence through differences in the fitness cost they impose on the host. On their side, it is also possible that MI-integrases show differential properties, such as distinct rates of activity or differences in the range of *attC*-sites recognized. Broader substrate recognition has indeed been observed to be the case for IntI1 compared with VchIntIA and IntIPstQ from *Pseudomonas stutzeri* (37, 132). It is tempting to speculate that the success of class 1 integrons could be due, at least partly, to IntI1 recognizing a broader variety of *attC* sites and conferring access to a larger portion of cassette-encoded functions. Regarding the influence of integron characteristics in their persistence, it could be argued that MI are stabilized by high antibiotic pressure alone. Nevertheless, the picture we have of CIs suggests that the stability of integrons is probably assured through both regulation of integrase activity by host regulatory networks (98), and control over *attC* site folding (SSB flattening the folded *attC* sites to avoid their instability and/or their unviability, integrase stabilizing them, supercoiling, horizontal gene transfer and replication favoring their folding).

Genesis of Integron Cassettes

We now have a good understanding of the recombination reactions catalyzed by integrases, and an increasing knowledge of the connection of the integron with the host. Still, some major questions remain unanswered, such as the *de novo* creation of cassettes. In cassettes, the general absence of promoters, together with the paucity of pseudogenes or noncoding sequence, has been interpreted to be representative of an RNA origin (133). In this model, the transcriptional terminator of the gene of interest is fused to an *attC* site through homologous recombination and further retrotranscribed into a DNA cassette. This is performed by a Group IIC-*attC* intron, a type of intron with affinity for palindromic sequences including *attC* sites, as well as retrotranscriptase activity (133). Unfortunately, direct evidence supporting this model remains scarce, and these elements have seldom been found within integrons (133, 134). Although the rationale for the RNA origin of cassettes is interesting, some arguments against this theory are also solid. For instance, in this scenario, what mechanism would allow for the creation of cassettes that do encode promoters, such as TA cassettes? Indeed, if the RNA origin of cassettes were to be true, at least one additional mechanism would be needed to produce cassettes from DNA so that promoters are present. Also, and bearing in mind the millions of years of evolution of integrons and the subtleties of their intertwining with the host machinery, it seems unlikely that a function as important as the creation of cassettes relies on an independent entity that is found in a very low percentage of integrons. Instead, if Group IIC-*attC* introns were cassette generators, one would expect them to be found as part of the constant platform of integrons. This should be especially the case in CIs in which *attC* sites show high levels of identity, a feature suggesting a link of some kind between cassette creation and the host. This is the case, as we have already mentioned, of the VCRs within the SI of *V. cholerae*, a species in which the presence of Group IIC-*attC* introns has not been reported. All in all, the creation of cassettes remains a subject of the utmost importance for understanding integrons, for which an undoubted model is not available.

Acknowledgments. *This work was supported by the Institut Pasteur, the Centre National de la Recherche Scientifique (CNRS-UMR3525), the European Union Seventh Framework Programme (FP7-HEALTH-2011-single-stage), the "Evolution and Transfer of Antibiotic Resistance" (EvoTAR) and the French Government's Investissement d'Avenir program Laboratoire d'Excellence "Integrative Biology of Emerging Infectious Diseases" (ANR-10-LABX-62-IBEID) and the French National Research Agency (ANR-12-BLAN-DynamINT). JAE is supported by the Marie Curie Intra-European Fellowship for Career Development (FP-7-PEOPLE-2011-IEF).*

The authors acknowledge Dr Jason Bland and Henry Kemble for critical reading of the review.

Citation. Escudero JA, Loot C, Nivina A, Mazel D. 2014. The integron: adaptation on demand. Microbiol Spectrum 2(6):MDNA3-0019-2014.

References

1. **Stokes HW, Hall RM.** 1989. A novel family of potentially mobile DNA elements encoding site-specific gene-integration functions: integrons. *Mol Microbiol* **3:** 1669–1683.
2. **Jové T, Da Re S, Denis F, Mazel D, Ploy MC.** 2010. Inverse correlation between promoter strength and excision activity in class 1 integrons. *PLoS Genet* **6:**e1000793.
3. **Levesque C, Brassard S, Lapointe J, Roy PH.** 1994. Diversity and relative strength of tandem promoters for the antibiotic-resistance genes of several integrons. *Gene* **142:**49–54.
4. **Collis CM, Hall RM.** 1995. Expression of antibiotic resistance genes in the integrated cassettes of integrons. *Antimicrob Agents Chemother* **39:**155–162.
5. **Collis CM, Hall RM.** 1992. Gene cassettes from the insert region of integrons are excised as covalently closed circles. *Mol Microbiol* **6:**2875–2885.
6. **Rowe-Magnus DA, Guerout AM, Ploncard P, Dychinco B, Davies J, Mazel D.** 2001. The evolutionary history of chromosomal super-integrons provides an ancestry for multiresistant integrons. *Proc Natl Acad Sci U S A* **98:** 652–657.
7. **Cambray G, Guerout AM, Mazel D.** 2010. Integrons. *Annu Rev Genet* **44:**141–166.
8. **Nemergut DR, Robeson MS, Kysela RF, Martin AP, Schmidt SK, Knight R.** 2008. Insights and inferences about integron evolution from genomic data. *BMC genomics* **9:**261.
9. **Cambray G, Mazel D.** 2008. Synonymous genes explore different evolutionary landscapes. *PLoS Genet* **4**(11): e1000256.
10. **Boucher Y, Labbate M, Koenig JE, Stokes HW.** 2007. Integrons: mobilizable platforms that promote genetic diversity in bacteria. *Trends Microbiol* **15:**301–309.
11. **Mazel D.** 2006. Integrons: agents of bacterial evolution. *Nature RevMicrobiol* **4:**608–620.
12. **Mazel D, Dychinco B, Webb VA, Davies J.** 1998. A distinctive class of integron in the *Vibrio cholerae* genome. *Science* **280:**605–608.
13. **Ogawa A, Takeda T.** 1993. The gene encoding the heat-stable enterotoxin of *Vibrio cholerae* is flanked by 123-base pair direct repeats. *Microbiol Immunol* **37:** 607–616.

14. **Barker A, Clark CA, Manning PA.** 1994. Identification of VCR, a repeated sequence associated with a locus encoding a hemagglutinin in *Vibrio cholerae* O1. *J Bacteriol* **176:**5450–5458.

15. **Rowe-Magnus DA, Guerout AM, Biskri L, Bouige P, Mazel D.** 2003. Comparative analysis of super-integrons: engineering extensive genetic diversity in the Vibrionaceae. *Genome Res* **13:**428–442.

16. **Mitsuhashi S, Harada K, Hashimoto H, Egawa R.** 1961. On the drug-resistance of enteric bacteria. 4. Drug-resistance of Shigella prevalent in Japan. *Jpn J Exp Med* **31:**47–52.

17. **Martinez E, de la Cruz F.** 1988. Transposon Tn21 encodes a RecA-independent site-specific integration system. *Mol Gen Genet* **211:**320–325.

18. **Arakawa Y, Murakami M, Suzuki K, Ito H, Wacharotayankun R, Ohsuka S, Kato N, Ohta M.** 1995. A novel integron-like element carrying the metallo-beta-lactamase gene blaIMP. *Antimicrob Agents Chemother* **39:**1612–1615.

19. **Collis CM, Kim MJ, Partridge SR, Stokes HW, Hall RM.** 2002. Characterization of the Class 3 integron and the site-specific recombination system it determines. *J Bacteriol* **184:**3017–3026.

20. **Xu H, Davies J, Miao V.** 2007. Molecular characterization of class 3 integrons from *Delftia* spp. *J Bacteriol* **189:**6276–683.

21. **Ramírez MS, Piñeiro S, Centrón D.** 2010. Novel insights about class 2 integrons from experimental and genomic epidemiology. *Antimicrob Agents Chemother* **54:**699–706.

22. **Hochhut B, Lotfi Y, Mazel D, Faruque SM, Woodgate R, Waldor MK.** 2001. Molecular analysis of antibiotic resistance gene clusters in *Vibrio cholerae* O139 and O1 SXT constins. *Antimicrob Agents Chemother* **45:**2991–3000.

23. **Sørum H, Roberts MC, Crosa JH.** 1992. Identification and cloning of a tetracycline resistance gene from the fish pathogen *Vibrio salmonicida*. *Antimicrob Agents Chemother*. **36:**611–615.

24. **Naas T, Mikami Y, Imai T, Poirel L, Nordmann P.** 2001. Characterization of In53, a class 1 plasmid- and composite transposon-located integron of *Escherichia coli* which carries an unusual array of gene cassettes. *J Bacteriol* **183:**235–249.

25. **Partridge SR, Tsafnat G, Coiera E, Iredell JR.** 2009. Gene cassettes and cassette arrays in mobile resistance integrons. *FEMS Microbiol Rev* **33:**757–784.

26. **Fluit AC, Schmitz FJ.** 2004. Resistance integrons and super-integrons. *Clin Microbiol Infect* **10:**272–288.

27. **Labbate M, Case RJ, Stokes HW.** 2009. The integron/gene cassette system: an active player in bacterial adaptation. *Methods Mol Biol* **532:**103–125.

28. **Martin C, Timm J, Rauzier J, Gomez-Lus R, Davies J, Gicquel B.** 1990. Transposition of an antibiotic resistance element in mycobacteria. *Nature* **345:**739–743.

29. **Nandi S, Maurer JJ, Hofacre C, Summers AO.** 2004. Gram-positive bacteria are a major reservoir of Class 1 antibiotic resistance integrons in poultry litter. *Proc Natl Acad Sci USA* **101:**7118–7122.

30. **Nesvera J, Hochmannová J, Pátek M.** 1998. An integron of class 1 is present on the plasmid pCG4 from gram-positive bacterium *Corynebacterium glutamicum*. *FEMS Microbiol Lett* **169:**391–395.

31. **Shi L, Zheng M, Xiao Z, Asakura M, Su J, Li L, Yamasaki S.** 2006. Unnoticed spread of class 1 integrons in gram-positive clinical strains isolated in Guangzhou, China. *Microbiol Immunol* **50:**463–467.

32. **Hocquet D, Llanes C, Thouverez M, Kulasekara HD, Bertrand X, Plésiat P, Mazel D, Miller SI.** 2012. Evidence for induction of integron-based antibiotic resistance by the SOS response in a clinical setting. *PLoS Pathog* **8:**e1002778.

33. **Gravel A, Fournier B, Roy PH.** 1998. DNA complexes obtained with the integron integrase IntI1 at the attI1 site. *Nucleic Acids Res* **26:**4347–4355.

34. **Partridge SR, Recchia GD, Scaramuzzi C, Collis CM, Stokes HW, Hall RM.** 2000. Definition of the attI1 site of class 1 integrons. *Microbiology* **146:**2855–2864.

35. **Nield BS, Holmes AJ, Gillings MR, Recchia GD, Mabbutt BC, Nevalainen KM, Stokes HW.** 2001. Recovery of new integron classes from environmental DNA. *FEMS Microbiol Lett* **195:**59–65.

36. **Elsaied H, Stokes HW, Kitamura K, Kurusu Y, Kamagata Y, Maruyama A.** 2011. Marine integrons containing novel integrase genes, attachment sites, attI, and associated gene cassettes in polluted sediments from Suez and Tokyo Bays. *ISME J* **5:**1162–1177.

37. **Biskri L, Bouvier M, Guérout AM, Boisnard S, Mazel D.** 2005. Comparative study of class 1 integron and *Vibrio cholerae* superintegron integrase activities. *J Bacteriol* **187:**1740–1750.

38. **Stokes HW, O'Gorman DB, Recchia GD, Parsekhian M, Hall RM.** 1997. Structure and function of 59-base element recombination sites associated with mobile gene cassettes. *Mol Microbiol* **26:**731–745.

39. **Hall RM, Brookes DE, Stokes HW.** 1991. Site-specific insertion of genes into integrons: role of the 59-base element and determination of the recombination cross-over point. *Mol Microbiol* **5:**1941–1959.

40. **Francia MV, Zabala JC, de la Cruz F, García Lobo JM.** 1999. The IntI1 integron integrase preferentially binds single-stranded DNA of the attC site. *J Bacteriol* **181:**6844–6849.

41. **Johansson C, Kamali-Moghaddam M, Sundstrom L.** 2004. Integron integrase binds to bulged hairpin DNA. *Nucleic Acids Res* **32:**4033–4043.

42. **Bouvier M, Demarre G, Mazel D.** 2005. Integron cassette insertion: a recombination process involving a folded single strand substrate. *EMBO J* **24:**4356–4367.

43. **Bouvier M, Ducos-Galand M, Loot C, Bikard D, Mazel D.** 2009. Structural features of single-stranded integron cassette attC sites and their role in strand selection. *PLoS Genetics* **5:**e1000632.

44. **MacDonald D, Demarre G, Bouvier M, Mazel D, Gopaul DN.** 2006. Structural basis for broad DNA-specificity in integron recombination. *Nature* **440:**1157–1162.

45. **Loot C, Bikard D, Rachlin A, Mazel D.** 2010. Cellular pathways controlling integron cassette site folding. *EMBO J* **29:**2623–2634.

46. **Nunes-Düby SE, Kwon HJ, Tirumalai RS, Ellenberger T, Landy A.** 1998. Similarities and differences among 105 members of the Int family of site-specific recombinases. *Nucleic Acids Res* **26:**391–406.

47. **Boyd EF, Almagro-Moreno S, Parent MA.** 2009. Genomic islands are dynamic, ancient integrative elements in bacterial evolution. *Trends Microbiol* **17:**47–53.

48. **Grindley N, Whiteson K, Rice P.** 2006. Mechanisms of site-specific recombination. *Annu Rev Biochem* **75:** 567–605.

49. **Messier N, Roy PH.** 2001. Integron integrases possess a unique additional domain necessary for activity. *J Bacteriol* **183:**6699–6706.

50. **Demarre G, Frumerie C, Gopaul DN, Mazel D.** 2007. Identification of key structural determinants of the IntI1 integron integrase that influence attC × attI1 recombination efficiency. *Nucleic Acids Res* **35:**6475–6489.

51. **Gravel A, Messier N, Roy PH.** 1998. Point mutations in the integron integrase IntI1 that affect recombination and/or substrate recognition. *J Bacteriol* **180:**5437–5442.

52. **Johansson C, Boukharta L, Eriksson J, Aqvist J, Sundström L.** 2009. Mutagenesis and homology modeling of the Tn21 integron integrase IntI1. *Biochemistry* **48:**1743–1753.

53. **Val ME, Bouvier M, Campos J, Sherratt D, Cornet F, Mazel D, Barre FX.** 2005. The single-stranded genome of phage CTX is the form used for integration into the genome of *Vibrio cholerae*. *Mol Cell* **19:**559–566.

54. **Loot C, Ducos-Galand M, Escudero JA, Bouvier M, Mazel D.** 2012. Replicative resolution of integron cassette insertion. *Nucleic Acids Res* **40:**8361–8370.

55. **Loot C, Parissi V, Escudero JA, Amarir-Bouhram J, Bikard D, Mazel D.** 2014. The integron integrase efficiently prevents the melting effect of *Escherichia coli* single-stranded DNA-binding protein on folded attC sites. *J Bacteriol* **196:**762–771.

56. **Collis CM, Grammaticopoulos G, Briton J, Stokes HW, Hall RM.** 1993. Site-specific insertion of gene cassettes into integrons. *Mol Microbiol* **9:**41–52.

57. **Collis CM, Recchia GD, Kim MJ, Stokes HW, Hall RM.** 2001. Efficiency of recombination reactions catalyzed by class 1 integron integrase IntI1. *J Bacteriol* **183:**2535–2542.

58. **Baharoglu Z, Bikard D, Mazel D.** 2010. Conjugative DNA transfer induces the bacterial SOS response and promotes antibiotic resistance development through integron activation. *PLoS Genet* **6:**e1001165.

59. **Collis CM, Hall RM.** 1992. Site-specific deletion and rearrangement of integron insert genes catalyzed by the integron DNA integrase. *J Bacteriol* **174:**1574–1585.

60. **Gestal AM, Stokes HW, Partridge SR, Hall RM.** 2005. Recombination between the dfrA12-orfF-aadA2 cassette array and an aadA1 gene cassette creates a hybrid cassette, aadA8b. *Antimicrob Agents Chemother* **49:**4771–4774.

61. **Hansson K, Sköld O, Sundström L.** 1997. Non-palindromic attI sites of integrons are capable of site-specific recombination with one another and with secondary targets. *Mol Microbiol* **26:**441–453.

62. **Roy Chowdhury P, Merlino J, Labbate M, Cheong EY, Gottlieb T, Stokes HW.** 2009. Tn6060, a transposon from a genomic island in a Pseudomonas aeruginosa clinical isolate that includes two class 1 integrons. *Antimicrob Agents Chemother* **53:**5294–5296.

63. **González-Zorn B, Catalan A, Escudero JA, Domínguez L, Teshager T, Porrero C, Moreno MA.** 2005. Genetic basis for dissemination of armA. *J Antimicrob Chemother* **56:**583–585.

64. **Recchia GD, Hall RM.** 1995. Plasmid evolution by acquisition of mobile gene cassettes: plasmid pIE723 contains the aadB gene cassette precisely inserted at a secondary site in the incQ plasmid RSF1010. *Mol Microbiol* **15:**179–187

65. **Segal H, Francia MV, Lobo JM, Elisha G.** 1999. Reconstruction of an active integron recombination site after integration of a gene cassette at a secondary site. *Antimicrob Agents Chemother* **43:**2538–2541.

66. **Jacquier H, Zaoui C, Sanson-le Pors MJ, Mazel D, Berçot B.** 2009. Translation regulation of integrons gene cassette expression by the attC sites. *Mol Microbiol* **72:**1475–1486.

67. **Shultzaberger RK, Bucheimer RE, Rudd KE, Schneider TD.** 2001. Anatomy of Escherichia coli ribosome binding sites. *J Mol Biol* **313:**215–228.

68. **Hanau-Berçot B, Podglajen I, Casin I, Collatz E.** 2002. An intrinsic control element for translational initiation in class 1 integrons. *Mol Microbiol* **44:**119–130.

69. **Bissonnette L, Champetier S, Buisson JP, Roy PH.** 1991. Characterization of the nonenzymatic chloramphenicol resistance (*cmlA*) gene of the In4 integron of Tn1696: similarity of the product to transmembrane transport proteins. *J Bacteriol* **173:**4493–4502.

70. **Stokes HW, Hall RM.** 1991. Sequence analysis of the inducible chloramphenicol resistance determinant in the Tn1696 integron suggests regulation by translational attenuation. *Plasmid* **26:**10–19.

71. **Biskri L, Mazel D.** 2003. Erythromycin esterase gene ere(A) is located in a functional gene cassette in an unusual class 2 integron. *Antimicrob Agents Chemother* **47:**3326–3331.

72. **da Fonseca ÉL, Vicente AC.** 2012. Functional characterization of a Cassette-specific promoter in the class 1 integron-associated qnrVC1 gene. *Antimicrob Agents Chemother* **56:**3392–3394.

73. **Szekeres S, Dauti M, Wilde C, Mazel D, Rowe-Magnus DA.** 2007. Chromosomal toxin-antitoxin loci can diminish large-scale genome reductions in the absence of selection. *Mol Microbiol* **63:**1588–1605.

74. **Guérout AM, Iqbal N, Mine N, Ducos-Galand M, Van Melderen L, Mazel D.** 2013. Characterization of the phd-doc and ccd toxin-antitoxin cassettes from *Vibrio* superintegrons. *J Bacteriol* **195:**2270–2283.

75. **Elsaied H, Stokes HW, Nakamura T, Kitamura K, Fuse H, Maruyama A.** 2007. Novel and diverse integron

integrase genes and integron-like gene cassettes are prevalent in deep-sea hydrothermal vents. *Environ Microbiol* 9:2298–2312.

76. **Stokes HW, Holmes AJ, Nield BS, Holley MP, Nevalainen KM, Mabbutt BC, Gillings MR.** 2001. Gene cassette PCR: sequence-independent recovery of entire genes from environmental DNA. *Appl Environ Microbiol* 67:5240–5246.
77. **Koenig JE, Boucher Y, Charlebois RL, Nesbø C, Zhaxybayeva O, Bapteste E, Spencer M, Joss MJ, Stokes HW, Doolittle WF.** 2008. Integron-associated gene cassettes in Halifax Harbour: assessment of a mobile gene pool in marine sediments. *Environ Microbiol* 10:1024–1038.
78. **Gillings MR, Holley MP, Stokes HW.** 2009. Evidence for dynamic exchange of qac gene cassettes between class 1 integrons and other integrons in freshwater biofilms. *FEMS Microbiol Lett* 296:282–288.
79. **Holmes AJ, Gillings MR, Nield BS, Mabbutt BC, Nevalainen KM, Stokes HW.** 2003. The gene cassette metagenome is a basic resource for bacterial genome evolution. *Environ Microbiol* 5:383–394.
80. **Gillings M, Boucher Y, Labbate M, Holmes A, Krishnan S, Holley M, Stokes HW.** 2008. The evolution of class 1 integrons and the rise of antibiotic resistance. *J Bacteriol* 190:5095–5100.
81. **Gillings MR, Holley MP, Stokes HW, Holmes AJ.** 2005. Integrons in Xanthomonas: a source of speciesgenome diversity. *Proc Natl Acad Sci USA* 102:4419–4424.
82. **Rapa RA, Labbate M.** 2013. The function of integron-associated gene cassettes in *Vibrio* species: the tip of the iceberg. *Front Microbiol* 4:385.
83. **Gerdes K, Christensen SK, Lobner-Olesen A.** 2005. Prokaryotic toxin-antitoxin stress response loci. *Nat Rev Microbiol* 3:371–382.
84. **Yamaguchi Y, Park JH, Inouye M.** 2011. Toxin-antitoxin systems in bacteria and archaea. *Annu Rev Genet* 45: 61–79.
85. **Van Melderen L, Saavedra De Bast M.** 2009. Bacterial toxin-antitoxin systems: more than selfish entities? *PLoS Genetics* 5:e1000437.
86. **Sberro H, Leavitt A, Kiro R, Koh E, Peleg Y, Qimron U, Sorek R.** 2013. Discovery of functional toxin/antitoxin systems in bacteria by shotgun cloning. *Molecular Cell* 50:136–148.
87. **Tsafnat G, Copty J, Partridge SR.** 2011. RAC: Repository of Antibiotic resistance Cassettes. Database: the journal of biological databases and curation. 2011:bar054.
88. **Moura A, Soares M, Pereira C, Leitão N, Henriques I, Correia A.** 2009. INTEGRALL: a database and search engine for integrons, integrases and gene cassettes. *Bioinformatics* 25:1096–1098.
89. **Joss MJ, Koenig JE, Labbate M, Polz MF, Gillings MR, Stokes HW, Doolittle WF, Boucher Y.** 2009. ACID: annotation of cassette and integron data. *BMC Bioinformatics* 10:118.
90. **Rowe-Magnus DA, Guérout AM, Mazel D.** 1999. Super-integrons. *Res Microbiol* 150:641–651.
91. **Chowdhury N, Asakura M, Neogi SB, Hinenoya A, Haldar S, Ramamurthy T, Sarkar BL, Faruque SM, Yamasaki S.** 2010. Development of simple and rapid PCR-fingerprinting methods for Vibrio cholerae on the basis of genetic diversity of the superintegron. *J Appl Microbiol* 109:304–312.
92. **Feng L, Reeves PR, Lan R, Ren Y, Gao C, Zhou Z, Cheng J, Wang W, Wang J, Qian W, Li D, Wang L.** 2008. A recalibrated molecular clock and independent origins for the cholera pandemic clones. *PloS One* 3:e4053.
93. **Labbate M, Boucher Y, Joss MJ, Michael CA, Gillings MR, Stokes HW.** 2007. Use of chromosomal integron arrays as a phylogenetic typing system for *Vibrio cholerae* pandemic strains. *Microbiology* 153:1488–1498.
94. **Erill I, Campoy S, Barbé J.** 2007. Aeons of distress: an evolutionary perspective on the bacterial SOS response. *FEMS Microbiol Rev* 31:637–656.
95. **Guerin E, Cambray G, Sanchez-Alberola N, Campoy S, Erill I, Da Re S, Gonzalez-Zorn B, Barbe J, Ploy MC, Mazel D.** 2009. The SOS response controls integron recombination. *Science* 324:1034.
96. **Cambray G, Sanchez-Alberola N, Campoy S, Guerin E, Da Re S, González-Zorn B, Ploy MC, Barbe J, Mazel D, Erill I.** 2011. Prevalence of SOS-mediated control of integron integrase expression as an adaptive trait of chromosomal and mobile integrons. *Mob DNA* 2:6.
97. **Harms K, Starikova I, Johnsen PJ.** 2013. Costly Class-1 integrons and the domestication of the the functional integrase. *Mobile Genetic Elements* 3:e24774.
98. **Starikova I, Harms K, Haugen P, Lunde T, Primicerio R, Samuelsen Ø, Nielsen KM, Johnsen PJ.** 2012. A trade-off between the fitness cost of functional integrases and long-term stability of integrons. *PLoS pathogens* 8:e1003043.
99. **Baharoglu Z, Mazel D.** 2011. *Vibrio cholerae* triggers SOS and mutagenesis in response to a wide range of antibiotics: a route towards multiresistance. *Antimicrob Agents Chemother* 55:2438–2441.
100. **Baharoglu Z, Krin E, Mazel D.** 2013. RpoS plays a central role in the SOS induction by sub-lethal aminoglycoside concentrations in *Vibrio cholerae. PLoS Genetics* 9:e1003421.
101. **Gutierrez A, Laureti L, Crussard S, Abida H, Rodríguez-Rojas A, Blázquez J, Baharoglu Z, Mazel D, Darfeuille F, Vogel J, Matic I.** 2013. β-Lactam antibiotics promote bacterial mutagenesis via an RpoS-mediatedreduction in replication fidelity. *Nature Commun* 4:1610.
102. **Baharoglu Z, Krin E, Mazel D.** 2012. Connecting environment and genome plasticity in the characterization of transformation-induced SOS regulation and carbon catabolite control of the *Vibrio cholerae* integron integrase. *J Bacteriol* 194:1659–1667.
103. **Krin E, Cambray G, Mazel D.** 2014. The superintegron integrase and the cassette promoters are co-regulated in *Vibrio cholerae. PLoS One* 9:e91194.

104. **Guérin E, Jové T, Tabesse A, Mazel D, Ploy MC.** 2011. High-level gene cassette transcription prevents integrase expression in class 1 integrons. *J Bacteriol* **193:**5675–5682.

105. **Vinué L, Jové T, Torres C, Ploy MC.** 2011. Diversity ofclass 1 integron gene cassette Pc promoter variants inclinical *Escherichia coli* strains and description of a new P2 promoter variant. *Int J Antimicrob Agents* **38:** 526–529.

106. **Moura A, Jové T, Ploy MC, Henriques I, Correia A.** 2012. Diversity of gene cassette promoters in class 1 integrons from wastewater environments. *Appl Environ Microbiol* **78:**5413–5416.

107. **Bikard D, Loot C, Baharoglu Z, Mazel D.** 2010. Folded DNA in action: hairpin formation and biological functions in prokaryotes. *Microbiol Mol BiolRev* **74:** 570–588.

108. **Trinh TQ, Sinden RR.** 1991. Preferential DNA secondary structure mutagenesis in the lagging strand of replication in *E. coli*. *Nature* **352:**544–547.

109. **Prudhomme M, Attaiech L, Sanchez G, Martin B, Claverys JP.** 2006. Antibiotic stress induces genetic transformability in the human pathogen *Streptococcus pneumoniae*. *Science* **313:**89–92.

110. **Moore ME, Lam A, Bhatnagar S, Solnick JV.** 2014. Environmental determinants of transformation efficiency in *Helicobacter pylori*. *J Bacteriol* **196:**337–344.

111. **Pearson CE, Zorbas H, Price GB, Zannis-Hadjopoulos M.** 1996. Inverted repeats, stem-loops, and cruciforms: significance for initiation of DNA replication. *J Cell Biochem* **63:**1–22.

112. **Lodge JK, Kazic T, Berg DE.** 1989. Formation of supercoiling domains in plasmid pBR322. *J Bacteriol* **171:**2181–2187.

113. **Pruss GJ, Drlica K.** 1986. Topoisomerase I mutants: the gene on pBR322 that encodes resistance to tetracycline affects plasmid DNA supercoiling. *Proc Natl Acad Sci USA* **83:**8952–8956.

114. **Liu LF, Wang JC.** 1987. Supercoiling of the DNA template during transcription. *Proc Natl Acad Sci USA* **84:** 7024–7027.

115. **Balke VL, Gralla JD.** 1987. Changes in the linking number of supercoiled DNA accompany growth transitions in *Escherichia coli*. *J Bacteriol* **169:**4499–4506.

116. **Jaworski A, Higgins NP, Wells RD, Zacharias W.** 1991. Topoisomerase mutants and physiological conditions control supercoiling and Z-DNA formation *in vivo*. *J Biol Chem* **266:**2576–2581.

117. **Ferrándiz MJ, Martín-Galiano AJ, Schvartzman JB, de la Campa A.** 2010. The genome of *Streptococcus pneumoniae* is organized in topology-reacting gene clusters. *Nucleic Acids Res* **38:**3570–3581.

118. **Majchrzak M, Bowater RP, Staczek P, Parniewski P.** 2006. SOS repair and DNA supercoiling influence the genetic stability of DNA triplet repeats in *Escherichia coli*. *J Mol Biol* **364:**612–624.

119. **Champion K, Higgins NP.** 2007. Growth rate toxicity phenotypes and homeostatic supercoil control differentiate *Escherichia coli* from *Salmonella enterica* serovar *Typhimurium*. *J Bacteriol* **189:**5839–5849.

120. **Collins J, Volckaert G, Nevers P.** 1982. Precise and nearly-precise excision of the symmetrical inverted repeats of Tn5; common features of recA-independent deletion events in *Escherichia coli*. *Gene* **19:**139–146.

121. **Meyer RR, Laine PS.** 1990. The single-stranded DNA-binding protein of *Escherichia coli*. *Microbiol Rev* **54:**342–380.

122. **Roy R, Kozlov AG, Lohman TM, Ha T.** 2009. SSB protein diffusion on single-stranded DNA stimulates RecA filament formation. *Nature* **461:**1092–1097.

123. **Dubois V, Debreyer C, Quentin C, Parissi V.** 2009. In vitro recombination catalyzed by bacterial class 1 integron integrase IntI1 involves cooperative binding and specific oligomeric intermediates. *PloS One* **4:**e5228.

124. **Dubois V, Debreyer C, Litvak S, Quentin C, Parissi V.** 2007. A new in vitro strand transfer assay for monitoring bacterial class 1 integron recombinase IntI1 activity. *PLoS One* **2:**e1315.

125. **Tanaka T, Mizukoshi T, Sasaki K, Kohda D, Masai H.** 2007. *Escherichia coli* PriA protein, two modes of DNA binding and activation of ATP hydrolysis. *J Biol Chem* **282:**19917–19927.

126. **Grompone G, Ehrlich SD, Michel B.** 2003. Replication restart in gyrB Escherichia coli mutants. *Mol Microbiol* **48:**845–854.

127. **Rowe-Magnus DA.** 2009. Integrase-directed recovery of functional genes from genomic libraries. *Nucleic Acids Res* **37:**e118.

128. **Sureshan V, Deshpande CN, Boucher Y, Koenig JE, Midwest Center for Structural G, Stokes HW, Harrop SJ, Curmi PM, Mabbutt BC.** 2013. Integron gene cassettes: a repository of novel protein folds with distinct interaction sites. *PloS One* **8:**e52934.

129. **Gestal AM, Liew EF, Coleman NV.** 2011. Natural transformation with synthetic gene cassettes: new tools for integron research and biotechnology. *Microbiology* **157:**3349–3360.

130. **Bikard D, Julié-Galau S, Cambray G, Mazel D.** 2010. The synthetic integron: an in vivo genetic shuffling device. *Nucleic Acids Res* **38:**e153.

131. **Baquero F, Coque TM, de la Cruz F.** 2011. Ecology and evolution as targets: the need for novel eco-evo drugs and strategies to fight antibiotic resistance. *Antimicrob Agents Chemother* **55:**3649–3660.

132. **Holmes AJ, Holley MP, Mahon A, Nield B, Gillings M, Stokes HW.** 2003. Recombination activity of a distinctive integron-gene cassette system associated with *Pseudomonas stutzeri* populations in soil. *J Bacteriol* **185:** 918–928.

133. **Leon G, Roy PH.** 2009. Potential role of group IIC-attC introns in integron cassette formation. *J Bacteriol* **191:** 6040–6051.

134. **Quiroga C, Centrón D.** 2009. Using genomic data to determine the diversity and distribution of target site motifs recognized by class C-attC group II introns. *J Mol Evol* **68:**539–549.

Mobile DNA, 3rd Edition
Nancy L. Craig, Michael Chandler, Martin Gellert, Alan M. Lambowitz, Phoebe A. Rice and Suzanne Sandmeyer

doi:10.1128/microbiolspec.MDNA3-0056-2014

Caroline Midonet[1]
Francois-Xavier Barre[1]

7 Xer Site-Specific Recombination: Promoting Vertical and Horizontal Transmission of Genetic Information

INTRODUCTION

It was Barbara McClintock who first described the problems of segregation arising from the circularity of chromosomes during her studies on maize variegation (1). The importance of this observation, which could have passed as a mere oddity at the time because of the linear nature of chromosomes in Eukaryota, was only realized after the demonstration of the circular nature of the *Escherichia coli* chromosome by François Jacob and Elie Wollman in the 1960s (2). Since then, the wealth of information gained by genomic studies has shown that circular chromosomes are the norm in Bacteria and Archaea.

DNA circularity can result in the formation of two major topological threats for the segregation of genetic information at the time of cell division: sister chromosome catenation and concatenation [Fig. 1(A)]. Replication of the double-stranded DNA helix introduces one catenation link per helical turn (3). In *E. coli*, catenanes are largely unlinked by the action of a Type II topoisomerase, Topo IV, which transports one double-stranded DNA segment through another after having introduced a double strand break in one of the two segments and then seals back the cleaved segment (Fig. 1(A), [4]). Sister chromosome concatenation, which is more classically referred to as chromosome dimer formation, results from odd numbers of crossovers due to homologous recombination between sister chromatids (Fig. 1(A), [5]). In *E. coli*, chromosome dimers are resolved by the addition of a crossover at a specific locus, *dif*, by a pair of chromosomally encoded tyrosine recombinases, XerC and XerD, which reestablishes the parity of strand exchanges (6, 7, 8, 9). XerCD-*dif* are also able to unlink catenanes in a

[1]Institute for Integrative Biology of the Cell (I2BC), Université Paris Saclay, CEA, CNRS, Université Paris Sud, 1 avenue de la Terrasse, 91198 Gif sur Yvette, France.

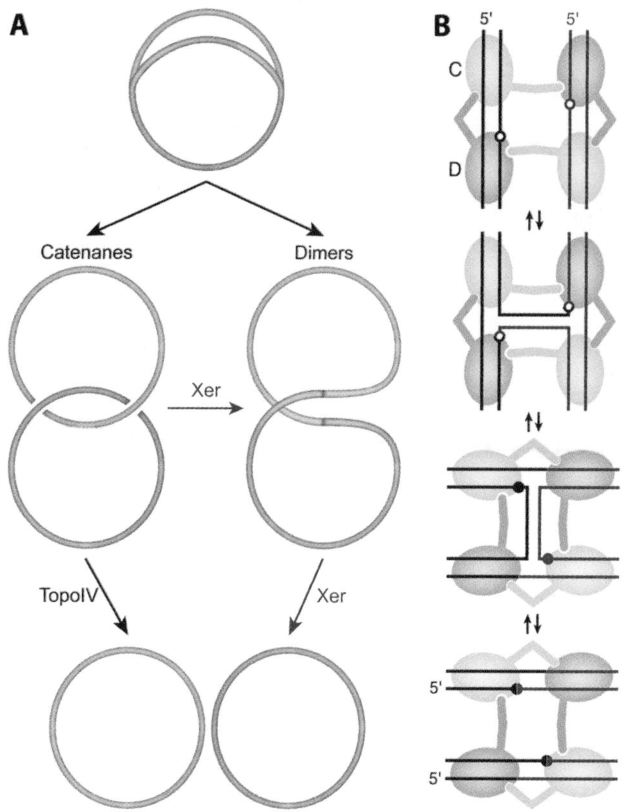

C

	XerC		XerD
dif1	5′-AGTGCGCATTA TCACGCGTAAT	TGTATG ACATAC	TTATGTTAAAT AATACAATTTA-5′
dif2	5′-AATGCGCATTA TTACGCGTAAT	CGTGCG GCACGC	TTATGTTAAAT AATACAATTTA-5′
difG	5′-AGTGCGCATTA TCACGCGTAAT	GGTATA CCATAT	TTATGTTAAAT AATACAATTTA-5′
core cer	5′-GGTGCGTACAA CCACGCATGTT	TTAAGGGA AATTCCCT	TTATGGTAAAT AATACCATTTA-5′
core psi	5′-GGTGCGCGCAA CCACGCGCGTT	GATCCA CTAGGT	TTATGTTAAAC AATACAATTTG-5′
attP[ET]	5′-AgTGCGTATTA TtACGCATAAT	Tgt CA AtgcgGT	TTATGTTgaGG AATACAAtgCC
attP[VGJ]	5′-ACTTCGCATTA TGAAGCGTAAT	TGTCGGC ACAGCCG	TTATGGTAAAA TTATCCATTTT-5′
attP[TLC]	5′-AGTGCGCATTA TCACGCGTAAT	TGTATG ACATAC	TAGAGAAAGTG ATCTCTTTCAC-5′

stepwise manner, which can compensate for a partial loss of the activity of Topo IV (Fig. 1(A), [10, 11, 12]). In addition to *E. coli*, chromosome dimer resolution has been studied in several Bacteria, including *Bacillus subtilis*, *Campylobacter jejuni*, *Caulobacter crescentus*, *Haemophilus influenzae*, *Helicobacter pylori*, *Lactococcus lactis*, *Staphylococcus aureus*, *Vibrio cholerae*, and *Xanthomonas campestris*, and in two Archaea, *Pyrococcus abyssi* and *Sulfolobus solfataricus* (13, 14, 15, 16, 17, 18, 19, 20, 21, 22, 23, 24, 25, 26). Evidence of its importance for the fitness of cells has been directly obtained in *B. subtilis*, *C. crescentus*, *E. coli*, *H. pylori*, *S. solfataricus*, and *V. cholerae* (13, 14, 15, 16, 17, 27, 28, 29).

Other than its importance for the evolutionary success of organisms harboring circular chromosomes, research on Xer recombination has been motivated by three characteristics. First, two recombinases are most often used, XerC and XerD. Each of them catalyzes the exchange of a specific pair of strands between recombining sites, with a Holliday Junction (HJ) as an essential reaction intermediate. As a result, two pathways of reaction can be defined depending on whether XerC- or XerD-catalysis initiates the reaction [Fig. 1(B)]. This feature makes it an excellent model for studying the molecular interplay among the proteins during site-specific recombination. Second, XerC and XerD are exploited by numerous mobile genetic elements: some

Figure 1 Chromosome dimer resolution and sister chromosome decatenation. (A) Topological maintenance of bacterial chromosomes. Violet and pink circles represent sister circular chromosomes. Topo IV resolves catenanes. The Xer machinery resolves both catenanes and dimers. (B) Xer recombination. Light and dark grey ovoid shapes depict the C-terminal domains of the XerC and XerD tyrosine recombinases, respectively. The N-terminal domains are omitted for clarity. Tails have been added to indicate the C-terminal interactions of the recombinases. Red and black lines indicate the two strands of the recombining sites. Full and empty circles represent the XerC and XerD cleavage points, respectively. Horizontal and vertical substrates are proficient for XerC and XerD-strand exchanges, respectively. (C) Consensus sequence obtained from the alignment of the *dif* sites of 715 bacterial chromosomes. The XerC and XerD recognition sites are underlined. Double-stranded DNA sequence of *V. cholerae dif1*, *dif2* and *difG*, of the core *cer* and *psi* plasmid sites and of the three types of attachment sites observed in the genome of integrative mobile elements exploiting Xer are shown below (ET: El Tor CTX; VGJ: VGJ phage; TLC: TLC satellite phage). XerC and XerD process the top and bottom strands, respectively. Bases differing from the consensus are shown in red. Lower case letters indicate the absence of conventional Watson–Crick pairing interactions.
doi:10.1128/microbiolspec.MDNA3-0056-2014.f1

plasmids use it to maintain the monomeric state of their genome and some lysogenic phages use it to integrate their genome into the genome of their host. Indeed, the Xer recombinases were initially discovered because of their implication in the stability of the *Escherichia coli* ColE1 plasmid (7, 30). Exploitation of Xer recombination by plasmids and integrative mobile elements relies on the extreme versatility of the Xer recombination machinery, which can recombine sites with very different structures and/or sequences (Fig. 1(C), [31]). The third interesting feature of Xer recombination is that despite the diversity of sites that XerC and XerD can recombine, their actions remain under very strict cellular and/or topological control.

THE XER SITE-SPECIFIC RECOMBINATION MACHINERY

The Xer Recombinases

Although they share only 38% of their identity, *E. coli* XerC and XerD are the closest relatives to each other in the tyrosine recombinase family. Orthologs of *E. coli* XerC and XerD are readily identified in the genome of most bacteria that carry circular chromosomes (13, 32). When this is the case, a *dif* sequence can almost always be predicted on each of the circular chromosomes of the bacterium based on its homology to *E. coli dif* (Fig. 1(C), [13, 32]). The XerC gene, the XerD gene, and *dif* are invariably encoded in different regions of the genome, which suggests that the Xer machinery appeared early in the evolution of bacteria (13, 32, 33, 34). However, the XerCD-*dif* dimer resolution machinery is not universal: in the *streptococci* and the *lactococci*, chromosome dimer resolution depends on a single Xer-like tyrosine recombinase, XerS (25); a subgroup of the ε-proteobacteria, including *Helicobacter pylori*, uses a single Xer-like tyrosine recombinase, XerH (33); a single tyrosine recombinase is also used in Archaea, XerA in *P. abyssi* (22) and SSQ0395 in *S. solfataricus* (15). These recombinases do not derive from each other, but can all be grouped with XerC and XerD in the phylogenetic tree of tyrosine recombinases. It is probable that the XerS and XerH machineries recently replaced the conventional XerCD-*dif* machinery in the bacteria that harbor them. Indeed, several ε-proteobacteria still possess a conventional XerCD-*dif* machinery (33). In addition, both XerC, XerD, and XerS have been found encoded in the genome of *Lactococcus helveticus* (32). Finally, the genes encoding for XerH, XerS, and XerA are immediately adjacent to their target site on the chromosome, which raises the possibility that they could have been acquired by horizontal gene transfer, via the integration of a mobile element (22, 25, 33).

The use of two recombinases is unusual but not unique among tyrosine recombinases. For instance, two recombinases mediate an inversion gene switch that regulates the expression of Type I fimbriae (35). The Tn554 conjugative transposon also uses two related tyrosine recombinases (36, 37).

Despite high sequence divergence, the crystal structure of XerD is strikingly similar to that of Cre (38). The predicted secondary structure of XerC fits well with the known structure of XerD. It is therefore very likely that the structure of XerC is similar to that of XerD. The XerC and XerD proteins can be divided into a small N terminal and a large C-terminal domain that make a C-shaped clamp into which the DNA is bound. Like Cre, and in contrast to Flp, a functional catalytic unit is contained within a single recombinase molecule. Indeed, each recombinase can by itself act as a site-specific type I topoisomerase by relaxing DNA containing its binding site (39). However, this activity is prevented when the two recombinases are bound together (39). This is likely due to the cyclic interactions that regulate and coordinate the two pairs of strand exchanges during recombination (Fig. 1(B), [40, 41, 42]).

The Recombination Mechanism

Recombination occurs within a nucleoprotein complex consisting of the two recombining sites and a pair of each of XerC and XerD. By analogy to Cre/*loxP* recombination, cyclic interactions made via the extension of the extreme C-terminus of the recombinases [the donor region, depicted as little tails in Fig. 1(B)] are thought to hold together the nucleoprotein complex. In these interactions, the donor region of each recombinase contacts the neighboring partner recombinase close to its active site (the acceptor region). The synaptic complex is asymmetric, the donor region of two of the recombinases contacting the acceptor region of the partner recombinases in *cis*. By analogy to Cre/*loxP*, this is thought to involve a more extended conformation of the donor recombinases, which activates their catalytic tyrosine (depicted in Fig. 1(B) by the addition of an open and filled circle at the position where the XerD and XerC recombinases will cleave their target strands, respectively). Each pair of activated recombinases catalyzes the cleavage of a specific pair of strands immediately 3′ to their binding sites. It generates recombinase/DNA covalent phosphorotyrosyl linkages on one side of the cleaved strands and free 5′-hydroxyl extremities on the other side. Following cleavage, a few bases from each of the two liberated 5′-hydroxyl extremities melt from their complementary strand and attack the recombinase/DNA covalent phosphorotyrosyl linkage of their recombining partner. Subsequent ligation requires the stabilization of strand invasion by Watson–Crick or Wobble base pairing interactions (43). Small movements of DNA and proteins in the resulting HJ allow a change in conformation that reciprocally activates the second pair of strand exchanges (depicted by a 90° rotation of the XerC and XerD binding sites in Fig. 1(B) and in the rest of the figures of this chapter). This model is supported by both genetic and biochemical ensemble experiments on various chromosome, plasmid, and synthetic recombination sites, as well as synthetic HJ intermediates (44, 45, 46, 47, 48, 49, 50). It is also supported by the direct observation of the predicted conformational changes during the recombination of two *dif* sites in single molecule experiments (51).

CHROMOSOME DIMER RESOLUTION

Cell Cycle Regulation

Decatenation and chromosome dimer resolution are highly regulated processes. They depend on two main factors: cell division and the organization of the chromosome.

During cell proliferation, the timing of DNA synthesis, chromosome segregation, and cell division must be coordinated to ensure the stable inheritance of the genetic material. In eukaryotes, this is achieved by checkpoint mechanisms that separate these processes in time. No such strict temporal separation exists in bacteria, in which these processes are concomitant. Replication can even be initiated more than once in a single cell, giving rise to siblings that inherit already partially duplicated chromosomes (52). This notably allows for shorter doubling times than the time required for the replication of the entire genome (53). The length of time required between the end of a new round of replication and the next cell scission event, D, is one of the key factors limiting the doubling time of the cell population [Fig. 2(A)]. Cell scission takes place at the end of the D period, however the assembly of the cell division apparatus can start much earlier [Fig. 2(A)]. Two steps can be defined in the process: (i) mid-cell assembly of a ring of early cell division proteins, including the tubulin-like FtsZ protein, and (ii) recruitment of late cell division proteins such as FtsI, which is involved in the remodeling of the rigid cell wall. Using thermosensitive mutants of early and late cell division proteins, it was shown that Xer recombination at *dif* happens only after the assembly of the complete cell division apparatus (54, 55, 56). Using cephalexin, an antibiotic that inhibits cell constriction without disrupting the cell division machinery, it was further shown that *dif* recombination occurs shortly before, or commensurate with, septum closure (Fig. 2(A), [54, 55, 56]). Likewise, even though Topo IV removes precatenanes behind the replication forks during the course of replication, the majority of topoisomerase IV activity is restricted to the D period of the cell cycle (57, 58).

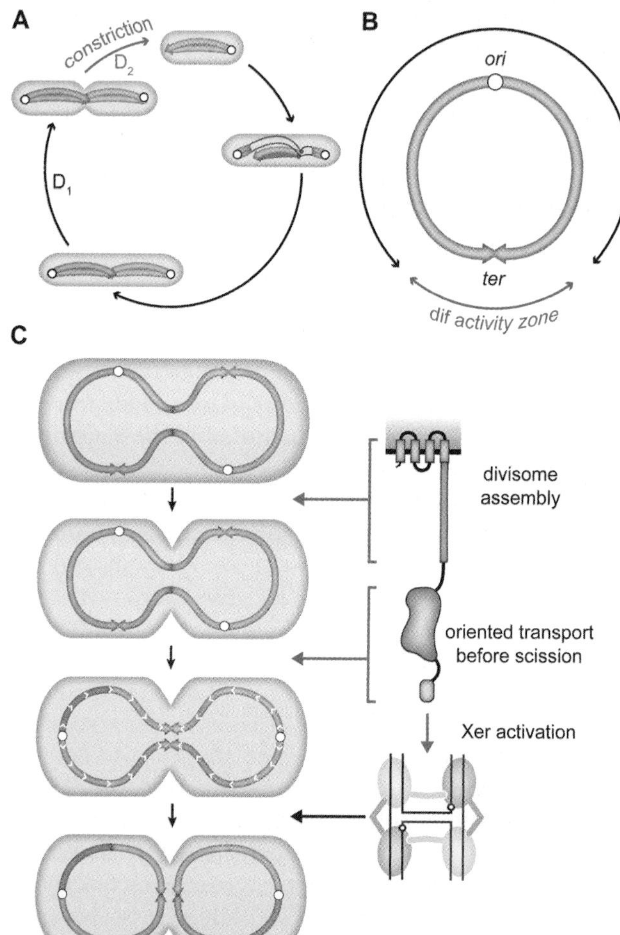

Figure 2 Spatial and temporal control of chromosome dimer resolution. (A) Temporal restriction of Xer recombination during the bacterial cell cycle. White disk: origin of replication region; Converging arrows: terminus of replication region. The two sister chromatids are depicted as pink and purple tubes. (B) Spatial restriction of Xer recombination along bacterial chromosomes. The *dif* activity zone corresponds to the region in which *dif* can still resolve dimers if displaced. (C) FtsK controls Xer recombination. Violet and pink circles represent bacterial sister circular chromosomes. White arrows indicate the KOPS motifs and their orientation. *dif* sites are shown as red and black lines. The FtsK protein is drawn in blue.
doi:10.1128/microbiolspec.MDNA3-0056-2014.f2

Bacterial chromosomes harbor a single origin of bidirectional replication. When they are circular, replication ends in a region opposite the origin of replication, the terminus region, which defines two replichores of approximately equivalent lengths [depicted as arrows in Fig. 2(B)]. There is an asymmetry between the nucleotide compositions of the leading strand and the lagging strand, which results in an inversion in GC-skew, i.e., the richness of G over C and T over A, between the two replichores of bacterial chromosomes. The position at which the two replication forks most frequently collide on the circular chromosome can be determined by computing the GC-skew (59). Analysis of hundreds of bacterial genomes demonstrated that *dif* sites were invariably located at a short distance from the termination point determined by GC-skew profiling (32, 60). The *dif* site was found to be a hot spot of TopoIV activity on the *E. coli* chromosome, linking decatenation to the replichore organization of bacterial chromosomes (61). In *E. coli*, chromosomal inversion and *dif*-displacement studies further showed that *dif* needs to be located in the terminus region of the chromosome to be effective for chromosome dimer resolution (62, 63, 64). In addition, excision of small DNA cassettes inserted in the chromosome between two directly-repeated *dif* sites was found to be limited to the terminus region of chromosomes in *E. coli* and *V. cholerae* (13, 27, 56, 65). However, there is no direct coupling between the recombination activity of XerCD at *dif* and replication: cells in which termination is displaced away from *dif* remain proficient in chromosome dimer resolution as long as *dif* is kept at the inversion of polarity of the two replichores (66, 67, 68).

It was the discovery of the function of a cell division protein, FtsK, in chromosome segregation that led to our understanding of the temporal restriction of *dif* recombination to the time of cell division and of its spatial restriction to the terminus region of the chromosomes. Most of our knowledge on FtsK came from the study of its role in *E. coli* and from the study of *B. subtilis* SpoIIIE, a homolog implicated in sporulation. *B. subtilis* and *E. coli* each possess nucleoid occlusion machineries (69, 70). Nevertheless, septum formation can initiate over partially segregated chromosomes during vegetative cell division in both species, at least when a chromosome dimer has been created by homologous recombination (65, 71, 72). In addition, it invariably occurs during sporulation in *B. subtilis* (73). The correct distribution of the genetic material then requires the respective activities of SpoIIIE and FtsK (Fig. 2C). The two proteins share a common structural organization (Fig. 2(C), [74]). This includes an integral domain at the amino-terminus ($FtsK_N$), which is anchored to the inner membrane by 4 transmembrane segments (75), a linker domain ($FtsK_L$), which lacks any evolutionarily conserved features, and a conserved RecA-type ATPase fold at the C-terminus ($FtsK_C$, [76]). In *E. coli*, $FtsK_N$ and $FtsK_L$ are implicated in the stabilization of the cell division apparatus, and the presence of at least one of these two domains is essential for the assembly of the cell division apparatus (77). $FtsK_C$ assembles into hexameric motors that use the energy of ATP to translocate on DNA (78, 79). As FtsK motors are anchored at the site of division by $FtsK_N$ and $FtsK_L$, it results in the mobilization of chromosomal DNA (28, 80).

FtsK translocation on chromosomal DNA is not random. Using a combination of genetics, bioinformatics, and biochemical experiments at the single molecule level, specific DNA motifs that dictate the orientation of translocation of FtsK were identified on the *E. coli* chromosome, the KOPS (FtsK oriented polar sequence) (81, 82, 83, 84). Identical motifs orient the activity of *V. cholerae* FtsK (13). KOPS are small, highly repeated, polar motifs that point from the origin of replication of bacterial chromosomes towards their *dif* site in their terminus region (13, 82, 85). As a result, FtsK motors pump the origin-proximal regions of a chromosome dimer away from the division site and the zone of convergence of the KOPS to mid-cell [Fig. 2(C)]. The sequence of KOPS is not conserved in all bacteria. The motifs that orient the activities of *B. subtilis* SpoIIIE and *L. lactis* FtsK differ from the *E. coli* and *V. cholerae* KOPS and between each other, which suggests that different polar DNA motifs have been co-opted during the course of evolution to serve as FtsK oriented polar sequences (85, 86). Many studies have focused on the mechanism of translocation of FtsK and on the mode of action of the KOPS (see (87, 88) for a review). The end result of the action of FtsK is that the two *dif* sites of a dimer are brought together if they are correctly located in the zone of convergence of the KOPS but separated if they are located elsewhere on the chromosome, which explains the *dif activity zone* [Fig. 2(B) and 2 (C)]. FtsK was also shown to stimulate the activity of Topo IV, which might explain in part the increase in Topo IV activity in the D period of the cell cycle (89, 90). Impeding FtsK translocation has varying effects on fitness depending on the bacterial species. In *E. coli*, $FtsK_C$ participates in the orderly segregation of the terminus region of sister chromatids, whether monomeric or dimeric (91). However, it is not essential. It is only required when a chromosome dimer has been formed, which under laboratory growth conditions has been estimated to occur in ~15% and 10% of cell divisions in *E. coli* and *V. cholerae*, respectively (65, 92). In contrast, $FtsK_C$ is essential in *Caulobacter crescentus*, which is attributed to its role in chromosome decatenation (93).

A direct interaction between FtsK and XerD is necessary for the addition of a crossover between *dif* sites (78, 94). This is the limiting factor in Xer recombination at *dif* (65, 78, 95, 96). FtsK does not take charge of the segregation of the bulk of the DNA, but only of the terminus regions of chromosomes (56, 91). As a result, Xer recombination is normally limited to *dif* sites located in this region, which explains the absence of excision of *dif* cassettes inserted elsewhere on chromosomes (56, 65, 66, 92) and the low efficiency of excision of *dif* cassettes on plasmids (65, 97). It also explains why recombination at chromosomal *dif* sites is limited to the time of division since this is the only period when FtsK needs to take charge of chromosomal DNA. However, hexamers of FtsK and SpoIIIE form before any visible sign of cell constriction (98, 99). Therefore, it remains to be understood why in *E. coli* and *V. cholerae* most recombination events only occur after the initiation of constriction and not directly after the recruitment of FtsK to mid-cell (54, 56). In addition, it remains unclear why there is over a 10-fold drop in the frequency of chromosomal tandem *dif* cassette excision when homologous recombination is abolished in *E. coli* (54, 65, 92). It was long postulated that monomeric chromosomes were segregated away from mid-cell before the activation of the FtsK translocation, thereby limiting Xer recombination to chromosome dimers. However, microscopic observations suggest that, even though FtsK-dependent DNA translocation is not essential in *E. coli*, FtsK normally takes charge of the orderly segregation of the terminus region of all the chromosomes, whether concatenated or not (91). In addition, *dif* cassette excision was found to be independent from chromosome dimer formation in *V. cholerae*, further suggesting that the influence of RecA might be linked to a specificity of the regulation of the activity and/or production of the *E. coli* recombinases (56). In favor of this hypothesis, Xer recombination between plasmid-borne *dif* sites is also slightly stimulated by chromosome dimer formation in *E. coli* (97, 100).

Control of Catalysis

Most of what we know about the control of Xer recombination at *dif* by FtsK has been gained from the study of the *E. coli* Xer recombination machinery. By default, the XerD recombinases are inactive and the XerC recombinases promote a low level of HJ formation *in vitro* and *in vivo* (Fig. 3(A), [65]). However, these HJs could only be observed *in vivo* in the presence of a DNA intercalating and crosslinking agent, psoralen (65). Their formation is favored *in vitro* by high glycerol concentrations and the presence of a DNA intercalating agent, ethidium bromide (65). This recombination pathway is a dead-end. HJs resulting from XerC-catalysis cannot be converted into a crossover by XerD-catalysis (44). In addition, no product can be created by replication across them *in vivo*, suggesting that they are unstable and very transient (65, 78). Indeed, HJs resulting from XerC-catalysis are rapidly eliminated by reverse reactions (Fig. 3(A), [44, 45, 49, 50, 65, 78]). During chromosome dimer resolution, the XerC-first futile recombination cycle is broken by a direct interaction with FtsK, which triggers the exchange of a first pair of strands by XerD-catalysis (Fig. 3(B), [13, 51, 78, 96, 101]). Structural and functional studies indicate that the 62-amino acid C-terminal fragment of FtsK, FtsKγ, interacts directly with the XerD C-terminus (94, 95). Indeed, FtsKγ can activate XerCD-*dif* recombination in the absence of the translocase domain, when it is fused to XerCD or added in isolation (102). Based on the structure of the Cre/*loxP* synaptic complex, it is not possible to switch a synapse in which the XerC recombinases are in a suitable conformation to be active into a configuration suitable for XerD catalysis without breaking the cyclic interactions between the recombinases [Fig. 1(B)]. Single molecule studies demonstrated that XerCD/*dif* synaptic complexes readily form *in vitro*, ruling out models in which FtsK activated XerD prior to synapsis and in which FtsK actively remodeled complexes initially poised to undergo XerC catalysis (51, 103). *In vitro* dissection of the reaction further indicated that the HJs created by XerD catalysis rapidly isomerized into a conformation suitable for XerC-strand exchange (51). Correspondingly, HJs created by XerD catalysis are rapidly converted into product by XerC-catalysis *in vitro* and *in vivo* (Fig. 3(B), [13, 51, 78, 96, 101]). As XerD catalysis is impeded by coordinated reciprocal switches in recombinase activity, HJs created by XerD catalysis are readily detectable without chemical treatment when XerC catalysis is inhibited and are stable enough to be converted into product by replication (13, 45, 78).

Work in *H. influenzae*, *V. cholerae* and *B. subtilis* suggest that the mechanism by which FtsK controls Xer recombination at *dif* is conserved in bacteria harboring two Xer recombinases (13, 72, 95, 104). Even more interestingly, the FtsK control seems to have been kept in *L. lactis* even though its XerCD machinery was replaced by a single tyrosine recombinase, XerS (105).

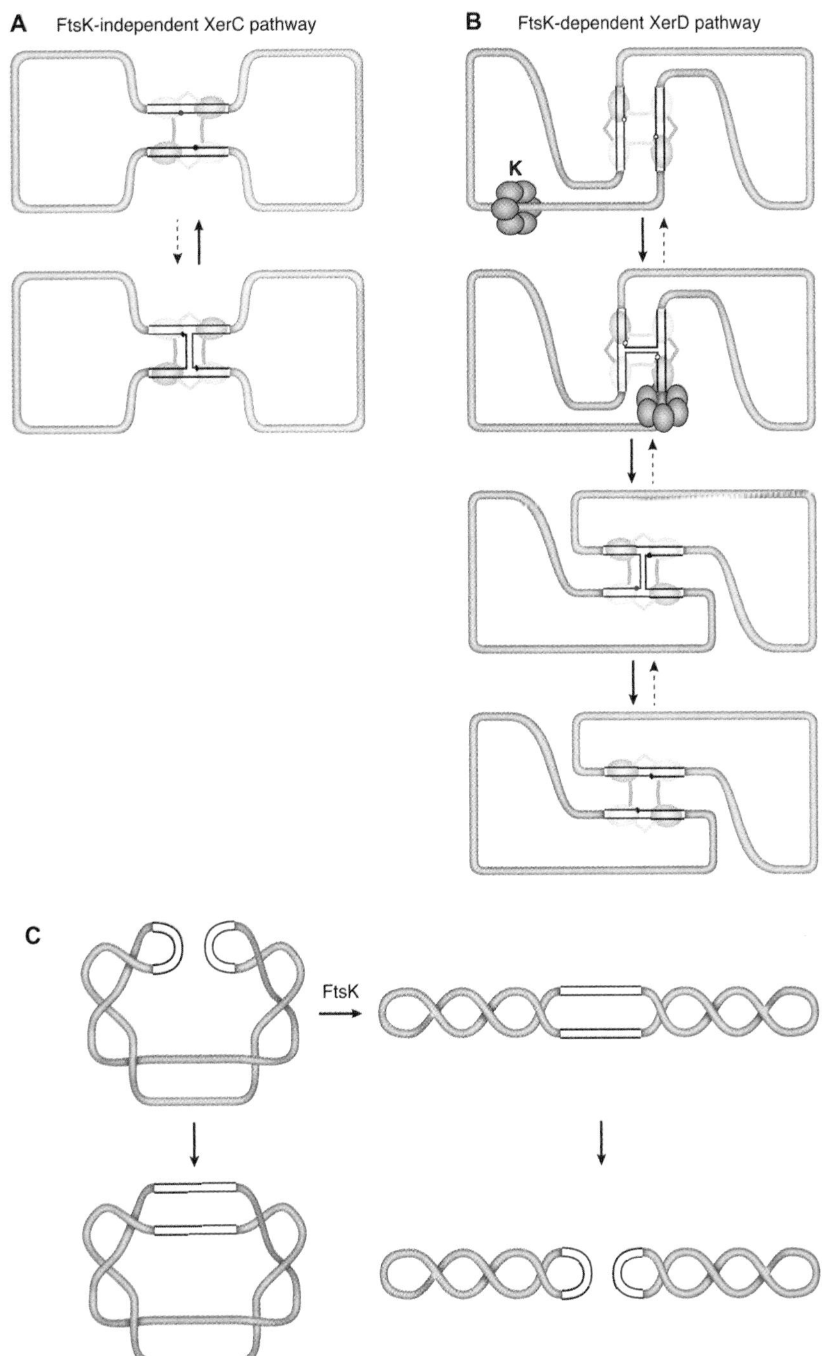

Figure 3 Chromosome dimer resolution. (A) Dead-end FtsK-independent XerC pathway of recombination between *dif* sites. (B) Chromosome dimer resolution pathway. (C) Topological control of Xer recombination. doi:10.1128/microbiolspec.MDNA3-0056-2014.f3

Topological Control

FtsK translocation is not absolutely required to activate Xer recombination at *dif* (102, 105). However, the ATPase activity of FtsK considerably enhances the efficiency of the reaction *in vitro* (51, 78, 81, 96, 101). In addition, products of FtsK-dependent Xer recombination on supercoiled circular substrates containing directly repeated *dif* sites have a simple fixed topology. The only products that could be detected were free circles, as if the two *dif* sites are always brought together by a slithering mechanism (Fig. 3(C), [78]). In contrast, in the absence of translocation, FtsK-dependent Xer recombination of supercoiled circular substrates harboring tandem *dif* sites produces catenanes, as expected from random collision events (24, 102). FtsK can also simplify the products of Cre/*loxP* recombination (10). The capacity of chromosome dimer resolution machinery to remove catenation links in a stepwise manner is probably related to this property of FtsK (11, 12). The mechanism by which FtsK might help bring together *dif* sites in a specific topological configuration is not understood yet. Some clues to the answer might reside in the observation that although FtsK and SpoIIIE can strip proteins off the DNA (101, 106, 107), FtsK translocation is stopped by XerCD-*dif* (107, 108).

PLASMID DIMER RESOLUTION

Multimerization of circular plasmids reduces the number of independently segregating units, which leads to failures in their vertical transmission from mother cell to daughter cells (109, 110, 111, 112, 113). Hence, many plasmids and phages encode a tyrosine recombinase that serves to maintain the monomeric state of their genome. This is notably the case for phage P1, which encodes the Cre recombinase. However, several plasmids do not possess their own dedicated dimer resolution machinery and exploit the Xer recombinases of their host (30, 114, 115, 116, 117, 118). These plasmids do not carry a *dif* site and resolution of plasmid multimers is independent of FtsK. Indeed, addition of a *dif* site to a plasmid rather leads to its concatenation inside the cell (97).

Topological Regulation

The ColE1 and pSC101 dimer resolution systems have been characterized in great detail (112, 116). ColE1 and pSC101 carry a complex recombination site composed of a core *dif*-like sequence and ~150 bp of accessory sequences, *cer* and *psi*, respectively (Fig. 1(C) and Fig. 4(A), [119, 120, 121]). In contrast to Xer recombination at *dif*, Xer recombination at *cer* and *psi* senses the topological connectivity between the sites such that they recombine only if they are directly repeated on the same DNA molecule (122). This is due to a "topological filter," which ensures recombination selectivity, i.e., that multimers are resolved rather than formed by Xer recombination. A direct consequence of the topological filter is that Xer recombination yields products with a specific topology (122): recombination between two *psi* sites produces four-noded catenanes, in which the two component circles are wrapped around each other

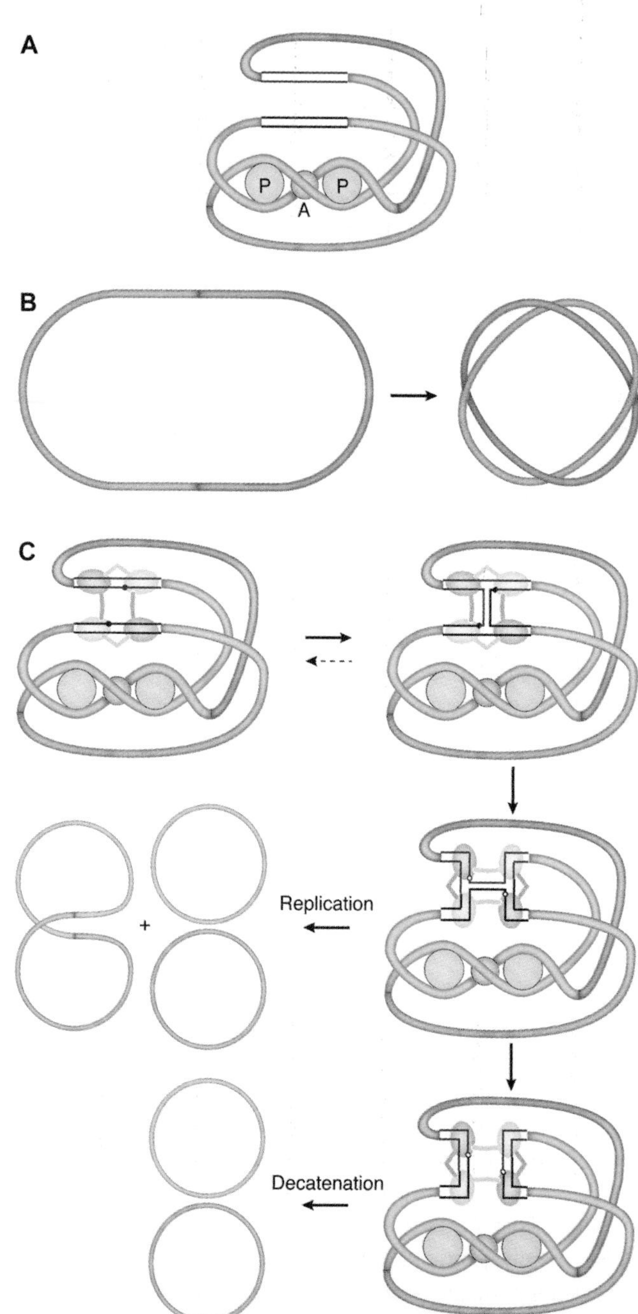

Figure 4 Plasmid dimer resolution. (A) Schematic representation of the topological filter. Yellow circles represent accessory proteins. P: PepA; A: ArgR or phosphorylated ArcA; Green tubes: accessory sequences. (B) Topology of the products of Xer recombination at *cer* and *psi* multimer four-node catenanes. (C) The topological filter controls Xer catalysis for plasmid dimer resolution.
doi:10.1128/microbiolspec.MDNA3-0056-2014.f4

exactly four times (Fig. 4(B), four-noded catenane); recombination between two *cer* sites produces an equivalent structure with an HJ.

In the *cer* and *psi* recombination reactions, the two product circles are wrapped around each other in a right-handed fashion and the sites in these circles are in antiparallel orientations (122). It was demonstrated using the mathematical tangle theory that this specific topology could only be produced by a single 3-dimensional model of the recombination reaction (Fig. 4(A), [123, 124]). The *cer* and *psi* accessory sequences are binding sites for two host proteins, PepA and ArgR (125, 126, 127) and PepA and phosphorylated ArcA (128), respectively. All three proteins are known transcription factors with specific DNA binding properties. ArgR binds as a hexamer to a single ArgR box within *cer*; PepA binds as a hexamer to two synapsed *cer* accessory sequences on both sides of the ArgR binding site (Fig. 4(A), [127]). PepA and phosphorylated ArcA bind to the *psi* accessory sequences at places similar to those bound by PepA and ArgR in *cer* (Fig. 4(A), [128]). Based on the X-ray structure of PepA and on the analysis of mutants of PepA, a molecular model could be built showing how two hexamers of PepA and one hexamer of ArgR could bind to the accessory sequences of two *cer* sites (129, 130). Atomic force microscopy now supports this model (131).

In contrast to recombination at *cer* and *psi*, recombination between *loxP* sites requires only the Cre protein and gives products of different topologies depending on the reaction conditions and the degree of supercoiling of the substrate (132, 133). Complex product topologies are created after synapsis by random collision (Fig. 3(C), [134, 135]). However, the addition of PepA was sufficient to alter the topology of Cre recombination at hybrid sites consisting of accessory sequences from *cer* or *psi* adjacent to a *loxP* core (136). These results show that the accessory sequences and the accessory proteins are sufficient to create a topological filter and thus ensure topological selectivity.

Control of Catalysis

The pathway of recombination followed by the Xer machinery at *cer* and *psi* is reversed compared to the pathway of recombination at *dif*: the two XerC recombinases of the nucleoprotein synaptic complex catalyze the first pair of strand exchanges in the reaction (Fig. 4(C), [7, 137]). It was partly because of these observations that the XerC-first pathway was proposed to be the default pathway of recombination in the early models on the control of Xer recombination at *dif* (Fig. 3(A) and 3(B), [45, 65]). However, in the same year when the role of FtsK in promoting a first pair of strand exchanges by XerD catalysis at *dif* was demonstrated, it was shown that the accessory sequences and the accessory proteins determined the order of strand exchanges at *psi*: a plasmid containing two inverted-core *psi* sites recombined with a reversed order of strand exchange, but with unchanged product topology (121). This result suggests that the topological constraints imposed by the accessory sequences and the accessory proteins switch the synaptic complex into a configuration suitable for XerC-strand exchanges and that they switch the resulting HJ intermediate into a configuration suitable for XerD-strand exchanges. This result also has structural implications: the recombinase–core nucleoprotein complex must form with the recombinase C-terminal domains facing the accessory proteins and sequences to explain both the order of strand exchanges and the topology of the products observed with normal and inverted *psi* sites (121).

The recombination reaction doesn't proceed further than the formation of an HJ at *cer* (138). This is probably linked to the structure of the *cer* core site, in which the central region contains 8 bp instead of the canonical 6 bp overlap region of most tyrosine recombinase sites [Fig. 1(C)]. The HJ intermediate is not processed by the host RuvC HJ resolvase (138). Instead, it is proposed that initiation of replication on the HJ-containing circular molecule leads to the production of two plasmid monomers and one plasmid dimer [Fig. 4(C)]. This also seems to occur at *psi* when the catalytic activity of XerD is inactivated (45, 121). In agreement with this model, the HJ intermediates resulting from XerC-catalysis at *cer* and *psi* are stable enough to be detected without any chemical trap, in contrast to the HJs resulting from XerC-catalysis at *dif*. It is probable that the topological constraints imposed by the accessory proteins and sequences permit the stabilization of the HJs by switching them into a conformation unsuitable for XerC-catalysis (and reciprocally suitable for XerD-catalysis in the case of *psi*).

Cell Cycle Regulation

An advantage of the mechanism by which plasmids exploit the Xer machinery of their host to maintain their monomeric state is to escape the spatial and temporal control normally exerted on Xer recombination. Moreover, plasmids are able to control the cell cycle of their host to allow more time for their replication and/or for multimer resolution before the cell divides. The RepA protein of plasmid pSC101 controls the cell division of *E. coli* to permit the repopulation of cells with plasmids when copy number falls and thus limit variation in plasmid copy number (139). ColE1 encodes a short, untranslated RNA called Rcd (regulator of cell division) that delays cell division to allow more time for multimer resolution (140, 141, 142). Rcd and its promoter are found within the accessory sequences of *cer* and plasmid concatenation triggers its expression (143). The exact mechanism by which Rcd inhibits cell division might be linked to the production of indole, a small molecule that affects the transmembrane potential of bacteria (144, 145, 146, 147).

INTEGRATIVE MOBILE ELEMENTS EXPLOITING XER

Many mobile genetic elements encode a tyrosine recombinase for the integration/excision of their genome into the genome of their host [Fig. 5(A)]. This is notably the case for lysogenic phages, such as phage λ. However, some integrative mobile elements do not encode their own recombinase but use the Xer machinery of their host to integrate at *dif*. These elements are referred to as integrative mobile elements exploiting Xer (IMEXs). IMEXs are found integrated in the genome of many bacteria, including Enterobacteriaceae like *E. coli* and *Yersinia pestis*, Xanthomonadaceae like *X. campestris* and *Xylella fastidiosa*, Neisseriaceae like *Neisseria gonorrhoeae* and *Neisseria meningitidis*, and Vibrionaceae like *V. cholerae* (31, 148, 149, 150, 151, 152, 153, 154, 155).

Initial work on IMEXs was motivated by their implication in the evolution of human pathogens. In particular, the genes encoding cholera toxin, the principal

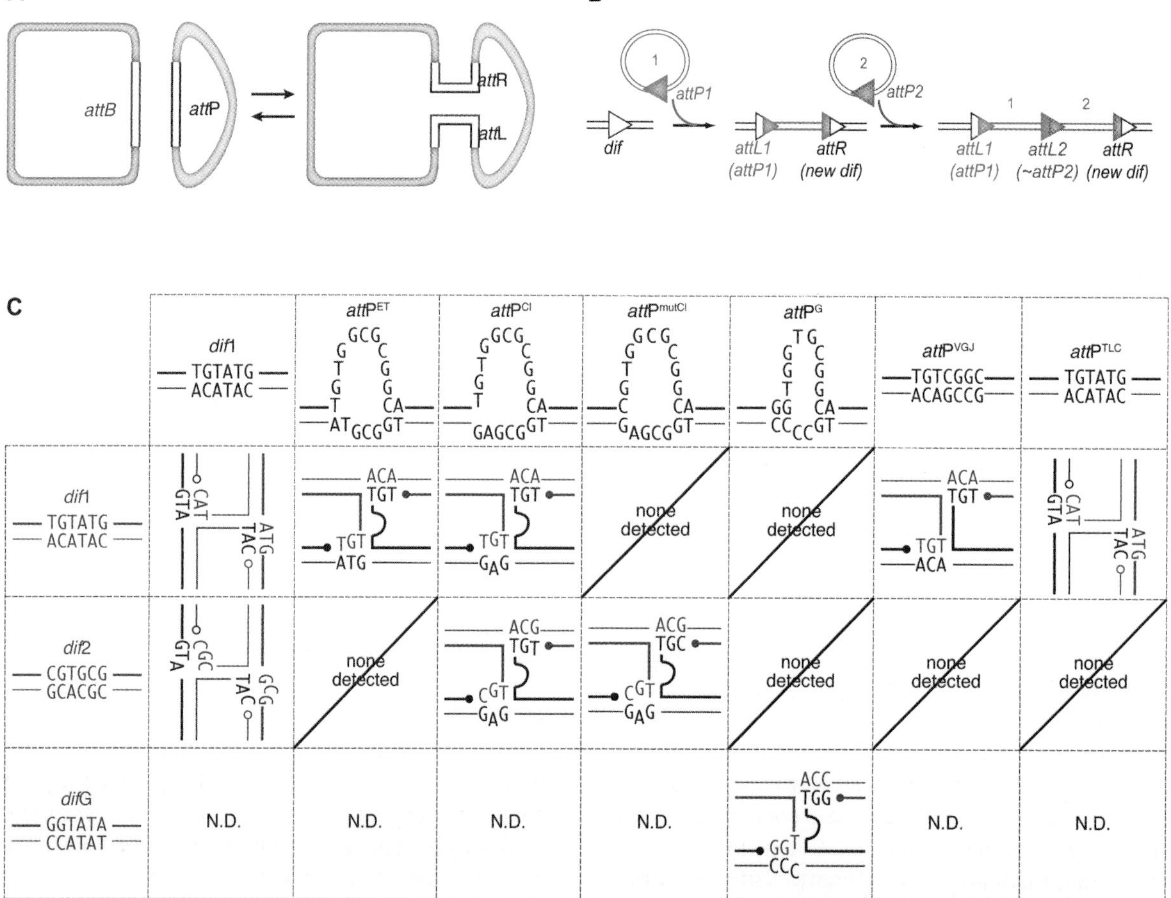

Figure 5 IMEX integration depends on little homology with *dif*. (A) Schematic of the integration/excision of mobile elements into the genome of their host. (B) IMEX integration generates a new *dif* site, which allows for multiple successive integration events. (C) IMEX integration depends on limited homology. The distance separating the two bases of a base pair indicates the quality of the base pair interactions that are formed. N.D.: not determined
doi:10.1128/microbiolspec.MDNA3-0056-2014.f5

virulence factor of the diarrhea-causing Gram-negative bacterium *Vibrio cholerae*, are encoded by one of the first phages to be identified as an IMEX, CTXϕ (152, 153). Several other IMEXs use the Xer machinery of *V. cholerae*. In each case, a functional *dif* site is invariably recreated after their integration, at the *attR* junction [Fig. 5(B)]. The new *dif* site is composed of the XerD-side of the previous *dif* site and of the XerC-side of the *attP* site of the element (Fig. 5(B), [156, 43, 157, 158]). This implies that the IMEX *attP* sites must carry a conserved XerC-binding site. *dif* restoration is important for the fitness of the host, and thus for the success of the IMEX. In addition, it permits multiple successive integration events [Fig. 5(B)]. Correspondingly, analysis of more than 150 complete genomes of clinical and environmental *V. cholerae* isolates revealed that most of them harbored large IMEX arrays (159, 160). Molecular interactions between the different IMEXs that compose these arrays are implicated in the ecological cycle that permits the constant and rapid acquisition of new cholera toxin gene variants, as observed in the ongoing 7th cholera pandemic (157, 158, 159, 160, 161, 162, 163, 164).

Xer Recombination Between Nonhomologous Sites

The bacterial attachment site of a given integrative mobile element, *attB*, must match the attachment site of the element, *attP*, and should be in a highly conserved genomic region of the host genome (165, 166). In this regard, exploitation of the Xer machinery is advantageous since chromosomal *dif* sites are highly conserved (13, 32, 33). The simplest solution for an element to become an IMEX would be to harbor a *dif* site. Indeed, *E. coli* and *V. cholerae* XerC and XerD can promote the efficient integration of plasmids harboring a *dif* site into their chromosomal counterpart in a FtsK-dependent manner (65, 158). However, such integration events are highly unstable (56, 65). All the IMEXs described to date possess an attachment site with limited homology to *dif*, as illustrated with the different *V. cholerae* IMEXs [Fig. 1(C) and Fig. 5(C)]. The genome of *V. cholerae* is divided on two circular chromosomes, chromosome I and II. Both chromosomes harbor a unique *dif* site. *V. cholerae dif* sites are polymorphous. They can be grouped into three classes based on the sequence of their central region, *dif1*, *dif2*, and *difG* (Fig. 1(C) and 5(C), [13, 43]). Chromosome II always carries a *dif2* site whereas chromosome I generally harbors a *dif1* site. Each of the different *V. cholerae* IMEXs specifically targets a subset of these sites (43, 157, 158). This implies some homology between the central regions of the *attP* site and the targeted *dif* site. However, the amount of homology that is required is limited (Fig. 5(C), [43, 157, 158]): a single Watson–Crick strand on one side of the exchanged overlap regions and a G–T Wobble base pair on the other side are sufficient to stabilize the exchange of a pair of strands [Fig. 5(C)]. From this point of view, the integration/excision reactions of IMEXs are markedly different from plasmid and chromosome dimer resolution reactions. The ability of XerCD to recombine sites with very little homology is further highlighted by the observation of rare *dif1*–*dif2* chromosome fusion events (167).

Strategies for the Control of Integration and of Excision

It is critically important for temperate phages to control their timely excision from the host genome, such as when the survival of the host is compromised. In most cases, this is ensured via a virally encoded accessory protein, the excisionase. In the absence of the excisionase, the phage recombinase recombines *attP* and *attB* to produce *attL* and *attR*. In the presence of the excisionase, it recombines *attL* and *attR* to produce *attB* and *attP* again. To date no excisionases have been characterized in the genome of IMEXs. This is apparently compensated by smart life cycle strategies.

Three very different types of IMEX *attP* sites have been identified so far, which correspond to three categories of IMEXs with distinct life cycle strategies [Fig. 1(C) and Fig. 6, [31, 148]). These three IMEX categories are referred to as the CTX-, VGJ- and TLC-families, according to their most well-studied representative in *V. cholerae*, CTXϕ (152, 153), VGJϕ (168) and TLCϕ (162, 163).

The CTXϕ Life Cycle

CTXϕ is a filamentous phage. Its double-stranded DNA (dsDNA) replicative form carries two *dif*-like sites in inverted orientation (169). None of them can be recombined with any of the *V. cholerae dif* sites because the central region of the first one has no homology with the central regions of the *V. cholerae dif* sites and because the central region of the other one is too long to permit XerC and XerD cyclic interactions (156). Instead the attachment site of CTXϕ, $attP^{CTX}$, consists of the stem of a hairpin that is generated by the folding of the region encompassing these two sites in the single-stranded DNA (ssDNA) phage genome: the one which is packaged in the phage particles and which is

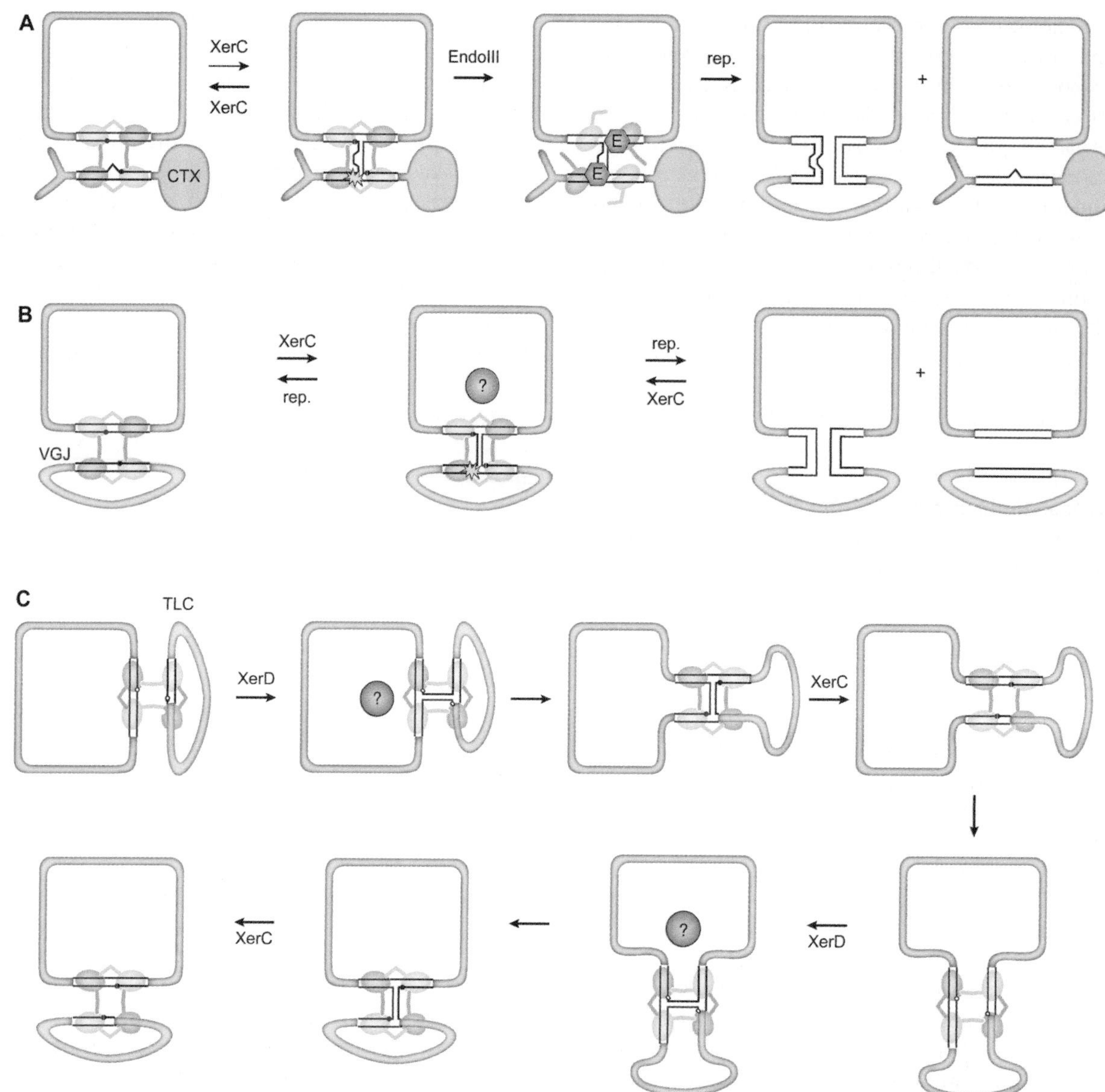

Figure 6 Diversity of the IMEX integration pathways. (A) Irreversible integration of CTX-type elements. The single-stranded DNA and double-stranded DNA forms of the element are represented in pink. The host genome is shown in purple. The incapacity of XerD to perform strand exchanges is indicated in yellow. Orange hexagons labeled with the letter E are EndoIII. (B) Integration/excision of VGJ-type elements. The double-stranded DNA replicative form of the element is shown in pink. The host genome is shown in purple. A yellow explosion indicates the impossibility for XerD to perform strand exchanges. The orange circle labeled with a question mark indicates a putative unknown integration factor. (C) Integration/excision pathway of TLC-type elements. The double-stranded DNA form of the element is shown in pink. The host genome is shown in purple. The Blue circle labeled with a question mark represents an unknown factor that could permit the binding of XerD and its catalytic activation. doi:10.1128/microbiolspec.MDNA3-0056-2014.f6

generated by rolling circle replication [Fig. 1(C)]. Integration is catalyzed by the XerC recombinases, which promote the formation of an HJ intermediate [Fig. 5(C) and 6(A), [43, 156]). The lack of homology on the XerD-side of the overlap regions of *attP*^{CTX} and *dif* prevents any potential XerD-mediated strand exchange. Correspondingly, CTXϕ-integration does not depend on the catalytic activity of XerD, which only plays a structural role in the formation of the synaptic complex (43). The HJ intermediate is probably converted into product by replication [Fig. 6(A)]. The active form of *attP*^{CTX} is masked in the double-stranded genome of prophage. As a result, the integration of CTXϕ is intrinsically irreversible (156). However, free CTXϕ genome copies can be produced by a process analogous to rolling circle replication after the integration of a second IMEX harboring the same integration/replication machinery, such as the RS1 satellite phage (164). Production of these new extrachromosomal copies of the genome of CTXϕ is under the control of SOS, at least in part because the SOS repressor, LexA, regulates the expression of the phage nickase, RstA (170, 171).

The VGJϕ Life Cycle

Like CTXϕ, VGJϕ is a filamentous phage. However, its attachment site, *attP*^{VGJ}, is found in its dsDNA replicative form. As in the case of CTXϕ, VGJϕ-integration results from a single pair of strand exchanges between this site and its chromosomal target, which is catalyzed by XerC (Fig. 5(C), [157]). The resulting HJ is probably converted into product by replication (Fig. 6(B), [157]). However, a major difference with the integration of CTXϕ is that a functional phage attachment site is recreated at the *attL* junction. As a result, a new round of Xer recombination can excise integrated copies of VGJϕ: XerC can promote the exchange of a pair of strands between the phage attachment site that is recreated at *attL* and the *dif* site that is recreated at *attR*; replication of the HJ intermediate leads to a free copy of the dsDNA genome of VGJϕ, one sister chromosome free of the prophage and one sister chromosome in which the prophage is still present (157). Correspondingly, IMEXs of the VGJ-type are only very rarely found integrated in the genome of clinical and environmental strains (159, 160). How do these cycles of integration and excision participate in the life cycle of VGJϕ? The dsDNA replicative form of the phage, which is normally extremely stable, is easily lost in cells in which Xer recombination is abolished, suggesting that the integration/excision cycles participate in the vertical inheritance of the phage genome and in the steady state production of new viral particles (157). As a consequence, VGJϕ production escapes the SOS control, which is markedly different from most other lysogenic phages (157).

The TLCϕ Life Cycle

TLCϕ is a satellite phage that exploits phages of the VGJϕ IMEX family for dissemination (162). As in the case of VGJϕ, its attachment site, *attP*^{TLC}, is found in its dsDNA replicative form. However, the overlap region of *attP*^{TLC} is fully homologous with the overlaps of its target *dif* site in contrast to *attP*^{CTX} and *attP*^{VGJ} [Fig. 5(B)]. In addition, it lacks a *bona fide* XerD-binding DNA arm [Fig. 1(B)]. Correspondingly, *attP*^{TLC} is a poor XerD binding substrate (158). Nevertheless, a pair of strand exchanges catalyzed by XerD initiates TLCϕ-integration, the resulting HJ intermediate being converted into product by XerC-catalysis (Fig. 6(C), [158]). A functional phage attachment site is recreated at the *attL* junction so that the reaction is fully reversible: XerD can promote the exchange of a pair of strands between the phage attachment site that is recreated at *attL* and the *dif* site that is recreated at *attR*, the resulting HJ intermediate being converted into product by XerC-catalysis (Fig. 6(C), [158]).

Control of Catalysis

The propensity of HJs resulting from XerC-catalysis to be eliminated by reverse reactions should severely compromise the chances for replication to finalize the integration process of CTXϕ and VGJϕ (44, 45, 65, 172). In the case of CTXϕ, Endo III, a ubiquitous base excision repair enzyme, seems to dislodge the Xer recombinases from the HJs once they are formed, which prevents XerC from reversing the reaction (172). This action is crucial to the life cycle of CTXϕ because it increases the chances for success of independent integration events, which facilitates the formation of tandem CTXϕ arrays (172). However, it is not an absolute requirement. *attP*^{CTX}/*dif* HJs are stable enough to be processed by replication in the absence of Endo III (172). Likewise, no other host or phage factor seems to be involved in the integration of VGJϕ, suggesting that *attP*^{VGJ}/*dif* HJs are stable enough to be processed by replication (172). In contrast, the *dif*/*dif* HJs that result from XerC-strand exchanges cannot be converted into product by replication [Fig. 3(A)]. These observations suggest that the unconventional central region of *attP*^{CTX} and *attP*^{VGJ} might help stabilize their HJ integration intermediates.

The integration/excision strategy of TLCϕ came as a surprise because all other mobile elements so far characterized, whether plasmids or IMEXs, exploited the XerC-first pathway of recombination (see above). In this respect, the TLCϕ integration pathway is similar to the chromosome dimer resolution pathway. However, TLCϕ integration and excision were both found to be independent of FtsK, whereas chromosome dimer resolution depends on the activation of XerD-catalysis by FtsK (158). A second puzzling aspect of the exploitation of Xer by TLCϕ is that integration seems to be much more efficient than excision, even though they rely on the same Xer reaction pathway (158). The factor(s) that permit the activation of XerD-catalysis in the TLCϕ reactions and help favor its integration remain to be explored.

CONCLUDING REMARKS

Two related tyrosine recombinases, XerC and XerD, are encoded in the genome of most bacteria where they serve to resolve dimers of circular chromosomes by the addition of a crossover at a specific site, *dif*. The most striking feature of Xer recombination is possibly the variety of the mechanisms that permit their exploitation as chromosome and plasmid dimer resolvases as well as mobile genetic element integrases.

The versatility of exploitation of Xer recombination seems intimately connected to the many possibilities that exist to coordinate and control the strand exchange reactions catalyzed by XerC and XerD. In particular, four recombination pathways can be considered, depending on whether recombination is initiated by XerC- or XerD-catalysis, and whether it ends with a second pair of strand exchange or with the conversion of the HJ intermediate into product by replication. Each of these different recombination pathways has been observed. The molecular bases of their control are being deciphered. This includes the reciprocal control of catalysis that the two Xer partner recombinases exert over each other, the topological modifications imposed by accessory sequences and accessory factors, and the sequence of the recombination sites.

A difficulty in our progress towards the molecular characterization of the Xer recombination pathways has long been the lack of structural information on the different intermediates of the reactions. However, it is safe to assume that Xer recombinases more amenable to structural studies will be found with the characterization of Xer machineries from more and more organisms. In addition, the development of new biophysical techniques to track the recombination reaction at the single molecule level promises future great advances in the field.

Acknowledgments. *We would like to acknowledge support from the Agence Nationale pour la Recherche [ANR-11-BLAN-O2401], the Fondation Bettencourt Schueller [2012 Coup d'Elan award], and the European Research Council under the European Community's Seventh Framework Programme [FP7/2007-2013 Grant Agreement no. 281590].*

Citation. Midonet C, Barre F-X. 2014. Xer site-specific recombination: promoting vertical and horizontal transmission of genetic information. Microbiol Spectrum 2(6):MDNA3-0056-2014.

References

1. **McClintock B.** 1932. A Correlation of Ring-Shaped Chromosomes with Variegation in Zea Mays. *Proc Natl Acad Sci U S A* **18:**677–681.
2. **Jacob F, Wollman E.** 1961. *Sexuality and the Genetics of Bacteria.* Academic Press, New York.
3. **Watson JD, Crick FHC.** 1953. Genetical implications of the structure of deoxyribonucleic acid. *Nature* **171:** 964–967.
4. **Adams DE, Shekhtman EM, Zechiedrich EL, Schmid MB, Cozzarelli NR.** 1992. The role of topoisomerase IV in partitioning bacterial replicons and the structure of catenated intermediates in DNA replication. *Cell* **71:** 277–288.
5. **Barre FX, Søballe B, Michel B, Aroyo M, Robertson M, Sherratt D.** 2001. Circles: the replication-recombination-chromosome segregation connection. *Proc Natl Acad Sci U S A* **98:**8189–8195.
6. **Blakely G, Colloms S, May G, Burke M, Sherratt D.** 1991. Escherichia coli XerC recombinase is required for chromosomal segregation at cell division. *New Biol* **3:** 789–798.
7. **Blakely G, May G, McCulloch R, Arciszewska LK, Burke M, Lovett ST, Sherratt DJ.** 1993. Two related recombinases are required for site-specific recombination at dif and cer in E. coli K12. *Cell* **75:**351–361.
8. **Kuempel PL, Henson JM, Dircks L, Tecklenburg M, Lim DF.** 1991. dif, a recA-independent recombination site in the terminus region of the chromosome of Escherichia coli. *New Biol* **3:**799–811.
9. **Clerget M.** 1991. Site-specific recombination promoted by a short DNA segment of plasmid R1 and by a homologous segment in the terminus region of the Escherichia coli chromosome. *New Biol* **3:**780–788.
10. **Ip SC, Bregu M, Barre FX, Sherratt DJ.** 2003. Decatenation of DNA circles by FtsK-dependent Xer site-specific recombination. *EMBO J* **22:**6399–6407.
11. **Grainge I, Bregu M, Vazquez M, Sivanathan V, Ip SC, Sherratt DJ.** 2007. Unlinking chromosome catenanes in vivo by site-specific recombination. *EMBO J* **26:**4228–4238.
12. **Shimokawa K, Ishihara K, Grainge I, Sherratt DJ, Vazquez M.** 2013. FtsK-dependent XerCD-dif recombination unlinks replication catenanes in a stepwise manner. *Proc Natl Acad Sci U S A* **110:**20906–20911.
13. **Val M-E, Kennedy SP, El Karoui M, Bonne L, Chevalier F, Barre F-X.** 2008. FtsK-dependent dimer resolution on multiple chromosomes in the pathogen Vibrio cholerae. *PLoS Genet* **4**(9):e1000201.
14. **Debowski AW, Carnoy C, Verbrugghe P, Nilsson H-O, Gauntlett JC, Fulurija A, Camilleri T, Berg DE, Marshall BJ, Benghezal M.** 2012. Xer recombinase and genome integrity in Helicobacter pylori, a pathogen without topoisomerase IV. *PloS One* **7:**e33310.
15. **Duggin IG, Dubarry N, Bell SD.** 2011. Replication termination and chromosome dimer resolution in the archaeon Sulfolobus solfataricus. *EMBO J* **30:**145–153.

16. **Jensen RB.** 2006. Analysis of the terminus region of the Caulobacter crescentus chromosome and identification of the dif site. *J Bacteriol* **188:**6016–6019.
17. **Britton RA, Grossman AD.** 1999. Synthetic lethal phenotypes caused by mutations affecting chromosome partitioning in Bacillus subtilis. *J Bacteriol* **181:**5860–5864.
18. **Sciochetti SA, Piggot PJ, Sherratt DJ, Blakely G.** 1999. The ripX locus of Bacillus subtilis encodes a site-specific recombinase involved in proper chromosome partitioning. *J Bacteriol* **181:**6053–6062.
19. **Sciochetti SA, Piggot PJ, Blakely GW.** 2001. Identification and Characterization of the dif Site from Bacillus subtilis. *J Bacteriol* **183:**1058–68.
20. **Leroux M, Rezoug Z, Szatmari G.** 2013. The Xer/dif site-specific recombination system of Campylobacter jejuni. *Mol Genet Genomics* **288:**495–502.
21. **Yen M-R, Lin N-T, Hung C-H, Choy K-T, Weng S-F, Tseng Y-H.** 2002. oriC region and replication termination site, dif, of the Xanthomonas campestris pv. campestris 17 chromosome. *Appl Environ Microbiol* **68:**2924–2933.
22. **Cortez D, Quevillon-Cheruel S, Gribaldo S, Desnoues N, Sezonov G, Forterre P, Serre M-C.** 2010. Evidence for a Xer/dif system for chromosome resolution in archaea. *PLoS Genet* **6:**e1001166.
23. **Serre M-C, El Arnaout T, Brooks MA, Durand D, Lisboa J, Lazar N, Raynal B, van Tilbeurgh H, Quevillon-Cheruel S.** 2013. The carboxy-terminal αN helix of the archaeal XerA tyrosine recombinase is a molecular switch to control site-specific recombination. *PloS One* **8:**e63010.
24. **Neilson L, Blakely G, Sherratt DJ.** 1999. Site-specific recombination at dif by Haemophilus influenzae XerC. *Mol Microbiol* **31:**915–926.
25. **Le Bourgeois P, Bugarel M, Campo N, Daveran-Mingot ML, Labonte J, Lanfranchi D, Lautier T, Pages C, Ritzenthaler P.** 2007. The Unconventional Xer Recombination Machinery of Streptococci/Lactococci. *PLoS Genet* 3:e117.
26. **Chalker AF, Lupas A, Ingraham K, So CY, Lunsford RD, Li T, Bryant A, Holmes DJ, Marra A, Pearson SC, Ray J, Burnham MK, Palmer LM, Biswas S, Zalacain M.** 2000. Genetic characterization of gram-positive homologs of the XerCD site-specific recombinases. *J Mol Microbiol Biotechnol* **2:**225–233.
27. **Pérals K, Cornet F, Merlet Y, Delon I, Louarn JM.** 2000. Functional polarization of the Escherichia coli chromosome terminus: the dif site acts in chromosome dimer resolution only when located between long stretches of opposite polarity. *Mol Microbiol* **36:**33–43.
28. **Bigot S, Corre J, Louarn J, Cornet F, Barre FX.** 2004. FtsK activities in Xer recombination, DNA mobilization and cell division involve overlapping and separate domains of the protein. *Mol Microbiol* **54:**876 886.
29. **Hendricks EC, Szerlong H, Hill T, Kuempel P.** 2000. Cell division, guillotining of dimer chromosomes and SOS induction in resolution mutants (dif, xerC and xerD) of Escherichia coli. *Mol Microbiol* **36:**973–981.
30. **Colloms SD, Sykora P, Szatmari G, Sherratt DJ.** 1990. Recombination at ColE1 cer requires the Escherichia coli xerC gene product, a member of the lambda integrase family of site-specific recombinases. *J Bacteriol* **172:**6973–6980.
31. **Das B, Martínez E, Midonet C, Barre F-X.** 2013. Integrative mobile elements exploiting Xer recombination. *Trends Microbiol* **21:**23–30.
32. **Kono N, Arakawa K, Tomita M.** 2011. Comprehensive prediction of chromosome dimer resolution sites in bacterial genomes. *BMC Genomics* **12:**19. doi: 10.1186/1471-2164-12-19.
33. **Carnoy C, Roten CA.** 2009. The dif/Xer recombination systems in proteobacteria. *PLoS One* **4:**e6531.
34. **Recchia GD, Sherratt DJ.** 1999. Conservation of Xer site-specific recombination genes in bacteria. *Mol Microbiol* **34:**1146–8.
35. **Klemm P.** 1986. Two regulatory fim genes, fimB and fimE, control the phase variation of type 1 fimbriae in Escherichia coli. *EMBO J* **5:**1389–1393.
36. **Bastos MC, Murphy E.** 1988. Transposon Tn554 encodes three products required for transposition. *EMBO J* **7:**2935–2941.
37. **Murphy E, Huwyler L, de Freire Bastos M do C.** 1985. Transposon Tn554: complete nucleotide sequence and isolation of transposition-defective and antibiotic-sensitive mutants. *EMBO J* **4:**3357–3365.
38. **Subramanya HS, Arciszewska LK, Baker RA, Bird LE, Sherratt DJ, Wigley DB.** 1997. Crystal structure of the site-specific recombinase, XerD. *EMBO J* **16:**5178–87.
39. **Cornet F, Hallet B, Sherratt DJ.** 1997. Xer recombination in Escherichia coli. Site-specific DNA topoisomerase activity of the XerC and XerD recombinases. *J Biol Chem* **272:**21927–21931.
40. **Gopaul DN, Duyne GD.** 1999. Structure and mechanism in site-specific recombination. *Curr Opin Struct Biol* **9:**14–20.
41. **Gopaul DN, Guo F, Van Duyne GD.** 1998. Structure of the Holliday junction intermediate in Cre-loxP site-specific recombination. *EMBO J* **17:**4175–4187.
42. **Guo F, Gopaul DN, Van Duyne GD.** 1999. Asymmetric DNA bending in the Cre-loxP site-specific recombination synapse. *Proc Natl Acad Sci U S A* **96:**7143–7148.
43. **Das B, Bischerour J, Val M-E, Barre F-X.** 2010. Molecular keys of the tropism of integration of the cholera toxin phage. *Proc Natl Acad Sci U S A* **107:**4377–4382.
44. **Arciszewska LK, Baker RA, Hallet B, Sherratt DJ.** 2000. Coordinated control of XerC and XerD catalytic activities during Holliday junction resolution. *J Mol Biol* **299:**391–403.
45. **Hallet B, Arciszewska LK, Sherratt DJ.** 1999. Reciprocal control of catalysis by the tyrosine recombinases XerC and XerD: an enzymatic switch in site-specific recombination. *Mol Cell* **4:**949–959.
46. **Ferreira H, Sherratt D, Arciszewska L.** 2001. Switching catalytic activity in the XerCD site-specific recombination machine. *J Mol Biol* **312:**45–57.

47. **Ferreira H, Butler-Cole B, Burgin A, Baker R, Sherratt DJ, Arciszewska LK.** 2003. Functional analysis of the C-terminal domains of the site-specific recombinases XerC and XerD. *J Mol Biol* **330:**15–27.

48. **Spiers AJ, Sherratt DJ.** 1999. C-terminal interactions between the XerC and XerD site-specific recombinases. *Mol Microbiol* **32:**1031–1042.

49. **Arciszewska LK, Grainge I, Sherratt DJ.** 1997. Action of site-specific recombinases XerC and XerD on tethered Holliday junctions. *EMBO J* **16:**3731–3743.

50. **Arciszewska LK, Sherratt DJ.** 1995. Xer site-specific recombination in vitro. *EMBO J* **14:**2112–2120.

51. **Zawadzki P, May PFJ, Baker RA, Pinkney JNM, Kapanidis AN, Sherratt DJ, Arciszewska LK.** 2013. Conformational transitions during FtsK translocase activation of individual XerCD-dif recombination complexes. *Proc Natl Acad Sci U S A* **110:**17302–17307.

52. **Kuzminov A.** 2013. The chromosome cycle of prokaryotes. *Mol Microbiol* **90:**214–227.

53. **Cooper S, Helmstetter CE.** 1968. Chromosome replication and the division cycle of Escherichia coli B/r. *J Mol Biol* **31:**519–540.

54. **Kennedy SP, Chevalier F, Barre FX.** 2008. Delayed activation of Xer recombination at dif by FtsK during septum assembly in Escherichia coli. *Mol Microbiol* **68:** 1018–1028.

55. **Steiner WW, Kuempel PL.** 1998. Cell division is required for resolution of dimer chromosomes at the dif locus of Escherichia coli. *Mol Microbiol* **27:**257–268.

56. **Demarre G, Galli E, Muresan L, David A, Paly E, Possoz C, Barre F-X.** 2014. Differential management of the replication terminus regions of the two Vibrio cholerae chromosomes during cell division. *PLoS Genet* **10**(9)**:**e1004557. doi: 10.1371/journal.pgen.1004557. eCollection 2014.

57. **Espeli O, Levine C, Hassing H, Marians KJ.** 2003. Temporal regulation of topoisomerase IV activity in E. coli. *Mol Cell* **11:**189–201.

58. **Wang X, Reyes-Lamothe R, Sherratt DJ.** 2008. Modulation of Escherichia coli sister chromosome cohesion by topoisomerase IV. *Genes Dev* **22:**2426–2433.

59. **Lobry JR.** 1996. Asymmetric substitution patterns in the two DNA strands of bacteria. *Mol Biol Evol* **13:** 660–665.

60. **Hendrickson H, Lawrence JG.** 2007. Mutational bias suggests that replication termination occurs near the dif site, not at Ter sites. *Mol Microbiol* **64:**42–56.

61. **Hojgaard A, Szerlong H, Tabor C, Kuempel P.** 1999. Norfloxacin-induced DNA cleavage occurs at the dif resolvase locus in Escherichia coli and is the result of interaction with topoisomerase IV. *Mol Microbiol* **33:** 1027–1036.

62. **Cornet F, Louarn J, Patte J, Louarn JM.** 1996. Restriction of the activity of the recombination site dif to a small zone of the Escherichia coli chromosome. *Genes Dev* **10:**1152–1161.

63. **Kuempel P, Hogaard A, Nielsen M, Nagappan O, Tecklenburg M.** 1996. Use of a transposon (Tndif) to obtain suppressing and nonsuppressing insertions of the dif resolvase site of Escherichia coli. *Genes Dev* **10:** 1162–1171.

64. **Tecklenburg M, Naumer A, Nagappan O, Kuempel P.** 1995. The dif resolvase locus of the Escherichia coli chromosome can be replaced by a 33-bp sequence, but function depends on location. *Proc Natl Acad Sci U S A* **92:**1352–1356.

65. **Barre FX, Aroyo M, Colloms SD, Helfrich A, Cornet F, Sherratt DJ.** 2000. FtsK functions in the processing of a Holliday junction intermediate during bacterial chromosome segregation. *Genes Dev* **14:**2976–2988.

66. **Deghorain M, Pages C, Meile JC, Stouf M, Capiaux H, Mercier R, Lesterlin C, Hallet B, Cornet F.** 2011. A defined terminal region of the E. coli chromosome shows late segregation and high FtsK activity. *PLoS One* **6:** e22164.

67. **Lesterlin C, Pages C, Dubarry N, Dasgupta S, Cornet F.** 2008. Asymmetry of chromosome Replichores renders the DNA translocase activity of FtsK essential for cell division and cell shape maintenance in Escherichia coli. *PLoS Genet* **4:**e1000288.

68. **Lesterlin C, Mercier R, Boccard F, Barre FX, Cornet F.** 2005. Roles for replichores and macrodomains in segregation of the Escherichia coli chromosome. *EMBO Rep* **6:**557–562.

69. **Bernhardt TG, de Boer PA.** 2005. SlmA, a nucleoid-associated, FtsZ binding protein required for blocking septal ring assembly over Chromosomes in E. coli. *Mol Cell* **18:**555–564.

70. **Wu LJ, Errington J.** 2004. Coordination of cell division and chromosome segregation by a nucleoid occlusion protein in Bacillus subtilis. *Cell* **117:**915–925.

71. **Kaimer C, Schenk K, Graumann PL.** 2011. Two DNA translocases synergistically affect chromosome dimer resolution in Bacillus subtilis. *J Bacteriol* **193:**1334–1340.

72. **Biller SJ, Burkholder WF.** 2009. The Bacillus subtilis SftA (YtpS) and SpoIIIE DNA translocases play distinct roles in growing cells to ensure faithful chromosome partitioning. *Mol Microbiol* **74:**790–809.

73. **Wu LJ, Errington J.** 1994. Bacillus subtilis spoIIIE protein required for DNA segregation during asymmetric cell division. *Science* **264:**572–575.

74. **Barre FX.** 2007. FtsK and SpoIIIE: the tale of the conserved tails. *Mol Microbiol* **66:**1051–1055.

75. **Dorazi R, Dewar SJ.** 2000. Membrane topology of the N-terminus of the escherichia coli FtsK division protein. *FEBS Lett* **478:**13–18.

76. **Massey TH, Mercogliano CP, Yates J, Sherratt DJ, Lowe J.** 2006. Double-stranded DNA translocation: structure and mechanism of hexameric FtsK. *Mol Cell* **23:**457–469.

77. **Dubarry N, Possoz C, Barre FX.** 2010. Multiple regions along the Escherichia coli FtsK protein are implicated in cell division. *Mol Microbiol* **78:**1088–1100.

78. **Aussel L, Barre FX, Aroyo M, Stasiak A, Stasiak AZ, Sherratt D.** 2002. FtsK is a DNA motor protein that activates chromosome dimer resolution by switching the catalytic state of the XerC and XerD recombinases. *Cell* **108:**195–205.

79. **Saleh OA, Perals C, Barre FX, Allemand JF.** 2004. Fast, DNA-sequence independent translocation by FtsK in a single-molecule experiment. *EMBO J* **23:**2430–2439.

80. **Dubarry N, Barre FX.** 2010. Fully efficient chromosome dimer resolution in Escherichia coli cells lacking the integral membrane domain of FtsK. *EMBO J* **29:** 597–605.

81. **Bigot S, Saleh OA, Cornet F, Allemand JF, Barre FX.** 2006. Oriented loading of FtsK on KOPS. *Nat Struct Mol Biol* **13:**1026–1028.

82. **Bigot S, Saleh OA, Lesterlin C, Pages C, El Karoui M, Dennis C, Grigoriev M, Allemand JF, Barre FX, Cornet F.** 2005. KOPS: DNA motifs that control E. coli chromosome segregation by orienting the FtsK translocase. *EMBO J* **24:**3770–3780.

83. **Pease PJ, Levy O, Cost GJ, Gore J, Ptacin JL, Sherratt D, Bustamante C, Cozzarelli NR.** 2005. Sequence-directed DNA translocation by purified FtsK. *Science* **307:**586–590.

84. **Levy O, Ptacin JL, Pease PJ, Gore J, Eisen MB, Bustamante C, Cozzarelli NR.** 2005. Identification of oligonucleotide sequences that direct the movement of the Escherichia coli FtsK translocase. *Proc Natl Acad Sci U S A* **102:**17618–17623.

85. **Nolivos S, Touzain F, Pages C, Coddeville M, Rousseau P, El Karoui M, Le Bourgeois P, Cornet F.** 2012. Coevolution of segregation guide DNA motifs and the FtsK translocase in bacteria: identification of the atypical Lactococcus lactis KOPS motif. *Nucleic Acids Res* **40:**5535–45.

86. **Ptacin JL, Nollmann M, Becker EC, Cozzarelli NR, Pogliano K, Bustamante C.** 2008. Sequence-directed DNA export guides chromosome translocation during sporulation in Bacillus subtilis. *Nat Struct Mol Biol* **15:** 485–493.

87. **Demarre G, Galli E, Barre F-X.** 2013. The FtsK Family of DNA Pumps. *Adv Exp Med Biol* **767:**245–262.

88. **Crozat E, Grainge I.** 2010. FtsK DNA translocase: the fast motor that knows where it's going. *ChemBioChem* **11:**2232–2243.

89. **Espeli O, Lee C, Marians KJ.** 2003. A physical and functional interaction between Escherichia coli FtsK and topoisomerase IV. *J Biol Chem* **278:**44639–44644.

90. **Bigot S, Marians KJ.** 2010. DNA chirality-dependent stimulation of topoisomerase IV activity by the C-terminal AAA+ domain of FtsK. *Nucleic Acids Res* **38:** 3031–3040.

91. **Stouf M, Meile J-C, Cornet F.** 2013. FtsK actively segregates sister chromosomes in Escherichia coli. *Proc Natl Acad Sci U S A* **110:**11157–11162.

92. **Pérals K, Capiaux H, Vincourt JB, Louarn JM, Sherratt DJ, Cornet F.** 2001. Interplay between recombination, cell division and chromosome structure during chromosome dimer resolution in Escherichia coli. *Mol Microbiol* **39:**904–913.

93. **Wang SC, West L, Shapiro L.** 2006. The bifunctional FtsK protein mediates chromosome partitioning and cell division in Caulobacter. *J Bacteriol* **188:**1497–1508.

94. **Yates J, Zhekov I, Baker R, Eklund B, Sherratt DJ, Arciszewska LK.** 2006. Dissection of a functional interaction between the DNA translocase, FtsK, and the XerD recombinase. *Mol Microbiol* **59:**1754–1766.

95. **Yates J, Aroyo M, Sherratt DJ, Barre FX.** 2003. Species specificity in the activation of Xer recombination at dif by FtsK. *Mol Microbiol* **49:**241–249.

96. **Massey TH, Aussel L, Barre F-X, Sherratt DJ.** 2004. Asymmetric activation of Xer site-specific recombination by FtsK. *EMBO Rep* **5:**399–404.

97. **Recchia GD, Aroyo M, Wolf D, Blakely G, Sherratt DJ.** 1999. FtsK-dependent and -independent pathways of Xer site-specific recombination. *EMBO J* **18:**5724–5734.

98. **Bisicchia P, Steel B, Mariam Debela MH, Löwe J, Sherratt D.** 2013. The N-terminal membrane-spanning domain of the Escherichia coli DNA translocase FtsK hexamerizes at midcell. *mBio* **4:**e00800–00813.

99. **Fiche J-B, Cattoni DI, Diekmann N, Langerak JM, Clerte C, Royer CA, Margeat E, Doan T, Nöllmann M.** 2013. Recruitment, assembly, and molecular architecture of the SpoIIIE DNA pump revealed by superresolution microscopy. *PLoS Biol* **11:**e1001557.

100. **Barre F-X, Sherratt DJS.** 2002. Xer Site-Specific Recombination: Promoting Chromosome Segregation, p 149–161. *In* Craig NL, Craigie R, Gellert M, Lambowitz A (ed), *Mobile DNA II*, ASM Press, Washington, DC.

101. **Bonne L, Bigot S, Chevalier F, Allemand JF, Barre FX.** 2009. Asymmetric DNA requirements in Xer recombination activation by FtsK. *Nucleic Acids Res* **37:** 2371–2380.

102. **Grainge I, Lesterlin C, Sherratt DJ.** 2011. Activation of XerCD-dif recombination by the FtsK DNA translocase. *Nucleic Acids Res* **39:**5140–5148.

103. **Diagne CT, Salhi M, Crozat E, Salomé L, Cornet F, Rousseau P, Tardin C.** 2014. TPM analyses reveal that FtsK contributes both to the assembly and the activation of the XerCD-dif recombination synapse. *Nucleic Acids Res* **42:**1721–1732.

104. **Kaimer C, Gonzalez-Pastor JE, Graumann PL.** 2009. SpoIIIE and a novel type of DNA translocase, SftA, couple chromosome segregation with cell division in Bacillus subtilis. *Mol Microbiol* **74:**810–825.

105. **Nolivos S, Pages C, Rousseau P, Le Bourgeois P, Cornet F.** 2010. Are two better than one? Analysis of an FtsK/Xer recombination system that uses a single recombinase. *Nucleic Acids Res* **38:**6477–6489.

106. **Marquis KA, Burton BM, Nollmann M, Ptacin JL, Bustamante C, Ben-Yehuda S, Rudner DZ.** 2008. SpoIIIE strips proteins off the DNA during chromosome translocation. *Genes Dev* **22:**1786–1795.

107. **Lee JY, Finkelstein IJ, Crozat E, Sherratt DJ, Greene EC.** 2012. Single-molecule imaging of DNA curtains reveals mechanisms of KOPS sequence targeting by the DNA translocase FtsK. *Proc Natl Acad Sci U S A* **109:** 6531–6536.

108. **Graham JE, Sivanathan V, Sherratt DJ, Arciszewska LK.** 2009. FtsK translocation on DNA stops at XerCD-dif. *Nucleic Acids Res* **38:**72–81.

109. **Boe L, Tolker-Nielsen T.** 1997. Plasmid stability: comments on the dimer catastrophe hypothesis. *Mol Microbiol* **23:**247–253.

110. **Field CM, Summers DK.** 2011. Multicopy plasmid stability: revisiting the dimer catastrophe. *J Theor Biol* **291:**119–127.

111. **Summers DK, Beton CW, Withers HL.** 1993. Multicopy plasmid instability: the dimer catastrophe hypothesis. *Mol Microbiol* **8:**1031–1038.

112. **Summers DK, Sherratt DJ.** 1984. Multimerization of high copy number plasmids causes instability: ColE1 encodes a determinant essential for plasmid monomerization and stability. *Cell* **36:**1097–1103.

113. **Austin S, Ziese M, Sternberg N.** 1981. A novel role for site-specific recombination in maintenance of bacterial replicons. *Cell* **25:**729–736.

114. **Bui D, Ramiscal J, Trigueros S, Newmark JS, Do A, Sherratt DJ, Tolmasky ME.** 2006. Differences in resolution of mwr-containing plasmid dimers mediated by the Klebsiella pneumoniae and Escherichia coli XerC recombinases: potential implications in dissemination of antibiotic resistance genes. *J Bacteriol* **188:**2812–2820.

115. **Pallecchi L, Riccobono E, Sennati S, Mantella A, Bartalesi F, Trigoso C, Gotuzzo E, Bartoloni A, Rossolini GM.** 2010. Characterization of small ColE-like plasmids mediating widespread dissemination of the qnrB19 gene in commensal enterobacteria. *Antimicrob Agents Chemother* **54:**678–682.

116. **Cornet F, Mortier I, Patte J, Louarn JM.** 1994. Plasmid pSC101 harbors a recombination site, psi, which is able to resolve plasmid multimers and to substitute for the analogous chromosomal Escherichia coli site dif. *J Bacteriol* **176:**3188–3195.

117. **Tolmasky ME, Colloms S, Blakely G, Sherratt DJ.** 2000. Stability by multimer resolution of pJHCMW1 is due to the Tn1331 resolvase and not to the Escherichia coli Xer system. *Microbiology* **146:**581–589.

118. **Tran T, Andres P, Petroni A, Soler-Bistué A, Albornoz E, Zorreguieta A, Reyes-Lamothe R, Sherratt DJ, Corso A, Tolmasky ME.** 2012. Small plasmids harboring qnrB19: a model for plasmid evolution mediated by site-specific recombination at oriT and Xer sites. *Antimicrob Agents Chemother* **56:**1821–1827.

119. **Summers DK, Sherratt DJ.** 1988. Resolution of ColE1 dimers requires a DNA sequence implicated in the three-dimensional organization of the cer site. *EMBO J* **7:**851–858.

120. **Blake JA, Ganguly N, Sherratt DJ.** 1997. DNA sequence of recombinase-binding sites can determine Xer site-specific recombination outcome. *Mol Microbiol* **23:**387–398.

121. **Bregu M, Sherratt DJ, Colloms SD.** 2002. Accessory factors determine the order of strand exchange in Xer recombination at psi. *EMBO J* **21:**3888–3897.

122. **Colloms SD, Bath J, Sherratt DJ.** 1997. Topological selectivity in Xer site-specific recombination. *Cell* **88:** 855–864.

123. **Bath J, Sherratt DJ, Colloms SD.** 1999. Topology of Xer recombination on catenanes produced by lambda integrase. *J Mol Biol* **289:**873–883.

124. **Vazquez M, Colloms SD, Sumners DW.** 2005. Tangle analysis of Xer recombination reveals only three solutions, all consistent with a single three-dimensional topological pathway. *J Mol Biol* **346:**493–504.

125. **Stirling CJ, Colloms SD, Collins JF, Szatmari G, Sherratt DJ.** 1989. xerB, an Escherichia coli gene required for plasmid ColE1 site-specific recombination, is identical to pepA, encoding aminopeptidase A, a protein with substantial similarity to bovine lens leucine aminopeptidase. *EMBO J* **8:**1623–1627.

126. **Stirling CJ, Szatmari G, Stewart G, Smith MC, Sherratt DJ.** 1988. The arginine repressor is essential for plasmid-stabilizing site-specific recombination at the ColE1 cer locus. *EMBO J* **7:**4389–4395.

127. **Alen C, Sherratt DJ, Colloms SD.** 1997. Direct inter action of aminopeptidase A with recombination site DNA in Xer site-specific recombination. *EMBO J* **16:** 5188–5197.

128. **Colloms SD, Alen C, Sherratt DJ.** 1998. The ArcA/ArcB two-component regulatory system of Escherichia coli is essential for Xer site-specific recombination at psi. *Mol Microbiol* **28:**521–530.

129. **Sträter N, Sherratt DJ, Colloms SD.** 1999. X-ray structure of aminopeptidase A from Escherichia coli and a model for the nucleoprotein complex in Xer site-specific recombination. *EMBO J* **18:**4513–4522.

130. **Reijns M, Lu Y, Leach S, Colloms SD.** 2005. Mutagenesis of PepA suggests a new model for the Xer/cer synaptic complex. *Mol Microbiol* **57:**927–941.

131. **Minh PNL, Devroede N, Massant J, Maes D, Charlier D.** 2009. Insights into the architecture and stoichiometry of Escherichia coli PepA*DNA complexes involved in transcriptional control and site-specific DNA recombination by atomic force microscopy. *Nucleic Acids Res* **37:**1463–1476.

132. **Abremski K, Hoess R, Sternberg N.** 1983. Studies on the properties of P1 site-specific recombination: evidence for topologically unlinked products following recombination. *Cell* **32:**1301–1311.

133. **Abremski K, Hoess R.** 1985. Phage P1 Cre-loxP site-specific recombination. Effects of DNA supercoiling on catenation and knotting of recombinant products. *J Mol Biol* **184:**211–220.

134. **Adams DE, Bliska JB, Cozzarelli NR.** 1992. Cre-lox recombination in Escherichia coli cells. Mechanistic differences from the in vitro reaction. *J Mol Biol* **226:** 661–673.

135. **Kilbride E, Boocock MR, Stark WM.** 1999. Topological selectivity of a hybrid site-specific recombination system with elements from Tn3 res/resolvase and bacteriophage P1 loxP/Cre. *J Mol Biol* **289:**1219–1230.

136. **Gourlay SC, Colloms SD.** 2004. Control of Cre recombination by regulatory elements from Xer recombination systems. *Mol Microbiol* **52:**53–65.

137. **Colloms SD, McCulloch R, Grant K, Neilson L, Sherratt DJ.** 1996. Xer-mediated site-specific recombination in vitro. *EMBO J* **15:**1172–1181.

138. **McCulloch R, Coggins LW, Colloms SD, Sherratt DJ.** 1994. Xer-mediated site-specific recombination at cer

generates Holliday junctions in vivo. *EMBO J* **13:** 1844–1855.

139. **Ingmer H, Miller C, Cohen SN.** 2001. The RepA protein of plasmid pSC101 controls Escherichia coli cell division through the SOS response. *Mol Microbiol* **42:** 519–526.

140. **Patient ME, Summers DK.** 1993. ColE1 multimer formation triggers inhibition of Escherichia coli cell division. *Mol Microbiol* **9:**1089–1095.

141. **Sharpe ME, Chatwin HM, Macpherson C, Withers HL, Summers DK.** 1999. Analysis of the ColE1 stability determinant Rcd. *Microbiology (Reading U K)* **145**(Pt 8): 2135–2144.

142. **Balding C, Blaby I, Summers D.** 2006. A mutational analysis of the ColE1-encoded cell cycle regulator Rcd confirms its role in plasmid stability. *Plasmid* **56:**68–73.

143. **Chatwin HM, Summers DK.** 2001. Monomer-dimer control of the ColE1 P(cer) promoter. *Microbiology (Reading U K)* **147:**3071–3081.

144. **Chant EL, Summers DK.** 2007. Indole signalling contributes to the stable maintenance of Escherichia coli multicopy plasmids. *Mol Microbiol* **63:**35–43.

145. **Chimerel C, Field CM, Piñero-Fernandez S, Keyser UF, Summers DK.** 2012. Indole prevents Escherichia coli cell division by modulating membrane potential. *Biochim Biophys Acta* **1818:**1590–1594.

146. **Field CM, Summers DK.** 2012. Indole inhibition of ColE1 replication contributes to stable plasmid maintenance. *Plasmid* **67:**88–94.

147. **Gaimster H, Cama J, Hernández-Ainsa S, Keyser UF, Summers DK.** 2014. The indole pulse: a new perspective on indole signalling in Escherichia coli. *PloS One* **9:** e93168.

148. **Das B, Bischerour J, Barre FX.** 2011. Molecular mechanism of acquisition of the cholera toxin genes. *Indian J Med Res* **133:**195–200.

149. **Dai H, Chow TY, Liao HJ, Chen ZY, Chiang KS.** 1988. Nucleotide sequences involved in the neolysogenic insertion of filamentous phage Cf16-v1 into the Xanthomonas campestris pv. citri chromosome. *Virology* **167:**613–620.

150. **Simpson AJ, Reinach FC, Arruda P, Abreu FA, Acencio M, Alvarenga R, Alves LM, Araya JE, Baia GS, Baptista CS, Barros MH, Bonaccorsi ED, Bordin S, Bove JM, Briones MR, Bueno MR, Camargo AA, Camargo LE, Carraro DM, Carrer H, Colauto NB, Colombo C, Costa FF, Costa MC, Costa-Neto CM, Coutinho LL, Cristofani M, Dias-Neto E, Docena C, El-Dorry H, Facincani AP, Ferreira AJ, Ferreira VC, Ferro JA, Fraga JS, Franca SC, Franco MC, Frohme M, Furlan LR, Garnier M, Goldman GH, Goldman MH, Gomes SL, Gruber A, Ho PL, Hoheisel JD, Junqueira ML, Kemper EL, Kitajima JP, Krieger JE, Kuramae EE, Laigret F, Lambais MR, Leite LC, Lemos EG, Lemos MV, Lopes SA, Lopes CR, Machado JA, Machado MA, Madeira AM, Madeira HM, Marino CL, Marques MV, Martins EA, Martins EM, Matsukuma AY, Menck CF, Miracca EC, Miyaki CY, Monteriro-Vitorello CB, Moon DH, Nagai MA, Nascimento AL, Netto LE, Nhani A, Nobrega FG, Nunes LR, Oliveira MA, de Oliveira MC, de Oliveira RC, Palmieri DA, Paris A, Peixoto BR, Pereira GA, Pereira HA, Pesquero JB, Quaggio RB, Roberto PG, Rodrigues V, de MR AJ, de Rosa VE, de Sa RG, Santelli RV, Sawasaki HE, da Silva AC, da Silva AM, da Silva FR, da Silva WA, da Silveira JF.** 2000. The genome sequence of the plant pathogen Xylella fastidiosa. The Xylella fastidiosa Consortium of the Organization for Nucleotide Sequencing and Analysis. *Nature* **406:**151–159.

151. **Dillard JP, Seifert HS.** 2001. A variable genetic island specific for Neisseria gonorrhoeae is involved in providing DNA for natural transformation and is found more often in disseminated infection isolates. *Mol Microbiol* **41:**263–277.

152. **Waldor MK, Mekalanos JJ.** 1996. Lysogenic conversion by a filamentous phage encoding cholera toxin. *Science* **272:**1910–1914.

153. **Huber KE, Waldor MK.** 2002. Filamentous phage integration requires the host recombinases XerC and XerD. *Nature* **417:**656–659.

154. **Gonzalez MD, Lichtensteiger CA, Caughlan R, Vimr ER.** 2002. Conserved filamentous prophage in Escherichia coli O18:K1:H7 and Yersinia pestis biovar orientalis. *J Bacteriol* **184:**6050–6055.

155. **Derbise A, Chenal-Francisque V, Pouillot F, Fayolle C, Prevost MC, Medigue C, Hinnebusch BJ, Carniel E.** 2007. A horizontally acquired filamentous phage contributes to the pathogenicity of the plague bacillus. *Mol Microbiol* **63:**1145–1157.

156. **Val M-E, Bouvier M, Campos J, Sherratt D, Cornet F, Mazel D, Barre F-X.** 2005. The single-stranded genome of phage CTX is the form used for integration into the genome of Vibrio cholerae. *Mol Cell* **19:**559–566.

157. **Das B, Bischerour J, Barre F-X.** 2011. VGJphi integration and excision mechanisms contribute to the genetic diversity of Vibrio cholerae epidemic strains. *Proc Natl Acad Sci U S A* **108:**2516–2521.

158. **Midonet C, Das B, Paly E, Barre F-X.** 2014. XerD-mediated FtsK-independent integration of TLCϕ into the Vibrio cholerae genome. *Proc Natl Acad Sci U S A* **pii:**201404047. [Epub ahead of print]

159. **Mutreja A, Kim DW, Thomson NR, Connor TR, Lee JH, Kariuki S, Croucher NJ, Choi SY, Harris SR, Lebens M, Niyogi SK, Kim EJ, Ramamurthy T, Chun J, Wood JL, Clemens JD, Czerkinsky C, Nair GB, Holmgren J, Parkhill J, Dougan G.** 2011. Evidence for several waves of global transmission in the seventh cholera pandemic. *Nature* **477:**462–465.

160. **Chun J, Grim CJ, Hasan NA, Lee JH, Choi SY, Haley BJ, Taviani E, Jeon YS, Kim DW, Brettin TS, Bruce DC, Challacombe JF, Detter JC, Han CS, Munk AC, Chertkov O, Meincke L, Saunders E, Walters RA, Huq A, Nair GB, Colwell RR.** 2009. Comparative genomics reveals mechanism for short-term and long-term clonal transitions in pandemic Vibrio cholerae. *Proc Natl Acad Sci U S A* **106:**15442–15447.

161. **Campos J, Martinez E, Marrero K, Silva Y, Rodriguez BL, Suzarte E, Ledon T, Fando R.** 2003. Novel type of

specialized transduction for CTX phi or its satellite phage RS1 mediated by filamentous phage VGJ phi in Vibrio cholerae. *J Bacteriol* **185:**7231–7240.

162. **Hassan F, Kamruzzaman M, Mekalanos JJ, Faruque SM.** 2010. Satellite phage TLCphi enables toxigenic conversion by CTX phage through dif site alteration. *Nature* **467:**982–985.

163. **Rubin EJ, Lin W, Mekalanos JJ, Waldor MK.** 1998. Replication and integration of a Vibrio cholerae cryptic plasmid linked to the CTX prophage. *Mol Microbiol* **28:**1247–1254.

164. **Moyer KE, Kimsey HH, Waldor MK.** 2001. Evidence for a rolling-circle mechanism of phage DNA synthesis from both replicative and integrated forms of CTXphi. *Mol Microbiol* **41:**311–323.

165. **Campbell AM.** 1992. Chromosomal insertion sites for phages and plasmids. *J Bacteriol* **174:**7495–7499.

166. **Reiter WD, Palm P, Yeats S.** 1989. Transfer RNA genes frequently serve as integration sites for prokaryotic genetic elements. *Nucleic Acids Res* **17:**1907–1914.

167. **Val M-E, Kennedy SP, Soler-Bistué AJ, Barbe V, Bouchier C, Ducos-Galand M, Skovgaard O, Mazel D.** 2014. Fuse or die: how to survive the loss of Dam in Vibrio cholerae. *Mol Microbiol* **91:**665–678.

168. **Campos J, Martinez E, Suzarte E, Rodriguez BL, Marrero K, Silva Y, Ledon T, del Sol R, Fando R.** 2003. VGJ phi, a novel filamentous phage of Vibrio cholerae, integrates into the same chromosomal site as CTX phi. *J Bacteriol* **185:**5685–5696.

169. **McLeod SM, Waldor MK.** 2004. Characterization of XerC- and XerD-dependent CTX phage integration in Vibrio cholerae. *Mol Microbiol* **54:**935–947.

170. **Quinones M, Kimsey HH, Waldor MK.** 2005. LexA cleavage is required for CTX prophage induction. *Mol Cell* **17:**291–300.

171. **Quinones M, Kimsey HH, Ross W, Gourse RL, Waldor MK.** 2006. LexA represses CTXphi transcription by blocking access of the alpha C-terminal domain of RNA polymerase to promoter DNA. *J Biol Chem* **281:** 39407–39412.

172. **Bischerour J, Spangenberg C, Barre F-X.** 2012. Holliday junction affinity of the base excision repair factor Endo III contributes to cholera toxin phage integration. *EMBO J* **31:**3757–3767.

Mobile DNA, 3rd Edition
Nancy L. Craig, Michael Chandler, Martin Gellert, Alan M. Lambowitz, Phoebe A. Rice and Suzanne Sandmeyer

doi:10.1128/microbiolspec.MDNA3-0020-2014

Margaret M. Wood[1]
Jeffrey F. Gardner[2]

8 The Integration and Excision of CTnDOT

INTRODUCTION

Bacteroides spp. are one of the more prevalent members of the human colonic microbiota, representing approximately 40% of the bacterial community (1). *Bacteroides* spp. are normally in symbiosis with their human hosts. Although they are usually harmless members of the gut microbiota, they can become opportunistic pathogens if released from the colon (2, 3). This most commonly occurs due to surgery, trauma or disease such as gangrenous appendicitis or malignancies (4). Among anaerobic bacteria, *Bacteroides* spp. are the pathogens most commonly isolated from clinical samples, including blood (2). The treatment of *Bacteroides* infections has become more challenging as they have acquired a variety of genes that encode resistances to antibiotics. In the 1970s, only 20 to 30% of *Bacteroides* spp. clinical isolates were resistant to tetracycline. By the 1990s, the prevalence of tetracycline resistance had increased to 80% (5). This increase in tetracycline resistance can be attributed to the presence of integrative and conjugative elements (ICEs) that encode antibiotic resistance genes.

ICEs, formerly referred to as conjugative transposons, have been increasingly implicated in the dissemination of antibiotic resistance in *Bacteroides* spp. and other members of the human colonic microbiota. The best characterized ICE in *Bacteroides* spp. is the 65 kb CTnDOT. CTnDOT was discovered as a mobile element in the chromosome of a *Bacteroides* spp. strain from a local clinical isolate (Shoemaker N, Salyers A, personal communication).

Like other ICEs, CTnDOT encodes the genes necessary for its excision and transfer to recipient cells via conjugation, as well as integration into the recipient cell chromosome. Fig. 1 shows the genes relevant to the integration, excision and transfer of CTnDOT. The *intDOT* gene encodes the integrase (IntDOT) required for both the integration and excision of CTnDOT. The excision operon contains the *xis2c*, *xis2d*, *orf3*, and *exc* genes. With the exception of *orf3*, the genes contained in this operon are (as its name implies) involved in CTnDOT excision. However, Xis2c and Xis2d have also been identified as positive regulators of the *tra* genes that encode proteins involved in conjugation (6). In addition, Xis2d and Exc also appear to stimulate expression of the *mob* operon that encodes the relaxase and coupling proteins required for mobilization (7). CTnDOT also encodes two antibiotic resistance genes, *tetQ* and *ermF* (8, 9).

[1]Simmons College, 300 The Fenway, Boston, MA 02115; [2]Department of Microbiology, University of Illinois at Urbana-Champaign, 601 S Goodwin Avenue, Urbana, IL 61801.

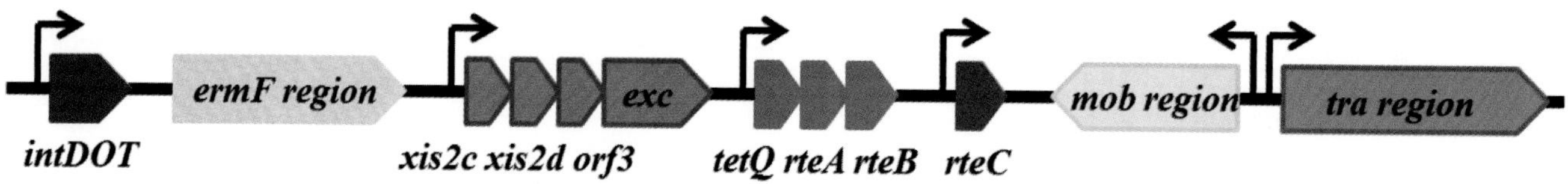

Figure 1 A schematic of CTnDOT. The genes *xis2c*, *xis2d*, *orf3*, and *exc* (shown in blue) are part of the excision operon. The genes shown in orange (*tetQ*, *rteA*, and *rteB*) are involved in regulating CTnDOT excision. doi:10.1128/microbiolspec.MDNA3-0020-2014.f1

The regulation of CTnDOT excision is complex and highly coordinated. One of the novel features of the CTnDOT element is that excision and conjugative transfer are stimulated in the presence of low levels of tetracycline (10). Upon exposure to tetracycline, the *tetQ-rteA-rteB* operon (shown in orange in Fig. 1) is expressed via a translational attenuation mechanism (11, 12). The *tetQ* gene encodes a ribosomal protection protein, and *rteA* and *rteB* encode a two-component regulatory system. Exactly what RteB is sensing in the environment is not yet known, although it is likely not tetracycline (13). RteB is the transcriptional activator of another regulatory protein, RteC. The function of RteC is to activate expression of the excision operon which contains the *xis2c*, *xis2d*, *orf3*, and *exc* genes (14, 15). RteC binds to two inverted repeat half-sites upstream of the excision operon promoter (16, 17).

After activation of the excision operon CTnDOT excises from the *Bacteroides* chromosome and forms a closed circular intermediate (Fig. 2). The intermediate is nicked at *oriT* and a single strand is transferred to a recipient cell via conjugation. The DNA is replicated in the recipient cell to form the double stranded circular form, which then integrates into the *Bacteroides* chromosome (18, 19). CTnDOT is stably maintained even in the absence of selection, perhaps due to the complexity of its regulatory system (19). In addition, no *Bacteroides* strains have thus far been identified that lack ICEs (19). Many of the ICEs contained in *Bacteroides* are cryptic and do not carry antibiotic resistance genes. However, they can potentially acquire resistance determinants which could facilitate their transfer to other bacterial species (19). CTnDOT can also mobilize other mobile elements, including conjugative plasmids and mobilizable transposons (20, 21, 22, 23). In the case of mobilizable transposon NBU1, the CTnDOT-encoded RteB activates expression of *orf2x*, one of the genes required for NBU1 excision (Moon K, unpublished results).

Originally, CTnDOT was called a conjugative transposon because it was thought that the element integrated into random target sites after transfer by conjugation. However, it was subsequently discovered that CTnDOT was site-selective because it integrates into a few target sites (24, 25). DNA sequencing of the *intDOT* gene suggested that it was a member of the tyrosine recombinase family which also includes lambda integrase (see below) (24, 26, 27, 28).

Since CTnDOT utilizes site-selective recombination during integration into and excision from the *Bacteroides* chromosome, it was initially assumed that the processes would mechanistically resemble that of bacteriophage lambda. However, recent studies have demonstrated that CTnDOT recombination differs significantly in several respects from that of the lambda system.

IntDOT

The enzyme responsible for catalyzing the integration and excision of CTnDOT is IntDOT, a tyrosine recombinase. The expression of IntDOT is constitutive, and IntDOT and a host factor are sufficient for integrative recombination (24). IntDOT was identified as a member of the tyrosine recombinase family because it contained five of six residues of a characteristic RK(H/K)R(H/W)Y motif, including a catalytic tyrosine found in tyrosine recombinases (24, 26, 27, 28, 29). However, alignments with other tyrosine recombinases (including lambda Int) demonstrated that IntDOT has a serine (S259) in place of the putative first arginine (24, 27). Substitution of S259 to alanine or arginine had no effect on *in vitro* integration, cleavage or ligation activities (30). This amino acid substitution, along with other experimental results that will be detailed later in this chapter, was among the first indicators that IntDOT may have a unique active site structure.

The reactions catalyzed by tyrosine recombinases (including IntDOT) are described in detail in the chapters by Makkuni Jayaram, Anca Segall, Gregory Van Duyne, and Art Landy. The reactions involve the binding of two monomers of the enzyme to each of the two DNA substrates. In the case of CTnDOT integration, the substrates are the joined ends of the excised element (*attDOT*) and the target sequence in the *Bacteroides* chromosome (*attB*) (24, 31). In CTnDOT excision, the substrates are called *attR* and *attL* (Fig. 2) (24, 32, 33). Initially, two monomers on each site are active.

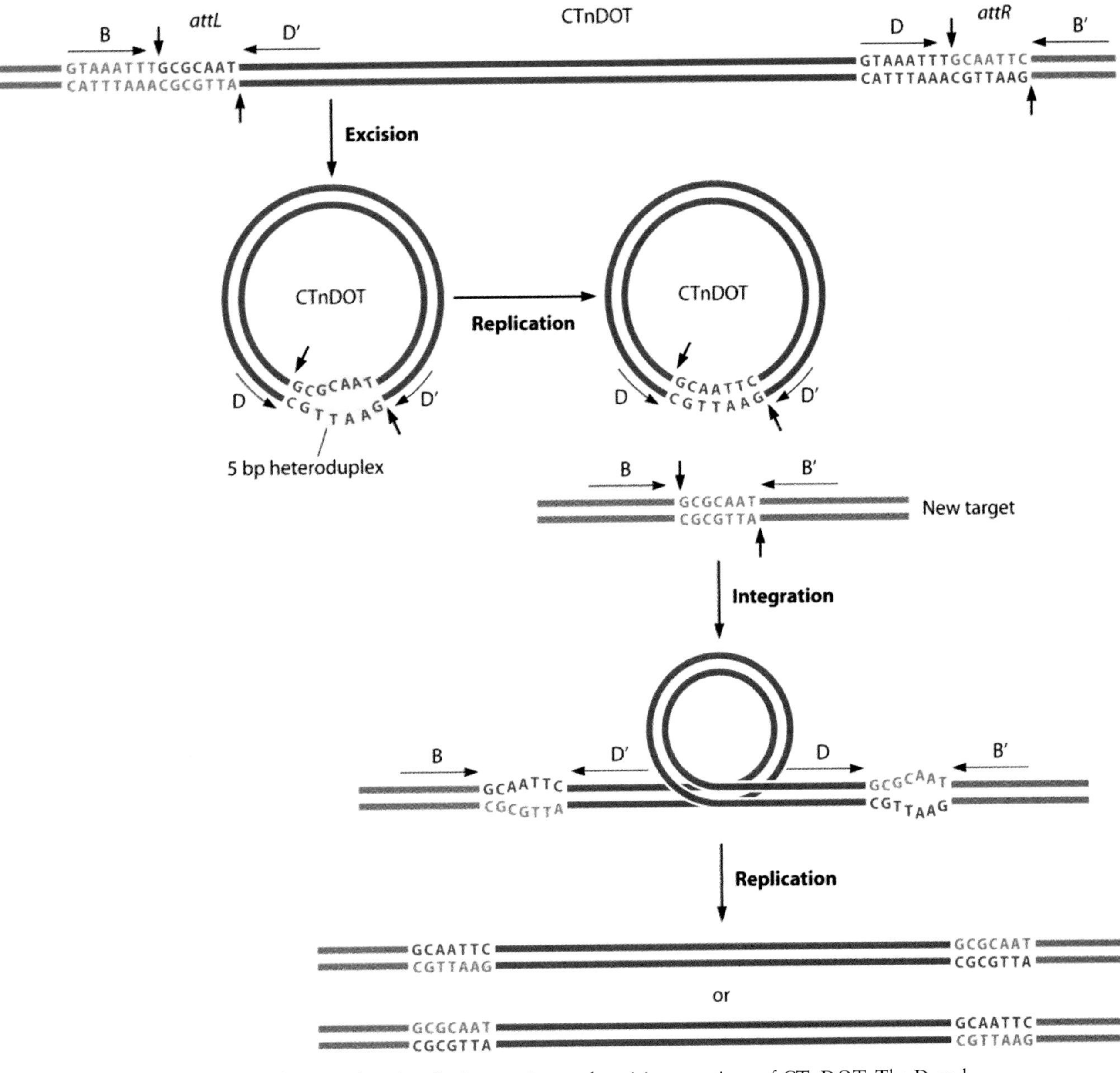

Figure 2 A diagram showing the integration and excision reactions of CTnDOT. The D and D′ core-type sites are located on CTnDOT (red), while B and B′ core-type sites are located in the bacterial chromosome (purple). IntDOT makes staggered cuts 7 bp apart (vertical arrows) on *attL* and *attR* to excise CTnDOT. The element forms a closed circular intermediate containing a 5 bp heteroduplex known as the coupling sequence. The heterology is likely resolved following conjugative transfer into a recipient cell. During integration into the bacterial chromosome, IntDOT makes staggered cuts 7 bp apart on the *attDOT* site of the circular intermediate as well as the *attB* target sequence, and CTnDOT integrates into the bacterial chromosome. The 5 bp heteroduplexes that form following integration are resolved by DNA replication or repair in the recipient cell.
doi:10.1128/microbiolspec.MDNA3-0020-2014.f2

The active monomers each cleave one strand of DNA to form a 3′-phosphotyrosyl bond with the DNA releasing free 5′-OH groups. Next, the 5′-OH groups undergo a ligation reaction by performing a nucleophilic attack on the partner phosphotyrosine linkage, forming a Holliday junction (HJ). A conformational change activates the other pair of monomers, and another round of cleavage, strand exchange, and ligation occurs. The end result is a recombinant with attachment sites containing DNA from each substrate (34, 35).

IDENTIFYING RESIDUES IMPORTANT FOR IntDOT FUNCTION

Like lambda Int and several other tyrosine recombinases, IntDOT is a heterobivalent DNA binding protein that contains three different domains (30, 36, 37). The core-binding (CB) and catalytic (CAT) domains bind to core-type sites (D and D′ on *attDOT* and B and B′ on *attB*) (Fig. 2) that flank the region of cleavage and strand exchange (overlap sequence). The arm-binding (N) domain binds to distal arm-type sites (R1′, R1, R2, R2′, L1, L2; see below) (30, 37, 38, 39). Initially an *in vivo* screen for recombination was developed to identify randomly generated mutants deficient in integration (37, 40). Mutants were isolated with substitutions in each of the IntDOT domains and tested for deficiencies in DNA binding, cleavage, ligation, and HJ resolution. The locations of the IntDOT substitutions isolated through random mutagenesis are shown in Fig. 3.

Four mutants with substitutions in the N domain were isolated. Two mutants (R13C and S38N) contain substitutions located in the N domain and 2 additional substitutions (V95M and G101R) occurred in the region where the N domain meets the CB domain (Fig. 3) (37). All four mutants showed reduced ability to form protein/DNA complexes in gel shift assays but displayed wild-type levels of *in vitro* cleavage and ligation (37). This result suggests that the N domain is involved in interactions with the DNA (37).

The V95M protein had an additional unique phenotype. While the R13C, S38N, and G101R mutant proteins were able to resolve HJs, the V95M protein could not (40). This result was interesting, since the methionine substitution in the V95M protein is located in the N domain that binds arm-type sites. The phenotype of the V95M mutant suggested that the N domain may interact with the CB and/or CAT domains during HJ resolution. The V95M substitution is located in a putative alpha helix (H2, Fig. 3) that corresponds to the coupler region of lambda Int. An alignment of the lambda Int coupler and the putative IntDOT coupler is shown in Fig. 4. The lambda Int coupler is required for cooperative Int interactions necessary for arm-type site binding (41), and it is possible that the IntDOT coupler performs a similar function (37). The lambda Int coupler has also been shown to be involved in HJ resolution via a protein–protein interaction (42). Lee *et al.* (42) performed a charge switch experiment in lambda Int in which they reversed charges of two residues in the same protein. One residue was in the coupler and the other residue was in the N domain. They demonstrated that a substitution of D71 in the coupler or R30 in the

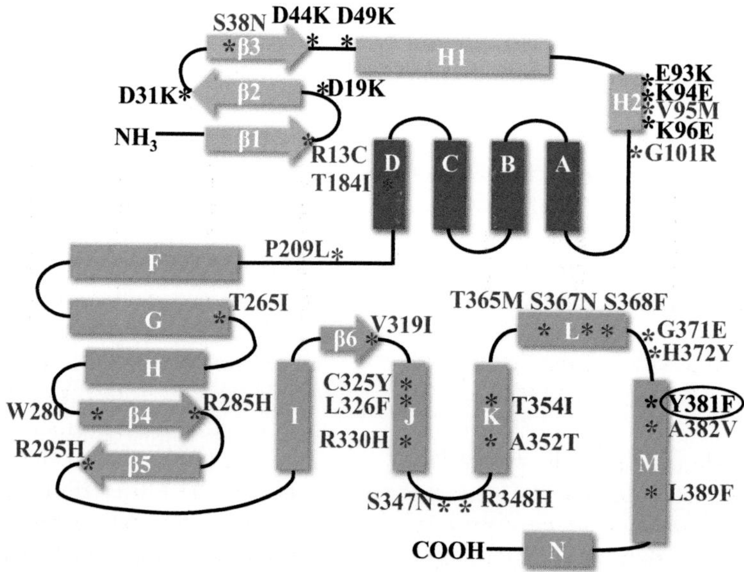

Figure 3 The location of IntDOT substitution mutants isolated from hydroxylamine random mutagenesis experiments (red) or site-directed mutagenesis (black). The circled residue is the catalytic tyrosine, which was mutated to phenylalanine via site-directed mutagenesis. The N domain is shown in purple, the CB domain is shown in dark blue, and the CAT domain is shown in light blue. doi:10.1128/microbiolspec.MDNA3-0020-2014.f3

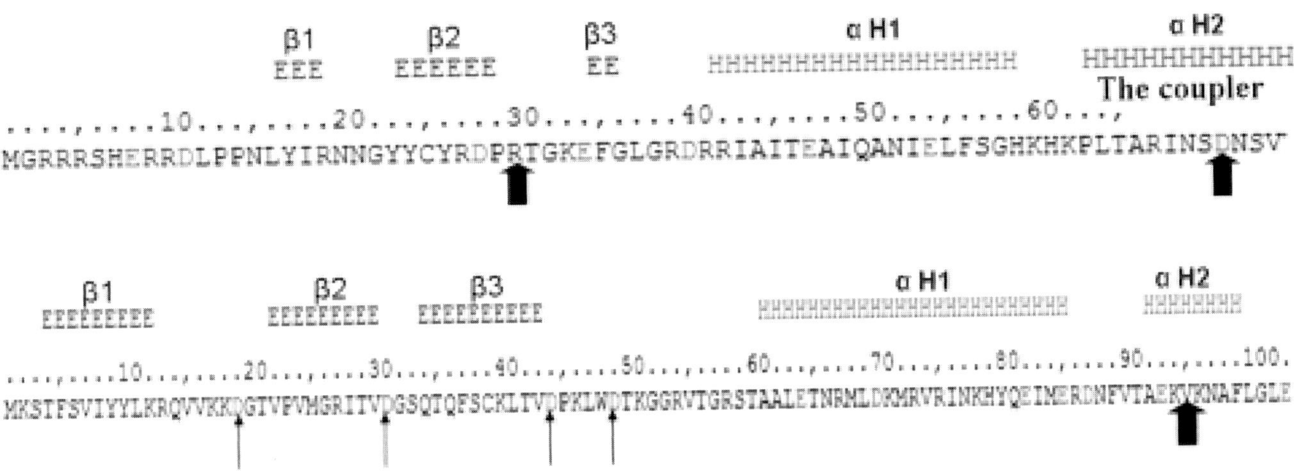

Figure 4 Alignment of the amino acid sequences and secondary structures of the N domains of lambda Int (top) and IntDOT (bottom). The helix α H2 is the lambda Int coupler. The bold arrows pointing to the lambda Int sequence denote residues R30 and D71 that form a protein–protein interaction. The V95 residue of IntDOT (indicated by a bold arrow on the bottom sequence) is located in the putative IntDOT coupler. Thinner arrows denote aspartic acid residues that were mutated to lysines to examine possible charge interactions between IntDOT monomers. doi:10.1128/microbiolspec.MDNA3-0020-2014.f4

N domain of lambda Int abolished HJ resolution. However, a protein containing both the D71R and R30D substitutions regained HJ resolution activity due to the restoration of an intermolecular ion bridge. They concluded that D71 and R30 make an intermolecular ion pair.

Since the V95 residue is in the alpha helix of IntDOT that is analogous to the lambda Int coupler helix, substitutions of charged residues in the coupler helix in the IntDOT helix were made. In contrast to the results found for lambda Int, it appears that an ion pair interaction is not involved in resolution of HJs by IntDOT. To ascertain whether charge interactions are similarly involved in IntDOT HJ resolution, the three charged residues (Fig. 3 and 4; E93, K94, and K96) in the putative IntDOT coupler region were mutated to the reverse-charge residues and tested for DNA binding, cleavage, ligation, and HJ resolution activity (40). The E93K protein showed wild-type activity in all the assays. The other two mutant proteins (K94E and K96E) were defective for HJ resolution. However, they were also defective in DNA binding, thus making it difficult to draw a conclusion about the effect of the charge reversal on HJ resolution (40). In addition, mutant proteins that contained substitutions of each of the four negatively charged residues in the N domain (Fig. 3 and 4; D19K, D30K, D44K, and D49K) were able to resolve HJs. In sum, it appears that the IntDOT coupler is involved in HJ resolution. However, unlike lambda Int, there is no evidence that IntDOT uses a charge interaction between the coupler and the N domain of a neighboring monomer.

Two mutants were isolated with substitutions in the IntDOT CB domain (T184I and P209L) (Fig. 3), and both showed defects in DNA binding, cleavage, and ligation activities (37). This result demonstrates that substitutions in the CB domain can have dramatic effects on catalytic activity. Additional CB residues that are likely important for IntDOT catalytic function were identified using a predicted three-dimensional model developed through homology modeling (30). Site-directed mutagenesis was utilized to construct alanine substitutions at 15 representative residues in the CB domain. Interestingly, several of these CB domain mutants (T139A, H143A, N183A, and T194A) were defective in *in vitro* cleavage but could still perform *in vitro* ligation (30). With the exception of tyrosine recombinases with substitutions of the catalytic tyrosine, these are the only ones that are defective in cleavage but retain ligation activity. The homology modeling predicted that T139, H143, and N183 interact with the cleaved strand (Fig. 5) (30). The location of N183 is striking because it is predicted to interact with DNA adjacent to the point of cleavage. T194 is predicted to interact with the uncleaved strand. Malanowska *et al.* (30) speculated that the IntDOT conformation may differ between cleavage and ligation, and that some residues, like N183, H143, and T194, are required only in the cleavage-activated conformation.

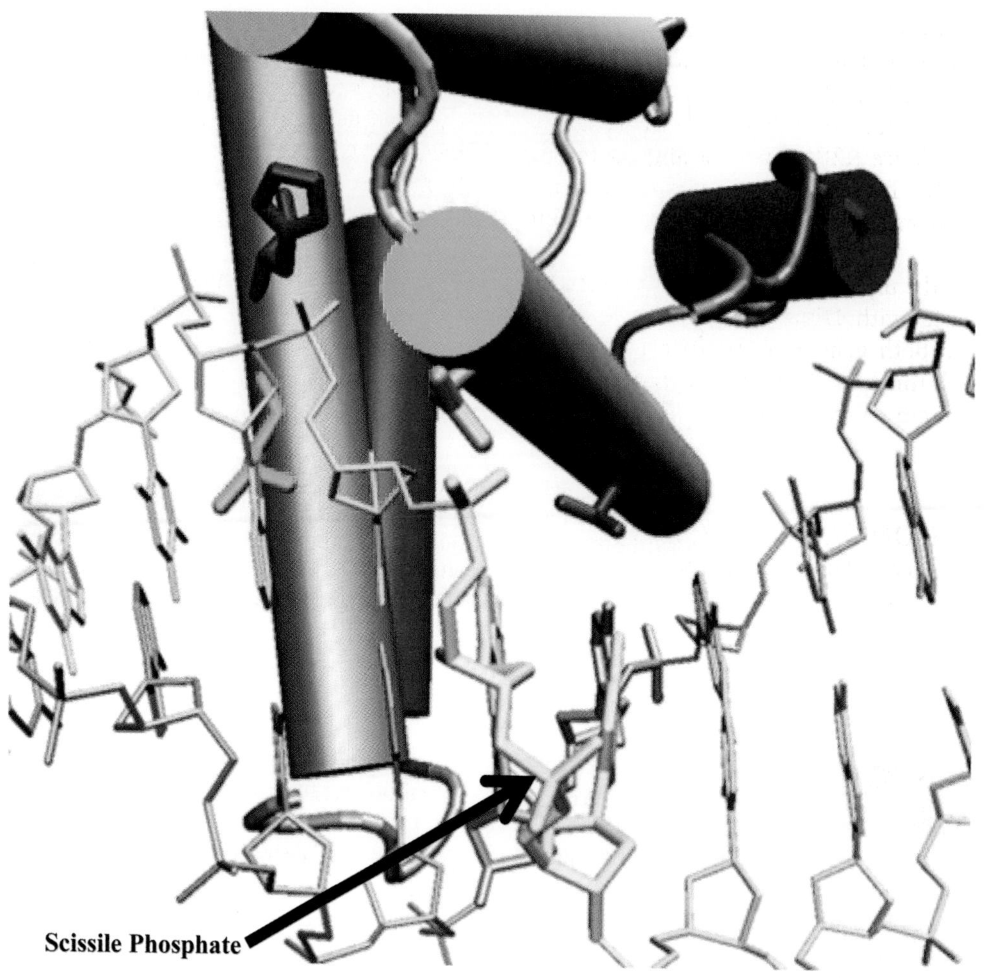

Figure 5 Locations of amino acids that, when mutated, yielded cleavage-defective phenotypes. N183 is shown in cyan, H143 is shown in blue, T183 is shown in green, and T194 is shown in purple. N183, H143, and T183 interact with the cleaved DNA strand, shown in yellow. T194 is proximal to the uncleaved strand, shown in gray. The scissile phosphate is indicated with a bold black arrow. doi:10.1128/microbiolspec.MDNA3-0020-2014.f5

Several mutants isolated using the *in vivo* screen had substitutions located in the CAT domain. One of the interesting mutants isolated was A382V (Fig. 3), which is located right next to the catalytic tyrosine (Y381). IntDOT containing the A382V substitution was defective in *in vivo* recombination but showed wild-type levels of *in vitro* DNA binding, cleavage, and ligation (37). It is possible that the positioning of the catalytic tyrosine is altered in the A382V mutant while the interactions of IntDOT with DNA are unaffected (37). Another mutant, L389F (Fig. 3), contains a substitution located in a predicted C-terminal tail of IntDOT. The IntDOT tail likely interacts with other IntDOT monomers during recombination (37). Like A382V, this mutant is defective in *in vivo* recombination but DNA binding, cleavage, and ligation were unaffected (37). In lambda Int, the deletion of the last eight amino acids in the C-terminal tail resulted in deficiencies in recombination but increased topoisomerase activity, suggesting that the C-terminus is important for regulating catalytic activity (43). The IntDOT C-terminal tail may serve a similar function.

One of the unique features of IntDOT is the missing arginine residue in the catalytic motif. As mentioned above, IntDOT has a serine residue (S259) where an arginine is predicted to be located (27). One of the CAT domain mutants isolated in the hydroxylamine-generated mutant screen, R285H (Fig. 3), was defective for *in vitro* cleavage and ligation in addition to *in vivo* recombination (37). Mutating R285 to an alanine, aspartate, or lysine residue also resulted in severe deficiencies in recombination, cleavage, and ligation, although

DNA binding was not affected (37). The IntDOT homology model suggests that R285 can enter the active site and interact with bound core-type site DNA similar to the Arg I residues of other tyrosine recombinases (Fig. 6) (37). Since R285 does not align with Arg I residues, the IntDOT active site appears to be structurally different from active sites of other tyrosine recombinases.

Work is currently under way to isolate crystals of IntDOT complexed with DNA, and it will be interesting to combine both genetic and structural data to generate a model for how IntDOT monomers form protein–protein interactions, bind DNA, and perform catalysis.

THE ROLE OF HOMOLOGY IN CTnDOT INTEGRATIVE RECOMBINATION

The altered architecture of the IntDOT active site is not the only feature that distinguishes it from other tyrosine recombinases. CTnDOT integrates site-selectively into at least six *attB* sites in the *Bacteroides* chromosome (24) while most other tyrosine recombinases promote site-specific recombination reactions between two unique sites containing at least 7 bp of sequence identity. One consequence of the site-selectivity of IntDOT is that it recombines sites that contain different overlap sequences (Fig. 7A). All six known *attB* sites, as well as *attDOT*, contain a conserved TTTGC in the top strand. The three thymines are located in the B core-type site and the G and C (known as the GC dinucleotide (31)) are located in the overlap sequence (Fig. 7A) (24, 31). The strand exchanges at the top strands occur between the third T and the G in the top strands of the partner sites, and the strand exchanges at the bottom strands occur 7 bp away. Because the 5 bp between the GC dinucleotide and the sites of the second strand exchange in the overlap sequences are usually different, a heteroduplex is formed (27, 31, 44). The 5 bp of heterology that are formed is known as the coupling sequence. The exact sequence of the GC dinucleotide is not important for the integration reaction. As long as the 2 bp of identity are present between *attDOT* and *attB* at the sites of strand exchanges, wild-type levels of *in vitro* integration are observed. For example, when the GC dinucleotide was changed to AT or CG in only one of the recombination substrates (*attDOT* or *attB*), *in vitro* integration was barely detectable. However, if a complementary mutation was constructed in the partner site, such that both *attDOT* and *attB* contained identical dinucleotides, wild-type levels of *in vitro* integration were restored (44). Low levels of *in vitro* integration were also detected when only 1 bp of homology existed between *attDOT* and *attB* (31). This demonstrated that the CTnDOT integration reaction was dependent on at least 1 bp of homology between *att* sites. This is significantly different from what has been seen in other systems. Most tyrosine recombinases, including lambda Int, have strict requirements for sequence identity in the overlap sequence. For example, a mutation adjacent to one of the cleavage sites in the lambda overlap sequence significantly decreased integration frequency. However, if the same mutation was made in the overlap of the partner attachment site, recombination was restored (45).

Malanowska *et al.* (27) showed that the first strand exchanges during the integration reaction occur at the sites of identity in the top strand overlap regions. They used cleavage assays performed with nicked *attB* substrates to demonstrate that the top strands of *attDOT* and *attB* adjacent to the G of the GC dinucleotide are exchanged first during recombination with the wild-type substrates (Fig. 7A). To further elucidate the role

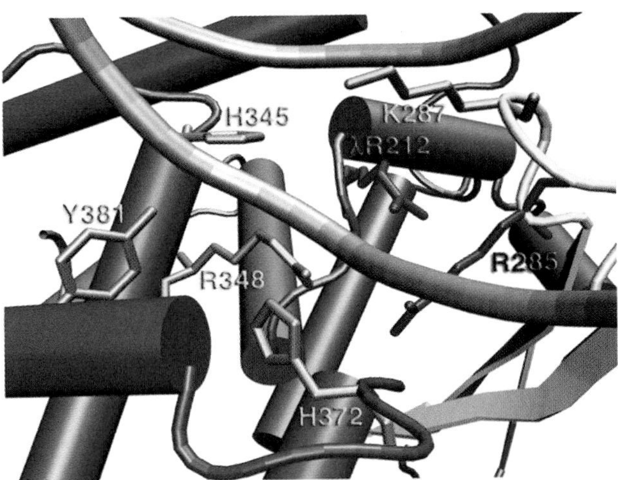

Figure 6 Predicted amino acid residues and active site of IntDOT. The DNA backbone of core-type site DNA is shown in gray. The protein backbone is shown in blue and yellow. Residues 279 to 295, which contain the active site residues and interact with DNA, are represented in yellow. Alpha helices are shown as cylinders, beta sheets are shown as arrows, and regions lacking predicted secondary structure are shown as tubes. The side chains for the predicted active site residues are shown in orange except R285, which is shown in red. The "Arg I" residue from lambda Int (R212) has been superimposed on the IntDOT structure and is shown in purple. R212 aligns with S259 in IntDOT, however S259 is not involved in catalysis. The model suggests that R285 and R212 enter the active site from different positions on the peptide backbone but interact with the similar regions of DNA.
doi:10.1128/microbiolspec.MDNA3-0020-2014.f6

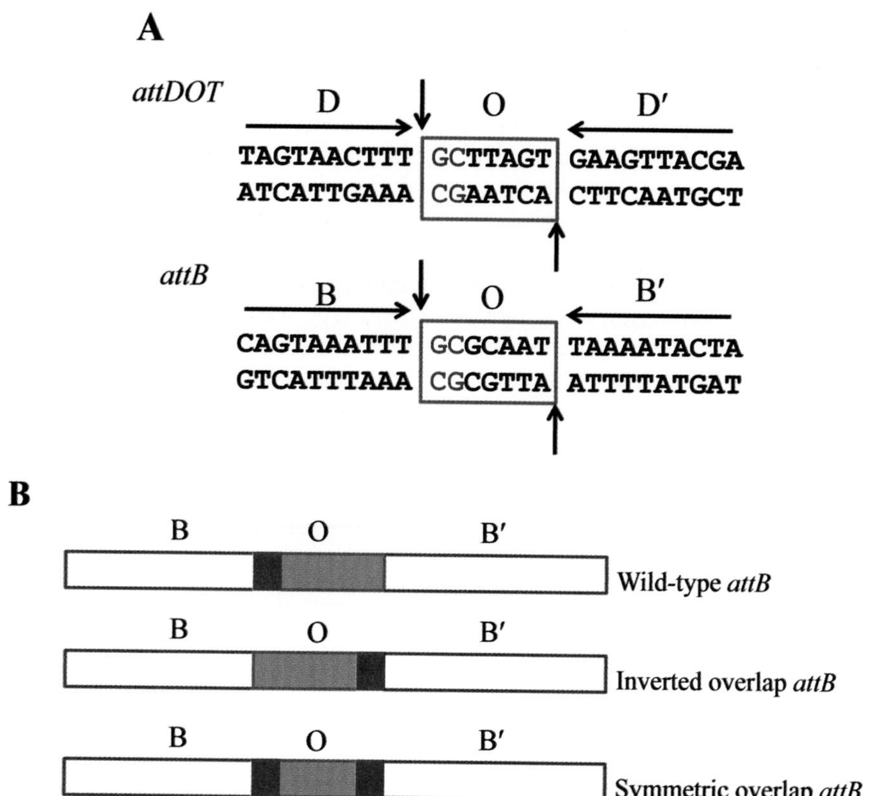

Figure 7 (A) The core-type sites and overlap regions of wild-type *attDOT* and *attB*. The horizontal arrows denote the core-type sites D and D′ on *attDOT* and B and B′ on *attB*. The blue boxes designate the overlap regions, also marked with "O." The GC dinucleotides are shown in red. The vertical arrows indicate the cleavage sites. (B) Simplified schematics of the wild-type, inverted overlap and symmetric *attB* substrates utilized in homology studies. The blue boxes indicate the heterology in the overlap regions, and the red boxes indicate positioning of the GC dinucleotides in the substrates.
doi:10.1128/microbiolspec.MDNA3-0020-2014.f7

of sequence identity during the IntDOT reaction, experiments were performed with *attB* substrates containing substitutions adjacent to the cleavage site in the bottom strand. In one *attB* substrate an inverted GC dinucleotide was placed next to the bottom strand cleavage site adjacent to the B′ core-type site (Fig. 7B, inverted overlap *attB*). This substrate recombined with wild-type *attDOT* at nearly wild-type levels, despite the absence of the GC dinucleotide at the first site of cleavage and strand exchange in the top strand of *attB* (31). Further work showed that, during recombination with the wild-type *attDOT* and the inverted *attB* substrates, the first strand exchanges occur between the top strand of *attDOT* and the bottom strand of *attB*. If an *attB* substrate is constructed with the GC dinucleotides next to both sites of strand exchange in *attB* (Fig. 7B, symmetric overlap *attB*), the top and bottom strands of *attB* are exchanged at approximately an equal frequency (31). Taken together, these results indicate that IntDOT first catalyzes homology-dependent exchanges at the sites of the GC dinucleotides, followed by homology-independent exchanges (31, 44). The location of the GC dinucleotide (or homology with the partner *attDOT* site) appears to dictate the order of cleavage and strand exchange in the CTnDOT system, irrespective of the location of the arm-type sites relative to the core. This is in contrast to the lambda system where locations of the arm-type sites in relation to the core dictate the order of strand exchanges (46).

The working model for CTnDOT integration is the following. First, four IntDOT monomers and the *Bacteroides* host factor assemble an intasome on *attDOT*. This is similar to the initial step proposed for lambda integration (47). Second, the intasome undergoes synapsis with a partner *attB* site. The complex is cleaved by two IntDOT protomers bound to *attDOT* and *attB*. If the partner sites are cleaved next to the GC dinucleotide in each site, the reaction proceeds and the HJ intermediate is formed. The second strand exchanges occur 7 bp away to form the recombinant products. However, if synapsis occurs with the sites in the opposite orientation, the ligation reaction is aborted and the reaction is reversed to form substrate.

ACCESSORY FACTORS INVOLVED IN THE CTnDOT EXCISION REACTION

Xis2c and Xis2d are required for excision (14, 15, 33). They are small, basic proteins that resemble other recombination directionality factors (RDFs) such as Xis from the bacteriophage lambda system (33, 48, 49). The role of Orf3 in CTnDOT excision or conjugative transfer (if any) has not been elucidated, as a phenotype for an *orf3* deletion has not been detected (6, 10, 14). Exc is a type 1A topoisomerase (similar to *E. coli* DNA topoisomerase III) and utilizes an active site tyrosine to relax negatively supercoiled DNA in the presence of Mg^{++} (50, 51). However, the topoisomerase function of Exc is not required for CTnDOT excision. If the catalytic tyrosine is mutated to a phenylalanine, excision remains unaffected (33, 51). As described below, Exc stimulates excision under certain conditions and may play a structural role in the excision reaction (15, 33). In addition to the proteins encoded by the excision operon, IntDOT and a *Bacteroides* host factor (BHFa) are also required for CTnDOT excision (14; Ringwald K, unpublished results).

ELUCIDATING THE ROLE OF Exc

One of the major questions about CTnDOT excision is the role of Exc in the reaction. Initially, the *exc* gene was shown to be required for *in vivo* excision. When the gene was deleted, excision was not detectable *in vivo* (14, 15). However, Exc did not affect the efficiency of recombination in an *in vitro* intermolecular excision assay (15). In addition, excision frequencies with the intermolecular reaction were inefficient, usually less than 5% (15, 32, 39). The original *in vitro* excision assay involved recombination between the *attL* and *attR* sites on two different plasmids (15). Thus, this intermolecular assay utilizes the sites used in the excision reaction but the reaction does not mimic the topology of the DNA containing the sites in the *in vivo* excision reaction, where *attL* and *attR* are on the same DNA molecule. Subsequently, Keeton and Gardner created an intramolecular *in vitro* excision assay with both *attL* and *attR* sites on a single plasmid (33). Several substrates were constructed that contained varying distances between the *attL* and *attR* sites and different or identical overlap sequences. Exc was not absolutely required for the intramolecular *in vitro* excision reaction (33). Depending on the substrate used, Exc was shown to stimulate the excision frequency by up to five-fold, although the exact extent of the stimulation was dependent on both the length of DNA between the *attL* and *attR* sites and the exact sequences of the overlap sequences. Optimal excision was obtained when Exc was present and the *attL* and *attR* sites contained identical overlap sequences. However, the magnitude of Exc stimulation of recombination between *attL* and *attR* sites was the greatest when the overlap sequences were different (33).

Although the precise role of Exc during CTnDOT excision is not known, it is unlikely that Exc interacts specifically with DNA because Exc does not appear to bind DNA in electrophoretic mobility shift assays (Hopp C, unpublished results). Instead, Exc may participate in protein–protein interactions with IntDOT, BHFa and/or Xis2c and Xis2d, thus facilitating excisive recombination (33). As described below, Exc may also play a role in resolving HJs during excisive recombination.

HOLLIDAY JUNCTION RESOLUTION BY IntDOT

Tyrosine recombinases form HJ intermediates after the first strand exchanges. Since the overlap regions of recombination sites of most tyrosine recombinase systems are identical, it was originally thought that the first strand exchanges occurred at one end of the overlap region and the HJ intermediate formed branch migrated to the other end where the second strand exchanges between the bottom strands could occur (34, 52). The branch migration would require homology or sequence identity in the overlap region. However, it was discovered that the lambda Int and Flp systems did not utilize simple branch migration to resolve synthetic HJs (53, 54). It was proposed that a series of isomerization steps occur after the first set of strand exchanges. A "strand swapping" isomerization model was proposed where two sequential swaps of three bases occur between the two partners (55). The homology-sensitive step was proposed to be at the annealing step before the ligation step could occur. Because the overlap region of a HJ formed by IntDOT contains heterology (Fig. 2), the final ligation step is at sites that contain mismatches with the partner strand (44). Thus, IntDOT differs from other tyrosine recombinases because it can perform the final ligation reactions when the sites of ligation contain mismatches.

However, IntDOT cannot resolve synthetic HJ intermediates to products when the intermediate contains mismatches in the overlap region. Synthetic HJs were created *in vitro* with either homologous (identical) overlap sequences (Fig. 8A) or overlaps with 5 bp of heterology in the overlap sequences (mismatched) (Fig. 8B). The latter substrate is similar to the HJ that would be encountered after the first strand exchanges occur during a reaction *in vivo* (40). These substrates contained only the core-type sites and lacked the arm-type sites. Thus, interactions of IntDOT with arm-type sites cannot occur. Each synthetic HJ was incubated with IntDOT in order to determine whether it could be resolved by IntDOT and whether there would be a bias in the direction of resolution to either substrates or products. The HJ containing identical overlap sequences was resolved by IntDOT into both products (*attR* and *attL*; Fig. 8A, black arrows) and substrates (*attDOT* and *attB*; Fig. 8A, gray arrows) (40). Because this HJ has identical overlap sequences, a strand undergoing ligation can form a Watson–Crick basepair with the partner strand at the site of ligation at either end of the overlap sequence. Thus the HJ can be resolved to both substrates and products. Surprisingly, however, the synthetic HJ containing mismatches in the overlap sequences was only resolved back to the *attDOT* and *attB* substrates (40). In this case, the ligation reaction works only when the strand undergoing ligation contains the 2 bp GC dinucleotide and can form base pairs with the complementary GC sequence at the site of ligation to form substrates (Fig. 8B, black arrows). Formation of products (*attL* and *attR*) would require ligations where the strands are mismatched. Because the HJs did not contain arm-type sites, it was proposed that IntDOT interactions with arm-type sites allow the reaction to bypass the heterology in the overlap region (40). It was also proposed that IntDOT and BHFa form an intasome on *attDOT* which undergoes synapsis with *attB* and subsequently performs the first strand exchange. This complex is able to bypass the heterology in the overlap region and perform the second set of strand exchanges to form products (40).

THE EFFECT OF HOMOLOGY ON EXCISIVE RECOMBINATION

As described above, IntDOT tolerates heterology in the overlap region during the integration and excision reactions. The integration reaction works equally well whether the overlap sites are the same or different in *attDOT* and *attB* (Laprise J, unpublished results). This observation suggests that the conversion of the HJ intermediate to a conformation that allows resolution to products is not affected by heterology in the overlap sequence. However, an unexpected result occurred when the *in vitro* intramolecular excision reaction was performed with *attL* and *attR* substrates with identical or different overlap sequences. It was expected that the same observation would be made for the excision reaction with *attL* and *attR* sites that contain identical or different *attL* and *attR* overlap sequences. However, excision substrates containing *attL* and *attR* sites with identical overlap sequences recombined 10- to 30-fold more efficiently than substrates with mismatches in the overlap regions of the sites in the absence of Exc (33). When Exc was added to reactions when *attL* and *attR* contained identical overlap sequences, excision frequencies were nearly 100%. However, the strongest enhancement effect with Exc was seen with excision substrates containing heterologous overlap sequences. Exc stimulated recombination between these substrates by up to five-fold (33). The mechanism behind this stimulation is unknown. One possibility is that Exc participates in a protein–protein interaction in the *attL* or *attR* intasome that somehow stimulates the ligation reaction when the sites have mismatches at the sites of ligation. The fact that sites with identical overlap sequences result in more efficient excision also highlights a major difference between the integration and excision reactions of CTnDOT.

THE ROLES OF THE IntDOT ARM-TYPE SITES IN CTnDOT INTEGRATION AND EXCISION

Several tyrosine recombinases, such as the lambda, L5, NBU1, and HP1 integrases, perform directional reactions. These recombinases are typically heterobivalent DNA binding proteins that interact with two classes of binding sites. The N domain interacts with arm-type sites, which flank the sites of recombination, while the CB and CAT domains interact with the core-type sites at the sites of strand exchange (56, 57, 58, 59). Arm-type sites have different consensus sequences that bind a monomer of the recombinase (39, 60, 61, 62). Arm-type sites play an architectural role in recombination reactions where they help form intasomes. For example, a monomer of recombinase can potentially interact simultaneously with an arm-type site and a core-type site on the same DNA molecule (intramolecular interactions) or different DNA molecules (intermolecular interactions) in an intasome (63, 64, 65). They can also interact cooperatively with each other or with accessory factors. For example, in the lambda system, four

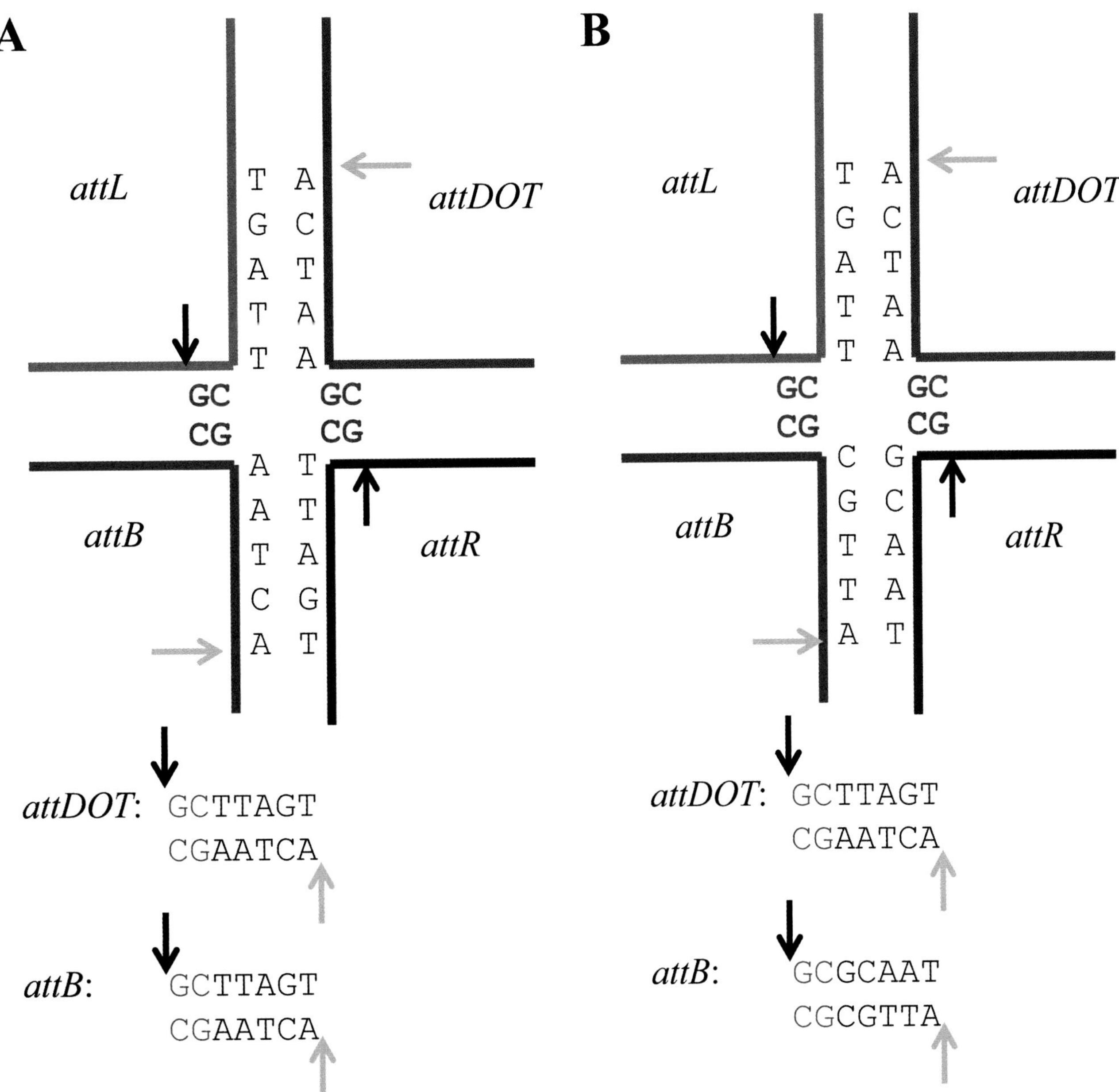

Figure 8 Synthetic Holliday junction substrates. (A) A Holliday junction containing core-type sites (not shown) and identical (homologous) overlap sequences shown with the GC dinucleotide denoted in red. The site of first cleavage and strand exchange is denoted with black arrows, while the site of second strand cleavage and strand exchange is indicated with gray arrows. The *attL* site is shown with a purple line, the *attR* is shown with a gray line, the *attB* is denoted with a blue line, and *attDOT* is denoted with a red line. Resolution at the sites of the black arrows form the *attDOT* and *attB* substrates while resolution at the sites of the gray arrows forms the *attL* and *attR* products. (B) A synthetic Holliday junction with two bp of homology at the GC dinucleotide and mismatches in the rest of the overlap sequences. Resolution occurs only at the sites of the black arrows forming the *attDOT* and *attB* substrates. doi:10.1128/microbiolspec.MDNA3-0020-2014.f8

of the five arm-type sites (P1, P′1, P′2, and P′3) are required for the integration reaction, while the P2, P′1, and P′2 arm-type sites are required for the excision reaction (66, 37, 68). Xis, an accessory factor required for excision in the lambda system, bends the DNA and interacts cooperatively with an Int monomer bound to the adjacent P2 arm-type site (69, 70, 71, 72, 73). Xis also indirectly blocks Int from binding to the P1 arm-type site, preventing an interaction required for the integration reaction (74).

DNase I footprinting and site-directed mutagenesis identified six CTnDOT arm-type sites that bind IntDOT: R1′, R1, R2, R2′, L1, and L2 (Fig. 9) (38, 39). Each arm-type site was mutated and tested in an *in vitro* integration competition assay. The results showed that mutations in either the R1′ or L1 arm-type sites reduced recombination to an undetectable level, indicating that these sites are required for integrative recombination (39). Interestingly, an *attDOT* site with mutations in the R1, R2, R2′, or L2 arm-type sites showed wild-type levels of integration. However, some *attDOT* sites with mutations in two arm-type sites were unable to undergo integrative recombination. Substrates with mutations in the R1 and R2 or R1 and R2′ arm-type sites did not undergo integrative recombination at a detectable level (39). These results suggest that an IntDOT monomer binds cooperatively at the R1 arm-type site with another monomer at either the R2 or R2′ arm-type sites (39). Because five arm-type sites appear to play a role in the integration reaction, it is possible that five IntDOT monomers may be bound to *attDOT* in the integrative intasome. Since only four monomers are likely to be necessary for catalysis, the fifth monomer could be playing an architectural role in forming the intasome (39).

By analogy to the lambda system, IntDOT and BHFa likely assemble an intasome on *attDOT* that undergoes synapsis with a naked *attB* site (47). The IntDOT proteins bound to D and D′ as well as to the R1′, R1, R2, and R2′ arm-type sites form the integrative intasome (39, 75). Further work will be necessary to determine the three-dimensional structure of the integrative intasome and the intramolecular and intermolecular contacts made by IntDOT.

When arm-type site mutants were tested in the *in vitro* intramolecular excision assay, the R1′, R1, and L1 arm-type sites were shown to be essential for CTnDOT excision (Fig. 9) (48). It is somewhat surprising that the same arm-type sites were shown to be important for both the integration and excision of CTnDOT. Again, future work will be required to determine the structures of the *attL* and *attR* intasomes as well as the synaptic complex.

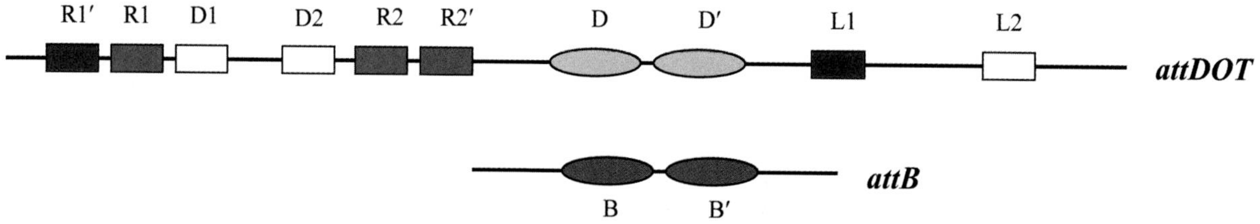

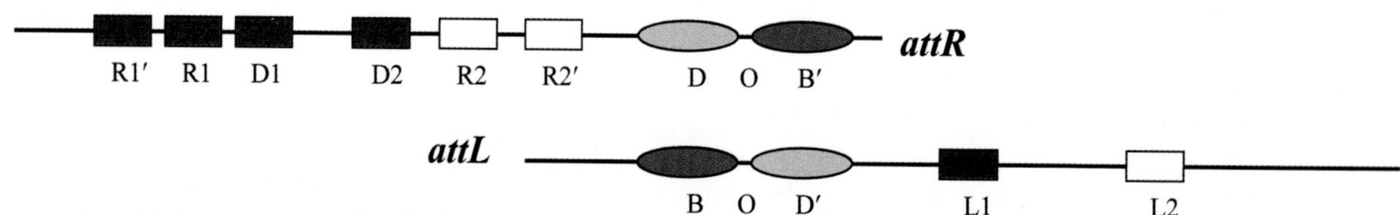

Figure 9 The IntDOT arm-type sites and the Xis2d binding sites. Boxes denote arm-type sites and Xis2d binding sites, and ovals denote core-type sites. Binding sites shown in red are required for recombination. Blue boxes denote arm-type sites that have a stimulatory effect on the integration reaction. White boxes indicate binding sites that are not required for the given reaction. Figure not drawn to scale.
doi:10.1128/microbiolspec.MDNA3-0020-2014.f9

Xis2c AND Xis2d INTERACTIONS DURING EXCISION

Along with IntDOT, the other CTnDOT-encoded proteins required for excision are Xis2c and Xis2d. Both Xis2c and Xis2d are small and basic, and contain helix-turn-helix motifs. Electrophoretic mobility shift assay (EMSA) analyses demonstrated that both proteins interact specifically with *attR* (48). Footprinting analyses identified two Xis2d binding sites, denoted D1 and D2, on *attR* (Fig. 9). The D1 and D2 sites are approximately 20 bp apart (48). A mutation in the D1 site had the most drastic effect on *in vitro* excision, reducing excision frequency by approximately five-fold. Mutating both D1 and D2 reduced *in vitro* excision frequency to background levels (48), showing the importance of Xis2d binding for excisive recombination. It is possible that binding of Xis2d dimers at D1 and D2 forms a DNA loop.

It is likely that a protein–protein interaction between IntDOT bound to the R1 arm-type site and Xis2d bound to the nearby D1 site is important for the excision reaction (Fig. 9). EMSAs with a substrate containing the R1′ and R1 arm-type sites as well as the D1 and D2 sites showed that IntDOT forms two different complexes with Xis2d. One complex was believed to contain one Xis2d dimer bound to the D1 site and one IntDOT monomer bound to the R1 arm-type site, while the other complex may contain two Xis2d dimers bound to the D1 and D2 and one IntDOT monomer bound to the R1 arm-type site (48). These results also suggested that IntDOT and Xis2d bind cooperatively to *attR*.

Xis2c shifts *attR* in an EMSA. However, interpretation of the results was not straightforward and the binding sites have not been identified conclusively by footprint analyses (48). It is possible that it interacts with the DNA between the D1 and D2 sites in the *attR* intasome.

CONCLUSIONS

The tyrosine family of recombinases contains members that are highly diverse. Experiments on the CTnDOT system have revealed some properties that highlight this diversity. For example, IntDOT catalyzes a chemical reaction characteristic of tyrosine recombinases. However, the observation that IntDOT lacks a conserved arginine residue indicates that the active site architecture likely differs from that of other family members. Presumably an arginine residue is donated to the catalytic pocket from another region of the protein.

Unlike virtually all of the other well-studied tyrosine recombinases, IntDOT does not require sequence identity in the overlap regions of partner sites. During integrative recombination the enzyme requires some sequence identity at the sites of the first strand exchanges but can perform ligation reactions during the second exchanges where the bases undergoing ligation cannot pair with complementary bases in the partner strand. Thus IntDOT must be able to accommodate heteroduplex DNA in the HJ intermediate formed after the first strand exchanges while the other enzymes cannot form recombinants if the overlap sequences contain heterology. Interestingly, the IntN1 enzyme encoded by mobilizable transposon NBU1 can catalyze a reaction where the first strand exchanges are homology-independent (76). Thus it is possible that experiments with other members of the tyrosine recombinase family could expand further the flexibility of DNA interactions that occur during the reactions.

The excision reaction of CTnDOT is also different from those of other tyrosine recombinase systems because it is so complex. Most systems that display directionality, like the lambda system, utilize a single accessory factor encoded by the element itself. The CTnDOT excision reaction utilizes three CTnDOT-encoded proteins (Xis2c, Xis2d, and Exc) encoded in an operon that is regulated by an intricate signal cascade. In addition, the Exc protein appears to play a role in the excision reaction that is absent in other characterized systems because it may participate in the resolution of the HJ intermediate to products.

There is still much to learn about the CTnDOT system. How does the active site of IntDOT differ from the active sites of other tyrosine recombinases? Why are there differences in the role of homology in this system's integration and excision reactions, and how does Exc influence the resolution reaction? Finally, the structures of the integrative and excisive intasomes and the synaptic complexes await future investigations.

Acknowledgments. *This article is dedicated to the memory of Professor Abigail A. Salyers who pioneered studies on CTnDOT and other conjugative elements.*

Citation. Wood MM, Gardner JF. 2014. The integration and excision of CTnDOT. Microbiol Spectrum **2**(6):MDNA3-0020-2014.

References

1. **Costello EK, Lauber CL, Hamady M, Fierer N, Gordon JI, Knight R.** 2009. Bacterial community variation in human body habitats across space and time. *Science* **326:** 1694–1697.
2. **Falagas ME, Siakavellas E.** 2000. Bacteroides, Prevotella, and Porphyromonas species: a review of antibiotic resistance and therapeutic options. *Int J Antimicrob Agents* **15:**1–9.

3. **Smith CJ, Callihan DR.** 1992. Analysis of rRNA restriction fragment length polymorphisms from Bacteroides spp. and Bacteroides fragilis isolates associated with diarrhea in humans and animals. *J Clin Microbiol* **30:** 806–812.
4. **Wexler HM.** 2007. Bacteroides: the good, the bad, and the nitty-gritty. *Clin Microbiol Rev* **20:**593–621.
5. **Shoemaker NB, Vlamakis H, Hayes K, Salyers AA.** 2001. Evidence for extensive resistance gene transfer among Bacteroides spp. and among Bacteroides and other genera in the human colon. *Appl Environ Microbiol* **67:**561–568.
6. **Keeton CM, Park J, Wang GR, Hopp CM, Shoemaker NB, Gardner JF, Salyers AA.** 2013. The excision proteins of CTnDOT positively regulate the transfer operon. *Plasmid* **69:**172–179.
7. **Waters JL, Wang GR, Salyers AA.** 2013. Tetracycline-related transcriptional regulation of the CTnDOT mobilization region. *J Bacteriol* **195:**5431–5438.
8. **Whittle G, Hund BD, Shoemaker NB, Salyers AA.** 2001. Characterization of the 13-kilobase ermF region of the Bacteroides conjugative transposon CTnDOT. *Appl Environ Microbiol* **67:**3488–3495.
9. **Nikolich MP, Shoemaker NB, Salyers AA.** 1992. A Bacteroides tetracycline resistance gene represents a new class of ribosome protection tetracycline resistance. *Antimicrob Agents Chemother* **36:**1005–1012.
10. **Whittle G, Shoemaker NB, Salyers AA.** 2002. Characterization of genes involved in modulation of conjugal transfer of the Bacteroides conjugative transposon CTnDOT. *J Bacteriol* **184:**3839–3847.
11. **Wang Y, Shoemaker NB, Salyers AA.** 2004. Regulation of a Bacteroides operon that controls excision and transfer of the conjugative transposon CTnDOT. *J Bacteriol* **186:**2548–2557.
12. **Wang Y, Rotman ER, Shoemaker NB, Salyers AA.** 2005. Translational control of tetracycline resistance and conjugation in the Bacteroides conjugative transposon CTnDOT. *J Bacteriol* **187:**2673–2680.
13. **Salyers AA, Shoemaker NB, Stevens AM, Li LY.** 1995. Conjugative transposons: an unusual and diverse set of integrated gene transfer elements. *Microbiol Rev* **59:** 579–590.
14. **Cheng Q, Sutanto Y, Shoemaker NB, Gardner JF, Salyers AA.** 2001. Identification of genes required for excision of CTnDOT, a Bacteroides conjugative transposon. *Mol Microbiol* **41:**625–632.
15. **Sutanto Y, DiChiara JM, Shoemaker NB, Gardner JF, Salyers AA.** 2004. Factors required in vitro for excision of the Bacteroides conjugative transposon, CTnDOT. *Plasmid* **52:**119–130.
16. **Moon K, Shoemaker NB, Gardner JF, Salyers AA.** 2005. Regulation of excision genes of the Bacteroides conjugative transposon CTnDOT. *J Bacteriol* **187:**5732–5741.
17. **Park J, Salyers AA.** 2011. Characterization of the Bacteroides CTnDOT regulatory protein RteC. *J Bacteriol* **193:**91–97.
18. **Whittle G, Salyers AA.** 2002. Bacterial transposons—an increasingly diverse group of elements, p 385–427. *In* Streips UN, Yasbin RE (ed), *Modern microbial genetics*. John Wiley & Sons, Inc, New York, NY.
19. **Salyers AA, Gardner JF, Shoemaker NB.** 2013. Excision and Transfer of Bacteroides Conjugative Integrated Elements, p 246–249. *In* Roberts AP, Mullaney P (ed), *Bacterial Integrative Mobile Genetic Elements*. Landes Bioscience, London, UK.
20. **Shoemaker NB, Getty C, Guthrie EP, Salyers AA.** 1986. Regions in Bacteroides plasmids pBFTM10 and pB8-51 that allow Escherichia coli-Bacteroides shuttle vectors to be mobilized by IncP plasmids and by a conjugative Bacteroides tetracycline resistance element. *J Bacteriol* **166:**959–965.
21. **Valentine PJ, Shoemaker NB, Salyers AA.** 1988. Mobilization of Bacteroides plasmids by Bacteroides conjugal elements. *J Bacteriol* **170:**1319–1324.
22. **Li LY, Shoemaker NB, Salyers AA.** 1993. Characterization of the mobilization region of a Bacteroides insertion element (NBU1) that is excised and transferred by Bacteroides conjugative transposons. *J Bacteriol* **175:** 6588–6598.
23. **Li LY, Shoemaker NB, Wang GR, Cole SP, Hashimoto MK, Wang J, Salyers AA.** 1995. The mobilization regions of two integrated Bacteroides elements, NBU1 and NBU2, have only a single mobilization protein and may be on a cassette. *J Bacteriol* **177:**3940–3945.
24. **Cheng Q, Paszkiet BJ, Shoemaker NB, Gardner JF, Salyers AA.** 2000. Integration and excision of a Bacteroides conjugative transposon, CTnDOT. *J Bacteriol* **182:**4035–4043.
25. **Bedzyk LA, Shoemaker NB, Young KE, Salyers AA.** 1992. Insertion and excision of Bacteroides conjugative chromosomal elements. *J Bacteriol* **174:**166–172.
26. **Argos P, Landy A, Abremski K, Egan JK, Haggard-Ljungquist E, Hoess RH, Kahn ML, Kalionis B, Narayana SV, Pierson LS III, Sternberg N, Leong JM.** 1986. The integrase family of site-specific recombinases: regional similarities and global diversity. *EMBO J* **5:**433–440.
27. **Malanowska K, Salyers AA, Gardner JF.** 2006. Characterization of a conjugative transposon integrase, IntDOT. *Mol Microbiol* **60:**1228–1240.
28. **Nunes-Duby SE, Kwon HJ, Tirumalai RS, Ellenberger T, Landy A.** 1998. Similarities and differences among 105 members of the Int family of site-specific recombinases. *Nucleic Acids Res* **26:**391–406.
29. **Esposito D, Scocca JJ.** 1997. The integrase family of tyrosine recombinases: evolution of a conserved active site domain. *Nucleic Acids Res* **25:**3605–3614.
30. **Malanowska K, Cioni J, Swalla BM, Salyers A, Gardner JF.** 2009. Mutational analysis and homology-based modeling of the IntDOT core-binding domain. *J Bacteriol* **191:**2330–2339.
31. **Laprise J, Yoneji S, Gardner JF.** 2010. Homology-dependent interactions determine the order of strand exchange by IntDOT recombinase. *Nucleic Acids Res* **38:**958–969.
32. **DiChiara JM, Salyers AA, Gardner JF.** 2005. In vitro analysis of sequence requirements for the excision reaction of the Bacteroides conjugative transposon, CTnDOT. *Mol Microbiol* **56:**1035–1048.

33. **Keeton CM, Gardner JF.** 2012. Roles of Exc Protein and DNA Homology in the CTnDOT Excision Reaction. *J Bacteriol* **194:**3368–3376.

34. **Rajeev L, Malanowska K, Gardner JF.** 2009. Challenging a paradigm: the role of DNA homology in tyrosine recombinase reactions. *Microbiol Mol Biol Rev* **73:**300–309.

35. **Van Duyne GD.** 2002. A structural view of tyrosine recombinase site-specific recombination, p 93–117. *In* Craig NL, Craigie R, Gellert M, Lambowitz AM (ed), *Mobile DNA II*. ASM Press, Washington, DC.

36. **Azaro MA, Landy A.** 2002. Lambda integrase and the lambda Int family, p 119–148. *In* Craig NL, Craigie R, Gellert M, Lambowitz AM (ed), *Mobile DNA II*. ASM Press, Washington, DC.

37. **Kim S, Swalla BM, Gardner JF.** 2010. Structure-function analysis of IntDOT. *J Bacteriol* **192:**575–586.

38. **Dichiara JM, Mattis AN, Gardner JF.** 2007. IntDOT interactions with core- and arm-type sites of the conjugative transposon CTnDOT. *J Bacteriol* **189:**2692–2701.

39. **Wood MM, Dichiara JM, Yoneji S, Gardner JF.** 2010. CTnDOT integrase interactions with attachment site DNA and control of directionality of the recombination reaction. *J Bacteriol* **192:**3934–3943.

40. **Kim S, Gardner JF.** 2011. Resolution of Holliday junction recombination intermediates by wild-type and mutant IntDOT proteins. *J Bacteriol* **193:**1351–1358.

41. **Warren D, Lee SY, Landy A.** 2005. Mutations in the amino-terminal domain of lambda-integrase have differential effects on integrative and excisive recombination. *Mol Microbiol* **55:**1104–1112.

42. **Lee SY, Radman-Livaja M, Warren D, Aihara H, Ellenberger T, Landy A.** 2005. Non-equivalent interactions between amino-terminal domains of neighboring lambda integrase protomers direct Holliday junction resolution. *J Mol Biol* **345:**475–485.

43. **Kazmierczak RA, Swalla BM, Burgin AB, Gumport RI, Gardner JF.** 2002. Regulation of site-specific recombination by the C-terminus of lambda integrase. *Nucleic Acids Res* **30:**5193–5204.

44. **Malanowska K, Yoneji S, Salyers AA, Gardner JF.** 2007. CTnDOT integrase performs ordered homology-dependent and homology-independent strand exchanges. *Nucleic Acids Res* **35:**5861–5873.

45. **Bauer CE, Gardner JF, Gumport RI.** 1985. Extent of sequence homology required for bacteriophage lambda site-specific recombination. *J Mol Biol* **181:**187–197.

46. **Kitts PA, Nash HA.** 1988. Bacteriophage lambda site-specific recombination proceeds with a defined order of strand exchanges. *J Mol Biol* **204:**95–107.

47. **Richet E, Abcarian P, Nash HA.** 1988. Synapsis of attachment sites during lambda integrative recombination involves capture of a naked DNA by a protein-DNA complex. *Cell* **52:**9–17.

48. **Keeton CM, Hopp CM, Yoneji S, Gardner JF.** 2013. Interactions of the excision proteins of CTnDOT in the attR intasome. *Plasmid* **70:**190–200.

49. **Waters JL, Salyers AA.** 2013. Regulation of CTnDOT conjugative transfer is a complex and highly coordinated series of events. *mBio* **4:**e00569-00513.

50. **Champoux JJ.** 2001. DNA topoisomerases: structure, function, and mechanism. *Annu Rev Biochem* **70:**369–413.

51. **Sutanto Y, Shoemaker NB, Gardner JF, Salyers AA.** 2002. Characterization of Exc, a novel protein required for the excision of Bacteroides conjugative transposon. *Mol Microbiol* **46:**1239–1246.

52. **Weisberg RA, Enquist LW, Foeller C, Landy A.** 1983. Role for DNA homology in site-specific recombination. The isolation and characterization of a site affinity mutant of coliphage lambda. *J Mol Biol* **170:**319–342.

53. **Lee J, Jayaram M.** 1995. Role of partner homology in DNA recombination. Complementary base pairing orients the 5′-hydroxyl for strand joining during Flp site-specific recombination. *J Biol Chem* **270:**4042–4052.

54. **Nunes-Duby SE, Azaro MA, Landy A.** 1995. Swapping DNA strands and sensing homology without branch migration in lambda site-specific recombination. *Curr Biol* **5:**139–148.

55. **Voziyanov Y, Pathania S, Jayaram M.** 1999. A general model for site-specific recombination by the integrase family recombinases. *Nucleic Acids Res* **27:**930–941.

56. **Moitoso de Vargas L, Pargellis CA, Hasan NM, Bushman EW, Landy A.** 1988. Autonomous DNA binding domains of lambda integrase recognize two different sequence families. *Cell* **54:**923–929.

57. **Esposito D, Thrower JS, Scocca JJ.** 2001. Protein and DNA requirements of the bacteriophage HP1 recombination system: a model for intasome formation. *Nucleic Acids Res* **29:**3955–3964.

58. **Wood MM, Rajeev L, Gardner JF.** 2013. Interactions of NBU1 IntN1 and Orf2x proteins with attachment site DNA. *J Bacteriol* **195:**5516–5525.

59. **Pena CE, Lee MH, Pedulla ML, Hatfull GF.** 1997. Characterization of the mycobacteriophage L5 attachment site, attP. *J Mol Biol* **266:**76–92.

60. **Ross W, Landy A.** 1982. Bacteriophage lambda int protein recognizes two classes of sequence in the phage att site: characterization of arm-type sites. *Proc Natl Acad Sci U S A* **79:**7724–7728.

61. **Ross W, Landy A.** 1983. Patterns of lambda Int recognition in the regions of strand exchange. *Cell* **33:**261–272.

62. **Ross W, Landy A, Kikuchi Y, Nash H.** 1979. Interaction of int protein with specific sites on lambda att DNA. *Cell* **18:**297–307.

63. **Kim S, Landy A.** 1992. Lambda Int protein bridges between higher order complexes at two distant chromosomal loci attL and attR. *Science* **256:**198–203.

64. **Lee EC, MacWilliams MP, Gumport RI, Gardner JF.** 1991. Genetic analysis of Escherichia coli integration host factor interactions with its bacteriophage lambda H′ recognition site. *J Bacteriol* **173:**609–617.

65. **Moitoso de Vargas L, Kim S, Landy A.** 1989. DNA looping generated by DNA bending protein IHF and the two domains of lambda integrase. *Science* **244:**1457–1461.

66. **Bauer CE, Hesse SD, Gumport RI, Gardner JF.** 1986. Mutational analysis of integrase arm-type binding sites of bacteriophage lambda. Integration and excision involve distinct interactions of integrase with arm-type sites. *J Mol Biol* **192:**513–527.

67. **Numrych TE, Gumport RI, Gardner JF.** 1990. A comparison of the effects of single-base and triple-base changes in the integrase arm-type binding sites on the site-specific recombination of bacteriophage lambda. *Nucleic Acids Res* **18:**3953–3959.

68. **Hazelbaker D, Azaro MA, Landy A.** 2008. A biotin interference assay highlights two different asymmetric interaction profiles for lambda integrase arm-type binding sites in integrative versus excisive recombination. *J Biol Chem* **283:**12402–12414.

69. **Bushman W, Yin S, Thio LL, Landy A.** 1984. Determinants of directionality in lambda site-specific recombination. *Cell* **39:**699–706.

70. **Cho EH, Gumport RI, Gardner JF.** 2002. Interactions between integrase and excisionase in the phage lambda excisive nucleoprotein complex. *J Bacteriol* **184:**5200–5203.

71. **Numrych TE, Gumport RI, Gardner JF.** 1992. Characterization of the bacteriophage lambda excisionase (Xis) protein: the C-terminus is required for Xis-integrase cooperativity but not for DNA binding. *EMBO J* **11:** 3797–3806.

72. **Swalla BM, Cho EH, Gumport RI, Gardner JF.** 2003. The molecular basis of co-operative DNA binding between lambda integrase and excisionase. *Mol Microbiol* **50:**89–99.

73. **Thompson JF, de Vargas LM, Skinner SE, Landy A.** 1987. Protein-protein interactions in a higher-order structure direct lambda site-specific recombination. *J Mol Biol* **195:**481–493.

74. **Moitoso de Vargas L, Landy A.** 1991. A switch in the formation of alternative DNA loops modulates lambda site-specific recombination. *Proc Natl Acad Sci U S A* **88:** 588–592.

75. **Laprise J, Yoneji S, Gardner JF.** 2013. IntDOT interactions with core sites during integrative recombination. *J Bacteriol* **195:**1883–1891.

76. **Rajeev L, Segall A, Gardner J.** 2007. The bacteroides NBU1 integrase performs a homology-independent strand exchange to form a holliday junction intermediate. *J Biol Chem* **282:**31228–31237.

Mobile DNA, 3rd Edition
Nancy L. Craig, Michael Chandler, Martin Gellert, Alan M. Lambowitz, Phoebe A. Rice and Suzanne Sandmeyer

doi:10.1128/microbiolspec.MDNA3-0047-2014

Reid C. Johnson[1]

9 Site-specific DNA Inversion by Serine Recombinases

INTRODUCTION

Reversible site-specific inversions of DNA segments occur within the genomes of many bacteria and their phages. These reactions are catalyzed by a dedicated recombinase and do not employ the use of the general recombination-repair machinery. In some cases, additional host regulatory proteins also perform critical functions, and DNA superstructure can play a profound role. In general, site-specific DNA inversions occur at a low frequency and regulate gene expression by coupling alternative protein coding segments to a fixed promoter or by switching the orientation of a promoter with respect to coding region(s). In this manner, a small subset of the population becomes preadapted to a potential change in the environment.

The two major classes of enzymes mediating site-specific recombination include a diverse family of tyrosine recombinases and a more conserved family of serine recombinase, as named from their catalytic active site residues (1). This chapter focuses on the Fis/enhancer-dependent DNA invertase subfamily of serine recombinases. The DNA invertases are closely related to resolvases, which primarily mediate deletions, and as a group, these are often referred to as small serine recombinases. Serine recombinases are relatively easily identified by conserved signature motifs over the active site regions within their catalytic domains. As of the beginning of 2014, there were tens of thousands of entries in the sequence database that exhibit >30% identity over at least part of the catalytic domains of small serine recombinases such as the Hin DNA invertase.

The chapter begins by reviewing the biology of well characterized Fis/enhancer-dependent DNA inversion reactions. A more detailed discussion on biological aspects of these reactions plus other site-specific DNA inversion systems is provided in Mobile DNA II Chapter 13 (2). The multiple DNA inversion reactions mediated by small serine recombinases in *Bacteriodes*, which were discovered after publication of the Mobile DNA II chapter, are also briefly discussed here.

The emphasis of most of the chapter is on recent developments concerning mechanistic studies of the Fis/enhancer-dependent inversion reactions, focusing on the Hin- and Gin-catalyzed reactions. Two defining features of these serine DNA invertase systems are discussed in detail. One is the requirement for the enhancer regulatory element that functions in *cis*, but in a largely distance- and orientation-independent manner with respect to the sites of recombination. The enhancer specifies two binding sites for the Fis protein, which was initially discovered because of its activity in the inversion reactions – hence its full name Factor for inversion

[1]Department of Biological Chemistry, David Geffen School of Medicine at UCLA, Los Angeles, CA 90095-1737.

stimulation. Fis is now known as a general nucleoid-associated protein that controls many different DNA reactions.

A second defining feature is the strict specificity for promoting intramolecular inversions over deletions or intermolecular recombination reactions. This directionality of recombination is mediated by the Fis/enhancer element in concert with DNA supercoiling. Thus, a major focus of research in the field has been to elucidate the steps involved in formation of the active recombination complex, called an invertasome. The Fis/enhancer element functions as a molecular scaffold in the assembly of the invertasome, whereby inactive DNA invertase dimers are remodeled into an active tetramer that breaks, exchanges, and ligates the DNA strands into the inverted orientation.

Research on DNA invertases and resolvases has elucidated our current understanding of the subunit rotation reaction that mediates the exchange of DNA strands by serine recombinases. Although the translocation of subunits after double strand DNA cleavages would seem like a perilous reaction because of the potential for chromosome breaks, recent evidence provides a compelling case for this unique mechanism of DNA recombination [see also chapters by Rice and Stark (3, 4)].

At the end of the chapter, recent progress on identifying specific residues on serine DNA invertases that control key conformational transitions during the reaction is discussed. Mutations in these residues often enable efficient recombination without the requirement for the Fis/enhancer element, and thus, without control of directionality.

BIOLOGY AND BASIC FEATURES OF SERINE DNA INVERTASE REACTIONS

Hin-Catalyzed Site-Specific DNA Inversion Controlling Flagellar Phase Variation in *Salmonella*

Antigenic variation, whereby clonally-derived populations of *Salmonella* are converted to an alternate antigenic form, was first described by Andrewes in 1922 (5). Subsequent genetic studies on phase variation in *Salmonella* by the laboratories of Stocker, Lederberg, and Iino showed that the variable H antigen was the result of alternate expression of two unlinked flagellin genes originally called *H1* and *H2*, but renamed *fliC* and *fljB*, respectively (6, 7, 8, 9). Stocker, and more recently Kutsukake et al., measured switching rates between the two flagellins in rich media cultures to be in the range of 10^{-4} to 10^{-5} per cell per generation (9, 10). Gillen and Hughes reported 2- to 20-fold higher switching rates using chromosomal *lac* reporters (11). All of these studies have shown a 2.5- to 30-fold bias in favor of switching from FljB to FliC expression over the reverse direction, but the molecular basis for this difference remains unknown. The low rate of switching is consistent with the postulated role of flagellar phase variation in escaping a host immune response. A clonally-derived population of *Salmonella* will express primarily one flagellin type, enabling the few members of the population that have switched to be insensitive to antibody generated by the host against the dominant flagellin.

The flagellar phase variation system is present in many but not all of the thousands of serovars of *Salmonella enterica* subsp. *enteric*, including the intensively studied serovar Typhimurium (*Salmonella* Typhimurium). Studies have shown that *Salmonella* Typhimurium cells expressing either flagellin are equally efficient at invading mouse intestinal epithelial cells and colonizing Peyers patches, but strains genetically locked into expressing only FliC are more virulent than FljB phase-locked strains (12). The FliC phase-locked cells were found to be more efficient at colonizing the spleen during later stages of infection. Moreover, infecting wild-type bacteria that were initially in the FljB phase tended to have switched to the FliC phase when spleens were analyzed 2 weeks post infection, whereas infecting FliC cells remained in the FliC phase.

Molecular analyses in the latter 1970s by Silverman, Zieg, and Simon first demonstrated that the "H-controlling region" adjacent to the *H2/fljB* flagellin gene in the *Salmonella* Typhimurium chromosome contains an inverting segment of DNA. Restriction mapping, together with the formation of hybridization bubbles between DNA molecules cloned from strains expressing the different flagellin forms, provided the physical proof for inversion of a ~1 kb DNA segment (13, 14). The region can be transferred into *Escherichia coli* where it also undergoes inversion. Genetic and sequence analyses showed that the *hin* (H inversion) gene, coding for a 190 amino acid residue protein that is responsible for the inversion reaction, was contained within the 993 bp invertible segment (Fig. 1A) (15, 16, 17). The segment is bounded by two imperfectly palindromic 26 bp recombination sites *hixL* and *hixR* (Fig. 1A and 2) and it contains a σ^{28}-dependent promoter that initiates transcription 28 bp upstream of *hixR* (18). In the *fljB* ON orientation, transcription from this promoter reads through *hixR*, and translation of the FljB flagellin begins just five bases after the end of the

A. Hin (*S. enterica* Typhimurium)

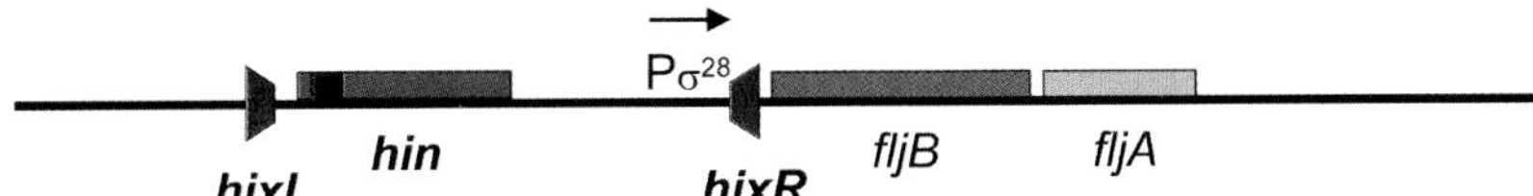

B. Gin (Phage Mu)

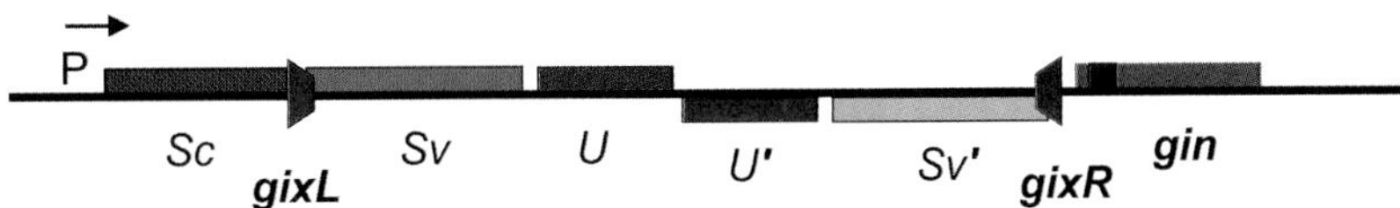

C. Cin (Phage P1)

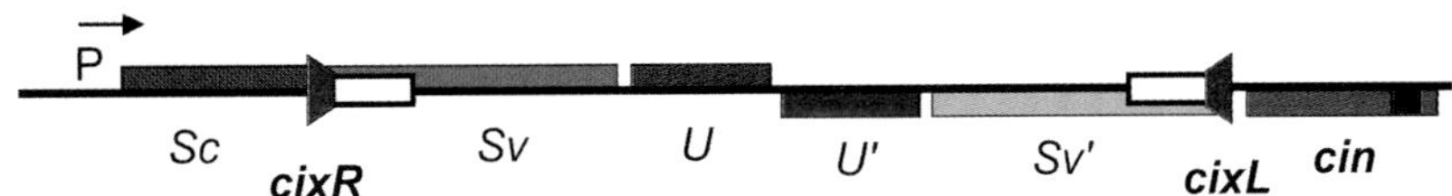

D. Min (p15B plasmid/phage)

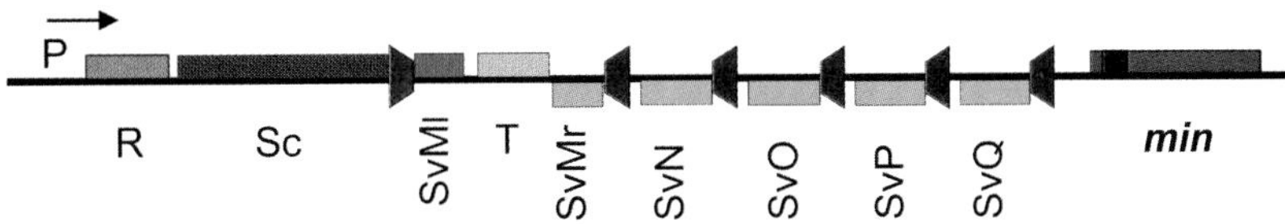

Figure 1 Genetic organization of Fis/enhancer-dependent DNA inversion systems. (A) The Hin system controlling flagellin synthesis in *Salmonella enterica* Typhimurium. *fljB* codes for one of the alternatively expressed flagellins, and *fljA* is a repressor of the *fliC* flagellin gene located elsewhere on the chromosome. (B, C) The Gin and Cin systems from phages Mu and P1, which control phage host range. *Sv* and *Sv´* gene segments are alternatively fused in-frame to the *Sc* gene segment. The S and U genes encode tail fiber proteins. (D) The complex Min locus from the p15B plasmid. The different *Sv* gene segments are alternatively fused to the *Sc* gene segment. In each case, the recombination sites are colored red, and the positions of the DNA invertases (red-orange) with their associated recombinational enhancer segments (blue) are denoted. P designates a promoter with the *S.* Typhimurium *fljB* and *fljA* genes being transcribed by the sigma 28 form of RNA polymerase.
doi:10.1128/microbiolspec.MDNA3-0047-2014.f1

hixR sequence. *hixR* has diverged from recombination site sequences of other Fis/enhancer-dependent DNA inversion systems (Fig. 2), probably because the locus must also encode the *fljB* ribosome binding site. The *fljA* gene (originally called *rH1*, repressor of H1), whose protein product is responsible for preventing synthesis of the alternative flagellin FliC, is co-transcribed with and begins 68 nt downstream of *fljB* (Fig. 1A) (19). FljA binds to the *fliC* mRNA to both inhibit its translation and promote its degradation (20, 21, 22). In the opposite or *fljBA* OFF orientation, the absence of *fljA* transcription enables the FliC flagellin, which is encoded elsewhere on the chromosome, to be expressed. The closest gene, *iroB*, which is annotated as a glycosyl transferase family protein, begins about 650 bp from the σ^{28}-dependent promoter in the *fljBA* OFF orientation.

A purified *in vitro* system supporting Hin-catalyzed DNA inversion was reported in 1986 (23). In addition to the Hin recombinase, two host factors were found to be required for efficient inversion. Factor I was identified as the nucleoid-associated protein HU, and Factor II was renamed Fis and later also classified as a nucleoid

```
                  -13                  -1 +1                 +13
Consensus          T T - T C - - A A A C C  A A  G G T T T - - G A - A A

hixL           t g g T T C T T G A A A A C C  A A  G G T T T T T G A T A A a g c
hixR           a a a T T T T C C T T T T G G  A A  G G T T T T T G A T A A c c a
gixL           c g t T T C C T G T A A A C C  G A  G G T T T T G G A T A A a c a
gixR           c g t T T C C T G T A A A C C  G A  G G T T T T G G A T A A t g g
cixL           g a g T T C T C T T A A A C C  A A  G G T T T A G G A T T G a a a
cixR           g a g T T C T C T T A A A C C  A A  G G T A T T G G A T A A c a g
mixMr'N'       g c c T T C C C C T A A A C C  A A  C G T T T T T A T G C C g c c
mixR'MI"       g c c T T C C C C C A A A C C  A A  G G T A A T C A A G A A c g c
```

Figure 2 Fis/enhancer-dependent serine DNA invertase recombination sites. DNA sequences of recombination sites from the systems depicted in Fig. 1 are listed along with the consensus sequence at the top. Sequences matching the consensus are highlighted in cyan. doi:10.1128/microbiolspec.MDNA3-0047-2014.f2

protein (24, 25, 26). Along with the *hixL* and *hixR* recombination sites, a third *cis*-acting sequence was found to be required (27). In the Hin inversion system, this 69 bp recombinational enhancer locus is encoded within the amino-terminal coding segment of Hin (residues 8 to 30), and thus, located within the invertible segment about 100 bp from the center of the *hix* site (Fig. 1A). However, it can function effectively from 10 bp closer to over 4 kb from the closest *hix* site, both inside and outside of the invertible segment and in either orientation (27, 28). The enhancer contains two binding sites for the Fis protein (29). Very low rates of inversion can occur on substrates lacking an enhancer when Fis is present, but inversion is nearly undetectable without Fis (23, 25). DNA supercoiling is absolutely essential for the inversion reaction (23, 30). The functions of these components in the inversion reaction will be discussed in detail below.

The primary determinant that is believed to be responsible for limiting the frequency of inversion in *Salmonella* ($<10^{-4}$ per cell per generation) is the very low synthesis of Hin because ectopic increases in Hin synthesis result in coordinate increases in DNA inversion rates (31). LacZ reporter experiments have provided evidence for transcription initiating in the 74 bp segment between the *hix* recombination site and start of the *hin* gene (R.C. Johnson, unpublished). Experiments designed to test whether Hin binding to *hix* or to a low affinity Hin binding site between *hix* and the start of the *hin* gene have not provided evidence for regulation of *hin* transcription under laboratory culturing conditions. Likewise, there is no evidence that Fis regulates *hin* expression. Translation initiation signals do not appear to limit Hin protein synthesis. Fis levels, which are very low under slow growth and late exponential-stationary phase conditions (18, 32), and DNA supercoiling densities influence inversion rates (33). Supercoiling densities in *Salmonella* Typhimurium are reported to be high in the gut of infected mice but decrease within macrophages, which is a major route used to enter the blood stream to cause systemic infection (34).

Fis/Enhancer-Dependent Serine Invertase Systems Controlling Phage Host Range

A large number of DNA inversion systems that are associated with phage or phage-like genetic elements are closely related to Hin. Among these, the Gin- and Cin-catalyzed inversion reactions from phages Mu and P1, respectively, are the best characterized (Fig. 1B, C). Gin catalyzes inversion of a 3015 bp G segment within the Mu genome, and Cin catalyzes inversion of a ~4.2 kb segment within the P1 genome (35, 36). Both of these invertible segments contain genes encoding alternative tail fiber proteins that enable the phage to adsorb to different bacterial hosts, thereby increasing the phage host range (37, 38). In the case of phage Mu, a constant N-terminal coding segment of the S gene, which is encoded immediately adjacent to the invertible segment, is linked to its two different C-terminal coding segments, Sv or Sv′, depending on the orientation (Fig. 1B) (39, 40, 41, 42, 43). The U/U' genes are also alternatively expressed depending on the orientation. The gene encoding Gin is located immediately outside of the invertible segment on the side opposite to the Sc gene fragment (44). The organization of the phage P1 C-inversion region is analogous to the Mu system with one important difference: the *cin* gene is oriented oppositely to the *gin* gene relative to the invertible segments (Fig. 1C) (36, 45, 46). Fis, together with recombinational enhancer sequences that are located within the N-terminal end of the respective recombinase genes, are required for efficient DNA inversion in both systems (26, 47, 48, 49, 50, 51). Thus, whereas the *hin* enhancer is located within the invertible segment beginning about 100 bp from *hixL/R*, the *gin* enhancer is located outside of the invertible segment beginning about 90 bp from *gixR* and the *cin* enhancer is located outside of the invertible segment beginning about 500 bp from *cixL*. The different organizations provided an initial clue that the enhancer sequences could function at variable positions and orientations relative to the recombination sites, which was also experimentally confirmed for the phage systems (47, 48, 52).

Like Hin, Gin-catalyzed inversion occurs only rarely under native *in vivo* conditions. Inversion rates measured after a single round of lytic phage growth are about 10^{-6} per phage particle (53). Conversely, induction of a Mu lysogen after long term growth gives rise to equal numbers of progeny containing their G segments in either orientation (54). This is advantageous to the phage as lytic infections will primarily result in progeny capable of attacking the same host species, but phage induced from long-term lysogenic growth will have a broad host range. Gin expression is limited by both low transcription and poor translation initiation signals (35, 55, 56). Transcription of *gin* is believed to initiate within the *gix* recombination site and thus be subject to autoregulation.

Whereas most DNA inversion reactions by serine recombinases involve single locus inversions, more complex systems have been reported (2, 57). The best analyzed is the Min system from p15B, a 94 kb plasmid from *E. coli* 15T$^-$ that exhibits considerable homology with phage P1 (58, 59, 60). There are six different 26 bp recombination sites within the 3.5 kb locus (Figs 1D and 2). Restriction and sequence analyses are consistent with all being active for supporting inversion to generate 240 different sequence combinations. Transformation of individual recombinant plasmids containing the locus showed that an equilibrium population representing most or all of the 240 isomeric forms could be obtained after 40 generations of growth (60). The different rearrangements give rise to six different alternative 3′ ends of a tail fiber-like gene fused in-frame to a constant 5′ region.

The extended group of related phage, cryptic prophage, and plasmid serine DNA invertase genes that are associated with tail fiber gene expression typically share greater than 75% amino acid identity over the entire protein sequence (2). Hin is somewhat more distantly related, sharing 68% identity (82% similarity) with Gin (Fig. 3) and 66% identity (80% similarity) with Cin. Nevertheless, the phage DNA invertases can invert the H invertible segment as well as function on each other's substrates (60, 61, 62, 63, 64). Kutsukake et al. reported that a serine DNA invertase in *Salmonella* Typhimurium called Fin that is associated with the phage P2-like Fels-2 prophage can function to promote flagellar phase variation at 1 to 3% of the *hin*$^+$ rate, although it is only 63% identical to the Hin amino acid sequence (10).

Specificity for DNA Inversion by the Fis/Enhancer-Dependent Serine Invertase Reactions

A hallmark of the Fis/enhancer-dependent DNA invertase reactions is that they only efficiently catalyze inversions between recombination sites located in *cis*. As discussed further below, the overall palindromic recombination sites have an orientation that is specified solely by the central nonpalindromic two base pairs. In all cases, the active recombination sites are oriented in an inverted configuration with respect to each other as defined by this dinucleotide sequence. Where tested, the products of recombination reactions between sites oriented in a directly repeated configuration are deletions, but the rate of this reaction is less than a few percent of the inversion reaction (53, 65, 66, 67). In the case of the *min* locus, deletions were detected but occurred at <0.1% the frequency of inversions (60). An extremely low rate of deletions is especially critical for maintaining the integrity of the *min* locus, which has five of its recombination sites in a directly repeated orientation. Likewise, unlike many other site-specific recombination systems, intermolecular reactions by serine DNA invertases occur at vanishingly low rates (53, 65, 66, 68). As elaborated below, the requirements for Fis, the enhancer, and DNA supercoiling combine to impart the strict directionality to the serine DNA invertase reactions.

Multiple Inversions Catalyzed by Serine DNA Recombinases in *Bacteriodes*

The abundant human intestinal symbiont *Bacteriodes* sp. are loaded with mobile DNA elements, and in particular, short DNA segments containing promoters that undergo reversible inversion (69, 70). The genome of

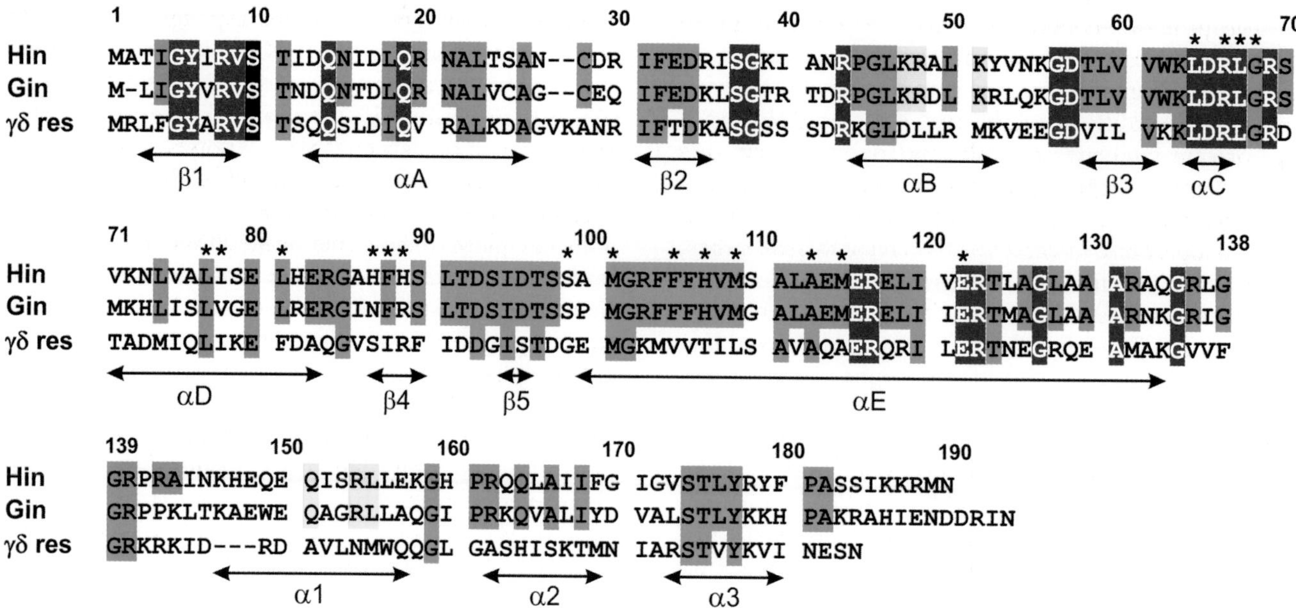

Figure 3 Amino acid sequence alignment of the Hin and Gin DNA invertases together with γδ resolvase. Highly conserved residues among all serine recombinases are highlighted in magenta. The active site serine residue is shaded black. Highly conserved residues among Fis/enhancer-dependent DNA invertases are in cyan; many of these are conserved among the small serine recombinases. Yellow highlighted residues within the α-helix B region are those that contact enhancer DNA and within α-helix 1 of the DNA binding domain (Hin residues 139 to 190) are those that interact with Fis. Residue numbering is according to Hin. Asterisks denote residues where single substitutions result in strong gain-of-function activities, often resulting in Fis/enhancer-independence (65, 109, 185). Secondary structure designations are from γδ resolvase (residues 1 to 138; PDB: 1GDT) and Hin (residues 139 to 190; PDB: 1IJW). An alignment that includes additional Fis/enhancer-dependent DNA invertases along with other small serine recombinases is given in reference 2.
doi:10.1128/microbiolspec.MDNA3-0047-2014.f3

Bacteriodes fragilis contains 30 potential DNA invertases; most are members of the tyrosine recombinase family, but three are serine recombinases. Unlike most site-specific recombination systems, individual recombinases in *Bacteriodes* promote inversions at many loci that are distributed throughout the chromosome. These inversion reactions are thought to enable the bacterium to continually adapt to the host environment and avoid immune surveillance, thereby enabling a long-term commensal relationship with their host.

A plasmid-encoded serine recombinase in *B. fragilis* called FinB (*fragilis* invertase B) catalyzes inversion of promoter regions upstream of seven polysaccharide biosynthesis genes (71, 72). Fin is quite similar to Hin (49% identical, 72% similar by Clustal O), and the sequences of the 32 bp inverted repeats flanking the promoter segments resemble the *hix* sequence. The same strain of *B. fragilis* also contains a 197 amino acid residue serine recombinase called Mpi (multiple promoter invertase) that promotes inversion of 13 different promoter elements that are responsible for expression of seven related capsular polysaccharide genes plus six other unidentified genes (73). Each of these invertible segments are flanked by related 20 to 22 bp recombination sites. Mpi is clearly a serine recombinase of similar overall structure as the invertase/resolvase fold but is less related to Hin in primary sequence (27% identical, 55% similar).

The Comstock group has also shown that the Gfi recombinase (glycoprotein family invertase) catalyzes inversions at promoters controlling seven different glycoprotein genes in *Bacteroides distasonis* (74). While these promoters are primarily maintained in the OFF orientation, all seven were found to invert at a low frequency under *in vitro* culturing conditions. Moreover, evidence for inversion into the ON orientation at some of the loci was obtained by PCR analysis of human stool samples, confirming glycoprotein phase variation in the organism's native environment. Gfi is one of four serine recombinases present in *B. distasonis*.

The promoter-containing invertible segments in all the *Bacteroides* systems are extremely short with most being less than 225 bp and one being only 139 bp. Thus, it would not be unexpected if a host DNA bending protein facilitates assembly of recombination sites into synaptic complexes. It will be interesting to learn if a regulatory system analogous to the Fis/enhancer element is operating in these reactions and whether there are mechanisms to ensure that recombination only occurs between sites flanking a specific promoter to avoid large rearrangements of the chromosome.

RELATIONSHIP OF FIS/ENHANCER-DEPENDENT DNA INVERTASES WITH OTHER SERINE RECOMBINASES

The Fis/enhancer-dependent serine DNA invertase genes typically contain 185 to 200 residues that exhibit over 60% sequence identity over the entire protein, with many of the phage-associated DNA invertases being >75% identical to each other (Fig. 3) (2). Their domain organization consists of a ~100 amino acid residue catalytic domain connected to a long oligomerization helix followed by a 50 to 60 amino acid helix-turn-helix DNA binding domain (DBD) (Fig. 4) (1). It was first recognized in 1980 that the sequence of DNA invertases resembled the TnpR repressors/resolvases of transposons Tn3 and γδ (Tn1000) (75). The resolvases, which primarily catalyze deletion reactions, have the same protein architecture and share up to about 35% sequence identity (55% similarity) with DNA invertases over the entire polypeptide and 41% identity (62% similarity) over the catalytic domain + oligomerization helix (Figs 3 and 4). These recombinases thus became the founding members of the DNA invertase/resolvase class, which together with more distantly related subgroups, constitute the serine recombinase superfamily (Fig. 4) (1, 76). Unlike DNA invertases, resolvases do not utilize Fis or a remote enhancer sequence. However, they do assemble elaborate supercoiling-dependent synaptic complexes, which employ additional resolvase subunits and sometimes DNA bending proteins like HU that perform critical DNA architectural and regulatory functions. They also share the same enzymology of DNA exchange (1, 3).

The "small" serine recombinases exhibiting the resolvase/invertase protein architecture cannot be easily classified into enzymes that are specific for promoting deletions or inversions based on their overall amino acid sequence. For example, the ISXc5 recombinase from *Xanthomonas campestri* catalyzes only deletions, yet is 55% identical to Hin and only 38% identical to γδ resolvase over the catalytic domain and oligomerization helix regions (77). Nevertheless, Fis/enhancer-dependent DNA invertases may be recognized by the presence of three conserved residues that contact Fis, which are located within helix 1 of their DNA binding domains, together with several basic residues in the helix B region of the catalytic domain, which contact the enhancer DNA (Fig. 3 and section "The dimer-tetramer remodeling reaction: role of the helix B region") (78). Some small serine recombinases have been shown to catalyze both DNA inversions and deletions. Examples include the β-recombinases (~32/34% identical to Hin/Gin, respectively) and Tn552 BinR (47/48% identical to Hin/Gin) (79, 80, 81, 82, 83). Neither of

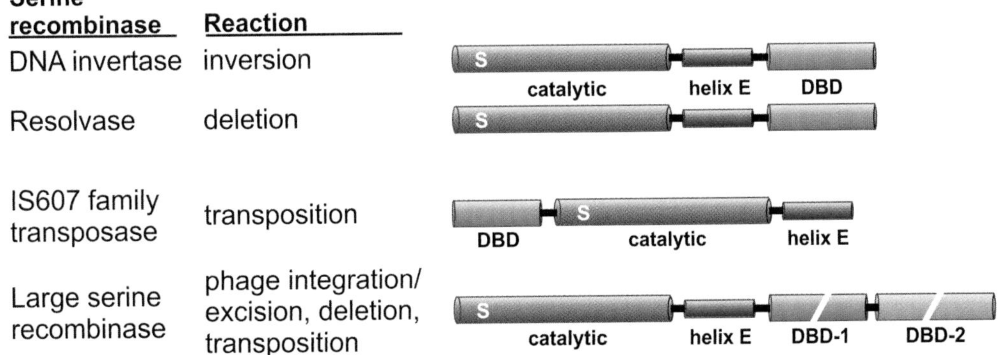

Figure 4 Serine recombinase subfamilies. The domain architectures of serine recombinase subfamily members denoting the ~100 amino acid residue catalytic domain containing the active site serine (S), the oligomerization helix E, and the DNA binding domain (DBD). The DNA binding regions of large serine recombinases can be quite large (300 to 450 residues) and consist of two discrete domains. The DNA invertases and resolvases are often grouped together as small serine recombinases. doi:10.1128/microbiolspec.MDNA3-0047-2014.f4

these utilize a Fis-like protein, as expected since they are active in gram (+) bacteria that do not contain Fis, but the β-recombinase requires a DNA bending protein related to HU (84, 85).

A large class of serine recombinases, which are known as "large serine recombinases" and include the serine integrases, mediate a diversity of reactions in bacteria, including developmentally-controlled deletions and phage integration and excision from bacterial host chromosomes (Fig. 4) (76, 86, 87). These recombinases share the catalytic domain and oligomerization helix that is homologous with the resolvase/invertase subgroup but have a much larger 300-550 residue C-terminal domain. The C-terminal extension is folded into two subdomains that participate in DNA binding and contribute to synapsis (88, 89, 90). Many of the serine integrases have been shown to require recombination directionality factors or Xis-like proteins to promote the excision/deletion reaction (87). Some members of this class catalyze an excision-integration reaction that translocates the intervening DNA to a new location within the host DNA, and thus, are considered transposases. Another class of serine recombinases, the IS607 class, also function as transposases (91). These are distinguished by their small winged-helix DNA binding domain being located N-terminal to the catalytic domain (Fig. 4) (92). Although less is currently known about the mechanistic and structural details of reactions catalyzed by members of these classes, the pathways and regulation of synaptic complex formation appear to be very different from those of the resolvases/invertase group. However, fundamental features regarding the enzymology of recombination, including the subunit rotation mechanism for DNA exchange, are believed to be conserved among all the serine recombinases.

MECHANISM OF FIS/ENHANCER-DEPENDENT SERINE DNA INVERTASE REACTIONS

General Properties of Fis/Enhancer-Dependent DNA Inversion Reactions *In Vitro*

Of the serine DNA invertase systems, the Hin- and Gin-catalyzed reactions have been studied most intensively *in vitro*. Reactions employing purified Hin or Gin and Fis proceed optimally at 37° in a buffered salt solution with a divalent cation like Mg^{2+} (23, 27, 47, 93, 94). The DNA substrate must be supercoiled, consistent with the sensitivity of the in vivo reaction to novobiocin (31). The Hin reaction also requires HU for efficient inversion on the wild-type substrate (discussed in section "DNA looping and the role of HU and other DNA bending proteins") (23). Under standard inversion conditions, the *in vitro* Hin reaction is essentially dependent upon Fis. An extremely inefficient reaction occurs in the absence of Fis when up to 15% ethylene glycol or glycerol is added; this basal reaction without Fis is about 1% of the rate of the reaction with Fis. Gin-mediated DNA inversion is also strongly stimulated by Fis but may exhibit a higher Fis-independent activity than Hin (93, 94). A variety of RNA molecules such as rRNA or synthetic polycytidylic or polyuridylic acid or even long chain polyanions like polyglutamic acid were found to strongly stimulate the activity of partially purified Hin preparations, but the stimulatory effect is much less in reactions employing highly purified native or refolded Hin (23). The basis of the stimulation by RNA has not been determined; possibilities include stabilizing Hin or titrating an inhibitor.

Early preparations of Hin supported recombination in the presence of 10 mM EDTA (23, 30). However, highly purified native or refolded Hin preparations require a divalent cation or spermidine for complete inversion but not for the initial DNA cleavage step. Mg^{2+}, Ca^{2+}, or spermidine are similarly effective, but reactions with Mn^{2+} are poorer. Gin is reported to catalyze inversion in 10 mM EDTA (95) but is strongly stimulated by Mg^{2+} or Ca^{2+} and weakly by Mn^{2+} (94, 96, 97). The mechanism behind the requirement for divalent cations or polyamines, particularly for the DNA ligation step, is not known, but they are not believed to be directly functioning in DNA chemistry. Both Hin and Gin exhibit a robust substrate-specific supercoiled DNA relaxing activity that is also dependent on divalent cations or polyamines (30, 93, 98).

Hin and Gin promote multiple inversions on supercoiled plasmids to generate an equilibrium state where both DNA orientations are equally represented (23, 30, 93, 96, 97). Therefore, starting with one orientation the product population never exceeds 50% inversion because of the reverse reaction. Reactions with highly purified Hin are rapid, recombining the DNA substrate to near equilibrium within 1 to 2 min. Hin turns over between DNA substrate molecules very inefficiently but can catalyze multiple reactions utilizing different recombination sites on the same DNA molecule (R.C. Johnson, unpublished). The inefficient turnover between substrates is at least partially due to the slow Hin-DNA dissociation rate (99). Deletion rates *in vitro* on substrates with directly-oriented recombination sites are typically <1% of inversion rates between sites in their native "inverted" orientation (67), and intermolecular reactions are undetectable, making the specificity for the inversion reaction *in vitro* even greater than measured *in vivo*.

The DNA-Cleaved Invertasome Intermediate

Hin generates inversion products without the appearance of readily detectable intermediates under standard reaction conditions. However, the combination of high concentrations of ethylene glycol (30% for Hin and up to 50% for Gin) and EDTA in place of Mg^{2+} stalls the reaction in an intermediate nucleoprotein complex where both recombination sites contain double strand cuts within their centers (100, 101). Formation of these "cleavage" complexes is recognized by the appearance of the vector backbone and invertible segment following rapid quenching with a protein denaturant such as SDS or HCl. Each of the four cleaved DNA ends contains a recombinase protomer covalently associated with the 5′ end and a two nucleotide protruding 3′ end terminating with a hydroxyl (100). The nondenatured cleavage complex represents a true reaction intermediate since it can be chased into ligated inversion products within seconds by adding back Mg^{2+} and diluting the ethylene glycol to ≤5%. Thus, DNA exchange occurs through double strand breaks over the central two base pairs of the 26 bp recombination sites (Fig. 2).

Unlike Hin, incubation of Gin with radiolabeled fragments generates a low level of site-specific DNA cleavages at the same respective locations within *gix* sites (98). Two dimensional thin layer chromatography of the covalent Gin-DNA complex after acid hydrolysis demonstrated that Gin was attached to the DNA 5′ end via a phosphoserine linkage, as observed earlier with γδ resolvase (102). Mutagenesis studies of Gin and Hin are consistent with the active site serine being at residue 9 in Gin and residue 10 in Hin, within one of the most conserved regions present in all serine recombinases (Fig. 3) (98, 103).

Electron microscopy of crosslinked Hin cleavage complexes revealed up to ~25% in an "invertasome" structure containing the two *hix* sites looped onto the enhancer (Fig. 5) (104). Similar invertasome structures have also been observed from Gin reactions (105). Formation of Hin invertasomes required Hin, the two *hix* sites, Fis and the enhancer on a supercoiled DNA plasmid. Antibody probing confirmed that both Hin and Fis were present in the tripartite complex. Significantly, no such structures were observed from reactions starting with relaxed DNA. Moreover, single-looped molecules containing only one *hix* site associated with the enhancer were not observed, even when the enhancer was missing one of the two Fis binding sites (104).

Nucleoprotein structures assembled from complete reactions but consisting of only the two *hix* sites paired together were also observed by electron microscopy at high frequency after protein crosslinking (104). Most of these paired-*hix* structures probably represent complexes where the enhancer segment was released from an assembled invertasome during sample preparation because the majority of them contained Fis covalently crosslinked to the Hin complex by antibody probing. Glutaraldehyde crosslinked paired-*hix* structures were also obtained in reactions without Fis or the enhancer. The formation of these complexes do not require DNA supercoiling and can even form with Hin mutants or disulfide-linked Hin dimers that cannot generate active

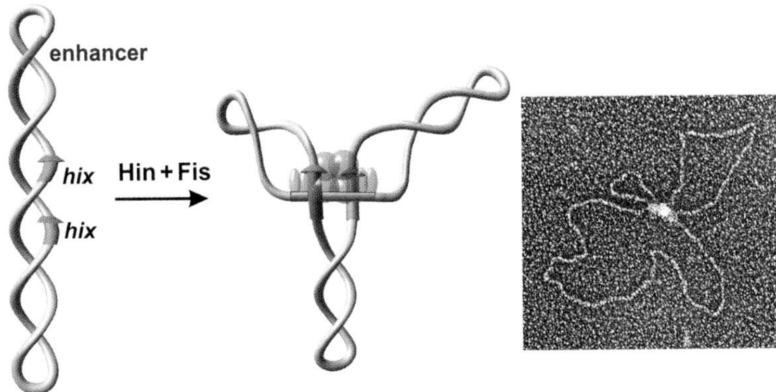

Figure 5 The Hin invertasome. Electron micrograph of an invertasome together with a schematic drawing of the structure. The invertasome structure was stabilized by crosslinking and supercoils removed prior to spreading onto the grid and low angle platinum shadowing (104). Hin subunits are rendered as translucent spheres; Fis subunits are rendered as ovals. *Hix* sites are depicted as arrows and the enhancer segment is blue.
doi:10.1128/microbiolspec.MDNA3-0047-2014.f5

synaptic complexes in the presence of Fis (106) (R.C. Johnson and co-workers, unpublished). Thus, the functional relevance of crosslinked paired-*hix* complexes formed by the wild-type enzyme without an operating Fis/enhancer system is questionable.

The imaging of invertasome intermediate complexes, combined with the requirements for their assembly, suggest a pathway for inversion in which the two recombination sites assemble into chemically active synaptic complexes at the enhancer in a supercoiling-dependent reaction (Figs 5 and 6). Stoichiometry calculations based on activity measurements or radiolabeling indicate that two dimers of Hin and two dimers of Fis are present in active synaptic complexes (104, 107). As discussed further below (section "The specificity for DNA inversion"), the supercoiling-directed assembly will orient the recombination sites in a configuration which specifies that the product of the DNA exchange reaction will be inversion of the intervening DNA.

The Serine DNA Invertase Dimer and Recombination Site Binding

The serine invertase dimer

Hin chromatographs as a dimer by gel filtration in the presence of 0.5 to 1 M NaCl and 0.1% Triton X-100 or 4 mM CHAPS (108) (Y. Chang and R.C. Johnson, unpublished). Dimeric interactions between subunits are highly sensitive to the zwitterionic detergent CHAPS; Hin behaves as a monomer in solution and cooperative DNA binding to *hix* is abrogated in the presence of high CHAPS concentrations (106, 108, 109). Gin is reported to chromatograph on Sephacryl S-200 as a monomer (94) but binds highly cooperatively as a dimer to *gix* sites (110). As noted above, Gin dimers exhibit weak DNA cleavage activity, but Hin dimers are chemically silent.

All evidence indicates that the dimeric structures of DNA invertases match very closely to those of resolvases (111, 112). A folded globular catalytic domain, which appears to be highly conserved among all serine recombinases, extends over the N-terminal ~100 amino acid residues (Figs 3, 4, and 7). Within the catalytic domain are two particularly highly conserved regions, one around the active site serine (Hin residue 10) and the other around arginines at Hin residues 66 and 69 (Fig. 3), which are proximal to each other in the inactive dimer structures. Invariant arginines at Hin residues 8, 66, 69 are proposed to perform direct chemical roles in the cleavage-ligation reactions within the remodeled active tetramers of serine recombinases (113, 114). As expected, mutations of each of these arginines, along with Ser10, Arg43, and Asp65, which are also believed to participate in catalysis, abolish DNA cleavage by Hin (103, 109, 115, 116) (R.C. Johnson, unpublished). Of particular significance is the fact that the serine hydroxyl nucleophiles from each subunit are separated by >35 Å and are not close to the scissile phosphate in the dimer structures of resolvase (111, 112). Thus, a major reorganization of the recombinase

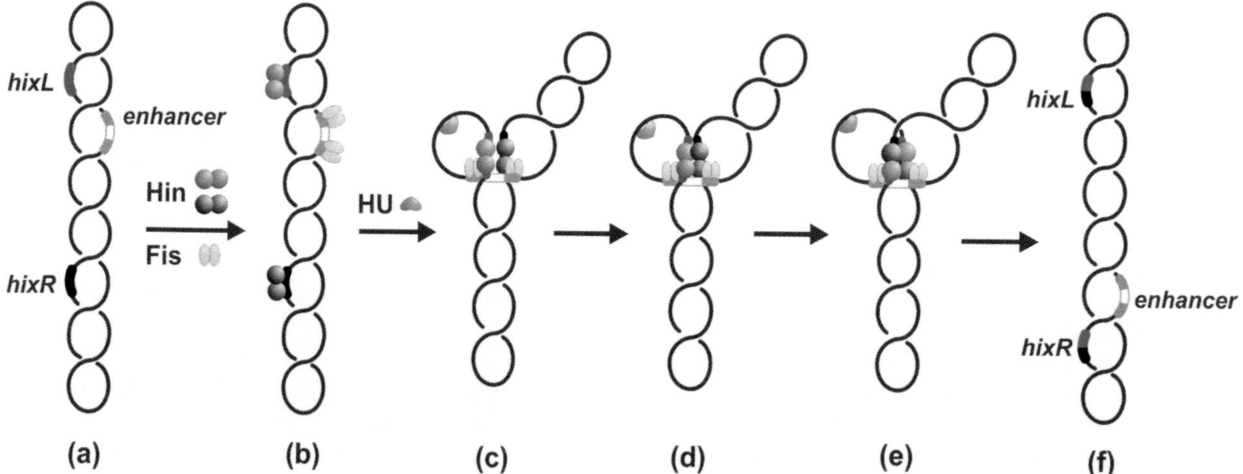

Figure 6 DNA inversion reaction pathway by Fis/enhancer-dependent serine invertases. In step (c) Hin dimers bound to *hixL* and *hixR* are associated with the Fis-bound enhancer at the base of a branch on supercoiled DNA. Formation of the Hin tetramer (d) generates an enzyme active for double strand cleavage and subunit rotation (e). Ligation and resolution of the complex (f) results in inversion of the DNA segment between recombination sites.
doi:10.1128/microbiolspec.MDNA3-0047-2014.f6

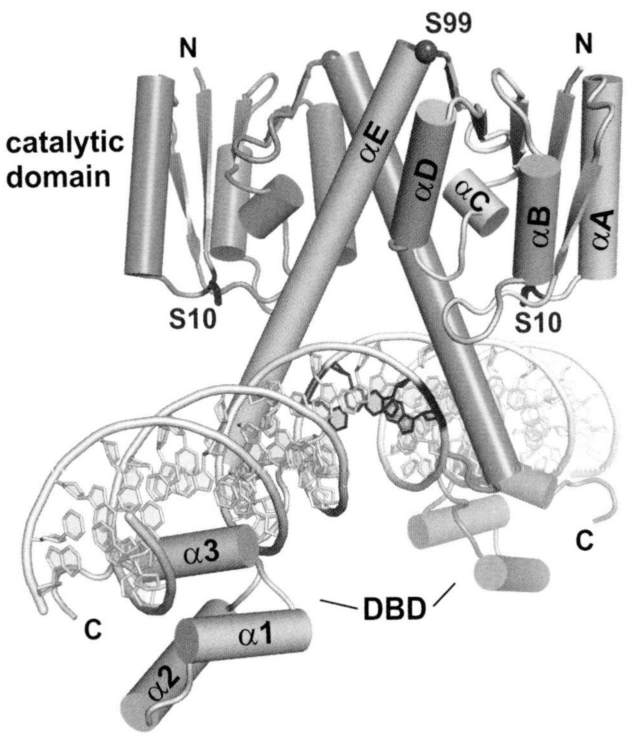

-13 -1+1 +13
TTCTTGAAAACCAAGGTTTTTGATAA
AAGAACTTTTGGTTCCAAAAACTATT

Figure 7 The serine DNA invertase dimer. Model of the Hin dimer bound to *hixL* derived from the catalytic domain and oligomerization helix E of γδ resolvase (PDB: 1GDT) linked to the DNA binding domain (DBD) of Hin (PDB: 1IJW) is shown. The Ser10 active site residue and core nucleotides where DNA exchange occurs are colored red. Hinge residue Ser99 (Cα) is rendered as a dark red sphere. The sequence of *hixL* showing the Hin cleavage sites (arrows) and core nucleotides (red) is given below.
doi:10.1128/microbiolspec.MDNA3-0047-2014.f7

quaternary structure, together with adjustments surrounding the active site pockets, is required prior to any DNA chemistry.

The catalytic domain is followed by a long amphipathic α-helical region referred to as helix E from the γδ resolvase secondary structure designations (Figs 3, 4, and 7). For Hin, modeling and experimental data are consistent with helix E extending from residues 100 to 134 (117, 118), though residues at the C-terminal end of helix E that are not in contact with the catalytic domain may be unstructured when not bound to DNA. Helix E residues between Met101 and Glu122 participate in most of the interactions between subunits in the dimer and undergo extensive remodeling to form much of the synaptic tetramer interface, including the connections between subunits of the newly formed rotating dimer and subunits across the rotating interface (discussed below). The "hinge" residue connecting the catalytic domain to helix E, Ser99 in Hin and Ser97/98 in Gin, is believed to play a critical role in regulating the quaternary transition between the dimers and the tetramer (sections "The dimer-tetramer remodeling reaction: role of the helix B region" and "The hinge residue Ser99") (109, 119, 120). A four residue linker connects the oligomerization helix to the C-terminal DNA binding domain (DBD), which begins at Gly139 in Hin (121, 122).

The Hin DNA Binding Domain and Recognition of the Recombination Site

X-ray crystal structures of the monomeric Hin DBD bound to the half-site that is identical between *hixL* and *hixR* reveal the details of sequence recognition of the recombination site (Fig. 8) (99, 122). At the N-terminal end, Gly139-Arg140 engage in extensive van der Waals interactions and make a series of hydrogen bonds with nucleotides along the floor of the minor groove of an A-tract sequence. The structure fits well with extensive mutagenesis and chemical probing studies which have shown that A/T or T/A base pairs, or more specifically the absence of a purine 2-amino group, at bp ±(5 and 6) are of critical importance (123, 124). Unlike most A-tract structures, however, the minor groove over this region is not highly compressed. Residues 146 to 180 fold into a helix-turn-helix (HTH) motif in which helix 1 and 2 are in a near antiparallel configuration, unlike classical phage repressor HTH structures but resembling homeodomains. However, the Hin recognition helix 3 is only eight residues long, which is much shorter than homeodomains and more similar to repressors. The overall structure of the Hin DBD is similar to that of γδ and Sin resolvases, although the lengths of helix 1 differ (111, 112).

Sequence recognition within the major groove involves direct contacts between only two amino acids on helix 3 and three base pairs at positions ±(9 to 11) (Fig. 8) (99, 122, 123). X-ray structures of the Hin DBD bound to the wild-type and four mutant *hix* sites reveal that in addition to direct base contacts by Arg178 and Ser174, two ordered water molecules engage in bridging contacts that are critical for *hix* binding. The importance of these waters is illustrated by a T/A to C/G substitution at position 11; this mutation decreases binding 25-fold but the only significant change in the structure is the loss of the two ordered water molecules (99).

A

-13 -1+1 +13

TTCTTGAAAACCAAGGTTTTTGATAA

AAGAACTTTTGGTTCCAAAAACTATT

binding mutations: gg ccc
cc t

B

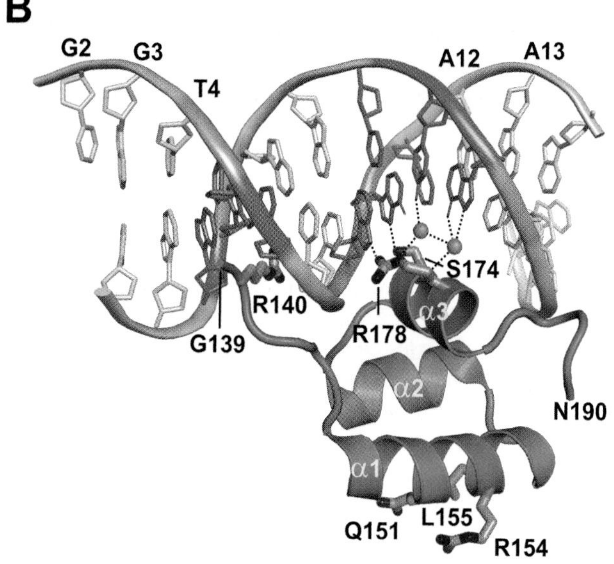

C

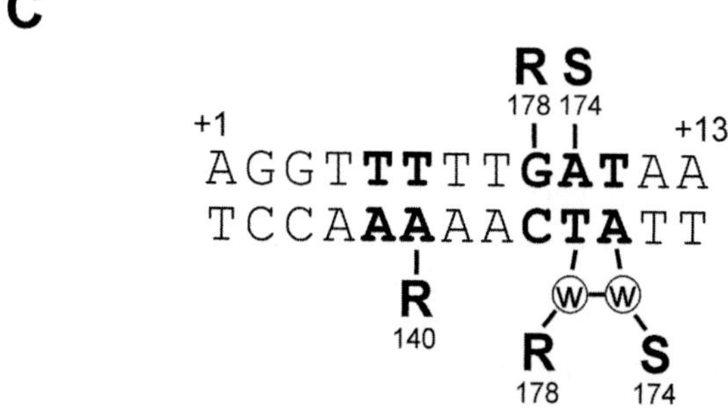

Figure 8 The Hin DNA binding domain bound to the conserved *hix* half-site located within the invertible segment. (A) Sequence of *hixL* with the locations of severe DNA binding mutations (99, 123). (B) Hin DBD-DNA complex (PDB: 1IJW). Side chains of residues Arg140, Ser174, and Arg178 along with the two ordered water molecules (cyan spheres) that make critical bridging contacts between Hin and DNA atoms are shown. Conserved residues Gln151, Arg154, and Leu155 on α-helix 1 that contact Fis are also shown. (C) Schematic representation of sequence-specific contacts, including water bridged hydrogen bonding.
doi:10.1128/microbiolspec.MDNA3-0047-2014.f8

Binding of the Serine Invertase Dimer to the Recombination Site

The Hin dimer binds to the intact *hix* site with about 10-fold higher affinity than the isolated DNA binding domains (99, 121). In the case of Hin-*hixR* and Gin-*gix* where the sequence of one of the half sites is suboptimal, cooperative interactions are important for binding (108, 110). The C-terminal ends of the E helices traverse the minor groove on opposite sides of the DNA immediately adjacent to the two bp crossover region and are modeled to make a series of DNA contacts beginning with Hin Arg123 that are important for function (Fig. 7). Arg69 and Arg43 within the catalytic domain may indirectly influence DNA binding by the dimer, as evidenced by the ability of substitutions at these positions to suppress the effects of particular *hix* binding site mutations (116). The orientation of the recombination sites is determined solely by the identity of the nucleotides within the 2 bp crossover region (67). In native DNA invertase recombination sites, the sequence of the core nucleotides is never symmetrical (e.g., Fig. 2). Hin can efficiently bind and recombine *hix* sites containing different core sequences, including the symmetrical sequence A/T-T/A. Although a 2 bp core sequence separating the half sites is essential for recombination, Hin can still bind to a *hix* site with one or even three nucleotides introduced into the center (108, 125). Gel electrophoresis experiments imply that the DNA is bent within the Hin and Gin complexes (106, 110), but the DNA curvature in the Hin-*hix* complex is estimated to be much less than the ~60° bend angle present in the γδ resolvase-*res* site I crystals (111). Indeed, a recent low resolution X-ray structure of the Gin-*gix* complex models the DNA with only a small curvature (126).

The Fis/Enhancer Regulatory Element

The most distinguishing feature of inversion reactions catalyzed by the serine DNA invertase family is the requirement for a remote recombinational enhancer and its binding partner Fis. Reactions performed without an enhancer but with the Fis protein exhibit very low inversion rates *in vivo* and *in vitro* (23, 27, 47). The ~1% inversion rate that occurs without an enhancer under standard *in vitro* conditions is dependent upon Fis binding to DNA and thus is believed to reflect the prolific nonspecific DNA binding properties of Fis. Reaction rates without Fis are reduced >1000-fold, but as noted above, a detectable basal Fis-independent reaction occurs *in vitro* in the presence of ethylene glycol (25, 51). Enhancer sequences can be functionally exchanged between DNA inversion systems (48, 60).

Location of Recombinational Enhancers

The native distances from the center of the closest recombination site to the minimal boundary of the enhancer elements range from 89 bp (Gin) to 481 bp (Cin). As depicted in Fig. 1, the enhancer can be located inside the invertible segment (Hin) or outside the invertible segment (Gin and Cin), and the DNA specifying enhancer activity also functions as the coding sequence for residues 8/9-29/30 of the respective DNA invertase. The Hin enhancer has been shown to function effectively in either orientation and when located from 89 bp to over 4 kb from the nearest *hix* site, though a measureable decrease in inversion rates is evident when the enhancer is repositioned many kbs from a *hix* site (27). The *hin* enhancer does not function when positioned very close to a recombination site and its activity varies with the helical repeat of supercoiled DNA over distances extending out to at least 150 bp. This is consistent with a physical interaction between the two sites through protein-mediated DNA looping (see section "DNA looping and the role of HU and other DNA bending proteins") (28, 117).

Kanaar et al. reported a situation where the *gin* enhancer can function even though it is not strictly in *cis* with respect to the *gix* recombination site (127). Two multiply interlinked supercoiled DNA molecules containing the enhancer on one DNA circle and one or both *gix* sites on the second DNA circle support a remarkably robust recombination reaction. By contrast, similar experiments with singly catenated circles exhibit no activity, indicating that physical proximity of the three sites is insufficient for assembling an active recombination complex. Moreover, supplying the enhancer in *trans* on a short linear fragment at high concentrations is ineffective (101) (M. Haykinson and R.C. Johnson, unpublished), unlike the enhancer-dependent Mu transposition reaction where similar experiments reveal enhancer activity (128). These properties of enhancers in the DNA inversion systems are consistent with the obligatory assembly of the tripartite invertasome at an interwound DNA branch on a plectonemically supercoiled DNA molecule.

Architecture of Enhancers and Fis Binding Sites

The *hin* enhancer sequence is minimally contained within a 63 bp segment. An additional 3 bp of nonspecific sequence on each end are required for Fis binding making the total length 69 bp (27, 129, 130). Deletion analysis of the *cin* enhancer gives similar boundaries (48). The Fis binding sites on each end of the enhancer are separated by 47 bp between their centers, which would position the Fis dimers on nearly opposite sides of the DNA helix separated by approximately 4.5 helical turns of DNA (Fig. 9C). Inactivation of either of the Fis binding sites abolishes all enhancer activity. The spacing between Fis binding sites is critical for activity (131) (Y. Chang, M.M. McLean, and R.C. Johnson, unpublished). Addition of one bp is well tolerated, but a +2 insertion or -1 or -2 deletion severely decreases and a 5 bp insertion inactivates enhancer function. Ten bp insertions retain significant activity, but enhancers containing two or more DNA helical turns or a deletion of 10 bp are not functional. Recently, a modest effect of sequence in the segment between the Fis binding sites in the *hin* enhancer has been reported (78). The effect is localized to two A/T-rich patches separated by 10 bp where basic residues from the Hin catalytic domain interact (Fig. 9C and see below). The A/T-rich sequences, which face Hin in the invertasome structure, may form narrower and more negatively charged minor grooves and/or induce curvature into the DNA that facilitates assembly of the invertasome by enhancing Hin-enhancer and/or Fis-Hin interactions, respectively.

DNA sequences of Fis binding sites are highly diverse but contain a 15 bp core "recognition" segment that usually has G/C and C/G base pairs at its boundaries combined with an A/T-rich region in the center (Fig. 9A) (49, 129, 132 to 134). High affinity Fis binding sites often contain pyrimidine-purine dinucleotide steps at the ±(3 to 4) positions, which correspond to locations of modest bending in Fis-DNA complexes (130). In all studied enhancers, one of the Fis binding sites (corresponding to site I or the *hix* proximal site in the *hin* enhancer) is considerably poorer than the other, in part due to a T/A instead of the preferred G/C bp on one end. This binding site overlaps the coding region for the highly conserved patch around the active site serine. The difference in binding sites appears irrelevant for enhancer function *in vitro* as a reconstructed enhancer containing two high affinity site II Fis binding sites functions nearly indistinguishably from the wild-type enhancer in the Hin system (29).

The Fis Protein Structure

Fis is a homodimer of 98 amino acid residue polypeptides. X-ray crystal structures reveal an ellipsoid core containing four α-helices beginning at Pro26 (Fig. 9B) (135, 136). Helices A and B from each subunit assemble into a four helix bundle and helices C and D of each subunit fold into a helix-turn-helix DNA binding unit stabilized by the C-terminal ends of helix B from the same polypeptide. A unique feature of the Fis dimer structure is that each DNA recognition helix D is separated

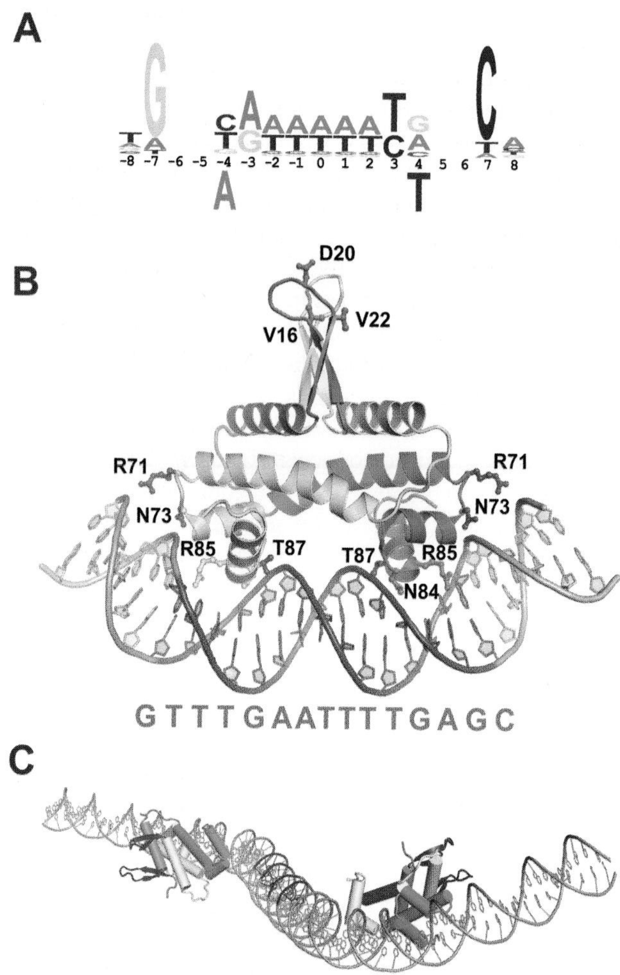

Figure 9 Fis and the recombinational enhancer. (A) Fis binding motif derived from footprinting, mutagenesis, genome-wide ChIP, and X-ray crystallography (see reference 130). Bases below the numbering are strongly inhibitory for binding. (B) Structure of the Fis dimer bound to a high affinity DNA segment (Fis residues 10 to 98; PDB: 3IV5). The sequence of the 15 bp core between ±7 (colored brown) is given below. Arg85 contacts the conserved guanines at the borders of the core sequence; Asn84 contacts the DNA backbone and often the base at ±4 and is responsible for the inhibitory effect of a thymine at this position (panel A). A subset of other important residues making DNA backbone contacts are colored grey. The Arg71 side chains, which are poorly resolved in most structures of DNA complexes, are shown oriented towards DNA. Bending of the flanking DNA segments varies depending on the DNA sequence. The triad of residues (Val16, Asp20, and Val22) near the tips of the mobile β-hairpin arms that contact DNA invertases are denoted for the cyan colored subunit. (C) Model of the *hin* enhancer. The two Fis dimers are docked onto the *hin* enhancer DNA sequence. The Fis β-hairpin arms are highlighted in red. The A/T-rich DNA segments contacted by the helix B regions of the DNA invertase tetramer in the invertasome are colored magenta (78).
doi:10.1128/microbiolspec.MDNA3-0047-2014.f9

by about 25 Å as opposed to the 32 to 34 Å separation that is typically found among helix-turn-helix DNA binding proteins. This separation is maintained in the DNA-bound complex and is accommodated by a severe narrowing of the central minor groove as described below. Another unusual feature of the Fis structure is the presence of mobile β-hairpin "arm" motifs at the N-terminal end of each subunit (residues 12 to 25) that protrude over 20 Å from the protein core (137). In many crystal forms, one or both of the β-hairpin arms are poorly resolved due to their mobility, but five different crystal forms, including Fis-DNA co-crystals, resolve most or all of the peptide chain of the β-hairpin arms (130, 137, 138) (S. Stella and R.C. Johnson, unpublished). As expected, the arms are positioned somewhat differently in these crystals due to crystal lattice interactions. Although the two arms are well separated from each other in the crystal structures, direct disulfide linkages between residues in the two arms that are separated in crystals by up to ~20 Å rapidly form when they are substituted with cysteines (137). The residue-specific cysteine crosslinking data provides strong support for the structure of the β-hairpin arm motifs and for their mobility in solution.

Mutagenesis and protein crosslinking studies have mapped the sites of Fis-Hin interaction within the invertasome to be between residues near the tip of the β-hairpin arm of Fis and helix 1 of the Hin DBD (78, 137, 139, 140). Extensive mutagenesis and chemical modification studies over the Fis β-hairpin arm identified the critical residues for activation of Hin-catalyzed DNA inversion to be Val16, Asp20, and Val22 (Fig. 9B). The critical feature of residues 16 and 22 is their branched chain character; a subset of polar residues and even a leucine at residue 20 retain some activity (78). Loss of the Asn17 side chain, which stabilizes the conformation of the β-hairpin loop, also compromises Fis-activation of inversion (137). In addition, mutations at Fis residues Gln33 and Lys36 within helix A negatively impact Hin inversion and so may also directly or indirectly influence invertasome formation. Only one of the Fis arms on each dimer bound to the enhancer is required, as demonstrated by efficient DNA inversion promoted by heterodimers containing only one functional β-hairpin arm (103).

Fis Binding to DNA and Model of Enhancer

Crystal structures of Fis bound to over 20 different DNA sequences with varying affinities have been determined (Fig. 9B) (130, 141) (S.P. Hancock, S. Stella, and R.C. Johnson, unpublished). Fis covers a 21 bp DNA segment in the crystals, and most of the direct contacts over this region are to the backbone of the DNA. The most important base-specific contact is by Arg85 on helix D to the consensus guanines at the edges of the 15 bp core. The Asn84 side chain is also hydrogen bonded to a base in most structures, but its most significant role is to exclude a thymine at position ±3 due to a clash with the 5-methyl group (Fig. 9A, B). The DNA helix axis exhibits overall curvatures of around 65° in the various crystal structures with most of the bending being over the major groove interfaces. Solution biochemistry experiments imply variable amounts of additional DNA bending occur over the regions flanking the 15 bp core sequence; the sequence of the flanking DNA together with Arg71 on Fis modulate this dynamic bending (129, 142) (S. Hancock, S. Stella, and R.C. Johnson, unpublished). The 5 to 7 bp A/T-rich segment at the center of the binding site is relatively straight but contains a highly compressed minor groove (130, 141). The narrow minor groove is critical for Fis binding because it enables the closely spaced recognition helices of the dimer to insert into the adjacent major grooves.

The Fis-DNA crystal structures enabled construction of a molecular model for the Fis-bound *hin* enhancer segment (Fig. 9C) (78). The DNA adopts an S-shaped structure with the β-hairpin arms from the two dimers oriented oppositely from one another. The S-shaped structure of the Fis-bound enhancer segment is supported by gel electrophoresis studies on spacing mutants (131, 142). As discussed below, the Fis-bound enhancer functions in the initial stages of the reaction as a scaffold to juxtapose serine invertase dimers for remodeling into the tetramer.

Functions of Fis in Other DNA Reactions and in Chromosome Compaction

Subsequent to the discovery of Fis as the factor required to activate DNA inversion reactions, Fis was shown to perform many other functions in the cell (24, 143). These include regulation of transcription and replication reactions, in addition to other specialized recombination reactions such as integration and excision of phage λ and Tn5 transposition. With respect to transcription, Fis can positively regulate promoter activity by interacting with RNA polymerase through its αCTD or σ subunits and negatively regulate transcription by competitively binding with RNAP or gene-specific activators (144). Transcriptome studies have reported significant changes in expression of over 20% of *E. coli* genes by the loss of Fis, but the correlation of Fis binding by genome-wide chromatin immunoprecipitation experiments and gene expression changes indicates that many of these effects are indirect (133, 134).

Fis is one of a small group of abundant nucleoid-associated proteins that also include HU, IHF, H-NS, and StpA. All of these proteins perform specific regulatory functions and are thought to contribute to chromosome compaction by means of DNA bending and/or stabilization of DNA loops (145, 146). Single-DNA molecule experiments have demonstrated moderate global DNA compaction by Fis-induced bending, as predicted from the structural and biochemical data, plus robust DNA condensation activity through trapping of DNA loops (147, 148). Under rapid growth conditions, Fis is one of the most abundant DNA binding proteins in *E. coli*; however, cellular Fis levels are much lower under poor growth conditions and levels of Fis in stationary phase cells are very low (18, 32, 149). Genome-wide studies reveal that Fis binds singly or in clusters every 3 to 6 kb on average throughout the chromosome and it has been proposed that interactions between tracts of bound Fis protein may contribute to the dynamic looped-domain structure of the bacterial nucleoid (133, 134). In addition to local effects on chromosome structure, Fis also has been reported to regulate expression of DNA gyrase and topoisomerase I and thereby modulate DNA supercoiling densities on a global scale (150, 151). Because assembly of active invertasomes is dependent upon supercoiling-induced DNA branching, these effects by Fis can also indirectly impact the efficiency of the inversion reaction *in vivo*.

Clear Fis homologs are found within many members of the gammaproteobacteria class. Members of the *Enterobacteriales* and related orders have conserved N-terminal β-hairpin arm sequences, but the sequence over this region becomes increasingly diverged and then partially or totally missing in Fis homologs from more distantly related gammaproteobacteria members. Distant Fis homologs are present in the betaproteobacteria class, but these are typically missing the N-terminal β-hairpin region that activates serine invertases.

DNA Looping and the Role of HU and Other DNA Bending Proteins

Formation of invertasomes requires DNA looping between the recombination sites and the enhancer element, and in many systems, one of the DNA loops is relatively small. For the Hin reaction, the distance between the center of *hixL* and edge of the enhancer is 99 bp. DNA inversion *in vitro* is very inefficient on substrates with distances less than 87 bp, even in the presence of the DNA bending protein HU (28). However, DNA-cleaved invertasomes can assemble with as little as 56 bp between the two elements. The larger loop size required for the complete inversion reaction probably reflects the winding of DNA strands during the DNA exchange step (discussed below). With separations between 56 and 125 bp, the formation of DNA-cleaved invertasomes strongly correlates with the helical repeat of DNA, which was calculated from the activities of 38 different spacing substrates to be 11.2 bp/DNA turn (28). This value is similar to the helical repeat measured by others for supercoiled DNA (152, 153), and it reflects the linking number deficit of supercoiled DNA relative to the 10.5 average for linear DNA.

Early *in vitro* Hin reactions with the wild-type substrate displayed up to a 10-fold stimulation of inversion rates by HU (23, 28), though more recent measurements employing different preparations of Hin exhibit less dependence. However, inversion reactions on substrates with shorter spacers are virtually dependent upon HU both *in vitro* and *in vivo*. Hin catalyzed inversions in *hupAB*$^+$ cells can occur at low rates with *hixL*-enhancer segments as short as 56 bp, implying that DNA in vivo is more flexible than in a buffered salt solution with Mg^{2+} and HU *in vitro* (28). The additional flexibility *in vivo* may be in part mediated by spermidine, as *in vitro* reactions have shown that long chain polyamines such as spermidine and spermine can enhance Hin invertasome assembly in the absence of HU. Nevertheless, the helical repeat derived from inversion activities of the spacer substrates in vivo also was calculated to be 11.2 bp/DNA turn, the same as determined *in vitro* in buffered salt solutions with Mg^{2+} and HU (28).

Hillyard et al. observed fewer than 4% Hin-mediated inversion events on the chromosome of *hupAB* mutant *Salmonella* Typhimurium strains as compared to the wild type (154). Using plasmid substrates, Haykinson et al. found that *E. coli hupAB* (HU deficient) mutants generated low but significant rates of Hin-catalyzed inversion with the wild-type *hix*-enhancer spacing but extremely low inversion was observed with shorter spacer lengths (28). As with *in vitro* reactions, plasmids with long spacers (e.g., 703 bp to the closest *hix* site) supported inversion rates in *hupAB* mutants that were less than 2-fold different from the wild-type parent. Wada et al. reported a strong dependence on HU for inversions catalyzed *in vivo* by the Hin, Gin, and Pin DNA invertases (155). HU has been reported to not affect *in vitro* inversion rates by Gin, however, even though there is only 89 bp separation between the *gixL* site and the *gin* enhancer (26). As expected, HU does not have a significant effect on Cin-catalyzed DNA inversion because the enhancer is located nearly 500 bp from the closest *cix* site (50, 155).

The *in vivo* phenotypes of *hupAB* mutants and the inability of *hupAB* mutant extracts to complement HU-deficient inversion reactions *in vitro*, imply that HU is the primary protein in *E. coli* responsible for DNA looping activity in the Hin reaction. Surprisingly, mammalian nuclear extracts are more active than *hupAB*$^+$ *E. coli* extracts for supplying activities that complement HU-deficient Hin-catalyzed inversion reactions (28). The proteins responsible are the abundant HMGB1 and HMGB2 chromatin-associated proteins (156). Likewise, the *S. cerevisiae* HMGB homologs Nhp6A and Nhp6B promote Hin invertasome assembly (157). Like HU, the nonspecifically binding HMGB proteins have robust DNA bending activity, enabling ligation of DNA microcircles that are much smaller than the persistence length of DNA. Eukaryotic HMGB proteins can induce efficient formation of DNA microcircles down to 66 bp, one helical turn shorter than the smallest circle capable of ligation in the presence of HU. Likewise, HMGB proteins are more active than HU in promoting assembly of Hin invertasomes with short *hix*-enhancer segments and can enable invertasomes to be assembled on substrates containing *hix*-enhancer segments 10 bp shorter than possible by HU. Significant stimulation of invertasome assembly occurs with only a few molecules of HU or HMGB proteins per DNA substrate, although maximum amounts of invertasomes are assembled with sufficient protein to bind every 150 to 300 bp (28, 157).

Assembly and Molecular Structure of the Invertasome

Overview of Invertasome Assembly Pathway and DNA Exchange

The critical regulatory step of the inversion reaction is the formation of the active tetramer at the enhancer. The overall pathway is summarized below and illustrated with molecular models in Fig. 10 (see also Fig. 6); experimental evidence for the individual steps is discussed in subsequent sections. The first step is the looping of recombinase dimers bound to the two recombination sites into the Fis-bound enhancer segment (Fig. 10A). Formation of this tripartite complex occurs at the base of a supercoiled DNA branch, which provides conformational energy to generate a specific geometry of crossing DNA strands (two negative nodes, see also Fig. 11 and below). The complex is stabilized by protein interactions between Fis dimers bound on each end of the enhancer and residues on helix 1 of one of the subunits on each recombinase dimer. The catalytic domains of each dimer are thereby positioned next to each other for isomerization into the active tetramer. The massive conformational rearrangement that generates the tetramer breaks most of the original contacts between subunits of the dimer, which are largely mediated by residues from the E helices that are oriented parallel to each other (Figs 7 and 10A). Transition into the tetramer creates an extensive set of new contacts between synapsed subunits, which largely involve residues of the E helices that are oriented antiparallel to each other (Fig. 10B). During the final stage of the remodeling reaction, basic residues from the B helices on the catalytic domains of the same subunits clamp onto the enhancer DNA between the two Fis binding sites (Fig. 10C). During this stage, the subunits of the original dimer become separated from each other by a relatively flat and completely hydrophobic interface. This interface enables the top subunit pair as drawn in Fig. 10D to F to rotate about the bottom subunit pair, which is fixed onto the enhancer through its connections with the Fis/enhancer segment. Upon full assembly of the tetramer, each subunit cleaves its respective DNA strand by forming the serine ester bond with the 5′ phosphodiester by a still rather mysterious process that probably involves additional coupled conformation changes. The rotation of one synapsed pair of subunits about the other translocates the covalently linked DNA strands to the recombinant orientation. Reversal of the phosphoserine linkage through attack by the free 3′ OH end (ligation) restores the DNA phosphodiester backbone.

Fis-Hin Contacts

The connections between the mobile β-hairpin arms of Fis and helix 1 of the Hin DBD were initially identified within DNA-cleaved invertasomes by site-directed protein-protein crosslinks utilizing reagents containing linker lengths as short as 4.4 Å (78). The crosslinkers targeted a cysteine introduced at Fis residue 19 or 21 between the critical Hin-activating residues Val16, Asp20, and Val22 within the β-hairpin loop (Fig. 9B) and either native lysines on helix 1 (Hin residues 146 or 158) or cysteines introduced at solvent-exposed residues

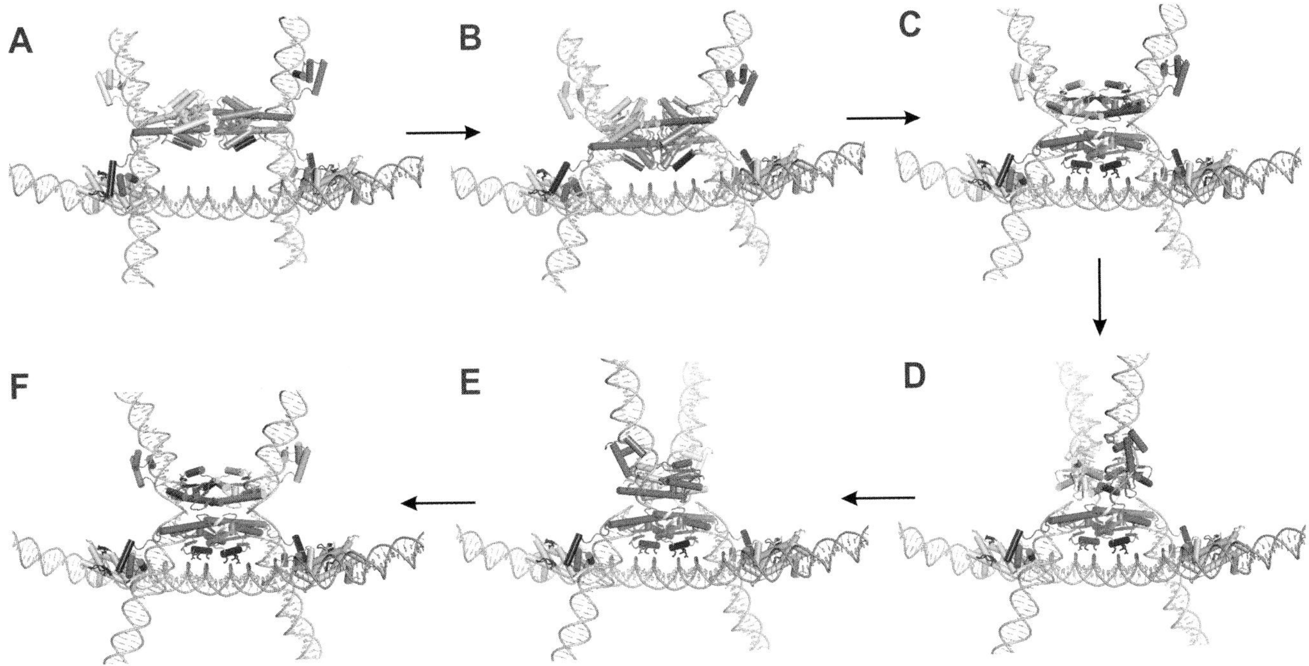

Figure 10 Assembly of the Hin invertasome and subunit rotation. (A) *Hix*-Hin dimers associated at the Fis-bound enhancer. Fis dimers are gold with their β-hairpin arms colored magenta. Hin α-helix B and α-helix 1 are colored red. (B) Pre-activated Hin tetramer intermediate (based on 3BVP) and (C) post-cleavage tetramer (based on 1ZR4). Side chains of residues from helix B that contact enhancer DNA are denoted. (D, E) Partial rotations (50° and 90°, respectively) of the top synaptic subunit pair and (F) complete subunit rotation to mediate the exchange of DNA strands. A movie depicting the assembly of the invertasome and DNA exchange by subunit rotation is provided in Video 2 of reference (78), from which these images are taken. Details of the models are described in references 78, 109, and 117. doi:10.1128/microbiolspec.MDNA3-0047-2014.f10

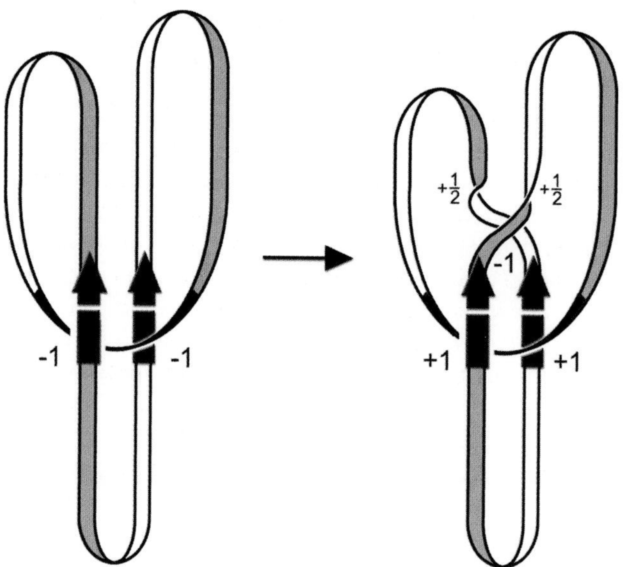

Figure 11 DNA topological changes during the inversion reaction. The starting complex (invertasome) between the two recombination sites (arrows) and the enhancer at the base of a plectonemic branch traps two negative DNA nodes (DNA depicted as a ribbon without supercoiling). Double strand cleavages and DNA exchange by the equivalent of a 180° clockwise rotation create a negative node but also introduce two half turns of helical twist that cancel the negative node. The recombinant configuration of DNA strands changes the trapped nodes to a positive sign resulting in an overall linking number change of +4. Node signs are determined by directionally tracing the entire path of the DNA molecule (193). By convention, a node is defined as negative when the DNA strand in front is pointed upwards and the strand underneath crosses in a rightward direction. A positive node is when the strand underneath crosses in a leftward direction. (Figure is modified from reference 105).
doi:10.1128/microbiolspec.MDNA3-0047-2014.f11

between 146 and 151. Mutant analysis of residues on helix 1 then identified Arg154 and Leu155 as being most critical for activation by Fis, with contributions by Gln151 and possibly Lys158 (Fig. 8B). Notably, residues 151, 154, and 155 are completely conserved among members of the DNA invertase subfamily that are known to be regulated by Fis (Fig. 3). Fis-Hin crosslinking experiments in which individual Hin subunits were radiolabeled demonstrated that only the bottom subunits of the tetramer (colored blue and green in Fig. 10) are associated with Fis. Moreover, Fis could be crosslinked to inactive covalently locked Hin dimers containing a disulfide bond across the dimer interface. Subsequent reduction of the disulfide bond enabled the Fis-linked Hin dimers to remodel into a tetramer that is active for DNA cleavage. This key result supports a model in which the Fis dimers initially contact inactive Hin dimers and maintain this contact during the dimer-tetramer remodeling reaction (Fig. 10A, B, C). Hin tetramers crosslinked to Fis are also competent for DNA ligation, providing evidence that Hin normally remains associated with the Fis/enhancer element through all the chemical and mechanical steps of the reaction (78).

Structure of the Invertase Tetramer

Initial evidence for the quaternary arrangement of subunits in the Hin tetramer came from experiments employing a strong hyperactive Hin mutant H107Y that can assemble stable DNA-cleaved synaptic complexes on oligonucleotides substrates without the Fis/enhancer (158). Cysteines were substituted around the protein surface targeting specific interfaces and site-directed crosslinking experiments employing bis-maleimide reagents with different spacer lengths were performed. Crosslinks between residues located proximal to the N-terminal end of the E helix (residues 94, 99, and 101) of subunits originating from different dimers implied that this region of the catalytic domain is within the synaptic interface of the tetramer and that the catalytic domains from the originating dimers are positioned orthogonally to each other. In this configuration, the DNA crossover sites are far apart from each other on opposite sides of the synaptic tetramer. A similar overall architecture for synaptic tetramers of resolvase were also proposed based on molecular and biophysical data, including small angle scattering of X-rays and neutrons (159, 160, 161).

The Steitz laboratory has determined several X-ray crystal structures of reaction intermediates of γδ resolvase tetramers bound to DNA (PDB: 1ZR2, 1ZR4, and 2GM4; 3.9, 3.4, and 3.5 Å resolution, respectively) (119, 162). These were generated using hyperactive mutants of resolvase with the protein used for the 2GM4 crystals being a chimera containing Hin residues 94 to 103 substituted for the analogous residues of resolvase. This segment includes the "flexible joint" between the catalytic domain and oligomerization helix E, in which a serine (Hin residue 99) replaces a glycine (resolvase residue 101). Each of the tetrameric structures represents a post cleavage complex where the 5′ ends of the cleaved DNA are covalently joined to each subunit at serine 10. Surprisingly, the cleaved DNA ends of each recombination site are separated by about 30 Å, nearly the distance between the two active site serines in the structure of the inactive resolvase dimer bound to DNA (PDB: 1GDT) (111). The post cleavage tetramers contain a flat and hydrophobic interface between synapsed subunit pairs that can support subunit rotation (discussed further in the section "Structural nature of the rotation interface").

A tetrameric crystal structure of a Gin hyperactive mutant containing the catalytic domains plus the first 26 residues of the E helices has also been reported (PDB: 3UJ3; 3.7 Å resolution) (120). Interestingly, this structure was crystallized in ethylene glycol that promotes stable tetramer formation in Hin. Each Gin subunit closely resembles γδ resolvase tetrameric subunits (RMSD over common backbone atoms = 1.3 to 1.5 Å; see Fig. 15 legend below). The synapsed subunit pairs are again separated by a flat and completely hydrophobic interface (see Fig. 15B below), but they are in a conformation that is equivalent to a 26° rotation of synapsed subunit pairs relative to the resolvase tetramer structures. A high resolution tetrameric crystal structure, also grown in ethylene glycol, of the Sin resolvase catalytic domains plus part of the E helices from a hyperactive mutant has also been determined (PDB: 3PKZ; 1.86 Å resolution) (114). The Sin tetramer exhibits yet a different rotational conformation, with the synapsed subunit pairs rotated 35 to 45° relative to those in the resolvase tetramers. Even though DNA is not present in the complex, the three Sin tetramer structures present in the asymmetric unit of the crystal currently provide the best view of the active site of a serine recombinase in a conformation poised for catalysis. The active site serines together with other key catalytic residues are organized around sulfate ions in the crystal, and docking of the bent uncleaved DNA of the γδ resolvase dimer structure positions the scissile phosphodiester bonds over the sulfates. Rice and coworkers suggest that this rotational state may represent the tetramer conformation that is active for cleavage and ligation (114).

Evidence for a Structural Intermediate in the Formation of the Active Tetramer

Two additional X-ray structures provide evidence that the dimer-tetramer remodeling may occur through a defined intermediate step referred to here as a "pre-activated" tetramer (Fig. 10B). One structure is a tetramer complex of the TP901 serine integrase catalytic domain plus N-terminal portion of its E helix (PDB: 3BVP; 2.1 Å resolution) (163). In the TP901 integrase structure, the four E helices are associated in a conformation similar to the tetrameric structures described above, but the catalytic domains are in an incompatible position for DNA chemistry and have not fully rotated about the hinge immediately prior to the E helices to form the flat interface between rotating dimers. A tetramer structure of another hyperactive γδ resolvase mutant (PDB: 2GM5; 2.1 Å resolution) appears to be a hybrid containing one dimer resembling the pre-activated TP901 conformation and one dimer matching that in the post cleavage resolvase tetramer (162). This asymmetric tetrameric structure may be relevant to the phenotypes of some mutant Hin proteins, particularly those with changes in the E helix region that exhibit uncoupling of double strand cleavages between the two *hix* sites (106, 109). Whereas wild-type Hin promotes near concerted double strand cuts at both *hix* sites, these mutants generate substantial amounts of double strand cuts at only one *hix* site. No Hin mutant, however, has been obtained that uncouples the cleavages of the two DNA strands at a single *hix* site to generate a single strand break; the highly concerted nature of the double strand cleavage reaction implies a coupled conformational change by the subunits of an original dimer. Site-directed crosslinks obtained from stable uncleaved mutant tetramers (section "Key residues controlling conformational changes: locations and properties of gain-of-function mutations that often lead to Fis/enhancer-independence") are consistent with a pre-activated tetramer structure (M.M. McLean, G. Dhar, and R.C. Johnson, unpublished).

The Dimer-Tetramer Remodeling Reaction: Role of the Helix B Region

Remodeling of inactive invertase dimers into the active tetramer is likely to be initiated by coincident scissor-like opening of the two juxtaposed dimer interfaces to expose their hydrophobic surfaces. During formation of the initial pre-activated tetramer, the E helices from synapsed subunits slide along each other over about 15 residues. In the later phase of assembly into the active tetramer, further rotation of the catalytic domains about the hinge located at Hin Ser99 (Gly101 in γδ resolvase) (109, 119, 162) associates the D helices of subunits that form rotating pairs (118) and positions Hin residues Lys47, Arg48, and Lys51 in the B helices of the two enhancer proximal subunits in contact with enhancer DNA between the bound Fis dimers (Fig. 10B, C) (78). Experimental evidence indicates that tetramers with rotating subunit pairs covalently linked across their D helices are competent for DNA cleavage, subunit rotation, and ligation, indicating that the DNA chemical and mechanical steps of the reaction occur within the fully assembled tetramer [(118) and M.M. McLean and R.C. Johnson, unpublished data].

Attractive electrostatic forces between the basic residues in the B helices and the enhancer DNA are postulated to help drive the latter conformational changes to the active tetramer structure (78). In the initial pre-activated tetramer, the B helices from the enhancer proximal Hin subunits are modeled to be oriented at about a 50° angle with respect to the enhancer DNA (Fig. 10B). In the active tetramer, Lys47, Arg48, and Lys51 from both subunits are positioned against the enhancer DNA between the bound Fis dimers (Fig. 10C). Unlike other serine recombinases, all of the Fis/enhancer-dependent serine invertases have basic residues at these positions, although arginines or lysines are found at positions 48 and 51 (Fig. 3). Loss of one of the helix B basic residues has a relatively small effect on Hin-catalyzed inversion rates, but substitution of two of these residues with alanine reduces Fis-activated inversion rates ≥20-fold but does not affect Fis-independent inversion by Hin hyperactive mutants. The locations where the B helices contact enhancer DNA were mapped by coupling the chemical nuclease FeBABE to Hin residues near the C-terminal end of the B helix (78). The hydroxyl radicals cleaved within A/T-rich segments in the center of the enhancer where the DNA minor grooves face the Hin tetramer (Figs 9C and 10C). Substitution of these segments with G/C-rich sequences has a demonstrable but small effect on inversion rates. The modest importance of base sequence, combined with the conservation of either arginine or lysine at the equivalent of residues 48 and 51 within Fis/enhancer-dependent invertases, suggests that nonspecific electrostatic forces are operating during this step.

DNA Topological Changes Introduced by Reactions Initiated within the Invertasome

DNA topological analyses of recombinant products generated by the Hin and Gin have provided crucial structural and mechanistic insights into the reaction. As discussed below, these studies established the starting geometry of DNA strands within the invertasome and provided the first experimental evidence for the subunit rotation model for DNA exchange. They also first revealed the role of the enhancer in regulating the processivity of the subunit rotation reaction and identified the importance of DNA superstructure in mediating the bias for intramolecular DNA inversion over other recombinant products.

A Single DNA Inversion Reaction Results in a Linking Number Change of +4

To determine the change in DNA linking number that accompanies a single DNA exchange reaction resulting in inversion, reactions were performed on unique negatively supercoiled plasmid topoisomers. High resolution electrophoresis in agarose gels containing chloroquine revealed that the overwhelming majority of inversion products have lost four negative supercoils and are unknotted. This was first demonstrated for Gin under standard inversion reaction conditions (164, 165) and was later shown for Hin under conditions ensuring only a single recombination reaction, which was necessary because of the robust DNA relaxing activity (166). The +4 linking number change was interpreted by Kanaar and Cozzarelli as having been generated by the following mechanism (Fig. 11) (105, 164). Two negative nodes (DNA crossings) are entrapped within the initiating recombination complex, consistent with the geometry of DNA strands at the base of a plectonemic supercoiled branch. DNA exchange by double strand breaks at the two recombination sites followed by a 180° clockwise rotation of each duplex above the crossing enhancer DNA and then ligation introduces a negative crossover node, but is also accompanied by a half turn of twist on each duplex thereby cancelling out the crossover node. However, because recombination switches the geometry of the entrapped nodes from −2 to +2, the linking number changes by +4. Thus, the energetically favorable reaction, in which there is a net loss of four negative supercoils together with the release of twist, is the clockwise rotation of DNA strands. Because the recombinase subunits are covalently linked to the DNA ends by the serine phosphodiester bond, it was proposed that the subunits must also undergo the equivalent 180° clockwise rotation within the invertasome complex.

Additional DNA Rotations Generate Specific DNA Knots

DNA knot nodes in plasmid DNA can be resolved by agarose gel electrophoresis after nicking the DNA to remove supercoils. Under standard *in vitro* conditions, Hin and Gin generate very few knotted products because most reactions ligate after a single 180° rotation (Fig. 12) (164, 167). Likewise, a comparison of knotted to inverted products of Hin-catalyzed reactions *in vivo* under conditions where topoisomerase IV, the major unknotting enzyme in *E. coli*, is blocked indicated that <5% of *in vivo* recombination reactions generate knots (166, 168). By contrast, Hin reactions performed on DNA substrates where one of the *hix* sites contains a mutation within the central 2 bp crossover region exhibit robust knotting. Reactions on these substrates are unable to ligate in the recombinant (inverted) orientation after a single 180° rotation because the 2 bp overlap region cannot base pair (Fig. 13) (166, 167). A second 180° rotation restores base pairing and enables ligation, but the two processive rotations generate a

3-noded knot (trefoil) with the invertible segment repositioned into the parental orientation (Fig. 12). Most of the products from a single reaction on a substrate with mutant core nucleotides are trefoils without inversion. However, occasionally the reaction fails to ligate after two rotations and undergoes a third rotation where it again cannot ligate over the mismatched crossover region and thus undergoes a fourth rotation that can support ligation back into the parental orientation; the four subunit rotations starting in the invertasome structure generate a 5-noded twist family knot (95, 101, 167). A distinguishing feature of Hin reactions on crossover mutant substrates is the absence of 4- and 6-noded twist knots (note that 6-noded compound knots can be generated from two independent reactions that each introduces three nodes). Reactions with wild-type recombination sites, however, do form twist knots with even numbers of nodes, albeit rarely. Knot formation *in vitro* by Gin has been most extensively studied in reactions employing low Mg^{2+} and low salt concentrations, conditions that inhibit ligation and destabilize the Fis/enhancer-bound invertasome complex (95). The stereostructures of knots generated by Hin and Gin have been determined by electron microscopy (95, 101, 167). In all cases for Hin and the vast majority of cases for Gin, the structures of the knots, including the chirality of the nodes, are compatible with only clockwise rotations starting with the −2 topology of DNA strands present in the starting invertasome. The structures of the DNA knots thus provide additional strong support for assembly of the invertasome at a DNA branch and the subunit rotation model for DNA exchange.

Short DNA lengths between one of the recombination sites and the enhancer inhibit processive rotations because of torsional constraints introduced by the multiple DNA windings (Fig. 12). This has been demonstrated for Hin both *in vitro* and *in vivo* where substrates containing mutant crossover regions were shown to produce few knots when the separation between *hixL* and the enhancer is the wild-type length of 99 bp as compared to substrates with much longer spacings (166, 167). Reactions on the mutant substrates forming invertasomes with short loops accumulate significant amounts of broken DNA molecules because the mismatched DNA ends are held in a nonligatible configuration after a single exchange (100). The low frequency knots that are generated under these conditions tend to be highly complex because of release of the Fis/enhancer from the invertasome complex

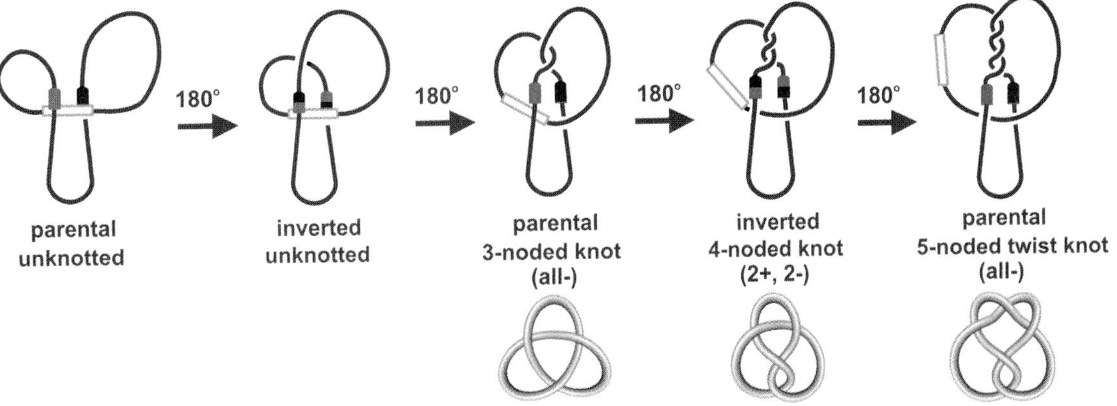

Figure 12 DNA topological changes during processive recombination by serine DNA invertases. Initially, the recombination sites (green and blue) are assembled at the enhancer DNA (brown) in the invertasome structure. The initial assembly depicts short and long loops between the recombination sites and enhancer as found in the Hin and Gin systems. The first DNA exchange by the equivalent of 180° rotations of DNA strands covalently linked to recombinase subunits generates inversion; these molecules have lost four negative supercoils (not shown) but no knot nodes are introduced. Subsequent processive rounds of DNA exchange generate knots of increasing complexity with the orientation of the invertible segment alternating between parental and inverted; these have all lost two negative supercoils. Multiple windings of DNA (processive DNA exchanges) are torsionally restricted as long as the enhancer remains associated with the recombination sites, but release of the enhancer removes this constraint. The structures of the knots, which have been confirmed by electron microscopy, are depicted below.
doi:10.1128/microbiolspec.MDNA3-0047-2014.f12

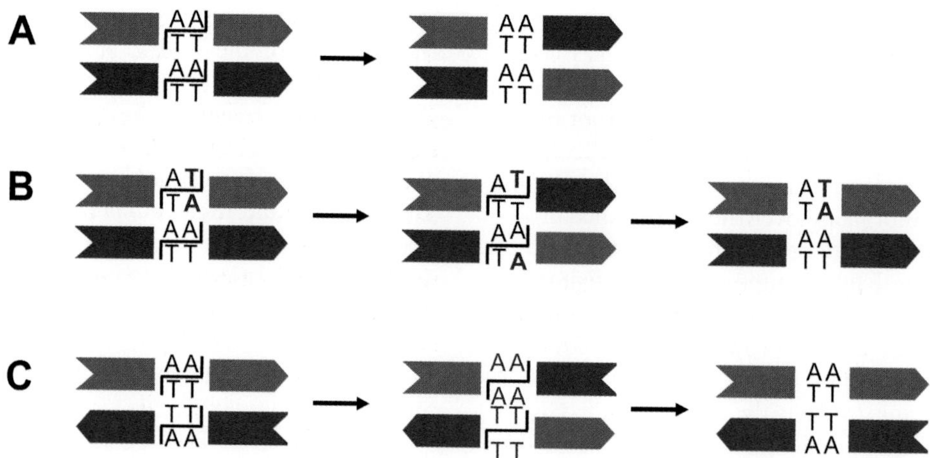

Figure 13 Synapsis and DNA exchange of recombination sites. (A) Synapsis of wild-type recombination sites in a parallel configuration as in the standard invertasome complex. DNA cleavage followed by exchange and ligation inverts the intervening DNA segment. (B) Synapsis where one recombination site contains a mutation (red) within the core nucleotides. DNA cleavage and a single exchange result in unpaired core nucleotides that cannot ligate. A second exchange is required for ligation, which reorients the intervening DNA segment into the starting (parental) configuration and generates a 3-noded knot (Fig. 12). (C) Synapsis of wild-type recombination sites that are oriented in a directly repeated configuration. Most of the time the sites align in an antiparallel configuration within an invertasome structure (see Fig. 16), and the core nucleotides cannot base pair after the first DNA exchange. Ligation can occur after a second exchange back into the parental orientation creating a 3-noded knot as in panel B.
doi:10.1128/microbiolspec.MDNA3-0047-2014.f13

with the consequent loss of torsional constraints (see section D.10). Knotted DNA molecules generated by at least eight processive exchanges have been observed from Hin and Gin reactions (95, 167). It has been suggested that Gin can also recombine DNA via alternative pathways that are topologically distinct from the primary pathway, especially when the loop sizes are short (101).

Evidence for Subunit Rotation at the Protein Level

Time-resolved site-directed crosslinking experiments using Hin mutants with strategically placed cysteines have provided direct evidence for subunit rotation during the recombination reaction. Initially, these experiments employed tetramers assembled with the Hin-H107Y Fis/enhancer-independent mutant in covalently-linked cleavage complexes with radiolabeled linear DNA substrates of different lengths in order to tag individual subunits (158). Crosslinks formed within cleavage complexes assembled using proteins containing cysteines located shortly prior to (Hin residues 94 or 99) or within the N-terminal end of helix E (Hin residue 101) link *trans*-diagonally located subunits (Fig. 14). Crosslinks formed early in the reaction are between subunits originating from synapsed dimers, but after a few seconds, subunits from the original dimers become crosslinked because of subunit translocation within the tetramer. The Hin tetramer models based on the resolvase post-cleavage structures suggest that even the starting configuration of subunits would require a small amount of rotational movement in order for the sulfhydryl groups to be within the 8 Å crosslinking distance for residues 94 (Fig. 14) and 99 (117).

Strikingly, robust crosslinking also occurs between helically phased cysteines on the C-terminal third of helix E (117, 118, 158). These residues are located up to 70 Å away from each other in the Hin tetramer model (Fig. 14). In order for the cysteines across the rotational interface to be in position for crosslinking, one pair of synapsed subunits would be required to rotate 90±15° in either the clockwise or counterclockwise direction. Nevertheless, subunits from the original dimers (requires counterclockwise rotation) or from the synapsed dimers (requires clockwise rotation) crosslink within seconds of Hin cleaving both *hix* sites.

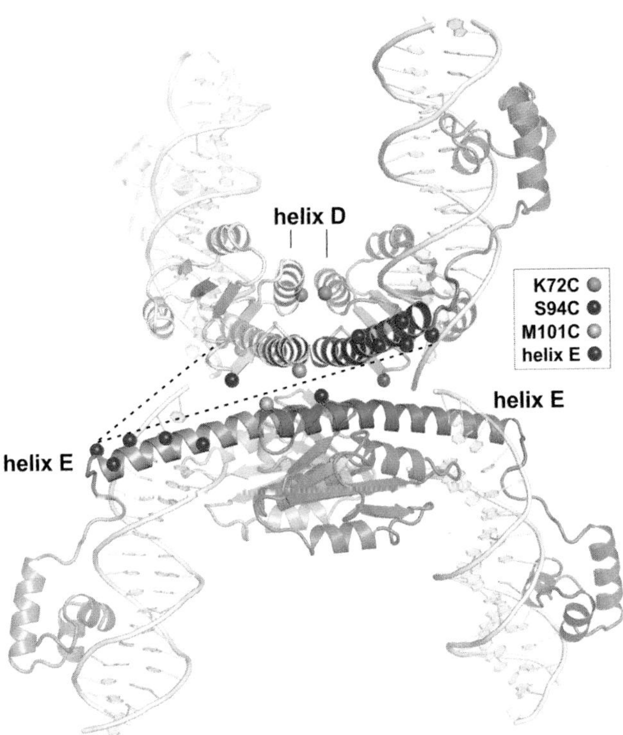

Figure 14 Site-directed crosslinking demonstrating subunit rotation within active Hin tetramers. Model of the Hin tetramer based on resolvase tetramer structures. The view is of the rotation interface with residues converted to cysteine for site-directed crosslinking rendered as colored spheres (Cβ atom). Rotations refer to the top subunit pair (gold and purple) relative to a fixed bottom subunit pair. M101C residues (orange spheres) from the gold and blue *trans*-diagonally located subunits are within range for crosslinking with a bis-thio reactive reagent containing an 8 Å spacer but becomes optimally positioned for crosslinking after a ~20° clockwise (cw) rotation. Initial crosslinks at residue 101 are between gold and blue subunits, but with time, purple and blue cross-linked subunit products that are the result of subunit exchange become equally represented. Crosslinking between S94C residues (magenta spheres) requires either a counter-clockwise (ccw) rotation of at least 20° (optimal between 40 to 70°) to link *trans*-diagonally positioned gold and blue subunits or a clockwise rotation of about 270° to link purple and blue subunits from an original dimer. Fis/enhancer-activated reactions on supercoiled DNA primarily form 94-94 cross-links between purple and blue subunits (cw rotation), whereas crosslinks between gold and blue subunits (ccw rotation) are overrepresented initially in Fis/enhancer-independent reactions but then both products become equally represented. Red spheres mark residues within the C-terminal third of helix E that support robust crosslinking even though they can be up to 70 Å from their nearest partner; rotations of 90±15° in the cw or ccw direction are required to generate crosslinks. Fis/enhancer-activated reactions on supercoiled DNA primarily form crosslinks between purple and green subunits (cw rotation) containing helix E cysteines, whereas Fis/enhancer-independent reactions without DNA supercoiling form cross-links between gold and green subunits (ccw) or purple and green subunits (cw) with the same kinetics. Dashed lines highlight the positions of Q134C residues at the C-terminal ends of helix E. The kinetics of crosslinking between residues at positions 101, 94, and the C-terminal end of helix E provide strong support for bidirectional subunit rotation within Fis/enhancer-independent reactions and for primarily single round cw rotation in Fis/enhancer-dependent reactions on negatively supercoiled DNA. Crosslinks between helix D residues like K72C (green spheres) readily form upon full assembly of the tetramer and do not block subunit rotation or DNA ligation. Details are provided in references 117, 118, and 158. doi:10.1128/microbiolspec.MDNA3-0047-2014.f14

Clockwise and counterclockwise rotation products exhibit nearly equivalent rates of crosslinking suggesting that both directions occur equally efficiently within the Hin tetramer complex on linear DNA. Bidirectional rotation within DNA invertase tetramers is also consistent with earlier topological analyses of reactions on relaxed plasmid DNA by a Fis-independent Gin mutants (169). Hin tetramers crosslinked between residues on the C-terminal end of the E helices are unable to support DNA ligation because the rotating subunit pairs are covalently locked about half way through a complete rotation leaving the DNA ends far apart from each other. As noted above, crosslinking between helix D residues 72 (Fig. 14) or 76 covalently links the synapsed rotating pairs of subunits with no effect on the subunit rotation and ligation steps (118).

Crosslinking reactions on supercoiled DNA using Fis/enhancer-dependent Hin recombinases generate a different kinetic profile than those with DNA fragments or relaxed plasmid DNA (117). The profiles are consistent with the vast majority of complexes undergoing a rapid but single clockwise translocation of subunits (Fig. 14). Most complexes remained in a singly rotated state for at least 20 min, even with mutant substrates where the crossover regions cannot base pair (Fig. 13B). Crosslinking reactions have also been performed on positively supercoiled DNA preparations. Although the Hin-catalyzed inversion reactions are less efficient than on negatively supercoiled DNA, a strong bias for rotation in the counterclockwise direction was observed. These results, combined with the evidence for equivalent rates of bidirectional rotations on linear fragments or relaxed plasmids, indicate that DNA supercoiling is determining the direction of subunit rotation in reactions initiated from invertasomes. As noted above, the clockwise direction of rotation is energetically favorable on native negatively supercoiled DNA because it is accompanied by a loss of supercoils and helical twist.

Structural Nature of the Rotation Interface

The rotation interface separating the synapsed subunit pairs is almost exclusively composed of residues from the first half of the E helices. Hin and Gin share only about 25% sequence identity with γδ resolvase over this surface. Nevertheless, the molecular character of the rotating interfaces of Gin (120), Hin (117), and γδ resolvase (119, 162) are very similar (Fig. 15A, B, C). In each case, the interface surface is circular (diameter 20 to 25 Å), relatively flat, exclusively aliphatic, and is surrounded by mostly acidic residues that are somewhat conserved between the invertases and resolvases. The interface surface areas of different rotational conformers vary with the greatest surface overlap where the cleaved DNA ends are linearly aligned (0 to 10° clockwise rotation) and where the E-helices between rotating pairs are aligned in a parallel/antiparallel configuration (~100° clockwise rotation) (Fig. 15) (117, 119). The overall buried surface areas, combined with shape complementarity and van der Waals interactions between specific surface residues, may function as energetic barriers or pause sites in the rotation reaction. A pause in subunit rotation over the DNA-aligned conformer may facilitate the ligation step. Rotational pausing at helix E-aligned conformers may contribute to the robust crosslinking efficiencies observed between cysteines on the C-terminal third of the E helices (117, 118). Crosslinking reactions performed in high and low ethylene glycol concentrations suggest that high ethylene glycol favors Hin tetramers to be in the helix E-aligned rotational conformer over the DNA-aligned conformer, providing a partial explanation for why ligation is inhibited and the tetramer intermediate is stabilized by high ethylene glycol (117).

Factors Regulating Processivity of Subunit Exchange

The protein crosslinking experiments together with the topological changes in the recombinant products determined both *in vitro* and *in vivo* on wild-type substrates demonstrate that mechanisms exist to limit subunit rotations to a single exchange. Nevertheless, additional exchanges can occur with high efficiency under certain conditions, the most biologically important is when sites synapse in the incorrect orientation or pseudosites synapse that have different nucleotides at their crossover site (e.g., Fig. 13B, C). The identity of the 2 bp crossover sequence upon the first rotation is the only check after the initial DNA binding step to ensure that the reaction is between the correct DNA sites that are synapsed into the correct orientation. The ability to undergo additional forward subunit exchanges enables ligation back into the parental DNA sequence, thereby avoiding a chromosome break. Rare examples have been reported from *in vivo* reactions between pseudo-sites where ligation must have occurred over mismatched crossover regions to generate sequence conversions (170, 171).

One major mechanism inhibiting multiple rotations of subunits is the interaction between the Fis/enhancer element and the recombinase complex in the invertasome structure (101, 117, 166, 167). The short DNA loop that is trapped with the native Hin substrate inhibits the multiple DNA windings that occur coincidently with each subunit exchange (Fig. 12). Thus, formation of knotted products on substrates with sequence differences at the crossover dinucleotides requires an expanded length of DNA between the enhancer and *hixL*. Crosslinking reactions also demonstrate that the length of the DNA loop between the closest *hix* site and the enhancer controls the ability of the subunits to undergo multiple exchanges (117). Conditions that destabilize the connections between the Fis/enhancer element and recombinase complex once the active invertasome has formed result in processive subunit exchanges. These include (i) mutations in the contact sites between Hin and Fis (78, 166), (ii) mutations in the helix B region of Hin that contacts the enhancer DNA (78), and (iii) positioning of the enhancer close to, but not in helical phase with, a *hix* site (117). In each case, reactions generate an increased frequency of higher-order knots, and where tested by crosslinking, generate a random mix of products that are indicative of multiple subunit exchanges even when the shortest loop size is ≤150 bp.

Even when the lengths of the DNA segments separating the enhancer from the recombination sites are sufficiently large such that multiple subunit rotations would not be expected to be torsionally constrained by DNA windings, most Hin reactions generate unknotted inversion products indicative of a single exchange (166, 167). For example, <3% of *in vivo* reactions on plasmid substrates in which the shortest loop size would be ~700 bp were estimated to have undergone multiple exchanges. Likewise, reactions on substrates with mutant crossover regions primarily ligate after two exchanges and only very rarely exceed four exchanges. These features suggest that additional mechanisms exist to restrict subunit rotations such that ligation will occur after exchange if the overlapping nucleotides can base pair.

Additional support for a mechanism limiting subunit rotations comes from single-DNA molecule experiments (B. Xiao, R.C. Johnson, and J.F. Marko,

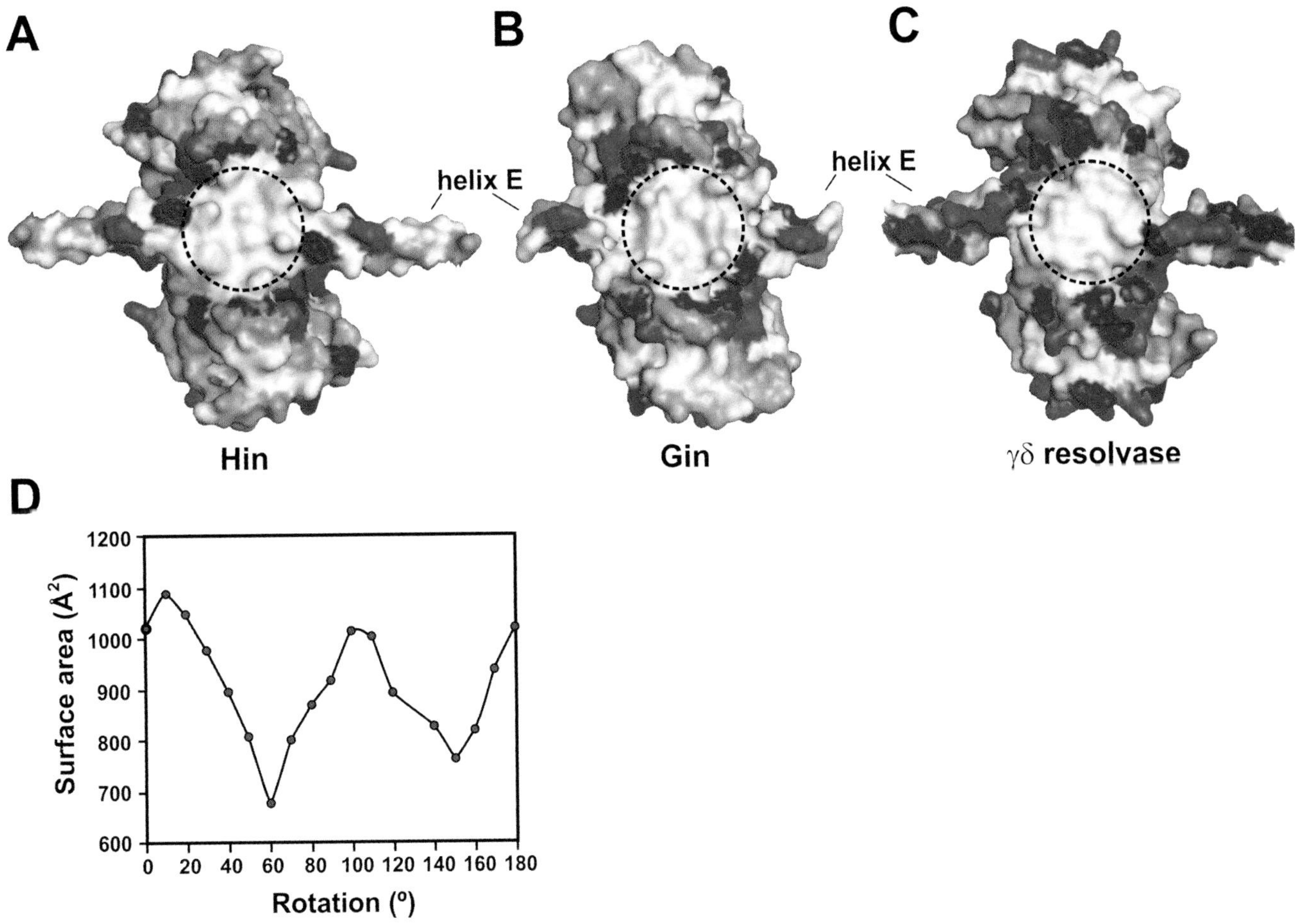

Figure 15 The subunit rotation interface. Surfaces of rotating subunit pairs from the (A) Hin model (residues 2 to 134 based on PDB: 2GM4), (B) Gin X-ray structure (residues 2 to 125, PDB: 3UJ3), and (C) γδ resolvase X-ray structure (residues 1 to 132, PDB: 2GM4) are shown after alignment. Hydrophobic residues are colored yellow, acidic residues are red, basic residues are blue, and polar residues are green. A 1.6 Å probe was used to render the surfaces. Dashed circles demarking the rotating interface have a diameter of ~20 Å. (D) Surface area overlap calculated for different clockwise rotational conformers from the Hin model; γδ resolvase gives a very similar pattern (117, 119). Rotations of around 0 to 10° correspond to conformers where the DNA ends are in-line for ligation and conformers around 100° have the E-helices between dimer pairs in a parallel/antiparallel configuration. The Hin models are based on γδ resolvase structures (shown here based on 2GM4); comparison of subunit structures with those from resolvase tetramers (2GM4 or 1ZR4) give RMSD values of <0.7 Å over the peptide backbone (residues 1 to 120). Subunits from the Gin tetramer structure (3UJ3) exhibit RMSD values of 1.3 to 1.5 Å over the peptide backbone atoms from residues 3 to 120 and 1.1 to 1.4 Å over just the catalytic domains from the Hin models or γδ resolvase tetramers (1ZR4 or 2GM4). Much of the difference between Gin and resolvase structures or Hin models is over poorly resolved loops connecting β1 to αA and β2 to αB. doi:10.1128/microbiolspec.MDNA3-0047-2014.f15

unpublished). Reactions employing a Fis/enhancer-independent Hin mutant were performed between two linear DNA duplexes containing a *hix* site that were each attached to a glass surface on one end and a paramagnetic bead on the other. The DNA molecules were coiled about each other by winding of the beads by a magnet. Under conditions inhibiting ligation, the Hin mutant unwound the braided DNA molecules in discrete steps whereby unwinding events were separated by time periods of seconds to minutes. The profiles are unlike those observed in similar experiments employing the serine integrase Bxb1, where an apparently ungated reaction results in full relaxation of the braided DNA molecules (172). The single-DNA molecule experiments provide further support for the subunit rotation mechanism and strongly imply that an intrinsic feature of the Hin tetramer structure mediates pausing of the rotation reaction.

Rotational pausing over the surface overlap maxima between rotating subunit pairs where the DNA ends are linearly aligned for ligation is one mechanism that may contribute to gating of the subunit rotation reaction. Consistent with this idea, certain Hin mutations in large hydrophobic residues within the rotating interface are defective in ligation (Y. Chang and R.C. Johnson, unpublished). Li et al. (119) have suggested that charge interactions between two arginines on the C-terminal end of the helix E on one subunit and two aspartic acids in the catalytic domain near the beginning of helix E on the subunit across the rotating interface may pause the rotation of γδ resolvase tetramers at the ligation-ready conformer. The Hin and Gin sequence is compatible with potential formation of one of these ion pairs, Hin residues Asp93-Arg123. The properties of Hin mutations at these residues imply that both play important roles early in the inversion reaction, but there is currently no evidence for a role of these residues in controlling subunit rotation (115, 118) (Y. Chang, S. Merickel. and R.C. Johnson, unpublished).

DNA Supercoiling and Branching

DNA supercoiling plays an essential role in the Fis/enhancer-dependent reaction by localizing the three interacting sites at the base of a DNA branch with the correct −2 topology. Solution experiments and even assays employing proteins bound to relaxed DNA plasmids have failed to provide evidence for meaningful Fis-Hin interactions (104) (R.C. Johnson, unpublished). Simulations by Vologodskii and coworkers predict that supercoiling increases the local concentration of two sites about 100-fold and three sites at a branch by more than 1000-fold over nonsupercoiled DNA (173). Thus, the conformational energy present in plectonemically supercoiled DNA is probably required to stabilize interactions between DNA-bound invertases and the Fis/enhancer element, which are otherwise too weak to form. The ability of hyperactive mutants of Hin and Gin to efficiently recombine DNA in Fis/enhancer-independent reactions without supercoiling indicates that supercoiling is not required for the protein conformational changes, chemical steps, or potential topological steps like DNA unwinding.

Electron microscopy and long-chain DNA modeling studies generally agree that naked supercoiled DNA is branched every 2±1 kb (173, 174, 175, 176). The DNA duplexes within branched molecules are under constant thermal fluctuation and can readily undergo slithering motions to align distant sites (176, 177, 178). The extent of branching does not vary significantly over physiological levels of supercoiling and may actually decrease at high supercoiling densities (σ) (174, 175, 179). Formation of active invertasomes begins at $\sigma \sim -0.025$ and increases up to $\sigma \sim -0.06$ (33) (I. Moskowitz and R.C. Johnson, unpublished). This correlates with the reduction in the superhelix diameter over this range of σ (175); likewise, a decrease in the separation of DNA duplexes at a branch point with increasing σ would also be expected. Computer simulations give a superhelix diameter of about 100 Å at $\sigma = -0.06$ where invertasome assembly is most efficient (174). In the invertasome model, the separation between *hix* sites where they cross the enhancer is about 75 Å. Hin reactions employing mutant enhancers that contain an additional helical turn of DNA, and thus would exhibit a separation between *hix* sites of around 110 Å, remain remarkably efficient (131).

Once Hin dimers are localized together at the enhancer, it is possible that supercoiling contributes to remodeling into the active tetrameric structure utilizing energy in the underwound supercoiled DNA. Alternatively, energy associated with the mobility of the β-hairpin arms of Fis may assist in the dimer-tetramer remodeling. A prominent role for such forces, however, seem unlikely given that enhancers with additional 10 bp separation between Fis sites function within 30 to 50% of the activity of wild-type enhancers (131). Moreover, Fis/enhancer activation proceeds efficiently when the mobility of the β-hairpin arms is restricted due to disulfide crosslinks between arm residues 15 or 18 (137).

Invertasomes could assemble via several potential pathways. One is a near simultaneous collision of the three sites slithered into supercoiled DNA branch. A second pathway involves sequential looping of individual recombinase bound recombination sites into the Fis/enhancer. *In vivo* assays employing transcriptional repression reporters have been interpreted as providing evidence for single loops between a Fis binding site and a *hix* site, but recruitment of pseudo-*hix* sites in these experiments, which are observed *in vitro*, cannot be ruled out (115, 180). A third pathway posits an initial interaction between recombinase dimers bound to recombination sites in a manner that entraps a branch, followed by joining of the Fis/enhancer on the branched DNA to the paired-recombination site complex. *In vivo* data from transcriptional repression assays are also consistent with formation of paired-*hix* complexes (115, 180). In addition, glutaraldehyde readily crosslinks paired-*hix* complexes (104), but there is little current *in vitro* evidence that paired-*hix* complexes are precursors for invertasome formation and not the result of nonspecific crosslinking.

DNA modeling studies by Halford and co-workers have concluded that there is only a 10-fold difference in the rates of 3-site as compared with 2-site synapses on supercoiled DNA. They modeled three sites on a 4.2 kb supercoiled DNA converging to a branch point within 0.1 to 1 s (181). Tripartite Hin invertasomes have been measured to assemble within seconds under certain conditions (106). Taken together, the available experimental data combined with DNA modeling studies appears most compatible with a near 3-site collision at a supercoiled branch being the primary pathway for invertasome assembly.

The Specificity for DNA Inversion

The conformational energy of supercoiled DNA also mediates the large bias for promoting inversions over deletions between appropriately oriented recombination sites in *cis*. Formation of deletions by Hin is completely dependent upon Fis, but only about half of the Gin-catalyzed deletion reactions go through a Fis-dependent pathway (169). Significantly, all of the rare deletion products generated by Hin are catenated circles and in all analyzed cases were found to be linked only once (67). Alignment of directly repeated recombination sites for deletion within an invertasome-like structure at a plectonemic branch demands that an additional loop must be introduced into the DNA (Fig. 16), which would be energetically costly. This alignment therefore occurs only rarely, but the product of a single clockwise rotation is the observed singly interlinked catenane. The Hin-catalyzed deletion reaction is highly dependent upon HU, which fits with the additional DNA bending that is required to assemble the deletion complex.

Directly repeated recombination sites overwhelmingly prefer to synapse in the standard invertasome configuration, but this generates an antiparallel orientation of sites (Fig. 16). An invertasome complex assembled with directly repeated recombination sites results in unpaired crossover nucleotides that cannot ligate after a single subunit rotation, and therefore, must undergo an even number of rotations to enable ligation back into the parental orientation (Fig. 13C). In Hin- and Gin-catalyzed reactions, both the robust rate of formation and structures of the knots produced by substrates with recombination sites oriented for deletion mimic the knotting

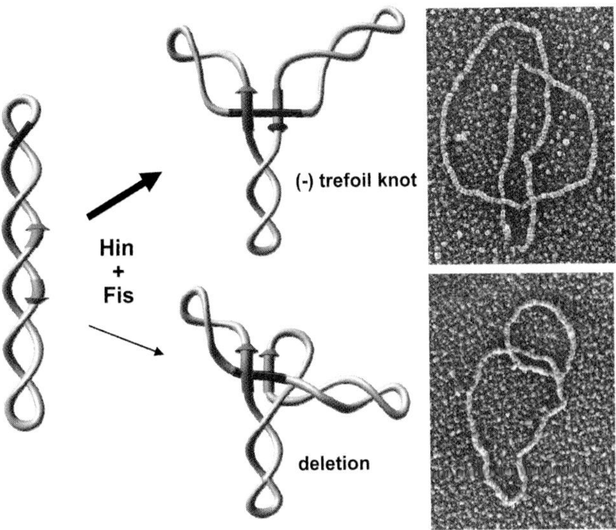

Figure 16 Products formed on substrates containing directly repeated recombination sites. Most synaptic complexes assemble in an invertasome structure as diagrammed in the top pathway, but the recombination sites are in an antiparallel orientation. The unpaired core nucleotides after a single DNA exchange prevent ligation (Fig. 13C). Ligation after two exchanges results in a knot containing three negative nodes (Fig. 12), as shown in the electron micrograph of a trefoil generated by Hin. In the bottom pathway, which occurs rarely, the recombination sites assemble in parallel orientation that requires an additional DNA loop. This structure is energetically unfavorable on a negatively supercoiled substrate. A single DNA exchange results in a singly linked catenated deletion product, as shown in the micrograph. The DNA molecules were coated with RecA protein to facilitate visualization of the DNA nodes.
doi:10.1128/microbiolspec.MDNA3-0047-2014.f16

behavior of substrates with mutant core sites in the standard inverted orientation (67, 95).

Further support for the supercoiling-directed alignment of recombination sites into active complexes comes from experiments employing sites in which the 2 bp crossover regions between recombining sites are internally symmetric (e.g., A/T-T/A). Recombination reactions between such sites overwhelmingly generate inversions and only rarely deletions, regardless of their relative orientations on the DNA molecule (67). These results confirm that site orientation is solely specified by the two nucleotides at the crossover region and that the configuration of sites that will generate inversion upon DNA exchange is preferentially assembled within the active synaptic complex (invertasome). Recombination site orientation by other serine recombinase reactions is also exclusively determined by the crossover dinucleotide sequence and tested only at the ligation step after DNA exchange (182, 183, 184).

Key Residues Controlling Conformational Changes: Locations and Properties of Gain-of-Function Mutations That Often Lead to Fis/Enhancer-Independence

Identification and properties of Fis/enhancer-independent mutants

Gain-of-function mutants that exhibit altered regulatory properties have been isolated in the Hin, Gin, and Cin serine invertases. The mutations identify important interactions within or between subunits that control the conformational rearrangements that occur during the dimer to tetramer remodeling reaction. As discussed above, strong Fis/enhancer-independent mutants of Hin and Gin have been invaluable for structural and mechanistic studies where regulatory controls imparted by the Fis/enhancer element and DNA supercoiling are not relevant. Moreover, comparison of the Fis/enhancer-dependent verses -independent reactions has provided valuable insights into the functions of the Fis/enhancer control element and DNA topological forces involved in regulating the specificity for inversion.

The first reported mutants that could function without the Fis/enhancer regulatory element were in Gin (H107Y, M115V, and weakly F105V) and Cin (H107Y and R72H) (Hin residue numbering is used throughout this discussion) (65, 169, 185). H107Y and M115V are also among the strongest Hin hyperactivating mutations, but changes in over 20 additional residues have been shown to confer altered regulatory properties (Fig. 3 and not shown) (106, 107, 109). Hin gain-of-function mutants have been isolated by genetic screens for mutants that (i) catalyze inversion in *fis* mutant hosts, (ii) activate the cellular SOS DNA damage response, or (iii) suppress inactive or weakly active Hin mutants, or by site-directed mutagenesis based on structure modeling or information from resolvase systems.

A hallmark of strong hyperactive mutants is their newfound ability to catalyze deletions and intermolecular fusions, in addition to promoting inversions, all without the requirement for Fis or the enhancer (65, 107, 169, 185). *In vitro* analyses also demonstrate that they no longer require DNA supercoiling and can synapse and recombine short oligonucleotide substrates containing their cognate recombination site sequence. They are further activated by the Fis/enhancer system provided the DNA is supercoiled, and under these conditions, recombination through an invertasome complex dominates (109, 169) (J. Heiss, Y. Chang, and R.C. Johnson, unpublished).

DNA Topological Properties of Fis-Independent Reactions

Comprehensive topological studies on reactions catalyzed by the strong Fis-independent Gin-M115V and Hin-H107Y mutants have been performed with similar conclusions (101, 169) (S. Merickel, J. Heiss, and R.C. Johnson, unpublished). Nearly all reactions on negatively supercoiled DNA in the presence of Fis initiate from a synaptic complex trapping two negative nodes indicative of the invertasome structure. However, unlike the wild-type enzymes, most reactions undergo multiple subunit rotations to generate a series of increasingly complex knotted products. The more processive nature of the subunit exchange reaction is probably due to the mutations stabilizing the active tetramer conformation (see below). In the absence of Fis, Gin-M115V and Hin-H107Y also preferentially promote recombination through a −2 synaptic complex to give a linking number change (ΔLK) of +4 after the first subunit exchange (169) and generate knotting profiles on substrates with wild type or mismatched recombination sites that reflect processive exchanges (101) (S. Merickel, J. Heiss, and R. Johnson, unpublished). However, unlike the wild-type enzymes, a diversity of other synaptic complex configurations also generates recombinant products. These include a substantial number of simple synapses between recombination sites on apposed supercoiled strands where no topological change occurs upon DNA exchange (ΔLK=0 complex) and a lower number of complexes that trap from 3 to at least 12 (-) supercoiled nodes via random intramolecular collision pathways. Many of the diverse product structures generated by Gin-M115V have been confirmed by electron microscopy (101). Productive synapses on nonsupercoiled substrates occur with equal efficiencies with recombination sites aligned in a parallel or antiparallel orientation (Fig. 13A, C, respectively) (107, 169). Thus, in the absence of the Fis/enhancer system, the hyperactive mutants have lost the strict control of recombination site synapsis, and this enables all types of recombination products to occur, including those from intermolecular reactions. The overrepresentation of intramolecular recombination by synaptic complexes that trap two (-) nodes presumably reflects the propensity for branching within plectonemically supercoiled DNA and could also reflect the contribution of helix B-DNA interactions in promoting tetramer assembly at a branch.

Residues Controlling Conformational Changes Required for Active Tetramer Assembly

The locations of regulatory mutations and the biochemical activities of the mutant proteins have identified key interactions between residues that control steps of the dimer-tetramer remodeling reaction. These include intrasubunit and intersubunit interactions that stabilize the dimer conformation, contacts that occur during the remodeling process, and contacts that are of particular importance in stabilizing the active tetramer conformation (109). Individual residues can participate in more than one step, and some of the hyperactivating mutations appear to gain new contacts that stabilize the preactivated or activated tetramer. As might be expected, there is considerable overlap of positions where mutations generate hyperactive serine invertases and resolvases (186, 187, 188, 189). A subset of the characterized DNA invertase hyperactivating mutations are discussed below focusing on those that have the largest effects and reveal particularly important features controlling the reaction. The proposed functions of the residues rely on structural models and morphed intermediates of Hin (78, 109), as discussed above, and the Gin tetramer structure (120).

Helix E Residues Involved in Intermolecular Interactions

Most of the strongest hyperactivating mutations involve residues on helix E, which undergoes extensive structural changes during the dimer-to-tetramer conversion. Whereas most mutations within helix E block steps after DNA binding, mutations in seven different residues can result in hyperactivating mutations. One of the strongest is H107Y that has been isolated in Hin, Gin, and Cin (65, 106, 185) and studied intensively in Hin (107, 118, 158). The histidine imidazole ring is predicted to be hydrogen bonded and in van der Waals contact with residues of the partner subunit on the surface of the dimer (Fig. 17A). The hyperactivating tyrosine side chain is predicted to clash or reorient into the solvent, consistent with the mild destabilization of the dimer interface by the mutant (109). Upon formation of the initial pre-activated tetramer, the mutant tyrosine rings between subunits of newly synapsed pairs are predicted to adopt a pi-stacked conformation (Fig. 17B), thereby stabilizing the tetrameric synaptic complex as experimentally observed. A glutamine to arginine substitution at the analogous position of the Sin recombinase also leads to a strong hyperactivating mutation, and a crystal structure of this mutant shows the arginine side chains from subunits of each rotating pair in a pi-stacked configuration (114). In the low resolution Gin tetramer X-ray structure, the native imidazole side chain is modeled to be packed against the helix E backbone (120). It seems likely that the larger tyrosine, with its greater pi-stacking potential (190, 191), adopts the alternate conformation. A phenylalanine at this position confers a milder hyperactive phenotype, but all other substitutions tested inactivate the invertase (65, 106, 185, 192).

Methionine 115 to valine creates another very robust Fis/enhancer-independent mutant that has been studied extensively (65, 101, 109, 120, 169). During formation of the tetramer, the Met115 side chain is modeled to move out of a cleft in the dimer partner subunit, slide along the dimer interface, and insert into a new pocket organized by residues from the N-terminal end of helix E of the synaptic subunit partner to both stabilize the synaptic subunit pair and contribute to the rotating interface of the tetramer (109, 120) (see Movie 2 in reference 109). Hin mutations within residues of the dimer partner cleft (V71 and Leu64) and synaptic subunit pocket (M101) also lead to hyperactivity (109). In Hin, the M115V mutation strongly destabilizes dimeric interactions and results in efficient formation of DNA-cleaved tetrameric synaptic complexes. In contrast, M115I generates tetrameric synaptic complexes that are inefficiently cleaved and less able to support complete recombination without Fis/enhancer-activation. The mutant isoleucine side chain may impede latter rearrangements to the fully active tetramer conformation or alter the tetramer structure for the cleavage step.

The Helix E Phenylalanine Triad and Phe88 Catalytic Domain Pocket

An invariant triad of phenylalanines between residues 104 to 106 on helix E of serine invertases play a critical role in controlling the conformational changes required for formation of the active tetramer (109). In the dimer, Phe105 is inserted into a hydrophobic pocket of the catalytic domain of its own subunit (Fig. 17C). This pocket is organized by Phe88 and a series of surrounding hydrophobic residues. Hin hyperactivating mutations have been isolated in six of these, and in all cases, they are changed to a smaller hydrophobic side chain. During tetramer formation, the Phe105 side chain rotates out and the Phe106 side chain inserts into this pocket (Fig. 17D, E) (see Movie 3 in reference 109). The Phe105 ring is modeled to become stacked on the Phe105 ring across the rotating interface from the diagonally oriented subunit (Fig. 17B). In the Gin tetramer, which is rotated by 26°, the Phe105 rings have separated (120). Substitutions at Phe105 in Hin generate a

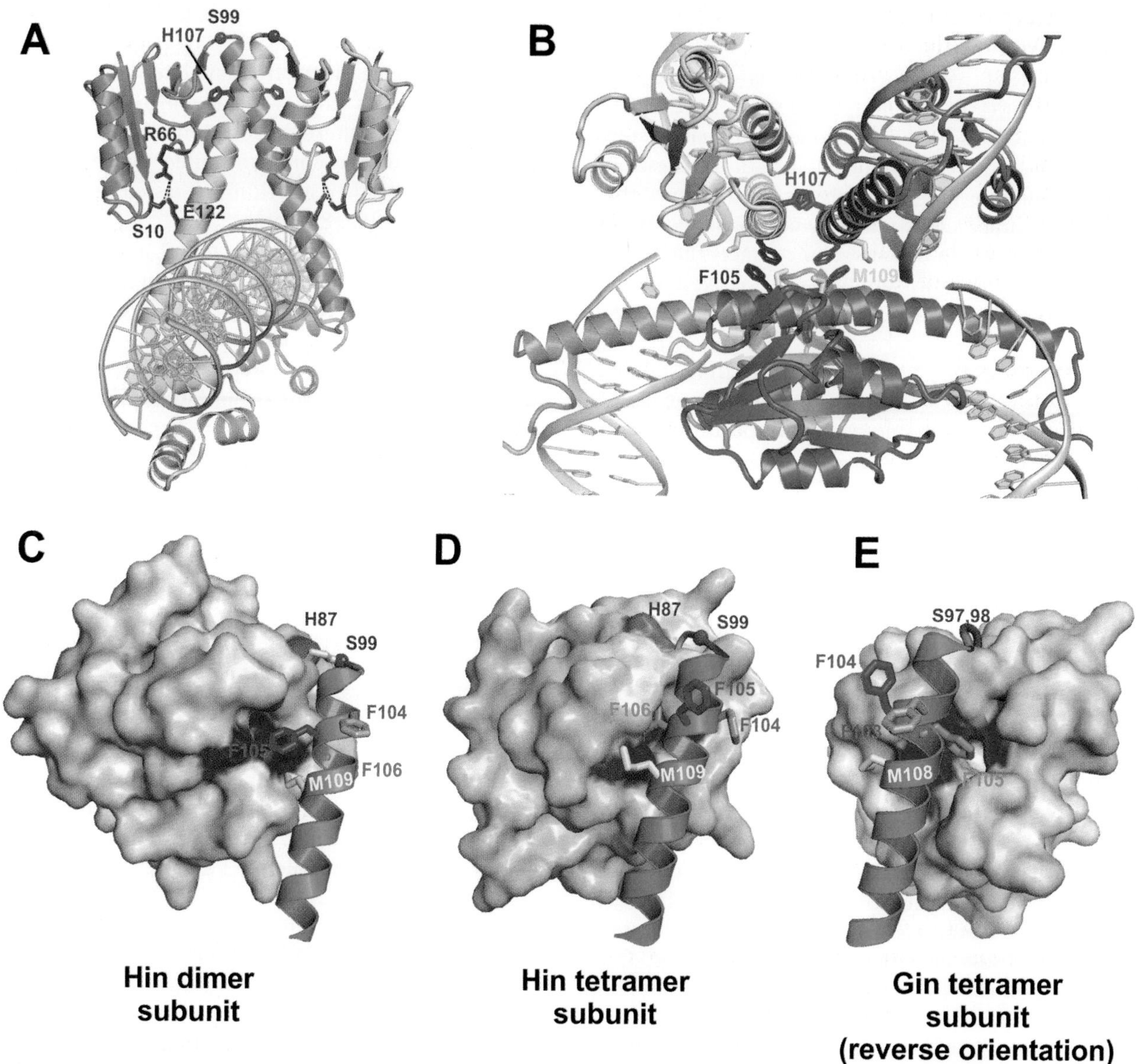

Figure 17 Residues regulating conformational transitions. (A) Hin dimer model depicting some of the residues where mutations have been isolated that control the dimer-tetramer transition. (B) Hin tetramer model (based on PDB: 2GM4) highlighting helix E residues Phe105 and Met109, whose side chains are located within the rotation interface and His107, whose side chain is predicted to stabilize the synaptic interface. (C) Subunit from the Hin dimer model showing helix E residues Phe104, Phe105, Phe106 (behind helix E), and Met109. The surface of the catalytic domain is also shown highlighting the pocket organized by residues surrounding Phe88. Residues around the Phe88 pocket where mutations lead to Fis/enhancer-independence are colored red. (D) Subunit from the Hin tetramer model rendered similarly as in panel C. Note that the side chains of Phe105 and Met109 have rotated away from the catalytic domain and are now within the rotation interface (panel B). Phe104 rotates away from the partner dimer subunit and becomes associated with the synaptic and rotation interfaces. Certain mutations at His87 (colored maroon) lead to strong Fis/enhancer-independence, and different substitutions of the hinge residue Ser99 generate various phenotypes (see text). (E) Subunit from the Gin tetramer X-ray structure (PDB: 3UJ3) in the reverse orientation as shown in panel D. Numbering of residues in Gin is one less than Hin. doi:10.1128/microbiolspec.MDNA3-0047-2014.f17

wide range of phenotypes ranging from inactivity to strong hyperactivity (Y. Chang and R.C. Johnson, unpublished). Most of the active mutants have defects in ligation, consistent with a prominent role of Phe105 in stabilizing the DNA-aligned rotational conformer. All Phe106 substitutions are defective for steps subsequent to DNA binding, except for weak inversion activity by tyrosine (192) (Y. Chang and R.C. Johnson, unpublished). The Phe104 side chain ring switches from the dimer interface to the synaptic and rotating interfaces and all substitutions of this residue tested are also severely defective in post DNA binding steps (106, 192).

Met109 on Helix E

Met109 on helix E is an invariant residue among serine invertases that is predicted to also engage in intrasubunit van der Waals contacts with its catalytic domain in the dimer (Fig. 17C). During tetramer formation, the Met109 side chain is repositioned into the rotating interface (Fig. 17B, D, E). Most substitutions at residue 109 in Hin severely interfere with recombination, but a large subset generate cleaved DNA molecules, even in the absence of Fis, that are competent for subunit rotation but defective in ligation (106) (Y. Chang and R.C. Johnson, unpublished). The Fis-independent cleavage activities of Met109 mutations can be rationalized by destabilization of the dimer, whereas the ligation defects presumably reflect perturbations of the rotation interface by the mutant side chains.

Additional Regulatory Dimeric Contacts: Residues Ser10, Arg66, and Glu122

Hin gain-of-function mutants point to another set of interactions between conserved residues that appear to stabilize the inactive dimer conformation. Mutations at Glu122 on helix E and at Arg66 within the active site region lead to efficient tetramer formation without Fis (109). The side chains of Glu122, Arg66, and the active site Ser10 have the potential to engage in hydrogen bonding between dimer subunits that would need to be broken very early during the remodeling process (Fig. 17A). One side of the γδ resolvase dimer crystal structure has the equivalent glutamic acid on helix E hydrogen bonded to the active site serine (111), and mutations of this glutamic acid also exhibit altered regulatory properties (119, 160, 162, 186, 189). Arg66 is believed to play a chemical role in the DNA cleavage/ligation reaction, consistent with the synaptic tetramers formed by Hin-R66D being inactive for DNA cleavage (109, 113, 114). The phenotypes of different substitutions at Hin residue 122 range from no activity after dimer DNA binding to efficient formation of uncleaved or cleaved tetramers. The diversity can be rationalized in light of the large number of transient interactions residue 122 is predicted to make during the transition to the tetramer.

The Hinge Residue Ser99

Serine 99 at the junction between the catalytic domain and helix E is modeled to be at the position where the pivoting and rotation occurs between these structural units during the dimer-tetramer remodeling (Fig. 17A, C, D, E) (109, 119). In some serine invertases, this residue is a threonine, and in γδ/Tn3 resolvase it is a glycine, but a serine (along with an aspartic acid to tyrosine change at the next residue) encourages formation of resolvase tetramers (189). Serine 99 has unfavorable peptide backbone angles in the Hin dimer models and thus may facilitate remodeling to the tetramer where the backbone angles are in the normal range. Consistent with this idea, Hin-S99G remains as an inactive dimer even under Fis/enhancer stimulation (109). Alternatively, Hin Ser99 mutants with lysine and arginine substitutions generate small numbers of uncleaved synaptic tetramers in reactions without Fis that exhibit slightly altered electrophoretic migrations from DNA-cleaved tetramers. These mutations synergize with other hyperactivating mutations to promote robust synaptic tetramer formation, but the resulting complexes remain blocked for DNA cleavage. These properties suggest that certain substitutions at position 99 facilitate early steps of tetramer formation, but cannot complete assembly of catalytically competent tetramers in the absence of Fis.

Acknowledgments. *I thank John Marko (Northwestern University) for discussions and comments on the manuscript. I am very grateful to the present and past members of my laboratory for their many contributions to work presented in this chapter. Research on recombination systems in my laboratory has been supported by the National Institutes of Health (GM038509).*

Citation. Johnson RC. 2014. Site-specific DNA inversion by serine recombinases. Microbiol Spectrum 2(6):MDNA3-0047-2014.

References

1. **Grindley ND, Whiteson KL, Rice PA.** 2006. Mechanisms of site-specific recombination. *Annu Rev Biochem* 75:567–605.
2. **Johnson RC.** 2002. Bacterial site-specific DNA inversion systems, p 230–271. *In* Craig NL, Craigie R, Gellert M, Lambowitz AM (ed), *Mobile DNA II*. ASM Press, Washington DC.
3. **Rice PA.** Serine Resolvases. *In* Craig NL (ed), *Mobile DNA III*. ASM Press, Washington, DC, in press.

4. **Stark WM.** 2014. The Serine Recombinases. *Microbiolspec* 2(6): doi: 10.1128/microbiolspec.MDNA3-0046-2014.
5. **Andrewes FW.** 1922. Studies in group-agglutination - The *Salmonella* group and its antigenic structure. *J Pathol Bacteriol* **25:**505–521.
6. **Iino T.** 1969. Genetics and chemistry of bacterial flagella. *Bacteriol Rev* **33:**454–475.
7. **Lederberg J, Edwards P.** 1953. Serotypic recombination in *Salmonella*. *J Immunol* **71:**323–340.
8. **Lederberg J, Iino T.** 1956. Phase variation in *Salmonella*. *Genetics* **41:**743–757.
9. **Stocker BAD.** 1949. Measurement of the rate of mutation of flagellar antigenic phase in *Salmonella typhimurium*. *J Hyg* **47:**398–413.
10. **Kutsukake K, Nakashima H, Tominaga A, Abo T.** 2006. Two DNA invertases contribute to flagellar phase variation in *Salmonella enterica* serovar Typhimurium strain LT2. *J Bacteriol* **188:**950–957.
11. **Gillen KL, Hughes KT.** 1991. Negative regulatory loci coupling flagellin synthesis to flagellar assembly in *Salmonella typhimurium*. *J Bacteriol* **173:**2301–2310.
12. **Ikeda JS, Schmitt CK, Darnell SC, Watson PR, Bispham J, Wallis TS, Weinstein DL, Metcalf ES, Adams P, O'Connor CD, O'Brien AD.** 2001. Flagellar phase variation of *Salmonella enterica* serovar Typhimurium contributes to virulence in the murine typhoid infection model but does not influence *Salmonella*-induced enteropathogenesis. *Infect Immun* **69:**3021–3030.
13. **Zieg J, Silverman M, Hilmen M, Simon M.** 1977. Recombinational switch for gene expression. *Science* **196:** 170–172.
14. **Zieg J, Hilmen M, Simon M.** 1978. Regulation of gene expression by site-specific inversion. *Cell* **15:**237–244.
15. **Zieg J, Simon M.** 1980. Analysis of the nucleotide sequence of an invertible controlling element. *Proc Natl Acad Sci U S A* **77:**4196–4200.
16. **Silverman M, Simon M.** 1980. Phase variation: genetic analysis of switching mutants. *Cell* **19:**845–854.
17. **Silverman M, Zieg J, Mandel G, Simon M.** 1981. Analysis of the functional components of the phase variation system. *Cold Spring Harb Symp Quant Biol* **45:** 17–26.
18. **Osuna R, Lienau D, Hughes KT, Johnson RC.** 1995. Sequence, regulation, and functions of *fis* in *Salmonella typhimurium*. *J Bacteriol* **177:**2021–2032.
19. **Silverman M, Zieg J, Simon M.** 1979. Flagellar-phase variation: isolation of the *rh1* gene. *J Bacteriol* **137:** 517–523.
20. **Aldridge PD, Wu C, Gnerer J, Karlinsey JE, Hughes KT, Sachs MS.** 2006. Regulatory protein that inhibits both synthesis and use of the target protein controls flagellar phase variation in *Salmonella enterica*. *Proc Natl Acad Sci U S A* **103:**11340–11345.
21. **Yamamoto S, Kutsukake K.** 2006. FljA-mediated post-transcriptional control of phase 1 flagellin expression in flagellar phase variation of *Salmonella enterica* serovar Typhimurium. *J Bacteriol* **188:**958–967.
22. **Bonifield HR, Hughes KT.** 2003. Flagellar phase variation in *Salmonella enterica* is mediated by a post-transcriptional control mechanism. *J Bacteriol* **185:** 3567–3574.
23. **Johnson RC, Bruist MF, Simon MI.** 1986. Host protein requirements for *in vitro* site-specific DNA inversion. *Cell* **46:**531–539.
24. **Finkel SE, Johnson RC.** 1992. The Fis protein: it's not just for DNA inversion anymore. *Mol Microbiol* **6:** 3257–3265.
25. **Johnson RC, Ball CA, Pfeffer D, Simon MI.** 1988. Isolation of the gene encoding the Hin recombinational enhancer binding protein. *Proc Natl Acad Sci U S A* **85:** 3484–3488.
26. **Koch C, Kahmann R.** 1986. Purification and properties of the *Escherichia coli* host factor required for inversion of the G segment in bacteriophage Mu. *J Biol Chem* **261:**15673–15678.
27. **Johnson RC, Simon MI.** 1985. Hin-mediated site-specific recombination requires two 26 bp recombination sites and a 60 bp recombinational enhancer. *Cell* **41:**781–791.
28. **Haykinson MJ, Johnson RC.** 1993. DNA looping and the helical repeat *in vitro* and *in vivo*: effect of HU protein and enhancer location on Hin invertasome assembly. *EMBO J* **12:**2503–2512.
29. **Bruist MF, Glasgow AC, Johnson RC, Simon MI.** 1987. Fis binding to the recombinational enhancer of the Hin DNA inversion system. *Genes Dev* **1:**762–772.
30. **Johnson RC, Bruist MB, Glaccum MB, Simon MI.** 1984. *In vitro* analysis of Hin-mediated site-specific recombination. *Cold Spring Harb Symp Quant Biol* **49:** 751–760.
31. **Bruist MF, Simon MI.** 1984. Phase variation and the Hin protein: *in vivo* activity measurements, protein overproduction, and purification. *J Bacteriol* **159:**71–79.
32. **Ball CA, Osuna R, Ferguson KC, Johnson RC.** 1992. Dramatic changes in Fis levels upon nutrient upshift in *Escherichia coli*. *J Bacteriol* **174:**8043–8056.
33. **Lim HM, Simon MI.** 1992. The role of negative supercoiling in Hin-mediated site-specific recombination. *J Biol Chem* **267:**11176–11182.
34. **Ó Cróinín T, Carroll RK, Kelly A, Dorman CJ.** 2006. Roles for DNA supercoiling and the Fis protein in modulating expression of virulence genes during intracellular growth of *Salmonella enterica* serovar Typhimurium. *Mol Microbiol* **62:**869–982.
35. **Koch C, Mertens G, Rudt F, Kahmann R, Kanaar R, Plasterk RH, van de Putte P, Sandulache R, Kamp D.** 1987. The invertible G segment, p 75–91. *In* Symonds N, Toussaint A, van de Putte P, Howe MM (ed), *Phage Mu*, 0 ed. Cold Spring Harbor Laboratory, New York, NY.
36. **Hiestand-Nauer R, Iida S.** 1983. Sequence of the site-specific recombinase gene Cin and of its substrates serving in the inversion of the C segment of bacteriophage P1. *EMBO J* **2:**1733–1740.
37. **Chow LT, Bukhari AI.** 1976. The invertible DNA segments of coliphages Mu and P1 are identical. *Virology* **74:**242–248.

38. **Toussaint A, Lefebvre N, Scott JR, Cowan JA, de Bruijn F, Bukhari AI.** 1978. Relationships between temperate phages Mu and P1. *Virology* **89:**146–161.

39. **Grundy FJ, Howe MM.** 1984. Involvement of the invertible G segment in bacteriophage Mu tail fiber biosynthesis. *Virology* **134:**296–317.

40. **Howe MM, Schumm JW, Taylor AL.** 1979. The S and U genes of bacteriophage Mu are located in the invertible G segment of Mu DNA. *Virology* **92:**108–124.

41. **Giphart-Gassler M, Plasterk RH, van de Putte P.** 1982. G inversion in bacteriophage Mu: a novel way of gene splicing. *Nature* **297:**339–342.

42. **Kamp D, Kahmann R, Zipser D, Broker TR, Chow LT.** 1978. Inversion of the G DNA segment of phage Mu controls phage infectivity. *Nature* **271:**577–580.

43. **van de Putte P, Cramer S, Giphart-Gassler M.** 1980. Invertible DNA determines host specificity of bacteriophage Mu. *Nature* **286:**218–222.

44. **Plasterk RH, Brinkman A, van de Putte P.** 1983. DNA inversions in the chromosome of *Escherichia coli* and in bacteriophage Mu: relationship to other site-specific recombination systems. *Proc Natl Acad Sci U S A* **80:** 5355–5358.

45. **Iida S, Meyer J, Kennedy KE, Arber W.** 1982. A site-specific, conservative recombination system carried by bacteriophage P1. Mapping the recombinase gene *cin* and the cross-over sites *cix* for the inversion of the C segment. *EMBO J* **1:**1445–1453.

46. **Iida S, Huber H, Hiestand-Nauer R, Meyer J, Bickle TA, Arber W.** 1984. The bacteriophage P1 site-specific recombinase Cin: recombination events and DNA recognition sequences. *Cold Spring Harb Symp Quant Biol* **49:**769–777.

47. **Kahmann R, Rudt F, Koch C, Mertens G.** 1985. G inversion in bacteriophage Mu DNA is stimulated by a site within the invertase gene and a host factor. *Cell* **41:** 771–780.

48. **Huber HE, Iida S, Arber W, Bickle TA.** 1985. Site-specific DNA inversion is enhanced by a DNA sequence element in *cis*. *Proc Natl Acad Sci U S A* **82:**3776–3780.

49. **Hubner P, Arber W.** 1989. Mutational analysis of a prokaryotic recombinational enhancer element with two functions. *EMBO J* **8:**577–585.

50. **Haffter P, Bickle TA.** 1987. Purification and DNA-binding properties of FIS and Cin, two proteins required for the bacteriophage P1 site-specific recombination system, cin. *J Mol Biol* **198:**579–587.

51. **Koch C, Vandekerckhove J, Kahmann R.** 1988. *Escherichia coli* host factor for site-specific DNA inversion: cloning and characterization of the *fis* gene. *Proc Natl Acad Sci U S A* **85:**4237–4241.

52. **Kanaar R, van Hal JP, van de Putte P.** 1989. The recombinational enhancer for DNA inversion functions independent of its orientation as a consequence of dyad symmetry in the Fis-DNA complex. *Nucleic Acids Res* **17:**6043–6053.

53. **Plasterk RH, Ilmer TA, Van de Putte P.** 1983. Site-specific recombination by Gin of bacteriophage Mu: inversions and deletions. *Virology* **127:**24–36.

54. **Symonds N, Coelho A.** 1978. Role of the G segment in the growth of phage Mu. *Nature* **271:**573–574.

55. **Kahmann R, Rudt F, Mertens G.** 1984. Substrate and enzyme requirements for *in vitro* site-specific recombination in bacteriophage Mu. *Cold Spring Harb Symp Quant Biol* **49:**285–294.

56. **Plasterk RH, van de Putte P.** 1984. Inversion of DNA *in vivo* and *in vitro* by Gin and Pin proteins. *Cold Spring Harb Symp Quant Biol* **49:**295–300.

57. **Komano T.** 1999. Shufflons: multiple inversion systems and integrons. *Annu Rev Genet* **33:**171–191.

58. **Iida S, Sandmeier H, Hubner P, Hiestand-Nauer R, Schneitz K, Arber W.** 1990. The Min DNA inversion enzyme of plasmid p15B of *Escherichia coli* 15T-: a new member of the Din family of site-specific recombinases. *Mol Microbiol* **4:**991–997.

59. **Sandmeier H, Iida S, Hubner P, Hiestand-Nauer R, Arber W.** 1991. Gene organization in the multiple DNA inversion region *min* of plasmid p15B of *E. coli* 15T-: assemblage of a variable gene. *Nucleic Acids Res* **19:** 5831–5838.

60. **Sandmeier H, Iida S, Meyer J, Hiestand-Nauer R, Arber W.** 1990. Site-specific DNA recombination system Min of plasmid p15B: a cluster of overlapping invertible DNA segments. *Proc Natl Acad Sci U S A* **87:** 1109–1113.

61. **Kamp D, Kahmann R.** 1981. The relationship of two invertible segments in bacteriophage Mu and *Salmonella typhimurium* DNA. *Mol Gen Genet* **184:** 564–566.

62. **Kutsukake K, Iino T.** 1980. Inversions of specific DNA segments in flagellar phase variation of *Salmonella* and inversion systems of bacteriophages P1 and Mu. *Proc Natl Acad Sci U S A* **77:**7338–7341.

63. **Kutsukake K, Nakao T, Iino T.** 1985. A gene for DNA invertase and an invertible DNA in *Escherichia coli* K-12. *Gene* **34:**343–350.

64. **van de Putte P, Plasterk R, Kuijpers A.** 1984. A Mu Gin complementing function and an invertible DNA region in *Escherichia coli* K-12 are situated on the genetic element *e14*. *J Bacteriol* **158:**517–522.

65. **Klippel A, Cloppenborg K, Kahmann R.** 1988. Isolation and characterization of unusual Gin mutants. *EMBO J* **7:**3983–3989.

66. **Scott TN, Simon MI.** 1982. Genetic analysis of the mechanism of the *Salmonella* phase variation site specific recombination system. *Mol Gen Genet* **188:** 313–321.

67. **Moskowitz IP, Heichman KA, Johnson RC.** 1991. Alignment of recombination sites in Hin-mediated site-specific DNA recombination. *Genes Dev* **5:**1635–1645.

68. **Kennedy KE, Iida S, Meyer J, Stalhammar-Carlemalm M, Hiestand-Nauer R, Arber W.** 1983. Genome fusion mediated by the site specific DNA inversion system of bacteriophage P1. *Mol Gen Genet* **189:**413–421.

69. **Cerdeno-Tarraga AM, Patrick S, Crossman LC, Blakely G, Abratt V, Lennard N, Poxton I, Duerden B, Harris B, Quail MA, Barron A, Clark L, Corton C, Doggett J, Holden MT, Larke N, Line A, Lord A, Norbertczak H,**

Ormond D, Price C, Rabbinowitsch E, Woodward J, Barrell B, Parkhill J. 2005. Extensive DNA inversions in the *B. fragilis* genome control variable gene expression. *Science* 307:1463–1465.

70. Kuwahara T, Yamashita A, Hirakawa H, Nakayama H, Toh H, Okada N, Kuhara S, Hattori M, Hayashi T, Ohnishi Y. 2004. Genomic analysis of *Bacteroides fragilis* reveals extensive DNA inversions regulating cell surface adaptation. *Proc Natl Acad Sci U S A* 101:14919–14924.

71. Patrick S, Parkhill J, McCoy LJ, Lennard N, Larkin MJ, Collins M, Sczaniecka M, Blakely G. 2003. Multiple inverted DNA repeats of *Bacteroides fragilis* that control polysaccharide antigenic variation are similar to the *hin* region inverted repeats of *Salmonella typhimurium*. *Microbiology* 149:915–924.

72. Krinos CM, Coyne MJ, Weinacht KG, Tzianabos AO, Kasper DL, Comstock LE. 2001. Extensive surface diversity of a commensal microorganism by multiple DNA inversions. *Nature* 414:555–558.

73. Coyne MJ, Weinacht KG, Krinos CM, Comstock LE. 2003. Mpi recombinase globally modulates the surface architecture of a human commensal bacterium. *Proc Natl Acad Sci U S A* 100:10446–10451.

74. Fletcher CM, Coyne MJ, Bentley DL, Villa OF, Comstock LE. 2007. Phase-variable expression of a family of glycoproteins imparts a dynamic surface to a symbiont in its human intestinal ecosystem. *Proc Natl Acad Sci U S A* 104:2413–2418.

75. Simon M, Zieg J, Silverman M, Mandel G, Doolittle R. 1980. Phase variation: evolution of a controlling element. *Science* 209:1370–1374.

76. Smith M, Thorpe H. 2002. Diversity in the serine recombinases. *Mol Microbiol* 44:299–307.

77. Liu CC, Huhne R, Tu J, Lorbach E, Droge P. 1998. The resolvase encoded by *Xanthomonas campestris* transposable element ISXc5 constitutes a new subfamily closely related to DNA invertases. *Genes Cells* 3:221–233.

78. McLean MM, Chang Y, Dhar G, Heiss JK, Johnson RC. 2013. Multiple interfaces between a serine recombinase and an enhancer control site-specific DNA inversion. *Elife* 2:e01211.

79. Canosa I, Lurz R, Rojo F, Alonso JC. 1998. beta Recombinase catalyzes inversion and resolution between two inversely oriented *six* sites on a supercoiled DNA substrate and only inversion on relaxed or linear substrates. *J Biol Chem* 273:13886–13891.

80. Janniere L, McGovern S, Pujol C, Petit MA, Ehrlich SD. 1996. *In vivo* analysis of the plasmid pAM beta 1 resolution system. *Nucleic Acids Res* 24:3431–3436.

81. Rojo F, Alonso JC. 1994. The beta recombinase from the *Streptococcal* plasmid pSM19035 represses its own transcription by holding the RNA polymerase at the promoter region. *Nucleic Acids Res* 22:1855–1860.

82. Rowland SJ, Dyke KG. 1989. Characterization of the staphylococcal beta-lactamase transposon Tn552. *EMBO J* 8:2761–2773.

83. Rowland SJ, Dyke KG. 1990. Tn552, a novel transposable element from *Staphylococcus aureus*. *Mol Microbiol* 4:961–975.

84. Alonso JC, Gutierrez C, Rojo F. 1995. The role of chromatin-associated protein Hbsu in beta-mediated DNA recombination is to facilitate the joining of distant recombination sites. *Mol Microbiol* 18:471–478.

85. Alonso JC, Weise F, Rojo F. 1995. The *Bacillus subtilis* histone-like protein Hbsu is required for DNA resolution and DNA inversion mediated by the beta recombinase of plasmid pSM19035. *J Biol Chem* 270:2938–2945.

86. Smith MC, Brown WR, McEwan AR, Rowley PA. 2010. Site-specific recombination by phiC31 integrase and other large serine recombinases. *Biochem Soc Trans* 38:388–394.

87. Smith MC. Phage-encoded Serine Integrases and Other Large Serine Recombinases. *In* Craig NL (ed), *Mobile DNA III*. ASM Press, Washington, DC, in press.

88. Van Duyne GD, Rutherford K. 2013. Large serine recombinase domain structure and attachment site binding. *Crit Rev Biochem Mol Biol* 48:476–491.

89. Rutherford K, Yuan P, Perry K, Sharp R, Van Duyne GD. 2013. Attachment site recognition and regulation of directionality by the serine integrases. *Nucleic Acids Res* 41:8341–8356.

90. Rutherford K, Van Duyne GD. 2014. The ins and outs of serine integrase site-specific recombination. *Curr Opin Struct Biol* 24:125–131.

91. Kersulyte D, Mukhopadhyay AK, Shirai M, Nakazawa T, Berg DE. 2000. Functional organization and insertion specificity of IS607, a chimeric element of *Helicobacter pylori*. *J Bacteriol* 182:5300–5308.

92. Boocock MR, Rice PA. 2013. A proposed mechanism for IS607-family serine transposases. *Mob DNA* 4:24.

93. Kanaar R, van de Putte P, Cozzarelli NR. 1986. Purification of the Gin recombination protein of *Escherichia coli* phage Mu and its host factor. *Biochim Biophys Acta* 866:170–177.

94. Mertens G, Fuss H, Kahmann R. 1986. Purification and properties of the DNA invertase Gin encoded by bacteriophage Mu. *J Biol Chem* 261:15668–15672.

95. Kanaar R, Klippel A, Shekhtman E, Dungan JM, Kahmann R, Cozzarelli NR. 1990. Processive recombination by the phage Mu Gin system: implications for the mechanisms of DNA strand exchange, DNA site alignment, and enhancer action. *Cell* 62:353–366.

96. Plasterk RH, Kanaar R, van de Putte P. 1984. A genetic switch *in vitro*: DNA inversion by Gin protein of phage Mu. *Proc Natl Acad Sci U S A* 81:2689–2692.

97. Mertens G, Hoffmann A, Blocker H, Frank R, Kahmann R. 1984. Gin-mediated site-specific recombination in bacteriophage Mu DNA: overproduction of the protein and inversion *in vitro*. *EMBO J* 3:2415–2421.

98. Klippel A, Mertens G, Patschinsky T, Kahmann R. 1988. The DNA invertase Gin of phage Mu: formation of a covalent complex with DNA via a phosphoserine at amino acid position 9. *EMBO J* 7:1229–1237.

99. **Chiu TK, Sohn C, Dickerson RE, Johnson RC.** 2002. Testing water-mediated DNA recognition by the Hin recombinase. *EMBO J* **21:**801–814.
100. **Johnson RC, Bruist MF.** 1989. Intermediates in Hin-mediated DNA inversion: a role for Fis and the recombinational enhancer in the strand exchange reaction. *EMBO J* **8:**1581–1590.
101. **Crisona NJ, Kanaar R, Gonzalez TN, Zechiedrich EL, Klippel A, Cozzarelli NR.** 1994. Processive recombination by wild-type Gin and an enhancer-independent mutant. Insight into the mechanisms of recombination selectivity and strand exchange. *J Mol Biol* **243:** 437–457.
102. **Reed RR, Moser CD.** 1984. Resolvase-mediated recombination intermediates contain a serine residue covalently linked to DNA. *Cold Spring Harb Symp Quant Biol* **49:**245–249.
103. **Merickel SK, Haykinson MJ, Johnson RC.** 1998. Communication between Hin recombinase and Fis regulatory subunits during coordinate activation of Hin-catalyzed site-specific DNA inversion. *Genes Dev* **12:**2803–2816.
104. **Heichman KA, Johnson RC.** 1990. The Hin invertasome: protein-mediated joining of distant recombination sites at the enhancer. *Science* **249:**511–517.
105. **Kanaar R, Cozzarelli NR.** 1992. Roles of supercoiled DNA structure in DNA transactions. *Cur Opin Struct Biol* **2:**369–379.
106. **Haykinson MJ, Johnson LM, Soong J, Johnson RC.** 1996. The Hin dimer interface is critical for Fis-mediated activation of the catalytic steps of site-specific DNA inversion. *Curr Biol* **6:**163–177.
107. **Sanders ER, Johnson RC.** 2004. Stepwise dissection of the Hin-catalyzed recombination reaction from synapsis to resolution. *J Mol Biol* **340:**753–766.
108. **Glasgow AC, Bruist MF, Simon MI.** 1989. DNA-binding properties of the Hin recombinase. *J Biol Chem* **264:**10072–10082.
109. **Heiss JK, Sanders ER, Johnson RC.** 2011. Intrasubunit and intersubunit interactions controlling assembly of active synaptic complexes during Hin-catalyzed DNA recombination. *J Mol Biol* **411:**744–764.
110. **Mertens G, Klippel A, Fuss H, Blocker H, Frank R, Kahmann R.** 1988. Site-specific recombination in bacteriophage Mu: characterization of binding sites for the DNA invertase Gin. *EMBO J* **7:**1219–1227.
111. **Yang W, Steitz TA.** 1995. Crystal structure of the site-specific recombinase gammadelta resolvase complexed with a 34 bp cleavage site. *Cell* **82:**193–207.
112. **Mouw KW, Rowland SJ, Gajjar MM, Boocock MR, Stark WM, Rice PA.** 2008. Architecture of a serine recombinase-DNA regulatory complex. *Mol Cell* **30:** 145–155.
113. **Olorunniji FJ, Stark WM.** 2009. The catalytic residues of Tn3 resolvase. *Nucleic Acids Res* **37:**7590–7602.
114. **Keenholtz RA, Rowland SJ, Boocock MR, Stark WM, Rice PA.** 2011. Structural basis for catalytic activation of a serine recombinase. *Structure* **19:**799–809.
115. **Nanassy OZ, Hughes KT.** 1998. *In vivo* identification of intermediate stages of the DNA inversion reaction catalyzed by the *Salmonella* Hin recombinase. *Genetics* **149:**1649–1663.
116. **Adams CW, Nanassy O, Johnson RC, Hughes KT.** 1997. Role of arginine-43 and arginine-69 of the Hin recombinase catalytic domain in the binding of Hin to the *hix* DNA recombination sites. *Mol Microbiol* **24:** 1235–1247.
117. **Dhar G, Heiss JK, Johnson RC.** 2009. Mechanical constraints on Hin subunit rotation imposed by the Fis/enhancer system and DNA supercoiling during site-specific recombination. *Mol Cell* **34:** 746–759.
118. **Dhar G, McLean MM, Heiss JK, Johnson RC.** 2009. The Hin recombinase assembles a tetrameric protein swivel that exchanges DNA strands. *Nucleic Acids Res* **37:**4743–4756.
119. **Li W, Kamtekar S, Xiong Y, Sarkis GJ, Grindley ND, Steitz TA.** 2005. Structure of a synaptic gamma delta resolvase tetramer covalently linked to two cleaved DNAs. *Science* **309:**1210–1215.
120. **Ritacco CJ, Kamtekar S, Wang J, Steitz TA.** 2013. Crystal structure of an intermediate of rotating dimers within the synaptic tetramer of the G-segment invertase. *Nucleic Acids Res* **41:**2673–2682.
121. **Bruist MF, Horvath SJ, Hood LE, Steitz TA, Simon MI.** 1987. Synthesis of a site-specific DNA-binding peptide. *Science* **235:**777–780.
122. **Feng JA, Johnson RC, Dickerson RE.** 1994. Hin recombinase bound to DNA: the origin of specificity in major and minor groove interactions. *Science* **263:** 348–355.
123. **Hughes KT, Gaines PC, Karlinsey JE, Vinayak R, Simon MI.** 1992. Sequence-specific interaction of the *Salmonella* Hin recombinase in both major and minor grooves of DNA. *EMBO J* **11:**2695–2705.
124. **Sluka JP, Horvath SJ, Glasgow AC, Simon MI, Dervan PB.** 1990. Importance of minor-groove contacts for recognition of DNA by the binding domain of Hin recombinase. *Biochemistry* **29:**6551–6561.
125. **Hughes KT, Youderian P, Simon MI.** 1988. Phase variation in *Salmonella*: analysis of Hin recombinase and *hix* recombination site interaction *in vivo*. *Genes Dev* **2:** 937–948.
126. **Ritacco CJ, Steitz TA, Wang J.** 2014. Exploiting large non-isomorphous differences for phase determination of a G-segment invertase-DNA complex. *Acta Crystallogr D Biol Crystallogr* **70:**685–693.
127. **Kanaar R, van de Putte P, Cozzarelli NR.** 1989. Gin-mediated recombination of catenated and knotted DNA substrates: implications for the mechanism of interaction between *cis*-acting sites. *Cell* **58:**147–159.
128. **Surette MG, Chaconas G.** 1992. The Mu transpositional enhancer can function in trans: requirement of the enhancer for synapsis but not strand cleavage. *Cell* **68:** 1101–1108.
129. **Pan CQ, Finkel SE, Cramton SE, Feng JA, Sigman DS, Johnson RC.** 1996. Variable structures of Fis-DNA complexes determined by flanking DNA-protein contacts. *J Mol Biol* **264:**675–695.

130. **Stella S, Cascio D, Johnson RC.** 2010. The shape of the DNA minor groove directs binding by the DNA-bending protein Fis. *Genes Dev* **24:**814–826.

131. **Johnson RC, Glasgow AC, Simon MI.** 1987. Spatial relationship of the Fis binding sites for Hin recombinational enhancer activity. *Nature* **329:**462–465.

132. **Shao Y, Feldman-Cohen LS, Osuna R.** 2008. Functional characterization of the *Escherichia coli* Fis-DNA binding sequence. *J Mol Biol* **376:**771–785.

133. **Kahramanoglou C, Seshasayee AS, Prieto AI, Ibberson D, Schmidt S, Zimmermann J, Benes V, Fraser GM, Luscombe NM.** 2011. Direct and indirect effects of H-NS and Fis on global gene expression control in *Escherichia coli*. *Nucleic Acids Res* **39:**2073–2091.

134. **Cho BK, Knight EM, Barrett CL, Palsson BO.** 2008. Genome-wide analysis of Fis binding in *Escherichia coli* indicates a causative role for A-/AT-tracts. *Genome Res* **18:**900–910.

135. **Yuan HS, Finkel SE, Feng JA, Kaczor-Grzeskowiak M, Johnson RC, Dickerson RE.** 1991. The molecular structure of wild-type and a mutant Fis protein: relationship between mutational changes and recombinational enhancer function or DNA binding. *Proc Natl Acad Sci U S A* **88:**9558–9562.

136. **Kostrewa D, Granzin J, Koch C, Choe HW, Raghunathan S, Wolf W, Labahn J, Kahmann R, Saenger W.** 1991. Three-dimensional structure of the *E. coli* DNA-binding protein FIS. *Nature* **349:**178–180.

137. **Safo MK, Yang WZ, Corselli L, Cramton SE, Yuan HS, Johnson RC.** 1997. The transactivation region of the Fis protein that controls site-specific DNA inversion contains extended mobile beta-hairpin arms. *EMBO J* **16:**6860–6873.

138. **Cheng YS, Yang WZ, Johnson RC, Yuan HS.** 2000. Structural analysis of the transcriptional activation region on Fis: crystal structures of six Fis mutants with different activation properties. *J Mol Biol* **302:**1139–1151.

139. **Osuna R, Finkel SE, Johnson RC.** 1991. Identification of two functional regions in Fis: the N-terminus is required to promote Hin-mediated DNA inversion but not lambda excision. *EMBO J* **10:**1593–1603.

140. **Koch C, Ninnemann O, Fuss H, Kahmann R.** 1991. The N-terminal part of the *E. coli* DNA binding protein FIS is essential for stimulating site-specific DNA inversion but is not required for specific DNA binding. *Nucleic Acids Res* **19:**5915–5922.

141. **Hancock SP, Ghane T, Cascio D, Rohs R, Di Felice R, Johnson RC.** 2013. Control of DNA minor groove width and Fis protein binding by the purine 2-amino group. *Nucleic Acids Res* **41:**6750–6760.

142. **Perkins-Balding D, Dias DP, Glasgow AC.** 1997. Location, degree, and direction of DNA bending associated with the Hin recombinational enhancer sequence and Fis-enhancer complex. *J Bacteriol* **179:**4747–4753.

143. **Johnson RC, Johnson LM, Schmidt JW, Gardner JF.** 2005. Major nucleoid proteins in the structure and function of the *Escherichia coli* chromosome, p 65–132. *In* Higgins NP (ed), *The bacterial chromosome*. ASM Press, Washington DC.

144. **Browning DF, Grainger DC, Busby SJ.** 2010. Effects of nucleoid-associated proteins on bacterial chromosome structure and gene expression. *Curr Opin Microbiol* **13:**773–780.

145. **Luijsterburg MS, Noom MC, Wuite GJ, Dame RT.** 2006. The architectural role of nucleoid-associated proteins in the organization of bacterial chromatin: A molecular perspective. *J Struct Biol* **156:**262–272.

146. **Dillon SC, Dorman CJ.** 2010. Bacterial nucleoid-associated proteins, nucleoid structure and gene expression. *Nat Rev Microbiol* **8:**185–195.

147. **Skoko D, Yan J, Johnson RC, Marko JF.** 2005. Low-force DNA condensation and discontinuous high-force decondensation reveal a loop-stabilizing function of the protein Fis. *Phys Rev Lett* **95:**208101.

148. **Skoko D, Yoo D, Bai H, Schnurr B, Yan J, McLeod SM, Marko JF, Johnson RC.** 2006. Mechanism of chromosome compaction and looping by the *Escherichia coli* nucleoid protein Fis. *J Mol Biol* **364:**777–798.

149. **Ishihama A, Kori A, Koshio E, Yamada K, Maeda H, Shimada T, Makinoshima H, Iwata A, Fujita N.** 2014. Intracellular concentrations of 65 species of transcription factors with known regulatory functions in *Escherichia coli*. *J Bacteriol* **196:**2718–2727.

150. **Schneider R, Travers A, Kutateladze T, Muskhelishvili G.** 1999. A DNA architectural protein couples cellular physiology and DNA topology in *Escherichia coli*. *Mol Microbiol* **34:**953–964.

151. **Weinstein-Fischer D, Elgrably-Weiss M, Altuvia S.** 2000. *Escherichia coli* response to hydrogen peroxide: a role for DNA supercoiling, topoisomerase I and Fis. *Mol Microbiol* **35:**1413–1420.

152. **Richardson SM, Boles TC, Cozzarelli NR.** 1988. The helical repeat of underwound DNA in solution. *Nucleic Acids Res* **16:**6607–6616.

153. **Bellomy GR, Record MTJ.** 1990. Stable DNA loops *in vivo* and *in vitro*: Roles in gene regulation at a distance and in biophysical characterization of DNA. *Prog Nucleic Acid Res Mol Biol* **39:**81–127.

154. **Hillyard DR, Edlund M, Hughes KT, Marsh M, Higgins NP.** 1990. Subunit-specific phenotypes of *Salmonella typhimurium* HU mutants. *J Bacteriol* **172:**5402–5407.

155. **Wada M, Kutsukake K, Komano T, Imamoto F, Kano Y.** 1989. Participation of the *hup* gene product in site-specific DNA inversion in *Escherichia coli*. *Gene* **76:**345–352.

156. **Paull TT, Haykinson MJ, Johnson RC.** 1993. The non-specific DNA-binding and -bending proteins HMG1 and HMG2 promote the assembly of complex nucleoprotein structures. *Genes Dev* **7:**1521–1534.

157. **Paull TT, Johnson RC.** 1995. DNA looping by *Saccharomyces cerevisiae* high mobility group proteins NHP6A/B. Consequences for nucleoprotein complex assembly and chromatin condensation. *J Biol Chem* **270:**8744–8754.

158. **Dhar G, Sanders ER, Johnson RC.** 2004. Architecture of the Hin synaptic complex during recombination:

the recombinase subunits translocate with the DNA strands. *Cell* **119:**33–45.

159. **Leschziner AE, Grindley NDF.** 2003. The architecture of the gamma delta resolvase crossover site synaptic complex revealed by using constrained DNA substrates. *Mol Cell* **12:**775–781.

160. **Sarkis GJ, Murley LL, Leschziner AE, Boocock MR, Stark WM, Grindley ND.** 2001. A model for the gamma delta resolvase synaptic complex. *Mol Cell* **8:** 623–631.

161. **Nollmann M, He J, Byron O, Stark WM.** 2004. Solution structure of the Tn3 resolvase-crossover site synaptic complex. *Mol Cell* **16:**127–137.

162. **Kamtekar S, Ho RS, Cocco MJ, Li W, Wenwieser SV, Boocock MR, Grindley NDF, Steitz TA.** 2006. Implications of structures of synaptic tetramers of gamma delta resolvase for the mechanism of recombination. *Proc Natl Acad Sci U S A* **103:**10642–10647.

163. **Yuan P, Gupta K, Van Duyne GD.** 2008. Tetrameric structure of a serine integrase catalytic domain. *Structure* **16:**1275–1286.

164. **Kanaar R, van de Putte P, Cozzarelli NR.** 1988. Gin-mediated DNA inversion: product structure and the mechanism of strand exchange. *Proc Natl Acad Sci U S A* **85:**752–756.

165. **Kahmann R, Mertens G, Klippel A, Brauer B, Rudt R, Koch C.** 1987. The mechanism of G inversion, p 681–689. *In* McMacken R, Kelly TJ (ed), *DNA replication and recombination, UCLA symposium on molecular and cellular biology*, vol. **47**, Alan R. Liss, Inc., New York, NY.

166. **Merickel SK, Johnson RC.** 2004. Topological analysis of Hin-catalysed DNA recombination *in vivo* and *in vitro*. *Mol Microbiol* **51:**1143–1154.

167. **Heichman KA, Moskowitz IP, Johnson RC.** 1991. Configuration of DNA strands and mechanism of strand exchange in the Hin invertasome as revealed by analysis of recombinant knots. *Genes Dev* **5:**1622–1634.

168. **Deibler RW, Rahmati S, Zechiedrich EL.** 2001. Topoisomerase IV, alone, unknots DNA in *E. coli*. *Genes Dev* **15:**748–761.

169. **Klippel A, Kanaar R, Kahmann R, Cozzarelli NR.** 1993. Analysis of strand exchange and DNA binding of enhancer-independent Gin recombinase mutants. *EMBO J* **12:**1047–1057.

170. **Iida S, Hiestand-Nauer R.** 1986. Localized conversion at the crossover sequences in the site-specific DNA inversion system of bacteriophage P1. *Cell* **45:** 71–79.

171. **Iida S, Hiestand-Nauer R.** 1987. Role of the central dinucleotide at the crossover sites for the selection of quasi sites in DNA inversion mediated by the site-specific Cin recombinase of phage P1. *Mol Gen Genet* **208:** 464–468.

172. **Bai H, Sun M, Ghosh P, Hatfull GF, Grindley ND, Marko JF.** 2011. Single-molecule analysis reveals the molecular bearing mechanism of DNA strand exchange by a serine recombinase. *Proc Natl Acad Sci U S A* **108:** 7419–7424.

173. **Vologodskii A, Cozzarelli NR.** 1996. Effect of supercoiling on the juxtaposition and relative orientation of DNA sites. *Biophys J* **70:**2548–2556.

174. **Vologodskii AV, Levene SD, Klenin KV, Frank-Kamenetskii M, Cozzarelli NR.** 1992. Conformational and thermodynamic properties of supercoiled DNA. *J Mol Biol* **227:**1224–1243.

175. **Boles TC, White JH, Cozzarelli NR.** 1990. Structure of plectonemically supercoiled DNA. *J Mol Biol* **213:** 931–951.

176. **Marko JF.** 1997. The internal 'slithering' dynamics of supercoiled DNA. *Physica A* **244:**263–277.

177. **Oram M, Marko JF, Halford SE.** 1997. Communications between distant sites on supercoiled DNA from non-exponential kinetics for DNA synapsis by resolvase. *J Mol Biol* **270:**396–412.

178. **Marko JF.** 2001. Short note on the scaling behavior of communication by 'slithering' on a supercoiled DNA. *Physica A* **296:**289–292.

179. **Vologodskii AV, Cozzarelli NR.** 1994. Conformational and thermodynamic properties of supercoiled DNA. *Annu Rev Biophys Biomol Struct* **23:**609–643.

180. **Lee SY, Lee HJ, Lee H, Kim S, Cho EH, Lim HM.** 1998. *In vivo* assay of protein-protein interactions in Hin-mediated DNA inversion. *J Bacteriol* **180:**5954–5960.

181. **Sessions RB, Oram M, Szczelkun MD, Halford SE.** 1997. Random walk models for DNA synapsis by resolvase. *J Mol Biol* **270:**413–425.

182. **Stark WM, Grindley NDF, Hatfull GF, Boocock MR.** 1991. Resolvase-catalysed reactions between *res* sites differing in the central dinucleotide of subsite I. *EMBO J* **10:**3541–3548.

183. **Ghosh P, Kim AI, Hatfull GF.** 2003. The orientation of mycobacteriophage Bxb1 integration is solely dependent on the central dinucleotide of *attP* and *attB*. *Mol Cell* **12:**1101–1111.

184. **Mandali S, Dhar G, Avliyakulov NK, Haykinson MJ, Johnson RC.** 2013. The site-specific integration reaction of Listeria phage A118 integrase, a serine recombinase. *Mobile DNA* **4:**2.

185. **Haffter P, Bickle TA.** 1988. Enhancer-independent mutants of the Cin recombinase have a relaxed topological specificity. *EMBO J* **7:**3991–3996.

186. **Burke ME, Arnold PH, He J, Wenwieser SV, Rowland SJ, Boocock MR, Stark WM.** 2004. Activating mutations of Tn3 resolvase marking interfaces important in recombination catalysis and its regulation. *Mol Microbiol* **51:**937–948.

187. **Rowland SJ, Boocock MR, McPherson AL, Mouw KW, Rice PA, Stark WM.** 2009. Regulatory mutations in Sin recombinase support a structure-based model of the synaptosome. *Mol Microbiol* **74:**282–298.

188. **Arnold PH, Blake DG, Grindley NDF, Boocock MR, Stark WM.** 1999. Mutants of Tn3 resolvase which do not require accessory binding sites for recombination activity. *EMBO J* **18:**1407–1414.

189. **Olorunniji FJ, He J, Wenwieser SV, Boocock MR, Stark WM.** 2008. Synapsis and catalysis by activated

Tn3 resolvase mutants. *Nucleic Acids Res* **36:**7181–7191.

190. **McGaughey GB, Gagne M, Rappe AK.** 1998. pi-Stacking interactions. Alive and well in proteins. *J Biol Chem* **273:**15458–15463.

191. **Brocchieri L, Karlin S.** 1994. Geometry of interplanar residue contacts in protein structures. *Proc Natl Acad Sci U S A* **91:**9297–9301.

192. **Lee HJ, Lee SY, Lee HM, Lim HM.** 2001. Effects of dimer interface mutations in Hin recombinase on DNA binding and recombination. *Mol Genet Genomics* **266:** 598–607.

193. **Cozzarelli NR, Krasnow MA, Gerrard SP, White JH.** 1984. A topological treatment of recombination and topoisomerases. *Cold Spring Harb Symp Quant Biol* **49:**383–400.

Mobile DNA, 3rd Edition
Nancy L. Craig, Michael Chandler, Martin Gellert, Alan M. Lambowitz, Phoebe A. Rice and Suzanne Sandmeyer

doi:10.1128/microbiolspec.MDNA3-0045-2014

Phoebe A. Rice[1]

Serine Resolvases

10

INTRODUCTION

The serine resolvases are a group of recombinases that, in their native contexts, resolve large fused replicons into smaller separated ones (1). Serine resolvases and the closely related invertases were the first serine recombinases to be studied in detail, and much of our understanding of the serine recombinase mechanism is owed to those early studies. Resolvases and invertases have also served as paradigms for understanding how DNA topology can be harnessed to regulate recombination reactions (2–5). Like other serine recombinases, the resolvases have a largely modular structure. In the resolvase case, the conserved catalytic domain is followed by a DNA-binding domain that is a simple helix-turn-helix similar to that found in many prokaryotic repressors (6). This modularity, combined with a wealth of structural and biochemical data, has made them good targets for engineering chimeric recombinases with designer sequence specificity (7, 8).

This chapter will focus on the current understanding of the mechanism of serine resolvases, with a focus on how structural studies have informed (and sometimes confounded) that understanding. For a broader view of serine recombinases, including an in-depth discussion of the large body of data addressing the mechanism of strand exchange, the reader is referred to the chapter by W.M. Stark. Additionally, the chapter by R.C. Johnson describes the invertases, which are quite closely related to the resolvases, and that by M.C.M. Smith describes the large serine recombinases, which have a different and much larger DNA-binding domain—also reviewed in reference (9). Although different groups of serine recombinases have different biological roles and regulatory properties, the strong conservation of the catalytic domain means that many of the lessons learned from the resolvase group can be directly applied to the rest of the serine recombinase family.

The generally accepted and well-supported mechanism of serine-recombinase-mediated strand exchange involves an unusual rotation of half of an entire complex relative to the other half. Briefly, formation of an active tetramer brings together the two dimer-bound crossover site DNAs (Figure 1). Formation of this tetramer is a key step that is regulated in different ways in different serine recombinase systems: see below for resolvases and chapters by RC Johnson and MCM Smith for other systems. Within the tetramer, the hydroxyl group of each subunit's active site serine then attacks a particular DNA phosphate group, displacing the 3′ hydroxyl and creating a covalent protein–DNA linkage. Once double-strand breaks are formed, one half of the DNA-bound tetramer can rotate relative to the other. Although thermal energy is sufficient to drive rotation within an active tetramer, DNA supercoiling can provide an additional driving force and can favor rotation in one direction over the other. Once rotation has aligned the broken DNA ends with new partners, the free 3′ hydroxyls attack the phosphoserine linkages

[1]Department of Biochemistry & Molecular Biology, The University of Chicago, Chicago, IL.

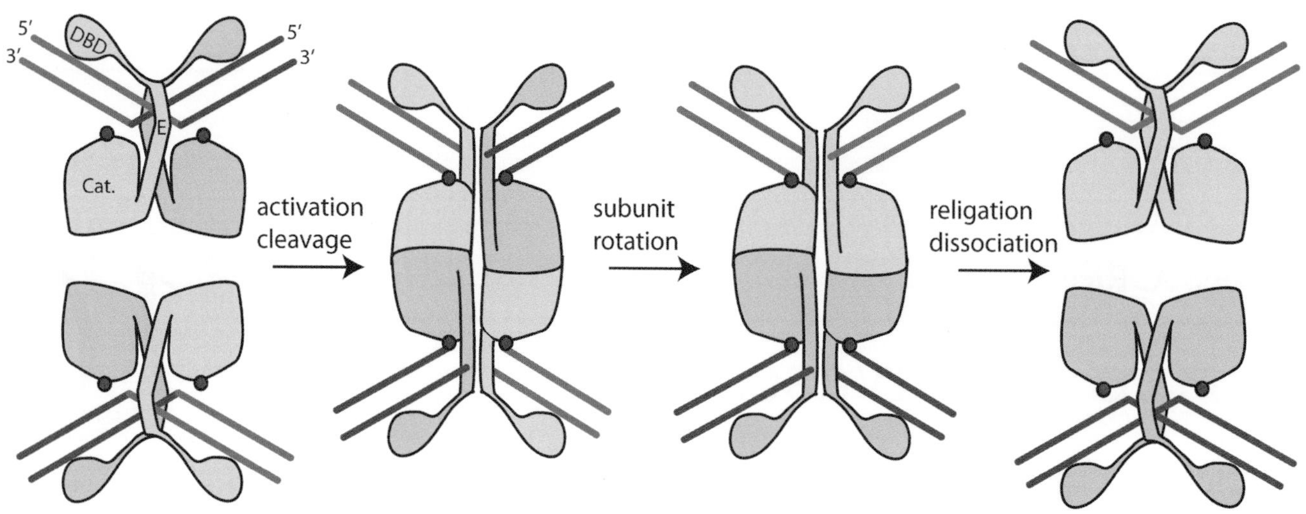

Figure 1 Cartoon of serine resolvase-mediated strand exchange. The wild-type protein initially binds crossover sites as an inactive dimer. Upon activation (see text), the catalytic domains (labeled "Cat.") form a tetramer that synapses the two partner sites. Within the tetramer, the active site serines (red dots) attack the DNA, creating double strand breaks with 5′ phosphoserine linkages and 2-nucleotide 3′ overhangs. Two subunits and the DNA segments covalently linked to them can then rotate relative to the other two. A 180° rotation aligns the broken ends for re-ligation in the recombinant configuration. Much of both the dimer and tetramer interface is contributed by helix E.
doi:10.1128/microbiolspec.MDNA3-0045-2014.f1

to reseal the DNA. The DNA can be re-ligated without the use of high-energy cofactors such as ATP because the chemical energy of the broken phosphodiester bond is stored in the phosphoserine linkage. However, as the net change in bond energy between substrate and product is zero, other factors are needed to tip the equilibrium towards product.

DNA resolvases play at least two biological roles. First, the transposition pathway of certain replicative transposons such as Tn3, γδ, and Tn4430 (see chapters by F. Dyda & A. B. Hickman and B. Hallet) leads to a "cointegrate" product in which the donor and target DNA replicons are fused, with a copy of the transposon at each junction (Figure 2A) (10). These elements generally encode site-specific recombinases, usually but not always of the serine family, that "resolve" the cointegrate product into two replicons, each bearing a copy of the transposon. Second, circular replicons (plasmids or entire bacterial chromosomes) can become dimerized as an accident of replication. When a collapsed replication fork is re-assembled in a RecA-dependent pathway, there are two possibilities for resolving the ensuing Holliday junction. In one case, the final product of replication will be two daughter circles, and in the other case the product will be one large fused circle (Figure 2B) (11). Although bacterial chromosomes (and certain plasmids) use tyrosine recombinases to resolve these dimers (see chapter by FX Barre), some plasmids encode their own serine resolvase systems. The functionally best-characterized of these is the β recombinase of *Streptococcus pyogenes* (12–14), whereas the closely related Sin recombinase, which is encoded by many staphylococcal plasmids, has been more extensively characterized biochemically and structurally (15, 16). Both transposon and plasmid resolvases act at sites termed "*res*" which include regulatory elements as well as the crossover sites.

The regulatory mechanism of serine resolvases exploits DNA topology to sense the relative location and orientation of the two partner *res* sites. Efficient recombination by wild-type (WT) resolvases occurs only between *res* sites that lie within the same circular DNA molecule and are in direct rather than inverted orientation. This requirement ensures that resolvases only catalyze the resolution of cointegrates (or dimers) and not inversion of the DNA segment between their cognate sites, or even intermolecular reactions that would create larger fused replicons. Formation of an active tetramer that synapses the two crossover sites occurs only within a larger complex termed a "synaptosome" that contains two copies of the full cognate *res* site for that resolvase, each one bound by multiple proteins. Although *res* sites for different serine resolvases vary in detail (Figure 3A), the topology of the DNA (when tested) is constant:

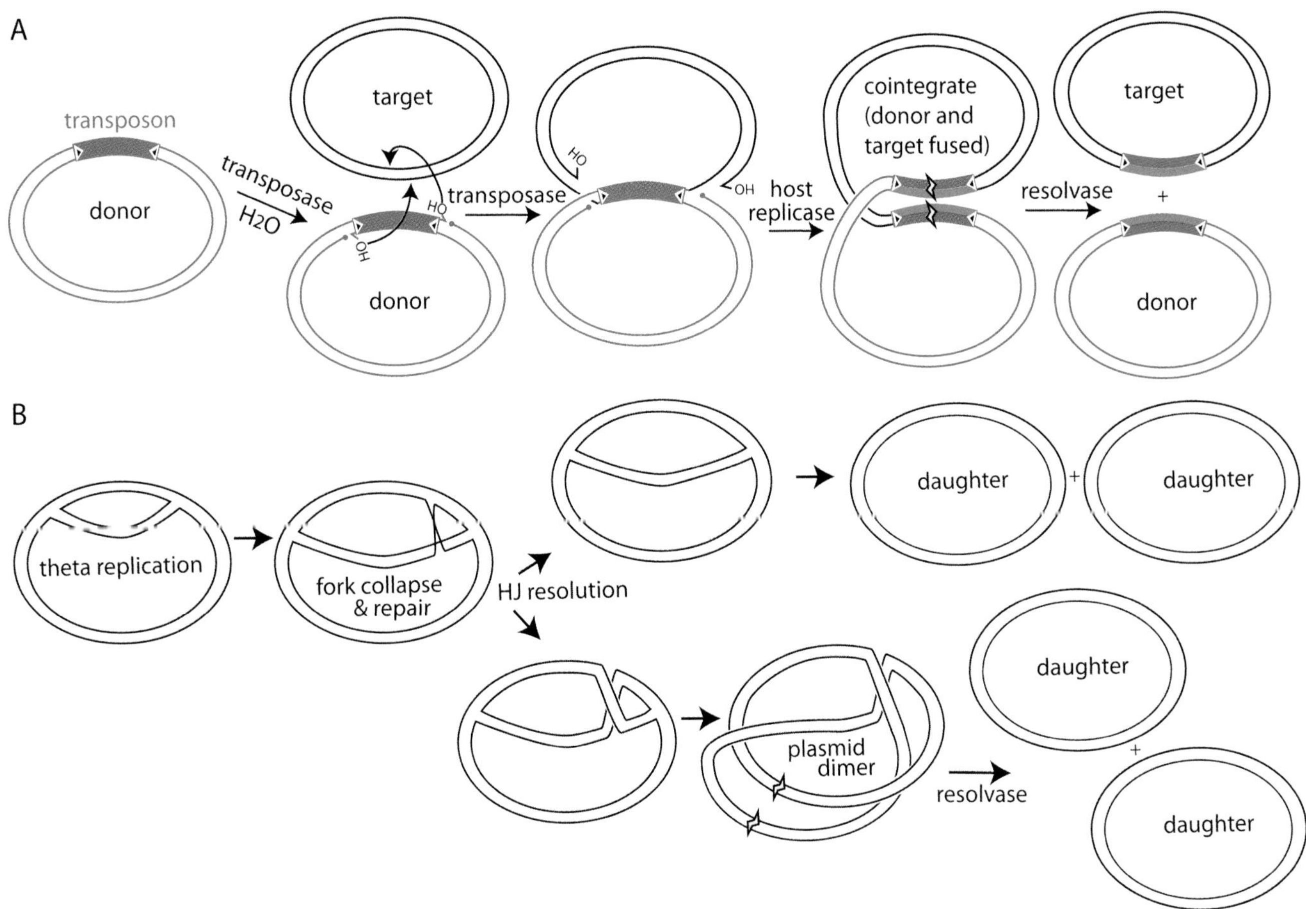

Figure 2 Biological roles for resolvases. (A) Some resolvases (e.g., γδ and Tn3) are encoded by replicative transposons. Their transposition creates a branched intermediate (center panel) that is processed by the host replication and repair machinery (new DNA strands are in blue) to yield a "cointegrate" (fourth panel) in which both the donor and recipient replicons are fused. Resolvase action at a *res* site within the transposon (yellow) resolves the cointegrate into the original donor and the recipient that now carries a copy of the transposon. (B) Some resolvases (e.g., β and Sin) are encoded by plasmids. Rescue of a stalled replication fork by a homologous recombination-mediated pathway can lead to a Holliday Junction (HJ) behind the rescued fork (second panel). Depending on which pair of strands is cut to resolve the Holliday Junction, replication results in two daughter circles (upper branch) or in a plasmid dimer (lower branch). Action of the plasmid-encoded resolvase at a *res* site (yellow) converts this dimer into two daughter circles. In both examples, the product circles are initially linked as catenanes (not shown for simplicity; see Figure 3) that are later separated by a host type II topoisomerase.
doi:10.1128/microbiolspec.MDNA3-0045-2014.f2

the protein–protein and protein–DNA interactions within the synaptosome trap 3 supercoiling nodes (Figure 3B) reviewed in references (1, 17). Supercoiling therefore stabilizes the complex, and in fact the protein–protein interactions appear to be tuned to render stable synaptosome formation dependent on supercoiling (18). Requiring synaptosome formation before catalysis accomplishes several things, reviewed in reference (2). First, it greatly favors intramolecular over intermolecular recombination, because the energy barrier to trapping three DNA–DNA crossings between two *res* sites on the same supercoiled plasmid is much lower than that for trapping three crossings between two separate plasmids. Second, the synaptosome aligns the two crossover sites properly such that recombination will produce resolved rather than inverted products. Third, the topology of the synaptosome is such that the 180° rotation accompanying strand exchange

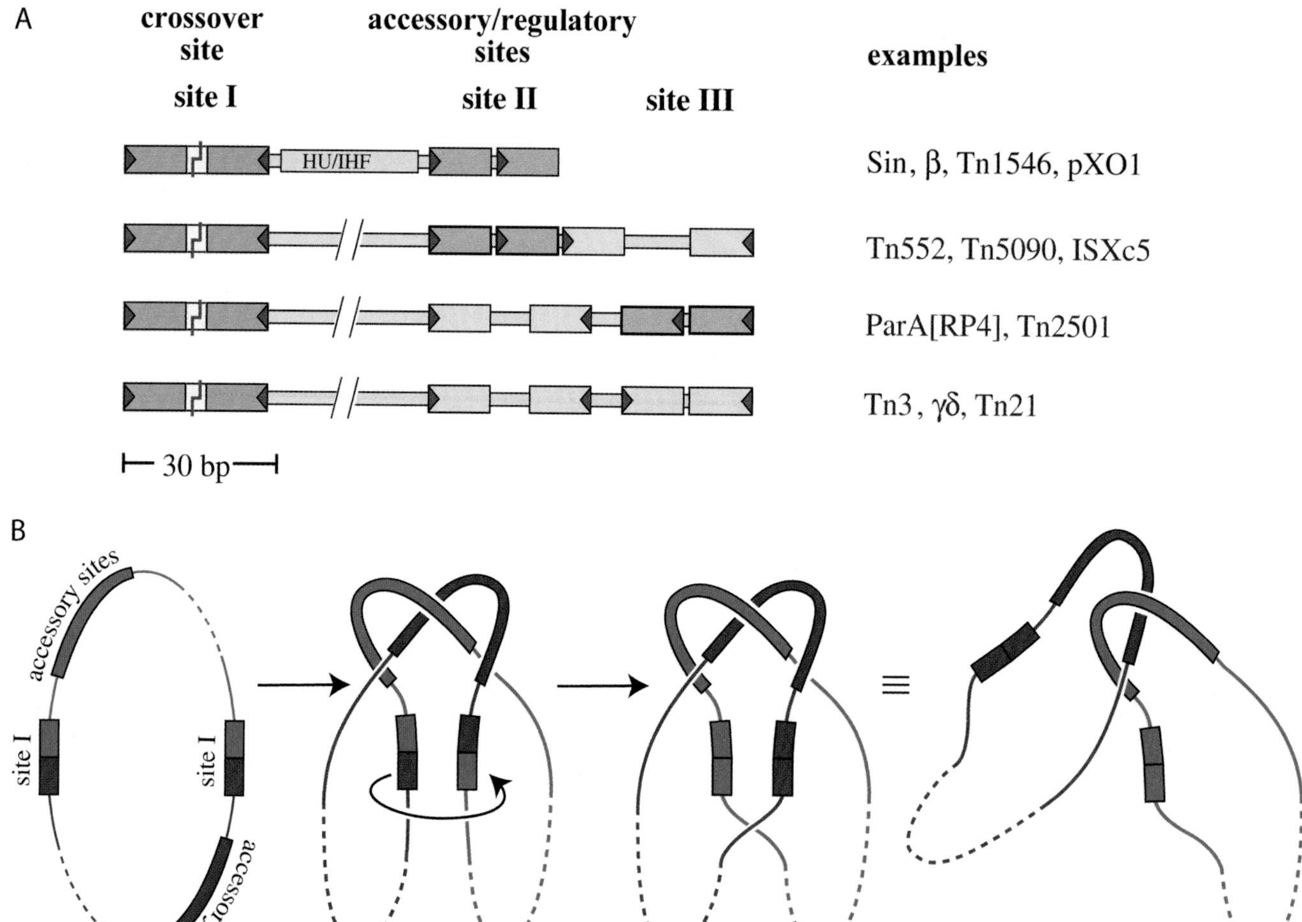

Figure 3 *Res* sites and the topology of the synaptosome. (A) Examples of serine resolvase *res* sites. Specific recognition sequences (~12 bp each) for individual resolvase subunits are shown as colored boxes: green for the crossover site (always an inverted repeat), purple for accessory sites that form direct repeats, and yellow for accessory sites that form inverted repeats. The recombinase dimers bound to the accessory sites are catalytically inactive, and these sites always differ from the crossover site in the length of their central spacers and/or the relative orientation of their half-sites. Sin and related resolvases require a DNA bending protein as well as additional recombinase subunits. The coding sequence for the resolvase protein is usually adjacent to its cognate *res* site. Figure adapted from reference (60) with permission. (B) Resolvase synaptosomes trap 3 supercoiling nodes. The resolvase and accessory proteins (if any) bound to each of the cognate *res* sites form a complex (the "synaptosome") that traps 3 dsDNA-over-dsDNA crossings and juxtaposes the two site Is. Synaptosome formation activates the site I-bound resolvase subunits, which then introduce double-strand breaks. A 180° rotation of the bottom two subunits (as drawn) realigns the broken ends, which are then re-ligated. In a negatively supercoiled substrate a right-handed rotation is favored because it introduces a + supercoiling node ($\Delta Lk = +1$) that cancels one of the pre-existing (–) nodes trapped in the synaptosome and because it allows rewinding of each duplex by a half turn (ΔTw for each = +½). The remaining two crossings trapped by the synaptosome are no longer intramolecular and instead catenate the two daughter circles (which can be separated by a host type II topoisomerase).
doi:10.1128/microbiolspec.MDNA3-0045-2014.f3

causes a change in linking number of +4. This prediction was verified by careful experimental measurements (19). (For a useful primer on DNA topology and the changes predicted by different recombination mechanisms see (20), and for a more advanced mathematical treatment, see (21).) Assuming a physiological supercoiling density of roughly −0.025 and a 20-kb substrate, relaxing four superhelical turns corresponds to a ΔG of about −9 kcal/mol (22). For comparison, that is slightly more than the standard free energy of ATP hydrolysis, and so a considerable, although unconventional, driving force. The product circles are still topologically linked as a two-noded catenane, but that is a problem that any type II topoisomerase can handle. Additional rotations after the first 180° are possible, but create knots (23, 24). Finally, within the product catenane the architecture of the synaptosome is no longer stabilized by supercoiling as it was in the substrate, and thus the complex is more likely to dissociate. Regulation of the serine invertases is conceptually similar, although the details differ (see the chapter by RC Johnson).

STRUCTURAL BIOLOGY OF SERINE RESOLVASES

Serine resolvases have been the target of structural studies for three decades, and the combined results of these efforts provide an increasingly detailed three-dimensional view of the reaction pathway. A brief overview of the surprisingly convoluted history of using serine resolvase structures to understand function is presented here. The first structure, of the catalytic domain of WT γδ resolvase, reported in 1990, was as puzzling as it was informative (25). DNA-binding and footprinting studies had suggested that the WT protein forms a dimer in solution, and that dimer–dimer interactions form the synaptosome (reviewed in reference (26)). Although multiple protein–protein interfaces were seen in the crystal, additional biochemical data were needed to determine which mediated the solution dimer, which represented contacts formed between dimers within the synaptosome, and which were probably irrelevant in solution (27, 28).

The resolvase catalytic domain contains a small α/β fold containing the catalytic serine followed by a long helix (helix E) that is sometimes now considered a separate connector domain (Figure 4A). Although the overall fold has been noted to show similarities to that of 5′ → 3′ exonucleases and topoisomerases of the TOPRIM fold, the active site is unrelated to either of those (29, 30). In the solution dimer, the E helices of two subunits dock against each other and into a shallow groove on the partner's surface. However, this dimer places the two active site serines much too far apart to attack the scissile phosphate groups in B-form DNA. Furthermore, because the serines lie on flexible solvent-exposed loops, they do not appear to be activated for nucleophilic attack. It was also unclear how two such dimers could dock together to form a recombination-competent tetramer, or, given the partially interdigitated nature of the interface, how subunit exchange could occur.

A similar catalytic domain dimer was seen in the structure of the WT γδ resolvase bound to crossover site DNA (Figure 4A) (31). That structure was very informative regarding resolvase–DNA contacts: it showed that while the most stringent aspects of protein–DNA recognition are, as expected, accomplished by the helix-turn-helix DNA-binding domain (32), additional contacts are made by an extension of helix E and the subsequent linker, which bind in the minor groove near the center of the site. The insertion of helix E into the central minor groove widens it, inducing it to bend away from the catalytic domains (a feature that recurs in the complexes with accessory sites described below). However, despite the DNA distortions seen in the resolvase–site I complex, the scissile phosphates were still not docked near the catalytic serines, and the mechanisms of catalysis and strand exchange remained puzzling.

Other lines of evidence also suggested that the conformation of the isolated WT dimer must be significantly different from that of the catalytically active species formed within the synaptosome. Constitutively active mutants of resolvases (and invertases) had been selected that, unlike the WT proteins, would recombine isolated crossover sites (and many more have been selected since that time) (33–41). Such mutants form tetramers rather than dimers in solution (36, 42–45). The positions of the activating mutations map to the dimer interface region: helix E, the turn leading into it, and the surfaces that it packs against. It had already been noticed when comparing multiple structures that the WT dimer is quite flexible, and nuclear magnetic resonance data showed that helix E is not integral to the folding of the catalytic domain (43, 46, 47). Thus, it seemed likely that activation might involve a rearrangement of helix E relative to the rest of the catalytic domain.

Despite the foreshadowing described above, the magnitude of the activating conformational change, finally revealed in 2005, was unexpected (Figures 4B, C and 5A) (48, 49). The breakthrough structure used an activated mutant of γδ resolvase that crystallized as a

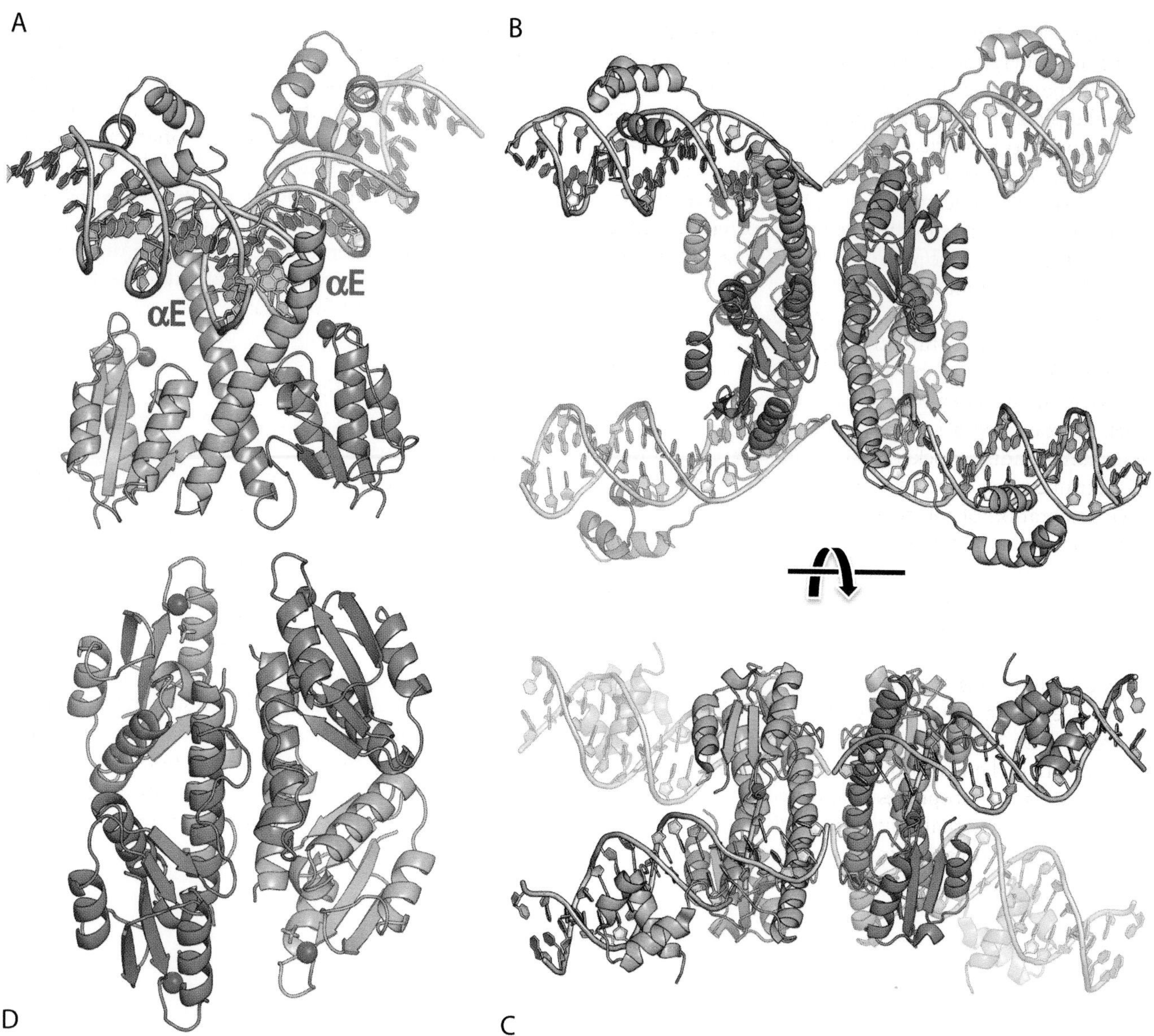

Figure 4 Serine resolvase structures. (A) Wild-type γδ resolvase bound to crossover site (or "site I") DNA. The active site serine residues are marked with red spheres (PDBid 1gdt; (31)). (B) Activated mutant γδ resolvase with crossover site DNA in the covalent protein–DNA intermediate state. Note that each catalytic domain has undergone major conformational changes in the transition from dimer to tetramer (PDBid 2gm4; (48, 49)). (C) The same structure as in B, rotated by ~90° about a horizontal axis. (D) Activated Sin resolvase tetramer catalytic domain tetramer. Sulfate ions that mark the binding pockets for the scissile phosphate are shown as sticks (PDBid 3pkz) (50).
doi:10.1128/microbiolspec.MDNA3-0045-2014.f4

tetramer, with bound DNA in the double-strand break form and with each active site serine residue covalently linked to the 5′ end of a DNA segment. Within each subunit, the catalytic domain core had rotated and tilted with respect to helix E. Furthermore, the old dimer-mediating protein–protein contacts were completely rearranged. The mutations resulting in constitutive activation appeared to both destabilize the dimeric form and to stabilize the tetrameric form. In the center of the tetramer, holding the broken DNA ends together, was a hydrophobic but unusually flat interface. At last one could picture rotation of two subunits relative to the other two! However, the active site itself appeared to have undergone a post-catalytic conformational change: the displaced 3′OH groups were far (> 10 Å) from the phosphoserine linkages and not all of the residues known to be critical for catalysis pointed into the active site. As a result, it remained difficult to propose a structure-based model of the phosphotransfer reactions.

In 2011, a high-resolution structure of an activated Sin resolvase catalytic domain provided a picture of the serine recombinase active site apparently fully assembled for catalysis (50). In this structure, the catalytic domain core is further tilted relative to Helix E (Figures 4D and 5B), and all of the catalytically important side chains assemble into a hydrogen-bonded network surrounding a sulfate ion that marks the scissile phosphate-binding pocket. The overall architecture of this tetramer also appears "ready for cleavage": the relative orientation of the two halves differs by about 35° from that seen in the γδ tetramer, and that rotation brings pairs of active sites closer together. Duplex DNA taken from the inactive WT γδ dimer structure could be docked as a rigid body onto the activated Sin tetramer such that the scissile phosphate groups superimposed on the sulfate ions bound in the Sin active site. This docking exercise also implied that the bending of the DNA may be relatively constant throughout the reaction, and that the catalytic domains simply swing in to meet the scissile phosphates when necessary.

Although more than one structure was obtained for activated γδ resolvase tetramers they all adopted the same overall rotation angle. However, as described above, an activated Sin resolvase tetramer adopted a different angle. A third rotational state was later seen in the structure of an activated Gin invertase tetramer (Figure 6) (51), and most recently a different activated Sin mutant crystallized with three independent tetramers in the asymmetric unit: two similar to that seen previously for Sin, and one in a fourth rotational state (C.S. Trejo and P.A. Rice, unpublished data). These structures illustrate that serine recombinase tetramers can indeed adopt multiple rotational states.

MOLECULAR MODELS FOR THE SYNAPTOSOME

The history of modeling the synaptosomes within which resolution occurs is similarly long and convoluted. Early work focused on Tn3 and γδ resolvases, which are 81% identical and have similar *res* sites. The overall path of the DNA was defined by elegant *in vitro* studies of the Tn3 system showing that within the synaptosome, wrapping of the two *res* sites about one another traps three supercoiling nodes (see chapter by WM Stark). Early work also identified residues that make inter-dimer contacts in the crystal and are important in assembling the synaptosome (28). However, because the full Tn3/γδ synaptosome contains 12 copies of the same protein (Figure 3A), deciphering exactly which pairs of subunits those important contacts are between has been challenging (36, 52, 53). Modeling

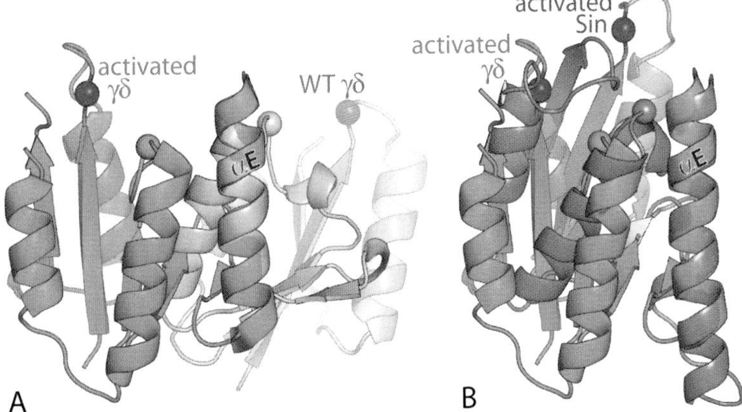

Figure 5 Intramolecular conformational changes upon activation. (A) Superposition, using the E-helices as guides, of one subunit from an activated, DNA-bound γδ resolvase tetramer (green) in a post-cleavage state with one from an inactive wild-type γδ dimer (light green) (PDBids 2gm4 and 2rsl (46, 49)). Red spheres mark the α carbons of the active site serines (S10 for γδ, S9 for Sin); green and blue spheres those of the probable general acid R71 (γδ)/R69 (Sin). (B) Similar superposition of the same activated γδ subunit as in (A), and one subunit from an activated Sin tetramer that appeared to be in the cleavage-ready state (PDBid 3pkz, (50)). doi:10.1128/microbiolspec.MDNA3-0045-2014.f5

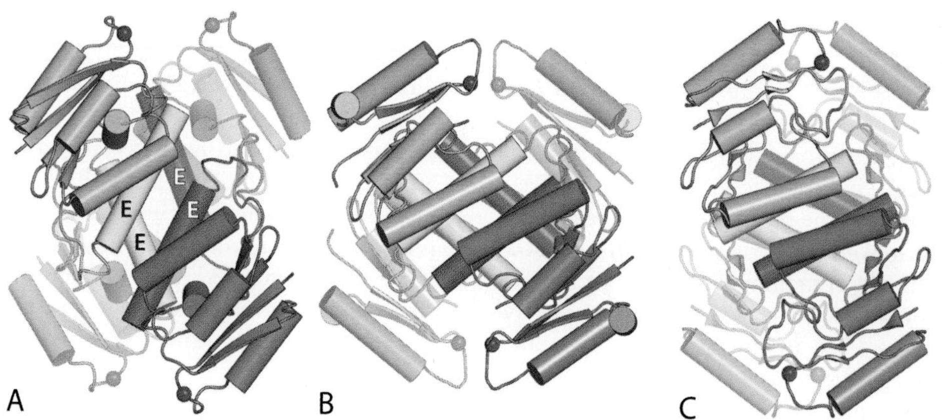

Figure 6 Structures of serine recombinase tetramers in three different rotational states. Colors are as in Figure 4, except that the E-helices are highlighted in yellow (green subunits) and magenta (blue subunits). (A) Activated γδ resolvase tetramer; (B) activated Sin resolvase tetramer; (C) activated Gin invertase tetramer. PDBids 2gm4, 3pkz, and 3uj3, respectively (49–51). doi:10.1128/microbiolspec.MDNA3-0045-2014.f6

has been further complicated by the fact that the central spacer of the three dimer-binding sites of *res* varies by nearly a helical turn, and while biochemical data confirm that complexes with sites II and III are highly asymmetric (54–56), structural data have only been published for γδ resolvase bound to site I. Nevertheless, a number of increasingly sophisticated models for the Tn3/γδ synaptosome have been proposed over the years (19, 36, 53, 57–59).

Sin's *res* site is shorter and its synaptosome proved simpler to model (Figure 3A) (60). The topology of the Sin synaptosome is the same as for Tn3 and γδ resolvases, but the *res* site contains only two Sin dimer-binding sites separated by a gap that is synergistically bound by a DNA bending protein (38). The activated γ δ resolvase–site I complex provides a good model for that of Sin because of the overall similarity of the proteins (~33% sequence identity and nearly identical folds) and the geometry of their site Is. In the natural Sin host *Staphylococcus aureus*, the DNA bending protein is most likely the highly conserved nucleoid-associated protein HU (originally named as Histone-like protein from strain U13). Structures are available for HU in complex with DNA, but its lack of sequence specificity complicates precise modeling (61). Fortunately, HU can be replaced with the structurally similar but sequence-specific *Escherichia coli* IHF ("Integration Host Factor") protein if the WT site I–site II spacer DNA sequence is replaced with an IHF binding site in the correct register (62, 63). Hence for modeling the synaptosome, the only missing structure was that of the Sin–site II complex.

Sin's site II is a direct rather than inverted repeat of the two half-sites. The structure of a WT Sin–site II complex showed how the dimeric protein can bind a direct repeat (Figure 7A). Although the catalytic domains form an approximately two-fold symmetric dimer similar to that of γδ resolvase, there is a break in one of the E-helices that allows major re-orientation of the DNA-binding domain (64). The extension of the other E helix binds in the minor groove, bending the DNA in the same way that γδ resolvase does. Nevertheless, when the γδ resolvase–site I complex is superimposed on the Sin–site II complex, the DNAs are almost mutually perpendicular. The observation that two Sin–site II complexes can pair even in the absence of the other synaptosome components could be explained by previously unsuspected protein–protein contacts between the DNA-binding domains seen in the crystal. The importance of these new contacts in site II–site II interactions was independently established by an elegant genetic screen (39, 64). The direct repeat arrangement of the binding sites perfectly orients the two pairs of DNA binding domains to make synergistic interactions, which explains why Sin–site I complexes do not form similar binding-domain-mediated tetramers.

Docking of the activated γδ-site I, IHF-DNA, and Sin-site II structures created a model of the Sin synaptosome that matches the known topology (Figure 7) (64). It also arranges the site I- and II-bound proteins such that a rotation of the site II-bound catalytic domains about helix E's hinge point would lead to dimer–dimer contacts mediated by a patch of residues known to be

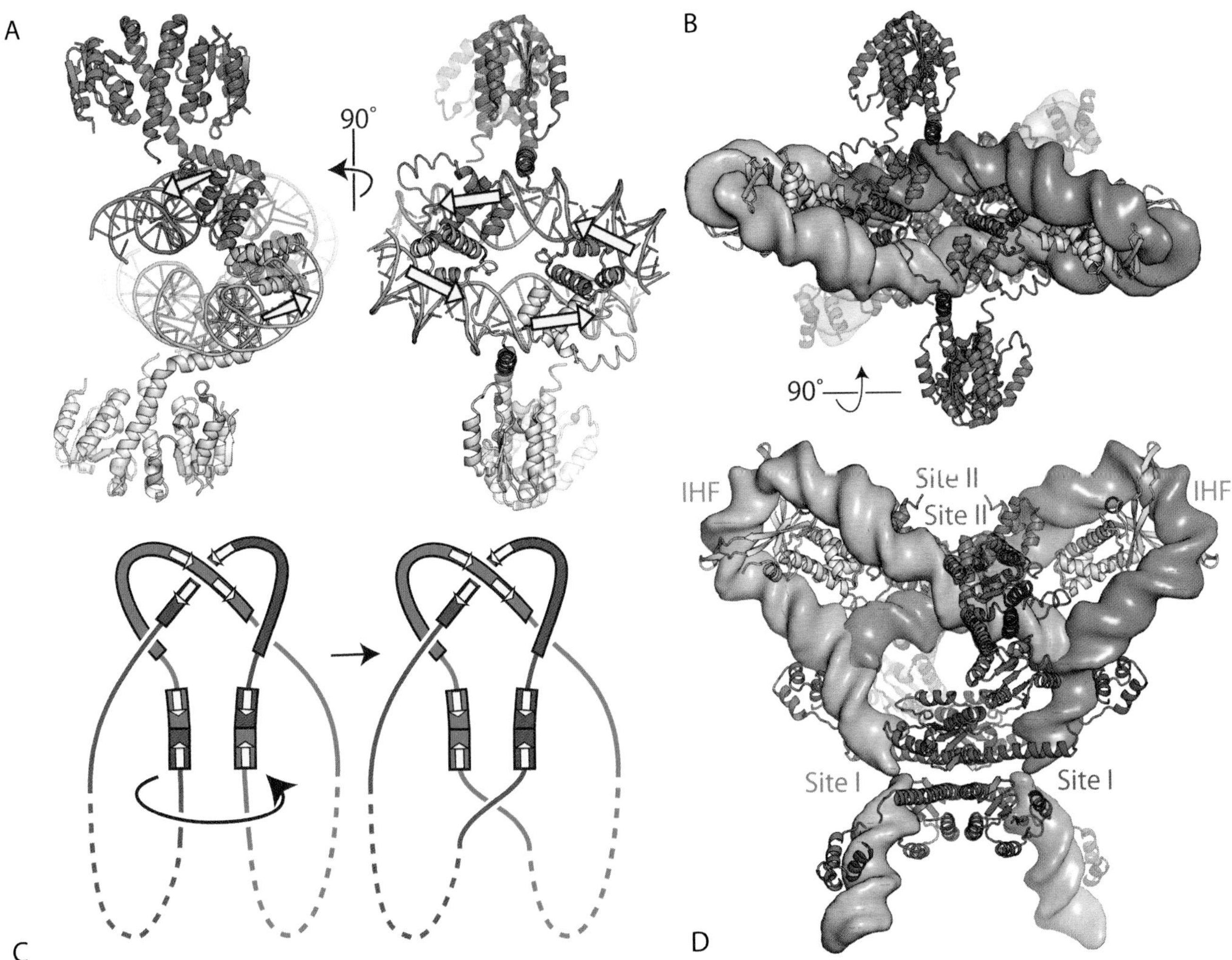

Figure 7 Modeling the Sin synaptosome. (A) Two orthogonal views of the structure of a tetramer of WT Sin resolvase bound to its cognate site II. In the left panel, the catalytic domains of the lower dimer are oriented in a similar way to those of wild-type γδ resolvase in Figure 4(A). The dimer–dimer contacts are mediated by the DNA-binding domains (PDBid 2r0q (64)). Yellow arrows show the orientations of the individual monomer-binding sites in the DNA. (B) Model for the synaptosome, created by rigid-body docking together of a symmetrized version of an activated γδ resolvase tetramer–DNA structure (blue and green proteins), two copies of an IHF–DNA complex structure (pink proteins), and the Sin–site II structure (purple proteins). The DNAs are shown as smoothed green and blue surfaces. The view is the same as for the right hand panel of (A). PDBids 1zr4, 1ihf, and 2r0q were used (48, 64, 84). (C) Cartoon similar to that in Figure 3B showing the expected synaptosome topology. Yellow arrows mark the orientations of the half-sites within each dimer-binding site. (D) Second view of the synaptosome model, rotated 90° about a horizontal axis. doi:10.1128/microbiolspec.MDNA3-0045-2014.f7

important in stabilizing the synaptosome (and analogous to those that make inter-dimer contacts in Tn3 and γδ resolvases). The model was supported by careful analysis of the combined effects of mutations in this patch, mutations that interfere with DNA-binding domain-mediated interactions and activating mutations (39).

How similar are other serine resolvase synaptosomes to Sin's? A general pattern among diverse *res* sites can be seen in Figure 3A: Site I, where the crossover occurs, is always an inverted repeat, and it is always followed by two additional "accessory" resolvase-binding and/or bending protein-binding sites. The additional resolvase binding sites vary in the relative orientation and spacing of their half-sites, but they always differ from site I. The only other resolvase with a Sin-like res site that has been studied is β recombinase, and it is likely to form a Sin-like synaptosome (14, 65, 66). However, under some conditions β can also catalyze inversions, implying that the requirements for catalytic activation may be less stringent than those for Sin (14).

In contrast to Sin and β, Tn3 resolvase appears to have found a different solution for constructing a three-noded synaptosome. A recent Tn3 resolvase–site III complex structure (S.P. Montaño and P.A. Rice, unpublished data) shows that its short central spacer allows only one E-helix to bind in the minor groove, giving the complex an overall asymmetry that is different yet again from the Sin–site II complex but that agrees well with previous footprinting data (54). It appears that the variations in the additional (noncrossover) resolvase-binding sites are tailored to induce particular geometries in the resulting protein–DNA complex. A Tn3 synaptosome model has been constructed using this new Tn3–site III structure, the activated γδ–site I structure, and SAXS-supported modeling of the Tn3–site II complex. The model also uses the inter-dimer contacts seen in crystals of Tn3 and γδ resolvases and new data that finally deconvolutes exactly which pairs of subunits interact (S.-J. Rowland, M.R. Boocock and W.M. Stark, unpublished data). Interestingly, while Sin and Tn3 use the same patch of protein to make those contacts, they differ in which subunits within the synaptosomes those contacts link. It appears that these systems have converged upon different ways to assemble three-noded synaptosomes because that topology is perfect for channeling the recombination reaction to produce only resolution products (rather than inversion or integration products).

ACTIVATION AND CATALYSIS

How does incorporation into a synaptosome activate the site-I bound proteins? It must at least transiently tip the conformational equilibrium from dimer to tetramer. By bringing the two site I-bound dimers into close proximity, the synaptosome can favor the tetramer simply by applying mass action. The local concentration of site-I-bound resolvase in the synaptosome can be roughly estimated as that of four subunits in a cube 130 Å on each side, which is 30 mM. *In vitro*, WT Sin bound to isolated crossover sites is normally inactive, but slight catalytic activity begins to be detectable if the concentration is raised to even 10 μM (45). In addition, the protein–protein contacts between crossover- and accessory-site-bound subunits may be activating as well as structurally stabilizing. The residues involved lie on the opposite side of the catalytic domain from helix E, and formation of these contacts may induce and/or stabilize rotation of the catalytic domain core relative to helix E, thus favoring the conformation found in the tetramer (Figure 8). This mechanism would explain why, even though the overall arrangement of accessory proteins is different in the Sin and Tn3 synaptosomes, the same patch of side chains is used to make these contacts. Furthermore, an overlapping patch is important in activating the serine invertase Hin, albeit through protein–DNA rather than protein–protein contacts (67).

Once activated, how do serine recombinases catalyze DNA cleavage and re-ligation? The chemistry of phosphotransfer reactions has been extensively reviewed elsewhere (68, 69). Serine recombinases are unusual, although not unique, in that they do not require divalent cations such as Mg^{2+} but instead rely only on amino acid side chains to provide catalytic power (19, 70). Biochemical studies, in conjunction with the evolving structural data, have identified a set of residues that are important for activity and proposed roles for them (42, 71–74). As shown in Figure 9, the "inner circle" surrounding the scissile phosphate is comprised of three arginines. Arginine is often found in phosphotransferase

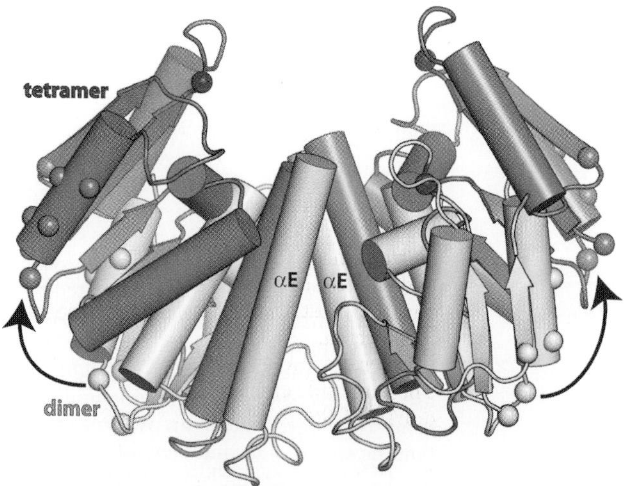

Figure 8 Interdimer contacts within the synaptosome may affect the dimer–tetramer equilibrium. A wild-type Sin dimer (yellow) is superimposed on two subunits from the activated Sin tetramer (blue). Red spheres mark the α carbons of the catalytic serines, and other spheres mark the positions of side chains whose mutation interferes with interdimer contacts in Sin and γδ resolvases and with catalytic domain–DNA contacts in the related Hin invertase (28, 39, 67).
doi:10.1128/microbiolspec.MDNA3-0045-2014.f8

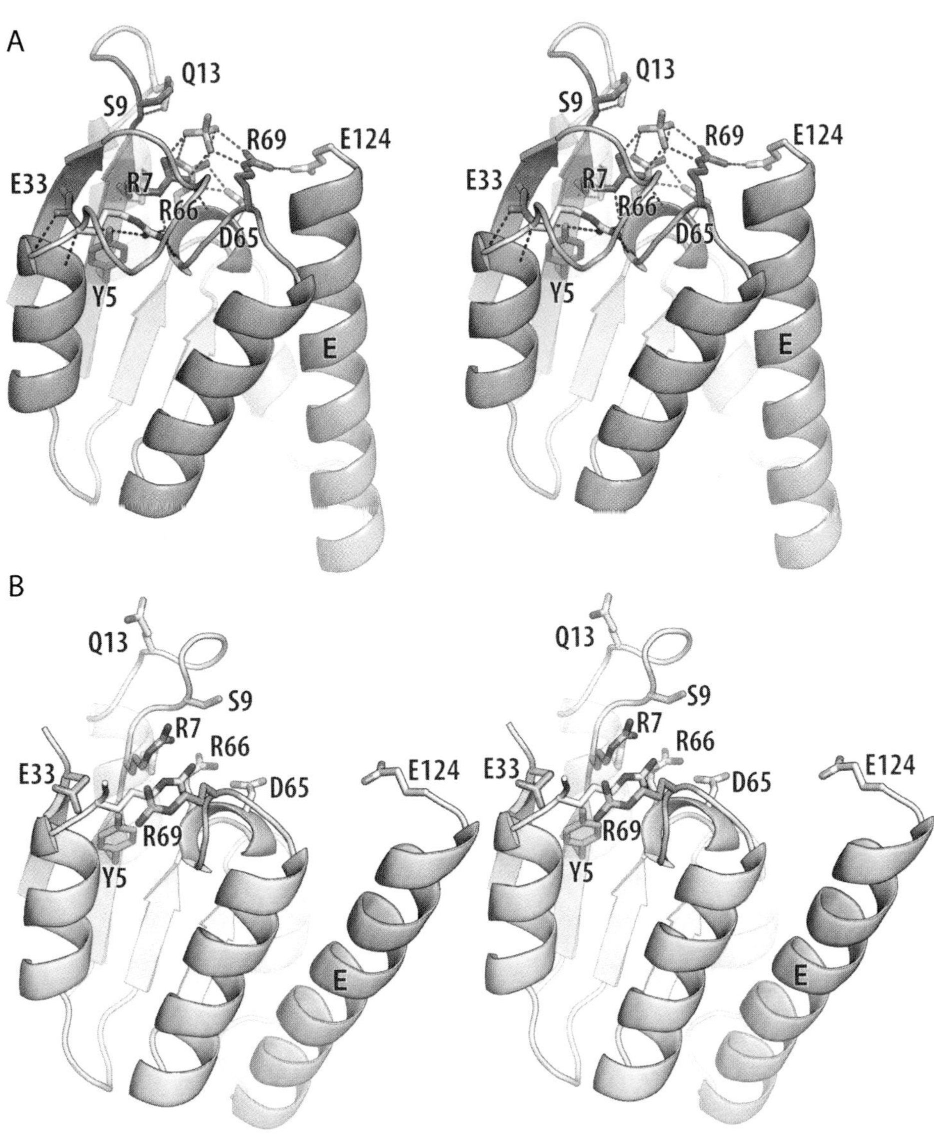

Figure 9 Details of the Sin active site. (A) Stereo view of one subunit from an activated Sin tetramer (PDBid 3pkz (50)). A sulfate ion marks the scissile-phosphate binding pocket, and side chains important for catalysis are shown as sticks. Those residues whose mutation had the most deleterious effect on the rate of DNA cleavage by Tn3 resolvase are shown in magenta, shading to white for those whose mutation had more moderate effects (74). (B) Stereo view of one subunit from a site II-bound wild-type Sin dimer (PDBid 2r0q (64)). The same side chains are shown, similarly shaded from a dark color to white. doi:10.1128/microbiolspec.MDNA3-0045-2014.f9

active sites, and may aid catalysis in several ways: it can localize the scissile phosphate through bidentate hydrogen bonds, it can stabilize additional partial negative charge on the transition state, and if properly oriented, it may preferentially stabilize the geometry of the transition state as well.

What is surprising about the serine recombinase active site is the lack of residues commonly associated with general acid–base catalysis such as histidine. The high intrinsic pKa of arginine makes it an unusual candidate for that role, but there is precedent (75). Furthermore, the clustering of several arginines in close proximity may alter the equilibrium between the positively charged and neutral forms, thus lowering their pKas. In the Sin active site, R7 is well-positioned to abstract a proton from the nucleophilic serine as it attacks, R66 to hold the scissile phosphate in place and stabilize the transition state, and R69 and R66 to donate a proton to the DNA 3′ oxygen as it leaves. Evidence that R69 is important in stabilizing the leaving group (during the cleavage reaction) and likely to be the general acid was supplied by experiments replacing the 3′O with an S, which is predicted to be a better leaving group in the absence of a general acid (76). Mutants of R69, but not of other arginines, were partially rescued by the 3′S (although it should be noted that R66 mutants were simply inactive with all substrates). Careful analysis of the kinetic effects of R to A versus R to K changes also showed that the residues in this positively charged cluster probably do collaborate to lower one another's pKas (76). In this respect, the catalytic center should be considered as a cooperative network.

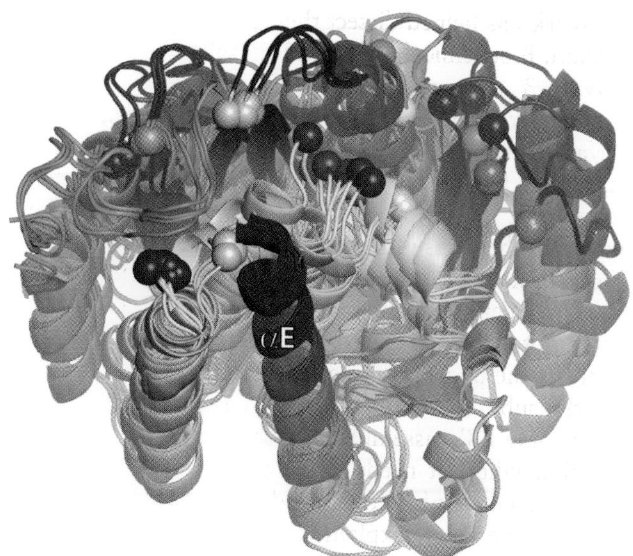

Figure 10 Superposition of all currently published serine resolvase catalytic domain structures, aligned using their E helices as guides. Each structure is shaded from blue (N-ter) to red (end of catalytic domain), and all unique copies from the asymmetric unit for each structure were used (two to four per structure). Colored spheres mark the positions of the active site serines. From right to left (roughly), they are: light blue, wild-type (WT) Sin from the Sin–site II structure (PDBid 2r0q); dark blue, WT γδ determined without DNA (PDBid 2rsl); green, WT γδ from the γδ–site I complex structure (PDBid 1gdt); yellow, activated γδ from a structure without DNA (PDBid 2gm5); orange and red, two different activated variants of γδ from structures with covalently linked crossover site DNA (PDBids 1zr4 and 2gm4, respectively), and pink, activated Sin tetramer (determined without DNA; PDBid 1pkz). Black and Gray spheres mark the Cα positions of the probable general acid R71 (γδ)/R69 (Sin). Those that cluster towards the left are in structures of activated mutants, whereas those that cluster towards the right are from WT structures.
doi:10.1128/microbiolspec.MDNA3-0045-2014.f10

FUTURE QUESTIONS

Several decades of careful study have drawn an increasingly detailed picture of these unusual enzymes. Armed with buffets of designer variants and numerous structures, the field is poised to branch into new directions. From a basic science viewpoint, serine resolvases provide a rich system for investigating protein dynamics as well as the enzymology of phosphotransfer reactions. From a medical viewpoint, serine resolvases are underappreciated functional components of many antibiotic resistance-bearing transposons and plasmids. Finally, from a biotech viewpoint, they can be useful genetic engineering tools.

The subunit rotation hypothesis, once rather puzzling at the molecular level, is now well-supported by many lines of evidence, from topological to structural. However, the detailed dynamics of this molecular swivel have not been examined. Does it move smoothly or are there significant energy minima along the way? Is there an intrinsic energy minimum at 180° that would cause pausing when re-ligation should occur, or is the only stop-the-merry-go-round signal the pairing of the 2-nucleotide overhangs? Although describing the protein conformational changes as simply dimer to tetramer is convenient, the superposition of structures shown in Figures 5B and 10 suggests that reality is more complicated. With the caveat that structures of different proteins (Sin vs. γδ) are used to compare different states, it appears that the ready-to-cleave tetramer conformation differs from the post-cleavage one. Figure 10 also highlights the flexibility of the inactive dimer (note the spread of blue and green spheres marking the position of its active site serine). Much, but not all, of this conformational variability reflects the motion of helix E. That helix and the region of protein it packs against are unusually rich in methionine residues, which are the most flexible of the large hydrophobic amino acids. Methionine-rich binding pockets are also used by the signal recognition particle and the chaperone DnaK to make favorable interactions with hydrophobic but variable substrates (77–79). While more structures, especially at high resolution, that trap additional rotational states would be very useful, even now the system seems ripe for computational exploration of its dynamics.

Although the residues that are critical for catalysis have been identified and recent structural and biochemical work has helped dissect their roles, several questions remain. Biochemical data link R69 with protonating the leaving 3′ oxygen during the DNA cleavage reaction, but is there a general base that accepts a proton from S9 as it attacks? R7 is nicely poised to do so in the structure, but there is no direct experimental evidence to support its playing that role. It is reasonable to expect that the pKas of the active site arginines are shifted towards neutral, which could make them more efficacious as general acid–base catalysts, but their pKas have not been measured. Another question is, what protects the phosphoserine linkage from hydrolysis during the subunit rotation step? It may be that the active site is only fully assembled at the 0° and 180° points of rotation where it needs to be catalytically competent. The role effect of divalent and multivalent cations on the enzyme is unexplained in detail. While serine recombinases do not require Mg^{2+} to catalyze phosphotransfer reactions, the phosphoserine-linked intermediate tends to accumulate in its absence (80, 81). No structures have revealed binding pockets for Mg^{2+}, and it can be replaced with Ca^{2+} or polyvalent cations such as spermidine (19, 70). Do these cations facilitate pairing of the 2-nucleotide overhangs before re-ligation through charge neutralization, perhaps stabilizing the catalysis-ready conformation, or do they affect some other stage of the strand exchange process? Finally, if the 2-nucleotide overhangs do not form Watson : Crick pairs after rotation, the complex simply rotates another 180° to realign the original partners, but how exactly is that base pairing checked (23, 82)?

The extensive structural and biochemical data regarding small serine recombinases (resolvases and invertases) have made them excellent targets for engineering. As discussed in more detail in the chapter by W.M. Stark, genomic rearrangements can be targeted to specific sequences by fusing the catalytic domain of an activated serine resolvase with the DNA binding domain of choice (7, 8). These simple, largely modular recombinases currently lack the directionality shown by large serine recombinases (83). Adding directionality to these designer-specificity recombinases will be an interesting protein engineering challenge for the future.

Acknowledgments. *I thank Martin Boocock, Sally Rowland and Marshall Stark for comments on the manuscript, decades of informative discussion, and help with Figure 3A, and Caitlin S. Trejo and Nancy Craig for comments on the manuscript. Macromolecular structure figures were generated using The PyMOL Molecular Graphics System, Version 1.7 Schrödinger, LLC. (http://www.pymol.org/)*

Citation. Rice P. 2014. Serine resolvases. Microbiol Spectrum 3(2):MDNA3-0045-2014

References

1. **Grindley NDF, Whiteson KL, Rice PA.** 2006. Mechanisms of site-specific recombination. *Annu Rev Biochem* **75:**567–605.
2. **Rice PA, Mouw KW, Montaño SP, Boocock MR, Rowland S-J, Stark WM.** 2010. Orchestrating serine resolvases. *Biochem Soc Trans* **38:**384–387.
3. **Boocock MR, Brown JL, Sherratt DJ.** 1986. Structural and catalytic properties of specific complexes between Tn3 resolvase and the recombination site res. *Biochem Soc Trans* **14:**214–216.
4. **Wasserman SA, Dungan JM, Cozzarelli NR.** 1985. Discovery of a predicted DNA knot substantiates a model for site-specific recombination. *Science* **229:**171–174.
5. **Merickel SK, Johnson RC.** 2004. Topological analysis of Hin-catalysed DNA recombination *in vivo* and *in vitro*. *Mol Microbiol* **51:**1143–1154.
6. **Garvie CW, Wolberger C.** 2001. Recognition of specific DNA sequences. *Mol Cell* **8:**937–946.
7. **Akopian A, He J, Boocock MR, Stark WM.** 2003. Chimeric recombinases with designed DNA sequence recognition. *Proc Natl Acad Sci USA* **100:**8688–8691.
8. **Mercer AC, Gaj T, Fuller RP, Barbas CF.** 2012. Chimeric TALE recombinases with programmable DNA sequence specificity. *Nucleic Acids Res* **40:**11163–11172.
9. **Rutherford K, Van Duyne GD.** 2014. The ins and outs of serine integrase site-specific recombination. *Curr Opin Struct Biol* **24:**125–131.
10. **Hickman AB, Chandler M, Dyda F.** 2010. Integrating prokaryotes and eukaryotes: DNA transposases in light of structure. *Crit Rev Biochem Mol Biol* **45:**50–69.
11. **Lesterlin C, Barre F-X, Cornet F.** 2004. Genetic recombination and the cell cycle: what we have learned from chromosome dimers. *Mol Microbiol* **54:**1151–1160.
12. **Rojo F, Alonso JC.** 1994. A novel site-specific recombinase encoded by the *Streptococcus pyogenes* plasmid pSM19035. *J Mol Biol* **238:**159–172.
13. **Ceglowski P, Boitsov A, Karamyan N, Chai S, Alonso JC.** 1993. Characterization of the effectors required for stable inheritance of *Streptococcus pyogenes* pSM19035-derived plasmids in *Bacillus subtilis*. *Mol Gen Genet* **241:**579–585.
14. **Canosa I, López G, Rojo F, Boocock MR, Alonso JC.** 2003. Synapsis and strand exchange in the resolution and DNA inversion reactions catalysed by the beta recombinase. *Nucleic Acids Res* **31:**1038–1044.
15. **Paulsen IT, Gillespie MT, Littlejohn TG, Hanvivatvong O, Rowland SJ, Dyke KG, Skurray RA.** 1994. Characterisation of sin, a potential recombinase-encoding gene from *Staphylococcus aureus*. *Gene* **141:**109–114.
16. **Shearer JES, Wireman J, Hostetler J, Forberger H, Borman J, Gill J, Sanchez S, Mankin A, Lamarre J, Lindsay JA, Bayles K, Nicholson A, O'Brien F, Jensen SO, Firth N, Skurray RA, Summers AO.** 2011. Major families of multiresistant plasmids from geographically and epidemiologically diverse staphylococci. *G3 Bethesda Md* **1:**581–591.

17. **Stark WM, Boocock MR, Sherratt DJ.** 1989. Site-specific recombination by Tn3 resolvase. *Trends Genet* **5:** 304–309.

18. **Watson MA, Boocock MR, Stark WM.** 1996. Rate and selectively of synapsis of res recombination sites by Tn3 resolvase. *J Mol Biol* **257:**317–329.

19. **Stark WM, Sherratt DJ, Boocock MR.** 1989. Site-specific recombination by Tn3 resolvase: topological changes in the forward and reverse reactions. *Cell* **58:** 779–790.

20. **Cozzarelli NR, Krasnow MA, Gerrard SP, White JH.** 1984. A topological treatment of recombination and topoisomerases. *Cold Spring Harb Symp Quant Biol* **49:** 383–400.

21. **Sumners DW, Ernst C, Spengler SJ, Cozzarelli NR.** 1995. Analysis of the mechanism of DNA recombination using tangles. *Q Rev Biophys* **28:**253–313.

22. **Vologodskii AV, Cozzarelli NR.** 1994. Conformational and thermodynamic properties of supercoiled DNA. *Annu Rev Biophys Biomol Struct* **23:**609–643.

23. **Stark WM, Grindley ND, Hatfull GF, Boocock MR.** 1991. Resolvase-catalysed reactions between res sites differing in the central dinucleotide of subsite I. *EMBO J* **10:**3541–3548.

24. **Stark WM, Boocock MR.** 1994. The linkage change of a knotting reaction catalysed by Tn3 resolvase. *J Mol Biol* **239:**25–36.

25. **Sanderson MR, Freemont PS, Rice PA, Goldman A, Hatfull GF, Grindley ND, Steitz TA.** 1990. The crystal structure of the catalytic domain of the site-specific recombination enzyme gamma delta resolvase at 2.7 A resolution. *Cell* **63:**1323–1329.

26. **Grindley ND, Reed RR.** 1985. Transpositional recombination in prokaryotes. *Annu Rev Biochem* **54:**863–896.

27. **Hughes RE, Rice PA, Steitz TA, Grindley ND.** 1993. Protein-protein interactions directing resolvase site-specific recombination: a structure–function analysis. *EMBO J* **12:** 1447–1458.

28. **Hughes RE, Hatfull GF, Rice P, Steitz TA, Grindley ND.** 1990. Cooperativity mutants of the gamma delta resolvase identify an essential interdimer interaction. *Cell* **63:** 1331–1338.

29. **Yang W.** 2010. Topoisomerases and site-specific recombinases: similarities in structure and mechanism. *Crit Rev Biochem Mol Biol* **45:**520–534.

30. **Artymiuk PJ, Ceska TA, Suck D, Sayers JR.** 1997. Prokaryotic 5′-3′ exonucleases share a common core structure with gamma-delta resolvase. *Nucleic Acids Res* **25:** 4224–4229.

31. **Yang W, Steitz TA.** 1995. Crystal structure of the site-specific recombinase gamma delta resolvase complexed with a 34 bp cleavage site. *Cell* **82:**193–207.

32. **Rimphanitchayakit V, Hatfull GF, Grindley ND.** 1989. The 43 residue DNA binding domain of gamma delta resolvase binds adjacent major and minor grooves of DNA. *Nucleic Acids Res* **17:**1035–1050.

33. **Klippel A, Cloppenborg K, Kahmann R.** 1988. Isolation and characterization of unusual gin mutants. *EMBO J* **7:** 3983–3989.

34. **Haffter P, Bickle TA.** 1988. Enhancer-independent mutants of the Cin recombinase have a relaxed topological specificity. *EMBO J* **7:**3991–3996.

35. **Haykinson MJ, Johnson LM, Soong J, Johnson RC.** 1996. The Hin dimer interface is critical for Fis-mediated activation of the catalytic steps of site-specific DNA inversion. *Curr Biol* **6:**163–177.

36. **Sarkis GJ, Murley LL, Leschziner AE, Boocock MR, Stark WM, Grindley ND.** 2001. A model for the gamma delta resolvase synaptic complex. *Mol Cell* **8:**623–631.

37. **Burke ME, Arnold PH, He J, Wenwieser SVCT, Rowland S-J, Boocock MR, Stark WM.** 2004. Activating mutations of Tn3 resolvase marking interfaces important in recombination catalysis and its regulation. *Mol Microbiol* **51:**937–948.

38. **Rowland SJ, Boocock MR, Stark WM.** 2005. Regulation of Sin recombinase by accessory proteins. *Mol Microbiol* **56:**371–382.

39. **Rowland S-J, Boocock MR, McPherson AL, Mouw KW, Rice PA, Stark WM.** 2009. Regulatory mutations in Sin recombinase support a structure-based model of the synaptosome. *Mol Microbiol* **74:**282–298.

40. **Heiss JK, Sanders ER, Johnson RC.** 2011. Intrasubunit and intersubunit interactions controlling assembly of active synaptic complexes during Hin-catalyzed DNA recombination. *J Mol Biol* **411:**744–764.

41. **Arnold PH, Blake DG, Grindley ND, Boocock MR, Stark WM.** 1999. Mutants of Tn3 resolvase which do not require accessory binding sites for recombination activity. *EMBO J* **18:**1407–1414.

42. **Olorunniji FJ, He J, Wenwieser SVCT, Boocock MR, Stark WM.** 2008. Synapsis and catalysis by activated Tn3 resolvase mutants. *Nucleic Acids Res* **36:**7181–7191.

43. **Nöllmann M, He J, Byron O, Stark WM.** 2004. Solution structure of the Tn3 resolvase-crossover site synaptic complex. *Mol Cell* **16:**127–137.

44. **Dhar G, McLean MM, Heiss JK, Johnson RC.** 2009. The Hin recombinase assembles a tetrameric protein swivel that exchanges DNA strands. *Nucleic Acids Res* **37:** 4743–4756.

45. **Mouw KW, Steiner AM, Ghirlando R, Li N-S, Rowland S-J, Boocock MR, Stark WM, Piccirilli JA, Rice PA.** 2010. Sin resolvase catalytic activity and oligomerization state are tightly coupled. *J Mol Biol* **404:**16–33.

46. **Rice PA, Steitz TA.** 1994. Refinement of gamma delta resolvase reveals a strikingly flexible molecule. *Struct Lond Engl* 1993 **2:**371–384.

47. **Pan B, Deng Z, Liu D, Ghosh S, Mullen GP.** 1997. Secondary and tertiary structural changes in gamma delta resolvase: comparison of the wild-type enzyme, the I110R mutant, and the C-terminal DNA binding domain in solution. *Protein Sci Publ Protein Soc* **6:**1237–1247.

48. **Li W, Kamtekar S, Xiong Y, Sarkis GJ, Grindley NDF, Steitz TA.** 2005. Structure of a synaptic gammadelta resolvase tetramer covalently linked to two cleaved DNAs. *Science* **309:**1210–1215.

49. **Kamtekar S, Ho RS, Cocco MJ, Li W, Wenwieser SVCT, Boocock MR, Grindley NDF, Steitz TA.** 2006. Implications of structures of synaptic tetramers of gamma delta

resolvase for the mechanism of recombination. *Proc Natl Acad Sci USA* **103:**10642–10647.

50. **Keenholtz RA, Rowland S-J, Boocock MR, Stark WM, Rice PA.** 2011. Structural basis for catalytic activation of a serine recombinase. *Struct Lond Engl* **19:** 799–809.

51. **Ritacco CJ, Kamtekar S, Wang J, Steitz TA.** 2013. Crystal structure of an intermediate of rotating dimers within the synaptic tetramer of the G-segment invertase. *Nucleic Acids Res* **41:**2673–2682.

52. **Soultanas P, Oram M, Halford SE.** 1995. Site-specific recombination at res sites containing DNA-binding sequences for both Tn21 and Tn3 resolvases. *J Mol Biol* **245:**208–218.

53. **Murley LL, Grindley ND.** 1998. Architecture of the gamma delta resolvase synaptosome: oriented heterodimers identity interactions essential for synapsis and recombination. *Cell* **95:**553–562.

54. **Mazzarelli JM, Ermácora MR, Fox RO, Grindley ND.** 1993. Mapping interactions between the catalytic domain of resolvase and its DNA substrate using cysteine-coupled EDTA-iron. *Biochemistry (Mosc)* **32:** 2979–2986.

55. **Blake DG, Boocock MR, Sherratt DJ, Stark WM.** 1995. Cooperative binding of Tn3 resolvase monomers to a functionally asymmetric binding site. *Curr Biol CB* **5:** 1036–1046.

56. **Nöllmann M, Byron O, Stark WM.** 2005. Behavior of Tn3 resolvase in solution and its interaction with res. *Biophys J* **89:**1920–1931.

57. **Krasnow MA, Cozzarelli NR.** 1983. Site-specific relaxation and recombination by the Tn3 resolvase: recognition of the DNA path between oriented res sites. *Cell* **32:** 1313–1324.

58. **Rice PA, Steitz TA.** 1994. Model for a DNA-mediated synaptic complex suggested by crystal packing of gamma delta resolvase subunits. *EMBO J* **13:**1514–1524.

59. **Leschziner AE, Grindley NDF.** 2003. The architecture of the gammadelta resolvase crossover site synaptic complex revealed by using constrained DNA substrates. *Mol Cell* **12:**775–781.

60. **Rowland S-J, Stark WM, Boocock MR.** 2002. Sin recombinase from *Staphylococcus aureus*: synaptic complex architecture and transposon targeting. *Mol Microbiol* **44:**607–619.

61. **Swinger KK, Lemberg KM, Zhang Y, Rice PA.** 2003. Flexible DNA bending in HU-DNA cocrystal structures. *EMBO J* **22:**3749–3760.

62. **Rowland S-J, Boocock MR, Stark WM.** 2006. DNA bending in the Sin recombination synapse: functional replacement of HU by IHF. *Mol Microbiol* **59:**1730–1743.

63. **Swinger KK, Rice PA.** 2004. IHF and HU: flexible architects of bent DNA. *Curr Opin Struct Biol* **14:**28–35.

64. **Mouw KW, Rowland S-J, Gajjar MM, Boocock MR, Stark WM, Rice PA.** 2008. Architecture of a serine recombinase-DNA regulatory complex. *Mol Cell* **30:** 145–155.

65. **Alonso JC, Weise F, Rojo F.** 1995. The *Bacillus subtilis* histone-like protein Hbsu is required for DNA resolution and DNA inversion mediated by the beta recombinase of plasmid pSM19035. *J Biol Chem* **270:** 2938–2945.

66. **Rojo F, Alonso JC.** 1995. The beta recombinase of plasmid pSM19035 binds to two adjacent sites, making different contacts at each of them. *Nucleic Acids Res* **23:** 3181–3188.

67. **McLean MM, Chang Y, Dhar G, Heiss JK, Johnson RC.** 2013. Multiple interfaces between a serine recombinase and an enhancer control site-specific DNA inversion. *eLife* **2:**e01211.

68. **Lassila JK, Zalatan JG, Herschlag D.** 2011. Biological phosphoryl-transfer reactions: understanding mechanism and catalysis. *Annu Rev Biochem* **80:**669–702.

69. **Cassano AG, Anderson VE, Harris ME.** 2004. Understanding the transition states of phosphodiester bond cleavage: insights from heavy atom isotope effects. *Biopolymers* **73:**110–129.

70. **Castell SE, Jordan SL, Halford SE.** 1986. Site-specific recombination and topoisomerization by Tn21 resolvase: role of metal ions. *Nucleic Acids Res* **14:**7213–7226.

71. **Reed RR, Moser CD.** 1984. Resolvase-mediated recombination intermediates contain a serine residue covalently linked to DNA. *Cold Spring Harb Symp Quant Biol* **49:** 245–249.

72. **Boocock MR, Zhu X, Grindley ND.** 1995. Catalytic residues of gamma delta resolvase act in cis. *EMBO J* **14:** 5129–5140.

73. **Leschziner AE, Boocock MR, Grindley ND.** 1995. The tyrosine-6 hydroxyl of gamma delta resolvase is not required for the DNA cleavage and rejoining reactions. *Mol Microbiol* **15:**865–870.

74. **Olorunniji FJ, Stark WM.** 2009. The catalytic residues of Tn3 resolvase. *Nucleic Acids Res* **37:**7590–7602.

75. **Guillén Schlippe YV, Hedstrom L.** 2005. A twisted base? The role of arginine in enzyme-catalyzed proton abstractions. *Arch Biochem Biophys* **433:**266–278.

76. **Keenholtz RA, Mouw KW, Boocock MR, Li N-S, Piccirilli JA, Rice PA.** 2013. Arginine as a general acid catalyst in serine recombinase-mediated DNA cleavage. *J Biol Chem* **288:**29206–29214.

77. **Bernstein HD, Poritz MA, Strub K, Hoben PJ, Brenner S, Walter P.** 1989. Model for signal sequence recognition from amino-acid sequence of 54K subunit of signal recognition particle. *Nature* **340:**482–486.

78. **Keenan RJ, Freymann DM, Walter P, Stroud RM.** 1998. Crystal structure of the signal sequence binding subunit of the signal recognition particle. *Cell* **94:** 181–191.

79. **Zhu X, Zhao X, Burkholder WF, Gragerov A, Ogata CM, Gottesman ME, Hendrickson WA.** 1996. Structural analysis of substrate binding by the molecular chaperone DnaK. *Science* **272:**1606–1614.

80. **Reed RR, Grindley ND.** 1981. Transposon-mediated site-specific recombination in vitro: DNA cleavage and protein–DNA linkage at the recombination site. *Cell* **25:** 721–728.

81. **Bai H, Sun M, Ghosh P, Hatfull GF, Grindley NDF, Marko JF.** 2011. Single-molecule analysis reveals the molecular bearing mechanism of DNA strand exchange by a serine recombinase. *Proc Natl Acad Sci USA* **108:**7419–7424.

82. **McIlwraith MJ, Boocock MR, Stark WM.** 1997. Tn3 resolvase catalyses multiple recombination events without intermediate rejoining of DNA ends. *J Mol Biol* **266:** 108–121.

83. **Colloms SD, Merrick CA, Olorunniji FJ, Stark WM, Smith MCM, Osbourn A, Keasling JD, Rosser SJ.** 2014. Rapid metabolic pathway assembly and modification using serine integrase site-specific recombination. *Nucleic Acids Res* **42:**e23.

84. **Rice PA, Yang S, Mizuuchi K, Nash HA.** 1996. Crystal structure of an IHF-DNA complex: a protein-induced DNA U-turn. *Cell* **87:**1295–1306.

Mobile DNA, 3rd Edition
Nancy L. Craig, Michael Chandler, Martin Gellert, Alan M. Lambowitz, Phoebe A. Rice and Suzanne Sandmeyer

doi:10.1128/microbiolspec.MDNA3-0059-2014

Margaret C. M. Smith[1]

11 Phage-encoded Serine Integrases and Other Large Serine Recombinases

INTRODUCTION

Conservative site-specific recombination systems are ubiquitous in bacteria, where they play important roles in the horizontal transfer of genetic information, genome stability, and the control of gene expression. The outcomes of site-specific recombination are DNA integration, DNA excision (sometimes referred to as resolution), and DNA inversion. The systems comprise a recombinase, the sequence specific DNA substrates, and any accessory factors that are required for control. There are two evolutionarily and mechanistically different families of site-specific recombinases, the tyrosine and serine recombinases (1). All types of recombination outcomes are mediated by recombinases from both families. In the serine recombinase family there is a clear division between the resolvase/invertases and the enzymes that mediate integration/excision. The resolvase/invertases are approximately 180 to 200 amino acid proteins and are increasingly referred to as the small serine recombinases. The (pro)phage-encoded serine integrases, the transposases, such as those from clostridial ICE elements, Tn*4451* and Tn*5397*, and the recombinases from the staphylococcal cassette chromosomes (SCC) elements are between 400 and 700 amino acids and are collectively known as the large serine recombinases (LSRs) (2). The first LSRs to be studied in *in vitro* recombination systems were the integrases from the *Streptomyces* phage, ϕC31 (3) and mycobacteriophage Bxb1 (4). Understanding the mechanism of the LSRs, however, was greatly hindered by the lack of structural information. In a breakthrough paper, Rutherford *et al.* published the structure of the large C-terminal domain (CTD) of a serine integrase bound to one half site of one substrate (5). This work has led to a step change in our understanding of the mechanism of the serine integrases (6, 7).

BIOLOGICAL ROLES OF LSRs

LSRs are not evenly distributed throughout the eubacteria. Phages that infect Gram-positive bacteria, in particular the Actinobacteria and Firmicutes, are rich in serine integrases. If one searches for the signature "recombinase" domain in Pfam (pfam7508; pfam.xfam.org), the majority of hits are in the Firmicutes, many in mobile elements encoding clinically relevant antibiotic resistances (8, 9, 10). In the Gram-negatives, large serine recombinases are found sporadically, mostly in

[1]Department of Biology, University of York, York, United Kingdom.

environmental organisms. By contrast, the small serine recombinases are common in both Gram-positive and Gram-negative bacteria.

ROLE OF SERINE INTEGRASES IN PHAGE LIFE CYCLES

When temperate phages enter the lysogenic life cycle, the phage genome often becomes integrated into the host DNA. First described by Allan Campbell (11) working with phage λ, the integration requires an attachment site in the host chromosome, the *attB* site and a phage attachment site, *attP* on the circular phage genome (Fig. 1A). Campbell proposed that a single crossover occurs between the two attachment sites and results in the integration of the phage genome, which is then flanked by two new attachment sites, *attL* and *attR*. Excision is the recombination between *attL* and *attR* that regenerates *attP* on the excised phage genome and *attB* in the host chromosome. In order for the phage to properly control entry and exit from lysogeny, integration and excision must be tightly controlled. To do this, phages encode accessory proteins, called the recombination directionality factors, RDFs or excisionase, Xis (12). Although phage integration and excision can be driven by tyrosine or serine integrases, the mechanisms of the integrases and accessory proteins and the organization of the attachment sites differ fundamentally between the two types of systems (2).

The attachment sites used by the serine integrases comprise the binding sites for their cognate integrase; no host or accessory factors bind the attachment sites (3, 7, 13). The *attP* and *attB* sites, whilst sharing little sequence identity with each other, both contain imperfect inverted repeats and the *attL* and *attR* sites contain half sites from *attP* and *attB* (2).

AttB sites used by the serine integrases tend to be located intragenically and, in most cases, integration of DNA into these sites appears to disrupt their target genes (Table 1). However, some integration sites are sufficiently close to the 5′ or 3′ ends of the genes that a sequence within the *attP* site might compensate for the loss of translational signals; for example, SCC*mec* elements regenerate a functional 3′ end of the *rmlH* gene (14). ϕC31 integrase can tolerate minor species-specific sequence variations in *attB* but integration activity drops by orders of magnitude if the cognate *attB* site is deleted (15). Secondary or pseudo-*attB* sites with limited sequence identity to *attB* can be used for integration if the cognate *attB* is absent (15). The *Mycobacterium* prophage ϕRv1 was found in different locations in two *M. tuberculosis* genomes integrated into a repetitive

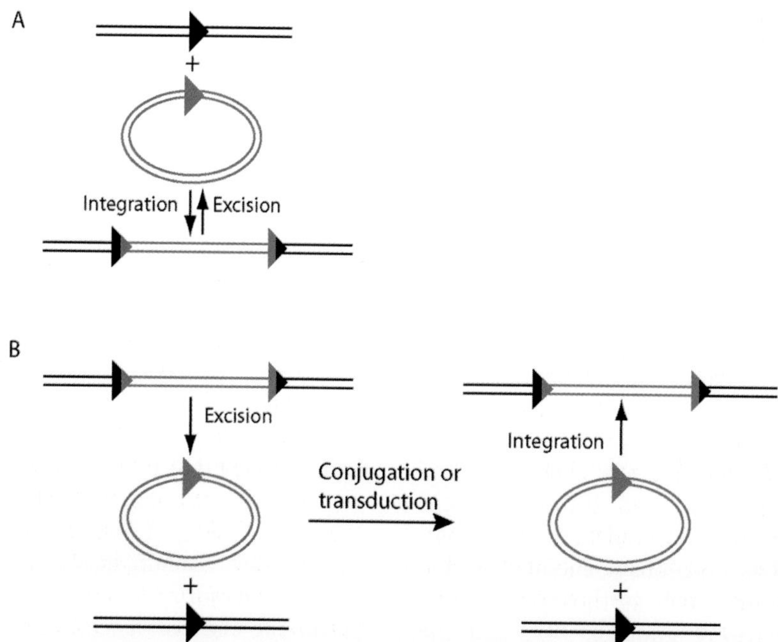

Figure 1 Role of large serine recombinases in (A) phage integration and excision and (B) movement by mobile elements. Black and blue double lines represent the host and phage/mobile element DNA, respectively. Triangles are the *attP*/*attTn*/*attSCC* sites (blue) or host *attB* sites (black). The hybrid *attL* and *attR* sites are mixed. Figure adapted from Smith (79) with permission from Elsevier Press. doi:10.1128/microbiolspec.MDNA3-0059-2014.f1

Table 1 Location of *attB* sites

Phage	Host	Locus *attB*	Reference
A118	*Listeria monocytogenes*	*comK* homologue	(27)
LI prophage	*L. innocua*	*comK* homologue	(5)
Bxb1	*Mycobacterium smegmatis*	*groEL1*, 3′ end	(4)
ϕC31	*Streptomyces lividans*	SCO3798, pirin homologue	(69)
ϕRV	*M. tuberculosis*	Rv1587c in REP13E12	(65)
TP901	*Lactococcus cremoris 901-1*	5′ end *orf125* (accession Y15043), homologue of *comGC* from *B. subtilis*	(80)
ϕBT1	*S. coelicolor*	SCO4848, encoding putative membrane protein	(81)
Peaches	*M. smegmatis*	MSmeg_5156, putative DNA repair gene	(18)
Bxz2	*M. smegmatis*	MSmeg_5156, putative DNA repair gene	(18)
R4	*S. parvulus*	Putative acyl CoA synthetase-encoding gene	(82)
TG1	*S. cattleya*	SAV4517, putative *dapC*, encoding N-succinyl diaminopimelate aminotransferase	(83)
FC1	*Enterococcus faecalis*	Putative *radC*	(84)
MR11	*Staphylococcus aureus*	Mercury reductase gene	(85)
TndX	*Clostridium difficile*	*C. difficile* homologue of PivNM-2	(34)
TnpX	*C. perfingens*	Various	(31)
SpoIVCA	*Bacillus subtilis*	*sigK*	(21)
CcrAB	*S. aureus*	*rlmH*	(17)
SV1	*S. venezuelae*	SCO4383, putative 4-coumarate CoA ligase gene[a]	(86)
BL3	*Brochothrix thermosphacta*	3′ end RNA methyltransferase gene, similar to *B. mycoides* NZ_CM000744	(87)
SPBc	*B. subtilis*	*ypqP*, putative capsular polysaccharide biosynthesis gene	(88)
Wbeta	*B. anthracis*	NP_844065, encoding hypothetical protein	(89)
ϕK38	*S. avermitilis*	SAV1570, putative ABS transporter gene	AB251919, submitted by Ikeda (2006)
ϕ370.1	*Streptococcus pyogenes SF370*	5′ end *pep*DNP_268945	(90)

[a]Mapped in *S. coelicolor*.

element, REP12E12. *M. tuberculosis* and *M. bovis* BCG have seven copies of REP13E12 and ϕRv1 can integrate efficiently into three almost identical copies; as the sequences diverge, efficiency of recombination decreases (16).

It is not surprising that phages or mobile elements that encode very similar integrases and *attP* sites use the same *attB* sites, for example, the SCC*mec* elements or phage A118 and *Listeria innocua* (LI) prophage (5, 17) (Table 1). Surprisingly, mycobacteriophages Peaches and Bxz2, which have integrases and *attP* sites that have diverged significantly (59% and 50% identical, respectively) use the same *attB* site (18). Bxz2 and Peaches are not able to swap *attP* sites, as only Bxz2 integrase can use both *attP* sites for recombination and DNA binding. These two integrases provide an insight into how integrases might coevolve with their attachment sites.

DOMESTICATION OF PROPHAGE EXCISION BY THE HOST

LSRs were first described in relation to their roles in heterocyst formation in the cyanobacterium *Anabaena* and in *Baciillus subtilis* sporulation (19, 20, 21). Both types of events only occur in terminally differentiated cells types, thus ensuring that the prophage is not lost in the vegetatively proliferating cells. In *Anabaena*, the response to nitrogen limiting conditions is to develop specialized cells, the heterocysts, in which atmospheric dinitrogen is reduced to ammonia. The nitrogen fixation depends on the nitrogenase complex, which is only expressed and active in the anaerobic microenvironment of the heterocysts. Remarkably, there are two large, independent genome rearrangements that take place to generate intact copies of *nifD* and *fdxN*, encoding a subunit of nitrogenase and ferredoxin, respectively (22). An 11 kbp insertion in *nifD* is excised by the action of a tyrosine recombinase, XisA and a 55 kbp insertion is removed from the nearby *fdxN* gene by a large serine recombinases, XisF (19, 23). Expression of the *xisF* gene in *Escherichia coli* containing a compatible reporter plasmid encoding DNA from the endpoints of the 55 kbp insertion yielded a low level of recombination activity and the products were consistent with a reciprocal site-specific recombination reaction (19). XisF requires two accessory proteins, XisH and XisI, for excision of the 55 kbp element, consistent with the dependence of prophages on RDFs to activate the excision reaction (24).

SpoIVCA in *B. subtilis* was also identified as a protein related to the resolvases and is essential for excision of a 42 kbp prophage-like element, *skin* (20, 21). As in prophage excision, the excised DNA element was shown to be a circular (25). In the bacterial chromosome, the recombination resulted in the precise joining of two gene fragments, *spoIVCB* and *spoIIIC*, to form an intact open reading frame (ORF) encoding the mothercell specific sigma factor, SigmaK (21, 25). The recombination sites were shown to resemble *attL* and *attR* sites (2, 26). The recombinase, SpoIVCA, binds to these regions protecting about 50 bp from DNAseI digestion. Expression of SpoIVCA is dependent on a small regulatory protein encoded by *spoIIID* and it has been proposed that SpoIIID might also be required as an accessory protein in excision (21, 25).

Domestication of prophage excision to regulate gene expression also occurs in nonterminally differentiated cell types. In *Listeria monocytogenes*, prophages including A118 and ϕ10403S (27, 28), are frequently inserted within *comK*, which in *B. subtilis* encodes a regulator of several genes expressed late in the competence pathway that leads to DNA uptake (29). The late expressed *com* genes encode the proteins required for the assembly of the DNA uptake apparatus, the pseudopilus. *L. monocytogenes* is an intracellular pathogen that is initially taken up by mammalian cells in the vacuole but the bacterium must escape this compartment and enter the cytosol to spread and multiply. ComK-dependent expression of the pseudopilus is essential for *L. monocytogenes* 10403S to escape from the vacuole and the intact *comK* gene is generated only after excision of ϕ10403S (28). ComK in *Listeria* is therefore an important virulence factor. Rabinovich *et al.* also demonstrated that, although the excised prophage is transcribed and mRNA for capsid and tail genes could be detected, phage genes for host lysis and DNA packaging were uninduced, suggestive of an abortive life cycle (28).

ROLE OF SERINE INTEGRASES IN THE MOVEMENT OF TRANSPOSONS AND THE SCC

The mobilisable element, Tn*4451*, in *Clostridium perfringens* and the conjugative element, Tn*5397*, from *Clostridium difficile* encode chloramphenicol and tetracycline resistance mechanisms, respectively. Both encode LSRs that are used for element excision and integration (30, 31, 32). TnpX in Tn*4451* excises the transposon generating a circular intermediate that, at the joint, forms the attachment site (*attTn*). After conjugation of Tn*4451* to a new host, *attTn* will recombine with one of several targets sites in the new host's chromosome (Fig. 1B). Lyras and Rood showed that this circular intermediate generates a strong promoter within *attTn* that reads into the *tnpX* gene such that after the circular intermediate has been delivered into the recipient, there is a burst of TnpX expression to mediate integration (33). The genetic organization of Tn*5397* precludes the same strategy for regulation of expression and integration of this element appears to be mostly site-specific (9, 32, 34). BLASTp searching with TnpX and TndX indicate that both transposases are present throughout *Clostridium spp.* and related genera.

SCC elements are prevalent mobile elements in *Staphylococcus* and *Enterococcus* spp. Many of the SCC elements encode two recombinases, CcrA and CcrB, and they form two distinct clades in a phylogenetic tree of LSRs. The SCC elements were first observed as they contain the *mecA* gene, encoding for a variant penicillin binding protein with low affinity for methicillin and hence conferring resistance to this antibiotic. The two recombinases are both required for excision and integration of the cassette (14, 35, 36). CcrA has a specific role in the recognition of the host *attB* site (14) as it recognizes a specific half of the *attB* site and it may enable the SCC element to insert into a site that is not an inverted repeat (14). Another unusual feature of the SCC elements it that the recombinase genes, *ccrA* and *ccrB*, are located distant from their recombination substrates (17). Excision and reintegration of the elements also occurs through a circular intermediate (35, 37) (Fig. 1B). Recently, transfer of the SCC*mec* element from cell to cell was shown to occur by transduction (38).

OVERVIEW OF THE PROPERTIES OF LARGE SERINE RECOMBINASES; CONSERVATIVE SITE-SPECIFIC RECOMBINATION AND SITE-SELECTIVITY

To date, 16 LSRs have been studied at the biochemical level and they have similar properties (Table 1) (6, 13). LSRs have the N-terminal domain (NTD), also called the catalytic domain, that is highly conserved in the serine recombinases. The LSRs act by binding to their DNA substrates as dimers and bring the sites together by protein–protein interactions to form a tetrameric synaptic complex (1, 39, 40, 41). Activation of the nucleophilic serine in each of the four subunits results in DNA cleavage to give 2nt 3′overhangs and transient phosphoseryl bonds to the recessed 5′ ends (Fig. 2). Reaction intermediates in LSR-mediated recombination can be observed in nondenaturing gels (42, 43, 44, 45). Complexes that are dependent on both recombination substrates being present have been shown to contain either the covalently bound integrase to cleaved DNA or noncovalently bound synaptic tetramers with substrates (42, 44, 45). The precise position of cleavage with Bxb1 was mapped demonstrating the formation of the 3′ dinucleotide overhang that lies exactly where the crossover site had been mapped by mutagenesis (42). In addition, the use of methylphosphonate substitutions in *attB* placed either at the scissile phosphate or at adjacent positions close to the crossover dinucleotide greatly inhibited double strand cleavage, suggesting that cleavage must be concerted, a hallmark of the serine recombinase mechanism (42).

Facilitated by the flat interface within the tetrameric synaptic complex, DNA strand exchange occurs by subunit rotation (discussed below). The 3′ dinucleotide overhangs base pair with the recessed 5′ bases and the 3′ OH attacks the phosphoseryl bond in the reverse of the cleavage reaction to join the recombinant half sites (1). The transient covalent attachment between each

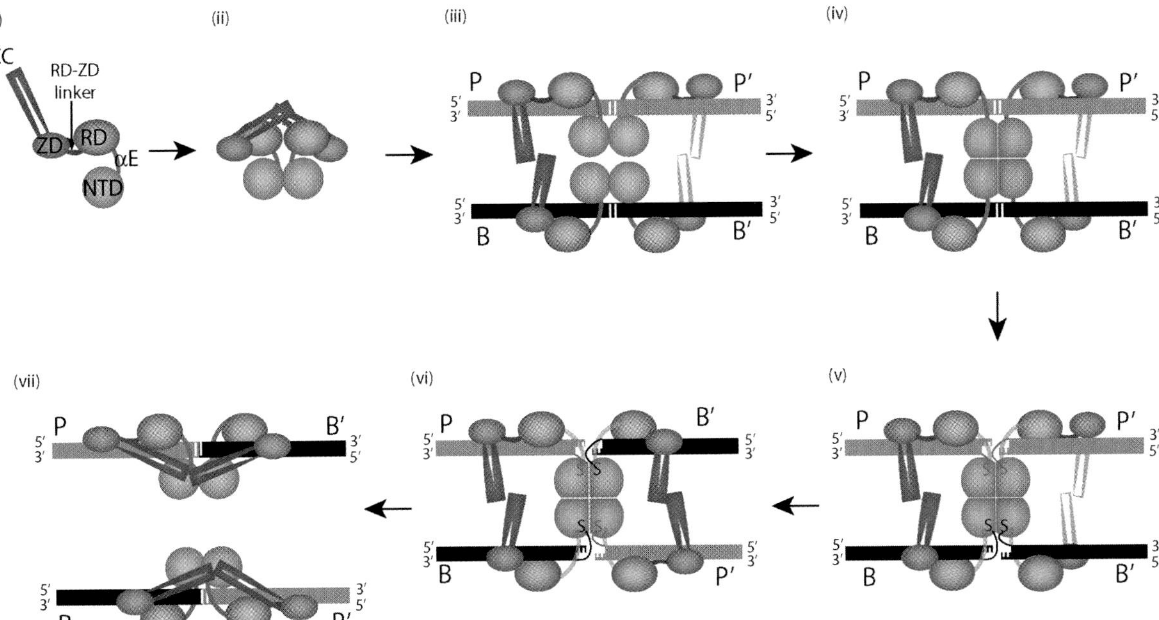

Figure 2 The recombination pathway by large serine recombinases (LSRs). (i) Diagram of the domain structure of the LSRs; the N-terminal domain (NTD; green) is the catalytic domain; αE (green line) is the long alpha helix that connects the NTD to the recombinase domain (RD; red). The short linker between the RD and the zinc ribbon domain (ZD) is indicated by the arrow. The coiled-coil (CC) motif is depicted by rectangular projections. (ii) The LSRs in the unbound state are likely to be compact and globular (6). (iii) The LSRs binding to *attP* (gray) and *attB* (black). The relative positions of the ZD domains are different on *attP* and *attB* and these have consequences for the positions of the CC motifs. The CC motifs are either flared and dark, indicating they are projecting out of the paper or not flared and light, where they project into the paper. The scissile phosphates are located flanking the two nucleotide crossover sites shown as two white vertical lines. Bound to *attP* and *attB*, the CC to CC motifs can begin to interact and initiate synapsis. (iv) A synaptic complex is formed that requires interactions between the CC motifs as well as through the NTD tetrameric interface (41). Conformational changes occur in the NTDs to generate a flat interface for subunit rotation. (v) DNA cleavage occurs with concomitant formation of the phosphoseryl bonds. (vi) Subunit rotation swaps half sites, in this case B′ and P′. (vii) Joining of the recombinant products leads to a closed conformation of LSRs on the *attL* and *attR* sites and conformation changes in the NTDs. Figure adapted from Rutherford and van Duyne (6) with permission from Elsevier Press.
doi:10.1128/microbiolspec.MDNA3-0059-2014.f2

recombinase subunit and the DNA backbone is an important reaction intermediate that conserves the high energy phosphodiester bond and negates any requirement for an additional energy source. The reactions are therefore termed conservative, as an energy source is not required. Moreover, recombination is reciprocal as no DNA is lost or gained in the reaction.

Conditions for *in vitro* recombination assays with LSRs usually consist of a simple buffer, with no requirement for divalent cations (3). Recombination rates and conversion efficiencies vary; ϕC31 and Bxb1 integrases can convert approximately 80% of substrates to products within 30 min, whereas ϕRv1 integrase can take an overnight incubation to reach this level of efficiency (16, 42, 46). The recombination sites are usually less than 50 bp in length and are referred to as "simple" because they bind only the recombinases, no other DNA binding or host-encoded proteins are required (42, 47). Where minimal sites have been precisely mapped, the *attP* sites are usually about 10 bp longer than the *attB* sites (6).

The arrangements and orientations of the sites determine the outcome of recombination (3). For instance, *attP* and *attB* can recombine when they are on different molecules, as is the case when they are located on the phage and host genomes to generate the integrated prophage, but they can also recombine, often with greater efficiency, if they are placed on the same molecule. In an intramolecular reaction, the outcome depends on the relative orientation of the sites; if the sites are in a head-to-tail orientation, deletion occurs and if they are head-to-head, inversion occurs (see below for an explanation of how polarity is determined). Similarly, with the *attL* and *attR* sites, their location and orientations determine the outcomes of recombination. Note that in studies of integrases with model substrates, the terms "integration" and "excision" are used to describe the *attP* × *attB* and *attL* × *attR* reactions, respectively, irrespective of whether the outcome is DNA integration or excision. Despite this topological promiscuity, there is strict control over the combinations of sites that can and cannot recombine (3, 43, 48). The default reaction by integrase in the absence of other proteins is *attP* and *attB* recombination to generate *attL* and *attR*; no other pairs of site recombine efficiently. Only when the RDF is present can integrase mediate *attL* versus *attR* recombination (although both TP901-1 and ϕBT1 can mediate a low level of detectable excision) (49, 50) and only under exceptional circumstances can two *attL* or two *attR* sites recombine (51, 52, 53). Site-selectivity and directional control are entirely consistent with the biology of phage integration and excision; a lysis/lysogeny decision can be made without expression of early phage genes but the induction back into the lytic cycle will necessarily need early phage gene expression, including the gene encoding the RDF.

SERINE INTEGRASES: DOMAIN STRUCTURE AND FUNCTION

The integrase from LI prophage is 98% identical to the *Listeria monocytogenes* phage, A118 integrase, and both integrases use almost identical attachment sites (5, 27). A structural model of LI CTD in a complex with a 25 bp A118 *attP* half site was generated by X-ray crystallography (5). This structure and the TP901-1 integrase NTD show that the LSRs are comprised of three major structural domains and functionally important motifs that connect them (5, 41) (Fig. 2 and Fig. 3). The three domains are the catalytic NTD, the "recombinase" domain (RD) and the zinc ribbon domain (ZD). The NTD and RD are linked by the long αE helix and the RD and ZD are connected by a short linker (5). The RD and ZD are collectively termed the CTD and it is from the large size of the LSR CTD, ranging in size from 319 to about 550 amino acids, that they earn their name (2).

THE CONSERVED N-TERMINAL DOMAIN

The LSR NTDs share the conserved residues and the functions of the small serine recombinase NTDs (39, 40, 41). Mutation of the catalytic serine in TnpX, ϕC31, Bxb1, and R4 abolishes recombination (31, 42, 44, 54). The structure of the NTD of TP901-1 integrase shows that the overall fold and shape of the LSR NTD strongly resembles that for γδ resolvase, and Sin recombinases (40, 41, 55). Although TP901-1 NTD is a tetramer in the X-ray crystal structure, it, along with isolated NTDs from Bxb1, A118 and TnpX, is a dimer in solution (41, 43, 56, 57). The dimer interface (I-II in Yuan *et al.* [41]) is formed largely by the crossing of αE, reminiscent of the dimer interface in γδ resolvase (41). The tetramer interface in the TP901 NTD structure is the putative synaptic interface formed by three different secondary structure elements including the base of αE at Met118 and Phe122. The equivalent amino acids in γδ (aas 96 to 105) and Tn*3* resolvases are believed to be part of a flexible linker region (1, 6). In γδ, as the tetramer interface forms, the crossing angle of the αE helices changes, weakening the I-II interactions and resulting in a flat interface that would provide a barrier free rotation of subunits to bring about strand exchange (39). Conformational changes

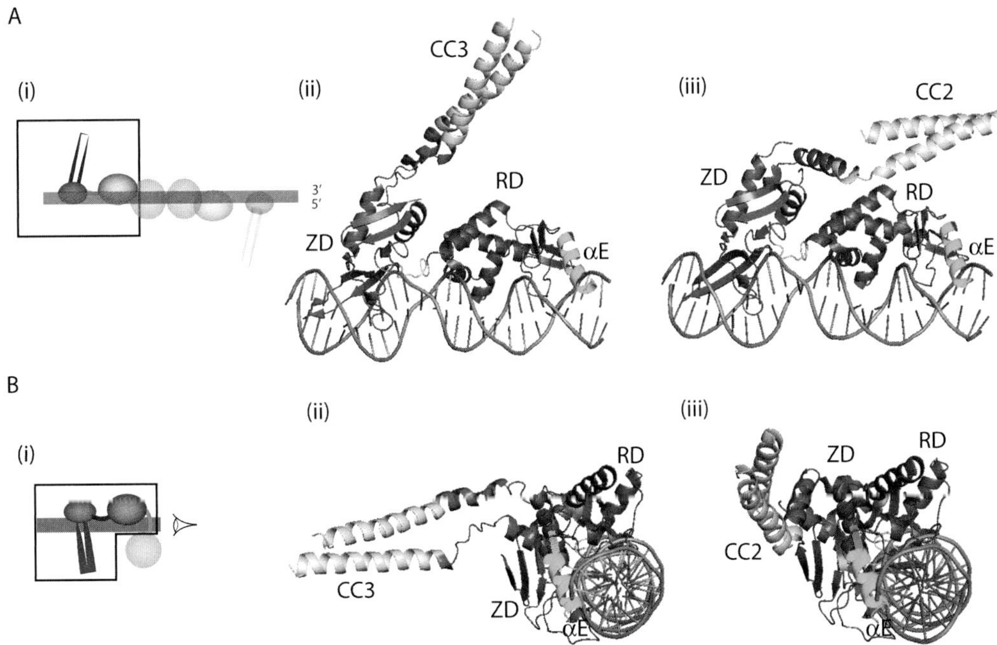

Figure 3 Structures of the *Listeria innocua* integrase C-terminal domain (CTD) bound to the A118 *attP* half site. Four views are shown to illustrate two different trajectories of the coiled-coil (CC) motif. (A) (i) A schematic of a dimer of a large serine recombinase bound to an *attP* site. The structures shown in (ii) and (iii) relate to the boxed area in (i) and display two different trajectories of the CC motif as described in Rutherford *et al.* (5). (B) (i) The schematic indicates that the views in (ii) and (iii), which show the same structures as in (A)(ii) and (A)(iii), are looking down through the DNA. Domains are color-coded: green is the C-terminal end of αE from the N-terminal domain (NTD), Red is the recombinases domain (RD), blue is the zinc ribbon domain (ZD) and the light blue region within ZD is the CC motif. The NTDs, which are absent in the structures, would connect to the αE helix (green) to bind to the opposite side of the DNA from the CTDs. Figures were constructed using the PDP file 4KIS. doi:10.1128/microbiolspec.MDNA3-0059-2014.f3

during the dimer to tetramer transition are also associated with activation of DNA cleavage. Similar conformational changes are important in the LSRs (6, 58). Substitutions at V129, a residue that is predicted to lie the flexible linker at base of the αE helix in ϕC31 integrase, affect either the stability of the synapse (IntV129A) or the ability of the synaptic complex to cleave the substrates (V129G) (58). The formation of the tetrameric interface and the associated conformational changes induced within the NTDs thus appear to be conserved in the serine recombinases. In all serine recombinases, there are barriers that need to be overcome to generate the active tetramers. In the LSRs, the CTDs play a major role in overcoming these barriers.

THE LARGE C-TERMINAL DOMAIN

Despite high diversity between recombinases, sequence alignments indicate that the secondary structure features elucidated by the LI integrase structure are generally conserved, arguing that all the LSRs will have similar overall structure (5). Length and amino acid sequence diversity is particularly noticeable between secondary structural elements (6). Indeed, ϕC31 integrase is predicted to have a large number of loops between the conserved secondary structure elements (born out by the partial structure shown in PDB entry 4BQQ). Despite the loops, the only protease sensitive site identified in ϕC31 integrase is between the NTD and CTD, suggesting that the structure in solution is globular and compact, an idea proposed by van Duyne and Rutherford (6) (Fig. 2).

The structure of LI integrase confirmed the presence of zinc coordinated by four cysteines in the ZD (5). Zinc had been shown to be present in ϕC31 integrase and that recombination and DNA binding activity were lost if integrase was preincubated in the presence of EDTA (59). ϕC31 integrase appears to be more sensitive to EDTA than other LSRs, possibly a consequence of a sequence variation in the vicinity of the coordinating cysteines in ϕC31 integrase (6, 57). The greatest surprise from the LI structure showed that embedded within the ZD is an antiparallel coiled coil (CC) that projects out from the more globular structure of the ZD. The CC motif can adopt at least four different trajectories relative to the ZD and RD domains indicative of the presence of a hinge region within the ZD (Fig. 3). Previous to the structure elucidation, the CC motif in C31 integrase was shown to be involved in the control of integration and excision and in protein–protein interactions between CTDs (48, 60).

TnpX (707aa) has, in addition to the three canonical LSR domains, a C-terminal domain containing 170 aa located beyond the ZD (56, 61). A derivative of TnpX lacking the last 110 amino acids, $TnpX_{1\text{-}597}$ was functional for excision, insertion, transposition, dimerization, and DNA binding. Full length TnpX forms more stable dimers and is more active in a deletion assay *in vitro* compared to its truncated form. Van Duyne and Rutherford suggest that this additional domain in TnpX is a duplicated zinc ribbon domain that provides additional DNA and protein–protein contacts (6). A compilation of the other putative LSRs with additional functional domains can be found in Pfam (pfam.xfam.org) (62).

THE LARGE CTD IN SERINE INTEGRASES: DNA BINDING

The work by Rutherford *et al.* has revealed for the first time not only the structure of a serine integrase CTD but also how it recognizes and contacts the *attP* half site (5). The structure supports biochemical data obtained from work with A118, Bxb1, ϕC31, and TnpX, and is therefore, likely to reflect the overall DNA binding mechanism for all the LSRs. The insights obtained from the LI CTD structure explain many of the generic properties of the LSRs, including observations pertaining to site-selectivity and directionality. A detailed review by van Duyne and Rutherford considers all of the structural and biochemical evidence to date of LSR DNA binding to its attachment sites and how binding might control the assembly of the synapse (6).

Recognition of the *attP* Site

The CTD from LI integrase makes multiple interactions along the entire length of the 25 bp A118 *attP* half site in both the major and minor grooves. The domains of LI integrase CTD contact distinct regions of the *attP* half site with RD (including the αE) contacting base pairs 0 to 12, the RD/ZD linker contacts bases 13 to 15 and the ZD domain base pairs 16 to 24 (Fig. 4). Correspondingly, sequence motifs within these contacted regions of the *attP* sites can be identified and are referred to as RD, ZD or RD-ZD linker binding motifs. The extensive interactions all along the A119 *attP* half site match both DNAseI footprinting and the minimal length determination for A118 *attP* site using a linear fragment (50 bp) (57).

Binding by the αE, The RD Domain and the RD-ZD Linker

The 13 bp at the center of the A118 *attP* half site is contacted by the C-terminal part of the αE and the RD domain (5) (Fig. 4 and Fig. 5). This conserved domain within the LSRs contains three helices (αF, αG and αH) at its core and resembles the three helix bundle of the CTD of the small serine recombinases (Fig. 3 and Fig. 5). Major groove contacts occur via the αH that makes direct contacts to bp 9 to 11. Minor groove contacts are made by aa within the RD-ZD linker region and from conserved aa in αE to bp close to the center of the *attP* site.

In a model of the LI integrase CTD bound to a full *attP* site, two RD motifs are located symmetrically around the crossover dinucleotide at base pairs 0 to −12 for the left arm and 0 to +12 for the right arm (Fig. 4 and Fig. 5). Rutherford *et al.* propose that the RD motif also recognizes this motif in *attB* (at base pairs 0 to ±12 bp) (5). This extrapolation is supported by the fact that most attachment site pairs (the exceptions are the SCC*mec* sites) share more sequence identity in the central 0 to ±12 bp than in their outer, flanking regions (2). Mutation scanning studies with Bxb1 *attP* and ϕC31 *attB* sites indicated that the central 24 bp in the attachment sites are recognized as a core sequence (63, 64). Although most of the positions are not particularly sensitive to mutations, the RD motif sequence appears to be enriched within the pseudo-*attB* sites recognized by ϕC31 integrase (15). Exceptionally, in both Bxb1 *attP* and ϕC31 *attB*, there are mutations close to the crossover dinucleotides that can strongly reduce recombination affecting integrase binding, synapsis and/or cleavage (Fig. 4) (63, 64). It seems plausible that interactions close to the crossover dinucleotide might have a significant impact on the ability of the integrase subunits to undergo the conformational changes required to activate cleavage and that failure to complete these changes might result in unstable DNA binding (5, 63, 64).

Interactions by the ZD to Attachment Sites

The majority of contacts by the ZD to the ZD motif are in the major groove and are mediated by a hairpin

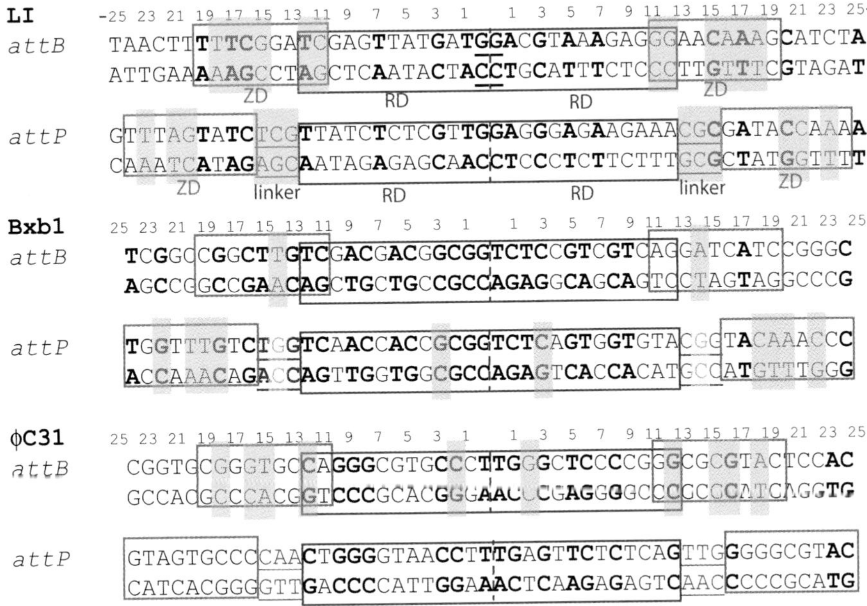

Figure 4 Binding motifs in the attachment sites for *Listeria innocua* (LI) integrase, Bxb1, and ϕC31 integrases. Numbering of the bases is outwards from the crossover dinucleotide (00) to the left (minus) and the right (plus). The red and blue boxes indicate the recombinase domain (RD) and zinc ribbon domain (ZD) binding motifs, respectively. The three base pairs recognized by the RD-ZD linker in the *attP* sites are shown in pink. Highlighted orange are bases that are mutational sensitive and yellow is the discriminatory base described in Singh *et al.* (64). doi:10.1128/microbiolspec.MDNA3-0059-2014.f4

(β11–β12) held by the coordinated zinc atom and by a loop (between β12 and αJ) (5) (Fig. 3 and Fig. 5). Contacts here were anticipated by mutations in the equivalent region in ϕC31 integrase that reduced DNA binding to variable extents (59). The ZD motifs lie in the outer flanks of the *attP* site and mutations in A118 *attP* site at 20, 21 and 23 are defective for LI integrase activity (5). Equivalent mutations in *attP* from Bxb1 (bp 18 to 21 and called the Flank-L and Flank-R sequences) were also defective for binding (64) (Fig. 4). Singh *et al.* recognized that there was no counterpart to the Flank/ZD motif sequences at the same positions in *attB*. However, Rutherford *et al.* provide a compelling case that the *attB* sites, in fact, do have the ZD motifs, recognizable by sequence similarity to *attP* ZD motifs, but they are positioned 5 bp towards the center of the *attB* site compared to their locations in *attP* (5) (Fig. 4). Mutations in the A118 *attB* site confirm that the most sensitive bases are 15 to 18, exactly 5 bp inwards from the equivalent bp in the *attP* site. Mutation scanning of the ϕC31 *attB* site also showed that the most mutationally sensitive base pairs in *attB* for recombination activity are located at bp 15 and 16 where changes resulted in very slow synapsis and recombination rates, but no significant decrease in DNA binding affinities (63). In the light of the structural information it seems likely that binding affinities could be maintained by compensatory contacts to mutant *attB* sites but these could also alter the positions of the CC motifs, which as described below, would affect synapse formation (6).

RD-ZD Linker Interactions

Rutherford *et al.* identified a 3 bp motif in the *attP* site that is contacted in the minor groove by an eight aa residue linker, the RD-ZD linker (5). The linker motif is well conserved in both arms in many of the LSR *attP* sites (Fig. 4) (6). A consequence of the relocation of the ZD motif 5 bp towards the center of the *attB* site, relative to the *attP* site, is that the linker motif and the ZD motif have to overlap in the bps 11 and 12 (5). Mutation scanning of the A118 *attB* site shows that mutations here reduce DNA binding and recombination whilst mutations in equivalent positions in *attP* are hardly affected. Similarly in ϕC31, *attB* double symmetrical mutations at positions ±12 are defective for recombination (63).

Further independent insight into the relationship between *attP* and *attB* was obtained by Singh *et al.* (64). Remarkably, only a few base pairs in *attP* need to be changed to their *attB* counter parts to lose *attP* identity

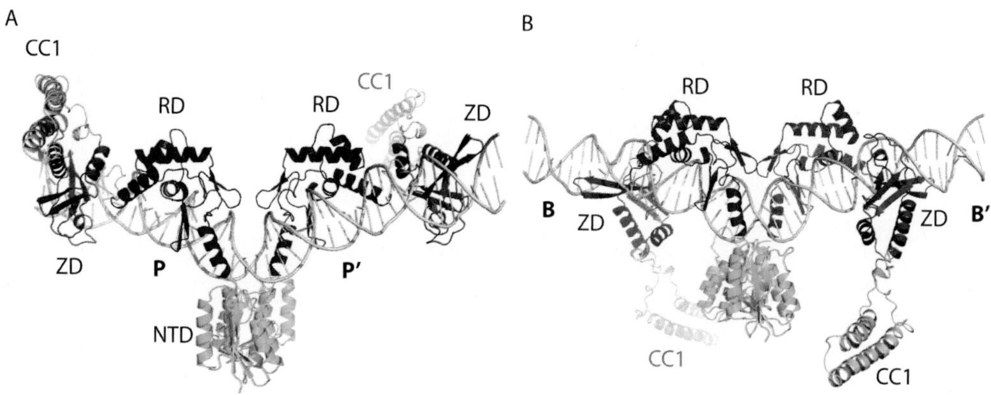

Figure 5 Structural models of a large serine recombinase bound to a full *attP* site (A) and a full *attB* site (B) adapted from Rutherford *et al.* (5) with permission from Oxford University Press and redrawn by Dr. Greg van Duyne. The domains are colored using the same scheme as in Fig. 3. ZD, zinc ribbon domain; RD, recombinases domain; NTD, N-terminal domain. doi:10.1128/microbiolspec.MDNA3-0059-2014.f5

and gain a moderately active *attB* site. The key residues include a so-called discriminator base at position 14 and a base in the ZD motif at position 20 (Fig. 4). With the benefit of the LI integrase CTD structural information, these changes disrupt the RD-ZD linker interaction (bp at position 20) and (at position 14) weaken the interaction to the ZD motif (6). The mutation in the ZD-RD linker motif actually causes the linker region to resemble the ZD binding motif at the correct position for the *attB* site (6). Overall, we now have a much clearer understanding of how *attP* and *attB* are related to each other and how they could both be recognized by their cognate LSRs (5).

FORMATION OF THE SYNAPSE IS THE KEY STEP IN THE CONTROL OF DIRECTIONALITY

Synaptic complexes containing a presumed tetramer of integrase, assembled by the interactions between dimers bound to a pair of attachment site substrates, can be observed experimentally (42, 44, 45, 47). Although LSRs generally bind to all their attachment sites with approximately equal affinities, LSRs do not synapse any pair of attachment sites other than *attP* and *attB* in the absence of their RDFs, reflecting the site-selectivity of integrase for integration. We now have insight into how the assembly of the synaptic complex could be controlled by the participating attachment sites. There are at least three important protein–protein interactions occurring to generate the synaptic complex, the dimer and tetramer interfaces in the NTDs (already discussed), and the CC interactions in the CTDs.

Protein–Protein Interactions between CTDs at Synapsis Determine Site-Selectivity

The isolated CTDs from ϕC31, Bxb1, Peaches, Bxz2, and A118 bind to their cognate attachment sites (18, 43, 56, 57, 60). When bound to *attP* and *attB* the isolated ϕC31 CTD can assemble synaptic complexes and displays the same site-selectivity of full-length (catalytically inactive) integrase. Thus, the CTDs bound to the *attP* and *attB* sites may be interacting to generate tetramers and, supported by observations that the CC motif fused to maltose binding protein (MBP) could oligomerize, it was proposed that tetramer interactions were mediated through the CC motif (60).

Explanations for how the CTD interactions are mediated have been proposed based on the LI integrase CTD bound to the *attP* half site (5). Within the ZD domain, the CC motif projects outwards from the more globular part of the domain in different trajectories (e.g., Fig. 3). Rutherford *et al.* constructed a model of an intact LI integrase bound to full *attP*, *attB* and *attL/R* sites, using as a guide the γδ resolvase/site1 complex (5, 40). In the model, the ZD domains located in integrase dimers bound to either *attP* and *attB* are located on opposite faces of the DNA, and correspondingly, the CC motifs project outwards from the DNA in opposite directions (Fig. 5 and Fig. 6). As described above, ZD domains are predicted to bind *attB* sites 5 bp towards the center of the attachment site compared to its location on the *attP* site. The consequence of the relocation of the ZD domains in *attB* is that CC domains project outwards from half a turn further towards the centre of the attachment site in *attB* compared with *attP* (Fig. 5). Thus, when a model synaptic complex is assembled based on the LI CTD-*attP* structure, the CC motifs in the integrase subunits bound to *attP* and *attB* are now close enough to interact (Fig. 6). Conversely, if the modeled LI integrase dimers bound to two *attP* sites or two *attB* sites are brought together in a hypothetical synaptic complex, the CC motifs in both structures project outwards in opposite directions and cannot interact (5). Deletion of the CC motif from LI integrase results in a recombination defective phenotype in intermolecular reactions where efficient synapsis is required (5). However, in intramolecular reactions loss of the CC motif leads to loss of recombination site-selectivity, confirming the importance of the CC motif in the control of directionality. The models by Rutherford *et al.* offer compelling explanations of how different conformations of integrase bound to *attP* and *attB* can enable synapsis.

WHY SERINE INTEGRASES ONLY MEDIATE INTEGRATION IN THE ABSENCE OF THE RDF

Models of LI integrase bound to *attL* and *attR* were also generated by Rutherford *et al.* (5). A hypothetic synaptic tetramer model of LI integrase bound to *attL* and *attR* permits CC to CC interactions between integrase subunits. However, despite the apparent ease at which the *attL* and *attR* bound synaptic tetramer could form, we know that native phage integrases cannot generate this complex in the absence of the RDF (13). Rutherford explain this paradox by proposing that the CC motifs from the two adjacently bound integrase subunits on *attL* or *attR* are sufficiently close to enable auto-inhibitory CC to CC interactions, if they assume a fairly closed trajectory by reaching across the RD domains towards the center of each attachment site (Figs 2, 3, and 6). These interactions between adjacently bound CTDs would then explain the cooperativity displayed by ϕC31, Bxz2, and Peaches CTDs binding to *attL* and *attR* (60, 64). Rutherford *et al.* also suggested that similar interactions might also be occurring between integrase subunits bound to *attB*, as they would be close enough to stretch across the RD domains and wind around the helix to come into contact and form a rather closed conformation compared with the integrase-*attP* complex.

These models are reinforced by the experiments by Rowley *et al.* who concluded that ϕC31 integrase bound to *attL* and *attR* had an auto-inhibitory activity that could be partially removed by mutation (48). Rowley *et al.* identified amino acid substitutions in the CTD that led to recombination promiscuity or "hyperactivity," that is, the ability to recombine *attL* and *attR* without the RDF. All the mutations lie close to or within the CC motif. Two mutant integrases, IntY475H and IntL460P, were severely defective in integration but showed significant activity in *attL* × *attR* recombination (compared to wild type integrase) in the absence of the RDF. These mutations abolish the CC to CC interactions in experimental assays, and this would explain both reduction in the ability for integrase to synapse *attP* and *attB* and a reduction in the auto-inhibitory interactions between the CC motifs in adjacently bound CTDs on *attL* and *attR* (60). IntE449K, however, is nearly as active in integration as native integrase and has a high level of activity in an intramolecular assay for *attL* and *attR* (much less so for intermolecular activity) (48). Unlike native integrase, Int E449K can synapse *attL* and *attR* (in addition to *attB* and *attP*) but the isolated CTDE449K could no longer generate the *attP*/B or *attL*/R tetramer complexes in EMSAs (60). The location of IntE449 is at the N-terminal base of the CC motif in a putative hinge region and if affecting the range of trajectories, might weaken CC to CC interactions in both auto-inhibition and synapsis.

HOW RDFS SWITCH THE DIRECTIONALITY OF INTEGRASE TO FAVOR EXCISION

The first two RDFs to be shown to activate excision were small proteins encoded by TP901-1 *orf7* (64aa) and Rv1584c (73aa), respectively (16, 49, 65). The TP901-1 *orf7* gene was unusual as it was not located adjacent to the integrase (encoded by *orf1* in TP901-1) and this was a distinct departure from the canonical arrangement of *int* and *xis* in the tyrosine integrase systems (49). Rv1584c, encoding the ϕRv1 Xis, is located adjacent to the *int* gene (Rv1586c) and encodes a protein with low level identity to a characterized Xis (gp36) for a tyrosine integrase (from phage L5) (65). Despite this, RDFs for serine integrases do not act with the same mechanism as those for tyrosine integrases (16).

The RDFs from Bxb1 (gp47) and from ϕC31 (gp3) are much larger proteins than Xis, 255aa in gp47 and 244 aa in gp3. These RDFs and that from TP901-1 are expressed early in the phage developmental cycles (49, 52, 53). Neither gp47 nor gp3 bind to any of the attachment sites used by their cognate integrases and the minimal sites for *attL* × *attR* recombination are less than 50 bp. RDFs from Rv1, Bxb1 and ϕC31 inhibit integration as well as activate excision (16, 52, 53).

ϕC31 gp3 binds integrase in solution and when integrase is bound to all four of its attachment sites (53).

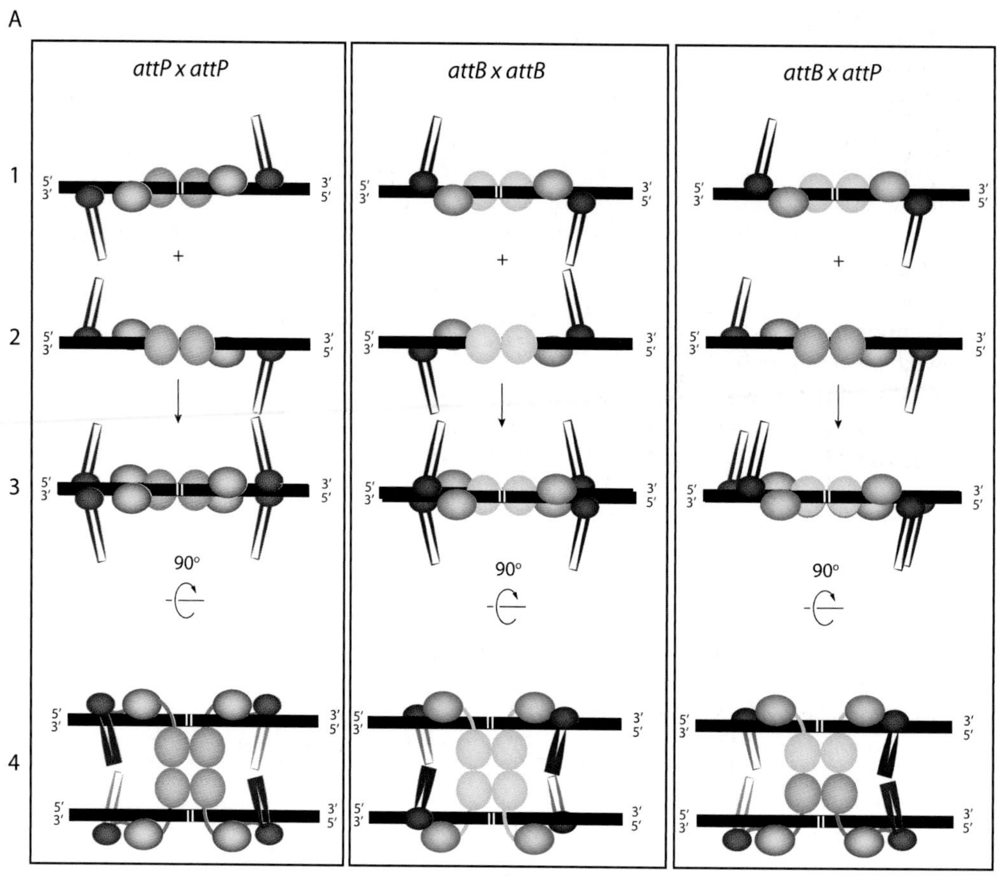

B *attL x attR* - complementary interactions with parallel alignment

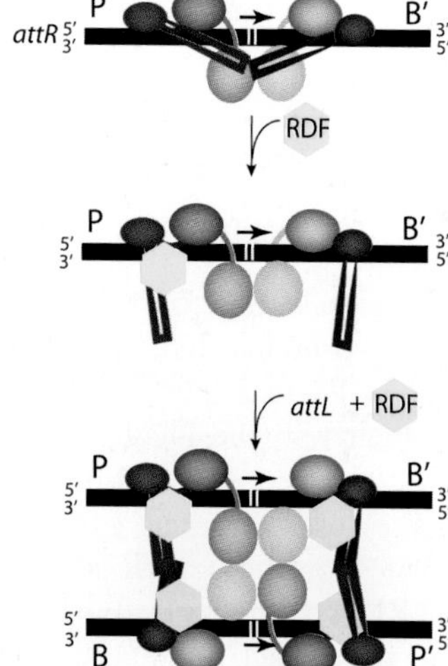

attL x attR - noncomplementary interactions with antiparallel alignment

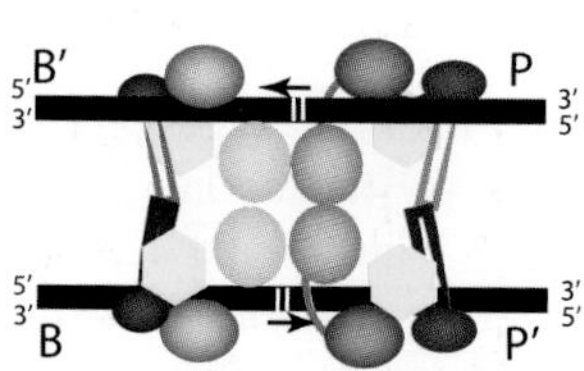

The presence of gp3 stimulates the formation of the *attL* × *attR* synaptic complex and inhibits or destabilizes the *attB* × *attP* synaptic complex. Gp3 acts stoichiometrically with integrase, consistent with its activity mediated through binding to integrase. Using the hyperactive integrase mutant, IntE449K, the mobilities of both integration and excision synaptic complexes with and without gp3 could be compared and in both types of synapse gp3 appears to remain bound. Kinetic analysis of a time course of excision using wild type integrase, gp3 and substrates showed that the initial rate of reaction is dependent on the order of addition of the components. Notably, if integrase is in excess and allowed to bind to the substrates *attL* and *attR* before the addition of gp3, the initial rate is slow compared to a reaction in which integrase and gp3 are incubated before the addition of substrates. Khaleel *et al.*, therefore, proposed that gp3 remodels integrase before the *attL*/R synapse can be made (53) (Fig. 6B).

In the light of the LI integrase-attachment site models, Rutherford and van Duyne proposed that the RDFs are likely to have a generic mechanism that displaces the auto-inhibitory activity of the CC motifs in the integrase-*attL* and integrase-*attR* complexes to enable the formation of the excisive synaptic complex (5, 7) (Fig. 6B). Binding of the Bxb1 RDF, gp47 to integrase bound to *attP* and *attB* is more stable than to *attL* or *attR* where there could be competition between RDF binding and the auto-inhibitory CC to CC interactions (52). Rutherford and van Duyne also suggested a possible mechanism through which the RDFs inhibit the integrative synapse, that is, the RDFs could stabilizing the possible auto-inhibitory activity of the CC motifs in the Int-*attB* complex (7).

The RDFs from Bxb1, ϕC31, TP901-1 and ϕRv1 are unrelated to each other in sequence and RDF genes cannot be predicted on the basis of their genetic loci (other than being early expressed phage genes) (49, 52, 53, 65). Identifying RDF genes is therefore a challenge. However, some RDFs are easily identified. ϕBT1 is a close relative of ϕC31 and contains a homologue to gp3 ($gp3^{BT1}$) (66). $gp3^{BT1}$ is active as an RDF for both ϕBT1 and for ϕC31, a cross-reactivity that is remarkable given the divergence between the two integrases (ϕBT1 and ϕC31 integrases share only approximately 29% aa identity). Homologues to ϕC31 gp3 and Bxb1 gp47 are present in other phages with serine integrases, but also in phages that encode tyrosine integrases (52, 53). ϕC31 gp3 and Bxb1 gp47 are both essential genes for lytic growth indicating that RDFs have two functions in the phage life cycle (52, 66). The mechanism of action of Bxb1 gp47 and ϕC31 (and ϕBT1) gp3 proteins are very similar but the proteins are apparently

Figure 6 Site-selectivity explained by the geometry of the CC motifs bound to the attachment sites. (A) The three panels show the hypothetical assembly of synaptic complexes by integrase bound to two *attP* sites (left panel), two *attB* sites (middle panel), and an *attP* and *attB* site (right panel). The integrase subunits are colored red if bound to P or P′ and blue on B or B′. Line 1 in each panel shows an integrase dimer bound to an attachment site and the dimer is viewed from the perspective of zinc ribbon domain (ZD) and recombinase domain (RD) and looking down towards the N-terminal domain (NTD) bound to the opposite face of the DNA (black line). Line 2 is a second dimer bound to an attachment site but viewed the other way, that is, from the NTD and looking down towards the RD and ZD domains underneath. Line 3 shows what happens when the dimer from line 1 is superimposed on the dimer from line 2 to generate an integrase tetramer and line 4 is where this complex is rotated by 90°. In line 4, the CC motifs are flared and dark where they project out of the page and pale and thin where they project into the page. The CC motifs project in opposite directions from dimers bound to two *attP* sites or two *attB* sites but are proposed to be close enough to interact between integrase dimers bound to *attP* and *attB* site. (B) Possible pathway for assembly of the excision synapse with so-called complementary interactions between the integrase subunits is shown on the left (48). The *attL* and *attR* sites are proposed to be in a closed conformation with respect to the CC motif in the absence of the recombination directionality factor (RDF). Addition of the RDF might bind to the ZD domain to change the trajectory of the CC motifs so that they are in a more open conformation. As the CC motifs in the dimers bound to *attL* and *attR* project in the same direction (here projecting out of the page), there is an opportunity for them to interact, but only in the presence of the RDF. If one of the *attL* or *attR* sites aligns in the opposite orientation as shown by the tetramer synapse on the right, the CC motifs cannot interact as they are projecting in different directions, possibly explaining noncomplementary interaction and the bias against this type of synaptic complex in excision (48). doi:10.1128/microbiolspec.MDNA3-0059-2014.f6

structurally unrelated. The emergence of these two protein families as RDFs is likely to be independent evolutionary events in which integrases recruit phage proteins for RDF function (53).

MOBILE ELEMENT LSRs AND ATTACHMENT SITES

Generally, the LSRs from mobile elements display less stringent site-selectivity than the canonical phage-encoded serine integrase systems, which raises questions on how their synaptic complexes might assemble. In the case of the transposases, TnpX and TndX, from Tn*4451* and Tn*5397*, respectively, no accessory factors are required for excision or integration (32, 67). TndX can mediate *attTn* × *attB*, *attL* or *attR* recombination but not *attTn* × *attTn*. Put more generally, TndX can mediate a recombination between *attTn* and any site with only one *attB* arm. Mechanistically, this pattern of site-selectivity could be explained if the CC motif interactions are only required between two of the DNA bound subunits, those bound to an *attTn* arm and an *attB* arm, rather than between all four as in the phage integrases.

The SCC*mec* elements use two recombinases, CcrA and CcrB for both integration (*attSCC* × *attB*) and excision (*attL* × *attR*) (35). The nature of the recombination sites and how they are recognized by the Ccr proteins is fundamentally different from the phage-encoded systems (14, 36). The *attB* site for SCC*mec* elements lies within the conserved gene *rlmH* (encoding a 23S rRNA pseudouridine methyl-transferase) at the 3′ end and entry of the SCC*mec* element regenerates the 3′ end and the nearby stop codon (14, 35). The left arm of *attB* is entirely *rlmH* and is called the *attB*rlmH arm. CcrA specifically binds to sites that contain the *attB*rmlH arm, that is, *attB* and *attL* (14). The right arm of *attB* contains a conserved 14 bp sequence that is also present as an imperfect inverted repeat in *attSCC* and *attR* (14). CcrB binds well to *attB*, *attSCC*, and *attR* but not to the *attB* rmlH arm (14, 36). In recombination, CcrB can recombine sites that exclude the *attB*rmlH arm (SCC or *attR* × SCC or *attR*). In the presence of CcrA and CcrB, the recombination repertoire expands to include recombination between any pairing of sites except those that contain more than one *attB* rlmH arm (*attB* or *attL* × *attB* or *attL*). It has been proposed that CcrA might be refractory for synapsis, which could explain why the only pairs of sites that cannot recombine are those with two CcrA binding sites (14). Explanations of how permissive sites can synapse and recombine in the context of the roles of the CC motifs in these proteins will be an interesting contrast to those proposed for the phage integrases.

THE CENTRAL DINUCLEOTIDE IN THE INTEGRATION SUBSTRATES DETERMINES SITE POLARITY

The outcome of site-specific recombination reactions is dependent upon the orientation of the recombination sites with respect to each other in the case where sites are intramolecular (previously discussed). In an intermolecular reaction such as phage integration, the orientation of the *attB* site will determine the orientation of the integrated prophage with respect to markers flanking the *attB* site. There are clear implications if the control over orientation fails, for example, inversion instead of prophage excision. Thus, the recombination sites need to possess polarity, referred to as left and right arms or P/B and P′/B′ arms. A left arm (P or B) must always be joined to a right arm (P′ or B′) for the substrates not to become scrambled (Fig. 7). In the LSRs, undergoing integration, that is, using *attP* and *attB* (or *attTn* and *attB*), the polarity of the attachment sites is determined by the nature of the dinucleotide at the crossover site (34, 42, 45, 46).

In integration, the *attP* and *attB* sites are imperfect inverted repeat sequences that flank an identical minimal sequence of 2 nucleotides (nts), the crossover sequence. The DNA cleavage of the recombination substrates generates a 3′ overhang at these two nucleotides that must base pair in the recombinants prior to joining. Mutations in this dinucleotide in one substrate and not the other still allows for initiation of the recombination pathway but joining of the recombination products is prevented (42, 46). Mutation of the dinucleotide to the reverse complement leads to complete reversal of the polarity of the site. For example, in the case of ϕC31 *attP* and *attB* sites, the central 2 nts are 5′TT but if one site is changed to 5′AA (e.g., $attB^{AA}$) the polarity of this site is switched leading to aberrant recombination products in which the left arms are joined and the right arms are joined (Fig. 7) (46). If an $attB^{AA}$ is integrated into the host chromosome, phage integration occurs in the opposite orientation compared to the wild type *attB* site. This confirmed that the dinucleotide is the only determinant of site polarity *in vivo* as well as *in vitro*. In both Bxb1 and ϕC31 integration, palindromic dinucleotides in both *attP* and *attB* substrates result in complete loss of polarity such that, in intramolecular reactions, deletions occur with equal frequency to inversions (42, 46).

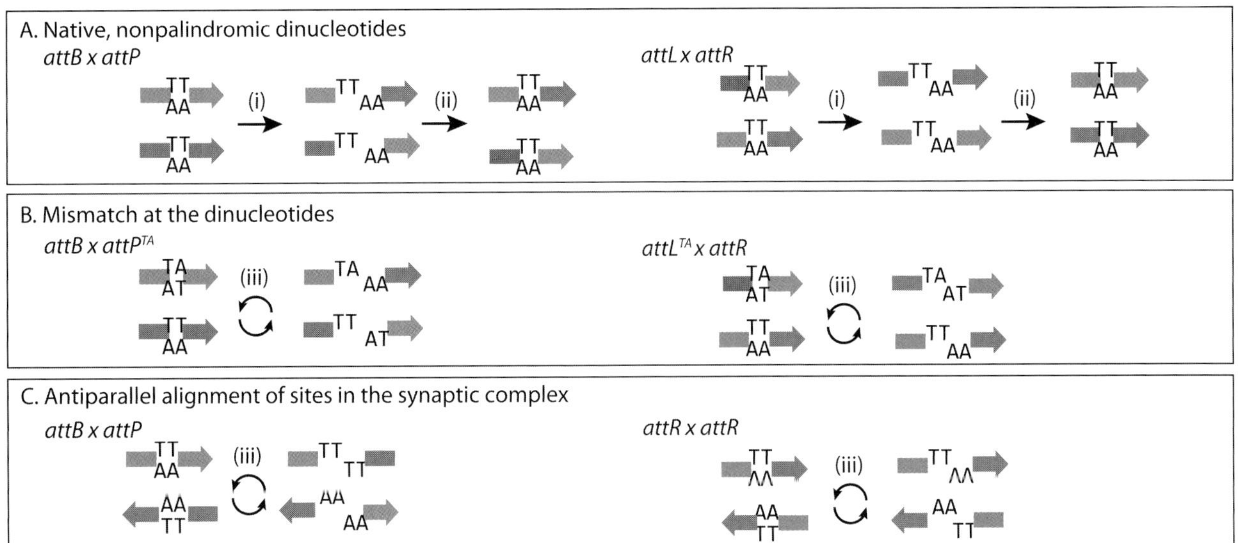

Figure 7 The role of the central dinucleotide in site polarity. The central dinucleotide determines the polarity of the *attP* and *attB* sites and the identity of the *attL* and *attR* sites. (A) The wild type recombination sites for ϕC31 integrase have a nonpalindromic 5′TT at the crossover dinucleotides that after cleavage and strand exchange by subunit rotation [indicated by arrow (i)] can base pair in the recombinants allowing joining of the DNA backbone [indicated by arrow (ii)]. (B) Any sites that have a mismatch at the dinucleotide cannot join after DNA cleavage and one round of strand exchange by subunit rotation. Strand exchange can either reverse or the rotation iterates to regenerate the substrates [indicated by arrows (iii)]. The recombination pathway can then begin again. (C) Fifty percent of synaptic complexes assemble with *attP* and *attB* in an antiparallel orientation such that cutting of the substrates yields dinucleotides at the crossover that cannot base pair in the recombinants leading to reversal or iteration of strand exchange (iii). In excision, *attL* and *attR* do not normally assemble an antiparallel synapse as this would require noncomplementary interactions between integrases bound to two P type (or two B type) half sites. However, *attR* × *attR* (or *attL* × *attL*) can form a synapse with the permitted complementary interactions by integrase subunits but joining of the products is prevented.
doi:10.1128/microbiolspec.MDNA3-0059-2014.f7

To understand how the dinucleotide is solely responsible for site polarity in integration, it is useful to think again about how integrase brings the site together in a synapse. The synapse can assemble with the *attP* and *attB* sites in two different alignments, parallel and antiparallel (Fig. 7) (42, 46). In both types of synapse, the same CC to CC interactions can be made between an integrase subunit bound to an *attB* type arm and an *attP* type arm. On formation of the synapse, DNA cleavage and strand exchange then takes place and, only when the half sites have been exchanged and are poised to form recombinants, does the proofreading step for site polarity occur, that is, whether the base pairing can occur between the overhangs. Only when the synapse is assembled with two sites in the parallel arrangement are the staggered breaks compatible. The *attP* and *attB* sites are therefore functionally symmetrical. Moreover, the sequences of the 2 bp at the crossover are not specifically recognized by integrases.

THE CENTRAL DINUCLEOTIDE IN THE EXCISION SUBSTRATES DETERMINES *attL* VERSUS *attR* IDENTITY

Similar experiments to the above have been used to study the excision substrates, *attL* and *attR*, specifically questioning their asymmetry and how this relates to the recombinational outcomes (48, 51, 53). As integrase alone does not mediate excision, these studies were performed either with a hyperactive integrase, such as IntE449K (48), or integrase in conjunction with its RDF (51, 53), either way, the properties of integrases are the similar. When the crossover dinucleotides are palindromic, the favored outcome for recombination with *attL* and *attR* is the correct outcome, that is, left arm to right arm joining to form *attP* and *attB* (Fig. 7). Insights into the nature of the *attL* and *attR* synapse from Rutherford *et al.* can explain this observation; the CC to CC motif interactions can occur between subunits located on an *attP* arm and an *attB* arm but not between subunits both bound to the *attP* arms or the *attB* arms (5, 7) (Fig. 6B). These interactions were anticipated and termed, respectively, complementary and noncomplementary interactions within the synapse (48, 51). A synaptic complex with complementary integrase interactions was strongly favored over noncomplementary interactions, with only the former yielding the correct *attP* and *attB* products.

Ghosh *et al.* noticed that the bias in the synapse structure in the excision reaction could explain how *attL* × *attL* and *attR* × *attR* reactions were prevented (51). These recombination substrates have the same asymmetry and complementary integrase interactions can occur if the two *attL* (or two *attR*) sites come together to synapse. Using FRET, Ghosh *et al.* provided direct evidence that the synapse containing the complementary integrase subunit interactions was strongly favored over the noncomplementary synapse using two *attR* sites (51) (Fig. 6B). Recombination between two wild type *attL* or two *attR* sites is largely prevented, as the overhangs generated will be unable to base pair after strand exchange, preventing product formation (Fig. 7). This two–step proofreading process ensures that only *attL* and *attR* sites recombine to give the expected outcome after recombination. Supporting this hypothesis is that two *attL* (or two *attR*) sites can recombine if the central dinucleotide is palindromic and the outcome is an aberrant product where P is joined to P and B′ to B′ (P′/P′ and B/B with two *attR* sites) (48, 51, 53).

STRAND EXCHANGE IN THE LSRs OCCURS BY SUBUNIT ROTATION

A remarkable property of the serine recombinases is that after DNA cleavage, the only forces that hold the recombining complex together are noncovalent, protein–protein interactions (Fig. 2). This complex is also believed to be dynamic, with supercoiling driving the rotation of two subunits of DNA bound recombinases 180° with respect to the other pair of subunits to bring about strand exchange (1). Mutation of a single bp at the crossover dinucleotide in γδ or Tn*3* resolvase or Hin invertase leads to iteration of the subunit rotation (i.e., two rotations of 180°) to reform the substrates as the mismatch in the recombinant arrangement of half sites strongly inhibits joining of the DNA ends.

The simple requirements of the LSR mediated integration reaction has permitted the use of a single molecule approach to study the phenomenon of subunit rotation. Bai *et al.* tethered DNA containing Bxb1 substrate(s) to a microscope cover slip at one end and to a paramagnetic bead at the other. Magnetic fields were used to control the tension on the DNA and the DNA linking number. As the DNA linking number is unwound, plectonemic twists spontaneously form and this shortens the distance between the bead and the cover slip. Subunit rotation can release the supercoiling and lengthen the distance between the bead and the tether. To initiate recombination and the formation of an open rotationally active complex, the tethered DNA contains an *attP* site and the *attB* is either provided on an oligonucleotide or in a second DNA strand that can be intertwined with the *attP* strand to generate plectonemes. In both experiments, all of the supercoiled DNA relaxes in a single step implying multiple rotations by the Bxb1 complex. When a mismatched *attB* site is used, products are prevented from forming and the cleaved recombination complex could remain "rotationally open" for many minutes. This work is significant as its shows using a direct assay that strand exchange is iterative, consistent with subunit rotation model.

Olorunniji *et al.* used a different experimental set up to show that strand exchange is normally gated after a single 180° rotation (68). Olorunniji *et al.* imposed a topological filter on the reaction driven by ϕC31 integrase using hybrid Tn3 *res/att* sites located in the same plasmid in head-to-tail orientation (68). The crossover sites from *res*, site I, were replaced by *attP* (to form *phesP*) or *attB* (*phesB*) and the *res* accessory sites II and III retained. Reactions with integrase, but without Tn3 resolvase, gave rise to deletions but contained multiple different topologies indicative of random collision at synapsis. Reactions with integrase and resolvase led to the majority of products being 2-noded catenanes, indicative of a single 180° rotation within a defined synapse topology. If the *phes* sites have a mismatch at the crossover site, a large amount of cleaved substrate is observed, indicative of stalling in the rotationally open state (68). Addition of the RDF to such reactions greatly reduces the amount of cleaved DNA, presumably by licensing a second 180° rotation to reform the substrates. The authors propose that the gating activity observed after strand exchange in an *attP* × *attB* reaction could relate to the proposed auto-inhibitory, CC to CC interactions that are likely to form when integrase is bound to *attL* and *attR* (5, 7).

APPLICATIONS OF PHAGE INTEGRASES, PROBLEMS REMAINING AND FUTURE PROSPECTS

As soon as the phage-encoded serine integrases were discovered, they were being incorporated into vectors for genetic engineering (54, 69, 70). Integrating vectors, that is, suicide plasmids encoding an *int* gene and an *attP* site, are used widely in microbes in which they either seek out an endogenous *attB* site for recombination or use one that is ectopically introduced (71, 72). For cloning DNA in *Streptomyces* spp., integrating vectors are the vectors of choice as they are stable, single (or sometimes double) insertions at a defined position. It quickly became clear from the biochemical properties of ϕC31 integrase in an *in vitro* system, that these relatively simple recombination systems could be exported into other organisms, in particular mammals, flies, plants and most model systems (72, 73, 74). For the most efficient integrations, the prior insertion of a docking site (either *attP* or *attB*) is required but inefficient ectopic recombination is also observed into sites that resemble an attachment site (72, 73).

More recently, applications of phage integrases have taken hold in synthetic biology, specifically in metabolic pathway assembly and in memory and counting devices. The attraction of using serine integrases in DNA assembly procedures is due to the efficiency of *in vitro* recombination (about 80% conversion of substrate to products), and the highly directional nature of the integrases. Moreover, for joining several fragments flanked by compatible *attP* and *attB* sites, the order of the fragments can be imposed by the dinucleotides at the crossover sites (75, 76). Colloms *et al.* also made use of the ϕC31 RDF to remove fragments of an assembled pathway and replace them with variants and this adds versatility that has not been realized in other assembly procedures.

Serine integrases have also been applied in the generation of *in vivo* counting modules and logic gates (77, 78). These applications also use the RDFs to control the inversion switch of a counting module. The switches are vertically inheritable and stable over many generations until a reset signal is imposed. These modules could be incorporated into organisms that sense their environment, record information and respond accordingly in a predetermined manner.

This brief description of the applications of serine integrases provides examples of how they are deployed widely in biology (see review by Fogg *et al.* [72]). Limitations are currently imposed by the few complete systems that are available for deployment, and not all of these are fully orthologous. Our understanding of the mechanisms of LSRs are also limited to only a few, especially the insights we can gain from structural information. The most immediate mechanistic questions concern how the RDFs control the formation of synaptic complexes, the nature of the proposed auto-inhibitory complexes, and how the RDFs might remodel these complexes. The CC to CC interactions are essential for an efficient intermolecular recombination reactions, but what are the contacts made between them? Are there other contacts between the CC motifs and other domains of the LSRs? Finally, how can we engineer integrases and/or their recombination sites for new and novel applications?

Acknowledgments. *MCMS thanks Dr. Greg van Duyne for redrawing the structural models shown in Fig. 5. Research on ϕC31 integrase and other LSRs in the laboratory of MCMS is funded by the Biotechnology and Biological Science Research Council, UK.*

Citation. Smith MCM. 2014. Phage-encoded serine integrases and other large serine recombinases. Microbiol Spectrum 3(1):MDNA3-0059-2014.

References

1. **Grindley NDF, Whiteson KL, Rice PA.** 2006. Mechanisms of site-specific recombination. *Annu Rev Biochem* 75:567–605.
2. **Smith MCM, Thorpe HM.** 2002. Diversity in the serine recombinases. *Mol Microbiol* **44:**299–307.
3. **Thorpe HM, Smith MCM.** 1998. *In vitro* site-specific integration of bacteriophage DNA catalyzed by a recombinase of the resolvase/invertase family. *Proc Natl Acad Sci USA* **95:**5505–5510.
4. **Kim AI, Ghosh P, Aaron MA, Bibb LA, Jain S, Hatfull GF.** 2003. Mycobacteriophage Bxb1 integrates into the *Mycobacterium smegmatis groEL1* gene. *Mol Microbiol* 50:463–473.
5. **Rutherford K, Yuan P, Perry K, Sharp R, Van Duyne GD.** 2013. Attachment site recognition and regulation of directionality by the serine integrases. *Nucleic Acids Res* **41:**8341–8356.
6. **Van Duyne GD, Rutherford K.** 2013. Large serine recombinase domain structure and attachment site binding. *Crit Rev Biochem Molec Biol* **48:**476–491.
7. **Rutherford K, Van Duyne GD.** 2014. The ins and outs of serine integrase site-specific recombination. *Curr Opin Struct Biol* **24:**125–131.
8. **Adams V, Lyras D, Farrow KA, Rood JI.** 2002. The clostridial mobilisable transposons. *Cell Mol Life Sci* **59:** 2033–2043.
9. **Mullany P, Roberts AP, Wang H.** 2002. Mechanism of integration and excision in conjugative transposons. *Cell Mol Life Sci* **59:**2017–2022.
10. **Hanssen AM, Ericson Sollid JU.** 2006. SCC*mec* in staphylococci: genes on the move. *FEMS Immunol Med Microbiol* **46:**8–20.
11. **Campbell AM.** 1962. Episomes, p 101–145. *In* Caspari EW, Thoday JM (ed), *Advances in Genetics*. Academic Press, New York, NY.
12. **Lewis JA, Hatfull GF.** 2001. Control of directionality in integrase-mediated recombination: examination of recombination directionality factors (RDFs) including Xis and Cox proteins. *Nucleic Acids Res* **29:**2205–2216.

13. **Smith MCM, Brown WR, McEwan AR, Rowley PA.** 2010. Site-specific recombination by phiC31 integrase and other large serine recombinases. *Biochem Soc Trans* **38:**388–394.

14. **Misiura A, Pigli YZ, Boyle-Vavra S, Daum RS, Boocock MR, Rice PA.** 2013. Roles of two large serine recombinases in mobilizing the methicillin-resistance cassette SCC*mec*. *Mol Microbiol* **88:**1218–1229.

15. **Combes P, Till R, Bee S, Smith MCM.** 2002. The streptomyces genome contains multiple pseudo-attB sites for the ϕC31-encoded site-specific recombination system. *J Bacteriol* **184:**5746–5752.

16. **Bibb LA, Hancox MI, Hatfull GF.** 2005. Integration and excision by the large serine recombinase phiRv1 integrase. *Mol Microbiol* **55:**1896–1910.

17. **Ito T, Katayama Y, Asada K, Mori N, Tsutsumimoto K, Tiensasitorn C, Hiramatsu K.** 2001. Structural comparison of three types of staphylococcal cassette chromosome mec integrated in the chromosome in methicillin-resistant *Staphylococcus aureus*. *Antimicrob Agents Chemother* **45:**1323–1336.

18. **Singh S, Rockenbach K, Dedrick RM, VanDemark AP, Hatfull GF.** 2014. Cross-talk between diverse serine integrases. *J Mol Biol* **426:**318–331.

19. **Carrasco CD, Ramaswamy KS, Ramasubramanian TS, Golden JW.** 1994. Anabaena *xisF* gene encodes a developmentally regulated site-specific recombinase. *Genes Dev* **8:**74–83.

20. **Sato T, Samori Y, Kobayashi Y.** 1990. The *cisA* cistron of *Bacillus subtilis* sporulation gene *spoIVC* encodes a protein homologus to a site-sepcific recombinase. *J Bacteriol* **172:**1092–1098.

21. **Stragier P, Kunkel B, Kroos L, Losick R.** 1989. Chromosomal rearrangement generating a composite gene for a developmental transcription factor. *Science* **243:**507–512.

22. **Golden JW, Mulligan ME, Haselkorn R.** 1987. Different recombination site specificity of two developmentally regulated genome rearrangements. *Nature* **327:**526–529.

23. **Carrasco CD, Buettner JA, Golden JW.** 1995. Programmed DNA rearrangement of a cyanobacteria *hupL* gene in heterocysts. *Proc Natl Acad Sci USA* **92:**791–795.

24. **Ramaswamy KS, Carrasco CD, Fatma T, Golden JW.** 1997. Cell-type specificity of the *Anabaena fdxN*-element rearrangement requires *xisH* and *xisI*. *Mol Microbiol* **23:** 1241–1249.

25. **Kunkel B, Losick R, Stragier P.** 1990. The *Bacillus subtilis* gene for the development transcription factor sigma K is generated by excision of a dispensable DNA element containing a sporulation recombinase gene. *Genes Dev* **4:**525–535.

26. **Popham DL, Stragier P.** 1992. Binding of the *Bacillus subtilis spoIVCA* product to the recombination sites of the element interrupting the sigma-K encoding gene. *Proc Natl Acad Sci USA* **89:**5991–5995.

27. **Loessner MJ, Inman RB, Lauer P, Calendar R.** 2000. Complete nucleotide sequence, molecular analysis and genome structure of bacteriophage A118 of *Listeria monocytogenes*: implications for phage evolution. *Mol Microbiol* **35:**324–340.

28. **Rabinovich L, Sigal N, Borovok I, Nir-Paz R, Herskovits AA.** 2012. Prophage excision activates Listeria competence genes that promote phagosomal escape and virulence. *Cell* **150:**792–802.

29. **Claverys JP, Prudhomme M, Martin B.** 2006. Induction of competence regulons as a general response to stress in gram-positive bacteria. *Annu Rev Microbiol* **60:**451–475.

30. **Bannam TL, Crellin PK, Rood JI.** 1995. Molecular genetics of the chloramphenicol-resistance transposon Tn*4451* from *Clostridium perfringens*: the TnpX site-specific recombinase excises a circular transposon molecule. *Mol Microbiol* **16:**535–551.

31. **Crellin PK, Rood JI.** 1997. The resolvase/invertase domain of the site-specific recombinase TnpX is functional and recognizes a target sequence that resembles the junction of the circular form of the *Clostridium perfringens* transposon Tn*4451*. *J Bacteriol* **179:**5148–5156.

32. **Wang H, Mullany P.** 2000. The large resolvase TndX is required and sufficient for integration and excision of derivatives of the novel conjugative transposon Tn*5397*. *J Bacteriol* **182:**6577–6583.

33. **Lyras D, Rood JI.** 2000. Transposition of Tn*4451* and Tn*4453* involves a circular intermediate that forms a promoter for the large resolvase, TnpX. *Mol Microbiol* **38:**588–601.

34. **Wang H, Smith MCM, Mullany P.** 2006. The conjugative transposon Tn*5397* has a strong preference for integration into its *Clostridium difficile* target site. *J Bacteriol* **188:**4871–4878.

35. **Katayama Y, Ito T, Hiramatsu K.** 2000. A new class of genetic element, staphylococcus cassette chromosome mec, encodes methicillin resistance in *Staphylococcus aureus*. *Antimicrob Agents Chemother* **44:**1549–1555.

36. **Wang L, Archer GL.** 2010. Roles of CcrA and CcrB in excision and integration of staphylococcal cassette chromosome mec, a *Staphylococcus aureus* genomic island. *J Bacteriol* **192:**3204–3212.

37. **Ito T, Katayama Y, Hiramatsu K.** 1999. Cloning and nucleotide sequence determination of the entire mec DNA of pre-methicillin-resistant *Staphylococcus aureus* N315. *Antimicrob Agents Chemother* **43:**1449–1458.

38. **Scharn CR, Tenover FC, Goering RV.** 2013. Transduction of staphylococcal cassette chromosome mec elements between strains of *Staphylococcus aureus*. *Antimicrob Agents Chemother* **57:**5233–5238.

39. **Li W, Kamtekar S, Xiong Y, Sarkis GJ, Grindley ND, Steitz TA.** 2005. Structure of a synaptic gamma-delta resolvase tetramer covalently linked to two cleaved DNAs. *Science* **309:**1210–1215.

40. **Yang W, Steitz TA.** 1995. Crystal-structure of the site-specific recombinase gamma-delta resolvase complexed with a 34bp cleavage site. *Cell* **82:**193–207.

41. **Yuan P, Gupta K, Van Duyne GD.** 2008. Tetrameric structure of a serine integrase catalytic domain. *Structure* **16:**1275–1286.

42. **Ghosh P, Kim AI, Hatfull GF.** 2003. The orientation of mycobacteriophage Bxb1 integration is solely dependent on the central dinucleotide of *attP* and *attB*. *Mol Cell* **12:** 1101–1111.

43. **Ghosh P, Pannunzio NR, Hatfull GF.** 2005. Synapsis in phage Bxb1 integration: selection mechanism for the correct pair of recombination sites. *J Mol Biol* **349:** 331–348.

44. **Smith MCA, Till R, Brady K, Soultanas P, Thorpe H, Smith MCM.** 2004. Synapsis and DNA cleavage in ϕC31 integrase-mediated site-specific recombination. *Nucleic Acids Res* **32:**2607–2617.

45. **Zhang L, Wang L, Wang J, Ou X, Zhao G, Ding X.** 2010. DNA cleavage is independent of synapsis during *Streptomyces* phage ϕBT1 integrase-mediated site-specific recombination. *J Mol Cell Biol* **2:**264–275.

46. **Smith MCA, Till R, Smith MCM.** 2004. Switching the polarity of a bacteriophage integration system. *Mol Microbiol* **51:**1719–1728.

47. **Thorpe HM, Wilson SE, Smith MCM.** 2000. Control of directionality in the site-specific recombination system of the Streptomyces phage ϕC31. *Mol Microbiol* **38:** 232–241.

48. **Rowley PA, Smith MCA, Younger E, Smith MCM.** 2008. A motif in the C-terminal domain of ϕC31 integrase controls the directionality of recombination. *Nucleic Acids Res* **36:**3879–3891.

49. **Breuner A, Brondsted L, Hammer K.** 1999. Novel organisation of genes involved in prophage excision identified in the temperate lactococcal bacteriophage TP901-1. *J Bacteriol* **181:**7291–7297.

50. **Zhang L, Ou X, Zhao G, Ding X.** 2008. Highly efficient *in vitro* site-specific recombination system based on *Streptomyces* phage ϕBT1 integrase. *J Bacteriol* **190:**6392–6397.

51. **Ghosh P, Bibb LA, Hatfull GF.** 2008. Two-step site selection for serine-integrase-mediated excision: DNA-directed integrase conformation and central dinucleotide proofreading. *Proc Natl Acad Sci USA* **105:**3238–3243.

52. **Ghosh P, Wasil LR, Hatfull GF.** 2006. Control of phage Bxb1 excision by a novel recombination directionality factor. *PLoS Biol* **4:**e186.

53. **Khaleel T, Younger E, McEwan AR, Varghese AS, Smith MCM.** 2011. A phage protein that binds ϕC31 integrase to switch its directionality. *Mol Microbiol* **80:**1450–1463.

54. **Matsuura M, Noguchi T, Yamaguchi D, Aida T, Asayama M, Takahashi H, Shirai M.** 1996. The sre gene (ORF469) encodes a site-specific recombinase responsible for integration of the R4 phage genome. *J Bacteriol* **178:**3374–3376.

55. **Mouw KW, Rowland SJ, Gajjar MM, Boocock MR, Stark WM, Rice PA.** 2008. Architecture of a serine recombinase-DNA regulatory complex. *Mol Cell* **30:** 145–155.

56. **Adams V, Lucet IS, Lyras D, Rood JI.** 2004. DNA binding properties of TnpX indicate that different synapses are formed in the excision and integration of the Tn*4451* family. *Mol Microbiol* **53:**1195–1207.

57. **Mandali S, Dhar G, Avliyakulov NK, Haykinson MJ, Johnson RC.** 2013. The site-specific integration reaction of *Listeria* phage A118 integrase, a serine recombinase. *Mobile DNA* **4:**2.

58. **Rowley PA, Smith MCM.** 2008. Role of the N-terminal domain of ϕC31 integrase in *attB-attP* synapsis. *J Bacteriol* **190:**6918–6921.

59. **McEwan AR, Raab A, Kelly SM, Feldmann J, Smith MCM.** 2011. Zinc is essential for high-affinity DNA binding and recombinase activity of ϕC31 integrase. *Nucleic Acids Res* **39:**6137–6147.

60. **McEwan AR, Rowley PA, Smith MCM.** 2009. DNA binding and synapsis by the large C-terminal domain of ϕC31 integrase. *Nucleic Acids Res* **37:**4764–4773.

61. **Lucet IS, Tynan FE, Adams V, Rossjohn J, Lyras D, Rood JI.** 2005. Identification of the structural and functional domains of the large serine recombinase TnpX from *Clostridium perfringens*. *J Biol Chem* **280:**2503–2511.

62. **Finn RD, Bateman A, Clements J, Coggill P, Eberhardt RY, Eddy SR, Heger A, Hetherington K, Holm L, Mistry J, Sonnhammer EL, Tate J, Punta M.** 2014. Pfam: the protein families database. *Nucleic Acids Res* **42:**D222–D230.

63. **Gupta M, Till R, Smith MCM.** 2007. Sequences in *attB* that affect the ability of ϕC31 integrase to synapse and to activate DNA cleavage. *Nucleic Acids Res* **35:**3407–3419.

64. **Singh S, Ghosh P, Hatfull GF.** 2013. Attachment site selection and identity in Bxb1 serine integrase-mediated site-specific recombination. *PLoS Genet* **9:**e1003490.

65. **Bibb LA, Hatfull GF.** 2002. Integration and excision of the *Mycobacterium tuberculosis* prophage-like element, ϕRv1. *Mol Microbiol* **45:**1515–1526.

66. **Zhang L, Zhu B, Dai R, Zhao G, Ding X.** 2013. Control of directionality in *Streptomyces* phage ϕBT1 integrase-mediated site-specific recombination. *PloS One* **8:**e80434.

67. **Lyras D, Adams V, Lucet I, Rood JI.** 2004. The large resolvase TnpX is the only transposon-encoded protein required for transposition of the Tn*4451/3* family of integrative mobilizable elements. *Mol Microbiol* **51:**1787–1800.

68. **Olorunniji FJ, Buck DE, Colloms SD, McEwan AR, Smith MCM, Stark WM, Rosser SJ.** 2012. Gated rotation mechanism of site-specific recombination by ϕC31 integrase. *Proc Natl Acad Sci USA* **109:**19661–19666.

69. **Kuhstoss S, Rao RN.** 1991. Analysis of the integration function of the streptomycete bacteriophage phic31. *J Mol Biol* **222:**897–908.

70. **Kuhstoss S, Richardson MA, Rao RN.** 1991. Plasmid cloning vectors that integrate site-specifically in *Streptomyces* spp. *Gene* **97:**143–146.

71. **St-Pierre F, Cui L, Priest DG, Endy D, Dodd IB, Shearwin KE.** 2013. One-step cloning and chromosomal integration of DNA. *ACS Synth Biol* **2:**537–541.

72. **Fogg PC, Colloms S, Rosser S, Stark M, Smith MCM.** 2014. New applications for phage integrases. *J Mol Biol* **426:**2703–2716.

73. **Groth AC, Calos MP.** 2004. Phage integrases: biology and applications. *J Mol Biol* **335:**667–678.

74. **Groth AC, Olivares EC, Thyagarajan B, Calos MP.** 2000. A phage integrase directs efficient site-specific integration in human cells. *Proc Natl Acad Sci USA* **97:** 5995–6000.

75. **Zhang L, Zhao G, Ding X.** 2011. Tandem assembly of the epothilone biosynthetic gene cluster by *in vitro* site-specific recombination. *Sci Rep* **1**:141.

76. **Colloms SD, Merrick CA, Olorunniji FJ, Stark WM, Smith MCM, Osbourn A, Keasling JD, Rosser SJ.** 2013. Rapid metabolic pathway assembly and modification using serine integrase site-specific recombination. *Nucleic Acids Res* **42**:e23.

77. **Bonnet J, Subsoontorn P, Endy D.** 2012. Rewritable digital data storage in live cells via engineered control of recombination directionality. *Proc Natl Acad Sci USA* **109**: 8884–8889.

78. **Bonnet J, Yin P, Ortiz ME, Subsoontorn P, Endy D.** 2013. Amplifying genetic logic gates. *Science* **340**:599–603.

79. **Smith MCM.** 2013. Conservative site-specific recombination, p 555–561. *In* Lennarz W (ed), *The Encyclopedia of Biological Chemistry*, **vol 1**. Academic Press, Waltham, MA.

80. **Christiansen B, Johnsen MG, Stenby E, Vogensen FK, Hammer K.** 1994. Characterization of the lactococcal temperate phage TP901-1 and its site-specific integration. *J Bacteriol* **176**:1069–1076.

81. **Gregory MA, Till R, Smith MCM.** 2003. Integration site for *Streptomyces* phage ϕBT1 and the development of novel site-specific integrating vectors. *J Bacteriol* **185**: 5320–5323.

82. **Shirai M, Nara H, Sato A, Aida T, Takahashi H.** 1991. Site-specific integration of the actinophage R4 genome into the chromosome of *Streptomyces parvulus* upon lysogenization. *J Bacteriol* **173**:4237–4239.

83. **Morita K, Yamamoto T, Fusada N, Komatsu M, Ikeda H, Hirano N, Takahashi H.** 2009. The site-specific recombination system of actinophage TG1. *FEMS Microbiol Lett* **297**:234–240.

84. **Park MO, Lim KH, Kim TH, Chang HI.** 2007. Characterization of site-specific recombination by the integrase MJ1 from enterococcal bacteriophage ϕFC1. *J Microbiol Biotechnol* **17**:342–347.

85. **Rashel M, Uchiyama J, Ujihara T, Takemura I, Hoshiba H, Matsuzaki S.** 2008. A novel site-specific recombination system derived from bacteriophage ϕMR11. *Biochem Biophys Res Commun* **368**:192–198.

86. **Fayed B, Younger E, Taylor G, Smith MCM.** 2014. A novel *Streptomyces* spp. integration vector derived from the *S. venezuelae* phage, SV1. *BMC Biotechnol* **14**:51.

87. **Kilcher S, Loessner MJ, Klumpp J.** 2010. Brochothrix thermosphacta bacteriophages feature heterogeneous and highly mosaic genomes and utilize unique prophage insertion sites. *J Bacteriol* **192**:5441–5453.

88. **Lazarevic V, Dusterhoft A, Soldo B, Hilbert H, Mauel C, Karamata D.** 1999. Nucleotide sequence of the *Bacillus subtilis* temperate bacteriophage SPβc2. *Microbiology* **145**(Pt 5):1055–1067.

89. **Fouts DE, Rasko DA, Cer RZ, Jiang L, Fedorova NB, Shvartsbeyn A, Vamathevan JJ, Tallon L, Althoff R, Arbogast TS, Fadrosh DW, Read TD, Gill SR.** 2006. Sequencing *Bacillus anthracis* typing phages gamma and cherry reveals a common ancestry. *J Bacteriol* **188**: 3402–3408.

90. **Canchaya C, Desiere F, McShan WM, Ferretti JJ, Parkhill J, Brussow H.** 2002. Genome analysis of an inducible prophage and prophage remnants integrated in the *Streptococcus pyogenes* strain SF370. *Virology* **302**: 245–258.

Mobile DNA, 3rd Edition
Nancy L. Craig, Michael Chandler, Martin Gellert, Alan M. Lambowitz, Phoebe A. Rice and Suzanne Sandmeyer

doi:10.1128/microbiolspec.MDNA3-0023-2014

Kerri Kobryn[1]
George Chaconas[2]

Hairpin Telomere Resolvases

12

HAIRPIN TELOMERE RESOLVASES

Hairpin telomere resolvases (also known as protelomerases) have emerged as a unique solution to the end replication problem (1, 2). These enzymes promote the formation of covalently closed hairpin ends on linear DNA molecules in some phage (3, 4, 5), bacterial plasmids and bacterial chromosomes (6, 7, 8, 9). Telomere resolvases are mechanistically related to tyrosine recombinases and type IB topoisomerases and are also believed to play a role in the genome plasticity that characterizes *Borrelia* species. Fig. 1 shows the reaction pathway for replication of linear DNA molecules with covalently closed hairpin telomeres. Duplication of the DNA molecule results in replicated telomeres (*rTel*, also referred to as dimer junctions) that are recognized and processed in a DNA breakage and reunion reaction promoted by a hairpin telomere resolvase. The reaction products are covalently closed hairpin telomeres at both ends of linear monomeric DNA molecules. At this writing telomere resolvases have been purified from three phage and seven bacterial species: *E. coli* phage N15 (3), *Klebsiella oxytoca* phage φKO2, *Yersinia enterocolitica* phage PY54 (5), *Agrobacterium tumefaciens* (8), the Lyme spirochete *Borrelia burgdorferi* (6), the relapsing fever borreliae *B. hermsii, B. parkeri, B. recurrentis, B. turicatae,* and the avian spirochete *B. anserina* (7). The *B. burgdorferi* enzyme, ResT (**Re**solvase of **T**elomeres) has been the most extensively studied at the biochemical level (6, 7, 10, 11, 12, 13, 14, 15, 16, 17, 18, 19, 20, 21, 22, 23) and is the primary focus of this review, with properties of the other enzymes noted (3, 4, 5, 8, 24). Structural studies of the *Klebsiella* phage φKO2 (25) and the *Agrobacterium* (26) resolvases have been reported and have shed additional light on reaction mechanisms and on differences between the resolvases from different organisms.

REACTION MECHANISM – SIMILARITY TO TYPE IB TOPOISOMERASES AND TYROSINE RECOMBINASES

ResT and other telomere resolvases/protelomerases constitute a new class of DNA breakage and rejoining enzymes since they catalyze a unique reaction in which a replicated telomere (*rTel*) substrate is cleaved six base pairs apart (4, 6, 8) on opposite strands at the center of the substrate, giving rise to covalently closed hairpin telomeres after the 6 nt 5′-overhangs have been folded back to effect strand rejoining.

Despite the unique hairpin products produced by the telomere resolution reaction, the reaction has key similarities to the reactions catalyzed by the type IB

[1]Department of Microbiology and Immunology, College of Medicine, University of Saskatchewan, Saskatoon, SK S7N 5E5, Canada; [2]Department of Biochemistry & Molecular Biology and Department of Microbiology, Immunology & Infectious Diseases, Snyder Institute, The University of Calgary, Calgary, AB T2N 4N1, Canada.

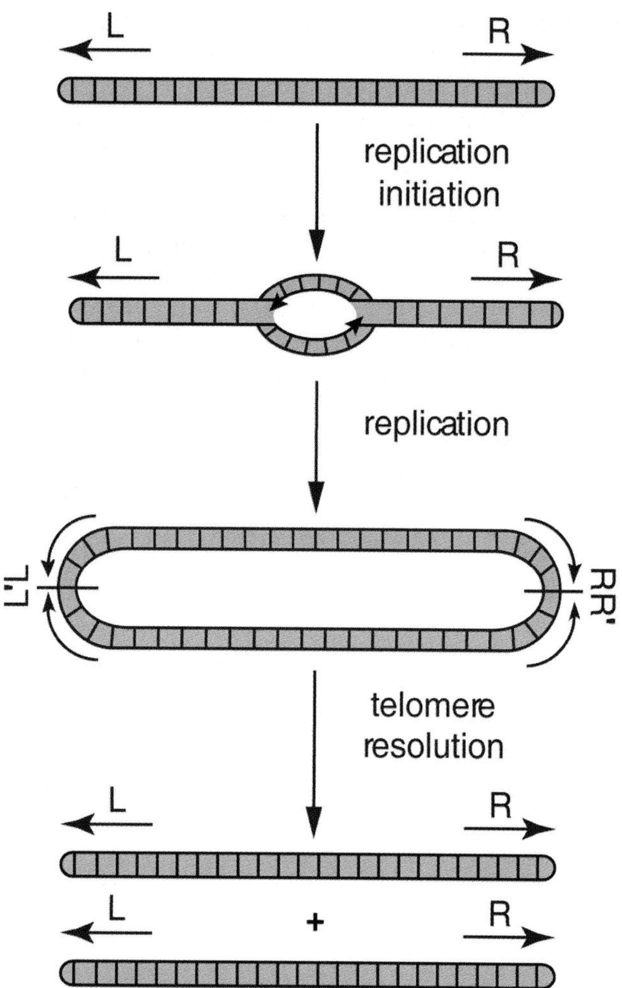

Figure 1 Hairpin telomere resolution as a solution to the end replication problem. Replication of a linear molecule results in the formation of dimer junctions or replicated telomeres (L′L, RR′) that are processed by telomere resolution, a unique type of DNA breakage and reunion reaction. Telomere resolution results in the formation of hairpin telomeres at the ends of the linear DNA molecule and separates the dimer replication intermediate into monomeric products of DNA replication. See text for further details. This figure is adapted from reference 6 and reprinted from reference 1. doi:10.1128/microbiolspec.MDNA3-0023-2014.f1

topoisomerase and tyrosine recombinase enzyme families. The principal similarity lies in the fact that the DNA breakage and reunion reactions proceed as a chemically isoenergetic two-step transesterification without a requirement for divalent metal ions or high-energy cofactors. This is made possible by a shared mode of phosphoryl transfer in which DNA cleavage is effected by attack on the scissile phosphate(s) by an active site tyrosine accompanied by formation of a transient protein–DNA covalent complex that stores the phosphodiester bond energy; the cleaved DNA strand(s) are resealed by attack on the 3′-phosphotyrosine linkage by the 5′-OH group of the cleaved strand(s) to form the final products releasing the enzyme.

The products of the reactions of these enzyme families vary by virtue of differences in the number of DNA strands cleaved and in the identity of the strands that are rejoined. The type IB topoisomerases cleave a single DNA strand that is rejoined after duplex rotation or strand passage to change levels of DNA supercoiling; the type IB topoisomerases act as monomers (27). Tyrosine recombinases cleave four DNA strands in two reaction steps and rejoin the cleaved strands to new partners to produce recombinant products. Consequently, the tyrosine recombinases act as tetramers that form a synaptic complex of the reaction sites; strand exchange occurs in the context of the synaptic complex (28, 29). The reaction of telomere resolvases lies midway in this spectrum; two DNA strands are cleaved and after strand foldback the 5′-overhanging strands are rejoined to the opposite strand to form two hairpin telomeres. Because of these differences telomere resolvases act as dimers on a single *rTel* reaction site (22, 25, 26). In some ways telomere resolution may be thought of as a stripped down site-specific recombination reaction that occurs with only one reaction site rather than two (4, 14).

Similar to the tyrosine recombinases, it is known that ResT requires interprotomer communication to initiate reaction chemistry. ResT is inactive on a half-site embedded in a plasmid and requires oligonucleotide half-sites to assemble into a 'cross-axis complex' in which a dimer of ResT brings the two half-sites into an arrangement that mimics the structure of the *bona fide rTel* substrate (Fig. 2) for reaction chemistry to proceed (14, 16). Formation of the cross-axis complex requires conformational changes in ResT to unmask the telomere binding determinants in the catalytic core and/or the C-terminal domain of ResT, which are partially occluded by the N-terminal domain of the protein (18). Unexpectedly, positive DNA supercoiling helps to overcome this barrier, promoting cooperative binding of ResT to *rTel*s and subsequent cross-axis complex formation (16); negative supercoiling inhibits the reaction. The positive DNA supercoiling likely promotes positive DNA rotation in the center of the *rTel* to facilitate the cooperative interactions required for cross-axis complex formation to occur.

Tyrosine recombinases have a strict temporal order and a half-of-sites activity scheme in which only one protomer in the dimer bound to each reaction site can be active at a time, such that the reactions transit

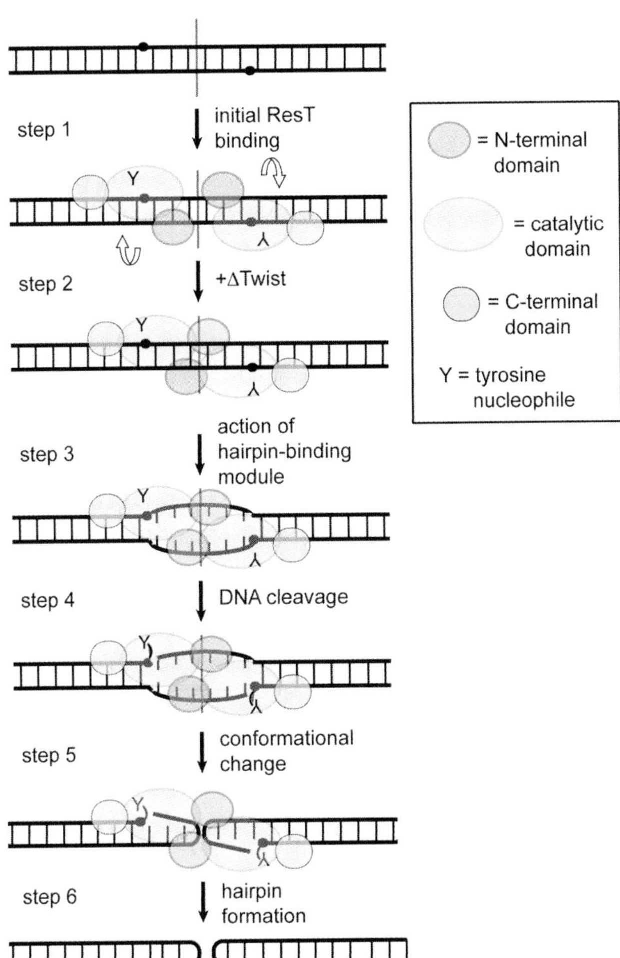

Figure 2 ResT promotes telomere resolution via a two-step transesterification. (1) The telomere resolvase, ResT, binds to a replicated telomere (*rTel*) substrate, which corresponds to the dimer junction L′L or RR′ shown in Fig. 1 (11). The black dots denote the scissile phosphates and the vertical bar in the center of *rTel* the axis of 180-degree rotational symmetry. (2) Positive supercoiling is believed to facilitate the cooperative formation of a cross-axis complex where communication between ResT protomers bound on both sides of the symmetry axis occurs (14, 16). (3) The action of the "hairpin binding module," a region with sequence similarity to a motif found in cut-and-paste transposases, induces a DNA distortion that facilitates (4) DNA cleavage, the first transesterification event (12). (5) A conformational change then occurs to juxtapose the free 5′-OH groups to the 3-phosphotyrosyl enzyme intermediates on the opposite strands. (6) Nucleophilic attack of the phosphotyrosyl linkage by the 5′-OH groups, the second transesterification step, results in phosphodiester bond formation to stabilize the hairpins generated during step 5. This figure and legend are adapted from reference 43 and reference 21, and reprinted from reference 2.
doi:10.1128/microbiolspec.MDNA3-0023-2014.f2

through the Holliday junction (HJ) intermediate without producing double-strand breaks in the substrate DNAs (30, 31, 32). Tyrosine recombinases will usually regenerate substrate if they cannot complete a reaction cycle (33, 34). In contrast to this, ResT activity on the two sides of the *rTel* must be simultaneous, or nearly so, in order that the 5′-OH nucleophiles for the second pair of transesterifications (hairpin formation) are both available at the same time (14). Because of these features it was of interest to know whether the reaction chemistry of DNA cleavage and strand transfer in telomere resolution and in the reverse reaction of telomere fusion were concerted on the two halves of the substrate, linking reaction completion on the two sides, or whether the reaction chemistry on the two half-sites could be uncoupled from each other. Using substrates modified so as to inhibit either DNA cleavage or DNA strand transfer on one half-site but not the other, it was determined that execution of both chemical steps on the two halves of the reaction substrate can be uncoupled from each other (14). This combination of simultaneous action on the two halves of the substrate with the freedom to uncouple the completion of the DNA cleavage or strand transfer reactions on the two sides of the substrate was somewhat surprising since it implied that in certain conditions a hairpin telomere might be formed on one side of the substrate at the expense of a ResT-capped double-strand break on the other half of the *rTel*.

A TYROSINE RECOMBINASE-LIKE ACTIVE SITE

A weak similarity of putative telomere resolvases to tyrosine recombinases was noted in early sequence alignments (6, 35). A detailed study of the putative active site residues for ResT was made based on these alignments and on alignments among confirmed or suspected telomere resolvases (7, 13). Similar to type IB topoisomerases and tyrosine recombinases, ResT can cleave substrates modified at the scissile phosphates with 5′-bridging phosphorothiolates but cannot use the liberated 5′-sulfhydryl groups as nucleophiles for the subsequent strand transfer by virtue of the lower pKa of sulfhydryl compared to hydroxyl groups (33, 36, 37, 38). These modified substrates can therefore be used as suicide substrates to accumulate the covalent ResT-DNA complex. ResT cleavage of such a suicide substrate, followed by affinity purification of tryptic peptides attached to the DNA, allowed the identification of the active site tyrosine nucleophile as Y335 (see Fig. 3) by tandem mass spectroscopy (13).

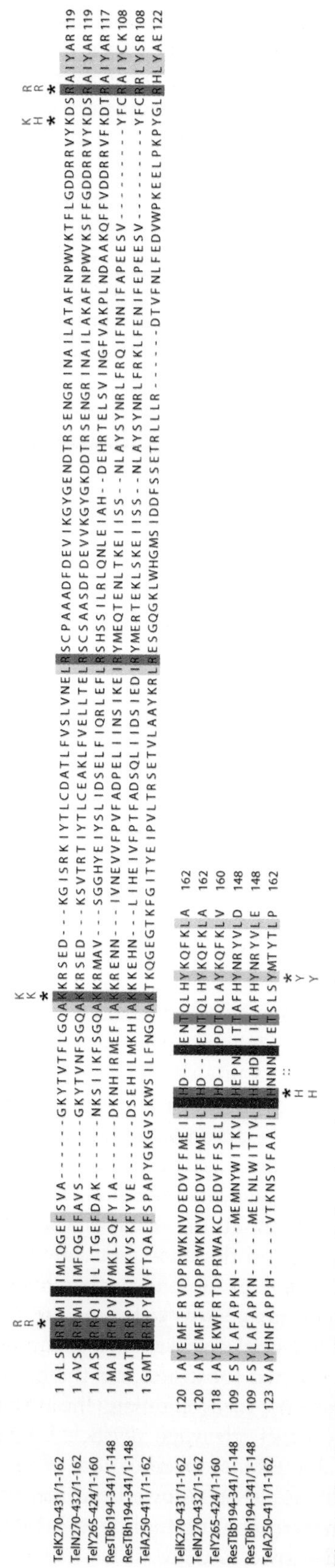

Figure 3 Sequence alignment of the domains carrying the active site residues of purified telomere resolvases. An alignment is shown for the telomere resolvases from the phages N15, TelN (3), φKO2, TelK (4), PY54, TelY (5) and the bacterial resolvases from *Agrobacterium tumefaciens*, TelA (8), the Lyme spirochete *Borrelia burgdorferi*, ResTBb (6) and the relapsing fever species *Borrelia hermsii*, ResTBh (7). Several other purified hairpin resolvases from other *Borrelia* species are not included in the lineup. The catalytic residues are indicated by asterisks and the active site tyrosine by a red asterisk. The double colon indicates the position of the proline in *B. burgdorferi* ResT that confers permissiveness for Type 2 telomeres (7). The corresponding active site residues for the tyrosine recombinase family are indicated above and below the alignment in red and those for type IB topoisomerases in blue. doi:10.1128/microbiolspec.MDNA3-0023-2014.f3

Tyrosine recombinases use a constellation of additional conserved active site residues (RKHRH/W) thought to be involved in transition state stabilization and in general acid/base chemistry for the protonation/deprotonation of leaving groups and nucleophiles, respectively (29, 39, 40, 41). Alignments imply an active site composition of RKYRH for ResT in addition to the nucleophilic tyrosine. Examination of mutants of ResT in the corresponding positions revealed that mutation of most of these residues produces a resolvase that can still bind DNA with normal affinity but has dramatically reduced telomere resolution proficiency. Another interesting feature for ResT is that the final residue in the putative active site is a highly conserved histidine but was found to be dispensable for the reaction (13). This is reminiscent of the situation in the Flp recombinase in which this residue is a tryptophan that plays a mainly structural rather than a catalytic role (40). Moreover, no evidence was obtained for any of the expected active site residues or additional conserved basic residues queried to play a role in leaving group protonation (13). This was investigated using the approach of rescue of DNA cleavage defects of putative general acid mutants by reaction with substrates containing a 5′-bridging phosphorothiolate substitution for the scissile phosphate (13, 33, 36, 37, 38). This result could indicate water-mediated chemistry (42) or an additional critical role for the suspected general acid residue (K224) in the reaction, precluding rescue solely by lowering the pKa of the leaving group.

There are now crystal structures of two hairpin telomere resolvases from a *Klebsiella oxytoca* phage with a linear plasmid prophage (TelK) and from *Agrobacterium tumefaciens,* a bacterium that possesses a linear as well as a circular chromosome (TelA). The structures of TelK and TelA were solved in complex with cleaved *rTel* and hairpin product, respectively. The active site of TelK is present as a mimic of the transition state of the DNA cleavage step and shows an active site that is a hybrid of that found in type IB topoisomerases and tyrosine recombinases with the RKHRH constellation being composed of an RKKRH constellation instead. The structure of the active site supports the proposed two-step transesterification mechanism inferred for telomere resolution (25). The active site of TelA, obtained as a transition state mimic of strand transfer/ligation, shows the expected arrangement of active site residues with the third active site residue represented by a tyrosine, as it is in ResT (13, 26). This position has been implicated in studies on Flp as important for protonation of the tyrosine leaving group in the second transesterification (41). Protonation of tyrosine is less dependent upon enzyme mediation than that of the 5′-OH leaving group; this is consistent with the less severe defect of the ResT (Y293F) mutant in telomere resolution than mutation of the other active site residues (13).

TELOMERE SUBSTRATES

A ubiquitous feature of the substrates for hairpin telomere resolvases (3, 4, 5, 8, 11, 24, 43) is their 180-degree rotational symmetry (see L′L and RR′, Fig. 1). The reaction is concerted in that both halves of the substrate are required before reaction chemistry will ensue (3, 14, 16, 24). Negative supercoiling of the substrate is a potent inhibitor of ResT (6, 16) and the reaction is dramatically stimulated by positive supercoiling, discussed further in a later section (16). In contrast, TelA can process a negatively supercoiled substrate, although not as efficiently as one that is relaxed (8).

An interesting feature of *Borrelia* species is their segmented genomes with a linear plasmid and multiple linear plasmids. The prototype *B. burgdorferi* strain B31 contains almost two dozen replicons, most of which are linear (44, 45). This strain contains 19 unique but related hairpin telomeres (Fig. 4) necessitating a relaxed ability for substrate utilization by ResT that is not found in the other characterized hairpin resolvases. The 19 unique telomeres display initial rates of reaction *in vitro* that vary by over 160-fold (20). The hairpin telomeres have an invariant box 3 that is recognized by the large proteolytic ResT fragment ($ResT_{164–449}$) that carries the active site. The hairpin telomeres have been divided into three groups (Type 1, 2 and 3) based upon the position or absence of box 1 sequences, which are bound by the N-terminal domain ($ResT_{1–163}$), which is delivered to the box 1 location by specific binding of $ResT_{164–449}$ to box 3 (18). Type 2 telomeres have their box 1 offset by three base pairs in the direction of the axis of symmetry in a replicated telomere relative to Type 1 telomeres. Type 1 and Type 2 telomeres are both processed efficiently. Type 3 telomeres are lacking a box 1 and in general are processed much less efficiently than Type 1 and Type 2 telomeres. Three of the Type 3 telomeres were surprisingly, completely unreactive *in vitro*. The *in vivo* activity of the three *in vitro* unreactive telomeres was confirmed, pointing to a fundamental difference in reaction conditions in the living organism. This might be the presence of stimulatory proteins or a difference in DNA topology, such as positive supercoiling, that potentiates the reactivity of these Type 3 telomeres. Interestingly, the Type 2 telomeres display differential reactivity by ResT from Lyme spirochetes versus relapsing fever spirochetes (17). Through the use of chimeric and mutant ResT proteins the determinant conferring permissive substrate usage was localized to a single proline residue in the catalytic region that was absent in the relapsing fever enzymes but present in ResT from Lyme spirochetes (7).

REACTION REVERSAL AND GENOME PLASTICITY

Another unusual feature in the genus *Borrelia* is the ongoing genome plasticity in the telomeric and subtelomeric regions of the linear chromosome and plasmids (44, 46, 47, 48, 49). Extensions of linear plasmid sequences can be found on the ends of some chromosomes and other linear plasmids (Fig. 5). In addition, the subterminal regions of many linear plasmids are a patchwork of repeated sequences found on other linear plasmids. Sequencing of recombination breakpoints indicates that these genome rearrangements result from nonhomologous recombination events, however, no molecular mechanism for their generation was proposed (44, 46, 47, 48). A hint as to a possible molecular mechanism comes from the demonstration that telomere resolution by ResT is a reversible reaction (15). In the reverse reaction, referred to as telomere fusion, two hairpin telomeres on different DNA molecules are covalently linked to generate a replicated telomere structure (reversal of the reaction depicted in Fig. 2). However, completely unrelated DNA molecules can be fused, generating chimeric plasmids or chromosomes with plasmid extensions. In the overwhelming majority of cases, these replicon fusions would be quickly resolved by the forward reaction, separating the fused replicons. However, at a very low frequency, the appearance of a mutation in the fused telomere would preclude resolution, resulting in a stabilized telomere fusion. Subsequent deletions would be expected to occur to remove one of the two sets of replication/maintenance functions on the replicon consisting of two fused plasmids. Such deletions could be quite large, removing large chunks of plasmid DNA, and the stabilized fusions would be frozen in time and available to participate in subsequent telomere fusion and recombination events in the future. This type of processive mechanism can explain the patchwork of repeated sequences found in the subtelomeric regions of most of the linear *Borrelia* plasmids and the plasmid extensions found on some *B. burgdorferi* chromosomes (Fig. 5). Once established, the patchwork of repeats would also be available for homologous recombination reactions driven by the RecA pathway or perhaps by the strand exchange activity of ResT (23) discussed in a later section. For further discussion see references 1, 2, 15, 23.

Type	Telomere	Sequence (box 1, box 3)	Initial Rate (fmol/min)
2	lp28-4R	AATTTATTATCTTTTAGTATAATGA	48.8 ± 2.5
2	lp36R	TATTTATTATCTTTTAGTATAATGA	43.0 ± 7.8
1	ChromR	ATATAATTTTTAATTAGTATAGAAT	41.7 ± 7.3
2	lp28-1R	TATTTATTATCTTTTAGTATATATA	41.4 ± 0.9
1	lp28-1L	ATATAATTTTTAATTAGTATAGATA	40.4 ± 7.4
2	lp28-2L, 36L	AATTTATTATTAATTAGTATAAATA	39.3 ± 13.8
1	lp17L, 28-3L, 28-4L, 21L	ATATAATTTTTATTAGTATAGAGT	33.7 ± 6.1
2	lp28-3R	ATATTATTATTACTTAGTATAAATT	29.2 ± 6.8
1	lp56R, lp17R	ATATAATATTTATTTAGTACAAAGT	28.4 ± 2.8
2	lp38R	AATTTATTATCTTTTAGTATAATAG	27.4 ± 0.1
2	lp56L	AATTTATTATCTTTTAGTATAATGC	26.5 ± 5.4
3	lp21R	AAATTAGTTTTTTTAGTATAAAGC	26.5 ± 6.1
2	ChromL	TAAATATAATTTAATAGTATAAAAA	25.6 ± 0.4
3	lp25L	TATAAATTTTTAAATAGTATAGTTA	18.3± 3.1
3	lp25R	AAATATTTTTTTATTAGTATAGAGA	13.6± 2.1
3	lp28-2R	ATACAATTATTAATTAGTATAGAAA	4.7± 2.2
3	lp38L	AATTTACAATTTTTTAGTATAAAAA	<0.3
3	lp54L	TGAAGATAATCTATTAGTATACTAA	<0.3
3	lp54R	TAATAAGAGTTTATTAGTATACTAA	<0.3

Figure 4 *B. burgdorferi* telomere sequence alignment. Telomere sequences are arranged in descending order, according to the initial rate of telomere resolution. The initial rate, expressed in fmol/min is shown in the right-hand column and telomere sequences are aligned with the hairpins (or symmetry axis in the replicated telomeres) at the left end. The telomeres shown are half of the actual replicated telomere substrates used in the telomere resolution reactions. The colored boxes labeled 1 and 3 refer to previously identified regions of sequence homology (11), with some modifications. The original box 1 sequence, TATAAT is indicated by a light blue box, while the newly identified box 1 sequence, TATTAT is shown in dark blue. The homology box 3 region has been expanded from the five nucleotide sequence TAGTA to the eight nucleotide sequence TTAGTATA. The telomere sequences of lp17L, lp17R, lp21R, lp28-1R, lp56R ChromL and ChromR have been reported previously (45, 47, 48, 70, 71, 72). Reprinted from reference 20.
doi:10.1128/microbiolspec.MDNA3-0023-2014.f4

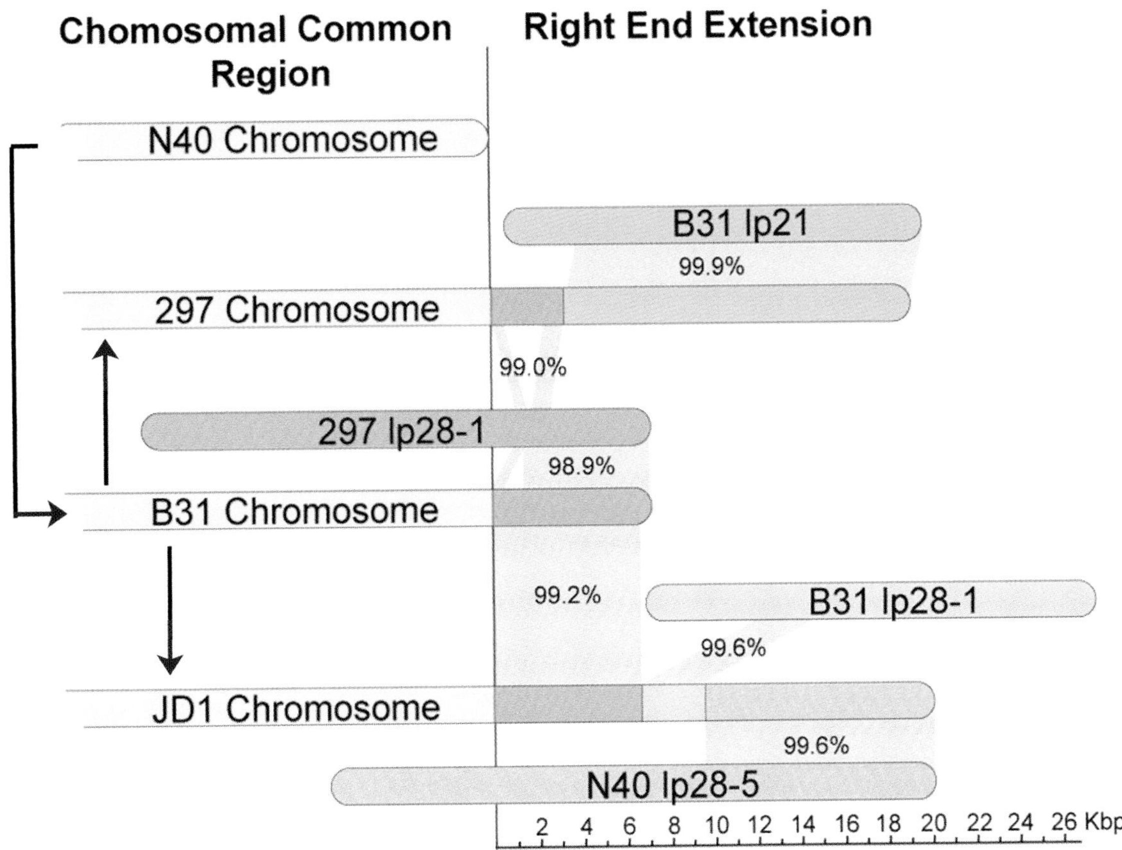

Figure 5 Telomere exchanges believed to be mediated by ResT promoted telomere fusions. The proposed mechanism for telomere exchange between linear plasmids and the right end of the *B. burgdorferi* chromosome is a two-step process. The first is a telomere fusion event and the second is a deletion or other type of mutation to inactivate or remove the newly fused telomere and prevent its resolution and to remove competing replication maintenance functions. The telomere fusion event is promoted by reversal of the ResT reaction such that two hairpin telomeres from different molecules are fused to generate a single DNA molecule carrying a replicated telomere. A deletion removing the telomere resolution site might be specifically targeted to the fused telomere by incomplete joining in the reverse reaction, to leave a ResT molecule covalently linked at a nick in the telomere; such covalent protein–DNA complexes are known to be foci for the formation of deletions and other chromosomal aberrations (73). Alternatively, a deletion could be derived from palindrome instability induced by passage of a replication fork through the inverted repeat of the fused telomere (74, 75). *B. burgdorferi* chromosome extensions that may have arisen by ResT-mediated telomere fusions followed by deletion formation (see references 1, 2, 15). The B31 chromosome appears to have arisen from fusion of an lp28-1 plasmid with the N40 chromosome. Subsequently, a single fusion of the B31 chromosome with lp21, followed by sequence deletion would have generated the 297 chromosome. Similarly, two rounds of fusion/deletion of the B31 chromosome, first with an lp28-1 and subsequently with lp28-5, would have generated the JD1 chromosome. The sequence relatedness of the chromosomes and plasmids shown were reported by references 44, 46, 47, 48. This figure is slightly modified from reference 46. doi:10.1128/microbiolspec.MDNA3-0023-2014.f5

DOMAIN DIVERSITY IN HAIRPIN TELOMERE RESOLVASES

As noted above and in Fig. 3, hairpin telomere resolvases all share a catalytic core domain with an active site similar to that of type IB topoisomerases and tyrosine recombinases. However, these enzymes appear to have a multidomain structure in which N- and C-terminal domain elaborations surrounding a central catalytic domain can be unrelated to each other in the different resolvases and likely have divergent functions (Fig. 6). For ResT, partial chymotrypsin cleavage, domain boundary mapping by mass spectroscopy and independent expression of the identified domains has uncovered two-proteolytic domains: an N-terminal domain ($ResT_{1-163}$) and a C-terminal domain, ($ResT_{164-449}$) (18). $ResT_{1-163}$ terminates just after the hairpin-binding module that corresponds to the long linker α-helix that connects the N-terminal and catalytic domains of TelA and TelK (12, 18, 25, 26). In ResT, this domain possesses nonspecific DNA binding activity apart from the hairpin-binding module. In the context of the full-length resolvase $ResT_{1-163}$ binds the box 1 sequences found near the center of the *rTel* substrate (see Fig. 4). $ResT_{164-449}$ encompasses the tyrosine recombinase-like catalytic domain (164 to 340) as well as an extension (341 to 449) that is likely an independently folded domain (18). The determinants of telomere recognition, mediated by the distally located box 3 to 5 sequences of the *rTel*, are found in $ResT_{164-449}$ (18). The TelK hairpin resolvase possesses an extension of the C-terminal domain (the stirrup) that is required for telomere resolution (25). The stirrup domain makes site-specific contacts with distal portions of the *rTel*, stabilizing a large distributed bend made in the *rTel* that contributes to hairpin formation (25). The stirrup is completely absent in the TelA resolvase (26). TelA, instead, possesses a short C-terminal extension that contributes to dimerization contacts (see Fig. 6). TelK and TelA have a shared N-terminal core subdomain (N-core) that forms the linker and top of a C-clamp shaped embrace of the substrate DNA. TelK contains a large insertion in the N-core domain, called the muzzle, which makes extensive interactions required for dimerization and is essential for telomere resolution (25). TelA has an additional N-terminal subdomain in addition to the N-core that is dispensable for telomere resolution (26).

An intriguing aspect of the domain arrangement of ResT is the observation that the determinants of telomere recognition harboured in the $ResT_{164-449}$ domain are partially masked in the full-length resolvase, either by steric hindrance from $ResT_{1-163}$ or by a conformational change in the full-length enzyme (18). The conserved box 3 sequence (TTAGTATA) that ResT recognizes in the *rTel* is present in thousands of copies in the genome; the autoinhibition of telomere binding may act to down regulate ResT binding and activity until a genuine cross-axis complex can be formed at *bona fide* replicated telomere substrates (18).

REQUIREMENT FOR THE HAIRPIN-BINDING MODULE OF ResT FOR DNA CLEAVAGE

In addition to the tyrosine recombinase-like active site, ResT employs a distinct part of the protein located at the end of the N-terminal domain (positions 139 to 159) called the hairpin-binding module (50, 51), to license DNA cleavage by the catalytic domain. A hairpin-binding module is not present in TelK or TelA. A similarity of this part of ResT to the hairpin-binding module of IS4 transposases was noted and investigated (12). The hairpin-binding module in these transposases is composed of a hydrophobic pocket and an α-helix with charged residues. The hairpin-binding module stabilizes the transient hairpin intermediate that arises during transposon excision. The hydrophobic pocket stacks on a flipped-out base that changes the normal trajectory of the cleaved DNA strand so that it can form the hairpin intermediate. A series of electrostatic interactions between the charged residues on the α-helix and the DNA backbone further stabilizes the hairpin intermediate (50, 51). Mutants in the transposase hairpin-binding module have defects in hairpin formation (52, 53, 54, 55). Mutation of the analogous positions in ResT resulted in variants defective in telomere resolution, most notably in the analogue of the hydrophobic pocket, and cold-sensitive variants when charged residues were mutated. These mutants were defective for DNA cleavage but were readily and specifically rescued by introduction of heteroduplex DNA at the central two base pairs in the center of the *rTel* (12). The rescued reactions on heteroduplexed *rTel*s proceeded to hairpin telomere formation rather than just rescuing the DNA cleavage reaction. This implies that the DNA distortions in the *rTel* induced by the ResT hairpin-binding module, which are mimicked by the heteroduplex, license DNA cleavage but also promote subsequent hairpin formation (12). The DNA distortion induced by the hairpin-binding module is distinct from that which promotes cross-axis formation since the central nick that promotes cross-axis complex formation does not rescue hairpin-binding module mutants (12). The precise nature of the distortion and the role played by individual sidechains within the hairpin-binding module remain to be elucidated. In contrast to the requirement for a ResT region involved in hairpin formation, for DNA cleavage TelA mutations in the N-core domain in key linker helix residues that appear to direct strand refolding (Y201A and R205A) are reported as cleavage competent but defective for hairpin formation (26). However, the observation that these mutants require days or weeks of incubation with suicide half-site substrates for cleavage points to a possible additional role for these residues in early interactions with the substrate DNA to allow cleavage.

HAIRPIN FORMATION/STRAND FOLDBACK

In the model of hairpin telomere formation based on the TelK-cleaved *rTel* complex it was inferred that torsion stored in the distributed out-of-plane bend in the DNA site drives dimer dissolution and spontaneous strand foldback after product release (25). This model was motivated by considerations of steric clash between hairpins that might form in the context of the extensively interlocked dimer (Fig. 7) and by the lack of contacts involving TelK and the DNA between the scissile phosphates. This model was also supported by the observation that deletion of the stirrup subdomain, important for forming and stabilizing the substrate DNA bend, produced a TelK variant that was cleavage competent but defective in hairpin formation (25).

In contrast, biochemical studies on ResT demonstrated that hairpin telomere formation must occur in the context of the dimer of ResT bound to the *rTel*, implying an active role for the telomere resolvase in the strand foldback required for hairpin telomere formation (22). These experiments made use of bead immobilized telomere resolution assays in which the *rTel* substrate was linked via one strand to paramagnetic beads by a 3′-biotin moiety. In this assay system, telomere resolution produces hairpin telomeres of distinct size: one remains attached to the bead and one hairpin is released into the supernatant fluid. Bead and supernatant fractions are separated by bead pull down followed by gel analysis of the fractions. This experimental set-up allowed for investigation of issues of product release. Fortuitous and deliberately engineered *rTel* substrates that showed a large bias in the speed and efficiency of hairpin telomere formation on the two halves of the substrate in free reactions were assayed in bead immobilized form. It was discovered that hairpin telomere product was not released into the supernatant until both hairpins had been formed (22). These results indicated an active role for ResT in hairpin formation. The observed hairpin product release for ResT is in contrast to the binding of the hairpin telomeres within a dimeric product complex observed for the TelA hairpin resolvase (8, 26).

The TelA-hairpin telomere product complexes, as well as structures of TelA with suicide substrates that prevent hairpin formation, argued for an active enzyme-mediated strand foldback mechanism of telomere formation (26). Structures of the product complex feature a dimer of TelA with the hairpin telomeres folded into an extremely compact conformation; steric clash between the hairpins is avoided by a difference in the helix trajectory in the two hairpins, breaking the 2-fold symmetry of the complex (Fig. 7). TelA makes extensive stabilizing interactions with the hairpin turnaround, includingvwith residues in the linker α-helix that comprises the end of the N-core domain, a region of TelA that occupies the same relative position in TelA as the hairpin-binding module of ResT. Structures with suicide substrates that prevent hairpin formation identified a "refolding intermediate" in which the refolding DNA strands assume an open conformation featuring extrahelical bases in a linearly stacked arrangement. The crystal structure also reveals Hoogstein base-pairing across the symmetry axis and interactions with key hairpin-interacting residues that stack an extrahelical base and otherwise constrain the possible trajectory of the refolding strands (26).

The overall structure of the N-terminal core and catalytic domain of TelA that was present in the structures was found be very similar to that of TelK. Moreover, the feature of the dimer complex being offset by a "jog" in the area that corresponds to the helix between the scissile phosphates in the TelK-cleaved substrate complex was also present (see Fig. 7). The α-helical linker of the N-core domain blocks the helical path that substrate DNA would have to assume to fit into the TelA dimer complex. This argues that the substrate DNA would likely have to be severely distorted or unwound during initial interactions with the resolvase. Such a distortion, believed to be made by the hairpin-binding module in ResT, allows productive engagement of the catalytic domains for DNA cleavage, blocks resealing of parental strands and promotes subsequent hairpin formation (12, 14).

HOLLIDAY JUNCTION FORMATION BY ResT

A further dramatic similarity of ResT to the tyrosine recombinases was uncovered under conditions known to inhibit telomere resolution, including low incubation temperature or negative DNA supercoiling in the

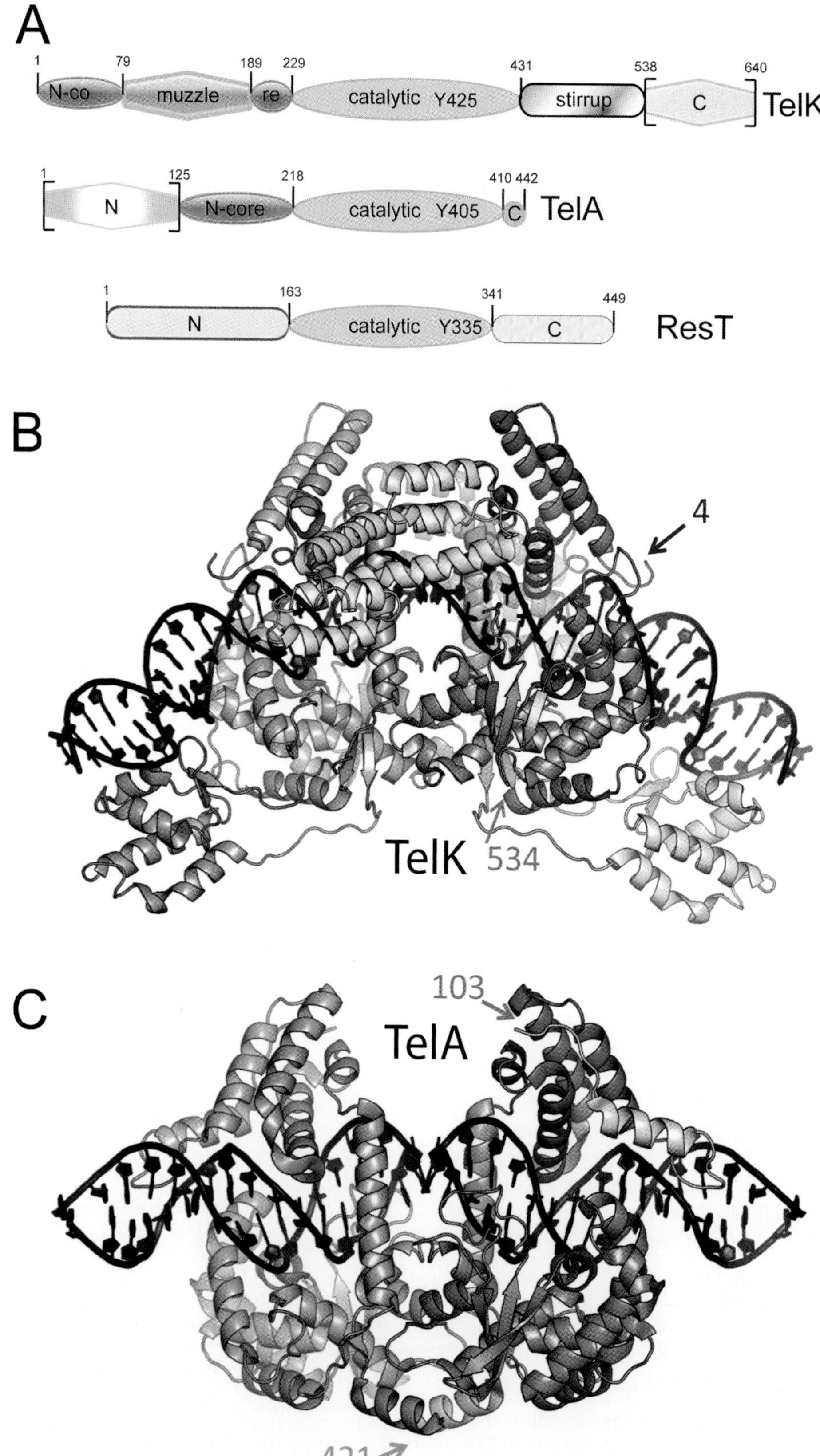
A
1
79
189
229
431
538
640
N-co
muzzle
re
catalytic Y425
stirrup
C
TelK
1
125
218
410 442
N
N-core
catalytic Y405
C
TelA
1
163
341
449
N
catalytic Y335
C
ResT
B
4
TelK
534
C
103
TelA
421

substrate. Under these conditions ResT can synapse *rTel*s together and catalyze the formation of a HJ (21). The simplest tyrosine recombinases are typified by Flp and Cre. They produce inversion or deletion recombinants from reaction sites composed of simple inverted repeat sequences that direct recombinase binding, coupled with sequence asymmetry in the central region of the reaction site between the scissile phosphates. The central sequence asymmetry directs strand exchange between sites synapsed in an anti-parallel orientation and directs a preferred order of strand cleavage and exchange that promotes the formation of recombinant products (56, 57, 58, 59, 60). Replicated telomere substrates for ResT lack such directional cues and consequently the HJs formed from symmetric *rTel*s lacked directionality (21). Artificially asymmetrizing *rTel* sites so that they progressively appeared more like recombination sites resulted in progressive inhibition of telomere resolution with a corresponding increase in directional HJ formation. Since replicated telomeres *in vivo* will always be symmetric, it is unlikely that ResT catalyzes site-specific recombination *in vivo*. It is simpler to posit that this activity reflects the evolutionary origin of telomere resolvases as ancestral phage integrases or cellular plasmid/chromosome dimer resolvases (see references 2, 21). Consistent with this idea is the close structural similarity of the catalytic domain of TelK and TelA to that of λ integrase (25, 26). This contention is also supported by experiments demonstrating a one-step linearization of the *E. coli* chromosome without deleterious effects on cell growth (61). The chromosome was linearized by introduction of the phage N15 replicated telomere near the *E. coli* replication terminus, coupled with the expression of the N15 telomere resolvase, TelN.

The discovery of ResT's ability to synapse *rTel*s and to catalyze HJ formation raised the interesting possibility that telomere resolution might occur in the context of a synapse of *rTel*s and that HJ formation is a possible side-reaction that occurs when hairpin formation is inhibited in some fashion. This possibility had some attractive features since it offered a mechanism for linking telomere resolution to cell division, similar to the dynamics of chromosome dimer resolution catalyzed by the Xer/*dif* system (34, 62, 63). Synapsis-dependent models of telomere resolution were tested and definitively ruled out by assays with replicated telomeres immobilized at subsaturating conditions on streptavidin-coated paramagnetic beads (22).

A POSSIBLE ROLE FOR ResT IN HOMOLOGOUS RECOMBINATION OF LINEAR REPLICONS

Recently, an unexpected activity of ResT was uncovered: a vigorous single-strand DNA annealing activity (SSA) and the ability to promote DNA strand exchange

Figure 6 The domain structure of hairpin telomere resolvases. (**A**) The domain composition of the telomere resolvases from *Klebsiella* phage KO2 (TelK), *Agrobacterium* (TelA) and *Borrelia* species (ResT) are shown (see also text). All telomere resolvases have a central catalytic domain with active site residues similar those of tyrosine recombinases and type IB topoisomerses (see Fig. 3 and text). In addition, hairpin resolvases carry divergent N-terminal domains; they may also carry a divergent C-terminal region or it may be absent as for TelA. Domains represented by the same shape and color scheme are structurally related. Domains delimited by brackets represent protein sequences dispensable for telomere resolution *in vitro* (8, 25). Also shown for each hairpin telomere resolvase is the tyrosine nucleophile (Y). The numbers above the graphics refer to the amino acid numbers of the proteins. Precise domain boundaries have been adjusted from those previously reported based upon structural alignments. (**B**) and (**C**) A structural view of the domain organization of TelK and TelA, respectively. The domains are represented by the same color scheme used in (**A**) on one monomer in each dimer. The N-core domain, represented in blue, is composed of a helical bundle that forms the top of the C-clamp embrace the resolvases make with the substrate DNA and the long linker α-helix that connects the top and bottom of the C-clamp. TelK has a large insertion in the N-core domain called the muzzle; this is represented in yellow. The shared catalytic domain is represented in red. TelK has an additional C-terminal domain essential for telomere resolution called the stirrup; this domain is represented in grey. Beyond the shared catalytic domain TelA lacks an additional domain like the stirrup but instead has a short segment that contributes to dimerization contacts; this short C-terminal extension is represented in green. The arrows with residue numbers indicate the position in the structure of the first and last resolvable residues present in the structures. The structures presented were generated with The PyMOL Molecular Graphics System, Version 1.7 Schrödinger, LLC. (http://www.pymol.org/) using PDB ID 2v6e for TelK (25) and 4e0g for TelA (26). doi:10.1128/microbiolspec.MDNA3-0023-2014.f6

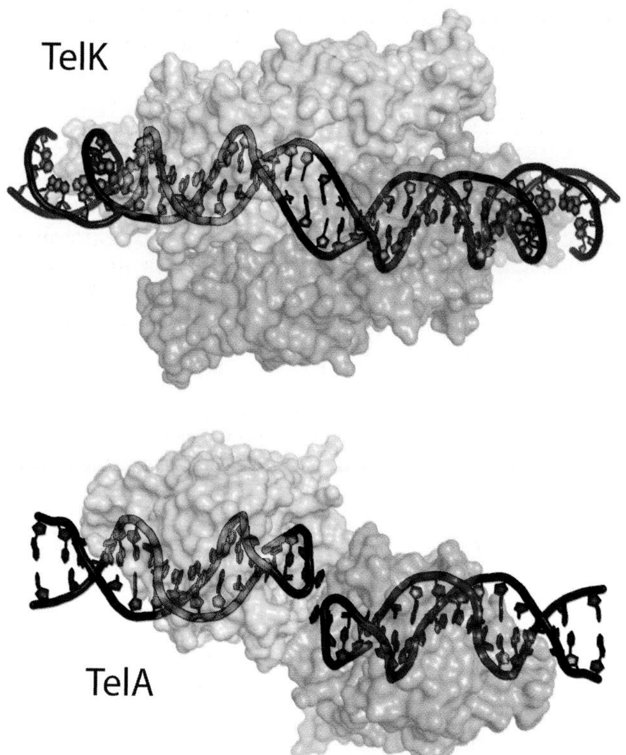

Figure 7 TelK and TelA-DNA complexes showing displacement of the helical axes at the dimer interface by 7.5 and >10 angstroms, respectively. The structures presented were generated with The PyMOL Molecular Graphics System, Version 1.7 Schrödinger, LLC. (http://www.pymol.org/) using PDB ID 2v6e for TelK (25) and 4e0g for TelA (26). doi:10.1128/microbiolspec.MDNA3-0023-2014.f7

between single-stranded donor DNAs and homologous partial duplex target DNAs (23). Interestingly, the ResT N-terminal domain, $ResT_{1-163}$ was found to promote SSA and strand exchange on its own (23). The properties of these reactions promoted by ResT are consistent with a mode of action similar to single-strand annealing proteins (SSAPs) typified by phage λ Beta of the λ Red recombination system and of Rad52, the eukaryotic SSAP that participates in RecA/Rad51-dependent and -independent homologous recombination reactions (64). ResT was found to be able to support SSA between complementary or partially complementary strands of up to 106 nt in length and to promote strand exchange through up to 63 bp of duplex DNA. Strand exchange was found to be sensitive to heterology between the ssDNA donor and the exchanging segment of the partial duplex target, giving rise to double-flap complexes that have strand exchanged beyond the heterology but that have stalled shortly thereafter. Such events, if they occur *in vivo,* would generate short-patch recombinants. Such short-patch recombinants typify the gene conversion events found in switching at the *B. burgdorferi vlsE* locus to promote antigenic variation (65). The mechanism of switching at *vlsE* remains uncharacterized and the only host factor known to participate in the reaction is the RuvAB branch migrase (66, 67). It is tempting to speculate that ResT might play a role in recombination at *vlsE*. Alternatively, ResT could play a possible role in homologous recombination reactions or perhaps in some, as yet, unappreciated role in hairpin telomere biology. Evidence has been presented that *B. burgdorferi* telomeres are recombinogenic (2, 10) and the possibility exists that the strand exchange activity that resides in $ResT_{1-163}$ may play a role in these recombination events.

IN VIVO EFFECTS OF A CONDITIONAL *resT* MUTANT

Our current understanding of the telomere resolution process began with the demonstration that the process occurred *in vivo* to generate the hairpin telomeres in *B. burgdorferi* (10). Thereafter, a large body of data has been generated through biochemical experiments with purified proteins and DNA substrates and by structural studies. However, at least for ResT, there is not a complete congruence of *in vivo* and *in vitro* reaction properties. In particular, some telomere substrates that are active *in vivo* show no detectable activity *in vitro* (20). In addition, Type 2 telomeres cannot be processed *in vitro* by relapsing fever telomere resolvases but are efficiently utilized *in vivo* (17). It is reasonable to postulate that the reaction *in vivo* is stimulated by other proteins or by substrate topology, such as positive supercoiling (16). Moreover, we know little about whether telomere resolvases interact with other proteins within the cell and how they may be involved with regulation of other cellular processes such as replication initiation, elongation and cell division.

As one might expect, the *resT* gene on cp26 of *B. burgdorferi* cannot be disrupted (68) unless a complementing *resT* gene has been added to the cell (17). To further study the *in vivo* activities of this essential gene, a strain with conditional expression of *resT* was constructed (69). The *resT* gene was cloned under the control of an IPTG-inducible expression system and the endogenous gene was disrupted. The protein was found to be expressed at about 15,000 monomers per cell, which is on the high side for a replication protein, suggesting that it may have other functions. Removal of IPTG from the culture allowed determination of the half-life of ResT to be 16 hours. ResT depletion resulted in growth arrest of the spirochetes at 48 hours. Filamentous forms did not appear, as might be expected if telomere resolution was linked to cell division. DNA replication did not continue after growth arrest, suggesting that ResT might interact directly or indirectly with factors controlling the initiation or elongation of DNA synthesis. Simple dimeric replication intermediates of linear plasmids were only visible at 24 to 48 hours after IPTG washout after which time they morphed into higher molecular weight forms. These more complex molecules may result from additional replication initiation events or stalled replication. Further studies will be required to elucidate the *in vivo* activities and interactions of ResT.

Acknowledgments. *The authors would like to thank Phoebe Rice for structural comparisons to define domain boundaries and for generating Fig. 6(B) and (C) and Fig. 7. Work in the authors' laboratories was supported by grant MOP 53086 from the Canadian Institutes of Health Research to G.C. (http://www.cihr-irsc.gc.ca/e/193.html), by grant MOP 79344 from the Canadian Institutes of Health Research to K.K., by grant RGPIN 326797-2011 from the Natural Sciences & Engineering Research Council of Canada to K.K. (http://www.nserc-crsng.gc.ca/index_eng.asp) and by grant 2570 from the Saskatchewan Health Research Fund to K.K. (http://shrf.ca/). G.C. holds a Canada Research Chair in the Molecular Biology of Lyme Borreliosis (http://www.chairs-chaires.gc.ca/home-accueil-eng.aspx) and a Scientist Award from Alberta Innovates – Health Solutions (http://www.ahfmr.ab.ca/).*

Citation. Kobryn K, Chaconas G. 2014. Hairpin telomere resolvases. Microbiol Spectrum **2**(6):MDNA3-0023-2014.

References

1. **Chaconas G.** 2005. Hairpin telomeres and genome plasticity in *Borrelia*: all mixed up in the end. *Mol Microbiol* **58:**625–635.
2. **Chaconas G, Kobryn K.** 2010. Structure, function, and evolution of linear replicons in Borrelia. *Annu Rev Microbiol* **64:**185–202.
3. **Deneke J, Ziegelin G, Lurz R, Lanka E.** 2000. The protelomerase of temperate *Escherichia coli* phage N15 has cleaving-joining activity. *Proc Natl Acad Sci U S A* **97:**7721–7726.
4. **Huang WM, Joss L, Hsieh T, Casjens S.** 2004. Protelomerase uses a topoisomerase IB/Y-recombinase type mechanism to generate DNA hairpin ends. *J Mol Biol* **337:**77–92.
5. **Hertwig S, Klein I, Lurz R, Lanka E, Appel B.** 2003. PY54, a linear plasmid prophage of *Yersinia enterocolitica* with covalently closed ends. *Mol Microbiol* **48:**989–1003.
6. **Kobryn K, Chaconas G.** 2002. ResT, a telomere resolvase encoded by the Lyme disease spirochete. *Mol Cell* **9:** 195–201.
7. **Moriarty TJ, Chaconas G.** 2009. Identification of the determinant conferring permissive substrate usage in the telomere resolvase, ResT. *J Biol Chem* **284:**23293–23301.
8. **Huang WM, DaGloria J, Fox H, Ruan Q, Tillou J, Shi K, Aihara H, Aron J, Casjens S.** 2012. Linear chromosome-generating system of *Agrobacterium tumefaciens* C58: protelomerase generates and protects hairpin ends. *J Biol Chem* **287:**25551–25563.
9. **Ramirez-Bahena MH, Vial L, Lassalle F, Diel B, Chapulliot D, Daubin V, Nesme X, Muller D.** 2014. Single acquisition of protelomerase gave rise to speciation of a large and diverse clade within the Agrobacterium/Rhizobium supercluster characterized by the presence of a linear chromid. *Mol Phylogenet Evol* **73:** 202–207.
10. **Chaconas G, Stewart PE, Tilly K, Bono JL, Rosa P.** 2001. Telomere resolution in the Lyme disease spirochete. *EMBO J* **20:**3229–3237.
11. **Tourand Y, Kobryn K, Chaconas G.** 2003. Sequence-specific recognition but position-dependent cleavage of two distinct telomeres by the *Borrelia burgdorferi* telomere resolvase, ResT. *Mol Microbiol* **48:**901–911.
12. **Bankhead T, Chaconas G.** 2004. Mixing active site components: A recipe for the unique enzymatic activity of a telomere resolvase. *Proc Natl Acad Sci U S A* **101:** 13768–13773.
13. **Deneke J, Burgin AB, Wilson SL, Chaconas G.** 2004. Catalytic residues of the telomere resolvase ResT: a pattern similar to, but distinct from tyrosine recombinases and type IB topoisomerases. *J Biol Chem* **279:**53699–53706.
14. **Kobryn K, Burgin AB, Chaconas G.** 2005. Uncoupling the chemical steps of telomere resolution by ResT. *J Biol Chem* **280:**26788–26795.
15. **Kobryn K, Chaconas G.** 2005. Fusion of hairpin telomeres by the *B. burgdorferi* telomere resolvase ResT: Implications for shaping a genome in flux. *Mol Cell* **17:** 783–791.
16. **Bankhead T, Kobryn K, Chaconas G.** 2006. Unexpected twist: harnessing the energy in positive supercoils to control telomere resolution. *Mol Microbiol* **62:**895–905.
17. **Tourand Y, Bankhead T, Wilson SL, Putteet-Driver AD, Barbour AG, Byram R, Rosa PA, Chaconas G.** 2006. Differential telomere processing by *Borrelia* telomere resolvases *in vitro* but not *in vivo*. *J Bacteriol* **188:** 7378–7386.
18. **Tourand Y, Lee L, Chaconas G.** 2007. Telomere resolution by *Borrelia burgdorferi* ResT through the collaborative efforts of tethered DNA binding domains. *Mol Microbiol* **64:**580–590.
19. **Lefas G, Chaconas G.** 2009. High-throughput screening identifies three inhibitor classes of the telomere resolvase from the Lyme disease spirochete. *Antimicrob Agents Chemother* **53:**4441–4449.
20. **Tourand Y, Deneke J, Moriarty TJ, Chaconas G.** 2009. Characterization and *in vitro* reaction properties of 19 unique hairpin telomeres from the linear plasmids of the Lyme disease spirochete. *J Biol Chem* **284:** 7264–7272.
21. **Kobryn K, Briffotaux J, Karpov V.** 2009. Holliday junction formation by the *Borrelia burgdorferi* telomere resolvase, ResT: implications for the origin of genome linearity. *Mol Microbiol* **71:**1117–1130.

22. **Briffotaux J, Kobryn K.** 2010. Preventing broken borrelia telomeres: Rest couples dual hairpin telomere formation to product release. *J Biol Chem* **285:**41010–41018.
23. **Mir T, Huang SH, Kobryn K.** 2013. The telomere resolvase of the Lyme disease spirochete, *Borrelia burgdorferi*, promotes DNA single-strand annealing and strand exchange. *Nucleic Acids Res* **41:**10438–10448.
24. **Deneke J, Ziegelin G, Lurz R, Lanka E.** 2002. Phage N15 telomere resolution: Target requirements for recognition and processing by the protelomerase. *J Biol Chem* **277:** 10410–10419.
25. **Aihara H, Huang WM, Ellenberger T.** 2007. An interlocked dimer of the protelomerase TelK distorts DNA structure for the formation of hairpin telomeres. *Mol Cell* **27:**901–913.
26. **Shi K, Huang WM, Aihara H.** 2013. An enzyme-catalyzed multistep DNA refolding mechanism in hairpin telomere formation. *PLoS Biol* **11:**e1001472.
27. **Shuman S.** 1998. Vaccinia virus DNA topoisomerase: a model eukaryotic type IB enzyme. *Biochim Biophys Acta* **1400:**321–337.
28. **Van Duyne GD.** 2002. A structural view of tyrosine recombinase site-specific recombination, p 93–117. *In* Craig NL, Craigie R, Gellert M, Lambowitz AM (ed), *Mobile DNA II.* ASM Press, Washington, DC.
29. **Grindley ND, Whiteson KL, Rice PA.** 2006. Mechanisms of site-specific recombination. *Annu Rev Biochem* **75:** 567–605.
30. **Lee J, Tonozuka T, Jayaram M.** 1997. Mechanism of active site exclusion in a site-specific recombinase: role of the DNA substrate in conferring half-of-the-sites activity. *Genes Dev* **11:**3061–3071.
31. **Voziyanov Y, Pathania S, Jayaram M.** 1999. A general model for site-specific recombination by the integrase family recombinases. *Nucleic Acids Res* **27:**930–941.
32. **Conway AB, Chen Y, Rice PA.** 2003. Structural plasticity of the Flp-Holliday junction complex. *J Mol Biol* **326:** 425–434.
33. **Burgin AB Jr, Nash HA.** 1995. Suicide substrates reveal properties of the homology-dependent steps during integrative recombination of bacteriophage lambda. *Curr Biol* **5:**1312–1321.
34. **Barre FX, Aroyo M, Colloms SD, Helfrich A, Cornet F, Sherratt DJ.** 2000. FtsK functions in the processing of a Holliday junction intermediate during bacterial chromosome segregation. *Genes Dev* **14:**2976–2988.
35. **Rybchin VN, Svarchevsky AN.** 1999. The plasmid prophage N15: a linear DNA with covalently closed ends. *Mol Microbiol* **33:**895–903.
36. **Krogh BO, Shuman S.** 2000. Catalytic mechanism of DNA topoisomerase IB. *Mol Cell* **5:**1035–1041.
37. **Krogh BO, Shuman S.** 2002. Proton relay mechanism of general acid catalysis by DNA topoisomerase IB. *J Biol Chem* **277:**5711–5714.
38. **Burgin AB.** 2001. Synthesis and use of DNA containing a 5′-bridging phosphorothioate as a suicide substrate for type I DNA topoisomerases. *Methods Mol Biol* **95:** 119–128.
39. **Van Duyne GD.** 2001. A structural view of *cre-loxp* site-specific recombination. *Annu Rev Biophys Biomol Struct* **30:**87–104.
40. **Chen Y, Rice PA.** 2003. The role of the conserved Trp330 in Flp-mediated recombination. Functional and structural analysis. *J Biol Chem* **278:**24800–24807.
41. **Whiteson KL, Chen Y, Chopra N, Raymond AC, Rice PA.** 2007. Identification of a potential general acid/base in the reversible phosphoryl transfer reactions catalyzed by tyrosine recombinases: Flp H305. *Chem Biol* **14:** 121–129.
42. **Davies DR, Mushtaq A, Interthal H, Champoux JJ, Hol WG.** 2006. The structure of the transition state of the heterodimeric topoisomerase I of Leishmania donovani as a vanadate complex with nicked DNA. *J Mol Biol* **357:**1202–1210.
43. **Kobryn K.** 2007. The linear hairpin replicons of Borrelia burgdorferi, p 117–140. *In* Meinhardt F, Klassen R (ed), *Microbial Linear Plasmids*. Springer, Berlin Heidelberg.
44. **Casjens S, Palmer N, van Vugt R, Huang WH, Stevenson B, Rosa P, Lathigra R, Sutton G, Peterson J, Dodson RJ, Haft D, Hickey E, Gwinn M, White O, Fraser CM.** 2000. A bacterial genome in flux: the twelve linear and nine circular extrachromosomal DNAs in an infectious isolate of the Lyme disease spirochete *Borrelia burgdorferi. Mol Microbiol* **35:**490–516.
45. **Fraser CM, Casjens S, Huang WM, Sutton GG, Clayton R, Lathigra R, White O, Ketchum KA, Dodson R, Hickey EK, Gwinn M, Dougherty B, Tomb JF, Fleischmann RD, Richardson D, Peterson J, Kerlavage AR, Quackenbush J, Salzberg S, Hanson M, van Vugt R, Palmer N, Adams MD, Gocayne J, Weidman J, Utterback T, Watthey L, McDonald L, Artiach P, Bowman C, Garland S, Fujii C, Cotton MD, Horst K, Roberts K, Hatch B, Smith HO, Venter JC.** 1997. Genomic sequence of a Lyme disease spirochaete, *Borrelia burgdorferi. Nature* **390:**580–586.
46. **Casjens SR, Mongodin EF, Qiu WG, Luft BJ, Schutzer SE, Gilcrease EB, Huang WM, Vujadinovic M, Aron JK, Vargas LC, Freeman S, Radune D, Weidman JF, Dimitrov GI, Khouri HM, Sosa JE, Halpin RA, Dunn JJ, Fraser CM.** 2012. Genome stability of Lyme disease spirochetes: comparative genomics of *Borrelia burgdorferi* plasmids. *PloS One* **7:**e33280.
47. **Huang WM, Robertson M, Aron J, Casjens S.** 2004. Telomere exchange between linear replicons of *Borrelia burgdorferi. J Bacteriol* **186:**4134–4141.
48. **Casjens S, Murphy M, DeLange M, Sampson L, van Vugt R, Huang WM.** 1997. Telomeres of the linear chromosomes of Lyme disease spirochaetes: nucleotide sequence and possible exchange with linear plasmid telomeres. *Mol Microbiol* **26:**581–596.
49. **Terekhova D, Iyer R, Wormser GP, Schwartz I.** 2006. Comparative genome hybridization reveals substantial variation among clinical isolates of *Borrelia burgdorferi* sensu stricto with different pathogenic properties. *J Bacteriol* **188:**6124–6134.
50. **Davies DR, Goryshin IY, Reznikoff WS, Rayment I.** 2000. Three-dimensional structure of the Tn5 synaptic complex transposition intermediate. *Science* **289:**77–85.

51. **Lovell S, Goryshin IY, Reznikoff WR, Rayment I.** 2002. Two-metal active site binding of a Tn5 transposase synaptic complex. *Nat Struct Biol* **9:**278–281.

52. **Ason B, Reznikoff WS.** 2002. Mutational analysis of the base flipping event found in Tn5 transposition. *J Biol Chem* **277:**11284–11291.

53. **Allingham JS, Wardle SJ, Haniford DB.** 2001. Determinants for hairpin formation in Tn10 transposition. *EMBO J* **20:**2931–2942.

54. **Bischerour J, Chalmers R.** 2007. Base-flipping dynamics in a DNA hairpin processing reaction. *Nucleic Acids Res* **35:**2584–2595.

55. **Bischerour J, Chalmers R.** 2009. Base flipping in tn10 transposition: an active flip and capture mechanism. *PloS One* **4:**e6201.

56. **Ghosh K, Lau CK, Gupta K, Van Duyne GD.** 2005. Preferential synapsis of loxP sites drives ordered strand exchange in Cre-loxP site-specific recombination. *Nat Chem Biol* **1:**275–282.

57. **Hoess RH, Wierzbicki A, Abremski K.** 1986. The role of the loxP spacer region in P1 site-specific recombination. *Nucleic Acids Res* **14:**2287–2300.

58. **Lee L, Chu LC, Sadowski PD.** 2003. Cre induces an asymmetric DNA bend in its target loxP site. *J Biol Chem* **278:**23118–23129.

59. **Lee L, Sadowski PD.** 2003. Sequence of the loxP site determines the order of strand exchange by the Cre recombinase. *J Mol Biol* **326:**397–412.

60. **Senecoff JF, Cox MM.** 1986. Directionality in FLP protein-promoted site-specific recombination is mediated by DNA-DNA pairing. *J Biol Chem* **261:**7380–7386.

61. **Cui T, Moro-oka N, Ohsumi K, Kodama K, Ohshima T, Ogasawara N, Mori H, Wanner B, Niki H, Horiuchi T.** 2007. *Escherichia coli* with a linear genome. *EMBO Rep* **8:**181–187.

62. **Aussel L, Barre FX, Aroyo M, Stasiak A, Stasiak AZ, Sherratt D.** 2002. FtsK Is a DNA motor protein that activates chromosome dimer resolution by switching the catalytic state of the XerC and XerD recombinases. *Cell* **108:**195–205.

63. **Bigot S, Saleh OA, Lesterlin C, Pages C, El Karoui M, Dennis C, Grigoriev M, Allemand JF, Barre FX, Cornet F.** 2005. KOPS: DNA motifs that control E. coli chromosome segregation by orienting the FtsK translocase. *EMBO J* **24:**3770–3780.

64. **Kuzminov A.** 1999. Recombinational repair of DNA damage in *Escherichia coli* and bacteriophage lambda. *Microbiol Mol Biol Rev* **63:**751–813.

65. **Coutte L, Botkin DJ, Gao L, Norris SJ.** 2009. Detailed analysis of sequence changes occurring during *vlsE* antigenic variation in the mouse model of *Borrelia burgdorferi* infection. *PLoS Pathog* **5:**e1000293.

66. **Dresser AR, Hardy P-O, Chaconas G.** 2009. Investigation of the role of DNA replication, recombination and repair genes in antigenic switching at the vlsE locus in *Borrelia burgdorferi*: an essential role for the RuvAB branch migrase. *PLoS Pathog* **5:**e1000680.

67. **Lin T, Gao L, Edmondson DG, Jacobs MB, Philipp MT, Norris SJ.** 2009. Central role of the Holliday junction helicase RuvAB in *vlsE* recombination and infectivity of *Borrelia burgdorferi*. *PLoS Pathog* **12:**e1000679.

68. **Byram R, Stewart PE, Rosa P.** 2004. The essential nature of the ubiquitous 26-kilobase circular replicon of *Borrelia burgdorferi*. *J Bacteriol* **186:**3561–3569.

69. **Bandy NJ, Salman-Dilgimen A, Chaconas G.** 2014. Construction and characterization of a *B. burgdorferi* strain with conditional expression of the essential telomere resolvase, ResT. *Journal of Bacteriology* **196:** 2396–2404.

70. **Hinnebusch J, Barbour AG.** 1991. Linear plasmids of *Borrelia burgdorferi* have a telomeric structure and sequence similar to those of a eukaryotic virus. *J Bacteriol* **173:**7233–7239.

71. **Hinnebusch J, Bergstrom S, Barbour AG.** 1990. Cloning and sequence analysis of linear plasmid telomeres of the bacterium *Borrelia burgdorferi*. *Mol Microbiol* **4:** 811–820.

72. **Zhang JR, Hardham JM, Barbour AG, Norris SJ.** 1997. Antigenic variation in Lyme disease borreliae by promiscuous recombination of VMP-like sequence cassettes. *Cell* **89:**275–285.

73. **Froelich-Ammon SJ, Osheroff N.** 1995. Topoisomerase poisons: harnessing the dark side of enzyme mechanism. *J Biol Chem* **270:**21429–21432.

74. **Pinder DJ, Blake CE, Lindsey JC, Leach DR.** 1998. Replication strand preference for deletions associated with DNA palindromes. *Mol Microbiol* **28:**719–727.

75. **Leach DR, Okely EA, Pinder DJ.** 1997. Repair by recombination of DNA containing a palindromic sequence. *Mol Microbiol* **26:**597–606.

Mobile DNA, 3rd Edition
Nancy L. Craig, Michael Chandler, Martin Gellert, Alan M. Lambowitz, Phoebe A. Rice and Suzanne Sandmeyer

doi:10.1128/microbiolspec.MDNA3-0008-2014

Nicolas Carraro[1]
Vincent Burrus[1]

13 Biology of Three ICE Families: SXT/R391, ICE*Bs1*, and ICE*St1*/ICE*St3*

INTRODUCTION

Nomenclature

In the early 1980s, conjugative transposons were defined as large DNA segments of bacterial chromosomes capable of "intercellular transposition," i.e., fragments able to move from the chromosome of a donor bacterium to the chromosome of a recipient bacterium during cell-to-cell contact. All these mobile genetic elements were found in pathogenic low GC Gram-positive bacteria, conferred antibiotic resistance properties, and were often capable of integrating into a large array of different sites (for review, see references 1, 2, 3, 4). Characterization of the molecular mechanism allowing integration into and excision from the chromosome revealed that conjugative transposons such as Tn*916* do not encode a DDE transposase, but rather a site-specific tyrosine recombinase. Fundamental differences in the molecular mechanism of DNA strand exchanges catalyzed by transposases and site-specific tyrosine recombinases, and subsequent identification of conjugative mobile elements integrating into a unique site of the bacterial chromosome in both Gram-positive and Gram-negative bacteria exposed the inadequacy of the naming "conjugative transposons." In fact, at the time the confusion in the scientific community was such that, in some instances, related elements were mislabeled as conjugative plasmids, R factors, or integrating conjugative plasmids (5). Two nomenclatures proposed to replace the obsolete term by a more adequate nomenclature: constin, an acronym that stands for conjugative, self-transmissible, integrating element, and ICE, an acronym for integrative and conjugative element (5, 6). Over the years the term ICE gained a broader acceptance among many authors to describe elements found in both Gram-positive and Gram-negative bacteria, so this term is used hereafter instead of conjugative transposon.

Scope

Several previous excellent reviews have focused on ICEs found in *Bacteroides* or in pathogenic low GC Gram-positive bacteria, with a strong emphasis on Tn*916* (1, 2, 3, 4, 5, 7, 8, 9). Owing to space constraints, the goals of this chapter will not be to provide

[1]Département de Biologie, Faculté des Sciences, Université de Sherbrooke, Sherbrooke, QC, J1K 2R1 Canada.

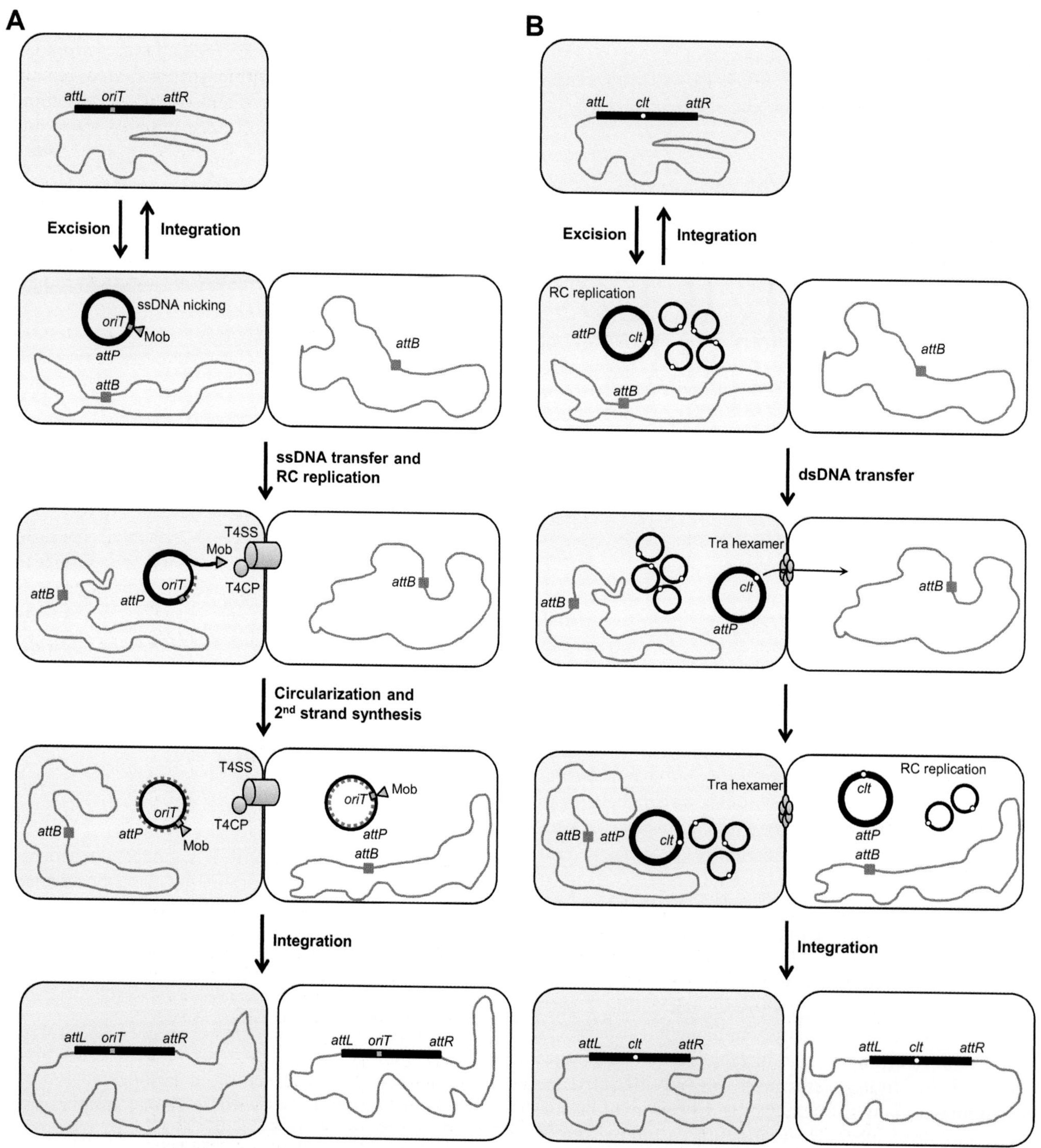
A
attL oriT attR
Excision
Integration
ssDNA nicking
oriT
Mob
attP
attB
attB
ssDNA transfer and
RC replication
T4SS
Mob
T4CP
attB
attP
oriT
attB
Circularization and
2nd strand synthesis
T4SS
oriT
Mob
attB
attP
Mob
T4CP
oriT
attP
attB
Integration
attL oriT attR
attL oriT attR
B
attL clt attR
Excision
Integration
RC replication
attP
clt
attB
attB
dsDNA transfer
Tra hexamer
clt
attB
attP
attB
Tra hexamer
RC replication
clt
attB
attP
clt
attP
attB
Integration
attL clt attR
attL clt attR

an exhaustive review of all of the families of ICEs. Excellent reviews that cover the ICEs of the Tn*916*/Tn*1545* family, the *Bacteroides* conjugative transposons, the Tn*1549* and Tn*4371* families, and ICE*clc* have already been published elsewhere (10). Here, we have chosen to focus on three ICE families (SXT/R391, ICE*Bs1*, ICE*St1*/ICE*St3*) for which extensive developments in the understanding of their biology have been obtained in the past decade.

General Mechanism of Dissemination

ICEs usually remain integrated into and are replicated as part of their host's chromosome. Upon induction by environmental or intracellular signals, ICEs can excise from the chromosome, usually by site-specific recombination mediated by a tyrosine or a serine recombinase or, in a few atypical cases, by a transposition mechanism mediated by a DDE transposase (11, 12, 13, 14). The resulting circular covalently closed ICE molecule then serves as a substrate for the conjugative machinery that translocates the ICE DNA from the donor to the recipient cell. Like for conjugative plasmids, translocation of DNA between the two mating partners can proceed either using single-strand DNA (ssDNA) or double-strand DNA (dsDNA), depending on the type of conjugative apparatus encoded by the conjugative element (15).

The most widespread strategy used by ICEs of both Gram-negative and Gram-positive bacteria relies on the translocation of an ssDNA molecule through a type IV secretion system (T4SS) (Fig. 1A) (15, 16, 17). T4SSs are complex machineries that usually involve the secretion and assembly of an extracellular pilus (18, 19). The assembly and/or activity of the secretion channel are thought to be energized by the ATPase activity of a VirB4-like subunit, which is a key component of T4SS. Biochemical processing of the circular ICE molecule is initiated at the origin of transfer (*oriT*), an ICE *cis*-acting locus, recognized and bound by a DNA relaxase and other ICE- and host-encoded auxiliary proteins assembled together as a nucleoprotein complex called the relaxosome (20, 21). The DNA relaxase mediates a strand-specific cleavage within the *oriT* and remains covalently attached to the 5′ end of the nicked strand. Unwinding of the DNA molecule coupled to rolling-circle (RC) replication facilitate the 5′ to 3′ translocation of the relaxase-bound ssDNA to the recipient cell. The coupling protein of conjugative T4SS (T4CP), a VirD4-like subunit, acts as a docking site for the relaxosome and its ATPase activity energizes the translocation of the ssDNA across the donor and recipient cell membranes (18, 20, 21). Once in the recipient, the relaxase ligates the extremities of the linear transferred ssDNA molecule and its complementary strand is synthesized.

The second strategy is exclusively documented for ICEs found in *Actinobacteria* (AICEs) and relies on a mechanism of translocation of dsDNA molecules (Fig. 1B) (17, 22). Since transfer of a dsDNA molecule is not a conservative mechanism, RC replication of the excised AICE initiated by a dedicated replication initiator protein (Rep) is required prior to its transfer to the recipient cell, likely to prevent loss in donor cells. RC-replication Rep proteins are usually related to the MOB_C family of DNA relaxases (23). The conjugative machinery used to translocate dsDNA to the recipient cell is structurally much simpler than the T4SS used for transfer of ssDNA. TraB, also known as TraSA or Tra, a single protein subunit resembling the prokaryotic septal DNA translocator protein FtsK, is sufficient for this process (24). Although the molecular aspects of dsDNA transfer have not yet been elucidated for AICEs, the

Figure 1 General models of ssDNA and dsDNA conjugative transfer of ICEs. (A) In the donor cell, the ICE excises from the chromosome by site-specific recombination between the *attL* and *attR* attachment sites. Following excision, the relaxase (Mob) recognizes the origin of transfer (*oriT*) and cleaves the strand, thereby becoming covalently bound to the 5′ end of the nicked strand. The single-stranded nucleoprotein complex is displaced by RC replication and interacts with the type IV coupling protein (T4CP), which energizes the translocation of the relaxase-bound ssDNA through the T4SS. Once in the recipient cell, the relaxase ligates the ssDNA molecule and the complementary strand is synthesized prior to integration into the chromosome by site-specific recombination between the *attP* and *attB* sites. The same process is also generally thought to occur in the donor cell. (B) Like ICEs, AICEs excise from the chromosome by site-specific recombination. Prior to transfer the excised circular AICE undergoes RC replication. The FtsK-like transfer protein Tra recognizes the AICE *clt* and mediates the translocation of the double-stranded AICE DNA by forming a hexameric pore structure and by hydrolyzing ATP. As for ICEs, the AICE integrates into the chromosome by site-specific recombination. Alternatively, RC replication can occur in the recipient prior to integration into the chromosome.
doi:10.1128/microbiolspec.MDNA3-0008-2014.f1

process is thought to be similar to the transfer of the conjugative plasmid pSVH1 of *Streptomyces venezuelae* (24). A *cis*-acting locus for transfer (*clt*) is thought to be recognized and bound by TraB. This protein is capable of assembling as a hexameric pore within the membrane, thereby allowing the ATP-dependent translocation of the AICE DNA into the recipient cell.

Regardless of the mode of transfer, once in the recipient cell, the ICE integrates by site-specific recombination into the chromosome to be vertically inherited in the recipient progeny. Reintegration of the ICE in the chromosome of the donor is usually presumed to occur after transfer, although the fate of the donor cells has rarely been investigated and often remains unknown. In the case of ICE*clc* of *Pseudomonas putida*, subpopulations of potential donor cells engaging into the transfer competent state have been shown to undergo fewer and slower division than non-transfer competent cells and eventually lyse (25).

Overview of the Modular Structure of ICEs

ICEs are composed of functional modules that correspond to DNA segments and genes or group of genes involved in specific key steps of their maintenance and dissemination. Four essential modules compose the basic ICE structure: the integration/excision module which ensures intracellular mobility functions, the replication/DNA processing module, the DNA secretion module and the regulation module (Fig. 2) (16, 17). The integration/excision module carries the *attP* attachment site as well as the gene coding for the integrase, either a serine or tyrosine recombinase. Tyrosine recombinase genes are often associated with a small gene coding for a recombination directionality factor (RDF) helping the integrase to mediate the excision step. The replication/DNA processing module consists of either an *oriT* and cognate relaxase- and T4CP-coding genes (ssDNA transfer) or an origin of replication (*ori*) and associated Rep gene (dsDNA transfer) (17, 23). In the case of ssDNA transfer, the associated DNA secretion module is always a cluster of genes coding for a T4SS. In the case of dsDNA transfer, this module is replaced by a gene coding for an FtsK-like Tra protein and a *clt* sequence. Finally, the regulation module is important to control the activity of the other modules and varies in complexity. Additional auxiliary modules are often present within ICEs and code for diverse functions, such as resistance to antibiotics or heavy metals, alternative catabolic pathways, establishment of symbiosis, or virulence, that can increase the fitness of their bacterial host in a given environmental context (5, 8, 9).

THE SXT/R391 FAMILY

SXT and R391 were, respectively, isolated from *Vibrio cholerae* O139 and *Providencia rettgeri* (6, 26). SXT was originally identified as a self-transmissible, chromosomally integrated genetic element conferring resistance to sulfamethoxazole, trimethoprim, and streptomycin in the first non-O1 serogroup of *V. cholerae* to cause

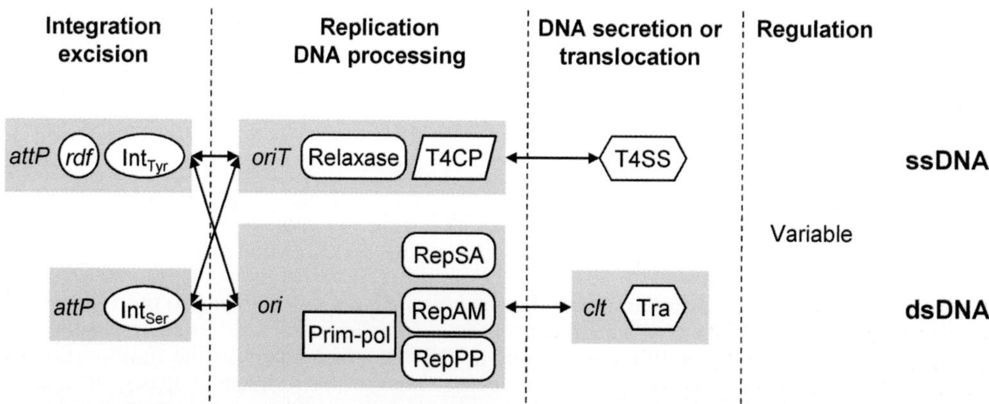

Figure 2 Schematic representation of the modular organization of ICEs and of the typical functional protein signatures associated with each module depending on the mode of DNA transfer (ssDNA vs dsDNA). Possible combinations of integration/excision, replication, and conjugative transfer modules are represented. Int$_{Tyr}$, tyrosine recombinase; Int$_{Ser}$, serine recombinase; T4CP, type IV coupling protein (VirD4-like protein); T4SS, type IV secretion system; Tra, FtsK-like DNA translocation protein; RepSA, RepAM, and RepPP, replication initiator proteins; Prim-pol, bifunctional DNA primase/polymerase. The regulation module is extremely variable between families of ICEs.
doi:10.1128/microbiolspec.MDNA3-0008-2014.f2

epidemic cholera in 1993 in India and Bangladesh (6, 27). Together with other "R factors," R391 was initially identified in the early 1970s as a plasmid conferring resistance to kanamycin and was defined as the prototypical representative of the "IncJ" incompatibility group of plasmids, which included other "R factors" such as R997 from an Indian *Proteus mirabilis* isolate as well as the so-called plasmids pMERPH from *Shewanella putrefaciens* and pJY1 from *V. cholerae* O1 (for a review, see reference 28). More than two decades after the discovery of R391, an accumulation of evidence based on the comparison of integration sites and conservation of a large set of nearly identical genes confirmed that SXT as well as all of the "IncJ elements" belong to the same family of ICEs, referred to as the SXT/R391 family (29, 30, 31, 32).

SXT/R391 ICEs are major contributors to the spread of multidrug resistance in *V. cholerae* (27, 31, 32). In fact, most SXT/R391 ICEs found in clinical isolates confer resistance to sulfamethoxazole and trimethoprim, two commonly used antibiotics. Furthermore, several of these ICEs have been shown to code for functional diguanylate cyclases that participate in bis-(3′,5′)-cyclic dimeric guanosine monophosphate (c-di-GMP) signaling in *V. cholerae*, which inhibits cell motility and enhances biofilm formation (33). SXT/R391 ICEs seem to have emerged in Asian *V. cholerae* clinical isolates in the early 1990s, subsequently spread in environmental and clinical strains in Asia and Africa, and have recently been found in all isolates recovered from cholera patients in Haiti (32, 34, 35, 36). SXT/R391 ICEs are also naturally occurring in non-*Vibrio* pathogens (37, 38, 39).

Genetic Organization

The genetic organization of SXT/R391 ICEs is shown in Fig. 3. SXT/R391 ICEs range from 79 to 110 kb in size and consist of a scaffold of genes and sequences that share more than 95% identity at the nucleotide level (30, 31, 37, 40, 41). Among the conserved genes are *xis* and *int* that catalyze the excision and integration, a mutagenic DNA repair system (*rumAB* (42)), five clusters of transfer genes coding for the mating bridge, synthesis of the pilus (T4SS), and initiation of DNA transfer (*oriT-mobI*, *mob*, *mpf1*, *mpf2* and *mpf3*), the main repressor gene *setR*, and the transcriptional activator genes *setCD*. The conserved core of genes also includes a large group of genes, most with unknown function, that include a λ Red-like homologous recombination system (*bet*/*exo*). The conserved gene core of SXT/R391 ICEs is related to the conserved core of IncA/C conjugative plasmids, which are widely

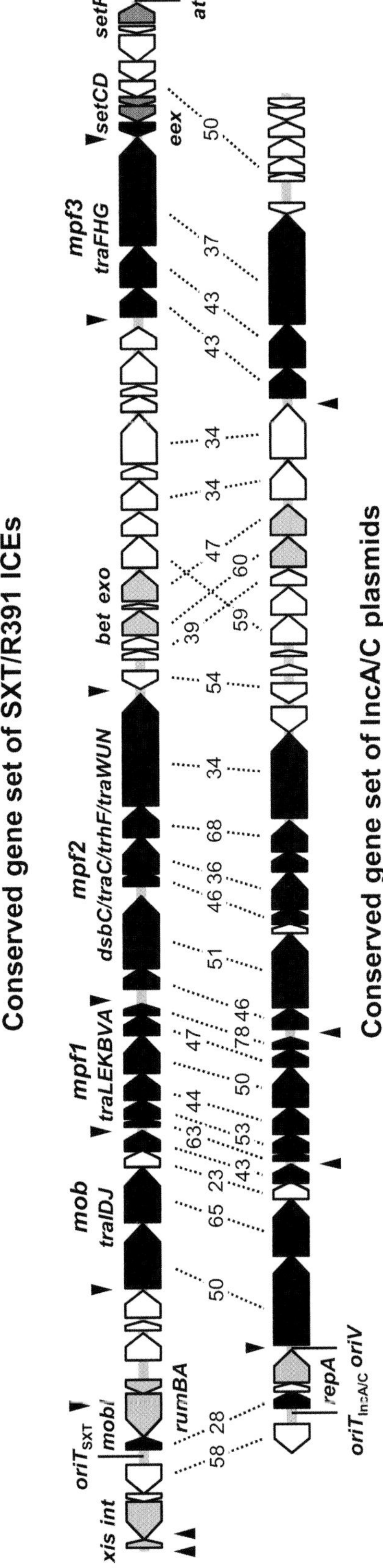

Figure 3 Comparison of the linear genetic maps of the conserved genes of SXT/R391 ICEs and IncA/C conjugative plasmids. Alignment of the conserved genes of the ICE SXT and IncA/C conjugative plasmid pIP1202. ORFs are color coded as follows: black, DNA processing and mating pair formation; dark gray, genes involved in regulation; light gray, genes involved in replication, recombination, or repair; white, genes of unknown function. Numbers shown in the middle represent % identity between the orthologous proteins encoded by SXT and pIP1202 (Genbank AY055428.1 and NC_009141, respectively). The positions of insertions of variable DNA in SXT/R391 ICEs and IncA/C plasmids are marked by arrowheads. The positions of the origins of transfer (*oriT*), origin of replication (*oriV*), and site-specific attachment site (*attP*) are indicated. doi:10.1128/microbiolspec.MDNA3-0008-2014.f3

distributed among several species of pathogenic *Enterobacteriaceae* and *Vibrionaceae* (31, 43, 44). Every predicted transfer protein encoded by SXT has a homolog encoded by the IncA/C plasmid pIP1202 of *Yersinia pestis*. In addition, the *tra* and *mob* genes of these two families of mobile elements are perfectly syntenic, which likely reflects their common ancestry (Fig. 3).

The conserved core of SXT/R391 ICEs is disrupted by insertions of variable DNA coding for diverse functions such as antibiotic resistance, secondary messenger production, toxin/antitoxin modules, or restriction/modification systems conferring resistance to bacteriophage infection (31, 33, 41).

Integration and Excision

Regardless of the host from which they were originally isolated, all the members of the SXT/R391 family integrate into the 5′ end of the gene *prfC*, which encodes the peptide chain release factor 3 (RF3) (29, 31, 45, 46). Integration of SXT occurs by site-specific recombination between two nearly identical 17-bp sequences, which likely represent the core of the attachment sites *attP* on the circular form of SXT and *attB* on the chromosome. The 17 bp of the *attB* site corresponds to amino acids 18 to 23 of RF3. Consequently, the integration of SXT disrupts *prfC*, which ends shortly after the 17-bp sequence inside SXT on the *attL* side. However, the *attR* side of SXT provides a new promoter, a ribosome binding site, and a new coding sequence that restore the reading frame of *prfC*. In an *E. coli* mutant lacking *prfC*, SXT integrates into different secondary integration sites, with a marked preference for the 5′ end of *pntB*, a gene coding for the β subunit of the pyridine nucleotide transhydrogenase (47). SXT transfers less efficiently to an *E. coli* recipient cell lacking *prfC* (~100 reduction), likely because suboptimal base pairing between the core of *attP* and alternative *attB* sites reduces the efficiency of integration. This observation suggests that integration of SXT into the recipient cell's chromosome becomes the rate-limiting step of transfer in a Δ*prfC* strain.

Integrative recombination of SXT/R391 ICEs between the *attP* and *attB* sites is mediated by a site-specific tyrosine recombinase (Int_{SXT}) that belongs to the P4 family. Chromosomal excision of SXT/R391 ICEs results from the site-specific recombination between *attL* and *attR* sites. It is also mediated by Int_{SXT} and is facilitated by the RDF Xis, a 64-amino acid basic protein that contains a predicted helix-turn-helix DNA-binding motif (47). Excision of SXT from the chromosome of *E. coli* was estimated to occur in ca. 1% of the cell population in non-induced conditions, which is about 100 times as high as the frequency of transfer (47). Therefore, excision is not a rate-limiting step of SXT transfer. The host encoded nucleoid associated protein IHF (integration host factor), which is necessary for integration and excision of the bacteriophage λ in *E. coli*, is not required for SXT integration into or excision from the *V. cholerae* chromosome (48).

Interestingly, an atypical genetic element sharing most of its structural genes with SXT/R391 ICEs has been identified in a natural isolate of *V. cholerae* O37 (31, 49). This element, ICE*Vch*Ban8, carries a different integration module coding for Int_{Ban8} and is integrated into a $tRNA^{Ser}$-coding gene, a locus usually occupied by the pathogenicity island VPI-2. Although the ability of ICE*Vch*Ban8 to transfer has not been demonstrated, it seems to be able to excise by site-specific recombination and form the circular intermediate required for transfer (49).

Conjugative Transfer

Dissemination of SXT/R391 ICEs by conjugation results from the activation of no less than five gene clusters coding for DNA processing and mating pair formation (Fig. 3). The two gene clusters *oriT-mobI* and *traIDJ* (*mob*) code for the DNA processing function that prepare the DNA for translocation by the T4SS encoded by the three operon-like gene clusters *traLEKBVA* (*mpf1*), *dsbC/traC/trhF/traWUN* (*mpf2*), and *traFHG* (*mpf3*). Transfer is initiated at *oriT* by the relaxase TraI as well as by the auxiliary mobilization factor MobI. TraI belongs to the MOB_{H1} family of relaxases, which are predicted to be atypical HD phosphohydrolases (23). Relaxases of the MOB_{H1} family also include TraI proteins encoded by conjugative plasmids of the IncA/C, IncP-7, IncHI, and IncT groups. Currently, the mechanism by which TraI nicks the DNA at *oriT* remains unclear, since no relaxase of the MOB_{H1} family has been characterized at the biochemical level and experimental data about the MOB_{H1}-type relaxosomes remain scarce (23). Furthermore, while the exact role of MobI in this process remains unknown, SXT/R391 ICEs are unable to transfer in its absence (50). Interestingly, IncA/C plasmids also encode a distant MobI ortholog that has been shown to be absolutely required for transfer (31, 51). Despite the complete lack of homology between the respective *oriT*s of SXT/R391 ICEs and IncA/C plasmids, the *oriT* locus and the *mobI* gene orthologs are adjacent in both of them (Fig. 3).

In the laboratory, transfer of SXT often results in the formation of exconjugants harboring multiple copies of SXT integrated into *prfC* in a tandem fashion both in *E. coli* and *V. cholerae*. Although the mechanism responsible for this behavior remains obscure, these tandem arrays do not persist in *recA*+ strains and are rapidly brought down to a singleton after only a few generations (52). Natural isolates harboring SXT/R391 ICEs are devoid of such arrays, suggesting that their formation is infrequent, unstable, or prevented in natural settings (52). Exclusion of ICE entry likely contributes to reducing the possibility of tandem array formation by preventing redundant acquisition of multiple identical ICEs by conjugation. Cells containing SXT exclude transfer of a second copy of SXT by reducing the efficiency of transfer by approximately 30-fold (53). Entry exclusion was shown to be mediated by the cytoplasmic portions of the two inner membrane proteins TraG and Eex in the donor and recipient cells, respectively (54). Complex topological rearrangements of the T4SS are likely at play during mating to enable cytoplasmic moieties of inner membrane proteins in two different cells to interact and mediate exclusion. All members of the SXT/R391 family do not exclude each other. In fact, two variants of the exclusion system coexist within the family: the S and R groups (55). SXT is the prototypical member of the S group and excludes the entry of all other SXT/R391 ICEs bearing an *eexS* allele but not those bearing an *eexR* allele. Conversely, R391 is the prototypical element of the R group. It excludes itself and all other SXT/R391 ICEs bearing an *eexR* allele but not those bearing an *eexS* allele. Remarkably, every ICE bearing an S group *eex* gene also bears an S group *traG* gene and similarly for ICEs of the R group (55). The comparison of the variants of *traG/eex* pairs and the construction of chimeric mutants allowed the specificity of exclusion in TraG to be pinpointed to only three amino acids at position 606–608, TDD or TGD for $TraG_R$ and PGE for $TraG_S$ (53, 55). In the 143 amino acids Eex protein, specificity determination was attributed to the last 56 C-terminal residues which exhibit the highest divergence (41% identity) (55). The reasons for the emergence and persistence of two exclusion groups in a single family of ICEs are unknown.

Regulation of Excision and Conjugative Transfer

Conjugative transfer of SXT is strongly regulated, and mostly repressed in normal conditions, typically occurring at frequencies below 10^{-4} per donor from *E. coli* and 10^{-7} per donor from *V. cholerae* (6, 56). The reasons for this striking difference of regulation are unknown but could be the result of a difference in the fraction of the cell population spontaneously expressing the SOS response between *V. cholerae* and *E. coli* (57). Indeed, exposure to DNA-damaging agents that are known to induce the SOS response (UV light, mitomycin C, ciprofloxacin) have been shown to stimulate SXT conjugative transfer at frequencies up to 10^{-2} per donor (58). Pairing the induction of the conjugative transfer of SXT/R391 ICEs with the host SOS response is reminiscent of the activation of the lytic pathway of bacteriophage λ and explains early observations reporting that, unlike most ICEs, the transfer of SXT and related elements requires *recA* in the donor cells (6, 58, 59), the key derepressor of the SOS response (60, 61). At least three ICE-encoded conserved genes regulate the excision and conjugative transfer of SXT: *setR*, *setC*, and *setD* (Fig. 3). *setR* encodes a λ cI-like transcriptional repressor which binds to four operators sites between the divergently transcribed genes *s086* and *setR* in noninducing conditions (56, 62). *s086* codes for a putative Cro-like regulator of unknown function. SetR represses expression from the promoter P_L, which drives the expression of an operon that includes *s086*, *setC*, and *setD*, as well as its own expression from the phage lambda promoter P_R (58, 62). Autorepression of SetR is thought to maintain a low level of the repressor, thereby allowing a quick response to inducing stimuli, notably when DNA damages stimulate the co-protease activity of RecA, which in return would facilitate the autocleavage and inactivation of SetR, thereby alleviating the repression of *setC* and *setD*. The proteins SetC and SetD are very distant orthologs of *E. coli*'s activators FlhC and FlhD, respectively (56). In *E. coli* FlhC associates with FlhD in an $FlhD_2C_2$ heterotetrameric complex that induces the transcription of genes involved in the biogenesis of cell surface flagella to mediate bacterial cell motility. SetC and SetD are thought to assemble in a similar complex to activate the expression of various genes or groups of genes in SXT, including *traLEKBVA* and *traFHG*, the *int* and *xis* genes, the *mosAT* toxin-antitoxin system, and the *bet/exo* RecA-independent homologous recombination system (47, 56, 58, 63, 64, 65). Interestingly, the SetCD transcriptional activator also targets genes that are located outside of SXT/R391 ICEs; it stimulates the expression of int_{MGI} and triggers the expression of *rdfM*, two genes encoded by mobilizable genomic islands (MGIs) (see below) (66, 67).

Evolution and Plasticity

The conserved gene core of SXT/R391 ICEs seems to be subjected to frequent recombination events within specific intergenic regions (Fig. 3) that appear to serve as hotspots for the integration of variable DNA (31).

These hotspots are likely insertion regions that do not jeopardize the viability of the ICE, because they do not disrupt a gene or operon essential for transfer or maintenance and do not alter negatively the regulation of *tra* gene expression by SetCD, thereby allowing the preservation of a suitable protein dosage required for the assembly of the T4SS. Analysis of the DNA content of these hotspots in several ICEs revealed the occurrence of inter-ICE exchanges of variable DNA, thereby suggesting that in natural conditions two different SXT/R391 ICEs can coexist in the same cell and recombine to generate a hybrid ICE with new properties contributing to their evolution. In fact, like SXT tandem arrays forming during conjugation, SXT and R391 have been shown to be able to integrate into *prfC* in the same cell in tandem arrays (29). Remarkably, the transfer from cells bearing such tandem arrays led to the isolation of up to 7% of exconjugant colonies bearing fully functional hybrid ICEs generated by homologous recombination (52, 63). Although RecA has been shown to contribute to this process (ca. 4% of the hybrids), the ICE-encoded λ Red-like homologous recombination system *bet*/*exo* was also shown to be involved. Both RecA and Bet/Exo act synergistically to promote hybrid ICE formation from ICE tandem arrays (63). Since *bet* and *exo* are part of a SetCD-activated operon, their activity promotes ICE plasticity in response to DNA damages (63, 64).

SXT/R391 ICE-Driven Mobilization of Genomic Islands: the MGIs

SXT is capable of mobilizing in *trans* the mobilizable plasmid CloDF13 with high efficiency and can also mobilize the broad-host-range plasmid RSF1010, although at a much reduced frequency (68). In contrast, SXT is unable to mobilize ColE1 and ColE3 plasmids. SXT is also capable of mobilizing chromosomal DNA located up to 1.5 Mb downstream of *prfC* in an Hfr-like manner between strains of *E. coli* (66, 68).

Furthermore, identification of the *oriT* of SXT allowed the identification of a novel family of MGIs mimicking this locus and enabling them to hijack the conjugative machinery encoded by SXT/R391 ICEs to mediate their own transfer (Fig. 4) (50, 66). MGIs range between 18 and 33 kb in size and are found in several bacterial species thriving in aquatic environment, such as *Vibrio* sp., *Alteromonas* sp., *Pseudoalteromonas*, and *Methylophaga*, with some of them naturally harboring SXT/R391 ICEs (66, 69). MGIs share a small conserved core of four genes and the *oriT*-like sequence that ensures their basic maintenance and mobilization function. These four genes are *cds4* and *cds8*, which code for putative proteins of unknown function, and int_{MGI} and *rdfM* which promote integration and excision of MGIs. The gene int_{MGI} codes for a tyrosine recombinase which mediates the integration of MGIs into the 3′ end of *E. coli*'s *yicC*, a gene of unknown function, and its orthologs in all the known natural hosts of MGIs (66, 69). Integration of MGIs into the chromosome of a naive host is independent of the co-transfer of an SXT/R391 ICE as int_{MGI} is constitutively expressed at a low level, thereby favoring MGI dissemination regardless of the ability of the ICE to co-establish in the recipient cell (67). Int_{MGI} also mediates the excision of MGIs, but only in the presence of the RDF RdfM (66, 67). In a host devoid of an ICE of the SXT/R391 family, *rdfM* is not expressed and MGIs are unable to excise from the host's chromosome. *rdfM* expression is triggered by the ICE-encoded transcriptional activator SetCD (67). Hence, MGI excision is responsive to DNA damage and occurs exclusively when an SXT/R391 ICE is present. This checkpoint mechanism likely stabilizes the genomic island in the progeny of the host cell by preventing untimely excision that could result in the loss of the element in a fraction of the cell population after cell division. Besides the four conserved genes, MGIs also carry a cargo of variable genes, most of them coding for diverse type I and type III restriction/modification systems as well as diverse toxin/antitoxin systems (69).

Remarkably, mobilization of genomic islands by SXT/R391 ICEs considerably increases the availability of chromosomal DNA for exchange between bacteria. Indeed, MGIs extend the range of chromosomal sequences that these ICEs can transfer in an Hfr manner by providing an *oriT*-like sequence from which chromosomal DNA transfer can be initiated remotely from the ICE integration site (66). An ICE from *V. fluvialis* integrated into the 5′ end of *prfC* in *E. coli* was shown to mediate a transfer in *trans* of up to 1 Mb of chromosomal DNA located upstream of *yicC*, through the recognition of the *oriT*-like sequence of an MGI integrated into the 3′ end of this gene.

ICE*Bs1* OF *BACILLUS SUBTILIS*

ICE*Bs1* is integrated at the 3′ end of the *trnS-leu2* locus in the chromosome of most, if not all, strains of the low GC Gram-positive bacterium *B. subtilis* (70, 71) (Fig. 5). Although ICE*Bs1* has been extensively studied over the past decade, its role in the biology of *B. subtilis* and the adaptive traits that this element might confer to its host remain unknown. Nevertheless, the

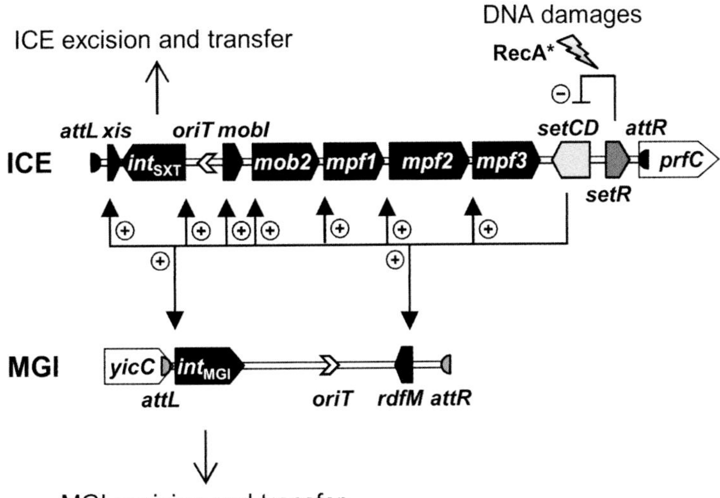

Figure 4 Model of ICE-mediated activation and mobilization of an MGI. DNA-damaging agents trigger the SOS response, alleviating the SetR-mediated repression of *setCD*. The transcriptional activator SetCD activates the expression of *mpf* and *mob* genes of the ICE. SetCD also directly activates the expression of *int* of the ICE and *int*$_{MGI}$ and *rdfM*, while Int catalyzes the excision of the ICE, and Int$_{MGI}$ and RdfM mediate the excision of the MGI. Expression of the *mpf* operons leads to the formation of the mating pore that will connect the donor and recipient cells, and deliver the DNA. Specific *mob* genes produce the mobilization proteins (TraI and MobI) that will recognize, bind to, and cleave the *oriT*s on both the ICE and MGI. The nicked DNA bound to TraI is directed to the mating pore and translocated to the recipient cell. doi:10.1128/microbiolspec.MDNA3-0008-2014.f4

ICE*Bs1* biology remains the most thoroughly studied among Gram-positive bacteria, providing valuable information regarding the global lifestyle of nearly all ICE families, especially vis-à-vis replication of the excised form of the element.

Genetic Organization

With Tn*916* from *Enteroccocus faecalis*, ICE*Bs1* is one of the smallest ICEs discovered. Only 20,451-bp large, ICE*Bs1* carries 24 genes organized in three main functional modules (Fig. 5). The genes belonging to the conjugation module are part of a single operon that codes for both the DNA processing and mating-pair formation. Most of the predicted or experimentally demonstrated transfer proteins encoded by ICE*Bs1* have homologs encoded by conjugative plasmids found in the *Firmicutes*. Four of these proteins are homologous to VirB1, VirB4, VirB11, and VirD4 proteins, which belong to inner membrane part of the T4SS of conjugative plasmids found in Gram-negative bacteria (72, 73, 74). Interestingly, the genes encoding the site-specific recombination functions and allowing intracellular mobility of ICE*Bs1* are not grouped in the same region. The integrase gene *int* belongs to the regulation module, whereas the RDF gene *xis* is transcribed as part of the conjugation module (Fig. 5).

Integration and Excision

ICE*Bs1* integration occurs by site-specific recombination within two long direct repeats of 60-bp sequences that constitute the core of the *attP* site carried by the extrachromosomal form of the element and the *attB* site located in the 3′ end of the *trnS-leu2* locus (75). Alternatively, in a recipient strain lacking the primary integration site, ICE*Bs1* maintains itself by site-specific recombination into at least 15 secondary integration sites (75, 76). However, in such circumstances, ICE*Bs1* excision becomes deficient and cell viability is impaired likely because of single-stranded replication initiated within ICE*Bs1* interfering with chromosome stability (76). Comparison of these recombination sites highlighted a conserved 17-bp sequence encompassing two 5-bp inverted repeats separated by 7 bp, which is the target for the recombination. When integrated, ICE*Bs1* is flanked by the *attL* and *attR* sites, targets of the site-specific recombination leading to ICE*Bs1* excision (Fig. 5). ICE*Bs1* site-specific excision and integration both require the integrase-encoding gene *int* localized near the *attL* site (Fig. 5) (75). The integrase of ICE*Bs1* belongs to the tyrosine recombinase family and is distantly related to the integrase of Tn*916* (75). ICE*Bs1* excision from the chromosome occurs in 0.005% of the cell population under normal growth conditions

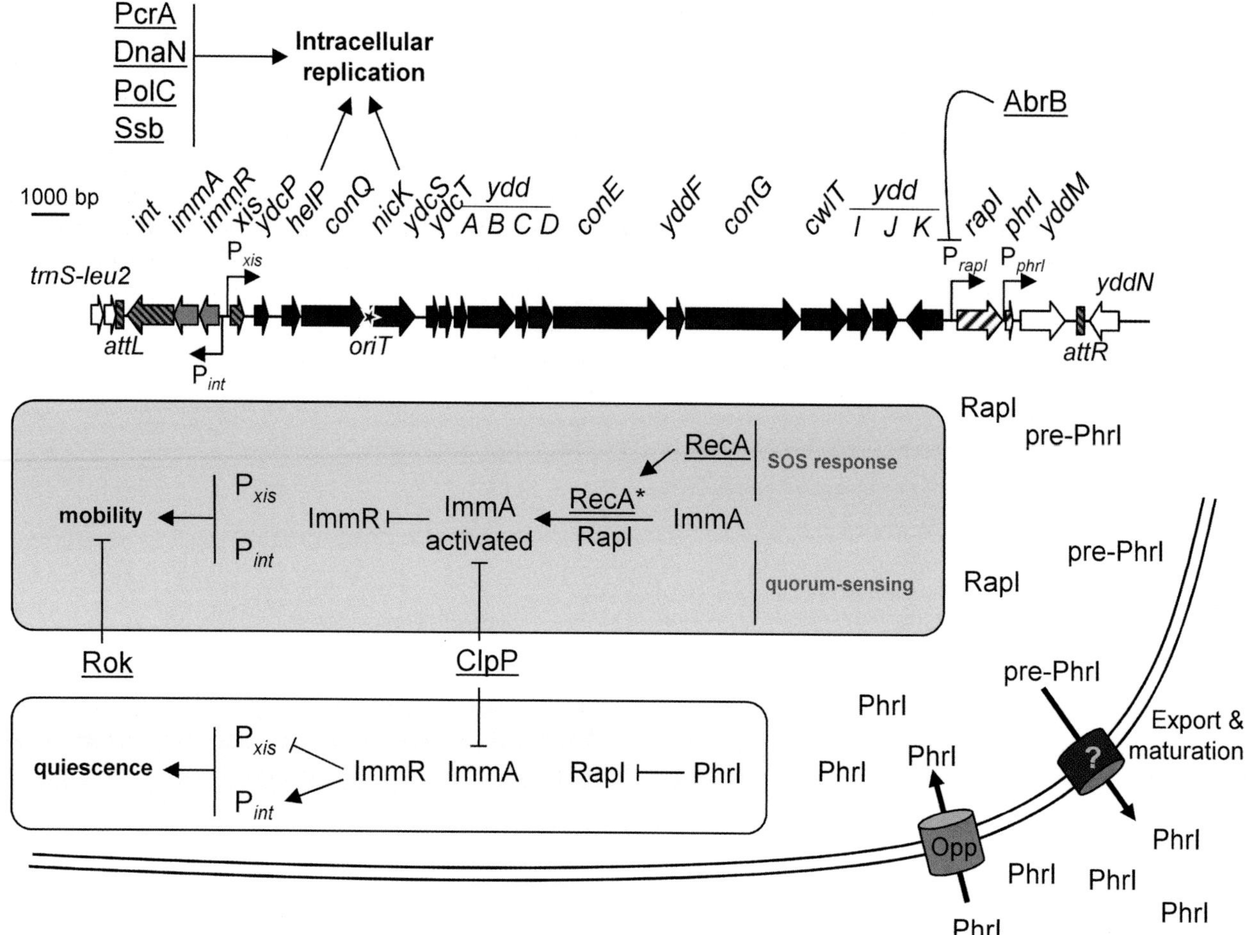

Figure 5 Genetic organization of the integrated ICE*Bs1*. ORFs are symbolized by arrowed boxes with their name above and color coded as follow: black, DNA processing and mating pair formation; dark gray, regulation; dashed light gray, recombination; dashed black and white, quorum sensing; white, genes of unknown function or that do not belong to ICE*Bs1*. The promoters are indicated by angled arrows. The position of the origin of transfer (*oriT*) and the site-specific attachment sites (*attL* and *attR*) are indicated and represented by a black star and gray rectangles, respectively. Transporters are represented by cylinders. Name of proteins are written with a capital and the host's factors are underlined. The quorum sensing system of ICE*Bs1* produces both the phosphate RapI and the prepeptide PhrI (pre-PhrI). Pre-PhrI is exported and maturated by an unknown transporter. When the mature pentapeptide PhrI reaches a threshold concentration in the extracellular environment, it is imported by the oligopeptide permease Opp. In the cell, PhrI inhibits the phosphatase RapI. Regulation of ICE*Bs1*. ICE*Bs1* activation pathway is encompassed in a gray round-angled box. Both the phosphatase. The RapI and the activation of RecA in RecA* during the SOS response activate the protease ImmA. ImmA site-specifically cleaves the transcriptional regulator ImmR, allowing the transcription from P_{xis}, thereby activating ICE*Bs1* excision and transfer. The pathway leading to ICE*Bs1* quiescence as an integrated element is depicted in a white round-angled box. In the absence of ImmA activation, i.e., without induction of the SOS response or after RapI inactivation by the pentapeptide PhrI, ImmR represses P_{xis} and activates P_{int}, leading to ICE*Bs1* quiescence. Besides RecA, the dynamics of ICE*Bs1* involves host factors, such as the transcriptional regulators AbrB and Rok, or the protease ClpP, that directly or indirectly inhibit ICE*Bs1* mobility. Intracellular replication of ICE*Bs1* initiated at *oriT* requires the ICE-encoded relaxase NicK and the helicase processivity factor HelP, as well as the host-encoded helicase PcrA, DNA polymerase subunit PolC, processivity clamp DnaN, and single-strand DNA binding protein Ssb.
doi:10.1128/microbiolspec.MDNA3-0008-2014.f5

(75). It is mediated by the integrase and triggered by the expression the RDF gene *xis*. Although *int* seems to be constitutively expressed, the expression of *xis* is repressed from its σ^A-recognized promoter P_{xis} and depends on a complex regulatory system described below (77, 78).

Conjugative Transfer

Intraspecific conjugative transfer of ICE*Bs1* in *B. subtilis* occurs at a frequency of about 10^{-2} transconjugants per recipient cell devoid of ICE*Bs1* and drops by about 50-fold when ICE*Bs1* is already integrated at its primary integration site (77). ICE*Bs1* can also transfer toward other low GC Gram-positive species such as *B. anthracis* and *B. licheniformis* as well as *Listeria monocytogenes*. The conjugative transfer of ICE*Bs1* mainly depends on the expression of the operon that encodes the RDF, which also codes for the entire DNA processing and T4SS machinery (Fig. 5).

After excision from the chromosome, the transfer of ICE*Bs1* is initiated at *oriT*, which corresponds to a 800-bp fragment overlapping the 3′ end of *conQ* and the 5′ end of the relaxase-encoding gene *nicK* (75) (Fig. 5). Nicking by the relaxase NicK occurs prior to single-strand transfer of ICE*Bs1* between two inverted repeats within a 24-bp GC-rich sequence located within *nicK*, close to its start codon. NicK is related to the relaxase TecH (Orf20) encoded by Tn*916*. TecH-mediated nicking at *oriT* of Tn*916* requires the presence of the integrase encoded by Tn*916* to confer sequence specificity to TecH (79). Unlike TecH, the relaxase NicK of ICE*Bs1* does not seem to require any ICE-encoded auxiliary protein to mediate specific nicking at *oriT* (75). Finally, *yddM* is predicted to encode a putative helicase that, together with the helicase processivity factor HelP, could be involved in extrachromosomal DNA replication, which occurs during conjugative transfer (80).

After the DNA has been processed, transfer of the single-strand molecule occurs through the mating apparatus. *B. subtilis* grows in chains of several connected cells in which ICE*Bs1* can rapidly spread by conjugation, thereby invading the whole chain from a single transconjugant (81). The VirB4-like protein ConE, an ATPase of the HerA/FtsK superfamily, is required for the transfer of ICE*Bs1* (72, 82). Specific localization of ConE at the cell poles might be a key factor in the intrachain transfer of ICE*Bs1*. ConQ is an FtsK/SpoIIIE protein also required for ICE*Bs1* transfer and likely acts as the T4CP (83). Although its precise role remains unknown, ConG is predicted to be a membrane-associated protein and has been experimentally shown to be necessary for ICE*Bs1* transfer (Fig. 5) (81). *cwlT* encodes a bifunctional murein hydrolase that contains a muramidase and a DL-endopeptidase domain (84). The *N*-acetylmuramidase domain of CwlT is necessary for the conjugation of ICE*Bs1* as it is implicated in cell-wall processing prior to the assembly of the T4SS (85). The peptidase domain of CwlT is not essential, but it enhances ICE*Bs1* conjugative transfer. Finally, the remaining genes of the conjugation module encode proteins that have no predicted function or known functional domain.

Besides its own conjugation, ICE*Bs1* was shown to mobilize plasmids devoid of dedicated mobilization functions. The mechanism involves a functional mating pair and the T4CP ConQ encoded by ICE*Bs1*, the RC origin of replication of the plasmids acting as an *oriT*, and its replicative relaxase acting as a conjugative relaxase (83).

ImmR/ImmA as a Central Regulatory System of ICE*Bs1*

The regulation of ICE*Bs1* transfer is now well understood, in spite of its complexity. ICE*Bs1* is tightly regulated by the central ICE-encoded regulatory system ImmR/ImmA which is itself regulated by the host SOS response and quorum sensing (75, 77, 86).

Several prophages, like TP901-1 and ϕ105, do not encode a cI regulator but rather a cI-like regulator (87, 88). Such regulators are not related to cI and are devoid of autopeptidase activity. The only well characterized cI-like regulator, ImmR, is encoded by ICE*Bs1* (Fig. 5) (78, 87). ImmR regulates its own expression and represses the expression of the RDF Xis from P_{xis}, thereby repressing the expression of the whole operon encoding the T4SS and DNA processing functions (78). A constant intracellular level of ImmR protein is maintained by the dose-dependent regulation of ImmR on its own promoter, leading to its autorepression when the amount of intracellular ImmR is high (78). ImmR is associated with ImmA, a protease encoded by the same operon that allows the derepression of the mobility of ICE*Bs1* (Fig. 5) (87). Activated by the SOS response *via* an unknown mechanism, ImmA directly interacts with ImmR and cleaves it site specifically (86, 87). Cleavage of ImmR by ImmA could result from the stimulation of ImmA proteolytic activity but also from its stabilization (86). Interestingly, in addition to keeping ICE*Bs1* quiescent, the transcriptional regulator ImmR can also act as an immunity factor. Expression of *immR* in recipient cells prevents the establishment of an incoming ImmR-regulated element by repressing the expression and/or inhibiting the activity of its integrase (78).

In *B. subtilis*, many cellular processes are regulated by Rap-Phr quorum-sensing systems. Although the pathways that they regulate can be very different, such systems have a common mode of action (89). ICE*Bs1* excision and transfer depend upon the phosphatase RapI and its cognate pentapeptide PhrI, both encoded by the same locus (Fig. 5) (77). The prepeptide (pre-PhrI) encoded by *phrI* is maturated and exported by an unknown transporter. The mature active form of PhrI corresponds to the five C-terminal amino acids of the prepeptide. When the extracellular concentration of PhrI reaches a threshold, PhrI is imported in the cell by the oligopeptide permease Opp (Fig. 5). Once internalized in the cell, PhrI inhibits the activity of RapI and, indirectly, ICE*Bs1* mobility. RapI is responsible for the activation of the protease ImmA and triggers the derepression of ICE*Bs1* (87). In a population in which the majority of the cells are devoid of ICE*Bs1*, the extracellular concentration of PhrI is low, which does not allow the repression of the element. As the population of cells carrying ICE*Bs1* increases, the buildup of extracellular PhrI triggers its import in the donor cells, thereby leading to the repression of RapI and thus of ICE*Bs1*. This strategy allows a tight regulation and optimizes the dissemination of ICE*Bs1* when the conditions are favorable, i.e., when the surrounding cells are devoid of ICE*Bs1*.

ICE*Bs1* and its host genome are interconnected. On one hand, ImmR was shown to bind diverse loci in the chromosome of *B. subtilis*, which could have a significant impact on host gene expression, and consequently on the fitness of the organism (78). On the other hand, several studies have reported the implication of host factors in the modulation of ICE dynamics, in both Gram-positive and Gram-negative bacteria (48, 90, 91). At least three host factors are known to influence ICE*Bs1* mobility. First, several genes of *B. subtilis* are repressed under favorable growth conditions by a transcriptional regulator called AbrB, which becomes inactive under low nutriment and/or high cell-density conditions (Fig. 5) (92). Transfer of ICE*Bs1* has been shown to increase in an *abrB* null mutant. AbrB targets the promoter region of *rapI* and represses its expression, thereby repressing excision and transfer of ICE*Bs1* in the exponential growth phase (77). Second, Rok is a transcriptional regulator that represses the establishment of natural competence in *B. subtilis* (Fig. 5) (93). A global analysis revealed that Rok targets AT-rich sequences, which is usually the hallmark of DNA acquired by horizontal gene transfer (94). Rok binds regions located at the extremities of ICE*Bs1*, in the recombination/regulation locus and the quorum sensing locus (94). Deletion of *rok* leads to a 4-fold increase of ICE*Bs1* excision. Finally, the proteasome could also participate in ICE*Bs1* regulation as the amount of ImmA significantly increases in a *clpP* mutant (Fig. 5) (86).

Replication of ICE*Bs1*

The principal mechanism of vertical inheritance of ICEs relies on their integration into and replication with the host chromosome. Nevertheless, several studies showed or strongly suggested that at least some ssDNA-transferring ICEs may be capable of extrachromosomal replication as plasmid-like molecules in both Gram-positive and Gram-negative bacteria (91, 95, 96, 97).

Several observations suggested that ICE*Bs1* was capable of extrachromosomal replication, such as the relative stability of a Δ*immR* mutant and the relocalization of the replisome during ICE*Bs1* conjugative transfer (72, 78). Further studies led to a precise model of RC replication of the activated extrachromosomal circular form of ICE*Bs1* (98). ICE*Bs1* replication is unidirectional, initiated at *oriT*, and requires the relaxase NicK (Fig. 5). Therefore, NicK initiates not only the intercellular replication of ICE*Bs1* during the conjugative transfer but also its extrachromosomal plasmid-like replication. The *oriT* locus plays the dual role of origin of transfer and origin of replication. ICE*Bs1* replication also requires host-encoded factors such as the helicase PcrA, the DNA polymerase subunit PolC, the processivity clamp DnaN, and the ssDNA binding protein Ssb (Fig. 5) (98). Furthermore, ICE*Bs1* replication involves the ICE-encoded ssDNA binding helicase processivity factor HelP, which stimulates the activity of PcrA (Fig. 5) (80). HelP is not only essential for intracellular replication but also for the transfer of ICE*Bs1*. The replication module is an intrinsic part of the intercellular mobility module of ICE*Bs1*. Hence, extrachromosomal replication could be a common feature of most ICEs (99). Although extrachromosomal replication is not necessary for conjugative transfer, it participates in the stability of ICE*Bs1* by preventing the loss of the excised form of the ICE and/or by allowing a single donor to transfer ICE*Bs1* to multiple recipient cells at once (98).

THE ICE*St1*/ICE*St3* FAMILY

ICE*St1* was first described as a 35-kb polymorphic region in the chromosome of the lactic acid bacterium *Streptococcus thermophilus* CNRZ368 (100). Further investigations revealed that this region contains an ICE integrated in the 3′ end of *fda*, a gene coding for the 1,6-bisphosphate aldolase (101). *fda* is also the target for the integration of another closely related ICE named ICE*St3*, initially identified in *S. thermophilus* CNRZ385 (102). Despite the challenge to genetically manipulate *S. thermophilus*, ICE*St1* and ICE*St3* have been studied over the past decade for their conjugative properties, their regulation, and the adaptive traits that they confer to their host as well as for their evolution by accretion–mobilization (91, 102, 103, 104, 105, 106). Diverse ICEs of streptococci are closely related to ICE*St1* and ICE*St3* and thus belong to the same family based on the strong similarities of their respective regulation and conjugation modules (91).

Genetic Organization

ICE*St1* and ICE*St3* are two closely related ICEs of 35 kb (34,734 bp) and 28 kb (28,098 bp), respectively. They share a common organization in two main regions: a core region that encompasses the genes and sequences necessary for their intra- and intercellular mobility, and a highly variable region delimited by the *attL* site and the regulation module (Fig. 6) (70, 101, 102). The core region of ICE*St1* and ICE*St3* is subdivided into three modules: the nearly identical conjugation and recombination modules that share 95% identity at the nucleotide level and a more variable regulation module (Fig. 6) (102). The conjugation and recombination modules are part of a single 14.6-kb polycistronic operon transcribed from the P_{cr} promoter (91). The regulation module of ICE*St1* and ICE*St3* is composed of three shared genes named *arp1*, *arp2*, and *orfQ* and the variable genes *orfP* and *orfR* in ICE*St1*, and *orf385A*, *orf385B*, and Δ*orf385C* in ICE*St3* (Fig. 6). The regulation modules of the two elements share a common organization with two operons transcribed from the P_{arp2} and P_{orfQ} promoters, separated by a rho-independent transcriptional terminator (91). Although they share a common organization, ICE*St1* and ICE*St3* regulation modules each exhibit specific features. The two operons of ICE*St3* are co-transcribed, whereas those of ICE*St1* are not. Furthermore, ICE*St3* carries a stationary phase-specific promoter, P_{arp2s}, that is inactive in ICE*St1* (Fig. 6) (91).

The variable region of both ICEs carries genes that encode putative or demonstrate accessory functions, and proteins of unknown function (Fig. 6). The variable region of ICE*St1* encodes functions conferring resistance to bacteriophage infection, i.e., a type II restriction-modification system named Sth368I and a putative Abi resistance system (70, 106). The variable region of ICE*St3* harbors two ORFs, *orf385F* and *orf385G*, encoding two putative methyltransferases that may belong to a type II restriction-modification system but no gene coding for an associated restriction enzyme has been found (102).

Integration and Excision

Regardless of the host genome in which ICE*St1* and ICE*St3* are integrated, both ICEs seem to always target the 3′ end of *fda* (105). Integration and excision occur by site-specific recombination between two 27-bp direct repeats that constitute the core of the *attB* and *attP* or *attL* and *attR* attachment sites, respectively. The tyrosine integrase Int, which belongs to the ϕLC3 integrase subfamilly, catalyzes both excision and integration, while the RDF Xis encoded by the same operon is needed for excision (103, 105). Excision in the donor cell and integration in the recipient cell are not the rate-limiting step for the transfer of these ICEs as both occur at a rate that is at least 1,000-fold higher than the frequency of transfer (91, 105).

Conjugative Transfer

ICE*St3* is a genuine ICE that transfers at 3.6×10^{-7} transconjugants per recipient cell without any inducing treatment. In contrast, only one transconjugant of ICE*St1* was ever obtained (105). Since *S. thermophilus* is naturally competent for DNA transformation under certain conditions and conjugative transfer experiments were carried out without Dnase, it is not clear whether ICE*St1* is still a functional ICE or was acquired by transformation (107). Nevertheless, ICE*St3* is also able to transfer to other species like *S. pyogenes* as well as to and from *E. faecalis* (105). Intraspecific transfer of ICE*St3* between two strains of *S. thermophilus* is significantly reduced if the *attB* site of the recipient cells is occupied by either ICE*St3* or a synthetic minielement (CIMEL_3catR_3) (103). Exconjugants resulting from such conjugation assays carry unstable tandem arrays that undergo frequent loss of one copy of one or the other element (2×10^{-4} loss per cell and per generation) (103).

The conjugation module of ICE*St1* and ICE*St3* has been experimentally shown to be functional, but no further investigation has been carried out to characterize precisely each individual component (105). Nevertheless, the conjugation module of these ICEs is more or less related to numerous ICEs, some of them having been studied in depth (91, 102). For example, the DNA processing machinery is thought to involve the putative relaxase OrfJ which is related to the one of ICE*Bs*1 (NicK) and to Tn*916* (TecH), and the location of the putative associated *oriT* seems to be conserved among ICE*St1*, ICE*St3*, ICE*Bs1*, and Tn*916* (70, 79, 108). The gene coding for the predicted conjugative proteins of ICE*St1* and ICE*St3* are syntenic with the orthologous

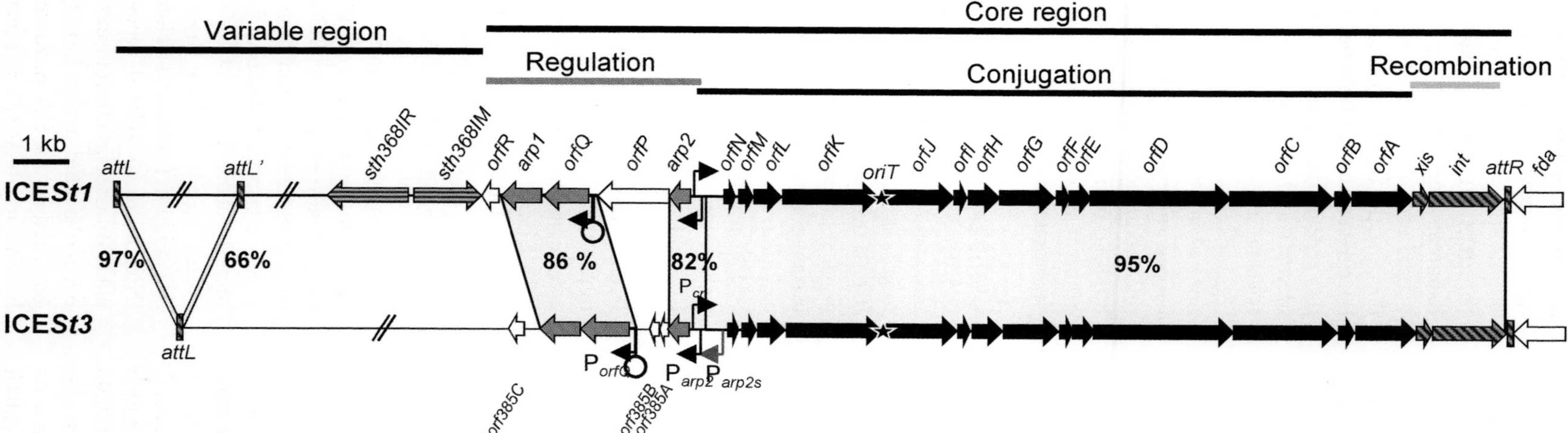

Figure 6 Genetic organization of integrated ICE*St1* and ICE*St3*. The organization of the ICEs is indicated by lines delimiting the variable and the core regions, which encompass the regulation, conjugation, and recombination modules. ORFs are symbolized by arrowed boxes with their name above and color coded as follow: black, DNA processing and mating pair formation; dark gray, regulation; dashed light gray, recombination; horizontal dashed gray and black, restriction-modification; white, genes of unknown function or that do not belong to the ICE. The promoters and rho-independent terminators are indicated by angled arrows and stem-loops. The positions of the putative origin of transfer (*oriT*) and the site-specific attachment sites (*attL* and *attR*) are indicated and represented by black stars and gray rectangles, respectively. The light gray areas indicate related sequences with the percentage of nucleotide identity. doi:10.1128/microbiolspec.MDNA3-0008-2014.f6

genes of ICE*Bs1* and Tn*916*, suggesting a common ancestry (70, 102).

Regulation

The regulation module of ICE*St1* and ICE*St3* is organized as two operons that carry three conserved ORFs that encode putative regulators. *arp1* encodes a cI regulator, functionally similar to *setR* in SXT (Fig. 6) (102, 104). *arp2* codes for a protein that is homologous to the ImmR regulator of ICE*Bs1*, whereas *orfQ* codes for a putative ImmA-like metalloprotease (Fig. 6). Interestingly, although the genes *immR* and *immA* of ICE*Bs1* and those coding for related protein in prophages belong to a unique operon, *arp2* and *orfQ* belong to two distinct operons. In addition, ICEs of the ICE*St1*/ICE*St3* family are the only elements that code for both cI and ImmR regulators, suggesting that a complex pathway in which the two types of regulators can interact to govern the regulation of these ICEs. Although there is no precise model for their regulation, the functional organization of ICEs of the ICE*St1*/ICE*St3* family strongly suggests that their mobility depends on the activation and/or derepression of the promoter P_{cr} (91). Preliminary results suggest that *arp1* codes for a repressor that represses ICE*St1* and ICE*St3* conjugative transfer (104). In agreement with the presence of cI and ImmR regulators, mitomycin C treatment of donor cells triggers the derepression of the conjugation and recombination modules, thereby leading to the excision of both ICE*St1* and ICE*St3* (91). Although conjugative transfer of ICE*St1* remains undetectable under mitomycin C induction, the transfer of ICE*St3* was shown to be derepressed by 25-fold (105). Using excision as a reporter to monitor the activity of ICE*St1* and ICE*St3*, Carraro *et al.* (91) showed that ICE*St1* and ICE*St3* are more active in a stationary growth phase. Finally, ICE*St3* dynamics was shown to be influenced by unknown host factors as excision and transfer have been shown to be differentially modulated in different *S. thermophilus* strains (91).

Replication

DNA damage strongly induces excision of ICE*St1* and ICE*St3*. Although mitomycin C induction does not alter the ICE*St1* copy number per cell, this treatment leads to ICE*St3* excision in more than 90% of the cell population and triggers the multicopy state of ICE*St3*, which reaches about ten extrachromosomal copies per cell (91). As for ICE*Bs1*, the replication of ICE*St3* could involve the putative relaxase OrfJ, which is distantly related to NicK of ICE*Bs1*. Moreover, based on the synteny of the conjugation modules of ICE*Bs1*, Tn*916*, and ICE*St1*/ICE*St3*, *orfL* could encodes a functional homolog of the helicase processivity factor HelP of ICE*Bs1* and its distant homolog Orf22 in Tn*916* (70, 80). Nevertheless, no relevant homology or domain has been detected.

Evolution by Accretion–Mobilization

CIME302 is a 13-kb element integrated at the 3′ end of *fda* in the chromosome of *S. thermophilus* CNRZ302 that probably derived from ICE*St1* following a major deletion extending from the variable region to most of the core region, but retaining almost identical *attR* and *attL* sites (102). Thus, this defective element is neither conjugative nor *trans*-mobilizable. Such nonautonomous elements seem to be widespread in *S. thermophilus* genomes (102). Analysis of the ICE*St1* structure and sequence revealed an internal *attL*′ site in the variable region that corresponds to 22 bp of the 27 bp of the direct repeat (102). Recombination between *attL*′ and *attR* leads to the excision of ICE*St2*, leaving a *cis*-mobilizable element (CIME) called CIME368 integrated in the 3′ end of *fda*. CIME368 is mobilizable by ICE*St2* following integration of ICE*St2* in the left recombination site *attB*′ and subsequent excision of the whole composite structure by recombination between *attL* and *attR*. This mobilization mechanism has been called accretion–mobilization and further investigation revealed that it could be a major driver for evolution of ICEs (Fig. 7) (102, 103). Evolution of autonomous and mobilizable conjugative elements has been well described by Bellanger *et al.* (109) in their recent review and has also been observed in *S. agalactiae*, suggesting that it could be a paradigm for the evolution of genomic islands (109, 110).

Other ICEs of Streptococci

Several other ICEs or putative ICEs have been detected and/or characterized in various streptococci genomes. For example, the analysis of the region of difference 2 (RD2) from *S. pyogenes* revealed that it is a functional ICE distantly related to ICE*St1* and ICE*St3* (95, 111). RD2 is also activated by DNA-damaging agents and replicates in this condition. Sequence analyses also revealed a large variety of putative mobile elements integrated in various loci of numerous *S. agalactiae* genomes (112). At least one of these elements, *ICE_515_tRNA*^{Lys}, is a genuine ICE that self-transfers, can *cis*-mobilize a related element, and encodes a new CAMP factor implicated in the cohemolytic activity of the pathogenic strain of *S. agalactiae* (110, 112, 113). Finally, the ICE 89K of *S. suis* has been well characterized for its diversity and functionality (114, 115).

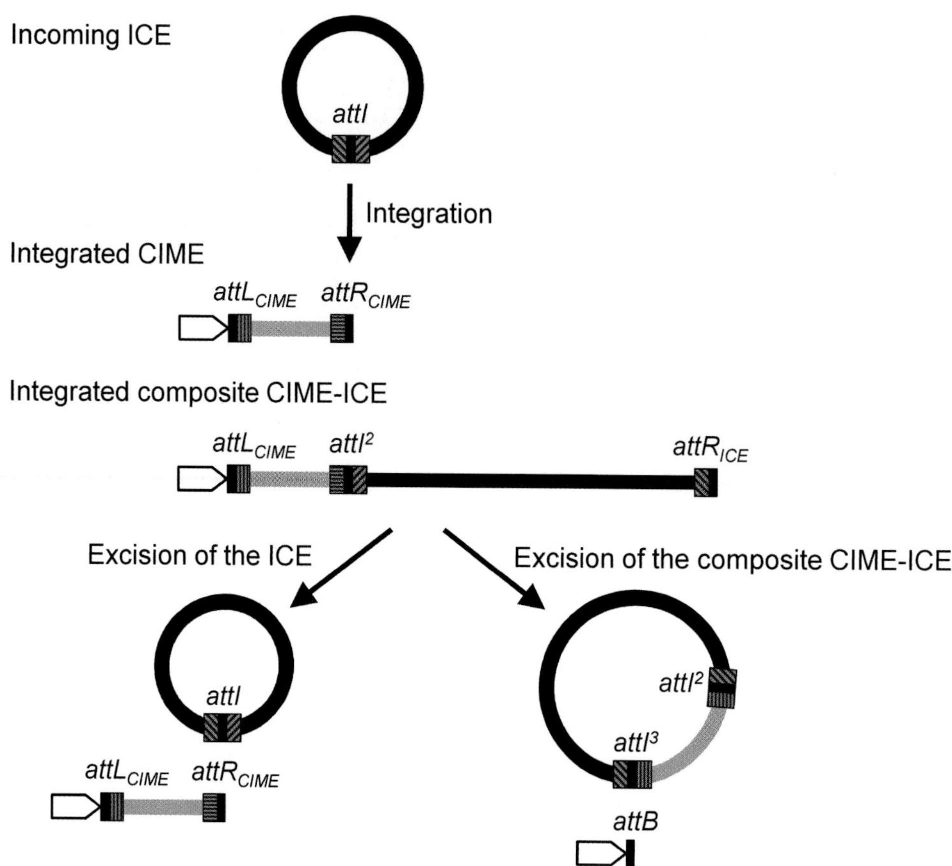

Figure 7 Model for accretion–mobilization of related ICEs and CIMEs. The ICE and the CIME are schematized in black and gray, respectively. The attachment sites (*att*) are symbolized by dashed rectangles with the direct repeats in black. The integration site is located at the 3′ end of a hypothetical gene represented by a white arrowed box. For accretion, an incoming ICE resulting from acquisition by conjugation integrates by site-specific recombination between its *attI* site and the closely related *attR* (or *attL*) site of a nonautonomous CIME. The resulting CIME–ICE composite structure carries the *attL* of the CIME ($attL_{CIME}$), the *attR* of the ICE ($attL_{ICE}$), and an internal $attI_2$ site. The subsequent recombination between $attL_{CIME}$ and $attR_{ICE}$ leads to the *cis*-mobilization of the CIME by the ICE, whereas the recombination between $attI_2$ and $attR_{ICE}$ leads to the excision of the ICE, each element being able to conjugate toward a recipient cell.
doi:10.1128/microbiolspec.MDNA3-0008-2014.f7

CONCLUDING REMARKS

Currently, ICEs play a crucial role in the mobilization and dissemination of multidrug-resistant genes among pathogenic microorganisms. ICEs have been primarily identified because of and studied for that same reason. However, the large prevalence in virtually all bacterial genomes of ICEs or ICE-like sequences suggests that they have been around and have been circulating among bacteria largely before the beginning of the antibiotic era, likely contributing extensively to the evolution and shaping of bacterial genomes (8, 9, 15, 16, 17). This review has covered a sample of three ICE families representing a minuscule fraction of the fascinating bestiary of mobile genetic elements that are concealed in bacterial chromosomes. The ability of SXT/R391 ICEs to drive the mobilization of chromosomal DNA and genomic islands likely represents the tip of the iceberg of the potential of horizontal gene transfer mediated by ICEs. For most ICEs, this potential remains unexplored and ought to be investigated. Exciting future directions concern the impact that genes carried by ICEs can have on the behavior of the host. Besides antibiotic resistance and transfer functions, most genes carried by ICEs found in Gram-negative bacteria have no predicted function. Furthermore, ICE genes coding for transcriptional activators and repressors could potentially alter the host gene expression. Finally, ICE genes coding for enzymes catalyzing c-di-GMP second-messenger synthesis or degradation can drastically alter the regulatory pathways triggering the transition between a free-living motile lifestyle to a biofilm lifestyle. From there, it is obvious that, despite recent important progress, still relatively little is known about ICEs, their biology, and the extent of their impact on the evolution and behavior of their hosts.

Acknowledgments. *We thank Alain Lavigueur for his insightful comments during the preparation of this manuscript. V.B. holds a Canada Research Chair in molecular bacterial genetics.*

Citation. Carraro N, Burrus V. 2014. Biology of three ICE families: SXT/R391, ICEBs1, and ICESt1/ICESt3. Microbiol Spectrum 2(6):MDNA3-0008-2014.

References

1. **Salyers AA, Shoemaker NB, Stevens AM, Li LY.** 1995. Conjugative transposons: an unusual and diverse set of integrated gene transfer elements. *Microbiol Rev* **59:** 579–590.
2. **Scott JR, Churchward GG.** 1995. Conjugative transposition. *Annu Rev Microbiol* **49:**367–397.
3. **Clewell DB, Flannagan SE, Jaworski DD.** 1995. Unconstrained bacterial promiscuity: the Tn*916*-Tn*1545* family of conjugative transposons. *Trends Microbiol* **3:** 229–336.
4. **Rice LB.** 1998. Tn*916* family conjugative transposons and dissemination of antimicrobial resistance determinants. *Antimicrob Agents Chemother* **42:**1871–1877.
5. **Burrus V, Pavlovic G, Decaris B, Guédon G.** 2002. Conjugative transposons: the tip of the iceberg. *Mol Microbiol* **46:**601–610.
6. **Waldor MK, Tschäpe H, Mekalanos JJ.** 1996. A new type of conjugative transposon encodes resistance to sulfamethoxazole, trimethoprim, and streptomycin in *Vibrio cholerae* O139. *J Bacteriol* **178:**4157–4165.
7. **Roberts AP, Mullany P.** 2011. Tn*916*-like genetic elements: a diverse group of modular mobile elements conferring antibiotic resistance. *FEMS Microbiol Rev* **35:** 856–71.
8. **Wozniak RAF, Waldor MK.** 2010. Integrative and conjugative elements: mosaic mobile genetic elements enabling dynamic lateral gene flow. *Nat Rev Microbiol* **8:** 552–563.
9. **Burrus V, Waldor MK.** 2004. Shaping bacterial genomes with integrative and conjugative elements. *Res Microbiol* **155:**376–386.
10. **Waters JL, Salyers AA.** 2013. Regulation of CTnDOT conjugative transfer is a complex and highly coordinated series of events. *MBio* **4:**e00569.
11. **Guérillot R, Da Cunha V, Sauvage E, Bouchier C, Glaser P.** 2013. Modular evolution of TnGBSs, a new family of integrative and conjugative elements associating insertion sequence transposition, plasmid replication, and conjugation for their spreading. *J Bacteriol* **195:** 1979–1990.
12. **Brochet M, Da Cunha V, Couvé E, Rusniok C, Trieu-Cuot P, Glaser P.** 2009. Atypical association of DDE transposition with conjugation specifies a new family of mobile elements. *Mol Microbiol* **71:**948–959.
13. **Marenda M, Barbe V, Gourgues G, Mangenot S, Sagne E, Citti C.** 2006. A new integrative conjugative element occurs in *Mycoplasma agalactiae* as chromosomal and free circular forms. *J Bacteriol* **188:**4137–4141.
14. **Dordet Frisoni E, Marenda MS, Sagné E, Nouvel LX, Guérillot R, Glaser P, Blanchard A, Tardy F, Sirand-Pugnet P, Baranowski E, Citti C.** 2013. ICEA of *Mycoplasma agalactiae*: a new family of self-transmissible integrative elements that confers conjugative properties to the recipient strain. *Mol Microbiol* **89:**1226–1239.
15. **Bordeleau E, Ghinet MG, Burrus V.** 2012. Diversity of integrating conjugative elements in Actinobacteria: coexistence of two mechanistically different DNA-translocation systems. *Mob Genet Elements* **2:**119–124.
16. **Guglielmini J, Quintais L, Garcillán-Barcia MP, de la Cruz F, Rocha EPC.** 2011. The repertoire of ICE in prokaryotes underscores the unity, diversity, and ubiquity of conjugation. *PLoS Genet* **7:**e1002222.
17. **Ghinet MG, Bordeleau E, Beaudin J, Brzezinski R, Roy S, Burrus V.** 2011. Uncovering the prevalence and diversity of integrating conjugative elements in Actinobacteria. *PLoS One* **6:**e27846.
18. **Alvarez-Martinez CE, Christie PJ.** 2009. Biological diversity of prokaryotic type IV secretion systems. *Microbiol Mol Biol Rev* **73:**775–808.
19. **Fronzes R, Christie PJ, Waksman G.** 2009. The structural biology of type IV secretion systems. *Nat Rev Microbiol* **7:**703–714.
20. **De la Cruz F, Frost LS, Meyer RJ, Zechner EL.** 2010. Conjugative DNA metabolism in Gram-negative bacteria. *FEMS Microbiol Rev* **34:**18–40.
21. **Smillie C, Garcillán-Barcia MP, Francia MV, Rocha EPC, de la Cruz F.** 2010. Mobility of plasmids. *Microbiol Mol Biol Rev* **74:**434–452.
22. **Te Poele EM, Bolhuis H, Dijkhuizen L.** 2008. Actinomycete integrative and conjugative elements. *Antonie Van Leeuwenhoek* **94:**127–143.
23. **Garcillán-Barcia MP, Francia MV, de la Cruz F.** 2009. The diversity of conjugative relaxases and its application in plasmid classification. *FEMS Microbiol Rev* **33:** 657–687.
24. **Vogelmann J, Ammelburg M, Finger C, Guezguez J, Linke D, Flötenmeyer M, Stierhof Y-D, Wohlleben W, Muth G.** 2011. Conjugal plasmid transfer in *Streptomyces* resembles bacterial chromosome segregation by FtsK/SpoIIIE. *EMBO J* **30:**2246–2254.
25. **Reinhard F, Miyazaki R, Pradervand N, van der Meer JR.** 2013. Cell differentiation to "mating bodies" induced by an integrating and conjugative element in free-living bacteria. *Curr Biol* **23:**255–259.
26. **Coetzee JN, Datta N, Hedges RW.** 1972. R factors from *Proteus rettgeri*. *J Gen Microbiol* **72:**543–552.

27. **Hochhut B, Lotfi Y, Mazel D, Faruque SM, Woodgate R, Waldor MK.** 2001. Molecular analysis of antibiotic resistance gene clusters in *Vibrio cholerae* O139 and O1 SXT constins. *Antimicrob Agents Chemother* **45:** 2991–3000.

28. **Pembroke JT, MacMahon C, McGrath B.** 2002. The role of conjugative transposons in the *Enterobacteriaceae*. **59:**2055–2064.

29. **Hochhut B, Beaber JW, Woodgate R, Waldor MK.** 2001. Formation of chromosomal tandem arrays of the SXT element and R391, two conjugative chromosomally integrating elements that share an attachment site. *J Bacteriol* **183:**1124–1132.

30. **Beaber JW, Burrus V, Hochhut B, Waldor MK.** 2002. Comparison of SXT and R391, two conjugative integrating elements: definition of a genetic backbone for the mobilization of resistance determinants. **59:**2065–2070.

31. **Wozniak RAF, Fouts DE, Spagnoletti M, Colombo MM, Ceccarelli D, Garriss G, Déry C, Burrus V, Waldor MK.** 2009. Comparative ICE genomics: insights into the evolution of the SXT/R391 family of ICEs. *PLoS Genet* **5:**e1000786.

32. **Burrus V, Marrero J, Waldor MK.** 2006. The current ICE age: biology and evolution of SXT-related integrating conjugative elements. *Plasmid* **55:**173–183.

33. **Bordeleau E, Brouillette E, Robichaud N, Burrus V.** 2010. Beyond antibiotic resistance: integrating conjugative elements of the SXT/R391 family that encode novel diguanylate cyclases participate to c-di-GMP signalling in *Vibrio cholerae. Environ Microbiol* **12:**510–523.

34. **Chin C-S, Sorenson J, Harris JB, Robins WP, Charles RC, Jean-Charles RR, Bullard J, Webster DR, Kasarskis A, Peluso P, Paxinos EE, Yamaichi Y, Calderwood SB, Mekalanos JJ, Schadt EE, Waldor MK.** 2011. The origin of the Haitian cholera outbreak strain. *N Engl J Med* **364:**33–42.

35. **Reimer AR, Van Domselaar G, Stroika S, Walker M, Kent H, Tarr C, Talkington D, Rowe L, Olsen-Rasmussen M, Frace M, Sammons S, Dahourou GA, Boncy J, Smith AM, Mabon P, Petkau A, Graham M, Gilmour MW, Gerner-Smidt P.** 2011. Comparative genomics of *Vibrio cholerae* from Haiti, Asia, and Africa. *Emerg Infect Dis* **17:**2113–2121.

36. **Ceccarelli D, Spagnoletti M, Cappuccinelli P, Burrus V, Colombo MM.** 2011. Origin of *Vibrio cholerae* in Haiti. *Lancet Infect Dis* **11:**262.

37. **Osorio CR, Marrero J, Wozniak RAF, Lemos ML, Burrus V, Waldor MK.** 2008. Genomic and functional analysis of ICE*Pda*Spa1, a fish-pathogen-derived SXT-related integrating conjugative element that can mobilize a virulence plasmid. *J Bacteriol* **190:**3353–3361.

38. **Badhai J, Kumari P, Krishnan P, Ramamurthy T, Das SK.** 2013. Presence of SXT integrating conjugative element in marine bacteria isolated from the mucus of the coral *Fungia echinata* from Andaman Sea. *FEMS Microbiol Lett* **338:**118–123.

39. **Harada S, Ishii Y, Saga T, Tateda K, Yamaguchi K.** 2010. Chromosomally encoded blaCMY-2 located on a novel SXT/R391-related integrating conjugative element in a *Proteus mirabilis* clinical isolate. *Antimicrob Agents Chemother* **54:**3545–3550.

40. **Pembroke JT, Piterina AV.** 2006. A novel ICE in the genome of *Shewanella putrefaciens* W3-18-1: comparison with the SXT/R391 ICE-like elements. *FEMS Microbiol Lett* **264:**80–88.

41. **Balado M, Lemos ML, Osorio CR.** 2013. Integrating conjugative elements of the SXT/R391 family from fish-isolated Vibrios encode restriction-modification systems that confer resistance to bacteriophages. *FEMS Microbiol Ecol* **83:**457–467.

42. **Kulaeva OI, Wootton JC, Levine AS, Woodgate R.** 1995. Characterization of the *umu*-complementing operon from R391. *J Bacteriol* **177:**2737–2743.

43. **Welch TJ, Fricke WF, McDermott PF, White DG, Rosso M-L, Rasko DA, Mammel MK, Eppinger M, Rosovitz MJ, Wagner D, Rahalison L, Leclerc JE, Hinshaw JM, Lindler LE, Cebula TA, Carniel E, Ravel J.** 2007. Multiple antimicrobial resistance in plague: an emerging public health risk. *PLoS One* **2:**e309.

44. **Fricke WF, Welch TJ, McDermott PF, Mammel MK, LeClerc JE, White DG, Cebula TA, Ravel J.** 2009. Comparative genomics of the IncA/C multidrug resistance plasmid family. *J Bacteriol* **191:**4750–4757.

45. **Hochhut B, Waldor MK.** 1999. Site-specific integration of the conjugal *Vibrio cholerae* SXT element into *prfC*. *Mol Microbiol* **32:**99–110.

46. **McGrath BM, Pembroke JT.** 2004. Detailed analysis of the insertion site of the mobile elements R997, pMERPH, R392, R705 and R391 in *E. coli* K12. *FEMS Microbiol Lett* **237:**19–26.

47. **Burrus V, Waldor MK.** 2003. Control of SXT integration and excision. *J Bacteriol* **185:**5045–5054.

48. **McLeod SM, Burrus V, Waldor MK.** 2006. Requirement for *Vibrio cholerae* integration host factor in conjugative DNA transfer. *J Bacteriol* **188:**5704–5711.

49. **Taviani E, Spagnoletti M, Ceccarelli D, Haley BJ, Hasan NA, Chen A, Colombo MM, Huq A, Colwell RR.** 2012. Genomic analysis of ICE*Vch*Ban8: an atypical genetic element in *Vibrio cholerae. FEBS Lett* **586:** 1617–1621.

50. **Ceccarelli D, Daccord A, Rene M, Burrus V.** 2008. Identification of the origin of transfer (oriT) and a new gene required for mobilization of the SXT/R391 family of integrating conjugative elements. *J Bacteriol* **190:** 5328–5338.

51. **Carraro N, Sauvé M, Matteau D, Lauzon G, Rodrigue S, Burrus V.** 2014. Development of pVCR94ΔX from *Vibrio cholerae*, a prototype for studying multidrug resistant IncA/C conjugative plasmids. *Front Microbiol* **5:**44.

52. **Burrus V, Waldor MK.** 2004. Formation of SXT tandem arrays and SXT-R391 hybrids. *J Bacteriol* **186:** 2636–2645.

53. **Marrero J, Waldor MK.** 2005. Interactions between inner membrane proteins in donor and recipient cells limit conjugal DNA transfer. *Dev Cell* **8:**963–970.

54. **Marrero J, Waldor MK.** 2007. Determinants of entry exclusion within Eex and TraG are cytoplasmic. *J Bacteriol* **189:**6469–6473.

55. **Marrero J, Waldor MK.** 2007. The SXT/R391 family of integrative conjugative elements is composed of two exclusion groups. *J Bacteriol* **189:**3302–3305.

56. **Beaber JW, Hochhut B, Waldor MK.** 2002. Genomic and functional analyses of SXT, an integrating antibiotic resistance gene transfer element derived from *Vibrio cholerae*. *J Bacteriol* **184:**4259–4269.

57. **McCool JD, Long E, Petrosino JF, Sandler HA, Rosenberg SM, Sandler SJ.** 2004. Measurement of SOS expression in individual *Escherichia coli* K-12 cells using fluorescence microscopy. *Mol Microbiol* **53:** 1343–1357.

58. **Beaber JW, Hochhut B, Waldor MK.** 2004. SOS response promotes horizontal dissemination of antibiotic resistance genes. *Nature* **427:**72–74.

59. **McGrath BM, O'Halloran JA, Pembroke JT.** 2005. Pre-exposure to UV irradiation increases the transfer frequency of the IncJ conjugative transposon-like elements R391, R392, R705, R706, R997 and pMERPH and is *recA*+ dependent. *FEMS Microbiol Lett* **243:** 461–465.

60. **Butala M, Zgur-Bertok D, Busby SJW.** 2009. The bacterial LexA transcriptional repressor. **66:**82–93.

61. **Patel M, Jiang Q, Woodgate R, Cox MM, Goodman MF.** 2010. A new model for SOS-induced mutagenesis: how RecA protein activates DNA polymerase V. *Crit Rev Biochem Mol Biol* **45:**171–184.

62. **Beaber JW, Waldor MK.** 2004. Identification of operators and promoters that control SXT conjugative transfer. *J Bacteriol* **186:**5945–5949.

63. **Garriss G, Waldor MK, Burrus V.** 2009. Mobile antibiotic resistance encoding elements promote their own diversity. *PLoS Genet* **5:**11.

64. **Garriss G, Poulin-Laprade D, Burrus V.** 2013. DNA damaging agents induce the RecA-independent homologous recombination functions of integrating conjugative elements of the SXT/R391 family. *J Bacteriol.* **195:** 1991–2003.

65. **Wozniak RAF, Waldor MK.** 2009. A toxin-antitoxin system promotes the maintenance of an integrative conjugative element. *PLoS Genet* **5:**e1000439.

66. **Daccord A, Ceccarelli D, Burrus V.** 2010. Integrating conjugative elements of the SXT/R391 family trigger the excision and drive the mobilization of a new class of Vibrio genomic islands. *Mol Microbiol* **78:**576–588.

67. **Daccord A, Mursell M, Poulin-Laprade D, Burrus V.** 2012. Dynamics of the SetCD-regulated integration and excision of genomic islands mobilized by integrating conjugative elements of the SXT/R391 family. *J Bacteriol* **194:**5794–5802.

68. **Hochhut B, Marrero J, Waldor MK.** 2000. Mobilization of plasmids and chromosomal DNA mediated by the SXT element, a constin found in *Vibrio cholerae* O139. *J Bacteriol* **182:**2043–2047.

69. **Daccord A, Ceccarelli D, Rodrigue S, Burrus V.** 2013. Comparative analysis of mobilizable genomic islands. *J Bacteriol* **195:**606–614.

70. **Burrus V, Pavlovic G, Decaris B, Guédon G.** 2002. The ICESt1 element of *Streptococcus thermophilus* belongs to a large family of integrative and conjugative elements that exchange modules and change their specificity of integration. *Plasmid* **48:**77–97.

71. **Earl AM, Losick R, Kolter R.** 2007. *Bacillus subtilis* genome diversity. *J Bacteriol* **189:**1163–1170.

72. **Berkmen MB, Lee CA, Loveday E-K, Grossman AD.** 2010. Polar positioning of a conjugation protein from the integrative and conjugative element ICE*Bs1* of *Bacillus subtilis*. *J Bacteriol* **192:**38–45.

73. **Grohmann E.** 2010. Conjugative transfer of the integrative and conjugative element ICE*Bs1* from *Bacillus subtilis* likely initiates at the donor cell pole. *J Bacteriol* **192:**23–25.

74. **Grohmann E, Muth G, Espinosa M.** 2003. Conjugative plasmid transfer in gram-positive bacteria. *Microbiol Mol Biol Rev* **67:**277–301.

75. **Lee CA, Auchtung JM, Monson RE, Grossman AD.** 2007. Identification and characterization of *int* (integrase), *xis* (excisionase) and chromosomal attachment sites of the integrative and conjugative element ICE*Bs1* of *Bacillus subtilis*. *Mol Microbiol* **66:**1356–1369.

76. **Menard KL, Grossman AD.** 2013. Selective pressures to maintain attachment site specificity of integrative and conjugative elements. *PLoS Genet* **9:**e1003623.

77. **Auchtung JM, Lee CA, Monson RE, Lehman AP, Grossman AD.** 2005. Regulation of a *Bacillus subtilis* mobile genetic element by intercellular signaling and the global DNA damage response. *Proc Natl Acad Sci U S A* **102:**12554–12559.

78. **Auchtung JM, Lee CA, Garrison KL, Grossman AD.** 2007. Identification and characterization of the immunity repressor (ImmR) that controls the mobile genetic element ICE*Bs1* of *Bacillus subtilis*. *Mol Microbiol* **64:** 1515–1528.

79. **Rocco JM, Churchward G.** 2006. The integrase of the conjugative transposon Tn*916* directs strand- and sequence-specific cleavage of the origin of conjugal transfer, *oriT*, by the endonuclease Orf20. *J Bacteriol* **188:**2207–2213.

80. **Thomas J, Lee CA, Grossman AD.** 2013. A conserved helicase processivity factor is needed for conjugation and replication of an integrative and conjugative element. *PLoS Genet* **9:**e1003198.

81. **Babic A, Berkmen MB, Lee CA, Grossman AD.** 2011. Efficient gene transfer in bacterial cell chains. *MBio* **2:** e00027.

82. **Iyer LM, Makarova KS, Koonin EV, Aravind L.** 2004. Comparative genomics of the FtsK-HerA superfamily of pumping ATPases: implications for the origins of chromosome segregation, cell division and viral capsid packaging. *Nucleic Acids Res* **32:**5260–5279.

83. **Lee CA, Thomas J, Grossman AD.** 2012. The *Bacillus subtilis* conjugative transposon ICE*Bs1* mobilizes

plasmids lacking dedicated mobilization functions. *J Bacteriol* **194:**3165–3172.

84. **Fukushima T, Kitajima T, Yamaguchi H, Ouyang Q, Furuhata K, Yamamoto H, Shida T, Sekiguchi J.** 2008. Identification and characterization of novel cell wall hydrolase CwlT: a two-domain autolysin exhibiting *N*-acetylmuramidase and DL-endopeptidase activities. *J Biol Chem* **283:**11117–11125.

85. **Dewitt T, Grossman AD.** 2014. The bifunctional cell wall hydrolase CwlT is needed for conjugation of the integrative and conjugative element ICE*Bs1* in *Bacillus subtilis* and *B. anthracis*. *J Bacteriol.* **196:**1588–1596.

86. **Bose B, Grossman AD.** 2011. Regulation of horizontal gene transfer in *Bacillus subtilis* by activation of a conserved site-specific protease. *J Bacteriol* **193:**22–29.

87. **Bose B, Auchtung JM, Lee CA, Grossman AD.** 2008. A conserved anti-repressor controls horizontal gene transfer by proteolysis. *Mol Microbiol* **70:**570–582.

88. **Madsen PL, Johansen AH, Hammer K, Brøndsted L.** 1999. The genetic switch regulating activity of early promoters of the temperate lactococcal bacteriophage TP901-1. *J Bacteriol* **181:**7430–7438.

89. **Perego M, Brannigan JA.** 2001. Pentapeptide regulation of aspartyl-phosphate phosphatases. *Peptides* **22:**1541–7.

90. **Bringel F, Van Alstine GL, Scott JR.** 1991. A host factor absent from *Lactococcus lactis* subspecies *lactis* MG1363 is required for conjugative transposition. *Mol Microbiol* **5:**2983–2993.

91. **Carraro N, Libante V, Morel C, Decaris B, Charron-Bourgoin F, Leblond P, Guédon G.** 2011. Differential regulation of two closely related integrative and conjugative elements from *Streptococcus thermophilus*. *BMC Microbiol* **11:**238.

92. **Strauch MA, Ballar P, Rowshan AJ, Zoller KL.** 2005. The DNA-binding specificity of the *Bacillus anthracis* AbrB protein. *Microbiology* **151:**1751–1759.

93. **Albano M, Smits WK, Ho LTY, Kraigher B, Mandic-Mulec I, Kuipers OP, Dubnau D.** 2005. The Rok protein of *Bacillus subtilis* represses genes for cell surface and extracellular functions. *J Bacteriol* **187:**2010–2019.

94. **Smits WK, Grossman AD.** 2010. The transcriptional regulator Rok binds A+T-rich DNA and is involved in repression of a mobile genetic element in *Bacillus subtilis*. *PLoS Genet* **6:**e1001207.

95. **Sitkiewicz I, Green NM, Guo N, Mereghetti L, Musser JM.** 2011. Lateral gene transfer of streptococcal ICE element RD2 (region of difference 2) encoding secreted proteins. *BMC Microbiol* **11:**65.

96. **Dimopoulou ID, Russell JE, Mohd-Zain Z, Herbert R, Crook DW.** 2002. Site-specific recombination with the chromosomal tRNA(Leu) gene by the large conjugative Haemophilus resistance plasmid. *Antimicrob Agents Chemother* **46:**1602–1603.

97. **Kiewitz C, Larbig K, Klockgether J, Weinel C, Tümmler B.** 2000. Monitoring genome evolution *ex vivo*: reversible chromosomal integration of a 106 kb plasmid at two tRNA(Lys) gene loci in sequential *Pseudomonas aeruginosa* airway isolates. *Microbiology* **146:**2365–2373.

98. **Lee CA, Babic A, Grossman AD.** 2010. Autonomous plasmid-like replication of a conjugative transposon. *Mol Microbiol* **75:**268–279.

99. **Grohmann E.** 2010. Autonomous plasmid-like replication of *Bacillus* ICE*Bs1*: a general feature of integrative conjugative elements? *Mol Microbiol* **75:**261–263.

100. **Roussel Y, Bourgoin F, Guédon G, Pébay M, Decaris B.** 1997. Analysis of the genetic polymorphism between three *Streptococcus thermophilus* strains by comparing their physical and genetic organization. *Microbiology* **143:**1335–1343.

101. **Burrus V, Roussel Y, Decaris B, Guédon G.** 2000. Characterization of a novel integrative element, ICESt1, in the lactic acid bacterium *Streptococcus thermophilus*. *Appl Environ Microbiol* **66:**1749–1753.

102. **Pavlovic G, Burrus V, Gintz B, Decaris B, Guédon G.** 2004. Evolution of genomic islands by deletion and tandem accretion by site-specific recombination: ICESt1-related elements from *Streptococcus thermophilus*. *Microbiology* **150:**759–774.

103. **Bellanger X, Morel C, Gonot F, Puymege A, Decaris B, Guédon G.** 2011. Site-specific accretion of an integrative conjugative element together with a related genomic island leads to *cis* mobilization and gene capture. *Mol Microbiol* **81:**912–925.

104. **Bellanger X, Morel C, Decaris B, Guédon G.** 2007. Derepression of excision of integrative and potentially conjugative elements from *Streptococcus thermophilus* by DNA damage response: implication of a cI-related repressor. *J Bacteriol* **189:**1478–1481.

105. **Bellanger X, Roberts AP, Morel C, Choulet F, Pavlovic G, Mullany P, Decaris B, Guédon G.** 2009. Conjugative transfer of the integrative conjugative elements ICE*St1* and ICE*St3* from *Streptococcus thermophilus*. *J Bacteriol* **191:**2764–2775.

106. **Burrus V, Bontemps C, Decaris B, Guédon G.** 2001. Characterization of a novel type II restriction-modification system, Sth368I, encoded by the integrative element ICE*St1* of *Streptococcus thermophilus* CNRZ368. *ApplEnviron Microbiol* **67:**1522–1528.

107. **Fontaine L, Boutry C, de Frahan MH, Delplace B, Fremaux C, Horvath P, Boyaval P, Hols P.** 2010. A novel pheromone quorum-sensing system controls the development of natural competence in *Streptococcus thermophilus* and *Streptococcus salivarius*. *J Bacteriol* **192:**1444–1454.

108. **Jaworski DD, Clewell DB.** 1995. A functional origin of transfer (oriT) on the conjugative transposon Tn916. *J Bacteriol* **177:**6644–6651.

109. **Bellanger X, Payot S, Leblond-Bourget N, Guédon G.** 2014. Conjugative and mobilizable genomic islands in bacteria: evolution and diversity. *FEMS Microbiol Rev* **38:**720–760.

110. **Puymège A, Bertin S, Chuzeville S, Guédon G, Payot S.** 2013. Conjugative transfer and *cis*-mobilization of a genomic island by an integrative and conjugative element of *Streptococcus agalactiae*. *J Bacteriol* **195:**1142–1151.

111. **Green NM, Zhang S, Porcella SF, Nagiec MJ, Barbian KD, Beres SB, LeFebvre RB, Musser JM.** 2005.

Genome sequence of a serotype M28 strain of group a streptococcus: potential new insights into puerperal sepsis and bacterial disease specificity. *J Infect Dis* **192:** 760–770.

112. **Brochet M, Couvé E, Glaser P, Guédon G, Payot S.** 2008. Integrative conjugative elements and related elements are major contributors to the genome diversity of *Streptococcus agalactiae. J Bacteriol* **190:** 6913–6917.

113. **Chuzeville S, Puymège A, Madec J-Y, Haenni M, Payot S.** 2012. Characterization of a new CAMP factor carried by an integrative and conjugative element in *Streptococcus agalactiae* and spreading in *Streptococci. PLoS One* **7:**e48918.

114. **Li M, Shen X, Yan J, Han H, Zheng B, Liu D, Cheng H, Zhao Y, Rao X, Wang C, Tang J, Hu F, Gao GF.** 2011. GI-type T4SS-mediated horizontal transfer of the 89K pathogenicity island in epidemic *Streptococcus suis* serotype 2. *Mol Microbiol* **79:**1670–1683.

115. **Wu Z, Li M, Wang C, Li J, Lu N, Zhang R, Jiang Y, Yang R, Liu C, Liao H, Gao GF, Tang J, Zhu B.** 2011. Probing genomic diversity and evolution of *Streptococcus suis* serotype 2 by NimbleGen tiling arrays. *BMC Genomics* **12:**219.

Programmed Rearrangements III

Mobile DNA, 3rd Edition
Nancy L. Craig, Michael Chandler, Martin Gellert, Alan M. Lambowitz, Phoebe A. Rice and Suzanne Sandmeyer

doi:10.1128/microbiolspec.MDNA3-0041-2014

David B. Roth[1]

14 V(D)J Recombination: Mechanism, Errors, and Fidelity

GENERATION OF ANTIGEN-RECEPTOR DIVERSITY: A DOUBLE-EDGED SWORD

The realization, now more than half a century ago, that B cells can generate antibodies to an astounding variety of chemical structures sparked intense interest in the "generation of diversity" question (reviewed in reference 1). The correct solution to this puzzle turned out to be both surprising and simple: the exons encoding the antigen-binding portions of the receptor (the so-called variable regions) are assembled by chromosomal breakage and rejoining in developing lymphocytes (2). Immunoglobulins and T-cell receptors are composed of two polypeptide chains, each of which contributes to the antigen-binding domain. The exons encoding the antigen-binding domains are assembled from so-called V (variable), D (diversity), and J (joining) gene segments by "cut and paste" DNA rearrangements. This process, termed V(D)J recombination, chooses a pair of segments, introduces double-strand breaks adjacent to each segment, deletes (or, in selected cases, inverts) the intervening DNA, and ligates the segments together (Fig. 1). Rearrangements occur in an ordered fashion, with D-to-J joining proceeding before a V segment is joined to the rearranged D-J segments. This process of combinatorial assembly—choosing one segment of each type from several (sometimes many) possibilities—is the fundamental engine driving antigen-receptor diversity in mammals. Diversity is tremendously amplified by the characteristic variability at the junctions (loss or gain of small numbers of nucleotides) between the various segments. This process leverages a relatively small investment in germline coding capacity into an almost limitless repertoire of potential antigen-binding specificities.

This elegant process does, however, have a potential downside. A system that must break chromosomal DNA several times in order to generate a functional antigen receptor gene—many millions of times over the lifetime of an organism—creates significant opportunities for error. The necessity of enforcing a high degree of fidelity in V(D)J recombination has been recognized for decades (reviewed in reference 3). Aberrant V(D)J recombination events do occur, and they can be life-threatening, underlying the genesis of common lymphoid neoplasms (4, 5, 6, 7), as discussed below. Recent genome-wide analyses of lymphoid neoplasms have revealed V(D)J recombination-driven oncogenic events and have intensified interest in the regulatory mechanisms responsible for ensuring fidelity during V(D)J recombination. This chapter reviews the basic aspects of V(D)J recombination, the mechanisms responsible for aberrant rearrangements, and the types of

[1]Department of Pathology and Laboratory Medicine and the Center for Personalized Diagnostics, Perelman School of Medicine of the University of Pennsylvania, Philadelphia, Pennsylvania 19104.

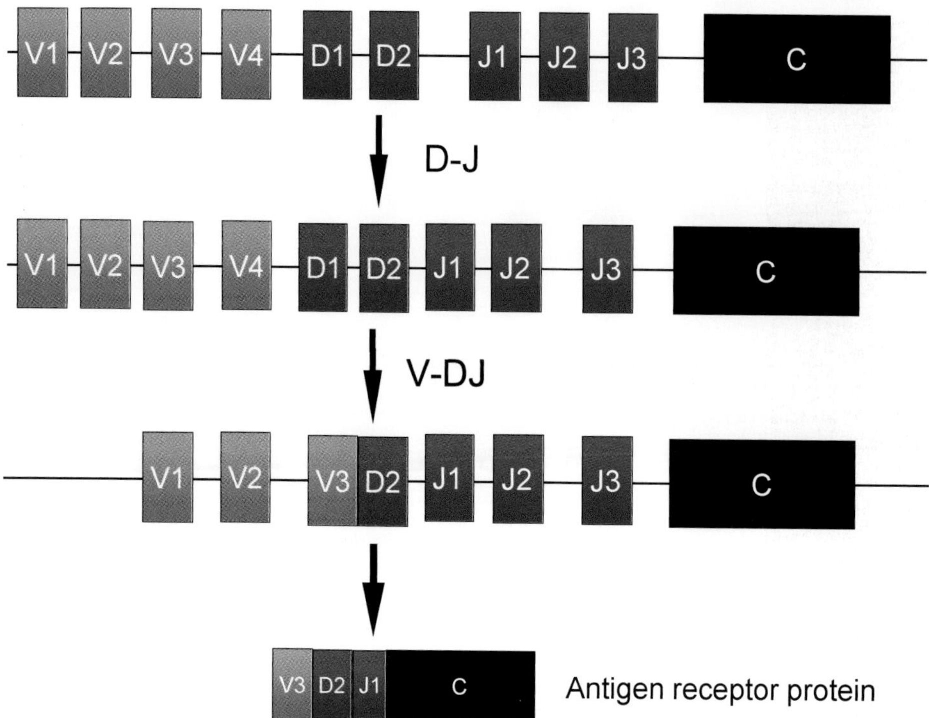

Figure 1 Antigen receptor variable exons are assembled by V(D)J recombination. Assembly of a complete variable exon occurs in two steps (in the case of an Ig heavy chain gene or a T-cell receptor β or δ gene), as shown. First, a D and a J segment are chosen from among several possibilities and are brought together to form a D-J rearrangement. A V region is then selected and joined with the D-J rearrangement to form a complete VDJ exon. Ig light chain genes and T-cell receptor α and γ genes rearrange in a single step, involving V-J recombination, as D segments are absent from these loci. C denotes the constant region exon. doi:10.1128/microbiolspec.MDNA3-0041-2014.f1

events uncovered in recent analyses of tumor genomes. Recent advances in understanding the mechanisms responsible for safeguarding genomic integrity during V(D)J recombination are also discussed.

THE NORMAL V(D)J RECOMBINATION MECHANISM

This section briefly overviews the normal mechanism of V(D)J recombination (for more details, see references 8 and 9). The V(D)J recombinase recognizes conserved DNA sequence elements, termed recombination signal sequences (RSSs), located adjacent to each V, D, and J coding segment. RSSs consist of conserved heptamer and nonamer elements, separated by 12 or 23 nucleotides of less conserved "spacer" sequence (Fig. 2). Efficient recombination only occurs between RSSs with different spacer lengths (the "12/23 rule"). Additional restrictions are imparted at some antigen-receptor loci by other DNA sequence features (the so-called "beyond 12/23 rule") (10). The RSSs are the only DNA segments required to allow V(D)J recombination to occur on artificial substrates, and their relative orientation determines whether the reaction proceeds by inversion or by deletion (Fig. 3) (11). An additional outcome, occasionally observed at antigen receptor loci, is formation of a "hybrid joint," in which a coding segment is joined to an RSS (12) (Fig. 3). Hybrid joints do not contribute to antigen-receptor diversity, nor do they appear to play a role in oncogenic transformation.

Nucleotide sequences of natural RSSs display considerable variability. Those with sequences closest to the consensus support the most efficient recombination (13). The first three nucleotides of the heptamer (closest to the coding flank) show the highest sequence conservation and are critical for recombination, whereas the remaining heptamer positions are much less important (13) (Fig. 2). The nonamer sequence conforms less closely to the consensus, with only a few highly conserved positions (particularly the A/T tract), and the nonamer is dispensable under certain conditions *in vitro* (13, 14). Spacer length is critical and can be changed

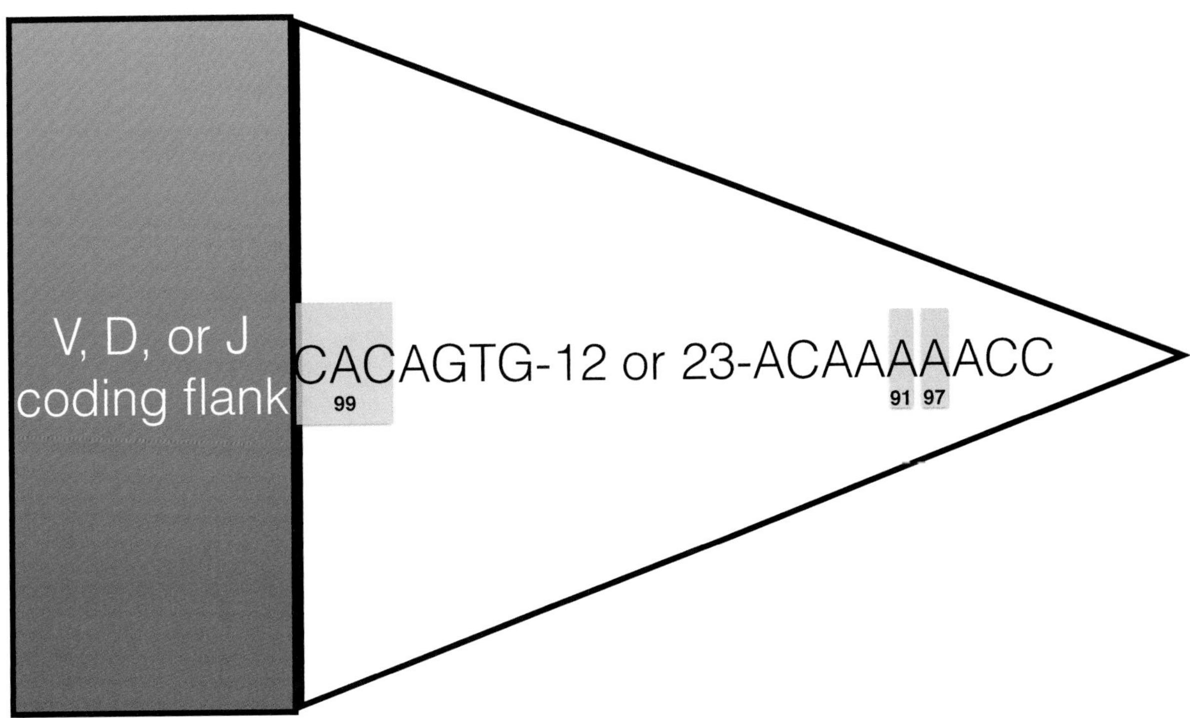

Figure 2 Consensus RSS. The consensus RSS is shown, with the heptamer abutting the coding flank. The most highly conserved positions of the heptamer and nonamer are shaded in red, with conservation (%) given below. Sequence conservation data are from (13).
doi:10.1128/microbiolspec.MDNA3-0041-2014.f2

successfully only in increments that preserve the helical spacing of the nonamer and heptamer elements (13). Other nucleotide sequence features can influence recombination efficiency, most notably the sequence of the coding segment adjacent to the heptamer (the coding flank) (14, 15). This reflects a requirement for DNA distortion during DNA cleavage (14, 15, 16, 17).

The V(D)J recombinase consists of two lymphoid-specific proteins, RAG1 and RAG2 (18), which work together with a non-lymphoid-specific DNA bending factor, HMG1A or HMG1B (19), to carry out DNA cleavage. The *RAG1* and *RAG2* genes are located quite close to each other in all species examined, and their open reading frames are generally encoded in single exons. These observations led to speculation that the V(D)J recombinase may have evolved from an ancestral prokaryotic transposase (20). Indeed, the mechanism of DNA cleavage by RAG1/2 (one-step transesterification) (21) is shared with a class of bacterial transposases, and the RAG proteins can catalyze *bona fide* transposition events (22, 23). Definitive evidence that RAG1/2 indeed evolved from an ancestral transposase remains elusive (24).

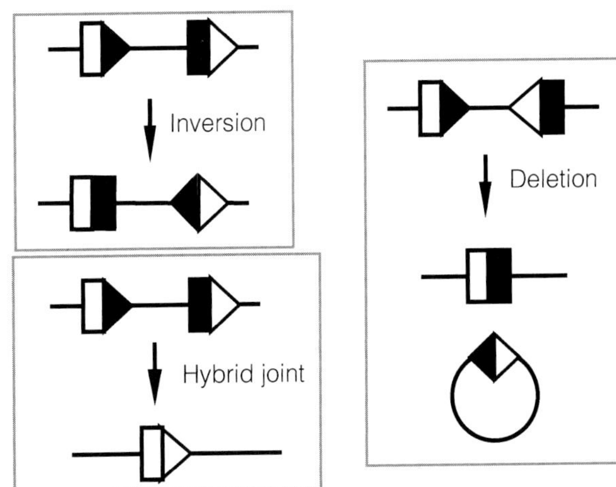

Figure 3 Products of V(D)J recombination. Inversional and deletional recombination are shown in the top portion of the figure. Whether recombination proceeds in a deletional or inversional manner is specified by the relative orientation of the two RSSs. Hybrid joint formation is shown at the bottom of the figure and involves an inappropriate joining of a coding end to a signal end. The reciprocal hybrid joint product, in this case an excised circle, is not shown.
doi:10.1128/microbiolspec.MDNA3-0041-2014.f3

The functional anatomy of the RAG1 and RAG2 proteins has been reviewed recently (8). Mutational studies have defined the minimally functional regions of both proteins. The "core" region of murine RAG1 is comprised of amino acids 384 to 1008 of the 1040-amino acid protein (25, 26) and is sufficient to catalyze V(D)J recombination, albeit with some abnormal features (27). Core RAG1 contains elements important for binding to the nonamer, as well as amino acids required for catalysis of cleavage. Neither specific DNA binding nor catalytic activities have been attributed to RAG2, leading to the view that RAG1 is the catalytic component of the recombinase, with RAG2 serving as an essential cofactor with some regulatory activities (described below). The essential "core" region of RAG2 historically has been defined as amino acids 1 to 383 (of 527) (28, 29). Recent work has shown that the minimal region extends only to amino acid 360 (30), closely coinciding with the predicted six-bladed β-propeller structure (31). This core domain is connected to the C-terminal domain via a flexible acidic hinge. The C terminus, while dispensable for recombination, is important for optimal recombination (32) and for enforcing the proper order of recombination events in developing lymphocytes (33). In its absence, aberrant recombination events are observed (33, 34, 35). The C terminus is also important for maintaining genomic stability (36, 37, 38), as is the acidic hinge (30). Within the C terminus reside a plant homeodomain (PHD) capable of recognizing histone H3K4 trimethylation (39, 40) and a cell-cycle-regulated protein degradation signal (41). These elements are discussed in more detail below.

Efficient cleavage of a DNA substrate requires only RAG1, RAG2, a divalent metal ion, and HMGB1 or HMGB2 (19, 42). Cleavage proceeds via a two-step mechanism (Fig. 4). First, a nick is introduced between the RSS and the coding flank, and the resulting 3′OH group is then used to attack the opposite strand by trans esterification, forming a hairpin coding end and a blunt signal end. This second step is similar to trans-esterification reactions catalyzed by the HIV integrase and by bacterial transposases (21). Whereas nicking can occur independently on either RSS, in the presence of the physiological divalent metal ion Mg^{2+}, trans-esterification requires assembly of a synaptic complex including both a 12- and a 23-RSS (43), providing a molecular basis for the 12/23 rule. After cleavage, the four DNA ends remain associated with the RAG proteins in a post-cleavage complex (Fig. 5), which retains the signal ends more stably than the coding ends (43, 45, 46). This complex is important for the proper rejoining of the broken DNA ends (47, 48, 49) and

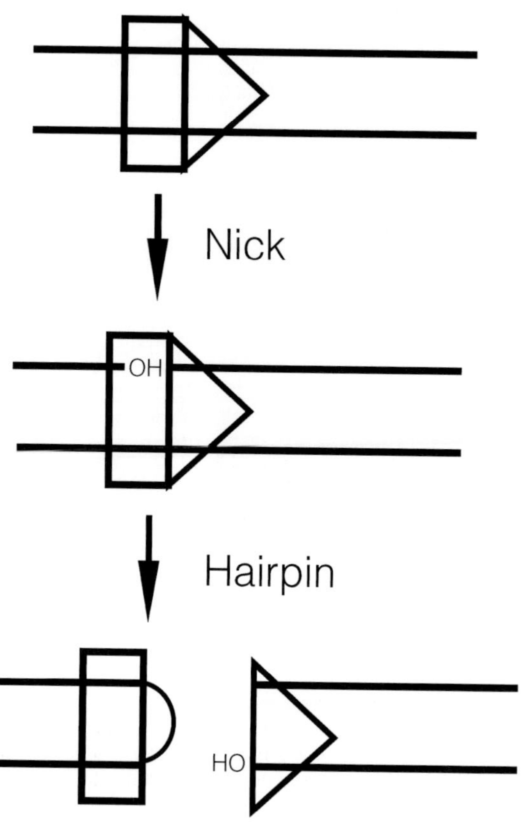

Figure 4 The biochemistry of cleavage. Cleavage occurs at the junction between the heptamer and the adjoining coding flank, and occurs in two steps, as described in the text. doi:10.1128/microbiolspec.MDNA3-0041-2014.f4

shepherds the ends to the classical nonhomologous end-joining (cNHEJ) pathway (45). This function, which prevents access of the ends to the low fidelity, translocation-prone alternative nonhomologous end-joining machinery (aNHEJ) joining pathway (50), is hypothesized to be important for maintenance of genomic stability during V(D)J recombination (30, 36), as discussed below. The C terminus of RAG2 contributes to the stability of the post-cleavage signal end complex (30, 50). Flexibility of the acidic hinge is also important: mutations (including some nucleotide sequence polymorphisms identified in humans) that reduce the negative charge destabilize the RAG-signal end complex and reduce genomic stability in pre-B-cell lines (30).

A characteristic feature of V(D)J recombination is the asymmetric processing of the signal and coding ends. Coding ends are joined with slight variations (small deletions, short insertions), whereas signal ends are generally joined with little or no end processing, so that the majority of the signal joints are perfect heptamer-to-heptamer fusions (51). This asymmetry

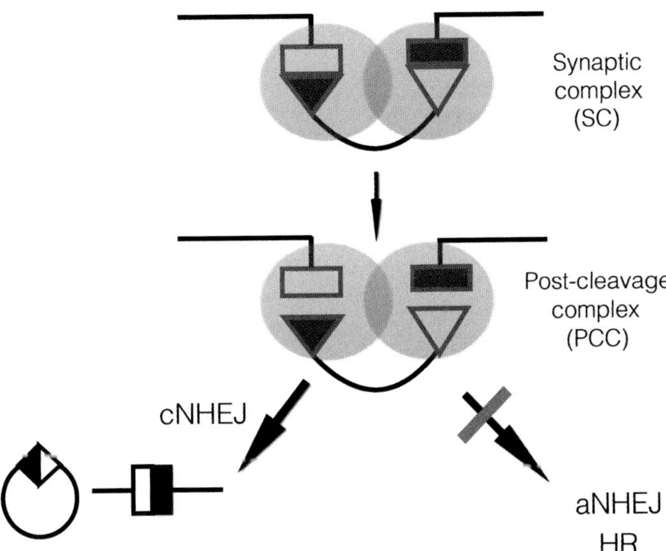

Figure 5 V(D)J recombination overview. Recombination is thought to be initiated by binding of the RAG proteins to a single 12-RSS (not shown), which then captures the 23-RSS to form a synaptic complex (44). RAG1/2 complexes are shown as shaded circles. Double-strand break formation generates a DNA–protein complex, the post-cleavage complex, which then helps to control the "shepherding" of the broken DNA ends to the classical nonhomologous end-joining machinery (cNHEJ; left), preventing the ends from accessing other repair mechanisms such as the alternative nonhomologous end-joining machinery (aNHEJ) or homologous recombination (HR) (right).
doi:10.1128/microbiolspec.MDNA3-0041-2014.f5

may partly be explained by the requirement for additional processing of the coding ends, which are covalently sealed. Other factors may also contribute, including the differential stability of the RAG-coding end and RAG-signal end post-cleavage complexes. Hairpin opening occurs through the action of the Artemis endonuclease, which often cuts off-axis, resulting in short, single-stranded extensions that can give rise to palindromic insertions (P nucleotides) at the coding joint. These are never observed in signal joints, which are formed by blunt end-to-end joining. Formation of ends with single-stranded extensions may also increase opportunities for loss of nucleotides from the coding ends. Another source of extra nucleotides is provided by terminal deoxynucleotidyl transferase, which adds short, nontemplated, GC-rich inserts (N regions) to coding joints (and occasionally to signal joints). The frequent presence of such "microscopic" junctional alterations at coding joints provides a powerful mechanism for amplifying the diversity of antigen-binding sites in T-cell receptor and immunoglobulin molecules.

RAG-generated DNA ends are normally joined by the cNHEJ (reviewed in reference 52). Inactivation of any of the key components of cNHEJ (e.g. Ku70/80, DNA ligase IV, XRCC4) results in a severe impairment of joining. The few junctions formed under these conditions are often (but not always) abnormal, showing excessive deletions, the frequent presence of microhomologies and the occasional presence of abnormally long stretches of extra nucleotides. These features have been considered characteristic of alternative joining pathways, collectively termed aNHEJ (52), although, as discussed below, they are not always observed. aNHEJ is error-prone in two senses: the tendency toward formation of abnormal junctions and an increased propensity for forming gross genomic rearrangements such as chromosome translocations (53, 54, 55).

As noted above, the "shepherding" function of the RAG post-cleavage complex prevents the coding and signal ends from accessing aNHEJ. This was demonstrated by the observation that certain RAG mutants allow much higher levels of aNHEJ with artificial substrates in cultured cells, in both the presence and the absence of functional cNHEJ (45, 50). This may be important for preserving genomic stability, as discussed below.

V(D)J RECOMBINATION ERRORS

As noted above, the rearranging gene strategy that so successfully generates antigen-receptor diversity comes with a price: the potential for generating deleterious genomic rearrangements. Indeed, chromosomal rearrangements involving antigen-receptor loci were reported in both B- and T-cell neoplasms shortly after the discovery of V(D)J recombination (4, 56, 57). With the advent of next-generation sequencing technologies, the genomic landscapes of these malignancies are being studied with increasingly fine resolution. Aberrant events identified in lymphoid neoplasms include chromosome translocations, relatively small (kilobase to megabase) inversions and deletions (5, 6, 58), and reinsertion of excised fragments bounded by signal ends (59). Recent work has highlighted the importance of deletions that affect numerous genes implicated or suspected in tumorigenesis (7, 38, 60, 61) (M. Mijuskovic, S. Lewis, D. Roth, unpublished data). These deletions are RAG mediated, as they occur between pairs of sequences closely resembling RSSs, they often follow the 12/23 rule, and reciprocal signal joints have been detected (M. Mijuskovic, S. Lewis, D. Roth, unpublished data). Thus, different types of V(D)J recombination errors play important roles in initiating oncogenic transformation.

Aberrant V(D)J recombination events observed in lymphoid neoplasms fall into two broad conceptual categories: errors in target recognition (Fig. 6) and errors in joining (Fig. 7). The first type consists of recognition of one authentic RSS and one DNA sequence fortuitously resembling an RSS (termed a "cryptic RSS" or cRSS). Given the relatively small size of RSS sequences and that recombination does not require strict adherence of this sequence to consensus heptamer/nonamer sequences, it is not surprising that cRSSs capable of supporting recombination are present approximately once

a. **Recognition error**: RSS and cRSS *in trans* (translocation)

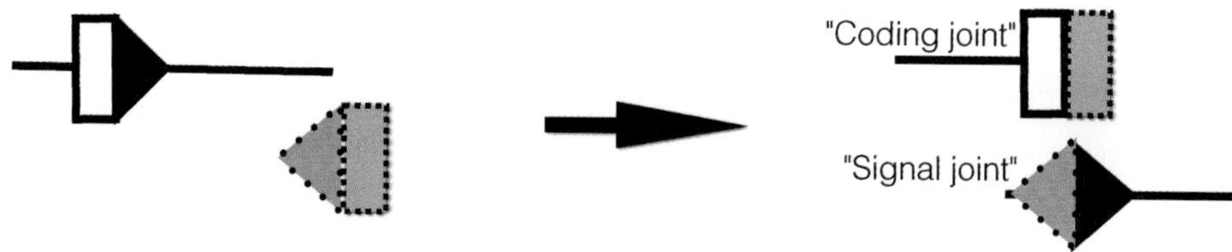

b. **Recognition error**: cRSS pair *in trans* (translocation)

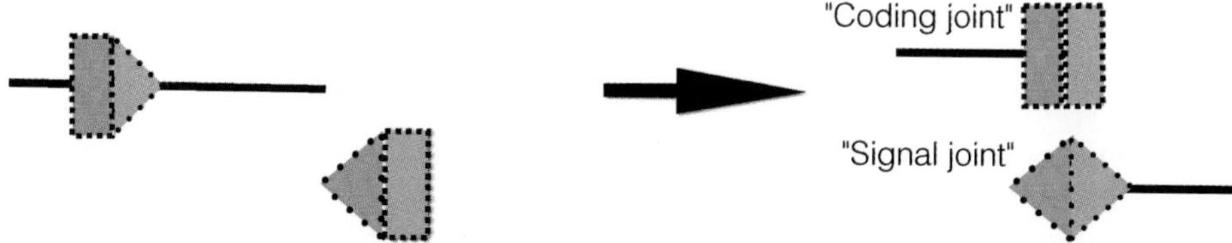

c. **Recognition error**: cRSS pair *in cis* (deletion)

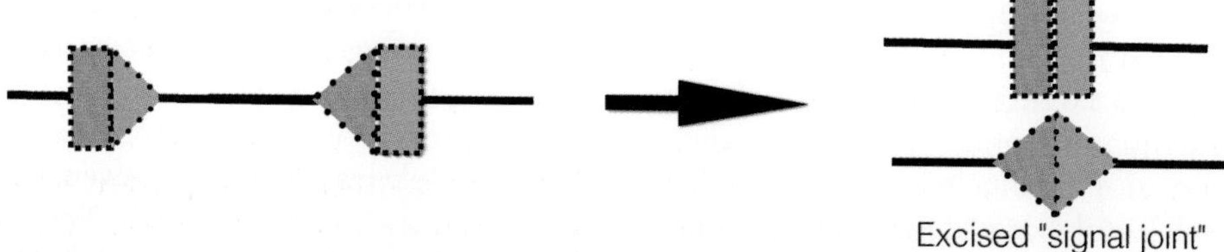

Figure 6 V(D)J recombination: recognition errors. Three types of recognition error are shown. (a) Recombination occurs between an authentic RSS (black triangle) with its associated coding flank (white box) and a cryptic RSS (cRSS; green triangle) with its associated coding flank (orange box), located on a separate DNA molecule. Recombination produces a *trans* rearrangement, with a pseudo-coding joint and a pseudo-signal joint. (b) The recombinase recognizes a pair of cRSSs located on separate DNA molecules. These recombine, generating a reciprocal chromosome translocation. The two products bear a pseudo-coding joint and a pseudo-signal joint. (c) The recombinase recognizes a pair of cRSSs located on the same DNA molecule and generates a deletion, forming a pseudo-coding joint (retained on the chromosome) and an excised circle containing a pseudo-signal joint.
doi:10.1128/microbiolspec.MDNA3-0041-2014.f6

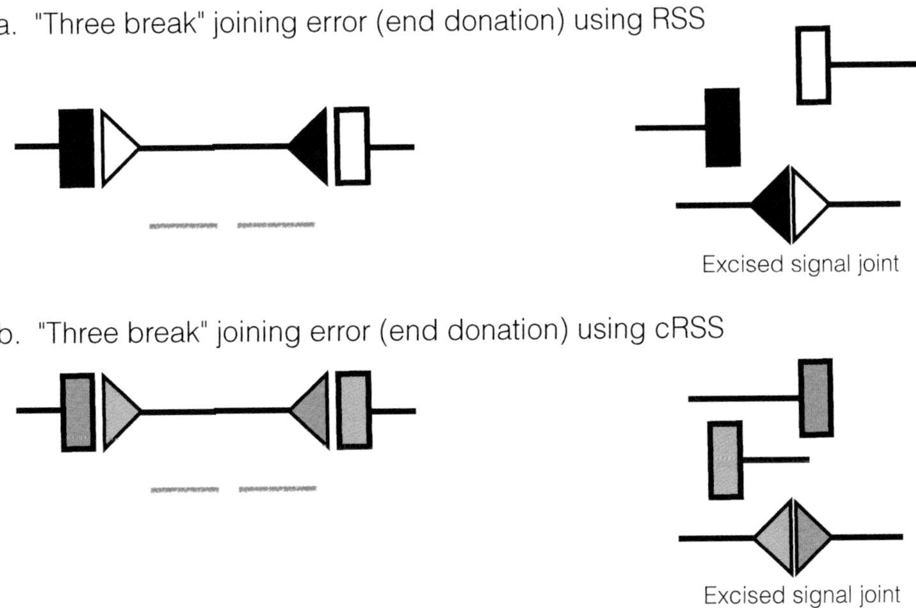

Figure 7 V(D)J recombination: joining errors. Two versions of a three-break event (end donation) are shown. (a) An event occurring between a normal V(D)J recombination event involving authentic RSSs and a chromosome break generated by some other means (break in the red DNA molecule). (b) A similar event, this time involving a V(D)J recombination event involving a pair of cRSSs. doi:10.1128/microbiolspec.MDNA3-0041-2014.f7

per kilobase in random DNA sequences (62). Perhaps the first example of such events was provided by cytogenetic analyses of human lymphoid neoplasms, which revealed chromosome translocations involving authentic RSSs at antigen-receptor loci and cRSSs adjacent to proto-oncogenes (4, 56) (Fig. 6a). These events can cause inappropriate expression of the target gene due to, for example, the presence of transcriptional regulatory elements from the antigen-receptor loci. Recombination events involving a cRSS/RSS pair can also deregulate oncogenes through amplification, likely through a breakage–fusion-bridge mechanism (63).

Events also occur between pairs of cRSSs. These can occur *in trans*, generating a chromosome translocation (Fig. 6b) as in T-cell acute lymphoblastic leukemia (ALL) cases involving translocations between T-cell receptor gene segments and the SCL locus (3) , or *in cis*, generating a deletional "coding joint" and an excised "signal joint" Fig. 6c). Interestingly, although one might expect events between cRSSs to also generate deletional "signal joints" retained in the chromosome, or inversion events, these are rarely observed (7) (M. Mijuskovic, unpublished data). Deletional recombination between cRSS pairs generates recurrent deletions at the SIL/SCL locus (64) and in *Notch1*, *Izkf1*, *PTEN*, and other critical genes in lymphoid neoplasms in humans and in mice (5, 7, 38, 58, 60, 61, 65) (Fig. 6c). These are now thought to be major drivers of oncogenic transformation in lymphocytes. Another type of target recognition error, less commonly observed, involves RAG-mediated cleavage at non-B-form DNA structures. This type of error has been implicated in oncogenic rearrangements joining the Bcl-2 major breakpoint region to an authentic RSS at the immunoglobulin heavy chain locus (66).

Errors in joining involve events that join a RAG-mediated double-strand break to a broken DNA end created by a non-RAG-mediated mechanism. The events observed in lymphoid neoplasms generally involve three DNA breaks and are referred to as "end donation" (67) or "type 2" events (68). These can involve a pair of breaks made during an apparently normal V(D)J recombination event, which are then mistakenly joined to another break generated by another mechanism (Fig. 7a). These can generate chromosome translocations or insertions of signal-ended fragments into another chromosomal location (59, 68). Similar events can involve a combination of recognition and joining errors: cleavage at a pair of cRSSs followed by joining to a non-RAG-mediated double-strand break (Fig. 7b). Although joining errors may involve the normal cNHEJ mechanism, it is thought that such events may be favored by the use of error-prone aNHEJ mechanisms (50, 63), which are known to favor the formation of translocations (53, 54).

RAG-mediated transposition events cannot be conveniently classified as errors in recognition or in joining, as in this case the initial cleavage event, occurring at a pair of authentic RSSs, is followed by RAG-mediated integration of the excised signal-ended fragment at another genomic location. These events, while observed *in vitro* (22, 23) in artificial systems in cultured cells (69, 70) and at the hypoxanthine artificiphosphoribosyl transferase (HPRT) locus in human peripheral T cells (71), have not yet been demonstrated definitively in lymphoid neoplasms. It should be noted, however, that certain types of transposition events generate products that would not be recognizable as having been derived from transposition (72).

HOW IS FIDELITY PRESERVED DURING V(D)J RECOMBINATION?

Given the fact that V(D)J recombination occurs in many millions of lymphocytes each day and that a variety of V(D)J recombination errors generate oncogenic lesions in lymphoid neoplasms, it seems logical to suppose that mechanisms exist to maintain the fidelity of the process. Perhaps the most basic of these is to ensure that the recombinase is active only in the appropriate target cells and only during the appropriate developmental stages. Indeed, expression of RAG1 and RAG2 is carefully limited in a cell- and developmental stage-specific fashion. Bypassing these controls by introducing *RAG1* and *RAG2* transgenes under the control of strong promoters causing constitutive expression during lymphocyte development and in extralymphoid tissues results in a spectrum of phenotypes (including lymphopenia, growth retardation, and early death) reminiscent of DNA damage deficiency syndromes (73). An additional temporal control is provided by cell-cycle-specific protein degradation of the RAG2 protein, mediated by phosphorylation of a threonine (T490) located in the dispensable C terminus (41). Disabling this feature (via a T490A mutation) results in accelerated lymphomagenesis in p53-deficient mice (37). Autoubiquitylation of RAG1 may also play a regulatory role (74, 75).

Choice of the joining pathway used to repair RAG-generated breaks also appears to be important in maintaining fidelity. Mice lacking a functional cNHEJ pathway exhibit accelerated lymphomagenesis in the absence of p53 (76, 77), with complex chromosome translocations (mediated by aNHEJ) accompanied by gene amplification (63). Choice of joining pathway appears to involve the RAG post-cleavage complex, as mutations in RAG1 or RAG2 that destabilize the complex allow the ends to be joined by alternative pathways, including homologous recombination and aNHEJ (30, 45, 50). Further work has shown that mutations in the nonessential C terminus of RAG2 lead to genomic instability, accompanied by chromosomal aberrations (30, 36). To test the hypothesis that the C-terminal mutations caused oncogenic transformation by encouraging joining errors mediated by aNHEJ, lymphomas from two mouse models lacking the RAG2 C terminus were examined by whole-genome sequencing. Scant evidence for oncogenic aberrant joining events was observed. Instead, most genomic lesions that could be linked to potentially oncogenic events were deletions between pairs of cRSSs (38, 60). These data suggest that interstitial deletions may be more important drivers of RAG-mediated oncogenesis, at least in some systems, than gross chromosomal aberrations and are in agreement with recent studies of B- and T-cell ALL in humans (7, 61).

In the case of the RAG2 C-terminal mutants, it is not yet clear whether increased access of RAG-mediated DNA breaks to aNHEJ plays a role in the observed genomic instability. A severe RAG2 C-terminal truncation that allows high levels of aNHEJ in cultured cells (50) increases access of RAG-mediated breaks to alternative joining mechanisms, as shown by rescue of joining in a Ku80/RAG2 double mutant. This observation supports the idea that the post-cleavage complex enforces pathway choice (38). Recognizing a particular junction as having arisen from aNHEJ in this system is complicated, however, because junctional features considered characteristic of aNHEJ were rarely observed, even in the double mutants (in which junctions must have formed by a cNHEJ-independent process) (38). These data are consistent with previous analysis of rare junctions isolated from Ku80-deficient mice (78). Together, these data indicate that aNHEJ is not always distinguishable on the basis of junction structures, and provide support for the suggestion that aNHEJ may actually consist of several distinct pathways (55, 79), only some of which generate junctional "signatures." Thus, caution must be observed when inferring the involvement of aNHEJ in aberrant recombination events in cancer genomes by sequence features alone.

As noted above, the 140+ amino acids in the RAG2 C terminus, largely conserved throughout evolution, contain several known or suspected regulatory elements. These may play important roles in maintaining the fidelity of V(D)J recombination. Clear evidence implicates the cell-cycle-regulated phosphorylation of T490 in suppressing persistence of broken DNA ends through the cell cycle and in suppressing lymphomagenesis. The PHD domain, which recognizes trimethylated histone H3K4, may play a role in limiting recognition of cRSSs located outside the antigen-receptor loci and/or in downregulating RAG cleavage activity in the absence of this histone modification (80). It should be noted, however, that these potential regulatory activities do not prevent the generation of deletions between cRSS pairs in known or suspected oncogenes and tumor-suppressor genes, as these are observed in thymic lymphomas of p53-deficient mice bearing wild-type RAG2 (38, 60) and in T- and B-cell ALL genomes from patients who presumably bear wild-type RAG alleles (7, 61). Other potential regulatory mechanisms involving the C terminus include its ability to inhibit RAG-mediated transposition (81, 82) and suppression of bi-allelic cleavage at antigen-receptor loci (83).

Clearly, V(D)J recombination fidelity is also enforced by non-RAG-specific mechanisms. These include ATM (84, 85), p53 (36, 38, 76, 77), and phosphorylated histone H2AX (86), as well as other aspects of the DNA damage response (9, 52). It is relatively straightforward to imagine how DNA damage-response factors may act to limit errors in joining, such as limiting the persistence of broken DNA ends or the survival of cells bearing persistent broken ends. How these factors might limit errors in recognition, such as deletions between pairs of cRSSs or the survival of cells bearing such events, is less obvious. Investigating these regulatory mechanisms provides an interesting focus for future research.

Citation. Roth DB. 2014. V(D)J Recombination: Mechanism, errors, and fidelity. Microbiol Spectrum 2(6):MDNA3-0041-2014.

References

1. **Brandt VL, Roth DB.** 2008. G.O.D.'s Holy Grail: discovery of the RAG proteins. *J Immunol* **180:**3:m.
2. **Hozumi N, Tonegawa S.** 1976. Evidence for somatic rearrangement of immunoglobulin genes coding for variable and constant regions. *Proc Natl Acad Sci U S A* **73:** 362862632.
3. **Onozawa M, Aplan PD.** 2012. Illegitimate V(D)J recombination involving nonantigen receptor loci in lymphoid malignancy. *Genes Chromosomes Cancer* **51:**5252n35.
4. **Tsujimoto Y, Gorham J, Cossman J, Jaffe E, Croce CM.** 1985. The t(14;18) chromosome translocations involved in B-cell neoplasms result from mistakes in VDJ joining. *Science* **229:**139099393.
5. **Mullighan CG, Miller CB, Radtke I, Phillips LA, Dalton J, Ma J, White D, Hughes TP, Le Beau MM, Pui CH, Relling MV, Shurtleff SA, Downing JR.** 2008. BCR-ABL1 lymphoblastic leukaemia is characterized by the deletion of Ikaros. *Nature* **453:**1101314.
6. **Zhang J, Ding L, Holmfeldt L, Wu G, Heatley SL, Payne-Turner D, Easton J, Chen X, Wang J, Rusch M, Lu C, Chen SC, Wei L, Collins-Underwood JR, Ma J, Roberts KG, Pounds SB, Ulyanov A, Becksfort J, Gupta P, Huether R, Kriwacki RW, Parker M, McGoldrick DJ, Zhao D, Alford D, Espy S, Bobba KC, Song G, Pei D, Cheng C, Roberts S, Barbato MI, Campana D, Coustan-Smith E, Shurtleff SA, Raimondi SC, Kleppe M, Cools J, Shimano KA, Hermiston ML, Doulatov S, Eppert K, Laurenti E, Notta F, Dick JE, Basso G, Hunger SP, Loh ML, Devidas M, Wood B, Winter S, Dunsmore KP, Fulton RS, Fulton LL, Hong X, Harris CC, Dooling DJ, Ochoa K, Johnson KJ, Obenauer JC, Evans WE, Pui CH, Naeve CW, Ley TJ, Mardis ER, Wilson RK, Downing JR, Mullighan CG.** 2012. The genetic basis of early T-cell precursor acute lymphoblastic leukaemia. *Nature* **481:**1575163.
7. **Papaemmanuil E, Rapado I, Li Y, Potter NE, Wedge DC, Tubio J, Alexandrov LB, Van Loo P, Cooke SL, Marshall J, Martincorena I, Hinton J, Gundem G, van Delft FW, Nik-Zainal S, Jones DR, Ramakrishna M, Titley I, Stebbings L, Leroy C, Menzies A, Gamble J, Robinson B, Mudie L, Raine K, O'Meara S, Teague JW, Butler AP, Cazzaniga G, Biondi A, Zuna J, Kempski H, Muschen M, Ford AM, Stratton MR, Greaves M, Campbell PJ.** 2014. RAG-mediated recombination is the predominant driver of oncogenic rearrangement in ETV6-RUNX1 acute lymphoblastic leukemia. *Nat Genet* **46:**1161:25.
8. **Schatz DG, Swanson PC.** 2011. V(D)J recombination: mechanisms of initiation. *Annu Rev Genet* **45:**1676:02.
9. **Helmink BA, Sleckman BP.** 2012. The response to and repair of RAG-mediated DNA double-strand breaks. *Annu Rev Immunol* **30:**1757:02.
10. **Bassing CH, Alt FW, Hughes MM, D'Auteuil M, Wehrly TD, Woodman BB, Gartner F, White JM, Davidson L, Sleckman BP.** 2000. Recombination signal sequences restrict chromosomal V(D)J recombination beyond the 12/23 rule. *Nature* **405:**5838586.
11. **Hesse JE, Lieber MR, Gellert M, Mizuuchi K.** 1987. Extrachromosomal DNA substrates in pre-B cells undergo inversion or deletion at immunoglobulin V-(D)-J joining signals. *Cell* **49:**7757:83.
12. **Lewis SM, Hesse JE, Mizuuchi K, Gellert M.** 1988. Novel strand exchanges in V(D)J recombination. *Cell* **55:** 109909107.
13. **Hesse JE, Lieber MR, Mizuuchi K, Gellert M.** 1989. V(D)J recombination: a functional definition of the joining signals. *Genes Dev* **3:**105305061.
14. **Ramsden DA, McBlane JF, van Gent DC, Gellert M.** 1996. Distinct DNA sequence and structure requirements for the two steps of V(D)J recombination signal cleavage. *EMBO J* **15:**319719206.
15. **Cuomo CA, Mundy CL, Oettinger MA.** 1996. DNA sequence and structure requirements for cleavage of V(D)J recombination signal sequences. *Mol Cell Biol* **16:** 568368690.
16. **Kale SB, Landree MA, Roth DB.** 2001. Conditional RAG-1 mutants block the hairpin formation step of V(D)J recombination. *Mol Cell Biol* **21:**4595:66.
17. **Bischerour J, Lu C, Roth DB, Chalmers R.** 2009. Base flipping in V(D)J recombination: insights into the mechanism of hairpin formation, the 12/23 rule, and the coordination of double-strand breaks. *Mol Cell Biol* **29:** 588988899.

18. **Oettinger MA, Schatz DG, Gorka C, Baltimore D.** 1990. RAG-1 and RAG-2, adjacent genes that synergistically activate V(D)J recombination. *Science* **248:**151751523.

19. **van Gent DC, Hiom K, Paull TT, Gellert M.** 1997. Stimulation of V(D)J cleavage by high mobility group proteins. *EMBO J* **16:**266566670.

20. **Thompson CB.** 1995. New insights into V(D)J recombination and its role in the evolution of the immune system. *Immunity* **3:**5313m39.

21. **van Gent DC, Mizuuchi K, Gellert M.** 1996. Similarities between initiation of V(D)J recombination and retroviral integration. *Science* **271:**159259594.

22. **Agrawal A, Eastman QM, Schatz DG.** 1998. Transposition mediated by RAG1 and RAG2 and its implications for the evolution of the immune system. *Nature* **394:**7444451.

23. **Hiom K, Melek M, Gellert M.** 1998. DNA transposition by the RAG1 and RAG2 proteins: a possible source of oncogenic translocations. *Cell* **94:**4636:70.

24. **Litman GW, Rast JP, Fugmann SD.** 2010. The origins of vertebrate adaptive immunity. *Nat Rev Immunol* **10:** 5434:53.

25. **Silver DP, Spanopoulou E, Mulligan RC, Baltimore D.** 1993. Dispensable sequence motifs in the RAG-1 and RAG-2 genes for plasmid V(D)J recombination. *Proc Natl Acad Sci U S A* **90:**610010104.

26. **Sadofsky MJ, Hesse JE, McBlane JF, Gellert M.** 1994. Expression and V(D)J recombination activity of mutated RAG-1 proteins. *Nucleic Acids Res* **22:**550.

27. **Dudley DD, Sekiguchi J, Zhu C, Sadofsky MJ, Whitlow S, DeVido J, Monroe RJ, Bassing CH, Alt FW.** 2003. Impaired V(D)J recombination and lymphocyte development in core RAG1-expressing mice. *J Exp Med* **198:** 143943450.

28. **Cuomo CA, Oettinger MA.** 1994. Analysis of regions of RAG-2 important for V(D)J recombination. *Nucleic Acids Res* **22:**181081814.

29. **Sadofsky MJ, Hesse JE, Gellert M.** 1994. Definition of a core region of RAG-2 that is functional in V(D)J recombination. *Nucleic Acids Res* **22:**180580809.

30. **Coussens MA, Wendland RL, Deriano L, Lindsay CR, Arnal SM, Roth DB.** 2013. RAG23s acidic hinge restricts repair-pathway choice and promotes genomic stability. *Cell Rep* **4:**8707178.

31. **Callebaut I, Mornon JP.** 1998. The V(D)J recombination activating protein RAG2 consists of a six-bladed propeller and a PHD fingerlike domain, as revealed by sequence analysis. *Cell Mol Life Sci* **54:**8808:91.

32. **Akamatsu Y, Monroe R, Dudley DD, Elkin SK, Gartner F, Talukder SR, Takahama Y, Alt FW, Bassing CH, Oettinger MA.** 2003. Deletion of the RAG2 C terminus leads to impaired lymphoid development in mice. *Proc Natl Acad Sci U S A* **100:**120920214.

33. **Curry JD, Schlissel MS.** 2008. RAG2's non-core domain contributes to the ordered regulation of V(D)J recombination. *Nucleic Acids Res* **36:**575075762.

34. **Talukder SR, Dudley DD, Alt FW, Takahama Y, Akamatsu Y.** 2004. Increased frequency of aberrant V(D)J recombination products in core RAG-expressing mice. *Nucleic Acids Res* **32:**453953549.

35. **Curry JD, Schulz D, Guidos CJ, Danska JS, Nutter L, Nussenzweig A, Schlissel MS.** 2007. Chromosomal reinsertion of broken RSS ends during T cell development. *J Exp Med* **204:**229329303.

36. **Deriano L, Chaumeil J, Coussens M, Multani A, Chou Y, Alekseyenko AV, Chang S, Skok JA, Roth DB.** 2011. The RAG2 C terminus suppresses genomic instability and lymphomagenesis. *Nature* **471:**1191123.

37. **Zhang L, Reynolds TL, Shan X, Desiderio S.** 2011. Coupling of V(D)J recombination to the cell cycle suppresses genomic instability and lymphoid tumorigenesis. *Immunity* **34:**1636:74.

38. **Gigi V, Lewis S, Shestova O, Mijuskovic M, Deriano L, Meng W, Luning Prak ET, Roth DB.** 2014. RAG2 mutants alter DSB repair pathway choice in vivo and illuminate the nature of vy choice in NHEJE. *Nucleic Acids Res* **42:**6352eic Ac.

39. **Liu Y, Subrahmanyam R, Chakraborty T, Sen R, Desiderio S.** 2007. A plant homeodomain in RAG-2 that binds hypermethylated lysine 4 of histone H3 is necessary for efficient antigen-receptor-gene rearrangement. *Immunity* **27:**5616:71.

40. **Matthews AG, Kuo AJ, Ramon-Maiques S, Han S, Champagne KS, Ivanov D, Gallardo M, Carney D, Cheung P, Ciccone DN, Walter KL, Utz PJ, Shi Y, Kutateladze TG, Yang W, Gozani O, Oettinger MA.** 2007. RAG2 PHD finger couples histone H3 lysine 4 trimethylation with V(D)J recombination. *Nature* **450:**110610110.

41. **Li Z, Dordai DI, Lee J, Desiderio S.** 1996. A conserved degradation signal regulates RAG-2 accumulation during cell division and links V(D)J recombination to the cell cycle. *Immunity* **5:**5757m89.

42. **van Gent DC, McBlane JF, Ramsden DA, Sadofsky MJ, Hesse JE, Gellert M.** 1996. Initiation of V(D)J recombinations in a cell-free system by RAG1 and RAG2 proteins. *Curr Top Microbiol Immunol* **217:**1170.

43. **Hiom K, Gellert M.** 1997. A stable RAG1-RAG2-DNA complex that is active in V(D)J cleavage. *Cell* **88:**658:2.

44. **Jones JM, Gellert M.** 2002. Ordered assembly of the V (D)J synaptic complex ensures accurate recombination. *EMBO J* **21:**4162–4171.

45. **Lee GS, Neiditch MB, Salus SS, Roth DB.** 2004. RAG proteins shepherd double-strand breaks to a specific pathway, suppressing error-prone repair, but RAG nicking initiates homologous recombination. *Cell* **117:**1717784.

46. **Wang G, Dhar K, Swanson PC, Levitus M, Chang Y.** 2012. Real-time monitoring of RAG-catalyzed DNA cleavage unveils dynamic changes in coding end association with the coding end complex. *Nucleic Acids Res* **40:**608208096.

47. **Qiu JX, Kale SB, Yarnell Schultz H, Roth DB.** 2001. Separation-of-function mutants reveal critical roles for RAG2 in both the cleavage and joining steps of V(D)J recombination. *Mol Cell* **7:**77:e7.

48. **Yarnell Schultz H, Landree MA, Qiu JX, Kale SB, Roth DB.** 2001. Joining-deficient RAG1 mutants block V(D)J recombination in vivo and hairpin opening in vitro. *Mol Cell* **7:**65:e5.

49. **Tsai CL, Drejer AH, Schatz DG.** 2002. Evidence of a critical architectural function for the RAG proteins in

end processing, protection, and joining in V(D)J recombination. *Genes Dev* **16:**193493949.

50. **Corneo B, Wendland RL, Deriano L, Cui X, Klein IA, Wong SY, Arnal S, Holub AJ, Weller GR, Pancake BA, Shah S, Brandt VL, Meek K, Roth DB.** 2007. Rag mutations reveal robust alternative end joining. *Nature* **449:** 4838986.
51. **Lewis S, Gifford A, Baltimore D.** 1985. DNA elements are asymmetrically joined during the site-specific recombination of kappa immunoglobulin genes. *Science* **228:** 6777885.
52. **Deriano L, Roth DB.** 2013. Modernizing the nonhomologous end-joining repertoire: alternative and classical NHEJ share the stage. *Annu Rev Genet* **47:**4333:55.
53. **Yan CT, Boboila C, Souza EK, Franco S, Hickernell TR, Murphy M, Gumaste S, Geyer M, Zarrin AA, Manis JP, Rajewsky K, Alt FW.** 2007. IgH class switching and translocations use a robust non-classical end-joining pathway. *Nature* **449:**4787982.
54. **Simsek D, Brunet E, Wong SY, Katyal S, Gao Y, McKinnon PJ, Lou J, Zhang L, Li J, Rebar EJ, Gregory PD, Holmes MC, Jasin M.** 2011. DNA ligase III promotes alternative nonhomologous end-joining during chromosomal translocation formation. *PLoS Genet* **7:**e1002080.
55. **Boboila C, Alt FW, Schwer B.** 2012. Classical and alternative end-joining pathways for repair of lymphocyte-specific and general DNA double-strand breaks. *Adv Immunol* **116:**1169.
56. **Haluska FG, Finver S, Tsujimoto Y, Croce CM.** 1986. The t(8; 14) chromosomal translocation occurring in B-cell malignancies results from mistakes in V-D-J joining. *Nature* **324:**1585461.
57. **Kagan J, Finan J, Letofsky J, Besa EC, Nowell PC, Croce CM.** 1987. -Chain locus of the T-cell antigen receptor is involved in the t(10;14) chromosome translocation of T-cell acute lymphocytic leukemia. *Proc Natl Acad Sci U S A* **84:**454354546.
58. **Haydu JE, De Keersmaecker K, Duff MK, Paietta E, Racevskis J, Wiernik PH, Rowe JM, Ferrando A.** 2012. An activating intragenic deletion in NOTCH1 in human T-ALL. *Blood* **119:**521121214.
59. **Vanura K, Montpellier B, Le T, Spicuglia S, Navarro JM, Cabaud O, Roulland S, Vachez E, Prinz I, Ferrier P, Marculescu R, Jager U, Nadel B.** 2007. In vivo reinsertion of excised episomes by the V(D)J recombinase: a potential threat to genomic stability. *PLoS Biol* **5:**e43.
60. **Mijuskovic M, Brown SM, Tang Z, Lindsay CR, Efstathiadis E, Deriano L, Roth DB.** 2012. A streamlined method for detecting structural variants in cancer genomes by short read paired-end sequencing. *PLoS One* **7:**e48314.
61. **Mendes RD, Sarmento LM, Cante-Barrett K, Zuurbier L, Buijs-Gladdines JG, Povoa V, Smits WK, Abecasis M, Yunes JA, Sonneveld E, Horstmann MA, Pieters R, Barata JT, Meijerink JP.** 2014. PTEN micro-deletions in T-cell acute lymphoblastic leukemia are caused by illegitimate RAG-mediated recombination events. *Blood* **124:**567:do-d.
62. **Lewis SM, Agard E, Suh S, Czyzyk L.** 1997. Cryptic signals and the fidelity of V(D)J joining. *Mol Cell Biol* **17:**312512136.
63. **Zhu C, Mills KD, Ferguson DO, Lee C, Manis J, Fleming J, Gao Y, Morton CC, Alt FW.** 2002. Unrepaired DNA breaks in p53-deficient cells lead to oncogenic gene amplification subsequent to translocations. *Cell* **109:**8111921.
64. **Aplan PD, Lombardi DP, Ginsberg AM, Cossman J, Bertness VL, Kirsch IR.** 1990. Disruption of the human SCL locus by "illegitimate" V-(D)-J recombinase activity. *Science* **250:**142642429.
65. **Ashworth TD, Pear WS, Chiang MY, Blacklow SC, Mastio J, Xu L, Kelliher M, Kastner P, Chan S, Aster JC.** 2010. Deletion-based mechanisms of Notch1 activation in T-ALL: key roles for RAG recombinase and a conserved internal translational start site in Notch1. *Blood* **116:**545545464.
66. **Raghavan SC, Swanson PC, Wu X, Hsieh CL, Lieber MR.** 2004. A non-B-DNA structure at the Bcl-2 major breakpoint region is cleaved by the RAG complex. *Nature* **428:**88283.
67. **Lewis SM.** 1994. The mechanism of V(D)J joining: lessons from molecular, immunological, and comparative analyses. *Adv Immunol* **56:**276:50.
68. **Marculescu R, Le T, Simon P, Jaeger U, Nadel B.** 2002. V(D)J-mediated translocations in lymphoid neoplasms: a functional assessment of genomic instability by cryptic sites. *J Exp Med* **195:**85958.
69. **Chatterji M, Tsai CL, Schatz DG.** 2006. Mobilization of RAG-generated signal ends by transposition and insertion in vivo. *Mol Cell Biol* **26:**155855568.
70. **Reddy YV, Perkins EJ, Ramsden DA.** 2006. Genomic instability due to V(D)J recombination-associated transposition. *Genes Dev* **20:**157557582.
71. **Messier TL, O'Neill JP, Hou SM, Nicklas JA, Finette BA.** 2003. In vivo transposition mediated by V(D)J recombinase in human T lymphocytes. *EMBO J* **22:**138138388.
72. **Roth DB, Craig NL.** 1998. VDJ recombination: a transposase goes to work. *Cell* **94:**4111:14.
73. **Barreto V, Marques R, Demengeot J.** 2001. Early death and severe lymphopenia caused by ubiquitous expression of the Rag1 and Rag2 genes in mice. *Eur J Immunol* **31:** 376376772.
74. **Jones JM, Gellert M.** 2003. Autoubiquitylation of the V(D)J recombinase protein RAG1. *Proc Natl Acad Sci U S A* **100:**15446545451.
75. **Yurchenko V, Xue Z, Sadofsky M.** 2003. The RAG1 N-terminal domain is an E3 ubiquitin ligase. *Genes Dev* **17:**5818:85.
76. **Difilippantonio MJ, Zhu J, Chen HT, Meffre E, Nussenzweig MC, Max EE, Ried T, Nussenzweig A.** 2000. DNA repair protein Ku80 suppresses chromosomal aberrations and malignant transformation. *Nature* **404:** 5101414.
77. **Gao Y, Ferguson DO, Xie W, Manis JP, Sekiguchi J, Frank KM, Chaudhuri J, Horner J, DePinho RA, Alt FW.** 2000. Interplay of p53 and DNA-repair protein XRCC4 in tumorigenesis, genomic stability and development. *Nature* **404:**8979400.
78. **Bogue MA, Wang C, Zhu C, Roth DB.** 1997. V(D)J recombination in Ku86-deficient mice: distinct effects on coding, signal, and hybrid joint formation. *Immunity* **7:**37:m7.

79. **Lieber MR.** 2010. The mechanism of double-strand DNA break repair by the nonhomologous DNA end-joining pathway. *Annu Rev Biochem* **79:**1818:11.

80. **Grundy GJ, Yang W, Gellert M.** 2010. Autoinhibition of DNA cleavage mediated by RAG1 and RAG2 is overcome by an epigenetic signal in V(D)J recombination. *Proc Natl Acad Sci U S A* **107:**22487242492.

81. **Elkin SK, Matthews AG, Oettinger MA.** 2003. The C-terminal portion of RAG2 protects against transposition in vitro. *EMBO J* **22:**193193938.

82. **Swanson PC, Volkmer D, Wang L.** 2004. Full-length RAG-2, and not full-length RAG-1, specifically suppresses RAG-mediated transposition but not hybrid joint formation or disintegration. *J Biol Chem* **279:**403403044.

83. **Chaumeil J, Micsinai M, Ntziachristos P, Roth DB, Aifantis I, Kluger Y, Deriano L, Skok JA.** 2013. The RAG2 C-terminus and ATM protect genome integrity by controlling antigen receptor gene cleavage. *Nat Commun* **4:**2231.

84. **Bredemeyer AL, Huang CY, Walker LM, Bassing CH, Sleckman BP.** 2008. Aberrant V(D)J recombination in ataxia telangiectasia mutated-deficient lymphocytes is dependent on nonhomologous DNA end joining. *J Immunol* **181:**262062625.

85. **Mahowald GK, Baron JM, Mahowald MA, Kulkarni S, Bredemeyer AL, Bassing CH, Sleckman BP.** 2009. Aberrantly resolved RAG-mediated DNA breaks in Atm-deficient lymphocytes target chromosomal breakpoints in cis. *Proc Natl Acad Sci U S A* **106:**18339838344.

86. **Celeste A, Difilippantonio S, Difilippantonio MJ, Fernandez-Capetillo O, Pilch DR, Sedelnikova OA, Eckhaus M, Ried T, Bonner WM, Nussenzweig A.** 2003. H2AX haploinsufficiency modifies genomic stability and tumor susceptibility. *Cell* **114:**3717483.

Mobile DNA, 3rd Edition
Nancy L. Craig, Michael Chandler, Martin Gellert, Alan M. Lambowitz, Phoebe A. Rice and Suzanne Sandmeyer

doi:10.1128/microbiolspec.MDNA3-0037-2014

Joyce K. Hwang*,[1]
Frederick W. Alt[1]
Leng-Siew Yeap*,[1]

15 Related Mechanisms of Antibody Somatic Hypermutation and Class Switch Recombination

OVERVIEW AND INTRODUCTION

Immunoglobulin genes, B cell receptors and antibodies

The B cell receptor (BCR) is expressed on the B lymphocyte cell surface where it serves as a receptor for foreign antigens (1). The BCR is comprised of two immunoglobulin (Ig) heavy (IgH) chains encoded by the *IgH* heavy chain locus and two Ig light (IgL) chains encoded by, for a given BCR, either the *Igκ* or *Igλ* (collectively referred to as *IgL*) light chain loci (Fig. 1). These three *Ig* loci lie on different chromosomes in both humans and mice. While there are certain differences in organization, the overall strategies for *Ig* gene diversification in mice and humans are very much the same (2, 3), so this review will focus mainly on the mouse. The amino-terminal portions of the IgH and IgL chains have a highly variable amino acid sequence from species to species of antibody and are called variable (V) regions. The IgH and IgL variable regions interact to generate the antigen-binding portion of the BCR/antibody. The carboxy-terminal end of IgH and IgL chains have only a few variations in their sequences and thus are called constant (C) regions.

The antigen-independent generation of an extremely large population of B cells in which individual cells express BCRs with unique antigen-binding specificity is of fundamental importance for vertebrates to generate effective humoral adaptive immune responses, as it enables B cells to recognize and respond to an enormous variety of foreign antigens. In this context, *IgH* and *IgL* variable region exons are not encoded in the germline, but rather are assembled during early B cell development prior to antigen exposure in the fetal liver and bone marrow by the V(D)J recombination process (2). V(D)J recombination generates an *IgH* $V_H(D)J_H$

[1]Howard Hughes Medical Institute, Program in Cellular and Molecular Medicine, Boston Children's Hospital, and Department of Genetics, Harvard Medical School, Boston, MA 02115. *Equal contribution.

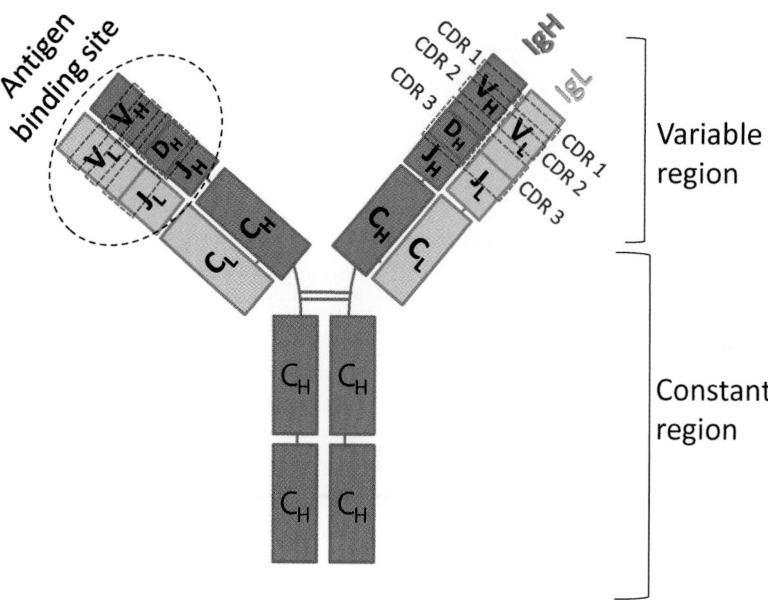

Figure 1 Antibody structure. The BCR is comprised of two immunoglobulin (Ig) heavy (IgH) chains encoded by the *IgH* heavy chain locus and two Ig light (IgL) chains. The rectangles represent Ig domains that constitute the structural units of the immunoglobulin heavy and light chains. The variable regions are assembled through V(D)J recombination of V_H, D_H, and J_H gene segments on the heavy chain and V_L and J_L gene segments on the light chain. Complementarity-determining regions (CDRs) are indicated as regions in dashed red boxes: CDR 1 and 2 are encoded in the V_H or V_L gene segments, and CDR 3 is encoded by the V_H D_H J_H junctional region or V_L and J_L junctional region. The heavy and light chain variable regions form the antigen-binding site. The constant region determines the class and effector function of the antibody molecule.
doi:10.1128/microbiolspec.MDNA3-0037-2014.f1

variable region exon by assembling different combinations of the numerous *IgH* variable (V_H) segments, diversity (D) segments, and joining (J_H) segments that lie within a 1 to 3 Mb region at the 5′ end of the *IgH* locus. V(D)J recombination assembles an *IgL* V_LJ_L variable region exon from *Igκ* or *Igλ* V segments and J segments (2).

V(D)J recombination is initiated by the lymphocyte-specific RAG1 and RAG2 ("RAG") endonuclease that recognizes conserved recombination signal sequences (RSS) that flank the V, D, and J segments (4). RAG cleaves between the RSSs and the coding sequences of a pair of involved segments, generating a pair of blunt RSS double strand break (DSB) ends that are later joined to each other and a pair of hair-pinned coding DSB ends that are processed and joined to each other (4) by the general cellular classical nonhomologous end-joining (C-NHEJ) DSB repair pathway (5, 6). Coding ends are often further diversified before they are joined, including the *de novo* additions of N nucleotides by the terminal deoxynucleotidyl transferase (Tdt), another lymphocyte-specific factor involved in V(D)J recombination (7). The combinatorial diversity arising from the numerous V, D, and J segments, as well as the junctional diversity that arises from junctional diversification during joining the segments, generates an enormous repertoire of primary variable region exons (8). Within the IgH and IgL variable regions there are three regions that show "hypervariability" separated by much less variable "framework" regions (FWR). As they are involved in antigen contact, these three hypervariable regions are termed complementarity-determining regions (CDRs) (9). CDR1 and CDR2 are encoded in the different germline V_H and V_L gene segments. The most diverse portion of the primary variable region exon is CDR3, which is generated through combinatorial assortment of V, D, and J sequences and from junctional diversification mechanisms (10).

Transcription of fully assembled *IgH* and *IgL* chain genes is initiated from the promoter of the V segment used in the V(D)J exon and continues through downstream exons that encode the C regions of IgH and IgL chains (11). The mouse *IgH* locus contains 8 sets of exons that encode different C_H regions (sometimes termed "C_H genes") within the approximately 200 kb region downstream of the J_H segment and lying in the order 5′-VDJ-Cμ-Cδ-Cγ3-Cγ1-Cγ2b-Cγ2a-Cε-Cα 3′ (12) (Fig. 2a). The set of C_H exons expressed with the V region exon determines the class of the BCR/antibody (e.g. IgM, IgG, IgE, IgA). Within the *IgH* locus, there are several B cell specific enhancers, for example iEμ, which lies between the J_H segments and the Cμ exons and a 30 kb IgH 3′ regulatory region ("IgH 3′ RR"), which is downstream of the Cα exons (11, 13) (Fig. 2a). Initially, the *IgH* variable region exon is transcribed in association with the immediately downstream Cμ exons, and in some cells, Cδ exons. Alternative RNA splicing of these primary IgH transcripts leads to differential expression of Cμ and Cδ and also to differential expression of C_H sequences that specify whether the IgH chain is expressed as membrane-bound BCR or is secreted as an antibody (14). Thus, prior to antigenic stimulation, resting B cells express IgM (or IgD).

During B cell development, V(D)J recombination generally occurs first at the *IgH* locus in progenitor (pro) B cells (2, 6). In this regard, developing B cells generate D to J_H rearrangements on both *IgH* alleles and then append V_H segments to pre-existing DJ_H rearrangements (15). If the first V_H to DJ_H is in frame (productive), the resulting IgH μ heavy chain protein generates a signal that feeds back to prevent V_H to DJ_H joining on the other DJ_H rearranged allele and to promote development to the precursor (pre) B cell stage. The resulting pre B cells will have a productive V(D)J *IgH* allele and a "frozen" DJ_H intermediate allele. If, due to junctional diversification, the first V_H to DJ_H rearrangement is out of frame, the cell can move on and

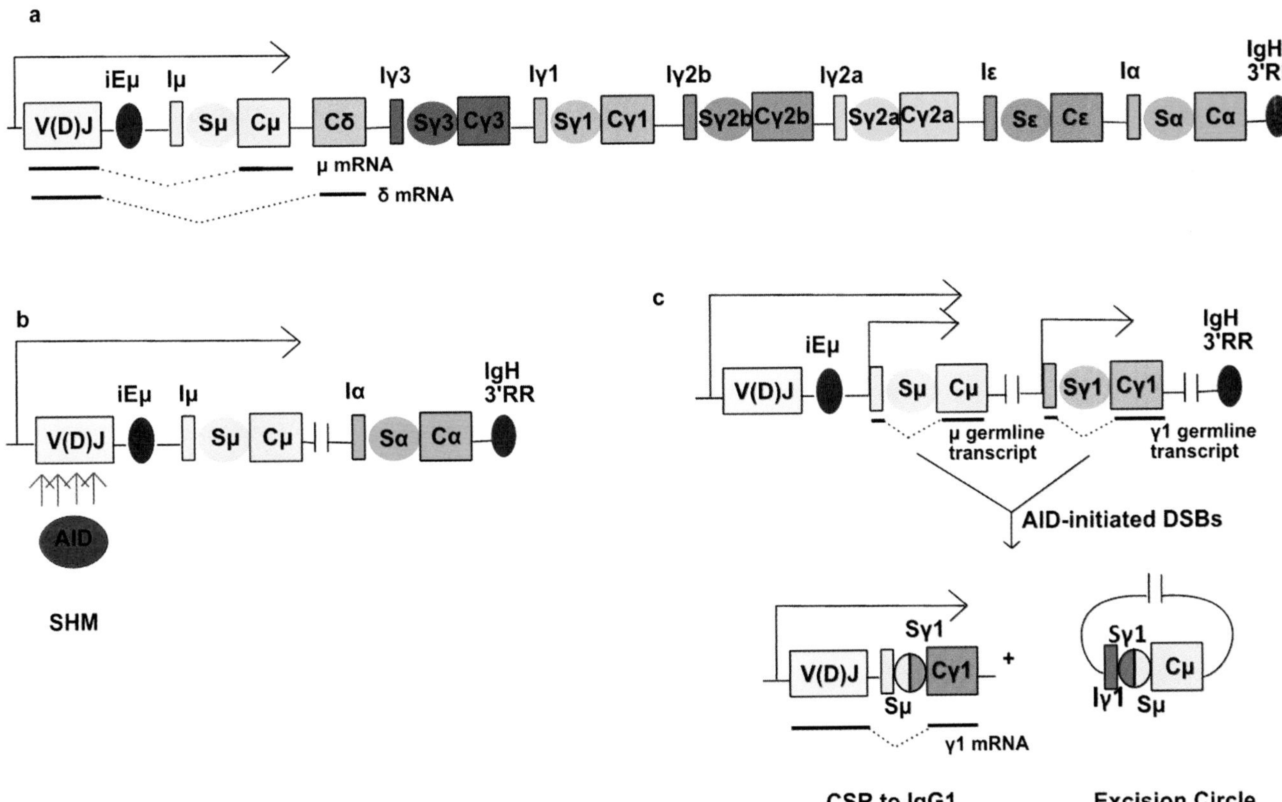

Figure 2 Genomic alterations of the *IgH* locus. **a.** Organization of the IgH constant (C) region. Each C region is preceded by a switch (S) region and a noncoding "I" exon. Blue oval between V(D)J exon and Iμ represents IgH intronic enhancer (iEμ). Blue oval downstream of Cα represents IgH 3′ regulatory region (IgH 3′RR). μ and δ mRNAs are shown below the corresponding genes. Dashed line represents spliced transcript. **b.** AID generates point mutations and/or DNA double strand breaks (DSBs) at the V(D)J exon during somatic hypermutation (SHM). **c.** AID-initiated DSBs in Sμ and Sγ1 result in CSR to IgG1. μ and γ1 germline transcripts are initiated from promoters upstream of the corresponding I exons.
doi:10.1128/microbiolspec.MDNA3-0037-2014.f2

append a V_H segment to the second DJ_H allele which, if productive, will again promote development to the pre B stage with the resulting pre B cells having an in-frame productive V(D)J rearrangement and out-of-frame nonproductive V(D)J rearrangement. Because about two thirds of V to DJ_H rearrangements are nonproductive, about 40% of normal B cells have two V(D)J rearrangements (one productive and one nonproductive). This "feedback" mechanism for the control of V_H to DJ_H rearrangement is thought to have evolved to ensure mono-specificity of B cell clones in the context of the phenomenon of "allelic exclusion" (see reference 16 for details). Precursor B cells rearrange *IgL* genes and if they form a productive IgL rearrangement leading to an IgL chain that pairs with μ heavy chain, they then express the complete IgM molecule on their surface as the BCR (2, 6). These newly generated IgM^+ B cells then migrate into the periphery and survey the secondary lymphoid organs, including spleen, lymph nodes and Peyer's patches, for cognate antigen that binds their BCR.

Overview of SHM, CSR and the Role of AID

An encounter with cognate antigen in the secondary lymphoid organs, usually in the context of a T-dependent immune response, can activate mature B cells. Activation can lead to the generation of B cells that secrete their BCR as a secreted antibody. Antigen-dependent B cell activation can also lead to the two somatic processes of genomic rearrangement that enhance the efficacy of the antibody response against specific antigen: SHM further diversifies the variable region exon and alters the affinity of the BCR for antigen (9) (Fig. 2b), while CSR switches the C_H region exon used and alters the antibody's antigen elimination function (17) (Fig. 2c). SHM occurs in the germinal centers (GCs) (18), specialized compartments of secondary lymphoid organs, while CSR can occur inside or outside of the GCs (19, 20). Both SHM and CSR are initiated by activation-induced cytidine deaminase (AID) (21, 22). AID is a small (24 kDa) protein that deaminates cytidine residues on single-stranded DNA (ssDNA), usually in the context of preferred sequence substrate motifs (9). Both SHM and CSR require transcription, both to promote specific AID targeting and also to contribute to formation of requisite ssDNA substrates. Both processes also co-opt activities of normal cellular base excision repair (BER) and mismatch repair (MMR) to convert AID cytidine deamination lesions to mutational and/or DSB outcomes. Each of these AID-associated processes will be discussed in depth in following sections.

During SHM, AID deaminates cytosine residues in *IgH* and *IgL* V(D)J exons and the deamination products are processed through specific repair pathways into predominantly point mutations, as well as a low frequency of small insertions and deletions (9, 23, 24). SHM produces nucleotide substitutions at all four bases, with a bias towards transitions over transversions such that approximately two thirds of nucleotide substitutions are transitions (25). The mutation frequency over the rearranged variable region exon as a whole is approximately 10^{-3} mutations per base pair per generation, with the highest levels found within complementarity-determining regions (CDRs) (9, 26, 27). In the context of the GC reaction, B cells with SHMs that increase antigen-binding affinity of their BCR are positively selected and those with SHMs that decrease affinity or inactivate the receptor are negatively selected via rounds of SHM, clonal expansion and affinity-based selection; in this manner SHM leads to affinity maturation of the antibody response (18).

For CSR, AID deaminates cytosine residues in long (1 to 10 kb), repetitive, noncoding, switch (S) regions that lie just upstream of each set of C_H exons (except Cδ exons, which do not undergo traditional CSR) (17, 28). Deamination products at donor and acceptor S regions are processed to DNA DSBs, as well as point mutations (17, 28). CSR is completed when AID-initiated DSBs generated in two participating S regions are fused to delete intervening DNA including the Cμ exons (12, 28). A switch in expression from IgM to different IgH classes such as IgG, IgE, and IgA occurs when Cμ exons are replaced with one of the sets of downstream C_H exons (e.g. Cγ, Cε, or Cα exons). Each antibody class is specialized for certain pathogen-elimination functions. For example, IgG promotes phagocytosis of antibody-coated particles, IgE triggers mast cell degranulation, and IgA defends against pathogens at mucosal surfaces (29, 30, 31). Thus, CSR alters an antibody's effector function to one that may be better suited for a given pathogen-elimination response, while maintaining the same variable region exon and thus antibody-binding specificity.

AID FUNCTIONS THROUGH CYTIDINE DEAMINATION OF TARGET DNA

AID was discovered by a subtractive hybridization approach that employed a mouse B cell line stimulated to undergo CSR from IgM to IgA (32). AID knock-out mice were found to be specifically defective for SHM and CSR (21). Likewise, contemporaneous studies of human patients with an autosomal recessive form of hyper-IgM syndrome, characterized by high levels of IgM in the serum and profound defects in IgH CSR and SHM, showed that they had AID mutations (22). These two types of studies showed that AID is required for both SHM and for IgH CSR and, thus, can be considered a master regulator of peripheral antibody diversification. Subsequent studies further showed that AID also is required for the variable region exon DSBs that initiate the gene conversion process that diversifies chicken antibody repertoires (33).

AID Target Sequences for SHM

AID deaminates cytidines to uridines in ssDNA (34, 35, 36, 37) preferentially deaminating cytidines in the context of "hotspots" described by the consensus motif DGYW (WRCH on the complementary strand, D = A/G/T, Y = C/T, W = A/T, H = T/C/A, R = A/G) (38). DGYW motifs are very abundant in the tandem repeat units of S regions, with a high density of DGYW motifs a conserved feature among the S regions of species from frogs to mammals (39). V region exons contain a lower density of DGYW motifs than S regions, but their DGYW frequency is still mildly enriched compared to bulk genomic DNA (39). As genomic DNA usually is in duplex form, transcription-based mechanisms have evolved to generate the requisite ssDNA substrates for AID (this is discussed in more detail in subsequent sections).

Mammalian and frog S regions are highly enriched for AGCT motifs, a palindromic variant of the canonical DGYW, which provide AID hotspot motifs on both DNA strands and contribute to DSB generation (see below). DGYW motifs are considered favored mutational hotspots, but not perfect predictors of mutability, since identical DGYW motifs within a given sequence undergo different levels of mutation (40). In addition, DGYW motifs are ubiquitous throughout the genome, but only undergo AID-initiated mutations in a subset of genes and, in most cases, at frequencies orders of magnitude less than at *Ig* gene targets (38, 41, 42). Thus, additional targeting mechanisms are important, including substrate sequence context beyond the DGYW motif, and higher level mechanisms including transcription (described in later sections). In addition, there is evidence that differential repair of AID cytidine deamination lesions also can influence final mutation and DSB outcome (9, 42).

AID-initiated lesions are processed by normal repair pathways to yield mutations and DSBs

The point mutations and DSBs that occur during SHM and CSR are generated in two steps (9). In the first step, AID deaminates cytidines to uridines (U) in V(D)J exons during SHM, or in S regions during CSR to produce uracil:guanine (U/G) mismatches. The second step involves error-prone resolution of the U/G lesion by co-opted BER and/or MMR pathway activities (Fig. 3). Normally the BER and MMR pathways repair such lesions in an error-free manner. How activities of the BER and MMR pathways that evolved to maintain genome fidelity are coerced into contributing to generating mutations and DSBs downstream of AID lesions is understood only in part. Below we will describe current knowledge of enzymatic processes involved in the two steps leading from AID-generated U/G mismatches to mutations or DSBs, starting with a description of the normal BER and MMR pathways.

BER repairs damaged bases by the following general steps: (1) recognition and excision of a damaged base (e.g. uracil) from the DNA backbone by an initiating DNA glycosylase (e.g. uracil-DNA-glycosylase) to create an abasic site; (2) cleavage of the DNA backbone at the abasic site by an apurinic/apyridimic (AP) endonuclease, generating a ssDNA nick adjacent to the abasic site; (3) processing of the nick to a single-nucleotide gap; (4) filling in of the gap by DNA polymerase β; and finally (5) sealing of the nick by DNA ligase 1 or DNA ligase 3 (43). An alternative to this one-nucleotide short-patch form of BER is long-patch BER. In long-patch BER, after nicking of the DNA by AP endonuclease, DNA polymerase $\beta/\delta/\varepsilon$ displaces and polymerizes an approximately 2 to 10 bp long tract of DNA (43, 44). The displaced strand is removed by activity catalyzed by the flap structure-specific endonuclease 1 (FEN1), and a remaining nick sealed by DNA ligase 1 (43, 44). MMR functions primarily in repair of base–base mismatches by a process involving (1) recognition of the mismatch by the MSH2-MSH6 heterodimer; (2) recruitment of a complex of MLH1 and PMS2 (MutLα); (3) excision of the patch of DNA surrounding the mismatch by exonuclease-1 (Exo1) to generate a gap; (4) gap-filling by DNA polymerase δ bound to PCNA; and (5) ligation by DNA ligase 1 to seal the nick (45, 46).

During SHM, replication over the initiating U/G lesion can produce transition (purine > purine or pyrimidine > pyrimidine) mutations at C/G base pairs (Fig. 3a) (9). In addition, the uracil can also be excised by UNG of the BER pathway which leads to an abasic site. Replication over the abasic site can lead to both transversions (purine > pyrimidine or pyrimidine > purine) as well as transitions at initiating C/G base pairs (Fig. 3b) (9). Thus, UNG deficiency produces mainly transition mutations at C/G base pairs (9, 25).

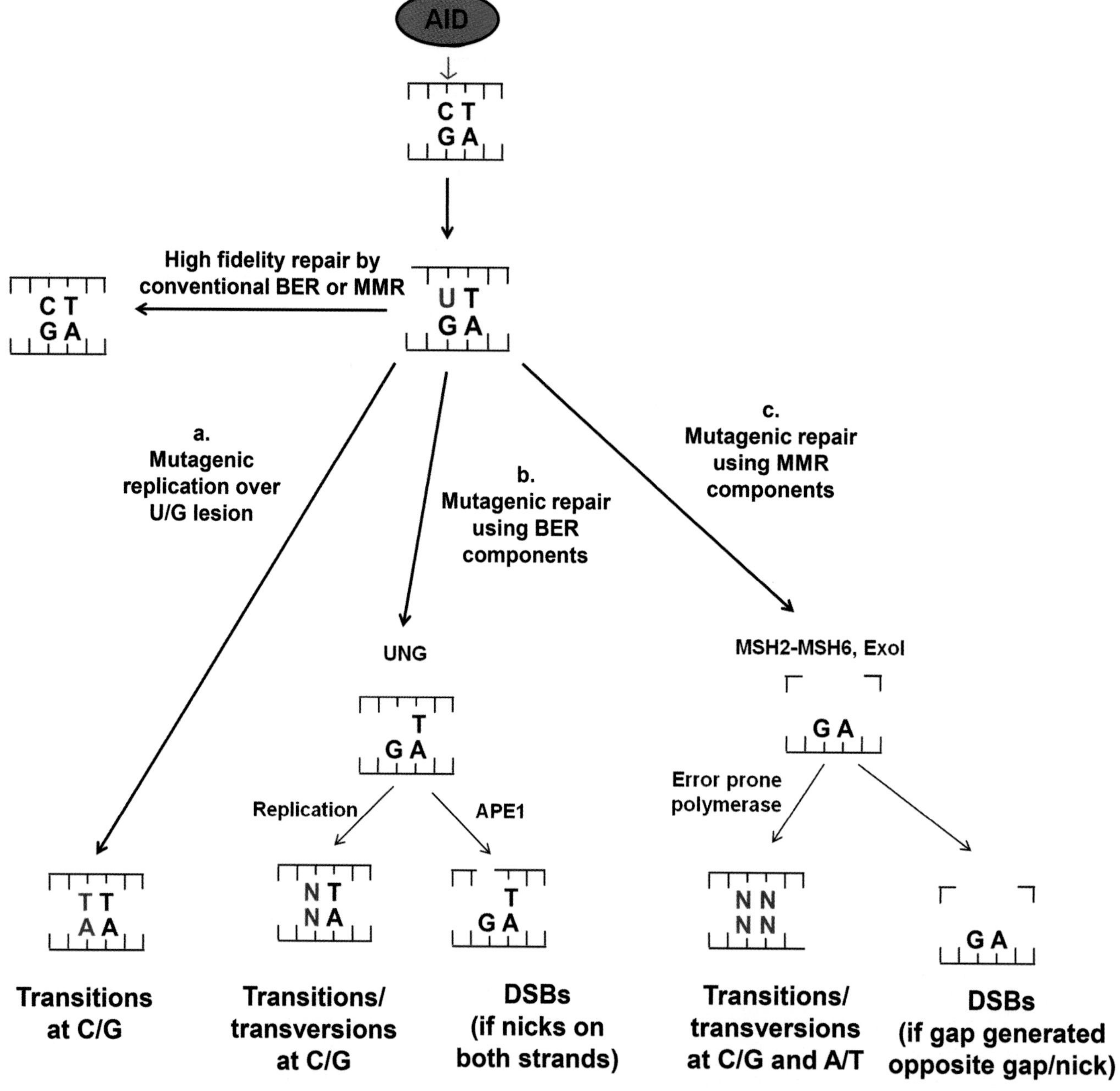

Figure 3 Mechanisms of AID cytidine deamination in SHM and CSR. AID deaminates cytidine (C) to uridines (U). The U/G lesion may be repaired with high fidelity (i.e. to C/G) by conventional base excision repair (BER) or mismatch repair (MMR). Mutagenic outcomes during SHM and CSR are generated by the following processes. **a.** Replication over the U/G lesion produces transition mutations at C/G base pairs. **b.** Uracil-DNA-Glycosylase (UNG) of the BER pathway excises the U creating an abasic site. Replication over the abasic site generates transition and transversion mutations at C/G base pairs. N indicates any nucleotide A,G,C, or T. AP endonuclease 1 (APE1) may create a nick at the abasic site. Nicks on both DNA strands may lead to DSBs. **c.** MSH2-MSH6 of the mismatch repair pathway recognize the U/G mismatch. Exo1 excises the patch of DNA containing the mismatch. Error-prone polymerase resynthesizes the patch leading to spreading of mutations to A/T base pairs. Overlapping gaps may lead to DSBs.
doi:10.1128/microbiolspec.MDNA3-0037-2014.f3

As AID only deaminates C's, the mutagenic BER processes described above cannot account for SHM at A and T residues. Rather, SHMs at A's and T's depends largely on components of the MMR pathway (47, 48). The MSH2-MSH6 heterodimer recognizes the U/G mismatch (49, 50), Exo1 then excises the patch of DNA containing the mismatch (51), and an error-prone polymerase such as polymerase η resynthesizes the patch (52, 53). Thus, mutations are "spread" from the C/G sites of deamination to nearby A/T sites.

The generation of DSBs during CSR also employs the activities of the BER and MMR pathways (Fig. 3b and c) (9, 20). Following the excision of uracil by UNG, AP endonuclease 1 (APE1) may create a nick at the abasic site (20, 44) (Fig. 3b). Adjacent nicks on opposite strands, for example in the context of AGCT motifs, may be sufficient to generate a DSB, particularly if target motifs are very dense as they are in S regions (54). Overlapping gaps generated during MMR can also lead to DSBs during CSR (55, 56) (Fig. 3c). Accordingly, while BER-deficiency or MMR-deficiency alone reduces CSR, combined BER and MMR deficiencies (e.g. UNG and MSH2 deficiency) abrogate CSR (9, 55). AID-initiated SHMs also accompany the DSBs that are generated in CSR-activated B cells, but in this case only mutations at C/G residues are found with little or no spreading to A/T residues. Such differential targeting/outcomes of AID activity during SHM and CSR will be discussed in more depth below.

UNG and MMR double deficiency, in addition to ablating CSR, also eliminates both C/G transversion mutations and spreading of mutations in the context of V region SHM, leaving only C/G transition mutations and also eliminates C/G transversion mutations in S regions during CSR (50, 55, 57). Thus in the absence of BER and MMR, V and S regions exhibit only the footprint of AID deamination, strongly supporting the two-step model in which AID deamination is followed by processing of the resulting lesions by BER and MMR to DSBs, transversion mutations at C and G residues, and transition and transversion mutations at A and T residues.

SOMATIC HYPERMUTATION

SHM Occurs During Germinal Center Responses

The process of somatic hypermutation (SHM) introduces point mutations in the assembled V regions of *IgH* and *IgL* genes of mature activated B cells (9). SHM takes place in the GC, a specialized structure found in B cell follicles of peripheral lymphoid organs (e.g lymph nodes and spleen) where rapidly proliferating B cells accumulate after primary immunization (18). Consistent with the role of AID in SHM, GC B cells express high levels of AID (32). In the GC, antigen-activated B cells, usually with the help of T cells, undergo multiple rounds of SHM and Darwinian-like selection for clones with high-affinity antigen binding followed by clonal expansion, leading to the evolution of B cells that express BCRs with increased affinity for the antigen. This process of affinity maturation is fundamental to the production of high-affinity antibodies to particular pathogens (18, 58).

SHM Targets in the Endogenous *IgH* and *IgL* Loci

Both productive and nonproductive *IgH* and *IgL* variable region exon alleles of B cells undergoing a GC response are subject to SHM (59, 60); however, only mutations in productive alleles affect the BCR and influence the fate of GC B cells that bind the antigen (58). GC B cells with SHMs that decrease BCR affinity for antigen or lead to auto-reactivity are not selected. In addition, B cells in which SHMs alter residues necessary for normal BCR functions (e.g. certain framework residues necessary for proper folding) are lost as BCR tonic (e.g. ligand-independent) signaling is required for survival (61). Given the strict selection for and against SHMs on the productive V(D)J alleles, SHMs on the nonproductive allele, which are not selected for or against, are considered a better indicator of intrinsic mutational patterns that are not biased by antigen selection (62, 63).

SHMs concentrate prominently within the CDRs of the V region exon (9). This accumulation does not appear to be merely due to selection since nonproductive V exons and passenger V transgenes (that provide a transcribed V(D)J exon that is not involved in BCR expression) also show preferential accumulation of SHMs in CDRs (40, 63). Preferential targeting of SHM to CDRs may reflect evolutionary pressure to direct SHM to parts of the V region that bind antigen and away from the intervening FWR, which are important for the Ig's structural integrity (64). However, the mechanism by which SHMs preferentially accumulate in the CDRs is not known. Differential AID targeting is a likely possibility, with CDRs of V_H exons containing somewhat more AID hotspots than other regions of the V exon (40). In this regard, although various codons can encode a given amino acid, CDRs preferentially use those that contain AID hotspot motifs (65, 66). However, it is notable that in nonproductively rearranged V exons, identical AID hotspot motifs (e.g. AGCT) mutate more when located in the CDRs as compared to when located in FWRs (40, 67). Thus, the underlying sequence of CDRs or their flanking regions, beyond AID target motifs, also may have a role in recruiting AID activity for SHM. In addition, the location of the CDRs in terms of their distance from the transcriptional start site (68) and/or other aspects of the overall structure of the V region exon could play a role in directing AID activity. Finally, the possibility of differential repair of AID deamination lesions might also contribute (42). Clearly, there is much to be learned about how AID is targeted within V exons during SHM.

The question of whether or how specific features of V exon sequences contribute to AID targeting must take into account findings that non-Ig sequences, including β-globin, chloramphenicol acetyltransferase and Ig Cκ sequences, are apparently robust substrates when inserted in place of the V exon in transgenic passenger alleles (69, 70, 71). However, the degree of SHM of these transgenes, while sometimes approaching that of the *bona fide* V exon, is quite dependent on integration site and copy-numbers (69). Thus, it is possible that features of the integration site or chromatin structural alterations associated with tandem transgene arrays could influence AID targeting in unknown ways. In this regard, targeting of such non-Ig sequences as single copies in place of an endogenous V exon will be required to conclusively address the degree to which these sequences undergo SHM relative to V exon sequences in a physiological endogenous setting. If non-Ig sequences mutate as well as V sequences in the endogenous V location, it would imply that there is nothing absolutely specific to the V exon sequence that targets high levels of SHM. In this case, the question would be why and how SHMs are focused on the CDRs, which may imply evolution of V exon to suppress SHM at FWRs versus CDRs. If non-Ig sequences do not mutate as well as V sequences when expressed in single copies in the normal physiological location, it would support the notion that CDRs evolved to specifically support AID-initiated SHMs.

Mutational Versus Deletional Outcomes During SHM

SHM, in contrast to CSR, is generally considered to involve predominantly point mutations and much less frequently DSBs (9). However, deletions have been found at relatively high frequency in some studies of nonproductive V exons (24) and passenger V exons (72). Such deletions generally would result from DSBs, which lead to deletions either through resection or by joining to another DSB in the same V exon (73). Such internal deletions are frequent in S regions in accord with their high DSB frequency (see below). DSBs can also lead to insertions, which have also been found in V exons in association with SHM (24). Together, deletions and insertions are often generically referred to as "indels" (74). DSBs and associated indels must occur at some frequency during SHM but are likely mostly selected against in productive V exon alleles, since they could disrupt reading frame or overall V region structure. The wide variability of the levels of indels found in V exons that have undergone SHM in different experiments (24, 63, 75, 76) could reflect most samples coming from productive alleles, the possibility that different V exon sequences have different propensities to undergo DSBs, limitations of sample size, or other factors. High-throughput sequencing of the Ig variable region exons from HIV-1 infected patients that produce rare broadly neutralizing antibodies have revealed that certain of these broadly neutralizing antibodies are extensively mutated and harbor very frequent indels (77, 78). How these anti-HIV broadly neutralizing antibodies accumulate such high levels of SHM and indels during affinity maturation is still speculative (78). An important question is whether AID-induced SHMs in some unmutated or affinity matured V exons can generate new sequences that further promote or direct DSBs and SHMs.

Mechanisms that Target SHM to Specific Variable Region Exon Targets

The mechanisms by which AID is targeted to its substrates is of great interest given the potentially deleterious consequences of AID's mutagenic activity. Off-target AID activities can activate oncogenes via mutations or translocations and, thereby, contribute to cancer (79, 80). In this regard, transcription has been shown to be a key factor for targeting AID to V exons (9, 26, 81, 82) as well as to S regions (see below). Correspondingly, deleting the V promoter eliminates SHM (83). In addition, non-Ig promoters can support SHM at least to some degree (83, 84, 85), suggesting that the V promoter *per se* may not direct AID targeting; but rather that such targeting is provided by transcription in general. Consistent with a key role for transcription, the spatial distribution of SHMs in a V exon is influenced by distance from the transcription start site (TSS), with the TSS defining the 5′ boundary of SHM (9, 76) and SHM frequency decreasing with distance from the TSS (68).

Ig enhancers, which are known transcriptional regulators (13, 82), promote SHM within transgenic V(D)J substrates (84, 86, 87). However, such enhancers, including the IgH intronic enhancer (iEμ), intronic Igκ enhancer (iEκ) and 3′ Igκ enhancer were deleted in mice and found to not be required for SHM (87, 88, 89). The difference between the transgene and endogenous findings may reflect redundancy of tested enhancer elements with other enhancers or other types of elements in the endogenous setting (82). The 30 kb IgH 3′RR contains a number of different enhancers and deletion of several of them in the endogenous locus can abolish germline C_H transcription and CSR to most C_Hs (91; see below), without affecting V(D)J transcription or SHM (92). However, recent studies showed that complete deletion of the 30 kb IgH 3′RR in the endogenous locus completely abolishes germline C_H transcription and CSR (93) and also severely impairs SHM (94). However, the impairment of SHM in the absence of the 30 kb IgH 3′RR was accompanied by only marginal reduction in transcription (94). These studies imply that the full IgH 3′RR contains elements that may impact AID targeting during SHM via mechanisms beyond transcription (94), as has also been suggested by mutation targeting studies in chicken *Igκ* ((95), see below). Another possibility is that the type of transcription (e.g. sense versus anti-sense) is important.

AID must be directed to its intended *Ig* gene substrate sequences versus other sequences to maintain specificity and reduce its potential off-target mutational activity. In addition, once at a target sequence, AID must gain access to a requisite ssDNA substrate. Transcription and transcription-related mechanisms have been implicated in facilitating both of these steps. In this context, a number of AID-associated proteins have been described that may contribute to these activities (96, 97). Transcriptional stalling has been implicated in directing AID to its targets (71, 81). The transcription associated factor Suppressor of Ty5 homolog (Spt5) associates with both AID (98) and RNA pol II (Pol II) (99, 100). Co-localization of Pol II, AID and Spt5 on genomic sites is predictive of AID-induced mutated sites, suggesting that Spt5 may target AID to genomic loci by recruiting AID to sites of stalled Pol II (98). The factors that lead to Pol II stalling in AID targets are unknown.

How AID Gains Access to the ssDNA Substrate Following Targeting

Following recruitment to targets, AID must gain access to the ssDNA template. Purified AID deaminates the nontemplate strand of mammalian S regions transcribed *in vitro* because mammalian S regions loop out the ssDNA nontemplate strand in the form of R loops upon transcription (35; see below). Purified AID does not deaminate T7-transcribed V exon substrates, which do not form R loops (101). However, serine 38 (S38)-phosphorylated AID in association with replication protein A (RPA) deaminates V exon substrates *in vitro* but only on the nontemplate strand, suggesting that RPA may assist in stabilizing ssDNA template for AID deamination and/or enhance further AID recruitment (101). Consistent with this model, disrupting the S38 phosphorylation of AID dramatically reduces SHM in GC B cells and CSR in activated B cells *in vivo* (102, 103). Both strands of DNA are targeted for SHM. Thus, while transcriptional stalling may target AID, and co-factors such as RPA may stabilize ssDNA substrates, a mechanism must exist to provide AID with access to the template strand, which may be masked by nascent RNA transcripts. The RNA exosome has been implicated in this role as it allows AID to robustly deaminate both strands of T7 transcribed substrates *in vitro* and is required for normal levels of CSR *in vivo* (104). The RNA exosome is an evolutionarily conserved exonuclease that processes nuclear RNA precursors and degrades RNA in the nucleus and cytoplasm (105). A working model suggests that once AID is brought to a target via stalled Pol II and Spt5, the RNA exosome displaces or degrades the nascent RNA, thus making the template strand available for deamination, which may *in vivo* be further augmented by RPA association (104) (Fig. 4b). Negative supercoiling is an additional mechanism that has been proposed to make both DNA strands available as ssDNA substrates (106, 107).

Factors that Promote AID Activity at "On-" and "Off-Target" Genes

As transcription occurs in a large number of genes in activated B cells, transcription alone cannot explain occurrence of SHM (and CSR) specifically within *Ig* loci and also within just a limited number of "off-target" genes, including potential B cell oncogenes, that undergo SHM at lower levels (42, 108, 109, 110). Thus many transcribed genes in B cells do not appear to undergo SHM. In this regard, unknown aspects of sequence context of AID targeting motifs in V(D)J exon, type of transcription (see above), aspects of specific chromatin structure, or *cis* elements beyond those involved in transcription may contribute to focusing SHM to Ig substrates and also contribute to AID

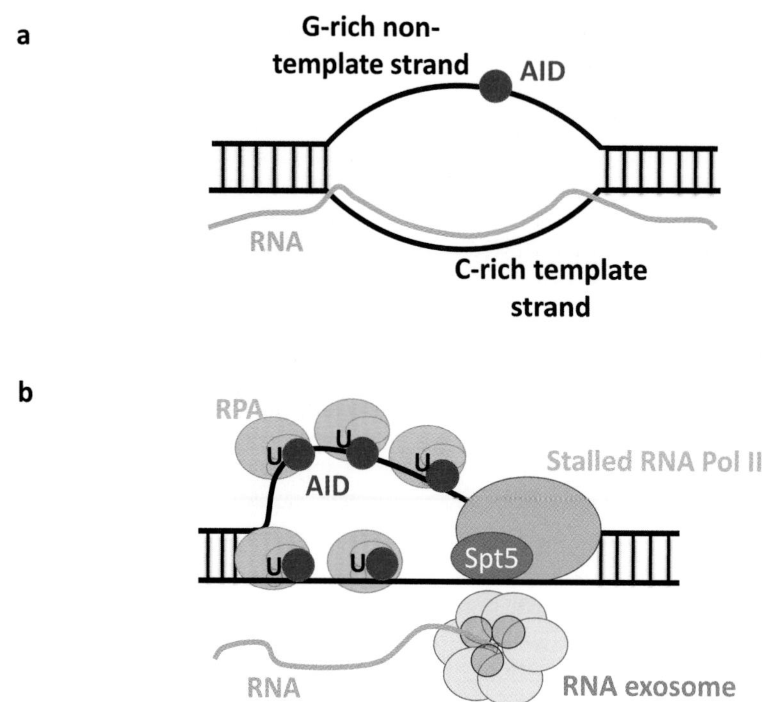

Figure 4 Transcriptional targeting of AID. **a.** R-loop structure. An R loop forms from G-rich RNA transcribed from the C-rich template strand forming a stable RNA-DNA hybrid with the C-rich template strand and looping out the G-rich nontemplate strand as ssDNA. **b.** A working model suggests that once AID is brought to a target via stalled Pol II and Spt5, the RNA exosome displaces or degrades the nascent RNA, thus making the template strand available for deamination, which may in vivo be further augmented by RPA association. Figure adapted from reference 104.
doi:10.1128/microbiolspec.MDNA3-0037-2014.f4

mutation to particular off-targets. Recent studies of the DT40 chicken B cell line identified within the chicken *Igκ* gene a diversification activator (DIVAC) that targets SHM to the V exon (95, 111, 112). The DIVAC contains multiple, redundant transcription factor binding motifs yet enhances SHM without stimulating a major increase in transcription (at least as measured by steady-state transcript levels), suggesting that it may target SHM by a transcription-independent mechanism (95). Notably, various mouse and human Ig enhancers also served as very strong DIVACs in this system, suggesting possible functional redundancy between enhancers and DIVACs in promoting SHM in the chicken *Igκ* locus, by mechanisms that do not involve influences on transcription *per se* (95). If so, however, then there still must be additional elements that function in this context in mammalian cells since mouse iEμ and mouse iEκ functioned as DIVACs in the DT40 assay (95), but their deletion does not markedly impair SHM in mouse GC B cells (88, 90).

IgH CLASS SWITCH RECOMBINATION

Overview

During class switch recombination (CSR), AID introduces cytidine deamination C to U mismatch lesions into large repetitive S regions that flank the upstream portion of C_H genes. These AID-initiated lesions are then converted into DSBs by co-opted BER and MMR activities (Fig. 3b and c). Deletional end-joining, usually via C-NHEJ of the upstream end of a DSB in Sμ to the downstream end of a DSB in a downstream S region, juxtaposes the V(D)J exon and downstream C_H gene to effect CSR. Whether there are mechanisms that promote such deletional joining versus the inversional joining alternative is unknown (6). The primary IgH class switching event involves switching from IgM to the various IgGs, IgE, and IgA, in which DSBs in the "donor" Sμ is joined to "acceptor" Sγ, Sε, or Sα. CSR also can in some cases occur successively. For example, an initial CSR event from Sμ to Sγ1 can generate a hybrid Sμ/Sγ1 donor S region that can subsequently successively undergo CSR with Sε to generate an IgE-producing B cell (30). Downstream "CSR" events between Sγ and Sε can also occur in cells in which Sμ has been truncated (113); in theory such a pathway could also contribute to successive CSR but the physiological significance of such downstream CSR events in normal B cells remains to be determined. The targeting of AID to S-region sequences shares at least some common mechanistic aspects with AID targeting during SHM. A primary mechanistic feature of AID targeting for both SHM and CSR is the requirement for transcription, which again both contributes to directing AID to its target S region sequence and which contributes to providing AID with access to ssDNA template.

Targeting of Specific IgH CSR Events via Differential Activation of Transcription in Various C_H Genes

Each set of C_H exons is part of a "germline" transcription unit referred to as a C_H gene, even though its transcribed RNA does not encode a known protein product (17) (Fig. 2). In these germline C_H genes, transcription is initiated from a cytokine activation-specific promoter upstream of a noncoding "I" (for "intervening") exon, continues through the associated S region, and terminates downstream of the C_H exons (17, 114). Different cytokines secreted by T cells and other immune cells can stimulate transcription from different I region promoters and, thereby, transcriptionally direct CSR to the C_H region most appropriate for a given type of pathogen or setting of pathogen infection (17, 20). As an example of such regulation, B cells activated through the CD40/CD40L pathway of B and T cell interaction that are also exposed to interleukin-4 (IL-4) (a cytokine secreted by T helper cells) will activate germline transcription from the IL-4 inducible Iγ1 and Iε promoters, and thereby direct CSR to IgG1 and IgE (114, 115, 116).

Induction of such germline C_H transcription occurs on both productive and nonproductive alleles (117), frequently leading to CSR to the same C_H gene on both alleles (118, 119), showing that CSR is a directed recombinational process. Deletion of the I exon abolishes germline C_H transcription from the associated I promoter and, thus, eliminates CSR to the corresponding S regions (120, 121). Moreover, replacement of I promoters with heterologous promoters ectopically targets CSR under B cell activation conditions in which it would not normally occur, implicating a direct role for transcription in directing CSR (122, 123, 124, 125). In addition to the cytokine and/or activation of specific I region promoters, differential regulation of transcription from these promoters is modulated by the IgH 3′RR located just downstream of the *IgH* locus (126).

The 30 kb region of IgH 3′RR contains multiple DNase I hypersensitive sites that correspond to enhancers (13). Combined deletion of one subset of these enhancers severely impairs germline C_H transcription from all I region promoters, except the Iγ1 promoter, and, correspondingly, severely impairs CSR to the transcriptionally inhibited C_H genes (91, 126). Deletion of the entire 30 kb region deletes additional enhancers and abrogates CSR to all C_H genes, including Cγ1, confirming that the IgH 3′RR is a master regulator of CSR, and implicating differential activity of elements within it in controlling CSR (93) and, as described earlier SHM (94). The IgH 3′RR has been implicated in differentially regulating CSR to different C_H genes by a "promoter competition" mechanism, by which certain activated I region promoters compete with and exclude interaction of other I promoters with the IgH 3′RR (126, 127). For example, LPS treatment of B cells induces germline transcription of Cγ2b and Cγ3 and CSR to these C_H genes. However, including IL-4 along with LPS in the treatment activates the Iγ1 and Iε promoters which are proposed to inhibit germline transcription from Iγ2b and Iγ3 promoters via competition for the IgH 3′RR, resulting in inhibition of CSR to Cγ2b and Cγ3 and activation of CSR to Cγ1 and Cε (127). Such regulation is consistent with the finding that the IgH 3′RR forms chromosomal loops with activated I region promoters (128, 129).

Targeting AID Activity Within S Regions

Once AID is targeted to S regions via transcriptionally related mechanisms, other features of targeting may overlap at least in part with those discussed for SHM, but there also are notable differences that are discussed below. Mammalian S regions are 1 to 10 kb long sequences composed primarily of tandem repetitive units (39). The deletion of S regions and their replacement with a sequence that lacks S-region features severely impairs CSR to the associated C_H exon, demonstrating that the S region plays a specific role in CSR (117, 130, 131). Mouse and human Sμ, Sα, and Sε are comprised of pentameric repeats, while the Sγ regions are comprised of 49 to 52 bp long repeats, that are enriched in smaller repeats including the AGCT and other DGYW targeting motifs (3, 39). In addition, mammalian S regions are C/G-rich and G-rich on the nontemplate strand (39). In contrast to mammalian S regions, the Sμ of *Xenopus* (frog) is A/T-rich and contains a high density of DGYW sequences, in particular AGCT motifs

(39, 132). As mentioned earlier, the palindromic nature of the AGCT motif may make it an optimal substrate for generating DSB breaks by providing AID hotspot motifs on both DNA strands (130, 131), with the high density of such motifs in S regions thereby promoting DSBs.

Transcription through mammalian S regions generates stable R loops *in vivo* (133). Such R loops result from G-rich RNA transcribed from the C-rich nontemplate S-region strand forming a stable RNA-DNA hybrid with the C-rich DNA template strand and looping out the G-rich nontemplate strand as ssDNA (134, 135, 136, 137, 138) (Fig. 4a). In biochemical experiments, AID robustly deaminates the nontemplate strand of a T7 transcribed S-region substrate that forms an R loop but not a C-rich substrate that does not form an R loop (35). R-loop formation during CSR in B cells is also abolished by inversion of the S-region sequence, which decreases, but does not eliminate, CSR (117). Based on such findings, the formation of R loops in mammalian S regions has been suggested to have evolved to enhance AID access to S regions (35, 138). However, as mentioned, inversion experiments show that R-loop formation is not absolutely required for S-region function in CSR (117). Likewise, *Xenopus* Sμ regions, which are A/T-rich and do not form R loops support substantial CSR when substituted for mouse S regions in activated B cells (131). We note, though, that it remains possible that R-loop formation in mammalian S regions evolved to enhance some other CSR function, for example by playing a role in S-region synapsis (see below).

As for SHM, AID targeting in CSR also appears to involve Pol II stalling, as revealed by accumulation of Pol ll in transcribed S regions (139, 140). R loops generated in the transcribed S region may enhance stalling (139, 140). Spt5 also has been implicated in recruiting AID in the context of stalled Pol II in S regions during CSR (98). *In vitro*, RPA facilitates AID access to the nontemplate strand of T7 transcribed *Xenopus* Sμ by association with S38-phosphorylated AID (101) and also binds R loop forming S regions in a S38-phosphorylated-AID-dependent manner (101, 141, 142). Thus, the prevention of AID S38-phosphorylation by an S38A mutation severely impairs RPA association and CSR in mouse B cells without affecting AID catalytic activity, supporting the notion that RPA interaction with S38-phosphorylated AID is important for CSR *in vivo* (102, 103, 143). The S38-phosphorylation of AID by PKA (143) may be important for CSR in a feedback loop mechanism that involves RPA and downstream repair factors (141, 144). Within transcribed mammalian S regions, template DNA is likely shielded from AID activity by stably hybridized RNA in the form of R loops (133). In this regard, the RNA exosome could function to displace/degrade the nascent RNA, facilitating targeting to the template strand (104) (Fig. 4b). *In vitro*, at least, the RNA exosome facilitates AID access to both DNA strands of non-R-loop-forming transcribed substrates (104, 145). Whether it plays such a role *in vivo* is unknown (104). Finally, RPA may stabilize the ssDNA substrates and augment AID activity in a phosphorylation-dependent manner (101, 143).

Evolution of CSR from SHM

Evolutionarily, SHM precedes CSR, with SHM emerging in early jawed vertebrates and CSR emerging in amphibians (146, 147). It has been proposed that *Xenopus* Sμ evolved to employ mechanisms utilized for AID targeting during SHM of variable region exons, which are not C/G-rich and which do not form R loops when transcribed (131). In this context, the AGCT-rich *Xenopus* Sμ region may have evolved via duplication of AGCT motif-dense CDR regions of V exons and in that context would target AID by SHM-like mechanisms. The novel features of mammalian S regions (C/G-richness, R-loop formation, much higher AGCT content) may have evolved to further enhance AID targeting or other aspects of CSR (e.g. synapsis, see below; 148). Further understanding of how S-region structure contributes to CSR may be illuminated by studying the divergent S regions of other species which may have evolved alternative or additional solutions for optimizing S-region substrates for CSR. For example, with respect to base composition, duck Sμ is C/G-rich but G:C content is equal between the two strands, while the putative duck Sα has an almost even distribution of the four bases on both strands with only a minor enrichment of G on one strand (39, 149). Assays of the abilities of these divergent S regions to support AID targeting and/or CSR in mammalian B cells could provide new insights into the S-region elements and the types of functions they support.

Joining AID-Initiated S-Region DSBs to Complete CSR: Overview

Productive CSR requires the upstream end of a DSB in Sμ to join to the downstream end of a DSB in a downstream S region that lies between 60 and 160 kb away, depending on the targeted S region. AID deaminates multiple cytosines within a given targeted S region (57); this may lead to multiple DSBs, with Sμ thought to be a particularly strong target. In addition to joining to

DSBs in other S regions, S-region DSBs may be directly joined back together, be joined back together following end resection, or be joined to another DSB within the same S region (Fig. 5). The latter two outcomes result in internal S-region deletions (ISDs). In addition, DSBs generated in an S region may participate in chromosomal translocations by joining to other non-S-region DSBs on the *cis* chromosome or to DSBs on other chromosomes (80). Yet, in activated B cell populations *in vitro*, CSR joins can occur in up to 50% or more cells over a 4-day period (102), raising the question of the nature of the mechanisms that promote CSR events over substantial distances within the *IgH* locus.

Factors That Promote S-Region Synapsis During CSR

For two DSBs in different genomic locations to be joined, they must be simultaneously broken and physically juxtaposed (synapsed). Thus, joining of two separate DSBs will be influenced by the frequency of DSBs at each site (which reflects both frequency of generation and time of persistence) and by the frequency with which the DSB target sequences are synapsed (6, 150). During V(D)J recombination, appropriate pairs of RSSs are likely synapsed by stochastic mechanisms that are enhanced by locus contraction. Then, RAG1/2 binds to and cleaves the synapsed pairs of RSS ends and subsequently holds them in a post-cleavage complex in which the appropriate ends are joined (e.g. coding end to coding end and RSS end to RSS end) exclusively by C-NHEJ (4, 5, 6). In contrast, AID can clearly act on S regions in the absence of their synapsis, as evidenced by frequent ISDs and other experiments (151). Thus, the question arises as to how joining between DSBs in two separate S regions occurs at sufficient levels to yield physiologic levels of CSR, as opposed to just being rejoined or joined as ISDs.

Studies of recombinational IgH class switching in the absence of S regions or AID have provided insight. Specifically, B cells in which Sμ and Sγ1 are replaced

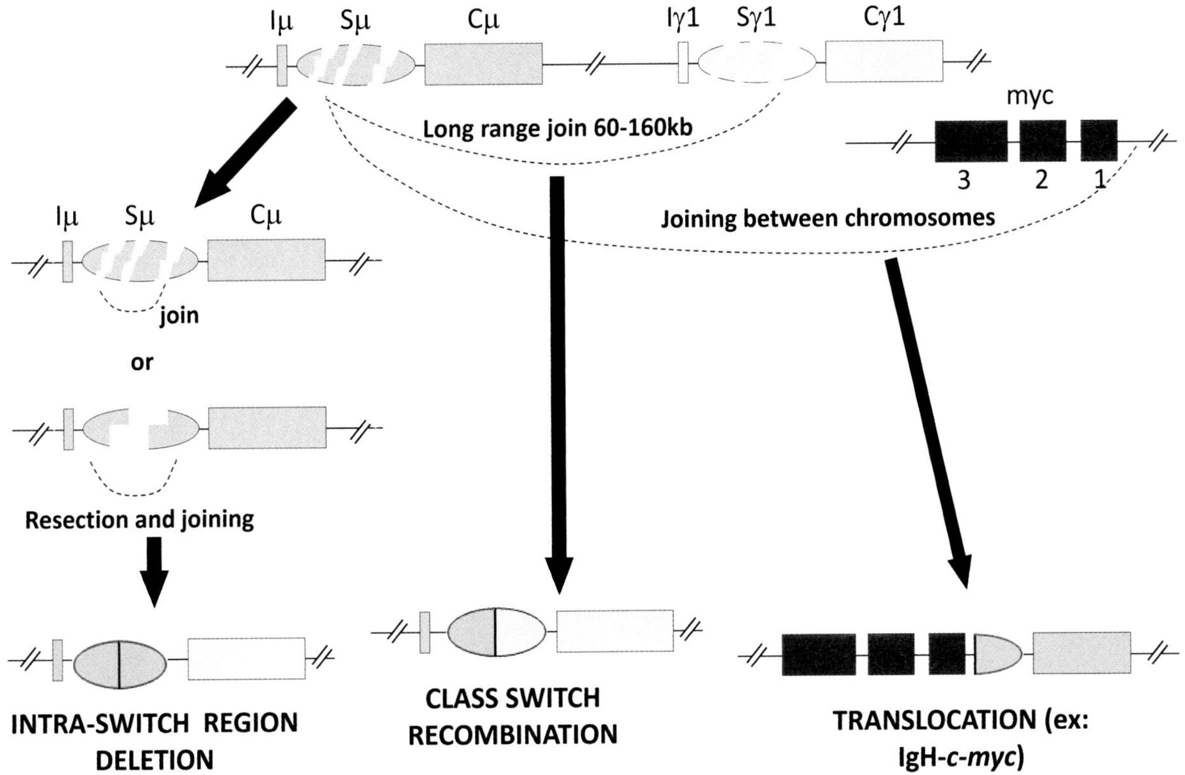

Figure 5 Outcomes of DSBs in S regions. DSBs within a S region may be directly joined back together or be joined back together following end resection, leading to intra-switch region deletions. Alternatively, a DSB generated in one S region may join to a DSB in another S region over a long-range (60 to 160 kb), which may lead to CSR. In addition, DSBs generated in an S region may participate in chromosomal translocations by joining to other non-S-region DSBs on the *cis* chromosome or to DSBs on other chromosomes.
doi:10.1128/microbiolspec.MDNA3-0037-2014.f5

with yeast I-SceI meganuclease target sites can undergo I-SceI-dependent recombinational class switching to IgG1 at levels approaching those of the wild type (WT) (152, 153). High-throughput analyses of joining frequencies between these widely separated IgH locus I-SceI-generated DSBs indicated that high level joining frequencies also occur in other cell types (e.g. T cells); in addition, similarly separated I-SceI or Cas-9/gRNA separated DSBs joined at high frequency in other chromosomal locations (e.g. around the *c-myc* gene) (153). Indeed, high-throughput DSB joining assays showed that DSBs within so-called megabase or submegabase topological "domains" (154, 155, 156, 157) are joined at surprisingly high frequency in several tested sites across the genome (6, 150).

The occurrence of sequences, such as of different S regions, within such megabase domains would generate a much greater chance of them being transiently synapsed by diffusion or related mechanisms (153). Thus, if two sequences in such a domain also were frequently broken, they would have a higher probability of being joined when broken while synapsed, thereby driving frequent joining (6, 153) (Fig. 6). CSR has been speculated to have evolved to exploit this general property of gene organization in chromatin (6, 17, 153). In addition, *IgH*-locus-specific mechanisms may further promote S-region synapsis in activated B cells; such mechanisms might involve the organization of the locus into specific loops via interactions of activated I-region promoters with enhancers and or other types of intralocus interactions (158). While AID has no known properties of stabilizing the synapsis of two S-region DSBs similar to RAG holding the DSBs in a post-cleavage complex, other factors including those of the DNA damage response regulated by the ataxia telangiectasia mutated kinase (ATM), as discussed below, may contribute such stabilization activities that increase the duration of S-region DSB synapsis to further promote CSR (97, 159, 160) (Fig. 6).

Contribution of the ATM-Dependent DNA Damage Response to Both Synapsis and Joining During CSR

DSBs, including AID-dependent S-region DSBs during CSR, activate the ATM-dependent DNA damage

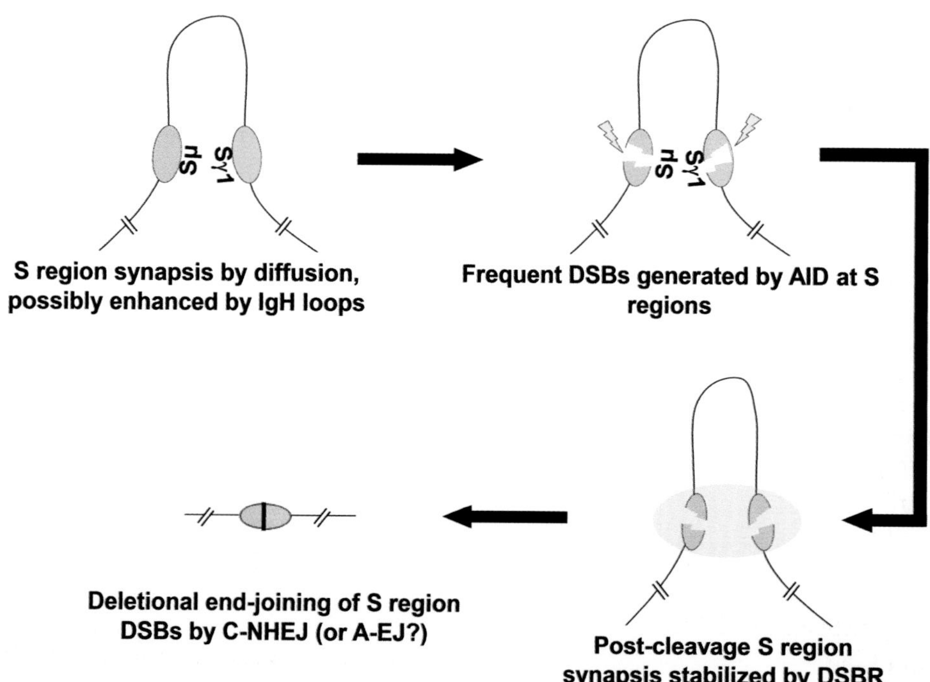

Figure 6 Synapsis and end-joining. The roles of synapsis and tethering in promoting long-range joining are shown. We propose that S regions are synapsed by diffusion, and that synapsis is possibly enhanced by proximity of S regions resulting from chromatin organization into megabase/submegabase domains. Post-cleavage, synapsis may be maintained by general DSB response (DSBR) factors, promoting the joining of S-region DSB ends by classical nonhomologous end-joining (C-NHEJ) and possibly alternative end-joining (A-EJ).
doi:10.1128/microbiolspec.MDNA3-0037-2014.f6

response. In the ATM-dependent DNA damage response, ATM phosphorylates substrates that mediate cell cycle checkpoints, such as p53, and DNA repair factors, including the proteins histone H2AX, mediator of DNA damage checkpoint protein 1 (MDC1), the nibrin (Nbs1) component of the Mre11-Rad50-Nbs1 complex (MRN), and tumor suppressor p53 binding protein 1 (53BP1) (160). ATM-dependent double-strand-break response factors (DSBR factors) form foci that extend megabases along chromatin flanking DSB sites and provide docking sites for protein complexes that bind and tether broken DNA ends (159, 161). DSBs in two different S regions are well within the range of overlapping DSBR foci, which might contribute to stabilizing the synapsis of broken S regions (159). Thus, deficiency for ATM or for several of its substrates, including H2AX, MDC1, and 53BP1 reduce CSR while increasing AID-dependent IgH breaks and translocations (162, 163).

Of the ATM substrates implicated in CSR, 53BP1 appears to have a specialized and especially critical role in CSR. In this regard, CSR in 53BP1-deficient B cells is profoundly reduced (164, 165). In addition, while overall levels of genome instability in CSR-activated 53BP1-deficient B cells are similar to those of ATM- or H2AX-deficient B cells, most of the instability arising in the context of 53BP1 deficiency occurs at the *IgH* locus, in contrast to the more wide spread genomic instability found in the context of ATM or H2AX deficiency (162, 163). Proposed roles for 53BP1 include preventing resection of DSB ends (166, 167), influencing repair pathway choice between C-NHEJ and alternative end-joining (A-EJ) (166), and promoting long-range joining (168). In the latter context, one potential function of 53BP1 would be to stabilize synapsed S regions once they are broken. ATM-dependent phosphorylation of 53BP1 recruits Rif1 to DSBs where it counters resection of DNA breaks that would inhibit CSR (169, 170, 171). The ATM-dependent DSB response also contributes directly to the joining of CSR DSBs via the C-NHEJ pathway discussed next (172).

End-Joining of Synapsed DSB Ends to Complete IgH CSR

DSBs in two synapsed S regions are end-joined to complete CSR. Most CSR DSBs appear to be joined by C-NHEJ, one of the two major known DSB joining pathways. C-NHEJ DSB end-joining may generate "direct" (or "blunt") DSB joins, which involves ligating blunt DNA ends (either generated as blunt ends or blunted by end-processing) or, alternatively, C-NHEJ can employ base pairing interactions between short (usually 1 to 4 bp) stretches of homology ("micro-homologies", MH) present in single-strand overhangs (173, 174). Consistent with C-NHEJ, most CSR junctions are direct or use short MHs (175, 176). C-NHEJ and a number of C-NHEJ factors were originally discovered based on the exclusive role of C-NHEJ for joining during V(D)J recombination (177). During C-NHEJ, DSBs are bound by the Ku70 and Ku80 end-binding complex (Ku), which promotes end ligation carried out by the XRCC4/ligase 4 (Lig4) complex (173). These factors Ku70, Ku80, XRCC4, and Lig4 are considered the "core" C-NHEJ factors as they are required for joining of all types of broken ends during C-NHEJ (174). Ku also recruits DNA-PKcs which appears to function both in end synapsis (178) and also by recruiting other factors, most notably the Artemis endonuclease, that process certain types of ends before they can be joined by core C-NHEJ factors (173).

Although V(D)J recombination and, thus, mature B cell development is strictly dependent on C-NHEJ (177, 179, 180, 181, 182), this developmental requirement can be circumvented by providing developing B cells with preassembled IgH and IgL variable region exons in their germline ("HL mice"). Germline inactivation of Ku, XRCC4 or Lig4 in HL mice, or conditional inactivation of the latter two factors in mature B cells, demonstrated that CSR can occur at up to 40% of normal levels in the absence of core C-NHEJ factors (175, 183, 184). In such C-NHEJ deficient B cells, unjoined AID-initiated DSBs progress into AID-dependent *IgH* locus chromosomal breaks and translocations (73, 175). DNA-PKcs and Artemis deficiencies generally have a milder effect on CSR levels (185, 186, 187, 188), consistent at least in part with major roles for these factors in processing certain types of ends but not others for C-NHEJ. In addition, Artemis-independent functions of DNA-PKcs, for example DSB end synapsis, can be provided by functionally redundant activities of the ATM kinase (189). Despite their more modest contribution to CSR joining than core C-NHEJ factors, the roles of DNA-PKcs and Artemis in CSR are clear since in the absence of either factor, activated B cells accumulate significantly increased levels of IgH locus chromosomal breaks and translocations (190).

CSR in the absence of C-NHEJ is carried out by one or more alternative end-joining pathways (5, 174). A-EJ has been described in many ways, including being considered MH-dependent. Yet, depending on context, A-EJ in mammalian cells is not necessarily MH-dependent (174). Thus, perhaps the best working definition is any end-joining that occurs in the absence of the core C-NHEJ factors (174, 191). In the absence of XRCC4 or Lig4, nearly all CSR junctions have MHs at their junctions in contrast to only 50% of CSR junctions having MHs in WT B cells (175). Notably, a fraction of these MHs in CSR junctions from C-NHEJ deficient cells are longer than those typically associated with C-NHEJ (174, 175, 176). The use of long MH may be influenced by sequence context, with ends from more related S regions having a greater chance to produce long MHs to support joining by A-EJ (175, 192). Currently, however, the degree to which A-EJ contributes to normal CSR in the presence of intact C-NHEJ remains to be determined.

A-EJ during CSR likely represents more than one pathway. These pathways include a Ku-independent pathway, as Ku-deficient cells also have reduced yet substantial CSR that is somewhat less MH-dependent than that of XRCC4 or Lig4 deficient B cells, and a Lig4-independent pathway that uses a different ligase downstream of Ku and other C-NHEJ components, as B cells lacking both Ku and Lig4 undergo CSR similarly to B cells lacking Ku (184). Factors thus far implicated in A-EJ during CSR include components of other known DNA repair pathways, including poly (ADP-ribose) polymerase 1 (PARP-1), which may provide a DSB end recognition function, Mre11 and the C-terminal-binding protein 1-interacting protein (CtIP), which may be involved in end-processing, and X-ray repair cross-complementing protein 1 (XRCC1), ligase 1, and ligase 3, which may be involved in end-joining (174). A-EJ, including the potential role of this pathway in CSR and translocations, has been recently reviewed in depth (174).

PERSPECTIVE

Elucidating mechanisms that promote differential targeting and outcome of AID activity in SHM and CSR is a major ongoing question. While many advances have been made, the question of how AID cytidine deamination activity is specifically targeted during SHM and CSR remains in substantial part unanswered. As outlined above, transcription of target sequences is required for CSR and SHM. Yet transcription *per se* is not sufficient to explain the specificity of AID targeting since most transcribed genes in CSR-activated or GC B cells are not subject to detectable AID deamination. In this context, the non-*Ig* loci that are recurrent targets of lower level AID targeting in CSR-activated or GC B cells also show no readily apparent similarity with S regions and V(D)J exons at the sequence level (193). A longstanding mystery involves the differential targeting of AID to V exons during the GC SHM response versus to S regions during CSR. Thus, AID acts on the S regions in CSR-activated B cells, but not (detectably) on the adjacent V(D)J exons, even though they are actively transcribed (194, 195). Likewise, some GC B cells have robust SHM within their *IgH* V exons in the absence of CSR (196, 197, 198). How AID achieves such specificity with respect to physiological targets is unknown. Another unanswered question is how the mutational outcome of AID activity is targeted to the three CDRs within variable region exons and the relative contributions of cellular selection versus actual targeting. Clearly, in all of these contexts, there are likely, yet to be defined, sequence motifs that may couple with unique transcription features of particular genes and contribute to make them AID targets. The chicken DIVAC elements, which function to enhance SHM through unknown mechanisms, are one potential example (95, 111, 112).

Once AID cytidine deamination is targeted, various mechanisms may lead to differential outcomes of this potentially mutagenic activity during SHM and CSR. In this regard, the level of AID targeting to a particular sequence is usually estimated based on the accumulation of mutations or DSBs and rearrangements/translocations. However, it remains possible that, for at least some sequences, the level at which such genomic alterations are found at particular sequences may be influenced by their predisposition to undergo high fidelity versus mutagenic repair outcomes of the AID-generated cytidine deamination lesions (42). It also is generally considered that AID targeting can preferentially lead to point mutational versus DSB outcomes, respectively, during SHM and CSR. However, more rigorous analyses need to be performed to determine the extent to which these generalizations actually apply, since AID activity at S regions clearly can lead to point mutations and AID activity on V regions can lead to DSBs. Beyond this potential caveat, sequences likely play an important role in the DSB outcome of AID activity in S regions, with a prime example being the abundance in S regions of the palindromic AGCT motif that could promote DSBs by leading to AID targeting on both DNA strands (6, 9, 26). In addition, other aspects of target sequences, differential repair pathways, or SHM versus CSR specific co-factors that favor the generation of point mutation versus DSBs may also play a role.

Another question that is not fully answered is how AID activity leads to C/G mutations in cell lines (199, 200, 201, 202, 203) or in CSR-activated B cells (194) versus spreading of the C/G mutations to A/T base pairs in GC B cells (204, 205). Error-prone DNA polymerases during BER and MMR in GC B cells have been implicated in the spreading process that generates mutations at A/T base pairs (206, 207), raising the possibility that differential expression of these enzymes or other related factors in GC B cells could contribute to SHM spreading (208). However, since V(D)J exons are not AID targets in B cells activated in culture for CSR, this notion has not yet been tested directly.

Acknowledgments. *We thank Feilong Meng, Ming Tian, and Vipul Kumar for helpful comments. This work was supported by NIH R01 AI077595 (to FA) and a fellowship from the Cancer Research Institute (to LSY). FWA is an Investigator of the Howard Hughes Medical Institute.*

Citation. Hwang JK, Alt FW, Yeap L-S. 2014. Related mechanisms of antibody somatic hypermutation and class switch recombination. Microbiol Spectrum 3(1):MDNA3-0037-2014

References

1. **Harwood NE, Batista FD.** 2008. New insights into the early molecular events underlying B cell activation. *Immunity* **28**(5):609–619.
2. **Cobb RM, Oestreich KJ, Osipovich OA, Oltz EM.** 2006. Accessibility control of V(D)J recombination. *Adv Immunol* **91**:45–109.
3. **Pan-Hammarstrom Q, Zhao Y, Hammarstrom L.** Class switch recombination: a comparison between mouse and human. *Adv Immunol* **93**:1–61.
4. **Schatz DG, Swanson PC.** 2011. V(D)J recombination: mechanisms of initiation. *Annu Rev Genet* **45**:167–202.
5. **Deriano L, Roth DB.** 2013. Modernizing the nonhomologous end-joining repertoire: alternative and classical NHEJ share the stage. *Annu Rev Genet* **47**:433–455.
6. **Alt FW, Zhang Y, Meng FL, Guo C, Schwer B.** 2013. Mechanisms of programmed DNA lesions and genomic instability in the immune system. *Cell* **152**(3):417–429.
7. **Alt FW, Baltimore D.** 1982. Joining of immunoglobulin heavy chain gene segments: implications from a chromosome with evidence of three D-JH fusions. *Proc Natl Acad Sci USA* **79**(13):4118–4122.
8. **Davis MM, Bjorkman PJ.** 1988. T-cell antigen receptor genes and T-cell recognition. *Nature* **334**(6181):395–402.
9. **Di Noia JM, Neuberger MS.** 2007. Molecular mechanisms of antibody somatic hypermutation. *Annu Rev Biochem* **76**:1–22.
10. **Stewart AK, Schwartz RS.** 1994. Immunoglobulin V regions and the B cell. *Blood* **83**(7):1717–1730.
11. **Roy AL, Sen R, Roeder RG.** 2011. Enhancer-promoter communication and transcriptional regulation of Igh. *Trends Immunol* **32**(11):532–539.
12. **Muramatsu M, Nagaoka H, Shinkura R, Begum NA, Honjo T.** 2007. Discovery of activation-induced cytidine deaminase, the engraver of antibody memory. *Adv Immunol* **94**:1–36.
13. **Pinaud E, Marquet M, Fiancette R, Peron S, Vincent-Fabert C, Denizot Y, Cogne M.** 2011. The IgH locus 3′ regulatory region: pulling the strings from behind. *Adv Immunol* **110**:27–70.
14. **Chen K, Cerutti A.** 2010. New insights into the enigma of immunoglobulin D. *Immunol Rev* **237**(1):160–179.
15. **Alt FW, Yancopoulos GD, Blackwell TK, Wood C, Thomas E, Boss M, Coffman R, Rosenberg N, Tonegawa S, Baltimore D.** 1984. Ordered rearrangement of immunoglobulin heavy chain variable region segments. *EMBO J* **3**(6):1209–1219.
16. **Mostoslavsky R, Alt RW, Rajewsky K.** 2004. The lingering enigma of the allelic exclusion mechanism. *Cell* **118**(5):539–544.
17. **Chaudhuri J, Basu U, Zarrin A, Yan C, Franco S, Perlot T, Vuong B, Wang J, Phan RT, Datta A, Manis J, Alt FW.** 2007. Evolution of the immunoglobulin heavy chain class switch recombination mechanism. *Adv Immunol* **94**:157–214.
18. **Victora GD, Nussenzweig MC.** 2012. Germinal centers. *Annu Rev Immunol* **30**:429–457.
19. **Fagarasan S, Kawamoto S, Kanagawa O, Suzuki K.** 2010. Adaptive immune regulation in the gut: T cell-dependent and T cell-independent IgA synthesis. *Annu Rev Immunol* **28**:243–273.
20. **Stavnezer J, Guikema JE, Schrader CE.** 2008. Mechanism and regulation of class switch recombination. *Annu Rev Immunol* **26**:261–292.
21. **Muramatsu M, Kinoshita K, Fagarasan S, Yamada S, Shinkai Y, Honjo T.** 2000. Class switch recombination and hypermutation require activation-induced cytidine deaminase (AID), a potential RNA editing enzyme. *Cell* **102**(5):553–563.
22. **Revy P, Muto T, Levy Y, Geissmann F, Plebani A, Sanal O, Catalan N, Forveille M, Dufourcq-Labelouse R, Gennery A, Tezcan I, Ersoy F, Kayserili H, Ugazio AG, Brousse N, Muramatsu M, Notarangelo LD, Kinoshita K, Honjo T, Fischer A, Durandy A.** 2000. Activation-induced cytidine deaminase (AID) deficiency causes the autosomal recessive form of the Hyper-IgM syndrome (HIGM2). *Cell* **102**(5):565–575.
23. **Wilson PC, de Bouteiller O, Liu YJ, Potter K, Banchereau J, Capra JD, Pascual V.** 1998. Somatic hypermutation introduces insertions and deletions into immunoglobulin V genes. *J Exp Med* **187**(1):59–70.
24. **Goossens T, Klein U, Kuppers R.** 1998. Frequent occurrence of deletions and duplications during somatic hypermutation: implications for oncogene translocations and heavy chain disease. *Proc Natl Acad Sci USA* **95**(5):2463–2468.
25. **Rada C, Williams GT, Nilsen H, Barnes DE, Lindahl T, Neuberger MS.** 2002. Immunoglobulin isotype switching is inhibited and somatic hypermutation perturbed in UNG-deficient mice. *Curr Biol* **12**(20):1748–1755.
26. **Maul RW, Gearhart PJ.** 2010. AID and somatic hypermutation. *Adv Immunol* **105**:159–191.
27. **Peled JU, Kuang FL, Iglesias-Ussel MD, Roa S, Kalis SL, Goodman MF, Scharff MD.** 2008. The biochemistry of somatic hypermutation. *Annu Rev Immunol* **26**:481–511.
28. **Matthews AJ, Zheng S, DiMenna LJ, Chaudhuri J.** 2014. Regulation of immunoglobulin class-switch

recombination: choreography of noncoding transcription, targeted DNA deamination, and long-range DNA repair. *Adv Immunol* **122:**1–57.

29. **Woof JM, Kerr MA.** 2006. The function of immunoglobulin A in immunity. *J Pathol* **208**(2):270–282.
30. **Wu LC, Zarrin AA.** 2014. The production and regulation of IgE by the immune system. *Nat Rev Immunol* **14** (4):247–259.
31. **Nimmerjahn F, Ravetch JV.** 2008. Fcgamma receptors as regulators of immune responses. *Nat Rev Immunol* **8** (1):34–47.
32. **Muramatsu M, Sankaranand VS, Anant S, Sugai M, Kinoshita K, Davidson NO, Honjo T.** 1999. Specific expression of activation-induced cytidine deaminase (AID), a novel member of the RNA-editing deaminase family in germinal center B cells. *J Biol Chem* **274**(26): 18470–18476.
33. **Arakawa H, Hauschild J, Buerstedde JM.** 2002. Requirement of the activation-induced deaminase (AID) gene for immunoglobulin gene conversion. *Science* **295** (5558):1301–1306.
34. **Bransteitter R, Pham P, Scharff MD, Goodman MF.** 2003. Activation-induced cytidine deaminase deaminates deoxycytidine on single-stranded DNA but requires the action of RNase. *Proc Natl Acad Sci USA* **100**(7): 4102–4107.
35. **Chaudhuri J, Tian M, Khuong C, Chua K, Pinaud E, Alt FW.** 2003. Transcription-targeted DNA deamination by the AID antibody diversification enzyme. *Nature* **422**(6933):726–730.
36. **Dickerson SK, Market E, Besmer E, Papavasiliou FN.** 2003. AID mediates hypermutation by deaminating single stranded DNA. *J Exp Med* **197**(10):1291–1296.
37. **Pham P, Bransteitter R, Petruska J, Goodman MF.** 2003. Processive AID-catalysed cytosine deamination on single-stranded DNA simulates somatic hypermutation. *Nature* **424**(6944):103–107.
38. **Rogozin IB, Diaz M.** 2004. Cutting edge: DGYW/WRCH is a better predictor of mutability at G:C bases in Ig hypermutation than the widely accepted RGYW/WRCY motif and probably reflects a two-step activation-induced cytidine deaminase-triggered process. *J Immunol* **172**(6):3382–3384.
39. **Hackney JA, Misaghi S, Senger K, Garris C, Sun Y, Lorenzo MN, Zarrin AA.** 2009. DNA targets of AID evolutionary link between antibody somatic hypermutation and class switch recombination. *Adv Immunol* **101:** 163–189.
40. **Dorner T, Brezinschek HP, Brezinschek RI, Foster SJ, Domiati-Saad R, Lipsky PE.** 1997. Analysis of the frequency and pattern of somatic mutations within nonproductively rearranged human variable heavy chain genes. *J Immunol* **158**(6):2779–2789.
41. **Rogozin IB, Kolchanov NA.** 1992. Somatic hypermutagenesis in immunoglobulin genes. II. Influence of neighbouring base sequences on mutagenesis. *Biochim Biophys Acta* **1171**(1):11–18.
42. **Liu M, Duke JL, Richter DJ, Vinuesa CG, Goodnow CC, Kleinstein SH, Schatz DG.** 2008. Two levels of protection for the B cell genome during somatic hypermutation. *Nature* **451**(7180):841–845.
43. **Krokan HE, Bjoras M.** 2013. Base excision repair. *Cold Spring Harbor Perspect Biol* **5**(4): a012583.
44. **Robertson AB, Klungland A, Rognes T, Leiros I.** 2009. DNA repair in mammalian cells: Base excision repair: the long and short of it. *Cell Mol Life Sci* **66**(6):981–993.
45. **Jiricny J.** 2006. The multifaceted mismatch-repair system. *Nat Rev Mol Cell Biol* **7**(5):335–346.
46. **Saribasak H, Gearhart PJ.** 2012. Does DNA repair occur during somatic hypermutation? *Semin Immunol* **24**(4):287–292.
47. **Neuberger MS, Di Noia JM, Beale RC, Williams GT, Yang Z, Rada C.** 2005. Somatic hypermutation at A.T pairs: polymerase error versus dUTP incorporation. *Nat Rev Immunol* **5**(2):171–178.
48. **Neuberger MS, Rada C.** 2007. Somatic hypermutation: activation-induced deaminase for C/G followed by polymerase eta for A/T. *J Exp Med* **204**(1):7–10.
49. **Wiesendanger M, Kneitz B, Edelmann W, Scharff MD.** 2000. Somatic hypermutation in MutS homologue (MSH)3-, MSH6-, and MSH3/MSH6-deficient mice reveals a role for the MSH2-MSH6 heterodimer in modulating the base substitution pattern. *J Exp Med* **191**(3):579–584.
50. **Shen HM, Tanaka A, Bozek G, Nicolae D, Storb U.** 2006. Somatic hypermutation and class switch recombination in Msh6(-/-)Ung(-/-) double-knockout mice. *J Immunol* **177**(8):5386–5392.
51. **Bardwell PD, Woo CJ, Wei K, Li Z, Martin A, Sack SZ, Parris T, Edelmann W, Scharff MD.** 2004. Altered somatic hypermutation and reduced class-switch recombination in exonuclease 1-mutant mice. *Nat Immunol* **5** (2):224–229.
52. **Zeng X, Winter DB, Kasmer C, Kraemer KH, Lehmann AR, Gearhart PJ.** 2001. DNA polymerase eta is an A-T mutator in somatic hypermutation of immunoglobulin variable genes. *Nat Immunol* **2**(6):537–541.
53. **Delbos F, Aoufouchi S, Faili A, Weill JC, Reynaud CA.** 2007. DNA polymerase eta is the sole contributor of A/T modifications during immunoglobulin gene hypermutation in the mouse. *J Exp Med* **204**(1):17–23.
54. **Petersen-Mahrt SK, Harris RS, Neuberger MS.** 2002. AID mutates E. coli suggesting a DNA deamination mechanism for antibody diversification. *Nature* **418** (6893):99–103.
55. **Rada C, Di Noia JM, Neuberger MS.** 2004. Mismatch recognition and uracil excision provide complementary paths to both Ig switching and the A/T-focused phase of somatic mutation. *Mol Cell* **16**(2):163–171.
56. **Chahwan R, Edelmann W, Scharff MD, Roa S.** 2012. AIDing antibody diversity by error-prone mismatch repair. *Semin Immunol* **24**(4):293–300.
57. **Xue K, Rada C, Neuberger MS.** 2006. The in vivo pattern of AID targeting to immunoglobulin switch regions deduced from mutation spectra in msh2-/- ung-/- mice. *J Exp Med* **203**(9):2085–2094.
58. **Neuberger MS.** 2008. Antibody diversification by somatic mutation: from Burnet onwards. *Immunol Cell Biol* **86**(2):124–132.

59. **Pech M, Hochtl J, Schnell H, Zachau HG.** 1981. Differences between germ-line and rearranged immunoglobulin V kappa coding sequences suggest a localized mutation mechanism. *Nature* **291**(5817):668–670.

60. **Roes J, Huppi K, Rajewsky K, Sablitzky F.** 1989. V gene rearrangement is required to fully activate the hypermutation mechanism in B cells. *J Immunol* **142**(3):1022–1026.

61. **Lam KP, Kuhn R, Rajewsky K.** 1997. In vivo ablation of surface immunoglobulin on mature B cells by inducible gene targeting results in rapid cell death. *Cell* **90**(6): 1073–1083.

62. **Betz AG, Neuberger MS, Milstein C.** 1993. Discriminating intrinsic and antigen-selected mutational hotspots in immunoglobulin V genes. *Immunol Today* **14**(8): 405–411.

63. **Betz AG, Rada C, Pannell R, Milstein C, Neuberger MS.** 1993. Passenger transgenes reveal intrinsic specificity of the antibody hypermutation mechanism: clustering, polarity, and specific hot spots. *Proc Natl Acad Sci USA* **90**(6):2385–2388.

64. **Jolly CJ, Wagner SD, Rada C, Klix N, Milstein C, Neuberger MS.** 1996. The targeting of somatic hypermutation. *Semin Immunol* **8**(3):159–168.

65. **Wagner SD, Milstein C, Neuberger MS.** 1995. Codon bias targets mutation. *Nature* **376**(6543):732.

66. **Kepler TB.** 1997. Codon bias and plasticity in immunoglobulins. *Mol Biol Evol* **14**(6):637–643.

67. **Foster SJ, Dorner T, Lipsky PE.** 1999. Somatic hypermutation of VkappaJkappa rearrangements: targeting of RGYW motifs on both DNA strands and preferential selection of mutated codons within RGYW motifs. *Eur J Immunol* **29**(12):4011–4021.

68. **Rada C, Milstein C.** 2001. The intrinsic hypermutability of antibody heavy and light chain genes decays exponentially. *EMBO J* **20**(16):4570–4576.

69. **Yelamos J, Klix N, Goyenechea B, Lozano F, Chui YL, Gonzalez Fernandez A, Pannell R, Neuberger MS, Milstein C.** 1995. Targeting of non-Ig sequences in place of the V segment by somatic hypermutation. *Nature* **376**(6537):225–229.

70. **Azuma T, Motoyama N, Fields LE, Loh DY.** 1993. Mutations of the chloramphenicol acetyl transferase transgene driven by the immunoglobulin promoter and intron enhancer. *Int Immunol* **5**(2):121–130.

71. **Peters A, Storb U.** 1996. Somatic hypermutation of immunoglobulin genes is linked to transcription initiation. *Immunity* **4**(1):57–65.

72. **Bross L, Fukita Y, McBlane F, Demolliere C, Rajewsky K, Jacobs H.** 2000. DNA double-strand breaks in immunoglobulin genes undergoing somatic hypermutation. *Immunity* **13**(5):589–597.

73. **Boboila C, Jankovic M, Yan CT, Wang JH, Wesemann DR, Zhang T, Fazeli A, Feldman L, Nussenzweig A, Nussenzweig M, Alt FW.** 2010. Alternative end-joining catalyzes robust IgH locus deletions and translocations in the combined absence of ligase 4 and Ku70. *Proc Natl Acad Sci USA* **107**(7):3034–3039.

74. **Briney BS, Willis JR, Crowe EF Jr.** 2012. Location and length distribution of somatic hypermutation-associated DNA insertions and deletions reveals regions of antibody structural plasticity. *Genes Immun* **13**(7):523–529.

75. **Weiss U, Zoebelein R, Rajewsky K.** 1992. Accumulation of somatic mutants in the B cell compartment after primary immunization with a T cell-dependent antigen. *Eur J Immunol* **22**(2):511–517.

76. **Lebecque SG, Gearhart PJ.** 1990. Boundaries of somatic mutation in rearranged immunoglobulin genes: 5′ boundary is near the promoter, and 3′ boundary is approximately 1 kb from V(D)J gene. *J Exp Med* **172**(6): 1717–1727.

77. **Wu X, Zhou T, Zhu J, Zhang B, Georgiev I, Wang C, Chen X, Longo NS, Louder M, McKee K, O'Dell S, Perfetto S, Schmidt SD, Shi W, Wu L, Yang Y, Yang ZY, Yang Z, Zhang Z, Bonsignori M, Crump JA, Kapiga SH, Sam NE, Haynes BF, Simek M, Burton DR, Koff WC, Doria-Rose NA, Connors M, Program NCS, Mullikin JC, Nabel GJ, Roederer M, Shapiro L, Kwong PD, Mascola JR.** 2011. Focused evolution of HIV-1 neutralizing antibodies revealed by structures and deep sequencing. *Science* **333**(6049):1593–1602.

78. **Mascola JR, Haynes BF.** 2013. HIV-1 neutralizing antibodies: understanding nature's pathways. *Immunol Rev* **254**(1):225–244.

79. **Kuppers R, Dalla-Favera R.** 2001. Mechanisms of chromosomal translocations in B cell lymphomas. *Oncogene* **20**(40):5580–5594.

80. **Gostissa M, Alt FW, Chiarle R.** 2011. Mechanisms that promote and suppress chromosomal translocations in lymphocytes. *Annu Rev Immunol* **29**:319–350.

81. **Storb U.** 2014. Why does somatic hypermutation by AID require transcription of its target genes? *Adv Immunol* **122**:253–277.

82. **Odegard VH, Schatz DG.** 2006. Targeting of somatic hypermutation. *Nat Rev Immunol* **6**(8):573–583.

83. **Fukita Y, Jacobs H, Rajewsky K.** 1998. Somatic hypermutation in the heavy chain locus correlates with transcription. *Immunity* **9**(1):105–114.

84. **Betz AG, Milstein C, Gonzalez-Fernandez A, Pannell R, Larson T, Neuberger MS.** 1994. Elements regulating somatic hypermutation of an immunoglobulin kappa gene: critical role for the intron enhancer/matrix attachment region. *Cell* **77**(2):239–248.

85. **Tumas-Brundage K, Manser T.** 1997. The transcriptional promoter regulates hypermutation of the antibody heavy chain locus. *J Exp Med* **185**(2):239–250.

86. **Sharpe MJ, Milstein C, Jarvis JM, Neuberger MS.** 1991. Somatic hypermutation of immunoglobulin kappa may depend on sequences 3′ of C kappa and occurs on passenger transgenes. *EMBO J* **10**(8):2139–2145.

87. **Terauchi A, Hayashi K, Kitamura D, Kozono Y, Motoyama N, Azuma T.** 2001. A pivotal role for DNase I-sensitive regions 3b and/or 4 in the induction of somatic hypermutation of IgH genes. *J Immunol* **167** (2):811–820.

88. **Inlay MA, Gao HH, Odegard VH, Lin T, Schatz DG, Xu Y.** 2006. Roles of the Ig kappa light chain intronic and 3′ enhancers in Igk somatic hypermutation. *J Immunol* **177**(2):1146–1151.

89. **van der Stoep N, Gorman JR, Alt FW.** 1998. Reevaluation of 3′Ekappa function in stage- and lineage-specific rearrangement and somatic hypermutation. *Immunity* **8** (6):743–750.

90. **Perlot T, Alt FW, Bassing CH, Suh H, Pinaud E.** 2005. Elucidation of IgH intronic enhancer functions via germ-line deletion. *Proc Natl Acad Sci USA* **102**(40): 14362–14367.

91. **Pinaud E, Khamlichi AA, Le Morvan C, Drouet M, Nalesso V, Le Bert M, Cogne M.** 2001. Localization of the 3′ IgH locus elements that effect long-distance regulation of class switch recombination. *Immunity* **15**(2): 187–199.

92. **Morvan CL, Pinaud E, Decourt C, Cuvillier A, Cogne M.** 2003. The immunoglobulin heavy-chain locus hs3b and hs4 3′ enhancers are dispensable for VDJ assembly and somatic hypermutation. *Blood* **102**(4):1421–1427.

93. **Vincent-Fabert C, Fiancette R, Pinaud E, Truffinet V, Cogne N, Cogne M, Denizot Y.** 2010. Genomic deletion of the whole IgH 3′ regulatory region (hs3a, hs1,2, hs3b, and hs4) dramatically affects class switch recombination and Ig secretion to all isotypes. *Blood* **116**(11): 1895–1898.

94. **Rouaud P, Vincent-Fabert C, Saintamand A, Fiancette R, Marquet M, Robert I, Reina-San-Martin B, Pinaud E, Cogne M, Denizot Y.** 2013. The IgH 3′ regulatory region controls somatic hypermutation in germinal center B cells. *J Exp Med* **210**(8):1501–1507.

95. **Buerstedde JM, Alinikula J, Arakawa H, McDonald JJ, Schatz DG.** 2014. Targeting of somatic hypermutation by immunoglobulin enhancer and enhancer-like sequences. *PLoS Biol* **12**(4):e1001831.

96. **Xu Z, Zan H, Pone EJ, Mai T, Casali P.** 2012. Immunoglobulin class-switch DNA recombination: induction, targeting and beyond. *Nat Rev Immunol* **12**(7):517–31.

97. **Daniel JA, Nussenzweig A.** 2013. The AID-induced DNA damage response in chromatin. *Mol Cell* **50**(3): 309–321.

98. **Pavri R, Gazumyan A, Jankovic M, Di Virgilio M, Klein I, Ansarah-Sobrinho C, Resch W, Yamane A, Reina San-Martin B, Barreto V, Nieland TJ, Root DE, Casellas R, Nussenzweig MC.** 2010. Activation-induced cytidine deaminase targets DNA at sites of RNA polymerase II stalling by interaction with Spt5. *Cell* **143**(1): 122–133.

99. **Wada T, Takagi T, Yamaguchi Y, Ferdous A, Imai T, Hirose S, Sugimoto S, Yano K, Hartzog GA, Winston F, Buratowski S, Handa H.** 1998. DSIF, a novel transcription elongation factor that regulates RNA polymerase II processivity, is composed of human Spt4 and Spt5 homologs. *Genes Dev* **12**(3):343–356.

100. **Hartzog GA, Wada T, Handa H, Winston F.** 1998. Evidence that Spt4, Spt5, and Spt6 control transcription elongation by RNA polymerase II in Saccharomyces cerevisiae. *Genes Dev* **12**(3):357–369.

101. **Chaudhuri J, Khuong C, Alt FW.** 2004. Replication protein A interacts with AID to promote deamination of somatic hypermutation targets. *Nature* **430**(7003): 992–998.

102. **Cheng HL, Vuong BQ, Basu U, Franklin A, Schwer B, Astarita J, Phan RT, Datta A, Manis J, Alt FW, Chaudhuri J.** 2009. Integrity of the AID serine-38 phosphorylation site is critical for class switch recombination and somatic hypermutation in mice. *Proc Natl Acad Sci USA* **106**(8):2717–2722.

103. **McBride KM, Gazumyan A, Woo EM, Schwickert TA, Chait BT, Nussenzweig MC.** 2008. Regulation of class switch recombination and somatic mutation by AID phosphorylation. *J Exp Med* **205**(11):2585–2594.

104. **Basu U, Meng FL, Keim C, Grinstein V, Pefanis E, Eccleston J, Zhang T, Myers D, Wasserman CR, Wesemann DR, Januszyk K, Gregory RI, Deng H, Lima CD, Alt FW.** 2011. The RNA exosome targets the AID cytidine deaminase to both strands of transcribed duplex DNA substrates. *Cell* **144**(3):353–363.

105. **Houseley J, LaCava J, Tollervey D.** 2006. RNA-quality control by the exosome. *Nat Rev Mol Cell Biol* **7**(7): 529–539.

106. **Shen HM, Storb U.** 2004. Activation-induced cytidine deaminase (AID) can target both DNA strands when the DNA is supercoiled. *Proc Natl Acad Sci USA* **101** (35):12997–13002.

107. **Longerich S, Basu U, Alt F, Storb U.** 2006. AID in somatic hypermutation and class switch recombination. *Curr Opin Immunol* **18**(2):164–174.

108. **Pasqualucci L, Migliazza A, Fracchiolla N, William C, Neri A, Baldini L, Chaganti RS, Klein U, Kuppers R, Rajewsky K, Dalla-Favera R.** 1998. BCL-6 mutations in normal germinal center B cells: evidence of somatic hypermutation acting outside Ig loci. *Proc Natl Acad Sci USA* **95**(20):11816–11821.

109. **Pasqualucci L, Neumeister P, Goossens T, Nanjangud G, Chaganti RS, Kuppers R, Dalla-Favera R.** 2001. Hypermutation of multiple proto-oncogenes in B-cell diffuse large-cell lymphomas. *Nature* **412**(6844):341–346.

110. **Shen HM, Peters A, Baron B, Zhu X, Storb U.** 1998. Mutation of BCL-6 gene in normal B cells by the process of somatic hypermutation of Ig genes. *Science* **280** (5370):1750–1752.

111. **Kohler KM, McDonald JJ, Duke JL, Arakawa H, Tan S, Kleinstein SH, Buerstedde JM, Schatz DG.** 2012. Identification of core DNA elements that target somatic hypermutation. *J Immunol* **189**(11):5314–5326.

112. **Blagodatski A, Batrak V, Schmidl S, Schoetz U, Caldwell RB, Arakawa H, Buerstedde JM.** 2009. A cis-acting diversification activator both necessary and sufficient for AID-mediated hypermutation. *PLoS Genet* **5**(1):e1000332.

113. **Zhang T, Franklin A, Boboila C, McQuay A, Gallagher MP, Manis JP, Khamlichi AA, Alt FW.** 2010. Downstream class switching leads to IgE antibody production by B lymphocytes lacking IgM switch regions. *Proc Natl Acad Sci USA* **107**(7):3040–3045.

114. **Lutzker S, Rothman P, Pollock R, Coffman R, Alt FW.** 1988. Mitogen- and IL-4-regulated expression of germline Ig gamma 2b transcripts: evidence for directed heavy chain class switching. *Cell* **53**(2):177–184.

115. **Rothman P, Chen YY, Lutzker S, Li SC, Stewart V, Coffman R, Alt FW.** 1990. Structure and expression of germ line immunoglobulin heavy-chain epsilon transcripts: interleukin-4 plus lipopolysaccharide-directed switching to C epsilon. *Mol Cell Biol* **10**(4):1672–1679.

116. **Esser C, Radbruch A.** 1989. Rapid induction of transcription of unrearranged S gamma 1 switch regions in activated murine B cells by interleukin 4. *EMBO J* **8**(2): 483–488.

117. **Shinkura R, Tian M, Smith M, Chua K, Fujiwara Y, Alt FW.** 2003. The influence of transcriptional orientation on endogenous switch region function. *Nat Immunol* **4** (5):435–441.

118. **Radbruch A, Sablitzky F.** 1983. Deletion of Cmu genes in mouse B lymphocytes upon stimulation with LPS. *EMBO J* **2**(11):1929–1935.

119. **Radbruch A, Muller W, Rajewsky K.** 1986. Class switch recombination is IgG1 specific on active and inactive IgH loci of IgG1-secreting B-cell blasts. *Proc Natl Acad Sci USA* **83**(11):3954–3957.

120. **Jung S, Rajewsky K, Radbruch A.** 1993. Shutdown of class switch recombination by deletion of a switch region control element. *Science* **259**(5097):984–987.

121. **Zhang J, Bottaro A, Li S, Stewart V, Alt FW.** 1993. A selective defect in IgG2b switching as a result of targeted mutation of the I gamma 2b promoter and exon. *EMBO J* **12**(9):3529–3537.

122. **Bottaro A, Lansford R, Xu L, Zhang J, Rothman P, Alt FW.** 1994. S region transcription per se promotes basal IgE class switch recombination but additional factors regulate the efficiency of the process. *EMBO J* **13**(3): 665–674.

123. **Seidl KJ, Bottaro A, Vo A, Zhang J, Davidson L, Alt FW.** 1998. An expressed neo(r) cassette provides required functions of the 1gamma2b exon for class switching. *Int Immunol* **10**(11):1683–1692.

124. **Lorenz M, Jung S, Radbruch A.** 1995. Switch transcripts in immunoglobulin class switching. *Science* **267** (5205):1825–1828.

125. **Qiu G, Harriman GR, Stavnezer J.** 1999. Ialpha exon-replacement mice synthesize a spliced HPRT-C-(alpha) transcript which may explain their ability to switch to IgA. Inhibition of switching to IgG in these mice. *Int Immunol* **11**(1):37–46.

126. **Cogne M, Lansford R, Bottaro A, Zhang J, Gorman J, Young F, Cheng HL, Alt FW.** 1994. A class switch control region at the 3′ end of the immunoglobulin heavy chain locus. *Cell* **77**(5):737–747.

127. **Seidl KJ, Manis JP, Bottaro A, Zhang J, Davidson L, Kisselgof A, Oettgen H, Alt FW.** 1999. Position-dependent inhibition of class-switch recombination by PGK-neor cassettes inserted into the immunoglobulin heavy chain constant region locus. *Proc Natl Acad Sci USA* **96**(6):3000–3005.

128. **Wuerffel R, Wang L, Grigera F, Manis J, Selsing E, Perlot T, Alt FW, Cogne M, Pinaud E, Kenter AL.** 2007. S-S synapsis during class switch recombination is promoted by distantly located transcriptional elements and activation-induced deaminase. *Immunity* **27**(5):711–722.

129. **Yan Y, Pieretti J, Ju Z, Wei S, Christin JR, Bah F, Birshtein BK, Eckhardt LA.** 2011. Homologous elements hs3a and hs3b in the 3′ regulatory region of the murine immunoglobulin heavy chain (Igh) locus are both dispensable for class-switch recombination. *J Biol Chem* **286**(31):27123–27131.

130. **Han L, Masani S, Yu K.** 2011. Overlapping activation-induced cytidine deaminase hotspot motifs in Ig class-switch recombination. *Proc Natl Acad Sci USA* **108** (28):11584–11589.

131. **Zarrin AA, Alt FW, Chaudhuri J, Stokes N, Kaushal D, Du Pasquier L, Tian M.** 2004. An evolutionarily conserved target motif for immunoglobulin class-switch recombination. *Nat Immunol* **5**(12):1275–1281.

132. **Mussmann R, Courtet M, Schwager J, Du Pasquier L.** 1997. Microsites for immunoglobulin switch recombination breakpoints from Xenopus to mammals. *Eur J Immunol* **27**(10):2610–2619.

133. **Yu K, Chedin F, Hsieh CL, Wilson TE, Lieber MR.** 2003. R-loops at immunoglobulin class switch regions in the chromosomes of stimulated B cells. *Nat Immunol* **4**(5):442–451.

134. **Reaban ME, Griffin JA.** 1990. Induction of RNA-stabilized DNA conformers by transcription of an immunoglobulin switch region. *Nature* **348**(6299):342–344.

135. **Daniels GA, Lieber MR.** 1995. RNA:DNA complex formation upon transcription of immunoglobulin switch regions: implications for the mechanism and regulation of class switch recombination. *Nucleic Acids Res* **23** (24):5006–5011.

136. **Tian M, Alt FW.** 2000. Transcription-induced cleavage of immunoglobulin switch regions by nucleotide excision repair nucleases in vitro. *J Biol Chem* **275**(31): 24163–24172.

137. **Mizuta R, Iwai K, Shigeno M, Mizuta M, Uemura T, Ushiki T, Kitamura D.** 2003. Molecular visualization of immunoglobulin switch region RNA/DNA complex by atomic force microscope. *J Biol Chem* **278**(7):4431–4434.

138. **Roy D, Yu K, Lieber MR.** 2008. Mechanism of R-loop formation at immunoglobulin class switch sequences. *Mol Cell Biol* **28**(1):50–60.

139. **Rajagopal D, Maul RW, Ghosh A, Chakraborty T, Khamlichi AA, Sen R, Gearhart PJ.** 2009. Immunoglobulin switch mu sequence causes RNA polymerase II accumulation and reduces dA hypermutation. *J Exp Med* **206**(6):1237–1244.

140. **Wang L, Wuerffel R, Feldman S, Khamlichi AA, Kenter AL.** 2009. S region sequence, RNA polymerase II, and histone modifications create chromatin accessibility during class switch recombination. *J Exp Med* **206**(8): 1817–1830.

141. **Vuong BQ, Lee M, Kabir S, Irimia C, Macchiarulo S, McKnight GS, Chaudhuri J.** 2009. Specific recruitment of protein kinase A to the immunoglobulin locus regulates class-switch recombination. *Nat Immunol* **10**(4): 420–426.

142. **Yamane A, Resch W, Kuo N, Kuchen S, Li Z, Sun HW, Robbiani DF, McBride K, Nussenzweig MC, Casellas R.**

2011. Deep-sequencing identification of the genomic targets of the cytidine deaminase AID and its cofactor RPA in B lymphocytes. *Nat Immunol* **12**(1):62–69.

143. **Basu U, Chaudhuri J, Alpert C, Dutt S, Ranganath S, Li G, Schrum JP, Manis JP, Alt FW.** 2005. The AID antibody diversification enzyme is regulated by protein kinase A phosphorylation. *Nature* **438**(7067):508–511.

144. **Vuong BQ, Herrick-Reynolds K, Vaidyanathan B, Pucella JN, Ucher AJ, Donghia NM, Gu X, Nicolas L, Nowak U, Rahman N, Strout MP, Mills KD, Stavnezer J, Chaudhuri J.** 2013. A DNA break- and phosphorylation-dependent positive feedback loop promotes immunoglobulin class-switch recombination. *Nat Immunol* **14**(11):1183–1189.

145. **Keim C, Kazadi D, Rothschild G, Basu U.** 2013. Regulation of AID, the B-cell genome mutator. *Genes Dev* **27** (1):1–17.

146. **Cannon JP, Haire RN, Rast JP, Litman GW.** 2004. The phylogenetic origins of the antigen-binding receptors and somatic diversification mechanisms. *Immunol Rev* **200**:12–22.

147. **Stavnezer J, Amemiya CT.** 2004. Evolution of isotype switching. *Semin Immunol* **16**(4):257–275.

148. **Larson ED, Duquette ML, Cummings WJ, Streiff RJ, Maizels N.** 2005. MutSalpha binds to and promotes synapsis of transcriptionally activated immunoglobulin switch regions. *Curr Biol* **15**(5):470–474.

149. **Lundqvist ML, Middleton DL, Hazard S, Warr GW.** 2001. The immunoglobulin heavy chain locus of the duck. Genomic organization and expression of D, J, and C region genes. *J Biol Chem* **276**(50):46729–46736.

150. **Zhang Y, McCord RP, Ho YJ, Lajoie BR, Hildebrand DG, Simon AC, Becker MS, Alt FW, Dekker J.** 2012. Spatial organization of the mouse genome and its role in recurrent chromosomal translocations. *Cell* **148**(5): 908–921.

151. **Dudley DD, Manis JP, Zarrin AA, Kaylor L, Tian M, Alt FW.** 2002. Internal IgH class switch region deletions are position-independent and enhanced by AID expression. *Proc Natl Acad Sci USA* **99**(15):9984–9989.

152. **Zarrin AA, Del Vecchio C, Tseng E, Gleason M, Zarin P, Tian M, Alt FW.** 2007. Antibody class switching mediated by yeast endonuclease-generated DNA breaks. *Science* **315**(5810):377–381.

153. **Gostissa M, Schwer B, Chang A, Dong J, Meyers RM, Marecki GT, Choi VW, Chiarle R, Zarrin AA, Alt FW.** 2014. IgH class switching exploits a general property of two DNA breaks to be joined in cis over long chromosomal distances. *Proc Natl Acad Sci USA* **111**(7): 2644–2649.

154. **Dixon JR, Selvaraj S, Yue F, Kim A, Li Y, Shen Y, Hu M, Liu JS, Ren B.** 2012. Topological domains in mammalian genomes identified by analysis of chromatin interactions. *Nature* **485**(7398):376–380.

155. **Nagano T, Lubling Y, Stevens TJ, Schoenfelder S, Yaffe E, Dean W, Laue ED, Tanay A, Fraser P.** 2013. Single-cell Hi-C reveals cell-to-cell variability in chromosome structure. *Nature* **502**(7469):59–64.

156. **Naumova N, Imakaev M, Fudenberg G, Zhan Y, Lajoie BR, Mirny LA, Dekker J.** 2013. Organization of the mitotic chromosome. *Science* **342**(6161):948–953.

157. **Nora EP, Lajoie BR, Schulz EG, Giorgetti L, Okamoto I, Servant N, Piolot T, van Berkum NL, Meisig J, Sedat J, Gribnau J, Barillot E, Bluthgen N, Dekker J, Heard E.** 2012. Spatial partitioning of the regulatory landscape of the X-inactivation centre. *Nature* **485**(7398):381–385.

158. **Kenter AL, Feldman S, Wuerffel R, Achour I, Wang L, Kumar S.** 2012. Three-dimensional architecture of the IgH locus facilitates class switch recombination. *Ann N Y Acad Sci* **1267**:86–94.

159. **Bassing CH, Alt FW.** 2004. H2AX may function as an anchor to hold broken chromosomal DNA ends in close proximity. *Cell Cycle* **3**(2):149–153.

160. **Nussenzweig A, Nussenzweig MC.** 2010. Origin of chromosomal translocations in lymphoid cancer. *Cell* **141**(1):27–38.

161. **Rogakou EP, Boon C, Redon C, Bonner WM.** 1999. Megabase chromatin domains involved in DNA double-strand breaks in vivo. *J Cell Biol* **146**(5):905–916.

162. **Franco S, Gostissa M, Zha S, Lombard DB, Murphy MM, Zarrin AA, Yan C, Tepsuporn S, Morales JC, Adams MM, Lou Z, Bassing CH, Manis JP, Chen J, Carpenter PB, Alt FW.** 2006. H2AX prevents DNA breaks from progressing to chromosome breaks and translocations. *Mol Cell* **21**(2):201–214.

163. **Ramiro AR, Jankovic M, Callen E, Difilippantonio S, Chen HT, McBride KM, Eisenreich TR, Chen J, Dickins RA, Lowe SW, Nussenzweig A, Nussenzweig MC.** 2006. Role of genomic instability and p53 in AID-induced c-myc-Igh translocations. *Nature* **440**(7080): 105–109.

164. **Manis JP, Morales JC, Xia Z, Kutok JL, Alt FW, Carpenter PB.** 2004. 53BP1 links DNA damage-response pathways to immunoglobulin heavy chain class-switch recombination. *Nat Immunol* **5**(5):481–487.

165. **Ward IM, Reina-San-Martin B, Olaru A, Minn K, Tamada K, Lau JS, Cascalho M, Chen L, Nussenzweig A, Livak F, Nussenzweig MC, Chen J.** 2004. 53BP1 is required for class switch recombination. *J Cell Biol* **165** (4):459–464.

166. **Bothmer A, Robbiani DF, Feldhahn N, Gazumyan A, Nussenzweig A, Nussenzweig MC.** 2010. 53BP1 regulates DNA resection and the choice between classical and alternative end joining during class switch recombination. *J Exp Med* **207**(4):855–865.

167. **Bothmer A, Robbiani DF, Di Virgilio M, Bunting SF, Klein IA, Feldhahn N, Barlow J, Chen HT, Bosque D, Callen E, Nussenzweig A, Nussenzweig MC.** 2011. Regulation of DNA end joining, resection, and immunoglobulin class switch recombination by 53BP1. *Mol Cell* **42**(3):319–329.

168. **Difilippantonio S, Gapud E, Wong N, Huang CY, Mahowald G, Chen HT, Kruhlak MJ, Callen E, Livak F, Nussenzweig MC, Sleckman BP, Nussenzweig A.** 2008. 53BP1 facilitates long-range DNA end-joining during V(D)J recombination. *Nature* **456**(7221):529–533.

169. **Chapman JR, Barral P, Vannier JB, Borel V, Steger M, Tomas-Loba A, Sartori AA, Adams IR, Batista FD, Boulton SJ.** 2013. RIF1 is essential for 53BP1-dependent nonhomologous end joining and suppression of DNA double-strand break resection. *Mol Cell* **49**(5):858–871.

170. **Di Virgilio M, Callen E, Yamane A, Zhang W, Jankovic M, Gitlin AD, Feldhahn N, Resch W, Oliveira TY, Chait BT, Nussenzweig A, Casellas R, Robbiani DF, Nussenzweig MC.** 2013. Rif1 prevents resection of DNA breaks and promotes immunoglobulin class switching. *Science* **339**(6120):711–715.

171. **Zimmermann M, Lottersberger F, Buonomo SB, Sfeir A, de Lange T.** 2013. 53BP1 regulates DSB repair using Rif1 to control 5′ end resection. *Science* **339**(6120): 700–704.

172. **Kumar V, Alt FW, Oksenych V.** 2014. Functional overlaps between XLF and the ATM-dependent DNA double strand break response. *DNA Repair* **16**:11–22.

173. **Lieber MR.** 2010. The mechanism of double-strand DNA break repair by the nonhomologous DNA end-joining pathway. *Annu Rev Biochem* **79**:181–211.

174. **Boboila C, Alt FW, Schwer B.** 2012. Classical and alternative end-joining pathways for repair of lymphocyte-specific and general DNA double-strand breaks. *Adv Immunol* **116**:1–49.

175. **Yan CT, Boboila C, Souza EK, Franco S, Hickernell TR, Murphy M, Gumaste S, Geyer M, Zarrin AA, Manis JP, Rajewsky K, Alt FW.** 2007. IgH class switching and translocations use a robust non-classical end-joining pathway. *Nature* **449**(7161):478–482.

176. **Stavnezer J, Bjorkman A, Du L, Cagigi A, Pan-Hammarstrom W.** 2010. Mapping of switch recombination junctions, a tool for studying DNA repair pathways during immunoglobulin class switching. *Adv Immunol* **108**:45–109.

177. **Taccioli GE, Rathbun G, Oltz E, Stamato T, Jeggo PA, Alt FW.** 1993. Impairment of V(D)J recombination in double-strand break repair mutants. *Science* **260**(5105): 207–210.

178. **Meek K, Gupta S, Ramsden DA, Lees-Miller SP.** 2004. The DNA-dependent protein kinase: the director at the end. *Immunol Rev* **200**:132–141.

179. **Nussenzweig A, Chen C, da Costa Soares V, Sanchez M, Sokol K, Nussenzweig MC, Li GC.** 1996. Requirement for Ku80 in growth and immunoglobulin V(D)J recombination. *Nature* **382**(6591):551–555.

180. **Frank KM, Sekiguchi JM, Seidl KJ, Swat W, Rathbun GA, Cheng HL, Davidson L, Kangaloo L, Alt FW.** 1998. Late embryonic lethality and impaired V(D)J recombination in mice lacking DNA ligase IV. *Nature* **396** (6707):173–177.

181. **Gao Y, Sun Y, Frank KM, Dikkes P, Fujiwara Y, Seidl KJ, Sekiguchi JM, Rathbun GA, Swat W, Wang J, Bronson RT, Malynn BA, Bryans M, Zhu C, Chaudhuri J, Davidson L, Ferrini R, Stamato T, Orkin SH, Greenberg ME, Alt FW.** 1998. A critical role for DNA end-joining proteins in both lymphogenesis and neurogenesis. *Cell* **95**(7):891–902.

182. **Gu Y, Seidl KJ, Rathbun GA, Zhu C, Manis JP, van der Stoep N, Davidson L, Cheng HL, Sekiguchi JM, Frank K, Stanhope-Baker P, Schlissel MS, Roth DB, Alt FW.** 1997. Growth retardation and leaky SCID phenotype of Ku70-deficient mice. *Immunity* **7**(5):653–665.

183. **Soulas-Sprauel P, Le Guyader G, Rivera-Munoz P, Abramowski V, Olivier-Martin C, Goujet-Zalc C, Charneau P, de Villartay JP.** 2007. Role for DNA repair factor XRCC4 in immunoglobulin class switch recombination. *J Exp Med* **204**(7):1717–1727.

184. **Boboila C, Yan C, Wesemann DR, Jankovic M, Wang JH, Manis J, Nussenzweig A, Nussenzweig M, Alt FW.** 2010. Alternative end-joining catalyzes class switch recombination in the absence of both Ku70 and DNA ligase 4. *J Exp Med* **207**(2):417–427.

185. **Manis JP, Tian M, Alt FW.** 2002. Mechanism and control of class-switch recombination. *Trends Immunol* **23** (1):31–39.

186. **Cook AJ, Oganesian L, Harumal P, Basten A, Brink R, Jolly CJ.** 2003. Reduced switching in SCID B cells is associated with altered somatic mutation of recombined S regions. *J Immunol* **171**(12):6556–6564.

187. **Bosma GC, Kim J, Urich T, Fath DM, Cotticelli MG, Ruetsch NR, Radic MZ, Bosma MJ.** 2002. DNA-dependent protein kinase activity is not required for immunoglobulin class switching. *J Exp Med* **196**(11): 1483–1495.

188. **Kiefer K, Oshinsky J, Kim J, Nakajima PB, Bosma GC, Bosma MJ.** 2007. The catalytic subunit of DNA-protein kinase (DNA-PKcs) is not required for Ig class-switch recombination. *Proc Natl Acad Sci USA* **104**(8):2843–2848.

189. **Callen E, Jankovic M, Wong N, Zha S, Chen HT, Difilippantonio S, Di Virgilio M, Heidkamp G, Alt FW, Nussenzweig A, Nussenzweig M.** 2009. Essential role for DNA-PKcs in DNA double-strand break repair and apoptosis in ATM-deficient lymphocytes. *Mol Cell* **34** (3):285–297.

190. **Franco S, Murphy MM, Li G, Borjeson T, Boboila C, Alt FW.** 2008. DNA-PKcs and Artemis function in the end-joining phase of immunoglobulin heavy chain class switch recombination. *J Exp Med* **205**(3):557–564.

191. **Zhang Y, Gostissa M, Hildebrand DG, Becker MS, Boboila C, Chiarle R, Lewis S, Alt FW.** 2010. The role of mechanistic factors in promoting chromosomal translocations found in lymphoid and other cancers. *Adv Immunol* **106**:93–133.

192. **Pan-Hammarstrom Q, Jones AM, Lahdesmaki A, Zhou W, Gatti RA, Hammarstrom L, Gennery AR, Ehrenstein MR.** 2005. Impact of DNA ligase IV on nonhomologous end joining pathways during class switch recombination in human cells. *J Exp Med* **201**(2):189–194.

193. **Duke JL, Liu M, Yaari G, Khalil AM, Tomayko MM, Shlomchik MJ, Schatz DG, Kleinstein SH.** 2013. Multiple transcription factor binding sites predict AID targeting in non-Ig genes. *J Immunol* **190**(8):3878–3888.

194. **Nagaoka H, Muramatsu M, Yamamura N, Kinoshita K, Honjo T.** 2002. Activation-induced deaminase (AID)-directed hypermutation in the immunoglobulin

Smu region: implication of AID involvement in a common step of class switch recombination and somatic hypermutation. *J Exp Med* **195**(4)**:**529–534.

195. **Liu M, Schatz DG.** 2009. Balancing AID and DNA repair during somatic hypermutation. *Trends Immunol* **30** (4)**:**173–181.

196. **Klein U, Rajewsky K, Kuppers R.** 1998. Human immunoglobulin (Ig)M+IgD+ peripheral blood B cells expressing the CD27 cell surface antigen carry somatically mutated variable region genes: CD27 as a general marker for somatically mutated (memory) B cells. *J Exp Med* **188**(9)**:**1679–1689.

197. **Rosner K, Winter DB, Kasmer C, Skovgaard GL, Tarone RE, Bohr VA, Gearhart PJ.** 2001. Impact of age on hypermutation of immunoglobulin variable genes in humans. *J Clin Immunol* **21**(2)**:**102–115.

198. **van Es JH, Meyling FH, Logtenberg T.** 1992. High frequency of somatically mutated IgM molecules in the human adult blood B cell repertoire. *Eur J Immunol* **22** (10)**:**2761–2764.

199. **Martin A, Bardwell PD, Woo CJ, Fan M, Shulman MJ, Scharff MD.** 2002. Activation-induced cytidine deaminase turns on somatic hypermutation in hybridomas. *Nature* **415**(6873)**:**802–806.

200. **Sale JE, Neuberger MS.** 1998. TdT-accessible breaks are scattered over the immunoglobulin V domain in a constitutively hypermutating B cell line. *Immunity* **9**(6)**:**859–869.

201. **Denepoux S, Razanajaona D, Blanchard D, Meffre G, Capra JD, Banchereau J, Lebecque S.** 1997. Induction of somatic mutation in a human B cell line in vitro. *Immunity* **6**(1)**:**35–46.

202. **Bachl J, Wabl M.** 1996. An immunoglobulin mutator that targets G.C base pairs. *Proc Natl Acad Sci USA* **93** (2)**:**851–855.

203. **Sale JE, Bemark M, Williams GT, Jolly CJ, Ehrenstein MR, Rada C, Milstein C, Neuberger MS.** 2001. In vivo and in vitro studies of immunoglobulin gene somatic hypermutation. *Philos Trans R Soc B* **356**(1405)**:** 21–28.

204. **Milstein C, Neuberger MS, Staden R.** 1998. Both DNA strands of antibody genes are hypermutation targets. *Proc Natl Acad Sci USA* **95**(15)**:**8791–8794.

205. **Wagner SD, Neuberger MS.** 1996. Somatic hypermutation of immunoglobulin genes. *Annu Rev Immunol* **14:** 441–457.

206. **Reynaud CA, Delbos F, Faili A, Gueranger Q, Aoufouchi S, Weill JC.** 2009. Competitive repair pathways in immunoglobulin gene hypermutation. *Philos Trans R Soc B* **364**(1517)**:**613–619.

207. **Diaz M, Lawrence C.** 2005. An update on the role of translesion synthesis DNA polymerases in Ig hypermutation. *Trends Immunol* **26**(4)**:**215–220.

208. **Ouchida R, Ukai A, Mori H, Kawamura K, Dolle ME, Tagawa M, Sakamoto A, Tokuhisa T, Yokosuka T, Saito T, Yokoi M, Hanaoka F, Vijg J, Wang JY.** 2008. Genetic analysis reveals an intrinsic property of the germinal center B cells to generate A:T mutations. *DNA Repair* **7**(8)**:**1392–1398.

Mobile DNA, 3rd Edition
Nancy L. Craig, Michael Chandler, Martin Gellert, Alan M. Lambowitz, Phoebe A. Rice and Suzanne Sandmeyer

doi:10.1128/microbiolspec.MDNA3-0012-2014

Meng-chao Yao[1]
Ju-lan Chao[1]
Chao-yin Cheng[1]

16

Programmed Genome Rearrangements in *Tetrahymena*

INTRODUCTION

This chapter reviews recent studies on the remarkable phenomenon of programmed DNA rearrangements in ciliated protozoa, focusing primarily on the species *Tetrahymena thermophila*. The phenomenon occurs widely among ciliates, a diverse group of single-celled eukaryotes. It varies significantly in mechanistic detail among the species that have been described, chiefly *Tetrahymena*, *Paramecium*, *Euplotes*, *Stylonychia*, and *Oxytricha*. Readers are referred to other chapters in this volume for studies in *Paramecium* and *Oxytricha*. *Tetrahymena* displays perhaps the simplest version of these DNA rearrangements and is the easiest to grow and manipulate in the laboratory, hence offering excellent opportunities for in-depth understanding. Since the publication of *Mobile DNA II* (1), significant progress has brought about fundamental changes in our understanding. Among other things, clear links have now been established between these processes and RNA interference (RNAi) and transposon domestication. This chapter will concentrate on progress made during this period, with a brief summary of earlier work to provide an introduction.

"Chromatin diminution" was first reported by Boveri in 1887 (2) and has since been observed in a number of diverse eukaryotic groups including ciliates, nematodes, crustaceans, and vertebrates. This developmental process eliminates a large portion of the inherited genome from all somatic nuclei of an organism. The phenomenon represents a dramatic departure from the rule of genome constancy that is the foundation of organismic development and differentiation, and raises questions regarding its regulatory mechanisms, biological roles, and evolution origin. Answers to these questions are beginning to emerge, and ciliates, being the simplest group, have been providing the bulk of information.

Programmed genome rearrangements were first reported in ciliates in 1971 in the spirotrichous ciliate *Stylonychia*, in which a drastic reduction in somatic DNA composition (as determined by CsCl density gradients) and molecular size were observed (3, 4). DNA renaturation kinetics studies of the distantly related

[1]Institute of Molecular Biology, Academia Sinica, Taipei, Taiwan.

Tetrahymena soon followed (5), revealing a 10 to 20% reduction in sequence complexity in the somatic genome. These findings expanded the phenomenon of programmed genome rearrangements in organisms beyond the Ascarid group, and established ciliates as model organisms for its molecular analysis. In the following decades, studies of the molecular details of this process ensued, some of which are summarized below.

TETRAHYMENA LIFE CYCLE

Tetrahymena lives freely in fresh water. It was among the earliest protozoa to be mass cultured and studied biochemically. The species *Tetrahymena thermophila* was chosen for genetic and molecular studies in the mid-20th century (6). Like most ciliates it displays nuclear dualism. A *Tetrahymena* cell contains two nuclei: a large polyploid macronucleus and a small diploid micronucleus, which separately carry out the somatic and germline functions (Fig. 1). *Tetrahymena* cells multiply through binary fission, typically in axenic peptone medium in the laboratory. During this vegetative growth phase, the micronucleus divides by mitosis, but the macronucleus divides by amitosis, an unusual division process without apparent chromosome condensation or spindle formation (7). RNA transcription occurs exclusively in the macronucleus, whereas the micronucleus remains condensed and silent throughout this phase. Upon starvation, *T. thermophila* cells of different mating types pair and initiate the sexual phase of the life cycle. In the laboratory, this process can be induced to occur in large quantities and with good synchrony. During this time, the micronucleus enters an elaborate meiotic division process that includes dramatic nuclear elongation and active transcription (Fig. 2). It produces four haploid nuclei; three of them are degraded and the remaining one divides once to generate two haploid gametic nuclei. The mating partners exchange one gametic nucleus, and the two gametic nuclei from the opposite partners fuse to form a diploid zygotic nucleus in each cell, thus completing the sexual process of genetic exchange. The zygotic nucleus then initiates a series of nuclear events to generate the new nuclei for the following vegetative growth phase. It divides twice mitotically to give rise to precursors for two new macronuclei, one new micronucleus, and a fourth nucleus that degenerates. During this time, the old macronucleus goes through a degenerating process that involves genes related to apoptosis (8) and autophagy (9). The mating pair separates and rests in this state with two new macronuclei and one new micronucleus per cell ("2 MAC 1 MIC"). The whole process takes no more than 20 h to complete. In the laboratory, the stages of mating are usually represented by hours after mixing starved cells of different mating types (hours post-mixing, hpm). The cell resumes growth when nutrients become available. The first cell division occurs with the division of the micronucleus and the distribution of the two macronuclei to restore the state of two nuclei per cell.

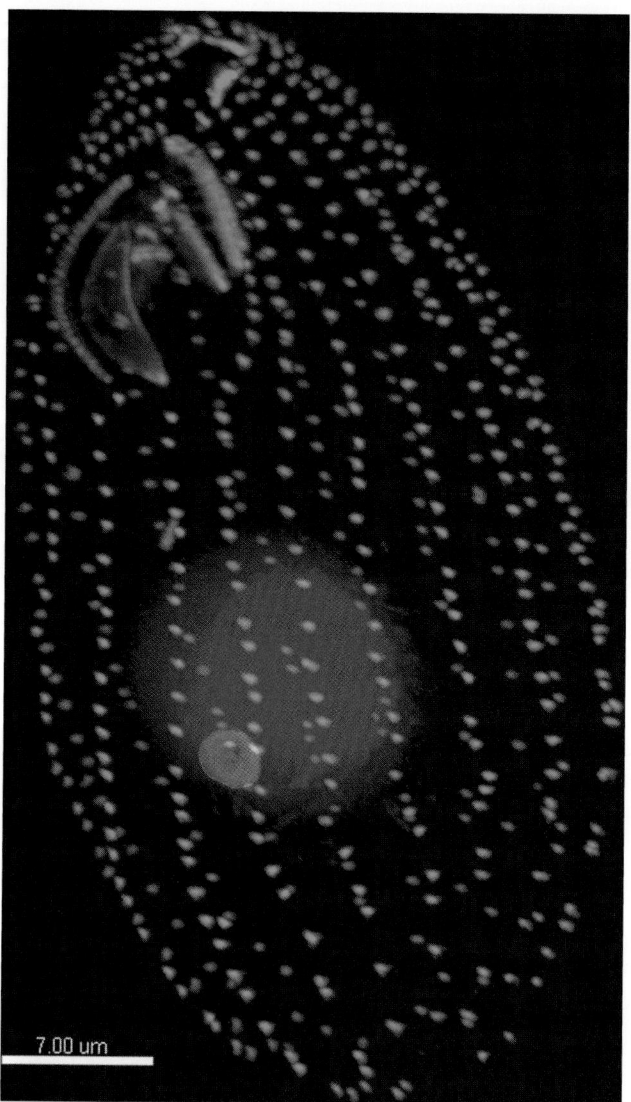

Figure 1 An image of *Tetrahymena thermophila* during vegetative growth. The fixed cell was labeled with antibodies against centrin (green) to show basal bodies on the cell cortex and a modified histone (red) to show the micronucleus. The macronucleus and micronucleus were stained with 4′,6-diamidino-2-phenylindole (DAPI; blue). The image was captured using Delta Vision.
doi:10.1128/microbiolspec.MDNA3-0012-2014.f1

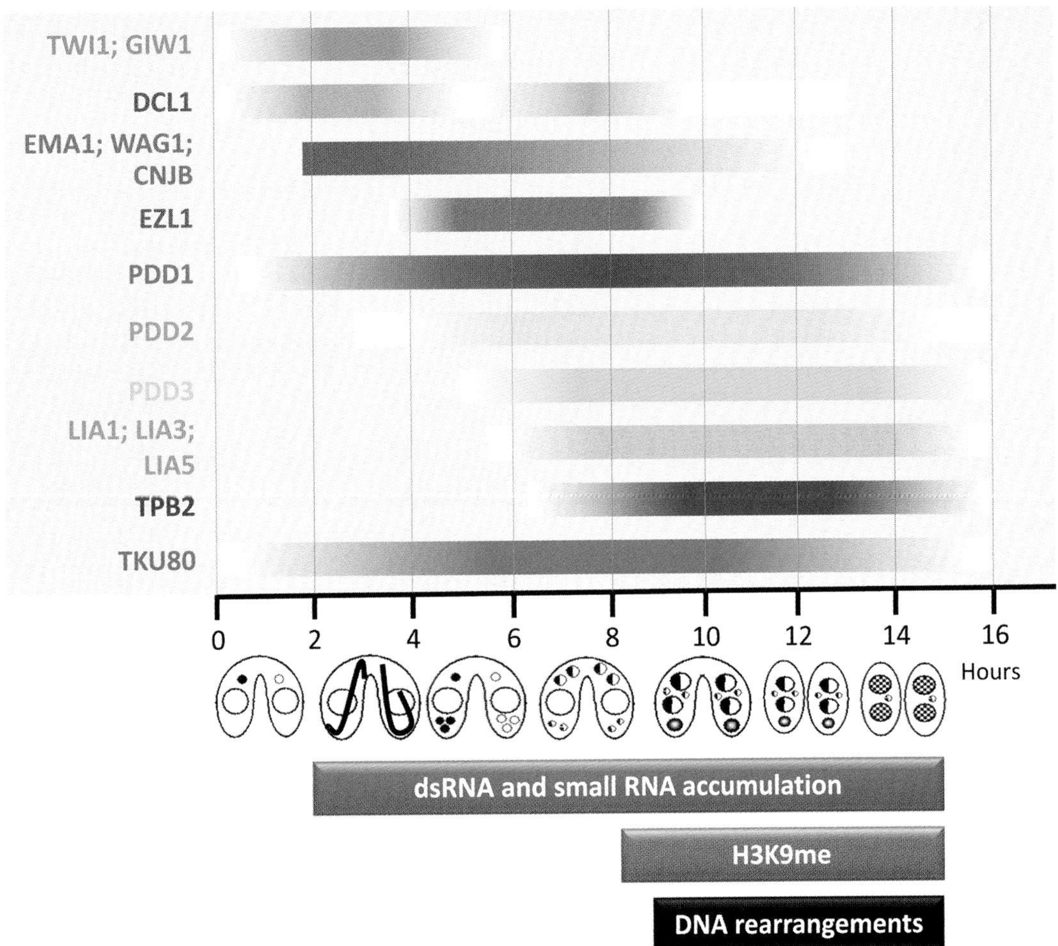

Figure 2 Cytological and molecular events during *Tetrahymena* conjugation. Various cytological events during mating progression (from 0 to 16 hours after cells are mixed) are represented by drawings of cells and nuclei. Expressions of key genes during this process are indicated above and activities of DNA and RNA are indicated below the cells.
doi:10.1128/microbiolspec.MDNA3-0012-2014.f2

OVERVIEW OF *TETRAHYMENA* DNA REARRANGEMENTS

New macronuclear development initiates at around 8 hpm, and the rearrangement of the genome occurs mostly at around 12 to 14 hpm (10). During the first three decades of studies, a number of genome-altering events were discovered, including chromosome breakage and telomere formation, selective amplification of the rRNA gene, and internal DNA deletion and elimination. These are described briefly below.

Chromosome Breakage and Telomere Formation

Chromosome breakage in *Tetrahymena* was discovered as the initiation step of rRNA gene amplification (11). It was later found to occur at roughly 200 sites in the genome, breaking the five pairs of chromosomes down into smaller fragments with new telomeres (12), which were maintained in the mature macronucleus throughout the vegetative life. Recent macronuclear genome sequence analysis has confirmed this configuration and suggested the presence of 187 macronuclear DNA molecules (13). A well-conserved 15-nucleotide (nt) sequence, CBS (for chromosome breakage sequence), marks the breakage sites and is necessary and sufficient for breakage to occur (12, 14, 15). The same sequence is found in all *Tetrahymena* species analyzed (16), suggesting a mechanism arising before the species had diverged. Breakage is highly efficient in *T. thermophila*, with no exceptions reported. After breakage, CBS and about 20 bp of DNA on both sides are lost, and the new ends are capped with new telomeres. Thus, the process reduces chromosome size drastically with the loss of relatively little DNA. The protein(s) that recognizes CBS and breaks chromosomes has not yet been found. It has been speculated that a homing endonuclease may have been domesticated to serve this function (17), but direct evidence is still lacking.

Ribosomal RNA Gene Amplification

Soon after the new macronucleus begins to enlarge, its DNA content increases through endoduplication. Most genomic sequences are replicated roughly equally, but the ribosomal RNA gene (rDNA) is selectively amplified (18). It was the first programmed gene-altering event known for *Tetrahymena*. *T. thermophila* contains a single copy of rDNA in the germline genome (19), which is unique among eukaryotes and rare among organisms analyzed. This copy is released from the chromosome through breakage at the CBS that is present in both ends; it becomes a head-to-head dimer (a palindrome) with new telomeres added to both ends and is selectively amplified. In the mature macronucleus, there are roughly 200-fold more copies of the rDNA than other sequences (18). The DNA ploidy level of the macronucleus has been determined by cytochemical methods. At the G1 stage, it contains about 23 times as much DNA as the diploid micronucleus and has been referred to as being 45C (20). Since the macronucleus has a reduced genome size (by 34%, see below), there are about 68 copies of most genomic sequences and about 13,000 copies of the rDNA in each macronucleus at the G1 stage. Palindrome formation appears to occur through an intramolecular recombination process guided by a pair of 42-bp inverted repeats that are present at one end of the rDNA (21). A similar process was later found to also occur in the yeast *Saccharomyces cerevisiae* (22) and in Chinese hamster ovary cells (23), and thus is probably widespread among eukaryotes. It could be a major contributing mechanism for gene amplification in cancer cells (24).

Internal DNA Deletion and Elimination

DNA renaturation kinetics studies revealed the loss of 10 to 20% of germline sequences from the macronucleus (5). The molecular basis for this was not known until about a decade later when cloning, sequencing, and hybridization methods became available. Focusing on separate genomic regions, two studies uncovered the deletion of specific DNA segments that occurred reproducibly as a programmed process (25, 26). One of these regions contains the M- (for middle) and R- (for right) element, which are relatively small in size (0.6 and 1.1 kb, respectively), and has been used extensively for analysis since then as a model of DNA deletion. By analyzing 20 randomly chosen DNA segments, a rough estimate was made that put the total number of deletion segments at 5,850 or more for the genome, with an average size of 2 to 3 kb to account for the total amount of DNA eliminated (26). Impressively, this is not too far from the actual number and sizes, which it is now possible to measure using micronuclear genomic sequence data (see below). In the following two decades, a number of deletion elements (or IESs for internal eliminated sequence) were cloned and characterized, which revealed several common features (27, 28, 29, 30, 31, 32, 33, 34). First, they were all simple deletions not accompanied by any additional sequence change at the junction. Secondly, the IES were mostly noncoding sequences but also included transposons and their degenerated remnants. No cellular genes were found in these studies and no shared sequence features could be identified. Thirdly, all IESs resided in noncoding regions including spacers and introns, and none was found to interrupt coding sequences. Fourthly, the majority of these elements had invariable boundaries, although some clearly had variable boundaries. These studies revealed a remarkable process that cuts, deletes, and rejoins DNA precisely at thousands of specific genomic locations, all within a period of approximately 2 h (12 to 14 hpm) during nuclear differentiation. At the same time, they have offered little information as to how the specificity of deletion is determined and how the process may affect cellular functions. Solving this mystery remains an interesting challenge. Recent genome sequence studies (35) (http://www.broadinstitute.org/annotation/genome/Tetrahymena) have offered a near-complete picture of all deletion elements, which basically confirms these general features with possible exceptions. Based on the draft genome sequences released, about 34% of the micronuclear genome is eliminated from the macronucleus (reducing the genome from 157 to 104 Mb), including more than 10,000 IESs. Some IESs may occur within coding regions and some genes may reside within IESs, although these exceptions are yet to be confirmed. The rich sequence information offers opportunities for further understanding, which awaits detailed analysis.

Since the publication of *Mobile DNA II* in 2002, major advances have been made in our understanding of DNA deletion in *Tetrahymena* and *Paramecium*. The process is shaping up as a specialized form of RNAi that has evolved through the domestication of transposases to delete all transposons from the somatic genome. In the following sections, we will review these lines of progress in *Tetrahymena* and offer our views on its mechanisms, functions, and evolution.

LINKS TO NONCODING RNA

Transcription of Internal Eliminated Sequences

Perhaps the most significant discovery made on ciliate DNA deletion in the past decade is the link to RNAi (Fig. 3). *Tetrahymena* IESs contain mainly noncoding sequences, yet they are transcribed in an unusual manner. The first observation was reported in 2001, soon after the landmark publication of RNAi in nematodes (36). In studying the well-characterized M-element, it was found through Northern blot hybridization that the element was transcribed from both strands to produce heterogeneous-sized RNAs averaging several hundred nucleotides in length that have no defined ends or poly(A) tails (37). They occur only during conjugation, starting shortly after pairing begins (2 hpm), reaching peak levels just after meiosis (4 to 5 hpm), and continuing through the middle stages of macronuclear development (12 hpm). Similar transcription was also found for the R-element, the repetitive element 2512, and several additional elements in later studies (38). The exact timing of transcription differs among elements, with some occurring only after the new macronucleus begins to develop. It was thought that double-stranded RNA (dsRNA) transcription occurred in most IESs but not in sequences to be retained in the macronucleus (known as MDSs for macronucleus-destined sequences),

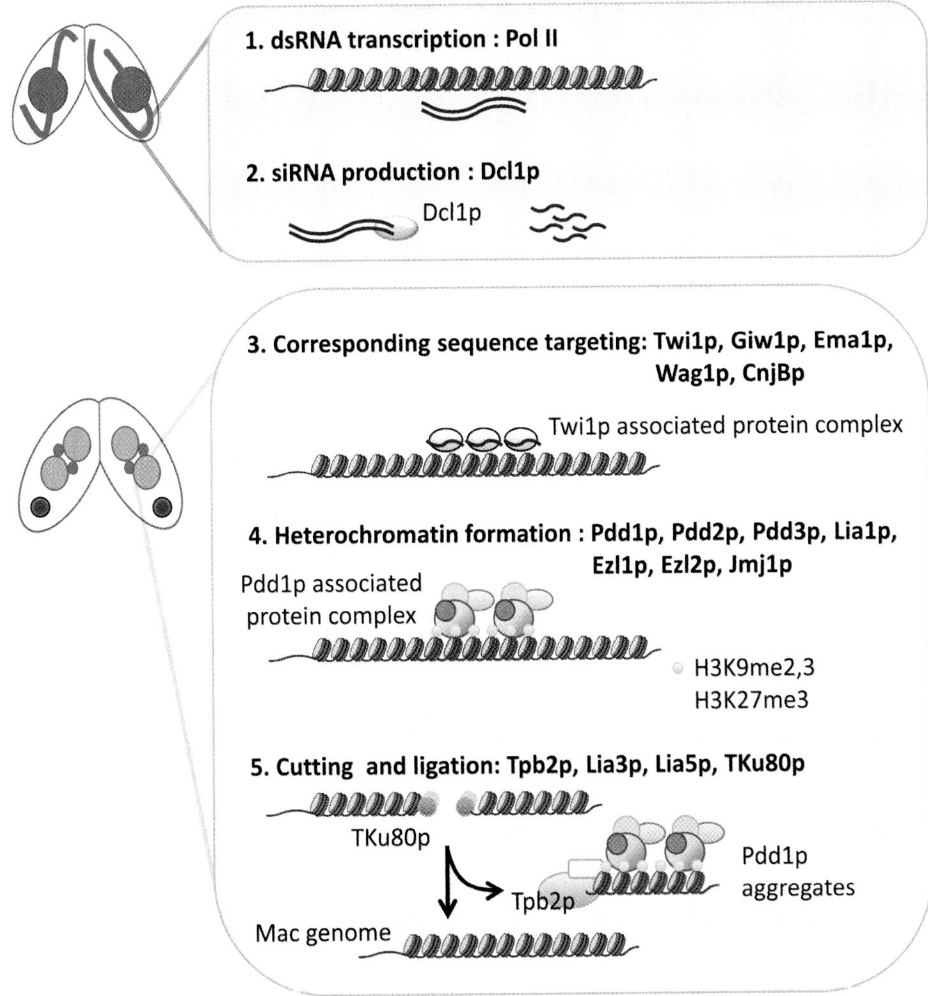

Figure 3 Key steps of programmed DNA deletion in *Tetrahymena*. The steps of programmed DNA deletion are summarized. The earlier two events occur in meiotic nuclei and the later events in developing macronuclei, as indicated by the drawings of cells to the left. Key proteins involved in each step are also indicated.
doi:10.1128/microbiolspec.MDNA3-0012-2014.f3

but definitive evidence has been hard to obtain partly due to the low and varying abundances of these transcripts (37). High-throughput sequencing analysis may provide additional insights but has not yet been reported. Analysis of small RNA (sRNA), which presumably is derived from this dsRNA, however, has supported this notion (see later sections). The importance of this transcription to DNA deletion was supported by actinomycin D treatment studies. Incubation with actinomycin D for 3 h during these stages (starting every half hour between 4.5 and 7 hpm) produced progeny with partial defects in M-element deletion (37). Although subject to other interpretations, the study offered a strong support for the role of dsRNA in DNA deletion. This unusual form of transcription is probably carried out by polymerase II, since a subunit, *RPB3* (*CNJC*), is highly expressed during meiosis and the protein is localized to the meiotic nucleus (39). Direct evidence is yet to be obtained to prove this point.

sRNA and dsRNA

The discovery of RNAi in eukaryotes (36) opened up new possibilities, and a study of Piwi proteins in *Tetrahymena* provided the first clear link to this process. Mochizuki *et al.* (40) discovered the presence of a new class of sRNA expressed specifically during *Tetrahymena* conjugation. This sRNA is around 27 nt (26 to 31 nt) in size and is highly abundant (detectable even by ethidium bromide staining after electrophoresis). It appears early during conjugation and is present until late stages (2 to 16 hpm). A *piwi*-related gene (*TWI1*), expressed early during conjugation, is necessary for the maintenance of this sRNA. Twi1p first appears in the cytoplasm, then mainly in the old macronucleus, and finally only in the new macronucleus (Figs 2 and 4). In somatic knockout strains, conjugation is arrested at the stage with two new macronuclei (MAC) and two micronuclei (MIC) ("two-MAC two-MIC"), typical of most mutants blocked in DNA rearrangements. Deletions of the M- and R-elements are completely blocked and chromosome breakage is partially affected at the site (Tt819) examined. This study provided a strong link for two main players of RNAi, namely a Piwi protein and sRNA, with DNA deletion, and led logically to the conclusion that DNA deletion involved the RNAi machinery. Based on these and earlier data, the authors went on to propose a provocative model for DNA deletion, in which sRNA (which they named scanning RNA or scnRNA) was generated from the entire genome and "scanned" the old macronuclear genome to move all similar sequences, and the remaining "germline-specific" sRNA was then used to direct DNA

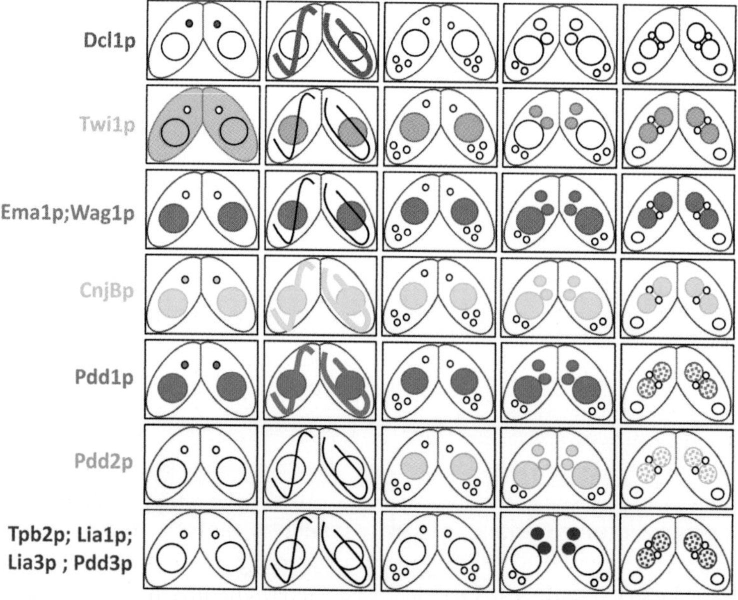

Figure 4 Localization of key proteins during conjugation. The panels summarize the localization patterns of various proteins in different cellular compartments during conjugation. Genes for the proteins involved are listed to the left, with each row indicating the progression of mating (from left to right). Proteins with similar patterns are shown in the same row. The compartments in which proteins have been detected (using antibodies or peptide tags) are colored. doi:10.1128/microbiolspec.MDNA3-0012-2014.f4

deletion. This model is appealing in proposing a role for sRNAs in controlling DNA deletion and explaining how the sequence specificity of DNA deletion is achieved. It has dominated the field for the past decade. We will return to it again later to examine its current supports.

The involvement of dsRNA in DNA deletion was demonstrated directly by injecting dsRNA into conjugating cells (41). The RNAs were 500 bp or longer and were produced *in vitro* from cloned macronuclear DNA templates of either coding or noncoding regions. They were injected into the cytoplasm around the time of macronuclear development, and the progeny cells produced, at rates nearing 50%, showed deletion of the targeted region in the macronuclear genome. Unlike endogenous deletions, these induced deletions had variable boundaries, some reaching beyond the targeted region by several kilobases. This study has an interesting resemblance to the original RNAi experiments in *Caenorhabditis elegans* (36), except that the consequence is DNA deletion, perhaps the most extreme form of gene silencing. It provided a direct support for the involvement of RNAi in DNA deletion, and showed that dsRNA alone was sufficient to trigger this process.

Dicer Proteins

These two studies firmly established the link between DNA deletion and RNAi, and prompted studies of other RNAi players in *Tetrahymena* including Dicer, Piwi, and related proteins, which are summarized briefly here. The *Tetrahymena* genome contains three orthologs of Dicer genes, *DCR1*, *DCR2* and *DCL1*. Only *DCL1* is expressed exclusively during conjugation. Genetic knockout strains lacking *DCL1* were able to initiate mating but did not produce scnRNA and appeared to accumulate the heterogeneous-sized dsRNA. These cells were arrested at the two-MAC two-MIC stage and failed to carry out DNA deletion and chromosome breakage at the sites examined (42, 43). Thus, *DCL1* is the Dicer gene responsible for generating scnRNA. Tagged Dcl1p appeared mainly in meiotic nuclei, which apparently were the compartments in which dsRNA was processed into scnRNA. However, it was not detected in macronculear anlagen (42), raising the question of where dsRNA was processed at these stages (8 to 16 hpm) (Figs 2 and 4). Studies of *DCR1* and *DCR2* revealed no obvious link to scnRNA production. *DCR1* is not essential and no clear function has been assigned to it (42, 44). *DCR2* is essential for growth. The encoded protein cleaves dsRNA to produce 23 to 24 nt sRNA *in vitro* (45) and is likely responsible for making this class of endogeneous sRNA in *Tetrahymena* (46). The *Tetrahymena* genome contains a single gene, *RDR1*, for the RNA-dependent RNA polymerase, which is essential for growth (45). The encoded protein physically interacts with Dcr2p and has been implicated in 23 to 24 nt sRNA production. Its role in scnRNA production, if any, is not known.

Piwi-Interacting Proteins

Four Twi1p -interacting proteins have been identified through co-immunoprecipitation experiments (38, 47, 48). Giw1p has no known conserved function. It is required for the localization of Twi1p to the nucleus and the ability to remove one strand of the sRNA. Ema1p is likely a DExH box RNA helicase. It is thought to be involved in the binding of Twi1p to chromatin. Immunoprecipitation experiments suggest that it is involved in the association between scnRNA and long noncoding RNA. Wag1p and CnjBp are two GW repeat proteins. All four proteins are expressed preferentially (*CNJB*) or specifically (the other three) during conjugation and localize to the parental macronuclei early and the developing macronuclei late during conjugation (Figs 2 and 4). In addition, Ema1p showed strong cytoplasmic localization throughout the stages and CnjBp was found mainly in the meiotic nucleus. These genes are required for the completion of DNA deletion either singly (*GIW1* and *EMA1*) or in combination (*WAG1* and *CNJB*). Interestingly, *EMA1* mutants were blocked in the deletion of some (M and Tlr1) but not other (R and Cal) elements tested, revealing IES subgroups that differed in deletion control. Although the underlying mechanism remains unclear, these studies began to offer the basic information necessary toward a comprehensive understanding of scnRNA actions in *Tetrahymena*.

CHROMATIN MARKINGS

Chromodomain Proteins

The link to RNAi helps explain the dynamics of chromatin during DNA deletion. Analogous to small interfering RNA (siRNA) and Piwi-interacting RNA (piRNA) in other organisms, the scnRNA in *Tetrahymena* is thought to target chromatin through sequence homology and brings about changes that turn it into "heterochromatin." The association between heterochromatin and DNA deletion was known before the RNA link was discovered. Earlier work identified several nuclear proteins that were synthesized preferentially during macronuclear development (49). One of the most prominent, p65, was partially sequenced and the gene identified and named *PDD1*. It is related to the conserved heterochromatin protein HP1, with two chromodomains and a chromo shadow domain (50). The protein is expressed only during conjugation and is localized mainly to the parental macronucleus at early stages and to the developing macronucleus at late stages. In the developing macronucleus, Pdd1p first appears homogenously, becomes punctate (10 hpm), and finally forms a few large aggregates with doughnut-like appearances under the electron microscope before it disappears completely (14 hpm) (Figs 2 and 4). Through *in situ* hybridization, micronucleus-specific sequences have been shown to have similar distributions (51). A colocalization study indeed found Pdd1p present in aggregates of IESs in the developing macronucleus (50). Thus, IES chromatin appears to be highly dynamic and forms heterochromatin-like aggregates before being eliminated, with Pdd1p as an integral part of this process. Gene knockout studies indeed showed that *PDD1* was essential for DNA deletion. Conjugation was arrested at the two-MAC two-MIC stage in strains lacking the macronuclear copies of this gene. Remarkably, tethering Pdd1p to a sequence not normally deleted in a special plasmid system (52) could cause it to be efficiently deleted, arguing for a key role for this protein. These studies set the foundation for all following studies of genes involved in DNA rearrangements. Two other related genes, *PDD2* and *PDD3* (also a chromodomain protein) were found to have similar properties (53, 54, 55) (Fig. 5). A recent study (56) analyzed *PDD1* through substitutions and deletions of various parts and determined that the two chromodomains and the chromo shadow domain are all essential (or highly important) with distinct functions. The first chromodomain can be substituted with a DNA-binding motif, implying a main role in DNA targeting. The other two domains are likely engaged in histone modifications and Pdd1p aggregation separately.

Histone Modifications

Other heterochromatin marks have also been investigated. The first study addressed the importance of histone deacetylation using the inhibitor trichostatin A, which partially blocked DNA deletion when applied during conjugation (57). Di- and trimethylation of histone 3 lysine 9 (H3K9me2,3) are perhaps the most

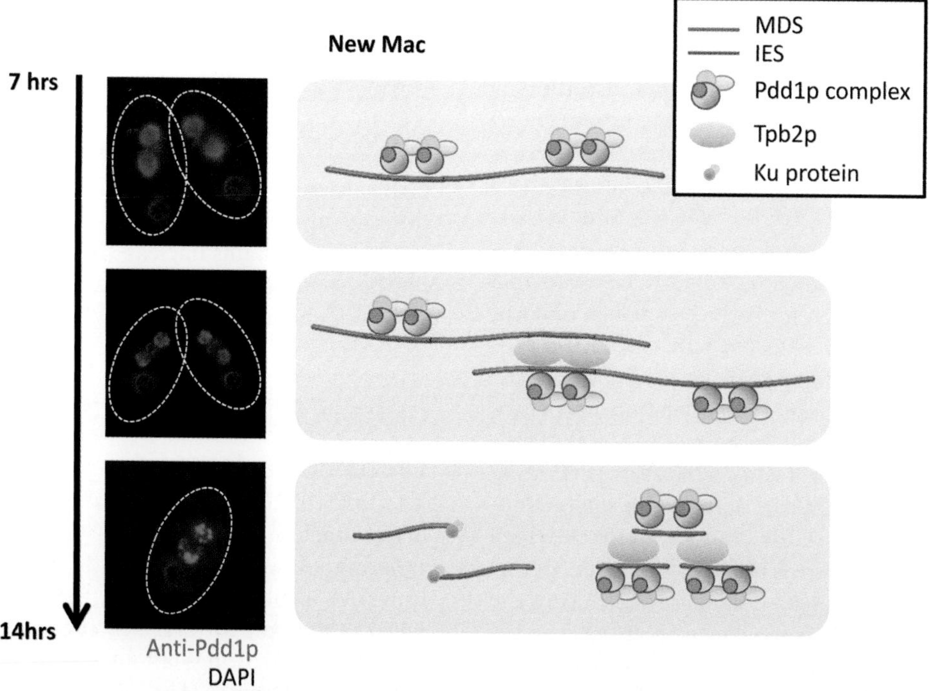

Figure 5 Formation of the Pdd1p aggregates during DNA deletion. The chromodomain protein Pdd1p and other proteins that participate in the formation of the large heterochromatin aggregates are shown. The proteins are guided to IESs through sRNAs and eventually form large aggregates with excised IESs. The left panels show micrographs of mating cells in three successive stages stained with antibodies against Pdd1p (green) and DAPI (blue). doi:10.1128/microbiolspec.MDNA3-0012-2014.f5

consistent marks of heterochromatin in eukaryotes. These modifications appear in developing macronuclei and disappear after DNA deletion has occurred (52) (Figs 2 and 4). Chromatin immunoprecipitation (ChIP) experiments have shown that they are associated specifically with selected IESs at this stage (52), although their overall genomic distributions have yet to be determined. Mutating both copies of the H3 genes in the genome to prevent K9 methylation (K9Q) led to failure in DNA deletion and conjugation arrest (at the two-MAC two-MIC stage), establishing the importance of these modifications. In these mutant lines, scnRNA continued to accumulate, supporting the argument that H3K9 methylation acted downstream of scnRNA accumulation (58). In this regard, it is interesting to note that in *DCL1* mutants, H3K9 methylation still occurs in developing macronuclei but is no longer enriched in IESs.

Another heterochromatin mark, trimethylation of histone 3 lysine 27 (H3K27me3) also plays important roles. Unlike H3K9me2,3, H3K27me3 occurs in both growing and mating cells and likely in all types of nuclei. In developing macronuclei, it has a distribution pattern very similar to those of H3K9me2,3 and Pdd1p, suggesting a possible link to DNA deletion. Three methylase genes related to this modification are present in the genome, and one of them, *EZL1*, is expressed only during conjugation. Mutants of *EZL1* lack H3K27me3 in parental and developing macronuclei and cannot carry out DNA deletion, indicating an essential role in this process. H3K27me3 occurs before, and likely facilitates, H3K9 methylation, since H3K9me2,3 does not occur in *EZL1* mutants but H3K27me3 occurs normally in H3K9Q mutants (59). It should be noted that, unlike H3K9 methylation, which occurs only in developing macronuclei, H3K27me3 also occurs in other nuclei at other stages and could have additional roles. A demethylase gene, *JMJ1*, has been studied in *Tetrahymena*. It is partially responsible for the removal of H3K27me3 in the developing macronucleus. It has only a minor role in DNA deletion since knockdown mutants showed only slight defects in DNA deletion but were unable to complete macronucleus development beyond DNA deletion (60).

Other Chromatin Proteins

Four other proteins have been identified for their associations with heterochromatin-like structures during DNA deletion (61). They were found through a screen of a cDNA library for proteins localized specifically in developing macronuclei during DNA deletion. Five novel proteins were identified, and four of them were found to localize to the Pdd1p structure. Three of these proteins, Lia1p, Lia3p and Lia5p, have been further analyzed (62, 63) and all appear to be important for DNA deletion. *LIA1* and *LIA5* are essential for DNA deletion and the completion of conjugation, and *LIA3* appears to affect boundary determination of some IESs (Figs 2 and 4). Further analysis of these novel proteins will likely expand our understanding of heterochromatin in eukaryotes.

These studies indicate that scnRNA leads to heterochromatin formation in *Tetrahymena*, involving conserved proteins and histone modifications that are shared by other siRNA-directed transcriptional gene silencing processes (Fig. 3). H3K27 methylation likely occurs before and promotes K9 methylation. Pdd1p (and probably Pdd2p, Pdd3p, and Lia1p) binds to these moieties and eventually causes deletion to occur. It probably also has a role in re-enforcing these modifications. In addition, Pdd1p has an unknown role in scnRNA production, since scnRNA is not produced in its absence (40).

ROLE OF PARENTAL MACRONUCLEI AND THE RNA SCANNING MODEL

Nuclear dualism provides special opportunities for interactions between nuclei. This is especially true during conjugation, when three different types of nuclei, each with its own fate, are present in the same cytoplasm. The interaction between the parental and progeny somatic genomes offers an intriguing aspect of ciliate DNA rearrangements. Early genetic work in *Paramecium* demonstrated the effect of the parental macronucleus on progeny macronucleus differentiation (64, 65), which was validated and further elucidated with the discovery of DNA rearrangements in this organism (66, 67). Similar effects are known to occur in *Tetrahymena* and *Oxytricha* as well, implying a common inheritance strategy accompanying nuclear dualism. In *Tetrahymena*, the first evidence was derived from experiments that put high-copy-number plasmids containing M- or R-elements in the parental macronucleus, which led to partial inhibition of the respective IES during subsequent conjugation (68, 69). Due to its sequence specificity, this effect argued for an RNA-mediated cross-talk between these two nuclei (1). The discovery of a link to RNAi immediately suggested a role for scnRNA in this cross-talk. The "scanning RNA" model was proposed (40), which argued that the entire genome was transcribed to produce dsRNA during meiosis, which was processed into sRNA and transported into the parental macronuclei to "scan" the genome. All identical sequences were removed or "subtracted" and the remaining sRNA, which now contained sequences present only in the germline genome (and not the somatic genome), was transported into the developing macronucleus to direct chromatin changes and DNA deletion. It provided a simple and elegant solution for the establishment of sequence specificity in DNA deletion and offered a nice explanation for how IESs inserted in parental macronuclei could inhibit the deletion of the same sequences in the developing macronucleus.

The RNA scanning model can be tested through some of its molecular assumptions and predictions. The followings are three of the basic requirements of the mechanism. First, the entire genome should be transcribed into dsRNA and processed into siRNAs. Secondly, these siRNAs must enter the parental macronucleus and then move into the developing macronucleus. Thirdly, there is a molecular mechanism to "subtract" siRNAs of similar sequences, thus reducing the sequence complexity after leaving the parental macronucleus. During the past decade, attempts to gather support have met with mixed results. The movements of siRNAs through various nuclear compartments have been difficult to measure directly. However, the appearance of the Piwi protein Twi1p, which is known to bind to this species of sRNA, has been followed, and largely agreed with the expected patterns. Whether the change in protein localization is actually due to movement or to degradation and new synthesis will require additional studies (40, 70).

Another major issue that has been studied is scnRNA abundance throughout conjugation. Several earlier studies offered partial support, although they were not conclusive (40, 43, 70). With the release of the micronuclear genome sequence, high-throughput sequence analysis of sRNA become possible and has now been reported (71). The result of this comprehensive analysis is not in total agreement with the scanning model. First, sRNA was not derived from the entire genome, and secondly, although sRNA abundances did change through conjugation, they were not at the level expected from the model. The study determined sRNA sequences from stages of conjugation before, during, and after the proposed scanning process (3, 4.5, 6 and 8 hpm). It appeared that most (~80%) of the sRNA (26 to 32 nt) at the earliest stage measured (presumably before any "scanning" could have occurred) was derived from the ~25% of the genome that can be uniquely assigned to IESs. The sequences to be retained (MDSs) were also transcribed but to a much lower extent (producing 15% of the sRNA). This result suggested a biased transcription, and not RNA scanning, for the enrichment of sRNAs from IESs. Using a special genetic strain (nullisomic for chromosome 4 in the micronucleus), it was clear that this sRNA was derived from the micronuclear (and not the macronuclear) compartment. This bias was seen also in mutants of *TWI1* and *EMA1*, which should not have carried out "scanning" as proposed. Clearly, sRNA production was not equal over the entire genome, which presumably was the result of selective transcription. Nonetheless, a significant proportion of sRNA was indeed derived from MDSs. Their roles are not clear and will need to be explained. The relative amounts of sRNA at a later stage (8 hpm) appeared to change, with slight increases for those from IESs and significant decreases (about 2-fold) for those from MDSs. This decrease was confirmed in selected loci by hybridization. *EMA1* also appeared to be important for some of these decreases. This change indicated a dynamic regulation of sRNA abundance. Whether it was due to "subtraction" by the macronucleus or other unknown processes has yet to be determined.

The model made one clear prediction that is testable, but the results have not been supportive so far. If a sequence is artificially removed from the parental macronucleus, the corresponding micronuclear sequence should now appear as a "micronuclear-specific" sequence. Since no "subtraction" of the corresponding scnRNA sequence can be performed, the sequence should be subjected to deletion. Thus, all somatic knockout strains (and other strains with somatic deletions) should generate offspring with the same somatic sequences deleted. In other words, somatic deletion should be heritable in *Tetrahymena*. This is a clear and unambiguous prediction, and is indeed reported for at least one gene in *Paramecium* (72). However, it has never been seen in *Tetrahymena*.

If the scanning model (in its original form) cannot be validated, one will need modifications or alternative explanations for the influence of the parental macronucleus on DNA deletion as well as the mechanism for selecting a sequence for deletion. One suggestion is that selective transcription of dsRNA could be the primary force for determining a sequence for deletion (73, 74). How the transcription is regulated is not known and will be the key to understanding this process. The parental macronucleus could play a secondary role to augment this sequence selection. If the same sequence is also present in the macronucleus, it will be transcribed, but the transcripts may not be processed into sRNA for the lack of *DCL1* in this compartment. This RNA may interact with the sRNA made from the micronuclear genome (similar to the scanning model idea) and prevent it from carrying out deletion (73).

DNA CUTTING AND TRANSPOSON DOMESTICATION

The link to RNAi provides an attractive explanation for the evolution of DNA deletion. Like other eukaryotes, ciliates may have evolved RNAi machineries to defend against invading genetic elements. *Tetrahymena* has a 23 nt siRNA-mediated, post-transcriptional gene silencing process against endogenous and exogenous RNAs (44, 46). The 27 nt sRNA pathway of DNA deletion, on the other hand, shares features with RNAi processes of transcriptional gene silencing and may have evolved from it. One key step in this evolution would be the acquisition of DNA cutting activities to remove the silenced chromatin. The recent discovery of a domesticated transposase may have provided this crucial link. *Tetrahymena* contains a number of transposons in the genome (31, 35, 75, 76), which are all removed from the macronucleus, mostly through DNA deletion. Interestingly, two genes (*TPB1* and *TPB2*) that show high similarity to piggyBac transposases and contain the catalytic DDD motif are retained in the macronucleus. These genes lack other transposon features and are expressed specifically during mid- to late stages of conjugation (6 to 14 hpm) (Fig. 2). Both encode proteins roughly twice as long as most piggyBac transposases, probably through fusions with additional exons at the N terminus (for *TPB1*) or C terminus (for *TPB2*) (77). Further studies showed that *TPB2* indeed participated in DNA deletion. Green fluorescent protein-tagged proteins colocalized with the chromodomain protein Pdd1p to form distinct aggregates (Fig. 4). Genetic knockdown strains were unable to carry out DNA deletion or chromosome breakage and arrested at the two-MAC two-MIC stage, typical of mutants defective in DNA rearrangements. Proteins expressed in *Escherichia coli* showed weak but specific endonuclease activities to generate ends with 4-nt 5′ overhangs, a transient feature of M-element deletion (78). These results suggested strongly that Tpb2p participated directly in DNA cutting, and was likely the key enzyme that turned the RNAi process into the DNA deletion process (Fig. 3). Tpb2p probably lost its original specificity for the transposon terminal inverted repeats and became targeted to heterochromatin through the acquired C-terminal sequence. Site-directed mutagenesis studies showed that the zinc-finger domain in the C-terminal half was indeed important for Tpb2p localization (79). A *TPB2* ortholog, *PGM1*, has been found in *Paramecium* and likely serves a similar role (80). Thus, the domestication process probably occurred before these two species split. *PGM1* also contains an extra C-terminal half with a zinc-finger domain, although it shares little sequence identity with that of *TPB2*, suggesting rapid evolution of this sequence. Another gene, *LIA5*, has recently been described in *Tetrahymena* (61, 63). It has some sequence similarity with the IS4 family of transposons that includes *TPB2* but lacks the conserved DDD motif. It also contains a zinc-ribbon domain in the C-terminal half. It is essential for the formation of the Pdd1p aggregates and DNA deletion but is not colocalized with Pdd1p. *LIA5* is probably related to *TPB2* in evolution and function, which further illustrates the importance of transposase domestication in this process, in a way similar to the domestication of *RAG1* during the evolution of the adaptive immune system of vertebrates. *TPB1*, on the other hand, is not essential for most DNA deletion and the completion of conjugation. It is more divergent from *TPB2* than *PGM1*, and no ortholog has been found in *Paramecium*. It plays some non-essential role during conjugation, but its nature is not yet clear (77).

The discovery of *PGM1* in *Paramecium* and *TPB2* in *Tetrahymena* offered a nice explanation for the evolution of DNA deletion. This transposase gene could have been domesticated in the ancestor species, acquired the ability to target all heterochromatin, and modified its activity from transposition to deletion. It thus turned transcriptional gene silencing of RNAi into DNA deletion and heterochromatic DNA into IESs. PiggyBac transposase may not be the only transposase captured for this activity in ciliates. The transposase of a presumably active Mariner family transposon, *TBE1*, is required for DNA deletion in the distantly related ciliate *Oxytricha* (81). Thus, DNA deletion may have evolved independently in different branches of ciliates, or may have continued to evolve by adopting new transposases for this role after its initial establishment.

NONHOMOLOGOUS END JOINING AND DNA DELETION

After DNA cutting to remove IESs, the broken ends generated are ligated back together, apparently involving the cellular DNA double-strand break repair machinery (Fig. 3). Ku70 and Ku80 proteins are conserved in eukaryotes for repairing double-strand breaks through the nonhomologous end-joining (NHEJ) pathway, presumably by binding and protecting these ends (82, 83). *Tetrahymena* contains two *KU70* and one *KU80* (*TKU80*) orthologous genes. An analysis of *TKU80* indicated that it was not essential for growth but was essential for completing conjugation. Mutants of *TKU80* could not complete DNA deletion and were arrested at the two-MAC two-MIC stage of conjugation. In these mutants, however, cleaved IESs were detected through their circularized junctions, indicating the occurrence of DNA cutting. Without *TKU80*, the developing macronucleus accumulated broken DNA ends (as revealed by terminal deoxynucleotidyl transferase dUTP nick end-labeling assays) and gradually lost most of its DNA, presumably through degradation from these ends (84). TKu80p is not localized to but is required for the formation of Pdd1p aggregates, indicating that DNA cutting is not dependent on the formation of these aggregates (Figs 4 and 5). In mammalian immune systems, Ku is also required for completing V(D)J recombination, but this process differs from IES deletion in that ligation of the deleted segments (the signal ends) also requires Ku (85, 86).

Although aggregates of Pdd1p are the most prominent structural features associated with DNA deletion, their role is not entirely clear. They contain common heterochromatin features including histone H3 modifications and chromodomain proteins, as well as the domesticated transposase Tpb2p (Fig. 5). Their DNA contents include repetitive sequences of IESs and possibly all IESs. Since IESs are widely distributed in the genome, one wonders if they have already been cleaved before appearing in these aggregates. The *TKU80* studies mentioned above suggest that these aggregates are not required for cleavage, raising the possibility that they instead are involved in repairing the cleaved ends, a step that would normally require Ku proteins. In this regard, it is interesting to note that some "aggregates" can be induced to form by UV treatment in *LIA5* mutants, which normally do not form any such aggregates. This result suggested a strong link to a DNA damage response (63).

SETTING DELETION BOUNDARIES

In *Tetrahymena*, essentially all IESs are located within noncoding regions including introns (35). Precision of deletion thus is probably not as critical for somatic gene expression as it is in *Paramecium*, *Euplotes*, and *Oxytricha*, in which many IESs interrupt coding regions. Nonetheless, DNA deletion is still a largely precise process in *Tetrahymena*. Initial analysis of the M- and R-elements revealed that, among dozens of independent progeny lines analyzed, the rejoined junctions had exactly the same sequences, indicating the precision of deletion at the single-nucleotide level. The few exceptions had junctions that differed by only a few nucleotides (27, 87). This precision is probably a common feature, although some IESs clearly have nonprecise junctions that vary by several hundred nucleotides or more (29, 34). It is likely that the majority of IESs in the genome are deleted precisely, like the M- and R-elements, but a small proportion has highly variable junctions. This raises the question of how the deletion junction is determined. There are at least two issues. First, if the cutting enzyme (presumably Tpb2p) is targeted to heterochromatin as is currently thought, how does it locate the two ends of an IES for cutting? And secondly, how is the single-nucleotide precision achieved through a guiding system that targets nucleosomes?

DNA deletion can be induced to occur anywhere in the genome through dsRNA injection. In this case, no pre-existing *cis*-acting sequence is required. DNA deletion can also occur to the *aph* gene of *E. coli* inserted at any genomic location. Unlike the M- and R-elements, these induced deletions have boundaries that vary by hundreds to thousands of nucleotides (41, 88). Presumably, these dsRNAs (or derived sRNAs) guide chromatin marking over the entire region to be deleted, which in turn attracts Tpb2p to carry out deletion. We suggest that Tpb2p is targeted to the two ends of the element by recognizing the borders and not the entire "heterochromatin" (Fig. 6). After cutting, the IES would be released as one piece, allowing the production of the circularized junctions observed. Junction variations can be explained if the heterochromatin region is imprecise, being somewhat variable in different chromatids.

Earlier studies revealed flanking sequences that set the boundaries of the M- and R-elements. By introducing specifically modified elements into conjugating cells in a rDNA plasmid, *cis*-acting sequences important for deletion have been determined. A 10-bp motif (5′-AAAAAGGGGG-3′) located about 45 bp away from both ends of the M-element was found to be crucial for setting deletion boundaries (89, 90). Without it, the boundary became variable. Inserting this sequence within the M-element induced new boundaries to form at about 45 bp downstream. Thus, this sequence is necessary and sufficient to set a deletion boundary at a specific distance from it. The R-element also appears to have a pair of flanking sequences that set its boundaries (91). They are AT-rich sequences, although their exact identities have not been determined. It has been suggested that there are many families of IESs, each defined by a particular flanking sequence that they share (91).

We suggest that flanking regulatory sequences set boundaries for heterochromatin, similar to chromatin boundary elements in other eukaryotes (92, 93). Establishing heterochromatin boundaries helps to provide defined target sites for Tpb2p (and other nucleases) binding and cutting. This interaction could set the Tpb2p

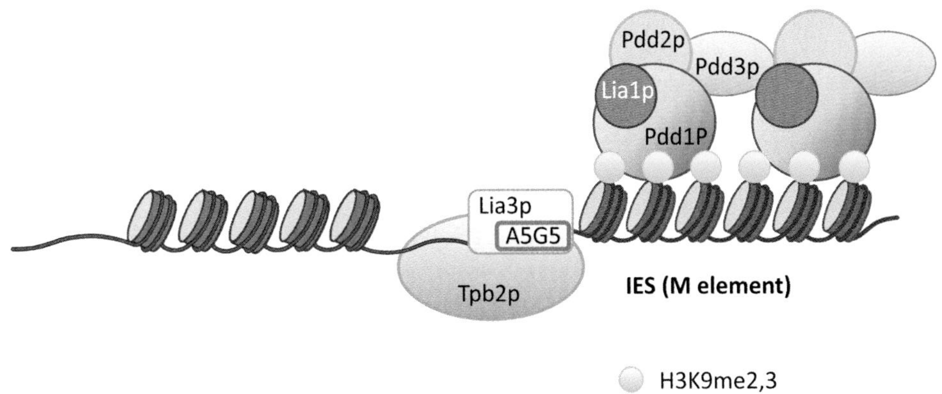

Figure 6 Setting deletion boundaries. The cartoon shows a possible heterochromatin structure with associated proteins over the IES to the right and the neighboring euchromatin to the left. The domesticated transposase Tpb2p is targeted to the junction for DNA cutting. In the example represented here, the M-element is shown to have a flanking regulatory sequence (5′-AAAAAGGGGG-3′) that is recognized by Lia3p to set the heterochromatin boundary and specify the cutting site. doi:10.1128/microbiolspec.MDNA3-0012-2014.f6

binding target to a small region, perhaps no more than a few dozen base pairs, within which Tpb2p will cut at a specific nucleotide due to its binding preference. Several proteins including CTCF have been known to help set chromatin boundaries in various eukaryotes (94, 95). We suggest that these or related proteins are involved here. These, or proteins that interact with them, could recognize and bind the flanking regulatory sequence to set the deletion boundary. A recent study has revealed that Lia3p possessed the expected properties for setting the boundaries of the M-element (C. M. Carle and D. L. Chalker, personal communication) (Fig. 6). Even though *LIA3* mutants were able to carry out DNA deletion, they lost the boundary specificities for M- and a few other elements that contain the flanking 5′-AAAAAGGGGG-3′ motif. *In vitro*, the protein is able to bind to this sequence, likely through the formation of a G quadruplex.

If IESs were derived from invading genetic elements such as transposons, they would be without boundary sequences when first inserted into the genome and would be deleted without precise boundaries. Those inserted within or very close to coding regions would have detrimental effects and could not be kept. Those inserted some distance away would still be harmful until *cis*-acting sequences had evolved to set boundaries and prevent chance deletions of neighboring genes. Thus, we suggest that flanking regulatory sequences evolved later, and that IESs with variable junctions are probably younger IESs (or located in regions poor in coding sequences, such as heterochromatic regions).

MATING-TYPE DETERMINATION INVOLVES A NEW TYPE OF DNA REARRANGEMENT

Although chromatin diminution affects a large part of the genome, specific phenotypic roles had not been found in association with any specific rearrangement until recently when mating-type determination mechanisms were characterized. In *Tetrahymena* and *Paramecium*, studies over the past four decades have hinted at the involvement of DNA rearrangements in mating-type determination. *T. thermophila* has seven mating types determined by a single locus with several known alleles. A progeny cell can be any one of these seven types, regardless of its parental mating types, and different subclones derived from a single mating pair, while having identical germline genomes, can express different mating types. Once determined, the mating type is very stable and has never been known to change. These features suggest the involvement of somatic DNA rearrangements (96). Cloning and analysis of this gene should help reveal the nature of this event, and this was finally achieved with the aid of genome sequence data (97). In the macronucleus, the *MAT* locus is composed of a pair of divergently transcribed genes (*MTA* and *MTB*) with different sequences in cells of different mating types, presumably as the result of DNA rearrangements. Sequencing of the micronuclear genome revealed the nature of these rearrangements. The germline locus is around 91 kb. At its two ends are the C-terminal halves of *MTA* and *MTB*, which are shared (and thus invariable) among all mating types, and the internal region contains an array of the different

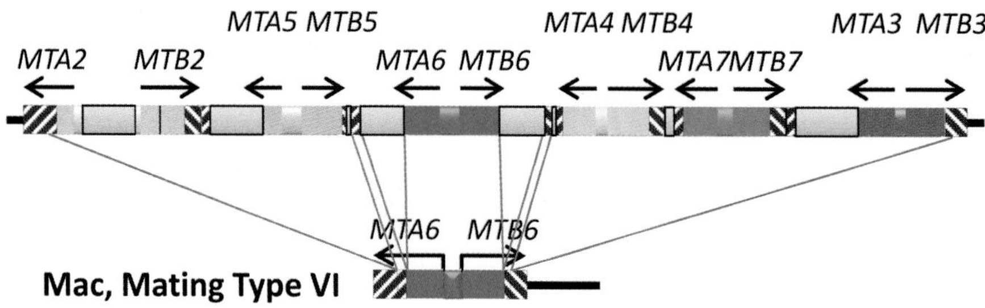

Figure 7 Mating-type determination through homologous recombination. The top shows a germline configuration of the *MAT* locus and the bottom shows the corresponding locus in the macronucleus of a mating type VI cell. The connecting lines indicate the two types of rearrangements required: deletion of IESs and removal of the other five mating-type sequences, presumably by recombination between homologous sequences (marked by similarly shaded areas) (modified from 97). doi:10.1128/microbiolspec.MDNA3-0012-2014.f7

N-terminal halves of all potential mating-type genes. For the allele that was sequenced (the B allele), there are six mating-type genes (all except type I) arranged in the order 2-5-6-4-7-3, in addition to a number of IESs and spacers (Fig. 7). To generate a particular mating-type gene pair during conjugation, all the IESs and all except one of the six partial gene pairs in the internal region are deleted. The remaining partial pair is joined to the "constant" C-terminal regions at both ends to generate the complete gene pair. This is a remarkable rearrangement that is apparently subject to strict regulations. It is interesting to note that the rearrangement occurs by recombination between the constant regions and the internal regions in sections that share sequence homology for hundreds of nucleotides. This site-specific homologous recombination is distinctive from IES deletion, which also occurs. How it is regulated to produce only one complete gene pair per chromosome remains an interesting problem to solve. The study nicely explains the determination of mating type in *Tetrahymena* and reveals a new DNA rearrangement mechanism distinctively different from IES deletion and chromosome breakage. It involves homologous recombination at specific ectopic sites, which potentially could also occur in other genomic regions.

CONCLUDING REMARKS

The past decade has witnessed transforming progress in our understanding of ciliate genome rearrangements. The link to RNAi has provided a plausible evolutionary explanation, suggested likely functional roles, and revealed crucial details in mechanism for DNA deletion. DNA deletion can now be viewed as a genome defense system against invading genetic elements including transposons. It has evolved from the RNAi process through the domestication of transposases and has turned the transcription silencing mechanism into the DNA deletion process. This intriguing phenomenon illustrates the powerful forces of transposons in shaping genomes and driving evolution.

Many interesting questions still remain. For instance, how are sequences recognized for deletion? Since dsRNA is the initiating event, control of its transcription becomes the key issue. Are there *cis*-acting sequences to be recognized, and how could this mechanism detect a foreign sequence? Relative little is known about the control of noncoding RNA transcription in general. Studies in *Tetrahymena* may help to provide some insights. Also, how does the old macronucleus interfere with DNA deletion in a sequence-specific manner? Little molecular detail is known, but RNA most likely plays a key role. Does the sequence recognition involve RNA–RNA or RNA–DNA interactions? This mysterious process may reveal new biological principles. Finally, how do sRNAs find their chromatin targets? In transcriptional silencing systems, it is generally thought that pairing with nascent transcripts guides sRNAs to their DNA targets (98). Little is known in ciliates in this regard. Since injecting dsRNA is sufficient to induce deletion at any locus, transcription of that locus may not be required, raising questions regarding a targeting mechanism that depends on nascent transcripts. Could sRNA directly target DNA? This appears to be the case in the CRISPR system in bacteria (99, 100) and remains a distinct possibility here. Questions also remain regarding processes other than DNA deletion. How has chromosome breakage evolved in *Tetrahymena* and what endonucleases are used in DNA cutting? What regulates the site-specific homologous recombination process of the *MAT* locus? These and other questions continue to make *Tetrahymena* a fertile ground in which to uncover mechanisms that remodel the genome.

Massive somatic genome rearrangements such as chromatin diminution occur sporadically in diverse groups of organisms, including limited species of nematodes, crustaceans, and vertebrates, and most species of ciliates. This diversity suggests independent evolutionary origins, which have likely resulted in mechanisms that differ in details. Recent genome sequencing studies in *Ascaris* did not reveal the involvement of internal DNA deletion or sRNA in chromatin diminution (101). In lamprey, sRNA also has not been reported in association with chromatin diminution (102). Nonetheless, the prevalence of transposons and the wide occurrences of RNAi offer excellent opportunities for the independent evolution of similar processes in diverse groups. Further investigation should clarify this possibility.

Among these organisms, ciliates are particularly interesting. While global genome rearrangements occur only in isolated groups in other phyla, they occur in all ciliate species studied. Moreover, multiple forms of rearrangements appear to occur within a single ciliate species. In *Tetrahymena*, in addition to the well-studied DNA deletion process, other programmed processes including chromosome breakage, ribosomal gene dimerization and amplification, and homologous recombination have also been found. This complexity might also be present in other ciliates as well. Ciliates appear to be hotbeds of programmed somatic DNA rearrangements, and we think nuclear dualism may have played a major role. Having evolved two nuclei in the same cell that are separately responsible for gene expression (the macronucleus) and high-fidelity genome transmission (the micronucleus), the organism can tailor these two compartments differently for their respective roles. Programmed DNA rearrangements provide the scissor and glue to restructure the somatic genome for optimal gene expression. This ability may allow the germline genome to accumulate changes that facilitate genome transmission without hampering gene expression. Presumably, similar changes can also evolve in non-dividing tissues of metazoans. The fly salivary glands and mammalian red blood cells perhaps provide such examples.

Tetrahymena may have the simplest version of genome rearrangements among ciliates studied. It has relatively modest levels of DNA elimination and chromosome fragmentation and probably the simplest form of DNA deletion. It is also the easiest to grow and manipulate in the laboratory. It should continue to offer the insights needed for a thorough understanding of this remarkable phenomenon of genome dynamics.

Acknowledgments. *We thank members of the Yao laboratory for inputs and comments. This work was supported by Academia Sinica and a grant (NSC 102-2311-B-001 -023 -MY3) from the Ministry of Science and Technology (Taiwan) to M.C.Y.*

Citation. YaoM-C, Chao J-L, Cheng C-Y. 2014. Programmed genome rearrangements in Tetrahymena. Microbiol Spectrum 2(6):MDNA3-0012-2014.

References

1. **Yao M-C, Duharcourt S, Chalker DL.** 2002. Genome-wide rearrangements of DNA in ciliates, p 730–758. *In* Craig NL, Craigie R, Martin G, Lambowitz AM (ed), *Mobile DNA II.* ASM Press, Washington, DC.
2. **Boveri T.** 1887. Uber Differenzierung der Zellkerne wahrend der Furchung des Eies von *Ascaris megalocephala. Ann Anat* **2:**688–693.
3. **Ammermann D, Steinbruck G, von Berger L, Hennig W.** 1974. The development of the macronucleus in the ciliated protozoan *Stylonychia mytilus. Chromosoma* **45:** 401–429.
4. **Prescott DM, Murti KG, Bostock CJ.** 1973. Genetic apparatus of *Stylonychia* sp. *Nature* **242:**576, 597–600.
5. **Yao MC, Gorovsky MA.** 1974. Comparison of the sequences of macro- and micronuclear DNA of *Tetrahymena pyriformis. Chromosoma* **48:**1–18.
6. **Elliott AM, Nanney DL.** 1952. Conjugation in *Tetrahymena. Science* **116:**33–34.
7. **Cervantes MD, Coyne RS, Xi X, Yao MC.** 2006. The condensin complex is essential for amitotic segregation of bulk chromosomes, but not nucleoli, in the ciliate *Tetrahymena thermophila. Mol Cell Biol* **26:**4690–4700.
8. **Akematsu T, Endoh H.** 2010. Role of apoptosis-inducing factor (AIF) in programmed nuclear death during conjugation in *Tetrahymena thermophila. BMC Cell Biol* **11:**13.
9. **Liu ML, Yao MC.** 2012. Role of ATG8 and autophagy in programmed nuclear degradation in *Tetrahymena thermophila. Eukaryot Cell* **11:**494–506.
10. **Austerberry CF, Allis CD, Yao MC.** 1984. Specific DNA rearrangements in synchronously developing nuclei of Tetrahymena. *Proc Nat Acad Sci U S A* **81:**7383–7387.
11. **Yao MC.** 1981. Ribosomal RNA gene amplification in Tetrahymena may be associated with chromosome breakage and DNA elimination. *Cell* **24:**765–774.

12. **Yao MC, Zhu SG, Yao CH.** 1985. Gene amplification in *Tetrahymena thermophila*: formation of extrachromosomal palindromic genes coding for rRNA. *Mol Cell Biol* **5:**1260–1267.

13. **Eisen JA, Coyne RS, Wu M, Wu D, Thiagarajan M, Wortman JR, Badger JH, Ren Q, Amedeo P, Jones KM, Tallon LJ, Delcher AL, Salzberg SL, Silva JC, Haas BJ, Majoros WH, Farzad M, Carlton JM, Smith RK Jr, Garg J, Pearlman RE, Karrer KM, Sun L, Manning G, Elde NC, Turkewitz AP, Asai DJ, Wilkes DE, Wang Y, Cai H, Collins K, Stewart BA, Lee SR, Wilamowska K, Weinberg Z, Ruzzo WL, Wloga D, Gaertig J, Frankel J, Tsao CC, Gorovsky MA, Keeling PJ, Waller RF, Patron NJ, Cherry JM, Stover NA, Krieger CJ, del Toro C, Ryder HF, Williamson SC, Barbeau RA, Hamilton EP, Orias E.** 2006. Macronuclear genome sequence of the ciliate *Tetrahymena thermophila*, a model eukaryote. *PLoS Biol* **4:**e286.

14. **Yao MC, Zheng K, Yao CH.** 1987. A conserved nucleotide sequence at the sites of developmentally regulated chromosomal breakage in Tetrahymena. *Cell* **48:**779–788.

15. **Fan Q, Yao MC.** 2000. A long stringent sequence signal for programmed chromosome breakage in *Tetrahymena thermophila*. *Nucleic Acids Res* **28:**895–900.

16. **Coyne RS, Yao MC.** 1996. Evolutionary conservation of sequences directing chromosome breakage and rDNA palindrome formation in tetrahymenine ciliates. *Genetics* **144:**1479–1487.

17. **Fan Q, Yao M.** 1996. New telomere formation coupled with site-specific chromosome breakage in *Tetrahymena thermophila*. *Mol Cell Biol* **16:**1267–1274.

18. **Yao MC, Kimmel AR, Gorovsky MA.** 1974. A small number of cistrons for ribosomal RNA in the germinal nucleus of a eukaryote, *Tetrahymena pyriformis*. *Proc Nat Acad Sci U S A* **71:**3082–3086.

19. **Yao MC, Gall JG.** 1977. A single integrated gene for ribosomal RNA in a eucaryote, *Tetrahymena pyriformis*. *Cell* **12:**121–132.

20. **Woodard J, Kaneshiro E, Gorovsky MA.** 1972. Cytochemical studies on the problem of macronuclear subnuclei in Tetrahymena. *Genetics* **70:**251–260.

21. **Yasuda LF, Yao MC.** 1991. Short inverted repeats at a free end signal large palindromic DNA formation in Tetrahymena. *Cell* **67:**505–516.

22. **Butler DK, Yasuda LE, Yao MC.** 1996. Induction of large DNA palindrome formation in yeast: implications for gene amplification and genome stability in eukaryotes. *Cell* **87:**1115–1122.

23. **Tanaka H, Tapscott SJ, Trask BJ, Yao MC.** 2002. Short inverted repeats initiate gene amplification through the formation of a large DNA palindrome in mammalian cells. *Proc Natl Acad Sci U S A* **99:**8772–8777.

24. **Tanaka H, Yao MC.** 2009. Palindromic gene amplification —an evolutionarily conserved role for DNA inverted repeats in the genome. *Nat Rev Cancer* **9:**216–224.

25. **Callahan RC, Shalke G, Gorovsky MA.** 1984. Developmental rearrangements associated with a single type of expressed α-tubulin gene in Tetrahymena. *Cell* **36:**441–445.

26. **Yao MC, Choi J, Yokoyama S, Austerberry CF, Yao CH.** 1984. DNA elimination in Tetrahymena: a developmental process involving extensive breakage and rejoining of DNA at defined sites. *Cell* **36:**433–440.

27. **Austerberry CF, Yao MC.** 1987. Nucleotide sequence structure and consistency of a developmentally regulated DNA deletion in *Tetrahymena thermophila*. *Mol Cell Biol* **7:**435–443.

28. **Austerberry CF, Yao MC.** 1988. Sequence structures of two developmentally regulated, alternative DNA deletion junctions in *Tetrahymena thermophila*. *Mol Cell Biol* **8:**3947–3950.

29. **Chau MF, Orias E.** 1996. Developmentally programmed DNA rearrangement in *Tetrahymena thermophila*: isolation and sequence characterization of three new alternative deletion systems. *Biol Cell* **86:**111–120.

30. **Chilcoat ND, Turkewitz AP.** 1997. In vivo analysis of the major exocytosis-sensitive phosphoprotein in Tetrahymena. *J Cell Biol* **139:**1197–1207.

31. **Heinonen TY, Pearlman RE.** 1994. A germ line-specific sequence element in an intron in *Tetrahymena thermophila*. *J Biol Chem* **269:**17428–17433.

32. **Huvos PE, Wu M, Gorovsky MA.** 1998. A developmentally eliminated sequence in the flanking region of the histone H1 gene in *Tetrahymena thermophila* contains short repeats. *J Eukaryot Microbiol* **45:**189–197.

33. **Katoh M, Hirono M, Takemasa T, Kimura M, Watanabe Y.** 1993. A micronucleus-specific sequence exists in the 5′-upstream region of calmodulin gene in *Tetrahymena thermophila*. *Nucleic Acids Res* **21:**2409–2414.

34. **Patil NS, Hempen PM, Udani RA, Karrer KM.** 1997. Alternate junctions and microheterogeneity of Tlr1, a developmentally regulated DNA rearrangement in *Tetrahymena thermophila*. *J Eukaryot Microbiol* **44:**518–522.

35. **Fass JN, Joshi NA, Couvillion MT, Bowen J, Gorovsky MA, Hamilton EP, Orias E, Hong K, Coyne RS, Eisen JA, Chalker DL, Lin D, Collins K.** 2011. Genome-scale analysis of programmed DNA elimination sites in *Tetrahymena thermophila*. *G3 (Bethesda)* **1:**515–522.

36. **Fire A, Xu S, Montgomery MK, Kostas SA, Driver SE, Mello CC.** 1998. Potent and specific genetic interference by double-stranded RNA in *Caenorhabditis elegans*. *Nature* **391:**806–811.

37. **Chalker DL, Yao MC.** 2001. Nongenic, bidirectional transcription precedes and may promote developmental DNA deletion in *Tetrahymena thermophila*. *Genes Dev* **15:**1287–1298.

38. **Aronica L, Bednenko J, Noto T, DeSouza LV, Siu KW, Loidl J, Pearlman RE, Gorovsky MA, Mochizuki K.** 2008. Study of an RNA helicase implicates small RNA-noncoding RNA interactions in programmed DNA elimination in Tetrahymena. *Genes Dev* **22:**2228–2241.

39. **Mochizuki K, Gorovsky MA.** 2004. RNA polymerase II localizes in *Tetrahymena thermophila* meiotic micronuclei when micronuclear transcription associated with genome rearrangement occurs. *Eukaryot Cell* **3:**1233–1240.

40. **Mochizuki K, Fine NA, Fujisawa T, Gorovsky MA.** 2002. Analysis of a piwi-related gene implicates small RNAs in genome rearrangement in Tetrahymena. *Cell* **110:**689–699.

41. **Yao MC, Fuller P, Xi X.** 2003. Programmed DNA deletion as an RNA-guided system of genome defense. *Science* **300:**1581–1584.

42. **Malone CD, Anderson AM, Motl JA, Rexer CH, Chalker DL.** 2005. Germ line transcripts are processed by a Dicer-like protein that is essential for developmentally programmed genome rearrangements of *Tetrahymena thermophila*. *Mol Cell Biol* **25:**9151–9164.

43. **Mochizuki K, Gorovsky MA.** 2005. A Dicer-like protein in Tetrahymena has distinct functions in genome rearrangement, chromosome segregation, and meiotic prophase. *Genes Dev* **19:**77–89.

44. **Howard-Till RA, Yao MC.** 2006. Induction of gene silencing by hairpin RNA expression in *Tetrahymena thermophila* reveals a second small RNA pathway. *Mol Cell Biol* **26:**8731–8742.

45. **Lee SR, Collins K.** 2007. Physical and functional coupling of RNA-dependent RNA polymerase and Dicer in the biogenesis of endogenous siRNAs. *Nat Struct Mol Biol* **14:**604–610.

46. **Lee SR, Collins K.** 2006. Two classes of endogenous small RNAs in *Tetrahymena thermophila*. *Genes Dev* **20:**28–33.

47. **Bednenko J, Noto T, DeSouza LV, Siu KW, Pearlman RE, Mochizuki K, Gorovsky MA.** 2009. Two GW repeat proteins interact with *Tetrahymena thermophila* argonaute and promote genome rearrangement. *Mol Cell Biol* **29:**5020–5030.

48. **Noto T, Kurth HM, Kataoka K, Aronica L, DeSouza LV, Siu KW, Pearlman RE, Gorovsky MA, Mochizuki K.** 2010. The Tetrahymena argonaute-binding protein Giw1p directs a mature argonaute-siRNA complex to the nucleus. *Cell* **140:**692–703.

49. **Madireddi MT, Davis MC, Allis CD.** 1994. Identification of a novel polypeptide involved in the formation of DNA-containing vesicles during macronuclear development in Tetrahymena. *Dev Biol* **165:**418–431.

50. **Madireddi MT, Coyne RS, Smothers JF, Mickey KM, Yao MC, Allis CD.** 1996. Pdd1p, a novel chromodomain-containing protein, links heterochromatin assembly and DNA elimination in Tetrahymena. *Cell* **87:**75–84.

51. **Yokoyama RW, Yao MC.** 1982. Elimination of DNA sequences during macronuclear differentiation in *Tetrahymena thermophila*, as detected by in situ hybridization. *Chromosoma* **85:**11–22.

52. **Taverna SD, Coyne RS, Allis CD.** 2002. Methylation of histone H3 at lysine 9 targets programmed DNA elimination in Tetrahymena. *Cell* **110:**701–711.

53. **Nikiforov MA, Gorovsky MA, Allis CD.** 2000. A novel chromodomain protein, Pdd3p, associates with internal eliminated sequences during macronuclear development in *Tetrahymena thermophila*. *Mol Cell Biol* **20:**4128–4134.

54. **Nikiforov MA, Smothers JF, Gorovsky MA, Allis CD.** 1999. Excision of micronuclear-specific DNA requires parental expression of Pdd2p and occurs independently from DNA replication in *Tetrahymena thermophila*. *Genes Dev* **13:**2852–2862.

55. **Smothers JF, Mizzen CA, Tubbert MM, Cook RG, Allis CD.** 1997. Pdd1p associates with germline-restricted chromatin and a second novel anlagen-enriched protein in developmentally programmed DNA elimination structures. *Development* **124:**4537–4545.

56. **Schwope RM, Chalker DL.** 2013. Mutations in Pdd1 reveal distinct requirements for its chromo- and chromoshadow domains in directing histone methylation and heterochromatin elimination. *Eukaryot Cell* **13:**190–201.

57. **Duharcourt S, Yao MC.** 2002. Role of histone deacetylation in developmentally programmed DNA rearrangements in *Tetrahymena thermophila*. *Eukaryot Cell* **1:** 293–303.

58. **Liu Y, Mochizuki K, Gorovsky MA.** 2004. Histone H3 lysine 9 methylation is required for DNA elimination in developing macronuclei in Tetrahymena. *Proc Nat Acad Sci U S A* **101:**1679–1684.

59. **Liu Y, Taverna SD, Muratore TL, Shabanowitz J, Hunt DF, Allis CD.** 2007. RNAi-dependent H3K27 methylation is required for heterochromatin formation and DNA elimination in Tetrahymena. *Genes Dev* **21:**1530–1545.

60. **Chung PH, Yao MC.** 2012. *Tetrahymena thermophila* JMJD3 homolog regulates H3K27 methylation and nuclear differentiation. *Eukaryot Cell* **11:**601–614.

61. **Yao MC, Yao CH, Halasz LM, Fuller P, Rexer CH, Wang SH, Jain R, Coyne RS, Chalker DL.** 2007. Identification of novel chromatin-associated proteins involved in programmed genome rearrangements in Tetrahymena. *J Cell Sci* **120:**1978–1989.

62. **Rexer CH, Chalker DL.** 2007. Lia1p, a novel protein required during nuclear differentiation for genome-wide DNA rearrangements in *Tetrahymena thermophila*. *Eukaryot Cell* **6:**1320–1329.

63. **Shieh AW, Chalker DL.** 2013. LIA5 is required for nuclear reorganization and programmed DNA rearrangements occurring during Tetrahymena macronuclear differentiation. *PLoS One* **8:**e75337.

64. **Sonneborn TM.** 1977. Genetics of cellular differentiation: stable nuclear differentiation in eucaryotic unicells. *Annu Rev Genet* **11:**349–367.

65. **Sonneborn TM, Schneller MV.** 1979. A genetic system for alternative stable characteristics in genomically identical homozygous clones. *Dev Genet* **1:**21–46.

66. **Meyer E, Duharcourt S.** 1996. Epigenetic programming of developmental genome rearrangements in ciliates. *Cell* **87:**9–12.

67. **Meyer E, Duharcourt S.** 1996. Epigenetic regulation of programmed genomic rearrangements in Paramecium aurelia. *J Eukaryot Microbiol* **43:**453–461.

68. **Chalker DL, Fuller P, Yao MC.** 2005. Communication between parental and developing genomes during Tetrahymena nuclear differentiation is likely mediated by homologous RNAs. *Genetics* **169:**149–160.

69. **Chalker DL, Yao MC.** 1996. Non-Mendelian, heritable blocks to DNA rearrangement are induced by loading

the somatic nucleus of *Tetrahymena thermophila* with germ line-limited DNA. *Mol Cell Biol* **16:**3658–3667.

70. **Mochizuki K, Gorovsky MA.** 2004. Conjugation-specific small RNAs in Tetrahymena have predicted properties of scan (scn) RNAs involved in genome rearrangement. *Genes Dev* **18:**2068–2073.

71. **Gao S, Liu Y.** 2012. Intercepting noncoding messages between germline and soma. *Genes Dev* **26:**1774–1779.

72. **Garnier O, Serrano V, Duharcourt S, Meyer E.** 2004. RNA-mediated programming of developmental genome rearrangements in *Paramecium tetraurelia*. *Mol Cell Biol* **24:**7370–7379.

73. **Yao MC, Chao JL.** 2005. RNA-guided DNA deletion in Tetrahymena: an RNAi-based mechanism for programmed genome rearrangements. *Annu Rev Genet* **39:** 537–559.

74. **Schoeberl UE, Kurth HM, Noto T, Mochizuki K.** 2012. Biased transcription and selective degradation of small RNAs shape the pattern of DNA elimination in Tetrahymena. *Genes Dev* **26:**1729–1742.

75. **Cherry JM, Blackburn EH.** 1985. The internally located telomeric sequences in the germ-line chromosomes of Tetrahymena are at the ends of transposon-like elements. *Cell* **43:**747–758.

76. **Wells JM, Ellingson JL, Catt DM, Berger PJ, Karrer KM.** 1994. A small family of elements with long inverted repeats is located near sites of developmentally regulated DNA rearrangement in *Tetrahymena thermophila*. *Mol Cell Biol* **14:**5939–5949.

77. **Cheng CY, Vogt A, Mochizuki K, Yao MC.** 2010. A domesticated piggyBac transposase plays key roles in heterochromatin dynamics and DNA cleavage during programmed DNA deletion in *Tetrahymena thermophila*. *Mol Biol Cell* **21:**1753–1762.

78. **Saveliev SV, Cox MM.** 1996. Developmentally programmed DNA deletion in *Tetrahymena thermophila* by a transposition-like reaction pathway. *EMBO J* **15:** 2858–2869.

79. **Vogt A, Mochizuki K.** 2013. A domesticated PiggyBac transposase interacts with heterochromatin and catalyzes reproducible DNA elimination in Tetrahymena. *PLoS Genet* **9:**e1004032.

80. **Baudry C, Malinsky S, Restituito M, Kapusta A, Rosa S, Meyer E, Betermier M.** 2009. PiggyMac, a domesticated piggyBac transposase involved in programmed genome rearrangements in the ciliate *Paramecium tetraurelia*. *Genes Dev* **23:**2478–2483.

81. **Nowacki M, Higgins BP, Maquilan GM, Swart EC, Doak TG, Landweber LF.** 2009. A functional role for transposases in a large eukaryotic genome. *Science* **324:** 935–938.

82. **Lieber MR.** 2010. The mechanism of double-strand DNA break repair by the nonhomologous DNA end-joining pathway. *Annu Rev Biochem* **79:**181–211.

83. **Downs JA, Jackson SP.** 2004. A means to a DNA end: the many roles of Ku. *Nat Rev Mol Cell Biol* **5:** 367–378.

84. **Lin IT, Chao JL, Yao MC.** 2012. An essential role for the DNA breakage-repair protein Ku80 in programmed DNA rearrangements in *Tetrahymena thermophila*. *Mol Biol Cell* **23:**2213–2225.

85. **Jeggo PA, Taccioli GE, Jackson SP.** 1995. Menage a trois: double strand break repair, V(D)J recombination and DNA-PK. *BioEssays* **17:**949–957.

86. **Finnie NJ, Gottlieb TM, Blunt T, Jeggo PA, Jackson SP.** 1995. DNA-dependent protein kinase activity is absent in *xrs-6* cells: implications for site-specific recombination and DNA double-strand break repair. *Proc Nat Acad Sci U S A* **92:**320–324.

87. **Austerberry CF, Snyder RO, Yao MC.** 1989. Sequence microheterogeneity is generated at junctions of programmed DNA deletions in *Tetrahymena thermophila*. *Nucleic Acids Res* **17:**7263–7272.

88. **Liu Y, Song X, Gorovsky MA, Karrer KM.** 2005. Elimination of foreign DNA during somatic differentiation in *Tetrahymena thermophila* shows position effect and is dosage dependent. *Eukaryot Cell* **4:**421–431.

89. **Godiska R, Yao MC.** 1990. A programmed site-specific DNA rearrangement in *Tetrahymena thermophila* requires flanking polypurine tracts. *Cell* **61:**1237–1246.

90. **Godiska R, James C, Yao MC.** 1993. A distant 10-bp sequence specifies the boundaries of a programmed DNA deletion in Tetrahymena. *Genes Dev* **7:**2357–2365.

91. **Chalker DL, La Terza A, Wilson A, Kroenke CD, Yao MC.** 1999. Flanking regulatory sequences of the Tetrahymena R deletion element determine the boundaries of DNA rearrangement. *Mol Cell Biol* **19:**5631–5641.

92. **Barkess G, West AG.** 2012. Chromatin insulator elements: establishing barriers to set heterochromatin boundaries. *Epigenomics* **4:**67–80.

93. **Tamaru H.** 2010. Confining euchromatin/heterochromatin territory: jumonji crosses the line. *Genes Dev* **24:**1465–1478.

94. **Merkenschlager M, Odom DT.** 2013. CTCF and cohesin: linking gene regulatory elements with their targets. *Cell* **152:**1285–1297.

95. **Phillips JE, Corces VG.** 2009. CTCF: master weaver of the genome. *Cell* **137:**1194–1211.

96. **Orias E.** 1981. Probable somatic DNA rearrangements in mating type determination in *Tetrahymena thermophila*: a review and a model. *Dev Genet* **2:**185–202.

97. **Cervantes MD, Hamilton EP, Xiong J, Lawson MJ, Yuan D, Hadjithomas M, Miao W, Orias E.** 2013. Selecting one of several mating types through gene segment joining and deletion in *Tetrahymena thermophila*. *PLoS Biol* **11:**e1001518.

98. **Castel SE, Martienssen RA.** 2013. RNA interference in the nucleus: roles for small RNAs in transcription, epigenetics and beyond. *Nat Rev Genet* **14:**100–112.

99. **Westra ER, Swarts DC, Staals RH, Jore MM, Brouns SJ, van der Oost J.** 2012. The CRISPRs, they are a-changin′: how prokaryotes generate adaptive immunity. *Annu Rev Genet* **46:**311–339.

100. **Deveau H, Garneau JE, Moineau S.** 2010. CRISPR/Cas system and its role in phage-bacteria interactions. *Annu Rev Microbiol* **64:**475–493.

101. Jex AR, Liu S, Li B, Young ND, Hall RS, Li Y, Yang L, Zeng N, Xu X, Xiong Z, Chen F, Wu X, Zhang G, Fang X, Kang Y, Anderson GA, Harris TW, Campbell BE, Vlaminck J, Wang T, Cantacessi C, Schwarz EM, Ranganathan S, Geldhof P, Nejsum P, Sternberg PW, Yang H, Wang J, Gasser RB. 2011. *Ascaris suum* draft genome. *Nature* 479:529–533.

102. Smith JJ, Kuraku S, Holt C, Sauka-Spengler T, Jiang N, Campbell MS, Yandell MD, Manousaki T, Meyer A, Bloom OE, Morgan JR, Buxbaum JD, Sachidanandam R, Sims C, Garruss AS, Cook M, Krumlauf R, Wiedemann LM, Sower SA, Decatur WA, Hall JA, Amemiya CT, Saha NR, Buckley KM, Rast JP, Das S, Hirano M, McCurley N, Guo P, Rohner N, Tabin CJ, Piccinelli P, Elgar G, Ruffier M, Aken BL, Searle SM, Muffato M, Pignatelli M, Herrero J, Jones M, Brown CT, Chung-Davidson YW, Nanlohy KG, Libants SV, Yeh CY, McCauley DW, Langeland JA, Pancer Z, Fritzsch B, de Jong PJ, Zhu B, Fulton LL, Theising B, Flicek P, Bronner ME, Warren WC, Clifton SW, Wilson RK, Li W. 2013. Sequencing of the sea lamprey (*Petromyzon marinus*) genome provides insights into vertebrate evolution. *Nat Genet* 45:415–421, 421e1–2.

Mobile DNA, 3rd Edition
Nancy L. Craig, Michael Chandler, Martin Gellert, Alan M. Lambowitz, Phoebe A. Rice and Suzanne Sandmeyer

doi:10.1128/microbiolspec.MDNA3-0035-2014

Mireille Bétermier[1]
Sandra Duharcourt[2]

17 Programmed Rearrangement in Ciliates: *Paramecium*

INTRODUCTION

Ciliates belong to a monophyletic group of unicellular eukaryotes within the Alveolate branch (1). The species that have been used as model organisms are free-living organisms, but parasitic or endosymbiotic ciliates have also been characterized (2, 3). A handful of ciliates have been studied, and common features could be deduced (reviewed in 4). They all carry motile and sensory cilia at the cell surface that allow swimming, food uptake, and the sensing of environmental signals. They present a characteristic nuclear dimorphism and undergo spectacular, genome-wide programmed rearrangements during development. The study of the mechanisms and regulation pathways underlying genome rearrangements has revealed a great diversity in the strategies used by different ciliates (5). The present chapter will focus on *Paramecium*, a widespread group of species that can be found on all continents. The sequences of the somatic genomes of two species, *Paramecium tetraurelia* and *Paramecium caudatum*, have been published recently (6, 7, 8). *P. tetraurelia*, which belongs to the *Paramecium aurelia* group of sibling species (9), is by far the most extensively studied species at the genomic level, and will be the main focus of the present chapter.

Nuclear Dimorphism

Paramecium exhibits two functionally distinct types of nuclei that coexist in its cytoplasm at all stages of its life cycle. This unique nuclear dimorphism actually reflects the physical separation of the germline and somatic functions of chromosomes (10). *P. tetraurelia* harbors two identical diploid micronuclei (MICs), which divide by mitosis at each cell division. Under laboratory conditions, *Paramecium* grows vegetatively when fed on bacteria inoculated in plant infusions (11). During vegetative growth, the MICs are not transcribed and appear to be dispensable (12, 13). They are referred to as the germline nuclei because they undergo meiosis during the sexual processes and transmit the genetic information to the progeny. The single, highly polyploid somatic macronucleus (MAC, $800n$) is responsible for gene transcription and governs the cell phenotype. During vegetative growth, the MAC divides in an unconventional manner (14), splitting in two halves without evidence of either chromosome condensation or formation of a classical mitotic spindle, which led to the hypothesis that MAC chromosomes do not carry centromeres. Even though its presence is essential for cell survival, the MAC is fragmented and

[1]Institute of Integrative Biology of the Cell (I2BC), CNRS, CEA, Université Paris Sud, 1 avenue de la Terrasse, 91198 Gif-sur-Yvette cedex, France; [2]Institut Jacques Monod, CNRS, UMR 7592, Université Paris Diderot, Sorbonne Paris Cité, Paris, F-75205, France.

lost during sexual processes, and a new somatic MAC has to differentiate from a copy of the zygotic nucleus in each cell of the next sexual generation.

Upon mild starvation, *Paramecium* reproduces sexually through conjugation between reactive partners from compatible mating types. In each species of the *P. aurelia* complex, two mating types—O and E—have been reported (15). At the cytological level, the cellular and nuclear reorganization that takes place during conjugation has been particularly well described in *P. tetraurelia* (Fig. 1A) (16). Mating starts with MIC meiosis, which takes place in each conjugating partner and gives rise to eight haploid nuclei. While MICs undergo meiosis, the parental MAC is fragmented into ~30 pieces, in which replication stops rapidly but transcriptional activity is maintained (17). Following meiosis, one haploid gametic nucleus, most likely selected randomly, moves to a specific compartment, the paroral cone, close to the interaction surface with the mating partner, and divides once to yield two genetically identical gametic nuclei. Fertilization involves the reciprocal exchange of one migratory gametic nucleus between the two partners. During karyogamy, fusion between the incoming migratory nucleus and the resident gametic nucleus gives rise to the diploid zygotic nucleus. The zygotic nucleus undergoes two successive mitotic divisions: the first takes place in the paroral cone of each mating partner, while the second takes place along the longitudinal axis of each cell, right after the separation of exconjugants. The latter division drives polar localization of the four mitotic products: the longitudinal orientation of mitotic spindles drives the positioning of two nuclei at the cell posterior pole, which triggers their differentiation into new MACs, while the two nuclei at the anterior pole will become the new MICs (16). The exact nature of the signal(s) that drive(s) MIC and MAC determination is not clear. Grandchamp and Beisson noticed that the second division of the zygotic nucleus is accompanied by a transient longitudinal contraction of the exconjugants, which become ~30% shorter in length. Their unpublished data (Fig. 1B) suggested that cell contraction activates asymmetrically located membrane mechanoreceptor channels (specific for Ca^{2+} at the anterior pole and for K^{+} at the posterior pole), which may form an intracellular ionic gradient between the two cell poles: Ca^{2+} at the anterior pole would favor MIC determination, while K^{+} at the posterior pole would trigger MAC determination (S. Grandchamp and J. Beisson, personal communication). Following nuclear determination, the two future new MACs adopt a central position in the cell where they start their differentiation in each exconjugant. This process, also called MAC development, involves massive endoduplication of the genome from $2n$ to $800n$, during a period extending over two cell cycles (17). Four to five synchronous replication peaks take place during the first cell cycle, concomitant with large-scale programmed genome rearrangements (see below). At the first cell division (or caryonidal division), the two MICs divide through mitosis, and one developing new MAC segregates into each daughter cell. Replication in the new MAC switches to a more continuous mode after the first cell division, until the final ploidy is reached (17).

The Peculiar Genetics of *P. aurelia*

P. aurelia species are ideally suited for genetic analyses because of their two alternative modes of sexual reproduction. In conjugation, reciprocal exchange of one of the gametic nuclei between two cells of complementary mating types, followed by karyogamy, produces F1 cells that are always genetically identical, whatever the genotypes of the parental cells. The presence of different phenotypes in the two genetically identical F1 cells immediately reveals non-Mendelian characters. Known cases of maternally inherited phenotypes have been described, including the inheritance of mating-type determination in *P. tetraurelia* (18, 19). In addition to conjugation, *P. aurelia* species undergo autogamy, a self-fertilization process in which the two genetically identical gametic nuclei in each cell fuse within the same cell. This results in an entirely homozygous zygotic genome, which, in addition to the extremely low mutation rate observed in *P. tetraurelia* (20), considerably facilitates genomic analyses.

STRUCTURE OF THE GERMLINE AND SOMATIC GENOMES

The Somatic Genome: Breakthrough from Genome Sequencing Programs

Although they both differentiate from identical mitotic copies of the zygotic nucleus, the MIC and MAC exhibit striking differences in their chromosome organization and genome structure (Fig. 2) (21).

According to pulse-field electrophoresis, MAC chromosomes are linear molecules of variable size ranging from 50 kbp to 1 Mbp (22). The ends of MAC chromosomes are capped by a mixture of G_3T_3 and G_4T_2 telomeric repeats, which are added by an error-prone telomerase at heterogeneous positions within a telomere-addition region that extends over ~1 kb (23, 24). An additional level of heterogeneity has been attributed

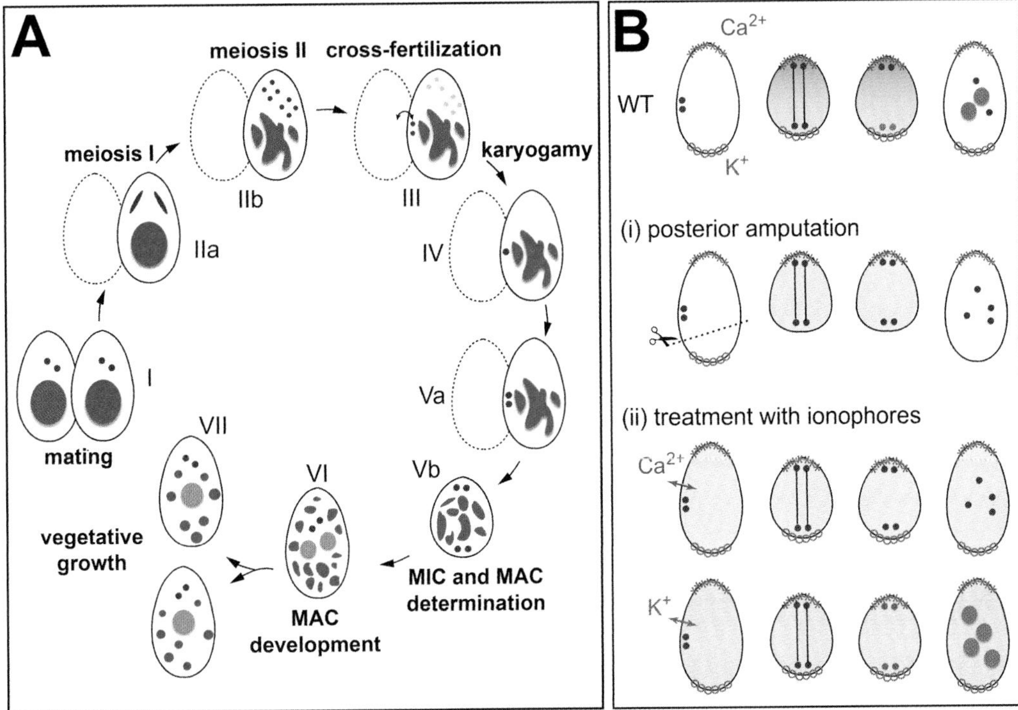

Figure 1 MIC and MAC determination during the *P. tetraurelia* sexual cycle. (A) Nuclear reorganization during the *P. tetraurelia* sexual cycle. Conjugation occurs between cells with compatible mating types. Nuclear reorganization events take place in both mating partners, but details are represented in only one cell for clarity, with the parental MAC in blue. Following mating of two reactive cells (I), MIC meiosis starts, while the MAC remains intact (IIa). At meiosis II, eight haploid gametic nuclei are produced and the MAC begins its fragmentation process (IIb). In each partner, one meiotic product divides once to give rise to two identical gametic nuclei, and fertilization takes place through reciprocal exchange of one gametic nucleus (III). The remaining seven meiotic products are degraded. In each conjugating cell, a diploid zygotic nucleus is formed through the fusion of a resident and a migratory haploid nucleus (IV). The zygotic nucleus undergoes two mitotic divisions (Va and Vb), and exconjugants separate between the first and second divisions. During the second division of the zygotic nucleus, exconjugants shorten dramatically in size (Vb), which triggers the determination of two new MICs at the anterior cell pole (black) and two new MACs at the posterior pole. Programmed genome rearrangements take place in the developing new MACs (brown in VI). At the first cell division, the new MICs divide mitotically, and each of the two developing new MACs segregates into a daughter cell (VII), where it continues to amplify the rearranged genome to a final ploidy of ~800*n*. (B) MIC and MAC determination during cell contraction. To simplify the figure, the old MAC is not represented. Mechanical stimulation experiments have indicated that K^+ mechanoreceptor channels (brown circles) are mostly located at the posterior pole while Ca^{2+} mechanoreceptor channels (red stars) are at the anterior pole (114). For wild-type cells (WT), an attractive hypothesis would be that the transient (~15 min) shortening of exconjugants exerts a pressure on the membrane, which activates mechanoreceptor channels and increases membrane permeability for Ca^{2+} at the anterior pole (where MICs are determined) and K^+ at the posterior pole (where MACs are determined). Two unpublished experiments confirm that a pre-existing determinant, already present at the cell posterior pole before the polar positioning of the products of the second mitotic division of the zygotic nucleus, drives MAC determination during cell contraction (S. Grandchamp and J. Beisson, personal communication). (i) Amputation of the cell posterior part right before the shortening of exconjugants results in a large excess of progeny with four MICs. (ii) Manipulating the Ca^{2+} or K^+ intracellular concentration using specific ionophores strongly disturbs nuclear determination: Ca^{2+} ionophores induce an excess of MICs, while K^+ ionophores trigger an excess of MACs.
doi:10.1128/microbiolspec.MDNA3-0035-2014.f1

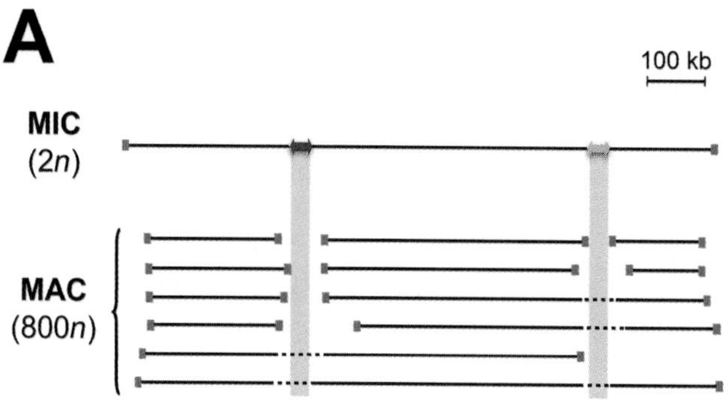

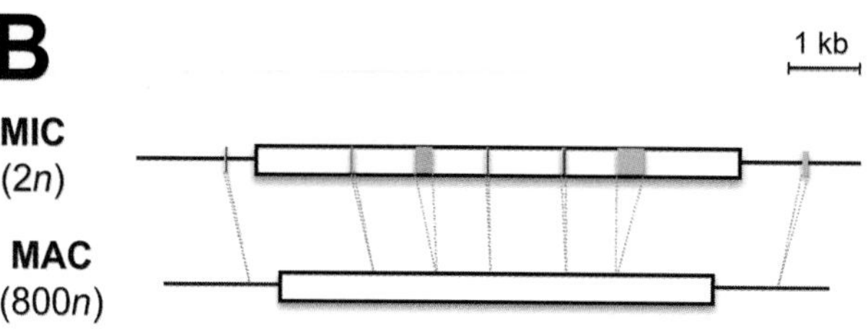

Figure 2 Programmed genome rearrangements during MAC development. (A) General organization of MIC and MAC chromosomes and chromosome fragmentation. A representative MIC chromosome is shown on top, with repeated germline sequences (e.g., transposons) drawn as colored double-headed arrows. The exact structure of MIC telomeres (green boxes) is at present not known. During MAC development, each MIC chromosome is amplified ~400-fold. Programmed heterogeneous elimination of repeated DNA (represented by vertical grey boxes) is associated with two alternative genome rearrangements: chromosome fragmentation and telomere addition to new MAC chromosome ends (blue boxes), or imprecise joining of the two chromosome arms that flank the eliminated region (dotted line). (B) Precise excision of internal eliminated sequences (IESs). The map represents a gene (white box) and its flanking regions. IESs are drawn as colored boxes and their precise excision is represented by dotted lines. Note that the scale is different from that used in (A).
doi:10.1128/microbiolspec.MDNA3-0035-2014.f2

to the presence of several telomere-addition regions, distant by several kilobase pairs from each other, at the ends of MAC chromosomes (25, 26, 27). Thanks to joint efforts of the international community of *Paramecium* laboratories (28, 29, 30), a draft assembly of the MAC genome of *P. tetraurelia* was released in 2006 (7) and a polished version in 2012 (6). One haploid equivalent of the MAC genome consists of 72 Mbp of 72% AT-rich DNA, distributed into 188 scaffolds ranging from 45 kbp to 1 Mbp, ~150 of which are terminated by telomeric repeats and therefore represent full-length somatic chromosomes. With regard to gene content, it has been known for a long time that *Paramecium* does not use the universal genetic code and translates UGA and UAG stop codons into glutamine (31, 32). Taking this into account, annotation of the MAC genome sequence revealed a strikingly high coding density (78%), with short intergenic regions (350 bp on average) and very little repeated DNA, except for a few segmental duplications (6, 7). Around 40,000 genes were annotated in the somatic genome of *P. tetraurelia*, as a consequence of three successive whole-genome duplications (7). Sequencing of the MAC genome from other species of the *P. aurelia* complex indicated that the most recent whole-genome duplication immediately preceded the burst of this group of species (7, 33). Inside genes, around 90,000 tiny introns (between 20 and 34 nt in length) were predicted and largely confirmed through expressed sequence tag sequencing (34) and high-throughput cDNA sequencing (O. Arnaiz, personal communication). *Paramecium* introns appear to be under selective pressure for the elimination of incorrectly spliced mRNAs through nonsense-mediated decay: indeed, $3n$ introns without in-frame stop codons (relative to their upstream exon) are significantly underrepresented among the population of all introns (34).

A Glimpse of the Organization of MIC Chromosomes

The exact number of MIC chromosomes in *P. tetraurelia* is not known. An early cytological study based on azure A staining was performed on three other species of the *P. aurelia* group (35). Depending on the strain and the species, each MIC contains between 30 and 60 pairs of small poorly characterized germline chromosomes, barely distinguishable under the microscope during the first metaphase of meiosis. It should be noted that tiny chromosomes probably would have escaped detection under these experimental conditions, and accurate counting of MIC chromosomes will require the use of higher-resolution techniques. Since then, unraveling the structure of germline chromosomes has been the subject of intensive work. In 1992, J. Preer and colleagues used a cell fractionation procedure to purify vegetative MICs from *P. tetraurelia* and constructed the first MIC phage library. This library enabled them to compare the MIC and MAC versions of a limited number of genomic loci, leading to the landmark discovery that germline-specific interstitial sequences, called internal eliminated sequences (IESs), are absent from the MAC genome (36, 37). Thanks to the MIC library, a few dozen IESs were identified inside genes or in intergenic regions: IESs are flanked by two

TA dinucleotides in the MIC, while a single TA is found at their excision site in the MAC (reviewed in 38). As the purification of pure MIC DNA remained a technical challenge, several tricks were used to gain further insight into the genome-wide distribution of IESs. Analysis of individual MAC sequence reads allowed the identification of several thousand polymorphic "TA indels," present in only a fraction of reads for each given locus and representing putative IES excision errors, such as IES retention on some MAC copies of the locus or erroneous excision of genomic regions (39, 40). Significant progress was made in 2012, following the discovery of the *PiggyMac* gene (*PGM*), which encodes a domesticated transposase required to initiate programmed genome rearrangements, including IES excision (41). Following Pgm depletion, the germline genome is amplified to high ploidy levels in the developing new MAC (see below), thus providing a source of genomic "*PGM* DNA" enriched for non-rearranged sequences and suitable for next-generation sequencing (6). The complexity of *PGM* DNA was found to be around 100 Mbp, consistent with previous estimates of the complexity of the MIC genome (10). This indicates that 25 to 30 Mbp of MIC-restricted sequences are eliminated from the somatic MAC. Alignment of *PGM* DNA reads with the reference MAC genome allowed the identification of a genome-wide set of ~45,000 single-copy IESs (6). Strikingly, 47% of all genes are interrupted by at least one IES in the germline genome, which makes precise excision of IESs essential for the reconstitution of functional somatic genes.

Based on the estimated number of MIC chromosomes (30 to 60 pairs, which represents a minimal value) and the estimated complexity of the haploid germline genome (100 Mbp), the average size of MIC chromosomes would be around ~1 to 2 Mbp at most. This size range is significantly larger than that of MAC chromosomes (~150 kbp to 1 Mbp, based on MAC DNA assembly [7]). The difference can be attributed to chromosome fragmentation, which takes place during MAC development. Indeed, molecular analysis of a couple of fragmentation loci indicates that chromosome fragmentation does not depend upon the presence of a specific DNA sequence motif but is associated with imprecise elimination of large germline regions (over several kilobase pairs) located downstream of telomere addition sites (6, 27). Telomere addition at MAC chromosome ends or the formation of internal deletions through end joining were proposed to be alternative products of the same elimination event (27, 39, 42). A body of molecular evidence indicates that the eliminated germline regions located downstream of chromosome fragmentation sites harbor multicopy transposable elements and other repeated elements (6, 27, 41, 43).

All information available so far indicates that, during MAC development, *P. tetraurelia* discards 25 to 30 Mbp of its germline genome, representing a wide diversity of sequences (long or short, single-copy or repeated DNA, eliminated in a precise or heterogeneous manner). It is noteworthy that all IESs taken together represent 3 Mbp of germline DNA, which corresponds to only 10% of total eliminated DNA (6). Due to the presence of repeated DNA, the remaining germline eliminated sequences, which thus represent 90% of all eliminated DNA (22 to 27 Mbp), could not be assembled properly from the *PGM* DNA sequence reads. These reads could not be mapped to genomic regions collinear with MAC chromosomes. Therefore, these sequences most likely are included in chromosome fragmentation regions. Only the complete assembly and annotation of the germline genome sequence will provide a complete view of the structure of MIC chromosomes. Of particular interest will be the identification of transposable elements, germline-specific genes, and other chromosome features (discussed in 44).

Characteristic Features of IESs

Among the different types of eliminated sequences that have been described in *Paramecium*, IESs have been by far the most extensively studied. Analysis of the genome-wide set of 45,000 IESs (6, 45) confirmed the conclusions drawn from the study of the first IESs that were identified in the MIC phage library (reviewed in 38 and 46). IESs are invariably flanked by one TA dinucleotide at each end, and only a loosely conserved 8-bp consensus sequence (5′-TAYAGYNR-3′), reminiscent of the inverted repeats found at the ends of *Tc1/mariner* transposons (47), could be deduced from the nucleotide sequence of their ends. Variations in this consensus sequence may exist according to IES size (46) or to specific requirements for efficient excision (48; see below).

In the mid-1990s, L. Klobutcher and G. Herrick proposed their "invade, bloom, abdicate, and fade" (IBAF) model, according to which *Paramecium* IESs have derived from ancestral cut-and-paste transposons, which would have invaded the germline genome and decayed over time through internal deletions and base substitutions, while being kept under selection pressure for being precisely excised from the somatic genome (49). Interestingly, *Tc1/mariner* transposons integrate preferentially into TA dinucleotides, which they duplicate on each side of the newly integrated copy (50): conservation of the ancestral target site duplication at IES boundaries indicates that the TA dinucleotide has become essential for IES excision. In agreement with the IBAF hypothesis, IESs from the genome-wide set are mostly short, single-copy, noncoding sequences, with 93% ranging between 26 and 150 bp long (6). The analysis of whole-genome duplications, which allow dating the arrival of individual IESs at particular genomic loci, confirmed that the shortest IESs also tend to be the oldest. Among the largest IESs (over 500 bp long), clearly recognizable portions of multicopy *Tc1 mariner* transposons were identified, providing further support to a transposon origin for at least a fraction of extant IESs. In parallel with the transposon origin of IESs, the system has also been co-opted to excise genomic fragments as IESs, providing an additional layer for the regulation of genome plasticity and gene expression (18).

Although IES excision is highly precise and efficient overall, excision errors may occur, as witnessed by the identification of IES excision polymorphisms (39). Evidence also exists in some cases for the use of alternative excision boundaries, which represent a fraction of the so-called TA indels (our unpublished observations; see also 6 and 51). When localized inside genes, these errors may be incorporated into transcripts, yielding nonfunctional mRNAs. Thus, even though IESs are excised from DNA and introns are spliced from RNA, interesting parallels may be drawn in *P. tetraurelia* between these two classes of elements. Both classes carry little sequence information at their boundaries, but their elimination is essentially precise, which raises the question of their recognition. In addition, exactly as was shown for introns, IESs and TA indels appear to be under selection pressure for recognition by mRNA quality-control systems, such as nonsense-mediated decay: 3*n* IESs (or TA indels) that would not introduce an in-frame premature stop codon are significantly under-represented among the population (6, 39).

THE IES EXCISION TOOLBOX

Paramecium: A Powerful Model for the Mechanistic Study of Programmed Genome Rearrangements

The molecular mechanisms involved in programmed genome rearrangements in *Paramecium* have been particularly well studied for IES excision, essentially because of the precision of the reaction at the nucleotide level. The search for essential genes involved in IES excision started from a detailed molecular analysis of DNA intermediates produced *in vivo* during the course of MAC development and benefited greatly from the availability of the fully annotated somatic genome (7) and of the transcriptome during sexual processes (52). All genome-wide data have been integrated into a user-friendly database, ParameciumDB, which is continuously curated by the community (53, 54). The development of a simple and efficient RNA interference (RNAi) technique, now used routinely in *Paramecium* laboratories (55), has been of considerable help for the functional analysis of candidate genes, in combination with next-generation sequencing to monitor genome-wide the effect of each RNAi on IES excision (6, 43, 56). Finally, the use of green fluorescent protein (GFP) translational fusions has allowed the localization of each corresponding candidate protein to be monitored (Fig. 3). Several essential components of the core IES excision machinery have now been identified (Table 1), providing further support to a mechanistic connection between programmed DNA elimination and cut-and-paste transposition.

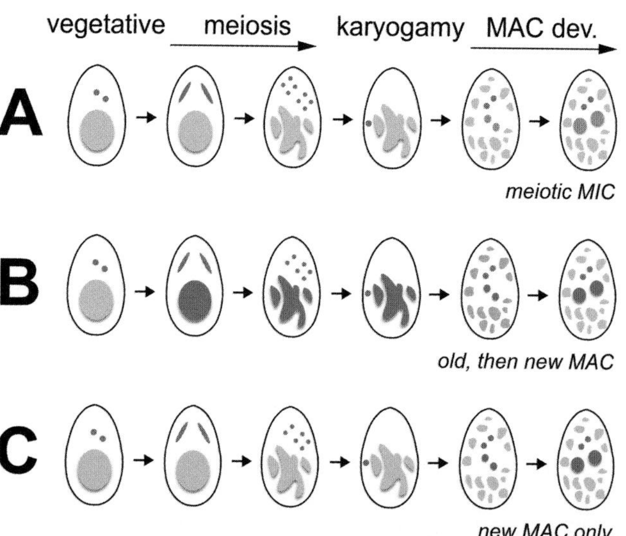

Figure 3 Major localization patterns of proteins involved in MAC development. All localization patterns were defined using GFP translational fusions expressed from transgenes that were microinjected into the vegetative MAC before induction of sexual processes (conjugation or autogamy; see Table 1 for the list of the genes that were tested). Based on GFP fluorescence (green), three major localization patterns were observed: (A) specific localization in the MICs at meiosis I; (B) transfer of GFP fluorescence from the old to the new MAC; and (C) exclusive localization in the new developing MAC. doi:10.1128/microbiolspec.MDNA3-0035-2014.f3

Table 1 Genes involved in genome rearrangements in the new developing MAC

Name	Putative function	Induction[a]	Nuclear localization[b]	RNAi or mutant phenotype[c]	Reference
DCL2	Dicer-like ribonuclease III scnRNA biogenesis	Early	A	Phenotype of Dcl2/Dcl3-depleted cells: Retention of a subset of IESs[d]	(56, 91, 92)
DCL3	Dicer-like ribonuclease III scnRNA biogenesis	Early	ND[f]	No scnRNAs[d] Inhibition of chromosome fragmentation/ transposon elimination[e]	
EZL1	H3K27 and H3K9 histone methyl transferase	Early	B	Retention of a subset of IESs[d] No H3K27me3, no H3K9me3 Inhibition of chromosome fragmentation/ transposon elimination[e]	(43)
NOWA1/2	GW repeat protein RNA binding	Early/intermediate	B	Retention of a subset of IESs[e] Inhibition of chromosome fragmentation/transposon elimination[e]	(85)
PTIWI09/01	Piwi protein scnRNA accumulation	Early	B	Retention of all IESs[e] No scnRNA accumulation[e] Inhibition of chromosome fragmentation/ transposon elimination[e]	(92)
PTMB.220	DEAH-box RNA helicase	Early	B	Retention of a subset of IESs[e] Inhibition of chromosome fragmentation/ transposon elimination[e]	(115)[g]
SUMO I/II/III	protein modifier	Early	B	Retention of all IESs[e]	(116)
DCL5	Dicer-like ribonuclease III iesRNA biogenesis	intermediate induction	C	Retention of a subset of IESs[d] No effect on transposon elimination[e] No iesRNAs[d]	(43, 56)
DIE5a/b	unknown	Intermediate	C	Retention of all IESs[e]	(117)
KU70a	Pgm partner C-NHEJ	Intermediate	C	Retention of all IESs[e] Inhibition of chromosome fragmentation/ transposon elimination[e]	(74)
KU80c	Pgm partner C-NHEJ	Intermediate	C	Retention of all IESs[e] Inhibition of chromosome fragmentation/ transposon elimination[e] No DSB[e]	(74)
LIG4a/b	DNA ligase IV C-NHEJ	Early	C	DSB accumulation at IES ends[e] No repair	(66)
PGM	domesticated transposase DNA cleavage	Intermediate	C	Retention of all IESs[d] Inhibition of chromosome fragmentation/ transposon elimination[e] No DSB[e]	(41)
UBA2	SUMO activating E1 enzyme	Intermediate	C	Retention of all IESs[e]	(116)
DNA-PKcs	C-NHEJ	Early/intermediate	ND	Retention of a subset of IESs[e] Lower DSB repair efficiency[e]	[h]
mtF	nd	ND	ND	Retention of a subset of IESs[e]	(118)
XRCC4	Lig4 partner	Early	ND	DSB accumulation at IES ends[e] No repair	(66)

[a]Transcriptome analysis from microarray hybridization data (52). Early, induction peak during meiosis; intermediate, induction peak during MAC development; intermediate induction, transcription starts during MAC development and increases gradually until late stages.
[b]Localization of GFP fusions expressed following transgene microinjection into the vegetative parental MAC. A, meiotic MIC; B, old then new MAC; C, new MAC only (see Fig. 3). For each pair of ohnologs, only the first gene of the pair (as indicated in the table) was fused to GFP.
[c]When only a subset of IESs is affected in RNAi-treated or mutant cells, the exact identity and number of sensitive IESs and their levels of retention differ largely from one gene to the other (see text). Data on chromosome fragmentation and/or transposon elimination are indicated only for those genes for which this particular type of genome rearrangement was examined.
[d]Whole-genome next-generation sequencing data.
[e]Data obtained for a limited number of elements (PCR, Southern blot hybridization, LM-PCR, analysis of NGS reads).
[f]ND, not determined.
[g]J. K. Nowak, personal communication.
[h]S. Malinsky, personal communication.

Insights from the Molecular Analysis of IES Excision Intermediates

Molecular characterization of IES excision intermediates produced during MAC development revealed that IESs are first amplified within the bulk of the germline genome during three to four endoduplication cycles, before excision starts (57). IES excision is essentially completed when caryonidal division takes place, although the excision machinery may still be active after this particular cell division (58). Ligation-mediated polymerase chain reaction (LM–PCR) experiments further established that IES excision is initiated by staggered double-strand cleavages centered on each flanking TA, generating DNA double-strand breaks (DSBs) with 4-nt 5′ overhangs (59). The conserved flanking TAs are essential for DNA cleavage, as confirmed by the inhibitory effect of a point mutation within one TA at either end of a given IES on IES excision (60, 61, 62, 63). Other positions within the consensus sequence may also be important (64, 65). Molecular analysis of a small set of IESs showed that programmed DSBs are detected over different periods of time depending on the IES (41, 63), suggesting that excision of particular IESs may be programmed to be completed early or late with regard to MAC development. Understanding whether DNA replication influences IES excision will require genome-wide monitoring of programmed genome rearrangements relative to DNA amplification.

Following DNA cleavage, IESs are released as linear molecules (66) and circularized in a second step to form precisely closed and supercoiled minicircles that do not replicate and are actively degraded at later stages during MAC development (46, 57, and M. Bétermier, unpublished data). At each chromosomal excision site, IES removal leaves a DSB that needs to be repaired efficiently and precisely to ensure that functional somatic chromosomes are assembled correctly. During both chromosome repair and IES circularization, the currently available model (Fig. 4A) proposes that broken DNA ends are aligned through the pairing of the TA dinucleotides that are present on each 5′ overhang, and undergo limited 5′ and 3′ processing before they are joined in a highly precise manner (59).

piggyBac to the Rescue

The establishment of an evolutionary connection between *P. tetraurelia* IESs and *Tc/mariner* transposons raised a puzzling issue with regard to the origin of the endonuclease responsible for the cleavage of IES boundaries. The original IBAF hypothesis proposed that exaptation of a *Tc/mariner* transposase gene allowed ancestral transposons to decay and give rise to extant IESs. However, as discussed previously (46), the transposition of *Tc/mariner* transposons leaves a characteristic footprint at the donor site from which the transposon is excised before integrating at another target site (50), making it unlikely that a *Tc/mariner*-related protein allows the precise excision of *Paramecium* IESs.

It was possible to unravel this puzzle following a survey of the somatic genome annotation, which allowed identification of the gene encoding PiggyMac (Pgm), a domesticated PiggyBac transposase that is expressed specifically during MAC development and localizes exclusively in the new developing MACs, at the time when IES excision starts (41). When compared with PiggyBac transposases, Pgm harbors at least three distinct domains (Fig. 4B): a putative catalytic domain containing a triad of aspartic acids (DDD) typical of PiggyBac transposases, a short cysteine-rich region similar to the PHD finger found at the C terminus of the PiggyBac transposases (67), and a long C-terminal coiled-coil extension, which seems to be an innovation of ciliates (68). Strikingly, the DSBs characterized at IES ends have the same geometry as those catalyzed *in vitro* by PiggyBac transposases, with 4-nt 5′-protruding ends carrying a central TA (41, 67). Conservation of the DDD catalytic triad makes Pgm a good candidate for DNA cleavage during IES excision, although its catalytic activity still has to be demonstrated *in vitro*. Consistent with this hypothesis, no DSBs are detected at IES ends in the new developing MACs of cells silenced for expression of the PGM gene, while DNA endoduplication occurs normally (41). Supporting the hypothesis that Pgm is the catalytic subunit of the IES excision complex, the conserved DDD triad found in Tpb2p, the homologous protein from the ciliate *Tetrahymena thermophila*, was shown to be essential for *in vitro* DNA cleavage activity (68, 69).

PGM RNAi results in IES retention in the genome of the new developing MACs, causing subsequent massive cell death in the progeny. As described above, next-generation sequencing of the DNA of Pgm-depleted autogamous cells led to the identification of a reference set of 45,000 IESs (6). This set probably includes all IESs present in the MAC-collinear fraction of the germline genome, as indicated by the sequencing of Preer's phage library, but whether Pgm is responsible for excision of all IESs needs to be confirmed by the sequencing of purified MIC DNA.

The discovery of Pgm revealed that the IES excision system in *Paramecium* has brought together two distinct transposon families: *Tc/mariner* (DNA substrates) and *piggyBac* (endonuclease). Pgm was also shown to be involved in the heterogeneous elimination of

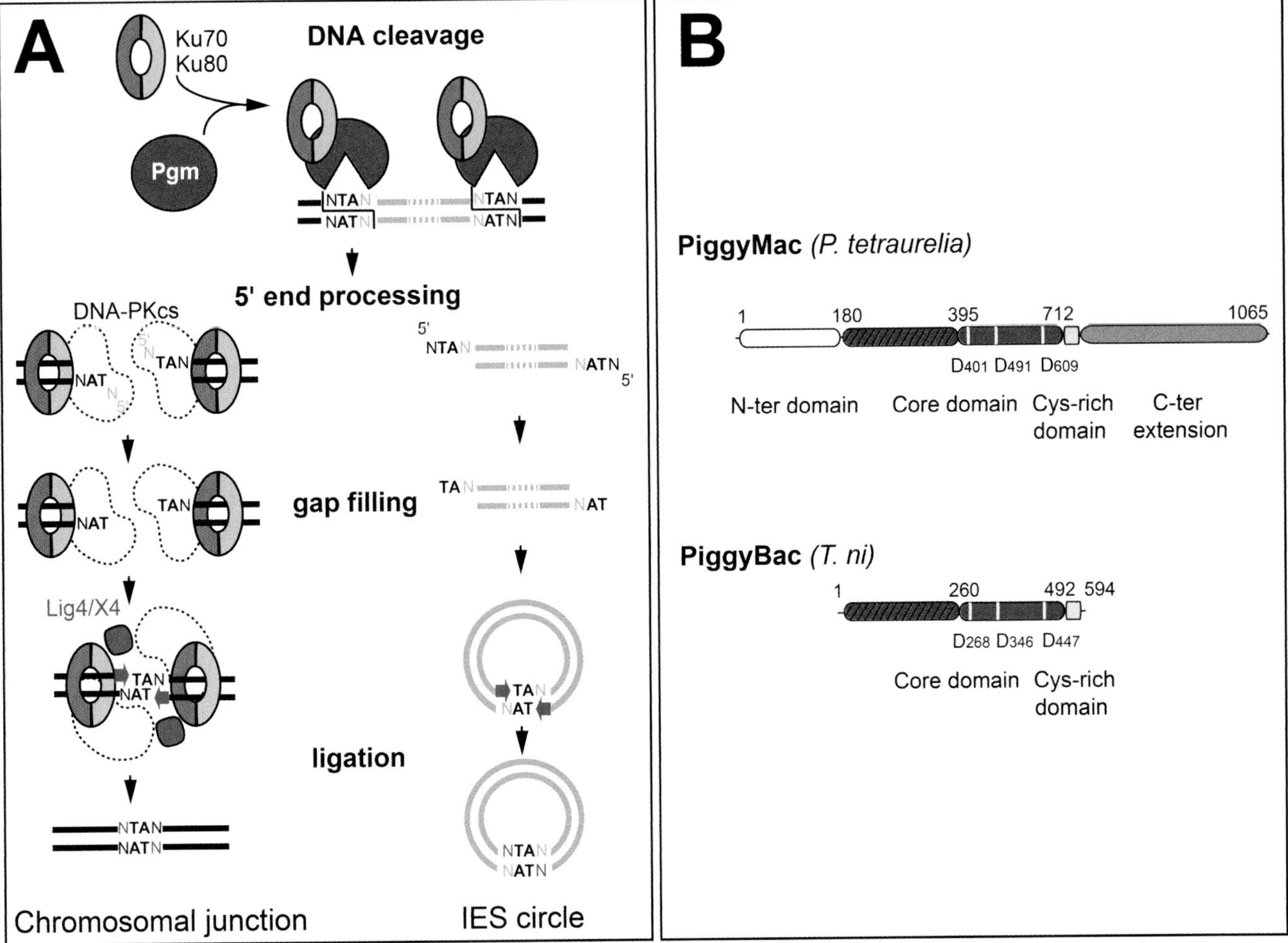

Figure 4 IES excision in *P. tetraurelia.* (A) Molecular mechanism and protein actors involved in IES excision. Both DNA strands are represented, with the IES in orange and flanking MAC-destined DNA in black. In the absence of any information about stoichiometry, the Pgm complex is represented by a red ball. Activation of DNA cleavage is thought to involve a physical interaction between Pgm and a development-specific Ku heterodimer (in grey). At each IES excision site (left), Ku is immediately positioned on the resulting DSBs and recruits all other actors of the classical nonhomologous end-joining pathway. In a first step, 5′ processing of the 4-nt overhangs, mediated by a nuclease that remains to be identified, results in the removal of the 5′-terminal nucleotide from each overhang. A gap-filling step involves addition of one nucleotide to each 3′-recessed end before the ligase IV–Xrcc4 (Lig4/X4) complex closes the junction. The linear excised IES (right) is thought to be circularized through a similar pathway. (B) Domain organization of the PiggyMac domesticated transposase (Pgm) and comparison with the PiggyBac transposase isolated from *Trichoplusia ni*. The boundaries of each domain are indicated (numbers refer to amino acid positions). The core transposase domain is in red, with the conserved catalytic aspartic acid triad (DDD) drawn as vertical bars and an upstream domain conserved in PiggyBac transposases represented by a hatched red box. The cysteine-rich domain that is proposed to fold into a putative PHD finger is in pink, and the C-terminal coiled-coil extension in purple. An additional N-terminal domain (in white) is found in the *Paramecium* protein.
doi:10.1128/microbiolspec.MDNA3-0035-2014.f4

transposons and other repeated DNA that is associated with chromosome fragmentation (41), but whether this role is direct or indirect still has to be established. During IES excision, the formation of a transpososome-like complex is supported by genetic evidence indicating that the DNA cleavage step requires crosstalk between the two ends of each IES (63). Analysis of the genome-wide set of IESs revealed a remarkable periodicity in the size distribution of these sequences: 35% of IESs are found in a first peak (26 to 32 bp in length) and, up to 150 bp, regularly spaced peaks are conspicuous every 10 to 11 bp, which coincides with the phasing of the DNA double helix (6). Intriguingly, one peak (centered at 35 to 36 bp) is largely under-represented in the distribution, suggesting the existence of a "forbidden" size range. The biological significance of both the 10-bp periodicity and the forbidden peak will require further investigation. As discussed previously (6), the distribution of IES sizes may reflect strong constraints on the assembly of the excision complex, such as the bridging of IES ends through the binding of Pgm subunits that might interact directly for very short elements (26 to 28 bp), or the formation of a DNA loop—perhaps assisted by accessory DNA bending factors—for larger IESs (44 to 46 bp). Major issues will also have to be addressed regarding the molecular mechanisms involved in IES excision, in particular what directs the precision of DNA cleavage at IES boundaries and whether DNA hairpins similar to those introduced by piggyBac transposases are formed during DNA cleavage.

Association of the Classical Nonhomologous End-Joining Pathway (C-NHEJ) with the Pgm Endonuclease Couples DNA Cleavage and Precise DSB Repair

When compared with cut-and-paste transposition, IES excision should rather be viewed as a "cut-and-close" reaction, during which the excision site is repaired precisely and the excised DNA is destined for degradation. As discussed in (45), precise and efficient DSB repair at IES excision sites ensures that MAC chromosomes are assembled in the right order and open reading frames are reconstituted with high precision. Using RNAi-mediated functional analysis of candidate genes, the ligase IV–Xrcc4 complex, an essential actor of the C-NHEJ pathway (70), was shown to carry out DSB repair at chromosomal IES excision sites and during the circularization of excised IESs (66). Interestingly, right before ligation, gap filling of recessed 3′ ends was inhibited in cells depleted for ligase IV (Fig. 4), suggesting that recruitment or activation of C-NHEJ DNA polymerases is coupled to the final ligation step (66).

The demonstration that the ligase IV–Xrcc4 specialized ligation complex carries out DSB repair at IES excision sites made *Paramecium* a relevant model to support the notion that the NHEJ pathway can be intrinsically precise, depending on the geometry of the broken ends that have to be joined (discussed in 71). Given the high number of IESs (one every 1 to 2 kbp along the germline genome), a major issue is how the NHEJ repair machinery is efficiently recruited to IES boundaries to carry out the joining of MAC-destined ends following Pgm-dependent cleavage. During general C-NHEJ-mediated DSB repair, broken ends are first bound by the Ku heterodimer and protected against extensive resection (reviewed in 72). Ku recruits the DNA-dependent protein kinase DNA-PKcs to form the DNA-PK complex, which stabilizes the bridging of broken ends at the DSB, regulates end resection, and activates downstream C-NHEJ actors (73). *Paramecium* encodes two closely related Ku70 and three Ku80 homologs (66), and one homolog of DNA-PKcs (S. Malinsky, personal communication). Using a combination of reverse genetics and molecular approaches, it was established that a development-specific Ku70/Ku80 heterodimer is required *in vivo* for DNA cleavage at IES ends and interacts physically with Pgm in heterologous cell extracts (74). Depletion of DNA-PKcs results in the persistence of free chromosomal ends at abnormally late stages during MAC development for a subset of IESs, consistent with the expectation that DNA-PKcs is involved in DSB repair at IES excision sites (S. Malinsky, personal communication). Intriguingly, nonexcised copies of DNA-PKcs-dependent IESs persist in the replicating chromosomes upon DNA-PKcs depletion, suggesting a possible defect in DNA cleavage. Thus, although the details of the Ku–Pgm interaction and the mechanisms involved in Ku-mediated activation of Pgm-dependent DNA cleavage need to be analyzed, integration of DSB repair proteins into the catalytically active IES excision machinery provides an ideal way to channel cleaved IES boundaries towards precise C-NHEJ repair right after DNA cleavage.

EPIGENETIC CONTROL OF PROGRAMMED DNA ELIMINATION IN *PARAMECIUM*

The molecular mechanisms underlying the specific recognition of germline eliminated sequences have been a long-standing issue. Clearly the excision complex does not simply target a nucleotide sequence motif. Even though a loosely conserved consensus sequence is found at IES ends, it does not contain sufficient information to explain the pattern of excision across the genome. Indeed, many IESs match the consensus poorly, while other sequences with perfect matches to the consensus are not excised. The identification of the Pgm endonuclease did not provide an answer to this question, because IESs are not related to *piggyBac* transposons and do not carry terminal inverted repeats that could be sequence-specific binding sites for Pgm. Likewise, the transposon remnants that have been annotated in the MIC genome are not related to *piggyBac* transposons. RNA-mediated mechanisms were shown to participate in the epigenetic regulation of genome rearrangements and may therefore contribute to the recognition of eliminated sequences (reviewed in 75 and 76).

Homology-Dependent Maternal Control of Genome Rearrangements

An early clue that genome rearrangements are regulated in a homology-dependent manner came from studies in *Paramecium* showing that modifying the DNA content of the maternal macronucleus can profoundly alter the rearrangement profile in the new macronucleus of progeny after sexual events.

The first described example of maternal inheritance of a rearrangement profile came from a study of the d48 epimutant, in which a surface antigen gene is reproducibly deleted from the MAC genome although its germline MIC genome is wild type (25). The gene deletion in the MAC genome is inherited through sexual generations in a non-Mendelian manner as shown by genetic analyses (77). This alternative rearrangement profile is transmitted from the maternal MAC genome to the new MAC genome of the sexual progeny through the cytoplasmic lineage. Introduction of the missing gene in the MAC of d48 epimutant was shown to restore the wild-type rearrangement profile in the new MAC genome of sexual progeny (78, 79, 80). This provided the demonstration that presence of the gene in the maternal MAC genome is needed to maintain the gene in the new MAC genome of sexual progeny. Other examples of MAC deletions of nonessential genes were also shown to be maternally inherited (81, 82). As for d48, reintroduction of the missing gene by transformation of the maternal MAC rescued the wild-type rearrangement profile in the new MAC genome of sexual progeny (83), supporting the idea that the new MAC only maintains sequences that are present in the maternal MAC.

A similar regulation by the maternal MAC was observed for IES excision. Introduction of an IES sequence in the maternal MAC led to the retention of the homologous zygotic IES sequence in the new MAC of sexual progeny, but the excision of all other IESs occurred normally (84). Once induced by transformation, the retention of this IES in MAC chromosomes can further inhibit excision of the homologous germline IES in subsequent sexual generations. This is a perfect example of epigenetic inheritance because one particular IES is no longer excised, although the germline genome remains entirely wild type and the initial trigger for retention is absent. This sequence-specific maternal inhibition of IES excision was observed for a third of the IESs that were tested (5/13), which were called maternally controlled IESs (mcIESs) (48). Among this set of mcIESs, only one was submitted to epigenetic inheritance. A quantitative analysis of IES excision inhibition in a population of post-autogamous cells provided evidence that the higher the IES copy number in the maternal MAC, the higher the IES retention rate in the new MAC. Yet, excision inhibition is a stochastic process in each individual developing MAC, because a wide variability is observed among isolated clones derived from a single transformed clone. At the population level, the efficiency of inhibition varies from 0 to 100% among IESs, for a given IES copy number in the maternal MAC, suggesting that IESs differ in some intrinsic properties (48). Yet no obvious characteristics that could distinguish mcIESs from non-mcIESs were identified, even though a variant TATT consensus was found at the ends of some mcIESs. The classification of IESs into mcIESs or non-mcIESs categories, according to whether inhibition was observed, may be misleading as it does not reflect the gradient of quantitative differences observed between IESs. A putative RNA-binding protein, Nowa1, was shown to be essential for transposon elimination and IES excision (85). Because all known mcIESs require Nowa1 for excision, it was then proposed that Nowa-dependent IESs are under maternal control, although this has to be tested at a genome-wide level.

Because identical germline genomes can produce alternative rearrangement profiles in the zygotic genome—deletion/retention of a gene (or of an IES)—the fate of germline sequences appears to be not solely determined by *cis*-acting determinants but controlled, at least to some extent, by the DNA content of the maternal somatic MAC, which is still present in the cytoplasm during the development of the new one. These observations raised the idea that reproduction of pre-existing rearrangement profiles was achieved by a global comparison of germline and somatic genomes, which would eventually lead to the removal of all germline sequences that were absent from the maternal somatic MAC genome (reviewed in 86).

An RNA-Dependent Mechanism is Involved in the Maternal Control of Genome Rearrangements

An important breakthrough was the identification of two classes of noncoding RNA (ncRNA) molecules with antagonistic functions that were shown to participate in the epigenetic regulation of genome rearrangements (Fig. 5).

A class of noncoding transcripts, produced by the maternal somatic genome, appears to protect homologous sequences against elimination. In *Paramecium*, reverse transcription-PCR experiments have provided evidence for the existence of ncRNAs produced at low levels from both strands by the somatic MAC genome. RNAi-targeted degradation of IES-containing transcripts prevented maternal inhibition of IES excision in

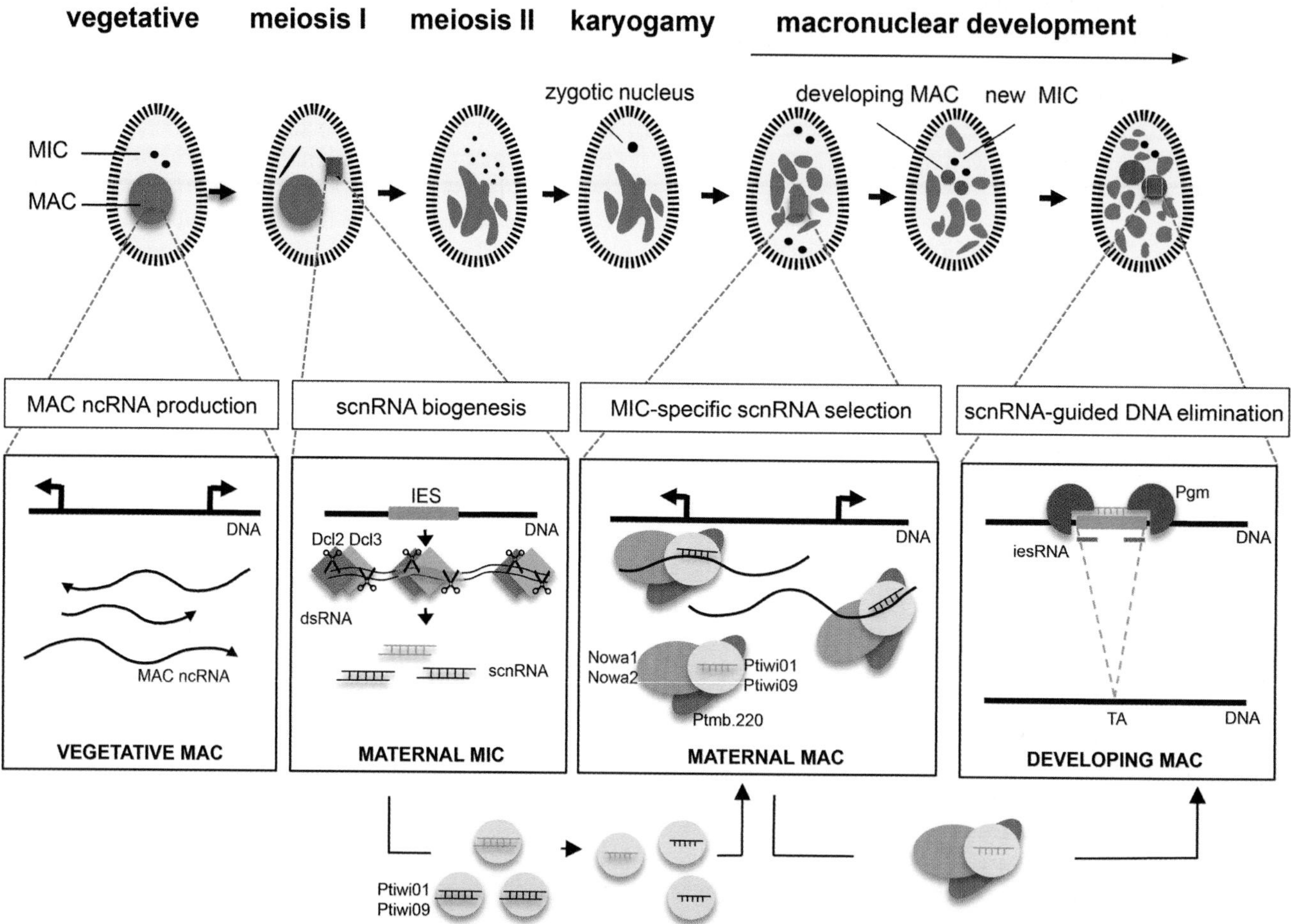

Figure 5 The genome-scanning model. The progression of nuclear events during autogamy is represented on top. The corresponding steps of the scanning model are schematized at the bottom. (i) MAC noncoding transcription. Noncoding transcripts are constitutively produced from the entire MAC genome. (ii) Biogenesis of scnRNAs. Early in meiosis I, 25-nt scnRNA duplexes are produced by the Dicer-like proteins Dcl2 and Dcl3 from the whole MIC genome. These duplexes are then transported to the cytoplasm where they are loaded onto Ptiwi01–Ptiwi09 complexes, leading to removal of the "passenger" strand. (iii) MIC-specific scnRNA selection. Ptiwi–scnRNA complexes are transported to the maternal MAC. Homologous pairing between scnRNAs and MAC noncoding transcripts, enhanced by the putative helicase Ptmb.220 and assisted by the putative RNA-binding proteins Nowa1–Nowa2, leads to selection of MIC-specific scnRNAs that are then transported to the developing MAC. (iv) DNA elimination. Recognition of scnRNA-homologous sequences in the developing MAC, enhanced by IES-specific small RNAs, guides recruitment of the endonuclease PiggyMac, allowing elimination of MIC-specific sequences.
doi:10.1128/microbiolspec.MDNA3-0035-2014.f5

the new macronucleus after sexual events (87). Similarly, RNAi-mediated silencing of a nonessential gene during vegetative growth triggered deletion of the homologous gene in the new MAC genome after sexual reproduction (83). Targeting a small region of the gene is sufficient to delete the entire gene, suggesting that RNAi-mediated silencing leads to the degradation of the transcripts covering the entire gene, although these are not necessarily mRNAs. This is consistent with earlier studies showing that the ability of a maternal transgene to rescue MAC deletions of the homologous gene in sexual progeny does not require the expression of a full-length, functional mRNA (88, 89). Thus, in order to be maintained in the developing MAC, a sequence needs not only to be present but also to be transcribed in the maternal MAC. Based on these observations, it was proposed that somatic ncRNAs protect homologous zygotic sequences against DNA elimination during MAC development. This would rely on generalized transcription of the maternal somatic genome, which has not yet been investigated genome-wide in *Paramecium*. In support of this hypothesis, increasing evidence of pervasive transcription has been provided in various organisms (90).

A second class of ncRNAs is composed of small RNAs, called scnRNAs, that resemble the metazoan piRNAs. scnRNAs are produced from the maternal germline genome by a meiosis-specific RNAi pathway and promote elimination of homologous sequences in the developing new MAC. Production of scnRNAs, approximately 25 nt in length, requires two Dicer-like proteins, Dcl2 and Dcl3 (56, 91). Sequencing of scnRNAs reveals a 5′-UNG signature and indicates that scnRNAs are cleaved from double-stranded precursors to yield RNA duplexes with 2-nt 3′ overhangs at both ends (91). The scnRNA population is highly complex and does not correspond exclusively to MIC-specific sequences. Indeed, scnRNAs are produced from a large fraction of the germline genome (18, 56, 91), suggesting that the entire micronuclear genome is transcribed and produces scnRNAs. Knockdown of both *Paramecium DCL2* and *DCL3* results in the disappearance of scnRNAs, DNA elimination defects, and death in the sexual progeny (56, 91). Thus, Dicer-dependent scnRNAs are required for the elimination of germline-specific sequences. Further support of the role of scnRNAs in the DNA elimination process came from the functional analysis of two Piwi proteins, Ptiwi01 and Ptiwi09, necessary for scnRNA accumulation and DNA elimination (92). The most direct evidence that scnRNAs can trigger DNA elimination in the developing MAC was obtained by injection of 25-nt synthetic RNA duplexes that mimic the structure of scnRNAs (87).

If, as suggested by the evidence above, scnRNAs are produced from the entire germline genome and are capable of promoting elimination of homologous DNA in the zygotic MAC, one needs to explain how the specificity of scnRNA induced deletions is achieved. The "genome-scanning" model posits that the highly complex population of scnRNAs is filtered out by base-pairing interactions with noncoding transcripts produced by the maternal somatic genome, resulting in the selective inactivation of those able to find a perfect match, and thus in the selection of MIC-specific scnRNAs (Fig. 5). How MIC-specific scnRNA selection is achieved is not yet clear. High-throughput sequencing supports the idea that scnRNAs are progressively enriched in MIC-restricted sequences during MAC development (18, 56), and favors the hypothesis that the scanning procedure leads to the selective degradation of MAC scnRNAs. Once selected by this "scanning" procedure, which likely occurs in the maternal MAC, MIC-specific scnRNAs would be exported to the developing zygotic MAC to target homologous sequences, thereby recruiting the excision machinery. The idea that scnRNAs travel from the maternal MAC to the developing MAC is supported by the observation that GFP fusions of Nowa1 and Ptiwi09 accumulate in the maternal MAC before the end of meiosis of the MIC and are later found in the developing MAC (Fig. 3 and Table 1).

The "genome-scanning" model was initially designed to account for the epigenetic inheritance of genome rearrangements, and it indeed provides a satisfactory explanation for all known cases of maternal inheritance: imprecise DNA elimination events and induced deletions, as well as regulation of mcIES excision. This process also accounts for the maternal inheritance of mating types in *P. tetraurelia*. The mating types, O and E, are not genetically determined in the MIC but are maternally inherited and controlled by the maternal MAC (19, 93). A recent study demonstrated that mating-type determination is achieved by an alternative rearrangement of the *mtA* gene during MAC development: in mating-type O clones, the *mtA* promoter is excised during MAC development by the scnRNA pathway, preventing expression of the *mtA* gene, while in mating-type E clones, the nonexcision of the *mtA* promoter drives the expression of the transmembrane protein mtA (18). The scnRNA pathway, very much like the piRNA pathway in metazoans, appears to be a genome defense mechanism that allows the removal from the somatic genome of transposons and their relics but can also regulate cellular genes and mediate epigenetic inheritance of phenotypic polymorphisms.

However, a "dark side" to this mechanism would in theory be unveiled during conjugation. Let us consider a cross between cells harboring two allelic IESs that have diverged to the point that scnRNAs produced by one allele can no longer recognize the other allele. Because of maternal inheritance of genome rearrangements, it is assumed that there is no significant exchange of cytoplasm between conjugating cells. Therefore, scnRNAs are independently sorted in each conjugating cell. Thus, in each exconjugant, scnRNAs homologous to the resident allele will be unable to excise the IES from the allele it has received from its conjugating partner, resulting in IES retention in the new MAC. Retention of an IES can lead to disruption of an essential gene and death of the sexual progeny. Given that IESs appear to evolve rapidly and that 16% of them have been acquired since the last whole-genome duplication (6), such mating incompatibilities may not be so rare. Since the *P. aurelia* group of 15 morphologically identical but sexually incompatible species arose shortly after the last whole-genome duplication (7, 33), it would be interesting to test whether the scnRNA pathway has played a role in driving speciation.

Is the scnRNA Pathway the Tip of the Iceberg for Control of IES Excision?

Depletion of the Dicer-like proteins Dcl2 and Dcl3 impairs the excision of all known mcIESs. In agreement with the "genome-scanning" model, scnRNAs are thus required for the excision of mcIESs (43, 56, 91). However, high-throughput sequencing indicates that only a small fraction of IESs, less than 10%, are retained in the new MAC after co-depletion of Dcl2 and Dcl3, suggesting that the number of IESs subject to maternal control might be smaller than anticipated based on previous studies (48). In fact, *DCL2* and *DCL3* knockdowns have revealed that the vast majority of IESs are correctly excised in the complete absence of scnRNAs. Therefore, excision of non-mcIESs may not depend on scnRNAs (43, 56). Yet the Piwi proteins Ptiwi01 and Ptiwi09 are required for the excision of the small set of IESs that have been tested so far (92). Even though genome-wide studies of the effects of Ptiwi01 and Ptiwi09 depletions are needed to reach a definitive conclusion, all IESs appear to rely on Ptiwi01/09 proteins for their efficient excision. One possibility is that excision of non-mcIESs relies on other small RNAs, which are not produced by Dcl2 and Dcl3 but are still loaded onto Ptwi01 and Ptiwi09.

Recently, a novel class of 26 to 30 nt long, IES-specific small RNAs, called iesRNAs, was reported (56). They accumulate during late MAC development and require the Dicer-like protein Dcl5 for their biogenesis. Dcl5 depletion partially impairs the excision of about 10% of IESs (43), but no lethality is observed in the sexual progeny, likely because only a partial retention is observed for each affected IES. Dcl5-dependent IESs do not correspond to those retained after Dcl2 and Dcl3 co-depletion, and the precise role of iesRNAs in IES excision remains to be elucidated. Other sRNAs may still remain to be discovered, because many Piwi proteins with unknown functions are found in the MAC genome, and some are expressed during sexual events and MAC development (92).

What Drives the DNA Cleavage Machinery to its Target Sites?

DNA elimination in the developing MAC relies on accurate recruitment of the excision machinery to MIC-specific sequences. How this is achieved remains to be discovered. According to the scanning model, once selected, MIC-specific scnRNAs would be exported to the developing zygotic MAC and used to target homologous sequences, thereby defining the eliminated sequences and recruiting the excision machinery. As observed in small RNA-guided heterochromatin formation in other organisms (94), scnRNAs would promote the deposition of specific chromatin modifications at homologous loci in the developing MAC (Figs 5 and 6A). In *T. thermophila*, the scnRNA pathway is conserved, and scnRNA-mediated trimethylation of histone H3 on lysine 9 and lysine 27 (H3K9me3 and H3K27me3) (95, 96, 97, 98) is thought to guide recruitment of the Tpb2p endonuclease (69). As in *Tetrahymena*, imprecise elimination of long, repetitive germline sequences in *P. tetraurelia* relies on a putative lysine histone methyltransferase, Ezl1, responsible for H3K27 and H3K9 trimethylation in the developing MAC (43).

As discussed in (75), in the case of IESs, the situation is radically different because excision of these 45,000 DNA segments must require a marking mechanism of considerable precision, allowing the demarcation of these very short, numerous, interspersed germline sequences from adjacent retained somatic sequences. Indeed, the vast majority of IESs are shorter than 150 bp and one-third of them are between 26 and 30 bp in length. IESs are thus not even as long as the DNA wrapped around a single nucleosome. Yet high-throughput sequencing showed that Ezl1-dependent H3K27/K9 methylation is required for the precise excision of about 70% of IESs (43). Ezl1 might not trigger heterochromatin formation

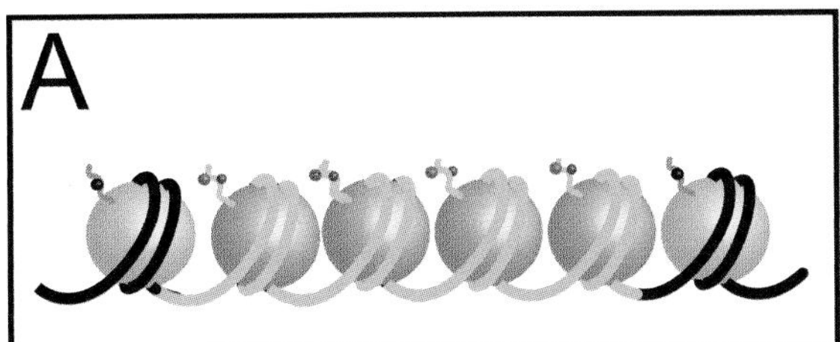

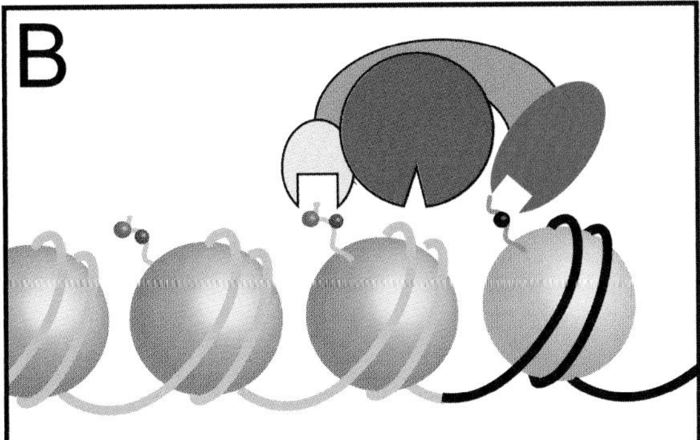

Figure 6 Models for IES recognition by the Pgm complex. (A) Eliminated sequences are defined by a specific chromatin organization (in orange). According to the scanning model, scnRNAs would promote the deposition of specific chromatin modifications at homologous loci in the developing MAC that distinguish them from MAC retained sequences (in black) (B) The excision machinery is recruited at IES ends via chromatin modifications. The Pgm complex (in red) might be specifically targeted to IES ends due to its association with chromatin transition zones that mark the boundary between the internal side of IESs and their flanking regions. The Pgm endonuclease might have the capacity to recognize histone marks loaded on eliminated sequences through its putative PHD finger and might interact with other chromatin-interacting proteins (in blue) via its putative coiled-coil domain. (C) Precise Pgm-dependent DNA cleavages occur at IES boundaries. Precision of DNA cleavages might be determined by a combination of factors: the DNA sequence found at IES ends (in particular the TA dinucleotide), DNA accessibility, and small RNAs that might guide DNA modifications or/and the Pgm endonuclease to its TA cleavage sites. doi:10.1128/microbiolspec.MDNA3-0035-2014.f6

on IESs but instead might act locally and methylate histone H3 on one or a few nucleosomes that encompass the IES sequence. Precise mapping of histone marks along the genome will be needed to determine whether chromatin modifications are specifically associated with IESs.

How chromatin modifications might recruit the excision machinery is not yet clear. The Pgm endonuclease might have the capacity to recognize the marks loaded on eliminated sequences. Indeed, Pgm, like its *Tetrahymena* counterpart, contains a putative PHD finger, potentially able to directly bind histone marks (41, 69). However, other chromatin-interacting proteins might bridge histone modifications to the Pgm endonuclease via its putative coiled-coil domain (Fig. 6B). In *Tetrahymena*, conjugation-specific chromodomain proteins Pdd1p and Pdd3p can bind H3K27me3 and form higher-order heterochromatic structures at the nuclear periphery (98, 99). Clearly, the interplay with chromatin features requires further investigation in *Paramecium*, and in particular the involvement of other histone modifications and higher-order chromatin conformation needs to be explored.

One important question is to understand the mechanisms by which chromatin structure may position the excision machinery for precise DNA cleavages. The Pgm complex may be specifically targeted to IES ends due to its association with chromatin transition zones that mark the boundary between the internal side of IESs and their flanking regions (Fig. 6B). Strikingly, iesRNAs preferentially map to IES ends and may participate in the definition of IES boundaries (56). Where the DNA cleavages occur precisely may be determined, at least in part, by the DNA sequence found at IES ends, in particular the absolute requirement for the TA dinucleotide. As previously discussed (75), access to DNA cleavage sites internal to a nucleosome might be possible through the action of chromatin-remodeling complexes that could promote DNA "breathing" on the nucleosome, nucleosome sliding, or nucleosome eviction (Fig. 6C). In the case of short IESs, targeting might involve a particular nucleosome positioning, whereby nucleosome-free IESs would leave DNA segments accessible to the excision machinery. One possibility would be that sRNAs (scnRNAs, iesRNAs, or others) trigger DNA modifications that position the excision complex right at its cleavage site (discussed in 75). Another, nonexclusive hypothesis is that small RNAs directly guide the Pgm complex to its TA cleavage sites (Fig. 6C), as described for the RNA-dependent endonuclease Cas9 of the bacterial CRISPR system (100).

CONCLUSION

Recent advances in ciliate research have provided a remarkable example of the impact of domesticated transposases on genome plasticity. For other organisms, increasing evidence already supported the notion that genome rearrangements take place in somatic tissues during development, cell differentiation, or aging. Some rearrangements were proposed to involve the activation of transposon-related elements, as reported in mouse and *Drosophila* brain (101, 102), and/or the programmed elimination of variable amounts of germline DNA (103, 104, 105, 106, 107). Although the significance of somatic mosaicism remains poorly understood, genome rearrangements may contribute to plasticity and regulate gene expression by removing germline-restricted genes or allowing the assembly of functional genes. The molecular mechanisms triggering genome rearrangements are also largely unknown, except for a very few cases. The best-documented example is the assembly of immunoglobulin genes in vertebrates, which takes place during lymphocyte differentiation and generates the diversity of the acquired immune response (108). The demonstration that the RAG1 component of the RAG1/RAG2 endonuclease responsible for V(D)J recombination is related to a *Transib* transposase provided the first evidence of transposase domestication in a cellular recombination system (109). Genome sequencing programs have revealed the existence of numerous other domesticated transposases in various organisms, but the function of these proteins has largely remained elusive (110, 111, 112, 113). The study of programmed genome rearrangements in *Paramecium* represents a unique opportunity to unravel the role of domesticated *piggyBac* transposases in genome plasticity, in a system in which DNA sites are not recognized through conspicuous sequence-specific protein–DNA interactions. Understanding how ncRNAs guide PiggyMac to its cleavage sites within the chromatin context of the developing new MAC will certainly provide new insights into the mechanisms that drive maternal inheritance of variant rearrangement patterns across sexual generations.

Acknowledgments. *We thank Jean Cohen and Janine Beisson for transmitting their invaluable knowledge of the biology of* P. tetraurelia, *and all members of our laboratories and those of Jean Cohen, Laurent Duret, Eric Meyer, and Linda Sperling for long-standing collaborations and stimulating discussions. We are grateful to Janine Beisson, Jean Cohen, and Linda Sperling for critical reading of our manuscript, to Simone Grandchamp, Janine Beisson and Jacek Nowak for allowing us to cite their unpublished results, and to Sophie Malinsky and Olivier Arnaiz for sharing their data prior to publication. Work in our two laboratories was supported by intramural funding from the CNRS, and by grants from the Agence Nationale de la Recherche (ANR 2010-BLAN-1603 and ANR-12-BSV6-0017). M.B. was supported by a grant from the Fondation ARC (#SFI20121205487) and S.D. by an ATIP Plus grant from the CNRS. We acknowledge the support of the CNRS GDRE "*Paramecium *Genome Dynamics and Evolution" and the European COST network "Ciliates as model systems to study genome evolution, mechanisms of non-Mendelian inheritance, and their roles in environmental adaptation network" (COST Action BM1102).*

Citation. Betermier M., Duharcourt S. 2014. Programmed rearrangement in ciliates: Paramecium. Microbiol Spectrum 2(6):MDNA3-0035-2014.

References

1. **Derelle R, Lang BF.** 2012. Rooting the eukaryotic tree with mitochondrial and bacterial proteins. *Mol Biol Evol* **29:**1277–1289.
2. **Coyne RS, Hannick L, Shanmugam D, Hostetler JB, Brami D, Joardar VS, Johnson J, Radune D, Singh I, Badger JH, Kumar U, Saier M, Wang Y, Cai H, Gu J, Mather MW, Vaidya AB, Wilkes DE, Rajagopalan V, Asai DJ, Pearson CG, Findly RC, Dickerson HW, Wu M, Martens C, Van de Peer Y, Roos DS, Cassidy-Hanley DM, Clark TG.** 2011. Comparative genomics of the pathogenic ciliate *Ichthyophthirius multifiliis*, its free-living relatives and a host species provide insights into adoption of a parasitic lifestyle and prospects for disease control. *Genome Biol* **12:**R100.
3. **Moon-van der Staay SY, van der Staay GW, Michalowski T, Jouany JP, Pristas P, Javorsky P, Kisidayova S, Varadyova Z, McEwan NR, Newbold CJ, van Alen T, de Graaf R, Schmid M, Huynen MA, Hackstein JH.** 2014. The symbiotic intestinal ciliates and the evolution of their hosts. *Eur J Protistol* **50:** 166–173.
4. **Simon M, Plattner H.** 2014. Unicellular eukaryotes as models in cell and molecular biology: critical appraisal of their past and future value. *Int Rev Cell Mol Biol* **309:**141–198.
5. **Vogt A, Goldman AD, Mochizuki K, Landweber LF.** 2013. Transposon domestication versus mutualism in ciliate genome rearrangements. *PLoS Genet* **9:**e1003659.
6. **Arnaiz O, Mathy N, Baudry C, Malinsky S, Aury JM, Denby-Wilkes C, Garnier O, Labadie K, Lauderdale BE, Le Mouel A, Marmignon A, Nowacki M, Poulain J, Prajer M, Wincker P, Meyer E, Duharcourt S, Duret L, Bétermier M, Sperling L.** 2012. The *Paramecium* germline genome provides a niche for intragenic parasitic DNA: Evolutionary dynamics of internal eliminated sequences. *PloS Genetics* **8:**e1002984.
7. **Aury JM, Jaillon O, Duret L, Noel B, Jubin C, Porcel BM, Segurens B, Daubin V, Anthouard V, Aiach N, Arnaiz O, Billaut A, Beisson J, Blanc I, Bouhouche K, Camara F, Duharcourt S, Guigo R, Gogendeau D, Katinka M, Keller AM, Kissmehl R, Klotz C, Koll F, Le Mouel A, Lepère G, Malinsky S, Nowacki M, Nowak JK, Plattner H, Poulain J, Ruiz F, Serrano V, Zagulski M, Dessen P, Bétermier M, Weissenbach J,**

Scarpelli C, Schachter V, Sperling L, Meyer E, Cohen J, Wincker P. 2006. Global trends of whole-genome duplications revealed by the ciliate *Paramecium tetraurelia*. *Nature* **444:**171–178.

8. **McGrath CL, Gout JF, Doak TG, Yanagi A, Lynch M.** 2014. Insights into three whole-genome duplications gleaned from the *Paramecium caudatum* genome sequence. *Genetics* **197:**1417–1428.
9. **Catania F, Wurmser F, Potekhin AA, Przybos E, Lynch M.** 2009. Genetic diversity in the *Paramecium aurelia* species complex. *Mol Biol Evol* **26:**421–431.
10. **Beale GH, Preer JR Jr.** 2008. *Paramecium: Genetics and Epigenetics*. CRC Press, Boca Raton, FL.
11. **Beisson J, Bétermier M, Bré MH, Cohen J, Duharcourt S, Duret L, Kung C, Malinsky S, Meyer E, Preer JR Jr., Sperling L.** 2010. *Paramecium tetraurelia*: the renaissance of an early unicellular model. *CSH Protocols* **2010:**pdb emo140.
12. **Brygoo Y, Sonneborn TM, Keller AM, Dippell RV, Schneller.** 1980. Genetic analysis of mating type differentiation in *Paramecium tetraurelia*. II. Role of the micronuclei in mating-type determination. *Genetics* **94:** 951–959.
13. **Tam LW, Ng SF.** 1986. The role of the micronucleus in stomatogenesis in sexual reproduction of *Paramecium tetraurelia*: laser microbeam irradiation of the micronucleus. *J Cell Sci* **86:**287–303.
14. **Tucker JB, Beisson J, Roche DL, Cohen J.** 1980. Microtubules and control of macronuclear 'amitosis' in *Paramecium*. *J Cell Sci* **44:**135–151.
15. **Sonneborn MT.** 1975. *Paramecium aurelia*, p 469–594. *In* King CR (ed), *Handbook of Genetics: Plants, Plant Viruses and Protists*, **vol. 2.** Plenum Press, New York.
16. **Grandchamp S, Beisson J.** 1981. Positional control of nuclear differentiation in *Paramecium*. *Dev Biol* **81:** 336–341.
17. **Berger DJ.** 1973. Nuclear differentiation and nucleic acid synthesis in well-fed exconjugants of *Paramecium aurelia*. *Chromosoma* **42:**247–268.
18. **Singh DP, Saudemont B, Guglielmi G, Arnaiz O, Gout JF, Prajer M, Potekhin A, Przybos E, Aubusson-Fleury A, Bhullar S, Bouhouche K, Lhuillier-Akakpo M, Tanty V, Blugeon C, Alberti A, Labadie K, Aury JM, Sperling L, Duharcourt S, Meyer E.** 2014. Genome-defence small RNAs exapted for epigenetic mating-type inheritance. *Nature* **509:**447–452.
19. **Sonneborn MT.** 1977. Genetics of cellular differentiation: stable nuclear differentiation in eucaryotic unicells. *Annu Rev Genet* **11:**349–367.
20. **Sung W, Tucker AE, Doak TG, Choi E, Thomas WK, Lynch M.** 2012. Extraordinary genome stability in the ciliate *Paramecium tetraurelia*. *Proc Natl Acad Sci U S A* **109:**19339–19344.
21. **Prescott MD.** 1994. The DNA of ciliated protozoa. *Microbiol Rev* **58:**233–267.
22. **Caron F, Meyer E.** 1989. Molecular basis of surface antigen variation in paramecia. *Annu Rev Microbiol* **43:** 23–42.
23. **McCormick-Graham M, Romero DP.** 1996. A single telomerase RNA is sufficient for the synthesis of variable telomeric DNA repeats in ciliates of the genus *Paramecium*. *Mol Cell Biol* **16:**1871–1879.
24. **McCormick-Graham M, Haynes WJ, Romero DP.** 1997. Variable telomeric repeat synthesis in *Paramecium tetraurelia* is consistent with misincorporation by telomerase. *EMBO J* **16:**3233–3242.
25. **Forney JD, Blackburn EH.** 1988. Developmentally controlled telomere addition in wild-type and mutant paramecia. *Mol Cell Biol* **8:**251–258.
26. **Amar L, Dubrana K.** 2004. Epigenetic control of chromosome breakage at the 5′ end of *Paramecium tetraurelia* gene A. *Eukaryot Cell* **3:**1136–1146.
27. **Le Mouël A, Butler A, Caron F, Meyer E.** 2003. Developmentally regulated chromosome fragmentation linked to imprecise elimination of repeated sequences in *Paramecium*. *Eukaryot Cell* **2:**1076–1090.
28. **Dessen P, Zagulski M, Gromadka R, Plattner H, Kissmehl R, Meyer E, Bétermier M, Schultz JE, Linder JU, Pearlman RE, Kung C, Forney J, Satir BH, Van Houten JL, Keller AM, Froissard M, Sperling L, Cohen J.** 2001. *Paramecium* genome survey: a pilot project. *Trends Genet* **17:**306–308.
29. **Sperling L, Dessen P, Zagulski M, Pearlman RE, Migdalski A, Gromadka R, Froissard M, Keller AM, Cohen J.** 2002. Random sequencing of *Paramecium* somatic DNA. *Eukaryot Cell* **1:**341–352.
30. **Zagulski M, Nowak JK, Le Mouel A, Nowacki M, Migdalski A, Gromadka R, Noel B, Blanc I, Dessen P, Wincker P, Keller AM, Cohen J, Meyer E, Sperling L.** 2004. High coding density on the largest *Paramecium tetraurelia* somatic chromosome. *Curr Biol* **14:**1397–1404.
31. **Caron F, Meyer E.** 1985. Does *Paramecium primaurelia* use a different genetic code in its macronucleus? *Nature* **314:**185–188.
32. **Preer JR Jr, Preer LB, Rudman BM, Barnett AJ.** 1985. Deviation from the universal code shown by the gene for surface protein 51A in *Paramecium*. *Nature* **314:** 188–190.
33. **McGrath CL, Gout JF, Johri P, Doak TG, Lynch M.** 2014. Differential retention and divergent resolution of duplicate genes following whole-genome duplication *Genome Res* **24:**1665–1675.
34. **Jaillon O, Bouhouche K, Gout JF, Aury JM, Noel B, Saudemont B, Nowacki M, Serrano V, Porcel BM, Segurens B, Le Mouel A, Lepère G, Schachter V, Bétermier M, Cohen J, Wincker P, Sperling L, Duret L, Meyer E.** 2008. Translational control of intron splicing in eukaryotes. *Nature* **451:**359–362.
35. **Jones WK.** 1956. *Nuclear differentiation in* Paramecium. *PhD thesis*, University of Wales, Aberystwyth, UK.
36. **Preer LB, Hamilton G, Preer JR.** 1992. Micronuclear DNA from *Paramecium tetraurelia*: serotype 51 *A* gene has internally eliminated sequences. *J Protozool* **39:** 678–682.
37. **Steele CJ, Barkocy-Gallagher GA, Preer LB, Preer JR Jr.** 1994. Developmentally excised sequences in micro-

nuclear DNA of *Paramecium. Proc Natl Acad Sci U S A* **91**:2255–2259.

38. **Bétermier M.** 2004. Large-scale genome remodelling by the developmentally programmed elimination of germ line sequences in the ciliate *Paramecium. Res Microbiol* **155**:399–408.
39. **Duret L, Cohen J, Jubin C, Dessen P, Gout JF, Mousset S, Aury JM, Jaillon O, Noel B, Arnaiz O, Bétermier M, Wincker P, Meyer E, Sperling L.** 2008. Analysis of sequence variability in the macronuclear DNA of *Paramecium tetraurelia*: a somatic view of the germline. *Genome Res* **18**:585–596.
40. **Catania F, McGrath CL, Doak TG, Lynch M.** 2013. Spliced DNA sequences in the *Paramecium* germline: their properties and evolutionary potential. *Genome Biol Evol* **5**:1200–1211.
41. **Baudry C, Malinsky S, Restituito M, Kapusta A, Rosa S, Meyer E, Bétermier M.** 2009. PiggyMac, a domesticated *piggyBac* transposase involved in programmed genome rearrangements in the ciliate *Paramecium tetraurelia. Genes Dev* **23**:2478–2483.
42. **Amar L.** 1994. Chromosome end formation and internal sequence elimination as alternative genomic rearrangements in the ciliate *Paramecium. J Mol Biol* **236**: 421–426.
43. **Lhuillier-Akakpo M, Frapporti A, Denby Wilkes C, Matelot M, Vervoort M, Sperling L, Duharcourt S.** Local effect of Enhancer of zeste-like reveals cooperation of epigenetic and cis-acting determinants for zygotic genome rearrangements. *PLoS Genet* **10**:e1004665.
44. **Sperling L.** 2011. Remembrance of things past retrieved from the *Paramecium* genome. *Res Microbiol* **162**: 587–597.
45. **Dubois E, Bischerour J, Marmignon A, Mathy N, Régnier V, Bétermier M.** 2012. Transposon invasion of the *Paramecium* germline genome countered by a domesticated PiggyBac transposase and the NHEJ pathway. *Int J Evol Biol* **2012**:436196.
46. **Gratias A, Bétermier M.** 2001. Developmentally programmed excision of internal DNA sequences in *Paramecium aurelia. Biochimie* **83**:1009–1022.
47. **Klobutcher LA, Herrick G.** 1995. Consensus inverted terminal repeat sequence of *Paramecium* IESs: resemblance to termini of Tc1-related and *Euplotes* Tec transposons. *Nucleic Acids Res* **23**:2006–2013.
48. **Duharcourt S, Keller AM, Meyer E.** 1998. Homology-dependent maternal inhibition of developmental excision of internal eliminated sequences in *Paramecium tetraurelia. Mol Cell Biol* **18**:7075–7085.
49. **Klobutcher LA, Herrick G.** 1997. Developmental genome reorganization in ciliated protozoa: the transposon link. *Progr Nucleic Acid Res Mol Biol* **56**:1–62.
50. **Plasterk RH, Izsvak Z, Ivics Z.** 1999. Resident aliens: the Tc1/*mariner* superfamily of transposable elements. *Trends Genet* **15**:326–332.
51. **Dubrana K, Le Mouël A, Amar L.** 1997. Deletion endpoint allele-specificity in the developmentally regulated elimination of an internal sequence (IES) in *Paramecium. Nucleic Acids Res* **25**:2448–2454.
52. **Arnaiz O, Gout JF, Bétermier M, Bouhouche K, Cohen J, Duret L, Kapusta A, Meyer E, Sperling L.** 2010. Gene expression in a paleopolyploid: a transcriptome resource for the ciliate *Paramecium tetraurelia. BMC Genomics* **11**:547.
53. **Arnaiz O, Cain S, Cohen J, Sperling L.** 2007. ParameciumDB: a community resource that integrates the *Paramecium tetraurelia* genome sequence with genetic data. *Nucleic Acids Res* **35**:D439–D444.
54. **Arnaiz O, Sperling L.** 2011. ParameciumDB in 2011: new tools and new data for functional and comparative genomics of the model ciliate *Paramecium tetraurelia. Nucleic Acids Res* **39**:D632–D636.
55. **Galvani A, Sperling L.** 2002. RNA interference by feeding in *Paramecium. Trends Genet* **18**:11–12.
56. **Sandoval PY, Swart EC, Arambasic M, Nowacki M.** 2014. Functional diversification of Dicer-like proteins and small RNAs required for genome sculpting. *Dev Cell* **28**:174–188.
57. **Bétermier M, Duharcourt S, Seitz H, Meyer E.** 2000. Timing of developmentally programmed excision and circularization of *Paramecium* internal eliminated sequences. *Mol Cell Biol* **20**:1553–1561.
58. **Ku M, Mayer K, Forney JD.** 2000. Developmentally regulated excision of a 28-base-pair sequence from the *Paramecium* genome requires flanking DNA. *Mol Cell Biol* **20**:8390–8396.
59. **Gratias A, Bétermier M.** 2003. Processing of double-strand breaks is involved in the precise excision of *Paramecium* IESs. *Mol Cell Biol* **23**:7152–7162.
60. **Matsuda A, Mayer KM, Forney JD.** 2004. Identification of single nucleotide mutations that prevent developmentally programmed DNA elimination in *Paramecium tetraurelia. J Eukaryot Microbiol* **51**:664–669.
61. **Mayer KM, Forney JD.** 1999. A mutation in the flanking 5′-TA-3′ dinucleotide prevents excision of an internal eliminated sequence from the *Paramecium tetraurelia* genome. *Genetics* **151**:597–604.
62. **Ruiz F, Krzywicka A, Klotz C, Keller A, Cohen J, Koll F, Balavoine G, Beisson J.** 2000. The *SM19* gene, required for duplication of basal bodies in *Paramecium*, encodes a novel tubulin, eta-tubulin. *Curr Biol* **10**:1451–1454.
63. **Gratias A, Lepère G, Garnier O, Rosa S, Duharcourt S, Malinsky S, Meyer E, Bétermier M.** 2008. Developmentally programmed DNA splicing in *Paramecium* reveals short-distance crosstalk between DNA cleavage sites. *Nucleic Acids Res* **36**:3244–3251.
64. **Haynes WJ, Ling KY, Preston RR, Saimi Y, Kung C.** 2000. The cloning and molecular analysis of pawn-B in *Paramecium tetraurelia. Genetics* **155**:1105–1117.
65. **Mayer KM, Mikami K, Forney JD.** 1998. A mutation in *Paramecium tetraurelia* reveals function and structural features of developmentally excised DNA elements. *Genetics* **148**:139–149.
66. **Kapusta A, Matsuda A, Marmignon A, Ku M, Silve A, Meyer E, Forney J, Malinsky S, Bétermier M.** 2011. Highly precise and developmentally programmed genome assembly in *Paramecium* requires ligase IV-dependent end joining. *PloS Genetics* **7**:e1002049.

67. **Mitra R, Fain-Thornton J, Craig NL.** 2008. *piggyBac* can bypass DNA synthesis during cut and paste transposition. *EMBO J* **27:**1097–1109.

68. **Cheng CY, Vogt A, Mochizuki K, Yao MC.** 2010. A domesticated piggyBac transposase plays key roles in heterochromatin dynamics and DNA cleavage during programmed DNA deletion in *Tetrahymena thermophila*. *Mol Biol Cell* **21:**1753–1762.

69. **Vogt A, Mochizuki K.** 2013. A domesticated PiggyBac transposase interacts with heterochromatin and catalyzes reproducible DNA elimination in *Tetrahymena*. *PLoS Genet* **9:**e1004032.

70. **Lieber MR.** 2010. The mechanism of double-strand DNA break repair by the nonhomologous DNA end-joining pathway. *Annu Rev Biochem* **79:**181–211.

71. **Bétermier M, Bertrand P, Lopez BS.** 2014. Is non-homologous end-joining really an inherently error-prone process? *PLoS Genet* **10:**e1004086.

72. **Symington LS, Gautier J.** 2011. Double-strand break end resection and repair pathway choice. *Annu Rev Genet* **45:**247–271.

73. **Davis AJ, Chen BP, Chen DJ.** 2014. DNA-PK: A dynamic enzyme in a versatile DSB repair pathway. *DNA Repair* **17:**21–29.

74. **Marmignon A, Bischerour J, Silve A, Fojcik C, Dubois E, Arnaiz O, Kapusta A, Malinsky S, Bétermier M.** 2014. Ku-mediated coupling of DNA cleavage and repair during programmed genome rearrangements in the ciliate *Paramecium tetraurelia*. *PLoS Genetics* **10:** e1004552.

75. **Coyne RS, Lhuillier-Akakpo M, Duharcourt S.** 2012. RNA-guided DNA rearrangements in ciliates: is the best genome defence a good offence? *Biol Cell* **104:** 309–325.

76. **Duharcourt S, Lepère G, Meyer E.** 2009. Developmental genome rearrangements in ciliates: a natural genomic subtraction mediated by non-coding transcripts. *Trends Genet* **25:**344–350.

77. **Epstein LM, Forney JD.** 1984. Mendelian and non-Mendelian mutations affecting surface antigen expression in *Paramecium tetraurelia*. *Mol Cell Biol* **4:** 1583–1590.

78. **Jessop-Murray H, Martin LD, Gilley D, Preer JR Jr, Polisky B.** 1991. Permanent rescue of a non-Mendelian mutation of *Paramecium* by microinjection of specific DNA sequences. *Genetics* **129:**727–734.

79. **Koizumi S, Kobayashi S.** 1989. Microinjection of plasmid DNA encoding the A surface antigen of *Paramecium tetraurelia* restores the ability to regenerate a wild-type macronucleus. *Mol Cell Biol* **9:**4398–4401.

80. **You Y, Aufderheide K, Morand J, Rodkey K, Forney J.** 1991. Macronuclear transformation with specific DNA fragments controls the content of the new macronuclear genome in *Paramecium tetraurelia*. *Mol Cell Biol* **11:** 1133–1137.

81. **Meyer E.** 1992. Induction of specific macronuclear developmental mutations by microinjection of a cloned telomeric gene in *Paramecium primaurelia*. *Genes Dev* **6:**211–222.

82. **Meyer E, Butler A, Dubrana K, Duharcourt S, Caron F.** 1997. Sequence-specific epigenetic effects of the maternal somatic genome on developmental rearrangements of the zygotic genome in *Paramecium primaurelia*. *Mol Cell Biol* **17:**3589–3599.

83. **Garnier O, Serrano V, Duharcourt S, Meyer E.** 2004. RNA-mediated programming of developmental genome rearrangements in *Paramecium tetraurelia*. *Mol Cell Biol* **24:**7370–7379.

84. **Duharcourt S, Butler A, Meyer E.** 1995. Epigenetic self-regulation of developmental excision of an internal eliminated sequence in *Paramecium tetraurelia*. *Genes Dev* **9:**2065–2077.

85. **Nowacki M, Zagorski-Ostoja W, Meyer E.** 2005. Nowa1p and Nowa2p: novel putative RNA binding proteins involved in *trans*-nuclear crosstalk in *Paramecium tetraurelia*. *Curr Biol* **15:**1616–1628.

86. **Chalker DL, Meyer E, Mochizuki K.** 2013. Epigenetics of ciliates. *Cold Spring Harb Perspect Biol* **5:**a017764.

87. **Lepère G, Bétermier M, Meyer E, Duharcourt S.** 2008. Maternal noncoding transcripts antagonize the targeting of DNA elimination by scanRNAs in *Paramecium tetraurelia*. *Genes Dev* **22:**1501–1512.

88. **Kim CS, Preer JR Jr, Polisky B.** 1994. Identification of DNA segments capable of rescuing a non-Mendelian mutant in *Paramecium*. *Genetics* **136:**1325–1328.

89. **You Y, Scott J, Forney J.** 1994. The role of macronuclear DNA sequences in the permanent rescue of a non-Mendelian mutation in *Paramecium tetraurelia*. *Genetics* **136:**1319–1324.

90. **Morris KV, Mattick JS.** 2014. The rise of regulatory RNA. *Nat Rev Genet* **15:**423–437.

91. **Lepère G, Nowacki M, Serrano V, Gout JF, Guglielmi G, Duharcourt S, Meyer E.** 2009. Silencing-associated and meiosis-specific small RNA pathways in *Paramecium tetraurelia*. *Nucleic Acids Res* **37:**903–915.

92. **Bouhouche K, Gout JF, Kapusta A, Bétermier M, Meyer E.** 2011. Functional specialization of Piwi proteins in *Paramecium tetraurelia* from post-transcriptional gene silencing to genome remodelling. *Nucleic Acids Res* **39:**4249–4264.

93. **Nanney DL.** 1957. Mating-type inheritance at conjugation in variety 4 of *Paramecium aurelia*. *J Protozool* **4:**89–95.

94. **Castel SE, Martienssen RA.** 2013. RNA interference in the nucleus: roles for small RNAs in transcription, epigenetics and beyond. *Nat Rev Genet* **14:**100–112.

95. **Liu Y, Mochizuki K, Gorovsky MA.** 2004. Histone H3 lysine 9 methylation is required for DNA elimination in developing macronuclei in *Tetrahymena*. *Proc Natl Acad Sci U S A* **101:**1679–1684.

96. **Liu Y, Taverna SD, Muratore TL, Shabanowitz J, Hunt DF, Allis CD.** 2007. RNAi-dependent H3K27 methylation is required for heterochromatin formation and DNA elimination in *Tetrahymena*. *Genes Dev* **21:**1530–1545.

97. **Malone CD, Anderson AM, Motl JA, Rexer CH, Chalker DL.** 2005. Germ line transcripts are processed by a Dicer-like protein that is essential for developmentally programmed genome rearrangements of *Tetrahymena thermophila*. *Mol Cell Biol* **25:**9151–9164.

98. **Taverna SD, Coyne RS, Allis CD.** 2002. Methylation of histone H3 at lysine 9 targets programmed DNA elimination in *Tetrahymena*. *Cell* **110:**701–711.

99. **Smothers JF, Madireddi MT, Warner FD, Allis CD.** 1997. Programmed DNA degradation and nucleolar biogenesis occur in distinct organelles during macronuclear development in *Tetrahymena*. *J Eukaryot Microbiol* **44:** 79–88.

100. **Wiedenheft B, Sternberg SH, Doudna JA.** 2012. RNA-guided genetic silencing systems in bacteria and archaea. *Nature* **482:**331–338.

101. **Perrat PN, DasGupta S, Wang J, Theurkauf W, Weng Z, Rosbash M, Waddell S.** 2013. Transposition-driven genomic heterogeneity in the *Drosophila* brain. *Science* **340:**91–95.

102. **Reilly MT, Faulkner GJ, Dubnau J, Ponomarev I, Gage FH.** 2013. The role of transposable elements in health and diseases of the central nervous system. *J Neurosci* **33:**17577–17586.

103. **Smith JJ, Antonacci F, Eichler EE, Amemiya CT.** 2009. Programmed loss of millions of base pairs from a vertebrate genome. *Proc Natl Acad Sci U S A* **106:** 11212–11217.

104. **Smith JJ, Baker C, Eichler EE, Amemiya CT.** 2012. Genetic consequences of programmed genome rearrangement. *Curr Biol* **22:**1524–1529.

105. **Sun C, Wyngaard G, Walton DB, Wichman HA, Mueller RL.** 2014. Billions of basepairs of recently expanded, repetitive sequences are eliminated from the somatic genome during copepod development. *BMC Genomics* **15:**186.

106. **Wang J, Mitreva M, Berriman M, Thorne A, Magrini V, Koutsovoulos G, Kumar S, Blaxter ML, Davis RE.** 2012. Silencing of germline-expressed genes by DNA elimination in somatic cells. *Dev Cell* **23:**1072–1080.

107. **Shibata A, Moiani D, Arvai AS, Perry J, Harding SM, Genois MM, Maity R, van Rossum-Fikkert S, Kertokalio A, Romoli F, Ismail A, Ismalaj E, Petricci E, Neale MJ, Bristow RG, Masson JY, Wyman C, Jeggo PA, Tainer JA.** 2014. DNA double-strand break repair pathway choice is directed by distinct MRE11 nuclease activities. *Mol Cell* **53:**7–18.

108. **Alt FW, Zhang Y, Meng FL, Guo C, Schwer B.** 2013. Mechanisms of programmed DNA lesions and genomic instability in the immune system. *Cell* **152:** 417–429.

109. **Kapitonov VV, Jurka J.** 2005. RAG1 core and V(D)J recombination signal sequences were derived from Transib transposons. *PLoS Biol* **3:**e181.

110. **Aziz RK, Breitbart M, Edwards RA.** 2010. Transposases are the most abundant, most ubiquitous genes in nature. *Nucleic Acids Res* **38:**4207–4217.

111. **Feschotte C, Pritham EJ.** 2007. DNA transposons and the evolution of eukaryotic genomes. *Annu Rev Genet* **41:**331–368.

112. **Sinzelle L, Izsvak Z, Ivics Z.** 2009. Molecular domestication of transposable elements: from detrimental parasites to useful host genes. *Cell Mol Life Sci* **66:** 1073–1093.

113. **Volff JN.** 2006. Turning junk into gold: domestication of transposable elements and the creation of new genes in eukaryotes. *Bioessays* **28:**913–922.

114. **Machemer H, Ogura A.** 1979. Ionic conductances of membranes in ciliated and deciliated *Paramecium*. *J Physiol* **296:**49–60.

115. **Nowak JK, Gromadka R, Juszczuk M, Jerka-Dziadosz M, Maliszewska K, Mucchielli MH, Gout JF, Arnaiz O, Agier N, Tang T, Aggerbeck LP, Cohen J, Delacroix H, Sperling L, Herbert CJ, Zagulski M, Bétermier M.** 2011. Functional study of genes essential for autogamy and nuclear reorganization in *Paramecium*. *Eukaryot Cell* **10:**363–372.

116. **Matsuda A, Forney JD.** 2006. The SUMO pathway is developmentally regulated and required for programmed DNA elimination in *Paramecium tetraurelia*. *Eukaryot Cell* **5:**806–815.

117. **Matsuda A, Shieh AW, Chalker DL, Forney JD.** 2010. The conjugation-specific Die5 protein is required for development of the somatic nucleus in both *Paramecium* and *Tetrahymena*. *Eukaryot Cell* **9:**1087–1099.

118. **Meyer E, Keller AM.** 1996. A mendelian mutation affecting mating-type determination also affects developmental genomic rearrangements in *Paramecium tetraurelia*. *Genetics* **143:**191–202.

Mobile DNA, 3rd Edition
Nancy L. Craig, Michael Chandler, Martin Gellert, Alan M. Lambowitz, Phoebe A. Rice and Suzanne Sandmeyer

doi:10.1128/microbiolspec.MDNA3-0025-2014

V. Talya Yerlici[1]
Laura F. Landweber[2]

18

Programmed Genome Rearrangements in the Ciliate *Oxytricha*

INTRODUCTION

Ciliates are microbial eukaryotes with separate germline and somatic nuclei. The DNA-rich somatic macronucleus forms by differentiation of a copy of the diploid, zygotic germline micronucleus during sexual reproduction. The distinctive genome architectures of ciliates make them attractive model systems to study a wide range of key biological phenomena. These include complex genome rearrangements on a massive scale, a diverse range of noncoding RNA pathways, and several examples of non-Mendelian inheritance. In particular, ciliates belonging to the subclass *Stichotrichia*, such as the genus *Oxytricha*, display the most exaggerated form of genome remodeling, stitching together somatic chromosomes from precursor gene segments, all under the epigenetic control of novel noncoding RNA pathways.

By contrast with these processes that actually reorder and recombine elements of germline DNA, many eukaryotes use RNA splicing to stitch together exons that are not linear in the genome. Such rearrangements at the RNA level can produce chimeric or permuted RNAs that reorder endogenous genetic material. Examples include the circularly permuted nuclear tRNA genes in the red alga *Cyanidioschyzon merolae* (1, 2, 3) and *trans*-splicing in both metazoa and protists, such as the *bursicon* gene in mosquitoes (4), at least two *Giardia lamblia* genes with pre-mRNAs transcribed from distinct chromosomal regions (5), the unusually spliced mitochondrial *cox3* of Dinoflagellates (6), and spliced leader *trans*-splicing in *Caenorhabditis elegans* and kinetoplastid protists (7, 8, 9), plus alternative splicing of 3′-UTRs in mammals (10) and countless other elaborate examples of alternative splicing (11). Numerous chimeric RNAs in humans have attracted attention due to their possible association with cancer (12, 13). One such example is the heterogeneously *trans*-spliced mRNA of the human transcription factor Sp1, which may alter the regulation of various genes (14, 15). Importantly, there is also evidence for the translation of such chimeric mRNAs in humans (16).

Although less common than rearrangement at the RNA level, programmed genomic remodeling at the

[1]Department of Molecular Biology, Princeton University, Princeton, NJ 08544; [2]Department of Ecology and Evolutionary Biology, Princeton University, Princeton, NJ 08544.

DNA level is highly relevant to humans, since it underlies adaptive immunity. For example, V(D)J recombination (17) and class-switch recombination allow B and T cells to express a range of antigen receptors through the reshuffling of germline DNA sequences to produce genes for novel antigen receptors (18, 19, 20). Aberrant genomic rearrangements are also a hallmark of cancer (21, 22, 23, 24). Similar to fusion proteins encoded by chimeric mRNAs, *trans*-spliced DNA segments may also encode malfunctioning proteins.

The process of genomic remodeling in *Oxytricha* also bears similarity to the developmentally regulated processes of chromosome elimination and diminution in metazoa, including nematodes, copepods, insects, jawless fish and marsupials (25). In the nematode *Parascaris univalens* and copepod *Mesocyclops edax*, the eliminated sequences account for ~90% of the germline genome as part of genome differentiation (26, 27). Similarly in Lamprey, an order of jawless fish, 20% of the germline sequences are eliminated during somatic differentiation. This may be an economical way to exclude genes from the soma that only function in the germline (28, 29). Sciarid flies go through a similar process of chromosome elimination during development, discarding entire chromosomes from their somatic genome (30).

The extensive genome remodeling in ciliates, while sharing some ostensible features with all these other types of genomic rearrangements, also illustrates the great power of noncoding RNA-mediated pathways. Small RNAs, together with long noncoding RNAs, in *Oxytricha* mediate the transgenerational transfer of information necessary to decrypt the germline genome during development of the soma. Surprisingly, ciliates have shown that RNAi-related pathways play a major part in genome remodeling (31, 32) in addition to their conventional role in gene silencing (33, 34). Moreover, these RNAi-related pathways have evolved to take on orthogonally different functions in the different branches of the ciliate clade (32).

Therefore, an emerging area of ciliate biology has been to study this intriguing form of non-Mendelian inheritance, which even permits mutations that accumulate in the soma to transmit to the sexual offspring. Epigenetic inheritance is ubiquitous among ciliates, due to the nature of their sexual life cycle (Fig. 1) (35). This allows maternal effects to transfer epigenetic information for decrypting the germline after each round of sexual conjugation. This review will explore in detail the radical genome architecture of *Oxytricha* and the evidence for the RNA-guided mechanisms underlying its elaborate process of genome remodeling.

BACKGROUND

Nuclear Dualism and the Life Cycle of Ciliates

In ciliates, the germline-soma differentiation normally associated with multicellular eukaryotes manifests in an unusual way. Each single celled organism contains a germline diploid micronucleus (MIC) and a somatic DNA-rich macronucleus (MAC) (Fig. 1). During vegetative growth, cells divide asexually through binary fission (Fig. 1) as the MIC undergoes mitosis while the MAC replicates through amitosis. In the absence of spindle fibers, the MAC chromosomes may segregate randomly. During the asexual life cycle, the germline MIC is transcriptionally silent and all transcription for maintenance of the vegetative cell takes place in the somatic MAC (36).

Ciliates also undergo non-replicative conjugation under certain conditions, such as starvation in the laboratory setting (Fig. 1). Conjugation leads to the formation of pairs between compatible mating types (Fig. 1). This initiates meiosis and an ensuing cascade of events leading to the formation of a new zygotic nucleus that will become the new germline, while a copy of the germline will develop into the new somatic nucleus after major processing. This differentiation occurs during each round of sexual conjugation.

After completion of meiosis, one copy of a haploid MIC from each cell exchanges between the two mating partners to fuse with an endogenous haploid MIC, forming a new, diploid, zygotic nucleus. Subsequently, the new zygotic nucleus undergoes mitosis, as the old maternal MAC degrades. One or more zygotic nuclei will be maintained as the germline, while others will differentiate into a new MAC (Fig. 1). The developing MAC is called the "anlage" (plural anlagen). The numbers of each type of nuclei in a cell vary between different ciliate genera, up to hundreds in *Urostyla* (36).

The parental and newly developing macronuclei coexist in the same cell (Fig. 1), allowing for the direct transmission of cytoplasmic, surface, or other epigenetic factors from parent to offspring (35, 37). Compared to the nuclear differentiation process in Oligohymenophorea, the class to which the well-studied *Tetrahymena* and *Paramecium* belong, the genome remodeling process in stichotrichous ciliates, which include *Oxytricha trifallax*, is much more complex (Fig. 2).

Oxytricha trifallax and its Distinct Genome Architectures

The *Oxytricha trifallax* macronuclear genome contains over 16,000 unique gene-sized chromosomes. Averaging

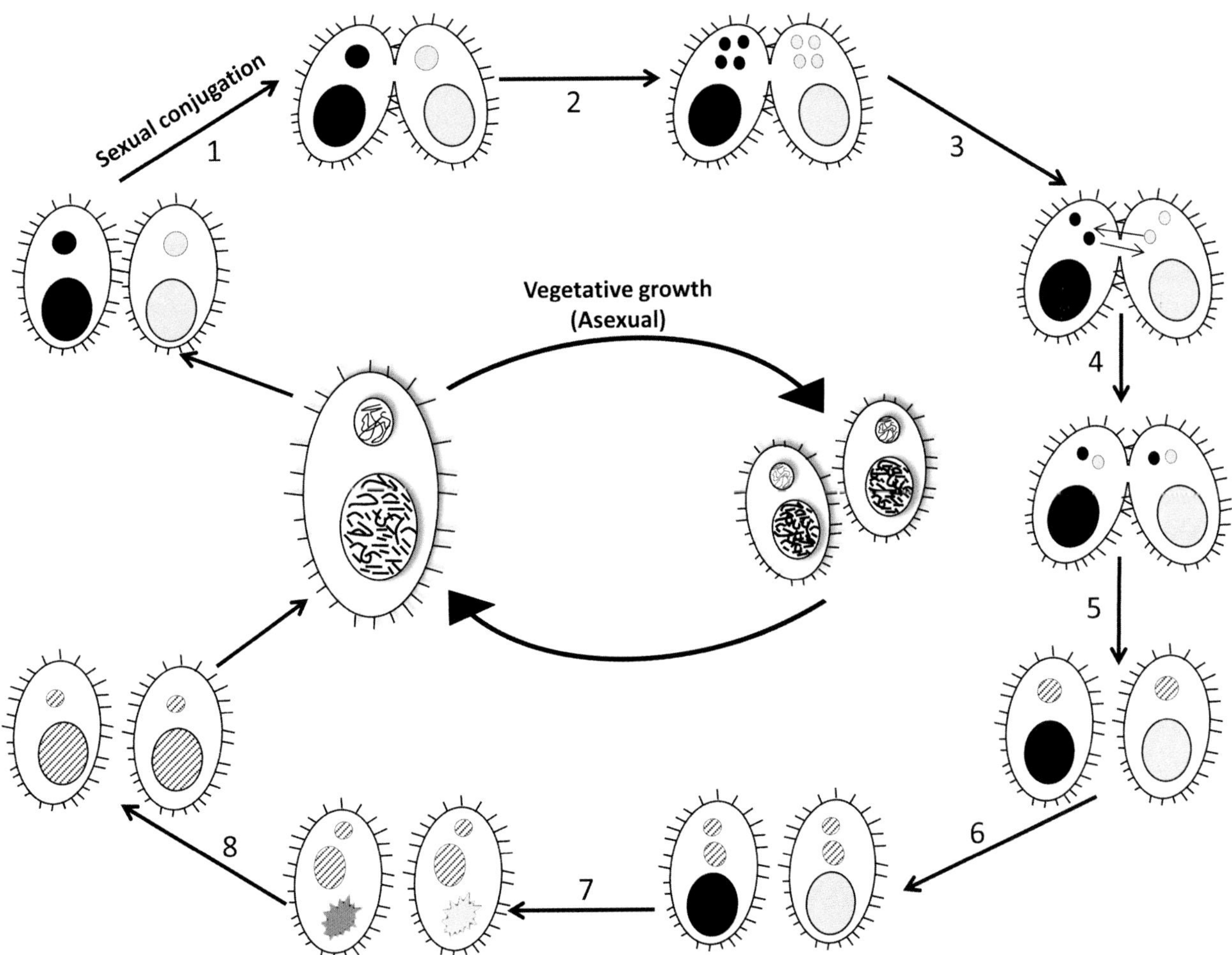

Figure 1 *Oxytricha* vegetative and sexual life cycles. *Oxytricha* cells contain one somatic macronucleus and two identical, germline, micronuclei. For simplicity only one MIC is shown. *Oxytricha* divides asexually through binary fission producing clonal offspring. (1) Under starvation conditions, two cells of different mating type form pairs. (2) The diploid MIC undergoes meiosis and produces four haploid gametes. (3) Three of the haploid gametes degrade while the one remaining divides mitotically. (4) The pair exchange a copy each of these haploid micronuclei. (5) The haploid micronuclei fuse, producing a new, diploid, zygotic nucleus shown with cross hatching. This produces exconjugants with identical zygotic genomes. (6) The zygotic nucleus divides mitotically, producing two identical zygotic genomes. (7) One copy of the zygotic genome differentiates into a new soma, while the old maternal soma degrades. The developing MAC at this stage is called the "anlage" (plural anlagen). The other zygotic nucleus will maintain the new germline. doi:10.1128/microbiolspec.MDNA3-0025-2014.f1

just 3.2 kb each with short telomeres, they are so small that they are sometimes called "nanochromosomes." Each encodes just 1 to 8 genes, with ~90% encoding only a single gene (38). In contrast to the MAC, the germline MIC harbors all the precursor macronuclear genomic sequences, plus a vast quantity of germline-limited DNA, on a set of ~120 long archival chromosomes (36).

Stichotrichous ciliates, such as *Oxytricha* and *Stylonychia*, differ from the oligohymenophorean ciliates *Tetrahymena* and *Paramecium* (Fig. 2) in important ways, such as discarding nearly all of their germline genome during formation of the somatic nucleus (compare over 90% in *Oxytricha* (39) and possibly 98% in *Stylonychia* (36) to ~33% in *Tetrahymena* (40, 41)) and the possession of over 200,000 scrambled genic segments in the germline that must be unscrambled to produce functional genes in the soma (39, 42). Somatic nanochromosomes in *Oxytricha* assemble from approximately 5–10% of the micronuclear genome

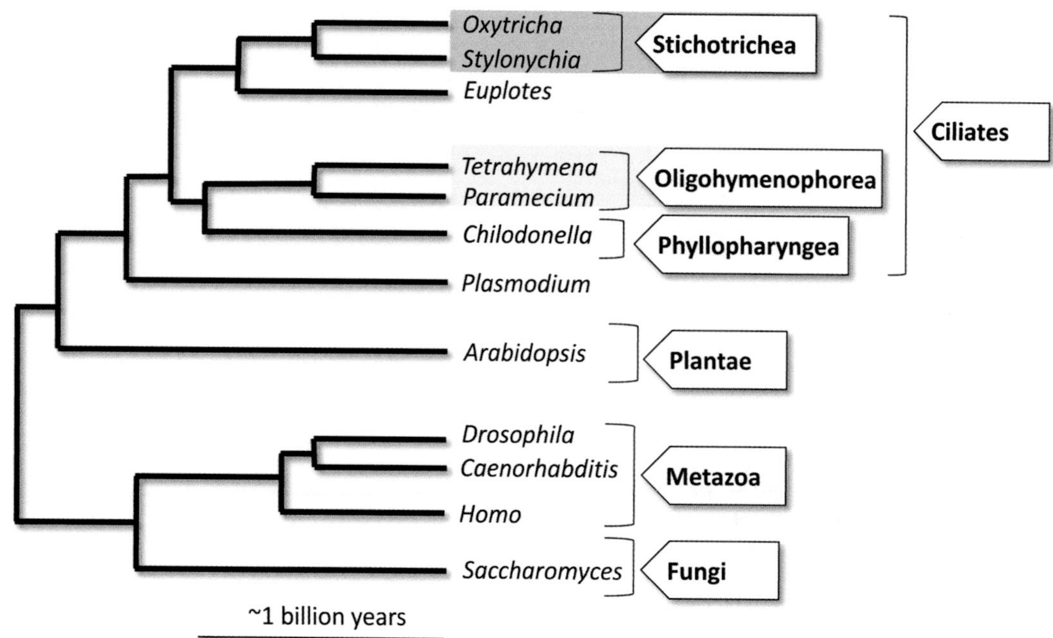

Figure 2 The inferred evolutionary relationship of *Oxytricha* to other well-studied ciliate lineages and representative eukaryotic genera. Figure modified from Bracht et al. (77). Branch order, branch lengths and scale bar are based on estimates in Parfrey et al. (83) and Chang et al. (141). doi:10.1128/microbiolspec.MDNA3-0025-2014.f2

after a sequence of events that includes chromosome polytenization, massive DNA elimination and chromosome diminution, DNA descrambling, chromosome fragmentation—likely coupled to telomere addition—and, finally, chromosome amplification. Genome differentiation begins with polytenization, where 15- to 32-fold amplification of the zygotic MIC genome produces giant polytene chromosomes (36, 43, 44). The first wave of amplification is followed by destruction of all germline-restricted DNA, including intragenic spacer sequences (internal **e**liminated**s**equences, or IESs) that interrupt macronuclear precursor loci (Fig. 3) (45), and intergenic DNA in the form of satellite repeats, transposons, and even hundreds of MIC-limited protein-coding genes (39). In stichotrichous ciliates and in a few other distant ciliates, such as *Chilodonella* (46), the ligation of DNA segments to reconstruct functional genes in the new MAC (**m**acronuclear-**d**estined **s**equences, or MDSs) may require DNA unscrambling (45, 47, 48) (Fig. 3). In *Oxytricha trifallax*, approximately 20% of the 18,400 non-redundant genes have MDSs that are scrambled in the MIC and all require decryption by various combinations of DNA translocations or inversions during genome remodeling (39). Inverted MDSs map to the reverse orientation on the opposite strand in the germline precursor locus in the MIC (Fig. 3).

IESs are commonly flanked by 2 to 20 base-pair regions of microhomology called "pointers", short, precise matches at the 3′ end of the *n*th MDS and the 5′ end of (*n*+1)st MDS (Fig. 3) (47). Several lines of evidence (49, 50) suggest a strong mechanistic preference for recombination between pointers, with one copy of the pointer retained in the MAC DNA. Because they can be as short as two base-pairs (or possibly a one base-pair perfect repeat, considering mismatches (50)—see "Epigenetic regulation of genome rearrangement in *Oxytricha*" below), pointers are not sufficient to guide MDS unscrambling on their own, though pointers that map between scrambled MDSs do contain more information and are typically longer than the nonscrambled pointers that form the junctions between consecutive MDSs (11 versus five base-pairs, respectively (39)). Chromosome fragmentation and telomere addition together yield nanochromosomes that are much smaller than the archival MIC chromosomes. Even though *Tetrahymena* fragments DNA at a well conserved 15-nucleotide chromosome breakage site (Cbs) (51, 52), and *Euplotes crassus* uses a different motif (*Euplotes* Cbs, E–Cbs) (53) to fragment chromosomes and to regulate IES excision (54, 55), no such motif has yet been identified in *Oxytricha*. Telomerase, originally discovered in ciliates (56), adds *de novo* telomeres to the ends of the first and last MDS. Alternative chromosome

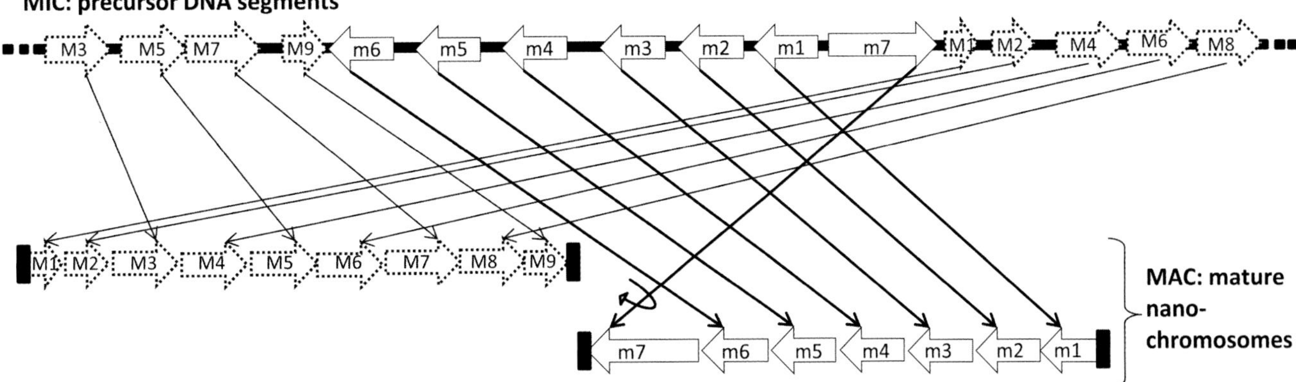

Figure 3 Schematic illustration of gene unscrambling in stichotrichous ciliates. Black bars represent IESs to be eliminated. These IESs separate the MDSs (shown as white arrows) in the micronuclear precursor region. MDS labeled M1-9 and drawn with dotted lines are the precursors for one somatic nanochromosome, whereas MDSs drawn with solid lines (m1-7) assemble into a separate nanochromosome. The numbers represent the order in which the MDSs appear in the mature MAC nanochromosomes. Mature nanochromosomes in the MAC form following IES excision, MDSs descrambling and chromosome fragmentation. Addition of *de novo* telomeres (black tall rectangles) at the boundaries of the first and last MDS produces mature nanochromosomes.
doi:10.1128/microbiolspec.MDNA3-0025-2014.f3

fragmentation, or alternative addition of telomeres to different sites, can produce nanochromosomes of variable length and with different 5′ or 3′ ends. This sometimes generates nanochromosome isoforms with different coding gene contents (38, 57, 58). Approximately 10% of *Oxytricha* nanochromosomes are alternatively fragmented, and most of these cases display a weaker pyrimidine to purine transition that may be important for chromosome breakage or telomere addition (38, 39). Amplification of telomere-capped nanochromosomes produces an average macronuclear chromosome copy number of 1900 (36).

Epigenetically Programmed, Small RNA-Mediated Genome Rearrangement in Oligohymenophorea

In the late 1980s and early 1990s, several pioneering experiments in *Paramecium*, as well as *Tetrahymena*, demonstrated a powerful, epigenetic influence of the maternal soma on the offspring in the next sexual generation (59, 60, 61, 62, 63, 64, 65, 66). More recent experiments, summarized in detail elsewhere (52, 67, 68; and see *Tetrahymena* and *Paramecium* chapters in this volume), led to a widely accepted epigenetic model which accounts for some aspects of non-Mendelian inheritance in *Tetrahymena* and *Paramecium*, particularly the regulation of DNA deletions by small noncoding RNAs. Briefly, in this "scan RNA" model, MIC transcripts are processed into small RNAs that range in size among Oligohymenophorea (~28 and 25 nucleotides in *Tetrahymena* and *Paramecium*, respectively). After scanning the parental macronuclear genome to subtract MDS sequences, the remaining scnRNAs mark MIC-limited regions in the anlagen for elimination by RNAi-related pathways (31, 69, 70, 71, 72). Recent experiments indicate involvement of multiple, parallel sRNA pathways, including "iesRNAs" that exclusively target IESs in *Paramecium* (73). Together, these collections of sRNAs are reminiscent of the extensively studied maternal mRNAs that drive the early steps of development in *Drosophila* embryos (74). But unlike the maternal effects of mRNAs in *Drosophila* and the scnRNAs in Oligohymenophorea that mark DNA sequences for deletion, the parental generation in *Oxytricha* supplies the daughter exconjugants with an enormous set of both small RNAs that mark DNA sequences in the anlagen for *retention* and long noncoding RNAs that provide the information to rebuild the soma.

ROLE OF SMALL RNAS IN *OXYTRICHA* GENOME REARRANGEMENT

Oxytricha, like the oligohymenophorean ciliates, produces a huge quantity of conjugation-specific small RNAs (sRNAs), so abundant that they can be seen on an ethidium bromide-stained acrylamide gel without radiolabeling (32, 75). Total RNA extracts from early conjugating pairs indicate an overwhelming enrichment of 27-nucleotide long 5′U sRNAs, which are absent in vegetative cells. Deep sequencing of these sRNAs demonstrated that they map bidirectionally to the entire MAC genome, including pointer recombination junctions that are only formed in the MAC, and that these sRNAs are depleted in MIC-limited segments (32, 75) (Fig. 4A). This is opposite to the scnRNAs and iesRNAs in *Tetrahymena* and *Paramecium* where the sRNAs are ultimately specific for primarily MIC-limited segments and IESs (73, 76), and not MAC-destined DNA. This suggests that a sign change occurred earlier in the evolutionary history of the lineage of these distant ciliate classes, separated by over one billion years of evolution (77). Furthermore, immunoprecipitation with an antibody specific to Otiwi1, the *Oxytricha* ortholog of a Piwi protein (78, 79), demonstrates an association between Otiwi1 and *Oxytricha*'s conjugation specific 27-nucleotide sRNAs. Hence, they are labeled Piwi-interacting RNAs, or piRNAs (32). Immunofluorescence staining suggests that Otiwi1 may participate in the transportation of piRNAs from the maternal MAC to the anlagen during early genome differentiation, a step that is essential for exconjugant viability and programmed macronuclear development, since knockdown of Otiwi1 results in nonviable offspring.

The contrast is striking between the model for piRNA guided-genome rearrangement in *Oxytricha*

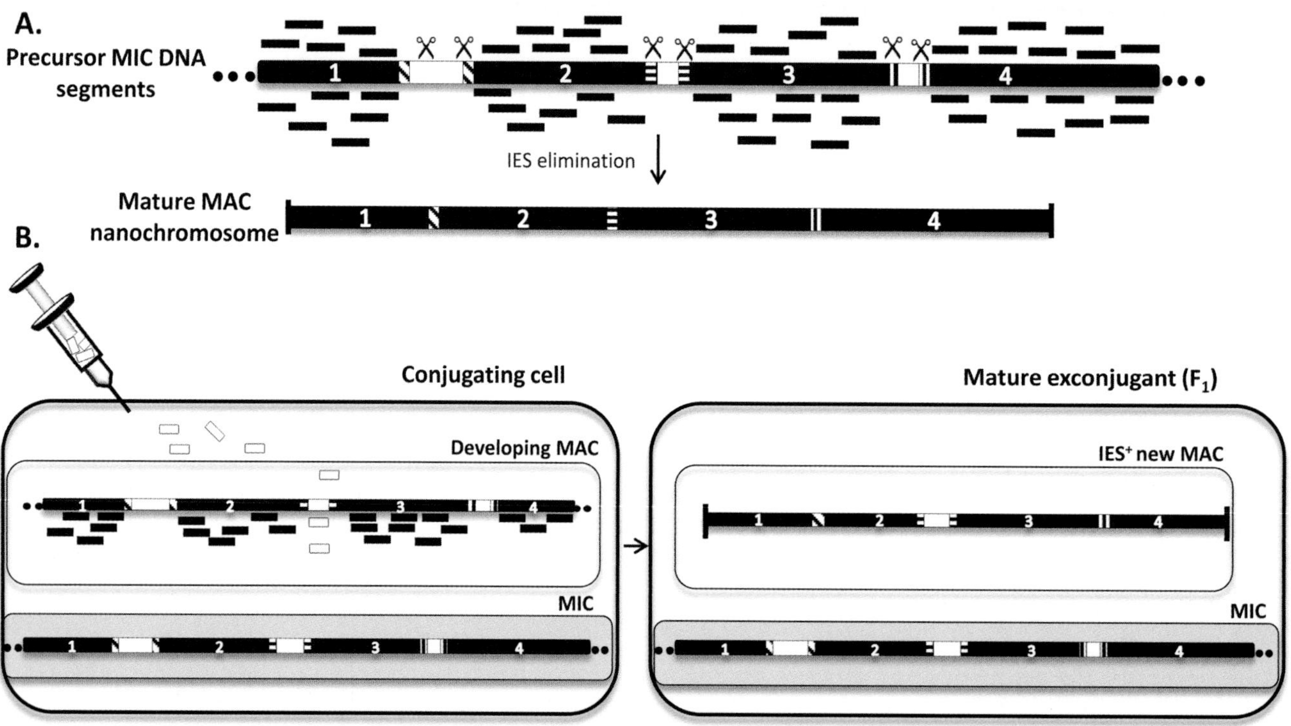

Figure 4 *Oxytricha* piRNAs mark MDSs for retention during genome rearrangement. A. 27-nucleotide piRNAs (short black bars) derive from RNA copies of the maternal somatic genome. These piRNAs mark and protect MDSs for retention. DNA segments to be eliminated, such as IESs (shown as white rectangles between MDSs in the precursor DNA segment), lack corresponding piRNAs. Direct repeats (pointers) flanking the IESs are shown as small, patterned vertical rectangles. Identical pointers that join two consecutive MDSs are marked with the same pattern. The excision machinery (represented as scissors) can cleave the regions not protected by piRNAs and permit recombination between pointers. This leads to IES elimination, with retention of one copy of the pointer between MDSs in the mature nanochromosome. Dots at the ends of precursor loci in the micronucleus and developing macronucleus represent sequences that continue beyond the regions shown. Mature, macronuclear nanochromosomes terminate in telomeres (thin, vertical, black rectangles). B. Microinjection into conjugating cells of synthetic 27-nucleotide piRNAs (short white bars) complementary to IESs that are normally deleted in wild type cells, leads to retention of the IES in the mature MAC of the exconjugants and subsequent sexual offspring (32). This creates IES^+ strains. doi:10.1128/microbiolspec.MDNA3-0025-2014.f4

and the scan RNA model in Oligohymenophorea, in which scnRNAs mark IESs via an RNAi-related pathway for elimination. The piRNAs in *Oxytricha* that originate from the maternal MAC must mark and possibly tile the hundreds of thousands of MDSs that are retained in the anlagen, leaving most of the MIC genome, including all MIC-limited sequences (e.g. IESs and transposons), unmarked. The nature of this mark is unknown, but potential epigenetic marks are heterochromatin formation through the methylation of DNA or histones (see "Epigenetic marks for retained vs. deleted sequences" below); however, the small size of the tiniest MDSs suggests that it would be difficult for the marks to be on nucleosomes. Whether directly, by a physical block, or indirectly, via epigenetic marks, piRNAs may ultimately prevent DNA excision within MDSs by antagonizing transposases, part of the proposed catalytic machinery responsible for DNA elimination (80) (see "Genome remodeling mechanisms and machinery" below).

As experimental proof of this model in *Oxytricha*, microinjection into conjugating cells of synthetic piRNAs complementary to IESs that are normally deleted in wild type cells leads to retention of the IES in the new mature MAC (32) (Fig. 4B). Furthermore, the retention of the IES in the new MAC is stable across many asexual as well as sexual generations. In a backcross between the F1 cells from such an experiment and a wild type strain, Fang et al. (32) specifically demonstrated the biological production of new piRNAs that target the retained IES (but that differ in sequence from the injected piRNA), supporting the epigenetic model in Fig. 4). Hence, even the transient availability of piRNAs during early development can influence heritable changes in the soma that transfer across multiple sexual generations, similar to the transgenerational epigenetic programming effect of piRNAs in *C. elegans* (81, 82).

The orthogonality between the two sRNA pathways within ciliates is less surprising when one considers the deep-rootedness of the ciliate lineage among eukaryotes, and that the evolutionary distance between *Oxytricha* and *Tetrahymena* is as divergent as that between humans and fungi (Fig. 1) (83). The opposite piRNA pathways also make economic sense for the respective organisms, because in all cases the piRNAs mark the minority class of the germline, either for deletion (33% in *Tetrahymena* (40)) or retention (5 to 10% in *Oxytricha*). Furthermore, because some *Oxytricha* IESs are smaller than 27 nucleotides, these would be problematic if sRNAs had to exclusively mark eliminated sequences, like the iesRNAs of *Paramecium* (73).

Along the same lines, it is also notable that both MDSs and IESs in *Oxytricha* can be smaller than the length of one nucleosome unit, suggesting that epigenetic marks are more likely to occur directly on the DNA. Accordingly, cytosine methylation has been noted as one developmental mark for deletion in *Oxytricha* (84). Additional distinguishing features in the respective sRNA pathways may explain *Tetrahymena*'s tolerance for imprecise IES excision. For example, *Tetrahymena* might lack the RNA-guided DNA proofreading available in *Oxytricha* (85) (see "RNA-guided DNA repair in *Oxytricha* and broader implications of RNA templating" below). As a result of its imprecision, IESs in *Tetrahymena* usually reside within introns or intergenic regions and only very rarely disrupt open reading frames (86, 87). Like *Oxytricha*, IES excision in *Paramecium* is mostly precise; however, unlike *Oxytricha*, the ends of *Paramecium* IESs possess a highly conserved TA pointer and uniform sequence motifs at both IES ends (52). This may constrain them from possessing enough information to support gene unscrambling and explain the lack of scrambled genes in *Paramecium*. (*Tetrahymena*, on the other hand, has no published cases of scrambled genes, because its IESs are mostly intergenic, with the exception of approximately 10 cases that interrupt exons (87).) Curiously, *Stylonychia*, another stichotrichous ciliate with a complex, scrambled germline genome like *Oxytricha* (Fig. 2), may discard close to 98% of its MIC DNA during MAC development (36). An earlier paper suggested that its sRNAs are more consistent with the scan RNA model (88); however, that inference was based only on hybridization, with no sequence analysis, and numerous other factors could have affected these early results.

LONG NONCODING MATERNAL RNA TEMPLATES

MDS Unscrambling

The scan RNA model that extends to *Tetrahymena* and *Paramecium* genome rearrangement would be insufficient to program the precise DNA segment reordering and occasional inversion of MDSs in *Oxytricha* (Fig. 3). Additionally, piRNAs are not likely candidates for guiding DNA rearrangements either, since microinjection of synthetic piRNAs has been unable to reprogram MDS order or to template RNA-guided DNA proofreading (32). Finally, the pointers, themselves, do not contain enough information to unambiguously reorder and invert MDSs, because they can be as small as 2 base-pairs, which would occur often by chance (39).

This conclusion, in addition to the observation of a surprising number of errors at MDS-MDS junctions early during early rearrangement (49), suggested a need for error correction, possibly via a DNA or RNA template. Prescott et al. (89) and, later, Angeleska et al. (90) proposed a specific epigenetic model for RNA-guided DNA recombination acting via maternal templates. Nowacki et al. (85) offered several lines of evidence that RNA templates derived from the maternal macronucleus do guide assembly of the new macronuclear chromosomes (Fig. 5A, 5B). Either RNAi against specific, endogenous RNA templates (Fig. 5C) or microinjection of new, alternative DNA or RNA templates (Fig. 5D) resulted in aberrant rearrangements in the progeny. RNAi to reduce the concentration of available templates in the cell stalled DNA rearrangement (Fig. 5C), whereas injection of foreign templates programmed new MDS patterns in the targeted gene (Fig. 5D), respectively. Presumably, the injection of DNA templates (complete telomere-containing synthetic nanochromosomes) permitted their transcription to RNA *in vivo*, to mimic the effect of injecting RNA directly. These experiments historically provided the first opportunity to parse the specific steps in this complex process and helped develop *Oxytricha* as a powerful model system to study programmed genome rearrangement. Moreover, as for piRNAs, the reprogramming effect of the foreign templates propagated over multiple sexual generations. These data strongly supported the epigenetic models for sequence-dependent comparison between two genomes (the parental and developing MAC) within a single cell. More broadly, these experiments revealed a new mechanism for epigenetic wiring and rewiring of cellular programs, and an RNA-mediated pathway for transmission of an acquired epigenetic state (Fig. 6).

Chromosome Copy Number Regulation

In addition to programming the DNA rearrangement pattern during genome differentiation, the maternal RNA templates described in the previous section also establish DNA copy number after conjugation through an unknown epigenetic mechanism. Gene dosage and DNA copy number are critical for many aspects of cell biology, including the developmental regulation of gene expression; additionally, increases in DNA copy number can contribute to acquisition of new function by gene or segment duplication (91). Furthermore, defects in the regulation of gene dosage may contribute to diseases, possibly including cancer (92, 93), autism (94) and disorders associated with trisomy (95). Classic examples for regulating gene expression through DNA dosage include the *chorion* locus in *Drosophila* (96), the circular rRNA genes in *Xenopus* (97), and the *Tetrahymena* rDNA locus, which forms differentially amplified palindromic segments in the MAC that originate from a single *rDNA* chromosomal copy during rearrangement (98, 99). Many other ciliate species, including the stichotrichous ciliates *Oxytricha* and *Stylonychia*, also have an overamplified rDNA chromosome. However, the more general relationship in *Oxytricha* between nanochromosome copy number and gene expression levels is surprisingly modest, probably constrained by the limited amount of copy number variation relative to gene expression variation (38).

Two groups demonstrated that *Oxytricha* and *Stylonychia* can regulate whole nanochromosome copy number based on the abundance of the corresponding maternal RNA templates during each round of sexual conjugation (100, 101). This discovery offered a novel biological role for long, noncoding RNAs. Both Nowacki et al. (100) and Heyse et al. (101) showed that changes in the available quantity of maternal RNA templates affect DNA copy number in the exconjugants following conjugation. Microinjection of two-telomere-containing RNA copies corresponding to different nanochromosomes resulted in an increase in the chromosome copy number in the sexual progeny, whereas RNAi against the endogenous templates for these chromosomes led to a decrease in nanochromosome copy number. The F2 progeny in the first experiment also displayed increased copy number, demonstrating once again that somatically acquired traits triggered by exposure to a noncoding RNA may propagate across sexual generations, and established this phenomenon in two groups of stichotrichous ciliates. Point substitutions that marked the synthetic RNA templates did not transfer to the exconjugants, confirming amplification of the endogenous chromosomes, rather than reverse transcription or copying of the injected templates. Two possible models for RNA-template-mediated regulation of DNA copy number could entail an influence of the level of maternal RNA templates on the timing and/or efficiency of either DNA rearrangement (100) or endoreplication after rearrangement (101), leading to a testable prediction that there will be concordant levels of amplification among the maternal and exconjugant nanochromosomes, as well as the maternal RNA templates.

RNA-GUIDED DNA REPAIR IN *OXYTRICHA* AND BROADER IMPLICATIONS OF RNA TEMPLATING

The role of the maternal RNA templates is not even limited to programmed genome remodeling and regulation

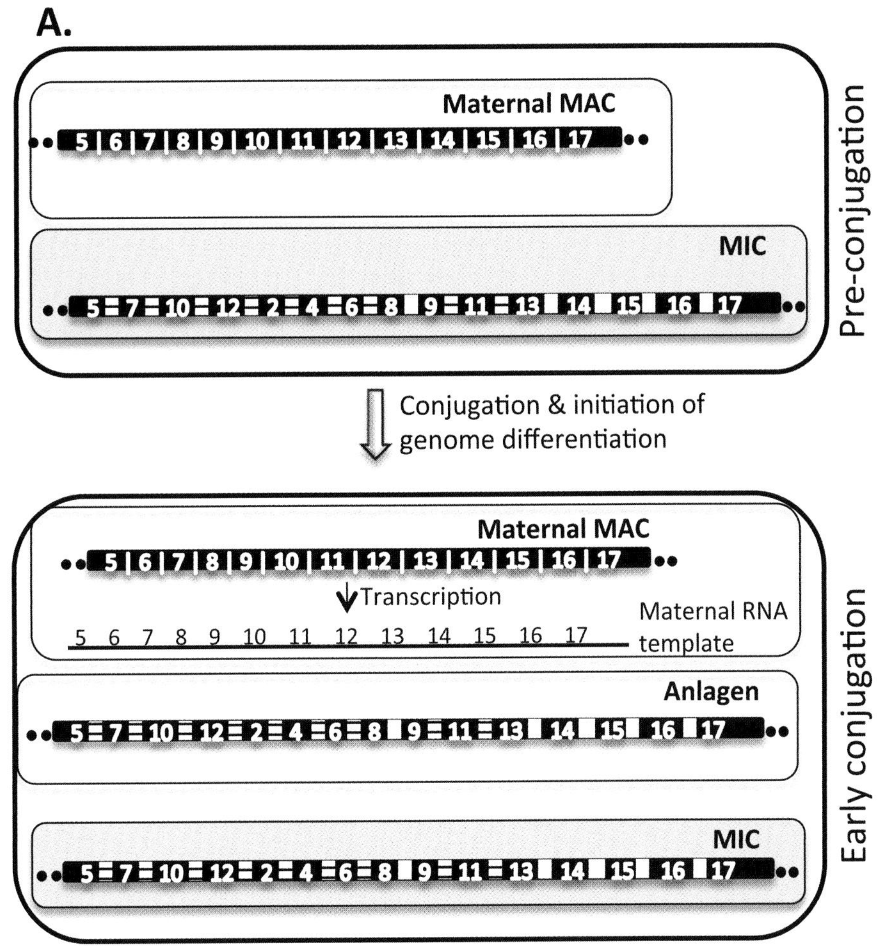

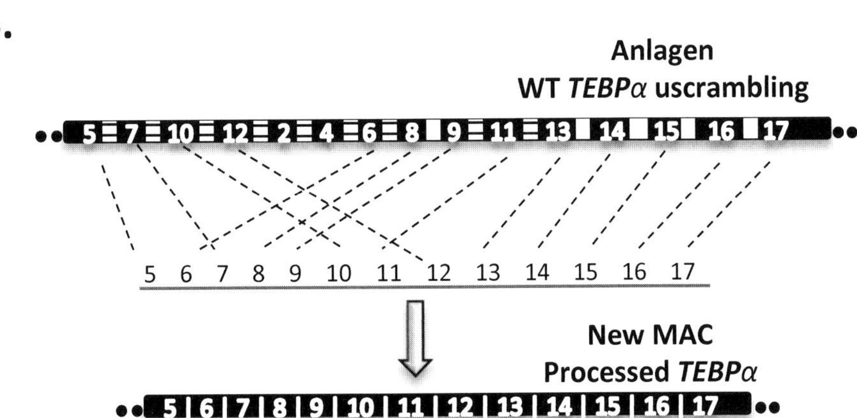

Figure 5 *continues on next page*

C.

Anlagen
Defective *TEBPα* unscrambling
TEBPα RNA template levels reduced by RNAi

5 7 10 12 2 4 6 8 9 11 13 14 15 16 17

5 6 7 8 9 10 11 12 13 14 15 16 17

New MAC
Aberrantly processed *TEBPα*

i. 5 7 10 12 2 4 6 8 | 9 11 13 | 14 | 15 | 16 | 17

ii. 5 7 10 12 2 4 6 8 9 11 13 | 14 | 15 | 16 17

iii. 5 | 6 | 7 | 8 | 9 11 13 14 15 16 | 17

iv. 5 7 10 12 2 4 6 8 | 9 11 | 12 | 13 | 14 | 15 | 16 | 17

v. 5 | 2 | 3 | 4 | 5 | 6 | 7 | 8 | 9 | 10 | 11 | 12 | 13 | 14 | 15 | 16 | 17

vi. 5 | 6 | 7 | 8 | 9 |10|11|12 1 | 2 | 3 | 4 | 5 | 6 | 7 | 8 | 9 |10|11|12|13|14|15|16|17

D.

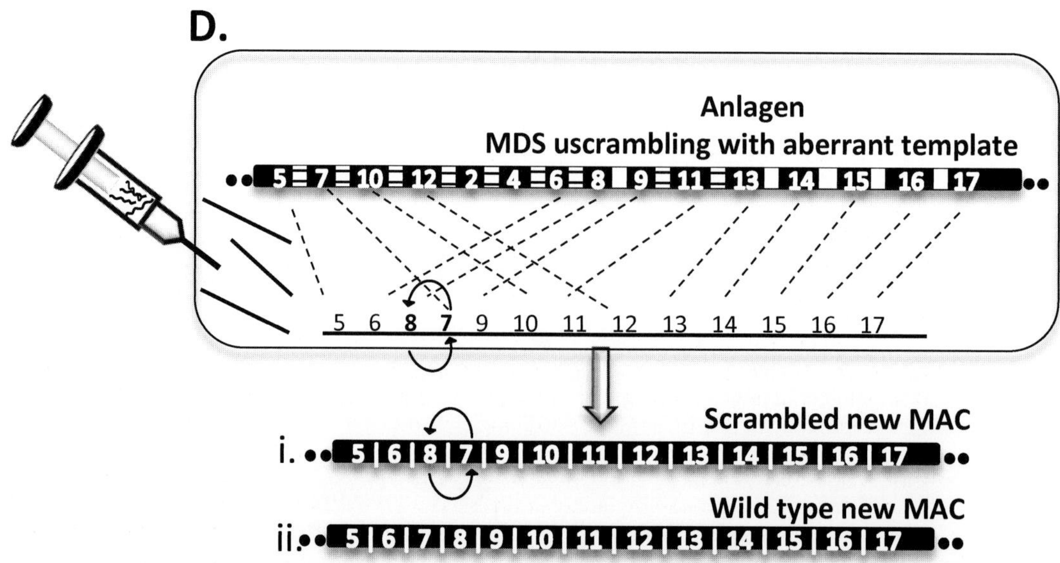

Figure 5 *continued*

of gene dosage, but also takes on a third level in DNA repair. Nowacki et al. (85) demonstrated the transfer of point mutations close to the recombination junctions, or pointers, from the RNA template to the developing MAC. This occurred despite the absence of corresponding DNA substitutions in the germline, confirming the epigenetic nature of this transmission. For example, a C-to-T substitution four base-pairs away from the end of an MDS transferred faithfully from the template, implicating RNA-directed DNA repair, as had previously been seen in yeast (102). The epigenetic transfer of the somatic point substitution was inherited across at least three sexual generations in the MAC, even in the absence of the corresponding mutation in the germline (Wang and Landweber, unpublished), similar to the RNA-mediated epigenetic inheritance of the three observed phenomena of IES retention, alternative MDS order, and chromosome copy number. *Oxytricha* was the first organism to offer experimental proof of the transient existence of an RNA-templated cache of genetic information passed across generations (85); however, the epigenetic inheritance of genomic variation through transiently available, and hence elusive, RNA templates may be more widespread than previously thought (102). On an evolutionary time scale, such events might influence the pace of adaptation to changing environments by permitting the propagation of fitter somatic variants, while preserving the opportunity to revert to wild type.

RNA-templated DNA repair may also have broader implications for diseases, such as cancer. Chromosomal rearrangements, as well as *trans*-spliced RNAs, can produce fusion proteins that facilitate cell growth, as discussed in the introduction to this chapter (103, 104, 105, 106). An additional consequence of *de novo* chimeric RNA formation in the absence of a corresponding DNA translocation may be to facilitate chromosomal rearrangements, with the *trans*-spliced or chimeric RNA acting as a template, similar to the action of RNA template-guided DNA recombination in *Oxytricha* (85, 90).

EPIGENETIC MARKS FOR RETAINED VS. DELETED SEQUENCES

The physical detection of piRNAs (see "Role of small RNAs in genome rearrangement" above) precedes the onset of genome rearrangement in *Oxytricha* (32). This apparent absence of piRNAs during later development suggests that piRNAs may mark the precursor MIC DNA segments for retention through various epigenetic modifications. These could be alterations to the

Figure 5 Genome differentiation in *Oxytricha* is mediated by long noncoding maternal RNA templates. A. Nowacki et al. (85) demonstrated the transient presence of two-telomere-containing, bidirectional RNA transcripts of maternal nanochromosomes in conjugating cells. Panel A shows part of the *O. trifallax* TEBPα (Telomere End Binding Protein *α*) locus between segments 5 and 17 and the corresponding region of the RNA template (long thin lines). B. These templates may guide rearrangement in the anlagen by interacting with the MIC precursor DNA. Numbered black rectangles represent MDSs in the linear MAC order. White rectangles indicate nonscrambled IESs (interrupting consecutive MDSs in the MAC) and striped rectangles are scrambled IESs that map between nonconsecutive MDSs. C. RNAi targeting the maternal *TEBPα* RNA templates (wavy line) leads to gross rearrangement defects in the corresponding gene (85). A PCR assay of the region between MDS 5 to 17 revealed the presence of molecules that are longer than wild type MAC sequences because they often retain unspliced IESs, with a strong bias towards retaining IESs between scrambled segments (examples i, ii, iv). Consistent with studies of earlier timepoints in the rearrangement cascade (49), example (ii) only eliminated a subset of the nonscrambled IESs in this region, and examples (i), (iv), (v) and (vi) eliminated *all* nonscrambled IESs. In addition, cases (i) and (ii) stalled before any reordering. The presumed decrease in template abundance also leads to accumulation of partially unscrambled molecules, such as (iii), (iv) and (v). Case (iv) correctly repaired MDS 12 between MDS 11 and 13, but with an unexpected duplication of MDS 12, retaining a copy in its scrambled location as well. The duplications in (iv) and (vi) indicate that intermolecular DNA rearrangement may be tolerated. Aberrantly spliced junctions, marked by thin vertical ovals, occur at cryptic repeats, instead of endogenous pointers. D. Microinjection of synthetic *TEBPα* templates with the order of segments 7 and 8 transposed (indicated with curly arrows) produces new, scrambled MAC nanochromosomes with the reprogrammed order. Wild type, unscrambled nanochromosomes coexist in the cell with the permuted chromosomes, because the endogenous wild type templates were not destroyed in this experiment (85).
doi:10.1128/microbiolspec.MDNA3-0025-2014.f5

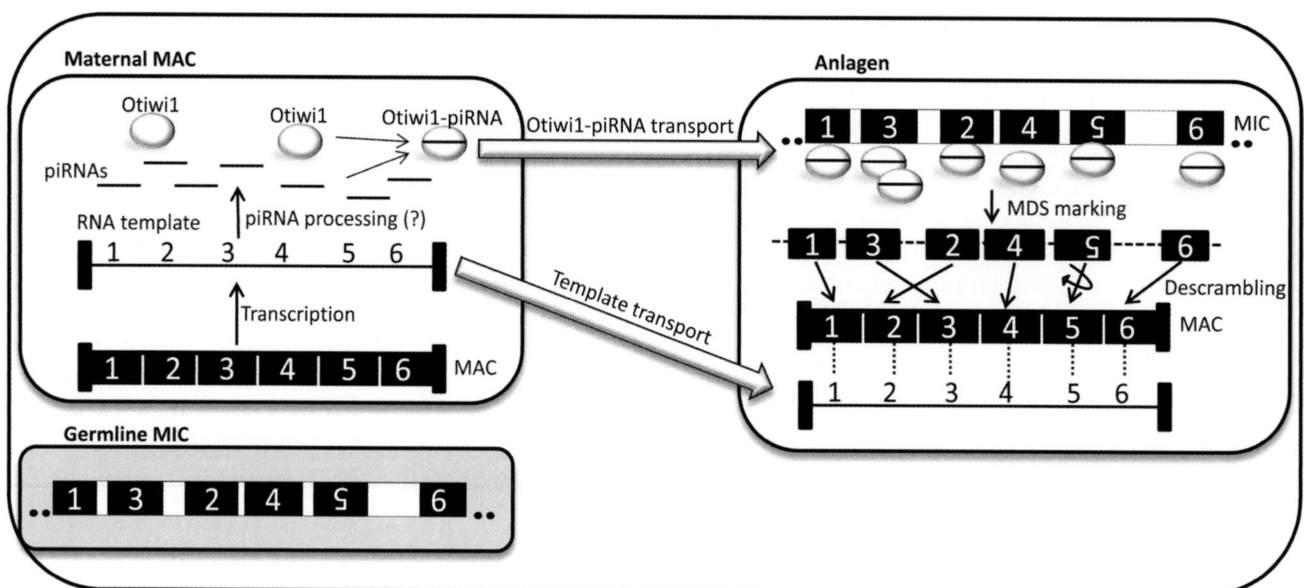

Figure 6 A model for noncoding RNA-mediated genome rearrangement in *Oxytricha*. Figure modified from Bracht et al. (77). The bidirectional transcription of the maternal MAC genome produces whole-chromosome transcripts which includes telomeres (long thin line flanked by black vertical rectangles). Either this maternal RNA template or other MAC transcripts are processed into 27-nucleotide piRNAs (short line segments) that interact with the *Oxytricha* Piwi protein, Otiwi1 (drawn as white ovals). Otiwi1 transports the piRNAs to the anlagen, in conjunction with the template RNAs. Here, piRNAs interact with the MIC precursor DNA to differentiate MDSs (black boxes) from IESs (white intervening rectangles). The template RNA then guides MDS descrambling, which, together with chromosome fragmentation and *de novo* telomere addition, forms mature MAC nanochromosomes. doi:10.1128/microbiolspec.MDNA3-0025-2014.f6

chromatin configuration, analogous to the heterochromatic regions discarded during rearrangement in *Stylonychia* and *Euplotes* (107, 108). Heterochromatin formation is a hallmark of RNAi silencing and studies in yeast indicate that heterochromatin may form as a result of histone H3 lysine-9 (H3K9) methylation directed by complementary small interfering RNAs (109, 110). Similarly, in *Tetrahymena*, H3K9 methylation occurs prior to DNA elimination and targets IESs for excision (111). Furthermore, scnRNAs are required for IESs to be marked by histone methylation, and DNA elimination is significantly reduced in the absence of such methylation (112, 113). This provides direct evidence for the role of histone methylation and heterochromatin formation in genome remodeling in *Tetrahymena*. However, *Oxytricha*'s orthogonally different piRNA pathway presents challenges because, instead, 1) piRNAs mark sequences for retention rather than deletion, and 2) both MDSs and IESs may be shorter than one nucleosomal unit, whereas the smallest IES in *Tetrahymena* is 194 base-pairs long (87). No stable epigenetic mark has yet been discovered for *Oxytricha* IESs so far. However, recent evidence suggests development-specific H3 variants may influence the fate of genome segments during *Stylonychia* genome rearrangement. The acetylated form of one variant specifically appears in association with *Stylonychia* MDSs, while another variant's expression and deposition depends on the piRNA pathway (114).

DNA methylation offers another epigenetic modification that can lead to heterochromatin formation (115, 116). Similar to H3K9 methylation in *Tetrahymena*, cytosine methylation may lead to heterochromatin formation and induce DNA excision in stichotrichous ciliates (117). In *Stylonychia*, *de novo* cytosine methylation occurs on transposon-like elements during differentiation of the anlagen (118). Therefore, DNA methylation may be required for the programmed removal of such MIC-limited segments from the anlagen. Similarly, very extensive cytosine methylation and hydroxymethylation has been observed in *Oxytricha* and may play a role in genome rearrangement by marking DNA for elimination and later degradation (84). This study identified cytosine methylation and hydroxymethylation in MIC-limited transposons and satellite repeats, as well as aberrant DNA rearrangement products and old MAC nanochromosomes that are destined for degradation during the later stages of nuclear development. Another candidate modification could be adenine methylation, which was reported to increase during *Tetrahymena* genome rearrangement (119, 120).

GENOME REMODELING MECHANISMS AND MACHINERY

Cis-Acting Sequence Features

In many ciliates, except *Tetrahymena* (86, 87), IESs frequently interrupt coding regions and hence must be precisely excised (121). *Paramecium* and *Euplotes crassus* IESs have highly conserved 5′-TA-3′ dinucleotide repeats demarcating their IESs, and this is part of a larger, conserved inverted repeat (41, 52, 122). Following excision, one copy of the TA pointer is retained in the new MDS junction (121), as for the more diverse sequence pointers in *Oxytricha*. While the presence of *cis*-acting sequences and retention of a TA pointer are conserved properties between the very distantly related ciliates *Paramecium* and *Euplotes*, their excision by-products containing the released IESs have drastically different structures. Longer *Paramecium* IESs (>200 base-pairs) circularize post excision (123), retaining a single TA precisely at the circle junction, whereas in *Euplotes* the observed circular by-products of both IES excision and deletion of transposon-like Tec elements retain both copies of the TA pointer, separated by a 10 base-pair heteroduplex region containing sequence flanking the IES (52, 124, 125). This supports the convergent evolutionary acquisition of two different mechanisms for deleting DNA sequences that map between a TA pointer. In the absence of any clear, conserved motifs flanking IESs in the stichotrichs *Oxytricha* and *Stylonychia*, the recruitment of longer and highly diverse direct repeats at MDS-MDS junctions may support the ability of this lineage to reorder complex scrambled genes, a phenomenon that appears absent from *Paramecium* and *Euplotes crassus*. Accordingly, the circularized excision by-products of many IESs in *Oxytricha* typically display imprecisely joined sequences at complex circular junctions (50) (also see below).

Intriguingly, the set of all two base-pair pointers in *Oxytricha* display a strong preference for the TA dinucleotide (39) similar to *Paramecium* and *Euplotes*, consistent with a similar but parallel evolutionary origin and common mechanistic constraints acting on all three of these independently acquired DNA splicing mechanisms (126). Furthermore, the set of all three base-pair pointers in *Oxytricha* is enriched in 5′-ANT-3′ (39), which is the characteristic target site duplication of the transposon-like telomere-bearing elements (TBEs) in *Oxytricha* (127). This suggests that IESs of this kind could be relics of ancient transposons from this family that mutated beyond sequence recognition, but retained simple features in the pointers that conform to the requirements or preferences of the machinery that catalyzes their excision (52, 128) (see "Transposases" below). Alternatively, pointers may have evolved to mimic the ends of transposons to facilitate their excision by transposases.

Both *Oxytricha* IESs and TBE transposons (127) appear often to be excised as DNA circles (129, 50). However, unlike the circular IES by-products in *Paramecium* and *Euplotes*, the *Oxytricha* IES circle junction sequences are more often highly imprecise, in contrast to the accurately spliced junctions between MDSs in mature macronuclear nanochromosomes (50). The circular junctions usually form near one of the endogenous pointers, but do not necessarily retain one complete copy of the pointer at the circle junction. Circularization often favors nearby cryptic pointers, which are direct repeats in the vicinity of at least one of the real pointers at the correct MDS borders (50). While aberrantly spliced MDS junctions at cryptic pointers (49) may be repaired by maternal RNA templates to yield precise MDS junctions in the mature macronuclear molecules, the by-products of IES excision would lack an RNA template for DNA repair, and this absent step can explain their observed sequence heterogeneity. Accordingly, the circularized IESs appear to be methylated, presumably as a tag for their destruction (50).

Transposases

The sequence similarity between the TA or ANT repeats flanking IESs and the preferred target sequences of ciliate transposons is compelling. The TA dinucleotide flanking *Paramecium* IESs, as well as flanking both IESs and Tec elements in *Euplotes crassus*, is strikingly similar to the termini of transposons in the Tc1/*mariner* superfamily (130, 131 and reviewed in 126 and 132). This similarity suggested a possible recruitment of transposases for DNA elimination during genome rearrangement in ciliates (128). Indeed, in *Tetrahymena* (*Tetrahymena piggyBac*-like transposase 2, TPB2) (133) and *Paramecium* (PiggyMac, Pgm) (134), domesticated *piggyBac*-like transposases participate in DNA elimination by inducing DNA double stranded breaks (*piggyBac* transposons reviewed in 135, 126). These *piggyBac* transposases with a DDD catalytic domain may localize to heterochromatin, possibly with the additional help of the *cis*-acting signals discussed earlier, to cleave DNA, leaving behind a four nucleotide overhang at the 5′ end (136).

In contrast to the domesticated *piggyBac* transposase genes in the macronuclear genomes of oligohymenophorean ciliates, *Oxytricha* bears thousands of active transposase genes within the Tc1/*mariner* superfamily TBE transposons in the MIC, and the encoded transposase proteins are implicated in the DNA elimination and rearrangement pathway during nuclear development (80). The *Oxytricha* macronuclear genome assembly does contain a modest number of somatic genes with transposase-like domains (38), including Phage_integrase, DDE_Tnp_IS1595 and MULE Pfam domains. Proteins containing the latter two domains offer additional candidates for a functional role in *Oxytricha*'s genome remodeling, because they are highly expressed in a development-specific manner, contain the DDE catalytic motif, and also lack the characteristic terminal inverted repeats—indicating possible domestication—similar to the *piggyBac* transposases (38). However, their expression is predominantly later than the TBE transposases, and this may implicate them in a different DNA elimination pathway. The possible participation of multiple catalysts in the excision machinery may also explain the lack of conserved motifs marking IESs for excision. It is possible that different transposases are responsible for excising different types of IESs, as part of the intricate orchestration of the process that eliminates most of the germline genome and then reorganizes the hundreds of thousands of pieces that remain.

Herrick et al.'s initial description of the MIC-limited TBE transposons (127) identified three closely related types which contain 17 base-pair terminal telomeric sequences (C4A4) flanked by 5′-ANT-3′ direct repeats. Full-length TBE transposons actually encode three proteins, including a 57 kDa zinc finger kinase and a 22 kDa protein (129, 137, 138, 139), both of unknown function, and notably a transposase with a DDE catalytic motif (137). The transposase proteins in the three transposon types, TBE1, 2, and 3, share >83% similarity at the amino acid level and appear to be under purifying selection (80), which further supported the hypothesis that they may play a biological role in genome differentiation (139). More recently, RNAi against all three groups of *Oxytricha* TBE transposase genes, exclusively expressed during rearrangement, leads to the aberrant or stalled elimination of not only the TBE transposon itself but also of IESs, as well as incorrect or incomplete MDS reordering (80). The knockdown experiments show accumulation of high-molecular-weight DNA, but significantly, the mutant phenotype is only observed after knockdown of all three TBE transposase types, indicating that their encoded transposase proteins may act redundantly in the DNA elimination process and the thousands of transposase genes may collectively encode a massive quantity of transposase protein needed for such a tremendous number of DNA elimination steps during development.

One pressing question regarding the evolution of this mutualistic relationship between ciliates and active transposons is why *Oxytricha* recruited an army of active, germline-limited transposases for DNA elimination, whereas in the distant clade of *Tetrahymena*/*Paramecium* the excision machinery was domesticated. This emphasizes the likelihood of independent origins, although domestication may have been a later evolutionary step after recruitment of active transposases. In the latter scenario, *Oxytricha* could represent a more primitive state prior to transposase domestication in other lineages (126).

Germline-Limited Genes

The TBE transposase genes are truly germline-limited, playing their roles early in development and then "expiring" in the soma but retained in the archival germline. With the exception of these transposase genes, there have been no other extensive studies of germline-limited genes—including the two other genes encoded in the TBE transposons, though both appear to be under purifying selection, suggesting that they also play a role (139, Schwartz and Landweber, unpublished). Until the discovery of active germline transposases, *Oxytricha*'s micronucleus had generally been considered transcriptionally silent. However, the recent sequencing of *Oxytricha*'s germline micronuclear genome revealed a surprising protein-coding content of hundreds of MIC-limited protein-coding genes, including additional genes encoding DDE_Tnp_IS1595 protein domains (39). All are expressed only during conjugation, according to RNA-sequence data (38), and proteomic surveys confirmed translation of some of these genes, so they are not just predicted gene products. Complete gene elimination may seem an unconventional, but is an absolute way to abrogate gene expression for proteins that are only required during early genome remodeling, analogous to genome diminution and the programmed loss of germline-specific genes from the soma of *Ascaris* and Lamprey (28, 29). Such extreme gene expression regulation would certainly prevent the ectopic expression of genes with germline-specific roles that could otherwise contribute to uncontrolled cell growth (140), and hence may be important for prevention of disease states or the maintenance of genome integrity. The MIC-limited gene products in *Oxytricha* may provide important proteins required for nuclear differentiation, bridging the period of growth between destruction of the old MAC and nascent gene expression from the developing MAC. Future studies of *Oxytricha*'s rich set of germline-limited genes have the potential to reveal fundamental aspects of *Oxytricha*'s life cycle and its division of labor between the two genomes, in addition to revealing more missing components of the excision apparatus that are no longer needed in vegetative cells and clandestinely, if not permanently, silenced by elimination, only to be resuscitated again by another round of sexual division.

CONCLUSION

Investigating genome remodeling in *Oxytricha* has uncovered many new and fundamental discoveries in RNA biology and explored the biological limits of DNA processing in a model eukaryote. Its extremely exaggerated genome plasticity and diversity of noncoding RNA pathways, plus unorthodox inheritance patterns owing to its sexual life cycle, together make *Oxytricha* a compelling model for studying all these phenomena that relate to genome integrity. Though many aspects of its biology seem, at first glance, to defy conventional genetics, it can also be viewed as an extended examination into the limits of genomic architecture and nucleic acid biodiversity on our planet.

Many unanswered questions remain, including the biogenesis of *Oxytricha* piRNAs and how they mark MDSs for retention. Furthermore, many of the components of the remodeling machinery have not yet been identified, including the source of proofreading from an RNA template. Similarly, the mechanism of gene expression regulation in *Oxytricha*, with such a tiny quantity of noncoding DNA in the macronucleus, is unknown. Future work should also examine the possible roles of other transposase-related genes in the macronucleus, as well as the two additional protein-coding genes (zinc finger kinase and 22 kDa unknown protein) present on the TBE transposons themselves. And finally, what are the roles of the MIC-limited protein-coding genes? The answers to these questions over the next several years will greatly improve our understanding of the genome dynamics of *Oxytricha* and relatives, near and far, including many eukaryotes that harbor more conventional twists on this creative RNA-regulated genome biology.

Acknowledgments. *We sincerely thank Mariusz Nowacki, Estienne C. Swart, and John R. Bracht for comments on the manuscript and valuable discussion. This work was supported by National Institute of General Medical Sciences Grant GM59708 to LFL.*

Citation. Yerlici VT, Landweber LF. 2014. Programmed genome rearrangements in the ciliate *Oxytricha*. Microbiol Spectrum **2**(6):MDNA3-0025-2014.

References

1. **Soma A, Onodera A, Sugahara J, Kanai A, Yachie N, Tomita M, Kawamura F, Sekine Y.** 2007. Permuted tRNA genes expressed via a circular RNA intermediate in *Cyanidioschyzon merolae*. *Science* **318:**450–453.
2. **Soma A, Sugahara J, Onodera A, Yachie N, Kanai A, Watanabe S, Yoshikawa H, Ohnuma M, Kuroiwa H, Kuroiwa T, Sekine Y.** 2013. Identification of highly-disrupted tRNA genes in nuclear genome of the red alga, *Cyanidioschyzon merolae* 10D. *Sci Rep* **3:**2321.
3. **Landweber LF.** 2007. Why genomes in pieces? *Science* **318:**405–407.
4. **Robertson HM, Navik JA, Walden KKO, Honegger HW.** 2007. The bursicon gene in mosquitoes: An unusual example of mRNA *trans*-splicing. *Genetics* **176:**1351–1353.
5. **Blumenthal T.** 2011. Split genes: Another surprise from *Giardia*. *Curr Biol* **21:**R162–163.
6. **Jackson CJ, Waller RF.** 2013. A widespread and unusual RNA *trans*-splicing type in dinoflagellate mitochondria. *PLoS One* **8:**e56777.
7. **Blumenthal T.** 2002. A global analysis of *Caenorhabditis elegans* operons. *Nature* **417:**851–854.
8. **Gopal S, Awadalla S, Gassterland T, Cross GAM.** 2005. A computational investigation of kinetoplastid *trans*-splicing. *Genome Biol* **6:**R951–R9511.
9. **Lasda EL, Blumenthal T.** 2011. *Trans*-splicing. *Wiley Interdiscip Rev RNA* **2:**417–34.
10. **Zhang C, Xie Y, Martignetti JA, Yeo TT, Massa SM, Longo FM.** 2003. A candidate chimeric mammalian mRNA transcript is derived from distinct chromosomes and is associated with nonconsensus splice junction motifs. *DNA Cell Biol* **22:**303–315.
11. **Moreira S, Breton S, Burger G.** 2012. Unscrambling genetic information at the RNA level. *Wiley Interdiscip Rev RNA* **3:**213–228.
12. **Li H, Wang J, Mor G, Sklar J.** 2008. A neoplastic gene fusion mimics *trans*-splicing of RNAs in normal human cells. *Science* **321:**1357–1361.
13. **Herai RH, Yamagishi MEB.** 2010. Detection of human interchromosomal *trans*-splicing in sequence databanks. *Brief Bioinform* **11:**198–209.
14. **Takahara T, Kanazu SI, Yanagisawa S, Akanuma H.** 2000. Heterogeneous Sp1 mRNAs in human HepG2 cells include a product of homotypic *trans*-splicing. *J Biol Chem* **275:**38067–38072.
15. **Takahara T, Kasahara D, Mori D, Yanagisawa S, Akanuma H.** 2002. The *trans*-spliced variants of Sp1 mRNA in rat. *Biochem Biophys Res Commun* **298:** 156–162.

16. **Frenkel-Morgenstern M, Lacroix V, Ezkurdia I, Levin Y, Gabashvili A, Prilusky J, del Pozo A, Tress M, Johnson R, Guigo R, Valencia A.** 2012. Chimeras taking shape: Potential functions of proteins encoded by chimeric RNA transcripts. *Genome Res* **22:**1231–1242.

17. **Gellert M.** 2002. V(D)J Recombination, p 705–729. *In* Craig NL, Craigie R, Gellert M, Lambowitz AM (ed), *Mobile DNA II*, 2nd ed. ASM Press, Washington, DC.

18. **Schatz DG, Ji Y.** 2011. Recombination centres and the orchestration of V(D)J recombination. *Nat Rev Immunol* **11:**251–263.

19. **Rooney S, Chaudhuri J, Alt FW.** 2004. The role of the non-homologous end-joining pathway in lymphocyte development. *Immunol Rev* **200:**115–131.

20. **Chaudhuri J, Alt FW.** 2004. Class-switch recombination: Interplay of transcription, DNA deamination and DNA repair. *Nat Rev Immunol* **4:**541–552.

21. **Mani RS, Chinnaiyan AM.** 2010. Triggers for genomic rearrangements: Insights into genomic, cellular and environmental influences. *Nat Rev Genet* **11:**819–829.

22. **Jones MJK, Jallepalli PV.** 2012. Chromothripsis: Chromosomes in crisis. *Dev Cell* **23:**908–917.

23. **Forment JV, Kaidi A, Jackson SP.** 2012. Chromothripsis and cancer: Causes and consequences of chromosome shattering. *Nat Rev Cancer* **12:**663–670.

24. **Zhang CZ, Leibowitz ML, Pellman D.** 2013. Chromothripsis and beyond: Rapid genome evolution from complex chromosomal rearrangements. *Genes Dev* **27:**2513–2330.

25. **Kloc M, Zagrodzinska B.** 2001. Chromatin elimination – an oddity or a common mechanism in differentiation and development? *Differentiation* **68:**84–91.

26. **Müller F, Tobler H.** 2000. Chromatin diminution in the parasitic nematodes *Ascaris suum* and *Parascaris univalens*. *Int J Parasitol* **30:**391–399.

27. **McKinnon C, Drouin G.** 2013. Chromatin diminution in the copepod *Mesocyclops edax*: Elimination of both highly repetitive and nonhighly repetitive DNA. *Genome* **56:**1–8.

28. **Smith JJ, Baker C, Eichler EE, Amemiya CT.** 2012. Genetic consequences of programmed genome rearrangement. *Curr Biol* **22:**1524–1529.

29. **Streit A.** 2012. Silencing by throwing away: A role for chromatin diminution. *Dev Cell* **23:**918–919.

30. **Goday C, Esteban MR.** 2001. Chromosome elimination in sciarid flies. *BioEssays* **23:**242–250.

31. **Mochizuki K, Fine NA, Fujisawa T, Gorovsky MA.** 2002. Analysis of a *piwi*-related gene implicates small RNAs in genome rearrangement in *Tetrahymena*. *Cell* **110:**689–699.

32. **Fang W, Wang X, Bracht JR, Nowacki M, Landweber LF.** 2012. Piwi-interacting RNAs protect DNA against loss during *Oxytricha* genome rearrangement. *Cell* **151:** 1243–1255.

33. **Wilson RC, Doudna JA.** 2013. Molecular mechanisms of RNA interference. *Annu Rev Biophys* **42:**217–239.

34. **Ruiz F, Vayssié L, Klotz C, Sperling L, Madeddu L.** 1998. Homology-dependent gene silencing in *Paramecium*. *Mol Biol Cell* **9:**931–943.

35. **Nowacki M, Landweber LF.** 2009. Epigenetic inheritance in ciliates. *Curr Opin in Microbiol* **12:**638–643.

36. **Prescott DM.** 1994. The DNA of ciliated protozoa. *Microbiol Rev* **58:**233–267.

37. **Aufderheide KJ, Frankel J, Williams NE.** 1980. Formation and positioning of surface-related structures in protozoa. *Microbiol Rev* **44:**252–302.

38. **Swart EC, Bracht JR, Magrini V, Minx P, Chen X, Zhou Y, Khurana JS, Goldman AD, Nowacki M, Schotanus K, Jung S, Fulton RS, Ly A, McGrath S, Haub K, Wiggins JL, Storton D, Matese JC, Parsons L, Chang WJ, Bowen MS, Stover NA, Jones TA, Eddy SR, Herrick GA, Doak TG, Wilson RK, Mardis ER, Landweber LF.** 2013. The *Oxytricha trifallax* macronuclear genome: A complex eukaryotic genome with 16,000 tiny chromosomes. *PLoS Biology*, **11:**e1001473.

39. **Chen X, Bracht JR, Goldman AD, Dolzhenko E, Clay DM, Swart EC, Perlman DH, Doak TG, Stuart A, Amemiya CT, Sebra RP, Landweber LF.** 2014. The architecture of a scrambled genome reveals massive levels of genomic rearrangement during development. *Cell* **158:**1187–1198.

40. **Coyne RS, Stover NA, Miao W.** 2012. Whole genome studies of *Tetrahymena*. *Methods Cell Biol* **109:**53.

41. **Jahn CL, Klobutcher LA.** 2002. Genome remodeling in ciliated protozoa. *Annu Rev Microbiol* **56:**489–520.

42. **Prescott DM.** 2000. Genome gymnastics: Unique modes of DNA evolution and processing in ciliates. *Nat Rev Genet* **1:**191–198.

43. **Spear BB, Lauth MR.** 1976. Polytene chromosomes of *Oxytricha*: Biochemical and morphological changes during macronuclear development in a ciliated protozoan. *Chromosoma* **54:**1–13.

44. **Ammermann D, Steinbrück G, von Berger L, Hennig W.** 1974. The development of the macronucleus in the ciliated protozoan *Stylonychia mytilus*. *Chromosoma* **45:**401– 429.

45. **Prescott DM.** 1999. The evolutionary scrambling and developmental unscrambling of germline genes in hypotrichous ciliates. *Nucleic Acids Res* **27:**1243–1250.

46. **Katz LA, Kovner AM.** 2010. Alternative processing of scrambled genes generates protein diversity in the ciliate *Chilodonella uncinata*. *J Exp Zool B Mol Dev Evol* **314:**480.

47. **Greslin AF, Prescott DM, Oka Y, Loukin SH, Chappell JC.** 1989. Reordering of nine exons is necessary to form a functional actin gene in *Oxytricha nova*. *Proc Natl Acad Sci USA* **86:**6264–6268.

48. **Prescott DM, Greslin AF.** 1992. Scrambled actin I gene in the micronucleus of *Oxytricha nova*. *Dev Genet* **13:** 66–74.

49. **Möllenbeck MY, Zhou Y, Cavalcanti ARO, Jönsson F, Higgins BP, Chang WJ, Juranek S, Doak TG, Rozenberg G, Lipps HJ, Landweber LF.** 2008. The pathway to detangle a scrambled gene. *PLoS One* **3:**e2330.

50. **Bracht JR, Higgins BP, Wang K, Angeleska A, Dolzhenko E, Fang W, Chen X, Landweber LF.** Oxytricha*: A model of genome catastrophe and recovery*. Submitted.

51. **Yao MC, Yao CH, Monks B.** 1990. The controlling sequence for site-specific chromosome breakage in *Tetrahymena*. *Cell* **63:**763–772.

52. **Yao MC, Duharcourt S, Chalker DL.** 2002. Genome-wide rearrangement of DNA in ciliates, p 730–760. *In* Craig NL, Craigie R, Gellert M, Lambowitz AM (ed), *Mobile DNA II*, 2nd ed. ASM Press, Washington, DC.

53. **Baird SE, Klobutcher LA.** 1989. Characterization of chromosome fragmentation in two protozoans and identification of a candidate fragmentation sequence in *Euplotes crassus*. *Genes Dev* **3:**585–597.

54. **Klobutcher LA, Gygax SE, Podoloff JD, Vermeesch JR, Price CM, Tebeau CM, Jahn CL.** 1998. Conserved DNA sequences adjacent to chromosome fragmentation and telomere addition sites in *Euplotes crassus*. *Nucleic Acids Res* **26:**4230–4240.

55. **Klobutcher LA.** 1999. Characterization of *in vivo* developmental chromosome fragmentation intermediates in *E. crassus*. *Mol Cell* **4:**695–704.

56. **Greider CW, Blackburn EH.** 1985. Identification of a specific telomere terminal transferase activity in *Tetrahymena* extracts. *Cell* **43:**405–413.

57. **Herrick G, Hunter D, Williams K, Kotter K.** 1987. Alternative processing during development of a macronuclear chromosome family in *Oxytricha fallax*. *Genes Dev* **1:**1047–1058.

58. **Williams KR, Doak TG, Herrick G.** 2002. Telomere formation on macronuclear chromosomes of *Oxytricha trifallax* and *O. fallax*: alternatively processed regions have multiple telomere addition sites. *BMC Genet* **3:**16.

59. **Jessop-Murray H, Martin LD, Gilley D, Preer-Jr JR, Polisky B.** 1991. Permanent rescue of a non-Mendelian mutation of paramecium by microinjection of specific DNA sequences. *Genetics* **129:**727–734.

60. **Harumoto T.** 1986. Induced change in a non-Mendelian determinant by transplantation of macronucleoplasm in *Paramecium tetraurelia*. *Mol Cell Biol* **6:**3498–3501.

61. **Koizumi S, Kobayashi S.** 1989. Microinjection of plasmid DNA encoding the A surface antigen of *Paramecium tetraurelia* restores the ability to regenerate a wild-type macronucleus. *Mol Cell Biol* **9:**4398–4401.

62. **You Y, Aufderheide K, Morand J, Rodkey K, Forney J.** 1991. Macronuclear transformation with specific DNA fragments controls the content of the new macronuclear genome in *Paramecium tetraurelia*. *Mol Cell Biol* **11:** 1133–1137.

63. **Kim CS, Preer-Jr JR, Polisky B.** 1994. Identification of DNA segments capable of rescuing a non-Mendelian mutant in *Paramecium*. *Genetics* **136:**1325–1328.

64. **You Y, Scott J, Forney J.** 1994. The role of macronuclear DNA sequences in the permanent rescue of a non-Mendelian mutation in *Paramecium tetraurelia*. *Genetics* **136:**1319–1324.

65. **Duharcourt S, Butler A, Meyer E.** 1995. Epigenetic self-regulation of developmental excision of an internal eliminated sequence on *Paramecium tetraurelia*. *Genes Dev* **9:**2065–2077.

66. **Chalker DL, Yao MC.** 1996. Non-Mendelian, heritable blocks to DNA rearrangement are induced by loading the somatic nucleus of *Tetrahymena thermophila* with germ line-limited DNA. *Mol Cell Biol* **16:**3658–3667.

67. **Mochizuki K.** 2010. DNA rearrangements directed by noncoding RNAs in ciliates. *Wiley Interdiscip Rev RNA* **1:**376–387.

68. **Schoeberl UE, Mochizuki K.** 2011. Keeping the soma free of transposons: Programmed DNA elimination in ciliates. *J Biol Chem* **286:**37045–37052.

69. **Chalker DL, Yao MC.** 2001. Nongenic, bidirectional transcription precedes and may promote developmental DNA deletion in *Tetrahymena thermophila*. *Genes Dev* **15:**1287–1298.

70. **Lepère G, Bétermier M, Meyer E, Duharcourt S.** 2008. Maternal noncoding transcripts antagonize the targeting of DNA elimination by scanRNAs in *Paramecium tetraurelia*. *Genes Dev* **22:**1501–1512.

71. **Lepère G, Nowacki M, Serrano V, Gout JF, Guglielmi G, Duharcourt S, Meyer E.** 2009. Silencing-associated and meiosis-specific small RNA pathways in *Paramecium tetraurelia*. *Nucleic Acids Res* **37:**903–915.

72. **Bouchouche K, Gout JF, Kapusta A, Bétermier M, Meyer E.** 2011. Functional specialization of Piwi proteins in *Paramecium tetraurelia* from post-transcriptional gene silencing to genome remodelling. *Nucleic Acids Res* **39:**4249–4264.

73. **Sandoval PY, Swart EC, Arambasic M, Nowacki M.** 2014. Functional diversification of dicer-like proteins and small RNAs required for genome sculpting. *Dev Cell* **28:**174–88.

74. **Schier AF.** 2007. The maternal-zygotic transition: Death and birth of RNAs. *Science* **316:**406–407.

75. **Zahler AM, Neeb ZT, Lin A, Katzman S.** 2012. Mating of the stichotrichous ciliate *Oxytricha trifallax* induces production of a class of 27 nt small RNAs derived from the parental macronucleus. *PLoS One* **7:**e42371.

76. **Schoeberl UE, Kurth HM, Noto T, Mochizuki K.** 2012. Biased transcription and selective degradation of small RNAs shape the pattern of DNA elimination in *Tetrahymena*. *Genes Dev* **26:**1729–1742.

77. **Bracht JR, Fang W, Goldman AD, Dolzhenko E, Stein EM, Landweber LF.** 2013. Genomes on the edge: Programmed genome instability in ciliates. *Cell* **152:** 406–416.

78. **Thomson T, Lin H.** 2009. The biogenesis and function of PIWI proteins and piRNAs: Progress and prospect. *Annu Rev Cell Dev Biol* **25:**355–376.

79. **Ross RJ, Weiner MM, Lin H.** 2014. PIWI proteins and PIWI-interacting RNAs in the soma. *Nature* **505:** 353–259.

80. **Nowacki M, Higgins BP, Maquilan GM, Swart EC, Doak TG, Landweber LF.** 2009. A functional role for transposases in a large eukaryotic genome. *Science* **324:** 935–938.

81. **Ashe A, Sapetschnig A, Weick EM, Mitchell J, Bagijn MP, Cording AC, Doebley AL, Goldstein LD, Lehrbach NJ, Le Pen J, Pintacuda G, Sakaguchi A, Sarkies P, Ahmed S, Miska EA.** 2012. piRNAs can trigger a multigenerational epigenetic memory in the germline of *C. elegans*. *Cell* **150:**88–99.

82. **Shirayama M, Seth M, Lee HC, Gu W, Ishidate T, Conte D Jr., Mello CC.** 2012. piRNAs initiate an epigenetic memory of nonself RNA in the *C. elegans* germline. *Cell* **150:**65.

83. **Parfrey LW, Lahr DJG, Knoll AH, Katz LA.** 2011. Estimating the timing of early eukaryotic diversification with multigene molecular clocks. *Proc Natl Acad Sci USA* **108:**13624–13629.

84. **Bracht JR, Perlman DH, Landweber LF.** 2012. Cytosine methylation and hydroxymethylation mark DNA for elimination in *Oxytricha trifallax*. *Genome Biol (Online Edition)* **13:**R99.

85. **Nowacki M, Vijayan V, Zhou Y, Schotanus K, Doak TG, Landweber LF.** 2008. RNA-mediated epigenetic programming of a genome-rearrangement pathway. *Nature* **451:**153–158.

86. **Austerberry CF, Yao MC.** 1987. Nucleotide sequence structure and consistency of a developmentally regulated DNA deletion in *Tetrahymena thermophila*. *Mol Cell Biol* **7:**435– 443.

87. **Fass JN, Joshi NA, Couvillion MT, Bowen J, Gorovsky MA, Hamilton EP, Orias E, Hong K, Coyne RS, Eisen JA, Chalker DL, Lin D, Collins K.** 2011. Genome-scale analysis of programmed DNA elimination sites in *Tetrahymena thermophila*. *G3* **1:**515–522.

88. **Juranek SA, Rupprecht S, Postberg J, Lipps HJ.** 2005. snRNA and heterochromatin formation are involved in DNA excision during macronuclear development in stichotrichous ciliates. *Eukaryot Cell* **4:**1934–1941.

89. **Prescott DM, Ehrenfeucht A, Rozenberg G.** 2003. Template-guided recombination for IES elimination and unscrambling of genes in stichotrichous ciliates. *J Theor Biol* **222:**323–330.

90. **Angeleska A, Jonoska N, Saito M, Landweber LF.** 2007. RNA-guided DNA assembly. *J Theor Biol* **248:**706–720.

91. **Korbel JO, Kim PM, Chen X, Urban AE, Weissman S, Snyder M, Gerstein MB.** 2008. The current excitement about copy-number variation: How it relates to gene duplications and protein families. *Curr Opin Struct Biol* **18:**366–374.

92. **Hyman E, Kauraniemi P, Hautaniemi S, Wolf M, Mousses S, Rozenblum E, Ringnér M, Sauter G, Monni O, Elkahloun A, Kallioniemi OP, Kallioniemi A.** 2002. Impact of DNA amplification on gene expression patterns in breast cancer. *Cancer Res* **62:**6240–6245.

93. **Pollack JR, Sørlie T, Perou CM, Rees CA, Jeffrey SS, Lonning PE, Tibshirani R, Botstein D, Børresen-Dale AL, Brown PO.** 2002. Microarray analysis reveals a major direct role of DNA copy number alteration in the transcriptional program of human breast tumors. *Proc Natl Acad Sci USA* **99:**12963–12968.

94. **Sebat J, Lakshmi B, Malhotra D, Troge J, Lese-Martin C, Walsh T, Yamrom B, Yoon S, Krasnitz A, Kendall J, Leotta A, Pai D, Zhang R, Lee YH, Hicks J, Spence SJ, Lee AT, Puura K, Lehtimäki T, Ledbetter D, Gregersen PK, Bregman J, Sutcliffe JS, Jobanputra V, Chung W, Warburton D, King MC, Skuse D, Geschwind DH, Gilliam TC, Ye K, Wigler M.** 2007. Strong Association of *de novo* copy number mutations with autism. *Science* **316:**445–449.

95. **FitzPatrick DR.** 2005. Transcriptional consequences of autosomal trisomy: Primary gene dosage with complex downstream effects. *Trends Genet* **21:**249–253.

96. **Spradling AC.** 1981. The organization and amplification of two chromosomal domains containing *Drosophila* chorion genes. *Cell* **27:**193–201.

97. **Hourcade D, Dressler D, Wolfson J.** 1973. The amplification of ribosomal RNA genes involves a rolling circle intermediate. *Proc Natl Acad Sci USA* **70:** 2926–2930.

98. **Yao MC, Blackburn E, Gall JG.** 1979. Amplification of the rRNA genes in *Tetrahymena*. *ColdSpring Harbor Symposia on Quantitative Biology* **43:**1293–1296.

99. **Ward JG, Blomberg P, Hoffman N, Yao MC.** 1997. The intranuclear organization of normal, hemizygous and excision-deficient rRNA genes during developmental amplification in *Tetrahymena thermophila*. *Chromosoma* **106:**233–242.

100. **Nowacki M, Haye JE, Fang W, Vijayan V, Landweber LF.** 2010. RNA-mediated epigenetic regulation of DNA copy number. *Proc Natl Acad Sci USA* **107:**22140–22144.

101. **Heyse G, Jönsson F, Chang WJ, Lipps HJ.** 2010. RNA-dependent control of gene amplification. *Proc Natl Acad Sci USA* **107:**22134–22139.

102. **Storici F, Bebenek K, Kunkel TA, Gordenin DA, Resnick MA.** 2007. RNA-templated DNA repair. *Nature* **447:**338–341.

103. **Tomlins SA, Rhodes DR, Perner S, Dhanasekaran SM, Mehra R, Sun XW, Varambally S, Cao X, Tchinda J, Kuefer R, Lee C, Montie JE, Shah RB, Pienta KJ, Rubin MA, Chinnaiyan AM.** 2005. Recurrent fusion of *TMPRSS2* and ETS transcription factor genes in prostate cancer. *Science* **310:**644–648.

104. **Mani RS, Chinnaiyan AM.** 2010. Triggers for genomic rearrangements: Insights into genomic, cellular and environmental influences. *Nat Rev Genet* **11:**819–829.

105. **Li H, Wang J, Mor G, Sklar J.** 2008. A neoplastic gene fusion mimics *trans*-splicing of RNAs in normal human cells. *Science* **321:**1357–1361.

106. **Fang W, Wei Y, Kang Y, Landweber LF.** 2012. Detection of a common chimeric transcript between human chromosomes 7 and 16. *Biol Direct* **7:**49.

107. **Meyer GF, Lipps HJ.** 1980. Chromatin elimination in the hypotrichous ciliate *Stylonychia mytilus*. *Chromosoma* **77:**285–297.

108. **Jahn CL.** 1999. Differentiation of chromatin during DNA elimination in *Euplotes crassus*. *Mol Biol Cell* **10:** 4217–4230.

109. **Volpe TA, Kidner C, Hall IM, Teng G, Grewal SIS, Martienssen RA.** 2002. Regulation of heterochromatic silencing and histone H3 lysine-9 methylation by RNAi. *Science* **297:**1833–1837.

110. **Verdel A, Jia S, Gerber S, Sugiyama T, Gygi S, Grewal SIS, Moazed D.** 2004. RNAi-mediated targeting of heterochromatin by the RITS complex. *Science* **303:**672–676.

111. **Taverna SD, Coyne RS, Allis CD.** 2002. Methylation of histone H3 at lysine 9 targets programmed DNA elimination in *Tetrahymena*. *Cell* **110:**701–711.

112. **Liu Y, Mochizuki K, Gorovsky MA.** 2004. Histone H3 lysine 9 methylation is required for DNA elimination in developing macronuclei in *Tetrahymena*. *Proc Natl Acad Sci USA* **101:**1679–1684.

113. **Liu Y, Taverna SD, Muratore TL, Shabanowitz J, Hunt DF, Allis CD.** 2007. RNAi-dependent H3K27 methylation is required for heterochromatin formation and DNA elimination in *Tetrahymena*. *Genes Dev* **21:**1530–1545.

114. **Forcob S, Bulic A, Jönsson F, Lipps HJ, Postberg J.** 2014. Differential expression of histone H3 genes and selective association of the variant H3.7 with a specific sequence class in *Stylonychia* macronuclear development. *Epigenetics Chromatin* **7:**4.

115. **Law JA, Jacobsen SE.** 2010. Establishing, maintaining and modifying DNA methylation patterns in plants and animals. *Nat Rev Genet* **11:**204–220.

116. **Onodera Y, Haag JR, Ream T, Nunes PC, Pontes O, Pikaard CS.** 2005. Plant nuclear RNA polymerase IV mediates siRNA and DNA methylation-dependent heterochromatin formation. *Cell* **120:**613–622.

117. **Bracht JR.** 2014. Beyond transcriptional silencing: Is methylcytosine a widely conserved eukaryotic DNA elimination mechanism? *BioEssays* **36:**346–352.

118. **Juranek S, Wieden HJ, Lipps HJ.** 2003. *De novo* cytosine methylation in the differentiating macronucleus of the stichotrichous ciliate *Stylonychia lemnae*. *Nucleic Acids Res* **31:**1387–1391.

119. **Blackburn EH, Pan WC, Johnson CC.** 1983. Methylation of ribosomal RNA genes in the macronucleus of *Tetrahymena thermophila*. *Nucleic Acids Res* **11:** 5131–5145.

120. **Hattman S.** 2005. DNA-[adenine] methylation in lower eukaryotes. *Biochemistry* **70:**550–558.

121. **Tausta SL, Turner LR, Buckley LK, Klobutcher LA.** 1991. High fidelity developmental excision of Tec1 transposons and internal eliminated sequences in *Euplotes crassus*. *Nucleic Acids Res* **19:**3229–3236.

122. **Arnaiz O, Mathy N, Baudry C, Malinsky S, Aury JM, Wilkes CD, Garnier O, Labadie K, Lauderdale BE, Le Mouël A, Marmignon A, Nowacki M, Poulain J, Prajer M, Wincker P, Meyer E, Duharcourt S, Duret L, Bétermier M, Sperling L.** 2012. The *Paramecium* germline genome provides a niche for intragenic parasitic DNA: Evolutionary dynamics of internal eliminated sequences. *PLoS Genetics* **8:**e1002984.

123. **Bétermier M, Duharcourt S, Seitz H, Meyer E.** 2000. Timing of developmentally programmed excision and circularization of *Paramecium* internal eliminated sequences. *Mol Cell Biol* **20:**1553–1561.

124. **Tausta SL, Klobutcher LA.** 1989. Detection of circular forms of eliminated DNA during macronuclear development in *E. crassus*. *Cell* **59:**1019–1026.

125. **Jaraczewski JW, Jahn CL.** 1993. Elimination of Tec elements involves a novel excision process. *Genes Dev* **7:**95–105.

126. **Vogt A, Goldman AD, Mochizuki K, Landweber LF.** 2013. Transposon domestication versus mutualism in ciliate genome rearrangements. *PLoS Genetics* **9:**e1003659.

127. **Herrick G, Cartinhour S, Dawson D, Ang D, Sheets R, Lee A, Williams K.** 1985. Mobile elements bounded by C4A4 telomeric repeats in *Oxytricha fallax*. *Cell* **43:** 759–768.

128. **Klobutcher LA, Herrick G.** 1997. Developmental genome reorganization in ciliated protozoa: the transposon link. *Prog Nucleic Acid Res Mol Biol* **56:**1–62.

129. **Williams K, Doak TG, Herrick G.** 1993. Developmental precise excision of *Oxytricha trifallax* telomere-bearing elements and formation of circles closed by a copy of the flanking target duplication. *EMBO J* **12:**4593–4601.

130. **Jacobs ME, Klobutcher LA.** 1996. The long and the short of developmental DNA deletion in *Euplotes crassus*. *J Eukaryot Microbiol* **43:**442–452.

131. **Klobutcher LA, Herrick G.** 1995. Consensus inverted terminal repeat sequence of *Paramecium* IESs: resemblance to termini of Tc1-related and *Euplotes* Tec transposons. *Nucleic Acids Res* **23:**2006–2013.

132. **Plasterk RHA, van Luenen HGAM.** 2002. The Tc1/*mariner* family of transposable elements, p 519–532. *In* Craig NL, Craigie R, Gellert M, Lambowitz AM (ed), *Mobile DNA II*, 2nd ed. ASM Press, Washington, DC.

133. **Cheng CY, Vogt A, Mochizuki K, Yao MC.** 2010. A domesticated *piggyBac* transposase plays key roles in heterochromatin dynamics and DNA cleavage during programmed DNA deletion in *Tetrahymena thermophila*. *Mol Biol Cell* **21:**1753–1762.

134. **Baudry C, Malinsky S, Restituito M, Kapusta A, Rosa S, Meyer E, Bétermier M.** 2009. *PiggyMac*, a domesticated *piggyBac* transposase involved in programmed genome rearrangements in the ciliate *Paramecium tetraurelia*. *Genes Dev* **23:**2478–2483.

135. **Robertson HM.** 2002. Evolution of DNA transposons in eukaryotes, p 1093–1110. *In* Craig NL, Craigie R, Gellert M, Lambowitz AM (ed), *Mobile DNA II*, 2nd ed. ASM Press, Washington, DC.

136. **Mitra R, Fain-Thornton J, Craig NL.** 2008. *PiggyBac* can bypass DNA synthesis during cut and paste transposition. *EMBO J* **27:**1097–1109.

137. **Doak TG, Doerder FP, Jahn CL, Herrick G.** 1994. A proposed superfamily of transposase genes: Transposon-like elements in ciliated protozoa and a common "D35E" motif. *Proc Natl Acad Sci USA* **91:**942–946.

138. **Doak TG, Witherspoon DJ, Doerder FP, Williams K, Herrick G.** 1997. Conserved features of TBE1 transposons in ciliated protozoa. *Genetica* **101:**75–86.

139. **Witherspoon DJ, Doak TG, Williams KR, Seegmiller A, Seger J, Herrick G.** 1997. Selection on the protein-coding genes of the TBE1 family of transposable elements in the ciliates *Oxytricha fallax* and *O. trifallax*. *Mol Biol Evol* **14:**696–706.

140. **Janic A, Mendizabal L, Llamazares S, Rossell D, Gonzalez C.** 2010. Ectopic expression of germline genes drives malignant brain tumor growth in *Drosophila*. *Science* **330:**1824–1827.

141. **Chang WJ, Bryson PD, Liang H, Shin MK, Landweber LF.** 2005. The evolutionary origin of a complex scrambled gene. *Proc Natl Acad Sci USA* **102:**15149–1554.

Mobile DNA, 3rd Edition
Nancy L. Craig, Michael Chandler, Martin Gellert, Alan M. Lambowitz, Phoebe A. Rice and Suzanne Sandmeyer

doi:10.1128/microbiolspec.MDNA3-0016-2014

Richard McCulloch[1]
Liam J. Morrison[1,2]
James P.J. Hall[1,3]

19 DNA Recombination Strategies During Antigenic Variation in the African Trypanosome

WHAT IS ANTIGENIC VARIATION?

One of the most powerful drivers of evolutionary change is the process of adaptation and counter-adaptation by interacting species (1). The so-called "arms race" between parasites and their hosts is a prime example of such reciprocal coevolution: host adaptations that reduce or attempt to remove parasites select for parasite adaptations that enable evasion of host defences. Elaborate, powerful and sometimes elegant mechanisms of host immunity and parasite infectivity are thought to have arisen from many iterations of this process. A case in point is the mammalian adaptive immune system, perhaps one of the more complex host defence mechanisms detailed to date, which uses directed DNA rearrangements, mutagenesis and selection during the development of T and B immune cells to generate vast numbers of genes encoding immunoglobulin receptors capable of recognizing the huge range of antigens in infecting pathogens (2). Parasites, on the other hand, have evolved various means of evading adaptive immunity. One such mechanism of immune evasion that is widely recorded among viruses and bacterial and eukaryotic pathogens is antigenic variation. Because parasite killing often depends on a match between circulating host immunity and parasite antigen, individual parasites that no longer express that antigen variant, but instead express an antigenically different variant in its place, survive and can proliferate. However, this advantage tends to be short-lived because immune responses will develop against the different antigen in turn. Hence, members of parasite lineages inhabiting an immunocompetent host are repeatedly being selected for antigenic novelty over the course of infection.

In contrast to depending on general processes of mutation to generate this novelty, antigenic variation *sensu stricto* is applied to cases where it is believed that pathogens have developed mechanisms to facilitate evolvability; that is, selection has acted on a "higher order" phenotype, and features and mechanisms have evolved to more effectively generate and express adaptive variation to evade adaptive immunity. Evolvability is likely

[1]Wellcome Trust Centre for Molecular Parasitology, Institute of Infection, Immunity and Inflammation, University of Glasgow, Glasgow, UK; [2]Roslin Institute, University of Edinburgh, UK; [3]Department of Biology, University of York, York, UK.

to be an adaptation in only a quite restricted set of circumstances (3), but it is increasingly clear that the generally high degree of relatedness between co-infecting pathogens and the intense selective pressures imposed by host immunity can favour evolvability in antigenically variant pathogens (4). In shaping an antigenic variation phenotype, selection is likely to act on individual pathogens by favouring individuals that restrict the number of different variants they express (5), and on a lineage of clonally related pathogens by favouring those lineages that express a range of antigenically distinct variants (6). Because adaptive immunity is a selective force interacting with a range of organisms, common features of antigenic variation have convergently evolved in many phylogenetically disparate pathogens. These include (7, 8): the possession of a "family" of silent antigen genes, either explicitly encoded as variant genes or implicitly encoded by pseudogenes or gene fragments; monoallelic expression, which is that a single antigen gene is expressed by an individual at a time; and genetic or epigenetic strategies to "switch" exclusive antigen gene expression to another, previously silent member of the antigen gene family. Beyond these features, we can add that antigenic variation typically occurs at high rates, above background mutation, and that the process is stochastic and preemptive of immune recognition of the expressed antigen. These latter features are shared with other processes of phenotypic change (8–10), such as phase variation, but may not be inviolate in antigenic variation. For instance, though antigenic variation in African trypanosomes occurs at rates up to 10^{-2} events/division it can be reversibly reduced 1,000-fold by prolonged growth (11); in bacterial spirochetes of the genus *Borrelia* antigenic variation occurs when the pathogens infect mammals, but is not observed when grown *in vitro* (12). Irrespective of these variations, when interacting with the selective forces of the immune system, antigenic variation manifests as a pathogen population continually switching "forwards" through different antigen variants, as individuals expressing "old" variants (either by not switching, or by "switching back") are neutralized by immunity.

ANTIGENIC VARIATION IN TRYPANOSOMES

'African' Salivarian trypanosomes of the *Trypanosoma brucei* clade represent one of the best-studied systems of antigenic variation. *Trypanosoma brucei* are kinetoplastid (protozoan flagellate) parasites that are usually transmitted between mammal hosts by insect vectors (tsetse flies, genus *Glossina*) and cause disease in humans (human African trypanosomiasis, or sleeping sickness; caused by *T. b. gambiense* and *T. b. rhodesiense*) and animals (animal African trypanosomiasis, or Nagana; caused by all *T. brucei* trypanosomes as well as by *Trypanosoma congolense* and *Trypanosoma vivax*) in numerous foci across sub-Saharan Africa (13). Besides *T. brucei*, several other "species" of Salivarian trypanosomes have been studied in the context of antigenic variation. *T. b. evansi* and *T. b. equiperdum*, both non-human infective derivatives of *T. brucei* adapted to mechanical and sexual transmission (14), respectively, possess many if not all of the features of *T. brucei* antigenic variation. *Trypanosoma congolense* and *T. vivax* are more distantly related to *T. brucei* and are known to undergo antigenic variation but the details, particularly at the molecular and mechanistic levels, are less well understood in these species (15, 16).

The key biochemical and genetic features of trypanosome antigenic variation have been known for around 40 years. Pioneering studies by K. Vickerman demonstrated that trypanosomes survive in mammals through changes in a dense surface "coat" (17, 18), which was revealed in the mid-1970s to be composed of variant surface glycoprotein (VSG) (19). Rapidly thereafter, the genes encoding the VSG were identified and the molecular basis of *VSG* gene switching was established (20). Antigenic variation in *T. brucei* was discussed by P. Borst in the previous version of this book, *Mobile DNA II* (7), and therefore we will attempt in this chapter to update that report by reviewing the findings that have been published since then. For more detailed discussion of specific aspects of *T. brucei* antigenic variation, the reader is referred to several recent reviews (21–27).

Transmission by hematophagous biting insects tends to select for parasite mechanisms that prolong presence in the blood, because it is those parasites that are most likely to be transmitted (28). Salivarian trypanosomes, unlike other chronic pathogens such as *Plasmodium* or the American trypanosome *T. cruzi*, do not invade host cells, so their residence in the blood leaves them exposed for the entirety of their existence in the mammalian host, were it not for ~5×10^6 VSG dimers covering their entire surface. The VSG coat acts as a physical barrier, blocking immune effectors from other elements of the parasite surface (29). Individual VSG dimers can move across the parasite surface, and the coat is constantly turning over through endocytosis and exocytosis occurring in the flagellar pocket at the posterior of the parasite (30) and a limited degree of shedding (31). For a long time it was believed that the thickness and depth of the VSG coat was key to physically shielding all other surface molecules, as well as the parasite surface itself, from immune effectors. It now appears that this is not the whole story because some invariant surface molecules are predicted to protrude from the VSG layer (32), and the coat of *T. vivax*, which appears considerably less dense than that of *T. brucei* (33), is able to resist innate immune effectors. It is currently unclear whether these findings will change the paradigm of VSG's putative sole function as a physical barrier—at any rate, all attempts to use vaccines to direct immune responses against alternative and perhaps more productive targets have so far not been promising (34).

Each trypanosome cell expresses a single *VSG*, and it is against the encoded VSG that host immunity mounts a most vigorous attack, primarily an antibody response capable of destroying parasites by activating complement and by recruiting phagocytes (35). In fact, VSG appears to function solely as an antigen and does not perform any important biochemical activity, meaning that it is remarkably free to vary in sequence, with different functional VSGs sharing as little as 20% primary sequence identity (36). Still, there are clearly some limitations to VSG structure because all *VSG*s identified to date share strong features of secondary structure (37), in particular long α helices, which presumably stand perpendicular to the plasma membrane and contribute to the intact coat's depth and density (26, 38).

African trypanosome antigenic variation comes about when individual trypanosomes "switch" VSG, resulting in diversity emerging in the expanding lineage; a contingency that preempts the immune responses being mounted by the host. Below, we will discuss the molecular events that underpin switching, and how these contribute to antigenic variation and the survival strategies of African trypanosomes.

THE GENOMIC COMPONENTS OF ANTIGENIC VARIATION IN *TRYPANOSOMA BRUCEI*

Antigenic variation in *T. brucei* is remarkable for the huge levels of elaboration in the genome resources devoted to the process, and potentially therefore in the machinery that acts upon these resources to execute the reactions involved. In contrast with other organisms, where antigenic variation among surface antigen genes appears normally to be executed either by transcriptional or recombination mechanisms, *T. brucei* employs both strategies. Moreover, the sites of *VSG* transcription have evolved to drive the expression of not merely the VSG antigen but also to co-express many other proteins, and the number of silent *VSG* genes that act as "donors" for generating and expressing new VSG variants by recombination is of an unparalleled size. As the available evidence suggests that transcription-based and recombination-based VSG switching reactions are mechanistically distinct, these are considered in turn below.

VSG genes are expressed in the mammal from multigenic, telomeric transcription sites termed bloodstream expression sites (BES), as shown in Fig. 1. Multigenic transcription is not unusual in *T. brucei* or in related kinetoplastids, as virtually all RNA polymerase (Pol) II transcribed protein-coding genes are expressed in this way, with extensive *trans*-splicing and coupled polyadenylation acting to generate mature mRNAs from precursor RNAs (39, 40). However, the promoter that drives *VSG* transcription is unusual in that it is recognized by RNA Pol I (41), which also conventionally transcribes *rRNA* genes in the genome. Despite this, the BES promoter displays limited sequence homology with the *rRNA* promoter (42) or, indeed, with the promoter of *procyclin* genes, which are also RNA Pol I transcribed (41) and encode surface coat proteins found in the tsetse fly. In fact, though each promoter appears to bind similar, somewhat diverged, RNA Pol I transcription factors (43), the subnuclear sites of *rRNA* and BES transcription are distinct: *T. brucei* possess a conventional nucleolus for *rRNA* transcription, but some RNA Pol I is recruited to the active BES in a separate, discrete nuclear location termed the expression site body (44, 45). Whether or not BES promoter sequence specificity contributes to expression site body recruitment, or whether it reflects further adaptations for antigenic variation remains unknown.

Changes in the identity of the single transcribed BES provides one route for antigenic variation because *T. brucei* possesses not one BES, but many, with different *VSG*s occupying each BES. Whether the number of BES in the genome is fixed remains somewhat uncertain, as does the extent to which they vary in sequence composition during evolution. This is because the BES are telomeric and frequently large (many 10s of kbp), meaning that they are underrepresented in the bacterial clones used in published whole-genome *T. brucei* sequencing strategies (46, 47) and are frequently not found as assembled entities. For this reason, Hertz-Fowler et al. (48) specifically targeted the BES for cloning and sequencing by a transformation-associated recombination approach, and this remains the most complete survey of BES number and content to date.

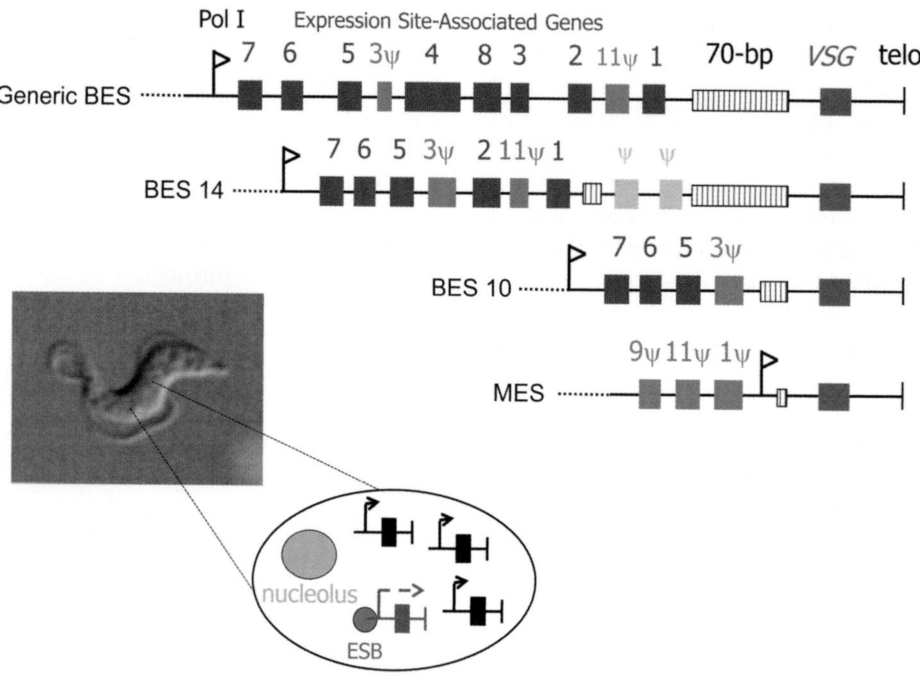

Figure 1 Architecture and singular transcription of variant surface glycoprotein (*VSG*) gene expression sites in *Trypanosoma brucei*. The four line diagrams show cartoon representations of telomeric *VSG* expression sites. The top diagram shows a generic bloodstream expression site (BES), while the two diagrams below display examples of variant BES (48) in which *VSG* pseudogenes (ψ, peach box) are found (BES 14) or where there has been loss of several expression site associated genes (*ESAG*s; dark blue box) or pseudogenes (light blue box) (BES 10). The final line diagram shows a *VSG* expression site (MES) used in metacyclic form *T. brucei*, which are found in the tsetse; here, the RNA polymerase I (Pol I) promoter (flag) does not drive expression of ESAGs, as it does in the BES, but only the *VSG* (red box), which in all cases is found adjacent to the telomere (telo; vertical line). Upstream of the MES promoter, several *ESAG* pseudogenes have been described, suggesting that these sites were derived from the BES. Arrays of 70-bp DNA repeats in the BES and MES are shown (hatched box), which always appear to be upstream of *VSG* genes or pseudogenes. Only one BES or MES is actively transcribed at a time in a single cell. A bloodstream form *T. brucei* cell is shown, in which the nucleus is diagrammed. The single active BES (red, extended arrow denotes transcription) is shown associated with the expression site body (ESB, small green circle), which is spatially distinct from the nucleolus (large green circle), though both subnuclear structures are sites of RNA Pol I transcription. Silent BES (three are shown in black; truncated arrow denotes limited transcription) do not associate with the ESB or nucleolus. doi:10.1128/microbiolspec.MDNA3-0016-2014.f1

Fourteen distinct BES were found by this approach in *T. b. brucei* strain Lister 427, slightly less than estimates made by a different approach in the same strain (49) and by the same method in a different *T. b. brucei* strain (50). The BES are undoubtedly a focus for recombination (see below) and rearrangements occurring between them, which could lead to changes in number, would be consistent with subtelomeres being a rich source of gene diversification in many organisms (51–55). Nonetheless, gene and sequence feature order within the BES appears to be rather conserved and to follow the general model of BES structure shown in Fig. 1 (48). In all cases, the *VSG* is found most proximal to the telomere repeats. Moreover, despite the wealth of *VSG* pseudogenes in the *T. brucei* genome (see below), the telomere-proximal *VSG*s described were all functional, with any *VSG* pseudogenes more distal to the telomere. Upstream of the *VSG* there is invariably a stretch of 70 bp repeats, which appear to be uniquely associated with the huge majority (~90%) of *VSG*s throughout the genome (37). Individual 70-bp repeats within arrays show considerable size variation and sequence complexity, including (TRR) repeats and GT- and AT-rich elements (56). The triplet repeat component has been shown to have a propensity to become non-H bonded (57) and may promote recombination, at least in bacterial plasmids and when transcribed (58); features that may tally with these elements being able to provide upstream homology during *VSG* recombination (see below).

BES are not just sites of VSG transcription, as each also contains a variable suite of Expression Site-Associated Genes (*ESAG*s). *ESAG*s are invariably found upstream of the 70-bp repeats, meaning that the size of the BES (from telomere to promoter) can be up to ~60 kbp and can include up to 13 genes (48). The order of *ESAG*s is broadly conserved between BES, but this is complicated by *ESAG* duplications, the presence of some *ESAG*s in only a minority of BES, *ESAG* pseudogenes, and the fact that some BES have suffered truncations in which *ESAG*s (but not *VSG*s) are lost. These variations suggest that recombination processes in the BES may not be limited to the *VSG*. An alternative suggestion is that variations in *ESAG* composition between BES may be an adaptation to different *T. brucei*–host species interactions, given the promiscuity of *T. brucei* in terms of infecting different mammalian species (59, 60). Testing this hypothesis is complicated by our limited understanding of ESAG function, though many *ESAG*s encode confirmed or predicted cell surface receptors (61). It seems plausible that RNA Pol I has been co-opted for BES transcription to allow for high level expression (43), so contributing to the generation of the 10^7 VSG proteins needed for the bloodstream form coat; indeed, the VSG accounts for a very large proportion of the bloodstream form mRNA population (62). Nonetheless, a need for high level expression does not readily explain the co-expression of the *ESAG*s with the *VSG*, as *ESAG* mRNAs appear not to be notably abundant (62). Bitter et al. (63) showed that when *T. brucei* is grown in culture media containing serum from different mammalian hosts, parasites are selected that have switched to transcribe a different BES. Switching correlates with expression of a different heterodimeric transferrin receptor (TfR) encoded by *ESAG6* and *ESAG7*, the genes that are most proximal to the BES promoter and in which sequence variation has been detected (64). Hence, it is proposed that multiple BES provide a range of TfRs that enable *T. brucei* to bind the variant transferrin molecules found in different host mammals that the parasite infects, thereby maximizing iron uptake in the face of anti-TfR antibodies (65). This hypothesis has been questioned (66, 67), and attempts to evaluate whether levels of *ESAG* sequence variation correlate with *Trypanosoma* species' host range have been somewhat unclear (50). Nonetheless, *in vitro* selection for BES switches is also been seen when *T. b. gambiense* is grown in differing host serum (68). Moreover, *ESAG4* genes, which encode novel, variable receptor-like adenylate cyclases, have been correlated with controlling the early immune response of the host (69), and BES-specific expression of a VSG-related serum resistance-associated factor allows *T. b. rhodesiense* to survive lysis by human serum components (70), suggesting that ESAGs may provide a variety of host-specific adaptations.

We clearly have much to learn about ESAG function, but this is central to any discussion of the role of multiple BES in antigenic variation. It is clear that *T. brucei* has evolved strategies to ensure expression of only one BES at a time and can execute a switch in transcription status from one BES to another, thereby changing the VSG coat (discussed below). Have these mechanisms evolved to serve immune evasion or do they underlie host adaptation, or both? Evolutionary pressures appear to have selected for drastic measures to secure increased numbers of BES in the *T. brucei* genome, since some are found in aneuploid "intermediate" chromosomes that are structurally distinct from the core genome, which is composed of 11 predominantly diploid "megabase" chromosomes (71) (Fig. 2). Moreover, *T. brucei* has adapted the BES to generate distinct *VSG* expression sites (metacyclic ES; MES) (Fig. 1) that are used to express a VSG coat in metacyclic stage parasites in the tsetse fly; a preadaptation for survival in the mammal for these infective stages (8, 72–74). However, recent trypanosome genome sequencing studies suggest that *ESAG*-rich BES may be a *T. brucei*-specific feature. In *T. vivax* and *T. congolense*, which also use antigenic variation for survival, homologues of most *T. brucei ESAG*s can be found, but they are not detectably linked to *VSG* in BES and may even provide distinct functions (75). Examining how *T. vivax* and *T. congolense* prosper in their mammal hosts will reveal much about the relationship between VSG and ESAG co-expression and the role of transcriptional switching in antigenic variation.

Despite their complexity in structure, and their potential occupation of all telomeres in the megabase chromosomes, the *VSG* expression sites (both BES and MES) provide only a minor component of the *T. brucei* armoury of genes for antigenic variation. The total number of *VSG*s in the *T. brucei* genome is enormous: the sequence of *T. b. brucei* strain TREU927 suggests 1,000 to 2,000 *VSG*s (37, 46), of which around half have been annotated (76). Remarkably, this "archive" represents around 20% of the coding capacity of the

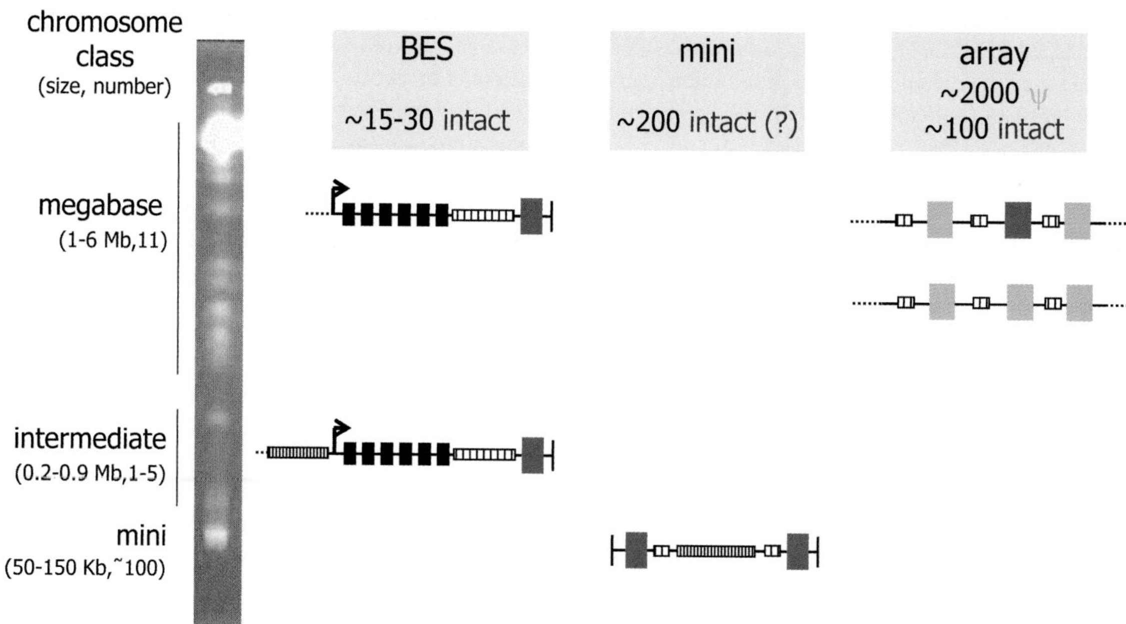

Figure 2 The variant surface glycoprotein (*VSG*) gene archive in *Trypanosoma brucei*. Whole *T. brucei* chromosomes are shown separated by pulsed field gel electrophoresis and stained with ethidium bromide. To the left of the gel, the positions of the megabase chromosomes, intermediate chromosomes and minichromosomes that comprise the nuclear genome are indicated, including the size and number of the different chromosome classes. To the right of the gel, the different loci in which *VSG*s are found are indicated (bloodstream expression site (BES), mini, array), including the number of *VSG*s in each locus type and whether they are functional (intact, red box) or are pseudogenic (ψ, peach box). BES denotes *VSG*s in expression sites that are used in the mammalian bloodstream and are found in the megabase and intermediate chromosomes. Mini denotes *VSG*s found in the minichromosomes, and array denotes *VSG*s found in the subtelomeres of the megabase chromosomes. In each case the presence or absence of a number of sequence features in addition to the *VSG* is shown: the telomere (vertical line), 70-bp repeats (widely hatched box), expression site-associated genes (black box), the RNA Pol I promoter (arrow) and 177-bp repeats (narrow hatched box). doi:10.1128/microbiolspec.MDNA3-0016-2014.f2

genome, an investment in variable antigens dwarfing that of other pathogens that employ antigenic variation. The *VSG*s outside the BES and MES—that is, the huge majority—are silent, as they are found in two locus types in which transcription has rarely been detected (Fig. 2). The first of these locus types encompasses *VSG*s found adjacent to the telomeres in ~200 minichromosomes: linear molecules ~30 to 150 kbp in size that are related to intermediate chromosomes in possessing central 177-bp repeats that may provide stability during cell division (71, 77, 78). Though it has been reported that a minichromosomal *VSG* can be expressed *in situ* (79), the lack of BES-related features other than 70-bp repeats in the minichromosomes that have been examined to date suggests that they evolved to provide a repertoire of silent, telomeric *VSG*s that are easily activated; indeed, in all cases examined to date the minichromosomes contain intact, functional *VSG*s. The sequence of the whole repertoire of minichromosomes has not yet been reported for any trypanosome strain, and this approach is needed to evaluate whether the above generalizations about these molecules are widely held. It is also possible that such sequencing will determine how these minichromosomes evolved: for instance, did this part of the *VSG* archive arise from the BES or MES?

The other locus type comprises huge arrays of *VSG*s found in the subtelomeres of the megabase chromosomes (46) (Fig. 2). These *VSG*s have been referred to as "basic copy", "chromosomal-internal" or "array" VSGs, reflecting the fact that they are substrates for switching by recombination and differ from BES *VSG*s in their genomic context. Like the minichromosome *VSG*s, virtually all array *VSG*s are flanked by 70-bp repeats (37), though typically the number of repeats is lower than is found in the BES. However, genome sequencing has revealed that most array *VSG*s appear distinct from telomere-adjacent *VSG*s, because only a very small fraction (~5%) of this part of the archive encodes intact, functional VSGs (37, 46). Most of the array *VSG*s (and therefore most of the archive) are pseudogenes in which the open reading frame is interrupted by stop codons or frameshifts (~65%), or where the gene is truncated (~20%) and lacks one of the two VSG protein domains. The remainder of the array *VSG*s (~10%) appear atypical, encoding products with predicted folding or posttranslational modifications that differ from known VSGs. Another, distinct, set of ~30 *VSG*-related genes appear not to contribute to antigenic variation, because they are transcribed in both bloodstream and procyclic form *T. brucei* cells, all of them lack associated 70-bp repeats and they localize where the chromosome cores and *VSG* array-containing subtelomeres abut (37). Nonetheless, the demonstration that most of the *VSG* archive is composed of pseudogenes has huge implications for how we consider the mechanisms, dynamics and functions of *VSG* antigenic variation (see below).

Sequence features of the *VSG*s and their subtelomeric location appear to promote diversification of the archive. The sizes of individual megabase chromosomes vary considerably between *T. brucei* isolates, a variation that extends to considerable size differences between chromosome homologues within a cell (80). Though some of this variation can be attributed to changes in gene and sequence copy number throughout the genome, most is accounted for by expansions and contractions in the *VSG* arrays (81). In part, this variation is due to partitioning the core and the subtelomere, allowing ectopic recombination in the latter and limiting this in the former, where it could have deleterious effects (27, 51). Beyond subtelomeric recombination, it is possible that specific mechanisms are in place to further promote diversification: mutagenic processes favoured by higher-order selection (82). Though *VSG*s share very limited sequence homology, and the most related genes are dispersed in the subtelomeres rather than being adjacent to each other (46), around a third of *VSG*s are in families of at least two genes that share around ~75% nucleotide identity (37). Hence, duplication of *VSG*s may be a prevalent process, and appears to be associated with gene conversions extending from the 70-bp repeats to regions of C-terminal homology. Whether other dispersed sequences, such as ingi transposable elements, are also involved is less clear. Smaller-scale mutations have also been inferred from *VSG* sequence comparisons within a single archive (83), though whether these are related to *VSG* gene conversion or have a different source is unknown. It is clear that we have only just begun to examine the processes that shape and diversify the *VSG* archive, and do not fully understand the events that operate during replication, over the course of an infection and through the life cycle. Understanding this will determine the extent to which archives compare or diverge in extant trypanosome isolates (47, 84). Just as importantly, we do not yet understand if and how these largely silent processes within the archive intersect with our growing understanding of how *VSG*s are recombined into the BES.

A BRIEF DISCUSSION OF TRANSCRIPTIONAL *VSG* SWITCHING

As discussed above, while the evolutionary selection for multiple *VSG* BES is still being debated, the multiplicity of these sites provides a route for *T. brucei* antigenic variation. Since this book is primarily devoted to DNA rearrangement processes, we will provide only a brief overview of transcriptional *VSG* switching; more detailed discussion can be found in the previous version of this book (*Mobile DNA II*; chapter authored by Piet Borst: Chapter 40) (7) and in several excellent reviews (23, 25, 85–89). It is clear that *T. brucei* possesses an active mechanism for ensuring that only a single VSG is expressed: attempts to select for *T. brucei* cells in which two or more BES are actively transcribed have failed (90, 91), but this is not because a mixed VSG coat is toxic (92, 93). The nature of this counting mechanism remains uncertain, however. A range of factors have been detailed whose loss through mutation or RNAi elevates transcription from the silent BES. These factors act in diverse functions, including telomere (94), chromatin (95–98) or nuclear envelope (99) activities. Despite this, perturbation of these activities shares a common outcome: though the silent BES become transcribed, the levels of this upregulation never reach transcription rates seen in the active BES. It is possible, then, that these factors influence processes that follow from, rather than dictate, singular BES expression. To date, the expression site body (Fig. 1) remains the best candidate for establishing monoallelic control (45), but its composition and association with the BES remains elusive. Indeed, the mechanistic focus of transcriptional control is still to be clarified. The observation that some transcripts can in fact be detected from silent BES, even in the absence of the above perturbations, suggests that selective RNA Pol I elongation may be the key (100). However, other studies have shown that RNA Pol I occupation levels are higher at the promoter of the active BES, suggesting that selective transcription initiation may also contribute (101).

Beyond the mechanism(s) for BES counting, the process of transcriptional switching between BES is arguably even more poorly understood. Only one factor has so far been described that influences the transfer of active transcription from one BES to another. Mutation of one of two related histone methyltransferases, DOT1B, causes ~10-fold elevated levels of transcription from silent BES, similar to the effects seen for the factors above. However, in DOT1B mutants transcription derepression is found throughout the transcription unit and, strikingly, loss of DOT1B appears to delay the time *T. brucei* takes to switch its VSG coat (102), potentially because normally unstable switch intermediates (90) are stabilized. How stabilization might occur remains uncertain (103), in part because the trigger and timing of transcriptional *VSG* switching is unknown. DNA replication and segregation may provide a window for transcriptional change, which would be consistent with observations that RNAi targeting of factors involved in nuclear replication and chromatid cohesion functions can elevate levels of silent *VSG* expression and switching (104–107). However, whether directed or random events can precipitate a transcriptional switch remains unknown. Indeed, though this switch reaction does not involve DNA rearrangements, it remains possible that it shares repair-related or replication-related functions that act in recombination-based switching (see below). Such a link would explain why mutation of core homologous recombination (HR) factors not only impairs VSG switching by recombination (see below), but also VSG switching by transcription (108–110), and why multiple treatments that cause DNA damage or replication arrest in *T. brucei* can elevate silent BES expression (111).

ACTIVATION OF INTACT *VSGs* INVOLVES HOMOLOGOUS RECOMBINATION

Activation of silent *VSGs* by their recombination into the BES is the primary route by which antigenic variation occurs, and recombination is the only route that can access all the *VSGs* in the archive for the formation of new VSG coats. Three recombination reactions have been described (Figs. 3, 4 and 5), which to some extent reflect different features of *VSGs* localized at telomeres and in subtelomeric arrays, but our understanding of how interrelated these reactions are is incomplete. What seems increasingly clear, however, is that *VSG* location and whether or not the gene is intact or pseudogenic strongly influences the timing of its activation (see below).

One recombination pathway is *VSG* gene conversion, where a copy is generated of a silent, functional *VSG* and replaces the BES-resident *VSG* (Fig. 3). This reaction is capable of activating functional *VSGs* throughout the archive. To generate a wholly new VSG, the extent of sequence copied must extend beyond the *VSG* open reading frame. Upstream, the 70-bp repeats can provide homology for both telomeric (BES, MES or or minichromosomal) and array *VSGs* (112), though gene conversions from a silent BES are frequently seen where homology is further upstream within the *ESAGs* (113, 114), even encompassing the *VSG* promoter (48). Downstream, gene conversion of array genes can end in 3′ coding or noncoding parts of the *VSG* (115), while conversion of telomeric *VSGs* can extend to the chromosome end (116, 117). A second pathway is reciprocal *VSG* recombination (118), where chromosome ends are exchanged by crossover without sequence loss (Fig. 5): a silent *VSG* moves into the BES and is activated, while the previously active *VSG* moves to the other chromosome end. This reaction, though readily detected (119), can only occur between telomere-adjacent *VSGs*, and so seems a minor pathway of *VSG* switching. The third recombination pathway, segmental gene conversion (SGC), is distinct from the two above in that it does not activate intact *VSGs*, but instead pieces together segments of multiple pseudogenes to generate novel, functional *VSG* "mosaics" (120–122) (Figs 3 and 4). This has substantial implications. First, the substrates used in SGC are different from activation of intact *VSGs*, in that at least one "end" of the reaction must be engaged through sequence homology between recombining *VSG* open reading frames, rather than through flanking homology. Since sequence homology between any two *VSGs* is normally very limited (26, 38), this raises questions about the nature, efficiency and substrate preferences of the recombination reaction; for instance, can RAD51-directed HR act in this process (see below)? Second, segmental *VSG* conversion has the capacity to generate many-fold larger numbers of VSG coats than merely the number of *VSG* genes in the archive (123) and, in this sense, has a comparable purpose and amplification of scale to the rearrangements used to generate mature immunoglobulin genes (120). Recent studies have built upon the genome sequence of *T. brucei* to show that segmental *VSG* conversion comes to predominate as *T. brucei* infections progress and that VSG diversity appears enormous (37, 124). These studies have confirmed far-sighted

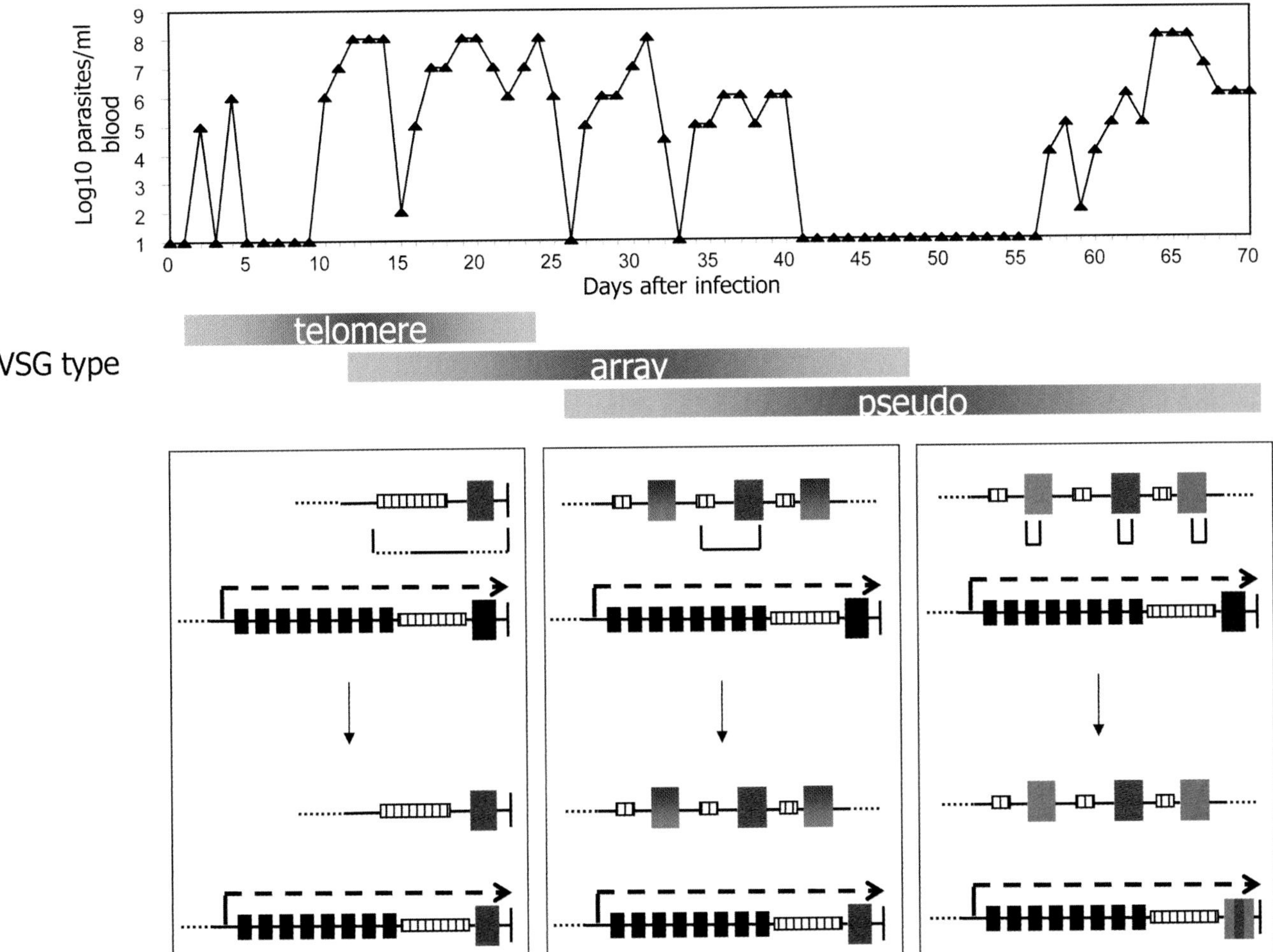

Figure 3 Hierarchy of variant surface glycoprotein (*VSG*) gene switching by recombination during infections by *Trypanosoma brucei*. The graph depicts the log of the number of *T. brucei* cells in a cow up to 70 days after infection (day 0). The schematic below details the timing of activation, by switching, of the different *VSGs* found in the genome (*VSG* type): silent telomeric *VSG*s (telomere) are activated more frequently than intact, subtelomeric array *VSG*s (array), which in turn are activated more frequently than *VSG* pseudogenes (pseudo). Gene conversion is the most frequent route for the above activation events, and the features associated with gene conversion of each *VSG* type are diagrammed. The *VSG* expressed before a switch (blue box) is transcribed (dotted arrow) from a bloodstream expression site (BES), in which the *VSG* is adjacent to the telomere (vertical line) and flanked upstream by 70-bp repeats (hatched box) and expression site associated genes (*ESAG*s; black boxes). The amount of sequence copied during *VSG* gene conversion is shown. For telomeric *VSG*s the sequence copied normally encompasses the *VSG* open reading frame (red box) and extends upstream to the 70-bp repeats, but also can extend further upstream into the *ESAG*s if the silent *VSG* is in an inactive BES; the downstream conversion limit may be the end of the *VSG*, but can also extend to the telomere from either a minichromosome *VSG* or inactive BES. Gene conversion of an intact subtelomeric array *VSG* is more limited in the range of sequence copied. In segmental *VSG* gene conversion parts of multiple, normally nonfunctional *VSG* pseudogenes (orange, red or brown boxes) are combined to generate a novel mosaic *VSG*; though this is shown to occur in the BES, it is not known if this is the location of gene assembly. Note also, the *VSG* pseudogene donors are shown for convenience as a contiguous array; in fact, segmental gene conversions using adjacent genes have never been observed.
doi:10.1128/microbiolspec.MDNA3-0016-2014.f3

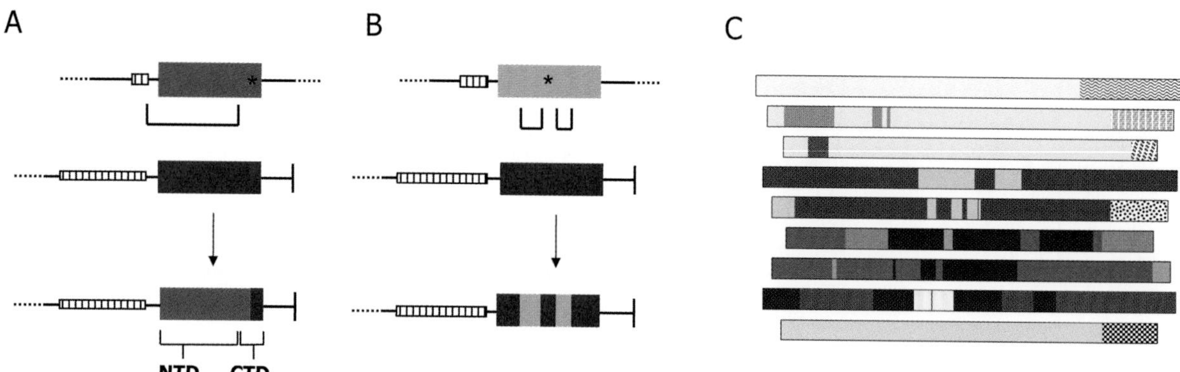

Figure 4 Complexity of variant surface glycoprotein (*VSG*) mosaics formed by segmental gene conversion in *Trypanosoma brucei*. (A) Where the 3′ boundary of segmental gene conversion occurs within the coding sequence of the *VSG* (3′ donation), part or all of the previously expressed C-terminal-domain-(CTD) -encoding region of *VSG* is retained, allowing the expression of a large contingent of silent *VSG*s (red box) that contain frameshifts or stop codons towards their 3′ ends (frameshift or premature stop codon indicated by an asterisk); as in Fig. 3, the recipient *VSG* (blue) is shown in the bloodstream expression site (BES) and the extent of conversion is indicated (NTD denotes N-terminal domain). Donors of *VSG*s formed in this way were found to share little sequence similarity over their whole sequence. (B) Mosaic *VSG*s can allow (partial) expression of pseudogene *VSG*s. Donors of *VSG*s (pink box) formed in this way share relatively high levels of sequence similarity (73% identity at the nucleotide level). (C) Segmental gene conversion yields diverse products: the diagram shows nine different *VSG*s detected during chronic infections (124); different donors are indicated in different colours, with 3′ donors indicated by hatching.
doi:10.1128/microbiolspec.MDNA3-0016-2014.f4

predictions by Kamper and Barbet, which predated whole genome sequencing and suggested that mosaic VSG formation is the key to infection chronicity and, most likely, even longer-term parasite–host interaction (125, 126).

To date, genetic studies to dissect the recombination machinery and sequences that act in *VSG* switching (Fig. 5) have been limited to understanding the activation of intact *VSG*s (106, 108, 110, 114, 117, 127–129). This stems from the fact that these studies have used *in vivo* or *in vitro* selection strategies that detect the most frequently activated *VSG*s; the lack of detectable mosaic *VSG* formation in these studies may suggest that SGC is less efficient than intact *VSG* conversion, though a mechanistic switch during an infection cannot be currently ruled out. A number of these studies make it clear that switching of intact *VSG*s is largely catalyzed by HR. In other words, though antigenic variation is a specialized recombination reaction targeting *VSG*s, it is executed by a nonspecific, general repair pathway rather than a specifically evolved *VSG* recombination reaction. The same conclusion has been reached regarding pilin antigenic variation in *Neisseria gonorrhoeae*, which occurs by gene conversion of silent, nonfunctional *pilS* genes to a *pilE* expression locus and where similar gene knockout approaches have been adopted (130, 131). A role for HR in VSG switching was first demonstrated by generating *T. brucei* mutants of RAD51; loss of this central enzyme of HR (132) significantly impairs *VSG* switching (110). This phenotype is similar, but not identical, to the effect that loss of RecA (the bacterial homologue of Rad51) has on *N. gonorrhoeae* antigenic variation. In *N. gonorrhoeae*, RecA mutants are effectively unable to switch their pili by recombination (133), whereas *T. brucei* VSG switching, though reduced approximately 10-fold, can still be detected in RAD51 mutants, including by gene conversion (110). This appears to suggest greater flexibility in the recombination pathways that can lead to antigenic variation in *T. brucei*. Indeed, *N. gonorrhoeae* antigenic variation appears to be a rather defined form of HR, because it only uses a sub-pathway of RecA-dependent HR initiated by a RecF-like machinery (134), and not RecBCD (135).

Further HR-related factors have been tested for their contribution to *T. brucei* antigenic variation, confirming the role of HR but suggesting some specialization. RAD51 activity *in vivo* is mediated by a number of factors (132), including BRCA2 and a collection of Rad51-related proteins termed Rad51 paralogs. BRCA2

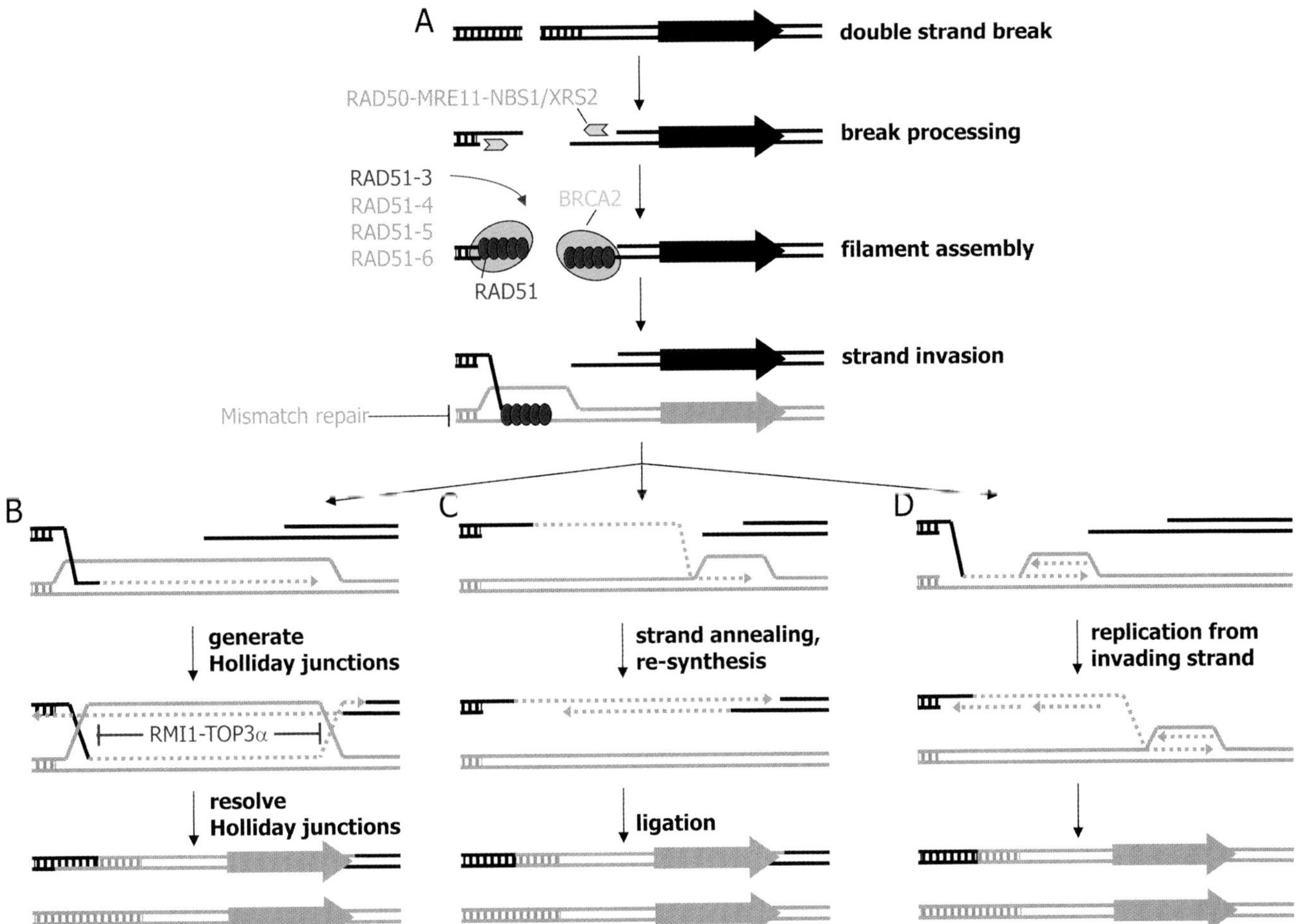

Figure 5 Models for variant surface glycoprotein (*VSG*) recombination during antigenic variation in *Trypanosoma brucei*. (A) Recombination is shown initiated by a DNA double-strand break (DSB) in the 70 bp repeats (hatched box) upstream of the *VSG* (black arrow) in the bloodstream expression site (BES) (*ESAG*s and promoter are not shown). Only factors that have been examined for a role in *VSG* switching are indicated; those shown in color have been found to act, while those for whom no evidence of a role in *VSG* switching has been found are shown in gray. DSB processing to reveal 3′ single-stranded ends is, in part, catalyzed by MRE11-RAD50-XRS2/NBS1 (MRX), generating a substrate on which RAD51 forms a nucleoprotein filament; note, however, that a further exonuclease (not shown) normally acts with MRX of both ends of the DSB are processed. RAD51 function is mediated by a number of factors: BRCA2 influences RAD51 filament dynamics, while the detailed roles of RAD51 paralogs (RAD51-3, RAD51-4, RAD51-5 and RAD51-6 in *T. brucei*) are unclear. RAD51 catalyzes repair by homology-dependent invasion of the single-stranded end into intact DNA (gray lines), containing a silent *VSG* (gray arrow). Mismatch repair constrains homologous recombination to act only on sufficiently homologous sequences. Three pathways for DSB repair have been described and may contribute to *VSG* switching. (B) DSB repair; here, newly synthesized DNA is copied from the intact DNA duplex and remains base-paired, generating Holliday junction structures whose enzymatic resolution can lead to gene conversion with (not shown) or without (shown) crossover of flanking sequence. In *T. brucei*, RMI1-TOP3 has been shown to suppress crossover, by perhaps acting on the Holliday junctions. (C) Synthesis-dependent strand annealing; here, newly synthesized DNA is displaced from the intact duplex and reanneals with homologous sequence at the DSB, allowing synthesis of the other strand. Break-induced replication is shown in (D); in this mechanism, an origin-independent replication fork forms on the strand invasion intermediate allowing replication to the chromosome end.
doi:10.1128/microbiolspec.MDNA3-0016-2014.f5

interacts with Rad51 via conserved BRC repeats and a distinct, less-conserved motif that is found towards the C-terminus of the mammalian protein, and these interactions appear to guide the formation and disassembly of Rad51 nucleoprotein filaments on broken DNA (136) (Fig. 5). Mutation of BRCA2 in *T. brucei* impairs VSG switching to a very similar extent to RAD51 mutants (109). In *T. brucei*, but not in related African trypanosomes or more distantly related trypanosomatids, BRCA2 possesses a remarkable expansion in the number of BRC repeats relative to most eukaryotes, suggesting a need for pronounced interaction with RAD51. However, this expansion appears not to be an adaptation for gene conversion of intact *VSG*s, because cells expressing BRCA2 engineered to have only one BRC repeat still support wild-type levels of such switching (109). To date, BRC repeat number has only been demonstrated to be important in the efficiency of RAD51 relocalization after induced DNA damage (137), but in what way this might relate to antigenic variation remains elusive. Nonetheless, there are intriguing parallels between BRCA2 function in *T. brucei* and RecX in *N. gonorrhoeae*. Loss of RecX in *N. gonorrhoeae* impairs pilin switching (138) and, though RecX is unrelated in sequence to BRCA2, it acts in a related way, modulating RecA nucleoprotein filament formation (139, 140). Hence, it seems likely that in both pathogens antigen gene recombination is controlled in ways not yet understood through the size or stability of the filaments formed by the related recombinases. The functions of Rad51 paralogs remain somewhat mysterious. These factors are related to Rad51 but, unlike a meiosis-specific cousin of the HR recombinase termed Dmc1, there is no evidence that they catalyze recombination. Instead, they appear to aid Rad51-dependent HR (141) and show some overlap with BRCA2 function (142, 143), though potentially also broader functions in genome maintenance. The numbers of Rad51 paralogs vary widely in eukaryotes, with five in mammals but only one in *Caenorhabditis elegans*. Among single-celled eukaryotes, trypanosomes appear to possess a relatively large repertoire of four proteins (108, 144). In *T. brucei* each RAD51 paralog acts in DNA recombination and repair. However, mutants of only one RAD51 paralog, RAD51–3, show a pronounced impairment in intact *VSG* switching, suggesting some specificity in the HR reaction(s) used. In contrast with the above factors that promote VSG switching, one HR enzyme complex has been shown to suppress the reaction. Mutation of TOP3a or RMI1, which interact and are probably two components of the *T. brucei* RTR (RecQ/Sgs1-Top3/TOPO3a- Rmi1/BLAP75/18) complex described in yeast and mammals, results in increased levels of VSG switching, as well as recombination elsewhere in the genome (117, 145). Though the extent of elevated VSG switching is surprisingly different in the two mutants, each shares common mechanisms that are, in part, RAD51-dependent: increased levels of crossovers within the *ESAG*s and increased *VSG* gene conversion delimited by the 70-bp repeats (Fig. 5). Again, comparison of these findings with *N. gonorrhoeae* is revealing, since RecQ has been implicated in pilin antigenic variation (146, 147). However, loss of *N. gonorrhoeae* RecQ or mutation of the expanded number of helicase domains in the bacterial protein do not result in an elevation of antigenic variation rates, as in *T. brucei*, but in a reduction. *Neisseria gonorrhoeae* RecQ acts to unwind guanine quartet (or G4) structures that initiate pilin gene switching (148–150), meaning that the different phenotypes in *T. brucei* might indicate a distinct strategy for antigenic variation initiation. However, eukaryotes typically possess more than one RecQ-like factor, and indeed no *T. brucei* RECQ factors have so far been examined, so there is the potential for redundancy in this aspect of HR-directed VSG switching.

Despite the above broad association of HR with switching of intact *VSG*s, not all factors that might have been predicted to act behave in the expected ways. A key factor in the early steps of eukaryotic DNA double-strand break (DSB) repair is the MRE11 complex (Fig. 5), composed of MRE11, RAD50 and Xrs2/NBS1 in eukaryotes, with roles in processing DSBs, activating the DNA damage response through the kinase ATM, bridging DNA molecules and suppressing non-HR repair (151). MRE11 mutants in *T. brucei* display DNA damage repair and genome instability phenotypes (152, 153), including gross chromosomal rearrangements due to loss of subtelomeric *VSG*s, an instability also seen in bloodstream form *T. brucei* BRCA2 mutants (109, 137). Despite this, loss of MRE11 has no detectable effect on VSG switching, questioning whether DSBs can initiate the reaction (see below) or whether loss of MRE11 alleviates suppression of alternative DSB repair reactions that can be upregulated and affect switching. Mismatch repair (MMR) recognizes and repairs base mismatches, which can occur during replication and during HR between DNA molecules that are diverged in sequence (Fig. 5). In this latter context, MMR suppresses HR (154), including in *T. brucei* (155, 156). Again, however, *T. brucei* mutations of MSH2 or MLH1 (central factors in MMR) do not detectably alter the frequency of *VSG* switching. Given that *VSG* recombination appears to frequently use highly variable flanking sequences for homology, such as the 70-bp repeats, this observation is perhaps surprising, and remains unexplained. For instance, these data might implicate an MMR-independent HR reaction (157) in VSG switching. Alternatively, MMR mutation might be neutral if the repair machinery functions not merely to survey for poorly matched *VSG*s during the downstream strand exchange steps of HR, but also acts in the initiation of the process, perhaps akin to the contribution of MMR to immunoglobulin rearrangements and transcription-associated triplet repeat expansions (158). Whatever the explanation, the lack of a detectable effect of MMR mutation may warrant further investigation, as these findings mirrored initial observations in *N. gonorrhoeae* antigenic variation, where mutants of MutS, the bacterial ortholog of MSH2, also appeared to show unaltered pilin switching (159). However, later analysis showed that MutS mutants do result in increased pilin switching, explained by altered recombination tract lengths (160). *Trypanosoma brucei* MMR may yet be revealed to contribute to *VSG* donor selection, but the far greater numbers of available silent *VSG*s mean that any effects of MMR mutation may not be seen by simply measuring switch efficiency.

A further complexity in *VSG* switching is the relatively modest effect of any single HR-associated gene mutation detailed to date. A consistent finding from many studies is that in the absence of RAD51-dependent HR, *VSG* switching by recombination continues to operate, albeit at reduced levels (110, 145). To date, the nature of this RAD51-independent HR is unknown. DSBs in eukaryotes can be repaired by at least three routes: HR, nonhomologous end-joining (NHEJ), and a still poorly understood pathway (or pathways) termed alternative end-joining or microhomology-mediated end-joining (MMEJ) (161). It seems likely that NHEJ can be discounted as a route for *VSG* gene conversion; not only does this appear mechanistically incompatible, but mutants of the Ku heterodimer that recognize DSBs in this reaction have no effect on *VSG* switching (162, 163). In fact, no study to date has successfully detected NHEJ in *T. brucei*, and it may well be absent, like in some other eukaryotes (164), due to the absence of key factors (165). In contrast, MMEJ is readily detected in *T. brucei*, and may have assumed the role of NHEJ in DSB repair (165–168). Testing whether MMEJ provides a route for *VSG* gene conversion in the absence of RAD51-dependent HR is hampered by the lack of clarity in the machinery involved, but it is increasingly clear that this reaction is highly flexible, for instance contributing to many forms of structural genome variation and acting in immunoglobulin rearrangements (161). Beyond MMEJ, some evidence suggests that bonafide HR occurs in *T. brucei* in the absence of RAD51, since homologous integration following transformation is readily detected in RAD51 mutants (166). Moreover, transformation assays have suggested that two HR reactions might operate in *T. brucei*, one that has a minimal homology length requirement of ~100 bp and another, less efficient reaction that needs around ~30 bp of homology (157). To date, however, the machinery needed for an HR reaction of such a short length is unknown. Gene conversion during antigenic variation that is independent of the HR recombinase is not unique to *T. brucei*: *Borrelia burgdorferi* is a tick-borne spirochete where recombination of silent *vls* gene segments into the single *vlsE* expression site is unaffected by mutation of RecA (169), though it does require RuvAB, a helicase that processes Holliday junction HR intermediates (170, 171). Like *T. brucei*, how homology-directed strand transfer initiates in *B. burgdorferi* without RecA/RAD51 is unknown, though the potential involvement of a novel, telomere-directed resolvase would be very interesting (172).

HOW IS *T. BRUCEI* VSG SWITCHING BY RECOMBINATION INITIATED?

The high rate of VSG switching suggests that antigenic variation must be promoted by some process(es) to elevate the frequency of coat changes above general mutation levels. In common with programmed rearrangements such a mating type switching in yeast and immunoglobulin gene maturation, DSBs in the BES have long been suspected as likely initiating lesions in *T. brucei*, promoting *VSG* recombination through HR (59, 173) (Fig. 5). Boothroyd et al. provided the first evidence that DSBs might indeed act in this way: by controllably expressing an endonuclease, IsceI, and localizing its recognition sequence within the BES, it was possible to show that an experimentally-induced DSB elevates VSG switching through gene conversion (127). Moreover, elevation of switching was dependent on localizing the IsceI target sequence adjacent to the 70-bp repeats and upstream of the *VSG*, and was abolished if the 70-bp repeats were removed from the BES. Further evidence for the role of DNA breaks was found using ligation-mediated PCR, which suggested that such lesions were readily detectable in the 70-bp repeats within the actively transcribed BES, but were less detectable in a silent BES or, indeed, within the *ESAG*s of the active BES (127). These data appear to implicate the 70-bp repeats as the culprit for promoting the formation of DNA breaks, perhaps even DSBs, and that this role is associated with, or seen more readily, when the BES is traversed by RNA Pol I. This is an attractive model, but several questions and areas of uncertainty remain (174).

The foremost question promoted by the above model, is how might the 70-bp repeats promote the formation of DSBs? One possibility is that they are the target of an endonuclease (173). To date, such an enzyme has escaped detection, perhaps because it is novel or its activity or expression is tightly regulated. Endonuclease-independent break formation might also be considered, such as seen by Activation-Induced (Cytidine) Deaminase (AID) targeting of the repair machinery during immunoglobulin class switching (175). Indeed, given the association between the breaks and BES expression, transcription-associated breakage, also mediated by repair such as has been proposed for triplet repeat expansion, is possible (158). Alternatively, the 70-bp repeats, perhaps due to the structurally unstable (TRR) motif, could stall DNA replication and promote HR. Certain DNA sequences, termed fragile sites, are regions of elevated genomic instability, and at least some these are associated with stalling of replication (176, 177). Indeed, the capacity of replication stalling to induce rearrangements has been harnessed in other organisms to promote locus-specific genetic changes: in *Schizosaccharomyces pombe* a specific replication-blocking lesion promotes *mat* switching by HR (178), and replication barriers are needed for gene amplification in *Tetrahymena* macronuclear development (179). As tractable as such mechanisms are to experimental investigation, the centrality of the 70-bp repeats for switching and the suggestion that it is DSBs that form on these sequences, is not supported by all data. Though deletion of the 70-bp repeats impedes the elevation of *VSG* switching after IsceI targeting, removal of the repeats from the active BES in the absence of IsceI appears not to affect *VSG* switching rate (114, 127). Moreover, though a later study confirmed that IsceI-mediated DSB formation can activate switching in a BES location-dependent fashion, strict localization of breaks within the active BES and around the 70-bp repeats was not confirmed, and instead ligation-mediated PCR suggested that all BES are fragile adjacent to the telomere (128). Finally, as noted above, the lack of a role for MRE11 in *VSG* switching is perplexing, given its central function in DSB detection and repair (152). One explanation for these inconsistencies, as we have argued throughout, is that *VSG* recombination cannot be explained by a single mechanism, or by a sequence to direct that mechanism, and the system acts flexibly through several routes. Such flexibility may be compatible with the suggestion that telomere length, and not the 70-bp repeats, is the determinant of VSG switching (180, 181). This suggestion was promoted by studies in telomerase-deficient *T. brucei*, which gradually lose telomere repeats over time (182). In cells with critically short telomeres, *VSG* switching rate is higher than in the same cells with longer telomeres (181). How gene conversion is promoted in these cells is unclear, given that IsceI-mediated removal of the telomere tract does not elicit this response (128), though short telomeres might precipitate subtelomeric breaks or switching might relate to a process of telomere stabilization (183). How significant such telomere processes are in telomerase-proficient cells is unclear, but it may be one route among many for *VSG* switching.

A larger area of uncertainty regarding the model of DSB induction of *VSG* switching lies in selection of the *VSG* donors that provide the substrates for recombination. As we have discussed, *T. brucei* has a large, diverse archive of *VSG*s found in BES, minichromoomes and in subtelomeric arrays (Fig. 2). Despite this, the overwhelming majority of *VSG* switching events that were characterized following IsceI DSB induction used BES *VSG*s as donors (~85%); minichromosomal *VSG*s, despite their ~10-fold numerical superiority, were used infrequently (~15%) and array *VSG*s, whether intact or pseudogene, were not detected (127). This may reflect the hierarchy of switching events that are seen during long-term infections (see below), suggesting that there is a probability spectrum of donor preference and it is immune selection that allows the emergence of less frequently accessed genes in the archive. Alternatively, it is possible that activation of some *VSG*s may be initiated by a distinct route. Answers to such questions will be informed by in-depth profiling of *VSG*s expressed during infections, but require detailed assays of switch rates and mechanisms of distinct *VSG* types for a definitive answer.

SEGMENTAL GENE CONVERSION PRODUCES FUNCTIONAL VSG SURFACE COATS

The SGC in *VSG* switching appears to be a remarkably flexible reaction, judging by its products (Fig. 4). In some instances SGC can be simple: in cases where the 3′ boundary of gene conversion occurs within the coding sequence, SGC causes only the part of the gene encoding the exposed N-terminal domain of the antigen to be exchanged ("3′ donation") (124), a process similar to the exchange of antigen variable region "cassettes" by pathogens such as *Anaplasma marginale* (184), *Treponema pallidum* (185), and *Mycoplasma genitalum* (186–188). In addition, SGC can form a "mosaic" *VSG*, exchanging segments across the gene and introducing variation into the antigenically-important N-terminal domain. Mosaic *VSG*s identified from chronic infections can involve some quite complex rearrangements, involving up to four donors and more than ten segments (124). Both patterns of SGC—mosaicism and 3′ donation—appear to have been facilitated by homology: donors in both cases shared at least 85% nucleotide identity at the boundary of conversion, and the donors for mosaic *VSG*s were at least 73% identical across their N-terminal domain-encoding regions (124).

Segmentally converted *VSG*s identified from infections share the structural features of other expressed *VSG*s: they are not markedly different in length to intact *VSG*s, they possess a similar domain structure, and appear to form functional surface coats (124, 189). This is significant, because unlike gene conversion of intact, full-length *VSG*s, both 3′ donation and mosaicism readily use segments of pseudogene *VSG*s. Most likely, the lack of incorporation of the mutations that would impair the VSG coat or its expression reflects selection: those cells that generate *VSG* mosaics in which mutations are present are rapidly killed. Death may be immune-mediated, or may result from a cessation of protein synthesis following detection of nonfunctional VSG expression (190191). Mosaic VSGs tend to appear only relatively later in infection (after 3 weeks in mice, although the kinetics of antigenic variation are likely to vary significantly between hosts) (37, 124). The "late" appearance of mosaic VSGs suggests either that they have a lower probability of becoming expressed than intact VSGs, perhaps because the SGC events are less efficient and tend not to occur earlier in infection, and/or that trypanosomes expressing mosaic VSGs grow more slowly. The former is more likely, as previous experimental studies found poor correlation between trypanosome growth rate and expressed VSGs (192). However, these studies did not examine mosaic VSGs constructed during an infection, which may be suboptimally expressed compared with intact VSGs. This latter point may also explain why the products of SGC identified to date have been fairly conservative, with the replacement of part of one gene with the homologous region from another and with no dramatic changes to the features or structure of VSG: more radical variants may occur, but impose such a growth defect on their expressers that they are outcompeted.

The molecular mechanisms underlying *T. brucei* SGC remain mysterious. It is unknown whether *VSG* SGC products are rare consequences of "normal" recombination processes that act during intact *VSG* gene conversion, or whether the reaction involves a distinct, dedicated machinery. Some clues can be gleaned from the homology between mosaics and their donors: similarity is required, but perfect identity is not. Classical HR in *T. brucei* seems too demanding to explain this process, being relatively intolerant of base mismatches and requiring extensive regions of homology (157). The RAD51-independent MMEJ pathway, requiring only ~5 to 15 bp of homology and tolerant of mismatches (166–168), appears a more promising candidate, but the machinery involved in this process is unknown. Just as importantly, it is unclear if the form of strand transfer and break repair in MMEJ is capable of catalyzing *VSG* SGC; for instance, are mosaic *VSG*s generated in a stepwise fashion (perhaps over a number of cell cycles) involving successive gene conversions (37), or can they result from a single construction event? Understanding where this process occurs would be very revealing: are the array *VSG* pseudogenes "assembly intermediates" that arise from ongoing subtelomeric recombination, or might the reaction be rapid enough to allow assembly within the active BES without leading to protracted expression of nonfunctional VSG protein (87)? Loss of array *VSG*s during growth of *T. brucei* BRCA2 and MRE11 mutants hints at recombination within the subtelomeres, but the nature and scale of such events is unclear (109, 137, 152).

WHAT IS THE FUNCTION OF *VSG* SEGMENTAL GENE CONVERSION?

The finding that *VSG* SGC predominates in chronic infections raises many questions. Most important is: what is the function of this reaction in general—and mosaic VSGs in particular—in trypanosome antigenic variation? There are several different ways to answer this question (193). Clearly SGC contributes to antigenic variation by allowing the expression of a different VSG, but as discussed above, *T. brucei* has several other mechanisms for switching between intact *VSG*s, and SGC poses the potentially lethal risk of causing expression of a dysfunctional *VSG* (191). Therefore, we want to ask: what are the advantages of expressing *VSG*s by SGC, *rather than as full-length genes*? What aspect(s) of SGC (if any) have made it a selected trait, causing trypanosomes with this phenotype to have an edge over their competitors, and what aspects are simply consequences of SGC?

Three theories have been advanced for the function of SGC and of mosaic VSGs, none of which are mutually exclusive. Each theory appeals to the ability of SGC to increase the diversity phenotype, either by increasing the number of variants possible (combinatorial variation, expression of pseudogenes) or by increasing the evenness of diversity over time (hierarchy).

Combinatorial Variation

The first explanation for SGC in antigenic variation is that it increases the effective VSG repertoire by introducing combinatorial variation, allowing either prolonged infection within a host, reinfection or superinfection of a previously-infected host, or both (194); the sheer size of the *VSG*, and particularly the pseudogene *VSG*, repertoire suggests that this aspect has been a strong selective force. Specifically, combinatorial variation proposes that x donors can interact by segmental gene conversion to produce y antigenically distinct variants, where $x < y$. Many different patterns of *VSG* SGC between the same donors have been identified in individual infections, demonstrating the ability of SGC to introduce combinatorial genetic variation. Combinatorial variation as an explanation for SGC has gained substantial support in other antigenically variant pathogens, in particular *A. marginale* and *B. burgdorferi* (195), and it is well known that the mammalian immune system harnesses combinatorial variation during B-cell maturation, generating perhaps 10^{10} different antibody idiotypes within a single host (196, 197). As an explanation for trypanosome SGC, combinatorial variation has considerable appeal. However, the data are somewhat more ambiguous. Mosaic *VSG*s identified to date use donors with relatively high levels of identity, and as a consequence mosaic *VSG*s tend to be antigenically similar: four mosaic *VSG*s constructed from four different donors isolated by Barbet et al. (189) were found to be cross-reactive using polyclonal rabbit antisera ($x = 4$, $y = 1$; $x \nless y$), and of the five mosaic *VSG*s constructed from four different donors isolated by Hall et al. (124), only one was antigenically distinct from the others when tested with polyclonal mouse antiplasma ($x = 4$, $y = 2$; $x \nless y$). In the latter case, related mosaics were antigenically distinct, but not as a consequence of combinatorial variation *per se*, as antigenic variation could be explained by differences between the donors. It is possible that the acute immune responses used in these tests are not representative of those mounted during a complex natural infection—indeed, examination of chronic infection showed that antibody responses against the segmentally generated *Anaplasma* surface antigen MSP2 were highly variable, with poor maintenance of antibody against individual gene segments (198). Neither may these responses be representative of the immunological memory of previously infected host; indeed, there is evidence of trypanosomiasis causing substantial immune dysfunction in laboratory models of infection, though the effects in natural hosts are less well understood (34). Either way, the data collected to date do not show SGC during VSG expression to be a particularly efficacious means of *generating* antigenic novelty. Furthermore, the most frequently identified SGC event in chronic infections was 3′ donation (124), which introduced combinatorial genetic variation into an antigenically unimportant region of VSG (although it is possible that mosaicism and 3′ donation are acted on by different selective pressures). Whether combinatorial variability is the specific property of SGC that caused it to be selected is therefore unclear.

Hierarchy

For many years it has been recognized that there is a semi-predictable order to the different VSGs that predominate as an infection progresses, referred to as the "expression hierarchy" (8, 199–202). An advantage of this phenotype, as opposed to one in which each *VSG* has an equal probability of being activated, is that infection is prolonged because the host cannot mount responses against all the antigen variants at once. SGC has been proposed to contribute to hierarchy in two ways: (i) by generating "strings" of mosaics (that is, the progressive accumulation of segments in the expressed *VSG* over time has an implicit order to it) (203); and (ii) by causing pseudogenes to be activated at low probability, and therefore causing them to be expressed "late" in the hierarchy (204). With regards the first mechanism, there is good theoretical support (6) and some experimental evidence for a progressive increase in mosaic *VSG* complexity over time (124, 189), consistent with SGC in other pathogens (205, 206). With regards the second mechanism, there is an obvious benefit to the expression of pseudogenes (see below) and a consequence of SGC is that pseudogenes are less likely to be activated and therefore tend not to appear until later in infection. However, the resulting hierarchy being the selected feature of SGC implies that trypanosome clones with delayed expression of particular genes (call them $T_{delayed}$) had an advantage over their peers ($T_{not\ delayed}$), which could not be the case if the competition were occurring within an infection. For $T_{not\ delayed}$, on expressing the antigen variant relatively earlier, would induce immunity that would also neutralize $T_{delayed}$ on expressing the antigen variant relatively later.

The benefit of the $T_{delayed}$ phenotype would be seen only if selection were occurring at a between-infection level, which is not impossible, but less likely to occur than selection at the between-individual level (194, 207).

Expression of Pseudogenes

An obvious beneficial effect of SGC is that it allows the expression of pseudogene *VSG*s, which would otherwise be junk, greatly increasing the potential for antigenic variation. Yet, one might suppose that if trypanosomes were being selected for increased archive size, the most straightforward adaptation would have been an increased number of intact archive *VSG*s, rather than the evolution of an archive of pseudogenes activated by risky SGC. This supposition does not take account of how selection is likely to act on the *VSG* archive, however. The strength of selection for function on an individual *VSG* gene in the archive is likely to be weak: an individual *VSG* is seldom expressed, and it usually requires to be copied for expression (208). On the other hand, genetic drift is likely to be strong: the *VSG* archive is subtelomeric, a genomic region subject to high mutation rates (that probably cause a beneficial increase in archive diversity) (51–55); and over the course of its lifecycle *T. brucei* goes through dramatic population bottlenecks (209). A consequence is that intact archive *VSG*s are likely to be under constant risk of becoming pseudogenes. In fact, like other multigene families, the *VSG* archive is probably subject to a process of birth-and-death evolution, where *VSG*s are "born" by gene duplication, and eventually "die" by the acquisition of mutations (210). Under these dynamics, the ability to use a "dead" *VSG*—which has acquired pseudogenizing mutations during the process of evolution—represents an obvious benefit of SGC that would give an individual trypanosome clone an advantage over its competitors within an individual infection. Consequently, expressing the archive in segments would further relax the strength of selection for intact genes, possibly explaining the high proportion of pseudogenes among *VSG*s when compared with other multigene families. The most severe selective bottleneck in the lifecycle of *T. brucei* is migration to the tsetse salivary glands, with perhaps only a handful of individuals surviving the journey (209)—a journey that seems to be absent from the lifecycles of *T. congolense* and *T. vivax* (which colonize the mouthparts) (211). Intriguingly, the proportion of pseudogenes in the *VSG* archives of *T. congolense* and *T. vivax* is much lower than in *T. brucei* (84), consistent with genetic drift driving pseudogenization of the *T. brucei* archive.

Comparative analyses may help to resolve these questions. It remains to be seen, for example, whether *T. congolense* and *T. vivax* use SGC to express VSG to the extent that *T. brucei* does. Finding that they do, in spite of their lower pseudogene content, would suggest that the primary function of SGC is in generating combinatorial variation and in the production of strings of mosaics, whereas finding that they do not would be more consistent with SGC performing a role primarily associated with pseudogene expression and archive evolution.

CAN VSG SWITCHING BE EXPLAINED BY A SINGLE RECOMBINATION MODEL?

Description of the locations of *VSG*s in the *T. brucei* genome, as well as the demonstration that the genes are flanked by blocks of homology (notably, the 70-bp repeats) and can be telomeric, prompted the proposal of HR mechanism models to encapsulate the process of *VSG* switching. Synthesis-dependent strand-annealing (SDSA) was initially favoured, as this process does not generate crossovers and would therefore avoid potentially lethal translocations of subtelomeric array *VSG*s into the BES (173, 212). More recently, it has been suggested that break-induced replication might be a better model for the activation of telomeric *VSG*s, where gene conversions can extend to include the telomeric repeats (8, 180). These suggestions are, at least in part, supported by the finding that loss of HR factors impairs switching. However, as we have increasing knowledge about the factors involved in *VSG* switching, it appears increasingly likely that no single mechanism is used, and instead *T. brucei* antigenic variation can exploit a number of recombination pathways (Fig. 5). Multiple lines of evidence support such reaction flexibility. First, mutation of RAD51-dependent HR impairs, but does not abolish, VSG recombination (108–110, 157). Second, though the 70-bp repeats make an important contribution to VSG switching, their removal is not an impediment (114, 127). Third, though induction of a DSB promotes VSG switching, the reaction(s) induced may not use all available *VSG* donors (127). Finally, no experiments to date have examined the machinery involved in *VSG* mosaic formation, and it is unclear if this can be accounted for by BES-focused recombination models.

Given the above, it seems plausible that all the mechanisms shown in Fig. 5 can be used in VSG switching. SDSA remains the most robustly supported: it is catalyzed by RAD51 and associated factors, is able to account for recombination of any intact *VSG* in the archive, and can lead to crossover events. Break-induced replication is kinetically distinct from SDSA (213) and differs mechanistically in that it uses a processive DNA replication fork established from one end of a DSB (214). Such a reaction could only account for switching of telomeric *VSGs*, but appears highly plausible, since it could more readily explain long-range gene conversions that span much of the BES (48, 215), can be catalyzed without Rad51 (216–218) and, in yeast, has been shown to provide a route for telomere stabilization (213, 215). Nonetheless, an experiment that shows by genetic or mechanistic dissection that break-induced replication is distinct from SDSA in *VSG* switching is lacking. Elevated switching by crossover in *T. brucei* RTR mutants (117, 145) provides evidence not only that reciprocal recombination can occur but also that it is normally suppressed between BES. However, the formation and resolution of Holliday junction intermediates has not been demonstrated, such as by testing for a role of Holliday junction resolvases (219), which have been shown to act in both *N. gonorrhoeae* and *B. burgdorferi* antigenic variation (131, 170, 171). It should be noted that it remains unclear whether any of the above reactions can account for *VSG* SGC, and this will be an important next phase of investigation, given the importance of this reaction to the parasite.

THE INTERPLAY BETWEEN DIFFERENTIATION AND VSG SWITCHING

A classical characteristic of trypanosome infections, shared with many other pathogens that undergo antigenic variation, is the establishment of very chronic infections; for example, cattle experimental infections with trypanosomes can last hundreds of days to years. Bearing in mind that these organisms are permanently extracellular in the mammalian host, and therefore constantly exposed to the multi-faceted assault of the host's immune system (in particular, with respect to VSGs, the adaptive immune response) the ability to employ the armoury of the VSG expression system is clearly an essential part of establishing chronicity. However, a critical component that must be considered when analyzing how *VSG* archive use and switching mechanisms influence *in vivo* dynamics, is the ability of the trypanosome to self-regulate its growth by terminally differentiating from the long slender multiplicative life cycle stage to the short stumpy stage, which is preadapted for transmission to the tsetse fly. While the exact identity of the cAMP-sensitive molecular trigger that mediates this differentiation, termed the "stumpy identification factor" (220, 221), remains elusive, there have been significant recent advances in identifying the mechanisms underlying the process (222–224). It is becoming clear that the interplay between the differentiation and antigenic variation is even more critical than previously considered in shaping the infection dynamics of trypanosome infections (225, 226); one cannot consider the influence of *VSG* archive utilization on chronicity without taking into account the population effects of differentiation, particularly as recent data suggest that differentiation has a profound impact upon the proportion of the population in which switching is occurring.

Estimates of the contribution of differentiation have been explicitly incorporated into previous modelling analyses of *in vivo* infection dynamics (6, 227, 228), but these have been limited by relatively low-resolution measurements (e.g. microscopy-based morphological analysis) and parameter estimations based upon a linear density-dependent effect. The recent identification of stumpy-specific expression markers, in particular the surface-expressed carboxylate transporter Protein Associated with Differentiation 1 (PAD1) (222), has allowed much finer resolution analysis of the role that differentiation plays in determining the infection kinetics (229). This work quantified relative PAD1 expression during chronic (30-day) infections in mice, and generated a mathematical model to analyze the role of differentiation within infection using this much more refined input. These data revealed that stumpy forms predominate to a much greater degree than previously thought, and this is particularly marked in the chronic stages of infection; modeling predictions were validated by morphological analysis with only approximately 15% of trypanosomes being long and slender by day 15 of infection. This has a significant impact upon our understanding of the utilization of the *VSG* archive, as only long slender trypanosomes switch VSGs. The work further suggested that the early period of infection, when long slender forms predominate, is where most of the population will be able to switch VSGs, and the authors suggest that this is a period of initial adaptation to the new environment enabling the trypanosome to overcome, for example, antibodies present due to previous infections or coinfections. After this period of putative rapid switching and adaptation, short stumpy forms predominate and only a minority of the population are able to switch to a new variant for the remainder of the infection timescale (230). These observations emphasize the requirement for further analysis of the chronic stages of infection, and in particular how the observation that only a small proportion of trypanosomes are switching VSGs may potentially influence the mechanistic processes during infection. Analyzing this interrelationship between *VSG* switching and differentiation may provide clues to any mechanistic switch between the utilization of intact *VSG*s early in infection and the predominance of SGC and mosaic *VSG* formation later in infection. However, it also raises the question of why the repertoire of *VSG*s in *T. brucei* is relatively so enormous: if only a fraction of cells are undergoing switching during the majority of the infection lifetime, what selective pressures have resulted in formation and retention of ~2000 *VSG* genes?

WHAT DO WE KNOW ABOUT ANTIGENIC VARIATION IN OTHER SPECIES OF TRYPANOSOME?

Antigenic variation as a phenotype is expressed and has been studied in all three species of African trypanosome (*T. brucei*, *T. congolense*, and *T. vivax*), and findings from the more laboratory-tractable and therefore intensively studied *T. brucei* have been largely assumed to translate to the other two species. Genomic analyses have challenged this assumption, revealing significant differences between the *VSG* archives in the three species (75, 84). Whole genome comparisons revealed that in *T. congolense* there is no equivalent of the *T. brucei* a-VSG subfamily; in contrast, there are two b-VSG subfamilies and a much greater range of distinct C-terminal domains (15–20 CTD types, each of which are found to be associated with particular *VSG* subtypes, in contrast to the single CTD found in *T. brucei*). This finding suggests that there is much greater structural heterogeneity in *T. congolense* VSGs, and that the *T. brucei* common CTD has evolved through transfer from one subfamily to the other, perhaps due to *VSG* switching. *Trypanosoma vivax* appears to possess the most structurally diverse *VSG* repertoire, with a-*VSG*s and b-*VSG*s, as well as two further types. Jackson and colleagues also compared the phylogenies of *VSG* families, which suggests differences in the role that recombination has played in shaping the archives in the three species. The data suggest that in *T. brucei* recombination occurs across the whole repertoire of *VSG*s, whereas in *T. congolense* recombination is focused within *VSG* clades, and in *T. vivax* there is relatively little evidence of *VSG* recombination. Finally, given our previous discussion of the increasing weight given to SGC, mosaic gene formation and use of the pseudogene repertoire in the chronic (or post-acute) stages of *T. brucei* infections, it is notable that the proportion of *VSG*s that are pseudogenes is markedly lower in either *Trypanosoma congolense* (21% and 29% of the two *VSG* subfamilies) or *T. vivax* (15% and 27%of two of the four *VSG* subfamilies) compared with *T. brucei* (69% and 72% of a- and b-VSGs, respectively). These findings raise intriguing questions about the different selective pressures that have shaped the size and organization of the different *VSG* archives, whether these relate to any mechanistic differences in *VSG* usage, and how these divergent systems express a similarly efficient antigenic variation phenotype that results in establishment of chronic infections and onwards transmission. It is clear that much insight will be gained from examining VSG expression diversity and the prevalent switch mechanisms in all three trypanosome species.

CONCLUSIONS

The 10 years that have passed since the publication of *Mobile DNA II* have seen substantial progress in our understanding of antigenic variation in African trypanosomes. In part these advances have stemmed from increasingly sophisticated tools for manipulation of, in particular, *T. brucei*, but they have also been accelerated by sequencing the genomes of the African trypanosomes and related kinetoplastid parasites. This has firmly connected *VSG* switching by recombination with general repair of genome damage through HR, aligning this strategy with antigenic variation as it operates in bacterial pathogens. A major shift in our understanding of the strategy of *VSG* switching is the realization that segmental gene conversion, frequently piecing together gene segments from multiple pseudogenes, allows for enormous coat diversity and is a major element of how antigenic variation promotes trypanosome survival; again, this aligns antigen switching in the African trypanosome with many other pathogens. The full extent of VSG diversity in trypanosomes, and how it is generated, remains unclear, but ever-improving next-generation strategies for analysis of DNA, RNA and protein content in all cells will provide routes to answer these questions. The same approaches are likely to allow us to identify putative trypanosome-specific features of antigenic variation.

Acknowledgments. *We thank the Wellcome Trust, the Royal Society, the BBSRC, the MRC and GALVmed/DfID for the predominant funding of our work over the last 10 years. We are grateful to many colleagues for their discussions over this time, but wish to particularly acknowledge Dave Barry, Piet Borst and Andy Tait for their guidance and support as each of us began our studies in trypanosomes.*

Citation. McCulloch R, Morrison LJ, Hall JPJ. 2014. DNA recombination strategies during antigenic variation in the african trypanosome. Microbiol Spectrum 2(6):MDNA3-0016-2014.

References

1. **Brockhurst MA, Koskella B.** 2013. Experimental coevolution of species interactions. *Trends Ecol Evol* **28:** 367–375.
2. **Hirano M, Das S, Guo P, Cooper MD.** 2011. The evolution of adaptive immunity in vertebrates. *Adv Immunol* **109:**125–157.
3. **Sniegowski PD, Murphy HA.** 2006. Evolvability. *Curr Biol* **16:**R831–R834.
4. **Graves CJ, Ros VI, Stevenson B, Sniegowski PD, Brisson D.** 2013. Natural selection promotes antigenic evolvability. *PLoS Pathog* **9:**e1003766.
5. **Nuismer SL, Otto SP.** 2005. Host–parasite interactions and the evolution of gene expression. *PLoS Biol* **3:**e203.
6. **Gjini E, Haydon DT, Barry JD, Cobbold CA.** 2010. Critical interplay between parasite differentiation, host immunity, and antigenic variation in trypanosome infections. *Am Nat* **176:**424–439.
7. **Borst P.** 2002. Antigenic Variation in Eukaryotic Parasites, p 953–971. *In* Craig NL, Berg DE (ed), *Mobile DNA II*. ASM Press, Washington.
8. **Barry JD, McCulloch R.** 2001. Antigenic variation in trypanosomes: enhanced phenotypic variation in a eukaryotic parasite. *Adv Parasitol* **49:**1–70.
9. **Deitsch KW, Moxon ER, Wellems TE.** 1997. Shared themes of antigenic variation and virulence in bacterial, protozoal, and fungal infections. *Microbiol Mol Biol Rev* **61:**281–293.
10. **Deitsch KW, Lukehart SA, Stringer JR.** 2009. Common strategies for antigenic variation by bacterial, fungal and protozoan pathogens. *Nat Rev Microbiol* **7:**493–503.
11. **Turner CM.** 1997. The rate of antigenic variation in fly-transmitted and syringe-passaged infections of *Trypanosoma brucei*. *FEMS Microbiol Lett*, **153:**227–231.
12. **Norris SJ.** 2006. Antigenic variation with a twist—the *Borrelia* story. *Mol Microbiol* **60:**1319–1322.
13. **Barrett MP, Burchmore RJ, Stich A, Lazzari JO, Frasch AC, Cazzulo JJ, Krishna S.** 2003. The trypanosomiases. *Lancet* **362:**1469–1480.
14. **Lai DH, Hashimi H, Lun ZR, Ayala FJ, Lukes J.** 2008. Adaptations of *Trypanosoma brucei* to gradual loss of kinetoplast DNA: *Trypanosoma equiperdum* and *Trypanosoma evansi* are petite mutants of *T. brucei*. *Proc Natl Acad Sci USA* **105:**1999–2004.
15. **Barbet AF, McGuire TC.** 1978. Crossreacting determinants in variant-specific surface antigens of African trypanosomes. *Proc Natl Acad Sci USA* **75:**1989–1993.
16. **Barry JD.** 1986. Antigenic variation during *Trypanosoma vivax* infections of different host species. *Parasitology* **92:**51–65.
17. **Vickerman K.** 1978. Antigenic variation in trypanosomes. *Nature* **273:**613–617.
18. **Vickerman K, Luckins AG.** 1969. Localization of variable antigens in the surface coat of *Trypanosoma brucei* using ferritin conjugated antibody. *Nature* **224:**1125–1126.
19. **Cross GA.** 1975. Identification, purification and properties of clone-specific glycoprotein antigens constituting the surface coat of *Trypanosoma brucei*. *Parasitology* **71:**393–417.
20. **Borst P, Cross GA.** 1982. Molecular basis for trypanosome antigenic variation. *Cell* **29:**291–303.
21. **Morrison LJ, Marcello L, McCulloch R.** 2009. Antigenic variation in the African trypanosome: molecular mechanisms and phenotypic complexity. *Cell Microbiol* **11:**1724–1734.
22. **Horn D.** 2009. Antigenic variation: extending the reach of telomeric silencing. *Curr Biol* **19:**R496–R498.
23. **Rudenko G.** 2011. African trypanosomes: the genome and adaptations for immune evasion. *Essays Biochem* **51:**47–62.
24. **Horn D, McCulloch R.** 2010. Molecular mechanisms underlying the control of antigenic variation in African trypanosomes. *Curr Opin Microbiol* **13:**700–705.
25. **Glover L, Hutchinson S, Alsford S, McCulloch R, Field MC, Horn D.** 2013. Antigenic variation in African trypanosomes: the importance of chromosomal and nuclear context in VSG expression control. *Cell Microbiol* **15:**1984–1993.
26. **Higgins MK, Carrington M.** 2014. Sequence variation and structural conservation allows development of novel function and immune evasion in parasite surface protein families. *Protein Sci* **23:**354–365.
27. **Barry JD, Hall JP, Plenderleith L.** 2012. Genome hyperevolution and the success of a parasite. *Ann NY Acad Sci* **1267:**11–17.
28. **Barbour AG, Restrepo BI.** 2000. Antigenic variation in vector-borne pathogens. *Emerg Infect Dis* **6:**449–457.
29. **Schwede A, Jones N, Engstler M, Carrington M.** 2011. The VSG C-terminal domain is inaccessible to antibodies on live trypanosomes. *Mol Biochem Parasitol* **175:**201–204.
30. **Engstler M, Pfohl T, Herminghaus S, Boshart M, Wiegertjes G, Heddergott N, Overath P.** 2007. Hydrodynamic flow-mediated protein sorting on the cell surface of trypanosomes. *Cell* **131:**505–515.
31. **Seyfang A, Mecke D, Duszenko M.** 1990. Degradation, recycling, and shedding of *Trypanosoma brucei* variant surface glycoprotein. *J Protozool* **37:**546–552.
32. **Higgins MK, Tkachenko O, Brown A, Reed J, Raper J, Carrington M.** 2013. Structure of the trypanosome haptoglobin-hemoglobin receptor and implications for nutrient uptake and innate immunity. *Proc Natl Acad Sci USA* **110:**1905–1910.
33. **Greif G, Ponce de LM, Lamolle G, Rodriguez M, Pineyro D, Tavares-Marques LM, Reyna-Bello A, Robello C, Alvarez-Valin F.** 2013. Transcriptome analysis of the bloodstream stage from the parasite *Trypanosoma vivax*. *BMC Genomics* **14:**149.
34. **La GF, Magez S.** 2011. Vaccination against trypanosomiasis: can it be done or is the trypanosome truly the

ultimate immune destroyer and escape artist? *Hum Vaccin* 7:1225–1233.

35. **Guirnalda P, Murphy NB, Nolan D, Black SJ.** 2007. Anti-*Trypanosoma brucei* activity in Cape buffalo serum during the cryptic phase of parasitemia is mediated by antibodies. *Int J Parasitol* **37:**1391–1399.

36. **Blum ML, Down JA, Gurnett AM, Carrington M, Turner MJ, Wiley DC.** 1993. A structural motif in the variant surface glycoproteins of *Trypanosoma brucei*. *Nature* **362:**603–609.

37. **Marcello L, Barry JD.** 2007. Analysis of the VSG gene silent archive in *Trypanosoma brucei* reveals that mosaic gene expression is prominent in antigenic variation and is favored by archive substructure. *Genome Res* **17:** 1344–1352.

38. **Metcalf P, Blum M, Freymann D, Turner M, Wiley DC.** 1987. Two variant surface glycoproteins of *Trypanosoma brucei* of different sequence classes have similar 6 A resolution X-ray structures. *Nature* **325:**84–86.

39. **Daniels JP, Gull K, Wickstead B.** 2010. Cell biology of the trypanosome genome. *Microbiol Mol Biol Rev* **74:** 552–569.

40. **Siegel TN, Gunasekera K, Cross GA, Ochsenreiter T.** 2011. Gene expression in *Trypanosoma brucei*: lessons from high-throughput RNA sequencing. *Trends Parasitol* **27:**434–441.

41. **Gunzl A, Bruderer T, Laufer G, Schimanski B, Tu LC, Chung HM, Lee PT, Lee MG.** 2003. RNA polymerase I transcribes procyclin genes and variant surface glycoprotein gene expression sites in *Trypanosoma brucei*. *Eukaryot Cell* **2:**542–551.

42. **Zomerdijk JC, Ouellette M, Ten Asbroek AL, Kieft R, Bommer AM, Clayton CE, Borst P.** 1990. The promoter for a variant surface glycoprotein gene expression site in *Trypanosoma brucei*. *EMBO J* **9:**2791–2801.

43. **Brandenburg J, Schimanski B, Nogoceke E, Nguyen TN, Padovan JC, Chait BT, Cross GA, Gunzl A.** 2007. Multifunctional class I transcription in *Trypanosoma brucei* depends on a novel protein complex. *EMBO J* **26:**4856–4866.

44. **Chaves I, Zomerdijk J, Dirks-Mulder A, Dirks RW, Raap AK, Borst P.** 1998. Subnuclear localization of the active variant surface glycoprotein gene expression site in *Trypanosoma brucei*. *Proc Natl Acad Sci U S A* **95:** 12328–12333.

45. **Navarro M, Gull K.** 2001. A pol I transcriptional body associated with VSG mono-allelic expression in *Trypanosoma brucei*. *Nature* **414:**759–763.

46. **Berriman M, Ghedin E, Hertz-Fowler C, Blandin G, Renauld H, Bartholomeu DC, Lennard NJ, Caler E, Hamlin NE, Haas B, Böhme U, Hannick L, Aslett MA, Shallom J, Marcello L, Hou L, Wickstead B, Alsmark UC, Arrowsmith C, Atkin RJ, Barron AJ, Bringaud F, Brooks K, Carrington M, Cherevach I, Chillingworth TJ, Churcher C, Clark LN, Corton CH, Cronin A, Davies RM, Doggett J, Djikeng A, Feldblyum T, Field MC, Fraser A, Goodhead I, Hance Z, Harper D, Harris BR, Hauser H, Hostetler J, Ivens A, Jagels K, Johnson D, Johnson J, Jones K, Kerhornou AX, Koo H, Larke N, Landfear S, Larkin C, Leech V, Line A, Lord A, Macleod A, Mooney PJ, Moule S, Martin DM, Morgan GW, Mungall K, Norbertczak H, Ormond D, Pai G, Peacock CS, Peterson J, Quail MA, Rabbinowitsch E, Rajandream MA, Reitter C, Salzberg SL, Sanders M, Schobel S, Sharp S, Simmonds M, Simpson AJ, Tallon L, Turner CM, Tait A, Tivey AR, Van Aken S, Walker D, Wanless D, Wang S, White B, White O, Whitehead S, Woodward J, Wortman J, Adams MD, Embley TM, Gull K, Ullu E, Barry JD, Fairlamb AH, Opperdoes F, Barrell BG, Donelson JE, Hall N, Fraser CM, Melville SE, El-Sayed NM.** 2005. The genome of the African trypanosome *Trypanosoma brucei*. *Science* **309:**416–422.

47. **Jackson AP, Sanders M, Berry A, McQuillan J, Aslett MA, Quail MA, Chukualim B, Capewell P, MacLeod A, Melville SE, Gibson W, Barry JD, Berriman M, Hertz-Fowler C.** 2010. The genome sequence of *Trypanosoma brucei* gambiense, causative agent of chronic human african trypanosomiasis. *PLoS Negl Trop Dis*, **4:**e658.

48. **Hertz-Fowler C, Figueiredo LM, Quail MA, Becker M, Jackson A, Bason N, Brooks K, Churcher C, Fahkro S, Goodhead I, Heath P, Kartvelishvili M, Mungall K, Harris D, Hauser H, Sanders M, Saunders D, Seeger K, Sharp S, Taylor JE, Walker D, White B, Young R, Cross GA, Rudenko G, Barry JD, Louis EJ, Berriman M.** 2008. Telomeric expression sites are highly conserved in *Trypanosoma brucei*. *PLoS ONE* **3:**e3527.

49. **Navarro M, Cross GA.** 1996. DNA rearrangements associated with multiple consecutive directed antigenic switches in *Trypanosoma brucei*. *Mol Cell Biol* **16:** 3615–3625.

50. **Young R, Taylor JE, Kurioka A, Becker M, Louis EJ, Rudenko G.** 2008. Isolation and analysis of the genetic diversity of repertoires of VSG expression site containing telomeres from *Trypanosoma brucei gambiense*, *T. b. brucei* and *T. equiperdum*. *BMC Genomics* **9:**385.

51. **Barry JD, Ginger ML, Burton P, McCulloch R.** 2003. Why are parasite contingency genes often associated with telomeres? *Int J Parasitol* **33:**29–45.

52. **Linardopoulou EV, Williams EM, Fan Y, Friedman C, Young JM, Trask BJ.** 2005. Human subtelomeres are hot spots of interchromosomal recombination and segmental duplication. *Nature* **437:**94–100.

53. **Fan C, Zhang Y, Yu Y, Rounsley S, Long M, Wing RA.** 2008. The subtelomere of *Oryza sativa* chromosome 3 short arm as a hot bed of new gene origination in rice. *Mol Plant* **1:**839–850.

54. **Brown CA, Murray AW, Verstrepen KJ.** 2010. Rapid expansion and functional divergence of subtelomeric gene families in yeasts. *Curr Biol* **20:**895–903.

55. **Moraes Barros RR, Marini MM, Antonio CR, Cortez DR, Miyake AM, Lima FM, Ruiz JC, Bartholomeu DC, Chiurillo MA, Ramirez JL, da Silveira JF.** 2012. Anatomy and evolution of telomeric and subtelomeric regions in the human protozoan parasite *Trypanosoma cruzi*. *BMC Genomics* **13:**229.

56. **Shah JS, Young JR, Kimmel BE, Iams KP, Williams RO.** 1987. The 5′ flanking sequence of a *Trypanosoma brucei*

variable surface glycoprotein gene. *Mol Biochem Parasitol* **24:**163–174.

57. **Ohshima K, Kang S, Larson JE, Wells RD.** 1996. TTA. TAA triplet repeats in plasmids form a non-H bonded structure. *J Biol Chem* **271:**16784–16791.
58. **Pan X, Liao Y, Liu Y, Chang P, Liao L, Yang L, Li H.** 2010. Transcription of AAT*ATT triplet repeats in *Escherichia coli* is silenced by H-NS and IS1E transposition. *PLoS One* **5:**e14271.
59. **Borst P, Rudenko G, Blundell PA, van Leeuwen F, Cross MA, McCulloch R, Gerrits H, Chaves IM.** 1997. Mechanisms of antigenic variation in African trypanosomes. *Behring Inst Mitt*1–15.
60. **Pays E, Lips S, Nolan D, Vanhamme L, Perez-Morga D.** 2001. The VSG expression sites of *Trypanosoma brucei*: multipurpose tools for the adaptation of the parasite to mammalian hosts. *Mol Biochem Parasitol* **114:**1–16.
61. **McCulloch R, Horn D.** 2009. What has DNA sequencing revealed about the VSG expression sites of African trypanosomes? *Trends Parasitol* **25:**359–63.
62. **Siegel TN, Hekstra DR, Wang X, Dewell S, Cross GA.** 2010. Genome-wide analysis of mRNA abundance in two life-cycle stages of *Trypanosoma brucei* and identification of splicing and polyadenylation sites. *Nucleic Acids Res* **38:**4946–4957.
63. **Bitter W, Gerrits H, Kieft R, Borst P.** 1998. The role of transferrin-receptor variation in the host range of *Trypanosoma brucei*. *Nature* **391:**499–502.
64. **van Luenen HG, Kieft R, Mussmann R, Engstler M, ter Riet B, Borst P.** 2005. Trypanosomes change their transferrin receptor expression to allow effective uptake of host transferrin. *Mol Microbiol* **58:**151–165.
65. **Gerrits H, Mussmann R, Bitter W, Kieft R, Borst P.** 2002. The physiological significance of transferrin receptor variations in *Trypanosoma brucei*. *Mol Biochem Parasitol* **119:**237–247.
66. **Steverding D.** 2006. On the significance of host antibody response to the *Trypanosoma brucei* transferrin receptor during chronic infection. *Microbes Infect* **8:** 2777–2782.
67. **Salmon D, Paturiaux-Hanocq F, Poelvoorde P, Vanhamme L, Pays E.** 2005. *Trypanosoma brucei*: growth differences in different mammalian sera are not due to the species-specificity of transferrin. *Exp Parasitol* **109:**188–194.
68. **Cordon-Obras C, Cano J, Gonzalez-Pacanowska D, Benito A, Navarro M, Bart JM.** 2013. *Trypanosoma brucei gambiense* Adaptation to Different Mammalian Sera Is Associated with VSG Expression Site Plasticity. *PLoS ONE* **8:**e85072.
69. **Salmon D, Vanwalleghem G, Morias Y, Denoeud J, Krumbholz C, Lhommé F, Bachmaier S, Kador M, Gossmann J, Dias FB, De Muylder G, Uzureau P, Magez S, Moser M, De Baetselier P, Van Den Abbeele J, Beschin A, Boshart M, Pays E.** 2012. Adenylate cyclases of *Trypanosoma brucei* inhibit the innate immune response of the host. *Science* **337:**463–466.
70. **Xong HV, Vanhamme L, Chamekh M, Chimfwembe CE, Van den AJ, Pays A, Van Meirvenne N, Hamers R, De Baetselier P, Pays E.** 1998. A VSG expression site-associated gene confers resistance to human serum in *Trypanosoma rhodesiense*. *Cell* **95:**839–846.
71. **Wickstead B, Ersfeld K, Gull K.** 2004. The small chromosomes of *Trypanosoma brucei* involved in antigenic variation are constructed around repetitive palindromes. *Genome Res* **14:**1014–1024.
72. **Ginger ML, Blundell PA, Lewis AM, Browitt A, Gunzl A, Barry JD.** 2002. Ex Vivo and In Vitro Identification of a Consensus Promoter for VSG Genes Expressed by Metacyclic-Stage Trypanosomes in the Tsetse Fly. *Eukaryot Cell* **1:**1000–1009.
73. **Sharma R, Gluenz E, Peacock L, Gibson W, Gull K, Carrington M.** 2009. The heart of darkness: growth and form of *Trypanosoma brucei* in the tsetse fly. *Trends Parasitol* **25:**517–524.
74. **Kolev NG, Ramey-Butler K, Cross GA, Ullu E, Tschudi C.** 2012. Developmental progression to infectivity in *Trypanosoma brucei* triggered by an RNA-binding protein. *Science* **338:**1352–1353.
75. **Jackson AP, Allison HC, Barry JD, Field MC, Hertz-Fowler C, Berriman M.** 2013. A cell-surface phylome for African trypanosomes. *PLoS Negl Trop Dis* **7:**e2121.
76. **Marcello L, Menon S, Ward P, Wilkes JM, Jones NG, Carrington M, Barry JD.** 2007. VSGdb: a database for trypanosome variant surface glycoproteins, a large and diverse family of coiled coil proteins. *BMC Bioinformatics* **8:**143.
77. **Weiden M, Osheim YN, Beyer AL, Van der Ploeg LH.** 1991. Chromosome structure: DNA nucleotide sequence elements of a subset of the minichromosomes of the protozoan *Trypanosoma brucei*. *Mol Cell Biol* **11:** 3823–3834.
78. **Akiyoshi B, Gull K.** 2014. Discovery of Unconventional Kinetochores in Kinetoplastids. *Cell* **156:**1247–1258.
79. **Rothwell V, Aline R Jr, Parsons M, Agabian N, Stuart K.** 1985. Expression of a minichromosomal variant surface glycoprotein gene in *Trypanosoma brucei*. *Nature* **313:**595–597.
80. **Melville SE, Gerrard CS, Blackwell JM.** 1999. Multiple causes of size variation in the diploid megabase chromosomes of African trypanosomes. *Chromosome Res* **7:** 191–203.
81. **Callejas S, Leech V, Reitter C, Melville S.** 2006. Hemizygous subtelomeres of an African trypanosome chromosome may account for over 75% of chromosome length. *Genome Res* **16:**1109–1118.
82. **MacLean RC, Torres-Barcelo C, Moxon R.** 2013. Evaluating evolutionary models of stress-induced mutagenesis in bacteria. *Nat Rev Genet* **14:**221–227.
83. **Gjini E, Haydon DT, Barry JD, Cobbold CA.** 2012. The Impact of Mutation and Gene Conversion on the Local Diversification of Antigen Genes in African Trypanosomes. *Mol Biol Evol* **29:**3321–3331.
84. **Jackson AP, Berry A, Aslett M, Allison HC, Burton P, Vavrova-Anderson J, Brown R, Browne H, Corton N, Hauser H, Gamble J, Gilderthorp R, Marcello L, McQuillan J, Otto TD, Quail MA, Sanders MJ,**

van Tonder A, Ginger ML, Field MC, Barry JD, Hertz-Fowler C, Berriman M. 2012. Antigenic diversity is generated by distinct evolutionary mechanisms in African trypanosome species. *Proc Natl Acad Sci USA* **109:** 3416–3421.

85. **Borst P, Ulbert S.** 2001. Control of VSG gene expression sites. *Mol Biochem Parasitol* **114:**17–27.

86. **Borst P.** 2002. Antigenic variation and allelic exclusion. *Cell* **109:**5–8.

87. **Pays E.** 2005. Regulation of antigen gene expression in *Trypanosoma brucei*. *Trends Parasitol* **21:**517–520.

88. **Navarro M, Penate X, Landeira D.** 2007. Nuclear architecture underlying gene expression in *Trypanosoma brucei*. *Trends Microbiol* **15:**263–270.

89. **Schwede A, Carrington M.** 2010. Bloodstream form Trypanosome plasma membrane proteins: antigenic variation and invariant antigens. *Parasitology* **137:**2029–2039.

90. **Chaves I, Rudenko G, Dirks-Mulder A, Cross M, Borst P.** 1999. Control of variant surface glycoprotein gene-expression sites in *Trypanosoma brucei*. *EMBO J* **18:** 4846–4855.

91. **Ulbert S, Chaves I, Borst P.** 2002. Expression site activation in *Trypanosoma brucei* with three marked variant surface glycoprotein gene expression sites. *Mol Biochem Parasitol* **120:**225–235.

92. **Baltz T, Giroud C, Baltz D, Roth C, Raibaud A, Eisen H.** 1986. Stable expression of two variable surface glycoproteins by cloned *Trypanosoma equiperdum*. *Nature* **319:**602–604.

93. **Munoz-Jordan JL, Davies KP, Cross GA.** 1996. Stable expression of mosaic coats of variant surface glycoproteins in *Trypanosoma brucei*. *Science* **272:**1795–1797.

94. **Yang X, Figueiredo LM, Espinal A, Okubo E, Li B.** 2009. RAP1 is essential for silencing telomeric variant surface glycoprotein genes in *Trypanosoma brucei*. *Cell* **137:**99–109.

95. **Denninger V, Fullbrook A, Bessat M, Ersfeld K, Rudenko G.** 2010. The FACT subunit TbSpt16 is involved in cell cycle specific control of VSG expression sites in *Trypanosoma brucei*. *Mol Microbiol* **78:**459–474.

96. **Povelones ML, Gluenz E, Dembek M, Gull K, Rudenko G.** 2012. Histone H1 Plays a Role in Heterochromatin Formation and VSG Expression Site Silencing in *Trypanosoma brucei*. *PLoS Pathog* **8:**e1003010.

97. **Alsford S, Horn D.** 2012. Cell-cycle-regulated control of VSG expression site silencing by histones and histone chaperones ASF1A and CAF-1b in *Trypanosoma brucei*. *Nucleic Acids Res* **40:**10150–10160.

98. **Narayanan MS, Rudenko G.** 2013. TDP1 is an HMG chromatin protein facilitating RNA polymerase I transcription in African trypanosomes. *Nucleic Acids Res* **41:**2981–2992.

99. **DuBois KN, Alsford S, Holden JM, Buisson J, Swiderski M, Bart JM, Ratushny AV, Wan Y, Bastin P, Barry JD, Navarro M, Horn D, Aitchison JD, Rout MP, Field MC.** 2012. NUP-1 Is a large coiled-coil nucleoskeletal protein in trypanosomes with lamin-like functions. *PLoS Biol* **10:**e1001287.

100. **Vanhamme L, Poelvoorde P, Pays A, Tebabi P, Van Xong H, Pays E.** 2000. Differential RNA elongation controls the variant surface glycoprotein gene expression sites of *Trypanosoma brucei*. *Mol Microbiol* **36:** 328–340.

101. **Nguyen TN, Muller LS, Park SH, Siegel TN, Gunzl A.** 2013. Promoter occupancy of the basal class I transcription factor A differs strongly between active and silent VSG expression sites in Trypanosoma brucei. *Nucleic Acids Res* **42:**3164–3176.

102. **Figueiredo LM, Janzen CJ, Cross GA.** 2008. A histone methyltransferase modulates antigenic variation in African trypanosomes. *PLoS Biol* **6:**e161.

103. **Stockdale C, Swiderski MR, Barry JD, McCulloch R.** 2008. Antigenic variation in *Trypanosoma brucei*: joining the DOTs. *PLoS Biol* **6:**e185.

104. **Landeira D, Bart JM, Van Tyne D, Navarro M.** 2009. Cohesin regulates VSG monoallelic expression in trypanosomes. *J Cell Biol* **186:**243–254.

105. **Tiengwe C, Marcello L, Farr H, Dickens N, Kelly S, Swiderski M, Vaughan D, Gull K, Barry JD, Bell SD, McCulloch R.** 2012. Genome-wide Analysis Reveals Extensive Functional Interaction between DNA Replication Initiation and Transcription in the Genome of *Trypanosoma brucei*. *Cell Rep* **2:**185–197.

106. **Benmerzouga I, Concepcion-Acevedo J, Kim HS, Vandoros AV, Cross GA, Klingbeil MM, Li B.** 2013. *Trypanosoma brucei* Orc1 is essential for nuclear DNA replication and affects both VSG silencing and VSG switching. *Mol Microbiol* **87:**196–210.

107. **Kim HS, Park SH, Gunzl A, Cross GA.** 2013. MCM-BP is required for repression of life-cycle specific genes transcribed by RNA polymerase I in the mammalian infectious form of Trypanosoma brucei. *PLoS ONE* **8:** e57001.

108. **Dobson R, Stockdale C, Lapsley C, Wilkes J, McCulloch R.** 2011. Interactions among *Trypanosoma brucei* RAD51 paralogues in DNA repair and antigenic variation. *Mol Microbiol* **81:**434–456.

109. **Hartley CL, McCulloch R.** 2008. Trypanosoma brucei BRCA2 acts in antigenic variation and has undergone a recent expansion in BRC repeat number that is important during homologous recombination. *Mol Microbiol* **68:**1237–1251.

110. **McCulloch R, Barry JD.** 1999. A role for RAD51 and homologous recombination in *Trypanosoma brucei* antigenic variation. *Genes Dev* **13:**2875–2888.

111. **Sheader K, te VD, Rudenko G.** 2004. Bloodstream form-specific up-regulation of silent vsg expression sites and procyclin in *Trypanosoma brucei* after inhibition of DNA synthesis or DNA damage. *J Biol Chem* **279:** 13363–13374.

112. **Liu AY, Van der Ploeg LH, Rijsewijk FA, Borst P.** 1983. The transposition unit of variant surface glycoprotein gene 118 of *Trypanosoma brucei*. Presence of repeated elements at its border and absence of promoter-associated sequences. *J Mol Biol* **167:**57–75.

113. **Pays E, Van Assel S, Laurent M, Dero B, Michiels F, Kronenberger P, Matthyssens G, Van Meirvenne N,**

Le Ray D, Steinert M. 1983. At least two transposed sequences are associated in the expression site of a surface antigen gene in different trypanosome clones. *Cell* **34:** 359–369.

114. **McCulloch R, Rudenko G, Borst P.** 1997. Gene conversions mediating antigenic variation in Trypanosoma brucei can occur in variant surface glycoprotein expression sites lacking 70- base-pair repeat sequences. *Mol Cell Biol* **17:**833–843.

115. **Bernards A, Van der Ploeg LH, Frasch AC, Borst P, Boothroyd JC, Coleman S, Cross GA.** 1981. Activation of trypanosome surface glycoprotein genes involves a duplication-transposition leading to an altered 3′ end. *Cell* **27:**497–505.

116. **de Lange T, Kooter JM, Michels PA, Borst P.** 1983. Telomere conversion in trypanosomes. *Nucleic Acids Res* **11:**8149–8165.

117. **Kim HS, Cross GA.** 2010. TOPO3alpha influences antigenic variation by monitoring expression-site-associated VSG switching in *Trypanosoma brucei. PLoS Pathog* **6:** e1000992.

118. **Pays E, Guyaux M, Aerts D, Van Meirvenne N, Steinert M.** 1985. Telomeric reciprocal recombination as a possible mechanism for antigenic variation in trypanosomes. *Nature* **316:**562–564.

119. **Rudenko G, McCulloch R, Dirks-Mulder A, Borst P.** 1996. Telomere exchange can be an important mechanism of variant surface glycoprotein gene switching in *Trypanosoma brucei. Mol Biochem Parasitol* **80:** 65–75.

120. **Thon G, Baltz T, Giroud C, Eisen H.** 1990. Trypanosome variable surface glycoproteins: composite genes and order of expression. *Genes Dev* **4:**1374–1383.

121. **Roth C, Bringaud F, Layden RE, Baltz T, Eisen H.** 1989. Active late-appearing variable surface antigen genes in Trypanosoma equiperdum are constructed entirely from pseudogenes. *Proc Natl Acad Sci USA* **86:** 9375–9379.

122. **Thon G, Baltz T, Eisen H.** 1989. Antigenic diversity by the recombination of pseudogenes. *Genes Dev* **3:** 1247–1254.

123. **Roth C, Jacquemot C, Giroud C, Bringaud F, Eisen H, Baltz T.** 1991. Antigenic variation in *Trypanosoma equiperdum. Res Microbiol* **142:**725–730.

124. **Hall JP, Wang H, Barry JD.** 2013. Mosaic VSGs and the scale of *Trypanosoma brucei* antigenic variation. *PLoS Pathog* **9:**e1003502.

125. **Barbet AF, Kamper SM.** 1993. The importance of mosaic genes to trypanosome survival. *Parasitol Today* **9:** 63–66.

126. **Kamper SM, Barbet AF.** 1992. Surface epitope variation via mosaic gene formation is potential key to long-term survival of *Trypanosoma brucei. Mol Biochem Parasitol* **53:**33–44.

127. **Boothroyd CE, Dreesen O, Leonova T, Ly KI, Figueiredo LM, Cross GA, Papavasiliou FN.** 2009. A yeast-endonuclease-generated DNA break induces antigenic switching in *Trypanosoma brucei. Nature* **459:** 278–281.

128. **Glover L, Alsford S, Horn D.** 2013. DNA break site at fragile subtelomeres determines probability and mechanism of antigenic variation in african trypanosomes. *PLoS Pathog* **9:**e1003260.

129. **Aitcheson N, Talbot S, Shapiro J, Hughes K, Adkin C, Butt T, Sheader K, Rudenko G.** 2005. VSG switching in *Trypanosoma brucei*: antigenic variation analysed using RNAi in the absence of immune selection. *Mol Microbiol* **57:**1608–1622.

130. **Cahoon LA, Seifert HS.** 2011. Focusing homologous recombination: pilin antigenic variation in the pathogenic *Neisseria. Mol Microbiol* **81:**1136–1143.

131. **Vink C, Rudenko G, Seifert HS.** 2011. Microbial antigenic variation mediated by homologous DNA recombination. *FEMS Microbiol Rev* **36:**917–948.

132. **San Filippo J, Sung P, Klein H.** 2008. Mechanism of Eukaryotic Homologous Recombination. *Annu Rev Biochem* **77:**229–257.

133. **Koomey M, Gotschlich EC, Robbins K, Bergstrom S, Swanson J.** 1987. Effects of recA mutations on pilus antigenic variation and phase transitions in *Neisseria gonorrhoeae. Genetics* **117:**391–398.

134. **Mehr IJ, Seifert HS.** 1998. Differential roles of homologous recombination pathways in *Neisseria gonorrhoeae* pilin antigenic variation, DNA transformation and DNA repair. *Mol Microbiol* **30:**697–710.

135. **Helm RA, Seifert HS.** 2009. Pilin antigenic variation occurs independently of the RecBCD pathway in *Neisseria gonorrhoeae. J Bacteriol* **191:**5613–5621.

136. **Roy R, Chun J, Powell SN.** 2012. BRCA1 and BRCA2: different roles in a common pathway of genome protection. *Nat Rev Cancer* **12:**68–78.

137. **Trenaman A, Hartley C, Prorocic M, Passos-Silva DG, van den Hoek M, Nechyporuk-Zloy V, Machado CR, McCulloch R.** 2013. *Trypanosoma brucei* BRCA2 acts in a life cycle-specific genome stability process and dictates BRC repeat number-dependent RAD51 subnuclear dynamics. *Nucleic Acids Res* **41:**943–960.

138. **Stohl EA, Seifert HS.** 2001. The recX gene potentiates homologous recombination in *Neisseria gonorrhoeae. Mol Microbiol* **40:**1301–1310.

139. **Gruenig MC, Stohl EA, Chitteni-Pattu S, Seifert HS, Cox MM.** 2010. Less is more: *Neisseria gonorrhoeae* RecX protein stimulates recombination by inhibiting RecA. *J Biol Chem* **285:**37188–37197.

140. **Cardenas PP, Carrasco B, Defeu SC, Cesar CE, Herr K, Kaufenstein M, Graumann PL, Alonso JC.** 2012. RecX facilitates homologous recombination by modulating RecA activities. *PLoS Genet* **8:**e1003126.

141. **Suwaki N, Klare K, Tarsounas M.** 2011. RAD51 paralogs: roles in DNA damage signalling, recombinational repair and tumorigenesis. *Semin Cell Dev Biol* **22:** 898–905.

142. **Jensen RB, Ozes A, Kim T, Estep A, Kowalczykowski SC.** 2013. BRCA2 is epistatic to the RAD51 paralogs in response to DNA damage. *DNA Repair (Amst)* **12:** 306–311.

143. **Chun J, Buechelmaier ES, Powell SN.** 2013. Rad51 paralog complexes BCDX2 and CX3 act at different

stages in the BRCA1-BRCA2-dependent homologous recombination pathway. *Mol Cell Biol* **33:**387–395.

144. **Proudfoot C, McCulloch R.** 2005. Distinct roles for two RAD51-related genes in *Trypanosoma brucei* antigenic variation. *Nucleic Acids Res* **33:**6906–6919.

145. **Kim HS, Cross GA.** 2011. Identification of *Trypanosoma brucei* RMI1/BLAP75 Homologue and Its Roles in Antigenic Variation. *PLoS One* **6:**e25313.

146. **Killoran MP, Kohler PL, Dillard JP, Keck JL.** 2009. RecQ DNA helicase HRDC domains are critical determinants in *Neisseria gonorrhoeae* pilin antigenic variation and DNA repair. *Mol Microbiol* **71:**158–171.

147. **Cahoon LA, Manthei KA, Rotman E, Keck JL, Seifert HS.** 2013. *Neisseria gonorrhoeae* RecQ helicase HRDC domains are essential for efficient binding and unwinding of the pilE guanine quartet structure required for pilin antigenic variation. *J Bacteriol* **195:**2255–2261.

148. **Cahoon LA, Seifert HS.** 2009. An alternative DNA structure is necessary for pilin antigenic variation in *Neisseria gonorrhoeae*. *Science* **325:**764–767.

149. **Cahoon LA, Seifert HS.** 2013. Transcription of a cis-acting, noncoding, small RNA is required for pilin antigenic variation in *Neisseria gonorrhoeae*. *PLoS Pathog* **9:**e1003074.

150. **Kuryavyi V, Cahoon LA, Seifert HS, Patel DJ.** 2012. RecA-binding pilE G4 sequence essential for pilin antigenic variation forms monomeric and 5′ end-stacked dimeric parallel G-quadruplexes. *Structure* **20:**2090–2102.

151. **Stracker TH, Petrini JH.** 2011. The MRE11 complex: starting from the ends. *Nat Rev Mol Cell Biol* **12:** 90–103.

152. **Robinson NP, McCulloch R, Conway C, Browitt A, Barry JD.** 2002. Inactivation of Mre11 Does Not Affect VSG Gene Duplication Mediated by Homologous Recombination in *Trypanosoma brucei*. *J Biol Chem* **277:** 26185–26193.

153. **Tan KS, Leal ST, Cross GA.** 2002. *Trypanosoma brucei* MRE11 is non-essential but influences growth, homologous recombination and DNA double-strand break repair. *Mol Biochem Parasitol* **125:**11–21.

154. **Jiricny J.** 2013. Postreplicative mismatch repair. *Cold Spring Harb Perspect Biol* **5:**a012633.

155. **Bell JS, McCulloch R.** 2003. Mismatch repair regulates homologous recombination, but has little influence on antigenic variation, in *Trypanosoma brucei*. *J Biol Chem* **278:**45182–45188.

156. **Bell JS, Harvey TI, Sims AM, McCulloch R.** 2004. Characterization of components of the mismatch repair machinery in *Trypanosoma brucei*. *Mol Microbiol* **51:** 159–173.

157. **Barnes RL, McCulloch R.** 2007. Trypanosoma brucei homologous recombination is dependent on substrate length and homology, though displays a differential dependence on mismatch repair as substrate length decreases. *Nucleic Acids Res* **35:**3478–3493.

158. **Slean MM, Panigrahi GB, Ranum LP, Pearson CE.** 2008. Mutagenic roles of DNA "repair" proteins in antibody diversity and disease-associated trinucleotide repeat instability. *DNA Repair (Amst)* **7:**1135–1154.

159. **Hill SA, Davies JK.** 2009. Pilin gene variation in *Neisseria gonorrhoeae*: reassessing the old paradigms. *FEMS Microbiol Rev* **33:**521–530.

160. **Criss AK, Bonney KM, Chang RA, Duffin PM, LeCuyer BE, Seifert HS.** 2010. Mismatch correction modulates mutation frequency and pilus phase and antigenic variation in *Neisseria gonorrhoeae*. *J Bacteriol* **192:**316–325.

161. **Ottaviani D, Lecain M, Sheer D.** 2014. The role of microhomology in genomic structural variation. *Trends Genet* **30:**85–94.

162. **Conway C, McCulloch R, Ginger ML, Robinson NP, Browitt A, Barry JD.** 2002. Ku is important for telomere maintenance, but not for differential expression of telomeric VSG genes, in African trypanosomes. *J Biol Chem* **277:**21269–21277.

163. **Janzen CJ, Lander F, Dreesen O, Cross GA.** 2004. Telomere length regulation and transcriptional silencing in KU80-deficient *Trypanosoma brucei*. *Nucleic Acids Res* **32:**6575–6584.

164. **Gill EE, Fast NM.** 2007. Stripped-down DNA repair in a highly reduced parasite. *BMC Mol Biol* **8:**24.

165. **Burton P, McBride DJ, Wilkes JM, Barry JD, McCulloch R.** 2007. Ku Heterodimer-Independent End Joining in *Trypanosoma brucei* Cell Extracts Relies upon Sequence Microhomology. *Eukaryot Cell* **6:**1773–1781.

166. **Conway C, Proudfoot C, Burton P, Barry JD, McCulloch R.** 2002. Two pathways of homologous recombination in *Trypanosoma brucei*. *Mol Microbiol* **45:** 1687–1700.

167. **Glover L, McCulloch R, Horn D.** 2008. Sequence homology and microhomology dominate chromosomal double-strand break repair in African trypanosomes. *Nucleic Acids Res* **36:**2608–2618.

168. **Glover L, Jun J, Horn D.** 2011. Microhomology-mediated deletion and gene conversion in African trypanosomes. *Nucleic Acids Res* **39:**1372–1380.

169. **Liveris D, Mulay V, Sandigursky S, Schwartz I.** 2008. *Borrelia burgdorferi* vlsE antigenic variation is not mediated by RecA. *Infect Immun* **76:**4009–4018.

170. **Dresser AR, Hardy PO, Chaconas G.** 2009. Investigation of the genes involved in antigenic switching at the vlsE locus in *Borrelia burgdorferi*: an essential role for the RuvAB branch migrase. *PLoS Pathog* **5:**e1000680.

171. **Lin T, Gao L, Edmondson DG, Jacobs MB, Philipp MT, Norris SJ.** 2009. Central role of the Holliday junction helicase RuvAB in vlsE recombination and infectivity of *Borrelia burgdorferi*. *PLoS Pathog* **5:**e1000679.

172. **Mir T, Huang SH, Kobryn K.** 2013. The telomere resolvase of the Lyme disease spirochete, *Borrelia burgdorferi*, promotes DNA single-strand annealing and strand exchange. *Nucleic Acids Res* **41:**10438–10448.

173. **Barry JD.** 1997. The relative significance of mechanisms of antigenic variation in African trypanosomes. *Parasitol Today* **13:**212–218.

174. **Barry D, McCulloch R.** 2009. Molecular microbiology: a key event in survival. *Nature* **459:**172–173.

175. **Keim C, Kazadi D, Rothschild G, Basu U.** 2013. Regulation of AID, the B-cell genome mutator. *Genes Dev* **27:**1–17.

176. **Durkin SG, Glover TW.** 2007. Chromosome fragile sites. *Annu Rev Genet* **41:**169–192.

177. **Ozeri-Galai E, Lebofsky R, Rahat A, Bester AC, Bensimon A, Kerem B.** 2011. Failure of origin activation in response to fork stalling leads to chromosomal instability at fragile sites. *Mol Cell* **43:**122–131.

178. **Klar AJ.** 2007. Lessons learned from studies of fission yeast mating-type switching and silencing. *Annu Rev Genet* **41:**213–236.

179. **Yakisich JS, Kapler GM.** 2006. Deletion of the *Tetrahymena thermophila* rDNA replication fork barrier region disrupts macronuclear rDNA excision and creates a fragile site in the micronuclear genome. *Nucleic Acids Res* **34:**620–634.

180. **Dreesen O, Li B, Cross GA.** 2007. Telomere structure and function in trypanosomes: a proposal. *Nat Rev Microbiol* **5:**70–75.

181. **Hovel-Miner GA, Boothroyd CE, Mugnier M, Dreesen O, Cross GA, Papavasiliou FN.** 2012. Telomere Length Affects the Frequency and Mechanism of Antigenic Variation in *Trypanosoma brucei*. *PLoS Pathog* **8:**e1002900.

182. **Dreesen O, Li B, Cross GA.** 2005. Telomere structure and shortening in telomerase-deficient *Trypanosoma brucei*. *Nucleic Acids Res* **33:**4536–4543.

183. **Dreesen O, Cross GA.** 2006. Telomerase-independent stabilization of short telomeres in *Trypanosoma brucei*. *Mol Cell Biol* **26:**4911–4919.

184. **Meeus PF, Brayton KA, Palmer GH, Barbet AF.** 2003. Conservation of a gene conversion mechanism in two distantly related paralogues of *Anaplasma marginale*. *Mol Microbiol* **47:**633–643.

185. **Giacani L, Molini BJ, Kim EY, Godornes BC, Leader BT, Tantalo LC, Centurion-Lara A, Lukehart SA.** 2010. Antigenic variation in *Treponema pallidum*: TprK sequence diversity accumulates in response to immune pressure during experimental syphilis. *J Immunol* **184:**3822–3829.

186. **Iverson-Cabral SL, Astete SG, Cohen CR, Totten PA.** 2007. mgpB and mgpC sequence diversity in *Mycoplasma genitalium* is generated by segmental reciprocal recombination with repetitive chromosomal sequences. *Mol Microbiol* **66:**55–73.

187. **Ma L, Jensen JS, Myers L, Burnett J, Welch M, Jia Q, Martin DH.** 2007. *Mycoplasma genitalium*: an efficient strategy to generate genetic variation from a minimal genome. *Mol Microbiol* **66:**220–236.

188. **Burgos R, Wood GE, Young L, Glass JI, Totten PA.** 2012. RecA mediates MgpB and MgpC phase and antigenic variation in *Mycoplasma genitalium*, but plays a minor role in DNA repair. *Mol Microbiol* **85:**669–683.

189. **Barbet AF, Myler PJ, Williams RO, McGuire TC.** 1989. Shared surface epitopes among trypanosomes of the same serodeme expressing different variable surface glycoprotein genes. *Mol Biochem Parasitol* **32:**191–199.

190. **Sheader K, Vaughan S, Minchin J, Hughes K, Gull K, Rudenko G.** 2005. Variant surface glycoprotein RNA interference triggers a precytokinesis cell cycle arrest in African trypanosomes. *Proc Natl Acad Sci USA* **102:**8716–8721.

191. **Smith TK, Vasileva N, Gluenz E, Terry S, Portman N, Kramer S, Carrington M, Michaeli S, Gull K, Rudenko G.** 2009. Blocking variant surface glycoprotein synthesis in Trypanosoma brucei triggers a general arrest in translation initiation. *PLoS ONE* **4:**e7532.

192. **Aslam N, Turner CM.** 1992. The relationship of variable antigen expression and population growth rates in *Trypanosoma brucei*. *Parasitol Res* **78:**661–664.

193. **Wouters A.** 2005. The function debate in philosophy. *Acta Biotheor* **53:**123–151.

194. **Marcello L, Barry JD.** 2007. From silent genes to noisy populations-dialogue between the genotype and phenotypes of antigenic variation. *J Eukaryot Microbiol* **54:**14–17.

195. **Palmer GH, Bankhead T, Lukehart SA.** 2009. 'Nothing is permanent but change'- antigenic variation in persistent bacterial pathogens. *Cell Microbiol* **11:**1697–1705.

196. **Glanville J, Zhai W, Berka J, Telman D, Huerta G, Mehta GR, Ni I, Mei L, Sundar PD, Day GM, Cox D, Rajpal A, Pons J.** 2009. Precise determination of the diversity of a combinatorial antibody library gives insight into the human immunoglobulin repertoire. *Proc Natl Acad Sci USA* **106:**20216–20221.

197. **Ueti MW, Tan Y, Broschat SL, Castaneda Ortiz EJ, Camacho-Nuez M, Mosqueda JJ, Scoles GA, Grimes M, Brayton KA, Palmer GH.** 2012. Expansion of variant diversity associated with a high prevalence of pathogen strain superinfection under conditions of natural transmission. *Infect Immun* **80:**2354–2360.

198. **Zhuang Y, Futse JE, Brown WC, Brayton KA, Palmer GH.** 2007. Maintenance of antibody to pathogen epitopes generated by segmental gene conversion is highly dynamic during long-term persistent infection. *Infect Immun* **75:**5185–5190.

199. **Gray AR.** 1965. Antigenic variation in a strain of *Trypanosoma brucei* transmitted by *Glossina morsitans* and *G. palpalis*. *J Gen Microbiol* **41:**195–214.

200. **Van MN, Janssens PG, Magnus E.** 1975. Antigenic variation in syringe passaged populations of *Trypanosoma (Trypanozoon). brucei*. 1. Rationalization of the experimental approach. *Ann Soc Belg Med Trop* **55:**1–23.

201. **Capbern A, Giroud C, Baltz T, Mattern P.** 1977. [*Trypanosoma equiperdum*: antigenic variations in experimental trypanosomiasis of rabbits]. *Exp Parasitol* **42:**6–13.

202. **Morrison LJ, Majiwa P, Read AF, Barry JD.** 2005. Probabilistic order in antigenic variation of *Trypanosoma brucei*. *Int J Parasitol* **35:**961–972.

203. **Barry JD, Marcello L, Morrison LJ, Read AF, Lythgoe K, Jones N, Carrington M, Blandin G, Bohme U, Caler E, Hertz-Fowler C, Renauld H, El Sayed N, Berriman M.** 2005. What the genome sequence is revealing about trypanosome antigenic variation. *Biochem Soc Trans* **33:**986–989.

204. **Pays E.** 1989. Pseudogenes, chimaeric genes and the timing of antigen variation in African trypanosomes. *Trends Genet* **5:**389–391.

205. **Futse JE, Brayton KA, Knowles DP Jr, Palmer GH.** 2005. Structural basis for segmental gene conversion in

generation of *Anaplasma marginale* outer membrane protein variants. *Mol Microbiol* 57:212–221.

206. **Coutte L, Botkin DJ, Gao L, Norris SJ.** 2009. Detailed analysis of sequence changes occurring during vlsE antigenic variation in the mouse model of Borrelia burgdorferi infection. *PLoS Pathog* 5:e1000293.

207. **West SA, Griffin AS, Gardner A, Diggle SP.** 2006. Social evolution theory for microorganisms. *Nat Rev Microbiol* 4:597–607.

208. **Van Dyken JD, Wade MJ.** 2010. The genetic signature of conditional expression. *Genetics* **184:**557–570.

209. **Oberle M, Balmer O, Brun R, Roditi I.** 2010. Bottlenecks and the maintenance of minor genotypes during the life cycle of *Trypanosoma brucei*. *PLoS Pathog* **6:** e1001023.

210. **Nei M, Rooney AP.** 2005. Concerted and birth-and-death evolution of multigene families. *Annu Rev Genet* **39:**121–152.

211. **Peacock L, Cook S, Ferris V, Bailey M, Gibson W.** 2012. The life cycle of *Trypanosoma (Nannomonas) congolense* in the tsetse fly. *Parasit Vectors* 5:109.

212. **Borst P, Rudenko G, Taylor MC, Blundell PA, van Leeuwen F, Bitter W, Cross M, McCulloch R.** 1996. Antigenic variation in trypanosomes. *Arch Med Res* **27:** 379–388.

213. **Lydeard JR, Jain S, Yamaguchi M, Haber JE.** 2007. Break-induced replication and telomerase-independent telomere maintenance require Pol32. *Nature* **448:**820–823.

214. **Donnianni RA, Symington LS.** 2013. Break-induced replication occurs by conservative DNA synthesis. *Proc Natl Acad Sci USA* **110:**13475–13480.

215. **Malkova A, Ira G.** 2013. Break-induced replication: functions and molecular mechanism. *Curr Opin Genet Dev* **23:**271–279.

216. **Malkova A, Signon L, Schaefer CB, Naylor ML, Theis JF, Newlon CS, Haber JE.** 2001. RAD51-independent break-induced replication to repair a broken chromosome depends on a distant enhancer site. *Genes Dev* **15:** 1055–1060.

217. **Signon L, Malkova A, Naylor ML, Klein H, Haber JE.** 2001. Genetic requirements for RAD51- and RAD54-Independent Break-Induced Replication Repair of a Chromosomal Double-Strand Break. *Mol Cell Biol* **21:** 2048–2056.

218. **Davis AP, Symington LS.** 2004. RAD51-dependent break-induced replication in yeast. *Mol Cell Biol* **24:** 2344–2351.

219. **Wyatt HD, Sarbajna S, Matos J, West SC.** 2013. Coordinated actions of SLX1-SLX4 and MUS81-EME1 for Holliday junction resolution in human cells. *Mol Cell* 52:234–247.

220. **Reuner B, Vassella E, Yutzy B, Boshart M.** 1997. Cell density triggers slender to stumpy differentiation of *Trypanosoma brucei* bloodstream forms in culture. *Mol Biochem Parasitol* 90:269–280.

221. **Vassella E, Reuner B, Yutzy B, Boshart M.** 1997. Differentiation of African trypanosomes is controlled by a density sensing mechanism which signals cell cycle arrest via the cAMP pathway. *J Cell Sci* **110:**2661–2671.

222. **Dean S, Marchetti R, Kirk K, Matthews KR.** 2009. A surface transporter family conveys the trypanosome differentiation signal. *Nature* **459:**213–217.

223. **Macgregor P, Matthews KR.** 2012. Identification of the regulatory elements controlling the transmission stage-specific gene expression of PAD1 in Trypanosoma brucei. *Nucleic Acids Res* **40:**7705–7717.

224. **Mony BM, Macgregor P, Ivens A, Rojas F, Cowton A, Young J, Horn D, Matthews K.** 2014. Genome-wide dissection of the quorum sensing signalling pathway in *Trypanosoma brucei*. *Nature* **505:**681–685.

225. **Matthews KR.** 2011. Controlling and coordinating development in vector-transmitted parasites. *Science* **331:** 1149–1153.

226. **Rico E, Rojas F, Mony BM, Szoor B, Macgregor P, Matthews KR.** 2013. Bloodstream form pre-adaptation to the tsetse fly in Trypanosoma brucei. *Front Cell Infect Microbiol* 3:78.

227. **Tyler KM, Higgs PG, Matthews KR, Gull K.** 2001. Limitation of *Trypanosoma brucei* parasitaemia results from density-dependent parasite differentiation and parasite killing by the host immune response. *Proc Biol Sci* **268:**2235–2243.

228. **Lythgoe KA, Morrison LJ, Read AF, Barry JD.** 2007. Parasite-intrinsic factors can explain ordered progression of trypanosome antigenic variation. *Proc Natl Acad Sci USA* **104:**8095–8100.

229. **Macgregor P, Savill NJ, Hall D, Matthews KR.** 2011. Transmission stages dominate trypanosome within-host dynamics during chronic infections. *Cell Host Microbe* 9:310–318.

230. **Macgregor P, Szoor B, Savill NJ, Matthews KR.** 2012. Trypanosomal immune evasion, chronicity and transmission: an elegant balancing act. *Nat Rev Microbiol* **10:**431–438.

Mobile DNA, 3rd Edition
Nancy L. Craig, Michael Chandler, Martin Gellert, Alan M. Lambowitz, Phoebe A. Rice and Suzanne Sandmeyer

doi:10.1128/microbiolspec.MDNA3-0022-2014

Laura A. Kirkman[1,2]
Kirk W. Deitsch[2]

20 Recombination and Diversification of the Variant Antigen Encoding Genes in the Malaria Parasite *Plasmodium falciparum*

INTRODUCTION

Antigenic variation is of great importance for the success and survival of various pathogens ranging from trypanosomes to bacteria, fungi, and the focus of this paper, *Plasmodium falciparum*, the most virulent of the human malaria parasites (1). For each pathogen, the pressure to diversify surface proteins exposed to the immune system is counterbalanced by the need to preserve function, which in the case of *P. falciparum* is the maintenance of binding capacity to receptors on vasculature endothelial cells (2). Each pathogen has developed a systematic method to diversify surface proteins while balancing these strong but opposing selection pressures. This typically involves the generation of DNA sequence modifications to the genes that encode the surface proteins in ways that generate diversity without compromising function. These changes are created using the particular complement of DNA recombination and repair pathways present within the pathogen. Due to the critical nature of maintaining DNA integrity, DNA repair pathways are highly conserved across species from bacteria to mammals and components of most pathways can be readily identified in various organisms (3), making DNA recombination/repair a subject of interest both for evolutionary biologists as well as for those interested in host–pathogen interactions.

With the increasing availability of whole genome sequences from an ever increasing number of organisms, it is becoming clear that presence and preference for the use of specific DNA repair pathways does in fact vary considerably amongst different species. A well described example is the preference for which mechanism is utilized to repair DNA double-strand breaks (DSBs);

[1]Department of Internal Medicine, Division of Infectious Diseases, Weill Medical College of Cornell University, New York, NY, 10065;
[2]Department of Microbiology and Immunology, Weill Medical College of Cornell University, New York, NY, 10065; Kirk W. Deitsch 1300 York Avenue, Box 62, New York, NY 10065.

higher eukaryotes show a pronounced propensity to use nonhomologous end joining (NHEJ) while the yeast *Saccharomyces cerevisiae* exhibits a strong preference for homologous recombination (HR), though both repair pathways are complete and present in these organisms (4). Recently it was discovered that some lineages of very divergent lower eukaryotes have lost components of DNA repair pathways (5, 6, 7, 8), resulting in strong propensities toward certain types of DNA repair products and thus potentially influencing how these genomes change over time. Since DNA rearrangements play a crucial role in the diversification of important surface antigens, the availability and preferential use of different DNA repair pathways should be considered a key component of the process of antigenic variation. While DNA recombination and repair have long been considered to be intimately tied to antigenic variation in organisms like African trypanosomes (9, 10), they are just recently becoming appreciated for malaria parasites. The unique complement of repair pathways present in malaria parasites are thus likely to be tied to pathogenesis as well as the generation of antigenic diversity.

ANTIGENIC VARIATION AND IMMUNE EVASION BY MALARIA PARASITES

P. falciparum is a eukaryote parasite with a complex life cycle involving both a human host and a mosquito vector (Fig. 1). The pathogenic stage of the parasite's life cycle involves invasion and asexual replication within red blood cells, a process that takes approximately 48 hours and results in the release of 20 to 30 daughter merozoites from each infected cell. These merozoites go on to invade new red blood cells, thus establishing a cyclic infection. Occasionally parasites differentiate into sexual stages (gametocytes) that when taken up by a mosquito further differentiate into male and female gametes, mate, and after several complex stage transformations, form new infectious cells (sporozoites) that migrate to the mosquito salivary gland, ready to establish a new infection with the next blood meal. It is only during the brief period in the mosquito midgut when the parasites are diploid and undergo meiotic recombination. With the mosquito's next blood meal the now haploid sporozoites leave the salivary gland, enter the human blood stream and make their way to the liver where they transform to the liver stage of the life cycle. For the remainder of the life cycle the parasites remain in the haploid state. Within the hepatocytes the parasites undergo many rounds of DNA replication and division in a process termed merogony. Once the liver stage is completed, the parasites exit the liver and reestablish the erythroctyic cycle.

As with all pathogens, the ability to maintain a chronic infection requires the organisms to avoid clearance by the immune response of the infected individual. Modulation of the host immune system, stage transitions between dormant and active states, and antigenic variation are all important mechanisms that pathogens use for persistence. The antibody response is a major component of metazoan immunity and enables host organisms to recognize and destroy foreign antigens. Thus many infectious organisms have evolved specialized pathways devoted specifically to this problem. The significance of avoiding immune recognition is also reflected in how the genomes of pathogens have evolved. For example, African trypanosomes have dedicated 10 to 30% of their genomes to the multicopy gene families implicated in antigenic variation (11).

In the case of *Plasmodium*, being surrounded by the membrane of a host cell that is largely metabolically inactive renders the parasites almost entirely hidden from circulating antibodies, thus enabling them to mostly avoid this arm of the immune response. However, in the case of *P. falciparum*, growth within the red cell leads to a distorted shape and increased rigidity of the infected cell, properties that will result in its destruction within the spleen (12). To avoid splenic clearance, *P. falciparum* places the adhesive protein *Plasmodium falciparum* erythrocyte protein 1 (PfEMP1) on the red cell surface where it binds to receptors on the vascular endothelium, thereby sequestering the parasites from the peripheral circulation and avoiding movement through the spleen (13). The adhesive properties of the infected cells are also a primary virulence factor and are thought to be the cause of much of the pathogenesis associated with *P. falciparum* infection, including severe disease syndromes such as pregnancy-associated and cerebral malaria (14). The placement of parasite encoded proteins on the infected cell surface stimulates antibody production against these surface antigens, leading to recognition and clearance of the infected cells. Thus variant PfEMP1 surface proteins are directly at the interface of host–parasite interactions, virulence, and the persistent nature of malaria infections.

Because PfEMP1 and other antigens expressed on the surface of *P. falciparum* infected red cells stimulate a strong antibody response, the parasites must continuously vary expression to antigenically distinct forms in order to maintain a chronic infection. Similar to many other infectious organisms, the genomes of malaria parasites contain large, multicopy gene families in which each individual gene encodes an antigenically

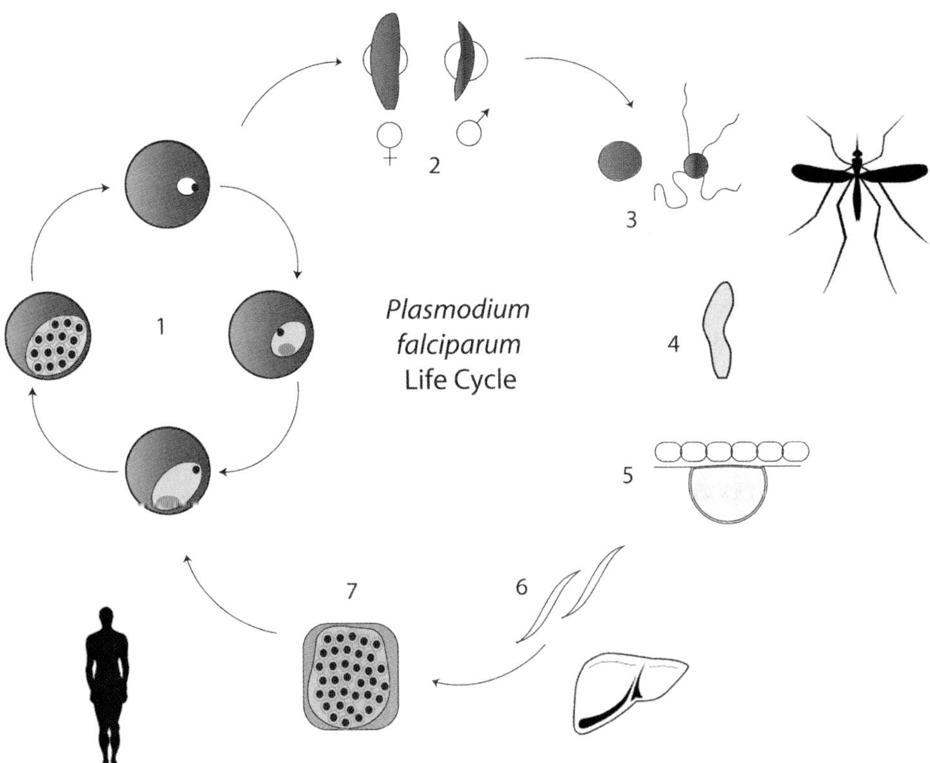

Figure 1 Schematic representation of the *Plasmodium* life cycle highlighting major points of morphological transition and replication. (1) Asexual replication within the erythrocytes of the mammalian host. (2) Male and female gametocytes circulating within the blood stream of the mammalian host prior to being acquired by the mosquito vector during a blood meal. (3) Male and female parasites as they leave the red blood cells prior to fusion and sexual division. The "rounded up" female is shown on the left and the male undergoing exflagellation is shown on the right. (4) The motile ookinete that crosses the gut wall of the blood fed mosquito. (5) An oocyst that forms after the ookinete crosses the gut wall. The parasite undergoes numerous asexual divisions at this point, giving rise to thousands of sporozoites. (6) Sporozoites that infect the salivary glands of the mosquito and are injected into a mammalian host during a subsequent blood meal. They then travel to and invade cells within the liver. (7) A liver cell infected with asexually replicating parasites. These parasites leave the liver and infect circulating red blood cells, thus completing the cycle. Figure adapted with permission from reference 83.
doi:10.1128/microbiolspec.MDNA3-0022-2014.f1

distinct form of the surface protein (15). In *P. falciparum*, the best studied is the *var* gene family, which includes ~60 members, each encoding a different form of PfEMP1 (16, 17, 18). However *P. falciparum* also has three other multicopy gene families, called *rifin* (135 members), *stevor* (~35 members) and *Pfmc-2TM* (~15 members) (19, 20, 21). Genes from all three families are arranged in clusters or tandem arrays, primarily within the subtelomeric regions of most chromosomes. *var* genes are highly variant and different approaches to classify members of this gene family have identified three main subfamilies based on 5′ upstream noncoding regions and the presence of particular functional domains within the encoded proteins (22, 23). These subfamilies have been termed A, B, and C and expression of members of each different subfamily has been tied to different degrees of disease severity (24, 25, 26, 27, 28, 29). Each *var* gene is complete, including a promoter and flanking regulatory regions, and genes are activated or silenced *in situ*, without recombination or movement to alternative positions in the genome (30). This is in contrast to antigenic variation in African trypanosomes or *Babesia bovis*, in which silent genes are activated by transposition into a specific chromosomal expression site (31, 32, 33). Only one *var* gene is actively transcribed at a time while the remainder are maintained in a silent state (34, 35). Thus antigenic variation results from switches in which gene is expressed, a process that has been shown to be controlled epigenetically (36).

HYPERVARIABILITY WITHIN THE *var* GENE FAMILY

Surveys of *var* gene sequences from different parasite isolates have found that these genes display an extraordinary degree of diversity. While such studies detected extreme *var* gene diversity on a global scale, within a given location, especially one with lower transmission rates, the variability can be more limited (37, 38). By comparing the degree of *var* gene diversity with that found within the rest of the genome, it is possible to gain insights into how the large multicopy gene families are maintained differently. Studies on parasites from areas of South America have been particularly informative because transmission rates are generally low or associated with clonal outbreaks, therefore the overall degree of genome heterogeneity is low. In addition, in certain areas malaria was close to being eradicated, thus upon reemergence the parasite genomes appear to have undergone selective sweeps (39, 40). For example, a study of genome sequences from 14 *P. falciparum* isolates collected in the Peruvian Amazon, an area notable for low transmission and a historic bottleneck from previous eradication efforts, detected a relatively genetically homogeneous population. Comparative microarrays found that genes annotated as "metabolic processes" displayed an average polymorphic probe frequency of only 2.6%, confirming the general lack of genetic diversity of these closely related parasites within the majority of the genome. In contrast, genes associated with antigenic variation were found to be hypervariable with a 60.2% average polymorphic probe frequency (41).

Given the relative degree of sequence conservation throughout the rest of the genome, the hypervariability displayed by *var* genes suggests that they must be subject to much higher frequencies of mutation or recombination. Bioinformatic comparisons of large sequence sets of laboratory lines noted that *var* gene sequences appear to "shuffle," predominantly between members of the same subfamily, creating mosaic sequences (42). Further, *var* genes display a segmental organization that is reflected in "homology blocks" within the encoded PfEMP1 proteins. Relatively conserved domains are separated by hypervariable regions, potentially facilitating recombination between semihomologous genes (43). While *var* genes in general are immensely diverse, the degree of variation is not uniform, with the most sequence diversity being observed at the 3′ end of exon 1. In contrast, exon 2, introns, and the 5′ and 3′ noncoding regions are relatively conserved [Fig. 2(A)] (16, 22, 23).

It has also been proposed that the preferential recombination between *var* genes of the same subfamily is in part facilitated by the location of the genes within the subtelomeric regions of the chromosomes, with each subfamily displaying a particular orientation (transcribed either toward or away from the chromosome end; [Fig. 2(B)] [42]). Like many other eukaryotes, the telomeres of *P. falciparum* are tethered to the nuclear periphery and arranged in clusters or "bouquets," an arrangement that brings members of each subfamily into alignment and potentially facilitates recombination [Fig. 2(C)] (44, 45). Evidence of *var* gene reassortment has highlighted the likely predominance of gene conversion as a mechanism driving diversification. The resulting composites of partially conserved homology blocks are thought to balance the need for continuous generation of diversity with the constraints of maintaining functional cytoadhesive structures within the encoded forms of PfEMP1.

var GENE RECOMBINATION – MEIOSIS VERSUS MITOSIS

While there is clear evidence that *var* genes are subject to increased rates of recombination and diversification, the mechanisms underlying this characteristic are poorly understood. During transmission by the mosquito vector, parasites undergo an obligate meiotic stage and corresponding sexual recombination, providing a potential opportunity for *var* gene shuffling and diversification. Previous studies have highlighted that the meiotic recombination rate of *var* and other clustered multicopy gene families is higher than that observed for the rest of the genome. Specifically, close examination of the progeny from experimental genetic crosses found that recombination between *var* genes is more frequent than expected when compared to the overall rate of meiotic recombination (40). In addition, examples of meiotic gene conversion events between *var* genes have been documented (45, 46). A recent paper proposed that DNA secondary structures within *var* coding regions, specifically potential hairpins formed during DNA replication, might play a role as recombination inducers (46). A probable homologue of Spo11, the endonuclease responsible for generating the DNA DSBs that initiate sexual recombination between homologous chromosomes, has been identified and is

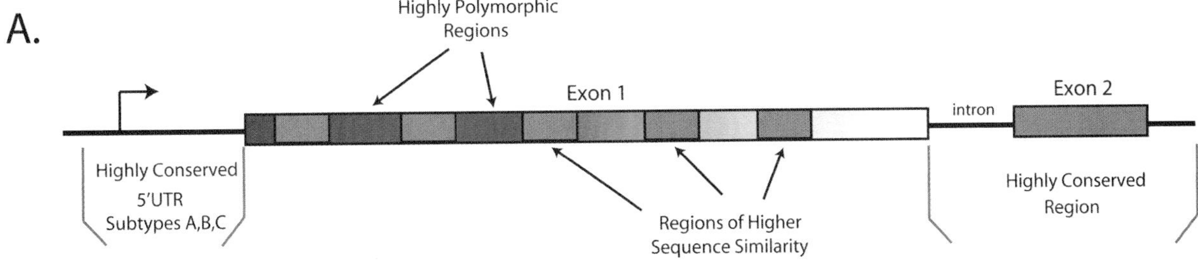

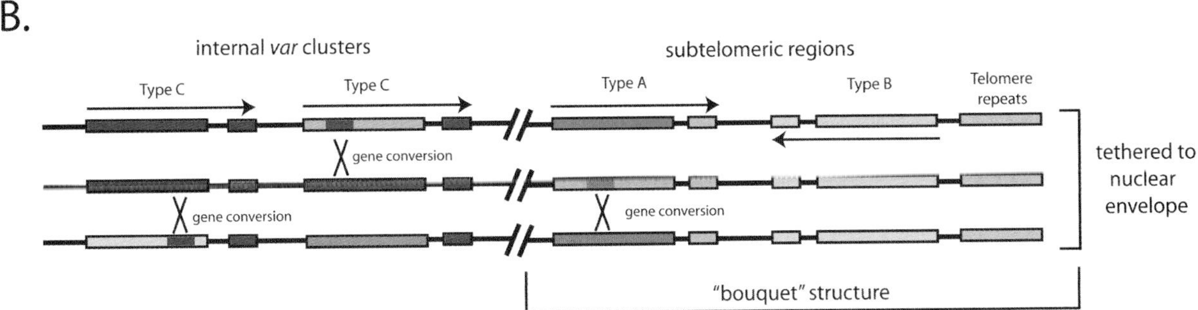

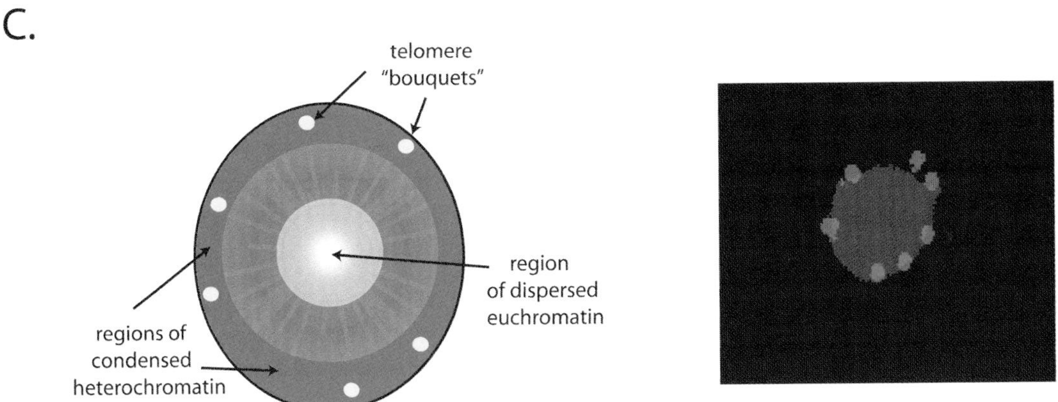

Figure 2 Structure and genomic arrangement of the *var* gene family in *P. falciparum*. (A) Schematic showing the two exon structure of all *var* genes. Note that the 5′ UTR, intron, exon 2 and 3′ UTR are all highly conserved (gray). The sequence of the 5′ UTR and upstream regulatory domains can be classified into three basic types called A, B and C. Exon 1 encodes the polymorphic portion of PfEMP1 and is arranged as an alternating series of highly diverse sequences separating regions of higher similarity. Overall the greatest degree of sequence diversity is found at the 3′ end of exon 1 as represented by the color gradient. Exon 2 encodes a highly conserved region of PfEMP1 that is not exposed to the immune system. (B) General chromosomal arrangement of *var* genes, with type C genes typically found in tandem arrangement in the internal regions of the chromosomes while types A and B genes are located next to the telomere repeats. The telomeres are known to cluster into "bouquets" which align the genes in a way that is proposed to facilitate recombination and gene conversion events. (C) Illustration (left) showing the subnuclear localization of the telomere bouquets (yellow spots) that are found near the nuclear envelope within regions of dense heterochromatin (dark blue). Regions of less dense euchromatin are typically found near the center of the nucleus (light blue). Fluorescent *in situ* hybridization showing the location of the *var* gene clusters within the parasite's nucleus. A probe that hybridizes to the conserved exon 2 of *var* genes is shown in yellow while the nuclear DNA is stained with DAPI (4′,6-diamidino-2-phenylindole) and is shown in blue. Image in Fig. 2(B) modified with permission from reference 84. Image in Fig. 2(C) modified with permission from reference 85. doi:10.1128/microbiolspec.MDNA3-0022-2014.f2

expressed in the gametocyte to ookinete stage as expected (Plasmodb.org; PF3D7_1217100). Accessory proteins known to play a role in meiotic recombination have not been identified but might simply be too divergent to be recognized by ordinary comparative analysis. It has also been observed that *var* gene recombination during meiosis can be ectopic (between gene copies in different chromosomal positions) (46, 45), promoting an even greater degree of potential recombination products.

In addition to meiotic recombination during transition through the mosquito, it is also possible that *var* genes could undergo recombination and diversification during asexual replication within the human host. While each individual erythrocytic stage parasite is initially haploid, after invasion of a red blood cell they undergo repeated genome replications and nuclear divisions in a process called schizogony, resulting in the release of 20 to 30 haploid daughter cells. These repeated rounds of replication provide ample opportunities for the occasional formation of DSBs as the DNA strands are copied. In addition, the mammalian immune response is known to include the production of numerous substances that can cause DNA damage, for example reactive oxygen and nitrogen species (47, 48, 49). Repair of DNA DSBs often includes mechanisms of HR, which when applied to multicopy gene families with large tracts of homologous sequences, could easily result in the generation of hybrid or mosaic genes. If such recombination happens at even low frequencies, given the large numbers of parasites present within an infected individual (up to 10^5 parasites per μl of blood), the probability is high that new *var* genes are being generated over the course of a single infection.

To investigate the frequency of mitotic recombination, studies mapped chromosomal regions from cloned laboratory lines grown *in vitro* as asexual parasites for prolonged periods of time. The most common finding has been large scale deletions of subtelomeric and telomeric sequences at the chromosome ends (50, 20). Because of their location within subtelomeric domains, these deletions often included the loss of members of multicopy gene families. Similar deletions have also been noted in field isolates, thus the most frequently observed structural variation is related to genetic loss (51, 52, 53, 54). The most thorough evaluation of genetic changes resulting from long-term asexual replication was performed with clones of laboratory lines grown *in vitro* for up to a year, then recloned and analyzed by both microarray hybridization and whole genome sequencing (55). This study indeed detected several recombination events leading to the formation of chimeric *var* genes. Three events were characterized in detail and likely represented examples of break induced recombination (BIR), with the hybrid sequence extending from a break site within the *var* gene through the end of the chromosome. No small gene conversion events were observed in this study, although evidence for small (~100 bp) gene conversions was found in a separate study of closely related parasite lines (56). Taken together, there is cumulative evidence for both mitotic and meiotic recombination between *var* gene family members, thus driving the continuous diversification and hypervariability that is a hallmark of these genes.

THE ROLE OF DNA REPAIR PATHWAYS IN GENERATING ANTIGENIC DIVERSITY

Recombination between DNA strands invariably initiates at the site of a DNA break. How the break is repaired therefore dictates the nature of the resulting repair product. In the case of large, variant gene families like *var*, the DNA repair process appears to be skewed toward recombination between family members, thereby generating the high degree of diversity observed when *var* gene complements are compared between different parasite lines. Given the important role that DNA repair likely plays in generating *var* gene diversity, a closer examination of the DNA repair pathways utilized by malaria parasites to maintain genome integrity is warranted. In eukaryotic organisms, DNA DSBs are repaired by two basic pathways: NHEJ, in which the broken DNA ends are ligated without the involvement of DNA sequences from elsewhere in the genome (57), and HR, which utilizes sequences homologous to those surrounding the break to serve as either a template to guide repair or as a site of exchange between the two DNA strands (58). NHEJ can be further classified into two distinct pathways, canonical (C-NHEJ) and alternative (A-NHEJ) (59, 60, 61). These three pathways rely on different molecular machinery and generate different types of repair products, thus the propensity of an organism to use one pathway versus another can have a profound influence on genome evolution, and in particular on how large, repetitive, and hyper-recombinogenic regions of the genome are maintained and diversify.

Pathogens have developed varied ways to maintain genome integrity in the face of DNA damage due to both metabolic stress and immune pressure, and mechanisms of re-assortment of the parasite surface proteins are in part dictated by DNA repair pathways active and present in the pathogen. The repair of DSB via NHEJ occurs in organisms ranging from some bacteria (although not in *E. coli*, for example) to mammals, indicating that this type of repair has been conserved during evolution (57). Most of the major factors involved were identified in the mammalian system, reflecting the major contribution of NHEJ to cell survival in the face of DNA damage. However, in diverse lower eukaryotes it appears there are many exceptions to the standard DNA repair pathways described in model organisms. By searching an organism's genome to identify genes encoding required components for the different repair pathways and by examining the products of repair, it is possible to infer how different evolutionary lineages have adapted their DNA repair pathways to suit their specific needs. For example, the DNA repair proteins Ku70/80 are thought to be required for C-NHEJ, but the genes encoding these proteins cannot be identified in the genomes of the protozoan parasites *Trichomonas vaginalis* and *Encephalitozoon cuniculi*, and only Ku70 is found in the gut parasite *Entamoeba histolytica* (62, 6). Similarly, the kinetoplast parasites *Trypanosoma brucei*, *Trypanosoma cruzi* and *Leishmania sp.* all possess Ku70/80, but appear to be missing other key components required for C-NHEJ, including DNA ligase IV and XRCC4/Lif1 (63). In these organisms, the Ku proteins appear to be involved in telomere maintenance rather than DSB repair (64). The loss of key components of C-NHEJ and the resulting shift toward the use of HR to repair DNA breaks are predicted to lead to a higher frequency of gene conversion events, the resulting product when an organism utilizes a homologous sequence within the genome as a template for repair (65). In *T. brucei*, gene conversion from alternative or even pseudogenes is important in the creation of mosaic *vsg* antigens (66), and similar mechanisms have been proposed for the tick borne parasite *Babesia bovis* (32), a parasite that also appears to lack components of the C-NHEJ pathway.

There has been discussion in the literature regarding the presence or absence of C-NHEJ in *Plasmodium* (15, 7, 67). Early analysis of the complete genome sequence noted that the several key components of the pathway could not be identified. Subsequent analyses have also failed to identify the Ku proteins, Ligase 4, Artemis, DNA-PKcs, XRCC4 or Cernunnos/XLF, all key components of C-NHEJ. Remarkably, in the closely related Apicomplexan parasite *Toxoplasma gondii*, these components were readily identified and genetic manipulations found that NHEJ was by far the dominant pathway utilized to repair DNA DSBs (8). In fact, NHEJ was so dominant that the Ku proteins had to be knocked out, thus disabling the pathway, to enable genetic manipulations that rely on HR (68, 69, 70). This stark difference in DNA repair pathways in closely related organisms suggests a selective advantage for utilizing one type of repair and the resulting effect this has on the evolution of the genome. For example, the relatively recent loss of C-NHEJ in the *Plasmodium* lineage suggests a selective advantage of skewing repair toward HR.

A closer examination of the phylogenetic tree shows that amongst the Alveolates, the presence of C-NHEJ (as defined by identifiable Ku proteins) is variable (Fig. 3). Ku proteins are easily identified by using Psi blast (http://blast.ncbi.nlm.nih.gov) or Hmmer (http://hmmer.janelia.org) with the yeast or human Ku 70 core sequence as query, or alternatively using the prokaryotic core domains (71). With this strategy Ku proteins (and Ligase IV) were identified for the Alveolates *Perkinsus marinus* and *Paramecium tetraurelia*. However within the Apicomplexan parasites, the same search strategy only identified orthologs of these proteins in the Coccidian branch, which includes *Toxoplasma*, *Eimeria* and *Neospora*. A similar approach for core components of HR (Rad 51, Ligase I, Mre11) easily identified this pathway as intact across all the Alveolates. Interestingly, both *Cryptosporidia* and the hematozoa (*Plasmodium*, *Babesia*, *Theileria*) do not possess any of the components required for C-NHEJ, implying that the pathway was lost twice while being maintained by the *Coccidia*.

The remarkable differences in the DNA repair pathways found in these relatively closely related organisms raises interesting evolutionary questions, in particular why both the *Cryptosporidia* and hematozoa lineages appear to have lost the C-NHEJ pathway. *Cryptosporidia* are distantly related to the *Coccidia* and have undergone substantial reductive evolution, including the loss of both the plastid (a chloroplast like organelle distinct to Apicomplexan parasites) and mitochondrial genomes, and thus the loss of C-NHEJ might be related to this general loss of genomic material (72). Toxoplasma, the best studied representative of the *Coccidia*, is an intracellular parasite that evades the immune system primarily by transitioning to a latent, intracellular cyst stage. No hyper-recombinogenic, large multicopy gene families involved in antigenic variation have been identified. In contrast, multicopy gene families with potential roles in antigenic variation have been identified in most hematozoa and it is tempting to posit that the hematozoa evolved to favor HR (and ultimately lost C-NHEJ) as part of their strategy for maintaining and diversifying these gene families. Gene conversion through HR preserves open reading frames and maintains functional domains intact, characteristics that are key to preserving large gene families in which virtually all copies are functional.

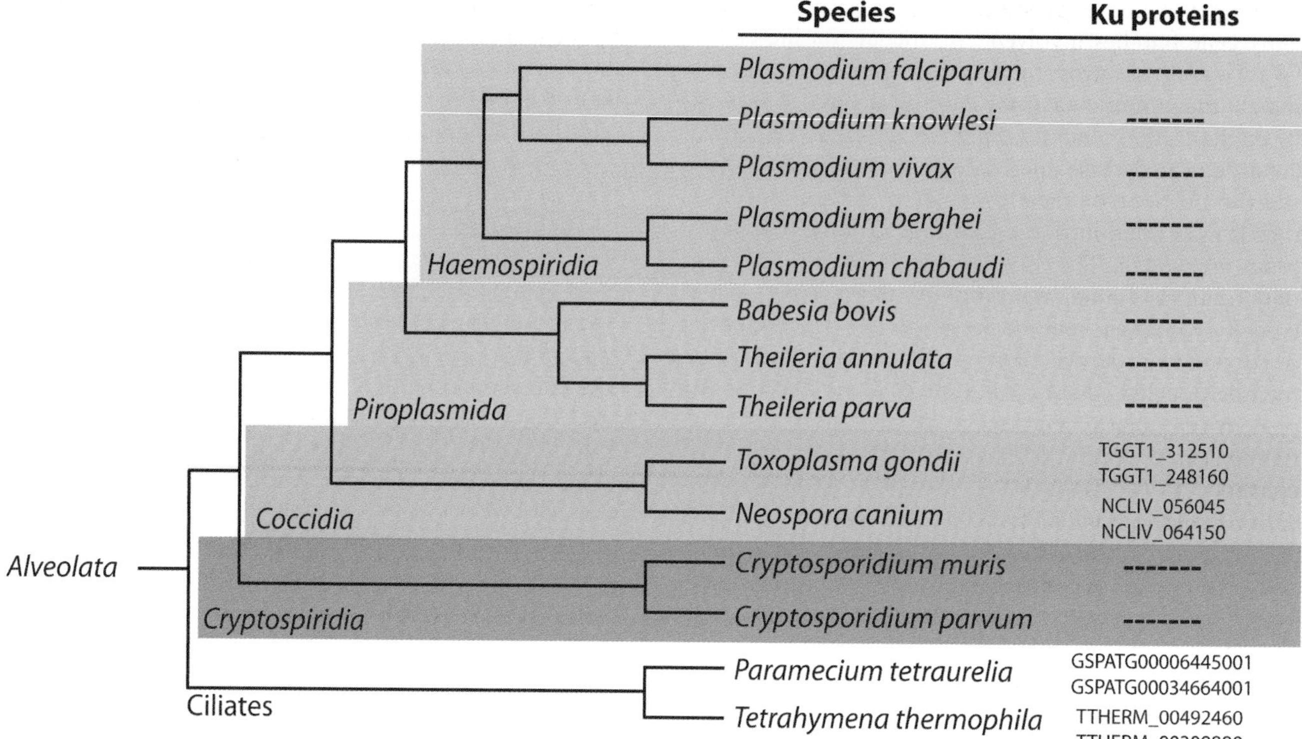

Figure 3 Phylogenetic tree of the Apicomplexan lineage. The different colors highlight several groups of obligate, Apicomplexan parasites. The *Haemosporidia* (orange) include parasites that cause malaria in humans (*P. falciparum* and *P. vivax*), nonhuman primates (*P. knowlesi*) and rodents (*P. berghei* and *P. chabaudi*). The *Piroplasmida* (green) are parasites transmitted by ticks that infect either red blood cells (*Babesia bovis*) or white blood cells (*T. annulata* and *T. parva*). The *Coccidia* (blue) are cyst forming parasites that do not undergo antigenic variation. The *Cryptospiridia* (purple) also form cysts and have a highly reduced genome. The ciliates, as exemplified by *P. tetraurelia* and *T. thermophila*, are non-parasitic, free living organisms. The ability of the organisms to utilize the C-NHEJ pathway for DSB repair is inferred by the presence or absence of proteins of the Ku family. Note that, of the Apicomplexans, only the *Coccidia* have retained these genes within their genomes. The annotations numbers for Ku70/80 are given for both the *Coccidia* and the Ciliates. The phylogenetic relationships are shown as described in reference 86.
doi:10.1128/microbiolspec.MDNA3-0022-2014.f3

THE POTENTIAL ROLE OF CHROMATIN STRUCTURE IN *var* GENE DIVERSIFICATION

Unlike some other pathogens where antigenic variation is mediated by the transposition of silent genes or gene fragments into an active expression site (for example African trypanosomes or *Babesia bovis*) (31, 32, 33), antigenic variation in *P. falciparum* results from the *in situ* activation of previously silent *var* genes without recombination (30). Activation of a silent gene is always coupled to silencing of the previously active gene, resulting in strict mutually exclusive expression. The molecular mechanisms underlying this tightly coordinated activation and silencing process are only partly understood, however it has become clear in recent years that epigenetic modifications to the chromatin surrounding the genes play a definitive role. Modifications to histone H3 at the lysine 9 position (H3K9) appear to play a prominent role, with acetylation of this residue (H3K9ac) associated with the single active gene and trimethylation (H3K9me3) found at the silent remaining members of the family (73, 74). In addition, trimethylation of the lysine at the 36th position of the same histone (H3K36me3) is found at all *var* genes and has been shown to be important for maintaining mutually exclusive expression (75, 76).

An unusual aspect of the histone modifications associated with *var* genes (H3K9me3 and H3K36me3) is that they are almost exclusively localized to genes that undergo variable expression, in particular the multicopy gene families involved in antigenic variation (77, 78). This is in contrast to most eukaryotic organisms in which these marks are found associated with genes throughout the genome. Malaria parasites appear to have dedicated these specific epigenetic marks to very narrow portions of their genomes, and the chromatin that is assembled in these regions is different than that found elsewhere. The multicopy gene families are also only found in limited regions of the chromosomes, specifically within the subtelomeric domains or within defined tandem arrays within internal chromosomal regions. Further, these chromosomal regions are tethered to the nuclear envelope and thereby anchored at the nuclear periphery, localized within distinct spots or clusters (79, 80) [Fig. 2(C)]. Collectively, these observations suggest that malaria parasites have segregated the hyper-recombinogenic regions of their genome away from the typical euchromatin that surrounds genes required for "housekeeping" functions, both along the linear length of the chromosomes and also spatially within the nucleoplasm. This model raises the interesting possibility that the unique chromatin structure found at the multicopy gene families that is known to play a role in transcriptional regulation might also be important for controlling the accessibility of templates for DNA DSB repair. If so, this would have major implications for recombination between genes and therefore the ongoing diversification of *var* genes.

There is growing evidence from model organisms consistent with the hypothesis that higher order chromosomal architecture exerts just as profound an influence on DNA repair as it does on nuclear processes like transcription and DNA replication (81, 82). Some differences in chromatin structure can be accounted for by chromosomal position, for example proximity to centromeres or telomeres, an observation directly relevant to recombination of multicopy genes in *P. falciparum* given their arrangement in mostly subtelomeric and several internal clusters. The A, B and C groups of *var* genes mentioned previously are distinguishable by chromosomal location as well as by 5′ UTR sequence and domain architecture, with A and B subtypes more frequently arranged in the subtelomeric regions and C types more often located in internal tandem arrays (22, 23). The aforementioned "bouquets" specifically align the different *var* subtypes (A genes with A genes, B genes with B genes, etc.) thus potentially facilitating the steady divergence of these families over time.

In model organisms, DSBs occurring in areas of heterochromatin are repaired with much slower kinetics than those in euchromatic regions (81). In addition, telomeric DSB repair is regulated differently than breaks found in more central chromosomal locations, in part due to the role of DNA repair enzymes in maintaining telomeric ends. For example, the addition of telomeric repeats to a plasmid integrated into an internal site dramatically decreased repair of DSBs regardless of which repair pathway, HR or NHEJ, was used (82). To repair a DSB regardless of the surrounding chromatin structure, the same general machinery is required, however in a state of heterochromatin an additional set of enzymes and histone modifiers are necessary (81). This might be particularly relevant for DSB repair at *var* genes due to the unique histone modifications and chromatin structure found only at these particular chromosomal regions. The possibility that the unique chromatin environment found at *var* genes influences the DNA repair pathways used to repair DSBs might help to explain recent data reported on DNA repair in *P. falciparum*. Examination of *var* gene sequences from large datasets provides ample evidence for frequent gene conversion events driving *var* gene diversification (42, 56). However, experimental induction of DSBs at a non-*var* site in the parasite's genome found that sequence divergence greater than 2% resulted in exclusive utilization of a repair pathway resembling A-NHEJ, and no gene conversion events were detected (67). Since *var* gene sequences typically diverge by much greater than 2%, DSBs that occur within *var*-associated regions of the genome must be handled differently if HR induced gene conversions are indeed the driving force behind *var* gene recombination.

CONCLUSIONS

The forces driving recombination and diversification of *var* and other multicopy gene families in *P. falciparum* remain poorly understood. The domain architecture of the gene coding regions, the chromatin structure found at these chromosomal regions, location of the genes along the length of the chromosome and within the nuclear space are all likely to be key players. To date descriptions of *var* gene recombination events, either those detected in the progeny of a genetic cross or arising in lab isolates grown over time, remain limited to the extent that it is difficult to account for the actual degree of diversity observed in the field. Recombination during meiosis or through BIR/gene conversion during mitotic replication are likely to contribute to overall *var* diversification, but our understanding of the molecular mechanism that influences these processes is far behind what is known in other pathogenic organisms including trypanosomes or various bacterial species.

Acquisition of large amounts of data has also been limited by difficulties in sequencing through the most AT-rich genome yet described, and assembly of repetitive gene families poses particular difficulties for the current generation of high-throughput genome sequencing platforms that rely on massive numbers of short sequence reads. This problem has been significant enough that most genome sequences to date have excluded the multicopy gene families and focused on the core of the genome. Similar analyses using microarrays have been limited to studying closely related parasites to which the array was designed. Advancing sequencing technologies that yield much longer read lengths and an improved ability to assemble closely related, repetitive sequences will solve many of these problems. Combined with a better understanding of nuclear structure, mechanisms of DNA repair, and knowledge about how parasites handle DNA stress will combine to greatly enlighten our concept of "mobile DNA" in malaria parasites.

Citation. Kirkman LA, Deitsch KW. 2014. Recombination and diversification of the variant antigen encoding genes in the malaria parasite Plasmodium falciparum. Microbiol Spectrum 2(6):MDNA3-0022-2014.

References

1. **Deitsch KW, Lukehart SA, Stringer JR.** 2009. Common strategies for antigenic variation by bacterial, fungal and protozoan pathogens. *Nat Rev Microbiol* 7:493–503.
2. **Smith JD, Deitsch KW.** 2012. Antigenic Variation, Adherence and Virulence, p 338–361. *In* Sibley LD, Howlett BJ, Heitman J (ed), *Evolution of Virulence in Eukaryotic Microbes*, Wiley-Blackwell, Hoboken, NJ.
3. **Kass EM, Jasin M.** 2010. Collaboration and competition between DNA double-strand break repair pathways. *FEBS Lett* **584**:3703–3708.
4. **Moynahan ME, Jasin M.** 2010. Mitotic homologous recombination maintains genomic stability and suppresses tumorigenesis. *Nat Rev Mol Cell Biol* **11**:196–207.
5. **Gill EE, Becnel JJ, Fast NM.** 2008. ESTs from the microsporidian Edhazardia aedis. *BMC Genomics* **9**:296.
6. **Lopez-Camarillo C, Lopez-Casamichana M, Weber C, Guillen N, Orozco E, Marchat LA.** 2009. DNA repair mechanisms in eukaryotes: Special focus in Entamoeba histolytica and related protozoan parasites. *Infect, Genet Evol* **9**:1051–1056.
7. **Aravind L, Iyer LM, Wellems TE, Miller LH.** 2003. Plasmodium biology: Genomic gleanings. *Cell* **115**:771–785.
8. **Smolarz B, Wilczynski J, Nowakowska D.** 2014. DNA repair mechanisms and Toxoplasma gondii infection. *Arch Microbiol* **196**:1–8.
9. **Machado CR, Augusto-Pinto L, McCulloch R, Teixeira SM.** 2006. DNA metabolism and genetic diversity in Trypanosomes. *Mutat Res* **612**:40–57.
10. **Bhattacharyya MK, Norris DE, Kumar N.** 2004. Molecular players of homologous recombination in protozoan parasites: implications for generating antigenic variation. *Infect Genet Evol* **4**:91–98.
11. **Donelson JE.** 2003. Antigenic variation and the African trypanosome genome. *Acta Trop* **85**:391–404.
12. **David PH, Hommel M, Miller LH, Udeinya IJ, Oligino LD.** 1983. Parasite sequestration in *Plasmodium falciparum* malaria: Spleen and antibody modulation of cytoadherence of infected erythrocytes. *Proc Natl Acad Sci U S A* **80**:5075–5079.
13. **Scherf A, Lopez-Rubio JJ, Riviere L.** 2008. Antigenic variation in Plasmodium falciparum. *Annu Rev Microbiol* **62**:445–470.
14. **Miller LH, Baruch DI, Marsh K, Doumbo OK.** 2002. The pathogenic basis of malaria. *Nature* **415**:673–679.
15. **Gardner MJ, Hall N, Fung E, White O, Berriman M, Hyman RW, Carlton JM, Pain A, Nelson KE, Bowman S, Paulsen IT, James K, Eisen JA, Rutherford K, Salzberg SL, Craig A, Kyes S, Chan MS, Nene V, Shallom SJ, Suh B, Peterson J, Angiuoli S, Pertea M, Allen J, Selengut J, Haft D, Mather MW, Vaidya AB, Martin DM, Fairlamb AH, Fraunholz MJ, Roos DS, Ralph SA, McFadden GI, Cummings LM, Subramanian GM, Mungall C, Venter JC, Carucci DJ, Hoffman SL, Newbold C, Davis RW, Fraser CM, Barrell B.** 2002. Genome sequence of the human malaria parasite Plasmodium falciparum. *Nature* **419**:498–511.
16. **Su X, Heatwole VM, Wertheimer SP, Guinet F, Herrfeldt JV, Peterson DS, Ravetch JV, Wellems TE.** 1995. A large and diverse gene family (*var*) encodes 200–350 kD proteins implicated in the antigenic variation and cytoadherence of *Plasmodium falciparum*-infected erythrocytes. *Cell* **82**:89–100.
17. **Baruch DI, Pasloske BL, Singh HB, Bi X, Ma XC, Feldman M, Taraschi TF, Howard RJ.** 1995. Cloning the *P. falciparum* gene encoding PfEMP1, a malarial variant antigen and adherence receptor on the surface of parasitized human erythrocytes. *Cell* **82**:77–87.
18. **Smith JD, Chitnis CE, Craig AG, Roberts DJ, Hudson-Taylor DE, Peterson DS, Pinches R, Newbold CI, Miller LH.** 1995. Switches in expression of *Plasmodium falciparumvar* genes correlate with changes in antigenic and cytoadherent phenotypes of infected erythrocytes. *Cell* **82**:101–110.
19. **Cheng Q, Cloonan N, Fischer K, Thompson J, Waine G, Lanzer M, Saul A.** 1998. Stevor and rif are Plasmodium falciparum multicopy gene families which potentially encode variant antigens. *Mol Biochem Parasitol* **97**:161–176.
20. **Lavazec C, Sanyal S, Templeton TJ.** 2006. Hypervariability within the Rifin, Stevor and Pfmc-2TM superfamilies in Plasmodium falciparum. *Nucleic Acids Res* **34**:6696–6707.
21. **Sam-Yellowe TY, Florens L, Johnson JR, Wang T, Drazba JA, Le Roch KG, Zhou Y, Batalov S, Carucci DJ, Winzeler EA, Yates JR III.** 2004. A Plasmodium gene family encoding Maurer's cleft membrane proteins: structural properties and expression profiling. *Genome Res* **14**:1052–1059.

22. **Kraemer SM, Smith JD.** 2003. Evidence for the importance of genetic structuring to the structural and functional specialization of the Plasmodium falciparum var gene family. *Mol Microbiol* **50:**1527–1538.
23. **Lavstsen T, Salanti A, Jensen ATR, Arnot DE, Theander TG.** 2003. Sub-grouping of Plasmodium falciparum 3D7 var genes based on sequence analysis of coding and non--coding regions. *Malar J* **2:**27.
24. **Rottmann M, Lavstsen T, Mugasa JP, Kaestli M, Jensen AT, Muller D, Theander T, Beck HP.** 2006. Differential expression of var gene groups is associated with morbidity caused by Plasmodium falciparum infection in Tanzanian children. *Infect Immun* **74:**3904–3911.
25. **Falk N, Kaestli M, Qi W, Ott M, Baea K, Cortes A, Beck HP.** 2009. Analysis of Plasmodium falciparum var genes expressed in children from Papua New Guinea. *J Infect Dis* **200:**347–356.
26. **Kaestli M, Cockburn IA, Cortes A, Baea K, Rowe JA, Beck HP.** 2006. Virulence of malaria is associated with differential expression of Plasmodium falciparum var gene subgroups in a case-control study. *J Infect Dis* **193:** 1567–1574.
27. **Lavstsen T, Turner L, Saguti F, Magistrado P, Rask TS, Jespersen JS, Wang CW, Berger SS, Baraka V, Marquard AM, Sequin-Orlando A, Willerslev E, Gilbert MTP, Lusingu J, Theander TG.** 2012. P. falciparum erythrocyte membrane protein 1 domain cassettes 8 and 13 are associated with severe malaria in children. *Proc Natl Acad Sci U S A* **109:**E1791–1800.
28. **Avril M, Tripathi AK, Brazier AJ, Andisi C, Janes JH, Soma VL, Sullivan DJ, Bull PC, Stins MF, Smith JD.** 2012. A restricted subset of var genes mediates adherence of Plasmodium falciparum infected erythrocytes to brain endothelial cells. *Proc Natl Acad Sci U S A* **109:**E1782–1790.
29. **Claessens A, Adams Y, Ghumra A, Lindergard G, Buchan CC, Andisi C, Bull PC, Mok SC, Gupta AP, Wang CW, Turner L, Arman M, Raza A, Bozdech Z, Rowe JA.** 2012. A subset of Group A-like var genes encode the malaria parasite ligands for binding to human brain endothelial cells. *Proc Natl Acad Sci U S A* **109:** E1772–1781.
30. **Scherf A, Hernandez-Rivas R, Buffet P, Bottius E, Benatar C, Pouvelle B, Gysin J, Lanzer M.** 1998. Antigenic variation in malaria: *in situ* switching, relaxed and mutually exclusive transcription of *var* genes during intra-erythrocytic development in *Plasmodium falciparum*. *EMBO J* **17:**5418–5426.
31. **Allred DR, Carlton JM, Satcher RL, Long JA, Brown WC, Patterson PE, O'Connor RM, Stroup SE.** 2000. The ves multigene family of B. bovis encodes components of rapid antigenic variation at the infected erythrocyte surface. *Mol Cell* **5:**153–162.
32. **al Khedery B, Allred DR.** 2006. Antigenic variation in Babesia bovis occurs through segmental gene conversion of the ves multigene family, within a bidirectional locus of active transcription. *Mol Microbiol* **59:**402–414.
33. **Pays E.** 2005. Regulation of antigen gene expression in Trypanosoma brucei. *Trends Parasitol* **21:**517–520.
34. **Dzikowski R, Frank M, Deitsch K.** 2006. Mutually Exclusive Expression of Virulence Genes by Malaria Parasites Is Regulated Independently of Antigen Production. *PLoS Pathog* **2:**e22.
35. **Voss TS, Healer J, Marty AJ, Duffy MF, Thompson JK, Beeson JG, Reeder JC, Crabb BS, Cowman AF.** 2006. A var gene promoter controls allelic exclusion of virulence genes in Plasmodium falciparum malaria. *Nature* **439:** 1004–1008.
36. **Lopez-Rubio JJ, Riviere L, Scherf A.** 2007. Shared epigenetic mechanisms control virulence factors in protozoan parasites. *Curr Opin Microbiol* **10:**560–568.
37. **Barry AE, Leliwa-Sytek A, Tavul L, Imrie H, Migot-Nabias F, Brown SM, McVean GA, Day KP.** 2007. Population Genomics of the Immune Evasion (var) Genes of Plasmodium falciparum. *PLoS Pathog* **3:**e34.
38. **Albrecht L, Castiñeiras C, Carvalho BO, Ladeia-Andrade S, Santos dS, Hoffmann EH, dalla Martha RC, Costa FT, Wunderlich G.** 2010. The South American Plasmodium falciparum var gene repertoire is limited, highly shared and possibly lacks several antigenic types. *Gene* **453:** 37–44.
39. **Ord RL, Tami A, Sutherland CJ.** 2008. ama1 genes of sympatric Plasmodium vivax and P. falciparum from Venezuela differ significantly in genetic diversity and recombination frequency. *PLoS One* **3:**e3366.
40. **Mu J, Awadalla P, Duan J, McGee KM, Joy DA, McVean GA, Su XZ.** 2005. Recombination hotspots and population structure in Plasmodium falciparum. *PLoS Biol* **3:**e335.
41. **Dharia NV, Plouffe D, Bopp SE, Gonzalez-Paez GE, Lucas C, Salas C, Soberon V, Bursulaya B, Kochel TJ, Bacon DJ, Winzeler EA.** 2010. Genome scanning of Amazonian Plasmodium falciparum shows subtelomeric instability and clindamycin-resistant parasites. *Genome Res* **20:**1534–1544.
42. **Kraemer SM, Kyes SA, Aggarwal G, Springer AL, Nelson SO, Christodoulou Z, Smith LM, Wang W, Levin E, Newbold CI, Myler PJ, Smith JD.** 2007. Patterns of gene recombination shape var gene repertoires in Plasmodium falciparum: comparisons of geographically diverse isolates. *BMC Genomics* **8:**45.
43. **Bull PC, Buckee CO, Kyes S, Kortok MM, Thathy V, Guyah B, Stoute JA, Newbold CI, Marsh K.** 2008. Plasmodium falciparum antigenic variation. Mapping mosaic var gene sequences onto a network of shared, highly polymorphic sequence blocks. *Mol Microbiol* **68:**1519–1534.
44. **Therizols P, Fairhead C, Cabal GG, Genovesio A, Olivo-Marin JC, Dujon B, Fabre E.** 2006. Telomere tethering at the nuclear periphery is essential for efficient DNA double strand break repair in subtelomeric region. *J Cell Biol* **172:**189–199.
45. **Freitas-Junior LH, Bottius E, Pirrit LA, Deitsch KW, Scheidig C, Guinet F, Nehrbass U, Wellems TE, Scherf A.** 2000. Frequent ectopic recombination of virulence factor genes in telomeric chromosome clusters of P. falciparum. *Nature* **407:**1018–1022.

46. **Sander AF, Lavstsen T, Rask TS, Lisby M, Salanti A, Fordyce SL, Jespersen JS, Carter R, Deitsch KW, Theander TG, Pedersen AG, Arnot DE.** 2014. DNA secondary structures are associated with recombination in major Plasmodium falciparum variable surface antigen gene families. *Nucleic Acids Res* **42:**2270–2281.

47. **Becker K, Tilley L, Vennerstrom JL, Roberts D, Rogerson S, Ginsburg H.** 2004. Oxidative stress in malaria parasite-infected erythrocytes: host-parasite interactions. *Int J Parasitol* **34:**163–189.

48. **Prada J, Kremsner PG.** 1995. Enhanced production of reactive nitrogen intermediates in human and murine malaria. *Parasitol Today* **11:**409–410.

49. **Barzilai A, Yamamoto K.** 2004. DNA damage responses to oxidative stress. *DNA Repair* **3:**1109–1115.

50. **Mattei D, Scherf A.** 1994. Subtelomeric chromosome instability in Plasmodium falciparum: short telomere-like sequence motifs found frequently at healed chromosome breakpoints. *Mutat Res* **324:**115–120.

51. **Houze S, Hubert V, Le Pessec G, Le Bras J, Clain J.** 2011. Combined deletions of pfhrp2 and pfhrp3 genes result in Plasmodium falciparum malaria false-negative rapid diagnostic test. *J Clin Microbiol* **49:**2694–2696.

52. **Koita OA, Doumbo OK, Ouattara A, Tall LK, Konare A, Diakite M, Diallo M, Sagara I, Masinde GL, Doumbo SN, Dolo A, Tounkara A, Traore I, Krogstad DJ.** 2012. False-negative rapid diagnostic tests for malaria and deletion of the histidine-rich repeat region of the hrp2 gene. *Am J Trop Med Hyg* **86:**194–198.

53. **Maltha J, Gamboa D, Bendezu J, Sanchez L, Cnops L, Gillet P, Jacobs J.** 2012. Rapid diagnostic tests for malaria diagnosis in the Peruvian Amazon: impact of pfhrp2 gene deletions and cross-reactions. *PLoS One* **7:**e43094.

54. **Kumar N, Pande V, Bhatt RM, Shah NK, Mishra N, Srivastava B, Valecha N, Anvikar AR.** 2013. Genetic deletion of HRP2 and HRP3 in Indian Plasmodium falciparum population and false negative malaria rapid diagnostic test. *Acta Trop* **125:**119–121.

55. **Bopp SE, Manary MJ, Bright AT, Johnston GL, Dharia NV, Luna FL, McCormack S, Plouffe D, McNamara CW, Walker JR, Fidock DA, Denchi EL, Winzeler EA.** 2013. Mitotic evolution of Plasmodium falciparum shows a stable core genome but recombination in antigen families. *PLoS Genet* **9:**e1003293.

56. **Frank M, Kirkman L, Costantini D, Sanyal S, Lavazec C, Templeton TJ, Deitsch KW.** 2008. Frequent recombination events generate diversity within the multi-copy variant antigen gene families of Plasmodium falciparum. *Int J Parasitol* **38:**1099–1109.

57. **Hefferin ML, Tomkinson AE.** 2005. Mechanism of DNA double-strand break repair by non-homologous end joining. *DNA Repair* **4:**639–648.

58. **Johnson RD, Jasin M.** 2001. Double-strand-break-induced homologous recombination in mammalian cells. *Biochem Soc Trans* **29:**196–201.

59. **Mladenov E, Iliakis G.** 2011. Induction and repair of DNA double strand breaks: the increasing spectrum of non-homologous end joining pathways. *Mutat Res* **711:** 61–72.

60. **Fattah F, Lee EH, Weisensel N, Wang Y, Lichter N, Hendrickson EA.** 2010. Ku regulates the non-homologous end joining pathway choice of DNA double-strand break repair in human somatic cells. *PLoS Genet* **6:** e1000855.

61. **Yu AM, McVey M.** 2010. Synthesis-dependent microhomology-mediated end joining accounts for multiple types of repair junctions. *Nucleic Acids Res* **38:**5706–5717.

62. **Gill EE, Fast NM.** 2007. Stripped-down DNA repair in a highly reduced parasite. *BMC Mol Biol* **8:**24.

63. **Burton P, McBride DJ, Wilkes JM, Barry JD, McCulloch R.** 2007. Ku heterodimer-independent end joining in Trypanosoma brucei cell extracts relies upon sequence microhomology. *Eukaryotic Cell* **6:**1773–1781.

64. **Conway C, McCulloch R, Ginger ML, Robinson NP, Browitt A, Barry JD.** 2002. Ku is important for telomere maintenance, but not for differential expression of telomeric VSG genes, in African trypanosomes. *J Biol Chem* **277:**21269–21277.

65. **Glover L, Jun J, Horn D.** 2011. Microhomology-mediated deletion and gene conversion in African trypanosomes. *Nucleic Acids Res* **39:**1372–1380.

66. **Barry JD, Marcello L, Morrison LJ, Read AF, Lythgoe K, Jones N, Carrington M, Blandin G, Bohme U, Caler E, Hertz-Fowler C, Renauld H, El Sayed N, Berriman M..** 2005. What the genome sequence is revealing about trypanosome antigenic variation. *Biochem Soc Trans* **33:** 986–989.

67. **Kirkman LA, Lawrence EA, Deitsch KW.** 2014. Malaria parasites utilize both homologous recombination and alternative end joining pathways to maintain genome integrity. *Nucleic Acids Res* **42:**370–379.

68. **Fox BA, Falla A, Rommereim LM, Tomita T, Gigley JP, Mercier C, Cesbron-Delauw MF, Weiss LM, Bzik DJ.** 2011. Type II Toxoplasma gondii KU80 knockout strains enable functional analysis of genes required for cyst development and latent infection. *Eukaryotic Cell* **10:** 1193–1206.

69. **Fox BA, Ristuccia JG, Gigley JP, Bzik DJ.** 2009. Efficient gene replacements in Toxoplasma gondii strains deficient for nonhomologous end joining. *Eukaryotic Cell* **8:** 520–529.

70. **Huynh MH, Carruthers VB.** 2009. Tagging of endogenous genes in a Toxoplasma gondii strain lacking Ku80. *Eukaryotic Cell* **8:**530–539.

71. **Aravind L, Koonin EV.** 2001. Prokaryotic homologs of the eukaryotic DNA-end-binding protein Ku, novel domains in the Ku protein and prediction of a prokaryotic double-strand break repair system. *Genome Res* **11:** 1365–1374.

72. **Abrahamsen MS, Templeton TJ, Enomoto S, Abrahante JE, Zhu G, Lancto CA, Deng M, Liu C, Widmer G, Tzipori S, Buck GA, Xu P, Bankier AT, Dear PH, Konfortov BA, Spriggs HF, Iyer L, Anantharaman V, Aravind L, Kapur V.** 2004. Complete genome sequence of the apicomplexan, Cryptosporidium parvum. *Science* **304:**441–445.

73. **Chookajorn T, Dzikowski R, Frank M, Li F, Jiwani AZ, Hartl DL, Deitsch KW.** 2007. Epigenetic memory at malaria virulence genes. *Proc Natl Acad Sci U S A* **104:** 899–902.

74. **Lopez-Rubio JJ, Gontijo AM, Nunes MC, Issar N, Hernandez RR, Scherf A.** 2007. 5′ flanking region of var genes nucleate histone modification patterns linked to phenotypic inheritance of virulence traits in malaria parasites. *Mol Microbiol* **66:**1296–1305.

75. **Jiang L, Mu J, Zhang Q, Ni T, Srinivasan P, Rayavara K, Yang W, Turner L, Lavstsen T, Theander TG, Peng W, Wei G, Jing Q, Wakabayashi Y, Bansal A, Luo Y, Ribeiro JM, Scherf A, Aravind L, Zhu J, Zhao K, Miller LH.** 2013. PfSETvs methylation of histone H3K36 represses virulence genes in Plasmodium falciparum. *Nature* **499:** 223–227.

76. **Ukaegbu UE, Kishore SP, Kwiatkowski DL, Pandarinath C, Dahan-Pasternak N, Dzikowski R, Deitsch KW.** 2014. Recruitment of PfSET2 by RNA polymerase II to variant antigen encoding loci contributes to antigenic variation in P. falciparum. *PLoS Pathog* **10:**e1003854.

77. **Lopez-Rubio JJ, Mancio-Silva L, Scherf A.** 2009. Genome-wide analysis of heterochromatin associates clonally variant gene regulation with perinuclear repressive centers in malaria parasites. *Cell Host Microbe* **5:**179–190.

78. **Flueck C, Bartfai R, Volz J, Niederwieser I, Salcedo-Amaya AM, Alako BT, Ehlgen F, Ralph SA, Cowman AF, Bozdech Z, Stunnenberg HG, Voss TS.** 2009. Plasmodium falciparum heterochromatin protein 1 marks genomic loci linked to phenotypic variation of exported virulence factors. *PLoS Pathog* **5:**e1000569.

79. **Marty AJ, Thompson JK, Duffy MF, Voss TS, Cowman AF, Crabb BS.** 2006. Evidence that Plasmodium falciparum chromosome end clusters are cross-linked by protein and are the sites of both virulence gene silencing and activation. *Mol Microbiol* **62:**72–83.

80. **Figueiredo LM, Freitas-Junior LH, Bottius E, Olivo-Marin JC, Scherf A.** 2002. A central role for Plasmodium falciparum subtelomeric regions in spatial positioning and telomere length regulation. *EMBO J* **21:** 815–824.

81. **Goodarzi AA, Jeggo P, Lobrich M.** 2010. The influence of heterochromatin on DNA double strand break repair: Getting the strong, silent type to relax. *DNA Repair* **9:** 1273–1282.

82. **Miller D, Reynolds GE, Mejia R, Stark JM, Murnane JP.** 2011. Subtelomeric regions in mammalian cells are deficient in DNA double-strand break repair. *DNA Repair* **10:**536–544.

83. **Hakimi MA, Deitsch KW.** 2007. Epigenetics in Apicomplexa: control of gene expression during cell cycle progression, differentiation and antigenic variation. *Curr Opin Microbiol* **10:**357–362.

84. **Kirkman LA, Deitsch KW.** 2012. Antigenic variation and the generation of diversity in malaria parasites. *Curr Opin Microbiol* **15:**456–462.

85. **Dzikowski R, Templeton TJ, Deitsch K.** 2006. Variant antigen gene expression in malaria. *Cell Microbiol* **8:** 1371–1381.

86. **DeBarry J, Fatumo S, Kissinger J.** 2013. The apicomplexan genomic landscape: the evolutionary context of Plasmodium, p 17–35. *In* Carlton JM, Perkins SL, Deitsch KW (ed), *Malaria Parasites: Comparative Genomics, Evolution and Molecular Biology*. Caister Academic Press, Norfolk, UK.

Mobile DNA, 3rd Edition
Nancy L. Craig, Michael Chandler, Martin Gellert, Alan M. Lambowitz, Phoebe A. Rice and Suzanne Sandmeyer

doi:10.1128/microbiolspec.MDNA3-0015-2014

Kyle P. Obergfell[1]
H. Steven Seifert[1]

21 Mobile DNA in the Pathogenic *Neisseria*

INTRODUCTION

The majority of species in the genus *Neisseria* are commensal bacteria that colonize mucosal surfaces. The two pathogenic species, *Neisseria gonorrhoeae* (the gonococcus) and *Neisseria meningitidis* (the meningococcus), are the causative agent of gonorrhea and the primary cause of bacterial meningitis in young adults, respectively. Both organisms are strict human pathogens with no known environmental reservoirs that have evolved from commensal organisms within the human population (1). The study of the *Neisseria* is important for public health reasons, but also provides a defined system to study evolution of two highly related organisms that cause distinct diseases. One unique aspect of the pathogenic *Neisseria* is the presence of sophisticated genetic systems that contribute to pathogenesis. The processes of DNA transformation and pilin antigenic variation will be discussed in this chapter.

NATURAL DNA TRANSFORMATION

There is a diverse set of more than 80 identified naturally transformable bacterial species that are able to recognize free DNA in the environment, import it across the envelope and recombine exogenous DNA with resident DNA molecules (2). Unlike the majority of naturally competent species, *Neisseria* sp. are constitutively competent, capable of transformation at all phases of growth (3, 4). Natural transformation is the primary means of horizontal gene transfer (HGT) in *Neisseria* with a documented flow of information among both commensal and pathogenic members of the genus (5, 6). Similar to the vast majority of Gram-negative bacteria, transformation in *Neisseria* is dependent on a type IV pilus (Tfp) complex (7). Pathogenic *Neisseria* often undergo HGT with recombination occurring so frequently that there is a marked inability to establish stable clonal lineages (8). These genomic signatures suggest that mixed infections are common and this idea has been supported by certain studies (9, 10). Although *N. meningitidis* maintains limited lineage structure, both species tend toward linkage equilibrium, and this frequent genetic exchange is thought to contribute to the rapid spread of antibiotic resistance among *N. gonorrhoeae* strains (11–14). HGT has led to clinical isolates of *N. gonorrhoeae* exhibiting resistance to multiple antibiotics (15). Though no single strain has accumulated all of the resistance markers, there are resistant lineages to every currently

[1]Northwestern University Feinberg School of Medicine, Department of Microbiology and Immunology, Chicago, IL 60611.

recommended therapy (15). The threat of untreatable gonorrhea has earned *N. gonorrhoeae* a spot on the Center for Disease Control's list of superbugs and its highest threat level, which is only invoked for three organisms (16–18).

The Type IV Pilus

Type IV pili are critical virulence factors for many pathogens and also promote interactions of nonpathogens with their environments. Tfp are long, thin fibers that undergo dynamic cycles of extension and retraction and mediate twitching motility, cellular adherence, microcolony formation, and natural transformation in both *N. gonorrhoeae* and *N. meningitidis* (19–23). The expression of *Neisseria* Tfp correlates directly with transformation efficiency, an observation that has been expanded to many Gram-negative species (3). In the presence of excess DNA, highly piliated strains of *N. gonorrhoeae* can achieve transformation efficiencies more than a million times higher than strains lacking the major Tfp pilin PilE (24). There is a complex Tfp assembly apparatus present in the bacterial envelope that is responsible for pilus expression and associated functions (Fig. 1). Many of the Tfp complex proteins are required for transformation. Among the Tfp complex proteins that have a defined function, PilD is a periplasmic protease responsible for processing PilE into the mature form that can be assembled into the

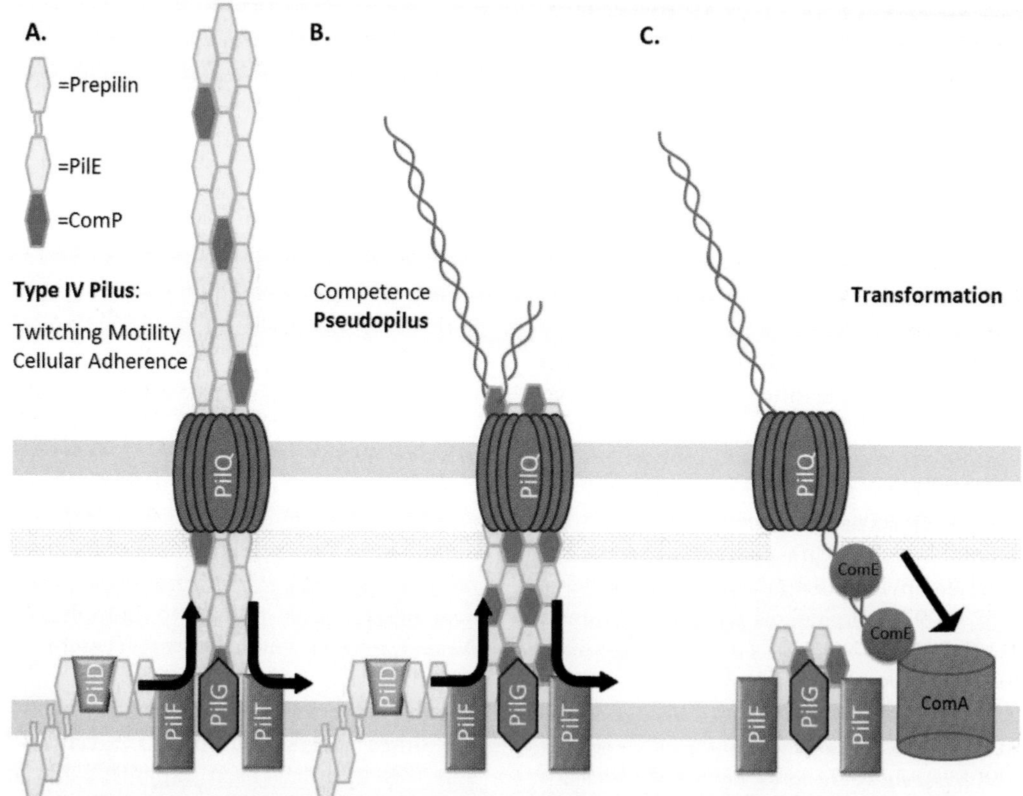

Figure 1 Type IV pilus and DNA uptake. (A) Type IV pilus—the Tfp is a several micron long, 60 Å wide fiber anchored in the inner membrane by PilG that extends through the PilQ secretin pore. Composed mainly of the major pilin PilE (pilin), which is processed by a dedicated protease, PilD. The PilF and PilT NTPases mediate extension and retraction of the pilus through polymerization and depolymerization of the pilin subunits. (B) Competence pseudopilus—hypothesized pseudopilus that could mediate transformation. Uses the type IV pilus complex including the PilQ pore but is not an extended fiber. Possible localization of ComP to the pseudopilus could mediate specific DNA binding. (C) DNA uptake model—retraction of the (pseudo)pilus mediated by PilT brings the initial length of DNA into the periplasm. DNA is then bound by a protein or protein complex possibly containing ComE, which mediates import of the remaining length of DNA into the periplasm. The inner membrane protein ComA facilitates DNA entry into the cytoplasm.
doi:10.1128/microbiolspec.MDNA3-0015-2014.f1

pilus fiber (23). PilD is required for pilus expression and transformation competence. Mutations in Tfp structural proteins including the inner membrane protein PilG and the pore-forming secretin PilQ abrogate transformation (25, 26). Additionally, transformation is dependent on the two cytoplasmic NTPases PilF and PilT (23, 27). PilF is thought to power pilus extension while PilT is required for pilus retraction and twitching motility. The requirement for many of the Tfp complex proteins in transformation has led to the generally accepted hypothesis that the pilus fiber mediates the binding and initial uptake of DNA into the periplasm as well as transport of DNA through the outer membrane during transformation (28–30). This has never been conclusively shown, and it is an open question whether Tfp or a pilus-like apparatus (pseudopilus) is actually responsible for DNA uptake across the outer membrane (31).

Contributing to this uncertainty are the observations that small amounts and altered forms of pilin are sufficient for transformation. Pilin (*pilE* gene product) can exist in variant forms such as S-pilin (a short or secreted form) and L-pilin (long form). S-pilin results from a cleavage event and produces a soluble form that is secreted from the cell by an unknown mechanism (32). Some S-pilin variants display intermediate piliation phenotypes but wild-type levels of transformation (33, 34). L-pilin variants result from a duplication of coding sequences in *pilE* that produces an oversized pilin monomer that cannot be assembled into pilus fibers, yet only reduces transformation efficiencies 35-fold (32, 33). Alongside investigations into pilin variants, a 2003 study showed that when the level of pilin expression was reduced to the point where observable pili are extremely rare in a population of cells, the cells still exhibit considerable transformation efficiencies (24). The competence in these pilus-deficient gonococci was still dependent on PilT and PilQ, leading to the hypothesis that extended Tfp are not necessary for transformation; rather a pseudopilus apparatus, using the type IV pilus complex of proteins, is sufficient for transformation (Fig. 1).

DNA Uptake

Regardless of whether extended Tfp or pseudopili are responsible for specific binding of extracellular DNA, retraction of the Tfp or pseudopilus is likely not sufficient to account for the difficulties of transporting DNA across the outer membrane. The PilQ pore is only 6 nm in width, the same width as the predicted Tfp structure (35). This leaves no room for concomitant transport of other substrates, much less the doubled up DNA structure that would result from binding at a mid-strand site. Taking into consideration that lengths of DNA several times longer than the cell are routinely transformed, a single pilus retraction event would not bring the entire DNA molecule into the periplasm (36). Therefore, DNA transport across the outer membrane requires a mechanism more complicated than pilus elaboration, DNA binding, and a single retraction event. Successive cycles of pilus extension and retraction could be responsible for pulling long DNA molecules across the outer membrane, but evidence from studies of the Gram-positive bacterium *Bacillus subtilis* suggests that the import of DNA is processive and occurs at a constant velocity (37). This could be explained by the cooperative retraction of several pili in succession but there is no experimental evidence to support this hypothesis (38).

An alternative hypothesis revolves around the gonococcal protein ComE. Although not surface localized, ComE is required for DNA uptake and transformation and binds DNA nonspecifically (39). Four copies of ComE are encoded in the gonococcal genome and deletion of individual *comE* genes results in an additive negative effect on transformation. A recent study in *Vibrio cholerae* showed that the ComE ortholog of *V. cholerae*, ComEA, is required for DNA uptake into the periplasm and that ComEA binding of DNA can potentially prevent retrograde transport of DNA through the PilQ pore (40). The authors propose that ComEA is responsible for pulling DNA into the periplasm through a ratcheting mechanism reliant on nonspecific DNA binding and entropic forces similar to what has been proposed to drive double-stranded DNA (dsDNA) transport through the nuclear pore complex in eukaryotic cells (41). Although it remains to be seen if ComE plays an analogous role in *Neisseria*, these observations form an alternative working model in which DNA is bound extracellularly by the (pseudo-)pilus (Fig. 1). Retraction of the (pseudo-)pilus pulls the initial length of DNA into the periplasm. Nonspecific binding of a complex possibly containing ComE then mediates import of the remaining length of DNA into the periplasm.

DNA Uptake Sequence

Although constitutively competent and lacking any apparent regulation of transformation, *Neisseria* species preferentially transform self-DNA (42, 43). This is accomplished through repeat sequences spread across the genome that aid in efficient uptake and transformation of self-DNA. The initially identified DNA uptake sequence (DUS) is a 10-base sequence (DUS10 5′-GCCGTCTGAA), but an extended 12-mer DUS

(DUS12 5′-ATGCCGTCTGAA) occurs at > 75% of DUS10 locations and slightly increases transformation efficiencies over the 10-mer DUS (42, 44, 45). Remarkably, the 10-base sequence occurs about once every kilobase of the genome, a frequency a thousand times higher than chance predicts, and is often located as an inverted repeat in putative Rho-independent transcriptional terminators (42, 45). It has been suggested that DUS are more often found within DNA repair genes due to a role in genome maintenance, but this analysis does not take into account that DUS can act over distances of several kilobases (46). It is possible that the enrichment of DUS in core genes (genes shared among all isolates) may indicate slow accumulation of the sequences with the oldest (most essential) genes accumulating the highest proportion of DUS (47). The 10-base DUS sequence was identified for its ability to competitively inhibit transformation (42) and to enhance both DNA uptake and transformation when added to a previously untransformable plasmid (43). DUS-mediated transformation enhancement is both strain and strand specific in *N. gonorrhoeae*; DUS containing DNA only enhances transformation 20-fold in strain FA1090 whereas it increases efficiencies 150-fold in strain MS11 (48). Because the DUS is non-palindromic, the two divergent single-stranded DUS were investigated for their relative effect on transformation and given identifiers Watson (5′-ATGCCGTCTGAA) and Crick (5′-TTCAGACGGCAT). The single-stranded Watson DUS12 increased transformation efficiencies with single-stranded DNA (ssDNA) 180-fold to 470-fold while the single-stranded Crick DUS12 only enhanced transformation seven-fold over non-DUS-containing ssDNA (49). Notably, even with the Watson DUS12, ssDNA was two-fold to 24-fold less efficient than DUS12 containing dsDNA in transformation assays. These data suggest that there may be different uses for the double-stranded and single-stranded uptake sequences.

Although the Tfp and its components were exhaustively investigated for a role in specific binding of the DUS, the mediator of self-DNA recognition was not identified until a landmark 2013 study by the Pelicic laboratory (39, 50–57). This study identified the type IV minor pilin ComP as the DUS receptor in *N. meningitidis* (54). The study investigated ComP because of its high level of sequence conservation among several *Neisseria* species (99%) suggesting an important function. They showed that ComP was the only known pilin component that bound dsDNA and had higher affinity for DUS-containing DNA. Through mutational analysis, it was also shown that an electronegative stripe on ComP that is predicted to be surface exposed on Tfp is responsible for the DNA binding ability of ComP (54). The DUS is conserved in both *N. meningitidis* and *N. gonorrhoeae*, but different members of the *Neisseria* genus and the broader *Neisseriaceae* family have slightly different DUS, termed dialects (58). The efficiency of the DUS dialects for transformation of *N. meningitidis* was shown to be dependent on the expression of the cognate ComP protein (59). Notably, while the inner bases of the DUS are most critical for allowing high transformation efficiency and DNA binding by the ComP protein, not all residues that were important for full transformation efficiency were important for ComP binding (59). This result suggests that the DUS may act in a different step during transformation independent of ComP and could rely on the single-stranded Watson DUS, shown to function in transformation (49).

Processing and Recombination

There are three gonococcal genes—*comL*, *tpc*, and *comA*—important for transformation competence that are proposed to act during the transport of DNA into the cytoplasm (60–62). ComL and Tpc are both localized to the periplasm and have been implicated in DNA transport across the peptidoglycan layer, possibly through the creation of localized breaks in the cell wall (60, 61). ComA displays homology with ComEC of *B. subtilis*, which is a polytopic membrane protein localized to the inner membrane that delivers ssDNA into the cytoplasm (62, 63). Whether ComA also transports dsDNA into the cytoplasm is an unanswered question.

Following import into the cell, DNA is subject to restriction modification (64). *N. gonorrhoeae* encodes a large array of methyltransferases and their corresponding endonucleases that form an effective restriction barrier to plasmid DNA, making plasmids 1,000-fold less transformable than chromosomal loci (65, 66). *N. meningitidis* is able to restrict transforming DNA through a unique CRISPR/Cas system that is not present in the sequenced gonococcal isolates (67). Clustered, regularly interspaced, short palindromic repeat (CRISPR) loci confer sequence-specific adaptive immunity based on the ability of CRISPR-associated (Cas) protein complexes to cleave incoming DNA (68). Although generally thought to be an adaptation to protect against phage invasion and foreign plasmid conjugation, the type II CRISPR/Cas system of *N. meningitidis* was the first system shown to naturally prevent transformation (67). The meningococcal CRISPR/Cas pathway is the most streamlined system characterized to date and encodes sequences that may be able to restrict the transfer of certain virulence determinants among cells of different lineages. This may explain the increased ability of the meningococcus to form semi-clonal lineages in comparison to the gonococcus, which lacks a CRISPR/Cas system.

Following entry into the cytoplasm, transforming DNA is integrated into the chromosome through homologous recombination mediated by the major recombinase, RecA, which can bind ssDNA, find the complement strand in a homologous DNA duplex, and catalyze D-loop formation (69, 70). RecA activity is limited by RecX, which facilitates more efficient recombination (71). Investigations into the role of DNA repair pathways in transformation showed the RecF-like pathway that mediates ssDNA gap repair in *Escherichia coli* is not involved in transformation-mediated homologous recombination, but mutations in the RecBCD pathway show a 10-fold to 100-fold decrease in transformation efficiency (72, 73). The RecBCD pathway is involved in dsDNA break repair in *E. coli* and would be predicted to only act on dsDNA transformation substrates (74). It is likely that transformation proceeds mainly through ssDNA intermediates, and the observation that the majority of transformation is independent of the RecBCD pathway confirms the ssDNA dependence. However, the amount of transformation that is RecBCD-dependent suggests that either there is some dsDNA transported into the cytoplasm or that a portion of the transported ssDNA is converted to dsDNA and then used for recombination. A role for DprA nuclease, which is involved in DNA transformation in several bacterial species, has not been reported (2). Finally, mutation of PriA helicase, which helps restart stalled replication forks in *E. coli*, lowers transformation efficiency. The helicase may act directly on the D-loop produced by RecA-mediated ssDNA invasion of the duplex, or alternatively, may process a different intermediate formed during the recombination process that requires replication restart (75, 76). There are no other naturally transformable organisms reported where PriA has been tested for a role in transformation processes, so it is unclear if this is a unique processing requirement in the *Neisseria*. The details of DNA processing and recombination during transformation in the *Neisseria* are still open to further investigation.

Delivery of DNA for DNA Transformation—Gonococcal Genetic Island and Type IV Secretion

One of the major issues in genetic transfer is the source of transforming DNA. It has been argued that using DNA from dead bacteria may promote the acquisition of alleles that are at a selective disadvantage (77). This viewpoint ignores the fact that even if a genetic change occurs that causes death of a bacterial cell, the negative allele is only one gene in a chromosome of 500 to 8,000 genes that can be transferred. In addition, most competent bacteria exhibit a natural level of autolysis that provides DNA from cells that are not under negative selection (78). The gonococcus, however, has a unique way of providing donor DNA for transformation. Over 75% of investigated gonococcal strains carry a genomic island approximately 57 kb in size encoding a type IV secretion system (T4SS) that secretes DNA into the extracellular environment (79, 80). While named the gonococcal genetic island (GGI) because of its discovery in *N. gonorrhoeae*, the GGI has also been found in 17.5% of meningococcal strains, although the functionality of the encoded T4SS varies among these strains (81, 82). The GGI displays many characteristics of a horizontally acquired genomic island; including a G+C content different from the rest of the chromosome and a paucity of DNA uptake sequences relative to the rest of the genome (80). Like many other horizontally acquired genomic islands, the GGI has short direct repeats on both ends (80).

Based on the GGI being flanked by the direct repeat sequences *difA*, a consensus *dif* site, and *difB*, a *dif* site varying by four mismatches, it is thought that the acquisition of the GGI was mediated by the *dif*-recognizing gonococcal recombinase XerCD (80, 83). Under laboratory conditions, the GGI undergoes XerCD mediated excision once every 10^6 cells after 18 hours of growth (84). Notably, when the non-consensus *difB* site is mutated to restore the consensus *dif* sequence, the excision rate increases to once per 1,000 cells (84). These observations have led to the conclusion that the nonconsensus sequence of the *difB* site acts to limit the rate of GGI excision (85). While the maintenance of the GGI and its encoded T4SS in 80% of gonococcal strains indicates a survival advantage, it is possible that infrequent excision of the GGI is maintained, perhaps to help propagate the GGI. The mutations in the *difB* site allow for this to happen, but also significantly decrease the frequency of GGI excision, which would favor maintenance. Interestingly, many *N. meningitidis* strains carry more mutations in the *difB* site than the gonococcal versions, which may explain why the GGI is maintained in the meningococcus despite carrying mutations that render the T4SS inactive (83).

The GGI consists of 62 open reading frames, 23 of which show significant homology to T4SS proteins (80). The encoded T4SS is similar to other F-like (named for the prototypical conjugative plasmid) T4SS, but in contrast to typical conjugative elements, the GGI secretes chromosomal DNA into the extracellular milieu in a contact-independent manner (86). Nuclease sensitivity experiments show that the secreted DNA is single-stranded and blocked at the 5′ end, presumably by a bound relaxase molecule (84). Unlike the prototypical T4SS, the secreted DNA is exposed to the extracellular environment rather than being transferred directly between cells (79). This idea is supported by the observation that the encoded TraA conjugative pilin subunit is not required for T4S (85). The secreted DNA, however, is competent for transformation and has been shown to be a better transformation substrate than DNA released by autolysis (79). This conclusion stands in contrast to the observation that ssDNA is a less efficient substrate than dsDNA for transformation (49). While protein secretion has yet to be shown, it has been postulated that the increased efficiency of transformation by T4S ssDNA over autolysis-released dsDNA is due to proteins bound to the secreted DNA (85).

Although the gonococcal T4SS is nonconjugative, inferences can be made about the structure of the gonococcal T4SS based on the structure revealed by cryo-electron tomography for the conjugative plasmid pKM101 and gonococcal experimental data (87, 88) (Fig. 2). In the conjugation apparatus of pKM101, the core complex spans both the inner and outer membranes with polymerized TraK forming the secretin in the outer membrane. Interacting with TraK in a one-to-one-to-one ratio are TraV, thought to stabilize the pore complex, and TraB (87, 88). Dillard and colleagues have established that TraK and TraV interact in gonococci and that TraK is surface exposed (85). In the pKM101 aparatus, TraB spans both the outer and inner membranes to connect the core complex and provide a channel for substrate translocation, and the same is predicted to be true in gonococci (87). The inner membrane complex is likely composed of the N-terminal domain of TraB, an ATPase TraC, a coupling protein TraD, and TraG (89, 90).

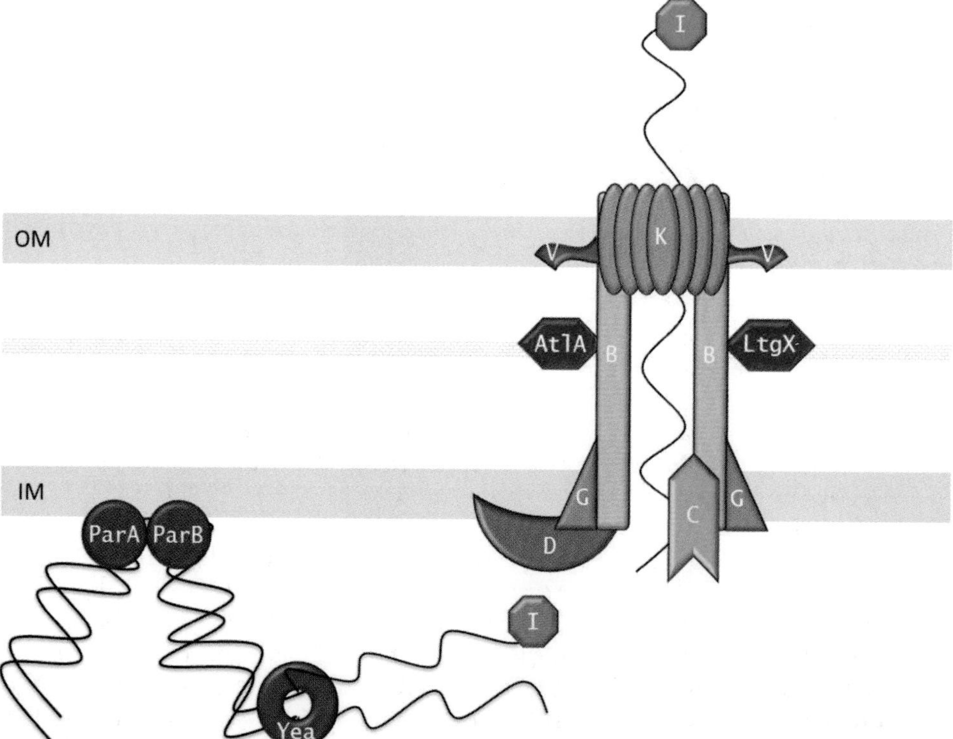

Figure 2 Type IV secretion system model. ParA and ParB recruit the chromosomal DNA to the type IV secretion system. TraI relaxase nicks the DNA at the *oriT* site and the DNA is unwound most likely by the Yea helicase. The resulting single-stranded DNA, possibly still bound by TraI, is then secreted through the type IV secretion complex into the extracellular milieu in a contact-independent manner. The inner membrane complex is predicted to consist of TraG, TraD, and TraC with TraB spanning both the inner and outer membranes to form a channel for the DNA. The transglycosylases AtlA and LtgX create localized breaks in the peptidoglycan to allow the system to assemble. The outer membrane complex consists of TraB, TraK, and TraV. doi:10.1128/microbiolspec.MDNA3-0015-2014.f2

Although shown to stabilize mating pairs, aid in pilus production, and act in entry exclusion in homologous systems; TraG plays a novel, uncharacterized role in gonococcal T4S (90). Among sequenced *N. gonorrhoeae* strains, three different alleles of TraG are found (79). The shortest allele was found to be nonfunctional for the secretion of DNA but it could still have a role in the secretion of proteins or other substrates (90). Like other T4SS, the GGI is dependent on lytic transglycosylases for secretion, presumably to produce localized breaks in the peptidoglycan to allow for secretion complex assembly (91). Unique to the gonococcal system is that two lytic transglycosylases, AtlA and LtgX, are required for T4S (91).

Besides the structural components of the T4SS, secretion of DNA is thought to occur through the action of ParA, ParB, TraI, Yea, and TraD (Fig. 2). ParA and ParB are partitioning proteins responsible for segregation of chromosomes and plasmid DNA during replication (92). ParA, specifically the DNA binding Walker Box domain, has been shown to be required for DNA secretion leading to the hypothesis that ParA and ParB are responsible for recruitment of the DNA to the secretion complex and the processing enzymes (80). TraI is a GGI-encoded relaxase, a class of proteins that nick DNA at a specific recognition site *oriT* (84, 93). TraI is required for DNA secretion and mutation of the putative catalytic tyrosine residues either blocks or severely hinders DNA secretion into the medium (84). While TraI does not appear to use the typical histidine-rich motif for metal coordination, an HD phosphohydrolase domain may substitute for metal ion coordination. Additionally, gonococcal TraI contains a predicted unique N-terminal amphiphatic helix, which may allow association with the inner membrane. TraI interacts with the DNA at the single *oriT* site on the GGI. This site is an inverted repeat located near the *traI* gene and is required for DNA secretion. While the GGI encodes the only *oriT* site, the site can still support DNA secretion if moved to a distant location on the chromosome (84). This leads to the model that DNA secretion begins with a TraI-induced nick at *oriT* and delivery of the nicked strand to the secretion complex by the required coupling protein TraD (94). Secretion then continues via unwinding of the chromosome by the GGI-encoded helicase Yea. As the DNA is unwound for secretion, strand replacement synthesis is used to regenerate the chromosome and secretion proceeds along the length of the chromosome with the efficiency decreasing the further the DNA is from the original *oriT*.

Whether the GGI (and the encoded T4SS) has a direct role in gonococcal pathogenesis remains unclear. Certain versions of the GGI (containing *atlA* and the *sac-4* allele of *traG*) are found more often in isolates that cause disseminated gonococcal infection (79). Additionally, the GGI T4SS is active during intracellular infection and can compensate for an inactive iron transport complex (95). These data suggest that the T4SS either acts as an import apparatus or exports proteins that can then be subsequently used to import iron into the bacterial cell. Finally, it has been established that secretion of ssDNA is required for the initial stages of biofilm formation, which is presumed to help with colonization (96). Although these studies establish possible roles for the GGI in *N. gonorrhoeae* virulence, the lack of an appropriate infection model makes it difficult to characterize the relative contribution of the system in causing disease. Perhaps the most interesting remaining question is what role the T4SS plays in protein secretion. The increased ability to transform with secreted DNA and the Ton-independent iron acquisition data suggest the possibility that the GGI mediates protein secretion, but this hypothesis still requires direct experimental support.

The fundamental question of what benefit HGT provides to any organism remains controversial (2). Several doubts have been raised about whether HGT is the evolutionary basis of competence because of the unpredictable results of genetic transfer with critics supporting a model where competence evolved as a nutrient acquisition system (77). An *in silico* study, however, reports that even if extracellular DNA comes from dead cells with more deleterious alleles than the recipient cell, HGT can actually allow a population to escape the predicted irreversible accumulation of harmful alleles (97). This in turn suggests that HGT may be important for long-term genomic maintenance rather than having deleterious effects. Another possibility is that transformation competence allows a species to have a larger gene pool to draw upon (98). Although this simulative study does not end the controversy, there is no doubt that natural transformation plays a significant role in genetic exchange and nutrient uptake and thus is an important area for continued investigation. With the HGT-mediated spread of antibiotic resistance, the specter of untreatable gonorrheal infections is a looming reality.

ANTIGENIC VARIATION

Introduction

As obligate human pathogens, *N. gonorrhoeae* and *N. meningitidis* are constantly under immune surveillance.

One vital mechanism for immune avoidance is antigenic variation, by which a pathogenic organism constantly modifies surface-exposed immunogenic molecules. This strategy can result in prolonged colonization or allow for re-infection of a previously infected host who has a potentially effective immune response that is made ineffectual by the variation—reviewed in ref. (99). Phase variation is a related but distinct process of phenotypic variation where a cell switches between defined phases of expression, either between ON and OFF phases or between two variant forms. Both phase and antigenic variation processes generally occur at rates higher than the normal mutation rate of the organism. Phase variation systems have the ability to reversibly switch between the phases and are usually mediated by polynucleotide repeat variation, invertible elements, or differential methylation—reviewed in ref. (100). In contrast, antigenic variation systems have the ability to stochastically express many different forms of a gene product and can be based on multigene phase variation or a recombination-based diversity generation system (see next section). Both phase and antigenic variation systems can have functional as well as immune surveillance consequences.

The two pathogenic *Neisseria* species express three antigenically or phase variable major surface determinants: the opacity (Opa) outer membrane proteins, which act as adhesins; lipooligosaccharide (LOS), which decorates the outer membrane and is also involved in host interactions; and Tfp (101). The commensal organisms express some of these molecules but do not undergo the variation processes. Both LOS and Opa proteins undergo phenotypic variation through an ON/OFF phase variation mechanism, mediated by slipped-strand mispairing of tandem repeats in multiple genes (102–104). In the case of the Opa gene family, there are 4–13 individual genes that phase vary ON and OFF independently through changes in a pentamer repeat found in the signal sequence coding region of each gene (102). It is likely that the Opa protein antigenic variation process is mainly used to create functional variants to promote interactions with different human cellular receptors (105). Five of the LOS biosynthetic glycosyltransferases are phase variable due to polynucleotide repeats in promoter or coding regions that when altered turn each gene ON or OFF (104, 106). The LOS structure is then defined by which set of these five genes is expressed in combination with the eight invariant biosynthetic genes. Pilus antigenic variation, in contrast, uses a complex, programmed homologous recombination system to express antigenically distinct peptide sequences on the Tfp. Even though homologous recombination-based systems for antigenic variation are found in both prokaryotic and eukaryotic pathogens, the pilin antigenic variation in *Neisseria* has become a model system for these types of diversity generation systems.

Pilin Antigenic Variation

Pilin antigenic variation is mediated by non-reciprocal recombination events, where a sequence from a silent pilin copy is donated to the pilin expression locus but does not change in the reaction (Fig. 3). Gonococci typically carry around 18 silent pilin copies in four to six *pilS* loci (107, 108). Lacking a promoter, ribosome binding site, and the coding sequence for the N-terminal α helix of pilin, the silent copies are not expressed (109–111). The *pilE* gene encodes a 5′ constant region followed by a semi-variable (SV) region, two highly conserved *cys* regions (containing the two-disulfide bond forming cysteines) sandwiched around the hypervariable loop, and finally the hypervariable tail. The amount of conservation and diversity in the different regions is due directly to the sequence variation in the silent copies (109). The hypervariable regions of *pilE* correspond to the coding regions that display the largest diversity among the silent copies. Highlighting the functional role of pilin variation in creating antigenic diversity, the variable regions correspond to the surface and antibody-exposed areas of pilin in the pilus fiber (112, 113). During pilin antigenic variation a portion of a *pilS* copy is transferred into the expressed *pilE*. Because the amount of transferred sequence can range from a single nucleotide to the entire *pilS* copy and sequence can be donated from multiple *pilS* copies along the length of *pilE*, this process can result in a remarkably large set of expressed pilin sequences (Fig. 3A–C) (114).

The molecular process of antigenic variation results in a wide range of functional consequences. Noted as early as the 1960s, gonococcal colonies can exhibit a visible phase variation due to expression or lack thereof of pili (115). Some pilus phase variation is due to ON/OFF switching of PilC or *pilE* deletion, but antigenic variation events that introduce a premature stop codon from a silent copy or a nonfavorable combination of silent copies can also prevent pilus expression (116–119). While not fully understood, it is likely that such nonfavorable combinations of silent copies result in a pilin molecule that is unable to efficiently assemble into a pilus fiber. This is supported by the observation that strains with different *pilE* coding sequences exhibit different levels of piliation (118). Although the lack of a suitable animal model means experimental data are

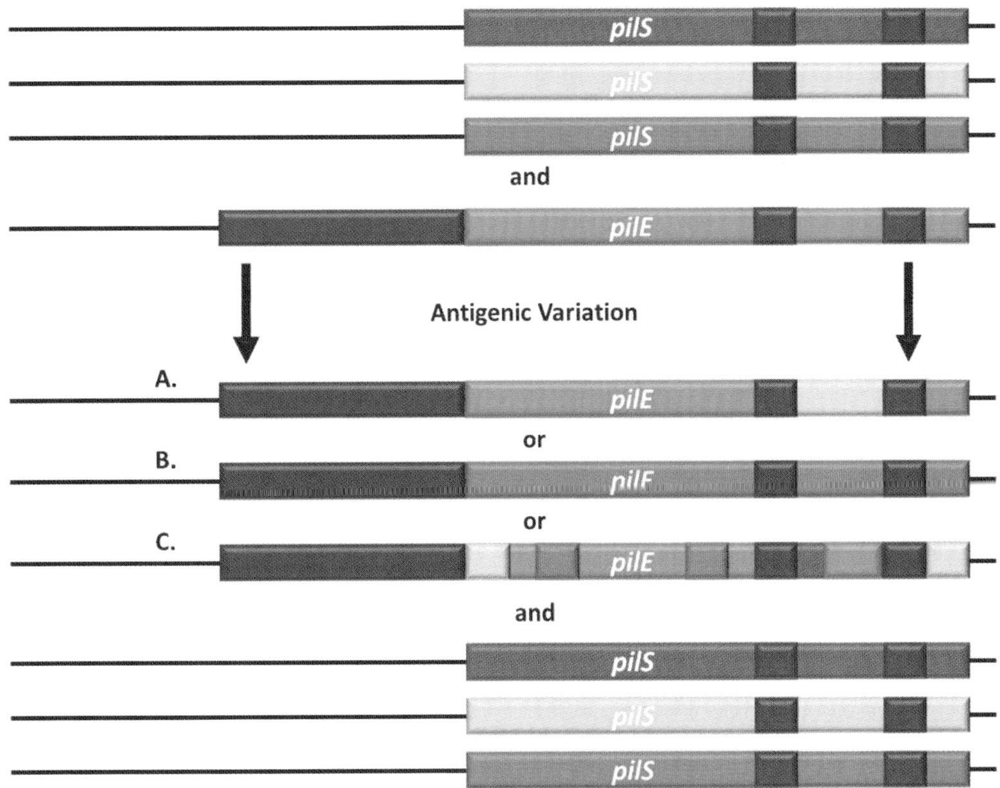

Figure 3 Molecular description of antigenic variation. The *pilE* and *pilS* loci have regions of sequence microhomology (grey) and variability (colored). Sequence from a nonexpressed *pilS* loci copy is transferred into the expression locus with the *pilS* sequence not changing. Recombination can occur (A) in just a section of the gene resulting in a *pilE-pilS* hybrid, (B) across the entire *pilS* gene resulting in an entirely new variable region of *pilE*, or (C) multiple times with different silent copies resulting in a new *pilE* sequence containing information from different silent copies throughout the variable regions. doi:10.1128/microbiolspec.MDNA3-0015-2014.f3

lacking, it is likely that phase variation is critical for multiple reasons as all gonococcal isolates have this ability and multiple avenues to achieve it. Antigenic variation can also alter the sites of post-translational modification of the pilus, which has been implicated in a variety of biological processes in *N. gonorrhoeae* including cellular adhesion and host-cell activation (120, 121). In *N. meningitidis*, changes in the glycosylation status of the pilus can enhance transit across epithelial barriers, a critical step in pathogenesis (122). In addition, variation of the exposed pilin residues has been implicated in controlling host-cell response in *N. meningitidis* (123). Engineered pilin variants demonstrated that the C-terminal domain of pilin is critical for colonization promoting host-cell interactions, and that different pilin sequences conferred different host-cell specificities (123). These findings have direct implications on the pathogenesis of *N. meningitidis* and underscore the importance of the diversity generation systems of the *Neisseria* to provide both functional and immune avoidance capabilities to the organism.

Trans-Acting Factors Important for Pilin Antigenic Variation

A series of broad and directed genetic screens have identified many factors important for pilin antigenic variation, although the majority of their roles have been inferred from orthologous proteins rather than direct biochemical characterization (124–126). The first protein identified as critical to antigenic variation was RecA as mutations in the *recA* gene decreased pilus phase variation by 100-fold to 1,000-fold (119). These results not only established RecA as a mediator of pilin antigenic variation but also demonstrated that antigenic variation is mediated by a homologous recombination-based process (119). *E. coli* RecA can complement a gonococcal RecA mutant and has similar biochemical properties (127, 128). Interestingly, expression of the *E. coli* RecA in *N. gonorrhoeae* resulted in increased antigenic variation and this increased frequency was due to the co-transcription of the *E. coli* RecX protein (129). This work led to the discovery that in the gonococcus, RecX is required for efficient pilin antigenic variation and that *E. coli* RecX is a negative regulator of RecA filamentation (71, 73, 129). Another protein that modulates RecA polymerization and activity named RdgC is also involved in promoting pilin antigenic variation presumably by also modulating RecA activity (130, 131).

The RecF-like recombination pathway has a central role in pilin antigenic variation (*Neisseria* do not encode a *recF* ortholog). In *E. coli*, the RecF pathway uses proteins both pathway-specific (RecF, RecR, RecO, RecQ) and nonspecific (RecA, RecN). The pathway is mainly responsible for repairing single-stranded gaps in DNA, although it can repair dsDNA breaks when the primary RecBCD pathway is inactive (132). In *N. gonorrhoeae*, mutational analysis of the pathway revealed that RecQ and RecJ are required for some but not all antigenic variation events, but only the RecOR recombinase is required for all pilin antigenic variation (72, 133). RecQ involvement was shown to be dependent on the helicase activity of two of the three HRDC domains at the C-terminus (134). The Rep protein, a 3′–5′ helicase in *E. coli*, is also required for some events, and it is not known whether RecQ and Rep are partially redundant to one another (135, 136). RecJ is a 5′–3′ single-strand exonuclease, and this activity suggests that single-strand end resection is involved (133). RecR and RecO form a necessary recombinase that acts with RecA in the process of antigenic variation (125). RecN, whose role is not limited to the RecF pathway in *E. coli*, was not found to play a role in antigenic variation (125, 133). The identification of the RecF-like pathway as being necessary for pilin antigenic variation suggests that there is a gapped intermediate that is required for one step of the recombination process, but the molecular characteristics of the intermediate have not been described.

The RecBCD recombination pathway was reported to not be involved in pilin antigenic variation, but subsequent studies published that a *recD* mutant showed an increased frequency of antigenic variation and that a *recB* mutant was deficient for pilin antigenic variation in one strain of *N. gonorrhoeae*, but not another (72, 137, 138). All of these studies relied on assays that scored only a subset of potential variants and did not always account for the greatly reduced growth rate of the *recB*, *recC*, and *recD* mutants. A less biased sequencing assay, used to detect all pilin antigenic variation events, conclusively showed that insertional mutations in *recB*, *recC*, and *recD* in both strains MS11 and FA1090 resulted in DNA repair phenotypes but not a pilin antigenic variation phenotype (139). This study confirmed that the impaired growth of the *recB* and *recD* mutants resulted in a shift in the frequency of some, but not all, donor silent copies—possibly explaining the reason for the contrasting conclusion of a role for the RecBCD pathway in pilin antigenic variation.

Both the RuvABC and RecG Holliday Junction processing pathways are required for pilin antigenic variation (125, 140). Mutations in either pathway prevent pilin antigenic variation, leading to the hypothesis that there are distinct substrates acted on by these pathways. Importantly, double mutants in *recG* and *ruvA*, *ruvB*, or *ruvC* created a partial synthetic lethality with an increase in *pilE* deletions in the surviving bacteria. This lethality could be rescued by a *recA* mutation as well as several other mutations that prevent antigenic variation (126, 140). This led to the hypothesis that antigenic variation involves two Holliday junctions, explaining the phenotype of the single mutations, and that the double *recG* and *ruvA*, *ruvB*, or *ruvC* mutations prevents the reversal of one or both of the Holliday junctions, so locking the cell into a lethal antigenic variation intermediate. Preventing antigenic variation through processes such as *recA* mutation or *pilE* deletion prevents formation of the unresolvable lethal intermediate and rescues the synthetic lethality (140).

Required DNA Sequences and Structures

Several *cis*-acting sites have been identified as contributing to pilin antigenic variation. The 63 bp Sma/Cla repeat is located downstream of all pilin loci and often carries *Sma*I and *Cla*I restriction endonuclease sites (110). Because of the similarity of the Sma/Cla sequence to recombinase binding sites, the Sma/Cla site was investigated, and it was found that deletion of the site downstream of the expressed pilin locus in strain MS11 reduced antigenic variation using a semi-quantitative assay that only detected transfer from two donor silent copies (141). This observation has not been reported for other strains and remains unexplained. The *cys2* region of pilin is another conserved sequence involved in antigenic variation. Two studies have implicated the importance of *cys2* in antigenic variation as well as the spacing between the *cys1* and *cys2* region (142, 143). There are two likely roles of these conserved regions in antigenic variation. First, the sequence conservation of the *cys2* region to that of the silent copies could serve as a shared region of homology to drive the recombination process. Second, the *cys2* region might serve as a binding site for *trans*-acting factors required for antigenic variation. Whatever the reason, it is clear that conserved sequences have a role in pilin antigenic variation and more investigation is needed to identify the critical sequences.

A breakthrough in the understanding of the mechanisms allowing pilin antigenic variation came from identification of a *cis*-acting DNA structure in the region of DNA immediately upstream of *pilE*. Transposon insertions isolated in a genetic screen prevented antigenic variation without disrupting pilin expression or an obvious open reading frame (125). A targeted mutagenesis of the genomic region carrying these transposon insertions identified 12 G-C base pairs that prevented antigenic variation when individually mutated (Fig. 4A and B) (126). The sequence conforms to the definition of a guanine quadruplex (G4) forming motif (Fig. 4B), which can adopt a four-stranded, square planar structure using nontraditional Hoogsteen base pairing (Fig. 4C). Further investigation confirmed that the sequence forms a G4 structure *in vitro* and the individual mutations that abrogate antigenic variation also prevent G4 formation *in vitro*. Additionally, the point mutations that prevent G4 formation also decreased the detection of single-stranded nicks in both the G4 structure and in the C-rich strand opposite the G4. These nicks are proposed to be required for the initiation of recombination and the subsequent gene conversion (126). Nuclear magnetic resonance analysis defined the *pilE* G4 structure, showing that the sequence forms a three-layer, all-parallel stranded monomeric G4 with single residue double-chain-reversal loops (Fig. 4C) (144). Importantly, the *pilE* G4 structure (but not two other G4-forming sequences present in the *N. gonorrhoeae* genome) binds to RecA and does so with an affinity similar to that of RecA for ssDNA (144). A G4 structure present on an ssDNA substrate can stimulate RecA-mediated strand exchange *in vitro* (144). Together these data suggest that the G4 serves to recruit RecA to the *pilE* locus and possibly serves as a nucleation site for RecA filamentation. Finally, the two extra HRDC domains of RecQ that were shown to be required for antigenic variation are also necessary to unwind the *pilE* G4 *in vitro*, suggesting that the effect of RecQ mutation on antigenic variation is due to the reduced capacity of the helicase to bind and unwind the G4 structure (134, 145).

Although the G4 structure was shown to be required for antigenic variation, it was unclear what initiated G4 formation because the dsDNA must be melted to allow the G4 structure to form. Directed mutational analysis identified a promoter downstream of the G4-forming sequence that is required for pilin antigenic variation (146). Transcription of a small RNA (sRNA) from this promoter was confirmed by 5′-RACE to start within the second set of Gs within the G4-forming sequence (Fig. 4A). Expression of the sRNA at a distal chromosomal site did not restore antigenic variation in a promoter mutant, showing that the sRNA was *cis*-acting. Together these data suggest that it is transcription of the sRNA at the G4 sequence and not some downstream role of the sRNA that initiates antigenic variation (146). Throughout the investigation of pilin antigenic variation, a variety of

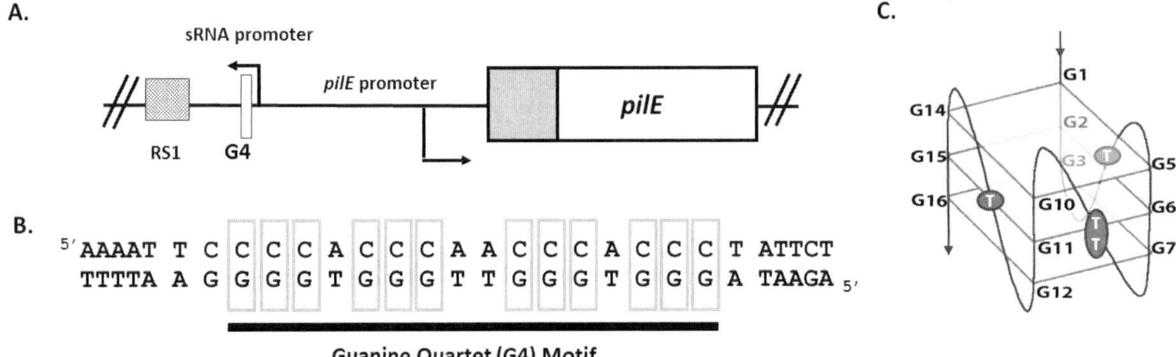

Figure 4 The *pilE* guanine quartet (G4). (A) Gene map showing the location of the *pilE*-associated G4-forming sequence and the sRNA promoter required for antigenic variation at the *pilE* locus. (B) The sequence upstream of *pilE* that forms a G4. Mutation of the boxed guanine residues leads to loss of antigenic variation implicating the G4 in antigenic variation. (C) The parallel G4 structure of the *pilE* G4 as solved by nuclear magnetic resonance analysis. doi:10.1128/microbiolspec.MDNA3-0015-2014.f4

models have been proposed to explain the phenomenon. In light of the most recent data, the feasibility of each model can be re-evaluated.

Antigenic Variation Models

Pilin antigenic variation is a gene conversion event, i.e., an apparent nonreciprocal recombination process that has been mainly studied in eukaryotic cells that possess two copies of their chromosomes. It was therefore proposed that the efficient DNA transformation system of the *Neisseria* could be used to allow for gene conversion from *pilS* to *pilE*, if the donor molecule was from a different cell than the recipient (33, 147). While there are data supporting or refuting transformation as a means to allow pilin antigenic variation, it is now generally accepted that the majority of pilin antigenic variation events do not involve transformation. (33, 147–149). With transformation ruled out as the main source for antigenic variation, all remaining proposed models of antigenic variation involve intracellular recombination and necessitate at least two chromosomes. In most bacteria, two chromosomes only exist after replication, but it has been shown that both *N. gonorrhoeae* and *N. meningitidis* are polyploid, most likely diploid, and that *Neisseria lactamica*, which does not undergo pilin antigenic variation, has a single copy of its chromosome (150, 151). It has yet to be directly demonstrated that pilin antigenic variation relies on diploid homozygous chromosomes, but it remains a compelling hypothesis.

One of the first models of antigenic variation proposed was the mini-cassette theory (109). It postulated that there were seven defined mini-cassettes of variable sequence interspersed among regions of homology that were used to affect the segmental recombination that defines pilin antigenic variation. This model was discounted by sequencing data that established that antigenic variation events can change as few as one nucleotide or incorporate an entire *pilS* copy and can occur anywhere in the variable sequences where microhomology occurs between the recombining silent and expressed genes (32, 152).

There are three main models of pilin antigenic variation that have been proposed. The unequal crossing-over model (or RecBCD-mediated double-chain-break repair model) (138, 153) (Fig. 5A) proposes that a double-strand break occurs in the *pilE* locus, and after RecBCD-mediated end resection, the single-stranded 3′ overhang invades the homologous *pilS* donor locus, presumably through RecA. The 3′ end of the invading strand is extended by polymerase while the displaced donor strand is used as a template to repair the non-invading strand. Holliday junction resolution results in the donor DNA sequence replacing the recipient *pilE* sequence. Although not conclusively ruled out, the data unambiguously showing that RecBCD plays no role in antigenic variation make the unequal crossing-over model less likely (139).

The successive half crossing-over model also proposes an initiating double-strand break at the recipient *pilE* locus, but could also be initiated by a gapped substrate (Fig. 5B) (72, 153). RecJ catalyzed 3′ end resection at the *pilE* double-strand break or nick would provide a substrate for RecA, RecX, and RecOR to mediate recombination with a donor *pilS* copy on a sister chromosome. This recombination event would create a *pilE–pilS* intermediate and link the sister chromosomes. A second half crossing-over event between the *pilS* region of the *pilE–pilS* 3′ intermediate and the original *pilE* locus would result in the original *pilE* locus containing a hybrid *pilE–pilS* sequence and the destruction of the donor *pilS* chromosome. As a result of the first half crossing-over event the original *pilE* locus will not be intact, requiring an as yet uncharacterized tethering mechanism to keep the loose *pilE* end of the dsDNA break in close proximity to the donor *pilS* locus.

The hybrid intermediate model is a variation of the half crossing-over model that evolved from experimental observations (Fig. 5C) (142, 154). This model proposes that a recombination event occurs between *pilE* and *pilS* in a region of shared microhomology similar to that of the half crossing-over model but between genes located on the same chromosome. The crossover event would result in a circular *pilE–pilS* hybrid intermediate with the chromosomal sequences that existed between the two recombining loci also carried on the episomal circle and the resultant loss of the donor chromosome. The hybrid intermediate then requires two recombination events with a recipient *pilE* on a sister chromosome. One recombination event would occur in the extensive region of upstream homology and may use the homologous recombination factors while the second event occurs in a region of microhomology within the *pilE* coding sequence. Resolution of the resulting double Holliday junction would create a new *pilE* sequence without any changes outside of the variable regions of *pilE*. Although *pilE–pilS* hybrids can be isolated from the gonococcus and have been shown to undergo recombination at the *pilE* locus more readily than *pilS* loci, a hybrid intermediate consistent with all experimental data has yet to be defined (154).

While the recombination event(s) that allow gene conversion in a bacterial chromosome remain undefined, the known antigenic variation proteins and the discovery of the G4 and *cis*-acting noncoding sRNA

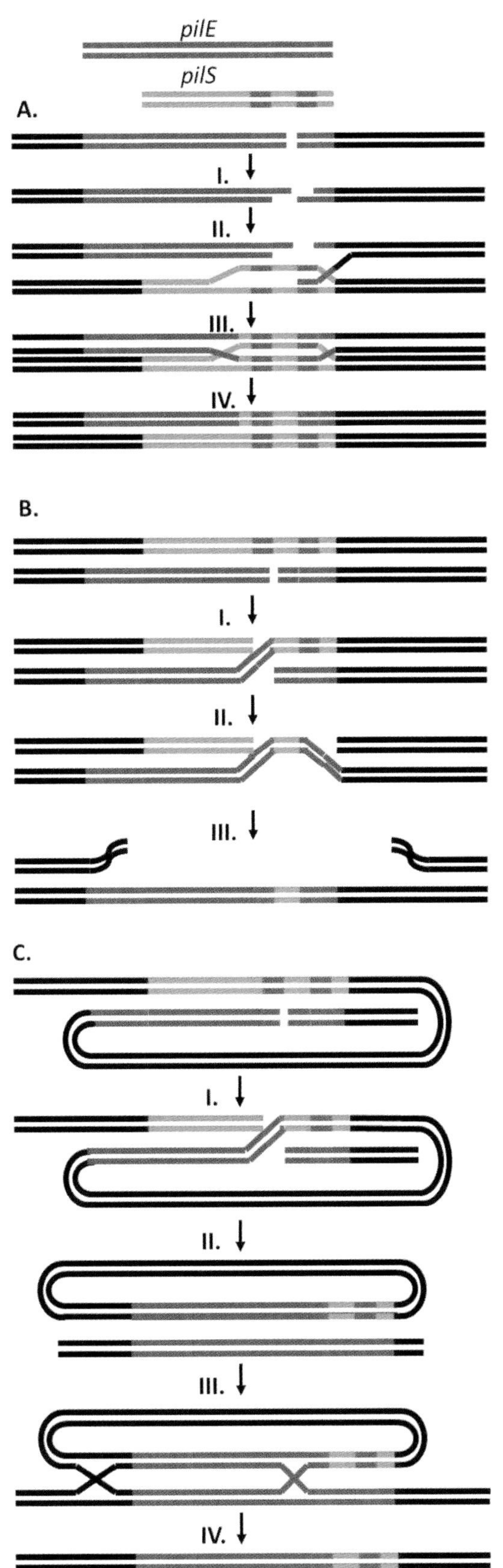

allows for the formation of a speculative, working model of the key events in antigenic variation (Fig. 6). Initiation of antigenic variation occurs with transcription of the noncoding sRNA. The process of transcription melts the dsDNA and the occlusion of the C-rich strand by the formation of a DNA:RNA intermediate allows for the formation of the G4. The activation energy required for G4 formation may be lowered by an as yet unknown protein. The fact that G4 unwinding by RecQ (and possibly Rep) is required for pilin antigenic variation suggests that the G4 is resolved by RecQ either during or after antigenic variation. Formation of the G4 leads to local nicking of the DNA, possibly induced by a stalled replication fork on the leading strand. The nicked substrate is probably processed by RecJ endonuclease and either the RecQ or Rep helicases. Based on RecA affinity for the *pilE* G4 structure, RecA may be recruited to the G4 structure to initiate RecA filamentation. A RecOR-assisted, RecA-mediated homologous pairing between the processed *pilE* and a *pilS* copy would create the half crossing-over intermediate and the second half crossing-over reaction. If the hybrid intermediate model is correct, these factors could be involved with the initial hybrid

Figure 5 Proposed recombination pathways. (A) Unequal crossing-over model—a dsDNA break occurs at the *pilE* locus and (I) the 5′ ends are resected by RecBCD to leave 3′ overhangs. (II) A single 3′ end mediated by RecA, invades the *pilS* locus forming a D-loop. (III) The 3′ ends are extended by DNA polymerase using the *pilS* gene as a template. (IV) Resolution of the double Holliday junctions results in a new *pilE* sequence without altering the donor *pilS* sequence. (B) Successive half crossing-over model–recombination begins with a dsDNA break or single-stranded gap in *pilE* in a region of homology. (I) An RecA and RecOR mediated half crossing-over event occurs linking the *pilE* and a *pilS* locus on a sister chromosome. (II) A second half crossing-over event occurs in another region of microhomology downstream of the first event between the *pilE:pilS* hyrbid and the original *pilE* locus. (III) This recombination event leads to a new sequence at the *pilE* locus and destruction of the donor chromosome. (C) Hybrid intermediate model—similar to the half crossing-over model, recombination initiates with a double-stranded break or single-stranded gap at *pilE* and (I) a half crossing-over event with a donor *pilS* on the same chromosome. (II) This results in a *pilE:pilS* hybrid intermediate and the loss of the donor chromosome. (III) The hybrid intermediate then undergoes two recombination events with the recipient *pilE* on a different chromosome. The first recombination event would occur in the extensive region of homology upstream of the genes while the second even would use microhomology within the variable regions of the genes. (IV) Resolution of the Holliday junction intermediates leads to a new *pilE* sequence on the recipient chromosome.
doi:10.1128/microbiolspec.MDNA3-0015-2014.f5

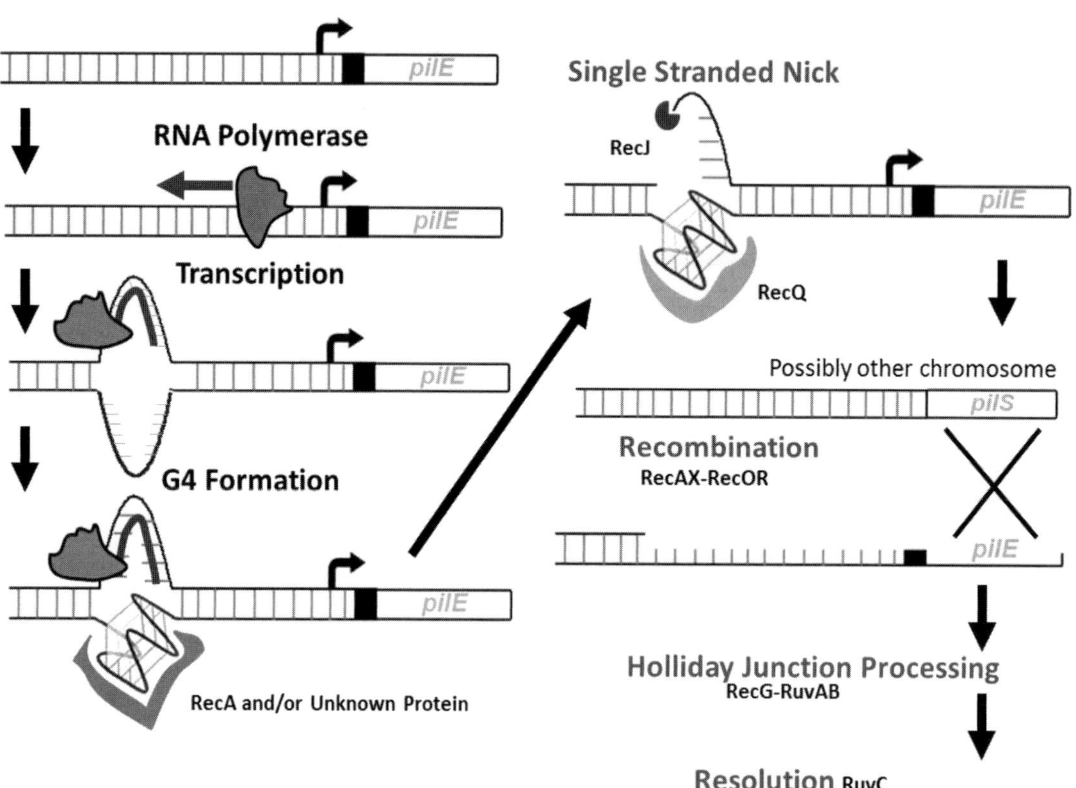

Figure 6 Proposed antigenic variation initiation pathway. Transcription initiation at the sRNA upstream of *pilE* melts the DNA allowing the G4 structure to form. An unknown protein likely binds the G4 to stabilize the structure. A single-stranded nick may occur on the strand opposite the G4 due to a stalled replication fork. RecQ could unwind the G4 structure. RecJ resects the 5′ nicked end allowing RecA to mediate recombination, possibly enhanced by binding the G4 structure, with RecOR using regions of homology between *pilE* and the donor *pilS*, presumably through a recombination mechanism detailed in Fig. 5. RecG and RuvABC then process and resolve the recombination intermediate.
doi:10.1128/microbiolspec.MDNA3-0015-2014.f6

intermediate formation or recombination of the intermediate with the recipient *pilE*. The presence of microhomology at the ends of many pilin antigenic variation recombination tracts suggests that there may be an annealing process involved, but the identity of the protein that promotes annealing is unknown. Regardless, the models need to account for the requirement for both RuvABC and RecG and an intermediate that cannot be resolved if both pathways are inactivated. Though the actual nature of the recombination events is still unclear, the proposed working model provides predictions for conducting future studies.

CONCLUSIONS

The pathogenic *Neisseria* undergo transformation with remarkable frequency and efficiency, and this genetic exchange is critical to allow the pathogens to survive as human-restricted organisms. Additionally, the system of antigenic variation allows for high-frequency productive gene conversion events without adverse effects on genomic stability. Although pilin antigenic variation has undergone the most thorough investigation of any antigenic diversity generating system, implicating specific DNA sequences and structures, as well as a multitude of DNA repair proteins, the detailed mechanisms of recombination are largely unresolved. Continued exploration of the molecular mechanisms will provide understanding of how directed recombination can occur as well as increasing our knowledge of a system thought to allow *N. gonorrhoeae* to escape immune detection, re-infect core populations, and cause the failure of attempted gonococcal vaccines.

Citation. Obergfell KP, Seifert HS. 2014. Mobile DNA in the pathogenic Neisseria. Microbiol Spectrum 3(1):MDNA3-0015-2014.

References

1. **Virji M.** 2009. Pathogenic *Neisseriae*: surface modulation, pathogenesis and infection control. *Nat Rev Microbiol* **7:**274–286.
2. **Johnston C, Martin B, Fichant G, Polard P, Claverys JP.** 2014. Bacterial transformation: distribution, shared mechanisms and divergent control. *Nat Rev Microbiol* **12:**181–196.
3. **Sparling PF.** 1966. Genetic transformation of *Neisseria gonorrhoeae* to streptomycin resistance. *J Bacteriol* **92:** 1364–1371.
4. **Biswas GD, Sox T, Blackman E, Sparling PF.** 1977. Factors affecting genetic transformation of *Neisseria gonorrhoeae*. *J Bacteriol* **129:**983–992.
5. **Koomey M.** 1998. Competence for natural transformation in *Neisseria gonorrhoeae*: a model system for studies of horizontal gene transfer. *APMIS Suppl* **84:**56–61.
6. **Sox TE, Mohammed W, Blackman E, Biswas G, Sparling PF.** 1978. Conjugative plasmids in *Neisseria gonorrhoeae*. *J Bacteriol* **134:**278–286.
7. **Chen I, Dubnau D.** 2004. DNA uptake during bacterial transformation. *Nat Rev Microbiol* **2:**241–249.
8. **Smith JM, Smith NH, O'Rourke M, Spratt BG.** 1993. How clonal are bacteria? *Proc Natl Acad Sci U S A* **90:** 4384–4388.
9. **Martin IM, Ison CA.** 2003. Detection of mixed infection of *Neisseria gonorrhoeae*. *Sex Transm Infect* **79:**56–58.
10. **Lynn F, Hobbs MM, Zenilman JM, Behets FM, Van Damme K, Rasamindrakotroka A, Bash MC.** 2005. Genetic typing of the porin protein of *Neisseria gonorrhoeae* from clinical noncultured samples for strain characterization and identification of mixed gonococcal infections. *J Clin Microbiol* **43:**368–375.
11. **Gibbs CP, Meyer TF.** 1996. Genome plasticity in *Neisseria gonorrhoeae*. *FEMS Microbiol Lett* **145:**173–179.
12. **Hobbs MM, Seiler A, Achtman M, Cannon JG.** 1994. Microevolution within a clonal population of pathogenic bacteria: recombination, gene duplication and horizontal genetic exchange in the opa gene family of *Neisseria* meningitidis. *Mol Microbiol* **12:**171–180.
13. **Snyder LA, Davies JK, Saunders NJ.** 2004. Microarray genomotyping of key experimental strains of *Neisseria gonorrhoeae* reveals gene complement diversity and five new neisserial genes associated with Minimal Mobile Elements. *BMC Genomics* **5:**23.
14. **Buckee CO, Jolley KA, Recker M, Penman B, Kriz P, Gupta S, Maiden MC.** 2008. Role of selection in the emergence of lineages and the evolution of virulence in *Neisseria meningitidis*. *Proc Natl Acad Sci U S A* **105:** 15082–15087.
15. **Goire N, Lahra MM, Chen M, Donovan B, Fairley CK, Guy R, Kaldor J, Regan D, Ward J, Nissen MD, Sloots TP, Whiley DM.** 2014. Molecular approaches to en hance surveillance of gonococcal antimicrobial resistance. *Nat Rev Microbiol* **12:**223–229.
16. **Kirkcaldy RD, Ballard RC, Dowell D.** 2011. Gonococcal resistance: are cephalosporins next? *Curr Infect Dis Rep* **13:**196–204.
17. **Lewis DA.** 2010. The Gonococcus fights back: is this time a knock out? *Sex Trans Inf* **86:**415–421.
18. **Prevention CfDCa.** 2013. *Antibiotic resistance threats in the United States, 2013*. CDC, Atlanta.
19. **Craig L, Pique ME, Tainer JA.** 2004. Type IV pilus structure and bacterial pathogenicity. *Nat Rev Microbiol* **2:**363–378.
20. **Merz AJ, So M, Sheetz MP.** 2000. Pilus retraction powers bacterial twitching motility. *Nature* **407**(6800): 98–102.
21. **Swanson J.** 1973. Studies on gonococcus infection. IV. Pili: their role in attachment of gonococci to tissue culture cells. *J Exp Med* **137:**571–589.
22. **Dietrich M, Bartfeld S, Munke R, Lange C, Ogilvie LA, Friedrich A, Meyer TF.** 2011. Activation of NF-kappaB by *Neisseria gonorrhoeae* is associated with microcolony formation and type IV pilus retraction. *Cell Microbiol* **13:**1168–1182.
23. **Freitag NE, Seifert HS, Koomey M.** 1995. Characterization of the *pilF-pilD* pilus-assembly locus of *Neisseria gonorrhoeae*. *Mol Microbiol* **16:**575–586.
24. **Long CD, Tobiason DM, Lazio MP, Kline KA, Seifert HS.** 2003. Low-level pilin expression allows for substantial DNA transformation competence in *Neisseria gonorrhoeae*. *Infect Immun* **71:**6279–6291.
25. **Drake SL, Koomey M.** 1995. The product of the *pilQ* gene is essential for the biogenesis of type IV pili in *Neisseria gonorrhoeae*. *Mol Microbiol* **18:**975–986.
26. **Tonjum T, Freitag NE, Namork E, Koomey M.** 1995. Identification and characterization of *pilG*, a highly conserved pilus-assembly gene in pathogenic *Neisseria*. *Mol Microbiol* **95:**451–464.
27. **Wolfgang M, Lauer P, Park HS, Brossay L, Hebert J, Koomey M.** 1998. *pilT* mutations lead to simultaneous defects in competence for natural transformation and twitching motility in piliated *Neisseria gonorrhoeae*. *Mol Microbiol* **29:**321–330.
28. **Biswas GD, Lacks SA, Sparling PF.** 1989. Transformation-deficient mutants of piliated *Neisseria gonorrhoeae*. *J Bacteriol* **171:**657–664.
29. **Aas FE, Wolfgang M, Frye S, Dunham S, Lovold C, Koomey M.** 2002. Competence for natural transformation in *Neisseria gonorrhoeae*: components of DNA binding and uptake linked to type IV pilus expression. *Mol Microbiol* **46:**749–760.
30. **Berry JL, Cehovin A, McDowell MA, Lea SM, Pelicic V.** 2013. Functional analysis of the interdependence between DNA uptake sequence and its cognate ComP receptor during natural transformation in *Neisseria* species. *PLoS Genet* **9:**19.
31. **Chen I, Dubnau D.** 2003. DNA transport during transformation. *Front Biosci* **8:**s544–556.
32. **Haas R, Schwarz H, Meyer TF.** 1987. Release of soluble pilin antigen coupled with gene conversion in *Neisseria gonorrhoeae*. *Proc Natl Acad Sci U S A* **84:**9079–9083.
33. **Gibbs CP, Reimann BY, Schultz E, Kaufmann A, Haas R, Meyer TF.** 1989. Reassortment of pilin genes in *Neisseria gonorrhoeae* occurs by two distinct mechanisms. *Nature* **338**(6217):651–652.

34. **Long CD, Madraswala RN, Seifert HS.** 1998. Comparisons between colony phase variation of *Neisseria gonorrhoeae* FA1090 and pilus, pilin, and S-pilin expression. *Infect Immun* **66**(5):1918–1927.

35. **Collins RF, Davidsen L, Derrick JP, Ford RC, Tonjum T.** 2001. Analysis of the PilQ secretin from *Neisseria meningitidis* by transmission electron microscopy reveals a dodecameric quaternary structure. *J Bacteriol* **183**(13):3825–3832.

36. **Hamilton HL, Dillard JP.** 2006. Natural transformation of *Neisseria gonorrhoeae*: from DNA donation to homologous recombination. *Mol Microbiol* **59:**376–385.

37. **Maier B, Chen I, Dubnau D, Sheetz MP.** 2004. DNA transport into *Bacillus subtilis* requires proton motive force to generate large molecular forces. *Nat Struct Mol Biol* **11:**643–649.

38. **Burton B, Dubnau D.** 2010. Membrane-associated DNA transport machines. *Cold Spring Harbor Persp Biol* **2:**a000406.

39. **Chen I, Gotschlich EC.** 2001. ComE, a competence protein from *Neisseria gonorrhoeae* with DNA-binding activity. *J Bacteriol* **183:**3160–3168.

40. **Seitz P, Pezeshgi Modarres H, Borgeaud S, Bulushev RD, Steinbock LJ, Radenovic A, Dal Peraro M, Blokesch M.** 2014. ComEA Is Essential for the Transfer of External DNA into the Periplasm in Naturally Transformable *Vibrio cholerae* Cells. *PLoS Genet* **10**(1):2.

41. **Salman H, Zbaida D, Rabin Y, Chatenay D, Elbaum M.** 2001. Kinetics and mechanism of DNA uptake into the cell nucleus. *Proc Natl Acad Sci U S A* **98**(13):7247–7252.

42. **Goodman SD, Scocca JJ.** 1988. Identification and arrangement of the DNA sequence recognized in specific transformation of *Neisseria gonorrhoeae*. *Proc Natl Acad Sci U S A* **85:**6982–6986.

43. **Elkins C, Thomas CE, Seifert HS, Sparling PF.** 1991. Species-specific uptake of DNA by gonococci is mediated by a 10-base-pair sequence. *J Bacteriol* **173:**3911–3913.

44. **Ambur OH, Frye SA, Tonjum T.** 2007. New functional identity for the DNA uptake sequence in transformation and its presence in transcriptional terminators. *J Bacteriol* **189:**2077–2085.

45. **Smith HO, Gwinn ML, Salzberg SL.** 1999. DNA uptake signal sequences in naturally transformable bacteria. *Res Microbiol* **150:**603–616.

46. **Davidsen T, Rodland EA, Lagesen K, Seeberg E, Rognes T, Tonjum T.** 2004. Biased distribution of DNA uptake sequences towards genome maintenance genes. *Nucleic Acids Res* **32:**1050–1058.

47. **Treangen TJ, Ambur OH, Tonjum T, Rocha EP.** 2008. The impact of the neisserial DNA uptake sequences on genome evolution and stability. *Genome Biol* **9:**2008–2009.

48. **Duffin PM, Seifert HS.** 2010. DNA uptake sequence-mediated enhancement of transformation in *Neisseria gonorrhoeae* is strain dependent. *J Bacteriol* **192:**4436–4444.

49. **Duffin PM, Seifert HS.** 2012. Genetic transformation of *Neisseria gonorrhoeae* shows a strand preference. *FEMS Microbiol Lett* **334:**44–48.

50. **Mathis LS, Scocca JJ.** 1984. On the role of pili in transformation of *Neisseria gonorrhoeae*. *J Gen Microbiol* **130:**3165–3173.

51. **Dorward DW, Garon CF.** 1989. DNA-binding proteins in cells and membrane blebs of *Neisseria gonorrhoeae*. *J Bacteriol* **171:**4196–4201.

52. **Wolfgang M, van Putten JP, Hayes SF, Koomey M.** 1999. The *comP* locus of *Neisseria gonorrhoeae* encodes a type IV prepilin that is dispensable for pilus biogenesis but essential for natural transformation. *Mol Microbiol* **31:**1345–1357.

53. **Aas FE, Lovold C, Koomey M.** 2002. An inhibitor of DNA binding and uptake events dictates the proficiency of genetic transformation in *Neisseria gonorrhoeae*: mechanism of action and links to Type IV pilus expression. *Mol Microbiol* **46:**1441–1450.

54. **Cehovin A, Simpson PJ, McDowell MA, Brown DR, Noschese R, Pallett M, Brady J, Baldwin GS, Lea SM, Matthews SJ, Pelicic V.** 2013. Specific DNA recognition mediated by a type IV pilin. *Proc Natl Acad Sci U S A* **110:**3065–3070.

55. **Assalkhou R, Balasingham S, Collins RF, Frye SA, Davidsen T, Benam AV, Bjoras M, Derrick JP, Tonjum T.** 2007. The outer membrane secretin PilQ from *Neisseria meningitidis* binds DNA. *Microbiology* **153:**1593–1603.

56. **Lang E, Haugen K, Fleckenstein B, Homberset H, Frye SA, Ambur OH, Tonjum T.** 2009. Identification of neisserial DNA binding components. *Microbiology* **155:**852–862.

57. **Benam AV, Lang E, Alfsnes K, Fleckenstein B, Rowe AD, Hovland E, Ambur OH, Frye SA, Tonjum T.** 2011. Structure-function relationships of the competence lipoprotein ComL and SSB in meningococcal transformation. *Microbiology* **157:**1329–1342.

58. **Frye SA, Nilsen M, Tonjum T, Ambur OH.** 2013. Dialects of the DNA uptake sequence in *Neisseriaceae*. *PLoS Genet* **9:**18.

59. **Berry JL, Cehovin A, McDowell MA, Lea SM, Pelicic V.** 2013. Functional Analysis of the Interdependence between DNA Uptake Sequence and Its Cognate ComP Receptor during Natural Transformation in *Neisseria* Species. *PLoS Genet* **9:**e1004014.

60. **Fussenegger M, Facius D, Meier J, Meyer TF.** 1996. A novel peptidoglycan-linked lipoprotein (ComL) that functions in natural transformation competence of *Neisseria gonorrhoeae*. *Mol Microbiol* **19:**1095–1105.

61. **Fussenegger M, Kahrs AF, Facius D, Meyer TF.** 1996. Tetrapac (*tpc*), a novel genotype of *Neisseria gonorrhoeae* affecting epithelial cell invasion, natural transformation competence and cell separation. *Mol Microbiol* **19:**1357–1372.

62. **Chaussee MS, Hill SA.** 1998. Formation of single-stranded DNA during DNA transformation of *Neisseria gonorrhoeae*. *J Bacteriol* **180:**5117–5122.

63. **Draskovic I, Dubnau D.** 2005. Biogenesis of a putative channel protein, ComEC, required for DNA uptake: membrane topology, oligomerization and formation of disulphide bonds. *Mol Microbiol* **55:**881–896.

64. **Sox TE, Mohammed W, Sparling PF.** 1979. Transformation-derived *Neisseria gonorrhoeae* plasmids with altered structure and function. *J Bacteriol* **138:**510–518.
65. **Stein DC, Gunn JS, Radlinska M, Piekarowicz A.** 1995. Restriction and modification systems of *Neisseria gonorrhoeae*. *Gene* **157:**19–22.
66. **Eisenstein BI, Sox T, Biswas G, Blackman E, Sparling PF.** 1977. Conjugal transfer of the gonococcal penicillinase plasmid. *Science* **195:**998–1000.
67. **Zhang Y, Heidrich N, Ampattu BJ, Gunderson CW, Seifert HS, Schoen C, Vogel J, Sontheimer EJ.** 2013. Processing-independent CRISPR RNAs limit natural transformation in *Neisseria meningitidis*. *Mol Cell* **50:** 488–503.
68. **Barrangou R.** 2013. CRISPR-Cas systems and RNA-guided interference. *Wiley Interdiscip Rev RNA* **4:**267–278.
69. **Koomey JM, Falkow S.** 1987. Cloning of the *recA* gene of *Neisseria gonorrhoeae* and construction of gonococcal *recA* mutants. *J Bacteriol* **169:**790–795.
70. **Chen Z, Yang H, Pavletich NP.** 2008. Mechanism of homologous recombination from the RecA-ssDNA/dsDNA structures. *Nature* **453**(7194)**:**489–494.
71. **Gruenig MC, Stohl EA, Chitteni-Pattu S, Seifert HS, Cox MM.** 2010. Less is more: *Neisseria gonorrhoeae* RecX protein stimulates recombination by inhibiting RecA. *J Biol Chem* **285:**37188–37197.
72. **Mehr IJ, Seifert HS.** 1998. Differential roles of homologous recombination pathways in *Neisseria gonorrhoeae* pilin antigenic variation, DNA transformation and DNA repair. *Mol Microbiol* **30:**697–710.
73. **Stohl EA, Seifert HS.** 2001. The *recX* gene potentiates homologous recombination in *Neisseria gonorrhoeae*. *Mol Microbiol* **40:**1301–1310.
74. **Dillingham MS, Kowalczykowski SC.** 2008. RecBCD enzyme and the repair of double-stranded DNA breaks. *Microbiol Mol Biol Rev* **72:**642–671.
75. **Kline KA, Seifert HS.** 2005. Mutation of the *priA* gene of *Neisseria gonorrhoeae* affects DNA transformation and DNA repair. *J Bacteriol* **187:**5347–5355.
76. **Marians KJ.** 2000. PriA-directed replication fork restart in *Escherichia coli*. *Trends Biochem Sci* **25:**185–189.
77. **Mell JC, Redfield RJ.** 2014. Natural competence and the evolution of DNA uptake specificity. *J Bacteriol* **31:**31.
78. **Lewis K.** 2000. Programmed death in bacteria. *Microbiol Mol Biol Rev* **64:**503–514.
79. **Dillard JP, Seifert HS.** 2001. A variable genetic island specific for *Neisseria gonorrhoeae* is involved in providing DNA for natural transformation and is found more often in disseminated infection isolates. *Mol Microbiol* **41:**263–277.
80. **Hamilton HL, Dominguez NM, Schwartz KJ, Hackett KT, Dillard JP.** 2005. *Neisseria gonorrhoeae* secretes chromosomal DNA via a novel type IV secretion system. *Mol Microbiol* **55:**1704–1721.
81. **Snyder LA, Jarvis SA, Saunders NJ.** 2005. Complete and variant forms of the 'gonococcal genetic island' in *Neisseria meningitidis*. *Microbiology* **151:**4005–4013.
82. **Woodhams KL, Benet ZL, Blonsky SE, Hackett KT, Dillard JP.** 2012. Prevalence and detailed mapping of the gonococcal genetic island in *Neisseria meningitidis*. *J Bacteriol* **194:**2275–2285.
83. **Dominguez NM, Hackett KT, Dillard JP.** 2011. XerCD-mediated site-specific recombination leads to loss of the 57-kilobase gonococcal genetic island. *J Bacteriol* **193:**377–388.
84. **Salgado-Pabón W, Jain S, Turner N, Van Der Does C, Dillard JP.** 2007. A novel relaxase homologue is involved in chromosomal DNA processing for type IV secretion in *Neisseria gonorrhoeae*. *Mol Microbiol* **66:** 930–947.
85. **Ramsey ME, Woodhams KL, Dillard JP.** 2011. The Gonococcal Genetic Island and Type IV Secretion in the Pathogenic *Neisseria*. *Front Microbiol* **2**(61)**:**00061.
86. **Bhatty M, Laverde Gomez JA, Christie PJ.** 2013. The expanding bacterial type IV secretion lexicon. *Res Microbiol* **164:**620–639.
87. **Chandran V, Fronzes R, Duquerroy S, Cronin N, Navaza J, Waksman G.** 2009. Structure of the outer membrane complex of a type IV secretion system. *Nature* **462**(7276)**:**1011–1015.
88. **Fronzes R, Schafer E, Wang L, Saibil HR, Orlova EV, Waksman G.** 2009. Structure of a type IV secretion system core complex. *Science* **323**(5911)**:**266–268.
89. **Fronzes R, Christie PJ, Waksman G.** 2009. The structural biology of type IV secretion systems. *Nat Rev Microbiol* **7:**703–714.
90. **Kohler PL, Chan YA, Hackett KT, Turner N, Hamilton HL, Cloud-Hansen KA, Dillard JP.** 2013. Mating pair formation homologue TraG is a variable membrane protein essential for contact-independent type IV secretion of chromosomal DNA by *Neisseria gonorrhoeae*. *J Bacteriol* **195:**1666–1679.
91. **Chan YA, Hackett KT, Dillard JP.** 2012. The lytic transglycosylases of *Neisseria gonorrhoeae*. *Microb Drug Resist* **18:**271–279.
92. **Leonard TA, Moller-Jensen J, Lowe J.** 2005. Towards understanding the molecular basis of bacterial DNA segregation. *Philos Trans R Soc Lond B Biol Sci* **360** (1455)**:**523–535.
93. **Grinter NJ.** 1981. Analysis of chromosome mobilization using hybrids between plasmid RP4 and a fragment of bacteriophage lambda carrying IS1. *Plasmid* **5:**267–276.
94. **Salgado-Pabon W, Du Y, Hackett KT, Lyons KM, Arvidson CG, Dillard JP.** 2010. Increased expression of the type IV secretion system in piliated *Neisseria gonorrhoeae* variants. *J Bacteriol* **192:**1912–1920.
95. **Zola TA, Strange HR, Dominguez NM, Dillard JP, Cornelissen CN.** 2010. Type IV secretion machinery promotes ton-independent intracellular survival of *Neisseria gonorrhoeae* within cervical epithelial cells. *Infect Immun* **78:**2429–2437.
96. **Zweig MA, Schork S, Koerdt A, Siewering K, Sternberg C, Thormann K, Albers SV, Molin S, van der Does C.** 2013. Secreted single-stranded DNA is involved in the initial phase of biofilm formation by *Neisseria gonorrhoeae*. *Environ Microbiol* **3:**1462–2920.

97. **Takeuchi N, Kaneko K, Koonin EV.** 2013. Horizontal Gene Transfer Can Rescue Prokaryotes from Muller's Ratchet: Benefit of DNA from Dead Cells and Population Subdivision. *G3* **17**(113):009845.

98. **Baumdicker F, Hess WR, Pfaffelhuber P.** 2012. The infinitely many genes model for the distributed genome of bacteria. *Genome Biol Evol* **4:**443–456.

99. **Vink C, Rudenko G, Seifert HS.** 2012. Microbial antigenic variation mediated by homologous DNA recombination. *FEMS Microbiol Rev* **36:**917–948.

100. **van der Woude MW.** 2011. Phase variation: how to create and coordinate population diversity. *Curr Opin Microbiol* **14:**205–211.

101. **Kline KA, Sechman EV, Skaar EP, Seifert HS.** 2003. Recombination, repair and replication in the pathogenic *Neisseriae*: the 3 R's of molecular genetics of two human-specific bacterial pathogens. *Mol Microbiol* **50:**3–13.

102. **Stern A, Brown M, Nickel P, Meyer TF.** 1986. Opacity genes in *Neisseria gonorrhoeae*: control of phase and antigenic variation. *Cell* **47:**61–71.

103. **Danaher RJ, Levin JC, Arking D, Burch CL, Sandlin R, Stein DC.** 1995. Genetic basis of *Neisseria gonorrhoeae* lipooligosaccharide antigenic variation. *J Bacteriol* **177:** 7275–7279.

104. **Jennings MP, Hood DW, Peak IR, Virji M, Moxon ER.** 1995. Molecular analysis of a locus for the biosynthesis and phase-variable expression of the lacto-*N*-neotetraose terminal lipopolysaccharide structure in *Neisseria meningitidis*. *Mol Microbiol* **18:**729–740.

105. **Bos MP, Hogan D, Belland RJ.** 1999. Homologue scanning mutagenesis reveals CD66 receptor residues required for neisserial Opa protein binding. *J Exp Med* **190:**331–340.

106. **Gotschlich EC.** 1994. Genetic locus for the biosynthesis of the variable portion of *Neisseria gonorrhoeae* lipooligosaccharide. *J Exp Med* **180:**2181–2190.

107. **Meyer TF, Mlawer N, So M.** 1982. Pilus expression in *Neisseria gonorrhoeae* involves chromosomal rearrangement. *Cell* **30:**45–52.

108. **Hamrick TS, Dempsey JA, Cohen MS, Cannon JG.** 2001. Antigenic variation of gonococcal pilin expression in vivo: analysis of the strain FA1090 pilin repertoire and identification of the *pilS* gene copies recombining with *pilE* during experimental human infection. *Microbiology* **147:**839–849.

109. **Haas R, Meyer TF.** 1986. The repertoire of silent pilus genes in *Neisseria gonorrhoeae*: evidence for gene conversion. *Cell* **44:**107–115.

110. **Haas R, Veit S, Meyer TF.** 1992. Silent pilin genes of *Neisseria gonorrhoeae* MS11 and the occurrence of related hypervariant sequences among other gonococcal isolates. *Mol Microbiol* **6:**197–208.

111. **Segal E, Hagblom P, Seifert HS, So M.** 1986. Antigenic variation of gonococcal pilus involves assembly of separated silent gene segments. *Proc Natl Acad Sci U S A* **83:**2177–2181.

112. **Craig L, Volkmann N, Arvai AS, Pique ME, Yeager M, Egelman EH, Tainer JA.** 2006. Type IV pilus structure by cryo-electron microscopy and crystallography: implications for pilus assembly and functions. *Mol Cell* **23:**651–662.

113. **Forest KT, Bernstein SL, Getzoff ED, So M, Tribbick G, Geysen HMX, Deal CD, Tainer JA.** 1996. Assembly and antigenicity of the *Neisseria gonorrhoeae* pilus mapped with antibodies. *Infect Immun* **64:**644–652.

114. **Criss AK, Kline KA, Seifert HS.** 2005. The frequency and rate of pilin antigenic variation in *Neisseria gonorrhoeae*. *Mol Microbiol* **58:**510–519.

115. **Kellogg DS Jr, Peacock WL Jr, Deacon WE, Brown L, Pirkle DI.** 1963. *Neisseria gonorrhoeae*. I. Virulence Genetically Linked to Clonal Variation. *J Bacteriol* **85:** 1274–1279.

116. **Jonsson AB, Nyberg G, Normark S.** 1991. Phase variation of gonococcal pili by frameshift mutation in *pilC*, a novel gene for pilus assembly. *EMBO J* **10:**477–488.

117. **Segal E, Billyard E, So M, Storzbach S, Meyer TF.** 1985. Role of chromosomal rearrangement in *N. gonorrhoeae* pilus phase variation. *Cell* **40:**293–300.

118. **Hagblom P, Segal E, Billyard E, So M.** 1985. Intragenic recombination leads to pilus antigenic variation in *Neisseria gonorrhoeae*. *Nature* **315**(6015):156–158.

119. **Koomey M, Gotschlich EC, Robbins K, Bergstrom S, Swanson J.** 1987. Effects of *recA* mutations on pilus antigenic variation and phase transitions in *Neisseria gonorrhoeae*. *Genetics* **117:**391–398.

120. **Jennings MP, Jen FE, Roddam LF, Apicella MA, Edwards JL.** 2011. *Neisseria gonorrhoeae* pilin glycan contributes to CR3 activation during challenge of primary cervical epithelial cells. *Cell Microbiol* **13:**885–896.

121. **Marceau M, Forest K, Beretti JL, Tainer J, Nassif X.** 1998. Consequences of the loss of O-linked glycosylation of meningococcal type IV pilin on piliation and pilus-mediated adhesion. *Mol Microbiol* **27:**705–715.

122. **Chamot-Rooke J, Mikaty G, Malosse C, Soyer M, Dumont A, Gault J, Imhaus AF, Martin P, Trellet M, Clary G, Chafey P, Camoin L, Nilges M, Nassif X, Dumenil G.** 2011. Posttranslational modification of pili upon cell contact triggers *N. meningitidis* dissemination. *Science* **331**(6018):778–782.

123. **Miller F, Phan G, Brissac T, Bouchiat C, Lioux G, Nassif X, Coureuil M.** 2014. The Hypervariable Region of Meningococcal Major Pilin PilE Controls the Host Cell Response via Antigenic Variation. *mBio* **5:**01024–13.

124. **Mehr IJ, Seifert HS.** 1997. Random shuttle mutagenesis: gonococcal mutants deficient in pilin antigenic variation. *Mol Microbiol* **23:**1121–1131.

125. **Sechman EV, Rohrer MS, Seifert HS.** 2005. A genetic screen identifies genes and sites involved in pilin antigenic variation in *Neisseria gonorrhoeae*. *Mol Microbiol* **57:**468–483.

126. **Cahoon LA, Seifert HS.** 2009. An alternative DNA structure is necessary for pilin antigenic variation in *Neisseria gonorrhoeae*. *Science* **325**(5941):764–767.

127. **Stohl EA, Blount L, Seifert HS.** 2002. Differential cross-complementation patterns of *Escherichia coli* and *Neisseria gonorrhoeae* RecA proteins. *Microbiology* **148:**1821–1831.

128. **Stohl EA, Gruenig MC, Cox MM, Seifert HS.** 2011. Purification and characterization of the RecA protein from *Neisseria gonorrhoeae*. *PloS one* **6:**0017101.

129. **Stohl EA, Brockman JP, Burkle KL, Morimatsu K, Kowalczykowski SC, Seifert HS.** 2003. *Escherichia coli* RecX inhibits RecA recombinase and coprotease activities in vitro and in vivo. *J Biol Chem* **278:**2278–2285.

130. **Mehr IJ, Long CD, Serkin CD, Seifert HS.** 2000. A homologue of the recombination-dependent growth gene, *rdgC*, is involved in gonococcal pilin antigenic variation. *Genetics* **154:**523–532.

131. **Drees JC, Chitteni-Pattu S, McCaslin DR, Inman RB, Cox MM.** 2006. Inhibition of RecA protein function by the RdgC protein from *Escherichia coli*. *J Biol Chem* **281:**4708–4717.

132. **Hiom K.** 2009. DNA Repair: Common Approaches to Fixing Double-Strand Breaks. *Curr Biol* **19:**R523–R525.

133. **Skaar EP, Lazio MP, Seifert HS.** 2002. Roles of the *recJ* and *recN* genes in homologous recombination and DNA repair pathways of *Neisseria gonorrhoeae*. *J Bacteriol* **184:**919–927.

134. **Killoran MP, Kohler PL, Dillard JP, Keck JL.** 2009. RecQ DNA helicase HRDC domains are critical determinants in *Neisseria gonorrhoeae* pilin antigenic variation and DNA repair. *Mol Microbiol* **71:**158–171.

135. **Lane HE, Denhardt DT.** 1974. The *rep* mutation. III. Altered structure of the replicating *Escherichia coli* chromosome. *J Bacteriol* **120:**805–814.

136. **Kline KA, Seifert HS.** 2005. Role of the Rep helicase gene in homologous recombination in *Neisseria gonorrhoeae*. *J Bacteriol* **187:**2903–2907.

137. **Chaussee MS, Wilson J, Hill SA.** 1999. Characterization of the *recD* gene of *Neisseria gonorrhoeae* MS11 and the effect of *recD* inactivation on pilin variation and DNA transformation. *Microbiology* **145:**389–400.

138. **Hill SA, Woodward T, Reger A, Baker R, Dinse T.** 2007. Role for the RecBCD recombination pathway for *pilE* gene variation in repair-proficient *Neisseria gonorrhoeae*. *J Bacteriol* **189:**7983–7990.

139. **Helm RA, Seifert HS.** 2009. Pilin antigenic variation occurs independently of the RecBCD pathway in *Neisseria gonorrhoeae*. *J Bacteriol* **191:**5613–5621.

140. **Sechman EV, Kline KA, Seifert HS.** 2006. Loss of both Holliday junction processing pathways is synthetically lethal in the presence of gonococcal pilin antigenic variation. *Mol Microbiol* **61:**185–193.

141. **Wainwright LA, Pritchard KH, Seifert HS.** 1994. A conserved DNA sequence is required for efficient gonococcal pilin antigenic variation. *Mol Microbiol* **13:** 75–87.

142. **Howell-Adams B, Wainwright LA, Seifert HS.** 1996. The size and position of heterologous insertions in a silent locus differentially affect pilin recombination in *Neisseria gonorrhoeae*. *Mol Microbiol* **22:**509–522.

143. **Howell-Adams B, Seifert HS.** 1999. Insertion mutations in *pilE* differentially alter gonococcal pilin antigenic variation. *J Bacteriol* **181:**6133–6141.

144. **Kuryavyi V, Cahoon LA, Seifert HS, Patel DJ.** 2012. RecA-binding *pilE* G4 sequence essential for pilin antigenic variation forms monomeric and 5′ end-stacked dimeric parallel G-quadruplexes. *Structure* **20:**2090–2102.

145. **Cahoon LA, Manthei KA, Rotman E, Keck JL, Seifert HS.** 2013. The *Neisseria gonorrhoeae* RecQ helicase HRDC domains are essential for efficient binding and unwinding of the *pilE* guanine quartet structure required for pilin Av. *J Bacteriol* **195:**2255–2261.

146. **Cahoon LA, Seifert HS.** 2013. Transcription of a *cis*-acting, noncoding, small RNA is required for pilin antigenic variation in *Neisseria gonorrhoeae*. *PLoS Pathog* **9**(1):e1003074.

147. **Seifert HS, Ajioka RS, Marchal C, Sparling PF, So M.** 1988. DNA transformation leads to pilin antigenic variation in *Neisseria gonorrhoeae*. *Nature* **336**(6197): 392–395.

148. **Swanson J, Morrison S, Barrera O, Hill S.** 1990. Piliation changes in transformation-defective gonococci. *J Exp Med* **171:**2131–2139.

149. **Zhang QY, DeRyckere D, Lauer P, Koomey M.** 1992. Gene conversion in *Neisseria gonorrhoeae*: evidence for its role in pilus antigenic variation. *Proc Natl Acad Sci U S A* **89:**5366–5370.

150. **Tobiason DM, Seifert HS.** 2006. The obligate human pathogen, *Neisseria gonorrhoeae*, is polyploid. *PLoS Biol* **4**(6).

151. **Stabler RA, Marsden GL, Witney AA, Li Y, Bentley SD, Tang CM, Hinds J.** 2005. Identification of pathogen-specific genes through microarray analysis of pathogenic and commensal *Neisseria* species. *Microbiology* **151:** 2907–2922.

152. **Seifert HS, Wright CJ, Jerse AE, Cohen MS, Cannon JG.** 1994. Multiple gonococcal pilin antigenic variants are produced during experimental human infections. *J Clin Invest* **93:**2744–2749.

153. **Kobayashi I.** 1992. Mechanisms for gene conversion and homologous recombination: the double-strand break repair model and the successive half crossing-over model. *Adv Biophys* **28:**81–133.

154. **Howell-Adams B, Seifert HS.** 2000. Molecular models accounting for the gene conversion reactions mediating gonococcal pilin antigenic variation. *Mol Microbiol* **37:** 1146–1158.

Mobile DNA, 3rd Edition
Nancy L. Craig, Michael Chandler, Martin Gellert, Alan M. Lambowitz, Phoebe A. Rice and Suzanne Sandmeyer

doi:10.1128/microbiolspec.MDNA3-0038-2014

Steven J. Norris[1]

22 *vls* Antigenic Variation Systems of Lyme Disease *Borrelia*: Eluding Host Immunity through both Random, Segmental Gene Conversion and Framework Heterogeneity

INTRODUCTION

Antigenic variation is defined as a hereditable, reversible variation in an antigenic structure that occurs during the course of infection at a rate higher than would be expected for standard recombination or mutation mechanisms. Many bacterial and protozoal pathogens have developed antigenic variation systems in which surface antigens can be continually altered as a means of evading the constant onslaught of adaptive antibody and T cell responses (1). In 1997, an elaborate antigenic variation system was identified in *Borrelia burgdorferi* B31 (2). Because of sequence similarity between this system and the previously characterized variable major protein (VMP) system of relapsing fever bacteria, it was termed the VMP-like sequence (*vls*) locus. Its expression site, called *vls* Expressed (*vlsE*), undergoes remarkable sequence variation involving segmental gene conversion events from *vls* silent cassettes. This review describes what is currently known about the structure, properties, role in host–pathogen interactions, recombination process and evolution of the *vls* system.

LYME BORRELIOSIS

Lyme borreliosis (LB; also called Lyme disease) is a multistage, tick-transmitted infection caused by spirochetes in the genus *Borrelia*. *B. burgdorferi* is the principal human pathogen in North America, whereas *B. garinii*, *B. afzelii*, and *B. burgdorferi* all give rise to Lyme borreliosis in Euroasia (3, 4, 5). These organisms are transmitted by hard-bodied ticks of the genus *Ixodes*; *Ixodes scapularis* and *I. pacificus* are the transmitting ticks in North America, whereas *I. ricinus* and *I. persulcatus* are most active in Europe and Asia,

[1]Department of Pathology and Laboratory Medicine, UTHealth Medical School, PO Box 20708, Houston, TX 77225-0708.

respectively. *B. spielmanii, B. bissettii* and *B. valaisiana* have also been associated with rare cases of human infections (6). There are many additional Lyme *Borrelia* species that are not known to cause human disease. All of the Lyme *Borrelia* species are referred to collectively as *Borrelia burgdorferi* sensu lato (in a broad sense), whereas *B. burgdorferi* sensu stricto (in a strict sense) refers only to the type species of the group. Relapsing fever *Borrelia* (including *B. hermsii*, *B. crocidurae*, and *B. recurrentis*), although related to Lyme *Borrelia*, cause an entirely different disease transmitted by soft-bodied, fast-feeding *Ornithodoros* ticks.

B. burgdorferi and other Lyme *Borrelia* survive by contiguous transmission between *Ixodes* ticks and susceptible mammalian hosts. Infection of humans occurs through the bite of an infected tick (usually at the nymphal stage), causing a localized infection and a resulting expanding red rash called erythema migrans (Table 1). The spirochetes multiply locally, but even at these early stages of infection are able to penetrate blood vessels and lymphatics and thereby disseminate to other tissues. The erythema migrans lesion will eventually clear. However, most patients will go on to develop disseminated symptoms, including a variety of musculoskeletal, neurologic, and cardiovascular manifestations. Months to years later, persistent infection causes Lyme arthritis, which is the most prominent late symptom in North American patients infected with *B. burgdorferi*. *B. garinii* infection tends to cause neurologic signs, whereas most cases of the skin lesion acrodermatitis chronica atrophicans (ACA) are caused by *B. afzelii*. The manifestations shown in Table 1 really form a continuum, and the disease properties differ greatly from patient to patient. Lyme *Borrelia* are present at high concentrations only in erythema migrans skin lesions, and otherwise are typically present in small numbers and can be distributed to almost any tissue. The organisms produce no known toxins; rather, pathogenesis appears to be primarily due to the induction of inflammatory reactions in the infected mammalian host (7). During the transitions between the tick and mammalian hosts, Lyme *Borrelia* undergo massive changes in gene expression (8), resulting in concomitant shifts in the proteins required for survival and growth in the arthropod or warm-blooded animal environments.

Table 1 Stages of Lyme borreliosis.[a]

Localized (days to weeks post infection)
Erythema migrans skin lesion
Headache, malaise, fatigue, muscle and joint pain
Disseminated (weeks to months post infection)
Secondary annular skin lesions
Neuroborreliosis – meningitis, facial palsy, radiculoneuritis
Migratory musculoskeletal pain
Atrioventricular nodal heart block
Lymphocytoma[b]
Eye manifestations
Persistent (months to years post infection)
Migratory arthritis of large joints
Neuroborreliosis – meningitis, encephalitis, facial palsy, radiculoneuropathy, polyneuritis
Atrioventricular nodal heart block
Acrodermatitis chronica atrophicans skin lesions[b]

[a]Information from Steere et al. (3) and Müllegger et al. (5).
[b]Found in Eurasia.

While it is not known how long humans can be infected with Lyme *Borrelia*, viable spirochetes can be cultured from almost any tissue of experimentally infected mice throughout their lifetimes; thus it is likely that, without treatment, humans can carry viable organisms for years. Lyme *Borrelia* thus fall in a category of persistent, nontoxigenic pathogens that also includes the syphilis spirochete *Treponema pallidum* subsp. *pallidum* (9).

Persistence requires mechanisms for evading host immune responses, particularly the adaptive immune response. Immune evasion mechanisms that have been described in Lyme *Borrelia* include complement regulator-acquiring surface proteins (CRASPs), which bind factor H and factor H-like protein 1 (FHL-1) and thus inhibit the activation of the complement cascade (10, 11). Another mechanism involves the downregulation of the antigenic tick phase-associated outer surface lipoproteins OspA and OspB, as well as OspC, which is required for survival of *B. burgdorferi* during the early phase of mammalian infection (12, 13).

vls SYSTEM OF *B. BURGDORFERI* B31

The B31 strain of *B. burgdorferi* was isolated from *I. scapularis* (then called *I. dammini*) ticks collected on Shelter Island, New York, and was the first strain of Lyme *Borrelia* described (14) . In early studies of this organism, it was determined that its genome consisted of a linear chromosome and multiple linear and circular plasmids ranging in size from 56 kbp to 5 kbp. Many of these plasmids were easily lost during *in vitro* culture, and absence of some of these plasmids correlated with loss of infectivity in animal models (15, 16, 17, 18, 19, 20, 21, 22, 23). The availability of the genome sequence of *B. burgdorferi* B31 showed that the organism contained 12 linear and 9 circular plasmids (with a total of over 600 kbp) as well as a 972-kbp linear chromosome (24, 25). The linear replicons of *Borrelia* were found to have covalently closed, hairpin telomeres in which a 5′–3′ bond is present between the positive and negative strands (25, 26, 27). The replication of these linear molecules occurs through the formation of a circular intermediate that undergoes asymmetric single-stranded cleavage by the plasmid-encoded telomere resolvase ResT at specific sites near the telomere, followed by separation of the two plasmid copies, snap-back hybridization and ligation (28).

The *vls* system was first discovered prior to the availability of the genomic sequence as the result of a subtractive hybridization study aimed at identifying sequences that were present in low-passage, infectious clones of *B. burgdorferi* B31 but absent in a high-passage, noninfectious B31 clone (2). A single recombinant plasmid called pJRZ53 was found to encode an amino acid sequence with a low but significant sequence identity with the *B. hermsii* antigenic variation protein Vlp17. The pJRZ53 insert hybridized with several restriction fragments of lp28-1, a 28-kbp linear plasmid. Cloning of the 10-kbp region containing the *vls* sequences into a lambda phage vector followed by sequencing revealed the presence of a single telomeric copy of the expression site *vlsE* as well as a contiguous array of 15 sequences that shared 90.0% to 96.1% nucleotide sequence identity and 76.9% to 91.4% encoded amino acid sequence identity with the central "cassette" region of the *vlsE* gene (Fig. 1A). The 15 unexpressed (silent) cassette regions are 474 to 594 bp in length and are demarcated by 17-bp direct repeats found also at either end of the cassette region of *vlsE*. They are present in a head-to-tail arrangement, forming a long contiguous open reading frame interrupted only by one stop codon (in cassette *vls11*) and two frame shifts (in cassettes *vls14* and *vls16*). Alignment of the nucleotide or encoded amino acid sequences of the *vls* cassettes reveals the presence of 6 variable regions (VRs) interspersed among 7 relatively invariant regions (IRs) (Fig. 1B). In the variable regions, up to 6 different

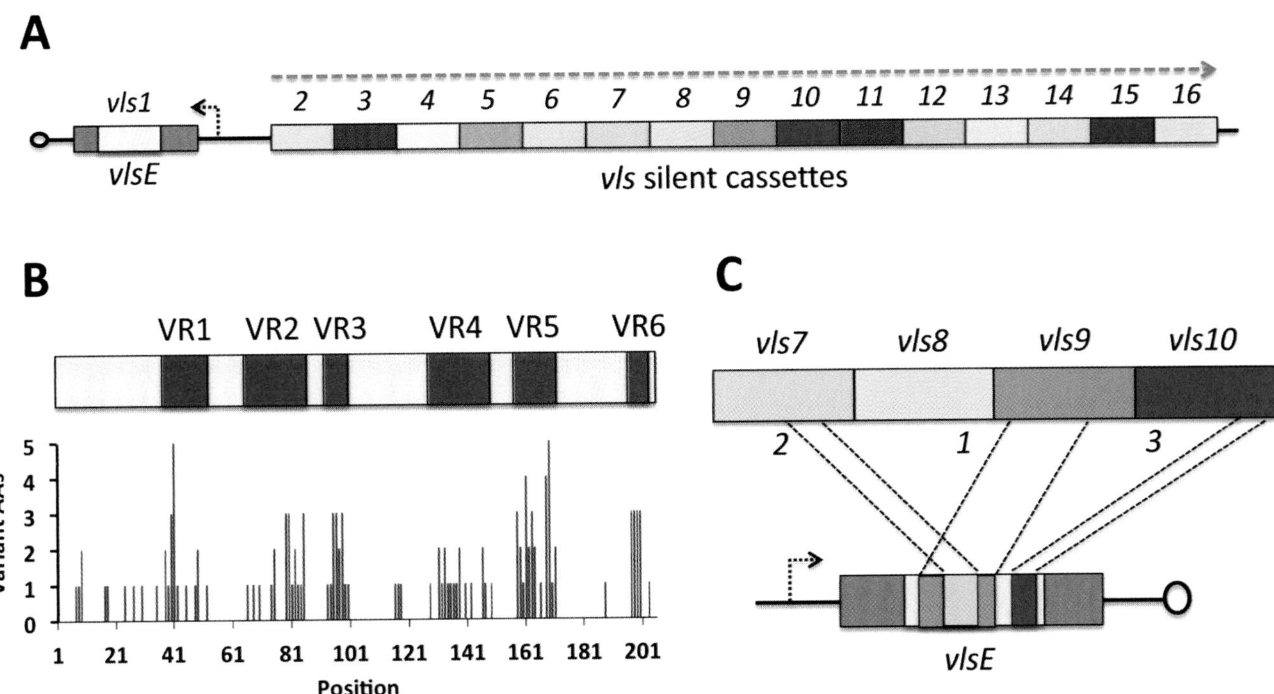

Figure 1 Characteristics of the *vls* locus of *B. burgdorferi* B31. (A) Arrangement of the *vlsE* expression site and the 15 silent cassettes near the telomere of the linear plasmid lp28-1. The promoter for *vlsE* is indicated by the short arrow and the orientation of the silent cassettes is shown by the large arrow. (B) The cassette regions contain six variable regions (VR1 through VR6) separated by relatively invariant regions. The graph indicates the number of different amino acids encoded by the silent cassettes at each codon in the aligned sequences. (C) Unidirectional, random, segmental recombination occurs sequentially during mammalian infection, as indicated by this hypothetical example of sequential recombination between *vlsE* and silent cassettes 9, 7, and 10. doi:10.1128/microbiolspec.MDNA3-0038-2014.f1

amino acids can be encoded in some positions of the aligned sequence. Nearly all indels are in multiples of three base pairs, indicating the importance of maintaining an intact open reading frame. Some of the silent cassettes are truncated or have long internal deletions; these regions were excluded from the analysis shown in Fig. 1B, because they do not appear to participate in *vlsE* recombination events.

The *vlsE* gene itself encodes a 36-kDa protein with a lipoprotein leader sequence; further studies verified that the VlsE product is lipidated and localized on the outer surface of the outer membrane (2). The gene has a consensus σ^{70} promoter region, and primer extension analysis showed that transcription is initiated with nearly equivalent efficiency at two adjacent thymidine residues at 13 and 14 bp downstream from the beginning of the −10 sequence (29). Interestingly, inverted repeats that include portions of the *vlsE* promoter region and the 5′ end of the first silent cassette are predicted to form a 51-bp stem–loop structure; this feature is likely involved in *vlsE* transcription and/or recombination.

To date, no *vlsE* recombination events have been reported in *in vitro* cultures of Lyme *Borrelia* or in infected ticks (2, 30, 31, 32). In contrast, *vlsE* sequence changes have been detected as early as 4 days postinoculation of C3H/HeN mice with *B. burgdorferi*; by 28 days of infection, few (if any) organisms expressing the 'parental' *vlsE* sequence are present (30, 33). Similar results have been obtained with infected rabbits, with multiple *vlsE* sequence changes observed at 2 weeks, the earliest postinoculation time point in these experiments (34). In contrast to relapsing fever *Borrelia*, in which the typical pattern is the sequential outgrowth of clones expressing a single VMP type, mammalian infection with Lyme *Borrelia* results in the outgrowth of a myriad of clones each expressing a different VlsE variant. Indeed, the degree of sequence variation is such that it is difficult to find two clones with the same *vlsE* sequence at time points beyond 4 weeks postinoculation, even in the same tissue biopsy (2, 30, 33).

Given the presence of 15 silent cassettes that represent variants of the *vlsE* cassette region sequence, the initial hypothesis was that each of these silent cassettes could replace the *vlsE* cassette, resulting in 15 different variants. Instead, the sequencing of *vlsE* cassette regions from over 1,400 *B. burgdorferi* clones has revealed that sequence variation occurs through the replacement of segments of DNA in the expression site with segments of the corresponding regions of the silent cassettes (Fig. 1C) (2, 30, 33). These segments can range in size from a few base pairs to nearly the entire length of the *vlsE* cassette region (33). The recombinations represent gene conversion events, in that the sequences of the silent cassettes are not altered (35), as will be discussed further in the context of recombination mechanisms. *vlsE* gene conversion events appear to occur continuously during the course of mammalian infection, resulting in a mosaic of overlapping recombinations that are difficult to decipher. An Excel spreadsheet-based Visual Basic program has been developed to help determine which silent cassettes could have been the source of sequence replacements (33). Because the silent cassettes have many regions of sequence identity, it is often not possible to attribute a given recombination event to a single silent cassette; multiple cassettes could have served as potential 'donors'. For the same reason, the boundaries of the recombination usually cannot be determined with confidence, so recombination events are defined as minimal recombination regions (including the nucleotides that differ from the parental sequence) and maximal recombination regions (encompassing the minimal region plus the surrounding nucleotides that are identical between the parental sequence and the silent cassette donor) (33). The sequence variation is restricted to the *vlsE* cassette region, in that 5′ and 3′ ends of the gene do not undergo sequence changes (35).

vls SYSTEMS OF OTHER LYME *BORRELIA*

vls sequences have been identified in every Lyme *Borrelia* organism for which a complete or draft genomic sequence is available, indicating that this antigenic variation system is ubiquitous among all Lyme disease *Borrelia*. After the initial description of the *B. burgdorferi* B31 *vls* system, the identification of *vls* sequences in recombinant DNA libraries derived from *B. burgdorferi* 297 (36), *B. garinii* IP90 (37), and *B. garinii* A87SA (38) were reported. A comprehensive analysis of *vls* sequences of *B. garinii* IP90 and *B. afzelii* ACA-1 revealed the near-complete sequence of the silent cassettes of these two strains (39) . The intact *vlsE* genes of these two strains were not isolated, but partial sequences were obtained by the cloning of cDNA products and by PCR analysis (39).

All of the 26 Lyme *Borrelia* strains from which extensive plasmid DNA sequences have been obtained (24, 40, 41, 42, 43, 44, 45) contain *vls* sequences, greatly increasing the available information regarding the characteristics of *vls* systems. Shotgun sequencing of genomes has yielded nearly complete sequences of *vls* silent cassette arrays. However, *vlsE* sequences are typically missing from genomic sequences, probably because of their location near a telomere or predicted stem–loop structures and resulting poor cloning and sequencing efficiency.

Currently, complete or nearly complete *vlsE* sequences are available in GenBank for 16 Lyme *Borrelia* strains. In addition, the *B. burgdorferi* 29805 silent cassette *vlsS1* is contiguous with a region homologous to the 5′ constant (non-cassette) region of *vlsE*, so it is also included in this comparison. The phylogenetic tree of the predicted VlsE amino acid sequences with the *B. hermsii* Vlp sequences as outliers (Fig. 2) reveals the conservation and diversity among the VlsE sequences. VlsE proteins of the German *B. burgdorferi* isolates PKa2, PAbe, PBoe, and PBre all exhibit a high sequence identity/similarity with the United States isolate B31 (83 to 93% and 90 to 95%, respectively), with nearly all the sequence differences existing in the variable regions. This result indicates that a closely related clade of *B. burgdorferi* clones exists in the North American and European continents, such that the VlsE sequence differences are essentially the same as are observed among B31 antigenic variants (33). Indeed, the availability of several genomic sequences from this clade

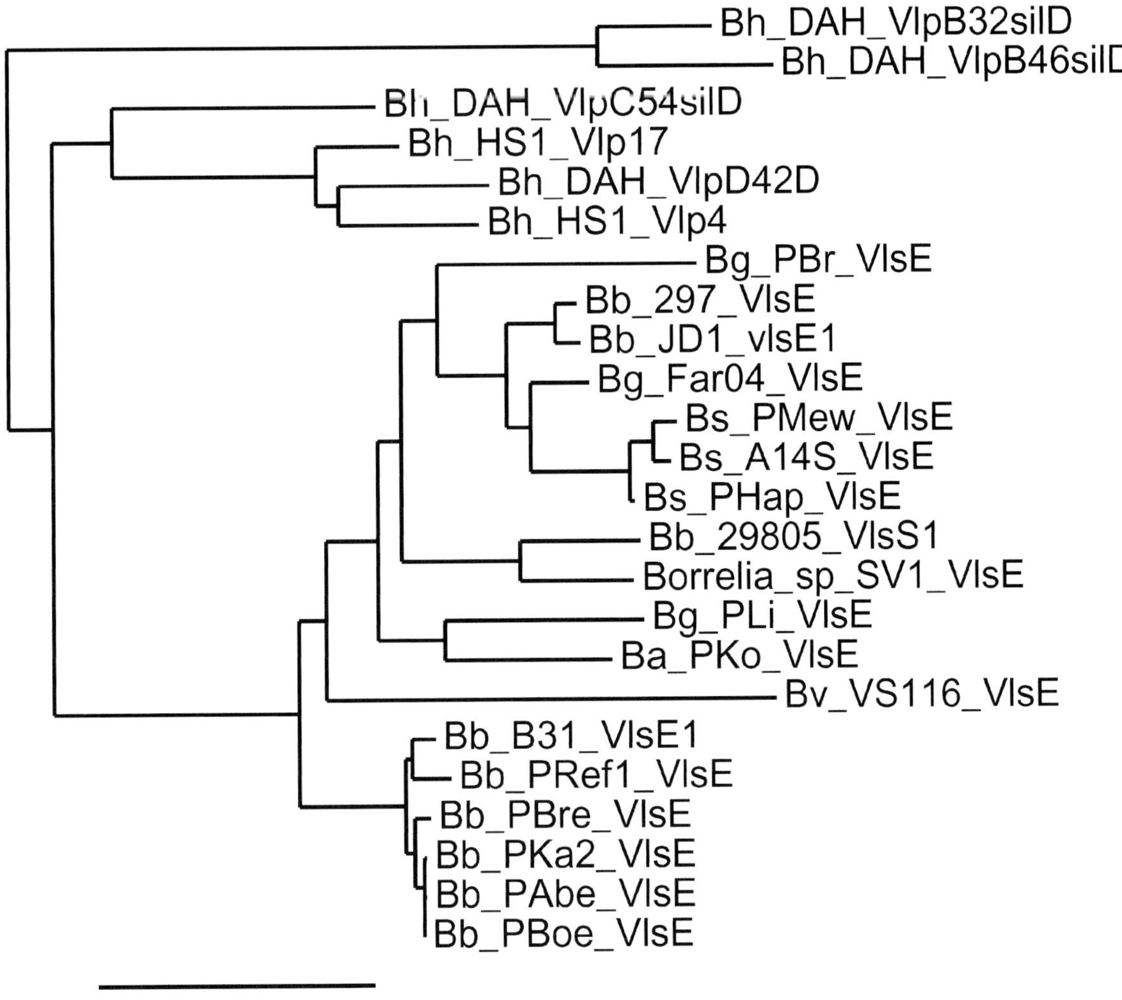

Figure 2 Dendrogram depicting the conservation and diversity of VlsE amino acid sequences and their relatedness to Vlp proteins from relapsing fever organisms. Representative Vlp proteins from *B. hermsii* (Bh) strains DAH and HS1 are clustered into two groups at the top of the dendrogram. The B31 strain of *B. burgdorferi* (Bb) and closely related European strains form a distinct grouping at the bottom, whereas the remaining Lyme disease *Borrelia* VlsE sequences exhibit considerable diversity. Additional species abbreviations: Bg = *B. garinii*, Bs = *B. spielmanii*, Ba = *B. afzelii*, and Bv = *B. valaisiana*.
doi:10.1128/microbiolspec.MDNA3-0038-2014.f2

indicate a pan-genomic clonality, consistent with the recent geographic spread of this group of Lyme *Borrelia* (43). In contrast, the VlsE sequences of North American *B. burgdorferi* isolates B31 and 297 exhibit a high degree of divergence with only 46% identity and 53% similarity. A similar level of divergence is observed between the Lyme *Borrelia* species (Fig. 2). Again using *B. burgdorferi* B31 for comparison, the *B. garinii*, *B. afzelii*, and *B. spielmanii* strains have only 35 to 49% identity and 41 to 59% similarity, indicative of a high level of diversification in "framework" regions outside of the variable regions. *B. valaisiana* VlsE was the most divergent from the B31 sequence (33% identity, 41% similarity). A very similar pattern of diversification between strains (not shown here) is evident in the much larger group of available *vls* silent cassette sequences, and also when nucleotide sequences are used for comparison.

The degree of diversity among *vlsE* sequences is higher than that of any other orthologous gene group in the Lyme *Borrelia* strains examined. *ospC* is the next most divergent, with the lowest OspC identity and similarity among available full-length sequences being 68% and 79% respectively (between *B. garinii* TCLSK and *B. valaisiana* VS116). The high diversity in *vlsE* (and *ospC*) is indicative of a strong selective pressure, most likely driven by mammalian host adaptive immune responses as discussed later in this chapter.

The sequences of only three intact *vls* systems (including both *vlsE* and the *vls* silent cassettes) are currently available (Fig. 3). In each of these, the arrangement is similar to the B31 system, with *vlsE* and the silent cassettes facing in opposite directions and separated by a short DNA segment. This region includes inverted repeats that could form a stem–loop structure; however, these repeats are comprised of different sequences in each case. For example, the possible stem–loop structure in strain B31 involves portions of the *vlsE* promoter on one side and a segment upstream of the first *vls* silent cassette region on the other. All three strains lack an intact promoter for the silent cassette region. However, both *B. burgdorferi* JD1 and *B. garinii* Far04 have sequences identical to the 5′ end of the *vlsE* open reading frame that are contiguous with the first silent cassette. In JD1, the region complementary with *vlsE* lacks a start codon but includes a 47-nucleotide (47-nt) region of identity (with 2 nt differences) with the 5′ end of *vlsE* and a portion of the first *vls* cassette. The Far04 silent cassette region likewise has a 476-nt region of identity to the 5′ end of *vlsE* and part of the cassette region, but also encompasses the ribosome binding site and –10 sequence. The B31 strain silent cassette region has a shorter segment of sequence identical to the 5′ end of *vlsE*. The 'loop' region between the complementary 'stem' sequences is 300 nt and 299 nt in JD1 and Far04, respectively, but is only 6 nt long in the B31 sequence. The significance and potential functionality of the duplication of the 5′ end of *vlsE* is not known, but the fact that it is conserved to a varying extent in each organism indicates that its preservation is favored and that 'purifying' recombinations may maintain the structure. Of course, there are many additional regions of sequence identity between *vlsE* and the silent cassettes that likely participate in the *vlsE* gene conversion events.

Many of the silent cassettes in both *B. garinii* Far04 and *B. burgdorferi* JD1 are separated by frame shifts (Fig. 3), and this pattern is found in the *vls* loci of other Lyme *Borrelia* organisms as well. Thus the presence of a nearly contiguous open reading frame in the *vls* cassette region as seen in *B. burgdorferi* B31 is not required for *vlsE* recombination and locus functionality. The *B. garinii* Far04 *vls* locus is found close to a telomere, as is the B31 locus; the *vlsE* sequence in JD1 is incomplete, so the nature of the downstream sequence is not known. *vls* silent cassette regions have been identified in every other complete or near-complete genomic sequence from Lyme *Borrelia* organisms, currently including 2 *B. afzelii*, 13 *B. burgdorferi*, 4 *B. garinii*, 2 *B. spielmanii*, and 1 *B. valaisiana* strains (data not shown). In all of these cases, the silent cassette arrangements are similar to those shown in Fig. 3. As mentioned previously, the lack of a contiguous sequence containing *vlsE* appears to be due to either its telomeric location or the inverted repeats, which interferes with sequencing reactions. Overall, the general properties of *vls* loci in Lyme *Borrelia* are well conserved, although substantial sequence diversity is present.

VlsE STRUCTURE AND THE LOCALIZATION OF VARIABLE REGIONS

Determination of the three-dimensional structure of *B. burgdorferi* B31 VlsE (46) revealed one of the most remarkable examples of immune evasion through the optimized localization of variable epitopes. As mentioned previously, VlsE is a surface-localized lipoprotein, and is thereby anchored to the outer membrane surface by lipid moieties covalently linked to the N-terminal cysteine residue of the processed protein. Although the unit cell of the crystal structure consists of four monomers with two side-by-side pairs (46), VlsE appears to exist as a monomer rather than a dimer in its native state. The N- and C-terminal sequences

Borrelia burgdorferi B31, lp28-1

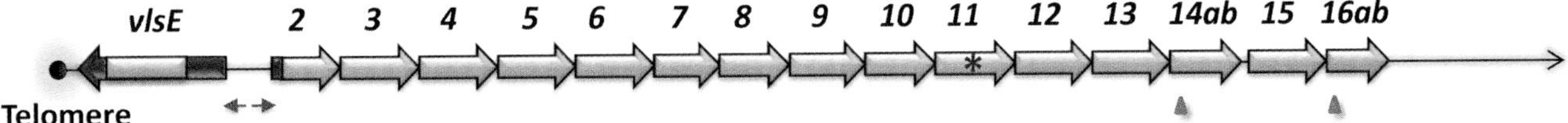

Borrelia garinii Far04, lp28-1

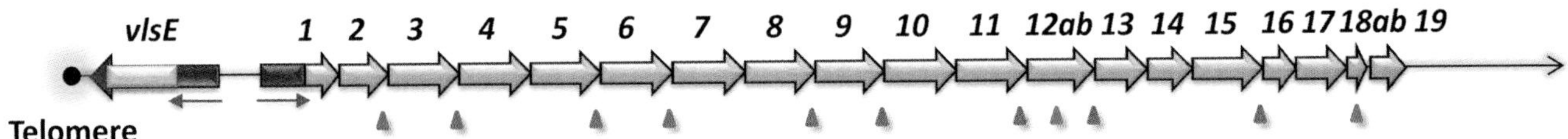

Borrelia burgdorferi JD1, lp28-1

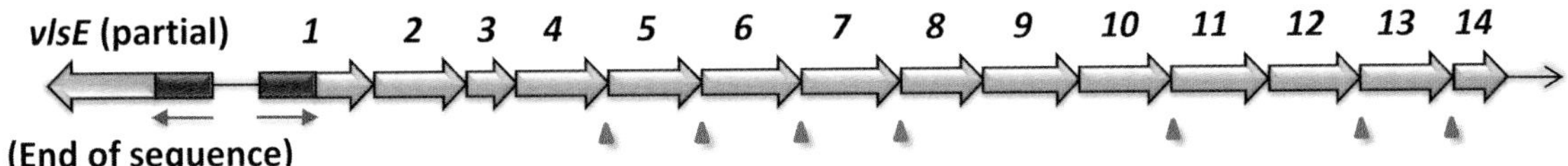

Frameshifts within cassettes (disruptive)

Frameshifts between cassettes (non-disruptive)

*** Stop codon**

Figure 3 Arrangement of *vlsE* and the silent cassettes in *B. burgdorferi* B31, *B. garinii* Far04, and *B. burgdorferi* JD1. Cassette region sequences are shown in green, whereas flanking 5′ and 3′ *vlsE* sequences and homologous 5′ sequences at the beginning of the silent cassette arrays are shown in red. Arrows indicate the locations of inverted repeats. Arrowheads correspond with frameshifts, with the blue arrowheads indicating frameshifts between silent cassettes and red arrowheads showing those located within silent cassettes (as also indicated by the *a* and *b* designations of the ORFs before and after the frameshift). The B31 silent cassette 11 contains a stop codon (red asterisk).
doi:10.1128/microbiolspec.MDNA3-0038-2014.f3

and the relatively invariant regions within the cassette form α-helices that constitute the structural framework of VlsE (Fig. 4A). Within this framework, the six VRs constitute random coil regions that form a 'dome' on the membrane distal surface of the protein. In the space-filling model (Fig. 4B), these variable regions essentially cover the top of the protein. Therefore, the region of the polypeptide that is most likely to interact with antibodies is precisely the area that undergoes rapid sequence variation during the course of mammalian infection.

It is not known to what degree the relatively invariant lateral surfaces of this elongated protein are exposed to the external fluid phase in the intact organism. The surface lipoproteins may form a contiguous layer during infection that would prevent antibodies from accessing the lateral surfaces. Such a 'forest' of outer surface proteins has been documented by cryo-electron tomography (cryo-ET) in *in vitro* cultured *B. burgdorferi* (47), but under these conditions the lipoproteins OspA, OspB, and OspC are highly expressed. During mammalian infection, OspA and OspB are dramatically downregulated, OspC appears to be transiently expressed, and VlsE expression is thought to increase (29, 48, 49); however, to date, the surface topology of mammalian host-adapted *Borrelia* has not been examined in detail by cryo-ET or other means. It has been shown that antibodies against the invariant regions, particularly the highly immunogenic IR6 region, do not react with VlsE

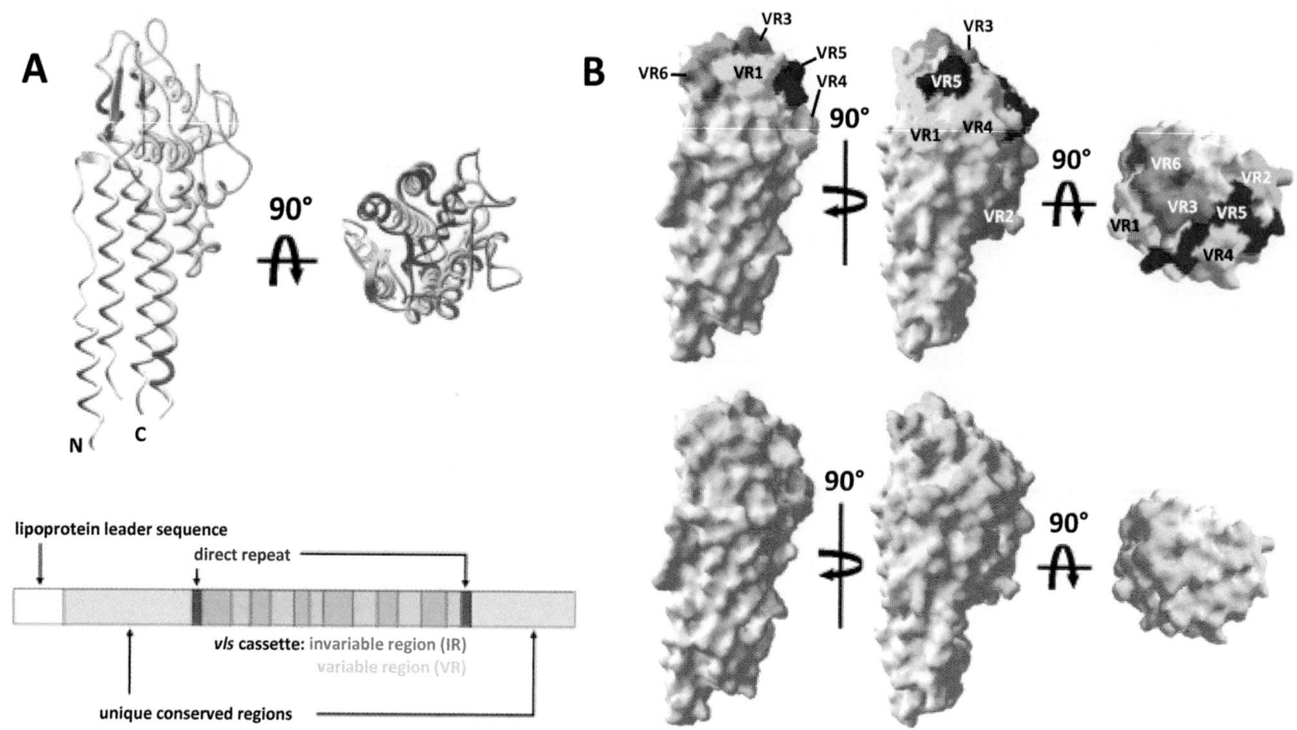

Figure 4 Localization of the VRs in the three-dimensional structure of the *B. burgdorferi* B31 allele VlsE1. (A) Ribbon diagram showing the abundance of alpha helices and the location of the cassette VR (yellow) and IRs (blue). Amino acids encoded by the direct repeats at either end of the cassette region are shown in red. The protein is anchored to the outer membrane by lipid moieties associated with the N-terminal cysteine. A schematic of the cassette region and flanking sequences is shown below the 3D structure. (B) Space-filling models indicating the locations of VRs 1 through 6. The VRs cover most of the membrane distal surface of the protein. The bottom panel shows the locations of variable amino acid residues in yellow. Modified from Eicken et al. (46) with permission.
doi:10.1128/microbiolspec.MDNA3-0038-2014.f4

in intact organisms, although they will bind to recombinant VlsE in solution (37, 50). IR6 forms a compact α-helix that is 'buried' underneath the variable region random coil structures (46), and thus may lack the surface exposure necessary to permit antibody binding. Recombinant VlsE is difficult to crystallize; it is possible that a significant amount is misfolded and thus may have exposed IR6 available for antibody reactions.

ROLE IN PATHOGENICITY

The earliest indication that the *vls* system was required for full infectivity was that loss of the encoding plasmid (a 28-kb linear plasmid, lp28-1, in most Lyme *Borrelia*) resulted in an intermediate infectivity phenotype in experimentally infected mice (15, 18, 20, 21, 22, 23). Needle inoculation of immunocompetent C3H/HeN mice with lp28-1$^-$ clones of *B. burgdorferi* strains B31 or 297 results in a transient infection in which organisms can be cultured from joints and, occasionally, other tissues for up to two weeks postinoculation (but not thereafter). Anti-*B. burgdorferi* antibody responses occur in the animals infected with the lp28-1$^-$ clones, consistent with the establishment of sufficient numbers of spirochetes to induce B-cell responses. Surprisingly, lp28-1$^-$ clones cause long-term infections in severe combined immunodeficiency (SCID) mouse strains (21, 30, 51). *Trans*-complementation with *vlsE* alone on a shuttle vector does not alter the infectivity phenotype of a lp28-1$^-$ clone, indicating that VlsE expression is not sufficient to restore full infectivity (51). Loss of lp28-1 does not affect the colonization of *I. scapularis* ticks (52). Taken together, these results show that lp28-1 carries virulence determinant(s) that protect Lyme *Borrelia* against adaptive immune responses in mammalian hosts.

More definitive evidence of the role of the *vls* system in mammalian infection was provided by elegant studies reported by Bankhead and Chaconas in 2007 (53). Using a telomere resolvase-mediated targeted deletion approach, they were able to delete the entire *vls* locus from the right end of lp28-1 in *B. burgdorferi* B31. As a control, they also generated mutants lacking genes (*BBF01-BBF19*) from the left end of lp28-1, upstream from the region required for plasmid replication. The two *Δvls* mutant clones tested exhibited the same intermediate infectivity phenotype in C3H/HeN mice as the control lp28-1$^-$ strain, whereas the mutants lacking the left end of lp28-1 were fully infectious. Expression of VlsE by itself in the *Δvls* background was not sufficient to restore full infectivity. These results clearly indicate that an intact *vls* system is required for persistent infection.

Antibody Responses Against VlsE

Immune evasion through antigenic variation systems is thought to act through the alteration of surface-exposed epitopes so that antibodies induced against one form of the antigen will not react with subsequent variants. An unexpected finding was that VlsE induces a robust antibody response in human Lyme borreliosis patients and naturally and experimentally infected animals (2, 37, 54, 55, 56, 57, 58, 59, 60). The antibody responses are predominantly against the conserved, non-variable regions of VlsE, particularly the IR6 region (also called C6) within the *vls* cassette (61, 62); serologic reactivity has also been demonstrated against the N-terminal and C-terminal regions outside the central cassette (63). However, McDowell et al. (64) established by VlsE variant cross-absorption studies that the variable regions of VlsE also induce antibody responses. Antibodies against conserved regions are typically cross-reactive with other VlsE sequences, and either full-length VlsE or the C6 peptide have been incorporated into many serological diagnostic tests for Lyme borreliosis in humans or dogs (55, 56, 58, 59).

In studies conducted to date, immunization with recombinant forms of VlsE or the C6 peptide has not resulted in the protection of mice against infection with *B. burgdorferi* expressing the homologous VlsE variant (65) (Norris SJ, unpublished data). Also, *B. burgdorferi* strains with a functional *vls* system that are experimentally modified to constitutively express an invariant form of VlsE cause persistent infection (51, 53). It could be expected that *Borrelia* expressing invariant VlsE would be eliminated by antibodies against that allele, but that is not the case. Liang et al. (37) have suggested that IR6 may serve as a "decoy" epitope that diverts the immune response away from protective responses to other epitopes on VlsE or other antigens. Although the recombinant B31 VlsE or a single C6 peptide is effective in detecting antibodies induced in humans and dogs infected with a variety of Lyme *Borrelia* species and strains, Baum et al. (66) found that the white-footed mouse *Peromyscus leucopus*, a natural host of *B. burgdorferi* in North America, often produces a weak response to infection with several different *B. burgdorferi* strains. They suggest that coevolution of the pathogen and host has resulted in the elimination of sequences that induce protective anti-VlsE responses. Overall, much remains to be understood about the nature of the immune response against VlsE and host–pathogen interactions.

VlsE EXPRESSION

As is the case for many Lyme *Borrelia* proteins, the expression of VlsE appears to be highly regulated during the organism's infectious cycle (31, 48, 67). When *B. burgdorferi* is acquired by tick larvae, it continues to express VlsE for approximately 96 hours, after which the percentage of VlsE-positive spirochetes decreases (48). VlsE is expressed at very low levels by the spirochete in the tick in between feedings. Disparate results were obtained regarding VlsE level changes occurring when an infected tick acquires a blood meal from mice. In two studies, the expression increased significantly within 24 hours, in a manner similar to the increase in OspC expression (31, 67); in a third report, the percent of VlsE-positive organisms remained <10% during feeding, but increased to ~90% in the skin of the mice at 72 hours after feeding was initiated (48). Piesman et al. (68) found that *vlsE* transcript levels were higher in *B. burgdorferi* in the salivary glands as compared to the levels in the midgut of feeding ticks, suggesting that the salivary gland environment is stimulatory toward *vlsE* expression. Similarly, Kοči et al. (67) observed *vlsE* transcript increases during tick feeding in *I. ricinus* ticks infected with *B. afzelii*. *In vitro* studies are contradictory, with opposite changes in VlsE expression occurring when *B. burgdorferi* were incubated at pH 7 or pH 8 at 34°C (31, 48). However, the reports were in agreement that VlsE expression was high at pH 7.5 to 8 at 23°C, conditions thought to mimic the tick midgut.

Hudson et al. (29) determined that *vlsE* is transcribed at higher levels in *B. burgdorferi* B31 in the presence of endothelial cells than in cell-free liquid culture, suggesting that the presence of these cells or their products triggers an upregulation of *vlsE* expression. No *vlsE* recombination was detected in these studies.

The *vlsE* promoter region has predicted –35 and –10 sequences typical of RpoD recognition sites (Fig. 5A). The transcriptional initiation site was mapped by Bykowski et al. (48) by primer extension and was found to be located at two adjacent T residues 6 nt downstream of the –10 sequence. One can speculate that the predicted stem–loop structure in the B31 has regulatory effects on *vlsE* expression. However, as men-

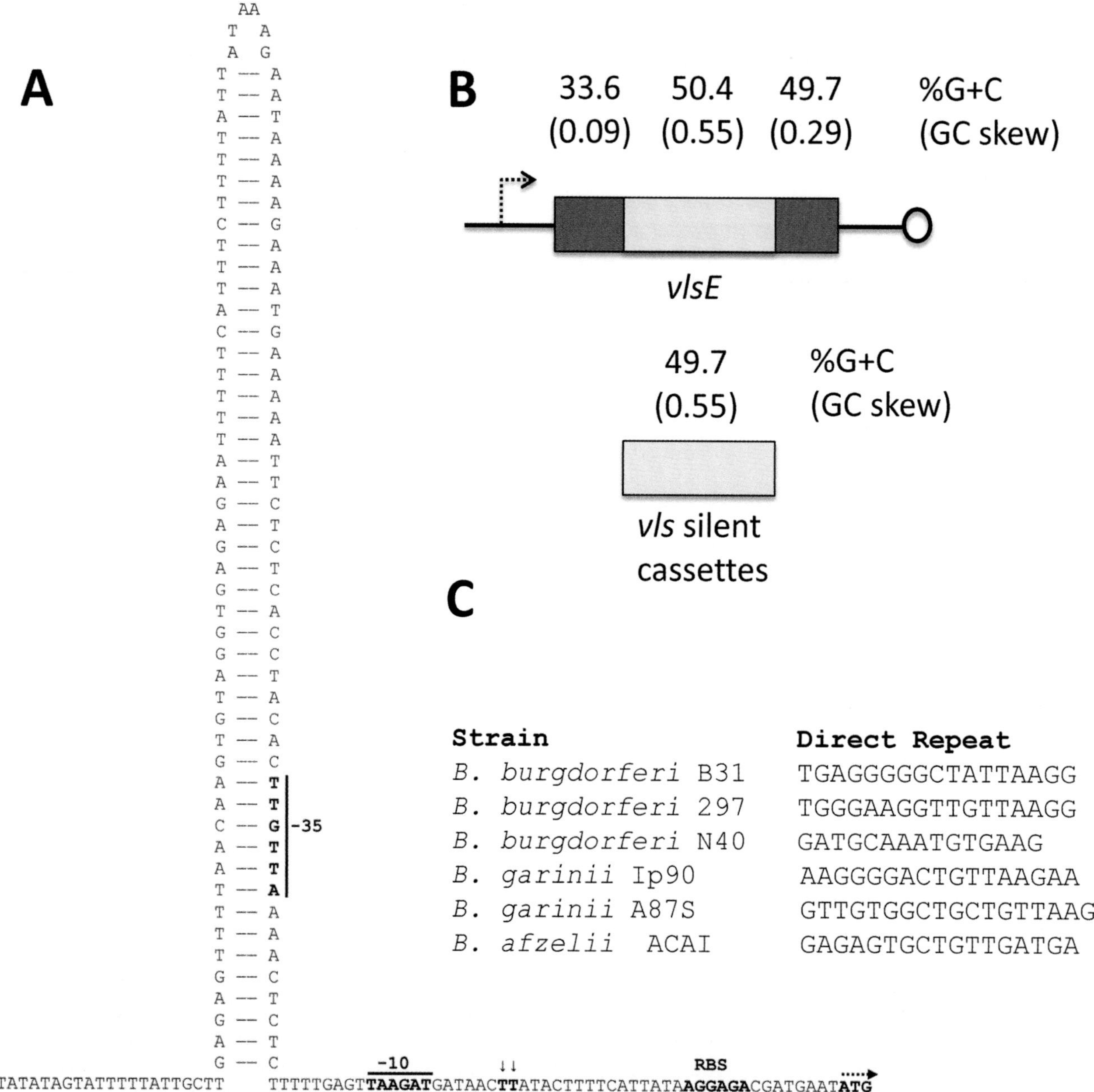

Figure 5 Properties of the *vls* locus potentially involved in recombination and regulation of transcription. (A) Inverted repeat between *vlsE* and the *vls* silent cassettes of *B. burgdorferi* B31 (48; Norris SJ, Howell JK, unpublished data). The predicted stem–loop structure includes the –35 region of the *vlsE* promoter. (B) G+C content and GC skew of *vlsE* and the *vls* silent cassettes in *B. burgdorferi* B31. (C) Lack of conservation of the *vls* direct repeat sequences in Lyme disease *Borrelia* suggests a limited role of these regions in *vlsE* recombination (38). Representative sequences are shown.
doi:10.1128/microbiolspec.MDNA3-0038-2014.f5

tioned previously, the location and nature of inverted repeats is variable in the few organisms in which the intact sequence of this region is available. Factors that bind to the promoter region have not as yet been reported. Jutras et al. (69) used electrophoretic mobility shift assays (EMSAs) to demonstrate that SpoVG binds to a dsDNA region of the *B. burgdorferi* B31 *vlsE* open reading frame; it is thus possible that this factor may affect transcription after initiation (see further discussion of SpoVG below).

Crother et al. (49, 70) performed novel studies on the expression of *B. burgdorferi* lipoproteins during mouse or rabbit infection, in which the tissue specimens were extracted by Triton X114 phase partitioning, which preferentially extracts lipoproteins. In the first study (70), SCID mice were inoculated to increase the yield of organisms, and infected joint, heart, ear, and skin tissues were utilized. The extracted proteins were detected by immunoblot analysis with polypeptide-specific or anti-*B. burgdorferi* polyspecific antisera. In these experiments, VlsE was the most abundant borrelial protein recovered from all tissues except the heart. In the heart, the levels were low throughout the course of infection. In the second study, rabbits were infected by intradermal injection, and rabbits were sacrificed and the skin sites analyzed on days 5, 7, 9, 11, 14, and 21 postinoculation (PI). By this methodology, VlsE protein levels increased dramatically on day 7 PI, and remained high throughout the experiment. Conversely, OspC levels peaked on day 7 PI and decreased rapidly thereafter. When VlsE was analyzed using VlsE-specific antiserum and two-dimensional gel electrophoresis, it appeared as a series of spots over a broad pH range; this may reflect the presence of multiple variant VlsE proteins with different isoelectric points (pI's). In addition, low molecular weight isoforms of VlsE were also present, but it was unclear whether these were degradation products or truncated forms. The results of this study provided evidence that OspC levels are downregulated and VlsE levels are upregulated during the course of experimental rabbit infection.

Further studies by Liang et al. (71, 72) examined transcript levels of *vlsE*, *ospC* and other lipoprotein genes during the course of *B. burgdorferi* infection in normal mice, B-cell deficient mice, SCID mice, and SCID mice treated with sera from infected animals. In SCID mice, transcript levels of most lipoprotein genes, including *vlsE* and *ospC*, remain at high levels; the same results were obtained with *vlsE* and *ospC* in B-cell-deficient mice (72). In contrast, *ospC* transcripts were decreased below detectable levels and *vlsE* transcript levels are reduced in immunocompetent mice at day 33 to 40 PI. Treatment of SCID mice with sera from *B. burgdorferi* infected mice resulted in dramatic decreases (up to 76-fold) in *ospC* mRNA levels and increases in *vlsE* transcripts (up to 44-fold) relative to SCID mice treated with prebleed mouse sera (72). Similar results were obtained with a monoclonal antibody against OspC, indicating that anti-OspC antibodies were principally responsible for this effect. Interestingly, the effects varied according to the tissues sampled; for example, neither of the antibody preparations had a significant effect on *vlsE* transcript levels in joint tissue.

Overall, studies to date indicate that *vlsE* transcription is dramatically affected by antibody responses, most likely through the direct or indirect effects of OspC transcript or protein expression levels. It is not known currently if these effects are due to the selection of OspC-underexpressing cells, or a generalized downregulation of OspC expression in the infecting population. Studies by Anguita et al. (73) suggest that the inflammatory process (in particular, interferon-γ expression) may also affect spirochete tissue adaptation and *vlsE* recombination. Clearly, additional studies are needed to gain a better understanding of this interesting and novel phenomenon.

THE *vlsE* RECOMBINATION PROCESS

Gene conversion clearly represents the major mechanism of *vlsE* sequence variation, although some have suggested that point mutations may also play a role (74). In terms of the recombination results, *vlsE* recombination most closely resembles that in the *Neisseria gonorrhoeae pilE* system and the *var* system of *Plasmodium falciparum*, in that the region of DNA replacement is variable in both length and region; in each case, the sequence changes appear to be restricted to a given area (e.g. the cassette region in *vlsE*). The *vls* cassette regions uniformly exhibit high GC content (~50%) and marked GC skew relative to the rest of the *B. burgdorferi* sensu lato genomes, and the existence of a G quadraplex (G4) structure similar to that in neisserial *pilE* loci has been suggested (75). Beyond these similarities, little is known about the *vlsE* recombination process. It does not require the extensive array of DNA recombination and repair proteins that are involved in *pilE* sequence variation (76, 77), nor has the involvement of a small RNA (as recently described in *N. gonorrhoeae* [78]) been implicated. Unlike *Neisseria* species and most other antigenic variation systems described, *vlsE* recombination is not detectable during *in vitro* culture (or during tick infection), indicating that the process is activated in some manner by conditions in the mammalian host. However, to date, the factor(s) contributing to this activation have not been identified, limiting the study of *vlsE* recombination to animal infection experiments. Analysis of the recombination process is further hampered by the recalcitrance of the *vls* locus to genetic manipulation, other than its elimination through upstream insertion of telomere sequences (53).

As a result, the present understanding of *vlsE* recombination requirements is limited to the aspects listed below. These points are addressed in greater detail later in the chapter.

Observations regarding the mechanisms of *vlsE* recombination. (i) The *cis* arrangement of *vlsE* and the *vls* silent cassettes appears to be required for recombination to occur.

(ii) Inverted repeats with the potential of forming stem–loop structures are present between *vlsE* and the *vls* cassette arrays in the three strains in which intact *vls* region sequences have been obtained, but these inverted repeats vary in sequence, location and length.

(iii) The uniform existence of high GC content and GC skew in *vls* cassette sequences indicates the involvement of these features in recombination.

(iv) Among an extensive list of DNA recombination and repair gene products examined to date, only RuvAB (a Holliday junction resolvase) and potentially MutS (77) have a substantial effect on the *vlsE* recombination process (76, 77).

(v) The regulatory protein SpoVG binds to a sequence in the *vlsE* reading frame upstream of the cassette region of *B. burgdorferi* B31 (69); however, this sequence is not uniformly present in other Lyme disease *Borrelia vlsE* genes, and the potential effects of SpoVG on *vlsE* expression and recombination are currently unknown.

(vi) *vlsE* expression and recombination are markedly upregulated during infection of mammalian hosts, but the mechanism(s) of these effects are not known.

Cis Arrangement of *vlsE* and *vls* Silent Cassettes

Complete (or nearly complete) *vls* loci sequences have only been determined for three *Borrelia* strains: *B. burgdorferi* B31, *B. burgdorferi* JD1, and *B. garinii* Far04. In each of these cases, the reading frames of *vlsE* and the silent cassettes are arranged in opposite directions, pointing away from one another (Fig. 3). The silent cassette region also begins with a region highly homologous to the 5′ end of the *vlsE* open reading frame; the promoter, however, is not retained. This region may be a remnant of the initial duplication event that gave rise to the first silent cassette. As described previously, the region at the junction of *vlsE* and the *vlsE* cassettes in each strain contains an inverted repeat of 51, 476 and 476 nt respectively, although these differ in location and composition. In the B31 strain, the inverted repeat is in the region between *vlsE* and the *vls* silent cassettes, with inclusion of the –35 sequence of the *vlsE* promoter (Fig. 5A; (29, 48); Howell JK, Norris SJ, unpublished data). In the JD1 and Far04 strains, the inverted repeats encompass the 5′ ends of the open reading frames of *vlsE* and the first silent cassette, ending at the sequence differences within the cassette regions. As mentioned previously, the genes at the far end of lp28-1 in *B. burgdorferi* B31 are not required for *vlsE* sequence changes, in that removal of *BBF01-BBF19* from lp28-1 did not interfere with *vlsE* recombination or infectivity of the clone in immunocompetent mice (53).

Based on the limited sampling of intact *vls* loci, the *cis* 'head-to-head' arrangement of *vlsE* and the silent cassettes may be required for *vlsE* recombination. No sequence changes were detected when mice were infected with *B. burgdorferi* B31 harboring an intact *vlsE* gene in a shuttle vector (51), suggesting the importance of the *cis* arrangement. A possible model is that *vlsE* "doubles back" onto the silent cassette, promoting close proximity of the donor and recipient regions and thereby strand invasion as part of the gene conversion process. It is conceivable that the inverted repeats could form hairpin loops and in some way facilitate the doubling back. However, one would expect the nearest silent cassettes to recombine more frequently with *vlsE* in this scenario, but that is not the case (33). If site-directed mutagenesis of the *vls* locus becomes feasible, then the potential roles of the head-to-head arrangement and inverted repeat regions in *vlsE* recombination could be examined more thoroughly.

High G+C Content, GC Skew, and Potential G4 Structures

The *vlsE* cassette region and the *vls* silent cassettes have a remarkably high G+C content and GC skew compared to the remainder of the *B. burgdorferi* B31 genome (2, 39, 75) (Fig. 5B). At 49.7% compared with the 28.2% value for the overall genome, the average percent G+C is 20.5% higher in the silent cassette regions. Although the 3′ end of the *vlsE* also has a high G+C content, the coding strand GC skew (G-C/G+C) is preserved only in the cassette region. The cassette region GC skew of 0.55 is much higher than that in the leading strand of DNA replication (e.g. 0.18 in the *B. burgdorferi* B31 chromosome). This pattern is conserved in all of the *vls* sequences that have been characterized to date, despite significant sequence divergence (as low as 40% identity and 49% similarity). The preservation of this pattern is therefore likely to be important in the promotion of *vlsE* recombination.

In *N. gonorrhoeae*, the segmental recombination of *pilS* cassette sequences into the *pilE* pilus protein expression site requires the formation of a G-quadruplex (G4) structure in a region just upstream of the *pilE* open reading frame (79, 80). This conformational change involves the formation of intrachain G–G hydrogen bonds in neighboring G-rich regions to create a complex of four antiparallel segments. It is promoted by the transcription of a small RNA that overlaps with the G4 region (78).

The high GC content and skew in the *vls* cassette region suggests that similar secondary structure changes may be involved in *vlsE* recombination. In 2013, Walia and Chaconas (75) explored the possible occurrence of G4 structure in *vls* sequences. First, they noted that PCR amplification of *B. burgdorferi* B31 *vlsE* resulted in an unexpected lower molecular weight band in which the cassette region and one of the 17-bp direct repeats were missing. This phenomenon is apparently due to aberrant base pairing between the two identical direct repeats during the amplification process. Thus it is not thought to occur during *vlsE* recombinations; rather, it indicates unusual base-pairing properties of the 17-bp repeat, which includes a stretch of five G residues. In solution, the 17-bp region of *B. burgdorferi* B31 (or a shorter 14-bp oligonucleotide that also encompasses the G_5 stretch) forms a stable, higher order complex in the presence of high concentrations of KCl, conditions that favor G4 formation (75). The occurrence of hydrogen bonding between the G-rich regions was indicated by protection against methylation by methylsulfoxide. It should be noted, however, that direct repeats at the ends of *vlsE* cassette regions are variable both between and within strains; their lack of conservation argues against a sequence-specific role of these regions (e.g. recognition by a protein involved in the recombination process). Walia and Chaconas (75) further reported that stretches of 3 or more G's are found at very high frequencies in the coding strand of the *vlsE* cassette regions of *B. burgdorferi* strains B31, N40, and JD1; indeed, this is a common property of all known *vls* sequences. Whereas all of these findings support the potential formation of G4 or other intrastrand secondary structures in *vls* sequences, to date no G4 structures have been demonstrated either conceptually (based on sequence) or biophysically using native *vls* regions. This represents a promising area of future research.

DNA Repair and Recombination Proteins

Dresser et al. (76) and Lin et al. (77) examined a comprehensive list of *B. burgdorferi* B31 genes encoding predicted DNA repair and recombination proteins for their potential role in *vlsE* recombination. Mutations in these genes were introduced by either site-directed mutagenesis (76) or random transposon mutagenesis (77). Surprisingly, only a few genes were found to have a marked effect on *vlsE* sequence changes during mouse infection (and hence also on survival of the *B. burgdorferi* strains in immunocompetent mice). Both studies identified *ruvA* and *ruvB* as important genes in the recombination process, in that the rate of accumulation and diversity of *vlsE* sequence changes were dramatically reduced in clones in which these genes were mutated (76, 77). RuvA and RuvB form a Holliday junction branch migrase. Holliday junctions are mobile junctions between two homologous DNA duplexes, in which strand exchange results in base pairing between homologous regions and formation of a structure with four double-stranded branches. RuvAB branch migrases promote the release and reformation of the base pairs, resulting in migration of the branch point in a zipper-like manner. This branch migration process may be important in the random 'selection' of regions of the silent cassette sequences for exchange with the parental *vlsE* sequences. In most bacterial systems, the endonuclease RuvC is required to resolve the Holliday junction and complete homologous recombination. *Borrelia* lack an obvious *ruvC* homolog; it possible that putative prophage nucleases on the multiple cp32 plasmids could fulfill this function (76). In the Lin et al. study (77), the mutation of *mutS* also appeared to have a relatively small effect on the rate of *vlsE* sequence variation during infection. The other genes examined had no obvious effect on *vlsE* recombination; this list includes genes encoding the recombination proteins RecA, RecG and RecJ, repair proteins MutL and MutS2, general DNA replication/processing proteins NucA, Mag, Mfd, Nth, SbcC, SbcD, and PriA, plasmid-encoded putative recombinases BBD20 and BBG32, predicted DNA methyltransferase BBE29, and Rep helicase. The finding that the single strand DNA-binding protein RecA is not necessary for *vlsE* sequence changes was also determined previously by Liveris et al. (81). The apparent lack of involvement of this long list of genes differs from the *N. gonorrhoeae pilE* system, in which many proteins of the common recombination pathways (including RecA) are required (see references 75 and 80). This pattern, coupled with the induction of *vlsE* recombination during mammalian infection, suggests that this process may utilize as yet undiscovered recombinase(s) or other proteins (or RNAs) that become active in the mammalian environment and are specific for the *vls* sequences.

Potential Role of SpoVG

It is likely that proteins or small RNAs that specifically recognize *vls* sequences are involved in the recombination process. SpoVG was initially described as a cytoplasmic protein that is required for normal sporulation, vegetative cell structure, and cell division in *Bacillus subtilis*. However, homologs of this protein are found in most eubacteria. In a search for proteins that bind to *vlsE* DNA sequences, Jutras et al. (69) identified a *B. burgdorferi* homolog of SpoVG that binds to an 18-bp sequence within the 5′ region of the B31 *vlsE* gene, just upstream of the cassette region. The predicted SpoVG sequence is highly conserved in all *Borrelia* species, and also has high similarity with SpoVG sequences in other bacteria. However, the B31 *vlsE* sequence to which SpoVG binds is not well conserved in other Lyme disease *Borrelia* outside the B31 'clade'. Further study is thus necessary to determine if the observed binding activity is related to a conserved mechanism among *B. burgdorferi* sensu lato for regulation of *vlsE* expression or recombination.

Induction of vlsE Expression and Recombination

As mentioned previously, *vlsE* recombination has not been detected consistently in *in vitro* cultures or in infected ticks, yet can be detected as early as three days after the initiation of mammalian infection (30, 33, 34). The fact that the recombination can be detected as soon as organisms can be cultured, combined with its occurrence in SCID animals, argues against these results being due to immune selection in the infected animals, although it is clear that the immune response accelerates the clearance of the parental clone used for inoculation (33). Thus recombination appears to be induced in the mammalian environment by unknown mechanisms that appear to be unrelated to temperature, CO_2 concentration, presence of serum components, or other simple explanations. This situation differs from that of relapsing fever organisms and *N. gonorrhoeae*, in which recombination of VMP genes and *pilE* occur readily during *in vitro* culture. *vlsE* sequence changes have been detected by incubation of *B. burgdorferi* with mouse or rabbit tissue explants using sensitive PCR procedures, but it is as yet unclear whether these exceedingly rare events are occurring at rates higher than in standard *in vitro* cultures (Edmondson DG, Norris SJ, unpublished data). As described previously, *vlsE* transcription and VlsE protein expression is increased during mammalian infection, and it is possible that gene expression facilitates the recombination process. Attempts to mutate *vlsE* in its native lp28-1 location have thus far been unsuccessful, but would be useful in determining whether transcription and/or translation contribute to the occurrence of cassette region sequence changes, as well as in identifying important *cis* elements.

EVOLUTION OF THE *vls* SYSTEM

Lyme borreliosis (LB) and relapsing fever (RF) *Borrelia* are closely related, with nearly complete synteny across the chromosomes and a high degree of homology among the protein products. Plasmid structure and content in these groups differ considerably and likely account in part for the biological differences observed, including the colonization and transmission by *Ixodes* versus *Ornithodoros* ticks and distinct patterns of clinical manifestations. However, recent characterizations of *B. miyamotoi* and strain LB-2001 genome sequences indicate a blurring of these phenotypic and genotypic lines, in that both organisms are firmly positioned within the RF group genetically but are transmitted by *Ixodes* ticks (82). Overall, the phylogeny of *Borrelia* species clearly indicate that they are monophyletic, i.e. arose from a common primordial ancestor distinct from those that gave rise to other major groups of spirochetes, including the *Spirochaeta/Treponema/Sphaerochaeta* group, the *Leptospira/Leptonema/Turneriella* group, and *Brachyspira*.

All Lyme disease organisms characterized to date have *vls* antigenic variation systems, and all relapsing fever organisms have variable major protein (VMP) antigenic variation systems, which feature a single expression site in which either variable large protein (*vlp*) or variable small protein (*vsp*) open reading frames can be inserted. The two systems differ fundamentally, in that *vlsE* sequence changes arise from replacement with random segments from silent cassettes, whereas most recombinations in the VMP system involve replacement of the expression site gene with a nearly complete open reading frame from one of ~40 *vlp* and *vsp* promoterless gene copies (83). These recombination events predominantly utilize two sites termed upstream homology sequences (UHS) and downstream homology sequences (DHS) that are shared among nearly all of the 'donor' *vlp* and *vsp* copies. This feature is remarkable, in that the *vlp* and *vsp* sequences are otherwise essentially unrelated to each other.

Despite this diversity in the LB and RF recombination mechanisms, the *vlp* and *vsp* RF lipoprotein gene families each have relatives in LB organisms. *vlsE* has a moderate level of sequence homology with the *vlp* genes; for example, comparison of the originally described variant *vlsE1* with *B. hermsii vlp17* reveals nucleotide sequence identity of 58.8% and deduced amino acid sequence identity and similarity of 37.4% and 58.8%, respectively (2). In addition to the multiple *vsp* copies, RF organisms also contain a *vsp*-related 'variable tick protein' (*vtp*) gene that does not participate in *vlp*/*vsp* recombination but is preferentially expressed in the tick environment. Interestingly, LB organisms have a prominent *vsp* homolog encoding outer surface protein C (*ospC*), which is predominantly expressed during the transmission of *Borrelia* from the tick to the mammal and in the mammal during the early stages of infection. Therefore the *vlp* and *vsp* gene families each must have evolved from single primordial *vlp* and *vsp* genes in the common *Borrelia* ancestor and then undergone duplication and remarkable sequence and functional diversification during the development of the separate LB and RF *Borrelia* lineages. This scenario is shown diagrammatically in Fig. 6.

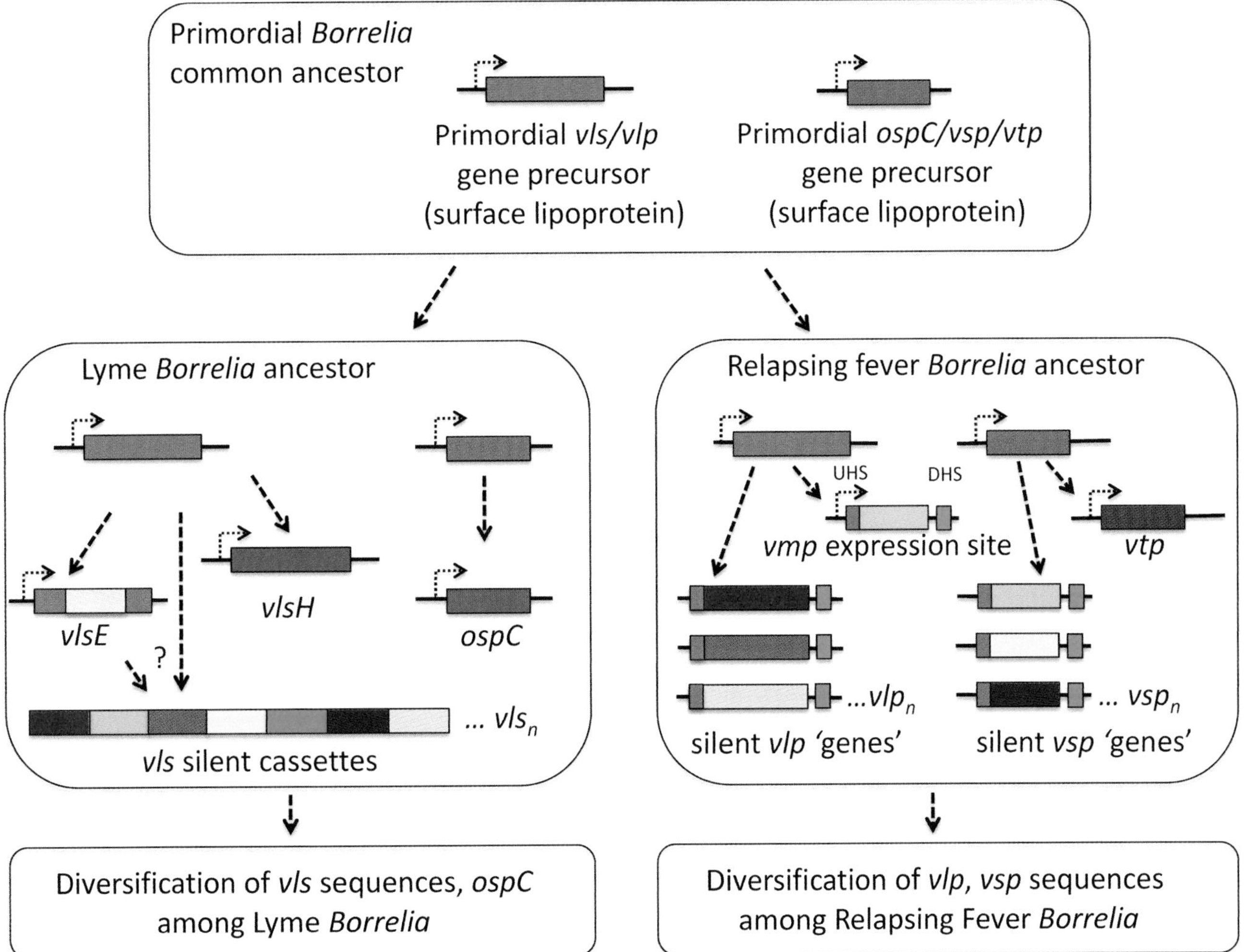

Figure 6 Model for the evolution of *vlp*- and *vsp*-related genes in *Borrelia*. It is hypothesized that a common ancestor of LB and RF organisms contained single copies of primordial *vlp* and *vsp* homologs. Following the divergence of LB and RF groups, these precursor genes duplicated and developed into the *vls* system and *ospC* in a primordial LB organism, and into the VMP system and *vtp* in a primordial RF ancestor. Each of these antigenic variation systems and related surface proteins continued to evolve and diverge under the pressure of immune selection. doi:10.1128/microbiolspec.MDNA3-0038-2014.f6

Is there any evidence of a common *vlp* precursor? Surprisingly, there is a paralogous gene in LB *Borrelia* that may be a remnant of this ancient ancestral gene. In *B. burgdorferi* B31, it is a fragmented pseudogene. However, in several strains, including *B. burgdorferi* JD1, the reading frame is intact; some strains, including *B. burgdorferi* WI91-23 and *B. garinii* PBr, have two intact copies. These genes, located on 28-kb, 38-kb, or 36-kb plasmids, often have been mistakenly annotated as *vlsE* in genomic sequences; it is suggested that they be referred to as *vls* homologs or *vlsH*. In JD1, the lp28-6 gene BbuJD1_Z01 encodes a 40.6-kDa lipoprotein that is 28% identical and 43% similar to B31 VlsE1. One conserved region is located within IR6, indicating that this α-helical sequence is structurally important in both VlsE and Vlp proteins (46). This region of VlsE is also highly immunogenic and is used as a diagnostic antigen in immunologic tests for Lyme disease. Several strains, including *B. burgdorferi* B31, N40, 29805, and 64b, have an identical frameshift following the 56th codon of *vlsH*; this occurrence may indicate ancestral inheritance of this damaged gene from a common progenitor. The *vlsH* genes and their products have not been studied; it would be of interest to examine their expression and other properties.

Both OspC and VlsE are surface-localized lipoproteins that are thought to be expressed sequentially during mammalian infection, with OspC being the predominant surface protein in the first days or weeks of infection followed by replacement by VlsE (84). Likely due to the constant assault of the host antibody response, these proteins are the most heterogeneous of all LB *Borrelia* polypeptides. The natural hosts of LB *Borrelia* (such as the white-footed mouse *Peromyscus leucopus*) are continually exposed to different strains of *Borrelia* delivered through multiple tick bites. Thus antibodies against OspC and the 'framework' portions of VlsE from prior infections would select for organisms that expressed different epitopes, promoting the fixation of mutations that change the antigenic structure without destroying the proteins' structural integrity. This constant immune pressure thus resulted in strains expressing increasingly heterogeneous versions of this surface protein. In the case of VlsE, this selection not only drove the development of the *vls* antigenic variation system, but also accelerated the evolution of diverse framework regions of this protein.

CONCLUSIONS

The *vls* system represents one of the most elaborate and elegant antigenic variation systems in bacterial pathogens. Its consistent presence in LB organisms indicates its importance in the tenuous survival of these spirochetes during their tightrope act of continual transmission between mammalian and arthropod hosts. The intense immune selection not only drove the parallel but divergent evolution of two different antigenic variation systems in RF and LB *Borrelia,* but also has promoted VlsE framework divergence within different LB species and strains. In addition, the patterns of expression of OspC in LB spirochetes and Vtp in RF organisms could be considered forms of phase variation to evade the immune response; OspC heterogeneity provides yet another example of the power of antibody reactions in the selection of antigenic variants. Together, these elaborate adaptations promote survival of Lyme borreliosis organisms for months to years in mammalian hosts, thus assuring the passage of future generations to arthropods and back again to sylvan mammals and, as accidental hosts, humans.

Acknowledgments. *Research reported in this publication was supported by the National Institute of Allergy and Infectious Diseases of the National Institutes of Health under Award Number R01 AI037277. The content is solely the responsibility of the author and does not necessarily represent the official views of the National Institutes of Health.*

Citation. Norris SJ. 2014. vls antigenic variation systems of lyme disease borrelia: eluding host immunity through both random, segmental gene conversion and framework heterogeneity. Microbiol Spectrum **2**(6):MDNA3-0038-2014.

References

1. **van der Woude MW, Bäumler AJ.** 2004. Phase and antigenic variation in bacteria. *Clin Microbiol Rev* **17:** 581–611.
2. **Zhang JR, Hardham JM, Barbour AG, Norris SJ.** 1997. Antigenic variation in Lyme disease borreliae by promiscuous recombination of VMP-like sequence cassettes. *Cell* **89:**275–285.
3. **Steere AC, Coburn J, Glickstein L.** 2004. The emergence of Lyme disease. *J Clin Invest* **113:**1093–1101.
4. **Stanek G, Strle F.** 2008. Lyme disease – European perspective. *Infect Dis Clinics North Amer* **22:**327–339.
5. **Müllegger RR, Glatz M.** 2008. Skin manifestations of Lyme borreliosis: diagnosis and management. *Am J Clin Dermatol* **9:**355–368.
6. **Stanek G, Reiter M.** 2011. The expanding Lyme Borrelia complex—clinical significance of genomic species? *Clin Microbiol Infect* **17:**487–493.
7. **Norris SJ, Coburn J, Leong JM, Hu LT, Höök M.** 2010. Pathobiology of Lyme disease *Borrelia*, p 299–331. *In* Samuels DS, Radolf JD (ed), Borrelia: *Molecular and Cellular Biology*. Caister Academic Press, Hethersett, Norwich, UK.
8. **Skare JT, Carroll JA, Yang XF, Samuels DS, Akins DR.** 2010. Gene regulation, transcriptomics, and proteomics,

p 67–102. *In* Samuels DS, Radolf JD (ed), Borrelia*: Molecular and Cellular Biology*. Caister Academic Press, Hethersett, Norwich, UK.

9. **Radolf JD, Lukehart SA.** 2006. *Pathogenic Treponema: Molecular and Cellular Biology*. Caister Academic Press, Hethersett, Norwich, UK.
10. **Kraiczy P, Stevenson B.** 2013. Complement regulator-acquiring surface proteins of *Borrelia burgdorferi*: Structure, function and regulation of gene expression. *Ticks Tick-borne Dis* **4:**26–34.
11. **de Taeye SW, Kreuk L, van Dam AP, Hovius JW, Schuijt TJ.** 2013. Complement evasion by *Borrelia burgdorferi*: it takes three to tango. *Trends Parasitol* **29:**119–128.
12. **Tilly K, Bestor A, Dulebohn DP, Rosa PA.** 2009. OspC-independent infection and dissemination by host-adapted *Borrelia burgdorferi*. *Infect Immun* **77:**2672–2682.
13. **Tilly K, Krum JG, Bestor A, Jewett MW, Grimm D, Bueschel D, Byram R, Dorward D, Vanraden MJ, Stewart P, Rosa P.** 2006. *Borrelia burgdorferi* OspC protein required exclusively in a crucial early stage of mammalian infection. *Infect Immun* **74:**3554–3564.
14. **Burgdorfer W, Barbour AG, Hayes SF, Benach JL, Grunwaldt E, Davis JP.** 1982. Lyme disease, a tick-borne spirochetosis? *Science* **216:**1317–1319.
15. **Labandeira-Rey M, Baker E, Skare J.** 2001. Decreased infectivity in *Borrelia burgdorferi* strain B31 is associated with loss of linear plasmid 25 or 28-1. *Infect Immun* **69:** 446–455.
16. **Barbour AG.** 1988. Plasmid analysis of *Borrelia burgdorferi*, the Lyme disease agent. *J Clin Microbiol* **26:** 475–478.
17. **Johnson RC, Marek N, Kodner C.** 1984. Infection of Syrian hamsters with Lyme disease spirochetes. *J Clin Microbiol* **20:**1099–1101.
18. **Norris SJ, Howell JK, Garza SA, Ferdows MS, Barbour AG.** 1995. High- and low-infectivity phenotypes of clonal populations of *in vitro*-cultured *Borrelia burgdorferi*. *Infect Immun* **63:**2206–2212.
19. **Xu Y, Johnson RC.** 1995. Analysis and comparison of plasmid profiles of *Borrelia burgdorferi* sensu lato strains. *J Clin Microbiol* **33:**2679–2685.
20. **Xu Y, Kodner C, Coleman L, Johnson RC.** 1996. Correlation of plasmids with infectivity of *Borrelia burgdorferi* sensu stricto type strain B31. *Infect Immun* **64:**3870–3876.
21. **Labandeira-Rey M, Seshu J, Skare JT.** 2003. The absence of linear plasmid 25 or 28-1 of *Borrelia burgdorferi* dramatically alters the kinetics of experimental infection via distinct mechanisms. *Infect Immun* **71:**4608–4613.
22. **Grimm D, Eggers CH, Caimano MJ, Tilly K, Stewart PE, Elias AF, Radolf JD, Rosa PA.** 2004. Experimental assessment of the roles of linear plasmids lp25 and lp28-1 of *Borrelia burgdorferi* throughout the infectious cycle. *Infect Immun* **72:**5938–5946.
23. **Purser JE, Norris SJ.** 2000. Correlation between plasmid content and infectivity in *Borrelia burgdorferi*. *Proc Natl Acad Sci USA* **97:**13865–13870.
24. **Fraser CM, Casjens S, Huang WM, Sutton GG, Clayton R, Lathigra R, White O, Ketchum KA, Dodson R, Hickey EK, Gwinn M, Dougherty B, Tomb JF, Fleischmann RD, Richardson D, Peterson J, Kerlavage AR, Quackenbush J, Salzberg S, Hanson M, van Vugt R, Palmer N, Adams MD, Gocayne J, Weidman J, Utterback T, Watthey L, McDonald L, Artiach P, Bowman C, Garland S, Fujii C, Cotton MD, Horst K, Roberts K, Hatch B, Smith HO, Venter JC.** 1997. Genomic sequence of a Lyme disease spirochaete, *Borrelia burgdorferi*. *Nature* **390:**580–586.
25. **Casjens S, Palmer N, van Vugt R, Huang WM, Stevenson B, Rosa P, Lathigra R, Sutton G, Peterson J, Dodson RJ, Haft D, Hickey E, Gwinn M, White O, Fraser CM.** 2000. A bacterial genome in flux: the twelve linear and nine circular extrachromosomal DNAs in an infectious isolate of the Lyme disease spirochete *Borrelia burgdorferi*. *Mol Microbiol* **35:**490–516.
26. **Hinnebusch J, Bergstrom S, Barbour AG.** 1990. Cloning and sequence analysis of linear plasmid telomeres of the bacterium *Borrelia burgdorferi*. *Mol Microbiol* **4:**811–820.
27. **Hinnebusch J, Barbour AG.** 1991. Linear plasmids of *Borrelia burgdorferi* have a telomeric structure and sequence similar to those of a eukaryotic virus. *J Bacteriol* **173:**7233–7239.
28. **Kobryn K, Chaconas G.** 2002. ResT, a telomere resolvase encoded by the Lyme disease spirochete. *Mol Cell* **9:** 195–201.
29. **Hudson CR, Frye JG, Quinn FD, Gherardini FC.** 2001. Increased expression of *Borrelia burgdorferi vlsE* in response to human endothelial cell membranes. *Mol Microbiol* **41:**229–239.
30. **Zhang JR, Norris SJ.** 1998. Kinetics and *in vivo* induction of genetic variation of *vlsE* in *Borrelia burgdorferi*. *Infect Immun* **66:**3689–3697.
31. **Indest KJ, Howell JK, Jacobs MB, Scholl-Meeker D, Norris SJ, Philipp MT.** 2001. Analysis of *Borrelia burgdorferi vlsE* gene expression and recombination in the tick vector. *Infect Immun* **69:**7083–7090.
32. **Nosbisch LK, de Silva AM.** 2007. Lack of detectable variation at *Borrelia burgdorferi vlsE* locus in ticks. *J Med Entomol* **44:**168–170.
33. **Coutte L, Botkin DJ, Gao L, Norris SJ.** 2009. Detailed analysis of sequence changes occurring during *vlsE* antigenic variation in the mouse model of *Borrelia burgdorferi* infection. *PLoS Pathog* **5:**e1000293.
34. **Embers ME, Liang FT, Howell JK, Jacobs MB, Purcell JE, Norris SJ, Johnson BJ, Philipp MT.** 2007. Antigenicity and recombination of VlsE, the antigenic variation protein of *Borrelia burgdorferi*, in rabbits, a host putatively resistant to long-term infection with this spirochete. *FEMS Immunol Med Microbiol* **50:**421–429.
35. **Zhang JR, Norris SJ.** 1998. Genetic variation of the *Borrelia burgdorferi* gene *vlsE* involves cassette-specific, segmental gene conversion. *Infect Immun* **66:**3698–3704.
36. **Kawabata H, Myouga F, Inagaki Y, Murai N, Watanabe H.** 1998. Genetic and immunological analyses of Vls (VMP-like sequences) of *Borrelia burgdorferi*. *Microb Pathog* **24:**155–166.
37. **Liang FT, Alvarez AL, Gu Y, Nowling JM, Ramamoorthy R, Philipp MT.** 1999. An immuno-

dominant conserved region within the variable domain of VlsE, the variable surface antigen of *Borrelia burgdorferi*. *J Immunol* **163:**5566–5573.

38. **Wang G, van Dam AP, Dankert J.** 2001. Analysis of a VMP-like sequence (*vls*) locus in *Borrelia garinii* and Vls homologues among four *Borrelia burgdorferi* sensu lato species. *FEMS Microbiol Lett* **199:**39–45.

39. **Wang D, Botkin DJ, Norris SJ.** 2003. Characterization of the *vls* antigenic variation loci of the Lyme disease spirochaetes *Borrelia garinii* Ip90 and *Borrelia afzelii* ACAI. *Mol Microbiol* **47:**1407–1417.

40. **Casjens SR, Fraser-Liggett CM, Mongodin EF, Qiu WG, Dunn JJ, Luft BJ, Schutzer SE.** 2011. Whole genome sequence of an unusual *Borrelia burgdorferi* sensu lato isolate. *J Bacteriol* **193:**1489–1490.

41. **Casjens SR, Mongodin EF, Qiu WG, Dunn JJ, Luft BJ, Fraser-Liggett CM, Schutzer SE.** 2011. Whole-genome sequences of two *Borrelia afzelii* and two *Borrelia garinii* Lyme disease agent isolates. *J Bacteriol* **193:**6995–6996.

42. **Casjens SR, Mongodin EF, Qiu WG, Luft BJ, Schutzer SE, Gilcrease EB, Huang WM, Vujadinovic M, Aron JK, Vargas LC, Freeman S, Radune D, Weidman JF, Dimitrov GI, Khouri HM, Sosa JE, Halpin RA, Dunn JJ, Fraser CM.** 2012. Genome stability of Lyme disease spirochetes: comparative genomics of *Borrelia burgdorferi* plasmids. *PLoS One* **7:**e33280.

43. **Mongodin EF, Casjens SR, Bruno JF, Xu Y, Drabek EF, Riley DR, Cantarel BL, Pagan PE, Hernandez YA, Vargas LC, Dunn JJ, Schutzer SE, Fraser CM, Qiu WG, Luft BJ.** 2013. Inter- and intra-specific pan-genomes of *Borrelia burgdorferi* sensu lato: genome stability and adaptive radiation. *BMC Genomics* **14:**693.

44. **Schutzer SE, Fraser-Liggett CM, Casjens SR, Qiu WG, Dunn JJ, Mongodin EF, Luft BJ.** 2011. Whole-genome sequences of thirteen isolates of *Borrelia burgdorferi*. *J Bacteriol* **193:**1018–1020.

45. **Schutzer SE, Fraser-Liggett CM, Qiu WG, Kraiczy P, Mongodin EF, Dunn JJ, Luft BJ, Casjens SR.** 2012. Whole-genome sequences of *Borrelia bissettii*, *Borrelia valaisiana*, and *Borrelia spielmanii*. *J Bacteriol* **194:**545–546.

46. **Eicken C, Sharma V, Klabunde T, Lawrenz MB, Hardham JM, Norris SJ, Sacchettini JC.** 2002. Crystal structure of Lyme disease variable surface antigen VlsE of *Borrelia burgdorferi*. *J Biol Chem* **277:**21691–21696.

47. **Liu J, Lin T, Botkin DJ, McCrum E, Winkler H, Norris SJ.** 2009. Intact flagellar motor of *Borrelia burgdorferi* revealed by cryo-electron tomography: evidence for stator ring curvature and rotor/C-ring assembly flexion. *J Bacteriol* **191:**5026–5036.

48. **Bykowski T, Babb K, von Lackum K, Riley SP, Norris SJ, Stevenson B.** 2006. Transcriptional regulation of the *Borrelia burgdorferi* antigenically variable VlsE surface protein. *J Bacteriol* **188:**4879–4889.

49. **Crother TR, Champion CI, Whitelegge JP, Aguilera R, Wu XY, Blanco DR, Miller JN, Lovett MA.** 2004. Temporal analysis of the antigenic composition of *Borrelia burgdorferi* during infection in rabbit skin. *Infect Immun* **72:**5063–5072.

50. **Liang FT, Nowling JM, Philipp MT.** 2000. Cryptic and exposed invariable regions of VlsE, the variable surface antigen of *Borrelia burgdorferi* sl. *J Bacteriol* **182:**3597–3601.

51. **Lawrenz MB, Wooten RM, Norris SJ.** 2004. Effects of vlsE complementation on the infectivity of *Borrelia burgdorferi* lacking the linear plasmid lp28-1. *Infect Immun* **72:**6577–6585.

52. **Grimm D, Tilly K, Bueschel DM, Fisher MA, Policastro PF, Gherardini FC, Schwan TG, Rosa PA.** 2005. Defining plasmids required by Borrelia burgdorferi for colonization of tick vector Ixodes scapularis (Acari: Ixodidae). *J Med Entomol* **42:**676–684.

53. **Bankhead T, Chaconas G.** 2007. The role of VlsE antigenic variation in the Lyme disease spirochete: persistence through a mechanism that differs from other pathogens. *Mol Microbiol* **65:**1547–1558.

54. **Lawrenz MB, Hardham JM, Owens RT, Nowakowski J, Steere AC, Wormser GP, Norris SJ.** 1999. Human antibody responses to VlsE antigenic variation protein of *Borrelia burgdorferi*. *J Clin Microbiol* **37:**3997–4004.

55. **Bacon RM, Biggerstaff BJ, Schriefer ME, Gilmore RD Jr, Philipp MT, Steere AC, Wormser GP, Marques AR, Johnson BJ.** 2003. Serodiagnosis of Lyme disease by kinetic enzyme-linked immunosorbent assay using recombinant VlsE1 or peptide antigens of *Borrelia burgdorferi* compared with 2-tiered testing using whole-cell lysates. *J Infect Dis* **187:**1187–1199.

56. **Marangoni A, Moroni A, Accardo S, Cevenini R.** 2008. *Borrelia burgdorferi* VlsE antigen for the serological diagnosis of Lyme borreliosis. *Eur J Clin Microbiol Infect Dis* **27:**349–354.

57. **Goettner G, Schulte-Spechtel U, Hillermann R, Liegl G, Wilske B, Fingerle V.** 2005. Improvement of Lyme borreliosis serodiagnosis by a newly developed recombinant immunoglobulin G (IgG) and IgM line immunoblot assay and addition of VlsE and DbpA homologues. *J Clin Microbiol* **43:**3602–3609.

58. **Ledue TB, Collins MF, Young J, Schriefer ME.** 2008. Evaluation of the LIAISON(R) Borrelia burgdorferi Test, a Recombinant VlsE-based Chemiluminescence Immunoassay for the Diagnosis of Lyme Disease. *Clin Vaccine Immunol* **15:**1796–1804.

59. **Levy SA, O'Connor TP, Hanscom JL, Shields P.** 2003. Evaluation of a canine C6 ELISA Lyme disease test for the determination of the infection status of cats naturally exposed to Borrelia burgdorferi. *Vet Ther* **4:**172–177.

60. **Magnarelli LA, Stafford KC 3rd, Ijdo JW, Fikrig E.** 2006. Antibodies to whole-cell or recombinant antigens of *Borrelia burgdorferi*, *Anaplasma phagocytophilum*, and *Babesia microti* in white-footed mice. *J Wildl Dis* **42:** 732–738.

61. **Liang FT, Philipp MT.** 1999. Analysis of antibody response to invariable regions of VlsE, the variable surface antigen of *Borrelia burgdorferi*. *Infect Immun* **67:** 6702–6706.

62. **Liang FT, Philipp MT.** 2000. Epitope mapping of the immunodominant invariable region of *Borrelia burgdorferi* VlsE in three host species. *Infect Immun* **68:**2349–2352.

63. **Liang FT, Bowers LC, Philipp MT.** 2001. C-terminal invariable domain of VlsE is immunodominant but its antigenicity is scarcely conserved among strains of Lyme disease spirochetes. *Infect Immun* **69:**3224–3231.
64. **McDowell JV, Sung SY, Hu LT, Marconi RT.** 2002. Evidence that the variable regions of the central domain of VlsE are antigenic during infection with Lyme disease spirochetes. *Infect Immun* **70:**4196–4203.
65. **Liang F, Jacobs M, Philipp M.** 2001. C-terminal invariable domain of VlsE may not serve as target for protective immune response against *Borrelia burgdorferi. Infect Immun* **69:**1337–1343.
66. **Baum E, Hue F, Barbour AG.** 2012. Experimental infections of the reservoir species *Peromyscus leucopus* with diverse strains of *Borrelia burgdorferi*, a Lyme disease agent. *mBio* **3:**e00434-00412.
67. **Koči J, Derdakova M, Peterkova K, Kazimirova M, Selyemova D, Labuda M.** 2006. *Borrelia afzelii* gene expression in *Ixodes ricinus* (Acari: *Ixodidae*) ticks. *Vector Borne Zoonotic Dis* **6:**296–304.
68. **Piesman J, Zeidner NS, Schneider BS.** 2003. Dynamic changes in *Borrelia burgdorferi* populations in *Ixodes scapularis* (Acari: *Ixodidae*) during transmission: studies at the mRNA level. *Vector Borne Zoonotic Dis* **3:** 125–132.
69. **Jutras BL, Chenail AM, Rowland CL, Carroll D, Miller MC, Bykowski T, Stevenson B.** 2013. Eubacterial SpoVG homologs constitute a new family of site-specific DNA-binding proteins. *PLoS One* **8:**e66683.
70. **Crother TR, Champion CI, Wu XY, Blanco DR, Miller JN, Lovett MA.** 2003. Antigenic composition of *Borrelia burgdorferi* during infection of SCID mice. *Infect Immun* **71:**3419–3428.
71. **Liang FT, Nelson FK, Fikrig E.** 2002. Molecular adaptation of *Borrelia burgdorferi* in the murine host. *J Exp Med* **196:**275–280.
72. **Liang FT, Yan J, Mbow ML, Sviat SL, Gilmore RD, Mamula M, Fikrig E.** 2004. *Borrelia burgdorferi* changes its surface antigenic expression in response to host immune responses. *Infect Immun* **72:**5759–5767.
73. **Anguita J, Thomas V, Samanta S, Persinski R, Hernanz C, Barthold SW, Fikrig E.** 2001. *Borrelia burgdorferi*-induced inflammation facilitates spirochete adaptation and variable major protein-like sequence locus recombination. *J Immunol* **167:**3383–3390.
74. **Sung SY, McDowell JV, Marconi RT.** 2001. Evidence for the contribution of point mutations to *vlsE* variation and for apparent constraints on the net accumulation of sequence changes in *vlsE* during infection with Lyme disease spirochetes. *J Bacteriol* **183:**5855–5861.
75. **Walia R, Chaconas G.** 2013. Suggested role for G4 DNA in recombinational switching at the antigenic variation locus of the Lyme disease spirochete. *PLoS One* **8:**e57792.
76. **Dresser AR, Hardy PO, Chaconas G.** 2009. Investigation of the genes involved in antigenic switching at the *vlsE* locus in *Borrelia burgdorferi*: an essential role for the RuvAB branch migrase. *PLoS Pathog* **5:**e1000680.
77. **Lin T, Gao L, Edmondson DG, Jacobs MB, Philipp MT, Norris SJ.** 2009. Central role of the Holliday junction helicase RuvAB in *vlsE* recombination and infectivity of *Borrelia burgdorferi. PLoS Pathog* **5:**e1000679.
78. **Cahoon LA, Seifert HS.** 2013. Transcription of a cis-acting, noncoding, small RNA is required for pilin antigenic variation in *Neisseria gonorrhoeae. PLoS Pathog* **9:** e1003074.
79. **Cahoon LA, Seifert HS.** 2009. An alternative DNA structure is necessary for pilin antigenic variation in *Neisseria gonorrhoeae. Science* **325:**764–767.
80. **Cahoon LA, Seifert HS.** 2011. Focusing homologous recombination: pilin antigenic variation in the pathogenic *Neisseria. Mol Microbiol* **81:**1136–1143.
81. **Liveris D, Mulay V, Sandigursky S, Schwartz I.** 2008. *Borrelia burgdorferi vlsE* antigenic variation is not mediated by RecA. *Infect Immun* **76:**4009–4018.
82. **Barbour AG.** 2014. Phylogeny of a relapsing fever *Borrelia* species transmitted by the hard tick *Ixodes scapularis. Infect Genet Evol* **27:**551–558.
83. **Dai Q, Restrepo BI, Porcella SF, Raffel SJ, Schwan TG, Barbour AG.** 2006. Antigenic variation by *Borrelia hermsii* occurs through recombination between extragenic repetitive elements on linear plasmids. *Mol Microbiol* **60:** 1329–1343.
84. **Tilly K, Bestor A, Rosa PA.** 2013. Lipoprotein succession in *Borrelia burgdorferi*: similar but distinct roles for OspC and VlsE at different stages of mammalian infection. *Mol Microbiol* **89:**216–227.

Mobile DNA, 3rd Edition
Nancy L. Craig, Michael Chandler, Martin Gellert, Alan M. Lambowitz, Phoebe A. Rice and Suzanne Sandmeyer

doi:10.1128/microbiolspec.MDNA3-0013-2014

Cheng-Sheng Lee[1]
James E. Haber[1]

23 Mating-type Gene Switching in *Saccharomyces cerevisiae*

INTRODUCTION

The budding yeast *Saccharomyces cerevisiae* propagates vegetatively either as *MAT***a** or *MAT*α haploids or as *MAT***a**/*MAT*α diploids created by conjugation of the opposite haploid types (Fig. 1). Mating type is determined by two different alleles of the mating-type (*MAT*) locus.

Like many other fungi, budding yeast has acquired the capacity to convert some cells in a colony from one haploid mating type to the other (Fig. 1) by a process called homothallism. The subsequent mating of cells to the opposite mating type enables these homothallic organisms to self-diploidize. The diploid state provides yeast with a number of evolutionarily advantageous strategies unavailable to haploids, most notably greater resistance to radiation and the ability to undergo meiosis and spore formation under nutritionally limiting conditions. Mating-type gene switching in *S. cerevisiae* is a highly choreographed process that has taught us much about many aspects of gene regulation, chromosome structure, and homologous recombination.

The *MAT* locus lies in the middle of the right arm of chromosome 3, ~100 kb from both the centromere and the telomere. The two mating-type alleles, *MAT*α and *MAT***a**, differ by ~700 bp of sequences designated Yα and Y**a**, respectively (Fig. 2A). Yα and Y**a** contain the promoters and most of the open reading frames for proteins that regulate many aspects of the cell's sexual activity (for reviews, see 1, 2, 3, 4, 5, 6). The *MAT* locus is divided into five regions (W, X, Y, Z1, and Z2) on the basis of sequences that are shared between *MAT* and the two cryptic copies of mating-type sequences located at *HML*α and *HMR***a** (Fig. 2B). *HML*α and *HMR***a** are maintained in an untranscribed, heterochromatic configuration with a highly ordered nucleosome structure (7, 8, 9, 10, 11, 12, 13). *HML*α and *HMR***a** serve as donors during the recombinational process that allows a *MAT***a** cell to switch to *MAT*α or vice versa.

FUNCTIONS AND REGULATION OF MAT PROTEINS

*MAT*α encodes two proteins, Matα1 and Matα2. Matα1 pairs with a constitutively expressed protein, Mcm1, to activate transcription of a set of α-specific genes (1, 14, 15), including those encoding the mating pheromone, α-factor, and Ste3, a *trans*-membrane receptor of the opposite mating pheromone, **a**-factor. The **a**- and α-factor mating pheromones trigger G1 arrest of the budding yeast expressing the opposite mating type and

[1]Department of Biology and Rosenstiel Basic Medical Sciences Research Center Brandeis University, Waltham, MA 02454-9110.

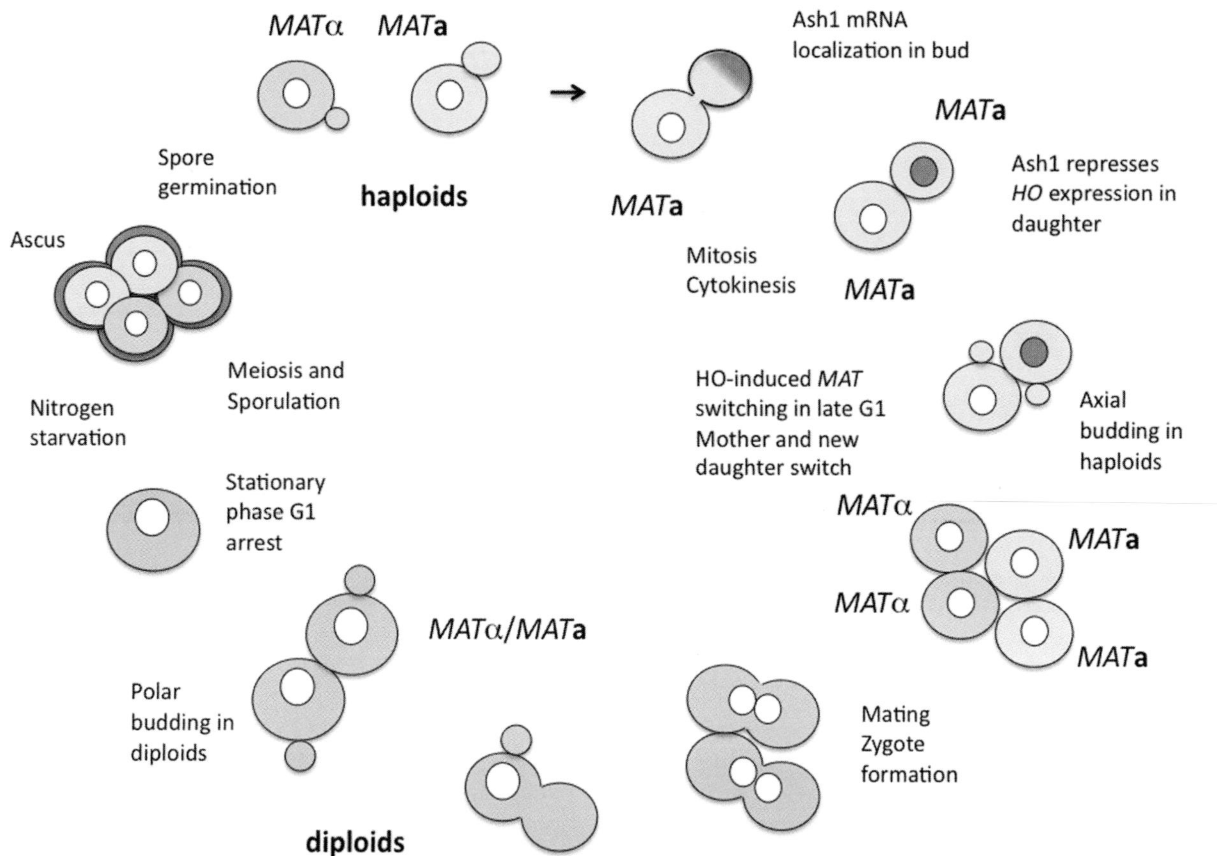

Figure 1 Homothallic life cycle of *S. cerevisiae*. A homothallic *MAT*a (pink) mother cell and its new daughter can switch to *MAT*α (light blue). This lineage is established by the asymmetric partitioning of the mRNA encoding the Ash1 repressor of *HO* gene expression in daughter cells (light green). These cells can conjugate to form a zygote that gives rise to *MAT*a/*MAT*α diploids (lilac), in which *HO* gene expression is repressed. Under nitrogen starvation, diploids undergo meiosis and sporulation to produce four haploid spores (two *MAT*a and two *MAT*α) in an ascus. The spores germinate and grow vegetatively and can repeat the homothallic cycle. Heterothallic cells have stable mating types and grow vegetatively until they exhaust their nutrients and enter stationary phase. Used with permission from the Genetics Society of America. doi:10.1128/microbiolspec.MDNA3-0013-2014.f1

facilitate conjugation, thus ensuring that the zygote will contain two unreplicated nuclei. *MAT*α2 encodes a homeodomain helix–turn–helix protein that acts with Mcm1 to form a repressor that binds to a 31-bp symmetric site with Mcm1 in the center and Matα2 at the ends (16). Matα2–Mcm1 represses **a**-specific genes, including those that produce **a**-factor (*MF*a*1* and *MF*a*2*) and Ste2, a *trans*-membrane receptor of α-factor. Repression also requires the action of Tup1 and Ssn6 proteins (17, 18, 19, 20).

*MAT*a is not required to turn on **a**-specific genes, which are constitutively expressed in the absence of *MAT*α2. However, Mata1 is required, along with Matα2, to form a very stable corepressor that turns off a set of haploid-specific genes. Because of this transcriptional repression by the **a**1–α2 repressor, *MAT*a/*MAT*α diploids acquire several important features: First, they are nonmating because the **a**1–α2 repressor turns off transcription of *MAT*α1, the activator of α-specific genes, while **a**-specific genes are repressed by *MAT*α*2* (5, 21, 22, 23, 24, 25). Secondly, they are able to initiate meiosis and spore formation because *RME1*, repressor of meiosis 1, is turned off. Thirdly, they have a bipolar budding pattern (26) because of the suppression of Axl1, which is required, in conjunction with a number of other gene products (Bud3, Bud4, Bud5, and Bud10), for axial budding (Fig. 1) (27, 28). Fourthly, nonhomologous end joining (NHEJ) is turned off by **a**1–α2 repression of the *NEJ1* gene and by the partial repression of another NHEJ component, *LIF1*

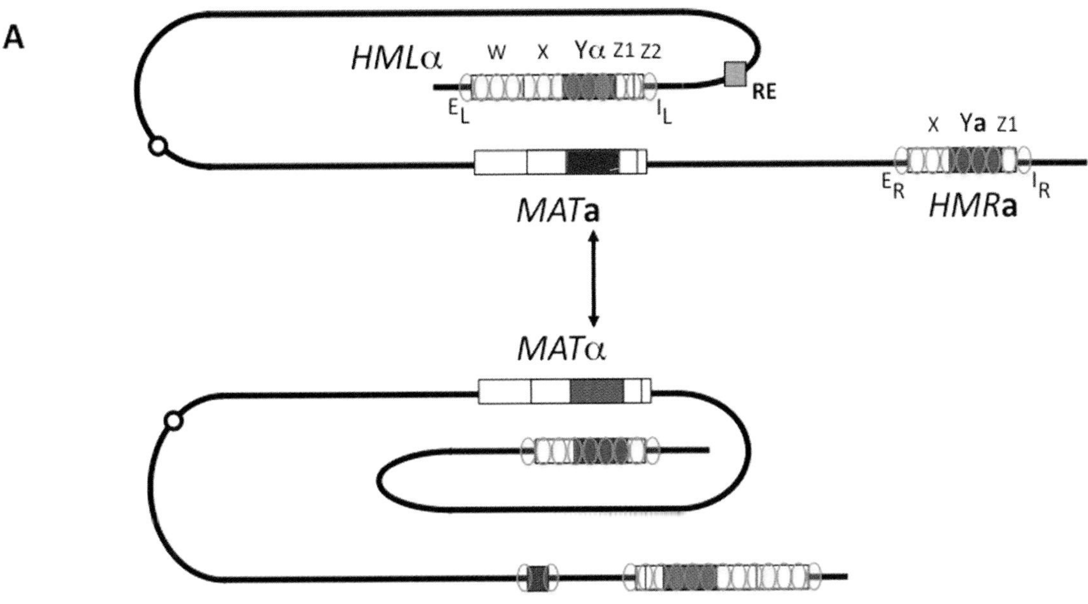

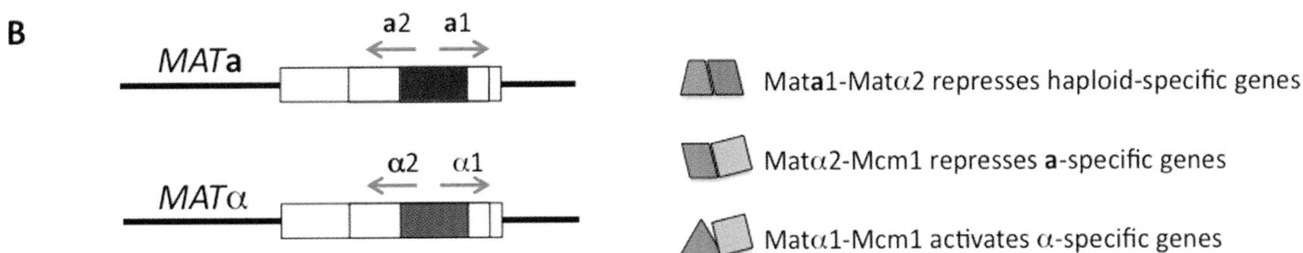

Figure 2 (A) Arrangement of *MAT*, *HML*, and *HMR* on chromosome 3. The gene conversion between *MAT***a** and *MAT*α is illustrated. Both *HML* and *HMR* could be transcribed but are silenced by the creation of short regions of heterochromatin (ordered nucleosomes are presented as blue circles) by the interaction of silencing proteins with flanking *cis*-acting silencer E and I sequences. The recombination enhancer (RE), located 17 kb centromere-proximal to *HML*, acts to promote the usage of *HML* as the donor in *MAT***a** cells. (B) Control of mating type-specific genes. Transcription of **a**- and α-regulatory genes at *MAT* occurs from a bidirectional promoter. The Mcm1 protein, in combination with Matα1 and Matα2, activates the transcription of α-specific genes or represses **a**-specific genes, respectively, while a Mata1–Matα2 repressor turns off haploid-specific genes. Mata2 has no known function. doi:10.1128/microbiolspec.MDNA3-0013-2014.f2

(29, 30, 31, 32), so that DNA double-strand breaks (DSBs) can only be repaired by HR. It is possible that NHEJ is repressed to prevent end joining in meiosis when more than 100 DSBs are created. Last, the expression of the *HO* endonuclease gene is suppressed by the **a**1–α2 repressor, once cells of opposite mating type conjugate to form a diploid.

The phenotypic switch from *MAT*α to *MAT***a** is quite rapid—within a single cell division. Hence, it is not surprising that Mata1, Matα1, and Matα2 transcription regulators are quite rapidly turned over, degraded by ubiquitin-mediated proteolysis by the proteasome (33, 34). In contrast, the **a**1–α2 corepressor is much more stable (24).

MAT switching depends on four phenomena: (i) a cell lineage pattern that ensures that only half of the cells in a population switch at any one time, to ensure that there will be cells of both mating types in close proximity; (ii) the presence of two unexpressed (silenced) copies of mating-type sequences that act as donors during *MAT* switching; (iii) the programmed creation of a site-specific DSB at *MAT* that results in the replacement of Ya or Yα sequences; and (iv) a remarkable mechanism that regulates the selective use of the two donors (donor preference).

CELL LINEAGE AND CONTROL OF *HO* GENE EXPRESSION

MAT switching provided a powerful early model to study the determination of cell lineage. Only half of the cells in a colony are able to switch mating type in any one cell division (Fig. 1). To be more specific, any haploid cell that has previously divided is capable of switching *MAT*, while new daughter cells cannot (35). Thus, a germinating *MAT***a** spore will produce a *MAT***a** daughter, and then in the next generation the original mother cell will switch to *MAT*α along with its second daughter, while the first daughter (and its daughter) remain *MAT***a**. The axial budding pattern of haploids places two *MAT***a** cells immediately adjacent to two *MAT*α cells and they readily conjugate, forming *MAT***a**/*MAT*α diploids in which the *HO* endonuclease gene is turned off so that further mating-type switching is repressed.

Nasmyth's laboratory (36, 37, 38) first demonstrated that the control of this lineage pattern depends on the asymmetric expression of the *HO* endonuclease gene, which is restricted to mother cells that have divided at least once. Control of *HO* expression depends on the localization of Ash1 mRNA to the daughter cell prior to cell division (39). The Ash1 repressor protein, localized only in the daughter cell, suppresses the expression of Swi5 transcription factor (Fig. 1) (40, 41), which is required for HO expression (36), thus restricting *HO* expression to the mother cell in the next G1 stage of the cell cycle. Since Ash1 mRNA localization was first discovered, more than 20 other mRNAs have been shown to show similar localization in *Saccharomyces* (42, 43).

In addition to the spatial control, *HO* transcription is confined to a narrow window in the cell cycle, after the cell passes "start" in the G1 stage of the cell cycle. The cell division kinase, Cdk1 (Cdc28), is required for *HO* expression and becomes active only after the start. Cdc28 combines with G1 cyclins to activate the Swi4–Swi6 transcription factor required for transcription of the *HO* gene to initiate *MAT* switching. In addition, the HO endonuclease protein is quite unstable, with a half-life of only 10 min, so that mother cells experience a brief pulse of endonuclease activity. The HO endonuclease is rapidly degraded by the ubiquitin-mediated SCF protein degradation complex (44, 45).

HO is a member of the LAGLIDADG family of site-specific endonucleases (reviewed in 46); it recognizes a degenerate 24-bp sequence that spans the *MAT*-Y/Z border (47, 48). A haploid yeast has three possible targets for HO: the *MAT* locus, *HML*α, and *HMR***a**, but only the *MAT* locus is accessible under normal conditions. Thus, combining all these controls, there is a single programmed DSB inflicted at the *MAT* locus only in mother cells and prior to the initiation of DNA replication.

SILENCING OF *HML* AND *HMR*: A MODEL FOR GENE SILENCING BY HETEROCHROMATIN

The presence of intact but unexpressed copies of mating-type genes at *HML* and *HMR* implied that these two loci have to be maintained in an unusual, silent configuration. Both *HML* and *HMR* are surrounded by a pair of related but distinct silencer sequences, designated *HML*-E, *HML*-I, *HMR*-E, and *HMR*-I (Fig. 3). Each of these elements contains several binding domains for different regulators, including the origin recognition complex proteins (ORC), the Rap1 transcription factor, and the transcription factor Abf1 (49). These *cis*-acting elements interact, directly or indirectly, with several *trans*-acting factors to repress transcription of these genes. The most critical roles in this process are played by four silent information regulator (Sir) proteins, discussed below. Together, these gene products and *cis*-acting sequences create short regions (~3 kb) of heterochromatin, in which the DNA sequences of *HML*

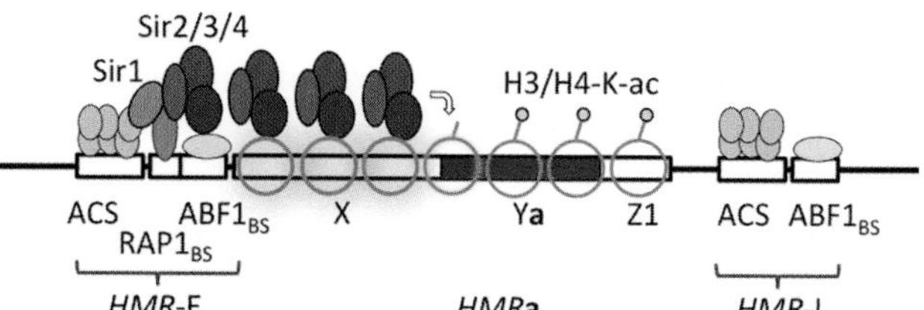

Figure 3 Silencing of *HMR* and *HML*. The processive process of silencing establishment from *HMR*-E is illustrated. Proteins (ORC proteins, Rap1, and Abf1, all in gray) bound to the three elements [autonomously replicating sequence consensus sequence (ACS), Rap1-binding site, and Abf1-binding site] of the *HMR*-E silencer recruit Sir1 that in turn recruits the Sir2–Sir3–Sir4 complex. The NAD$^+$- dependent HDAC Sir2 deacetylates lysines on the N-terminal tails of histones H3 and H4, which allows the Sir3–Sir4 complex to bind and stabilize the position of the nucleosome. Sir2 can then deactylate the next nucleosome and silencing spreads further. The progressing spread of silencing in the simplified figure is shown only in one direction and from one of the two silencers. doi:10.1128/microbiolspec.MDNA3-0013-2014.f3

and *HMR* are found as a highly ordered nucleosome structure (7, 8, 9). These heterochromatic regions are transcriptionally silent for both PolII- and PolIII-transcribed genes (10, 11) and resistant to cleavage by several endogenously expressed endonucleases, including the HO endonuclease (12, 13). Silencing is weakened if the distance between the two silencer sequences is increased. The reader is encouraged to consult more detailed reviews of *HML* and *HMR* silencing for more details (13, 50, 51, 52, 53, 54, 55, 56, 57, 58).

In addition to the action of the Sir proteins and *cis*-acting silencers, the extent of silencing appears to be strengthened by the location of both *HMR* and *HML* relatively near chromosome ends (telomeres) that also exhibit gene silencing (59, 60, 61, 62). There are some important differences in the silencing of *HML* and *HMR*. *HMR* appears to be more silenced than *HML*, because several mutations, including the histone H4-K16N mutation, which strongly affect *HML* silencing have little effect on *HMR* silencing (63) (J. E. Haber, unpublished data). Moreover, whereas overexpression of HO endonuclease will create some cleavage at *HML*, *HMR* is much more resistant to such cutting (12, 64). Interestingly, if both *HML* and *HMR* are inserted ~60 kb from a telomere on chromosome 6, both *HML* and *HMR* show equivalent silencing defects with a histone H4-K16N mutation (65), indicating that there are "booster" sequences in the vicinity of *HMR* that make it much more silent. Furthermore, silencing of *HMR* requires passage through S phase (but not replication per se), while "substantial silencing of *HML*α" could be established without passage through S phase (66). The spread of silencing to adjacent regions is restricted by several boundary elements (67, 68, 69).

The establishment and maintenance of silencing requires four Sir proteins. Mutations in these genes lead to coexpression of both *HML*α and *HMR*a in a haploid cell, thus producing a nonmating phenotype normally seen in *MAT*a/*MAT*α dipoloids (35, 70, 71, 72). A deletion of any of these—*SIR2*, *SIR3*, and *SIR4*—completely abolished silencing, while loss of *SIR1* had a less extreme phenotype. Sir1 interacts with the ORC complex at silencers to recruit the rest of the Sir complex. In the absence of Sir1, silencing—already established—can be epigenetically maintained for several generations without it (73). The keystone of these silencing proteins is the NAD$^+$-dependent Sir2 histone deacetylase, which is responsible for deacetylating a number of lysines on the N-terminal tails of histones H3 and H4 (74). Sir3 protein directly binds to various domains on nucleosomes, including H4-K16 and H3-K79 (75). Methylation of histone H3-K79 by the Dot1 methyltransferase correlates with the unsilenced chromatin state (76). Sir3 also binds with Sir4 (77), which in turn interacts with yKu70 (78), which is important in telomere associations with the nuclear periphery (79). Regulation of acetylation of the N-terminal tails of histones H3 and H4 is directly implicated in silencing, first by mutations that replace the four evolutionarily conserved lysine residues (59, 80, 81, 82, 83). More direct evidence came from the fractionation of chromatin in terms of the state of acetylation of histone H4-K16 (84), showing that *HML* and *HMR* are preferentially recovered in the hypoacetylated fraction. Deletion of the first 32 amino acids of the N-terminal tail of histone H3 unsilences *HML* but not *HMR*, whereas deletion of the N-terminal 16 amino acids of H4 unsilences both loci (85). A number of other proteins also contribute to silencing, including the acetyltransferase Esa1 and Sas2 protein (86).

It should be noted that silencing also occurs adjacent to yeast telomeres, and many of the genes involved in *HML*/*HMR* gene silencing also play a role in telomeric silencing (reviewed in 13, 50, 51, 53, 54, 55, 56, 57, 58, 87). At telomeres, there are no specific silencer sequences, but the telomere-associated Rap1 protein interacts with both Sir3 and Sir4 (88, 89, 90). Moreover telomere termini are also enriched in yKu70–yKu80, which also recruit Sir4 (91, 92). Telomeric silencing can extend for more than 10 kb (93), but strong silencing is confined to the first 1–2 kb (94). Several mutations that strongly affect telomeric silencing (e.g., *yku70*Δ) have either no effect on *HM* loci or have an effect only with a partially debilitated *HMR*-E sequence, suggesting that *HM* loci are more strongly silenced than telomeres (88, 89, 95).

FIRST MODELS OF *MAT* SWITCHING

Early studies of *MAT* switching recognized the existence of two additional key loci that were required for the replacement of *MAT* alleles: *HML* and *HMR* (96, 97, 98). A remarkably insightful hypothesis by Oshima and Takano (1971) suggested that these loci were the seat of controlling elements that could transpose to *MAT* and activate opposite mating-type alleles (99). Coupled with the key experiments of Hawthorne (1963), these ideas led Herskowitz' laboratory (100, 101) to suggest a specific version of the transposition model known as the "cassette model" in which an unexpressed copy of Yα sequences was located at *HML* (*HML*α) and unexpressed Ya sequences were found at *HMR*a. These sequences could be transposed to the *MAT* locus, where they would be expressed. In these early models, there was no suggestion that *MAT* switching involved homologous recombination; rather, a site-specific duplicative transposition was imagined. Subsequent studies (102, 103, 104, 105) confirmed that there were indeed two additional copies of mating-type information at *HML* and *HMR*. Most laboratory strains carry *HML*α and *HMR***a**, but natural variants exist that carry the opposite configuration: *HML***a** and *HMR*α (96, 106, 107). One early surprise in the molecular analysis of *MAT*, *HML*, and *HMR* was that the two donor cassettes did not carry simply the Y**a** and Yα donor sequences that could be "played" in the cassette player of the *MAT* locus, but that they were in fact intact, complete copies of mating-type genes carrying their own bidirectional promoters (Fig. 2). However, these genes were not transcribed. The two unexpressed cassettes differ in the extent of homology they share with *MAT*. *HMR*, *HML*, and *MAT* all share two regions flanking the Y sequences, termed X and Z1. *HML* and *MAT* share additional sequences, termed W and Z2 (Fig. 2).

We now know that during switching there is no change in either donor sequence; that is, *MAT* switching does not involve a reciprocal exchange of Y**a** and Yα sequences but rather a copying of the sequences from either *HML*α or *HMR***a** to replace the original *MAT* allele (108). This asymmetric recombination event is termed gene conversion. The idea that *HML* and *HMR* could repeatedly serve as donors during *MAT* switching provided an explanation for an early observation of Hawthorne (1963) that a mutant *MAT*α cell could be replaced by *MAT***a**, which then switched to a wild-type *MAT*α allele. Subsequent "healing" and "wounding" experiments were carried out in which mutations at *MAT* were corrected by recombination with the donor or in which a mutation at the donor was introduced into the *MAT* locus (100, 109, 110). In some cases, the replacement of *MAT* information included not only the Y region but at least part of the flanking X and Z1 regions as well that were shared by *MAT* and its two donors (110, 111).

In the 37 years since the cassette model was articulated, the *MAT* switching system and other HO-induced DSBs have been the object of intense study, to learn both about gene silencing and about the multiple mechanisms of DSB repair by homologous recombination, nonhomologous end joining, and new telomere addition (112, 113, 114, 115, 116, 117, 118, 119, 120, 121, 122, 123, 124, 125). Some related studies have been done by inserting small HO endonuclease recognition sites at other locations and from the induction of other site-specific endonucleases, most notably I-SceI (119, 123, 126, 127, 128, 129, 130, 131, 132, 133, 134, 135, 136, 137). Additional information about DSB repair has been gleaned from the analysis of DSB-induced recombination in meiotic cells (reviewed in 138, 139, 140, 141). By and large, the results are sequence independent; however, some interesting aspects particular to *MAT* switching are noted below.

MAT SWITCHING: A MODEL FOR HOMOLOGOUS RECOMBINATION

The conversion of one mating type to the other involves the replacement at the *MAT* locus of Y**a** or Yα by a gene conversion induced by an HO endonuclease-mediated DSB (142, 143). The process is highly directional, in that the sequences at *MAT* are replaced by copying new sequences from either *HML*α or *HMR***a**, while the two donor loci remain unchanged. Directional gene conversion reflects the fact that the HO endonuclease cannot cleave its recognition sequence at either *HML* or *HMR*, as these sites are apparently occluded by nucleosomes in silenced DNA. Thus, the *MAT* locus is cleaved and becomes the recipient in this gene conversion process. In cells deleted for *SIR2*, *SIR3*, or *SIR4*, where *HML* or *HMR* is unsilenced, HO can readily cut these loci and they become recipients (144, 145).

Normally, the *HO* gene is tightly regulated to be expressed only in haploid mother cells and only at the G1 stage of the cell cycle (36); however, the creation of a galactose-inducible *HO* gene made it possible to express *HO* at all stages of the cell cycle and in all cells (146). This made it possible to deliver a DSB to all cells simultaneously and to follow the appearance of intermediates and final products by physical analysis of DNA extracted at times after *HO* induction (12, 147, 148). An example of Southern blot analysis of *MAT***a** switching to *MAT*α is shown in Fig. 4. Physical monitoring of recombination at *MAT* has yielded much of what we know about DSB-induced mitotic recombination (reviewed in 3, 113, 115, 149, 150).

The overall process of *MAT***a** switching to *MAT*α is illustrated in Fig. 5. Following HO cleavage of *MAT***a**, the ends are resected in a 5′-to-3′ direction, creating a 3′-ended single-stranded DNA (ssDNA) tail that assembles a filament of the Rad51 recombinase protein. This protein–DNA complex engages in a search for a homologous sequence (in this case *HML*α) with which repair can be effected. The homology search culminates in strand exchange in which the ssDNA base pairs with the complementary sequence in the donor, creating a displacement loop, or D-loop. The 3′ end of the invading

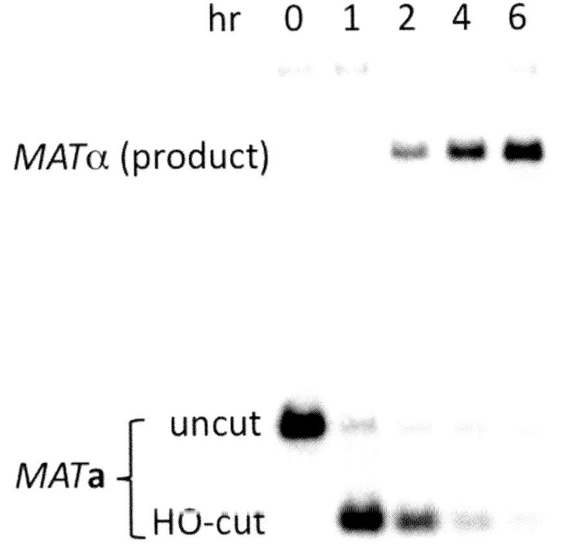

Figure 4 Physical monitoring of *MAT* switching. Southern blot analysis of *Sty*I-digested DNA after galactose induction of the HO endonuclease. The probe detects sequences just distal to *MAT*- Z1/Z2 and shows a difference in the size of the *Sty*I restriction fragments of *MAT*a and *MAT*α. In this experiment, an ade3::GAL::HO strain carrying *HML*α *MAT*a*hmr*Δ was used. The cleavage of *MAT*a into a smaller HO-cut segment was followed by the appearance of the *MAT*α product. (Figure modified from 149).
doi:10.1128/microbiolspec.MDNA3-0013-2014.f4

strand is then used as a primer to initiate copying of one strand of the donor locus, and the newly copied strand is displaced until it can anneal with homologous sequences on the opposite end of the DSB. The 3′-ended nonhomologous tail of the Y region is clipped off and the new 3′ end is used to prime a second strand of DNA synthesis, completing the replacement of *MAT*a by *MAT*α. This process is known as the synthesis-dependent strand-annealing (SDSA) mechanism, in which all the newly copied DNA is found at the recipient locus while the donor remains unaltered (151). Each of these steps is discussed in more detail below.

HO Cleavage

The HO endonuclease cleaves a degenerate recognition site of 24 bp *in vitro* (47), although sites of 117 bp down to 33 bp are generally used when the HO recognition site is inserted at other locations. A site with only 21 bp results in inefficient single-strand nicking that, by replication, can be converted to a DSB (152). Single base-pair mutations in the recognition site, such as *MAT*-inc (inconvertible) or *MAT*-stk (stuck), abolish or greatly reduce switching (153, 154). HO cutting generates 4 bp, 3′-overhanging ends (143). *In vivo*, the DSB is processed by several 5′-to-3′ exonucleases to create long 3′-ended tails (148). As discussed more fully below, the 3′ end is remarkably resistant to exonucleolytic removal. There do not seem to be any 3′-to-5′ exonucleases that act on a 3′-overhanging end, but protection of the ssDNA tail depends on binding of the ssDNA-binding protein complex, RPA (155).

MAT switching, induced by a galactose-regulated HO endonuclease, is a surprisingly slow process, requiring more than 1 h to complete, independent of the time during the cell cycle (12, 147, 148). It is possible that *MAT* switching may be more rapid when HO is expressed from its normal promoter in G1, only in mother cells, and in a coordinated fashion with other genes (142). However, additional experiments using HO whose expression is restricted to the G1 phase of the cell cycle (36) showed similar slow kinetics, as did expressing a conditional allele of HO under normal cell-cycle control (M. Yamaguchi, M. Gartenberg, and J. E. Haber, unpublished data).

5′-to-3′ Resection of the DSB Ends

5′-to-3′ resection proceeds rather slowly, at a rate of ~4 kb h^{-1} (~1 nucleotide s^{-1}) (156, 157). In a strain where there is no repair of the DSB, resection will continue at roughly this rate for more than 24 h, while if the DSB is being repaired, the slowly proceeding resection proteins most likely are displaced by the repair DNA synthesis machinery.

It is possible that resection is downregulated once recombination intermediates are established. Southern blot analysis of the rate of 5′-to-3′ degradation first implicated the trio of interacting proteins, Mre11, Rad50, and Xrs2 (MRX complex) (158, 159, 160). MRX somehow associates with Sae2, although a direct interaction has not been demonstrated, and together these proteins appear to do the initial 5′-to-3′ resection at HO-induced DSB ends. Mre11 has 3′-to-5′ exonuclease activity, and both Mre11 and Sae2 have endonuclease activity (161, 162, 163); however, mutation of the nuclease motifs of Mre11 has only a modest effect on resection (161, 162, 164). The nuclease activity of Mre11 is required in mitotic cells to cleave hairpin ends (165, 166) and in meiosis to remove the Spo11 protein from DSB ends (but in both of these cases, Sae2 is also necessary). Deleting Sae2 more significantly retards resection (167). However, while MRX–Sae2 appears to get resection started, extensive resection depends on two competing pathways of resection, one comprising the 5′-to-3′ exonuclease Exo1 and the other consisting of the Sgs1–Top3–Rmi1 (STR) helicase complex coupled to the nuclease function of Dna2 (which itself has nuclease

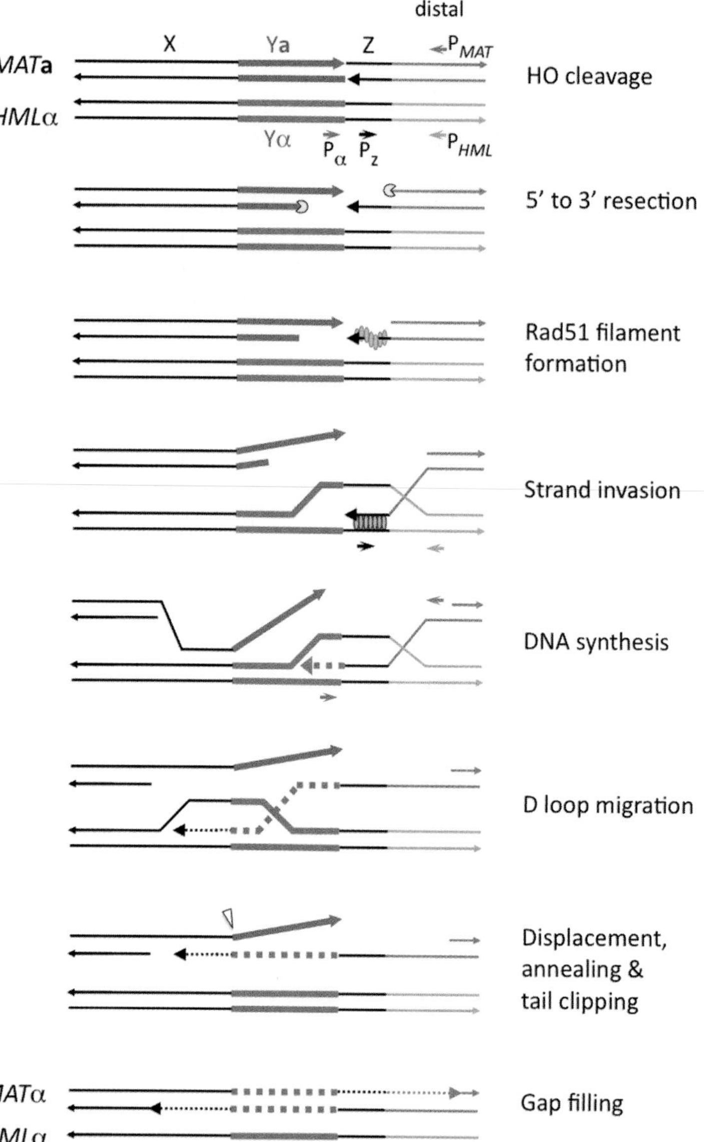

Figure 5 Mechanism of *MAT* switching. Key steps in the switching of *MAT*a to *MAT*α by a synthesis-dependent strand-annealing (SDSA) mechanism. An HO-induced DSB is resected by 5′-to-3′ exonucleases or helicase endonucleases to produce a 3′-end single-stranded DNA (ssDNA) tail, on which assembles a Rad51 filament (shown only on one side of the DSB). The Rad51::ssDNA complex engages in a search for homology. Strand invasion of *MAT*-Z into the homologous *HML*-Z can be detected by anti-Rad51 chromatin immunoprecipitation followed by quantitative PCR using the primer pair P_z and P_{HML}. Strand invasion can form an interwound (plectonemic) joint molecule (D-loop) that can assemble DNA replication factors to copy the Yα sequences, which can be monitored by a primer extension assay using the primer pair P_α and P_{MAT}. The D-loop is thought to migrate as DNA synthesis proceeds. Unlike normal replication, the newly copied strand is postulated to dissociate from the template and, when sufficiently extended, anneal with the second end, still blocked from forming a plectonemic structure by the long nonhomologous Ya tail. The single-stranded tail is clipped off once strand annealing occurs by the Rad1– Rad10 flap endonuclease, so that the new 3′ end can also be used as a primer to fill in the gap. Consequently, all newly synthesized DNA is found at the *MAT* locus, while the donor is unaltered. However, a small fraction of DSB repair events apparently proceed by a different repair mechanism involving the formation of a double Holliday junction. doi:10.1128/microbiolspec.MDNA3-0013-2014.f5

activity that is not relevant for this process) (157, 168, 169, 170). The activity of Exo1 is impaired by the binding of yKu70 and yKu80 proteins, which are important in NHEJ (171, 172).

When both Sgs1 and Exo1 are deleted, there is very limited resection, which appears to depend on MRX and Sae2. This result suggests that MRX acts first and hands off resection to Exo1 or the STR–Dna2 complex, but this idea is contradicted by the long-observed fact that deletion of MRX proteins only reduces resection by about 50% in cycling cells. However, in G2-arrested cells, deletion of Rad50 eliminates nearly all resection (173). These observations suggest that there must be an alternative resection activity that is absent in G2-arrested cells. Moreover, in G1 cells prior to the activation of Cdk1 at the start point of the cell cycle, there is nearly no resection (174, 175). Inhibition of Cdk1 at other points in the cell cycle also blocks resection (174, 175). This control is exerted through Cdk1's phosphorylation of both Sae2 and Dna2 (168, 176). Taken together, it seems that the modest inhibition of resection in cycling cells by deleting MRX or Sae2 argues that MRX–Sae2 does not act as the obligate gatekeeper of resection.

Resection of course must plow through chromatin, and it is not yet clear how these complexes accomplish the necessary chromatin remodeling. Deleting the Arp8 subunit of the Ino80 chromatin-remodeling complex has—in some hands—a modest effect on resection. However, a much more profound inhibition of resection is seen when the Swi2/Snf2 homolog, Fun30, is deleted (177, 178, 179). Fun30 is an ATPase that has been shown to displace a positioned nucleosome *in vitro* (180, 181). Resection of HO-cut ends is also reduced by the phosphorylation of histone H2A (called γ-H2AX) by the Mec1 and Tel1 checkpoint kinases (homologs of mammalian ATR and ATM). Interestingly, Fun30 associates less strongly *in vitro* with nucleosomes carrying H2A-S129E, mimicking γ-H2AX (179).

Recruitment of Rad51 Recombinase and the Search for Homology

As long 3′ tails are generated, they are bound by the Rad51 recombination protein, which facilitates a search for homologous regions, to initiate recombination (Fig. 5). Chromatin immunoprecipitation (ChIP) experiments have shown that once ssDNA is generated, it is first bound by the ssDNA-binding protein complex, RPA, which is then displaced by Rad51 (182). The loading of Rad51 depends completely on the Rad52 protein (183, 184). In the absence of the Rad55 and Rad57 proteins, which are known as Rad51 paralogs, Rad51 filament assembly is slow and apparently incomplete, and *MAT* switching fails to occur. In other DSB-mediated repair events where the donor is not silenced, recombination also fails in the absence of Rad55 or Rad57 (183, 184).

ChIP, using an anti-Rad51 antibody, allows one to visualize the kinetics of Rad51 loading onto ssDNA. The same approach permits visualization of the synapsis between the *MAT* DSB and the donor, as Rad51 will associate with both the invading *MAT* strand and the *HML*α duplex DNA. This step takes ~15 min after appearance of Rad51 assembly at the DSB (149, 183, 184). The time to achieve pairing between *MAT* and *HML* has also been seen microscopically by examining green fluorescent protein (GFP)-tagged LacO and TetR arrays situated close to *HML* and *MAT*, respectively (185, 186). It should be noted that the time to pair with donor sequences located interchromosomally is significantly longer than what occurs between *HML* and *MAT* (187). The relatively rapid encounter between these two loci is undoubtedly aided by the *cis*-acting recombination enhancer (RE), located ~17 kb centromere-proximal to *HML*, which will be discussed in detail below. It is striking that the amount of homology shared by *MAT* and its donors is quite small, especially on the Z side, which seems to initiate copying of the donor. *MAT* and *HMR* only share 230 bp, while *MAT* and *HML* share 327 bp. In contrast, there is much more extensive homology on the W/X side, but this lies beyond the nonhomologous Y sequences, so this DSB end cannot directly initiate new DNA synthesis. Thus, the efficiency of repair is dictated by the smaller Z side. We will discuss the mechanism of donor preference in detail below, but suffice it to say that one can set up an experiment in which *MAT* will normally switch with *HMR* as a partner and *HML* is the "wrong donor." By artificially increasing the size of the homology on the Z side of *HML* from 327 to 650 to 1800 bp, one can significantly increase its use as a donor in competition with *HMR* (188). It has been estimated that encounters between the Rad51 filament and the preferred donor will happen on average four times before some irreversible step will lead to the completion of recombination (188). The location of a donor in the genome also seems to play an important role in homology searching. Donors located in the regions showing higher contact probability, determined by genome-wide chromosome configuration capture, are used more efficiently to repair a DSB on a different chromosome (C.-S. Lee, R. Wang, and J. E. Haber, unpublished data). Once resection exposes both the X and Z homologous regions flanking the DSB, they can each synapse with their homologous sites in *HML* (149). However, for the left end of the DSB to act as a primer, the nonhomologous Ya tail has to be removed. The nonhomologous tail is clipped off by the Rad1–Rad10 endonuclease (189, 190). In other HO-induced events (and probably in *MAT* switching), removal of the nonhomologous tail also requires a number of other proteins, including Msh2–Msh3, Saw1, and Slx4 (191, 192). This "clippase" acts apparently only after annealing of the resected end with the newly copied complementary strand to create a branched, annealed structure with a 3′-ended tail. This step also requires the action of either of the two DNA damage-responsive protein kinases, Mec1 or Tel1, to phosphorylate Slx4 (193). Although a branched structure with the nonhomologous tail might form when *MAT*-X synapses with the donor, tail clipping preferentially occurs later, apparently after the annealing of the elongated first strand in the *MAT*-X region (149).

During strand invasion, any mismatch in the synapse could be corrected by mismatch repair machinery. McGill *et al.* (111) used artificial restriction sites inserted at different places in the X and Z regions and showed a highly biased transfer of markers from the donor to *MAT*. In the absence of mismatch repair, a single base-pair mutation only 8 bp from the 3′ end of the HO cut, in the Z region, was most often retained during switching (154). A kinetic analysis further demonstrated that mismatch correction occurred soon after the strand invaded the donor locus (194). That mutant sequence in the invading Z DNA was corrected to the genotype of the donor is the most direct *in vivo* demonstration of the idea that mismatch repair will preferentially correct a mismatch adjacent to a nick (in this case, the 3′ end of an invading strand) (195, 196).

In addition to anti-Rad51 ChIP, strand invasion has been examined with a polymerase chain reaction (PCR)-based nucleosome protection assay (149). As noted before, *HML* has very highly positioned nucleosomes, whereas there is little order to nucleosomes at *MAT*. At the time of synapsis, there is a transient reduction in protection of nucleosomes at the site of strand invasion. By delaying the initiation of new DNA synthesis, it was possible to detect the change more reliably. This study produced two interesting results. First, the region of reduced nucleosome protection extended to several nucleosomes in the nonhomologous Y region prior to the initiation of new DNA synthesis, suggesting that the D-loop is extended beyond the 3′ end of the invading strand, perhaps by helicases. It has not been determined yet whether the extended and apparently open region would bind RPA. Interestingly, a mutation of the largest subunit of RPA, *rfa1-t11* (K45E), has normal Rad51 filament formation but fails to engage in synapsis with donors (182). This result could mean that RPA is needed to stabilize the D-loop.

The other important finding was that nucleosome rearrangement at *HML* did not occur in a *rad54Δ* mutant. Rad54 is a Swi2/Snf2 homolog that has been shown to engage in chromatin remodeling *in vitro* (197). Deletion of Rad54 did not abolish strand invasion as measured by the association of Rad51 with the donor (183), but this association apparently is distinct from the full chromatin rearrangement necessary to complete DSB repair. Without Rad54, there is no primer extension and new DNA synthesis. One possible explanation for this result is that the kind of association of *MAT* and *HML* strands in the absence of Rad54 is a *paranemic* joint in which the invading strand does not intertwine with the donor duplex, whereas with Rad54 a *plectonemic*, interwound structure is formed with the displaced strand in an extended D-loop (Fig. 5).

Copying the Donor Sequences

The initiation of new DNA synthesis can be detected by a PCR-based primer extension assay, using two primers— one complementary to sequences distal to *MAT* and one in the Yα region of the donor (148). PCR amplification can take place only after the invasion of the 3′-ended single strand into the Z region of the donor locus and the copying of at least 50 nt, thus creating a recombination intermediate that covers both primers. This step occurs ~15–20 min after synapsis between *MAT* and *HML* is observed by ChIP and 15–30 min prior to the completion of gene conversion, as monitored by both Southern blotting and a second PCR assay, detecting the time when the donor Y sequences are joined to the proximal side of *MAT* (148, 149).

Physical analysis has also made it possible to analyze conditional lethal mutants to ask which DNA replication enzymes are required for *MAT* switching. Unlike break-induced replication, where only one end of a DSB can invade homologous sequences (reviewed in 198), only a fraction of the proteins necessary for origin-dependent DNA replication are also required for *MAT* switching (136). In this process, which appears to involve the elongation of one strand and then the elongation of the second strand, primase and DNA polymerase α, the lagging strand DNA polymerase, are not required, while both DNA polymerases δ and ε appear to act either sequentially or redundantly (199). The proliferating cell nuclear antigen (PCNA) clamp is required, but the GINS–Cdc45–Mcm helicase complex is dispensable (199). DNA synthesis during *MAT* switching does not need most of the loading factors required at an origin for normal replication, including the ORC proteins, Cdt1 and Cdc6 (200). However, the Dpb11–Sld2–Sld3 proteins are required (149). These proteins have been shown to be part of a preloading complex at origins (201), but how they would work when DNA copying is not dependent on an autonomously replicating sequence or on other early-functioning proteins is unknown. Three other mutations that impair break-induced replication do not block *MAT* switching or other gene conversion events: (i) deletion of the nonessential DNA polymerase δ subunit, Pol32; (ii) inactivation of the Pif1 helicase; and (iii) the dominant *pol30-FF248, 249AA* mutation of PCNA (119, 136, 202, 203).

The fact that DNA synthesis during gene conversion does not use all the processivity factors employed in normal replication may explain why gene conversion is much more susceptible to mutation of the replicated sequences. Taking advantage of several features of *MAT* switching, it was possible to select for mutations that arose during gene conversion (204). The majority of mutations were base-pair substitutions, but the rest (~40%) represented various types of template jumps, as if the DNA polymerase was less processive than would be found during normal replication. Surprisingly, all types of mutation representing template jumps were eliminated in a strain with a proofreading defect in DNA polymerase δ. We surmised, on the basis of some *in vitro* studies of a similar proofreading mutation (205), that the proofreading-defective mutant enzyme is in fact less prone to dissociate from the template. This result argues strongly that DNA polymerase δ is a major player in *MAT* switching. However, there was also evidence that DNA polymerase ε was active, since a proofreading-defective mutant of DNA polymerase ε resulted in the appearance of +1 frameshifts. The appearance of these mutations in gene conversion was apparently independent of the mismatch repair system and insensitive to the error-prone DNA polymerase ζ or another translesion DNA polymerase, DNA polymerase η. However, DNA polymerase ζ does seem to be principally responsible for mutations in ssDNA outside the regions of shared homology that must be filled in after the DSB is repaired (206, 207).

Direct evidence that *MAT* switching proceeds by an SDSA mechanism was obtained by Ira *et al.* (151), who used "heavy–light" density transfer methods to analyze the location of new DNA synthesis during *MAT* switching. These experiments confirmed that all the newly copied DNA is in the recipient *MAT* locus, while the donor remains unaltered.

Completion of Switching

One of the other striking aspects of *MAT* switching is that it is very rarely accompanied by crossing over, which produces lethal outcomes in *MAT* switching (208, 209). Crossovers are not expected when the SDSA mechanism is used, because there is no stable single or double Holliday junction that would be cleaved to produce crossovers. Moreover, such crossovers would be lethal between *MAT* and either *HML* or *HMR*. On the basis of ectopic recombination studies in which HO induces a gene conversion between *MAT***a** on chromosome 5 and an uncuttable *MAT***a**-inc allele on chromosome 3 (and where the normal donors are deleted), it seems that crossovers, which are viable in this scenario and can be easily scored, are prevented predominantly by the action of two helicases, Sgs1 (with its partners Top3 and Rmi1) and Mph1 (210, 211). The Sgs1 complex acts as a dissolvase to remove double Holliday junctions that would possibly become crossovers if removed by resolvases. Mph1 appears to ensure that the SDSA pathway is used rather than the alternative DSB repair mechanism that is much more prevalent in meiosis.

Finally, it is worth noting that *MAT* switching represents a case of gap repair rather than strictly break repair; that is, the regions of homology located by the two DSB ends are separated on the donor template by ~700 bp. In a related study, using HO-cleaved *LEU2* sequences, Jain *et al.* (123) found that break repair and gap repair, when the gap was larger than ~2 kb, are surprisingly different repair processes. Break repair occurs with relatively rapid kinetics, whereas there is a delay of hours in new DNA synthesis when there is a long gap. This delay is quite similar to that seen in break-induced replication, where only one end of the DSB is homologous to a template and repair can only occur by assembling a complete replication fork. Consequently, long gap repair and break-induced replication depend on Pol32 and are impaired by *pol30-FF248, 249AA*, while break repair (and *MAT* switching) is unimpaired. Apparently the two ends of the DSB need to be in contact with each other at the time of strand invasion; they must pair close enough to each other, and in the proper orientation, to permit some signal to be propagated. We termed the assessment of the nature of the strand invasion—the difference between break repair and gap repair—as a manifestation of a *recombination execution checkpoint*. *MAT* switching appears to have a small enough gap to be treated as break repair.

DONOR PREFERENCE

During *MAT* switching, there is an elaborate mechanism that gives yeast the ability to choose between its two donors. *MAT***a** cells use *HML*α as the template to repair the DSB ~85–90% of the time, while *MAT*α cells preferentially choose *HMR***a**. This preference makes sense in that it leads to a productive switch to the opposite mating type. However, donor selection is not dictated by the Y**a** or Yα content of the donors: a strain with reversed silent information (*HML***a** *MAT***a** *HMR*α) still chooses *HML*~85–90% of the time (212, 213). Weiler and Broach (213) showed that replacing the entire *HML* region including its silencers with a cloned *HMR* locus did not change donor preference, suggesting that the *location* of the donor, not the sequence differences between *HML* and *HMR*, directs donor selection. There must therefore be one or more *cis*-acting sequences, outside of the donors themselves, that activate or repress one or both donors, depending on mating type.

*MAT*α's choice of *HMR* over *HML* occurs independently of the *MAT*α1 gene but is strongly dependent on *MAT*α2, the gene that acts as a repressor of **a**-specific genes (100, 214, 215, 216, 217, 218). *MAT***a** donor preference does not depend on a functional *MAT***a**1 gene (219). *HMR* appears to be used as the default locus and all the active regulation is in making *HML* more accessible (Fig. 6). Thus, a *MAT***a** cell deleted for its preferred donor, *HML*, can easily use *HMR*, but some of *MAT*α cells die when their only choice of a donor is *HML* (216, 220, 221). Cells experiencing an unrepaired DSB become arrested at a G2/M checkpoint (222), which should theoretically have allowed cells time to locate a donor and repair the DSB by gene conversion, but as DSB ends are resected, other recombination events—for example, between several Ty retrotransposons located 30 kb proximal to *MAT* and Tys on other chromosomes—may lead to the death of cells (223).

Identification of a Recombination Enhancer

Wu and Haber (220) identified the key *cis*-acting element responsible for donor preference (now called the recombination enhancer or RE) by creating a series of truncations and internal deletions. A 2.5-kb deletion located 17 kb proximal to *HML* completely reversed donor preference, so that a *MAT***a** cells now used *HML* only 10% of the time instead of 90% (Fig. 6). Deletion of this sequence also abolished *MAT***a** donor preference for donors located 41, 62, and 92 kb from the left end. Deletion of this region had no effect on *MAT*α cells, which continued to use *HMR* most of the time.

Further mapping of RE was accomplished by inserting subfragments of the smallest deletion back into the chromosome. This led initially to the identification of a 700-bp RE that restored *MAT***a** donor preference almost to wild-type levels. Subsequent analysis has narrowed down the most important sequences to ~250 bp, although full activity resides in a region of ~400 bp. This further refinement was accomplished by showing that a syntenic region in *Saccharomyces carlsbergensis* and in *Saccharomyces bayanus* (but not in more distant

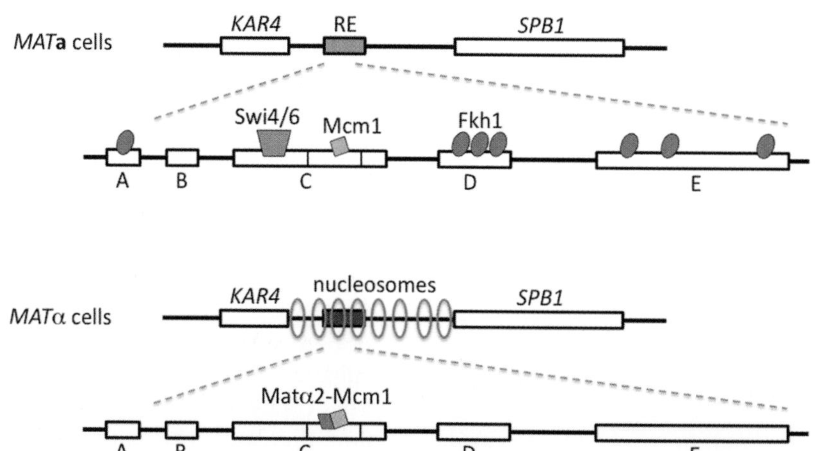

Figure 6 Protein binding to consensus elements in the RE. In *MAT***a** cells, Mcm1 binding facilitates the binding of Swi4–Swi6 and multiple copies of Fkh1. These proteins are important for the ability of RE in promoting *HML* usage in *MAT***a** cells. In *MAT*α cells, binding of the Matα2–Mcm1 repressor to a 31-bp conserved operator, shared by **a**-specific genes, leads to the formation of highly positioned nucleosomes between the two flanking genes and excludes binding of Fkh1 or Swi4–Swi6. doi:10.1128/microbiolspec.MDNA3-0013-2014.f6

species such as *Saccharomyces servazzii*) contained an active RE that would substitute for the *S. cerevisiae* RE (224, 225). By comparing the divergent sequences of these REs, Wu *et al.* (224) defined five well-conserved subdomains, named A–E (Fig. 6). Deletion analysis indicated that subdomains A, C, D, and E are all important for donor preference, while subdomain B is not. Further analysis revealed that each of the subdomains A, D, and E contains one or more binding sites for the Fkh1 transcription regulator (225). As we will see below, this protein plays a central role in the activity of RE. A key finding was that one could replace the entire RE with multimers of only subdomain A, D, or E to mimic RE activity. Most strikingly, four copies of the 21-bp subdomain A was sufficient to raise *HML* usage from ~10% in the absence of any RE to 65%. The activity of the 4×A construct depends on Fkh1 (225).

Role of a Mat2–Mcm1 Operator in Turning RE Off in *MAT* Cells

RE lies in perhaps the largest "empty" region of the yeast genome, ~2.5 kb with no open reading frames or regulatory sequences associated with the flanking genes, *KAR4* and *SPB1*. In *MAT*a, the region is "open" and binds a number of proteins, whereas in *MAT*α cells, highly positioned nucleosomes cover the entire 2.5-kp intergenic region harboring RE (224, 226). The 90-bp subdomain C, which has no Fkh1-binding site but is important for donor preference, harbors a conserved 31-bp consensus Matα2–Mcm1 binding site. In *MAT*α cells, α2–Mcm1 repressor binding, in conjunction with the corepressor Tup1, alters chromatin structure and thus turns off RE (217). Mutation of the Matα2-binding sites is sufficient to alter donor preference in *MAT*α, increasing *HML* usage from ~10 to ~50% (217, 224). RE contains no open reading frame, but indeed there are two "sterile" transcripts of the RE region that are transcribed in *MAT*a but not in *MAT*α (218). However, these two noncoding RNA transcripts are not important for RE activity (85, 220, 224). It is interesting to note that the 2.5-kb region between *KAR4* and *SPB1* contains a second α2–Mcm1 binding site centromere-proximal to the defined RE element. This second repressor site may have a role in repressing *HML* usage in *MAT*α cells.

Activation of RE in *MAT*a Cells Depends on the Mcm1 Protein

A 2-bp mutation that eliminates Mcm1 binding in the Matα2–Mcm1 operator sequence completely inactivates RE and reduces *HML* usage from ~90 to 10–20% in *MAT*a cells (224). Similarly, a single amino acid substitution mutation in *MCM1* (*mcm1-R89A*), which reduces Mcm1 binding, also showed a reduction in *HML* usage (224). Surprisingly, in *MAT*a cells, where there is no Matα2 protein, the mutant RE with the 2-bp operator mutation that prevents Mcm1 binding has an array of positioned nucleosomes that is similar to what is seen in normal *MAT*α cells (224). Apparently, other sequences within RE can, at least in *MAT*a cells, organize a phased nucleosome structure in the absence of Matα2–Mcm1 binding.

Clearly, Mcm1 binding is critical in the activation of the normal RE sequence, but it is also evident that, since multimers of domain A, D, or E (which lack Mcm1-binding sites) are sufficient to promote preferential use of *HML*, the importance of Mcm1 is in regulating other chromatin features of the normal RE.

RE Binds Fkh1 and Swi4–Swi6

Binding of Fkh1 to several domains of RE was confirmed by ChIP against a functional epitope-tagged Fkh1–haemagglutinin construct. Fkh1 bound only in *MAT*a, and preferentially in the G2/M stages of the cell cycle (227). Deletion of Fkh1 markedly reduced *HML* usage in *MAT*a from ~85 to ~35%, which has more *HML* usage than when RE was fully deleted. This result suggests that there must be other proteins involved in the action of the complete RE (227). In contrast, when a 4×A construct was used in place of the whole RE, deleting Fkh1 dropped *HML* usage to the same level as deleting RE (227). Further inspection revealed that subdomain C contains a Swi4–Swi6-regulated cell cycle box (SCB) that binds the G1-to-S-phase cell-cycle regulators Swi4–Swi6. When the SCB was mutated or if Swi6 was deleted, *HML* usage dropped, again to ~35% (227). Swi4–Swi6 and Fkh1 are not in the same pathway in the regulation of RE activity because a deletion of SCB, coupled with *fkh1*Δ, further reduced *HML* usage to ~15%.

Other Donor Preference Mutations

Two other *trans*-acting factors have been shown to play less decisive roles in donor preference. Screening directly for donor preference mutations has not been very productive, largely because mutations that affect HO expression and *cis*-acting mutations that reduce HO cleavage tend to interfere with the evaluation of donor choice scored at the colony level. Only *chl1*Δ has emerged in this way. Deletion of the *CHL1* gene reduces *HML* usage in *MAT*a cells from 80 to 60% but has no effect on *MAT*α (215). *CHL1* encodes a nuclear protein with putative helicase activity that is implicated in the establishment and maintenance of sister-chromatid cohesion (228). Deletion of *CHL1* causes a large increase in both the loss and gain of chromosomes by mitotic chromosome nondisjunction (229, 230). The other factors implicated in donor preference are the yKu70 and yKu80 proteins. The Ku complex plays many roles in chromosome architecture and in DSB repair. These proteins are required for the predominant mechanism of nonhomologous end joining that can rejoin the 4-bp 3′-overhanging DSB ends created by HO, but they are also critical in associating telomeres with the nuclear periphery and in ensuring the full activity of telomerase. When Ku proteins are deleted, telomeric regions are delocalized from the periphery, but, paradoxically, it seems that *HML* is more frequently associated with the nuclear periphery in a *yku70*Δ mutant (231). This might explain why, when either Ku protein is deleted, there is an ~10% reduction in *HML* usage in *MAT***a** but no discernible effect on *MAT*α. It seems that Chl1 and yKu70–yKu80 may all act in the Swi4–Swi6 pathway, as double mutants SCBΔ *chl1*Δ and SCBΔ *yku80*Δ continue to use *HML* ~30% of the time, whereas *fkh1*Δ *chl1*Δ and *fkh1*Δ *yku80*Δ both resemble *fkh1*Δ SCBΔ (227). Telomeres are also tethered by the interaction of the Sir4 silencing protein with Esc1 (232, 233); however, deletion of *ESC1* had no important defect in donor preference, and a *yku70*Δ *esc1*Δ double mutant also did not exhibit a significant perturbation of donor preference (E. Coïc, unpublished data). Hence, it is not clear that anchoring RE or *HML* to the nuclear periphery plays any meaningful role in this regulation.

RE Acts Over a Long Distance and is Portable

When *HML* was deleted and either *HML* or *HMR* was inserted at other chromosome 3 locations, the donor could be activated at several sites along the entire left arm in *MAT***a** cells, although the efficiency decreased as the donor was moved further from RE (219, 220). Conversely, RE itself can be moved to sites along the left arm and still stimulate the use of *HML* (234). RE can also stimulate *HML* usage when *MAT* is moved to a different chromosome (221). Moreover, RE also works with non-*MAT* sequences. In a competition assay, an HO-induced DSB in a *leu2* sequence on chromosome 5 can be repaired with a *LEU2* gene placed near RE or a second, competing donor located ~100-kb centromere-proximal on chromosome 5. The presence of RE increased the use of the adjacent interchromosomal *LEU2* donor from ~15 to ~50% (85). The increased usage of the RE-adjacent donor was reflected in a more rapid appearance of the association of the Rad51 recombination protein with the donor sequences, as measured by ChIP (85), suggesting that RE accelerates the search for homology and/or the conversion of an initial encounter between the DSB end and the donor into a stable recombination intermediate.

Further evidence that RE is portable has come from two additional experiments. First, insertion of a second copy of RE near *HMR* in *MAT***a** cells increased the usage of *HMR* from ~10 to ~50% (234). Secondly, when *MAT***a**, *HML*, *HMR*, and RE were moved from chromosome 3 to the larger chromosome 5 in roughly the same configuration, the use of *HML* increased from ~40 to ~90% with the help of RE (234).

RE Affects Other Recombination Events

The effect of RE is not restricted to recombination induced by HO endonuclease. If *HML* is replaced with an allele of the *leu2* gene and a second *leu2* allele was placed elsewhere on chromosome 3—or even on another chromosome—the rate of Leu$^+$ spontaneous recombination was ~10 times higher in *MAT***a** cells than in *MAT*α in an RE-dependent manner (216). There was no significant mating-type-dependent difference when a similar experiment was done with one *leu2* allele in place of *HMR*, thus supporting the conclusion that donor preference was effected through changes in the left arm of the chromosome, with *HMR* being a more passive, although efficient, participant.

How Does RE Work?

There is no obvious difference in the positioning of nucleosomes over *HML* in *MAT***a** versus *MAT*α strains (8), suggesting that RE does not make *HML* more accessible by altering the silencing of *HML*. Furthermore, the silencing of *HML* does not seem to interfere with the recombination with *MAT*, and donor preference is unchanged if the donor regions are unsilenced but contain alterations so that they cannot be cleaved by HO (188). Although *HML* and *HMR* have been shown to reside preferentially near the nuclear periphery, it appears that this tethering does not prevent the donors from engaging *MAT* in the center of the nucleus (231).

One attractive idea to explain the role of RE is that it changes the localization or the higher-order folding of the entire left arm of chromosome 3 to make *HML* more flexible and available in locating and pairing with the recipient site in *MAT***a** cells. Several approaches suggest that there are differences in chromosome arrangement in the two mating types, but these changes, in the absence of a DSB, do not seem to explain donor preference. First, chromosome segments near *MAT*, *HML*, and *HMR* can be fluorescently tagged by binding LacI–GFP or TetR–GFP (or some other color) to LacO or TetO arrays and thus the distance between these elements can be estimated. In one such study of fixed cells, however, *HML* was not closer to *MAT***a** than *HMR* (185). More detailed studies using fluorescent techniques (I. Lassadi and K. Bystricky, personal communication) or by sequencing-based chromosome conformation capture (termed 5C) (J. Dekker, personal communication) suggest that in *MAT*α cells the left arm of the chromosome is more extended and perhaps more associated with the nuclear periphery. If RE is deleted in *MAT*α cells the arm is more "crumpled" and more like the configuration seen in *MAT***a** cells. These data suggest that the inactive RE, bound by the Matα2–Mcm1 repressor, actually impairs left arm usage in *MAT*α cells, so that RE would positively affect usage in *MAT***a** but might also negatively affect *HML* usage in *MAT*α.

A Simple Model of the Mechanism Underlying the Fkh1-Regulated Donor Preference

When RE was replaced with four copies of the LexA operator, *HML* usage was very low, but the use of the left donor could be markedly increased by expressing a LexA–Fkh1 fusion protein (85). LexA–Fkh1 was then truncated so that it contained only the first 230 amino acids, comprising the forkhead-associated (FHA) domain. This construct was shown to completely restore *HML* usage (Fig. 7), whereas other fusions carrying the transcription regulatory domain near the C terminus had no activity. In addition, mutation of a conserved arginine residue, which is required for the phosphothreonine-binding activity of FHA domains, abolished LexA–FHA activity (Fig. 7). These results suggested that Fkh1 binds to RE in *MAT***a** cells and that its FHA domain physically interacts with proteins containing a phosphothreonine residue, around the DSB, and thus tethers *HML* close to *MAT*. Consistent with this model, ChIP experiments showed the association of LexA–FHA with *MAT* only after a DSB was induced, even in strains where *HML* itself was deleted (85). Whereas phosphorylation of histone H2A (γ-H2AX) spreads more than 50 kb on either side of the DSB, FHA binding is confined to only a few kilobases around the break. However, the binding partner of the FHA domain remains to be determined. Neither of the two DSB-dependent phosphorylations of histones, H2A-S129 or H4-S1, which spread around the *MAT* locus, is responsible for donor preference. Moreover, donor preference is not influenced in the mutants deleted for N-terminal tails of histones H3 and H4 or proteins known to associate with the DSB ends, including Mre11 and Sae2. As for the identity of the damage-dependent protein kinase that phosphorylates the threonine residue, casein kinase II but not the ATR and ATM homologs, Mec1 and Tel1, is involved in the donor preference regulation (85).

A simple model to explain donor preference is presented in Fig. 8. The cluster of Fkh1 proteins bound at RE in *MAT***a** (but not *MAT*α) comes into contact with DSB-induced, casein kinase II-dependent phosphorylated threonines in proteins that bind near the DSB ends. This association effectively tethers the nearby *HML* locus close to *MAT* and facilitates its use in *MAT* switching. This tethering brings *HML* from its location 200 kb from *MAT* to within about 20 kb of the DSB, while *HMR* remains 100-kb distant, and can account for the strong preferential use of this donor. It is still not clear whether there are any other constraints preventing *HML* use in *MAT*α cells or whether there are any facilitating sequences that aid *HMR* usage. Of course, this is hardly the entire story since RE also binds Swi4–Swi6 in domain C, in a cell-cycle-dependent fashion. The activity of the entire RE is likely to be substantially more elaborate.

UNANSWERED QUESTIONS

Although *MAT* switching is probably the best-understood example of repair of a DSB by homologous recombination, there are still many aspects of the process that are not well understood. Recent studies have shown that a broken chromosome explores a larger fraction of the nuclear volume (235, 236), but how this affects repair is still not clear. The manner in which the Rad51 filament locates its partner is also a matter of active investigation (237, 238). The roles of several chromatin remodelers in gaining access to heterochromatic donor sequences also remain to be elucidated, and the overlapping roles of several DNA polymerases still need to be untangled. Indeed, recent evidence has shown that the repair process itself is surprisingly mutagenic (204), suggesting that the repair DNA synthesis process is quite different from the normally processive and accurate chromosomal replication. Finally, although it is evident that RE brings *HML* into proximity of the DSB, the relevant target sites (phosphorylated residues on proteins near the DSB) have yet to be identified. As some of these questions are answered, there will be new issues raised by the continuing advances in our ability to study molecular events *in vivo* in ever greater resolution.

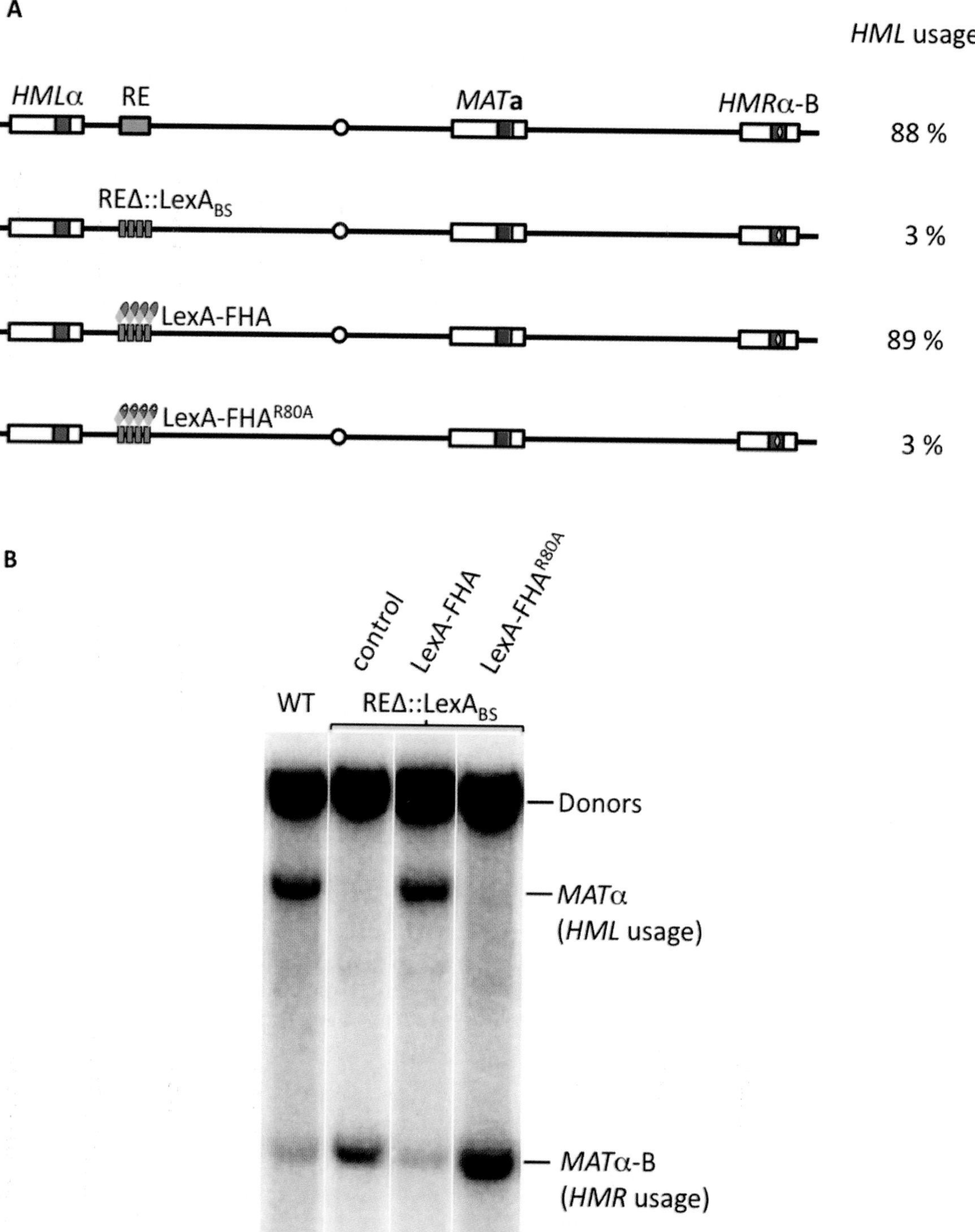

Figure 7 Role of the recombination enhancer in *MAT***a** donor preference. (A) Arrangement of *HML*α, *MAT***a**, and *HMR*α–*Bam*HI (*HMR*α*-B*). When RE is replaced by four LexA-binding sites (LexA$_{BS}$), *HML* usage is strongly impaired. Expression of LexA–FHA (the phosphothreonine-binding domain of Fkh1) fusion protein completely rescues *HML* usage, while expression of the mutant LexA–FHA^{R80A}, which has lost phosphothreonine-binding activity, fails to rescue it. (B) Southern blot data after induction of switching showing the proportion of *Bam*HI-digested *MAT*α or *MAT*α*-B* DNA in the strains depicted above. (Figure modified from 85.) doi:10.1128/microbiolspec.MDNA3-0013-2014.f7

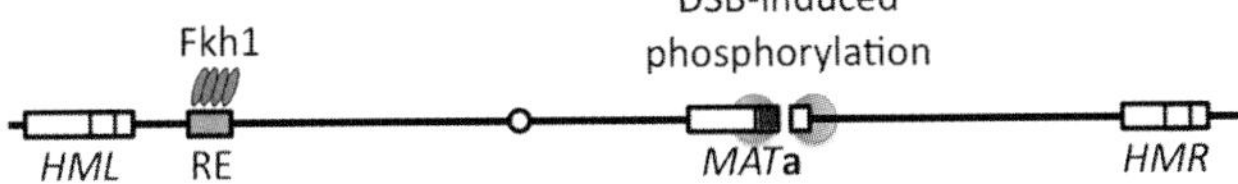

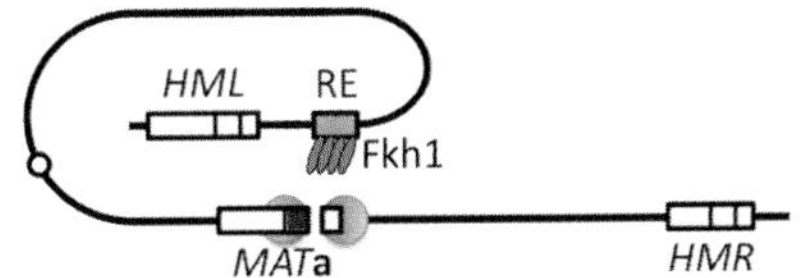

Figure 8 Model for Fkh1-regulated donor preference. A cluster of Fkh1 bound to RE in *MAT*a cells can associate with phosphothreonine residues that are located near the DSB and created by casein kinase II, and possibly other kinases, in response to the DSB. The association of Fkh1 and the DSB, which has been demonstrated by ChIP, tethers *HML* within ~20 kb of the DSB ends and facilitates its use over *HMR*, located 100 kb away.
doi:10.1128/microbiolspec.MDNA3-0013-2014.f8

Citation. Lee C-S, Haber JE. 2014. Mating-type gene switching in saccharomyces cerevisiae. Microbiol Spectrum 3(1): MDNA3-0013-2014.

References

1. **Klar AJ.** 1987. Determination of the yeast cell lineage. *Cell* **49:**433–435.
2. **Herskowitz I.** 1988. Life cycle of the budding yeast *Saccharomyces cerevisiae*. *Microbiol Rev* **52:**536–553.
3. **Haber JE.** 2006. Transpositions and translocations induced by site-specific double-strand breaks in budding yeast. *DNA Repair* **5:**998–1009.
4. **Haber JE.** 2012. Mating-type genes and MAT switching in *Saccharomyces cerevisiae*. *Genetics* **191:**33–64.
5. **Strathern JN.** 1988. Control and execution of mating type switching in *Saccharomyces cerevisiae*, p 445–464. *In* Kucherlapati R, Smith GR (ed), *Genetic Recombination*. ASM Press, Washington, DC.
6. **Haber JE.** 2007. Decisions, decisions: donor preference during budding yeast mating-type switching, p 159–170. *In* Heitman J, Kronstad JW, Taylor JW, Casselton LA (ed), *Sex in Fungi: Molecular Determination and Evolutionary Implications*. ASM Press, Washington, DC.
7. **Nasmyth KA.** 1982. Molecular genetics of yeast mating type. *Annu Rev Genet* **16:**439–500.
8. **Weiss K, Simpson RT.** 1998. High-resolution structural analysis of chromatin at specific loci: *Saccharomyces cerevisiae* silent mating type locus HMLα. *Mol Cell Biol* **18:**5392–5403.
9. **Ravindra A, Weiss K, Simpson RT.** 1999. High-resolution structural analysis of chromatin at specific loci: *Saccharomyces cerevisiae* silent mating-type locus HMRa. *Mol Cell Biol* **19:**7944–7950.
10. **Brand AH, Breeden L, Abraham J, Sternglanz R, Nasmyth K.** 1985. Characterization of a "silencer" in yeast: a DNA sequence with properties opposite to those of a transcriptional enhancer. *Cell* **41:**41–48.
11. **Schnell R, Rine J.** 1986. A position effect on the expression of a tRNA gene mediated by the SIR genes in *Saccharomyces cerevisiae*. *Mol Cell Biol* **6:**494–501.
12. **Connolly B, White CI, Haber JE.** 1988. Physical monitoring of mating type switching in *Saccharomyces cerevisiae*. *Mol Cell Biol* **8:**2342–2349.
13. **Loo S, Rine J.** 1994. Silencers and domains of generalized repression. *Science* **264:**1768–1771.
14. **Hagen DC, Bruhn L, Westby CA, Sprague GF Jr.** 1993. Transcription of α-specific genes in *Saccharomyces cerevisiae*: DNA sequence requirements for activity of the coregulator α1. *Mol Cell Biol* **13:**6866–6875.
15. **Bruhn L, Sprague GF Jr.** 1994. MCM1 point mutants deficient in expression of α-specific genes: residues important for interaction with α1. *Mol Cell Biol* **14:** 2534–2544.
16. **Smith DL, Johnson AD.** 1992. A molecular mechanism for combinatorial control in yeast: MCM1 protein sets the spacing and orientation of the homeodomains of an α2 dimer. *Cell* **68:**133–142.
17. **Keleher CA, Passmore S, Johnson AD.** 1989. Yeast repressor α 2 binds to its operator cooperatively with yeast protein Mcm1. *Mol Cell Biol* **9:**5228–5230.
18. **Herschbach BM, Arnaud MB, Johnson AD.** 1994. Transcriptional repression directed by the yeast α 2 protein in vitro. *Nature* **370:**309–311.
19. **Patterton HG, Simpson RT.** 1994. Nucleosomal location of the STE6 TATA box and Matα2p-mediated repression. *Mol Cell Biol* **14:**4002–4010.
20. **Smith RL, Johnson AD.** 2000. Turning genes off by Ssn6–Tup1: a conserved system of transcriptional repression in eukaryotes. *Trends Biochem Sci* **25:**325–330.
21. **Jensen R, Sprague GF Jr, Herskowitz I.** 1983. Regulation of yeast mating-type interconversion: feedback control of HO gene expression by the mating-type locus. *Proc Nat Acad Sci U S A* **80:**3035–3039.
22. **Goutte C, Johnson AD.** 1988. a1 protein alters the DNA binding specificity of α2 repressor. *Cell* **52:**875–882.
23. **Li T, Stark MR, Johnson AD, Wolberger C.** 1995. Crystal structure of the MATa1/MAT α 2 homeodomain heterodimer bound to DNA. *Science* **270:**262–269.
24. **Johnson PR, Swanson R, Rakhilina L, Hochstrasser M.** 1998. Degradation signal masking by heterodimerization of MATα2 and MATa1 blocks their mutual destruction by the ubiquitin-proteasome pathway. *Cell* **94:**217–227.
25. **Tan S, Richmond TJ.** 1998. Crystal structure of the yeast MATα2/MCM1/DNA ternary complex. *Nature* **391:**660–666.
26. **Chant J.** 1996. Generation of cell polarity in yeast. *Curr Opin Cell Biol* **8:**557–565.
27. **Chant J, Pringle JR.** 1991. Budding and cell polarity in *Saccharomyces cerevisiae*. *Curr Opin Genet Dev* **1:** 342–350.
28. **Lord M, Inose F, Hiroko T, Hata T, Fujita A, Chant J.** 2002. Subcellular localization of Axl1, the cell type-specific regulator of polarity. *Curr Biol* **12:**1347–1352.

29. **Frank-Vaillant M, Marcand S.** 2001. NHEJ regulation by mating type is exercised through a novel protein, Lif2p, essential to the ligase IV pathway. *Genes Dev* **15:** 3005–3012.

30. **Kegel A, Sjostrand JO, Astrom SU.** 2001. Nej1p, a cell type-specific regulator of nonhomologous end joining in yeast. *Curr Biol* **11:**1611–1617.

31. **Ooi SL, Shoemaker DD, Boeke JD.** 2001. A DNA microarray-based genetic screen for nonhomologous end-joining mutants in *Saccharomyces cerevisiae*. *Science* **294:**2552–2556.

32. **Valencia M, Bentele M, Vaze MB, Herrmann G, Kraus E, Lee SE, Schar P, Haber JE.** 2001. NEJ1 controls non-homologous end joining in *Saccharomyces cerevisiae*. *Nature* **414:**666–669.

33. **Laney JD, Hochstrasser M.** 2003. Ubiquitin-dependent degradation of the yeast Matα2 repressor enables a switch in developmental state. *Genes Dev* **17:**2259–2270.

34. **Laney JD, Mobley EF, Hochstrasser M.** 2006. The short-lived Matα2 transcriptional repressor is protected from degradation in vivo by interactions with its corepressors Tup1 and Ssn6. *Mol Cell Biol* **26:**371–380.

35. **Haber JE, George JP.** 1979. A mutation that permits the expression of normally silent copies of mating-type information in *Saccharomyces cerevisiae*. *Genetics* **93:** 13–35.

36. **Nasmyth K.** 1987. The determination of mother cell-specific mating type switching in yeast by a specific regulator of HO transcription. *EMBO J* **6:**243–248.

37. **Breeden L, Nasmyth K.** 1987. Cell cycle control of the yeast HO gene: *cis-* and *trans-* acting regulators. *Cell* **48:**389–397.

38. **Nasmyth K, Stillman D, Kipling D.** 1987. Both positive and negative regulators of HO transcription are required for mother-cell-specific mating-type switching in yeast. *Cell* **48:**579–587.

39. **Long RM, Singer RH, Meng X, Gonzalez I, Nasmyth K, Jansen RP.** 1997. Mating type switching in yeast controlled by asymmetric localization of ASH1 mRNA. *Science* **277:**383–387.

40. **Bobola N, Jansen RP, Shin TH, Nasmyth K.** 1996. Asymmetric accumulation of Ash1p in postanaphase nuclei depends on a myosin and restricts yeast mating-type switching to mother cells. *Cell* **84:**699–709.

41. **Sil A, Herskowitz I.** 1996. Identification of asymmetrically localized determinant, Ash1p, required for lineage-specific transcription of the yeast HO gene. *Cell* **84:**711–722.

42. **Shepard KA, Gerber AP, Jambhekar A, Takizawa PA, Brown PO, Herschlag D, DeRisi JL, Vale RD.** 2003. Widespread cytoplasmic mRNA transport in yeast: identification of 22 bud-localized transcripts using DNA microarray analysis. *Proc Nat Acad Sci U S A* **100:** 11429–11434.

43. **Jambhekar A, McDermott K, Sorber K, Shepard KA, Vale RD, Takizawa PA, DeRisi JL.** 2005. Unbiased selection of localization elements reveals *cis*-acting determinants of mRNA bud localization in *Saccharomyces cerevisiae*. *Proc Nat Acad Sci U S A* **102:**18005–18010.

44. **Kaplun L, Ivantsiv Y, Bakhrat A, Raveh D.** 2003. DNA damage response-mediated degradation of Ho endonuclease via the ubiquitin system involves its nuclear export. *J Biol Chem* **278:**48727–48734.

45. **Kaplun L, Ivantsiv Y, Bakhrat A, Tzirkin R, Baranes K, Shabek N, Raveh D.** 2006. The F-box protein, Ufo1, maintains genome stability by recruiting the yeast mating switch endonuclease, Ho, for rapid proteasome degradation. *Isr Med Assoc J* **8:**246–248.

46. **Haber JE, Wolfe KH.** 2005. Evolution and function of HO and VDE endoncucleases in fungi, p 161–175. *In* Belfort B, Derbyshire V, Stodddard B, Wood D (ed), *Homing Endonucleases and Inteins*. Springer-Verlag, New York.

47. **Nickoloff JA, Chen EY, Heffron F.** 1986. A 24-base-pair DNA sequence from the MAT locus stimulates intergenic recombination in yeast. *Proc Nat Acad Sci U S A* **83:**7831–7835.

48. **Nickoloff JA, Singer JD, Heffron F.** 1990. In vivo analysis of the *Saccharomyces cerevisiae* HO nuclease recognition site by site-directed mutagenesis. *Mol Cell Biol* **10:**1174–1179.

49. **McNally FJ, Rine J.** 1991. A synthetic silencer mediates SIR-dependent functions in *Saccharomyces cerevisiae*. *Mol Cell Biol* **11:**5648–5659.

50. **Laurenson P, Rine J.** 1992. Silencers, silencing, and heritable transcriptional states. *Microbiol Rev* **56:**543–560.

51. **Sherman JM, Pillus L.** 1997. An uncertain silence. *Trends Genet* **13:**308–313.

52. **Astrom SU, Rine J.** 1998. Theme and variation among silencing proteins in *Saccharomyces cerevisiae* and *Kluyveromyces lactis*. *Genetics* **148:**1021–1029.

53. **Grunstein M.** 1998. Yeast heterochromatin: regulation of its assembly and inheritance by histones. *Cell* **93:** 325–328.

54. **Lustig AJ.** 1998. Mechanisms of silencing in *Saccharomyces cerevisiae*. *Curr Opin Genet Dev* **8:**233–239.

55. **Stone EM, Pillus L.** 1998. Silent chromatin in yeast: an orchestrated medley featuring Sir3p [corrected]. *BioEssays* **20:**30–40.

56. **Rusche LN, Kirchmaier AL, Rine J.** 2003. The establishment, inheritance, and function of silenced chromatin in *Saccharomyces cerevisiae*. *Annu Rev Biochem* **72:** 481–516.

57. **McConnell KH, Muller P, Fox CA.** 2006. Tolerance of Sir1p/origin recognition complex-dependent silencing for enhanced origin firing at HMRa. *Mol Cell Biol* **26:** 1955–1966.

58. **Hickman MA, Froyd CA, Rusche LN.** 2011. Reinventing heterochromatin in budding yeasts: Sir2 and the origin recognition complex take center stage. *Eukaryot Cell* **10:**1183–1192.

59. **Thompson-Stewart D, Karpen GH, Spradling AC.** 1994. A transposable element can drive the concerted evolution of tandemly repetitious DNA. *Proc Nat Acad Sci U S A* **91:**9042–9046.

60. **Shei GJ, Broach JR.** 1995. Yeast silencers can act as orientation-dependent gene inactivation centers that

respond to environmental signals. *Mol Cell Biol* **15:** 3496–3506.

61. **Maillet L, Boscheron C, Gotta M, Marcand S, Gilson E, Gasser SM.** 1996. Evidence for silencing compartments within the yeast nucleus: a role for telomere proximity and Sir protein concentration in silencer-mediated repression. *Genes Dev* **10:**1796–1811.
62. **Marcand S, Buck SW, Moretti P, Gilson E, Shore D.** 1996. Silencing of genes at nontelomeric sites in yeast is controlled by sequestration of silencing factors at telomeres by Rap 1 protein. *Genes Dev* **10:**1297–1309.
63. **Bird AW, Yu DY, Pray-Grant MG, Qiu Q, Harmon KE, Megee PC, Grant PA, Smith MM, Christman MF.** 2002. Acetylation of histone H4 by Esa1 is required for DNA double- strand break repair. *Nature* **419:**411–415.
64. **Wolner B, Peterson CL.** 2005. ATP-dependent and ATP-independent roles for the Rad54 chromatin remodeling enzyme during recombinational repair of a DNA double strand break. *J Biol Chem* **280:**10855–10860.
65. **Thompson JS, Johnson LM, Grunstein M.** 1994. Specific repression of the yeast silent mating locus HMR by an adjacent telomere. *Mol Cell Biol* **14:**446–455.
66. **Ren J, Wang CL, Sternglanz R.** 2010. Promoter strength influences the S phase requirement for establishment of silencing at the *Saccharomyces cerevisiae* silent mating type loci. *Genetics* **186:**551–560.
67. **Ishii K, Arib G, Lin C, Van Houwe G, Laemmli UK.** 2002. Chromatin boundaries in budding yeast: the nuclear pore connection. *Cell* **109:**551–562.
68. **Donze D, Adams CR, Rine J, Kamakaka RT.** 1999. The boundaries of the silenced HMR domain in *Saccharomyces cerevisiae*. *Genes Dev* **13:**698–708.
69. **Dhillon N, Raab J, Guzzo J, Szyjka SJ, Gangadharan S, Aparicio OM, Andrews B, Kamakaka RT.** 2009. DNA polymerase epsilon, acetylases and remodellers cooperate to form a specialized chromatin structure at a tRNA insulator. *EMBO J* **28:**2583–2600.
70. **Klar AJ, Fogel S, Macleod K.** 1979. *MAR1*—a Regulator of the *HMa* and *HMα* loci in *Saccharomyces cerevisiae*. *Genetics* **93:**37–50.
71. **Rine J, Strathern JN, Hicks JB, Herskowitz I.** 1979. A suppressor of mating-type locus mutations in *Saccharomyces cerevisiae*: evidence for and identification of cryptic mating-type loci. *Genetics* **93:**877–901.
72. **Rine J, Herskowitz I.** 1987. Four genes responsible for a position effect on expression from HML and HMR in *Saccharomyces cerevisiae*. *Genetics* **116:**9–22.
73. **Pillus L, Rine J.** 1989. Epigenetic inheritance of transcriptional states in *S. cerevisiae*. *Cell* **59:**637–647.
74. **Imai S, Armstrong CM, Kaeberlein M, Guarente L.** 2000. Transcriptional silencing and longevity protein Sir2 is an NAD-dependent histone deacetylase. *Nature* **403:**795–800.
75. **Armache KJ, Garlick JD, Canzio D, Narlikar GJ, Kingston RE.** 2011. Structural basis of silencing: Sir3 BAH domain in complex with a nucleosome at 3.0 Å resolution. *Science* **334:**977–982.
76. **Takahashi YH, Schulze JM, Jackson J, Hentrich T, Seidel C, Jaspersen SL, Kobor MS, Shilatifard A.** 2011. Dot1 and histone H3K79 methylation in natural telomeric and HM silencing. *Mol Cell* **42:**118–126.
77. **Moazed D, Kistler A, Axelrod A, Rine J, Johnson AD.** 1997. Silent information regulator protein complexes in *Saccharomyces cerevisiae*: a SIR2/SIR4 complex and evidence for a regulatory domain in SIR4 that inhibits its interaction with SIR3. *Proc Nat Acad Sci U S A* **94:** 2186–2191.
78. **Tsukamoto Y, Kato J, Ikeda H.** 1996. Hdf1, a yeast Ku-protein homologue, is involved in illegitimate recombination, but not in homologous recombination. *Nucleic Acids Res* **24:**2067–2072.
79. **Taddei A, Hediger F, Neumann FR, Bauer C, Gasser SM.** 2004. Separation of silencing from perinuclear anchoring functions in yeast Ku80, Sir4 and Esc1 proteins. *EMBO J* **23:**1301–1312.
80. **Megee PC, Morgan BA, Mittman BA, Smith MM.** 1990. Genetic analysis of histone H4: essential role of lysines subject to reversible acetylation. *Science* **247:** 841–845.
81. **Park EC, Szostak JW.** 1990. Point mutations in the yeast histone H4 gene prevent silencing of the silent mating type locus HML. *Mol Cell Biol* **10:**4932–4934.
82. **Fisher-Adams G, Grunstein M.** 1995. Yeast histone H4 and H3 N-termini have different effects on the chromatin structure of the GAL1 promoter. *EMBO J* **14:** 1468–1477.
83. **Hecht A, Laroche T, Strahl-Bolsinger S, Gasser SM, Grunstein M.** 1995. Histone H3 and H4 N-termini interact with SIR3 and SIR4 proteins: a molecular model for the formation of heterochromatin in yeast. *Cell* **80:** 583–592.
84. **Braunstein M, Sobel RE, Allis CD, Turner BM, Broach JR.** 1996. Efficient transcriptional silencing in *Saccharomyces cerevisiae* requires a heterochromatin histone acetylation pattern. *Mol Cell Biol* **16:**4349–4356.
85. **Li J, Coic E, Lee K, Lee CS, Kim JA, Wu Q, Haber JE.** 2012. Regulation of budding yeast mating-type switching donor preference by the FHA domain of Fkh1. *PLoS Genet* **8:**e1002630.
86. **Lafon A, Chang CS, Scott EM, Jacobson SJ, Pillus L.** 2007. MYST opportunities for growth control: yeast genes illuminate human cancer gene functions. *Oncogene* **26:**5373–5384.
87. **Gasser SM, Cockell MM.** 2001. The molecular biology of the SIR proteins. *Gene* **279:**1–16.
88. **Moretti P, Freeman K, Coodly L, Shore D.** 1994. Evidence that a complex of SIR proteins interacts with the silencer and telomere-binding protein RAP1. *Genes Dev* **8:**2257–2269.
89. **Wotton D, Shore D.** 1997. A novel Rap1p-interacting factor, Rif2p, cooperates with Rif1p to regulate telomere length in *Saccharomyces cerevisiae*. *Genes Dev* **11:** 748–760.
90. **Mishra K, Shore D.** 1999. Yeast Ku protein plays a direct role in telomeric silencing and counteracts inhibition by rif proteins. *Curr Biol* **9:**1123–1126.
91. **Roy R, Meier B, McAinsh AD, Feldmann HM, Jackson SP.** 2004. Separation-of- function mutants of yeast

Ku80 reveal a Yku80p–Sir4p interaction involved in telomeric silencing. *J Biol Chem* **279:**86–94.

92. **Ribes-Zamora A, Mihalek I, Lichtarge O, Bertuch AA.** 2007. Distinct faces of the Ku heterodimer mediate DNA repair and telomeric functions. *Nat Struct Mol Biol* **14:**301–307.

93. **Strahl-Bolsinger S, Hecht A, Luo K, Grunstein M.** 1997. SIR2 and SIR4 interactions differ in core and extended telomeric heterochromatin in yeast. *Genes Dev* **11:**83–93.

94. **Rusche LN, Lynch PJ.** 2009. Assembling heterochromatin in the appropriate places: a boost is needed. *J Cell Physiol* **219:**525–528.

95. **Vandre CL, Kamakaka RT, Rivier DH.** 2008. The DNA end-binding protein Ku regulates silencing at the internal HML and HMR loci in *Saccharomyces cerevisiae. Genetics* **180:**1407–1418.

96. **Takahashi T, Saito H, Ikeda Y.** 1958. Heterothallic behavior of a homothallic strain in Saccharomyces yeast. *Genetics* **43:**249–260.

97. **Takano I, Oshima Y.** 1967. An allele specific and a complementary determinant controlling homothallism in *Saccharomyces oviformis. Genetics* **57:**875–885.

98. **Santa Maria J, Vidal D.** 1970. Segregación anormal del "mating type" en Saccharomyces. *Inst Nac Invest Agron Conf* **30:**1–21.

99. **Oshima Y, Takano I.** 1971. Mating types in Saccharomyces: their convertibility and homothallism. *Genetics* **67:**327–335.

100. **Hicks J, Strathern JN.** 1977. Interconversion of mating type in *S. cerevisiae* and the cassette model for gene transfer. *Brookhaven Symp Biol*233–242.

101. **Hicks J, Strathern J, Herskowitz I.** 1977. The cassette model of mating-type interconversion, p 457–462. *In* Bukhari A, Shapiro J, Adhya S (ed), *DNA Insertion Elements, Plasmids and Episomes*. Cold Spring Harbor Laboratory Press, Cold Spring Harbor, NY.

102. **Nasmyth KA, Tatchell K.** 1980. The structure of transposable yeast mating type loci. *Cell* **19:**753–764.

103. **Strathern JN, Spatola E, McGill C, Hicks JB.** 1980. Structure and organization of transposable mating type cassettes in Saccharomyces yeasts. *Proc Nat Acad Sci U S A* **77:**2839–2843.

104. **Astell CR, Ahlstrom-Jonasson L, Smith M, Tatchell K, Nasmyth KA, Hall BD. 1981. The sequence of the DNAs coding for the mating-type loci of *Saccharomyces cerevisiae*. *Cell* 27:**15–23.

105. **Tatchell K, Nasmyth KA, Hall BD, Astell C, Smith M.** 1981. In vitro mutation analysis of the mating-type locus in yeast. *Cell* **27:**25–35.

106. **Naumov GI, Tolstorukov II.** 1971. [Discovery of an unstable homothallic strain of *Saccharomyces cerevisiae* var. *elipsoideus*.] *Nauchnye Doki Vyss Shkoly Biol Nauki* **9:**92–94 (in Russian).

107. **Tolstorukov II, Naumov GI.** 1973. [Comparative genetics of yeasts. XI. A genetic study of autodiploidization in natural homothallic strains of Saccharomyces]. *Nauchnye Doki Vyss Shkoly Biol Nauki* **117:**111–115 (in Russian).

108. **Hicks J, Strathern JN, Klar AJ.** 1979. Transposable mating type genes in *Saccharomyces cerevisiae. Nature* **282:**478–473.

109. **Klar AJ, Fogel S, Radin DN.** 1979. Switching of a mating-type a mutant allele in budding yeast *Saccharomyces cerevisiae. Genetics* **92:**759–776.

110. **Sprague GF Jr, Rine J, Herskowitz I.** 1981. Homology and non-homology at the yeast mating type locus. *Nature* **289:**250–252.

111. **McGill C, Shafer B, Strathern J.** 1989. Coconversion of flanking sequences with homothallic switching. *Cell* **57:** 459–467.

112. **Rattray AJ, Symington LS.** 1995. Multiple pathways for homologous recombination in *Saccharomyces cerevisiae. Genetics* **139:**45–56.

113. **Paques F, Haber JE.** 1999. Multiple pathways of recombination induced by double- strand breaks in *Saccharomyces cerevisiae. Microbiology and Molecular Biology Reviews* **63:**349–404.

114. **Aylon Y, Kupiec M.** 2004. DSB repair: the yeast paradigm. *DNA repair* **3:**797–815.

115. **Krogh BO, Symington LS.** 2004. Recombination proteins in yeast. *Annu Rev Genet* **38:**233–271.

116. **Daley JM, Palmbos PL, Wu D, Wilson TE.** 2005. Nonhomologous end joining in yeast. *Annu Rev Genet* **39:** 431–451.

117. **McEachern MJ, Haber JE.** 2006. Break-induced replication and recombinational telomere elongation in yeast. *Annu Rev Biochem* **75:**111–135.

118. **Sung P, Klein H.** 2006. Mechanism of homologous recombination: mediators and helicases take on regulatory functions. *Nat Rev Mol Cell Biol* **7:**739–750.

119. **Lydeard JR, Jain S, Yamaguchi M, Haber JE.** 2007. Break-induced replication and telomerase-independent telomere maintenance require Pol32. *Nature* **448:**820–823.

120. **Li X, Heyer WD.** 2008. Homologous recombination in DNA repair and DNA damage tolerance. *Cell Res* **18:** 99–113.

121. **McVey M, Lee SE.** 2008. MMEJ repair of double-strand breaks (director's cut): deleted sequences and alternative endings. *Trends Genet* **24:**529–538.

122. **San Filippo J, Sung P, Klein H.** 2008. Mechanism of eukaryotic homologous recombination. *Annu Rev Biochem* **77:**229–257.

123. **Jain S, Sugawara N, Lydeard J, Vaze M, Tanguy Le Gac N, Haber JE.** 2009. A recombination execution checkpoint regulates the choice of homologous recombination pathway during DNA double-strand break repair. *Genes Dev* **23:**291–303.

124. **Heyer WD, Ehmsen KT, Liu J.** 2010. Regulation of homologous recombination in eukaryotes. *Annu Rev Genet* **44:**113–139.

125. **Schwartz EK, Heyer WD.** 2011. Processing of joint molecule intermediates by structure-selective endonucleases during homologous recombination in eukaryotes. *Chromosoma* **120:**109–127.

126. **Rudin N, Haber JE.** 1988. Efficient repair of HO-induced chromosomal breaks in *Saccharomyces*

cerevisiae by recombination between flanking homologous sequences. *Mol Cell Biol* **8**:3918–3928.

127. **Nickoloff JA, Singer JD, Hoekstra MF, Heffron F.** 1989. Double-strand breaks stimulate alternative mechanisms of recombination repair. *J Mol Biol* **207**:527–541.

128. **Ray A, Machin N, Stahl FW.** 1989. A DNA double chain break stimulates triparental recombination in *Saccharomyces cerevisiae*. *Proc Nat Acad Sci U S A* **86**: 6225–6229.

129. **Plessis A, Perrin A, Haber JE, Dujon B.** 1992. Site-specific recombination determined by I-SceI, a mitochondrial group I intron-encoded endonuclease expressed in the yeast nucleus. *Genetics* **130**:451–460.

130. **McGill CB, Shafer BK, Derr LK, Strathern JN.** 1993. Recombination initiated by double-strand breaks. *Curr Genet* **23**:305–314.

131. **Liefshitz B, Parket A, Maya R, Kupiec M.** 1995. The role of DNA repair genes in recombination between repeated sequences in yeast. *Genetics* **140**:1199–1211.

132. **Weng YS, Whelden J, Gunn L, Nickoloff JA.** 1996. Double-strand break-induced mitotic gene conversion: examination of tract polarity and products of multiple recombinational repair events. *Curr Genet* **29**:335–343.

133. **Inbar O, Kupiec M.** 1999. Homology search and choice of homologous partner during mitotic recombination. *Mol Cell Biol* **19**:4134–4142.

134. **Wilson TE.** 2002. A genomics-based screen for yeast mutants with an altered recombination/end-joining repair ratio. *Genetics* **162**:677–688.

135. **Storici F, Durham CL, Gordenin DA, Resnick MA.** 2003. Chromosomal site-specific double-strand breaks are efficiently targeted for repair by oligonucleotides in yeast. *Proc Nat Acad Sci U S A* **100**:14994–14999.

136. **Lydeard JR, Lipkin-Moore Z, Sheu YJ, Stillman B, Burgers PM, Haber JE.** 2010. Break-induced replication requires all essential DNA replication factors except those specific for pre-RC assembly. *Genes Dev* **24**: 1133–1144.

137. **Marrero VA, Symington LS.** 2010. Extensive DNA end processing by exo1 and sgs1 inhibits break-induced replication. *PLoS Genetics* **6**:e1001007.

138. **Kleckner N.** 1996. Meiosis: how could it work? *Proc Nat Acad Sci U S A* **93**:8167–8174.

139. **Borner GV, Kleckner N, Hunter N.** 2004. Crossover/noncrossover differentiation, synaptonemal complex formation, and regulatory surveillance at the leptotene/zygotene transition of meiosis. *Cell* **117**:29–45.

140. **Keeney S, Neale MJ.** 2006. Initiation of meiotic recombination by formation of DNA double-strand breaks: mechanism and regulation. *Biochem Soc Trans* **34**: 523–525.

141. **Longhese MP, Bonetti D, Guerini I, Manfrini N, Clerici M.** 2009. DNA double- strand breaks in meiosis: checking their formation, processing and repair. *DNA repair* **8**:1127–1138.

142. **Strathern JN, Klar AJ, Hicks JB, Abraham JA, Ivy JM, Nasmyth KA, McGill C.** 1982. Homothallic switching of yeast mating type cassettes is initiated by a double-stranded cut in the MAT locus. *Cell* **31**:183–192.

143. **Kostriken R, Strathern JN, Klar AJ, Hicks JB, Heffron F.** 1983. A site-specific endonuclease essential for mating-type switching in *Saccharomyces cerevisiae*. *Cell* **35**:167–174.

144. **Klar AJ, Hicks JB, Strathern JN.** 1981. Irregular transpositions of mating-type genes in yeast. *Cold Spring Harb Symp Quant Biol* **45**:983–990.

145. **Miyazaki T, Bressan DA, Shinohara M, Haber JE, Shinohara A.** 2004. In vivo assembly and disassembly of Rad51 and Rad52 complexes during double-strand break repair. *EMBO J* **23**:939–949.

146. **Jensen RE, Herskowitz I.** 1984. Directionality and regulation of cassette substitution in yeast. *Cold Spring Harb Symp Quant Biol* **49**:97–104.

147. **Raveh D, Hughes SH, Shafer BK, Strathern JN.** 1989. Analysis of the HO-cleaved MAT DNA intermediate generated during the mating type switch in the yeast *Saccharomyces cerevisiae*. *Mol Gen Genet* **220**: 33–42.

148. **White CI, Haber JE.** 1990. Intermediates of recombination during mating type switching in *Saccharomyces cerevisiae*. *EMBO J* **9**:663–673.

149. **Hicks WM, Yamaguchi M, Haber JE.** 2011. Real-time analysis of double-strand DNA break repair by homologous recombination. *Proc Nat Acad Sci U S A* **108**: 3108–3115.

150. **Haber JE.** 1995. In vivo biochemistry: physical monitoring of recombination induced by site-specific endonucleases. *BioEssays* **17**:609–620.

151. **Ira G, Satory D, Haber JE.** 2006. Conservative inheritance of newly synthesized DNA in double-strand break-induced gene conversion. *Mol Cell Biol* **26**: 9424–9429.

152. **Cortes-Ledesma F, Aguilera A.** 2006. Double-strand breaks arising by replication through a nick are repaired by cohesin-dependent sister-chromatid exchange. *EMBO Reports* **7**:919–926.

153. **Weiffenbach B, Haber JE.** 1981. Homothallic mating type switching generates lethal chromosome breaks in *rad52* strains of *Saccharomyces cerevisiae*. *Mol Cell Biol* **1**:522–534.

154. **Ray BL, White CI, Haber JE.** 1991. Heteroduplex formation and mismatch repair of the "stuck" mutation during mating-type switching in *Saccharomyces cerevisiae*. *Mol Cell Biol* **11**:5372–5380.

155. **Chen H, Lisby M, Symington LS.** 2013. RPA coordinates DNA end resection and prevents formation of DNA hairpins. *Mol Cell* **50**:589–600.

156. **Fishman-Lobell J, Rudin N, Haber JE.** 1992. Two alternative pathways of double- strand break repair that are kinetically separable and independently modulated. *Mol Cell Biol* **12**:1292–1303.

157. **Zhu Z, Chung WH, Shim EY, Lee SE, Ira G.** 2008. Sgs1 helicase and two nucleases Dna2 and Exo1 resect DNA double-strand break ends. *Cell* **134**:981–994.

158. **Sugawara N, Haber JE.** 1992. Characterization of double-strand break-induced recombination: homology requirements and single-stranded DNA formation. *Mol Cell Biol* **12**:563–575.

159. **Ivanov EL, Sugawara N, White CI, Fabre F, Haber JE.** 1994. Mutations in XRS2 and RAD50 delay but do not prevent mating-type switching in *Saccharomyces cerevisiae. Mol Cell Biol* **14:**3414–3425.

160. **Tsubouchi H, Ogawa H.** 1998. A novel *mre11* mutation impairs processing of double-strand breaks of DNA during both mitosis and meiosis. *Mol Cell Biol* **18:** 260–268.

161. **Bressan DA, Olivares HA, Nelms BE, Petrini JH.** 1998. Alteration of N-terminal phosphoesterase signature motifs inactivates *Saccharomyces cerevisiae* Mre11. *Genetics* **150:**591–600.

162. **Lee SE, Bressan DA, Petrini JH, Haber JE.** 2002. Complementation between N- terminal *Saccharomyces cerevisiae mre11* alleles in DNA repair and telomere length maintenance. *DNA Repair* **1:**27–40.

163. **Nicolette ML, Lee K, Guo Z, Rani M, Chow JM, Lee SE, Paull TT.** 2010. Mre11–Rad50–Xrs2 and Sae2 promote 5′ strand resection of DNA double-strand breaks. *Nat Struct Mol Biol* **17:**1478–1485.

164. **Moreau S, Ferguson JR, Symington LS.** 1999. The nuclease activity of Mre11 is required for meiosis but not for mating type switching, end joining, or telomere maintenance. *Mol Cell Biol* **19:**556–566.

165. **Lobachev K, Vitriol E, Stemple J, Resnick MA, Bloom K.** 2004. Chromosome fragmentation after induction of a double-strand break is an active process prevented by the RMX repair complex. *Curr Biol* **14:** 2107–2112.

166. **Yu J, Marshall K, Yamaguchi M, Haber JE, Weil CF.** 2004. Microhomology-dependent end joining and repair of transposon-induced DNA hairpins by host factors in *Saccharomyces cerevisiae. Mol Cell Biol* **24:** 1351–1364.

167. **Clerici M, Mantiero D, Lucchini G, Longhese MP.** 2005. The *Saccharomyces cerevisiae* Sae2 protein promotes resection and bridging of double strand break ends. *J Biol Chem* **280:**38631–38638.

168. **Huertas P, Cortes–Ledesma F, Sartori AA, Aguilera A, Jackson SP.** 2008. CDK targets Sae2 to control DNA-end resection and homologous recombination. *Nature* **455:**689–692.

169. **Mimitou EP, Symington LS.** 2008. Sae2, Exo1 and Sgs1 collaborate in DNA double- strand break processing. *Nature* **455:**770–774.

170. **Niu H, Chung WH, Zhu Z, Kwon Y, Zhao W, Chi P, Prakash R, Seong C, Liu D, Lu L, Ira G, Sung P.** 2010. Mechanism of the ATP-dependent DNA end-resection machinery from *Saccharomyces cerevisiae*. *Nature* **467:** 108–111.

171. **Mimitou EP, Symington LS.** 2010. Ku prevents Exo1 and Sgs1-dependent resection of DNA ends in the absence of a functional MRX complex or Sae2. *EMBO J* **29:**3358–3369.

172. **Shim EY, Chung WH, Nicolette ML, Zhang Y, Davis M, Zhu Z, Paull TT, Ira G, Lee SE.** 2010. *Saccharomyces cerevisiae* Mre11/Rad50/Xrs2 and Ku proteins regulate association of Exo1 and Dna2 with DNA breaks. *EMBO J* **29:**3370–3380.

173. **Diede SJ, Gottschling DE.** 2001. Exonuclease activity is required for sequence addition and Cdc13p loading at a de novo telomere. *Curr Biol* **11:**1336–1340.

174. **Aylon Y, Liefshitz B, Kupiec M.** 2004. The CDK regulates repair of double-strand breaks by homologous recombination during the cell cycle. *EMBO J* **23:** 4868–4875.

175. **Ira G, Pellicioli A, Balijja A, Wang X, Fiorani S, Carotenuto W, Liberi G, Bressan D, Wan L, Hollingsworth NM, Haber JE, Foiani M.** 2004. DNA end resection, homologous recombination and DNA damage checkpoint activation require CDK1. *Nature* **431:**1011–1017.

176. **Chen X, Niu H, Chung WH, Zhu Z, Papusha A, Shim EY, Lee SE, Sung P, Ira G.** 2011. Cell cycle regulation of DNA double-strand break end resection by Cdk1-dependent Dna2 phosphorylation. *Nat Struct Mol Biol* **18:**1015–1019.

177. **Chen X, Cui D, Papusha A, Zhang X, Chu CD, Tang J, Chen K, Pan X, Ira G.** 2012. The Fun30 nucleosome remodeller promotes resection of DNA double-strand break ends. *Nature* **489:**576–580.

178. **Costelloe T, Louge R, Tomimatsu N, Mukherjee B, Martini E, Khadaroo B, Dubois K, Wiegant WW, Thierry A, Burma S, van Attikum H, Llorente B.** 2012. The yeast Fun30 and human SMARCAD1 chromatin remodellers promote DNA end resection. *Nature* **489:** 581–584.

179. **Eapen VV, Sugawara N, Tsabar M, Wu WH, Haber JE.** 2012. The *Saccharomyces cerevisiae* chromatin remodeler Fun30 regulates DNA end resection and checkpoint deactivation. *Mol Cell Biol* **32:**4727–4740.

180. **Neves-Costa A, Will WR, Vetter AT, Miller JR, Varga–Weisz P.** 2009. The SNF2- family member Fun30 promotes gene silencing in heterochromatic loci. *PLoS One* **4:**e8111.

181. **Awad S, Ryan D, Prochasson P, Owen–Hughes T, Hassan AH.** 2010. The Snf2 homolog Fun30 acts as a homodimeric ATP-dependent chromatin-remodeling enzyme. *J Biol Chem* **285:**9477–9484.

182. **Wang X, Haber JE.** 2004. Role of Saccharomyces single-stranded DNA-binding protein RPA in the strand invasion step of double-strand break repair. *PLoS Biol* **2:**E21.

183. **Sugawara N, Wang X, Haber JE.** 2003. In vivo roles of Rad52, Rad54, and Rad55 proteins in Rad51-mediated recombination. *Mol Cell* **12:**209–219.

184. **Wolner B, van Komen S, Sung P, Peterson CL.** 2003. Recruitment of the recombinational repair machinery to a DNA double-strand break in yeast. *Mol Cell* **12:** 221–232.

185. **Bressan DA, Vazquez J, Haber JE.** 2004. Mating type-dependent constraints on the mobility of the left arm of yeast chromosome III. *J Cell Biol* **164:**361–371.

186. **Houston PL, Broach JR.** 2006. The dynamics of homologous pairing during mating type interconversion in budding yeast. *PLoS Genet* **2:**e98.

187. **Kim JA, Hicks WM, Li J, Tay SY, Haber JE.** 2011. Protein phosphatases Pph3, Ptc2, and Ptc3 play redundant

roles in DNA double-strand break repair by homologous recombination. *Mol Cell Biol* **31**:507–516.

188. **Coic E, Martin J, Ryu T, Tay SY, Kondev J, Haber JE.** 2011. Dynamics of homology searching during gene conversion in *Saccharomyces cerevisiae* revealed by donor competition. *Genetics* **189**:1225–1233.

189. **Fishman-Lobell J, Haber JE. 1992. Removal of nonhomologous DNA ends in double-strand break recombination: the role of the yeast ultraviolet repair gene RAD1.** *Science* **258**:480–484.

190. **Lyndaker AM, Goldfarb T, Alani E.** 2008. Mutants defective in Rad1–Rad10–Slx4 exhibit a unique pattern of viability during mating-type switching in *Saccharomyces cerevisiae*. *Genetics* **179**:1807–1821.

191. **Colaiacovo MP, Paques F, Haber JE.** 1999. Removal of one nonhomologous DNA end during gene conversion by a RAD1- and MSH2-independent pathway. *Genetics* **151**:1409–1423.

192. **Li F, Dong J, Pan X, Oum JH, Boeke JD, Lee SE.** 2008. Microarray-based genetic screen defines SAW1, a gene required for Rad1/Rad10-dependent processing of recombination intermediates. *Mol Cell* **30**:325–335.

193. **Toh GW, Sugawara N, Dong J, Toth R, Lee SE, Haber JE, Rouse J.** 2010. Mec1/Tel1-dependent phosphorylation of Slx4 stimulates Rad1–Rad10-dependent cleavage of non-homologous DNA tails. *DNA Repair* **9**: 718–726.

194. **Haber JE, Ray BL, Kolb JM, White CI.** 1993. Rapid kinetics of mismatch repair of heteroduplex DNA that is formed during recombination in yeast. *Proc Nat Acad Sci U S A* **90**:3363–3367.

195. **Porter SE, White MA, Petes TD.** 1993. Genetic evidence that the meiotic recombination hotspot at the HIS4 locus of *Saccharomyces cerevisiae* does not represent a site for a symmetrically processed double-strand break. *Genetics* **134**:5–19.

196. **Leung W, Malkova A, Haber JE.** 1997. Gene targeting by linear duplex DNA frequently occurs by assimilation of a single strand that is subject to preferential mismatch correction. *Proc Nat Acad Sci U S A* **94**: 6851–6856.

197. **Jaskelioff M, Van Komen S, Krebs JE, Sung P, Peterson CL.** 2003. Rad54p is a chromatin remodeling enzyme required for heteroduplex DNA joint formation with chromatin. *J Biol Chem* **278**:9212–9218.

198. **Anand RP, Lovett ST, Haber JE.** 2013. Break-induced DNA replication. *Cold Spring Harb Perspect Biol* **5**: a010397.

199. **Wang X, Ira G, Tercero JA, Holmes AM, Diffley JF, Haber JE.** 2004. Role of DNA replication proteins in double-strand break-induced recombination in *Saccharomyces cerevisiae*. *Mol Cell Biol* **24**:6891–6899.

200. **Holmes AM, Haber JE.** 1999. Double-strand break repair in yeast requires both leading and lagging strand DNA polymerases. *Cell* **96**:415–424.

201. **Muramatsu S, Hirai K, Tak YS, Kamimura Y, Araki H.** 2010. CDK-dependent complex formation between replication proteins Dpb11, Sld2, Pol ε, and GINS in budding yeast. *Genes Dev* **24**:602–612.

202. **Chung WH, Zhu Z, Papusha A, Malkova A, Ira G.** 2010. Defective resection at DNA double-strand breaks leads to de novo telomere formation and enhances gene targeting. *PLoS Genet* **6**:e1000948.

203. **Wilson MA, Kwon Y, Xu Y, Chung WH, Chi P, Niu H, Mayle R, Chen X, Malkova A, Sung P, Ira G.** 2013. Pif1 helicase and Polδ promote recombination-coupled DNA synthesis via bubble migration. *Nature* **502**: 393–396.

204. **Hicks WM, Kim M, Haber JE.** 2010. Increased mutagenesis and unique mutation signature associated with mitotic gene conversion. *Science* **329**:82–85.

205. **Stith CM, Sterling J, Resnick MA, Gordenin DA, Burgers PM.** 2008. Flexibility of eukaryotic Okazaki fragment maturation through regulated strand displacement synthesis. *J Biol Chem* **283**:34129–40.

206. **Holbeck SL, Strathern JN.** 1997. A role for REV3 in mutagenesis during double- strand break repair in *Saccharomyces cerevisiae*. *Genetics* **147**:1017–1024.

207. **Yang Y, Sterling J, Storici F, Resnick MA, Gordenin DA.** 2008. Hypermutability of damaged single-strand DNA formed at double-strand breaks and uncapped telomeres in yeast *Saccharomyces cerevisiae*. *PLoS Genet* **4**:e1000264.

208. **Haber JE, Rogers DT, McCusker JH.** 1980. Homothallic conversions of yeast mating-type genes occur by intrachromosomal recombination. *Cell* **22**:277–289.

209. **Klar AJ, Strathern JN.** 1984. Resolution of recombination intermediates generated during yeast mating type switching. *Nature* **310**:744–748.

210. **Ira G, Malkova A, Liberi G, Foiani M, Haber JE.** 2003. Srs2 and Sgs1–Top3 suppress crossovers during double-strand break repair in yeast. *Cell* **115**:401–411.

211. **Prakash R, Satory D, Dray E, Papusha A, Scheller J, Kramer W, Krejci L, Klein H, Haber JE, Sung P, Ira G.** 2009. Yeast Mph1 helicase dissociates Rad51-made D-loops: implications for crossover control in mitotic recombination. *Genes Dev* **23**:67–79.

212. **Klar AJ, Hicks JB, Strathern JN.** 1982. Directionality of yeast mating-type interconversion. *Cell* **28**:551–561.

213. **Weiler KS, Broach JR.** 1992. Donor locus selection during *Saccharomyces cerevisiae* mating type interconversion responds to distant regulatory signals. *Genetics* **132**:929–942.

214. **Tanaka K, Oshima T, Araki H, Harashima S, Oshima Y.** 1984. Mating type control in *Saccharomyces cerevisiae*: a frameshift mutation at the common DNA sequence, X, of the *HML*α locus. *Mol Cell Biol* **4**: 203–211.

215. **Weiler KS, Szeto L, Broach JR.** 1995. Mutations affecting donor preference during mating type interconversion in *Saccharomyces cerevisiae*. *Genetics* **139**: 1495–1510.

216. **Wu X, Moore JK, Haber JE.** 1996. Mechanism of MATα donor preference during mating-type switching of *Saccharomyces cerevisiae*. *Mol Cell Biol* **16**:657–668.

217. **Szeto L, Broach JR.** 1997. Role of α2 protein in donor locus selection during mating type interconversion. *Mol Cell Biol* **17**:751–759.

218. **Szeto L, Fafalios MK, Zhong H, Vershon AK, Broach JR.** 1997. α2p controls donor preference during mating type interconversion in yeast by inactivating a recombinational enhancer of chromosome III. *Genes Dev* **11:** 1899–1911.

219. **Wu X, Haber JE.** 1995. MATa donor preference in yeast mating-type switching: activation of a large chromosomal region for recombination. *Genes Dev* **9:** 1922–1932.

220. **Wu X, Haber JE.** 1996. A 700 bp *cis*-acting region controls mating-type dependent recombination along the entire left arm of yeast chromosome III. *Cell* **87:** 277–285.

221. **Wu X, Wu C, Haber JE.** 1997. Rules of donor preference in Saccharomyces mating-type gene switching revealed by a competition assay involving two types of recombination. *Genetics* **147:**399–407.

222. **Sandell LL, Zakian VA.** 1993. Loss of a yeast telomere: arrest, recovery, and chromosome loss. *Cell* **75:** 729–739.

223. **VanHulle K, Lemoine FJ, Narayanan V, Downing B, Hull K, McCullough C, Bellinger M, Lobachev K, Petes TD, Malkova A.** 2007. Inverted DNA repeats channel repair of distant double-strand breaks into chromatid fusions and chromosomal rearrangements. *Mol Cell Biol* **27:**2601–2614.

224. **Wu C, Weiss K, Yang C, Harris MA, Tye BK, Newlon CS, Simpson RT, Haber JE.** 1998. Mcm1 regulates donor preference controlled by the recombination enhancer in Saccharomyces mating-type switching. *Genes Dev* **12:**1726–1737.

225. **Sun K, Coic E, Zhou Z, Durrens P, Haber JE.** 2002. Saccharomyces forkhead protein Fkh1 regulates donor preference during mating-type switching through the recombination enhancer. *Genes Dev* **16:**2085–2096.

226. **Weiss K, Simpson RT.** 1997. Cell type-specific chromatin organization of the region that governs directionality of yeast mating type switching. *EMBO J* **16:** 4352–4360.

227. **Coic E, Sun K, Wu C, Haber JE.** 2006. Cell cycle-dependent regulation of *Saccharomyces cerevisiae* donor preference during mating-type switching by SBF (Swi4/Swi6) and Fkh1. *Mol Cell Biol* **26:**5470–5480.

228. **Petronczki M, Chwalla B, Siomos MF, Yokobayashi S, Helmhart W, Deutschbauer AM, Davis RW, Watanabe Y, Nasmyth K.** 2004. Sister-chromatid cohesion mediated by the alternative RF-CCtf18/Dcc1/Ctf8, the helicase Chl1 and the polymerase- α-associated protein Ctf4 is essential for chromatid disjunction during meiosis II. *J Cell Sci* **117:**3547–3559.

229. **Liras P, McCusker J, Mascioli S, Haber JE.** 1978. Characterization of a mutation in yeast causing nonrandom chromosome loss during mitosis. *Genetics* **88:**651–671.

230. **Gerring SL, Spencer F, Hieter P.** 1990. The *CHL 1* (*CTF 1*) gene product of *Saccharomyces cerevisiae* is important for chromosome transmission and normal cell cycle progression in G_2/M. *EMBO J* **9:**4347–4358.

231. **Bystricky K, Van Attikum H, Montiel MD, Dion V, Gehlen L, Gasser SM.** 2009. Regulation of nuclear positioning and dynamics of the silent mating type loci by the yeast Ku70/Ku80 complex. *Mol Cell Biol* **29:**835–848.

232. **Gartenberg MR, Neumann FR, Laroche T, Blaszczyk M, Gasser SM.** 2004. Sir- mediated repression can occur independently of chromosomal and subnuclear contexts. *Cell* **119:**955–967.

233. **Andrulis ED, Zappulla DC, Ansari A, Perrod S, Laiosa CV, Gartenberg MR, Sternglanz R.** 2002. Esc1, a nuclear periphery protein required for Sir4-based plasmid anchoring and partitioning. *Mol Cell Biol* **22:**8292–8301.

234. **Coic E, Richard GF, Haber JE.** 2006. *Saccharomyces cerevisiae* donor preference during mating-type switching is dependent on chromosome architecture and organization. *Genetics* **173:**1197–1206.

235. **Miné–Hattab J, Rothstein R.** 2012. Increased chromosome mobility facilitates homology search during recombination. *Nat Cell Biol* **14:**510–517.

236. **Dion V, Kalck V, Horigome C, Towbin BD, Gasser SM.** 2012. Increased mobility of double-strand breaks requires Mec1, Rad9 and the homologous recombination machinery. *Nat Cell Biol* **14:**502–509.

237. **Renkawitz J, Lademann CA, Kalocsay M, Jentsch S.** 2013. Monitoring homology search during DNA double-strand break repair in vivo. *Mol Cell* **50:**261–272.

238. **Forget AL, Kowalczykowski SC.** 2012. Single-molecule imaging of DNA pairing by RecA reveals a three-dimensional homology search. *Nature* **482:**423–427.

Mobile DNA, 3rd Edition
Nancy L. Craig, Michael Chandler, Martin Gellert, Alan M. Lambowitz, Phoebe A. Rice and Suzanne Sandmeyer

doi:10.1128/microbiolspec.MDNA3-0003-2014

Amar J. S. Klar[1]
Ken Ishikawa[1]
Sharon Moore[1]

24 A Unique DNA Recombination Mechanism of the Mating/Cell-type Switching of Fission Yeasts: a Review

INTRODUCTION

Cells of the highly diverged *Schizosaccharomyces (S.) pombe* and *S. japonicus* fission yeasts exist in one of the two sex/mating types, called P (for plus) or M (for minus), specified by which allele, *M* or *P*, resides at *mat1* (Fig. 1). The fission yeasts have evolved an elegant mechanism for switching P or M information at *mat1* by a programmed DNA recombination event with a copy of one of the two silent mating-type genes residing nearby in the genome. The switching process is highly cell-cycle and generation dependent such that only one of four grandchildren of a cell switches mating type, and switching occurs in nearly half the cells of a population. Such a change of cell type is analogous to the stem-cell division found in higher eukaryotes whereby sister cells differ in their fate. Extensive studies of fission yeast established the natural DNA strand chirality at the *mat1* locus as the primary basis of asymmetric cell division. This asymmetry results from a unique site- and strand-specific epigenetic "imprint" at *mat1* installed in one of the two chromatids during DNA replication. The imprint is inherited by only one daughter cell, maintained for one cell cycle, and then used for initiating recombination during *mat1* replication. The progression through two replication cycles and two cell divisions leads to the "one-in-four" switching proportion among granddaughter cells. This mechanism of cell-type switching is considered to be unique to these two organisms, but determining the operation of such a mechanism in other organisms has not been possible for technical reasons. Thus, the validity of this mechanism for development in general remains untested. This review summarizes recent exciting developments in understanding the mechanism of *mat1* switching in fission yeasts and extends these observations to suggest how such a DNA strand-based

[1]National Cancer Institute at Frederick, Gene Regulation and Chromosome Biology Laboratory, National Cancer Institute at Frederick, P.O. Box B, Frederick, MD 21702-1201.

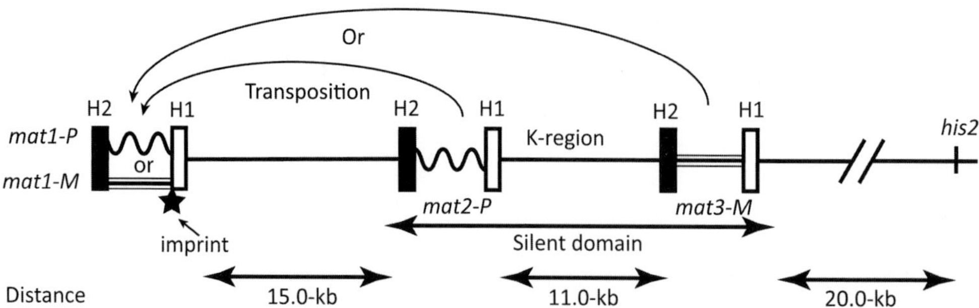

Figure 1 The mating-type region and *mat1*-switching process. The mating-type region comprises three genetic cassettes located on chromosome 2 at the indicated distances from each other in the order shown. The P-specific region, existing in both *mat1-P* and *mat2-P* loci, is 1,113-bp long and the M-specific sequence in *mat1-M* and *mat3-M* is 1,127-bp long (6). Recombination is facilitated by short homology boxes, which flank all cassettes. The cassette proximal homology box H1 is 135-bp long; the distal H2 element is 59-bp long. *mat1* switching occurs when genetic information copied from the *mat2* or *mat3* locus is unidirectionally transmitted to the *mat1* locus (curved horizontal arrows), where it displaces the previously existing *mat1* allele. A unique epigenetic imprint (star) found at *mat1* initiates the recombination leading to *mat1* switching. Whereas the *mat1* locus is transcriptionally active and dictates cell type, the *mat2*-K-*mat3* region is transcriptionally silenced by another kind of epigenetic mechanism (reviewed in reference 5). Thus, the cell type is defined only by the *mat1* locus even though genetic information for both mating types resides elsewhere in the chromosome. doi:10.1128/microbiolspec.MDNA3-0003-2014.f1

mechanism of cellular differentiation could also operate in diploid organisms. Although the analogous cell-type switching found in the *Saccharomyces (Sa.) cerevisiae* budding yeast by HO-endonuclease cleavage of the mating-type locus appears superficially similar to that of fission yeast, the mechanistic details are very different in these organisms. Thus, studies with diverse single-celled model yeast organisms have been helpful to appreciate how different paradigms of cellular differentiation have evolved. Considering that the ultimate basis of cellular differentiation in yeast is the double-helical structure of DNA, it is likely that such a mechanism operates in higher eukaryotes as well.

FISSION YEAST IS A LOWER EUKARYOTE IDEAL FOR BIOLOGICAL RESEARCH

S. pombe displays structural features that define eukaryotic organisms in that it contains a nucleus and mitochondria; however, unlike most other diploid eukaryotes, it exists primarily as a haploid, rod-shaped single cell. The cells are ~7 to 14-µm long and ~3 to 4-µm wide, nearly three times the size of *Escherichia coli* cells. The genome consists of three chromosomes with ~14.0 megabases of DNA, also about three times the size of the *E. coli* genome. Mitotically growing yeast cells elongate at the tips and divide by fission at the middle of the cell, and hence the name "fission yeast." Many biological processes, such as cell cycle, gametogenesis, meiosis, cellular differentiation, and cell biology, are evolutionarily conserved in eukaryotes. Yeasts are superb eukaryotic models, providing a well-established demonstration of classical Mendelian genetics as well as being an inexpensive and abundant source of material for biochemical analyses. For inclusive descriptions of earlier studies on yeast genetics, meiosis, and mating-type switching, previous review articles are available (1, 2, 3, 4, 5). This review primarily summarizes research on the cellular differentiation of yeast cells and highlights more recent advances for the mechanisms of DNA recombination employed for cellular differentiation, genome integrity, and gene silencing.

HAPLOID CELLS CONSIST OF P OR M CELL TYPE

The yeast cell exists as one of two sex/cell types, defined by which cassette of genetic information, *P* or *M*, resides at the *mat1* locus (1). These *mat1* alleles are composed of nonhomologous sequences of about 1.1 kilobases (kb) and each allele encodes two divergently transcribed transcripts (6). As these transcripts are induced by nutritional starvation, especially for nitrogen, cells do not express their mating type while growing in a rich medium. When mixed on a solid starvation (sporulation) medium, adjacent cells of opposite mating type fuse to produce a zygotic cell. Usually, the zygotic cell proceeds through conventional nuclear fusion and meiosis followed by sporulation, in which four haploid meiotic spore "segregants" are produced in an ascus, a zygotic cell sac. Mutational and molecular studies have demonstrated that the specific *mat1* transcripts are required for mating, meiosis, and sporulation (6). Two of the haploid spore segregants, or "ascospores," in each ascus are of the P cell type and the other two are of the M cell type. This 2:2 meiotic segregation pattern in each ascus establishes that a single locus with two alleles genetically specifies the cell's mating type (7).

In rare instances, the zygotic cell fails to sporulate, and remains in a diploid condition. Growing the culture in a rich medium that prohibits sporulation may perpetuate the diploid state, which is useful for genetic analyses. Once the diploid culture is transferred to a sporulation medium, cells with the *mat1-P*/*mat1-M* constitution proceed through meiosis and sporulation without further mating. As both *mat1* alleles are necessary for meiosis and sporulation, diploid cells homozygous for mating type (*mat1-P*/*mat1-P* and *mat1-M*/*mat1-M*) do not sporulate, but they can do so once they have become heterozygous because of *mat1* switching (described in the next section) or have mated with each other and then produce diploid ascospores (1).

The ascospores synthesize a starchy compound but mitotically growing cells do not. This property of yeast has provided a simple and powerful diagnostic tool to differentiate colonies that contain spores from those that do not: those with spores stain black when exposed to iodine vapors and nonsporulating colonies that lack starch fail to stain (8) (Fig. 2). This feature has become the single most useful tool in studies of the phenomenon of *mat1* switching and in the search for *cis*- and *trans*-acting factors that play a role in the switching process.

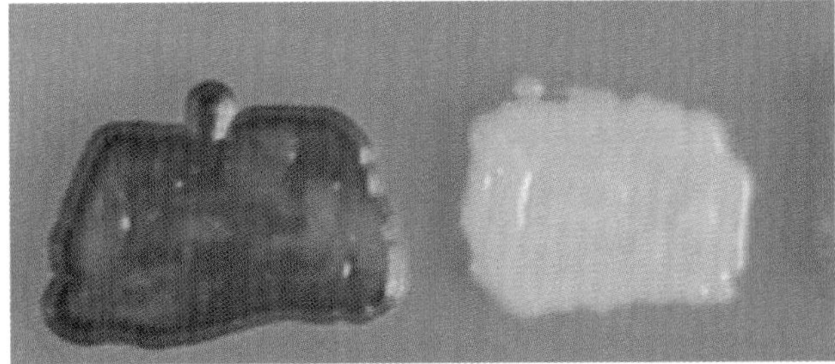

Figure 2 Staining phenotype of patches of cells growing on sporulation medium with iodine vapors. The patches are exposed to iodine vapors for about two minutes. Cell patches (or colonies, not shown) of the wild-type (*swi*$^+$) strain are composed of a mixture of P and M cells that engage in efficient mating and sporulation. Such patches stain black because they contain asci with spores that synthesize a starch-like compound that is stained by iodine vapors. The patch of *swi1*$^-$ cells fails to stain, indicating their switching defect. doi:10.1128/microbiolspec.MDNA3-0003-2014.f2

SWITCHING OCCURS BY THE STEM-CELL-LIKE ASYMMETRIC CELL DIVISIONS

A remarkable phenomenon exhibited by yeast cells is that the mating cell type is inherently unstable such that the two cell types readily interconvert by a phenomenon known as homothallism (9). Yeast populations that are capable of switching are designated h^{90} (for homothallic, 90% sporulation), but those incapable of efficient switching are called heterothallic. Heterothallic cells express very stable P (h^+) or M (h^-) cell types. Genetic and molecular analyses of heterothallic strains were fundamental in defining the structure of the mating-type locus and the mechanism of its switching (10). These studies showed that a heterothallic cell results from rare ($\sim 1 \times 10^{-5}$) and spontaneous mitotic recombination events that occur at the *mat1* locus. As the origin of the heterothallic state has been described in several review articles (1, 4, 5, 11) and as no new research has been conducted in this area recently, this aspect of mating-type research is not addressed here. Rather, this review highlights recent progress in *mat1* switching that reveals a mechanism of DNA recombination that is unusual in many respects compared to the canonical double-stranded DNA break repair mechanism operating elsewhere in this yeast and other eukaryotes.

Owing to the efficient *mat1*-switching process, a colony originating from a single cell of either mating type becomes a mixture of P and M cells in a 1:1 ratio (12). Usually, daughter/sister cells generated from a single cell can mate while growing on a poor growth medium, indicating that one of the sister cells has switched to the opposite mating type. Miyata and Miyata (13) observed an unusual pattern of switching in cell pedigrees. They found one zygote among four grandchildren of a cell in about 72 to 94% lineages and two zygotes were never observed. This pattern is now known as Miyata's one-in-four granddaughters switching rule and derives from the following switching restrictions imposed in cell pedigrees: (i) an unswitchable cell (for example, Pu, u for unswitchable [Fig. 3]) in ~80% of cell divisions produces one switchable daughter (Ps cell) while the sister cell is always unswitchable Pu; (ii) the Ps cell produces one switched Mu cell and one Ps cell in ~80% of cell divisions; and (iii) the Ps cell switches productively to the opposite mating type in ~80% of cell divisions. As a result of the last feature, chains of recurrent switching can be found by cell pedigree studies (14, 15). The same cell-lineage restrictions apply when cells switch from M to P. Thus, two consecutive asymmetric cell divisions are required to conform to the one-in-four pattern of switching (Fig. 3). This asymmetric cell division is very much analogous to the stem-cell asymmetric division reported in many eukaryotic cell divisions in which one of the two daughter cells changes fate, whereas the other maintains the fate exhibited by the parental cell. Remarkably, cellular differentiation decisions require two generations and this feature might be useful to help define mechanisms of cellular differentiation in multicellular organisms, such as vertebrates.

SWITCHING OCCURS VIA *mat1* ALLELE REPLACEMENT BY A PROGRAMMED RECOMBINATION MECHANISM

As individual cells can interconvert between the two cell types and *mat1* alleles contain different DNA sequence information, genetic information for both *mat1* alleles must reside in the genome (Fig. 1). However, only one of the two alleles occupies the *mat1* locus at a given time in a haploid cell. Before the *mat1* genes were molecularly cloned, classical genetic studies guided research for determining the mechanism of *mat1* switching. Perhaps influenced by the promoter flip-flop mechanism of genetic switches first discovered in prokaryotic organisms, interconversion of yeast *mat1* was initially proposed to be mediated by inversion of a shared promoter between the two mating-type genes of opposite allelic information residing in chromosome 2 at the *mat1* locus by recombination (16). However, after the mating-type gene replacement mechanism of *MAT* switching in the unrelated budding yeast *Sa. cerevisiae* was discovered (17, 18) (reviewed in reference 19), genetic analysis of two mutations that produced meiosis-deficient *mat1* alleles suggested that the gene-replacement model applied to switching in fission yeast as well (20, 21). These meiosis-defective mutations genetically defined the *mat2* and *mat3* loci, proposed as the source for copies of the genetic information that replaces the *mat1* allele by recombination (Fig. 1).

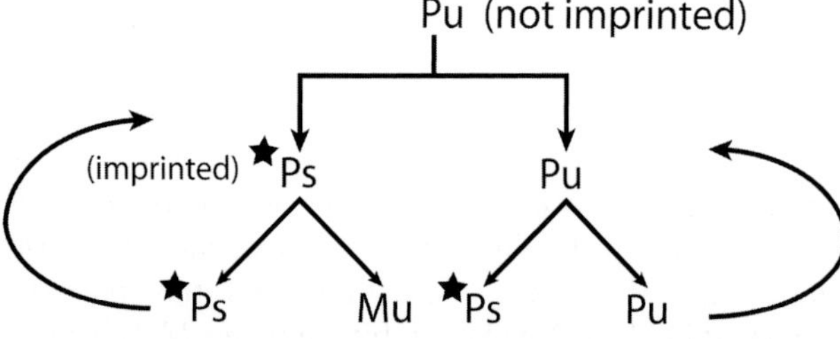

Figure 3 The one-in-four granddaughters switching pattern of yeast cells. The star indicates a cell that inherits the *mat1* imprint from the parent cell, making it switching competent so that it will produce one of the daughters that is switched. The same pattern is observed when cells switch from M to P. Pu, unswitchable cell; Ps, switchable cell.
doi:10.1128/microbiolspec.MDNA3-0003-2014.f3

A DNA FRAGILE SITE INITIATES RECOMBINATION FOR *mat1* SWITCHING

Southern analysis of DNA extracted from yeast cells (10, 22) revealed that 20 to 25% of *mat1* DNA exhibits a site-specific double-stranded DNA break (DSB) (Fig. 4). Analogous to the previously described similar finding in the *S. cerevisiae* cell-type switching system (23), the DSB in *S. pombe* was proposed to initiate recombination required for *mat1* switching. The association between the DSB and switching was supported by genetic analysis of three switching-defective mutants, *swi1* (*swi* for switch), *swi3*, and *swi7*, in which the switching deficiency was paralleled by a correspondingly reduced DSB level at the *mat1* locus (24, 25). The *swi1*, *-3*, and *-7* gene mutations played a fundamental role in deciphering the switching mechanism and help explain the biological basis of the switching pattern found in cell pedigrees (Fig. 3).

GENETIC ANALYSIS SUPPORTS THE STRAND-SPECIFIC IMPRINTING MECHANISM OF SWITCHING

Cell-pedigree experiments have established that two consecutive asymmetric cell divisions are required to generate the one-in-four pattern of switching (Fig. 3). Clearly, the first decision for a switch of the granddaughter cell must be made in the grandparental cell and the second one in one of its daughter cells. Why do some cells switch while their sisters never do? This question led to the suggestion that a single-stranded inheritable imprint could be installed during DNA replication in the grandparental cell, and this imprint is subsequently used in one daughter cell to induce recombination in one of the two chromatids during replication

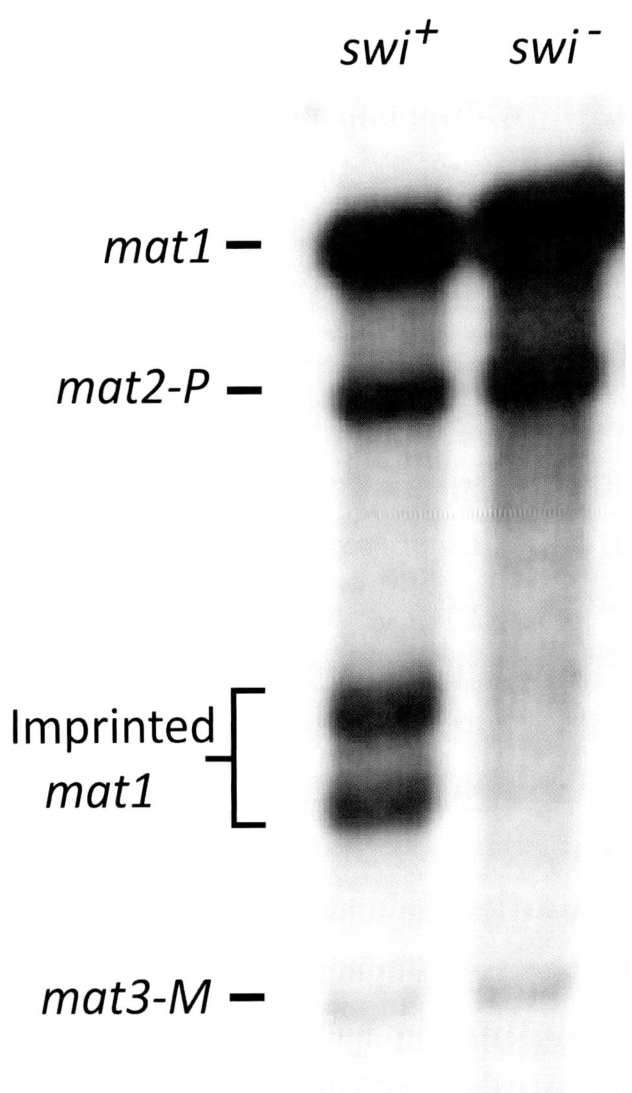

Figure 4 The imprint creates a fragile site in *mat1* DNA. Southern blot analysis of DNA extracted from a wild type and from an imprint-deficient *swi1*$^-$ yeast strain is shown. The DNA was digested with *Hin*dIII endonuclease, and the resulting blot was probed with a radiolabeled DNA fragment containing the *mat1-P* cassette. The intensity of the signal of each fragment reflects the extent of DNA sequence homology between the cassettes. The imprint causes a fragile site in DNA that results in a double-stranded break during conventional methods of DNA extraction. The imprint level is much reduced in the *swi1*$^-$ mutant.
doi:10.1128/microbiolspec.MDNA3-0003-2014.f4

(26). This scenario raises the problem of how to prohibit repair of the imprint for the length of an entire cell cycle. Compounding the problem is the observation that a constant level of DSB is maintained in cells regardless of the stage of the cell cycle. A molecular model was envisioned in which the Watson and Crick strands of DNA are nonequivalent in their ability to acquire the developmental potential for switching (26). That is, the double-helical structure of DNA might constitute the primary basis for cellular differentiation in this organism. To test this hypothesis genetically, another *mat1* gene in an inverted orientation was introduced by a DNA-mediated transformation adjacent to the indigenous *mat1* locus. According to the strand-specific imprinting/segregation model, two (cousin) cells among four grandchildren should switch as opposed to the one-in-four pattern found in standard strains that contain a single *mat1* gene. The predicted pattern of cousins switching was observed in ~32% of cell lineages having the inverted duplication (15). Additionally, Southern analysis showed that the DSB occurs in either one *mat1* gene or the other, but never in both genes on the same chromosome (26). Both of these results precisely conformed to the molecular and developmental predictions of the model. This conclusion required that the DNA sequence employed to make the *mat1* inverted duplication carried all the sequences necessary for switching and that the inverted gene did not interfere with the indigenous *mat1* locus's ability to switch. The two-in-four granddaughter cells switching result indicated a gain in switching proficiency according to the strand-segregation model in predicted ways. However, a control experiment with a direct *mat1* duplication could not be performed because the duplication is too efficiently removed by recombination at the *mat1* locus (26).

THE IMPRINT ALSO EFFICIENTLY INITIATES *mat1* GENE CONVERSION IN MEIOSIS

Analysis of strains engineered to be deleted for donor loci has significantly enhanced analyses of molecular events that occur at the *mat1* locus. The lack of donor loci "freezes" the switching mechanism and directs repair of the imprint by a different pathway. The two donor loci and the intervening K region were deleted by replacing this ~17.0-kb region with the *S. cerevisiae LEU2* gene (Fig. 1). The transgene was named *mat2/mat3Δ::LEU2* (7). The *mat1-P, mat2/mat3Δ::LEU2* cells exhibit only the P cell type and the *mat1-M, mat2/mat3Δ::LEU2* cells exhibit only the M cell type. Although these heterothallic strains do not switch mating type, they retain the usual level of DSB at *mat1*. Therefore, the DNA break can be repaired somehow without undergoing conventional *mat1* switching. It was initially hypothesized that repair might occur by recombination with the intact *mat1* gene of the sister chromatid (7). This suggestion was supported by the finding that donor-deleted cells die if defective in the DSB repair gene *rad22*, just as the h^{90}, *rad22* cells do (27, 28).

Having donor-deleted strains allowed investigators to ask if the imprint (normally used for *mat1* switching in mitosis) could act as a recombination "hotspot" to induce *mat1* gene conversion in meiosis as well. Crossing *mat1-M*, *mat2/mat3Δ::LEU2* cells with *mat1-P*, *mat2/mat3Δ::LEU2* cells followed by conventional meiotic tetrad analysis tested this idea. In these crosses, ~80% of the tetrads obtained were of the expected 2P:2M class type, which indicates conventional segregation of the *mat1* alleles, but ~10% of the tetrad genes converted *mat1-P* into *mat1-M* (1P:3M tetrad type), and another ~10% of the genes converted *mat1-M* into *mat1-P* (3P:1M tetrad type). Thus, ~20% of the asci underwent *mat1* gene conversion in meiosis, a high frequency of gene conversion compared to the level normally found at other loci (7). This result was interpreted to mean that the imprint normally used for conventional *mat1* switching (Fig. 3) could also efficiently induce meiotic gene conversion by recombination between *mat1* loci residing on chromosome 2 homologs. When *S. pombe* cells mate, the resulting zygotic cell does not divide further; instead, the zygote directly proceeds to meiosis and sporulation. This unique feature of the *S. pombe* life cycle was exploited to show that the imprint installed at the *mat1* locus induces *mat1* recombination in *cis* and segregates as an epigenetic entity, similar to a conventional Mendelian marker in mitosis and in meiosis.

swi1, *swi3*, AND *swi7* GENES FUNCTION TO INSTALL THE *mat1* *cis*-ACTING IMPRINT

Although a correlation had been observed between low switching efficiency and low DSB level in the *swi1*, *-3*, and *-7* mutants, further analysis was required to ascertain whether the lower level of DSB caused the switching deficiency, or whether both deficiencies stemmed from an upstream source (29). To test whether these *swi* gene factors function to form the *mat1* imprint and to determine whether the imprint is stable for one generation as stipulated by the one-in-four switching rule, crosses of donor-deleted strains were conducted in which one of the strains was *swi*$^-$ while the other was *swi*$^+$. Control crosses of *mat1-M*, *mat2/mat3Δ::LEU2* cells with *mat1-P*, *mat2/mat3Δ::LEU2* cells produced ~20% tetrads that were meiotically gene converted for both *mat1* alleles, but a cross between donor-deleted *mat1-P*, *swi3*$^-$ and donor-deleted *mat1-M swi*$^+$ strains generated an aberrant tetrad class of predominantly 3P:1M type (29). Conversely, a cross between donor-deleted *mat1-M swi3*$^-$ and donor-deleted *mat1-P swi*$^+$ strains generated a predominantly gene-converted class of 1P:3M type. Similar crosses with a *swi1*$^-$ or a *swi7*$^-$ parent generated meiotic *mat1* gene conversions in which only the allele contributed by the *swi*$^+$ parent converted. These results established several principles that further our understanding of the switching mechanism: (i) the competence for meiotic *mat1* gene conversion segregates in *cis* with the *mat1* locus; (ii) the *swi1*, *-3*,- and *-7* genes had already installed the competence in a fraction of mitotic cells that participated in the cross; (iii) the *swi*$^+$ gene present in the zygotic cell was incapable of conferring gene conversion potential to the *mat1* allele that was replicated in a respective *swi*$^-$ background; and (iv) because only tetrads with a 3:1 plus 1:3 ratio were observed, and none with a 0:4 ratio, only one of the two sister chromatids of the previously imprinted chromosome had undergone gene conversion. These results also established that chromosomally imprinted functions are installed by *swi* gene-encoded factors at least a generation before a gene conversion at *mat1* can occur. Such meiotic analysis of the *mat1* gene-conversion potential of individual chromosomes and their replication histories with respect to the *swi* gene constitution in mitotic cells was the key to defining the pattern of switching found in cell pedigrees (Fig. 3). These meiotic *mat1* gene conversion results, combined with the two-in-four granddaughters switching result of the *mat1*-inverted duplication, were central in deciphering this unusual mechanism of *mat1* switching by programmed recombination.

ORIENTATION OF *mat1* REPLICATION DICTATES THE SWITCHING PATTERN

Two-dimensional gel analysis of replication intermediates at the *mat1* locus had shown that the locus is replicated unidirectionally from centromere-distal origin(s) (3, 30). This is controlled by a polar replication termination site (*RTS1*) situated 0.7 kb to the centromere-proximal side of *mat1* that blocks DNA replication proceeding from the centromeric direction (31). Notably, imprinting/switching was drastically reduced when the *mat1* gene was inverted at its indigenous location in the chromosome or when the *RTS1* element was transplaced from its indigenous location to the centromere-distal side of the *mat1* gene. However, once both these rearrangements, each individually reducing switching drastically, were combined in *cis*, both imprinting and switching proficiency were restored. These results clearly indicate that the imprint is installed only during the lagging-strand synthesis in one cell cycle in a chromatid-specific fashion where it is maintained for the entire length of the cell cycle. The ensuing DNA replication of the imprinted strand by the leading-strand replication complex stalls to create a DSB that induces switching in the specific chromatid (29, 32). Another possibility to consider is that the inherent difference in the sequence of the two DNA strands alone might form the biological basis of differentiated sister cells. However, if this were true, then all cells could produce one switched and one unchanged daughter instead of only one switched cell in four grandchildren (Fig. 3). These studies defining the DNA replication intermediates further established the orientation of replication, and led to the strand-specific imprinting/segregation models (15, 26) that explained the pattern of switching in cell pedigrees. However, this analysis initially created a major inconsistency regarding the result obtained when the additional inverted *mat1* placed ~5.0 kb to the centromere-proximal side of the existing *mat1* was switching proficient (15, 26), whereas the inverted cassette at its indigenous location was switching deficient (30). The discovery of the *RTS1* element between the inverted *mat1* cassettes resolved this inconsistency (31). Indeed, the strand-specific imprinting/segregation model precisely predicted such a genetic outcome.

swi1, *swi3*, AND *swi7* FACTORS PLAY MULTIPLE ROLES IN IMPRINTING

The visualization of extracted DNA by Southern blot, probed with a *mat1* fragment and subsequent quantitation of the DSB, indicated that nearly one-quarter of the DNA is broken at a specific site and suggested that the DSB (Fig. 4) is maintained unrepaired through the entire cell cycle (10, 22). As a persistent reactive site, during the following cell cycle it induces recombination in one of the chromatids resulting from DNA replication of the imprinted chromosome. Although early molecular analyses of switching mechanisms relied on the level of the DSB, this model had to be modified upon the discovery that the DSB observed *in vitro* was an artifact of the DNA preparation procedure (30, 32). The DSB was not found in DNA purified by embedding cells in agarose plugs instead of the conventional method using phenol/chloroform extraction, vortexing, and RNase treatment (10, 22, 33). Furthermore, denaturing the DNA by formamide and formaldehyde indicated that the DNA strands at *mat1* were intact, but denaturing with sodium hydroxide produced a significant level of strand specific break at *mat1* (30). As the DSB is an artifact of the DNA extraction procedure because of the presence of a fragile site at the imprint, the level of the single-stranded nick is a better quantitation of the imprint. This method indicates that nearly 50% of the chromosomes are nicked, in agreement with the strand-segregation model (32, 34).

Imprinting requires several *mat1 cis*-acting sequences (Fig. 5): nearby replication origins, a Sap1 (switch activation protein 1)-binding site (SAS1) (35), a polar replication pause site (*MPS1*) (36), and the centromere proximal *RTS1* (31). The SAS1 site is about 150-bp centromere distal from the imprint site, the *MPS1* is situated near the imprint site, and the *RTS1* is about 0.7-kb centromere proximal to the *mat1* gene (31). Another recently described *cis*-acting element is a 104-bp "spacer region" that is essential for imprinting and lies proximal to the imprinting site (37).

At least six *trans*-acting proteins are necessary for imprinting as well (Fig. 5): Sap1 (38), Swi1, Swi3, and Swi7 (24), and the lysine-specific demethylases Lsd1

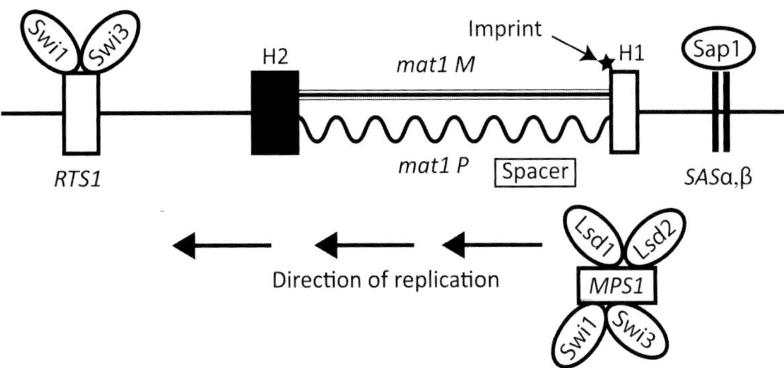

Figure 5 *mat1 cis*-acting sequence elements and *trans*-acting factors required for imprinting. See the text for details. doi:10.1128/microbiolspec.MDNA3-0003-2014.f5

and Lsd2 (39). The first *trans*-acting factor identified as essential for imprinting was the catalytic subunit of DNA polymerase α encoded by *swi7* (40). The *Swi1* protein sequence resembles a mammalian timeless protein (Tim) (36). *Swi3* is highly similar to the mammalian and mouse timeless interacting protein (Tipin) (41). Tim and Tipin are replisome-associated proteins that prevent replication-fork collapse (42). A complex including *Swi1* and *Swi3* (41) is required for replication-fork pausing by binding to the *MPS1* and for blocking replication forks of the wrong polarity at *RTS1*. Interestingly, a specific mutant allele of *swi1* indicates that different moieties of Swi1 function at *MPS1* and *RTS1* sites (36). Lsd1 and Lsd2 also facilitate imprinting by pausing replication at *MPS1* (39). The requirement for *swi7* (Polα), whose major function is synthesizing the RNA primer for the Okazaki fragment of the lagging strand, suggested the possibility that the primase function of the Polα/primase complex is responsible for imprint installation and that the imprint consists of ribonucleotides that are not removed from the Okazaki fragment primer by the replication machinery. Interestingly, *swi1* and *swi3* mutations compromised *MPS1* pausing, whereas a *swi7* mutation did not (36). Therefore, *MPS1* pausing might facilitate the placement of the postulated RNA primer at a specific position by lagging-strand replication.

Recent studies have extended the mechanism of installing the imprint (reviewed in references 39 and 43). The function of the *MPS1* site is to promote site-specific RNA priming at *mat1* by the primase/Polα complex at the imprint site so that the 5′ end of the stalled fork is dictated by the replication pause (37). Work from the Arcangioli laboratory suggests that the imprint is a single-stranded nick that contains unphosphorylated 3′-OH and 5′-OH termini (32, 44, 45). It is possible that different outcomes, such as alkali lability of the imprint, a strand-specific nick, and/or DSB, derive from inserted ribonucleotides forming the imprint. Regardless of whether the imprint consists of ribonucleotides and/or a nick, all biochemical studies on the imprinting mechanism have supported the strand-specific imprinting/segregation mechanism first proposed years ago from the genetic evidence (15, 26). Overall, several *cis*-acting sites spread around ~200 bp on both sides of the imprint site are required for imprinting. Such an arrangement suggests that a specific Okazaki fragment synthesized at the imprint site holds the key for imprinting. Accordingly, this yeast has adapted the usual DNA replication machinery to accomplish cellular differentiation by evolving an exquisitely regulated recombination mechanism at the *mat1* locus.

mat1 SWITCHING REPLACES BOTH DNA STRANDS

As no endonuclease activity has been identified that produces the nick or DSB at *mat1*, an alternative hypothesis is that when the replication fork synthesizing the leading strand reaches the imprint, a polar one-ended DSB is transiently formed (29). In donor-deleted cells, the DSB is repaired using the intact sister chromatid (7), but cells preferentially use the *mat2*/*mat3* donor loci, if available, for *mat1* switching. Supporting the sister chromatid hypothesis, sister chromatid recombination intermediates have been found by molecular methods (46). Interestingly, each recombination pathway uses a unique endonuclease to resolve the recombination intermediates; Mus81/Eme1 is used for sister chromatid-associated repair, and Swi9/Swi10 effect donor-associated repair (46).

The initiation step of repair remains unclear. One suggestion is that the broken end invades *mat2* or *mat3*, and extends the 3′-OH end by the copy-choice mechanism (29, 30). Another possibility is that repair is initiated by template switching through the replication–recombination coupled process (32, 34, 47, 48), similar to that of the canonical DSB repair mechanism (23, 49, 50). Following the fate of newly synthesized ^{13}C- and ^{15}N-labeled strands during switching showed that both *mat1* DNA strands are synthesized *de novo* when the opposite *mat1* information is substituted (51). However, when switching between two P cassettes with homologous sequences was tested, one at *mat1* and the other at *mat2*, each cassette marked with a different mutation, the wild-type *mat1-P* cassette was recreated through heteroduplex formation and repair (48). In switching between *mat1* and a donor, a single strand of the donor is synthesized and then transmitted to *mat1*, where it serves as a template for the complementary strand. An artificially introduced reversible imprinting mechanism shows that the imprint occurs in one cell cycle, remains stable throughout the next cell cycle, and causes switching in the following cell cycle (34). These molecular results support the strand-segregation model derived from previous genetic studies (15, 26).

DIRECTIONALITY OF SWITCHING

The donors that replace the *mat1* cassette are selected in a highly biased and counterintuitive manner. For example, *mat1-P* changes mostly by copying DNA from the nonhomologous *mat3-M*, and *mat1-M* from the nonhomologous *mat2-P* in ~90% of the switching events. Donor preference, called directionality of switching, is determined by location rather than by sequence of the donor loci (52). This was demonstrated using a strain with swapped genetic contents of the donors, *mat2-M* and *mat3-P*, referred to as h^{09}. The switching to opposite mating-type was very inefficient (~18%) in the h^{09} strain, which suggests that the directionality is working correctly and that most switches result in homologous replacement. Recently, it was reported that *cis*-elements adjacent to *mat2* and *mat3*, the Swi2-dependent recombination enhancer (SRE) adjacent to *mat2* (SRE2), and SRE3 at *mat3*, govern the directionality (53). When *mat2* and *mat3* are swapped, accompanied by their respective SRE2 and SRE3 elements, the cells switch as efficiently as wild type, indicating that the SRE elements are critical for determining the donor choice. How is the directionality determined by the location? There are two mechanisms identified for directionality; one is a cell-type specific distribution of a protein complex that mediates homologous recombination and the other is a chromosomal conformation to facilitate selective intrachromosomal interactions.

The recombination-promoting complex (RPC), containing Swi2 and Swi5, promotes the recombination between *mat1* and donor loci (54). Swi5 is also required for general recombination (28) and Swi2 recruits the recombination factor Rhp51 to the *mat* region (55). In P cells, Swi2 binding is mostly localized to the *mat3* region dictating *mat3-M* as the donor, whereas in M cells Swi2 binding is distributed throughout the *mat2/3* region (54). Swi2 binding to the *mat* region requires the SRE3, and it was proposed that in M cells Swi2 would reach to the *mat2* region by spreading from the SRE3 through a heterochromatin region between *mat2* and *mat3* via physical interaction with Swi6 (54). Two more proteins, Abp1 and Mc, are required for Swi2 binding to the *mat2* region, perhaps by promoting Swi2 spreading (56, 57, 58). The Mc protein, a transcription factor encoded within the M cassette, provides proper distribution of the RPC by positively regulating the expression of *swi2* and *swi5* genes in M cells (57). In addition, Mc stimulates expression of a shorter transcript (swi2S) of the *swi2* gene that facilitates *mat2* utilization in M cells (58). However, in conflict with the spreading hypothesis, another study suggests that Swi2 binds directly to the *mat2 cis*-acting element in M cells (53).

Heterochromatin in the *mat* region is another factor required for directionality (59). Defects of genes required for heterochromatin formation, *swi6*, *clr4*, *rik1*, *clr7*, and *clr8*, compromise directionality (52, 60, 61, 62). Therefore, heterochromatin may mediate Swi2 spreading and/or provide proper chromosomal conformation for promoting directionality.

mat1 IMPRINT IS A HOTSPOT OF MITOTIC RECOMBINATION BETWEEN HOMOLOGS

In addition to functioning in *mat1* switching, the imprint also creates a mitotic recombination hotspot in diploid cells that causes homozygosis of all the centromere-distal markers (Fig. 1). The efficiency of homozygosis loosely reflects the switching rate of a *mat1-cis*-acting mutation carried on one of the homologs, even though the other homolog carries the recombination-competent wild-type *mat1* allele (63, 64); thus, recombination only occurs when both *mat1* loci in the diploid are capable of switching (Fig. 6). This feature is in sharp contrast to the general recombination mechanism, which is limited by the frequency of the recombination-initiating event, such as the double-stranded chromosomal break (23, 49, 50), but once initiation has occurred in one of the chromosomes/chromatids, recombination and DNA repair ensue efficiently. This difference in recombination between imprint-associated recombination, which requires both *mat1* loci to be capable of switching, and general recombination, which is governed by the rate of the initiating event, led to the proposal of the selective chromatid recombination and arms-swapping model wherein chromatids that transiently and simultaneously carry DSBs and/or a single-stranded nick at *mat1* could recombine (Fig. 6). In this model, chromatid arms can recombine or segregate without recombination at an equal frequency and both without involving DNA repair. Thus, the hotspot recombination is proposed to occur through a pathway other than the conventional DSB repair model. In support of this hypothesis, the hotspot recombination is not reduced in the *swi5* mutant, encoding a function required for general mitotic recombination in yeast (65).

A recent study (65) has reported several unusual features of the mechanism of hotspot recombination: (i) recombination occurs in the S/G2 phase of the cell; (ii) recombination occurs at an appreciable rate in about 4.0% of cell divisions; (iii) one-half of recombination events causes homozygosis, while the other half only changes linkage of markers flanking the *mat1* locus; (iv) recombination occurs only between previously imprinted nonsister chromatids; and (v) recombination occurs only between chromatids that simultaneously contain DSBs/nick. These results support the arms-swapping model (Fig. 6) and advance our understanding of the mechanism of mitotic recombination at the *mat1* hotspot. Moreover, unique features, such as recombination occurring only in the S/G2 phase and only between specific nonsister chromatids, have helped define the mode of segregation of chromosome 2 strands

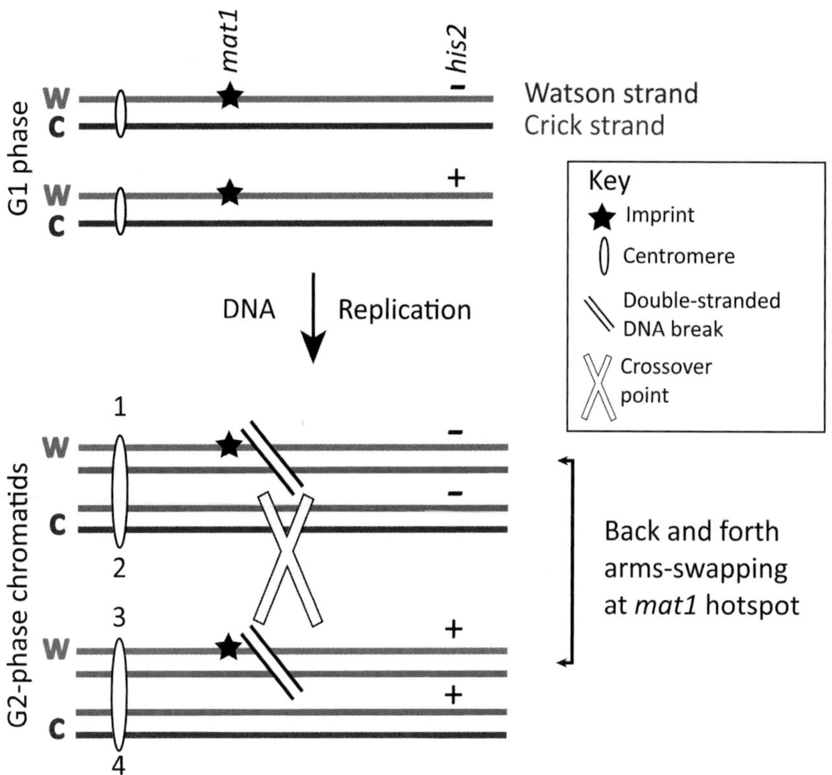

Figure 6 The selective chromatid recombination and arms-swapping model of the *mat1* hotspot recombination. Recombination occurs only between nonsister chromatids numbers 1 and 3, and only when they simultaneously have a DNA break/nick (\\) in the S/G2 phase of the cell. To depict specific strand distribution, template Watson (W) strands are colored blue and Crick (C) strands are colored red. Strands synthesized in the present replication cycle are indicated in black. The symbol X represents the crossover point at *mat1*. The figure is modified from reference 65. doi:10.1128/microbiolspec.MDNA3-0003-2014.f6

to daughter cells. By genetic analyses, it is argued that chromosome 2 strands of one homolog are segregated randomly and independently with respect to those of the second homolog during cell division. Overall, selective chromatid recombination occurs at the hotspot and it is followed by random chromatid distribution to daughter cells. It is generally thought that the DSB-induced recombination occurs without any chromatid bias in eukaryotes. Overall, hotspot recombination varies from the general DSB recombination/repair mechanism in several respects (discussed above).

S. JAPONICUS FISSION YEAST ALSO USES THE DNA STRAND-BASED MECHANISM TO GENERATE ASYMMETRIC CELL DIVISION

Chromosomally borne epigenetic differences between sister cells that change cell type have been demonstrated only in the *S. pombe* fission yeast. For technical reasons, it has been impossible to determine the existence of such a mechanism operating elsewhere in biology, especially during the embryonic development of multicellular organisms. Thus, it is easy to perceive that such a DNA-based mechanism does not operate in biology at large. However, unraveling the elegant system of *mat1* switching in *S. pombe* motivated the search for another experimentally favorable system that permits observation of a similar DNA strand-based mechanism of asymmetric cell division. With that goal in mind, a study was initiated with the recently sequenced fission yeast *S. japonicus* (44% GC content) (66), which is highly diverged from the well-studied *S. pombe* species (36% GC content) having only a 50% identity at the amino acid level in their protein orthologs. Remarkably, *S. japonicus* cells also switch cell/mating type after undergoing two consecutive cycles of asymmetric cell divisions and, as in *S. pombe*, only one of four granddaughter cells switches. The DNA strand-specific

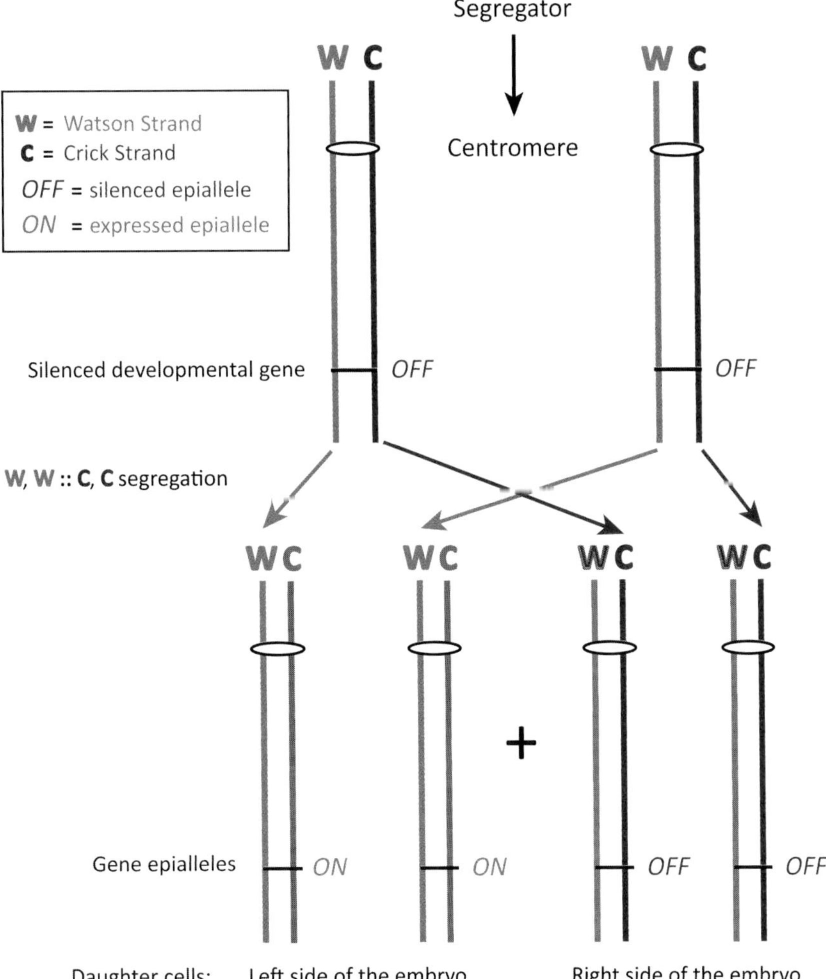

Figure 7 The SSIS model. The model (71, 72) predicts the evolution of two phenomena. One causes production of nonequivalent sister chromatids through epigenetic means to express differentially a developmentally important gene in the Watson (W) versus Crick (C) strand/chromatid-specific fashion. The other causes nonrandom segregation of all four chromatids belonging to a pair of homologous chromosomes, or sets of chromosomes, in mitosis. To depict the diagrammed nonrandom segregation pattern, termed "W,W::C,C," DNA strands are color-coded. The strands synthesized in the parent cell are depicted in black. A hypothetical *trans*-acting segregator factor is proposed to function by acting at the centromere to mediate selective chromatid segregation at a specific cell division in development to promote asymmetric cell division. Lateralized organs could be derived from the progeny of thus differentiated daughter cells. The selective chromatid segregation model has recently been invoked to explain the neurological disorder, congenital mirror hand movement [MIM 157600], which occurs in 50% of *rad51/RAD51* human subjects. In the absence of selective chromatid segregation, random segregation for a specific chromosome at a specific cell division is proposed to impact the laterality development of the brain hemispheres (76). Similarly, the observation of 50% penetrance of psychosis disorder development in individuals with chromosome 11 translocations is consistent with translocation-caused anomalies of the SSIS mechanism. (77). doi:10.1128/microbiolspec.MDNA3-0003-2014.f7

epigenetic imprint at *mat1* initiates the recombination event that is required for cellular differentiation (67). Therefore, the *S. pombe* and *S. japonicus* mating systems provide the first two examples in which the intrinsic chirality of the double-helical structure of DNA was found to confer asymmetric cell division. This conservation of the strand-specific mechanism of asymmetric cell division of *S. japonicus* is all the more interesting considering that the DNA sequences and genomic arrangement of cassettes are fundamentally different from those of *S. pombe* (67).

mat1-SWITCHING PARADIGM ADVANCED TO EXPLAIN ASYMMETRIC CELL DIVISION AND DEVELOPMENT IN HIGHER ORGANISMS

We learned from the fission-yeast studies that two consecutive asymmetric cell divisions are required to produce one switched granddaughter cell of a nonswitchable cell and that in both divisions the developmental program is advanced by the act of DNA replication itself. Also, DNA replication produces nonequivalent chromatids of the parental chromosome in the Watson versus Crick strand-specific fashion, and stable epigenetic states of gene expression are replicated along with the DNA and inherited both in mitosis and meiosis as conventional Mendelian factors (68, 69, 70). Could these lessons be exploited to explain cellular differentiation and development in higher eukaryotes?

In principle, differential gene regulation of developmentally important gene(s) may be accomplished by epigenetic entities somatically installed in a strand-specific fashion at specific stages in development (Fig. 7). This may occur by differential heterochromatin assembly and/or by altering cytosine base methylation of the relevant gene to turn off its transcription or to activate its transcription by disrupting the repressing epigenetic control. In this fashion, nonequivalent sister chromatids might be produced. However, for the hypothesized chromatid asymmetry to be a useful mechanism for cellular differentiation in diploid organisms, selective segregation of differentiated chromatids of both homologs is required. The somatic strand-specific imprinting and selective segregation (SSIS) model predicts the evolution of such a mechanism for producing developmentally nonequivalent sister cells in diploid organisms (Fig. 7). This model was advanced to explain visceral left–right axis laterality development in mice (71) and for human-brain hemisphere specialization (72). The SSIS model postulates selective chromatid segregation of specific chromosomes at a specific cell division to accomplish development during embryogenesis. Although the experiments involved are not straightforward, the selective chromatid segregation phenomenon appears to operate in mouse stem cells (73, 74) and in asymmetrically dividing *Drosophila* cells (75). For discovering aspects of the SSIS mechanism, we owe thanks to the model organisms of fission yeast for providing us with a paradigm possibly applicable to investigations of developmental biology in higher eukaryotes.

Acknowledgments. *It is a pleasure to summarize here the contributions of our mating-type research colleagues, with apologies to those whose work could not be cited because of space considerations. The Intramural Research Program of the National Cancer Institute, National Institutes of Health, supports this research.*

Citation. Klar AJS, Ishikawa K, and Moore S. 2014. A unique DNA recombination mechanism of the mating/cell-type switching of fission yeasts: a review. Microbiol Spectrum 2(5):MDNA3-0003-2014.

References

1. **Egel R.** 1989. Mating-type genes, meiosis and sporulation, p 31–74. *In* Nasim A, Young P, Johnson B (ed), *Molecular Biology of the Fission Yeast*. Academic Press, New York, NY.
2. **Gutz H, Schmidt H.** 1990. The genetic basis of homothallism and heterothallism in *Saccharomyces cerevisiae* and *Schizosaccharomyces pombe*. *Semin Dev Biol* **1:** 169–176.
3. **Dalgaard JZ, Klar AJS.** 2001. Does *S. pombe* exploit the intrinsic asymmetry of DNA synthesis to imprint daughter cells for mating-type switching? *Trends Genet* **17:** 153–157.
4. **Arcangioli B, Thon G.** 2004. Mating-type cassettes: structure, switching and silencing, p 129–147. *In* Egel R (ed), *The Molecular Biology of Schizosaccharomyces pombe*. Springer Verlag, Berlin, Germany.
5. **Klar AJS.** 2007. Lessons learned from studies of fission yeast mating-type switching and silencing. *Ann Rev Genet* **41:**213–236.
6. **Kelly M, Burke J, Smith M, Klar A, Beach D.** 1988. Four mating-type genes control sexual differentiation in the fission yeast. *EMBO J* **7:**1537–1547.
7. **Klar AJS, Miglio LM.** 1986. Initiation of meiotic recombination by double-strand DNA breaks in *S. pombe*. *Cell* **46:**725–731.
8. **Bresch C, Muller G, Egel R.** 1968. Genes involved in meiosis and sporulation of a yeast. *Mol Gen Genet* **102:** 301–306.
9. **Leupold U.** 1950. Die vererbung von homothallie and heterothallie bei *Schizosaccharomyces pombe*. *C R Trav Lab Carlsberg Ser Physiol* **24:**381–480.
10. **Beach DH, Klar AJS.** 1984. Rearrangements of the transposable mating-type cassettes of fission yeast. *EMBO J* **3:** 603–610.
11. **Singh J.** 1999. Mating type switching in fission yeast: a unique model for development and differentiation. *Curr Sci* **77:**1262–1272.
12. **Egel R.** 1977. Frequency of mating-type switching in homothallic fission yeast. *Nature* **266:**172–174.
13. **Miyata H, Miyata M.** 1981. Mode of conjugation in homothallic cells of *Schizosaccharomyces pombe*. *J Gen Appl Microbiol* **27:**365–371.

14. **Egel R, Eie E.** 1987. Cell lineage asymmetry for *Schizosaccharomyces pombe*: unilateral transmission of a high-frequency state of mating-type switching in diploid pedigrees. *Curr Genet* **3:**5–12.
15. **Klar AJS.** 1990. The developmental fate of fission yeast cells is determined by the pattern of inheritance of parental and grandparental DNA strands. *EMBO J* **9:**1407–1415.
16. **Egel R.** 1977. "Flip-flop" control and transposition of mating-type genes in fission yeast, p 446–455. *In* Bukhar AJ, Shapiro J, Adhya S (ed), *DNA Insertion Elements, Plasmids and Episomes*. Cold Spring Harbor Laboratory, Cold Spring Harbor, NY.
17. **Klar AJS, Fogel S.** 1979. Activation of mating type genes by transposition in *Saccharomyces cerevisiae*. *Proc Natl Acad Sci USA* **76:**4539–4543.
18. **Hicks J, Strathern JN, Klar AJ.** 1979. Transposable mating type genes in *Saccharomyces cerevisiae*. *Nature* **282:** 478–473
19. **Klar AJS.** 2010. The yeast mating-type switching mechanism: a memoir. *Genetics* **186:**443–449.
20. **Egel R.** 1984. The pedigree pattern of mating-type switching in *Schizosaccharomyces pombe*. *Curr Genet* **8:** 205–210.
21. **Egel R.** 1984. Two tightly linked silent cassettes in the mating type region of *Schizosaccharomyces pombe*. *Curr Genet* **8:**199–203.
22. **Beach DH.** 1983. Cell type switching by DNA transposition in fission yeast. *Nature* **305:**682–688.
23. **Strathern JN, Klar AJ, Hicks JB, Abraham JA, Ivy JM, Nasmyth KA, McGill C.** 1982. Homothallic switching of yeast mating type cassettes is initiated by a double-stranded cut in the *MAT* locus. *Cell* **31:**183–192.
24. **Egel R, Beach DH, Klar AJS.** 1984. Genes required for initiation and resolution steps of mating-type switching in fission yeast. *Proc Natl Acad Sci USA* **81:**3481–3485.
25. **Gutz H, Schmidt H.** 1985. Switching genes in *Schizosaccharomyces pombe*. *Curr Genet* **9:**325–331.
26. **Klar AJS.** 1987. Differentiated parental DNA strands confer developmental asymmetry on daughter cells in fission yeast. *Nature* **326:**466–470.
27. **Ostermann K, Lorentz A, Schmidt H.** 1993. The fission yeast *rad22* gene, having a function in mating-type switching and repair of DNA damages, encodes a protein homolog to Rad52 of Saccharomyces cerevisiae. *Nucleic Acids Res* **21:**5940–5944.
28. **Schmidt H, Kapitza P, Gutz H.** 1987. Switching genes in *Schizosaccharomyces pombe*: their influence on cell viability and recombination. *Curr Genet* **11:**303–308.
29. **Klar AJS, Bonaduce MJ.** 1993. The mechanism of fission yeast mating-type interconversion: evidence for two types of epigenetically inherited chromosomal imprinted events. *Cold Spring Harb Symp Quant Biol* **58:**457–465.
30. **Dalgaard JZ, Klar AJS.** 1999. Orientation of DNA replication establishes mating-type switching pattern in *S. pombe*. *Nature* **400:**181–184.
31. **Dalgaard JZ, Klar AJS.** 2001. A DNA replication-arrest site RTS1 regulates imprinting by determining the direction of replication at mat1 in *S. pombe*. *Genes Dev* **15:**2060–2068.
32. **Arcangioli B.** 1998. A site- and strand-specific DNA break confers asymmetric switching potential in fission yeast. *EMBO J* **17:**4503–4510.
33. **Nielsen O, Egel R.** 1989. Mapping the double-strand breaks at the mating-type locus in fission yeast by genomic sequencing. *EMBO J* **8:**269–276.
34. **Holmes AM, Kaykov A, Arcangioli B.** 2005. Molecular and cellular dissection of mating-type switching steps in *Schizosaccharomyces pombe*. *Mol Cell Biol* **25:**303–311.
35. **Arcangioli B, Klar AJS.** 1991. A novel switch-activating site (SAS1) and its cognate binding factor (SAP1) required for efficient mat1 switching in *Schizosaccharomyces pombe*. *EMBO J* **10:**3025–3032.
36. **Dalgaard JZ, Klar AJS.** 2000. *swi1* and *swi3* perform imprinting, pausing, and termination of DNA replication in *S. pombe*. *Cell* **102:**745–751.
37. **Sayrac S, Vengrova S, Godfrey EL, Dalgaard JZ.** 2011. Identification of a novel type of spacer element required for imprinting in fission yeast. *PLoS Genet* **7:**e1001328.
38. **Arcangioli B, Copeland TD, Klar AJ.** 1994. Sap1, a protein that binds to sequences required for mating-type switching, is essential for viability in *Schizosaccharomyces pombe*. *Mol Cell Biol* **14:**2058–2065.
39. **Holmes A, Roseaulin L, Schurra C, Waxin H, Lambert S, Zaratiegul M, Martienssen RA, Arcangioli B.** 2012. Lsd1 and lsd2 control programmed replication fork pauses and imprinting in fission yeast. *Cell Rep* **2:**1513–1520.
40. **Singh J, Klar AJS.** 1993. DNA polymerase-alpha is essential for mating-type switching in fission yeast. *Nature* **361:**271–273.
41. **Lee BS, Grewal SI, Klar AJ.** 2004. Biochemical interactions between proteins and mat1 cis-acting sequences required for imprinting in fission yeast. *Mol Cell Biol* **24:** 9813–9822.
42. **McFarlane RJ, Mian S, Dalgaard JZ.** 2010. The many facets of the Tim-Tipin protein families' roles in chromosome biology. *Cell Cycle* **9:**700–705.
43. **Dalgaard JZ.** 2012. Causes and consequences of ribonucleotide incorporation into nuclear DNA. *Trends Genet* **28:**592–597.
44. **Kaykov A, Arcangioli B..** 2004. A programmed strand-specific and modified nick in *S. pombe* constitutes a novel type of chromosomal imprint. *Curr Biol* **14:**1924–1928.
45. **Kaykov A, Holmes AM, Arcangioli B.** 2004. Formation, maintenance and consequences of the imprint at the mating-type locus in fission yeast. *EMBO J* **23:**930–938.
46. **Roseaulin L, Yamada Y, Tsutsui Y, Russell P, Iwasaki H, Arcangioli B.** 2008. Mus81 is essential for sister chromatid recombination at broken replication forks. *EMBO J* **27:**1378–1387.
47. **Arcangioli B, de Lahondes R.** 2000. Fission yeast switches mating type by a replication-recombination coupled process. *EMBO J* **19:**1389–1396.
48. **Yamada-Inagawa T, Klar AJS, Dalgaard JZ.** 2007. *Schizosaccharomyces pombe* switches mating type by the

synthesis-dependent strand-annealing mechanism. *Genetics* **177:**255–265.

49. **Resnick MA.** 1976. The repair of double-strand breaks in DNA: a model involving recombination. *J Theor Biol* **59:** 97–106.
50. **Szostak JW, Orr-Weaver TL, Rothstein RJ, Stahl FW.** 1983. The double-strand-break repair model for recombination. *Cell* **33:**25–35.
51. **Arcangioli B.** 2000. Fate of *mat1* DNA strands during mating-type switching in fission yeast. *EMBO Rep* **1:** 145–150.
52. **Thon G, Klar AJS.** 1993. Directionality of fission yeast mating-type interconversion is controlled by the location of the donor loci. *Genetics* **134:**1045–1054.
53. **Jakociunas T, Holm LR, Verhein-Hansen J, Trusina A, Thon G.** 2013. Two portable recombination enhancers direct donor choice in fission yeast heterochromatin. *PLoS Genet* **9:**e1003762.
54. **Jia S, Yamada T, Grewal SI.** 2004. Heterochromatin regulates cell type-specific long-range chromatin interactions essential for directed recombination. *Cell* **119:** 469–480.
55. **Akamatsu Y, Dziadkowiec D, Ikeguchi M, Shinagawa H, Iwasaki H.** 2003. Two different Swi5-containing protein complexes are involved in mating-type switching and recombination repair in fission yeast. *Proc Natl Acad Sci USA* **100:**15770–15775.
56. **Aguilar-Arnal L, Marsellach FX, Azorin F.** 2008. The fission yeast homologue of CENP-B, Abp1, regulates directionality of mating-type switching. *EMBO J* **27:** 1029–1038.
57. **Matsuda E, Sugioka-Sugiyama R, Mizuguchi T, Mehta S, Cui BC, Grewal SI.** 2011. A homolog of male sex-determining factor SRY cooperates with a transposon-derived CENP-B protein to control sex-specific directed recombination. *Proc Natl Acad Sci USA* **108:**18754–18759.
58. **Yu C, Bonaduce MJ, Klar AJS.** 2012. Going in the right direction: mating-type switching of Schizosaccharomyces pombe is controlled by judicious expression of two different swi2 transcripts. *Genetics* **190:**977–987.
59. **Grewal SI, Klar AJS.** 1997. A recombinationally repressed region between *mat2* and *mat3* loci shares homology to centromeric repeats and regulates directionality of mating-type switching in fission yeast. *Genetics* **146:**1221–1238.
60. **Ivanova AV, Bonaduce MJ, Ivanov SV, Klar AJS.** 1998. The chromo and SET domains of the Clr4 protein are essential for silencing in fission yeast. *Nat Genet* **19:** 192–195.
61. **Tuzon CT, Borgstrom B, Weilguny D, Egel R, Cooper JP, Nielsen O.** 2004. The fission yeast heterochromatin protein Rik1 is required for telomere clustering during meiosis. *J Cell Biol* **165:**759–765.
62. **Thon G, Hansen KR, Altes SP, Sidhu D, Singh G, Verhein-Hansen J, Bonaduce MJ, Klar AJS.** 2005. The Clr7 and Clr8 directionality factors and the Pcu4 cullin mediate heterochromatin formation in the fission yeast *Schizosaccharomyces pombe*. *Genetics* **171:**1583–1595.
63. **Angehrn P, Gutz H.** 1968. Influence of mating type on mitotic crossing-over in *Schizosaccharomyces pombe*. *Genetics* **60:**158.
64. **Egel R.** 1981. Mating-type switching and mitotic crossing-over at the mating-type locus in fission yeast. *Cold Spring Harb Symp Quant Biol* **45:**1003–1007.
65. **Klar AJS, Bonaduce MJ.** 2013. Unbiased segregation of fission yeast chromosome 2 strands to daughter cells. *Chromosome Res* **21:**297–309.
66. **Rhind N, Chen ZH, Yassour M, Thompson DA, Haas BJ, Habib N, Wapinski I, Roy S, Lin MF, Heiman DI, Young SK, Furuya K, Guo YB, Pidoux A, Chen HM, Robbertse B, Goldberg JM, Aoki K, Bayne EH, Berlin AM, Desjardins CA, Dobbs E, Dukaj L, Fan L, FitzGerald MG, French C, Gujja S, Hansen K, Keifenheim D, Levin JZ, Mosher RA, Muller CA, Pfiffner J, Priest M, Russ C, Smialowska A, Swoboda P, Sykes SM, Vaughn M, Vengrova S, Yoder R, Zeng QD, Allshire R, Baulcombe D, Birren BW, Brown W, Ekwall K, Kellis M, Leatherwood J, Levin H, Margalit H, Martienssen R, Nieduszynski CA, Spatafora JW, Friedman N, Dalgaard JZ, Baumann P, Niki H, Regev A, Nusbaum C.** 2011. Comparative functional genomics of the fission yeasts. *Science* **332:**930–936.
67. **Yu C, Bonaduce MJ, Klar AJS.** 2013. Defining the epigenetic mechanism of asymmetric cell division of *Schizosaccharomyces japonicus* yeast. *Genetics* **193:**85–94.
68. **Grewal SIS, Klar AJS.** 1996. Chromosomal inheritance of epigenetic states in fission yeast during mitosis and meiosis. *Cell* **86:**95–101.
69. **Thon G, Friis T.** 1997. Epigenetic inheritance of transcriptional silencing and switching competence in fission yeast. *Genetics* **145:**685–696.
70. **Klar AJS.** 1998. Propagating epigenetic states through meiosis: where Mendel's gene is more than a DNA moiety. *Trends Genet* **14:**299–301.
71. **Klar AJS.** 1994. A model for specification of the left-right axis in vertebrates. *Trends Genet* **10:**392–396.
72. **Klar AJS.** 2004. A genetic mechanism implicates chromosome 11 in schizophrenia and bipolar diseases. *Genetics* **167:**1833–1840.
73. **Armakolas A, Klar AJS.** 2006. Cell type regulates selective segregation of mouse chromosome 7 DNA strands in mitosis. *Science* **311:**1146–1149.
74. **Armakolas A, Klar AJS.** 2007. Left–right dynein motor implicated in selective chromatid segregation in mouse cells. *Science* **315:**100–101.
75. **Sauer S, Klar AJS.** 2013. Reply to "Chromosome-specific nonrandom sister chromatid segregation during stem-cell division." *Nature* **498:**254–256.
76. **Klar AJS.** 2014. Selective chromatid segregation mechanism invoked for the human congenital mirror hand movement disorder development by *RAD51* mutations: A hypothesis. *Int J Biol Sci* 2014; **10**(9):1018–1023. doi:10.7150/ijbs.9886. Available from http://www.ijbs.com/v10p1018.htm
77. **Klar AJS.** 2014. Selective Chromatid Segregation Mechanism Explains the Etiology of Chromosome 11 Translocation-Associated Psychotic Disorders: A Review. *J Neurol Disord* **2:**173. doi:10.4172/2329-6895.1000173

DNA-only Transposons

IV

Mobile DNA, 3rd Edition
Nancy L. Craig, Michael Chandler, Martin Gellert, Alan M. Lambowitz, Phoebe A. Rice and Suzanne Sandmeyer

doi:10.1128/microbiolspec.MDNA3-0034-2014

Alison B. Hickman[1]
Fred Dyda[1]

25 Mechanisms of DNA Transposition

In this chapter, we provide an overview of the fundamental concepts of DNA transposition mechanisms. Our aim is to emphasize basic themes and, in this effort, we will focus on specific illustrative cases rather than attempt an exhaustive review of the literature. We hope that the selected references will point the curious reader towards the landmark studies in the field as well as some of the most exciting recent results. We also direct the reader to other recent reviews (1–3).

DNA transposases are enyzmes that move discrete segments of DNA called transposons from one location in the genome (often called the donor site) to a new site without using RNA intermediates. DNA transposases are usually encoded by the mobile element itself (in which case they are "autonomous" transposons). However, some transposons are missing a self-encoded transposase yet have ends that can be recognized by a transposase encoded somewhere else in the genome, and thus are "non-autonomous" (Siguier *et al.*, this volume). Although logic suggests that all DNA transposons are moved by transposases, the term was originally reserved for those enzymes that do not require significant regions of homology between any part of the transposon and the sites to which they are moved, the so-called target (or insertion or integration) sites. As biology is not always neat and tidy, transposases can exhibit a spectrum of homology requirements and vagaries of terminology have arisen such that certain transposases are sometimes referred to as "resolvases" or by the generic term "recombinases".

From a mechanistic perspective, there are only a few ways in which transposases catalyze the required DNA strand breakage and rejoining reactions that comprise transposition (1), so from a structural point of view there are only a few different types of catalytic domain found in transposases. The catalytic domain topology, or its "fold", is a convenient way to classify DNA transposases although they have also historically been grouped according to whether or not their strand breakage mechanism involves a covalently-bound transposase/DNA intermediate. Other modes of classification include whether transposition proceeds through a replicative or non-replicative pathway, and whether transposition involves double- or single-stranded forms of DNA. These are useful distinctions, and it is worth noting that even within a single one of these categories, different DNA transposases can exhibit variations in their mechanisms (see also Siguier *et al.*, this volume).

[1]Laboratory of Molecular Biology, National Institute of Diabetes and Digestive and Kidney Diseases, National Institutes of Health, 5 Center Dr., Bethesda, MD 20892, USA.

It is this "similar-yet-distinct" property of transposition mechanisms that is part of their ongoing fascination.

CHEMISTRY OF DNA CLEAVAGE AND STRAND TRANSFER

There are four distinct types of catalytic nuclease domain folds (4) that are known to be used by DNA transposases to carry out the chemical reactions of transposition (Table 1). The most common is the so-called RNase H-like fold, sometimes also referred to as a DD(E/D) domain or the "retroviral integrase fold" as it has three catalytic acid residues at its active site (5–7). The second major type of catalytic domain is seen for those transposases that act on single-stranded DNA (ssDNA) and is known as an HUH domain (8). The serine transposases (such as those of IS*607*, Tn*5397*, and Tn*5541* (9, 10)) and tyrosine transposases (exemplified by those of CTnDOT and Tn*916* (11)) are predicted to have the same catalytic domain folds as serine and tyrosine site-specific recombinases, respectively. This last aspect is illustrated in Table 1: the four catalytic nuclease domains found in DNA transposases are also used by other enzymes that rearrange DNA such as retroviral integrases, invertases, resolvases, site-specific recombinases, and the RAG-1 recombinase involved in V(D)J recombination.

1. DNA Transposases with RNase H-like Catalytic Domains

At the core of RNase H-like transposases is an active site in which three catalytic acidic residues (DDE or DDD) coordinate two metal ions - no doubt Mg^{2+} *in vivo* - in order to activate either a water molecule or a 3′-OH group of a nucleotide for nucleophilic attack on a phosphodiester bond (Figure 1A). This type of active site was first structurally investigated and a chemical mechanism proposed in the context of the 3′-5′ exonuclease of the Klenow fragment of *E. coli* DNA polymerase I (12). More recently, the mechanism has been further investigated and described for RNase H from *Bacillus halodurans* (13, 14) and analyzed computationally with quantum mechanics/molecular mechanics (QM/MM) methods (15, 16).

There are two types of reactions catalyzed by RNase H-like active sites in transposases: (i) nucleophilic attack by an activated water molecule on a scissile phosphate at or close to a transposon end to break a DNA phosphodiester bond such that a 3′-OH group and a 5′-phosphate are generated at the cleavage site; and (ii) transesterification in which nucleophilic attack by the 3′-OH of a terminal nucleotide of a DNA strand rearranges the connectivity of DNA strands by simultaneously cleaving one strand while covalently joining one to another. This latter reaction can be used to join one DNA strand to its opposite strand to form a hairpin, to join one transposon end to the other end to form a circular intermediate, or to integrate a transposon into a new site. Each reaction occurs via an in-line S_N2 nucleophilic attack which occurs with inversion of the stereoconfiguration at the scissile phosphate being attacked (17, 18, 19). These two reactions, used in different combinations and on different DNA strands,

Table 1 Examples of proteins containing the four types of nuclease catalytic domains found in DNA transposases and other enzymes that rearrange DNA

	NUCLEASE DOMAIN			
FUNCTION	RNase H-like (DDE)	HUH	Phosphotyrosine site-specific recombinase	Phosphoserine site-specific recombinase
Transposase	• Tpase of most ISs • Tn5 • MuA • Mos1 • Hermes	• TnpA of IS*91* • TnpA of IS*Hp608* • ORFs of IS*CR*s • Helitrons	• CTnDOT • Tn916	• IS*607* • TnpX
Bacteriophage or viral integrase	HIV, PFV integrase		bacteriophage λ integrase	Bxb1, ΦC31 integrase
DNA invertase	Piv		Flp	Hin
DNA resolvase			• Cre • XerC/D	• Sin • γδ resolvase
Other	• RNase H • Some HJ resolvases • RAG1	• Rolling circle replication proteins • Some conjugative relaxases • Rep proteins of adeno-associated virus (AAV)	Type IB topoisomerases	

have been deployed by transposases to generate a plethora of reaction pathways for transposition.

One important characteristic shared by DNA transposases is that the hydrolysis of high energy cofactors is not required for any of the steps. (Although the *Drosophila* P element transposase contains a GTP-binding domain, GTP hydrolysis is not required for transposition (Majumdar & Rio, this volume). That said, transposition reactions often require supercoiling of either the transposon ends or the target DNA, in part because the potential energy stored in supercoils can drive the reaction forward when DNA is cleaved. At the same time, it has been demonstrated that in some cases hydrolysis of high energy cofactors is required for the disassembly of the final product protein-DNA complex (20): transposition reactions are often slow and proceed through several steps of assembling elaborate protein-DNA complexes in which the final assembly is so stable that taking it apart requires energy. It is possible that the process of assembling ever-more stable complexes ensures the directionality of the reaction.

Where it has been examined, evidence suggests that in the context of an active complex in which a transposase is bound to its transposon end DNA (also known as a "synaptic complex" or "transpososome"), a single active site is able to catalyze both types of reactions at one end of the transposon and does so sequentially (19, 21). In other words, there is no evidence that during the reactions at one transposon end, one transposase monomer catalyzes hydrolysis and a separate monomer performs strand transfer.

The only divalent metal ion species that appear to be able to support all the steps of transposition are Mg^{2+} and Mn^{2+}. Typically, reactions *in vitro* are faster and more robust in Mn^{2+}, although sometimes less accurate. Interestingly, Ca^{2+} generally does not support hydrolysis or transesterification and acts as an inhibitor (22) despite an ionic radius (1.14Å) that does not differ much from those of Mg^{2+} (0.86Å) and Mn^{2+} (0.81Å). It appears that, at least for hydrolysis, the reduced charge transfer from water to Ca^{2+} results in less effective activation of the nucleophile (22). However, it has been reported that Ca^{2+} can catalyze strand transfer in the case of the bacteriophage MuA transposase (23; see also Harshey, this volume), perhaps indicating that this step where the nucleophile is a 3′-OH group is less stringent.

The two metal ions that are coordinated by the catalytic acidic residues of the RNase H-like fold and the scissile phosphate (Figure 1A) serve to precisely position the reacting groups, activate the nucleophile, and stabilize the pentacovalent transition state that is presumed to exist as the reactions are believed to be associative. The two metal ions most likely adopt distinct roles at each step of the transposition reaction in which one activates the nucleophile and the other stabilizes the leaving group (24–27). This has been particularly well-characterized for the RNase H-like catalytic domain of the prototype foamy virus (PFV) integrase, a close relative of DNA transposases, which shows that in the presence of metal ions, there is a symmetric organization of the metal ions on either side of the scissile phosphate (where the metals M_A and M_B both coordinate the same oxygen of the scissile phosphate and each binds to one of the side chain carboxyl oxygens of a active site Asp residue (28; Engelman & Cherepanov, this volume). This suggests that this particular nuclease active site is adept at coordinating alternating nucleophilic attack from first one side and then another to choreograph the multiple steps that comprise transposition (19).

2. DNA Transposases with HUH Catalytic Domains

A large number of DNA transposases use an entirely different catalytic domain with an HUH nuclease fold (reviewed in (3)) to cut and join ssDNA (Figure 1B). The best characterized examples are the transposases of the prokaryotic IS*200* (see He *et al.* and Siguier *et al.*, this volume) and IS*91* families of insertion sequences. In the case of the IS*200* family, the transposition mechanism has been established experimentally in detail, and shown to require ssDNA forms of both the donor and target DNAs (29–35). HUH catalytic domains are used by a wide variety of proteins that occupy different biological niches, inititating processes such as plasmid rolling circle replication, the conjugative transfer of plasmids between cells, and the replication of parvoviruses such as the adeno-associated virus.

HUH nuclease domains use either one or two active site nucleophilic tyrosine residues (where the nucleophile is the OH group of the side chain) to cleave ssDNA through the formation of a 5′-phosphotyrosine covalent intermediate. If the 5′-phosphotyrosine linkage is subsequently attacked by a terminal 3′-OH group of another DNA strand, the covalent intermediate is said to be "resolved" i.e., the phosphotyrosine link is broken and the two DNA strands become connected in a strand transfer reaction. The name "HUH" refers to two conserved and catalytically required histidines (separated by a hydrophobic residue) that coordinate an single essential divalent metal ion cofactor (Figure 1B), which in all likelihood is Mg^{2+} in cells.

CLEAVAGE

STRAND TRANSFER

A. DD(E/D)

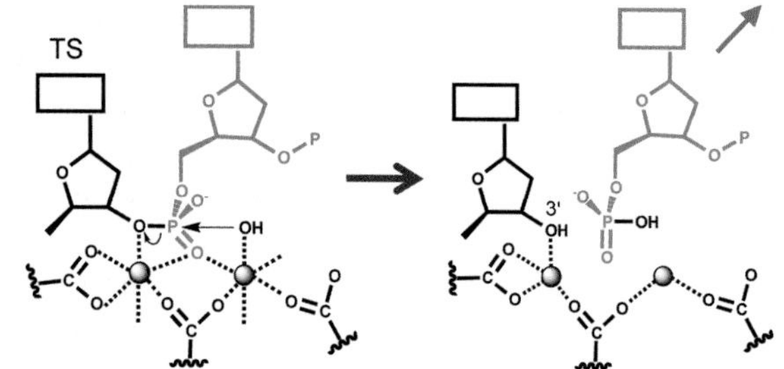

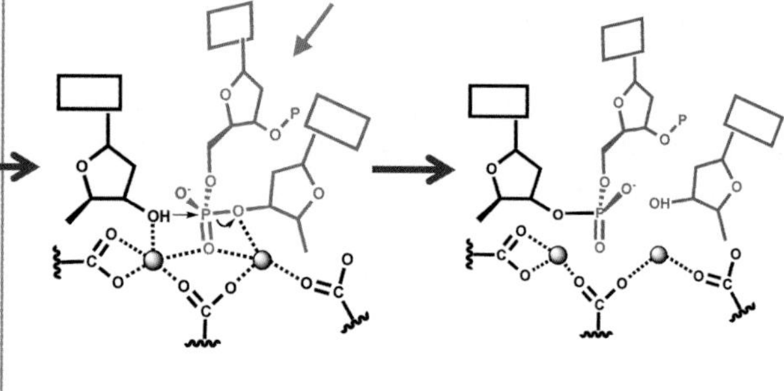

B. HUH

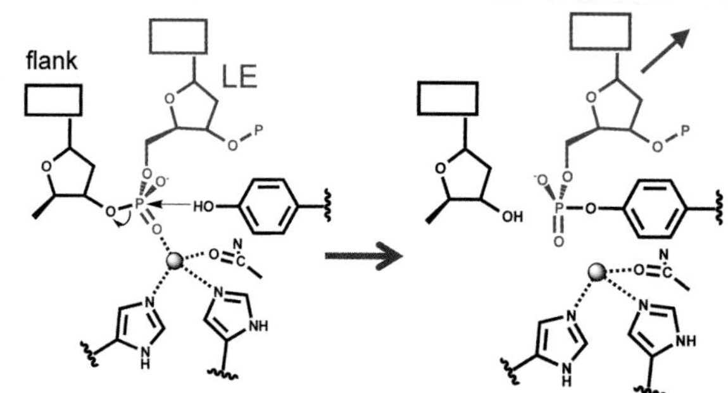

flank from RE

C. Ser transposase

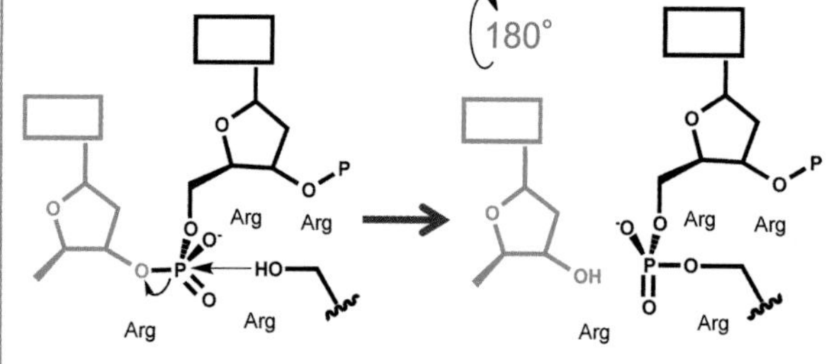

D. Tyr transposase

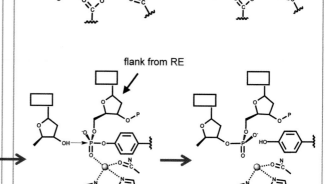

Mg^{2+} binds and polarizes the scissile phosphate, setting it up for nucleophilic attack by either tyrosine or a 3′-OH group (36–38). Some HUH transposases require only one active site tyrosine ("Y1 transposases") yet others need two closely spaced tyrosines ("Y2 transposases") to complete the cycle of strand cleavage and rejoining.

For some HUH nucleases, the chemical steps can be supported by a wide range of divalent metal ions, suggesting that the active site is relatively tolerant. For example, TrwC, the HUH relaxase of plasmid R388 which catalyzes the initial nicking reaction of conjugative DNA transfer, can cleave ssDNA in the presence of Mg^{2+}, Mn^{2+}, Ni^{2+}, Zn^{2+}, Ca^{2+}, or Cu^{2+} (39).

Beyond the IS*91* and IS*200* families, there are several recently discovered families of mobile elements which have associated proteins with HUH domains and which may turn out to fit the definition of DNA transposons. For example, the IS*CR*s, or Insertion Sequence Common Regions, are associated with antibody resistance genes and seem likely to be mobile elements with transposases resembling those of the IS*91* family, although their mobility has not yet been experimentally shown (40). More closely related to the IS*200* family of transposases are a group of $TnpA_{REP}$ proteins (41, 42) (also known as RAYTS (43)) that are associated with repetitive extragenic palindromic sequences (or REPs). REP sequences form hairpins and have been found scattered throughout many bacterial genomes (41, 43); it seems likely that the REP sequences and $TnpA_{REP}$ proteins are remnants of ssDNA transposons. Finally, the widespread eukaryotic helitron transposons (Thomas & Pritham, this volume) also appear to encode an HUH domain in their transposases but have not yet been demonstrated to be active in the myriad species in which they have been identified (44–46).

3. Serine Transposases

There are a number of bacterial transposons that encode a serine transposase. These include insertion sequences (ISs) such as IS*607* (47) which is proposed to use a circular intermediate of the IS to recombine with a target DNA (48) (Figures 1C & 2), and certain conjugative transposons such as Tn*5397* from *Clostridium difficile* and its relatives which also move using a circular intermediate (49, 50). Serine transposases are assumed to share many catalytic features with the serine site-specific recombinases such as resolvases and invertases (reviewed in (51) and elsewhere in this volume), yet exhibit a relaxed or practically-absent requirement for homology between the recombining sites (i.e., the abutted transposon ends in the circular intermediate and the target site). Nevertheless, specificity of target-site selection is a continuum; for instance, Tn*5397* displays a strong target site preference (52) whereas IS*607* has very little insertion specificity (53).

Serine recombinases are predicted to have the same catalytic core fold as the structurally characterized γδ resolvase (54, 55). They have been classified into four groups based on their domain organization and function (9) and, according to this classification, the Tn*5397* and Tn*4451* transposases are also known as "large serine" recombinases because - relative to γδ resolvase - they have an additional large C-terminal domain. The group of IS*607*-like transposases are distinguished from γδ resolvase by a reversal of the order of the DNA-binding domain and the catalytic domain within the primary sequence.

DNA cleavage by serine recombinases involves an active site serine nucleophile that attacks the scissile phosphate forming a covalent 5′-phosphoserine intermediate and a free 3′-OH group (Figure 1C). No divalent metal ion or any other cofactors are required, and key roles in catalysis are played by an array of arginine

Figure 1 Basic chemical reactions catalyzed by DNA transposases. (A) An RNase H-like active site, based on structures of PFV intasomes (28, 94, 95). The green DNA represents the cleaved dinucleotide, and orange is the target strand. Spheres indicate bound metal ions. (B) HUH nuclease active site acting on single-stranded DNA (based on PDB ID 2X06 of IS*Dra2* TnpA). Shown is the reaction that occurs at the transposon Left End (LE). After cleavage, the DNA flanking the LE (black) remains in the active site; upon exchange of α-helices between the two active sites of the dimeric transposase, the cleaved LE moves to the other monomer where it is joined to the cleaved RE to form a circular excised transposon (not shown). At the same time, the flanking DNA from the RE of the transposon switches active sites (as shown here in black) and subsequent joining results in a sealed donor backbone. (C) DNA cleavage catalyzed by a serine recombinase. The active serine is surrounded by many Arg residues. Upon 180° rotation of one dimer within a tetramer, one strand rotates out of the active site (green) while another rotates in (orange). (D) DNA cleavage catalyzed by a tyrosine recombinase. Crucial residues within the active site include a conserved RHR triad (for details, see also (181)).
doi:10.1128/microbiolspec.MDNA3-0034-2014.f1

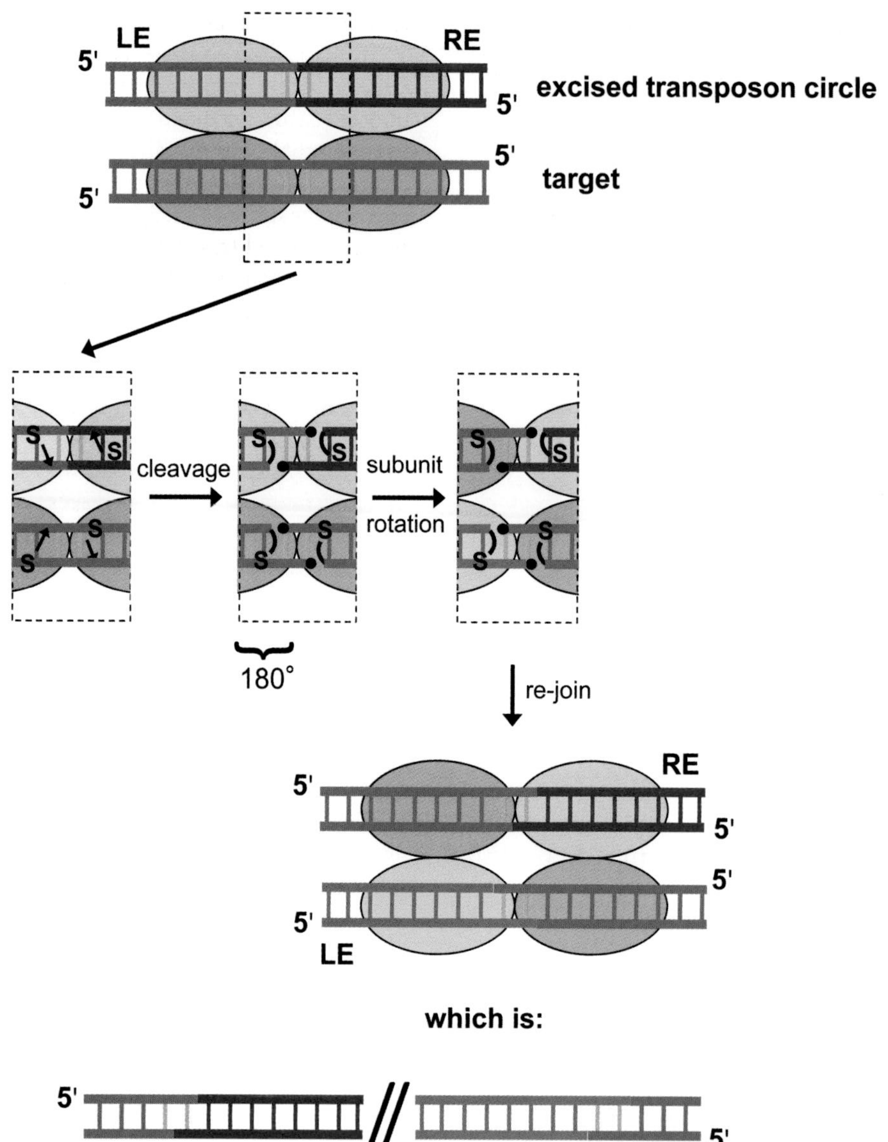

Figure 2 Proposed pathway for transposon circle integration into target DNA catalyzed by a serine transposase. At the top, a tetrameric assembly is shown bringing together the abutted Left End (LE; orange) and Right End (RE; red) of an excised circular transposon with a target DNA (green). The reactions in the dashed box show how four cleavage reactions in which each active site serine becomes covalently attached to one strand of DNA, followed by a 180 degree rotation of the left-most dimer, leads to a re-organization of the strands. Resolution of the four covalent intermediates results in an integrated transposon. doi:10.1128/microbiolspec.MDNA3-0034-2014.f2

residues at the active site (56, 57). The resolution of the covalent intermediate by a terminal 3′-OH group of another strand results in strand transfer.

Recombination by serine recombinases is understood to occur in the context of a tetrameric complex in which each subunit cleaves one of the four strands of the recombining DNA duplexes (Figure 2). In the tetramer, resolution of the phosphoserine linkages and strand transfer occur after a dramatic subunit rotation where one of the dimers of the tetramer rotates 180 degrees around the other (55).

4. Tyrosine Transposases

Most conjugative transposons (also known as Integrative Conjugative Elements, or ICEs; Figure 3) are mobilized by associated tyrosine transposases which are believed to be structurally and mechanistically related to the well-characterized site-specific tyrosine recombinases (51, 58, 59) such as Cre (van Duyne, this volume), Flp (Jayaram *et al.*, this volume; 60) and λ integrase (Landy, this volume) (3). The most extensively studied are the transposases of the Tn*916* family of conjugative transposons (61) and the CTnDOT (Wood & Gardner, this volume) conjugative transposon. These appear to have adapted a mechanism of site-specific recombination that proceeds through a Holliday junction to catalyze the transposition steps.

Tyrosine transposases cleave DNA using an active site tyrosine residue to attack the scissile phosphate and to form a covalent 3′-phosphotyrosine linkage (Figure 1D). This cleavage polarity is the opposite to that exhibited by the other three types of transposases. By analogy to tyrosine recombinases, a tetrameric complex is believed to assemble in which each subunit cleaves one of the four strands of the recombining sites (Figure 4). In the excision step, recombination between the two transposon ends (*attL* and *attR*) results in the formation of a free circular intermediate in which the ends are abutted (Figure 4A). To generate this intermediate, tyrosine transposases make staggered cuts at the transposon ends in which pairs of active site tyrosine residues are sequentially covalently bound (Figure 4A). The circular intermediate is then transferred into the recipient cell through conjugation. Once there, another recombination event takes place and the intermediate is integrated into a target at a *attB* site (Figure 4B).

Conjugative transposons display a spectrum of targeting specificity in their requirements for homology between the *attL*/*attR* junction and the *attB* site (11). Some insert essentially randomly, others display relatively strict specificity reflecting the site-specific recombinase machinery at work, and yet others occupy a middle ground with strict albeit usually very short sequence requirements for where they will integrate (11, 61, 62). While most conjugative transposons use tyrosine or serine transposases, it has recently been discovered that some conjugative transposons are mobilized by RNase H-like transposases (63, 64).

TRANSPOSITION PATHWAYS

1. RNase H-like Transposases

(i) Replicative Transposition

Replicative transposition couples transposition to extensive DNA replication and, in doing so, generates a second copy of the transposon at a new target site. This is an obvious mechanism that allows a mobile element to expand and proliferate within a genome. One of the best studied of these elements is bacteriophage Mu which uses replicative transposition to increase its copy number in infected cells (reviewed in (65) and Harshey, this volume). This mechanism is also employed by the Tn*3* family of transposons (Nicolas *et al.*, this volume). During phage Mu replicative transposition (66), the phage-encoded transposase, MuA, catalyzes two hydrolysis and transesterification reactions that occur sequentially on each transposon end with no intervening steps (Figure 5a): the first nucleophilic attack by water on the transposon end generates a 3′-OH group that is then used to attack the site for transposon insertion. This strand of the transposon end is therefore known as the "transferred strand". The end result are branched DNA structures at each transposon end; these are then substrates for a complex set of "handover" reactions between the transpososome and the replication fork (67, 68).

One key feature of replicative transposition is that it does not generate double-strand breaks (DSBs) at the transposon ends. In this sense, there is a mechanistic analogy with retroviral integration (see Engelman & Cherepanov, this volume): the donor DNA of retroviral integration is the blunt-ended linear product of reverse transcription and while integrase processes this by removing two nts from the viral transferred strand, no DSBs are needed. This mechanistic parallel is reflected in the close structural similarity between the catalytic domains of retroviral integrase (5) and the phage MuA transposase (6).

A different type of replicative transposition is carried out by the so-called "copy-and-paste" insertion sequences such as those of the large IS*3* family, and has been extensively studied for the representative element IS*911* (see Chandler *et al.*, this volume). Replication is required by these transposons to generate an excised transposon circle that is the substrate for the integration step (Figure 5b). After the initial generation of a 3′-OH group at one end of the transposon, this then attacks the same strand at the opposite end of the transposon to generate a "figure-of-eight" intermediate (i.e., what it looks like when the transposon is contained on a plasmid). This results in the joining of the two transposon ends by a ssDNA bridge. Replication by the host cell machinery then converts this intermediate to a transposon circle (69), which is the substrate for subsequent insertion into a new site. The integration step of IS*911* transposition (70) requires two more strand cleavage events at the junction of the joined transposon ends to generate the 3′-OH nucleophiles for insertion into a target.

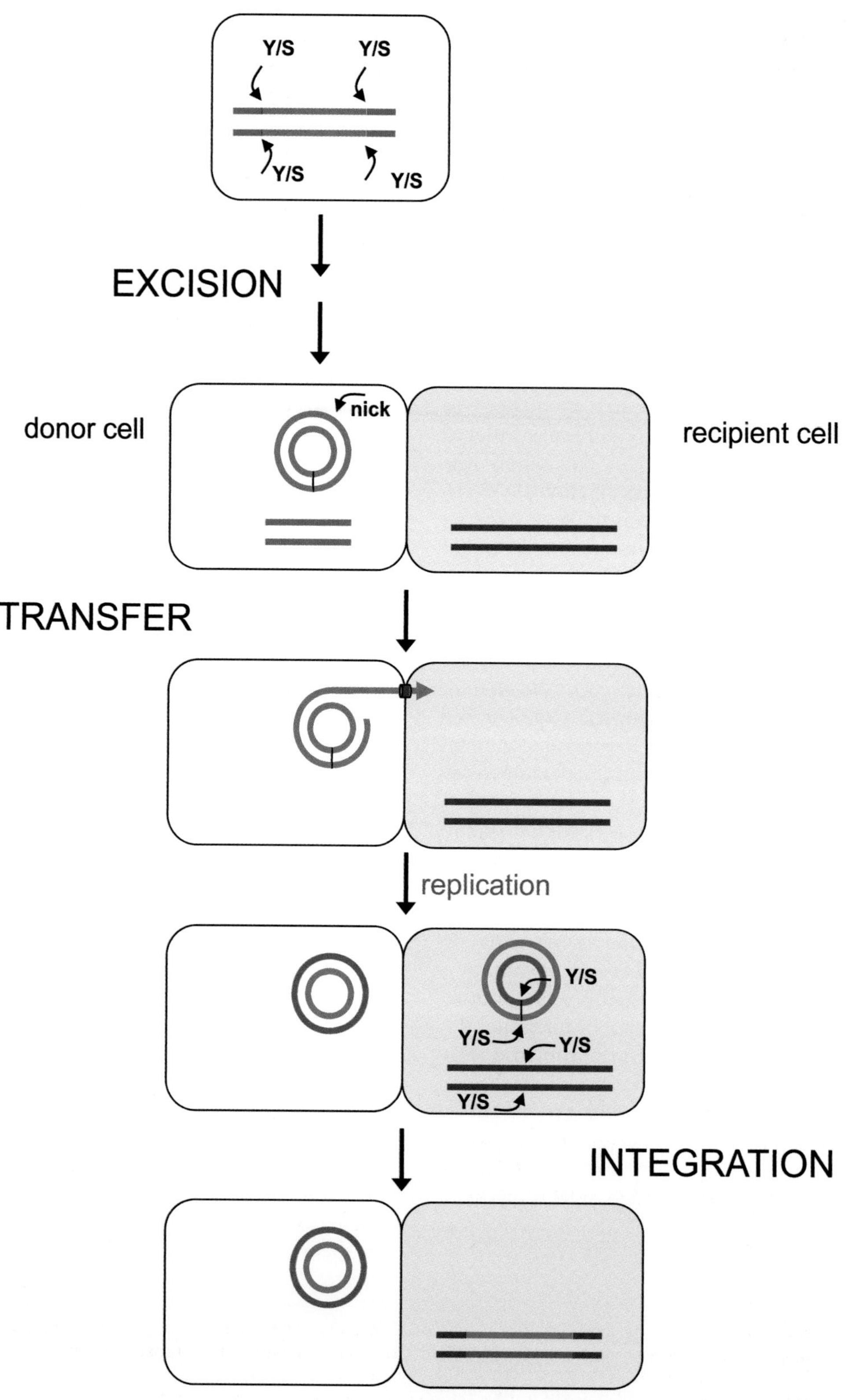
Y/S
Y/S
Y/S
Y/S
EXCISION
donor cell
nick
recipient cell
TRANSFER
replication
Y/S
Y/S
Y/S
Y/S
INTEGRATION

In the case of IS*911*, either end can be used to initiate transposition (71). In a remarkably clever twist, formation of an IS*911* transposon circle comcommitantly generates a strong promoter at the site of the junction between the two abutted transposon ends (72). This promoter drives expression of the transposase, high levels of which are thought to be particularly important for integration as it would increase the probability of rapid insertion of the transposon into a new site before loss of the transposon circle.

(ii) Cut-and-Paste Transposition: Excision

Many RNase H-like transposases catalyze "non-replicative" or "cut-and-paste" transposition. These enzymes must generate DSBs at their transposon ends, the prerequisite for liberating the mobile element from its donor site. After an initial hydrolysis reaction that cleaves one strand, different cut-and-paste transposases cleave the second strand and generate a DSB in a variety of different ways (73), differences that are also reflected in structural variations in the transposases (74). Discovering the details of second-strand processing is a particularly informative approach to understand transposition mechanisms.

After the initial nucleophilic attack by a water molecule to form the first 3′-OH group at a transposon end, one way in which the second strand break can be introduced is by another activated water molecule, as illustrated in Figure 5c for the Tc1/*mariner* pathway (see Tellier *et al.*, this volume). Thus, the generation of DSBs is simply a case of two sequential strand cleavage events on opposite strands at the same end (75–77). The eukaryotic P element also uses this pathway, but is unusual in that there is an atypically large offset in the position of cleavage events on the two strands: whereas the transferred strand is cleaved at the transposon-donor junction, the other is cleaved 17 bp into the transposon end (78).

Alternatively, the 3′-OH generated during the first strand cleavage can serve as the nucleophile to attack the opposite, second strand at the same transposon end, in which case a hairpin is formed. There are two possible variations on this step. If the first strand that is cleaved is the one that will eventually be the transferred strand, then the resulting 3′-OH is on the transposon end, and attack by this on the second strand generates a hairpin on the transposon end (Figure 5d); this is the case for the prokaryotic IS*4* family of transposases such as Tn*10* and Tn*5* (Haniford & Ellis, this volume; 79), as well as for the eukaryotic *piggyBac* transposases (Yusa, this volume; 80). For the transposition reaction to continue, the hairpin must be opened so that a 3′-OH group is available at the transposon end for the final transesterification step of integration, and this occurs by yet another nucleophilic attack by an activated water molecule.

In contrast, if the first cleavage reaction is on the "non-transferred strand", then the resulting free 3′-OH is on the flanking DNA, and the hairpin intermediate is also formed on the flanking donor DNA (Figure 5e). Examples of transposases that use this pathway include the eukaryotic *hAT* (Atkinson, this volume; 81) and CACTA transposases, as well as the related V(D)J RAG1/2 recombinase (Roth, this volume; 82) which almost certainly evolved from an ancient Transib transposon (83). In these cases, the hairpin on the flanking end DNA does not need to be opened up for the purposes of transposition, as the transferred strand 3′-OH is generated at the same time as hairpin formation. There are no known examples of prokaryotic transposases that form flanking hairpins during transposition.

It is clear from the accumulated experimental work over decades and on a variety of systems that, no matter how the second strand is processed, there is only one DD(E/D) active site involved at each transposon end, despite the observation that transpososomes always contain multiple transposase monomers. (One possible exception may be the P element with its 17-nucleotide staggered cuts; it has been proposed that the simplest model to explain this would be for two monomers of the tetrameric transposase to act on each end (84)). Mechanistic studies suggest that in the case of transposases that form transposon end hairpins, the cleavage and transesterification steps are accomplished with the transferred strand remaining in the transposase active site for all four chemical steps (19, 21), and major conformational rearrangements appear unnecessary. On the other hand, when the hairpin is formed on flanking DNA, the active site must

Figure 3 Pathway of conjugative transposition. Whether catalyzed by a serine or a tyrosine transposase, excision results in a circular intermediate in which the transposon ends are abutted. Only one of the strands of this intermediate is transferred to the recipient cell, and replication (new strands shown in blue) regenerates the double-stranded form in both cells. doi:10.1128/microbiolspec.MDNA3-0034-2014.f3

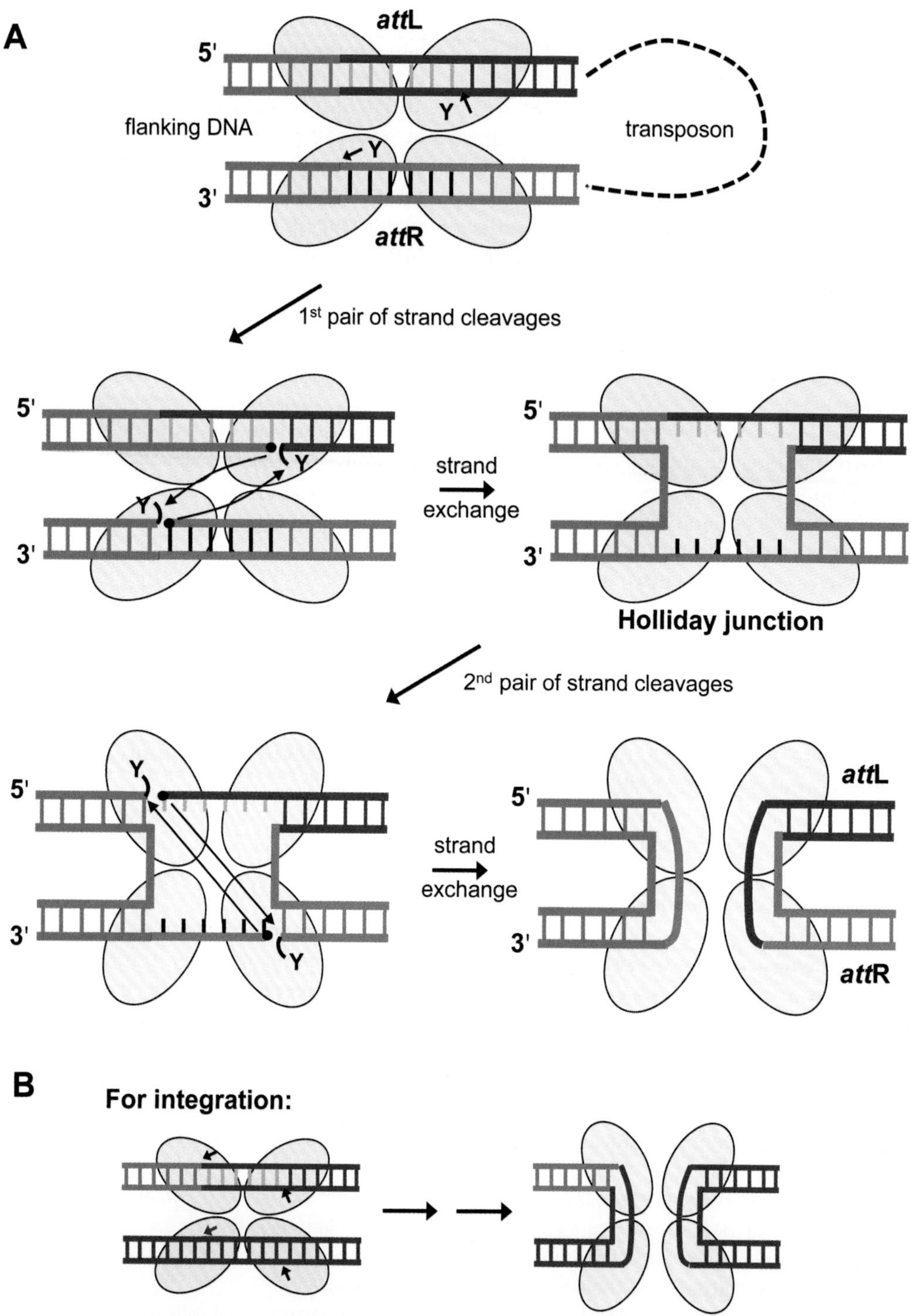

Figure 4 Proposed pathways of excision and integration by tyrosine transposases. (A) Transposon excision. (B) Transposon integration.
doi:10.1128/microbiolspec.MDNA3-0034-2014.f4

somehow switch from the non-transferred strand where the first nicking reaction takes place to the transferred strand to catalyze the subsequent transesterification reaction. Thus, it seems likely that significant conformational rearrangements will be needed for this type of reaction.

One cut-and-paste transposition system, Tn7 from *E. coli* (Peters, this volume), has solved the issue of DSB generation in a unique way. Tn7 encodes five transposition-related proteins (TnsA, B, C, D and E) with TnsA/TnsB forming the heteromeric transposase (85, 86). TnsB, which contains a predicted RNase H-like catalytic domain, cuts the transferred strand generating the 3′-OH but the non-transferred strand cut is by TnsA, which is a restriction endonuclease-like nuclease (87). In a clear demonstration of the deep mechanistic relationship between cut-and-paste transposition and DSBs, TnsA active site mutants that render the transposase unable to cut the non-transferred 5′ ends of Tn7 turn the system into a replicative transposon (88).

(iii) Insights from Structure

The stage of the transposition reaction where the transposon ends have been cleaved but the target DNA is not yet bound has been captured in three-dimensional structures (Figure 6) of three different cut-and-paste transposases bound to DNA (89, 90, 91; reviewed in (92)). The structures have been extremely valuable in illuminating several aspects of DNA transposition mechanisms relevant to excision. To date, only one transpososome - that of MuA - has been structurally characterized at a later step in the reaction in which its ends have been inserted into target DNA (93; Figure 6D). The retroviral intasome has also been structurally characterized at various stages along the integration pathway and is discussed in detail elsewhere (28, 94, 95; Engelman & Cherepanov, this volume; Figure 6B).

How does a transposase recognize its own DNA?

Clearly, all transposases must be able to specifically recognize both ends of their transposons. The types and organization of sequences at transposon ends that are important for transposase binding vary widely. In general, binding sites responsible for transposon end recognition do not extend to the very tip of the transposon where the cleavages occur but are slightly subterminal. Among the characterized systems, some transposases recognize a single stretch of DNA that contains enough basepairs to uniquely define a binding site (e.g., Tn5 (89)). Others have been shown to bind two different short sites that are close together using two distinct site-specific DNA binding domains within a single protein monomer (e.g., MuA (93, 96) and the *mariner* elements (90, 97)), or to bind multiple sites close to the transposon ends where each site binds a separate transposase monomer (e.g., Tn7 (98)).

The Tn5 transposase binds identical 19 bp sequences (known as Terminal Inverted Repeats (TIRs) or Inverted Terminal Repeats (ITRs)) that are found in reverse orientation at each transposon end. The crystal structure of the Tn5 transposase/TIR complex (Figure 6A) revealed that almost all of the 19 bp are contacted by protein (89), certainly sufficient for unique recognition of the ends. There are three distinct domains within the Tn5 transposase (99) (as there are for its close relative, the Tn*10* transposase (100)), and residues contributed by all of them participate in DNA binding.

In contrast, the Mos1 transposase uses two small site-specific DNA binding Helix-Turn-Helix (HTH) domains to recognize two distinct subterminal segments within its ends (Figure 6C). This is most likely a feature of all Tc/*mariner* transposases. HTH domains and their variants are encountered in a wide range of DNA binding proteins (101, 102), and are similarly employed by another structurally characterized transposase, bacteriophage MuA (93; Figure 6D). Many IS families and several eukaryotic transposon superfamilies have transposases with N-terminal HTH domains implicated in DNA binding (103–105).

Other transposons also feature two different binding sites at their ends (106). For example, analysis of the structure of the Hermes *hAT* transposase bound to its TIRs (91; Figure 6E) suggests a bipartite binding mode in which the TIRs are bound by multiple domains of the transposase whereas an N-terminal BED-finger domain (missing in the structure in Figure 6E; 107) recognizes short subterminal repeats which are a characteristic feature of *hAT* transposons (Atkinson, this volume).

How do transposases synapse their two transposon ends?

In their active forms, both the Tn5 and Mos1 transposase are dimeric, yet they arrive at this point through different assembly pathways. For Tn5, the transposase is a monomer in solution in the absence of DNA (108), as are most prokaryotic DNA transposases that contain an RNase H-like catalytic domain and act on double-stranded DNA, and protein dimerization is believed to occur after each end has been bound by one transposase monomer (79). The Tn5 transposase binds its transposon ends "in trans", which means that the active site that is engaged in processing one transposon end is part of the polypeptide chain that encodes the DNA binding domain(s) that binds the other. Thus, dimerization is concomitant with transpososome formation. In contrast, Mos1 is a dimer in solution prior to DNA binding (109), and *mariner* elements such as Mos1 sequentially capture their transposon ends (110) in a defined order as the relative binding affinites to the

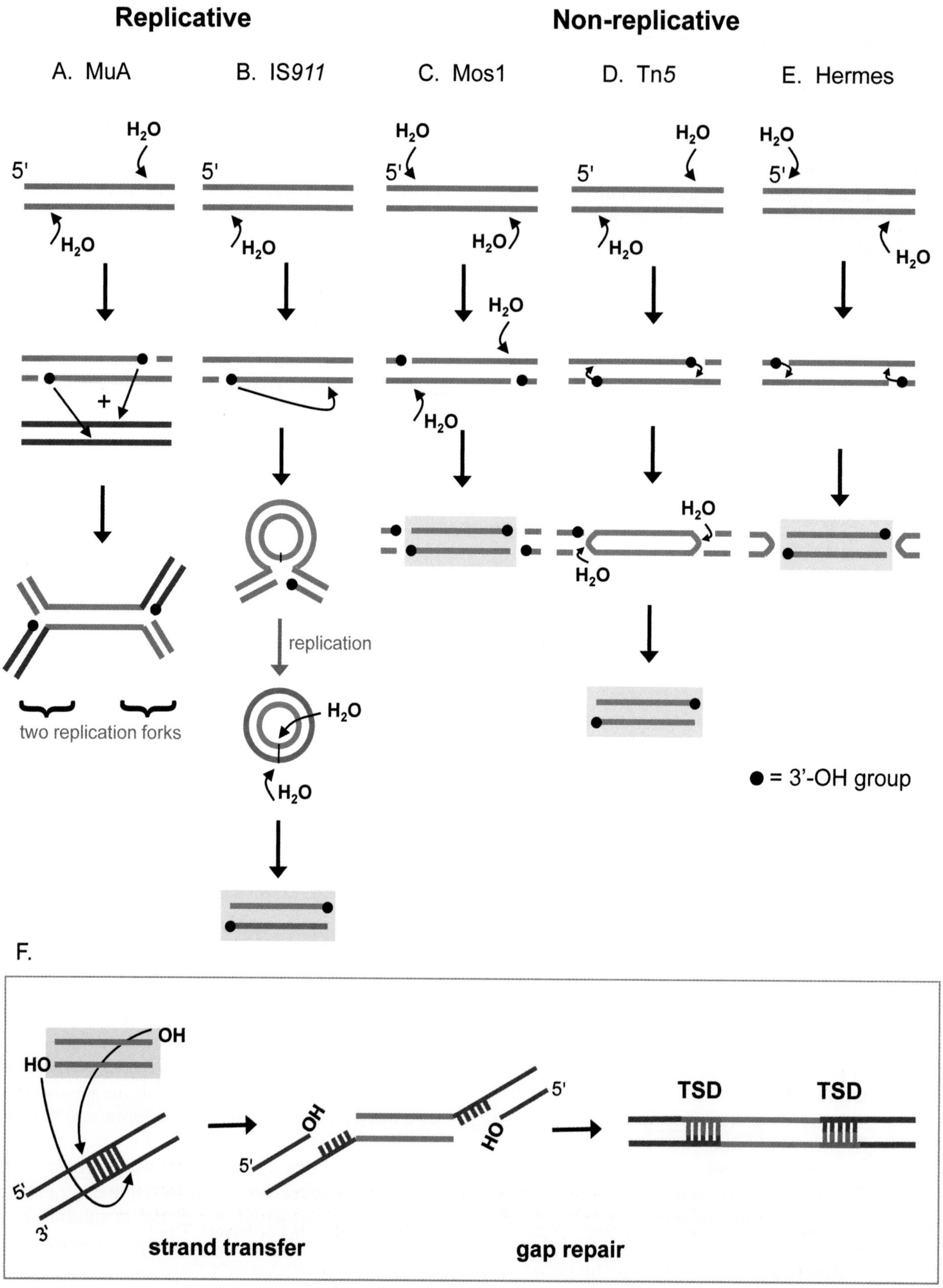

Replicative
Non-replicative
A. MuA
B. IS911
C. Mos1
D. Tn5
E. Hermes
H2O
5'
replication
two replication forks
● = 3'-OH group
F.
OH
HO
5'
3'
strand transfer
gap repair
TSD
TSD

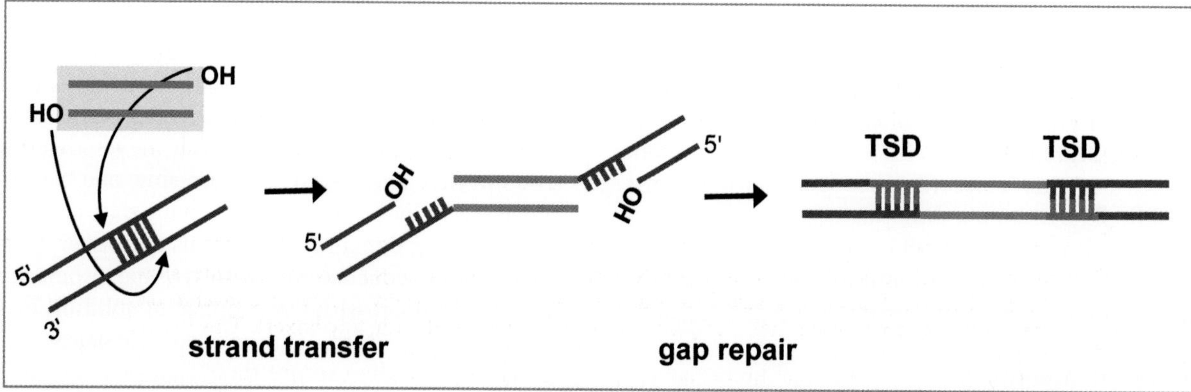

two ends are different (111, 112). The Mos1 transpososome structure reveals that catalysis is again in trans (90), suggesting a recurring regulation mechanism that would ensure that both ends are located and bound before any chemical reactions begin. Regardless of the assembly pathway, for these transpososomes, there is one transposase monomer per end and, at each, one active site per end appears to perform all of the chemical steps.

Transposases that bind to multiple asymmetrically-organized sites within their transposon ends tend to form transpososomes that contain more than two transposase protomers. For example, among the prokaryotic systems, the MuA transposase forms a series of distinct complexes with DNA - some involving DNA binding sites far removed from the bacteriophage ends - before it arrives at a tetrameric synaptic complex (Figure 6D) capable of initiating the chemical steps of transposition (66). The Tn7 transposase incorporates its ultimate target site into the synaptic complex before any chemical reactions begin (113, 114) and, after strand transfer, the Tn7 transpososome has been reported to contain one molecule of TnsD, at least 6 protomers of TnsB, and multiple copies of $TnsA_2C_2$ (115).

Among the eukaryotic transposons, the transposon ends of the *Hermes* transposon are typical of other *hAT* elements as its asymmetric ends are several hundred bp long and contain multiple apparently haphazardly arranged subterminal binding sites (91, 116). The active form of Hermes is a ring-shaped octamer in which eight N-terminal site-specific DNA binding domains are available to interact with these interior sites while presenting the two transposon TIRs to the catalytic sites of one of the dimers of the octameric assembly (91; Figure 6E). The P element transposase, which also has asymmetric ends (Majumdar & Rio, this volume), is reportedly tetrameric both prior to DNA binding and upon end synapsis (84). Sleeping Beauty, a resurrected vertebrate transposase of the Tc1/*mariner* family (117), is also proposed to form a tetrameric transpososome (118). It has been suggested that multimerization prior to DNA binding might be a way to down-regulate transposition activity (sometimes called over-production inhibition or OPI) (119).

(iv) Target Binding and Integration

In addition to binding its transposon ends, a DNA transposase must also bind the DNA into which it is going to insert its transposon, and the structures of the MuA transpososome and the PFV intasome bound to target DNA have been particularly instructive regarding this step of transposition (93, 95). For many transposition systems, target DNA binding is non-specific and the transposon can integrate essentially anywhere. On the other hand, a rare few integrate into specific sites such as Tn7 which integrates into a precise location downstream of the *E. coli glmS* gene (see Peters, this volume). This targeting by Tn7 is dependent on TnsD which site-specifically binds the target sequence (120, 121). However, Tn7 is very resourceful as it also encodes TnsE (122). When TnsE is incorporated in the Tn7 transpososome instead of TnsD, transposition is directed to chromosomal regions where replication is terminated and to DSBs (123). This target selection process is likely mediated by a direct physical interaction between TnsE and the β clamp replication processivity factor (124).

Other transposons exhibit varying degrees of target specificity. Among the RNase H-like transposases, the Tc/*mariner* transposases insert at TA sites (125) and the *piggyBac* transposases always insert into a TTAA sequence (126). Other RNase H-like transposases have preferred sites of integration. These involve distinct and often palindromic (127) patterns of base pairs (128–137), suggesting that these target site preferences might reflect some other property other than sequence specificity, for example perhaps DNA bendability. Indeed, target DNA has been repeatedly observed to be bent when bound by transpososomes (e.g., for Tn7 (138), Tn*10* (139), Mos1 (140), MuA (93)), and such bending has been proposed to be an effective mechanism to ensure the directionality of the reaction (93, 141). In the crystal structure of the MuA transpososome (Figure 6D), target DNA is bent through a total of 140° (93) whereas in the retroviral PFV intasome (Figure 6B), the target DNA is also bent but to less of an extent (95).

In another manifestation of the variability of transposition mechanisms, the point during the reaction at which target DNA is bound by the transpososome can

Figure 5 Transposition pathways for RNase H-like transposases. Arrows indicate sites of strand cleavage and the black dots indicate 3′-OH groups. Many pathways converge on essentially the same form of the excised transposon (highlighted with grey boxes). This linear intermediate is then integrated into target DNA as shown in (f). Target site duplications (TSDs) are generated when the cell repairs the gaps introduced by staggered strand transfer reactions. Adapted from (1). doi:10.1128/microbiolspec.MDNA3-0034-2014.f5

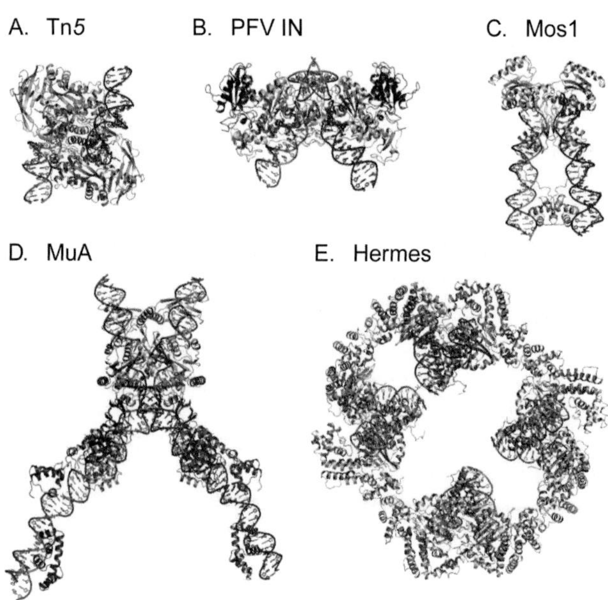

Figure 6 Transpososome structures for the RNase H-like DNA transposases. Cartoon representations of five transpososomes containing RNase H-like catalytic domains determined by X-ray crystallography. In all images, the catalytically active protomers acting on the two transposon ends are colored in orange and green, with the green catalytic domain acting on the DNA end shown in blue and the orange domain acting on the DNA end in red. Where target DNA is present, it is shown in grey. Inactive protomers (MuA, PFV IN and Hermes) are colored purple and magenta. The following PDB codes were used: (A) Tn*5*, 1MUH; (B) PFV IN, 4E7J; (C) Mos1, 3HOT; (D) MuA, 4FCY; (E) Hermes, 4D1Q. doi:10.1128/microbiolspec.MDNA3-0034-2014.f6

differ. For some transposases, target DNA cannot be bound until both transposon ends are cleaved and the flanking DNA has been released; this might be a consequence of using the same protein surface for binding both flanking and target DNA (142, 143). For other transposases such as that of Tn*7*, the requirement is precisely the opposite and target DNA must be bound before strand cleavage is initiated (113, 138).

Integration into target DNA (Figure 5f) occurs via two transesterification reactions involving the coordinated attack of the two free 3′-OH groups at the transposon ends on opposite strands of target DNA a defined distance (i.e., number of bp) apart. This staggered strand transfer generates short gaps at both sides of the integrated transposon that must be subsequently repaired, giving rise to target site duplications (TSDs). The length of the TSDs is a characteristic of each particular family of transposon, and is generally between 2 and 11 bp (Siguier *et al.*, this volume and (105)). In mechanistic terms, the constancy of TSD size reflects that integration into two target strands occurs in the context of a transpososome assembly with a fixed structural relationship between two active sites within the transpososome.

Sometimes, the TSDs reflect aspects of prior steps of the transposition pathway. For example, *piggyBac* excision results in a four bp 5′ overhang on each end that is 5′-TTAA (80), identical to its TTAA target site requirement and, due to a four bp offset in the sites of strand transfer into target, identical to its TSDs (126). Upon insertion, the overhangs basepair with the offsets generated by the four bp stagger in nucleophilic attacks (80), and there are no gaps that have to be filled but simply a bond to be formed. Thus, the TSDs are only temporary as, when *piggyBac* excises, the flanking DNA is restored to its original sequence without the need for any DNA synthesis. This property has made *piggyBac* a particularly attractive system for *in vivo* genomic applications: during a cycle of insertion and excision, it leaves no permanent genomic marks behind (144). Other transposons, notably *Sleeping Beauty* and *Tol2*, are also finding wide use for genomic manipulation experiments (145–147).

One final aspect of target site selection that should be mentioned is the notion of target immunity. Certain transposons such as phage Mu, Tn*7*, and those of the Tn*3* family possess the ability to distinguish self from non-self, and can avoid the suicidal step of integrating into themselves. In the case of Mu, immunity depends on a Mu-encoded ATPase, MuB (148, 149). For Tn*7*, the ATPase TnsC and the TnsB protein of the transposase work together to establish target immunity (150). Curiously, a similarly-functioning protein has not yet been identified for Tn*3* transposons (Nicolas *et al.*, this volume; 151) and the mechansim of target immunity is not currently understood for Tn*3* and its relatives.

(v) Host Proteins

Some RNase H-like transposases have been shown to require host proteins to carry out transposition, very often to bend or deform the transposon ends. This may be important when transposon binding sites are separated along the DNA yet need to be incorporated within the transpososome. Classical examples are the MuA, Tn*5*, and Tn*10* transposases which rely on highly expressed DNA bending proteins such as IHF ("Integration Host Factor") and HU ("Histone-like protein" from *E. coli* strain U93). For example, Mu transposition (Harshey, this volume) requires both IHF and HU; an IHF binding site is located within an enhancer sequence which is ~900 bp away from the phage left end, and HU is needed to assemble a catalytically active transpososome but is not needed thereafter (152). The structure of the MuA transpososome has provided a valuable starting point for modeling how HU may participate in the assembly process (93). The Tn*10* transposase also relies on IHF for transpososome assembly (153, 154). In contrast, the closely related Tn*5* transposase does not appear to require IHF whereas both Tn*5* and Tn*10* use the host protein H-NS (histone-like nucleoid-structuring protein) to assist proper transpososome assembly in roles that are different in detail (155, 156).

There are also examples of eukaryotic transposases that rely on a DNA binding protein, HMGB ("high mobility group box"), in a similar assisting role. Sleeping Beauty uses HMGB1 (157), probably due to two far-spaced binding sites on both transposon ends although this does not appear to be a general property of *mariner* transposases. HMGB plays a similar role in V(D)J recombination in the assembly of RAG1/2-RSS (Recombination Signal Sequence) complexes (158, 159). It would be surprising if HMGB did not participate in other as-yet biochemically uncharacterized eukaryotic transposition systems.

2. HUH Transposases

(i) Transposon End Recognition

Single-stranded DNA transposases of the IS*200*/IS*605* family mobilize their transposons when they become accessible in single-stranded form, for example on the lagging strands during replication or during certain types of DNA repair (160, 161). Other HUH DNA transposases such as those of the IS*91* family or helitrons may actively assist in generating ssDNA, either by recruiting a cellular helicase as has been suggested for IS*91* (162) or by encoding a helicase domain as proposed for helitrons (44), but the mechanisms of these "rolling circle" transposons (163) are far from being firmly established.

Recently, the mechanism of IS*200*/IS*605* transposition has been intensively studied through a series of genetic, biochemical, and structural experiments with the *Helicobacter pylori* IS*608* and *Deinococcus radiodurans* ISDra2 transposases (29–35; He *et al.*, this volume). One of the most important concepts to emerge is that neither the cleavage sites at the transposon ends nor the target site is recognized by a site-specific DNA binding domain of the transposase. Rather, the transposase mediates DNA-DNA interactions involving basepairing between two ssDNA regions, and this directs the appropriate scissile phosphate into the active site. This same mechanism also ensures site-specificity of integration, which precisely targets either a tetra- or pentanucleotide sequence (164, 165).

TnpA is an obligatory dimer even in the absence of DNA, and locates its transposon ends by binding DNA hairpins that are formed by palindromic sequences located subterminal to the transposon tips (30; Figure 7). Binding is neither strictly in cis or in trans, as both monomers contribute to the binding of both hairpins. Directly 5′ of the recognition hairpins are "guide sequences" that basepair with bases at the cleavage sites (32). This arrangement is provocatively reminiscent of the use of RNA guide sequences located 5′ of CRISPR hairpins in microbial immune systems (166). Figure 7 shows a model of how the IS*608* dimer would bind its two transposon ends. Each ssDNA end is folded into a distinct secondary structure similar to the types of complicated secondary structures formed by RNA, featuring two layers of base triplets and the subterminal hairpin. Upon cleavage at each end, the 5′-ends become covalently attached to the protein, one at each active site (Figure 1B). It is proposed that exchange of the 5′ ends between subunits through a dramatic conformational change involving the two α-helices bearing the active site tyrosine residues, followed by attack of the two free 3′ ends on the active site phosphotyrosines, leads to the generation of an excised single-stranded circle in which the two transposon ends are directly abutted (32). This is the substrate for subsequent insertion into a new location.

The IS*200* TnpA active sites are composite in the sense that, for cleavage, the HUH motif is provided by one monomer while the nucleophilic tyrosine is provided by the other. For strand transfer, the arrangement changes as the tyrosines covalently attached to DNA strands through a 5′-phosphotyrosine linkage move from one monomer to the other; strand transfer is therefore catalyzed by active sites that are now composed of residues from the same polypeptide chain.

Other ssDNA transposons have subterminal palindromic sequences capable of forming hairpins, suggesting that some aspects of end recognition may be conserved. For example, IS*91* has dissimilar hairpins at its two transposon ends, and they have been proposed to have distinct roles during rolling circle transposition (162). The IS*91* transposase is a Y2 transposase member of the HUH superfamily, and is likely a functional homolog of the plasmid ΦX174 rolling circle replication protein, gpA (163, 167). For the eukaryotic helitrons, alignments of reconstructed consensus sequences show that while a palindromic sequence followed by a highly conserved tetranucleotide at the transposon 3′ end is always present, the only common feature at

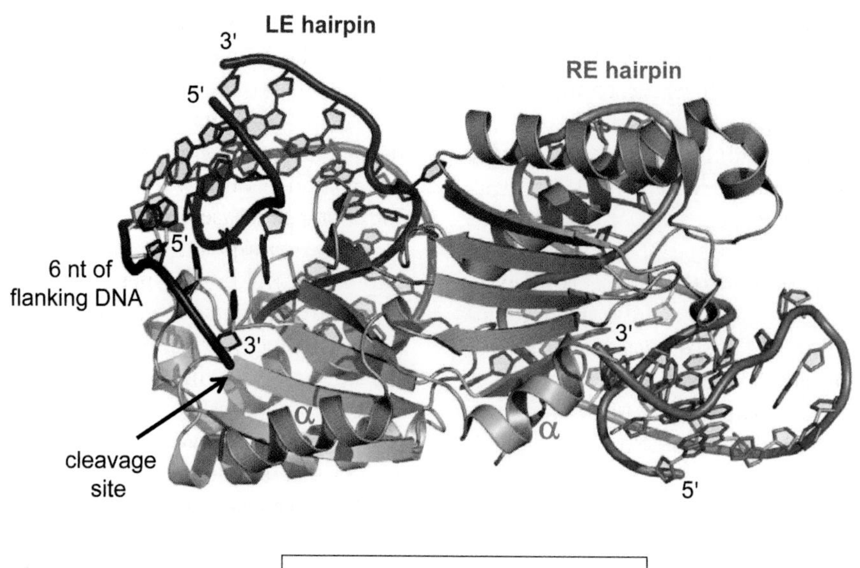

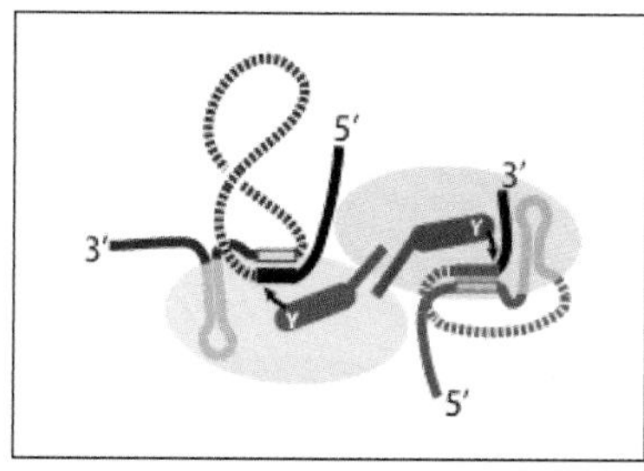

Figure 7 Transpososome of the HUH transposase TnpA of IS*608*, modelled as binding one Left End (LE; red) and one Right End (RE; blue). The PDB codes 2VJV and 2VJU were used. The inset shows the step of the reaction in the strand transfer and reset model for IS*608* transposition (see He *et al.*, this volume) to which the structure corresponds; note that the RE flank has not yet been observed crystallographically.
doi:10.1128/microbiolspec.MDNA3-0034-2014.f7

5′ transposon ends is a conserved dinucleotide (44). Thus, understanding how these transposases are able to recognize and cleave their two ends awaits experimental work to establish their mechanism.

(ii) Integration

The IS*200* DNA transposases integrate immediately 3′ to specific tetra- or pentanucleotide sequences dictated by the guide sequences located subterminal to the 5′ transposon end. For example, the target site requirement for the IS*608* transposase is 5′-TTAC. IS*91* also has a specific tetranucleotide target sequence, either 5′-CTTG or 5′-GTTC. Eukaryotic helitrons preferentially integrate between A and T nucleotides (44, 46), suggesting that a slightly different mechanism for target site recognition may be at work. None of these transposases generates TSDs upon integration.

For IS*200* transposases, an excised ssDNA transposon circle is inserted into a new target site through a mechanism that requires that the target site possess the same sequence as the original Left End cleavage site (reviewed in He *et al.*, this volume); thus, the DNA-DNA recognition step of initial cleavage is repeated but it is the single-stranded target that is directed into the active site rather than the uncleaved transposon-flank junction. It is proposed that integration occurs by a second set of strand cleavages, followed by another exchange of covalently attached 5′-ends between subunits of the transposase dimer. This reorganization of DNA segments results in an integrated transposon. One consequence of this mechanism is that the integration site-specificity can be manipulated at will by simply changing the bases comprising the guide sequence (168).

Thus, the key to strand transfer by HUH enzymes is that, upon nucleophilic attack on a phosphotyrosine intermediate, the covalent linkage to the active site tyrosine is resolved and the enzyme resumes its initial unbound state. It is this ability to cycle between covalent attachment, strand movement, and resolution that makes HUH enzymes particularly adept for repetitive processes such as generating multiple plasmid copies through rolling circle replication or breaking DNA bonds and rejoining them in a different configuration to comprise transposition (3).

Helitrons and IS*CR*s are notable in that most appear to have captured host genes (or fragments of genes). This may be a consequence of their proposed rolling circle-like mechanism which sporadically may not always terminate at the 5′ end of the element but continue beyond. When termination finally occurs, the mobilized DNA can now include sequences that were located beyond the authentic 5′ end. This capture may be an important contributor to the evolution of species where helitrons are particularly abundant, for example in maize (169, 170), and to the spread of antibiotic resistance genes by the IS*CR*s (40). Gene capture is likely a variation of one-ended transposition previously described for IS*91* (162).

GENOME REPAIR AFTER TRANSPOSON INSERTION

Once a transposon has been mobilized and re-integrated, there are two genomic sites that must be repaired: the empty donor site from which the transposon has excised, and the nicks and gaps that were introduced (if any) upon target site integration. Surprisingly little is known about these repair steps for most transposons. Among prokaryotic transposition systems, the donor site is believed to be repaired by homologous recombination as long as a sister chromosome is available i.e., after DNA replication but before cell division (171). This timing also leads to the regeneration of the transposon at the donor site, thereby leading to an increase in the number of transposon copies per cell. Gap repair has recently been investigated for the non-replicative pathway of Mu transposition and shown to involve proteins of both the replication restart and homologous recombination pathways (172).

Information on how donor sites are repaired in eukaryotic systems arises predominantly from studies on the P element (Majumdar & Rio, this volume) and the V(D)J recombination system (Roth, this volume). RAG1/2-mediated cleavage and excision leaves flanking hairpins at the sites of the DSBs (173), and the cellular Artemis complex is responsible for the initial step of repair which is hairpin opening (174). The NHEJ (nonhomologous end-joining) proteins such as the Ku70/80 heterodimer and DNA-PKcs are also involved (175). The *Drosophila* Ku70 homolog has been similarly implicated in the repair of gaps introduced by P element excision (176), and Ku70/80 is reported to be important in the proper execution of ciliate programmed genome rearrangements by the domesticated DNA transposase PiggyMac (182). It seems likely that the same proteins and repair pathways will be required for opening flanking hairpins generated by other eukaryotic transposases such as the *hAT* and CACTA family transposases.

MuA transposition has been particularly well-studied from the perspective of what happens to the transpososome itself once the chemical steps of transposition have been completed. The version of the MuA transpososome which remains bound to DNA once strand transfer is completed is exceedingly stable. To complete transposition, the MuA complex must be removed from the two branched junctions so that the host cell replication machinery can take over. This is accomplished by a host-encoded remodeling protein, ClpX, which is a member of the Clp/Hsp100 family of AAA+ ATPases (20, 177). ClpX recognizes a specific sequence at the C-terminus of MuA, and unfolds one particular subunit of the tetramer in an ATP-dependent reaction; this appears to destabilize the transpososome enough to allow functional replication forks to be assembled (178–180).

CONCLUDING REMARKS

The powerful combination of genetic, biochemical, and structural studies has illuminated many aspects of DNA recombination mechanisms. However, vast areas of great interest remain to be investigated. For example, although the field has a solid foundation for understanding of how catalysis is likely to proceed for the simpler RNase H-like transposases, we still do not have a clear, detailed sense of how protein structure mediates flanking hairpin formation and the evident need to act sequentially on first one strand and then the other. Similarly intriguing is the question of how certain transposases mediate the formation of circular intermediates. It remains unclear how serine and tyrosine transposases have taken the protein building blocks of site-specific recombinases and repurposed them for reactions that are considerably less stringent in terms of sequence specificity. Also, although at least 20 different superfamilies of eukaryotic transposases have been defined, few have yet been shown to be amenable to biochemical studies and we know very little about what distinguishes them and how this might relate to mechanism and structure. Notably, the helitron transposases are most curious as they surely must bear some functional and structural similarities to their well-characterized prokaryotic HUH relatives. Similarly intriguing are the

ISCRs, and as yet there is no direct experimental evidence regarding the mechanisms of either of these types of presumptively mobile elements. In all these areas, many important discoveries and answers to fundamental questions of structure and function await.

Acknowledgments. *This work was funded by the NIH Intramural Program of the National Institute of Diabetes and Digestive and Kidney Diseases (NIDDK). We thank Phoebe Rice for the inspiration for Table 1.*

Citation. Hickman AB., Dyda F. 2014. Mechanisms of DNA transposition. Microbiol Spectrum 3(2):MDNA3-0034-2014.

References

1. **Curcio MJ, Derbyshire KM.** 2003. The outs and ins of transposition: From Mu to kangaroo. *Nature Rev Mol Cell Biol* **4:**865–877.
2. **Montaño SP, Rice PA.** 2011. Moving DNA around: DNA transposition and retroviral integration. *Curr Opin Struct Biol* **21:**370–378.
3. **Chandler M, de la Cruz F, Dyda F, Hickman AB, Moncalian G, Ton-Hoang B.** 2013. Breaking and joining single-stranded DNA: the HUH endonuclease superfamily. *Nature Rev Microbiol* **11:**525–538.
4. **Yang W.** 2011. Nucleases: diversity of structure, function and mechanism. *Quart Rev Biophys* **44:**1–93.
5. **Dyda F, Hickman AB, Jenkins TM, Engelman A, Craigie R, Davies DR.** 1994. Crystal structure of the catalytic domain of HIV-1 integrase: Similarity to other polynucleotidyl transferases. *Science* **266:**1981–1986.
6. **Rice P, Mizuuchi K.** 1995. Structure of the bacteriophage Mu transposase core: A common structural motif for DNA transposition and retroviral integration. *Cell* **82:**209–220.
7. **Yuan YW, Wessler SR.** 2011. The catalytic domain of all eukaryotic cut-and-paste transposase superfamilies. *Proc Natl Acad Sci USA* **108:**7884–7889.
8. **Koonin EV, Ilyina TV.** 1993. Computer-assisted dissection of rolling circle DNA replication. *BioSystems* **30:** 241–268.
9. **Smith MCM, Thorpe HM.** 2002. Diversity in the serine recombinases. *Mol Microbiol* **44:**299–307.
10. **Smith MCM, Brown WRA, McEwan AR, Rowley PA.** 2010. Site-specific recombination by ΦC31 integrase and other large serine recombinases. *Biochem Soc Trans* **38:**388–394.
11. **Rajeev L, Malanowska K, Gardner JF.** 2009. Challenging a paradigm: the role of DNA homology in tyrosine recombinase reactions. *Microbiol Mol Biol Rev* **73:** 300–309.
12. **Beese LS, Steitz TA.** 1991. Structural basis for the 3′-5′ exonuclease activity of *Escherichia coli* DNA polymerase I: a two metal ion mechanism. *EMBO J* **10:**25–33.
13. **Nowotny M, Gaidamakov SA, Crouch RJ, Yang W.** 2005. Crystal structures of RNase H bound to an RNA/DNA hybrid: Substrate specificity and metal-dependent catalysis. *Cell* **121:**1005–1016.
14. **Nowotny M, Yang W.** 2006. Stepwise analyses of metal ions in RNase H catalysis from substrate destabilization to product release. *EMBO J* **25:**1924–1933.
15. **Rosta E, Woodcock HL, Brooks BR, Hummer G.** 2009. Artificial reaction coordinate "tunneling" in free-energy calculations: The catalytic reaction of RNase H. *J Comput Chem* **30:**1634–1641.
16. **Rosta E, Nowotny M, Yang W, Hummer G.** 2011. Catalytic mechanism of RNA backbone cleavage by ribonuclease H from quantum mechanics/molecular mechanics simulations. *J Am Chem Soc* **133:**8934–8941.
17. **Mizuuchi K, Adzuma K.** 1991. Inversion of the phosphate chirality at the target site of Mu DNA strand transfer: evidence for a one-step transesterification mechanism. *Cell* **66:**129–140.
18. **Engelman A, Mizuuchi K, Craigie R.** 1991. HIV-1 DNA integration: mechanism of viral DNA cleavage and DNA strand transfer. *Cell* **67:**1211–1221.
19. **Kennedy AK, Haniford DB, Mizuuchi K.** 2000. Single active site catalysis of the successive phosphoryl transfer steps by DNA transposases: Insights from phosphorothioate stereoselectivity. *Cell* **101:**295–305.
20. **Levchenko I, Luo L, Baker TA.** 1995. Disassembly of the Mu transposase tetramer by the ClpX chaperone. *Genes Dev* **9:**2399–2408.
21. **Bolland S, Kleckner N.** 1996. The three chemical steps of Tn10/IS10 transposition involve repeated utilization of a single active site. *Cell* **84:**223–233.
22. **Rosta E, Yang W, Hummer G.** 2014. Calcium inhibition of Ribonuclease H1 two-metal ion catalysis. *J Am Chem Soc* **136:**3137–3144.
23. **Savilahti H, Rice PA, Mizuuchi K.** 1995. The phage Mu transpososome core: DNA requirements for assembly and function. *EMBO J* **14:**4893–4903.
24. **Steitz TA, Steitz JA.** 1993. A general two-metal-ion mechanism for catalytic RNA. *Proc Natl Acad Sci USA* **90:**6498–6502.
25. **Stahley MR, Strobel SA.** 2005. Structural evidence for a two-metal-ion mechanism of group I intron splicing. *Science* **309:**1587–1590.
26. **Nowotny M.** 2009. Retroviral integrase superfamily: the structural perspective. *EMBO Reports* **10:**144–151.
27. **Nakamura T, Zhao Y, Yamagata Y, Hua YJ, Yang W.** 2012. Watching DNA polymerase η make a phosphodiester bond. *Nature* **487:**196–201.
28. **Hare S, Maertens GN, Cherepanov P.** 3′-processing and strand transfer catalysed by retroviral integrase *in crystallo*. *EMBO J* **31:**3020–3028.
29. **Ton-Hoang B, Guynet C, Ronning DR, Cointin-Marty B, Dyda F, Chandler M.** 2005. Transposition of ISHp*608*, member of an unusual family of bacterial insertion sequences. *EMBO J* **24:**3325–3338.
30. **Ronning DR, Guynet C, Ton-Hoang B, Perez ZN, Ghirlando R, Chandler M, Dyda F.** 2005. Active site sharing and subterminal hairpin recognition in a new class of DNA transposases. *Mol Cell* **20:**143–154.
31. **Guynet C, Hickman AB, Barabas O, Dyda F, Chandler M, Ton-Hoang B.** 2008. In vitro reconstitution of a

single-stranded transposition mechanism of IS*608*. *Mol Cell* **29:**302–312.

32. **Barabas O, Ronning DR, Guynet C, Hickman AB, Ton-Hoang B, Chandler M, Dyda F.** 2008. Mechanism of IS*200*/IS*605* family DNA transposases: Activation and transposon-directed target site selection. *Cell* **132:** 208–220.
33. **Hickman AB, James JA, Barabas O, Pasternak C, Ton-Hoang B, Chandler M, Sommer S, Dyda F.** 2010. DNA recognition and the precleavage state during single-stranded DNA transposition in *D. radiodurans*. *EMBO J* **29:**3840–3852.
34. **He S, Hickman AB, Dyda F, Johnson NP, Chandler M, Ton-Hoang B.** 2011. Reconstitution of a functional IS*608* single-strand transpososome: role of non-canonical base pairing. *Nucl Acids Res* **39:**8503–8512.
35. **He S, Guynet C, Siguier P, Hickman AB, Dyda F, Chandler M, Ton-Hoang B.** 2013. IS*200*/IS*605* family single strand transposition: mechanism of IS*608* strand transfer. *Nucl Acids Res* **41:**3302–3313.
36. **Hickman AB, Ronning DR, Kotin RM, Dyda F.** 2002. Structural unity among viral origin binding proteins: Crystal structure of the nuclease domain of adeno-associated virus Rep. *Mol Cell* **10:**327–337.
37. **Guasch A, Lucas M, Moncalián G, Cabezas M, Pérez-Luque R, Gomis-Rüth FX, de la Cruz F, Coll M.** 2003. Recognition and processing of the origin of transfer DNA by conjugative relaxase TrwC. *Nature Struct Biol* **10:**1002–1010.
38. **Datta S, Larkin C, Schildbach JF.** 2003. Structural insights into single-stranded DNA binding and cleavage by F factor TraI. *Struct* **11:**1369–1379.
39. **Boer R, Russi S, Guasch A, Lucas M, Blanco AG, Pérez-Luque R, Coll M, de la Cruz F.** 2006. Unveiling the molecular mechanism of a conjugative relaxase: The structure of TrwC complexed with a 27-mer DNA comprising the recognition hairpin and the cleavage site. *J Mol Biol* **358:**857–869.
40. **Toleman MA, Bennett PM, Walsh TR.** 2006. ISCR elements: Novel gene-capturing systems of the 21st century? *Microbiol Mol Biol Rev* **70:**296–316.
41. **Ton-Hoang B, Siguier P, Quentin Y, Onillon S, Marty B, Fichant G, Chandler M.** 2012. Structuring the bacterial genome: Y1-transposases associated with REP-BIME sequences. *Nucl Acids Res* **40:**3596–3609.
42. **Messing SAJ, Ton-Hoang B, Hickman AB, McCubbin AJ, Peaslee GF, Ghirlando R, Chandler M, Dyda F.** 2012. The processing of repetitive extragenic palindromes: the structure of a repetitive extragenic palindrome bound to its associated nuclease. *Nucl Acids Res* **40:**9964–9979.
43. **Nunvar J, Huckova T, Licha I.** 2010. Identification and characterization of repetitive extragenic palindromes (REP)-associated tyrosine transposases: implications for REP evolution and dynamics in bacterial genomes. *BMC Genomics* **11:**44.
44. **Kapitonov VV, Jurka J.** 2001. Rolling-circle transposons in eukaryotes. *Proc Natl Acad Sci USA* **98:** 8714–8719.
45. **Feschotte C, Wessler SR.** 2001. Treasures in the attic: Rolling circle transposons discovered in eukaryotic genomes. *Proc Natl Acad Sci USA* **98:**8923–8924.
46. **Pritham EJ, Feschotte C.** 2007. Massive amplification of rolling-circle transposons in the lineage of the bat *Myotis lucifugus*. *Proc Natl Acad Sci USA* **104:**1895–1900.
47. **Kersulyte D, Mukhopadhyay AK, Shirai M, Nakazawa T, Berg DE.** 2000. Functional organization and insertion specificity of IS*607*, a chimeric element of *Helicobacter pylori*. *J Bacteriol* **182:**5300–5308.
48. **Boocock MR, Rice PA.** 2013. A proposed mechanism for IS607-family serine transposases. *Mobile DNA* **4:**24.
49. **Bannam TL, Crellin PK, Rood JI.** 1995. Molecular genetics of the chloramphenicol-resistance transposon Tn*4451* from *Clostridium perfringens*: the TnpX site-specific recombinase excises a circular transposon molecule. *Mol Microbiol* **16:**535–551.
50. **Lyras D, Rood JI.** 2000. Transposition of Tn*4451* and Tn*4453* involves a circular intermediate that forms a promoter for the large resolvase, TnpX. *Mol Microbiol* **38:**588–601.
51. **Grindley NDF, Whiteson KL, Rice PA.** 2006. Mechanisms of site-specific recombination. *Annu Rev Biochem* **75:**567–605.
52. **Wang H, Smith MCM, Mullany P.** 2006. The conjugative transposon Tn*5397* has a strong preference for integration into its *Clostridium difficile* target site. *J Bacteriol* **188:**4871–4878.
53. **Kersulyte D, Kalia A, Zhang MJ, Lee HK, Subramaniam D, Kiuduliene L, Chalkauskas H, Berg DE.** 2004. Sequence organization and insertion specificity of the novel chimeric IS*Hp609* transposable element of *Helicobacter pylori*. *J Bacteriol* **186:**7521–7528.
54. **Sanderson MR, Freemont PS, Rice PA, Goldman A, Hatfull GF, Grindley NDF, Steitz TA.** 1990. The crystal structure of the catalytic domain of the site-specific recombination enzyme γδ resolvase at 2.7 Å resolution. *Cell* **63:**1323–1329.
55. **Li W, Kamtekar S, Xiong Y, Sarkis GJ, Grindley NDF, Steitz TA.** 2005. Structure of a synaptic γδ resolvase tetramer covalently linked to two cleaved DNAs. *Science* **309:**1210–1215.
56. **Keenholtz RA, Rowland SJ, Boocock MR, Stark WM, Rice PA.** 2011. Structural basis for catalytic activation of a serine recombinase. *Struct* **19:**799–809.
57. **Keenholtz RA, Mouw KW, Boocock MR, Li NS, Piccirilli JA, Rice PA.** 2013. Arginine as a general acid catalyst in serine recombinase-mediated DNA cleavage. *J Biol Chem* **288:**29206–29214.
58. **Hickman AB, Waninger S, Scocca JJ, Dyda F.** 1997. Molecular organization in site-specific recombination: The catalytic domain of bacteriophage HP1 integrase at 2.7Å resolution. *Cell* **89:**227–237.
59. **Kwon HJ, Tirumalai R, Landy A, Ellenberger T.** 1997. Flexibility in DNA recombination: Structure of the lambda integrase catalytic core. *Science* **276:**126–131.

60. **Chen Y, Rice PA.** 2003. New insight into site-specific recombination from Flp recombinase-DNA structures. *Annu Rev Biophys Biomol Struct* **32:**135–159.

61. **Roberts AP, Mullany P.** 2009. A modular master on the move: the Tn*916* family of mobile genetic elements. *Trends Microbiol* **17:**251–258.

62. **Waters JL, Salyers AA.** 2013. Regulation of CTnDOT conjugative transfer is a complex and highly coordinated series of events. *mBio* **4:**e00569–13.

63. **Brochet M, Da Cunha V, Couvé E, Rusniok C, Trieu-Cuot P, Glaser P.** 2009. Atypical association of DDE transposition with conjugation specifies a new family of mobile element. *Mol Microbiol* **71:**948–959.

64. **Guérillot R, Siguier P, Gourbeyre E, Chandler M, Glaser P.** 2014. The diversity of prokaryotic DDE transposases of the Mutator superfamily, insertion specificity, and association with conjugation machineries. *Genome Biol Evol* **6:**260–272.

65. **Harshey RM.** 2012. The Mu story: how a maverick phage moved the field forward. *Mobile DNA* **3:**21.

66. **Mizuuchi K.** 1992. Transpositional recombination: Mechanistic insights from studies of Mu and other elements. *Annu Rev Biochem* **61:**1011–1051.

67. **North SH, Kirtland SE, Nakai H.** 2007. Translation factor IF2 at the interface of transposition and replication by the PriA-PriC pathway. *Mol Microbiol* **66:**1566–1578.

68. **Jones JM, Nakai H.** 1999. Duplex opening by primosome protein PriA for replisome assembly on a recombination intermediate. *J Mol Biol* **289:**503–515.

69. **Duval-Valentin G, Marty-Cointin B, Chandler M.** 2004. Requirement of IS*911* replication before integration defines a new bacterial transposition pathway. *EMBO J* **23:**3897–3906.

70. **Ton-Hoang B, Polard P, Chandler M.** 1998. Efficient transposition of IS*911* circles *in vitro*. *EMBO J* **17:**1169–1181.

71. **Polard P, Chandler M.** 1995. An in vivo transposase-catalyzed single-stranded DNA circularization reaction. *Genes Dev* **9:**2846–2858.

72. **Ton-Hoang B, Bétermier M, Polard P, Chandler M.** 1997. Assembly of a strong promoter following IS*911* circularization and the role of circles in transposition. *EMBO J* **16:**3357–3371.

73. **Turlan C, Chandler M.** 2000. Playing second fiddle: second-strand processing and liberation of transposable elements from donor DNA. *Trends Microbiol* **8:**268–274.

74. **Hickman AB, Chandler M, Dyda F.** 2010. Integrating prokaryotes and eukaryotes: DNA transposases in light of structure. *Crit Rev Biochem Mol Biol* **45:**50–69.

75. **Dawson A, Finnegan DJ.** 2003. Excision of the *Drosophila* mariner transposon Mos1: Comparison with bacterial transposition and V(D)J recombination. *Mol Cell* **11:**225–235.

76. **Claeys Bouuaert C, Chalmers R.** 2010. Transposition of the human *Hsmar1* transposon: rate-limiting steps and the importance of the flanking TA dinucleotide in second strand cleavage. *Nucl Acids Res* **38:**190–202.

77. **Lampe DJ, Churchill MEA, Robertson HM.** 1996. A purified *mariner* transposase is sufficient to mediate transposition *in vitro*. *EMBO J* **15:**5470–5479.

78. **Beall EL, Rio DC.** 1997. *Drosophila* P-element transposase is a novel site-specific endonuclease. *Genes Dev* **11:**2137–2151.

79. **Steiniger-White M, Rayment I, Reznikoff WS.** 2004. Structure/function insights into Tn*5* transposition. *Curr Opin Struct Biol* **14:**50–57.

80. **Mitra R, Fain-Thornton J, Craig NL.** 2008. *piggyBac* can bypass DNA synthesis during cut and paste transposition. *EMBO J* **27:**1097–1109.

81. **Zhou L, Mitra R, Atkinson PW, Hickman AB, Dyda F, Craig NL.** 2004. Transposition of *hAT* elements links transposable elements and V(D)J recombination. *Nature* **432:**995–1001.

82. **Schatz DG, Swanson PC.** 2011. V(D)J recombination: Mechanisms of initiation. *Annu Rev Genet* **45:**167–202.

83. **Kapitonov VV, Jurka J.** 2005. RAG1 core and V(D)J recombination signal sequences were derived from *Transib* transposons. *PLoS Biol* **3:**e181.

84. **Tang M, Cecconi C, Bustamante C, Rio DC.** 2007. Analysis of P element transposase protein-DNA interactions during the early stages of transposition. *J Biol Chem* **282:**29002–29012.

85. **Biery MC, Lopata M, Craig NL.** 2000. A minimal system for Tn7 transposition: The transposon-encoded proteins TnsA and TnsB can execute DNA breakage and joining reactions that generate circularized Tn7 species. *J Mol Biol* **297:**25–37.

86. **Choi KY, Li Y, Sarnovsky R, Craig NL.** 2013. Direct interaction between the TnsA and TnsB subunits controls the heteromeric Tn7 transposase. *Proc Natl Acad Sci USA* **110:**E2038–E2045.

87. **Hickman AB, Li Y, Mathew SV, May EW, Craig NL, Dyda F.** 2000. Unexpected structural diversity in DNA recombination: The restriction endonuclease connection. *Mol Cell* **5:**1025–1034.

88. **May EW, Craig NL.** 1996. Switching from cut-and-paste to replicative Tn7 transposition. *Science* **272:**401–404.

89. **Davies DR, Goryshin IY, Reznikoff WS, Rayment I.** 2000. Three-dimensional structure of the Tn5 synaptic complex transposition intermediate. *Science* **289:**77–85.

90. **Richardson JM, Colloms SD, Finnegan DJ, Walkinshaw MD.** 2009. Molecular architecture of the Mos1 paired-end complex: The structural basis of DNA transposition in a eukaryote. *Cell* **138:**1096–1108.

91. **Hickman AB, et al.** 2014. Structural basis of *hAT* transposon end recognition by Hermes, an octameric DNA transposase from *Musca domestica*. *Cell* **158:**353–367.

92. **Dyda F, Chandler M, Hickman AB.** 2012. The emerging diversity of transpososome architectures. *Quart Rev Biophys* **45:**493–521.

93. **Montaño SP, Pigli YZ, Rice PA.** 2012. The Mu transpososome structure sheds light on DDE recombinase evolution. *Nature* **491:**413–417.

94. **Hare S, Gupta SS, Valkov E, Engelman A, Cherepanov P.** 2010. Retroviral intasome assembly and inhibition of DNA strand transfer. *Nature* **464:**232–236.

95. **Maertens GN, Hare S, Cherepanov P.** 2010. The mechanism of retroviral integration from X-ray structures of its key intermediates. *Nature* **468:**326–329.

96. **Schumacher S, Clubb RT, Cai M, Mizuuchi K, Clore GM, Gronenborn AM.** 1997. Solution structure of the Mu end DNA-binding Iβ subdomain of phage Mu transposase: modular DNA recognition by two tethered domains. *EMBO J* **16:**7532–7541.

97. **Watkins S, van Pouderoyen G, Sixma TK.** 2004. Structural analysis of the bipartite DNA-binding domain of Tc3 transposase bound to the transposon DNA. *Nucl Acids Res* **32:**4306–4312.

98. **Arciszewska LK, Craig NL.** 1991. Interaction of the Tn7-encoded transposition protein TnsB with the ends of the transposon. *Nucl Acids Res* **19:**5021–5029.

99. **Braam LAM, Reznikoff WS.** 1998. Functional characterization of the Tn*5* transposase by limited proteolysis. *J Biol Chem* **273:**10908–10913.

100. **Kwon D, Chalmers RM, Kleckner N.** 1995. Structural domains of IS*10* transposase and reconstitution of transposition activity from proteolytic fragments lacking an interdomain linker. *Proc Natl Acad Sci USA* **92:** 8234–8238.

101. **Wintjens R, Rooman M.** 1996. Structural classification of HTH DNA-binding domains and protein-DNA interaction modes. *J Mol Biol* **262:**294–313.

102. **Aravind L, Anantharaman V, Balaji S, Babu MM, Iyer LM.** 2005. The many faces of the helix-turn-helix domain:Transcription regulation and beyond. *FEMS Microbiol Rev* **29:**231–262.

103. **Rousseau P, Gueguen E, Duval-Valentin G, Chandler M.** 2004. The helix-turn-helix motif of bacterial insertion sequence IS*911* transposase is required for DNA binding. *Nucl Acids Res* **32:**1335–1344.

104. **Nagy Z, Szabó M, Chandler M, Olasz F.** 2004. Analysis of the N-terminal DNA binding domain of the IS*30* transposase. *Mol Microbiol* **54:**478–488.

105. **Feschotte C, Pritham EJ.** 2007. DNA transposons and the evolution of eukaryotic genomes. *Annu Rev Genet* **41:**331–368.

106. **Beall EL, Rio DC.** 1998. Transposase makes critical contacts with, and is stimulated by, single-stranded DNA at the P element termini *in vitro*. *EMBO J* **17:** 2122–2136.

107. **Aravind L.** 2000. The BED finger, a novel DNA-binding domain in chromatin-boundary-element-binding proteins and transposases. *Trends Biochem Sci* **25:**421–423.

108. **Braam LAM, Goryshin IY, Reznikoff WS.** 1999. A mechanism for Tn5 inhibition: Carboxyl-terminal dimerization. *J Biol Chem* **274:**86–92.

109. **Richardson JM, Dawson A, O'Hagan N, Taylor P, Finnegan DJ, Walkinshaw MD.** 2006. Mechanism of Mos1 transposition: insights from structural analysis. *EMBO J* **25:**1324–1334.

110. **Cuypers MG, Trubitsyna M, Callow P, Forsyth VT, Richardson JM.** 2013. Solution conformations of early intermediates in Mos1 transposition. *Nucl Acids Res* **41:**2020–2033.

111. **Augé-Gouillou C, Hamelin MH, Demattei MV, Periquet M, Bigot Y.** 2001. The wild-type conformation of the *Mos-1* inverted terminal repeats is suboptimal for transposition in bacteria. *Mol Genet Genomics* **265:** 51–57.

112. **Zhang L, Dawson A, Finnegan DJ.** 2001. DNA-binding activity and subunit interaction of the *mariner* transposase. *Nucl Acids Res* **29:**3566–3575.

113. **Bainton RJ, Kubo KM, Feng JN, Craig NL.** 1993. Tn7 transposition: Target DNA recognition is mediated by multiple Tn7-encoded proteins in a purified in vitro system. *Cell* **72:**931–943.

114. **Skelding Z, Sarnovsky R, Craig NL.** 2002. Formation of a nucleoprotein complex containing Tn7 and its target DNA regulates transposition initiation. *EMBO J* **21:** 3494–3504.

115. **Holder JW, Craig NL.** 2010. Architecture of the Tn7 posttransposition complex: an elaborate nucleoprotein structure. *J Mol Biol* **401:**167–181.

116. **Kim YJ, Hice RH, O'Brochta DA, Atkinson PW.** 2011. DNA sequence requirements for *hobo* transposable element transposition in *Drosophila melanogaster*. *Genetica* **139:**985–997.

117. **Ivics Z, Hackett PB, Plasterk RH, Izsvák Z.** 1997. Molecular reconstruction of *Sleeping Beauty*, a *Tc1*-like transposon from fish, and its transposition in human cells. *Cell* **91:**501–510.

118. **Izsvák Z, Khare D, Behlke J, Heinemann U, Plasterk RH, Ivics Z.** 2002. Involvement of a bifunctional, paired-like DNA-binding domain and a transpositional enhancer in *Sleepy Beauty* transposition. *J Biol Chem* **277:**34581–34588.

119. **Lohe AR, Hartl DL.** 1996. Autoregulation of *mariner* transposase activity by overproduction and dominant-negative complementation. *Mol Biol Evol* **13:**549–555.

120. **Waddell CS, Craig NL.** 1989. Tn7 transposition: Recognition of the *attTn7* target sequence. *Proc Natl Acad Sci USA* **86:**3958–3962.

121. **Chakrabarti A, Desai P, Wickstrom E.** 2004. Transposon Tn7 protein TnsD binding to *Escherichia coli* attTn7 DNA and its eukaryotic orthologs. *Biochem* **43:** 2941–2946.

122. **Peters JE, Craig NL.** 2001. Tn7: Smarter than we thought. *Nature Rev Mol Cell Biol* **2:**806–814.

123. **Peters JE, Craig NL.** 2000. Tn7 transposes proximal to DNA double-strand breaks and into regions where chromosomal DNA replication terminates. *Mol Cell* **6:** 573–582.

124. **Parks AR, Li Z, Shi Q, Owens RM, Jin MM, Peters JE.** 2009. Transposition into replicating DNA occurs through interaction with the processivity factor. *Cell* **138:**685–695.

125. **Plasterk RHA, Izsvák Z, Ivics Z.** 1999. Resident aliens: the Tc1/*mariner* superfamily of transposable elements. *Trends Genet* **15:**326–332.

126. **Fraser MJ, Cary L, Boonvisudhi K, Wang HH.** 1995. Assay for movement of Lepidopteran transposon IFP2

in insect cells using a baculovirus genome as a target DNA. *Virol* **211:**397–407.

127. **Linheiro RS, Bergman CM.** 2008. Testing the palindromic target site model for DNA transposon insertion using the *Drosophila melanogaster* P-element. *Nucl Acids Res* **36:**6199–6208.

128. **Halling SM, Kleckner N.** 1982. A symmetrical six-base-pair target site sequence determines Tn10 insertion specificity. *Cell* **28:**155–163.

129. **Davies CJ, Hutchison CA III.** 1995. Insertion site specificity of the transposon Tn3. *Nucl Acids Res* **23:** 507–514.

130. **Liao GC, Rehm EJ, Rubin GM.** 2000. Insertion site preferences of the P transposable element in *Drosophila melanogaster. Proc Natl Acad Sci USA* **97:**3347–3351.

131. **Shevchenko Y, Bouffard GG, Butterfield YSN, Blakesley RW, Hartley JL, Young AC, Marra MA, Jones SJM, Touchman JW, Green ED.** 2002. Systematic sequencing of cDNA clones using the transposon Tn5. *Nucl Acids Res* **30:**2469–2477.

132. **Vigdal TJ, Kaufman CD, Izsvák Z, Voytas DF, Ivics Z.** 2002. Common physical properties of DNA affecting target site selection of *Sleeping Beauty* and other Tc1/*mariner* transposable elements. *J Mol Biol* **323:** 441–452.

133. **Manna D, Deng S, Breier AM, Higgins NP.** 2005. Bacteriophage Mu targets the trinucleotide sequence CGG. *J Bacteriol* **187:**3586–3588.

134. **Liu S, Yeh CT, Ji T, Ying K, Wu H, Tang HM, Fu Y, Nettleton D, Schnable PS.** 2009. *Mu* transposon insertion sites and meiotic recombination events co-localize with epigenetic marks for open chromatin across the maize genome. *PLoS Genet* **5:**e1000733.

135. **Woodard LE, Li X, Malani N, Kaja A, Hice RH, Atkinson PW, Bushman FD, Craig NL, Wilson MH.** 2012. Comparative analysis of the recently discovered *hAT* transposon *TcBuster* in human cells. *PLoS ONE* **7:** e42666.

136. **Linheiro RS, Bergman CM.** 2012. Whole genome resequencing reveals natural target site preferences of transposable elements in *Drosophila melanogaster. PLoS ONE* **7:**e30008.

137. **Guo Y, Park JM, Cui B, Humes E, Gangadharan S, Hung S, FitzGerald PC, Hoe KL, Grewal SIS, Craig NL, Levin HL.** 2013. Integration profiling of gene function with dense maps of transposon integration. *Genetics* **195:**599–609.

138. **Kuduvalli PN, Rao JE, Craig NL.** 2001. Target DNA structure plays a critical role in Tn7 transposition. *EMBO J* **20:**924–932.

139. **Pribil PA, Haniford DB.** 2003. Target DNA bending is an important specificity determinant in target site selection in Tn*10* transposition. *J Mol Biol* **330:**247–259.

140. **Pflieger A, Jaillet J, Petit A, Augé-Gouillou C, Renault S.** 2014. Target capture during Mos1 transposition. *J Biol Chem* **289:**100–111.

141. **Cherepanov P, Maertens GN, Hare S.** 2011. Structural insights into the retroviral DNA integration apparatus. *Curr Opin Struct Biol* **21:**249–256.

142. **Sakai J, Kleckner N.** 1997. The Tn*10* synaptic complex can capture a target DNA only after transposon excision. *Cell* **89:**205–214.

143. **Gradman RJ, Ptacin JL, Bhasin A, Reznikoff WS, Goryshin IY.** 2008. A bifunctional DNA binding region in Tn*5* transposase. *Mol Microbiol* **67:**528–540.

144. **Yusa K, Zhou L, Li MA, Bradley A, Craig NL.** 2011. A hyperactive *piggyBac* transposase for mammalian applications. *Proc Natl Acad Sci USA* **108:**1531–1536.

145. **Claeys Bouuaert C, Chalmers RM.** 2010. Gene therapy vectors: the prospects and potentials of the cut-and-paste transposons. *Genetica* **138:**473–484.

146. **VandenDriessche T, Ivics Z, Izsvák Z, Chuah MKL.** 2009. Emerging potential of transposons for gene therapy and generation of induced pluripotent stem cells. *Blood* **114:**1461–1468.

147. **Copeland NG, Jenkins NA.** 2010. Harnessing transposons for cancer gene discovery. *Nature Rev Cancer* **10:**696–706.

148. **Adzuma K, Mizuuchi K.** 1988. Target immunity of Mu transposition reflects a differential distribution of Mu B protein. *Cell* **53:**257–266.

149. **Greene EC, Mizuuchi K.** 2002. Target immunity during Mu DNA transposition: Transpososome assembly and DNA looping enhance MuA-mediated disassembly of the MuB target complex. *Mol Cell* **10:**1367–1378.

150. **Stellwagen AE, Craig NL.** 1997. Avoiding self: two Tn7-encoded proteins mediate target immunity in Tn7 transposition. *EMBO J* **16:**6823–6834.

151. **Lambin M, Nicolas E, Oger CA, Nguyen N, Prozzi D, Hallet B.** 2012. Separate structural and functional domains of Tn*4430* transposase contribute to target immunity. *Mol Microbiol* **83:**805–820.

152. **Lavoie BD, Chaconas G.** 1993. Site-specific HU binding in the Mu transpososome: conversion of sequence-independent DNA-binding protein into a chemical nuclease. *Genes Dev* **7:**2510–2519.

153. **Chalmers R, Guhathakurta A, Benjamin H, Kleckner N.** 1998. IHF modulation of Tn10 transposition: Sensory transduction of supercoiling status via a proposed protein/DNA molecular spring. *Cell* **93:**897–908.

154. **Haniford DB.** 2006. Transpososome dynamics and regulation in Tn10 transposition. *Crit Rev Biochem Mol Biol* **41:**407–424.

155. **Whitfield CR, Wardle SJ, Haniford DB.** 2009. The global bacterial regulator H-NS promotes transpososome formation and transposition in the Tn5 system. *Nucl Acids Res* **37:**309–321.

156. **Liu D, Haniford DB, Chalmers RM.** 2011. H-NS mediates the dissociation of a refractory protein-DNA complex during Tn*10*/IS*10* transposition. *Nucl Acids Res* **39:**6660–6668.

157. **Zayed H, Izsvák Z, Khare D, Heinemann U, Ivics Z.** 2003. The DNA-bending protein HMGB1 is a cellular cofactor of *Sleeping Beauty* transposition. *Nucl Acids Res* **31:**2313–2322.

158. **van Gent DC, Hiom K, Paull TT, Gellert M.** 1997. Stimulation of V(D)J cleavage by high mobility group proteins. *EMBO J* **16:**2665–2670.

159. **Little AJ, Corbett E, Ortega F, Schatz DG.** 2013. Cooperative recruitment of HMGB1 during V(D)J recombination through interactions with RAG1 and DNA. *Nucl Acids Res* **41:**3289–3301.

160. **Ton-Hoang B, Pasternak C, Siguier P, Guynet C, Hickman AB, Dyda F, Sommer S, Chandler M.** 2010. Single-stranded DNA transposition is coupled to host replication. *Cell* **142:**398–408.

161. **Mennecier S, Servant P, Coste G, Bailone A, Sommer S.** 2006. Mutagenesis via IS transposition in *Deinococcus radiodurans*. *Mol Microbiol* **59:**317–325.

162. **Mendiola MV, Bernales I, de la Cruz F.** 1994. Differential roles of the transposon termini in IS*91* transposition. *Proc Natl Acad Sci USA* **91:**1922–1926.

163. **Garcillán-Barcia MP, Bernales I, Mendiola MV, de la Cruz F.** 2001. Single-stranded DNA intermediates in IS*91* rolling-circle transposition. *Mol Microbiol* **39:**494–501.

164. **Kersulyte D, Akopyants NS, Clifton SW, Roe BA, Berg DE.** 1998. Novel sequence organization and insertion specficity of IS*605* and IS*606*: chimaeric transposable elements of *Helicobacter pylori*. *Gene* **223:**175–186.

165. **Kersulyte D, Velapatiño B, Dailide G, Mukhopadhyay AK, Ito Y, Cahuayme L, Parkinson AJ, Gilman RH, Berg DE.** 2002. Transposable element IS*Hp608* of *Helicobacter pylori*: Nonrandom geographic distribution, functional organization, and insertion specificity. *J Bacteriol* **184:**992–1002.

166. **Pennisi E.** 2013. The CRISPR craze. *Science* **341:** 833–836.

167. **Garcillán-Barcia MP, de la Cruz F.** 2002. Distribution of IS*91* family insertion sequences in bacterial genomes: evolutionary implications. *FEMS Microbiol Ecol* **42:** 303–313.

168. **Guynet C, Achard A, Ton-Hoang B, Barabas O, Hickman AB, Dyda F, Chandler M.** 2009. Resetting the site: Redirecting integration of an insertion sequence in a predictable way. *Mol Cell* **34:**612–619.

169. **Du C, Fefelova N, Caronna J, He L, Dooner HK.** 2009. The polychromatic *Helitron* landscape of the maize genome. *Proc Natl Acad Sci USA* **106:**19916–19921.

170. **Yang L, Bennetzen JL.** 2009. Distribution, diversity, evolution, and survival of *Helitrons* in the maize genome. *Proc Natl Acad Sci USA* **106:**19922–19927.

171. **Hagemann AT, Craig NL.** 1993. Tn7 transposition creates a hotspot for homologous recombination at the transposon donor site. *Genetics* **133:**9–16.

172. **Jang S, Sandler SJ, Harshey RM.** 2012. Mu insertions are repaired by the double-strand break repair pathway of *Escherichia coli*. *PLoS Genet* **8:**e1002642.

173. **McBlane JF, van Gent DC, Ramsden DA, Romeo C, Cuomo CA, Gellert M, Oettinger MA.** 1995. Cleavage at a V(D)J recombination signal requires only RAG1 and RAG2 proteins and occurs in two steps. *Cell* **83:** 387–395.

174. **Ma Y, Pannicke U, Schwarz K, Lieber MR.** 2002. Hairpin opening and overhang processing by an Artemis/DNA-dependent protein kinase complex in nonhomologous end joining and V(D)J recombination. *Cell* **108:** 781–794.

175. **Malu S, Malshetty V, Francis D, Cortes P.** 2012. Role of non-homologous end joining in V(D)J recombination. *Immunol Res* **54:**233–246.

176. **Beall EL, Rio DC.** 1996. *Drosophila* IRBP/Ku p70 corresponds to the mutagen-sensitive *mus309* gene and is involved in P-element excision in vivo. *Genes Dev* **10:** 921–933.

177. **Mhammedi-Alaoui A, Pato M, Gama MJ, Toussaint A.** 1994. A new component of bacteriophage Mu replicative transposition machinery: the *Escherichia coli* ClpX protein. *Mol Microbiol* **11:**1109–1116.

178. **Abdelhakim AH, Sauer RT, Baker TA.** 2010. The AAA + ClpX machine unfolds a keystone subunit to remodel the Mu transpososome. *Proc Natl Acad Sci USA* **107:** 2437–2442.

179. **Kruklitis R, Welty DJ, Nakai H.** 1996. ClpX protein of *Escherichia coli* activates bacteriophage Mu transposase in the strand transfer complex for initiation of Mu DNA synthesis. *EMBO J* **15:**935–944.

180. **Burton BM, Baker TA.** 2003. Mu transpososome architecture ensures that unfolding by ClpX or proteolysis by ClpXP remodels but does not destroy the complex. *Chem Biol* **10:**463–472.

181. **Gibb B, Gupta K, Ghosh K, Sharp R, Chen J, Van Duyne GD.** 2010. Requirements for catalysis in the Cre recombinase active site. *Nucl Acids Res* **38:**5817–5832.

182. **Marmignon A, Bischerour J, Silve A, Fojcik C, Dubois E, Arnaiz O, Kapusta A, Malinsky S, Betermier M.** 2014. Ku-mediated coupling of DNA cleavage and repair during programmed genome rearrangements in the ciliate Paramecium tetraurelia. *PLoS Genet* **10:** e1004552.

Mobile DNA, 3rd Edition
Nancy L. Craig, Michael Chandler, Martin Gellert, Alan M. Lambowitz, Phoebe A. Rice and Suzanne Sandmeyer

doi:10.1128/microbiolspec.MDNA3-0030-2014

Patricia Siguier,[1] Edith Gourbeyre,[1] Alessandro Varani,[2]
Bao Ton-Hoang,[1] and Michael Chandler[1]

26 Everyman's Guide to Bacterial Insertion Sequences

INTRODUCTION

We have divided this review into two major sections. In one, we have attempted to present an overview of our current understanding of prokaryotic insertion sequences (IS), their diversity in sequence, in organization and in mechanism, their distribution and impact on their host genome, and their relation to their eukaryotic cousins. We discuss several IS-related transposable elements (TE) which have been identified since the previous edition of *Mobile DNA*. These include IS that use single-strand DNA intermediates and their related "domesticated" relations, insertion sequences with a common region (IS*CR*), and integrative conjugative elements (ICE), which use IS-related transposases (Tpases) for excision and integration. Several more specialized chapters in this volume include additional detailed information concerning a number of these topics. One of the major conclusions from this section is that the frontiers between the different types of TE are becoming less clear as more are identified. In the second part, we have provided a detailed description of the expanding variety of IS, which we have divided into families for convenience. We emphasize that there is no "quantitative" measure of the weight of each of the criteria we use to define a family. Our perception of these families continues to evolve and families emerge regularly as more IS are added. This section is designed as an aid and a source of information for consultation by interested specialist readers.

HISTORY

It is now over 40 years since the first IS were described. They were identified as short DNA segments found repeatedly associated with mutations in the *gal* operon and bacteriophage λ (1–3). Shortly after, it was established that IS were normal residents of the *Escherichia coli* chromosome (4) sometimes present in multiple copies. They were shown to be involved in generating deletions (5) and in activating gene expression (6). They were also identified as constituents of bacterial plasmids (7). At about the same time, it was observed that antibiotic resistance genes could also be transferred or "transposed" from one plasmid to another (8–10) and it was recognized that IS and "transposons" were both members of a group of genetic entities: transposable or mobile genetic elements (TE or MGE). This relationship between IS and transposons was reinforced by the observation that

[1]Laboratoire de Microbiologie et Génétique Moléculaires, CNRS, Toulouse, France; [2]Departamento de Tecnologia, Faculdade de Ciências Agrárias e Veterinárias de Jaboticabal, UNESP - Univ. Estadual Paulista, Jaboticabal, SP, Brazil.

different DNA segments carrying different genes could be translocated by two flanking IS (11, 12). It was also realized (13) that they might be related to the controlling elements discovered by genetic analysis of maize several decades previously (14).

However, in spite of the observation that IS can be present in some bacterial species in extremely high copy numbers (15, 16), little at the time prepared us for the subsequent recognition of the preponderant role they play in shaping genomes, of their extreme diversity and their widespread distribution (see reference 17).

WHAT IS AN IS?

The original definition of an IS was: a short, generally phenotypically cryptic, DNA segment encoding only the enzymes necessary for its transposition and capable of repeated insertion into many different sites within a genome using mechanisms independent of large regions of DNA homology between the IS and target (18, 19). Classical IS are between 0.7 and 2.5 kb in length, genetically compact with one or two open reading frames (*orfs*) which occupy the entire length of the IS and terminate in flanking imperfect terminal repeat sequences (IR). The *orfs* include the Tpase that catalyzes the DNA cleavages and strand transfers leading to IS movement and, in some cases, regulatory proteins. Their highly compact nature is illustrated by the fact that some IS have developed "recoding" strategies such as Programmed Ribosomal Frameshifting (involving ribosome slippage) and Programmed Transcriptional Realignment (involving RNA polymerase slippage) (Chandler et al., this volume; (20, 21)). These permit assembly of different functional protein domains effectively encoding two proteins of different function in one DNA segment. IS also often generate a short flanking directly repeated duplication (DR) of the target on insertion. These characteristics are not limited to prokaryotic IS but are also shared with most eukaryotic DNA transposons. However, for prokaryotic IS, this strict definition has been broadened over the years with the discovery of an increasing number of noncanonical derivatives and variants, some of which are described below. Moreover, as we learn more about diversity from sequenced genomes, classification is becoming more problematic because the large degree of MGE diversity is obscuring the borders between certain types of TE (see *Fuzzy Borders* section) (20).

Despite their abundance and diversity, the number of different chemical mechanisms used in TE movement is surprisingly limited and many quite divergent TE share a similar mechanism.

ISFINDER AND THE GROWING NUMBERS OF IS

Since 1998, IS have been centralized in the ISfinder database (www-is.biotoul.fr). This provides a basic framework for nomenclature and IS classification into related groups or families (22). Initially IS were each assigned a simple number (23). However, to provide information about their provenance, IS nomenclature rules were changed and now resemble those used for restriction enzymes: with the first letter of the genus followed by the first two letters of the species and a number (24) (e.g., IS*Bce1* for *Bacillus cereus*).

In 1977 only five IS (IS*1*, IS*2*, IS*3*, IS*4*, and IS*5*) had been identified (13). At the time of publication of the first edition of *Mobile DNA* this had risen to 50 (25); at the time of the second edition, there were more than 700 (26). Currently, ISfinder includes more than 4,000 different IS. This represents only a fraction of IS present in the public databases. Not only has the number of IS identified increased dramatically with the advent of high-throughput genome sequencing but examination of the public databases has shown that genes annotated as Tpases, the enzymes that catalyze TE movement (or proteins with related functions), are by far the most abundant functional class (17).

MAJOR IS GROUPS ARE DEFINED BY THE TYPE OF TRANSPOSASE THEY USE

Insertion sequences can be grouped into families but, in the first instance, the principal division in IS classification is based on the nature of their Tpases (Table 1). These can be divided into two major types based on the chemistry used in breaking and re-joining DNA during TE displacement: the DDE (and DEDD) and HUH enzymes.

DDE Transposases

DDE enzymes, so-called because of a conserved Asp, Asp, Glu triad of amino acids that coordinate essential metal ions, use OH (e.g., H_2O) as a nucleophile in a transesterification reaction (27) (Hickman and Dyda, this volume). They do not form covalent Tpase–DNA intermediates during the transposition process.

Insertion sequences with DDE enzymes are the most abundant type in the public databases. This is partly because the definition of an IS became implicitly coupled to the presence of a DDE Tpase, an idea probably reinforced by the similarity between Tpases of IS (and other prokaryotic and eukaryotic TE) and the retroviral integrases (28–30) particularly in the region including the catalytic site. More precisely, for these TE, the triad is DD(35)E in which the second D and E are separated by 35 residues. As more DDE Tpases were identified, the distance separating the D and E residues was found to vary slightly. However, for certain IS, this distance was significantly larger. In these cases, the Tpases include an "insertion domain" between the second D and E residues (31) with either α-helical or β-strand configurations. Although in most cases this is a prediction, it has been confirmed by crystallographic studies for the IS*50* (β-strand) (32) and Hermes (α–helical) (33) Tpases. The function of these "insertion domains" is not entirely clear.

Although DDE-type transposons share basic transposition chemistry, different TE vary in the steps leading to formation of an insertion intermediate that has shed flanking donor DNA (Figure 1) (Hickman and Dyda, this volume). They catalyze cleavage of a single DNA strand to generate a 3′ OH at the TE ends, which is subsequently used as a nucleophile to attack the DNA target phosphate backbone. This is known as the transferred strand. The variations are a result of the way in which the second (nontransferred) strand is processed (34, 35). There are several ways in which second-strand processing can occur (Figure 1).

For certain IS, the second strand is not cleaved but replication following transfer of the first strand fuses donor and target molecules to generate cointegrates with a directly repeated copy at each donor/target junction. This is known as replicative transposition (e.g., IS*6*, Tn*3*) or more precisely, target primed replicative transposition (Figure 1, first column).

In the other pathways, the flanking donor DNA can be shed in several different ways.

The nontransferred strand may be cleaved initially several bases within the IS before cleavage of the transferred strand (e.g., IS*630* and *Tc1*) (36, 37) (Figure 1, second column).

The 3′ OH generated by first-strand cleavage may be used to attack the second strand to form a hairpin structure at the IS ends liberating the IS from flanking DNA and subsequently hydrolyzed to regenerate the 3′ OH (conservative or cut-and-paste transposition; e.g., IS*4*; Haniford and Ellis, this volume) (Figure 1, third column).

The 3′ OH of the transferred strand from one IS end may attack the other to generate a donor molecule with a single-strand bridge that is then replicated to produce a double-strand transposon circle intermediate and regenerating the original donor molecule (Copy out-Paste in or more precisely donor primed transposon replication, e.g., IS*3*; Chandler et al, this volume) (Figure 1, fourth column).

Or finally, the 3′ OH at the flank of the nontransferred strand may attack the second strand to form a hairpin on the flanking DNA and a 3′ OH on the transferred strand (at present this has only been demonstrated for eukaryotic TE of the hAT family and in V(D)J recombination (38).

Clearly, many families produce double-strand circular intermediates but this does not necessarily mean that they all use the Copy out-Paste in donor primed transposon replication mechanism because a circle could be generated by excision involving recombination of both strands (see Hickman and Dyda, this volume). These differences are reflected in the different IS families.

DEDD Transposases

A similar type of Tpase, known as a DEDD Tpase, is related to the Holliday junction resolvase, RuvC (39, 40). These possess a similar predicted structural topology in their catalytic site and presumably have similar chemistry to the DDE enzymes. They are at present limited to a single IS family (IS*110*). They also exhibit a different order in their functional domains compared with most DDE Tpases: with the catalytic domain N-terminal to the DNA-binding domain. In addition, the associated IS do not have significant terminal IR and insertion does not necessarily generate direct target repeats. Hence, while the transposition chemistry may be similar to that of the DDE Tpases, the overall transposition mechanism is probably different.

HUH Transposases

Named after a conserved pair of His residues separated by a large hydrophobic amino acid (U), the HUH Tpases use tyrosine as a nucleophile and generate a transitory covalent 5′ tyrosine–DNA transposition intermediate—for review see (41) (Hickman and Dyda, this volume; He et al., this volume).

The TE encoding the second major type of Tpase, called HUH, have been identified more recently. HUH enzymes are widespread single-strand nucleases. They include Rep proteins involved in bacteriophage and plasmid rolling circle replication and relaxases or Mob proteins involved in conjugative plasmid transfer (41). They are limited to two prokaryotic (IS*91* and IS*200*/IS*605*; He et al., this volume) and one eukaryotic (helitron; Thomas and Pritham this volume) TE family. As Tpases, they are involved in presumed rolling circle transposition and also in single-strand transposition (see Hickman and Dyda, this volume; He et al., this volume). Their transposition chemistry is radically different to that of DDE group elements. It involves DNA

Table 1 Characteristics of insertion sequence families

Families	Sub-groups	Typical size-range	DR (bp)	Ends	IRs	No. of ORF	Frameshift	Catalytic residues	Comments	Mechanism
IS*1*	-	740–1,180	8–9	GGnnnTG	Y	2	ORFAB	DDE		copy-and-paste and cointegrate
	single ORF	800–1,200	0–9			1				
	IS*Mhu11*	900–4,600	0–10		Y	2	ORFAB			
IS1595	IS*Pna2*	1,000–1,150	8	GGCnnTG	Y	1		DDNK		copy-and-paste ?
	ISPna2+pass	1,500–2,600	8							
	IS*H4*	1,000	8	CGCTCTT		1		DDNK		
	IS*1016*	700–745	7–9	GGGgctg		1		DDEK		
	IS*1595*	900–1,100	8	CcTGATT		1		DDNK + ER4R7		
	IS*Sod11*	1,000–1,100	8	nnnGcnTATC		1		DDHK + ER4R7		
	IS*Nwi1*	1,080–1,200	8	ggnnatTAT		1		DDEK + ER4		
	ISNwi1+pass	1,750–4,750	8			1				
	IS*Nha5*	3,450–7,900	8	CGGnnTT		1		DDER/K		
IS*3*	IS*150*	1,200–1,600	3–4	TG	Y	2	ORFAB	DDE		copy-paste
	IS*407*	1,100–1,400	4	TG						
	IS*51*	1,000–1,400	3–4	TG						
	IS3	1,150–1,750	3–4	TGa/g						
	IS2	1,300–1,400	5	TG						
IS*481*	-	950–1,300	4–15	TGT	Y	1		DDE		copy-paste ?
ISNCY	IS*1202*	1,400–1,700	5	TGT	Y	1		DDEQ		
IS4	IS*10*	1,200–1,350	9	CT	Y	1		DDE	Hairpin intermediate	cut-and-paste
	IS*50*	1,350–1,550	8–9	C					Hairpin intermediate	
	IS*Pepr1*	1,500–1,600	7–8	-T-AA					?	
	IS4	1,400–1,600	10–13	-AAT					?	
	IS4*Sa*	1,150–1,750	8–10	CA					?	
	IS*H8*	1,400–1,800	10	CAT					?	
	IS231	1,450–5,400	10–12	CAT		1 or + *			* Passenger genes	
IS*701*	-	1,400–1,550	4		Y	1		DDE		
	IS*Aba11*				Y	1		DDE		
IS*H3*	-	1,225–1,500	4–5	C-GT	Y	1		DDE		
IS*1634*	-	1,500–2,000	5–6	C	Y	1		DDE		
IS*5*	IS903	950–1,150	9	GG	Y	1		DDE		
	IS*L2*	850–1,200	2–3			1				
	IS*H1*	900–1,150	8	-GC		1				
	IS*5*	1,000–1,500	4	Ga/g		1				
	IS*1031*	850–1,050	3	GAa/g		1				
	IS*427*	800–1,000	2–4	Ga/g		2	ORFAB			
IS*1182*	-	1,330–1,950	0–60		Y	1		DDE		
ISNCY	IS*Dol1*	1,600–1,900	6–7		Y	1		DDE		
IS6	-	700–900	8	GG	Y	1		DDE		co-integrate

IS*21*	-	1,750–2,600	4–8	TG	Y	2 *		DDE	* istB: transposition helper	
IS*30*	-	1,000–1,700	2–3		Y	1		DDE		copy-and-paste
IS*66*	-	2,000–3,000	8–9	GTAA	Y	3 *		DDE *	* TnpC has the DDE domain	
	IS*Bst12*	1,350–1,900	8–9	GTAA	Y	1		DDE		
IS*256*	-	1,200–1,500	8–9	Ga/g	Y	1		DDE		copy-paste
	IS*1249*	1,300	0–10	GG	Y	1		DDE		
	IS*C1250*	1,250	0–9	GG	Y	1		DDE		
IS*H6*	-	1,450	8	GGT	Y	1		DDE		
IS*Lre2*	-	1,500–2,000	9		Y	1		DDE		
IS*Kra4*	IS*Azba1*	1,400–2,900	0		Y	1 or + *		DDE	* Passenger genes	
	IS*Mich2*	1,250–1,400	8	GGG	Y	1 or 2	ORFAB	DDE		
	IS*Kra4*	1,400–3,700	9	GGG	Y	1 or + *		DDE	* Passenger genes	
IS*630*	-	1,000–1,400	2 *		Y	1 or 2	ORFAB	DDE	* Target site : often NTAN with duplication of the TA	cut-and-paste
IS*982*	-	1,000	3–9	AC	Y	1		DDE		
IS*1380*	-	1,550–2,000	4–5	CC	Y	1		DDE		
IS*As1*	-	1,200–1,500	8–10	CAGGG	Y	1				
IS*L3*	-	1,300–2,300	8	GG	Y	1				
Tn*3*	-	>3,000	0	GGGG	Y	>1		DDE		co-integrate
IS*Azo13*	-	1,250–2,200	0–4	Ga/g	Y	1				
IS*110*	-	1,200–1,550	0		N	1		DEDD		
	IS*1111*				Y *				* IRs not at the termini of the IS	
IS*91*	-	1,500–2,000	0		N	1		HUH/Y2	Target site GAAC ( ) CAAG	rolling circle
IS*200*/IS*605*	IS*200*	600–750	0		N	1 *		HUH/Y1	* TnpA	peel-and-paste
	IS*605*	1,300–2,000				2 *		HUH/Y1 **	* TnpA + TnpB ; ** Y1 on TnpA	
	IS*1341*	1,200–1,500				1 *			* TnpB	
IS*607*	-	1,700–2,500	0		N	2 *		Serine **	* TnpA + TnpB; ** TnpA	
IS*NCY*	IS*892*	1,600	0–8	CTAG	Y	2	ORFAB			
	IS*Lbi1*	1,400–1,500	5		Y	1				
	IS*Mae2*	1,400–2,400	9	CAG	Y	1				
	IS*Plu15*	800–1,000	0		N	1				
	IS*A1214*	1,000–1,200	8–12		Y	2				
	IS*C1217*	1,200	6–8	TAG	Y	1				
	IS*M1*	1,300–1,600	8–9		Y	1				

Abbreviations: DR, duplication repeat; IS, insertion sequence; ORF, open reading frame.

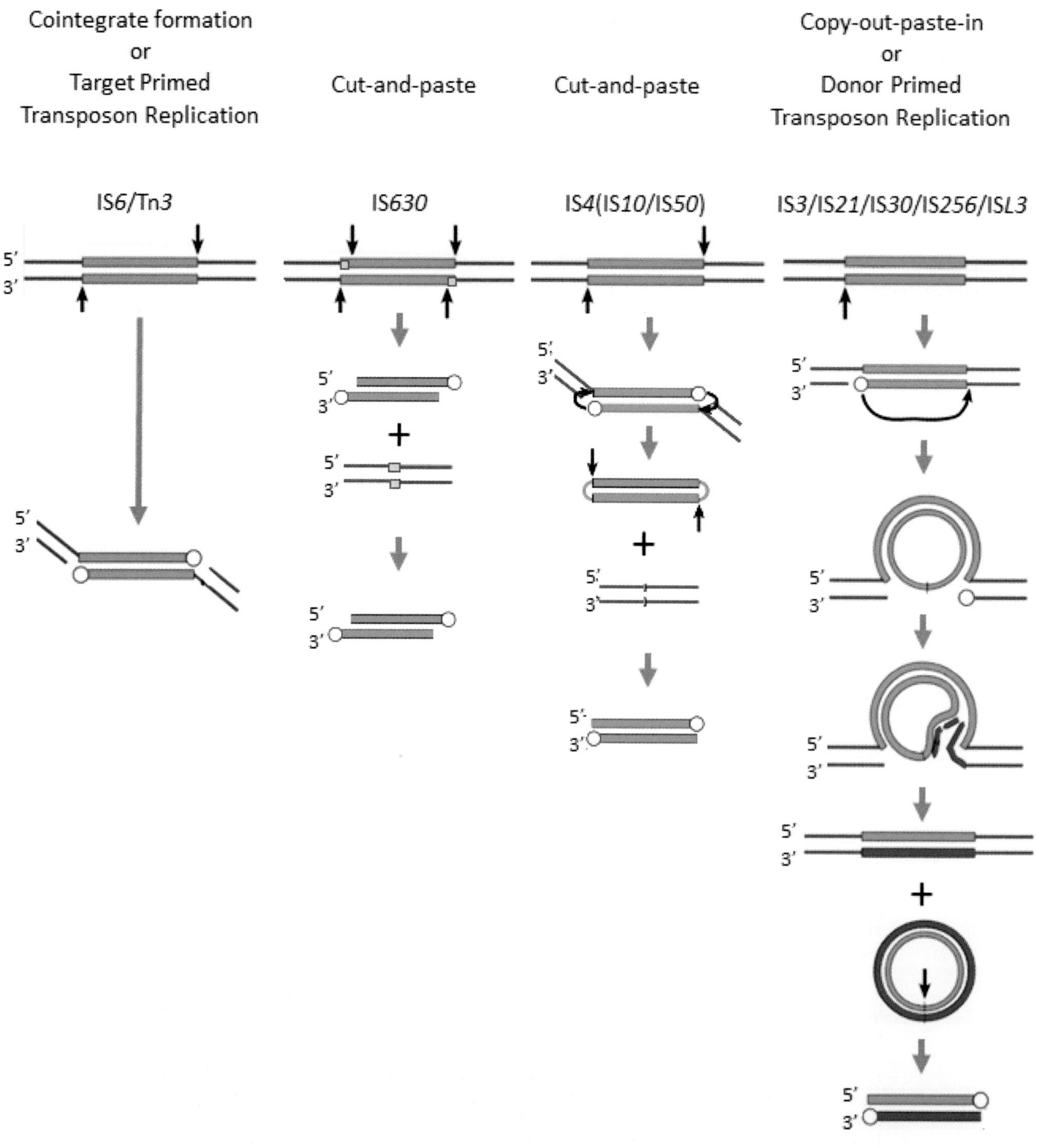
Second Strand Processing of DDE Transposases
Cointegrate formation
or
Target Primed
Transposon Replication
Cut-and-paste
Cut-and-paste
Copy-out-paste-in
or
Donor Primed
Transposon Replication
IS6/Tn3
IS630
IS4(IS10/IS50)
IS3/IS21/IS30/IS256/ISL3
5′
3′

cleavage using a tyrosine residue and transient formation of a 5′ phospho-tyrosine bond between the enzyme and its substrate DNA. In addition, the associated transposons have an entirely different organization and include subterminal secondary structures instead of IR (see *IS families* section below and He et al., this volume). There are two major HUH Tpase families: Y1 and Y2 enzymes depending on whether there is a single or two catalytic Y residues (41) (Dyda and Hickman, this volume). Although these enzymes use the same Y-mediated cleavage mechanism, IS*200*/IS*605* family Y1 transposases and IS*91* transposases appear to carry out the transposition process in quite different ways.

Serine Transposases

A third but minor type of Tpase resembles a site-specific serine recombinase and is at present limited to a single IS family, IS*607*. These presumably use the catalytic serine to generate a 5′ phospho-serine bond between the enzyme and its substrate DNA in a similar way to serine recombinases such as the Tn*3* resolvase (see chapters by: Stark; Rice; Smith; and Johnson, this volume).

Tyrosine Transposases

Finally, tyrosine site-specific recombinases of the bacteriophage integrase (Int) type are often associated with conjugative transposons (ICE) and are considered to be Tpases. However, at present there are no known IS that use this type of enzyme. These transposases presumably use a catalytic Y to generate a transitory intermediate with a 3′ phospho-tyrosine bond between the enzyme and its substrate DNA as do the site-specific recombinases (see chapters by Jayaram et al.; van Duyne; Landy; Carraro and Burrus; and Wood and Gardner, this volume) as suggested by early studies with Tn916 (42).

This nomenclature is clearly complex and stems directly from the history of the field. It has often led to confusion in genome annotations in the public databases. The reader is referred to Dyda and Hickman (this volume), He et al., (this volume) and to chapters in the section *Conservative Site-Specific Recombination* for more detailed descriptions of the overlapping issues of mechanism, function, and structure.

FUZZY BORDERS

With our increasing knowledge of mobilome diversity (the ensemble of mobile genetic elements including TE, ICE, Genomic Islands [GI], plasmids, phages, and integrons), the distinction between IS and other TE is becoming increasingly unclear. The major feature used to distinguish IS from transposons was that the former (Figure 2A) lack phenotypically detectable passenger genes (genes not involved in the transposition process) whereas the latter include one or more such genes (for antibiotic resistance, virulence and pathogenicity functions or genes permitting the use of unusual compounds). This is no longer the case (Figure 2). Many

Figure 1 Insertion sequence (IS) families with DDE transposases are distinguished by how the second ("nontransferred") strand is processed. IS are shown in green, flanking DNA in blue. Cleavage is shown as bold vertical arrows. 3′ OH residues are shown as red circles, replicated DNA is indicated in red. The first column shows initial cleavages which generate the 3′OH of the transferred strand and are subsequently used to attack target DNA (not shown) without prior liberation from the flanking donor DNA. Their transfer generates a forked molecule in which a donor and target strand are joined to the TE at each end and which provides a 3′ OH in the flanking target DNA that can prime replication of the transposable elements (TE). This might be called target primed transposon replication. TE of the Tn*3* and IS6 families transpose in this way. The second column shows a pathway adopted by the IS*630* family. Here, the nontransferred strand is cleaved two bases within the TE (light green square) before cleavage of the transferred strand, which generates the 3′ OH. Repair of the donor molecule would lead to inclusion of a noncomplementary 2-bp scar or footprint (light green square). This is a cut-and-paste mechanism without TE replication. The third column represents transposition using a hairpin intermediate in which the transferred strand is first cleaved and the resulting 3′ OH then attacks the opposite strand to form a hairpin at the TE ends liberating the TE from flanking donor DNA. This is then hydrolyzed to liberate the final transposition intermediate. This is a cut-and paste mechanism without TE replication. The fourth column shows a "copy out-paste" in mechanism adopted by a large number of IS families. It involves cleavage of one IS end and attack of the opposite end by the liberated 3′ OH, the TE then undergoes replication using the 3′ OH in the donor DNA, a process that might be called donor primed transposon replication. This generates a double-strand DNA transposon circle and regenerates the donor molecule. The circle then undergoes cleavage and insertion. Adapted from references 35 and 259.
doi:10.1128/microbiolspec.MDNA3-0030-2014.f1

examples have now been identified in which passenger genes are located within IS or in which TE with typical transposon structures are devoid of transposition proteins.

Transporter IS: IS and Relatives with Passenger Genes

Over the past few years, a number of TE have been identified that are very closely related to known IS but that carry passenger genes not directly involved in transposition. These are called transporter IS (tIS) (43) (Figure 2C). Passenger genes include transcription regulators (e.g., IS*Nha5*, members of the IS*1595* family), methyltransferases (e.g., IS*220*, IS*1380* family), and antibiotic resistance (e.g., IS*Cgl1*, IS*481* family) genes. They can include a significant amount of DNA with no clear coding capacity (e.g., IS*Bse1*, IS*Spo3*, and IS*Spo8*, IS*1595* family) and are longer than typical IS (e.g., IS*Causp2*, 7,915 bp, IS*1595* family). This has presumably delayed their identification. As the second IS end would occur at an unexpectedly distant position, they would resemble partial IS copies lacking a second end. They are never present in high numbers and often only in single copy, suggesting that their transposition activity may be limited.

IS Derivatives of Tn*3* Family Transposons

Another source of ambiguity for classification purposes occurs in the Tn*3* family (Nicolas et al., this volume) (Figure 2B). Tn*3* family members are quite variable. They include a number of diverse passenger genes that can represent entire operons, notably mercury resistance, or individual genes involved in antibiotic resistance, breakdown of halogenated aromatics or virulence (44). They often carry integron recombination platforms enabling them to incorporate additional resistance genes by recruiting integron cassettes (45). These are small DNA segments that carry promoterless passenger genes and integrate by site-specific recombination into the integron recombination platform. This platform provides an appropriate resident promoter to govern their expression (see Escudero et al., this volume). Members are characteristic: they have long relatively well-conserved IR and a particularly long Tpase (950 to 1,025 amino acids). They also encode a site-specific recombination ("resolution") system necessary for completion of their transposition (Nicolas et al., this volume). IS*1071*, composed of Tn*3*-like IR and Tpase gene but lacking both the site-specific recombination system and passenger genes, was identified many years ago (46). This clearly accords with the definition of an IS. Several other examples have now been identified (e.g., IS*Vsa19*, IS*Shfr9*, IS*Busp1*).

IS Related to ICE

The ICE, which were initially identified integrate and excise from their host chromosomes using a tyrosine-based enzyme related to phage integrases (47) (Figure 2B) (see chapters by Wood and Gardner and by Carraro and Burrus, this volume). These also carry genes permitting intercellular transfer, although derivatives exist that are not capable of autonomous transfer and are known as IMEs or CIMEs (integrative mobilizable elements; or *cis*-mobilizable elements) (48, 49). Some, known as GI, also include other types of passenger gene. Depending on the type of passenger genes, GI have been called pathogenicity islands or symbiotic islands.

More recent studies (50–53) have identified ICE with typical DDE Tpases. One group, TnGBS, initially found in Group B Streptococcus, has led to the identification of an entire family of typical insertion sequences, the IS*Lre2* family (52), whereas another shows a close relationship to IS*30* family members (53). The ICE that use typical DDE Tpases also include IR with sequences resembling those of the related IS. Hence it is becoming difficult to draw a distinction between certain GI, ICE and IS.

IS*91* and IS*CR*

A final example of the subtle line dividing IS and transposons is found in the IS*91*/IS*CR* group. IS*91* was identified some time ago (54) and carries a single Tpase *orf*. More recently, a group of related elements, IS*CR* (IS with a "common region") was described (reviewed in references 55 and 56). Although there has been no formal demonstration that these actually transpose, the CR is an *orf* that resembles the IS*91* family HUH Tpases (41). The major feature of IS*CR* elements is that they are associated with a diverse variety of antibiotic resistance genes and, particularly in the case of *Pseudomonas* IS*CR*, with aromatic degradation pathways, both upstream and downstream of the Tpase *orf*. This is therefore another example of a very particular group of IS derivatives that appear to include multiple passenger genes.

NON AUTONOMOUS IS DERIVATIVES

Many prokaryotic genomes are littered with IS fragments and small, non autonomous IS derivatives whose transposition can, in principal, be catalyzed in *trans* by the Tpase of a related complete IS (Figure 2C). These miniature inverted repeat transposable elements (MITE) were first identified in plants (57) and are related to IS with DDE Tpases. They are short (~300 bp), include terminal IR but no Tpase and generally generate flanking DR. Equivalent MITE-like structures called palindrome-associated transposable elements (PATE) (58) including the ends of IS*200*/IS*605* family members (see *IS families*, below; He et al., this volume) with their subterminal secondary structures have also been identified (Figure 2D). Both MITE and PATE probably derive from IS by internal deletion. The first MITE identified were related to eukaryotic Tc/mariner elements (distantly related to the bacterial IS*630* family) and IS*630*-related MITE were also the first described bacterial examples (59–61). MITE showing similarities to other IS families (e.g., IS*1*, IS*4*, IS*5*, IS*6*, and even Tn*3* family members) have since been identified in bacteria and archaea (62, 63).

Another group of IS derivatives related to MITE are called mobile insertion cassettes (MICs) (64) (Figure 2C). These, like MITE, are flanked by IR, do not include a Tpase gene, and generate flanking DR. They carry various coding sequences, and are present in relatively low copy number. The IS*231* subgroup of the large IS*4* family includes examples of many of these IS-derivatives (canonical IS, MITE, MIC and tIS) (65).

RELATIONSHIP BETWEEN IS AND EUKARYOTIC TE

In spite of their obvious similarities, there is often poor transfer of knowledge between studies of prokaryotic and eukaryotic TE. This artificial barrier is reflected in their nomenclature systems: Prokaryotic TE are named following the basic logic of bacterial genetics built on the initial Demerec rules (66); Eukaryotic TE, on the other hand, have more colorful names in keeping with the culture of nomenclature used in eukaryotic genetics. To a certain extent, this idiosyncratic nomenclature camouflages the diversity and relationships between members of the eukaryotic TE superfamilies and their prokaryotic cousins.

It is important to appreciate that the basic chemistry of transposition is identical for both prokaryotic and eukaryotic elements (Hickman and Dyda, this volume). Moreover, many eukaryotic DNA transposons have similar sizes and organization to those of prokaryotic IS and, as most do not carry additional "passenger" genes, they are not transposons in the prokaryotic sense and should strictly be considered as eukaryotic IS. The major differences lie in how Tpase expression and activity are regulated (67). One important difference is that most eukaryotic transposons are "insulated" by constraints of the nucleus (which physically separate the transposition process from that of Tpase expression) whereas those of prokaryotes are not, because prokaryotic transcription and translation are coupled. In addition, eukaryotic transposons are subject to a hierarchy of regulation via small RNAs (68, 69). In prokaryotes, it is possible that CRISPRs (clustered regularly interspaced short palindromic repeats) may impose some control at this level but, although it has been demonstrated that CRISPRs are active against mobile genetic elements and may regulate some endogenous gene expression (70), these are limited to plasmids and phage and to our knowledge have not yet been demonstrated to act on intracellular MGE such as IS and transposons.

In spite of these differences, a significant number of eukaryotic DNA TE are related to prokaryotic IS, and moreover, eukaryotic TE including passenger genes are now being identified (71). This reinforces the view that the borders between different types of TE are "fuzzier" than previously recognized.

IS IMPACT

There is an increasing body of data describing the impact of IS on their host genomes. Although outside the scope of this review, it is important to provide a brief overview here because this is a crucial feature of IS.

IS Expansion, Elimination and Genome Streamlining

Perhaps one of the most spectacular characteristics of IS populations is their capacity to expand within genomes and also participate in genome streamlining or trimming by facilitating DNA deletion. There are many striking examples of IS expansion (20). These are often observed in bacteria with recently adopted host-restricted lifestyles. Current opinion is that the nutritionally rich environment of the host reduces the requirement for many genes that are essential for free-living bacteria, allowing fixation of slightly deleterious mutations in the population by random stochastic IS transposition and concomitant increase in IS copy number.

Additionally, IS activity may also eliminate genes responsible for surface antigens and interfere with regulatory circuitry, providing increased protection against host defenses (72, 73). Fixation would be facilitated by successive population bottlenecks (74–77) and the effect would be more marked the more genetically isolated the bacterial population. Intracellular endosymbionts provide many examples of this. Accumulation of IS-generated endosymbiont pseudogenes in functions that can

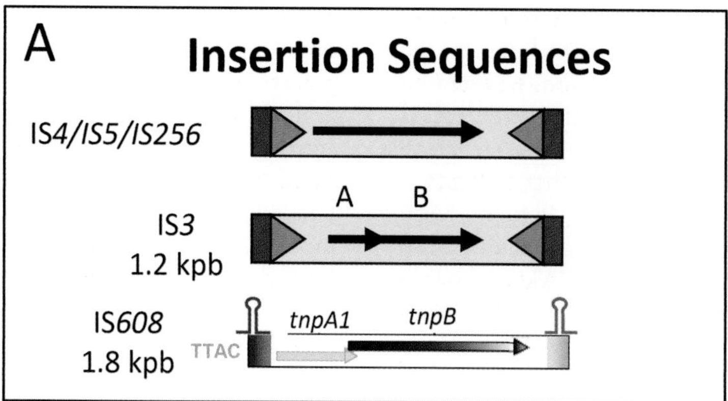
A
Insertion Sequences
IS4/IS5/IS256
A B
IS3
1.2 kpb
IS608
1.8 kpb
TTAC
tnpA1
tnpB

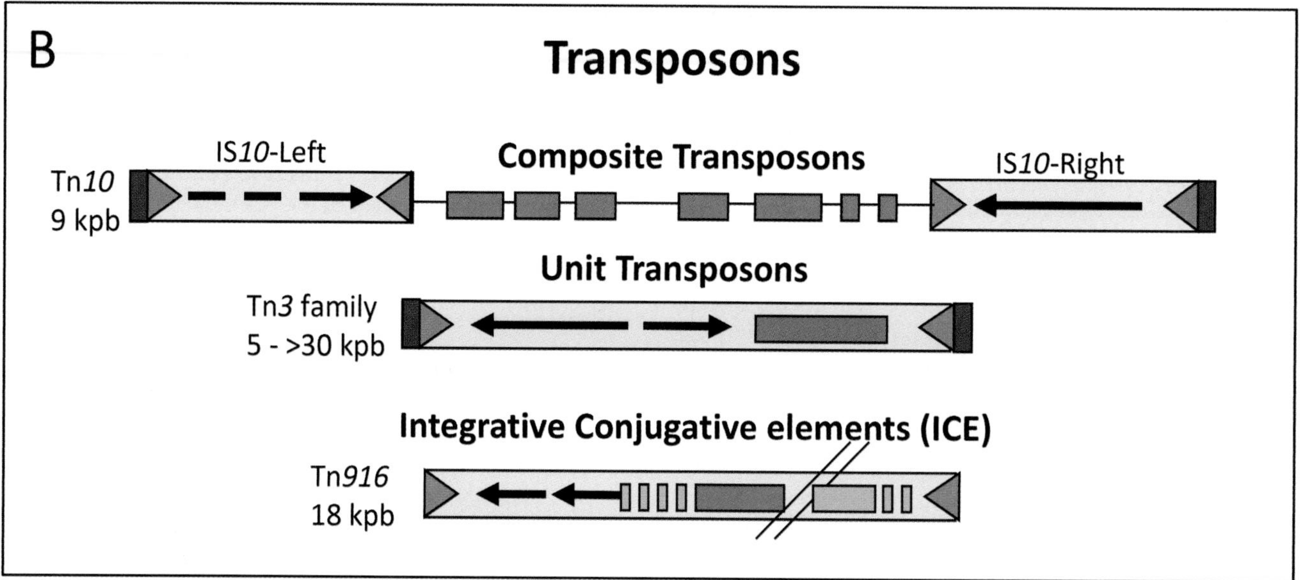
B
Transposons
Tn10
9 kpb
IS10-Left
Composite Transposons
IS10-Right
Unit Transposons
Tn3 family
5 - >30 kpb
Integrative Conjugative elements (ICE)
Tn916
18 kpb

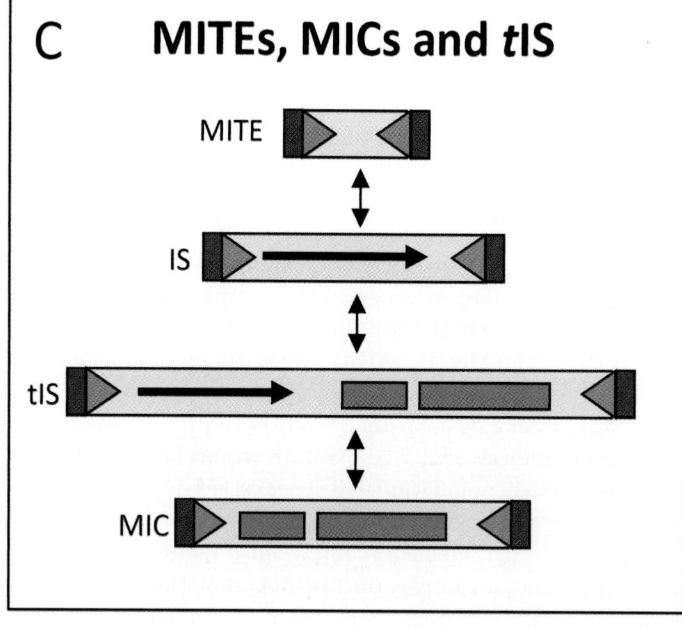
C
MITEs, MICs and tIS
MITE
IS
tIS
MIC

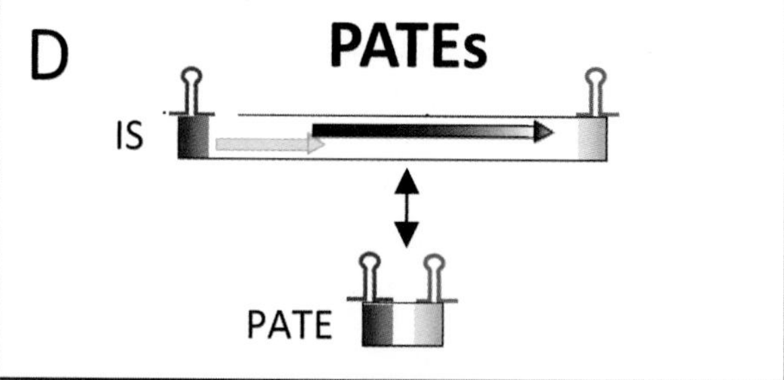
D
PATEs
IS
PATE

be supplied by the host would increase the host dependence of the bacterial population (78–80). The capacity of IS to generate deletions would also be expected to eventually lead to complete or partial elimination of the IS themselves. This is thought to have occurred in the IS-free ancient endosymbionts. Genetic isolation of the population is key to the process of IS elimination because reinfection could occur by lateral gene transfer from other bacterial populations, a phenomenon that is thought to have occurred several times in certain *Wolbachia* sp. (81).

IS and Gene Expression

Another important aspect of IS impact on their bacterial hosts is their ability to modulate gene expression. In addition to acting as vectors for gene transmission from one replicon to another in the form composite transposons (two IS flanking any gene; Figure 2B) and tIS (Figure 2C) and their ability to interrupt genes, it has been known for some time that IS can also activate gene expression (5, 82). This capacity has recently received much attention because of the increase in resistance to various antibacterials (83, 84), a worrying public health threat (85–87).

They can accomplish this in two ways: either by providing internal promoters whose transcripts escape into neighboring DNA (82, 88) or by hybrid promoter formation. Many IS carry –35 promoter components oriented towards the flanking DNA. In a number of cases this plays an important part in their transposition because a significant number of IS transpose using an excised transposon circle (see *Major IS groups are defined by the type of transposase they use* section, above) with abutted left and right ends. For these IS, the other end carries a −10 element oriented inwards towards the Tpase gene. Together with the −35, this generates a strong promoter on formation of the circle junction to drive the Tpase expression required for catalysis of integration (Chandler et al., this volume) (89–91). Hence, if integration occurs next to a resident −10 sequence, the IS −35 sequence can contribute to a hybrid promoter to drive expression of neighboring genes, see reference 92. At present this phenomenon had been reported to occur with over 30 different IS in at least 17 bacterial species (20, 93).

Target Choice

The influence of different IS on genome architecture will depend not only on their levels of activity but also on the type of target into which they insert. It was initially believed that TE show no or only low sequence specificity in their target choice. For example IS*630* and the eukaryotic Tc/*mariner* families (Tellier et al., this volume) both require a TA dinucleotide in the target (36, 37) whereas others such as the IS*200*/IS*605* and IS*91* families require short tetra- or penta-nucleotide sequences (94, 95). Yet others, such as IS*1* and IS*186* (of the IS*4* family), show some regional specificity (for AT–rich and GC-rich sequences, respectively) (96–98).

Although, from a global genome perspective, insertion may appear to occur without significant sequence specificity, accumulation of more statistically robust data has uncovered rather subtler insertion patterns. For example, there is some indication from the public databases suggesting that IS density is generally significantly higher in conjugative bacterial plasmids than in their host chromosomes with the exception of special cases in which the host has undergone IS expansion as described above. Such plasmids are major vectors in lateral gene transfer providing good delivery systems for TE. Some TE, including IS, appear to be attracted to replication forks (99, 100) and show a strong

Figure 2 Organization of different insertion sequence (IS) -related derivatives. IS with DDE transposases (Tpases) and their derivatives are shown as blue boxes, terminal inverted repeats as light blue triangles and flanking direct target repeats as red boxes. The Tpase *orf*s are shown as black horizontal arrows. Passenger genes are shown as orange boxes and transfer functions (in the case of ICE) are shown as purple boxes. The single-strand IS are indicated with their left (red) and right (blue) subterminal secondary structures indicated. (A) IS organization. From top to bottom: a typical IS with a single Tpase *orf*; an IS in which the Tpase reading frame is distributed over two reading phases and requires frameshifting for expression; and the organization of a typical member of the single-strand IS family IS*200*/IS*605*. (B) Different IS-related TE. From top to bottom: composite transposon Tn*10* with inverted flanking copies of IS*10* (note that the left IS*10* copy is not autonomously transposable); a unit transposon of the Tn*3* family; and an integrative conjugative element (ICE). (C) Relationship between IS, miniature inverted repeat transposable elements (MITE), transporter IS (tIS) and mobile insertion cassettes (MIC). (D) Generation of palindrome-associated transposable elements (PATE) from IS*200*/IS*605* family members. Adapted from references 20 and 43. doi:10.1128/microbiolspec.MDNA3-0030-2014.f2

orientation bias indicating strand preference at the fork (99–101). Moreover, in certain cases, insertion may target stalled replication forks (He et al., this volume). A link between replication (in this case, replication origins) and insertion has also now been observed for a eukaryotic TE: the P element of *Drosophila* (102).

Some transposons such as Tn*7* in *Escherichia coli* (103) and Tn*917* in *Bacillus subtilis* (104), *Enterococcus faecalis* (105) and *Streptococcus equi* (but not in *Listeria monocytogenes* or *Streptococcus suis*) (106) also show a preference for integration into the replication terminus region and sites of DNA breakage may also attract insertions (103). It remains to be seen whether any IS has adopted these types of target preference.

Potential topological characteristics or secondary structures are another feature that can attract certain TE. Changes in topology induced by the nucleoid protein, H-NS, for example, may explain the effects of H-NS mutants on the target choice of IS*903* and Tn*10* (IS*10*) (107) (Haniford and Ellis, this volume). Members of the IS*110*, IS*3*, and IS*4* families are examples of IS that insert into potential secondary structures such as repeated extragenic palindromes (He et al., this volume) (108–110), integrons (Escudero et al., this volume; 111, 112) or even the ends of other TE (113, 114).

In addition, IS*21*, IS*30*, and IS*911* have all been observed to insert close to sequences that resemble their own IR (115–118). Although these IS are members of different families, they have in common the formation of a dsDNA excised circular transposon intermediate with abutted left and right ends (Chandler et al., this volume). Insertion next to a resident "target" IR such that IR of the IS are abutted "head-to-head" presumably reflects the capacity of the Tpase to form a synaptic complex between one IR present in the transposon circle and the target IR. This type of structure is extremely active in transposition and will continue to generate genome rearrangements.

Other factors that may be influential in determining target choice are interactions (direct or indirect) with DNA-associated host proteins. For example, Tn*7* appears to be drawn to replication forks by interaction with the fork-associated β-clamp (119). Indeed, a recent large-scale analysis of IS insertions in bacterial chromosomes has pointed out the potential implication of sliding clamp in targeting and provided evidence for direct interaction of various transposases with the conserved replication processing factor (120). A more detailed exploration of these results would be useful. Another potential interaction, in this case with RNA polymerase, is suggested by the recent observation that the transposon TnGBS (an ICE from *Streptococcus agalactiae*) and members of the closely related IS*Lre2* family insert preferentially 15–17 bp upstream of σA promoters (50, 51). Targeting of upstream regions of transcription units has also been extensively documented for certain eukaryotic transposons (121).

Influence of Transposition Mechanisms

The way in which strand cleavages and transfers occur during transposition also affects the outcome of the transposition events and therefore impinges on genome structure. For example Tn*3* and IS*6* family members, both with DDE Tpases, generate fusions or cointegrates between the donor and target replicons by a process of replicative transposition, presumably by target primed transposon replication (Hickman and Dyda, this volume) (Figure 1 first column). However, in the event of intramolecular transposition, this type of mechanism is expected to give rise to inversions with a copy of the IS at each junction or inversions with a single IS copy remaining and a second copy segregating with a circularized deletion (122). Note that similar effects are also known to occur by homologous recombination between two inverted or directly repeated IS copies in a replicon. Other known mechanisms such as cut-and-paste, or Copy out-Paste in (donor primed transposon replication) (34) would not generate this type of genomic rearrangement but could contribute to genomic modifications in other ways such as "nearly precise excision" (123, 124) or by using alternative sequences that resemble their IR (125, 126).

Summary

The above considerations serve to provide an overview of the diversity of IS and their close TE relatives together with an outline of their behavior and its impact on prokaryotic genomes. Below we provide a guide to the present classification of IS and present a more detailed description of the individual IS families, their characteristics, differences, similarities, transposition mechanism (where known), and their distribution. There are clearly large disparities in our level of knowledge from family to family. Although we have tried to treat each family individually, we have grouped those that have proved to be related. We also include information concerning the identification of the different types of derivative elements such as MITES or tIS, which are related to given individual IS families.

IS FAMILIES

Insertion sequences in ISfinder are classified into families using a variety of characteristics (Table 1) (127) such as: transposition chemistry; length and sequence of the short imperfect terminal IR sequences (TIR in eukaryotes) carried by many IS at their ends; length and sequence of short flanking direct target DNA repeats; *orf* organization; and the nature of their target sequences. In many cases, these distinguishable differences are associated with different properties and behavior of the IS. An additional criterion is based on the presence/absence and order of various Tpase domains. Transposases are multidomain proteins. They include domains or modules involved in catalysis (e.g., DDE and HUH—see *Major IS groups are defined by the type of transposase they use* section, above), in sequence-specific binding to the TE ends (e.g., helix-turn-helix, HTH; zinc finger, ZF) and in multimerization (leucine zipper). The sequence specific binding module is generally located in the N-terminal region permitting folding and binding of the nascent peptide before completion of translation (see *IS3 and IS481 families* section, below). In addition, they may include specific signatures for interaction with host proteins (Peters, this volume).

At present, IS in ISfinder are grouped into 29 families and many of these are further divided into subgroups based on shared characteristics. Although there are very well-defined homogeneous families (such as IS*3*, IS*30*, and IS*256*) others have been redefined over time as more and more IS are identified (e.g., IS*4* and IS*5*).

It is important to note that the number of IS in a given ISfinder family does not necessarily reflect their relative abundance in nature. Some estimates of this have been made from time to time using different approaches, although the accuracy of these is difficult to assess (74, 128, 129). IS inclusion in ISfinder has not involved a systematic global search of the public databases.

We emphasize that this is a description of the ISfinder classifications at this time and that some groups will certainly emerge with further additions to the database and more detailed analysis.

IS WITH DDE TRANSPOSASES

IS*1* and IS*1595*: Two Related Families

IS*1*

(i) General

IS*1* was among the first bacterial insertion sequences to be identified (13). It is also one of the shortest (768 bp) and has been identified in over 40 different bacterial and archaeal species. IS*1* is a component of several compound transposons (12, 130) where it is present in direct or inverted orientation. It is also found in several conjugative plasmids flanking large regions carrying a number of antibiotic resistance genes (resistance determinant or r-det) (131) and can participate in homologous recombination to generate circular r-det forms or tandem multimers resulting in increased antibiotic resistance (132).

(ii) Organization

The family has been extensively reviewed previously (26, 127) (and references therein). Integration generates a 9 bp target DR but DR of 8, 10, and 14 bp also occasionally occur. The classic IS*1* includes 23 bp imperfect inverted repeats (IRL and IRR) and two partly overlapping *orfs* (*insA* and *insB'*) located in the 0 and −1 relative translational phases and expressed from a promoter partially located in IRL. Their integrity is essential for transposition. The *insA* product, InsA, binds both IR and regulates expression and probably transposition activity. The Tpase, InsAB', is a transframe fusion protein produced by programmed ribosomal frameshifting, between *insA* and *insB'* with typical frameshift signals. Programmed ribosomal frameshifting occurs at a frequency of about 1%. No InsB' protein species has been detected.

IS*1*-related derivatives carrying only a single *orf* have been identified. These tend to be longer (~1,000 bp) than the classic IS*1* with slightly longer Tpases due to an N-terminal extension (Table 1). They retain the characteristic IS*1* IR sequences (Table 1). They were first observed in the archaeal Sulfolobiales (IS*C1773a* and *b* and IS*Sto7*) where this arrangement appears to be the rule (63) but are not restricted to the Archaea. Several examples occur in Eubacteria (e.g., IS*Aba3*, *Acinetobacter baumannii*, and IS*Pa14*, *Pseudomonas aeruginosa*) (43). It is possible, as in the case of *dnaX* (133), that the upstream N-terminal protein is indeed produced but by frameshifting to generate the smaller derivative from the full-length protein.

Overall transposition activity appears to depend on the ratio of InsA/InsAB', serving to regulate activation of transposition by uncontrolled Tpase expression from external transcription. It had been suggested that a translational restart within the *insA* frame giving rise to an InsAB' protein with an N-terminal deletion generates the true Tpase (134). However, although the importance of this protein cannot be ruled out, the establishment of an *in vitro* IS*1* transposition system based on partially purified engineered InsAB' suggests that the translational restart product may not play a central role (135).

(iii) Mechanism

In vivo, direct visualization of 13 DNA species obtained following induction of IS*1* transposition and the kinetics of their appearance and disappearance clearly identified forms corresponding to reciprocal products of IS-mediated deletions as well as excised transposon circles. This suggests that IS*1* can transpose using both the cointegrate (target primed transposon replication) and Copy out-Paste in (donor primed transposon replication) pathways (136).

(iv) Transposase organization

Alignment of InsAB' from different members of the family confirmed the presence of a C-terminal DDE catalytic domain (137) and also revealed potential N-terminal ZF and HTH motifs (135, 138). Addition of 1,10-phenanthroline, which shows a high affinity for zinc, prevented binding to IS*1* IR, as did mutations in either of the two motifs whereas mutation of the DDE motif confirmed its importance in catalysis but not in binding (135, 138). All three motifs are observed in the longer Tpases with a single long reading frame.

However, members of the IS*Mhu11* subgroup lack the N-terminal ZF while retaining the HTH motif (43). They include a 30–120-residue C-terminal extension that is unrelated in different members of this group and the spacing between the second D and E residues is 40–60 amino acids longer. Three different organizations of IS*Mhu11* subgroup members were identified: examples with two *orf*s and a potential frameshift zone (IS*Mhu11*, IS*Mac25*, IS*Arch18*, and IS*Acma3*); a single example with additional noncoding DNA upstream of the Tpase *orf* (IS*Beg1*); and members that carry passenger genes (Table 1) generally with no known function, but often with other relatives in different bacteria. An exception is tIS*Sce1*, which includes *orf*s resembling a DNA methyltransferase, a possible sigma factor, and member of the HTH_XRE family of transcription regulators. However, only a single example of each type with passenger genes was identified, suggesting that these IS have low or no transposition activity.

More extensive comparisons have indicated that IS*1* is distantly related to another relatively newly recognized family, IS*1595*.

IS*1595*

(i) General

IS*1595* was identified in *Xanthomonas campestris* (43) and closely related IS (e.g., IS*Xo2*, IS*Xo5*, IS*Xo16*, and IS*Xca4*) are present in high copy number in other *Xanthomonas* species. The IS*1595* family is less homogeneous than the IS*1* family. BLAST analysis of ISfinder using IS*1595* as a query confirmed a distant relationship with IS*1* family members and with IS*1016* (139), a multicopy *Neisseria* element previously binned with the "unclassified" IS (IS*NYC*). Seven subgroups have been identified: IS*1595*, IS*1016*, IS*Pna2*, the halobacterial IS*H4* group (140), IS*Sod1*, IS*Nwi1* and IS*Nha5*. Members of the IS*Pna2*, IS*Nwi1* and IS*Nha5* subgroups may contain passenger genes or additional noncoding DNA (43). As in the case of IS*Mhu11*, only a single example of each member was identified, suggesting that these IS have low or no transposition activity. It is also related to IS*Sag10* from *Streptococcus agalactiae*. This was originally called MTnSag1 and thought to be a member of the IS*1* family (141) but now called tIS*Sag10* due to the presence of an *O*-lincosamide nucleotidyltransferase passenger gene. tIS*Sag10* also carries an origin of transfer and can be mobilized by transfer functions of Tn*916*. It can therefore also be defined as a non autonomous ICE, again underlining the increasing difficulty in TE classification.

(i) Organization

Most are flanked by 8 bp AT-rich DR and have a single Tpase *orf*. Like IS*1* family Tpases, they include an N-terminal ZF, an HTH motif, and a C-terminal catalytic motif with some exceptions. For example IS*1016* group members lack the N-terminal ZF but, as IS*1016* is present in multiple copies, it is probably active. The catalytic sites of all family members show group-specific variation particularly around the final E residue. More surprising is the apparent substitution of the final E residue for N or H in certain members. The exact nature of these possible noncanonical catalytic sites will require experimental determination.

It has been reported that the IS*1595* family (in particular IS*1016*) is related to the Merlin family of eukaryotic TE (142), especially at the level of the DDE motif and the position of an upstream HTH. They also have comparable lengths and similar IR. The Merlin Tpase is longer at the N-terminus than that of IS*1016* and more similar in size to the other members of the IS*1595* family although it does not exhibit the IS*1595* N-terminal ZF.

IS*3* and IS*481* Families

IS*3*

(i) General

The IS*3* family is one of the most coherent and largest IS families (143) (see Chandler et al., this volume). It is very widely distributed and at present ISfinder includes 554 different members from more than 270 bacterial species. The family is quite homogeneous in spite of its wide distribution in bacteria exhibiting a large range of

G+C contents and of the presence of members in hosts such as *Mycoplasma* with a nonuniversal genetic code (e.g., IS*1138*) or in bacteria that use stop codon read-through by insertion of the unusual amino acid selenocysteine (e.g., IS*Dvu3* from *Desulfovibrio vulgaris*). In the case of both copies of IS*1138*, which participates in high-frequency rearrangements of the *Mycoplasma pulmonis* chromosome, the Tpase *orf* carries 11 UGA codons, which are decoded as tryptophan (144).

Although most members had been limited to the eubacteria (63), an example, IS*Mco1*, has now been identified in the Archaea *Methanosaeta concilii*. As this Archaea is widespread in nature (145), it is possible that this represents a case of recent horizontal transfer. The presence of eight copies implies that IS*Mco1* is active in its archaeal host.

(ii) Organization

Members carry relatively well conserved IR and invariably express their Tpases as fusion proteins using programmed –1 frameshifting. All have characteristic terminal IR generally terminating with the dinucleotide 5′-CA-3′ and carrying a short block of GC-rich sequence. They generate 3 or 4 bp DR on insertion. Their Tpase, OrfAB, like those of the majority of IS*1* family members, is expressed as a fusion protein by programmed –1 programmed ribosomal frameshifting from two consecutive, partially overlapping reading frames with the second located in the –1 reading phase compared to the upstream frame (Chandler et al., this volume).

A clade carrying noncanonical ends has recently been identified. This is currently composed of 15 members. These IS include seven supplementary base pairs on each end flanking canonical IS*3* ends: a conserved stretch of five C residues is located 5′ to the left IR and a less conserved motif (CGG) is located 3′ to the right end. When these additional bases are taken into account every member of this clade exhibits a 4 bp DR characteristic of the IS*3* family (Table 1) (E. Gourbeyre, unpublished). This conclusion is supported by the presence of multiple IS copies (e.g., IS*Psy31*) and also by identification of "empty sites" (ISfinder). This clearly requires further experimental investigation.

Recently, an additional subgroup has been proposed, which includes IS*Ppy1* (146). However, all members belong to the IS*150* subgroup and their Tpases are not separated by our standard multiple alignment and Markov Cluster Analysis (MCL) analysis. Although they do exhibit some variation in the sequence of their terminal dinucleotides, similar variations are found for IS2 and members of other IS*3* subgroups.

(iii) Mechanism

IS*3* family members transpose using a Copy out-Paste in, donor primed transposon replication, mechanism that produces dsDNA circular intermediates (Figure 1). The recognition sequences and cleavage points have been well characterized experimentally in several IS*3* family members. *In vitro* transposition systems have been developed for several family members (Chandler et al., this volume). Bridged molecules are formed in an asymmetric manner by cleavage of one IS end to generate a 3′ OH, which then attacks the other end to generate a single-strand bridge between the IS ends. A copy of the IS appears to be replicated out as a circle leaving the other IS copy in the reconstituted donor DNA replicon (Chandler et al., this volume).

One member of this family, IS*911*, has been used to address the mechanism, common to many IS, in which the Tpase shows a preference for acting on the IS copy from which it was synthesized (*cis* activity). It was shown that this behavior depends on coupling between transcription and translation and results from co-translational binding of the nascent Tpase peptide to the IS ends (147).

IS*481*

Initially, IS*481* appeared to be an IS*3* family derivative that had been truncated for the N-terminal end of the Tpase and includes a C-terminal extension. The DDE active site domain and the IR (ending in 5′ TGT 3′) are similar to those of IS*3* family members. Their presence in high copy number in some species and the identification of at least 130 distinct but related IS from over 90 species strongly suggests that these represent a distinct transpositionally active family. Different members generate DR of between 4 and 15 bp. Moreover, certain members (e.g., IS*Sav7*) insert specifically into the tetranucleotide CTAG, which becomes the flanking DR and provides the UAG termination codon for the Tpase. In contrast to the vast majority of IS*3* family members, the IS*481* Tpase is not produced by frameshifting. There is no evidence for a leucine zipper as in IS3.

Some members include passenger genes including antibiotic resistance (Cm^R for IS*5564* and IS*Cgl1*), or potential transcriptional regulators (IS*Krh1*, IS*Pfr21*, IS*Sav7*). IS*481* itself has played a fundamental role in the evolution of the genomes of the Bordetellae where, in *Bordetella pertussis* it has undergone extensive amplification to several hundred copies with accompanying genome decay (72, 73).

These IS are distantly related to the eukaryotic Banshee transposon, which at present is restricted to the anaerobic flagellated protozoan *Trichomonas vaginalis* (Pritham per. comm.) (148). They share the highly conserved Pfam integrase core domain identified initially in the IS*3* family and retroviruses (28, 29). They also show a conserved 5′ TG 3′ tip to the IR, which is typical of this and other types of mobile element. It would be interesting to determine whether Banshee transposes using a dsDNA circular intermediate as do IS*3* family members.

IS*1202* group (IS*NCY*)

A small group including IS*1202* (149), composed of 10 IS which had been included in the IS*NCY* (not classified yet) group appears distantly related to IS*481*. Members are between 1,400 and 1,700 bp (except for IS*Kpn21*, which includes a passenger gene annotated as "hypothetical protein") with a Tpase *orf* of between 400 and 500 amino acids in a single reading frame. Their IR begin with TGT as do those of the IS*3* and IS*481* families. They generate DR of between 5 and, unusually, 27 bp.

They appear to have similarities at the level of their Tpases particularly in their DDE domains (e.g., IS*1202* has 39% amino acid similarity to IS*Pfr5* of the IS*481* family). They include a glutamine (Q) seven residues C-terminal to the conserved E instead of the characteristic K/R. Identification of additional IS will be necessary to clearly define this group.

The Redefined IS*4* and Related Families

The accumulation of additional related IS has permitted a more detailed analysis of the IS*4* family, which had already become heterogeneous displaying extremely elevated levels of internal divergence (150). Based on more than 200 IS*4*-related sequences from bacterial and archaeal genome sequences, seven subgroups and three emerging families (IS*701*, IS*H3*, and IS*1634*) were defined. Separation into three emerging families was principally due to variations in an important conserved YREK motif, a division that is supported by the IR sequences and the associated DR.

Members of all these encode a Tpase with an insertion domain rich in β-strand and located between the second D and the E of the DDE motif (31). That of the IS*50* Tpase (151) is the only example that has structural support (152), although bioinformatic analysis (31) indicated that IS*H3* (e.g., ISC*1359* and ISC*1439*), IS*701* (e.g., IS*701* and IS*Rso17;* (153)) and IS*1634* (e. g., IS*1634*, IS*Mac5*, IS*Plu4*; (154)) family members also exhibit a similar insertion domain.

IS*4*

The IS*4* family originally included a diverse collection of IS characterized by three conserved domains in the Tpase [N2, N3, and C1 containing D, D, and E respectively; (151)]. In addition to the conserved DDE triad, this family is defined by the presence of an additional tetrad YREK (150) which, in the case of IS*50*, is thought to be involved in coordination of a terminal phosphate group at the 5′ end of the cleaved IS (155) (Haniford and Ellis, this volume).

IS*701*

The IS*701* family was distinguished from the IS*4* family by a highly conserved 4 bp target duplication, 5′-YTAR-3′. MCL analysis also indicated that the Tpases form a defined and separate group and alignments indicated the absence of Y in the Tpase YREK motif. There are several clades within this family. A new clade, IS*Aba11*, was proposed as a new family based on five IS (156). Members of this group generate 5 bp target duplications (instead of 4 bp), exhibit conserved IR and include HHEK instead of YREK. However, additional examples (ISfinder) exhibiting the conserved IR did not universally contain HHEK and MCL cluster analysis did not strongly support the notion that IS*Aba11* constitutes a new family. At present, we have retained IS*Aba11* as a subgroup in the IS*701* family.

IS*H3*

The IS*H3* family is restricted to the Archaea. In roughly half of the 30 members identified, the Tpase lacked the K/R residue of the DDE motif whereas all except IS*Fac10* displayed a Y(2)R(3)E(3)R motif. A characteristic of this family is the presence of 5 bp DR flanked at one end by A and at the other by T.

IS*1634*

The IS*1634* family (previously IS*1549*) is characterized by large Tpases due to a β-strand insertion domain located between the conserved second D and E residues, which is 35 to 79 amino acids longer than that of IS*4* and members of the other related families. They generate 5 bp to 6 bp AT-rich DR and are present in both Archaea and bacteria (150). Certain members generate very long variable DR (e.g., IS*1634* from 17 to 478 bp, (154); ISC*sa8* from 16 to 131 bp; IS*Mhp1*, 80 bp).

The Redefined IS*5* and Related IS*1182* Families

Like the IS*4* family, growth in the number of identified members of the IS*5* (564 members) and IS*1182* families (142 members) has revealed that the two are related and has allowed a more detailed analysis and a separation into various subgroups and families The IS*5* family is now partitioned into six subgroups: IS*5*, IS*903*, IS*L2*, IS*H1*, IS*1031*, and IS*427*. Some of these may prove to be emerging families.

There is a distant relationship, about 30% similarity, with the Pif/Harbinger group of eukaryotic TE (148, 157).

IS*5*

Although the majority of members have a single Tpase *orf*, about 20% of family members may express Tpase by frameshifting because it is distributed between two translation phases similar to most of the IS*427* subgroup (82/116) (127). In these cases the frameshifting signals appear more appropriate for a programmed transcriptional realignment frameshift mechanism rather than for classical translation frameshifting (programmed ribosomal frameshifting) as there are no obvious downstream enhancement signals (21). Similar split reading frames have now been identified in several other subgroups: IS*1031* (13/65 members); IS*L2* (7/43); and few in the IS*5* subgroup (7/149). There is no experimental evidence that these frameshift signals are functional but many of the IS are in multiple copies, suggesting that the derivatives are active. In view of their diversity compared with families such as IS*3*, the subgroups will certainly be partitioned into additional groups as more IS are identified. At present, the IS*903* and the archaeal IS*H1* subgroups do not contain members with potential frameshifting.

In addition to their Tpases and the presence or absence of potential frameshifting, a further distinction between these elements resides in their target specificities. Certain IS*427* subgroup members and IS*1182* family members do not carry a termination codon for their Tpases but generate this on insertion into a specific target sequence, CTAG, which is duplicated on insertion. Other IS such as IS*1031*, duplicate a sequence TNA while others such IS*L2* appear to duplicate ANT.

IS*1182*

IS*1182* family members exhibit a diverse set of target specificities. Some duplicate 4 bp. These are of two types: those specific for CTAG and those that show no apparent target sequence specificity. Yet others target palindromic sequences. These are also of different types: some insert at the 3′ foot of a stem-loop and duplicate the entire structure whereas others insert 3′ of the loop and simply duplicate the loop (P. Siguier, E. Gourbeyre and M. Chandler, unpublished).

IS*Dol1* Group (IS*NCY*)

Another small group, IS*Dol1*, with 17 members from 15 bacterial species has recently emerged from the IS*NCY* "orphan" group. Members have a length of between 1,600 and 1,900 bp and generate DR of 6–7 bp.

IS*6*

(i) General

There are at present 130 family members in ISfinder from nearly 80 bacterial and archaeal species but this represents only a fraction of those present in the public databases. The family was named after the directly repeated insertion sequences in transposon Tn*6* (158) to standardize the various names that had been attributed to identical elements (e.g., IS*15*, IS*26*, IS*46*, IS*140*, IS*160*, IS*176*). Many are found as part of compound transposons (159–161) invariably as flanking direct repeats, a consequence of their transposition mechanism.

Recent activity resulting in horizontal dissemination is suggested by the observation that copies identical to *Mycobacterium fortuitum* IS*6100* occur in other bacteria: as part of a plasmid-associated catabolic transposon carrying genes for nylon degradation in *Arthrobacter* sp. (162), from the *Pseudomonas aeruginosa* plasmid R1003 (164), and within the *Xanthomonas campestris* transposon Tn*5393b* (164).

One member, IS*257*, has played an important role in sequestering a variety of antibiotic resistance genes in clinical methicillin-resistant *Staphylococcus aureus* isolates. It provides an outward oriented promoter that drives expression of genes located proximal to the left end. Moreover, both left and right ends appear to carry a −35 promoter component, which would permit formation of hybrid promoters on insertion next to a resident −10 element (165).

A single member, IS*Dsp3*, present in single copy in *Dehalococcoides* sp. BAV1 carries a passenger gene annotated as hypothetical protein.

IS*6* family elements are abundant in Archaea and cover almost all of the traditionally recognized archaeal lineages (methanogens, halophiles, thermoacidophiles, and hyperthermophiles) (63).

They form a monophyletic group related to bacterial IS from Firmicutes but can be further divided into three phylogenetic groups present in the halophiles, the sulfolobales, and the pyrococcales/methanosarcinales. The IR of the archaeal IS*6* members are variable compared with the bacterial members and generally terminate with 5′-GT or 5′-GA, as opposed to the 5′-GG found in bacteria. The large phylogenetic distribution of IS*6* family members in the Archaea and the monophyly of the IS*6* archaeal group suggest that these elements were ancestrally present in the Archaea rather than being recently acquired by lateral gene transfer from bacteria.

IS*26* is encountered with increasing frequency in plasmids of clinical importance where it is involved in expression of antibiotic resistance genes and plasmid rearrangements (166–171).

(ii) Organization

The putative Tpases are very closely related and show identity levels ranging from 40 to 94%. They generally range in length from 789 bp (IS*257*) to 880 bp (IS*6100*). However, a separate group represented by seven members are somewhat larger (approximately 1,200 bp) as a consequence of an N-terminal extension with a predicted ZF. Several members (e.g., IS*Rle39a*, IS*Rle39b*, and IS*Enfa1*) apparently require a frameshift for Tpase expression. It is at present unclear whether this is biologically relevant. However, alignment with similar sequences in the public databases suggests that IS*Enfa1* itself has an insertion of 10 nucleotides and is therefore unlikely to be active.

All carry short related (15–20 bp) terminal IR and generally create 8 bp DR. A single *orf* is transcribed from a promoter at the left end and stretches across almost the entire IS. In the case of IS*26* this is located within the first 82 bp of the left end and the intact *orf* is required for transposition activity.

(iii) Mechanism and insertion specificity

The predicted amino acid sequence of the Tpase exhibits a strong DDE motif. Translation products of this frame have been demonstrated for several members (e.g., IS*240*). Little is known about Tpase expression although transposition activity of IS*6100* in *Streptomyces lividans* is significantly increased when the element is placed downstream of a strong promoter.

Where analyzed, members of the IS*6* family give rise exclusively to replicon fusions (cointegrates) in which the donor and target replicons are separated by two directly repeated IS copies (e.g., IS*15D*, IS*26*, IS*257*, IS*1936*) (172). Transposition of these elements therefore presumably occurs in a replicative manner by target primed transposon replication. No known specific resolvase system such as that found in Tn*3*-related elements has been identified in this family and it is assumed that cointegrate resolution occurs via homologous recombination. It is for this reason that compound IS*6*-based transposons carry directly repeated flanking IS copies. Recent results suggest that the IS*6* family member IS*26* may transpose in an unusual manner (173), an observation which merits further investigation.

No marked target selectivity has been observed.

IS*21*

(i) General

The IS*21* family is fairly homogeneous. It has more than 140 members from about 80 bacterial species. IS*21* was discovered in plasmid R68 where it was subsequently observed to undergo a tandem duplication, which greatly facilitated the insertion of the plasmid into the *Pseudomonas* host genome. This led to the formation of Hfr strains (174, 175) and the subsequent demonstration that the *Pseudomonas* chromosome was genetically circular (176).

(ii) Organization

IS*21* family members encode two genes, the Tpase *istA* and a "helper" gene, *istB*, which exhibits some similarity to the DnaA replication initiator protein due to the presence of an ATP-binding motif, and often appears in BLAST searches of complete genomes. In many members, the termination codon of *istA* and the initiation codon of *istB* overlap, suggesting that IstB is produced by translational coupling. The IS*21* family terminal IR are complex and carry several tandem repeated sequences thought to be Tpase binding sites. Like a number of IS families, the ends of the element terminate with the dinucleotide 5′-CA-3′.

(iii) Mechanism

IS*21* family members transpose using a two-step mechanism by formation of a reactive junction, similar to those formed in the copy-paste mechanism of IS*3* and other families, in which two abutted IS*21* ends are separated by several nucleotides. This is consistent with the marked tendency of IS*21* to insert in, or close to, an IS*21* end. The reactive junction is subsequently integrated into the target DNA to give a DR of 4 to 8 bp. Integration is optimal when the distance separating the two ends in the junction is 4 bp. It is efficient with a 2 or 3 bp separation but inefficient with smaller or larger intervening sequences. IstA carries the DDE motif and is the Tpase. The molecular details of IstB activity are not known. Both IstA and IstB are required for efficient transposition whereas a product of an alternative translation initiation within the Tpase gene may facilitate integration. Using IstA-enriched *Escherichia coli* cell extracts, it was shown that this protein is responsible for 3′ end cleavage of IS*21* and of both ends in the IR junction (177).

IS*30*

(i) General

The IS*30* family currently comprises 94 members from over 70 bacterial species and an example, ISC*1041*, has also been found in the Archaea (178). IS*30*-like Tpases have also been found as integral parts of certain ICE from methicillin-resistant *Staphylococcus aureus* (53).

Members of this family are capable of activating neighboring genes by creation of a hybrid promoter on insertion next to a −10 promoter element (83, 179–181).

(ii) Organization

IS*30* family members have a single Tpase *orf* spanning almost their entire length. The Tpases show several well-conserved regions. One, in the N-terminus includes a potential helix–turn–helix (HTH) motif, which, in the case of IS*30*, is responsible for IR binding (182, 183). Another in the C-terminal region contains the DD(33)E motif. The terminal IR are between 20 and 30 bp and contain some conserved sequence signatures (183), and their tips show significant sequence variation although in most of the elements the IR have not been experimentally confirmed. Insertion generally generates DR of 2–3 bp but there are several exceptions in which the DR is between 12 and 32 bp (e.g., IS*1630*, IS*1470*, IS*658*, IS*L7*, IS*Lpl1*) (183).

(iii) Mechanism and insertion specificity

The founding member of the family, IS*30*, is the best characterized at the mechanistic level (117, 182–189) and an *in vitro* transposition system has been developed (190). This 1,221 bp long *Escherichia coli* element belongs to a growing class of IS known to transpose through an intermediate formed by abutting the IR, donor primed transposon replication. Here the IR are separated by 2 bp (179, 184, 186). Such an IR junction can be created by formation of a dimer of two directly repeated IS copies or by the formation of transposon circles. Both IS minicircles and dimers have been observed. IR–IR junctions have also been detected in some other IS*30* family members such as IS*18* (181), IS*4351* (191), and IS*1470* (193). A structure in which two IS*30* ends are linked by a single-strand bridge (forming a figure of eight structure on a circular plasmid, has been identified (190).

IS*30* also shows a preference for two distinct types of target sequence: "natural" hot spots, characterized by a 24 bp symmetric consensus, and, like many IS which transpose via a circular dsDNA intermediate, the IR of the element itself. A similar type of insertion specificity was observed for IS*1655* from *Neisseria meningitidis* (189).

IS66

(i) General

IS66 was first identified in the Ti plasmid pTi66 of *Agrobacterium tumefaciens* (193). The vast majority of IS*66* members originate from the Proteobacteria with several from the Bacteriodetes/Chlorobi and the Firmicutes. A second group of closely related IS, widely spread among both bacteria and Archaea are thought to represent a subgroup within the IS*66* family (194). The founder member, IS*Bst12*, originally isolated from *Bacillus stearothermophilus*, was described as a novel family (195), but identification of many additional examples suggests that the IS*Bst12* and IS*66* groups should be considered a single family. IS*Bst12* are found in Actinobacteria, Cyanobacteria, Deinococcus/Thermus, Firmicutes, and Planctomycetes as well as in Proteobacteria. They are also found in the Euryarchaeota phylum of Archaea (but have not yet been identified in the Crenarchaeota).

(ii) Organization

The IS*66* reference copy from a plasmid of the enteropathogenic *Escherichia coli* B171, IS*679*, (196) is defined by three *orfs*: *tnp*A, *tnp*B, and *tnp*C and relatively well conserved terminal IR of about 20–30 bp flanked by an 8 bp DR at their insertion sites. Orf *tnp*C is 1,572 bp and its predicted product includes a typical DDE motif. It also carries an insertion domain between the second D and the E of the DDE motif (e.g., IS*679*, IS*Psy5*, and IS*Mac8*) (31)

The role of the products of *tnpA* (651 bp) and *tnpB* (345 bp) is less clear. Mutation of each *orf* separately (by introduction of an in-frame deletion) reduced transposition by at least two orders of magnitude (196). The three frames are disposed in a pattern suggesting translational coupling: *tnpB* is in general in translational reading frame −1 compared with *tnpA* and in most cases the termination codon of *tnpA* and the initiation codon of *tnpB* overlap (ATGA). An initiation codon for *tnpC* occurs slightly downstream separated from *tnpB* by about 20 bp.

However, rather surprisingly, in the light of a requirement for all three *orfs* for transposition of the canonical IS*66* family member IS*679*, members of the IS*Bst12* group are devoid of *tnpA* and *tnpB* and carry only the *tnpC* reading frame. Although both IS*Bst12* and IS*66* members contain IR, which start with 5′-GTAA-3′, they are clearly distinguishable due to a single conserved A at bp 11 In IS*Bst12*, which is not conserved in IS*66*.

IS*66* members can be grouped into three classes based on their organization: those including all three *orfs*, A, B, and C transcribed in the same direction; those with additional passenger genes invariably present downstream of *orf*C and transcribed in the same direction; and those which lack *orf*A but retain both *orfs* B and C. Each of these organizations includes members with multiple copies, implying that they are active in transposition. In addition to the DDE catalytic domain (194), TnpC also exhibits a highly conserved CwAH-rR motif downstream of the second D residue, a relatively conserved CX2(C)X33CX2C motif characteristic of a ZF further upstream and a leucine-rich region which might form a leucine zipper necessary in multimerization of other Tpases (197), at the N-terminus.

(iii) Mechanism and insertion specificity

Nothing is known about the transposition mechanism of this group of IS and they exhibit no substantial target sequence specificity.

The IS*256* Cluster

Recently, a study of ICE elements identified examples from type B Streptococcus (TnGBS) (50) and Mycoplasma (199) that include a DDE type Tpase rather than the more common phage integrase-like gene. Using a cascade PSI-Blast approach not only revealed two new IS families (IS*Lre2* and IS*Kra4*) but established a distant relationship with the IS*256* and IS*H6* families (52).

Analysis of the N-terminal Tpase region (52) also identified two shared domains (N1 and N2). N2 corresponds to a potential HTH domain in the region of the IS*256* Tpase, which recognizes the terminal IR (199).

The cluster can be divided into five clades containing nine groups based on branching of the Tpases phylogenetic tree: two types of closely related TnGBS, TnGBS1, and TnGBS2, and IS*Lre2* (MULT3); the mycoplasma ICE; IS*256* (MULT1); IS*H6* (MULT2); IS*Azba1*, IS*Mich2*, IS*Kra4* (MULT4) (52).

There is a distant relationship with the Tpase of the eukaryotic Mutator TE and, like MuDR from *Zea mays*, many generate 8 bp to 9 bp target repeat on insertion. They have therefore been called MULE (for Mutator-like Elements). Like MuDr/Foldback, members of these groups carry a largely α-helical insertion domain (31) between the second D and E catalytic residues. This includes a conserved C/D(2)H signature present in the eukaryotic and prokaryotic IS (52, 200).

IS*256*

The IS*256* family can be subdivided into three groups: IS*256*, IS*1249*, and ISC*1250*. The classical IS*256* group has more than 180 members in both bacteria and Archaea. They are between 1,200 and 1,500 bp long with IR of 20–30 bp and generate DR of between 8 and 9 bp.

IS*256* itself was originally isolated as a component of the compound transposon Tn*4001* (201, 202). This family is quite homogeneous. Members carry related IR of between 24 and 41 bp, and most generate 8–9 bp duplications. A single long *orf* carrying a potential DDE motif with a spacing of 112 residues between the second D and E residues, together with a correctly placed K/R residue. The catalytic residues have been validated by mutagenesis (203). It was shown several years ago that the Tpase of IS*256* family elements shares some similarities with the eukaryotic Mutator element (204), a relationship that has been explored recently in more detail (205).

Members of this family transpose using an excised circular dsDNA transposon intermediate (203, 206).

IS*1249* group

There are more than 30 members confined at present to the actinobacteria and the firmicutes. They are about 1,300 bp in length with IR of about 26 bp and generally generate DR of 8 bp (with variations of between 0 and 10).

ISC*1250* group

At present there are only three members of this family in ISfinder. All are found in the Archaea *Sulfolobus solfataricus*.

IS*H6*

This group (MULT2) was originally observed uniquely in Archaea (63). There are 11 members of about 1,450 bp with highly conserved IR of 24–27 bp, DR of 8 bp and a single Tpase *orf* encoding a protein of 450 bp.

IS*Lre2*

There are 48 entries for IS*Lre2* family members in ISfinder. They are restricted at present to the bacteria. They are between 1,500 and 2,000 bp long, with IR from 15 to 29 bp and generate 9 bp DR. Together with the related TnGBS ICE, show strong target specificity and insert 13–17 bp upstream of σA promoters (50, 52) in oriented fashion with the right IS end (RE) proximal. PCR analysis has detected a transposon circle junction, as with the related ICE, suggesting that transposition may occur via a donor primed transposon replication process.

IS*Kra4*

This new family includes 83 members and is divided into three related groups: IS*Azba1*, IS*Mich2*, and IS*Kra4*.

(i) IS*Azba1*

There are presently 28 members of this group. They encode a Tpase of between 450 and 480 amino acids, are 1,400 to 2,900 bp long with IR of about 20 bp and no DR. Six (IS*Afe13*, IS*Cot1*, IS*Ec51*, IS*Kpn19*, IS*Sysp7*) carry an *orf* in addition to the Tpase and this specifies a protein related to serine-recombinases or resolvases. Four of these also include a third *orf* annotated as hypothetical protein. The fifth, IS*Afe13*, carries the Tpase, a resolvase and an alternative *orf* annotated as ORF-3-like from plasmid pRiA4b. Other proteins found in this family are annotated as being hypothetical or putative TnpR resolvases although no direct evidence for resolvase function is available. Eight other members simply encode the Tpase and the ORF-3 like protein. While IS*Cep1* includes the ORF-3-like protein and a third annotated as phage integrase or *xer*C/D.

(ii) IS*Mich2*

This includes 24 members which are presently limited to the cyanobacteria. Twenty two have a Tpase *orf* distributed between two reading phases whereas in the remaining two the Tpase forms a unique continuous *orf*. However, all show a potential but atypical frameshift motif, TTTTTT which could be involved in either programmed ribosomal frameshifting or programmed transcriptional realignment recoding. Further experimental analysis would be necessary to confirm or refute this. Members are between 1,250 and 1,400 bp long with a Tpase of 360 amino acids, IR of between 18 and 39 bp with 8 bp DR. Three members (IS*Cysp26*, IS*Mic1*, and IS*Mich2*) carry a passenger gene annotated as hypothetical protein.

(iii) IS*Kra4*

The 31 elements in this group range in size from 1,400 to 3,700 bp due to the presence in some of various passenger genes. They have IR of 18 to 31 bp and generate DR of 9 bp. Three carry passenger genes: IS*Ldr1*, a hypothetical protein and a reverse transcriptase; IS*Sri1*, a transcriptional regulator; and IS*Tni1*, a hypothetical protein. Six members may express their Tpases by frameshifting (five include a 7A motif and one with a motif, 5TC).

IS*630*

(i) General

There are over 160 members from over 80 bacterial and archaeal genomes.

(ii) Organization

Members are between 1,100 and 1,200 bp long with terminal IR and generally include a single *orf*. However, in about 90 members, the Tpase *orf* is distributed over two reading frames, suggesting that it may be produced as a fusion protein by frameshifting.

(iii) Mechanism

IS*630* transposition has been addressed *in vitro* (207) using IS*Y100* (IS*TcSa*) first identified in *Synechocystis* sp. PCC6803 (208). The Tpase was shown to specifically bind IS*Y100* IR using an N-terminal domain containing two potential HTH motifs. It is the only protein required for IS*Y100* excision and integration and introduces double-strand breaks on mini-IS*Y100* on a supercoiled DNA substrate. Tc*1*/*mariner* element transposition has also been extensively studied *in vitro* (37) and a Tpase structural model is available (209, 210). IS*630* Tpase cleaves exactly at the 3′ (transferred strand) IS ends and two nucleotides inside the 5′ (nontransferred strand) ends. Cleavage is less precise on linear substrates. Both single-end and, less frequently, double-end insertion occur *in vitro* in a TA-target-specific manner (208). Transposition does not involve a hairpin intermediate.

(iv) Insertion specificity

Family members show high target specificity inserting into and duplicating a TA dinucleotide with a preference for the sequence 5′-NTAN-3′ (211). As the cleavages of the nontransferred strand occur 2 nucleotides within the 5′ end of the IS, repair of the donor molecule after excision of the IS can result in a 2 bp scar at the excision site.

The IS*630* family is related to the Tc1/mariner family of eukaryotic TE particularly at the level of the DDE signature. There is also an N-terminal HTH motif. Moreover, IS*630* and the Tc1/mariner families target similar sequences, have similar DR and transposition of both involves cleavage of two nucleotides inside the 5′ ends (37).

IS*982*

(i) General

The IS*982* family has nearly 100 entries in ISfinder from over 40 bacterial and archaeal species. In the case of IS*Lpl4* from *Lactobacillus plantarum*, identical copies have been detected in *Leuconostoc mesenteroides*, *Oenococcus oeni*, and *Lactobacillus sakei* indicating horizontal gene transfer. At least two members, IS*982*B (212) and IS*1187* (83, 213, 214) can provide a −35 hexamer in their right IR capable of forming a hybrid promoter with a resident −10 and activating neighboring genes.

(ii) *Organization*

IS*982* family members are between 962 and 1,155 bp long and carry similar terminal IR of between 18 and 35 bp with conserved ends: 5′-ACCC-3′. They encode a single *orf* of between of 271 and 313 amino acids with a possible DDE motif but without a convincing conserved downstream K/R residue. Little is known about the transposition of these elements. They generate DR of 6 to 8 bp.

Although the Tpase of a majority of members occupies a single reading phase, there are several examples in which the gene is distributed over two phases. It has been reported that a +1 nucleotide insertion in the Tpase *orf* of an IS*Lpl4* from *Oenococcus oeni* may undergo programmed translational frameshifting at a low rate (215). Although this must be confirmed, it would represent the first functional case of +1 frameshifting in IS. The Tpase of archaeal element IS*Pfu3* is also distributed over two phases. IS*Pfu3* carries a potential transcriptional frameshift signal A7 (programmed transcriptional realignment) present in all five copies suggesting that IS*Pfu3* is active.

IS*1380*

This family is represented by 153 members from nearly 100 bacterial species in ISfinder. They show conserved ends terminating with CCt/c. Although the majority of their Tpases often include the canonical DDE(6)K/R, several members exhibit other residues in place of the K/R. These include DDE(6)Q or DDE(6)I. A subgroup, IS*942*, composed of 13 members all restricted to Bacteriodetes, include DDE(6)N. None of these differences appear to affect the predicted secondary structure. In addition, to the host-restricted IS*942* group, two other branches of the Tpase tree are restricted to the *Actinobacteria*. A single, poorly characterized NCBI database entry (WP_018034290) probably corresponds to an archaeal IS*1380* member and intriguingly, an *orf* (XP_002337507) with a 100% match to IS*Lsp5* (*Leptospirillum* sp.) has been identified in the genomic sequence of the black cottonwood tree, *Populus trichocarpa*. Tpases of this family include an insertion domain with a predominantly β-strand secondary structure (31).

IS*1380* itself (216) is present in high copy number in the *Acetobacter pasteurianus* NCI1380 genome and in several strains of acetic acid bacteria. At present the family contains two tIS present in a single copy: IS*Msm12* (*Mycobacterium smegmatis*; tetR + methyltransferase) and IS*Rop1* (*Rhodococcus opacus*; reverse transcriptase).

This IS family is distantly related to the eukaryotic PiggyBac TE family (see Yusa, this volume) which also includes an insertion domain largely in the form of β-strand (148).

IS*As1*

There are over 80 entries for this family in ISfinder from over 50 bacterial species. There are currently no archaeal members. IS*As1* family members are between 1,200 and 1,326 bp long and generally carry terminal IR of between 14 and 22 bp. A single *orf* of between 294 and 376 amino acids occupies almost the entire length. There are several conserved D and E residues. The putative Tpase of IS*1358* has been visualized using a phage T7 promoter-driven gene and that of ISAs*1* has been detected in *Escherichia coli* minicells. The family also includes "H-repeats", which form part of several so-called rearrangement hot spots (RHS) elements containing another repeated sequence, the H-rpt element (Hinc repeat). H-rpt display features of typical insertion sequences although no transposition activity has yet been detected. For the sake of clarity, the H-rpt DNA sequences B (RhsB), C1 to C3 (RhsC), E (RhsE), and min.5 as well as H-rptF were renamed ISEc*1* to ISEc7, respectively (127). The Tpases of this family include a β-strand insertion domain (31).

Little is known about the transposition properties of this family of elements. However, recent experiments with the *Vibrio cholerae* element IS*1358* have demonstrated that insertion generates 10 bp DR and that, in *Escherichia coli*, it undergoes simple insertion into a target plasmid, pOX38 (217).

IS*L3*

(i) General

There are more than 120 members from nearly 80 bacterial species. The family also includes archaeal members, particularly in the Methanomicrobia. A potential tIS derivative that includes a mercury-resistance operon has also been identified in a conjugative plasmid in *Enterococcus faecium* (218).

(ii) Organization

Members range in size from 1,186 bp to 1,553 bp, carry closely related IR of between 15 and 39 bp and generate DR of 8 bp. They generally have a single *orf* of between 400 and 440 amino acids, which shows good alignment and includes an α-helical insertion domain (31).

However, IS*1096* harbors two *orf*s: the upstream *orf* exhibits similarities to the IS*L3* family Tpases; the second, *tnp*R, a MerR-like transcription factor, is related to *orf*s from *Agrobacterium rhizogenes* and *Rhizobium* sp. plasmids. TnpR appears to regulate transposition activity of IS*Ppu12* (219).

In IS*1167*, the reading frame appears to be distributed between two consecutive *orf*s with a potential for translational coupling suggested by overlapping initiation and termination codons (ATGA). Small sequences (130 to 340 bp) related to the IR of IS*1167* have been detected in *Streptococcus sanguis*, *Streptococcus pneumoniae*, and *Streptococcus agalactiae* (3, 417).

(iii) Mechanism

Transposition of most of these elements has been demonstrated, but no detailed analysis of their transposition mechanism has yet appeared. IS*1411* from *Pseudomonas putida* forms a circular species with abutted IR separated by 5 bp (220). Transposon circles are also formed by IS*Pst9* (221) and an isoform of IS*Ppu12* (219) and IS*31831* forms DNA species with a size expected for an excised transposon (222). There is some evidence indicating that transposition of these IS can be induced by a form of zygotic induction following conjugative transfer (219).

No obvious target sequence specificity has yet been observed although there is some suggestion that there may be a preference for AT-rich regions.

Tn*3*

(i) General

his represents a large and highly homogeneous group in terms of their transposition enzymes and terminal IR (Nicolas et al., this volume). Many are complicated in structure and include multiple antibiotic resistances, virulence and other "accessory" genes. These are often carried by another type of transposable element, integron cassettes (45). The Tn*3* family is included here because certain family members resemble IS (e.g., IS*1071*, IS*101*, IS*Xc4*/IS*Xc5*) encoding only the Tpase flanked by IR.

(ii) Organization

The IR are generally 38 bp long, start with the sequence GGGG (occasionally GAGG) and terminate internally with TAAG. The Tn*3* family encodes large (>900 amino acids) DDE Tpases with an α-helical insertion domain (31). Classical Tn*3* family members also encode a site-specific recombinase. The major differences between members of this family are in the number and location of the many passenger genes and in the type of site-specific recombinase present.

(iii) Mechanism

The replicative transposition mechanism of this family involves formation of a cointegrate in which donor and recipient replicons are fused and separated at each junction with a directly repeated transposon copy (223). These structures must be "resolved", by recombination between the two transposon copies, to generate the donor and target replicons each retaining a single transposon copy. This is accomplished by a site-specific recombinase (resolvase), which acts at a unique DNA sequence, the Res site. Apart from passenger genes, the major difference between various Tn*3*-like elements is the nature of their resolvases (Nicolas et al., this volume).

IS*Azo13*

This family, represented by 37 members in ISfinder, emerged from the IS*NCY* orphan group. It is based on both Tpase and IR sequence similarities. Insertion generates a 3 bp AT-rich DR and the ends have a consensus GGa/g. Their Tpases are highly conserved with a probable DDE motif and an HTH motif at the N-terminus, which could function as a DNA binding do main. Two members encode two *orf*s with a possible programmed ribosomal frameshifting motif of 8 or 9 Å while the other members encode a unique *orf* which includes a triple lysine at the equivalent position (E. Gourbeyre unpublished).

IS WITH DEDD TRANSPOSASES

IS*110*

(i) General

At present, only a single IS family, IS*110*, is known to encode this type of DEDD enzyme, related to the RuvC Holliday junction resolvase (40). There are over 250 examples from nearly 130 bacterial and archaeal species.

The Tpase is closely related to the *Piv* and *MooV* invertases from *Moraxella lacunata*/*Moraxella bovis* (224, 225) and *Neisseria gonorrhoeae* (39, 226). *Piv* catalyzes inversion of a DNA segment permitting expression of a type IV pilin. However, the organization of IS*110* family members and the inversion systems are different. In the inversion systems, the recombinase is located outside the invertible segment, whereas it is located within the IS element (40).

The family includes two subgroups. It has been suggested that these may even represent two distinct families (114, 227): IS*110* and IS*1111*. Although their Tpases are very similar to those of the classical IS*110*, members of the second subgroup, IS*1111*, are distinguished from those of the IS*110* group principally by the presence of small (7 to 17 bp) subterminal IR. This would be the only family that is not defined by differences in the Tpase but by the nature of the IS ends.

However, the entire group of IS exhibit significant differences: some have subterminal IR whereas others do not; some appear to generate small DR while others do not; and the entire family show significant variations in their target preferences.

(ii) Organization

The organization of IS*110* family members is quite different from that of the DDE IS: they do not contain the typical terminal IR of the DDE IS and do not generally generate flanking target DR on insertion. This implies that their transposition occurs using a different mechanism to the DDE IS. They encode a single Tpase gene that spans the entire length of the IS. One characteristic which distinguishes IS*110* family members from all other elements whose Tpases exhibit a predicted RNase fold is that the predicted catalytic domain of their DEDD Tpases is located N-terminal to the DNA-binding domain (39, 228). In the DDE Tpases it is generally located upstream.

(iii) Mechanism

It has proved difficult to determine the activity of these Tpases in detail *in vitro*. Transposition of IS with DEDD Tpases may be unusual and involve HJ intermediates, which must be resolved using a RuvC-like mechanism. This type of recombination would be consistent with the close relationship between DEDD Tpases and the Piv/MooV invertases, which presumably resolve HJ structures during inversion (229). The difference in domain organization between the DEDD and DDE Tpases reinforces the idea that the two IS types possess a different transposition mechanism.

Members of this family produce double-strand circular transposon intermediates (e.g., IS*492*:(90), (230); IS*Pa11* (114); IS*Ec11*, (231); IS*117* (232, 233); IS*1383* (234)). However, it remains to be determined whether they use a copy-paste mechanism or whether the circular intermediate is formed by excision. For example, IS*492* was identified within the extracellular polysaccharide production (*eps*) gene of *Pseudoalteromonas atlantica* (90, 235). IS*492* clearly undergoes Tpase-dependent precise excision to regenerate a functional *eps* gene. Like many other IS that use double-strand circular intermediate, circle formation results in the assembly of a junction promoter from a −35 promoter element in the right end oriented outwards and a −10 promoter element in the left end oriented inwards (90). In the case of IS*1383*, the −10 component of the promoter appears to lie within the DNA sequence located between the abutted IS ends. This promoter presumably serves to drive Tpase expression for the final integration step (90, 234). For IS*1383*, amplification of a putative circle junction suggested that the abutted IS ends were separated by 10 bp composed of the 5 bp flanking each end in the original target site. For IS*492*, this is distance is 5 bp. The IS*492* copy within the *eps* gene is flanked by a 5-bp DR that is required for Tpase-dependent precise excision and for transposition of the IS (90).

(iv) Insertion specificity

Another characteristic of the IS*110* family is their particular insertion specificities. IS*492*, IS*Ptu2*, and IS*Spi5* recognize a 7 bp sequence, 5′-CTTGTTA-3′, and duplicate the first 5 bp (CTTGT). Excision of the IS regenerates the original target sequence (236). In the case of IS*492*, the flanking 5 bp were also essential for formation of the dsDNA circular form (90).

At least six different members of the IS*1111* subgroup (IS*Kpn4*, IS*Pa21*, IS*Pst6*, IS*UnCu1*, IS*Azvi12*, and IS*Pa25*) show a preference for *attC* sequences of integrons (111) whereas others appear to prefer repeated extragenic palindrome (REP) sequences (109). Both targets are capable of assuming hairpin-like secondary structures. The integron *attC* is central to integration of circular integron cassettes (45) whereas REP sequences are small repeated extragenic palindromic sequences often present in many hundreds of copies in bacterial genomes and which play a variety of structural and regulatory roles (He et al., this volume). There are at least seven examples (IS*621*, IS*Pa11*, IS*Ppu9*, IS*Ppu10*, IS*Rm19*, IS*Psy7*, and IS*1594*) from both the classical IS*110* and IS*1111* groups, which have been identified as insertions into REP sequences. These are two types of insertion. In type 1, the IS inserts at the same position in the REP whereas type 2 insertions occur adjacent to a REP. Only one IS*1111* member, IS*Psy7*, has been identified with type 2 insertion specificity (109). Certain IS that insert into REP (e.g., IS*621*, IS*Ppu9*, IS*Ppu10* and IS*1594*) generate a duplication of 2 bp.

Other IS of this family also appear to insert into conserved target sequences: IS*1533* occurs in 84 copies in *Leptospira borgpetersenii* and inserts into a partially conserved sequence (ttAGACAAAA[IS*1533*]TATCAG agcc-gtct–aaa); IS*Rfsp2* from *Roseiflexus* sp. RS-1, present in 40 copies in the host genome, is flanked by the sequence, CTCtGCGaaCGCtGCGc[IS*Rfsp2*]CTCtGCG Gtg. Yet other IS of this family appear to target the ends of other IS. Hence IS*1383* present in six copies in *Pseudomonas putida* plasmids, inserts into the sequence TTCAGATGGT[IS1383]ATAAG contained within the end of another IS of the IS*5* family, IS*1384* (227, 234); IS*4321* and IS*5075* (both members of the IS*1111* group) target the ends of the transposon Tn*21*; and IS*Sba8* (IS*1111*) inserts into a REP-like sequence located in the end of the IS*3* family member, IS*Sba5*.

Therefore, as a general rule, IS*110* family elements appear to insert in a sequence-specific and oriented way (236).

IS WITH HUH TRANSPOSASES

There are two major HUH Tpase families: Y1 and Y2 enzymes (41) according to whether they carry one or two Y residues involved in catalysis. One (Y1) is associated with the IS*200*/IS*605* family. Although these enzymes use the same Y-mediated cleavage mechanism, they appear to carry out the transposition process in quite different ways. The TE that use this type of Tpase do not terminate in IR, which often makes it difficult to determine their ends by bioinformatics analysis unless there are a number of identical copies in a genome.

IS*91*

The IS*91* family is fairly homogeneous. The canonical IS*91* identified three decades ago (54) carries only a single *orf*, encoding an HUH Y2 Tpase with an N-terminal zinc finger motif (237) (see also reference 41). Both Y residues are necessary for transposition (238). The IS does not terminate with extensive IR but includes some potential secondary structure. Several other family members (e.g., IS*Azo26*, IS*CARN110*, IS*Mno23*, IS*Sde12*, IS*Shvi3*, IS*Sod25*, and IS*Wz1*) include a second upstream *orf* related to the phage integrase/Y-recombinase family. Its role, if any, in transposition remains unclear.

More recently, a second related group, the IS*CR* (IS with a Common Region) was identified (55) associated with multiple antibiotic-resistance genes both upstream and downstream of the Tpase gene. They form a distinct class. Although their Tpases are closely related to the IS*91* Tpase, they carry a single catalytic Y residue. However, it has yet to be demonstrated that IS*CR* indeed transpose.

IS*91* is thought to transpose using a rolling circle-type mechanism (239) involving an initiation event at one IS end (ori, 3′ to the Tpase), polarized transfer of the IS strand into a target molecule and termination at the second end (ter, 5′ to the Tpase) (238). Sequestration of flanking genes is proposed to occur when the termination mechanism fails and rolling circle transposition extends into neighboring DNA, where it may encounter a second surrogate end (238). This has been estimated to occur at a frequency of about 1%. Indeed, *ori* is essential for activity whereas removal of *ter* reduces but does not eliminate transposition.

Insertion of IS*91* is oriented with *ori* adjacent to the 3′ of a specific tetranucleotide target (5′-CTTG or 5′-GTTC), which is essential for further transposition (238, 239). In the transposition model, displacement of an IS*91* active transposon strand would be driven by leading strand replication of the donor replicon from a 3′ OH generated by cleavage at *ori*. The original model proposed that the cleaved IS end is transferred to the target DNA and the IS is replicated "into" the target replicon. However, it is difficult to explain the occurrence of single- and double-strand IS*91* circles on this model and their role remains to be determined. Clearly, since IS*CR* may play an important role in the assembly and transmission of multiple antibiotic resistance (55, 240) it is important to address the transposition mechanism of these elements.

IS*91* and IS*CR* Tpases and organization are distantly related to the eukaryotic Helitrons (241) (Thomas and Pritham, this volume).

IS*200*/IS*605*

(i) General

The IS*200*/IS*605* family (242–244) (He et al., this volume) is divided into three major groups based on the presence or absence of two reading frames, *tnpA* (encoding a Y1 Tpase) and *tnpB* (whose exact role is unknown) (245). The IS*200* group includes *tnpA* only, the IS*1341* (246) group carries only *tnpB* (and as such its status as an autonomous TE remains questionable), while the third group, IS*605*, carries both *tnpA* and *tnpB*. These can be expressed divergently or sequentially and, in some cases overlap slightly, suggesting translational coupling. TnpB is not required for transposition (245, 247) but may have a role in regulation because expression of the protein under control of inducible promoter from IS*Dra2* reduces transposition in its original host, *Deinococcus radiodurans*, and in *Escherichia coli* (248). TnpB has also been identified in the IS*607* family (below) not only in prokaryotes but also in eukaryotes (71, 249) where it is sometimes associated with other TE.

(ii) Mechanism

The IS*200*/IS*605* transposition mechanism is well understood from a combination of genetic, biochemical and structural studies (94, 100, 245, 250–255). Briefly, it occurs using a single-strand "peel-and-paste" mechanism (He et al., this volume) in which a specific single transposon strand (the "top" strand) is excised to form a single-strand transposon circle. This then inserts into a single-strand target. Family members include subterminal secondary structures, which are recognized by TnpA. The cleavage site occurs a short distance 5′ to the left and

3′ to the right of the structure. These are not directly recognized by TnpA but form a complex set of interactions with the internal sequence permitting their cleavage. Transposition occurs by insertion of the left end 3′ to a specific tetra- or penta-nucleotide, which is also essential for excision and further transposition. Insertion does not generate DR and occurs preferentially into the lagging strand template of replication forks. This results in a clear orientation bias at the genome level reflecting the direction of replication of the target replicon. This can be detected in numerous bacterial genomes. It is also possible that this family targets stalled replication forks (He et al., this volume).

IS WITH SERINE TRANSPOSASES

The serine recombinase family includes three groups: the resolvase/integrase group (whose activity has been well characterized); the large serine recombinases; and the serine Tpases (256). For the two former groups, the catalytic domain is invariably N-terminal followed by the sequence-specific DNA-binding domain—a simple HTH for the resolvase/invertase group, or a much larger domain of unknown structure in the large serine recombinases (257); (see chapters in the section *Conservative Site-Specific Recombination*).

IS*607*

(i) General

IS*607* was first identified in *Helicobacter pylori* (258). Family members encode a Tpase related to serine site-specific recombinase which uses serine as a nucleophile for cleavage of the DNA strand (223). This family is linked to the IS*200*/IS*605* family (above) by the presence of a TnpB analogue. IS*607*-like elements have been observed in the Mimi virus and other nucleocytoplasmic large DNA viruses (NCLDV) (140). Interestingly it appears to be one of the rare prokaryotic IS identified in eukaryotic genomes (71, 249).

(ii) Mechanism

Little is known about IS*607* family transposition although it is thought that circular intermediates are involved (N.D.F. Grindley pers. comm. cited in reference 63). The enzyme presumably catalyzes similar cleavages and strand transfers as its site-specific serine recombinase cousins using a transitory 5′ phosphorserine covalent intermediate to excise a double-strand circular IS DNA copy. A mechanistic transposition model has been proposed based on Tpase structures from structural genomics studies and detailed knowledge of the general serine recombinase mechanism (256). This imagines a synaptic Tpase tetramer (as for classical serine recombinases) and explains the lack of IS*607* target specificity (258), unusual for this type of recombinase.

(iii) Organization

In contrast to the serine recombinases, the DNA-binding domain of the serine Tpases is located N-terminal to the catalytic domain (249, 256) in a similar way to that of DDE Tpases and may reflect a similar function: folding of the nascent peptide and co-translational binding to the IS ends (see *IS3 and IS481 families* section above)

ORPHAN IS

In addition to the major IS groupings, there are also several families for which the Tpase signature is not yet clear, either because there are a number of potential catalytic residues or because there are not a sufficient number of examples to define the highly conserved residues. In the former cases, definition of the important conserved residues will require experimental analysis. In the latter case, the IS are grouped as IS*NCY* (not classified yet), which includes small numbers of unclassified IS or orphans. Members of this group often emerge as families, or new distant groups of a known family, as more examples are added to the database. For example, both IS*1202 and ISDol1* have emerged as groups related to the IS*3*/IS*481* and IS*5* families, respectively.

Others include:

IS*892*: These are 1,600 bp long and represented by two sequences (with additional examples in the public databases). Their Tpases may be produced by frameshifting and include a Pfam MULE-like motif.

IS*Lbi1*: These are 1,400–1,500 bp long and represented by two sequences (with additional examples in the public databases) with a single *orf*, 30 bp IR and 5 bp DR.

IS*Mae2*: These are 1,400–2,400 bp long and represented by three sequences (with additional examples in the public databases) with a single *orf* carrying a potential DDE motif, IR and 9 bp DR. One of these, IS*Acif1*, includes a passenger gene of unknown function located downstream of the Tpase.

IS*Plu15*: These are 800–1,000 bp long and represented by two sequences (with additional examples in the public databases). They include a single *orf* and there are no apparent IR or DR, which makes definition of the IS ends difficult unless they are present in multiple copies.

IS*A1214*: These are archaeal-specific IS with a length of 1,000–1,200 bp. They are represented by five sequences with IR, with DR of 8 to 12 bp and a small *orf* upstream of the Tpase expressed in the opposite direction.

IS*C1217*: These archaeal-specific IS are 1,200 bp long and are represented by four sequences from the Sulfolobales with a single Tpase *orf*, IR, and DR of 6–8 bp.

IS*M1*: These archaeal-specific IS are 1,300–1,600 bp long and are represented by six sequences with low conservation and no clear DDE but with IR of about 24 bp, DR of 8 or 9 bp, and a single *orf*.

EUKARYOTIC TE AND PROKARYOTIC RELATIVES

Eukaryotic DNA transposons have been classified into "superfamilies" and, of those that have been analyzed in some detail, many have prokaryotic cousins. These relationships have been highlighted throughout this text. They are Tc1/mariner (IS*630*), Mutator(MuDR)/foldback (IS*256* and IS*Lre2*), PiggyBac (IS*1380*), PIF/Harbinger (IS*5*), Merlin (IS*1595*), Banshee (IS*3*/IS*481*), and Helitron (IS*91*). However, several elements with DDE transposases such as hAT, P, CACTA (En/Spm), Transib, Chapaev, Sola, Zator, and Ginger (148, 200) have yet to find prokaryotic cousins. It seems highly probable that, as the diversity of TE is explored more extensively aided by the massive accumulation of sequence data and the development of software designed to detect TE, more phylogenetic relationships between prokaryotic and eukaryotic TE will become evident.

CONCLUDING REMARKS

We have divided this review into two major sections. In one, we have attempted to present an overview of our current understanding of prokaryotic IS, their diversity in sequence, in organization and in mechanism, their distribution and impact on their host genome, and their relation to their eukaryotic cousins. We have discussed several IS-related TE that have been identified since the previous edition of *Mobile DNA*. This includes IS that use single-strand DNA intermediates and their related "domesticated" relations, and IS*CR* and ICE that use IS-related transposases for excision and integration. Several more specialized chapters in this volume include additional detailed information concerning a number of these topics. One of the major conclusions from this section is that the frontiers between the different types of TE are becoming less clear as more are identified.

In the second part, we have provided a detailed description of the expanding variety of IS, which we have divided into families for convenience. We emphasize that there is no "quantitative" measure of the weight of each of the criteria we use to define a family. Our perception of these families continues to evolve and families emerge regularly from the IS*NCY* "class" as more IS are added. This section is designed to be an aid and a source of information for consultation by interested specialist readers.

It is clear from this survey that there are many important and unanswered questions concerning both mechanistic and regulatory aspects of IS and the way in which they have and continue to spread and colonize their host genomes. An area of particular importance is in understanding the dynamics of TE-driven genome remodeling. A growing view is that the prokaryote genome is an ecological niche populated by a diverse collection of TE, in particular IS, including both endogenous elements and those acquired serendipitously by horizontal transfer. These form an integral part of a genomic landscape which is continuously modified by their activity. Understanding the evolutionary relationship between genomes and TE in this ecological niche and of the dynamics of TE-mediated genome remodeling is one of the fundamental challenges in genomics.

Citation. Siguier P, Gourbeyre E, Varani A, Ton-Hoang B, Chandler M. 2015. Everyman's guide to bacterial insertion sequences. Microbiol Spectrum 3(2):MDNA3-0030-2014.

References

1. **Fiandt M, Szybalski W, Malamy MH.** 1972. Polar mutations in lac, gal and phage lambda consist of a few IS-DNA sequences inserted with either orientation. *Mol Gen Genet* **119:**223–231.
2. **Hirsch HJ, Saedler H, Starlinger P.** 1972. Insertion mutations in the control region of the galactose operon of *E. coli*. II. Physical characterization of the mutations. *Mol Gen Genet* **115:**266–276.
3. **Hirsch HJ, Starlinger P, Brachet P.** 1972. Two kinds of insertions in bacterial genes. *Mol Gen Genet* **119:** 191–206.
4. **Saedler H, Heiss B.** 1973. Multiple copies of the insertion-DNA sequences IS1 and IS2 in the chromosome of *E. coli* K-12. *Mol Gen Genet* **122:**267–277.
5. **Reif HJ, Saedler H.** 1974. IS1 is Involved in Deletion Formation in the gal Region of *E. coli* K12. *Mol Gen Genet* **137:**17–28.
6. **Saedler H, Reif HJ, Hu S, Davidson N.** 1974. IS2, a genetic element for turn-off and turn-on of gene activity in *E. coli*. *Mol Gen Genet* **132:**265–289.

7. **Hu S, Ohtsubo E, Davidson N.** 1975. Electron microscopic heteroduplex studies of sequence relations among plasmids of *Escherichia coli*: structure of F13 and related F-primes. *J Bacteriol* **122:**749–763.

8. **Barth PT, Datta N, Hedges RW, Grinter NJ.** 1976. Transposition of a deoxyribonucleic acid sequence encoding trimethoprim and streptomycin resistances from R483 to other replicons. *J Bacteriol* **125:**800–810.

9. **Hedges RW, Jacob AE.** 1974. Transposition of ampicillin resistance from RP4 to other replicons. *Mol Gen Genet* **132:**31–40.

10. **Heffron F, Sublett R, Hedges RW, Jacob A, Falkow S.** 1975. Origin of the TEM-beta-lactamase gene found on plasmids. *J Bacteriol* **122:**250–256.

11. **Arber W, Iida S, Jutte H, Caspers P, Meyer J, Hanni C.** 1979. Rearrangements of genetic material in *Escherichia coli* as observed on the bacteriophage P1 plasmid. *Cold Spring Harb Symp Quant Biol* **43**(Pt 2)**:**1197–1208.

12. **So M, Heffron F, McCarthy BJ.** 1979. The *E. coli* gene encoding heat stable toxin is a bacterial transposon flanked by inverted repeats of IS1. *Nature* **277:**453–456.

13. **Nevers P, Saedler H.** 1977. Transposable genetic elements as agents of gene instability and chromosomal rearrangements. *Nature* **268:**109–115.

14. **McClintock B.** 1956. Controlling elements and the gene. *Cold Spring Harb Symp Quant Biol* **21:**197–216.

15. **Nyman K, Nakamura K, Ohtsubo H, Ohtsubo E.** 1981. Distribution of the insertion sequence IS1 in gram-negative bacteria. *Nature* **289:**609–612.

16. **Ohtsubo H, Nyman K, Doroszkiewicz W, Ohtsubo E.** 1981. Multiple copies of iso-insertion sequences of IS1 in *Shigella dysenteriae* chromosome. *Nature* **292:**640–643.

17. **Aziz RK, Breitbart M, Edwards RA.** 2010. Transposases are the most abundant, most ubiquitous genes in nature. *Nucleic Acids Res* **38:**4207–4217.

18. **Berg DE, Howe MM.** 1989. *Mobile DNA*. American Society for Microbiology, Washington DC.

19. **Craig NL, Craigie R, Gellert M, Lambowitz A.** 2002. *Mobile DNA II*. American Society of Microbiology, Washington DC.

20. **Siguier P, Gourbeyre E, Chandler M.** 2014. Bacterial insertion sequences: their genomic impact and diversity. *FEMS Microbiol Rev* **38:**865–891.

21. **Sharma V, Firth AE, Antonov I, Fayet O, Atkins JF, Borodovsky M, Baranov PV.** 2011. A pilot study of bacterial genes with disrupted ORFs reveals a surprising profusion of protein sequence recoding mediated by ribosomal frameshifting and transcriptional realignment. *Mol Biol Evol* **28:**3195–3211.

22. **Siguier P, Perochon J, Lestrade L, Mahillon J, Chandler M.** 2006. ISfinder: the reference centre for bacterial insertion sequences. *Nucleic Acids Res* **34:**D32–D36.

23. **Campbell A, Berg DE, Botstein D, Lederberg EM, Novick RP, Starlinger P, Szybalski W.** 1979. Nomenclature of transposable elements in prokaryotes. *Gene* **5:** 197–206.

24. **Mahillon J, Chandler M.** 2000. Insertion sequence nomenclature. *ASM News* **66:**324.

25. **Galas DJ, Chandler M.** 1989. Bacterial insertion sequences, p 109–162. *In* Berg D, Howe M (ed), *Mobile DNA*. American Society for Microbiology, Washington DC.

26. **Chandler M, Mahillon J.** 2002. Insertion sequences revisited, p 305–366. *In* Craig NL, Craigie R, Gellert M, Lambowitz A (ed), *Mobile DNA*, **vol II.** ASM Press, Washington DC.

27. **Mizuuchi K, Baker TA.** 2002. Chemical mechanisms for mobilizing DNA, p 12–23. *In* Craig NL, Craigie R, Gellert M, Lambowitz A (ed), *Mobile DNA*, **vol II.** ASM press, Washington DC.

28. **Fayet O, Ramond P, Polard P, Prere MF, Chandler M.** 1990. Functional similarities between retroviruses and the IS3 family of bacterial insertion sequences? *Mol Microbiol* **4:**1771–1777.

29. **Kulkosky J, Jones KS, Katz RA, Mack JP, Skalka AM.** 1992. Residues critical for retroviral integrative recombination in a region that is highly conserved among retroviral/retrotransposon integrases and bacterial insertion sequence transposases. *Mol Cell Biol* **12:**2331–2338.

30. **Khan E, Mack JP, Katz RA, Kulkosky J, Skalka AM.** 1991. Retroviral integrase domains: DNA binding and the recognition of LTR sequences [published erratum appears in Nucleic Acids Res 1991 Mar 25;19(6):1358]. *Nucleic Acids Res* **19:**851–860.

31. **Hickman AB, Chandler M, Dyda F.** 2010. Integrating prokaryotes and eukaryotes: DNA transposases in light of structure. *Crit Rev Biochem Mol Biol* **45:**50–69.

32. **Davies DR, Braam LM, Reznikoff WS, Rayment I.** 1999. The three-dimensional structure of a Tn5 transposase-related protein determined to 2.9-A resolution. *J Biol Chem* **274:**11904–11913.

33. **Hickman AB, Perez ZN, Zhou L, Musingarimi P, Ghirlando R, Hinshaw JE, Craig NL, Dyda F.** 2005. Molecular architecture of a eukaryotic DNA transposase. *Nat Struct Mol Biol* **12:**715–721.

34. **Curcio MJ, Derbyshire KM.** 2003. The outs and ins of transposition: from Mu to Kangaroo. *Nat Rev Mol Cell Biol* **4:**865–877.

35. **Turlan C, Chandler M.** 2000. Playing second fiddle: second-strand processing and liberation of transposable elements from donor DNA. *Trends Microbiol* **8:** 268–274.

36. **Feng X, Colloms SD.** 2007. *In vitro* transposition of ISY100, a bacterial insertion sequence belonging to the Tc1/mariner family. *Mol Microbiol* **65:**1432–1443.

37. **Plasterk RH.** 1996. The Tc1/mariner transposon family. *Curr Top Microbiol Immunol* **204:**125–143.

38. **Zhou L, Mitra R, Atkinson PW, Burgess Hickman A, Dyda F, Craig NL.** 2004. Transposition of hAT elements links transposable elements and V(D)J recombination. *Nature* **432:**995–1001.

39. **Choi S, Ohta S, Ohtsubo E.** 2003. A novel IS element, IS621, of the IS110/IS492 family transposes to a specific site in repetitive extragenic palindromic sequences in *Escherichia coli*. *J Bacteriol* **185:**4891–4900.

40. **Buchner JM, Robertson AE, Poynter DJ, Denniston SS, Karls AC.** 2005. Piv site-specific invertase requires a

DEDD motif analogous to the catalytic center of the RuvC Holliday junction resolvases. *J Bacteriol* **187:** 3431–3437.

41. **Chandler M, de la Cruz F, Dyda F, Hickman AB, Moncalian G, Ton-Hoang B.** 2013. Breaking and joining single-stranded DNA: the HUH endonuclease superfamily. *Nat Rev Microbiol* **11:**525–538.
42. **Taylor KL, Churchward G.** 1997. Specific DNA cleavage mediated by the integrase of conjugative transposon Tn916. *J Bacteriol* **179:**1117–1125.
43. **Siguier P, Gagnevin L, Chandler M.** 2009. The new IS1595 family, its relation to IS1 and the frontier between insertion sequences and transposons. *Res Microbiol* **160:**232–241.
44. **Liebert CA, Hall RM, Summers AO.** 1999. Transposon Tn21, Flagship of the Floating Genome. *Microbiol Mol Biol Rev* **63:**507–522.
45. **Mazel D.** 2006. Integrons: agents of bacterial evolution. *Nat Rev Microbiol* **4:**608–620.
46. **Nakatsu C, Ng J, Singh R, Straus N, Wyndham C.** 1991. Chlorobenzoate catabolic transposon Tn5271 is a composite class I element with flanking class II insertion sequences. *Proc Natl Acad Sci USA* **88:**8312–8316.
47. **Burrus V, Waldor MK.** 2004. Shaping bacterial genomes with integrative and conjugative elements. *Res Microbiol* **155:**376–386.
48. **Adams V, Lyras D, Farrow KA, Rood JI.** 2002. The clostridial mobilisable transposons. *Cell Mol Life Sci* **59:**2033–2043.
49. **Pavlovic G, Burrus V, Gintz B, Decaris B, Guedon G.** 2004. Evolution of genomic islands by deletion and tandem accretion by site-specific recombination: ICESt1-related elements from *Streptococcus thermophilus*. *Microbiology* **150:**759–774.
50. **Brochet M, Da Cunha V, Couve E, Rusniok C, Trieu-Cuot P, Glaser P.** 2009. Atypical association of DDE transposition with conjugation specifies a new family of mobile elements. *Mol Microbiol* **71:**948–959.
51. **Guerillot R, Da Cunha V, Sauvage E, Bouchier C, Glaser P.** 2013. Modular evolution of TnGBSs, a new family of integrative and conjugative elements associating insertion sequence transposition, plasmid replication, and conjugation for their spreading. *J Bacteriol* **195:**1979–1990.
52. **Guerillot R, Siguier P, Gourbeyre E, Chandler M, Glaser P.** 2014. The diversity of prokaryotic DDE transposases of the mutator superfamily, insertion specificity, and association with conjugation machineries. *Genome Biol Evol* **6:**260–272.
53. **Smyth DS, Robinson DA.** 2009. Integrative and sequence characteristics of a novel genetic element, ICE6013, in *Staphylococcus aureus*. *J Bacteriol* **191:**5964–5975.
54. **Diaz-Aroca E, de la Cruz F, Zabala JC, Ortiz JM.** 1984. Characterization of the new insertion sequence IS91 from an alpha-hemolysin plasmid of *Escherichia coli*. *Mol Gen Genet* **193:**493–499.
55. **Toleman MA, Bennett PM, Walsh TR.** 2006. ISCR Elements: Novel Gene-Capturing Systems of the 21st Century? *Microbiol Mol Biol Rev* **70:**296–316.
56. **Toleman MA, Walsh TR.** 2010. ISCR elements are key players in IncA/C plasmid evolution. *Antimicrob Agents Chemother* **54:**3534; author reply 3534.
57. **Feschotte C, Zhang X, Wessler S.** 2002. Miniature inverted repeat transposable elements and their relationship to established DNA transposons, p 1147–1158. *In* Craig NL, Craigie R, Gellert M, Lambowitz A (ed), *Mobile DNA*, **vol II**. ASM Press, Washington DC.
58. **Dyall-Smith ML, Pfeiffer F, Klee K, Palm P, Gross K, Schuster SC, Rampp M, Oesterhelt D.** 2011. *Haloquadratum walsbyi*: limited diversity in a global pond. *PLoS ONE* **6:**e20968.
59. **Correia FF, Inouye S, Inouye M.** 1988. A family of small repeated elements with some transposon-like properties in the genome of *Neisseria gonorrhoeae*. *J Biol Chem* **263:**12194–12198.
60. **Buisine N, Tang CM, Chalmers R.** 2002. Transposon-like Correia elements: structure, distribution and genetic exchange between pathogenic *Neisseria* sp. *FEBS Lett* **522:**52–58.
61. **Oggioni MR, Claverys JP.** 1999. Repeated extragenic sequences in prokaryotic genomes: a proposal for the origin and dynamics of the RUP element in *Streptococcus pneumoniae*. *Microbiology* **145**(Pt 10)**:**2647–2653.
62. **Brugger K, Redder P, She Q, Confalonieri F, Zivanovic Y, Garrett RA.** 2002. Mobile elements in archaeal genomes. *FEMS Microbiol Lett* **206:**131–141.
63. **Filee J, Siguier P, Chandler M.** 2007. Insertion sequence diversity in archaea. *Microbiol Mol Biol Rev* **71:**121–157.
64. **Chen Y, Braathen P, Léonard C, Mahillon J.** 1999. MIC231, a naturally occurring mobile insertion cassette from *Bacillus cereus*. *Mol Microbiol* **32:**657–668.
65. **De Palmenaer D, Vermeiren C, Mahillon J.** 2004. IS231-MIC231 elements from *Bacillus cereus sensu lato* are modular. *Mol Microbiol* **53:**457–467.
66. **Demerec M, Adelberg EA, Clark AJ, Hartman PE.** 1966. A proposal for a uniform nomenclature in bacterial genetics. *Genetics* **54:**61–76.
67. **Nagy Z, Chandler M.** 2004. Regulation of transposition in bacteria. *Res Microbiol* **155:**387–398.
68. **Dumesic PA, Madhani HD.** 2014. Recognizing the enemy within: licensing RNA-guided genome defense. *Trends Biochem Sci* **39:**25–34.
69. **Fedoroff NV.** 2012. Transposable elements, epigenetics, and genome evolution. *Science* **338:**758–767.
70. **Bikard D, Marraffini LA.** 2013. Control of Gene Expression by CRISPR-Cas systems. *F1000Prime Rep* **5:**47.
71. **Bao W, Jurka J.** 2013. Homologues of bacterial TnpB_IS605 are widespread in diverse eukaryotic transposable elements. *Mob DNA* **4:**12.
72. **Parkhill J, Sebaihia M, Preston A, Murphy LD, Thomson N, Harris DE, Holden MT, Churcher CM, Bentley SD, Mungall KL, Cerdeno-Tarraga AM, Temple L, James K, Harris B, Quail MA, Achtman M, Atkin R, Baker S, Basham D, Bason N, Cherevach I, Chillingworth T, Collins M, Cronin A, Davis P, Doggett J, Feltwell T, Goble A, Hamlin N, Hauser H,**

Holroyd S, Jagels K, Leather S, Moule S, Norberczak H, O'Neil S, Ormond D, Price C, Rabbinowitsch E, Rutter S, Sanders M, Saunders D, Seeger K, Sharp S, Simmonds M, Skelton J, Squares R, Squares S, Stevens K, Unwin L, et al. 2003. Comparative analysis of the genome sequences of *Bordetella pertussis*, *Bordetella parapertussis* and *Bordetella bronchiseptica*. *Nat Genet* **35**:32–40.

73. **Preston A, Parkhill J, Maskell DJ.** 2004. The Bordetellae: lessons from genomics. *Nat Rev Microbiol* **2**:379–390.

74. **Touchon M, Rocha EP.** 2007. Causes of insertion sequences abundance in prokaryotic genomes. *Mol Biol Evol* **24**:969–981.

75. **Gil R, Belda E, Gosalbes MJ, Delaye L, Vallier A, Vincent-Monegat C, Heddi A, Silva FJ, Moya A, Latorre A.** 2008. Massive presence of insertion sequences in the genome of SOPE, the primary endosymbiont of the rice weevil *Sitophilus oryzae*. *Int Microbiol* **11**:41–48.

76. **Plague GR, Dunbar HE, Tran PL, Moran NA.** 2008. Extensive proliferation of transposable elements in heritable bacterial symbionts. *J Bacteriol* **190**:777–779.

77. **Moran NA, Plague GR.** 2004. Genomic changes following host restriction in bacteria. *Curr Opin Genet Dev* **14**:627–633.

78. **Andersson JO, Andersson SG.** 1999. Insights into the evolutionary process of genome degradation. *Curr Opin Genet Dev* **9**:664–671.

79. **Lawrence JG, Hendrix RW, Casjens S.** 2001. Where are the pseudogenes in bacterial genomes? *Trends Microbiol* **9**:535–540.

80. **Mira A, Ochman H, Moran NA.** 2001. Deletional bias and the evolution of bacterial genomes. *Trends Genet* **17**:589–596.

81. **Cerveau N, Leclercq S, Leroy E, Bouchon D, Cordaux R.** 2011. Short- and long-term evolutionary dynamics of bacterial insertion sequences: insights from *Wolbachia* endosymbionts. *Genome Biol Evol* **3**:1175–1186.

82. **Glansdorff N, Charlier D, Zafarullah M.** 1981. Activation of gene expression by IS2 and IS3. *Cold Spring Harb Symp Quant Biol* **45**(Pt 1):153–156.

83. **Soki J, Eitel Z, Urban E, Nagy E.** 2013. Molecular analysis of the carbapenem and metronidazole resistance mechanisms of *Bacteroides* strains reported in a Europe-wide antibiotic resistance survey. *Int J Antimicrob Agents* **41**:122–125.

84. **Aubert D, Naas T, Heritier C, Poirel L, Nordmann P.** 2006. Functional characterization of IS1999, an IS4 family element involved in mobilization and expression of beta-lactam resistance genes. *J Bacteriol* **188**:6506–6514.

85. **Kieny M-P.** 2012. *The evolving threat of antimicrobial resistance: options for action*. World Health Organization, Geneva.

86. **McKenna M.** 2013. The Last Resort. *Nature* **499**: 394–396.

87. **Mole B.** 2013. Farming up trouble. *Nature* **499**: 398–400.

88. **Simons RW, Hoopes BC, McClure WR, Kleckner N.** 1983. Three promoters near the termini of IS10: pIN, pOUT, and pIII. *Cell* **34**:673–682.

89. **Ton-Hoang B, Bétermier M, Polard P, Chandler M.** 1997. Assembly of a strong promoter following IS911 circularization and the role of circles in transposition. *EMBO J* **16**:3357–3371.

90. **Perkins-Balding D, Duval-Valentin G, Glasgow AC.** 1999. Excision of IS492 requires flanking target sequences and results in circle formation in *Pseudoalteromonas atlantica*. *J Bacteriol* **181**:4937–4948.

91. **Duval-Valentin G, Normand C, Khemici V, Marty B, Chandler M.** 2001. Transient promoter formation: a new feedback mechanism for regulation of IS911 transposition. *EMBO J* **20**:5802–5811.

92. **Prentki P, Teter B, Chandler M, Galas DJ.** 1986. Functional promoters created by the insertion of transposable element IS1. *J Mol Biol* **191**:383–393.

93. **Depardieu F, Podglajen I, Leclercq R, Collatz E, Courvalin P.** 2007. Modes and modulations of antibiotic resistance gene expression. *Clin Microbiol Rev* **20**: 79–114.

94. **Guynet C, Achard A, Hoang BT, Barabas O, Hickman AB, Dyda F, Chandler M.** 2009. Resetting the site: redirecting integration of an insertion sequence in a predictable way. *Mol Cell* **34**:612–619.

95. **Mendiola MV, de la Cruz F.** 1989. Specificity of insertion of IS91, an insertion sequence present in alphahaemolysin plasmids of *Escherichia coli*. *Mol Microbiol* **3**:979–984.

96. **Galas DJ, Calos MP, Miller JH.** 1980. Sequence analysis of Tn9 insertions in the lacZ gene. *J Mol Biol* **144**: 19–41.

97. **Meyer J, Iida S, Arber W.** 1980. Does the insertion element IS1 transpose preferentially into A+T- rich DNA segments? *Mol Gen Genet* **178**:471–473.

98. **Sengstag C, Iida S, Hiestand-Nauer R, Arber W.** 1986. Terminal inverted repeats of prokaryotic transposable element IS186 which can generate duplications of variable length at an identical target sequence. *Gene* **49**: 153–156.

99. **Peters JE, Craig NL.** 2001. Tn7: smarter than we thought. *Nat Rev Mol Cell Biol* **2**:806–814.

100. **Ton-Hoang B, Pasternak C, Siguier P, Guynet C, Hickman AB, Dyda F, Sommer S, Chandler M.** 2010. Single-stranded DNA transposition is coupled to host replication. *Cell* **142**:398–408.

101. **Hu WY, Derbyshire KM.** 1998. Target choice and orientation preference of the insertion sequence IS903. *J Bacteriol* **180**:3039–3048.

102. **Spradling AC, Bellen HJ, Hoskins RA.** 2011. *Drosophila* P elements preferentially transpose to replication origins. *Proc Natl Acad Sci U S A* **108**:15948–15953.

103. **Peters JE, Craig NL.** 2000. Tn7 transposes proximal to DNA double-strand breaks and into regions where chromosomal DNA replication terminates. *Mol Cell* **6**: 573–582.

104. **Shi Q, Huguet-Tapia JC, Peters JE.** 2009. Tn917 Targets the Region Where DNA Replication Terminates in *Bacillus subtilis*, Highlighting a Difference in Chromosome Processing in the Firmicutes. *J Bacteriol* **191**: 7623–7627.

105. **Garsin DA, Urbach J, Huguet-Tapia JC, Peters JE, Ausubel FM.** 2004. Construction of an *Enterococcus faecalis* Tn917-mediated-gene-disruption library offers insight into Tn917 insertion patterns. *J Bacteriol* **186:** 7280–7289.

106. **Slater JD, Allen AG, May JP, Bolitho S, Lindsay H, Maskell DJ.** 2003. Mutagenesis of *Streptococcus equi* and *Streptococcus suis* by transposon Tn917. *Vet Microbiol* **93:**197–206.

107. **Swingle B, O'Carroll M, Haniford D, Derbyshire KM.** 2004. The effect of host-encoded nucleoid proteins on transposition: H-NS influences targeting of both IS903 and Tn10. *Mol Microbiol* **52:**1055–1067.

108. **Clement J-M, Wilde C, Bachellier S, Lambert P, Hofnung M.** 1999. IS1397 Is Active for Transposition into the Chromosome of *Escherichia coli* K-12 and Inserts Specifically into Palindromic Units of Bacterial Interspersed Mosaic Elements. *J Bacteriol* **181:**6929–6936.

109. **Tobes R, Pareja E.** 2006. Bacterial repetitive extragenic palindromic sequences are DNA targets for Insertion Sequence elements. *BMC Genomics* **7:**62.

110. **Wilde C, Escartin F, Kokeguchi S, Latour-Lambert P, Lectard A, Clement JM.** 2003. Transposases are responsible for the target specificity of IS1397 and ISKpn1 for two different types of palindromic units (PUs). *Nucleic Acids Res* **31:**4345–4353.

111. **Tetu SG, Holmes AJ.** 2008. A Family of Insertion Sequences That Impacts Integrons by Specific Targeting of Gene Cassette Recombination Sites, the IS1111-attC Group. *J Bacteriol* **190:**4959–4970.

112. **Post V, Hall RM.** 2009. Insertion sequences in the IS1111 family that target the attC recombination sites of integron-associated gene cassettes. *FEMS Microbiol Lett* **290:**182–187.

113. **Hallet B, Rezsohazy R, Delcour J.** 1991. IS231A from *Bacillus thuringiensis* is functional in *Escherichia coli*: transposition and insertion specificity. *J Bacteriol* **173:** 4526–4529.

114. **Partridge SR, Hall RM.** 2003. The IS1111 family members IS4321 and IS5075 have subterminal inverted repeats and target the terminal inverted repeats of Tn21 family transposons. *J Bacteriol* **185:**6371–6384.

115. **Loot C, Turlan C, Chandler M.** 2004. Host processing of branched DNA intermediates is involved in targeted transposition of IS911. *Mol Microbiol* **51:**385–393.

116. **Reimmann C, Haas D.** 1987. Mode of replicon fusion mediated by the duplicated insertion sequence IS21 in *Escherichia coli*. *Genetics* **115:**619–625.

117. **Olasz F, Farkas T, Kiss J, Arini A, Arber W.** 1997. Terminal inverted repeats of insertion sequence IS30 serve as targets for transposition. *J Bacteriol* **179:**7551–7558.

118. **Prere MF, Chandler M, Fayet O.** 1990. Transposition in *Shigella dysenteriae*: isolation and analysis of IS911, a new member of the IS3 group of insertion sequences. *J Bacteriol* **172:**4090–4099.

119. **Parks AR, Li Z, Shi Q, Owens RM, Jin MM, Peters JE.** 2009. Transposition into replicating DNA occurs through interaction with the processivity factor. *Cell* **138:** 685–695.

120. **Gomez MJ, Diaz-Maldonado H, Gonzalez-Tortuero E, Lopez de Saro FJ.** 2014. Chromosomal replication dynamics and interaction with the beta sliding clamp determine orientation of bacterial transposable elements. *Genome Biol Evol* **6:**727–740.

121. **Qi X, Daily K, Nguyen K, Wang H, Mayhew D, Rigor P, Forouzan S, Johnston M, Mitra RD, Baldi P, Sandmeyer S.** 2012. Retrotransposon profiling of RNA polymerase III initiation sites. *Genome Res* **22:** 681–692.

122. **Shapiro JA.** 1979. Molecular model for the transposition and replication of bacteriophage Mu and other transposable elements. *Proc Natl Acad Sci USA* **76:** 1933–1937.

123. **Ross DG, Swan J, Kleckner N.** 1979. Nearly precise excision: a new type of DNA alteration associated with the translocatable element Tn10. *Cell* **16:**733–738.

124. **Ross DG, Swan J, Kleckner N.** 1979. Physical structures of Tn10-promoted deletions and inversions: role of 1400 bp inverted repetitions. *Cell* **16:**721–731.

125. **Ohtsubo E, Zenilman M, Ohtsubo H.** 1980. Plasmids containing insertion elements are potential transposons. *Proc Natl Acad Sci U S A* **77:**750–754.

126. **Polard P, Seroude L, Fayet O, Prere MF, Chandler M.** 1994. One-ended insertion of IS911. *J Bacteriol* **176:** 1192–1196.

127. **Mahillon J, Chandler M.** 1998. Insertion sequences. *Microbiol Mol Biol Rev* **62:**725–774.

128. **Robinson DG, Lee MC, Marx CJ.** 2012. OASIS: an automated program for global investigation of bacterial and archaeal insertion sequences. *Nucleic Acids Res* **40:**e174.

129. **Wagner A, Lewis C, Bichsel M.** 2007. A survey of bacterial insertion sequences using IScan. *Nucleic Acids Res* **35:**5284–5293.

130. **MacHattie LA, Jackowski JB.** 1977. Physical Structure and Deletion Effects of the Chloramphenicol Resistance Element Tn9 in Phage Lambda, p 219–228. *In* Bukhari AI, Shapiro JA, Adhya SL (ed), *DNA Insertion Elements, Plasmids, and Episomes*. Cold Spring Harbour Laboratory, New York.

131. **Chandler M, Silver L, Lane D, Caro L.** 1979. Properties of an autonomous r-determinant from R100.1. *Cold Spring Harb Symp Quant Biol* **43**(Pt 2):1223–1231.

132. **Peterson BC, Rownd RH.** 1985. Recombination sites in plasmid drug resistance gene amplification. *J Bacteriol* **164:**1359–1361.

133. **Blinkowa AL, Walker JR.** 1990. Programmed ribosomal frameshifting generates the *Escherichia coli* DNA polymerase III gamma subunit from within the tau subunit reading frame. *Nucleic Acids Res* **18:**1725–1729.

134. **Matsutani S.** 1994. Genetic evidence for IS1 transposition regulated by InsA and the delta InsA-B'-InsB species, which is generated by translation from two alternative internal initiation sites and frameshifting. *J Mol Biol* **240:**52–65.

135. **Ton-Hoang B, Turlan C, Chandler M.** 2004. Functional domains of the IS1 transposase: analysis *in vivo* and *in vitro*. *Mol Microbiol* **53:**1529–1543.

136. **Turlan C, Chandler M.** 1995. IS1-mediated intramolecular rearrangements: formation of excised transposon circles and replicative deletions. *EMBO J* **14:**5410–5421.

137. **Ohta S, Tsuchida K, Choi S, Sekine Y, Shiga Y, Ohtsubo E.** 2002. Presence of a characteristic D-D-E motif in IS1 transposase. *J Bacteriol* **184:**6146–6154.

138. **Ohta S, Yoshimura E, Ohtsubo E.** 2004. Involvement of two domains with helix-turn-helix and zinc finger motifs in the binding of IS1 transposase to terminal inverted repeats. *Mol Microbiol* **53:**193–202.

139. **Parkhill J, Achtman M, James KD, Bentley SD, Churcher C, Klee SR, Morelli G, Basham D, Brown D, Chillingworth T, Davies RM, Davis P, Devlin K, Feltwell T, Hamlin N, Holroyd S, Jagels K, Leather S, Moule S, Mungall K, Quail MA, Rajandream MA, Rutherford KM, Simmonds M, Skelton J, Whitehead S, Spratt BG, Barrell BG.** 2000. Complete DNA sequence of a serogroup A strain of *Neisseria meningitidis* Z2491. *Nature* **404:**502–506.

140. **Filée J, Siguier P, Chandler M.** 2007. I am what I eat and I eat what I am: acquisition of bacterial genes by giant viruses. *Trends Genet* **23:**10–15.

141. **Achard A, Leclercq R.** 2007. Characterization of a small mobilizable transposon, MTnSag1, in *Streptococcus agalactiae*. *J Bacteriol* **189:**4328–4331.

142. **Feschotte C.** 2004. Merlin, a new superfamily of DNA transposons identified in diverse animal genomes and related to bacterial IS1016 insertion sequences. *Mol Biol Evol* **21:**1769–1780.

143. **Rousseau P, Normand C, Loot C, Turlan C, Alazard R, Duval-Valentin G, Chandler M.** 2002. Transposition of IS911, p 366–383. *In* Craig NL, Craigie R, Gellert M, Lambowitz A (ed), *Mobile DNA II*. American Society of Microbiology, Washington DC.

144. **Bhugra B, Dybvig K.** 1993. Identification and characterization of IS1138, a transposable element from *Mycoplasma pulmonis* that belongs to the IS3 family. *Mol Microbiol* **7:**577–584.

145. **Smith KS, Ingram-Smith C.** 2007. *Methanosaeta*, the forgotten methanogen? *Trends Microbiol* **15:**150–155.

146. **Petrova M, Shcherbatova N, Gorlenko Z, Mindlin S.** 2013. A new subgroup of the IS3 family and properties of its representative member ISPpy1. *Microbiology* **159:**1900–1910.

147. **Duval-Valentin G, Chandler M.** 2011. Cotranslational control of DNA transposition: a window of opportunity. *Mol Cell* **44:**989–996.

148. **Feschotte C, Pritham EJ.** 2007. DNA Transposons and the Evolution of Eukaryotic Genomes. *Annu Rev Genet* **41:**331–368.

149. **Morona JK, Guidolin A, Morona R, Hansman D, Paton JC.** 1994. Isolation, characterization, and nucleotide sequence of IS1202, an insertion sequence of *Streptococcus pneumoniae*. *J Bacteriol* **176:**4437–4443.

150. **De Palmenaer D, Siguier P, Mahillon J.** 2008. IS4 family goes genomic. *BMC Evol Biol* **8:**18.

151. **Rezsohazy R, Hallet B, Delcour J, Mahillon J.** 1993. The IS4 family of insertion sequences: evidence for a conserved transposase motif. *Mol Microbiol* **9:**1283–1295.

152. **Davies DR, Goryshin IY, Reznikoff WS, Rayment I.** 2000. Three-dimensional structure of the Tn5 synaptic complex transposition intermediate. *Science* **289:**77–85.

153. **Mazel D, Bernard C, Schwarz R, Castets AM, Houmard J, Tandeau de Marsac N.** 1991. Characterization of two insertion sequences, IS701 and IS702, from the cyanobacterium *Calothrix* species PCC 7601. *Mol Microbiol* **5:**2165–2170.

154. **Vilei EM, Nicolet J, Frey J.** 1999. IS1634, a Novel Insertion Element Creating Long, Variable-Length Direct Repeats Which Is Specific for *Mycoplasma mycoides* subsp. *mycoides* Small-Colony Type. *J Bacteriol* **181:**1319–1323.

155. **Klenchin VA, Czyz A, Goryshin IY, Gradman R, Lovell S, Rayment I, Reznikoff WS.** 2008. Phosphate coordination and movement of DNA in the Tn5 synaptic complex: role of the (R)YREK motif 10.1093/nar/gkn577. *Nucl Acids Res* **36:**5855–5862.

156. **Rieck B, Tourigny DS, Crosatti M, Schmid R, Kochar M, Harrison EM, Ou HY, Turton JF, Rajakumar K.** 2012. *Acinetobacter* insertion sequence ISAba11 belongs to a novel family that encodes transposases with a signature HHEK motif. *Appl Environ Microbiol* **78:**471–480.

157. **Zhang X, Jiang N, Feschotte C, Wessler SR.** 2004. PIF- and Pong-like transposable elements: distribution, evolution and relationship with Tourist-like miniature inverted-repeat transposable elements. *Genetics* **166:**971–986.

158. **Berg DE, Davies J, Allet B, Rochaix JD.** 1975. Transposition of R factor genes to bacteriophage lambda. *Proc Natl Acad Sci USA* **72:**3628–3632.

159. **Miriagou V, Carattoli A, Tzelepi E, Villa L, Tzouvelekis LS.** 2005. IS26-Associated In4-Type Integrons Forming Multiresistance Loci in Enterobacterial Plasmids. *Antimicrob Agents Chemother* **49:**3541–3543.

160. **Partridge SR, Zong Z, Iredell JR.** 2011. Recombination in IS26 and Tn2 in the Evolution of Multiresistance Regions Carrying blaCTX-M-15 on Conjugative IncF Plasmids from *Escherichia coli*. *Antimicrob Agents Chemother* **55:**4971–4978.

161. **Zhu Y-G, Johnson TA, Su J-Q, Qiao M, Guo G-X, Stedtfeld RD, Hashsham SA, Tiedje JM.** 2013. Diverse and abundant antibiotic resistance genes in Chinese swine farms. *Proc Natl Acad Sci USA* **110:**3435–3440.

162. **Kato K, Ohtsuki K, Mitsuda H, Yomo T, Negoro S, Urabe I.** 1994. Insertion sequence IS6100 on plasmid pOAD2, which degrades nylon oligomers. *J Bacteriol* **176:**1197–1200.

163. **Hall RM, Brown HJ, Brookes DE, Stokes HW.** 1994. Integrons found in different locations have identical 5′ ends but variable 3′ ends. *J Bacteriol* **176:**6286–6294.

164. **Sundin GW, Bender CL.** 1995. Expression of the strA-strB streptomycin resistance genes in *Pseudomonas syringae* and *Xanthomonas campestris* and character-

ization of IS6100 in *X. campestris*. *Appl Environ Microbiol* **61**:2891–2897.

165. **Simpson AE, Skurray RA, Firth N.** 2000. An IS257-derived hybrid promoter directs transcription of a tetA(K) tetracycline resistance gene in the *Staphylococcus aureus* chromosomal mec region. *J Bacteriol* **182**:3345–3352.

166. **Bertini A, Poirel L, Bernabeu S, Fortini D, Villa L, Nordmann P, Carattoli A.** 2007. Multicopy blaOXA-58 gene as a source of high-level resistance to carbapenems in *Acinetobacter baumannii*. *Antimicrob Agents Chemother* **51**:2324–2328.

167. **Zienkiewicz M, Kern-Zdanowicz I, Carattoli A, Gniadkowski M, Ceglowski P.** 2013. Tandem multiplication of the IS26-flanked amplicon with the bla (SHV-5) gene within plasmid p1658/97. *FEMS Microbiol Lett* **341**:27–36.

168. **Loli A, Tzouvelekis LS, Tzelepi E, Carattoli A, Vatopoulos AC, Tassios PT, Miriagou V.** 2006. Sources of diversity of carbapenem resistance levels in Klebsiella pneumoniae carrying blaVIM-1. *J Antimicrob Chemother* **58**:669–672.

169. **Doublet B, Praud K, Weill FX, Cloeckaert A.** 2009. Association of IS26-composite transposons and complex In4-type integrons generates novel multidrug resistance loci in *Salmonella* genomic island 1. *J Antimicrob Chemother* **63**:282–289.

170. **Nigro SJ, Farrugia DN, Paulsen IT, Hall RM.** 2013. A novel family of genomic resistance islands, AbGRI2, contributing to aminoglycoside resistance in *Acinetobacter baumannii* isolates belonging to global clone 2. *J Antimicrob Chemother* **68**:554–557.

171. **Cullik A, Pfeifer Y, Prager R, von Baum H, Witte W.** 2010. A novel IS26 structure surrounds blaCTX-M genes in different plasmids from German clinical *Escherichia coli* isolates. *J Med Microbiol* **59**:580–587.

172. **Trieu-Cuot P, Courvalin P.** 1985. Transposition behavior of IS15 and its progenitor IS15-delta: are cointegrates exclusive end products? *Plasmid* **14**:80–89.

173. **Harmer CJ, Moran RA, Hall RM.** 2014. Movement of IS26-associated antibiotic resistance genes occurs via a translocatable unit that includes a single IS26 and preferentially inserts adjacent to another IS26. *MBio* **5**: e01801–01814.

174. **Riess G, Holloway BW, Puhler A.** 1980. R68.45, a plasmid with chromosome mobilizing ability (Cma) carries a tandem duplication. *Genet Res* **36**:99–109.

175. **Willetts NS, Crowther C, Holloway BW.** 1981. The insertion sequence IS21 of R68.45 and the molecular basis for mobilization of the bacterial chromosome. *Plasmid* **6**:30–52.

176. **Watson JM, Holloway BW.** 1978. Chromosome mapping in *Pseudomonas aeruginosa* PAT. *J Bacteriol* **133**: 1113–1125.

177. **Berger B, Haas D.** 2001. Transposase and cointegrase: specialized transposition proteins of the bacterial insertion sequence IS21 and related elements. *Cell Mol Life Sci* **58**:403–419.

178. **Ammendola S, Politi L, Scandurra R.** 1998. Cloning and sequencing of ISC1041 from the archaeon Sulfolobus solfataricus MT-4, a new member of the IS30 family of insertion elements [In Process Citation]. *FEBS Lett* **428**:217–223.

179. **Dalrymple B.** 1987. Novel rearrangements of IS30 carrying plasmids leading to the reactivation of gene expression. *Mol Gen Genet* **207**:413–420.

180. **Rasmussen JL, Odelson DA, Macrina FL.** 1987. Complete nucleotide sequence of insertion element IS4351 from *Bacteroides fragilis*. *J Bacteriol* **169**:3573–3580.

181. **Rudant E, Courvalin P, Lambert T.** 1998. Characterization of IS18, an Element Capable of Activating the Silent aac(6′)-Ij Gene of *Acinetobacter* sp. 13 Strain BM2716 by Transposition. *Antimicrob Agents Chemother* **42**:2759–2761.

182. **Stalder R, Caspers P, Olasz F, Arber W.** 1990. The N-terminal domain of the insertion sequence 30 transposase interacts specifically with the terminal inverted repeats of the element. *J Biol Chem* **265**:3757–3762.

183. **Nagy Z, Szabó M, Chandler M, Olasz F.** 2004. Analysis of the N-terminal DNA binding domain of the IS30 transposase. *Mol Microbiol* **54**:478–488.

184. **Olasz F, Stalder R, Arber W.** 1993. Formation of the tandem repeat (IS30)2 and its role in IS30- mediated transpositional DNA rearrangements. *Mol Gen Genet* **239**:177–187.

185. **Olasz F, Farkas T, Stalder R, Arber W.** 1997. Mutations in the carboxy-terminal part of IS30 transposase affect the formation and dissolution of (IS30)2 dimer. *FEBS Lett* **413**:453–461.

186. **Kiss J, Olasz F.** 1999. Formation and transposition of the covalently closed IS30 circle: the relation between tandem dimers and monomeric circles. *Mol Microbiol* **34**:37–52.

187. **Kiss J, Szabo M, Olasz F.** 2003. Site-specific recombination by the DDE family member mobile element IS30 transposase. *Proc Natl Acad Sci USA* **100**: 15000–15005.

188. **Szeverenyi I, Nagy Z, Farkas T, Olasz F, Kiss J.** 2003. Detection and analysis of transpositionally active head-to-tail dimers in three additional Escherichia coli IS elements. *Microbiology* **149**:1297–1310.

189. **Kiss J, Nagy Z, Toth G, Kiss GB, Jakab J, Chandler M, Olasz F.** 2007. Transposition and target specificity of the typical IS30 family element IS1655 from *Neisseria meningitidis*. *Mol Microbiol* **63**:1731–1747.

190. **Szabó M, Kiss J, Nagy Z, Chandler M, Olasz F.** 2008. Sub-terminal Sequences Modulating IS30 Transposition *in Vivo* and *in Vitro*. *J Mol Biol* **375**:337–352.

191. **Hwa V, Shoemaker NB, Salyers AA.** 1988. Direct repeats flanking the Bacteroides transposon Tn4351 are insertion sequence elements. *J Bacteriol* **170**:449–451.

192. **Brynestad S, Granum PE.** 1999. Evidence that Tn5565, which includes the enterotoxin gene in Clostridium perfringens, can have a circular form which may be a transposition intermediate. *FEMS Microbiol Lett* **170**: 281–286.

193. **Machida Y, Sakurai M, Kiyokawa S, Ubasawa A, Suzuki Y, Ikeda JE.** 1984. Nucleotide sequence of the insertion sequence found in the T-DNA region of mutant Ti plasmid pTiA66 and distribution of its homologues in octopine Ti plasmid. *Proc Natl Acad Sci USA* **81:**7495–7499.

194. **Gourbeyre E, Siguier P, Chandler M.** 2010. Route 66: investigations into the organisation and distribution of the IS66 family of prokaryotic insertion sequences. *Res Microbiol* **161:**136–143.

195. **Egelseer EM, Idris R, Jarosch M, Danhorn T, Sleytr UB, Sara M.** 2000. ISBst12, a novel type of insertion-sequence element causing loss of S- layer-gene expression in *Bacillus stearothermophilus* ATCC 12980. *Microbiology* **146**(Pt 9):2175–2183.

196. **Han CG, Shiga Y, Tobe T, Sasakawa C, Ohtsubo E.** 2001. Structural and functional characterization of IS679 and IS66-family elements. *J Bacteriol* **183:**4296–4304.

197. **Haren L, Polard P, Ton-Hoang B, Chandler M.** 1998. Multiple oligomerisation domains in the IS911 transposase: a leucine zipper motif is essential for activity. *J Mol Biol* **283:**29–41.

198. **Dordet Frisoni E, Marenda MS, Sagné E, Nouvel LX, Guérillot R, Glaser P, Blanchard A, Tardy F, Sirand-Pugnet P, Baranowski E, Citti C.** 2013. ICEA of *Mycoplasma agalactiae*: a new family of self-transmissible integrative elements that confers conjugative properties to the recipient strain. *Mol Microbiol* **89:**1226–1239.

199. **Hennig S, Ziebuhr W.** 2010. Characterization of the Transposase Encoded by IS256, the Prototype of a Major Family of Bacterial Insertion Sequence Elements. *J Bacteriol* **192:**4153–4163.

200. **Yuan Y-W, Wessler SR.** 2011. The catalytic domain of all eukaryotic cut-and-paste transposase superfamilies. *Proc Natl Acad Sci USA* **108:**7884–7889.

201. **Lyon BR, Gillespie MT, Skurray RA.** 1987. Detection and characterization of IS256, an insertion sequence in *Staphylococcus aureus*. *J Gen Microbiol* **133:** 3031–3038.

202. **Lyon BR, May JW, Skurray RA.** 1984. Tn4001: a gentamicin and kanamycin resistance transposon in *Staphylococcus aureus*. *Mol Gen Genet* **193:**554–556.

203. **Loessner I, Dietrich K, Dittrich D, Hacker J, Ziebuhr W.** 2002. Transposase-dependent formation of circular IS256 derivatives in *Staphylococcus epidermidis* and *Staphylococcus aureus*. *J Bacteriol* **184:**4709–4714.

204. **Eisen JA, Benito MI, Walbot V.** 1994. Sequence similarity of putative transposases links the maize Mutator autonomous element and a group of bacterial insertion sequences. *Nucleic Acids Res* **22:**2634–2636.

205. **Hua-Van Al, Capy P.** 2008. Analysis of the DDE Motif in the Mutator Superfamily. *J Mol Evol* **67:**670–681.

206. **Prudhomme M, Turlan C, Claverys JP, Chandler M.** 2002. Diversity of Tn4001 transposition products: the flanking IS256 elements can form tandem dimers and IS circles. *J Bacteriol* **184:**433–443.

207. **Feng X, Bednarz AL, Colloms SD.** 2010. Precise targeted integration by a chimaeric transposase zinc-finger fusion protein. *Nucl Acids Res* **38:**1204–1216.

208. **Cassier-Chauvat C, Poncelet M, Chauvat F.** 1997. Three insertion sequences from the *Cyanobacterium synechocystis* PCC6803 support the occurrence of horizontal DNA transfer among bacteria. *Gene* **195:** 257–266.

209. **Dawson A, Finnegan DJ.** 2003. Excision of the Drosophila mariner transposon mos1. Comparison with bacterial transposition and v(d)j recombination. *Mol Cell* **11:**225–235.

210. **Richardson JM, Dawson A, O'Hagan N, Taylor P, Finnegan DJ, Walkinshaw MD.** 2006. Mechanism of Mos1 transposition: insights from structural analysis. *EMBO J* **25:**1324–1334.

211. **Tenzen T, Ohtsubo E.** 1991. Preferential transposition of an IS630-associated composite transposon to TA in the 5′-CTAG-3′ sequence. *J Bacteriol* **173:**6207–6212.

212. **Lopez de Felipe F, Magni C, de Mendoza D, Lopez P.** 1996. Transcriptional activation of the citrate permease P gene of Lactococcus lactis biovar diacetylactis by an insertion sequence- like element present in plasmid pCIT264. *Mol Gen Genet* **250:**428–436.

213. **Kato N, Yamazoe K, Han CG, Ohtsubo E.** 2003. New insertion sequence elements in the upstream region of cfiA in imipenem-resistant *Bacteroides fragilis* strains. *Antimicrob Agents Chemother* **47:**979–985.

214. **Podglajen I, Breuil J, Rohaut A, Monsempes C, Collatz E.** 2001. Multiple mobile promoter regions for the rare carbapenem resistance gene of *Bacteroides fragilis*. *J Bacteriol* **183:**3531–3535.

215. **de Las Rivas B, Marcobal A, Gomez A, Munoz R.** 2005. Characterization of ISLpl4, a functional insertion sequence in *Lactobacillus plantarum*. *Gene* **363:** 202–210.

216. **Takemura H, Horinouchi S, Beppu T.** 1991. Novel insertion sequence IS1380 from *Acetobacter pasteurianus* is involved in loss of ethanol-oxidizing ability. *J Bacteriol* **173:**7070–7076.

217. **Dumontier S, Trieu-Cuot P, Berche P.** 1998. Structural and Functional Characterization of IS1358 from *Vibrio cholerae*. *J Bacteriol* **180:**6101–6106.

218. **Davis IJ, Roberts AP, Ready D, Richards H, Wilson M, Mullany P.** 2005. Linkage of a novel mercury resistance operon with streptomycin resistance on a conjugative plasmid in *Enterococcus faecium*. *Plasmid* **54:**26–38.

219. **Christie-Oleza JA, Nogales B, Lalucat J, Bosch R.** 2010. TnpR encoded by an ISPpu12 isoform regulates transposition of two different ISL3-like insertion sequences in *Pseudomonas stutzeri* after conjugative interaction. *J Bacteriol* **192:**1423–1432.

220. **Kallastu A, Hõrak R, Kivisaar M.** 1998. Identification and Characterization of IS1411, a New Insertion Sequence Which Causes Transcriptional Activation of the Phenol Degradation Genes in *Pseudomonas putida*. *J Bacteriol* **180:**5306–5312.

221. **Christie-Oleza JA, Nogales B, Martin-Cardona C, Lanfranconi MP, Alberti S, Lalucat J, Bosch R.** 2008. ISPst9, an ISL3-like insertion sequence from *Pseudomonas stutzeri* AN10 involved in catabolic gene inactivation. *Int Microbiol* **11:**101–110.

222. **Vertes A, Asai Y, Inui M, Kobayashi M, Yukawa H.** 1995. The Corynebacterial insertion sequence IS31831 promotes formation of an excised transposon fragment. *Biotechnol Lett* **17:**1143–1148.

223. **Grindley NDF.** 2002. The movement of Tn3-like elements: transposition and cointegrate resolution, p 230–271. *In* Craig NL, Craigie R, Gellert M, Lambowitz A (ed), *Mobile DNA II*. ASM Press, Washington DC.

224. **Fulks KA, Marrs CF, Stevens SP, Green MR.** 1990. Sequence analysis of the inversion region containing the pilin genes of *Moraxella bovis*. *J Bacteriol* **172:**310–316.

225. **Rozsa FW, Meyer TF, Fussenegger M.** 1997. Inversion of *Moraxella lacunata* type 4 pilin gene sequences by a *Neisseria gonorrhoeae* site-specific recombinase. *J Bacteriol* **179:**2382–2388.

226. **Skaar EP, Lecuyer B, Lenich AG, Lazio MP, Perkins-Balding D, Seifert HS, Karls AC.** 2005. Analysis of the Piv recombinase-related gene family of *Neisseria gonorrhoeae*. *J Bacteriol* **187:**1276–1286.

227. **Lauf U, Muller C, Herrmann H.** 1999. Identification and characterisation of IS1383, a new insertion sequence isolated from *Pseudomonas putida* strain H [In Process Citation]. *FEMS Microbiol Lett* **170:**407–412.

228. **Tobiason DM, Buchner JM, Thiel WH, Gernert KM, Karls AC.** 2001. Conserved amino acid motifs from the novel Piv/MooV family of transposases and site-specific recombinases are required for catalysis of DNA inversion by Piv. *Mol Microbiol* **39:**641–651.

229. **Tobiason DM, Lenich AG, Glasgow AC.** 1999. Multiple DNA binding activities of the novel site-specific recombinase, Piv, from *Moraxella lacunata*. *J Biol Chem* **274:**9698–9706.

230. **Higgins BP, Carpenter CD, Karls AC.** 2007. Chromosomal context directs high-frequency precise excision of IS492 in *Pseudoalteromonas atlantica*. *Proc Natl Acad Sci USA* **104:**1901–1906.

231. **Prosseda G, Latella MC, Casalino M, Nicoletti M, Michienzi S, Colonna B.** 2006. Plasticity of the P junc promoter of ISEc11, a new insertion sequence of the IS1111 family. *J Bacteriol* **188:**4681–4689.

232. **Henderson DJ, Lydiate DJ, Hopwood DA.** 1989. Structural and functional analysis of the mini-circle, a transposable element of *Streptomyces coelicolor* A3(2). *Mol Microbiol* **3:**1307–1318.

233. **Smokvina T, Henderson DJ, Melton RE, Brolle DF, Kieser T, Hopwood DA.** 1994. Transposition of IS117, the 2.5 kb *Streptomyces coelicolor* A3(2) 'minicircle': roles of open reading frames and origin of tandem insertions. *Mol Microbiol* **12:**459–468.

234. **Muller C, Lauf U, Hermann H.** 2001. The inverted repeats of IS1384, a newly described insertion sequence from *Pseudomonas putida* strain H, represent the specific target for integration of IS1383. *Mol Genet Genom* **265:**1004–1010.

235. **Bartlett DH, Silverman M.** 1989. Nucleotide sequence of IS492, a novel insertion sequence causing variation in extracellular polysaccharide production in the marine bacterium *Pseudomonas atlantica*. *J Bacteriol* **171:** 1763–1766.

236. **Higgins BP, Popkowski AC, Caruana PR, Karls AC.** 2009. Site-specific insertion of IS492 in *Pseudoalteromonas atlantica*. *J Bacteriol* **191:**6408–6414.

237. **Mendiola MV, de la Cruz F.** 1992. IS91 transposase is related to the rolling-circle-type replication proteins of the pUB110 family of plasmids. *Nucleic Acids Res* **20:** 3521–3521.

238. **Garcillan-Barcia MP, Bernales I, Mendiola MV, De la Cruz F.** 2002. IS91 rolling circle transposition, p 891–904. *In* Craig NL, Craigie R, Gellert M, Lambowitz A (ed), *Mobile DNA*, **vol II**. ASM Press, Washington DC.

239. **del Pilar Garcillan-Barcia M, Bernales I, Mendiola MV, de la Cruz F.** 2001. Single-stranded DNA intermediates in IS91 rolling-circle transposition. *Mol Microbiol* **39:** 494–501.

240. **Toleman MA, Walsh TR.** 2011. Combinatorial events of insertion sequences and ICE in Gram-negative bacteria. *FEMS Microbiol Rev* **35:**912–935.

241. **Kapitonov VV, Jurka J.** 2007. Helitrons on a roll: eukaryotic rolling-circle transposons. *Trends Genet* **23:** 521–529.

242. **Lam S, Roth JR.** 1983. IS200: a Salmonella-specific insertion sequence. *Cell* **34:**951–960.

243. **Kersulyte D, Krishnan BR, Berg DE.** 1992. Nonrandom orientation of transposon Tn5supF insertions in phage lambda. *Gene* **114:**91–96.

244. **Kersulyte D, Velapatino B, Dailide G, Mukhopadhyay AK, Ito Y, Cahuayme L, Parkinson AJ, Gilman RH, Berg DE.** 2002. Transposable element ISHp608 of *Helicobacter pylori*: nonrandom geographic distribution, functional organization, and insertion specificity. *J Bacteriol* **184:**992–1002.

245. **Ton-Hoang B, Guynet C, Ronning DR, Cointin-Marty B, Dyda F, Chandler M.** 2005. Transposition of ISHp608, member of an unusual family of bacterial insertion sequences. *EMBO J* **24:**3325–3338.

246. **Murai N, Kamata H, Nagashima Y, Yagisawa H, Hirata H.** 1995. A novel insertion sequence (IS)-like element of the thermophilic bacterium PS3 promotes expression of the alanine carrier protein-encoding gene. *Gene* **163:**103–107.

247. **Kersulyte D, Akopyants NS, Clifton SW, Roe BA, Berg DE.** 1998. Novel sequence organization and insertion specificity of IS605 and IS606: chimaeric transposable elements of *Helicobacter pylori*. *Gene* **223:**175–186.

248. **Pasternak C, Dulermo R, Ton-Hoang B, Debuchy R, Siguier P, Coste G, Chandler M, Sommer S.** 2013. ISDra2 transposition in *Deinococcus radiodurans* is downregulated by TnpB. *Mol Microbiol* **88:**443–455.

249. **Gilbert C, Cordaux R.** 2013. Horizontal Transfer and Evolution of Prokaryote Transposable Elements in Eukaryotes. *Genome Biol Evol* **5:**822–832.

250. **Ronning DR, Guynet C, Ton-Hoang B, Perez ZN, Ghirlando R, Chandler M, Dyda F.** 2005. Active site sharing and subterminal hairpin recognition in a new class of DNA transposases. *Mol Cell* **20:**143–154.

251. **Guynet C, Hickman AB, Barabas O, Dyda F, Chandler M, Ton-Hoang B.** 2008. *In vitro* reconstitution of a

single-stranded transposition mechanism of IS608. *Mol Cell* **29:**302–312.

252. **Barabas O, Ronning DR, Guynet C, Hickman AB, Ton-Hoang B, Chandler M, Dyda F.** 2008. Mechanism of IS200/IS605 family DNA transposases: activation and transposon-directed target site selection. *Cell* **132:** 208–220.

253. **Hickman AB, James JA, Barabas O, Pasternak C, Ton-Hoang B, Chandler M, Sommer S, Dyda F.** 2010. DNA recognition and the precleavage state during single-stranded DNA transposition in *D. radiodurans*. *EMBO J* **29:**3840–3852.

254. **He S, Hickman AB, Dyda F, Johnson NP, Chandler M, Ton-Hoang B.** 2011. Reconstitution of a functional IS608 single-strand transpososome: role of non-canonical base pairing. *Nucleic Acids Res* **39:**8503–8512.

255. **He S, Guynet C, Siguier P, Hickman AB, Dyda F, Chandler M, Ton-Hoang B.** 2013. IS200/IS605 family single-strand transposition: mechanism of IS608 strand transfer. *Nucleic Acids Res* **41:**3302–3313.

256. **Boocock MR, Rice PA.** 2013. A proposed mechanism for IS607-family serine transposases. *Mob DNA* **4:**24.

257. **Van Duyne GD, Rutherford K.** 2013. Large serine recombinase domain structure and attachment site binding. *Crit Rev Biochem Mol Biol* **48:**476–491.

258. **Kersulyte D, Mukhopadhyay AK, Shirai M, Nakazawa T, Berg DE.** 2000. Functional organization and insertion specificity of IS607, a chimeric element of *Helicobacter pylori*. *J Bacteriol* **182:**5300–5308.

259. **Haren L, Ton-Hoang B, Chandler M.** 1999. Integrating DNA: transposases and retroviral integrases. *Annu Rev Microbiol* **53:**245–281.

Mobile DNA, 3rd Edition
Nancy L. Craig, Michael Chandler, Martin Gellert, Alan M. Lambowitz, Phoebe A. Rice and Suzanne Sandmeyer

doi:10.1128/microbiolspec.MDNA3-0031-2014

Michael Chandler,[1] Olivier Fayet,[1] Philippe Rousseau,[1] Bao Ton Hoang,[1] and Guy Duval-Valentin[1]

27

Copy-out–Paste-in Transposition of IS*911*: A Major Transposition Pathway

INTRODUCTION

The bacterial insertion sequence, IS*911*, is a member of the large IS*3* family. It transposes using a mechanism known as Copy-out–Paste-in. This is a major transposition pathway as judged by the number of transposable elements that use it. This pathway has not only been demonstrated to apply to various other members of the IS*3* insertion sequence family, IS*2* (1), IS*3* (2), and IS*150* (3), but has also been adopted by members of at least seven other large IS families: IS*1*, IS*21*, IS*30*, IS*256*, IS*110*, IS*Lre2*, IS*L3*, and their derivatives (see Siguier et al., this volume).

The various steps in Copy-out–Paste-in transposition are described in greater detail later in this review (*Transposition Pathway* section, and Figure 5). However, the essential features of this common pathway can be summarized as follows. It is essentially asymmetric: cleavage of one strand at one IS end generates a 3′OH group; this 3′OH is then used to attack the other IS end forming a single-strand bridge between both ends; the IS is replicated *in situ* from a 3′OH in the flanking donor DNA, which was generated by formation of the bridge; replication generates a double-strand circular IS copy with abutted ends (an active junction) and presumably regenerates the original donor molecule, which retains an IS copy. The closed circular transposon circle with abutted IS ends is then poised to integrate into a double-strand target DNA by single-strand cleavage at each end in the junction generating two 3′OH, which are then used to attack the two target strands.

In this chapter, we first review the origin and distribution of IS*911* itself (with supplementary information from other members of the IS*3* family where appropriate), its organization, the functional organization of its transposase, the regulation of expression and activity, the composition of the IS*911* transpososomes, nucleoprotein complexes involved in the Copy-out step (containing both transposon ends and transposase) and the Paste-in step (containing target DNA, transposon ends, and transposase) and, finally, a detailed description of the transposition pathway with some of the supporting data.

ISOLATION AND DISTRIBUTION

Insertion sequence IS*911* was isolated from a *Shigella dysenteriae* phage λ lysogen by spontaneous insertion

[1]Laboratoire de Microbiologie et Génétique Moléculaires, CNRS, Toulouse, France.

into the phage cI repressor gene. This resulted in a virulent phage whose lytic phenotype could be suppressed in an *Escherichia coli* strain able to produce the CI repressor protein (4). The element is present in multiple copies in the original host strain and in type strains of other *Shigella* species. Two vestigial copies, both interrupted by a copy of IS*30*, were also detected in the chromosome of *E. coli* K12 (5) and proved able to form transposition intermediates when supplied with IS*911* transposase (6). Entire or truncated IS*911* copies have also been identified in several *E. coli* virulence plasmids (7, 8), in pathogenicity islands of uropathogenic *E. coli* (9), in various other clinical isolates of *E. coli*, and in a large number of well-known and less well-known enterobacteria such as *Escherichia fergusonii*, *Chronobacter*, *Dickeya*, *Erwinia*, *Klebsiella*, *Pantoea*, *Shimwellia*, and *Yersinia*.

IS*911* is a member of the very widely distributed and one of the largest IS families: IS*3* (see Siguier *et al.*, this volume). Although this review is centered on studies of IS*911* as a model system for studying the IS*3* family, important insights and some differences have been revealed from studies of other family members such as IS2, IS*3*, and IS*150*. These will be described where appropriate below.

GENERAL ORGANIZATION

IS*911* (Figure 1A) is 1,250 bp long, bordered by imperfect 36-bp terminal inverted repeats (IRL and IRR) and generates 3-bp (sometimes 4-bp) target duplications on insertion (4). Like those of most family members and members of several other transposable element families, the IS*911* IRs terminate in a 5′-CA-3′ dinucleotide (Figure 2A). IS2 carries a natural mutation at one end which affects the activity of this end. These bases are important for the chemistry of transposition. Also like most other family members, IS*911* carries two consecutive and partially overlapping open reading frames, *orfA* and *orfB* under control of a weak promoter, p_{IRL}, partially located in IRL (Figure 2A and B). The 5′ end of *orfB* overlaps the 3′ end of *orfA* and occurs in reading phase −1 relative to *orfA* (Figure 1A). Complex inverted repeat sequences (Figure 2C) are located between co-ordinates 19 and 73 and include the −35 and −10 hexamers of p_{IRL}, the transcription start site and the ribosome-binding site for OrfA. This is thought to play a role at the mRNA level in preventing excess transposase expression resulting from external transcription. The full secondary structure would be present in transcripts initiated outside the IS, so sequestering the translation initiation signals, but only the 3′ part would be present if transcription initiates at p_{IRL}. In this case the translation initiation signals would be exposed. Initial studies (M.F. Prère and O. Fayet, unpublished) have shown that translation from the longer transcript is very low but that deletion of its 5′ end, to “liberate” the ribosome-binding site, indeed results in a significant increase in translation (Figure 2C). In the related IS2 element, a similar sequence appears to function as a DNA binding site for the OrfA protein, which represses promoter activity but further studies are necessary to confirm this (10).

In common with many other ISs of both the IS*3* and other families (e.g., IS*21* (11), IS*30* (12), IS*110* (13, 14)) the IS*911* IRR carries an outward-directed −35 promoter hexamer whereas IRL carries an inward-directed −10 promoter component (Figure 2A). These are assembled into a strong promoter, p_{Junc}, (Figure 2B), which serves to express high levels of the transposition proteins in one of its key transposition intermediates, an excised transposon circle (see *Transposition pathway* section below). It should be noted that transcription initiation from p_{Junc}, like that from impinging transcription, would also produce an RNA which could sequester the translation initiation signals but in a shorter and less stable stem loop structure (Figure 2C).

TRANSPOSITION PROTEINS: DNA SEQUENCE RECOGNITION, MULTIMERIZATION AND CATALYTIC DOMAINS

IS*911* expresses two major proteins (Figure 1B): OrfA, and the transposase, OrfAB (15). There is some evidence that OrfB may also be produced at low levels (15–17).

OrfA has a predicted molecular weight of 11.5 kDa. The OrfA sequence carries an α helix-turn-α helix (HTH) motif involved in sequence-specific binding to the terminal IRs of the transposase of the element OrfAB (see below) and a C-terminal leucine zipper (LZ) motif involved in protein multimerization (4, 15, 18).

Most IS*3* family members exhibit a similarly placed HTH signature (4, 19), whereas the LZ motif was identified in IS2, IS*150*, and IS*3* and appears to be conserved in the majority of known members (18).

OrfB is 299 residues long and has a predicted molecular weight of 34.6 kDa. Its TAA termination codon lies just within IRR and may be significant in regulation. The OrfB initiation codon is AUU and consequently initiation occurs only at low levels (15, 17) and is modulated by the level of initiation factor IF3 (16). It is possible that OrfB plays no direct role in IS*911* transposition chemistry but that its translation signals serve to modulate the programmed translational frameshifting required to generate a single transposase protein, OrfAB, from the two reading frames *orfA* and *orfB* (see *Frameshifting: transposase OrfAB as a fusion protein* section below). Expression of IS*911* genes leads to about five-fold less OrfB than OrfAB, mainly because initiation on the AUU codon is dampened by a downstream stem–loop structure also involved in the frameshifting event (16). OrfB has been observed in the case of IS*3* (20) (MF Prère and O. Fayet, unpublished), of IS*150* (21) and IS*3411*/IS*629* (22, 23). For IS*150*, the OrfB initiation codon is out of phase and expression requires a –1 frameshift after initiation (see *Frameshifting: transposase OrfAB as a fusion protein* section below). OrfB may be implicated in modulating transposition of IS*3* and IS*3411*/IS*629* in a way that has yet to be mechanistically defined. Sequence analysis suggests that it is probably synthesized by about 34% of the IS*3* family members through translational coupling: the stop codon of *orfA* is found to overlap with a potential start codon for *orfB* (e.g., AUGA or GUGA) in 134 out of 399 ISs. In IS*3* this leads to synthesis of an equal amount of OrfB and OrfAB (20). OrfB protein synthesis is therefore not a universal feature within the IS*3* family, as exemplified by its absence in IS*2* expression (24). The OrfB amino acid sequence shares significant similarities with retroviral integrases. In particular it was comparison of this sequence that contributed to defining the highly conserved amino acid triad DDE common to all IS*3* family members and to many of this type of phosphoryltransferase enzymes (25, 26). This constitutes part of the active site (for reviews see references 27–33). OrfB carries neither the HTH nor the LZ motif.

Although no structural information is available from crystallography, the roles of the HTH and LZ motifs have been investigated *in vivo* and *in vitro*.

The conserved N-terminal HTH motif is related to the LysR family of bacterial transcription factors and has a highly conserved tryptophan residue similar to that of certain homeodomain protein HTH motifs. Site-directed mutagenesis was used to probe the function of this motif using a series of *in vivo* and *in vitro* tests and demonstrated that the HTH domain is indeed important in directing transposase to bind IS*911* IR (34). The N-terminal helices of the related IS*2* transposase have also been shown to be involved in IR binding (35).

The LZ motif is composed of four heptameric units (Figure 1B) with a predicted coiled coil structure including potential buried inter-subunit hydrogen bonds across the dimer interface, to maintain the zipper in a dimeric state, and correctly placed residues with opposite charges potentially able to form characteristic inter-subunit salt-bridges to stabilize the dimeric structure (18). Mutation of specific critical residues in the OrfAB LZ, which lead to reduced levels of multimerization, resulted in reduction in the level of transposition intermediates *in vivo* and *in vitro* (18) (see *Transposition pathway* section below). Immunoprecipitation from extracts of cells expressing various C-terminal truncated OrfAB derivatives with a wild-type LZ and a C-terminal HA tag was used to assess multimerization. This demonstrated that OrfAB and OrfA could form both homomultimers as well as mixed OrfAB–OrfA multimers (18, 36). The critical mutations reduced or prevented multimer (dimer) formation.

These studies also showed that a poorly defined region, M, located between residues 109 and 135 (Figure 1B) and components in the catalytic domain of OrfAB (18) are also involved in its multimerization.

It is interesting that OrfAB and OrfA share three of their four heptads (Figure 1B). The last heptad of each differs in sequence due to the translational frameshift that occurs within the heptad in expression of OrfAB (see *Frameshifting: transposase OrfAB as a fusion protein* section below). This presumably results in different strengths of monomer–monomer interactions in the case of homo- and hetero-multimers, which may be involved in regulation of transposition.

For experimental purposes, production of OrfAB without necessitating a translational frameshift is obtained by introduction of a single additional base pair within the frameshift region (see *Frameshifting: transposase OrfAB as a fusion protein* section below) which artificially fuses the *orfA* and *orfB* frames and eliminates OrfA production (15). It was initially difficult to construct this mutant in the context of an entire IS*911* (i.e., with the two flanking IR). This is because constitutive expression of the OrfAB protein *in cis* results in exceptionally high transposition activity of the closely neighboring IS ends More recently this has been accomplished using a much longer artificial IS in which the ends are located at some distance from the transposase gene. Even in these conditions, the transposition frequency was exceptionally high and the transposon was unstable (37). A similar mutant in IS*3* results in a high frequency of adjacent deletions (20).

FRAMESHIFTING: TRANSPOSASE OrfAB AS A FUSION PROTEIN

OrfAB of IS*911* is assembled from *orfA* and *orfB* by a programmed –1 translational frameshift occurring near the 3′ end of *orfA*, as first demonstrated for the related

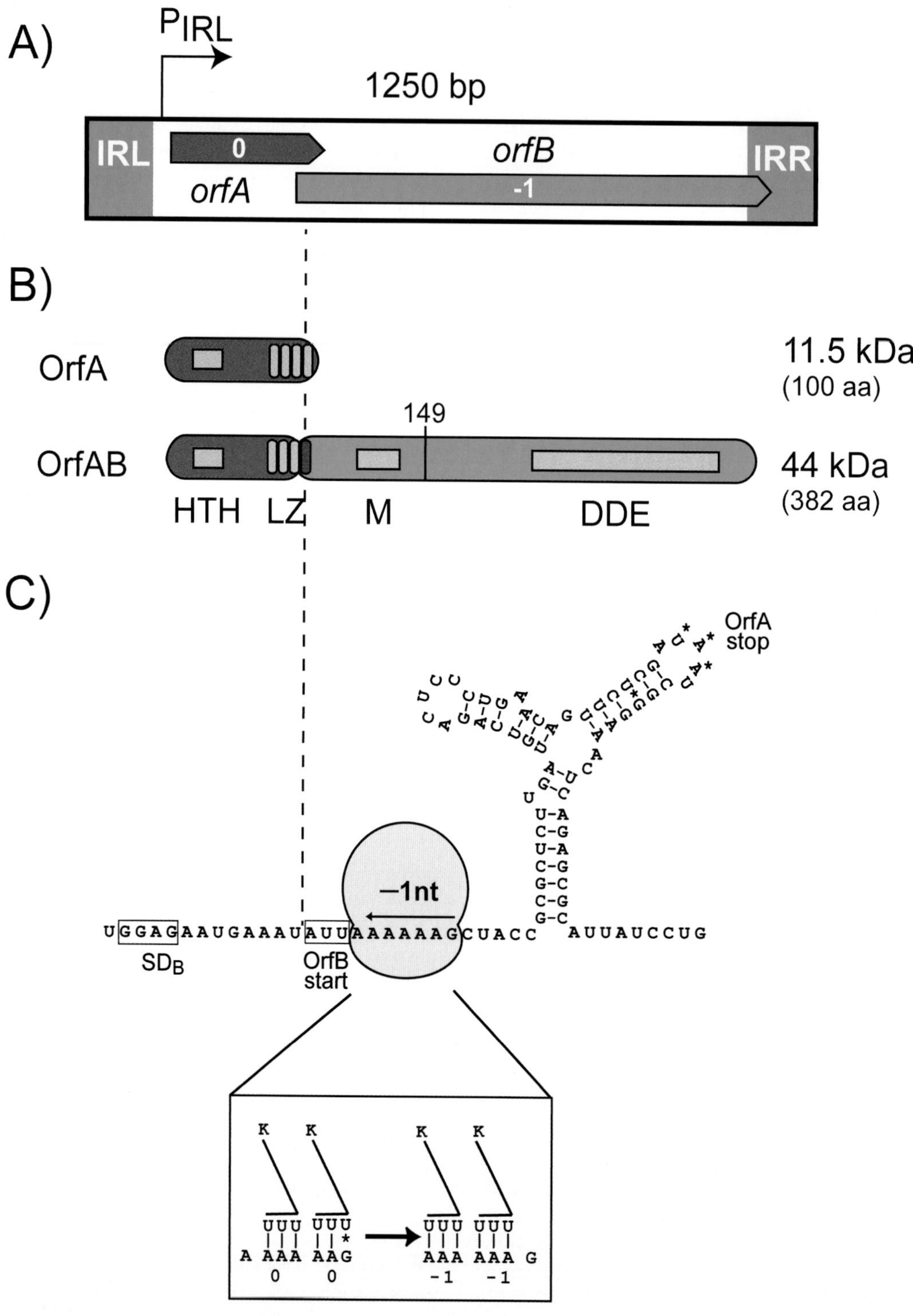
A)
PIRL
1250 bp
IRL
0
orfB
IRR
orfA
-1
B)
OrfA
11.5 kDa
(100 aa)
149
OrfAB
44 kDa
(382 aa)
HTH
LZ
M
DDE
C)
OrfA stop
−1nt
UGGAGAAUGAAAUAUUAAAAAAGCUACC
AUUAUCCUG
SDB
OrfB start
K K K K
UUU UUU UUU UUU
A AAA AAG AAA AAA G
0 0 -1 -1

IS*150* (21). The transframe protein combines the HTH motif of *orfA*, an LZ motif and the DD(35)E catalytic domain of *orfB* (18). OrfAB (382 amino acids) shares its 86 N-terminal amino acids with OrfA (100 amino acids) and its 296 C-terminal amino acids with OrfB (299 amino acids). Rephasing of the ribosome to generate OrfAB occurs on a group of "slippery" lysine codons with a frequency of about 15% (measured using systems driven by two different promoters; T7$p_{\phi 10}$ and p_{tac}). OrfA is therefore normally expressed at significantly higher levels than OrfAB. Frameshifting permits the combination of different functional protein domains. The frameshifting in IS*911* is similar to that used in some retroviruses to generate the pol-gag "polyprotein" (38) and in the *dnaX* gene of *E. coli* to synthesize γ, the subunit of DNA polymerase III (39). The relevant sequences involved in frameshifting in IS*911* are shown in Figure 1C. The group of slippery lysine codons is A AAA AAG and is directly preceded by the AUU OrfB initiation codon. As *E. coli* does not encode a $tRNA_{Lys}$ with a $_{3'}UUC_{5'}$ anti-codon for AAG, both lysine codons are decoded by the same $tRNA_{Lys}$ with a $_{3'}UUU_{5'}$ anti-codon (Figure 1C). Its pairing has been shown to be weaker when there is a G at the wobble position (40), probably because modifications of a base, U_{34}, increase the rigidity of the anti-codon (41). The presence of an upstream RBS (GGAG sequence) and a downstream secondary structure (Y-shaped stem–loop) has been shown experimentally to stimulate rephasing of the ribosome in the −1 direction. What drives frameshifting is probably the thermodynamically favorable re-pairing of the two $tRNA_{Lys}$ from codons AAA-AAG to codons AAA-AAA (39, 42). The stimulators likely have a mechanical effect bringing back in register the ribosome and the mRNA after tRNA slippage. Different groups of codons have been observed to allow rephasing of the ribosome (43) and, although the most common motif is A_6G, different members of the IS*3* family carry a variety of these (e.g., A_3G for IS*3*; for review see reference 44).

Two similarly located partially overlapping reading frames are present in IS*3*, IS*150* and IS*3411*(22) and indeed in most of the presently identified IS3 family members. The transposases, OrfAB, like that of IS*911*, are fusion products of the two *orfs* generated by a −1 translational frameshift. For IS*3*, frameshifting is also stimulated by a presumed H-type pseudoknot structure similar to those generally involved in viral recoding (45). In IS*3411*, −1 slippage on a U UUU motif requires a more convoluted form of pseudoknot structure formed by the pairing of an apical loop and an internal loop belonging to two hairpins located 65 nucleotides apart on the mRNA (22). Two similarly arranged *orfs* occur in IS2 and have been shown to encode OrfA and OrfAB equivalents only (10, 24). This organization is observed in most members of the IS*3* family but, beside the cases mentioned above, frameshifting has been analyzed in only a few other, less well-characterized,

Figure 1 Organization of IS*911*. (A) Genetic organization. The 1,250-bp IS*911* is shown as a box. The boxes at each end represent the left (IRL) and right (IRR) terminal inverted repeats. The two open reading frames, *orfA* (blue) and *orfB* (green) are positioned in relative reading phases 0 and −1, respectively, as indicated. The indigenous promoter, p_{IRL}, is shown. The region of overlap between *orfA* and *orfB*, which includes the frameshifting signals to produce OrfAB, lies within IS*911* coordinates 300 and 400. The precise point at which the frameshift occurs, within the last heptad of the LZ, is indicated by the vertical dotted line. (B) Structure function map of OrfAB and OrfA. HTH, a potential helix-turn-helix motif; LZ, a leucine zipper motif involved in homo- and hetero-multimerization of OrfAB and OrfA. Programmed translational frameshifting that fuses OrfA and OrfB to generate the transposase OrfAB occurs within the fourth heptad. The LZ of OrfA and OrfAB therefore differ in their fourth heptad. A second region, M, necessary for multimerization of OrfAB is shown, as is the catalytic core of the enzyme which carries a third multimerization domain. OrfA translation initiates at an AUG, terminates with UAA whereas OrfAB translation terminates within the right IR. The vertical line to the right of M shows the extent of the truncated transposase, OrfAB[1–149] described in the text. (C) Frameshifting window. The mRNA sequence around the programmed translational frameshifting window is presented. The boxed sequence GGAG is the potential ribosome-binding site located upstream of *orfB* whose potential translation would be initiated at the boxed AUU codon. A ribosome (not to scale) is shown covering a series of "slippery" codons (AAAAAAG). A downstream secondary structure is also shown with the UAA, OrfA translation termination codon. The ribosome-binding site, slippery codons and secondary structure all contribute to the efficiency of the programmed −1 frameshift. The box at the foot of this figure shows how the anti-codons of two $tRNA_{Lys}$ are thought to undergo re-pairing with their codons in the AAAAAAG motif. doi:10.1128/microbiolspec.MDNA3-0031-2014.f1

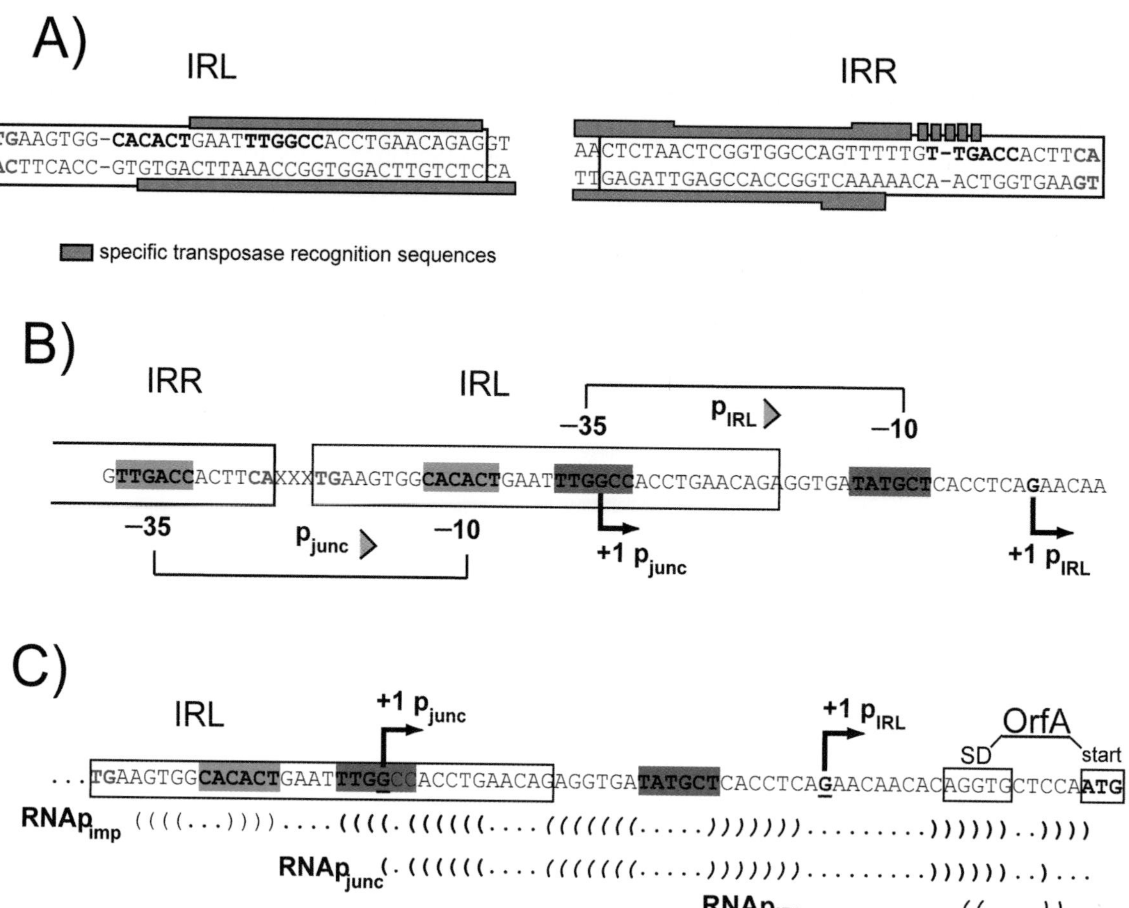

Figure 2 Important sequence features of IS*911*. (A) Organization of the IS*911* inverted repeat (IR). The nucleotide sequence of IRL and IRR is boxed. Grey horizontal bars above and below indicate the internal regions protected from DNaseI digestion by binding of OrfAB[1–149], a derivative of the 382-amino-acid OrfAB truncated for its catalytic domain. The dotted horizontal grey bar indicates partial protection. The dashes within the sequence indicate mismatches between the left and right ends. The −35 and −10 components of the indigenous promoter p_{IRL} (blue boxes) and of p_{junc} (green boxes) are shown. The conserved 5′TG tips are highlighted in red. (B) Organization of p_{junc}. The "junction" promoter assembled on circularization of IS*911* is shown as green boxes. The initiating transcript nucleotide (+1 p_{junc}), the indigenous p_{IRL} (blue boxes) and the initiating transcript nucleotide (+1 p_{IRL}) are also shown. The conserved 5′TG tips are highlighted in red. (C) Secondary structure at the left IS*911* end. The sequence of the "top" strand of IRL is shown together with the various transcription and translation signals. The symbols below are standard "dot–bracket" notations to indicate potential secondary structures formed with transcripts from top to bottom: from an external promoter, from p_{junc}, or from p_{IRL} respectively. The brackets shown in italic simply permit the reader to identify the apical stem of the secondary structure. doi:10.1128/microbiolspec.MDNA3-0031-2014.f2

elements (including IS*51*, IS*222*, IS*600*, IS*1133*, IS*1222*). The frequency of frameshifting appears to be quite variable from element to element: reported values are 15% for IS*911*, 50% for IS*150*, 6% for IS*3*, and 2% for IS*3411* (22). These values may not reflect the *in vivo* situation because they were not established by direct measurement of the amount of the OrfA and OrfAB proteins synthesized from an intact IS, but after modification of expression signals of the IS genes or after cloning the frameshift signals in a reporter system (15, 20, 21). However, the level of IS*911* circle formation measured by qPCR has indeed been shown to depend on frameshifting frequency *in vivo* (46). IS*911* copies from several clinical isolates that proved to contain variations in the frameshift region exhibited various reduced levels of frameshifting. When these were introduced into the model IS*911* they resulted in comparable reductions in circle formation. In addition, frameshifting is likely modulated by the physiological state of the host cells and by the environment: for example frameshifting decreases when temperature is raised or when ribosome density on the mRNA is increased (O. Fayet, unpublished).

TEMPERATURE SENSITIVITY OF TRANSPOSITION: A REGULATORY MECHANISM?

Formation of the first IS*911* transposition intermediate (figure-eight molecules; see *Overview of Transposition Pathway* section and Figure 5 below) is naturally temperature sensitive (6). Temperature sensitive transposition has been observed for other transposons such as IS*1* (47), Tn*3* (48), and Tn*951* (49). For IS*911*, very few figure-eight or transposon circle intermediates were observed at 42°C but levels were optimal at 30°C *in vivo*. Moreover, activity at 30°C was sufficiently high to detect circular copies of a single defective chromosomal IS*911* copy when OrfAB was provided *in trans* (6). Cell extracts containing OrfAB also proved somewhat temperature sensitive in figure-eight formation *in vitro*. Two partially temperature-resistant point mutants were obtained by selection following nitrous acid mutagenesis and showed increased activity compared with wild-type both *in vivo* and with OrfAB extracts *in vitro*. Both were located between the LZ motif and a second multimerization region, M, suggesting that this inter-domain region may be important for correct assembly of the transposition complex. A partial explanation for the temperature sensitivity came from the observation that OrfAB preparations included truncated OrfAB derivatives, in particular a species, OrfAB*, of approximately 16 kDa (around 149 amino acids), which were more prevalent at the higher temperature (50). Reconstruction experiments in which an engineered C-terminal deletion derivative, OrfAB[1–149] (Figure 1B), was supplied at the same time as full-length OrfAB, demonstrated a strong inhibition of transposition *in vivo* and *in vitro*. Partially temperature-resistant mutants were selected and found to have significantly reduced levels of truncated OrfAB derivatives, confirming this hypothesis (50).

Although this behavior might reveal a mechanism to regulate IS*911* transposition, it should be noted that all these studies involved overproduction of OrfAB. It seems possible that the high OrfAB levels themselves, either from the constitutive transframe protein or by high expression from the natural *orf* configuration, may lead to production of misfolded proteins, which are subject to cleavage. Further studies are essential to clarify this.

CO-TRANSLATIONAL BINDING

Certain prokaryotic IS transposases show a strong preference for acting on the element from which they are expressed rather than on other copies of the same element in the cell. This phenomenon of "cis" preference presumably serves to prevent general activation of several identical IS copies by any "accidental" (stochastic) transposase expression from a single IS. Several different IS such as IS*1*, IS*10*, IS*50*, IS*903*, and IS*911* (see reference 51 and references therein) exhibit this regulatory phenotype but "*cis*" preference may be the result of a combination of diverse mechanisms. Hence the Lon protease enhances "*cis*" preference of the IS*903* transposase (52) and, for IS*10*, it is influenced by translation levels, transposase mRNA half-life and translation efficiency (53, 54).

Another mechanism, co-translational binding based on tight coupling between prokaryotic transcription and translation, was proposed to explain the inability to complement a transposase mutant of IS*903* (55) and, more specifically, for Tn*5* (56). As its name indicates, co-translational binding would involve transposase binding to neighboring transposon DNA ends while the protein is being extruded from the ribosome and before it is released. It is therefore "tethered" to the transposable element from which it is being expressed.

Some full-length IS transposases bind weakly to their cognate IR but the isolated DNA-binding domain can bind more strongly. This has been observed for transposases of several elements including IS*1* (57) and IS*30* (58, 59) and has also been observed for that of IS*911*. Early studies using band shift assays demonstrated that full-length OrfAB binds the IRs only weakly and that OrfA binding was even lower or undetectable (36, 60). However, a truncated version of OrfAB, OrfAB[1–149], which is amputated for the C-terminal catalytic domain bound both ends avidly (18) (see also *Temperature sensitivity of transposition* section above). It is important to note that this implies that, in many *in vitro* systems, the majority of transposase is therefore likely to be inactive or only partially active because it would not bind stably to its substrate. The observations suggest that the C-terminal domain inhibits specific binding by the sequence-specific N-terminal DNA-binding domain, possibly by steric masking (Figure 3). This idea is consistent with the observation that IS*10* transposase activity is increased by partial denaturation (for example by treatment with low alcohol concentrations (61)). It is also consistent with the observation that the OrfAB protein of IS2 can bind the IS2 IRs when it carries a large GFP tag (35, 62).

One biological explanation for *cis* preference is that the nascent N-terminal domain might fold before completion of translation of the C-terminal domain and the nascent protein could initiate binding directly to the closest IS end. Once bound, it would no longer be sensitive to masking by the C-terminal domain. If binding fails to occur after translation of the N-terminal DNA-binding domain, continuing translation and folding of the C-terminal domain would then sterically mask the DNA-binding domain, resulting in an inactive protein. This implies that binding necessary for subsequent catalysis would occur only transitorily early in translation (Figure 3).

In vivo studies (37) demonstrated that OrfAB had a very strong *cis* preference (about 200-fold higher activity than *in trans*), that the strength of the *cis* effect depends on the distance of the transposase gene from the IS ends. Also modification of the translational frameshifting pause signal (see *Frameshifting: transposase OrfAB as a fusion protein* section above) has a strong influence on *cis* preference, presumably by delaying translation and folding of the C-terminal domain increasing the chance that the folded N-terminal domain will recognize and bind its target IR.

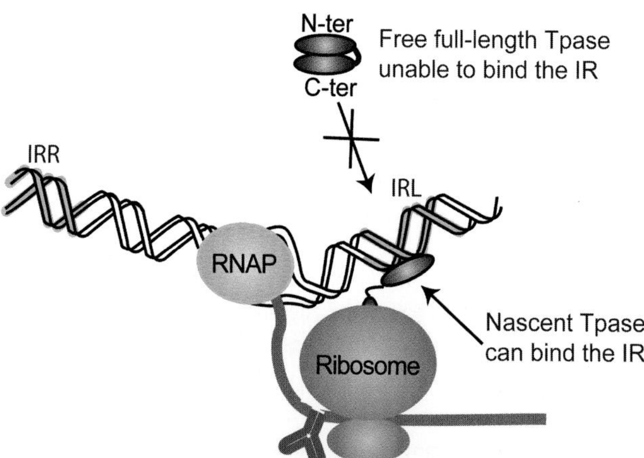

Figure 3 Co-translational binding model. This schematic, not to scale, shows the insertion sequence (IS) with its left (IRL) and right (IRR) ends in green. RNA polymerase, RNAP, is shown in pale green in the process of transcribing from the promoter p_{IRL}. The mRNA is shown in dark green with a ribosome (blue) paused at the secondary structure shown in Figure 1C. The nascent OrfAB peptide (brown) is shown binding to IRL while undergoing translation. Above is shown the full-length OrfAB in a folded configuration proposed to prevent its binding to the IR as a completed protein.
doi:10.1128/microbiolspec.MDNA3-0031-2014.f3

The model was further supported by *in vitro* analyses using ribosome display with a coupled *E. coli*-derived transcription–translation system together with size exclusion chromatography (37). It was demonstrated that an added IR bound nascent OrfAB derivatives while they are attached to the ribosome. Using C-terminal truncated OrfAB genes, ternary complexes containing mRNA, ribosome, and a nascent peptide specifically bound added IR copies if only the N-terminal 149 amino acids extended from the ribosome whereas a full-length Tpase exiting the ribosome did not. Moreover, OrfAB with a point mutation eliminating DNA binding (34) also failed to bind IRs under these co-translational conditions. Direct evidence of coupled translational binding was obtained using a staged coupled transcription/translation reaction: nascent OrfAB bound the IR before its synthesis was complete but not after. This clearly showed that OrfAB can efficiently bind the IR only before its complete translation.

Identification of co-translational binding as a regulatory mechanism raises some important questions concerning the dynamics of transpososome formation. To function, OrfAB must multimerize. Stable formation of the important synaptic complex containing both IS ends and the transposase requires a dimeric OrfAB (see *The IS911 transpososome* section below). An intriguing question arising directly from these results is therefore how OrfAB multimerizes to bind both ends of the IS: must two OrfAB monomers each bind one IS end and subsequently dimerize or, once one monomer is bound, does a second form the dimer before binding the second end? Moreover, OrfA is also translated, presumably from the same mRNA molecule and, although it does not bind IS*911* ends, does form heterodimers with OrfAB as well as homodimers. Additional studies will be necessary to unravel the temporal behavior of these proteins.

THE IS*911* TRANSPOSOSOME

A crucial checkpoint in transposition is the assembly of the "transpososome". This step is a general prerequisite for initiating DNA cleavage and the subsequent chemical steps in transposition for most elements that use DNA (rather than RNA) transposition intermediates. In this protein–DNA complex, both ends of the transposon are bridged by the transposase before it catalyzes the DNA strand cleavages and strand transfers necessary for transposon mobility (28, 63, 64). The transpososome adopts very precise architectures to accomplish these steps, and undergoes defined changes throughout the transposition process.

The overall IS*911* transposition pathway is a two-step process in which a closed circular copy of the IS is made by replication and is subsequently integrated (see below: Figure 4). This implies consecutive assembly of two types of transpososome: one required to pair the L and R transposon ends to begin the process leading to replication and circularization of the transposing IS copy (synaptic complex A; SCA) and the other (synaptic complex B; SCB) is required to bring the abutted transposon ends on the circular replicated copy into contact with the target DNA ensuring its integration (Figure 4A and B). Study of the synaptic complexes used the truncated OrfAB derivative because full-length OrfAB binds very poorly to the IS*911* ends.

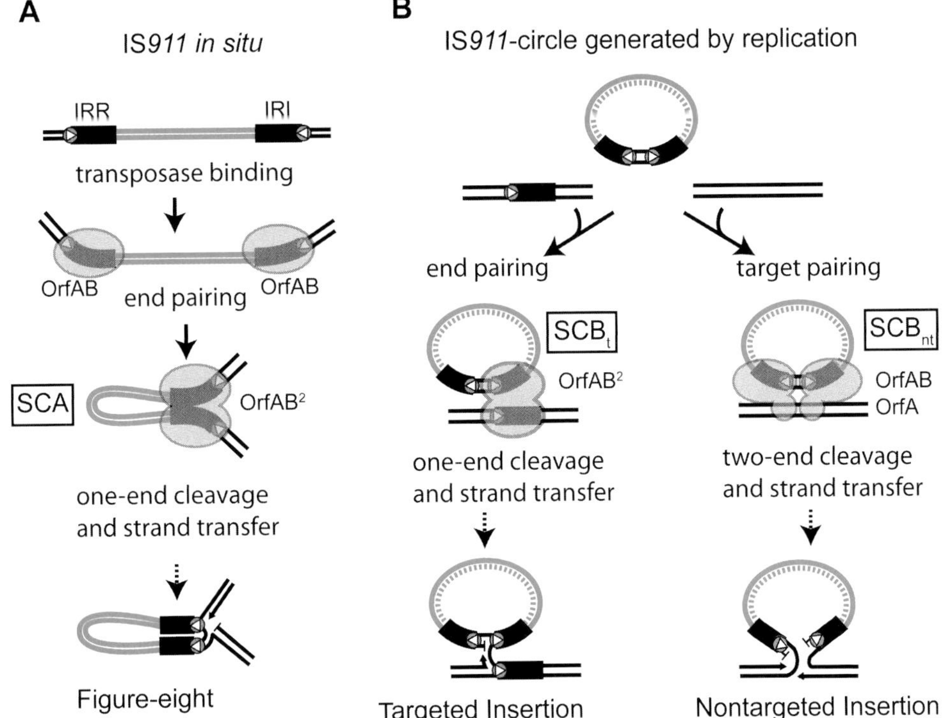

Figure 4 The IS*911* transpososome. The schematic shows the proposed configuration and composition of the different synaptic complexes (SCA and SCB) involved in different steps of the IS*911* transposition cycle. (A) The excision complex SCA. The tips of the insertion sequence (IS), which are not protected by the truncated transposase OrfAB[1–149] (Figure 2A) are shown as green circles containing an arrowhead. The inverted repeats (IR) are indicated by thick black lines and the IS as green lines. Full-length OrfAB, which is presumed to cover the entire IR, is shown bound as a monomer to each end and to introduce a small bend in the DNA. Dimerization creates SCA, resulting in pairing of both IRs and in the formation of a DNA loop which includes the IS. Finally, a cleavage and strand transfer event results in the formation of a single-strand bridge between the IRs. (B) The integration complex SCB. Symbols are as in (A). In the left hand column, the IS circle intermediate with its newly replicated strand (dotted line) is shown to form a complex between an IR in the circle and a second in the target to form SCB_t. Cleavage and strand transfer is shown to form a single-strand bridge between the two IRs. The RecG helicase is thought to intervene to drive strand migration before a second cleavage and strand transfer results in integration of the circle (89). This type of mechanism would explain the integration of the many different ISs observed to occur next to a resident IR in the target. The right-hand column shows an untargeted integration event that involves OrfA in addition to OrfAB. It should be emphasized that OrfA is known to interact with OrfAB. It also changes in some way OrfAB binding but it is not clear whether it remains in the complex. doi:10.1128/microbiolspec.MDNA3-0031-2014.f4

Excision Synaptic Complex SCA

OrfAB[1–149] forms a complex with two IR copies, the paired-end complex (PEC) (36). This was determined using a band shift assay and a mixture of IR of two different lengths (the so-called "long-short" experiment) in which three species (one with two long IR, one with two short IR and one with an IR of each length were observed equivalent to the SCA. An intact OrfAB[1–149] LZ was also found to be necessary for correct PEC/SCA formation (18, 36). At higher OrfAB [1–149] concentrations an additional species appeared, which was interpreted as a single-end complex (SEC) composed of one IR and OrfAB[1–149]. It is noteworthy that addition of OrfA appears to disturb both PEC/SCA and SEC and generates a fast migrating species whose composition remains to be determined but does not appear to contain OrfA itself (36).

DNaseI and copper phenanthroline footprinting revealed that OrfAB[1–149] protects a subterminal (internal) region of the IRs that includes two blocks of sequence conserved between the left (IRL) and right (IRR) ends (Figure 2A). DNA-binding assays *in vitro* and measurement of recombination activity *in vivo* of sequential deletion derivatives of the two IRs suggested a model in which the N-terminal region of OrfAB binds the conserved boxes in a sequence-specific manner and anchors the two IRs into the SCA. The external region of the inverted repeat was proposed to contact the C-terminal transposase domain carrying the catalytic site (60).

The stoichiometry of the IS*911* transposase in the formation of SCA was also studied using band shift approaches. Using tagged and untagged truncated forms of the transposase derivative OrfAB[1–149], it has been shown that SCA is formed by the pairing of two OrfAB[1–149]–IRR complexes. This suggests that SCA is composed of a dimer of transposase bridging to two IRs (65). The geometry of SCA has been studied at the single molecule level by atomic force microscopy using asymmetric IRR-carrying DNA fragments. It was shown that OrfAB[1–149] assembles two IRR copies in a parallel orientation (Figure 4A) (65).

Assembly of SCA has also been studied using a second single molecule approach: tethered particle motion (66) in which a DNA molecule is tethered to a glass support and its effective length is measured by observing the Brownian motion of a bead attached to its free end. OrfAB[1-149] binding to a single IR provoked a small shortening of the DNA, consistent with a DNA bend introduced by protein binding to the IR and was confirmed using an electrophoretic mobility shift assay. When two ends were present on the tethered DNA in their natural, inverted, configuration, OrfAB[149] not only provoked the short reduction in length but also generated species with greatly reduced effective length consistent with DNA looping between the ends and hence SCA formation (Figure 4A). Once formed, SCA was very stable and kinetic analysis in real-time suggested that passage from the bound unlooped to the looped state could involve another unlooped species of intermediate length in which OrfAB[1-149] is bound to both IRs. DNA carrying directly repeated IR also gave rise to the looped species but the level of the intermediate species was significantly enhanced. Its accumulation could reflect a less favorable SCA formation with directly repeated IR copies than with inverted IR. This is compatible with a model in which OrfAB binds separately to and bends each IR and protein–protein interactions then leads to SCA formation (Figure 4A) (67). Cleavage and strand transfer would then give rise to a species in which both IS ends are joined by a single-strand bridge, a figure-eight on a circular plasmid (Figures 4A and 5C) (see *Transposition pathway* section below).

Insertion Synaptic Complex SCB

Insertion of the circular transposon intermediate occurs in two ways: nontargeted (where insertion occurs with little sequence selectivity) and targeted (where insertion occurs next to a sequence resembling the IS end) (see *Transposition pathway* section below). In both cases, this is accomplished through a synaptic complex built on the abutted IR junction. The architecture of this complex, SCB, is probably different for each of the two types of integration event. SCB has not been characterized in such a precise way as SCA (Figure 4B). The two types of insertion synaptic complexes are called SCB_t and SCB_{nt}, for targeted and nontargeted synaptic complex, respectively (Figure 4B) (68). Little is known about the stoichiometry and geometry of these complexes.

SCB_{nt}, which is involved in the normal transposition insertion pathway is thought to differ from both SCA and SCB_t and to include the second IS*911* protein, OrfA. This second transposition protein binds nonspecifically to DNA and also interacts with OrfAB (see *Transposition proteins: DNA sequence recognition, multimerization and catalytic domains* section above) (18, 36), is proposed to direct an OrfAB–junction complex, the product of the replicative IS*911* excision (see below), to a randomly chosen target DNA to form SCB_{nt} (68, 69). This is based on the observation that integration of the transposon circle intermediate is greatly stimulated by preincubation of OrfAB and OrfA in an *in vitro* reaction (70).

OrfA does not appear to be directly involved in the targeted insertion pathway, as it presumably requires synapsis between an IR of the transposon circle and an IR-like sequence in the target. SCB_t may therefore show some similarities with SCA. Based on protein and DNA requirements for protein–DNA complex formation, as judged by band shift, and for transposition products, as judged by *in vitro* and *in vivo* transposition assays, it has been proposed that SCB_t is composed of a transposase dimer bridging a target DNA molecule carrying an IR and a DNA molecule carrying an IRR–IRR junction (IS*911* circle).

This IR targeted insertion explains how the original isolate of IS*911* might have occurred next to a sequence that strongly resembles an IR (4) and can also explain one ended insertion (71). In this regard IRR shows somewhat higher affinity than IRL. Note that if one of the two IR carried by the circle is omitted, SCB_t resembles SCA (Figure 4).

THE TRANSPOSITION PATHWAY

IS*911* is one of an increasing number of ISs and IS families known to transpose using a double-strand circular DNA intermediate through a Copy-out–Paste-in process (Figure 5). This represents a major transposition pathway that has yet to be widely recognized. Closely related pathways have been demonstrated for IS2 (1), IS*3* (2), and IS*150* (3) (all members of the IS3 family) and members of other IS families such as IS*1* (72), IS*21*, IS*30*, IS*256*, IS*110*, IS*Lre2*, IS*L3*, and their derivatives (see Siguier et al., this volume).

Briefly, this process involves: end pairing; cleavage at one IS end such that a liberated 3′OH attacks the second end to make an intact single strand (a figure of eight bridged by a single strand between left and right ends); replication to generate a double-strand IS DNA circle with abutted ends regenerating, at the same time, the original donor molecule. The transposase then breaks both strands, one at each end and joins them to the target.

Figure-Eight Formation

The key feature that differentiates Copy-out–Paste-in transposition from other mechanisms is the asymmetric cleavage at one IS end and its rejoining to itself at the 5′ end. This tethers both ends by a single-strand bridge and generates a figure-eight intermediate on a circular plasmid donor molecule.

The initial step in the pathway is recognition of the IR by OrfAB (which presumably occurs during its translation) and assembly of SCA (see above) to correctly position the DNA ends and the transposase catalytic site for the subsequent chemical steps (Figure 5A and B).

Like all known DDE transposase-catalyzed reactions (see Hickman and Dyda, this volume), IS*911* transposition proceeds by cleavage of a single strand at the transposon end generating a 3′OH. This then attacks a target phosphodiester bond in a strand transfer reaction. The particularity of this Copy-out–Paste-in mechanism is that initial cleavage occurs at only one transposon end, either left or right (Figure 5B).

This single liberated 3′OH directs strand transfer to the same strand three bases 5′ to the other end of the element. This generates a molecule in which a single transposon strand is circularized to produce a single-strand bridge generating a figure-eight structure on a circular plasmid donor molecule (Figure 5C), which can be easily observed *in vivo* (73).

The IR are joined by the single-stranded bridge and separated by three bases derived from flanking DNA from either the left or right end. The 3 (or 4) bp direct repeats flanking the original insertion are not required for further transposition (as also shown for IS*3* (74)) and an IS*911*-based transposon engineered to have different flanks generates a mixed population of figure-eight molecules with one or other flank sequence. Prevention of cleavage of one or other transposon end resulted in a homogeneous population that carries the three-nucleotide DNA flank associated with the mutant end confirming that the IRL can attack IRR and vice versa. The reaction can be viewed as a one-ended site-specific transposition event. This type of one ended attack occurs as the normal pathway for IS2. This IS carries a mutation at one end which prevents it from being cleaved and acting as an attacking donor strand.

These initial steps can be accomplished by OrfAB alone. However, it should be noted that in the presence of OrfA, no figure eight or IS circles could be detected by a simple gel assay *in vivo* although IS circles were found using a PCR approach (46). This suggests that OrfA may play a role in negatively regulating initiation of transposition. A similar conclusion has been reached for OrfA of IS*3* (75). Alternatively, OrfA may stimulate the disappearance of figure-eight and IS circles (see below) because no effect of OrfA was observed on figure-eight formation *in vitro* (76). Together with the fact that OrfAB is normally produced at low levels from a weak promoter (15), initiation of transposition to form the figure-eight intermediate may be stochastic.

Formation of the Circular Intermediate

A double-strand circular IS intermediate is then produced from the figure-eight intermediate by replication.

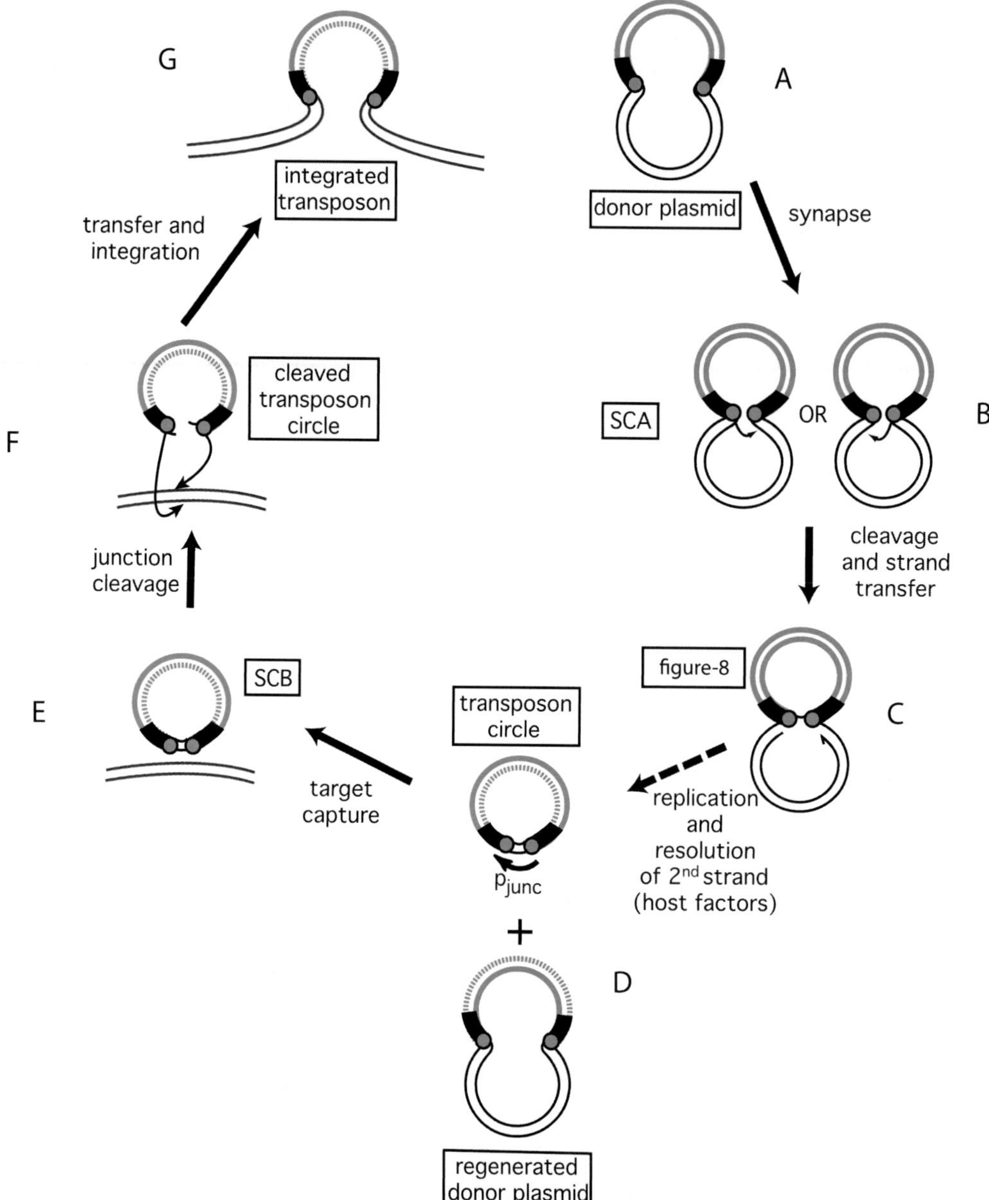

Figure 5 The IS*911* transposition cycle. The transposon is shown in green, the flanking donor DNA in black and the target DNA in blue. Transposon ends are shown as green filled circles. The small arrows shown in Figure 4 have been omitted for brevity. (A) Donor plasmid carrying the insertion sequence (IS). (B) Formation of the first synaptic complex SCA and cleavage of the left or right inverted repeat (IR) and attack of the other end. (C) Formation of a single-strand bridge to create a figure-eight molecule if the donor is a plasmid as shown here. (D) The products of IS-specific replication: the double strand circular IS transposition intermediate and the regenerated transposon donor plasmid. The replicated strand is shown as a green dotted line. (E) Formation of the second synaptic complex SCB and engagement of the target DNA (blue). (F) Cleavage of the IS circle and integration. (G) The newly integrated IS. doi:10.1128/microbiolspec.MDNA3-0031-2014.f5

This has been demonstrated by kinetic data (37, 73) which indicate that the figure-eight gives rise to the circular transposon form, which can easily be detected *in vivo* (77). The IRs in the IS circle are abutted and separated by 3 bp of DNA flanking the original insertion (Figure 5D). As in the case of the figure-eight molecules, a transposon engineered to have different flanks generates a mixed population of transposon circles with one or the other 3-bp flank located at the junction (77).

Further *in vivo* studies using a staged labeling protocol and a temperature-sensitive plasmid as transposon donor demonstrated that conversion from the figure-eight to the transposon circle occurs by semi-conservative replication where the circular intermediate is "copied out" leaving a copy in the transposon donor molecule (Figure 5D) (78). This is transposon-specific, requires OrfAB (presumably to generate the figure-eight and generate a 3′OH on the IS*911* DNA flank) and does not depend on replication from the donor plasmid origin of replication (78).

Using donor plasmids where cleavage of one or other IR was inactivated for cleavage would be expected to determine whether one or other of the 3′OH is used in transposon replication. This was tested using the Tus/*ter* system (79–82) (which blocks passage of a replication fork in an orientation-specific fashion) cloned into the transposon in either one or other orientation. In the presence of Tus protein, no transposon circles were observed if the orientation of the *ter* site was that expected to block replication from one or the other end (78). This demonstrates that replication occurs using the 3′OH of the donor flank. Circle formation can therefore be described as a donor-primed transposon replication mechanism.

At present, it is not known how OrfAB is removed and how this replication step is initiated or terminated to generate the final circles. It is possible that these processes involve host factors and mechanisms similar to those that operate in replicative transposition of bacteriophage Mu (see Harshey, this volume) (83, 84).

Integration of the Circular Intermediate

The circular IS intermediate is then cleaved to liberate a 3′OH, which then attacks target DNA for the final integration step. Indeed, it has also shown that the IR junction formed by IS circularization is very unstable in the presence of OrfAB and undergoes high levels of deletion and insertion *in vivo* (85) and *in vitro* (70).

Insertion of the transposon circles presumably requires further transposase synthesis. A remarkable consequence of transposon circle formation is that a strong promoter, p_{junc}, is assembled from a −35 hexamer contributed by IRR and a −10 hexamer contributed by IRL (Figure 2B; Figure 5D). The 3 (or more rarely 4) bp that separate IRL and IRR in the circle provide an ideal spacing between the −35 and −10 elements (85). The junction promoter, p_{junc}, is 30–50-fold stronger than the indigenous promoter, p_{IRL} (85), and more than two-fold stronger than lacUV5 (14). It is correctly placed to drive high levels of transposase synthesis and plays an active role in controlling IS*911* transposition. Inactivation of p_{junc} by mutagenesis strongly reduced IS*911* transposition *in vivo* when the transposase was expressed in its native configuration (14). Moreover, the truncated OrfAB derivative, OrfAB[1–149], which specifically binds IRR and IRL, reduced *in vivo* promoter activity 10-fold in a mutated junction resistant to cleavage. Full-length OrfAB, which binds the IR only weakly, and OrfA, which does not specifically bind the IR, had no effect (14). In the case of the related IS2, this junction promoter is required for transposition (86). Integration results in disassembly of p_{junc} providing a powerful feedback mechanism resulting in transient and controlled activation of integration only in the presence of the correct (circular) intermediate.

Circle junction formation brings both transposon ends together in an inverted orientation. This active junction must then participate in a second type of synaptic complex that includes target DNA (Figure 5E). Insertion can follow two pathways: nontargeted or targeted.

In the nontargeted pathway, insertion is relatively sequence independent. Two single-strand cleavages, one at each abutted IR, linearize the transposon circle, permitting the two liberated 3′OH groups to direct coordinated strand transfer (Figure 5F). The final step requires OrfAB but is greatly stimulated by OrfA and is sensitive to the ratio of OrfAB/OrfA (70). It is not known whether target capture occurs before or after cleavage of the circle junction although it has been observed that linear copies of IS*911* are produced from transposon circles *in vitro* and in the presence of high OrfAB levels *in vivo* and a pre-cleaved linear transposon was a robust substrate for integration *in vitro* (87). Based on kinetics and on the formation of the strong p_{junc} promoter, we favor a model in which the IS circles represent a reservoir of transposition intermediates and that linear forms are generated from the IS circles during the integration process. This has also been proposed for IS3 (74). Both IS circles and linear IS forms have also been identified for the related IS*150* (3, 74, 88).

In the targeted pathway, a synaptic complex is formed between an IR on the transposon circle and an IR-like sequence in the target. It seems probable that only a single IS end is cleaved and transferred to the "target" site. This process involves a target IS*911* end and strand transfer occurs between one cleaved end of the IS circle and the target IS end to create an *intermolecular* single-strand bridge rather than the *intramolecular* bridge of the figure-eight intermediate (Figure 4B left). Resolution of this structure implicates branch migration and replication from the donor plasmid (89). In this light, the RecG helicase is implicated in this targeted insertion process.

CONCLUSIONS

It is important to emphasize that the transposition mechanism uncovered for IS*911* appears to have been adopted by a considerable number of other IS families (e.g., IS*1*, IS*3*, IS*21*, IS*30*, IS*256*, IS*110*, IS*Lre2*, IS*L3* see Siguier *et al.*, this volume). This Copy-out–Paste-in mechanism therefore represents a major transposition pathway, which will undoubtedly prove to be more widespread than at present recognized.

Although it uses the same chemistry, the IS*911* transposition Copy-out–Paste-in mechanism is quite different from that of Cut-and-Paste or of Cointegration. The pathway has been relatively well defined both genetically and biochemically but a number of important questions remain to be answered. In contrast to ISs, which transpose using cut-and-paste or replicative transposition via cointegration (see Hickman and Dyda, this volume; Siguier et al., this volume) and in which cleavages occur at both ends concomitantly, IS*911* transposition is asymmetric. The product of cleavage at one IS*911* end, the 3′OH, is required to cleave the opposite end. Understanding the constraints within the transpososome that produce this asymmetric behavior awaits the results of structural studies, which have been hampered in the past by the presence of truncated forms in transposase preparations.

Replication is required to generate the circular IS intermediate. It is not yet clear how a replication fork is installed on the single-strand bridge that forms the figure-eight, how OrfAB may be removed and replaced by replication proteins, nor what replication proteins are involved. A second important unknown is how replication of the IS is terminated resulting in formation of the IS circle.

Another question central to IS*911* transposition is the role of OrfA. In its presence, the figure-eight and circle species are not detected *in vivo* but *in vitro*, it stimulates integration. It is possible that OrfA plays two roles: one in inhibiting accumulation of the transposition intermediates and a second in stimulating integration. It is clear that OrfA can indeed form heterodimers with OrfAB and affect the nature of the complexes formed by OrfAB and the IR. However, it does not bind the IR directly and, although OrfAB and OrfA interactions can be detected by co-immunoprecipitation (18, 36), it is not yet clear whether OrfAB and OrfA interact in a transitory way or form a more stable complex.

As underlined throughout, many aspects of this pathway have been documented for other members of the IS*3* family. Although presumably all members of the IS*3* family transpose using this replicative Copy-out–Paste-in mechanism, resulting in transposon circle formation, they may have adopted slightly different regulatory mechanisms. Although it seems likely that most of these produce IS circles by replication, it should be kept in mind that it is formally possible that some of the other IS families that produce circular intermediates (such as IS*607* family members, genomic islands or integrative conjugative elements) may use site-specific recombination that excises the entire IS from its donor replicon. Hence the presence of circular intermediates does not necessarily mean that they are intermediates in a Copy-out–Paste-in pathway.

Another fundamental question stems from the co-translational binding studies. This arises because stable SC formation requires a dimeric OrfAB. It is possible that dimerization also occurs co-translationally. *In vivo* under natural expression conditions, one OrfAB monomer would be expected to be synthesized for every 10–50 OrfA monomers (16), physically separating two OrfAB monomers. One solution to this problem would be that monomeric OrfAB molecules bind the two IRs independently and only subsequently form the full SCA. This would be consistent with the interpretation of tethered particle motion experiments that OrfAB[1–149] can occupy both IRs independently. To understand the relationship between co-translational binding and OrfAB multimerization it will be important to understand the dynamics of frameshifting *in vivo*.

Citation. **Chandler M, Fayet O, Rousseau P, Ton Hoang B, and Duval-Valentin G.** 2015. Copy-out-paste-in transposition of IS*911*: a major transposition pathway. Microbiol Spectrum 3(2):MDNA3-0031-2014.

References

1. **Lewis LA, Grindley ND.** 1997. Two abundant intramolecular transposition products, resulting from reactions initiated at a single end, suggest that IS2 transposes by an unconventional pathway. *Mol Microbiol* **25:**517–529.
2. **Ohtsubo E, Minematsu H, Tsuchida K, Ohtsubo H, Sekine Y.** 2004. Intermediate molecules generated by transposase in the pathways of transposition of bacterial insertion element IS3. *Adv Biophys* **38:**125–139.

3. **Haas M, Rak B.** 2002. *Escherichia coli* insertion sequence IS150: transposition via circular and linear intermediates. *J Bacteriol* **184:**5833–5841.
4. **Prere MF, Chandler M, Fayet O.** 1990. Transposition in *Shigella dysenteriae*: isolation and analysis of IS911, a new member of the IS3 group of insertion sequences. *J Bacteriol* **172:**4090–4099.
5. **Blattner FR, Plunkett G, Bloch CA, Perna NT, Burland V, Riley M, Collado-Vides J, Glasner JD, Rode CK, Mayhew GF, Gregor J, Davis NW, Kirkpatrick HA, Goeden MA, Rose DJ, Mau B, Shao Y.** 1997. The complete genome sequence of *Escherichia coli* K-12. *Science* **277:**1453–1474.
6. **Haren L, Bétermier M, Polard P, Chandler M.** 1997. IS911-mediated intramolecular transposition is naturally temperature sensitive. *Mol Microbiol* **25:**531–540.
7. **Schmidt H, Henkel B, Karch H.** 1997. A gene cluster closely related to type II secretion pathway operons of gram-negative bacteria is located on the large plasmid of enterohemorrhagic *Escherichia coli* O157 strains. *FEMS Microbiol Lett* **148:**265–272.
8. **Tobe T, Hayashi T, Han CG, Schoolnik GK, Ohtsubo E, Sasakawa C.** 1999. Complete DNA sequence and structural analysis of the enteropathogenic Escherichia coli adherence factor plasmid. *Infect Immun* **67:**5455–5462.
9. **Swenson DL, Bukanov NO, Berg DE, Welch RA.** 1996. Two pathogenicity islands in uropathogenic *Escherichia coli* J96: cosmid cloning and sample sequencing. *Infect Immun* **64:**3736–3743.
10. **Hu ST, Hwang JH, Lee LC, Lee CH, Li PL, Hsieh YC.** 1994. Functional analysis of the 14 kDa protein of insertion sequence 2. *J Mol Biol* **236:**503–513.
11. **Reimmann C, Moore R, Little S, Savioz A, Willetts NS, Haas D.** 1989. Genetic structure, function and regulation of the transposable element IS21. *Mol Gen Genet* **215:** 416–424.
12. **Dalrymple B.** 1987. Novel rearrangements of IS30 carrying plasmids leading to the reactivation of gene expression. *Mol Gen Genet* **207:**413–420.
13. **Perkins-Balding D, Duval-Valentin G, Glasgow AC.** 1999. Excision of IS492 requires flanking target sequences and results in circle formation in *Pseudoalteromonas atlantica*. *J Bacteriol* **181:**4937–4948.
14. **Duval-Valentin G, Normand C, Khemici V, Marty B, Chandler M.** 2001. Transient promoter formation: a new feedback mechanism for regulation of IS911 transposition. *EMBO J* **20:**5802–5811.
15. **Polard P, Prere MF, Chandler M, Fayet O.** 1991. Programmed translational frameshifting and initiation at an AUU codon in gene expression of bacterial insertion sequence IS911. *J Mol Biol* **222:**465–477.
16. **Prere MF, Canal I, Wills NM, Atkins JF, Fayet O.** 2011. The interplay of mRNA stimulatory signals required for AUU-mediated initiation and programmed -1 ribosomal frameshifting in decoding of transposable element IS911. *J Bacteriol* **193:**2735–2744.
17. **Rettberg CC, Pr, Gesteland RF, Atkins JF, Fayet O.** 1999. A Three-way junction and constituent stem-loops as the stimulator for programmed -1 frameshifting in bacterial insertion sequence IS911. *J Mol Biol* **286:**1365–1378.
18. **Haren L, Polard P, Ton-Hoang B, Chandler M.** 1998. Multiple oligomerisation domains in the IS911 transposase: a leucine zipper motif is essential for activity. *J Mol Biol* **283:**29–41.
19. **Fu R, Voordouw G.** 1998. ISD1, an insertion element from the sulfate-reducing bacterium *Desulfovibrio vulgaris* Hildenborough: structure, transposition, and distribution. *Appl Environ Microbiol* **64:**53–61.
20. **Sekine Y, Eisaki N, Ohtsubo E.** 1994. Translational control in production of transposase and in transposition of insertion sequence IS3. *J Mol Biol* **235:**1406–1420.
21. **Vogele K, Schwartz E, Welz C, Schiltz E, Rak B.** 1991. High-level ribosomal frameshifting directs the synthesis of IS150 gene products. *Nucleic Acids Res* **19:**4377–4385.
22. **Mazauric MH, Licznar P, Prere MF, Canal I, Fayet O.** 2008. Apical Loop-Internal Loop RNA Pseudoknots: a new type of stimulator of-1 translational frameshifting in bacteria. *J Biol Chem* **283:**20421–20432.
23. **Chen CC, Hu ST.** 2006. Two frameshift products involved in the transposition of bacterial insertion sequence IS629. *J Biol Chem* **281:**21617–21628.
24. **Hu ST, Lee LC, Lei GS.** 1996. Detection of an IS2-encoded 46-kilodalton protein capable of binding terminal repeats of IS2. *J Bacteriol* **178:**5652–5659.
25. **Fayet O, Ramond P, Polard P, Prere MF, Chandler M.** 1990. Functional similarities between retroviruses and the IS3 family of bacterial insertion sequences? *Mol Microbiol* **4:**1771–1777.
26. **Kulkosky J, Jones KS, Katz RA, Mack JP, Skalka AM.** 1992. Residues critical for retroviral integrative recombination in a region that is highly conserved among retroviral/retrotransposon integrases and bacterial insertion sequence transposases. *Mol Cell Biol* **12:**2331–2338.
27. **Hickman AB, Chandler M, Dyda F.** 2010. Integrating prokaryotes and eukaryotes: DNA transposases in light of structure. *Crit Rev Biochem Mol Biol* **45:**50–69.
28. **Montano SP, Rice PA.** 2011. Moving DNA around: DNA transposition and retroviral integration. *Curr Opin Struct Biol* **21:**370–378.
29. **Yuan Y-W, Wessler SR.** 2011. The catalytic domain of all eukaryotic cut-and-paste transposase superfamilies. *Proc Natl Acad Sci USA* **108:**7884–7889.
30. **Doak TG, Doerder FP, Jahn CL, Herrick G.** 1994. A proposed superfamily of transposase genes: transposon-like elements in ciliated protozoa and a common "D35E" motif. *Proc Natl Acad Sci USA* **91:**942–946.
31. **Polard P, Chandler M.** 1995. Bacterial transposases and retroviral integrases. *Mol Microbiol* **15:**13–23.
32. **Rezsohazy R, Hallet B, Delcour J, Mahillon J.** 1993. The IS4 family of insertion sequences: evidence for a conserved transposase motif. *Mol Microbiol* **9:**1283–1295.
33. **Rice P, Craigie R, Davies DR.** 1996. Retroviral integrases and their cousins. *Curr Opin Struct Biol* **6:**76–83.
34. **Rousseau P, Gueguen E, Duval-Valentin G, Chandler M.** 2004. The helix-turn-helix motif of bacterial insertion sequence IS911 transposase is required for DNA binding. *Nucleic Acids Res* **32:**1335–1344.

35. **Lewis LA, Astatke M, Umekubo PT, Alvi S, Saby R, Afrose J, Oliveira PH, Monteiro GA, Prazeres DM.** 2012. Protein-DNA interactions define the mechanistic aspects of circle formation and insertion reactions in IS2 transposition. *Mob DNA* **3:**1.
36. **Haren L, Normand C, Polard P, Alazard R, Chandler M.** 2000. IS911 transposition is regulated by protein–protein interactions via a leucine zipper motif. *J Mol Biol* **296:** 757–768.
37. **Duval-Valentin G, Chandler M.** 2011. Cotranslational control of DNA transposition: a window of opportunity. *Mol Cell* **44:**989–996.
38. **Brierley I.** 1995. Ribosomal frameshifting viral RNAs. *J Gen Virol* **76**(Pt 8):1885–1892.
39. **Tsuchihashi Z, Brown PO.** 1992. Sequence requirements for efficient translational frameshifting in the *Escherichia coli* DNAX gene and the role of an unstable interaction between tRNA(Lys) and an AAG lysine codon. *Genes Dev* **6:**511–519.
40. **Yokoyama S, Watanabe T, Murao K, Ishikura H, Yamaizumi Z, Nishimura S, Miyazawa T.** 1985. Molecular mechanism of codon recognition by tRNA species with modified uridine in the first position of the anticodon. *Proc Natl Acad Sci USA* **82:**4905–4909.
41. **Sundaram M, Durant PC, Davis DR.** 2000. Hypermodified nucleosides in the anticodon of tRNALys stabilize a canonical U-turn structure. *Biochemistry* **39:** 12575–12584.
42. **Licznar P, Mejlhede N, Prere MF, Wills N, Gesteland RF, Atkins JF, Fayet O.** 2003. Programmed translational -1 frameshifting on hexanucleotide motifs and the wobble properties of tRNAs. *EMBO J* **22:**4770–4778.
43. **Sharma V, Prere MF, Canal I, Firth AE, Atkins JF, Baranov PV, Fayet O.** 2014. Analysis of tetra- and heptanucleotides motifs promoting -1 ribosomal frameshifting in *Escherichia coli. Nucleic Acids Res* doi:10.1093/nar/gku386.
44. **Fayet O, Prère M-F.** 2010. Programmed ribosomal-1 frameshifting as a tradition: the bacterial transposable elements of the IS3 family, p 259–280. *In* Atkins JF, Gesteland R (ed), *Recoding: Expansion of Decoding Rules Enriches Gene Expression*, **vol 24**. Springer, New York and Heidelberg.
45. **Giedroc DP, Cornish PV.** 2009. Frameshifting RNA pseudoknots: structure and mechanism. *Virus Res* **139:** 193–208.
46. **Licznar P, Bertrand C, Canal I, Prere MF, Fayet O.** 2003. Genetic variability of the frameshift region in IS911 transposable elements from *Escherichia coli* clinical isolates. *FEMS Microbiol Lett* **218:**231–237.
47. **Reif HJ, Saedler H.** 1974. IS1 is involved in deletion formation in the gal region of *E. coli* K12. *Mol Gen Genet* **137:**17–28.
48. **Kretschmer PJ, Cohen SN.** 1979. Effect of temperature on translocation frequency of the Tn3 element. *J Bacteriol* **139:**515–519.
49. **Cornelis G.** 1980. Transposition of Tn951 (Tnlac) and cointegrate formation are thermosensitive processes. *J Gen Microbiol* **117:**243–247.
50. **Gueguen E, Rousseau P, Duval-Valentin G, Chandler M.** 2006. Truncated forms of IS911 transposase downregulate transposition. *Mol Microbiol* **62:**1102–1116.
51. **Nagy Z, Chandler M.** 2004. Regulation of transposition in bacteria. *Res Microbiol* **155:**387–398.
52. **Derbyshire KM, Kramer M, Grindley ND.** 1990. Role of instability in the cis action of the insertion sequence IS903 transposase. *Proc Natl Acad Sci USA* **87:**4048–4052.
53. **Jain C, Kleckner N.** 1993. IS10 mRNA stability and steady state levels in *Escherichia coli*: indirect effects of translation and role of rne function. *Mol Microbiol* **9:** 233–247.
54. **Jain C, Kleckner N.** 1993. Preferential cis action of IS10 transposase depends upon its mode of synthesis. *Mol Microbiol* **9:**249–260.
55. **Grindley ND, Joyce CM.** 1981. Analysis of the structure and function of the kanamycin- resistance transposon Tn903. *Cold Spring Harb Symp Quant Biol* **45**(Pt 1):125–133.
56. **Sasakawa C, Lowe JB, McDivitt L, Berg DE.** 1982. Control of transposon Tn5 transposition in Escherichia coli. *Proc Natl Acad Sci USA* **79:**7450–7454.
57. **Zerbib D, Jakowec M, Prentki P, Galas DJ, Chandler M.** 1987. Expression of proteins essential for IS1 transposition: specific binding of InsA to the ends of IS1. *EMBO J* **6:**3163–3169.
58. **Nagy Z, Szabó M, Chandler M, Olasz F.** 2004. Analysis of the N-terminal DNA binding domain of the IS30 transposase. *Mol Microbiol* **54:**478–488.
59. **Stalder R, Caspers P, Olasz F, Arber W.** 1990. The N-terminal domain of the insertion sequence 30 transposase interacts specifically with the terminal inverted repeats of the element. *J Biol Chem* **265:**3757–3762.
60. **Normand C, Duval-Valentin G, Haren L, Chandler M.** 2001. The terminal inverted repeats of IS911: requirements for synaptic complex assembly and activity. *J Mol Biol* **308:**853–871.
61. **Chalmers RM, Kleckner N.** 1994. Tn10/IS10 transposase purification, activation, and *in vitro* reaction. *J Biol Chem* **269:**8029–8035.
62. **Lewis LA, Astatke M, Umekubo PT, Alvi S, Saby R, Afrose J.** 2011. Soluble expression, purification and characterization of the full length IS2 Transposase. *Mob DNA* **2:**14.
63. **Dyda F, Chandler M, Hickman AB.** 2012. The emerging diversity of transpososome architectures. *Q Rev Biophys* **45:**493–521.
64. **Gueguen E, Rousseau P, Duval-Valentin G, Chandler M.** 2005. The transpososome: control of transposition at the level of catalysis. *Trends Microbiol* **13:**543–549.
65. **Rousseau P, Tardin C, Tolou N, Salome L, Chandler M.** 2010. A model for the molecular organisation of the IS911 transpososome. *Mob DNA* **1:**16.
66. **Pouget N, Dennis C, Turlan C, Grigoriev M, Chandler M, Salome L.** 2004. Single-particle tracking for DNA tether length monitoring. *Nucleic Acids Res* **32:**e73.
67. **Pouget N, Turlan C, Destainville N, Salome L, Chandler M.** 2006. IS911 transpososome assembly as analysed by

tethered particle motion. *Nucleic Acids Res* **34**:4313–4323.

68. **Rousseau P, Loot C, Guynet C, Ah-Seng Y, Ton-Hoang B, Chandler M.** 2007. Control of IS911 target selection: how OrfA may ensure IS dispersion. *Mol Microbiol* **63**: 1701–1709.
69. **Rousseau P, Loot C, Turlan C, Nolivos S, Chandler M.** 2008. Bias between the Left and Right Inverted Repeats during IS911 Targeted Insertion. *J Bacteriol* **190**:6111–6118.
70. **Ton-Hoang B, Polard P, Chandler M.** 1998. Efficient transposition of IS911 circles *in vitro*. *EMBO J* **17**: 1169–1181.
71. **Polard P, Seroude L, Fayet O, Prere MF, Chandler M.** 1994. One-ended insertion of IS911. *J Bacteriol* **176**:1192–1196.
72. **Turlan C, Chandler M.** 1995. IS1-mediated intramolecular rearrangements: formation of excised transposon circles and replicative deletions. *EMBO J* **14**:5410–5421.
73. **Polard P, Chandler M.** 1995. An in vivo transposase-catalyzed single-stranded DNA circularization reaction. *Genes Dev* **9**:2846–2858.
74. **Sekine Y, Aihara K, Ohtsubo E.** 1999. Linearization and transposition of circular molecules of insertion sequence IS3. *J Mol Biol* **294**:21–34.
75. **Sekine Y, Izumi K, Mizuno T, Ohtsubo E.** 1997. Inhibition of transpositional recombination by OrfA and OrfB proteins encoded by insertion sequence IS3. *Genes Cell* **2**: 547–557.
76. **Haren L.** 1998. *Relations Structure/Fonction des Facteurs Protéiques Exprimés par l'Elément Transposable Bacterien IS911* Université Paul Sabatier, Toulouse, France.
77. **Polard P, Prere MF, Fayet O, Chandler M.** 1992. Transposase-induced excision and circularization of the bacterial insertion sequence IS911. *EMBO J* **11**:5079–5090.
78. **Duval-Valentin G, Marty-Cointin B, Chandler M.** 2004. Requirement of IS911 replication before integration defines a new bacterial transposition pathway. *EMBO J* **23**:3897–3906.
79. **Bierne H, Ehrlich SD, Michel B.** 1994. Flanking sequences affect replication arrest at the *Escherichia coli* terminator TerB *in vivo*. *J Bacteriol* **176**:4165–4167.
80. **Hill TM, Marians KJ.** 1990. *Escherichia coli* Tus protein acts to arrest the progression of DNA replication forks *in vitro*. *Proc Natl Acad Sci USA* **87**:2481–2485.
81. **Kuempel PL, Pelletier AJ, Hill TM.** 1989. Tus and the terminators: the arrest of replication in prokaryotes. *Cell* **59**:581–583.
82. **Neylon C, Kralicek AV, Hill TM, Dixon NE.** 2005. Replication termination in *Escherichia coli*: structure and antihelicase activity of the Tus-Ter complex. *Microbiol Mol Biol Rev* **69**:501–526.
83. **Burton BM, Baker TA.** 2003. Mu transpososome architecture ensures that unfolding by ClpX or proteolysis by ClpXP remodels but does not destroy the complex. *Chem Biol* **10**:463–472.
84. **Nakai H, Doseeva V, Jones JM.** 2001. Handoff from recombinase to replisome: Insights from transposition. *Proc Natl Acad Sci USA* **98**:8247–8254.
85. **Ton-Hoang B, Bétermier M, Polard P, Chandler M.** 1997. Assembly of a strong promoter following IS911 circularization and the role of circles in transposition. *EMBO J* **16**:3357–3371.
86. **Lewis LA, Cylin E, Lee HK, Saby R, Wong W, Grindley ND.** 2004. The left end of IS2: a compromise between transpositional activity and an essential promoter function that regulates the transposition pathway. *J Bacteriol* **186**:858–865.
87. **Ton-Hoang B, Polard P, Haren L, Turlan C, Chandler M.** 1999. IS911 transposon circles give rise to linear forms that can undergo integration *in vitro*. *Mol Microbiol* **32**: 617–627.
88. **Sekine Y, Eisaki N, Ohtsubo E.** 1996. Identification and characterization of the linear IS3 molecules generated by staggered breaks. *J Biol Chem* **271**:197–202.
89. **Turlan C, Loot C, Chandler M.** 2004. IS911 partial transposition products and their processing by the *Escherichia coli* RecG helicase. *Mol Microbiol* **53**:1021–1033.

Mobile DNA, 3rd Edition
Nancy L. Craig, Michael Chandler, Martin Gellert, Alan M. Lambowitz, Phoebe A. Rice and Suzanne Sandmeyer

doi:10.1128/microbiolspec.MDNA3-0039-2014

S. He,[1] A. Corneloup,[1] C. Guynet,[1] L. Lavatine,[1] A. Caumont-Sarcos,[1] P. Siguier,[1] B. Marty,[1] F. Dyda,[2] M. Chandler,[1] and B. Ton Hoang[1]

28

The IS*200*/IS*605* Family and "Peel and Paste" Single-strand Transposition Mechanism

INTRODUCTION

Members of the widespread IS*200*/IS*605* bacterial insertion sequence (IS) family transpose using obligatory single-strand (ss) DNA intermediates. This distinguishes them from classical IS, which move via double-strand (ds) DNA intermediates (see Siguier et al., this volume). Members of this family also differ fundamentally from classic IS in their organization. They carry subterminal palindromic structures instead of inverted repeats at their ends (Figure 1A) and insert 3′ to specific AT-rich tetra- or penta-nucleotides without duplicating the target site. Importantly, the transposase, TnpA, does not share characteristics of the "DDE" enzymes of classical IS. It is a member of the "HuH" superfamily of enzymes including relaxases, Rep proteins of rolling circle replication (RCR) plasmids/single-stranded phages, bacterial and eukaryotic transposases of IS*91*/ISC*R*, and helitrons (see Thomas and Pritham, this volume) (1), which all catalyze cleavage and rejoining of ssDNA substrates. IS*200*, the founding member, was identified 30 years ago in *Salmonella typhimurium* (2) but there has been renewed interest for these elements since the identification of the IS*605* group in *Helicobacter pylori* (3, 4). Studies of two elements of this group, IS*608* from *H. pylori* and IS*Dra2* from the radiation-resistant *Deinococcus radiodurans*, have provided a detailed picture of their mobility (5–10).

This chapter presents an analysis of the organization and distribution of this IS family and summarizes studies of IS*608* (from *H. pylori*) and of IS*Dra2* (from *D. radiodurans*) as experimental model systems. It also addresses the probable domestication of IS*200*/IS*605* family transposases as enzymes involved in multiplication of chromosomal repeated extragenic palindromes (REP) and as potential homing endonucleases in Intron–IS chimeras (IStrons).

DISTRIBUTION AND ORGANIZATION

The family is widely distributed in prokaryotes: more than 153 distinct members have been identified so far. Of these, 89 are distributed over 45 genera and 61 species of eubacteria, and 64 are from archaea (ISfinder, http://www-is.biotoul.fr). The family can be divided

[1]Laboratoire de Microbiologie et Génétique Moléculaires, CNRS, Toulouse, France; [2]Laboratory of Molecular Biology, National Institute of Diabetes and Digestive and Kidney Diseases, NIH, Bethesda, MD.

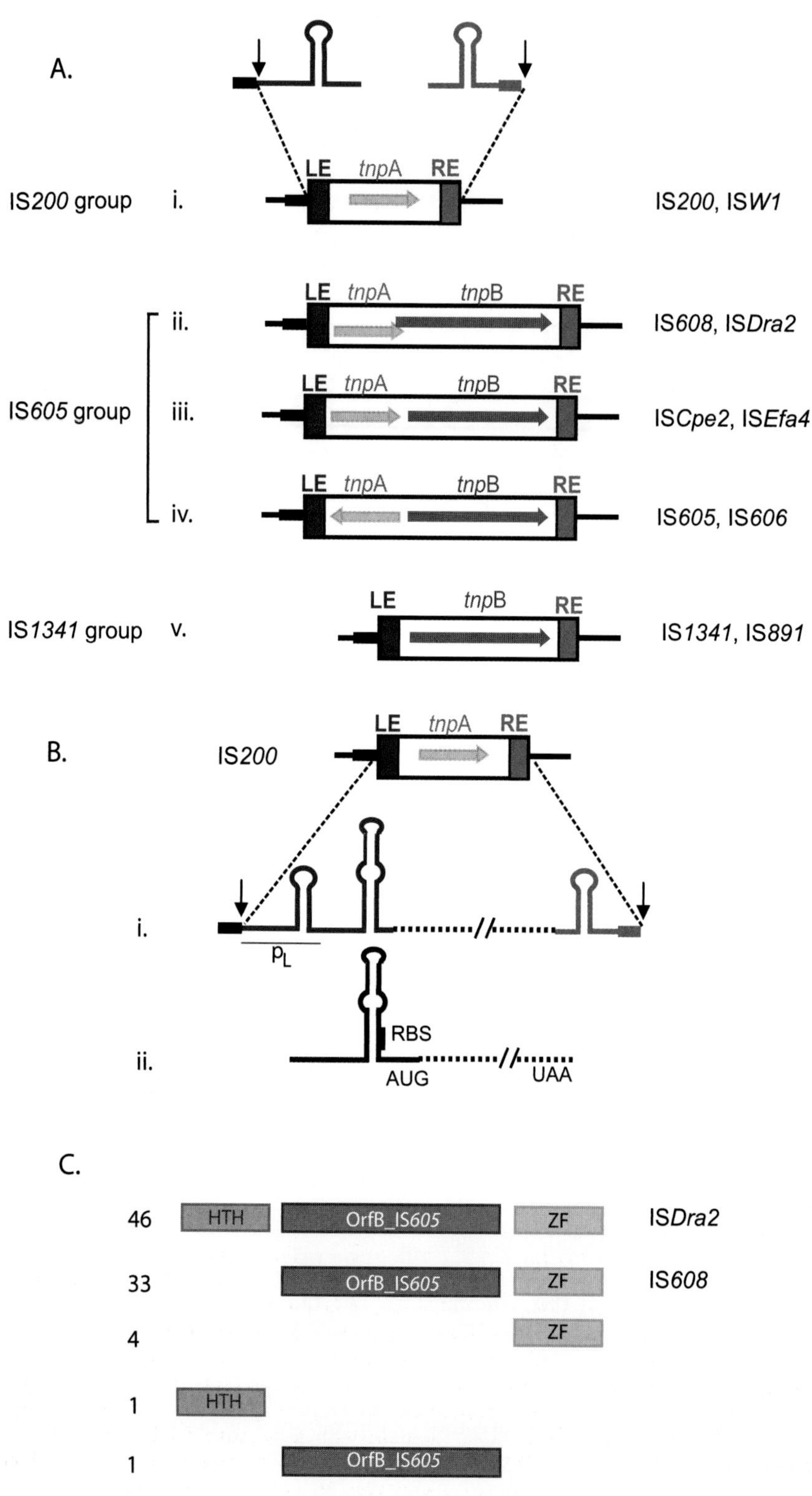
A.
LE
tnpA
RE
IS200 group
i.
IS200, ISW1
LE
tnpA
tnpB
RE
ii.
IS608, ISDra2
LE
tnpA
tnpB
RE
IS605 group
iii.
ISCpe2, ISEfa4
LE
tnpA
tnpB
RE
iv.
IS605, IS606
LE
tnpB
RE
IS1341 group
v.
IS1341, IS891
B.
IS200
LE
tnpA
RE
i.
pL
RBS
ii.
AUG
UAA
C.
46
HTH
OrfB_IS605
ZF
ISDra2
33
OrfB_IS605
ZF
IS608
4
ZF
1
HTH
1
OrfB_IS605

into three major groups based on the presence or absence and on the configuration of two genes: the transposase *tnpA*, sufficient to promote IS mobility *in vivo* and *in vitro*, and a second gene of unknown function, *tnpB*, which is not required for transposition activity (Figure 1A). These groups are: IS*200*, IS*605*, and IS*1341*.

The IS*200* Group

IS*200* group members encode only *tnpA*, and are present in Gram-positive and Gram-negative eubacteria and certain archaea (Figure 1Ai) (11, 12). Alignment of TnpA from various members shows that they are highly conserved but may carry short C-terminal tails of variable length and sequence. They can occur in relatively high copy number (e.g., > 50 copies of IS*1541* in *Yersinia pestis*) and are among the smallest known autonomous IS with lengths generally between 600 and 700 base pairs. Some members such as IS*W1* (from *Wolbachia* sp.) or IS*Prp13* (from *Photobacterium profundum*) are even shorter.

IS*200* (Figure 1B), the founding member of the family, was initially identified as an insertion mutation in the *Salmonella typhimurium* histidine operon (2). It is abundant in different *Salmonella* strains and has now also been identified in a variety of other enterobacteria such as *Escherichia*, *Shigella*, and *Yersinia* (see ISfinder). Different enterobacterial IS*200* copies have almost identical lengths of between 707 and 711 base pairs. Analysis of the ECOR (*Escherichia coli*) and SARA (*Salmonellae*) collections showed that the level of sequence divergence between IS*200* copies from these hosts is equivalent to that observed for chromosomally encoded genes from the same taxa (13). This suggests that IS*200* was present in the common ancestor of *E. coli* and *Salmonellae*.

In spite of their abundance, an enigma of IS*200* behavior has been its poor contribution to spontaneous mutation in its original *Salmonella* host: only very rare insertion events have been documented (12). One reason for these rare insertions could be the poor expression of the $tnpA_{IS200}$ gene from a weak promoter pL identified at the left IS end (LE) (Figure 1Bi) (14, 15).

Besides the characteristic major subterminal palindromes (14), presumed binding sites of the transposase at both LE and right end (RE) (see *Substrate recognition* section below), IS*200* also carries a potential supplementary interior stem–loop structure (Figure 1Bi). These two structures play a role in regulating IS*200* gene expression. The first (perfect palindrome at LE) overlaps the $tnpA_{IS200}$ promoter pL, can act as a bidirectional transcription terminator upstream of $tnpA_{IS200}$ and terminates up to 80% of transcripts (16) (Figure 1Bi). The second (interior stem–loop) (Figure 1Bi), at the RNA level can repress mRNA translation by sequestration of the ribosome binding site (Figure 1Bii). Experimental data suggested that the stem–loop is indeed formed *in vivo* and its removal by mutagenesis caused up to a 10-fold increase in protein production (16). Recent deep sequencing analysis has revealed another aspect in post-transcriptional regulation of IS*200* expression. A small anti-sense RNA to IS*200* transcript expression has been identified as a substrate of Hfq, an RNA chaperone involved in post-transcriptional regulation in numerous bacteria (17). Interestingly, anti-sense RNA and Hfq independently inhibit IS*200* transposase expression: knockout of both components resulted in a synergistic increase in transposase expression. Moreover, footprint data showed that Hfq binds directly to the 5′ part of the transposase transcript and blocks access to the ribosome binding site (Ellis MJ, Trussler RS and Haniford DB, personal communication).

Figure 1 Organization of IS*200*/IS*605* family. (A) Genetic organization. Left (LE) and right (RE) ends carrying the subterminal hairpin (HP) are presented as red and blue boxes, respectively (colour code retained throughout). Left and right cleavage sites (C_L and C_R) are presented as black and blue boxes respectively, where the black box also represents element-specific tetra-/pentanucleotide target site (T_S). The cleavage positions are indicated by small vertical arrows. Gray arrows: *tnpA* and *tnpB* open reading frames (*orfs*); (i) IS*200* group with *tnpA* alone; (ii) to (iv) IS*605* group with *tnpA* and *tnpB* in different configurations; (v) IS*1341* group with *tnpB* alone. (B) IS*200*. Secondary structures in the LE, adapted from reference 16: promoter (p_L), ribosome binding site (RBS), *tnpA* start and stop codons (AUG and UAA) are indicated. (i) DNA top strand with perfect palindromes at LE and RE in red and blue, interior stem–loop in black, (ii) RNA stem–loop structure in transcript originated from p_L. (C) Organization of TnpB protein and derivatives: putative N-terminal helix-turn helix motif (HTH), central OrfB_IS*605* domain with a putative DDE motif (Pfam) and C-terminal zinc finger motif (ZF) are shown. Numbers represent occurrence of corresponding variants among 85 analyzed sequences: 46 carry all the three domains (e.g., IS*Dra2*), 33 lack HTH motif (e.g., IS*608*), whereas others retain separate domains.
doi:10.1128/microbiolspec.MDNA3-0039-2014.f1

In spite of its very low transposition activity, an increase in IS*200* copy number was nevertheless observed during strain storage in stab cultures (2). However, the factors triggering this activity remain unknown (12). Transient high transposase expression leading to a burst of transposition was proposed to explain the observed high IS*200* (> 20) copy number in various hosts and in stab cultures (2).

Although regulatory structures similar to that observed in IS*200* (Figure 1B) were predicted in IS*1541*, another member of this group with 85% identity to IS*200*, this element can be detected in higher copy number (> 50) in *Salmonella* and *Yersinia* genomes. However, no detailed analysis of its transposition is available and since no *de novo* insertions have been experimentally documented and chromosomal copies appear stable in *Y. pestis* (18), it remains possible that IS*1541* also behaves like IS*200*. However, the regulatory structures are not systematically present in other IS*200* group members and understanding of control of transposase synthesis requires further study.

The IS*605* Group

IS*605* group members are generally longer (1.6–1.8 kb) due to the presence of a second open reading frame (*orf*), *tnpB* in addition to *tnpA*. Alignment of TnpA copies from this group indicated that although they do not form a separate clade from the IS*200* group TnpA, they generally carry the short C-terminal tail. The *tnpA* and *tnpB* *orf*s exhibit various configurations with respect to each other. They may be divergent (Figure 1Aiv: e.g., IS*605*, IS*606*) or expressed in the same direction with *tnpA* upstream of *tnpB*. In these latter cases, the *orf*s may be partially overlapping (Figure 1Aii: e.g., IS*608*, IS*Dra2*) or separate (Figure 1Aiii: e.g., IS*SCpe2*, IS*Efa4*). *tnpB* is also sometimes associated with another transposase that is a member of the S-transposases (e.g., IS*607*) (19, 20) (see Siguier et al., this volume). TnpB was not required for transposition of either IS*608* or IS*Dra2*.

Three related IS, IS*605*, IS*606*, and IS*608* (Figure 1A) have been identified in numerous strains of the gastric pathogen *H. pylori* (3, 4) and IS*605* appears to be involved in genomic rearrangements in various *H. pylori* isolates (21). The *H. pylori* elements transpose in *E. coli* at detectable frequencies in a standard “mating-out” assay using a derivative of the conjugative F plasmid as a target (3, 4).

The two best characterized members of this family are IS*608* and the closely related IS*Dra2* from *D. radiodurans*. Both have overlapping *tnpA* and *tnpB* genes (Figure 1Aii). Like other family members, insertion is sequence-specific: IS*608* inserts in a specific orientation with its left end 3′ to the tetranucleotide TTAC both *in vivo* and *in vitro* (3) whereas IS*Dra2* inserts 3′ to the pentanucleotide TTGAT (22). Interestingly IS*Dra2* transposition in its highly radiation resistant deinococcal host is strongly induced by irradiation (23) (see *Single strand DNAin vivo*). Their detailed transposition pathway has been deciphered by a combination of *in vivo* studies and *in vitro* biochemical and structural approaches (see also *Mechanism of IS200/IS605 single-strand DNA transposition*).

The IS*1341* Group

Elements of the third group, IS*1341*, are devoid of *tnpA* and carry only *tnpB* (Figure 1Av). The IS occurs in three copies in thermophilic bacterium PS3 (24). Multiple presumed full-length elements (including *tnpA* and *tnpB*) and closely related copies have been identified in other bacteria such as *Geobacillus* (see ISfinder). On the other hand, IS*891* from the cyanobacterium *Anabaena* is present in multiple copies on the chromosome. Moreover, it is thought to be mobile because a copy was observed to have inserted into a plasmid introduced into the strain (25). Another isolated *tnpB*-related gene, *gipA*, present in the *Salmonella* Gifsy-1 prophage was considered as a virulence factor because a *gipA* null mutation compromised *Salmonella* survival in a Peyer’s patch assay (26). Although no mobility function has been suggested for *gipA*, it is indeed bordered by structures characteristic of IS*200*/IS*605* family ends and closely related to *E. coli* IS*Ec42* (see ISfinder).

In spite of their presence in multiple copies, it is still unclear whether IS*1341* group members are autonomous IS or products of IS*605* group degradation and require TnpA supplied from a related IS in the same cell for transposition. Further work is clearly necessary to assess the transposition activity of this group.

IS Decay

Circumstantial evidence based on analysis of the IS database suggests that IS carrying both *tnpA* and *tnpB* genes may be unstable. Hence, although members of the IS*200* group are often present in high copy number in their host genomes, intact full-length IS*605* group members are invariably found in low copy number (Siguier P, unpublished) (see also *TnpB* section below). On the other hand, various truncated IS*605* group derivatives appear quite frequently. These forms seem to result from successive internal deletions and retain intact LE and RE copies. Sometimes, as in the case of IS*Soc3*, *orf* inactivation appears to have occurred by successive insertion/deletion of short sequences (indels) generating frameshifts and truncated proteins. For some IS (e.g., IS*Cco1*, IS*Tel2*, IS*Cysp14*, IS*Soc3*) degradation can be precisely reconstituted and each successive step can be validated by the presence of several identical copies (Siguier P, unpublished). This suggests that the degradation process is recent and that these derivatives are likely mobilized by TnpA supplied *in trans* by autonomous copies in the genome.

MECHANISM OF IS*200*/IS*605* SINGLE-STRAND DNA TRANSPOSITION

General Transposition Pathway

The transposition pathway of IS*200*/IS*605* family members is shown in Figure 2. Much of the biochemistry was elucidated using an IS*608* cell-free *in vitro* system, which recapitulates each step of the reaction. This requires purified $TnpA_{IS608}$ protein, single-strand IS*608* DNA substrates and divalent metal ions such as Mg^{2+} or Mn^{2+} (5, 7, 8). Similar and complementary results were also obtained with IS*Dra2* (9, 10, 27). The reactions are not only strictly dependent on ssDNA substrates but are also strand-specific: only the "top" strand (defined as the strand carrying the target sequence, T_S, 5′ to the IS; Figure 2A) is recognized and processed whereas the "bottom" strand is refractory (5, 7). Cleavage of the top strand at the left and right cleavage sites (T_S/C_L and C_R, note that T_S is also the left cleavage site C_L) (Figure 2B) leads to excision as a circular ssDNA intermediate with abutted left and right ends (transposon joint) (Figure 2C). This is accompanied by rejoining of the DNA originally flanking the excised strand (donor joint) (Figure 2C).

The transposon joint is then cleaved (Figure 2E) and integrated into a single-strand conserved element-specific target sequence (T_S) where the LE invariably inserts 3′ to T_S (Figure 2F). This target specificity is another unusual feature of IS*200*/IS*605* transposition. The target sequence is characteristic of the particular family member and, although it is not part of the IS, it is essential for further transposition because it is also the left end cleavage site C_L of the inserted IS (5) (see also *Cleavage site recognition* section below) and is therefore intimately involved in the transposition mechanism.

TnpA, Y1 Transposases, and Transposition Chemistry

IS*200*/IS*605* family transposases belong to the HuH enzyme superfamily. All contain a conserved amino-acid triad composed of Histidine (H)-bulky hydrophobic residue (u)-Histidine (H) (28) providing two of three ligands required for coordination of a divalent metal ion that localizes and prepares the scissile phosphate for nucleophilic attack. HuH proteins catalyze ssDNA breakage and joining with a unique mechanism. They all catalyze DNA strand cleavage using a transitory covalent 5′ phosphotyrosine enzyme–substrate intermediate and release a 3′ OH group (1).

The HuH enzyme family also includes other transposases of the IS*91*/IS*CR* and helitron families as well as proteins involved in DNA transactions essential for plasmid/virus rolling circle replication (Rep; not to be confused with the $TnpA_{REP}$/REP system described in the *Y1 transposase domestication* section below) and plasmid conjugation (Mob/relaxase).

IS*200*/IS*605* transposases are single-domain proteins containing a single catalytic tyrosine residue, called therefore Y1 transposase (see below). They use the tyrosine residue (Y127 for IS*608*) as a nucleophile to attack the phosphodiester link at the cleavage sites (vertical arrows in Figure 2A, D). Since cleavages at both IS ends occur on the same strand, the polarity of the reaction implies that the enzyme forms a covalent 5′-phosphotyrosine bond with the IS at LE producing a 3′-OH on the DNA flank and a 5′-phosphotyrosine bond at the RE flank producing a 3′-OH on RE itself (Figure 2B). The released 3′-OH groups then act as nucleophiles to attack the appropriate phosphotyrosine bond, resealing the DNA backbone in one case and generating an ssDNA transposon circle in the other (Figure 2C). The same polarity is applied to the integration step (Figure 2D–F). As an important mechanistic consequence of this chemistry, IS*200*/IS*605* transposition occurs without loss or gain of nucleotides. *In vitro*, the reaction requires only TnpA and does not require host cell factors.

TnpA Overall Structure

Crystal structures of Y1 transposases have been determined for three family members: IS*608* ($TnpA_{IS608}$) from *H. pylori* (6, 8), IS*Dra2* ($TnpA_{ISDra2}$) from *D. radiodurans* (10) and IS*C1474* from *Sulfolobus solfataricus* (29). In contrast to most characterized HuH enzymes, which are usually monomeric and have two catalytic tyrosines, Y1 transposases form obligatory dimers with two active sites (Figure 3A). In these structures, the two monomers dimerize by merging their β-sheets into one large central β-sheet sandwiched between α-helices. Each catalytic site is constituted by the HuH motif from one TnpA monomer (H64 and H66 in the case of $TnpA_{IS608}$) and a catalytic tyrosine residue (Y127)

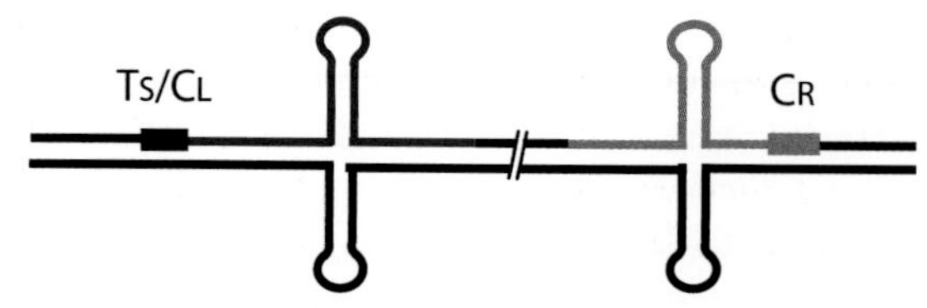
Ts/CL
CR

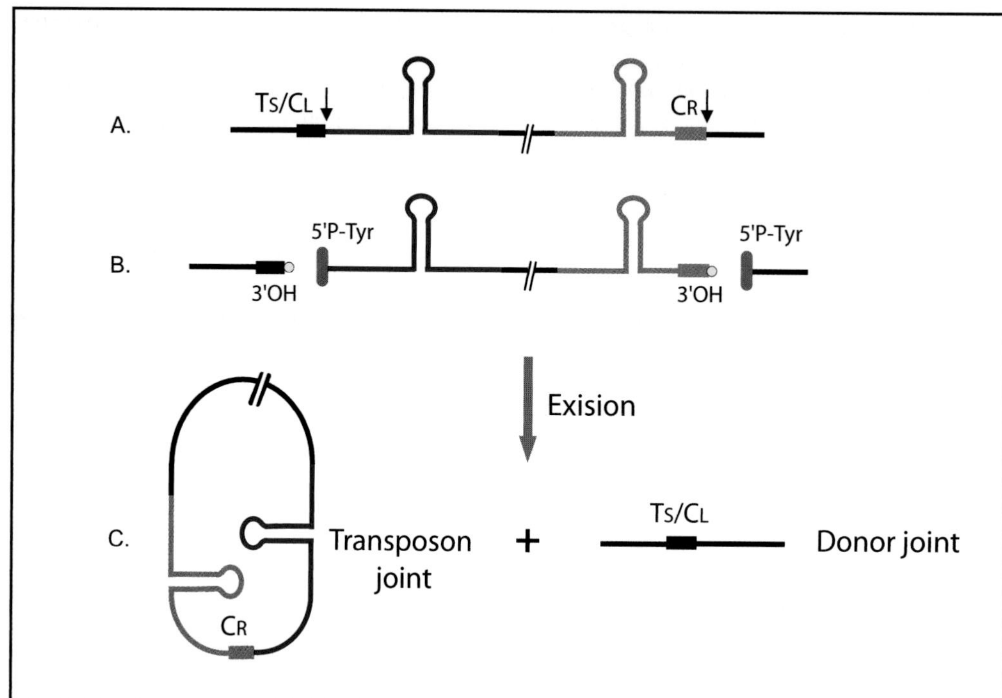
A.
Ts/CL
CR
B.
5'P-Tyr
3'OH
5'P-Tyr
3'OH
Exision
C.
Transposon joint
+
Ts/CL
Donor joint
CR

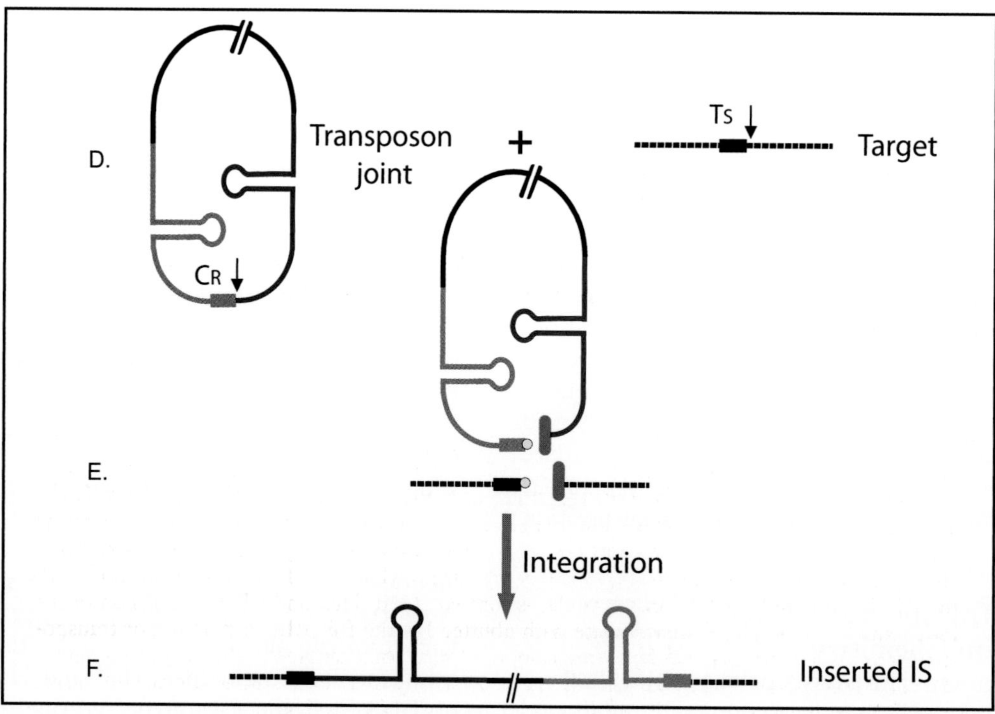
D.
Transposon joint
+
Ts
Target
CR
E.
Integration
F.
Inserted IS

located in the C-terminal αD helix tail of the other monomer (Figure 3A, C, D). This is joined to the body of the protein by a flexible loop (*trans* configuration, see the *Active site assembly and catalytic activation* and *Transposition cycle: the trans/cis rotational model* sections below).

The active sites of these TnpA enzymes are believed to adopt two functionally important conformations: the *trans* configuration described above (Figure 3A, C, D), in which each active site is composed of the HuH motif supplied by one monomer with the tyrosine residue supplied by the other, and the *cis* configuration, in which both motifs are contributed by the same monomer. The *trans* conformation is active during cleavage where tyrosine acts as nucleophile, whereas the *cis* conformation is thought to function during strand transfer where the 3′-OH is the attacking nucleophile (see *Transposition cycle: the trans/cis rotational model* section below). Only the *trans* configuration of $TnpA_{IS608}$ and $TnpA_{ISDra2}$ has been observed crystallographically (6, 8, 10) but the existence of the *cis* configuration is supported by biochemical data (30).

Single-Strand Transpososome

The key machinery for transposition is the higher order protein–DNA complex, the transpososome (or synaptic complex), which contains both transposase and two IS DNA ends with or without target DNA. Transpososome formation, stability and the temporal changes in configuration that occur during the transposition cycle have been relatively well characterized for $TnpA_{IS608}$ by crystallographic and biochemical approaches. Although for technical reasons it was not possible to obtain structures with both LE and RE hairpins together, co-crystal structures with either LE or RE showed that a TnpA dimer binds two subterminal DNA hairpins, suggesting that it could bind both LE and RE ends simultaneously. Binding sites for the hairpins are located on the same face of the TnpA dimer whereas the two catalytic sites are formed on the opposite surface (Figure 3A, B).

Substrate Recognition

A key feature of TnpA is that it is only active on one strand, the "top" strand. The IS*608* and IS*Dra2* ends carry subterminal imperfect hairpins. In addition to specific sequences on the loops, the irregularities on the hairpins help the enzyme to distinguish between "top" and "bottom" strands (6, 10) as described in detail below. The initial co-crystal structure was obtained with $TnpA_{IS608}$ and a 22-nucleotide (nt) imperfect RE hairpin (HP22) including its characteristic extrahelical T17 located mid-way along the DNA stem (Figures 3B, and 4A, B). In addition to a number of backbone contacts with HP22, $TnpA_{IS608}$ also shows several base-specific contacts, in particular with T10 in the loop and the extrahelical T17 (6) (Figures 3B and 4A, B). Exchange of T10 and neighboring T nucleotides in the loop abolished binding whereas exchange of T17 for an A significantly reduced but did not eliminate binding (31). Similar studies with $TnpA_{ISDra2}$ showed that it also recognizes a similarly located T in the hairpin loop of IS*Dra2* and that this is essential for binding (10). Instead of an extrahelical T, IS*Dra2* LE and RE include a bulge caused by two mismatched nucleotides (G and T) in the hairpin stem. These unpaired nucleotides are specifically recognized and stabilized by the protein. Again, mutation of the T (to C which, in this case, eliminates the bulge to generate a GC base pair in the stem) greatly reduces binding.

Although most members of the IS*605* group, which includes IS*608* and IS*Dra2*, have imperfect palindromes with extrahelical bases or bulges, some members of the IS*200* group (e.g., IS*200*, IS*1541*) include perfect hairpins. Whether base-specific interactions with the loop sequence are exclusively responsible for strand-specific activity of the corresponding transposase remains to be clarified.

Cleavage Site Recognition

The left (C_L/T_S) and right (C_R) IS*608* cleavage sites (TTAC[1] and TCAA[1] respectively, where [1] represents the point of cleavage) are located at some distance from the

Figure 2 IS*608* and transposition cycle. IS*608* organization. The left (LE) and right (RE) ends with subterminal hairpin (HP) are in red and blue, left and right cleavage sites (C_L/T_S and C_R) are represented by black and blue boxes, respectively. Excision. (A) TnpA activity: top strand (active strand) structures are recognized and cleaved by TnpA (vertical arrows). (B) Upon cleavage, a 5′ phosphotyrosine bond (green cylinder) is formed with LE and with the RE 3′ flank and 3′-OH (yellow circle) is formed at left flank and RE. (C) Excision of the IS*608* single-strand circle intermediate with abutted LE and RE (RE–LE junction or transposon joint) (C) accompanied by formation of donor joint retaining the target sequence. (D) The transposase catalyzes the cleavage of transposon joint and single-strand target (E) then integration (F). doi:10.1128/microbiolspec.MDNA3-0039-2014.f2

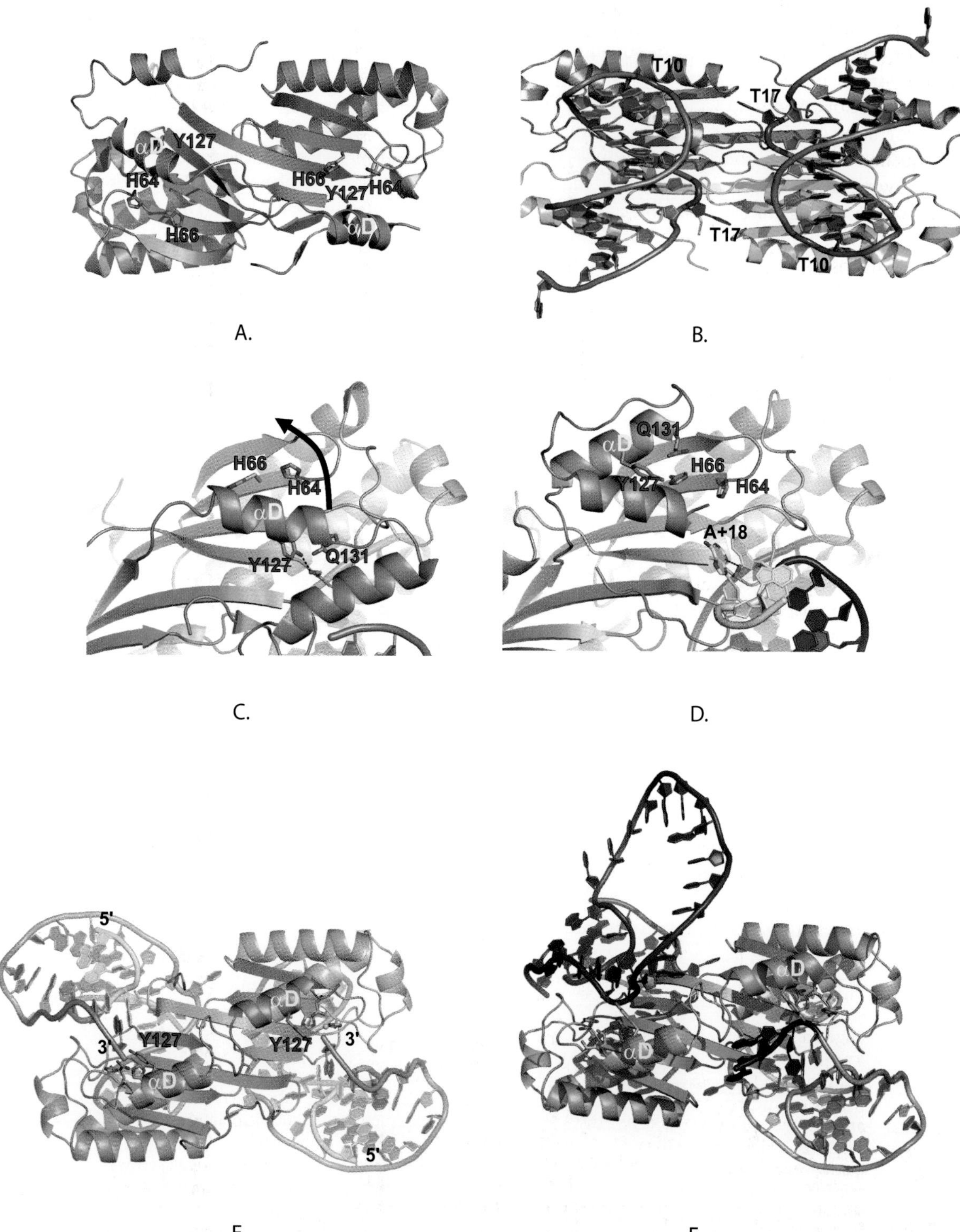
αD
Y127
H64
H66
H66
Y127
H64
αD
T10
T17
T17
T10
H66
H64
αD
Y127
Q131
αD
Q131
H66
Y127
H64
A+18
5'
αD
Y127
3'
3'
Y127
αD
5'
αD
αD

A.

B.

C.

D.

E.

F.

subterminal recognition hairpins (19 nt at LE and 10 nt at RE) (Figure 4A, B). The system is asymmetric because the two distinct cleavage sites are separated from the hairpins by linkers of different lengths and the C_L/T_S sequence does not form part of IS whereas C_R does.

Structural studies revealed that the cleavage sites are recognized in a unique way that does not involve direct sequence recognition by TnpA. Instead, an internal part of the IS sequence is co-opted to recognize different cleavage sites, allowing TnpA to catalyze both excision and integration of the element with a single DNA-binding domain.

Internal transposon sequences, the left (G_L) and right (G_R) tetranucleotide guide sequences, AAAG and GAAT, located 5′ to the foot of the hairpins (Figure 4A, B), recognize their respective cleavage sites by direct base interactions. These G_L/C_L and G_R/C_R interactions involve three of the four nucleotides of G_L and G_R. They include both canonical Watson–Crick interactions and, in the case of RE, noncanonical interactions resulting in base triplets (Figure 4B, C, bases joined by both regular and dotted lines, respectively) (8, 31). In the case of LE and the transposon joint, base triples (dotted lines) are suggested from biochemical data (31).

These interactions place the scissile phosphate precisely into the two active sites of $TnpA_{IS608}$ for nucleophilic attack by the catalytic Y127. Interestingly, the base-pairing patterns responsible for cleavage site recognition are similar at LE, RE, and the target site in spite of sequence differences (Figure 4A, B, D). As T_S is also C_L, this type of recognition not only explains the requirement for the T_S located at the left end of the inserted IS (Figures 2A and 4D) for further transposition, but also the target specificity. Upon integration, T_S is presumably recognized by the G_L present on the excised transposon joint. Note that the transposon joint contains only the LE guide sequence G_L but not the LE cleavage site C_L (Figures 2C and 4D).

Similar crystal structures were obtained with $TnpA_{ISDra2}$ (see also *Single strand DNA* in vivo section below) with a similar interaction network between the guide sequences and cleavage sites. The IS*Dra2* transpososome is structurally very similar to those of IS*608* despite only 34% sequence identity of the transposases. It is important to note that the target sequence in IS*Dra2* is a pentanucleotide instead of a tetranucleotide as in IS*608*. The fifth nucleotide in the IS*Dra2* sequence is, however, not involved in DNA–DNA interactions but in DNA–protein interactions (10).

The potential cleavage site recognition mode (i.e., the canonical interaction network between $C_{L,R}$ and $G_{L,R}$) is indeed well conserved throughout the family (Figure 4E, see also ISfinder).

This model has been validated *in vitro* and *in vivo* by showing that it is possible to modify cleavage sites by changing corresponding guide sequences. Moreover, in the case of IS*608*, modifications of G_L in the transposon joint generate predictable changes in insertion site specificity of the element (32).

Active Site Assembly and Catalytic Activation

Comparison of crystal structures of different TnpA protein–DNA complexes (6, 8, 29) revealed TnpA in both active and inactive configurations. In both the free $TnpA_{IS608}$ dimer and $TnpA_{IS608}$–DNA complexes bound to a "minimal" HP22 hairpin (which does not include the guide sequence), the catalytic tyrosine residue (Y127) points away from the HuH motif (H64 and H66) and therefore cannot act as a nucleophile (Figure 3A–C) (6). The enzyme is therefore in an inactive conformation. Binding to the appropriate substrate containing the 4-nt guide sequence 5′ to the hairpin foot (compare Figure 3C and D) triggers a change in TnpA configuration that permits assembly of functional active sites. A single A (A+18, Figure 3D) in the guide sequence present in both G_L and G_R does not

Figure 3 $TnpA_{IS608}$ structures (adapted from references 6 and 8). (A) Crystallographic structure of TnpA alone. The two monomers of the TnpA dimer are colored green and orange, respectively. Positions of helix αD and catalytic residues are shown. (B) Costructure TnpA–RE HP22. HP22 is shown in blue. The extrahelical T17 and the T located in the hairpin loop are indicated in red (6). Note that in the TnpA–HP22 costructure, binding sites for the hairpins are located on the same face of the TnpA dimer whereas the two catalytic sites are formed on the opposite surface (A, C–F). (C) Configuration of the active site in the TnpA–RE HP22. HP22 is shown in blue. Note that in A, B and C, TnpA is in the inactive conformation. The arrow shows the presumed rotation of the αD helix to activate the protein. (D) Configuration of the active site in the TnpA–LE HP26 costructure. LE HP26 is shown in red and the 5′ 4-nucleotide extension (G_L) in yellow). The base A+18 has displaced Y127 to activate the protein. (E) TnpA–RE35 complex. Interaction of G_R-C_R (in light and dark blue, respectively) positions the cleavage site within the catalytic site of the protein. (F) Modeled TnpA–LE–RE complex. LE, RE, and flanking sequences in red, blue, and black, respectively. doi:10.1128/microbiolspec.MDNA3-0039-2014.f3

A.

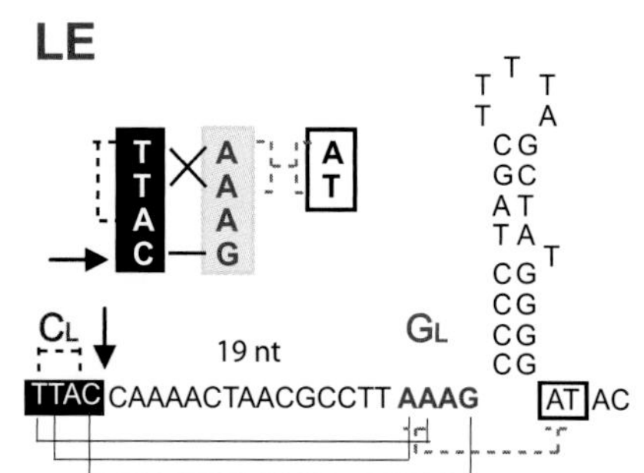
LE
TTAC
AAAG
AT
CL
19 nt
GL
TTAC CAAAACTAACGCCTT AAAG AT AC

B.

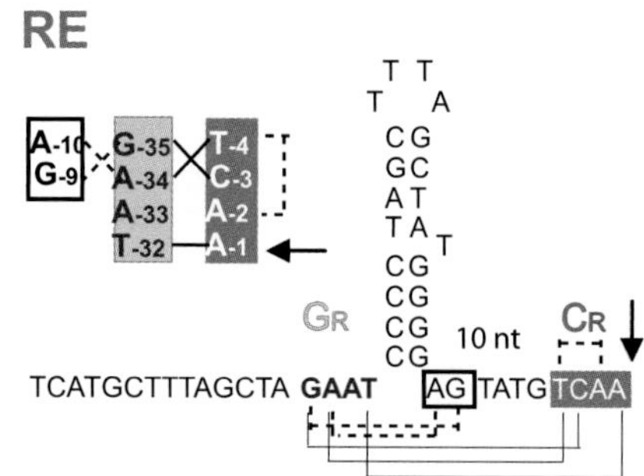
RE
A-10
G-9
G-35
A-34
A-33
T-32
T-4
C-3
A-2
A-1
GR
10 nt
CR
TCATGCTTTAGCTA GAAT AG TATG TCAA

C.

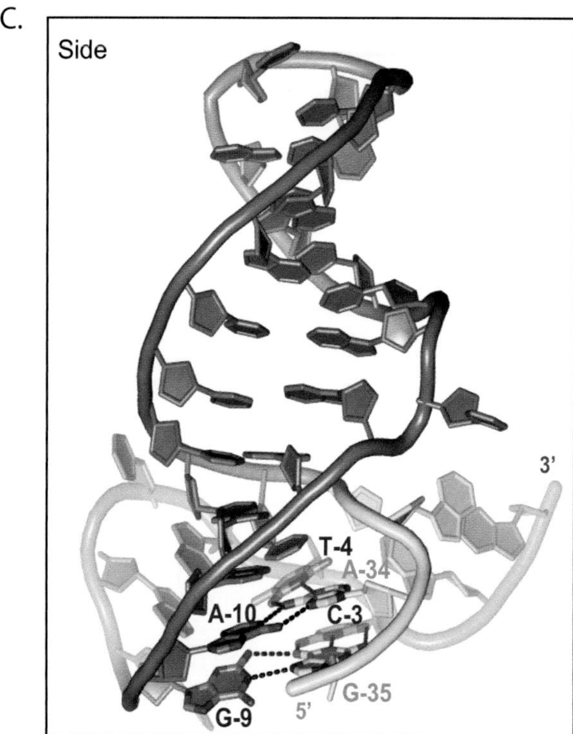
Side
3'
T-4
A-34
A-10
C-3
G-9
5'
G-35

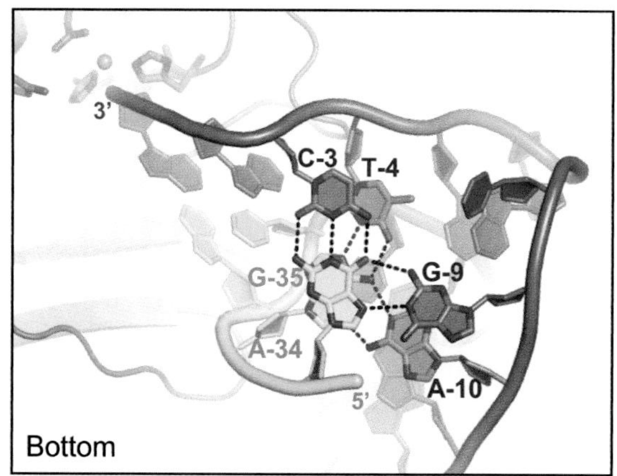
3'
C-3
T-4
G-35
G-9
A-34
5'
A-10
Bottom

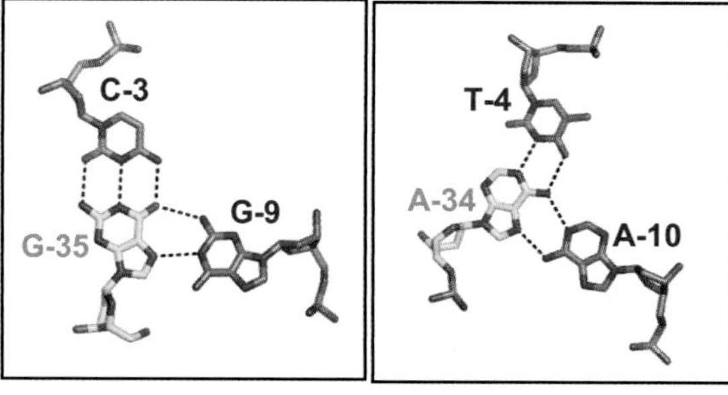
C-3
G-9
G-35
T-4
A-34
A-10

D.

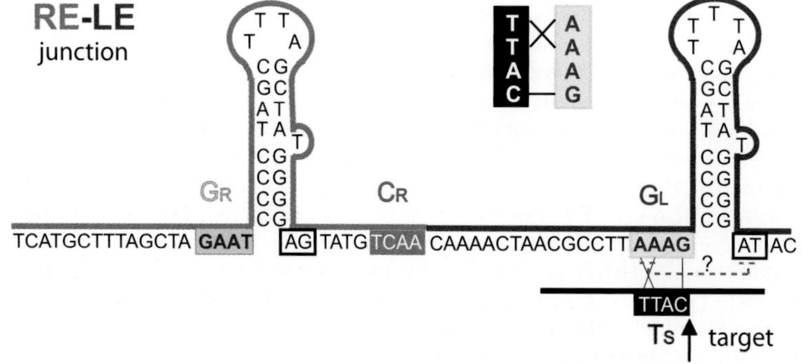
RE-LE
junction
TTAC
AAAG
GR
CR
GL
TCATGCTTTAGCTA GAAT AG TATG TCAA CAAAACTAACGCCTT AAAG AT AC
?
TTAC
Ts
target

E.

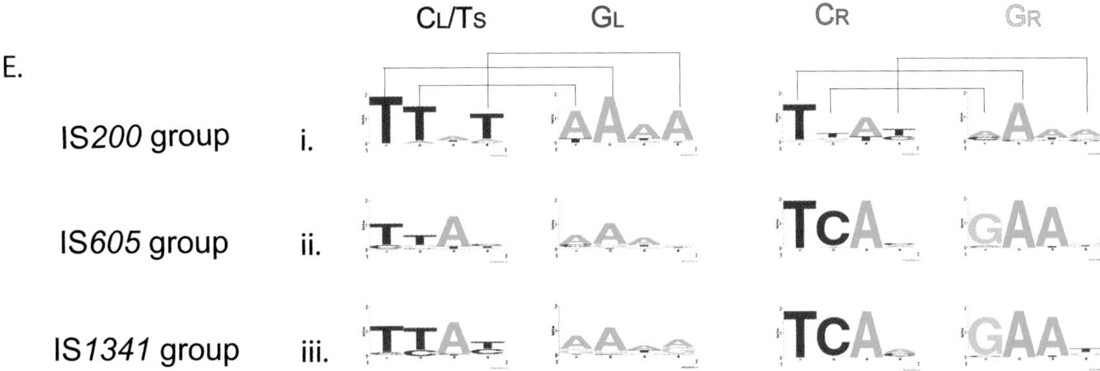

F.

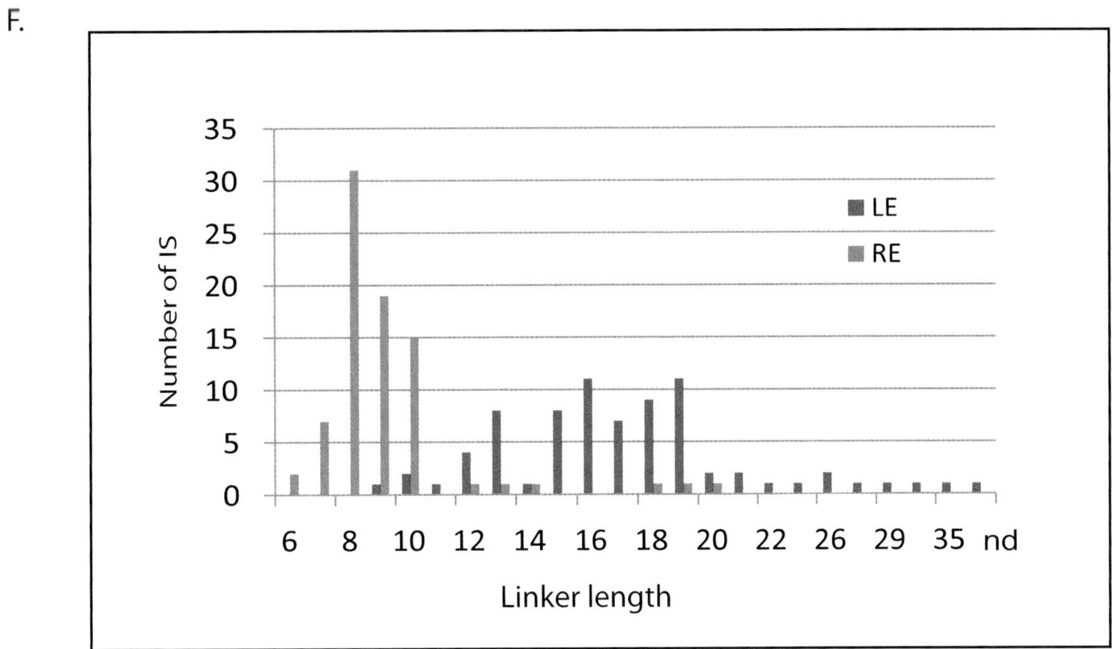

Figure 4 Recognition of cleavage sites. Schematic of the canonical and noncanonical base interactions in (A) left end (LE) and (B) right end (RE). The LE and RE are shown in red and blue. Cleavage sequences C_L or C_R are placed in black or dark blue boxes; guide sequences G_L and G_R are framed in pink and light blue, respectively. Two nucleotides at the 3′ foot of $HP_{L, R}$ involved in triplet formation are highlighted by bold and in black frame. Nucleotide sequences of LE and RE and the base paring within HP_L and HP_R are shown. The inset figures describe the interactions between the cleavage sequences and guide sequences. The filled lines indicate canonical base interactions and the dotted lines indicate additional noncanonical base interactions. (C) Structure of the co-complex $TnpA_{IS608}$–RE35 adapted from reference 8 showing the active site and the base pairs between C_R (TCAA, dark blue) and G_R (GAAT, light blue). The gray sphere is bound Mn^{2+}. Right: Two base triplets observed in the $TnpA_{IS608}$–RE35 complex. (D) Target recognition: single-strand transposon joint (RE–LE junction) and target Ts are presented. For simplicity only the recognition of the target cleavage site is indicated. (E) Cleavage sites recognition in the IS*200*/IS*605* family. Multiple sequence alignment of the cleavage sites and guide sequences using Weblogo (weblogo.berkeley.edu) was carried out on 38, 43 and 23 members of the IS*200* (i), the IS*605* (ii), and the IS*1341* (iii) groups, respectively. (F) Linker length distribution of LE and RE from 76 (red) and 80 (blue) different IS, respectively.
doi:10.1128/microbiolspec.MDNA3-0039-2014.f4

participate in base interactions with the cleavage site. On formation of the $C_{L(R)}/G_{L(R)}$ base interaction network, this single base penetrates the structure and forces the C-terminal αD helix carrying Y127 closer to the HuH motif, placing it in the correct position poised for catalysis (8) (Figure 3D). This movement also places a third amino acid (Q131 located at the C-terminal end of helix αD on the same face as Y127) in a position enabling it to function in conjunction with both H residues to complete the metal ion binding pocket. This movement is made possible by the fact that the αD helix is attached to the protein body by a flexible loop. This conformational change involving αD helix movement will be discussed below (see *Transposition cycle: the* trans/cis *rotational model* section below).

Transpososome Assembly and Stability

Excision requires assembly of a transpososome containing both LE and RE. However, it is technically difficult to generate crystallographically pure complexes of this type. Only crystal structures containing two LE or two RE were obtained. The excision transpososome was initially modeled using information obtained from the IS*608* LE–TnpA and RE–TnpA structures (Figure 3E) (8). However complexes containing both LE and RE have now been identified using a band shift assay and characterized biochemically (31).

A TnpA co-complex with either LE or RE can be titrated by addition of increasing quantities of the other end (RE or LE) to obtain a transpososome containing both LE and RE. This can be easily detected in a gel shift assay. Such species proved to be catalytically active because they could be removed from the gel and, when incubated with the essential divalent metal ion, robust reaction products could be detected in a denaturing gel (31).

This approach was used to monitor both transpososome formation and stability using oligonucleotides carrying point mutations in $G_{L,R}$ and $C_{L,R}$. Robust transpososome formation and cleavage activity requires much of the network of $G_{L,R}$ and $C_{L,R}$ interactions observed in the crystal structures (31) (schematized in Figure 4A, B). Although base triplets in the original LE cocrystal structure were not detected, because the LE substrate was too short (8), the biochemical data suggested that such interactions probably exist (gray dotted lines in Figure 4A). For example, the two nucleotides 3′ to the foot of the LE hairpin (at equivalent positions to triplet-forming bases in RE, Figure 4A) are required for robust synaptic complex formation and cleavage (31). This further implies that these base triplets might also be involved in target DNA capture (gray dotted lines in Figure 4D).

Base changes in G_L resulted in a predictable choice of target sequence (32). However, large differences in insertion frequencies were observed. The influence of the presumed noncanonical interactions in LE would provide an explanation for this variability because these were not taken into account in the choice of LE guide sequence.

In both IS*608* and IS*Dra2*, the extra-helical bases in the hairpin stem and nucleotides in the loop are also important for transpososome formation even in a context that includes both $G_{L,R}$ and $C_{L,R}$ (10, 31).

Transposition Cycle: The *trans/cis* Rotational Model

Transpososome assembly is followed by two critical chemical steps: cleavage and strand transfer. These are thought to be accomplished by a series of large changes in transpososome configuration. A detailed model has been proposed for the dynamics of the IS*608* transpososome during the transposition reactions (Figure 5) (8, 30). As described in the *TnpA overall structure* section (above), $TnpA_{IS608}$ could in principle assume two configurations: *trans* and *cis*. Switching between these two states would involve rotation of the two unconstrained flexible arms that join the αD helix to the protein body. The current model for IS*608* and IS*Dra2* transposition proposes that the strand transfer step involves rotation of these arms from the *trans* to the *cis* configuration: cleavage occurs while the enzyme is in the *trans* configuration. A *trans* to *cis* conformational change then occurs allowing strand transfer. The ground state of the IS*608* and IS*Dra2* transpososomes obtained from crystallography is the *trans* configuration. LE and RE binding and cleavage occur with the enzyme in its *trans* configuration (Figure 5B). This results in formation of the 5′ phosphotyrosine bond with LE liberating a 3′-OH on the flanking DNA and the 5′ phosphotyrosine bond with the RE DNA flank liberating a 3′-OH on the RE transposon end. Rotation of the two arms would displace LE towards the sequestered 3′-OH of RE and the RE flank towards the 3′-OH of the LE flank (Figure 5C) and position them so that both 3′-OH can attack the appropriate phosphodiester bond. This model is supported by several lines of indirect evidence from studies of IS*608*.

An initial piece of evidence concerns the length differences in the LE and RE "linker" (the distance between the hairpin foot and the cleavage site): this is only 10 nt for RE but 19 nt for LE (Figure 4A, B). The rotation model suggests that the longer LE linker may be required to provide sufficient length to rotate the

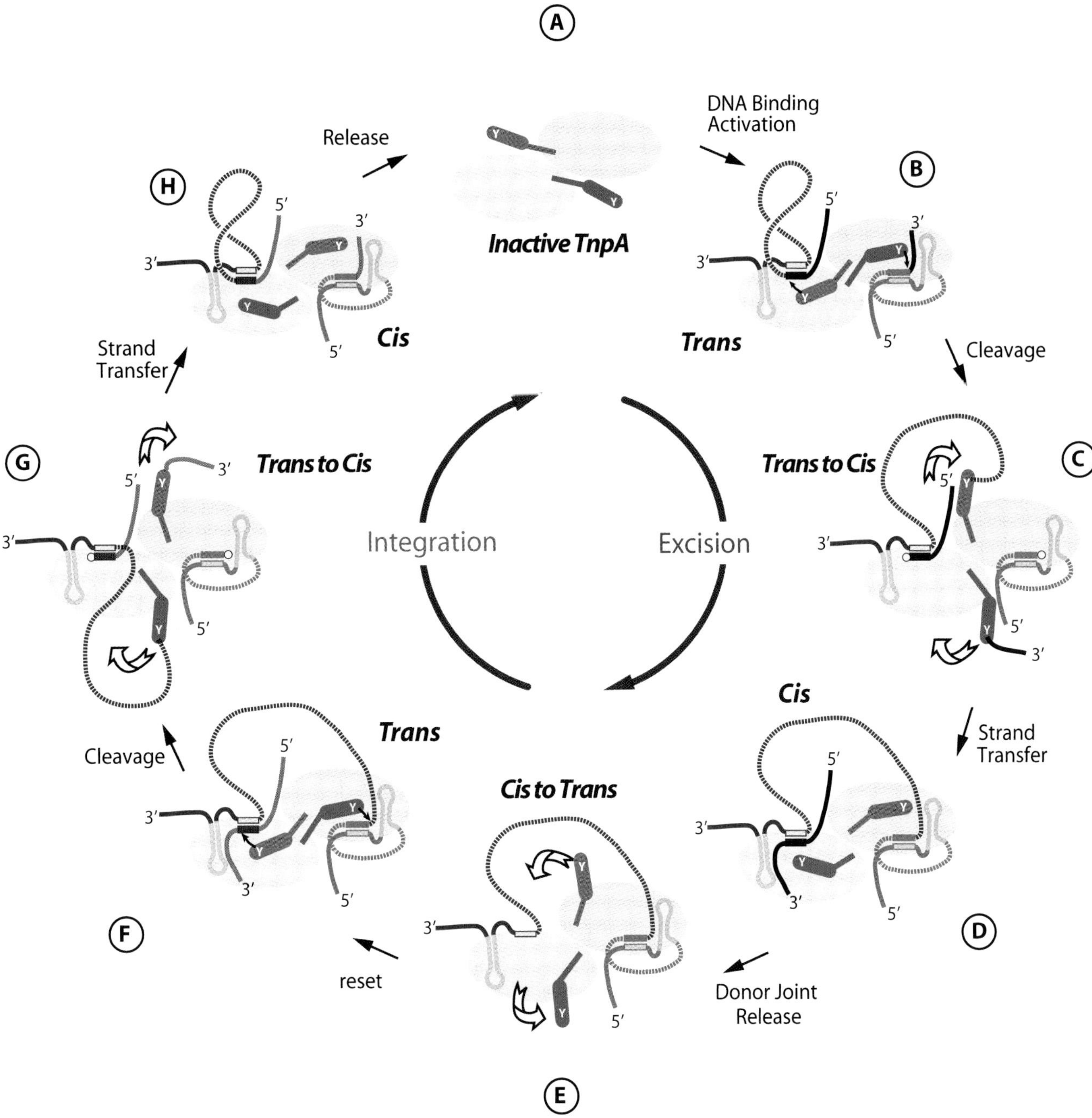

Figure 5 Strand transfer and reset model of IS*608* transpososome. (A) Inactive form of TnpA dimer in the absence of DNA (pale green, orange ovals and dark green and orange cylinders represent the body and the αD helices of two monomers, respectively). At the ends, dotted red and blue lines represent linkers at left end (LE) and right end (RE), light red and light blue boxes represent G_L and G_R, respectively. (B) Binding of a copy of LE and RE resulting in TnpA activation (catalytic sites *in trans*). (C) Cleavage of both ends forms a 5′ phosphotyrosine linkage between Y127 and LE on one αD helix (dark orange cylinders) and between Y127 and the RE flank on the other (dark green cylinders). 3′-OH groups are shown as yellow circles. Reciprocal rotation of both αD helices from *trans* to *cis* configuration are indicated by large arrows. (D) Strand transfer takes place to reconstitute the joined donor backbone (donor joint) and generate the RE–LE transposon junction at *cis* configuration. (E) Release of the donor joint and transition from *cis* to *trans* configuration. (F) Reset to the *trans* form and target site engagement. (G) Cleavage of the RE–LE junction and target and transition from *trans* to *cis* configuration. (H) Regeneration of the left and right transposon ends.
doi:10.1128/microbiolspec.MDNA3-0039-2014.f5

5′ LE phosphotyrosine bond to position it close to the immobile RE 3′-OH (Figure 5C, G). This would imply that LE linker length is critical for strand transfer. Indeed, sequential reduction in the length of the LE linker has a large effect on transposition frequency and excision *in vivo*. *In vitro*, it also had a somewhat larger effect on strand transfer than on cleavage (30), supporting the idea that the linker is important for mechanical movement. However, transpososome formation and stability was also observed to be affected with the shortest linkers. This presumably reflects steric barriers to $G_{L(R)}/C_{L(R)}$ interaction and supports the notion that these interactions are important in transpososome assembly. A survey of over 100 different IS from all three groups (35 from the IS*200* group; 47 from IS*605*, and 24 from IS*1341*) in the public databases has shown that the asymmetry of the IS*608* ends is conserved across the entire family: the left linker is always longer than the right (15–16 nt versus 8 nt) (Figure 4F) (30).

The second piece of evidence comes from the behavior of $TnpA_{IS608}$ heterodimers carrying point mutations in the HuH or catalytic Y. These were expressed and assembled *in vivo* and purified based on two different C-terminal affinity tags (one for each monomer). This permitted heterodimers to be distinguished from homodimers. A heterodimer with a combination of mutations that enforce a *trans*-active TnpA site (in which the wild-type HuH motif and Y127 belong to different TnpA monomers) is proficient for cleavage but not for rejoining. In contrast, a heterodimer with *cis*-active TnpA site (in which the wild-type HuH motif and Y127 belong to the same TnpA monomer) is proficient for rejoining but inactive in cleavage (30). This implies that all chemical reactions involved in cleavage occur in the *trans* site whereas the chemical reactions for strand transfer occur in the *cis* site. This strongly supports the rotational model.

A third piece of evidence comes from studies of the flexible arm that joins helix αD to the body of the protein and is proposed to play a pivotal role in the rotation. This flexibility may be facilitated by two glycine residues (G117 and G118). Mutation of these two residues did not affect strand cleavage but led to inhibition of strand transfer, suggesting that the two residues are required for achieving a *cis* configuration. The importance of these G residues is reflected in their conservation throughout the family (30).

Hence, while the *cis* configuration has not been observed crystallographically for these elements, its existence is strongly suggested by experimental data, supporting the *trans*/*cis* rotational model.

REGULATION OF SINGLE-STRAND TRANSPOSITION

Single-Strand DNA *in vivo*

The obligatory single-stranded nature of IS*200*/IS*605* transposition *in vitro* suggests that it is limited *in vivo* by the availability of its ssDNA substrates inside the cells and processes that produce ssDNA may stimulate transposition. We describe below a link between the transposition of these elements and the replication fork. Moreover, in the case of IS*Dra2*, ssDNA produced during reassembly of the *D. radiodurans* genome following irradiation results in stimulation of transposition (9, 33). Transcription or other processes leading to horizontal gene transfer such as transformation, conjugative transfer, or transduction with single-strand phages might also favor their mobility.

Replication Fork

The replication fork modulates the transposition of many transposable elements (Tn*7*, IS*903*, IS*10*, IS*50*, Tn*4430*, P element (34–39), Hallet B. personal communication). For IS*200*/IS*605* family members, the replication fork, in particular the lagging strand template, is an important source of ssDNA substrates for both excision and integration. Transposition can be considered to follow a “Peel-and-Paste” mechanism (Figure 6) where the IS excises or is “peeled” off as a single-strand circle from the lagging strand template of the donor molecule and then integrates, or is “pasted”, in a single-strand target at the replication fork.

(i) Excision

Excision of IS*608* is sensitive to the direction of replication across the element: it is more frequent when the active strand (top strand) is on the lagging-strand (discontinuous) template (Figure 6A) but is difficult to detect when it is on the leading-strand (continuous) (27). Moreover, excision *in vitro* requires that both ends are in single-strand form at the same time (5).

The length of ssDNA on the lagging-strand template depends on the initiation frequency of Okazaki fragment synthesis by the DnaG primase (40, 41). Transient inactivation of DnaG activity reduces this frequency and therefore increases the average length of ssDNA between Okazaki fragments; the IS608 excision frequency was found to increase under these conditions for *E. coli* carrying a *dnaGts* mutation, using a plasmid-based assay with IS*608* derivatives of different lengths, the excision frequency decreased strongly as IS length increased. In contrast, when DnaGts activity was reduced by growth under sublethal conditions,

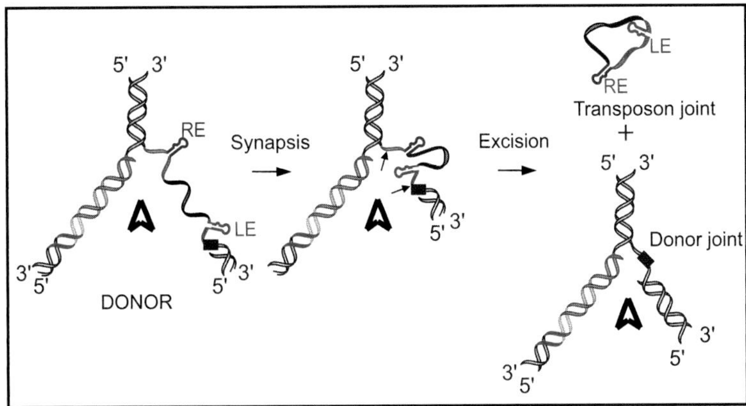

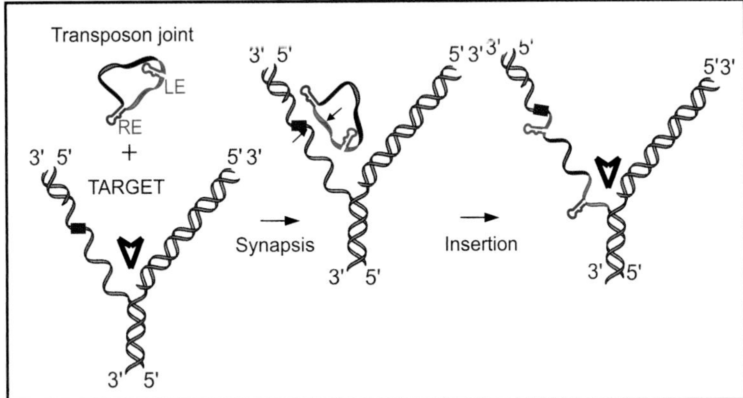

Figure 6 Transposition of the IS*200*/IS*605* family and the replication fork: "Peel and Paste" transposition mechanism. (A) Excision. Cartoon representing excision of the single-strand circular intermediate (transposon joint) from the lagging strand template of a donor plasmid. Arrow tip represents replication direction. (B) Integration. Integration of right end (RE)–left end (LE) transposon joint into single-strand target at the replication fork.
doi:10.1128/microbiolspec.MDNA3-0039-2014.f6

excision showed a much less pronounced length-dependence. This length-dependence might also contribute to the difference in copy numbers observed in the IS*200* and IS*605* groups (see *Distribution and organization* section above).

(ii) Integration

IS*608* integration is oriented (with its LE 3′ to a TTAC target site) and it requires an ssDNA target *in vitro* (3, 7). The close link between transposition and the replication fork is also illustrated by the integration bias, consistent with a preference for an ssDNA target on the lagging-strand template (Figure 6B). This was indeed found to be the case in *E. coli* for both plasmid and chromosome targets (27) (Lavatine et al., in preparation). As expected, the orientation of insertions into the *E. coli* chromosome was correlated with the direction of replication of each replicore and was consistent with integration into the lagging-strand template.

The orientation bias is not restricted to IS*608* and IS*Dra2*. An *in silico* analysis of a large number of bacterial genomes carrying copies of various family members revealed that most had a strong insertional bias consistent with the direction of replication (27). Moreover, in certain cases, elements that did not follow the orientation pattern could be correlated to genomic region that had undergone inversion or displacement, suggesting that, once they occur, insertions are quite stable. It seems possible that this type of genomic archaeology based on orientation patterns could be used to complement the study of bacterial genome evolution.

(iii) Stalled Replication Forks

Stalled replication forks appeared preferential targets for IS*608* insertion. In the experiments using the Tus/*ter* replication termination or operator/repressor system, replication fork arrest attracts IS*608* insertion (27). Transient blockage of a unidirectional replication fork by the Tus protein at the *ter* site resulted in preferential IS*608* insertion into the array of target sequences behind the stalled forks on the lagging strand but not on the leading strand. A similar result was obtained in the *E. coli* chromosome using the lacI/*lacO* and tetR/*tetO* repressor/operator roadblock systems (42) (Lavatine et al., in preparation). Moreover, a significant number of IS*608* insertions into the *E. coli* chromosome were localized in the highly transcribed *rrn* operons. This suggests that high transcription levels might affect replication fork progression (fork arrest by collision with RNA polymerase, R-loop formation, etc.) and could account for targeting the *rrn* operons. Hence, IS*608* insertions can be targeted to the stalled forks and this may well represent a major pathway for targeting transposition.

Genome Reassembly After Irradiation in *D. radiodurans*

Deinococcus radiodurans, arguably the most radiation-resistant organism known, has a remarkable capacity to survive the lethal effects of DNA-damaging agents, such as ionizing radiation, UV light, and desiccation. After exposure to high irradiation doses, the *D. radiodurans* chromosome, which is present in multiple copies per cell (43, 44), is shattered and degraded, but can be very rapidly reassembled in a process called extended synthesis dependent strand annealing (ESDSA). This involves resection of the multiple dsDNA fragments to generate extensive ssDNA segments, reannealing of complementary DNA, and reconstitution of the intact chromosome (23).

Mennecier et al. (33) analyzed the mutational profile in the *thyA* gene following irradiation. The majority of mutants were due to insertion of a single IS, IS*Dra2*, which is present in a single copy in the genome of the laboratory *D. radiodurans* strain. Furthermore, using a tailored genetic system, both IS*Dra2* excision and insertion efficiency was found to increase significantly following host cell irradiation (9). A PCR-based approach was used to follow irradiation-induced excision of the single genomic IS*Dra2* copy and reclosure of flanking sequences. Remarkably, these events are temporally closely correlated with the start of ESDSA. The signal that triggers IS*Dra2* transposition is likely the production of ssDNA intermediates generated during genome reassembly. Consistent with this, the requirement of ssDNA substrates for IS*Dra2*, as for IS*608*, was confirmed by *in vitro* studies of $\text{TnpA}_{\text{IS}\textit{Dra2}}$-catalyzed cleavage and strand transfer (9).

IS*Dra2* excision also depends on the direction of replication and is consistent with a requirement for the active strand to be located on the lagging-strand template in normally growing cells. However, this bias disappeared in irradiated *D. radiodurans* (27). As no apparent strand bias was observed in generating ssDNA during ESDSA, the lack of orientation bias in irradiated *D. radiodurans* suggests that ssDNA substrates are no longer limited to those rendered accessible during replication. This indicates that ssDNA sources are different in the contexts of vegetative replication and in genome reassembly.

TnpB

TnpA alone can carry out both the cleavage and joining steps *in vitro*. TnpB is encoded only by the IS*1341* and IS*605* groups and is not required for transposition of either IS*608* or IS*Dra2* in *E. coli* and *D. radiodurans*, respectively (3, 5, 9). TnpA activity has been extensively analyzed but few data are available concerning TnpB function. However, several observations suggest that TnpB might play a regulatory role in transposition of IS*200*/IS*605* family members.

Full-length TnpB is approximately 400 amino acids long. An overview of TnpB organization was obtained by comparing the entire ISfinder collection of 85 *tnpB* copies with the Pfam domain database. This revealed three major domains: an N-terminal putative helix-turn-helix, a longer and more variable central domain OrfB_IS*605* with a putative DDE motif (3) and a C-terminal zinc finger domain (45) (Figure 1C). The highest level of conservation is found at the C-terminal end of the protein, which includes a highly conserved zinc finger of the CPXCG type (46). Half of the analyzed TnpB copies including $\text{TnpB}_{\text{IS}\textit{Dra2}}$ but not $\text{TnpB}_{\text{IS}\textit{608}}$ contained all three domains, whereas only two did not include a zinc finger.

Pasternak et al. (45) observed that TnpB has an inhibitory effect on IS*Dra2* excision and insertion in its host, *D. radiodurans*, and on excision in *E. coli*, and that the integrity of its putative zinc finger motif is required for this effect. Understanding the exact molecular mechanism of TnpB activity will require further study. Accordingly, analysis of the ISfinder database revealed that IS*200*/IS*605* family elements exhibiting a high genomic copy number generally encode TnpA alone (ISfinder; Siguier P, unpublished).

Although the details of TnpB activity are unknown, the protein has been identified in both prokaryotes and eukaryotes. It is carried by members of the IS*607* family found both in prokaryotes and in eukaryotes and their viruses (47, 48) but is dispensable for IS*607* transposition in *E. coli* (19). TnpB analogues, known as Fanzor1 and Fanzor2, have also been identified in diverse eukaryotic transposable elements (49). TnpB/Fanzor proteins may function as transposition regulatory proteins *in vivo*.

Y1 TRANSPOSASE DOMESTICATION

There are many examples of eukaryotic transposases whose activities have been appropriated to perform various cellular functions (for review see references 50 and 51). However, the very few examples of this domestication for prokaryotic enzymes concern Y1 transposases.

$TnpA_{REP}$ and REP/BIME

Recently, a new clade of Y1 transposases ($TnpA_{REP}$) was found associated with REP/BIME sequences in structures called REPtrons (52, 53) (Figure 7A). In spite of their compact size, bacterial genomes carry many repetitive sequences, often important for genome function and evolution. Among them, the REP sequences are short DNA repeats of 20–40 bp that can form stem–loop structures preceded by a conserved tetranucleotide (GTAG or GGAG) (Figure 7Ai). REPs are found in intergenic regions in many bacterial species, particularly in proteobacteria, at high copy number (52, 54, 55) (Quentin Y, Fichant G, unpublished). There are nearly 590 copies in *E. coli* K12 (56) and up to 2,200 copies in *Pseudomonas* sp. GM79 (55). REPs can exist as individual units but can cluster in more complex structures called bacterial interspersed mosaic elements (BIME). These are composed of two individual REPs in inverse orientation (REP and iREP) separated by a short linker of variable length. BIME are often found in consecutive tandem copies (Figure 7Aii). Several roles have been attributed to these sequences including genome structuring, posttranscriptional regulation and genome plasticity. REPs are known to interact with protein partners such as integration host factor (57), DNA gyrase (58), and DNA polymerase I (59). REPs also increase mRNA stability and can act as transcriptional terminators (54, 60), or as targets for different IS (61) (ISfinder; Siguier et al., this volume). However, their origin and dissemination mechanisms are poorly understood.

Although more complex, REPtrons are reminiscent of IS*200* group members (Figure 7Aiii). However, REPtrons do not appear to be mobile and, in general, a single copy of a given REPtron coexists with numerous corresponding REP/BIME and genomes may harbor several distinct REPtrons (52, 55). It has therefore been suggested that REP/BIME represent a special type of nonautonomous transposable element mobilizable by $TnpA_{REP}$.

In vitro analysis of REPtrons

Analysis of *E. coli* REPtron activity *in vitro* has shown that, like $TnpA_{IS200/IS605}$, $TnpA_{REP}$ strictly requires single-stranded REP/BIME DNA substrates and is strand specific, only REP can be processed, whereas iREP are refractory to cleavage (62). Purified *E. coli* $TnpA_{REP}$ promotes single-stranded REP cleavage (in the linker sequences either 3′ or 5′ to the REP structure) and rejoining, and this activity requires the conserved tetranucleotide GTAG and the bulge in the middle of the REP stem (53, 63). Cleavage *in vitro* is less specific than that of $TnpA_{IS200/IS605}$ and occurs at a CT dinucleotide.

In contrast to $TnpA_{IS608}$ and $TnpA_{ISDra2}$, *E. coli* $TnpA_{REP}$ is a monomer in solution and in the crystal structure (63). Moreover, in the co-crystal structure, the short C-terminal tail is inserted into the active site blocking access to an ssDNA. It may therefore play a regulatory role in activity. Indeed C-terminal truncation of $TnpA_{REP}$ resulted in increased cleavage activity relative to the full-length protein *in vitro*. Biochemical and structural analysis suggested that the GTAG 5′ to the foot of the REP hairpin may play a similar role to the guide sequences $G_{L/R}$ in IS*200*/IS*605*. Moreover, structural data also highlighted numerous specific contacts between $TnpA_{REP}$ and GTAG, explaining its importance in the activity and clearly distinguishing $TnpA_{REP}$ from $TnpA_{IS200/IS605}$, which do not directly contact the guide sequences (see *Cleavage site recognition* section above). The way by which $TnpA_{REP}$ promotes REP/BIME proliferation through their host genomes remains to be determined.

IStrons

Another role for Y1 transposases was suggested by the identification of chimeric genetic elements widely distributed in the genome *of Clostridium difficile* (64) and *Bacillus cereus* group (65): IStrons. These combine functional and structural properties of group I introns at their 5′-end with those of an IS element at their 3′-end (Figure 7B). This 3′ part contains an IS*200*/IS*605*-related sequence including two full-length or truncated *orfs*, *tnpA* and *tnpB*, very similar to those found in IS*Dra2* (*D. radiodurans*) and IS*Cpe2* (*Clostridium perfringens*). IStrons are present at several loci in the same genome, indicating that this element is mobile and may move as a complete genetic unit. All IStron copies analyzed so far are inserted 3′ to the pentanucleotide TTGAT. *In vivo*, all variants can be efficiently and precisely

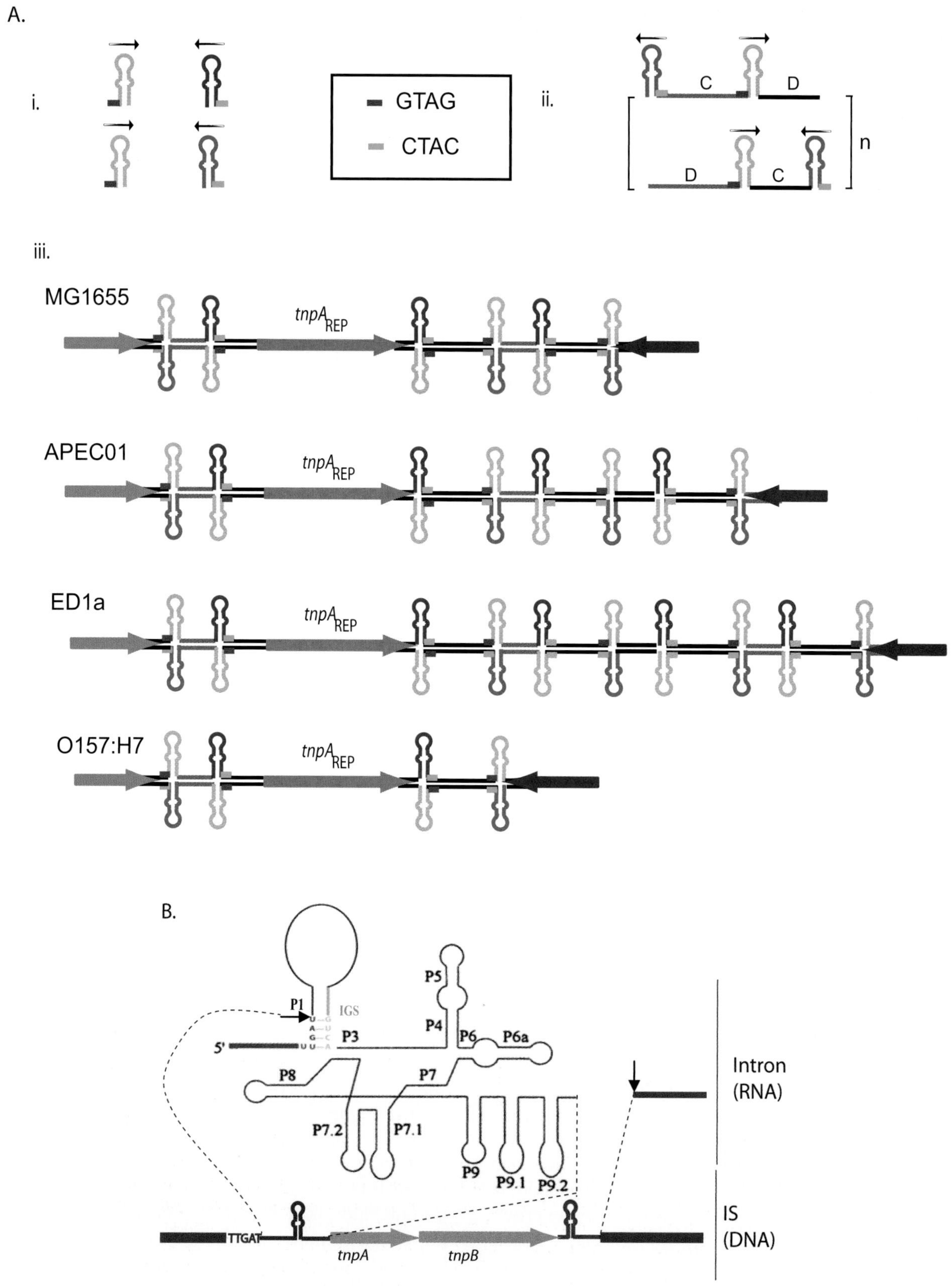

A.
i.
GTAG
CTAC
ii.
C
D
D
C
n
iii.
MG1655
tnpA REP
APEC01
tnpA REP
ED1a
tnpA REP
O157:H7
tnpA REP
B.
P1
IGS
5'
P3
P4
P5
P6
P6a
P7
P8
P7.2
P7.1
P9
P9.1
P9.2
Intron
(RNA)
TTGAT
tnpA
tnpB
IS
(DNA)

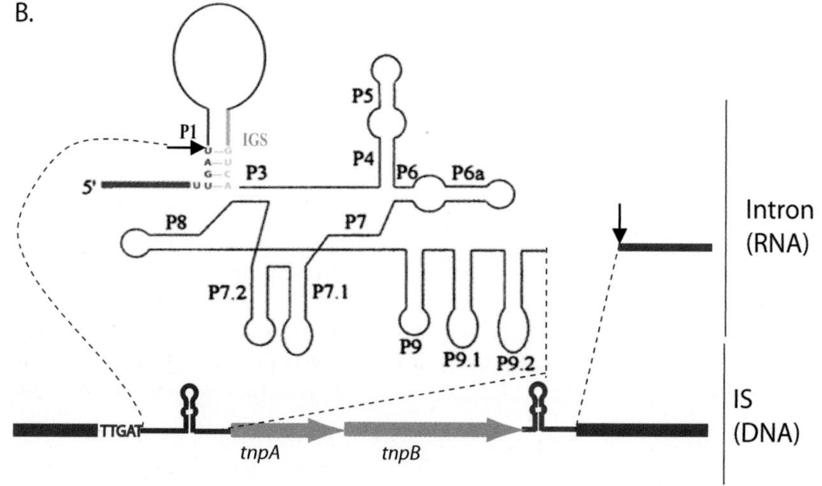

excised signifying that components necessary for ribozyme activity are present (64). Little is known about IStron behavior but the data suggest that IS components could mediate the spread of IStron whereas the intron component could assure splicing.

In vitro oligonucleotide-based assays using purified IStron transposase confirmed that at the DNA level, TTGAT is the LE cleavage site in excision and the target site, respectively (Caumont-Sarcos A, unpublished). At the RNA level, the same sequence is probably required in the splicing reaction (66). This would represent a novel type of intron invasion and transposition mechanism and provide a direct link between RNA and DNA worlds.

It is interesting to note that related IStrons have recently been identified that include components of the IS*607* family (67). These are characterized by a serine transposase together with a *tnpB* gene (19).

CONCLUSIONS AND PERSPECTIVES

The IS*200*/IS*605* family is distinguished from classical IS by their organization and their transposition pathway catalyzed by Y1 transposases. Not only is transposition strictly ssDNA-dependent but it is also asymmetric because of the characteristics discussed above: strand specificity, difference in left and right end organization and their activity and roles in the transposition pathway. Excision of the transposon as a circular ssDNA intermediate with abutted left and right ends is accompanied by rejoining of the DNA originally flanking the excised strand. The circle junction then inserts 3′ to a conserved element-specific sequence. This target specificity is another unusual feature of IS*200*/IS*605* transposition. Although the single-strand transposition mechanism is relatively well defined biochemically and structurally and transposition is primarily regulated by the availability of ssDNA *in vivo*, many regulatory aspects require further investigation.

Available data illustrate that members of the IS*200*/IS*605* family have evolved a mode of transposition that exploits the ssDNA at replication forks for their mobility. While the insertion orientation bias can be considered as a hallmark of the family, the detailed mechanism of replication fork targeting remains to be determined.

Moreover, the proposed single-strand transposition pathway raises questions concerning the enigma of IS*200* group abundance in various enterobacterial genomes and their apparent low transposition activity. Excision of a single-strand circular intermediate coupled with resealing of flanking sequence behind the replication fork would not increase global IS copy number in the genome. As chromosomal copies appeared stable, it is probable that these elements did not multiply from chromosomal copies bur rather arrived via horizontal transfer, excised and then integrated into the host chromosome. Their propagation remains to be assessed experimentally.

It will be important to address the control of transposase gene expression and how the level of transposase expression influences transposition, how TnpB expression is controlled and what is the molecular mechanism of TnpB activity.

In parallel, knowledge of IS*200*/IS*605* family distribution in bacterial genomes and metagenomes will provide an overview of the impact of horizontal transfer processes on the dissemination of these expanding classes of mobile genetic elements and their novel and distinct mechanism of transposition.

Acknowledgments. *We would like to thank O. Barabas, A. Hickman, D. Ronning, S. Messing, C. Pasternak, S. Sommer, Y. Quentin, and G. Fichant for fruitful collaboration. This work was supported by intramural CNRS (MC, BTH) funding and the intramural program of NIDDK (FD) and grants from the Agence National de la Recherche (France) Mobigen ANR-08-BLAN-0336 (MC), Mobising (ANR-12-BSV8-0009-01) (BTH).*

Citation. He S, Corneloup A, Guynet C, Lavatine L, Caumont-Sarcos A, Siguier P, Marty B, Dyda F, Chandler M,

Figure 7 Y1 transposase domestication. (A) *Escherichia coli* repeated extragenic palindromes (REP), bacterial interspersed mosaic elements (BIME) and REPtron. (i) Representation of two categories of REP structures in *E. coli*/*Shigella* with mismatches in the hairpin stem in orange and light blue, violet box represents the conserved tetranucleotide GTAG. Corresponding iREP structures in red and dark blue where green box represents the complementary tetranucleotide CTAC. (ii) Structure of BIME: REP and iREP separated by linkers C or D. BIME are frequently found as consecutive copies. (iii) Examples of REPtrons from some representative *E. coli* strains. *tnpA*$_{REP}$ is shown in gray, the flanking genes *yafL* and *fhiA* in green and in violet, respectively. Arrows represent the direction of transcription. (B) IStron: organization of IStron where Intron and IS parts are indicated. P1–P8 and IGS represents characteristic features of group I Introns. LE, RE, TTGAT target site and two *orf* of the IS part are indicated. doi:10.1128/microbiolspec.MDNA3-0039-2014.f7

Ton Hoang B. 2014. The IS*200*/IS*605* family and "peel and paste" single-strand transposition mechanism. Microbiol Spectrum 3(2):MDNA3-0039-2014.

References

1. **Chandler M, de la Cruz F, Dyda F, Hickman AB, Moncalian G, Ton-Hoang B.** 2013. Breaking and joining single-stranded DNA: the HUH endonuclease superfamily. *Nat Rev Microbiol* **11:**525–538.
2. **Lam S, Roth JR.** 1983. IS200: a Salmonella-specific insertion sequence. *Cell* **34:**951–960.
3. **Kersulyte D, Velapatino B, Dailide G, Mukhopadhyay AK, Ito Y, Cahuayme L, Parkinson AJ, Gilman RH, Berg DE.** 2002. Transposable element ISHp608 of *Helicobacter pylori*: nonrandom geographic distribution, functional organization, and insertion specificity. *J Bacteriol* **184:**992–1002.
4. **Kersulyte D, Akopyants NS, Clifton SW, Roe BA, Berg DE.** 1998. Novel sequence organization and insertion specificity of IS605 and IS606: chimaeric transposable elements of *Helicobacter pylori. Gene* **223:**175–186.
5. **Ton-Hoang B, Guynet C, Ronning DR, Cointin-Marty B, Dyda F, Chandler M.** 2005. Transposition of ISHp608, member of an unusual family of bacterial insertion sequences. *Embo J* **24:**3325–3338.
6. **Ronning DR, Guynet C, Ton-Hoang B, Perez ZN, Ghirlando R, Chandler M, Dyda F.** 2005. Active site sharing and subterminal hairpin recognition in a new class of DNA transposases. *Mol Cell* **20:**143–154.
7. **Guynet C, Hickman AB, Barabas O, Dyda F, Chandler M, Ton-Hoang B.** 2008. In vitro reconstitution of a single-stranded transposition mechanism of IS608. *Mol Cell* **29:**302–312.
8. **Barabas O, Ronning DR, Guynet C, Hickman AB, Ton-Hoang B, Chandler M, Dyda F.** 2008. Mechanism of IS200/IS605 family DNA transposases: activation and transposon-directed target site selection. *Cell* **132:**208–220.
9. **Pasternak C, Ton-Hoang B, Coste G, Bailone A, Chandler M, Sommer S.** 2010. Irradiation-induced *Deinococcus radiodurans* genome fragmentation triggers transposition of a single resident insertion sequence. *PLoS Genet* **6:** e1000799.
10. **Hickman AB, James JA, Barabas O, Pasternak C, Ton-Hoang B, Chandler M, Sommer S, Dyda F.** 2010. DNA recognition and the precleavage state during single-stranded DNA transposition in D. radiodurans. *Embo J* **29:**3840–3852.
11. **Devalckenaere A, Odaert M, Trieu-Cuot P, Simonet M.** 1999. Characterization of IS1541-like elements in *Yersinia enterocolitica* and *Yersinia pseudotuberculosis. FEMS Microbiol Lett* **176:**229–233.
12. **Beuzon CR, Chessa D, Casadesus J.** 2004. IS200: an old and still bacterial transposon. *Int Microbiol* **7:**3–12.
13. **Bisercic M, Ochman H.** 1993. The ancestry of insertion sequences common to *Escherichia coli* and *Salmonella typhimurium. J Bacteriol* **175:**7863–7868.
14. **Lam S, Roth JR.** 1986. Structural and functional studies of insertion element IS200. *J Mol Biol* **187:**157–167.
15. **Beuzon CR, Casadesus J.** 1997. Conserved structure of IS200 elements in *Salmonella. Nucleic Acids Res* **25:** 1355–1361.
16. **Beuzon CR, Marques S, Casadesus J.** 1999. Repression of IS200 transposase synthesis by RNA secondary structures. *Nucleic Acids Res* **27:**3690–3695.
17. **Sittka A, Lucchini S, Papenfort K, Sharma CM, Rolle K, Binnewies TT, Hinton JCD, Vogel J.** 2008. Deep sequencing analysis of small noncoding RNA and mRNA targets of the global post-transcriptional regulator, Hfq. *PLoS Genetics* **4:**e1000163.
18. **Odaert M, Devalckenaere A, Trieu-Cuot P, Simonet M.** 1998. Molecular characterization of IS1541 insertions in the genome of *Yersinia pestis. J Bacteriol* **180:**178–181.
19. **Kersulyte D, Mukhopadhyay AK, Shirai M, Nakazawa T, Berg DE.** 2000. Functional organization and insertion specificity of IS607, a chimeric element of *Helicobacter pylori. J Bacteriol* **182:**5300–5308.
20. **Boocock MR, Rice PA.** 2013. A proposed mechanism for IS607-family serine transposases. *Mob DNA* **4:**24.
21. **Akopyants NS, Clifton SW, Kersulyte D, Crabtree JE, Youree BE, Reece CA, Bukanov NO, Drazek ES, Roe BA, Berg DE.** 1998. Analyses of the cag pathogenicity island of *Helicobacter pylori. Mol Microbiol* **28:** 37–53.
22. **Islam SM, Hua Y, Ohba H, Satoh K, Kikuchi M, Yanagisawa T, Narumi I.** 2003. Characterization and distribution of IS8301 in the radioresistant bacterium *Deinococcus radiodurans. Genes Genet Syst* **78:**319–327.
23. **Zahradka K, Slade D, Bailone A, Sommer S, Averbeck D, Petranovic M, Lindner AB, Radman M.** 2006. Reassembly of shattered chromosomes in *Deinococcus radiodurans. Nature* **443:**569–573.
24. **Murai N, Kamata H, Nagashima Y, Yagisawa H, Hirata H.** 1995. A novel insertion sequence (IS)-like element of the thermophilic bacterium PS3 promotes expression of the alanine carrier protein-encoding gene. *Gene* **163:** 103–107.
25. **Bancroft I, Wolk CP.** 1989. Characterization of an insertion sequence (IS891) of novel structure from the cyanobacterium *Anabaena* sp. strain M-131. *J Bacteriol* **171:** 5949–5954.
26. **Stanley TL, Ellermeier CD, Slauch JM.** 2000. Tissue-specific gene expression identifies a gene in the lysogenic phage Gifsy-1 that affects *Salmonella enterica* serovar typhimurium survival in Peyer's patches. *J Bacteriol* **182:** 4406–4413.
27. **Ton-Hoang B, Pasternak C, Siguier P, Guynet C, Hickman AB, Dyda F, Sommer S, Chandler M.** 2010. Single-stranded DNA transposition is coupled to host replication. *Cell* **142:**398–408.
28. **Koonin EV, Ilyina TV.** 1993. Computer-assisted dissection of rolling circle DNA replication. *Biosystems* **30:** 241–268.
29. **Lee HH, Yoon JY, Kim HS, Kang JY, Kim KH, Kim DJ, Ha JY, Mikami B, Yoon HJ, Suh SW.** 2006. Crystal structure of a metal ion-bound IS200 transposase. *J Biol Chem* **281:**4261–4266.

30. **He S, Guynet C, Siguier P, Hickman AB, Dyda F, Chandler M, Ton-Hoang B.** 2013. IS200/IS605 family single-strand transposition: mechanism of IS608 strand transfer. *Nucleic Acids Res* **41:**3302–3313.
31. **He S, Hickman AB, Dyda F, Johnson NP, Chandler M, Ton-Hoang B.** 2011. Reconstitution of a functional IS608 single-strand transpososome: role of non-canonical base pairing. *Nucleic Acids Res* **39:**8503–8512.
32. **Guynet C, Achard A, Hoang BT, Barabas O, Hickman AB, Dyda F, Chandler M.** 2009. Resetting the site: redirecting integration of an insertion sequence in a predictable way. *Mol Cell* **34:**612–619.
33. **Mennecier S, Servant P, Coste G, Bailone A, Sommer S.** 2006. Mutagenesis via IS transposition in *Deinococcus radiodurans*. *Mol Microbiol* **59:**317–325.
34. **Parks AR, Li Z, Shi Q, Owens RM, Jin MM, Peters JE.** 2009. Transposition into replicating DNA occurs through interaction with the processivity factor. *Cell* **138:** 685–695.
35. **Hu WY, Derbyshire KM.** 1998. Target choice and orientation preference of the insertion sequence IS903. *J Bacteriol* **180:**3039–3048.
36. **Roberts D, Hoopes BC, McClure WR, Kleckner N.** 1985. IS10 transposition is regulated by DNA adenine methylation. *Cell* **43:**117–130.
37. **Yin JC, Krebs MP, Reznikoff WS.** 1988. Effect of dam methylation on Tn5 transposition. *J Mol Biol* **199:**35–45.
38. **Dodson KW, Berg DE.** 1989. Factors affecting transposition activity of IS50 and Tn5 ends. *Gene* **76:**207–213.
39. **Spradling AC, Bellen HJ, Hoskins RA.** 2011. *Drosophila* P elements preferentially transpose to replication origins. *Proc Natl Acad Sci U S A* **108:**15948–15953.
40. **Zechner EL, Wu CA, Marians KJ.** 1992. Coordinated leading- and lagging-strand synthesis at the *Escherichia coli* DNA replication fork. II. Frequency of primer synthesis and efficiency of primer utilization control Okazaki fragment size. *J Biol Chem* **267:**4045–4053.
41. **Wu CA, Zechner EL, Reems JA, McHenry CS, Marians KJ.** 1992. Coordinated leading- and lagging-strand synthesis at the Escherichia coli DNA replication fork. V. Primase action regulates the cycle of Okazaki fragment synthesis. *J Biol Chem* **267:**4074–4083.
42. **Lau IF, Filipe SR, Soballe B, Okstad OA, Barre FX, Sherratt DJ.** 2003. Spatial and temporal organization of replicating Escherichia coli chromosomes. *Mol Microbiol* **49:**731–743.
43. **Hansen MT.** 1978. Multiplicity of genome equivalents in the radiation-resistant bacterium *Micrococcus radiodurans*. *J Bacteriol* **134:**71–75.
44. **Harsojo, Kitayama S, Matsuyama A.** 1981. Genome multiplicity and radiation resistance in *Micrococcus radiodurans*. *J Biochem* **90:**877–880.
45. **Pasternak C, Dulermo R, Ton-Hoang B, Debuchy R, Siguier P, Coste G, Chandler M, Sommer S.** 2013. ISDra2 transposition in *Deinococcus radiodurans* is downregulated by TnpB. *Mol Microbiol* **88:**443–455.
46. **Krishna SS, Majumdar I, Grishin NV.** 2003. Structural classification of zinc fingers: survey and summary. *Nucleic Acids Res* **31:**532–550.
47. **Gilbert C, Cordaux R.** 2013. Horizontal transfer and evolution of prokaryote transposable elements in eukaryotes. *Genome Biol Evol* **5:**822–832.
48. **Filée J, Siguier P, Chandler M.** 2007. I am what I eat and I eat what I am: acquisition of bacterial genes by giant viruses. *Trends Genet* **23:**10–15.
49. **Bao W, Jurka J.** 2013. Homologues of bacterial TnpB_IS605 are widespread in diverse eukaryotic transposable elements. *Mob DNA* **4:**12.
50. **Volff JN.** 2006. Turning junk into gold: domestication of transposable elements and the creation of new genes in eukaryotes. *Bioessays* **28:**913–922.
51. **Vogt A, Goldman AD, Mochizuki K, Landweber LF.** 2013. Transposon domestication versus mutualism in ciliate genome rearrangements. *PLoS Genet* **9:**e1003659.
52. **Nunvar J, Huckova T, Licha I.** 2010. Identification and characterization of repetitive extragenic palindromes (REP)-associated tyrosine transposases: implications for REP evolution and dynamics in bacterial genomes. *BMC Genomics* **11:**44.
53. **Ton-Hoang B, Siguier P, Quentin Y, Onillon S, Marty B, Fichant G, Chandler M.** 2012. Structuring the bacterial genome: Y1-transposases associated with REP-BIME sequences. *Nucleic Acids Res* **40:**3596–3609.
54. **Rocco F, De Gregorio E, Di Nocera PP.** 2010. A giant family of short palindromic sequences in *Stenotrophomonas maltophilia*. *FEMS Microbiol Lett* **308:**185–192.
55. **Nunvar J, Licha I, Schneider B.** 2013. Evolution of REP diversity: a comparative study. *BMC Genomics* **14:**385.
56. **Gilson E, Bachellier S, Perrin S, Perrin D, Grimont PA, Grimont F, Hofnung M.** 1990. Palindromic unit highly repetitive DNA sequences exhibit species specificity within Enterobacteriaceae. *Res Microbiol* **141:**1103–1116.
57. **Boccard F, Prentki P.** 1993. Specific interaction of IHF with RIBs, a class of bacterial repetitive DNA elements located at the 3′ end of transcription units. *Embo J* **12:** 5019–5027.
58. **Espeli O, Boccard F.** 1997. In vivo cleavage of *Escherichia coli* BIME-2 repeats by DNA gyrase: genetic characterization of the target and identification of the cut site. *Mol Microbiol* **26:**767–777.
59. **Gilson E, Perrin D, Hofnung M.** 1990. DNA polymerase I and a protein complex bind specifically to *E. coli* palindromic unit highly repetitive DNA: implications for bacterial chromosome organization. *Nucleic Acids Res* **18:** 3941–3952.
60. **Khemici V, Carpousis AJ.** 2004. The RNA degradosome and poly(A) polymerase of *Escherichia coli* are required in vivo for the degradation of small mRNA decay intermediates containing REP-stabilizers. *Mol Microbiol* **51:** 777–790.
61. **Tobes R, Pareja E.** 2006. Bacterial repetitive extragenic palindromic sequences are DNA targets for Insertion Sequence elements. *BMC Genomics* **7:**62.
62. **Ton-Hoang B, Siguier P, Quentin Y, Onillon S, Marty B, Fichant G, Chandler M.** 2012. Structuring the bacterial genome: Y1-transposases associated with REP-BIME sequences. *Nucleic Acids Res* **40:**3596–3609.

63. **Messing SA, Ton-Hoang B, Hickman AB, McCubbin AJ, Peaslee GF, Ghirlando R, Chandler M, Dyda F.** 2012. The processing of repetitive extragenic palindromes: the structure of a repetitive extragenic palindrome bound to its associated nuclease. *Nucleic Acids Res* **40:**9964–9979.

64. **Braun V, Mehlig M, Moos M, Rupnik M, Kalt B, Mahony DE, von Eichel-Streiber C.** 2000. A chimeric ribozyme in *Clostridium difficile* combines features of group I introns and insertion elements. *Mol Microbiol* **36:**1447–1459.

65. **Tourasse NJ, Helgason E, Okstad OA, Hegna IK, Kolsto AB.** 2006. The *Bacillus cereus* group: novel aspects of population structure and genome dynamics. *J Appl Microbiol* **101:**579–593.

66. **Nielsen H, Johansen SD.** 2009. Group I introns: moving in new directions. *RNA Biol* **6:**375–383.

67. **Tourasse NJ, Stabell FB, Kolsto AB.** 2014. Survey of chimeric IStron elements in bacterial genomes: multiple molecular symbioses between group I intron ribozymes and DNA transposons. *Nucleic Acids Res* **42:**12333–12351.

Mobile DNA, 3rd Edition
Nancy L. Craig, Michael Chandler, Martin Gellert, Alan M. Lambowitz, Phoebe A. Rice and Suzanne Sandmeyer

doi:10.1128/microbiolspec.MDNA3-0002-2014

David B. Haniford[1]
Michael J. Ellis[1]

29 Transposons Tn*10* and Tn*5*

GENERAL BACKGROUND

Tn*10* and Tn*5* are composite bacterial transposons (Fig. 1). Both these transposons, as well as their respective IS elements (IS*10* and IS*50*), transpose by a nonreplicative cut-and-paste mechanism. Tn*10*/IS*10* was the first bacterial transposon shown to transpose by the cut-and-paste mechanism and as such provided an early model for this mode of transposition (1, 2). In cut-and-paste transposition the transposon is first excised from flanking donor DNA by a pair of transposase-catalyzed double-strand breaks at each transposon end after which the excised transposon is inserted into a target site. Host repair of the transposon-target DNA junction completes the transposition process and leaves a characteristic target-site duplication.

IS*10* and IS*50* are both members of the IS4 family of insertion sequences and their respective transposase proteins share an approximately 20% amino acid sequence identity (3). Both transposases catalyze four chemical steps at each of the two transposon ends. Remarkably, this involves repeated use of a single active site in each of the two transposase monomers that participate in the respective reactions (4, 5). All of these chemical steps, which include first-strand nicking, hairpin formation, hairpin resolution, and target-strand transfer, take place in the context of a higher order protein–DNA complex [referred to as a transpososome (or Tsome for short)] in which the two transposon ends are held together through a series of protein–DNA and protein–protein contacts.

One major difference between Tn*10* and Tn*5* involves the protein composition of the respective Tsomes. The outside end (OE) of Tn*10* contains a binding site for the integration host factor (IHF) (Fig. 2). IHF binding to the OE produces a 180° bend in the DNA which allows transposase to make both terminal and subterminal contacts with the OE (6). This appears to stabilize transposase binding to the OE DNA, as the formation of a stable Tsome is dependent on the presence of IHF (at least when the donor DNA containing the transposon lacks DNA supercoils) (7). The inside end (IE) of IS*10* does not contain an IHF binding site and, accordingly, the IS*10* Tsome (OE × IE) is an asymmetric structure containing one folded end produced by IHF binding (OE) and one unfolded end (IE). Interestingly, the Tn*10* Tsome (OE × OE) is also an asymmetric structure with one folded and one unfolded end, even though both OEs possess an equivalent IHF binding site (8). Once the Tsome is assembled, IHF can be removed without compromising the structural integrity of the complex. In fact, the mechanical force associated with IHF release and the concomitant end unfolding has been linked to the chemical steps in transposition (9). In addition, the position of the folded arm is known to influence target interactions. If one of the arms is refolded after full excision,

[1]Department of Biochemistry, University of Western Ontario, London, Ontario N6A 5C1, Canada.

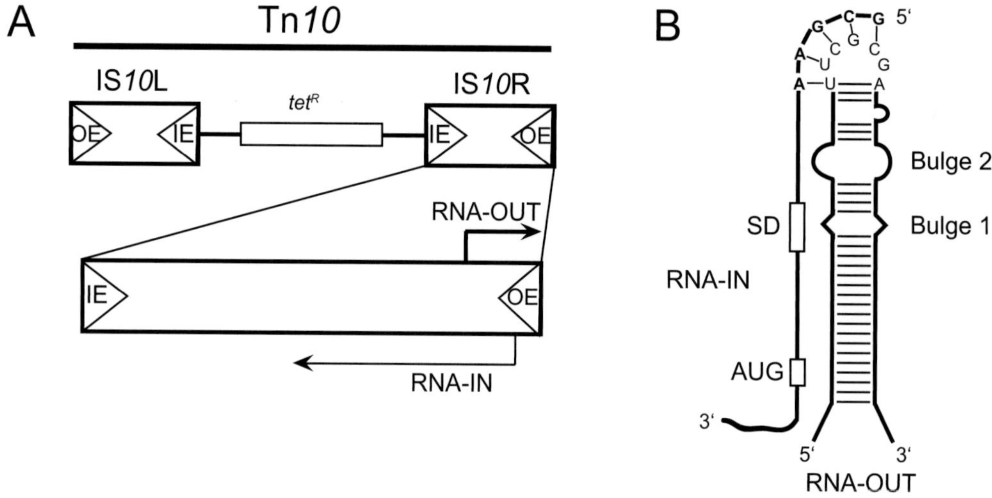

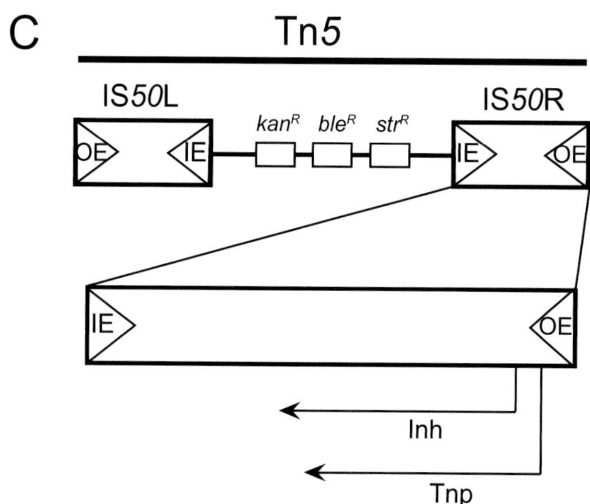

Figure 1 Structure of Tn*10* and Tn*5*. (**A**) Tn*10* is a composite transposon that encodes resistance determinants for tetracycline. The OEs and IEs of IS*10*-Right and IS*10*-Left are shown as well as the transcription units of IS*10*-Right. RNA-IN encodes the transposase and RNA-OUT encodes an antisense RNA. (**B**) Pairing of RNA-IN and RNA-OUT. RNA-OUT is a highly structured antisense RNA to the transposase RNA, RNA-IN. There are 35 nucleotides of perfect base complementarity between RNA-IN and RNA-OUT. Pairing initiates between the 5′ end of RNA-IN and the hairpin loop of RNA-OUT and full pairing results in the sequestration of both the Shine–Dalgarno (SD) sequence of RNA-IN and the start codon (AUG). Positions of the two main internal bulges in the stem are indicated. (**C**) Structure of Tn*5*. Tn*5* is a composite transposon that encodes resistance to kanamycin, bleomycin, and streptomycin. The OE and IE of IS*50*-Right and IS*50*-Left are shown as well as the transcription units in IS*50*-Right. Transcripts Tnp and Inh encode the transposase and inhibitor proteins, respectively. doi:10.1128/microbiolspec.MDNA3-0002-2014.f1

self-destructive intramolecular transposition events are favored over intermolecular events. In contrast, if the arms stay unfolded, intermolecular transposition events are favored over intramolecular events (Fig. 2) (10).

In contrast to Tn*10*/IS*10*, the DNA sequence requirements for Tn*5*/IS*50* Tsome formation are much simpler as Tsome formation does not require that transposases make contact with two separate end-sequence domains. All of the determinants necessary for transposase binding and Tsome formation appear to be present within the terminal 20 bp of the end sequences (11). Accordingly, there are no folded arms in the Tn*5*/

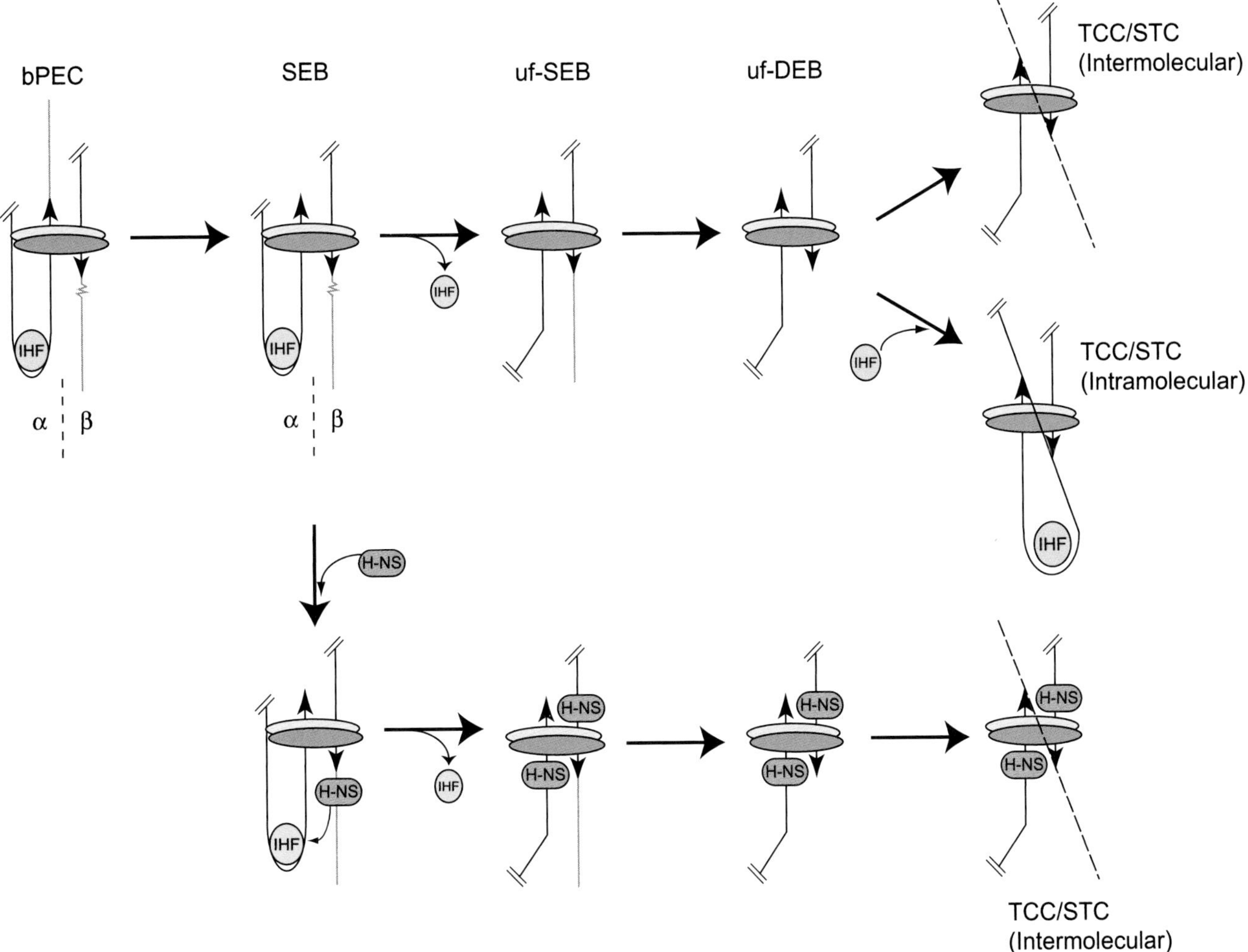

Figure 2 Tsome dynamics in Tn*10* transposition. The top line shows the initial Tsome with an IHF-folded arm (α-arm) transitioning to a single end break complex (αSEB) in which flanking donor DNA has been cleaved from one end and the α-arm remains folded, to an unfolded single end break complex (uf-SEB), and then to an unfolded double end break complex (uf-DEB) where flanking donor DNA has been cleaved from both transposon ends. At the branch point the uf-DEB can either capture a target DNA (TCC, target-capture complex) that is not part of the transposon and catalyze an intermolecular strand-transfer event (STC, strand-transfer complex) or the uf-DEB can rebind IHF to refold a transposon arm and undergo an intramolecular target-capture and strand-transfer event in which part of the transposon serves as the target DNA. The lower line shows the impact of H-NS on Tsome dynamics. An H-NS dimer is shown binding to the flanking DNA of the β-arm of the SEB that contains a distorted DNA structure (squiggly line in top panel). H-NS then facilitates the displacement of IHF (through an unknown mechanism) permitting the α-arm to unfold and subsequently additional H-NS dimers are recruited to the unfolded Tsome. H-NS binding within the transposon sequences is proposed to help maintain the Tsome in an unfolded form to both stabilize the fully cleaved unfolded Tsome and promote intermolecular target capture. Note that H-NS might first interact with the initial Tsome (denoted bPEC for historical reasons) instead of the α-SEB as shown. Transposon end sequences, arrows attached to black lines; flanking donor DNA, grey lines; target DNA, dashed lines; transposase, ovals. For clarity the two transposon ends are not joined (indicated by double dashes). doi:10.1128/microbiolspec.MDNA3-0002-2014.f2

IS*50* Tsome and the structure is symmetric. However, there is evidence, discussed below (H-NS and Tn*5* transposition), that another nucleoid-binding protein, H-NS, can bind to the Tn*5* Tsome (or transposase single-end complex) and influence the efficiency of Tsome assembly (12).

Both Tn*10* and Tn*5* are subject to multiple levels of regulation, most of which ensure that transposition frequencies are maintained at very low levels. Intrinsic factors, such as weak transposase promoters and protection against transposase production from read-through transcription, are common to both elements and serve to limit transposase expression (13, 14, 15). Other intrinsic factors are element specific, such as an antisense RNA system in Tn*10* that limits transposase translation (16), and the production of an inhibitor protein in Tn*5* that interferes with transposase dimerization and subsequent end binding (17) (Fig. 1). Extrinsic factors, including host-encoded proteins, also play key roles in limiting Tn*10* and Tn*5* transposition. For example, the expression of both transposase proteins is downregulated by Dam methylation of promoter elements in the respective transposase genes (18, 19). IHF and Fis (factor for inversion stimulation) have also been identified as regulators of Tn*5* transposition, although with the exception of the action of IHF in the Tn*10* system, the mechanisms of action for the other factors have not been well defined (20, 21, 22). In addition there is evidence that the transcriptional repressor LexA can repress Tn*5* transposase gene transcription (23). This latter observation may be particularly important given the evidence presented below that Hfq downregulates Tn*5* transposase expression at the transcriptional level.

The most recent comprehensive reviews of Tn*10* and Tn*5* were published in 2006 and 2008, respectively (24, 25). In the intervening period most of the research activity on these elements has been directed to furthering our understanding of how these transposons are regulated by their hosts. H-NS and Hfq are highly expressed proteins found in most enterobacteria and both have been implicated as regulators of Tn*10* and Tn*5* transposition. Interestingly, both proteins are known to play key roles in stress-response pathways and as such could link Tn*10* and/or Tn*5* transposition to the physiological state of cells (26, 27). This is an aspect of bacterial transposition that has not been thoroughly studied even though there are many (mostly anecdotal) examples of transposition systems responding to stress conditions. One exception is the IS*903* system, in which the response of this element to defects in purine metabolism has been worked out reasonably well (28). Outside of host regulation there has been some new activity directed at further defining molecular details of the Tn*5* and Tn*10* hairpin reactions. In addition, there have been some interesting uses of Tn*10* transposase RNAs in synthetic biology. Accordingly, this review will focus on these topics.

H-NS REGULATION OF TN*10* AND TN*5* TRANSPOSITION

Background

H-NS is a highly expressed nucleoid binding protein that is widely distributed among enterobacteria. It is a potent transcriptional repressor that regulates a large number of genes, including many that are important in stress adaptation and virulence. H-NS has structure-specific DNA binding activity, preferentially binding A-T rich sequences as a consequence of DNA binding determinants that are exquisitely sensitive to the shape of the minor groove of DNA (29). Given its well-defined role as a transcriptional repressor, it was unexpected that H-NS would be implicated as a positive regulator of Tn*10* transposition (30). It was perhaps even more surprising to find evidence that H-NS exerts its effects on Tn*10* transposition through the direct binding to Tn*10* Tsomes. At the time of the most recent review on Tn*10* transposition the impact of H-NS on Tsome dynamics was just beginning to be investigated. The initial studies provided evidence that H-NS binds selectively to both folded and unfolded forms of the Tsome (see Fig. 2 for a description of these Tsome forms). The flanking donor DNA was found to provide important determinants for H-NS binding to the folded Tsome and it was further shown that H-NS binding induced Tsome unfolding and increased the yield of strand-transfer products (STPs), apparently without influencing steps in transposon excision (31). In the sections below more recent results relating to H-NS function in Tn*10* (and Tn*5*) transposition are discussed.

H-NS-Tn*10* Tsome Interactions

Understanding how H-NS functions in Tn*10* transposition requires detailed knowledge of how it interacts with Tn*10* Tsomes. Footprinting studies with WT H-NS protein, as well as biochemical analysis of H-NS "separation of function" mutants, provided important insights into this problem. H-NS readily forms dimers in solution and higher order oligomers on DNA. It can bind DNA nonspecifically (low affinity) and in a structure-specific manner (high affinity) (26). P116S H-NS is a mutant form of H-NS that is defective in structure-specific DNA binding but that can still bind nonspecifically and form H-NS dimers (32, 33). Unlike WT H-NS, this form of H-NS did not bind the initial folded Tsome and did not stimulate strand transfer in a full reaction (34). Taken together with the requirement of flanking donor DNA for WT H-NS binding to the initial Tsome, the behavior of the P116S mutant described above is consistent with H-NS recognizing a distorted DNA structure in the flanking donor DNA as its primary binding determinant in the initial Tsome. Importantly, the formation of the initial Tsome is known to coincide with a structural deformation in the flanking DNA, as is the formation of the α-SEB, a form of Tsome in which flanking donor DNA has been removed from one end and the same end retains a folded configuration (8, 35) (Fig. 2).

Interestingly, P116S and a truncated form of H-NS containing only the first 64 residues both bound an unfolded Tsome (34). The latter form of H-NS does not retain significant DNA binding activity but can form dimers (33). Footprinting studies with P116S and WT H-NS revealed binding sites within the OE of the unfolded Tsome with one of these sites being close to where IHF normally binds. This raised the interesting possibility that H-NS binding within the core of the Tsome might help maintain a Tsome in an unfolded state. This may also have important implications for target recognition (see below). In addition, the finding that H-NS 1-64 retained Tsome binding capability further raised the possibility that a transposase protein provides determinants for H-NS binding. This was confirmed for WT H-NS in the context of an unfolded Tsome by protein–protein cross-linking analysis. Thus, it appears that H-NS recognizes both structural features in the DNA of the Tsome and transposase protein to bind this transposition complex. At this point it is not clear which determinants are bound by H-NS first. However, since P116S and H-NS 1-64 do not bind stably to the initial folded Tsome it is likely that H-NS first engages the flanking donor DNA of the Tsome to induce an unfolding event that subsequently exposes additional H-NS binding determinants (transposase protein and/or OE DNA) in the unfolded Tsome (34).

While H-NS is a highly abundant protein (~20,000 copies per cell in *Escherichia coli*), it also has many potential binding partners. Accordingly, for H-NS to have an impact on Tn*10* Tsome dynamics *in vivo*, the binding affinity of H-NS for Tsomes would have to be relatively strong. In fact, studies with an unfolded Tsome revealed that H-NS bound this complex with a K_d of ~0.3 nM (36). Prior to this work the H-NS–proU interaction ($K_d \approx 13$ nM) was the highest affinity H-NS–DNA interaction defined (37). The 40-fold higher affinity for the Tsome ranks the unfolded Tsome as the highest affinity H-NS binding partner defined to date.

H-NS may Function in a Postexcision Capacity in Tn*10* Transposition

In addition to H-NS, a divalent metal cation (Me^{2+}) binding to the active site of transposase can induce Tsome unfolding. Furthermore, there is evidence that Me^{2+} binding-induced unfolding promotes steps in transposon excision (8, 9). In contrast, H-NS did not have any obvious effects on steps in transposon excision. A possible explanation for this apparent paradox is that the primary role for H-NS in Tn*10* may be to stabilize unfolded Tsomes following donor cleavage. It is known that transitions from folded to unfolded Tsome forms coincide with the loss of Tsome signal in gel assays and a component of this is due to reduced Tsome stability (38). Conceptually, it is not difficult to imagine why unfolded forms of the Tsome would be less stable than folded forms as the former lack subterminal contacts with transposase. Interestingly, H-NS was shown to stabilize at least two different forms of the unfolded Tsome, including the initial uncleaved form and a form in which donor DNA is absent from one transposon end (35, 36). The ability of H-NS to stabilize a fully cleaved unfolded Tsome has not been tested, so at this point it can only be inferred that H-NS has the potential to stabilize this species.

Another way in which H-NS could function in a postexcision capacity is to ensure that IHF does not rebind to the unfolded fully cleaved Tsome (Fig. 2). Refolding of this complex has major implications with regard to the frequency of intra- versus intermolecular transposition events. Tn*10* is subject to a phenomenon called “target-site channeling” wherein the transposon ends are positioned within the fully cleaved Tsome in an orientation that favors self-destructive intramolecular transposition events (Fig. 2). This is a consequence of the subterminal region of the OE in a folded Tsome occupying the target site-binding cavity (6, 10). In transposition reactions with supercoiled plasmids containing Tn*10* derivatives, target-site channeling is manifested by the accumulation of a particular transposition product called an unknotted inversion circle (UKIC) (Fig. 3). In the equivalent reactions supplemented with H-NS, target-site channeling was reduced as evidenced by the reduction in the level of UKIC and an increase in the level of transposition products indicative of unconstrained target-site interactions, including intermolecular transposition events. Importantly, the antichanneling effect of H-NS was blocked by increasing the amount of IHF added to reactions, which is consistent with H-NS and IHF competing for binding sites within Tsomes (39). More generally, the capacity of H-NS to act as an antichanneling factor would be expected to increase the frequency of productive transposition events in the Tn*10* system; this is consistent with the observation that the frequency of Tn*10* transposition was reduced in strains of *E. coli* containing either an *hns* disruption or specific loss of function *hns* alleles, such as P116S (34). Notably there is precedent in the HIV system for the intasome interacting with a host factor (BAF-1) to limit intramolecular insertion events (40, 41). It will be interesting to see if there are

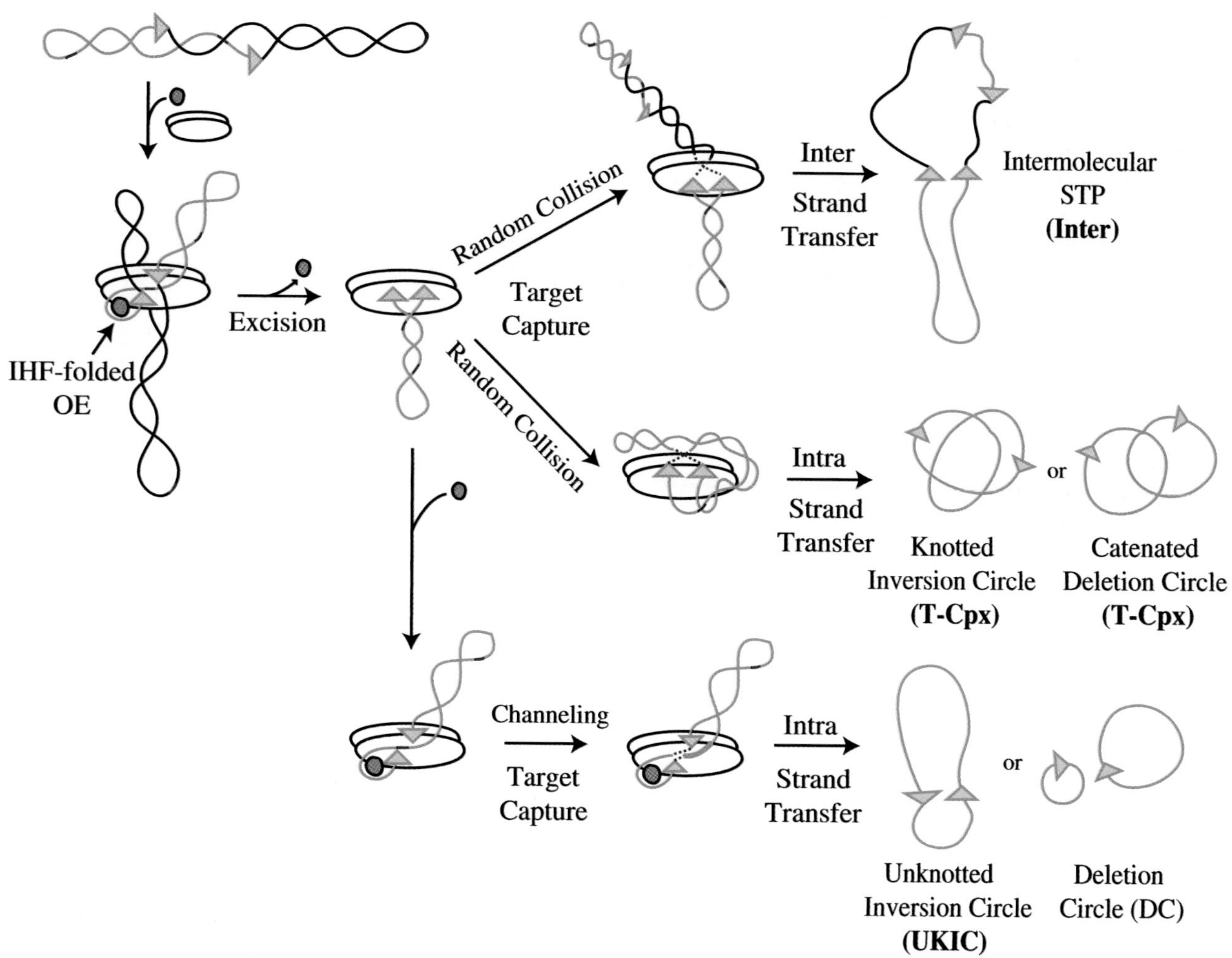

Figure 3 Strand-transfer products formed in Tn*10* transposition. If IHF (orange circle) maintains at least one OE in the folded form after excision (lower branch of diagram), insertion events are "channeled" into a target site (red line) located close to an OE within the transposon. Accordingly, intramolecular STPs are formed, and these products are "topologically simple" (UKIC) and unlinked deletion circles (DCs) because supercoiling nodes present in the excision product are not trapped between the OEs and the target site (dotted red line shows the recombination event; intrastrand strand transfer leads to the formation of DCs, whereas interstrand strand transfer leads to formation of ICs). If IHF does not remain bound to the OE and the OE unfolds, then the OE can interact with a target site through "random collision" (upper branches of diagram). If the target site is on a separate DNA molecule, intermolecular transposition occurs, whereas if the target site is within the transposon, intramolecular transposition occurs. The products of the latter include topologically complex (T-Cpx) DCs (catenated) and ICs (knotted). Green lines indicate transposon DNA; black lines, flanking donor DNA; ovals, transposase. Reprinted from the *Journal of Molecular Biology* (39) with permission from Elsevier. doi:10.1128/microbiolspec.MDNA3-0002-2014.f3

any mechanistic parallels in terms of how H-NS and BAF-1 limit autointegration events.

H-NS and Tn*5* Transposition

Although the Tn*5* Tsome does not contain a folded arm like the Tn*10* Tsome, the DNA contains a significant bend (centered at base pair 2) (42). As such a bend could provide determinants for H-NS binding, studies were performed to determine if the Tn*5* system might also be regulated by H-NS. Genetic studies revealed a drop in transposition of up to 6-fold for a Tn*5* derivative under conditions of *hns* deficiency, suggesting a positive regulatory role for H-NS in this system (12). It was subsequently shown that H-NS bound Tn*5* Tsomes with high specificity and three potential binding sites within the Tsome were identified by footprinting and mutational analyses. Interestingly, these sites were found within the terminal 20 bp of the transposon end sequence, a region in which transposase makes all of its contacts. Of the three potential binding sites, only one has an optimal sequence for structure-specific H-NS binding, a 5-bp A-T stretch from base pairs 8 to 12. In fact, *in silico* docking of an H-NS dimer with the Tn*5* Tsome revealed that an H-NS dimer could fit into the Tsome at this site without significant steric clashes. Cross-linking studies also revealed that H-NS could directly contact transposase specifically in the context of the Tsome, so that in the case of the Tn*10* system binding determinants for H-NS in the Tn*5* Tsome include both DNA and protein (43).

The precise role of H-NS in Tn*5* transposition remains to be established (see below), but it is important to note that the results of genetic studies are consistent with H-NS acting directly on transposition complexes, as opposed to acting indirectly through modulation of the gene expression of other factors that influence Tn*5* transposition. Mutations introduced into Tn*5* end sequences that blocked H-NS binding to Tsomes *in vitro* resulted in Tn*5* transposition becoming insensitive to the H-NS status of cells (43). With regard to a possible mechanism for H-NS function in Tn*5* transposition it has been suggested that H-NS may function in Tsome assembly. This is based on the findings that H-NS can both promote Tsome formation and bind to a single end-T ase complex *in vitro* (12).

REGULATION OF TN*10* AND TN*5* TRANSPOSITION BY Hfq

Background

Work on H-NS regulation of Tn*10* and Tn*5* transposition led indirectly to the discovery of Hfq as an additional regulator of both these systems. Hfq is an RNA-binding protein that functions primarily in the regulation of gene expression at the posttranscriptional level. By facilitating the pairing of small noncoding RNAs (sRNA) with coding RNAs and the recruitment of RNaseE to transcripts, Hfq can influence both the translation and stability of mRNAs (27). Hfq regulates *hns* gene expression by promoting the pairing of DsrA (a sRNA) and the *hns* mRNA. This pairing promotes rapid degradation of the *hns* mRNA and thus both DsrA and Hfq are negative regulators of *hns* gene expression (44). In an attempt to look at the impact of increasing H-NS protein levels on Tn*10* transposition, the transposition frequency of Tn*10* was measured in both *hfq* and *dsrA* disruption strains. The hypothesis was that increasing H-NS protein levels would increase the frequency of intermolecular Tn*10* transposition events. An increase in transposition of up to 80-fold was observed under conditions of *hfq* deficiency. However, *dsrA* deficiency did not have a significant impact on transposition. As DsrA is the mediator of posttranscriptional control of H-NS expression and Hfq is the accessory factor, it was inferred that the upregulation of Tn*10* transposition seen in *hfq* deficiency was unrelated to increased H-NS protein levels. That is, Hfq acts independent of H-NS to downregulate Tn*10* transposition (45).

Hfq Inhibits Tn*10* Transposase Expression at the Posttranscriptional Level

Hfq is considered a global regulator of gene expression as genome-wide expression studies have revealed that disruption of the *hfq* gene impacts on the expression of roughly one-third of *Salmonella* genes (46). Accordingly, the possibility was tested that Hfq might be inhibiting Tn*10* transposase expression. Experiments with transposase expression reporters provided evidence of this as the transposase expression increased substantially under conditions of *hfq* deficiency. Importantly, the increase in transposase expression was observed only in the situation where the reporter construct included both the native transposase promoter and sequences required for translational control. This is most consistent with Hfq acting as a posttranscriptional regulator of transposase expression (45).

Roughly 20 years prior to the work described above a potent posttranscriptional regulatory system had been described for the Tn*10* system. Tn*10* encodes an antisense RNA (RNA-OUT) that is perfectly complementary to 35 nucleotides of the transposase RNA (RNA-IN) (Fig. 1B). Pairing of the two RNAs, which occludes the ribosome binding site and start codon of RNA-IN, was shown to inhibit the translation of RNA-IN and decrease the stability of RNA-IN, thereby acting as a potent negative regulator of transposase expression and transposition (16, 47, 48). As Hfq typically functions in riboregulation by aiding in the pairing of RNA species, the possibility was considered that Hfq might play a role in the antisense pairing system of Tn*10*. Consistent with this possibility it was found that the impact of disrupting the *hfq* gene on Tn*10* transposition was greatly reduced (but not nullified) under conditions in which RNA-OUT was an ineffective inhibitor of transposase expression (i.e., when the transposon was present in the chromosome in a single copy). In addition, work *in vitro* demonstrated that Hfq could bind to both RNA-IN and RNA-OUT and accelerate the rate at which these molecules pair (45).

Transposase expression and transposition assays were also carried out under conditions in which possible synergy between RNA-OUT and Hfq could be assessed. Blocking both RNA-OUT and Hfq expression gave about a 10-fold increase in transposase expression and transposition relative to when either factor was blocked in isolation. If the negative regulatory effects of Hfq were restricted to its action in the antisense system, then the magnitude of the increase in either transposase expression or transposition in double and single mutants would be similar. As this was not the case, this experiment provided evidence that Hfq can also function independently of the antisense system to downregulate Tn*10*/IS*10* transposition (45). As described below, the results of *in vitro* Hfq-RNA-IN footprinting studies are at least consistent with the possibility that Hfq can bind directly to the translation initiation region of RNA-IN to inhibit the translation of this mRNA (Fig. 4) (49). Notably, evidence has been presented in a few other systems that Hfq can bind directly to the translation initiation region to inhibit translation (50, 51, 52). However, the RNA-OUT independent pathway for Hfq regulation of RNA-IN expression has not yet been studied in detail and it will be interesting to find out if other genetic determinants are required for this level of regulation.

In some respects it was surprising that the Tn*10* antisense system would have the need for a protein accessory factor. However, by binding sRNA–mRNA pairs Hfq can do more than simply increase the local concentration of potential pairing partners. A substantial body of work has shown that Hfq has the capacity to restructure RNA molecules (53, 54). So, despite the fact that there are 35 nucleotides of perfect complementarity between RNA-IN and -OUT, the potential for these molecules to form a stable hybrid might be limited by the fact that RNA-OUT is a highly structured RNA. In fact, RNase footprinting of Hfq–RNA-OUT complexes revealed that Hfq destabilized the stem of the RNA-OUT stem–loop structure (49). Notably, residues in the 5′ side of the stem are expected to basepair with the 5′ portion of RNA-IN. Previously, it was presumed, based on genetic studies, that preexisting discontinuities in the RNA-OUT stem would be sufficient to destabilize the stem for full pairing (55). Interestingly, one such discontinuity (bulge 2) appears to be a binding site for Hfq (Fig, 3). As Hfq was not previously implicated in the pairing reaction, the properties of mutations in the RNA-OUT stem previously used to probe aspects of RNA-IN/OUT pairing might be more complex than first appreciated.

In addition to restructuring RNA-OUT, Hfq can also restructure RNA-IN. Structure probing studies on the first 160 nucleotides of RNA-IN yielded evidence for a basepaired stem involving a subset of residues (nucleotides 25 to 36) expected to pair with RNA-OUT. The addition of Hfq to RNA-IN160 destabilized this stem, raising the possibility that Hfq-mediated restructuring of RNA-IN might also be important for RNA-IN/OUT pairing (49).

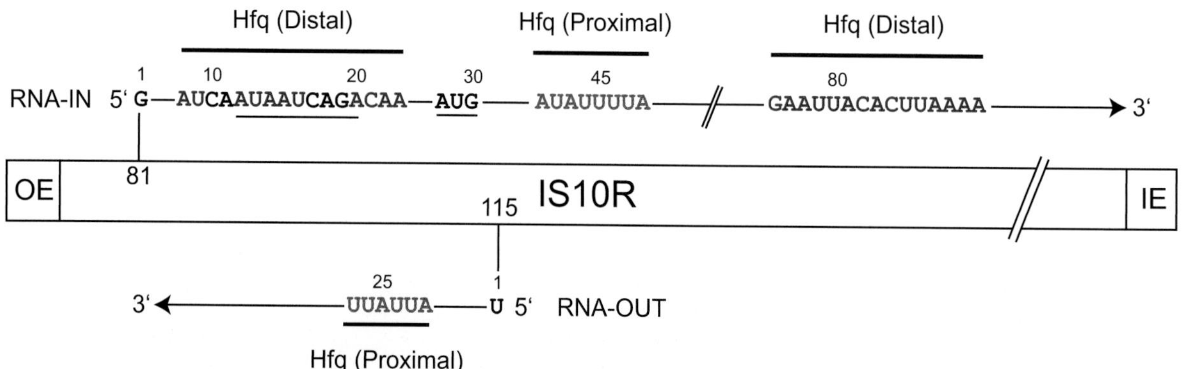

Figure 4 Hfq-binding sites within RNA-IN and RNA-OUT. A portion of IS*10*-Right is depicted along with partial sequences of the first 95 and 40 nucleotides of RNA-IN and RNA-OUT, respectively. The positions of Hfq-binding sites defined by *in vitro* footprinting are also shown (thick lines). Blue and red lettering distinguish sites thought to interact with the proximal and distal binding faces of Hfq, respectively. The SD and start codon of RNA-IN are underlined (thin lines). Note that the distal Hfq-binding site at the 5′ end of RNA-IN overlaps with the SD sequence. doi:10.1128/microbiolspec.MDNA3-0002-2014.f4

Details have been worked out regarding how Hfq interacts with RNA-IN and OUT. The results of a combination of footprinting and electrophoretic mobility shift assay analyses were consistent with the existence of three high-affinity Hfq binding sites within the first 160 nucleotides of RNA-IN and one high-affinity site within RNA-OUT (Fig. 4) (49). Hfq has three functionally distinct RNA binding surfaces referred to as the proximal, distal, and lateral sites (Fig. 5) (56, 57). In prototypical sRNA–mRNA pairs the body of the sRNA engages the proximal site and the body of the mRNA binds the distal site or both the distal and proximal sites (usually with a higher affinity for the former). The pairing (or seed) regions of both the sRNA and mRNA interact in the lateral site. Experiments with mutant forms of Hfq that were defective for either proximal or distal site binding (K56A and Y25A, respectively) were used to evaluate if RNA-IN/OUT pairings interact with Hfq, as is typical for an sRNA–mRNA pair. The results obtained were fully consistent with this. Moreover, it was found that both mutant forms of Hfq increased the expression of transposase and the frequency of IS*10* transposition, as would be expected if Hfq were to play an important role in RNA-IN/OUT pairing *in vivo* (49). Notably, RNA-IN and -OUT have not yet been tested for binding to the lateral surface of Hfq. To test further the importance of Hfq in RNA-IN/OUT pairing it will be important to establish if lateral site mutants fail to suppress transposase expression and transposition *in vivo*. The expectation is that these mutants will still be able to bind RNA-IN and RNA-OUT but will not suppress transposase expression and transposition because they cannot promote RNA-IN/OUT pairing.

One important caveat in the aforementioned *in vivo* studies is that since Hfq is a global regulator of gene expression, utilization of binding face mutants for

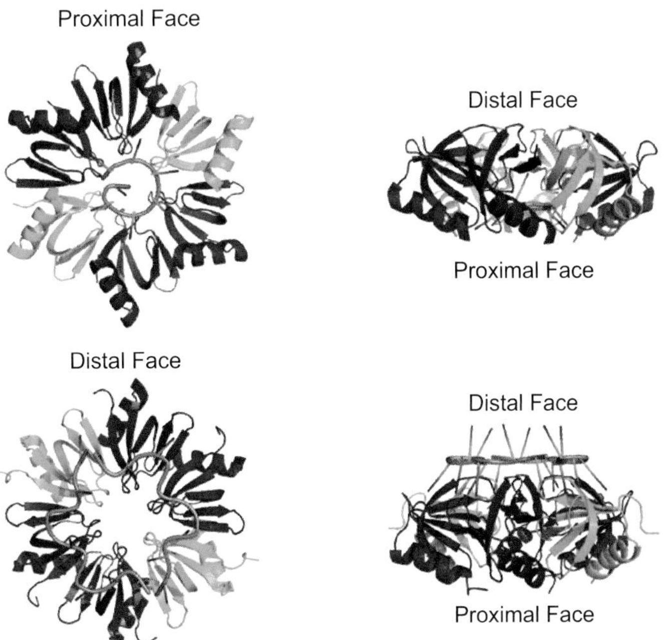

Figure 5 Structure and RNA-binding properties of Hfq. Hfq monomers assemble into a homohexameric ring-shaped complex with three distinct RNA-binding surfaces. The proximal surface (located close to the N-terminus) preferentially binds short U-rich stretches of RNA with each nucleotide binding to a pocket formed from adjacent protomers. The opposite face of the hexamer is termed the distal surface and binds longer RNA sequences that are purine rich. Each monomer contains three nucleotide-binding pockets termed the A, R, and N sites, which interact with adenines, purines, and the sugar-phosphate backbone, respectively. Each hexamer also possesses six additional RNA binding sites (lateral sites, not shown) positioned within each monomer between the distal and proximal sites. A lateral site can accommodate RNA extending from the proximal and distal sites. Hfq monomers are in red, blue, and green and RNA is in gold. Adapted from *Nature Reviews: Microbiology* (27) with permission from Macmillan Publishers.
doi:10.1128/microbiolspec.MDNA3-0002-2014.f5

determining the importance of Hfq in the IN–OUT pairing reaction could be complicated by the fact that the levels of other regulators of Tn*10* transposition may be affected in these mutant *hfq* strains. For this reason it will be advantageous to determine if mutating Hfq binding sites in RNA-IN and -OUT also cause increases in transposase expression and transposition. However, this important undertaking has its own complications because disrupting an Hfq binding site typically requires multiple nucleotide changes in an RNA, and changes in RNA structure and stability may end up being dominant factors in the properties of such mutants.

An intriguing possibility arising from the Hfq–RNA-IN/OUT pairing story is that other antisense RNAs might utilize Hfq as an accessory factor in their pairing reactions. The impact of this possibility could be quite significant given the recent recognition, based largely on "RNA seq" and microarray experiments, of the large number of potential antisense RNAs produced in bacteria (58). However, given the known role of Hfq in stabilizing sRNAs, it may be difficult to differentiate between the impact of Hfq on sRNA stability versus sRNA–mRNA pairing.

Hfq INHIBITS TN*5* TRANSPOSASE EXPRESSION AT THE TRANSCRIPTIONAL LEVEL

Detection of an antisense-independent Hfq regulatory pathway in the Tn*10* system led researchers to ask if a transposition system closely related to Tn*10* but without a known antisense regulatory system might also be subject to Hfq regulation. Tn*5* was chosen for study and the available evidence generated to date is fully consistent with Hfq acting as a potent negative regulator of this system. Transposition assays measuring transposition of the native Tn*5* positioned in the chromosome of *E. coli* revealed an increase in transposition events of approximately 10-fold under conditions of *hfq* deficiency. Consistent with this, transposase expression from a single copy Tn*5* reporter construct increased about 8-fold in *hfq* deficiency. However, unlike the Tn*10* situation, the "up-expression" phenotype was detected in both transcriptional and translation fusion reporters. In the former most of the 5′ UTR of the transposase transcript was not present. Thus, it would appear that in this system Hfq is acting to regulate negatively transcription of the transposase gene. This was supported by the finding that the steady-state level of the transposase transcript increased about 3-fold in *hfq* deficiency (59). Although Hfq is not typically associated with transcriptional control, there are some examples of *hfq* deficiency reducing transcript production independent of transcript degradation. For example, in the case of mRNAs RpsO, RpsB, and RpsT, which encode ribosomal proteins, it has been suggested that reduced transcript production in *hfq* deficiency might be due to Hfq binding to nascent transcripts during elongation to overcome pauses that otherwise might result in premature transcript release from the polymerase complex (60). As *hfq* deficiency caused an increase rather than a decrease in Tn*5* transposase transcript levels, an alternative mechanism of regulation must be considered in this case. More generally, it will be interesting to get a sense of how prevalent Hfq regulation of transposition systems is. As Hfq is a central player in stress-response pathways and there are many (mostly anecdotal) examples of bacterial transposition systems being responsive to stress, Hfq might turn out to be a very important regulatory component of numerous transposition systems.

DEVELOPMENTS IN TN*10* AND TN*5* HAIRPIN FORMATION/ RESOLUTION REACTIONS

A unique feature of the Tn*10* and Tn*5* excision reactions is that they take place through a DNA hairpin cleavage intermediate. It was originally proposed from work in the Tn*10* system that this mode of excision provided a solution to the general problem of how a single molecule of transposase with a single active site could make a flush double-strand break in a DNA molecule to release transposon from flanking donor DNA sequences (4, 61). Subsequently, the hairpin mechanism for transposon excision was discovered in a number of eukaryotic transposons that employ a cut-and-paste mechanism (62, 63, 64). Given the unusual nature of the hairpin cleavage reaction there was a lot of interest in understanding the molecular details of the reaction. The structure of the Tn*5* transpososome provided a crucial framework for understanding aspects of the hairpin reaction even though the crystallized complex can be considered a postexcision complex. First, this structure revealed an interesting deformation in the DNA structure that included a flipped-out base at the second residue of the nontransferred strand. Second, the structure provided clues as to the identity of amino acid residues in transposase that are likely to play key roles in the hairpin reaction (11).

A central issue in the hairpin reaction is how the 3′ OH of the transferred strand, exposed after first strand nicking, is able to carry out an in-line nucleophilic attack on the phosphodiester bond located directly across the helix some 16 to 18 angstroms away to generate the hairpin and simultaneously release the donor DNA from the transposon end. A deformation in the DNA structure involving the breaking of perhaps multiple base pairs in the vicinity of the transposon–donor junction would have to occur to bring together the aforementioned moieties. A bend in the DNA of the Tn*5* transpososome centered at base pair 2 had previously been identified and thus, prior to the detection of the flipped base in the crystal structure, it was already recognized that the structure of the junction region was substantially deformed. Finding that a flipped base was part of this deformed structure was somewhat surprising. In other systems in which base flipping had been seen, the production of an extrahelical base fits well with the biology of this system. For example, base flipping by M. HhaI exposes what otherwise would be an inaccessible base for methylation and removal of damaged or mismatched bases by DNA glycosylases is aided by the damaged or mismatched base becoming extrahelical (65, 66). It was less obvious what the function of the flipped-out base would be in the Tn*5* hairpin reaction. Of course, one possibility is that base flipping was simply part of the mechanism for providing the necessary strand separation for hairpin formation.

An elegant series of studies from the Chalmers lab first confirmed through structure probing and protein–DNA cross-linking studies that base flipping does, indeed, occur in both Tn*5* and Tn*10* systems and that it is critical for hairpin formation. They then went on to provide a sound biological rational for why production of an extrahelical base is a good fit for this class of reaction. In addition to contributing to the structural change in the junction that permits hairpin formation, it appears likely that the extrahelical base makes protein contacts that are necessary for the subsequent chemical step where the hairpin is reopened (or resolved) (67).

Briefly, in both Tn*5* and Tn*10* systems, base flipping appears to be driven by a combination of DNA bending and insertion of a "probe" residue into the helix (W323 in Tn*5* and M289 in Tn*10*). The flipped base is stabilized in its extrahelical position by insertion into a pocket where it stacks onto a tryptophan residue (W298 in Tn*5* and W265 in Tn*10*). In the case of Tn*10*, mutation of either the probe residue or the "stack" residue to alanine negatively affected both hairpin formation and resolution. Furthermore, when the flipped base was converted into an abasic residue so that no interaction between this residue and the stack amino acid could occur, there was a significant negative effect on hairpin formation and resolution implying that these steps require a base-specific contact (T-2) with W265 (67, 68). Further evidence of the importance of W265 in hairpin formation/resolution was provided by studies preceding the Chalmers work in which it was found that mutating W265 altered the fidelity of the hairpin formation and resolution reactions, an occurrence that certainly could be explained by the stacking residue participating in the guiding of the nontransferred strand into the active site for hairpin formation through its interaction with the extrahelical base (69).

IS*10* AND SYNTHETIC BIOLOGY

A major objective in the field of synthetic biology is to coordinate the expression of a set of (often heterologous) genes in an organism for the maximal production of a desired protein product. Typically, this requires modulation of the levels at which a set of genes within an operon are translated. In addition, if gene expression is linked to RNA processing, it may also be important to increase the stability of processed transcripts produced from an operon. Accordingly, custom-designed riboregulators have become an integral part of the discipline (70). Towards this end researchers have turned to modifying natural antisense systems rather than the *de novo* development of riboregulators to create novel orthogonal (high specificity with low off-target pairing) yet homogenous (similar kinetics and mechanism of pairing) RNA-based regulators of gene expression (71). These antisense RNAs can be designed to sense external metabolites as well as protein concentrations, and provide specific regulation of gene expression based on RNA:RNA interactions. The IS*10* antisense system has been extensively studied since 1983, and has a defined mechanism of pairing making it a convenient system for developing new riboregulators. Pairing begins between five nucleotides in the loop of RNA-OUT and the 5′ end of RNA-IN and ultimately results in 35 bp forming to inhibit translation initiation of RNA-IN (Fig. 1B) (48, 72).

Mutalik et al. mutated the 5-nt seed region in RNA-OUT and expressed these RNAs in the presence of a minimal RNA-IN–GFP reporter system to create a library of IS*10*-derived orthogonal regulators of translation (73). The goal of this work was to define the design parameters for constructing functionally homogenous yet orthogonal antisense regulators. By mutating the seed region to all possible combinations as well as introducing two exogenous nucleotides, this study produced 56 RNA-IN/OUT pairs. Half of these pairs were assayed for the *in vivo* repression of the RNA-IN–GFP translational fusion, and most cognate pairs showed >5-fold repression relative to no RNA-OUT. This work also tested the repression of RNA-OUT variants that were not perfectly complementary to RNA-IN (529 combinations) and showed that off-target repression could occur, with some RNA-OUT variants able to regulate multiple RNA-IN targets, and some RNA-IN targets repressed by multiple antisense RNAs. However, over 1000 "families" of compatible antisense pairs (i.e., <20% cross talk between family members) were identified ranging from three to seven family members. Ultimately, this work determined that the hybridization energy of the 5-bp seed region as well as the total hybridization energy of the RNA-IN/OUT pair were effective predictors of regulatory function. The authors speculate that the IS*10* antisense system is a convenient platform for designing new RNA-based translational regulators that could target any gene of interest, provided the regulatory RNA contains a similar structure to RNA-OUT, as well as satisfying the thermodynamic constraints.

The IS*10* antisense system has also been engineered to become responsive to an external ligand. Since the loop of RNA-OUT is critical for antisense pairing with RNA-IN, Qi et al. hypothesized that an engineered pseudoknot interaction between a ligand-sensing aptamer and RNA-OUT would result in an inducible translational regulator (74). The theophylline aptamer, which consists of a small RNA stem–loop structure with two internal loops that bind theophylline, was selected as the allosteric regulator of RNA-OUT. A chimeric noncoding RNA (ncRNA) was designed such that a pseudoknot would form between the loops of RNA-OUT and the aptamer in the absence of theophylline. This pseudoknot would sequester the loop nucleotides in RNA-OUT, disrupting antisense pairing. The addition of theophylline would prevent the pseudoknot from forming, allowing RNA-OUT to basepair with and prevent translation of RNA-IN. Expression of an RNA-IN–GFP translational fusion was repressed ~6-fold only in the presence of theophylline, and *in vitro* footprinting confirmed that the loop nucleotides of RNA-OUT become single stranded in the presence ligand.

RNA-IN/OUT pairing has also been used in conjunction with the pT181 transcriptional attenuator for RNA-based transcriptional regulation (75). Transcription attenuators form alternate RNA secondary structures within a leader transcript that either allow (ON) or block (OFF) elongation. RNA:RNA interactions between the attenuator and a regulatory ncRNA can stabilize the "OFF" state, resulting in an antisense RNA-regulated transcriptional switch. Owing to complex RNA pairing pathways, natural attenuators cannot easily be reprogrammed to alter specificity. Takahashi and colleagues designed a modular transcriptional attenuator using RNA-IN/RNA-OUT pairing for the antisense control of a pT181-derived attenuator. Since the pT181 attenuator stem–loop structure is essential for regulation, the authors reversed the natural roles of RNA-IN/RNA-OUT, resulting in a trans-acting RNA-IN (nt 1 to 40 of natural RNA-IN) base pairing to a chimeric RNA-OUT/pT181 attenuator. When the "top half" of RNA-OUT was included in the chimeric attenuator, the trans-acting RNA-IN was able to repress effectively the expression of a reporter gene by close to 4-fold. Furthermore, mutations to the loop region of RNA-OUT and the complementary positions on RNA-IN were introduced to produce a family of mutually orthogonal transcriptional attenuators all derived from the IS*10* antisense system.

The three studies discussed above use the IS*10* antisense system as a platform for designing new orthogonal RNA regulators of translation and transcription. However, the role of Hfq in facilitating antisense pairing was not considered. In particular, the RNA-IN/RNA-OUT regulated transcriptional attenuator was not functional until the sequences on RNA-OUT containing the putative Hfq binding site were included. However, this may only be a coincidence, as the authors suggested a complicated RNA structure forming between RNA-IN and the transcriptional terminator used in the expression construct was responsible for blocking antisense pairing.

CONCLUDING REMARKS

The study of transposons Tn*10* and Tn*5* has provided a wealth of information regarding steps in cut-and-paste transposition, Tsome dynamics and structure, and mechanisms used to regulate bacterial transposition. Current research on Tn*10* and Tn*5* is focusing mainly on newly discovered host regulators (H-NS and Hfq) of both transposons and attempts to define the mechanisms by which these regulators act to control transposition. In addition, there have been some clever applications of IS*10* antisense pairing in the emerging field of synthetic biology. These applications add to the long list of biotechnical applications (not discussed here) that have evolved out of detailed knowledge of both transposition systems (25, 76). With regard to the regulation of transposition it is intriguing that Tn*5* and Tn*10* are affected by proteins that are themselves considered global regulators of gene expression. This potentially provides a means of linking these transpositions systems to the physiological state of the cells they inhabit. It is also interesting that one of the factors,

Hfq, is a negative regulator of transposition while the other, H-NS, is a positive regulator of transposition. Accordingly, the two transposition systems could respond in either direction to changes in cell physiology. Although the original work on Tn*10* and Hfq focused on Hfq working through the antisense RNA system of Tn*10*, it may be more relevant (with regard to the possibility of Hfq playing a more general role in transposition systems) that Hfq can exert its regulatory effects on Tn*10* and Tn*5* independent of antisense RNAs. At this point it is not known if a broader collection of transposons will show evidence of Hfq regulation. On the one hand, the component insertion sequences of Tn*10* and Tn*5* are in the same IS family and so perhaps it is not surprising that they would be subject to similar regulatory mechanisms. On the other hand, Hfq-mediated riboregulation of bacterial transposition systems might turn out to be a common phenomenon, much like RNAi in eukaryotic systems. The simplest means of identifying other transposition systems involving Hfq regulation would be cloning *en masse* transposase genes into gene-expression reporters and looking for an impact on transposase expression under conditions of *hfq* deficiency.

It should be recognized that with regard to understanding regulatory pathways for transposition systems that involve Hfq we might only be seeing the tip of the iceberg. For example, in the Tn*5* system it is likely that Hfq is regulating the level of an as-yet unknown transcription factor or group of transcription factors. Presumably, genetic studies will provide a means of working out the details of this/these pathway(s). In the Tn*10* system, details of the antisense independent Hfq pathway remain to be elucidated. The bottom line is that studying some of the old mainstays of the "bacterial transposition world" continues to provide new insights into a class of genetic elements that has fascinated us for over 50 years.

Acknowledgments. *We thank members of the Haniford lab, Joseph Ross and Ryan Trussler, for helpful discussions and Alicia Haniford for proofreading the manuscript. All of the work contributed by my own lab was supported by funding from the Canadian Institutes of Health Research (MOP 11281).*

Citation. **Haniford D.** 2014 Transposons Tn*10* and Tn*5*. Microbiol Spectrum 3(1):MDNA3-0002-2014.

References

1. **Bender J, Kleckner N.** 1986. Genetic evidence that Tn10 transposes by a nonreplicative mechanism. *Cell* **45:** 801–815.
2. **Haniford DB, Chelouche AR, Kleckner N.** 1989. A specific class of IS*10* transposase mutants are blocked for target site interactions and promote formation of an excised transposon fragment. *Cell* **59:**385–394.
3. **Mahillon J, Chandler M.** 1998. Insertion sequences. *Microbiol Mol Biol Rev* **62:**725–774.
4. **Kennedy AK, Guhathakurta A, Kleckner N, Haniford DB.** 1998. Tn*10* transposition via a DNA hairpin intermediate. *Cell* **95:**125–134.
5. **Bhasin A, Goryshin IY, Reznikoff WS.** 1999. Hairpin formation in Tn*5* transposition. *J Biol Chem* **274:** 37021–37029.
6. **Chalmers R, Guhathakurta A, Benjamin H, Kleckner N.** 1998. IHF modulation of Tn*10* transposition: sensory transduction of supercoiling status via a proposed protein/DNA molecular spring. *Cell* **93:**897–908.
7. **Sakai J, Chalmers RM, Kleckner N.** 1995. Identification and characterization of a pre-cleavage synaptic complex that is an early intermediate in Tn*10* transposition. *EMBO J* **14:**4374–4383.
8. **Crellin P, Chalmers R.** 2001. Protein-DNA contacts and conformational changes in the Tn*10* transpososome during assembly and activation for cleavage. *EMBO J* **20:** 3882–3891.
9. **Crellin P, Sewitz S, Chalmers R.** 2004. DNA looping and catalysis; the IHF-folded arm of Tn*10* promotes conformational changes and hairpin resolution. *Mol Cell* **13:** 537–547.
10. **Benjamin HW, Kleckner N.** 1989. Intramolecular transposition by Tn*10*. *Cell* **59:**373–383.
11. **Davies DR, Goryshin IY, Reznikoff WS, Rayment I.** 2000. Three-dimensional structure of the Tn*5* synaptic complex transposition intermediate. *Science* **289:**77–85.
12. **Whitfield CR, Wardle SJ, Haniford DB.** 2009. The global bacterial regulator H-NS promotes transpososome formation and transposition in the Tn*5* system. *Nucleic Acids Res* **37:**309–321.
13. **Raleigh EA, Kleckner N.** 1986. Quantitation of insertion sequence IS*10* transposase gene expression by a method generally applicable to any rarely expressed gene. *Proc Natl Acad Sci U S A* **83:**1787–1791.
14. **Krebs MP, Reznikoff WS.** 1986. Transcriptional and translational initiation sites of IS50. Control of transposase and inhibitor expression. *J Mol Biol* **192:**781–791.
15. **Davis MA, Simons RW, Kleckner N.** 1985. Tn*10* protects itself at two levels from fortuitous activation by external promoters. *Cell* **43:**379–387.
16. **Simons RW, Kleckner N.** 1983. Translational control of IS*10* transposition. *Cell* **34:**683–691.
17. **Mahnke Braam LA, Goryshin IY, Reznikoff WS.** 1999. A mechanism for Tn*5* inhibition. carboxyl-terminal dimerization. *J Biol Chem* **274:**86–92.
18. **Roberts D, Hoopes BC, McClure WR, Kleckner N.** 1985. IS*10* transposition is regulated by DNA adenine methylation. *Cell* **43:**117–130.
19. **Yin JC, Krebs MP, Reznikoff WS.** 1988. Effect of dam methylation on Tn*5* transposition. *J Mol Biol* **199:**35–45.
20. **Signon L, Kleckner N.** 1995. Negative and positive regulation of Tn*10*/IS*10*-promoted recombination by IHF: two distinguishable processes inhibit transposition off of

multicopy plasmid replicons and activate chromosomal events that favor evolution of new transposons. *Genes Dev* **9:**1123–1136.

21. **Makris JC, Nordmann PL, Reznikoff WS.** 1990. Integration host factor plays a role in IS*50* and Tn*5* transposition. *J Bacteriol* **172:**1368–1373.
22. **Weinreich MD, Reznikoff WS.** 1992. Fis plays a role in Tn5 and IS50 transposition. *J Bacteriol* **174:**4530–4537.
23. **Kuan CT, Tessman I.** 1991. LexA protein of *Escherichia coli* represses expression of the Tn*5* transposase gene. *J Bacteriol* **173:**6406–6410.
24. **Haniford DB.** 2006. Transpososome dynamics and regulation in Tn10 transposition. *Crit Rev Biochem Mol Biol* **41:**407–424.
25. **Reznikoff WS.** 2008. Transposon Tn*5*. *Ann Rev Genet* **42:**269–286.
26. **Dorman CJ.** 2004. H-NS: a universal regulator for a dynamic genome. *Nat Rev Microbiol* **2:**391–400.
27. **Vogel J, Luisi BF.** 2011. Hfq and its constellation of RNA. *Nat Rev Microbiol* **9:**578–589.
28. **Coros AM, Twiss E, Tavakoli NP, Derbyshire KM.** 2005. Genetic evidence that GTP is required for transposition of IS*903* and Tn*552* in *Escherichia coli. J Bacteriol* **187:** 4598–4606.
29. **Navarre WW, McClelland M, Libby SJ, Fang FC.** 2007. Silencing of xenogeneic DNA by H-NS-facilitation of lateral gene transfer in bacteria by a defense system that recognizes foreign DNA. *Genes Dev* **21:**1456–1471.
30. **Swingle B, O'Carroll M, Haniford D, Derbyshire KM.** 2004. The effect of host-encoded nucleoid proteins on transposition: H-NS influences targeting of both IS*903* and Tn*10*. *Molec Microbiol* **52:**1055–1067.
31. **Wardle SJ, O'Carroll M, Derbyshire KM, Haniford DB.** 2005. The global regulator H-NS acts directly on the transpososome to promote Tn10 transposition. *Genes Dev* **19:**2224–2235.
32. **Spurio R, Falconi M, Brandi A, Pon CL, Gualerzi CO.** 1997. The oligomeric structure of nucleoid protein H-NS is necessary for recognition of intrinsically curved DNA and for DNA bending. *EMBO J* **16:**1795–1805.
33. **Badaut C, Williams R, Arluison V, Bouffartigues E, Robert B, Buc H, Rimsky S.** 2002. The degree of oligomerization of the H-NS nucleoid structuring protein is related to specific binding to DNA. *J Biol Chem* **277:** 41657–41666.
34. **Ward CM, Wardle SJ, Singh RK, Haniford DB.** 2007. The global regulator H-NS binds to two distinct classes of sites within the Tn*10* transpososome to promote transposition. *Molec Microbiol* **64:**1000–1013.
35. **Liu D, Haniford DB, Chalmers RM.** 2011. H-NS mediates the dissociation of a refractory protein–DNA complex during Tn*10*/IS*10* transposition. *Nucleic Acids Res* **39:**6660–6668.
36. **Wardle SJ, Chan A, Haniford DB.** 2009. H-NS binds with high affinity to the Tn*10* transpososome and promotes transpososome stabilization. *Nucleic Acids Res* **37:**6148–6160.
37. **Bouffartigues E, Buckle M, Badaut C, Travers A, Rimsky S.** 2007. H-NS cooperative binding to high-affinity sites in a regulatory element results in transcriptional silencing. *Nat Struct Molec Biol* **14:**441–448.
38. **Liu D, Crellin P, Chalmers R.** 2005. Cyclic changes in the affinity of protein–DNA interactions drive the progression and regulate the outcome of the Tn*10* transposition reaction. *Nucleic Acids Res* **33:**1982–1992.
39. **Singh RK, Liburd J, Wardle SJ, Haniford DB.** 2008. The nucleoid binding protein H-NS acts as an anti-channeling factor to favor intermolecular Tn*10* transposition and dissemination. *J Mol Biol* **376:**950–962.
40. **Lee MS, Craigie R.** 1998. A previously unidentified host protein protects retroviral DNA from autointegration. *Proc Natl Acad Sci U S A* **95:**1528–1533.
41. **Chen H, Engelman A.** 1998. The barrier-to-autointegration protein is a host factor for HIV type 1 integration. *Proc Natl Acad Sci U S A* **95:**15270–15274.
42. **York D, Reznikoff WS.** 1997. DNA binding and phasing analyses of Tn*5* transposase and a monomeric variant. *Nucleic Acids Res* **25:**2153–2160.
43. **Whitfield CR, Shilton BH, Haniford DB.** 2012. Identification of basepairs within Tn*5* termini that are critical for H-NS binding to the transpososome and regulation of Tn*5* transposition. *Mob DNA* **3:**7.
44. **Lease RA, Belfort M.** 2000. A trans-acting RNA as a control switch in *Escherichia coli*: DsrA modulates function by forming alternative structures. *Proc Natl Acad Sci U S A* **97:**9919–9924.
45. **Ross JA, Wardle SJ, Haniford DB.** 2010. Tn*10*/IS*10* transposition is downregulated at the level of transposase expression by the RNA-binding protein Hfq. *Molec Microbiol* **78:**607–621.
46. **Ansong C, Yoon H, Porwollik S, Mottaz-Brewer H, Petritis BO, Jaitly N, Adkins JN, McClelland M, Heffron F, Smith RD.** 2009. Global systems-level analysis of Hfq and SmpB deletion mutants in *Salmonella*: implications for virulence and global protein translation. *PloS One* **4:**e4809.
47. **Case CC, Simons EL, Simons RW.** 1990. The IS*10* transposase mRNA is destabilized during antisense RNA control. *EMBO J* **9:**1259–1266.
48. **Kittle JD, Simons RW, Lee J, Kleckner N.** 1989. Insertion sequence IS*10* anti-sense pairing initiates by an interaction between the 5′ end of the target RNA and a loop in the anti-sense RNA. *J Mol Biol* **210:**561–572.
49. **Ross JA, Ellis MJ, Hossain S, Haniford DB.** 2013. Hfq restructures RNA-IN and RNA-OUT and facilitates antisense pairing in the Tn*10*/IS*10* system. *RNA* **19:**670–684.
50. **Salvail H, Caron MP, Belanger J, Masse E.** 2013. Antagonistic functions between the RNA chaperone Hfq and an sRNA regulate sensitivity to the antibiotic colicin. *EMBO J* **32:**2764–2778.
51. **Desnoyers G, Masse E.** 2012. Noncanonical repression of translation initiation through small RNA recruitment of the RNA chaperone Hfq. *Genes Dev* **26:**726–739.

52. **Vytvytska O, Moll I, Kaberdin VR, von Gabain A, Blasi U.** 2000. Hfq (HF1) stimulates ompA mRNA decay by interfering with ribosome binding. *Genes Dev* **14:**1109–1118.

53. **Geissmann TA, Touati D.** 2004. Hfq, a new chaperoning role: binding to messenger RNA determines access for small RNA regulator. *EMBO J* **23:**396–405.

54. **Soper TJ, Doxzen K, Woodson SA.** 2011. Major role for mRNA binding and restructuring in sRNA recruitment by Hfq. *RNA* **17:**1544–1550.

55. **Case CC, Roels SM, Jensen PD, Lee J, Kleckner N, Simons RW.** 1989. The unusual stability of the IS*10* anti-sense RNA is critical for its function and is determined by the structure of its stem-domain. *EMBO J* **8:**4297–4305.

56. **Mikulecky PJ, Kaw MK, Brescia CC, Takach JC, Sledjeski DD, Feig AL.** 2004. *Escherichia coli* Hfq has distinct interaction surfaces for DsrA, rpoS and poly(A) RNAs. *Nat Struct Molec Biol* **11:**1206–1214

57. **Sauer E, Weichenrieder O.** 2011. Structural basis for RNA 3′-end recognition by Hfq. *Proc Natl Acad Sci U S A* **108:**13065–13070.

58. **Storz G, Vogel J, Wassarman KM.** 2011. Regulation by small RNAs in bacteria: expanding frontiers. *Molec Cell* **43:**880–891.

59. **McLellan CR.** 2012. *Regulation of the bacterial transposon Tn5: target deformation drives integration and the* Escherichia coli *host proteins H-NS and Hfq modulate transposition. Ph.D. dissertation* University of Western Ontario, London, Ontario, Canada.

60. **Le Derout J, Boni IV, Regnier P, Hajnsdorf E.** 2010. Hfq affects mRNA levels independently of degradation. *BMC Molec Biol* **11:**17.

61. **Kennedy AK, Haniford DB, Mizuuchi K.** 2000. Single active site catalysis of the successive phosphoryl transfer steps by DNA transposases: insights from phosphorothioate stereoselectivity. *Cell* **101:**295–305.

62. **Zhou L, Mitra R, Atkinson PW, Hickman AB, Dyda F, Craig NL.** 2004. Transposition of hAT elements links transposable elements and V(D)J recombination. *Nature* **432:**995–1001.

63. **Mitra R, Fain-Thornton J, Craig NL.** 2008. piggyBac can bypass DNA synthesis during cut and paste transposition. *EMBO J* **27:**1097–1109.

64. **Hencken CG, Li X, Craig NL.** 2012. Functional characterization of an active Rag-like transposase. *Nat Struct Molec Biol* **19:**834–836.

65. **Klimasauskas S, Kumar S, Roberts RJ, Cheng X.** 1994. HhaI methyltransferase flips its target base out of the DNA helix. *Cell* **76:**357–369.

66. **Hollis T, Ichikawa Y, Ellenberger T.** 2000. DNA bending and a flip-out mechanism for base excision by the helix-hairpin-helix DNA glycosylase, *Escherichia coli* AlkA. *EMBO J* **19:**758–766.

67. **Bischerour J, Chalmers R.** 2007. Base-flipping dynamics in a DNA hairpin processing reaction. *Nucleic Acids Res* **35:**2584–2595.

68. **Bischerour J, Chalmers R.** 2009. Base flipping in Tn*10* transposition: an active flip and capture mechanism. *PloS One* **4:**e6201.

69. **Allingham JS, Wardle SJ, Haniford DB.** 2001. Determinants for hairpin formation in Tn*10* transposition. *EMBO J* **20:**2931–2942.

70. **Callura JM, Dwyer DJ, Isaacs FJ, Cantor CR, Collins JJ.** 2010. Tracking, tuning, and terminating microbial physiology using synthetic riboregulators. *Proc Natl Acad Sci U S A* **107:**15898–15903.

71. **Chappell J, Takahashi MK, Meyer S, Loughrey D, Watters KE, Lucks J.** 2013. The centrality of RNA for engineering gene expression. *Biotechnol J* **8:**1379–1395.

72. **Ma C, Simons RW.** 1990. The IS*10* antisense RNA blocks ribosome binding at the transposase translation initiation site. *EMBO J* **9:**1267–1274.

73. **Mutalik VK, Qi L, Guimaraes JC, Lucks JB, Arkin AP.** 2012. Rationally designed families of orthogonal RNA regulators of translation. *Nat Chem Biol* **8:**447–454.

74. **Qi L, Lucks JB, Liu CC, Mutalik VK, Arkin AP.** 2012. Engineering naturally occurring trans-acting non-coding RNAs to sense molecular signals. *Nucleic Acids Res* **40:**5775–5786.

75. **Takahashi MK, Lucks JB.** 2013. A modular strategy for engineering orthogonal chimeric RNA transcription regulators. *Nucleic Acids Res* **41:**7577–7588.

76. **Kleckner N, Bender J, Gottesman S.** 1991. Uses of transposons with emphasis on Tn*10*. *Methods Enzymol* **204:**139–180.

Mobile DNA, 3rd Edition
Nancy L. Craig, Michael Chandler, Martin Gellert, Alan M. Lambowitz, Phoebe A. Rice and Suzanne Sandmeyer

doi:10.1128/microbiolspec.MDNA3-0010-2014

Joseph E. Peters[1]

30 Tn7

INTRODUCTION

The bacterial transposon Tn7 is distinguished by the levels of control it displays over when and where it directs transposition and its capacity to utilize different kinds of target sites. Over the 10 years since the second edition of Mobile DNA there have been many advances in our understanding of Tn7 (1). This chapter focuses on new findings since the previous edition and on areas not covered in other review articles on Tn7 (2, 3, 4). One significant finding over the past 10 years is the appreciation of the dissemination of Tn7, and related elements called Tn7-like elements that contain homologs of the Tn7 transposition proteins, in highly diverged bacteria adapted to a remarkable number of different environments (3, 5, 6). The success of these elements very likely stems from the control they have over the targets they select. The well-studied canonical Tn7 element stands as an important model system for understanding the regulation of transposition and provides insight into how Tn7-like elements and more-distantly related elements may function.

All mobile elements confront the same two fundamental problems: how to do as little harm to their host while still moving often enough to remain one step ahead of the inevitable occurrence of inactivating mutations. The apparent solution for Tn7 is to insert using two pathways of transposition that are each highly unlikely to harm the host, but yet maximize vertical and horizontal transfer of the element (Fig. 1). One pathway of transposition is directed into a single neutral site that is found in bacteria called its attachment site, or *attTn7*, a strategy that favors vertical transmission of the element to daughter cells (7, 8, 9). A second pathway directs transposition preferentially into mobile plasmids called conjugal plasmids and thereby facilitates horizontal transfer of the element to new hosts (10). This system involving two targeting pathways maximizes the distribution of Tn7 while minimizing the possibility of the element inserting into genes that are normally important or essential to the host.

Five transposon-encoded proteins, TnsA, TnsB, TnsC, TnsD, and TnsE (TnsABCDE), and *cis*-acting left and right end sequences mediate Tn7 transposition (Fig. 1) (11). TnsA and TnsB constitute the transposase, TnsD and TnsE are target specificity factors, and TnsC is a regulator protein that communicates between the transposase and targeting proteins. Arguably, the central feature that enables the control found with Tn7 hinges on a heteromeric TnsA and TnsB transposase, which only initiates transposition when an appropriate place to insert has been identified (12, 13, 14). Although all transposable elements have some level of target site specificity or bias, a limited number have been shown to control transposition such that they only initiate the process when certain specific targets are identified.

The TnsA protein appears to be a unique adaptation of Tn7 and its relatives. TnsA does not show homology to other proteins based on its primary amino acid

[1]Department of Microbiology, Cornell University, 175a Wing Hall, Ithaca, New York 14853.

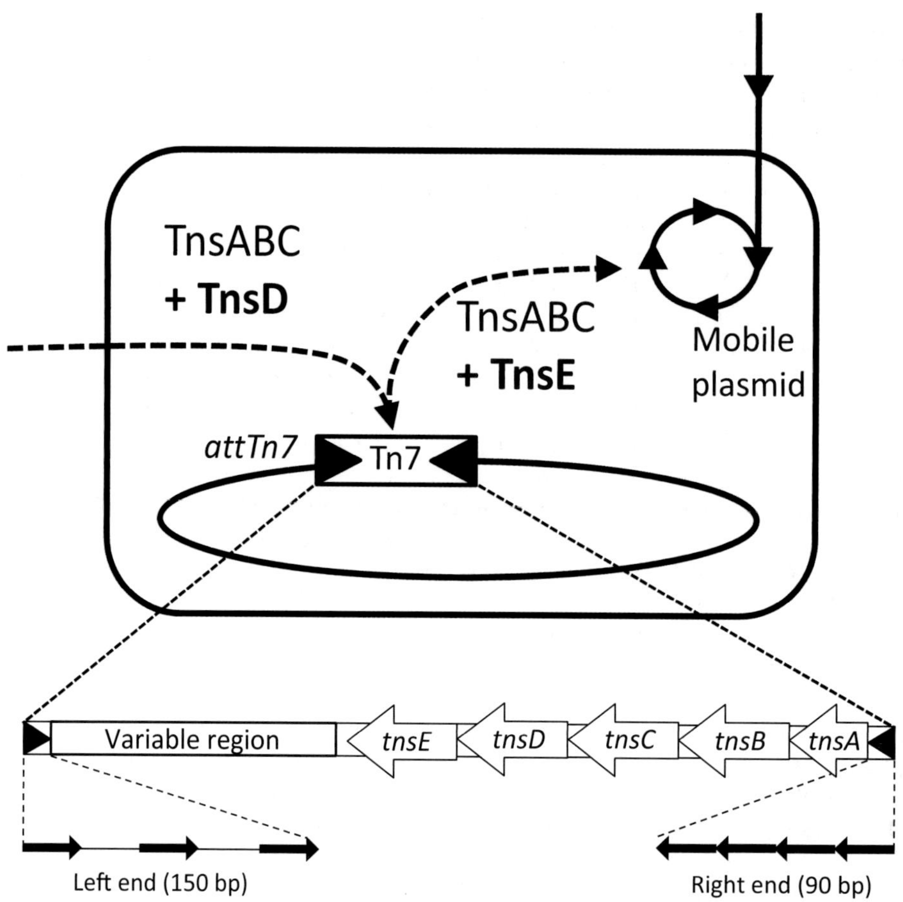

Figure 1 Tn7 encodes two pathways that recognize different types of target site. One pathway is directed by the TnsABC+TnsD proteins and directs insertions into a single site, called *attTn7*, found in bacteria. Insertion into the *attTn7* site does not appear to harm the bacterial host and likely maximizes vertical transmission of the element. A second pathway directed by the TnsABC+TnsE proteins directs insertions into plasmids capable of mobilizing between bacteria when they enter the host. Insertion into mobile plasmids probably facilitates the horizontal transfer of the element. Neither pathway is likely to inactivate host genes. The positions of the five Tn7 genes required for transposition, *tnsA, tnsB, tnsC, tnsD*, and *tnsE*, are shown including the "Variable region" that contains genes likely to benefit the element and/or the bacterial host. The right and left ends are indicated with black triangles and showing the layout of the seven 22-bp TnsB binding sites with black arrows.
doi:10.1128/microbiolspec.MDNA3-0010-2014.f1

sequence, but it is structurally related to restriction enzymes in the N-terminal portion of the protein and the H5 histone at its C-terminal region (12, 15). As explained in more detail below, having TnsA that cleaves at the 5′ ends of the element in addition to the TnsB transposase that liberates the element at the 3′ ends allows Tn7 to move by cut-and-paste transposition in a different way than other transposons. TnsB is a member of the retroviral integrase superfamily of transposases (16, 17). This large group of proteins includes other well-characterized bacterial DNA transposases like those found in Tn5, Tn10, and bacteriophage Mu. In addition to catalyzing breakage and joining events at the 3′ ends of the element during transposition, the TnsB protein is also responsible for recognizing the sequences that delineate the left and right ends of the element (Fig. 1) (18, 19). Together the co-dependent TnsA and TnsB proteins mediate complete excision of the element from a donor DNA and insertion into a preselected target site (20).

A core machinery of TnsA, TnsB, and TnsC is necessary but not naturally sufficient for Tn7 transposition without other Tn7-encoded proteins. Communication between the TnsA/TnsB transposase and the two dedicated target site selection proteins TnsD and TnsE is primarily mediated via the regulatory protein TnsC (2, 11, 21, 22). TnsD and TnsE will only signal the TnsA/TnsB transposase to initiate excision of the element through TnsC when they engage an appropriate target site. TnsD is a sequence-specific DNA binding protein that recognizes sequences at the *attTn7* site and activates high-frequency insertion into this site, while the TnsE protein is a structure-specific DNA binding protein that senses features of replication associated primarily with conjugal plasmids as they enter a cell (Fig. 1). Interestingly, Tn7 has an additional level of control beyond choosing where and when transposition occurs; it also controls the orientation of the resulting transposition events (by an unknown mechanism).

Tn7 was first isolated in *Escherichia coli* from an animal agriculture setting by the antibiotic resistance it confers (23, 24). Although Tn7 was regularly identified in hospital settings for decades it was unclear how widespread Tn7 was in the environment. The burgeoning databases of DNA sequence information from whole genomes or environmental samples now reveal that Tn7 and other elements with clear *tnsABCDE* genes, called Tn7-like elements, are extremely widespread in very different environments and from remarkably divergent bacteria. At this point it would be almost impossible to catalog the many hundreds of elements with obvious homologs to the five *tns* genes originally identified in Tn7 (3, 5). Even within the elements that are nearly identical to Tn7 in the left and right end sequences and *tns* genes there can be high levels of diversity in the so-called variable region found in the element (Fig. 1). In some Tn7-like elements diversity is further facilitated by the presence of an integron cassette system, which can include any number of different antibiotic resistance determinants (25, 26). Tn7-like elements that have diverged from Tn7 do not frequently have integron systems, but still display a broad variety of different families of gene products, as well as many genes of unknown function (3).

TOOLS FOR THE STUDY OF Tn7 TRANSPOSITION

As reviewed in detail elsewhere, *in vitro* reactions have been invaluable in establishing the molecular role of the Tns proteins and the progression of Tn7 transposition reactions (1, 13, 14, 20, 27, 28, 29). Our current understanding of Tn7 has also been made possible by a robust and adaptable genetic screen, called the papillation assay, for mutations that either increase or decrease the frequency of transposition (8, 30). The papillation assay used for monitoring transposition in a promoter capture assay where a synthetic miniTn7 element containing the left and right ends of Tn7 flanking the *lacZ* and *lacY* genes required for lactose import and utilization. However, in the construct used in this assay there is no promoter and when the miniTn7 element resides in a transcriptionally silent region of the chromosome the cells are phenotypically Lac^-. However, if the element transposes to new locations, insertion events in transcriptionally active regions of the chromosome will result in Lac^+ colonies that outgrow the lawn of otherwise Lac^- cells on appropriate indicator media. Mutants derived from genetic screens with this assay have been used with the cell-free reconstituted *in vitro* transposition reactions and other *in vitro* analyses to indicate important aspects of Tn7 transposition. Sensitive and quantitative *in vivo* transposition systems have also been important for assessing the effect of mutant Tn7 proteins and host mutations on the frequency and targeting of transposition (8, 11). More recently, bioinformatics analysis has also been of critical importance for helping identify important residues in the proteins for mutational analysis. This combination of experimental tools has allowed for a solid understanding of the basics of Tn7 transposition and the utility of these tools promises to help address the remaining questions with this adaptable element.

THE CORE TnsA, TnsB, AND TnsC MACHINERY

TnsA, TnsB, and TnsC are believed to be involved in all TnsAB interdependent transposition reactions. Tn7 is completely excised from a target DNA via double-stranded DNA breaks on both ends of the element. The heteromeric TnsAB transposase that makes these breaks is an important feature of Tn7, and work with this element provides a model system for understanding other elements with a heteromeric transposase. In the cut-and-paste mechanism found with Tn7, TnsA is responsible for cleaving on the 5′ ends of the transposon 3 bp outside the terminal bases of the element (although these bases are processed off by the host during repair after transposition) (Fig. 2) (12, 14, 20). TnsB is responsible for breaks at the 3′ ends that are directly joined to a target DNA (Fig. 2). These joining events to the target DNA that are carried out by TnsB occur at positions that are staggered by 5 bp. Repair of the 5-bp gaps at the ends of the element leads to the 5-bp target site duplication that is found with all Tn7 and Tn7-like elements investigated to date.

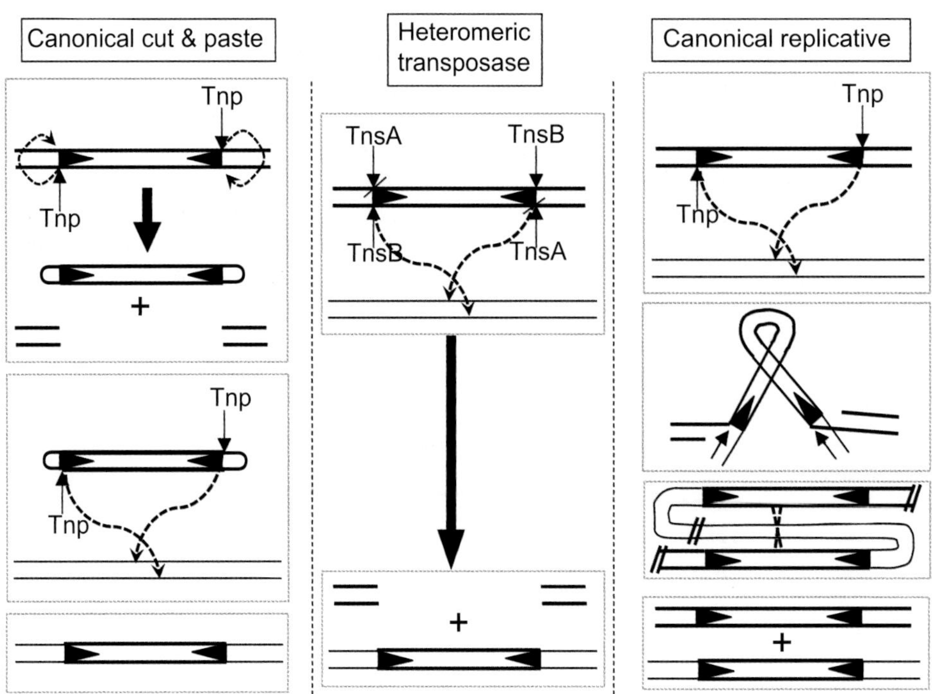

Figure 2 Mechanisms of DNA transposition. Canonical cut-and-paste transposition involves two separate reactions (left side). Canonical replicative transposition involves a direct joining reaction to the target DNA and extensive DNA replication, often along with resolution of a co-integrate (right side). Tn7 utilizes a heteromeric transposase that is capable of directly joining the broken ends of the element to a target DNA, but does not require extensive processing after transposition because it has a second protein, the TnsA endonuclease (center). The rectangle indicates the transposon element and the triangles indicate the *cis*-acting sequences at the ends of the element. (See the text for details.)
doi:10.1128/microbiolspec.MDNA3-0010-2014.f2

TnsB is also responsible for recognizing the left and right ends of the element (18, 19). The ends of Tn7 are similar, but not identical. The left end of Tn7 is about 150 bp long and contains three spaced binding sites for TnsB that are each 22 bp in length, whereas the right end of Tn7 is about 90 bp in length with four juxtaposed sites (Fig. 1). Analysis of the nucleoprotein complex showed that TnsB occupies each binding site in the complex that promotes transposition (31).

The molecular mechanism that allows Tn7 to control strictly the orientation of insertion events remains to be determined. Presumably this is somehow sensed by the configuration with which TnsB binds to the seven sites distributed between the left and right ends of the element and the interaction with TnsA. However, it is known from work with a core machinery composed of a mutant form of TnsC, TnsC^{A225V}, (i.e., TnsABC^{A225V}), which allows transposition without a target site selecting protein, that TnsA, TnsB, and TnsC are the only requirements for orientation specificity (32, 33, 34), a process reviewed in detail elsewhere (1).

Experiments with hybrid elements with various combinations of the left and right end sequence indicate that transposition can occur with two right ends, but not two left ends (18, 19). Although TnsB is the protein in the transposase that recognizes the left and right ends, TnsA greatly facilitates a process in which the ends are brought together to form the paired end complex that is required for transposition (35). As the paired end complex is only formed *in vitro* with two right ends or left and right ends (i.e., the same combinations needed for transposition), it has been suggested that the TnsA and TnsB interaction with certain configurations of the TnsB binding sites is an important regulatory step in the transposition process (35).

The Heteromeric Transposase and Multiple Targeting Pathways

The heteromeric transposase of Tn7 provides a unique mechanism for cut-and-paste transposition, but could it also be important for maintaining two target selection pathways? In many examples of cut-and-paste transposition, a transposase cuts on the 3′ ends of the element and removal of the element occurs via a hairpin intermediate found on the transposon or target DNA (Fig. 2, left side) (36, 37, 38, 39). The transposase then acts a second time to join the excised element to a target DNA. Presumably, given the success of these elements, there are advantages of regulating transposition independently of signals from the target DNA (Fig. 2, left side). Other types of elements directly join the ends of the element to the target DNA without the hairpin intermediate. For example, the process of replicative transposition involves direct joining to the target, where the new free 3′ ends in the target DNA are used to initiate DNA replication that proceeds across the entire element making a co-integrate structure between the donor and target DNAs that is often reconciled by a site-specific resolution system found in these elements (Fig. 2, right side). An advantage of the process of replicative transposition involves duplication of the elements, something essential for the lytic program of bacteriophage Mu.

In the case of Tn7, using a heteromeric transposase that couples excision of the element to target recognition, the TnsAB transposase completely liberates the element and directly joins the 3′ ends to the target DNA only when specific target sites are recognized (Fig. 2, center). This seemingly subtle requirement may have provided a unique opportunity to acquire the two specific targeting pathways with this cut-and-paste element. Of additional interest, there is evidence that different targeting proteins may exist that recognize other types of targets in other elements that are more distantly related to Tn7 and Tn7-like elements that also appear to use a heteromeric transposase (see below).

TnsD-MEDIATED TRANSPOSITION

Tn7 can choose between two different types of insertion site. The ability of Tn7 to direct transposition to the *attTn7* site found in bacteria probably provides multiple benefits to the element. Of primary importance is likely the ability to reduce greatly the chances of insertionally inactivating important or essential genes in the host. In support of the utility of this ability of Tn7 is the finding that nearly all Tn7 and Tn7-like elements identified using bioinformatics are found in the *attTn7* site of the host (3). The frequency of transposition into the *attTn7* site is also very high *in vivo* and *in vitro*, something that may be only naturally permissible with transposons that target a very specific site that is highly unlikely to inactivate host genes (8, 20, 40). Consistent with this idea, hyperactive transposase mutants can be isolated relatively easily in other transposon systems, suggesting that there is genetic pressure to maintain the frequency of transposition at a low level in elements that lack the site-specific targeting found with Tn7 (41, 42, 43, 44, 45, 46).

The TnsD protein functions with the core TnsABC machinery to direct transposition into the *attTn7* site and is reviewed in detail elsewhere (1). TnsD binds to a DNA sequence located in the 3′ end of the *glmS* gene (Fig. 3) (40). The *glmS* gene encodes the essential glucosamine 6-phosphate synthase (GlmS). GlmS catalyzes the formation of the glucosamine 6-phosphate product from a fructose 6-phosphate precursor in which ammonia is derived from glutamine. The product of this reaction, *N*-acetylglucosamine, is an essential building block of cell walls in bacteria. Although TnsD binds a conserved region of the GlmS coding region it does not direct transposition into the essential gene, but instead at a single position 25 bp downstream in the transcriptional terminator for the *glmS* gene (Fig. 3). Careful experimental analysis of the expression of *tns* genes has not been undertaken, but transcription from the upstream *glmUS* genes may contribute to the expression of the genes required for mobilization of the element, which provides cues to the metabolic state of the host (47). Sequences at the specific point of insertion do not appear to be important for the element, and only the adjacent sequence recognized by TnsD within glmS (3, 5). Recognition of a DNA sequence encoding an essential (and therefore conserved) gene, but not destroying it during transposition, has very likely contributed to the broad distribution of this element in highly diverged bacteria.

Complexes Involved in TnsABC+D Transposition

An elaborate preintegration nucleoprotein complex (also called a transpososome) that involves TnsA, TnsB, TnsC, and TnsD is assembled with the left and right ends of the element and the target *attTn7* DNA sequence to regulate transposition into this site. The analysis of this complex is facilitated *in vitro* through the use of Ca^{2+} in setting up the reaction, instead of the normal Mg^{2+} divalent metal. Using Ca^{2+} instead of Mg^{2+} supports the assembly of the complex without allowing the chemistry of transposition to initiate. Detection of the Ca^{2+} transpososome complex by gel electrophoresis requires the use of a protein cross-linker. However, the role of this complex could be confirmed with a remarkable experiment that indicated it was still

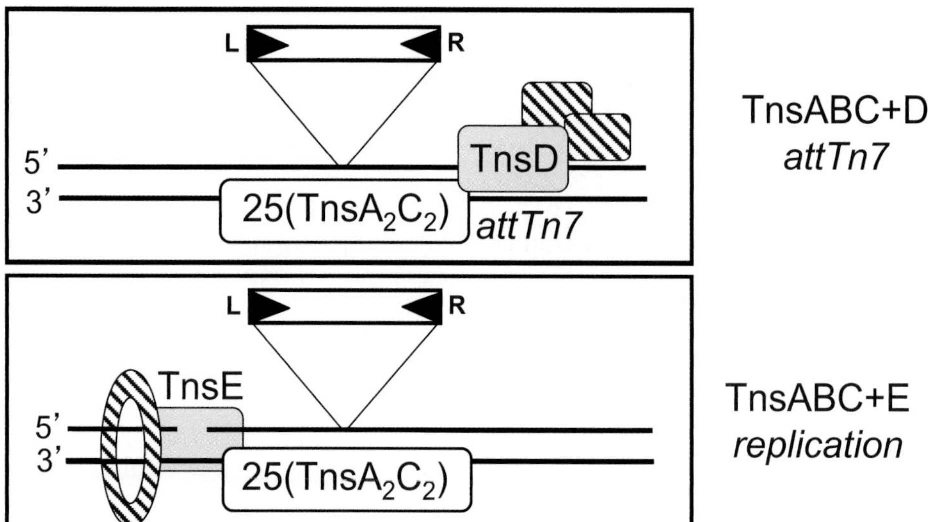

Figure 3 The relationship between the positions of Tns-protein and host-protein binding relative to the point of insertion and the orientation of the element. In TnsABC+TnsD-mediated transposition into the *attTn7* site, TnsD binds to the very C-terminal coding region of the *glmS* gene (not shown). Two host proteins, ACP and L29, help in this binding reaction (cross-hatched ovals). Approximately 25 $TnsA_2TnsB_2$ complexes are recruited to the *attTn7* site by TnsD, which appear to encompass a region across both sides of the point of insertion. Based on the findings with TnsD and the TnsABC+TnsE *in vitro* reaction, a plausible model for TnsE-mediated targeting of the lagging-strand template during DNA replication has the protein to the left of the point of insertion and the orientation of the element. TnsE interacts with the 3′ recessed end structure and the sliding clamp processivity factor protein (cross-hatched ring). TnsE may recruit a similar array of $TnsA_2TnsB_2$ complexes as found with the TnsABC+TnsD complexes. doi:10.1128/microbiolspec.MDNA3-0010-2014.f3

competent for bona fide transposition chemistry: isolation of a gel slice that contained this complex and incubation with Mg^{2+} resulted in the same products found with the standard solution-based transposition assay (48). Although formation of some of this preintegration complex can occur at low levels without TnsA, the TnsA protein stimulates the formation of much greater amounts of complex. The TnsABC+D complex (i.e., the complex that includes TnsA) is also much more stable than the TnsBC+D complex when challenged by incubation at elevated temperatures (48). These experiments indicate that there is a complex containing TnsABC+D and the end sequences and the *attTn7* target DNA in this transposition pathway prior to integration.

The complex involved in transposition becomes more stable after the element is inserted into target DNA and analysis of this complex reveals information about the number of proteins involved in the process. Following transposition, all of the components remain in the complex which indicates that TnsC and TnsD play structural roles in the nucleoprotein complex other than serving as "loaders" of the various components onto the target DNA (31). Interestingly, unlike the complex that is trapped in the pretransposition state because it is assembled in Ca^{2+}, the posttransposition Mg^{2+} complex does not need to be stabilized by cross-linking agents (31). Barring any unexpected effects of having Ca^{2+} instead of Mg^{2+} in the complex, this result implies that the release of the transposon ends and joining to the target DNA stabilizes the structure. A posttransposition complex that is more stable than the pretransposition complex would help drive the reaction in the correct direction, something also suggested with other systems (49, 50). Work with the posttransposition complex also suggests that the terminal TnsB-binding sites are more protected following transposition when compared with TnsB binding alone, supporting the idea that a tighter structure results after release of the element from the target DNA and integration into the *attTn7* site. The stability of this complex makes it likely that a presently unknown host factor or factors are required to disassemble the proteins following transposition, something that is known to be important in the bacteriophage Mu transposition system (51). Protection of the gaps at the ends of the element suggests that repair will not commence until this host-driven process occurs.

The posttransposition complex appears to encompass a region of about 350 bp of sequence including the left and right ends of the element and the region flanking the site of insertion. Examination of the transposome complex with Tns proteins tagged with fluorophores and radiolabeled *attTn7* DNA also gave important, and in some ways unexpected, estimates of the number of individual Tns proteins in the posttransposition complex. Assuming that there is one TnsD protein per *attTn7* site, about seven TnsB proteins are calculated to be in the posttransposition complex, consistent with the seven TnsB-binding sites found in the ends of the element. Unexpectedly, the $TnsA_2C_2$ complexes appear to be much more numerous, estimated at 25 $TnsA_2C_2$ complexes in the posttransposition transposome (31). Presumably, TnsA and TnsC are responsible for protecting the region flanking the insertion site in the footprint that was determined with the posttransposition complex (31). Only the TnsA and TnsB proteins positioned at the very ends of the element will carry out the chemistry involved in transposition, which indicates that an unknown signal must trigger activation of these proteins. If TnsB resides on the left and right end sequences and $TnsA_2C_2$ complexes reside in the *attTn7* target DNA, the only interface between TnsA and TnsB molecules would be at the transposon ends and possibly provides this signal.

Functional and Physical Organization of TnsA

The 273-amino-acid TnsA protein is the smallest of the transposition proteins in Tn7 and is responsible for cleaving at the 5′ ends of the element. Information provided by the crystal structure of TnsA was essential for unlocking its unexpected origins (15). Transposases from DNA transposition systems studied from a variety of bacteria (and including the TnsB protein of Tn7) fall into the retroviral integrase superfamily of transposases. It therefore came as a surprise when the crystal structure of TnsA indicated that it possessed the folds of an exonuclease similar to the restriction domain of the FokI restriction enzyme. This N-terminal portion of TnsA found at amino acid positions 77 to 168 defines the TnsA endonuclease N-terminal family (*Tn7_Tnp_TnsA_N*) (PF08722) that is part of the highly diverse PD-(D/E)XK phosphodiesterase superfamily (Fig. 4) (15, 52, 53). Mutations in TnsA identify the catalytic residues (E63, D114, K132) involved in coordinating the metals involved in transposition (Fig. 4). Interestingly, gain-of-activity and loss-of-activity mutations in TnsA that allow transposition without the target site selection proteins (54) map to the 65 to 81 amino acid region of TnsA that forms a portion of the alpha-helix that also includes the E63 active site residue (Fig. 4) (15).

Functional interaction between TnsA and TnsB was first evidenced in their co-dependence for transposition; no chemistry can be found with either protein alone (12, 14). The interdependence appears to be structural given that active site mutants in TnsA only effect 5′ end cleavage but not TnsB-mediated events and vice versa. Their interdependence is also shown with the finding that they can perform TnsA and TnsB chemistry under highly relaxed *in vitro* conditions in the presence of high glycerol and Mn^{2+} without any other Tns proteins (55). Mutants of TnsA and TnsB have also been isolated that allow transposition without other Tns proteins *in vivo* and *in vitro* suggesting they represent a functional unit (35, 54, 55). Of special interest are mutations in TnsA at Y180, S181, V182, and E185 that map to a surface-exposed helix in the C-terminus of the protein that appear to facilitate an interaction with TnsB (35); TnsA and TnsB can interact in solution as purified proteins and consistently loss-of-function $TnsA^{Y180A/S181A}$ mutants reduce this interaction and gain-of-function $TnsA^{Y180H/S181P}$ mutants increase this interaction. Additionally, TnsA stimulates the ability of TnsB to bind the Tn7 ends to form a paired end complex that contains both TnsA and TnsB; consistently, formation of this complex is reduced in a predictable fashion with a loss-of-activity mutant ($TnsA^{Y180A/S181A}$). The region 170 to 248 defines the TnsA endonuclease C terminal domain (*Tn7_Tnp_TnsA_C*) (pfam08721) (Fig. 4) (53). The C-terminal domain of the protein includes a helix-turn-helix domain and shows structural similarities with histone H5 (15, 56). The C-terminal domain of TnsA likely mediates interactions with TnsB, TnsC, and DNA (see below).

Not only do full-length TnsA and TnsC form a stable interaction, but also a cocrystal of TnsA and a C-terminal fragment of TnsC (504 to 555) could be used to define a hydrophobic pocket in TnsA that allows for this interaction (Fig. 4) (15, 56). Gross structural changes in TnsA are not found when the protein is bound to the fragment of TnsC. However, it was noted that TnsA contains two Mg^{2+} when crystalized alone and only one in the cocrystal (56). Significantly, although there are no major changes in the structure of TnsA in the cocrystal, changes in activity could be found with a $TnsA/TnsC^{495-555}$ complex. Although TnsA will not bind DNA alone, the $TnsA/TnsC^{495-555}$ complex will bind DNA in a sequence-independent manner (56) (see below). The interaction between TnsA and $TnsC^{504-555}$ as resolved by the cocrystal also

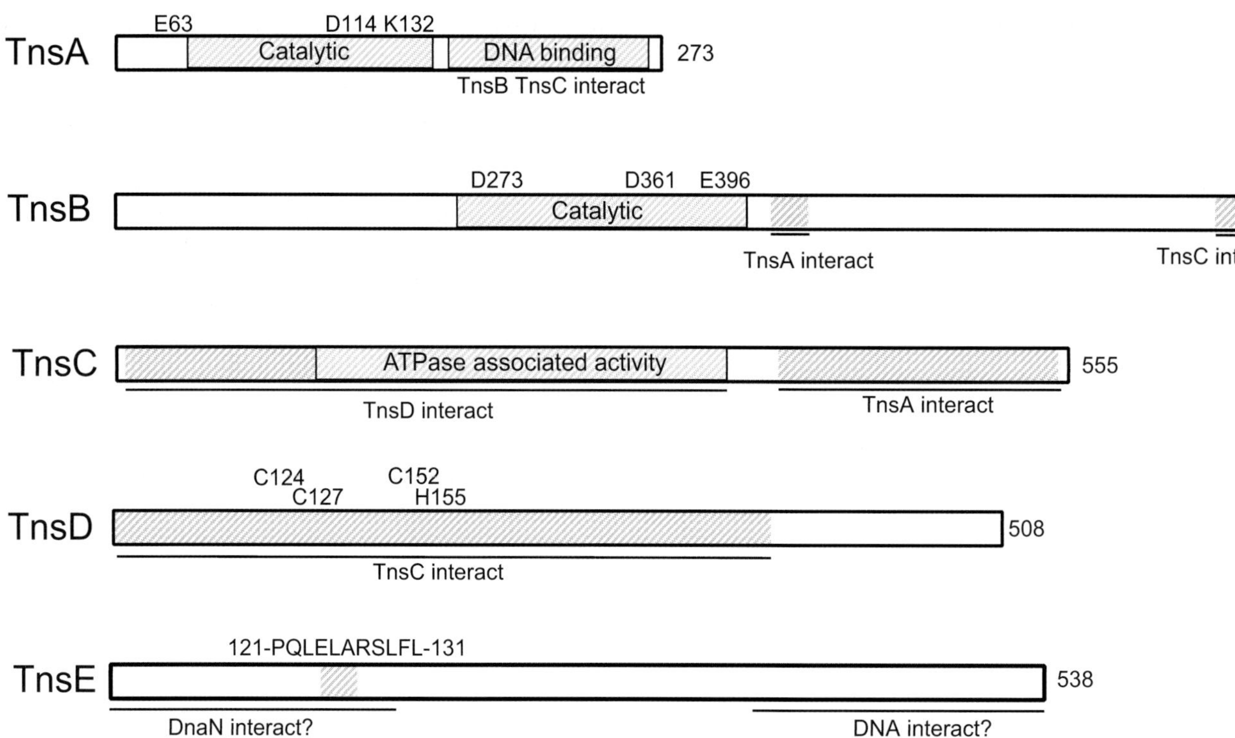

Figure 4 The relevant features of the Tns proteins of Tn7 and their total length. TnsA encodes a conserved N-terminal catalytic domain [77 to 168 amino acids (aa)] as indicated and the amino acids that coordinate the metal are shown (E63, D114, and K132). The conserved C-terminal region of TnsA is shown (170 to 248 aa) as is the region that interacts with TnsB, TnsC, and DNA. TnsB encodes a conserved catalytic domain (266 to 406 aa) and the amino acids that coordinate the metal are shown (D273, D361, and E396). The regions of TnsB that interact with TnsA (440 to 480 aa) and TnsC (662 to 702 aa) are indicated. TnsC contains a conserved domain from the AAA family (126 to 281 aa). The 1 to 293 region of TnsC interacts with TnsD and the 504 to 555 region of TnsC interacts with TnsA. TnsD has a CCCH zinc finger motif (C124, C127, C152, H155) and the region 1 to 309 interacts with TnsC. TnsE contains a sliding clamp interacting motif 121 to 131 aa and may generally act with DnaN across the N-terminus of the protein and DNA with the C-terminus of the protein based on TnsE gain-of-activity and loss-of-activity mutations. (See the text for details.) doi:10.1128/microbiolspec.MDNA3-0010-2014.f4

indicates that the catalytic residues remain available for cleavage in this structure.

Functional and Physical Organization of TnsB

The 702-amino-acid TnsB protein shows homology to other bacterial transposases in the retroviral integrase superfamily of transposases. The region from 266 to 406 aligns with the integrase core domain family (*rve*) (pfam00665) (53, 57), a region that coincides well with results from site-directed mutations that identified the residues that coordinate the metal, D273, D361, and E396 (Fig. 4) (12). As mentioned above, TnsB is responsible for recognizing the left and right ends of the element and *in vitro* TnsB will allow these ends to be paired in a process that is stimulated by TnsA. A TnsA–TnsB solution interaction with full-length or truncated constructs of TnsB was used as a tool to localize the region of TnsA interaction within the 440 to 480 region of TnsB (35). Supporting the putative contact region between the proteins, mutations in conserved residues in this 440–480 region, $TnsB^{D467A/L468A}$, produced a protein that did not show the TnsA-stimulated effects on end binding found with the wild type protein.

The most C-terminal 40 amino acids of TnsB are required for interaction with TnsC. While the $TnsB^{1-662}$ truncation can form a complex with the paired left and right ends of the element, this protein will not support transposition *in vivo* or *in vitro* and will not form the nucleoprotein complex shown to be an intermediate in the TnsABC+D transposition reaction (see above) (48). Although, technically, interactions with TnsD in the TnsABC+D complex could have been affected, this is essentially ruled out because the same result is found with a mutant TnsABC transposition reaction mediated with the TnsC^{A225V} mutant protein. TnsB mutants with small patches of alanine residues in the C-terminal 40 amino acids of TnsB will not form a productive interaction with TnsC. This region of TnsB is also responsible for mediating a process called target immunity; this process is found in Tn7 and some other elements where the transposition machinery actively inhibits a second insertion from occurring in a region in which an element already exists (see below). Certain specific mutations in the very C-terminus of the 702-amino-acid TnsB protein (TnsB^{P686S}, TnsB^{V689M}, and TnsB^{P690L}) are capable of bypassing target site immunity (58). Interactions between the most C-terminal portions of TnsB and TnsC show multiple similarities to those found with MuA and MuB in the bacteriophage Mu system (59, 60). Another link between the transposases of Tn7 and Mu comes from the surprising ability of TnsB to perform replicative transposition in a way that is similar to the MuA transposase; it has been shown that Tn7 will readily utilize the replicative transposition mechanism involving a co-integrate structure if the active site of TnsA is inactivated, abolishing cleavage at the 5′ ends of the element (14).

Functional and Physical Organization of TnsC

The 555-amino-acid TnsC protein has multiple functions as a regulator of Tn7 transposition; it acts as an "on" and "off" switch for the TnsA/TnsB transposase and as a communicator between the TnsA/TnsB transposase and the target site selection proteins. The ability of TnsC to control transposase activity involves nucleotide binding. Details of the role of the nucleotide binding state of TnsC are reviewed elsewhere (1, 61) and are reminiscent of the regulatory role carried out by MuB with bacteriophage Mu transposition. Productive TnsABC+D transposition *in vitro* requires ATP and will not occur in the presence of ADP (20). Constitutive (but untargeted) transposition is found *in vitro* in the presence of nonhydrolysable forms of ATP even in the absence of the normal target site selection proteins (i.e., TnsABC and AMP-PNP).

TnsC is a member of the large family of ATPase Associated Activity (AAA) proteins based on homology located at 126 to 281 (Fig. 4) (2, 53) and experimentally TnsC is found to display a weak ATPase activity (62). Gain-of-function mutations in TnsC that allow it to participate in transposition with TnsABC alone can be isolated and these mutations can be grouped into two classes (21). The gain-of-function TnsC^{A225V} mutant is the best-studied TnsC mutant of a class that can work without any target site selection protein (i.e., TnsABC^{A225V} only) and that can still respond to targeting signals from TnsD (targeting *attTn7*) or TnsE (targeting conjugal plasmids). The A225V change in TnsC^{A225V} is near one of the conserved ATP binding motifs and the TnsC^{A225V} protein is found to display reduced ATPase activity as compared to the wild-type protein (62). The ability to bind and hydrolyze nucleotides is critical in a process called target immunity in which transposition into regions near an existing copy of Tn7 is strongly inhibited (see below).

Multiple types of experiments indicate the region of TnsC that interacts with TnsA. Affinity-tagged TnsA will interact with full-length TnsC as well as a truncated TnsC$^{504-555}$ derivative, but a TnsC$^{1-503}$ truncation is incapable of this interaction (see above) (56). In addition, because the cleavage sites utilized by TnsA and TnsB during transposition are only separated by 3 bp they can also be modeled into the TnsA/TnsC$^{504-555}$ complex (56). Although this model suggests that the more extended region of TnsC from 495 to 501 could block access by TnsB , they could also identify a cluster of conserved positively charged residues from 495 to 501 in TnsC that can reorient, allowing these residues to interact with the DNA providing access for TnsB cleavage (56). Consistently, in follow-up experiments it was found that while neither TnsA alone nor the TnsA/TnsC$^{504-555}$ complex (nor TnsC$^{504-555}$, nor TnsC$^{495-555}$ alone) is capable of binding DNA, a TnsA/TnsC$^{495-555}$ complex binds DNA nonspecifically. Consistent with an important role for the 495 to 509 region of TnsC in DNA binding, the DNA binding ability of the TnsA/TnsC$^{495-555}$ complex is dependent on the basic lysine residues found in the TnsC 495 to 509 region (56).

TnsC plays an important role in interacting with the target DNA to identify potential insertion sites. TnsC is believed to recognize predominantly structural features of the target DNA, but protein:protein interactions with the target site selector protein TnsD are also found to be important (6, 33). Truncated derivatives of TnsC and TnsD were used to establish that the 1 to 293 region of TnsC is responsible for interacting with TnsD, as explained below.

Functional and Physical Organization of TnsD

The 508-amino-acid TnsD protein does not have obvious extended homologs outside Tn7 and Tn7-like elements. A recognizable zinc finger domain can be found in the first 170 amino acids with a characteristic CCCH motif (Cys^{124}, Cys^{127}, Cys^{152}, and His^{155}) (Fig. 4). Mutational analysis confirms the role of the characteristic CCCH residues for recognizing the *attTn7* site and their role in transposition (6). However, deletion derivatives that remove the zinc finger domain still bind DNA, which suggests multiple DNA binding motifs may reside throughout TnsD. Although the distribution of DNA binding function across the protein is atypical, it is not without precedent (63), and confirmation of this could help account for the ability of protein to form the prominent distortion in the *attTn7* target DNA that is important for recruiting TnsC and directing TnsABC +D transposition downstream of the *glmS* gene (33).

TnsC and TnsD form a stable complex with the *attTn7* DNA sequence that can be resolved using electromobility shift assays. Comparing the region of DNA occupied (i.e., protected in DNA footprint experiments) by TnsC and TnsD verses TnsD alone indicates that TnsC very likely extends to include the specific point of insertion utilized by the element (Fig. 3) (33). Full length TnsC and TnsD are shown to interact in a yeast two-hybrid assay and deletion derivatives of the proteins indicate that the 1–293 region of TnsC interacts with the 1–309 region of TnsD protein (Fig. 4). TnsD deletion derivatives that lack a C-terminal region (e.g. TnsD$^{1-309}$) show an enhanced interaction with full-length TnsC, which could indicate that allosteric structural changes occur with other components to help regulate the availability of TnsD to interact productively with TnsC.

Host proteins are also known to be important in the TnsABC+D pathway of transposition. Early work hinted at a role for host factors in experiments showing that cell lysates increased the ability of TnsD to bind to DNA with the *attTn7* sequence (40). Subsequent work revealed that TnsD binding to *attTn7* is stimulated by two host proteins, the essential acyl carrier protein (ACP) and a nonessential component of the large subunit of the ribosome L29 (64). Presumably, ACP and L29 could play structural and/or regulatory roles in TnsD-mediated transposition into the *attTn7* site. The regions of TnsD responsible for the interaction with ACP and L29 are unknown, as is whether these interactions are conserved or dispensable in other bacterial hosts.

Consequences of Site-Specific Transposition with TnsABC+D

It is not the sequence at the specific point of insertion in *attTn7*, but the distance from the TnsD binding site that determines the point of insertion (Fig. 3). One important implication of TnsD recognizing this highly conserved region of the *glmS* gene is that the site will be maintained even in diverged hosts. As described below, target immunity will inhibit another Tn7 element from inserting into the *attTn7* from a sister chromosome. However, target immunity is only active between highly similar Tn7-like elements (see below), which probably explains why multiple diverged Tn7-like elements can naturally be found in a single *attTn7* site (Fig. 5) (3, 5). The process of multiple Tn7-like elements inserting into an *attTn7* site and the inevitable erosion of these elements as they become inactivated plays an important evolutionary role in bacteria through the formation of fitness or pathogenicity genomic islands at the *attTn7* site (Fig. 5) (3, 5).

The TnsABC+D pathway of transposition has proved itself a very useful tool for inserting genetic information into bacteria at a single neutral chromosomal site (65, 66, 67, 68, 69, 70, 71, 72, 73, 74, 75). This is based on the high frequency of transposition occurring into this site and the conservation of this site across bacteria. Given that the *glmS* gene is found in a number of organisms across all three domains of life and the site recognized by TnsD is maintained, future possibilities for extending the utility of TnsD-mediated transposition as a tool in the other domains of life exists (6, 76).

TnsE-MEDIATED TRANSPOSITION

Tn7 is not only notable for its ability to direct insertion into a single site found in all bacteria, but also its ability to choose between different types of insertion sites in bacteria. The ability of Tn7 to recognize non-*attTn7* sites is through TnsE. It is easy to imagine why a transposon might be at an evolutionary disadvantage if it only recognized a single conserved site because it would not have an efficient mechanism for horizontal transfer into other bacteria (although low-frequency transposition into similar sites could help mitigate this [77]). The TnsE-mediated transposition pathway is well adapted for facilitating horizontal transfer.

TnsE-Mediated Transposition Preferentially Occurs Into Mobile Plasmids

Tn7 transposition with the TnsABC+E proteins appears adapted for mediating the spread of the element to other hosts through its ability to target transposition specifically into mobile plasmids, also known as conjugal plasmids, capable of moving between bacteria (10). The TnsE-mediated pathway does not recognize plasmids that are not capable of mobilizing between

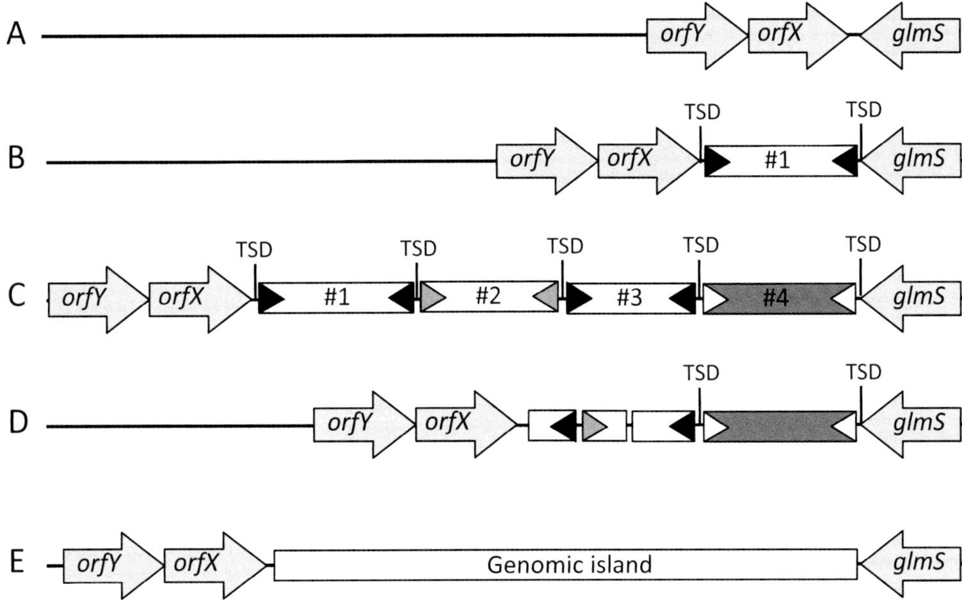

Figure 5 The TnsABC+D pathway directs transposition between *glmS* and a downstream gene. The downstream genes will differ by the particular genus and species (shown as *orfX* and *orfY*) (A). The TnsD protein recognizes sequences in *glmS* and directs insertions in one orientation at a single position downstream of *glmS* and this insertion will have the 5-bp target-site duplication (TSD) associated with this element (B). Additional Tn7-like elements that have diverged from the others can insert in series and can accumulate with new elements always inserting proximal to *glmS* at the exact same position and with the same target-site duplication (C). Over time the elements will pick up inactivating mutations and deletions that will contribute to the element eroding leading to a mix of functional and non-functional elements (D). Genomic islands can result when the transposase genes and ends are lost, but highly selected genes still reside in the *attTn7* site (E). (See the text for details.) doi:10.1128/microbiolspec.MDNA3-0010-2014.f5

bacteria, which indicates that a specific adaptation to recognize mobile plasmids exists. It was also shown that transposition into conjugal plasmids was stimulated when they entered cells and that the process is target-sequence independent. Transposition events are also directed into filamentous phage M13 possibility, indicating another mechanism to facilitate horizontal transfer of the element (78).

TnsE-Mediated Transposition Targets the Lagging-Strand Template during DNA Replication

Clues into how TnsE recognizes certain DNAs preferentially first came from the analysis of low-frequency TnsABC+E transposition events that can be isolated in the *E. coli* host chromosome. TnsE-mediated transposition occurs with a highly distinct distribution in the chromosome. There is a strong bias for transposition into the region of the chromosome where DNA replication terminates (79), but more strikingly transposition occurs with a single orientation in each of the two replicores of *E. coli*, the regions of the chromosome that are replicated by each replisome (80). The bias is explained by experiments indicating that transposition is targeted into the lagging-strand template during replication (see below).

Although a distinct regional preference is found with TnsE-mediated insertions, no preferred DNA sequence can be identified in DNA sequences utilized for insertion. Therefore, unlike TnsD, TnsE is not a sequence-specific DNA binding protein. TnsE binds DNA, but analysis of TnsE binding shows that the protein possess a structure-specific binding for 3′ recessed ends (80). Hyperactive mutations in TnsE that allow for much higher levels of transposition map to the C-terminal end of the protein and specifically enhance the ability to bind 3′ recessed end DNA structures, which indicates that this ability is important for target recognition and likely involves DNA binding in this region of TnsE (Fig. 4).

TnsE Interacts with 3′ Recessed Ends and the Sliding Clamp Protein

The minimal features that allow TnsE-mediated transposition to target the lagging-strand template during DNA replication involve TnsE interaction with a DNA structure with a 3′ recessed end and sliding clamp processivity factor (also called β-clamp or DnaN) (29). DNA polymerase and numerous other proteins that move along the bacterial chromosome during replication are tethered to the DNA via the sliding clamp. The expected distribution of sliding clamp proteins on the two template strands during DNA replication is very different. In theory, one clamp on the leading strand template could be sufficient to replicate an entire replicore. However, on the lagging-strand template replication is discontinuous and a new sliding clamp is loaded for DNA polymerase III and the old one abandoned (for a time) on the lagging-strand template. Interactions between DNA polymerase I and ligase with the sliding clamps on the lagging-strand template are thought to be important for processing the lagging-strand template. The ability of TnsE to interact with the sliding clamp provides a satisfying model for how TnsE could be recruited to the lagging-strand template. Interaction with the sliding clamp is dependent on a conserved clamp interaction sequence found at 121 to 131 in TnsE (Fig. 4). Single amino acid substitutions in the 121 to 131 region will both perturb the ability of TnsE to interact with the sliding clamp and disrupt TnsE-mediated transposition. The dissociation constant (K_D) of the interaction between TnsE and sliding clamp was found to be ~2.44 μM using surface plasmon resonance and this rather weak interaction could help explain how cells tolerate modest overexpression of TnsE.

Significantly, it was found that TnsE-mediated transposition could be reconstituted with purified Tns proteins on a circular target DNA containing a 20-bp gap. When this target plasmid was preloaded with sliding clamp the resulting insertions showed the same orientation bias as was found during DNA replication on the lagging-strand template *in vivo*. Transposition events found with the TnsABC+E *in vitro* reaction concentrated adjacent to the gap occurring in one orientation. This was significant because TnsABC+D insertions are found to occur at a single site in one orientation at a set distance from where TnsD binds to *attTn7*. Interestingly, if TnsE is recognizing the 3′ end of the 20-bp gap in the plasmid (as predicated by the preferred binding substrate *in vitro*) it would predict that TnsC is recruited differently with TnsE than is found with TnsD: TnsE binding occurs on the left side of the element instead of the right in the case of TnsD (Fig. 3).

Transposition with the TnsABC+E system *in vitro* only occurs at a low efficiency and also requires the use of high-activity mutants, which suggests that other requirements for transposition remain to be discovered. Other questions also remain to be answered. For example, nicked plasmids are not productive targets for TnsE-mediated transposition even when loaded with sliding clamp protein. This could indicate that some type of gap is required for TnsE-mediated transposition, which might also provide hints into the regional biases found with TnsE-mediated transposition. Gaps in the DNA can be argued to be more common when replication terminates or during DNA processing in recipient cells during conjugation, which could suggest a mechanism for regional biases (2). Other mobile elements have also been found to take advantage of features only found on the lagging-strand DNA template; Lambda Red recombination and the movement of multiple distinct types of mobile genetic elements preferentially or constitutively occur on the lagging-strand template (81, 82, 83, 84, 85, 86, 87). These findings are consistent with the idea that differences in processing make one strand more vulnerable during DNA replication (88).

TnsE-Mediated Transposition Targets Replication-Mediated Repair

TnsE-mediated transposition is stimulated by DNA double-strand breaks (79, 89). The mechanism used to induce the double-strand break does not appear to make a difference; UV light, phleomycin, mitomycin C, and even cut-and-paste transposon excision of another unrelated DNA transposon were found to induce specifically TnsE-mediated transposition (79, 89). Inducing a double-strand break at a specific position revealed that transposition is not targeted to the site of the break. Instead, TnsE-mediated transposition targets DNA replication induced for repair of the break. Of additional interest, replication events that were initiated for repair were only activated as targets at certain fixed small regions of the chromosome that could be hundreds of kilobytes from the original double-strand DNA break. Further work is needed to establish which features are recognized, but this work suggests that replication forks initiated for DNA repair are different to those initiated at the origin and that the effects of a break at a single position can have consequences at considerable distances from the original site of DNA damage.

Functional and Physical Organization of TnsE

The 538-amino-acid TnsE protein does not share any obvious homology with proteins other than those found in Tn7-like elements. TnsE initiates with an atypical valine start codon (80). Interaction with the sliding clamp processivity factor occurs with a conserved sliding clamp interaction sequence, 121-PQLELARSLFL-131 (Fig. 4) (29). One of the TnsE hyperactive mutations, $TnsE^{M371}$ (80), aggressively stimulates the SOS response in *E. coli* (A. R. Parks and J. P. Peters, unpublished observation), which could support that the N-terminal region of the protein that is important in interacting with the replication machinery. Hyperactive mutations that enhance the ability of TnsE to interact with DNA structures containing 3′ recessed ends are found in the C-terminal region, which suggests that this region of the protein is involved in DNA binding (80). TnsE likely multimerizes on DNA given that shifted and super-shifted products can be found with electromobility shift assays with 3′ recessed end structures (80). The ability of TnsE to multimerize is also consistent with the finding that a TnsE–TnsE interaction can be detected using surface plasmon resonance with a K_D of ~0.5 μM (A. R. Parks and J. P. Peters, unpublished observation). Hyperactive and loss-of-activity mutations that do not affect protein stability have been isolated across the entire coding sequence of TnsE and should provide insight into the structure and function of the TnsE protein in future work (Z. Li, Q. Shi, and J. P. Peters, unpublished observation).

Tn7 TARGET IMMUNITY INVOLVES INTERACTION BETWEEN TnsA, TnsB, AND TnsC

Target immunity is a process by which transposition of a second Tn7 element in a region already occupied by Tn7 is inhibited, a process found with a number of transposons (90, 91, 92, 93). Target immunity is a local effect in that transposition to distal sites in the cell is not affected. Target immunity can have different benefits depending on the type of transposon. With a number of different elements transposition is stimulated when the element itself is replicated. If transposition occurred into a nearby DNA, including a sister chromosome, damaging internal insertion could result. Even a local insertion could have adverse side effects because the homology that results could favor deletions of the intervening DNA sequence through host-mediated homologous recombination. Target immunity could also simply encourage the distribution of the element to other places in the cell that could enhance the spread of the element, including into other replicons like bacteriophage and plasmids [as well as from bacteriophage and plasmids into the chromosome (Fig. 6)].

With Tn7, target immunity is a basic function of the core machinery. In the case of Tn7 transposition in the TnsD pathway, target immunity very likely provides an important mechanism to control high-frequency transposition that could otherwise occur repeatedly into the *attTn7* site. Without this function, genome instability could result from multiple insertions accumulating in the *attTn7* site. Instability could also occur stemming from double-strand-break formation in the chromosome following excision of the element. The double-strand break that is found following excision of Tn7 from the donor site is known to be repaired by homologous recombination from the sister chromosome (92). TnsE-mediated transposition also displays target immunity: following transposition of the element into a conjugal plasmid, insertion of a second element will be inhibited (Fig. 6). Target immunity in the TnsE-mediated pathway may help prevent destabilizing effects that could result from having multiple elements in a plasmid.

A central problem in understanding the molecular foundation of immunity is reconciling the fact that the same proteins that act positively to encourage transposition can also act to discourage transposition at certain locations in the same cell. In the case of Tn7 the key players in the process of target immunity are TnsA, TnsB, and TnsC. A critical feature in the system is that TnsB has the capacity to discourage the ability of TnsC to interact productively on DNA to attract Tn7 transposition events. However, TnsB is only capable of displacing TnsC when it is at a sufficient local concentration, which is predicted to occur only in regions with one or more TnsB-binding sites. The target immunity process can be reconstituted *in vitro* in a reaction with two target plasmids, where one has a TnsB-binding site and one does not, and insertions are found to occur into the plasmid that is not bound by TnsB (61). This process is dependent of the inclusion of TnsB in the preincubation step of a staged reaction. TnsB is also able to redistribute TnsC from one molecule onto another in an *in vitro* reaction in a way that does not compromise the ability of TnsC to participate in transposition on these new DNAs.

The immunity process involves the ability of TnsC to hydrolyze ATP. This is based on the finding that immunity is not found in the presence of nonhydrolyzable analogs. Additionally, one class of TnsC mutant that no longer displays target immunity ($TnsC^{S401Y402}$) still allows transposition in the presence of ADP, unlike the wild-type protein (22, 61, 62). Mutations that show defects in target immunity can also be isolated in TnsB (21, 58). Target immunity bypass mutations localize to the very C-terminal region of TnsB (58). Specific

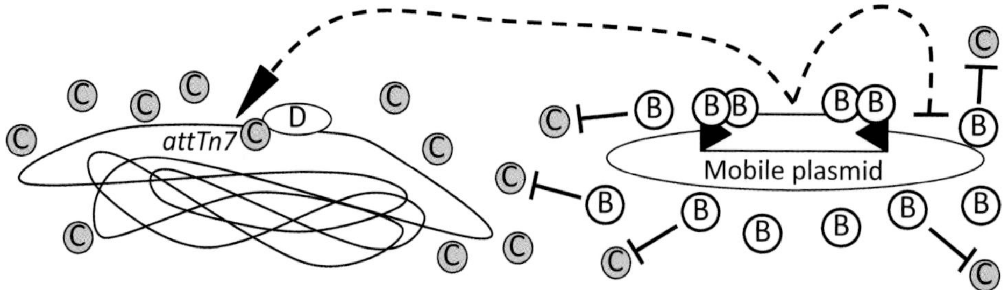

Figure 6 Target immunity inhibits addition copies of the element from occurring in the same site or region where one already exists. The target immunity process is mediated by the Tn7 proteins TnsA, TnsB, and TnsC. TnsB that is bound to the ends of the element, and presumably therefore at a higher concentration in this region, will not allow TnsC to form a productive target complex with TnsD and TnsE. This behavior is modeled to redistribute active TnsC to other sites where stable complexes can form with TnsA and the target-site selection proteins. (See the text for details.) doi:10.1128/microbiolspec.MDNA3-0010-2014.f6

alleles isolated genetically (TnsB^{P686S}, TnsB^{V689M}, and TnsB^{P690L}) are competent for transposition and allow immunity bypass, while alanine mutations in this same region result in transposition defective and TnsB–TnsC interaction defective proteins. Wild-type TnsB is capable of provoking disassembly of the TnsCD–*attTn7* complex, but not a TnsB mutant protein that bypasses immunity *in vivo* (TnsB^{V689M}). Interestingly, this activity can be independently traced to the very C-terminus of the protein because a short peptide of this sequence is capable of target-complex disassembly. TnsA has the capacity to stabilize a TnsC-bound target preventing TnsB from displacing the proteins, which could be important in the commitment to a specific target DNA for insertion.

UNEXPLORED PATHWAYS OF TRANSPOSITION IN ELEMENTS ENCODING HOMOLOGS TO THE TnsA, TnsB, and TnsC CORE MACHINERY

Bioinformatics analysis suggests that the TnsABC core machinery used by Tn7 and Tn7-like elements may also be used in other elements in unexplored targeting pathways. Transposons can carry out transposition with a single transposase as described above for canonical cut-and-paste transposition (Fig. 2). Replicative transposition occurs with elements that encoded a transposase and a regulator protein that helps identify transposon insertion sites, as in the case of bacteriophage Mu and Tn5090/Tn5053 elements (Fig. 2 and 7) (95, 96). Tn5090/Tn5053 completes the replicative transposition process using an element-encoded resolvase (*tniR*) and resolution site (*res*) (Fig. 7), while Mu packages the copies of its genome following replicative transposition. Tn5090/Tn5053 also encodes a target-site selector protein that targets sites of dimer resolution in plasmids (97).

Bioinformatics indicate that a great many elements encode homologs of TnsA, TnsB, and TnsC, but contain no homologs of the full length TnsD and TnsE proteins found in Tn7 and Tn7-like elements (B. Kapili and J. P. Peters, unpublished data). Constructing a tree using the conserved endonuclease domain of TnsA reveals elements distantly related to Tn7, like Tn6022 and Tn6230 that have uncharacterized mechanisms for target-site selection (see below) (Fig. 7 and 8).

Evidence for New Site-Specific Insertion Pathways

DNA sequence analysis identifies an uncharacterized group of elements that show homologs to the TnsABC machinery found in Tn7-like elements but appear to have a targeting pathway that recognizes a new putative attachment site (98). These nearly identical elements are found in the *yhiN* region of *Salmonella enterica* subsp. enterica serovar Senftenberg Strain A4-543 (98), *E. coli* O104:H4 str. 2011C-3493 (99), *E. coli* 55989 (NC_011748) (100), *Enterobacter cloacae* subsp. cloacae ATCC 13047, *Cronobacter sakazakii* ATCC BAA-894 (101), and *E. albertii* TW07627 (ABKX01000001.1). Transposition occurs at a position encoding the most C-terminal amino acids (2 to 5 in the current examples) of the YhiN protein producing a 5-bp target site duplication (M. T. Petassi and J. P. Peters, unpublished data). Insertion at this site replaces the terminal amino acids and adds four amino acids. The *yhiN* gene is a putative oxidoreductase with a FAD/NAD(P)-binding domain and is not essential in *E. coli*. As in the case of *attTn7*, the homology between the insertion sites is only found on the right side of the insertions site. We

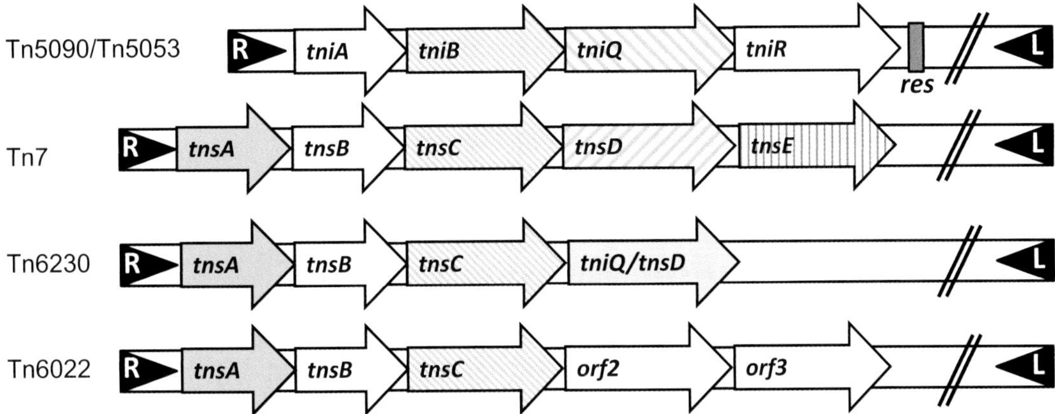

Figure 7 Comparison of various transposable elements. Representative transposons that utilize a transposase regulator protein (TnsC/TniB). The Tn5090/Tn5350 element undergoes replicative transposition where a resolvase (TniR) is used to resolve co-integrates at the *res* site after transposition and replication of the element. Tn5090/Tn5350 is compared with the known or putative heteromeric transposase elements Tn7, Tn6022, and Tn6230. (See the text for details.) doi:10.1128/microbiolspec.MDNA3-0010-2014.f7

name the 37 kb Tn7-like element in *S. enterica* subsp. enterica serovar Senftenberg as Tn6230 (Fig. 7).

Hinting at an interrelationship between Tn7, Tn5090/Tn5053, and Tn6230, downstream from the TnsC-like protein in Tn6230 is a protein with homology to TniQ of Tn5053 (~20% identity with the TniQ and containing the conserved family domain, PF06527) (102) and homology to TnsD from Tn7 (~20% identity with TnsD including possible matches in the alignment to the CCCH amino acids found in TnsD) (M. T. Petassi and J. P. Peters, unpublished data). Homologs to this TniQ/TnsD-like protein were previously called TnsD′ (3). No other target-site selection proteins are obvious among the ~30 other proteins found in the Tn6230 element. Among many other proteins, Tn6230 encodes metal resistance genes and a potential *hipAB* toxin/antitoxin system as associated with the formation of persister cells (102). An exciting possibility would be if the genes could allow persister-like resistance to a broad range of antibiotics. It is additionally interesting that, essentially, this exact element is found on conjugal plasmids in *S. enterica* subsp. enterica serovar Montevideo Strain S5-403 plasmid pS5-403-2 (98), *S. enterica* subsp. enterica serovar Heidelberg pSH111_227 (JN983042), *E. coli* APEC O1 plasmid pAPEC-O1-R (104), and *Serratia marcescens* pR478(98, 105). Further research is required to determine if transposition is directed by the TniQ/TnsD protein into the *yhiN* gene and/or plasmids.

An element first noted in *Acinetobacter baumannii* AYE also shows evidence of a new attachment site. The genome sequence of *A. baumannii* AYE indicates that it contains a single massive 86 kb resistance island within the *comM* gene with ~50 genes associated with antibiotic and biocide resistance (106). It was later appreciated that this element showed similarity to Tn7 and Tn5090, but also contained a gene in the position of TnsA containing the TnsA endonuclease N-terminal family domain (*Tn7_Tnp_TnsA_N*) (PF08722) (107). Although not confirmed, it seems very likely that this element also uses a heteromeric transposase like Tn7. Almost all *A. baumannii* genomes have a related element residing in an identical location in the *comM* gene flanked by the same 5-bp target site duplication (ACCGC). The sequence of the 5-bp target site duplication can be used to determine which transposon ends resulted from the same transposition event. Interestingly, the drug-susceptible *A. baumannii* strains AB307-0294 and D1279779 do not contain a similar element, which indicates it is not ancestral to all *A. baumannii* (108, 109).

As noted by others, the actual configuration of the element found in *comM* can be highly variable. However, the elements found at this position are flanked by the same target ACCGC site duplication. This observation could be consistent with transposition of the element into a preferred target site like *attTn7*, but a similar outcome would be expected if a chance insertion occurred once at this position and that further modifications occurred by homologous recombination within the element at this site. Further experiments will be needed to decide between these possibilities. Interestingly, *A. baumannii* strain AB0057 has a version of the element in *comM* and a second smaller element with essentially identical end sequences and encoding

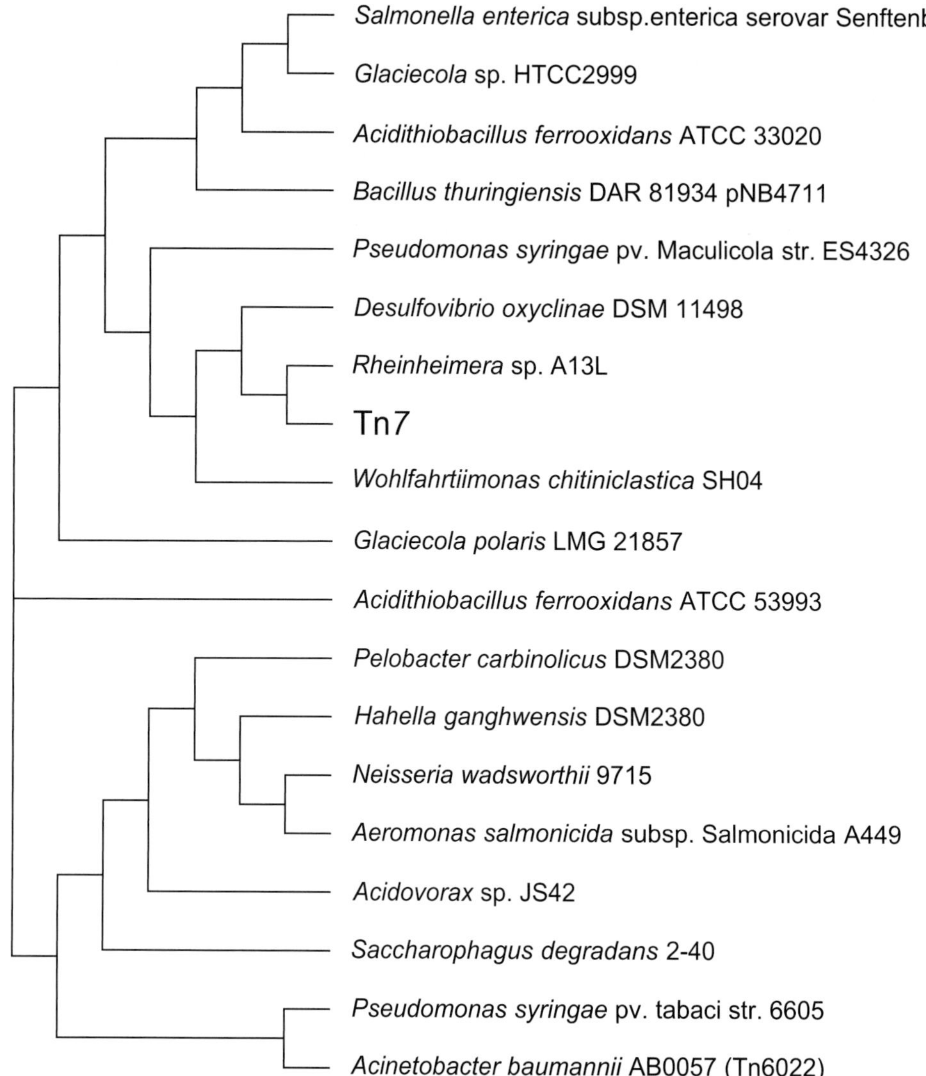

Figure 8 Phylogenetic tree of TnsA endonuclease transposase found with putative heteromeric transposase elements indicated by their host with Tn7, Tn6230, and a Tn6022 derivative (Tn6022) also indicated. A MUSCLE alignment file of extracted TnsA N-domain sequences (*Tn7_Tnp_TnsA_N)* (PF08722) with eight iterations was used to build the tree using the Jukes-Cantor genetic distance model and a Neighbor-Joining tree-building method (112, 113, 114) using the following sequences: *Acidithiobacillus ferrooxidans* ATCC 33020 (AAC21667) (115), *A. ferrooxidans* ATCC 53993 (YP_002220549), *Acidovorax* sp. JS42 (WP_011804647), *Ac. baumannii* AYE (WP_012300781) (116), *Aeromonas salmonicida* subsp. salmonicida A449 (WP_005317426) (117), *Bacillus cereus* ATCC10987 (WP_001129185) (118), *B. thuringiensis* DAR 81934 pNB4711 (WP_017762552) (119), *Desulfovibrio oxyclinae* DSM 11498 (WP_018123630), Tn7 *E. coli* (120), *Glaciecola polaris* LMG 21857 (WP_007103512), *Glaciecola* sp. HTCC2999 (WP_010179396), *Hahella ganghwensis* DSM 17046 (WP_020406004), *Idiomarina loihiensis* L2TR (WP_011235836) (121), *Neisseria wadsworthii* 9715 (WP_009116837), *Pelobacter carbinolicus* DSM 2380 (WP_011339779) (122), *Pseudomonas syringae* pv. tabaci str. 6605 (WP_016981931*), P. syringae* pv. Maculicola str. ES4326 (WP_007247747), *Rheinheimera* sp. A13L (WP_008897114) (123), *Saccharophagus degradans* 2-40 (WP_011466674) (124), *S. enterica* subsp. enterica serovar Senftenberg str. A4-543 (EHC89275) (125), *Shewanella baltica* OS155 (WP_011848289) (126), and *Wohlfahrtiimonas chitiniclastica* SH04 (WP_008315655). (See the text for details.) doi:10.1128/microbiolspec.MDNA3-0010-2014.f8

known transposition proteins along with other proteins of interest (108). This 17-kb element encodes resistance to carbapenems, the class D-type oxacillinase *bla-OXA-23* in the Tn2006 element among other genes and a derivative lacking this element is called Tn6022 (Fig. 7) (110, 111). This finding and other results indicate that the transposition machinery of the elements very likely has the capacity to direct transposition into sites to facilitate horizontal transfer. Similarly placed putative *tnsABC* genes in Tn6022 share the conserved domains indicated above for Tn7: TnsA(*Tn7_Tnp_TnsA_N*), TnsB(*rve*), and TnsC(*AAA*) (Fig. 7). In Tn6022 these genes are also known as *tniE/orf1*, *tniA*, and *tniB*, respectively (111). The other proteins in Tn6022 do not show homology to the target-site selecting proteins from Tn7, Tn6230, and Tn5090/Tn5053 (B. Kapili and J. P. Peters, unpublished data). Genes immediately 3′ of the regulator protein will be candidates as target-site selecting proteins for further study in this family of Tn6022-like elements (Fig. 7).

CONCLUSION

Extensive work with Tn7 has provided detailed molecular information on how a DNA transposon can target a single preferred site. This information provides insight into the formation of genetic islands in bacteria and the development of widely used tools for delivering genes stably into the chromosomes of diverse bacteria with high efficiency. Future work with the TnsE pathway will be important for understanding how Tn7-like elements have spread so successfully into diverse bacteria. It will be of great interest to see how the transposition system that has been characterized with Tn7 has been adapted for targeting pathways with other heteromeric transposase elements. Evidence suggests new attachment sites have evolved in *yhiN* and *comM* for related heteromeric transposase elements, but it remains unclear why these sites within genes are maintained and if additional attachment sites exist. New targeting systems recognizing *yhiN* and *comM* will also likely need mechanisms to facilitate horizontal transfer that remain to be discovered.

Acknowledgments. *I thank the members of the Peters' laboratory, Nancy L. Craig, and Mick Chandler for comments on the chapter and Qi Sun of the Cornell Computational Biology Service Unit for bioinformatics advice. Bennett Kapili provided the phylogenetic tree in Figure 8. Work in the lab is funded by the National Science Foundation (MCB-1244227).*

Citation. **Peters J.** 2014. Tn7. Microbiol Spectrum **2**(5): MDNA3-0010-2014.

References

1. **Craig NL.** 2002. Tn7, p 423–456. *In* Craig NL, Craigie R, Gellert M, Lambowitz AM (ed), *Mobile DNA II.* ASM Press, Washington, DC.
2. **Li Z, Craig NL, Peters JE.** 2012. Transposon Tn7, p 1–32. *In* Roberts A, Mullany P (ed), *Bacterial Integrative Mobile Genetic Elements.* Landes Bioscience, Austin, TX.
3. **Parks AR, Peters JE.** 2009. Tn7 elements: engendering diversity from chromosomes to episomes. *Plasmid* **61:** 1–14.
4. **Peters JE, Craig NL.** 2001. Tn7: smarter than we thought. *Nat Rev Mol Cell Biol* **2:**806–814.
5. **Parks AR, Peters JE.** 2007. Transposon Tn7 is widespread in diverse bacteria and forms genomic islands. *J Bacteriol* **189:**2170–2173.
6. **Mitra R, McKenzie GJ, Yi L, Lee CA, Craig NL.** 2010. Characterization of the TnsD-*attTn7* complex that promotes site-specific insertion of *Tn7*. *Mobile DNA* **1:**18.
7. **Gringauz E, Orle K, Orle A, Waddell CS, Craig NL.** 1988. Recognition of *Escherichia coliattTn7* by transposon Tn7: lack of specific sequence requirements at the point of Tn7 insertion. *J Bacteriol* **170:**2832–2840.
8. **McKown RL, Orle KA, Chen T, Craig NL.** 1988. Sequence requirements of *Escherichia coli* attTn7, a specific site of transposon Tn7 insertion. *J Bacteriol* **170:**352–358.
9. **Lichtenstein C, Brenner S.** 1982. Unique insertion site of Tn7 in *E. coli* chromosome. *Nature* **297:**601–603.
10. **Wolkow CA, DeBoy RT, Craig NL.** 1996. Conjugating plasmids are preferred targets for Tn7. *Genes Dev* **10:** 2145–2157.
11. **Waddell CS, Craig NL.** 1988. Tn7 transposition: two transposition pathways directed by five Tn7-encoded genes. *Genes Dev* **2:**137–149.
12. **Sarnovsky R, May EW, Craig NL.** 1996. The Tn7 transposase is a heteromeric complex in which DNA breakage and joining activities are distributed between different gene products. *EMBO J* **15:**6348–6361.
13. **Gary PA, Biery MC, Bainton RJ, Craig NL.** 1996. Multiple DNA processing reactions underlie Tn7 transposition. *J Mol Biol* **257:**301–316.
14. **May EW, Craig NL.** 1996. Switching from cut-and-paste to replicative Tn7 transposition. *Science* **272:**401–404.
15. **Hickman AB, Li L, Mathew SV, May EW, Craig NL, Dyda F.** 2000. Unexpected structural diversity in DNA recombination: the restriction endonuclease connection. *Mol Cell* **5:**1025–1034.
16. **Haren L, Ton-Hoang B, Chandler M.** 1999. Integrating DNA: transposases and retroviral integrases. *Annu Rev Microbiol* **53:**245–281.
17. **Hickman AB, Chandler M, Dyda F.** 2010. Integrating prokaryotes and eukaryotes: DNA transposases in light of structure. *Crit Rev Biochem Mol Biol* **45:**50–69.
18. **Arciszewska LK, Craig NL.** 1991. Interaction of the Tn7-encoded transposition protein TnsB with the ends of the transposon. *Nucleic Acids Res* **19:**5021–5029.

19. **Arciszewska LK, McKown RL, Craig NL.** 1991. Purification of TnsB, a transposition protein that binds to the ends of Tn7. *J Biol Chem* **266:**21736–21744.

20. **Bainton R, Gamas P, Craig NL.** 1991. Tn7 transposition *in vitro* proceeds through an excised transposon intermediate generated by staggered breaks in DNA. *Cell* **65:**805–816.

21. **Stellwagen A, Craig NL.** 1997. Gain-of-function mutations in TnsC, an ATP-dependent transposition protein which activates the bacterial transposon Tn7. *Genetics* **145:**573–585.

22. **Stellwagen AE, Craig NL.** 1998. Mobile DNA elements: controlling transposition with ATP-dependent molecular switches. *Trends Biochem Sci* **23:**486–490.

23. **Barth PT, Datta N, Hedges RW, Grinter NJ.** 1976. Transposition of a deoxyribonucleic acid sequence encoding trimethoprim and streptomycin resistances from R483 to other replicons. *J Bacteriol* **125:**800–810.

24. **Barth P, Datta N.** 1977. Two naturally occurring transposons indistinguishable from Tn7. *J Gen Microbiol* **102:**129–134.

25. **Hall RM, Brookes DE, Stokes HW.** 1991. Site-specific insertion of genes into integrons: role of the 59-base element and determination of the recombination crossover point. *Mol Microbiol* **5:**1941–1959.

26. **Sundstrom L, Roy PH, Skold O.** 1991. Site-specific insertion of three structural gene cassettes in transposon Tn7. *J Gen Microbiol* **173:**3025–3028.

27. **Bainton RJ, Kubo KM, Feng JN, Craig NL.** 1993. Tn7 transposition: target DNA recognition is mediated by multiple Tn7-encoded proteins in a purified *in vitro* system. *Cell* **72:**931–943.

28. **Biery MC, Steward F, Stellwagen AE, Raleigh EA, Craig NL.** 2000. A simple *in vitro* Tn7-based transposition system with low target site selectivity for genome and gene analysis. *Nucleic Acids Res* **28:**1067–1077.

29. **Parks AR, Li Z, Shi Q, Owens RM, Jin MM, Peters JE.** 2009. Transposition into replicating DNA occurs through interaction with the processivity factor. *Cell* **138:**685–695.

30. **Huisman O, Kleckner N.** 1987. A new generalizable test for detection of mutations affecting Tn10 transposition. *Genetics* **116:**185–189.

31. **Holder JW, Craig NL.** 2010. Architecture of the Tn7 posttransposition complex: an elaborate nucleoprotein structure. *J Mol Biol* **401:**167–181.

32. **Rao JE, Miller PS, Craig NL.** 2000. Recognition of triple-helical DNA structures by transposon Tn7. *Proc Natl Acad Sci USA* **97:**3936–3941.

33. **Kuduvalli P, Rao JE, Craig NL.** 2001. Target DNA structure plays a critical role in Tn7 transposition. *EMBO J* **20:**924–932.

34. **Rao JE, Craig NL.** 2001. Selective recognition of pyrimidine motif triplexes by a protein encoded by the bacterial transposon Tn7. *J Mol Biol* **307:**1161–1170.

35. **Choi KY, Li Y, Sarnovsky R, Craig NL.** 2013. Direct interaction between the TnsA and TnsB subunits controls the heteromeric Tn7 transposase. *Proc Natl Acad Sci USA* **110:**E2038–E2045.

36. **Kennedy A, Guhathakurta A, Kleckner N, Haniford DB.** 1998. Tn10 transposition via a DNA hairpin intermediate. *Cell* **95:**125–134.

37. **Bhasin A, Goryshin IY, Reznikoff WS.** 1999. Hairpin formation in Tn5 transposition. *J Biol Chem* **274:**37021–37029.

38. **Roth DB, Menetski JP, Nakajima PB, Bosma MJ, Gellert M.** 1992. V(D)J recombination: broken DNA molecules with covalently sealed (hairpin) coding ends in scid mouse thymocytes. *Cell* **70:**983–981.

39. **Zhou L, Mitra R, Atkinson PW, Hickman AB, Dyda F, Craig NL.** 2004. Transposition of hAT elements links transposable elements and V(D)J recombination. *Nature* **432:**995–1001.

40. **Bainton RJ, Kubo KM, Feng J-N, Craig NL.** 1993. Tn7 transposition: target DNA recognition is mediated by multiple Tn7-encoded proteins in a purified *in vitro* system. *Cell* **72:**931–943.

41. **Wiegand TW, Reznikoff WS.** 1992. Characterization of two hypertransposing Tn5 mutants. *J Gen Microbiol* **174:**1229–1239.

42. **Lampe DJ, Akerley BJ, Rubin EJ, Mekalanos JJ, Robertson HM.** 1999. Hyperactive transposase mutants of the Himar1 mariner transposon. *Proc Natl Acad Sci USA* **96:**11428–11433.

43. **Beall EL, Mahoney MB, Rio DC.** 2002. Identification and analysis of a hyperactive mutant form of *Drosophila* P-element transposase. *Genetics* **162:**217–227.

44. **Baus J, Liu L, Heggestad AD, Sanz S, Fletcher BS.** 2005. Hyperactive transposase mutants of the Sleeping Beauty transposon. *Mol Ther* **12:**1148–1156.

45. **Yusa K, Zhou L, Li MA, Bradley A, Craig NL.** 2011. A hyperactive piggyBac transposase for mammalian applications. *Proc Natl Acad Sci USA* **108:**1531–1536.

46. **Lazarow K, Du M-L, Weimer R, Kunze R.** 2012. A hyperactive transposase of the maize transposable element activator (Ac). *Genetics* **191:**747–756.

47. **Deboy RT, Craig NL.** 2000. Target site selection by Tn7: attTn7 transcription and target activity. *J Bacteriol* **182:**3310–3313.

48. **Skelding Z, Sarnovsky R, Craig NL.** 2002. Formation of a nucleoprotein complex containing Tn7 and its target DNA regulates transposition initiation. *EMBO J* **21:**3494–3504.

49. **Yanagihara K, Mizuuchi K.** 2003. Progressive structural transitions within Mu transpositional complexes. *Mol Cell* **11:**215–224.

50. **Gueguen E, Rousseau P, Duval-Valentin G, Chandler M.** 2005. The transpososome: control of transposition at the level of catalysis. *Trends Microbiol* **13:**543–549.

51. **Levchenko I, Luo L, Baker TA.** 1995. Disassembly of the Mu transposase tetramer by the ClpX chaperone. *Genes Dev* **9:**2399–2408.

52. **Steczkiewicz K, Muszewska A, Knizewski L, Rychlewski L, Ginalski K.** 2012. Sequence, structure and functional diversity of PD-(D/E)XK phosphodiesterase superfamily. *Nucleic Acids Res* **40:**7016–7045.

53. **Punta M, Coggill PC, Eberhardt RY, Mistry J, Tate J, Boursnell C, Pang N, Forslund K, Ceric G, Clements J, Heger A, Holm L, Sonnhammer ELL, Eddy SR, Bateman A, Finn RD.** 2012. The Pfam protein families database. *Nucleic Acids Res* **40:**D290–D301.

54. **Lu F, Craig NL.** 2000. Isolation and characterization of Tn7 transposase gain-of-function mutants: a model for transposase activation. *EMBO J* **19:**3446–3457.

55. **Biery M, Lopata M, Craig NL.** 2000. A minimal system for Tn7 transposition: the transposon-encoded proteins TnsA and TnsB can execute DNA breakage and joining reactions that generate circularized Tn7 species. *J Mol Biol* **297:**25–37.

56. **Ronning DR, Li Y, Perez ZN, Ross PD, Hickman AB, Craig NL, Dyda F.** 2004. The carboxy-terminal portion of TnsC activates the Tn7 transposase through a specific interaction with TnsA. *EMBO J* **23:**2972–2981.

57. **Dyda F, Hickman AB, Jenkins TM, Engelman A, Craigie R, Davies DR.** 1994. Crystal structure of the catalytic domain of HIV-1 integrase: similarity to other polynucleotidyl transferases. *Science* **266:**1981–1986.

58. **Skelding Z, Queen-Baker J, Craig NL.** 2003. Alternative interactions between the Tn7 transposase and the Tn7 target DNA binding protein regulate target immunity and transposition. *EMBO J* **22:**5904–5917.

59. **Wu Z, Chaconas G.** 1994. Characterization of a region in phage Mu transposase that is involved in interaction with the Mu B protein. *J Biol Chem* **269:**28829–28833.

60. **Levchenko I, Yamauchi M, Baker TA.** 1997. ClpX and MuB interact with overlapping regions of Mu transposase: implications for control of the transposition pathway. *Genes Dev* **11:**1561–1572.

61. **Stellwagen A, Craig NL.** 1997. Avoiding self: two Tn7-encoded proteins mediate target immunity in Tn7 transposition. *EMBO J* **16:**6823–6834.

62. **Stellwagen AE, Craig NL.** 2001. Analysis of gain of function mutants of an ATP-dependent regulator of Tn7 transposition. *J Mol Biol* **305:**633–642.

63. **Van Roey P, Waddling CA, Fox KM, Belfort M, Derbyshire V.** 2001. Intertwined structure of the DNA-binding domain of intron endonuclease I-TevI with its substrate. *EMBO J* **20:**3631–3637.

64. **Sharpe P, Craig NL.** 1998. Host proteins can stimulate Tn7 transposition: a novel role for the ribosomal protein L29 and the acyl carrier protein. *EMBO J* **17:**5822–5831.

65. **McKenzie GJ, Craig NL.** 2006. Fast, easy and efficient: site-specific insertion of transgenes into enterobacterial chromosomes using Tn7 without need for selection of the insertion event. *BMC Microbiol* **6:**39.

66. **Koch B, Jensen LE, Nybroe O.** 2001. A panel of Tn7-based vectors for insertion of the gfp marker gene or for delivery of cloned DNA into Gram-negative bacteria at a neutral chromosomal site. *J Microbiol Methods* **45:**187–195.

67. **Hahn G, Jarosch M, Wang JB, Berbes C, McVoy MA.** 2003. Tn7-mediated introduction of DNA sequences into bacmid-cloned cytomegalovirus genomes for rapid recombinant virus construction. *J Virol Methods* **107:**185–194.

68. **Berger I, Fitzgerald DJ, Richmond TJ.** 2004. Baculovirus expression system for heterologous multiprotein complexes. *Nat Biotechnol* **22:**1583–1587.

69. **Laitinen OH, Airenne KJ, Hytonen VP, Peltomaa E, Mahonen AJ, Wirth T, Lind MM, Makela KA, Toivanen PI, Schenkwein D, Heikura T, Nordlund HR, Kulomaa MS, Yla-Herttuala S.** 2005. A multipurpose vector system for the screening of libraries in bacteria, insect and mammalian cells and expression *in vivo*. *Nucleic Acids Res* **33:**e42.

70. **Choi KH, Gaynor JB, White KG, Lopez C, Bosio CM, Karkhoff-Schweizer RR, Schweizer HP.** 2005. A Tn7-based broad-range bacterial cloning and expression system. *Nat Methods* **2:**443–448.

71. **Crepin S, Harel J, Dozois CM.** 2012. Chromosomal complementation using Tn7 transposon vectors in enterobacteriaceae. *Appl Environ Microbiol* **78:**6001–6008.

72. **Damron FH, McKenney ES, Schweizer HP, Goldberg JB.** 2012. Construction of a broad-host-range Tn7-based vector for single-copy PBAD-controlled gene expression in Gram-negative bacteria. *Appl Environ Microbiol* **79:**718–721.

73. **Richards CA, Brown CE, Cogswell JP, Weiner MP.** 2000. The admid system: generation of recombinant adenoviruses by Tn7-mediated transposition in *E. coli*. *BioTechniques* **29:**146–154.

74. **Sibley MH, Raleigh EA.** 2012. A versatile element for gene addition in bacterial chromosomes. *Nucleic Acids Res* **40:**e19.

75. **Kvitko BH, McMillan IA, Schweizer HP.** 2013. An improved method for oriT-directed cloning and functionalization of large bacterial genomic regions. *Appl Environ Microbiol* **79:**4869–4878.

76. **Kuduvalli PN, Mitra R, Craig NL.** 2005. Site-specific Tn7 transposition into the human genome. *Nucleic Acids Res* **33:**857–863.

77. **Kubo KM, Craig NL.** 1990. Bacterial transposon Tn7 utilizes two classes of target sites. *J Bacteriol* **172:**2774–2778.

78. **Finn JA, Parks AR, Peters JE.** 2007. Transposon Tn7 directs transposition into the genome of filamentous bacteriophage M13 using the element-encoded TnsE protein. *J Bacteriol* **189:**9122–9125.

79. **Peters JE, Craig NL.** 2000. Tn7 transposes proximal to DNA double-strand breaks and into regions where chromosomal DNA replication terminates. *Mol Cell* **6:**573–582.

80. **Peters JE, Craig NL.** 2001. Tn7 recognizes target structures associated with DNA replication using the DNA binding protein TnsE. *Genes Dev* **15:**737–747.

81. **Ton-Hoang B, Pasternak C, Siguier P, Guynet C, Hickman AB, Dyda F, Sommer S, Chandler M.** Single-stranded DNA transposition is coupled to host replication. *Cell* **142:**398–408.

82. **Guynet C, Hickman AB, Barabas O, Dyda F, Chandler M, Ton-Hoang B.** 2008. *In vitro* reconstitution of a single-stranded transposition mechanism of IS608. *Mol Cell* **29:**302–312.

83. **Hu WY, Derbyshire KM.** 1998. Target choice and orientation preference of the insertion sequence IS903. *J Bacteriol* **180:**3039–3048.

84. **Zhong J, Lambowitz AM.** 2003. Group II intron mobility using nascent strands at DNA replication forks to prime reverse transcription. *EMBO J* **22:**4555–4565.

85. **Yu D, Ellis HM, Lee EC, Jenkins NA, Copeland NG, Court DL.** 2000. An efficient recombination system for chromosome engineering in *Escherichia coli. Proc Natl Acad Sci USA* **97:**5978–5983.

86. **Datsenko KA, Wanner BL.** 2000. One-step inactivation of chromosomal genes in *Escherichia coli* K-12 using PCR products. *Proc Natl Acad Sci USA* **97:**6640–6645.

87. **van Kessel JC, Hatfull GF.** 2008. Efficient point mutagenesis in mycobacteria using single-stranded DNA recombineering: characterization of antimycobacterial drug targets. *Mol Microbiol* **67:**1094–1107.

88. **Fricker A, Peters JE.** 2014. Vulnerabilities on the lagging-stand template: opportunities for mobile elements. *Ann Rev Genet* **48**, in press.

89. **Shi Q, Parks AR, Potter BD, Safir IJ, Luo Y, Forster BM, Peters JE.** 2008. DNA damage differentially activates regional chromosomal loci for Tn7 transposition in *Escherichia coli. Genetics* **179:**1237–1250.

90. **Hauer B, Shapiro JA.** 1984. Control of Tn7 transposition. *Mol Gen Genet* **194:**149–158.

91. **Reyes I, Beyou A, Mignotte-Vieux C, Richaud F.** 1987. Mini-Mu transduction: *cis*-inhibition of the insertion of Mud transposons. *Plasmid* **18:**183–192.

92. **Arciszewska LK, Drake D, Craig NL.** 1989. Transposon Tn7 *cis*-acting sequences in transposition and transposition immunity. *J Mol Biol* **207:**35–52.

93. **Robinson MK, Bennett PM, Richmond MH.** 1977. Inhibition of TnA translocation by TnA. *J Bacteriol* **129:**407–414.

94. **Hagemann AT, Craig NL.** 1993. Tn7 transposition creates a hotspot for homologous recombination at the transposon donor site. *Genetics* **133:**9–16.

95. **Kholodii GYG, Yurieva OVO, Lomovskaya OLO, Gorlenko ZZ, Mindlin SZS, Nikiforov VGV.** 1993. Tn5053, a mercury resistance transposon with integron's ends. *J Mol Biol* **230:**1103–1107.

96. **Radstrom P, Skold O, Swedberg G, Flensburg F, Roy PH, Sundstrom L.** 1994. Transposon Tn5090 of plasmid R751, which carries an integron, is related to Tn7, Mu, and the retroelements. *J Bacteriol* **176:**3257–3268.

97. **Minakhina S, Kholodii G, Mindlin S, Yurieva O, Nikiforov V.** 1999. Tn5053 family transposons are *res* site hunters sensing plasmidal *res* sites occupied by cognate resolvases. *Mol Microbiol* **33:**1059–1068.

98. **Moreno Switt AI, den Bakker HC, Cummings CA, Rodriguez-Rivera LD, Govoni G, Raneiri ML, Degoricija L, Brown S, Hoelzer K, Peters JE, Bolchacova E, Furtado MR, Wiedmann M.** 2012. Identification and characterization of novel *Salmonella* mobile elements involved in the dissemination of genes linked to virulence and transmission. *PLoS One* **7:**e41247.

99. **Ahmed SA, Awosika J, Baldwin C, Bishop-Lilly KA, Biswas B, Broomall S, Chain PSG, Chertkov O, Chokoshvili O, Coyne S, Davenport K, Detter JC, Dorman W, Erkkila TH, Folster JP, Frey KG, George M, Gleasner C, Henry M, Hill KK, Hubbard K, Insalaco J, Johnson S, Kitzmiller A, Krepps M, Lo C-C, Luu T, McNew LA, Minogue T, Munk CA, Osborne B, Patel M, Reitenga KG, Rosenzweig CN, Shea A, Shen X, Strockbine N, Tarr C, Teshima H, van Gieson E, Verratti K, Wolcott M, Xie G, Sozhamannan S, Gibbons HS, Threat Characterization C.** 2012. Genomic comparison of *Escherichia coli* O104:H4 isolates from 2009 and 2011 reveals plasmid, and prophage heterogeneity, including shiga toxin encoding phage stx2. *PloS One* **7:** e48228.

100. **Touchon M, Hoede C, Tenaillon O, Barbe V, Baeriswyl S, Bidet P, Bingen E, Bonacorsi S, Bouchier C, Bouvet O, Calteau A, Chiapello H, Clermont O, Cruveiller S, Danchin A, Diard M, Dossat C, Karoui ME, Frapy E, Garry L, Ghigo JM, Gilles AM, Johnson J, Le Bouguenec C, Lescat M, Mangenot S, Martinez-Jehanne V, Matic I, Nassif X, Oztas S, Petit MA, Pichon C, Rouy Z, Ruf CS, Schneider D, Tourret J, Vacherie B, Vallenet D, Medigue C, Rocha EPC, Denamur E.** 2009. Organised genome dynamics in the *Escherichia coli* species results in highly diverse adaptive paths. *PLoS Genetics* **5:**e1000344.

101. **Kucerova E, Clifton SW, Xia X-Q, Long F, Porwollik S, Fulton L, Fronick C, Minx P, Kyung K, Warren W, Fulton R, Feng D, Wollam A, Shah N, Bhonagiri V, Nash WE, Hallsworth-Pepin K, Wilson RK, McClelland M, Forsythe SJ.** 2010. Genome sequence of *Cronobacter sakazakii* BAA-894 and comparative genomic hybridization analysis with other *Cronobacter* species. *PloS One* **5:**e9556.

102. **Kholodii GY, Mindlin SZ, Bas IA, Yurieva OV, Minakhina SV, Nikiforov VG.** 1995. Four genes, two ends, and a *res* region are involved in transposition of Tn5053: a paradigm for a novel family of transposons carrying either a *mer* operon or an integron. *Mol Microbiol* **17:**1189–1200.

103. **Lewis K.** 2010. Persister cells. *Annu Rev Microbiol* **64:** 357–372.

104. **Johnson TJ, Wannemeuhler YM, Scaccianoce JA, Johnson SJ, Nolan LK.** 2006. Complete DNA sequence, comparative genomics, and prevalence of an IncHI2 plasmid occurring among extraintestinal pathogenic *Escherichia coli* isolates. *Antimicrob Agents Chemother* **50:** 3929–3933.

105. **Gilmour MW, Thomson NR, Sanders M, Parkhill J, Taylor DE.** 2004. The complete nucleotide sequence of the resistance plasmid R478: defining the backbone components of incompatibility group H conjugative plasmids through comparative genomics. *Plasmid* **52:**182–202.

106. **Fournier P-E, Vallenet D, Barbe V, Audic S, Ogata H, Poirel L, Richet H, Robert C, Mangenot S, Abergel C, Nordmann P, Weissenbach J, Raoult D, Claverie J-M.** 2006. Comparative genomics of multidrug resistance in *Acinetobacter baumannii. PLoS Genet* **2:**e7.

107. **Rose A.** 2010. TnAbaR1: a novel Tn7-related transposon in *Acinetobacter baumannii* that contributes to the

accumulation and dissemination of large repertoires of resistance genes. *Biosci Horizons* 3:40–48.

108. **Adams MD, Goglin K, Molyneaux N, Hujer KM, Lavender H, Jamison JJ, MacDonald IJ, Martin KM, Russo T, Campagnari AA, Hujer AM, Bonomo RA, Gill SR.** 2008. Comparative genome sequence analysis of multidrug-resistant *Acinetobacter baumannii*. *J Bacteriol* **190:**8053–8064.

109. **Farrugia DN, Elbourne LDH, Hassan KA, Eijkelkamp BA, Tetu SG, Brown MH, Shah BS, Peleg AY, Mabbutt BC, Paulsen IT.** 2013. The complete genome and phenome of a community- acquired *Acinetobacter baumannii*. *PLoS One* **8:**e58628.

110. **Lee H-Y, Chang R-C, Su L-H, Liu S-Y, Wu S-R, Chuang C-H, Chen C-L, Chiu C-H.** 2012. Wide spread of Tn2006 in an AbaR4-type resistance island among carbapenem-resistant *Acinetobacter baumannii* clinical isolates in Taiwan. *Int J Antimicrob Agents* **40:**163 167.

111. **Hamidian M, Hall RM.** 2011. AbaR4 replaces AbaR3 in a carbapenem-resistant *Acinetobacter baumannii* isolate belonging to global clone 1 from an Australian hospital. *J Antimicrob Chemother* **66:**2484–2491.

112. **Jukes TH, Cantor CR.** 1969. *Evolution of Protein Molecules*. Academic Press, New York.

113. **Saitou N, Nei M.** 1987. The neighbor-joining method: a new method for reconstructing phylogenetic trees. *Mol Biol Evol* **4:**406–425.

114. **Edgar RC.** 2004. MUSCLE: multiple sequence alignment with high accuracy and high throughput. *Nucleic Acids Res* **32:**1792–1797.

115. **Oppon JC, Sarnovsky RJ, Craig NL, Rawlings DE.** 1998. A Tn7-like transposon is present in the *glmUS* region of the obligately hemoautolithotrophic bacterium *Thiobacillus ferrooxidans*. *J Bacteriol* **180:**3007–3012.

116. **Vallenet D, Nordmann P, Barbe V, Poirel L, Mangenot S, Bataille E, Dossat C, Gas S, Kreimeyer A, Lenoble P, Oztas S, Poulain J, Segurens B, Robert C, Abergel C, Claverie J-M, Raoult D, Médigue C, Weissenbach J, Cruveiller S.** 2008. Comparative analysis of *Acinetobacters*: three genomes for three lifestyles. *PLoS One* **3:**e1805.

117. **Reith ME, Singh RK, Curtis B, Boyd JM, Bouevitch A, Kimball J, Munholland J, Murphy C, Sarty D, Williams J, Nash JH, Johnson SC, Brown LL.** 2008. The genome of *Aeromonas salmonicida* subsp. salmonicida A449: insights into the evolution of a fish pathogen. *BMC Genomics* **9:**427.

118. **Rasko DA, Ravel J, Okstad OA, Helgason E, Cer RZ, Jiang L, Shores KA, Fouts DE, Tourasse NJ, Angiuoli SV, Kolonay J, Nelson WC, Kolsto AB, Fraser CM, Read TD.** 2004. The genome sequence of *Bacillus cereus* ATCC 10987 reveals metabolic adaptations and a large plasmid related to *Bacillus anthracis* pXO1. *Nucleic Acids Res* **32:**977–988.

119. **Wang A, Pattemore J, Ash G, Williams A, Hane J.** 2013. Draft genome sequence of *Bacillus thuringiensis* strain DAR 81934, which exhibits molluscicidal activity. *Genome Announc* **1:**e0017512.

120. **Flores C, Qadri MI, Lichtenstein C.** 1990. DNA sequence analysis of five genes; tnsA, B, C, D and E, required for Tn7 transposition. *Nucleic Acids Res* **18:**901–911.

121. **Hou S, Saw JH, Lee KS, Freitas TA, Belisle C, Kawarabayasi Y, Donachie SP, Pikina A, Galperin MY, Koonin EV, Makarova KS, Omelchenko MV, Sorokin A, Wolf YI, Li QX, Keum YS, Campbell S, Denery J, Aizawa S, Shibata S, Malahoff A, Alam M.** 2004. Genome sequence of the deep-sea gamma-proteobacterium *Idiomarina loihiensis* reveals amino acid fermentation as a source of carbon and energy. *Proc Natl Acad Sci USA* **101:**18036–18041.

122. **Aklujkar M, Haveman SA, DiDonato R Jr, Chertkov O, Han CS, Land ML, Brown P, Lovley DR.** 2012. The genome of *Pelobacter carbinolicus* reveals surprising metabolic capabilities and physiological features. *BMC Genomics* **13:**690.

123. **Gupta HK, Gupta RD, Singh A, Chauhan NS, Sharma R.** 2011. Genome sequence of *Rheinheimera* sp. strain A13L, isolated from Pangong Lake, India. *J Bacteriol* **193:**5873–5874.

124. **Weiner RM, Taylor LE 2nd, Henrissat B, Hauser L, Land M, Coutinho PM, Rancurel C, Saunders EH, Longmire AG, Zhang H, Bayer EA, Gilbert HJ, Larimer F, Zhulin IB, Ekborg NA, Lamed R, Richardson PM, Borovok I, Hutcheson S.** 2008. Complete genome sequence of the complex carbohydrate-degrading marine bacterium, *Saccharophagus degradans* strain 2-40 T. *PLoS Genet* **4:**e1000087.

125. **den Bakker HC, Moreno Switt AI, Govoni G, Cummings CA, Ranieri ML, Degoricija L, Hoelzer K, Rodriguez-Rivera LD, Brown S, Bolchacova E, Furtado MR, Wiedmann M.** 2011. Genome sequencing reveals diversification of virulence factor content and possible host adaptation in distinct subpopulations of *Salmonella enterica*. *BMC Genomics* **12:**425.

126. **Caro-Quintero A, Auchtung J, Deng J, Brettar I, Hofle M, Tiedje JM, Konstantinidis KT.** 2012. Genome sequencing of five *Shewanella baltica* strains recovered from the oxic-anoxic interface of the Baltic Sea. *J Bacteriol* **194:**1236.

Mobile DNA, 3rd Edition
Nancy L. Craig, Michael Chandler, Martin Gellert, Alan M. Lambowitz, Phoebe A. Rice and Suzanne Sandmeyer

doi:10.1128/microbiolspec.MDNA3-0007-2014

Rasika M. Harshey[1]

Transposable Phage Mu

31

INTRODUCTION

Transposable phage Mu has played a historic role in the development of the mobile DNA element field (1). The very first paper that christened this phage after its <u>mu</u>tator properties (2) also drew attention to its ability to suppress the phenotypic expression of genes, and suggested that Mu resembled the "controlling elements" postulated by Barbara McClintock to regulate the mosaic color patterns of maize seeds (3). This bold postulate inspired equally insightful early experiments aimed at investigating its mobile properties (4, 5), and led to an influential model for transposition (6), which correctly predicted the cutting and joining steps of the Mu transposition reaction and their attendant DNA rearrangements. The high efficiency of the Mu reaction was responsible for the development of the first *in vitro* transposition system (7), which was critical for dissecting reaction chemistry as well as the function of several participating proteins (see references 8 and 9). This article focuses on the major developments in Mu transposition since this topic was last reviewed in Mobile DNA II, providing background information as necessary (9).

ONE TRANSPOSITION MECHANISM, TWO PATHWAYS FOR PRODUCT RESOLUTION

Mechanism

The mechanism of Mu transposition has been deciphered *in vitro* on both supercoiled and oligonucleotide substrates (10). Figure 1 shows transposition events in the context of *in vivo* substrates. Mu transposition has two distinct phases, which differ in donor substrate configuration and in the fate of the transposition products (8). During the infection phase, the Mu DNA injected into cells has a peculiar structure. This DNA is linear in the phage heads, and flanked by non-Mu host DNA acquired during packaging of integrated Mu replicas during the lytic cycle in a prior host. The number of base pairs of the host sequences that flank the left, or L, end of Mu is 60 to 150 kb, and 0.5 to 3 kb flank the right, or R, end (8). An injected phage protein N binds to the tip of the flanking DNA (FD), protecting the open ends from degradation while also converting the linear genome into a noncovalently closed supercoiled circle prior to integration into the host chromosome (drawn linear for clarity in Fig. 1A) (11, 12, 13). (Non-Mu flanking DNA is referred to as FD herein, irrespective of whether the donor substrate is phage, prophage, plasmid, or oligonucleotide.) During the lytic phase, Mu is part of a large covalently closed circular host genome (Fig. 1B). Thus, the donor Mu DNA configuration in the infection phase is different from that during the lytic phase. In both phases, the mechanism of Mu transposition is the same.

The transposase MuA initially generates a pair of water-mediated endonucleolytic cleavages on specific Mu–host phosphodiester bonds, producing 3′-OH nicks at Mu DNA ends (Fig. 1, Cleavage). In the subsequent strand-transfer (ST) step, the 3′-OH ends directly attack

[1]Department of Molecular Biosciences, Institute of Cellular and Molecular Biology, University of Texas at Austin, Austin, TX 78712.

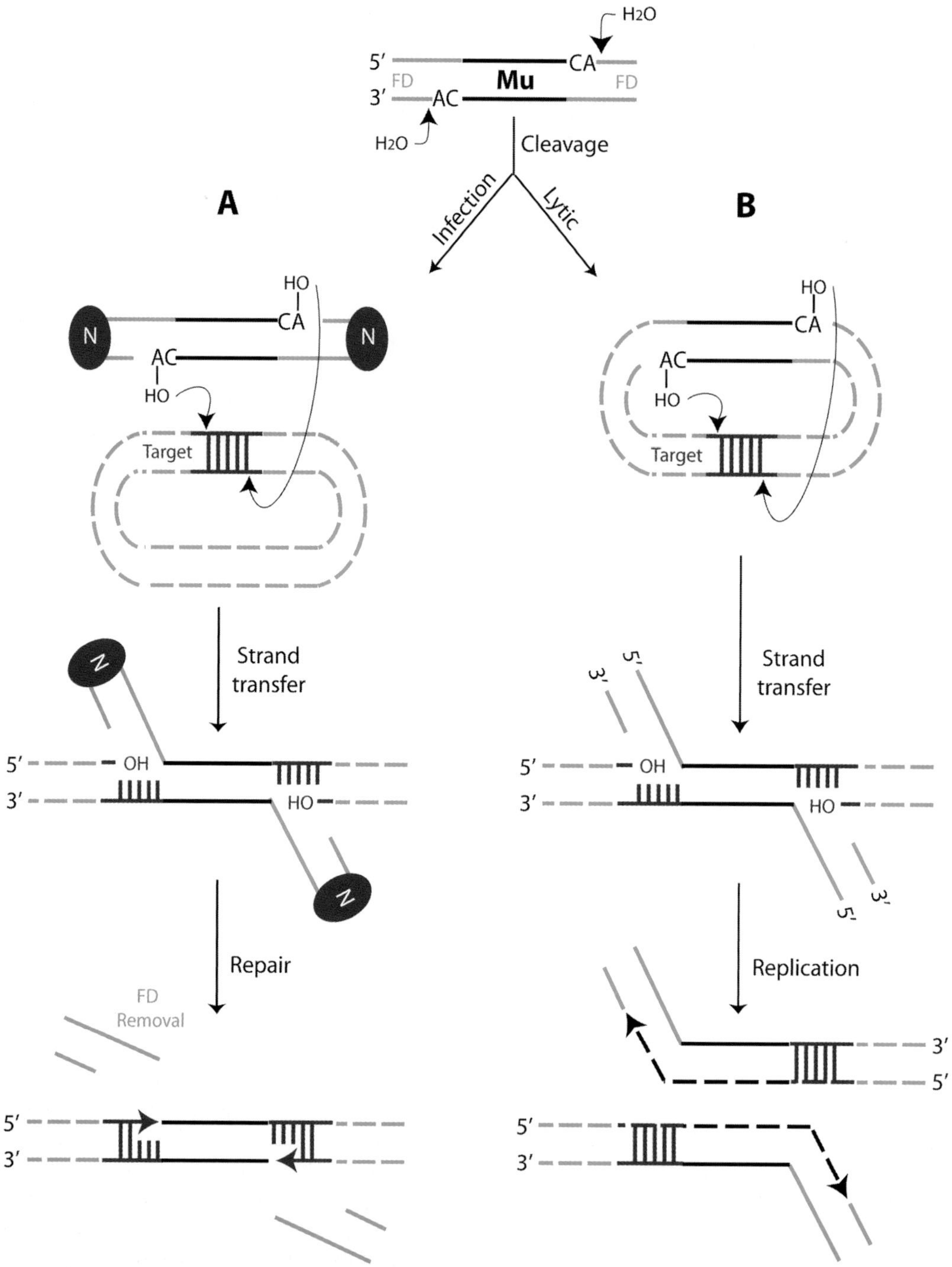

Figure 1 One transposition mechanism, two pathways for resolution of the ST intermediate. The chemical steps of cleavage and ST are the same in both the infection and lytic phases of transposition. (A) In the infection phase, the linear donor Mu genome is converted into a noncovalently closed circle, joined by the MuN protein (purple ovals, ends shown unjoined for clarity); *E. coli* genome is the target. The ST intermediate formed during intermolecular transposition is resolved by removal of the FD, and repair of the 5-bp gaps in the target by limited replication at the host–Mu junction. (B) In the lytic phase, Mu is part of the covalently closed circular *E. coli* genome. The ST intermediate formed during intramolecular transposition is resolved by replication across Mu.
doi:10.1128/microbiolspec.MDNA3-0007-2014.f1

phosphodiester bonds in the target DNA spaced 5 bp apart; this reaction is intermolecular in the infection phase (Fig. 1A) and intramolecular in the lytic phase (Fig. 1B). Mu ends join to 5′-Ps in the target, leaving 3′-OH nicks on the target. The MuB protein is essential for the efficient capture of the target, but plays critical roles at all stages of transposition by allosterically activating MuA (see below) (9). The cleavage and ST reactions, also called phosphoryl transfer reactions (14), are common to other DNA transposition systems including retroviral integration (15). These reactions take place within the same active site of MuA, which contains a structurally conserved "DDE" domain, so named for the three Mg^{2+}-binding carboxylate residues found in other transposases and recombinases (16). Divalent metal ions coordinated by the DDE residues are proposed to activate hydroxyl groups for nucleophilic attack on the reactive phosphodiester bonds in both steps of transposition (10, 17). These reactions proceed via bimolecular nucleophilic substitution (S_N2), a mechanism shared by metal-dependent nucleotidyl transferases and some nucleases (18, 19, 20). Crystal structures of the HIV-related prototype foamy virus (PFV) retroviral integrase assemblies (intasomes), whose phosphotransfer mechanism is similar to Mu, validate the S_N2 mechanism (21, 22, 23).

Two pathways for product resolution

Post-transposition, the branched ST intermediate product must be resolved (Fig. 1A and B). During the infection phase, the intermediate is resolved by FD removal/degradation and repair of the 5-bp gaps in the target by limited DNA replication, generating a simple insertion of Mu in the *Escherichia coli* genome. During the lytic phase, the intermediate is resolved by target-primed replication across the entire Mu genome. These product resolution pathways are referred to as repair or replication pathways herein. The replication pathway has been reconstituted *in vitro* and studied in some detail. The repair pathway has been studied only *in vivo*, and is in the early stages of characterization. The known steps in both pathways are described later.

DISINTEGRATION

A chemical reversal of the transposition/integration reaction is called disintegration. In such a reaction, the 3′-OHs created on the target DNA after ST would attack the phosphodiester bonds formed at the transposon-target junction during the forward reaction (Fig. 2A, red arrows), restoring the original target configuration and releasing the cleaved donor. Disintegration is normally not observed, despite the fact that the forward and reverse reactions are isoenergetic. The reverse reaction can, however, be demonstrated *in vitro* using altered substrates and reaction conditions. Reversal of ST was first described for HIV-1 integrase on oligonucleotide ST substrates (24), and subsequently reproduced in other systems (25, 26, 27, 28). The reported reaction is a pseudo reversal, also called foldback reversal, because the substrates employed resemble a ST joint that has been unpaired and flipped (or folded back), so that the target 3′-OHs attack the transposon-target joint in an unnatural *trans* configuration rather than the normal *cis* configuration, creating hairpin ends on the target (Fig. 2B). The disintegration reaction was also observed for Mu, on both oligonucleotide and plasmid substrates, and required high temperatures or altered metal ions (29, 30). Both true as well as pseudo reversal were observed for Mu (Fig. 2A and B), each showing distinct metal ion specificities indicative of different configurations of the reactive components within the active site. Ca^{2+} ions, which support ST of precleaved ends, also supported true reversal; these ions did not support pseudo reversal, suggesting that the metal ion binding pocket is similar in the forward and the true-reversal reactions, but different in the foldback reaction. When the transpososome was assembled on uncleaved Mu donor substrates, and the reaction proceeded through the 3′ Mu end cleavage and ST, true reversal required high temperatures (29, 30), suggesting that, in the normal course of events, the reactive groups are rearranged within the ST complex such that reversal is proscribed; high temperatures likely cause conformation changes that restore the pre-ST configuration of the active site (29). The higher stability of the ST complex compared to the cleaved complex (31) is consistent with the notion of structural transitions in the transpososomes (and hence the active sites) as the reaction proceeds forward. True reversal was also observed when mismatched bases were incorporated in the target (Fig. 2C, left), likely because of increased target flexibility around the mismatch, which allows the ST joint to explore normally disfavored spaces that promote reversal (29). Support for the conjecture that the reversal of ST is normally prevented because of misalignment of the reactive groups after ST comes from crystal structures of PFV intasomes captured at various stages of the reaction (32). The viral DNA–target DNA joint was observed to be ejected from the active site after ST, and thus no longer available for reversal (21).

Disintegration has also been observed on plasmid substrates at which only a single Mu end was allowed to undergo cleavage and ST (Fig. 2C, right) (33). Vulnerability of the single-ended joint to reversal is favored in the absence of the transposase-activator protein

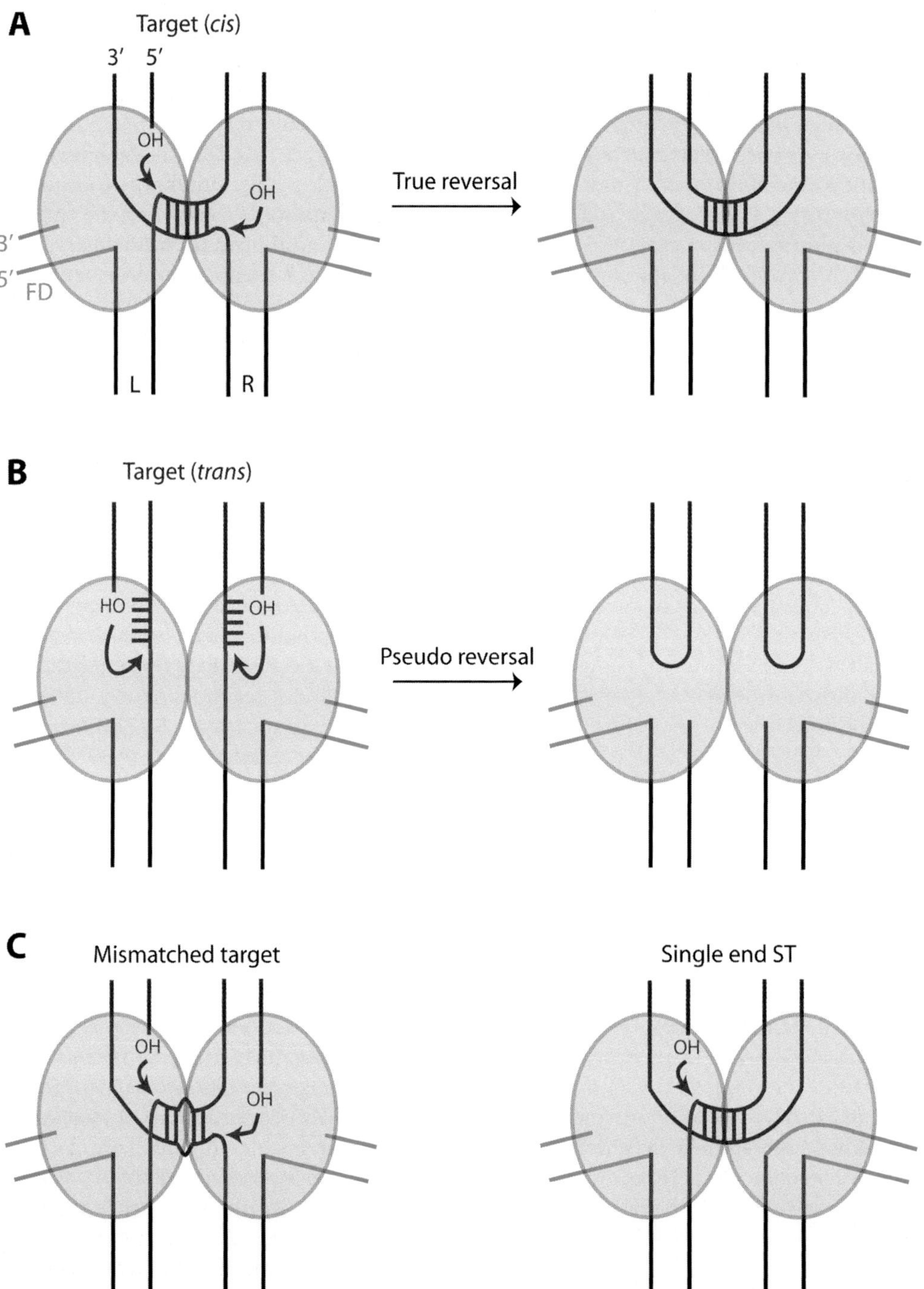

Figure 2 Disintegration: true and pseudo reversal. Ovals represent the transposase active sites. (A) True reversal refers to restoration of the original target configuration by a *cis* nucleophilic attack of the target 3′-OHs on the target–host junction generated during ST. L and R refer to the left and right ends of Mu. (B) In pseudo reversal, the target is rearranged (imagine unpairing the 5 bp in A, and flipping the DNA through 180°), so that the target nucleophiles attack the host–target junction in *trans*, resulting in hairpin products. (C) True reversal is more facile if the target carries a mismatch (left) indicated by an unpaired base pair, or if only a single end undergoes ST within the transpososome (right). See text for details. doi:10.1128/microbiolspec.MDNA3-0007-2014.f2

MuB, revealing a role for MuB in coordinating the configuration of both active sites so as to prevent reversal. On plasmid substrates, where the STs of both Mu ends are highly concerted, high-temperature disintegration (either true or pseudo) was observed at only one of the two ends (29), suggesting that both active sites cannot simultaneously adopt a reversal configuration.

TRANSPOSOSOME ASSEMBLY, ACTIVITY, AND STRUCTURE

Normal Versus Minimal DNA-Protein Requirements

Mu transposition requires MuA binding to sites at the left (L) and right (R) ends (also referred to as *att* ends), as well as at an enhancer (E) DNA segment located ~1 kb away from the L end on the Mu genome (Fig. 3A) (9). The L and R ends are asymmetric with respect to orientation and spacing of the three MuA-binding sites at each end (L1–L3 and R1–R3). There are three MuA-binding sites at E as well (O1–O3). Separate regions within MuA bind to the end and enhancer sites (Fig. 3B). Under physiological reaction conditions, transpososome assembly requires that both sets of DNA sites in their native configuration be present on supercoiled Mu donor DNA, along with the *E. coli* protein HU, which binds between L1 and L2 at the L end (Fig. 3A); the *E. coli* IHF protein, which binds between O1 and O2 at E, optimizes assembly. In the presence of the divalent metal ion Mg^{2+}, these components are sufficient for promoting the DNA cleavage reaction. ST requires MuB in addition (Fig. 3C). MuB is an AAA+ ATPase, which binds DNA nonspecifically (9, 34). The ATPase activity of MuB is stimulated by DNA and MuA. MuB not only captures target DNA and delivers it to the transpososome, but also its interactions with MuA optimize all stages of transpososome assembly (Fig. 4) (9, 34, 35, 36).

A minimal transposition system has been established *in vitro*, in which addition of the solvent Me_2SO (dimethyl sulfoxide) allows oligonucleotide substrates encoding only the R1–R2 subsites to undergo efficient pairing, cleavage, and ST by MuA alone, in the absence of DNA supercoiling, the enhancer, the N-terminal enhancer-binding MuA domain Iα, HU, MuB, and the C-terminal MuB/ClpX-binding MuA domain IIIβ (37). These permissive reaction conditions have greatly aided the dissection of the transposition reaction (9).

Assembly, Cleavage, Strand Transfer

The transposase MuA is a monomer in solution and does not assemble into a multimeric form unless bound to Mu ends (9). MuA subunits bound to all six L and R binding sites interact with the enhancer (Fig. 3A), and assemble into a transpososome. In a transpososome built from six MuA subunits, two subunits are loosely held; high salt treatment of the native transpososome yields a stable tetrameric core that is catalytically proficient. Within this tetrameric core, only two subunits, those bound to L1 and R1, catalyze cleavage and ST in *trans*, i.e., the subunit bound to L1 is responsible for catalysis at the R1 end and vice versa. The disintegration reaction also occurs in *trans* (33). Oligonucleotide substrates (R1–R2) also assemble a tetrameric MuA, and also catalyze the reaction in *trans* (9).

A variety of approaches have been used to dissect the configuration of the Mu ends within the two MuA active sites, the role of MuB in coordinating the reaction, and the role of metal ions. The results are summarized below ([i]–[vii]), and can be followed using Fig. 4 as a guide. Reaction chemistry occurs on the strands that end in CA (see Fig. 1), called the transferred strands; the opposite strands are the nontransferred strands. The terminal Mu base pair is designated +1, while the base pair immediately outside on the FD is −1. The role of the enhancer is discussed in a later section.

(i) Initial Engagement of Mu Ends and Stable Assembly

The rate-limiting step of Mu transposition is inferred to be a DNA melting event around the cleavage sites, not the subsequent reaction chemistry (37, 38, 39, 40). On supercoiled substrates, the free energy of supercoiling in the DNA outside the Mu ends (FD), but not inside Mu, is used to lower the activation barrier of this rate-limiting step, implying that the Mu ends pair and sequester the supercoils into separate Mu and FD DNA domains, before selectively using supercoiling energy stored in the FD domain (Fig. 4A) (38). The terminal CA dinucleotide, conserved in many transposable elements, including retroviruses (41), appears to be chosen at this position for its flexibility and ease of distortion (42, 43), rather than for its reaction chemistry (40, 43, 44). Mutant termini can assemble an unstable form of the left-end, enhancer and right-end (LER) transpososome (see "Enhancer and transpososome topology"), but fail to promote stable assembly (45). However, mismatched mutant termini assemble readily (40, 46), indicating that engagement of the melted DNA strands within the active site promotes larger conformational changes that allow MuA subunits to oligomerize into a stable complex (Fig. 4B). Assembly defects at +1 are more severe than those at at +2 or +3, and Mn^{2+} ions suppress these defects, suggesting a role for metal ions in modulating active-site

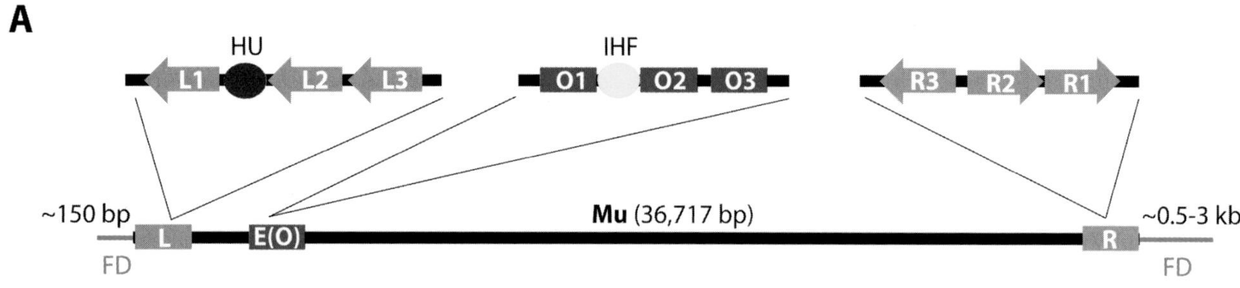

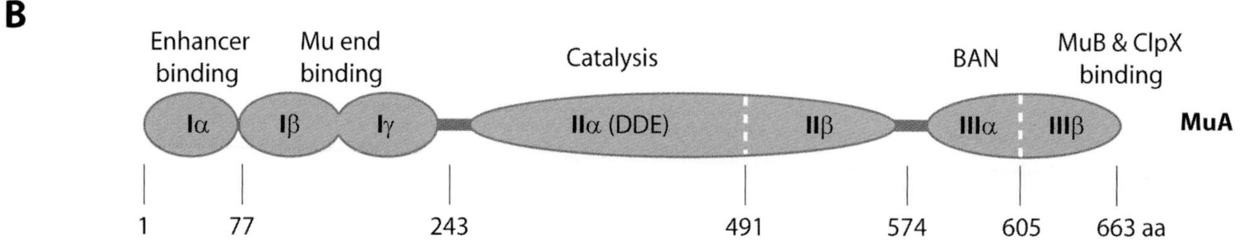

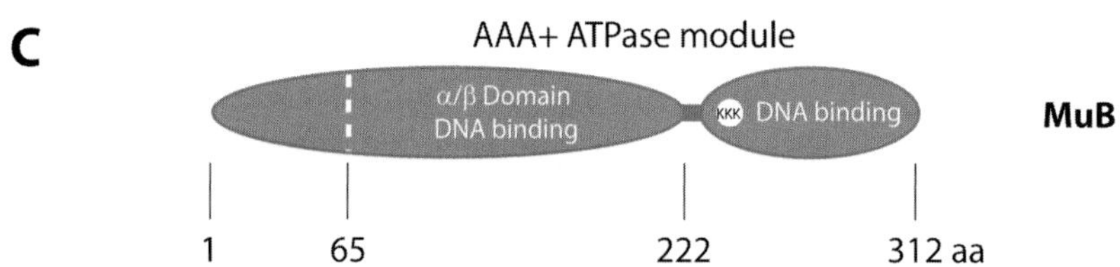

Figure 3 DNA and protein requirements for transposition. (A) Arrangement of MuA binding sites at the L (L1–L3) and R (R1–R3) ends of the Mu genome, and within the Mu enhancer E (O1–O3). E is positioned closer to the L end on the Mu genome, and is also labeled O because of its dual function as an operator that regulates lysis/lysogeny decision (8). HU and IHF bind within L and E as shown. FD on either side of the Mu genome is packaged into virions. (B) Domain and subdomain organization of MuA as assigned by partial proteolysis (140). NMR and crystal structures for the individual subdomains (except IIIβ) are available (9, 62, 141). The subdomain IIβ was observed in crystal structures (142), while IIIα and IIIβ were delineated by mutagenesis and functional studies (9). BAN stands for DNA-binding and nuclease function (122). See reference 143 for an insertion mutagenesis study across the domains. (C) Domain organization of MuB, as assigned by partial proteolysis (144). An NMR structure for the C-terminal domain is available (145). An AAA+ ATPase function spanning residues in both domains was deduced by bioinformatics, and supported by mutagenesis (34). Both domains also contribute to nonspecific DNA binding (9, 34, 35). A patch of positively charged residues (KKK) on the C-terminal domain likely interacts with MuA to trigger ATP hydrolysis (34, 35). The conformation of the hinge region between the domains is exquisitely modulated by ATP, DNA, and A protein, as judged by its sensitivity to proteolysis (146). doi:10.1128/microbiolspec.MDNA3-0007-2014.f3

conformation (45). On the nontransferred strand, the +1 T residue is essential for assembly even on mismatched substrates, with the R group at position 5 important for the initial contact but dispensable after cleavage of the transferred strand (Fig. 4B) (40, 46). Fully base-paired substrates require divalent metal ions for assembly (Mg^{2+}, Mn^{2+}, or Ca^{2+}), but this requirement is substantially reduced on substrates with mismatched termini (46), or if the FD strands are not complementary, or are very short, or the substrate is precleaved (37), implying that on fully base-paired substrates metal ions assist in the DNA opening observed around Mu ends (Fig. 4A and B) (39, 45, 47). MuB stimulates assembly on both plasmid and oligonucleotide substrates; the stimulation is independent of the presence of the FD on oligonucleotide substrates (36).

When concentrations of Mu DNA are limited, MuA bound to its end-recognition sites can capture and assemble non-Mu DNA sequences that resemble the Mu sequence; considerable variability in the placement of

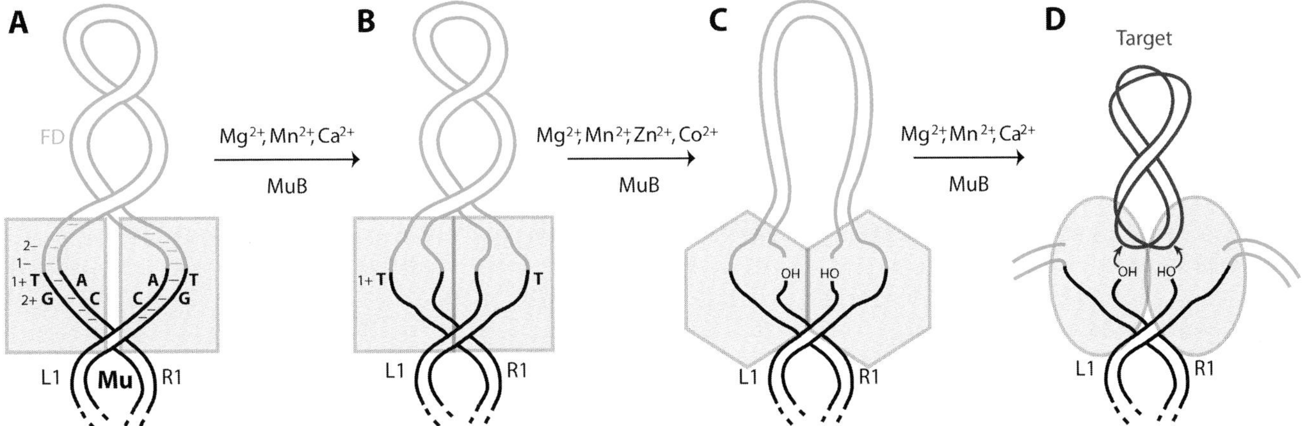

Figure 4 DNA–protein transitions within the MuA active sites. (A) Terminal base pairs at the Mu ends (L1 and R1 sites) and its adjoining FD are engaged within the MuA active sites (squares). Catalysis is in *trans*, i.e., the MuA subunit bound at R1 engages the L1 terminus and vice versa (9). This complex is not stable, and MuA has not tetramerized (indicated by a separation of the squares). (B) After Mu end synapsis, the free energy of supercoiling in the FD domain is used to melt several base pairs around the Mu–FD junction, concomitant with tetramerization of MuA (indicated by contiguous squares) (9). Mismatched substrates will tolerate any nucleotide at the terminal 2 bp, but require T at the +1 position on the nontransferred strand for stable MuA assembly. (C) Mismatched termini can cleave adjacent to any nucleotide at the +1 position on the transferred strand. The cleaved complex is more stable, indicated by a shape change to hexagons. (D) ST can occur on precleaved substrates even from an abasic site. This complex is the most stable, indicated by a shape change to the ovals. Each stage of transition (A–D) exhibits specific metal ion requirements and is regulated by MuB. See text for details. doi:10.1128/microbiolspec.MDNA3-0007-2014.f4

the MuA binding site with respect to the cleavage site appears to be tolerated by the transposition machinery (48, 49).

(ii) DNA Cleavage

As described above, the cleavage reaction is independent of nucleotide sequence when the terminal bases are unpaired. This reaction is also more efficient if the DNA flanking the Mu ends is not base paired (37). The MuA active sites accept hairpin substrates with different "loop" lengths for cleavage (50). Although Mg^{2+} is likely the biologically relevant cation, Mn^{2+} ions support both cleavage and ST in all the transposition systems studied. However, preassembled Mu substrates can be cleaved by Zn^{2+} and Co^{2+} (Fig. 4B and C) (39).

(iii) DNA ST

The wild-type nucleophile for ST is the 3′-OH of the terminal adenosine. However, Mu termini ending in a dideoxynucleotide allow target DNA hydrolysis by a water nucleophile (44). Remarkably, substrates that terminated with an abasic site, i.e., contained the terminal ribose and its 3′-OH, but were missing the adenine base, promoted efficient ST. Thus, it is not the terminal base, but rather the ribose (and/or the attached 5′-phosphate) that is the critical activating feature of the terminal nucleotide (Fig. 4C). It appears that the presence of the terminal nucleotide prevents the use of inappropriate nucleophiles, and offers an advantage to the attached 3′-OH to serve as the nucleophile.

Precleaved ends will perform ST in the presence of Ca^{2+} (Fig. 4D) (37). A true reversal of ST can also use Ca^{2+} (see "Disintegration"). Since a single active site carries out both cleavage and ST, the differential metal ion selectivity in cleavage (which is not supported by Ca^{2+}) versus ST must reflect either conformational differences between the active sites during the two steps or differences in the way the two distinct nucleophiles are activated.

(iv) Coupling Catalysis at the Paired Ends

The normal transposition reaction is highly concerted at both ends. When only one end carries a mutation in the terminal dinucleotide, stable assembly of both ends is blocked (51). Metal ions and MuB protein influence this tight coupling between the two active sites as judged by situations in which the coupling is lost. For example, both Mn^{2+} ions and MuB suppress the assembly and catalysis defects of mutations at one Mu end (9, 51, 52). So also, hairpin ends react in a concerted manner with

Mn^{2+} ions, but only produce single-end ST with Mg^{2+} (50). In the absence of MuB, the presence of FD in only one active site slows the ST of precleaved substrates in both active sites (36). MuB also suppresses mutations of G residues at the +2 position on the bottom strand, influencing the degree of concerted ST activity (46). Thus, MuB influences all steps of transpososome assembly and catalysis. The allosteric effect of MuB is apparently not effected through any specific MuA subunit (53).

(v) FD Configuration

As described above, several nucleotides adjacent to the Mu termini on the FD also undergo melting during assembly, and this single-stranded feature of the FD is important for DNA cleavage (Fig. 4B). After cleavage, the FD must be moved away to accommodate the target DNA (Fig. 4D); this is evident in the finding that in the absence of MuB, the presence of the FD in only one active site slowed one or more steps between DNA cleavage and joining in both sites, and that this slow step was not due to a change in the affinity of the transpososome for the target DNA (36). Thus, MuB interaction with MuA coordinates shifting the FD in both active sites, coupling this movement to a conformational change within the transpososome that positions the target DNA for coordinated ST.

(vi) Target Capture and DNA Conformation

Transpososomes assembled on plasmid donor Mu substrates can capture target DNA even prior to DNA cleavage, i.e., at the stage shown in Fig. 4A (54). Interestingly, on these substrates the terminal MuA binding site L1, but not R1, is required for target capture (47). Transpososomes assembled on R1–R2 oligonucleotide substrates lacking the terminal 3 bp are capable of target DNA capture, indicating that the target-interacting surface of the transpososome can be organized without the Mu-terminal nucleotides (Fig. 4D) (40). A single mismatch in the center of the 5-bp target recognition sequence promotes preferential use of the mismatched target, likely because it gives the target more flexibility for adopting the severely bent target configuration seen in the crystal structure (Fig. 5) (55), a feature found in other transpososome structures as well (21, 56). Given that transposition to nonmismatched sites is suppressed when mismatched sites are available suggests that the transposon–transposase complex samples a large number of potential target sites before ST.

(vii) Plasticity of the Active Site

As evidenced from the description above, the ability of a variety of DNA end configurations (fully paired, unpaired, hairpin) to be accommodated within the active site, of a variety of metal ions to support different stages of integration and disintegration, and of both half and full target sites to permit disintegration speaks to the remarkable plasticity of the active site (10).

Transpososome Structure

Electron Microscopy (EM)

The crystal structure of the dimeric Tn5 transposase in complex with DNA provided the earliest glimpse of a transposase active site (57). As first demonstrated for MuA (9), the Tn5 transposase subunits were observed to be arranged in *trans* for catalysis (57), a feature proving to be widespread among mobile elements (22, 58, 59, 60). The structure of a six-subunit Mu transpososome is expected to be more complex than that of a dimeric one. Although the structures of nearly all the individual domains of MuA were determined by nuclear magnetic resonance (NMR) or X-ray methods (Fig. 3B) (9), the entire complex proved difficult to crystallize. The first three-dimensional image of a tetrameric Mu transpososome assembled on precleaved R1–R2 oligonucleotide substrates was published by Yuan et al. who used scanning transmission electron microscopy (Fig. 5A) (61). Although the resolution of the reconstruction was relatively low (34 Å), it satisfied several biochemical observations, particularly the *trans* arrangement of the subunits for catalysis. A model of the complex resembled a large V (shown inverted in Fig. 5A), with the only significant protein–protein interactions occurring at the bottom of the V where the catalytic subunits were located, the rest of the complex apparently held together only by DNA–protein interactions. The latter observation was proposed to provide flexibility to the overall complex as it transitioned into increasingly stable states with each step of the reaction. Target DNA was modeled to fit into the cleft of the V in the 3D structure (Fig. 5A). The X-ray structure of the ST transpososome shows the target DNA also accommodated at the bottom, but on the other side of the V cleft (see below).

X-ray Crystallography

The crystal structure of the Mu ST transpososome assembled on R1–R2 subsites joined to target DNA is shown in Fig. 5B (62). A truncated version of MuA (residues 77 to 605) was used for assembly (Fig. 3B). Efficient ST of precleaved substrates was achieved by employing a target substrate with a single mismatch, described under "Target capture and DNA conformation" (above). The overall transpososome shape is consistent with the V shape described by scanning EM (Fig. 5A),

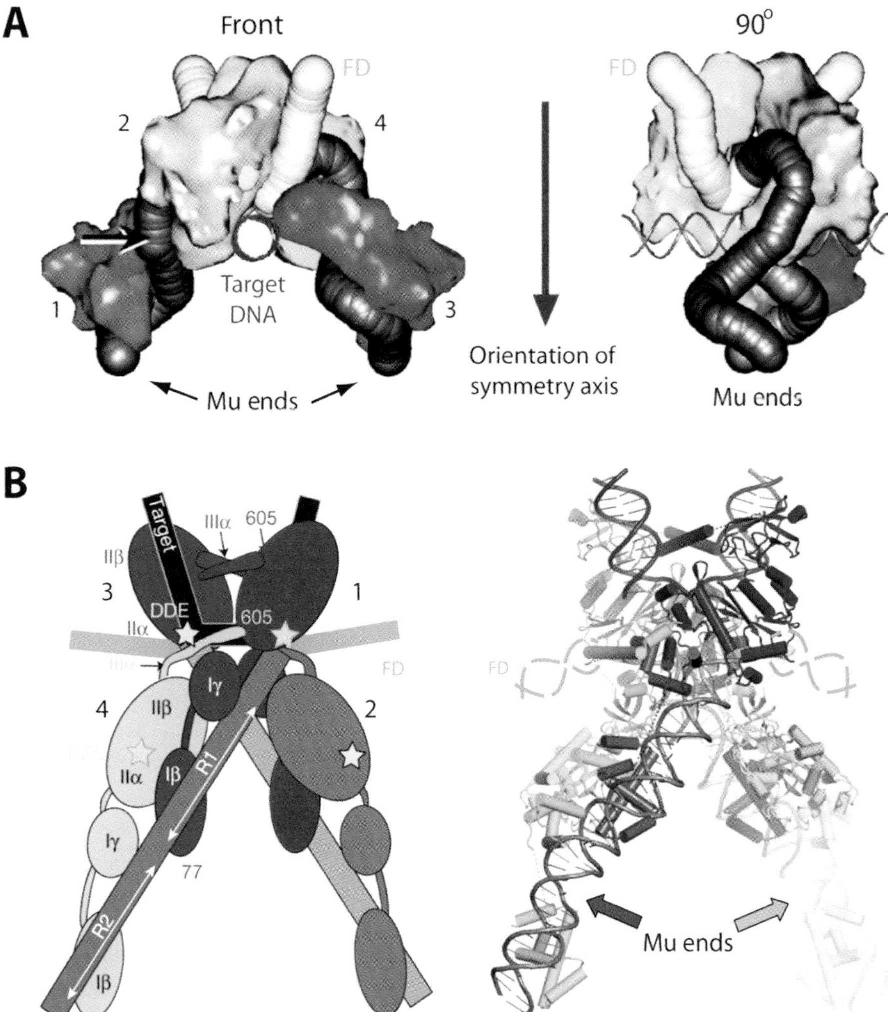

Figure 5 Mu transpososome structures assembled on oligonucleotide substrates. (A) Two views of the 3D reconstruction of images of a cleaved MuA tetramer bound to R1 and R2 ends, obtained by scanning transmission electron microscopy at cryotemperatures, combined with electron spectroscopic imaging of the DNA-phosphorus (61). Target DNA is modeled into the structure. Location of Mu ends (black tubes) and FD (gray tubes) is indicated. The image has been modified from the original in order to match the orientation of the X-ray image in B. (B) X-ray crystal structure of the Mu transpososome engaged with cleaved R1 and R2 Mu ends joined to target DNA (62). The MuA polypeptide in the crystal structure includes residues 77 to 605; it is missing the regulatory N (Iα)- and C (IIIβ)-terminal domains (see Fig. 3B). Left, schematic illustrating positions of the various MuA domains and DNA segments. Catalytic sites are marked as tan/yellow stars. Right, ribbon drawing, with the scissile phosphate groups shown as yellow spheres. The figure is modified to indicate position of the FD. In the crystal structure, the BAN region in domain IIIα (see Fig. 3B) of the R2-bound subunits (cyan and pink) makes contact with the FD near the Mu–FD junction; this region is associated with a nonspecific endonuclease activity (122). Images in A (61) and B (62) are adapted with permission from the Nature Publishing group.
doi:10.1128/microbiolspec.MDNA3-0007-2014.f5

where protein–protein interactions are mainly between the catalytic subunits at the bottom of the V. As expected, the DDE residues in domain IIα of subunits bound to R1 sites engage the Mu termini in *trans*; domain IIβ residues of these catalytic subunits make extensive contact with the target DNA. Domain II of subunits at the R2 sites also contact the opposite DNA but in a slightly different manner, and interact with DNA-binding domains of the R1 subunits. The IIIα domains of the subunits at R2 are positioned to be proximal to the FD on the opposite Mu termini; domain IIIα has an apparent role in repair of the ST intermediate in the infection phase (53, 63), the significance of which is discussed in Transition from Transposition to Repair. The target DNA is bent through a total of ~140°, a feature also seen in the PFV intasome (21). It is hypothesized that target bending may help render ST irreversible by straining the DNA conformation such that the ends snap away from the active site after ST (62), as seen in the PVF intasome (21).

ENHANCER AND TRANSPOSOSOME TOPOLOGY

Mu is unique among transposable elements in employing an enhancer (Fig. 3A). The E DNA segment is so named because it enhances transposition over 100-fold *in vivo* and is essential for the assembly of the transpososome on native supercoiled Mu donor substrates *in vitro* (9). All the end-binding sites (L1–L3 and R1–R3) participate in interactions with E (64, 65); MuA subunits bound to these L and R sites through their Iβγ domain are expected to make cross-bridging interactions with E through their Iα domain to form the three-segment LER complex (Fig. 6). HU, which binds in the spacer region between L1 and L2 and bends the DNA, and integration host factor (IHF), which binds E and also bends DNA, both play important roles in LER interactions (Fig. 3A) (9). Although the position of the enhancer can be varied, its orientation cannot, nor can it function when present outside the Mu ends on the same DNA molecule (9); however, it can function in *trans* when supplied in 50-fold molar excess (66), from where it maintains its topological specificity and crisscrossed R1–O1 and L1–O2 interactions (67).

The path of DNA through the LER complex was mapped using a methodology called "difference topology," where the topologically well-characterized site-specific recombinase Cre was used to seal the DNA crossings within the Mu synapse (68). Using a variety of approaches, including directed positioning of MuA subunits missing the E-binding Iα domain at individual end-binding sites, the following sequence of events and topology of LER interactions was deduced: an initial ER synapse (HU-independent) captures L (HU-dependent) to form an LER complex that has five supercoils trapped within it (Fig. 6A) (69, 70). The primacy of ER interactions has been confirmed using a gel electrophoresis assay, which shows, in addition, that at the L end the L1 site is the last to join the complex (47). Of the five DNA crossings held within the Mu synapse, two are between R and E, one between L and E, and two between L and R (Fig. 6B). The E–L crossing is thought to be trapped fortuitously, while the E–R crossings are specific; except for the E–L crossing, MuA can recreate the other four DNA crossings even on nicked circular DNA in the presence of Me_2SO (71). At the R end, R1–E interaction is essential for the initial steps in the assembly, R2–E interaction is not required, and R3–E interaction contributes to the native topology (72). The data on L2–E and L3–E interactions are not unequivocal. If these interactions do occur, either one is sufficient to support the assembly process. When stripped of the loosely bound subunits, the tetrameric MuA complex retains the two L–R crossings as well as the proximal E–R crossing (Fig. 6B, black dots with white circles), but loses the E–L and distal E–R crossings; the latter crossing is formed by the MuA subunit bound at R3 (Fig. 6B, black dot) (71, 72). The structure of the tetrameric Mu transpososome assembled without E and without the native arrangement of the L and R end sites shows only one L–R crossing (Fig. 5B) (62). A mathematical tangle analysis of the difference topology experiments concluded that the experimentally observed three-branched, five-crossing topological architecture of the Mu transpososome is the simplest solution, and thus biologically most likely (73, 74).

Footprinting experiments show that the Mu termini are not stably engaged in the active sites in the LER complex (Fig. 6A; see also Fig. 4A) (47). This was also deduced from the behavior of mutant termini, which stall at the LER stage as described in the section above. These experiments show, in addition, that the enhancer changes from an apparently strained to a less strained configuration as the LER transitions from an unstable to a stable state, and that HU-promoted engagement of the L1 site with ER is the last event that triggers transition into the stable form (Fig. 6A) (47); interestingly, in a complex assembled without L1, precleaved L1 can functionally join the complex when provided in *trans*. Although E is not required for reaction chemistry (9), it remains engaged with L and R ends even after ST is completed (Fig. 6A) (47, 70), suggesting that it may play an additional role post-ST. In the minimal reaction system with R1–R2 oligonucleotide substrates, E did

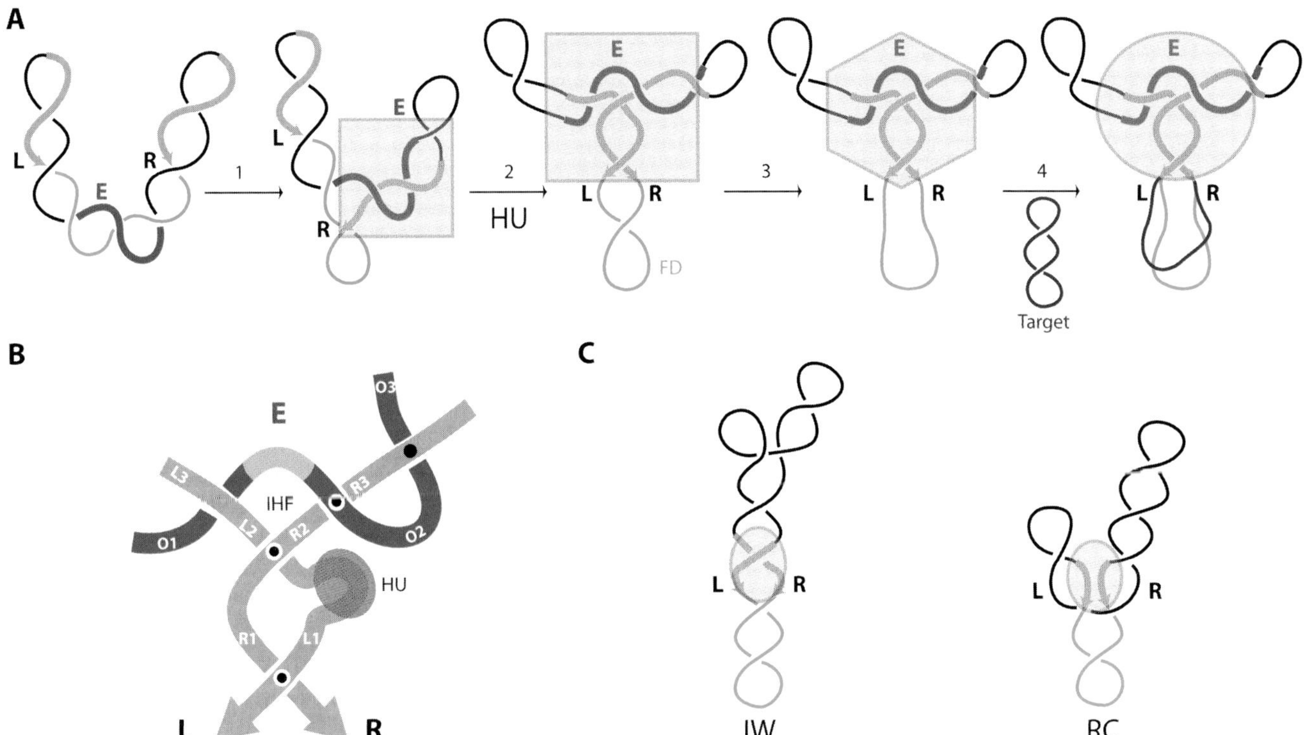

Figure 6 Interaction of MuA binding sites during transpososome assembly, and topology of the Mu DNA synapse. (A) 1. On a supercoiled DNA containing the native arrangement of L, E, and R segments, MuA-mediated interactions between E and R (square) trap two supercoil nodes within an initial ER synapse. 2. L is recruited into the ER complex with assistance from HU, and contributes one more crossing with E. The two L–R crossings within LER are fluid. 3. LER transitions into a stable complex (hexagon) which traps five supercoil nodes: two between E and R, one between E and L, and two between L and R. In this stable complex, the DNA around the Mu termini is first melted and then cleaved (see Fig. 4B, C). 4. The five-noded LER topology is maintained in the ST complex (oval), which is the most stable. (B) Contribution of the individual MuA-binding sites to the DNA topology. R–E interactions, particularly R1–O1, are essential in the initial stages of assembly, R2–E interactions are not required, and R3-E interactions contribute to the distal E–R crossing (black dot). Six MuA subunits (not shown) hold the five DNA crossings. The MuA tetramer retains two L–R and the proximal E–R DNA crossings (black dots with white circles). Of the two L–R crossings, the one between L1 and R1 is likely the one seen in the crystal structure (Fig. 5B). Placement of the second L–R crossing is arbitrary; see reference 72 for details. IHF binding and bending at E between O1 and O2 optimizes E interactions with L and R. HU binding and bending L between L1 and L2 delivers L1 to the ER complex (47). (C) Encounter and synapsis of Mu ends on supercoiled DNA. In the absence of E, the L and R ends can approach each other either by slithering to form a plectonemically interwrapped (IW) synapse or by random collision to form a random collision (RC) synapse; the presence of E channels synapsis toward the IW pathway (77). See text for details.
doi:10.1128/microbiolspec.MDNA3-0007-2014.f6

not enhance LR pairing, but accelerated transition to a stable form of the complex (75).

Enhancers in other site-specific recombination systems, such as the Tn3/γδ resolvase or Hin/Gin invertases, are highly selective with respect to orientation of the interacting sites and recombine through a very specific synapse topology (68). Enhancer-independent recombinases in these systems have lost the requirement for DNA supercoiling or for a specific orientation of the recombining sites. For Mu, two different enhancer-independent situations have been described. One involves an enhancer-independent transposase that, unlike the invertase and resolvase systems, does not relieve the dependence on DNA supercoiling or on the correct orientation of Mu ends (67). The other involves the addition of Me_2SO to the reaction, which does provide independence from constraints of substrate topology or site orientation (76). To determine the contribution of E to the interwrapping of Mu ends, the topology of the LR synapse was examined under the two enhancer-independent reaction conditions (77). Under the Me_2SO conditions, two topologically distinct arrangements of the ends were observed. In their normal relative orientation, L and R were either plectonemically interwrapped (IW) or aligned by random collision (RC) (Fig. 6C). Addition of the enhancer to this system channeled synapsis toward the IW pathway, showing that the enhancer imposes topological specificity on the synapse. When the ends were in the wrong relative orientation, synapsis occurred exclusively by the RC mode. In the second enhancer-independent condition, which retains the requirement for a specific orientation of Mu ends, synapsis of L and R was entirely via the IW synapse. This finding implies that the enhancer is not the only determinant of topological selectivity; the interaction of MuA with the Mu ends is also important. The mutant transposase has acquired independence from the enhancer but not from the orientation specificity of the ends. Thus, studies with this enhancer-independent Mu transposase have revealed that systems involving two-site interactions, and not necessarily three-site interactions, can also be subject to strict topological restrictions. Me_2SO conditions promote not only enhancer-independence, but independence from end orientation as well. If transposition can be supported by enhancer-independent pairing of the L and R ends, why does Mu use an enhancer? (For a discussion of this question, see reference 68.)

CENTRAL SGS SITE AND MU END PAIRING

Mu ends on plasmid substrates have no difficulty pairing, but those on a Mu prophage genome require a centrally located strong gyrase binding site (SGS) for efficient synapsis and Mu replication (78, 79). DNA gyrase bound at this site is proposed to promote the formation of a plectonemically interwound, supercoiled loop, with the site at the apex and the synapsed prophage ends at the base of the loop (Fig. 7A) (80, 81). Gyrase is known to protect and cleave within a ~100-bp region; if the cleaved region is defined as the "core" and the flanking sequences the "arms," then sequences in the right arm were delineated as an important feature of the SGS for Mu replication (82). It is speculated that the right arm may make favorable gyrase contacts, likely forming the T segment that is passed through the cleaved G segment during the supercoiling reaction. On plasmid substrates, SGS promotes highly processive supercoiling (81); this property of the Mu SGS has been exploited in studying gyrase structural dynamics using a single-molecule assay (83). Candidate SGS sequences obtained from five Mu-like prophages in different bacteria were not as proficient in supporting Mu replication, although some of them did support processive supercoiling on plasmid substrates (84).

The SGS site was seen to be important in maintaining the Mu prophage as a separate and stable chromosomal domain of *E. coli* (85). In the prophage, the two Mu ends are paired, segregating Mu into an independent chromosomal domain (Fig. 7B). The Mu domain configuration is assisted by MuB and several nucleoid-associated proteins (NAPs), and promotes low-level transcription from an early prophage promoter, which controls the expression of Mu *A* and *B*, as well as several genes not essential for phage growth, including a ligase and a Ku-like DNA repair function (86, 87). MuB might provide a NAP-like function (88). It is proposed that the Mu domain provides long-term survival benefits to both the prophage and the host: to the prophage in bestowing transposition-ready topological properties unique to the Mu reaction and to the host in contributing extraneous DNA housekeeping functions (85).

TARGET SITE SELECTION

Mu is the most promiscuous of known mobile elements. A consensus 5-bp target recognition site 5′-NY(G/C)RN-3′ reported in early experiments (89) was refined more recently as 5′-C-Py-(G/C)-Pu-G-3′ (90); this preference is MuA encoded, and is independent of the target capture protein MuB. *In vivo*, a preference for CGG as the central triplet was observed (91). In the transpososome crystal structure, MuA is seen to contact the target DNA over a 20 to 25-bp region (Fig. 5B) (62). Analysis of target sequences *in vitro* detected symmetrical base patterns spanning a ~23 to 24-bp region around the target pentamer, indicative not of a sequence preference, but possibly of a structural preference that might facilitate target deformation (90).

Mu does not generally display an orientation bias in target selection (91), although an exception was seen at the *E. coli bgl* locus *in vivo* (92). A DNA microarray analysis of target site selection during the Mu lytic phase in both *E. coli* and *Salmonella* found hot and cold spots throughout the genome, reflecting a >1,000-fold variation in target preference (93, 94). Transcription had a strong negative influence on transposition (93), although a direct relationship between transcription and transposition is unlikely (88).

MuB is essential for target capture *in vitro* (9). A whole-genome *E. coli* tiling array revealed that there

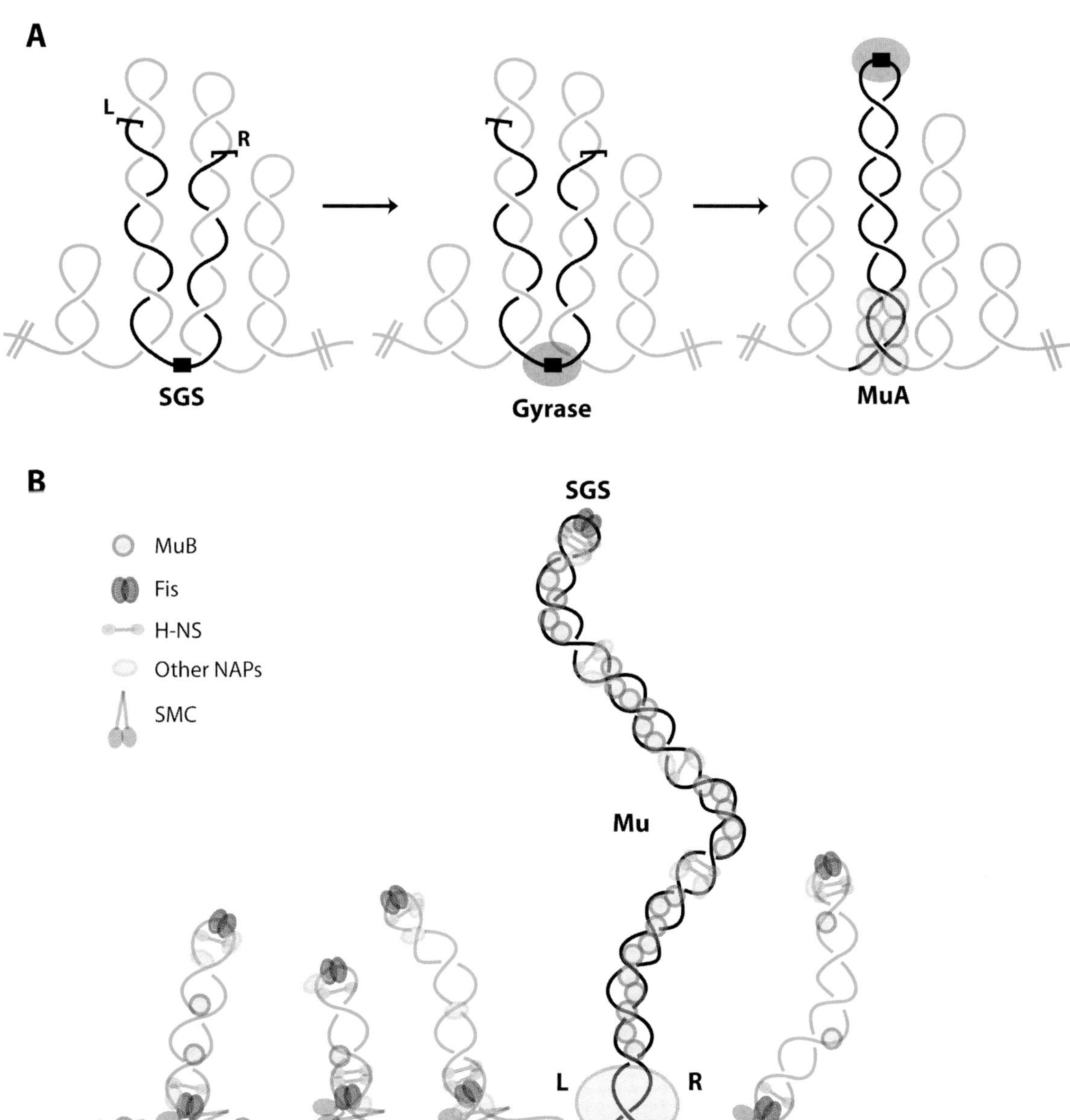

Figure 7 The central SGS site helps Mu prophage ends pair. (A) Plectonemically supercoiled domains of the *E. coli* nucleoid are shown carrying a copy of the Mu genome; L and R ends and the centrally located SGS are indicated. DNA gyrase binds at the SGS and initiates processive introduction of supercoils, leading to the extrusion of a novel nucleoid domain comprising the Mu genome in its entirety, aligning the L and R ends to promote transpososome (MuA) assembly. (B) A model for the structure of a Mu prophage and for Mu genome immunity. The model proposes that segregation of Mu into a separate domain, as shown in (A), is sealed by either the Mu transpososome assembled on the ends, or by nucleoid associated proteins (NAPs). Several NAPs are shown stabilizing this structure, hypothesized to promote the formation of MuB filaments. MuB, which itself has NAP-like properties, is proposed to provide immunity to self-integration. Fis and H-NS proteins may be expected to reside at the SGS and Mu ends because these proteins prefer A/T-rich regions. SMC proteins have been proposed to be involved in the creation of large topological loops by bridging two DNAs at the base of the stem of such loops. (A) Adapted from reference 81, and reprinted with permission from John Wiley and Sons. (B) Taken from references 85 and 106. doi:10.1128/microbiolspec.MDNA3-0007-2014.f7

were hot and cold MuB binding sites in the genome, and that Mu transposition was in the vicinity rather than within MuB-bound regions (88). MuB is an AAA+ ATPase and a nonspecific DNA-binding protein (Fig. 3C) (34), which hydrolyzes ATP for target selection; both activities are stimulated by MuA (9). MuB forms ATP-dependent helical filaments, with or without DNA (34, 95), and has been reported to exhibit a tendency to form larger filaments on A/T-rich DNA (95, 96). Single-particle EM imaging of MuB assembled on DNA found that MuB–ATP forms a solenoid-like filament, with DNA bound in the axial channel (34). The helical parameters of the MuB filament do not match those of the B-form DNA. Despite this mismatched symmetry between the protein and DNA, MuB does not deform bound DNA (34). Based on these and other findings, it is proposed that the MuB-imposed symmetry transiently deforms DNA at the boundary of the MuB filament and results in a bent DNA conformation favored by MuA for transposition (Fig. 8A,B) (34). Consistent with this model, Mu transposition was observed at the junction of A/T and non-A/T DNA *in vitro* (97) and in the vicinity rather than within MuB-bound regions *in vivo* (88). Two hot spots for Mu insertion near the *bgl* locus were also observed to be clustered at the borders of an A/T-rich segment (92). Interestingly, the majority of A/T-rich regions are unavailable for MuB binding *in vivo* because they are occupied by nucleoid proteins such as Fis, which have a similar binding preference (88, 97). Other strongly DNA-bound proteins also occlude Mu insertions (94).

MuB binds DNA nonspecifically (9). Transposition in the vicinity of a specific site was achieved by fusing MuB to the site-specific DNA-binding protein Arc repressor (98). The fusion variant could select target DNA independently of ATP hydrolysis, although ADP binding was required for the MuA-stimulating activity of MuB. Thus, the ATP-binding and MuA-regulated DNA-binding activity of MuB is not essential for target delivery, but activation of MuA by MuB requires nucleotide-bound MuB. Taken together, these results suggest that target delivery by MuB occurs as a consequence of the ability of nucleotide-bound MuB to stimulate MuA while simultaneously anchoring MuA to a selected target DNA. ATP hydrolysis has a different function, as discussed below.

TRANSPOSITION IMMUNITY

cis Immunity

Several bacterial transposons, including members of the Tn3 family, Tn7, and bacteriophage Mu, display transposition immunity (99, 100). These elements avoid insertion into DNA molecules that already contain a copy of the transposon (a phenomenon called *cis* immunity), and it is thought that this form of self-recognition must also provide protection against self-integration. *cis* immunity does not provide protection to the whole bacterial genome on which the transposon is resident, but can extend over large distances from the chromosomal site where the transposon is located or over an entire plasmid harboring the transposon.

In vitro studies with phage Mu provided the first molecular insights into the *cis*-immunity phenomenon (9). Using innovating fluorescence and microfluidic technology, single-molecule experiments show that MuA–MuB interaction, which stimulates the ATPase activity of MuB and promotes its dissociation from DNA, is the basis of the observed transposition immunity of mini-Mu plasmids *in vitro*; that is, MuB bound to DNA dissociates upon interaction in *cis* with MuA bound to the Mu ends, resulting in MuB-free DNA, which is a poor target for new insertions (Fig. 8C). The MuA–MuB interaction requires the formation of DNA loops between the MuA- and the MuB-bound DNA sites. Iterative loop formation/disruption cycles with intervening diffusional steps result in larger DNA loops, leading to preferential insertion of the transposon at sites distant from the transposon ends (101). MuB also dissociates upon interaction with MuA in *trans*, but the oligomeric state of MuA monomer, for example, when bound to ends versus multimer when assembled into an active transpososome, may distinguish interactions at the ends that underlie *cis* immunity from those that promote target capture and transposition in *trans* (Fig. 8C) (102, 103, 104). Support for a *cis*-immunity mechanism *in vivo*, which would remove MuB from the vicinity of the Mu genome, comes from studies using a 10-kb derivative of Mu (Mud), which was monitored for transposition into Tn10 elements placed at various distances from the Mud element on the *Salmonella typhimurium* chromosome (105). A gradient of insertion immunity was observed in both directions from the Mud insertion point; immunity was strongest around 5 kb outside the Mu ends.

Mu Genome Immunity

During the lytic phase, Mu amplifies its genome at least 100-fold. To produce viable progeny, Mu must avoid transposing into itself, a daunting task given that nearly half the host genome is composed of Mu sequences by the end of the lytic cycle and that Mu lacks target specificity. The *cis*-immunity mechanism, which is strongest around 5 kb outside the Mu ends, would not be expected to protect the 37-kb genome effectively. Indeed, the level of protection offered by this mechanism

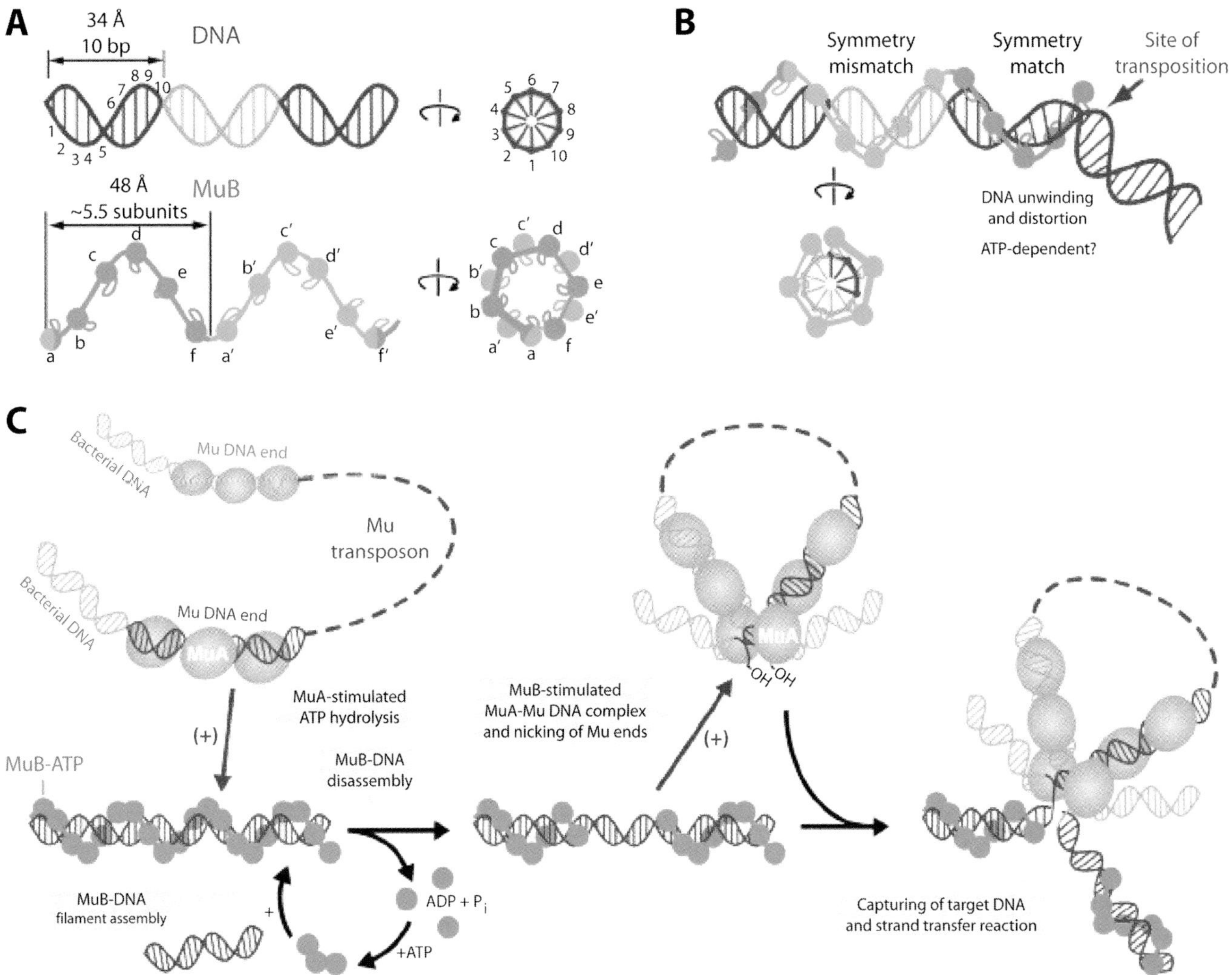

Figure 8 Model for MuB function in target capture and *cis* immunity. (A) The helical parameters of the MuB filament (represented as beads on a string) do not match those of B-form DNA. This results in a nucleoprotein complex with a symmetry mismatch. (B) Matching symmetry between MuB and DNA would require the DNA to be underwound and extended, which may occur at the boundary of the MuB filament with the help of MuA and possibly ATP hydrolysis. Deformed and bent DNA is a preferred target for transposition catalyzed by MuA. (C) A summary of interplay among MuA, MuB, and DNA during transposition. Upon ATP binding, MuB forms helical filaments on DNA. MuA bound to Mu DNA ends stimulates ATP hydrolysis by MuB and MuB dissociation from DNA, which generates MuB-free DNA regions. Reciprocally, MuB stimulates MuA to pair and nick Mu DNA ends at the junction with the flanking sequences. MuA and MuB together may induce the matching symmetry between MuB and DNA at the boundary of a MuB filament and thus DNA distortion, which leads to the target DNA capture and Mu transposition. Taken from reference 34, reprinted with permission from the *Proceedings of the National Academy of Sciences U S A*. doi:10.1128/microbiolspec.MDNA3-0007-2014.f8

is insufficient to explain the protection seen inside Mu (106). A new immunity mechanism labeled "Mu genome immunity" has been described, which protects actively replicating Mu from self-integration (106). Unlike the *cis*-immunity mechanism, which requires removal of MuB from DNA adjacent to Mu ends, the genome-immunity mechanism is associated with strong MuB binding within the Mu genome. Sharply different patterns of MuB binding were observed inside and outside Mu, suggesting that the Mu genome is segregated

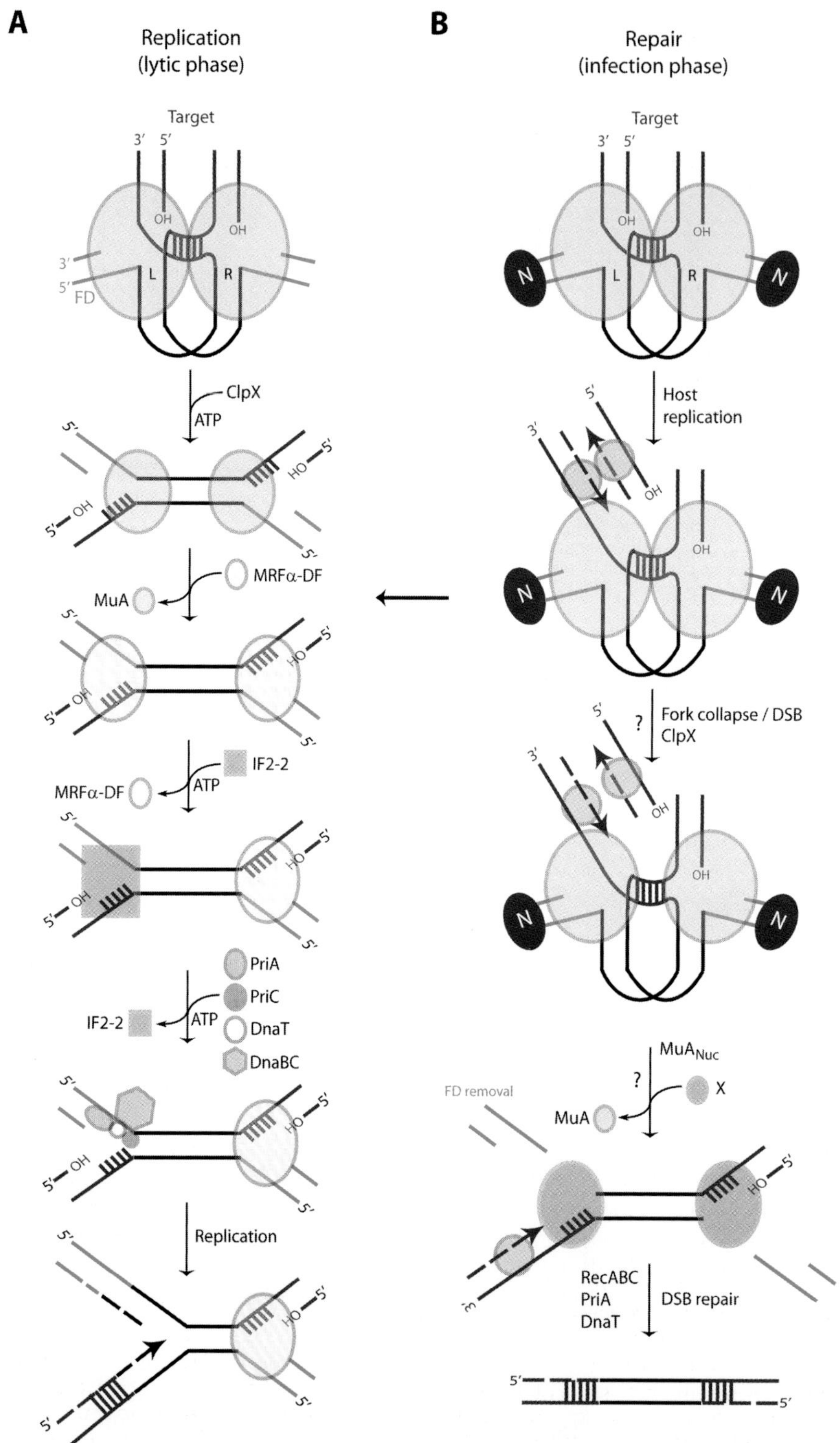
A
Replication
(lytic phase)
Target
3′ 5′
OH
OH
3′
5′
FD
L
R
ClpX
ATP
MRFα-DF
MuA
IF2-2
MRFα-DF
ATP
PriA
PriC
IF2-2
ATP
DnaT
DnaBC
Replication
B
Repair
(infection phase)
Target
3′ 5′
OH
OH
L
R
N
N
Host
replication
OH
Fork collapse / DSB
ClpX
?
MuA$_{Nuc}$
?
X
FD removal
MuA
RecABC
PriA
DnaT
DSB repair
5′
5′

into an independent chromosomal domain (Fig. 7). It is not clear why MuB would bind strongly within the Mu genome, which is not A/T rich, albeit A/T content was shown to be an unreliable predictor of MuB binding *in vivo*; MuB binding is expected to be modulated by host proteins *in vivo* (88, 97). A model for how the formation of an independent "Mu domain" might nucleate polymerization of MuB on the genome, forming a barrier against self-integration, has been proposed (Fig. 7B) (106). Mu-genome immunity might be functionally similar to the immunity conferred by the eukaryotic cellular barrier-to-autointegration factor (BAF) protein to HIV or murine leukemia virus retroviral genomes; BAF appears to play a dual role, compacting DNA reversibly to prevent autointegration on the one hand, while promoting intermolecular target capture on the other (see references in reference 106).

TRANSITION FROM TRANSPOSITION TO REPLICATION

As diagrammed in Fig. 1, the ST intermediate is resolved by replication through Mu during the lytic phase of Mu growth (107). To do so, the transpososome actively recruits the host restart replication machinery to a forked end generated by ST (Fig. 9A). In a highly choreographed series of events, the stable ST transpososome is first destabilized by the molecular chaperone ClpX, a member of the Clp/Hsp100 family of ATPases known to remodel and degrade multicomponent complexes (9, 108). On transpososomes assembled *in vitro*, the chaperone activity of ClpX selectively unfolds one or the other of the catalytic subunits so as to destabilize (but not destroy) the entire complex (Fig. 9A) (109, 110, 111). Interestingly, MuA residues exposed only in the MuA tetramer are important for enhanced recognition by ClpX, ensuring that the tetrameric complex is a high-priority substrate (111, 112).

The ClpX-destabilized transpososome is disassembled by an as-yet unidentified host factor (Mu replication factor [MRFα-DF]), which displaces the transpososome in an ATP-independent reaction, exchanging it for a truncated form of the translation initiation factor IF2 (IF2-2) (Fig. 9A) (113). The replication restart proteins (primosome) PriA, PriC, DnaT, and the DnaB–DnaC complex then promote the binding of the replicative helicase DnaB on the lagging strand template of the Mu fork. The PriA helicase activity is needed to displace IF2-2 and plays an important role in opening the DNA duplex for DnaB binding, which promotes assembly of DNA polymerase III holoenzyme to form the restart replisome (114).

TRANSITION FROM TRANSPOSITION TO REPAIR

The ST intermediate generated during the infection phase is repaired instead of being fully replicated, as diagrammed in Fig. 1 (115). This integration event is also called nonreplicative transposition because of limited replication at the Mu ends (116). The alternate fates of a similar ST intermediate in the lytic versus the infection phase could be due to different configurations of their FD (Fig. 1). Other differences during these two phases are the differential requirements for MuB and for the host gyrase. Although both MuA and MuB are required for efficient transposition, several MuB mutants that do not support lytic transposition are proficient for the integration of infecting Mu (117, 118). *In vitro*, these mutants retain the ability to stimulate MuA, but are defective in binding and delivery of target DNA to the transpososome. These data are puzzling because transposition during the infection phase also requires target capture. The requirement for the supercoiling enzyme DNA gyrase (Fig. 7A), essential during the lytic phase, is dispensable during the infection phase (119). These data are also puzzling, given the multiple roles of supercoiling in promoting binding of MuA and HU to Mu ends (9), in arranging a special topology of the interacting sites (Fig. 6), and in the essential role of supercoiling for DNA strand separation at the Mu termini (Fig. 4).

The transition from transposition to repair (Fig. 9B) appears to be as complex as the handoff of the transpososome to the replication restart machinery (Fig. 9A). There are two repair events associated with integration

Figure 9 Transition from transposition to replication or repair. The ST intermediate in the lytic versus infection phase differs primarily in the configuration of the FD (see Fig. 1). (A) The depicted order of events was established from *in vitro* experiments (107, 113). Mu replication is known to be unidirectional, primarily initiating at the L end (8). (B) Repair events are deduced from *in vivo* experiments (63, 120). Question marks signify that the order of these events is not as yet established. X is a hypothetical factor. The arrow from B to A indicates that infecting Mu can proceed directly to replication without FD removal, as seen in a domain III MuA mutant defective in FD removal (63). See text for details of both pathways. doi:10.1128/microbiolspec.MDNA3-0007-2014.f9

during the infection phase: filling the 5-bp target gaps and removing the N-linked FD. The first is likely done by the *E. coli* replisome (Pol III) because stable Mu insertions are not recovered in the absence of the machinery responsible for double-strand break (DSB) repair—RecA, RecB, RecC, PriA, and DnaT (120); arrival of the replisome at the gap would be expected to convert the gap into a DSB (Fig. 9B). Contrary to widely held assumptions that gaps left in the target after transposition are repaired by gap-filling polymerases, PolA (Pol I) appears to not be involved in filling the Mu gaps (120). Gap-filling repair is likely coordinated with the second repair event which removes FD, but the details are not known. FD removal has been monitored by PCR and other assays *in vivo* (115, 121). In the absence of ClpX, FD removal is delayed (63), suggesting that the requirement for ClpX is shared in the two pathways (Fig. 9B). MuA domain III residues, important for the cryptic nuclease activity MuA_{Nuc} (122) (Fig. 3B), are also required for FD removal (63). The requirement for ClpX and MuA_{Nuc} could be linked, in that ClpX might unmask the nuclease potential of domain III. However, such a mechanism must be suppressed in the lytic phase because removal of FD in this phase would affect the integrity of the entire chromosome (Fig. 1). Interestingly, domain III mutants that block FD removal allow replicative transposition (63), suggesting that there is a window of opportunity for FD removal, after which Mu replication can proceed by the restart pathway (Fig. 9, arrow from B to A).

MU-LIKE PHAGES

After the discovery of Mu by Larry Taylor (2), one other Mu-like phage called D108 was isolated from *E. coli* in the 1970s (123). Mu and D108 are fairly identical, except for their enhancer/operator sequences, and their respective binding regions in MuA and in the lysogenic repressor Rep. These differences have proved useful in understanding the contribution of E to transposition (65, 67, 124, 125). In the 1980s, several Mu-like phages were isolated from *Pseudomonas aeruginosa*; one of these, D3112, has been studied for its transposition properties in some detail (see reference 123). In the era of genomics, Mu-like prophages have been found in multiple distantly related species, indicating that they are widespread mobile genetic elements (126, 127). Of the mosaic Mu-like phages isolated from *P. aeruginosa* (128), *Rhodobacter capsulatus* (Rcap Mu) (129), and *Haemophilus parasuis* (130), the one from *Rhodobacter* (Rcap Mu) is similar to Mu in packaging host FD in phage particles, indicative of replicative transposition with little target site specificity. Many prophage sequences detected in Gram-negative bacteria have mosaic gene patterns, with only some Mu-like modules. Although the SGS sequences of some of these elements could substitute for the Mu SGS to varying degrees (see Central SGS Site and Mu End Pairing above) and the transposase gene from Hin-Mu in *H. influenzae* Rd was functional, as detected by *in vitro* reactions (131), there is no evidence that the majority of these more recently identified prophages are capable of active transposition. Complete and partial Mu-like prophages have been detected in several *Firmicutes* (132), opening the door for Mu-like genetic tools in Gram-positive bacteria.

MU AS A TOOL FOR GENE MANIPULATION

Since the mid-1980s, mini-Mu derivatives have been used extensively *in vivo* for insertional mutagenesis, for gene fusion and mapping, as well as for gene cloning and DNA sequencing strategies, including metabolic engineering (see references 9 and 133). More recently, the Mu enhancer element has been either supplied in *trans* or excised from mini-Mu vectors to control their transposition efficiency (reviewed in reference 133). Electroporation of *in vitro* assembled and cleaved Mu R1–R2 transpososomes has been used successfully for Mu integration in a variety of bacterial species, both Gram-positive and Gram-negative (134, 135), as well as in yeast and in mammalian genomes (136). Strong biases were seen in the target site distributions of the Mu insertion vectors in eukaryotic genomes, consistent with biases seen with other insertion vectors, illustrating the utility of Mu transpososome technology for gene transfer in eukaryotic cells. This technique has also been used to map integration sites directly using DNA barcoding and pyrosequencing (137). While MuA and MuB have been expressed in mammalian cells, integration of a transfected mini-Mu donor vector was from illegitimate recombination rather than from transposition, suggesting that Mu target capture complexes might promote nonhomologous recombination (138).

The superior target for Mu insertion presented by single nucleotide mismatches can be exploited to map genetic polymorphisms (55), as was demonstrated *in vitro* for a butterfly genome (139). Single nucleotide polymorphisms are an important resource for mapping human disease genes and have other biological applications as well (139).

SUMMARY

Transposable phage Mu has played a major role in elucidating the mechanism of movement of mobile DNA elements. The high efficiency of Mu transposition has facilitated a detailed biochemical dissection of the reaction mechanism, as well as of protein and DNA elements that regulate transpososome assembly and function. The deduced phosphotransfer mechanism involves in-line orientation of metal ion-activated hydroxyl groups for nucleophilic attack on reactive diester bonds, a mechanism that appears to be used by all transposable elements examined to date. A crystal structure of the Mu transpososome is available. Mu differs from all other transposable elements in encoding unique adaptations that promote its viral lifestyle. These adaptations include multiple DNA (enhancer, SGS) and protein (MuB, HU, IHF) elements that enable efficient Mu end synapsis, efficient target capture, low target specificity, immunity to transposition near or into itself, and efficient mechanisms for recruiting host repair and replication machineries to resolve transposition intermediates. MuB has multiple functions, including target capture and immunity. The SGS element promotes gyrase-mediated Mu end synapsis and the enhancer, aided by HU and IHF, participate in directing a unique topological architecture of the Mu synapse. The function of these DNA and protein elements is important during both lysogenic and lytic phases. Enhancer properties have been exploited in the design of mini-Mu vectors for genetic engineering. Mu ends assembled into active transpososomes have been delivered directly into bacterial, yeast and human genomes, where they integrate efficiently, and may prove useful for gene therapy.

Acknowledgments. *The principal support for Mu research in my lab has come from a long-running grant from the National Institutes of Health, and partial support by the Robert Welch foundation. I am grateful to Rudra Saha for assistance with the illustrations in this article.*

Citation. Harshey RM. 2014. Transposable phage Mu. Microbiol Spectrum **2**(5):MDNA3-0007-2014.

References

1. **Harshey RM.** 2012. The Mu story: how a maverick phage moved the field forward. *Mob DNA* **3**:21.
2. **Taylor AL.** 1963. Bacteriophage-induced mutations in *E. coli*. *Proc Natl Acad Sci USA* **50**:1043–1051.
3. **McClintock B.** 1950. The origin and behavior of mutable loci in maize. *Proc Natl Acad Sci USA* **36**:344–355.
4. **Ljungquist E, Bukhari AI.** 1977. State of prophage Mu DNA upon induction. *Proc Natl Acad Sci USA* **74**: 3143–3147.
5. **Faelen M, Huisman O, Toussaint A.** 1978. Involvement of phage Mu-1 early functions in Mu-mediated chromosomal rearrangements. *Nature* **271**:580–582.
6. **Shapiro JA.** 1979. Molecular model for the transposition and replication of bacteriophage Mu and other transposable elements. *Proc Natl Acad Sci USA* **76**: 1933–1937.
7. **Mizuuchi K.** 1983. *In vitro* transposition of bacteriophage Mu: a biochemical approach to a novel replication reaction. *Cell* **35**:785–794.
8. **Symonds N, Toussaint A, Van de Putte P, Howe MM.** 1987. *Phage Mu*. Cold Spring Harbor Laboratory, Cold Spring Harbor, NY.
9. **Chaconas G, Harshey RM.** 2002. Transposition of phage Mu DNA, p 384–402. *In* Craig NL, Craigie R, Gellert M, Lambowitz AM (ed), *Mobile DNA II*. ASM Press, Washington, DC.
10. **Mizuuchi K, Baker TA.** 2002. Chemical mechanisms for mobilizing DNA, p 12–23. *In* Craig NL, Craigie R, Gellert M, Lambowitz AM (ed), *Mobile DNA II*. ASM Press, Washington, DC.
11. **Harshey RM, Bukhari AI.** 1983. Infecting bacteriophage Mu DNA forms a circular DNA-protein complex. *J Mol Biol* **167**:427–441.
12. **Puspurs AH, Trun NJ, Reeve JN.** 1983. Bacteriophage Mu DNA circularizes following infection of *Escherichia coli*. *EMBO J* **2**:345–352.
13. **Gloor G, Chaconas G.** 1988. Sequence of bacteriophage Mu N and P genes. *Nucleic Acids Res* **16**:5211–5212.
14. **Lassila JK, Zalatan JG, Herschlag D.** 2011. Biological phosphoryl-transfer reactions: understanding mechanism and catalysis. *Annu Rev Biochem* **80**:669–702.
15. **Lewinski MK, Bushman FD.** 2005. Retroviral DNA integration–mechanism and consequences. *Adv Genet* **55**:147–181.
16. **Montano SP, Rice PA.** 2011. Moving DNA around: DNA transposition and retroviral integration. *Curr Opin Struct Biol* **21**:370–378.
17. **Kennedy AK, Haniford DB, Mizuuchi K.** 2000. Single active site catalysis of the successive phosphoryl transfer steps by DNA transposases: insights from phosphorothioate stereoselectivity. *Cell* **101**:295–305.
18. **Mizuuchi K, Adzuma K.** 1991. Inversion of the phosphate chirality at the target site of Mu DNA strand transfer: evidence for a one-step transesterification mechanism. *Cell* **66**:129–140.
19. **Nowotny M, Gaidamakov SA, Crouch RJ, Yang W.** 2005. Crystal structures of RNase H bound to an RNA/DNA hybrid: substrate specificity and metal-dependent catalysis. *Cell* **121**:1005–1016.
20. **Mizuuchi K.** 1992. Polynucleotidyl transfer reactions in transpositional DNA recombination. *J Biol Chem* **267**: 21273–21276.
21. **Maertens GN, Hare S, Cherepanov P.** 2010. The mechanism of retroviral integration from X-ray structures of its key intermediates. *Nature* **468**:326–329.
22. **Hare S, Gupta SS, Valkov E, Engelman A, Cherepanov P.** 2010. Retroviral intasome assembly and inhibition of DNA strand transfer. *Nature* **464**:232–236.
23. **Hare S, Maertens GN, Cherepanov P.** 2012. 3′-processing and strand transfer catalysed by retroviral integrase *in crystallo*. *EMBO J* **31**:3020–3028.

24. **Chow SA, Vincent KA, Ellison V, Brown PO.** 1992. Reversal of integration and DNA splicing mediated by integrase of human immunodeficiency virus. *Science* **255:**723–726.
25. **Jonsson CB, Donzella GA, Roth MJ.** 1993. Characterization of the forward and reverse integration reactions of the Moloney murine leukemia virus integrase protein purified from *Escherichia coli. J Biol Chem* **268:**1462–1469.
26. **Polard P, Ton-Hoang B, Haren L, Bétermier M, Walczak R, Chandler M.** 1996. IS911-mediated transpositional recombination *in vitro. J Mol Biol* **264:** 68–81.
27. **Beall EL, Rio DC.** 1998. Transposase makes critical contacts with, and is stimulated by, single-stranded DNA at the P element termini *in vitro. EMBO J* **17:**2122–2136.
28. **Melek M, Gellert M.** 2000. RAG1/2-mediated resolution of transposition intermediates: two pathways and possible consequences. *Cell* **101:**625–633.
29. **Au TK, Pathania S, Harshey RM.** 2004. True reversal of Mu integration. *EMBO J* **23:**3408–3420.
30. **Mizuuchi M, Rice PA, Wardle SJ, Haniford DB, Mizuuchi K.** 2007. Control of transposase activity within a transpososome by the configuration of the flanking DNA segment of the transposon. *Proc Natl Acad Sci USA* **104:**14622–14627.
31. **Surette MG, Buch SJ, Chaconas G.** 1987. Transpososomes: stable protein-DNA complexes involved in the *in vitro* transposition of bacteriophage Mu DNA. *Cell* **49:** 253–262.
32. **Cherepanov P, Maertens GN, Hare S.** 2011. Structural insights into the retroviral DNA integration apparatus. *Curr Opin Struct Biol* **21:**249–256.
33. **Lemberg KM, Schweidenback CT, Baker TA.** 2007. The dynamic Mu transpososome: MuB activation prevents disintegration. *J Mol Biol* **374:**1158–1171.
34. **Mizuno N, Dramicanin M, Mizuuchi M, Adam J, Wang Y, Han YW, Yang W, Steven AC, Mizuuchi K, Ramon-Maiques S.** 2013. MuB is an AAA+ ATPase that forms helical filaments to control target selection for DNA transposition. *Proc Natl Acad Sci USA* **110:**E2441–2450.
35. **Coros CJ, Sekino Y, Baker TA, Chaconas G.** 2003. Effect of mutations in the C-terminal domain of Mu B on DNA binding and interactions with Mu A transposase. *J Biol Chem* **278:**31210–31217.
36. **Williams TL, Baker TA.** 2004. Reorganization of the Mu transpososome active sites during a cooperative transition between DNA cleavage and joining. *J Biol Chem* **279:**5135–5145.
37. **Savilahti H, Rice PA, Mizuuchi K.** 1995. The phage Mu Transpososome core—DNA requirements for assembly and function. *EMBO J* **14:**4893–4903.
38. **Wang Z, Harshey RM.** 1994. Crucial role for DNA supercoiling in Mu transposition: a kinetic study. *Proc Natl Acad Sci USA* **91:**699–703.
39. **Wang Z, Namgoong SY, Zhang X, Harshey RM.** 1996. Kinetic and structural probing of the precleavage synaptic complex (type 0) formed during phage Mu transposition. Action of metal ions and reagents specific to single-stranded DNA. *J Biol Chem* **271:**9619–9626.
40. **Yanagihara K, Mizuuchi K.** 2003. Progressive structural transitions within Mu transpositional complexes. *Mol Cell* **11:**215–224.
41. **Lee I, Harshey RM.** 2003. Patterns of sequence conservation at termini of long terminal repeat (LTR) retrotransposons and DNA transposons in the human genome: lessons from phage Mu. *Nucleic Acids Res* **31:**4531–4540.
42. **El Hassan MA, Calladine CR.** 1997. Conformational characteristics of DNA: empirical classifications and a hypothesis for the conformational behaviour of dinucleotide steps. *Phil Trans R Soc Lond* **355:**43–100.
43. **Lee I, Harshey RM.** 2001. Importance of the conserved CA dinucleotide at Mu termini. *J Mol Biol* **314:**433–444.
44. **Goldhaber-Gordon I, Early MH, Baker TA.** 2003. The terminal nucleotide of the Mu genome controls catalysis of DNA strand transfer. *Proc Natl Acad Sci USA* **100:** 7509–7514.
45. **Watson MA, Chaconas G.** 1996. Three-site synapsis during Mu DNA transposition: A critical intermediate preceding engagement of the active site. *Cell* **85:**435–445.
46. **Lee I, Harshey RM.** 2003. The conserved CA/TG motif at Mu termini: T specifies stable transpososome assembly. *J Mol Biol* **330:**261–275.
47. **Kobryn K, Watson MA, Allison RG, Chaconas G.** 2002. The Mu three-site synapse: a strained assembly platform in which delivery of the L1 transposase binding site triggers catalytic commitment. *Mol Cell* **10:**659–669.
48. **Goldhaber-Gordon I, Williams TL, Baker TA.** 2002. DNA recognition sites activate MuA transposase to perform transposition of non-Mu DNA. *J Biol Chem* **277:** 7694–7702.
49. **Goldhaber-Gordon I, Early MH, Gray MK, Baker TA.** 2002. Sequence and positional requirements for DNA sites in a mu transpososome. *J Biol Chem* **277:**7703–7712.
50. **Saariaho AH, Savilahti H.** 2006. Characteristics of MuA transposase-catalyzed processing of model transposon end DNA hairpin substrates. *Nucleic Acids Res* **34:**3139–3149.
51. **Coros CJ, Chaconas G.** 2001. Effect of mutations in the Mu-host junction region on transpososome assembly. *J Mol Biol* **310:**299–309.
52. **Surette MG, Harkness T, Chaconas G.** 1991. Stimulation of the Mu A protein-mediated strand cleavage reaction by the Mu B protein, and the requirement of DNA nicking for stable type 1 transpososome formation. *In vitro* transposition characteristics of mini-Mu plasmids carrying terminal base pair mutations. *J Biol Chem* **266:** 3118–3124.
53. **Mariconda S, Namgoong SY, Yoon KH, Jiang H, Harshey RM.** 2000. Domain III function of Mu transposase analysed by directed placement of subunits within the transpososome. *J Biosci* **25:**347–360.

54. **Naigamwalla DZ, Chaconas G.** 1997. A new set of Mu DNA transposition intermediates: alternate pathways of target capture preceding strand transfer. *EMBO J* **16:** 5227–5234.

55. **Yanagihara K, Mizuuchi K.** 2002. Mismatch-targeted transposition of Mu: a new strategy to map genetic polymorphism. *Proc Natl Acad Sci USA* **99:**11317–11321.

56. **Craig NL, Craigie R, Gellert M, Lambowitz AM.** 2002. *Mobile DNA II.* ASM Press, Washington, DC.

57. **Steiniger-White M, Rayment I, Reznikoff WS.** 2004. Structure/function insights into Tn5 transposition. *Curr Opin Struct Biol* **14:**50–57.

58. **Barabas O, Ronning DR, Guynet C, Hickman AB, Ton-Hoang B, Chandler M, Dyda F.** 2008. Mechanism of IS200/IS605 family DNA transposases: activation and transposon-directed target site selection. *Cell* **132:** 208–220.

59. **Richardson JM, Colloms SD, Finnegan DJ, Walkinshaw MD.** 2009. Molecular architecture of the Mos1 paired-end complex: the structural basis of DNA transposition in a eukaryote. *Cell* **138:**1096–1108.

60. **Grandgenett D, Korolev S.** 2010. Retrovirus integrase-DNA structure elucidates concerted integration mechanisms. *Viruses* **2:**1185–1189.

61. **Yuan JF, Beniac DR, Chaconas G, Ottensmeyer FP.** 2005. 3D reconstruction of the Mu transposase and the Type 1 transpososome: a structural framework for Mu DNA transposition. *Genes Dev* **19:**840–852.

62. **Montano SP, Pigli YZ, Rice PA.** 2012. The Mu transpososome structure sheds light on DDE recombinase evolution. *Nature* **491:**413–417.

63. **Choi W, Harshey RM.** 2010. DNA repair by the cryptic endonuclease activity of Mu transposase. *Proc Natl Acad Sci USA* **107:**10014–10019.

64. **Allison RG, Chaconas G.** 1992. Role of the A protein-binding sites in the *in vitro* transposition of Mu DNA. A complex circuit of interactions involving the Mu ends and the transpositional enhancer. *J Biol Chem* **267:**19963–19970.

65. **Jiang H, Yang JY, Harshey RM.** 1999. Criss-crossed interactions between the enhancer and the att sites of phage Mu during DNA transposition. *EMBO J* **18:** 3845–3855.

66. **Surette MG, Chaconas G.** 1992. The Mu transpositional enhancer can function in *trans*: requirement of the enhancer for synapsis but not strand cleavage. *Cell* **68:** 1101–1108.

67. **Jiang H, Harshey RM.** 2001. The Mu enhancer is functionally asymmetric both in *cis* and in *trans*. Topological selectivity of Mu transposition is enhancer-independent. *J Biol Chem* **276:**4373–4381.

68. **Harshey RM, Jayaram M.** 2006. The Mu transpososome through a topological lens. *Crit Rev Biochem Mol Biol* **41:**387–405.

69. **Pathania S, Jayaram M, Harshey RM.** 2002. Path of DNA within the Mu transpososome. Transposase interactions bridging two Mu ends and the enhancer trap five DNA supercoils. *Cell* **109:**425–436.

70. **Pathania S, Jayaram M, Harshey RM.** 2003. A unique right end-enhancer complex precedes synapsis of Mu ends: the enhancer is sequestered within the transpososome throughout transposition. *EMBO J* **22:**3725–3736.

71. **Yin Z, Jayaram M, Pathania S, Harshey RM.** 2005. The Mu transposase interwraps distant DNA sites within a functional transpososome in the absence of DNA supercoiling. *J Biol Chem* **280:**6149–6156.

72. **Yin Z, Suzuki A, Lou Z, Jayaram M, Harshey RM.** 2007. Interactions of phage Mu enhancer and termini that specify the assembly of a topologically unique interwrapped transpososome. *J Mol Biol* **372:**382–396.

73. **Darcy IK, Chang J, Druivenga N, McKinney C, Medikonduri RK, Mills S, Navarra-Madsen J, Ponnusamy A, Sweet J, Thompson T.** 2006. Coloring the Mu transpososome. *BMC Bioinformatics* **7:**435.

74. **Darcy IK, Luecke J, Vazquez M.** 2009. Tangle analysis of difference topology experiments: applications to a Mu protein–DNA complex. *Algebr Geom Topol* **9:** 2247–2309.

75. **Mizuuchi M, Mizuuchi K.** 2001. Conformational isomerization in phage Mu transpososome assembly: effects of the transpositional enhancer and of MuB. *EMBO J* **20:**6927–6935.

76. **Mizuuchi M, Mizuuchi K.** 1989. Efficient Mu transposition requires interaction of transposase with a DNA sequence at the Mu operator: implications for regulation. *Cell* **58:**399–408.

77. **Yin Z, Harshey RM.** 2005. Enhancer-independent Mu transposition from two topologically distinct synapses. *Proc Natl Acad Sci USA* **102:**18884–18889.

78. **Pato ML, Banerjee M.** 2000. Genetic analysis of the strong gyrase site (SGS) of bacteriophage Mu: localization of determinants required for promoting Mu replication. *Mol Microbiol* **37:**800–810.

79. **Pato ML.** 2004. Replication of Mu prophages lacking the central strong gyrase site. *Res Microbiol* **155:**553–558.

80. **Pato ML, Banerjee M.** 1996. The Mu strong gyrase-binding site promotes efficient synapsis of the prophage termini. *Mol Microbiol* **22:**283–292.

81. **Oram M, Howells AJ, Maxwell A, Pato ML.** 2003. A biochemical analysis of the interaction of DNA gyrase with the bacteriophage Mu, pSC101 and pBR322 strong gyrase sites: the role of DNA sequence in modulating gyrase supercoiling and biological activity. *Mol Microbiol* **50:**333–347.

82. **Oram M, Travers AA, Howells AJ, Maxwell A, Pato ML.** 2006. Dissection of the bacteriophage Mu strong gyrase site (SGS): significance of the SGS right arm in Mu biology and DNA gyrase mechanism. *J Bacteriol* **188:**619–632.

83. **Basu A, Schoeffler AJ, Berger JM, Bryant Z.** 2012. ATP binding controls distinct structural transitions of Escherichia coli DNA gyrase in complex with DNA. *Nat Struct Mol Biol* **19:**538–546, S531.

84. **Oram M, Pato ML.** 2004. Mu-like prophage strong gyrase site sequences: analysis of properties required for promoting efficient Mu DNA replication. *J Bacteriol* **186:**4575–4584.

85. **Saha RP, Lou Z, Meng L, Harshey RM.** 2013. Transposable prophage Mu is organized as a stable chromosomal domain of *E. coli*. *PLoS Genet* **9:**e1003902.

86. **Paolozzi L, Ghelardini P.** 1992. A case of lysogenic conversion: modification of cell phenotype by constitutive expression of the Mu gem operon. *Res Microbiol* **143:** 237–243.

87. **d'Adda di Fagagna F, Weller GR, Doherty AJ, Jackson SP.** 2003. The Gam protein of bacteriophage Mu is an orthologue of eukaryotic Ku. *EMBO Rep* **4:**47–52.

88. **Ge J, Lou Z, Cui H, Shang L, Harshey RM.** 2011. Analysis of phage Mu DNA transposition by whole-genome *Escherichia coli* tiling arrays reveals a complex relationship to distribution of target selection protein B, transcription and chromosome architectural elements. *J Biosci* **36:**587–601.

89. **Mizuuchi M, Mizuuchi K.** 1993. Target site selection in transposition of phage Mu. *Cold Spring Harb Symp Quant Biol* **58:**515–523.

90. **Haapa-Paananen S, Rita H, Savilahti H.** 2002. DNA transposition of bacteriophage Mu. A quantitative analysis of target site selection *in vitro*. *J Biol Chem* **277:** 2843–2851.

91. **Manna D, Deng S, Breier AM, Higgins NP.** 2005. Bacteriophage Mu targets the trinucleotide sequence CGG. *J Bacteriol* **187:**3586–3588.

92. **Manna D, Wang X, Higgins NP.** 2001. Mu and IS1 transpositions exhibit strong orientation bias at the *Escherichia coli bgl* locus. *J Bacteriol* **183:**3328–3335.

93. **Manna D, Breier AM, Higgins NP.** 2004. Microarray analysis of transposition targets in *Escherichia coli*: the impact of transcription. *Proc Natl Acad Sci USA* **101:** 9780–9785.

94. **Manna D, Porwollik S, McClelland M, Tan R, Higgins NP.** 2007. Microarray analysis of Mu transposition in *Salmonella enterica*, serovar Typhimurium: transposon exclusion by high-density DNA binding proteins. *Mol Microbiol* **66:**315–328.

95. **Greene EC, Mizuuchi K.** 2004. Visualizing the assembly and disassembly mechanisms of the MuB transposition targeting complex. *J Biol Chem* **279:**16736–16743.

96. **Tan X, Mizuuchi M, Mizuuchi K.** 2007. DNA transposition target immunity and the determinants of the MuB distribution patterns on DNA. *Proc Natl Acad Sci USA* **104:**13925–13929.

97. **Ge J, Harshey RM.** 2008. Congruence of *in vivo* and *in vitro* insertion patterns in hot *E. coli* gene targets of transposable element Mu: opposing roles of MuB in target capture and integration. *J Mol Biol* **380:**598–607.

98. **Schweidenback CT, Baker TA.** 2008. Dissecting the roles of MuB in Mu transposition: ATP regulation of DNA binding is not essential for target delivery. *Proc Natl Acad Sci USA* **105:**12101–12107.

99. **Craig NL.** 1997. Target site selection in transposition. *Annu Rev Biochem* **66:**437–474.

100. **Lambin M, Nicolas E, Oger CA, Nguyen N, Prozzi D, Hallet B.** 2012. Separate structural and functional domains of Tn4430 transposase contribute to target immunity. *Mol Microbiol* **83:**805–820.

101. **Han YW, Mizuuchi K.** 2010. Phage Mu transposition immunity: protein pattern formation along DNA by a diffusion-ratchet mechanism. *Mol Cell* **39:**48–58.

102. **Greene EC, Mizuuchi K.** 2002. Direct observation of single MuB polymers: evidence for a DNA-dependent conformational change for generating an active target complex. *Mol Cell* **9:**1079–1089.

103. **Greene EC, Mizuuchi K.** 2002. Dynamics of a protein polymer: the assembly and disassembly pathways of the MuB transposition target complex. *EMBO J* **21:**1477–1486.

104. **Greene EC, Mizuuchi K.** 2002. Target immunity during Mu DNA transposition. Transpososome assembly and DNA looping enhance MuA-mediated disassembly of the MuB target complex. *Mol Cell* **10:**1367–1378.

105. **Manna D, Higgins NP.** 1999. Phage Mu transposition immunity reflects supercoil domain structure of the chromosome. *Mol Microbiol* **32:**595–606.

106. **Ge J, Lou Z, Harshey RM.** 2010. Immunity of replicating Mu to self-integration: a novel mechanism employing MuB protein. *Mob DNA* **1:**8.

107. **Nakai H, Doseeva V, Jones JM.** 2001. Handoff from recombinase to replisome: insights from transposition. *Proc Natl Acad Sci USA* **98:**8247–8254.

108. **Baker TA, Sauer RT.** 2012. ClpXP, an ATP-powered unfolding and protein-degradation machine. *Biochim Biophys Acta* **1823:**15–28.

109. **Burton BM, Williams TL, Baker TA.** 2001. ClpX-mediated remodeling of Mu transpososomes: selective unfolding of subunits destabilizes the entire complex. *Mol Cell* **8:**449–454.

110. **Burton BM, Baker TA.** 2003. Mu transpososome architecture ensures that unfolding by ClpX or proteolysis by ClpXP remodels but does not destroy the complex. *Chem Biol* **10:**463–472.

111. **Abdelhakim AH, Sauer RT, Baker TA.** 2010. The AAA+ ClpX machine unfolds a keystone subunit to remodel the Mu transpososome. *Proc Natl Acad Sci USA* **107:**2437–2442.

112. **Abdelhakim AH, Oakes EC, Sauer RT, Baker TA.** 2008. Unique contacts direct high-priority recognition of the tetrameric Mu transposase-DNA complex by the AAA+ unfoldase ClpX. *Mol Cell* **30:**39–50.

113. **North SH, Kirtland SE, Nakai H.** 2007. Translation factor IF2 at the interface of transposition and replication by the PriA-PriC pathway. *Mol Microbiol* **66:** 1566–1578.

114. **North SH, Nakai H.** 2005. Host factors that promote transpososome disassembly and the PriA-PriC pathway for restart primosome assembly. *Mol Microbiol* **56:** 1601–1616.

115. **Au TK, Agrawal P, Harshey RM.** 2006. Chromosomal integration mechanism of infecting mu virion DNA. *J Bacteriol* **188:**1829–1834.

116. **Harshey RM.** 1984. Transposition without duplication of infecting bacteriophage Mu DNA. *Nature* **311:**580–581.

117. **Chaconas G, Giddens EB, Miller JL, Gloor G.** 1985. A truncated form of the bacteriophage Mu B protein promotes conservative integration, but not replicative transposition, of Mu DNA. *Cell* **41:**857–865.

118. **Roldan LA, Baker TA.** 2001. Differential role of the Mu B protein in phage Mu integration vs. replication: mechanistic insights into two transposition pathways. *Mol Microbiol* **40:**141–155.

119. **Sokolsky TD, Baker TA.** 2003. DNA gyrase requirements distinguish the alternate pathways of Mu transposition. *Mol Microbiol* **47:**397–409.

120. **Jang S, Sandler SJ, Harshey RM.** 2012. Mu insertions are repaired by the double-strand break repair pathway of *Escherichia coli. PLoS Genet* **8:**e1002642.

121. **Chaconas G, Kennedy DL, Evans D.** 1983. Predominant integration end products of infecting bacteriophage Mu DNA are simple insertions with no preference for integration of either Mu DNA strand. *Virology* **128:** 48–59.

122. **Wu Z, Chaconas G.** 1995. A novel DNA binding and nuclease activity in domain III of Mu transposase: evidence for a catalytic region involved in donor cleavage. *EMBO J* **14:**3835–3843.

123. **DuBow MS.** 1987. Transposable Mu-like phages, p 201–213. *In* Symonds N, Toussaint A, Putte VDP, Howe MM (ed), *Phage Mu*. Cold Spring Harbor Laboratory, Cold Spring Harbor, NY.

124. **Yang JY, Jayaram M, Harshey RM.** 1995. Enhancer-independent variants of phage Mu transposase—enhancer-specific stimulation of catalytic activity by a partner transposase. *Genes Dev* **9:**2545–2555.

125. **Yang JY, Kim K, Jayaram M, Harshey RM.** 1995. A domain sharing model for active site assembly within the Mu A tetramer during transposition: the enhancer may specify domain contributions. *EMBO J* **14:**2374–2384.

126. **Morgan GJ, Hatfull GF, Casjens S, Hendrix RW.** 2002. Bacteriophage Mu genome sequence: analysis and comparison with Mu-like prophages in *Haemophilus, Neisseria* and *Deinococcus. J Mol Biol* **317:**337–359.

127. **Casjens S.** 2003. Prophages and bacterial genomics: what have we learned so far? *Mol Microbiol* **49:**277–300.

128. **Wang PW, Chu L, Guttman DS.** 2004. Complete sequence and evolutionary genomic analysis of the *Pseudomonas aeruginosa* transposable bacteriophage D3112. *J Bacteriol* **186:**400–410.

129. **Fogg PC, Hynes AP, Digby E, Lang AS, Beatty JT.** 2011. Characterization of a newly discovered Mu-like bacteriophage, RcapMu, in *Rhodobacter capsulatus* strain SB1003. *Virology* **421:**211–221.

130. **Zehr ES, Tabatabai LB, Bayles DO.** 2012. Genomic and proteomic characterization of SuMu, a Mu-like bacteriophage infecting *Haemophilus parasuis. BMC Genomics* **13:**331.

131. **Saariaho AH, Lamberg A, Elo S, Savilahti H.** 2005. Functional comparison of the transposition core machineries of phage Mu and *Haemophilus influenzae* Mu-like prophage Hin-Mu reveals interchangeable components. *Virology* **331:**6–19.

132. **Toussaint A.** 2013. Transposable Mu-like phages in Firmicutes: new instances of divergence generating retroelements. *Res Microbiol* **164:**281–287.

133. **Akhverdyan VZ, Gak ER, Tokmakova IL, Stoynova NV, Yomantas YA, Mashko SV.** 2011. Application of the bacteriophage Mu-driven system for the integration/amplification of target genes in the chromosomes of engineered Gram-negative bacteria. *Appl Microbiol Biotechnol* **91:**857–871.

134. **Lamberg A, Nieminen S, Qiao M, Savilahti H.** 2002. Efficient insertion mutagenesis strategy for bacterial genomes involving electroporation of *in vitro*-assembled DNA transposition complexes of bacteriophage Mu. *Appl Environ Microbiol* **68:**705–712.

135. **Pajunen MI, Pulliainen AT, Finne J, Savilahti H.** 2005. Generation of transposon insertion mutant libraries for Gram-positive bacteria by electroporation of phage Mu DNA transposition complexes. *Microbiology* **151:**1209–1218.

136. **Paatero AO, Turakainen H, Happonen LJ, Olsson C, Palomaki T, Pajunen MI, Meng X, Otonkoski T, Tuuri T, Berry C, Malani N, Frilander MJ, Bushman FD, Savilahti H.** 2008. Bacteriophage Mu integration in yeast and mammalian genomes. *Nucleic Acids Res* **36:**e148.

137. **Brady T, Roth SL, Malani N, Wang GP, Berry CC, Leboulch P, Hacein-Bey-Abina S, Cavazzana-Calvo M, Papapetrou EP, Sadelain M, Savilahti H, Bushman FD.** 2011. A method to sequence and quantify DNA integration for monitoring outcome in gene therapy. *Nucleic Acids Res* **39:**e72.

138. **Schagen FH, Rademaker HJ, Cramer SJ, van Ormondt H, van der Eb AJ, van de Putte P, Hoeben RC.** 2000. Towards integrating vectors for gene therapy: expression of functional bacteriophage MuA and MuB proteins in mammalian cells. *Nucleic Acids Res* **28:**E104.

139. **Orsini L, Pajunen M, Hanski I, Savilahti H.** 2007. SNP discovery by mismatch-targeting of Mu transposition. *Nucleic Acids Res* **35:**e44.

140. **Nakayama C, Teplow DB, Harshey RM.** 1987. Structural domains in phage Mu transposase: identification of the site-specific DNA-binding domain. *Proc Natl Acad Sci USA* **84:**1809–1813.

141. **Rice PA.** 2005. Visualizing Mu transposition: assembling the puzzle pieces. *Genes Dev* **19:**773–775.

142. **Rice P, Mizuuchi K.** 1995. Structure of the bacteriophage Mu transposase core: a common structural motif for DNA transposition and retroviral integration. *Cell* **82:**209–220.

143. **Rasila TS, Vihinen M, Paulin L, Haapa-Paananen S, Savilahti H.** 2012. Flexibility in MuA transposase family protein structures: functional mapping with scanning mutagenesis and sequence alignment of protein homologues. *PLoS One* **7:**e37922.

144. **Teplow DB, Nakayama C, Leung PC, Harshey RM.** 1988. Structure-function relationships in the transposition protein B of bacteriophage Mu. *J Biol Chem* **263:** 10851–10857.

145. **Hung LH, Chaconas G, Shaw GS.** 2000. The solution structure of the C-terminal domain of the Mu B transposition protein. *EMBO J* **19:**5625–5634.

146. **Leung PC, Harshey RM.** 1991. Two mutations of phage Mu transposase that affect strand transfer or interactions with B protein lie in distinct polypeptide domains. *J Mol Biol* **219:**189–199.

Mobile DNA, 3rd Edition
Nancy L. Craig, Michael Chandler, Martin Gellert, Alan M. Lambowitz, Phoebe A. Rice and Suzanne Sandmeyer

doi:10.1128/microbiolspec.MDNA3-0060-2014

Emilien Nicolas,[1,3] Michael Lambin,[1,2] Damien Dandoy,[1,2] Christine Galloy,[1]
Nathan Nguyen,[1] Cédric A. Oger,[1] and Bernard Hallet[1]

32 The Tn*3*-family of Replicative Transposons

INTRODUCTION

The ampicillin-resistance transposon Tn*3* is the archetype ("Tn*3*" being synonymous with "Tn*1*" or "Tn*2*"; (1)) of a large and widespread family of transposons with representatives in nearly all bacterial phyla including proteobacteria, firmicutes, and cyanobacteria. Family members are modular platforms allowing assembly, diversification, and redistribution of an ever-growing arsenal of antimicrobial resistance genes, thereby contributing along with other mobile genetic elements, to the emergence of multi-drug resistances at a rate that challenges the development of new treatments (2–4). They are also prevalent in horizontal transfer of large catabolic operons, allowing bacteria to metabolize various families of compounds, including industrial xenobiotic pollutants (5, 6).

A distinguishing feature contributing to the proliferation and evolutionary success of Tn*3*-family transposons is their replicative mode of transposition (7, 8) (Fig. 1). They transpose using a "copy-in" (or "paste-and-copy") mechanism in which replication of the transposon occurs during integration into the target. Intermolecular transposition generates a cointegrate intermediate, in which the donor and target molecules are fused by directly repeated transposon copies (Fig. 1). Cointegrate formation requires both the transposon-encoded transposase (TnpA), to cut and rejoin the transposon ends with the target DNA, and the cell replication machinery, to copy the complementary strands of the transposon. The transposase is also involved in a process termed target immunity whereby the transposon avoids inserting more than once into the same DNA molecule (7). This process is specific to each member of the family and is thought to act over large distances (up to several dozens of kilobases) within the genome.

Tn*3*-family transposons generally encode a DNA site-specific recombinase, or "resolvase", whose function is to resolve the cointegrate intermediate by catalyzing a reciprocal recombination reaction between the newly synthesized copies of the element (Fig. 1). Recombination takes place at a specific site, the "resolution site" (*res*) or "internal recombination site" (IRS) located inside the transposon (7). The reaction completes the transposition cycle, restoring the original donor molecule and producing a target molecule with a transposon copy (Fig. 1). The resolvase of most characterized Tn*3*-family transposons is a member of

[1]Institut des Sciences de la Vie, Université Catholique de Louvain, B-1348, Louvain-la-Neuve, Belgium; [2]GSK Biologicals, B-1300, Wavre, Belgium; [3]Laboratoire du Métabolisme de l'ARN FRS/FNRS-IBMM-CMMI-Université Libre de Bruxelles, B-6041 Charleroi-Gosselies, Belgium.

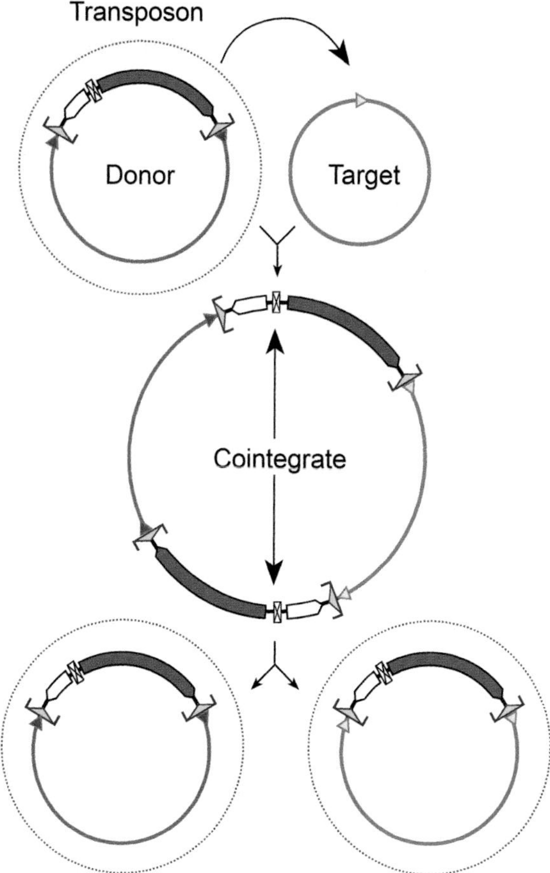

Figure 1 Overview of the replicative transposition cycle of Tn*3*-family transposons. Intermolecular transposition (curved arrow) from a donor (purple) to a target DNA molecule (blue) generates a cointegrate in which both molecules are fused together by directly repeated copies of the transposon. This step requires the transposase and the host replication machinery. The cointegrate is resolved by resolvase-mediated site-specific recombination (double arrow) between the duplicated copies of the transposon resolution site (boxed cross). Bracketed triangles are the terminal inverted repeats (IRs) of the transposon. The transposase and resolvase genes are represented by a purple and a white arrow, respectively. Small triangles show the short (usually 5-bp) direct repeats (DRs) that are generated upon insertion into the target. The red stippled circle indicates that a molecule that contains a copy of the transposon is immunized against further insertions due to target immunity.
doi:10.1128/microbiolspec.MDNA3-0060-2014.f1

the serine recombinase (S-recombinase) family, but in a few cases it is a member of the tyrosine recombinase (Y-recombinase) family (7, 9).

Although cointegrate resolution has been the subject of intense biochemical, topological, and structural studies over the past decades, much less is known regarding the transposase-catalyzed reactions in both the transposition and target immunity processes. After an overview of the biological diversity and significance of Tn*3*-family transposons, this chapter will provide an update of the mechanistic aspects of the transposition cycle with a special focus on the initial steps of the reaction.

MODULAR AND DYNAMIC STRUCTURE OF TN3-FAMILY TRANSPOSONS

Mobile genetic elements can be described as a juxtaposition of functional modules, which together provide each element with its own specificities (10). Members of the Tn*3* transposon family include three types of module (Fig. 2): the core transposition module, which is the mobility signature of the family; the cointegrate resolution module, which optimizes the transposition pathway by reducing the risk of generating aberrant replicon fusions and making the transposon less dependent on the host recombination functions; and various sets of cargo genes and operons that were assimilated, presumably because they proved to be useful for their host under certain conditions.

The Transposition Module: The ID of the Family

Autonomous Tn*3*-family members carry a transposition core module comprising the transposase gene (*tnpA*) and typical ~38-bp inverted repeats (IRs) at the transposon ends (Fig. 2). TnpA proteins are unusually large (from ~950 to ~1020 amino acids [aa]) compared to other transposases. A BLAST search performed with any Tn*3*-family transposase identifies only other members of the family. However, phylogenetic analysis of 924 full-length transposases from the Pfam database (http://pfam.xfam.org/) reveals five protein clusters sharing less than ~30% sequence identity (Fig. 3). One large cluster comprising 595 sequences can be further divided into three subgroups exemplified by the Tn*4430*, Tn*5393*, and Tn*21* transposases (Fig. 3). An alignment of 21 active transposases from the different clusters and separate bacterial phyla showed that only 15 residues, including the "DD-E" catalytic triad, are perfectly conserved (11) (see also *The transposase* section, below).

Remarkably, this phylogeny does not faithfully superimpose on that of the bacterial species from which the transposons were originally isolated, underlining the important level of horizontal transfer associated with the family. In contrast, there is a good correlation between the transposase phylogeny and the ~38-bp terminal transposon IRs, suggesting that both transposition module partners have coevolved to maintain specific and functional interactions (see *Interaction with the transposon ends* section, below).

Transposition Module-Only Tn3-Family Elements

Certain representatives from phylogenetically separate Tn3-family subgroups are distinctive because they contain only the transposition module, an organization that relates these elements to insertion sequences (ISs) (12) (see also the chapter by P. Siguier *et al.* in this volume) (Fig. 2). Cointegrate intermediates generated by these 'IS-like' Tn3-family members are thought to be resolved by RecA-dependent homologous recombination. The lack of a transposon-encoded resolution system may have facilitated formation of stable composite transposons that are typical of IS families, in which two copies of the element became associated with other genetic determinants to mediate their mobility as a single entity (12) (Fig. 2).

One of these, IS*1071*, was initially identified for its contribution to the 17-kb chlorobenzoate-catabolic *Comamonas testosteroni* (formerly *Alcaligenes* sp.) transposon, Tn*5271* (13). Similar associations were subsequently found between intact or truncated IS*1071* copies and different catabolic gene clusters from a variety of bacterial isolates (5, 14). Another "transposition module-only" Tn3-family transposon, IS*3000*, was identified adjacent to the extended-spectrum β-lactamase CTX-M-9 gene from enterobacteria, and more recently, on the multi-resistance *Klebsiella pneumoniae* plasmid pKp11-42 (15, 16).

Because of its minimal organization, IS*1071* was initially proposed as the archetype of an ancestral element from which the other Tn3-family transposons evolved by acquisition of a cointegrate resolution system (7, 13). However, the TnpA proteins of IS*1071* and IS*3000* belong to separate clusters within the Tn3-transposase family (Fig. 3), suggesting that these minimal structures have evolved separately.

The Cointegrate Resolution Modules

S-recombinases and Y-recombinases cleave and reseal DNA molecules at specific sequences using a serine or a tyrosine nucleophile, respectively. Both recombinase families are involved in a broad range of biological processes such as integration and excision of temperate bacteriophages, movement of different classes of mobile genetic elements and control of gene expression through programmed DNA rearrangements (17) (see also the section on "Conservative site-specific recombination" of this volume). In all of these types of biological process, there are clear cases where members of either family have been recruited to carry out similar functions. Cointegrate resolution systems of both recombinase families are for example closely related to those found on circular plasmids and chromosomes, the function of which is to reduce multimeric replicon forms and permit their segregation at the time of cell division (9) (Fig. 4 and 5).

Recombinases of both families mediate recombination between short (~28-bp to 30-bp) DNA sequences, the "core" sites, which typically contain inversely oriented recognition motifs separated by a central 2-bp (S-recombinases) or 6-bp to 8-bp (Y-recombinases) region (Fig. 6 and 7). The minimal core is usually insufficient to support efficient recombination. Recombination sites often contain extra DNA sequences to which additional recombinase subunits and/or accessory proteins bind to control the outcome of recombination. For transposon resolution systems, this is important to efficiently resolve cointegrate intermediates while avoiding detrimental DNA rearrangements (see *Convergent mechanisms to control the selectivity of recombination* section, below).

Cointegrate Resolution Modules Using an S-Recombinase

The S-resolvases

Tn3-family S-resolvases are typically small (~180 to 210 aa) comprising a relatively well conserved ~120-aa N-terminal catalytic domain connected to a short ~65-aa C-terminal DNA-binding domain (Fig. 6). However, the resolvases of some transposons (e.g., IS*Xc5* and Tn*5044* from *Xanthomonas campestris* or Tn*5063* and Tn*5046* from *Pseudomonas* sp.) are larger (~310 aa) due to a ~110-aa C-terminal extension (18–21) (Fig. 6). The role of this extension is not known. Its deletion abolished Tn*5044* resolvase recombination *in vivo* (18) but had no effect on ISXc5 resolvase activity (22). As previously observed (7), the S-resolvase phylogeny does not perfectly match that of the transposases, suggesting that cointegrate resolution modules were independently acquired or exchanged between transposons of the different subgroups after the divergence of their transposition module (Fig. 4).

Variations in S-resolvase *res* site organization

The transposon resolution site, *res*, is generally located immediately upstream of the resolvase gene (Fig. 6). The best-characterized Tn3-family *res* sites, those of Tn3 and γδ (also called Tn*1000*), are ~120 bp long and contain three subsites (I to III). Each subsite is

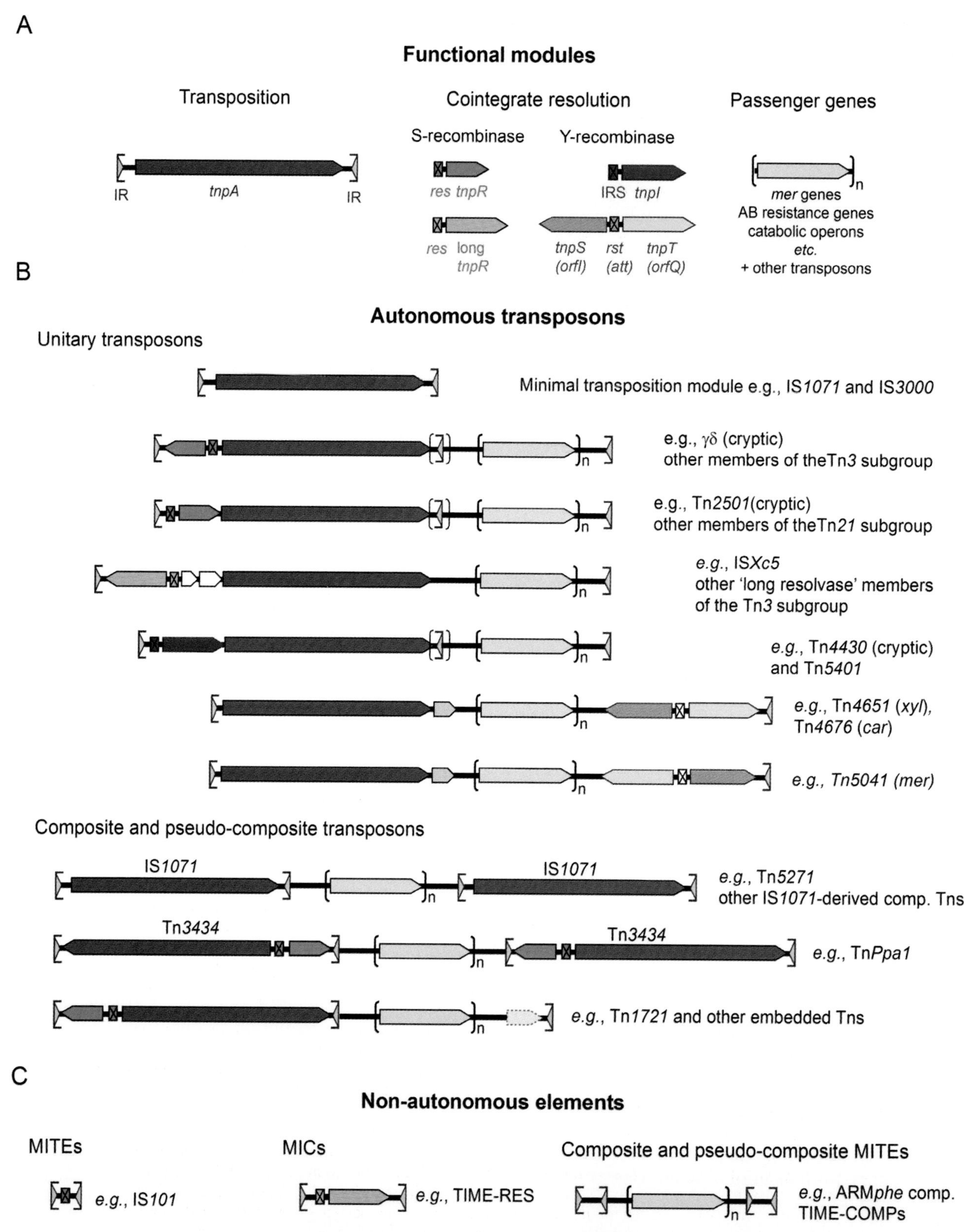
A
Functional modules
Transposition
IR
tnpA
IR
Cointegrate resolution
S-recombinase
res tnpR
res long tnpR
Y-recombinase
IRS tnpI
tnpS (orfI)
rst (att)
tnpT (orfQ)
Passenger genes
n
mer genes
AB resistance genes
catabolic operons
etc.
+ other transposons
B
Autonomous transposons
Unitary transposons
Minimal transposition module e.g., IS1071 and IS3000
e.g., γδ (cryptic)
other members of theTn3 subgroup
e.g., Tn2501(cryptic)
other members of theTn21 subgroup
e.g., ISXc5
other 'long resolvase' members
of the Tn3 subgroup
e.g., Tn4430 (cryptic)
and Tn5401
e.g., Tn4651 (xyl),
Tn4676 (car)
e.g., Tn5041 (mer)
Composite and pseudo-composite transposons
IS1071
IS1071
e.g., Tn5271
other IS1071-derived comp. Tns
Tn3434
Tn3434
e.g., TnPpa1
e.g., Tn1721 and other embedded Tns
C
Non-autonomous elements
MITEs
e.g., IS101
e.g., ΔxIS TIME family
MICs
e.g., TIME-RES
Composite and pseudo-composite MITEs
e.g., ARMphe comp.
TIME-COMPs
e.g., TnHad2

composed of inversely oriented 12-bp resolvase recognition motifs (Fig. 6). Site I is the recombination core site (or crossover site) at which strand exchange occurs; sites II and III are regulatory elements required for assembly of a topologically defined synaptic complex to control recombination directionality (see *Convergent mechanisms to control the selectivity of recombination* section, below).

The three-subsite organization of *res* is conserved in most Tn*3*-family transposons in spite of the differences in resolvase sequence specificity. There is some variation, however, in the spacer length between the core site I and the accessory sites II and III, ranging from four to seven integral helical turns (Fig. 6). Conservation of the helical phasing between sites I and II is important to properly align the core and regulatory resolvase subunits in the synaptic complex (23, 24).

A structurally different *res* site organization has been proposed for other transposons such as ISXc*5* in which the site II inverted repeats are replaced by direct repeats (7, 25). Tandem resolvase binding motifs were also identified in the putative *res* sites of several Tn*3*-family transposons from Gram-positive bacteria (e.g., Tn*1546* and TnXO*1*) (7, 25). These *res* sites also appear more compact with two resolvase binding sites instead of three (Fig. 6). This two-subsite organization was initially described for Sin and the β-recombinases of Gram-positive plasmids to which the Tn*1546* and TnXO*1* resolvases are only distant relatives (25, 26) (Fig. 4 and 6). Recombination mediated by these recombinases requires the host protein Hbsu (or any DNA bending protein like the nucleoid-associated protein HU, the integration host factor IHF, or high-mobility group of proteins [HMG] from eukaryotes) that binds the spacer segment between sites I and II, playing an architectural role in synaptic complex assembly (25, 27–31). Other serine recombinases involved in plasmid monomerization are more closely related to resolvases from cointegrate resolution systems (32) (Fig. 4). In some cases there is clear evidence that they were derived from transposons during recent evolution (25, 32).

Finally, related resolution systems are also found in a distinct class of transposable elements called "Mu-like transposons" for their resemblance to bacteriophage Mu. These include β-lactam resistance transposon Tn*552* from *Staphylococcus* sp. and transposons of the widespread Tn*5053*/Tn*402* family (33–35) (Fig. 4 and 6). Like Tn*3*-family transposons, these move by replicative transposition and generate cointegrates. The Tn*552* BinL resolvase is a close homolog of the plasmid pI9789 BinR recombinase, whereas the Tn*5053*/Tn*402* TniR recombinase is closer to the resolvase of classical Tn*3*-family members like Tn*163* and Tn*5393* (Fig. 4). TniR and BinL are also distant relatives of the C-terminally extended resolvases (e.g., those of ISXc*5* and Tn*5044*; Fig. 4) and the organization of the *resL* site of Tn*552*

Figure 2 The modular structure of Tn*3*-family transposons and derived elements. (A) Tn*3*-family transposons are constituted by the association of three classes of functional modules. The transposition module comprises the transposase gene (*tnpA*, purple arrow) and its associated ~38-bp inverted repeats (IRs, bracketed triangles). The cointegrate resolution module is made of a gene encoding a site-specific recombinase from the serine- (green) or tyrosine- (magenta) family and their cognate recombination site (boxed cross). Resolution modules working with a tyrosine recombinase of the TnpS/OrfI subgroup (pale magenta) contain an additional gene coding for an accessory recombination protein (TnpT/OrfQ). "Long *tnpR*" (pale green) refers to a resolvase gene that encodes a C-terminally extended member of the serine recombinase family. Passenger genes comprise a variety of phenotypic determinants and transposons that are specific to each element (gray arrows). (B) Autonomous transposons are characterized by a fully functional transposition module to mediate the mobility of the element *in cis*. The simplest elements (ISs) are solely constituted by the minimal transposition module. Most characterized transposons have a typical unitary structure in which the transposase gene and its associated modules are flanked by the IR ends. Unitary elements can associate to form composite transposons containing a pair of full-length elements flanking a specific genomic segment; or pseudo-composite transposons carrying an autonomous element on one side and an isolated IR end on the other side. (C) Nonautonomous elements are Tn*3*-family derivatives whose mobility requires a functional transposase to be provided in *Trans* by a related transposon. Miniature Inverted-repeats Transposable Elements (MITEs) are solely constituted by a pair of IR ends flanking a short DNA segment that sometime contains a cointegrate resolution site. Mobile Insertion Cassettes (MICs) are nonautonomous unitary elements carrying one or more passenger genes between the ends. MITEs can also associate with passenger genes and isolated ends to form composite and pseudo-composite mobilized structures. See the text for details and the indicated examples. doi:10.1128/microbiolspec.MDNA3-0060-2014.f2

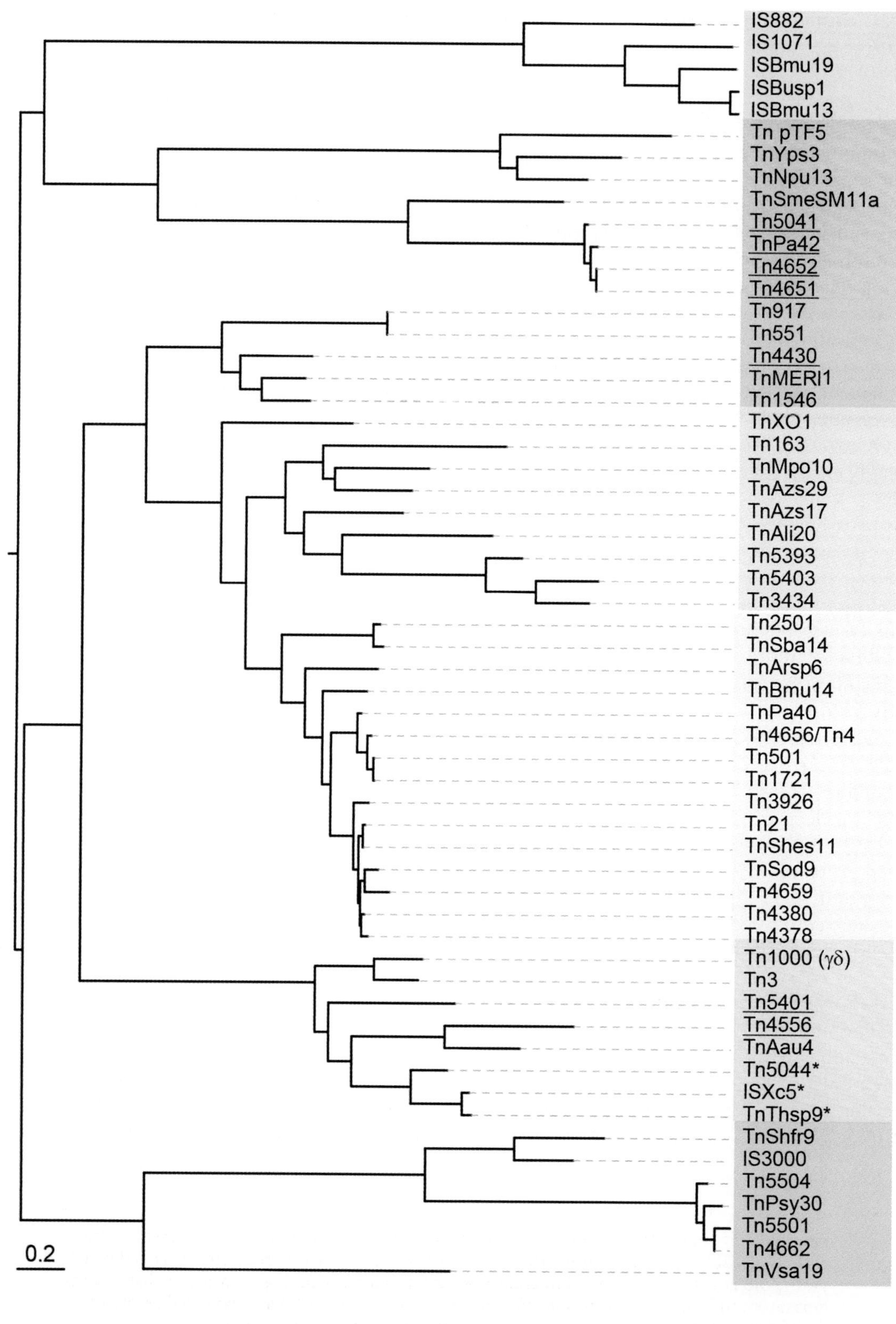
IS882
IS1071
ISBmu19
ISBusp1
ISBmu13
Tn pTF5
TnYps3
TnNpu13
TnSmeSM11a
Tn5041
TnPa42
Tn4652
Tn4651
Tn917
Tn551
Tn4430
TnMERI1
Tn1546
TnXO1
Tn163
TnMpo10
TnAzs29
TnAzs17
TnAli20
Tn5393
Tn5403
Tn3434
Tn2501
TnSba14
TnArsp6
TnBmu14
TnPa40
Tn4656/Tn4
Tn501
Tn1721
Tn3926
Tn21
TnShes11
TnSod9
Tn4659
Tn4380
Tn4378
Tn1000 (γδ)
Tn3
Tn5401
Tn4556
TnAau4
Tn5044*
ISXc5*
TnThsp9*
TnShfr9
IS3000
Tn5504
TnPsy30
Tn5501
Tn4662
TnVsa19
0.2
TnpA subgroups: IS1071 Tn4651 Tn4430 Tn163 Tn21 Tn3 IS3000

was proposed to resemble the resolution site of ISXc5 (7, 25) (Fig. 6).

Cointegrate Resolution Modules Using a Y-Recombinase

Phylogenetically Distinct Y-Resolvase Subgroups

Y-recombinases fall into well-separated subgroups based on sequence similarities, which correlate with different recombination site organization (Fig. 5 and 7). As for the S-resolvases, recombinases of the different subgroups are phylogenetically related to other Y-recombinases that play a role in stabilizing circular replicons (Fig. 5).

Recombinases of one group, exemplified by TnpI of *Bacillus thuringiensis* transposons Tn*4430* (285 aa) and Tn*5401* (306 aa) (36, 37) are distant homologs of the chromosomally encoded multifunctional XerCD recombinases whose primary function is to resolve chromosome dimers arising from homologous recombination before their segregation into daughter cells (38–40) (see also the chapter by C. Midonet and F.-X. Barre in this volume).

The TnpS/OrfI recombinases (323 aa and 351 aa, respectively) encoded by the *Pseudomonas* sp. Tn*4651*/Tn*5041* transposon subfamily (41–43) represent a second protein group more closely related to bacteriophage P1 Cre protein and to recombinases of a variety of large *Rhizobiacae* plasmids (9) (Fig. 5). Interestingly, TnpS-mediated or OrfI-mediated cointegrate resolution requires the product of a divergently oriented gene (TnpT or OrfQ) that shows no homology to any other characterized protein (41, 44).

Another putative Y-resolvase type identified in Tn*4556* from *Streptomyces fradiae* is more closely related to the ResD resolvase of *Escherichia coli* F factor than to the two other groups (7, 45) (Fig. 5). Finally, the defective *Pseudomonas putida* transposon Tn*4655* carries a unique site-specific recombination module composed of a recombination site (*attI*) and a Y-recombinase (TnpI) substantially larger (415 aa) than the other resolvases of the family (46). This system was proposed to have been initially involved in gene exchange between different replicons before incorporation as a bona fide cointegrate resolution system (46).

Variations in Y-resolvase *res* Site Organization

The IRS of Tn*4430* (116 bp) is located immediately upstream of *tnpI* (Fig. 7). It is composed of four TnpI-binding motifs with a common 9-bp sequence. Two 16-bp inverted motifs (IR1 and IR2) form the IRS core recombination site, whereas the adjacent 14-bp direct repeats (DR1 and DR2) are dispensable but required to control selectivity and directionality (36, 47) (Fig. 7) (see also *Variation on a theme: the TnpI/IRS recombination complex of Tn4430* section, below). The IRS of Tn*5401* exhibits a similar arrangement of TnpI binding motifs suggesting a similar mechanism to control cointegrate resolution (37, 48) (Fig. 7).

A completely different organization has been proposed for the Tn*4651*/Tn*5041* subfamily resolution site (designated "*rst*" for the resolution site associated with TnpS and TnpT) (41, 44). Deletion analysis located the minimal requirement for full *rst* activity to a 136-bp sequence between *tnpS* and *tnpT* (Fig. 7). DNA strand exchange takes place at a 33-bp inverted repeat (IRL-IRR), dividing *rst* into two arms (44). A similar 34-bp cross-over site was previously proposed for Tn*5041* (42). The accessory protein, TnpT, binds to another conserved 20-bp inverted repeat element (IR1–IR2) in the *tnpT*-proximal arm side of *rst* (44) (Fig. 7). However, it has yet to be determined how this protein collaborates with TnpS to mediate recombination.

Cryptic Transposons: Conduction and Formation of Compound Structures

A number of transposons belonging to different Tn3-family subgroups are phenotypically "cryptic" and encode no functions other than those associated with transposition and cointegrate resolution (Fig. 2). Replicative transposons of these elements can promote gene exchange by forming transient cointegrates between a conjugative plasmid and non-self-transmissible genetic material in a process known as "conduction" (49). This process led to the identification and initial characterization of several Tn3-family members (49–51).

Figure 3 Phylogenetic tree of the Tn3-family transposase proteins. The different clusters and subgroups identified within the family are boxed with different colors as indicated. Transposons that contain a cointegrate resolution module working with a tyrosine recombinase are underlined. Transposons that encode a "long", C-terminally extended resolvase of the serine-recombinase family are marked with an asterisk. The length of the branches is proportional to the average number of substitutions per residue.
doi:10.1128/microbiolspec.MDNA3-0060-2014.f3

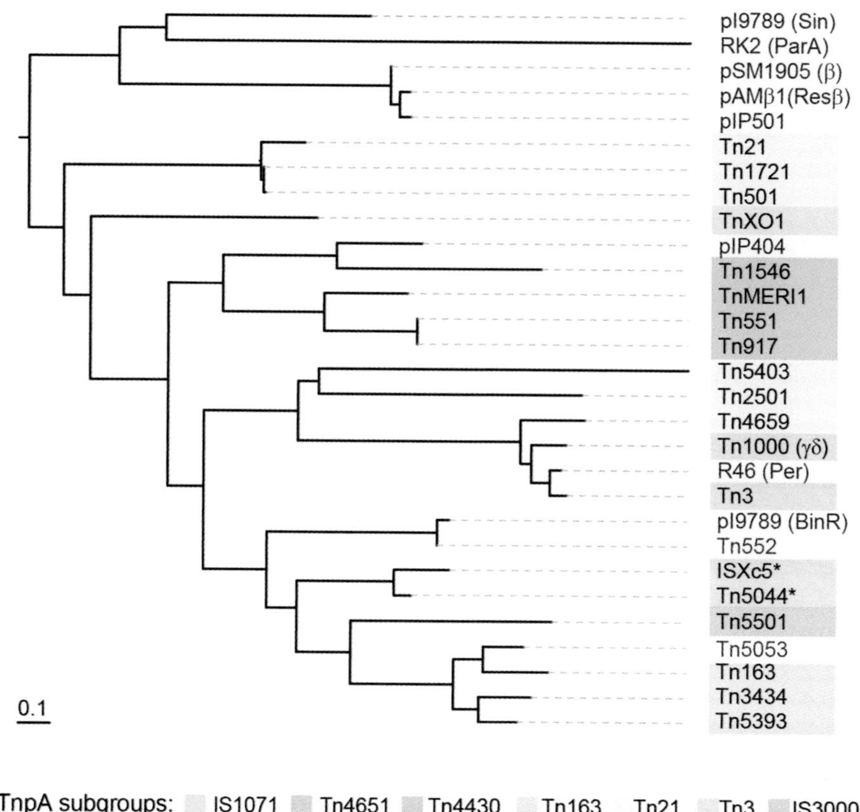

Figure 4 Phylogenetic tree of transposon-encoded and plasmid-encoded resolvases of the S-recombinase family. Transposons belonging to different subgroups are boxed with different colors as in Fig. 3. Transposons encoding a "long", C-terminally extended resolvase are marked by an asterisk. Plasmids are highlighted in blue with the name of the corresponding recombinase (when assigned) in brackets. "Mu-like" transposons are highlighted in magenta. The length of the branches is proportional to the average number of substitutions per residue. doi:10.1128/microbiolspec.MDNA3-0060-2014.f4

The process may potentially impact public health. For example, the cryptic transposon Tn*4430* from the large self-transmissible *B. thuringiensis* pXO12 plasmid mediates relatively high-frequency conduction of *Bacillus anthracis* virulence plasmids pXO1 and pXO2 among different *Bacillus cereus sensu lato* strains (52). More generally, Tn*4430* is thought to contribute, with other transposable elements, to diversification of the entomopathogenic specificity of *B. thuringiensis* serovars by promoting redistribution and subsequent recombination of the plasmid-encoded δ-endotoxin genes (53).

Cryptic transposons can form transpositionally active composite assemblies as exemplified by Tn*Ppa1*, a large genomic transposon from *Parococcus* spp (54, 55) (Fig. 2). Tn*Ppa1* is delineated by divergently oriented copies of the cryptic element Tn*3434* (3.7 kb). The mobilized Tn*Ppa1* core (37 kb) carries 34 open reading frames with a variety of predicted functions, such as housekeeping genes, different transporter families and transcriptional regulators (54, 55).

Passenger Genes and Operons

Tn*3*-family transposons are generally unitary, noncomposite structures (previously designated "class II" transposons) containing one or more passenger genes in addition to their transposition and cointegrate resolution modules (Fig. 2). The size of these elements varies from a few, to tens of kilobases depending on the number of genes they transport.

Antimicrobial Passenger Genes

Tn*3*-family members that confer resistance to antimicrobial compounds are found in different subgroups of the family. In the simplest cases, these elements encode one or more determinants active against a specific class of compounds, sometimes associated with additional

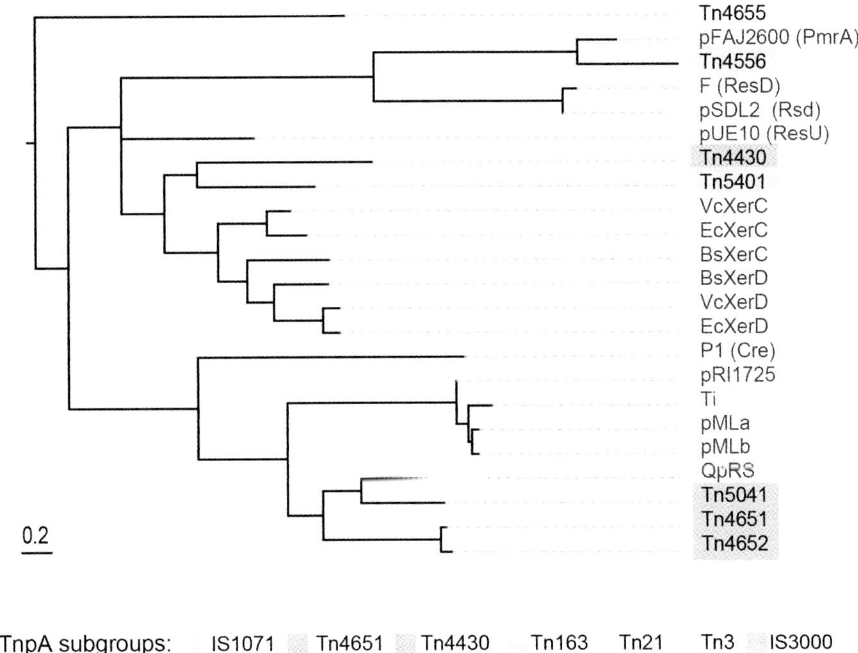

Figure 5 Phylogenetic tree of transposon-encoded and plasmid-encoded resolvases of the Y-recombinase family. Transposons belonging to different subgroups are boxed with different colors as in Fig. 3. Plasmids are highlighted in blue with the name of the corresponding recombinase (when assigned) in brackets. Representative and well-characterized XerC and D recombinases are from *Escherichia coli* (Ec), *Bacillus subtilis* (Bs) and *Vibrio cholerae* (Vc). The length of the branches is proportional to the average number of substitutions per residue. doi:10.1128/microbiolspec.MDNA3-0060-2014.f5

regulatory functions. Examples of these include the TEM-1 β-lactamase genes—Tn*3* and its derivatives (1), the inducible macrolide-lincosamide-streptogramin resistance gene *ermAM*—Tn*917* from *Enterococcus faecalis* (56), the aminoglycoside phosphotransefrase *strA* and *strB* genes—Tn*5393* (57) or the tetracycline efflux pump *tetA(A)* gene—Tn*1721* (58). Dissemination of vancomycin resistance by Tn*1546* (10.85 kb) from *Enterococcus faecium* involves a cluster of seven *van* genes. Five of these are required for the inducible synthesis of alternative peptidoglycan precursors that are refractory to the glycopeptide antibiotic (59).

Mercury-resistance transposons, which encode genetic determinants allowing detoxification of mercuric ions from the environment, are also found in different Tn*3* subfamilies. They are evolutionarily ancient and widely dispersed elements found in a variety of bacterial species and diverse environments (20, 60, 61). The mercury resistance operons (*mer*) comprise a variable number of genes depending on the element and its detoxification capabilities. In most cases, they include a metal-responsive regulator (*merR*), a transport system responsible for delivering mercuric ions into the cell (*merTP*), and a cytoplasmic mercuric reductase (*merA*) that converts toxic Hg^{2+} ions into less toxic reduced forms (61).

With the successive development of novel antibiotherapies, transposons initially active against only restricted ranges of compounds, some of which like mercury resistance transposons predated the antibiotic era, accumulated additional resistance genes, as the consequence of coselective pressure (see below). There are now many identified Tn*3* family representatives with different combinations of antibiotic resistance determinants, which contribute to the emergence and accelerated expansion of multi-resistant bacterial pathogens in both the clinical and community environments (3, 4, 61, 62).

Catabolic Passenger Genes

Tn*3*-family catabolic transposons are typically large (>40 kb) and enable the host cells to use xenobiotic compounds such as recalcitrant aromatic hydrocarbons (e.g., toluene, xylenes, and naphthalene) as a sole source of carbon and energy. This generally relies on complex clusters of genes that often comprise an "upper"

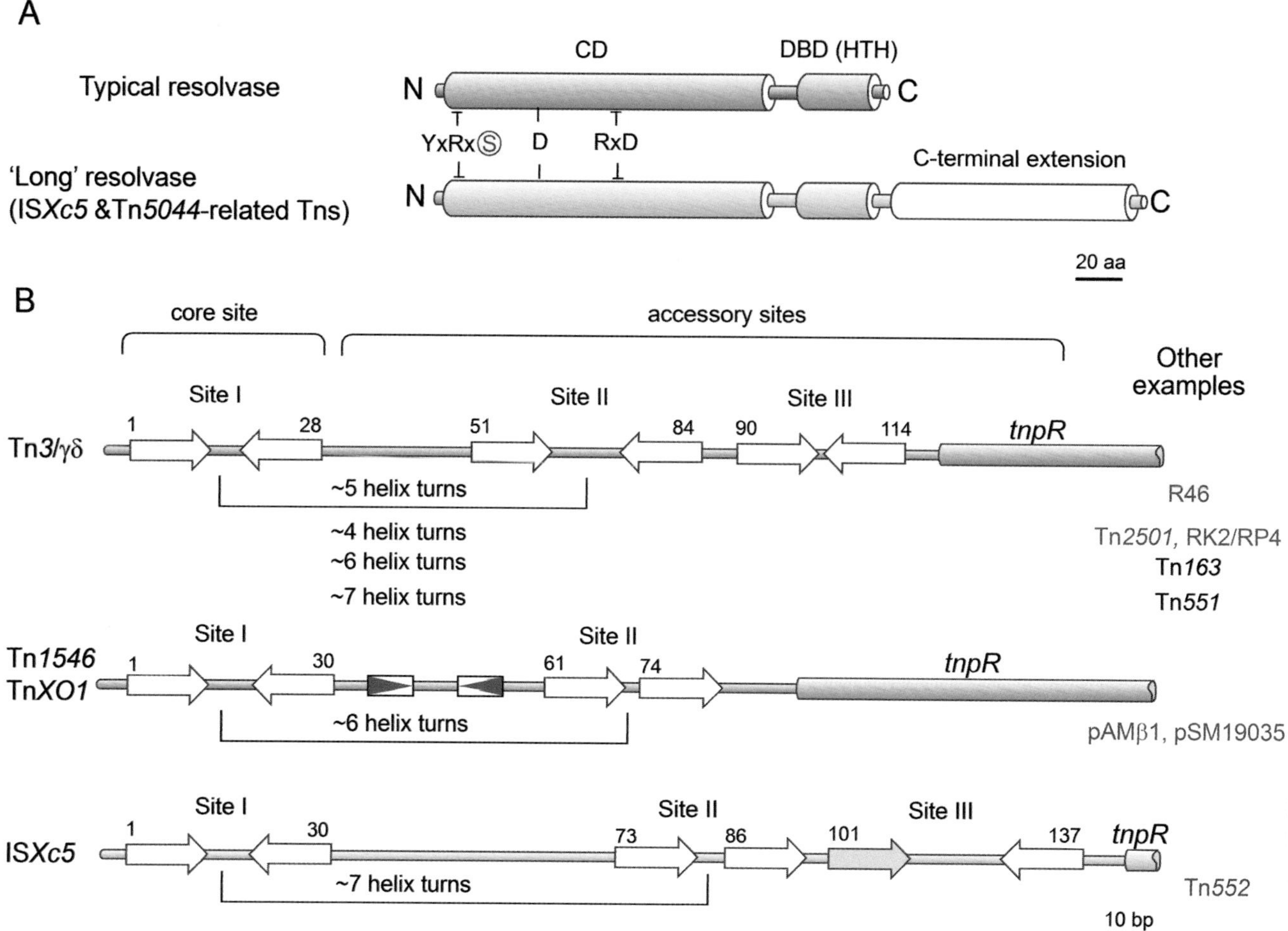

Figure 6 Cointegrate resolution modules working with a resolvase of the S-recombinase family. (A) Serine-resolvases are typically small proteins containing an N-terminal catalytic domain (CD) linked to a short helix-turn-helix (HTH) C-terminal DNA-binding domain (DBD). Resolvases of the "long" resolvase subgroup have a ~100-amino acid extension (white cylinder) at the C-terminus. Position of important catalytic residues (248) is indicated, with the active site serine (circled) highlighted in magenta. (B) Organization of the resolution sites *res*. Open arrows are the 12-bp resolvase binding motifs. Shaded arrows are for motifs with a poorer match to the consensus. Coordinates (in bp) of the first position of each motif are indicated above the recombination sites. Boxed triangles show the position of the putative Hbsu binding sites in the *res* site of Tn*1546* and Tn*XO1*. Organization of Tn*1546*, Tn*XO1* and IS*Xc5* res sites is as proposed by Rowland *et. al.* (25). Adapted from figure 6, p. 152 of reference (9). doi:10.1128/microbiolspec.MDNA3-0060-2014.f6

pathway operon that is responsible for the conversion of the initial substrate into oxidized intermediates, a "meta" operon that is required to degrade the resulting intermediates into central metabolites, and additional regulatory genes that are involved in the transcriptional activation of the whole pathway (5, 6).

Pathogenicity and Other Passenger Genes

Other functions associated with Tn*3*-family transposons include virulence factors from both animal and plant pathogens. For example, the *B. anthracis* virulence plasmid pXO1 transposon Tn*XO1* (8.7 kb) carries the *gerX*BAC operon responsible for spore germination during infection (63, 64) and the *Pseudomonas syringae* Tn*hopX1* promotes dispersal of HopX1, a conserved effector protein present in most plant pathogens with type III secretion systems (65). Recent bioinformatics analyses reveal that the contribution of Tn*3*-related elements to plant virulence is more general, being involved in the spread of different classes of secreted type

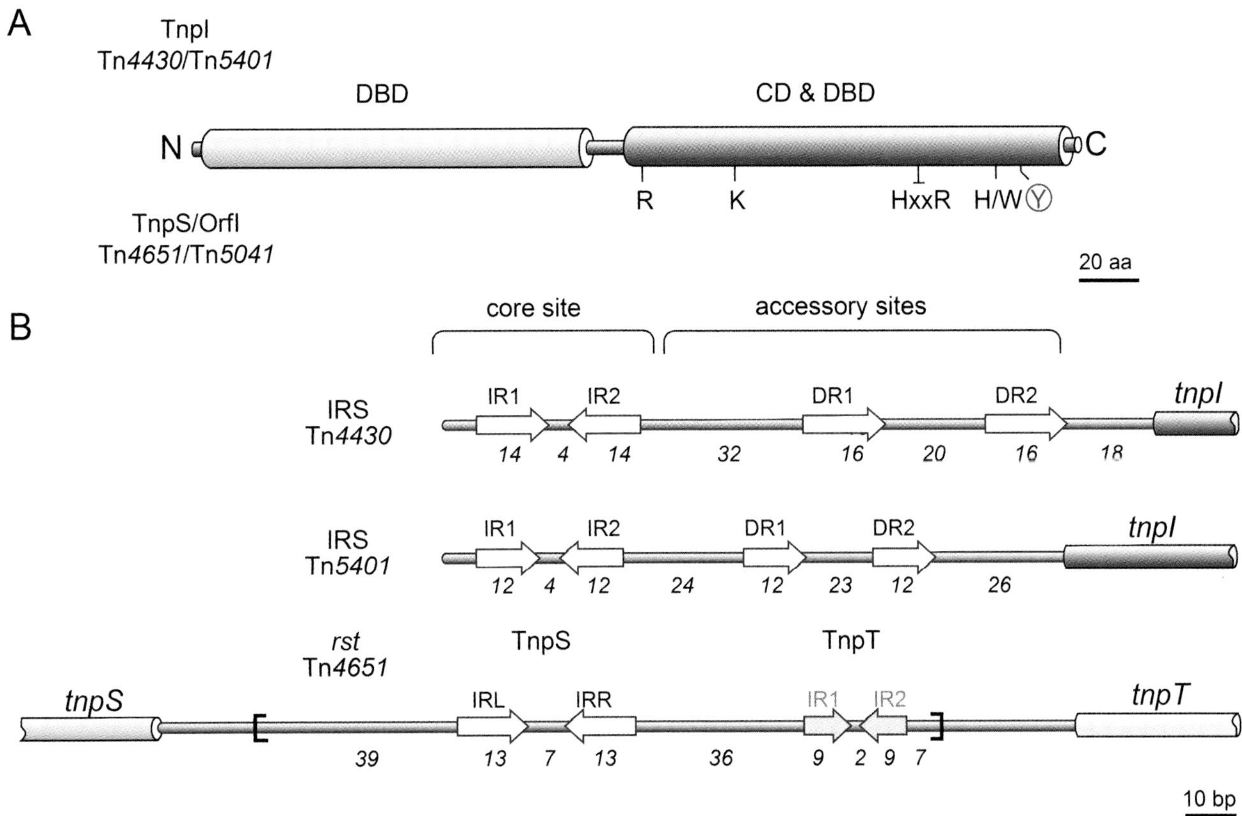

Figure 7 Cointegrate resolution modules working with a resolvase of the Y-recombinase family. (A) Typical two-domain structure of a tyrosine recombinase. The catalytic domain is in the C-terminal part of the protein. Positions of the conserved catalytic residues (RKHR^{H}/$_{W}$Y) are indicated, with the active site tyrosine (circled) highlighted in magenta. Both the N- and C-terminal domains contain the DNA-binding determinants of the protein (DBD). (B) Organization of the transposon recombination sites. Open arrows indicate the recombinase binding motifs. Shaded arrows in the *rst* site of Tn*4651* are the putative DNA recognition motifs for the auxiliary protein TnpT. Numbers indicate the length and the spacing between each motif (in bp) in the recombination site. Brackets in the *rst* site show the extent of the functional recombination site as determined by deletion analysis (44).
doi:10.1128/microbiolspec.MDNA3-0060-2014.f7

III effectors in the Xanthomonads (A. Varani and M. Chandler, personal communication). A few transposons from separate subgroups encode post-segregational killing toxin–antitoxin systems (59, 66, 67). This presumably helps to stabilize the transposon in a bacterial population.

The Dynamics of Tn3-Family Transposon Modular Organization: Module Capture and Reshuffling

It is often difficult to predict the exact mechanisms by which the Tn3 family building blocks have been assembled to generate their specific modular structures. Although each element has its own history, several general, nonexclusive scenarios can be drawn based on certain observations.

Creating and Sharing New Transposon Ends

Acquisition of new phenotypic traits may occur when the proximal end of a transposon inserted adjacent to a selectable marker gene is inactivated and a more distant "surrogate" IR sequence is recruited. This was suggested to occur during acquisition of the carbapenemase bla_{KPC} gene by the transposon Tn*4401* (68) and presumably in other single-trait Tn3-family transposons. In Tn3-related elements, a degenerate "ancestral" IR sequence is found immediately upstream of the bla_{TEM1} passenger gene (our unpublished observation); whereas in Tn*4401* the supernumerary internal IR end was inactivated by insertion of IS*Kpn7*, which also reinforced expression of the bla_{KPC} gene (68–70).

Unitary transposons may also be derived from composite transposons by inactivation of one of the two flanking transposon copies. This is the case for the tetracycline resistance transposon Tn*1721*, in which one element of the initial composite structure is inactivated by deletion leaving the other element (Tn*1722*) intact and capable of autonomous transposition (58, 71) (see also Fig. 2). In addition, Tn*1721*, like several other Tn*3*-family members, mediates one-ended transposition, a poorly understood transposition reaction catalyzed by the transposase at a single IR end that generates recombinant products containing one terminus of the transposon and variable lengths of adjacent DNA (72–76). This can directly provide single-end derivatives for the construction of new modular arrangements without requiring subsequent deletion events.

Combination of functional transposition modules is well illustrated by certain catabolic transposons that form large (up to ~90 kb) mosaic structures containing multiple mobile subentities (e.g., (46, 77–79)). This is viewed as an adaptive strategy allowing movement and reassortment of different combinations of catabolic functions within and between separate plasmids from environmental bacteria (6). A similar nested organization has been reported for other phenotypically distinct Tn*3*-family transposons such as the *Yersinia enterocolitica* lactose Tn*951* (80) or the mercury-resistance derivatives Tn*510* and Tn*511* (81). Some of these are transposition-defective and require an active copy of a related element to be mobilized in *trans* (46, 77, 80, 81).

Housing Other Transposable Elements

There are specific recombination systems that facilitate acquisition of new passenger genes. A particularly striking example is the recruitment of mobile integrons, site-specific recombination systems that use a Y-recombinase family integrase (IntI) to promote the exchange of specific gene cassettes (82–84) (see also the chapter by J. A. Escudero *et al.* in this volume). This led to the rapid and incremental diversification of antibiotic resistance genes in the Tn*21*-related mercury-resistance transposons and other Tn*3*-family members (85, 86). The evolution of multidrug-resistant transposons has been proposed to retrace the recent history of antimicrobial treatments (for discussion, see references 2, 3, 87, and 88).

A proposed founding event at the onset of multidrug resistance evolution was the incorporation of a specific integron and its associated cassettes into a "Mu-like" Tn*5053*/Tn*402*-family transposon (33), generating a specific class of "mobile" integrons known as "class 1" or "Class 1 In/Tn" (82, 87). Although this may have produced different variants of the ancestral Class 1 In/Tn (89, 90), the key to introduction and spread of one particular type of Class 1 integron among bacterial pathogens was presumably the presence of a gene (*qacE*) conferring resistance to quaternary ammonium compounds, which were commonly used as biocides before the antibiotic era. Concurrently, bacteria also accumulated mercury-resistance transposons, notably because of industrial enrichment of Hg^{2+} in the environment and the use of mercury-containing disinfectants in clinical practices (2, 3, 88). Since Tn*402* family transposons preferentially integrate within *res* sites (91–93), coselection of both types of mobile element could have led to the formation of recombinant transposons with Class 1 In/Tns inserted within or close to the *res* site of mercury-resistant Tn*3*-family members. Integration of different variants occurred not only at separate positions within different Tn*21*-subgroup members but also in non-mercury-resistance transposons such as Tn*1721* (85, 86). With the introduction of successive classes of therapeutic antibiotics, the integrons of these different transposons subsequently acted as a platform for collecting and redistributing new resistance cassettes from the reservoir, thereby accelerating their accumulation and dissemination among both established and emerging pathogens.

Passenger genes can also be incorporated into Tn*3*-family members in the form of composite transposons (e.g., TnHad2, (77)) or through the promiscuous transposition activity of members of certain IS families (e.g., IS*Ecp1* [IS*1380* family] or IS*CRs* [IS*91* family]) able to transduce variable lengths of DNA adjacent to their integration site (2, 85, 94–97). IS*Ecp1* and IS*CR1* are for example responsible for incorporation of variants of extended-spectrum $bla_{CTX\text{-}M}$ genes into Tn*21*-related elements (e.g., Tn*1722*), or into class 1 integrons with which they are associated (98–102).

Mixing by Recombination

Another major process driving Tn*3* family evolution is recombination between both related and unrelated transposons (85). This may occur by homologous recombination between common regions or by inter-transposon site-specific recombination at the *res* site (6, 20, 61, 85). Resolvase-mediated recombination between heterologous *res* sites has been inferred from sequence comparison (1, 20, 79, 103) or demonstrated experimentally (78, 104, 105). *Res* site recombination can occur both intra- or inter-molecularly and can also lead to inversions between adjacent transposons reflecting some promiscuity in the activity of certain S-resolvases, which are normally controlled by a mechanism of resolution selectivity (78, 105, 106) (see *Convergent mechanisms to control the selectivity of recombination* section, below). Similar activities were proposed for the TnpS/OrfI Y-recombinase of the Tn*4651*/Tn*5041* subgroup (43, 44) and for the atypical Tn*4656* TnpI/*attI* recombination system (46). Promiscuous recombination mediated by cointegrate resolution systems appears to be an ongoing process that constantly reshapes the Tn*3* family by exchanging functional module and passenger genes between its different members (6, 20, 61, 85).

MITEs and MICs: More Tricks in the Tn3-Family Toolbox

Besides the minimal transposable modules described as IS elements, early studies identified an even more compact transposon derivative, IS*101*, for its ability to promote cointegrate formation between its parental plasmid, pSC101, and other replicons (107) (Fig. 2). This 209-bp element consists of a pair of IR ends flanking a typical *res* site. IS*101* activity required the γδ transposase and resolvase in *trans* (108). IS*101* is a typical "MITE" (Miniature Inverted repeat Transposable Element). These are short nonautonomous elements found in both eukaryotes and prokaryotes (109, 110), which can be mobilized by active copies of related transposons. Several Tn*3*-family MITEs have been identified (111) and have been called "TIMEs" (Tn*3* family-derived Inverted repeat Miniature Elements) (67) (Fig. 2). Tn*5563*-derived TIMEs form a rather dispersed family of elements in *Pseudomonas* sp. and several other proteobacteria (67).

Another derivative, "RES-MITE", was identified that retains the IR ends and a cointegrate resolution module, but lacks the transposase (67). This element is a particular case of another class of nonautonomous transposable elements, an "MIC" (Mobile Insertion Cassette), composed of transposon ends flanking one or more genes not directly involved in transposition (112, 113) (Fig. 2). MIC mobility, like that of MITEs, also requires a source of cognate transposase and there is evidence that structurally equivalent elements contribute to the Tn*3* family outfit of tools, for example by promoting the dissemination of plant virulence factors in *Xanthomonas* spp (A. Varani and M. Chandler, personal communication).

THE MOLECULAR MECHANISM OF TRANSPOSITION

Different models were initially proposed for the Tn*3* family "copy-in" replicative transposition mechanism and that of other cointegrate-forming elements such as bacteriophage Mu, Mu-like transposons of the Tn*552*/Tn*5053* groups, and the IS*6* family (12, 34, 35, 114–117). In one of these models, verified for bacteriophage Mu (114), the transposase introduces single-strand nicks at both 3′ ends of the element, and then joins the ends to staggered phosphates of the target DNA (strand transfer) to generate an intermediate (the "Shapiro intermediate") in which the transposon DNA strands are connected to both the donor and target molecules through three-way branched structures resembling replication forks (Fig. 8). Recruitment of host replication factors leads to transposon duplication and the formation of short direct repeats flanking both copies of the element in the cointegrate (Fig. 8).

Studies on Mu have provided a detailed molecular picture of the different stages in transposition complex assembly and progression (the transpososome), and how it is then disassembled to recruit the host replication machinery (117–119) (see also the chapter by R. M. Harshey in this volume). However, the relevance of this model to other replicative transposable elements, including the Tn*3*-family, is currently not known.

The Transposase

During transposition, transposase must bind specifically to the transposon ends, interact with an appropriate target, and assemble a higher order nucleoprotein complex, the "transpososome", in which the transposition partners are brought together in a proper configuration to catalyze the DNA breakage and rejoining reactions. Additional accessory proteins often join the complex to control the outcome of transposition or to process reaction intermediates. As will be discussed below, Tn*3*-family transposases are also the specific determinant of target immunity, acting both as a "repulsive" signal to discard the transposon from immune targets and as a component of the transpososome that selectively responds to this signal.

A sequence alignment performed with a representative subset of transposases belonging to different Tn*3*-family subgroups revealed discrete regions of higher similarity (Fig. 9). Limited tryptic and α-chymotryptic proteolysis of the Tn*4430* transposase coupled to mass spectrometry showed that these conserved regions are included in separate structural domains (120).

The C-Terminal Catalytic Domain

The most conserved transposase region overlaps with a ~27-kDa proteolytic fragment roughly corresponding to the C-terminal third of the protein and contains 9 of the 15 perfectly conserved residues including the three acidic residues that constitute the "DD-E" catalytic triad

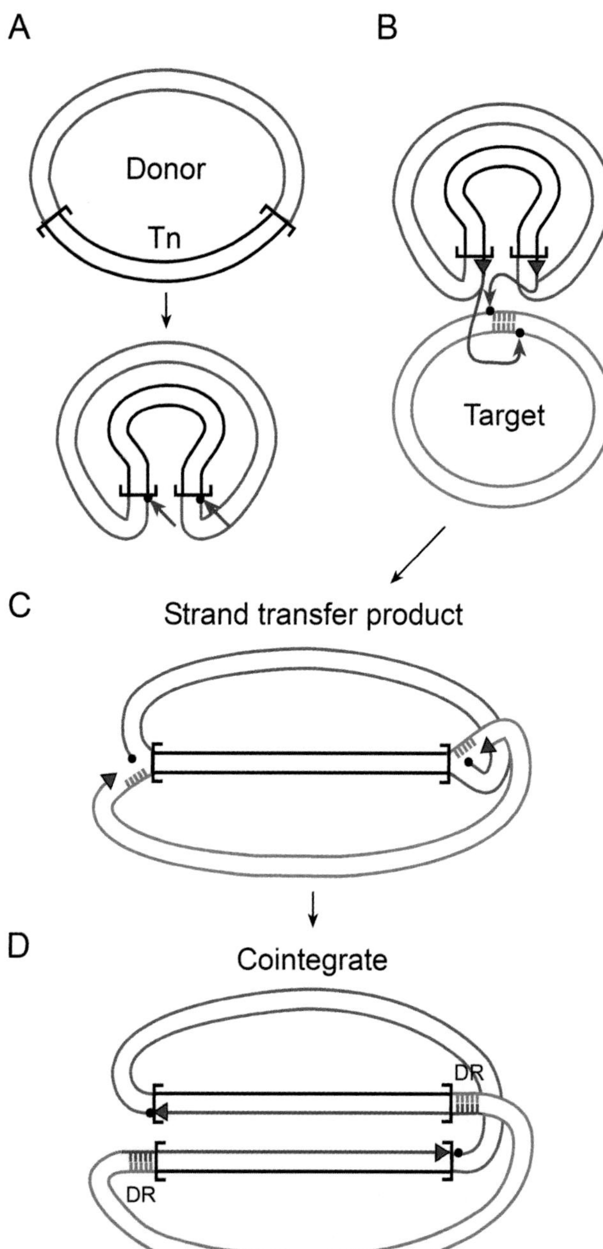

Figure 8 Model for copy-in replicative transposition. The diagram illustrates the case of intermolecular movement of a transposon (double black line delineated by brackets) between a donor molecule (purple) and a target molecule (blue). (A) Transposition starts when the transposase introduces specific single-strand nicks at both 3′ ends of the element, releasing 3′ OH groups in the donor (red triangles). (B) The transposon 3′ OH ends are then used as a nucleophile to attack separate phosphodiester bonds in both target DNA strands. (C) The reaction generates a strand transfer product (often called a "Shapiro intermediate") in which the transposon is linked to the target DNA through its 3′ ends, and to the donor DNA through its 5′ ends. The target phosphates (black dots) are usually staggered by five base pairs, which leaves 5-bp single-strand gaps at the junctions between the transposon and the target DNA in the strand transfer intermediate. (D) Replication initiates at the 3′ OH end(s) released by cleavage of the target to synthesize the complementary strands of the transposon (red line) and form the cointegrate. DNA synthesis also repairs the single-stranded gaps at the ends of the transposon generating directly repeated (DR) 5-bp target duplications that flank both copies of the element in the final cointegrate.
doi:10.1128/microbiolspec.MDNA3-0060-2014.f8

(7, 11) (Fig. 9). This motif is the signature of a superfamily of proteins, the DD-E/D polynucletotidyl transferases, that includes various nucleic-acid processing enzymes, as well as different transposase families and the retroviral integrases (121–123). These enzymes have a common "RNase H-like" fold of their catalytic core, in which the conserved triad provides the essential active site residues. Phosphodiester bond breakage and resealing reactions uses an Mg^{2+}-dependent transesterification mechanism with a water molecule or a 3′ OH group as a nucleophile (123, 124). The Tn*3*-family DD-E motif is surrounded by other highly conserved amino acid residues underlining the importance of this region. Mutation of the conserved Asp751 residue in the Tn*4430* transposase DD-E motif abolishes DNA cleavage and strand transfer *in vitro*, whereas the other activities (i.e., ability to bind and pair the transposon ends) are unaffected (Nicolas *et al.*, in preparation).

Secondary structure predictions suggest that the DD-E region exhibits the βββαβαβ-α topological arrangement typical of the RNase H fold, with the conserved catalytic residues occupying the expected positions within the predicted structure (120). This was confirmed by three-dimensional modeling using different protein threading tools, with the best fits being obtained with the catalytic core of HIV-1 integrase (Fig. 9) (M. Lambin and B. Hallet, unpublished data). However, the proposed Tn*3*-family RNAse H fold domain is predicted to contain an extra ~90-aa insertion between the last β strand of the central β sheet and the C-terminal α helix carrying the conserved Glu residue of the DD-E catalytic triad (120) (Fig. 9). This insert corresponds to one of the most conserved regions of the proteins, suggesting an important role in transposase activity. Insertion domains with different structures are found in an equivalent topological location of the Tn*5* and Hermes transposases where they are proposed to play a role in the formation of DNA hairpin intermediates for the excision of these elements (122, 125–127).

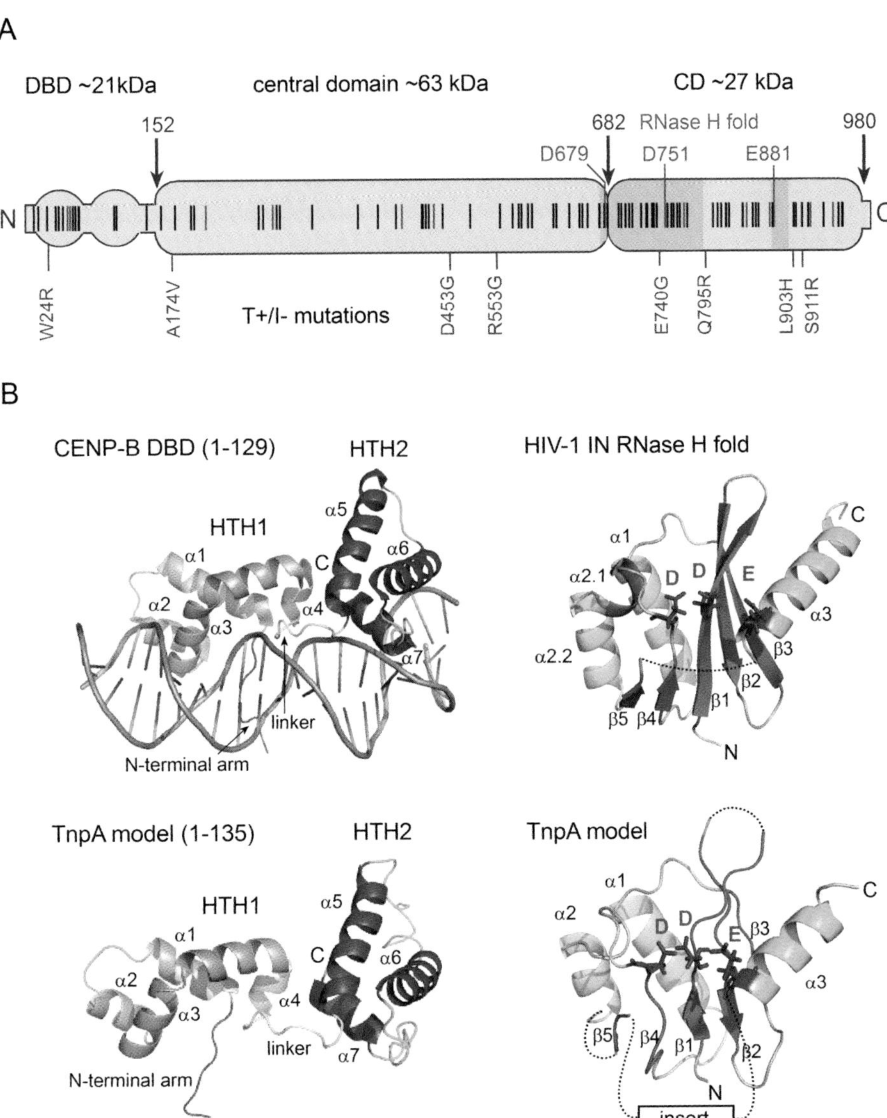

Figure 9 The Tn*3*-family transposase protein. (A) Structural organization of the transposase. The protein is depicted with three main domains based on limited proteolysis (vertical arrows) of the Tn*4430* transposase (987 amino acids) (120). The N-terminal DNA-binding domain (DBD) has a predicted bipartite structure analogous to that of the two-helix-turn-helix (HTH) DNA-binding domain of CENP-B. The C-terminal catalytic domain contains the predicted RNase H fold region (shaded in blue). Vertical bars correspond to highly conserved (>90%) amino acid residues in a representative subset of 21 transposases of the family, with the 15 perfectly conserved residues highlighted in red. Residues of the DD-E catalytic triad are indicated. Characterized mutations that selectively affect target immunity (T+/I– mutations) are reported below the diagram. (B) Structural models for the CENP-B-like DNA-binding domain (left panels) and the RNase H fold catalytic core (right panels). The actual structures of the CENP-B–DNA co-complex (128) and the HIV integrase RNase H fold (249) are shown on the top, and the predicted models derived for the Tn*4430* transposase are shown below. Nonstructured regions are represented by dashed lines. Positions of secondary structures and other critical structural elements are indicated and highlighted in different colors (see text for details). Nonstructured regions are represented by dashed lines. In the RNase H fold, the DD-E catalytic residues are shown in a stick configuration colored in red. The predicted location of the 90-amino acid insert in the putative RNase H fold of TnpA is indicated by a rectangle. doi:10.1128/microbiolspec.MDNA3-0060-2014.f9

The N-Terminal DNA-Binding Domain

Another highly conserved Tn*3*-family transposase segment lies at the N-terminus. Structural homology searches based on secondary structure predictions and fold recognition modeling revealed that this region has a high likelihood to form a bipartite DNA-binding domain analogous to that of the human centromeric protein CENP-B (128) (Fig. 9). As for the C-terminal RNase H fold, this prediction was verified for a consensus sequence derived from the different Tn*3*-family transposase subgroups, reflecting the high level of sequence similarity in this region (120) (Fig. 9). Modeling is consistent with previous results that identified the IR-binding domain at the N-terminus of the transposase (129, 130). Deletion of this region abolishes end binding by Tn*4430* TnpA *in vivo* and *in vitro*, while a 300-aa N-terminal fragment of the protein binds the IR end with the same affinity as the full-length transposase (E. Nicolas and B. Hallet, unpublished data).

The CENP-B DNA-binding domain consists of two helix-turn-helix motifs (HTH1 and HTH2) separated by a short, 10-aa linker (Fig. 9). Both motifs contact adjacent major DNA grooves making specific interactions with the target sequence. The N-terminal residues (residues 1 to 9) form a flexible arm that makes additional contacts with the phosphate backbone (128). These different structural components of the CENP-B binding domain appear to be conserved in the N-terminal domain of Tn*3*-family transposases based on secondary structure prediction and sequence comparison (120) (Fig. 9).

Intriguingly, the CENP-B DNA-binding domain is structurally homologous to that of Tc*1*/mariner transposon family and to the bipartite end-binding domain of bacteriophage Mu transposase, MuA (131–133). CENP-B is believed to have evolved from an ancestral transposase that has been "domesticated" during evolution to acquire a novel function (134). Confirming the presence of a CENP-B-like DNA-binding domain at the N-terminus of Tn*3*-family transposases would therefore provide an additional example illustrating the modular organization of the DD-E/D transposases, which presumably resulted from the recruitment of separate functional domains early during evolution (122, 123).

A Large Central Domain of Unknown Function

In the crystal structure of the Mos*1* transposase (a Tc*1*/mariner-family member) and MuA, the N-terminal DNA-binding domain is immediately fused to the RNase H fold domain by a short connector of a few amino-acid residues (131, 133). In transposons of the Tc*1*/mariner family, this linker region contributes to the transposase dimer interface and plays a role in reciprocal communication between transposase subunits during transposition (131, 135, 136). In Tn*3*-family transposases, the equivalent region of the protein is occupied by a large central domain, corresponding to a ~63-kDa proteolytic fragment of Tn*4430* TnpA (residue 152 to 682) (Fig. 9). This domain is predicted to have an all-alpha structure. Despite several patches of well-conserved residues, it shows no sequence or structural homology with other proteins (120). However, filter binding assays performed with the Tn*3* transposase suggested that separate subregions of the central domain (residue 243 to 283 and residue 439 to 505) exhibit nonspecific DNA-binding activity (130).

The Protein Determinants of Target Immunity

A possible function of the unique central TnpA domain might be in target immunity. Most transposons and ISs do not exhibit this property and this would explain why Tn*3*-family transposases are significantly larger than other transposases. A genetic screen was developed to identify Tn*4430* TnpA mutants that could bypass immunity by promoting high levels of transposition into an immune target (i.e., a target carrying a Tn*4430* IR end). The selected T^+/I^- (i.e., transposition-plus/Immunity-minus) mutations were clustered into five discrete regions belonging to separate domains. Although two clusters of mutations were identified within the large central domain, those that showed the strongest defect in immunity mapped in the C-terminus, beyond the DD-E motif. This scattering of the T^+/I^- mutations indicates that target immunity is likely to be a complex process involving multiple activities of the transposase (120) (see also *The mechanism of target immunity: 'schizophrenic-like' behavior of the transposase*, below).

Interaction with the Transposon Ends

Specific transposase recognition of the transposon ends is crucial to initiate transposition on the appropriate DNA substrates and to adequately position the partners for subsequent reactions steps within the transpososome.

The Conserved IR End

As for other transposable elements, the ~38-bp terminal IRs of Tn*3*-family transposons are composed of two functional domains (Fig. 10). The outermost highly conserved 5′-GGGG-3′ motif is essential for transposition, but is not required for stable TnpA binding. It contains the DNA cleavage site and is thought to be contacted by the transposase catalytic domain to carry out the chemical steps of transposition (137–142). TnpA binding to this region generally gives weak protection as determined by DNA footprinting, with some enhanced signals due to protein-induced DNA distortions around the cleavage site (140, 141, 143–148). For Tn*4430*, these signals are more pronounced upon formation of a paired-end complex, which corresponds to an activated intermediate of the transposition reaction (Nicolas *et al.*, in preparation).

The inner segment of the IR (position ~10 to 38) is the recognition domain with which the transposase makes the strongest and most specific interactions. This region also encodes the specific determinants to confer target immunity (138–140, 149, 150) (see also below). As shown in the alignment (Fig. 10), it can be further divided into two subdomains; a central AT-rich region (box B1, positions ~11 to 27) and a more conserved subdomain (box B2, position ~31 to 38) carrying the 5′-TAAG-3′ motif at the inward-facing boundary of the IR (Fig. 10). TnpA makes extended interactions with both recognition subdomains, protecting up to three adjacent major grooves (137, 141, 149). Mutations in the innermost box B2 subdomain showed the strongest defects in transposase binding and transposition activity (139, 140).

This binding pattern is consistent with the predicted CENP-B-like bipartite Tn3-family transposase DNA-binding domain (Fig. 9). In Tc*1*/Mariner-family transposases and MuA, a structurally equivalent domain recognizes sequence elements of about the same length (~22 to 26 bp) as Tn3-family transposases (~27 bp). Their HTH motifs interact with both sides of the recognition element making specific contacts within the major groove (131–133, 151). The basic linker connecting both HTHs forms a structure termed 'AT-hook' that makes additional interactions with the AT-rich stretch that separates both HTH binding sites (131, 132). It is tempting to propose that Tn3-family transposases bind the transposon ends in a similar way, with one predicted HTH motif interacting with box B2 of the IR and the other, together with the linker, interacting with the longer AT-rich box B1. Additional protein segments, such as the conserved N-terminal arm of the DNA-binding domain may also contribute to the interaction, as observed in the CENP–B/DNA complex (128) (see also Fig. 9).

Extended Ends: Role of Auxiliary Proteins

A number of Tn3-family transposons have significantly longer terminal inverted repeats than the canonical ~38-bp IR (18, 44, 48, 65, 146, 148, 152). For some of them, this was found to correlate with the presence of additional binding sites for accessory proteins in the subterminal regions of the transposon, providing a means to control transposition at the level of transpososome assembly.

The inner segment of the 53-bp Tn*5401* terminal inverted repeats includes an extra 12-bp binding motif for the TnpI recombinase (37, 48, 148). This motif is not included in the IRS and is therefore not required for cointegrate resolution (48). By contrast, TnpI binding to this motif is essential for the formation of a stable complex between the transposase and the transposon ends (148). At the same time, as was previously reported for the resolvase of other Tn3-family members (7), TnpI binding to the IRS down-regulates the expression of both *tnpI* and *tnpA*, keeping transposition activity at a low level (148). This antagonistic action of TnpI may be important to ensure efficient transposition under certain circumstances. In particular, bursts of TnpI and TnpA expression would occur when the transposon enters a new host cell, before transcriptional repression by TnpI becomes effective. Functional coupling between TnpI and TnpA activities may also be regarded as a safeguard mechanism to maintain the integrity of both the transposition and cointegrate resolution modules as the transposon multiplies and spreads.

Transposase binding to the ends of γδ and Tn*4652* is influenced by the host protein IHF (integration host factor). Since the cellular abundance of nucleoid-associated proteins like IHF varies according to the growth rate (153, 154), their contribution to transposition provides a means of communicating environment signals or the physiological state of the cells to the transpososome.

For γδ, transposase and IHF binding to the transposon ends is cooperative (145). This was found to primarily impact on target immunity (155) and to rescue transposition defects resulting from specific ends mutations (139). For Tn*4652*, IHF binding is a prerequisite for transposase binding and transposition (146). IHF was also found to up-regulate Tn*4652* transposition by stimulating the transcription of the *tnpA* gene (156). TnpA expression is also under strong dependence of the stationary-phase sigma factor σ^S (RpoS) (157). Together, these different mechanisms co-operate to restrict Tn*4652* transposition under carbon starvation conditions, when the cellular concentration of IHF is maximal and σ^S is induced. Consistently, binding of Tn*4652* transposase to the transposon ends is negatively affected by Fis (factor for inversion stimulation), which unlike IHF is maximally expressed in actively growing cells

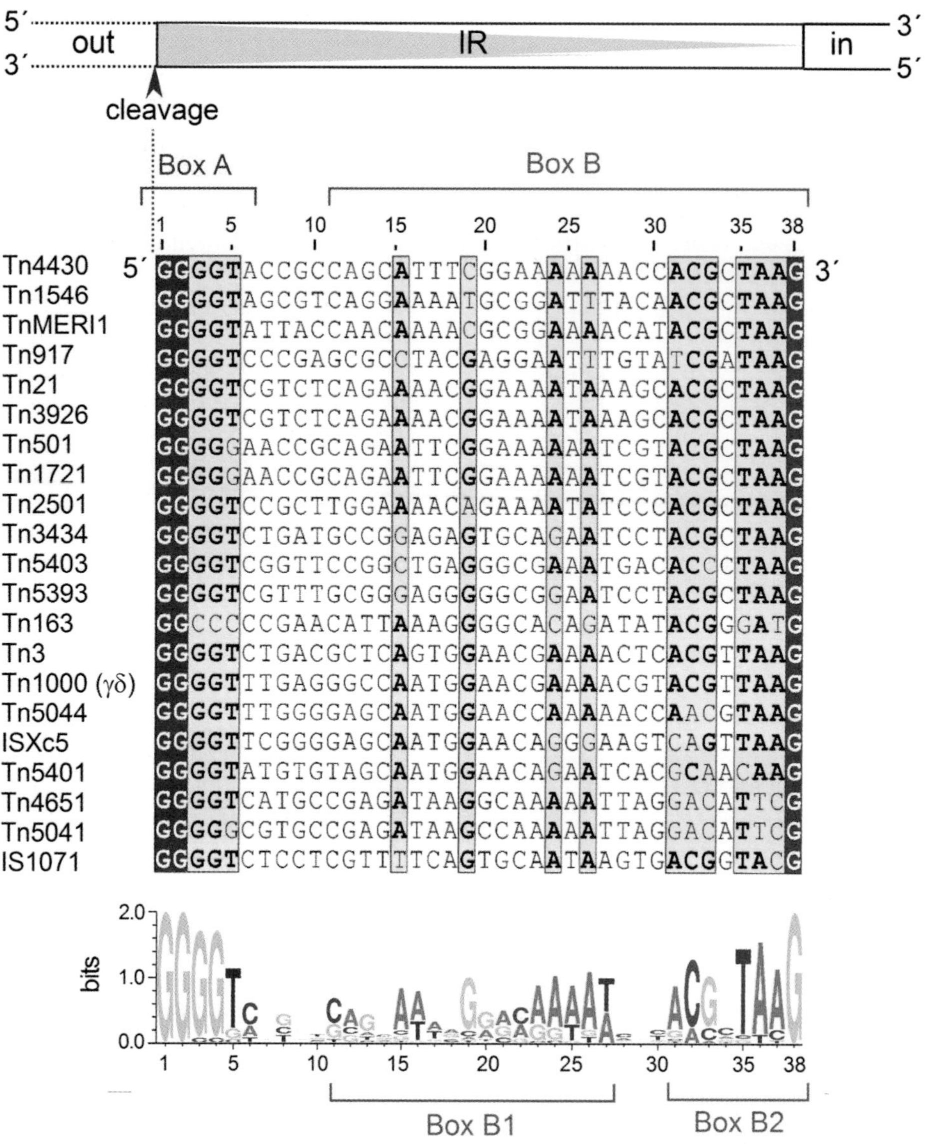

Figure 10 The Tn*3*-family transposons ends. The 38-bp terminal inverted repeat (IR) sequences of 21 representative transposons of the Tn*3* family are aligned, and perfectly or highly (>75%) conserved positions are boxed in red and yellow, respectively. The resulting consensus sequence is shown below the alignment, and a cartoon showing the orientation of the IR sequence (purple triangle) with respect to the inside (in) and outside (out) regions of the transposon is reported on the top. Position of the transposase 3′-end cleavage site is indicated by an arrow. Conserved regions corresponding to the external cleavage domain of the IR (Box A) and the internal transposase recognition domain (Box B) are indicated with brackets. The transposase recognition sequence is further subdivided into two conserved motifs (Box B 1 and Box B 2). doi:10.1128/microbiolspec.MDNA3-0060-2014.f10

(147). Fis binding to its cognate sites is proposed to outcompete IHF from the left end of Tn*4652*, thereby compromising the recruitment of the transposase. Transposition of Tn*4652* is also modulated by the *Pseudomonas putida* ColRS two-component system (158) and the small transposon-encoded protein TnpC, which appears to down-regulate the expression of the transposase at the post-transcriptional level (159).

If the requirement for specific host factors is important to correlate the rate of transposition with cell physiology, it may also restrict the host-range of transposons as was proposed for IS*1071* (152). This element is characterized by unusually long terminal IRs (110 bp) containing potential binding sites for host accessory proteins such as IHF (152).

Interaction with the Target

Target site choice has strong biological implications for both the host and the transposon because it determines whether transposition will have a beneficial outcome or not. Target site specificity also reflects the molecular mechanism whereby the transposition machinery interacts with the target DNA, offering an additional level of control on transposition (160).

DNA Sequence Specificity and Regional Preference

Tn3-family transposons are characterized by a low level of target specificity, which likely contributed to their wide dissemination. Early studies indicated that Tn3-related elements preferentially insert into specific AT-rich DNA regions (161–164). This was confirmed from large-scale analyses which revealed that the target sites of different family members conform to a similar, low-specificity consensus sequence centered on a degenerated 5-bp AT-rich core motif corresponding to the target duplications (i.e., 5′-TA[A/T]TA-3′ for Tn3, 5′-TATAA-3′ for Tn*917*, 5′-[T/A] [T/A]N[A/T] [A/T]-3′ for Tn*hopX1*, 5′-T [T/A] [T/A] [A/T] [A/T]-3′ for Tn*4652*, and 5′-[T/A]NNN[A/T]-3′, for Tn*4430*) immediately flanked by GC-rich positions (65, 165–168) (E. Nicolas and B. Hallet, unpublished data).The overall palindromic organization of this consensus likely reflects symmetrical interactions made by the transposition complex to transfer both transposon ends to the target DNA. G/C to A/T transition steps at the DNA cleavage sites may be required to distort or bend the DNA, making the target scissile phosphates more accessible for strand transfer as proposed for other transposable elements (169–171).

In all analyses, the target DNA sequence is clearly not sufficient to determine Tn3-family insertion specificity. Specific DNA sequences perfectly matching the consensus are poor targets, whereas other sites are targeted much more frequently than expected from their resemblance to the consensus (165, 166, 172). This suggests that integration of Tn3-family transposons obeys some other structural or dynamic properties of DNA.

Role of Target Replication

Recent work performed with Tn*4430* indicates that relatively subtle changes made in a target molecule may profoundly alter the insertion profile (173). In particular, integration of repeated copies of a pseudo-palindromic DNA sequence (i.e., three or five copies of the *E. coli lacO* operator) created strong insertion hot spots between the repeats and the unidirectional replication origin of the target plasmid (173) (E. Nicolas and B. Hallet, unpublished data). This altered regional preference did not change target sequence specificity, but was dependent on the direction of DNA replication with respect to the *lacO* repeats (E. Nicolas, C. Oger and B. Hallet, unpublished data). A plausible interpretation is that the palindromes interfere with replication forks progression in the hot spot region (e.g., through the formation of secondary structures in the DNA), thereby provoking the accumulation of preferred DNA target structures. Consistent with this, several Tn3-family transposons preferentially insert within or close to the replication origin of target plasmids (161, 174). On the other hand, Tn*917* transposition into the chromosome of *Enterococcus faecalis* and other Firmicutes showed a strong preference for the replication termination region (172, 175, 176). This region is subject to extensive processing, including numerous replication forks collapse and repair events (177)), which may provide the required transposition signal. Intriguingly, the Tn*4430* transposase binds *in vitro* with high affinity to branched DNA substrates that mimic a replication fork (E. Nicolas, C. Oger and B. Hallet, unpublished data).

Transposition of other transposons is also coupled to replication (178–185). Tn7 targets the lagging strand at a replication fork (178, 179, 185) using specific interactions between the Tn7-encoded auxiliary protein TnsE and the β-clamp processivity factor of the replisome (186) (see also the chapter by J. E. Peters in this volume). This results in a marked preference for inserting into conjugating plasmids and is regarded as a mechanism to promote horizontal dispersal. For other transposable elements such as the IS*200*/IS*608* family and Group II introns, coupling between DNA targeting and lagging-strand DNA synthesis is inherent to the transposition mechanism that requires single-stranded DNA for integration (183, 187).

The possibility that replicative transposons like those of the Tn3 family preferentially integrate into replication forks would provide a direct mechanism for recruiting the host replication machinery for the replicative transposition step. Indeed, experiments performed with conditionally replicating DNA molecules have shown that target replication is essential for Tn*1*/Tn3 transposition (188). Transposition into nonreplicating target DNA was essentially undetectable, whereas blocking replication of the donor molecule had no effect on transposition activity. Tn3-family transposons also prefer to insert into episomes such as phages and plasmids rather than into the chromosome, presumably because the replication dynamics of these different replicons is different (189, 190).

Control of Transposition by the Target: Target Immunity

Target immunity, also called transposition immunity, is a highly selective and *cis*-acting process by which certain classes of transposable elements avoid inserting into a DNA molecule that already contains a copy of themselves. A remarkable feature of this "repulsive" process is that it can act over extended genome regions that, depending on the element, range from ~20 kb to several dozen kilobases (138, 160, 191–193). Understanding this mechanism therefore has much to tell us about how DNA is organized within the cell and how distant sites in a genome communicate with each other.

The Mechanism of Target Immunity: "Schizophrenic-Like" Behavior of the Transposase

Target immunity was initially described for Tn*3* family members (194, 195) and subsequently for other transposable elements like Mu (196, 197) and Tn7 (198, 199). In all cases, a single transposon end or a subdomain of the transposon terminus capable of binding the transposase is sufficient to confer target immunity (196, 197, 199, 200), but the presence of multiple transposase binding sites usually strengthens the phenomenon (120, 192, 201, 202). For Tn*3*-family transposons, the DNA determinants of target immunity coincide with the internal ~27-bp to 30-bp recognition motif (box B) of the terminal IR (120, 138–140, 149, 150) (Fig. 10).

Experiments with Tn*4430* have shown that simply attracting the transposase to a nonspecific target molecule by fusing the protein to a heterologous DNA-binding domain is insufficient to make this molecule refractory to transposition (173). This suggests that establishment of immunity requires the formation of a specific complex between the transposase and a cognate site on the target (173).

The requirement for specific interactions between the transposase and the transposon ends is only part of the self-recognition mechanism that makes each transposon of the family immunized against itself, but not against other members of the family (195, 203). This was shown in "mixing" experiments where Tn*4430* and Tn*1* were introduced into the same cells; an immune target for Tn*4430* was not immune against Tn*1* and reciprocally, a target that was immunized by Tn*1* transposase was permissive for Tn*4430* (120). The transposase of Tn*3* family transposons therefore plays a dual role in the specificity of immunity by both communicating the immunity signal to the target and by interpreting this signal.

For Mu and Tn7, target immunity results from the interplay between the transposase (MuA and TnsAB, respectively) and an additional element-encoded protein (MuB and TnsC) that function as a molecular "match-maker" between the transpososome and the target (117, 118, 204–206) (see also the chapters by R. M. Harshey and J.E. Peters in this volume). Both proteins are DNA-dependent ATPases that bind to DNA in their ATP-bound form. By interacting with their partner transposase, MuB and TnsC stimulate transpososome assembly and DNA strand transfer (200, 207–210). As a result, MuB- or TnsC-bound DNA molecules are preferred substrates for Mu and Tn7 transposition, respectively (196, 200, 208, 211–213). Interaction with the transposase also stimulates ATP hydrolysis by MuB and TnsC, which triggers their dissociation from DNA and promotes their relocation to remote sites. As this preferentially takes place on DNA molecules that carry the end-bound MuA/TnsB proteins, displacement of MuB/TnsC makes those molecules less reactive, and hence immune against further integration (196, 200, 202, 214, 215).

For Tn*3*-family transposons, the transposase is the only element-encoded protein that takes part in the immunity process. However, this does not rule out the possibility that some additional cellular function or host protein contributes by modulating transposase action in the same way for all family members despite their specific sequence and self-recognition properties. The action of transposase on the target may either be "dissuasive", by selectively restricting access to immune DNA molecules, or "attractive", by directing the transposon toward permissive targets as is proposed for the MuB and TnsC target adaptors. Depending on their global effect on transposition, the Tn*4430* TnpA T^+/I^- mutations were proposed to either impair the ability of the transposase to establish immunity, or to bypass the immunity signal (whether positive or negative) by promoting efficient transposition into nonpermissive targets (120). Supporting the latter scenario, recent biochemical data obtained with selected T^+/I^- TnpA mutants indicate that they have promiscuous activities, suggesting a possible link between target immunity and transposase activation (Nicolas *et al.*, in preparation) (see also *Catalysis of the transposition reactions*, below).

Biological Roles of Target Immunity

Transposition immunity is traditionally presented as a strategy to favor intracellular and intercellular dispersal of transposable elements. For example, without immunity, Tn7 would accumulate into its high-affinity chromosome site *att*Tn7, precluding it from spreading among bacterial populations by horizontal transfer (204, 205). Another proposed role for transposition immunity is to avoid self-destruction by autointegration (i.e., integration into itself). This is particularly crucial for bacteriophage Mu, which multiplies its genome (38.5 kb) by undergoing multiple rounds of replicative transposition in a short period of time (216). Mu target immunity is effective over ~20–25 kb on either side of the genome ends (191, 217). This would be sufficient to protect Mu against self-integration while fully exploiting the available target DNA landscape for replication.

A serious drawback of the copy-in mechanism of replicative transposition is that the transposon remains tethered to the donor molecule through its 5′ ends after cleavage at the 3′ ends (Fig. 8). This could strongly restrict the ability of the element to reach a distant target within the cell. As a consequence, replicative transposons should exhibit a high propensity to integrate into close DNA regions belonging to the initial donor molecule. However, intramolecular transposition by the copy-in mechanism gives rise uniquely to inversions or deletions of adjacent DNA sequences, which can be detrimental for the host replicon (163, 218, 219). Therefore, the primary role of replicative transposon immunity would be to protect the host genome against such deleterious consequences of intramolecular transposition. The Tn*4430* T^+/I^- TnpA mutants provide a unique opportunity to test this possibility (120). Preliminary results indicate that these mutants mediate intramolecular transposition at a higher frequency than the wild-type transposase, correlating with their immunity defect (E. Nicolas, C. Oger and B. Hallet, unpublished data). It would be interesting to see whether this also correlates with a loss of fitness resulting from transposition-induced damages in the host replicon.

Catalysis of the Transposition Reactions

Our understanding of the molecular transposition mechanism of Tn3 family transposons has long remained scant because of the technical difficulties inherent in transposase purification for biochemical characterization. An *in vitro* transposition reaction based on TnpA-enriched cell extracts was developed for Tn3 and, as expected, TnpA was shown to introduce specific nicks at the 3′ ends of the transposon in an Mg^{2+}-dependent manner (220–222) (Fig. 8). This is consistent with TnpA being a member of the DD-E/D transposase superfamily (221, 222). However, the overall efficiency of the reaction was too low for the molecular characterization of transposition intermediates and products (220–222).

The Tn*4430* transposase was recently purified in an active and soluble form (120), allowing development of biochemical assays for the different transposition steps (from binding to the transposon ends to DNA cleavage and strand transfer; Nicolas *et al.*, in preparation). This was accomplished with the wild-type transposase and with selected immunity-deficient T^+/I^- mutants. T^+/I^- TnpA mutants are all deregulated to different extents. They spontaneously assemble an activated form of the transposition complex termed the paired-end complex in which two transposon ends are brought together by the transposase, and they exhibit higher DNA cleavage activity than the wild-type transposase. However, after cleavage, both the wild-type and mutant transposases appear to transfer the transposon ends to a DNA target with similar efficiencies. This suggests that transposition is controlled at an early stage of transpososome assembly, before initial cleavage, and that mutations that impair immunity have "unlocked" the transposase making it more prone to adopt an active configuration than wild-type TnpA (Nicolas *et al.*, in preparation).

Controlling transposition at the level of transpososome assembly is a recurrent strategy among transposable elements to ensure coordinated DNA cleavage and rejoining between appropriate partners. As the Tn*4430* T^+/I^- TnpA mutants were selected for their increased propensity to promote transposition into immune targets, it is reasonable to propose that formation of an activated paired-end complex by the wild-type transposase is normally conditioned by the target. This would provide a checkpoint allowing the transpososome to discriminate between permissive and nonpermissive integration sites. The molecular mechanism(s) that regulate(s) DNA targeting by Tn3-family transposons are not known, but as discussed above (see *Interaction with the target* section), an intriguing possibility is that they involve specific interactions between the transposase and cellular processes like DNA replication.

COINTEGRATE RESOLUTION BY SITE-SPECIFIC RECOMBINATION

Cointegrate resolution by Tn3-family S-resolvases has been the subject of several recent reviews (7, 9, 17, 31, 223, 224) (see also the chapters by W. M. Stark and P. A. Rice in this volume). Here, we will only give a brief summary of these studies, and focus on more recent work on resolvases of the Y-recombinase family.

S- and Y-Resolvases: Unrelated Mechanisms to Cut and Rejoin DNA Strands

S-recombinases and Y-recombinases cut and reseal DNA through the formation of a transient covalent protein–DNA intermediate involving their serine and tyrosine active site residue, respectively. For both recombinase families, DNA strand exchange takes place in the context of a tetrameric complex comprising four recombinase molecules bound onto the core recombination sites of both partners. However, the chemistry and the molecular choreography used by the S- and Y-recombinases to carry out recombination are different (17) (see also the chapters by M. Jayaram *et al.* and W. M. Stark in this volume).

S-Recombinases: Double-Strand Breaks and Rotation

S-recombinases cleave all four DNA strands concurrently (Fig. 11A). This generates double-strand breaks with a two-nucleotide 3′ OH overhang on one side of the break, and a recessed end to which the recombinase is attached via a 5′-phosphoseryl bond on the opposite side. The DNA strands are then exchanged by a 180° right-handed rotation of one pair of half-sites relative to the other. Base pairing between the exchanged overhangs is necessary to properly realign the DNA duplexes and orient the DNA ends for the ligation step (Fig. 11A). This step is the reversal of the cleavage reaction, with each 3′ OH group of the cleaved ends attacking the phosphoseryl bonds in the partner (7, 9, 17, 31, 223, 224).

There has been some debate in the literature concerning the molecular mechanism of rotation (7, 225–229). The model that currently conciliates most biochemical and structural data (the "subunit rotation" model) involves a complete 180° rotation of one half of the cleaved complex relative to the opposite half (227, 230). The crystal structure of an activated form of the γδ resolvase tetramer trapped in the act of strand exchange shows that both sides of the recombinase tetramer are held together by a flat hydrophobic interface that would make such a rotational motion between recombinase subunits thermodynamically favorable without causing complete dissociation of the complex (230, 231) (Fig. 11A).

Y-Recombinases: Sequential Strand Exchange and Isomerization

In contrast to S-recombinases, Y-recombinases sequentially exchange one pair of DNA strands at a time via the formation of a Holliday junction intermediate (Fig. 12A). Each strand exchange is a concerted process in which the 3′ phosphotyrosyl DNA–protein bond generated by DNA cleavage is subsequently attacked by the 5′ OH end of the partner strand. Exchange of both pairs of DNA strands implies that separate pairs of recombinase molecules in the complex are sequentially activated (Fig. 12A). Studies performed with different Y-recombinase family members have provided a detailed picture showing how this could be achieved at the molecular level (17, 232–234).

According to the model, the recombinase-bound recombination core sites are brought together in an antiparallel alignment, forming a synaptic complex with a pseudo-four-fold symmetry (Fig. 12A). The DNA is bent to expose one specific DNA strand from each duplex toward the center of the synapse for the first strand exchange. After cleavage, DNA strands are exchanged by extruding three or four nucleotides from the core central region, and by reannealing them to the partner strand to orient the cleaved 5′ OH ends for ligation. As for the S-recombinases, this "homology-testing" step ensures that appropriate DNA strands are exchanged during the reaction. The complex then isomerizes to activate the other pair of recombinase subunits for the second strand exchange (Fig. 12A).

For all Y-recombinases studied in detail, the activity of the core recombination complex is regulated by a cyclic network of allosteric interactions that places each recombinase subunit under the control of its two neighbors, so that diagonally opposed active sites of the tetramer are sequentially and reciprocally switched on and off during recombination (235–238). Isomerization of the tetramer involves relatively limited readjustments of DNA and proteins in the complex, and the mechanisms that promote these conformational changes to provide directionality to the strand exchange reaction appear to vary among different recombination systems (239).

Convergent Mechanisms to Control the Selectivity of Recombination

Although some promiscuity in resolvase-catalyzed recombination reactions may be useful from an evolutionary point of view to create new transposons (see *Mixing by recombination*, above), their primary biological function requires a preferential action on directly repeated copies of the *res* sites as they appear on a cointegrate. This is important to efficiently resolve cointegrate intermediates of the transposition process, while avoiding any other undesired DNA rearrangements.

The γδ/Tn*3* Resolvase Synaptsome Paradigm

Efficient S-resolvase-mediated recombination of Tn*3* and γδ only takes place if two full-length *res* sites are

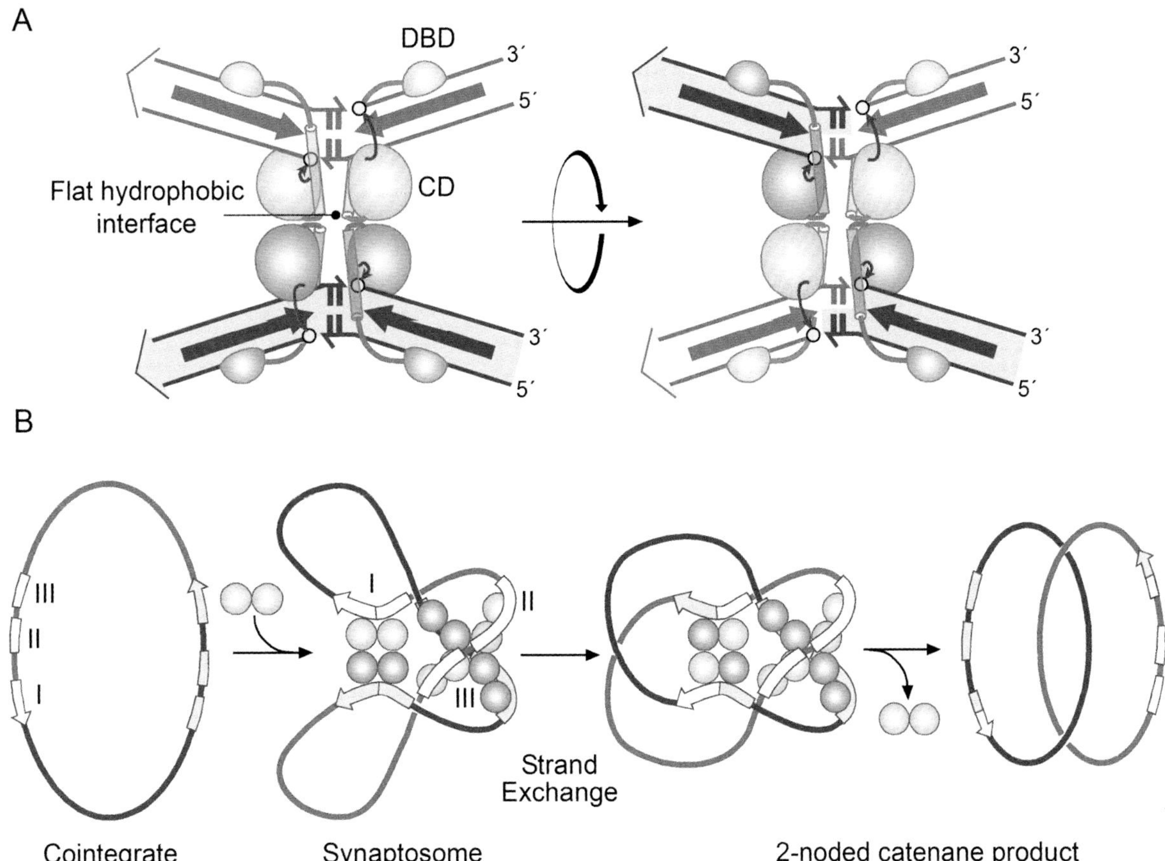

Figure 11 Mechanism of cointegrate resolution by resolvases of the S-recombinase family. (A) The rotational strand exchange reaction catalyzed by S-recombinases. Representation of the recombination complex is inspired from the structure of the synapse showing the activated γδ resolvase tetramer bound to paired core sites I of *res* (230, 231). The recombination sites are aligned in parallel. Blue arrows represent the 12-bp resolvase recognition motifs. The partner resolvase dimers are colored in pale and dark green with their catalytic domain (CD) lying at the inside of the synapse and their DNA-binding domain (DBD) at the outside. The four recombinase molecules have cleaved the DNA, generating double-strand breaks with phosphoseryl DNA–protein bonds at the 5′ ends (shown as yellow dots linked to red connectors) and free OH groups at the 3′ ends of the breaks (half-arrows). DNA strands are exchanged by 180° rotation of one pair of partner subunits with respect to the other around a flat hydrophobic interface within the tetramer. For the rejoining reaction, each free 3′ OH end attacks the phosphoseryl bond of the opposite DNA strand. (B) Topological selectivity in resolvase-mediated cointegrate resolution. Binding of resolvase dimers (green spheres) to sites I, II and III of the partner *res* sites results in the formation of a synaptosome in which the two *res* sites are inter-wrapped, trapping three negative crosses from the initial DNA substrate. This complex only readily forms if the starting *res* sites are in a head-to-tail configuration on a supercoiled DNA molecule. Strand exchange by right-handed 180° rotation as in (A) generates a two-node catenane product.
doi:10.1128/microbiolspec.MDNA3-0060-2014.f11

present in the appropriate head-to-tail orientation on a supercoiled DNA molecule. Resolvase binding to the three *res* subsites (Fig. 6) promotes assembly of a multi-subunit protein–DNA complex termed a synaptosome in which the two *res* sites are plectonemically inter-wrapped, trapping three negative supercoils from the initial DNA substrate (7, 17, 227) (Fig. 11B). The specific topology of this complex is proposed to act as a topological filter to ensure that the recombination sites are in the correct configuration (240). DNA wrapping around the synaptosome is facilitated by negative supercoiling. In contrast, formation of such a complex

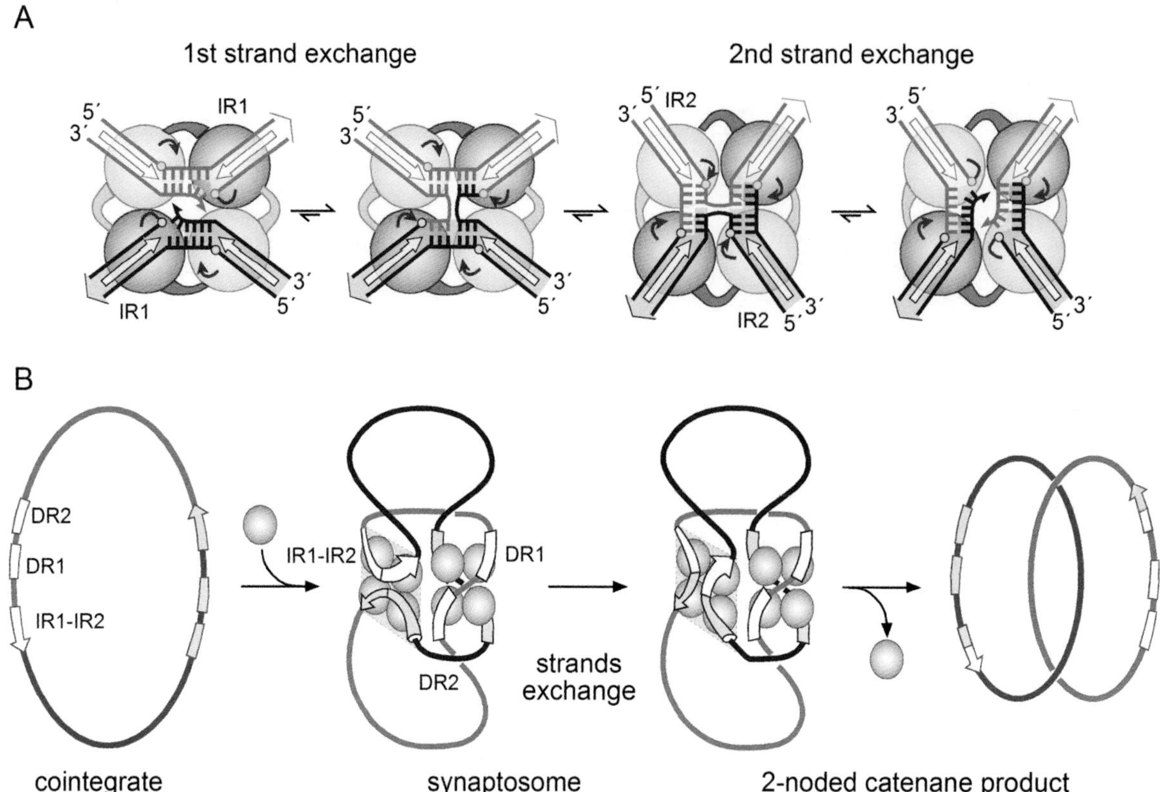

Figure 12 Mechanism of cointegrate resolution by the TnpI recombinase of Tn*4430*. (A) Ordered DNA strand exchange catalyzed by TnpI at the IR1–IR2 core site of the IRS. The TnpI tetramer bound to synapsed IR1–IR2 core sites is drawn according to the structure of the related Cre recombinase complexes (232). Only the C-terminal catalytic domain of the protein is shown for clarity. Each recombinase subunit is connected to its neighbors though a cyclic network of allosteric interactions that dictates its activation state during the consecutive steps of recombination. The recombination sites are brought together in an antiparallel configuration exposing one specific pair of DNA strands at the center of the synapse. In this configuration, the IR1-bound TnpI subunits (magenta) are activated to catalyze the first strand exchange and generate the Holliday junction (HJ) intermediate. The complex then isomerizes, which deactivates the IR1-bound TnpI subunits and activates the IR2-bound subunits (pink) for catalyzing the second strand exchange. For each strand exchange, the recombinase catalytic tyrosine (curved arrow) attacks the adjacent phosphate (yellow circle) to form a 3′ phosphotyrosyl protein–DNA bond, which is in turn attacked by the 5′ OH end (half-arrow) of the partner DNA strand. (B) Possible model for the topological organization of the TnpI/IRS recombination complex. TnpI binding to the DR1–DR2 accessory motifs of directly repeated IRSs generate a synaptic complex in which three DNA crosses are trapped. Proper antiparallel pairing of the IR1–IR2 core sites introduce a positive twist in the DNA so that strand exchange as in (A) generates a two-node catenane product.
doi:10.1128/microbiolspec.MDNA3-0060-2014.f12

between inversely oriented *res* sites, or between sites on separate DNA molecules would be topologically hindered (7, 17, 227, 240).

Synaptosome assembly and activation is a dynamic process during which selective interactions between adjacent resolvase dimers result in a succession of conformational transitions that progressively drive the complex towards its final, recombination-competent configuration (17). Reciprocal interactions between resolvase dimers bound to accessory sites II and III of both *res* sites are responsible for the establishment of the initial inter-wrapped structure in which three negative nodes are trapped (241, 242). Formation of this presynaptic complex is the key of the topological filter that dictates whether recombination will proceed by aligning the core recombination sites I for strand exchange. Productive parallel pairing of the crossover sites I involves specific interactions between the resolvase molecules bound to site I and site III, as well as additional interactions between the site I-bound dimers of both duplexes to form a synaptic interface within the tetramer (230, 231, 243, 244).

In addition to promoting spatial juxtaposition of the crossover sites, transient interactions between the regulatory and catalytic subunits of the complex are thought to play a more direct role in the directionality of strand exchange by inducing the required conformational change to bring resolvase from an inactive to an active configuration observed in structural studies (17, 31, 224, 230, 244). This activation step may be important to tip the recombination complex to an irreversible stage of the recombination reaction.

Because of the defined topology of the recombination complex and the rotational specificity of the strand exchange mechanism, recombination mediated by S-resolvases exclusively yields catenated molecules in which the two recombinant products are singly interlinked (Fig. 11B). The same product topology has been reported for other resolution systems using S-recombinases suggesting a common topological structure of the synaptic complex even if the organization of the recombination sites and the molecular architecture of the synaptosome are different (19, 25, 27, 30, 31) (see also the chapter by P. A. Rice in this volume).

Variation on a Theme: TnpI/IRS Recombination Complex of Tn*4430*

A mechanism of toplogical selectivity analogous to that described for the serine recombinases controls cointegrate resolution mediated by the Tn*4430* Y-recombinase, TnpI (47). As for S-resolvases, TnpI mediates recombination without additional host factors, acting both as a catalytic and as a regulatory component of the recombination complex. However, as opposed to most regulated site-specific recombinases, TnpI does not absolutely require the DR1 and DR2 accessory binding motifs of the resolution site IRS to be active (47) (see also Fig. 7). In the absence of DR1 and DR2, TnpI-mediated recombination at the IR1–IR2 core site is nonselective (or "unconstrained") giving rise to all possible DNA rearrangements *in vivo* (i.e., deletions, inversions, or intermolecular fusions) and to topologically complex products *in vitro* (47). In contrast, the DR1 and DR2 accessory motifs stimulate intramolecular recombination between directly repeated IRSs and generate exclusively two-node catenane products *in vitro* (47) (Fig. 12B).

More recent topological analyses indicate that TnpI binding to DR1 and DR2 results in the formation of a complex in which the accessory sequences are inter-wrapped approximately three times. As for the S-resolvases, formation of this complex acts as a checkpoint (i.e., topological filter) to ensure that the recombination sites are in a proper head-to-tail configuration. Changing the arrangement between the core site and the accessory motifs inhibits recombination by compromising productive core site pairing or forcing alignment in an incorrect configuration, while mutational inactivation of DR1 and DR2 increases the level of topologically unconstrained recombination arising from random collision of the recombination sites (47, 245) (C. Galloy, D. Dandoy and B. Hallet, unpublished data).

The formation of two-node catenanes as the unique products of recombination also implies that the topologically constrained synapse between the TnpI-bound DR1–DR2 sequences imposes a specific alignment of the IR1–IR2 core sites to carry out DNA strand exchange. This alignment is such that antiparallel pairing of the core sites introduces a positive twist in the synaptic complex, thereby compensating for one negative node trapped by the accessory sequences (Fig. 12B). Specific positioning of the core sites within the recombination complex correlates with a defined order of activation of the catalytic TnpI subunits within the tetramer (245). The most distant core subunits with respect to the accessory sites (i.e., the IR1-bound subunits in the wild-type IRS) initiate recombination by catalyzing the first strand exchange, while the proximal subunits (i.e., the IR2-bound subunits) resolve the Holliday junction intermediate by exchanging the second pair of strands (245) (Fig. 12B).

According to current models, directionality of strand exchange depends on the bending direction of the core sites as they are assembled in the synapse. There are two possible recombinase tetramer configurations in which diagonally opposed recombinase subunits are activated for catalysis (Fig. 12B). If one configuration is used to initiate recombination, then the opposite configuration of the complex will terminate the reaction following isomerization (245). In the TnpI/IRS system, the choice of starting recombination with one configuration of the synapse instead of the other is primarily dictated by the DR1 and DR2 accessory motifs. In their absence, unconstrained recombination catalyzed by TnpI at the IR1–IR2 core site takes place with both possible synapse configurations and no preferred order of strand exchange (245).

Based on structural data reported for other tyrosine recombinases, positioning of the TnpI core tetramer in the synaptic complex would orient the recombinase

C-terminal domains toward the regulatory region of the complex, while the N-terminal domains of the TnpI core subunits would point away from the accessory sequences (Fig. 12B). Selecting for this specific arrangement of the core complex may occur indirectly by imparting a specific path to the core sites, or more directly by requiring specific interactions between the core and accessory TnpI subunits of the complex. Imposing a specific pairing of the core recombinase complex to start recombination is important to avoid the formation of unproductive (i.e., parallel) synapses (245).

In addition to providing an architectural scaffold for assembly of the topologically and functionally selective synapse, the DR1–DR2 accessory sequences of Tn*4430* IRS also affect recombination directionality by acting at later steps in strand exchange (245). Their correct orientation in the recombination substrate stabilizes DNA cleavage and rejoining intermediates that could not be observed in reactions using the minimal IR1–IR2 core site alone (245). Cleavage of the first pair of DNA strands is proposed to release the free energy stored in the topologically constrained synapse to bring about conformational changes required to generate the Holliday junction intermediate and to promote its subsequent isomerization (245). This allosteric activity of the accessory components of the complex ensures efficient substrate conversion to products by preventing reversal of the reaction.

A topologically defined complex with a different molecular architecture promotes selectivity during resolution of plasmid multimers by the tyrosine recombinase XerCD from *E. coli* (39, 246). In this case, assembly of the topologically constrained synapse requires dedicated cellular proteins instead of extra recombinase molecules and the recombination product is a four-node catenane instead of a two-node catenane. It is presently unclear whether similar mechanisms of selectivity function in the TnpS–TnpT/*rst* (OrfQ-OrfI/*att5041*) cointegrate resolution system encoded by the Tn*4651*/Tn*5041* subgroup of Tn*3*-family transposons (41, 43, 44) (Fig. 7). Recombination mediated by the tyrosine recombinase TnpS at the *rst* resolution site of Tn*4651* is a relatively slow process that requires the accessory protein TnpT (44). However, it is not known whether TnpT binding to *rst* results in the formation of a topologically constrained synaptic complex as for other resolution systems. Likewise, further biochemical studies are required to decipher the molecular mechanisms that control recombination mediated by the PmrA-like tyrosine recombinase encoded by Tn*4556* (7) and by the TnpI/*attI* recombination system of Tn*4655* (247).

CONCLUSIONS AND PERSPECTIVES

Since their initial identification as the first antibiotic resistance transposons, isolation of new members of the Tn*3* family has continued to demonstrate their constant implication in the tit-for-tat race between bacterial pathogens and humans since the onset of antimicrobial therapies. However, both the collection and dissemination of antibiotic resistances among pathogens, and the emergence of bacterial isolates with new catabolic capabilities in polluted environments are clearly the result of recent adaptation to selective pressures imposed by human activities. In nature, Tn*3*-family transposons are likely associated with a much broader range of accessory functions, and their real impact on bacterial adaptability remains underestimated. Of particular interest is the recent finding that these transposons can promote the mobility of potentially highly versatile elements such as MITEs and MICs. Because of their minimal structure, these elements inevitably eluded classical genome annotation and thorough genomic surveys are therefore required to assess their contribution to bacterial diversification and phenotypic adaptation. Further comparative studies of both autonomous and nonautonomous elements are also necessary to decipher the mechanisms of functional module acquisition and reshuffling within and between separate subgroups of the family.

The biological relevance of Tn*3*-family transposons sharply contrasts with our current understanding of the molecular mechanisms that mediate and regulate their mobility. Most of what we known about the copy-in replicative transposition comes from studies on bacteriophage Mu, and it is presently not known whether the Mu paradigm applies to more "conventional" replicative transposons such as those of the Tn*3* family. Tn*3*-family TnpA proteins are only distantly related to other DDE-D transposases and contain a large and structurally unique central domain that is not found in other proteins. In contrast to Mu and Tn*7*, TnpA is the only transposon-specific protein involved in both transposition and target immunity, and the self-recognition mechanism whereby this single protein imposes and responds to the immunity signal remains an enigma. Finally, integration of Tn*3*-family transposons appears to depend on target DNA replication, suggesting a possible mechanism to synchronize transposition with DNA synthesis. These different aspects of the transposition mechanism raise a number of new working hypotheses that can now be addressed at the cellular and molecular levels by using newly developed genetic and biochemical tools. The prospect of these studies will be to provide an integrated view connecting transpososome assembly and activation with target immunity and DNA replication.

Acknowledgments. We are grateful to P. Siguier for her expertise in protein sequence comparison and for her help in building the phylogenetic trees. We also thank A. Varani and M. Chandler for sharing data before its publication. Work in the laboratory of B.H. is supported by grants from the Fonds National de la Recherche Scientifique (FNRS) and the Fonds Spéciaux de La Recherche (FSR) at UCL. M.L., D.D., C.G. and C.O held a FRIA fellowship and E.N. was research assistant at the FNRS. B.H. is honorary senior research associate at the FNRS.

Citation. Nicolas E, Lambin M, Dandoy D, Galloy C, Nguyen N, Oger CA, Hallet B. 2014. The Tn3-family of replicative transposons. Microbiol Spectrum 3(2):MDNA3-0060-2014.

References

1. **Partridge SR, Hall RM.** 2005. Evolution of transposons containing blaTEM genes. *Antimicrob Agents Chemother* **49:**1267–1268.
2. **Toleman MA, Walsh TR.** 2011. Combinatorial events of insertion sequences and ICE in Gram-negative bacteria. *FEMS Microbiol Rev* **35:**912–935.
3. **Stokes HW, Gillings MR.** 2011. Gene flow, mobile genetic elements and the recruitment of antibiotic resistance genes into Gram-negative pathogens. *FEMS Microbiol Rev* **35:**790–819.
4. **Nordmann P, Dortet L, Poirel L.** 2012. Carbapenem resistance in Enterobacteriaceae: here is the storm! *Trends Mol Med* **18:**263–272.
5. **Nojiri H, Shintani M, Omori T.** 2004. Divergence of mobile genetic elements involved in the distribution of xenobiotic-catabolic capacity. *Appl Microbiol Biotechnol* **64:**154–174.
6. **Tsuda M, Ohtsubo H, Yano H.** 2014. Mobile catabolic genetic elements in pseudomonads, p 83–103. *In* Nojiri H, Tsuda M, Fukuda M, Kamagata Y (ed), *Biodegradative Bacteria: How Bacteria Degrade, Survive, Adapt, and Evolve*. Springer, Japan.
7. **Grindley ND.** 2002. The movement of Tn3-like elements: transposition and cointegrate resolution, p 272–302. *In* Craig NL, Craigie R, Gellert M, Lambowitz AM (ed), *Mobile DNA II*. ASM Press, Washington D.C.
8. **Curcio MJ, Derbyshire KM.** 2003. The outs and ins of transposition: from mu to kangaroo. *Nat Rev Mol Cell Biol* **4:**865–877.
9. **Hallet B, Vanhooff V, Cornet F.** 2004. DNA-specific Resolution Systems, p 145–180. *In* Funnell BE, Phillips GJ (ed), *Plasmid Biology*. ASM Press, Washington D.C.
10. **Toussaint A, Merlin C.** 2002. Mobile elements as a combination of functional modules. *Plasmid* **47:**26–35.
11. **Yurieva O, Nikiforov V.** 1996. Catalytic center quest: comparison of transposases belonging to the Tn3 family reveals an invariant triad of acidic amino acid residues. *Biochem Mol Biol Int* **38:**15–20.
12. **Chandler M, Mahillon J.** 2002. Insertion sequences revisited, p 305–365. *In* Craig NL, Craigie R, Gellert M, Lambowitz AM (ed), *Mobile DNA II*. ASM Press, Washington D.C.
13. **Nakatsu C, Ng J, Singh R, Straus N, Wyndham C.** 1991. Chlorobenzoate catabolic transposon Tn5271 is a composite class I element with flanking class II insertion sequences. *Proc Natl Acad Sci USA* **88:**8312–8316.
14. **Dunon V, Sniegowski K, Bers K, Lavigne R, Smalla K, Springael D.** 2013. High prevalence of IncP-1 plasmids and IS1071 insertion sequences in on-farm biopurification systems and other pesticide-polluted environments. *FEMS Microbiol Ecol* **86:**415–431.
15. **Sabate M, Navarro F, Miro E, Campoy S, Mirelis B, Barbe J, Prats G.** 2002. Novel complex sul1-type integron in *Escherichia coli* carrying bla(CTX-M-9). *Antimicrob Agents Chemother* **46:**2656–2661.
16. **Mataseje LF, Boyd DA, Lefebvre B, Bryce E, Embree J, Gravel D, Katz K, Kibsey P, Kuhn M, Langley J, Mitchell R, Roscoe D, Simor A, Taylor G, Thomas E, Turgeon N, Mulvey MR.** 2014. Complete sequences of a novel blaNDM-1-harbouring plasmid from Providencia rettgeri and an FII-type plasmid from *Klebsiella pneumoniae* identified in Canada. *J Antimicrob Chemother* **69:**637–642.
17. **Grindley ND, Whiteson KL, Rice PA.** 2006. Mechanisms of site-specific recombination. *Annu Rev Biochem* **75:**567–605.
18. **Kholodii G, Yurieva O, Mindlin S, Gorlenko Z, Rybochkin V, Nikiforov V.** 2000. Tn5044, a novel Tn3 family transposon coding for temperature-sensitive mercury resistance. *Res Microbiol* **151:**291–302.
19. **Liu CC, Huhne R, Tu J, Lorbach E, Droge P.** 1998. The resolvase encoded by *Xanthomonas campestris* transposable element ISXc5 constitutes a new subfamily closely related to DNA invertases. *Genes Cells* **3:** 221–233.
20. **Mindlin S, Kholodii G, Gorlenko Z, Minakhina S, Minakhin L, Kalyaeva E, Kopteva A, Petrova M, Yurieva O, Nikiforov V.** 2001. Mercury resistance transposons of gram-negative environmental bacteria and their classification. *Res Microbiol* **152:**811–822.
21. **Yeo CC, Tham JM, Kwong SM, Yiin S, Poh CL.** 1998. Tn5563, a transposon encoding putative mercuric ion transport proteins located on plasmid pRA2 of *Pseudomonas alcaligenes*. *FEMS Microbiol Lett* **165:**253–260.
22. **Schneider F, Schwikardi M, Muskhelishvili G, Droge P.** 2000. A DNA-binding domain swap converts the invertase gin into a resolvase. *J Mol Biol* **295:**767–775.
23. **Salvo JJ, Grindley ND.** 1988. The gamma delta resolvase bends the res site into a recombinogenic complex. *EMBO J* **7:**3609–3616.
24. **Salvo JJ, Grindley ND.** 1987. Helical phasing between DNA bends and the determination of bend direction. *Nucleic Acids Res* **15:**9771–9779.
25. **Rowland SJ, Stark WM, Boocock MR.** 2002. Sin recombinase from *Staphylococcus aureus*: synaptic complex architecture and transposon targeting. *Mol Microbiol* **44:**607–619.
26. **Petit MA, Ehrlich D, Janniere L.** 1995. pAM beta 1 resolvase has an atypical recombination site and requires a histone-like protein HU. *Mol Microbiol* **18:** 271–282.

27. **Canosa I, Lopez G, Rojo F, Boocock MR, Alonso JC.** 2003. Synapsis and strand exchange in the resolution and DNA inversion reactions catalysed by the beta recombinase. *Nucleic Acids Res* **31:**1038–1044.
28. **Rowland SJ, Boocock MR, Stark WM.** 2006. DNA bending in the Sin recombination synapse: functional replacement of HU by IHF. *Mol Microbiol* **59:**1730–1743.
29. **Rowland SJ, Boocock MR, Stark WM.** 2005. Regulation of Sin recombinase by accessory proteins. *Mol Microbiol* **56:**371–382.
30. **Mouw KW, Rowland SJ, Gajjar MM, Boocock MR, Stark WM, Rice PA.** 2008. Architecture of a serine recombinase-DNA regulatory complex. *Mol Cell* **30:** 145–155.
31. **Rice PA, Mouw KW, Montano SP, Boocock MR, Rowland SJ, Stark WM.** 2010. Orchestrating serine resolvases. *Biochem Soc Trans* **38:**384–387.
32. **Dodd HM, Bennett PM.** 1987. The R46 site-specific recombination system is a homologue of the Tn3 and gamma delta (Tn1000) cointegrate resolution system. *J Gen Microbiol* **133:**2031–2039.
33. **Radstrom P, Skold O, Swedberg G, Flensburg J, Roy PH, Sundstrom L.** 1994. Transposon Tn5090 of plasmid R751, which carries an integron, is related to Tn7, Mu, and the retroelements. *J Bacteriol* **176:**3257–3268.
34. **Kholodii GY, Mindlin SZ, Bass IA, Yurieva OV, Minakhina SV, Nikiforov VG.** 1995. Four genes, two ends, and a res region are involved in transposition of Tn5053: a paradigm for a novel family of transposons carrying either a mer operon or an integron. *Mol Microbiol* **17:**1189–1200.
35. **Rowland SJ, Dyke KG.** 1989. Characterization of the staphylococcal beta-lactamase transposon Tn552. *EMBO J* **8:**2761–2773.
36. **Mahillon J, Lereclus D.** 1988. Structural and functional analysis of Tn4430: identification of an integrase-like protein involved in the co-integrate-resolution process. *EMBO J* **7:**1515–1526.
37. **Baum JA.** 1994. Tn5401, a new class II transposable element from *Bacillus thuringiensis*. *J Bacteriol* **176:** 2835–2845.
38. **Barre F-X, Sherratt DJ.** 2002. Xer site-specific recombination: promoting chromosome segregation, p 149–161. *In* Craig NL, Craigie R, Gellert M, Lambowitz AM (ed), *Mobile DNA II*. ASM Press, Washington D.C.
39. **Colloms SD.** 2013. The topology of plasmid-monomerizing Xer site-specific recombination. *Biochem Soc Trans* **41:**589–594.
40. **Das B, Martinez E, Midonet C, Barre FX.** 2013. Integrative mobile elements exploiting Xer recombination. *Trends Microbiol* **21:**23–30.
41. **Genka H, Nagata Y, Tsuda M.** 2002. Site-specific recombination system encoded by toluene catabolic transposon Tn4651. *J Bacteriol* **184:**4757–4766.
42. **Kholodii GY, Yurieva OV, Gorlenko Z, Mindlin SZ, Bass IA, Lomovskaya OL, Kopteva AV, Nikiforov VG.** 1997. Tn5041: a chimeric mercury resistance transposon closely related to the toluene degradative transposon Tn4651. *Microbiology* **143**(Pt 8):2549–2556.
43. **Kholodii G, Gorlenko Z, Mindlin S, Hobman J, Nikiforov V.** 2002. Tn5041-like transposons: molecular diversity, evolutionary relationships and distribution of distinct variants in environmental bacteria. *Microbiology* **148:**3569–3582.
44. **Yano H, Genka H, Ohtsubo Y, Nagata Y, Top EM, Tsuda M.** 2013. Cointegrate-resolution of toluene-catabolic transposon Tn4651: determination of cross-over site and the segment required for full resolution activity. *Plasmid* **69:**24–35.
45. **Siemieniak DR, Slightom JL, Chung ST.** 1990. Nucleotide sequence of *Streptomyces fradiae* transposable element Tn4556: a class-II transposon related to Tn3. *Gene* **86:**1–9.
46. **Sota M, Yano H, Ono A, Miyazaki R, Ishii H, Genka H, Top EM, Tsuda M.** 2006. Genomic and functional analysis of the IncP-9 naphthalene-catabolic plasmid NAH7 and its transposon Tn4655 suggests catabolic gene spread by a tyrosine recombinase. *J Bacteriol* **188:**4057–4067.
47. **Vanhooff V, Galloy C, Agaisse H, Lereclus D, Revet B, Hallet B.** 2006. Self-control in DNA site-specific recombination mediated by the tyrosine recombinase TnpI. *Mol Microbiol* **60:**617–629.
48. **Baum JA.** 1995. TnpI recombinase: identification of sites within Tn5401 required for TnpI binding and site-specific recombination. *J Bacteriol* **177:**4036–4042.
49. **Clark AJ, Warren GJ.** 1979. Conjugal transmission of plasmids. *Annu Rev Genet* **13:**99–125.
50. **Guyer MS.** 1978. The gamma delta sequence of F is an insertion sequence. *J Mol Biol* **126:**347–365.
51. **Palchaudhuri S, Maas WK.** 1976. Fusion of two F-prime factors in *Escherichia coli* studied by electron microscope heteroduplex analysis. *Mol Gen Genet* **146:**215–231.
52. **Green BD, Battisti L, Thorne CB.** 1989. Involvement of Tn4430 in transfer of *Bacillus anthracis* plasmids mediated by *Bacillus thuringiensis* plasmid pXO12. *J Bacteriol* **171:**104–113.
53. **Mahillon J, Rezsohazy R, Hallet B, Delcour J.** 1994. IS231 and other *Bacillus thuringiensis* transposable elements: a review. *Genetica* **93:**13–26.
54. **Mikosa M, Sochacka-Pietal M, Baj J, Bartosik D.** 2006. Identification of a transposable genomic island of *Paracoccus pantotrophus* DSM 11072 by its transposition to a novel entrapment vector pMMB2. *Microbiology* **152:**1063–1073.
55. **Dziewit L, Baj J, Szuplewska M, Maj A, Tabin M, Czyzkowska A, Skrzypczyk G, Adamczuk M, Sitarek T, Stawinski P, Tudek A, Wanasz K, Wardal E, Piechucka E, Bartosik D.** 2012. Insights into the transposable mobilome of *Paracoccus* spp. (Alphaproteobacteria). *PLoS One* **7:**e32277.
56. **Shaw JH, Clewell DB.** 1985. Complete nucleotide sequence of macrolide-lincosamide-streptogramin B-resistance transposon Tn917 in *Streptococcus faecalis*. *J Bacteriol* **164:**782–796.
57. **Chiou CS, Jones AL.** 1993. Nucleotide sequence analysis of a transposon (Tn5393) carrying streptomycin resistance genes in *Erwinia amylovora* and other gram-negative bacteria. *J Bacteriol* **175:**732–740.

58. **Allmeier H, Cresnar B, Greck M, Schmitt R.** 1992. Complete nucleotide sequence of Tn1721: gene organization and a novel gene product with features of a chemotaxis protein. *Gene* **111:**11–20.

59. **Arthur M, Molinas C, Depardieu F, Courvalin P.** 1993. Characterization of Tn1546, a Tn3-related transposon conferring glycopeptide resistance by synthesis of depsipeptide peptidoglycan precursors in *Enterococcus faecium* BM4147. *J Bacteriol* **175:**117–127.

60. **Mindlin S, Minakhin L, Petrova M, Kholodii G, Minakhina S, Gorlenko Z, Nikiforov V.** 2005. Present-day mercury resistance transposons are common in bacteria preserved in permafrost grounds since the Upper Pleistocene. *Res Microbiol* **156:**994–1004.

61. **Mindlin S, Petrova M.** 2013. Mercury resistance transposons, p 33–52. *In* Roberts AP, Mullany P (ed), *Bacterial Integrative Genetic Elements.* Landes Bioscience.

62. **Canton R, Coque TM.** 2006. The CTX-M beta-lactamase pandemic. *Curr Opin Microbiol* **9:**466–475.

63. **Van der Auwera G, Mahillon J.** 2005. TnXO1, a germination-associated class II transposon from *Bacillus anthracis. Plasmid* **53:**251–257.

64. **Okinaka RT, Cloud K, Hampton O, Hoffmaster AR, Hill KK, Keim P, Koehler TM, Lamke G, Kumano S, Mahillon J, Manter D, Martinez Y, Ricke D, Svensson R, Jackson PJ.** 1999. Sequence and organization of pXO1, the large *Bacillus anthracis* plasmid harboring the anthrax toxin genes. *J Bacteriol* **181:**6509–6515.

65. **Landgraf A, Weingart H, Tsiamis G, Boch J.** 2006. Different versions of *Pseudomonas syringae* pv. tomato DC3000 exist due to the activity of an effector transposon. *Mol Plant Pathol* **7:**355–364.

66. **Schluter A, Heuer H, Szczepanowski R, Poler SM, Schneiker S, Puhler A, Top EM.** 2005. Plasmid pB8 is closely related to the prototype IncP-1beta plasmid R751 but transfers poorly to *Escherichia coli* and carries a new transposon encoding a small multidrug resistance efflux protein. *Plasmid* **54:**135–148.

67. **Szuplewska M, Ludwiczak M, Lyzwa K, Czarnecki J, Bartosik D.** 2014. Mobility and generation of mosaic non-autonomous transposons by Tn3-derived inverted-repeat miniature elements (TIMEs). *PLoS One* **9:**e105010.

68. **Naas T, Cuzon G, Villegas MV, Lartigue MF, Quinn JP, Nordmann P.** 2008. Genetic structures at the origin of acquisition of the beta-lactamase bla KPC gene. *Antimicrob Agents Chemother* **52:**1257–1263.

69. **Cuzon G, Naas T, Nordmann P.** 2011. Functional characterization of Tn4401, a Tn3-based transposon involved in blaKPC gene mobilization. *Antimicrob Agents Chemother* **55:**5370–5373.

70. **Naas T, Cuzon G, Truong HV, Nordmann P.** 2012. Role of ISKpn7 and deletions in blaKPC gene expression. *Antimicrob Agents Chemother* **56:**4753–4759.

71. **Grinsted J, de la Cruz F, Schmitt R.** 1990. The Tn21 subgroup of bacterial transposable elements. *Plasmid* **24:**163–189.

72. **Avila P, Grinsted J, de la Cruz F.** 1988. Analysis of the variable endpoints generated by one-ended transposition of Tn21. *J Bacteriol* **170:**1350–1353.

73. **Heritage J, Bennett PM.** 1985. Plasmid fusions mediated by one end of TnA. *J Gen Microbiol* **131:**1130–1140.

74. **Motsch S, Schmitt R, Avila P, de la Cruz F, Ward E, Grinsted J.** 1985. Junction sequences generated by 'one-ended transposition'. *Nucleic Acids Res* **13:**3335–3342.

75. **Motsch S, Schmitt R.** 1984. Replicon fusion mediated by a single-ended derivative of transposon Tn1721. *Mol Gen Genet* **195:**281–287.

76. **Revilla C, Garcillan-Barcia MP, Fernandez-Lopez R, Thomson NR, Sanders M, Cheung M, Thomas CM, de la Cruz F.** 2008. Different pathways to acquiring resistance genes illustrated by the recent evolution of IncW plasmids. *Antimicrob Agents Chemother* **52:**1472–1480.

77. **Sota M, Endo M, Nitta K, Kawasaki H, Tsuda M.** 2002. Characterization of a class II defective transposon carrying two haloacetate dehalogenase genes from *Delftia acidovorans* plasmid pUO1. *Appl Environ Microbiol* **68:**2307–2315.

78. **Yano H, Garruto CE, Sota M, Ohtsubo Y, Nagata Y, Zylstra GJ, Williams PA, Tsuda M.** 2007. Complete sequence determination combined with analysis of transposition/site-specific recombination events to explain genetic organization of IncP-7 TOL plasmid pWW53 and related mobile genetic elements. *J Mol Biol* **369:**11–26.

79. **Yano H, Miyakoshi M, Ohshima K, Tabata M, Nagata Y, Hattori M, Tsuda M.** 2010. Complete nucleotide sequence of TOL plasmid pDK1 provides evidence for evolutionary history of IncP-7 catabolic plasmids. *J Bacteriol* **192:**4337–4347.

80. **Cornelis G, Sommer H, Saedler H.** 1981. Transposon Tn951 (TnLac) is defective and related to Tn3. *Mol Gen Genet* **184:**241–248.

81. **Petrovski S, Stanisich VA.** 2011. Embedded elements in the IncPbeta plasmids R772 and R906 can be mobilized and can serve as a source of diverse and novel elements. *Microbiology* **157:**1714–1725.

82. **Gillings MR.** 2014. Integrons: past, present, and future. *Microbiol Mol Biol Rev* **78:**257–277.

83. **Mazel D.** 2006. Integrons: agents of bacterial evolution. *Nat Rev Microbiol* **4:**608–620.

84. **Boucher Y, Labbate M, Koenig JE, Stokes HW.** 2007. Integrons: mobilizable platforms that promote genetic diversity in bacteria. *Trends Microbiol* **15:**301–309.

85. **Partridge SR.** 2011. Analysis of antibiotic resistance regions in Gram-negative bacteria. *FEMS Microbiol Rev* **35:**820–855.

86. **Liebert CA, Hall RM, Summers AO.** 1999. Transposon Tn21, flagship of the floating genome. *Microbiol Mol Biol Rev* **63:**507–522.

87. **Partridge SR, Tsafnat G, Coiera E, Iredell JR.** 2009. Gene cassettes and cassette arrays in mobile resistance integrons. *FEMS Microbiol Rev* **33:**757–784.

88. **Gillings MR, Stokes HW.** 2012. Are humans increasing bacterial evolvability? *Trends Ecol Evol* **27:**346–352.

89. **Labbate M, Roy CP, Stokes HW.** 2008. A class 1 integron present in a human commensal has a hybrid transposition module compared to Tn402: evidence of interaction with mobile DNA from natural environments. *J Bacteriol* **190:**5318–5327.

90. **Sajjad A, Holley MP, Labbate M, Stokes HW, Gillings MR.** 2011. Preclinical class 1 integron with a complete Tn402-like transposition module. *Appl Environ Microbiol* **77:**335–337.

91. **Minakhina S, Kholodii G, Mindlin S, Yurieva O, Nikiforov V.** 1999. Tn5053 family transposons are res site hunters sensing plasmidal res sites occupied by cognate resolvases. *Mol Microbiol* **33:**1059–1068.

92. **Petrovski S, Stanisich VA.** 2010. Tn502 and Tn512 are res site hunters that provide evidence of resolvase-independent transposition to random sites. *J Bacteriol* **192:**1865–1874.

93. **Kamali-Moghaddam M, Sundstrom L.** 2000. Transposon targeting determined by resolvase. *FEMS Microbiol Lett* **186:**55–59.

94. **Toleman MA, Bennett PM, Walsh TR.** 2006. ISCR elements: novel gene-capturing systems of the 21st century? *Microbiol Mol Biol Rev* **70:**296–316.

95. **Zhao WH, Hu ZQ.** 2013. Epidemiology and genetics of CTX-M extended-spectrum beta-lactamases in Gram-negative bacteria. *Crit Rev Microbiol* **39:**79–101.

96. **Poirel L, Lartigue MF, Decousser JW, Nordmann P.** 2005. ISEcp1B-mediated transposition of blaCTX-M in *Escherichia coli*. *Antimicrob Agents Chemother* **49:** 447–450.

97. **Garcillán-Barcia M, Bernales I, Mendiola M, de la Cruz F.** 2002. IS91 rolling circle transposition, p 891–904. *In* Craig NL, Craigie R, Gellert M, Lambowitz AM (ed), *Mobile DNA II*. ASM Press, Washington D.C.

98. **Poirel L, Decousser JW, Nordmann P.** 2003. Insertion sequence ISEcp1B is involved in expression and mobilization of a bla(CTX-M) beta-lactamase gene. *Antimicrob Agents Chemother* **47:**2938–2945.

99. **Zong Z, Yu R, Wang X, Lu X.** 2011. blaCTX-M-65 is carried by a Tn1722-like element on an IncN conjugative plasmid of ST131 *Escherichia coli*. *J Med Microbiol* **60:**435–441.

100. **Soler Bistue AJ, Martin FA, Petroni A, Faccone D, Galas M, Tolmasky ME, Zorreguieta A.** 2006. *Vibrio cholerae* InV117, a class 1 integron harboring aac(6′)-Ib and blaCTX-M-2, is linked to transposition genes. *Antimicrob Agents Chemother* **50:**1903–1907.

101. **Valverde A, Canton R, Galan JC, Nordmann P, Baquero F, Coque TM.** 2006. In117, an unusual In0-like class 1 integron containing CR1 and bla(CTX-M-2) and associated with a Tn21-like element. *Antimicrob Agents Chemother* **50:**799–802.

102. **Novais A, Canton R, Valverde A, Machado E, Galan JC, Peixe L, Carattoli A, Baquero F, Coque TM.** 2006. Dissemination and persistence of blaCTX-M-9 are linked to class 1 integrons containing CR1 associated with defective transposon derivatives from Tn402 located in early antibiotic resistance plasmids of IncHI2, IncP1-alpha, and IncFI groups. *Antimicrob Agents Chemother* **50:**2741–2750.

103. **Partridge SR, Hall RM.** 2004. Complex multiple antibiotic and mercury resistance region derived from the r-det of NR1 (R100). *Antimicrob Agents Chemother* **48:**4250–4255.

104. **Reed RR.** 1981. Resolution of cointegrates between transposons gamma delta and Tn3 defines the recombination site. *Proc Natl Acad Sci USA* **78:**3428–3432.

105. **Kholodii G.** 2001. The shuffling function of resolvases. *Gene* **269:**121–130.

106. **Michiels T, Cornelis G.** 1989. Site-specific recombinations between direct and inverted res sites of Tn2501. *Plasmid* **22:**249–255.

107. **Miller CA, Cohen SN.** 1980. F plasmid provides a function that promotes recA-independent site-specific fusions of pSC101 replicon. *Nature* **285:**577–579.

108. **Ishizaki K, Ohtsubo E.** 1985. Cointegration and resolution mediated by IS101 present in plasmid pSC101. *Mol Gen Genet* **199:**388–395.

109. **Filee J, Siguier P, Chandler M.** 2007. Insertion sequence diversity in archaea. *Microbiol Mol Biol Rev* **71:**121–157.

110. **Fattash I, Rooke R, Wong A, Hui C, Luu T, Bhardwaj P, Yang G.** 2013. Miniature inverted-repeat transposable elements: discovery, distribution, and activity. *Genome* **56:**475–486.

111. **Peters M, Heinaru E, Talpsep E, Wand H, Stottmeister U, Heinaru A, Nurk A.** 1997. Acquisition of a deliberately introduced phenol degradation operon, pheBA, by different indigenous *Pseudomonas* species. *Appl Environ Microbiol* **63:**4899–4906.

112. **Siguier P, Gourbeyre E, Chandler M.** 2014. Bacterial insertion sequences: their genomic impact and diversity. *FEMS Microbiol Rev* **38:**865–891.

113. **De Palmenaer D, Siguier P, Mahillon J.** 2008. IS4 family goes genomic 1. *Bmc Evolutionary Biology* **8:**18.

114. **Shapiro JA.** 1979. Molecular model for the transposition and replication of bacteriophage Mu and other transposable elements. *Proc Natl Acad Sci USA* **76:**1933–1937.

115. **Harshey RM, Bukhari AI.** 1981. A mechanism of DNA transposition. *Proc Natl Acad Sci USA* **78:**1090–1094.

116. **Galas DJ, Chandler M.** 1981. On the molecular mechanisms of transposition. *Proc Natl Acad Sci USA* **78:** 4858–4862.

117. **Chaconas G, Harshey RM.** 2002. Transposition of Phage Mu DNA, p 384–402. *In* Craig NL, Craigie R, Gellert M, Lambowitz AM (ed), *Mobile DNA II*. ASM Press, Washington D.C.

118. **Mizuuchi K.** 1992. Transpositional recombination: mechanistic insights from studies of mu and other elements. *Annu Rev Biochem* **61:**1011–1051.

119. **Madison KE, Abdelmeguid MR, Jones-Foster EN, Nakai H.** 2012. A new role for translation initiation factor 2 in maintaining genome integrity. *PLoS Genet* **8:** e1002648.

120. **Lambin M, Nicolas E, Oger CA, Nguyen N, Prozzi D, Hallet B.** 2012. Separate structural and functional domains of Tn4430 transposase contribute to target immunity. *Mol Microbiol* **83:**805–820.

121. **Nowotny M.** 2009. Retroviral integrase superfamily: the structural perspective. *EMBO Rep* **10:**144–151.

122. **Hickman AB, Chandler M, Dyda F.** 2010. Integrating prokaryotes and eukaryotes: DNA transposases in light of structure. *Crit Rev Biochem Mol Biol* **45:**50–69.

123. **Montano SP, Rice PA.** 2011. Moving DNA around: DNA transposition and retroviral integration. *Curr Opin Struct Biol* **21**:370–378.

124. **Yang W, Lee JY, Nowotny M.** 2006. Making and breaking nucleic acids: two-Mg2+-ion catalysis and substrate specificity. *Mol Cell* **22**:5–13.

125. **Davies DR, Mahnke BL, Reznikoff WS, Rayment I.** 1999. The three-dimensional structure of a Tn5 transposase-related protein determined to 2.9-A resolution. *J Biol Chem* **274**:11904–11913.

126. **Davies DR, Goryshin IY, Reznikoff WS, Rayment I.** 2000. Three-dimensional structure of the Tn5 synaptic complex transposition intermediate. *Science* **289**: 77–85.

127. **Hickman AB, Perez ZN, Zhou L, Musingarimi P, Ghirlando R, Hinshaw JE, Craig NL, Dyda F.** 2005. Molecular architecture of a eukaryotic DNA transposase. *Nat Struct Mol Biol* **12**:715–721.

128. **Tanaka Y, Nureki O, Kurumizaka H, Fukai S, Kawaguchi S, Ikuta M, Iwahara J, Okazaki T, Yokoyama S.** 2001. Crystal structure of the CENP-B protein-DNA complex: the DNA-binding domains of CENP-B induce kinks in the CENP-B box DNA. *EMBO J* **20**:6612–6618.

129. **Evans LR, Brown NL.** 1987. Construction of hybrid Tn501/Tn21 transposases in vivo: identification of a region of transposase conferring specificity of recognition of the 38-bp terminal inverted repeats. *EMBO J* **6**: 2849–2853.

130. **Maekawa T, Amemura-Maekawa J, Ohtsubo E.** 1993. DNA binding domains in Tn3 transposase. *Mol Gen Genet* **236**:267–274.

131. **Richardson JM, Colloms SD, Finnegan DJ, Walkinshaw MD.** 2009. Molecular architecture of the Mos1 paired-end complex: the structural basis of DNA transposition in a eukaryote. *Cell* **138**:1096–1108.

132. **Watkins S, van Pouderoyen G, Sixma TK.** 2004. Structural analysis of the bipartite DNA-binding domain of Tc3 transposase bound to transposon DNA. *Nucleic Acids Res* **32**:4306–4312.

133. **Montano SP, Pigli YZ, Rice PA.** 2012. The mu transpososome structure sheds light on DDE recombinase evolution. *Nature* **491**:413–417.

134. **Casola C, Hucks D, Feschotte C.** 2008. Convergent domestication of pogo-like transposases into centromere-binding proteins in fission yeast and mammals. *Mol Biol Evol* **25**:29–41.

135. **Claeys BC, Walker N, Liu D, Chalmers R.** 2014. Crosstalk between transposase subunits during cleavage of the mariner transposon. *Nucleic Acids Res* **42**: 5799–5808.

136. **Liu D, Chalmers R.** 2014. Hyperactive mariner transposons are created by mutations that disrupt allosterism and increase the rate of transposon end synapsis. *Nucleic Acids Res* **42**:2637–2645.

137. **Amemura-Maekawa J, Ohtsubo E.** 1991. Functional analysis of the two domains in the terminal inverted repeat sequence required for transposition of Tn3. *Gene* **103**:11–16.

138. **Kans JA, Casadaban MJ.** 1989. Nucleotide sequences required for Tn3 transposition immunity. *J Bacteriol* **171**:1904–1914.

139. **May EW, Grindley ND.** 1995. A functional analysis of the inverted repeat of the gamma delta transposable element. *J Mol Biol* **247**:578–587.

140. **Nissley DV, Lindh F, Fennewald MA.** 1991. Mutations in the inverted repeats of Tn3 affect binding of transposase and transposition immunity. *J Mol Biol* **218**:335–347.

141. **Wiater LA, Grindley ND.** 1991. Gamma delta transposase. Purification and analysis of its interaction with a transposon end. *J Biol Chem* **266**:1841–1849.

142. **Ichikawa H, Ikeda K, Amemura J, Ohtsubo E.** 1990. Two domains in the terminal inverted-repeat sequence of transposon Tn3. *Gene* **86**:11–17.

143. **New JH, Eggleston AK, Fennewald M.** 1988. Binding of the Tn3 transposase to the inverted repeats of Tn3. *J Mol Biol* **201**:589–599.

144. **Ichikawa H, Ikeda K, Wishart WL, Ohtsubo E.** 1987. Specific binding of transposase to terminal inverted repeats of transposable element Tn3. *Proc Natl Acad Sci USA* **84**:8220–8224.

145. **Wiater LA, Grindley ND.** 1988. Gamma delta transposase and integration host factor bind cooperatively at both ends of gamma delta. *EMBO J* **7**:1907–1911.

146. **Ilves H, Horak R, Teras R, Kivisaar M.** 2004. IHF is the limiting host factor in transposition of *Pseudomonas putida* transposon Tn4652 in stationary phase. *Mol Microbiol* **51**:1773–1785.

147. **Teras R, Jakovleva J, Kivisaar M.** 2009. Fis negatively affects binding of Tn4652 transposase by out-competing IHF from the left end of Tn4652. *Microbiology* **155**: 1203–1214.

148. **Baum JA, Gilmer AJ, Light Mettus AM.** 1999. Multiple roles for TnpI recombinase in regulation of Tn5401 transposition in *Bacillus thuringiensis*. *J Bacteriol* **181**: 6271–6277.

149. **Amemura J, Ichikawa H, Ohtsubo E.** 1990. Tn3 transposition immunity is conferred by the transposase-binding domain in the terminal inverted-repeat sequence of Tn3. *Gene* **88**:21–24.

150. **Wiater LA, Grindley ND.** 1990. Uncoupling of transpositional immunity from gamma delta transposition by a mutation at the end of gamma delta. *J Bacteriol* **172**: 4959–4963.

151. **Schumacher S, Clubb RT, Cai M, Mizuuchi K, Clore GM, Gronenborn AM.** 1997. Solution structure of the Mu end DNA-binding ibeta subdomain of phage Mu transposase: modular DNA recognition by two tethered domains. *EMBO J* **16**:7532–7541.

152. **Sota M, Yano H, Nagata Y, Ohtsubo Y, Genka H, Anbutsu H, Kawasaki H, Tsuda M.** 2006. Functional analysis of unique class II insertion sequence IS1071. *Appl Environ Microbiol* **72**:291–297.

153. **Dillon SC, Dorman CJ.** 2010. Bacterial nucleoid-associated proteins, nucleoid structure and gene expression. *Nat Rev Microbiol* **8**:185–195.

154. **Browning DF, Grainger DC, Busby SJ.** 2010. Effects of nucleoid-associated proteins on bacterial chromosome structure and gene expression. *Curr Opin Microbiol* **13:** 773–780.

155. **Wiater LA, Grindley ND.** 1990. Integration host factor increases the transpositional immunity conferred by gamma delta ends. *J Bacteriol* **172:**4951–4958.

156. **Horak R, Kivisaar M.** 1998. Expression of the transposase gene tnpA of Tn4652 is positively affected by integration host factor. *J Bacteriol* **180:**2822–2829.

157. **Ilves H, Horak R, Kivisaar M.** 2001. Involvement of sigma(S) in starvation-induced transposition of *Pseudomonas putida* transposon Tn4652. *J Bacteriol* **183:** 5445–5448.

158. **Horak R, Ilves H, Pruunsild P, Kuljus M, Kivisaar M.** 2004. The ColR-ColS two-component signal transduction system is involved in regulation of Tn4652 transposition in *Pseudomonas putida* under starvation conditions. *Mol Microbiol* **54:**795–807.

159. **Horak R, Kivisaar M.** 1999. Regulation of the transposase of Tn4652 by the transposon-encoded protein TnpC. *J Bacteriol* **181:**6312–6318.

160. **Craig NL.** 1997. Target site selection in transposition. *Annu Rev Biochem* **66:**437–474.

161. **Grinsted J, Bennett PM, Higginson S, Richmond MH.** 1978. Regional preference of insertion of Tn501 and Tn802 into RP1 and its derivatives. *Mol Gen Genet* **166:**313–320.

162. **Tu CP, Cohen SN.** 1980. Translocation specificity of the Tn3 element: characterization of sites of multiple insertions. *Cell* **19:**151–160.

163. **Heffron F.** 1983. Tn3 and its relatives, p 223–260. *In* Shapiro JA (ed), *Mobile Genetic Elements*. Academic Press Inc., New York.

164. **Liu L, Whalen W, Das A, Berg CM.** 1987. Rapid sequencing of cloned DNA using a transposon for bidirectional priming: sequence of the *Escherichia coli* K-12 avtA gene. *Nucleic Acids Res* **15:**9461–9469.

165. **Davies CJ, Hutchison CA III.** 1995. Insertion site specificity of the transposon Tn3. *Nucleic Acids Res* **23:** 507–514.

166. **Seringhaus M, Kumar A, Hartigan J, Snyder M, Gerstein M.** 2006. Genomic analysis of insertion behavior and target specificity of mini-Tn7 and Tn3 transposons in *Saccharomyces cerevisiae*. *Nucleic Acids Res* **34:**e57.

167. **Kumar A, Seringhaus M, Biery MC, Sarnovsky RJ, Umansky L, Piccirillo S, Heidtman M, Cheung KH, Dobry CJ, Gerstein MB, Craig NL, Snyder M.** 2004. Large-scale mutagenesis of the yeast genome using a Tn7-derived multipurpose transposon. *Genome Res* **14:** 1975–1986.

168. **Kivistik PA, Kivisaar M, Horak R.** 2007. Target site selection of *Pseudomonas putida* transposon Tn4652. *J Bacteriol* **189:**3918–3921.

169. **Hallet B, Rezsohazy R, Mahillon J, Delcour J.** 1994. IS231A insertion specificity: consensus sequence and DNA bending at the target site. *Mol Microbiol* **14:** 131–139.

170. **Pribil PA, Haniford DB.** 2003. Target DNA bending is an important specificity determinant in target site selection in Tn10 transposition. *J Mol Biol* **330:**247–259.

171. **Liu G, Geurts AM, Yae K, Srinivasan AR, Fahrenkrug SC, Largaespada DS, Takeda J, Horie K, Olson WK, Hackett PB.** 2005. Target-site preferences of Sleeping Beauty transposons. *J Mol Biol* **346:**161–173.

172. **Garsin DA, Urbach J, Huguet-Tapia JC, Peters JE, Ausubel FM.** 2004. Construction of an *Enterococcus faecalis* Tn917-mediated-gene-disruption library offers insight into Tn917 insertion patterns. *J Bacteriol* **186:** 7280–7289.

173. **Nicolas E, Lambin M, Hallet B.** 2010. Target immunity of the Tn3-family transposon Tn4430 requires specific interactions between the transposase and the terminal inverted repeats of the transposon. *J Bacteriol* **192:** 4233–4238.

174. **Sota M, Tsuda M, Yano H, Suzuki H, Forney LJ, Top EM.** 2007. Region-specific insertion of transposons in combination with selection for high plasmid transferability and stability accounts for the structural similarity of IncP-1 plasmids. *J Bacteriol* **189:**3091–3098.

175. **Slater JD, Allen AG, May JP, Bolitho S, Lindsay H, Maskell DJ.** 2003. Mutagenesis of *Streptococcus equi* and *Streptococcus suis* by transposon Tn917. *Vet Microbiol* **93:**197–206.

176. **Shi Q, Huguet-Tapia JC, Peters JE.** 2009. Tn917 targets the region where DNA replication terminates in *Bacillus subtilis*, highlighting a difference in chromosome processing in the firmicutes. *J Bacteriol* **191:**7623–7627.

177. **Rudolph CJ, Upton AL, Stockum A, Nieduszynski CA, Lloyd RG.** 2013. Avoiding chromosome pathology when replication forks collide. *Nature* **500:**608–611.

178. **Wolkow CA, Deboy RT, Craig NL.** 1996. Conjugating plasmids are preferred targets for Tn7. *Genes Dev* **10:** 2145–2157.

179. **Peters JE, Craig NL.** 2000. Tn7 transposes proximal to DNA double-strand breaks and into regions where chromosomal DNA replication terminates. *Mol Cell* **6:** 573–582.

180. **Peters JE, Craig NL.** 2001. Tn7 recognizes transposition target structures associated with DNA replication using the DNA-binding protein TnsE. *Genes Dev* **15:**737–747.

181. **Hu WY, Derbyshire KM.** 1998. Target choice and orientation preference of the insertion sequence IS903. *J Bacteriol* **180:**3039–3048.

182. **Swingle B, O'Carroll M, Haniford D, Derbyshire KM.** 2004. The effect of host-encoded nucleoid proteins on transposition: H-NS influences targeting of both IS903 and Tn10. *Mol Microbiol* **52:**1055–1067.

183. **Ton-Hoang B, Pasternak C, Siguier P, Guynet C, Hickman AB, Dyda F, Sommer S, Chandler M.** 2010. Single-stranded DNA transposition is coupled to host replication. *Cell* **142:**398–408.

184. **Gomez MJ, Diaz-Maldonado H, Gonzalez-Tortuero E, Lopez de Saro FJ.** 2014. Chromosomal replication dynamics and interaction with the beta sliding clamp determine orientation of bacterial transposable elements. *Genome Biol Evol* **6:**727–740.

185. **Fricker AD, Peters JE.** 2014. Vulnerabilities on the lagging-strand template: opportunities for mobile elements. *Annu Rev Genet* **48:**167–186.

186. **Parks AR, Li Z, Shi Q, Owens RM, Jin MM, Peters JE.** 2009. Transposition into replicating DNA occurs through interaction with the processivity factor. *Cell* **138:**685–695.

187. **Lambowitz AM, Zimmerly S.** 2011. Group II introns: mobile ribozymes that invade DNA. *Cold Spring Harb Perspect Biol* **3:**a003616.

188. **Muster CJ, Shapiro JA, MacHattie LA.** 1983. Recombination involving transposable elements: role of target molecule replication in Tn1 delta Ap-mediated replicon fusion. *Proc Natl Acad Sci USA* **80:**2314–2317.

189. **Kretschmer PJ, Cohen SN.** 1977. Selected translocation of plasmid genes: frequency and regional specificity of translocation of the Tn3 element. *J Bacteriol* **130:** 888–99.

190. **Muster CJ, Shapiro JA.** 1981. Recombination involving transposable elements: on replicon fusion. *Cold Spring Harb Symp Quant Biol* **45**(Pt 1):239–242.

191. **Manna D, Higgins NP.** 1999. Phage Mu transposition immunity reflects supercoil domain structure of the chromosome. *Mol Microbiol* **32:**595–606.

192. **Deboy RT, Craig NL.** 1996. Tn7 transposition as a probe of cis interactions between widely separated (190 kilobases apart) DNA sites in the *Escherichia coli* chromosome. *J Bacteriol* **178:**6184–6191.

193. **Huang CJ, Heffron F, Twu JS, Schloemer RH, Lee CH.** 1986. Analysis of Tn3 sequences required for transposition and immunity. *Gene* **41:**23–31.

194. **Robinson MK, Bennett PM, Richmond MH.** 1977. Inhibition of TnA translocation by TnA. *J Bacteriol* **129:** 407–414.

195. **Lee CH, Bhagwat A, Heffron F.** 1983. Identification of a transposon Tn3 sequence required for transposition immunity. *Proc Natl Acad Sci USA* **80:**6765–6769.

196. **Adzuma K, Mizuuchi K.** 1988. Target immunity of Mu transposition reflects a differential distribution of Mu B protein. *Cell* **53:**257–266.

197. **Darzins A, Kent NE, Buckwalter MS, Casadaban MJ.** 1988. Bacteriophage Mu sites required for transposition immunity. *Proc Natl Acad Sci USA* **85:**6826–6830.

198. **Hauer B, Shapiro JA.** 1984. Control of Tn7 transposition. *Mol Gen Genet* **194:**149–158.

199. **Arciszewska LK, Drake D, Craig NL.** 1989. Transposon Tn7. cis-Acting sequences in transposition and transposition immunity. *J Mol Biol* **207:**35–52.

200. **Stellwagen AE, Craig NL.** 1997. Avoiding self: two Tn7-encoded proteins mediate target immunity in Tn7 transposition. *EMBO J* **16:**6823–6834.

201. **Groenen MA, van de Putte P.** 1986. Analysis of the ends of bacteriophage Mu using site-directed mutagenesis. *J Mol Biol* **189:**597–602.

202. **Greene EC, Mizuuchi K.** 2002. Target immunity during Mu DNA transposition. Transpososome assembly and DNA looping enhance MuA-mediated disassembly of the MuB target complex. *Mol Cell* **10:**1367–1378.

203. **Arthur A, Nimmo E, Hettle S, Sherratt D.** 1984. Transposition and transposition immunity of transposon Tn3 derivatives having different ends. *EMBO J* **3:**1723–1729.

204. **Li Z, Craig NL, Peters JE.** 2013. Transposon Tn7, p 1–32. *In* Roberts AP, Mullany P (ed), *Bacterial Integrative Mobile Genetic Elements*. ASM Press, Austin, Texas.

205. **Peters JE, Craig NL.** 2001. Tn7: smarter than we thought. *Nat Rev Mol Cell Biol* **2:**806–814.

206. **Harshey RM.** 2012. The Mu story: how a maverick phage moved the field forward. *Mob DNA* **3:**21.

207. **Baker TA, Mizuuchi M, Mizuuchi K.** 1991. MuB protein allosterically activates strand transfer by the transposase of phage Mu. *Cell* **65:**1003–1013.

208. **Stellwagen AE, Craig NL.** 1997. Gain-of-function mutations in TnsC, an ATP-dependent transposition protein that activates the bacterial transposon Tn7. *Genetics* **145:**573–585.

209. **Naigamwalla DZ, Chaconas G.** 1997. A new set of Mu DNA transposition intermediates: alternate pathways of target capture preceding strand transfer. *EMBO J* **16:** 5227–5234.

210. **Skelding Z, Queen-Baker J, Craig NL.** 2003. Alternative interactions between the Tn7 transposase and the Tn7 target DNA binding protein regulate target immunity and transposition. *EMBO J* **22:**5904–5917.

211. **Mizuuchi M, Mizuuchi K.** 1993. Target site selection in transposition of phage Mu. *Cold Spring Harb Symp Quant Biol* **58:**515–523.

212. **Greene EC, Mizuuchi K.** 2002. Direct observation of single MuB polymers: evidence for a DNA-dependent conformational change for generating an active target complex. *Mol Cell* **9:**1079–1089.

213. **Tan X, Mizuuchi M, Mizuuchi K.** 2007. DNA transposition target immunity and the determinants of the MuB distribution patterns on DNA. *Proc Natl Acad Sci USA* **104:**13925–13929.

214. **Stellwagen AE, Craig NL.** 2001. Analysis of gain-of-function mutants of an ATP-dependent regulator of Tn7 transposition. *J Mol Biol* **305:**633–642.

215. **Han YW, Mizuuchi K.** 2010. Phage Mu transposition immunity: protein pattern formation along DNA by a diffusion-ratchet mechanism. *Mol Cell* **39:**48–58.

216. **Bukhari AI.** 1976. Bacteriophage mu as a transposition element. *Annu Rev Genet* **10:**389–412.

217. **Ge J, Lou Z, Harshey RM.** 2010. Immunity of replicating Mu to self-integration: a novel mechanism employing MuB protein. *Mob DNA* **1:**8.

218. **Bishop R, Sherratt D.** 1984. Transposon Tn1 intramolecular transposition. *Mol Gen Genet* **196:**117–122.

219. **Murata M, Uchida T, Yang Y, Lezhava A, Kinashi H.** 2011. A large inversion in the linear chromosome of *Streptomyces griseus* caused by replicative transposition of a new Tn3 family transposon. *Arch Microbiol* **193:** 299–306.

220. **Ichikawa H, Ohtsubo E.** 1990. *In vitro* transposition of transposon Tn3. *J Biol Chem* **265:**18829–18832.

221. **Maekawa T, Yanagihara K, Ohtsubo E.** 1996. A cell-free system of Tn3 transposition and transposition immunity. *Genes Cells* **1:**1007–1016.

222. **Maekawa T, Yanagihara K, Ohtsubo E.** 1996. Specific nicking at the 3′ ends of the terminal inverted repeat sequences in transposon Tn3 by transposase and an *E. coli* protein ACP. *Genes Cells* **1:**1017–1030.

223. **Olorunniji FJ, Stark WM.** 2010. Catalysis of site-specific recombination by Tn3 resolvase. *Biochem Soc Trans* **38:** 417–421.

224. **Stark WM, Boocock MR, Olorunniji FJ, Rowland SJ.** 2011. Intermediates in serine recombinase-mediated site-specific recombination. *Biochem Soc Trans* **39:**617–622.

225. **Merickel SK, Haykinson MJ, Johnson RC.** 1998. Communication between Hin recombinase and Fis regulatory subunits during coordinate activation of Hin-catalyzed site-specific DNA inversion. *Genes Dev* **12:**2803–2816.

226. **Rice PA, Steitz TA.** 1994. Model for a DNA-mediated synaptic complex suggested by crystal packing of gamma delta resolvase subunits. *EMBO J* **13:**1514–1524.

227. **Stark WM, Sherratt DJ, Boocock MR.** 1989. Site-specific recombination by Tn3 resolvase: topological changes in the forward and reverse reactions. *Cell* **58:** 779–790.

228. **Burke ME, Arnold PH, He J, Wenwieser SV, Rowland SJ, Boocock MR, Stark WM.** 2004. Activating mutations of Tn3 resolvase marking interfaces important in recombination catalysis and its regulation. *Mol Microbiol* **51:**937–948.

229. **McIlwraith MJ, Boocock MR, Stark WM.** 1997. Tn3 resolvase catalyses multiple recombination events without intermediate rejoining of DNA ends. *J Mol Biol* **266:**108–121.

230. **Li W, Kamtekar S, Xiong Y, Sarkis GJ, Grindley ND, Steitz TA.** 2005. Structure of a synaptic gammadelta resolvase tetramer covalently linked to two cleaved DNAs. *Science* **309:**1210–1215.

231. **Kamtekar S, Ho RS, Cocco MJ, Li W, Wenwieser SV, Boocock MR, Grindley ND, Steitz TA.** 2006. Implications of structures of synaptic tetramers of gamma delta resolvase for the mechanism of recombination. *Proc Natl Acad Sci USA* **103:**10642–10647.

232. **Van Duyne GD.** 2001. A structural view of cre-loxp site-specific recombination. *Annu Rev Biophys Biomol Struct* **30:**87–104.

233. **Chen Y, Rice PA.** 2003. New insight into site-specific recombination from Flp recombinase-DNA structures. *Annu Rev Biophys Biomol Struct* **32:**135–159.

234. **Radman-Livaja M, Biswas T, Ellenberger T, Landy A, Aihara H.** 2006. DNA arms do the legwork to ensure the directionality of lambda site-specific recombination. *Curr Opin Struct Biol* **16:**42–50.

235. **Biswas T, Aihara H, Radman-Livaja M, Filman D, Landy A, Ellenberger T.** 2005. A structural basis for allosteric control of DNA recombination by lambda integrase. *Nature* **435:**1059–1066.

236. **Guo F, Gopaul DN, Van Duyne GD.** 1997. Structure of Cre recombinase complexed with DNA in a site-specific recombination synapse. *Nature* **389:**40–46.

237. **Hallet B, Arciszewska LK, Sherratt DJ.** 1999. Reciprocal control of catalysis by the tyrosine recombinases XerC and XerD: an enzymatic switch in site-specific recombination. *Mol Cell* **4:**949–959.

238. **Chen Y, Narendra U, Iype LE, Cox MM, Rice PA.** 2000. Crystal structure of a Flp recombinase-Holliday junction complex: assembly of an active oligomer by helix swapping. *Mol Cell* **6:**885–897.

239. **Lee L, Sadowski PD.** 2005. Strand selection by the tyrosine recombinases. *Prog Nucleic Acid Res Mol Biol* **80:** 1–42.

240. **Stark WM, Boocock MR.** 1995. Topological selectivity in site-specific recombination, p 223–260. *In* Sherratt DJ (ed), *Mobile Genetic Elements.* IRL Press, Oxford, UK.

241. **Kilbride E, Boocock MR, Stark WM.** 1999. Topological selectivity of a hybrid site-specific recombination system with elements from Tn3 res/resolvase and bacteriophage P1 loxP/Cre. *J Mol Biol* **289:**1219–1230.

242. **Kilbride EA, Burke ME, Boocock MR, Stark WM.** 2006. Determinants of product topology in a hybrid Cre-Tn3 resolvase site-specific recombination system. *J Mol Biol* **355:**185–195.

243. **Sarkis GJ, Murley LL, Leschziner AE, Boocock MR, Stark WM, Grindley ND.** 2001. A model for the gamma delta resolvase synaptic complex. *Mol Cell* **8:**623–631.

244. **Murley LL, Grindley ND.** 1998. Architecture of the gamma delta resolvase synaptosome: oriented heterodimers identity interactions essential for synapsis and recombination. *Cell* **95:**553–562.

245. **Vanhooff V, Normand C, Galloy C, Segall AM, Hallet B.** 2010. Control of directionality in the DNA strand-exchange reaction catalysed by the tyrosine recombinase TnpI. *Nucleic Acids Res* **38:**2044–2056.

246. **Colloms SD, Bath J, Sherratt DJ.** 1997. Topological selectivity in Xer site-specific recombination. *Cell* **88:** 855–864.

247. **Sota M, Yano H, Ono A, Miyazaki R, Ishii H, Genka H, Top EM, Tsuda M.** 2006. Genomic and functional analysis of the IncP-9 naphthalene-catabolic plasmid NAH7 and its transposon Tn4655 suggests catabolic gene spread by a tyrosine recombinase. *J Bacteriol* **188:** 4057–4067.

248. **Olorunniji FJ, Stark WM.** 2009. The catalytic residues of Tn3 resolvase. *Nucleic Acids Res* **37:**7590–7602.

249. **Dyda F, Hickman AB, Jenkins TM, Engelman A, Craigie R, Davies DR.** 1994. Crystal structure of the catalytic domain of HIV-1 integrase: similarity to other polynucleotidyl transferases. *Science* **266:**1981–1986.

Mobile DNA, 3rd Edition
Nancy L. Craig, Michael Chandler, Martin Gellert, Alan M. Lambowitz, Phoebe A. Rice and Suzanne Sandmeyer

doi:10.1128/microbiolspec.MDNA3-0004-2014

Sharmistha Majumdar*,[1]
Donald C. Rio[1]

33 P Transposable Elements in *Drosophila* and other Eukaryotic Organisms

INTRODUCTION

P transposable elements are one of the best-studied eukaryotic mobile DNA elements in metazoans. These elements were initially discovered in the late 1960's because they cause a syndrome of genetic traits termed hybrid dysgenesis [1]. The molecular cloning and biochemical characterization of the P element transposition reaction have led to general insights regarding eukaryotic cut-and-paste-transposition. P elements have also facilitated many applications as genetic tools for molecular genetics in *Drosophila*.

This review will focus on the experiments and results that have taken place over the past decade related to understanding the mechanism, distribution, specificity and uses of P element transposition. It will also discuss studies aimed at providing insights into how P element transposition is controlled, how P elements rely upon their host cells to provide the functions necessary to aid their successful mobility and how the damage they incur to the genomes in which they reside is limited and repaired. For reviews about other aspects of P element biology, more historical perspectives on the invasion of P elements into *Drosophila* and the genetic inheritance patterns of hybrid dysgenesis and P cytotype regulation, the reader is referred to earlier review articles [2–10] and the previous review article in this ASM series [11].

I. P TRANSPOSABLE ELEMENTS IN *DROSOPHILA* AND OTHER EUKARYOTIC GENOMES

I.A. Structure and Protein Products of P Transposable Elements

Isolation and DNA sequence analysis of the P elements showed a wide variation in size in a typical P strain, from 0.5–2.9kb [12]. P strains carried 50–60 P element insertions, of which approximately one-third were full-length (2.9kb) whereas the remainder were internally-deleted in different ways [13]. All P elements have a canonical structure that includes 31bp terminal inverted repeats (TIR) and internal inverted repeats (IIR) of 11bp located about 100bp from the ends that interact with

[1]Department of Molecular and Cell Biology, University of California, Berkeley, Berkeley, CA 94720-3204. *Present address: Department of Biological Engineering, Indian Institute of Technology-andhinagar, Ahmedabad 382424, India, Email: sharmistham@iitgn.ac.in.

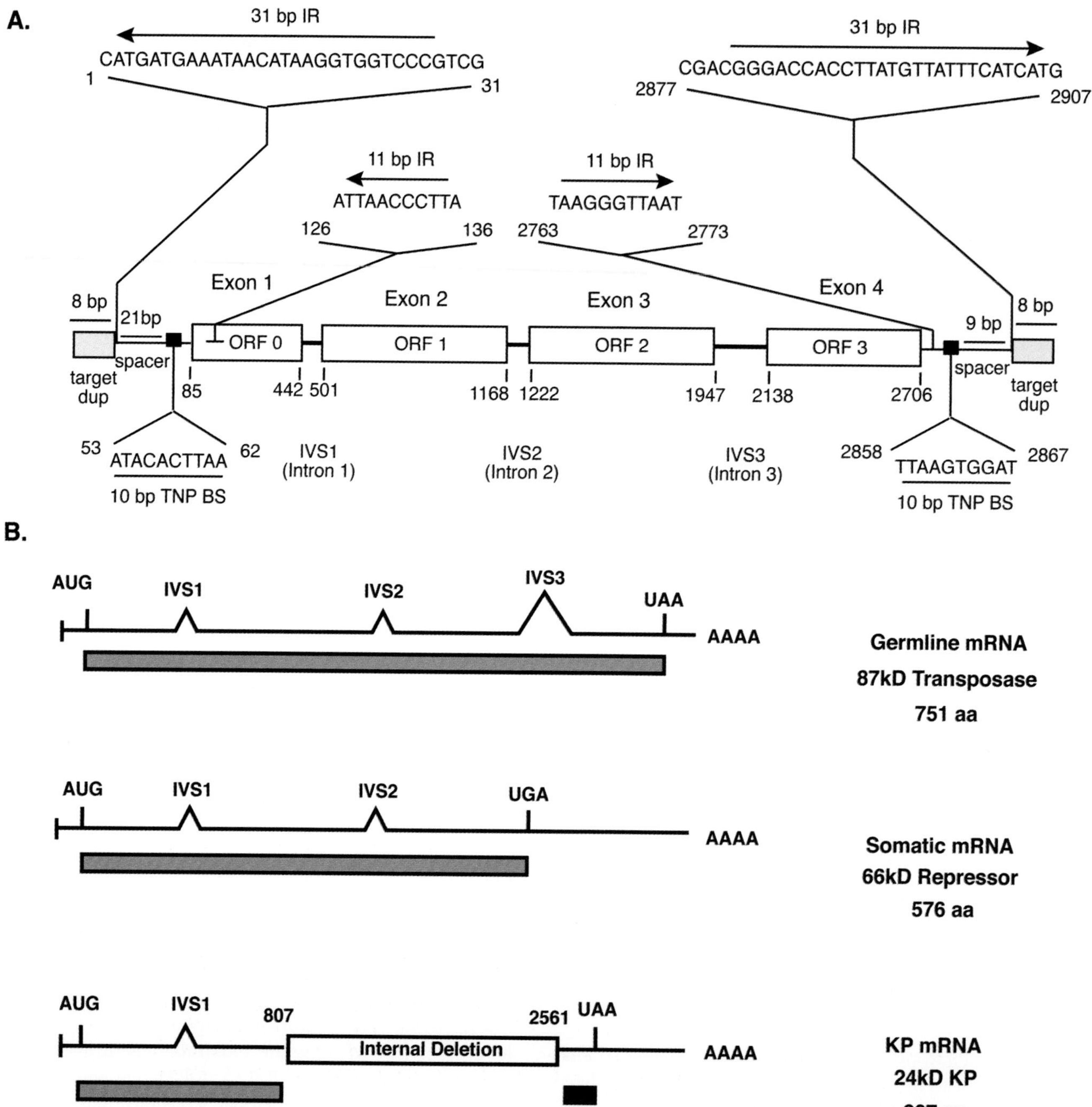
A.
31 bp IR
CATGATGAAATAACATAAGGTGGTCCCGTCG
1
31
31 bp IR
CGACGGGACCACCTTATGTTATTTCATCATG
2877
2907
11 bp IR
ATTAACCCTTA
126
136
11 bp IR
TAAGGGTTAAT
2763
2773
Exon 1
Exon 2
Exon 3
Exon 4
8 bp
21bp
9 bp
8 bp
ORF 0
ORF 1
ORF 2
ORF 3
spacer
target dup
spacer
target dup
85
442
501
1168
1222
1947
2138
2706
53
ATACACTTAA
62
10 bp TNP BS
IVS1 (Intron 1)
IVS2 (Intron 2)
IVS3 (Intron 3)
2858
TTAAGTGGAT
2867
10 bp TNP BS
B.
AUG
IVS1
IVS2
IVS3
UAA
AAAA
Germline mRNA
87kD Transposase
751 aa
AUG
IVS1
IVS2
UGA
AAAA
Somatic mRNA
66kD Repressor
576 aa
AUG
IVS1
807
2561
UAA
Internal Deletion
AAAA
KP mRNA
24kD KP
207 aa

the THAP domain of the transposase [14] (Fig. 1A). Between these two repeats, but distant from the site of DNA cleavage, are high-affinity binding sites for the P element transposase protein THAP DNA binding domain (see below; Fig. 1A; [15]).

The largest active, complete, P element analyzed is 2907 bp in length and contains four non-contiguous open reading frames [12] (Fig. 1B). All four open reading frames or exons are required for P element mobility and encode the 87kD transposase protein [16, 17]. These open reading frames are functionally joined together at the RNA level via pre-mRNA splicing [18]. The expression of transposase is normally restricted to germline cells because splicing of the P element third intron (IVS3) only occurs in the germline. In addition, it was shown that in somatic cells (and also to a large extent in the germline) the third intron (IVS3) is retained, producing a functional mRNA that encodes a 66kD protein [17]. The 66kD protein functions as a repressor of transposition and has been termed a Type I repressor [19–22] (Fig. 1B). In addition, some truncated proteins produced from the smaller, internally-deleted P elements present in natural P strains or engineered can act as repressors of transposition and are called Type II repressors [5, 14, 23–25]. In summary, P element transposition *in vivo* requires about 150bp of DNA in *cis* at each transposon end, including 31bp terminal inverted repeats and the high-affinity transposase binding sites. Upon insertion P elements create an 8bp duplication of target DNA and the elements can encode a transposase, as well as DNA binding repressors of transposition.

I.B. P Elements, P Element-Related THAP9 Genes and Active P Element "Transposase-Like" Genes in other Eukaryotic Genomes

Sequencing of the human genome revealed the presence of ~50 genes that were derived from DNA transposable elements [26]. One of these genes, termed THAP9, bears homology to the *Drosophila* P element transposase (Fig. 2 and [27, 28]). Further genome sequencing efforts over the past decade have led to the discovery of P element-related THAP9 genes or transposons in a variety of organisms in addition to humans, including other primates, zebrafish, *Xenopus* [29], *Ciona* [30], sea urchin, hydra [31] and the human pathogenic protozoan parasite, *Trichomonas vaginalis* [32]. Notably, the THAP9 gene is absent from rodents, rats and mice, and a defective copy is found in the chicken genome [33].

The human THAP9 gene is homologous (25% identical and 40% similar) to the *Drosophila* P element transposase throughout the entire length of the protein [27, 33]. The discovery of the human THAP9 gene suggests a possible invasion of P element-like transposons into vertebrates. The THAP9 gene appears most recently functional as a transposase in zebrafish (called Pdre2; [33], where there are obvious inverted repeat elements (called Pdre elements; [33, 34]) carrying 13bp

Figure 1 Features of the complete 2.9kb P element. A.) Sequence features of the 2.9kb P element. The four coding exons (ORF 0, 1, 2 and 3) are indicated by boxes with nucleotide numbers shown. The positions of the three introns (IVS 1, 2, 3) are indicated below. The DNA sequences of the 31bp terminal inverted repeats (TIR) and the 11bp internal inverted repeats (IIR) are shown, with corresponding nucleotide numbers shown above. The 8bp duplications of target site DNA are shown by boxes at the ends of the element. DNA binding sites for the transposase protein from the 5′ end (nt 48–68) and from the 3′ end (nt 2855–2871) that are bound by P element transposase [15]. The consensus 10bp transposase binding site is: 5′- AT(A/C)CACTTAA -3′. Distances of the beginning of the 10 bp core high affinity transposase binding sequence from the corresponding 31bp terminal repeat are indicated. Note that there are distinct spacer lengths between the 31bp repeats and the transposase binding sites, 21bp at the 5′ end and 9bp at the 3′ end, which are indicated. The sequence of the 11bp internal inverted repeats are also shown, which bind the P element THAP DNA binding domain [54]. Nucleotide numbers are from the 2907bp full-length P element sequence. B.) P element mRNAs and proteins. The 2.9kb P element and four exons (ORF 0, 1, 2 and 3) are shown at the top. The germline mRNA, in which all three introns are removed, encodes the 87kD transposase mRNA. The somatic mRNA, in which only the first two introns are removed (and which is also expressed in germline as well as somatic cells), encodes the 66kD repressor mRNA. Shown at the bottom is a KP element, which contains an internal deletion. This truncated element encodes a 24kD repressor protein. doi:10.1128/microbiolspec.MDNA3-0004-2014.f1

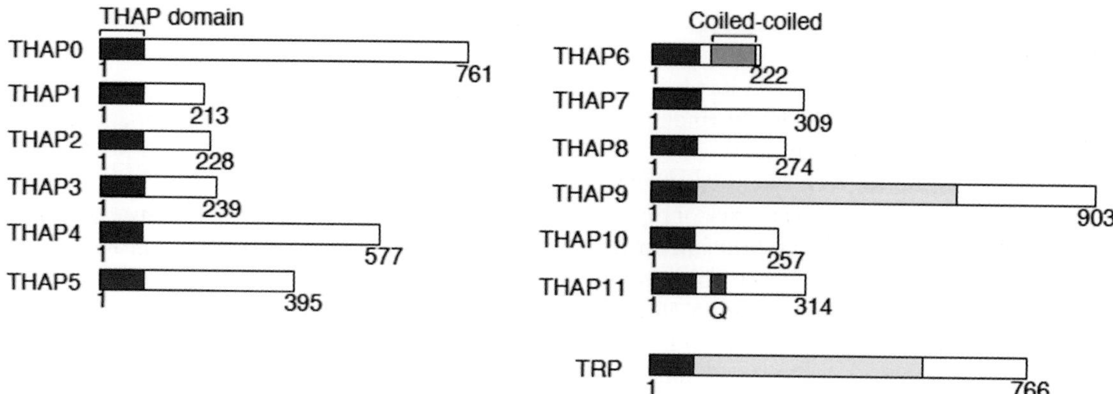

Figure 2 THAP domain-containing proteins in the human genome. Diagram of the 12 human THAP domain-containing proteins and *Drosophila* P element transposase. Note that the homology of human THAP9 and the Drosophila P element transposase extends the entire length of the protein, well beyond the N-terminal THAP DNA binding domain. Taken from [28]. doi:10.1128/microbiolspec.MDNA3-0004-2014.f2

terminal inverted repeats (TIRs) and 12bp sub-terminal inverted repeats (STIRs) with a ~400bp spacer and which carry direct 8bp target site duplications flanking the TIRs (Fig. 3A). In addition to the full-length Pdre2 THAP9 transposase-like gene, there are also multiple internally-deleted Pdre elements elsewhere in the zebrafish genome reminiscent of *Drosophila* P strains [33, 34].

Regarding the activity of the vertebrate THAP9 genes, it has recently been shown that the human THAP9 protein can mobilize *Drosophila* P elements in both *Drosophila* and human cells (Fig. 3B and 3C) [35]. These results indicate that human THAP9 is an active DNA recombinase that retains the catalytic activity to mobilize P transposable elements across species. However, the cellular function of human THAP9 is still unknown. It may be the case that the human THAP9 gene could encode a recombinase that acts on remote recombination signals elsewhere in the human genome, similar to Rad1/2 activity in V(D)J recombination. Gene expression profiling indicates that the human THAP9 gene is highly expressed in embryonic stem cells, testes and kidney.

In addition to THAP9, the human genome has 11 THAP domain-containing genes ([28]; Fig. 2). Many have been characterized as transcription factors that control the expression of genes involved in apoptosis, cell cycle regulation, stem cell pluripotency and epigenetic gene silencing. Human THAP domain family members have also been implicated in a variety of human diseases, including heart disease, torsional dystonia and cancer. The THAP DNA binding domain appears to be restricted to animals because no known or predicted THAP domain-containing genes have been found in plants, yeast, other fungi or bacteria.

I.C. P Elements as Tools for *Drosophila* Genetics and the *Drosophila* Genome Project

One of the most important uses of P elements since their discovery has been P-element-mediated germ line transformation. The method makes use of the fact that P elements normally only transpose in germline cells and that a P element carrying a foreign gene can be mobilized in *trans* using a source of transposase [36–38] (Fig. 4). The Berkeley Drosophila genome project (BDGP) has collectively developed strategies which efficiently use single P element mutagenesis [39, 40], leading to large-scale P element insertional mutation screens so that now 9440 or about two-thirds of the annotated protein-coding genes are tagged [41–43]. It was discovered that normally P elements transpose preferentially in *cis*, within about 50–100kb from their initial location, so-called "local hopping" [44, 45]. P elements can also undergo transposase-mediated excision and transposase-induced male recombination to generate deletions flanking an existing P element insertion [46]. These smaller deletions can be enlarged by performing P element excision crosses in a DmBLM/mus309 mutant genetic background (defective for repair of P element transposase-induced DNA double-strand breaks; more details in section II.F) [47–49]. P elements have also played a role in other molecular genetic methods, such as homologous gene targeting [50] and in the use of libraries of bacterial artificial chromosomes (bac) for transformation and recombineering [36, 38, 51]. Thus, P elements have continued to play important roles in the post-genome sequence era of *Drosophila* genetics and genomics.

II. MECHANISM OF P ELEMENT TRANSPOSITION

II.A. *Cis*-Acting DNA Sites Involved in *Drosophila* P Element Transposition

Experiments using P element-mediated transformation showed that the 31bp inverted repeats were required for transposition (Fig. 1A) [16, 52]. Extensive mutagenesis studies subsequently showed that both the terminal 31bp inverted repeats and the high affinity internal transposase binding sites are also required for transposition *in vivo* [53]. Additionally, the internal 11bp inverted repeats located at ~120bp from each transposon end function as transpositional enhancer elements (Fig. 1A) [53]. These 11bp internal inverted repeats can interact with the N-terminal THAP DNA binding domain of the P element transposase (Fig. 1A) [54]. Analysis of hybrid elements carrying tandem 5′ and 3′ ends showed that two different P element ends (a 5′ and 3′ end) must be paired for transposition to occur, indicating that the two P element ends are not equivalent [53]. In summary, the *cis*-acting DNA sites for P element transposition include 31bp terminal inverted repeats, 11bp sub-terminal inverted repeats and internal 10bp transposase binding sites, located between the terminal inverted repeats and sub-terminal inverted repeats. Interestingly, the zebrafish Pdre P-like elements also have both terminal and sub-terminal inverted repeats (see section, I.B., above and Fig. 3A). All P elements analyzed to date in *Drosophila*, as well as P element-related transposons in zebrafish, create 8bp direct duplications of target site DNA upon insertion [33, 55, 56].

II.B. The *Drosophila* P Element Transposase

Understanding the domain organization of the P element transposase had come from biochemical studies, genetic experiments and sequence comparisons of P elements from other *Drosophila* species and other eukaryotes with P element transposase-like THAP9 genes. The N-terminal THAP DNA binding domain is a C_2CH zinc binding motif with an adjacent basic region and is the site-specific DNA binding domain (Fig. 5A; see section II.D., below) that recognizes the internal transposase DNA binding sites and the 11bp sub-terminal inverted repeats [15, 54]. Adjacent to the THAP domain is a long coiled-coil region made up of a canonical leucine-zipper motif with heptad leucine/isoleucine repeats and an adjacent longer coiled-coil motif (up to residue 221; Fig. 5A). This type of coiled-coil region in the transposase protein, commonly found in THAP domain-containing proteins, allows protein dimerization, although this is not essential for high affinity site-specific DNA-protein recognition [54, 57].

Biochemical studies revealed a requirement for GTP as a cofactor during P element transposition [58], suggesting that the transposase would bind GTP. The central region of the transposase protein contains several sequence motifs found in the GTPase protein superfamily [59, 60] (Fig. 5A and 5B). Mutation of some of the key conserved residues abolished GTP binding *in vitro* and transposase activity *in vivo* [61]. Alteration of the most conserved NKXD guanine recognition motif to the NKXN motif xanthine recognition motif (D379N) altered the purine nucleotide requirement for transposase activity from guanosine to xanthosine triphosphate, both *in vitro* and *in vivo*, indicating that purine nucleotide cofactor binding is required for transposase activity [61] (Fig. 5B). It is know known that GTP plays a key role in synapsis of the two transposon ends during transposition [62, 63] (see section II.E., below).

The carboxyl-terminal region of the transposase protein, including exon 4 (ORF3), contains a high proportion of acidic amino acid residues (Fig. 5C). Mechanistically, the P element transposase belongs to the superfamily of polynucleotidyl transferases, including the transposases from bacteriophage Mu and other bacterial mobile elements, the retroviral integrases, the Holliday junction nuclease RuvC, the RNaseH superfamily [64, 65] and the RAG1 subunit of the V(D)J recombinase [66–68]. These enzymes generally use metal ion-mediated catalysis of phosphodiester bond hydrolysis and formation, where acidic amino acids in the protein active site serve to coordinate a divalent metal ion, usually magnesium [65, 69]. Because the P element-encoded 66kD repressor protein lacks catalytic activity and because of the high proportion of acidic amino acid residues in the C-terminal part of the protein, it seemed likely that acidic residues found in this region might constitute the catalytic domain of the protein. Mutagenesis studies of many acidic C-terminal residues have failed to identify a set of clear catalytic residues (D. Rio and colleagues, unpublished results). However, sequence- and structure-based (Phyre2) [70] alignments suggest a relationship to the hermes transposase, with putative RNaseH secondary structures and catalytic residues between the THAP domain and GTP binding regions at the N-terminal portion of the protein [64, 65, 71]. It may be the case that the GTP binding domain of the *Drosophila* P element transposase was inserted into a progenitor RNaseH-like catalytic domain [64, 65]. Recent exhaustive bioinformatics analyses, using all known P element-related protein sequences from the sequenced eukaryotic genomes in the Repbase

A **Zebrafish THAP9 transposon**

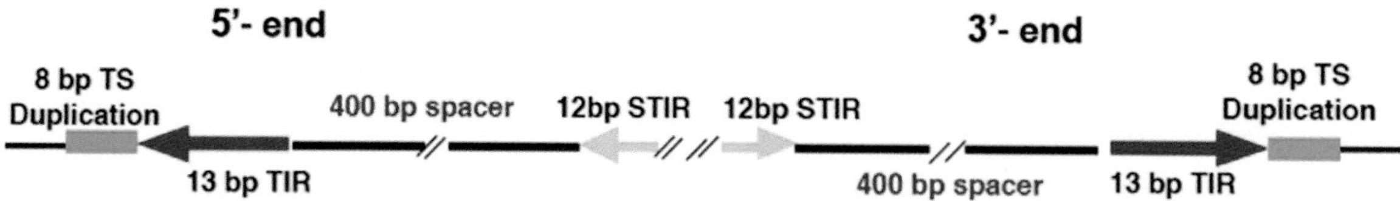

Drosophila P Element transposon

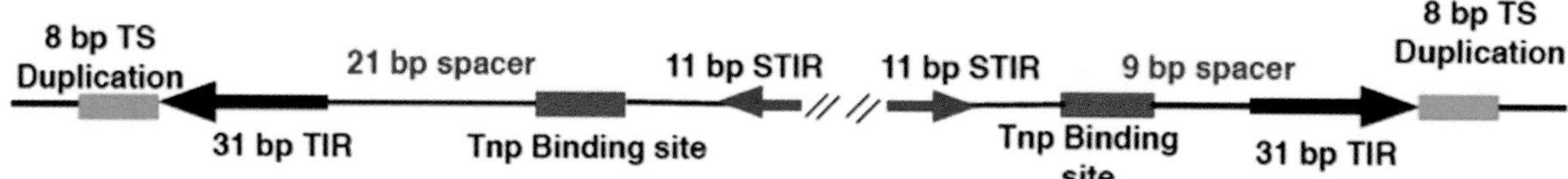

Figure 3 Organization of the zebrafish Pdre P element-like elements and activity of human THAP9 with Drosophila P element DNA. A.) Organization of the zebrafish Pdre inverted repeat elements. Indicated are the 8bp target site duplication (TSD), 13bp terminal inverted repeat (TIR) and 12bp internal inverted repeat (STR) [33]. B.) Assay for THAP9 transposition of Drosophila P element DNA in human cells HEK 293 cells. A P element vector (Cg4) carrying the G-418^R gene is transfected into human cells along with expression vectors for *Drosophila* P element transposase or human THAP9. Upon G-418 selection, individual colonies are assayed for novel DNA insertion sites [35]. C.) Colonies of human cells in which P elements have undergone transposition by *Drosophila* P element transposase or human THAP9 compared to a negative control plate [35].
doi:10.1128/microbiolspec.MDNA3-0004-2014.f3

Figure 3 *continues on next page*

database, revealed a conserved D, D, E-like motif common to all P element-transposase-related proteins known [71], including the THAP9 family (Fig. 5C). Thus, it appears that the P element transposase is a member of the DDE enzyme superfamily with a complex metal ion binding site configuration in the active site.

II.C. Biochemical Characterization of P Element Transposase

Purification and characterization of the P element transposase protein from *Drosophila* tissue culture cell nuclear extracts showed that the 87kD transposase protein binds to 10bp sites near each end of the P element, located between the terminal 31bp and internal 11bp inverted repeats [15] (Fig. 1A). The consensus transposase binding site is: 5′- AT(A/C)CACTTAA -3′. The two high-affinity transposase binding sites do not overlap the terminal 31bp inverted repeats which are known to be required for transposition *in vivo* [16, 53] (Fig. 1A). Following initial recognition of these binding sites by transposase, a GTP-dependent assembly (or synapsis) occurs to bring the two transposon ends together [62, 63] (see section II.E., below and Fig. 6B).

The purified P element transposase protein was used to develop *in vitro* assays to study the different steps of the transposition reaction: donor DNA cleavage and target DNA integration [58]. The *in vitro* reaction required wild type transposase DNA binding sites on the donor P element and linear pre-cleaved donor DNA could be used as a substrate, indicating that supercoiling of the donor DNA is not required for transposition. Importantly, 3′ hydroxyl groups at the P element ends were required for activity indicating that, like other transposition systems, a 3′-hydroxyl group is used as a nucleophile during strand transfer [58]. These studies demonstrated that P element transposition proceeds via a cut-and-paste mechanism.

Most surprisingly, the nucleoside triphosphate guanosine triphosphate (GTP) was discovered as a critical cofactor for P element transposition [58]. Non-hydrolyzable GTP analogs (GTP-γ-S, GMP-PNP and

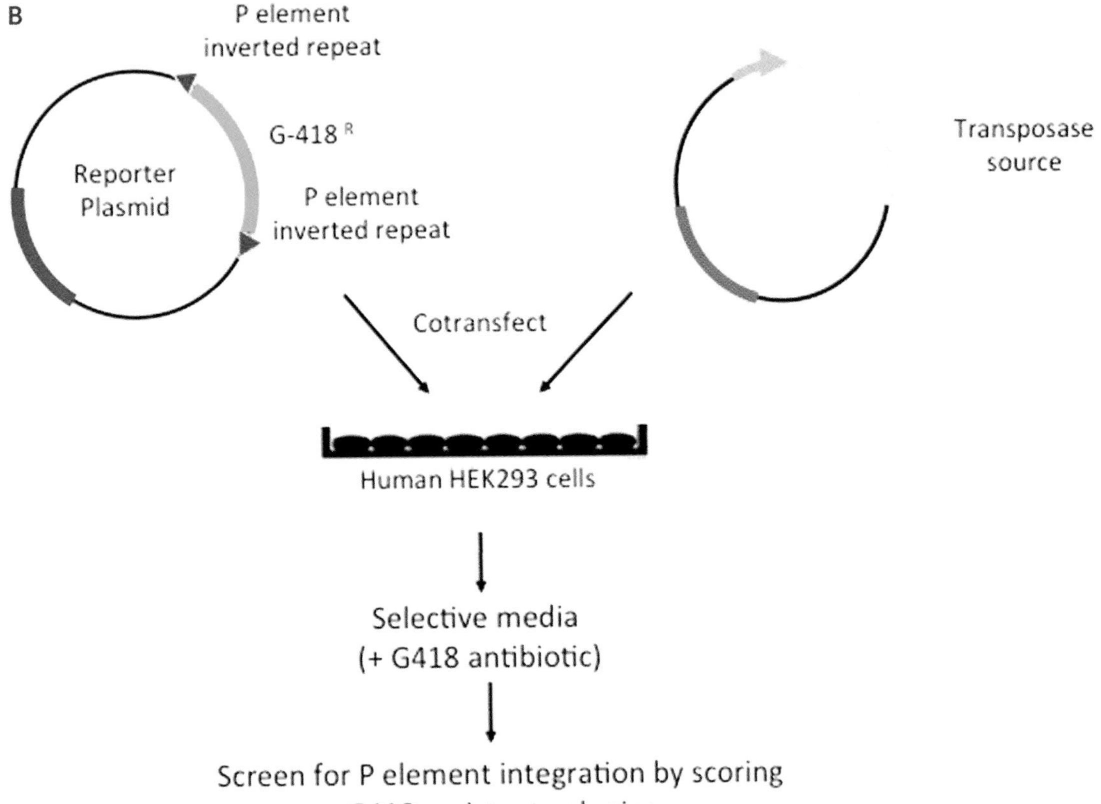

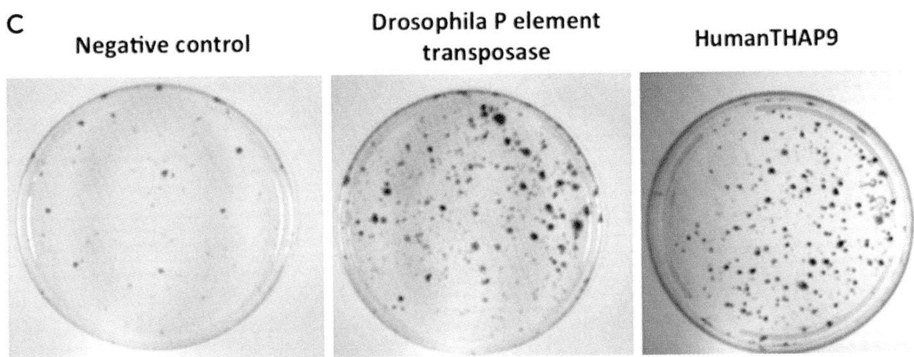

Figure 3 *continued*

GMP-PCP) showed levels of activity equivalent to those observed with normal GTP, indicating that hydrolysis of a high energy phosphoryl bond was not required for activity. This suggests that GTP plays an allosteric role by binding to P element transposase, in the same way that GTP modulates the conformation and activities of other GTP-binding proteins, such as ras and mammalian Ga subunits [59, 72] or the GAD family member dynamin [73]. The GTP requirement for the P element transposase protein is unique among this class of polynucleotidyl transferase proteins. Single-molecule imaging studies showed that GTP plays a critical role in initial synapsis of the transposon ends ([62, 63]; see section II.E., below).

Physical assays for the detection and analysis of reaction products and intermediates were developed using both radiolabeled and unlabeled P element DNAs leading to the detection of an excised P element transposon fragment, directly confirming that transposition occurs through double-strand DNA breaks at the

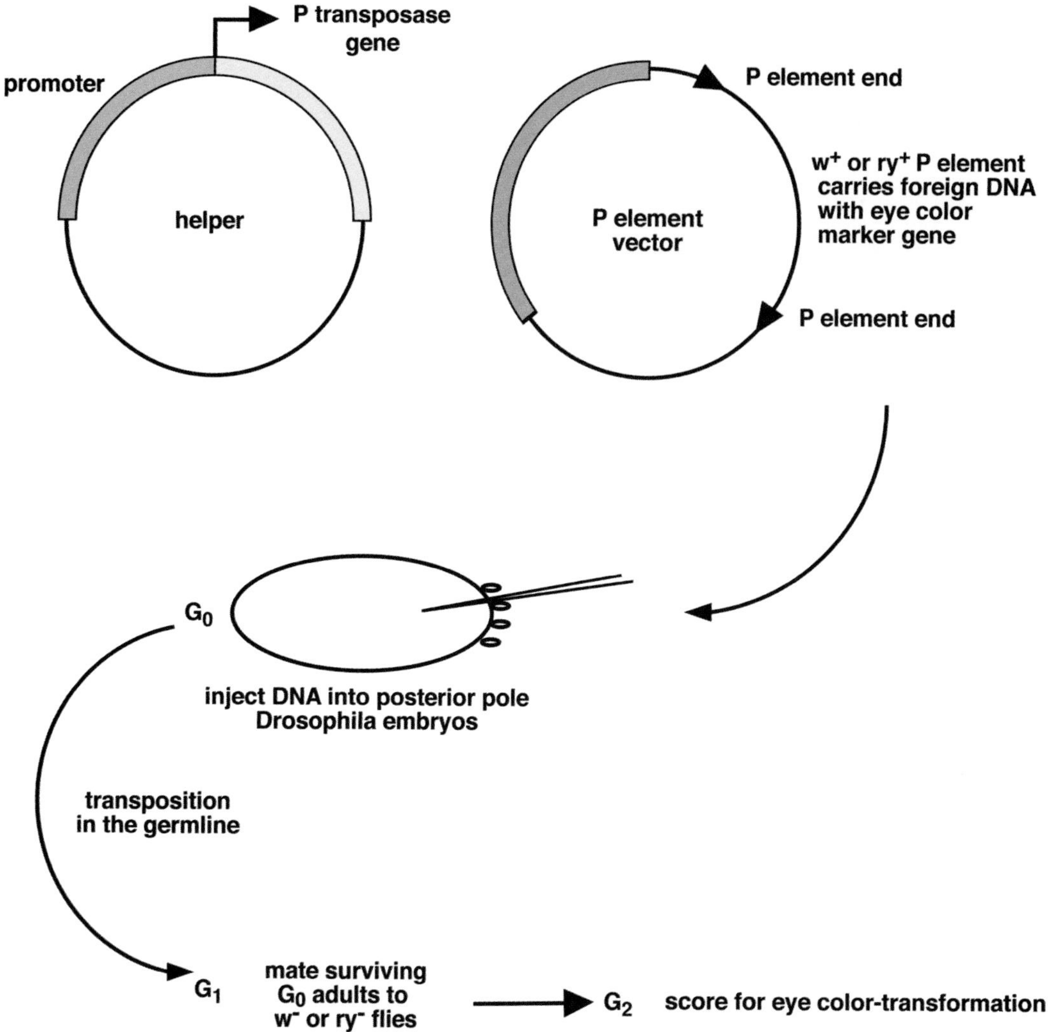

Figure 4 P element-mediated germline transformation. Outline of the method for germline transformation of *Drosophila* using P element vectors. Two plasmids, one encoding the P element transposase protein but lacking P element ends and the second plasmid carrying a foreign DNA segment and an eye color marker gene (w^+ or ry^+) within P element ends, are injected into the posterior pole of pre-blastoderm embryos. Once the transposase plasmid enters nuclei of presumptive germline cells and is expressed, it leads to transposition of the P element from the second plasmid into *Drosophila* germline chromosomes. Following development of the injected embryos (G_0 generation), the surviving adults are mated to w^- or ry^- flies (G_1 generation) and the progeny from this cross (G_2 generation) are scored for restoration of wild type eye color. The transformation frequency is typically ~20% of the fertile G_0 adults are carrying the transgene [37].
doi:10.1128/microbiolspec.MDNA3-0004-2014.f4

transposon termini [74]. DNA cleavage mapping experiments showed that P elements are excised from the donor DNA site by an unusual set of cleavages at the transposon ends [74]. The 3′ ends (bottom strands) of the transposon DNA are cleaved at the junction with the 8bp target site duplication (Fig. 6A). But surprisingly, the 5′ ends (top strands) were cleaved 17nt into the 31bp inverted repeats generating novel 17nt 3′ single-strand extensions on both the excised transposon and flanking donor DNAs (Fig. 6A). These transposase-mediated *in vitro* DNA cleavage sites are consistent with previously characterized *in vivo* P element excision products [47, 75]. This long 3′ single-strand extension may facilitate entry of cleaved donor DNA sites

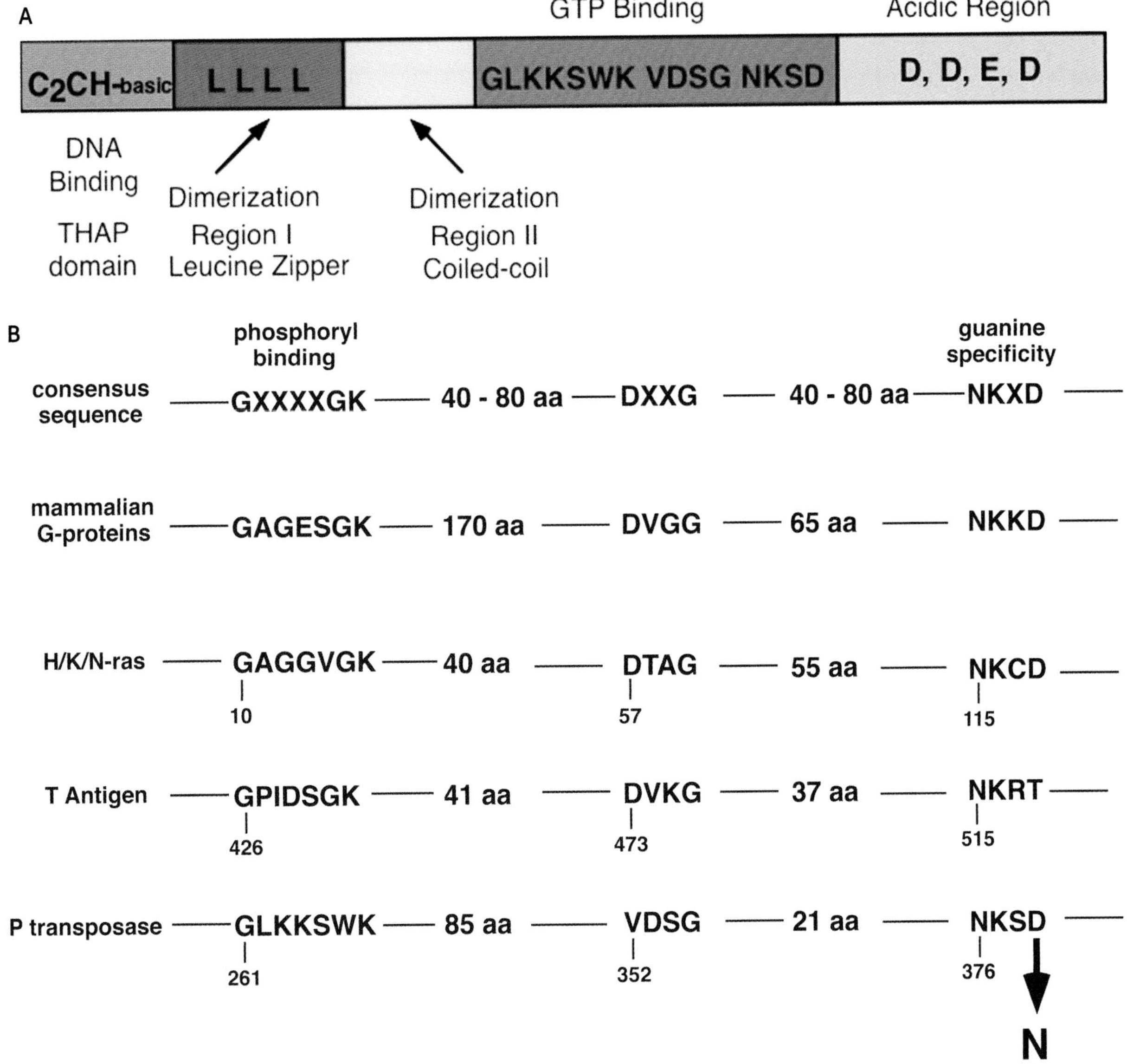

Figure 5 Domain organization of the *Drosophila* P element transposase protein. A.) Domains of P element transposase. The N-terminal region contains a C_2CH motif and basic region, called the THAP domain, involved in site-specific DNA binding. There are two dimerization regions adjacent to the N-terminal DNA binding domain: dimerization region I is a canonical leucine zipper motif and dimerization region II is C-terminal to the leucine zipper but does not resemble any known motif. The central part of the protein contains a GTP binding region, with some sequence motifs found in the GTPase superfamily [61]. Acidic residues are enriched at the C-terminus. B.) Similarities between P element transposase and GTPase superfamily members. Alignments of regions of P element transposase that bear some resemblance to known G proteins. The conserved motifs for phosphoryl binding and guanine specificity are indicated at the top. Amino acid numbers are given below for ras, T antigen and P element transposase. The residue D379 that when changed to N (aspartic acid to asparagine) switched the nucleotide specificity from guanosine to xanthosine in P element transposase is indicated at the bottom. Figure taken from [61]. C.) Sequence of the P element transposase with predicted secondary structural elements and putative catalytic signature residues. Taken from [71].
doi:10.1128/microbiolspec.MDNA3-0004-2014.f5

Figure 5 *continues on next page*

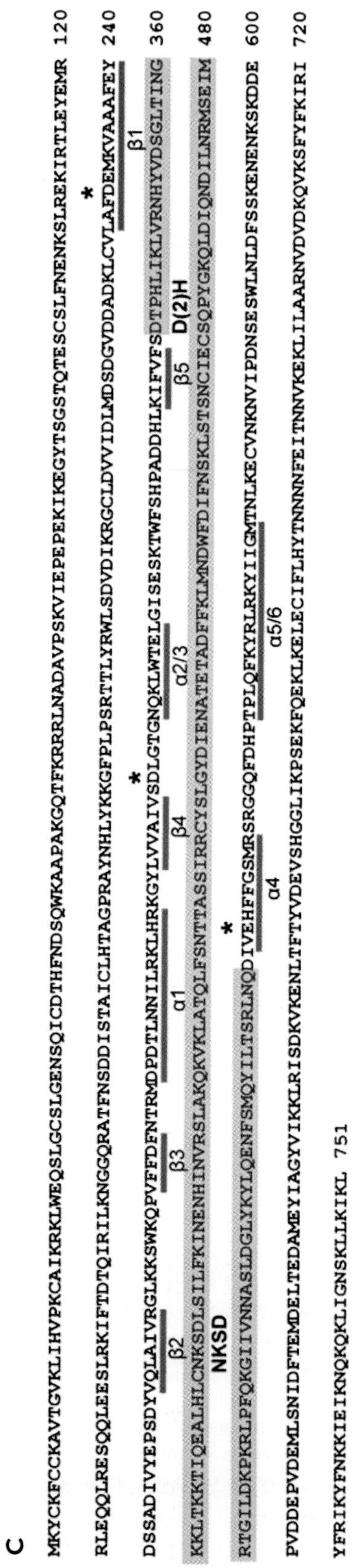

Figure 5 *continued*

into DNA repair pathways and could contribute to the irreversibility of the P element excision reaction.

Further studies of P element transposase protein *in vitro* used synthetic oligonucleotide substrates. First, these studies showed that the 17nt single-stranded 3′ extension is critical for strand transfer because a decrease in the length or mutation of the exposed single-stranded DNA extension caused a drastic reduction in strand transfer activity [76]. Second, chemically-modified oligonucleotide substrates were used to probe the protein-DNA contacts that were critical for strand transfer activity by P element transposase [76]. These experiments demonstrated that critical contacts between transposase and both the duplex and single-stranded regions of the substrate DNA are necessary for strand transfer activity *in vitro* [76]. In addition, there were sites in which DNA modification actually stimulates strand transfer, indicating that distortion of the substrate DNA may facilitate the chemistry of strand transfer, such as those observed with Mu transposase [77, 78] and HIV integrase [79]. Thus, while the initial donor DNA cleavage reaction occurs on duplex DNA, the strand transfer reaction uses a completely different substrate in which the single-stranded region at the P element ends are critical for activity indicating that there is a significant active site flexibility in the P element transposase protein during transposition.

Using oligonucleotide substrates that mimic a strand transfer intermediate carrying a pre-cleaved P element end joined to target DNA, the purified transposase protein can carry out disintegration [76], in a similar manner to the retroviral integrases [80], Mu transposase [81] and two other eukaryotic recombinases, the *C. elegans* Tc1 transposase [82] and the V(D)J RAG1/2 recombinase [83]. The ability of these proteins to perform disintegration may be significant in genomic surveillance because these disintegration reactions may serve to prevent chromosomal translocations and other rearrangements that might ensue following aberrant transposition events [83].

II.D. Protein-DNA Recognition by the THAP Family of C_2CH-Zinc-Coordinating DNA Binding Domains

THAP domains are a recently described family of zinc-coordinating DNA binding motifs, first recognized in the P element transposase [28]. The THAP domain is 80-90 amino acid residues, typically located at the amino terminus of the protein and contains a C_2CH (consensus: Cys-X_{2-4}-Cys-X_{35-50}-Cys-X_2-His) zinc-coordinating motif

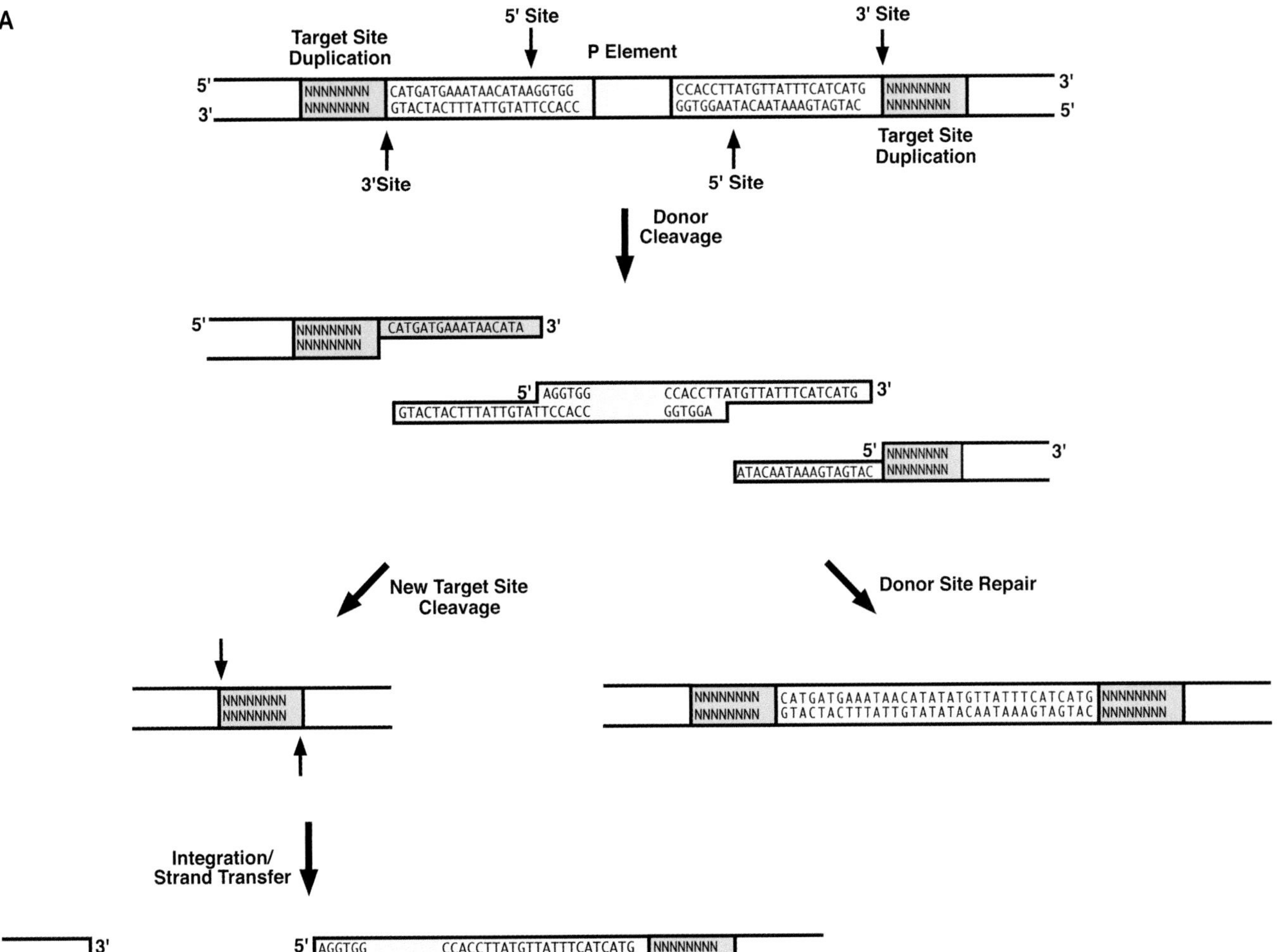

Figure 6 Pathway of DNA cleavage and joining and transposase-DNA assembly during P element transposition. A.) Shown at the top is the P element donor site with the target site duplications, 31bp inverted repeats and the 5′ and 3′ cleavage sites indicated. In the first step of transposition, donor cleavage occurs and both ends of the P element are cleaved. This novel DNA cleavage results in a 17nt single-strand extension on the P element transposon ends and leaves 17nt of single-stranded DNA from each P element inverted repeat attached to the flanking donor DNA cleavage site. Once transposon excision occurs, the donor site can be repaired via a non-homologous end joining (NHEJ) pathway (shown to the right) or via the synthesis-dependent strand annealing (SDSA) homology-dependent repair pathway (not shown) [157]. The excised P element then selects a target site and the strand transfer reaction integrates the P element into the donor site generating a gapped intermediate, which upon DNA repair completes integration creating a direct 8bp duplication of target DNA flanking the new P element insertion (bottom). Figure taken from [74]. B.) Synaptic and cleaved donor DNA intermediates detected by atomic force microscopy (AFM). Taken from [63]. C.) Single-end transposon binding of transposase in the absence of GTP. AFM imaging of P element DNA in the presence of transposase and in the absence of GTP. Taken from [62]. doi:10.1128/microbiolspec.MDNA3-0004-2014.f6

Figure 6 *continues on next page*

and other signature elements, including a C-terminal AVPTIF sequence. Although the THAP domain of THAP proteins share low primary sequence identity, recent structural studies have shown that there is strong conservation of the overall β–α–β protein fold and secondary structure elements. Structures of the P element transposase bound to its DNA site using X-ray diffraction [57] and of the human THAP1 protein bound to its DNA site using

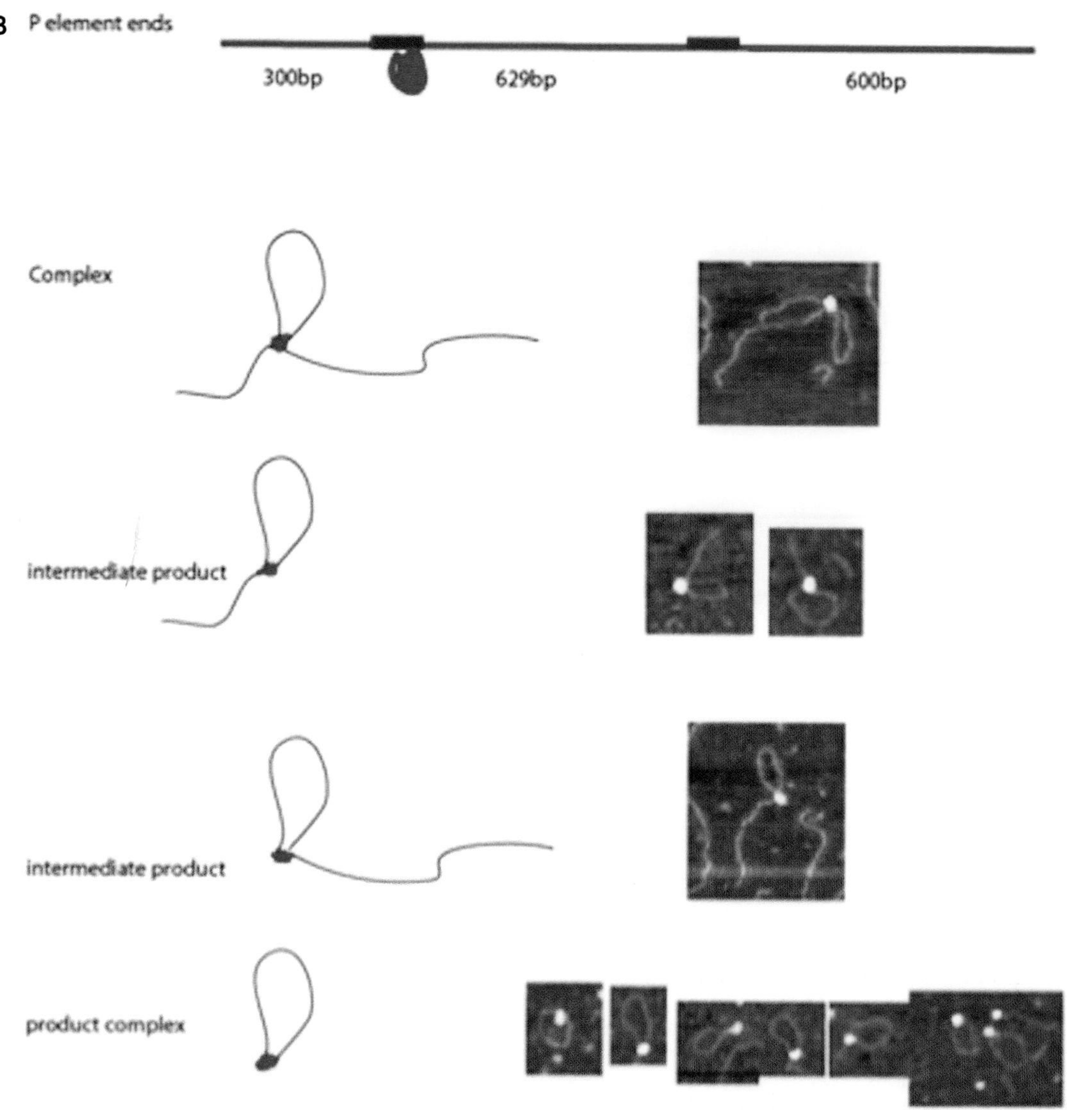

Figure 6 *continued*

NMR [84] have shown that THAP domains recognize their DNA sites in a bipartite manner, using two β-strands in the major groove and a basic C-terminal loop in the adjacent minor groove.

The X-ray crystal structure of the P element transposase THAP domain (DmTHAP) bound to DNA (Fig. 7 and 8) shows that His18 and Gln42 from the two β-strands at the N-terminus of DmTHAP, make a total of six direct contacts with DNA bases in the major groove of DNA and engage both strands of the DNA duplex (Fig. 7 and 8). Notably, the residues making the most central contacts with the DNA, including water-mediated hydrogen bonds to specific DNA bases in the major groove, show little (His18) or no (Gln42) sequence conservation within the human THAP protein family (Fig. 7C; [57]). Presumably, sequence variability at these positions, along with differences in the length and amino acid composition of the N-terminus, specifies the precise DNA sequences recognized by the THAP proteins through the major groove sub-site [57]. The variability of residue Gln42 (Fig. 7C), which interacts with several bases via multiple hydrogen bonds is unusual, since amino acid residues involved in multiple base-specific interactions are typically less variable [57]. This observation correlates well with the finding that the most conserved THAP

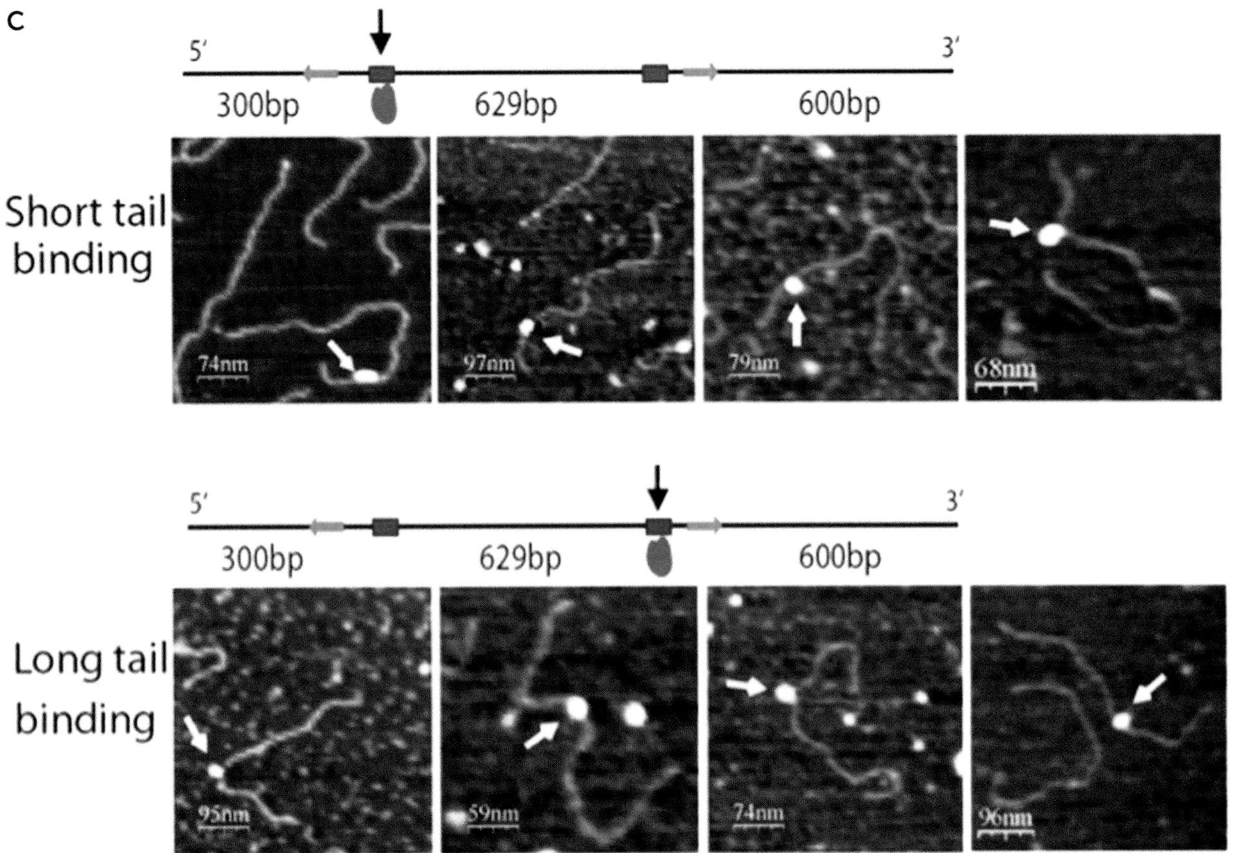

Figure 6 *continued*

domain residues play structural roles, namely all the invariant residues besides the C_2CH zinc-binding motif are involved in forming the hydrophobic core of the protein (Fig. 7A and 7C).

In addition to major groove DNA contacts, there are multiple residues in the loop 4 region of DmTHAP that are involved in adjacent minor groove DNA interactions (Fig. 7C and 8A and 8B). The residues corresponding to R65, R66 and R67 in DmTHAP, which are both contacting DNA (R65 and R67) and positioning the loop 4 region of the protein (R66), are different in human THAP9 (FK*R65*R*R67*LN in DmTHAP and GI*R*R*K*LK in human THAP9 (Fig. 7C). The minor groove binding residues may modulate DNA binding affinity, since the DmTHAP domain with its RRR motif has a higher affinity for its site than human THAP1, which has NKL in the corresponding position (Fig. 7C). The higher affinity DNA binding by the *Drosophila* P element transposase THAP domain compared to other THAP domain proteins [57], may have some role in the high frequency of P element mobility in *Drosophila*.

II.E. A GTP-Dependent Protein-DNA Assembly Pathway for *Drosophila* P Element Transposition

One of the unique and unexpected discoveries concerning P element transposition is its requirement for GTP as a cofactor. Biochemical studies indicated that GTP was not required for site-specific DNA binding by the P element transposase and that the purified transposase from *Drosophila* cells was a tetramer, that was unaffected by the presence or absence of GTP or DNA. Attempts to detect transposase-DNA complexes using native gel electrophoresis proved unsuccessful. However, single-molecule imaging using atomic force microscopy (AFM) of tranposase-DNA complexes has led to important insights into the role of GTP in the P element transposition pathway. First, GTP promotes formation of a stable synaptic complex between the transposase tetramer and the P element DNA (Fig. 6B; [63]). Second, time course reactions revealed the presence of both synaptic complexes and cleaved donor DNA complexes. These imaging experiments showed that while synapsis was fast (0–30 min.), cleavage of

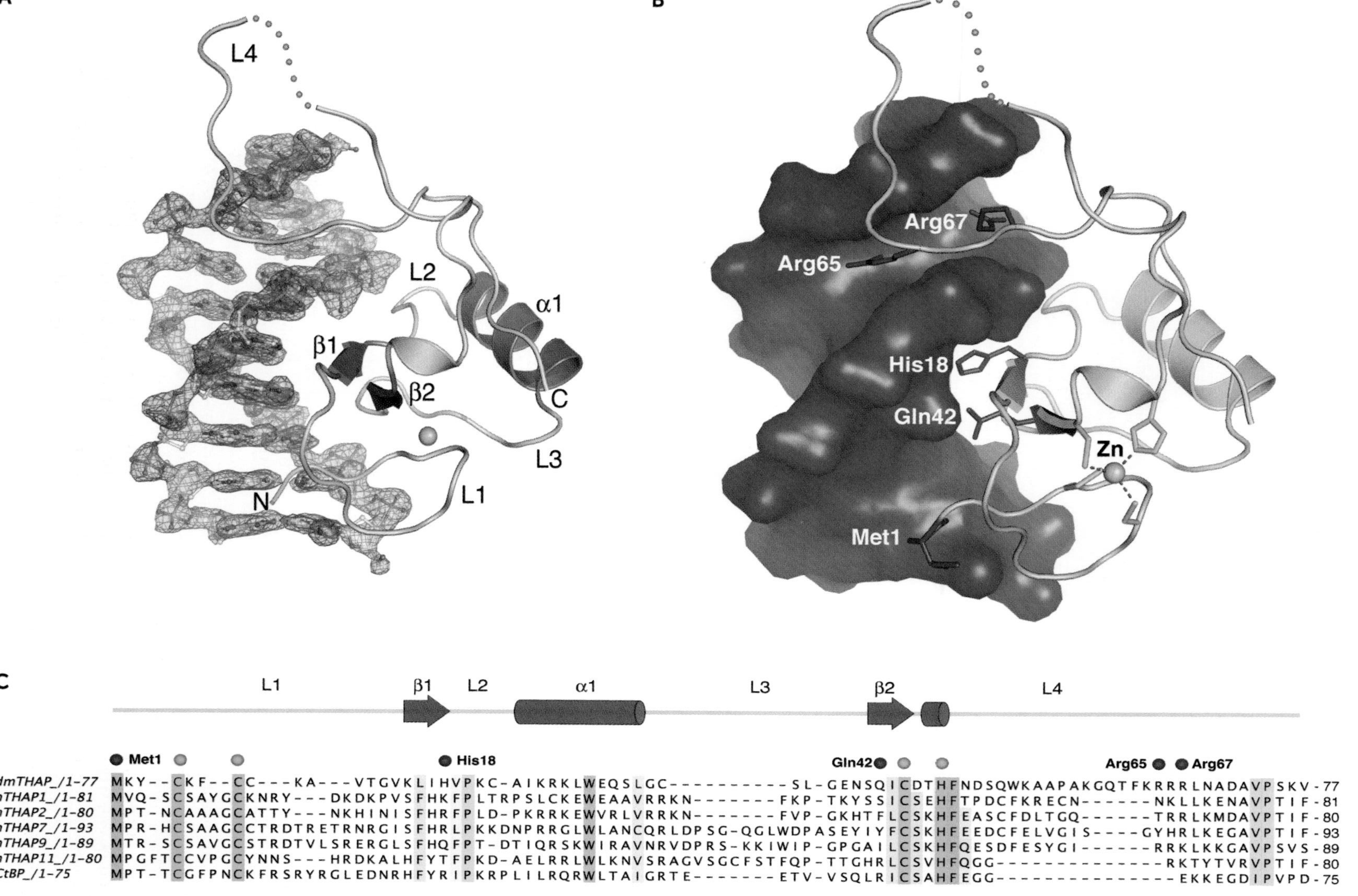

Figure 7 Overall structure of Drosophila P element transposase THAP domain (DmTHAP)-DNA complex. A.) The protein-DNA interface. Experimental electron density map of the DNA (blue mesh) is contoured at 1.5σ. DmTHAP is shown as a ribbon diagram and labeled by secondary structure, with the β–α–β motif highlighted in magenta. Zinc is shown as a green sphere. B.) Base-specific interactions in the major and minor groove. Interacting amino acids are shown as magenta sticks; DNA is shown in blue surface representation; zinc-coordinating residues are shown as green sticks. C.) Structure-based multiple sequence alignment of DmTHAP, human THAP1, 2, 7, 9 and 11, and *C. elegans* CtBP. Conserved residues are highlighted; zinc-coordinating C_2CH motif is highlighted in green and indicated by green circles; DNA-binding residues of DmTHAP are indicated by magenta circles and are labeled. The secondary structure diagram is shown for DmTHAP and labeled as in (A). Taken from [57]. doi:10.1128/microbiolspec.MDNA3-0004-2014.f7

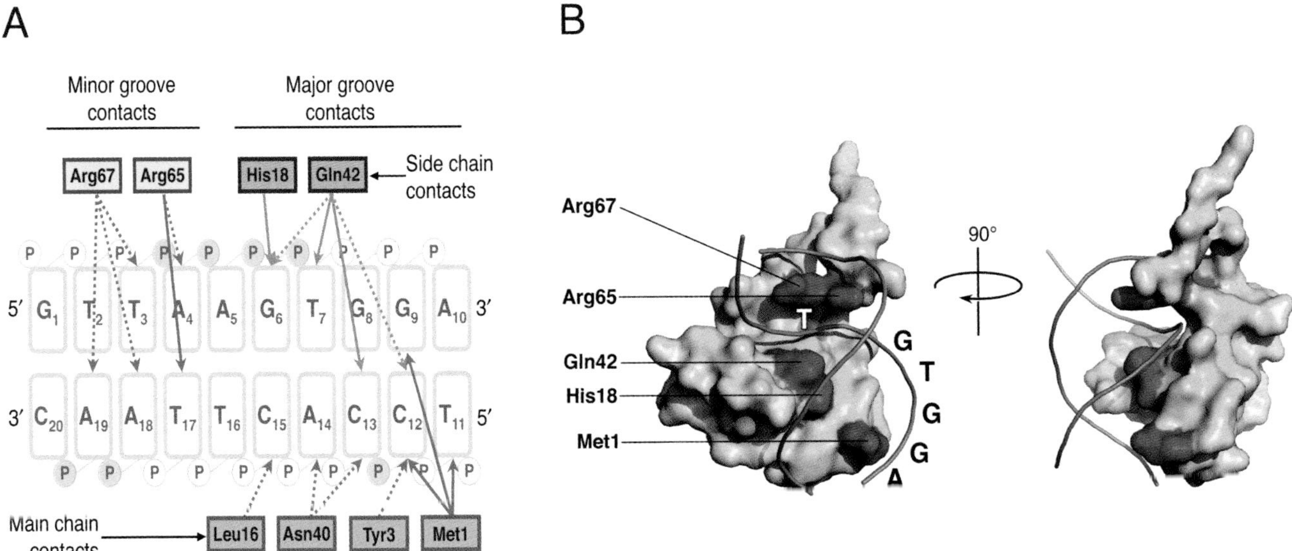

Figure 8 Base-specific DmTHAP-DNA contacts. A.) Schematic representation of all base-specific contacts in the major and minor groove. Direct contacts are shown as solid lines, base-specific water-mediated contacts are shown as dashed lines, interacting phosphates are highlighted yellow. B.) Surface representation of DmTHAP. Sequence specific DNA-binding residues are highlighted in magenta. DNA backbone is shown as lines with sub-site positions labeled. Taken from [57]. doi:10.1128/microbiolspec.MDNA3-0004-2014.f8

the donor DNA was slow (hours) and took place in a random, non-concerted manner in which one transposon end was cleaved and then later the second end was cleaved [62, 63]. Third, in the absence of GTP the tetrameric transposase only bound to one transposon end (Fig. 6C; [62]) and upon addition of GTP or non-hydrolyzable GTP analogs synaptic complexes formed [62, 63]. Thus, GTP promotes synapsis of the two transposon ends, a critical step in the P element transposition pathway, possibly by reorienting one of the THAP domains in the transposase tetramer.

II.F. Target Site Selection in *Drosophila* P Element Transposition

All of the P elements that were analyzed initially [12] and subsequently, including a recent set of 2266 insertions from the Berkeley *Drosophila* Genome Project (BDGP) showed that an 8bp duplication of target DNA occurs when P elements integrate [85]. Bioinformatic analysis of this data and additional data from the *Drosophila* genome project showed that the insertion sites tended to be GC-rich and, in fact, that there was a symmetric palindromic pattern of 14bp centered on the 8bp target duplication found at the insertion sites with a conserved consensus motif (Fig. 9) [55, 56, 85]. This pattern would make sense if the recognition of target site DNA required binding of two (or an even number of) subunits of the transposase protein in a synaptic complex juxtaposing the two P element ends with the insertion sites in the target DNA.

It was found that in cells P elements normally can transpose preferentially in *cis*, within about 50–100kb from the initial location, so-called "local hopping" [44, 45]. It has also been noted that P elements tend to insert near the 5′ ends of genes, near promoters, and a correlation was made to sites of binding of the DNA replication factor ORC [86], but this may simply reflect an open chromatin organization at *Drosophila* gene promoters.

II.G. DNA Repair Pathways Involved in P Element Transposition in *Drosophila*

Excision of a P element results in a double-strand DNA break, which can then be repaired via a gap repair process using a homologous chromosome or sister chromatid as a template [58, 87]. It was also shown that ectopic templates could be used for gap repair with about 30bp of homology and a dramatic preference for template use in *cis*, on the same chromosome, was observed [88–90]. Incomplete copying can also explain how internally-deleted P elements are generated in natural P strains. The generation of internally-deleted products are consistent with a model for the gap repair

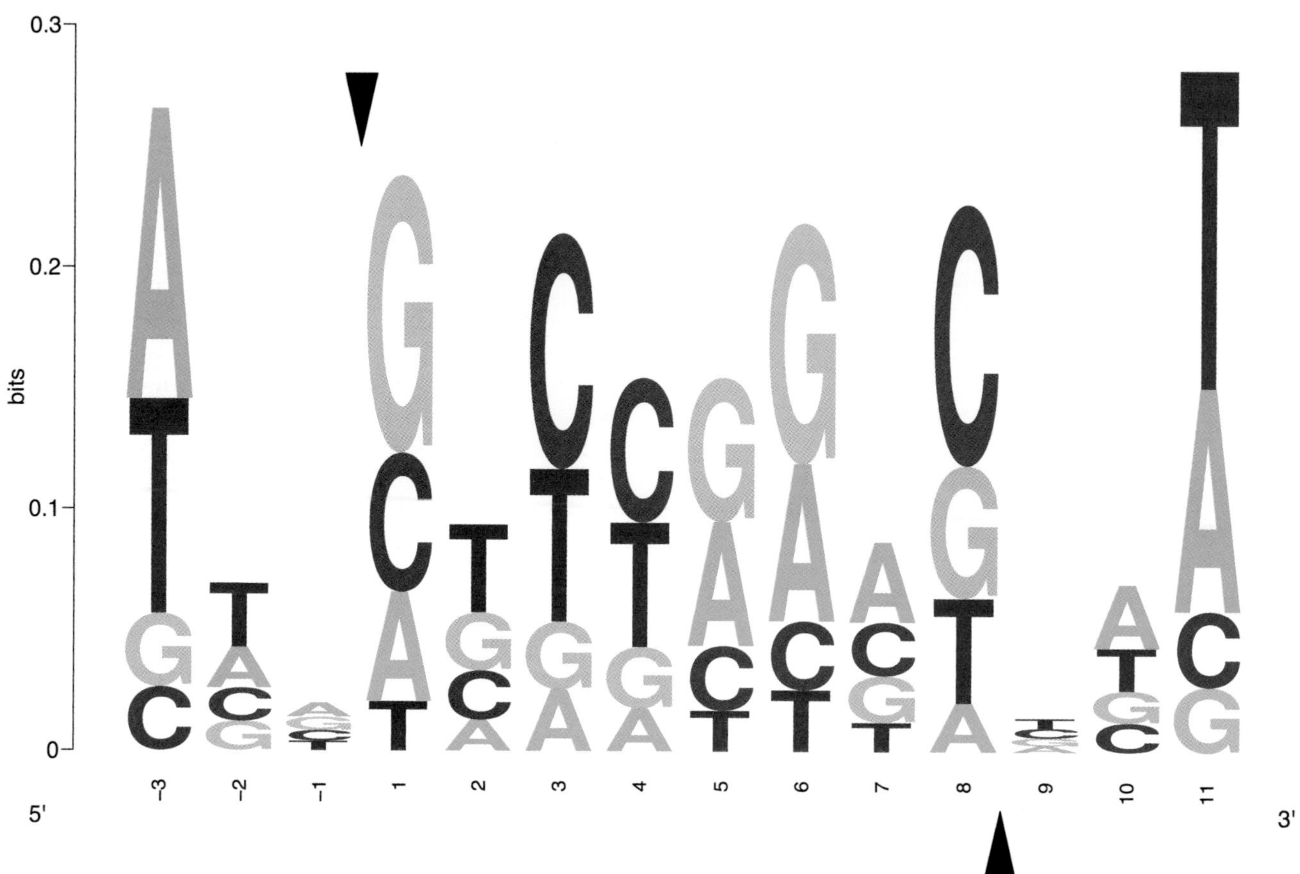

Figure 9 Consensus target site for P element integration. A 14bp palindromic motif deduced from analysis of > 20,000 P element insertions displayed as a position-specific scoring matrix (PSSM). Taken from [55, 56] and C. Bergman, personal communication. doi:10.1128/microbiolspec.MDNA3-0004-2014.f9

process occurring via a synthesis-dependent strand annealing (SDSA) mechanism [91, 92] first described for replication-linked recombination in bacteriophage T4 [93]. This template-directed DNA repair following P element excision suggests a way for P elements to increase in copy number since the gap repair process restores a new copy of the P element at the donor site, while the excised element transposes to a new location (Fig. 10).

Molecular and genetic studies have allowed testing of the involvement of different DNA repair pathways in the repair of P element-induced or other types of DNA breaks. The mus309 gene encodes the *Drosophila* homolog of the Bloom's DNA repair helicase (DmBLM) [94]. *Mus309* mutants are sterile and mutagen-sensitive [95]. They are also defective for repair of P element transposase-induced DNA double-strand breaks [47, 48, 91, 92, 94]. Other studies have shown that *mus309* mutants are defective for repair by the homology-directed SDSA pathway [91] and that the defects of *mus309* mutants can be rescued by transgenes encoding Ku70, a subunit of the heterodimeric Ku complex involved in the non-homologous end joining (NHEJ) pathway [47, 94]. Using RNA interference and plasmid-based assays, similar conclusions were reached regarding an interplay or competition between these two repair pathways [96]. More detailed genetic analyses have indicated differential requirements for DmBLM during development [97] and that in the absence of DmBLM there is an increase in the use of other DNA repair pathways, such as single-strand annealing (SSA) [98] or non-homologous end-joining (NHEJ) can then be used [99–101]. More recently, microhomology-dependent pathways for repair have been revealed in Drosophila [102]. A series of studies using transgenic reporters to detect alternate DNA repair pathways using either P element or I-Sce I nuclease-induced DNA breaks showed that different repair pathways can be

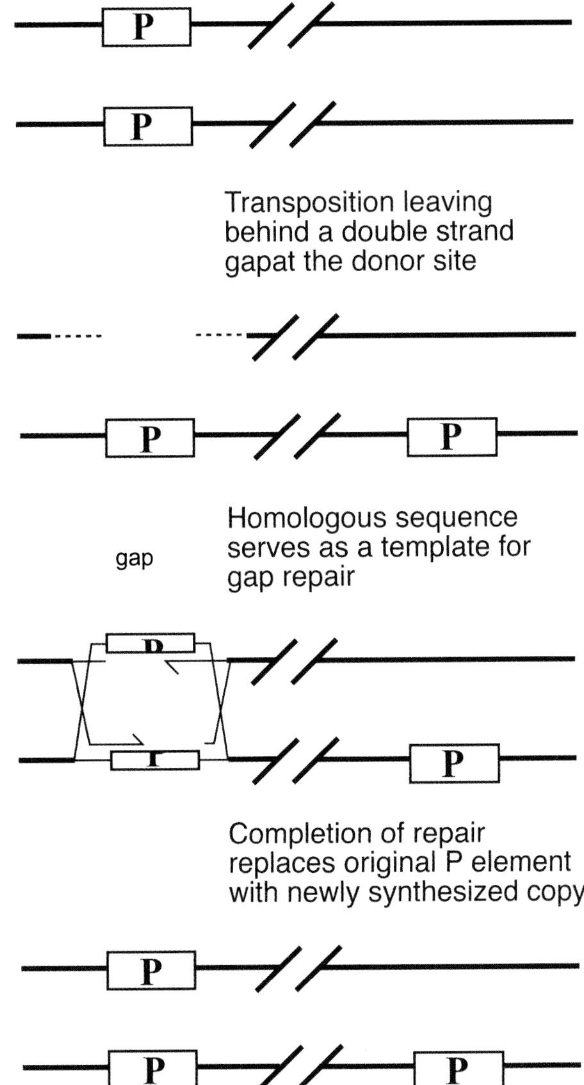

Figure 10 Homology-dependent gap repair following P element excision via the SDSA (synthesis-dependent strand annealing) pathway. Homologous chromosomes or sister chromatids, which after undergoing P element excision leave a double strand gap at the donor site. The homologous sequence then serves as a template for synthesis dependent strand annealing synthesis (SDSA) [91, 93, 157]. Completion of DNA repair replaces the original P element with a newly synthesized copy. If DNA synthesis during this gap repair process is incomplete, internal deletions of the P element would result.
doi:10.1128/microbiolspec.MDNA3-0004-2014.f10

used to repair these double-strand DNA breaks [99, 100]. Interestingly, when one DNA repair pathway is disrupted, others compensate to provide a means of genome stability. Thus, P elements are capable of efficiently using cellular DNA repair pathways.

III. REGULATION OF P ELEMENT TRANSPOSITION IN *DROSOPHILA*

III.A. Hybrid Dysgenesis, P Cytotype, the Piwi-Interacting (piRNA) Small RNA Pathway, Adaptation and Paramutation

Hybrid dysgenesis is the term used to describe a collection of symptoms including high rates of sterility, mutation induction, male recombination (which does not normally occur in *Drosophila*) and chromosomal abnormalities and rearrangements [2–8, 11]. Hybrid dysgenesis and its associated sterility and mutation induction are normally only observed in the germlines, but not in somatic tissues, of progeny from crosses in which males carrying P transposable elements (termed P or paternally-contributing strains) are mated to females that lack autonomously mobile P elements (termed M or maternally-contributing strains) (Fig. 11). Initial studies on the reciprocal cross effect in hybrid dysgenic crosses led to the description of two regulatory states based on genetic crosses between P and M strains. M strains were said to have an M cytotype state, which was permissive for P element mobility, whereas P strains were said to have P cytotype, a state which is restrictive for P element movement [103, 104]. These reciprocal cross experiments followed the repressive effect of P strain females for several generations [103, 105, 106]. The segregation pattern of P cytotype exhibited some aspects of strict maternal inheritance in that the cytotype of the great grandmother played a role in determining whether subsequent progeny displayed M or P cytotype. Studies of the regulatory effects of P cytotype have used several different genetic assays (Table I), which can differ in their tissue specificity and whether or not they directly assay P element mobility.

The study of one repressor-producing P strain derived from a wild population, called *Lk-P*(1A), has been particularly informative regarding the mechanism and genetics of P cytotype regulation [107]. *Lk-P*(1A) contains two full-length P elements near the telomere of the X chromosome at cytological position 1A, which lie in inverted orientation separated by ~5kb; these elements are integrated into sub-telomeric heterochromatic repeat sequences known as *TAS* repeats (for Telomere-Associated Sequences) [108, 109], which have been shown to be hotspots for P element insertion [110]. *Lk-P*(1A) exhibits a very strong P cytotype repressive effect characteristic of the regulatory properties observed with normal P strains. Studies on *Lk-P* (1A) used either of two genetic tests for P element cytotype control, gonadal dysgenic (GD) sterility or the

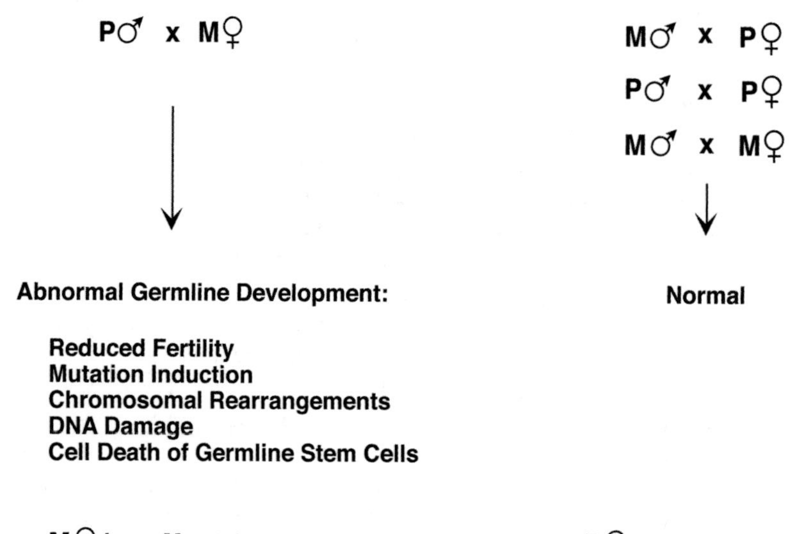

Figure 11 The genetics and symptoms of hybrid dysgenesis. The reciprocal crosses of hybrid dysgenesis are shown. Only when P strain males are mated to M strain females does abnormal germline development occur, due to high rates of P element transposition. Progeny from reciprocal M male by P female, P × P or M × M crosses are normal. M females give rise to eggs with a state permissive for P element transposition (M cytotype) whereas P females give rise to eggs with a state restrictive for P element transposition (P cytotype). doi:10.1128/microbiolspec.MDNA3-0004-2014.f11

singed-weak (sn^w) hypermutability assays (Table I) [108]. *Lk-P*(1A) displayed a P cytotype repression effect equal to that observed with several natural P stains carrying twenty to thirty times the number of P elements [108, 111]. The repressive properties of *Lk-P*(1A) were found to be strong in germline tissues and were maternally transmitted, but repression was observed only weakly in somatic tissues [108]. By contrast, a natural P strain, such as Harwich, displays the repressive properties of P cytotype strongly in both somatic and germline tissues [108]. More recent studies on Lk-P(1A), as well as other telomeric P element insertions [112–115], has shown that there is a connection to the maternal inheritance of P cytotype, the germline piRNA pathway [116–122] and the heterochromatin protein Su(var)205 (HP1) [121–123].

One of the most unusual features of hybrid dysgenesis is the multigenerational inheritance of the P element-repressive state known as P cytotype. One important connection that was made in this regard came from the analysis of small RNAs, known as piwi-interacting RNAs or piRNAs, that co-purified with the *Drosophila* germline-specific Argonaute family members, piwi, aubergine and Ago-3 [117, 124]. The application of small RNA cloning and high-throughput sequencing showed that these proteins bound pools of small RNAs that were derived largely from transposable elements, including P elements. These findings illuminated how sites in the genome, such as the flamenco locus which contains an endogenous I factor polyA retrotransposon, can act as piRNA-generating loci that generate large amounts of piRNAs that silence endogenous or exogenous I factor transposons. It is thought that small piRNAs, complementary to endogenous P element mRNA, are transmitted to P strain oocytes and serve to silence P elements introduced by P strain sperm [117]. The piRNA system also functions to control transposon mobility in the mouse germline [125]. More recent studies have revealed connections of piRNAs to deposition of repressive histone chromatin marks and reduced levels of gene expression [126].

One distinguishing feature of P strains isolated from the wild is the presence of 50–60 P element copies, about a third of which are full-length 2.9kb P elements. This evolution of a P strain has been recapitulated in population cages after transformation of M strains with single P elements, where the generation of P cytotype correlates with the accumulation of ~ 50 P elements [13, 127, 128]. More recently, high-throughput sequencing was used to follow restoration of fertility in

Table I Genetic assays for P cytotype repression

Assays	Tissues	References
singed-weak (sn^w) test	germline	[16, 52, 158]
gonadal dysgenic sterility	germline	[159]
singed female sterility (cytotype-dependent alleles)	germline	[22]
singed-weak (sn^w) bristle mosaics	soma	[135]
P[w⁺] white gene excision eye color mosaics	soma	[18]
P[w⁺] white gene transposition eye color mosaics	soma	[18]
modified *P[w⁺] white* gene expression	soma	[160]
suppression of Δ2-3 X Birm2 lethality	soma	[161]
singed bristle phenotype (cytotype-dependent alleles)	soma	[22]
vestigial wing phenotype (cytotype-dependent alleles)	soma	[22, 162]
P *[LacZ]* gene enhancer trap β-galactosidase expression	soma and/or germline	[111, 163]
Pre-P cytotype	germline	[164]
Trans-silencing	germline	[151, 165]
Combination effect	germline	[166]

sterile P-M dysgenic hybrids [129]. This restoration of fertility correlated with insertion of transposons into piRNA clusters resulting in the production of piRNAs directed to the P element transcript. Interestingly, this study also showed that P elements and other transposons were mobilized during hybrid dysgenesis. Thus, hybrid dysgenesis results in transient transposon activation and then insertion of elements at new locations leads to transposon silencing by the piRNA pathway and a concomitant restoration of fertility [129]. These findings can explain how the P elements in wild populations may have spread so quickly in nature.

Paramutation is defined as an epigenetic interaction between two alleles of a locus, through which one allele induces a heritable change in the other allele without modifying the DNA sequence. In the case of P elements, P cytotype could be induced by telomeric P elements or clusters of P element transgenes to cause a homology-dependent silencing termed the trans-silencing effect (TSE; [107]). Recent studies using genetic crosses and high-throughput sequencing of small RNAs showed that clusters of P element-derived transgenes can convert other, previously inactive homologous transgene clusters into strong mediators of TSE [130]. Interestingly, this TSE can be transmitted through 50 generations and occurs without any chromosome pairing between the paramutagenic and paramutated chromosomal loci [130]. This multigenerational paramutational effect is mediated by inheritance of maternal cytoplasm carrying piRNAs homologous to the P element transgenes and requires the aubergine gene product, which is involved in piRNA biogenesis, but not Dicer-2 which is involved in siRNA production. This landmark study provides a genetic basis for the multigenerational inheritance of P cytotype.

III.B. Control of the Tissue-Specificity of *Drosophila* P Element Transposition by Alternative RNA Splicing

One of the most startling findings regarding the regulation of P element transposition was the discovery that alternative pre-mRNA splicing was responsible for restricting expression of the P element transposase, and hence the entire syndrome of traits associated with hybrid dysgenesis, to germline cells [18]. It is now known that pre-mRNA splicing is a widely used regulatory mechanism for expanding proteomic diversity and increased splicing is correlated with organismal complexity [131]. The basic biochemical mechanism of pre-mRNA intron removal by the spliceosome is conserved in eukaryotic cells and alternative pre-mRNA splicing is mediated by RNA regulatory motifs called enhancers or silencers, which can be located in either introns or exons [132]. The tissue-specific splicing of the P element third intron (IVS3) has served as an important model system for investigating how alternative RNA splicing patterns are generated in distinct cell or tissue types and has led to detailed characterization of the first exonic splicing silencer (ESE) upstream of the P element third intron [133] (Fig. 12A).

Two lines of investigation of IVS3 splicing led to the conclusion that this control involves an inhibition of IVS3 splicing in somatic cells. First, a molecular genetic approach showed that mutations in the 5′ exon, upstream from IVS3, caused an activation of IVS3 splicing in somatic cells [18, 134, 135]. Specific sequence changes in the 5′ exon activated IVS3 splicing in somatic cells [134]. Second, a set of biochemical experiments showed that while IVS3 splicing occurred in mammalian cell splicing extracts, it was not observed in the *Drosophila* somatic cell extracts [136]. Titration of 5′ exon RNA into the somatic cell extract activated IVS3 splicing, suggesting that the action of *trans*-acting factors led to the inhibitory effect observed [136, 137]. It was noted that both IVS3 and the 5′ exon had a number of 5′ splice site-like sequences and that these pseudo-5′ splice sites might play a role in regulating IVS3 splicing [136, 137] (Fig. 12A). Another study using mammalian splicing extracts showed that mutations in

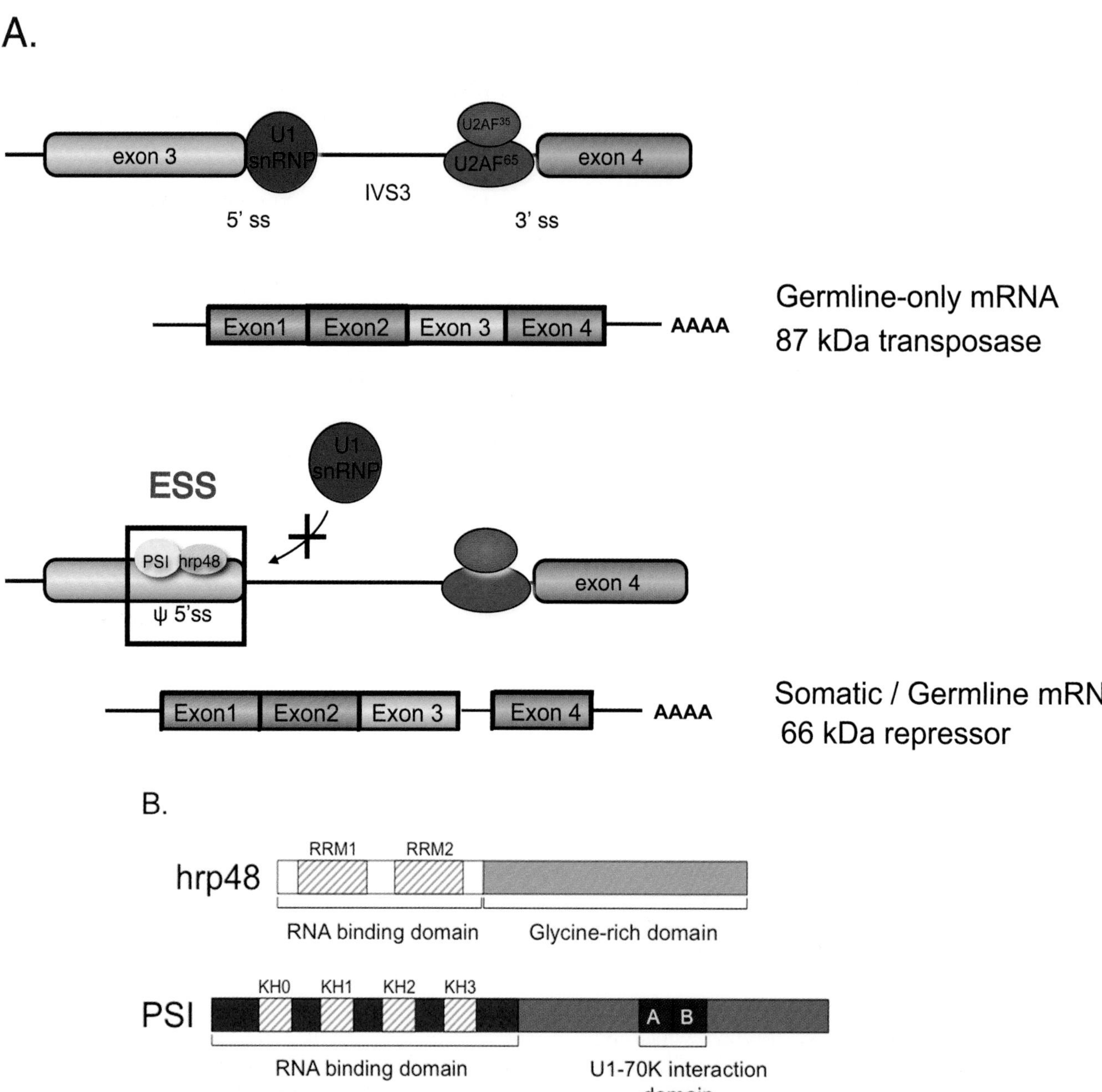

Figure 12 Model for somatic inhibition of IVS3 splicing and splicing factors involved. A.) U1 snRNP (small nuclear ribonucleoprotein particle) normally interacts with the IVS3 5′ splice site (5′ SS) during the early steps of intron recognition and spliceosome assembly. In somatic cells (and *in vitro*) this site is blocked [137]. Mutations in the upstream negative regulatory element lead to activation of IVS3 splicing *in vivo* [134] and *in vitro* [137, 138]. The F1 site is known to bind U1 snRNP [137, 144] and the F2 site is known to bind the hnRNP protein, hrp48 [139]. An RNA binding protein containing four KH-domains which is expressed highly in somatic cells, called PSI, has also been implicated in IVS3 splicing control [143]. B.) Diagram of the domain organization of PSI and hrp48. PSI contains four N-terminal KH-type RNA binding domains and a reiterated 100 amino segment (A and B domains) that interacts with the U1 snRNP 70K protein. Hrp48 contains two N-terminal RRM-type RNA binding domains and a low complexity $(RGG)_n$ glycine-rich C-terminal domain. doi:10.1128/microbiolspec.MDNA3-0004-2014.f12

the 5′ exon known to activate IVS3 splicing in *Drosophila*, also activated IVS3 splicing *in vitro*, suggesting the possible conservation of components [138]. Indeed, subsequent studies using *in vitro* splicing assays with *Drosophila* extracts showed that IVS3 splicing was activated by mutations in the 5′ exon [137]. RNA-protein interaction studies defined two elements in the 5′ exon termed F1 and F2, both of which bear sequence identity to 5′ splice sites and contain a reiterated sequence motif (AGNUUAAG) [137]. Mutations in the F1 and F2 sites that activate splicing *in vitro*, inhibit RNA-protein complex formation in nuclear extracts [137]. The F1 site binds U1 snRNP [137] and the F2 site binds hrp48 [139] (Fig. 12B), a *Drosophila* hnRNP protein similar to mammalian hnRNP A1 [140]. Interestingly, hnRNP A1 causes use of distal (upstream) 5′ splice sites *in vitro* [141] and binding site selection data indicate it binds to sites resembling 5′ splice sites [142]. These results led to a model in which RNA-protein and -snRNP interactions in the 5′ exon exonic splicing silencer (ESS) cause an inhibition of IVS3 splicing by blocking access of U1 snRNP to the normal IVS3 5′ splice site [137] (Fig. 12A).

Subsequent biochemical studies identified additional proteins that interact with the IVS3 5′ exon exonic splicing silencer. Proteins of 97kD and 50kD proteins were identified as PSI (P element somatic inhibitor) and hrp48, respectively (Fig. 12B). PSI contains four KH-type RNA binding motifs and two C-terminal direct ~100 amino acid repeats [143]. Conserved residues in these repeat motifs are involved in direct interaction between PSI and the U1 snRNP 70K protein [144, 145]. A similar motifs are found in a mammalian alternative splicing factor called KH-type splicing regulatory protein (KSRP) and a related family of proteins called FUSE-binding proteins (FBPs) [146]. PSI interacts with the IVS3 5′ exon RNA, but not with heterologous RNAs [143]. Preferred RNA binding sites and RNA binding by the individual PSI KH domains have been examined [147, 148]. PSI protein is expressed highly in somatic cells, but at low or undetectable levels in the female germline [143]. The hrp48 protein specifically binds to the F2 element in the 5′ exon [139] and is expressed in both the germline and soma [143]. Hrp48 contains two N-terminal RNP-CS type RNA binding domains and a C-terminal glycine-rich domain [140] (Fig. 12B). This structure is characteristic for this class of hnRNP proteins, termed 2XRBD-GLY, of which mammalian hnRNP A1 is a member [149]. Interestingly, these glycine-rich low complexity protein domains have been implicated in human diseases [150]. The expression pattern of hrp48 in both germline and soma suggests that hrp48 might also play a role in the inhibition of IVS3 splicing in the germline, where IVS3 is known to be inefficiently spliced [151]. Our model for splicing repression is consistent with the general idea that RNA binding proteins can direct spliceosome components to specific sites in pre-mRNAs to generate differential splicing patterns [132, 133].

Molecular genetic and genomic studies have addressed the roles of PSI and hrp48 in IVS3 splicing control *in vivo*. First, ectopic expression of PSI in germline cells caused a modest reduction in IVS3 splicing [152]. Anti-sense hammerhead ribozyme targeting of PSI mRNA in somatic cells resulted in activation of IVS3 splicing in the soma [152]. Both experiments suggest a role for PSI in the reduction of IVS3 splicing. Reduction of hrp48 levels in somatic cells using hypomorphic P element insertion alleles caused a small activation of IVS3 splicing in the soma [153]. These experiments also showed that hrp48 was encoded by an essential gene, and therefore must have additional functions involving cellular mRNAs. Genetic analysis of the PSI gene showed it to be essential and that the AB-repeat domain which interacts with U1 snRNP was essential for male fertility and normal courtship and mating behavior [154], suggesting that PSI must regulate RNA processing of other transcripts in somatic and male germline tissues. More recently, characterization of the general role of hrp48 in alternative splicing has been investigated using splice junction microarrays and RNA immunopurification procedures to demonstrate a general role for hrp48 as a splicing repressor protein [155, 156]. These studies indicate how effectively the P element has made use of cellular RNA binding proteins to control the tissue-specificity of P element RNA processing.

IV. SUMMARY AND CONCLUSION

P elements invaded *Drosophila* in the early 20th century and spread rapidly through wild populations to avert the deleterious effects of hybrid dysgenesis. The creation of mutational insertions allowed the molecular cloning of P elements and their use as tools for *Drosophila* molecular genetics. Perhaps the most unexpected finding regarding P elements comes from the genome sequencing efforts of the past decade and the realization that P element-like genes or transposons (THAP9) exist in a variety of other animals. The THAP DNA binding domain, found initially to be at the N-terminus of the P element transposase, is now one of the most common animal-specific zinc-coordinating site-specific DNA binding domains. Finally, the unique role of GTP in the P element reaction pathway, as an allosteric effector that promotes synapsis of the transposon ends shed light on how the complex protein-DNA assembly of the P element transpososome is initiated. We anticipate that much more is to be learned by continued study of this now-widespread family of eukaryotic transposons.

Acknowledgments. *We thank our colleagues in the Rio Lab, especially Mei Tang for their insights and critical discussions. We thank Yeon Lee and Lucas Horan for critical reading of the manuscript. We also thank James Berger and Artem Lyubimov for their efforts on the DmTHAP X-ray crystal structure. We thank a non-anonymous reviewer for helpful comments on the manuscript. Due to space limitations, we have not been able to cite all of the primary research articles related to each topic. We also express our gratitude for continued interest and funding from the NIH.*

Citation. Majumdar S, Rio DC. 2014. P Transposable elements in Drosophila and other eukaryotic organisms. Microbiol Spectrum 3(2):MDNA3-0004-2014.

References

1. **Hiraizumi Y.** 1971. Spontaneous recombination in Drosophila melanogaster males. *Proc Natl Acad Sci USA* **68** (2):268–270.
2. **Engels WR.** 1983. The P family of transposable elements in Drosophila. *Annual Review of Genetics* **17**:315–344.
3. **Engels WR.** 1989. P elements in Drosophila melanogaster, p 437–484. *In* Berg DE, Howe MM (ed), *Mobile DNA*. American Society for Microbiology, Washington, D.C.
4. **Engels WR.** 1992. The origin of P elements in Drosophila melanogaster. *Bioessays* **14**(10):681–686.
5. **Engels WR.** 1996. P elements in Drosophila. *Curr Top Microbiol Immunol* **204**:103–123.
6. **Engels WR.** 1997. Invasions of P elements. *Genetics* **145** (1):11–15.
7. **Kidwell MG.** 1992a. Horizontal transfer of P elements and other short inverted repeat transposons. *Genetica* **86**(1–3):275–286.
8. **Kidwell MG.** 1992b. Horizontal transfer. *Curr Opin Genet Dev* **2**(6):868–873.
9. **Rio DC.** 1990. Molecular mechanisms regulating Drosophila P element transposition. **24**:543–578.
10. **Rio DC.** 1991. Regulation of Drosophila P element transposition. *Trends in Genetics* **7**:282–287.
11. **Rio DC.** 2002. P Transposable Elements in Drosophila melanogaster, p 484–518. *In* Craig NL, Cragie R, Gellert M, Lambowitz AM (ed), *Mobile DNA II*. Amer Society for Microbiology, Washington, DC.
12. **O'Hare K, Rubin GM.** 1983. Structures of P transposable elements and their sites of insertion and excision in the Drosophila melanogaster genome. *Cell* **34**:25–35.
13. **O'Hare K, et al.** 1992. Distribution and structure of cloned P elements from the Drosophila melanogaster P strain pi 2. *Genet Res* **60**(1):33–41.
14. **Lee CC, Beall EL, Rio DC.** 1998. DNA binding by the KP repressor protein inhibits P element transposase activity in vitro. *EMBO J* **17**:4166–4174.
15. **Kaufman PD, Doll RF, Rio DC.** 1989. Drosophila P element transposase requires internal P element DNA sequences. *Cell* **38**:135–146.
16. **Karess RE, Rubin GM.** 1984. Analysis of P transposable element function in Drosophila. *Cell* **38**:135–146.
17. **Rio DC, Laski FA, Rubin GM.** 1986. Identification and immunochemical analysis of biologically active Drosophila P element transposase. *Cell* **44**:21–32.
18. **Laski FA, Rio DC, Rubin GM.** 1986. Tissue specificity of Drosophila P element transposition is regulated at the level of mRNA splicing. *Cell* **44**:7–19.
19. **Gloor GB, et al.** 1993. Type I repressors of P element mobility. *Genetics* **135**(1):81–95.
20. **Misra S, Rio DC.** 1990. Cytotype control of Drosophila P element transposition: the 66 kd protein is a repressor of transposase activity. *Cell* **62**(2):269–284.
21. **Misra S, et al.** 1993. Cytotype control of Drosophila melanogaster P element transposition: genomic position determines maternal repression. *Genetics* **135**(3):785–800.
22. **Robertson HM, Engels WR.** 1989. Modified P elements that mimic the P cytotype in Drosophila melanogaster. *Genetics* **123**(4):815–824.
23. **Black DM, et al.** 1987. KP elements repress P-induced hybrid dysgenesis in Drosophila melanogaster. *The EMBO Journal* **6**:4125–4135.
24. **Jackson MS, Black DM, Dover GA.** 1988. Amplification of KP elements associated with the repression of hybrid dysgenesis in Drosophila melanogaster. *Genetics* **120**:1003–1013.
25. **Simmons MJ, et al.** 2002. Regulation of P-element transposase activity in Drosophila melanogaster by hobo transgenes that contain KP elements. *Genetics* **161** (1):205–215.
26. **Lander ES, et al.** 2001. Initial sequencing and analysis of the human genome. *Nature* **409**(6822):860–921.
27. **Hagemann S, Pinsker W.** 2001. Drosophila P transposons in the human genome? *Mol Biol Evol* **18**(10): 1979–1982.
28. **Roussigne M, et al.** 2003. The THAP domain: a novel protein motif with similarity to the DNA-binding domain of P element transposase. *Trends Biochem Sci* **28** (2):66–69.
29. **Hellsten U, et al.** 2010. The genome of the Western clawed frog Xenopus tropicalis. *Science* **328**(5978):633–636.
30. **Kimbacher S, et al.** 2009. Drosophila P transposons of the urochordata Ciona intestinalis. *Mol Genet Genomics* **282**(2):165–172.
31. **Chapman JA, et al.** 2010. The dynamic genome of Hydra. *Nature* **464**(7288):592–596.
32. **Carlton JM, et al.** 2007. Draft genome sequence of the sexually transmitted pathogen Trichomonas vaginalis. *Science* **315**(5809):207–212.
33. **Hammer SE, Strehl S, Hagemann S.** 2005. Homologs of Drosophila P transposons were mobile in zebrafish but have been domesticated in a common ancestor of chicken and human. *Mol Biol Evol* **22**(4):833–844.

34. **Hagemann S, Hammer SE.** 2006. The implications of DNA transposons in the evolution of P elements in zebrafish (Danio rerio). *Genomics* **88**(5):572–579.

35. **Majumdar S, Singh A, Rio DC.** 2013. The human THAP9 gene encodes an active P-element DNA transposase. *Science* **339**(6118):446–448.

36. **Venken KJ, Bellen HJ.** 2007. Transgenesis upgrades for Drosophila melanogaster. *Development* **134**(20):3571–3584.

37. **Bachmann A, Knust E.** 2008. The use of P-element transposons to generate transgenic flies. *Methods Mol Biol* **420**:61–77.

38. **Venken KJ, Bellen HJ.** 2012. Genome-wide manipulations of Drosophila melanogaster with transposons, Flp recombinase, and PhiC31 integrase. *Methods Mol Biol* **859**:203–228.

39. **Cooley L, Kelley R, Spradling A.** 1988. Insertional mutagenesis of the Drosophila genome with single P elements. *Science* **239**(4844):1121–1128.

40. **Hummel T, Klambt C.** 2008. P-element mutagenesis. *Methods Mol Biol* **420**:97–117.

41. **Bellen HJ, et al.** 2011. The Drosophila gene disruption project: progress using transposons with distinctive site specificities. *Genetics* **188**(3):731–743.

42. **Bellen HJ, et al.** 2004. The BDGP gene disruption project: single transposon insertions associated with 40% of Drosophila genes. *Genetics* **167**(2):761–781.

43. **Spradling AC, et al.** 1995. Gene disruptions using P transposable elements: an integral component of the Drosophila genome project. *Proc Natl Acad Sci U S A* **92**(24):10824–10830.

44. **Tower J, et al.** 1993. Preferential transposition of Drosophila P elements to nearby chromosomal sites. *Genetics* **133**(2):347–359.

45. **Zhang P, Spradling AC.** 1993. Efficient and dispersed local P element transposition from Drosophila females. *Genetics* **133**(2):361–373.

46. **Adams MD, Sekelsky JJ.** 2002. From sequence to phenotype: reverse genetics in Drosophila melanogaster. *Nat Rev Genet* **3**(3):189–198.

47. **Beall EL, Rio DC.** 1996. Drosophila IRBP/Ku p70 corresponds to the mutagen-sensitive mus309 gene and is involved in P-element excision in vivo. *Genes Dev* **10:** 921–933.

48. **McVey M, et al.** 2004. Formation of deletions during double-strand break repair in Drosophila DmBlm mutants occurs after strand invasion. *Proc Natl Acad Sci U S A* **101**(44):15694–15699.

49. **Witsell A, et al.** 2009. Removal of the bloom syndrome DNA helicase extends the utility of imprecise transposon excision for making null mutations in Drosophila. *Genetics* **183**(3):1187–1193.

50. **Maggert KA, Gong WJ, Golic KG.** 2008. Methods for homologous recombination in Drosophila. *Methods Mol Biol* **420**:155–174.

51. **Venken KJ, et al.** 2006. P[acman]: a BAC transgenic platform for targeted insertion of large DNA fragments in D. melanogaster. *Science* **314**(5806):1747–1751.

52. **Spradling AC, Rubin GM.** 1982. Transposition of cloned P elements into Drosophila germ line chromosomes. *Science* **218**:341–347.

53. **Mullins MC, Rio DC, Rubin GM.** 1989. Cis-acting DNA sequence requirements for P-element transposition. *Genes Dev* **3**(5):729–738.

54. **Lee CC, Mul YM, Rio DC.** 1996. The Drosophila P-element KP repressor protein dimerizes and interacts with multiple sites on P-element DNA. *Mol Cell Biol* **16**:5616–5622.

55. **Linheiro RS, Bergman CM.** 2008. Testing the palindromic target site model for DNA transposon insertion using the Drosophila melanogaster P-element. *Nucleic Acids Res* **36**(19):6199–6208.

56. **Linheiro RS, Bergman CM.** 2012. Whole genome resequencing reveals natural target site preferences of transposable elements in Drosophila melanogaster. *PLoS One* **7**(2):e30008.

57. **Sabogal A, et al.** 2010. THAP proteins target specific DNA sites through bipartite recognition of adjacent major and minor grooves. *Nat Struct Mol Biol* **17**(1):117–123.

58. **Kaufman PD, Rio DC.** 1992. P element transposition in vitro proceeds by a cut-and-paste mechanism and uses GTP as a cofactor. *Cell* **69**(1):27–39.

59. **Bourne HR, Sanders DA, McCormick F.** 1990. The GTPase superfamily: a conserved switch for diverse cell functions. *Nature* **348**:125–132.

60. **Bourne HR, Sanders DA, McCormick F.** 1991. The GTPase superfamily: conserved structure and molecular mechanism. *Nature* **349**:117–127.

61. **Mul YM, Rio DC.** 1997. Reprogramming the purine nucleotide cofactor requirement of Drosophila P element transposase in vivo. *EMBO J* **16**(14):4441–4447.

62. **Tang M, et al.** 2007. Analysis of P element transposase protein-DNA interactions during the early stages of transposition. *J Biol Chem* **282**(39):29002–29012.

63. **Tang M, et al.** 2005. Guanosine triphosphate acts as a cofactor to promote assembly of initial P-element transposase-DNA synaptic complexes. *Genes Dev* **19** (12):1422–1425.

64. **Hickman AB, Chandler M, Dyda F.** 2010. Integrating prokaryotes and eukaryotes: DNA transposases in light of structure. *Crit Rev Biochem Mol Biol* **45**(1):50–69.

65. **Dyda F, Chandler M, Hickman AB.** 2012. The emerging diversity of transpososome architectures. *Q Rev Biophys* **45**(4):493–521.

66. **Fugmann SD, et al.** 2000. Identification of two catalytic residues in RAG1 that define a single active site within the RAG1/RAG2 protein complex. *Mol Cell* **5**(1):97–107.

67. **Kim DR, et al.** 1999. Mutations of acidic residues in RAG1 define the active site of the V(D)J recombinase. *Genes Dev* **13**(23):3070–3080.

68. **Landree MA, Wibbenmeyer JA, Roth DB.** 1999. Mutational analysis of RAG1 and RAG2 identifies three catalytic amino acids in RAG1 critical for both cleavage steps of V(D)J recombination. *Genes Dev* **13**(23): 3059–3069.

69. **Yang W, Lee JY, Nowotny M.** 2006. Making and breaking nucleic acids: two-Mg2+-ion catalysis and substrate specificity. *Mol Cell* **22**(1):5–13.

70. **Rice P.** pers. comm.

71. **Yuan YW, Wessler SR.** 2011. The catalytic domain of all eukaryotic cut-and-paste transposase superfamilies. *Proc Natl Acad Sci U S A* **108**(19):7884–7889.

72. **Wittinghofer A, Pai EF.** 1991. The structure of Ras protein: a model for a universal molecular switch. *Trends Biochem Sci* **16**:382–387.

73. **Gasper R, et al.** 2009. It takes two to tango: regulation of G proteins by dimerization. *Nat Rev Mol Cell Biol* **10**(6):423–429.

74. **Beall EL, Rio DC.** 1997. Drosophila P-element transposase is a novel site-specific endonuclease. *Genes Dev* **11**:2137–2151.

75. **Staveley BE, et al.** 1995. Protected P-element termini suggest a role for inverted-repeat-binding protein in transposase-induced gap repair in Drosophila melanogaster. *Genetics* **139**(3):1321–1329.

76. **Beall EL, Rio DC.** 1998. Transposase makes critical contacts with, and is stimulated by, single- stranded DNA at the P element termini in vitro. *Embo J* **17**(7):2122–2136.

77. **Savilahti H, Rice PA, Mizuuchi K.** 1995. The phage Mu transpososome core: DNA requirements for assembly and function. *Embo J* **14**(19):4893–4903.

78. **Montano SP, Pigli YZ, Rice PA.** 2012. The mu transpososome structure sheds light on DDE recombinase evolution. *Nature* **491**(7424):413–417.

79. **Scottoline BP, et al.** 1997. Disruption of the terminal base pairs of retroviral DNA during integration. *Genes Dev* **11**(3):371–382.

80. **Chow SA, et al.** 1992. Reversal of integration and DNA splicing mediated by integrase of human immunodeficiency virus. *Science* **255**(5045):723–726.

81. **Au TK, Pathania S, Harshey RM.** 2004. True reversal of Mu integration. *EMBO J* **23**(16):3408–3420.

82. **Vos JC, De Baere I, Plasterk RH.** 1996. Transposase is the only nematode protein required for in vitro transposition of Tc1. *Genes Dev* **10**(6):755–761.

83. **Melek M, Gellert M.** 2000. RAG1/2-mediated resolution of transposition intermediates: Two pathways and possible consequences. *Cell* **101**:625–633.

84. **Campagne S, et al.** 2010. Structural determinants of specific DNA-recognition by the THAP zinc finger. *Nucleic Acids Res* **38**(10):3466–3476.

85. **Liao GC, Rehm EJ, Rubin GM.** 2000. Insertion site preferences of the P transposable element in Drosophila melanogaster. *Proc Natl Acad Sci U S A* **97**(7):3347–3351.

86. **Spradling AC, Bellen HJ, Hoskins RA.** 2011. Drosophila P elements preferentially transpose to replication origins. *Proc Natl Acad Sci U S A* **108**(38):15948–15953.

87. **Engels WR, et al.** 1990. High-frequency P element loss in Drosophila is homolog-dependent. *Cell* **62**(3):515–525.

88. **Engels WR, Preston CR, Johnson-Schlitz DM.** 1994. Long-range cis preference in DNA homology search over the length of a Drosophila chromosome. *Science* **263**(5153):1623–1625.

89. **Dray T, Gloor GB.** 1997. Homology requirements for targeting heterologous sequences during P-induced gap repair in Drosophila melanogaster. *Genetics* **147**(2): 689–699.

90. **Keeler KJ, Gloor GB.** 1997. Efficient gap repair in Drosophila melanogaster requires a maximum of 31 nucleotides of homologous sequence at the searching ends. *Mol Cell Biol* **17**(2):627–634.

91. **Adams MD, McVey M, Sekelsky JJ.** 2003. Drosophila BLM in double-strand break repair by synthesis-dependent strand annealing. *Science* **299**(5604):265–267.

92. **McVey M, et al.** 2004. Evidence for multiple cycles of strand invasion during repair of double-strand gaps in Drosophila. *Genetics* **167**(2):699–705.

93. **Formosa T, Alberts BM.** 1986. DNA synthesis dependent on genetic recombination: characterization of a reaction catalyzed by purified bacteriophage T4 proteins. *Cell* **47**(5):793–806.

94. **Kusano K, Johnson-Schlitz DM, Engels WR.** 2001. Sterility of Drosophila with mutations in the Bloom syndrome gene–complementation by Ku70. *Science* **291** (5513):2600–2602.

95. **Boyd JB, et al.** 1981. Third-chromosome mutagen-sensitive mutants of Drosophila melanogaster. *Genetics* **97**(3–4):607–623.

96. **Min B, Weinert BT, Rio DC.** 2004. Interplay between Drosophila Bloom's syndrome helicase and Ku autoantigen during nonhomologous end joining repair of P element-induced DNA breaks. *Proc Natl Acad Sci U S A* **101**(24):8906–8911.

97. **McVey M, et al.** 2007. Multiple functions of Drosophila BLM helicase in maintenance of genome stability. *Genetics* **176**(4):1979–1992.

98. **Preston CR, Engels W, Flores C.** 2002. Efficient repair of DNA breaks in Drosophila: evidence for single-strand annealing and competition with other repair pathways. *Genetics* **161**(2):711–720.

99. **Preston CR, Flores CC, Engels WR.** 2006. Differential usage of alternative pathways of double-strand break repair in Drosophila. *Genetics* **172**(2):1055–1068.

100. **Johnson-Schlitz DM, Flores C, Engels WR.** 2007. Multiple-pathway analysis of double-strand break repair mutations in Drosophila. *PLoS Genet* **3**(4):e50.

101. **McVey M.** 2010. In vivo analysis of Drosophila BLM helicase function during DNA double-strand gap repair. *Methods Mol Biol* **587**:185–194.

102. **McVey M, Lee SE.** 2008. MMEJ repair of double-strand breaks (director's cut): deleted sequences and alternative endings. *Trends Genet* **24**(11):529–538.

103. **Kidwell MG, Kidwell JF.** 1975. Cytoplasm-chromosome interactions in Drosophila melanogaster. *Nature* **253**: 755–756.

104. **Kidwell MG, Kidwell JF, Sved JA.** 1977. Hybrid dysgenesis in Drosophila melanogaster: a syndrome of abberant traits including mutation, sterility and male recombination. *Genetics* **86**:813–833.

105. **Engels WR.** 1979a. Hybrid dysgenesis in Drosophila melanogaster: rules of inheritance of female sterility. *Genet Res Camb* **33**:219–236.

106. **Engels WR.** 1979b. Extrachromosomal control of mutability in Drosophila melanogaster. *Proc Natl Acad Sci U S A* **76**(8):4011–4015.

107. **Ronsseray S, et al.** 2003. Telomeric transgenes and trans-silencing in Drosophila. *Genetica* **117**(2–3):327–335.

108. **Ronsseray S, Lehmann M, Anxolabehere D.** 1991. The maternally inherited regulation of P elements in Drosophila melanogaster can be elicited by two P copies at cytological site 1A on the X chromosome. *Genetics* **129** (2):501–512.

109. **Ronsseray S, et al.** 1996. The regulatory properties of autonomous subtelomeric P elements are sensitive to a Suppressor of variegation in Drosophila melanogaster [published erratum appears in Genetics 1996 Nov;144 (3):1329]. *Genetics* **143**(4):1663–1674.

110. **Karpen GH, Spradling AC.** 1992. Analysis of subtelomeric heterochromatin in the Drosophila minichromosome Dp1187 by single P element insertional mutagenesis. *Genetics* **132**:737–753.

111. **Lemaitre B, Ronsseray S, Coen D.** 1993. Maternal repression of the P element promoter in the germline of Drosophila melanogaster: a model for the P cytotype. *Genetics* **135**(1):149–160.

112. **Stuart JR, et al.** 2002. Telomeric P elements associated with cytotype regulation of the P transposon family in Drosophila melanogaster. *Genetics* **162**(4):1641–1654.

113. **Simmons MJ, et al.** 2004. The P cytotype in Drosophila melanogaster: a maternally transmitted regulatory state of the germ line associated with telomeric P elements. *Genetics* **166**(1):243–254.

114. **Niemi JB, et al.** 2004. Establishment and maintenance of the P cytotype associated with telomeric P elements in Drosophila melanogaster. *Genetics* **166**(1):255–264.

115. **Thorp MW, Chapman EJ, Simmons MJ.** 2009. Cytotype regulation by telomeric P elements in Drosophila melanogaster: variation in regulatory strength and maternal effects. *Genet Res (Camb)* **91**(5):327–336.

116. **Aravin AA, Hannon GJ, Brennecke J.** 2007. The Piwi-piRNA pathway provides an adaptive defense in the transposon arms race. *Science* **318**(5851):761–764.

117. **Brennecke J, et al.** 2008. An epigenetic role for maternally inherited piRNAs in transposon silencing. *Science* **322**(5906):1387–1392.

118. **Reiss D, et al.** 2004. aubergine mutations in Drosophila melanogaster impair P cytotype determination by telomeric P elements inserted in heterochromatin. *Mol Genet Genomics* **272**(3):336–343.

119. **Josse T, et al.** 2007. Telomeric trans-silencing: an epigenetic repression combining RNA silencing and heterochromatin formation. *PLoS Genet* **3**(9):1633–1643.

120. **Todeschini AL, et al.** 2010. The epigenetic trans-silencing effect in Drosophila involves maternally-transmitted small RNAs whose production depends on the piRNA pathway and HP1. *PLoS One* **5**(6):e11032.

121. **Simmons MJ, et al.** 2007. Cytotype regulation by telomeric P elements in Drosophila melanogaster: interactions with P elements from M′ strains. *Genetics* **176**(4): 1957–1966.

122. **Simmons MJ, et al.** 2010. Maternal impairment of transposon regulation in Drosophila melanogaster by mutations in the genes aubergine, piwi and Suppressor of variegation 205. *Genet Res (Camb)* **92**(4):261–272.

123. **Haley KJ, et al.** 2005. Impairment of cytotype regulation of P-element activity in Drosophila melanogaster by mutations in the Su(var)205 gene. *Genetics* **171**(2): 583–595.

124. **Brennecke J, et al.** 2007. Discrete small RNA-generating loci as master regulators of transposon activity in Drosophila. *Cell* **128**(6):1089–1103.

125. **O'Donnell KA, Boeke JD.** 2007. Mighty Piwis defend the germline against genome intruders. *Cell* **129**(1): 37–44.

126. **Sienski G, Donertas D, Brennecke J.** 2012. Transcriptional silencing of transposons by Piwi and maelstrom and its impact on chromatin state and gene expression. *Cell* **151**(5):964–980.

127. **Daniels SB, Chovnick A, Kidwell MG.** 1989. Hybrid dysgenesis in Drosophila simulans lines transformed with autonomous P elements. *Genetics* **121**(2):281–291.

128. **Daniels SB, et al.** 1987. Genetic transformation of Drosophila melanogaster with an autonomous P element: phenotypic and molecular analyses of long-established transformed lines. *Genetics* **115**(4):711–723.

129. **Khurana JS, et al.** 2011. Adaptation to P element transposon invasion in Drosophila melanogaster. *Cell* **147** (7):1551–1563.

130. **de Vanssay A, et al.** 2012. Paramutation in Drosophila linked to emergence of a piRNA-producing locus. *Nature* **490**(7418):112–115.

131. **Nilsen TW, Graveley BR.** 2010. Expansion of the eukaryotic proteome by alternative splicing. *Nature* **463** (7280):457–463.

132. **Blencowe BJ.** 2006. Alternative splicing: new insights from global analyses. *Cell* **126**(1):37–47.

133. **Black DL.** 2003. Mechanisms of alternative pre-messenger RNA splicing. *Annu Rev Biochem* **72**:291–336.

134. **Chain AC, et al.** 1991. Identification of a cis-acting sequence required for germ line-specific splicing of the P element ORF2-ORF3 intron. **11**:1538–1546.

135. **Laski FA, Rubin GM.** 1989. Analysis of the cis-acting requirements for germ-line-specific splicing of the P-element ORF2-ORF3 intron. *Genes & Development* **3**:720–728.

136. **Siebel CW, Rio DC.** 1990. Regulated splicing of the Drosophila P transposable element third intron in vitro; somatic repression. *Science* **248**:1200–1208.

137. **Siebel CW, Fresco LD, Rio DC.** 1992. The mechanism of somatic inhibition of Drosophila P-element pre-mRNA splicing: multiprotein complexes at an exon pseudo-5′ splice site control U1 snRNP binding. *Genes Dev* **6**(8):1386–1401.

138. **Tseng JC, et al.** 1991. Splicing of the Drosophila P element ORF2-ORF3 intron is inhibited in a human cell extract. *Mech Dev* **35**(1):65–72.

139. **Siebel CW, Kanaar R, Rio DC.** 1994. Regulation of tissue-specific P-element pre-mRNA splicing requires the RNA-binding protein PSI. *Genes & Development* **8**: 1713–1725.

140. **Matunis EL, Matunis MJ, Dreyfuss G.** 1992. Characterization of the major hnRNP proteins from Drosophila melanogaster. *J Cell Biol* **116**(2):257–269.

141. **Mayeda A, Krainer AR.** 1992. Regulation of alternative pre-mRNA splicing by hnRNP A1 and splicing factor SF2. *Cell* **68**(2):365–375.

142. **Burd CG, Dreyfuss G.** 1994a. RNA binding specificity of hnRNP A1: significance of hnRNP A1 high- affinity binding sites in pre-mRNA splicing. *Embo J* **13**(5): 1197–1204.

143. **Siebel CW, Admon A, Rio DC.** 1995. Soma-specific expression and cloning of PSI, a negative regulator of P element pre-mRNA splicing. *Genes & Development* **9**:269–283.

144. **Labourier E, Adams MD, Rio DC.** 2001. Modulation of P-element pre-mRNA splicing by a direct interaction between PSI and U1 snRNP 70K protein. *Mol Cell* **8**(2): 363–373.

145. **Ignjatovic T, et al.** 2005. Structural basis of the interaction between P-element somatic inhibitor and U1-70k essential for the alternative splicing of P-element transposase. *J Mol Biol* **351**(1):52–65.

146. **Min H, et al.** 1997. A new regulatory protein, KSRP, mediates exon inclusion through an intronic splicing enhancer. *Genes Dev* **11**(8):1023–1036.

147. **Amarasinghe AK, et al.** 2001. An in vitro-selected RNA-binding site for the KH domain protein PSI acts as a splicing inhibitor element. *RNA* 7(9):1239–1253.

148. **Chmiel NH, Rio DC, Doudna JA.** 2006. Distinct contributions of KH domains to substrate binding affinity of Drosophila P-element somatic inhibitor protein. *RNA* **12**(2):283–291.

149. **Burd CG, Dreyfuss G.** 1994b. Conserved structures and diversity of functions of RNA-binding proteins. *Science* **265**(5172):615–621.

150. **Kim HJ, et al.** 2013. Mutations in prion-like domains in hnRNPA2B1 and hnRNPA1 cause multisystem proteinopathy and ALS. *Nature* **495**(7442):467–473.

151. **Roche SE, Schiff M, Rio DC.** 1995. P-element repressor autoregulation involves germ-line transcriptional repression and reduction of third intron splicing. *Genes Dev* **9**(10):1278–1288.

152. **Adams MD, Tarng RS, Rio DC.** 1997. The alternative splicing factor PSI regulates P-element third intron splicing in vivo. *Genes Dev* **11**:129–138.

153. **Hammond LE, et al.** 1997. Mutations in the hrp48 gene, which encodes a Drosophila heterogeneous nuclear ribonucleoprotein particle protein, cause lethality and developmental defects and affect P-element third-intron splicing in vivo. *Mol Cell Biol* **17**(12):7260–7267.

154. **Labourier E, et al.** 2002. The KH-type RNA-binding protein PSI is required for Drosophila viability, male fertility, and cellular mRNA processing. *Genes Dev* **16** (1):72–84.

155. **Blanchette M, et al.** 2005. Global analysis of positive and negative pre-mRNA splicing regulators in Drosophila. *Genes Dev* **19**(11):1306–1314.

156. **Blanchette M, et al.** 2009. Genome-wide analysis of alternative pre-mRNA splicing and RNA-binding specificities of the Drosophila hnRNP A/B family members. *Mol Cell* **33**(4):438–449.

Mobile DNA, 3rd Edition
Nancy L. Craig, Michael Chandler, Martin Gellert, Alan M. Lambowitz, Phoebe A. Rice and Suzanne Sandmeyer

doi:10.1128/microbiolspec.MDNA3-0033-2014

Michael Tellier[1]
Corentin Claeys Bouuaert[1]
Ronald Chalmers[1]

34 Mariner and the ITm Superfamily of Transposons

INTRODUCTION

The mariner elements belong to the ITm superfamily of cut-and-paste DNA-transposons. The acronym is derived from the IS630, Tc1, and mariner elements, which represent three major divisions within the grouping. The first member of the superfamily to be documented was Tc1 in *Caenorhabditis elegans* in 1983 (1). A few years later, Mos1 and IS630 were identified in *Drosophila mauritiana* and the bacterium *Shigella sonnei*, respectively (2, 3). A steady stream of ITm elements entered the literature in subsequent years. However, these were the tip of an iceberg and the depth and breadth of their phylogenetic distribution did not start to become apparent until 1993 when PCR experiments using mariner-specific primers revealed their presence in seven orders of insects (4). We now know that ITm is probably the most widespread superfamily of transposons in nature and that they are present in all branches of the tree of life (Fig. 1) (5).

The ITm transposons generally carry a single open reading frame (ORF) encoding the transposase [Fig. 2(A)]. The transposases share a common catalytic domain containing a triad of conserved aspartate and/or glutamate residues [Fig. 2(B)]. These coordinate two divalent metal ions in the active site, which resides within an RNase H-like structural fold. The ITm transposases also have an N-terminal domain for site-specific DNA-binding.

The transposase gene is flanked by a pair of inverted terminal-repeats (ITR) [Fig. 2(A)]. These can be as short as 20 to 30 bp, encoding a single transposase binding site. The longer ITRs may have additional binding sites and spacer regions, which differ between opposite ends of the transposon. Across the group as a whole, the ITRs are so divergent that no consensus can be discerned. However, the ITRs are always flanked by a TA dinucleotide, which is derived from the target site and duplicated during the integration step of the reaction. Apart from the evolutionary origin of the catalytic domain, this is the one feature that unites all members of the ITm family. The mechanistic significance of the invariant target site duplication will be considered in more detail below.

Phylogenetic analysis of the ITm catalytic domain reveals seven distinct lineages (Fig. 3). The most mechanistically-relevant nomenclature uses the identities of the residues in the DDE/D catalytic triad and the distances between them (6). The spacing between the first and second aspartate residues is variable. However, the distance between the second aspartate residue and the

[1]School of Life Sciences, University of Nottingham, QMC, Nottingham, NG7 2UH, UK.

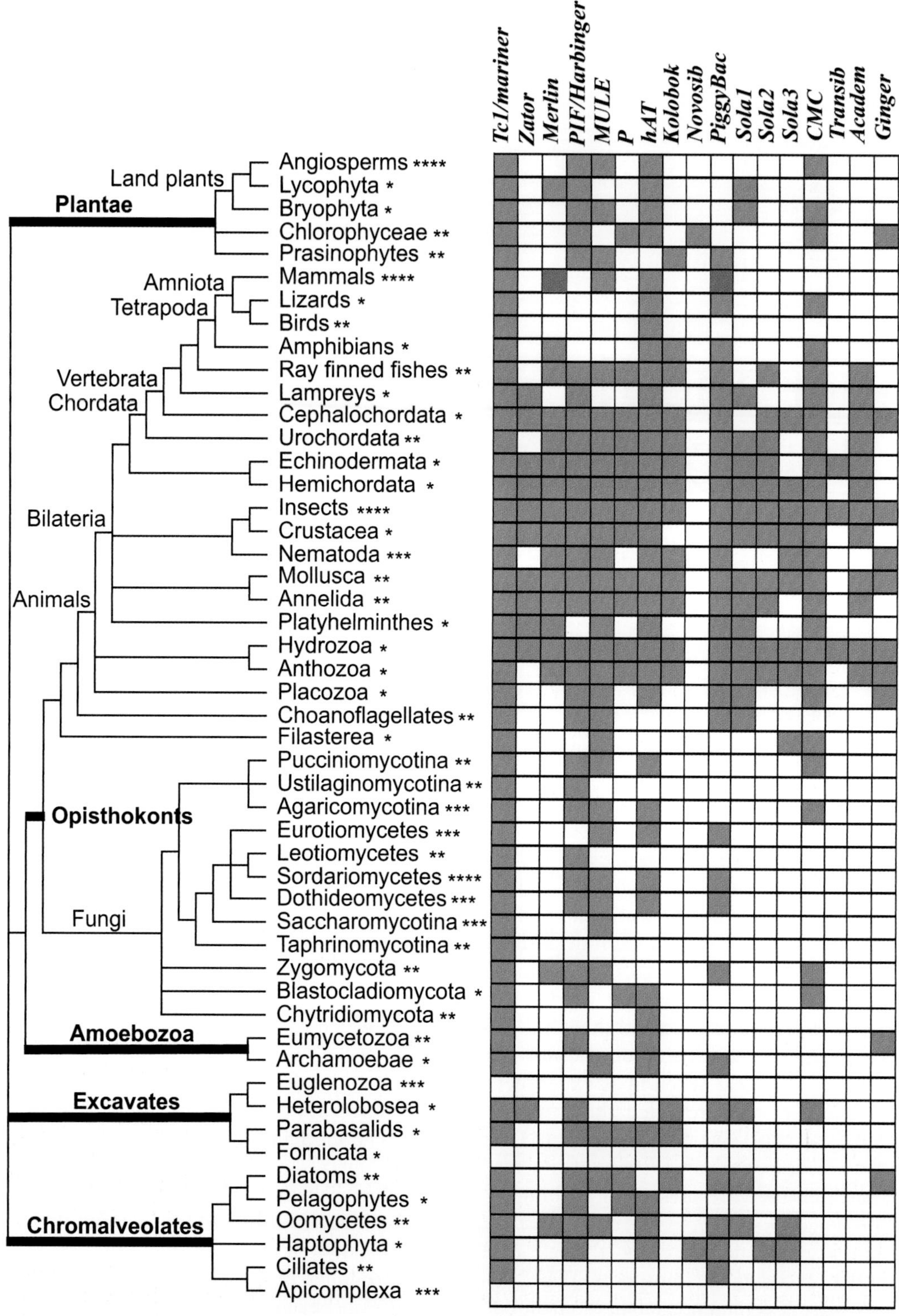

Figure 1 Taxonomic distribution of DNA transposons across the eukaryotic tree of life. Five eukaryotic super-groups are indicated with thickened lines. Asterisks indicate the numbers of genomes representing the branch: *, 1 genome; **, 2 to 5 genomes; ***, 6 to 10 genomes; ****, over 10 genomes. Figure reproduced from (5) with permission of the authors and the publisher. doi:10.1128/microbiolspec.MDNA3-0033-2014.f1

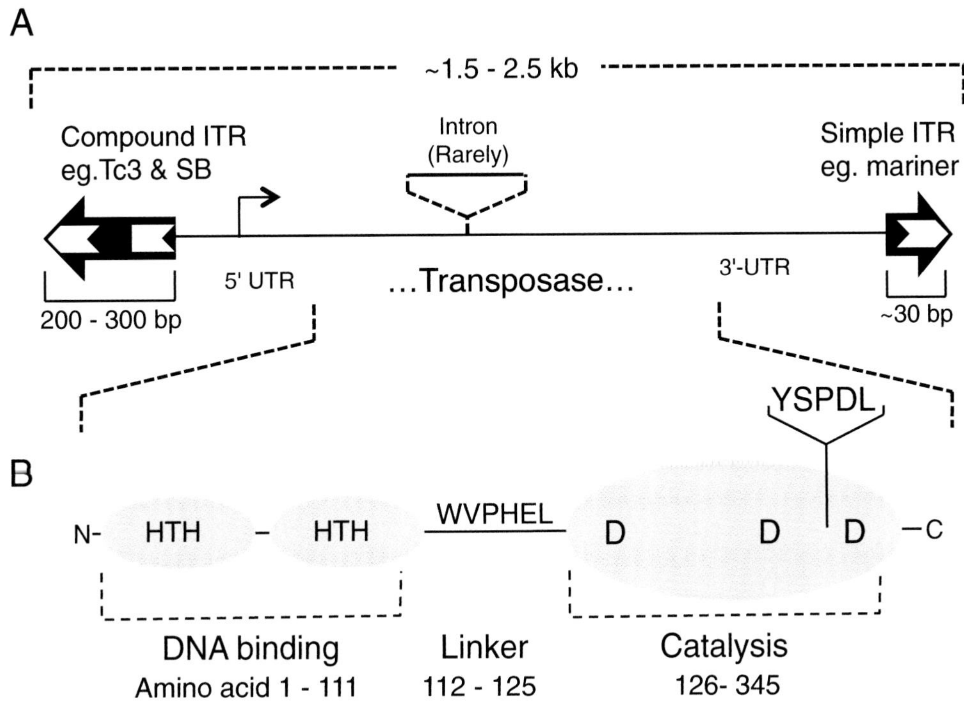

Figure 2 Organization of ITm transposons and mariner transposases. (A) ITm transposons are ~1.5 to 2.5 kb in length. In Tc3 and SB, ITRs have two transposase binding sites. In mariner, ITRs are ~30 bp and carry a single transposase binding site. ITm transposons have a single ORF, which codes for the transposase and is rarely interrupted by an intron. (B) Mariner transposases have an N-terminal DNA-binding domain with two helix-turn-helix (HTH) motifs and a C-terminal catalytic domain with three aspartate (D) residues that coordinate the catalytic metal ions. The DNA-binding and catalytic domains are joined by a linker region that includes the highly conserved stretch of amino acids, WVPHEL. YSPDL is another highly conserved motif that occurs just before the third aspartate. Both conserved motifs are important in communication between the active sites (see text for details). doi:10.1128/microbiolspec.MDNA3-0033-2014.f2

third aspartate or glutamate residue is sufficiently well conserved to distinguish the various lineages.

THE MARINER FAMILY

For the remainder of this review we will focus mainly on the mariner (ITmD34D) transposons because their mechanism has been examined in detail. They are generally about 1.5 kb long and encode an intron-less transposase of about 340 amino acids which is flanked by simple inverted repeats of about 30 bp (Fig. 2). Mariners are particularly widespread in higher animals where they are capable of horizontal transfer between species. This is evident from the incongruent phylo genies of the transposons and their hosts (7). Thus, closely related mariners may be found in distantly related hosts and vice versa. Furthermore, their distribution is patchy and they may be present or absent in closely related species.

The first mariner element to be studied *in vitro* was Himar1 (8). The transposase is a consensus sequence derived from six defunct elements cloned by PCR of the horn fly genome. This work established that the transposase was the only protein required for transposition, which is consistent with their wide phylogenetic distribution. In general, most transposons that have been tested are active in heterologous hosts, indicating that they are independent of specific host factors. Notable exceptions are Sleeping Beauty (SB, ITmD34E) and the P element, which may have a restricted host range owing to a host factor requirement. The second mariner element to be studied *in vitro* was the naturally active Mos1 (9). This was followed by *in vitro* analysis of Mboumar-9, Mcmar1 and Hsmar1 (10, 11, 12, 13, 14). However, most of the mechanistic information set out below has been derived from Mos1 and Hsmar1.

In vitro analysis revealed that the mariner transposons use a cut-and-paste mechanism. The transposase

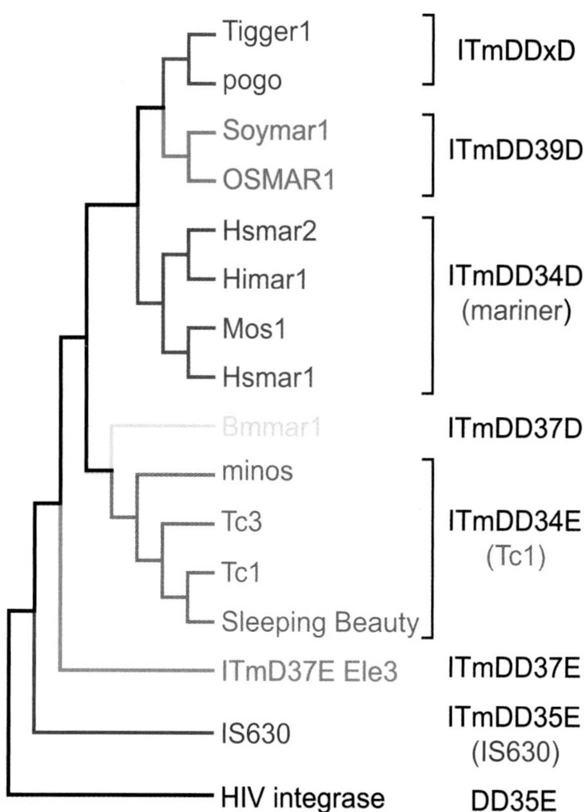

Figure 3 Phylogeny of the ITm transposons. ITm families are classified according to their triad of catalytic residues and the number of amino acids separating the second aspartate (D) and the third aspartate or glutamate (E). Sequences of the catalytic domains for 15 ITm transposases were extracted from Repbase, aligned with MUSCLE, and an unrooted tree was constructed with Mega 6.06 using HIV integrase as an outgroup. doi:10.1128/microbiolspec.MDNA3-0033-2014.f3

first makes staggered cuts at the end of the transposon [Fig. 4(A); (8, 15, 16)]. The 3′-end of the element, which is later transferred to the target, is cleaved precisely. Nicking at the 5′-end is imprecise, but is mostly staggered 3 bp within the element. Kinetic studies and analyses of gel-purified reaction intermediates demonstrated that the 5′-end is always cleaved before the 3′-end (15, 17). The order of cleavage events is also constrained and both 5′-ends of the transposon are cleaved before either of the 3′-ends (17).

TRANSPOSITION CHEMISTRY

Catalysis in mariner transposition is provided by a structural fold in the transposase that was first observed in RNase H from *Escherichia coli* (18, 19). This enzyme performs a hydrolysis reaction to nick the RNA strand in a DNA/RNA hybrid [Fig. 4(A)]. The active site contains two divalent metal ions, which are coordinated by three or four acidic residues (20). The transition state is illustrated in Fig. 4(C) using the three aspartate residues in the mariner active site as an example. The nucleophile, which in this case is the hydroxyl of water, makes an in-line attack on the scissile phosphate, displacing the bridging oxygen opposite and breaking the phosphodiester bond. The two metal-ion mechanism was first proposed in 1991 based on the structure of the 3′-5′ exonuclease domain of DNA polymerase I (21). The 'A' metal was proposed to activate the nucleophile and the 'B' metal to assist the 3′-oxyanion leaving group. Both metals also act as Lewis acids to stabilize the pentacovalent transition state by coordinating a nonbridging oxygen atom. Subsequent co-crystal structures between various enzymes and their substrates have refined the two metal-ion mechanism and tested its veracity, e.g., references 20, 22, 23, 24.

Cleavage and integration of a transposon end is an interesting enzymological problem because, like most nucleases, RNase H has a single active site, which acts on a single strand of nucleic acid. The mechanistic difficulties arise from the fact that the active site must act on at least three strands of DNA during cut-and-paste transposition: two at the transposon end and one at the target. The mechanisms of several other cut-and-paste transposons have been characterized in detail but some uncertainties still surround the mariner reaction, which appears to follow a distinctly different order of events. We will therefore begin by describing the mechanisms established in related transposons before setting out why the mechanism in mariner has remained slightly obscure. We will then describe some unpublished experiments that clarify the situation.

VARIATIONS ON THE TWO METAL-ION MECHANISM

The minimal catalytic requirements for a transposition reaction are exemplified by the retroviral integrases and bacteriophage Mu (Fig. 4(D), left). Hydrolysis introduces a nick at the 3′-end of the transposon (25, 26, 27). This is followed by a transesterification reaction, which integrates the 3′-end at the target site (28, 29). Hydrolysis and transesterification reactions are chemically identical. The only difference is that the 3′-hydroxyl at the end of the transposon replaces the hydroxyl of water as the nucleophile during the transesterification step. Overall, this type of transposition can therefore be described as a hydrolysis–transesterification (H–T) reaction.

Compared to the basic RNase H nicking reaction, the H–T reaction of the replicative transposons and HIV requires two mechanistic innovations. In RNase H, the A metal ion activates the nucleophile during hydrolysis [Fig. 4(C)]. However, if the 3′-hydroxy, which is the nucleophile in the transesterification reaction, was to be coordinated by the same metal ion it would have to move a considerable distance and rotate through 180°. This problem is solved by postulating that the metal ions swap roles so that the B metal ion activates the nucleophile for transesterification (23). Thus, the leaving group during the hydrolysis step becomes the nucleophile during integration (compare Fig. 4(C) with Fig. 4(D), left flowing top to bottom). In this elegant representation of the reaction the metal ions are designated as H and T according to whether they activate the nucleophile during hydrolysis or transesterification. The second innovation is to accommodate an additional strand of nucleic acid, represented by the target, in the active site during integration. The nicked 5′-end of the flanking DNA, which is still attached to the transposon by base pairing with the uncleaved strand, must move out of the way to allow the scissile phosphate in the target to take up position in the active site.

It is interesting to note that the RNase H fold seems quite flexible with respect to the identity of the nucleophilic group. In the case of HIV integrase it can accommodate water or glycerol or even the 3′-end of the dinucleotide that is removed in the cleavage step (28). Nevertheless, during integration the scissile phosphate of the target DNA must be accommodated in place of the scissile phosphate of the transposon end.

The two-step H–T mechanism is sufficient for the simplest types of transposition reaction. The replicative elements are never completely separated from their flanking donor sites, and the retroviruses are already linear and have only a few bp of flanking DNA left over from reverse transcription. In contrast, the cut-and-paste transposons require double strand breaks at both ends. This can be achieved via a DNA hairpin intermediate (Fig. 4(D), center left). As before, the initial hydrolysis is followed by a transesterification. However, transesterification is directed towards the opposite strand rather than the target DNA (30). The hairpin on the transposon end is resolved by a second hydrolysis, followed by a second transesterification, which mediates integration. The overall reaction can thus be designated H–T–H–T.

In the past we have referred to this as the "forward-hairpin" mechanism, which places the covalent intermediate proximal to the transposon end (31). The mechanism was first documented in two sets of experiments with Tn10. Firstly, it was shown that the transpososome contains a dimer of transposase and that all four phosphoryl transfer reactions at a given end are catalyzed by a single active site (32, 33). Next, stereochemical experiments with modified phosphorothioate substrates confirmed that the first three steps of the reaction proceed by direct in-line nucleophilic attacks (34). The proximal hairpin intermediate was later demonstrated in Tn5 and PiggyBac (35, 36).

A convincing rationale for the proximal-hairpin strategy of Tn10 is that it helps the active site to accommodate the intermediates at different stages of the reaction (Fig. 4(D), center left) (23). The H and T metal ions toggle back and forth, acting alternately as the Lewis acid to activate the nucleophile during the successive hydrolysis and transesterification steps. This calls for minimal reorganization of the active site and the bound nucleic acid(s) at the different stages of the reaction, since the leaving group in each hydrolysis reaction is used as a nucleophile in the next step without dissociating from the active site. Nevertheless, coordination of the scissile phosphodiester bond must be reestablished at each stage.

In the H–T–H–T reaction the hairpin intermediate is equivalent to the integration product in the H–T reaction. Although integration is isoenergetic, it is generally irreversible. Neither does the product succumb to further hydrolysis. Its stability is probably owing to the ejection of the new phosphodiester bond from the active site and the concomitant increase in the binding energy of the nucleoprotein complexes e.g., references 16, 24, 37. The mechanistic innovation in the H–T–H–T reaction is to reset the active site for the second hydrolysis, which resolves the hairpin.

Members of the hAT family of cut-and-paste transposons use an alternative "reverse-hairpin" strategy for cleavage (38). Here the first nick is on the other strand and liberates the 5′-end of the transposon (Fig. 4(D), center). As before, the A and B metal ions would take alternate roles, acting as the Lewis acid to activate the nucleophiles in the hydrolysis and hairpin reactions, respectively. The order of the steps for the reverse-hairpin reaction is therefore H–T–T. The mechanistic innovation is that the location of the hairpin in the flanking DNA means that it takes no further part in the reaction and can, in principle, be released from the complex. The active site does not therefore need to be reset for a second hydrolysis reaction. Nevertheless, it does have to capture the 3′-hydroxyl at the end of the transposon prior to the final integration step. This is illustrated as a 180° rotation of the active site and the transposon end with respect to each other (Fig. 4(D), center). Although

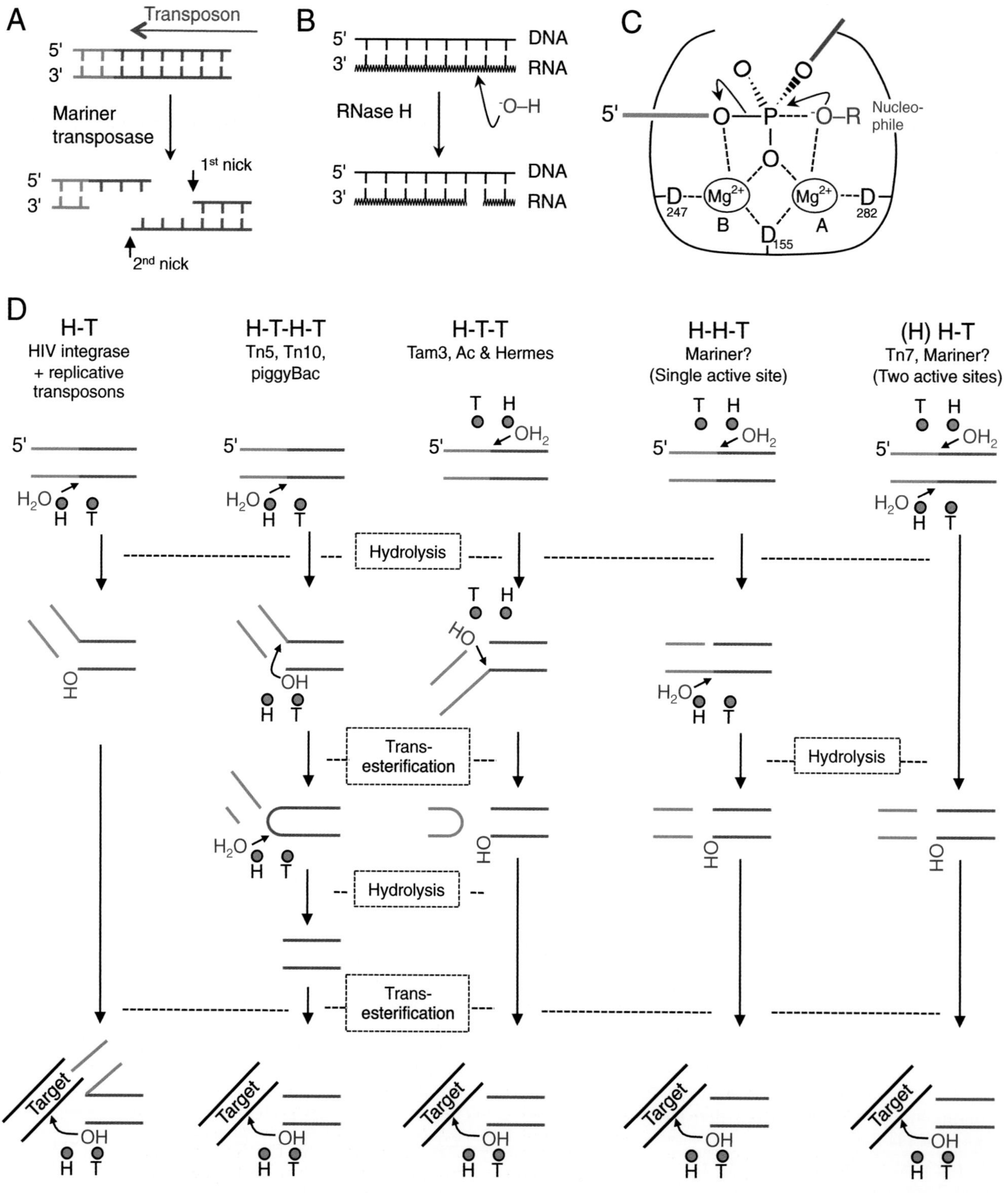
A
Transposon
5'
3'
Mariner transposase
1st nick
2nd nick
B
DNA
RNA
RNase H
-O–H
C
O
P
O–R
Nucleo-phile
D
247
Mg2+
B
Mg2+
A
D
282
D155
D
H-T
HIV integrase + replicative transposons
H-T-H-T
Tn5, Tn10, piggyBac
H-T-T
Tam3, Ac & Hermes
H-H-T
Mariner? (Single active site)
(H) H-T
Tn7, Mariner? (Two active sites)
T H
OH2
H2O
H T
Hydrolysis
HO
OH
Trans-esterification
Hydrolysis
Trans-esterification
Target
OH

perhaps inelegant, the saving grace of this strategy is that a reorganization of the complex is already required at this step to accept the scissile phosphate of the target.

The flanking hairpin intermediate was first proposed in 1986 by Enrico Coen and colleagues, based on the footprints left behind after excision of Tam3 in *Antirrhinum* (39). Their insight was to recognize that palindromic sequences in the footprints were evidence for the asymmetric resolution of a flanking hairpin during host-mediated repair of the empty donor site. The flanking hairpin was not directly demonstrated until 1992, when it was detected in V(D)J recombination (40).

In passing, it is perhaps worth noting that two further mechanistic innovations are required for V(D)J recombination. Firstly, the "transpososome" must retain the flanking hairpins and hand them off to the Artemis complex for resolution and end joining e.g., reference 41. Secondly, the "transpososome" should be defective for the integration step.

The polarity of the first nick in mariner transposition suggested that it should conform to the H–T–T strategy for double strand cleavage (Fig. 4(D), center right). However, a flanking hairpin product was not detected *in vitro* (15, 16). This suggests either that there is no hairpin intermediate or that it is resolved very quickly after it is formed. To distinguish these possibilities, the transposase was offered a prenicked substrate with a 3′-flanking dideoxynucleotide, lacking the nucleophilic hydroxyl group required for the hairpin step (15). Second-strand cleavage was still efficient, which is consistent with sequential hydrolysis reactions and the absence of a hairpin. Further confirmation was provided by efficient second-strand cleavage in the complete absence of a flanking top strand (17, 42). Curiously, although the presence of a flanking bottom strand is not a mechanistic prerequisite for mariner's sequential-hydrolysis strategy (Fig. 4(D), center right), its absence inhibits first-strand nicking (17, 42).

If mariner transposition is performed by a single active site it would have to be reset for a second hydrolysis without an intervening transesterification step (Fig. 4(D), center right). Similar to the H–T–T reaction, the top strand must move out of the way while the bottom strand, containing the 3′-end of the transposon, moves into the active site. However, this must be accomplished while the flanking DNA is still attached by one strand. Later we will describe how a flanking-DNA interaction involving the invariant TA dinucleotide is important in the progression between the two hydrolysis reactions. Finally, consistent with the anticipated mechanistic difficulties, the second hydrolysis reactions are the slowest steps during cleavage of mariner and Tn10 ends (12, 30).

A fourth strategy to generate a double strand break at a transposon end is to use a pair of active sites (Fig. 4(D), right). This strategy is used by the type II restriction endonucleases and Tn7. However, Tn7 is a special case because the transposase is a heterodimer

Figure 4 Nucleotidyl transfer reactions catalyzed by RNase-H type enzymes. (A) Double strand cleavage at the ends of mariner transposons. The first nick is usually recessed three nucleotides within the transposon DNA and exposes a 5′-phosphate on the end of the transposon. The second nick is precisely at the transposon end and exposes the 3′-hydroxyl, which is subsequently transferred to the target. (B) RNase H hydrolyzes the RNA strand in an RNA–DNA hybrid. (C) The transition state in the two metal-ion mechanism for nucleotidyl transfer reactions is illustrated. The RNase-H fold coordinates two divalent metal ions via three (or four) acidic residues. The figure illustrates the metal binding pocket of Hsmar1, which has three conserved aspartate (D) residues. The proposed transition state is shown with the nucleophile, R–O$^-$ attacking from the right in an in-line configuration with the leaving group (the 3′-OH of the transposon). Several different nucleophiles, including H_2O and glycerol, can be used in strand cleavage. The proposed role of the DDD triad in coordinating a pair of divalent metal ions and their role as Lewis acids and bases in promoting the reaction are indicated. (D) Several mechanisms of strand cleavage and joining reactions and the role of divalent metal ions in transposition are illustrated. DNA transposons can be classified according to the sequence of hydrolysis and transesterification reactions. Metal ions are designed H or T to indicate their role in activating the nucleophile in hydrolysis and transesterification reactions, respectively (see text for details). In the H–T–T mechanism illustrated for the hAT transposons (center) we have assumed that the reaction is performed by a single active site. If this is the case, the active site would have to be reorganized to capture the 3′-end of the transposon in preparation for the integration step. This is illustrated as a 180° rotation of the transposon end and the active site with respect to each other. If the mariner elements are also cleaved by a single active site a reorganization would be required between the two hydrolysis steps to exchange the top and bottom strands in the active site. See text for further details. doi:10.1128/microbiolspec.MDNA3-0033-2014.f4

(44). The top strand is cleaved by a subunit related to the type II restriction enzymes, while the bottom strand is cleaved and integrated by a DDE-family transposase (45). In Fig. 4(D) this is indicated as an (H) H–T reaction because the top strand is cleaved by a different active site and has no mechanistic connection to bottom strand cleavage or integration. This strategy has not been formally excluded for mariner transposition, as will be explained below.

MARINER NUCLEOPROTEIN COMPLEXES AND THE NUMBER OF ACTIVE SITES

Initial biochemical analysis of the mariner transpososome was difficult to interpret and yielded a confused picture of the reaction. The first electrophoretic mobility shift assays (EMSA) with the Mos1 transposase revealed a single protein–DNA complex (46). Subsequent analysis of Mos1, Himar1, and Hsmar1 revealed several additional complexes (16, 47, 48, 49). It was clear that two of the complexes contained a single transposon end. These became known as single-end complex (SEC) 1 and SEC2, which contain a monomer and a dimer of transposase, respectively (43, 48, 50). The paired-ends complex (PEC) was more difficult to study. Depending on which group was studying which transposon, there were either two forms of the complex, one form but only in the presence catalytic metal ions, or it was undetectable (15, 16, 48, 49).

Although the biochemical analyses were inconclusive, they seemed to suggest that the active multimer might be a tetramer (16, 48, 51). This would have fitted with the presence of multiple transposase binding sites in the ends of Sleeping Beauty and Tc1, with the tetrameric structure of the retroviral integrases, and with the fact that the mariner transposase is a dimer in free solution (42, 52, 53, 54, 55, 56). Another factor was that SEC2 seemed to be competent for 5′-cleavage (15, 16). It therefore appeared that SEC2 might represent half of a PEC and led to the suggestion that the 5′- and 3′-cleavage were performed before and after synapsis, respectively.

Three general models have been suggested to accommodate these views [Fig. 5(A)]: 1, a tetramer model; 2, a subunit exchange model; and 3, a dimer model. In the bottom part of the illustration, the grey glow indicates which monomer is active and when. There are also a number of obvious variants of each model that will not be discussed in detail. In the tetramer model for mariner transposition, the 5′-cleavage happens within SEC2 (Fig. 5(A), left). Tetramerization activates the other pair of monomers, which perform 3′-cleavages. Thus, the four DNA strands are cleaved by four different active sites, which corresponds to an (H)–H–T mechanism [see Fig. 4(D)]. As in Tn7, one subunit would be located at the transposon end by protein–DNA interactions, while the other is located by protein–protein interactions.

In the subunit exchange model, the 5′-cleavage also happens within SEC2 (Fig. 5(A), center). Synapsis is accompanied by the ejection of two subunits to form a dimeric PEC in which the 3′-cleavages are performed by the other pair of monomers. Cleavage thus involves four active sites, which corresponds to an (H)–H–T mechanism. To date there is no published experimental evidence that rules this out for mariner transposition.

In the third model, a dimer of transposase first binds a transposon to form SEC2 and then recruits a second, naked, transposon end to yield the transpososome within which all the cleavages take place (Fig. 5(A), right). Each active site performs two strand cleavages, which corresponds to an H–H–T mechanism. Two versions of the model are possible depending on whether the 5′- and 3′-cleavage events carried out by each active site are on the same or on different transposon ends.

Crystallographic analysis revealed that the Mos1 transposon ends are held together by a dimer of transposase in the post-cleavage intermediate (57). The two active sites in the dimer are sufficient to mediate the two transesterification reactions required for the subsequent integration step. However, this does not directly indicate the number of subunits required during the preceding cleavage step, which requires four phosphoryl transfer reactions. Nevertheless, the dimeric structure strongly challenges the tetramer model. Biochemical studies with Hsmar1 also challenged the notion that 5′-cleavage might take place in SEC2. In contrast to *in vitro* reactions with Mos1, catalysis in Hsmar1 transposition is dependent on prior synapsis of the ends (58). This put further constraints on the possible models for mariner transposition. The dimer model thus gained support in recent years and several groups, including us, have interpreted their studies in the context of this model. However, it should be noted that the dimer model has not yet been fully demonstrated and that some versions of the "subunit exchange" mechanism remain formally possible.

In an as-yet unpublished study, we have addressed the reaction order by defining the subunit architecture during cleavage and analyzing the distribution of products when active and inactive transposase subunits are mixed in various ratios. The results fit an H–H–T model in which double strand cleavage and integration are performed by a single transposase monomer at each transposon end (CCB and RC, in preparation). In the

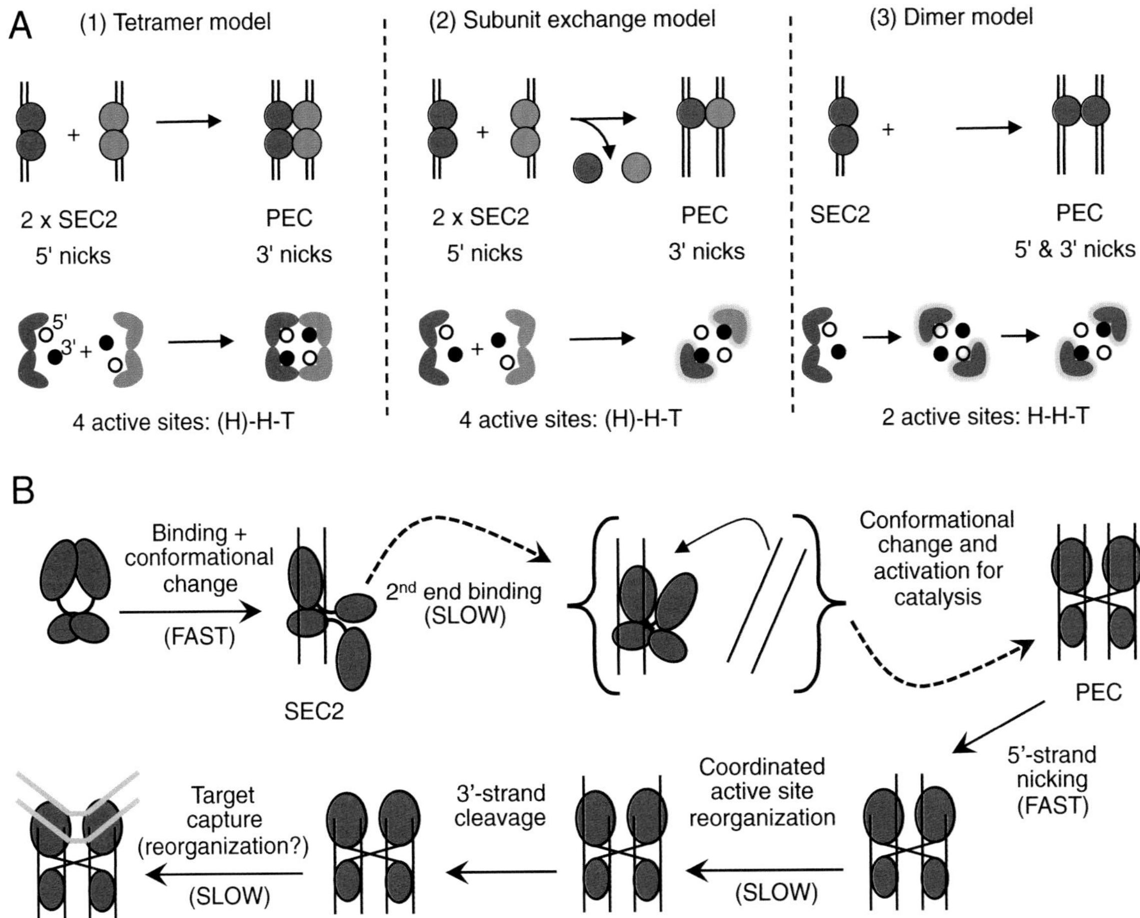

Figure 5 Models for the mechanism of mariner transposition. (A) Three proposed models for the arrangement of subunits in mariner cleavage: (1) tetramer; (2) subunit exchange; and (3) dimer models. Single-end complex 2 (SEC2) contains a transposase dimer and one transposon end. Paired-ends complex (PEC) contains two transposon ends and two or four subunits. The open and filled circles represent the DNA strand containing the 5′- and 3′-ends of the transposon, respectively, viewed down the axis of the double helix. (B) The nucleoprotein complexes deduced from biochemical analysis of Hsmar1 transposition are illustrated using the dimer model for cleavage. Binding of a transposase dimer to the first transposon end is fast. Recruitment of the second end within SEC2 is slow. Catalysis is within the context of the PEC. The 5′ strands of the transposon ends are cleaved first, followed by a structural change that is coordinated between ends. This is followed by cleavage of the 3′-ends and transposon integration. doi:10.1128/microbiolspec.MDNA3-0033-2014.f5

following sections we will therefore review the dynamics of mariner transposition in terms of the dimeric H–H–T model.

REACTION KINETICS AND TRANSPOSOSOME DYNAMICS

The Hsmar1 transposase, like other helix-turn-helix (HTH) proteins, binds rapidly to its recognition site [Fig. 5(B)] (50, 58). One would expect that binding of the second transposon end would be almost as fast. However, it is in fact very much slower (58). This was interpreted as the result of an allosteric conformational change induced by the disruption of the two-fold symmetry of the complex when it binds the first end. However, it is equally possible that the bound transposon end simply hinders access of the second end to the unoccupied DNA-binding domain. This phenomenon has been estimated to reduce the rate of synapsis by 10^5-fold (50). Later we will explain how this plays a role in regulating the rate of transposition *in vivo* and its topological selectivity.

Following synapsis, catalysis is initiated by 5′-cleavages of the transposon ends [Fig. 5(B)]. The rate of the 5′ nicks is difficult to measure because it is faster than the time needed to mix and stop an *in vitro* reaction ($t_{1/2}$ <30 s) (12, 50). The second hydrolysis reaction, which cleaves the 3′ end of the transposon, is much slower ($t_{1/2}$ = 5 to 10 min). However, there is also a mechanistic constraint. Both 5′ ends must be nicked before either of the 3′ ends (17). This suggests that the conformational change, which resets the active sites for the second hydrolysis, is coordinated across the transpososome.

The target capture and integration steps are also relatively slow. In a staged kinetic-analysis, where the PEC is preassembled on a supercoiled plasmid substrate prior to the addition of the catalytic metal ion, the excised transposon was detected at early time points and then chased into integration products (12).

DNA TOPOLOGY

Mariner transposons are sensitive to the topology of their substrates and targets. Negative supercoiling promotes synapsis, the rate limiting step of the reaction, by increasing the relative concentration of the transposon ends with respect to each other and by providing a favorable angular distribution when the inverted repeats meet in the plectosome (58). The reaction is much slower when the ends are present as direct repeats or when inverted repeats are positively supercoiled. The most stringent condition for synapsis is when the transposon ends are on different DNA molecules.

The increase in the rate of transposition afforded by negatively supercoiled inverted repeats over the other configurations has not been measured precisely, but it is many orders of magnitude (58). The strong bias towards one particular configuration of sites constitutes a topological filter. Topological filters have been observed in several other recombination systems such as the Tn3/ϒδ resolvases, the hin/gin/cin inversion systems, and phage Mu transposition (59, 60, 61, 62, 63). These are similarly dependent on negative supercoiling in the substrate. However, in addition to the two recombination sites, they require a third site, called an enhancer, to impart directionality.

In contrast, the topological filter in mariner does not require a third site. Directionality is provided by the allosterism in the developing transpososome, which slows synapsis by raising a kinetic barrier to the recruitment of the second transposon end. As explained above, this arises from the low number of degrees of freedom of the unoccupied DNA-binding domain in SEC2, which dictates that productive collision events are restricted to a narrow angle of approach. Although negative and positive supercoiling both increase the relative concentration of transposon ends, which will help to overcome the kinetic barrier by increasing the frequency of collision events, the favorable angular distribution is only provided by negatively supercoiled inverted-repeat substrates (58).

The topological filter could help to suppress genomic instability, which would arise from the promiscuous synapsis of transposon ends. For example, although Tn10 transposition is dependent on supercoiling there is no topological selectivity and ends in direct or inverted-repeat configuration are used almost equally well (64). In the Activator/Dissociation transposition system in maize, the promiscuous, noncanonical, synapsis of transposon ends causes large scale chromosomal rearrangements (65, 66). This is the mechanism responsible for the breakage–fusion–bridge cycles observed by McClintock (67).

Target site selection is also affected by DNA topology. Since the transpososome acquires its target by random collision, different sites are used according to their relative concentrations (68). Consequently, intramolecular target sites, within the transposon itself, are used more frequently than intermolecular sites, which are relatively dilute in *in vitro* reactions. However, increasing the concentration of intermolecular target sites increases their selection at the expense of intramolecular sites (68). Since the DNA concentration in the nucleus is about four orders of magnitude higher than in typical *in vitro* reactions, intermolecular target sites are used more frequently. Nevertheless, transposon inversion-circles, which are products of autointegration, have been detected in many systems including retroviruses, Tn10, SB, and PiggyBac (69, 70, 71, 72).

Negatively supercoiled targets are also strongly favored over relaxed or positively supercoiled targets (68). This raises an interesting paradox: whilst negative supercoiling favors synapsis and excision of the transposon, its retention within the transposon after excision will favor autointegration (12). Supercoiling is thus a positive and a negative regulator of transposition. The physiological significance of DNA topology on mariner transposition is unknown. However, excision and integration probably require open chromatin, which will be transiently supercoiled during replication, or episodes of nucleosome remodeling and transcription.

Although mariner and the ITm family transposons integrate at a TA dinucleotide target site, not all sites are used with equal frequency (68, 73). While the mechanism underlying the preferential selection of certain target sites remains unknown, this phenomenon shows that the transposon does not integrate into the first TA dinucleotide it encounters as it diffuses through space. Rather, like other site-specific DNA-binding proteins, it must first establish nonspecific interactions and then perform a one dimensional scan of the DNA. Presumably, some as yet unidentified structural feature of DNA then favors a particular integration site.

THE TA DINUCLEOTIDE AND THE TERMINAL RESIDUES

No consensus sequence can be discerned for ends of the ITm family transposons. However, they are all flanked by a 5′-TA dinucleotide, which is derived from the target site and duplicated during integration e.g., reference 74. When the nucleotide directly flanking the transposon end (position −1) is mutated, the kinetics of the first nick are identical to wild type (12, 17). However, 3′-cleavage is severely inhibited. This indicates that the mutation does not affect transpososome assembly or the initiation of catalysis: rather, it prevents the active site resetting for the second hydrolysis reaction (17). If we recall that the flanking 5′-strand is dispensable for 3′-cleavage, we can see that the crucial interactions are probably with the T residue opposite −1A. In contrast, none of the three substitutions at position −2T affect the kinetics of cleavage. Thus, whilst both residues in the TA dinucleotide are required for integration, only the innermost is important during excision.

Systematic mutagenesis of the residues between position +1 and +4 revealed that none of the changes significantly alter the kinetics of the first nick (17). First strand nicking must therefore be determined almost entirely by the location of the specific transposase binding sites with respect to the transposon end. This may explain why the first strand nicking activity in mariner tends to be imprecise. Finally, most of the +1 to +4 substitutions affect second strand nicking but none are as severe as the −1A substitutions (17).

THE LIFE-CYCLE AND REGULATION OF TRANSPOSITION

DNA transposons do not persist for long periods in a given eukaryotic genome and their survival requires frequent horizontal transfer into new hosts (75, 76). The life-cycle of a transposon might therefore be considered as the period between its first appearance in a virgin genome and its eventual extinction. Immediately following horizontal transfer, the transpositional activity of the element is under positive selection; an element with a fully active transposase, and perfect cognate transposon ends, will have a high rate of transposition. As time passes some copies will acquire mutations. However, the presence of multiple copies relaxes selection because a pool of freely diffusing transposase acts on all copies of the element, including those with defective transposase genes. Eventually, genetic drift is thought to poison the pool of active transposase by dominant-negative complementation leading to the extinction of the transposon (77, 78, 79, 80, 81).

Notwithstanding their inevitable demise in a given genome, the life-history of a transposon has many of the hallmarks of a classical host–parasite relationship. The host has adaptations, such as RNAi, to suppress the parasite, while the parasite has adaptations to spare the fitness of the host. Autoregulation of bacterial transposons has been studied in some detail (82, 83). However, until recently little was known about autoregulation in eukaryotes.

Transposition is an inherently exponential reaction because each new copy of the element is a source of further new copies. Any transposon active enough to found a genomic invasion therefore has the potential to cause a genomic meltdown. Experiments with the non-autonomous Mos1, responsible for the *peach* mutation in *Drosophila*, provided the first experimental evidence for autoregulation (84). The phenomenon, termed over-production inhibition (OPI), was revealed by a reduction in the frequency of excision when multiple copies of the transposase gene are present, or when transposase is over-expressed from a heat-shock promoter. OPI was later shown to affect other mariner elements, both *in vivo* and *in vitro* (12, 85, 86).

The key observation that helped reveal the mechanism of OPI was that the mariner transposases dimerize prior to transposon end binding (47). If synapsis in mariner transposition is by recruitment of a naked transposon end into SEC2, OPI will result from the progressive sequestration of the transposon ends as the transposase concentration rises [Fig. 6(A)]. *In vitro* and *in vivo* experiments with Hsmar1 confirmed this model, which was termed an assembly-site occlusion (ASO) mechanism (50).

The properties of the ASO mechanism were investigated using a computer model (50). This revealed that ASO, in its simplest form, as illustrated in Fig. 6(A), will not provide significant inhibition until the free transposase concentration reaches a significant fraction of its K_d for the transposon end [Fig. 6(B)]. This is because the transposase must search a much greater volume of the nucleus to find the first transposon end than

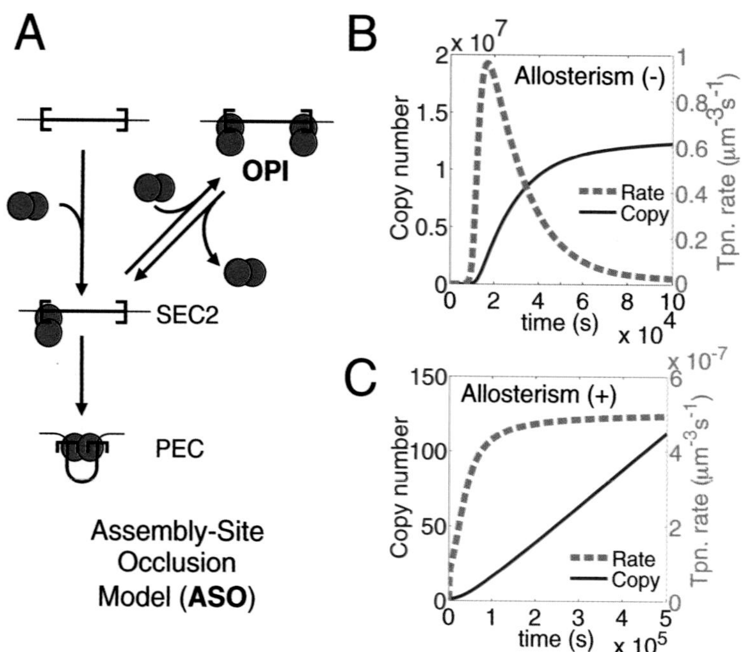

Figure 6 Autoregulation of mariner transposition by OPI. (A) Model of the ASO mechanism that underlies OPI in mariner transposition. PEC assembly is by recruitment of a naked transposon end. As the concentration of transposase rises, naked ends are sequestered leading to inhibition of the reaction. (B) Simulation of a genomic invasion by a mariner transposon. If the affinity of the transposase dimer for the first and second transposon ends are the same, OPI only starts to reduce the rate of transposition once there are a very large number of transposons contributing to the pool of transposase. The timescale of the simulation is very short because transposition events in the computer model are allowed to yield products instantly and the low rate of diffusion *in vivo*, owing to molecular crowding, has been ignored. See text for further details. (C) Simulation as in part (B) except that we account for allosterism, which slows synapsis by reducing the affinity of the developing transpososome for the second transposon end compared to the first. Part (B) reproduced from (50) available under Creative Commons License. doi:10.1128/microbiolspec.MDNA3-0033-2014.f6

to find the second, which can never be too far away owing to the continuity of the DNA connecting them. This simple version of the ASO mechanism is therefore ineffective until there are hundreds of thousands of transposons contributing to the pool of transposase (50). However, the allosterism in the developing transpososome slows the rate of synapsis, providing a corresponding increase in the inhibitory power of the ASO mechanism [Fig. 6(C)]. This means that a pseudo-steady-state rate of transposition is established early in a genomic invasion, when only a few copies of the transposon are present (17, 50).

It is worth noting that the ASO mechanism does not depend on the actual multimeric state of the transposase, only on the fact that the second transposon end is naked when it is recruited into the developing complex. This ensures that an increase in the transposase concentration will always lead to a reduction in the rate of transposition. Transposase dose–response experiments performed *in vivo* with SB and the distantly related PiggyBac transposon suggest that they may be regulated in the same way (50).

MITES

Because a freely diffusing pool of transposase acts on all copies of the transposon, autonomous transposons may be parasitized by miniature inverted repeat transposable elements (MITES). Some of these are simple deletion derivatives of their parental element, while others have a more complex genesis. Miniature elements are often very numerous and therefore appear to have a higher rate of transposition. In the case of the deletion derivatives, the shorter distance between transposon ends probably promotes synapsis and provides a lower probability of nonproductive autointegration. However, the more complex miniature elements may have sequence-features that enhance transposition e.g., reference 87. For example, the Osmar transposase mobilizes the Stowaway element very efficiently despite binding weakly to its ends (87). This is a counterintuitive finding that would make sense in the light of the ASO mechanism.

The human mariner transposon Hsmar1 is associated with an 80 bp derivative, which is known as Made1 in Repbase (88, 89, 90). Even though it is shorter than the persistence length of DNA, transposition is efficient *in vitro* (D. Liu and RC, to be presented elsewhere).

Nonautonomous and miniature elements are rare in bacteria. One reason for this is that the *cis*-action of some bacterial transposases restricts their activity to the encoding element e.g., references 83, 91. Nevertheless, the ITm family appears to have produced the Correia element in *Neisseria meningitidis* and *Neisseria gonorrhea* (92, 93). Although the element does not encode transposase, it encodes a strong binding site for a bacterial histone and a pair of outward facing promoters (94). Collectively, the hundreds of Correia elements have the potential to influence the expression of a large number of genes.

EXAPTATION

With the large number of whole genome sequences available it has become increasingly clear that transposons sometimes experience exaptation, or domestication. The key transition is when a transposon-derived sequence begins to perform a "useful" function and therefore comes under purifying selection, just like any other bona fide host gene or regulatory region e.g., reference 95, 96. A transposon may come under purifying selection simply because it has inserted near a gene and changed its expression either directly or by altering or adding regulatory elements. The transposase may also contribute functional domains to an expressed protein. In mariner, the best known example of this type is the human SETMAR protein, in which the Hsmar1 transposase is fused to a SET-domain protein-methylase (88). The transposase domain is 94% identical to the Hsmar1 consensus sequence. However, the third active site D is substituted with an N residue, which all but abolishes its transpositional activity (49). Nevertheless, structural analysis of SETMAR may provide insight into the transposition reaction (below).

Another notable example of exaptation is the internally eliminated sequences (IESs), which interrupt thousands of genes in ciliated protozoans. The IESs are highly degenerate, but they are all flanked by a TA dinucleotide, and many appear to be derived from an ITm transposon (97). Interestingly, IESs are excised by a domesticated PiggyBac transposon (98). This apparently loose relationship between short parasitic sequences and their cognate transposase is reminiscent of the Osmar–Stowaway relationship (above).

STRUCTURAL DYNAMICS AND REGULATION

Several crystal structures are available for DDE(D) transposases. The retroviral integrase has provided the most complete set, with structures representing pre- and postcleavage intermediates and target complexes (24, 99, 100). There are also structures for the postcleavage intermediates for Tn5 and Mos1 (57, 101). Phage Mu integrase provides a view of a replicative integration product (102). Collectively, the structures provide deep insights into the two metal-ion catalytic mechanism and the structural determinants of transpositional dynamics. However, we still lack structures poised for the hairpin step or the second hydrolysis.

The domain structure of the mariner transposases is illustrated in Fig. 2(A). It has a bipartite N-terminal DNA-binding domain connected to the catalytic domain by a proteolitically sensitive linker. Apart from the catalytic triad of D residues, the most highly conserved sequence motifs are WVPHEL and YSPDL (4). Although these motifs are less well conserved in other ITm transposons, the general arrangement is probably common. For example, the ITmD34E elements Tc3 and SB are similar in size to mariner and both have bipartite DNA-binding domains (103, 104). Other ITm transposases, such as Osmar, are significantly larger but appear to have a similar arrangement (105).

The crystal structure for the Mos1 postcleavage intermediate presents a classical *trans*-architecture, in which the transposase subunit bound to one transposon end contributes its catalytic domain to the opposite end [Fig. 7(A)] (57). It also helped to clarify a long-standing question about why the WVPHEL motif is conserved given its location in an interdomainal linker, which might be expected to have an unconstrained sequence. The structure revealed that the linker contributes a second level of complexity to the *trans*-architecture of the complex (Fig. 7A, B). An extended "clamp loop" feature emerges from the catalytic core of one transposase subunit and interacts with the transposon end and the WVPHEL motif on the other side of the complex. These interactions account for more than 70% of the dimer interface. On the opposite side, WVPHEL is in contact with the YSPDL motif, which forms part of the active site. This network of interactions connects the active sites of both subunits and provides a potential

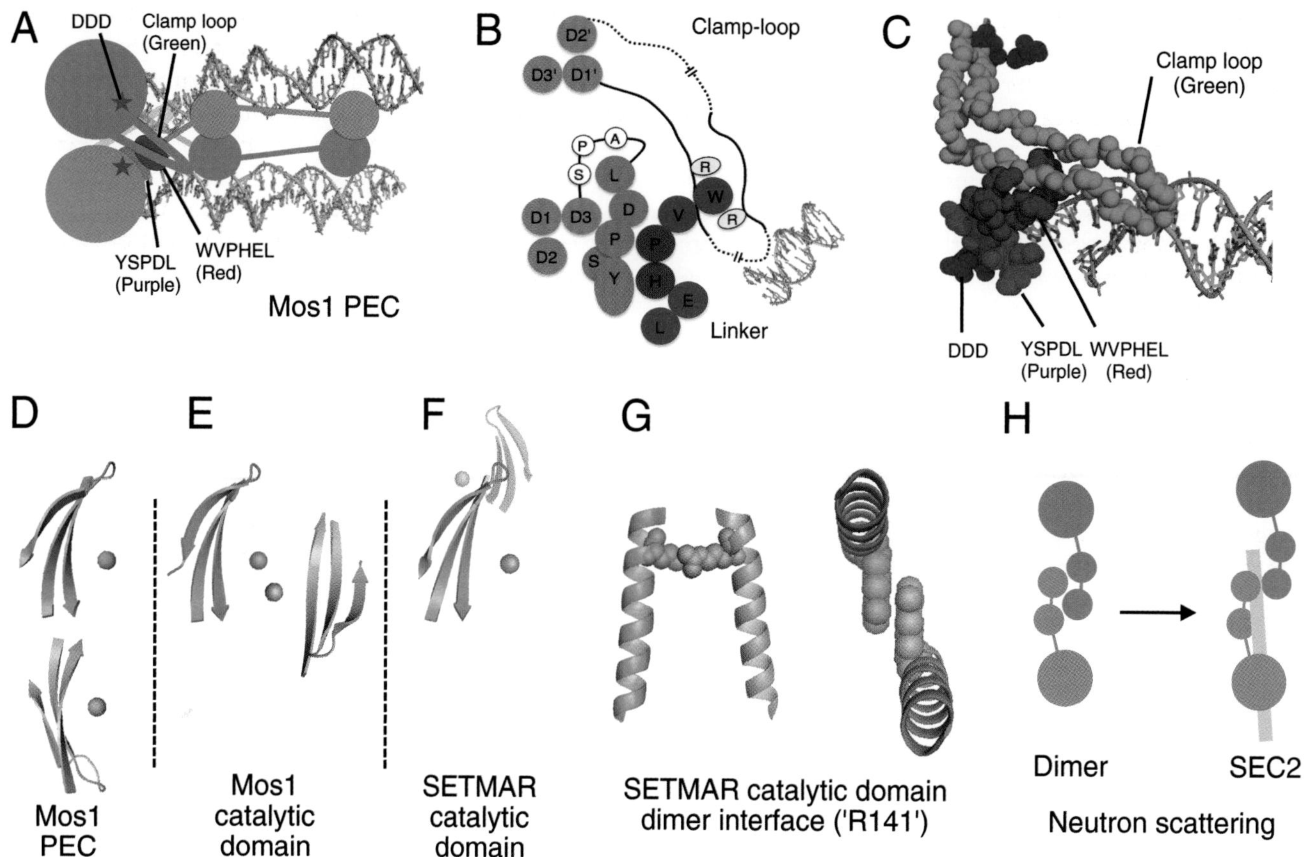

Figure 7 Structural features of mariner transposition intermediates. (A) A cartoon illustrating the *trans*-architecture of the Mos1 transpososome as visualized in the crystal structure of the postcleavage intermediate (57). Green and orange blobs, transposase; blue, active site; green, clamp loop; red, WVPHEL motif; purple, YSPDL motif. (B, C) Cartoon and space filling representation for the interactions between the clamp loop and the WVPHEL motif at the dimer interface. Coordinates from PDB HOT3. (D to F) The relationship between the catalytic domain of protein subunits in the crystal structures of the Mos1 PEC (PDB HOT3), the Mos1 catalytic domain (PDB 2F7T), and the SETMAR catalytic domain (PDB 3K9J). The three beta sheets forming the core of the RNase H fold are shown together with the third conserved aspartate residue from the active site. (G) One of the structural elements from the dimer interface in SETMAR (PDB 3K9J) is shown, highlighting residue R141 (numbering from Hsmar1). (H) A cartoon illustrating the elongation of the transposase dimer during first end binding as suggested by neutron scattering experiments. Part (A) reproduced from (17) available under Creative Commons License. Parts (B) and (C) reproduced from (83, 108) available under Creative Commons License. Part (H) reproduced from (111) available under Creative Commons License.
doi:10.1128/microbiolspec.MDNA3-0033-2014.f7

conduit for signals. It is worth noting that the clamp loop feature is inserted between the first and second strands of the core β-sheet. This is the same location as the much larger "insertion domains" in Tn5, RAG1 and Hermes transposases (101, 106, 107). The presence of an insertion domain seems to be related to the mechanism of second strand processing because it is absent in those enzymes which do not cleave the second strand, for example, RNase H, Mu transposase, and the retroviral integrases.

Biochemical analysis of Hsmar1 transposition showed that communication between subunits was important for two aspects of the reaction. Saturating mutagenesis of the WVPHEL motif showed that the vast majority of substitutions of the W, V, E, and L residues increase the rate of transposition (108). The underlying mechanism was an increase in the rate of synapsis. Presumably, the mutations increase the degrees of freedom of the unoccupied DNA-binding domain within SEC2, which increases the proportion of productive collisions. This also has the effect of relaxing the topological filter (83, 108). In addition, the mutants gained the ability to perform the 5′-nick prior to synapsis and some also acquired a nonspecific endonuclease activity. This suggests that a defect has arisen in the communication between opposite sides of the complex and in the control of catalysis.

The WVPHEL motif is also involved in the conformational change that resets the complex for the second hydrolysis. Within the transpososome the order of strand nicking is constrained and both 5′-nicks normally take place before either of the 3′-nicks (17). This suggests that the structural transition responsible for the exit of the cleaved strand from the active site, and the entry of the uncleaved strand, is coordinated between the partner subunits. One of the few hypoactive WVPHEL mutants (V119G) stalled after the 5′-nick, suggesting that it is unable to perform the transition at all (17). Furthermore, although the hyperactive mutants are able to complete the reaction, the coordination between the subunits is less robust than wild type. The WVPHEL motif thus coordinates catalysis across the complex.

Catalysis is also coordinated across the Tn10 transpososome (109). A defective transposon end was shown to prevent hydrolysis of the hairpin intermediate on one transposon end, which blocked initiation of catalysis on the other. This provides a parallel with the inhibition of 3′-end cleavage in the mariner V119G mutant. In both cases the blockage corresponds to resetting the active site for the mechanistically-difficult second hydrolysis reaction.

In mariner the most intimate subunit interactions are between the WVPHEL tryptophan residue and a pair of arginine residues in the clamp loop, which stack on its aromatic ring [Fig. 7(B), (C)]. Whilst all 16 of the W-substitutions tested were hyperactive, single and double alanine substitutions of the arginine residues were inactive (108). Likewise, no hyperactive mutations were identified at the P and H positions despite extensive screening, or in the YSPDL motif. This suggests that these residues do something over-and-above their regulatory function.

Since further structural intermediates for mariner transposition are lacking, it is worth considering two structures for the catalytic domain determined in the absence of DNA. One is for Mos1, while the other is derived from SETMAR (42, 110). The primary sequences of the catalytic domains from SETMAR and Hsmar1 transposase are 91% identical. Hsmar1 and Mos1 transposase are 37% identical with a single residue indel. This level of similarity in the primary sequence suggests that their three dimensional structures will be largely superimposable.

The relative spatial orientation of the catalytic cores in the Mos1 PEC is shown in Fig. 7(D). The orientation is significantly different in the apo-structure in the absence of DNA [Fig. 7(E)]. In this structure the dimer interface has a small area and it may therefore represent a crystal lattice interaction with no biological significance. The orientation of the catalytic cores in SETMAR is quite different from either of the Mos1 structures [Fig. 7(F)]. In this case the dimer interface is extensive. One element is shown in Fig. 7(G). The highlighted residue is R469, which is homologous to R142 in Mos1 and R141 in Hsmar1. In a random screen for hyperactive Hsmar1 mutations, R141L was identified as increasing the rate of transposition by 140-fold in an *in vivo* assay (D. Liu and RC, to be presented elsewhere). This suggests that the subunit interface observed in SETMAR may also be present in one of the transposition intermediates.

One way to interpret the hyperactivity of the R141L transposase is to postulate that the SETMAR interface is present in SEC2 and that the mutation promotes the conformational change that accompanies second end recruitment and activation for catalysis [see Fig. 5(B)]. However, the fact that the SETMAR dimer interface is located in the catalytic domain is inconsistent with a neutron scattering study of Mos1 (111). This study suggested that the transposase subunits have head-to-head interactions, with an extended structure that becomes even more extended in SEC2 [Fig. 7(H)].

Whether or not one or other of these alternative views is correct, it still raises questions about the mechanism responsible for hyperactivity in the WVPHEL mutants. In either case, the WVPHEL motif of one subunit is probably too far away from the clamp loop of its partner to engage in the interactions seen in the postcleavage PEC. Presumably, the WVPHEL motif must engage in different interactions at earlier stages of the reaction, and it is these that are disrupted by the mutations. This conclusion would fit with the lack of reciprocity between mutations in the WVPHEL motif and those in the clamp loop arginine residues, which do not yield hyperactive transposases (108). Thus, the hyperactive WVPHEL mutants, and the relaxation in the coupling of catalysis to synapsis, is probably owing to the loss of an as yet unrecognized intramolecular interaction.

FUTURE DIRECTIONS AND TECHNICAL PITFALLS

Cut-and-paste transposition is a complex reaction requiring between six and eight strand breaking and joining reactions, depending on whether the DNA strands are hydrolyzed directly or via a hairpin intermediate. Additional layers of complexity arise from the temporal progression and the coordination of events between opposite sides of the complex and from transposon regulatory mechanisms. Below, we describe a number of factors that are important to keep in mind when studying mariner elements.

First of all, many transposases are rather poorly active *in vitro*. In some cases, particularly in eukaryotes, this may be because the element studied is a defective copy. However, a low level of activity may also arise from bona fide regulatory functions. For example, the transposase of the bacterial element IS911 binds cotranslationally to the transposon end (91). In Tn5 the C-terminal end of the Tn5 transposase inhibits DNA binding by steric hindrance of the N-terminal domain. Mutation of this interaction was essential to develop a highly efficient *in vitro* system. However, *in vivo* this mechanism is important: it mediates the *cis*-biased activity of the transposase and controls the genomic invasion of the transposon (83, 112).

The Hsmar1 transposase is unusually active *in vitro*. But even here OPI and the ASO mechanism, which underlies autoregulation, complicate the interpretation of results (12). *In vitro*, OPI operates as soon as two transposase dimers are present in the reaction. This is because a free transposase will always find a naked transposon end before SEC2 can recruit the naked end. Any kinetic experiments that do not involve the preassembly of the PEC in noncatalytic conditions are dubious because the rate of the catalytic steps is underestimated by an inevitable fraction of substrates that are doubly occupied and inhibited by OPI. The ASO mechanism also leads to counterintuitive effects that have proved simple assumptions to be wrong (50). For example, under OPI conditions, increasing the affinity of the transposase for the transposon end may lower the activity, in contrast to what common sense dictates.

Band-shift assays can also provide misleading results because the species observed are not necessarily present in solution and may not correspond to an active species. For example, in an EMSA mariner transposases yield a complex containing a single transposon end (SEC1). However this is not an intermediate of the reaction. Instead it seems to arise when the PEC falls apart into two equal halves (50). All available evidence suggests that SEC1 is an artifact of the assay and plays no role in the reaction.

The study of mariners has revealed a number of differences between elements. For example, the initiation of catalysis in Hsmar1 is more tightly controlled than in other mariners studied to date (58). Whereas Mos1 and Himar1 may be capable of performing the 5′-nick within SEC2 (15, 16), Hsmar1 catalysis is almost completely dependent on synapsis of the transposon ends (58). This tight control of catalysis has facilitated biochemical analysis by providing a reaction that is largely free from the confounding effects of the nonspecific endonuclease activities (12). Whereas the Hsmar1 transposase is the reconstructed sequence from a transposon, which founded a successful genomic invasion (10), Himar1 is a consensus of defunct elements and Mos1 was recognized in laboratory populations where it happens to cause a visible phenotype in some backgrounds (2, 8). In Hsmar1, the tight coupling between catalysis and synapsis is therefore probably an adaptive feature of the reaction, which minimizes unproductive DNA damage. In contrast, the lax control of catalysis and the nonspecific endonuclease activities observed in Himar1 and Mos1 appear to be wholly detrimental to the transposition reaction and are probably therefore the result of genetic drift. Similar effects have been observed during mutational analysis of Hsmar1: several mutations, which alter amino acids in the dimer interface, uncouple catalysis from synapsis to various degrees (108).

We have made significant progress in understanding the biochemistry and dynamics of mariner transposition and we have a valuable structure for the post-cleavage intermediate. To conclude this review, we would like to propose some research directions that should keep the field busy for the years to come. Our immediate goals should be to obtain further structures for mariner intermediates and to extend *in vitro* analysis to other members of the ITm superfamily. Further exploring the effect of regulatory mechanisms *in vivo* in mariner and other DNA transposons will also be very interesting. Nevertheless, our current understanding of mariner regulatory mechanisms should readily provide guides to assist the design and optimization of new hyperactive transposase variants for biotechnological and medical applications (83).

Other avenues that remain to be explored include deciphering the effect of chromatinization and cellular events like transcription or replication and the role of DNA topology on transposition *in vivo*. The relationship between transposons and host defense mechanisms; the mechanisms responsible for the horizontal transfer of transposons; the function of domesticated transposases; and the role of transposons in the evolution of regulatory networks are also important directions to pursue in the future.

Citation. Tellier M, Claeys Bouuaert C, Chalmers R. 2014. Mariner and the ITm superfamily of transposons. Microbiol Spectrum 3(2):MDNA3-0033-2014.

References

1. **Emmons SW, Yesner L, Ruan KS, Katzenberk D.** 1983. Evidence for a transposon in Caenorhabditis elegans. *Cell* **32:**55–65.
2. **Jacobson JW, Medhora MM, Hartl DL.** 1986. Molecular structure of a somatically unstable transposable element in Drosophila. *Proc Natl Acad Sci U S A* **83:** 8684–8688.
3. **Matsutani S, Ohtsubo H, Maeda Y, Ohtsubo E.** 1987. Isolation and characterization of IS elements repeated in the bacterial chromosome. *J Mol Biol* **196:**445–455.
4. **Robertson HM.** 1993. The mariner transposable element is widespread in insects. *Nature* **362:**241–245.
5. **Yuan YW, Wessler SR.** 2011. The catalytic domain of all eukaryotic cut-and-paste transposase superfamilies. *Proc Natl Acad Sci U S A* **108:**7884–7889.
6. **Shao H, Tu Z.** 2001. Expanding the diversity of the IS630-Tc1-mariner superfamily: discovery of a unique DD37E transposon and reclassification of the DD37D and DD39D transposons. *Genetics* **159:**1103–1115.
7. **Robertson HM, Lampe DJ.** 1995. Distribution of transposable elements in arthropods. *Annu Rev Entomol* **40:** 333–357.
8. **Lampe DJ, Churchill ME, Robertson HM.** 1996. A purified mariner transposase is sufficient to mediate transposition in vitro. *EMBO J* **15:**5470–5479.
9. **Tosi LR, Beverley SM.** 2000. cis and trans factors affecting Mos1 mariner evolution and transposition in vitro, and its potential for functional genomics. *Nucleic Acids Res* **28:**784–790.
10. **Miskey C, Papp B, Mates L, Sinzelle L, Keller H, Izsvak Z, Ivics Z.** 2007. The ancient mariner sails again: transposition of the human Hsmar1 element by a reconstructed transposase and activities of the SETMAR protein on transposon ends. *Mol Cell Biol* **27:**4589–4600.
11. **Munoz-Lopez M, Siddique A, Bischerour J, Lorite P, Chalmers R, Palomeque T.** 2008. Transposition of Mboumar-9: identification of a new naturally active mariner-family transposon. *J Mol Biol* **382:**567–572.
12. **Claeys Bouuaert C, Chalmers R.** 2010. Transposition of the human Hsmar1 transposon: rate-limiting steps and the importance of the flanking TA dinucleotide in second strand cleavage. *Nucleic Acids Res* **38:**190–202.
13. **Renault S, Demattei MV, Lahouassa H, Bigot Y, Auge-Gouillou C.** 2010. In vitro recombination and inverted terminal repeat binding activities of the Mcmar1 transposase. *Biochemistry* **49:**3534–3544.
14. **Trubitsyna M, Morris ER, Finnegan DJ, Richardson JM.** 2014. Biochemical characterization and comparison of two closely related active mariner transposases. *Biochemistry* **53:**682–689.
15. **Dawson A, Finnegan DJ.** 2003. Excision of the Drosophila mariner transposon mos1. Comparison with bacterial transposition and v(d)j recombination. *Mol Cell* **11:**225–235.
16. **Lipkow K, Buisine N, Lampe DJ, Chalmers R.** 2004. Early intermediates of mariner transposition: catalysis without synapsis of the transposon ends suggests a novel architecture of the synaptic complex. *Mol Cell Biol* **24:**8301–8311.
17. **Claeys Bouuaert C, Walker N, Liu D, Chalmers R.** 2014. Crosstalk between transposase subunits during cleavage of the mariner transposon. *Nucleic Acids Res* **42:**5799–5808.
18. **Yang W, Hendrickson WA, Crouch RJ, Satow Y.** 1990. Structure of ribonuclease H phased at 2 A resolution by MAD analysis of the selenomethionyl protein. *Science* **249:**1398–1405.
19. **Katayanagi K, Miyagawa M, Matsushima M, Ishikawa M, Kanaya S, Ikehara M, Matsuzaki T, Morikawa K.** 1990. Three-dimensional structure of ribonuclease H from E. coli. *Nature* **347:**306–309.
20. **Nowotny M, Gaidamakov SA, Ghirlando R, Cerritelli SM, Crouch RJ, Yang W.** 2007. Structure of human RNase H1 complexed with an RNA/DNA hybrid: insight into HIV reverse transcription. *Mol Cell* **28:** 264–276.
21. **Beese LS, Steitz TA.** 1991. Structural basis for the 3′-5′ exonuclease activity of Escherichia coli DNA polymerase I: a two metal ion mechanism. *EMBO J* **10:**25–33.
22. **Yang W, Lee JY, Nowontny M.** 2006. Making and breaking nucleic acids: Two-Mg2+-ion catalysis and substrate specificity. *Mol Cell* **22:**5–13.
23. **Nowotny M, Gaidamakov SA, Crouch RJ, Yang W.** 2005. Crystal structures of RNase H bound to an RNA/DNA hybrid: substrate specificity and metal-dependent catalysis. *Cell* **121:**1005–1016.
24. **Hare S, Maertens GN, Cherepanov P.** 2012. 3′-processing and strand transfer catalysed by retroviral integrase in crystallo. *EMBO J* **31:**3020–3028.
25. **Craigie R, Fujiwara T, Bushman F.** 1990. The IN protein of Moloney murine leukemia virus processes the viral DNA ends and accomplishes their integration in vitro. *Cell* **62:**829–837.
26. **Roth MJ, Schwartzberg PL, Goff SP.** 1989. Structure of the termini of DNA intermediates in the integration of retroviral DNA: dependence on IN function and terminal DNA sequence. *Cell* **58:**47–54.
27. **Craigie R, Mizuuchi K.** 1985. Mechanism of transposition of bacteriophage Mu: structure of a transposition intermediate. *Cell* **41:**867–876.
28. **Engelman A, Mizuuchi K, Craigie R.** 1991. HIV-1 DNA integration: mechanism of viral DNA cleavage and DNA strand transfer. *Cell* **67:**1211–1221.
29. **Mizuuchi K, Adzuma K.** 1991. Inversion of the phosphate chirality at the target site of Mu DNA strand transfer: evidence for a one-step transesterification mechanism. *Cell* **66:**129–140.

30. **Kennedy AK, Guhathakurta A, Kleckner N, Haniford DB.** 1998. Tn10 transposition via a DNA hairpin intermediate. *Cell* **95:**125–134.

31. **Claeys Bouuaert C, Chalmers RM.** 2010. Gene therapy vectors: the prospects and potentials of the cut-and-paste transposons. *Genetica* **138:**473–484.

32. **Bolland S, Kleckner N.** 1996. The Three Chemical Steps of Tn10/IS10 Transposition Involve Repeated Utilization of a Single Active Site. *Cell* **84:**223–233.

33. **Sakai J, Chalmers RM, Kleckner N.** 1995. Identification and characterization of a pre-cleavage synaptic complex that is an early intermediate in Tn10 transposition. *EMBO J* **14:**4374–4383.

34. **Kennedy AK, Haniford DB, Mizuuchi K.** 2000. Single active site catalysis of the successive phosphoryl transfer steps by DNA transposases: insights from phosphorothioate stereoselectivity. *Cell* **101:**295–305.

35. **Bhasin A, Goryshin IY, Reznikoff WS.** 1999. Hairpin formation in Tn5 transposition. *J Biol Chem* **274:** 37021–37029.

36. **Mitra R, Fain-Thornton J, Craig NL.** 2008. piggyBac can bypass DNA synthesis during cut and paste transposition. *EMBO J* **27:**1097–1109.

37. **Stewart BJ, Wardle SJ, Haniford DB.** 2002. IHF-independent assembly of the Tn10 strand transfer transpososome: implications for inhibition of disintegration. *EMBO J* **21:**4380–4390.

38. **Zhou L, Mitra R, Atkinson PW, Hickman AB, Dyda F, Craig NL.** 2004. Transposition of hAT elements links transposable elements and V(D)J recombination. *Nature* **432:**995–1001.

39. **Coen ES, Carpenter R, Martin C.** 1986. Transposable elements generate novel spatial patterns of gene expression in Antirrhinum majus. *Cell* **47:**285–296.

40. **Roth DB, Menetski JP, Nakajima PB, Bosma MJ, Gellert M.** 1992. V(D)J recombination: broken DNA molecules with covalently sealed (hairpin) coding ends in scid mouse thymocytes. *Cell* **70:**983–991.

41. **Gellert M.** 2002. V(D)J recombination: rag proteins, repair factors, and regulation. *Annu Rev Biochem* **71:**101–132.

42. **Richardson JM, Dawson A, O'Hagan N, Taylor P, Finnegan DJ, Walkinshaw MD.** 2006. Mechanism of Mos1 transposition: insights from structural analysis. *EMBO J* **25:**1324–1334.

43. **Claeys Bouuaert C.** 2011. *The mechanism of mariner transposition. PhD Thesis*, University of Nottingham, Nottingham.

44. **Sarnovsky RJ, May EW, Craig NL.** 1996. The Tn7 transposase is a heteromeric complex in which DNA breakage and joining activities are distributed between different gene products. *EMBO J* **15:**6348–6361.

45. **Hickman AB, Li Y, Mathew SV, May EW, Craig NL, Dyda F.** 2000. Unexpected structural diversity in DNA recombination: the restriction endonuclease connection. *Mol Cell* **5:**1025–1034.

46. **Zhang L, Dawson A, Finnegan DJ.** 2001. DNA-binding activity and subunit interaction of the mariner transposase. *Nucleic Acids Res* **29:**3566–3575.

47. **Auge-Gouillou C, Brillet B, Germon S, Hamelin MH, Bigot Y.** 2005. Mariner Mos1 transposase dimerizes prior to ITR binding. *J Mol Biol* **351:**117–130.

48. **Auge-Gouillou C, Brillet B, Hamelin MH, Bigot Y.** 2005. Assembly of the mariner Mos1 synaptic complex. *Mol Cell Biol* **25:**2861–2870.

49. **Liu D, Bischerour J, Siddique A, Buisine N, Bigot Y, Chalmers R.** 2007. The human SETMAR protein preserves most of the activities of the ancestral Hsmar1 transposase. *Mol Cell Biol* **27:**1125–1132.

50. **Claeys Bouuaert C, Lipkow K, Andrews SS, Liu D, Chalmers R.** 2013. The autoregulation of a eukaryotic DNA transposon. *elife* **2:**e00668.

51. **Brillet B, Bigot Y, Auge-Gouillou C.** 2007. Assembly of the Tc1 and mariner transposition initiation complexes depends on the origins of their transposase DNA binding domains. *Genetica* **130:**105–120.

52. **Ivics Z, Hackett PB, Plasterk RH, Izsvak Z.** 1997. Molecular reconstruction of Sleeping Beauty, a Tc1-like transposon from fish, and its transposition in human cells. *Cell* **91:**501–510.

53. **Fischer SEJ, van Luenen H, Plasterk RHA.** 1999. Cis requirements for transposition of Tc1-like transposons in C. elegans. *Mol Gen Genet* **262:**268–274.

54. **Cherepanov P, Maertens G, Proost P, Devreese B, Van Beeumen J, Engelborghs Y, De Clercq E, Debyser Z.** 2003. HIV-1 integrase forms stable tetramers and associates with LEDGF/p75 protein in human cells. *J Biol Chem* **278:**372–381.

55. **Wang JY, Ling H, Yang W, Craigie R.** 2001. Structure of a two-domain fragment of HIV-1 integrase: implications for domain organization in the intact protein. *EMBO J* **20:**7333–7343.

56. **Deprez E, Tauc P, Leh H, Mouscadet JF, Auclair C, Brochon JC.** 2000. Oligomeric states of the HIV-1 integrase as measured by time-resolved fluorescence anisotropy. *Biochemistry* **39:**9275–9284.

57. **Richardson JM, Colloms SD, Finnegan DJ, Walkinshaw MD.** 2009. Molecular architecture of the Mos1 paired-end complex: the structural basis of DNA transposition in a eukaryote. *Cell* **138:**1096–1108.

58. **Claeys Bouuaert C, Liu D, Chalmers R.** 2011. A simple topological filter in a eukaryotic transposon as a mechanism to suppress genome instability. *Mol Cell Biol* **31:** 317–327.

59. **Benjamin HW, Matzuk MM, Krasnow MA, Cozzarelli NR.** 1985. Recombination site selection by Tn3 resolvase: topological tests of a tracking mechanism. *Cell* **40:** 147–158.

60. **Craigie R, Mizuuchi K.** 1986. Role of DNA topology in Mu transposition: mechanism of sensing the relative orientation of two DNA segments. *Cell* **45:**793–800.

61. **Jiang H, Harshey RM.** 2001. The Mu enhancer is functionally asymmetric both in cis and in trans. Topological selectivity of Mu transposition is enhancer-independent. *J Biol Chem* **276:**4373–4381.

62. **Johnson RC.** 1991. Mechanism of site-specific DNA inversion in bacteria. *Curr Opin Genet Dev* **1:**404–411.

63. **McLean MM, Chang Y, Dhar G, Heiss JK, Johnson RC.** 2013. Multiple interfaces between a serine recombinase and an enhancer control site-specific DNA inversion. *elife* **2:**e01211.
64. **Chalmers RM, Kleckner N.** 1996. IS10/Tn10 transposition efficiently accommodates diverse transposon end configurations. *EMBO J* **15:**5112–5122.
65. **English JJ, Harrison K, Jones JDG.** 1995. Aberrant transpositions of maize double Ds-like elements usually involve Ds ends on sister chromatids. *Plant Cell* **7:** 1235–1247.
66. **Zhang J, Zuo T, Peterson T.** 2013. Generation of tandem direct duplications by reversed-ends transposition of maize ac elements. *PLoS Genet* **9:**e1003691.
67. **McClintock B.** 1942. The Fusion of Broken Ends of Chromosomes Following Nuclear Fusion. *Proc Natl Acad Sci U S A* **28:**458–463.
68. **Claeys Bouuaert C, Chalmers R.** 2013. Hsmar1 transposition is sensitive to the topology of the transposon donor and the target. *PLoS One* **8:**e53690.
69. **Wang Y, Wang J, Devaraj A, Singh M, Jimenez Orgaz A, Chen JX, Selbach M, Ivics Z, Izsvak Z.** 2014. Suicidal autointegration of sleeping beauty and piggyBac transposons in eukaryotic cells. *PLoS Genet* **10:** e1004103.
70. **Chalmers R, Guhathakurta A, Benjamin H, Kleckner N.** 1998. IHF modulation of Tn10 transposition: sensory transduction of supercoiling status via a proposed protein/DNA molecular spring. *Cell* **93:**897–908.
71. **Benjamin HW, Kleckner N.** 1989. Intramolecular transposition by Tn10. *Cell* **59:**373–383.
72. **Shoemaker C, Hoffman J, Goff SP, Baltimore D.** 1981. Intramolecular integration within Moloney murine leukemia virus DNA. *J Virol* **40:**164–172.
73. **Crenes G, Moundras C, Demattei MV, Bigot Y, Petit A, Renault S.** 2010. Target site selection by the mariner-like element, Mos1. *Genetica* **138:**509–517.
74. **Lidholm DA, Lohe AR, Hartl DL.** 1993. The transposable element mariner mediates germline transformation in Drosophila melanogaster. *Genetics* **134:**859–868.
75. **Hellen EH, Brookfield JF.** 2013. Transposable element invasions. *Mobile Genet Elem* **3:**e23920.
76. **Hellen EH, Brookfield JF.** 2013. The diversity of class II transposable elements in mammalian genomes has arisen from ancestral phylogenetic splits during ancient waves of proliferation through the genome. *Mol Biol Evol* **30:**100–108.
77. **Lohe AR, Moriyama EN, Lidholm DA, Hartl DL.** 1995. Horizontal transmission, vertical inactivation, and stochastic loss of mariner-like transposable elements. *Mol Biol Evol* **12:**62–72.
78. **Hartl DL, Lohe AR, Lozovskaya ER.** 1997. Modern thoughts on an ancyent marinere: Function, evolution, regulation. *Annu Rev Genet* **31:**337–358.
79. **Le Rouzic A, Capy P.** 2005. The first steps of transposable elements invasion: parasitic strategy vs. genetic drift. *Genetics* **169:**1033–1043.
80. **Le Rouzic A, Boutin TS, Capy P.** 2007. Long-term evolution of transposable elements. *Proc Natl Acad Sci U S A* **104:**19375–19380.
81. **Hua-Van A, Le Rouzic A, Boutin TS, Filee J, Capy P.** 2011. The struggle for life of the genome's selfish architects. *Biol Direct* **6:**19.
82. **Nagy Z, Chandler M.** 2004. Regulation of transposition in bacteria. *Res Microbiol* **155:**387–398.
83. **Bouuaert CC, Tellier M, Chalmers R.** 2014. One to rule them all: A highly conserved motif in mariner transposase controls multiple steps of transposition. *Mobile Genet Elem* **4:**e28807.
84. **Lohe AR, Hartl DL.** 1996. Autoregulation of mariner transposase activity by overproduction and dominant-negative complementation. *Mol Biol Evol* **13:**549–555.
85. **Lampe DJ, Grant TE, Robertson HM.** 1998. Factors affecting transposition of the Himar1 mariner transposon in vitro. *Genetics* **149:**179–187.
86. **Clark KJ, Carlson DF, Leaver MJ, Foster LK, Fahrenkrug SC.** 2009. Passport, a native Tc1 transposon from flatfish, is functionally active in vertebrate cells. *Nucleic Acids Res* **37:**1239–1247.
87. **Yang G, Nagel DH, Feschotte C, Hancock CN, Wessler SR.** 2009. Tuned for transposition: molecular determinants underlying the hyperactivity of a Stowaway MITE. *Science* **325:**1391–1394.
88. **Robertson HM, Zumpano KL.** 1997. Molecular evolution of an ancient mariner transposon, Hsmar1, in the human genome. *Gene* **205:**203–217.
89. **Oosumi T, Belknap WR, Garlick B.** 1995. Mariner transposons in humans. *Nature* **378:**672.
90. **Morgan GT.** 1995. Identification in the human genome of mobile elements spread by DNA-mediated transposition. *J Mol Biol* **254:**1–5.
91. **Duval-Valentin G, Chandler M.** 2011. Cotranslational control of DNA transposition: a window of opportunity. *Mol Cell* **44:**989–996.
92. **Buisine N, Tang CM, Chalmers R.** 2002. Transposon-like Correia elements: structure, distribution and genetic exchange between pathogenic Neisseria sp. *FEBS Lett* **522:**52–58.
93. **Correia FF, Inouye S, Inouye M.** 1988. A family of small repeated elements with some transposon-like properties in the genome of Neisseria gonorrhoeae. *J Biol Chem* **263:**12194–12198.
94. **Siddique A, Buisine N, Chalmers R.** 2011. The transposon-like Correia elements encode numerous strong promoters and provide a potential new mechanism for phase variation in the meningococcus. *PLoS Genet* **7:**e1001277.
95. **Alzohairy AM, Gyulai G, Jansen RK, Bahieldin A.** 2013. Transposable elements domesticated and neofunctionalized by eukaryotic genomes. *Plasmid* **69:**1–15.
96. **Feschotte C.** 2008. Transposable elements and the evolution of regulatory networks. *Nat Rev Genet* **9:**397–405.
97. **Arnaiz O, Mathy N, Baudry C, Malinsky S, Aury JM, Wilkes CD, Garnier O, Labadie K, Lauderdale BE, Le Mouel A, Marmignon A, Nowacki M, Poulain J,**

Prajer M, Wincker P, Meyer E, Duharcourt S, Duret L, Betermier M, Sperling L. 2012. The Paramecium germline genome provides a niche for intragenic parasitic DNA: evolutionary dynamics of internal eliminated sequences. *PLoS Genet* **8**:e1002984.

98. **Baudry C, Malinsky S, Restituito M, Kapusta A, Rosa S, Meyer E, Betermier M.** 2009. PiggyMac, a domesticated piggyBac transposase involved in programmed genome rearrangements in the ciliate Paramecium tetraurelia. *Genes Dev* **23**:2478–2483.

99. **Hare S, Gupta SS, Valkov E, Engelman A, Cherepanov P.** 2010. Retroviral intasome assembly and inhibition of DNA strand transfer. *Nature* **464**:232–236.

100. **Maertens GN, Hare S, Cherepanov P.** 2010. The mechanism of retroviral integration from X-ray structures of its key intermediates. *Nature* **468**:326–329.

101. **Davies DR, Goryshin IY, Reznikoff WS, Rayment I.** 2000. Three-dimensional structure of the Tn5 synaptic complex transposition intermediate. *Science* **289**:77–85.

102. **Montano SP, Pigli YZ, Rice PA.** 2012. The mu transpososome structure sheds light on DDE recombinase evolution. *Nature* **491**:413–417.

103. **Watkins S, van Pouderoyen G, Sixma TK.** 2004. Structural analysis of the bipartite DNA-binding domain of Tc3 transposase bound to transposon DNA. *Nucleic Acids Res* **32**:4306–4312.

104. **Carpentier CE, Schreifels JM, Aronovich EL, Carlson DF, Hackett PB, Nesmelova IV.** 2014. NMR structural analysis of Sleeping Beauty transposase binding to DNA. *Protein Sci* **23**:23–33.

105. **Feschotte C, Osterlund MT, Peeler R, Wessler SR.** 2005. DNA-binding specificity of rice mariner-like transposases and interactions with Stowaway MITEs. *Nucleic Acids Res* **33**:2153–2165.

106. **Hickman AB, Perez ZN, Zhou L, Musingarimi P, Ghirlando R, Hinshaw JE, Craig NL, Dyda F.** 2005. Molecular architecture of a eukaryotic DNA transposase. *Nat Struct Mol Biol* **12**:715–721.

107. **Lu CP, Sandoval H, Brandt VL, Rice PA, Roth DB.** 2006. Amino acid residues in Rag1 crucial for DNA hairpin formation. *Nature Struct Mol Biol* **13**:1010–1015.

108. **Liu D, Chalmers R.** 2014. Hyperactive mariner transposons are created by mutations that disrupt allosterism and increase the rate of transposon end synapsis. *Nucleic Acids Res* **42**:2637–2645.

109. **Crellin P, Sewitz S, Chalmers R.** 2004. DNA looping and catalysis; the IHF-folded arm of Tn10 promotes conformational changes and hairpin resolution. *Mol Cell* **13**:537–547.

110. **Goodwin KD, He H, Imasaki T, Lee SH, Georgiadis MM.** 2010. Crystal structure of the human Hsmar1-derived transposase domain in the DNA repair enzyme Metnase. *Biochemistry* **49**:5705–5713.

111. **Cuypers MG, Trubitsyna M, Callow P, Forsyth VT, Richardson JM.** 2013. Solution conformations of early intermediates in Mos1 transposition. *Nucleic Acids Res* **41**:2020–2033.

112. **Reznikoff WS.** 2008. Transposon Tn5. *Annu Rev Genet* **42**:269–286.

LTR Retrotransposons

Mobile DNA, 3rd Edition
Nancy L. Craig, Michael Chandler, Martin Gellert, Alan M. Lambowitz, Phoebe A. Rice and Suzanne Sandmeyer

doi:10.1128/microbiolspec.MDNA3-0054-2014

Peter W. Atkinson[1]

hAT Transposable Elements 35

INTRODUCTION

hAT transposable elements are class II DNA transposons that are ancient in their origin. They are widespread across the plant and animal kingdoms and are found in all eukaryotes with the exception of ciliates, diatoms, and the protozoan *Trichomonas* (1). A survey of eight dicotyledons from five angiopsperm families and eight monocotyledons from two angiosperm families revealed that *hAT* elements comprised approximately 0.31% of the dicotyledon genomes (representing 6.4% of the total genomic DNA transposons) and 0.46% of the monocotyledon genomes (representing 8.2% of the total genomic DNA transposons) (2). This low abundance is countered by their apparent impact on angiosperm evolution and adaptation in which they have been estimated to contribute to approximately 20% of 65 examples of transposon-mediated alterations to gene function or creation (2). They comprise the most abundant superfamily of class II transposons found in humans, yet no active forms have been found in our species to date. Despite being very ancient in origin, phylogenetic trees constructed from the amino acid sequences of their transposases are often not completely congruent with those arising from sequence comparison of their chromosomal genes suggesting that other factors, such as horizontal transfer, may have played a role in the current distribution of these transposons.

hAT elements are typically less than 5 kb in length, with the sizes of their terminal inverted repeats (TIRs) varying from 5 to 27 bp with little conservation of sequences other than a preference for A and G at the second and fifth positions, respectively. Subterminal repeats are present in *hAT* transposons but their precise role in transposition is unclear. They likely play a role in positioning the transposase at the ends of the transposon (3, 4, 5, 6). Another defining characteristic of *hAT* elements is that, upon integration into a target DNA sequence, they generate 8 bp target site duplications (TSDs). *hAT* elements typically excise from the donor site imprecisely leaving small deletions and small repeated sequences of DNA flanking the transposon (7, 8). Analysis of the excision of the *Hermes hAT* element from the housefly *Musca domestica* showed that this is caused by the resolution of hairpin structures formed as reaction intermediates on this flanking DNA (9). The transposases of *hAT* elements contain a DDE motif derived from the RNaseH domain that constitutes the catalytic core domain of the enzyme (10, 11). The glutamate is located some linear distance from the second aspartate due to the presence of a large insertion domain (10). Other conserved residues are also present in the transposase (see below).

There have been several previous reviews on both *hAT* elements and individual members of this superfamily such as the *Ac/Ds* element of *Zea mays* and the *hobo* element of *Drosophila melanogaster*. The reader is referred to these for details about the status of this superfamily at the time of publication (7, 12, 13, 14,

[1]Department of Entomology and Institute for Integrative Genome Biology, University of California, Riverside, CA 92521.

15, 16, 17). Earlier reviews necessarily focused on chromosomal rearrangements associated with, or generated by, functional *hAT* elements such as *Ac* and *Tam3* from *Antirrhinum majus* and examined the molecular outcomes of excision and transposition of these transposons. The first transposable element identified, *Ac* of maize discovered by McClintock, is a member of the *hAT* superfamily, and understandably became a focus of ongoing studies in the many years since her initial discovery (18, 19). The *hobo* element was the first *hAT* transposon found in animals and remains one of the few active *hAT* elements characterized (20). Based on the amino acid sequence of their transposases, the *hAT* superfamily was initially divided into two families, *Ac* and *Buster* (21). While members of the *Ac* family are relatively widespread amongst kingdoms, the distribution of *Buster* family transposons appears to be confined to animals (21). A secondary consideration of this classification was the difference in 8 bp TSD consensus sequence generated between transposons of each family. That of *Ac* family members is 5′ nTnnnnAn 3′ while the sequence of *Buster* family members is 5′ nnnTAnnn 3′ (21). The functional basis of this difference remains unknown. Arensburger et al. (21) speculated on the presence of a third family of *hAT* elements centered on the *Tip100* transposons but declined to make this additional division due to the lack of transposon sequence information available. Several additional *Tip*-related transposons and their transposases have since been identified and sequenced, lending further support for this third family within the *hAT* superfamily (22, 23, 24). There is insufficient TSD information to determine the consensus TSD sequence generated by the integration of these transposons. The *Tip* family has members in the plant and animal kingdoms and so may be as widespread as the *Ac* family.

Whole genome sequencing combined with bioinformatics analysis has greatly increased the number of new *hAT* transposon sequences discovered. However, few full-length elements with intact TIR sequences and intact ORFs have so far been discovered. Our knowledge of how *hAT* elements actually function is thus restricted to the few active *hAT* transposons discovered, while more detailed molecular knowledge is confined to the single active *hAT* transposon for which there is a crystal structure of its transposase and its bound TIR, this being the *Hermes* transposon (25).

This review is not simply an update on the increasing catalog of *hAT* transposons, as that would be rendered obsolete within months. Rather it will focus on the structure and function of primarily active forms of transposons based on the most recent studies on the *Hermes* transposon for which the cocrystal structure of the transpososome was recently obtained (6).

THE DISTRIBUTION OF *hAT* ELEMENTS ACROSS KINGDOMS

hAT elements are widespread throughout the tree of life (1). As an illustration, approximately 160 selected genomes were examined for the presence of members of the 17 superfamilies of eukaryote cut and paste transposons using a TBLASTN search in which the DDE/D catalytic domain shared amongst many class II transposons was the query (11). Together with *Tc1/mariner*, *PIF/Harbinger*, *MULE*, and *piggyBac* elements, *hAT* elements are one of the most widespread transposon superfamilies present in plants, animals, fungi, and amoeba. As such, they are now relatively simple to identify and isolate but, as is the case with all transposons, it can be difficult to identify new active members of the superfamily as selection against active transposons by the host organism leads to increasing numbers of inactive forms. These can be elements with internal deletions or defective TIRs, or can be miniature inverted-repeat transposable elements (MITEs). However more than a dozen active *hAT* derivatives have been found and characterized. This provides different variants with the same basic structure and organization, potentially providing a basis for comparative and genetic and biochemical analysis.

The large number of *hAT* element sequences identified through the expanding database set has produced examples where horizontal transfer has been proposed in order to explain the distribution of these elements amongst distantly related species. A challenge presented by *hAT* elements is to understand how the basic components, such as its transposition machinery, the transposon DNA substrate itself, and the *hAT*-encoded transposase, which catalyzes its excision and integration, can facilitate transfer between sites and, at low frequencies, between species. *hAT* transposons can be used as gene vectors in biotechnology. Several of them can be introduced into new species for the purposes of genetic engineering, gene mapping and gene therapy. A question that immediately arises is whether an introduced, genetically modified *hAT* element will interact with any of the inactive *hAT* sequences already present in the genome (present in massive numbers in humans and plants for example) and this can be addressed through understanding how these transposons function *in vivo* and *in vitro*.

The circumstantial evidence for possible horizontal transfer of *hAT* elements was initially presented for the *hobo* element of *D. melanogaster* to explain its distribution between this species and *Drosophila simulans*, from which the authors proposed it invaded the *D. melanogaster* genome during the middle of the 20th century (26). Although the direction of transfer was later debated, the likelihood of its horizontal transfer between at least these two species and *Drosophila mauritiana* was supported by subsequent molecular analyses (27). The advent of whole genome sequencing combined with elegant bioinformatics tools enabled the distribution of *hobo* and newly discovered insect *hAT* elements to be determined at a far more detailed level in a greater range of *Drosophila* species as described below (28, 29).

The question of whether transkingdom horizontal transfer of *hAT* elements has occurred was initially addressed through both clustering analysis and by establishing phylogenies based on six conserved blocks in the transposase-coding region together with the sequences of 147 full-length elements as defined by the presence of probable TIRs and TDRs (12). Clustering analysis based on the conserved blocks examined indicated that they could be placed into six groups, with members of each group confined to a single kingdom (12). The conclusion was that while incongruence between phylogenies established for *hAT* transposons and the species they reside in were likely caused by horizontal transfer within kingdoms, there was no compelling evidence for the same occurring across kingdoms (12). Subsequent reexamination of the hAT transposases, which included the addition of the Buster and the space invader (SPIN) transposases discovered in the genome of the bushbaby *Otolemur garnettii* (30), grouped them into just two families with the *Buster* family containing members uniquely from the animal kingdom (21). The *Ac* family contained a subfamily consisting of transposons (such as *Ac* and *Tam3*) confined to plants and other members found in plants and fungi. This was further confirmed upon the identification of a third family of the *hAT* elements, the *Tip* family, which, like the *Ac* family has members in the plant and animal kingdoms (22, 23).

The *Tip* family is named after the *Tip100* element from the common morning glory, *Ipomoea purpurea*, in which it was discovered within an intron of a newly characterized gene, *CHS-D* involved in flower pigmentation (31). Subsequent investigation revealed that *Tip100* is present in three other species in the *Ipomoea* genus and that its distribution amongst these was consistent with vertical inheritance rather than with horizontal transfer (24). *Tip* transposons have recently been found in the genomes of *Drosophila buzzatii*, *Rhodnius prolixus*, and *Bombyx mori* representing a distribution across three orders of insects (22, 23). The high degree of similarity of the *BuT2* transposon combined with its patchy distribution across five species groups of the *Drosophila* and *Sophophora* subgenera suggest that it has been horizontally transferred within the genus *Drosophila* (22). Similarly, the high degree of similarity amongst the Tip transposases found in *Rhodnius* and *Bombyx* is highly indicative of horizontal transfer occurring at some stage between members of these two orders. Despite the patchy distribution of the *Tip* elements so far discovered in plant and insect species, there is no evidence of interkingdom horizontal transfer of these transposons.

Other *hAT* elements also show patchy distributions within the genus *Drosophila*. These are the proposed *harrow* subfamily of *Ac* family-*hATs* found in 10 species of *Drosophila*, spanning the *Sophophora* and *Drosophila* subgenera and the *hosimary* transposon that is found in some species of the subgenus *Sophomora* and in the distantly related species *Zaprionus indianus* (32, 33).

Recent evidence for the horizontal transfer of the piscean *hAT* transposons, *Tgf2* and *Tol2*, was obtained following the discovery of the *Tgf2* element in the goldfish, *Carassius auratus* (34, 35). *Tgf2* is mobile when introduced into the medaka fish, *Oryzias latipes*, and a phylogenetic tree constructed from the relationships based on a discontinuous 67 amino acid region of the Tgf2 transposase, which is conserved amongst the hAT transposases (see below), revealed that the distribution of *Tgf2* and the closely related *Tol2 hAT* element from the medaka fish were consistent with horizontal transfer (35). Supporting this is the very patchy distribution of *Tgf2* elements in genomic databases. It is present in medaka (in which the goldfish Tgf2 element is mobile) and in the Nile tilapia but absent in closely related species such as carp (35). The closely related *Tol2* transposon also shows a distribution inconsistent with inheritance by only vertical transmission. It is found in only two of 10 species of medaka with the copies of this transposon showing hardly any sequence divergence (36).

SPIN transposons, initially found in *O. garnettii* (30), are members of the *Buster* family of *hAT* elements and their relationship with other *Buster* elements has been previously described (21). A consensus sequence based on these transposons was used to interrogate genome databases resulting in the discovery of highly conserved *SPIN* elements in a range of mammalian species (30). However, the resulting distribution was inconsistent with the phylogeny of these species as *SPIN* is present in a diverse range of rodents, primates, bats, marsupials, lizards, and frogs but is completely absent in many species closely related to these (30). A subsequent analysis of reptilian genomes revealed highly related *SPINs* to be present throughout 14 families of reptiles leading to speculation that *SPIN* may have been horizontally transferred up to 13 separate times within reptiles (37).

A genome-wide analysis of the bats *Myotis lucifugus* and *Myotis austroriparius* revealed six families of *hAT* transposons with evidence of recent mobility based on the low level of sequence divergence of multiple copies across different insertion sites in the genome (38). This constituted the first report of a possibly active class II DNA transposon in a mammalian genome. However, no intact hAT transposase was found in these bat genomes suggesting that activity of bat *hATs* was most likely recent rather than current (38). Members from four groups of *hAT* elements (referred to as families by the authors) were found dispersed through species of mammals, reptiles, amphibians, and planarians (39). This discontinuous distribution is consistent with horizontal transfer. However, the authors noted that some species (for example *M. lucifugus*, the gray short-tailed opossum *Monodelphis domestica*, and the lizard *Anolis carolinensis*) seemed to possess multiple members of horizontally transferred *hATs*, including *SPIN* transposons (39). The basis for this perceived susceptibility to transposon invasion and subsequent retention of these sequences remains unknown.

hAT ELEMENTS IN BIOTECHNOLOGY

The ability of *hAT* transposons to cross species boundaries within kingdoms would indicate that they should be able to be harnessed in technological applications in which gene transfer into a new host is desired. Several *hAT* elements have indeed been used as genetic tools for this purpose in both plants and animals.

The *Ac* element of *Z. mays* was shown to be capable of mobility in tobacco (40) and has since been developed into a gene vector in a range of plant species, including tomato, lettuce, flax, barley, petunia, rice, *Arabidopsis thaliana*, sugar beet, and wheat (41, 42, 43, 44, 45, 46, 47, 48, 49). The ability of *Ac* to be used as an elegant genetic tool in this multitude of plant species has been described previously and will not be expanded upon here (15, 17). Improvements to the *Ac/Ds* system have involved the development of a binary delivery strategy similar to that used to introduce transposons into the genomes of other organisms. One vector contains the transposon into which a genetic marker and the gene or sequence of interest is inserted while the second plasmid contains the transposase placed under the control of an inducible promoter. Further refinements to this strategy that increases its efficiency typically involve the use of promoters that target the temporal and tissue specific expression of the transposase. For example, the ability of the *Ac/Ds* system to generate secondary transpositions of *Ds* in rice, in which the Ac transposase is very active, can be controlled by placing the transposase under the control of a chemically-inducible system, which activates transcription via a glucocorticoid binding domain/VP16 acidic activation domain/Gal4 DNA binding domain (50). Expression of the transposase can be simply induced upon introduction of the chemical dexamethasone through spraying or irrigation (50). In this study, the authors introduced a second modification of the *Ac/Ds* system that permitted stabilization of new *Ds* integrations in the rice genome by using the site-specific Cre/lox recombination system to remove the Ac transposase following *Ds* excision (50). This was achieved by placing the *Ds* element between the ubiquitin promoter used to drive expression of the Cre recombinase gene with the lox sites being placed either side of *Ac* transposase gene. Excision of the *Ds* element that presumably leads to its integration elsewhere in the rice genome and results in the reconstitution of the ubiquitin-Cre recombinase gene, which then results in removal of the Ac transposase, leading to a stable genetic line (50).

The generation of mutations in the transposase that increase its activity is another approach to increase efficiency. This has been achieved for the Ac transposase with four mutations, E249A, E336A, D459A, D545A, which resulted in close to a 100-fold increase in the ability of this transposase to excise the *Ds* element in yeast and in *Arabidopsis* (51). However, there was no significant increase in the integration frequency of *Ds* in either of these organisms (51). As described below, these four mutations are located in, or near, functionally important regions of the transposase, as revealed by the crystal structure of the Hermes transposase. This hyperactive transposase promotes elevated excision levels but has no effect on integration. However, the authors noted that this mutant transposase showed a reduced preference for promoting insertions into GC-rich regions of the genome. This indicates that it does impact insertion by exerting more subtle effects on target choice (51).

A functional copy of the *D. melanogaster hobo* transposon was used to genetically transform its host upon the discovery of a functional version of this transposon (52). The binary method of transformation was identical to that developed for the *P* transposon in that two plasmids were injected into preblastoderm insect embryos with one plasmid containing the transposon carrying a cloned genetic marker and the other plasmid containing the corresponding transposase gene placed under the control of an inducible promoter (53). This paradigm has essentially remained unchanged in the intervening years and it has been used to achieve the genetic transformation of other insect species using active *hAT* elements that were discovered subsequent to the *hobo* element. The examples of *Hermes* element-mediated transformation of insects and *Schizosaccharomyces pombe* are described below but this same technique has also been applied to the *TcBuster* and *AeBuster1 hAT* transposons to achieve genetic transformation of *D. melanogaster* (21). The frequencies of transformation of this species with both *hobo* and *Hermes* are comparable to those achieved with the *P*, *Mos1*, *Minos*, and *piggyBac* transposons. However, neither *hobo* nor *Hermes* have enjoyed widespread use as genetic tools in this model insect.

The *Hermes*, *SPIN*$_{ON}$ and *TcBuster* transposons have been examined for their insertion preference in *Saccharomyces cerevisiae* and HeLa cells, respectively (54, 55). An analysis of 175,600 insertions in yeast showed that *Hermes* preferentially inserted near the 5′ ends of genes and into regions of the genome not occupied by nucleosomes (54). *Hermes* also preferred to insert in GC-rich regions of the yeast genome indicating that this transposon is sensitive to both the base composition of DNA (which affects its ability to bend) and to the higher-order structure of chromatin although this has not been explored *in vitro* (54). *TcBuster* and *SPIN*$_{ON}$ also show a preference for insertion near the 5′ ends of genes and near genic regions that are actively transcribed in HeLa cells as reflected by their association with chromatin modifications consistent with transcription (55). This preference for insertion of *TcBuster* into active regions of transcription was also observed in similar studies performed in human HEK392 cells which, in addition, revealed that this transposon did not have an insertion preference for related *Tramp* or *Buster* sequences in the human genome (56). The efficiency of *TcBuster*-mediated gene transfer in HeLa cells is comparable to that achieved using the *piggyBac* element and this is also seen in HEK392 cells in which *TcBuster* was as efficient as the *Tol2* transposon for medicating gene transfer (55, 56). These data indicate that *TcBuster* and *SPIN*$_{ON}$ might be developed as efficient gene transfer tools in human cells.

The *Tol2* transposon from medaka fish (57) has been adapted as a gene transfer vector in murine embryonic stem cells, *D. melanogaster*, *Xenopus*, and the zebrafish, *Danio rerio* (58, 59, 60, 61, 62). Indeed, in zebrafish it was very quickly developed into an elegant gene and enhancer trap, which illustrated that this sophisticated genetic strategy that had been developed for plants and invertebrates could also be efficiently applied in vertebrates (63, 64). The efficiency of this transposon has led to many mutants being created through the integration of these modified *Tol2* transposons as is cataloged in the zTrap database (http://kakwakami.lab.nig.ac.jp/ztrap/) (65). Gene trapping using the *Tol2* transposon has also been achieved in chicken primordial stem cells in which it was found to preferentially insert into transcribed regions of chromosomes indicating that *Tol2* will be a useful gene vector for the genetic modification of chickens (66). The versatility of this transposon in identifying genes and promoters involved in zebrafish development may make it an attractive system for studying chicken and mouse development with obvious implications for the analysis of human development and disease (67, 68). Its application as a genetic tool in *Xenopus* is based on both its ability to genetically transform this amphibian but also to be remobilized following integration into the genome. This has been achieved by the subsequent injection of Tol2 transposase mRNA into stable transgenic lines and more recently by the crossing of transgenic lines containing the transposase and the transposon (69, 70). The preference of *Tol2* is for insertion into intergenic regions throughout the human genome and, to a lesser extent, into introns (71). It displays a significant preference for insertion near CpG islands and into regions of low gene density. Some 51% of insertions occur into repeated sequences with a predisposition to insert into LINE elements (71). Preference for integration near CpG islands was also seen in genome-wide profiling studies performed in human peripheral blood lymphocyte-derived T cells. However, in this genome, the *Tol2* element was also found to have an insertion preference near the transcription start sites of genes and near DNaseI hypersensitive sites (72). *Tol2* has recently been used for transgenesis of the marine annelid, *Platynereis dumerilii*, which also serves as a model for neuronal development. It was found to be more efficient at this task than the *mariner Mos1* transposon yet, for unknown reasons, *Tol2* became silenced in the germ-line (73).

THE CHEMICAL MECHANISM OF *hAT* ELEMENT EXCISION AND TRANSPOSITION

Initial insights into the mechanism of excision and transposition of *hAT* elements came principally from studies that examined the sequence structure of sites remaining after excision of the plant *hAT* transposons, *Tam3* and *Ac*, from *A. majus* and *Z. mays*, respectively (74, 75). These studies were the first to reveal the presence of short palindromic repeats of sequences immediately flanking the empty excision site. This led to a model in which a hairpin loop was generated on this flanking DNA and then nicked at a single site anywhere within the loop. The repair of the resulting asymmetric single stranded DNA by the nonhomologous end joining pathway then generated the short palindromic repeats. The same patterns of palindromic repeats were also seen upon excision of the *hobo* transposable element in assays performed in developing embryos of *D. melanogaster* (8). These functional data combined with comparative sequence data showing the sequence conservation of three regions within the three transposases indicated that these three transposons were members of the same transposon superfamily, named *hAT* after these three transposons (8, 76).

Discovery of the molecular mechanism confirming this model awaited the successful isolation of an active hAT transposase that could function in *in vitro* experiments and so would allow the chemical mechanism of excision and transposition to be dissected. To date, the isolation of active forms of the *Ac*, *Tam3*, and *hobo* transposases has proven elusive. An active form of a hAT transposase isolated from the housefly *hAT* transposon *Hermes* subsequently provided answers to the mechanism of *hAT* element excision and transposition and so, at this level, has become the archetype of this superfamily (6, 9, 77).

The *Hermes* element is 2,749 bp long, has imperfect TIRs 17 bp in length with A-T transitions at positions 9 and 10 of each end (78). It generates 8 bp TSDs of consensus sequence 5′ nTnnnnAn 3′ as do the *Ac*, *Tam3*, and *hobo* elements and is classified as being a member of the *Ac* family of *hAT* elements (79). The Hermes transposase is 612 aa in length and contains the conserved motifs that defines *hAT* element transposases (see below). *Hermes* activity was initially detected in plasmid-based excision assays in developing housefly embryos. These assays were performed in the absence of hobo transposase to determine whether this insect contained any enzymatic activity that could possibly recognize the *hobo* element ends and, in doing so, lead to its excision from a plasmid (8). These experiments indicated that the housefly did contain such an activity and, using degenerate primers designed to the conserved coding regions of the hobo, Ac and Tam3 transposases, the *Hermes* element was identified and reconstructed (25). Full-length copies of *Hermes* appear to be present in almost all global populations of housefly, although a more detailed picture of *Hermes* in its host will await the completion of the genome project for *M. domestica* (80).

Relatively rapidly, *Hermes* was shown to be active when introduced into developing embryos of a number of insect species such as *D. melanogaster*, *Bactrocera tryoni* (Queensland fruit fly), *Lucilia cuprina* (Australian sheep blowfly), *Ceratitus capitata* (Mediterranean fruit fly), *Aedes aegypti* (yellow fever mosquito), *Anopheles gambiae* (the principle mosquito vector of human malaria in Sub-Saharan Africa), and *Helicoverpa armigera* (79, 81, 82, 83). The choice of these insect species for examining *Hermes* activity was not accidental as all have a significant impact on human health or agriculture and their control by genetic manipulation using novel transposon tools is clearly important. These successes led to *Hermes* being used to genetically transform several insect species, including *D. melanogaster*, *C. capitata*, *Stomoxys calcitrans* (stable fly), the butterfly *Bicyclus anynana*, and *A. aegypti* (84, 85, 86, 87, 88, 89). It has subsequently been harnessed as a genetic tool in *S. pombe* and is active in *S. cerevisiae* (54, 90, 91, 92).

Like several other *hAT* elements, *Hermes* shares short subterminal repeated sequences, mutations in which alter transposition (3, 4, 93, 94, 95). Small direct repeats are typically distributed asymmetrically between the L and R ends of the transposon. This provides a potential structural basis for differential recognition of each end, probably by transposase binding. For example, the *Ac* element contains 12 copies of the repeat 5′ AAACGG 3′ within 300 bp of each end of the transposon and mutation of these eliminated transposase binding in gel-retardation experiments (4). Similar results have been obtained for the *Tag1* transposon of *A. thaliana* and these will be described in more detail below (96). Using Hermes transposase purified from *Escherichia coli*, it was initially shown that 80 bp of the L end and 110 bp of the R end enabled transposase binding (9). The L end showed higher affinity for binding than the R end did (9). However, it was subsequently shown that the Hermes transposase octamer can bind to as little as 16 bp of the L end TIR (6). Three copies of the 5′ GTGGC 3′ repeat are located within 300 bp of the L end, while two copies are located within 300 bp of the R end (9). For the L end, 91 bp is sufficient to support accurate excision of the transposon *in vitro*, while it remains unknown whether a fragment of corresponding length (containing both copies of 5′ GTGGC 3′) from the R end can also support strand cleavage and subsequent strand transfer to target DNA (9).

The mechanism of chemical cleavage at the ends of *Hermes* involve two chemical steps initiating with a single strand nick one nucleotide into the flanking DNA at the 5′ end of the element (Fig. 1) (9). The 3′ OH generated on this strand then executes a nucleophilic attack on the other strand immediately adjacent to the 3′ end of the *Hermes* element. This forms a labile hairpin loop intermediate on the flanking DNA and so generates a 3′OH on the transposon end that can then participate in a nucleophilic attack on the target DNA. How the 1 nt overhang on the 5′ strand is resolved remains unknown but it is removed prior to, or during, integration as only sequences delimited by the TIRs of the *Hermes* element are present at the new integration site (9).

This two-step reaction during excision presents an interesting quandary for the transposase in that it would seem to require a conformation change in the active site of the enzyme as it requires the strands to be nicked at a single end to be placed precisely and sequentially into the catalytic site.

While accurate excision and transposition of *Hermes* can be achieved *in vitro* using only 30 bp of end sequence, longer ends are needed for transposition of the transposon *in vivo*. A mini-*Hermes* element containing only precleaved ends containing only 30 bp of the L end and 30 bp of the R end was shown to be active *in vitro* as was a precleaved mini-*Hermes* element containing 30 bp of the left end present at both ends of the transposon (9). Accurate transposition was confirmed by transforming *E. coli* with plasmids generated by integration of the mini-*Hermes* element with DNA sequencing subsequently confirming the presence of 8 bp TSDs flanking the transposon at its new integration site.

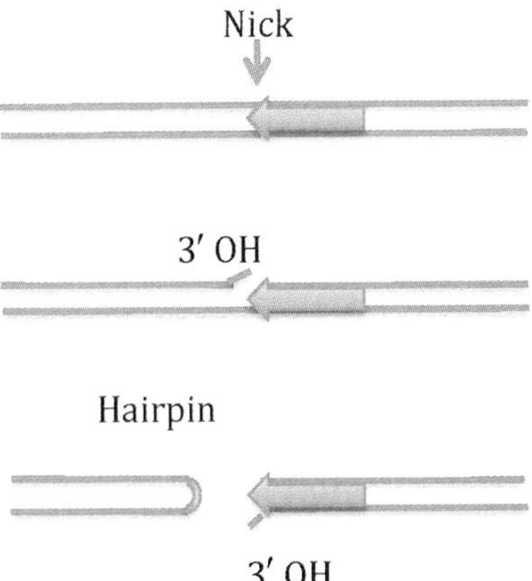

Figure 1 The mechanism of *hAT* element excision. The chemical mechanism of *hAT* element excision is shown based on studies undertaken with the *Hermes* element and transposase (9). The initial nick occurs on the nontransferred strand of the transposon that leads to the formation of an intermediate structure in which a hairpin loop is formed at the end of the flanking DNA with this second nick exposing the 3′OH of the terminal nucleotide on the transferred strand of the transposon. Therefore, it can undertake strand transfer to the target DNA molecule.
doi:10.1128/microbiolspec.MDNA3-0054-2014.f1

Thirty base pair ends are, however, insufficient for transposition of *Hermes* in *D. melanogaster* cell culture. This is presumed to reflect the likelihood of this transposon moving in the whole insect (6). As is seen with the related *hobo* element, the frequency of transposition *in vivo* increases with the lengths of the two ends, which, in both transposons, increases the number of subterminal repeats available for transposase binding. For *Hermes*, for which active transposase has been purified, the increase in transposition frequency as a function of increasing length of the ends correlates with an increase in transposase binding, at least in the case of the L end (6).

THE STRUCTURE OF *hAT* ELEMENT TRANSPOSASES

The general structure of hAT transposases is shown schematically in Fig. 2. The transposases so far characterized contain four shared domains, which are, reading from the N-terminal end, the BED domain, a second DNA binding domain, and the catalytic domain. An additional domain is inserted into, and bisects, the catalytic domain.

The crystal structure of the Hermes transposase bound with 16 bp of the L end of the transposon also provides context for earlier observations on conserved sequences within the hAT transposases. This reveals that the *Hermes* transposasome consists of a tetramer of four dimers to form an octameric ring. This complete structure is necessary for transposition *in vivo* but not *in vitro*.

This structure lacks the first 78 amino acids of the transposase and so omits the BED domain but contains multiple specific DNA binding domains throughout its structure as well as surfaces that interact nonspecifically with DNA. These sites of interaction between the transposase and DNA provide a frame of reference that enables a comparative analysis of active hAT transposases to be made to determine if specific function leads to conservation of sequences across the superfamily.

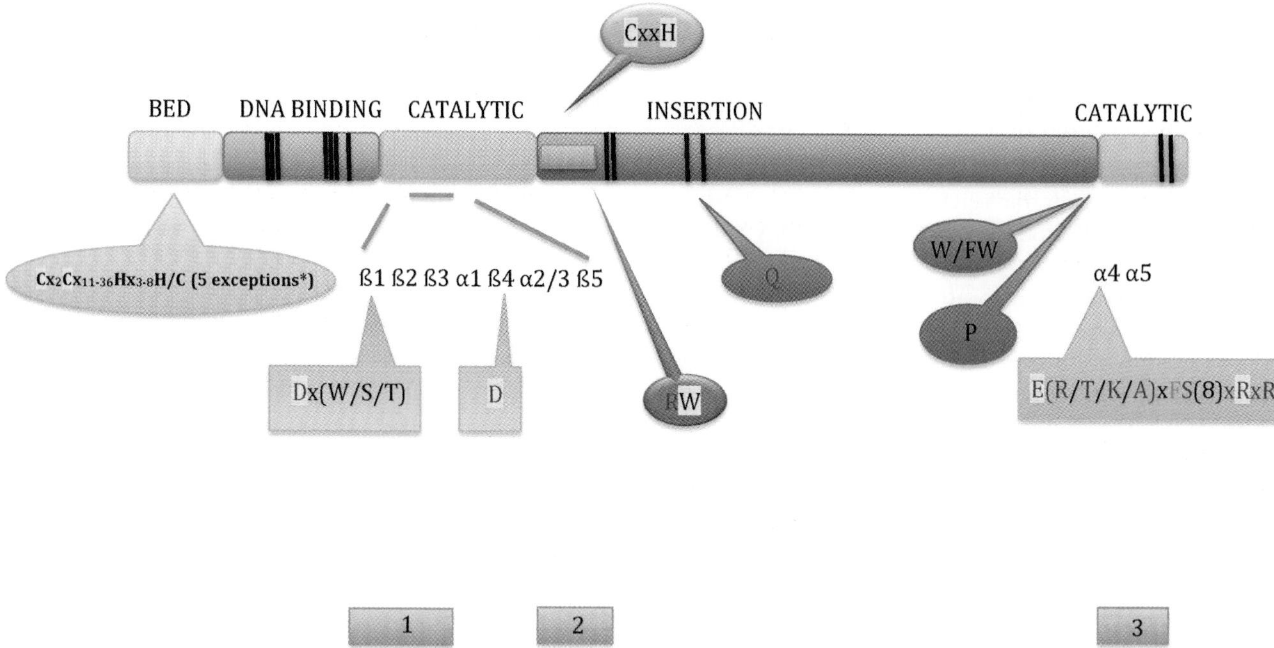

Figure 2 Conserved amino acids across hAT transposases. The four functional domains of hAT transposases are shown with conserved amino acids between 22 hAT transposases shown below. The DDE catalytic triad is shown in red and, based on the cocrystal structure of the Hermes transposase bound with the 16-mer L terminal inverted repeat (TIR), amino acids involved in DNA binding of the TIR, which are moderately conserved across the hAT transposases, are shown in blue. The relative positions of other amino acids that bind to the Hermes L TIR but are not conserved across the hAT transposases are shown by black bars. The position of the DNA binding cleft in the Hermes transposase is show by the blue bar located at the N-end of the insertion domain. Yellow-highlighted amino acids are very highly conserved across these transposases. The locations of the three conserved regions identified in the HFLI hobo transposase are shown underneath.
doi:10.1128/microbiolspec.MDNA3-0054-2014.f2

The first identification of sequence similarities between the Ac, Tam3, and hobo transposases revealed three regions of reasonably high conservation. We now know that they correspond to the first three β-strands of the RNaseH domain (the first containing the first catalytic carboxylate amino acid of the DDE motif), the α2/3-helices of the RNaseH domain containing a conserved CxxH motif and the conserved RW motif, and the previously named "*hAT* dimerization" domain at the C-terminus of the protein, which also contains the final carboxylate amino acid of the catalytic triad (97). These three regions were also identified in the transposase of the functional *hobo* transposon, HFL1 (98). The locations of these three regions are shown in Fig. 2.

Six conserved blocks between hAT transposases were subsequently identified using a bioinformatic analysis of 147 nonredundant *hAT* element sequences (12). These are shown in Fig. 3A in relation to 10 regions of 22 hAT transposases from members of the *Ac*, *Buster*, and *Tip* families. This alignment, using the multiple sequence alignment program T-Coffee (99), contains 17 hAT transposases of which there is evidence of current activity and five members of the recently confirmed *Tip* family, which are supplied for reference as the amino acid sequences of the transposases are, in large part, the basis for the classification of the *hAT* superfamily into three subfamilies. There is no experimental evidence supporting the activity of these five Tip transposases.

The six blocks (listed as A to F) identified by Rubin et al. (12) can now be identified as corresponding to: (A) the first β-strand of the RNAseH domain in which the first catalytic carboxylate amino acid is located; (B) the α2/3-helices of the same domain in which the highly conserved CxxH motif is located; (C) the highly conserved RW motif; (D) an α-helix toward the C-terminus of the transposase that contains two conserved and adjacent W residues in many of the transposases; (E) an α-helix containing the final catalytic carboxylate amino acid which forms part of the RNaseH domain; and (F) the final α-helix of the transposase (12).

Five of these conserved blocks (A, B, C, E, and F) are intimately involved in the chemical process of excision and in the positioning of the *Hermes* ends during excision and transposition, therefore, their relatively high level of conservation is consistent with their functional importance (6). Two (A and E) contain two of the three carboxylate amino acids (D, D) of the catalytic DDE triad while block C contains the RW motif, which, in the Hermes transposase, forms a hydrophobic pocket with H268 and is immediately adjacent to the DDE catalytic core following conformational changes during excision (6). By analogy with what is observed in *Tn5* excision, the tryptophan in the RW motif is believed to be critical for the binding and stabilizing of the hairpin formed on flanking DNA during excision (100). The CxxH motif in region B is present on the α2/3-helix of the RNaseH domain. In the Hermes transposase the histidine (H268) participates, as described above, in the formation of a hydrophobic pocket at the catalytic site (6). Mutation of H268 in the Hermes transposase severely reduces the catalytic activity of the enzyme (9). A similarly conserved histidine is also found in the same relative location, some 15 to 45 amino acids downstream of the second catalytic carboxylate. It is found in members of the *MULE* superfamily where it is present as C/DxxH, and in the *P* element in which it is present as DxxH , and the *Cacta/Mirage/Chapave* and *Transib* superfamilies in which it is present as CxxC (11). The Hermes transposase crystal structure reveals the CxxH motif to be located immediately adjacent to a structure present on the rim of the octamer, which has been proposed to be a target DNA binding cleft of the transposase (6). This cleft is formed, in part, by the same α-helix from adjacent monomers (the location of this region is shown by the blue bar in Fig. 2 and in Fig. 3B by the α-helices containing dark blue and orange sections). Mutations in some basic resides within this α-helix decrease transposition frequency of *Hermes* in *D. melanogaster* indicating these residues at least participate in the function of the transposase (6). At one end of this cleft, the CxxH motif lies close to the catalytic site, while the other end contains the RW motif that participates in the chemical mechanism of excision and so is proximal to the catalytic site. The intervening 48 amino acids constitute the proposed target binding cleft leading to the possibility that conformational changes occurring within the catalytic domain during excision are conveyed to the target binding cleft. A consequence of the arrangement of this region from two adjacent Hermes monomers may mean the excision at the L and R ends of the transposon in two separate catalytic domains may result in two separate conformation change signals to a single target DNA cleft.

Block F identified by Rubin et al. (12) contains the C-terminal α-helices, the penultimate one containing the third carboxylate amino acid of the catalytic domain and contains residues, which, based on the co-crystal structure of the Hermes transposase and the L end oligomer, interact with the L TIR and so position it within the catalytic site.

The only conserved motif identified by Rubin et al. (12) that does not yet have an assigned function in the transposition of *hAT* elements is the 543WW544 motif in block D (Fig. 3A). This is located adjacent to the small interface present in four locations of the Hermes transposase octamer. This small interface is important for transposase function as mutations that abolish it lead to Hermes transposase dimers that are active *in vitro* but inactive *in vivo* (6). This motif is highly conserved as WW in the *Ac* family of hAT transposases and as FW in the *Buster* family indicating that the presence of aromatic amino acids is important at this position. In the four Tip transposases examined F or W are present in the first position of this pair with only the BuT2 transposase having an aromatic amino acid at the second position. The function of this motif awaits identification.

It is notable that none of these conserved regions includes the two DNA binding domains of the hAT transposases that reside at the N-terminus of the molecule. While all contain a Zn-finger BED domain, the precise sequence of this varies amongst hAT transposases. The proposed model that the BED domain engages in binding to the many small subterminal repeats present at the *hAT* element ends, each peculiar to the particular *hAT* element, is consistent with this region not showing strong sequence conservation between the hAT transposases.

COMPARATIVE ANALYSIS OF THE FOUR DOMAINS OF hAT TRANSPOSASES

The BED Domain

The BED domain is present in all the hAT transposases so far examined (Fig. 3A). This includes the DNA binding domain of the Ac transposase (15), which aligns with the BED domain, albeit with a slightly different arrangement of the first two cysteines in the domain. Fig. 3A shows the alignment of the BED domains of the

```
A Catalytic triad:                                                                    D                          D
Tol2        47  CVL--C (19) HIER-MH          127 YIIQGLHPFST    163 RSK   191 TDCWT   234 EVLAS   258 TDS
Tgf2        83  CVL--C (19) HIER-MH          163 YIIQGLHPFST    199 RSK   227 TDCWT   270 EVLAS   294 TDS
Herves      26  CLY--C (16) HLNL-VH          115 LICKECLPFNL    151 SNA   180 SDGWT   222 RNIAN   246 TDN
Tam3        146 CLL--C (17) HLTA-KH          207 FIVONELPFSF    244 FRD   273 SDLWQ   316 DCIRD   340 LDN
Ac          168 CNFPNC (17) HLRT-SH          235 AIIMHEYPFNI    271 RKY   300 MDMWT   342 QRLSQ   264 LDN
nDart1      72  CRI--C (17) HAES-CA          135 LIARQDNPNLF    172 TRD   201 SDIWS   243 ENIAE   267 LDN
TCUP        101 CIH--C (17) HRDR-CP          165 LLARLEIPISL    202 TRD   232 SDIWS   274 QNIAD   298 LDN
Hobo        99  CRQ--C (16) HKCC-LT          149 WVVQDCRPFSA    196 SRK   227 VDMWT   270 ENILM   295 TDR
Homer       50  CRT--C (16) HPCC-RN          100 WVVEDCRPFSV    147 SRN   178 IDMWT   221 DNVLN   246 TDR
Hermes      51  CRK--C (16) HKCC-AS          101 WVVRDCRPFSA    148 SRK   179 IDLWT   222 ENIYK   247 TDR
TcBuster    77  CVI--C (16) HLDT-KH          147 RIAKQGEAYTI    193 SRR   222 MDEST   263 EEIFN   288 TDG
AeBuster1   75  CVI--C (16) HLET-KH          145 KIAKSGKAHTI    191 ARR   219 LDEST   259 DEIFD   284 TDG
SPINon      37  CVI--CC(15) HFDS-KH          108 RIARAMKPHTI    154 RRR   184 LDEST   225 RDVFD   250 TDG
Tag1        44  CAF--CH(13) HHLAG-VKGNTDAC   172 WFYDACIPMNA    211 RVG   242 ADGWK   280 ENLCN   303 TDS
Restless    109 CRL--C (23) HLRD-IH          195 WITHDNLPFRL    233 RSR   262 FDGWT   304 DNLAN   329 LDN
Toll        195 CFC--C (27) HEKSPEH          280 SLAVRNLALRG    341 NEL   368 LDCTL   412 QHLAS   437 YDN
Crypt1      98  CRL--C (23) HLEGRDNEGH       205 LVACCDIPLST    245 MRW   275 ADLWT   317 ENLQW   341 FDN
BuT2        33  CKY--C (11) HAKTAKH           91 YIA----THSS    126 I-I   154 IDEST   196 KTIVK   221 TDN
RP-hAT1     182 CFC--C (29) HEVSPLH          269 FLAEQNLAFRG    336 NEF   354 FDSTP   408 GEITD   433 YDN
hAT-4_BM    182 CFC--C (29) HEVSPLH          269 FLAKQNLAFRG    336 NEF   354 FDSTP   408 VEITD   433 YDN
hAT-34_HM   190 CFC--C (30) HESSDNH          278 FLAKQNLPRFG    338 NEF   366 FDSTP   410 LDLST   435 YDN
Ip          127 CLP--C (36) HIGKDLTSPH       221 WLAFQGCSFRG    278 KQI   305 VDSTP   346 STLKE   371 YDN
DOMAIN:     ---------- BED --------------    --------DNA BINDING -------  ------------CATALYTIC ----------
BLOCK:(12)                                                                -BLOCK A-

Tol2        325 KCACHL   382 LQLLRPNQTRWNSTFMAVD   449 DILQAETN
Tgf2        362 KCACHL   419 LQLLRPNQTRWNSTFMAVD   486 DILQAETN
Herves      265 PCFAHT   319 LKMTQEVSTRWNNSGYDML   382 NIVSAQKY
Tam3        369 RCICHI   421 KHIYLDVPHRWNNATYRML   484 KIMSSCYT
Ac          400 RCACHI   454 KGISYDVSTRWNSTYLMLR   520 ELLSGTGY
nDart1      306 RCACHI   349 RKFATDAEHRWNATYLMLK   418 LTLSHVYY
TCUP        334 RCACHI   387 RKFQLDMEVRWNSTYLMLK   455 VALSGVYY
Hobo        312 NCSSHL   357 TTLKSACRTRWNSNYKMMK   417 KKLQTSSS
Homer       263 NCSSHL   308 TSLKSQCPTRWNSNYGMIK   371 NKLQLCSS
Hermes      264 NCSSHL   309 SSLKSECRTRWNSTYTMLR   372 KELQTCSS
TcBuster    316 HCCLHR   369 KNLLLHTEVRWLSRGKVLT   433 LSLQGPNS
AeBuster1   312 HCSLHR   365 TSLLLHTEVRWLSRGKVLT   420 TYLQGSTS
SPINon      278 HCMIHR   331 NALLFHTEVRWLSRGKVKR   398 LSLQGPNA
Tag1        327 PCSAHC   377 REIIRPGETRFATTFIALQ   455 RICDADEK
Restless    356 RCLGHV   456 LRPIADNETRWNSRHRMMV   534 VRLQGNPK
Toll        455 PCGAHT   515 ITLKSWTETRWESKIKSIE   593 KLMQSPNM
Crypt1      377 RCWGHI   519 LRPVVAQETRWNSTEAEIS   586 KHHEGNA-
BuT2        249 KCVCHS   305 LKIPRVCETRWLSIEKAVS   379 KIFQTSTG
RP-hAT1     461 ACTNHF   511 QGVKRSVETRWSARAAAVT   589 KYLQIKGL
hAT-4_BM    461 ACTNHF   511 QGVKRSVETRWSARAAAVT   589 KYLQIKGL
hAT-34_HM   463 GCIDHS   513 ISVKRLSTTRWSAHYNAVK   591 IYLQTKGL
Ip          399 HCFAHR   470 GSLQRPGDTRWSSHLKSIS   548 QALQLQSQ
DOMAIN:     ------------------- INSERTION ----------------------
BLOCK:(12)  -BLOCK B-       ---- BLOCK C -----

Catalytic triad:                                      E
Tol2            583 --TRE--------SLLTFPAICSLSIKTNTPLPASAACERLFSTAGLLFSPKRARLDTNNFENQLLLKLNLRFYN
Tgf2            620 --TRE--------SLLTFPAMCSLSIKTNTPLPASAACERLFSTAGLLFSPKRARLDTNNFENQLLLKLNLRFYN
Herves          541 LWWKE--------HQVLYPSLYTLAMSTLCIPGTSVPCERLFSKAGQIYSEKRSRLAPKKLQEILFIQQN-----
Tam3            650 KWWRQ--------NESLTPVLARIARDLLSSQMSTVASERAFSAGHRVLTDARNRLKPGSVKFCMIWKDMLDQQY
Ac              689 SWWRG--------RVAEYPILTQIARDVLAIQVSTVASESAFSAGGRVVDPYRNRLGSEIVEALICTKDWVAASR
nDart1          600 GWWND--------HKITYPVLSKLARDVLTVPVSTVSSESAFSLCGRIIEDRRTTLRSDHVEMLLSVKDWELARQ
TCUP            651 HWWHQ--------HKLTYPVLSIMAKDILTVPVSTISSESTFSLTGRIIDERRRRLKSDVVEMLTCIKDWEDAQA
Hobo            589 EWWKN--------NANLYPQLSKLALKLLSIPASSAAAERVFSLAGNIITEKRNRLCPKSVDSLLFLHSYYKNLN
Homer           553 EWWQS--------NQSKYPHLSKFALQIHAIPASSAAAERSFSLAGNLITEKRNRIAPGSVDSLLFLNTYYKNYN
Hermes          542 EWWNL--------NSKKYPKLSKLALSLLSIPASSAASERTFSLAGNIITEKRNRIGQQTVDSLLFLNSFYKNFC
TcBuster        559 PFWIK--------LMDEFPEISKRAVKELMPFVTTYLCEKSFSVYVATKTKYRNRLDAEDDMRLQLTTIHPDIDN
AeBuster1       560 RFWIS--------VRHMYPCLYEEAVKFLLPFTTSYLCEAGFSEMVAIKTKYRNKLRLRLSPSLRLLTGIEIDVS
SPINon          524 KFWIK--------CLQSYPVLSETVLRLLLPFPTTYLCETGFSSLLVIKSKYRSRLVVEDDLRCALAKTAPRISD
Tag1            582 EWWRY--------FGHDAPNLQKMAIRILSQTASSSGCERNWCVFERIHTKKRNRLEHQRLNDLVFVHYNLRLQH
Restless        731 SWWQ--------EHEMEYPNLCRMATDLLSIPTMSAETERSFSSAGKMVSPLRTRLDRHTIGMAQGMRSWSREGI
Toll            762 SFMSD------KDLSEIYPNFWTALRIALTLPVTVAQAERSFSKLKLIKSYLRSTMSQERLTNLAVVSINHSVGE
Crypt1          865 VWWAA--------RKAEFPALSQMAFDVLSIPAMSSECERVFSQGKLTMSSQRRMKSSTLELLLCLKDWLKNGT
BuT2            533 PFWAEVLRYRDGAGTKRFSALASFALNLLSLPWSNAEVERVFSQMNIIKTKLRNSMANRTLNAILHIRYGLRLQK
RP-hAT1         756 KFIQQ------YRLENSVPNIVILLRIFLTISISVASCERSFSKLKLIKNYLRSTMSQLRLQNLAILSIEQEITN
hAT-4_BM        756 KFIQQ------YRLENSVPNIVILLRIFLTISISVASCERSFSKLKLIKNYLRSTMSQSRLQNLAILSIEQEITN
hAT-34_HM       763 TFITE------WDFTETLPMLSLTLRLFLTICVSVASCERSFSKLKLIKNYLRSTMGQSRLSDLAILYIESEFVK
Ip              719 RWLVKT------RKSNIYPLVFRVVTLVLTLPVSTATTERSFSAMNIVKTTLRNKMEDEFLSDCLLVYIEKQIAK
DOMAIN:             ------------------------- CATALYTIC -----------------------------------------
BLOCK:(12)           -BLOCK D-         --------- BLOCK E --------- ------ BLOCK F -------
```

Figure 3 *continues on next page*

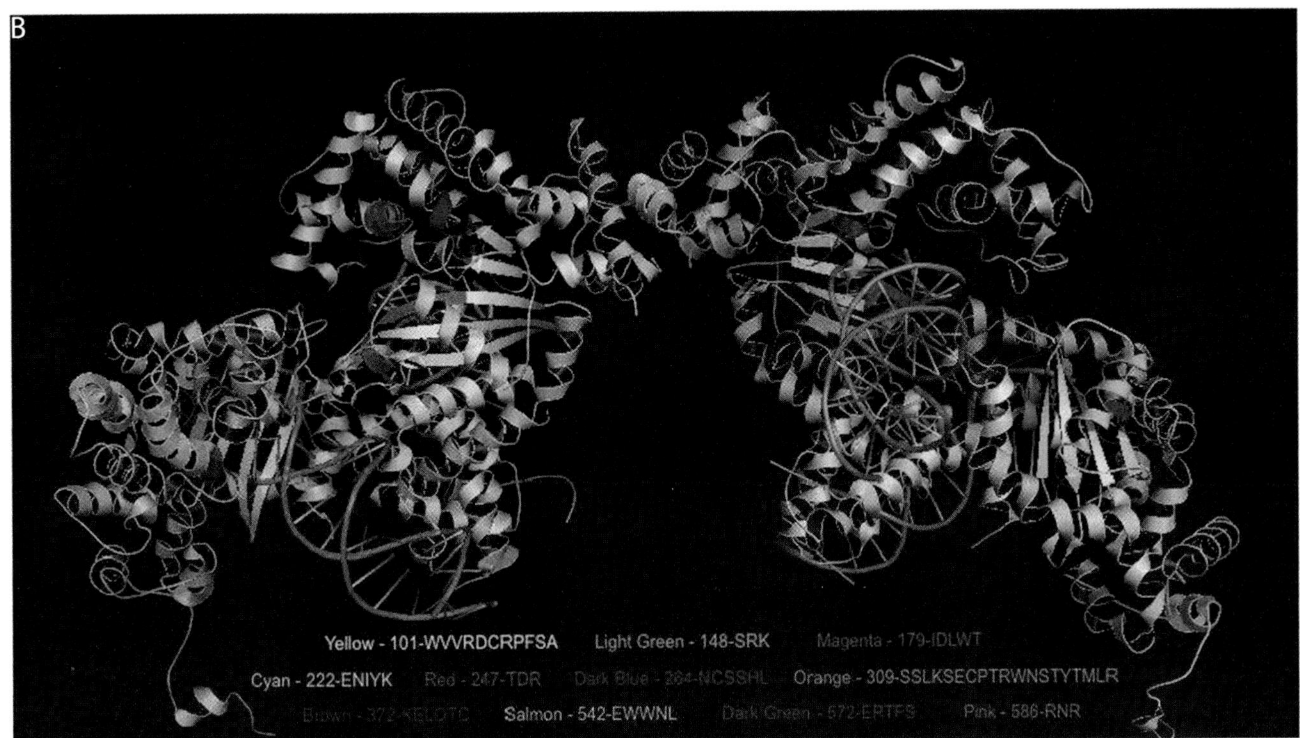

Figure 3 Comparison of hAT transposase sequences. (A) Consensus alignment of 22 hAT transposases. The alignment was obtained using the program M-Coffee, which uses multiple sequence alignment programs. The location of each sequence within its transposase is shown by the amino acid numbers. The location of the DDE motif is shown above the alignments, conserved resides are shown below the alignments along with the locations of the domains and the six conserved blocks originally identified by Rubin et al. (12). Amino acids identified as being critical for Hermes transposase activity and/or DNA binding from the Hermes transpososome cocrystal by Hickman et al. (6) are shown in red as are identical amino acids present in the same locations in the other transposases. The C and H amino acids proposed to constitute the BED domain are shown in blue. Accession numbers are: Tol2 (BAA87039), Tgf2 (AFC96942), Herves (AAS21248), Tam3 (CAA38906), Ac (CAA29005), nDart1 (BAI39457), TCUP (ABC59221.1), Hobo (A39652), Homer (AAD03082), Hermes (AAB60236), TcBuster (ABF20545), AeBuster1 (ABF20543), Tag1 (AAC25101), Restless (CAA93759), Tol1 (BAF64515), Crypt1 (AF283502), BuT2 (AF368884), and IpTip100 (BAA36225). (B) Relative location of the conserved regions identified in Fig. 3A on the Hermes transposase and *Hermes* terminal inverted repeat tetramer (6).
doi:10.1128/microbiolspec.MDNA3-0054-2014.f3

22 hAT transposases examined with evidence of activity in 17 of these from the *Ac* and *Buster* families together with five recently identified *Tip* family members, although no clear evidence for their current activity exists. As shown in Fig. 3A, the BED domain for all but five of the hAT transposases conforms to the structure $Cx_2Cx_{11\text{-}36}Hx_{3\text{-}8}$(H/C). The BED domain structure was initially described as $Cx_2Cx_nHx_{3\text{-}5}$(H/C) and is present in hAT transposases and genes that have evolved from them (101). The Ac transposase is exceptional in that it contains four amino acids between the two cysteines ($Cx_4Cx_{17}Hx_4H$) and the closely related Hermes, hobo, and Homer transposases have a cysteine replacing the final histidine ($Cx_2Cx_{16}HxCC$). The Tag1 transposase BED domain has the structure $Cx_2Cx_{13}HHx_{10}C$.

The BED domain is absent from the crystal structure of the Hermes transposase, therefore, its precise placement within the overall structure of it remains speculative, although a modeled placement of it within the crystal has been proposed (6). A model for its role in DNA binding has recently been proposed in which its inferred location in the octameric structure of the Hermes transposase facilitates binding of the eight copies of this domain to the subterminal sequences internal to the *Hermes* TIRs (6). The subterminal sequences that facilitate binding are thought to be the short repeated

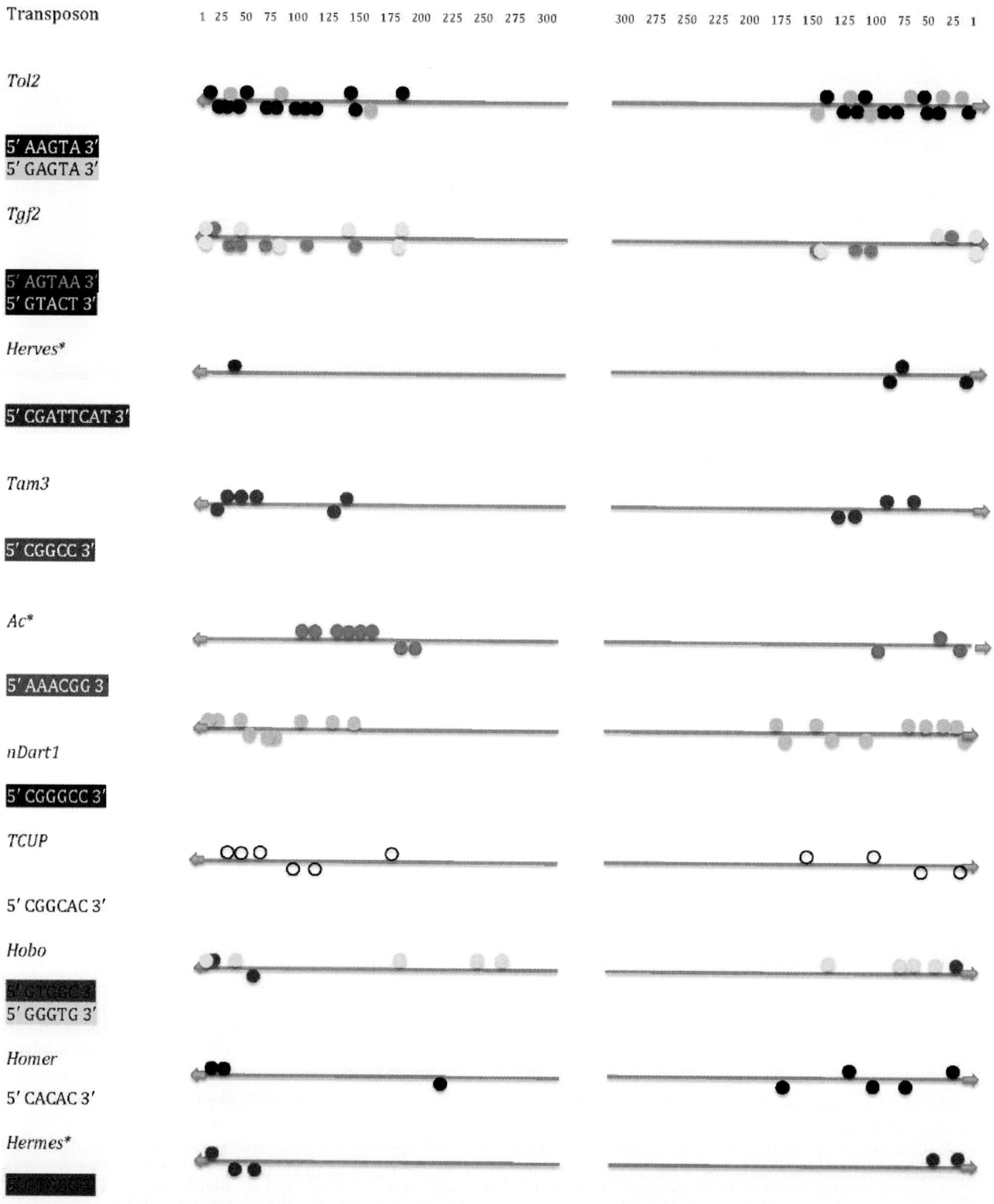

Transposon

1 25 50 75 100 125 150 175 200 225 250 275 300

300 275 250 225 200 175 150 125 100 75 50 25 1

Tol2

5′ AAGTA 3′

5′ GAGTA 3′

Tgf2

5′ AGTAA 3′

5′ GTACT 3′

*Herves**

5′ CGATTCAT 3′

Tam3

5′ CGGCC 3′

*Ac**

5′ AAACGG 3′

nDart1

5′ CGGGCC 3′

TCUP

5′ CGGCAC 3′

Hobo

5′ GGGTG 3′

Homer

5′ CACAC 3′

*Hermes**

Figure 4 *continues on next page*

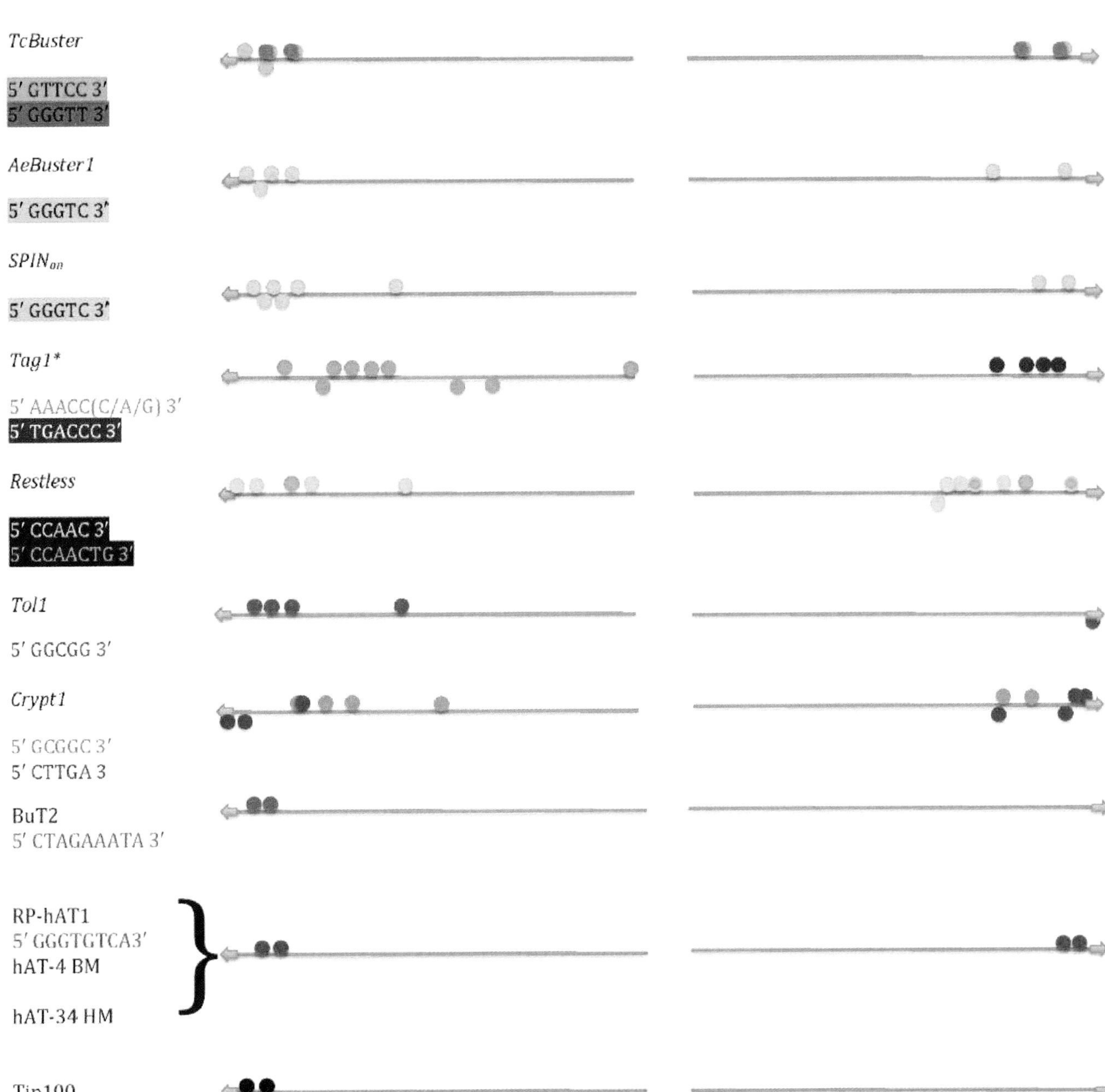

Figure 4 The organization of subterminal direct repeats in 17 active *hAT* transposon from the *Ac* and *Buster* families and five transposons from the *Tip* family. The relative positions of these repeats in the 300 bp at each end of the transposon is shown together with the sequences of these repeats that are located under the name of each transposon. Asterisks denote where biochemical studies using purified transposase have confirmed binding to these repeats. The orientation of the repeats on the top or bottom strand of the transposon is depicted by the position of the filled circle above or below the transposon ends. In some cases, the direct repeats overlap with the terminal inverted repeats. doi:10.1128/microbiolspec.MDNA3-0054-2014.f4

sequences that are present in the subterminal regions of *hAT* transposons and are a defining feature of this superfamily. For *Hermes*, these are 5′ GTGGC 3′ (4). As described above, these repeats are located in an asymmetrical way between the L and R transposon ends, which may explain two other features of *hAT* transposons; that they require both ends for *in vivo* activity and that, *in vivo*, the TIRs alone are insufficient for transposition (93, 94, 96).

Among the *hAT* family members, *Hermes* appears to carry the lowest number of small repeats. Other members, such as *Tol2*, can include up to 32 copies distributed between the R and L ends. The relative position of these subterminal repeats in the 17 active *hAT* transposons from the *Ac* and *Buster* families, together with the five *Tip* family transposons, are shown in Fig. 4. The small direct repeats shown were identified by the authors of the relevant papers or are identified from the present study. Several features are clear. First, there is a significant diversity in the sequences of these repeats between *hAT* transposons. Second, there is an asymmetry in their presence in the L and R ends of these transposons, which is more marked when fewer copies are present. Third, there is variation in the number, spacing, and orientation of these repeats between transposons with *Tol2* showing the greatest number of copies while transposons such as *Hermes* and *Herves* have the fewest. In many cases, these repeats are located very close to at least one of the TIRs, and in some cases, partially overlap with the more interior of the TIR sequences. The *Tip* elements appear to have longer direct repeats than are found in members of the *Ac* and *Buster* families, these being 5′ GGGTGTCA 3′ for the *hAT-34_HM*, *RP-hAT1* and *hAT-4_BM* transposons (23) and 5′ GGGGCCTGGC 3′ for *Tip100* element of *I. purpurea*. The presence of these longer direct repeats has led to the suggestion that these may be a defining feature of this family of *hAT* transposons (23). In both *Tip100* and *BuT2*, these longer repeats appear to be present only at the L end of each element (Fig. 4). In the absence of both biochemical studies confirming that these are the binding sites for active transposase and genetic evidence that alterations to them affect binding and/or transposition, their role remains speculative.

The DNA binding region of the Ac transposase was identified through using mobility shift assays in wild-type and mutant transposases as being between amino acids 158 and 239 at the N-terminal end of the transposase (102). A subsequent analysis of this region using gel retardation assays with a combination of transposase mutants and DNA oligomers containing either two copies of the 5′ AAACGG 3′ repeat or six copies of the 12 bp TIR indicated that this was a bipartite DNA binding region in which the the N-terminal region of the transposase appears to bind to the TIRs but not to the subterminal repeats, while the C-terminal half appeared essential for binding to both sequences (103). This region is located downstream from the N-terminal 102 aa of the transposase that can be deleted without affecting transposition (104). Despite the clear binding affinity of the transposase for these hexanucleotide repeats, *Ds* elements with only a low number of them are still capable of being mobilized by the Ac transposase, supporting earlier caution that they may not be the sole determiners of mobility (4). A recent investigation of the *Ac/Ds* family of maize illustrated the diversity of *Ds* elements with respect to the presence of these repeats (14). Several *Ds1* elements lacking repeats can still be mobilized by the Ac transposase leading to speculation that the sequences with sufficient similarity to these hexanucleotides and the TIRs are sufficient for transposition of these elements (14).

This region between aa 150 and 210 contains the BED domain of the Ac transposase and the delineation of it into two halves separates the CxxxxC and HxxxxH components of this domain. While neither of the cysteines were targeted by Becker and Kunze's mutagenesis strategy, the individual mutation of four lysines, each pair being either side of CxxxC, resulted in loss of binding to the TIR hexamer with little change to binding to the 5′ AAACGG 3′ repeat (103). In contrast, mutating each of the two histidines now known to be part of the BED domain, as well as basic amino acids either side of these resulted in loss of binding to both targets (97, 103). These data indicated that the BED domain may be capable of binding to both the TIRs and the subterminal repeats of *Ac* and perhaps other *hAT* transposons. This possibility is not excluded by the recent model arising from the analysis of the cocrystal structure of the Hermes transposase bound to the L end in which the modeled location of the placement of eight BED domains within the transposase octamer may place regions of the BED domain in close proximity to the more internal sequences of the *Hermes* TIRs (6). The second transposase DNA binding domain, which is also required for dimerization of Hermes transposase monomers within the octameric structure, is located close to the proposed location of the BED domain. Side chains from three amino acids (R107, F109, and S110) within this DNA binding domain interact with the sugar-phosphate backbone of the Hermes L TIR between bp 6 and 7 (6).

The Tag1 transposase DNA binding region that recognizes the asymmetrically distributed *TAG1* short repeats is also centered in the BED domain of the transposase (3). As stated above, this is an exceptional BED domain amongst the hAT transposases. Earlier studies showed that mutations in each of the three cysteines in this domain eliminated binding to the subterminal repeats of the transposon. However, any interactions with the TIRs were not examined, nor were mutations of the two histidine residues at the center of the domain (3). One interesting feature of the *Tag1* subterminal repeats is that their sequences differ between the L and R ends with the sequence 5′ AAACC (C/A/G) 3′ being present multiple times at the L end while 5′ TGACCC 3′ is present at the R end (96). That these are required for transposition *in vivo* is supported by a loss of transposition activity of a truncated *TAG1* derivative containing the TIRs but not the subterminal repeats (96).

The inability to purify hobo transposase has prevented a biochemical analysis of its target site. As is the case with the other *hAT* transposons, the presence of TIRs and subterminal sequences immediately interior to these are insufficient for transposition. Transposition frequency increases with increasing length of the transposon ends, consistent with a role for the 5′ GTGGC 3′ and 5′ GGGTG 3′ subterminal repeats in transposition (95). There is also evidence of cross-mobilization. At least in somatic tissue, the hobo and Hermes transposases can recognize and excise the *Hermes* and *hobo* elements respectively (105). While these transposons share the 5′ GTGGC 3′ subterminal sequence, they are also closely related both in terms of the sequences of their TIRs and their transposases. It is quite possible that the ability to cross-mobilize resides in several shared structural features of these transposons.

The DNA Binding Domain

The region identified originally as a DNA binding domain in hAT transposases has two distinct functions. The structure of the Hermes transposase shows it to be one of three regions of the Hermes monomer critical for dimerization (6, 77). It contains three α-helices that intertwine between adjacent monomers in the interior of the octameric oligomer (part of this region is colored yellow in Fig. 3B). As mentioned above, its precise position relative to the modeled BED domain is unknown but there is experimental evidence that regions of both domains are in close proximity in the Hermes transposase octamer (6). Two regions within this domain of the Hermes transposase interact with the L TIR. The side chains of R107 and S110 interact with the DNA backbone of the nontransferred strand between bp 6 and 7 and F109 interacts with the backbone of the same strand between bp 7 and 8, while E138, E139 and L141 coordinate a sodium ion that binds to the DNA backbone of the transferred strand between bp 11 and 12 (6).

In comparing these regions across the 17 *Ac* and *Buster* family transposases for which there is evidence of activity, as well as the five Tip transposases, there is little conservation of any of these amino acids other than a preference for phenylalanine at the corresponding position of 109 in the Hermes transposase in 11 other transposases from the *Ac* and *Tip* families. This is denoted in Fig. 3A as the first of two sequences within the DNA binding domain, commencing at amino acid 101 of the Hermes transposase. There is a low level of sequence conservation in this small region with a proline immediately preceding this phenylalanine being conserved in 15 of these transposases. Six residues upstream, the alignment shows preference for leucine, isoleucine, or valine. One of the four mutations that constitute the hyperactive form of the Ac transposase maps four amino acids downstream from F243 that corresponds to F109 in the Hermes transposase (51). The position of this 10 residue region within the Hermes transposase tetramer is shown in Fig. 3B. This region bridges two α-helices with the conserved P108 being on the loop between these. The amino acid residues, F109 and S110, are adjacent to the transposon TIR and are on a separate α-helix than the amino acids towards the N-terminal end of the sequences shown in Fig. 3A.

There is some indication of conservation of positively charged amino acids at the location corresponding to R149 in the Hermes transposase (shown as the second of two sequences within the DNA binding domain in Fig. 3A). The side chain of this amino acid has been found to interact with the C at bp 11 in the L TIR of *Hermes* (6). As shown in Fig. 3A, arginine is found in this corresponding position in 10 of the 22 transposases examined, while lysine is present in both the Ac and Tip100 transposases. Arginine is also present at the preceding site in seven of these transposases. This is consistent with this site being involved in DNA binding to the TIR of these transposases. In the Hermes transposase octamer, this three amino acid region is located on an α-helix some distance from the α-helix on which F109 and S110 are located and interacts with the other TIR (colored yellow in Fig. 3B).

The Catalytic Domain

The *hAT* element transposases contain a catalytic core consisting of a DDE motif that coordinates two divalent cations involved in catalyzing DNA cleavage immediately flanking the transposon and in the subsequent transfer of strands to the target molecule. This DDE catalytic domain is found in 11 of the 19 eukaryotic transposase superfamilies so far identified (11). As previously described in reference 10, the conserved core of the domain, known as an RNase H-like fold contains a mix of α- helices and β-strands in the order β1-β2-β3-α1-β4-α2/3-β5-α4-α5 (10). The catalytic triad is always present on topologically equivalent secondary structure features within the RNaseH-like fold with the primary aspartate on β1 (in Hermes this is D180), the second carboxylate amino acid located on or immediately after β4 (D248), and the final carboxylate amino acid on or immediately before α4 (E572). In the *Hermes* transposase, the final α-helices 4 and 5 are separated from the remainder of the catalytic domain by an insertion domain 287 amino acids in length. Amongst the hAT transposases, the DDE sequences are obviously highly conserved as the activity of the transposase is dependent on both their presence and their relative positions to each other within the active site of the enzyme. Amino acids flanking these three acidic amino acids also show high levels of conservation relative to the remainder of the transposase but some interesting differences remain, especially in the context of the recently studied cocrystal structure of the Hermes transposase and its L TIR.

Across all the 22 hAT transposases examined in this review, all retain the DDE residues at the corresponding locations to those known to form the catalytic core in the Hermes transposase (Fig. 3A). Mutations to each of the corresponding DDE residues in the Ac transposase (D301A, D367A, E719A) abolish catalytic activity of this enzyme (51). A tryptophan located two residues downstream from the first aspartate at 180 in the Hermes transposase has been proposed to correctly orientate D180 within the active site. W182 is also important as the W182A mutant is deficient in hairpin formation on flanking DNA during excision of the transposon (6). Moreover, this tryptophan, located two positions downstream from the first D of the catalytic triad, is conserved in the active members of the Ac family: *Tol2, Tgf2, Herves, Tam3, Ac, nDart1, TCUP, hobo, Hermes, TAG1, restless*, and *Crypt1*, which are found in plants, animals, and fungi. This suggests that it has undergone strong positive selection. One of the four mutations (E336A) that generated hyperactivity in the Ac transposase maps approximately midway between D301 and D367 in a region that is relatively nonconserved between these transposases, supporting a role for this region in the activity of the enzyme (51).

Conservation of leucine, valine, or isoleucine residues at a position corresponding to I224 of the Hermes transposase was also observed in all 22 hAT transposases examined (Fig. 3A). This region is located on an α-helix located towards the rim of the ocatamer away from both the 16 bp TIR end and neighboring monomers and its significance, if any, remains unknown (Fig. 3B).

The Insertion Domain

A defining aspect of *hAT* elements is the distance between the second D and E of the catalytic triad, which are separated by a large insertion domain that is variable in length but greater than approximately 250 amino acids (11). In the *Hermes* transposase, the length of the insertion domain is 287 amino acids and contains many α-helices, while in the related *hobo* element of *D. melanogaster*, it is of a similar size and contains a series TPE repeats which have no known function (52). Across its length, the insertion domain is perhaps one of the least conserved regions of the hAT transposases yet it plays a role in the correct positioning of the final glutamate residue of the catalytic triad in the active site (6, 77). Furthermore, there are other highly conserved residues located at the N-terminal end of the insertion domain in these transposases. This includes, for example, a CxxH motif at the immediate N-terminal end followed by a highly conserved RW motif approximately 40 to 50 amino acids downstream (Fig. 2). As described above, this histidine in the Hermes transposase (H268) is within the catalytic core following conformation changes that occur during excision of the transposon (Fig. 3B) (6). The CxxH motif is conserved across all of the 22 hAT transposases examined, indicating its importance for the functioning of the enzyme.

Immediately upstream from this motif in the Hermes transposase are two amino acid residues, S310 and K312, that interact with the backbone of the transferred strand between bp 5 and 6. These are not highly conserved amongst the 22 hAT transposases examined (Fig. 3A). The lysine corresponding to K312 of Hermes is found in the closely related hobo and Homer transposases and is present in some Tip transposases and the Tol1 transposase.

By analogy with the tryptophan found in the Tn*5* transposase that enables hairpin formation on the transposon end by stabilizing the hairpin formed as a reaction intermediate at the end of the transposon during excision, the RW motif conserved amongst the hAT transposases is proposed to stabilize the hairpin formed on the flanking DNA following cleavage of the nontransferred strand as the first step in the excision process (77, 100). This tryptophan is present in all hAT transposases examined except in the Tag1 transposase where it is replaced by phenylalanine, which presumably fulfills the same function (Fig. 3A). The arginine preceding the tryptophan in the RW motif is conserved in all of the 22 hAT transposases examined (Fig. 3A). In the Hermes transposase, this amino acid interacts with the DNA backbone on the transferred strand between bp 3 and 4 and presumably, together with W319, plays a significant role in properly positioning the transposon end and flanking DNA in the catalytic site. In the Ac transposase, the mutation D459A contributes to hyperactivity (51). This mutation is located four residues upstream from the RW motif indicating its proximity to the DNA immediately adjacent to the transposon end (Fig. 3A). Amongst the transposases examined, aspartate is only found in this position in the Ac, Tam3, nDart1, TCUP, and Restless transposases but, overall, there is strong conservation within this small region of the insertion domain across these transposases (Fig. 3A). The location of this region within the catalytic core of the Hermes transposase is shown in Fig. 3B.

Approximately 50 residues downstream from the RW motif of the Hermes transposase are three amino acids, K372, Q375, and S378, that interact with the *Hermes* L end (6). The side chain of K372 interacts with the second bp of the transposon end and lysine is found in this corresponding position eight of the 22 transposases examined across all three families. Eight different amino acids occur at this position in the remaining 14 transposases (Fig. 3A). The cocrystal structure of the Hermes transposase reveals that R573 also interacts with the second bp of the L end (6). In all of the transposases examined, this arginine residue immediately follows the final carboxylate residue of the catalytic triad and is found in all but three *Buster* elements (*AeBuster1*, *TcBuster*, and $SPIN_{on}$) in which it is replaced by lysine, alanine, or threonine, respectively (Fig. 3A).

The backbone of Q375 interacts with the initial bp of the transferred strand of the L end of *Hermes* and glutamine is found at this relative location in 14 of the 22 hAT transposases examined suggesting that the overall structure of this amino acid may play a role in its function (Fig. 3A). In three of the remaining transposases, a glutamine is located downstream within four residues of this location (Herves, Ac, and Tip100). The final of the four mutations that lead to hyperactivity in the Ac transposase, D545A, is located 19 resides downstream from this glutamine, suggesting that this region may contribute to the activity of the enzyme (51). In contrast, there is not strong conservation of either the lysine or serine corresponding to K372 or S378 in the Hermes transposase in these other hAT transposases. The side chain of S378 interacts with the DNA backbone of the nontransferred strand between bp 3 and 4, while the side chain of K372 interacts with highly conserved A found at the bp 2 in *hAT* elements (6). This region is located on an α-helix and loop immediately adjacent to the end of the L TIR of the *Hermes* element (Fig. 3B).

Based on the crystal structure of the Hermes transposase, the insertion domain plays a role in the oligomerization of the transposase as do other regions of this molecule (77). Indeed, there are three regions that participate in this. Two are located within the insertion domain (leading to protein–protein interactions with Hermes transposase monomers either side of a given transposase monomer). In the third, the previously discussed DNA binding domain between amino acids 79 and 150, three α-helices from each monomer intertwine to form a structure that also binds to the transposon DNA (6, 77).

The *hAT* Domain: A Continuation of the Catalytic Domain

The *hAT* domain was originally identified as a conserved motif located at the C-terminal of the Ac transposase, which was important for dimerization of the protein (106). Its involvement in the oligomerization of the Hermes transposase was confirmed by two-hybrid experiments (107). This, together with the fact that it is the most highly conserved large region amongst the hAT transposases led it to be identified as a "hAT C-terminal dimerization domain" or pfam05699. The crystal structure of the Hermes transposase reveals that this region is not, in fact, a domain, but plays a general role in maintaining the overall structure of the transposase (77). This explains earlier data based on two-hybrid analyses of the Hermes and Ac transposases in yeast that were misinterpreted leading to the conclusion that this domain was the primary site of dimerization of these enzymes (106, 107).

This region contains the final member of the catalytic triad of hAT transposases (E572 in the Hermes transposase) that is conserved in all hAT transposases examined (Fig. 3A). The arginine residue immediately following (R573 in Hermes) interacts with the Hermes TIR at bp 2 and is moderately conserved amongst the hAT transposases examined, being present in 17 of them (Fig. 3A). The next amino acid, phenylalanine (position F575 in Hermes), is found in 21 of the hAT

A

Transposon	2 5 11	L(bp)	Family
Tol2	cagaggtgtaaa	12	A
Tgf2	cagaggtg	8	A
Herves	tagagttgtgc	11	A
Tam3	taaagatgttgaa	13	A
Ac	tagggatgaaa	11	A
nDart1	tagaggtggccaaacgggc	19	A
TCUP	tatagttggccaa	12	A
Hobo	cagagaactgca	12	A
Homer	cagagatctgca	12	A
Hermes	cagagaacaacaacaag	17	A
TcBuster	cagtgttcttcaactg	16	B
AeBuster1	catagattcccaaact	16	B
SPIN$_{on}$	cagcggttctcaacct	16	B
TAG1	caatgttttcacgcccgacccg	22	A
Restless	cagagtgcgtaatc	14	A
Tol1	cagtagcggttcta	14	A
Crypt1	cagcgttccacacaagtcaag	21	B
BuT2	cagtgctgccaa	12	T
RP-hAT1	cagtggcgtaccta	14	T
hAT-4BM	cagtggcgtaccta	14	T
hAT-34HM	cagtggcgtaccta	14	T
Ip Tip100	caggggcggaggca	14	T

B

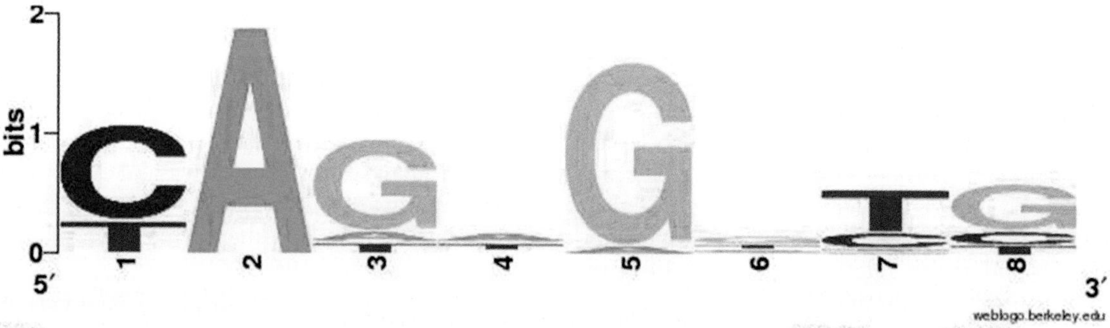

bits 0 1 2
5′ 1 2 3 4 5 6 7 8 3′
weblogo.berkeley.edu

DNA: R107 F109* S110

CAT 1:

INS: Q375* K372 R318 S310* S378 K312

CAT 2: R573 S576 K585 K605 N587 R588 R586

transposases examined with the exception of the Tag1 transposase in which it is replaced by tryptophan. In the Hermes transposase F575, stacked against W182, forms part of a pocket at the active site that is proposed be important for the correct positioning of the first aspartate of the catalytic triad (6). A comparison across these 22 transposases shows that the tryptophan corresponding to W182 is not as highly conserved as F575, perhaps indicating the more critical role for the later amino acid in this structure. The serine immediately following this phenyalanine is also found in 21 out of 22 transposases examined here, the exception again being the Tag1 transposase in which it is replaced by cysteine (Fig. 3A).

Eleven amino acids downstream is an arginine residue (R586 in the Hermes transposase), which is found in all 22 hAT transposases. This arginine is the second of a group of four consecutive amino acids that, in the Hermes transposase, interact with the minor groove of the DNA of the L TIR between the fourth and eighth base pairs. However, across these 22 transposases, it is the two arginines (positions 586 and 588 in the Hermes transposase) that show the strongest conservation (Fig. 3B). In the Hermes transposase, T593 interacts with the DNA backbone of the nontransferred strand between bp 8 and 9, however, this residue is not conserved amongst the other transposases.

Immediately upstream from this region is a conserved tryptophan (W545 in the Hermes transposase), which, in 11 of the 22 transposases examined, is immediately adjacent to another tryptophan residue (Fig. 3 – annotated as WW). However, the Tol2, Tgf2, Tol1, and Tip100 transposases do not carry the tryptophan corresponding Hermes W545 (Fig. 3A). This region is located away from the catalytic domain and from regions known to bind DNA and instead is in a location that enables it to possibly interact with neighboring Hermes dimers. Therefore, it may be important for the formation of the octamer (Fig. 3B). If so, its conservation amongst the transposases examined here may indicate that all have an oligomeric, ring-like quaternary structure. There is also strong conservation of a proline at the position corresponding to P552 of the Hermes transposase with this present in all but the BuT2 transposase (Fig. 3B). The function of this proline in maintaining the structure and function of the transposases remains unknown, however, it does define the border between the insertion and C-terminal catalytic domain.

THE ROLE OF THE TIRs

In a previous review of the *hAT* elements, Rubin et al. [12] concluded that a consensus sequence of (T/C)A (A/G)NG was present at the ends of the TIRs for all the 34 *hAT* elements they examined, not of all which were functional. Addition of functional *hAT* elements such as *Herves*, *TcBuster*, *AeBuster1*, and *SPIN$_{on}$* confirm this consensus sequence and add an additional preference for C at the 11th position and A at the 12th position of the TIRs for those *hAT* elements that have TIRs 11 bp or longer (Fig. 5).

Can this consensus sequence be interpreted with the respect to what is known about the binding of the 16 bp *Hermes* L end oligomer to the Hermes transposase? As described above, the first 5 bp of the L TIR interacts with several amino acids of the transposase. In particular, K372 and R573 interact with the highly conserved A at the second position of the TIR (6). R573 is conserved amongst the majority of the transposases within the *Ac* family but, in the Ac transposase, it is replaced by serine. (Fig. 3A). In the *Buster* family, threonine occupies this position in the active AeBuster1 and SPIN$_{on}$ transposases but in the active TcBuster transposase, lysine is present at this position. K372, located in the insertion domain, is not conserved amongst the hAT transposases (Fig. 3A). Within *hAT* TIRs, the most conserved nucleotides are A at position 2 and G at position 5 (Fig. 5) (78). The strong preference for A at position 2 is not reflected by a corresponding absolute conservation of the two amino acid residues known to interact with it. This suggests that other factors may also play a role in the conservation of this

Figure 5 Alignment of the left terminal inverted repeats (TIRs) of the 17 active transposons from the *Ac* and *Buster* families and the five from the *Tip* family. (A) Alignment showing the conservation of A2 and G5 amongst these transposons with a lower degree of conservation of C11 when the TIRs exceed 10 bp in length. (B) The weblogo generated from the first 8 bp of each of the TIRs in (A) with the amino acids that interact with these in the Hermes transpoase shown below (6). These are grouped according to their location in the DNA binding domain (DNA), the first catalytic domain (CAT 1), the insertion domain (INS), or the second catalytic domain (CAT 2) of the transposase. *Denotes that the interaction with the TIR is with the main chain of the amino acid. Amino acids in red interact with the nontransferred strand, those in black with the transferred strand. The nontransferred strand is shown in the weblogo. doi:10.1128/microbiolspec.MDNA3-0054-2014.f5

nucleotide pair. The lack of correlation between nucleotide conversation and conservation of those amino acids known to bind it is more marked for the G at position 5 at which, based on the Hermes cocrystal structure, N587 interacts (Fig. 5). As Fig. 3A shows, this amino acid is not well conserved amongst the 22 hAT transposases examined. However, as described above, it does reside within a small region that is relatively conserved with the preceding arginine (R586) present in all of these transposases (Fig. 3A). These are two of the four amino acids (S576, R586, N587, and R588) that interact with the minor groove of the L TIR between the third and eighth base pairs [6]. The lack of complete concordance between highly conserved nucleotides in the *hAT* TIRs and the amino acids in their transposases known to interact with them indicates that other factors influence the conservation of A2 and G5 on the nontransferred strand of *hAT* elements.

Within the insertion domain, only Q375 is conserved amongst the hAT transposases. In *Hermes*, this interacts with the terminal nucleotide pair of the TIR (Fig. 3A) (6). However, this conservation is not strong, Q being absent in this position in the Herves, Tam3, Ac, nDart1, TCUP, Tag1, and Crypt1 transposases (Fig. 3A). This amino acid residue is obviously immediately adjacent to the active site and the terminal nucleotide of the transposon (Fig. 3B). The lack of conservation might be surprising, however, it needs to be remembered that there is considerable variance in the final nucleotide of the 8 bp TSD flanking *hAT* elements. Depending on the family, consensus is seen only at the fourth and fifth (Buster family) or second and seventh (Ac family) base pairs of the TSD. Some aspects of the structure of the catalytic site of these transposases may therefore need to account for this variation. There is no detectable correlation between the presence of Q at this position and *hAT* element family (Fig. 3A). S378, which interacts with the third base pair of the *Hermes* L TIR, shows no conservation amongst the hAT transposases, however, R318, which also interacts with this nucleotide, is present in all the transposases and is immediately adjacent to the tryptophan proposed to stabilize the hairpin formed on flanking DNA as a reaction intermediate during excision (Fig. 3A). There is a strong preference for G at position 3 of the nontransferred strand with only the *Tam3*, *TCUP*, *AeBuster1*, and *Tag1* L TIRs having an alternate nucleotide present at this position (Figure 5). There is no correlation between the third nucleotide in this position in this transposon and the amino acid present in the position corresponding to S378 in their transposases (Fig. 3A).

As described above, there is some support for conservation of amino acids with positively charged side chains in the DNA binding domain tightly centered on the residue corresponding to R149 in the Hermes transposase in which it interacts with the C at bp 11 in the L TIR. Given the frequency of the presence of C in this position in TIRs that are greater than 11 bp in length, it is possible this correlates with the presence of this nucleotide in the TIR.

The close proximity of two (E249, D459) of the four mutations lead to hyperactivity of the Ac transposase to regions of the transposase implicated in interactions with sequences in the TIR (51). E249 is located five amino acids downstream from the amino acids corresponding to R107, F109 and S110 of the Hermes transposase that interacts with the sixth, seventh, and eighth nucleotides of the transferred strand of the Hermes TIR (Fig. 5). D459 is located four amino acids upstream from the conserved RW motif in the insertion domain that interacts with both flanking DNA and between the third and fourth nucleotides of the DNA backbone of the nontransferred strand (Fig. 5) (6).

DOMESTICATED hAT TRANSPOSASES

There are many examples of chromosomal genes that are derived from transposons and these have been reviewed previously (108, 109). Of the *hAT* transposons, the documented examples are *DAYSLEEPER* from *A. thaliana*, which is found in the nucleus and vesicles of cells in tissues undergoing development (110, 111), *gary*, which is expressed in the ears of rice and wheat but as yet has no known function(112), and *Gon-14* from *Caenorhabditis elegans* which is involved in larval development (113). Furthermore, there is the DREF gene from *D. melanogaster*, which is also involved in tissue morphogenesis and development (114), the P52rIPK gene from mammals, which acts as a repressor of phosphorylation of the initiation factor eIF-1alpha (115), and KIAA0543_ZnF826 from humans that has unknown function (116). In addition, there is GTF2IRD2, which is a transcription factor in humans (117), Lin-15B, which regulates the development of the vulva in *C. elegans* (118), and the ZBED genes found in vertebrates (116, 119, 120, 121, 122). The ZBED genes include the *Buster1* (*ZBED5*) gene of mammals, which is derived from the *Charlie hAT* elements and has an unknown function (116), and the *ZBED1* gene of mammals, which is expressed in dividing cells and regulates several ribosomal protein genes (119). In addition, there is the *ZBED4* gene, which is expressed in cones and Müller cells of the human retina but in mice shows a different localization within retinal cells and is also found in the heart, brain, liver, spleen, embryo, lung, testis, and ovary, with the highest expression levels detected in the thymus (123). There is also the *ZBED6* gene, which is a transcription factor that is highly conserved amongst eutherian mammals (124, 125).

The *ZBED6* gene was identified through a single nucleotide change in the intron of in the insulin-like growth factor 1 gene, which resulted from the artificial selection for lean growth in domesticated pigs over the past 60 years (125). This single transition increases muscle and heart size in domesticated pigs and, at the molecular level, reduces the interaction between this site and a repressor that was subsequently shown to be the ZBED6 protein (125). ZBED6 is highly expressed in muscles, brain, ovary, and testis, but also shows weaker expression beyond these tissues, and is a splice variant of the *Zch3h11a* gene, the other functions of which remain unknown (124). ZBED6 is highly conserved amongst placental mammals, with remnants remaining in marsupial genomes, indicating that it evolved subsequent to the split of the eutherian mammals from the marsupials (124). An analysis of ChIP-sequencing data obtained from the genomic binding sites of ZBED6 in the mouse genome revealed potential target sites in genes involved in development, the regulation of biological processes, transcriptional regulation, cell differentiation, morphogenesis, neurogenesis, intracellular signaling, and muscle development, with many of these having human orthologs that are associated with disease in humans, with the major class being developmental disorders (125).

The *DAYSLEEPER* gene is one member of a large family of domesticated genes found throughout the angiosperms (126). Orthologs to the *DAYSLEEPER* gene of *Arabidopsis* were found in rice and grapevine leading to a survey of its presence throughout the plant kingdom (126). Its distribution appears to be confined to the angiosperms in which it can be present in several copies, with most functional members of the family being found in the nucleus. The exception being a group of SLEEPER genes that lack the N-terminal sequences and which are found only in the cytoplasm (126). Interestingly, the rice and grapevine SLEEPER genes do not rescue the DAYSLEEPER mutant phenotype when placed into this *Arabidopsis* mutant but their expression does result in alterations to leaf number and shape and in alterations in the processes of flower and seed development (126). Members of this family contain the BED and hAT dimerization domains at the C-terminus of the protein.

The criterion that has been used to identify these domesticated genes has been their retention of one or more the domains that define the hAT transposases, these typically being one of the BED, catalytic, or hAT domains. Indeed, an analysis of genes containing the BED domain alone illustrates the degree to which this domain has populated genomes (101, 122). Some of the domesticated *hAT* genes listed by Sinzelle et al. (109) contain only this domain. In the analysis presented here, the focus is only on those domesticated *hAT* genes that retain these three domains and how the structure of the Hermes transposase may help explain their new functions in cells.

An M-Coffee-based alignment of 11 of these proteins is shown in Fig. 6A with the conserved amino acids shared between them shown in Fig. 6B. Comparing Fig. 6A with Fig. 3A shows that many of the amino acids critical for function of the active hAT transposases are conserved in the domesticated genes derived from them, however, none of them possess all of these critically important residues. For example, the *ZBED5* (*Buster1*) gene of humans contains a BED domain, the DDE motif, the glutamine (Q375 in the Hermes transposase) that interacts with the terminal base pairs of the *Hermes* TIR, the highly conserved phenylalanine from the C-terminal catalytic domain, and arginines corresponding to R586 and R588 of the Hermes transposase. Its CxxH motif is replaced by a CxxY motif (Fig. 6A, B). In this narrow context, it might be reasonable to suspect that *ZBED5* is capable of catalyzing some form of DNA recombination derived from transposition.

Allowing for the substitution of amino acids with the same chemical properties, across these 11 domesticated hAT proteins, the least conserved "essential" residues appear to be the second aspartate of the catalytic triad that is required for excision and transposition of the related transposases, the glutamine that lies immediate to the terminal nucleotide of the Hermes transposon (Q375 in Hermes) and the three arginines at the C-terminal end of the catalytic domain, which, as discussed above, interact with the sequences of the L TIR of *hAT* elements (Fig. 6A, B). One interpretation of these data is that most of these proteins have lost the ability to recognize and bind to sequences similar in structure to TIRs and have lost the ability to catalyze recombination. They appear to have retained the ability to bind DNA (through their BED domains) and the conservation of the WW motif may indicate that some exist as oligomers. They all retain the C of the CxxH domain with all but one retaining the W of the RW motif (Fig. 6B). Both are in regions, which, as described

A

```
Hermes               51 CRKC (16) HKCCAS    101 WVVRDC--RPFSA    179 IDLWT   247 TDRG--A-N
ZBED1                47 CRIC (15) HLEKNH    133 LICEGL--YPASI    198 TDMWR   270 NY--G-K-D
ZBED4               306 CIHC (20) HMWRAH    665 MIALDL--QPSYF    732 SGIWM   808 TD--N-A-S
ZBED5               132 CVLC (16) HLETKH    224 LLIKPCA------    277 LDESA   343 SD-AS-R-A
ZBED6               151 CNLC (21) HLQARH    417 MIVEDM--HPYNY    484 VDIWT   563 SD-NS-S-N
DAYSLEEPER           88 CKGC (16) HLKRH     172 MIIMHD--YPLHM    236 LDFWT   304 NHPA-SN-S
KIAA0543-ZnF862     161 CSAC (28) HAKSKAH   581 LLQSTG--TV---    628 LDSST   696 --TD--G-S
P52RIPK              58 CAKH (28) HLNNPH    206 CRINSG--EE---    264 TDDVV   330 YI-VS-S-S
DREF                 59 CIKC (15) HLQHRH    232 IVIKDL--RNVDS    300 FEMWV   360 NY----D-E
GON-14a              47 HFLC                259 FFIKSG--IPLRA    321 FDTT-   387 GK----A-E
GTF2IRD2            447 CLIC (14) HYQTNH    490 KKGLRKYLLGLSD    552 IDEIT   618 ST-GT-P-A
b-Gary              173 CIHC (13) HLSRH     300 MIILHA--YPFSI    385 ASIWT   461 VGDVRNNAN

CONSERVED
DOMAIN IN hATs:     -------- BED ---------  --- DNA BINDING ---  -------- CATALYTIC --------
```

```
Hermes              264 NCSSHL   309 SS--LKSECRTRWNST-YTMLR    372 KELQT--CSS
ZBED1               289 PCLGHT   345 -M--LVSNRVSWWGST-LAMLQ    409 EMLSA--SRY
ZBED4               826 QCFSHT   883 -H--LIQDVPSKWSTS-FHMLE    945 REMST--QMS
ZBED5               371 HCLLYR   425 -A--LLLNTEVRWLSR-GKVLR    489 LSMQG--KNV
ZBED6               582 PCFLHC   639 -N--LKQDETGHWIST-FYMLK    700 QKVSV--KTA
DAYSLEEPER          335 NCVART   389 KV--LSLDDQTQWNTT-YMMLV    453 STLQS--TGN
KIAA0543-ZnF862     723 HCVAHR   777 EIIRLKDLNAVRWVASRRRTLH    856 EVCQKEIVLI
P52RIPK             358 LCSSCA   418 -E--LKEICHSQWTGR-HDAFE    498 KNLQG--QTS
DREF                381 LCYVSV   427 ----MPT-YNEH-------FPW    477 DTLRG--EDI
GON-14a             406 ICFSQN   461 -E--YPEVDNGHWITT-LDFIT    523 KEMVD--VKA
GTF2IRD2            653 CCIIHP   704 -S--LLYYTEIKWLSR-GLVLK    768 ISLQG--HSQ
b-Gary              493 ACLDDI   546 ----CPQEDAKWWHKF-YFRLE    600 EVISC--PSS

CONSERVED
DOMAIN IN hATs:     -------------------- INSERTION ------------------------
```

```
Hermes               542 EWWNLNSKKYPKLSKLALSLLSIPASSAASERTFSLAG-NIITEKRNRIG
ZBED1                589 KWWSDRLALFPLLPKVLQKYWCVTATRVAPERLFGSAA-NVVSAKRNRLA
ZBED4               1102 TYWNLKKASWPGLSALAVRFLGCPPSIVPSEKLFNTPT-ENGSLGQSRLM
ZBED5                615 DFWSSLIQEYPSIARRAVRVLLPFATMHLCETGFSYYA-ATKTKYRKRLD
ZBED6                886 IYWQRKISIWPALTQVAIQYLSCPMCSWQSECIFTKNS-HFHPKQIMSLD
DAYSLEEPER           612 DWWKQNKLKYPTLSKMARDILSIPVSAAAFDYVFDMEP-REMDEYKTSLR
KIAA0543-ZnF862     1010 KNALAQHCRFPLLSKLMAVVVCVPISTSCCERGFKAMN-RIRTDERTKLS
P52RIPK              667 --HLPDIKFFPNVYALLKVLCILPVMKVENERYENGRK-RLKAYLRNTLT
DREF                 646 EWWLKMGHIYGTLRDLASLYHSVPGVVTLSFKKALRDQIYDFN-KRFMLT
GON-14a              654 DWWARHAGRFPRLYKYARELFIIPAFSIDAAYYLGEYG-ILTH-SINESD
GTF2IRD2             889 EFYKYLWGSYPKYKHHCAKILSMFGSTYICEQLFSIMK-LSKTKYCSQLK
b-Gary               764 NWWMNHTEKYPTLAAMAQDILAMPASAVQSEAAFSSSG-PVIPKHYSTLS

CONSERVED
DOMAIN IN hATs:     -------------------CATALYTIC--------------------------
```

Figure 6 *continues on next page*

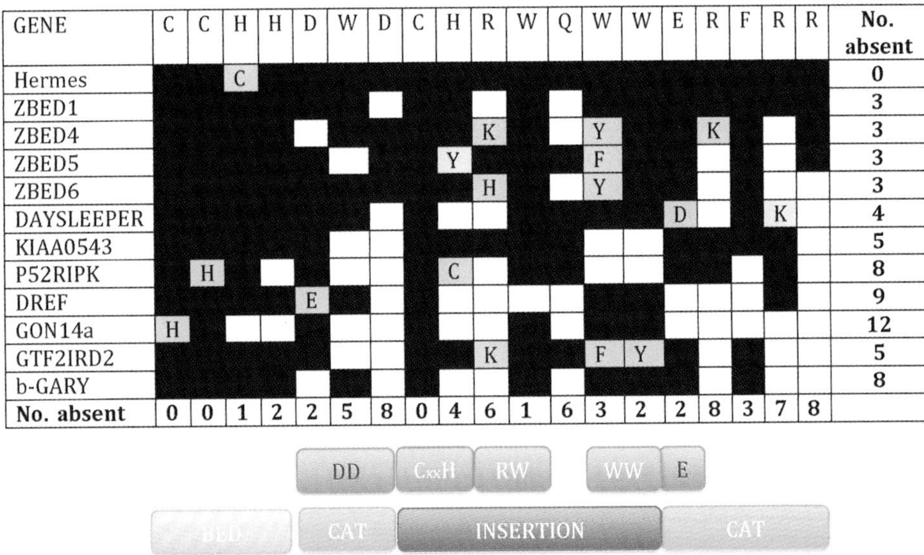

B

GENE	C	C	H	H	D	W	D	C	H	R	W	Q	W	W	E	R	F	R	R	No. absent
Hermes			C																	0
ZBED1																				3
ZBED4										K			Y			K				3
ZBED5									Y				F							3
ZBED6										H			Y							3
DAYSLEEPER															D			K		4
KIAA0543																				5
P52RIPK		H							C											8
DREF					E															9
GON14a	H																			12
GTF2IRD2										K			F	Y						5
b-GARY																				8
No. absent	0	0	1	2	2	5	8	0	4	6	1	6	3	2	2	8	3	7	8	

Figure 6 Amino acid sequence comparison of domesticated hAT transposases. (A) Consensus alignment of 11 domesticated genes derived from hAT transposases and the Hermes transposase. The location of each sequence within its transposase is shown by the amino acid numbers. The location of conserved resides are shown below the alignments along with the locations of the domains. Amino acids identified as being critical for Hermes transposase activity and/or DNA binding from the Hermes transpososome cocrystal by Hickman et al. (6) are shown in red as are identical amino acids present in the same locations in the other proteins. The C and H amino acids proposed to constitute the BED domain are shown in blue. Accession numbers are: ZBED1 (AAH15030), ZBED4 (NP_055653), ZBED5 (Q49AG3), ZBED6 (NP_001167579), DAYSLEEPER (Q9M2N5), KIAA0543_ZnF862 (060290), P52rIPK (O43422), DREF (BAA24727), GON-14a (CCD71205), GTF21RD2 (AAP14955), and b-Gary (CAJ32531). (B) Conserved key amino acids in domesticated genes derived from hAT transposases and the Hermes transposase. The amino acids are listed along the top of the table. Underneath is a diagram showing the domain they reside in. Black boxes indicate that the amino acid is present in this location in the domesticated gene. Gray boxes denote an amino acid is present that is chemically similar to the amino acid listed at the top of the table and this is shown in the box. Empty boxes indicate no conservation. The number of these amino acids missing from each protein is listed in the final column. The number missing at each location is listed in the final row.
doi:10.1128/microbiolspec.MDNA3-0054-2014.f6

above, interface with the catalytic domain and, together with the conservation of the first and last carboxylate amino acids of the catalytic triad, suggest that at least some components of the overall structure of the catalytic site might be retained.

The analysis of the SLEEPER genes from angiosperms identified three conserved regions: the BED domain, a previously unidentified region in the insertion domain, and the hAT domain (126). Interestingly, the previously unidentified conserved region identified in the insertion domain corresponds to the region that a proline (P437 in Hermes) is conserved amongst both the active transposases and is present in eight of the 11 domesticated transposases analyzed here.

CONCLUSIONS

The *hAT* elements offer a fascinating vehicle with which to explore the molecular biology of excision and integration of eukaryotic class II transposable elements. They are a very ancient family of transposons and so are abundant throughout many species of all kingdoms. They have diverged considerably, with the result that many different forms of them exist with the differences between then sufficient to allow them to be classified into different families. During evolution, there is very strong evidence that many have been horizontally transferred between species within kingdoms indicating their promiscuity. This has been harnessed in biotechnology where several hAT transposons from different kingdoms are routinely used as genetic tools in plants, insects, and vertebrates, with the clear potential of some being able to be applied in human gene therapy. Many different examples of *hAT* transposons becoming domesticated now exist with the intriguing similarities and differences between these proteins and their transposase counterparts leading to experimentally tractable questions about transposon evolution and transposase function. The diversity of this fascinating superfamily of transposons is matched by the number of questions that arise from their study.

Acknowledgments. *I thank Ms. Anna-Louse Doss for assistance with Fig. 3B, Dr. Z. Zhang for the peptide sequences of hAT transposases from* Rhodnius prolixus *and* Bombyx mori, *and colleagues for invaluable and interesting discussions throughout many years.*

Citation. Atkinson PW. 2014. *hAT* transposable elements. Microbiol Spectrum 3(2):MDNA3-0054-2014.

References

1. **Feschotte C, Pritham EJ.** 2007. DNA transposons and the evolution of eukaryotic genomes. *Ann Rev Genet* **41**:331–368.
2. **Oliver KR, McComb JA, Greene WK.** 2013. Transposable elements: powerful contributors to angiosperm evolution and diversity. *Genome Biol Evol* **5**:1886–1901.
3. **Mack AM, Crawford NM.** 2001. The Arabidopsis TAG1 transposase has an N-terminal zinc finger DNA binding domain that recognizes distinct subterminal motifs. *Plant Cell* **13**:2319–2331.
4. **Kunze R, Starlinger P.** 1989. The putative transposase of transposable element Ac from Zea mays L. interacts with subterminal sequences of Ac. *Embo J* **8**:3177–3185.
5. **Kahlon AS, Hice RH, O'Brochta DA, Atkinson PW.** 2011. DNA binding activities of the Herves transposase from the mosquito *Anopheles gambiae*. *Mobile DNA* **2**:9.
6. **Hickman AB, Ewis HE, Li X, Knapp JA, Laver T, Doss A-L, Tolun G, Steven AC, Grishaev A, Bax A, Atkinson PW, Craig NL, Dyda F.** 2014. Structural basis for transposon end recognition by *Hermes*, an octameric hAT DNA transposase from *Musca domestica*. *Cell* **158**: 353–367.
7. **Coen ES, Robbins TP, Almeida J, Hudson A, Carpenter R.** 1989. Consequences and mechanisms of transposition in *Antirrhinum majus*, p 413–436. *In* Berg D, Howe M (ed), *Mobile DNA*. American Society for Microbiology, Washington DC.
8. **Atkinson PW, Warren WD, O'Brochta DA.** 1993. The *hobo* transposable element of *Drosophila* can be cross-mobilized in houseflies and excises like the *Ac* element of maize. *Proc Natl Acad Sci USA* **83**:9693–9697.
9. **Zhou L, Mitra R, Atkinson PW, Hickman AB, Dyda F, Craig NL.** 2004. Transposition of *hAT* elements links transposable elements and V(D)J recombination. *Nature* **432**:995–1001.
10. **Hickman AB, Chandler M, Dyda F.** 2010. Integrating prokaryotes and eukaryotes: DNA transposases in light of structure. *Crit Rev Biochem Mol Biol* **45**:50–69.
11. **Yuan YW, Wessler SR.** 2011. The catalytic domain of all eukaryotic cut-and-paste transposon superfamilies. *Proc Natl Acad Sci USA* **108**:7884–7889.
12. **Rubin E, Lithwick G, Levy AA.** 2001. Structure and evolution of the hAT transposon superfamily. *Genetics* **158**:949–957.
13. **Kempken F, Windhofer F.** 2001. The hAT family: a versatile transposon group common to plants. fungi, animals, and man. *Chromosoma* **110**:1–9.
14. **Du C, Hoffman A, He L, Caronna J, Dooner HK.** 2011. The complete Ac/Ds transposon family of maize. *BMC Genomics* **12**:588.
15. **Lazarow K, Doll M-L, Kunze R.** 2013. Molecular biology of maize Ac/Ds elements: an overview, p 59–82. *In* Peterson T (ed), *Plant Transposable Elements: Methods and Protocols*, **vol. 1057**. Springer, New York, NY.
16. **Federoff NV.** 1989. Maize transposable elements, p 375–411. *In* Berg DE, Howe MM (ed), *Mobile DNA*. ASM Press, Washington DC.
17. **Kunze R, Weil CF.** 2002. The hAT and CACTA superfamilies of plant transposons, p 565–610. *In* Craig NL, Craigie R, Gellert M, Lambowitz AM (ed), *Mobile DNA II*. ASM Press, Washington DC.
18. **McClintock B.** 1947. Cytogenetic studies of maize and *Neurospora*. *Carnegie Institution of Washington Year Book* **46**:146–152.
19. **McClintock B.** 1948. Mutable loci in maize. *Carnegie Institution of Washington Year Book* **47**:155–169.
20. **McGinnis W, Shermoen AW, Beckendorf SK.** 1983. A transposable element inserted just 5′ to a *Drosophila* glue protein gene alters gene expression and chromatin structure. *Cell* **34**:75–84.
21. **Arensburger P, Hice RH, Zhou L, Smith RC, Tom AC, Wright JA, Knapp JA, O'Brochta DA, Craig NL, Atkinson PW.** 2011. Phylogenetic and functional characterization of the hAT transposon superfamily. *Genetics* **188**:45–57.
22. **Rossato DO, Ludwig A, Depra M, Loreto ELS, Ruiz A, Valenta VLS.** 2014. *BuT2* is a member of the third major group of *hAT* transposons and is involved in horizontal transfer events in the genus *Drosophila*. *Genome Biol Evol* **6**:352–365.
23. **Zhang HH, Shen YH, Xu HE, Liang HY, Han MJ, Zhang Z.** 2013. A novel *hAT* element in *Bombyx mori* and *Rhodnius prolixus*: its relationship with miniature repeat transposable elements (MITEs) and horizontal transfer. *Insect Mol Biol* **22**:584–596.
24. **Christoff A-P, Lerto ELS, Sepel LMN.** 2012. Evolutionary history of the Tip100 transposon in the genus Ipomoea. *Genet Mol Biol* **35**:460–465.
25. **Warren WD, Atkinson PW, Obrochta DA.** 1994. The Hermes transposable element from the house fly, *Musca domestica*, is a short inverted repeat-type element of the *hobo*, *Ac*, and *Tam3* (hAT) element family. *Genet Res* **64**:87–97.

26. **Pascual L, Periquet G.** 1991. Distribution of *hobo* transposable elements in natural populations of *Drosophila melanogaster*. *Mol Biol Evol* **8:**282–296.
27. **Simmons GM.** 1992. Horizontal transfer of hobo transposable elements within the *Drosophila melanogaster* species complex - evidence from DNA sequencing. *Mol Biol Evol* **9:**1050–1060.
28. **Ortiz MD, Loreto ELS.** 2009. Characterization of new *hAT* transposable elements in 12 *Drosophila* genomes. *Genetica* **135:**67–75.
29. **Ladeveze V, Chaminade N, Lemeunier F, Periquet G, Aulard S.** 2012. General survey of *hAT* transposon superfamily with highlight on *hobo* element in *Drosophila*. *Genetica* **140:**375–392.
30. **Pace JKI, Gilbert C, Clark MS, Feschotte C.** 2008. Repeated horizontal transfer of a DNA transposon in mammals and other tetrapods. *Proc Natl Acad Sci U S A* **105:**17023–17028.
31. **Habu Y, Histomi Y, Iida S.** 1998. Molecular characterization of the mutable *flaked* allele for flower variegation in the common morning glory. *Plant J* **16:**371–376.
32. **Mota NR, Ludwig A, da Silva Valente VL, Loreto ELS.** 2010. harrow: new Drosophila hAT transposons involved in horizontal transfer. *Insect Mol Biol* **19:**217–228.
33. **Depra M, Panzera Y, Ludwig A, Valenta VLS, Loreto ELS.** 2010. *hosimary*: a new *hAT* transposon group involved in horizontal transfer. *Mol Genet Genomics* **283:** 451–459.
34. **Zou S, Du X, Yuan J, Jiang X.** 2010. Cloning of goldfish *hAT* transposon *Tgf2* and its structure. *Hereditas* **32:**1–6.
35. **Jiang XY, Du XD, Tian YM, Shen RJ, Sun CF, Zou SM.** 2012. Goldfish transposase *Tgf2* presumably from recent horizontal transfer is active. *FASEB J* **26:**2743–2752.
36. **Koga A, Shimada A, Shima A, Sakaizumi M, Tachida H, Hori H.** 2000. Evidence for recent invasion of the medaka fish genome by the Tol2 transposable element. *Genetics* **155:**273–281.
37. **Gilbert C, Hernandez SS, Flores-Benabib J, Smith EN, Feschotte C.** 2012. Rampant horizontal transfer of *SPIN* transposons in squamate reptiles. *Mol Biol Evol* **29:**503–515.
38. **Ray DA, Pagan HJT, Thompson ML, Stevens RD.** 2006. Bats with *hATs*: Evidence for recent DNA transposon activity in genus *Myotis*. *Mol Biol Evol* **24:** 632–639.
39. **Novick P, Smith J, Ray D, Boissinot S.** 2010. Independent and parallel lateral transfer of DNA transposons in tetrapod genomes. *Gene* **449:**85–94.
40. **Baker B, Schell J, Lorz H, Fedoroff N.** 1986. Transposition of the maize controlling element *Activator* in tobacco. *Proc Natl Acad Sci U S A* **83:**4844–4848.
41. **Laufs J, Wirtz U, Kammann M, Matzeit V, Schaefer S, Schell J, Czernilofsky AP, Baker B, Gronenborn B.** 1990. Wheat dwarf virus Ac/Ds vectors: expression and excision of transposable elements introduced into various cereals by a viral replicon. *Proc Natl Acad Sci U S A* **87:**7752–7756.
42. **Izawa T, Miyazaki C, Yamamoto M, Terada R, Iida S, Shimamoto K.** 1991. Introduction and transposition of the maize transposable element *Ac* in rice (*Oryza sativa* L). *Mol Gen Genet* **227:**391–396.
43. **Chuck G, Robbins T, Nijjar C, Ralston E, Courtney-Gutterson N, Dooner HK.** 1993. Tagging and cloning of a petunia flower color gene with the maize transposable element activator. *Plant Cell* **5:**371–378.
44. **Finnegan EJ, Lawrence GJ, Dennis ES, Ellis JG.** 1993. Behaviour of modified *Ac* elements in flax callus and regenerated plants. *Plant Mol Biol* **22:**625–633.
45. **McElroy D, Louwerse JD, McElroy SM, Lemaux PG.** 1997. Development of a simple transient assay for Ac/Ds activity in cells of intact barley tissue. *Plant J* **11:**157–165.
46. **Yoder JI.** 1990. Rapid proliferation of the maize transposable element *Activator* in transgenic tomato. *Plant Cell* **2:**723–730.
47. **Yang CH, Ellis JG, Michelmore RW.** 1993. Infrequent transposition of Ac in lettuce, Lactuca sativa. *Plant Mol Biol* **22:**793–805.
48. **Lisson R, Hellert J, Ringleb M, Machens F, Kraus J, Hehl R.** 2010. Alternative splicing of the maize Ac transposase transcript in transgenic sugar beet (Beta vulgaris L.). *Plant Mol Biol* **74:**19–32.
49. **Babwah V, Waddell S.** 2002. Trans-activation of the maize transposable element, Ds, in Brassica napus. *Theor Appl Genet* **104:**1141–1149.
50. **Qu S, Jeon JS, Ouwerkekr PBF, Bellizzi M, Leach J, Ronald P, Wang GL.** 2009. Construction and application of efficient Ac-Ds transposon tagging vectors in rice. *J Integr Plant Biol* **51:**982–992.
51. **Lazarow K, Du M-L, Weimer R, Kunze R.** 2012. A hyperactive transposase of the maize transposable element *Activator* (*Ac*). *Genetics* **191:**747–756.
52. **Blackman RK, Macy M, Koehler D, Grimaila R, Gelbart WM.** 1989. Identification of a fully functional *hobo* transposable element and its use for germ line transformation of *Drosophila*. *EMBO J* **8:**211–217.
53. **Rubin GM, Spradling AC.** 1982. Genetic transformation of *Drosophila* with transposable element vectors. *Science* **218:**348–353.
54. **Gangadhrana S, Mularoni L, Fain-Thornton J, Wheelan SJ, Craig NL.** 2010. DNA transposon Hermes inserts into DNA in nucleosome-free regions *in vivo*. *Proc Natl Acad Sci U S A* **107:**21966–21972.
55. **Li X, Ewis HE, Hice RH, Malani N, Parker N, Zhou L, Feshotte C, Bushman FD, Atkinson PW, Craig NL.** 2013. A resurrected mammalian hAT transposable element and closely related insect element are highly active in human cell culture. *Proc Natl Acad Sci U S A* **110:** E478–E487.
56. **Woodward LE, Li X, Malani N, Kaja A, Hice RH, Atkinson PW, Bushman FD, Craig NL, Wilson MH.** 2012. Comparative analysis of the recently discovered hAT transposon TcBuster in human cells. *PLoS One* **7:** e42666.
57. **Hori H, Suzuki M, Inagaki H, Oshima T, Koga A.** 1998. An active Ac-like transposable element in teleost fish. *J Mar Biotechnol* **6:**206–207.

58. **Kawakami K, Noda T.** 2004. Transposition of the *Tol2* element, an *Ac*-like element from the Japanese Medaka fish *Oryzias latipes*, in mouse embryonic stem cells. *Genetics* **166:**895–899.

59. **Asakawa K, Suster ML, Mizusawa K, Nagayoshi S, Kotani T, Urasaki A, Kishimoto Y, Kawakami K.** 2008. Genetic dissection of neural circits by Tol2 transposon-mediated Gal4 gene and enhancer tranpping in zebrafish. *Proc Natl Acad Sci U S A* **105:**1255–1260.

60. **Asakawa K, Kawakami K.** 2009. The *Tol2*-mediated Gal4-UAS method for gene and enhancer trapping in zebrafish. *Methods* **49:**275–281.

61. **Urasaki A, Mito T, Noji S, Ueda R, Kawakami K.** 2008. Transposition of the vertebrate Tol2 transposable element in Drosophila melanogaster. *Gene* **425:** 64–68.

62. **Hamlet MR, Yergeau DA, Kuliyev E, Takeda M, Taira M, Kawakami M, Mead PE.** 2006. Tol2 transposon-mediated transgenesis in Xenopus tropicalis. *Genesis* **44:**438–445.

63. **Kawakami K, Takeda H, Kawakami N, Kobayashi M, Mishina M.** 2004. A transposon-mediated gene trap approach identifies developmentally regulated genes in zebrafish. *Developmental Cell* **7:**133–144.

64. **Scott EK, Mason L, Arrenberg AB, Ziv L, Goose NJ, Xiao T, Chi NC, K A, Kawakami K, Baier H.** 2007. Taregting neural circuitry in zebrafish using GAL4 enhancer trapping. *Nat Methods* **4:**332–326.

65. **Kawakami K, Abe G, Asada T, K A, Fukada R, Ito A, Lal P, Mouri N, Muto A, Suster ML, Takakubo A, Wada H, Yoshida M.** 2010. zTrap: zebrafish gene trap and enhancer trap database. *BMC Dev Biol* **10:**105.

66. **Macdonald J, Taylor L, Sherman A, Kawakami K, Takahashi Y, Sang HM, McGrew MJ.** 2012. Efficient gene modification and germ-line transmission of primordial germ cells using piggyBac and Tol2 transposons. *Proc Natl Acad Sci U S A* **109:**E1466–E1472.

67. **Mayasari NI, Mukougawa K, Shigeoka T, Kawaichi M, Ishida Y.** 2012. Mixture of differentially tagged Tol2 transposons accelerates conditional disruption of a broad spectrum of genes in mouse embryonic stem cells. *Nucleic Acids Res* **20:**e97.

68. **Freeman S, Chrysostomou E, Kawakami K, Takahashi Y, Daudet N.** 2012. Tol2-mediated gene transfer and in ovo electroporation of the optic placode: a powerful and versatile approach for invetifatin embryonic development and regeneration of the chicken inner-ear. *Methods Mol Biol* **916:**127–139.

69. **Yergeau DA, Kelley CM, Kuliyev E, Zhu H, Sater AK, Wells DE, Mead PE.** 2010. Remobilization of Tol2 transposons in Xenopus tropicalis. *BMC Dev Biol* **10:**11.

70. **Lane MA, Kimber M, Khokha MK.** 2013. Breeding based remobilization of Tol2 transposon in *Xenopus tropicalis*. *PLoS ONE* **8:**e76807.

71. **Meir YJ, Weirauch MT, Yang HS, Chung PC, Yu RK, Wu SC.** 2011. Genome-wide target profiling of piggyBac and Tol2 in HEK392: pros and cons for gene discovery and gene therapy. *BMC Biotechnol* **11:**28.

72. **Huang X, Guo H, Tammana S, Jung YC, E M, Bassi P, Cao Q, Tu ZJ, Kim YC, Ekker SC, Wu X, Wang SM, Zhou X.** 2010. Gene transfer efficiency and genome-wide integration profiling of Sleeping Beauty, Tol2, and piggyBac transposons in human primary T cells. *Mol Ther* **18:**1803–1813.

73. **Backfisch B, Kozin VV, Kirchmaier S, Tessmar-Raible K, Raible F.** 2014. Tools for gene-regulatory analyses in th marine annelid *Platynereis dumerilii*. *PLoS ONE* **9:** e93076.

74. **Coen ES, Carpenter R, Martin C.** 1986. Transposable elements generate novel spatial patterns of gene expression in Antirrhinum majus. *Cell* **47:**285–296.

75. **Weil CF, Kunze R.** 2000. Transposition of maize *Ac/Ds* transposable elements in the yeast *Saccharomyces cerevisiae*. *Nat Genet* **26:**187–190.

76. **Calvi BR, Hong TJ, Findley SD, Gelbart WM.** 1991. Evidence for a common evolutionary origin of inverted terminal repeat transposons in *Drosophila* and plants: *hobo*, *Activator* and *Tam3*. *Cell* **66:**465–471.

77. **Hickman AB, Perez ZN, Zhou L, Musingarimi P, Ghirlando R, Hinshaw JE, Craig NL, Dyda F.** 2005. Molecular architecture of a eukaryotic transposase. *Nat Struct Mol Biol* **12:**715–721.

78. **Warren WD, Atkinson PW, O'Brochta DA.** 1994. The *Hermes* transposable element from the housefly, *Musca domestica*, is a short inverted repeat-type element of the *hobo*, *Ac*, and *Tam3* (*hAT*) element family. *Genet Res* **64:**87–97.

79. **Sarkar A, Coates CJ, Whyard S, Willhoeft U, Atkinson PW, O'Brochta DA.** 1997. The *Hermes* element from *Musca domestica* can transpose in four families of cylorrhaphan flies. *Genetica* **99:**15–29.

80. **Subramanian RA, Cathcart LA, Krafsur ES, Atkinson PW, O'Brochta DA.** 2009. *Hermes* transposon distribution in *Musca domestica*. *J Hered* **100:**473–480.

81. **Sarkar A, Yardley K, Atkinson PW, James AA, O'Brochta DA.** 1997. Transposition of the *Hermes* element in embryos of the vector mosquito, *Aedes aegypti*. *Insect Biochem Mol Biol* **27:**359–363.

82. **Pinkerton AC, O'Brochta DA, Atkinson PW.** 1996. Mobility of *hAT* transposable elements in the Old World American bollworm, *Helicoverpa armigera*. *Insect Mol Biol* **5:**223–227.

83. **Zhao Y, Eggleston P.** 1998. Stable transformation of an *Anopheles gambiae* cell line mediated by the *Hermes* mobile genetic element. *Insect Biochem Mol Biol* **28:** 213–219.

84. **Lehane MJ, Atkinson PW, O'Brochta DA.** 2000. Hermes-mediated genetic transformation of the stable fly, *Stomoxys calcitrans*. *Insect Mol Biol* **9:**531–538.

85. **Michel K, Stamenova A, Pinkerton AC, Franz G, Robinson AS, Gariou-Papalexiou A, Zacharopoulou A, O'Brochta DA, Atkinson PW.** 2001. *Hermes*-mediated germ-line transformation of the Mediterranean fruit fly, *Ceratitis capitata*. *Insect Mol Biol* **10:**155–162.

86. **O'Brochta DA, Warren WD, Saville KJ, Atkinson PW.** 1996. *Hermes*, a functional non-drosophilid gene vector from *Musca domestica*. *Genetics* **142:**907–914.

87. **Marcus JM, Ramos DM, Monteiro A.** 2004. Germline transformation of the butterfly *Bicyclus anynana. Proc Biol Sci* **271**(Suppl 5):S263–S265.

88. **Pinkerton AC, Michel K, O'Brochta DA, Atkinson PW.** 2000. Green fluorescent protein as a genetic marker in transgenic *Aedes aegypti. Insect Mol Biol* **9:**1–10.

89. **Jasinskiene N, Coates CJ, Benedict MQ, Cornel AJ, Salazar-Rafferty C, James AA, Collins FH.** 1998. Stable, transposon-mediated transformation of the yellow fever mosquito, *Aedes aegypti*, using the *Hermes* element from the house fly. *Proc Natl Acad Sci U S A* **95:** 3743–3747.

90. **Evertts AG, Plymire C, Craig NL, Levin HL.** 2007. The Hermes transposon of Musca domestica is an efficient tool for the mutagenesis of Schizosaccharomyces pombe. *Genetics* **177:**2519–2523.

91. **Park JM, Evertts AG, Levin HL.** 2009. The *Hermes* transposon of *Musca domestica* and its use as a mutagen of *Schizosaccharomyces pombe. Methods* **49:**243–247.

92. **Guo Y, Park JM, Cui B, Humes E, Gangadhrana S, Hung S, FitzGerald PC, Hoe KL, Grewak SI, Craig NL, Levin HL.** 2013. Integration profiling if gene function with dense maps of transposon integration. *Genetics* **195:**599–609.

93. **Urasaki A, Morvan G, Kawakami K.** 2006. Functional dissection of the Tol2 transposable element identified the minimal cis-sequence and a highly repetive sequence in the subterminal region essential for transposition. *Genetics* **174:**639–649.

94. **Coupland G, Plum C, Chatterjee S, Post A, Starlinger P.** 1989. Sequences near the termini are required for transposition of the maize transposon Ac in transgenic tobacco plants. *Proc Natl Acad Sci U S A* **86:**9385–9388.

95. **Kim YJ, Hice RH, O'Brochta DA, Atkinson PW.** 2011. DNA sequence requirements for *hobo* transposable element transposition in *Drosophila melanogaster. Genetica* **139:**985–987.

96. **Liu D, Mack A, Wang R, Galli M, Belk J, Ketpura NI, Crawford NM.** 2001. Functional dissection of the cis-acting sequences of the *Arabidopsis* transposable element *Tag1* reveals dissimilar subterminal sequence and minimal spacing requirements for transposition. *Genetics* **157:**817–830.

97. **Feldmar S, Kunze R.** 1991. The ORFa protein, the putative transposase of maize transposable element Ac, has a basic DNA binding domain. *EMBO J* **10:**4003–4010.

98. **Calvi BR, Hong TJ, Findley SD, Gelbart WM.** 1991. Evidence for a common evolutionary origin of inverted repeat transposons in Drosophila and plants: hobo, Activator, and Tam3. *Cell* **66:**465–471.

99. **Di Tommaso P, Moretti S, Xenarios I, Orobitg M, Montanyola A, Chang JM, Taly JF, Notredame C.** 2011. T-Coffee: a web server for the mutiple sequence alignment of protein and RNA sequences using structural information and homology extension. *Nucleic Acids Res* **39:**W13–W17.

100. **Davies DR, Goryshin IY, Reznikoff WS, Rayment I.** 2000. Three-dimensional structure of the Tn5 synaptic complex transposition intermediate. *Science* **289:**77–85.

101. **Aravind L.** 2000. The BED finger, a novel DNA-binding domain in chromatin-boundary-element-binding proteins and transposase. *Trends Bicohem Sci* **25:**421–423.

102. **Feldmar S, Kunze R.** 1991. The ORFa protein, the putative transposase of maize transposable element *Ac*, has a basic DNA binding domain. *EMBO J* **10:**4003–4010.

103. **Becker HA, Kunze R.** 1997. Maize Activator transposase has a bipartite DNA binding domain that recognizes subterminal sequences and the terminal inverted repeats. *Mol Gen Genet* **254:**219–230.

104. **Kunze R, Behrens U, Courage-Franzkowiak U, Feldmar S, Kuhn S, Lutticke R.** 1993. Dominant transposition-deficient mutants of maize Activator (Ac) transposase. *Proc Natl Acad Sci U S A* **90:**7094–7098.

105. **Sundaraajan P, Atkinson PW, O'Brochta DA.** 1999. Transposable element interactions in insects: Cross mobilization of *hobo* and *Hermes. Insect Mol Biol* **8:** 359–368.

106. **Essers L, Adophs RH, Kunze R.** 2000. A highly conserved domain of the maize activator tranposases in involved in dimerization. *Plant Cell* **12:**211–224.

107. **Michel K, O'Brochta DA, Atkinson PW.** 2003. The C-terminus of the Hermes transposase contains a protein multimerization domain. *Insect Biochem Mol Biol* **33:**959–970.

108. **Volff JN.** 2006. Turning junk into gold: domestication of transposable elements and the creation of new genes in eukaryotes. *Bioessays* **28:**913–922.

109. **Sinzelle L, Izsvak Z, Ivics Z.** 2009. Molecular domestication of transposable elements: From detrimental parasites to useful host genes. *Cell Mol Life Sci* **66:**1073–1093.

110. **Bundock P, Hooykaas P.** 2005. An *Arabidopsis hAT*-like transposase is essential for plant development. *Nature* **436:**282–284.

111. **Knip M, Hiemstra S, Sietsma A, Castelein M, de Pater S, Hooykaas P.** 2013. DAYSLEEPER: a nuclear and vesicular-localized protein that is expressed in proliferating tissues. *BMS Plant Biology* **13:**211.

112. **Muehlbauer GJ, Bhau BS, Syed NH, Cho S, Marshall D, Patetron S, Buisine N, Chalhoub B, Flavell AJ.** 2006. A hAT superfamily transposase recruited by the cereal grass genome. *Mol Gen Genomics* **275:**553–563.

113. **Chesney MA, Kidd ARI, Kimble J.** 2006. gon-14 functions with class B and class C synthetic multivulva genes to control larval growth in *Caenorhabdidtis elegans. Genetics* **172:**915–928.

114. **Hirose F, Ohshima N, Shiraki M, Inoue YH, Taguchi O, Nishi Y, NMatsukage A, Yamaguchi M.** 2001. Ectopic expression of DREF induces DNA synthesis, apoptosis, and unusual morphogenesis in the Drosophila eye imaginal disc: Possible interactions with Polycomb and trithorax group proteins. *Mol Cell Biol* **21:** 7231–7242.

115. **Gale MJ, Blakely CM, Hopkins DA, Melville MW, Wambach M, Romano PR, Katze MG.** 1998. Regulatio of interferon-induced protein kinase PKR: modulation of P58IPK inhibitory function by a novel protein, P52rIPK. *Mol Cell Biol* **18:**859–871.

116. **Smit AFA.** 1999. Interspersed repeats and other mementos of transposable elements in mammalian genomes. *Curr Opin Genet Dev* **9:**657–663.

117. **Tipney HJ, Hinsley TA, Brass A, Metcalf K, Donnai D, Tassabehji M.** 2004. Isolation and characerization of GTF2IRD2, a novel fusion gene and member of the TFII-I family of transcription factors, deleted in Williams-Beuren syndrome. *Eur J Hum Genet* **12:**551–560.

118. **Robertson HM.** 2002. Evolution of DNA transposons in eukaryotes, p 1093–1110. *In* Craig NL, Cragie R, Gellert M, Lambowitz A (ed), *Mobile DNA II*. ASM Press, Washington DC.

119. **Yamashita D, Sano Y, Adachi Y, Okamoto Y, Osadai H, Takahashi T, Yamaguchi T, Osumi T, Hirose F.** 2007. hDREF regulates cell proliferation and expression of ribosomal protein genes. *Mol Cell Biol* **27:**2003–2013.

120. **Matsukage A, Hirose F, Yoo MA, Yamaguchi M.** 2008. The DRE/DREF transcriptional regulatory system: a master key for cell proliferation. *Biochim Biophys Acta* **1779:**81–89.

121. **Chen T, Li M, Ding Y, Zhang LS, Xi Y, Pan WJ, Tao DL, Wang JY, Li L.** 2009. Identification of zinc-finger BED domain-containing 3 (Zbed3) as a novel axin-interacting protein that activates Wnt/beta-catenin signaling. *J Biol Chem* **284:**6683–6689.

122. **Hayward A, Ghazal A, Andersson G, Andersson L, Jern P.** 2013. ZBED evolution: repeated utilization of DNA transposons as regulators of diverse host functions. *PLoS ONE* **8:**e59940.

123. **Saghizadeh M, Gribanova Y, Akhmedov NB, Farber DB.** 2011. ZBED4, a cone and Muller cell protein in human retina, has a different cellular expression in mouse. *Mol Vis* **17:**2011–2018.

124. **Andersson L, Andersson G, Hjalm G, Jiang L, Lindblad-Toh K, Lindroth AM, Markljung E, Nystrom AM, Rubin CJ, Sundstrom E.** 2010. ZBED6. The birth of a new transcription factor in the common ancestor of placental mammals. *Transcription* **1:**144–148.

125. **Markljung E, Jiang L, Jaffe JD, Mikkelson TS, Wallerman O, Larhammar M, Zhang X, Wang L, Saenz-Vash V, Gnirke A, Lindroth AM, Barres R, J Y, Stromberg S, De S, Ponten F, Lander ES, Carr SA, Zierath JR, Kullander K, Wadelius C, Lindblad-Toh K, Andersson G, Hjalm G, Andersson L.** 2009. ZBED6, a novel transcription factor derived from a domesticated DNA transposon regulates IGF2 expression and muscle growth. *PLoS Biol* **7:**e100256.

126. **Knip M, de Pater S, Hooykaas PJ.** 2012. The *SLEEPER* genes: a transposon-derived angiosperm-specific gene family. *BMC Plant Biol* **12:**192.

Mobile DNA, 3rd Edition
Nancy L. Craig, Michael Chandler, Martin Gellert, Alan M. Lambowitz, Phoebe A. Rice and Suzanne Sandmeyer

doi:10.1128/microbiolspec.MDNA3-0032-2014

Damon Lisch[1]

36

Mutator and *MULE* Transposons

INTRODUCTION

A distinguishing feature of transposable elements (TEs) is their propensity to induce mutations. Among the most mutagenic of all TEs are the *Mutator* elements of maize. Lines carrying large numbers of these elements can exhibit mutation frequencies 50 to 100 times that of background (1, 2). This is due to a very high transposition frequency, which can exceed one new insertion per element per generation (3, 4), as well as a propensity to insert into or near genes (5). Because they are so mutagenic, *Mutator* elements have been very useful in both forward and reverse genetic screens in maize and recent high-throughput methodologies have only made the system more so (6). However, in addition to its utility, the *Mutator* system has also provided important clues as to the consequences of unrestrained TE activity, and the means by which active TEs are controlled by their host. This chapter will provide a review of the biology, regulation, evolution and uses of this remarkable transposon system, with an emphasis on recent developments in our understanding of the ways in which this TE system is recognized and epigenetically silenced.

CLASSES OF *MUTATOR* ELEMENTS IN MAIZE

Like many TE systems, the *Mutator* system in maize includes both autonomous and nonautonomous members. All *Mutator* transposons share similar, ~215 bp terminal inverted repeats (TIRs), but each class of these transposons contains distinct internal sequences (7, 8). *MuDR* elements are the autonomous class of the *Mutator* element (9, 10, 11). They encode the proteins necessary for transposition of both themselves and the more numerous nonautonomous elements. Nonautonomous *Mutator* elements may be deletion derivatives of *MuDR* transposons, or they may be elements that share the same TIRs as *MuDR* elements but that flank unique internal sequences (7).

Autonomous *MuDR* Elements

MuDR elements carry two genes, *mudrA*, which encodes the putative transposase, MURA, and *mudrB*, which encodes a second protein, MURB, of unknown function (7) (Fig. 1A). Interestingly, while genes homologous to *mudrA* are extremely widespread and apparently ancient, those homologous to *mudrB* are restricted to the genus *Zea* (8, 12). Transcription of these two genes is convergent and is initiated from within the nearly identical TIRs (TIRA and TIRB, respectively) (Fig. 1B) (13). The two convergent transcripts are terminated 200 bp from each other. Interestingly, each transcript contains several direct repeats upstream of the site of polyadenylation, and the region between the two transcripts would exhibit a high degree of secondary structure if it were expressed in either direction (13). Presumably, these structural features inhibit the production of excessive antisense transcripts. However, despite this,

[1]Purdue University, West Lafayette, IN.

antisense transcripts due to read-through transcription or from deletion derivatives of *MuDR* have often been detected and do not appear to have negative effects on *Mutator* activity (13, 14, 15, 16). The *mudrA* transcript includes three introns, one of which is in the 5′ UTR (Fig. 1A). If all three introns are spliced, the result is an 823-amino acid polypeptide with a predicted molecular weight of 94 kD. Comparative analysis of *mudrA*-homologs in other species suggests that this is the predominant functional protein (7). The *mudrB* gene also includes three introns, only two of which are efficiently spliced, and encodes a 207-amino acid, 23 kD polypeptide.

MURA carries several conserved motifs consistent with its putative function, including a MULE (*Mu*-like element) transposase domain containing a catalytic DDE motif (part of the RNAseH fold), a SWIM zinc finger, and at least three nuclear localization signals (7, 17, 18) (Fig. 1A). DNAseI protection experiments indicate that this protein binds to a site within *Mutator* TIRs near the end of the elements (Fig. 1B) (19). Beyond that, little is known about the biochemistry of MURA activity, due in no small part to the fact that it is refractory to cloning in *E. coli*. Further, although there are published reports of antibodies raised to MURA, these antibodies lack specificity (they recognize protein in maize plants that lack activity), so their utility is questionable (20). A more detailed understanding of MURA function and biochemistry will require a sustained effort.

The function of MURB remains unclear. However, it is known that transgenic constructs or deletion derivatives containing only *mudrA* can cause somatic excisions of *Mutator* elements, but not germinally transmitted insertions (15, 21). Thus, it would appear that MURB is required for integration of *Mutator* elements, but not their excision. *In situ* analysis suggests that both *mudrA* and *mudrB* are expressed at highest levels in actively dividing cells and in pollen, consistent with the ob-

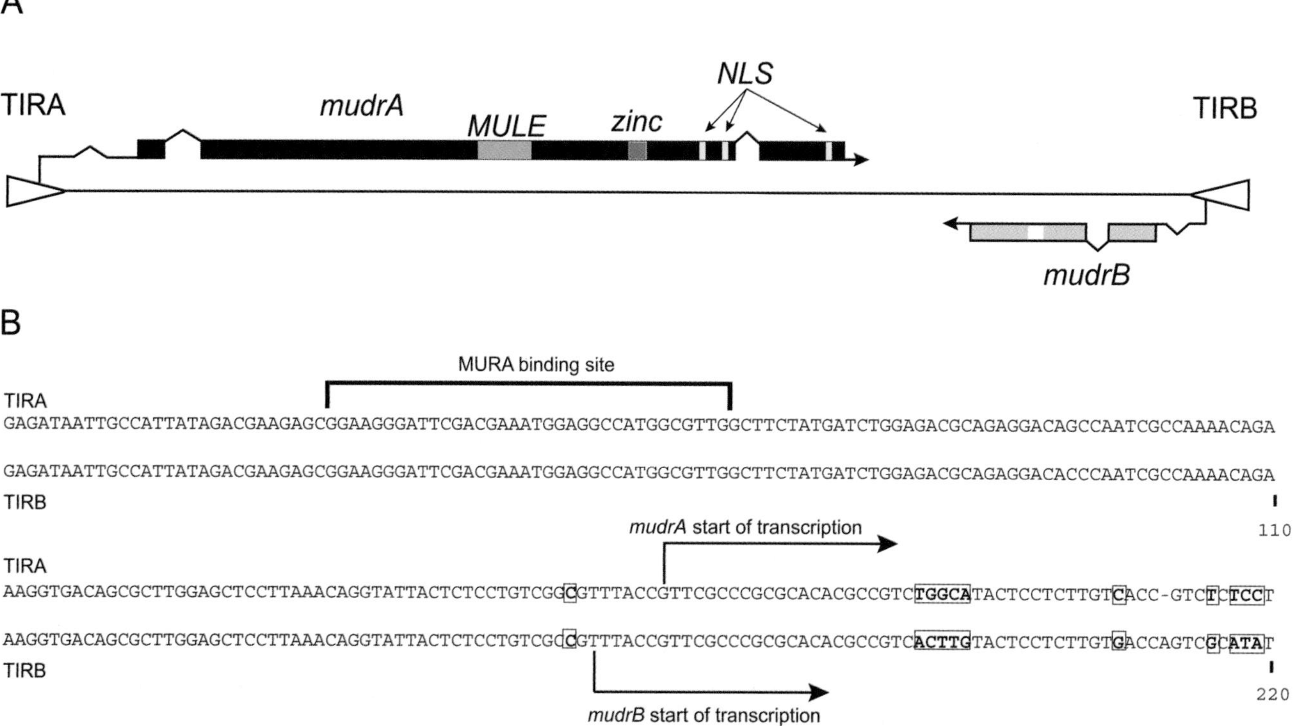

Figure 1 The *MuDR* transposon. (A) The structure of *MuDR*. Terminal inverted repeats (TIRs) at either end of the element are designated TIRA and TIRB. The two transcripts are indicated above and below. Exons are depicted as boxes. Introns are depicted as thin black lines. The third intron of *mudrB* is only infrequently spliced out. Domains identified within the MURA protein are as indicated. (B) The sequence of TIRA (top) and TIRB (bottom). The TIR sequences shows where the transposase binds and where *mudrA* and *mudrB* transcription are initiated are as shown. Note that TIRA and TIRB are identical for the first 158 nucleotides of each TIR. Differences between TIRA and TIRB are indicated by boxed residues. doi:10.1128/microbiolspec.MDNA3-0032-2014.f1

servation that the *MuDR* TIRs contain plant cell-cycle enhancer motifs and functionally defined pollen enhancers (16, 22, 23). Antibody analysis of MURB expression has shown that MURB protein is present at highest levels in actively dividing cells, particularly those in floral meristems, an observation that is consistent with *in situ* analysis of RNA expression (16, 23). Interestingly, MURB is undetectable in prepollen mother cells, suggesting that MURB expression is tightly regulated just prior to meiosis. In contrast, both *mudrB* RNA and MURB are readily detectable in pollen.

MuDR Deletion Derivatives

Like many other Class II (DNA) elements, *MuDR* elements are prone to produce deletion derivatives, sometimes at very high rates, and multiple copies of these derivatives have often been observed, many of which retain the promoter elements within the TIRs (15, 17). In most cases the derivatives contain deletions of *MuDR* sequences that include one copy of short repeat sequences as well as the region between those repeats (24). These deletions can occur during somatic development, resulting in clonal sectors of lost activity, or they can occur in the germ lineage, resulting in heritable transmission of these derivatives. A detailed analysis of one family segregating for a single *MuDR* element revealed that roughly one-third of the progeny carried independent deletion derivatives at the original position of that element (25). Some of the deletions were of particular interest. *MuDR-d107* has a deletion within *mudrA*. It produces a normal *mudrB* transcript and an internally deleted *mudrA* transcript. Although it can transpose in the presence of a fully functional *MuDR* element, it does not produce a functional transposase itself. These observations indicate that MURB is, by itself insufficient to cause any aspect of *Mutator* activity. In contrast, a deletion derivative that expresses only *mudrA* is competent to cause somatic excision of a reporter element, but is not sufficient to catalyze germinal insertions (15).

Finally, there are deletion derivatives that permit transcriptional read through from one *MuDR* gene into the other. *MuDR-d202*, for instance, has a deletion in the 3′ end of the *mudrA* gene. In addition to a slightly truncated *mudrA* transcript and a normal *mudrB* transcript, this deletion derivative produces a significant amount of transcript that initiates from *mudrB* and extends into *mudrA*, resulting in *mudrA* antisense transcript. The presence of this antisense *mudrA* transcript has, however, no obvious effect on activity. Similarly, there is evidence that full length *MuDR* elements, due to the convergent transcription of *mudrA* and *mudrB*, produce detectable amounts of antisense transcript that do not affect levels of activity, nor do constructs specifically designed to transcribe only antisense *mudrA* transcript (14, 16). Thus, it would appear that *MuDR* elements are refractive to epigenetic silencing via antisense transcript.

In addition to active *MuDR* elements, the maize genome is host to a number of *MuDR* homologs called *hMuDR* elements. These elements are very similar to *MuDR* elements, but do not produce functional protein, nor do they have any direct effect on *Mutator* activity (20). They do produce transcripts, but only at low levels, and most of that transcript is restricted to the nucleus (26). Given that these elements are also associated with 24 to 26nt small RNAs, it is likely that they are deeply silenced and refractive to reactivation by *MuDR* elements (27).

Nonautonomous *Mutator* Elements

To date, 16 classes of nonautonomous *Mutator* elements have been identified in *Mutator* active lines. There are currently 13 classes of these elements that have been fully sequenced in maize (Fig. 2). Those that are known to be transpositionally active are *Mu1*(or *Mu1.4*) (28, 29, 30), *Mu1.7* (31), *Mu3* (32), Mu4 (33, 34), *Mu7/rcy* (35) *Mu8* (36), and *Mu13* (37). Each of these has been fully sequenced. There are also at least three other active elements that have only been partially sequenced: *Mu10*, *Mu11*, and *Mu12* (34). *Mu14*, *Mu15*, *Mu16*, *Mu17*, *Mu18*, and *Mu19* are elements whose capacity to transpose has not been established, but whose structure and sequence suggest that they can (37). Although additional elements may be identified in the future, given that all *Mu*-active lines originated from a single line (2), and given the extensive sequencing that has been done in multiple derivatives of this line, it is likely that most or all active *Mutator* elements have now been identified. With the exception of deletion derivatives of *MuDR*, all nonautonomous *Mutator* elements in maize carry captured fragments of host genes, and are referred to generically as "*Pack-MULEs*" (7, 33, 38) (Fig. 2). As the captured sequences are very similar to the cognate host genes, it is likely that these sequences were captured, or transduplicated, relatively recently (8).

MUTATOR LINES

Standard *Mutator* Lines

For many years, *Mutator* activity was defined by the propensity for *Mutator* lines to generate large numbers of new mutations (39, 40). These lines are referred to

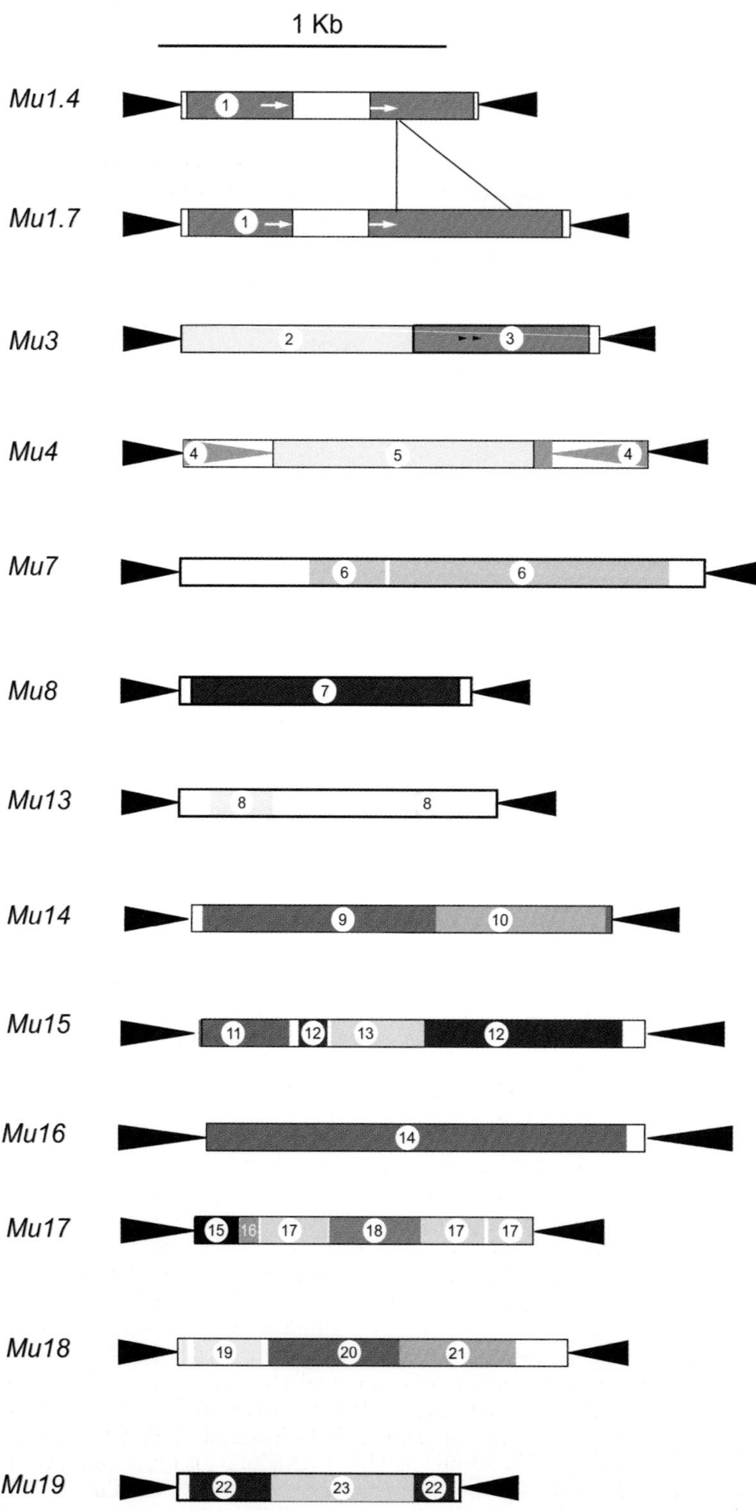
1 Kb
Mu1.4
1
Mu1.7
1
Mu3
2
3
Mu4
4
5
4
Mu7
6
6
Mu8
7
Mu13
8
8
Mu14
9
10
Mu15
11
12
13
12
Mu16
14
Mu17
15
16
17
18
17
17
Mu18
19
20
21
Mu19
22
23
22

as "Standard *Mutator* lines," or "Tagging lines." This effect is due to the presence of large numbers of actively transposing *Mutator* elements. Lines with a lower copy number of elements do not exhibit this form of *Mutator* activity even if they do carry actively transposing *Mutator* elements. A second way to monitor activity is to observe somatic excisions of a reporter element in, for instance, a color gene. If the color gene is expressed in the kernel, activity can be assessed by looking for the characteristic spots of revertant tissue (Fig. 3A). This assay does not depend on the copy number of *Mutator* elements, and is a reliable indicator of the presence of MURA transposase (albeit not of mutagenic potential) (9, 10).

Standard *Mutator* lines do not exhibit Mendelian inheritance of activity. Instead, inheritance is determined by variation in *MuDR* copy number as well as spontaneous silencing. Mutagenic potential is regularly lost in about 10% of the progeny of a standard *Mutator* line upon out crossing or self-fertilization. Interestingly, the frequency of lost activity in these lines increases following several rounds of self-fertilization (40). Spontaneous silencing is associated with methylation of both nonautonomous and autonomous *Mutator* elements, and it is largely irreversible (41, 42, 43).

In addition to the production of mutants, many standard *Mutator* lines exhibit a distinctive set of phenotypes referred to as "*Mutator* syndrome." These phenotypes include stunted growth, necrotic, often torn leaves low pollen production and poor seed set (2, 44). The fact that lines that share few of the same insertions exhibit similar phenotypes suggests that *Mutator* syndrome represents a generic effect of activity rather than the effects of specific insertions. Further, a comparison of RNA and protein in plants with *Mutator* activity and related individuals that had lost activity revealed dramatic changes in gene expression in response to *Mutator* activity (45). Interestingly, many of the differentially expressed genes encode proteins involved in stress response, suggesting that *Mutator* activity represents a chronic stressor.

Minimal *Mutator* Lines

As useful and interesting as the complex phenomenology of standard *Mutator* lines is, the lack of Mendelian genetics made analysis of regulation of the system difficult. This changed with the identification of lines that segregated for a single *MuDR* element. The lines, which are referred to as minimal *Mutator* lines, were derived by recurrent crosses to a line that lacked *Mutator* elements or any factors that inhibited *Mutator* activity. One such line contained a factor called *Cy*, and a reporter element in the *bz1* gene called *rcy* (46, 47). Subsequent molecular analysis revealed that Cy is a *MuDR* element, and *rcy* is a *Mu7* element (24, 47). This was the first published evidence that the *Mutator* system could be regulated by a single locus, in a fashion similar to that observed for systems regulated by *Ac* or

Figure 2 Structural features of nonautonomous *Mutator* elements in maize. Black triangles represent terminal inverted repeat (TIR) sequences. Shaded boxes represent captured host sequences, with each independent sequence indicated by a number. The cognate host genes are indicated here, along with the percent identity between the captured sequence and the host gene. *Mu1* and *Mu1.7*: (1) GRMZM2G117007, 94% identical (unknown function). *Mu3*: (2) GRMZM2G015352 (unknown function), 97% identical. (3) GRMZM2G542994 (putative mago nashi, protein), 94% identical. *Mu4*: (4) GRMZM2G177883 (putative receptor-like protein kinase 5 precursor), 97% identical. (5) GRMZM2G037164 (unknown function), 96% identical. *Mu7*: (6) GRMZM2G022945 (BRCA1 C Terminus domain containing protein), 98% identical. *Mu8*: (7) GRMZM2G315375 (P-glycoprotein 1), 99% identical. *Mu13*: (8) GRMZM2G317614 (putative nucleotide binding protein), 96% identical. *Mu14*: (9) GRMZM2G010000 (putative heat shock protein binding protein), 90% identical. (10) GRMZM2G120085 (subtilisin-like protease precursor), 94% identical. *Mu15*: (11) GRMZM2G181219 (unknown function), 96% identical. (12) AC196090.3 (putative xylem serine proteinase 1 precursor), 95% identical. (13) AC234154.1 (putative phospholipase A1), 95% identical. *Mu16*: (14) GRMZM2G001934 (putative receptor protein kinase TMK1 precursor), 95% identical. *Mu17*: (15) GRMZM2G029979 (TGACG-sequence-specific DNA-binding protein TGA-2), 97% identical (16) GRMZM2G331374 (unknown function), 100% identical (17) GRMZM2G148831 (unknown function), 97% identical. (18) GRMZM2G055809 (unknown function), 95% identical. (19) GRMZM2G126413 (VQ motif family protein), 87% identical. (20) GRMZM2G081406 (putative auxin response factor), 96% identical. (21) GRMZM2G152432 (putative calmodulin), 97% identical. (22) GRMZM2G106401 (putative xylem serine proteinase 1 precursor), 96% identical. (23) GRMZM2G116908 (unknown fuction), 96% identical.
doi:10.1128/microbiolspec.MDNA3-0032-2014.f2

Figure 3 Examples of somatic excision of *Mu1* from *a1-mum2* in the seed (A), the sheath (B) and the anthers (C). Note that in each case, reversion events are uniformly late. doi:10.1128/microbiolspec.MDNA3-0032-2014.f3

Spm. Analysis of *Cy* revealed that it transposed at a relatively low frequency, often to unlinked sites. A similar minimal line was independently derived by two groups (9, 48). In this case, a standard line carrying a *Mu1* element in the promoter of the *A1* gene (the *a1-mum2* allele) was crossed repeatedly to a line carrying a reference mutant allele of *A1* (*a1-rDt;sh2*), which lacked *Mutator* elements. Eventually, ears were identified that segregated for a single regulatory locus on chromosome 2L, later designated *MuDR(p1)*, for *MuDR* at position one. Repeated subsequent crosses to the *a1sh2* line, continually selecting for ears that segregated for a single regulatory locus gave rise to a line that carried a single *MuDR* element, and a single *Mu1* element at the reporter. Subsequent genetic and molecular analysis revealed that this was possible because *MuDR(p1)*, and the nonautonomous *Mutator* elements under its control, exhibits a relatively low germinal duplication frequency (48, 49). Thus, each back cross to *a1sh2* resulted in a reduced copy number of the nonautonomous elements.

MECHANISM OF *MUTATOR* ELEMENT TRANSPOSITION

There is very little direct biochemical evidence for the mechanism of *Mutator* element transposition. However, given the presence of a DDE motif, it is likely that excision and integration are similar that of a broad array of other members of this superfamily of Class II transposons, including the bacterial IS256 element (50, 51, 52). Significant advances in understanding the biochemistry of *Mutator* element transposition will require an amenable heterologous system. There are, however, some clues as to the mechanism of *Mutator* element transposition in maize based on the consequences of *Mutator* element activity in different tissues at different stages of development.

Possible Transposition Intermediates

IS256 employs a closed circular intermediate (53, 54) (Siguier et al., this volume). Interestingly, extrachromosomal *Mu1* circles have also been detected in maize (55). The presence of these circles is dependent on the presence of active *MuDR* elements, suggesting the MURA can catalyze their production. However, the junctions of these circles has not been determined, and it is not known if they are true transposition intermediates. Interestingly, a distantly related MULE, Transposon Ellen Dempsey (TED) (discussed below) also produces TE circles in the presence of the transposase (56). As with *Mutator* circles, the circles associated with TED are covalently closed and are more complex than simple end-to-end fusions. Those that were complete carried filler sequences not present at the donor site. Those that were truncated had deletions within the element, or of flanking sequences, or of either or both of the ends at the junctions. Interestingly, the internal deletions were flanked by imperfect direct repeats, similar to those observed within *MuDR* elements. As with

Mu, it remains unknown whether or not the circles represent true transposition intermediates.

Timing of Transposition

Mutator element transposition is tightly regulated, and the consequences of transposition vary dramatically depending on the cell lineage. In somatic tissue such as the aleurone, the vast majority of reversions occur in the last few cell divisions (57, 58) (Fig. 3). Given that a similar pattern of excision is observed when expression of *mudrA* is driven by a constitutive 35S cauliflower mosaic virus promoter, this pattern is unlikely to due to changes in expression of the native *MuDR* promoter during development (59). Direct cytological analysis of a modified transgenic variant of *Mu1* called *Rescue-Mutator* revealed that new insertions, as well as excisions, occur in somatic tissues. Importantly, each cell that carried a new insertion also lost the element at its original position, consistent with a simple cut-and-paste transposition reaction in these cells (21, 60). Further, analysis of large numbers of empty insertion sites in somatic tissues have revealed the presence of "footprints" consistent with repair of double-stranded gaps via nonhomologous end joining (NHEJ) (61, 62).

Germinally transmitted reversion events are exceedingly rare (39). In contrast, germinally transmitted insertions are frequent and often occur very late during development of the sprorophyte germinal lineage, or in the gametophyte (17, 63). Unlike insertions in somatic tissue, these insertions are not associated with a loss of the original element and are generally to unlinked sites (21, 64). Thus, the consequences of *Mutator* element activity vary dramatically depending on the tissue in which that activity occurs.

One way to explain these observations is to assume that transposition of *Mutator* elements always involves excision, but the repair of the resulting double-stranded gap varies depending on time and tissue (23, 24). If excision in the germinal lineage occurs exclusively in S1 subsequent to replication of the inserted element, then the sister chromatid would be available to mediate repair. The result would be an apparent replication of the element, as excised elements would be restored following excision (Fig. 4). In contrast, late during somatic development, excision events may occur prior to S1, or may be repaired primarily using nonhomologous end joining, a conclusion supported by analysis of footprint sequences in somatic tissues.

There are several lines of evidence supporting this model. Deletions within *MuDR* elements are consistent with strand slippage during template mediated repair, which results in a deleted region flanked by short repeats (24). While deletion events often occur late in the germinal lineage, they can also occur in earlier during development, resulting in clonal sectors of various sizes in which only a deletion derivative of a full length element is present (8, 49). In combination with evidence for clonal sectors in which elements have duplicated early during development, this observation suggests that excisions are happening prior to the last few cell divisions, but that those excisions are repaired successfully, resulting in duplication events, or unsuccessfully, resulting in deletion derivatives (9). Also consistent with gap repair associated with slippage, are insertions of filler DNA from nearby sequences into sites of excision in relatively rare early excision sectors (62). As with the deletion derivatives, these events appear to be mediated by short stretches of direct repeats and are presumably the result of strand slippage during gap repair. Finally, analysis of RAD51 mutants in maize strongly support a role for double-stranded gap repair in germinally transmitted *Mutator* transposition. RAD51 plays a central role in the Homologous Recombination pathway in maize (159, 160). In *Mutator* active plants that were mutant for RAD51, germinally transmitted deletions within *MuDR* elements and of sequences flanking *MuDR* element were 40 times more frequent than in wild type siblings. This suggests that germinally transmitted *Mutator* element replication requires homologous recombination mediated by RAD51.

Insertion Preferences

Like many TEs, *Mutator* elements do not insert randomly within the genome. Several high throughput experiments have yielded tens of thousands of novel insertions (5, 65, 66). Although a large majority of the maize genome is composed of retrotransposons, the vast majority of insertions are into low copy regions of the genome, in or near genes. The most comprehensive survey of 40,000 nonredundant insertions revealed a strong preference for the 5′ ends of genes and with recombinationally active regions (67). More specifically, the insertions showed a strong correlation with chromatin marks associated with open chromatin configuration and with single copy sequences. Thus, H3K9ac or H3K4me3 modifications were most associated with elevated frequencies of *Mutator* insertions and H3K27me3 and DNA methylation were least associated. Similar preferences for the 5′ ends of genes are also observed with other class II elements in both maize and other species, suggesting that preference for open chromatin around the transcriptional start site of genes is a common strategy for optimizing TE amplification. Overall, the target site duplications also showed an in-

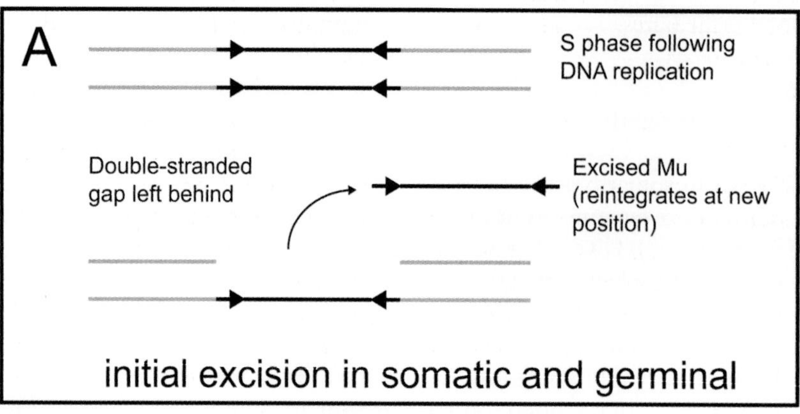
A
S phase following DNA replication
Double-stranded gap left behind
Excised Mu (reintegrates at new position)
initial excision in somatic and germinal

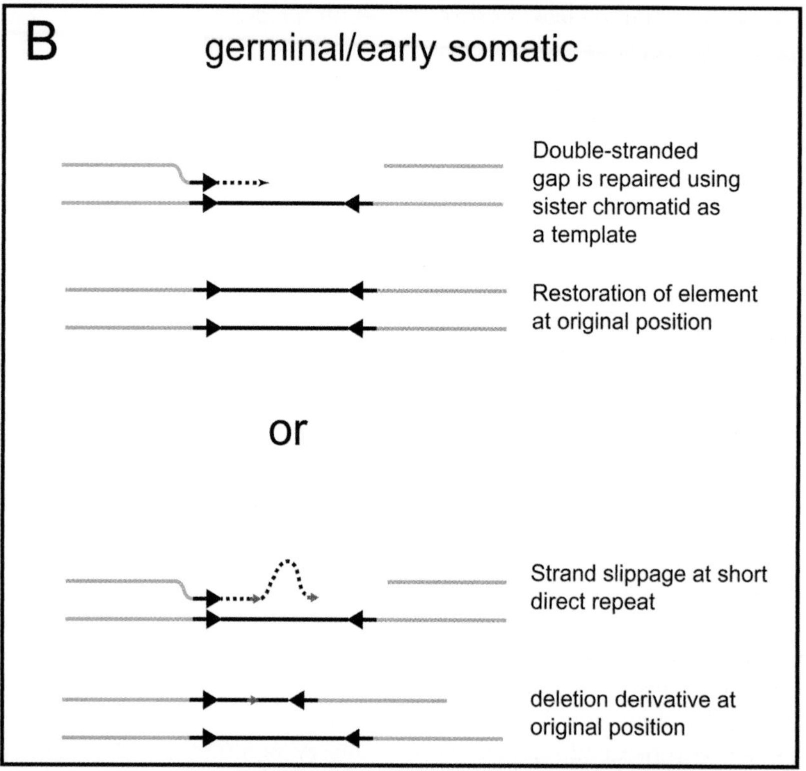
B
germinal/early somatic
Double-stranded gap is repaired using sister chromatid as a template
Restoration of element at original position
or
Strand slippage at short direct repeat
deletion derivative at original position

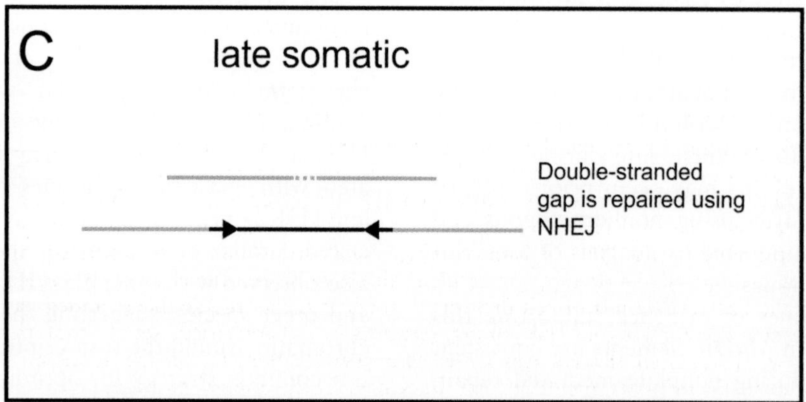
C
late somatic
Double-stranded gap is repaired using NHEJ

creased GC richness, coincident with the documented GC richness known to be found in the 5′ ends of genes in maize and other monocots (68).

SUPPRESSIBLE ALLELES

Suppressible insertion alleles are those in which the mutant phenotype is dependent on the presence or absence of the transposase (69). A wide variety of suppressible *Mutator* insertions have been identified, examples of which include *Les28* (42), *hcf106::Mu1* (70), *a1-mum2* (9), and mutant alleles of *rs1*, *lg3* (71), *kn1* (72, 73) and *rf2a* (74). The specific nature of the suppression effect depends the nature of the insertion, but in each case, it would appear that that the combination of the insertion and the transposase are required to alter gene expression. Thus, insertions that result in dominant ectopic expression, such as those in *Les28* or *Lg-Or422* only exhibit that ectopic expression in the presence of *Mutator* activity. Similarly, recessive insertion alleles such *hcf106::Mu1* or *a1-mum2* only express functional transcript in the absence of activity. The most parsimonious explanation for these observations is that the transposase binds to the ends of the inserted *Mutator* elements and blocks normal function, resulting in a mutant phenotype. Given that several of the suppressible mutations characterized to date have insertions in regulatory regions, and *Mutator* elements tend to insert into the 5′ region of genes, where regulatory motifs are often found it is likely that suppressible mutations are common in *Mutator* active lines. Because *Mutator* activity is often lost during introgression into various inbred backgrounds, many of these suppressible mutations have probably not been characterized, as the phenotype is lost when activity is lost. In most cases, given their instability, such mutations are not particularly useful. However, suppressible mutants can be useful in some cases. For instance, lethal suppressible mutations can be kept as homozygotes, and the phenotype can be uncovered by crossing activity back in. Further, because activity is often lost in clonal sectors, suppressible mutations can be used for somatic sector analysis (42, 75, 76).

The existence of suppressible mutations raises some interesting questions concerning the evolution of regulation of gene expression, as they suggest that regulation of gene expression can be determined by the presence of the transposase. Indeed, it was this effect of TEs on genes that Barbara McClintock found more intriguing than TE transposition (77). Although most suppressible mutations are unlikely to be selectively beneficial there are in fact many instances in which homologs of *mudrA* have in fact been integrated into regulatory networks (discussed below). In some of these cases, the initial step in that process may have been suppressible alleles that provided a selective advantage to the host.

POSITION EFFECTS

Because *MuDR(p1)* regularly transposed to new positions in the genome, it was possible to compare various positions with respect to various aspects of *Mutator* activity. The first effect observed was that when *MuDR (p1)* transposed to a new position, its transposition frequency doubled. This suggests that the low frequency of transposition of *MuDR* at position 1 was due to a *cis* effects. Subsequent analysis of the sequences flanking *MuDR(p1)* provides some clues as to why that might be. As discussed above, *Mutator* elements tend to transpose into single copy sequences. Often those are genes, but the maize genome, like that of many species, contains vast numbers of gene fragments that have been transduplicated by TEs (78). In the case of *MuDR(p1)*, *MuDR* is inserted next to a gene fragment that had been captured by a helitron and is flanked on the other side by a heavily methylated helitron terminus (Lisch, unpublished data). This may be responsible for the low level of transposition *MuDR* at this position. This also suggests a possible function for transduplicated sequences.

Figure 4 A model explaining the differences between late somatic and germinal *Mutator* element transposition. (A) In all tissues, *Mutator* element excision produces a double-stranded gap. What is hypothesized to vary is how that gap is repaired. (B) In germinal (and early somatic) lineages the gap is repaired using the sister chromatid, which requires that excision occurs primarily after DNA synthesis. Occasional strand slippage, mediated by short stretches of sequence homology, can result in deletions within the element. (C) In contrast, during the last few rounds of cell division in somatic tissue, the double-stranded gap is repaired using nonhomologous end joining, resulting in a characteristic set of "footprints." In each case, the excised element can insert at a new location, but in the germinal lineage, sister-chromatid-mediated repair restores an element at the original position.
doi:10.1128/microbiolspec.MDNA3-0032-2014.f4

Because they carry genic sequences, they may attract Class II elements like *Mutator* but the insertions do not have a fitness cost, as most transduplicated sequences are unlikely to be functional (79). If the resulting position effects also reduce (or even eliminate) subsequent transposition, then these tens of thousands of transduplicated genes may act as "traps" for TEs by soaking up and immobilizing potentially damaging TEs.

Although it has a reduced transition frequency, *MuDR(p1)* causes a typical frequency of somatic excision of the nonautonomous element at a1-mum2 (Fig. 3). A second, transposed copy of *MuDR(p1)*, *MuDR (p3)*, exhibits a dramatically lower excision frequency (64) (Fig. 5A). As with the reduced transposition of *MuDR(p1)*, this effect is due entirely to position; when this element transposes to a new position, the excision frequency of *Mu1* is restored to a more typical level (Fig. 5C). When this happens during development of the ear, the result is an ear sector in which heavily spotted kernels segregate (Fig. 5B). In the subsequent generation, *MuDR(p3)* and the newly transposed copy can be segregated away from each other and each retains its characteristic effect on *Mu1* excision (64).

EPIGENETIC SILENCING

A characteristic feature of standard *Mutator* lines is their propensity to undergo spontaneous silencing. This process results in inactivation of large numbers of *Mutator* elements simultaneously, and remains poorly understood. Spontaneous silencing involves the acquisition of cytosine methylation in *Mutator* elements of all classes, including *MuDR*, and the loss of *mudrA* and *mudrB* transcript (17, 80). Once it occurs, it is heritable and very stable. However, it does not appear to be a consequence of the production of a heritable factor that can go on to silence newly introduced *MuDR* elements. Thus, when activity is reintroduced to a line that had been silenced, subsequent rates of silencing are not increased (81). Indeed, when *MuDR(p1)* was introduced to silenced lines, activity segregated as a single Mendelian locus (64).

Mu Killer

A distinguishing feature of minimal *Mutator* lines is an absence of epigenetic silencing. As a rule, activity is lost in these lines due to simple genetic segregation, position effects, or deletions within *MuDR* elements (25). There is, however, a notable exception to this rule. In one family originally segregating for *MuDR(p1)*, there arose a single dominant locus that can heritably silence one or more *MuDR* elements. This locus, designated *Mu killer* (*Muk*) is a transposed copy of *MuDR(p1)* that underwent rearrangement, such that one end of the element (including TIRA and a portion of *mudrA*) was duplicated and inverted. In addition, portions of three genes were deleted (82). (Fig. 6A). This rearrangement resulted in a long (2.2 kb) inverted repeat sequence, flanked by the 5′ portions of two genes.

Although the normal promoter for *mudrA* is present and intact in *Muk*, it does not appear to be functional, perhaps as a consequence of epigenetic silencing. Instead, the dominant promoter for *Muk* expression is located in the 5′ adjacent gene, designated *Accomplice1* (GRMZM2G175065). The *Muk* transcript is initiated from within this promoter and it extends through the truncated *Accomplice1* gene, through the long inverted repeat of the *MuDR* sequence, and then is terminated and polyadenylated in sequences found upstream of the 3′ truncated gene (GRMZM2G175218). RNAase protection experiments reveal that the long hairpin transcript produced by *Muk* is double-stranded, as expected, and small RNA blots show that it is the source of small, mostly 22 nt siRNAs.

When plants carrying *MuDR* are crossed to those carrying *Muk* there is an immediate, uniform effect on somatic excision of the reporter when *Muk* is introduced via the female (Fig. 6C). In contrast, *Muk* has no effect on somatic excision of the reporter in the aleurone of the seed when it is introduced through the male. Regardless of the direction of cross, however, *MuDR* is effectively silenced early during development of F1 plants (83). If plants carrying both *MuDR* and *Muk* are crossed to a tester that carries only the reporter element, progeny seeds carrying only *MuDR* lack somatic excisions, and plants grown from those seeds lack transcripts from either *mudrA* or *mudrB*. Once silenced, these elements remain inactive for multiple generations (83).

Silencing of *mudrA* by *Muk* is associated with increased methylation in the TIR associated with *mudrA*, as well as enrichment in H3K27m2 and H3K9me2, two chromatin marks often associated with epigenetic silencing (84). All of these marks are stable over at least six generations propagation in the absence of *Muk*. Essentially, after a single generation of exposure to *Muk*, *MuDR* becomes indistinguishable from any other silenced maize TE.

Like *mudrA*, *mudrB* is heritably silenced by *Muk*. However, the developmental tragectory of that process is quite distinct. It does not appear to involve the production of small RNAs or homology to the *Muk* transcript, and only occurs when *mudrB* is in *cis* to mudA; deletion derivatives that express only *mudrB* are unaffected by

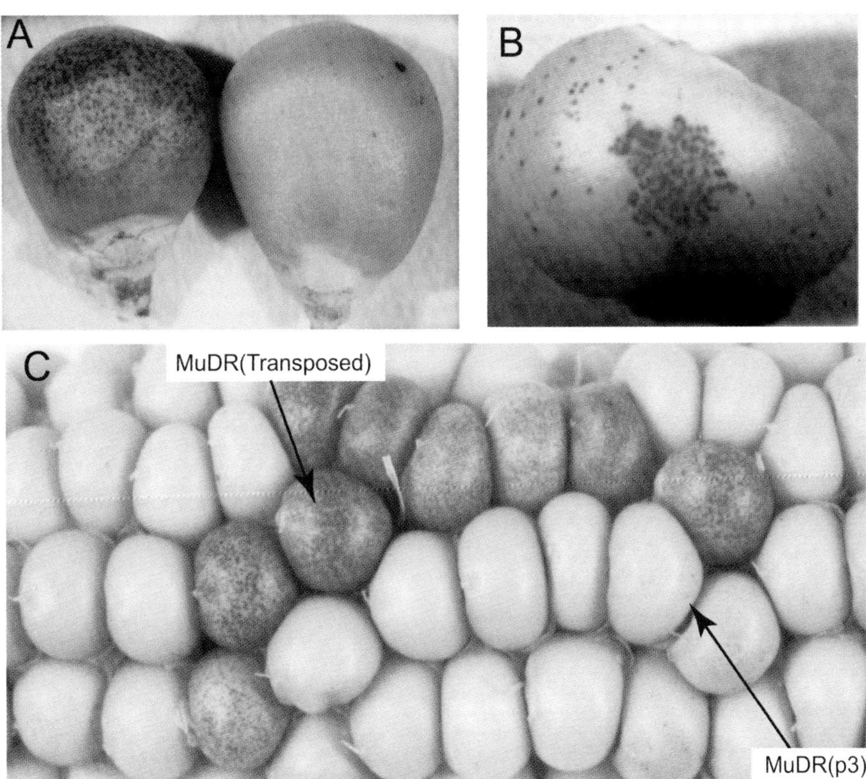

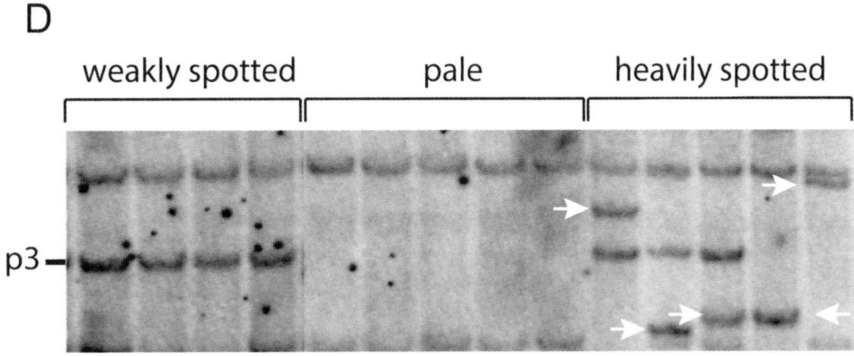

Figure 5 A striking example of a position effect on the capacity for a *MuDR* element to cause somatic excision of a reporter element. (A) Somatic excisions caused by *MuDR(p1)* (left) and *MuDR(p3)* (right). (B) An example of likely transposition of *MuDR(p3)* during somatic development, resulting in a kernel sector with an increased level of somatic excision. (C) A rare ear sector in which *MuDR(p3)* has undergone a duplication during development, resulting in an ear sector in which *MuDR(p3)* and transposed copies of *MuDR(p3)* segregate. (D) Southern blot analysis of more typical, single kernel *MuDR(p3)* duplication events. Weakly spotted, pale and heavily spotted kernels were picked from a single ear and their DNA was examined for evidence of *MuDR* transposition. Analysis of the weakly spotted and pale kernels show that segregation of *MuDR(p3)* correlates with the weak spotting phenotype. Analysis of the heavily spotted kernels shows that in each case, a new fragment, consistent with a transposition event, appeared. Note that in some cases both *MuDR(p3)* and a transposed copy were present, while in others, only the transposed copy is available, suggesting that these transposition events occurred prior to meiosis.
doi:10.1128/microbiolspec.MDNA3-0032-2014.f5

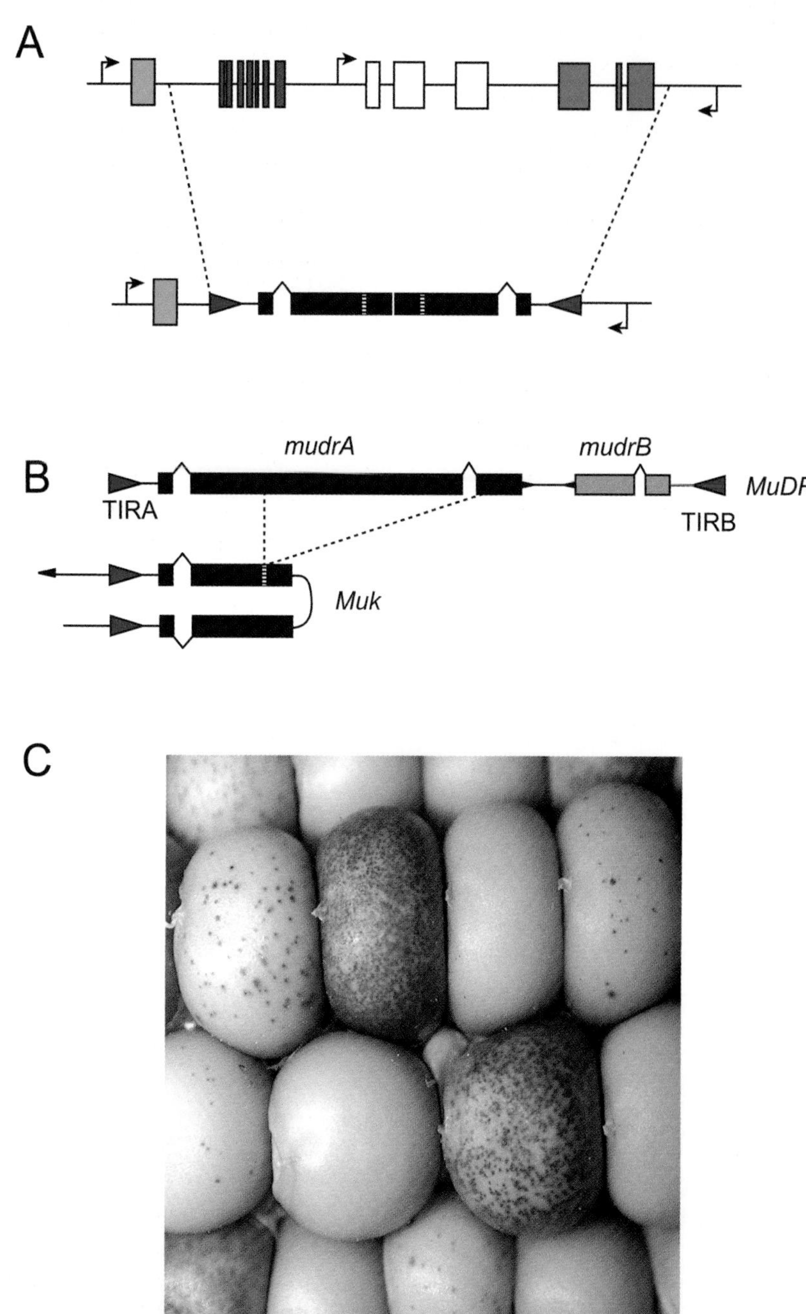

Figure 6 *Mu killer*. (A) The *Mu killer* (*Muk*) locus. *Mu killer* is a rearranged *MuDR* element, derived from the single *MuDR* present in the minimal line [*MuDR(p1)*]. The rearrangement that gave rise to *Muk* also caused the complet deletion of one gene and the deletion of portions of two other genes, as well as a duplication and inversion of a portion of the 5′ end of the *MuDR* element. Triangles represent the TIR. Boxes represent coding sequences. Transcriptional start sites for genes are as indicated. (B) The structure of the hairpin transcript derived from *Muk* relative to a *MuDR* element. (C) The effect of *Muk* on *MuDR* in the first generation after a plant carrying *Muk* is crossed to a plant carrying *MuDR*. Heavily spotted kernels are those that inherit only *MuDR*. Weakly spotted kernels are those that carry both *MuDR* and *Muk*. Pale kernels lack *MuDR*.
doi:10.1128/microbiolspec.MDNA3-0032-2014.f6

Muk, even in plants that carry both *Muk* and an intact *MuDR* in *trans* to the deletion derivative. In F1 plants carrying both *MuDR* and *Muk*, *mudrA* is heavily methylated and transcriptionally silenced by the immature ear stage (84). In contrast, *mudrB* continues to express normal levels of normally sized transcript in this organ. However, this transcript does not appear to be polyadenylated. By the next generation, it is entirely absent.

Developmental Progression of *Muk*-Induced Silencing

Although silencing of *MuDR* by *Muk* is highly efficient over the course of one generation, the process by which this occurs is unexpectedly complex. In part, this is due to features unique to the *Muk* locus, and in part, due to changes in epigenetic regulatory pathways during plant development.

Maize, like most plants, undergoes a transition from juvenile to adult growth. Maize produces a series of leaves sequentially from its vegetative meristem. The first few leaves produced in the seedling are called juvenile leaves. Subsequent leaves are called adult leaves. Juvenile and adult maize leaves are distinguished by a number of characteristics, including waxes, hairs and cell morphology (85) (Fig. 7A). Transition leaves are those that have sectors of both juvenile and adult morphology. It is only following the transition to adult growth that maize is reproductively competent.

In juvenile leaves of plants carrying both *MuDR* and *Muk*, TIRA is heavily methylated by the forth leaf (the last juvenile leaf). Remarkably, all methylation is absent in the fifth and sixth leaves (the transition leaves) (84). Methylation is then restored at a lower level in all subsequent leaves, as well as in the shoot apical meristem during adult development, and in young developing tassels and ears. The loss of methylation in the transition leaves is associated with a loss of repressive chromatin marks and a restoration of *mudrA* expression. Importantly, however, methylation of nonautonomous elements is not lost. This indicates that functional MURA (whose presence invariably results in hypomethylation of *Mutator* TIRs) is not present, and that the machinery necessary for maintaining *Mutator* TIR methylation is present in these transition leaves.

The loss of silencing of TIRA in transition leaves is associated with a transient loss of *Leafbladeless1* (*Lbl1*) transcript in those leaves. Analysis of *lbl1* mutants confirms that *Lbl1* is required for methylation of TIRA by *Muk* in juvenile leaves. Interestingly, a mutation that converts adult leaves to juvenile leaves, *Corngrass1* (*Cg1*) also shifts both the pattern of methylation of TIRA and expression of *lbl1*, suggesting that *Cg1* is epistatic to *lbl1*. *Lbl1* is a homolog of the Arabidopsis gene, *SUPRESSOR OF GENE SILENCING3 (SGS3)* (86). In arabidopsis, this gene works cooperatively with *RNA-DEPENDENT RNA POLYMERASE6* to produce double stranded RNA from a variety of templates, including transcript from sense-antisense gene pairs, viral RNA and transcripts involved in the tasiRNA pathway (87, 88). One target of that pathway is *AUXIN RESPONSE FACTOR3 (ARF3)*. In that case, a microRNA, mir390 is used to trigger the production of double-stranded RNAs, which are cleaved into phased tasiRNAs, some of which target *ARF3* (89, 90). In maize, *ARF3* expression is increased in transition leaves in which *lbl1* transcript is reduced, consistent with the known relationship between these genes (84). Similarly, a number of sense-antisense gene pairs in maize are also upregulated in both *lbl1* mutants and in transition leaves (Lisch, unpublished). These surprising findings suggests that a number of targets of the silencing pathway are coordinately released from repression in leaves just prior to reproductively competent adult growth. The function of this transient change in regulation of so many targets of the tasiRNA pathway is somewhat mysterious. However, it is tempting to speculate that it is a system that has evolved to briefly relax processing of targets of this pathway as a way to increase target RNAs. This idea is supported by the observation that expression of *Lbl1*, and methylation of TIRA, is restored as transition leaves grow larger. Thus, the loss of silencing of *MuDR* is transient, both during the development the plant as a whole, and during the development of individual leaves. This process is reminiscent of similar periods of relaxation of silencing at other stages of development. In arabidopsis pollen, for instance, silencing pathways are relaxed in the vegetative nucleus (which does not contribute to the next generation) (91, 92, 93, 94). This results in the production of *trans*-acting small RNAs that can reinforce silencing in the sperm cell nuclei, one of which will fertilize the egg (Fig. 7B). In fact, polyadenylated *mudrA* transcripts from silenced *MuDR* elements are readily detected in pollen, suggesting this also occurs in maize (91). Similarly, there is evidence that silencing is relaxed in the terminally differentiated endosperm (the starchy part of the seed), which is also thought to enhance silencing in the embryo, likely via movement of small RNAs (Fig. 7C) (95, 96, 97). Note that in each case, silencing is relaxed in a tissue that is adjacent to, but not part of, the germinal lineage. The overall theme is one in which information in the form of transcripts from otherwise silenced TEs is gathered in a tissue in which the consequences of TE activity is minimal in order to enhance silencing germinal lineages.

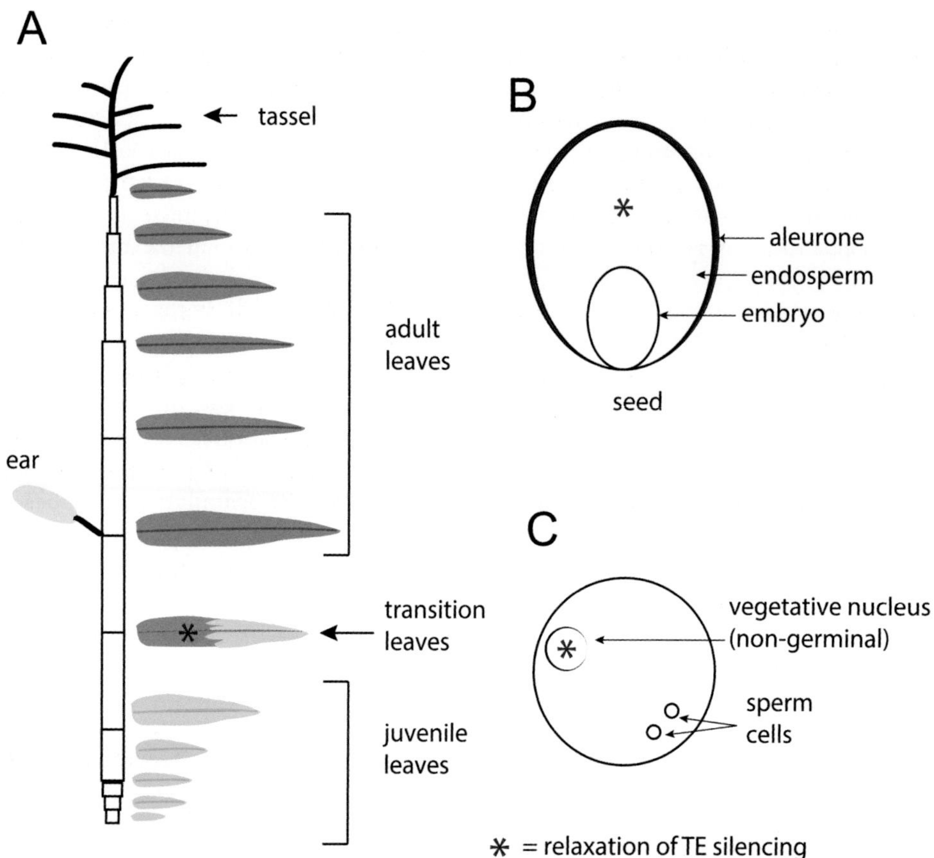

Figure 7 Maize development. (A) An illustration of the major components of an adult maize plant. Leaves are produced sequentially as the maize plant develops, with juvenile and adult leaves being distinguished by presence of epicuticular wax (a juvenile trait) and epidermal hairs (an adult trait). Transition leaves have patches of tissue with either adult or juvenile traits. Ears and tassels are only produced once adult leaves are produced. Each maize plant can be crossed as a male, or a female, or to itself. (B) A cartoon of a maize seed, showing the location of the embryo, endosperm (a terminally differentiated nutritive tissue), and the aleurone, which is the outer cell layer of the endosperm that is competent to express color. (C) A cartoon of a mature pollen grain, which contains three nuclei. The vegetative nucleus is responsible for the development of the pollen tube, and will not contribute to the next generation. Of the two sperm cells, one will fertilize the two polar nuclei to give rise to the triploid endosperm, and one of which will fertilize the egg cell to give rise to the embryo. In each part of the plant illustrated here, tissues and cells in which a relaxation of TE silencing has been observed are indicated by red asterisks.
doi:10.1128/microbiolspec.MDNA3-0032-2014.f7

Spontaneous Silencing

In addition to the directed heritable silencing initiated by *Muk*, standard lines exhibit high rates of spontaneous epigenetic silencing. This is often a progressive process, in which increasingly large and frequent clonal sectors contain methylated autonomous and nonautonomous *Mutator* elements (42). In cases in which those sectors include portions of developing ears, the result can be patches of kernels carrying inactive *Mutator* elements. Plants grown from these kernels remain inactive, as do progeny of those kernels. Interestingly, there is no evidence that the spontaneous loss of activity in these plants is due to the emergence of new "*Mu killers*." This is based on the observation that inactivated lines do not silence otherwise active lines, suggesting the absence of a heritable factor that is competent in inactivate *Mutator* elements (49, 81). However, lines that are self-fertilized for several generations exhibit an increased frequency of silencing, and there is evidence that these lines, unlike those in which activity was lost upon outcrossing, can inactivate otherwise active lines (40).

The cause of spontaneous silencing is not known. Within somatic sectors in which activity is spontaneously lost, there is no evidence of rearrangements of *MuDR* elements. There is evidence that standard lines undergoing silencing have an increased relative amount of nuclear localized *MuDR* and/or *hMuDR* transcript, but this is likely to be due to the loss of normal, polyadenylated transcript in the cytoplasm rather than in increase in the absolute quantity of nuclear transcript (26). Similarly, there is some evidence for small RNAs in these plants, but the quantity of these small RNAS (at least in the tissue examined) is not greater in silencing versus active plants.

Given that the most likely source of silencing information in standard lines are the *MuDR* elements, it may be that all *MuDR* elements produce some quantity of aberrant transcript, and at some level of expression, this aberrant transcript triggers a cascade of small RNA production via an RNA-dependent RNA polymerase such as RDR6. This scenario would be similar to that observed during silencing of the *Evade* retroelement in arabidopsis, in which silencing is tightly linked to increases in copy number of the element (98). In the case of *Evade*, this is due to the production of a small quantity of small RNAs that eventually trigger transitive silencing of the promoter. In *MuDR*, it could be that a similar source of small RNAs could be produced via read-through of the two convergently transcribed genes encoded by *MuDR*. It should be noted, however, that even a single *MuDR* element produces very large amounts of transcript, and that even 35-S driven antisense *mudrA* transcript does not appear to trigger *MuDR* silencing (14, 15).

A second possibility is that rearranged or aberrant elements are in fact produced at a high frequency late during somatic development. Any one of these may be insufficient to trigger silencing, but large numbers of them may eventually do so. As these events occur in somatic tissue, in order to trigger heritable silencing they would have to produce small RNAs that could be transferred to the meristem. Recent evidence from grafting experiments suggests that catalytically active small RNAs are in fact transported from leaves, and small RNA-mediated systemic silencing of viruses can protect the meristem from viroid infection (99, 100, 101). However, as we have seen with *Muk*, once silencing of an element is achieved it no longer requires the presence of the trigger.

Maintenance of Silencing

Once epigenetically silenced, *MuDR* elements are remarkably refractive to reactivation. There are, however some mutants and environmental treatments that can destabilize the silenced state of these elements.

Mop1 is the maize version of *RNA-DEPENDENT RNA POLYMERASE2* (27, 102). In maize, as in other plants, *RDR2* is required for the production of the vast majority of 24 nt small RNAs associated with RNA-directed DNA methylation (103). It is also required to maintain methylation at *Mutator* element TIRs. It is not, however, required for the initiation of silencing of *MuDR* by *Muk*, perhaps due to the fact that the *Muk* transcript is already a double-stranded hairpin and thus does not require the activity of an RNA polymerase (27, 82). Interestingly, *MuDR* is only reactivated after several generations in a *mop1* mutant background, despite immediate effects on *MuDR* TIR methylation. Further, it is only *mudrA* that becomes transcriptionally active; in one experiment, *mudrB* remained silenced after eight generations in the mutant background (104). Thus, although both *mudrA* and *mudrB* are silenced by *Muk*, only *mudrA* requires *mop1* for maintenance of that silenced state.

Interestingly, there are position effects with respect to the maintenance of silencing, one of which has been well documented. As discussed above, *MuDR(p3)* exhibited a reversible position effect with respect to its effect on the reporter element in *a1-mum2* (Fig. 5). As illustrated in the figure, subsequent transposition events can then be easily detected as heavily spotted kernels on an ear segregating mostly weakly spotted kernels. Although *Muk* has a strong effect on somatic excisions in the kernel in the first generation if it is present in the female parent, it has no effect in F1 kernels when it is present in the male parent (83). Thus, plants carrying *MuDR(p3)* could be crossed to plants homozygous for *Muk* and the resulting progeny kernels could be screened for transposed copies of *MuDR(p3)* by picking heavily spotted kernels.

Plants carrying those transposed copies could then be test crossed to determine the stability of silencing at these new positions. One position [*MuDR(p5)*], showed a strong propensity to become reactivated after the loss of *Muk* due to genetic segregation (105). When a reactivated *MuDR(p5)* element transposed to a new position, this propensity was lost, suggesting the epigenetic instability of *MuDR(p5)* was due to *cis*-acting sequences. These observations raise the intriguing possibility that plant genomes encode *cis*-acting sequences whose function is to erase previously established silencing.

MuDR(p5) is exceptional. In most cases, once silenced, either due to exposure to *Muk* or due to spontaneous silencing, *MuDR* elements remain that way for many generations. However, there are some environmental conditions that can result in reactivation. UV-B radiation, at conditions comparable to those at encountered 33% ozone depletion have been shown to reactivate *MuDR* expression and trigger excision of a reporter element (106). Subsequent experiments demonstrated that this increased expression was associated with decreases in modifications associated with epigenetic silencing, including H3K9 dimethylation, HP1 enrichment and DNA methylation (107). There is also evidence that low-dose ion implantation can reverse silencing and DNA methylation of *MuDR* elements, albeit infrequently (108).

USES OF THE *MUTATOR* SYSTEM

Because of its high mutagenic potential, the *Mutator* system has been used in forward screens for several decades (6, 109, 110). However, although new mutants were easy to obtain, isolating the insertion responsible for the mutation was laborious and often unsuccessful (110). More recently, high throughput sequencing of sequences flanking *Mutator* elements in active lines has dramatically improved the odds of doing so (111). Reverse genetic resources of *Mutator* element insertions are also available. TUSC, a resource provided by industry, has proved to be an invaluable source of new insertion alleles, with a 90% success rate and an average of three insertions in each targeted gene (6). A similar, publically financed project, the maize targeted mutagenesis population derived from 43,776 plants is also available (112). The advantage of this population is that *Mutator* activity was eliminated using a *Mu killer*-like locus, thus minimizing somatic insertions that are not heritably transmitted. Finally, 38,000 lines with sequenced indexed *Mutator* insertions are also now available (113). These lines lack active *MuDR* elements and are thus stable, and they have been recurrently backcrossed into the W22 inbred background, thereby minimizing confounding background variation. Collectively then, *Mutator* has been, and will continue to be, an invaluable source of new mutants.

OTHER ACTIVE *MULES* IN MAIZE

MuDR elements are the canonical representative of a supergroup of Class II elements referred to as *Mu*-Like Elements (*MULEs*). In addition to *MuDR*, maize contains at least two other distinct classes of active autonomous *MULEs*: *Jittery* and *TED* (56, 114). Phylogenetic analysis suggests each of these elements has been maintained as independent active element for many millions of years, as the *TED/MuDR* clade diverged prior to the rice/maize split, and the *Jittery/MuDR* clade diverged prior to the monocot/dicot split. Despite this, all three of these elements share characteristic features, including homology between the *mudrA* genes, long (~200 bp) TIRs, and 9 bp target site duplications upon integration.

There are, however, some interesting differences between these maize *MULEs*. Both *Jittery* and *TED* are very low copy elements, even in lines exhibiting activity, and neither element carries a *mudrB* homolog. *Jittery* is also unusual in that, unlike all other maize elements, Jittery elements excise precisely, and thus do no leave footprints. Interestingly, although it can excise from a reporter, it does not transpose. Obviously, as it is inserted into the reporter in the first place, it had to have been able to transpose at one time. Thus, it is likely that the *Jittery* that was isolated is deficient in some way, because either it is missing a MURB-like function, or because of a sequence alteration, such as that observed within one of the TIRS of the cloned *Jittery* element.

Unlike *Jittery*, *TED* is competent to transpose, and leaves characteristic footprints upon reversion. Like *Mutator* and Jittery elements, *TED* elements excise late during somatic development, resulting in a very uniform pattern of reversions (56). However, germinal reversions of *TED* induced alleles are 100-fold higher than the reversion frequency of *Mu1*-containing mutable alleles (161). Interestingly, the vast majority of these reversion events occur in the female gametophyte subsequent to meiosis in the linage that fertilizes the endosperm. In contrast, transposition defective deletion derivatives of *TED* arise frequently during meiosis, and are often associated with insertions of *TED* elements at new positions in the genome. These derivatives, like those of *MuDR*, appear to result from strand-slippage during template mediated repair subsequent to excision. These data suggest that the double-stranded gaps left following excision of *TED* elements are repaired differently during meiosis than in the haploid gametophyte. As with *Mutator* element, it would appear to be the response by the plant to a double-stranded gap produced by TE excision that appears to vary, rather than the mechanisms of TE transposition. *TED* elements, like *Mutator* elements transpose to unlinked sites and generally produce 9 bp TSDs on insertion.

EVOLUTION OF *MULE* ELEMENTS

Active *MULEs* in Other Species

MULEs are present in a wide variety of species, including in phytophthora, diatoms, entamoeba, invertebrates,

vertebrates, fungi, plants, trichomonas, and bacteria (51, 115) (Fig. 8). All autonomous *MULEs* (by definition) share the *MULE*-specific DDE integrase domain (pfam10551), and often also carry one or more zinc fingers (7, 52, 116). Most (but by no means all) have long terminal inverted repeats and produce target site duplications of 8 to 9 base pairs. A few putatively autonomous elements been demonstrated to be transpositionally active. This include those identified in rice (Os777 and *Os3378*) (117), *Fusarium oxysporum* (Hop) (118), and arabidopsis (*AtMu1, Hiun*) (119, 120). Of these, only *Hiun* has been unambiguously demonstrated to be an autonomous element.

Capture of Genes by Autonomous *MULEs*

In addition to *mudrA*-homologous sequences, *MULEs* can acquire additional genes, presumably because they enhance the fitness of the element. The first example of this was *mudrB*, which is only found in *MuDR* elements in the genus *Zea* (12). As of May 2014, there continues to be a conspicuous absence of sequences similar to *mudrB* detectable in any other genus (Lisch, unpublished). Presumably, *mudrB* was acquired by *MuDR* elements at some point, but there is no evidence for a cellular copy of *mudrB* in either maize or any other species. This suggests that either *mudrB* was horizontally transferred into a recent progenitor of maize from a species whose genome has not been sequenced, or that this gene was subject to high levels of positive selection, such that the cellular gene is no longer recognizably similar. The presence of distant relatives of *MuDR* in the maize genome that carry quite dissimilar *mudrA* and *mudrB* genes is consistent with the later scenario (Lisch, unpublished).

Interestingly autonomous *MULEs* in other species have captured other genes. *MULEs* in several species have captured a gene with strong similarity to a conserved domain found exclusively in ubiquitin-like protein-specific protease (ULP1-like genes) (121, 122). In arabidopsis, *MULEs* carrying this captured gene are of a class with short imperfect terminal inverted repeats (123). Many of the captured genes show signatures of purifying selection, and most of them encode potentially functional proteins, suggesting continued function. Like other plants, the arabidopsis genome also contains cellular versions of *ULP-1* proteases, presumably the source of the transduplicated copies within the *MULEs*. *MULEs* with transduplicated *ULP-1* genes have also been identified in melon and rice, although in these cases the *MULEs* are of the long perfect TIR variety, suggesting independent capture events. Also consistent with multiple independent capture events are the presence of individual clades of these *MULEs* that can carry one or two *ULP-1* genes in tandem or reversed orientation. In addition to *MULEs* in plants, *MULEs* carrying *ULP-1* have also been identified in hydra and lancelet (124, 125, 126).

Remarkably, *ULP-1* genes are also found in completely distinct DNA type elements. *Ginger* (for *Gypsy* INteGrasE Related) elements are a widespread superfamily of DNA type elements found in animals that are not related to *MULEs* (124). However, like some *MULEs*, some *Ginger* elements have captured *ULP-1* protease domains, and others have captured *OTU*, a protease domain related to *ULP-1*. Similarly, members of a third distinct superfamily of DNA type elements found in a wide variety of species, *Zisupton* elements, carry an *ULP-1* domain fused in *cis* to the transposase (127).

The function of these added protease related genes is not known, but they apparently provide a generic advantage to DNA type elements. As these elements invariably introduce DNA damage when excising, it has been speculated that they may be involved in activating factors in the DNA repair pathway via ubiquitinylation (124, 128).

A third instance in which a captured gene has a function for a *MULE* involves *Hiun (Hi)* (120). This *MULE* contains both a well-conserved transposase as well as a second gene, designated *VANC*. Normally, like most TEs, *Hi* is epigenetically silenced. However, when a plant containing silenced *Hi* elements was transformed with VANC, the silenced endogenous *Hi* elements were activated in *trans*. This process was associated with hypomethylation of the endogenous elements, suggesting that the VANC protein is competent to reverse epigenetic silencing of these elements. Interestingly, *Hi* is a member of a large and particularly successful family of *MULEs* in the genus *arabidopsis* that have short, degenerate TIRs, including the *Arnold* and *Vandal* subfamilies (123). All major subgroups of the non-LTR *MULEs*, but none of the TIR-*MULEs*, have members that encode VANC-like proteins, suggesting that this protein has provided a long-term benefit for these elements. Global analysis of patterns of methylation in plants suggest that the hypomethylation triggered by VANC is highly specific to *Hi* and its close relatives.

Horizontal Transfer of *MULEs*

One way that TEs can avoid epigenetic silencing is to transfer from a host that is competent to recognize them to one that cannot, because it lacks sequences such as *Muk* that can trigger silencing. Although horizontal

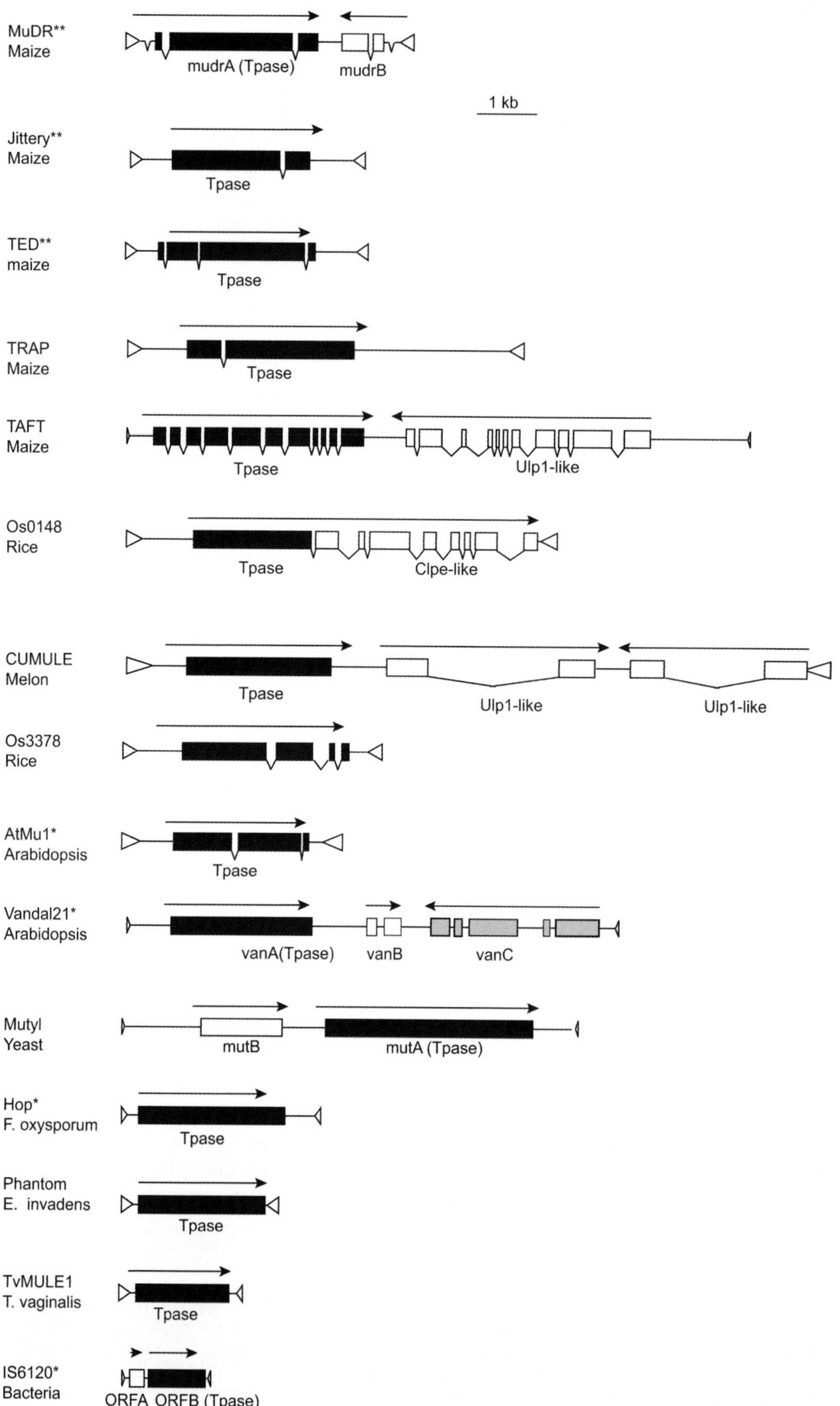
MuDR**
Maize
mudrA (Tpase)
mudrB
1 kb
Jittery**
Maize
Tpase
TED**
maize
Tpase
TRAP
Maize
Tpase
TAFT
Maize
Tpase
Ulp1-like
Os0148
Rice
Tpase
Clpe-like
CUMULE
Melon
Tpase
Ulp1-like
Ulp1-like
Os3378
Rice
AtMu1*
Arabidopsis
Tpase
Vandal21*
Arabidopsis
vanA(Tpase)
vanB
vanC
Mutyl
Yeast
mutB
mutA (Tpase)
Hop*
F. oxysporum
Tpase
Phantom
E. invadens
Tpase
TvMULE1
T. vaginalis
Tpase
IS6120*
Bacteria
ORFA ORFB (Tpase)

transfer of TEs is ubiquitous in bacteria (129) and common between animals (130), until recently there were no documented examples in plants. The first such case was transfer of one or more *MULE* between the ancestors of the genus *Setaria* and *Oryza* (131). Evidence for horizontal transfer was based on an unusually high degree of similarity between the *MULEs* in these species, a radical incongruity between the phylogenetic relationships between the *MULEs* and their hosts, and an absence of evidence for selection for protein function despite the high level of similarity. Similar analysis suggested a second horizontal transfer of a related *MULE* between *Oryza* and Old World (but not New World) bamboos. Since that study, additional investigations suggest that horizontal transfer of TEs may have been more common that was once thought (132, 133).

EVOLUTIONARY IMPACTS OF *MULE* TRANSPOSONS IN PLANTS

Like all TEs, *MULEs* are both shaped by and shape genomes in which they reside. In some plant species, *MULEs* appear to have had a significant impact on their hosts.

Pack-MULEs

Nonautonomous *MULEs* are a particularly diverse group of elements because many of them contain captured, or transduplicated, fragments of host genes. Because transduplication can be quite frequent, the result can be genomes that are littered with thousands of mobilized gene fragments. Further, many individual *Pack-MULEs* contain fragments of more than one gene, raising the possibility that new functional genes may arise from this process (134). The most comprehensive analysis of *MULE*-mediated transduplication was done in rice. In the genome of this species there are more than 3,000 *Pack-MULEs*, which contain fragments of 1,000 genes (38). Similarly, it is estimated that maize contains 276 *Pack-MULEs* in addition to those specifically mobilized by *MuDR* (135). *Pack-MULEs* have also been identified in arabidopsis, Lotus, and sugar cane, suggesting that transduplication is a common feature of at least some (but not all) families of *MULEs* (8). The gene fragments found in *Pack-MULEs* are generally very similar to their cognate host genes, suggesting that most of these transduplication events occurred relatively recently. Given the ubiquity of this process, and the fact that few individual transduplication events are shared by related species, it is likely that most transduplicated sequences are rapidly lost, and that *Mu*-induced transduplication is a frequent event in many plant species.

Most *Pack-MULEs* carry only fragments of one or more genes (Fig. 2) (8). Thus, the vast majority of the transduplication events that gave rise to these elements are unlikely to result in functional proteins (79). However, there is some evidence that a subset of rice *Pack-MULE* encoded genes are translated, suggesting potential functionality (136). Further, as many *Pack-MULEs* and other transduplicated sequences express transcript that is antisense relative to their cognate host genes, it is also been suggested that these elements may be a source of negative regulatory information (136, 137).

There is also evidence that gene fragment capture by *Pack-MULEs* may have contributed to the pronounced tendency genes in monocots to exhibit a GC gradient, with the 5′ end of genes have higher average GC content than the 3′ end (68). *Pack-MULEs* have a propensity to transduplicate GC regions of genes and they tend to insert into the 5′ ends of genes. In some cases, *Pack-MULEs* end up constituting the 5′ end of genes, presumably because after insertion, transcript initiates from within the element, a phenomenon that has been documented in both maize and rice (135, 138). This raises the intriguing possibility that the observed GC gradient in monocots is at least in part a consequence of millions of years of sequence capture and deposition of GC-rich sequences by *Pack-MULEs* into the 5′ end of genes. To the extent that 5′ GC richness has a generic effect on gene expression, this would suggest that these TEs might have had a profound and generic effect on gene expression in the monocots.

MULE Domestication

Although most TEs are present in genomes simply because they can replicate themselves, there are many documented instances in which TE sequences have been coopted and retained over longer periods because they

Figure 8 An illustration of the diversity of *MULEs* in a wide variety of species. In each case, triangles represent TIRs of various lengths. Boxes represent putative coding sequences. Black boxes represent putative transposases. All models are to scale. Names marked with one asterisks indicate elements that have been shown to be mobile. Those with two asterisks have been demonstrated to be autonomous.
doi:10.1128/microbiolspec.MDNA3-0032-2014.f8

provide a fitness advantage to their host. There are a wide variety of ways in which this can occur. For instance, TEs can add enhancer elements, new splice sites, or they can contribute to epigenetic regulation of genes via silencing pathways (139, 140). The most extreme example of this has been called molecular domestication, in which proteins encoded by TEs are exapted for a new function (141, 142). In plants, molecular domestication has resulted in several clades of *MULE* derived genes, some of which are known to act as transcription factors modulating light response in arabidopsis. The first instance of this was uncovered in a screen for mutants impaired in far-red light response. Two of these mutations were of the *FAR1* and *PHY3* genes, both of which are derived from a transposase in the *Jittery* subfamily of *MULEs* (143). Subsequent analysis revealed that these genes encode transcriptional factors that promote expression of a number of genes involved in various aspects of light response, including far-red-mediated seed deetolation, the circadian clock, chlorophyll biosynthesis, and chloroplast division (144, 145, 146, 147, 148, 149). They do so by activating genes whose promoters contain a FHY3/FAR1 binding site. (147, 148, 149, 150). Global analysis of binding of sites of FAR1 and PHY3 suggests that many other genes, and other pathways, may be influenced by these transposon derived transcription factors (148).

Phylogenetic analysis suggests that *FAR1* and *PHY3* are members of one of at least three distinct clades of genes with homology to *MULE* transposases, all of which appear to have been exapted prior to the monocot/dicot split. One such clade includes genes called MUSTANG elements (151). Although the function of these genes has not been established, at least some of them appear to be negatively modulated by phytohormones and by auxin, and their expression shows some tissue specificity (152). Further, mutations in some combinations of these genes have defects in growth, flowering and reproduction (153). Finally, a third clade of *MULE*-related genes has been identified in rice. Like *FAR1* and the *MUSTANG* genes, genes within this clade are, unlike *MULEs*, expressed ubiquitously and are often found at syntenic positions in related species (Lisch, unpublished data) (154).

CONCLUSIONS AND FUTURE DIRECTIONS

The *Mutator* system has proved to be a remarkable source of information concerning the biology and evolution of transposable elements. This is due to the fact that the autonomous element has been unambiguously identified, very large number of *de novo* insertions have been characterized, and the system can be easily silenced using a single dominant locus.

A more detailed understanding of the biochemistry of the transposition reaction will require a sustained effort to recreate a functioning system in an amenable system like yeast. Analysis of the function of MURB will be of particular interest, given its somewhat mysterious origin and function. As we know that a 35S-driven *mudrA* cDNA is sufficient to cause somatic excisions but not insertions of a reporter element, it will be interesting to examine a line that contains transgenic copies of both *mudrA* and *mudrB* to see if insertional activity can be restored in *trans* and whether or not both can function in a heterologous system. It will also be interesting to determine whether MURB interacts with proteins other than MURA, and whether constitutive expression of MURB has more general effects than those we see on *Mutator* element activity.

Although the means by which *mudrA* is silenced by *Muk* is reasonably well understood, its somewhat mysterious effects on *mudrB* are not. Why, for instance, does *Muk* trigger a loss of polyadenylation (but not transcription) in F1 plants carrying both *MuDR* and *Muk*, and why must *mudrB* be in *cis* to *mudrA* in order to be silenced by *Muk*? Similarly, we know almost nothing about spontaneous silencing. Presumably, this process involves the production of small RNAs, but no evidence for the production of a specific class of small RNAs associate with this process. It could be that the ratio of *MuDR* transcript and small RNAs produced by these *MuDR* elements at some low level reaches some kind of a threshold, as is observed in the *Evade* and *Onsen* retroelements in arabidopsis (155, 156). Alternatively, rearrangements in somatic tissue may produce a population of small RNAs that can be transported to the meristem, where they would trigger both somatic and germinal silencing. A test for this hypothesis could involve mutants that specifically affect the transport, but not the function, of small RNAs produced in leaves.

MULEs will continue to be a rich source of information concerning the evolution of TEs in general, and the role that they have played in shaping host evolution. Maize continues to be an excellent model because it is now host to at least three distinct, active autonomous *MULEs*. The similarities and differences between these elements can tell us how three versions of *MULEs* have evolved different strategies for long-term survival within the same genome. For instance, do *Jittery* and TED elements maintain a low duplication frequency as a strategy to avoid triggering the kind of silencing observed when *Mutator* elements reach high copy numbers? Under what circumstances, if any, would these other

MULEs rapidly increase their copy number, and what, if anything, will trigger silencing of these elements?

Given that we have already identified one instance of horizontal gene transfer, and new plant genomes are being sequenced at a rapid pace, it will also be interesting to see if horizontal transfer of these elements has been common. Conversely, we will also have the opportunity to determine how frequently molecular domestication of *MULEs*, or exaptation of any *MULE* sequence, has occurred. Those that have been identified were relatively ancient, but a comprehensive analysis of large numbers of related plant species may identify additional, more recent, instances. Plants are particularly amenable to this kind of analysis because of their high rates of pseudogene and TE deletion (157, 158). For instance, although sorghum and rice are only 10 million years diverged, TEs that lack a function for the host are rarely maintained at syntenic positions. Thus, any *MULE* sequence that is retained at the same position over a comparable distance between any two species will be an excellent candidate for exaptation.

One of the real advantages (and great joys) of studying TEs is the degree to which they can provide a window onto a wide variety of subjects, from the biochemistry of transposase function, to the physiology of TE silencing, to the evolution of genes and genomes. Because of their diversity, ubiquity, and sometimes very high levels of activity, *MULE*s are uniquely suited to serve as a model for understanding the ways in which TEs and their hosts engage in an ongoing coevolutionary conversation. As is illustrated in this chapter, this process can result in a remarkable continuum of interaction. From the rapid and mutagentic spread of *Mutator* elements in maize Standard *Mutator* lines that triggers efficient epigenetic silencing, to the domestication of *MULE* transposases that appear to have played a key role in the ability of flowering plants to sense and respond to light. Until recently, it was possible to gather fascinating but essentially anecdotal evidence for those interactions. With recent technological advances, it should be possible to gain a far more comprehensive view of the kinds of changes that currently active *MULEs* can cause, as well as those changes that have been contributed to plant adaptation over long periods of time.

Citation. Lisch D. 2014. *Mutator* and *MULE* transposons. Microbiol Spectrum 3(2):MDNA3-0032-2014

References

1. **Robertson DS.** 1978. Characterization of a mutator system in maize. *Mutat Res* **51**:21–28.
2. **Walbot V.** 1991. The *Mutator* transposable element family of maize. *Genet Eng (N Y)* **13**:1–37.
3. **Alleman M, Freeling M.** 1986. The *Mu* transposable elements of maize: evidence for transposition and copy number regulation during development. *Genetics* **112**:107–119.
4. **Walbot V, Warren C.** 1988. Regulation of *Mu* element copy number in maize lines with an active or inactive Mutator transposable element system. *Mol Gen Genet* **211**:27–34.
5. **Cresse AD, Hulbert SH, Brown WE, Lucas JR, Bennetzen JL.** 1995. *Mu1*-related transposable elements of maize preferentially insert into low copy number DNA. *Genetics* **140**:315–324.
6. **McCarty D, Meeley R.** 2009. Transposon resources for forward and reverse genetics in maize, p 561–584. *In* Bennetzen J, Hake S (ed), *Handbook of Maize: Genetics and Genomics*. Springer, Berlin.
7. **Lisch D.** 2002. Mutator transposons. *Trends Plant Sci* 7:498–504.
8. **Lisch D, Jiang H.** 2009. *Mutator* and *MULE* transposons, p 277–306. *In* Bennetzen J, Hake S (ed), *Handbook of Maize: Genetics and Genomics*. Springer, Berlin.
9. **Chomet P, Lisch D, Hardeman KJ, Chandler VL, Freeling M.** 1991. Identification of a regulatory transposon that controls the *Mutator* transposable element system in maize. *Genetics* **129**:261–270.
10. **Hershberger RJ, Warren CA, Walbot V.** 1991. Mutator activity in maize correlates with the presence and expression of the Mu transposable element *Mu9*. *Proc Natl Acad Sci U S A* **88**:10198–10202.
11. **Qin M, Robertson DS, Ellingboe AH.** 1991. Cloning of the mutator transposable element *MuA2*, a putative regulator of somatic mutability of the a1-Mum2 allele in maize. *Genetics* **129**:845–854.
12. **Lisch DR, Freeling M, Langham RJ, Choy MY.** 2001. *Mutator* transposase is widespread in the grasses. *Plant Physiol* **125**:1293–1303.
13. **Hershberger RJ, Benito MI, Hardeman KJ, Warren C, Chandler VL, Walbot V.** 1995. Characterization of the major transcripts encoded by the regulatory *MuDR* transposable element of maize. *Genetics* **140**:1087–1098.
14. **Kim SH, Walbot V.** 2003. Deletion derivatives of the *MuDR* regulatory transposon of maize encode antisense transcripts but are not dominant-negative regulators of mutator activities. *Plant Cell* **15**:2430–2447.
15. **Lisch D, Girard L, Donlin M, Freeling M.** 1999. Functional analysis of deletion derivatives of the maize transposon *MuDR* delineates roles for the MURA and MURB proteins. *Genetics* **151**:331–341.
16. **Joanin P, Hershberger RJ, Benito MI, Walbot V.** 1997. Sense and antisense transcripts of the maize *MuDR* regulatory transposon localized by *in situ* hybridization. *Plant Mol Biol* **33**:23–36.
17. **Walbot V, Rudenko GN.** 2002. *MuDR/Mu* transposable elements of maize, p 533–564. *In* Craig NL, Craigie R, Gellert M, Lambowitz AM (ed), *Mobile DNA II*. ASM Press, Washington DC.
18. **Ono A, Kim SH, Walbot V.** 2002. Subcellular localization of MURA and MURB proteins encoded by the maize *MuDR* transposon. *Plant Mol Biol* **50**:599–611.

19. **Benito MI, Walbot V.** 1997. Characterization of the maize *Mutator* transposable element MURA transposase as a DNA-binding protein. *Mol Cell Biol* **17:**5165–5175.

20. **Rudenko GN, Walbot V.** 2001. Expression and post-transcriptional regulation of maize transposable element *MuDR* and its derivatives. *Plant Cell* **13:**553–570.

21. **Raizada MN, Nan GL, Walbot V.** 2001. Somatic and germinal mobility of the *RescueMu* transposon in transgenic maize. *Plant Cell* **13:**1587–1608.

22. **Raizada MN, Benito MI, Walbot V.** 2001. The *MuDR* transposon terminal inverted repeat contains a complex plant promoter directing distinct somatic and germinal programs. *Plant J* **25:**79–91.

23. **Donlin MJ, Lisch D, Freeling M.** 1995. Tissue-specific accumulation of MURB, a protein encoded by *MuDR*, the autonomous regulator of the Mutator transposable element family. *Plant Cell* **7:**1989–2000.

24. **Hsia AP, Schnable PS.** 1996. DNA sequence analyses support the role of interrupted gap repair in the origin of internal deletions of the maize transposon, *MuDR*. *Genetics* **142:**603–618.

25. **Lisch D, Freeling M.** 1994. Loss of *Mutator* activity in a minimal line. *Maydica* **39:**289–300.

26. **Rudenko GN, Ono A, Walbot V.** 2003. Initiation of silencing of maize *MuDR/Mu* transposable elements. *Plant J* **33:**1013–1025.

27. **Woodhouse MR, Freeling M, Lisch D.** 2006. Initiation, establishment, and maintenance of heritable *MuDR* transposon silencing in maize are mediated by distinct factors. *PLoS Biol* **4:**e339.

28. **Strommer JN, Hake S, Bennetzen JL, Taylor WC, Freeling M.** 1982. Regulatory mutants of the maize *Adh1* gene caused by DNA insertions. *Nature* **300:**542–544.

29. **Bennetzen JL.** 1984. Transposable element *Mu1* is found in multiple copies only in Robertson's Mutator maize lines. *J Mol Appl Genet* **2:**519–524.

30. **Taylor LP, Chandler VL, Walbot V.** 1986. Insertion of 1.4 kb *Mu* elements into the *bronze1* gene of Zea mays L. *Maydica* **31:**31–45.

31. **Taylor LP, Walbot V.** 1987. Isolation and characterization of a 1.7-kb transposable element from a mutator line of maize. *Genetics* **117:**297–307.

32. **Oishi K, Freeling M.** 1983. The *Mu3* transposon in maize, p 289–292. *In* Nelson O (ed), *Plant Transposable Elements*. Plenum Press, New York, NY.

33. **Talbert LE, Chandler VL.** 1988. Characterization of a highly conserved sequence related to *mutator* transposable elements in maize. *Mol Biol Evol* **5:**519–529.

34. **Dietrich CR, Cui F, Packila ML, Li J, Ashlock DA, Nikolau BJ, Schnable PS.** 2002. Maize *Mu* transposons are targeted to the 5′ untranslated region of the *gl8* gene and sequences flanking *Mu* target-site duplications exhibit nonrandom nucleotide composition throughout the genome. *Genetics* **160:**697–716.

35. **Schnable P, Peterson PA.** 1989. Genetic evidence of a relationship between two maize transposable element systems: *Cy* and *Mutator*. *Mol Gen Genet* **215:**317–321.

36. **Fleenor D, Spell M, Robertson D, Wessler S.** 1990. Nucleotide sequence of the maize *Mutator* element, *Mu8*. *Nucleic Acids Res* **18:**6725.

37. **Tan BC, Chen Z, Shen Y, Zhang Y, Lai J, Sun SS.** 2011. Identification of an active new *mutator* transposable element in maize. *G3 (Bethesda)* **1:**293–302.

38. **Jiang N, Bao Z, Zhang X, Eddy SR, Wessler SR.** 2004. *Pack-MULE* transposable elements mediate gene evolution in plants. *Nature* **431:**569–573.

39. **Walbot V, Britt AB, Luehrsen K, McLaughlin M, Warren C.** 1988. Regulation of mutator activities in maize. *Basic Life Sci* **47:**121–135.

40. **Robertson DS.** 1986. Genetic studies on the loss of mu mutator activity in maize. *Genetics* **113:**765–773.

41. **Chandler VL, Walbot V.** 1986. DNA modification of a maize transposable element correlates with loss of activity. *Proc Natl Acad Sci U S A* **83:**1767–1771.

42. **Martienssen R, Baron A.** 1994. Coordinate suppression of mutations caused by Robertson's mutator transposons in maize. *Genetics* **136:**1157–1170.

43. **Walbot V.** 1986. Inheritance of mutator activity in Zea mays as assayed by somatic instability of the bz2-mu1 allele. *Genetics* **114:**1293–1312.

44. **Slotkin RK.** 2005. *The heritable epigenetic silencing of* mutator *transposons by* Mu killer. *Ph.D.* University of California at Berkeley, Berkeley, CA.

45. **Skibbe DS, Fernandes JF, Medzihradszky KF, Burlingame AL, Walbot V.** 2009. *Mutator* transposon activity reprograms the transcriptomes and proteomes of developing maize anthers. *Plant J* **59:**622–633.

46. **Schnable PS, Peterson PA.** 1988. The mutator-related *Cy* transposable element of Zea mays L. behaves as a near-Mendelian factor. *Genetics* **120:**587–596.

47. **Schnable PS, Peterson PA, Saedler H.** 1989. The bz-rcy allele of the Cy transposable element system of Zea mays contains a *Mu*-like element insertion. *Mol Gen Genet* **217:**459–463.

48. **Robertson DS, Stinard PS.** 1989. Genetic analyses of putative two-element systems regulating somatic mutability in *Mutator*-induced aleurone mutants of maize. *Dev Genet* **10:**482–506.

49. **Lisch D.** 1995. *Genetic and molecular characterization of the Mutator system in maize. Ph.D.* University of California at Berkeley, Berkeley, CA.

50. **Guerillot R, Siguier P, Gourbeyre E, Chandler M, Glaser P.** 2014. The diversity of prokaryotic DDE transposases of the mutator superfamily, insertion specificity, and association with conjugation machineries. *Genome Biol Evol* **6:**260–272.

51. **Eisen JA, Benito MI, Walbot V.** 1994. Sequence similarity of putative transposases links the maize *Mutator* autonomous element and a group of bacterial insertion sequences. *Nucleic Acids Res* **22:**2634–2636.

52. **Hua-Van A, Capy P.** 2008. Analysis of the DDE motif in the *Mutator* superfamily. *J Mol Evol* **67:**670–681.

53. **Loessner I, Dietrich K, Dittrich D, Hacker J, Ziebuhr W.** 2002. Transposase-dependent formation of circular

IS256 derivatives in *Staphylococcus epidermidis* and *Staphylococcus aureus*. *J Bacteriol* **184:**4709–4714.

54. **Prudhomme M, Turlan C, Claverys JP, Chandler M.** 2002. Diversity of *Tn4001* transposition products: the flanking IS256 elements can form tandem dimers and IS circles. *J Bacteriol* **184:**433–443.
55. **Sundaresan V, Freeling M.** 1987. An extrachromosomal form of the *Mu* transposons of maize. *Proc Natl Acad Sci U S A* **84:**4924–4928.
56. **Li Y, Harris L, Dooner HK.** 2013. *TED*, an autonomous and rare maize transposon of the *mutator* superfamily with a high gametophytic excision frequency. *Plant Cell* **25:**3251–3265.
57. **Levy AA, Britt AB, Luehrsen KR, Chandler VL, Warren C, Walbot V.** 1989. Developmental and genetic aspects of *Mutator* excision in maize. *Dev Genet* **10:**520–531.
58. **Levy AA, Walbot V.** 1990. Regulation of the timing of transposable element excision during maize development. *Science* **248:**1534–1537.
59. **Raizada MN, Walbot V.** 2000. The late developmental pattern of *Mu* transposon excision is conferred by a cauliflower mosaic virus 35S -driven MURA cDNA in transgenic maize. *Plant Cell* **12:**5–21.
60. **Yu W, Lamb JC, Han F, Birchler JA.** 2007. Cytological visualization of DNA transposons and their transposition pattern in somatic cells of maize. *Genetics* **175:**31–39.
61. **Britt AB, Walbot V.** 1991. Products of *Mu* Excision from the *bronze1* gene of Zea-mays. *J Cell Biochem Suppl* **15:**99.
62. **Doseff A, Martienssen R, Sundaresan V.** 1991. Somatic excision of the *Mu1* transposable element of maize. *Nucleic Acids Res* **19:**579–584.
63. **Robertson DS.** 1981. Mutator activity in maize: timing of its activation in ontogeny. *Science* **213:**1515–1517.
64. **Lisch D, Chomet P, Freeling M.** 1995. Genetic characterization of the Mutator system in maize: behavior and regulation of *Mu* transposons in a minimal line. *Genetics* **139:**1777–1796.
65. **Fernandes J, Dong Q, Schneider B, Morrow DJ, Nan GL, Brendel V, Walbot V.** 2004. Genome-wide mutagenesis of *Zea mays L.* using *RescueMu* transposons. *Genome Biol* **5:**R82.
66. **McCarty DR, Suzuki M, Hunter C, Collins J, Avigne WT, Koch KE.** 2013. Genetic and molecular analyses of UniformMu transposon insertion lines. *Methods Mol Biol* **1057:**157–166.
67. **Liu S, Yeh CT, Ji T, Ying K, Wu H, Tang HM, Fu Y, Nettleton D, Schnable PS.** 2009. *Mu* transposon insertion sites and meiotic recombination events co-localize with epigenetic marks for open chromatin across the maize genome. *PLoS Genet* **5:**e1000733.
68. **Carels N, Bernardi G.** 2000. Two classes of genes in plants. *Genetics* **154:**1819–1825.
69. **Fedoroff NV.** 2013. McClintock and epigenetics. *In* Fedoroff NV (ed), *Plant Transposons and Genome Dynamics in Evolution*. Wiley-Blackwell, Oxford.
70. **Martienssen R, Barkan A, Taylor WC, Freeling M.** 1990. Somatically heritable switches in the DNA modification of *Mu* transposable elements monitored with a suppressible mutant in maize. *Genes Dev* **4:**331–343.
71. **Girard L, Freeling M.** 2000. *Mutator*-suppressible alleles of rough sheath1 and liguleless3 in maize reveal multiple mechanisms for suppression. *Genetics* **154:**437–446.
72. **Lowe B, Mathern J, Hake S.** 1992. Active *Mutator* elements suppress the knotted phenotype and increase recombination at the *Kn1-O* tandem duplication. *Genetics* **132:**813–822.
73. **Greene B, Walko R, Hake S.** 1994. *Mutator* insertions in an intron of the maize *knotted1* gene result in dominant suppressible mutations. *Genetics* **138:**1275–1285.
74. **Cui X, Hsia AP, Liu F, Ashlock DA, Wise RP, Schnable PS.** 2003. Alternative transcription initiation sites and polyadenylation sites are recruited during *Mu* suppression at the *rf2a* locus of maize. *Genetics* **163:**685–698.
75. **Fowler JE, Meuhlbauer GJ, Freeling M.** 1996. Mosaic analysis of the liguleless3 mutant phenotype in maize by coordinate suppression of *mutator*-insertion alleles. *Genetics* **143:**489–503.
76. **Lisch D.** 2013. Regulation of the *Mutator* system of transposons in maize. *Methods Mol Biol* **1057:**123–142.
77. **McClintock B.** 1958. The suppressor-mutator system of control of gene action in maize. *Carnegie Inst Wash Yr Bk* **57:**415–429.
78. **Feschotte C, Pritham EJ.** 2009. A cornucopia of *Helitrons* shapes the maize genome. *Proc Natl Acad Sci U S A* **106:**19747–19748.
79. **Juretic N, Hoen DR, Huynh ML, Harrison PM, Bureau TE.** 2005. The evolutionary fate of *MULE*-mediated duplications of host gene fragments in rice. *Genome Res* **15:**1292–1297.
80. **Martienssen RA.** 1996. Epigenetic silencing of *Mu* transposable elements in maize, p 593–608. *In Cold Spring Harbor monograph series: epigenetic mechanisms of gene regulation*. Cold Spring Harbor Laboratory Press, Plainview, NY.
81. **Brown J, Sundaresan V.** 1992. Genetic study of the loss and restoration of *mutator* transposon activity in maize - evidence against dominant-negative regulator associated with loss of activity. *Genetics* **130:**889–898.
82. **Slotkin RK, Freeling M, Lisch D.** 2005. Heritable transposon silencing initiated by a naturally occurring transposon inverted duplication. *Nat Genet* **37:**641–644.
83. **Slotkin RK, Freeling M, Lisch D.** 2003. *Mu killer* causes the heritable inactivation of the *Mutator* family of transposable elements in *Zea mays*. *Genetics* **165:**781–797.
84. **Li H, Freeling M, Lisch D.** 2010. Epigenetic reprogramming during vegetative phase change in maize. *Proc Natl Acad Sci U S A* **107:**22184–22189.
85. **Walbot V, Evans MM.** 2003. Unique features of the plant life cycle and their consequences. *Nat Rev Genet* **4:**369–379.
86. **Nogueira FTS, Sarkar AK, Chitwood DH, Timmermans MCP.** 2006. Organ polarity in plants is specified through the opposing activity of two distinct small regulatory RNAs. *Cold Spring Harb Symp Quant Biol* **71:**157–164.

87. **Glick E, Zrachya A, Levy Y, Mett A, Gidoni D, Belausov E, Citovsky V, Gafni Y.** 2008. Interaction with host SGS3 is required for suppression of RNA silencing by tomato yellow leaf curl virus V2 protein. *Proc Natl Acad Sci USA* **105:**157–161.

88. **Kumakura N, Takeda A, Fujioka Y, Motose H, Takano R, Watanabe Y.** 2009. *SGS3* and *RDR6* interact and colocalize in cytoplasmic SGS3/RDR6-bodies. *FEBS Lett* **583:**1261–1266.

89. **Allen E, Xie Z, Gustafson AM, Carrington JC.** 2005. microRNA-directed phasing during *trans*-acting siRNA biogenesis in plants. *Cell* **121:**207–221.

90. **Pekker I, Alvarez JP, Eshed Y.** 2005. Auxin response factors mediate Arabidopsis organ asymmetry via modulation of KANADI activity. *Plant Cell* **17:**2899–2910.

91. **Slotkin RK, Vaughn M, Borges F, Tanurdzic M, Becker JD, Feijo JA, Martienssen RA.** 2009. Epigenetic reprogramming and small RNA silencing of transposable elements in pollen. *Cell* **136:**461–472.

92. **Creasey KM, Zhai J, Borges F, Van Ex F, Regulski M, Meyers BC, Martienssen RA.** 2014. miRNAs trigger widespread epigenetically activated siRNAs from transposons in Arabidopsis. *Nature* **508:**411–415.

93. **Borges F, Martienssen RA.** 2013. Establishing epigenetic variation during genome reprogramming. *RNA Biol* **10:**490–494.

94. **Calarco JP, Borges F, Donoghue MT, Van Ex F, Jullien PE, Lopes T, Gardner R, Berger F, Feijo JA, Becker JD, Martienssen RA.** 2012. Reprogramming of DNA methylation in pollen guides epigenetic inheritance via small RNA. *Cell* **151:**194–205.

95. **Gehring M, Bubb KL, Henikoff S.** 2009. Extensive demethylation of repetitive elements during seed development underlies gene imprinting. *Science* **324:**1447–1451.

96. **Hsieh TF, Ibarra CA, Silva P, Zemach A, Eshed-Williams L, Fischer RL, Zilberman D.** 2009. Genome-wide demethylation of Arabidopsis endosperm. *Science* **324:**1451–1454.

97. **Wollmann H, Berger F.** 2012. Epigenetic reprogramming during plant reproduction and seed development. *Curr Opin Plant Biol* **15:**63–69.

98. **Mari-Ordonez A, Marchais A, Etcheverry M, Martin A, Colot V, Voinnet O.** 2013. Reconstructing de novo silencing of an active plant retrotransposon. *Nat Genet* **45:**1029–1039.

99. **Di Serio F, Martinez de Alba AE, Navarro B, Gisel A, Flores R.** 2010. RNA-dependent RNA polymerase 6 delays accumulation and precludes meristem invasion of a viroid that replicates in the nucleus. *J Virol* **84:** 2477–2489.

100. **Molnar A, Melnyk CW, Bassett A, Hardcastle TJ, Dunn R, Baulcombe DC.** 2010. Small silencing RNAs in plants are mobile and direct epigenetic modification in recipient cells. *Science* **328:**872–875.

101. **Brosnan CA, Voinnet O.** 2011. Cell-to-cell and long-distance siRNA movement in plants: mechanisms and biological implications. *Curr Opin Plant Biol* **14:**580–587.

102. **Alleman M, Sidorenko L, McGinnis K, Seshadri V, Dorweiler JE, White J, Sikkink K, Chandler VL.** 2006. An RNA-dependent RNA polymerase is required for paramutation in maize. *Nature* **442:**295–298.

103. **Nobuta K, Lu C, Shrivastava R, Pillay M, De Paoli E, Accerbi M, Arteaga-Vazquez M, Sidorenko L, Jeong DH, Yen Y, Green PJ, Chandler VL, Meyers BC.** 2008. Distinct size distribution of endogeneous siRNAs in maize: Evidence from deep sequencing in the *mop1-1* mutant. *Proc Natl Acad Sci U S A* **105:**14958–14963.

104. **Woodhouse MR, Freeling M, Lisch D.** 2006. The *mop1* (*mediator of paramutation1*) mutant progressively reactivates one of the two genes encoded by the *MuDR* transposon in maize. *Genetics* **172:**579–592.

105. **Singh J, Freeling M, Lisch D.** 2008. A position effect on the heritability of epigenetic silencing. *PLoS Genet* **4:** e1000216.

106. **Walbot V.** 1999. UV-B damage amplified by transposons in maize. *Nature* **397:**398–399.

107. **Questa JI, Walbot V, Casati P.** 2010. *Mutator* transposon activation after UV-B involves chromatin remodeling. *Epigenetics* **5:**352–363.

108. **Qian Y, Cheng X, Liu Y, Jiang H, Zhu S, Cheng B.** 2010. Reactivation of a silenced minimal *Mutator* transposable element system following low-energy nitrogen ion implantation in maize. *Plant Cell Rep* **29:** 1365–1376.

109. **Candela H, Hake S.** 2008. The art and design of genetic screens: maize. *Nat Rev Genet* **9:**192–203.

110. **Walbot V, Questa J.** 2013. Using *MuDR/Mu* transposons in directed tagging strategies. *Methods Mol Biol* **1057:**143–155.

111. **Williams-Carrier R, Stiffler N, Belcher S, Kroeger T, Stern DB, Monde RA, Coalter R, Barkan A.** 2010. Use of Illumina sequencing to identify transposon insertions underlying mutant phenotypes in high-copy *Mutator* lines of maize. *Plant J* **63:**167–177.

112. **May BP, Liu H, Vollbrecht E, Senior L, Rabinowicz PD, Roh D, Pan X, Stein L, Freeling M, Alexander D, Martienssen R.** 2003. Maize-targeted mutagenesis: A knockout resource for maize. *Proc Natl Acad Sci U S A* **100:**11541–11546.

113. **Settles AM, Holding DR, Tan BC, Latshaw SP, Liu J, Suzuki M, Li L, O'Brien BA, Fajardo DS, Wroclawska E, Tseung CW, Lai J, Hunter CT 3rd, Avigne WT, Baier J, Messing J, Hannah LC, Koch KE, Becraft PW, Larkins BA, McCarty DR.** 2007. Sequence-indexed mutations in maize using the UniformMu transposon-tagging population. *BMC Genomics* **8:**116.

114. **Xu Z, Yan X, Maurais S, Fu H, O'Brien DG, Mottinger J, Dooner HK.** 2004. *Jittery*, a *Mutator* distant relative with a paradoxical mobile behavior: excision without reinsertion. *Plant Cell* **16:**1105–1114.

115. **Feschotte C, Pritham EJ.** 2007. DNA transposons and the evolution of eukaryotic genomes. *Annu Rev Genet* **41:**331–368.

116. **Babu MM, Iyer LM, Balaji S, Aravind L.** 2006. The natural history of the WRKY-GCM1 zinc fingers and

the relationship between transcription factors and transposons. *Nucleic Acids Res* **34**:6505–6520.

117. **Gao D.** 2012. Identification of an active *Mutator*-like element (*MULE*) in rice (Oryza sativa). *Mol Genet Genomics* **287**:261–271.

118. **Chalvet F, Grimaldi C, Kaper F, Langin T, Daboussi MJ.** 2003. *Hop*, an active *Mutator*-like element in the genome of the fungus *Fusarium oxysporum*. *Mol Biol Evol* **20**:1362–1375.

119. **Singer T, Yordan C, Martienssen RA.** 2001. Robertson's *Mutator* transposons in *A. thaliana* are regulated by the chromatin-remodeling gene *Decrease in DNA Methylation* (*DDM1*). *Genes Dev* **15**:591–602.

120. **Fu Y, Kawabe A, Etcheverry M, Ito T, Toyoda A, Fujiyama A, Colot V, Tarutani Y, Kakutani T.** 2013. Mobilization of a plant transposon by expression of the transposon-encoded anti-silencing factor. *EMBO J* **32**: 2407–2417.

121. **van Leeuwen H, Monfort A, Puigdomenech P.** 2007. *Mutator*-like elements identified in melon, Arabidopsis and rice contain ULP1 protease domains. *Mol Genet Genomics* **277**:357–364.

122. **Hoen DR, Park KC, Elrouby N, Yu Z, Mohabir N, Cowan RK, Bureau TE.** 2006. Transposon-mediated expansion and diversification of a family of ULP-like genes. *Mol Biol Evol* **23**:1254–1268.

123. **Yu Z, Wright SI, Bureau TE.** 2000. *Mutator*-like elements in *Arabidopsis thaliana*. Structure, diversity and evolution. *Genetics* **156**:2019–2031.

124. **Bao W, Kapitonov VV, Jurka J.** 2010. *Ginger* DNA transposons in eukaryotes and their evolutionary relationships with long terminal repeat retrotransposons. *Mob DNA* **1**:3.

125. **Bao W, Jurka J.** 2008. *MuDr*-type DNA transposons from Hydra magnipapillata. *Repbase Rep* **8**:2075–2075.

126. **Bao W, Jurka J.** 2009. *MuDr*-type DNA transposons from Branchiostoma floridae. *Repbase Rep* **9**:683–683.

127. **Bohne A, Zhou Q, Darras A, Schmidt C, Schartl M, Galiana-Arnoux D, Volff JN.** 2012. *Zisupton*–a novel superfamily of DNA transposable elements recently active in fish. *Mol Biol Evol* **29**:631–645.

128. **Bergink S, Jentsch S.** 2009. Principles of ubiquitin and SUMO modifications in DNA repair. *Nature* **458**: 461–467.

129. **Rocha EP.** 2013. Evolution. With a little help from prokaryotes. *Science* **339**:1154–1155.

130. **Schaack S, Gilbert C, Feschotte C.** 2010. Promiscuous DNA: horizontal transfer of transposable elements and why it matters for eukaryotic evolution. *Trends Ecol Evol* **25**:537–546.

131. **Diao X, Freeling M, Lisch D.** 2006. Horizontal transfer of a plant transposon. *PLoS Biol* **4**:e5.

132. **El Baidouri M, Carpentier MC, Cooke R, Gao D, Lasserre E, Llauro C, Mirouze M, Picault N, Jackson SA, Panaud O.** 2014. Widespread and frequent horizontal transfers of transposable elements in plants. *Genome Res* **24**:831–838.

133. **Roulin A, Piegu B, Fortune PM, Sabot F, D'Hont A, Manicacci D, Panaud O.** 2009. Whole genome surveys of rice, maize and sorghum reveal multiple horizontal transfers of the LTR-retrotransposon *Route66* in Poaceae. *BMC Evol Biol* **9**:58.

134. **Lisch D.** 2005. *Pack-MULEs*: theft on a massive scale. *Bioessays* **27**:353–355.

135. **Jiang N, Ferguson AA, Slotkin RK, Lisch D.** 2011. *Pack-Mutator*-like transposable elements (*Pack-MULEs*) induce directional modification of genes through biased insertion and DNA acquisition. *Proc Natl Acad Sci U S A* **108**:1537–1542.

136. **Hanada K, Vallejo V, Nobuta K, Slotkin RK, Lisch D, Meyers BC, Shiu SH, Jiang N.** 2009. The functional role of *pack-MULEs* in rice inferred from purifying selection and expression profile. *Plant Cell* **21**:25–38.

137. **Jiang SY, Christoffels A, Ramamoorthy R, Ramachandran S.** 2009. Expansion mechanisms and functional annotations of hypothetical genes in the rice genome. *Plant Physiol* **150**:1997–2008.

138. **Barkan A, Martienssen RA.** 1991. Inactivation of maize transposon Mu suppresses a mutant phenotype by activating an outward-reading promoter near the end of *Mu1*. *Proc Natl Acad Sci U S A* **88**:3502–3506.

139. **Lisch D.** 2013. How important are transposons for plant evolution? *Nat Rev Genet* **14**:49–61.

140. **Feschotte C.** 2008. Transposable elements and the evolution of regulatory networks. *Nat Rev Genet* **9**:397–405.

141. **Miller WJ, McDonald JF, Pinsker W.** 1997. Molecular domestication of mobile elements. *Genetica* **100**:261–270.

142. **Sinzelle L, Izsvak Z, Ivics Z.** 2009. Molecular domestication of transposable elements: from detrimental parasites to useful host genes. *Cell Mol Life Sci* **66**:1073–1093.

143. **Hudson ME, Lisch DR, Quail PH.** 2003. The *FHY3* and *FAR1* genes encode transposase-related proteins involved in regulation of gene expression by the phytochrome A-signaling pathway. *Plant J* **34**:453–471.

144. **Hudson M, Ringli C, Boylan MT, Quail PH.** 1999. The *FAR1* locus encodes a novel nuclear protein specific to phytochrome A signaling. *Genes Dev* **13**:2017–2027.

145. **Wang H, Deng XW.** 2002. Arabidopsis *FHY3* defines a key phytochrome A signaling component directly interacting with its homologous partner FAR1. *EMBO J* **21**: 1339–1349.

146. **Allen T, Koustenis A, Theodorou G, Somers DE, Kay SA, Whitelam GC, Devlin PF.** 2006. Arabidopsis *FHY3* specifically gates phytochrome signaling to the circadian clock. *Plant Cell* **18**:2506–2516.

147. **Li G, Siddiqui H, Teng Y, Lin R, Wan XY, Li J, Lau OS, Ouyang X, Dai M, Wan J, Devlin PF, Deng XW, Wang H.** 2011. Coordinated transcriptional regulation underlying the circadian clock in Arabidopsis. *Nat Cell Biol* **13**:616–622.

148. **Ouyang X, Li J, Li G, Li B, Chen B, Shen H, Huang X, Mo X, Wan X, Lin R, Li S, Wang H, Deng XW.** 2011. Genome-wide binding site analysis of FAR-RED ELON GATED HYPOCOTYL3 reveals its novel function in Arabidopsis development. *Plant Cell* **23**:2514–2535.

149. **Tang W, Wang W, Chen D, Ji Q, Jing Y, Wang H, Lin R.** 2012. Transposase-derived proteins FHY3/FAR1 interact with PHYTOCHROME-INTERACTING FACTOR1 to regulate chlorophyll biosynthesis by modulating HEMB1 during deetiolation in Arabidopsis. *Plant Cell* **24:**1984–2000.

150. **Lin R, Ding L, Casola C, Ripoll DR, Feschotte C, Wang H.** 2007. Transposase-derived transcription factors regulate light signaling in Arabidopsis. *Science* **318:** 1302–1305.

151. **Cowan RK, Hoen DR, Schoen DJ, Bureau TE.** 2005. *MUSTANG* is a novel family of domesticated transposase genes found in diverse angiosperms. *Mol Biol Evol* **22:**2084–2089.

152. **Kajihara D, de Godoy F, Hamaji TA, Blanco SR, Van Sluys MA, Rossi M.** 2012. Functional characterization of sugarcane mustang domesticated transposases and comparative diversity in sugarcane, rice, maize and sorghum. *Genet Mol Biol* **35:**632–639.

153. **Joly-Lopez Z, Forczek E, Hoen DR, Juretic N, Bureau TE.** 2012. A gene family derived from transposable elements during early angiosperm evolution has reproductive fitness benefits in *Arabidopsis thaliana*. *PLoS Genet* **8:**e1002931.

154. **Jiao Y, Deng XW.** 2007. A genome-wide transcriptional activity survey of rice transposable element-related genes. *Genome Biol* **8:**R28.

155. **Marsch-Martinez N.** 2011. A transposon-based activation tagging system for gene function discovery in Arabidopsis. *Methods Mol Biol* **754:**67–83.

156. **Hirochika H, Okamoto H, Kakutani T.** 2000. Silencing of retrotransposons in arabidopsis and reactivation by the *ddm1* mutation. *Plant Cell* **12:**357–369.

157. **Bennetzen JL, Wang H.** 2014. The contributions of transposable elements to the structure, function, and evolution of plant genomes. *Annu Rev Plant Biol* **65:** 505–530.

158. **Woodhouse MR, Schnable JC, Pedersen BS, Lyons E, Lisch D, Subramaniam S, Freeling M.** 2010. Following tetraploidy in maize, a short deletion mechanism removed genes preferentially from one of the two homologs. *PLoS Biol* **8:**e1000409.

159. **Franklin AE, McElver J, Sunjevaric I, Rothstein R, Bowen B, Cande WZ.** 1999. Three-dimensional microscopy of the Rad51 recombination protein during meiotic prophase. *Plant Cell* **11**(5):809–824.

160. **Li J, Harper LC, Golubovskaya I, Wang CR, Weber D, Meeley RB, McElver J, Bowen B, Cande WZ, Schnable PS.** 2007. Functional analysis of maize RAD51 in meiosis and double-strand break repair. *Genetics* **176**(3): 1469–1482.

161. **Bennetzen JL.** 1996. The Mutator transposable element system of maize. *Curr Top Microbiol Immunol* **204:** 195–229.

Mobile DNA, 3rd Edition
Nancy L. Craig, Michael Chandler, Martin Gellert, Alan M. Lambowitz, Phoebe A. Rice and Suzanne Sandmeyer

doi:10.1128/microbiolspec.MDNA3-0052-2014

Max Salganik[1,2]
Matthew L. Hirsch[1,3]
Richard Jude Samulski[1,4]

37 Adeno-associated Virus as a Mammalian DNA Vector

WILD-TYPE ADENO-ASSOCIATED VIRUS DISCOVERY TO VECTORIZATION

Adeno-associated Virus Discovery

In 1965, Atchinson and colleagues observed a small (25 nm) contaminant particle within electron micrographs of their adenovirus preparations (1). These contaminants were purified from adenovirus and applied to cells. They were shown to be nonautonomous, as particle production also required adenovirus coinfection (1). These defective, small particles were therefore named adeno-associated virus (AAV) which, even to this day (over 50 years later), remains one of the smallest viruses known to man. Consequently, AAV is a very simple virus with a protein capsid composed of 60 capsid subunits, and a ~4.7-kb single-stranded linear DNA genome that is framed by inverted terminal repeat sequences (ITRs) (2). Both polarities of the single-stranded genome are individually packaged at similar efficiencies (3). There are three AAV genes identified to date, which collectively mediate genome replication, site-specific integration, capsid production, and genome packaging (4–8). Twelve natural serotypes of AAV have been reported with many additional variants; however, the last 30 years have seen the most work done with AAV serotype 2 because of its amenability to cell culture.

In the case of wild-type AAV, the coordinated expression of viral genes involves complex self-regulatory feedback loops and appears to be mediated primarily by AAV gene products (9). The employment of these regulatory schemes controls the lysogenic or replicative phase of the viral life cycle which is induced by the presence of a helper virus. Although, AAV was originally described as an adenovirus contaminant, it was later determined that additional viruses, and cellular stress in general, also provide helper functions to induce the AAV replication phase decision (10). AAV is unique because it represents the only known case of site-specific integration in the human chromosome (5, 6); a process that is initiated when AAV infection occurs without a helper virus. In this process the AAV Rep protein interacts with the ITRs on the AAV genome and a very similar sequence (termed AAVS1) uniquely found on human chromosome 19 (reviewed in reference 10). It is thought that the endonuclease activity of the Rep protein mediates strand scission at AAVS1, facilitating host

[1]Gene Therapy Center; [2]Lineberger Comprehensive Cancer Center; [3]Department of Ophthalmology; [4]Department of Pharmacology, University of North Carolina, Chapel Hill, NC.

DNA polymerase strand switching, viral genome replication, and subsequent imprecise integration as both a monomer and concatamers in an approximate 1-kb region of AAVS1 (10, 11). The latent AAV genome remains primarily quiescent until favorable helper conditions shift viral gene expression in a manner that favors genome excision and entry into the replication phase (10). AAV particle escape relies on the disruption of cellular membranes by helper-virus-induced cell lysis and initiates a new round of the AAV life cycle.

ADENO-ASSOCIATED VIRUS COMPONENTS/VECTOROLOGY

Adeno-associated Virus Capsid

AAV virions have icosahedral symmetry with a triangulation number of 1 (T = 1), representing the simplest structure of all viruses. The capsid surface contains three axes of symmetry; two-fold, three-fold and five-fold. The two-fold axis is characterized by a valley-like depression and represents the thinnest capsid cross-section with the lowest number of contacts between capsid protein monomers in the AAV capsid. The three-fold axis is characterized by prominent spike-like protrusions and contains the known receptor binding sites (12, 13) as well as many of the identified antibody recognition sites (14, 15). Finally the five-fold axis is comprised of a low, flat pentamer with a prominent central pore that is surrounded by an elevated rim. This pore is the site of genome packaging (16) and is thought to be necessary for the externalization of N-terminal minor capsid protein motifs necessary for AAV infection (16, 17). Three AAV capsid proteins (VPs) have been identified that are transcribed from the P40 promoter (Figure 1). The P40 transcripts undergo slicing, with one major and one minor splice variant produced. The VPs of all AAVs share a common C-terminus, with VP1 (87 kDa), and VP2 (73 kDa) each having additional N-terminal sequences when compared with VP3 (62 kDa) (18, 19). VP3 represents nearly ~90% of the total protein of intact virions, whereas VP1 and VP2 together represent the remaining ~10% (18, 20, 21). Only the minor splice variant from the P40 transcript contains the VP1 translational start codon (AUG), so explaining its lower production levels compared with

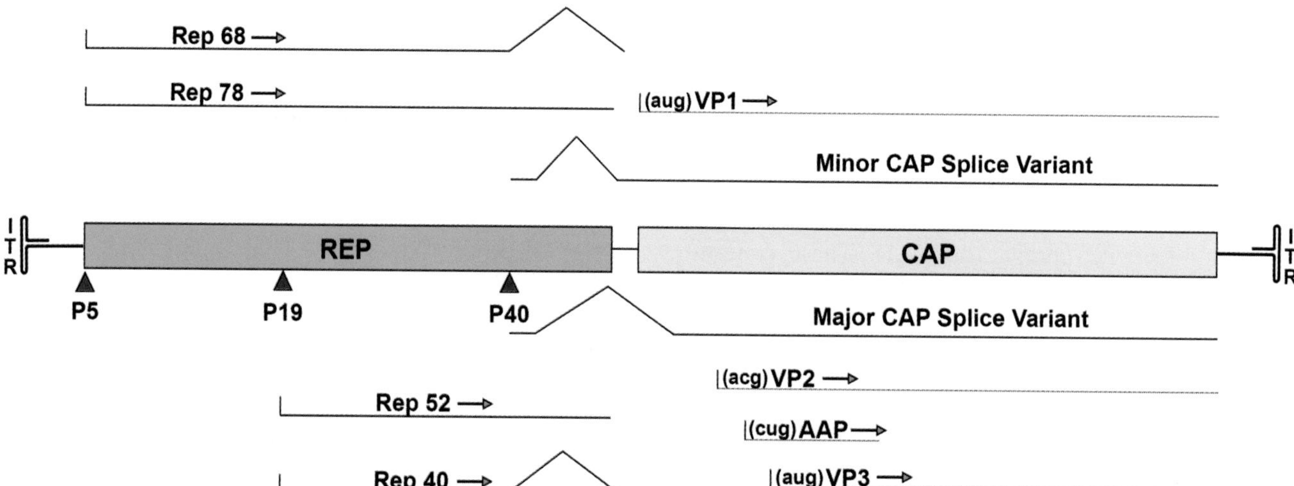

Figure 1 The adeno-associated virus (AAV) genome is a linear ~4.7-kb single-stranded DNA which is flanked by inverted terminal repeats. The genome contains three promoters that drive transcription of the viruses replication (REP), capsid (CAP) and assembly (AAP) genes. The first two promoters, p5 and p19, drive transcription of the large (78/68 kDa) and small (52/40 kDa) Rep proteins, respectively. In each case an alternate splice at the end of each transcript results in the smaller variant of each protein (Rep68 and Rep40). Transcription from the p40 promoter produces one large mRNA with an intron. A minor splice variant of this message contains the translational initiation codon (AUG) for the largest capsid protein (VP1), while the major splice variant truncates this sequence. The major splice variant contains a nontraditional translation initiation codon (ACG) at the start of VP2, which is often skipped for the downstream AUG of VP3, the smallest and most abundant capsid protein. The p40 transcripts also contain an alternate reading frame that encodes the assembly activating protein (AAP), which is translated from a nontraditional CUG.
doi:10.1128/microbiolspec.MDNA3-0052-2014.f1

VP3. Likewise, whereas the major splice variant contains translational initiation sites for VP2 and VP3, only the latter uses a traditional AUG whereas the former uses an ACG, which is frequently skipped; this accounts for VP2's lower production levels. Although VP3 alone is capable of assembling into a capsid, the presence of VP1 is critical for successful infection because of the enzymatic and targeting motifs in the 135-amino-acid N-terminal region that is unique to VP1. In contrast, VP2 has a much lower effect on infection, to the point that it is considered nonessential, and can tolerate the addition of large peptides to its N-terminus (22) (a fact further discussed in the *AAV receptors* section). To date, 12 serotypes (AAV1 to AAV12) and more than 100 variants have been isolated from human and nonhuman primate tissue samples (23). Among the 12 serotypes, AAV serotypes 1–9 are currently being developed as gene therapy vectors because of their broad tissue tropisms. A review of these serotypes and their receptor preferences is given in Table 1.

To date, the crystal structure of AAVs 1–9 have been determined using X-ray crystallography (24–36). Cocrystal structures of AAV serotypes 2, 5, and 6 with respective capsid receptors (e.g., heparin sulfate and sialic acid) have furthered the molecular understanding of AAV virus/receptor interactions (37–40). In solving the structures of intact AAV capsids, only the common C-terminal-most ~520 amino acids are typically observed (VP3 common region), with the remainder of the sequence (VP1 and VP2 unique regions) believed to either be disordered or in a nonicosahedral symmetry, which is assumed when solving these structures. When the AAV capsid structure is compared to other parvoviruses (adenovirus, parvovirus B19, canine parvovirus, feline parvovirus, minute virus of mice, and porcine parvovirus), the core of the protein is comprised of a remarkably conserved eight-stranded anti-parallel β-barrel motif among all parvoviruses. Structural differences between AAVs and other parvoviruses can primarily be mapped to their surface topology, which accounts for their binding to different receptors and the resulting tissue tropisms (Table 1). The surface loops equivalent to those making up the three-fold protrusions were found to be the most variable among serotypes. This variable region spans the center of the primary capsid sequence (residues ~440 to 600, AAV2 VP1 numbering); whereas residues located at the N- and C-termini are conserved (27). Variations within these loop regions are also thought to eliminate antibody recognition of serotypes AAV1 and AAV3–9 (41, 42). These observations highlight the conservation in capsid structure and yet illustrate the significant diversity of AAV serotype capsids and their ability to exploit multiple cell surface receptors, mediate successful intracellular trafficking with altered efficiencies/kinetics, and elicit specific humoral immunity.

Table 1 Known receptor usage for adeno-associated virus (AAV) serotypes 1–12

Serotype	Receptor/Co-Receptor
AAV1	α2–3/α2–6 N-linked SA (69, 178, 226)
AAV2	HSPG (70), FGFR1 (227), HGFR (228), Integrin $\alpha_V\beta_5/\alpha_5\beta_1$ (229, 230), 37/67 kDa LamR (231)
AAV3	HSPG (70), 37/67 kDa LamR (231), HGFR (232)
AAV4	α2–3 O-linked SA (71, 72)
AAV5	α2–3 N-linked SA (71, 72), PDGFR (233)
AAV6	HSPG, α2–3/ α2–6 N-linked SA (69, 234), EGFR (235)
AAV7	Undetermined
AAV8	37/67 kDa LamR (231)
AAV9	Galactose (73, 236), 37/67 kDa LamR (231)
AAVrh10	Undetermined
AAV11	Undetermined
AAV12	Undetermined

Note: Does not include mutated variants with alternate receptor usage.
FGFR1, fibroblast growth factor receptor; HGFR, hepatocyte growth factor receptor; HSPG, heparan sulfate proteoglycan; PDGFR, platelet-derived growth factor receptor; SA, sialic acid.

Adeno-associated Virus Genome

The 4.679-kb AAV2 genome displays the *rep*, *cap*, and *aap* genes in the sense orientation, flanked on either end by 145 nucleotide ITRs. There are three promoter sites (Figure 1), P5, P19, and P40. The first two promoters drive transcription of the large (78/68 kDa) and small (52/40 kDa) replication (Rep) proteins, with the large Rep proteins containing a unique N-terminal domain in addition to the sequence shared with the small Reps. The smaller variants of both the large and small Reps are produced by splicing the P5 and P19 transcripts (Figure 1). As previously mentioned, the P40 promoter is responsible for producing transcripts that are spliced and translated into VP1–3. Recently, an additional open reading frame involved in particle assembly, termed the assembly activating protein (AAP), was found nested within the *cap* gene (8) (Figure 3). AAP is thought to directly interact with the capsid protein's conserved C-terminus, serving as an assembly scaffold (8, 60).

The AAV ITRs can assume a T-shaped structure (43) and exist in two conformations termed "flip" or "flop" and generated during the rolling hairpin mechanism of replication (44). There are three salient features of the AAV ITR that are required for vector production: (i) a

16-nucleotide tetrameric repeat (GAGC) sequence that is specifically bound by the Rep68/78 proteins termed the Rep-binding element (RBE), (ii) a sequence within the ITR internal hairpin that serves to orient Rep68/78 towards its resolution site termed RBE' (45–49), and (iii) the terminal resolution sequence (trs), which is the site of the strand-specific Rep68/78 endonuclease activity (44, 50–52). The N-termini of the larger Rep proteins contain a DNA-binding domain, as well as strand-specific endonuclease activity, necessary for initiating replication during virion production and site-specific integration (46, 53–56). The C-terminal portion of Rep68/78 is required for multimerization and confers helicase activity which, in part, extrudes a putative nicking stem within the ITR sequence (54, 57, 58). This conformational change presents the trs loop structure to the endonuclease domain of RBE-bound Rep68/78. Rep then induces a strand-specific nick at the trs forming a 5′-phosphotyrosyl tether between Rep and the trs (52, 54, 59). In AAV genome replication, this allows for the resolution of the double-stranded intermediate and the subsequent replication of both ITRs (Figure 2). The smaller Rep proteins (Rep 52 and 40) also maintain these C-terminal properties. The temporal regulation of AAV gene expression is, in large part, controlled by Rep68/78 but is specific to the intracellular environment (reviewed in reference 10).

Adeno-associated Virus Vectorology

In 1982, the entire AAV2 genome was cloned into the pBR322 plasmid (pAAV). In that work it was demonstrated that transfection of pAAV into human cells, in the presence of adenovirus, resulted in the production of AAV2 particles (4). Shortly thereafter, it was shown that the viral genes (*rep*, *cap*, and *aap*) were not necessary in *cis* for genome replication and particle production, and were effectively complemented by another plasmid without the AAV ITRs (129). The key observation was that a transgenic sequence placed between the 145-nucleotide plasmid-borne ITRs could be used for AAV "vector" production in the presence of Rep, Cap, and AAP supplied in *trans* (61, 62). This result suggested that AAV capsids did not have genomic packaging signal sequences required for virion production. Additionally, at the time, these data made AAV vectors among the most efficient mobile contexts for human gene delivery, and did so in the complete absence of any viral genes (61, 62). Furthermore, in the absence of the Rep proteins, site-specific integration at AAVS1 did not occur and therefore, the majority of AAV vector genomes persisted as stable episomes with low levels of "random" integration (reviewed in reference 10).

The initial demonstrations of recombinant AAV (rAAV) production, however, were complicated by the presence of the now "contaminating" adenovirus particles, which initially appeared necessary for rAAV production because they provided the necessary "helper" functions. To overcome this limitation, the necessary adenoviral genes were cloned and supplied on a plasmid, so eliminating contaminating adenovirus particles from rAAV production (62, 63). Despite over 30 years of research on AAV vector production, this "helper-free" rAAV production strategy remains the most widely used method and relies, in its most basic embodiment, on transfection with three plasmids: (i) pRepCap (provides the viral genome excluding the ITRs, (ii) pITR-sequence* (provides a variable sequence to be packaged as the transgenic genome), (iii) pAdeno-help (provides the helper functions for vector genome replication) (Figure 3); reviewed elsewhere (64).

ADENO-ASSOCIATED VIRUS VECTOR TRANSDUCTION AND OPTIMIZATION

Adeno-associated Virus Receptors

Most of our current knowledge of AAV receptor usage comes from cell culture studies that have used a combination of biochemical and genetic tools. AAVs exploit several different types of glycans as primary attachment receptors and use a broad range of cell surface receptors to mediate cellular uptake. Receptor usage varies by serotype, though heparan sulfate proteoglycan, terminal galactose (Gal), and several linkage variants of sialic acid (SA) constitute the currently identified glycan repertoire for known AAVs (Table 1). The unique combination of specific glycan and entry receptor affinity is believed to be the primary cause for the different tissue tropisms seen among the serotypes, although there is evidence for post-entry restriction in some cell/tissue types (65–68). Structural, biochemical, and genetic studies have helped to identify the glycan-binding regions for AAV1–6 (69–72) and AAV9 (73). This has enabled a wide variety of experiments that have directly used this structural information in an attempt to alter the natural trafficking patterns of AAV. As an example, a single mutation (AAV2 585E) in the AAV2 heparin-binding footprint abolishes AAV2 heparin binding and results in a different transduction profile (cardiac versus hepatic tropism) *in vivo* (74). As a natural extension of such results, rational design studies of AAV capsids have looked at swapping the critical motifs for primary receptor binding sites between different serotypes and in one example, demonstrated that an AAV2/AAV8 chimera (AAV2i8) displays an altered transduction profile as the result of such a swap. AAV2i8 selectively transduces cardiac and whole-body skeletal muscles with high efficiency while losing liver tropism (75). Further studies integrating the AAV9 Gal binding residues into the AAV2 capsid (AAV2G9) have found that AAV2G9 has dual receptor function and exploits Gal and heparan sulfate receptors for infection (76). Of particular interest, AAV2G9 retains a similar tropism to AAV2 but confers more rapid onset and higher liver transgene expression in mice. The strategy of domain swaps continues to be actively used and is just one tool available for generating custom tropism.

Further efforts to tailor AAV tropism and to generate AAV capsids with unique properties have exploited two broad strategies: capsid modification with targeting molecules and generation of novel AAV capsids through directed evolution. Efforts in the former have included the insertion of targeting domains such as integrin-binding motifs (77), antibody domains (78), and other targeting peptides (79–83) into the outer variable loops of AAV. In each case, there were some promising results with novel tropism or the ability to infect previously refractory cell types. Unfortunately these successes were modest in their ability to infect new cells in culture and even more so *in vivo*, and more often resulted in viruses that were defective for infection. Subsequent work has shown that insertional modification is highly sequence specific (79, 84), suggesting that it was difficult to predict sites and peptide candidates for successful large capsid insertions. As AAV is a small and relatively simple virus it is not surprising that large insertions into the capsid proteins are often difficult to tolerate. An alternative strategy for retargeting has used the fact that VP2 is both nonessential and able to tolerate large additional sequences at its N-terminus, which would subsequently be displayed on the capsid surface (22, 85). Although this has been a successful strategy for decorating the capsid with a myriad of proteins such as GFP, the main pitfalls of this approach have traditionally been the retention of native capsid tropism, which competes with any retargeting molecule and low copy number of targeting molecules. This latter problem is the result of VP2 being a minor capsid protein, and while recent efforts (86, 87) have combined this VP2-based approach with detargeted vectors with some success, the problem of low copy number remains. The upper size limit for VP2-based tags is also not clearly defined and appears to be at least partially sequence dependent.

Conversely, directed evolution as a strategy has sought to generate new capsid properties without previous mechanistic knowledge. Early directed evolution strategies relied on phage display libraries to generate peptides that would bind to a desired target and that would then be inserted into the AAV capsid (88–93). Such strategies are an extension of the capsid insertion approach previously discussed and also had to contend with the fact that successful peptide capsid insertion is highly sequence specific. Hence, subsequent attempts transitioned to random peptide insertion followed by selection to ensure that peptide inserts would bind their molecular targets in the context of the capsid (83, 94–99). Others have taken a more holistic directed evolution approach by generating libraries of novel AAV sequences either by random mutagenesis (100–102) or shuffling of capsid sequences across multiple serotypes followed by selection on either a target cell population (99, 103–107) or *in vivo* (108–113). The latter approach in particular has been successful in yielding capsids with conserved and functional interiors and novel tropism. Although AAV offers a compelling platform for gene therapy, the rigorous and widely varying demands of a broad range of clinical applications will ensure that the efforts to augment and tailor the natural biology of AAV will continue.

Adeno-associated Virus Vector Trafficking/Uncoating

A broad overview of recombinant AAV transduction is provided in Figure 4. This is believed to be the same process as occurs with wild-type infection, with the exception of gene transcription, which in wild-type virus is controlled by the genome and in recombinant vectors is dictated by the promoter in a given expression cassette. AAV has traditionally been thought to exploit clathrin-mediated, dynamin-dependent endocytosis (114–118). However, a growing body of work suggests that this is only one of several entry pathways that different AAVs can use with evidence implicating the caveolin pathway (119, 120), as well as the clathrin-independent carrier mediated/glycosylphosphatidylinositol-anchored-protein-enriched endosomal compartment-associated (CLIC/GEEC) pathway (121, 122). There is evidence to suggest that different AAV serotypes may be able to use different entry pathways based on their native receptor tropism, which may help to explain the post-entry restriction seen in some cell types (65–68, 123). Furthermore, although ablation of native receptor binding often results in an attenuation of infection in cell culture (AAV2 non-heparan sulfate proteoglycan-binding mutants), the same effect is not always seen *in vivo*, where mutants often show altered tissue tropism instead of an outright defective phenotype (74, 103, 124). AAV trafficking post-entry

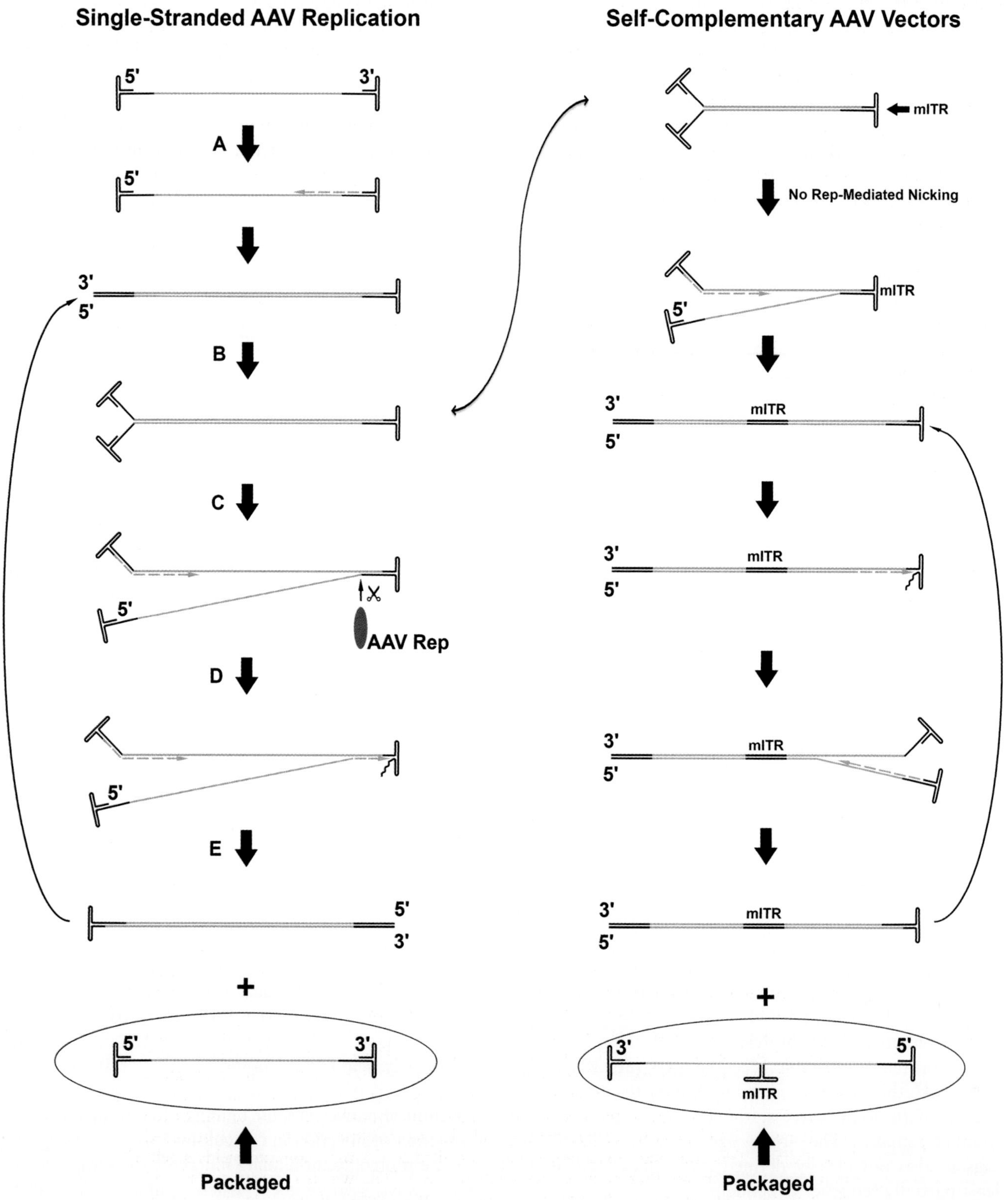

Single-Stranded AAV Replication

5' 3'

A

5'

3'
5'

B

C

5'

AAV Rep

D

5'

E

5'
3'

+

5' 3'

Packaged

Self-Complementary AAV Vectors

mITR

No Rep-Mediated Nicking

mITR
5'

3' mITR
5'

3' mITR
5'

3' mITR
5'

3' mITR
5'

+

3' 5'
mITR

Packaged

has been associated with a broad range of organelles from both the endosomal (88, 116, 121) and caveolin (120, 125) pathways, and appears to be influenced by a wide range of factors such as initial route of entry and initial infecting dose (126, 127). Despite the relative uncertainty over the exact trafficking route of AAV, all productive pathways share the common feature of leading to an acidified compartment. AAV trafficking through acidified endosomal compartments is required for infection. Either a pharmacological block of endosomal acidification (116, 121, 128, 129) or a bypass of the endosomal system via microinjection of AAV into the cytoplasm (3, 4, 88, 130–133), results in a failure to achieve vector transgene expression. An acidic environment has been shown to lead to structural transitions in the AAV capsid (85, 134), which activate an autocatalytic protease activity within the capsid (86, 134–136) and prime the capsid for externalization of the VP1 N-terminal domains (88, 135). The VP1 N-terminus is critical for infection and contains a phospholipase A2 domain (94, 137) along with a broad range of signaling motifs (100, 136, 137) including several nuclear localization sites (103, 104, 135, 137). Although mutation of the PLA2 domain is highly detrimental to infection and results in the perinuclear accumulation of virus, it also appears by itself to be insufficient to facilitate endosomal escape of the 3-MDa parvovirus capsid (108, 110, 112, 138, 139). It may however enable the virus to display the signaling motifs of the V1 N-terminus on the exterior of small vesicles, so enabling AAV to guide its own trafficking to the perinuclear region along the microtubule network while still contained in vesicles (114, 140, 141). Whatever the exact mechanism of endosomal escape, AAV does enter the cytoplasm where a portion of the infecting virions is subjected to phosphorylation and targeting for proteasome-mediated degradation (119, 142–144) The use of proteasome-inhibiting drugs (121, 140, 141) and surface tyrosine mutants (123, 135, 141) both result in notable increases in AAV transduction. AAV enters the nucleus as an intact capsid, most likely via active transport across the nuclear pore complex (103, 124, 141). The presence of the intact AAV capsid in the nucleus is necessary for productive infection (88, 124, 145, 146) and a carefully choreographed trafficking of the capsid to and from the nucleolus appears to influence the timing of uncoating and the success of vector-delivered transgene expression (125). There is also some evidence that genome uncoating does not involve the complete disassociation of the capsid and that the capsid may play a role in gene expression after genome uncoating (126).

Adeno-associated Virus Vector Genome Fate

Single-strand DNA vector genomes are uncoated within the nucleus and form discrete foci whose abundance directly correlates to the transduction efficiency (128, 147). The generation of a DNA molecule competent for transcription is dependent upon the genome concentration within the nucleus. When vector genomes are limiting, second-strand synthesis mediated by, minimally, DNA polymerase delta, is primed by the 3′ ITR giving rise to a duplexed, yet still single-stranded, genome (Figure 2) (3, 4, 130–133, 142, 146). Additionally, it has been demonstrated that AAV vector genomes, when present in the nucleus at high concentrations, can also form double-strand molecules via opposite polarity single-strand genome annealing (134, 148–150). Regardless of the exact mechanism of second-strand generation, it is apparent that target cell replication is not

Figure 2 Adeno-associated virus (AAV) genome replication is primed by the 3′ inverted terminal repeat (ITR), and primarily uses the host pol δ (A). The genome is replicated through the 5′ ITR (via strand displacement), yielding a double-stranded (dsDNA) intermediate (B). The newly generated 3′ ITR (produced when copying the 5′ ITR) then primes replication in the opposite direction (C). At the same time, the AAV Rep protein, bindings to the Rep-binding element (RBE) on the original 3′ ITR, and cuts the lower strand at the terminal resolution site (*trs*), generating a free 3′ end (C), which allows for the replication of the original 3′ ITR (D) and the production of a new single-stranded (ssDNA) genome as well as another dsDNA intermediate (E) which is fed back into the replication cascade. This replication mechanism results in the production of ssDNA genomes of both polarities, which are then packaged into the capsid. In the case of self-complementary (scDNA) vectors, one of the ITRs is mutanted (mITR) to remove the *trs*, and prevent resolution of the double-stranded intermediate. This resulting replication scheme (detailed in the right column) yields a self-complementary genome that at its center contains the mITR, which is never nicked and which serves as an intermolecular hinge. This hinge acts to form a duplex genome immediately upon uncoating and bypasses the typical need for second-strand synthesis that serves as a bottleneck for ssDNA vectors. doi:10.1128/microbiolspec.MDNA3-0052-2014.f2

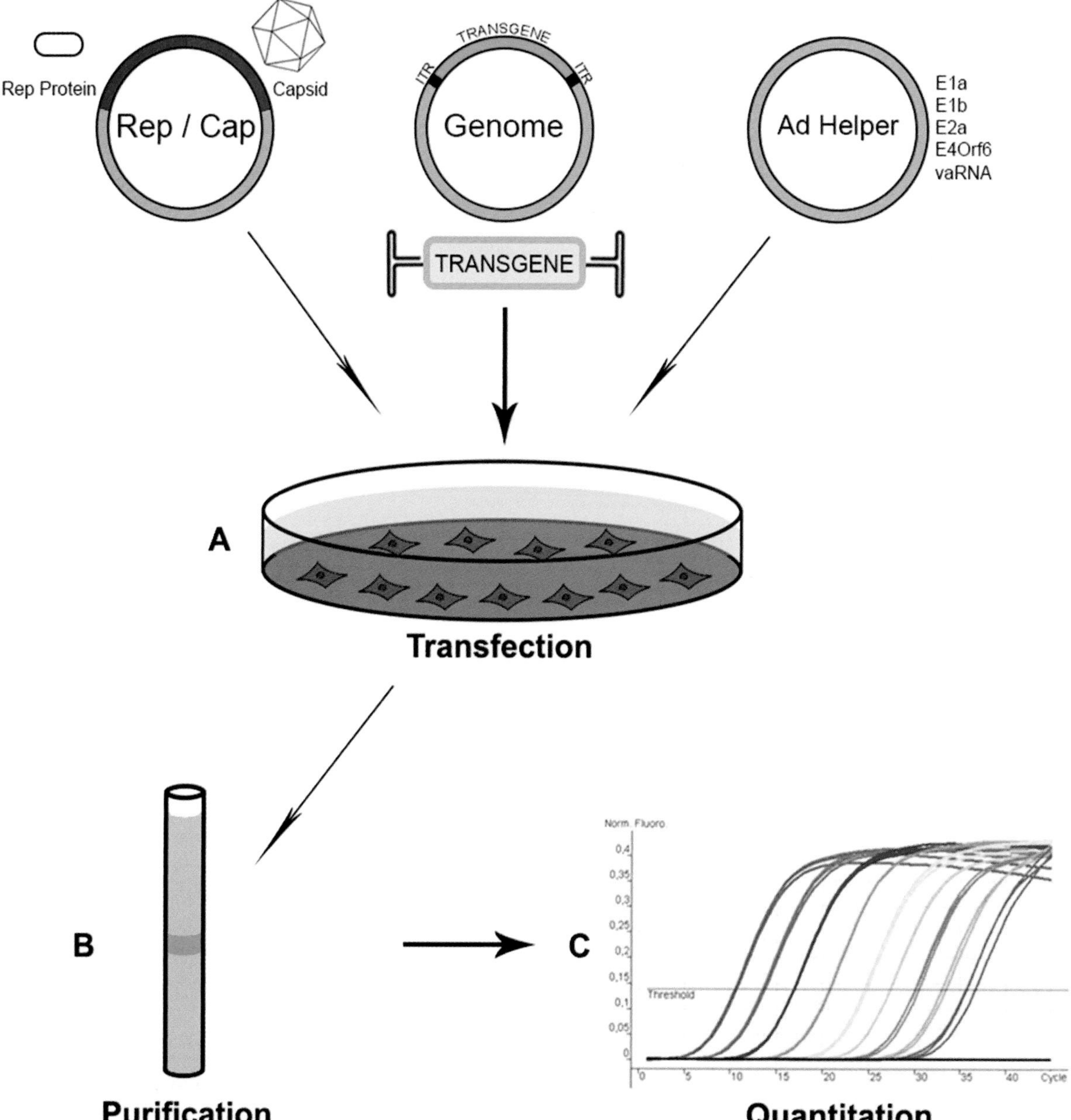

Figure 3 Adeno-associated virus (AAV) vector production. (A) Tradition AAV vector production begins with transfection of mammalian cells (commonly HEK 293) with three plasmids. The first provides the Cap proteins from the chosen AAV serotype in conjunction with Rep from AAV2. This plasmid lacks inverted terminal repeats (ITRs), ensuring that the Rep/Cap sequences are not packaged into AAV capsids and no replication-competent virus is made. The Genome plasmid contains the chosen transgene sequence flanked by ITRs, which are necessary for packaging and genome replication. In the case of ssDNA vectors, transgene cassettes up to 5 kbp can be packaged with high efficiency, whereas scDNA vectors can accommodate cassettes half that size. The third plasmid provides in *trans* the adenovirus genes that are necessary for AAV replication. It is common to see this plasmid and the Rep/Cap plasmid combined in a single large construct for simplified production. (B) After 48–72 hr cells are harvested and lysed. Vectors can be purified by either column chromatography or density gradient centrifugation, which can separate AAV from contaminating cellular proteins as well as separating empty capsids from genome-containing particles. It is not uncommon to see at least two purification steps employed for high purity vectors and column-based methods can vary depending on the serotype being purified. (C) Vector quantitation is most often performed by measuring DNase-resistant genomes by real-time PCR. In cases where verification is needed for the separation of full and empty capsids, this genome titering is combined with an ELISA-based assay for capsid protein. doi:10.1128/microbiolspec.MDNA3-0052-2014.f3

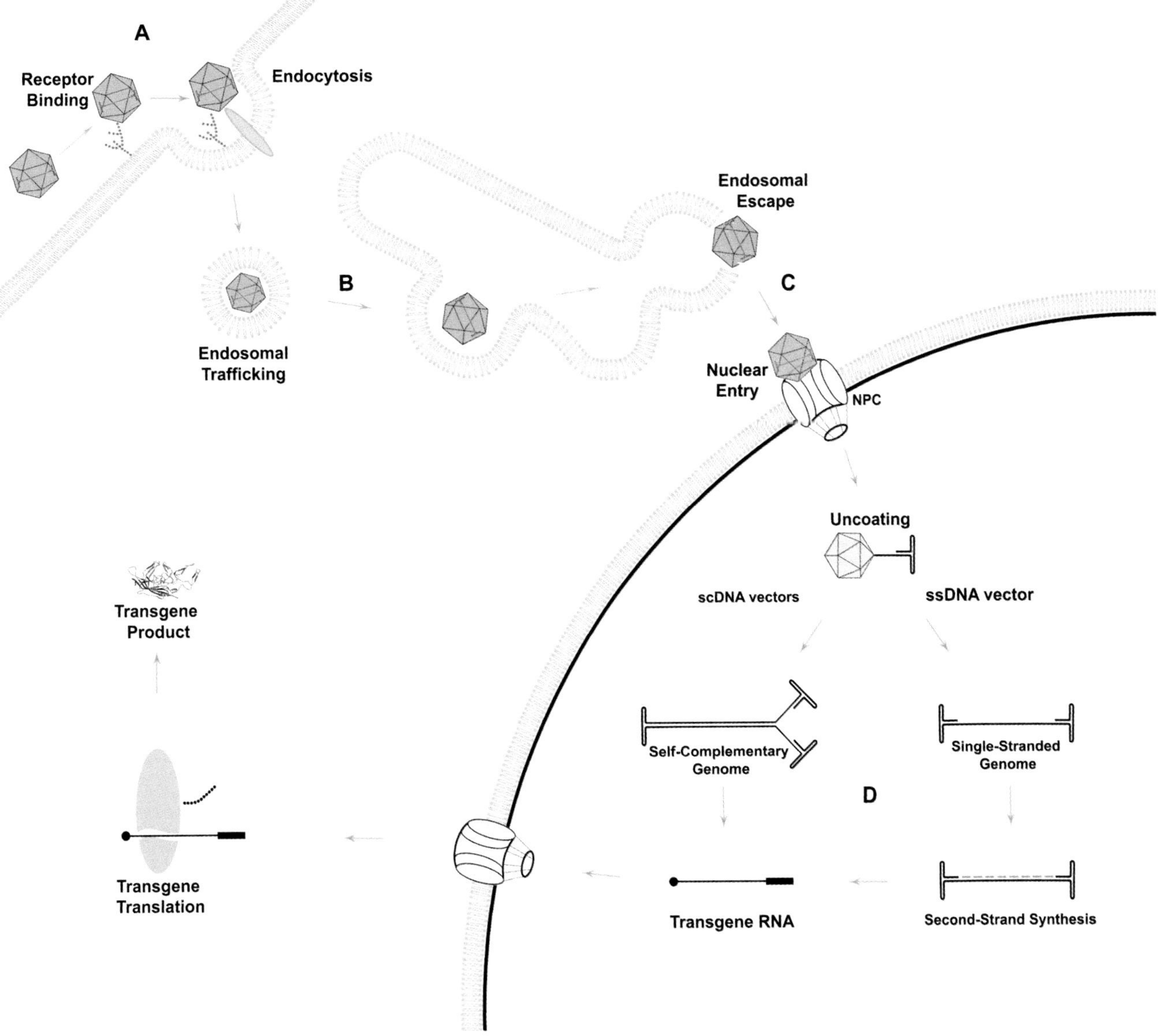

Figure 4 Adeno-associated virus (AAV) vector infectious pathway. (A) Infection is initiated by capsid binding to cell-surface receptors. In the case of AAV serotype 2 this is a two-step process involving binding to a primary glycan receptor followed by binding to a protein receptor that mediates endocytosis. Classical AAV2 internalization is believed to be a clathrin-mediated, dynamin-dependent process, although other pathways have been implicated. (B) AAV traffics through the endosomal system where it is exposed to a low-pH environment that triggers the externalization of the VP1/2 unique regions and activation of the autocatalytic capsid protease. Exposure to the low-pH environment is necessary for productive AAV infection. (C) Escape from the endosomal compartment is mediated by the VP1 PLA2 domain, and AAV uses several nuclear localization signaling motifs to traffic to the nucleus and enter, most likely via the nuclear pore complex (NPC). (D) AAV uncoats its genome in the nucleus. In the case of single-stranded DNA vectors, second-strand synthesis must first occur before transcription, creating a bottleneck in the transduction process. Self-complementary (sc) vectors use a single mutant inverted terminal repeat (ITR) in conjunction with both + and – strands to produce a genome that is double-stranded upon uncoating so bypassing second-strand synthesis, albeit at the cost of lower packaging capacity. After a double-stranded genome is formed transcription and translation are driven by the transgene cassette and are not virus-specific. doi:10.1128/microbiolspec.MDNA3-0052-2014.f4

necessary and that most, if not all, cell types appear to have the necessary functions to allow this aspect of AAV vector transduction. Following the generation of a duplexed DNA molecule, the vector genome primarily remains episomal and circularizes via intra- and intermolecular ITR linkages shortly after infection (134–136, 142, 151). Using circularization-dependent AAV vectors in dividing cell cultures, it was demonstrated that RecQ helicase family members, Mre11, Nbs1, and ATM are required for efficient vector genome circularization (135, 148, 151). In nondividing skeletal muscle fibers, ATM and DNA-PKcs were necessary for genome circularization wherease Nbs1 was not (137, 151, 152). Consistent with these *in vivo* data, elegant work performed in the mouse liver demonstrated that Artemis and DNA-PKcs process the AAV ITRs and promote the circularization/concatamerization of vector genomes, which is also supported by other reports (62, 136, 137, 153–156). Collectively, with regards to the generation of persistent AAV vector episomes, it appears that the circularization event is probably mediated at the ITRs by different DNA repair pathways; homology directed repair in dividing cells and nonhomologous end joining (NHEJ) in terminally differentiated tissues (135, 137, 157). One of the unique features of AAV vector transduction is the remarkably long-term transgene expression from the AAV vector episome (138, 139, 158). In general, these molecules are resistant to overt transcriptional silencing and have been demonstrated to produce the transgene product for over 4 years in humans following a single injection.

It is well appreciated that the rAAV transduction elicits a DNA damage response within the host cell, which influences the overall efficiency of transduction. In general, several factors have been identified that appear to inhibit, directly or indirectly, rAAV transduction. For instance, rAAV transduction in dividing cells deficient for ATM, a key regulator of double-strand break repair, is dramatically increased, an effect attributed to enhancement of second-strand synthesis (140, 141, 159). Interestingly, several forms of DNA stress (UV, hydroxyurea , mytomycin C) also increase AAV vector transduction (142, 143, 157), although not in the absence of ATM (140, 141, 155, 156), further supporting the notion that ATM inhibits rAAV transduction at the genome level. This supposed inhibitory effect, however, is not necessarily specific to ATM, as DNA-PKcs and Rad52 are also implicated in this process, indicating the involvement of both primary DNA repair pathways, homologous recombination (HR) and NHEJ (135, 141, 155, 156). In the absence of both Ku86 and Rad52, rAAV transduction increases, and consistently both Ku86 and Rad52 have been found to bind AAV genomes following vector transduction (137, 141). In addition, Mre11, a key sensor of damaged DNA normally considered part of the HR pathway, in conjunction with Nbs1 and Rad50, also appears to inhibit AAV transduction (145, 146, 160). A fascinating study in mice demonstrated that downregulation of DNA repair proteins, in particular the Mre11–Rad50–Nbs1 (MRN) complex, via terminal differentiation coincided with increased AAV vector transduction (146, 155, 156, 160). In addition to loss of DNA repair proteins involved in HR during terminal differentiation, cells also exited the cell cycle to a quiescent state, therefore complicating the assignment of discrete roles for particular proteins in the inhibition of rAAV transduction (61, 62, 146). In particular, the role of intact cell cycle checkpoints, and the growth phase in general, also appear to affect transduction and furthermore, confound the direct implication of cell cycle regulators, such as ATM and p53, as controllers of AAV vector transduction. As mentioned above, several genotoxic stresses that also arrest the cell cycle result in increased transduction (9, 147). The mechanism of this enhancement is not completely understood but could be explained by the sequestration of repair factors that are reported to inhibit rAAV transduction away from the AAV genome (such as the MRN complex). However, this explanation, and the actual role of the cell cycle in AAV transduction, is probably not attributable to a single event. For instance, exit from the cell cycle *in vivo* increases transduction, whereas an earlier report demonstrated that cells in S-phase are transduced much better (200-fold increase) compared with nondividing cells (142, 146, 158). Interestingly, in wild-type diploid cultured cells, AAV transduction has been reported to induce cell cycle arrests at both the G1 and G2/M checkpoints (148–150, 161, 162). However, it has been our observation that in cells defective for particular cell cycle checkpoints, transduction increases initially, perhaps due to a quicker entry into S-phase (142, 151, 161, 162). In fact, in the absence of normal checkpoint functions, such as in p53-deficient cells or human embryonic stem cells, it has been reported that AAV vector transduction results in cell death (148, 151, 163). The reasons for the rAAV mediated cytotoxicity remain unclear and are probably cell-type-dependent; however, ITRs have been implicated as stalled replication forks and/or perhaps forming G-quadruplex/quartets, both of which would induce DNA damage cascades from human telomeres (151, 152, 164).

Although the majority of AAV vector genomes persist as episomes, as with all exogenous DNA, a low percentage will undergo illegitimate host genome integration (62, 153–156, 164–166). One major factor influencing this largely undesirable event is whether or not the cells are actively replicating/dividing (157, 167). It has been reported in dividing cell cultures that illegitimate integration of AAV vector genomes occurs at DSBs within the host genome and that efficiencies can be as high as 1% of transduced cells (158, 166, 167). Furthermore, induced DSBs, either by general genotoxic stress or by a site-specific endonuclease, resulted in the illegitimate integration of rAAV genomes with microhomologies, deletions, and insertions present at the integration junctions (159, 168–171). In contrast, an early investigation in a nondividing bronchial epithelial cell line demonstrated no AAV vector integration despite transduction in over 90% of the cells (157, 158). In addition to cellular replication, AAV vector integration prefers CpG islands and demonstrates a modest inclination towards actively transcribed regions (155, 156, 172). Consistently, a bias for vector integration was demonstrated at ribosomal DNA repeats (155, 156, 172). A large analysis of rAAV integration events in multiple murine tissues found that 30% of all integration events occurred near DNA palindromes, which are known sites of chromosome instability (137, 173). Another report found that AAV vector integration in mice occurred, in all cases, in actively transcribed genes, further demonstrating the potential risk of oncogenesis following insertional mutagenesis (160, 174). Hence, it appears that AAV vector integration is not random, and the efficiency probably depends on the species, the cell type, and the phase of the cell cycle (155, 156, 160, 174). These collective reports imply that fragile DNA sequences and or replication-induced damage of the host chromosome both facilitate the incorporation of AAV vector DNA, and probably exogenous DNA in general.

ADENO-ASSOCIATED VIRUS VECTOR APPLICATIONS

Adeno-associated virus vectors have been employed as DNA delivery vehicles for diverse applications with a primary focus on gene addition and gene editing and strategies.

Gene Addition

The vectorization of AAV described above was the seminal discovery for mobile DNA applications in mammalian cells including clinical gene therapy (61, 62, 175), reviewed in references 9 and 175. Since then, the most popular use of AAV vectors has been gene addition strategies in which a gene, or more often cDNA, is placed within an expression cassette and packaged in the virion. The design and production of such vectors is straightforward when adherence to the AAV capsid packaging capacity (<5 kb) is respected. Key elements for these gene addition vectors include a promoter, cDNA or a gene, followed by a poly-adenylation sequence situated between the AAV ITRs (as depicted in Figure 2). Considering these elements, customization towards the specific application is possible (i.e. tissue-restricted promoters); however, as these sequences are not specific to AAV vectors they will not be discussed here. Regarding AAV vector-specific enhancements for gene addition applications, a pivotal advancement was spurred by the observation that transgenic cassettes less than half the size of the AAV genome (≤2.3 kb) packaged both monomer and dimer single-strand DNA species that had the capability to form a duplex vector genome through antiparallel single strand complementarity (158, 174, 176). Shortly thereafter, two reports demonstrated that these dimer species could be intentionally induced using a plasmid containing a wild-type ITR sequence, transgenic DNA (≤2.3 kb) and a truncated ITR in which the *trs* is deleted and therefore, is deficient for Rep-induced resolution of the typical double-stranded AAV genome intermediate (Figure 2) (161, 162, 177). Effectively, the deleted ITR functions as an intragenomic hinge, promoting the base pairing of antiparallel DNA arms capable of duplex formation (Figure 2), and as such were termed self-complementary AAV (41, 161, 162, 178). These vectors have consistently demonstrated earlier transgene onset and in general, a more than five-fold enhancement compared with single-strand AAV vector transduction (reviewed in reference 163). In hemophilia B patients, the increased transduction efficiency by self-complementary AAV vectors has effectively allowed therapeutic correction at lower administered doses, which importantly decreases immunological complications. On the other end of the spectrum, AAV large gene (>5 kb) transduction has been extensively studied (reviewed in reference 164). These approaches all rely on host-mediated assembly of partial transgene cassettes and are currently of three general types: (i) overlapping vectors, (ii) trans-splicing vectors, and (iii) fragment AAV vectors. Regarding the most efficient strategy for AAV large gene delivery, reports have been somewhat conflicting (164–166). A recent report demonstrated that the repair mechanism of trans-splicing AAV and fragment AAV vectors for large gene transduction is inherently different (167), suggesting that the observed differences may be attributed to the specific DNA repair pathways present in different tissues (166, 167). In general, all of the collective approaches can mediate large gene transduction *in vivo*, albeit at a significantly decreased efficiency when compared with intact AAV vectors.

Gene Editing

Another application of rAAV is as a tool for gene editing of specific sites in the human chromosome via HR. Notably, work primarily from the Russell laboratory has demonstrated that rAAV genomes are enhanced (up to 10,000-fold) for gene editing compared with other substrates, such as plasmid DNA whose efficiency of gene editing is approximately 1 in a million (168–171). In the process of rAAV-mediated gene editing, a repair sequence is used as the transgenic DNA that exhibits homologous arms to the target site with the desired modification preferably placed near the middle of the homology (158). Once inside the nucleus, the vector DNA may interact with chromosomal regions of homology, which stimulates repair via synthesis-dependent gene conversion and/or by direct strand incorporation (172). The reliance on the Rad51/54 pathway (172) suggests that this pathway is classical HR and, as such, would restrict AAV gene editing to cycling cells, which has been reported (173). Despite this academic understanding of HR, AAV gene editing has corrected a murine model of hereditary tyrosinemia type 1 (174). In this model, AAV gene-edited cells have a selective survival advantage, which is important to increase the relatively low correction efficiencies. As there appears to be a balance between HR and NHEJ, Paulk et al. elegantly demonstrated that genetic or drug inhibition of NHEJ increased AAV gene editing nearly 10-fold (174). AAV gene editing has also been reported as a potential treatment for epidermolysis bullosa simplex (175). In that work, AAV gene editing was coupled with a promoter trap clonal recovery strategy to correct the KRT14 mutation, a mutation in the basal epidermal keratin 14 protein, *ex vivo* in patient keratinocytes. The corrected cells were then expanded to levels relevant for skin transplant in these patients (175). These *in vivo* results suggest that AAV gene editing via HR occurs in the diseased liver and keratinocytes and have the ability to permanently correct disease mutations at the level of the host chromosome, especially when a selection scheme is employed (174, 176).

ADENO-ASSOCIATED VIRUS IN THE CLINIC

Adeno-associated virus is naturally replication deficient, and requires a helper virus to complete its life cycle. This fact, along with a lack of any known pathogenicity makes it an attractive candidate as a gene transfer vector for both clinical and research applications. The low immunogenic profile (177) and the ability to transencapsidate AAV2 ITRs with different serotype capsids (41, 178), in order to expand vector tissue targeting potential (179, 180), make AAV a versatile tool in the clinic. This versatility is enhanced by AAV's ability to infect both dividing and nondividing cells and to establish stable long-term gene expression (139, 181, 182), with a low risk of deleterious integration into the host genome (reviewed in reference 183). It is therefore not surprising that AAV has been in used in over one hundred clinical trials, for a broad range of conditions including systemic inherited monogenic diseases such as hemophilia (184, 185) and diseases in the central nervous system (186), eye (187–192), muscle (193), and heart (194). These trials have served to repeatedly highlight the safety, efficacy, and versatility of AAV as a clinical tool and its potential as a future widespread therapeutic platform.

The eye, with its immunoprivileged status (195) and plethora of established analytical tests (196), offers an attractive target for gene therapy. One of the prominent recent successes for gene therapy has been the AAV2-based treatment of Leber's congenital amaurosis (LCA). While LCA can be caused by mutation in several genes, it consistently results in the early onset of retinal degeneration and is the most prevalent cause of childhood blindness (197). AAV-based clinical trials have focused on a subset of LCA caused by loss of function mutations in the retinal pigment epithelium-specific protein 65kDa (*RPE65*) gene (187–192). In a normal eye, photosensitive pigments in the retina mediate the conversion of light to electric signals. One such pigment, 11-*cis*-retinal is converted to all-*trans*-retinal when exposed to light, which initiates a chemical cascade that ultimately leads to an electrical signal that can then be interpreted by the brain. RPE65 mediates the conversion of all-*trans*-retinal to 11-*cis*-retinal, so recycling this key pigment and allowing for the visual cycle to continue. The therapy was based on the subretinal delivery of an rAAV2 vector carrying a functional copy of *RPE65*, which restores the conversion of all-*trans*-retinal to 11-*cis*-retinal, and would allow for the repeated signaling from photoreceptors (198). So far 30 patients have received this treatment and have demonstrated lasting improvements in visual fields, nystagmus, dark-adapted perimetry, and mobility in low light (187–192). The success of these early trials, and the gene-therapy friendly attributes of the eye have led to eight more clinical trials being initiated to treat retinal diseases with rAAV, including a phase III trial targeting LCA (http://www.abedia.com/wiley).

Although the success of LCA clinical trails is both impressive and encouraging, currently there is only one clinically approved, commercial therapeutic that is based on AAV gene delivery. Glybera is an AAV1-based therapeutic that delivers a mutant lipoprotein lipase (LPL^{S447X}) gene via intramuscular injection, for the treatment of lipoprotein lipase deficiency (199), an inherited disease that often leads to hyperlipidemia, recurring pancreatitis, and liver failure. The LPL S447X variant is a truncation form of LPL, which naturally occurs in 20% of the population and is associated with increased lipolytic function and an anti-atherogenic lipid profile [reviewed in (200)], so making it ideal as an LPLD therapeutic. It received European Medical Agency (EMA) approval in 2012 for use in patients with genetically confirmed lipoprotein lipase deficiency who have experienced repeated pancreatitis attacks despite dietary restrictions. In clinical trials, Glybera showed a good safety profile and demonstrated efficacy in decreasing the frequency and severity of pancreatitis (199, 201–203). These results are admirable, but continuing work on vector optimization promises to deliver even more effective and targeted AAV-based therapeutics in the future. A glimpse of the progress that has been made in the vector development arena and its direct effect on clinical trial outcomes is afforded by the hemophilia B clinical trial at St. Jude.

ADENO-ASSOCIATED VIRUS CAPSID/TRANSGENE HUMAN IMMUNE RESPONSE

Cell-mediated Response

One property of AAV vectors that has been traditionally touted as an advantage has been low immunogenicity, as compared to other viral vectors such as adenovirus. The low efficiency with which AAVs transduce professional antigen-presenting cells such as macrophages and dendritic cells (67, 204, 205) was suspected as the reason behind their relatively low immunogenicity. That said, the production of cytotoxic T lymphocytes (CTLs) in response to capsid protein was documented in animal models transduced with wild-type AAV (206–208); however, such studies determined that the response was a byproduct of artificially overexpressed capsid protein and not relevant in the context of rAAV vectors. Even after initial human trials for hemophilia B revealed clearance of AAV-transduced hepatocytes by suspected anti-capsid CTLs (185), animal models once again failed to replicate the same phenomenon (209). These results highlighted the discrepancy between the inbred murine immune response and that of human patients, suggesting that better tools are necessary for characterizing anti-AAV immunity and evaluating any potential immunomodulatory strategies.

As hinted above, the first clinical trial using hepatic artery infusion of an AAV2-based clotting factor IX (F.IX) vector saw in patients who received the highest vector dose (2×10^{12} vector genomes/kg), clearance of vector-transduced hepatocytes. The same patients were able to initially produce therapeutic levels of F.IX but subsequently experienced an asymptomatic elevation of liver transaminases, accompanied by a drop in F.IX production (185). Follow-up studies revealed that the transaminitis was most likely due to AAV capsid-reactive CTL-killing of transduced hepatocytes. This initial study demonstrated therapeutic effects using AAV and showed that the relatively modest CTL activation appeared to be dose dependent, so leaving open the possibility that this issue could be addressed in the future by immunomodulation and increased vector efficiency (lower vector dose).

A second clinical trial, used peripheral vein infusion of AAV8, F.IX-expressing vector and was able to produce significantly higher, clinically relevant levels of transgene expression while using the same vector dosing schedule as the first trial (184). As in the previous trial, the two patients receiving the highest vector doses displayed a transient increase in liver enzymes, suggesting the onset of a capsid-targeted CTL response. As this possibility was anticipated, these patients were immunosuppressed with glucocorticoid therapy and saw a subsequent drop in liver enzymes and a continuation of therapeutic transgene production. This trial confirmed that the human anti-AAV CTL response was dose dependent and that it could potentially be managed with immunosuppressive therapy. Subsequent trials, such as the AAV1-based LPL (210) demonstrated similar dose-dependent CTL activation that appeared to be independent of route of administration.

Humoral Response

As AAV is ubiquitous in nature, the majority of the human population (95% of individuals) has been exposed to AAV2, with approximately 50% of the population having capsid neutralizing antibodies (NAbs) that can inhibit rAAV transduction (42, 184, 190, 211–218). The prevalence of NAbs in children is lower, ranging from 13 to 25% (216, 219). As a result, screening for NAbs is a prerequisite before enrollment in AAV2-based clinical trials, with high pre-existing titers being a parameter for exclusion. This criterion immediately limits the broad applicability of rAAV for disease treatment until further optimization generates the ability to evade NAbs in what may likely be a patient-specific manner (patient-specific AAV capsids). Although 11 additional AAV serotypes have been explored for gene therapy purposes, minimal cross-reactivity of NAbs has been demonstrated among these types in animals (218, 220–223). However, in humans, recent studies have shown that different degrees of NAb cross-reactivity exist between AAV2 and other types (213, 214, 220, 224). Collectively, there is a lower prevalence of NAbs against AAV1, -5, -6, -7, and -8 than against AAV2, supporting the use of other natural capsids as a means to evade AAV vector neutralization by NAbs (213, 214). Additionally, studies have aimed to map NAb epitopes to the capsid surface (14) in the hope that any conserved epitopes could be mutated away, generating serologically nonreactive vectors. While those efforts are ongoing, initial results, which have mapped multiple antibody epitopes to the three-fold protrusions of the AAV capsid, have suggested that such a mutagenesis based-strategy may result in capsids that are also compromised for infection. Alternately, a decoy strategy has been put forward, using empty vector particles, which are normally present in rAAV preparations and are typically purified out (225). This strategy relied on empty capsids as antibody decoys, and was shown to provide for an increase in transduction in the face of a challenge with NAbs. Although the strategy is simple and effective it does present the obvious shortcoming of introducing even more AAV capsid protein into patients; something that has repeatedly been shown to generate a CTL response as discussed above. As AAV continues to expand into the clinic, and more work is done to characterize the interaction between the vector and host immune systems, new strategies will inevitably emerge to meet the challenge presented by pre-existing NAbs and the anti-capsid CTL response.

Citation. Salganik M, Hirsch ML, Samulski RJ. 2014. Adeno-associated virus as a mammalian DNA vector. Microbiol Spectrum 3(2):MDNA3-0052-2014

References

1. **Atchison RW, Casto BC, Hammon WM.** 1965. Adenovirus-associated defective virus particles. *Science* **149:**754–756.
2. **Rose JA, Berns KI, Hoggan MD, Koczot FJ.** 1969. Evidence for a single-stranded adenovirus-associated virus genome: formation of a DNA density hybrid on release of viral DNA. *Proc Natl Acad Sci USA* **64:**863–869.
3. **Zhong L, Zhou X, Li Y, Qing K, Xiao X, Samulski RJ, Srivastava A.** 2008. Single-polarity recombinant adeno-associated virus 2 vector-mediated transgene expression *in vitro* and *in vivo* : mechanism of transduction. *Mol Ther* **16:**290–295.
4. **Samulski RJ, Berns KI, Tan M, Muzyczka N.** 1982. Cloning of adeno-associated virus into pBR322: rescue of intact virus from the recombinant plasmid in human cells. *Proc Natl Acad Sci USA* **79:**2077–2081.
5. **Kotin RM, Siniscalco M, Samulski RJ, Zhu XD, Hunter L, Laughlin CA, McLaughlin S, Muzyczka N, Rocchi M, Berns KI.** 1990. Site-specific integration by adeno-associated virus. *Proc Natl Acad Sci USA* **87:**2211–2215.
6. **Samulski RJ, Zhu X, Xiao X, Brook JD, Housman DE, Epstein N, Hunter LA.** 1991. Targeted integration of adeno-associated virus (AAV) into human chromosome 19. *EMBO J* **10:**3941–3950.
7. **Linden RM, Ward P, Giraud C, Winocour E, Berns KI.** 1996. Site-specific integration by adeno-associated virus. *Proc Natl Acad Sci USA* **93:**11288–11294.
8. **Sonntag F, Schmidt K, Kleinschmidt JA.** 2010. A viral assembly factor promotes AAV2 capsid formation in the nucleolus. *Proc Natl Acad Sci USA* **107:**10220–10225.
9. **Mitchell AM, Nicolson SC, Warischalk JK, Samulski RJ.** 2010. AAV's anatomy: roadmap for optimizing vectors for translational success. *Curr Gene Ther* **10:**319–340.
10. **McCarty DM, Young SM, Samulski RJ.** 2004. Integration of adeno-associated virus (AAV) and recombinant AAV vectors. *Annu Rev Genet* **38:**819–845.
11. **Young SM, Mccarty DM, Degtyareva N, Samulski RJ.** 2000. Roles of adeno-associated virus Rep protein and human chromosome 19 in site-specific recombination. *J Virol* **74:**3953–3966.
12. **Kern A, Schmidt K, Leder C, Müller OJ, Wobus CE, Bettinger K, Lieth Von der CW, King JA, Kleinschmidt JA.** 2003. Identification of a heparin-binding motif on adeno-associated virus type 2 capsids. *J Virol* **77:**11072–11081.
13. **Opie SR, Warrington KH, Agbandje-McKenna M, Zolotukhin S, Muzyczka N.** 2003. Identification of amino acid residues in the capsid proteins of adeno-associated virus type 2 that contribute to heparan sulfate proteoglycan binding. *J Virol* **77:**6995–7006.
14. **Gurda BL, Raupp C, Popa-Wagner R, Naumer M, Olson NH, Ng R, McKenna R, Baker TS, Kleinschmidt JA, Agbandje-McKenna M.** 2012. Mapping a neutralizing epitope onto the capsid of adeno-associated virus serotype 8. *J Virol* **86:**7739–7751.
15. **Gurda BL, Dimattia MA, Miller EB, Bennett A, McKenna R, Weichert WS, Nelson CD, Chen W-J, Muzyczka N, Olson NH, Sinkovits RS, Chiorini JA, Zolotutkhin S, Kozyreva OG, Samulski RJ, Baker TS, Parrish CR, Agbandje-McKenna M.** 2013. Capsid antibodies to different adeno-associated virus serotypes bind common regions. *J Virol* **87:**9111–9124.
16. **Bleker S, Sonntag F, Kleinschmidt JA.** 2005. Mutational analysis of narrow pores at the fivefold symmetry axes of adeno-associated virus type 2 capsids reveals a dual role in genome packaging and activation of phospholipase A2 activity. *J Virol* **79:**2528–2540.

17. **Kronenberg S, Böttcher B, Lieth von der CW, Bleker S, Kleinschmidt JA.** 2005. A conformational change in the adeno-associated virus type 2 capsid leads to the exposure of hidden VP1 N termini. *J Virol* **79:**5296–5303.
18. **Muzyczka N.** 1992. Use of adeno-associated virus as a general transduction vector for mammalian cells. *Curr Top Microbiol Immunol* **158:**97–129.
19. **Buller RM, Rose JA.** 1978. Characterization of adenovirus-associated virus-induced polypeptides in KB cells. *J Virol* **25:**331–338.
20. **Wistuba A, Kern A, Weger S, Grimm D, Kleinschmidt JA.** 1997. Subcellular compartmentalization of adeno-associated virus type 2 assembly. *J Virol* **71:**1341–1352.
21. **Rolling F, Samulski RJ.** 1995. AAV as a viral vector for human gene therapy. Generation of recombinant virus. *Mol Biotechnol* **3:**9–15.
22. **Warrington KH, Gorbatyuk OS, Harrison JK, Opie SR, Zolotukhin S, Muzyczka N.** 2004. Adeno-associated virus type 2 VP2 capsid protein is nonessential and can tolerate large peptide insertions at its N terminus. *J Virol* **78:**6595–6609.
23. **Gao G, Vandenberghe LH, Wilson JM.** 2005. New recombinant serotypes of AAV vectors. *Curr Gene Ther* **5:**285–297.
24. **Xie Q, Bu W, Bhatia S, Hare J, Somasundaram T, Azzi A, Chapman MS.** 2002. The atomic structure of adeno-associated virus (AAV-2), a vector for human gene therapy. *Proc Natl Acad Sci USA* **99:**10405–10410.
25. **DiMattia M, Govindasamy L, Levy HC, Gurda-Whitaker B, Kalina A, Kohlbrenner E, Chiorini JA, McKenna R, Muzyczka N, Zolotukhin S, Agbandje-McKenna M.** 2005. Production, purification, crystallization and preliminary X-ray structural studies of adeno-associated virus serotype 5. *Acta Crystallogr Sect F Struct Biol Cryst Commun* **61:**917–921.
26. **Xie Q, Ongley HM, Hare J, Chapman MS.** 2008. Crystallization and preliminary X-ray structural studies of adeno-associated virus serotype 6. *Acta Crystallogr Sect F Struct Biol Cryst Commun* **64:**1074–1078.
27. **Padron E, Bowman V, Kaludov N, Govindasamy L, Levy H, Nick P, McKenna R, Muzyczka N, Chiorini JA, Baker TS, Agbandje-McKenna M.** 2005. Structure of adeno-associated virus type 4. *J Virol* **79:**5047–5058.
28. **Kaludov N, Padron E, Govindasamy L, McKenna R, Chiorini JA, Agbandje-McKenna M.** 2003. Production, purification and preliminary X-ray crystallographic studies of adeno-associated virus serotype 4. *Virology* **306:**1–6.
29. **Nam H-J, Lane MD, Padron E, Gurda B, McKenna R, Kohlbrenner E, Aslanidi G, Byrne B, Muzyczka N, Zolotukhin S, Agbandje-McKenna M.** 2007. Structure of adeno-associated virus serotype 8, a gene therapy vector. *J Virol* **81:**12260–12271.
30. **Lerch TF, Xie Q, Ongley HM, Hare J, Chapman MS.** 2009. Twinned crystals of adeno-associated virus serotype 3b prove suitable for structural studies. *Acta Crystallogr Sect F Struct Biol Cryst Commun* **65:**177–183.
31. **Mitchell M, Nam H-J, Carter A, McCall A, Rence C, Bennett A, Gurda B, McKenna R, Porter M, Sakai Y, Byrne BJ, Muzyczka N, Aslanidi G, Zolotukhin S, Agbandje-McKenna M.** 2009. Production, purification and preliminary X-ray crystallographic studies of adeno-associated virus serotype 9. *Acta Crystallogr Sect F Struct Biol Cryst Commun* **65:**715–718.
32. **Miller EB, Gurda-Whitaker B, Govindasamy L, McKenna R, Zolotukhin S, Muzyczka N, Agbandje-McKenna M.** 2006. Production, purification and preliminary X-ray crystallographic studies of adeno-associated virus serotype 1. *Acta Crystallogr Sect F Struct Biol Cryst Commun* **62:**1271–1274.
33. **Quesada O, Gurda B, Govindasamy L, McKenna R, Kohlbrenner E, Aslanidi G, Zolotukhin S, Muzyczka N, Agbandje-McKenna M.** 2007. Production, purification and preliminary X-ray crystallographic studies of adeno-associated virus serotype 7. *Acta Crystallogr Sect F Struct Biol Cryst Commun* **63:**1073–1076.
34. **Lane MD, Nam H-J, Padron E, Gurda-Whitaker B, Kohlbrenner E, Aslanidi G, Byrne B, McKenna R, Muzyczka N, Zolotukhin S, Agbandje-McKenna M.** 2005. Production, purification, crystallization and preliminary X-ray analysis of adeno-associated virus serotype 8. *Acta Crystallogr Sect F Struct Biol Cryst Commun* **61:**558–561.
35. **Govindasamy L, Padron E, McKenna R, Muzyczka N, Kaludov N, Chiorini JA, Agbandje-McKenna M.** 2006. Structurally mapping the diverse phenotype of adeno-associated virus serotype 4. *J Virol* **80:**11556–11570.
36. **Walters RW, Agbandje-McKenna M, Bowman VD, Moninger TO, Olson NH, Seiler M, Chiorini JA, Baker TS, Zabner J.** 2004. Structure of adeno-associated virus serotype 5. *J Virol* **78:**3361–3371.
37. **O'Donnell J, Taylor KA, Chapman MS.** 2009. Adeno-associated virus-2 and its primary cellular receptor–Cryo-EM structure of a heparin complex. *Virology* **385:** 434–443.
38. **Levy HC, Bowman VD, Govindasamy L, McKenna R, Nash K, Warrington K, Chen W, Muzyczka N, Yan X, Baker TS, Agbandje-McKenna M.** 2009. Heparin binding induces conformational changes in Adeno-associated virus serotype 2. *J Struct Biol* **165:**146–156.
39. **Lerch TF, Xie Q, Chapman MS.** 2010. The structure of adeno-associated virus serotype 3B (AAV-3B): insights into receptor binding and immune evasion. *Virology* **403:**26–36.
40. **Ng R, Govindasamy L, Gurda BL, McKenna R, Kozyreva OG, Samulski RJ, Parent KN, Baker TS, Agbandje-McKenna M.** 2010. Structural characterization of the dual glycan binding adeno-associated virus serotype 6. *J Virol* **84:**12945–12957.
41. **Rabinowitz JE, Bowles DE, Faust SM, Ledford JG, Cunningham SE, Samulski RJ.** 2004. Cross-dressing the virion: the transcapsidation of adeno-associated virus serotypes functionally defines subgroups. *J Virol* **78:** 4421–4432.
42. **Wobus CE, Hügle-Dörr B, Girod A, Petersen G, Hallek M, Kleinschmidt JA.** 2000. Monoclonal antibodies against the adeno-associated virus type 2 (AAV-2) capsid: epitope mapping and identification of capsid domains involved in AAV-2-cell interaction and neutralization of AAV-2 infection. *J Virol* **74:**9281–9293.

43. **Horowitz ED, Rahman KS, Bower BD, Dismuke DJ, Falvo MR, Griffith JD, Harvey SC, Asokan A.** 2013. Biophysical and ultrastructural characterization of adeno-associated virus capsid uncoating and genome release. *J Virol* **87**:2994–3002.

44. **Cavalier-Smith T.** 1974. Palindromic base sequences and replication of eukaryote chromosome ends. *Nature* **250**:467–470.

45. **Snyder RO, Im DS, Ni T, Xiao X, Samulski RJ, Muzyczka N.** 1993. Features of the adeno-associated virus origin involved in substrate recognition by the viral Rep protein. *J Virol* **67**:6096–6104.

46. **Chiorini JA, Wiener SM, Owens RA, Kyöstiö SR, Kotin RM, Safer B.** 1994. Sequence requirements for stable binding and function of Rep68 on the adeno-associated virus type 2 inverted terminal repeats. *J Virol* **68**:7448–7457.

47. **Mccarty DM, Pereira DJ, Zolotukhin I, Zhou X, Ryan JH, Muzyczka N.** 1994. Identification of linear DNA sequences that specifically bind the adeno-associated virus Rep protein. *J Virol* **68**:4988–4997.

48. **Mccarty DM, Ryan JH, Zolotukhin S, Zhou X, Muzyczka N.** 1994. Interaction of the adeno-associated virus Rep protein with a sequence within the A palindrome of the viral terminal repeat. *J Virol* **68**:4998–5006.

49. **Weitzman MD, Kyöstiö SR, Kotin RM, Owens RA.** 1994. Adeno-associated virus (AAV) Rep proteins mediate complex formation between AAV DNA and its integration site in human DNA. *Proc Natl Acad Sci USA* **91**:5808–5812.

50. **Straus SE, Sebring ED, Rose JA.** 1976. Concatemers of alternating plus and minus strands are intermediates in adenovirus-associated virus DNA synthesis. *Proc Natl Acad Sci USA* **73**:742–746.

51. **Lusby E, Bohenzky R, Berns KI.** 1981. Inverted terminal repetition in adeno-associated virus DNA: independence of the orientation at either end of the genome. *J Virol* **37**:1083–1086.

52. **Snyder RO, Im DS, Muzyczka N.** 1990. Evidence for covalent attachment of the adeno-associated virus (AAV) rep protein to the ends of the AAV genome. *J Virol* **64**: 6204–6213.

53. **Im DS, Muzyczka N.** 1989. Factors that bind to adeno-associated virus terminal repeats. *J Virol* **63**:3095–3104.

54. **Im DS, Muzyczka N.** 1990. The AAV origin binding protein Rep68 is an ATP-dependent site-specific endonuclease with DNA helicase activity. *Cell* **61**:447–457.

55. **Owens RA, Weitzman MD, Kyöstiö SR, Carter BJ.** 1993. Identification of a DNA-binding domain in the amino terminus of adeno-associated virus Rep proteins. *J Virol* **67**:997–1005.

56. **Chiorini JA, Weitzman MD, Owens RA, Urcelay E, Safer B, Kotin RM.** 1994. Biologically active Rep proteins of adeno-associated virus type 2 produced as fusion proteins in Escherichia coli. *J Virol* **68**:797–804.

57. **Brister JR, Muzyczka N.** 1999. Rep-mediated nicking of the adeno-associated virus origin requires two biochemical activities, DNA helicase activity and transesterification. *J Virol* **73**:9325–9336.

58. **Wonderling RS, Kyöstiö SR, Owens RA.** 1995. A maltose-binding protein/adeno-associated virus Rep68 fusion protein has DNA–RNA helicase and ATPase activities. *J Virol* **69**:3542–3548.

59. **Snyder RO, Samulski RJ, Muzyczka N.** 1990. *In vitro* resolution of covalently joined AAV chromosome ends. *Cell* **60**:105–113.

60. **Naumer M, Sonntag F, Schmidt K, Nieto K, Panke C, Davey NE, Popa-Wagner R, Kleinschmidt JA.** 2012. Properties of the adeno-associated virus assembly-activating protein. *J Virol* **86**:13038–13048.

61. **Hermonat PL, Labow MA, Wright R, Berns KI, Muzyczka N.** 1984. Genetics of adeno-associated virus: isolation and preliminary characterization of adeno-associated virus type 2 mutants. *J Virol* **51**:329–339.

62. **Samulski RJ, Chang LS, Shenk T.** 1989. Helper-free stocks of recombinant adeno-associated viruses: normal integration does not require viral gene expression. *J Virol* **63**:3822–3828.

63. **Ferrari FK, Xiao X, McCarty D, Samulski RJ.** 1997. New developments in the generation of Ad-free, high-titer rAAV gene therapy vectors. *Nat Med* **3**:1295–1297.

64. **Grieger JC, Samulski RJ.** 2012. Adeno-associated virus vectorology, manufacturing, and clinical applications. *Meth Enzymol* **507**:229–254.

65. **Fisher KJ, Gao GP, Weitzman MD, DeMatteo R, Burda JF, Wilson JM.** 1996. Transduction with recombinant adeno-associated virus for gene therapy is limited by leading-strand synthesis. *J Virol* **70**:520–532.

66. **Ferrari FK, Samulski T, Shenk T, Samulski RJ.** 1996. Second-strand synthesis is a rate-limiting step for efficient transduction by recombinant adeno-associated virus vectors. *J Virol* **70**:3227–3234.

67. **Jooss K, Yang Y, Fisher KJ, Wilson JM.** 1998. Transduction of dendritic cells by DNA viral vectors directs the immune response to transgene products in muscle fibers. *J Virol* **72**:4212–4223.

68. **Thomas CE, Storm TA, Huang Z, Kay MA.** 2004. Rapid uncoating of vector genomes is the key to efficient liver transduction with pseudotyped adeno-associated virus vectors. *J Virol* **78**:3110–3122.

69. **Wu Z, Miller E, Agbandje-McKenna M, Samulski RJ.** 2006. Alpha2,3 and alpha2,6 N-linked sialic acids facilitate efficient binding and transduction by adeno-associated virus types 1 and 6. *J Virol* **80**:9093–9103.

70. **Summerford C, Samulski RJ.** 1998. Membrane-associated heparan sulfate proteoglycan is a receptor for adeno-associated virus type 2 virions. *J Virol* **72**:1438–1445.

71. **Kaludov N, Brown KE, Walters RW, Zabner J, Chiorini JA.** 2001. Adeno-associated virus serotype 4 (AAV4) and AAV5 both require sialic acid binding for hemagglutination and efficient transduction but differ in sialic acid linkage specificity. *J Virol* **75**:6884–6893.

72. **Walters RW, Yi SM, Keshavjee S, Brown KE, Welsh MJ, Chiorini JA, Zabner J.** 2001. Binding of adeno-associated virus type 5 to 2,3-linked sialic acid is required for gene transfer. *J Biol Chem* **276**:20610–20616.

73. **Shen S, Bryant KD, Brown SM, Randell SH, Asokan A.** 2011. Terminal N-linked galactose is the primary receptor

for adeno-associated virus 9. *J Biol Chem* **286**:13532–13540.

74. **Müller OJ, Leuchs B, Pleger ST, Grimm D, Franz W-M, Katus HA, Kleinschmidt JA.** 2006. Improved cardiac gene transfer by transcriptional and transductional targeting of adeno-associated viral vectors. *Cardiovasc Res* **70**:70–78.

75. **Asokan A, Conway JC, Phillips JL, Li C, Hegge J, Sinnott R, Yadav S, DiPrimio N, Nam H-J, Agbandje-McKenna M, McPhee S, Wolff J, Samulski RJ.** 2010. Reengineering a receptor footprint of adeno-associated virus enables selective and systemic gene transfer to muscle. *Nat Biotechnol* **28**:79–82.

76. **Shen S, Horowitz ED, Troupes AN, Brown SM, Pulicherla N, Samulski RJ, Agbandje-McKenna M, Asokan A.** 2013. Engraftment of a galactose receptor footprint onto adeno-associated viral capsids improves transduction efficiency. *J Biol Chem* **288**:28814–28823.

77. **Girod A, Ried M, Wobus C, Lahm H, Leike K, Kleinschmidt J, Deléage G, Hallek M.** 1999. Genetic capsid modifications allow efficient re-targeting of adeno-associated virus type 2. *Nat Med* **5**:1052–1056.

78. **Yang Q, Mamounas M, Yu G, Kennedy S, Leaker B, Merson J, Wong-Staal F, Yu M, Barber JR.** 1998. Development of novel cell surface CD34-targeted recombinant adenoassociated virus vectors for gene therapy. *Human Gene Ther* **9**:1929–1937.

79. **Grifman M, Trepel M, Speece P, Gilbert LB, Arap W, Pasqualini R, Weitzman MD.** 2001. Incorporation of tumor-targeting peptides into recombinant adeno-associated virus capsids. *Mol Ther* **3**:964–975.

80. **Loiler SA, Conlon TJ, Song S, Tang Q, Warrington KH, Agarwal A, Kapturczak M, Li C, Ricordi C, Atkinson MA, Muzyczka N, Flotte TR.** 2003. Targeting recombinant adeno-associated virus vectors to enhance gene transfer to pancreatic islets and liver. *Gene Ther* **10**:1551–1558.

81. **Rabinowitz JE, Xiao W, Samulski RJ.** 1999. Insertional mutagenesis of AAV2 capsid and the production of recombinant virus. *Virology* **265**:274–285.

82. **Shi W, Arnold GS, Bartlett JS.** 2001. Insertional mutagenesis of the adeno-associated virus type 2 (AAV2) capsid gene and generation of AAV2 vectors targeted to alternative cell-surface receptors. *Hum Gene Ther* **12**:1697–1711.

83. **Müller OJ, Kaul F, Weitzman MD, Pasqualini R, Arap W, Kleinschmidt JA, Trepel M.** 2003. Random peptide libraries displayed on adeno-associated virus to select for targeted gene therapy vectors. *Nat Biotechnol* **21**:1040–1046.

84. **Judd J, Wei F, Nguyen PQ, Tartaglia LJ, Agbandje-McKenna M, Silberg JJ, Suh J.** 2012. Random insertion of mCherry into VP3 domain of adeno-associated virus yields fluorescent capsids with no loss of infectivity. *Mol Ther Nucleic Acids* **1**:e54.

85. **Nam H-J, Gurda BL, McKenna R, Potter M, Byrne B, Salganik M, Muzyczka N, Agbandje-McKenna M.** 2011. Structural studies of adeno-associated virus serotype 8 capsid transitions associated with endosomal trafficking. *J Virol* **85**:11791–11799.

86. **Salganik M, Venkatakrishnan B, Bennett A, Lins B, Yarbrough J, Muzyczka N, Agbandje-McKenna M, McKenna R.** 2012. Evidence for pH-dependent protease activity in the adeno-associated virus capsid. *J Virol* **86**:11877–11885.

87. **Münch RC, Janicki H, Völker I, Rasbach A, Hallek M, Büning H, Buchholz CJ.** 2013. Displaying high-affinity ligands on adeno-associated viral vectors enables tumor cell-specific and safe gene transfer. *Mol Ther* **21**:109–118.

88. **Sonntag F, Bleker S, Leuchs B, Fischer R, Kleinschmidt JA.** 2006. Adeno-associated virus type 2 capsids with externalized VP1/VP2 trafficking domains are generated prior to passage through the cytoplasm and are maintained until uncoating occurs in the nucleus. *J Virol* **80**:11040–11054.

89. **White AF, Mazur M, Sorscher EJ, Zinn KR, Ponnazhagan S.** 2008. Genetic modification of adeno-associated viral vector type 2 capsid enhances gene transfer efficiency in polarized human airway epithelial cells. *Hum Gene Ther* **19**:1407–1414.

90. **Nicklin SA, Buening H, Dishart KL, de Alwis M, Girod A, Hacker U, Thrasher AJ, Ali RR, Hallek M, Baker AH.** 2001. Efficient and selective AAV2-mediated gene transfer directed to human vascular endothelial cells. *Mol Ther* **4**:174–181.

91. **White SJ, Nicklin SA, Büning H, Brosnan MJ, Leike K, Papadakis ED, Hallek M, Baker AH.** 2004. Targeted gene delivery to vascular tissue *in vivo* by tropism-modified adeno-associated virus vectors. *Circulation* **109**:513–519.

92. **Work LM, Büning H, Hunt E, Nicklin SA, Denby L, Britton N, Leike K, Odenthal M, Drebber U, Hallek M, Baker AH.** 2006. Vascular bed-targeted *in vivo* gene delivery using tropism-modified adeno-associated viruses. *Mol Ther* **13**:683–693.

93. **Yu C-Y, Yuan Z, Cao Z, Wang B, Qiao C, Li J, Xiao X.** 2009. A muscle-targeting peptide displayed on AAV2 improves muscle tropism on systemic delivery. *Gene Ther* **16**:953–962.

94. **Girod A, Wobus CE, Zádori Z, Ried M, Leike K, Tijssen P, Kleinschmidt JA, Hallek M.** 2002. The VP1 capsid protein of adeno-associated virus type 2 is carrying a phospholipase A2 domain required for virus infectivity. *J Gen Virol* **83**:973–978.

95. **Perabo L, Büning H, Kofler DM, Ried MU, Girod A, Wendtner CM, Enssle J, Hallek M.** 2003. *In vitro* selection of viral vectors with modified tropism: the adeno-associated virus display. *Mol Ther* **8**:151–157.

96. **Waterkamp DA, Müller OJ, Ying Y, Trepel M, Kleinschmidt JA.** 2006. Isolation of targeted AAV2 vectors from novel virus display libraries. *J Gene Med* **8**:1307–1319.

97. **Michelfelder S, Lee M-K, deLima-Hahn E, Wilmes T, Kaul F, Müller O, Kleinschmidt JA, Trepel M.** 2007. Vectors selected from adeno-associated viral display peptide libraries for leukemia cell-targeted cytotoxic gene therapy. *Exp Hematol* **35**:1766–1776.

98. **Michelfelder S, Kohlschütter J, Skorupa A, Pfennings S, Müller O, Kleinschmidt JA, Trepel M.** 2009. Successful

expansion but not complete restriction of tropism of adeno-associated virus by *in vivo* biopanning of random virus display peptide libraries. *PLoS ONE* **4:**e5122.

99. **Grimm D, Lee JS, Wang L, Desai T, Akache B, Storm TA, Kay MA.** 2008. *In vitro* and *in vivo* gene therapy vector evolution via multispecies interbreeding and retargeting of adeno-associated viruses. *J Virol* **82:**5887–5911.
100. **Popa-Wagner R, Porwal M, Kann M, Reuss M, Weimer M, Florin L, Kleinschmidt JA.** 2012. Impact of VP1-specific protein sequence motifs on adeno-associated virus type 2 intracellular trafficking and nuclear entry. *J Virol* **86:**9163–9174.
101. **Maheshri N, Koerber JT, Kaspar BK, Schaffer DV.** 2006. Directed evolution of adeno-associated virus yields enhanced gene delivery vectors. *Nat Biotechnol* **24:**198–204.
102. **Perabo L, Endell J, King S, Lux K, Goldnau D, Hallek M, Büning H.** 2006. Combinatorial engineering of a gene therapy vector: directed evolution of adeno-associated virus. *J Gene Med* **8:**155–162.
103. **Grieger JC, Snowdy S, Samulski RJ.** 2006. Separate basic region motifs within the adeno-associated virus capsid proteins are essential for infectivity and assembly. *J Virol* **80:**5199–5210.
104. **Johnson JS, Li C, DiPrimio N, Weinberg MS, McCown TJ, Samulski RJ.** 2010. Mutagenesis of adeno-associated virus type 2 capsid protein VP1 uncovers new roles for basic amino acids in trafficking and cell-specific transduction. *J Virol* **84:**8888–8902.
105. **Li W, Asokan A, Wu Z, Van Dyke T, DiPrimio N, Johnson JS, Govindaswamy L, Agbandje-McKenna M, Leichtle S, Redmond DE, McCown TJ, Petermann KB, Sharpless NE, Samulski RJ.** 2008. Engineering and selection of shuffled AAV genomes: a new strategy for producing targeted biological nanoparticles. *Mol Ther* **6:**1252–1260.
106. **Maguire CA, Gianni D, Meijer DH, Shaket LA, Wakimoto H, Rabkin SD, Gao G, Sena-Esteves M.** 2010. Directed evolution of adeno-associated virus for glioma cell transduction. *J Neurooncol* **96:**337–347.
107. **Ward P, Walsh CE.** 2009. Chimeric AAV Cap sequences alter gene transduction. *Virology* **386:**237–248.
108. **Farr GA, Zhang L-G, Tattersall P.** 2005. Parvoviral virions deploy a capsid-tethered lipolytic enzyme to breach the endosomal membrane during cell entry. *Proc Natl Acad Sci USA* **102:**17148–17153.
109. **Yang L, Jiang J, Drouin LM, Agbandje-McKenna M, Chen C, Qiao C, Pu D, Hu X, Wang D-Z, Li J, Xiao X.** 2009. A myocardium tropic adeno-associated virus (AAV) evolved by DNA shuffling and *in vivo* selection. *Proc Natl Acad Sci USA* **106:**3946–3951.
110. **Mani B, Baltzer C, Valle N, Almendral JM, Kempf C, Ros C.** 2006. Low pH-dependent endosomal processing of the incoming parvovirus minute virus of mice virion leads to externalization of the VP1 N-terminal sequence (N-VP1), N-VP2 cleavage, and uncoating of the full-length genome. *J Virol* **80:**1015–1024.
111. **Gray SJ, Blake BL, Criswell HE, Nicolson SC, Samulski RJ, McCown TJ, Li W.** 2010. Directed evolution of a novel adeno-associated virus (AAV) vector that crosses the seizure-compromised blood-brain barrier (BBB). *Mol Ther* **18:**570–578.
112. **Cohen S, Marr AK, Garcin P, Panté N.** 2011. Nuclear envelope disruption involving host caspases plays a role in the parvovirus replication cycle. *J Virol* **85:**4863–4874.
113. **Lisowski L, Dane AP, Chu K, Zhang Y, Cunningham SC, Wilson EM, Nygaard S, Grompe M, Alexander IE, Kay MA.** 2013. Selection and evaluation of clinically relevant AAV variants in a xenograft liver model. *Nature* **506**(7488)**:**382–366.
114. **Xiao P-J, Samulski RJ.** 2012. Cytoplasmic trafficking, endosomal escape, and perinuclear accumulation of adeno-associated virus type 2 particles are facilitated by microtubule network. *J Virol* **86:**10462–10473.
115. **Duan D, Li Q, Kao AW, Yue Y, Pessin JE, Engelhardt JF.** 1999. Dynamin is required for recombinant adeno-associated virus type 2 infection. *J Virol* **73:**10371–10376.
116. **Bartlett JS, Wilcher R, Samulski RJ.** 2000. Infectious entry pathway of adeno-associated virus and adeno-associated virus vectors. *J Virol* **74:**2777–2785.
117. **Uhrig S, Coutelle O, Wiehe T, Perabo L, Hallek M, Büning H.** 2012. Successful target cell transduction of capsid-engineered rAAV vectors requires clathrin-dependent endocytosis. *Gene Ther* **19:**210–218.
118. **Liu Y, Joo K-I, Wang P.** 2012. Endocytic processing of adeno-associated virus type 8 vectors for transduction of target cells. *Gene Ther* **20:**308–317.
119. **Zhong L, Li B, Jayandharan G, Mah CS, Govindasamy L, Agbandje-McKenna M, Herzog RW, Weigel-Van Aken KA, Hobbs JA, Zolotukhin S, Muzyczka N, Srivastava A.** 2008. Tyrosine-phosphorylation of AAV2 vectors and its consequences on viral intracellular trafficking and transgene expression. *Virology* **381:**194–202.
120. **Bantel-Schaal U, Braspenning-Wesch I, Kartenbeck J.** 2009. Adeno-associated virus type 5 exploits two different entry pathways in human embryo fibroblasts. *J Gen Virol* **90:**317–322.
121. **Douar AM, Poulard K, Stockholm D, Danos O.** 2001. Intracellular trafficking of adeno-associated virus vectors: routing to the late endosomal compartment and proteasome degradation. *J Virol* **75:**1824–1833.
122. **Nonnenmacher M, Weber T.** 2011. Adeno-associated virus 2 infection requires endocytosis through the CLIC/GEEC pathway. *Cell Host Microbe* **10:**563–576.
123. **Zhong L, Li B, Mah CS, Govindasamy L, Agbandje-McKenna M, Cooper M, Herzog RW, Zolotukhin I, Warrington KH, Weigel-Van Aken KA, Hobbs JA, Zolotukhin S, Muzyczka N, Srivastava A.** 2008. Next generation of adeno-associated virus 2 vectors: point mutations in tyrosines lead to high-efficiency transduction at lower doses. *Proc Natl Acad Sci USA* **105:**7827–7832.
124. **Nicolson SC, Samulski RJ.** 2014. Recombinant adeno-associated virus utilizes host cell nuclear import machinery to enter the nucleus. *J Virol* **88:**4132–4144.

125. **Johnson JS, Samulski RJ.** 2009. Enhancement of adeno-associated virus infection by mobilizing capsids into and out of the nucleolus. *J Virol* **83:**2632–2644.

126. **Salganik M, Aydemir F, Nam H-J, McKenna R, Agbandje-McKenna M, Muzyczka N.** 2014. Adeno-associated virus capsid proteins may play a role in transcription and second-strand synthesis of recombinant genomes. *J Virol* **88:**1071–1079.

127. **Ding W, Zhang LN, Yeaman C, Engelhardt JF.** 2006. rAAV2 traffics through both the late and the recycling endosomes in a dose-dependent fashion. *Mol Ther* **13:** 671–682.

128. **Cervelli T, Palacios JA, Zentilin L, Mano M, Schwartz RA, Weitzman MD, Giacca M.** 2008. Processing of recombinant AAV genomes occurs in specific nuclear structures that overlap with foci of DNA-damage-response proteins. *J Cell Sci* **121:**349–357.

129. **Hauck B, Zhao W, High K, Xiao W.** 2004. Intracellular viral processing, not single-stranded DNA accumulation, is crucial for recombinant adeno-associated virus transduction. *J Virol* **78:**13678–13686.

130. **Samulski RJ, Srivastava A, Berns KI, Muzyczka N.** 1983. Rescue of adeno-associated virus from recombinant plasmids: gene correction within the terminal repeats of AAV. *Cell* **33:**135–143.

131. **Hauswirth WW, Berns KI.** 1979. Adeno-associated virus DNA replication: nonunit-length molecules. *Virology* **93:**57–68.

132. **Ni TH, McDonald WF, Zolotukhin I, Melendy T, Waga S, Stillman B, Muzyczka N.** 1998. Cellular proteins required for adeno-associated virus DNA replication in the absence of adenovirus coinfection. *J Virol* **72:**2777–2787.

133. **Nash K, Chen W, McDonald WF, Zhou X, Muzyczka N.** 2007. Purification of host cell enzymes involved in adeno-associated virus DNA replication. *J Virol* **81:** 5777–5787.

134. **Nakai H, Storm TA, Kay MA.** 2000. Recruitment of single-stranded recombinant adeno-associated virus vector genomes and intermolecular recombination are responsible for stable transduction of liver *in vivo*. *J Virol* **74:**9451–9463.

135. **Choi VW, McCarty DM, Samulski RJ.** 2006. Host cell DNA repair pathways in adeno-associated viral genome processing. *J Virol* **80:**10346–10356.

136. **Song S, Laipis PJ, Berns KI, Flotte TR.** 2001. Effect of DNA-dependent protein kinase on the molecular fate of the rAAV2 genome in skeletal muscle. *Proc Natl Acad Sci USA* **98:**4084–4088.

137. **Inagaki K, Ma C, Storm TA, Kay MA, Nakai H.** 2007. The role of DNA-PKcs and artemis in opening viral DNA hairpin termini in various tissues in mice. *J Virol* **81:**11304–11321.

138. **Duan D, Sharma P, Yang J, Yue Y, Dudus L, Zhang Y, Fisher KJ, Engelhardt JF.** 1998. Circular intermediates of recombinant adeno-associated virus have defined structural characteristics responsible for long-term episomal persistence in muscle tissue. *J Virol* **72:**8568–8577.

139. **Xiao X, Li J, Samulski RJ.** 1996. Efficient long-term gene transfer into muscle tissue of immunocompetent mice by adeno-associated virus vector. *J Virol* **70:** 8098–8108.

140. **Sanlioglu S, Duan D, Engelhardt JF.** 1999. Two independent molecular pathways for recombinant adeno-associated virus genome conversion occur after UV-C and E4orf6 augmentation of transduction. *Hum Gene Ther* **10:**591–602.

141. **Zentilin L, Marcello A, Giacca M.** 2001. Involvement of cellular double-stranded DNA break binding proteins in processing of the recombinant adeno-associated virus genome. *J Virol* **75:**12279–12287.

142. **Russell DW, Alexander IE, Miller AD.** 1995. DNA synthesis and topoisomerase inhibitors increase transduction by adeno-associated virus vectors. *Proc Natl Acad Sci USA* **92:**5719–5723.

143. **Miller JL, Donahue RE, Sellers SE, Samulski RJ, Young NS, Nienhuis AW.** 1994. Recombinant adeno-associated virus (rAAV)-mediated expression of a human gamma-globin gene in human progenitor-derived erythroid cells. *Proc Natl Acad Sci USA* **91:**10183–10187.

144. **Li C, He Y, Nicolson S, Hirsch M, Weinberg MS, Zhang P, Kafri T, Samulski RJ.** 2013. Adeno-associated virus capsid antigen presentation is dependent on endosomal escape. *J Clin Invest* **123:**1390–1401.

145. **Schwartz RA, Palacios JA, Cassell GD, Adam S, Giacca M, Weitzman MD.** 2007. The Mre11/Rad50/Nbs1 complex limits adeno-associated virus transduction and replication. *J Virol* **81:**12936–12945.

146. **Lovric J, Mano M, Zentilin L, Eulalio A, Zacchigna S, Giacca M.** 2012. Terminal differentiation of cardiac and skeletal myocytes induces permissivity to AAV transduction by relieving inhibition imposed by DNA damage response proteins. *Mol Ther* **20:**2087–2097.

147. **Rahman SH, Bobis-Wozowicz S, Chatterjee D, Gellhaus K, Pars K, Heilbronn R, Jacobs R, Cathomen T.** 2013. The nontoxic cell cycle modulator indirubin augments transduction of adeno-associated viral vectors and zinc-finger nuclease-mediated gene targeting. *Hum Gene Ther* **24:**67–77.

148. **Raj K, Ogston P, Beard P.** 2001. Virus-mediated killing of cells that lack p53 activity. *Nature* **412:**914–917.

149. **Winocour E, Callaham MF, Huberman E.** 1988. Perturbation of the cell cycle by adeno-associated virus. *Virology* **167:**393–399.

150. **Fragkos M, Beard P.** 2011. Mitotic catastrophe occurs in the absence of apoptosis in p53-null cells with a defective G1 checkpoint. *PLoS ONE* **6:**e22946.

151. **Hirsch ML, Fagan BM, Dumitru R, Bower JJ, Yadav S, Porteus MH, Pevny LH, Samulski RJ.** 2011. Viral single-strand DNA induces p53-dependent apoptosis in human embryonic stem cells. *PLoS ONE* **6:**e27520.

152. **Fragkos M, Jurvansuu J, Beard P.** 2009. H2AX is required for cell cycle arrest via the p53/p21 pathway. *Mol Cell Biol* **29:**2828–2840.

153. **Miao CH, Snyder RO, Schowalter DB, Patijn GA, Donahue B, Winther B, Kay MA.** 1998. The kinetics of rAAV integration in the liver. *Nat Genet* **19:**13–15.

154. **Yang CC, Xiao X, Zhu X, Ansardi DC, Epstein ND, Frey MR, Matera AG, Samulski RJ.** 1997. Cellular recombination pathways and viral terminal repeat hairpin structures are sufficient for adeno-associated virus integration *in vivo* and *in vitro*. *J Virol* **71:**9231–9247.

155. **Miller DG, Trobridge GD, Petek LM, Jacobs MA, Kaul R, Russell DW.** 2005. Large-scale analysis of adeno-associated virus vector integration sites in normal human cells. *J Virol* **79:**11434–11442.

156. **Nakai H, Wu X, Fuess S, Storm TA, Munroe D, Montini E, Burgess SM, Grompe M, Kay MA.** 2005. Large-scale molecular characterization of adeno-associated virus vector integration in mouse liver. *J Virol* **79:**3606–3614.

157. **Flotte TR, Afione SA, Zeitlin PL.** 1994. Adeno-associated virus vector gene expression occurs in nondividing cells in the absence of vector DNA integration. *Am J Respir Cell Mol Biol* **11:**517–521.

158. **Hirata RK, Russell DW.** 2000. Design and packaging of adeno-associated virus gene targeting vectors. *J Virol* **74:**4612–4620.

159. **Miller DG, Petek LM, Russell DW.** 2004. Adeno-associated virus vectors integrate at chromosome breakage sites. *Nat Genet* **36:**767–773.

160. **Nakai H, Montini E, Fuess S, Storm TA, Grompe M, Kay MA.** 2003. AAV serotype 2 vectors preferentially integrate into active genes in mice. *Nat Genet* **34:** 297–302.

161. **Mccarty DM, Fu H, Monahan PE, Toulson CE, Naik P, Samulski RJ.** 2003. Adeno-associated virus terminal repeat (TR) mutant generates self-complementary vectors to overcome the rate-limiting step to transduction *in vivo*. *Gene Ther* **10:**2112–2118.

162. **Xiao X, Xiao W, Li J, Samulski RJ.** 1997. A novel 165-base-pair terminal repeat sequence is the sole cis requirement for the adeno-associated virus life cycle. *J Virol* **71:**941–948.

163. **McCarty DM.** 2008. Self-complementary AAV vectors; advances and applications. *Mol Ther* **16:**1648–1656.

164. **Hirsch ML, Agbandje-McKenna M, Samulski RJ.** 2010. Little vector, big gene transduction: fragmented genome reassembly of adeno-associated virus. *Mol Ther* **18:**6–8.

165. **Lai Y, Zhao J, Yue Y, Wasala NB, Duan D.** 2014. Partial restoration of cardiac function with ΔPDZ nNOS in aged mdx model of Duchenne cardiomyopathy. *Hum Mol Genet* **23:**3189–3199.

166. **Dyka FM, Boye SL, Chiodo VA, Hauswirth WW, Boye SE.** 2014. Dual adeno-associated virus vectors result in efficient *in vitro* and *in vivo* expression of an oversized gene, MYO7A. *Hum Gene Ther Meth* **25:**166–177.

167. **Hirsch ML, Li C, Bellon I, Yin C, Chavala S, Pryadkina M, Richard I, Samulski RJ.** 2013. Oversized AAV transductifon is mediated via a DNA-PKcs-independent, Rad51C-dependent repair pathway. *Mol Ther* **21:**2205–2216.

168. **Russell DW, Hirata RK.** 1998. Human gene targeting by viral vectors. *Nat Genet* **18:**325–330.

169. **Smithies O, Gregg RG, Boggs SS, Koralewski MA, Kucherlapati RS.** 1985. Insertion of DNA sequences into the human chromosomal beta-globin locus by homologous recombination. *Nature* **317:**230–234.

170. **Thomas KR, Capecchi MR.** 1987. Site-directed mutagenesis by gene targeting in mouse embryo-derived stem cells. *Cell* **51:**503–512.

171. **Russell DW, Hirata RK.** 2008. Human gene targeting favors insertions over deletions. *Hum Gene Ther* **19:** 907–914.

172. **Vasileva A, Linden RM, Jessberger R.** 2006. Homologous recombination is required for AAV-mediated gene targeting. *Nucleic Acids Res* **34:**3345–3360.

173. **Trobridge G, Hirata RK, Russell DW.** 2005. Gene targeting by adeno-associated virus vectors is cell-cycle dependent. *Hum Gene Ther* **16:**522–526.

174. **Paulk NK, Wursthorn K, Wang Z, Finegold MJ, Kay MA, Grompe M.** 2010. Adeno-associated virus gene repair corrects a mouse model of hereditary tyrosinemia *in vivo*. *Hepatology* **51:**1200–1208.

175. **Petek LM, Fleckman P, Miller DG.** 2010. Efficient KRT14 targeting and functional characterization of transplanted human keratinocytes for the treatment of epidermolysis bullosa simplex. *Mol Ther* **18:**1624–1632.

176. **Paulk NK, Wursthorn K, Haft A, Pelz C, Clarke G, Newell AH, Olson SB, Harding CO, Finegold MJ, Bateman RL, Witte JF, McClard R, Grompe M.** 2012. *In vivo* selection of transplanted hepatocytes by pharmacological inhibition of fumarylacetoacetate hydrolase in wild-type mice. *Mol Ther* **20:**1981–1987.

177. **Mueller C, Flotte TR.** 2008. Clinical gene therapy using recombinant adeno-associated virus vectors. *Gene Ther* **15:**858–863.

178. **Rabinowitz JE, Rolling F, Li C, Conrath H, Xiao W, Xiao X, Samulski RJ.** 2002. Cross-packaging of a single adeno-associated virus (AAV) type 2 vector genome into multiple AAV serotypes enables transduction with broad specificity. *J Virol* **76:**791–801.

179. **Ellis BL, Hirsch ML, Barker JC, Connelly JP, Steininger RJ, Porteus MH.** 2013. A survey of *ex vivo/in vitro* transduction efficiency of mammalian primary cells and cell lines with nine natural adeno-associated virus (AAV1-9) and one engineered adeno-associated virus serotype. *Virol J* **10:**74.

180. **Zincarelli C, Soltys S, Rengo G, Rabinowitz JE.** 2008. Analysis of AAV serotypes 1-9 mediated gene expression and tropism in mice after systemic injection. *Mol Ther* **16:**1073–1080.

181. **Podsakoff G, Wong KK, Chatterjee S.** 1994. Efficient gene transfer into nondividing cells by adeno-associated virus-based vectors. *J Virol* **68:**5656–5666.

182. **Berns KI, Pinkerton TC, Thomas GF, Hoggan MD.** 1975. Detection of adeno-associated virus (AAV)-specific nucleotide sequences in DNA isolated from latently infected Detroit 6 cells. *Virology* **68:**556–560.

183. **Tenenbaum L, Lehtonen E, Monahan PE.** 2003. Evaluation of risks related to the use of adeno-associated virus-based vectors. *Curr Gene Ther* **3:**545–565.

184. **Nathwani AC, Tuddenham EGD, Rangarajan S, Rosales C, McIntosh J, Linch DC, Chowdary P, Riddell**

A, Pie AJ, Harrington C, O'Beirne J, Smith K, Pasi J, Glader B, Rustagi P, Ng CYC, Kay MA, Zhou J, Spence Y, Morton CL, Allay J, Coleman J, Sleep S, Cunningham JM, Srivastava D, Basner-Tschakarjan E, Mingozzi F, High KA, Gray JT, Reiss UM, Nienhuis AW, Davidoff AM. 2011. Adenovirus-associated virus vector-mediated gene transfer in hemophilia B. *N Engl J Med* 365:2357–2365.

185. Manno CS, Pierce GF, Arruda VR, Glader B, Ragni M, Rasko JJ, Rasko J, Ozelo MC, Hoots K, Blatt P, Konkle B, Dake M, Kaye R, Razavi M, Zajko A, Zehnder J, Rustagi PK, Nakai H, Chew A, Leonard D, Wright JF, Lessard RR, Sommer JM, Tigges M, Sabatino D, Luk A, Jiang H, Mingozzi F, Couto L, Ertl HC, High KA, Kay MA. 2006. Successful transduction of liver in hemophilia by AAV-Factor IX and limitations imposed by the host immune response. *Nat Med* 12:342–347.

186. Janson C, McPhee S, Bilaniuk L, Haselgrove J, Testaiuti M, Freese A, Wang D-J, Shera D, Hurh P, Rupin J, Saslow E, Goldfarb O, Goldberg M, Larijani G, Sharrar W, Liouterman L, Camp A, Kolodny E, Samulski J, Leone P. 2002. Clinical protocol. Gene therapy of Canavan disease: AAV-2 vector for neurosurgical delivery of aspartoacylase gene (ASPA) to the human brain. *Hum Gene Ther* 13:1391–1412.

187. Testa F, Maguire AM, Rossi S, Pierce EA, Melillo P, Marshall K, Banfi S, Surace EM, Sun J, Acerra C, Wright JF, Wellman J, High KA, Auricchio A, Bennett J, Simonelli F. 2013. Three-year follow-up after unilateral subretinal delivery of adeno-associated virus in patients with Leber congenital Amaurosis type 2. *Ophthalmology* 120:1283–1291.

188. Bainbridge JWB, Smith AJ, Barker SS, Robbie S, Henderson R, Balaggan K, Viswanathan A, Holder GE, Stockman A, Tyler N, Petersen-Jones S, Bhattacharya SS, Thrasher AJ, Fitzke FW, Carter BJ, Rubin GS, Moore AT, Ali RR. 2008. Effect of gene therapy on visual function in Leber's congenital amaurosis. *N Engl J Med* 358:2231–2239.

189. Hauswirth WW, Aleman TS, Kaushal S, Cideciyan AV, Schwartz SB, Wang L, Conlon TJ, Boye SL, Flotte TR, Byrne BJ, Jacobson SG. 2008. Treatment of Leber congenital amaurosis due to RPE65 mutations by ocular subretinal injection of adeno-associated virus gene vector: short-term results of a phase I trial. *Hum Gene Ther* 19:979–990.

190. Maguire AM, High KA, Auricchio A, Wright JF, Pierce EA, Testa F, Mingozzi F, Bennicelli JL, Ying G-S, Rossi S, Fulton A, Marshall KA, Banfi S, Chung DC, Morgan JIW, Hauck B, Zelenaia O, Zhu X, Raffini L, Coppieters F, De Baere E, Shindler KS, Volpe NJ, Surace EM, Acerra C, Lyubarsky A, Redmond TM, Stone E, Sun J, McDonnell JW, Leroy BP, Simonelli F, Bennett J. 2009. Age-dependent effects of RPE65 gene therapy for Leber's congenital amaurosis: a phase 1 dose-escalation trial. *Lancet* 374:1597–1605.

191. Maguire AM, Simonelli F, Pierce EA, Pugh EN, Mingozzi F, Bennicelli J, Banfi S, Marshall KA, Testa F, Surace EM, Rossi S, Lyubarsky A, Arruda VR, Konkle B, Stone E, Sun J, Jacobs J, Dell'Osso L, Hertle R, Ma J-X, Redmond TM, Zhu X, Hauck B, Zelenaia O, Shindler KS, Maguire MG, Wright JF, Volpe NJ, McDonnell JW, Auricchio A, High KA, Bennett J. 2008. Safety and efficacy of gene transfer for Leber's congenital amaurosis. *N Engl J Med* 358:2240–2248.

192. Bennett J, Ashtari M, Wellman J, Marshall KA, Cyckowski LL, Chung DC, McCague S, Pierce EA, Chen Y, Bennicelli JL, Zhu X, Ying G-S, Sun J, Wright JF, Auricchio A, Simonelli F, Shindler KS, Mingozzi F, High KA, Maguire AM. 2012. AAV2 gene therapy readministration in three adults with congenital blindness. *Sci Transl Med* 4:120ra15.

193. Bowles DE, McPhee SWJ, Li C, Gray SJ, Samulski JJ, Camp AS, Li J, Wang B, Monahan PE, Rabinowitz JE, Grieger JC, Govindasamy L, Agbandje-McKenna M, Xiao X, Samulski RJ. 2012. Phase 1 gene therapy for Duchenne muscular dystrophy using a translational optimized AAV vector. *Mol Ther* 20:443–455.

194. Hajjar RJ, Zsebo K, Deckelbaum L, Thompson C, Rudy J, Yaroshinsky A, Ly H, Kawase Y, Wagner K, Borow K, Jaski B, London B, Greenberg B, Pauly DF, Patten R, Starling R, Mancini D, Jessup M. 2008. Design of a phase 1/2 trial of intracoronary administration of AAV1/SERCA2a in patients with heart failure. *J Card Fail* 14:355–367.

195. Barker SE, Broderick CA, Robbie SJ, Duran Y, Natkunarajah M, Buch P, Balaggan KS, MacLaren RE, Bainbridge JWB, Smith AJ, Ali RR. 2009. Subretinal delivery of adeno-associated virus serotype 2 results in minimal immune responses that allow repeat vector administration in immunocompetent mice. *J Gene Med* 11:486–497.

196. Jacobson SG, Cideciyan AV, Ratnakaram R, Heon E, Schwartz SB, Roman AJ, Peden MC, Aleman TS, Boye SL, Sumaroka A, Conlon TJ, Calcedo R, Pang J-J, Erger KE, Olivares MB, Mullins CL, Swider M, Kaushal S, Feuer WJ, Iannaccone A, Fishman GA, Stone EM, Byrne BJ, Hauswirth WW. 2012. Gene therapy for leber congenital amaurosis caused by RPE65 mutations: safety and efficacy in 15 children and adults followed up to 3 years. *Arch Ophthalmol* 130:9–24.

197. McClements ME, MacLaren RE. 2013. Gene therapy for retinal disease. *Transl Res* 161:241–254.

198. Cai X, Conley SM, Naash MI. 2009. RPE65: role in the visual cycle, human retinal disease, and gene therapy. *Ophthalm Genet* 30:57–62.

199. Gaudet D, Méthot J, Kastelein J. 2012. Gene therapy for lipoprotein lipase deficiency. *Curr Opin Lipidol* 23:310–320.

200. Ross CJD, Twisk J, Bakker AC, Miao F, Verbart D, Rip J, Godbey T, Dijkhuizen P, Hermens WTJMC, Kastelein JJP, Kuivenhoven JA, Meulenberg JM, Hayden MR. 2006. Correction of feline lipoprotein lipase deficiency with adeno-associated virus serotype 1-mediated gene transfer of the lipoprotein lipase S447X beneficial mutation. *Hum Gene Ther* 17:487–499.

201. Gaudet D, Méthot J, Déry S, Brisson D, Essiembre C, Tremblay G, Tremblay K, de Wal J, Twisk J, van den Bulk N, Sier-Ferreira V, van Deventer S. 2013. Efficacy and long-term safety of alipogene tiparvovec (AAV1-

LPLS447X) gene therapy for lipoprotein lipase deficiency: an open-label trial. *Gene Ther* **20:**361–369.

202. **Carpentier AC, Frisch F, Labbé SM, Gagnon R, de Wal J, Greentree S, Petry H, Twisk J, Brisson D, Gaudet D.** 2012. Effect of alipogene tiparvovec (AAV1-LPL (S447X)) on postprandial chylomicron metabolism in lipoprotein lipase-deficient patients. *J Clin Endocrinol Metab* **97:**1635–1644.

203. **Stroes ES, Nierman MC, Meulenberg JJ, Franssen R, Twisk J, Henny CP, Maas MM, Zwinderman AH, Ross C, Aronica E, High KA, Levi MM, Hayden MR, Kastelein JJ, Kuivenhoven JA.** 2008. Intramuscular administration of AAV1-lipoprotein lipase S447X lowers triglycerides in lipoprotein lipase-deficient patients. *Arterioscler Thromb Vasc Biol* **28:**2303–2304.

204. **Chuah M, VandenDriessche T.** 2007. Gene therapy for hemophilia "A" and "B": efficacy, safety and immune consequences. *Verh K Acad Geneeskd Belg* **69:**315–334.

205. **Zaiss A-K, Liu Q, Bowen GP, Wong NCW, Bartlett JS, Muruve DA.** 2002. Differential activation of innate immune responses by adenovirus and adeno-associated virus vectors. *J Virol* **76:**4580–4590.

206. **Chen J, Wu Q, Yang P, Hsu H-C, Mountz JD.** 2006. Determination of specific CD4 and CD8 T cell epitopes after AAV2- and AAV8-hF.IX gene therapy. *Mol Ther* **13:**260–269.

207. **Li C, Hirsch M, Asokan A, Zeithaml B, Ma H, Kafri T, Samulski RJ.** 2007. Adeno-associated virus type 2 (AAV2) capsid-specific cytotoxic T lymphocytes eliminate only vector-transduced cells coexpressing the AAV2 capsid *in vivo*. *J Virol* **81:**7540–7547.

208. **Li H, Murphy SL, Giles-Davis W, Edmonson S, Xiang Z, Li Y, Lasaro MO, High KA, Ertl HC.** 2007. Pre-existing AAV capsid-specific $CD8^+$ T cells are unable to eliminate AAV-transduced hepatocytes. *Mol Ther* **15:**792–800.

209. **Wang Z, Allen JM, Riddell SR, Gregorevic P, Storb R, Tapscott SJ, Chamberlain JS, Kuhr CS.** 2007. Immunity to adeno-associated virus-mediated gene transfer in a random-bred canine model of Duchenne muscular dystrophy. *Hum Gene Ther* **18:**18–26.

210. **Mingozzi F, Meulenberg JJ, Hui DJ, Basner-Tschakarjan E, Hasbrouck NC, Edmonson SA, Hutnick NA, Betts MR, Kastelein JJ, Stroes ES, High KA.** 2009. AAV-1-mediated gene transfer to skeletal muscle in humans results in dose-dependent activation of capsid-specific T cells. *Blood* **114:**2077–2086.

211. **Li C, DiPrimio N, Bowles DE, Hirsch ML, Monahan PE, Asokan A, Rabinowitz J, Agbandje-McKenna M, Samulski RJ.** 2012. Single amino acid modification of adeno-associated virus capsid changes transduction and humoral immune profiles. *J Virol* **86:**7752–7759.

212. **Gao G-P, Alvira MR, Wang L, Calcedo R, Johnston J, Wilson JM.** 2002. Novel adeno-associated viruses from rhesus monkeys as vectors for human gene therapy. *Proc Natl Acad Sci USA* **99:**11854–11859.

213. **Li C, Narkbunnam N, Samulski RJ, Asokan A, Hu G, Jacobson LJ, Manco-Johnson MJ, Monahan PE, Joint Outcome Study Investigators.** 2012. Neutralizing antibodies against adeno-associated virus examined prospectively in pediatric patients with hemophilia. *Gene Ther* **19:**288–294.

214. **Carlisle RC, Benjamin R, Briggs SS, Sumner-Jones S, McIntosh J, Gill D, Hyde S, Nathwani A, Subr V, Ulbrich K, Seymour LW, Fisher KD.** 2008. Coating of adeno-associated virus with reactive polymers can ablate virus tropism, enable retargeting and provide resistance to neutralising antisera. *J Gene Med* **10:**400–411.

215. **Georg-Fries B, Biederlack S, Wolf J, Hausen zur H.** 1984. Analysis of proteins, helper dependence, and seroepidemiology of a new human parvovirus. *Virology* **134:**64–71.

216. **Mendell JR, Campbell K, Rodino-Klapac L, Sahenk Z, Shilling C, Lewis S, Bowles D, Gray S, Li C, Galloway G, Malik V, Coley B, Clark KR, Li J, Xiao X, Samulski J, McPhee SW, Samulski RJ, Walker CM.** 2010. Dystrophin immunity in Duchenne's muscular dystrophy. *N Engl J Med* **363:**1429–1437.

217. **Jacobson EM, Huber A, Tomer Y.** 2008. The HLA gene complex in thyroid autoimmunity: from epidemiology to etiology. *J Autoimmun* **30:**58–62.

218. **Taneja V, David CS.** 2001. Lessons from animal models for human autoimmune diseases. *Nat Immunol* **2:** 781–784.

219. **Lee GK, Maheshri N, Kaspar B, Schaffer DV.** 2005. PEG conjugation moderately protects adeno-associated viral vectors against antibody neutralization. *Biotechnol Bioeng* **92:**24–34.

220. **Erles K, Sebökovà P, Schlehofer JR.** 1999. Update on the prevalence of serum antibodies (IgG and IgM) to adeno-associated virus (AAV). *J Med Virol* **59:**406–411.

221. **Scallan CD, Jiang H, Liu T, Patarroyo-White S, Sommer JM, Zhou S, Couto LB, Pierce GF.** 2006. Human immunoglobulin inhibits liver transduction by AAV vectors at low AAV2 neutralizing titers in SCID mice. *Blood* **107:**1810–1817.

222. **Grewal IS, Flavell RA.** 1998. CD40 and CD154 in cell-mediated immunity. *Annu Rev Immunol* **16:**111–135.

223. **Jiang H, Couto LB, Patarroyo-White S, Liu T, Nagy D, Vargas JA, Zhou S, Scallan CD, Sommer J, Vijay S, Mingozzi F, High KA, Pierce GF.** 2006. Effects of transient immunosuppression on adenoassociated, virus-mediated, liver-directed gene transfer in rhesus macaques and implications for human gene therapy. *Blood* **108:** 3321–3328.

224. **Nathwani AC, Gray JT, McIntosh J, Ng CYC, Zhou J, Spence Y, Cochrane M, Gray E, Tuddenham EGD, Davidoff AM.** 2007. Safe and efficient transduction of the liver after peripheral vein infusion of self-complementary AAV vector results in stable therapeutic expression of human FIX in nonhuman primates. *Blood* **109:**1414–1421.

225. **Mingozzi F, Anguela XM, Pavani G, Chen Y, Davidson RJ, Hui DJ, Yazicioglu M, Elkouby L, Hinderer CJ, Faella A, Howard C, Tai A, Podsakoff GM, Zhou S, Basner-Tschakarjan E, Wright JF, High KA.** 2013. Overcoming preexisting humoral immunity to AAV using capsid decoys. *Sci Transl Med* **5:**194ra92.

226. **Chen S, Kapturczak M, Loiler SA, Zolotukhin S, Glushakova OY, Madsen KM, Samulski RJ, Hauswirth WW, Campbell-Thompson M, Berns KI, Flotte TR, Atkinson MA, Tisher CC, Agarwal A.** 2005. Efficient transduction of vascular endothelial cells with recombinant adeno-associated virus serotype 1 and 5 vectors. *Hum Gene Ther* **16:**235–247.

227. **Qing K, Mah C, Hansen J, Zhou S, Dwarki V, Srivastava A.** 1999. Human fibroblast growth factor receptor 1 is a co-receptor for infection by adeno-associated virus 2. *Nat Med* **5:**71–77.

228. **Kashiwakura Y, Tamayose K, Iwabuchi K, Hirai Y, Shimada T, Matsumoto K, Nakamura T, Watanabe M, Oshimi K, Daida H.** 2005. Hepatocyte growth factor receptor is a coreceptor for adeno-associated virus type 2 infection. *J Virol* **79:**609–614.

229. **Summerford C, Bartlett JS, Samulski RJ.** 1999. AlphaVbeta5 integrin: a co-receptor for adeno-associated virus type 2 infection. *Nat Med* **5:**78–82.

230. **Asokan A, Hamra JB, Govindasamy L, Agbandje-McKenna M, Samulski RJ.** 2006. Adeno-associated virus type 2 contains an integrin alpha5beta1 binding domain essential for viral cell entry. *J Virol* **80:**8961–8969.

231. **Akache B, Grimm D, Pandey K, Yant SR, Xu H, Kay MA.** 2006. The 37/67-kilodalton laminin receptor is a receptor for adeno-associated virus serotypes 8, 2, 3, and 9. *J Virol* **80:**9831–9836.

232. **Ling C, Lu Y, Kalsi JK, Jayandharan GR, Li B, Ma W, Cheng B, Gee SWY, McGoogan KE, Govindasamy L, Zhong L, Agbandje-McKenna M, Srivastava A.** 2010. Human hepatocyte growth factor receptor is a cellular coreceptor for adeno-associated virus serotype 3. *Hum Gene Ther* **21:**1741–1747.

233. **Di Pasquale G, Davidson BL, Stein CS, Martins I, Scudiero D, Monks A, Chiorini JA.** 2003. Identification of PDGFR as a receptor for AAV-5 transduction. *Nat Med* **9:**1306–1312.

234. **Wu Z, Asokan A, Grieger JC, Govindasamy L, Agbandje-McKenna M, Samulski RJ.** 2006. Single amino acid changes can influence titer, heparin binding, and tissue tropism in different adeno-associated virus serotypes. *J Virol* **80:**11393–11397.

235. **Weller ML, Amornphimoltham P, Schmidt M, Wilson PA, Gutkind JS, Chiorini JA.** 2010. Epidermal growth factor receptor is a co-receptor for adeno-associated virus serotype 6. *Nat Med* **16:**662–664.

236. **Bell CL, Vandenberghe LH, Bell P, Limberis MP, Gao G-P, Van Vliet K, Agbandje-McKenna M, Wilson JM.** 2011. The AAV9 receptor and its modification to improve *in vivo* lung gene transfer in mice. *J Clin Invest* **121:**2427–2435.

Mobile DNA, 3rd Edition
Nancy L. Craig, Michael Chandler, Martin Gellert, Alan M. Lambowitz, Phoebe A. Rice and Suzanne Sandmeyer

doi:10.1128/microbiolspec.MDNA3-0042-2014

Zoltán Ivics[1]
Zsuzsanna Izsvák[2]

Sleeping Beauty Transposition 38

Sleeping Beauty KISSED BACK TO LIFE – A SHORT HISTORY

Members of the Tc1/*mariner* superfamily are probably the most widespread DNA transposons in nature (1). However, these elements appear to be transpositionally inactive in vertebrates due to the accumulation of mutations. In an attempt to isolate potentially active copies, we surveyed a number of fish genomes for the presence of Tc1-like elements from 11 different species. In summary, all the Tc1-like elements that we (2) and others (3, 4) described from the different fish species were defective copies carrying inactivating mutations that accumulated over long evolutionary times. Nevertheless, careful sequence analysis allowed us to predict a consensus sequence that would likely represent an active archetypal sequence. We have engineered this sequence by eliminating the inactivating mutations from the transposase open reading frame. The resurrected synthetic transposon was named *Sleeping Beauty* (*SB*), in analogy of the Grimm brothers' famous fairy tale. *SB* can be identical or closely related to an ancient transposon that once successfully invaded several fish genomes, in part by horizontal transmission between species (5). The resurrection of *SB* was the first demonstration that ancient transposable elements can be brought back to life. Before this work was published in 1997, there was no indication that any DNA-based transposon was active in vertebrates. *SB* not only represents the first DNA-based transposon ever shown to be active in cells of vertebrates, but the first functional gene ever reconstructed from inactive, ancient genetic material, for which an active, naturally occurring copy either does not exist or has not yet been isolated.

STRUCTURAL AND FUNCTIONAL COMPONENTS OF THE *Sleeping Beauty* TRANSPOSON

The *SB* transposon has a simple structure. In its natural form, it consists of a single gene encoding the transposase polypeptide, the enzymatic factor of transposition, which is flanked by terminal inverted repeats (IRs) containing binding sites for the transposase [Fig. 1(A)] (5). The transposase gene can be physically separated from the IRs, and replaced by other DNA sequences [Fig. 1(B)]. This is because the transposase can mobilize transposons in *trans*, as long as they retain the IRs. *SB* transposes through a conservative, cut-and-paste mechanism, during which the transposable element is excised from its original location by the transposase, and is integrated into a new location.

[1]Division of Medical Biotechnology, Paul Ehrlich Institute, Langen, Germany; [2]Max Delbrück Center for Molecular Medicine, Berlin, Germany.

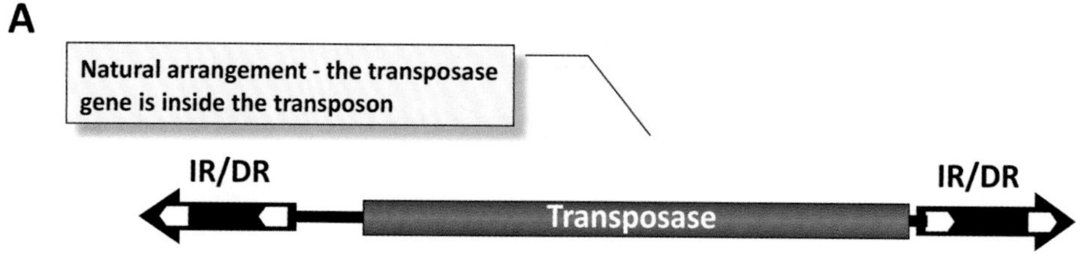

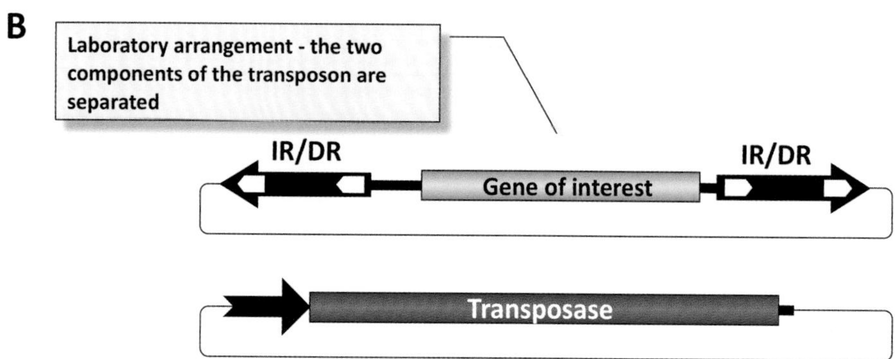

Figure 1 The *SB* transposon system. (A) Structure of the *SB* transposon. The central transposase gene (purple box) is flanked by terminal IRs (black arrows) that contain binding sites for the transposase (white arrows). (B) Gene transfer vector system based on *SB*. The transposase coding region can be replaced by a gene of interest (yellow box) within the transposable element. This transposon can be mobilized if a transposase source is provided in cells; for example, the transposase can be expressed from a separate plasmid vector containing a suitable promoter (black arrow). Reprinted from *Current Gene Therapy* (161) with permission from the publisher. doi:10.1128/microbiolspec.MDNA3-0042-2014.f1

THE TRANSPOSASE

The DNA-binding Domain

The overall domain structure of the transposase is conserved in the entire Tc1/*mariner* superfamily (1). Specific substrate recognition is mediated by an N-terminal, bipartite DNA-binding domain of the transposase (Fig. 2) (6, 7, 8). This DNA-binding domain has been proposed to consist of two helix-turn-helix (HTH) motifs, similar to the paired domain of some transcription factors in both amino acid sequence and structure (2, 8, 9). The modular paired domain has evolved versatility in binding to a range of different DNA sequences through various combinations of its subdomains (PAI+RED) (10). The origin of the paired domain is not clear, but phylogenetic analyses indicate that it might have been derived from an ancestral transposase (11).

The first of these HTH motifs has been crystallized in complex with double-stranded DNA corresponding to the termini of Tc3 transposons in *Caenorhabditis elegans* (12). The crystal structure indeed showed a HTH fold, and a dimer of transposase subunits bringing together the two DNA ends. The recently described NMR solution structure of the PAI subdomain of the *SB* transposase identified amino acid residues located in the second and third alpha helices forming the HTH motif to be involved in binding to DNA (Fig. 3) (13).

We found that a GRPR-like sequence (GRRR in *SB*) between the two HTH motifs is conserved in Tc1/*mariner* transposases (Fig. 2). The GRPR motif is similar to an AT-hook (6), and characteristic to homeodomain proteins (14). It mediates minor groove interactions with DNA in the case of the Hin invertase of *Salmonella* (15) and in the V(D)J recombination activating gene (RAG1) recombinase (see V(D)J chapter) (16).

Partially overlapping with the RED subdomain in the transposase is a bipartite nuclear localization signal (NLS) (Fig. 2), flanked by phosphorylation target sites of casein kinase II (2). Phosphorylation of these sites is a potential checkpoint in the regulation of transposition. The NLS indicates that these transposons, unlike

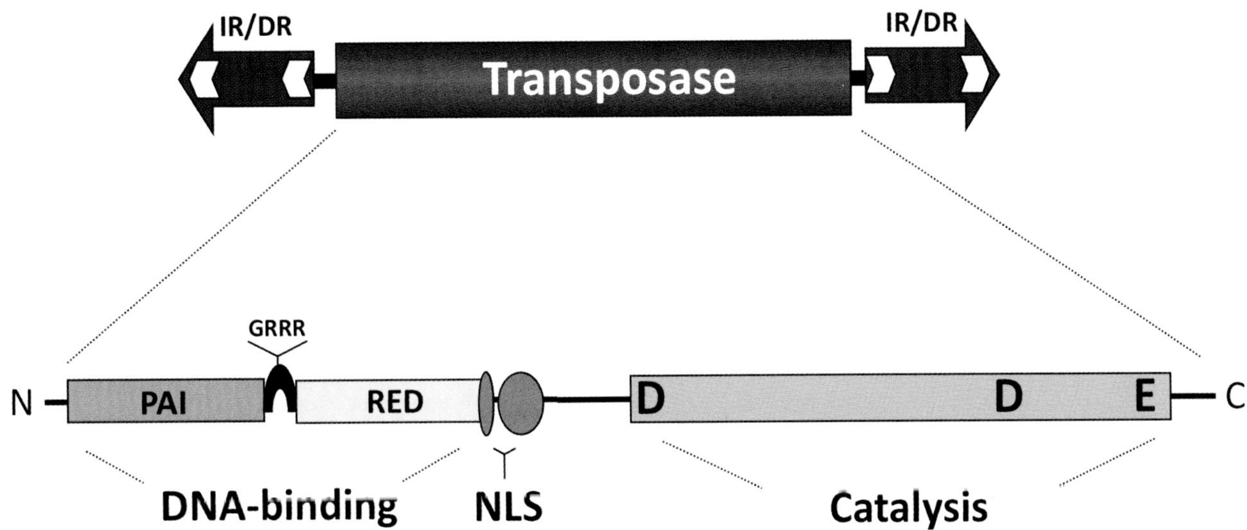

Figure 2 Structural and functional components of *SB*. On top, a schematic drawing of the transposon is shown. The terminal IR/DR (black arrows) contain two binding sites for the transposase (white arrows). The element contains a single gene encoding the transposase (purple box). The transposase has an N-terminal, bipartite, paired-like DNA-binding domain containing a GRRR AT-hook motif, an NLS, and a C-terminal catalytic domain. The DNA binding domain consists of a PAI and a RED subdomain containing helix-turn-helix DNA-binding motifs. The DDE amino acid triad is a characteristic signature of the catalytic domain that catalyzes the DNA cleavage and joining reactions. Reprinted from *Molecular Therapy* (164) with permission from the publisher.
doi:10.1128/microbiolspec.MDNA3-0042-2014.f2

murine retroviruses, can take advantage of the receptor-mediated transport machinery of host cells for nuclear uptake of their transposases.

The Catalytic Domain

The second major domain of the *SB* transposase has been referred to as the catalytic domain, because it is responsible for the DNA cleavage and joining reactions of transposition. The majority of known transposases, including *SB*, and retroviral integrases possess a well-conserved triad of amino acids, known as the aspartate–aspartate–glutamate, in short the DDE signature in their C-terminal catalytic domain (Fig. 2) (17, 18, 19, 20). These amino acids play an essential role in catalysis by coordinating, in general, two divalent cations necessary for activity. One metal ion acts as a Lewis acid, and stabilizes the transition state of the penta-coordinated phosphate, the other one acts as a general base and deprotonizes the incoming nucleophile during transesterification and strand transfer (21). The biologically relevant cation required for the catalytic steps of transposition is thought to be Mg^{++} (22, 23).

Site-directed mutagenesis of the DDE residues in the related Tc3 transposase confirmed that these three amino acids are indeed essential for all catalytic activities (24). Similarly, a DAE variant of the *SB* transposase is catalytically dead (25). Although the crystal structure of the *SB* transposase catalytic domain is yet to be solved, the available Mos1 *mariner* structure from *Drosophila mauritiana* (26) is assumed to be the closest to model the structure of the *SB* transposase. Intriguingly, crystallographic analyses of the catalytic domains of transposases and some other proteins whose functions are not obviously related to transposition, such as RNaseH (27) or RuvC (28), have revealed a remarkably similar overall fold. Besides similarities, there are important differences between *mariner* and *SB* transposases. First, while *SB* has the characteristic DDE motif in its catalytic domain, this motif in *mariner*s is DDD. Interestingly, a change of the exceptional third D of *mariner*, turning the DDD into the canonical DDE, inactivates the transposase (29), and the reciprocal substitution in the *SB* transposase (DDD) is also inactive (30). A conserved glycine-rich subdomain can be found within the catalytic domains of *SB* and Tc1-like transposases (5). This glycine-rich subdomain is not present in *mariner*s or retroviral integrases, and its function is yet to be determined. The emerging picture reinforces the notion of a common structural motif that catalyzes polynucleotidyl transfer reactions in diverse biological contexts

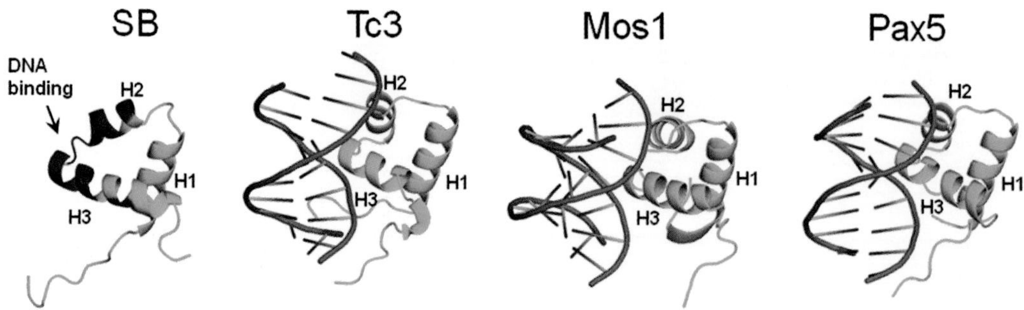

Figure 3 Structures of the PAI subdomain of the *SB* transposase and the DNA-bound N-terminal DNA-binding subdomains of the Tc3 and Mos1 *mariner* transposases and the Pax5 transcription factor. Residues on the second and third alpha-helices of the *SB* PAI subdomain are directly involved in DNA-binding. Reprinted from *Protein Science* (13) with permission from the publisher. doi:10.1128/microbiolspec.MDNA3-0042-2014.f3

(31, 32), and that the different specificities in binding to DNA might have evolved by the apparent acquisition of different DNA-binding domains in the evolution of DDE/D recombinases (18).

Molecular Evolution of the SB100X Hyperactive *Sleeping Beauty* Transposase

In evolutionary terms, the *SB* transposon was a successful element with the ability to colonize several fish genomes millions of years ago (2). However, even successful transposons have not been selected for the highest possible activity. On the contrary, there is strong selective pressure to avoid insertional mutagenesis of essential genes of their host. In an attempt to derive hyperactive transposase variants for advanced genetic engineering, amino acid substitutions spanning almost the entire *SB* transposase polypeptide have been screened for eliciting a change in catalytic activity. These amino acid replacements were conducted either by systematic alanine-scanning (33), by "importing" single amino acids or small (2 to 7 aa) blocks of amino acids from related transposases (34, 35, 36), and by rational replacement of selected amino acid residues based on charge (35). These approaches generated transposase variants with (i) no change in activity; (ii) reduced activity, or (iii) a relatively modest increase of transposition activity. The vast majority of the mutations have a neutral or a negative effect on transposition activity. The inactivating mutations generally map to the evolutionary conserved domains of the transposase. Surprisingly, some combinations of hyperactive mutations were found to result in a significant reduction of activity. Nevertheless, a strategy of identifying those hyperactive variants that acted in an additive or synergistic manner combined with a high throughput genetic screening of ~2,000 possible combinations yielded *SB* transposase variants with significantly enhanced activities (30). The most hyperactive version, *SB*100X, displays a ~100-fold hyperactivity when compared to the originally resurrected transposase (Fig. 4) (30). The hyperactivity of *SB*100X cannot be explained by altered transposase stability, nor by increased binding to the transposon IRs; instead, the particular combination of mutations in *SB*100X appears to affect the folding properties of the transposase (30). The use of the *SB*100X system yielded robust gene transfer efficiencies into human hematopoietic progenitors (30, 37), mesenchymal stem cells, muscle stem/progenitor cells (myoblasts), iPSCs (38), and T cells (39). These cells are relevant targets for stem cell biology and for regenerative medicine and gene- and cell-based therapies of complex genetic diseases. Moreover, the *SB*100X transposase enables highly efficient germline transgenesis in relevant mammalian models, including mice, rats, rabbits, and pigs (40, 41, 42).

The Inverted Repeats of the *Sleeping Beauty* Transposon

Similarly to most transposon ends, the IRs of *SB* are composed of two functional parts. The 2 to 3 terminal base pairs of the ends are the recombinationally active sequences involved in the cleavage and the strand transfer reactions. The other functional part is situated within the IRs and it ensures the sequence-specific positioning of the transposase on the transposon ends.

The IRs of *SB* are 200 to 250 bp long and carry a pair of transposase-binding sites within the ends of each IR, characterized by short, 15 to 20 bp direct repeats (DRs) (Fig. 1). This special organization of IRs is termed IR/DR (1, 43), and can be found in numerous elements in the Tc1 transposon family, including the *Minos*, *S*, *Paris* and *Bari* elements in various *Drosophila* species

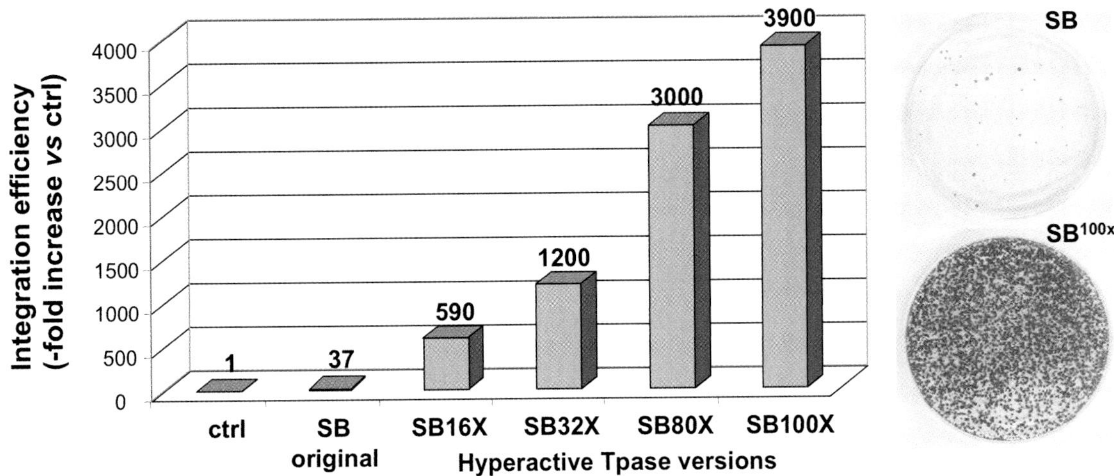

Figure 4 Comparison of different hyperactive versions of the *SB* transposase in transfected human HeLa cells. The chart shows the respective potential of transposase mutants to generate antibiotic-resistant cell colonies in human cell culture. The Petri dishes on the right show stained, antibiotic-resistant cell colonies obtained with the original *SB* transposase and with the *SB*100X hyperactive variant. doi:10.1128/microbiolspec.MDNA3-0042-2014.f4

(1, 44, 45, 46), *Quetzal* elements in mosquitos (47), at least three Tc1-like transposon subfamilies in fish (2), and *Txr*, *Eagle*, *Froggy*, and *Jumpy* transposons in *Xenopus* (48, 49). The spacing of about 200 bp between the outer and inner DRs is conserved in all elements within the IR/DR group, but the actual DNA sequences are not similar, suggesting convergent evolution of the IR/DR-type repeats. The IR/DR group differs significantly from Tc1 or the *mariner* elements, which are simpler and have repeats of less than 100 bp and a single transposase binding site per repeat (8, 50). The four DRs of *SB* are not identical; the outer ones are longer by two base pairs. The IRs are not identical either; the left IR contains a sequence motif called "half-DR" (HDR), which resembles the 3′-half of the transposase binding sites (6). A construct containing two left IRs transposes more efficiently than the wild-type transposon, but another version that has two right IRs has very poor mobility, indicating that the left and right IRs are functionally distinct (6). The multiple binding sites of the IR/DR elements likely impose control over the timing and specificity of the transposition reaction (see below).

THE MOLECULAR MECHANISM OF *Sleeping Beauty* TRANSPOSITION AND ITS REGULATION BY HOST-ENCODED FACTORS

The typical "cut-and-paste" transposition process of *SB* can be divided into at least four major steps: (1) binding of the transposase to its sites within the transposon IRs; (2) formation of a synaptic complex in which the two ends of the elements are paired and held together by transposase subunits; (3) excision from the donor site; (4) reintegration at a target site (Fig. 5).

The wide phylogenetic distribution of the Tc1/*mariner* family suggested no or weak host factor requirement of the transposition reaction. Supporting this assumption, the activity of Tc1 was reconstructed *in vitro*, and the reaction required only the presence of the transposon and the transposase (51). In addition, the nematode Tc3 element was demonstrated to jump in zebrafish; however, this *trans*-species transposition reaction was not efficient (52). Despite earlier assumptions, *SB* transposition turned out to be highly dependent on cellular host factors (described in detail in sections below) (53, 54, 55, 56) and became an excellent model system to study transposon–host interactions in higher eukaryotes (57). *SB* can transpose in a wide range of vertebrate cells from fish to human, although the efficiency of transposition varies significantly (58), suggesting that differential interactions between the transposon and host-encoded factors may affect activity and eventually limit the host range. Indeed, transposition of *SB* seems to be restricted to vertebrates, with the exception of a chordate, *Ciona intestinalis* (59). Importantly, the identified host factors of *SB* are evolutionarily conserved in vertebrates and support *SB* transposition from fish to human. In summary, the regulation of *SB* transposition is mediated both by transposon- and host-encoded factors. Thus, the *SB* transposon has an

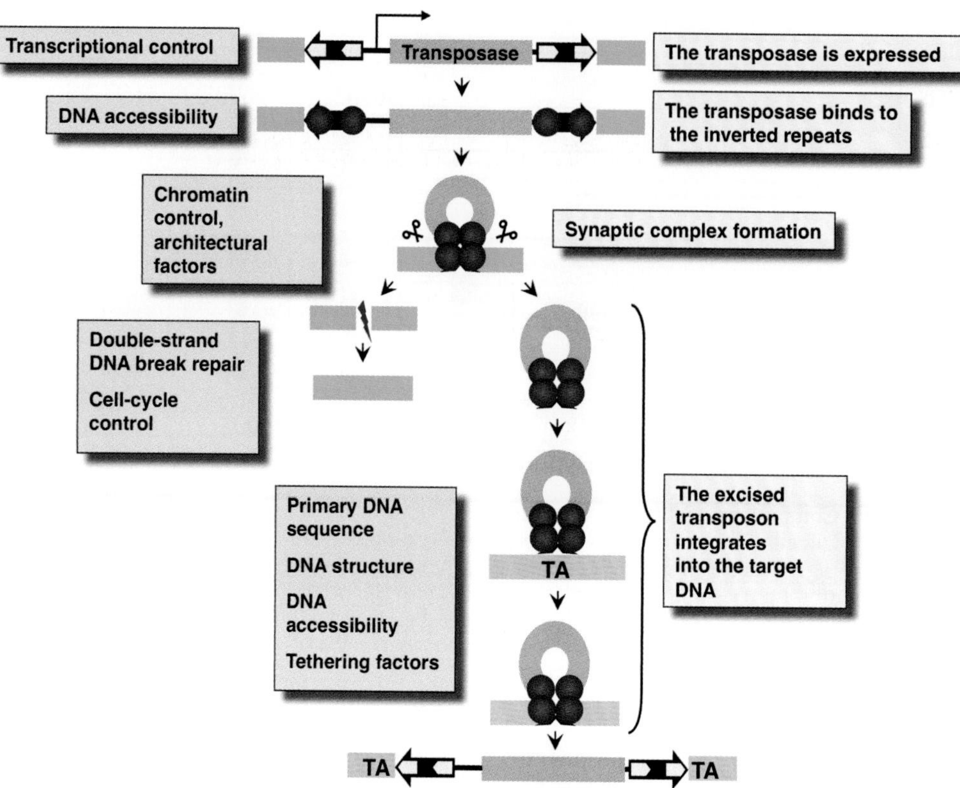

Figure 5 Mechanism and regulation of *SB* transposition. The transposable element consists of a gene encoding a transposase (orange box) bracketed by terminal IRs (solid black arrows) that contain binding sites of the transposase (white arrows) and flanking donor DNA (blue boxes). Transcriptional control elements in the 5′-UTR of the transposon drive transcription (arrow) of the transposase gene. The transposase (purple spheres) binds to its sites within the transposon IRs. Excision takes place in a synaptic complex, and separates the transposon from the donor DNA. The excised element integrates into a TA site in the target DNA (green box) that is duplicated and flanks the newly integrated transposon. On the right, the various steps of transposition are shown. On the left, mechanisms and host factors regulating each step of the transposition reaction are indicated. Reprinted from (57) with permission from the publisher. doi:10.1128/microbiolspec.MDNA3-0042-2014.f5

intimate relationship with the host that likely modulates transposition at every step of the transposition reaction (Fig. 5).

Transcriptional Regulation of the *Sleeping Beauty* Transposon

Some of the 5′-untranslated regions (UTRs) upstream of the initiation codon of the transposase gene contain promoter motifs (60), suggesting that they might have functions associated with control of transposition activity. However, previous studies did not reveal an internal promoter in the Tc1 element; instead they showed that the elements are transcribed by read-through transcription from *C. elegans* genes (61). The left IR of *SB* is separated from the transposase coding sequence by a 160-bp stretch of DNA [Fig. 6(A)] with no apparent function in the transposition reaction (35). As measured by transient luciferase reporter assays, transcription driven by the 5′-UTR of *SB* is ~18-fold higher than transcription of a promoter-less sequence, ~4.6-fold higher than transcription driven by a TATA-box minimal promoter, and ~2.5-fold higher than transcription driven by the 5′-UTR of the closely related *Frog Prince* (*FP*) transposon [Fig. 6(A)] (54). The 5′-UTR drives expression of the *SB* transposase at a level sufficient to detect *SB* transposition in a colony-forming transposition assay in HeLa cells [Fig. 6(A)]. The right IR can also drive expression towards the inside of the element, but at a lower efficiency than the 5′-UTR [Fig. 6(A)]. Convergent transcription of *SB* transposons raises the possibility of the formation of

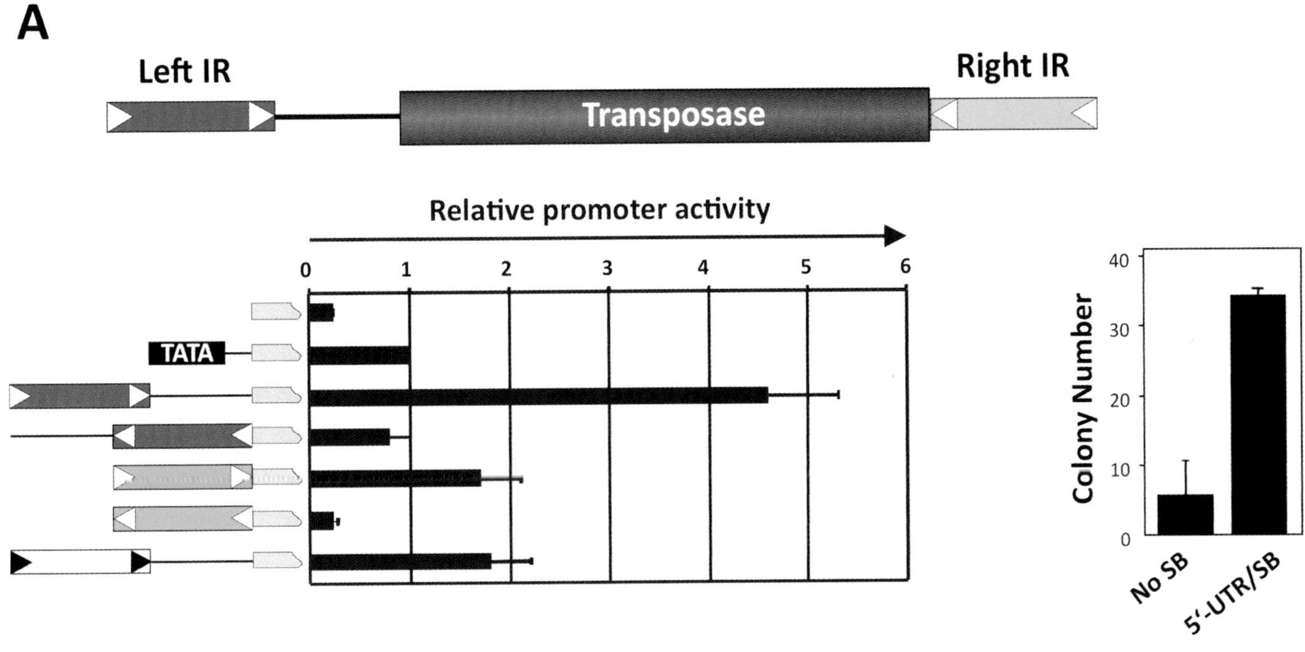

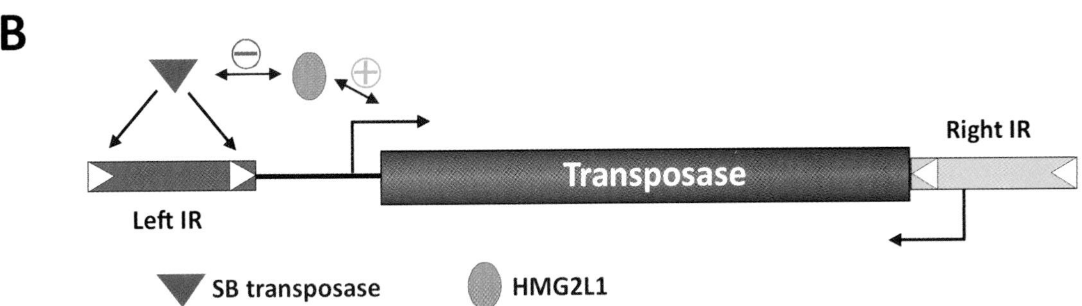

Figure 6 The UTRs of the *SB* transposon exhibit moderate, directional promoter activities. (A) Transcriptional activities residing within the *SB* transposon. On top, a schematic drawing of the transposon is shown. The terminal IRs contain two binding sites for the transposase (white arrows). The element contains a single gene encoding the transposase (purple box). Relative promoter activities as determined by transient luciferase assays in HeLa cells. Activity of a minimal promoter (TATA-box) control was arbitrarily set to value 1. Transposon sequences flanking the transposase gene were placed in front of a luciferase reporter gene in two possible orientations (in the case of the 5′-UTR, the luciferase gene precisely replaces the transposase coding region). Blue box: left IR of *SB*; green box: right IR of *SB*; beige box: left IR of *Frog Prince*; black lines connecting the IRs and the luciferase gene represent transposon sequences directly upstream of the transposase coding regions. The 5′-UTR of *SB* can drive transposase expression at a level sufficient for the detection of chromosomal transposition events in cultured cells. A neo-tagged *SB* transposon plasmid was cotransfected together with an *SB* expression construct, in which the transposase is expressed from the 5′-UTR of the transposon or with an empty cloning vector. The difference in numbers of G418-resistant cell colonies is evidence for transposition. (B) A model for transcriptional regulation of the *SB* transposase gene. In the wild-type, natural transposon, the central transposase gene (purple box) is flanked by UTRs that include the left and right inverted repeats (IRs, blue and green boxes, respectively) that contain binding sites for the transposase (white arrows). Arrows indicate the direction of transcription that is initiated within the UTRs. HMG2L1 upregulates, whereas *SB* transposase downregulates transcription from the 5′-UTR. Reprinted from *Molecular Therapy* (54) with permission from the publisher. doi:10.1128/microbiolspec.MDNA3-0042-2014.f6

transposon-specific double-stranded RNA molecules that may serve as triggers of transposon regulation by RNA interference (54, 62). Indeed, transgene expression from genomically integrated *SB* copies was enhanced in the presence of the p19 protein (derived from the tomato bushy stunt viruses) that suppresses RNA interference (63).

A cellular host factor, HMG2L1 (alias HMGXB4), was identified as a physical interacting partner of the *SB* transposase and shown to upregulate transcription from the 5′-UTR of *SB* by 10- to 15-fold (54) [Fig. 6(B)]. HMG2L1 is an HMG-box DNA-binding domain-containing protein that shares structural similarity with lymphocyte enhancer-binding factor 1 (LEF-1), sex-determining region Y (SRY), and SRY-related HMG-box protein 4 (SOX4) transcription factors (64). HMG-box transcription factors specifically bind their target DNA through their HMG-box domains, and regulate transcription of target genes (64). Indeed, in addition to its interaction with the transposase, HMG2L1 also interacts with the *SB* transposon DNA *in vivo*, as shown by chromatin immunoprecipitation (ChIP) experiments (54). Interestingly, co-expression of the *SB* transposase with HMG2L1 has a repressing effect on transcription by the 5′-UTR [Fig. 6(B)]. Thus, transposase expression in the context of the naturally occurring *SB* transposon is subject to negative feedback regulation, with the transposase acting as a transcriptional repressor. This model postulates a sensitive balance in the regulation of transposase expression that is calibrated by transposase concentrations in the cell: low concentrations allow more transposase to be made, whereas high concentrations lead to shutting off transposase expression.

Specific DNA-binding by the *Sleeping Beauty* Transposase

Similar to the DNA-binding domain of the transposase, the binding sites also have a bipartite structure in which the 3′-part of the binding site is recognized by the PAI subdomain, whereas the 5′-sequences interact with the RED subdomain of the transposase (6). Specificity of DNA-binding is predominantly determined by base-specific interactions mediated by the PAI subdomain (6).

All four binding sites within the IR/DR structure are required for *SB* transposition (58). The paired-like DNA-binding domain forms tetramers in complex with transposase binding sites (6). The inner DRs are more strongly bound by the transposase than the outer DRs (53, 65). Recent NMR data confirmed differential binding of the DNA-binding domain of the *SB* transposase to the inner and outer DRs (13). The PAI subdomain also binds to the HDR motif within the left IR of *SB* and mediates protein–protein interactions with other transposase subunits (6). Thus, the PAI subdomain is proposed to have at least three distinct functions: interaction with both the DRs and the HDR motif, and transposase oligomerization. In cooperation with the main DNA-binding domain, the GRRR motif was shown to function as an AT-hook, contributing to specific substrate recognition (6). Although part of the NLS is included in the RED subdomain, it does not appear to contribute to DNA recognition. Domain swapping experiments have shown that primary DNA-binding is not sufficient to determine specificity of the transposition reaction (6). Zebrafish *Tdr1* elements are closely related to *SB*, but are not mobilized by the *SB* transposase. Comparison of the transposase binding site sequences of *SB* and *Tdr1* elements revealed main differences in the 5′-half of the DRs. This sequence is contacted by the RED subdomain, indicating that the function of the RED is to enforce specificity at a later step in transposition. Substrate recognition of *SB* transposase is therefore sufficiently specific to prevent activation of transposons of closely related subfamilies.

Ordered Assembly of Synaptic Complexes and the role of HMGB1 in *Sleeping Beauty* Transposition

A uniform requirement among transposition reactions is the formation of a nucleoprotein complex, before transposon excision can take place. This very early step, synaptic complex assembly, is the process by which the two ends of the elements are paired and held together by transposase subunits (Fig. 5). The necessary factors that are required for synaptic complex assembly of *SB* include the complete IRs with four transposase binding sites, the HDR motif, and tetramerization-competent transposase. These tetrameric complexes form only if all the four binding sites are present and they are in the proper context. The HDR motif is important but not essential in transposition and therefore can be viewed as a transpositional enhancer that, together with the PAI subdomain of the transposase, stabilizes complexes formed by a transposase tetramer bound at the IR/DR. In contrast to the *Mu* transposase, where the two specificities of binding to the enhancer and to the recombination sites are encoded in two distinct domains (66), the paired-like region of *SB* transposase combines these two functions in a single protein domain. As described above, the *SB* transposase preferentially binds the inner DRs within the transposon IRs.

This suggests that the unequal strengths of transposase binding and the positions of the DRs within the IRs are required for ordered assembly of transposase-DNA complexes at the ends of the transposon that has a fundamental effect on the outcome of the transposition reaction.

The high mobility group protein HMGB1 is required for efficient *SB* transposition in mammalian cells (53). HMG family members have been shown to also be required for V(D)J recombination (67, 68). HMGB1 is an abundant, nonhistone, nuclear protein associated with eukaryotic chromatin and has the ability to bend DNA (69). *SB* transposition is significantly reduced in HMGB1-deficient mouse cells (53). This effect was complemented by expressing HMGB1 and HMGB2, but not with the more distantly related HMGA1 protein. Overexpression of HMGB1 in wild-type cells enhanced transposition, indicating that HMGB1 is a limiting factor of transposition (53). HMGs have low affinity for standard B-form DNA, and interactor proteins need to guide them to certain sites (69). *SB* transposase was found to interact with HMGB1 *in vivo* and to form a ternary complex with the transposase and transposon DNA (Fig. 7) (53), suggesting that the transposase may actively recruit HMGB1 to transposon DNA via protein–protein interactions. Considering the significant drop of transposition activity in HMGB1-deficient cells, the role of HMGB1 in transposition is a critical one.

HMGB1 was proposed to promote communication between DNA motifs within the transposon that are otherwise distant to each other, including the DRs, the transpositional enhancer, and the two IRs (53). However, as mentioned above, physical proximity of the DRs is not sufficient for *SB* transposition; a highly specific configuration of functional DNA elements within the IRs has a critical importance. It was also found that HMGB1 enhances transposase binding to both DRs, but its effect is significantly more pronounced at the inner sites (53). It appears, therefore, that the order of events that take place during the very early steps of transposition is binding of transposase molecules first to the inner sites and then to the outer sites. The pronounced effect of HMGB1 on binding of the transposase to the inner sites suggests that HMGB1 enforces ordered assembly of a catalytically active synaptic complex (Fig. 7). Indeed, interference with this sequence of events by replacing the outer transposase binding sites with the inner sites abolishes *SB* transposition (65). This ordered assembly process probably controls that cleavage at the outer sites occurs only if all the previous requirements have been fulfilled.

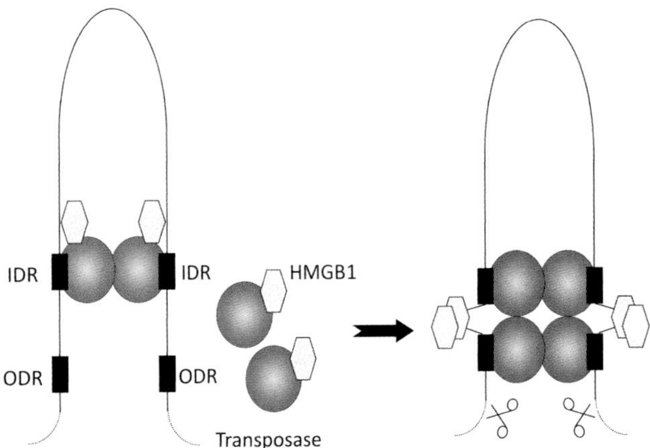

Figure 7 A model for the role of HMGB1 in *SB* synaptic complex formation. *SB* transposase (pink spheres) recruits HMGB1 (dotted hexagons) to the transposon IRs. First, HMGB1 stimulates specific binding of the transposase to the inner binding sites (IDRs). Once in contact with DNA, HMGB1 bends the spacer regions between the DRs, thereby assuring correct positioning of the outer sites (ODRs) for binding by the transposase. Cleavage (scissors) proceeds only if complex formation is complete. The complex includes the four binding sites (black boxes) and a tetramer of the transposase. Reprinted from *Nucleic Acids Res* (53) with permission from the publisher.
doi:10.1128/microbiolspec.MDNA3-0042-2014.f7

In summary, the IR/DR-type organization of IRs introduces a higher-level regulation into the transposition process. These elements might have evolved novel "built in regulatory checkpoints" to enforce synapsis prior to catalysis, thereby ensuring a higher level of accuracy and fidelity during the transposition process compared to transposons with simply structured IRs (53, 56, 70). The repeated transposase binding sites, their dissimilar affinity for the transposase, and the effect of HMGB1 to differentially enhance transposase binding to the inner sites are all important for a geometrically and timely orchestrated formation of synaptic complexes, which is a strict requirement for the subsequent catalytic steps of transposition. Such strictly regulated assembly of catalytically primed complexes could suppress unpaired reaction products or promiscuous synapses of distant ends of the transposon (70).

Regulation of Transposon Excision by DNA CpG Methylation

CpG methylation of chromosomal DNA leads to formation of heterochromatin and is known to decrease or inhibit transpositional activity of diverse transposons (71). Surprisingly, CpG methylation of the *SB* transposon was found to enhance transpositional activity in mouse embryonic stem (ES) cells (72). It was subsequently found that the enhancing effect of CpG methylation is not restricted to *SB* but, rather, seems to be an intrinsic feature associated with the characteristic IR/DR structure of the *SB*, *Frog Prince*, and *Minos* elements (73). At which step(s) of cut-and-paste transposition does the effect of CpG methylation manifest? Because (i) CpG methylation has no effect on *SB* transposase binding, (ii) CpG methylation induces the formation of a condensed chromatin structure (heterochromatin), and (iii) DNA compaction by protamine enhances transposition, a model was proposed in which CpG methylation and subsequent chromatin condensation aids synaptic complex formation. Because heterochromatin formation results in tight packaging of DNA and histones, DNA sites that are usually far away from each other – for example, the two transposase binding sites in the IRs – might be brought closer together (Fig. 8). The physical proximity of the inner and outer binding sites might assist the formation of transposase dimers as soon as they bind, thereby facilitating the formation of a catalytically active synaptic complex. Thus, similarly to the effect of HMGB1, conformational changes of the excising transposon may greatly in fluence the efficiency of transposition. Assuming that the transposase source is provided by a transcriptionally active element located in euchromatin, host-cell-induced CpG methylation/heterochromatin-based silencing of transposons can be offset by a higher transposition efficiency out of condensed chromatin,

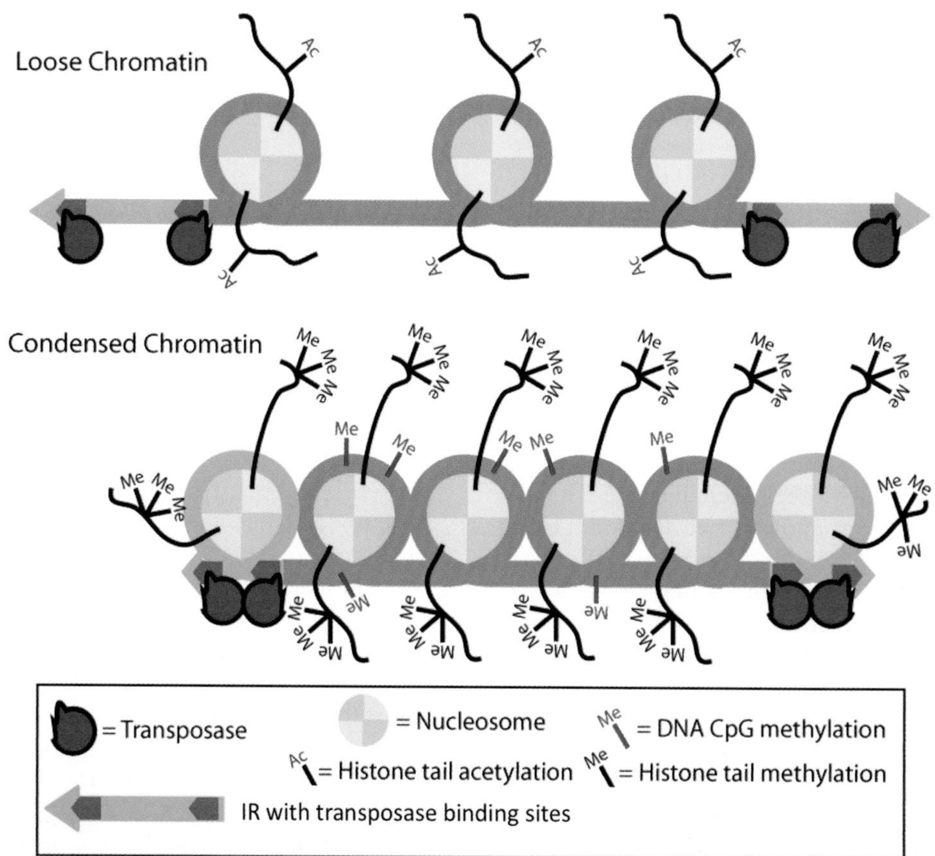

Figure 8 A model for the enhancing effect of a compact chromatin structure on *SB* transposition. Euchromatin contains DNA wrapped around nucleosomes in a "beads-along-a-string"-like conformation (upper panel). Transposase subunits bound within the transposon IRs are separated by 166 bp DNA. Heterochromatin (lower panel), characterized by DNA CpG methylation and specific histone tail modifications, e.g., trimethylated lysine 9 of histone H3, features a higher histone : DNA ratio. Positioning of a nucleosome between the transposase binding sites (small orange arrows) shortens the distance between these sites, thereby facilitating the formation of a transposase dimer per IR and subsequent assembly of the synaptic complex. Reprinted from *Mobile DNA* (73) with permission from the publisher. doi:10.1128/microbiolspec.MDNA3-0042-2014.f8

thereby constituting a potential mechanism of *SB* and other similar-structured transposons to escape CpG methylation-mediated silencing imposed by the host.

Excision of the *Sleeping Beauty* Transposon and Double-Strand Break Repair of Excision Sites

The key process of all transposon excision is the exposure of the 3′-OH groups of the transposon ends, which will later be used at the strand transfer reaction for integration (74). Every DNA strand cleavage in all transposition reactions is a transposase-catalyzed, Mg^{++}-dependent hydrolysis of the phosphodiester bonds of the DNA backbone, executed by a nucleophilic molecule. All the DDE recombinases catalyze similar chemical reactions (75), which begin with a single-strand nick that generates a free 3′-OH group. In the case of the first strand cleavage the nucleophile is H_2O (74). During cut-and-paste transposition, nicking of the element is followed by the cleavage of the complementary DNA strand. To catalyze second strand cleavage, DDE recombinases developed versatile strategies (76). Most DDE transposases use a single active site to cleave both strands of DNA at the transposon end via a DNA–hairpin intermediate (77, 78, 79, 80, 81). For example, in V(D)J recombination the single-strand nick is converted into a double-stranded break (DSB) by a transesterification reaction in which the free 3′-OH attacks the opposite strand, thereby creating a hairpin intermediate at the donor site (82, 83). Tn*5* and Tn*10* transposons also transpose *via* a hairpin intermediate, with the difference that the hairpin is on the transposon and not on the flanking DNA (84, 85). However, *mariner* (86) and *SB* (55) transposition does not proceed through a hairpin intermediate, and the exact mechanism of second-strand cleavage remains unknown. Thus, in the absence of a hairpin intermediate, *mariner* and *SB* transposases likely cleave the two strands of the DNA at each transposon end by sequential hydrolysis reactions.

Strand cleavage can occur at different positions relative to the transposon ends. The position of 5′-cleavage of the second strand required for the liberation of the element occurs directly opposite the 3′-cleavage site in V(D)J recombination (82) and for the bacterial Tn*5* (87) and Tn*10* elements (85) (thereby generating blunt ended products). In the case of the Tc1/*mariner* elements the non-transferred strand is cleaved a few nucleotides within the transposon [two nucleotides for the Tc1 and Tc3 elements (24, 50) and three nucleotides inside the element in the case of *mariner* and *SB* (Fig. 9) (50, 88)]. Thus, *SB* transposon excision leaves behind three-nucleotide-long 3′-overhangs (Fig. 9), which are processed by the DNA repair mechanisms of the cells leaving a transposon "footprint" at the transposon donor site (see below).

The DSBs generated by transposon excision are repaired either by the nonhomologous end joining pathway (NHEJ) or by homologous recombination (HR) (89, 90). The main factors that mediate NHEJ are a complex of DNA ligase IV and Xrcc4 and the DNA-dependent protein kinase (DNA-PK), a serine/threonine protein kinase (91, 92). DNA-PK consists of a catalytic subunit (DNA-PKcs) and a DNA-binding subunit termed Ku. The Ku heterodimer (composed of Ku70 and Ku80) binds to DNA ends and facilitates DSB repair by recruiting DNA-PKcs and additional factors such as the Xrcc4/DNA ligase IV complex to the site of damage.

The prominent pathway of repairing transposon excision sites in somatic mammalian cells is NHEJ, which generates transposon "footprints" that are identical to the first or last 2 to 4 nucleotides of the transposon in Tc1/*mariner* transposition (88, 93). NHEJ of the three-nucleotide-long 3′-overhangs left behind by *SB* excision generates a 3-bp footprint (Fig. 9) (55, 93). Factors of the NHEJ pathway of DSB repair, including Ku70 and DNA-PKcs, are required for *SB* transposition by repairing the transposon excision sites (55). NHEJ components have also been shown to be required for efficient retroelement integration and V(D)J recombination (92, 94). Ku70 physically interacts with the *SB* transposase (55), suggesting that it might be involved in shepherding excision site repair to NHEJ.

NHEJ and HR have overlapping roles in maintaining chromosomal integrity in vertebrate cells (95), and they can serve as alternative pathways for repair of the same DSB (96). Although the NHEJ pathway of DSB repair plays a dominant role in repair of transposon excision sites in somatic cells, the dependence of *SB* transposition on NHEJ factors is not absolute. In contrast to V(D)J recombination (92), HR can also be involved in excision site repair during *SB* transposition (55). Similarly, both NHEJ and HR play significant roles in the repair of DSBs generated by Tc1 excision in *C. elegans* (97), in P-element transposition in *Drosophila* (98, 99), and in Ty1 retrotransposon integration in yeast (100, 101). These observations support the view that DSBs generated by radiation, V(D)J recombination, retroviral integration, and DNA transposition require overlapping but different factors for repair. The interplay between the repair factors and the recombination machineries probably determines how mechanistically similar processes can produce different products.

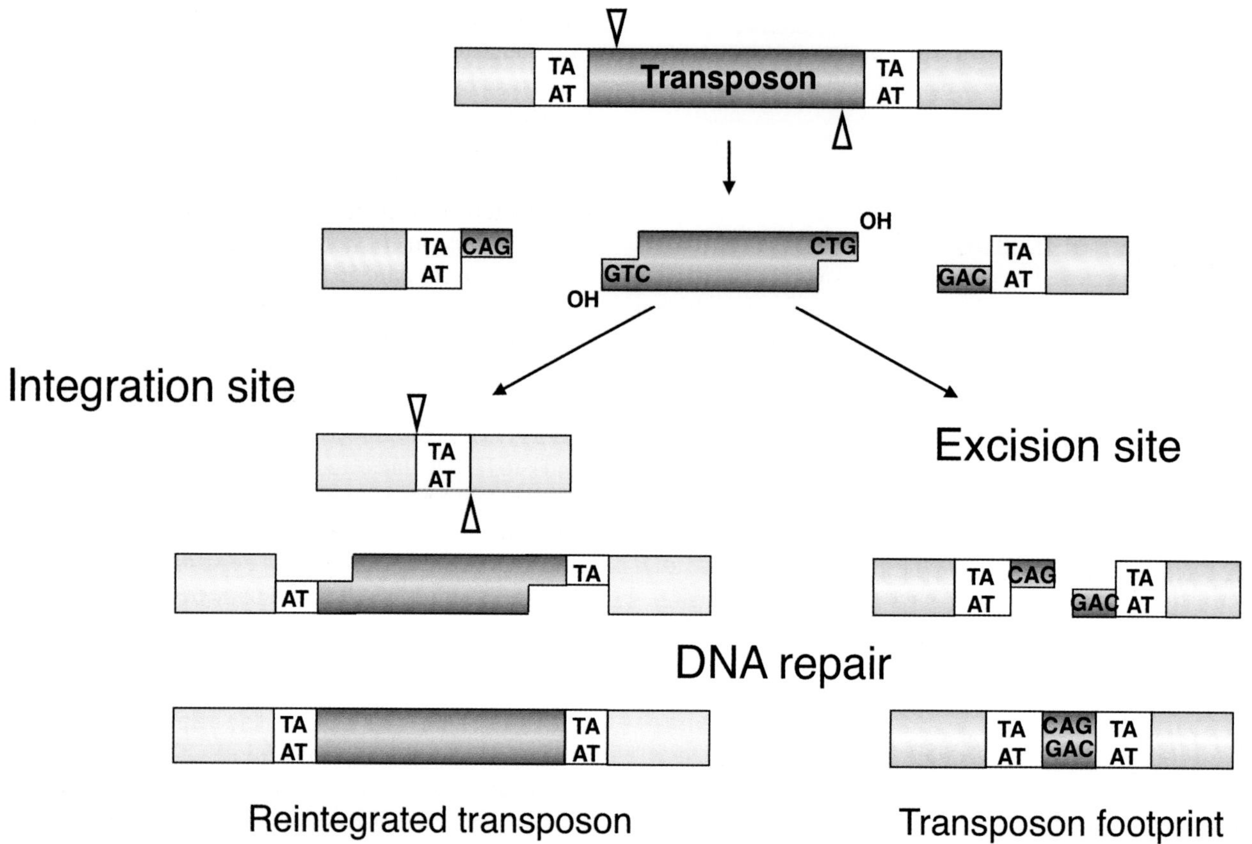

Figure 9 Molecular events during cut-and-paste transposition. The transposase initiates the excision of the transposon with staggered cuts and reintegrates it at a TA target dinucleotide. The single-stranded gaps at the integration site as well as the double-strand DNA breaks in the donor DNA are repaired by the host DNA repair machinery. After repair, the target TA is duplicated at the integration site, and a small footprint is left behind at the site of excision. The footprint is generated by the NHEJ pathway of DSB repair, and the central A:A mismatch is likely repaired by the mismatch repair system of the cell. Reprinted from *CMLS* (165) with permission from the publisher.
doi:10.1128/microbiolspec.MDNA3-0042-2014.f9

Sleeping Beauty Transposase Modulates Cell-Cycle Progression through Interaction with Miz-1

The Myc-interacting zinc finger protein 1 (Miz-1) transcription factor (102) was identified as an interactor of the *SB* transposase in a yeast two-hybrid screen (103). Miz-1 is a transcriptional regulator of the cyclin D1 gene (102), and it downregulates the cyclin D1 promoter resulting in slower cell growth. Decreased cellular levels of cyclin D1 prevent cells from entering the S phase, resulting in cell-cycle arrest in the G1 phase (104). Strikingly, through its physical association with Miz-1, the *SB* transposase seems to downregulate cyclin D1 expression in human cells, resulting in a cell-cycle slowdown (Fig. 10) (103).

The likely biological significance of our finding is that, by inducing a temporary G1 delay, the *SB* transposase potentiates the involvement of NHEJ to repair transposition-inflicted DNA damage (55). Indeed, a delay in the G1/S transition and S phase progression by cell-cycle checkpoints is thought to facilitate DNA repair to avoid replication and subsequent propagation of potentially hazardous mutations. In eukaryotic cells, the NHEJ and HR pathways are complementary but act at different stages of the cell-cycle: NHEJ is preferentially active in the G1 and early S phases (105), whereas HR is active in the late S and G2 phases (95). Accordingly, there is increasing evidence for a correlation between the particular pathway used for the repair of transposon-induced DNA damage and the cell-cycle

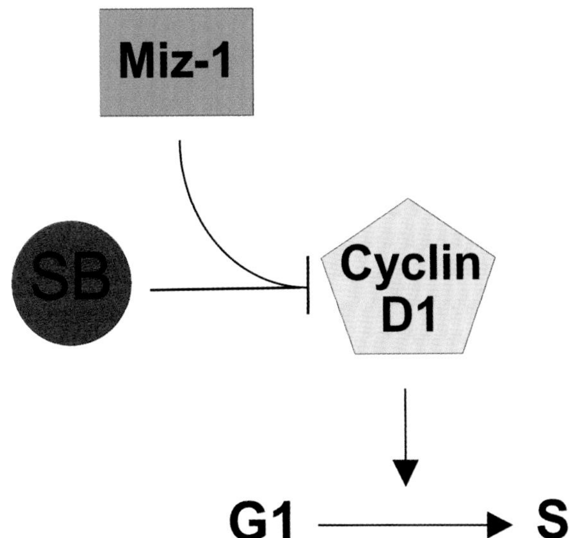

Figure 10 The *SB* transposase modulates cell-cycle progression through interaction with Miz-1. The *SB* transposase, through its interaction with Miz-1, downregulates cyclin D1 expression, which results in an inhibition of the G1/S transition of the cell-cycle. Reprinted from *PNAS* (103) with permission from the publisher.
doi:10.1128/microbiolspec.MDNA3-0042-2014.f10

stage where recombination occurs. This is nicely illustrated by gene rearrangements through V(D)J recombination, which is tightly linked to the G1 phase of the cell-cycle and to NHEJ (92, 106). As described above, DSBs generated by *SB* transposition are preferentially repaired by the NHEJ pathway (55, 107), and the *SB* transposase physically interacts with the Ku DNA-binding subunit of DNA-PK, a key component of the NHEJ machinery (55). The data suggest a model in which *SB* transposase induces a cyclin D1-dependent G1 slowdown in proliferating cells through interaction with Miz-1, thereby ensuring that transposon-induced DNA damage is repaired by NHEJ. In nature, preferential use of NHEJ for the repair of transposon-induced DSBs might help avoid homologous recombination events between dispersed copies of transposable elements in the genome, thereby assisting the maintenance of genomic stability.

Other parasitic genetic elements, including HIV-1 (108), Herpes simplex virus (109), cytomegalovirus (110), Epstein-Barr virus (111), Kaposi's sarcoma-associated herpesvirus (112), and mouse hepatitis virus (113), have also developed versatile strategies to perturb the cellular machinery to maximize their chance for survival and propagation. Thus, overriding the normal cell-cycle program seems to be a shared strategy of parasitic genetic elements.

Transposon Integration: Target Site Selection Properties of *Sleeping Beauty*

The second step of the transposition reaction is the transfer of the exposed 3′-OH transposon tip to the target DNA molecule by transesterification (Fig. 9). Similarly to the initial DNA cut, the strand transfer is executed by a nucleophilic attack. In this case, the 3′-OH groups of the already liberated transposon ends serve as a nucleophile that couples the element to the target, without previous target DNA cleavage. As a result, the transposon ends are covalently attached to staggered positions: one of the transposon ends joining to one of the target strands, the other end joining to a displaced position of the other target strand. Due to this staggered fashion of the strand joining reaction, and because the inserted element has 3′-overhangs, the integration is flanked by single-stranded gaps (Fig. 9). DNA repair at these gaps restores the terminal nucleotides of the inserted transposon and generates a characteristic duplication of the target site flanking the element that is called target site duplication (TSD) (Fig. 9). *SB* transposition almost exclusively occurs at TA dinucleotides (5, 114, 115, 116, 117, 118), and *SB* integrants therefore are flanked by TA TSDs (Fig. 9), which are molecular hallmarks of *SB* transposition. At very low frequencies, non-TA target sites were also found at insertions generated by *SB*100X (115, 119). About 75% of *SB* transposon excision events are coupled to chromosomal integration (93) and no extrachromosomal, excised molecules are readily detectable. The ~25% loss of excised transposons might be due to challenge of productive transposition by suicidal autointegration, i.e., when the transposon integrates into itself (see below).

The genome-wide insertion pattern of most transposons is non-random, showing characteristic preferences for insertion sites at the primary DNA sequence level and 'hotspots' and 'cold regions' on a genome-wide scale. Sequences responsible for target site selection of the bacterial Tn*10* transposon and retroviruses have been mapped to the core catalytic domain of the transposase or integrase, respectively (120, 121). However, despite the implication that the conserved catalytic DDE domain is responsible for locating the target site, no common pattern of integration emerges on the primary DNA sequence level. *SB* displays considerable specificity in target site selection at the primary DNA sequence level: in addition to the highly preferred TA, a palindromic AT-repeat consensus sequence in an AT-rich sequence context [Fig. 11(A)] with bendability and hydrogen bonding potential was found to constitute preferred target sites (122). It was shown that a characteristic deformation of the DNA sequence may be a recognition signal for target selection (118). This deformation, and the likelihood a particular TA will be targeted by *SB*, can be computationally predicted (118), which may allow a theoretical assessment of the likelihood of transposon insertions in particular genomic regions (123). These results indicate that a combination of particular physical properties generates a spatial optimum of the DNA for transposase interaction. This pattern of structural preference is conserved in the Tc1/*mariner* family and in other relatively randomly integrating transposons in the DDE recombinase family (122). However, these factors cannot be the only determinants of target site selection because the Tc1 and Tc3 elements have different insertion profiles in *C. elegans* (124). Therefore, it appears that there exist at least two levels of selection that together determine how favorable a particular DNA sequence is for transposon insertion. Physical properties of the DNA primarily specify a set of sequences in a genome that are in a spatial optimum to receive a transposon insertion, whereas the ability of the transposase to efficiently interact with such sequences specifies a subset within these sites where insertions occur.

In contrast to the considerable specificity at the primary DNA sequence level, *SB* integration can be considered fairly random on the genomic level (114, 115, 116, 117, 119, 122, 125). Roughly one-third of *SB* insertions in mouse and human cells occur in transcribed regions [Fig. 11(B)], and because genes cover about one-third of the genome, such frequency suggests neither preference for nor disfavoring of insertion into genes. *SB* shows no pronounced preference for inserting into transcription units or transcriptional regulatory regions of genes, and the vast majority of those insertions that occur in genes are located in introns (114, 115, 116, 117, 119, 122, 125). The transcriptional status of targeted genes apparently does not influence the integration profile of *SB* (125). This is in marked contrast to target site distributions of several other transposons including *Tol2* [Fig. 11(B)] (117, 126), *TcBuster* (115), *SPIN* (115), and *piggyBac* [Fig. 11(B)] (115, 117, 126, 127, 128). These all show significant difference from random insertion with respect to favored integration into genes and near chromatin marks characteristic of active transcription units (e.g., H3K27 acetylation and H3K4 monomethylation) and disfavored integration near marks characteristic of inactive chromatin (e.g., H3K27 trimethylation).

The random genomic distribution of *de novo SB* insertions can be observed when the transposon DNA is introduced into the nucleus by extrachromosomal gene delivery, including plasmid vectors (114, 115, 116, 117, 119, 122, 125), integration-deficient lentiviral vectors (IDLVs) (114), adenovirus vectors (129), herpesvirus vectors (130), and adeno-associated vectors (131). In these cases, transposition takes place from the extrachromosomal vector into the genome. However, target site selection properties of *SB* when launched from a chromosomal site are markedly different and are governed by "local hopping". Local hopping describes a phenomenon of chromosomal transposition in which transposons have a preference for landing into *cis*-linked sites in the vicinity of the donor locus. Local hopping seems to be a shared feature of "cut-and-paste" transposons. However, the actual extent of hopping to linked chromosomal sites and the interval of local transposition varies. For example, the P-element transposon of *Drosophila* prefers to insert within ~100 kb of the donor site at a rate ~50-fold higher than in regions outside that interval (132). Similarly, 30 to 80% of *SB* transposition events were found to re-insert locally on either side of the transposon donor locus (93, 133, 134, 135, 136, 137, 138, 139). In contrast to the P-element, *SB* seems to have a much larger local transposition interval (in the megabase range), but the targeted window for local reinsertion appears to be dependent on the donor locus and can range from <4 Mb (140) to 5 to 15 Mb (134). Local hopping of *SB* was also observed within 3 Mb of the donor locus in *Xenopus* (141).

The local hopping feature not only differs between different transposons, but a given transposon may show great variations in local hops in different hosts and in different donor loci even in the same host. For example, local hopping of the *Ac* element in tomato seems overall to be less prevalent than in maize (142, 143), and there are species-specific differences in its tendency for local hopping out of different transposon donor loci (144). This variation in local hopping of the same element could possibly be explained by varying affinity of the transposase for unknown, chromatin-associated factors in different hosts (145).

Self-disruptive Autointegration of *Sleeping Beauty*

In the process of productive transposition, the excised DNA molecule integrates into a new genomic location. However, in principle, the excised transposon molecule could reinsert, in a self-disruptive process, into its own genome. This suicidal transposition event is called autointegration, which has been observed with the bacterial systems Tn*10* (146) and *Mu* (147), with Ty1

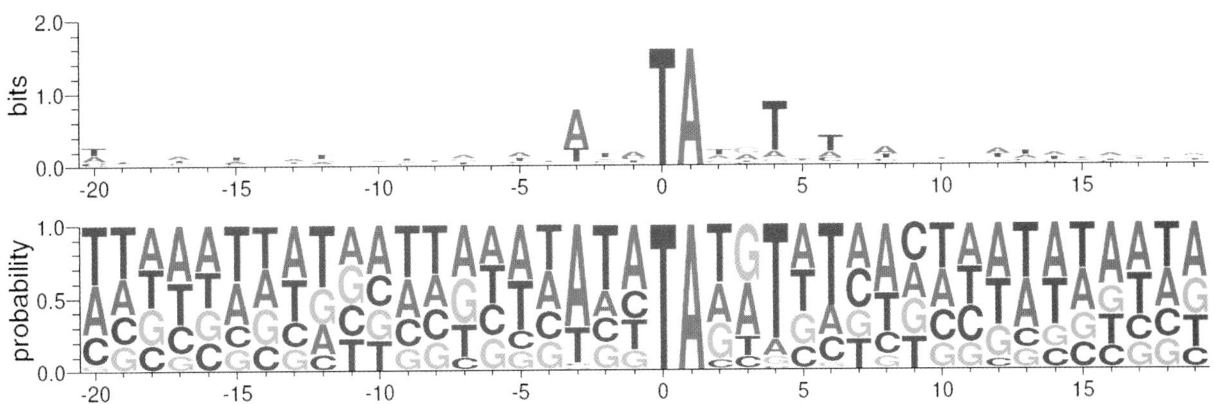

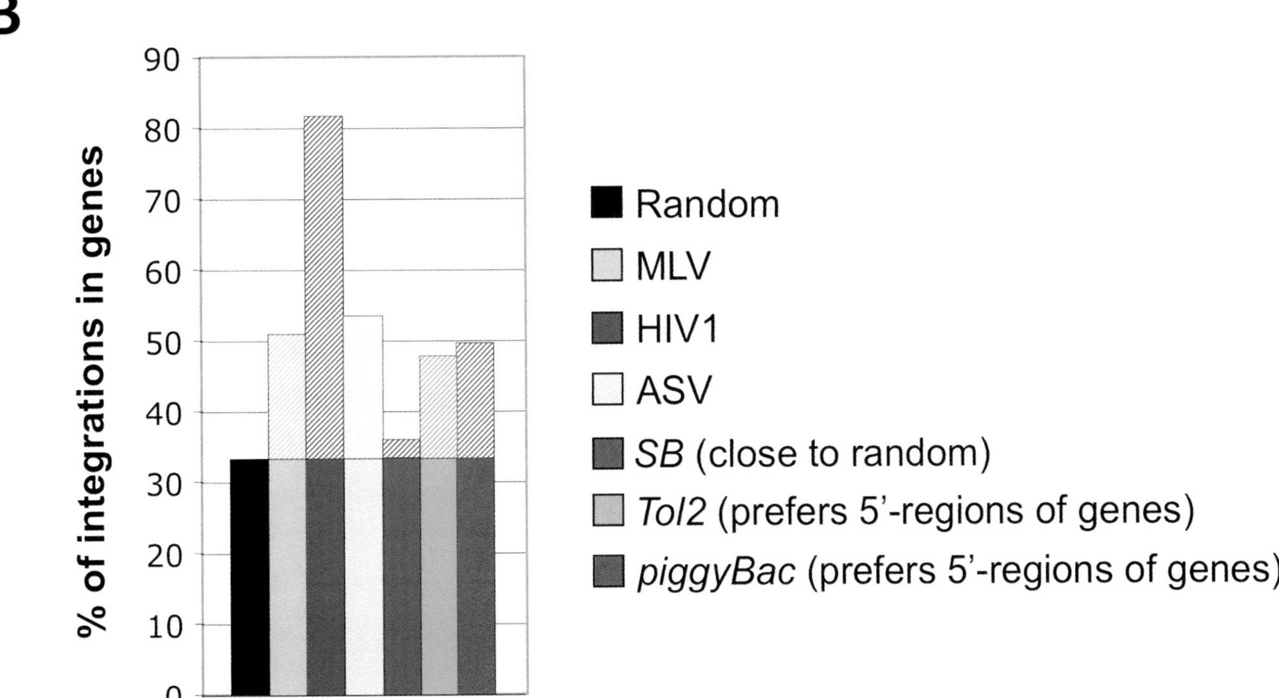

Figure 11 Genomic insertion preferences of *SB*. (A) Consensus sequence of *SB* insertion sites. Seqlogo analysis and nucleotide probability plot of *SB* insertion sites in HeLa cells. Twenty base pairs upstream and downstream of the TA target sites were analyzed. The y-axis represents the strength of the information, with 2 bits being the maximum for a DNA sequence. (B) Relative frequencies of insertions into genes by retroviruses and transposons. The top portions of the graphs indicate an over-representation of genic insertions as compared to random. Part (A) reprinted from *Molecular Therapy* (126) with permission from the publisher. Part (B) reprinted from *BioEssays* (170) with permission from the publisher. doi:10.1128/microbiolspec.MDNA3-0042-2014.f11

retrotransposons in yeast (148), and with retroviruses (149). Apparently, the *SB* transposon is also subject to autointegration (56), thereby compromising integration of a fraction of excised transposon molecules.

The conserved size of Tc1-like elements is between 1.6 and 1.8 kb. The efficacy of transposition usually correlates negatively with increasing size of the transposon (58, 150, 151, 152), and such an effect was also

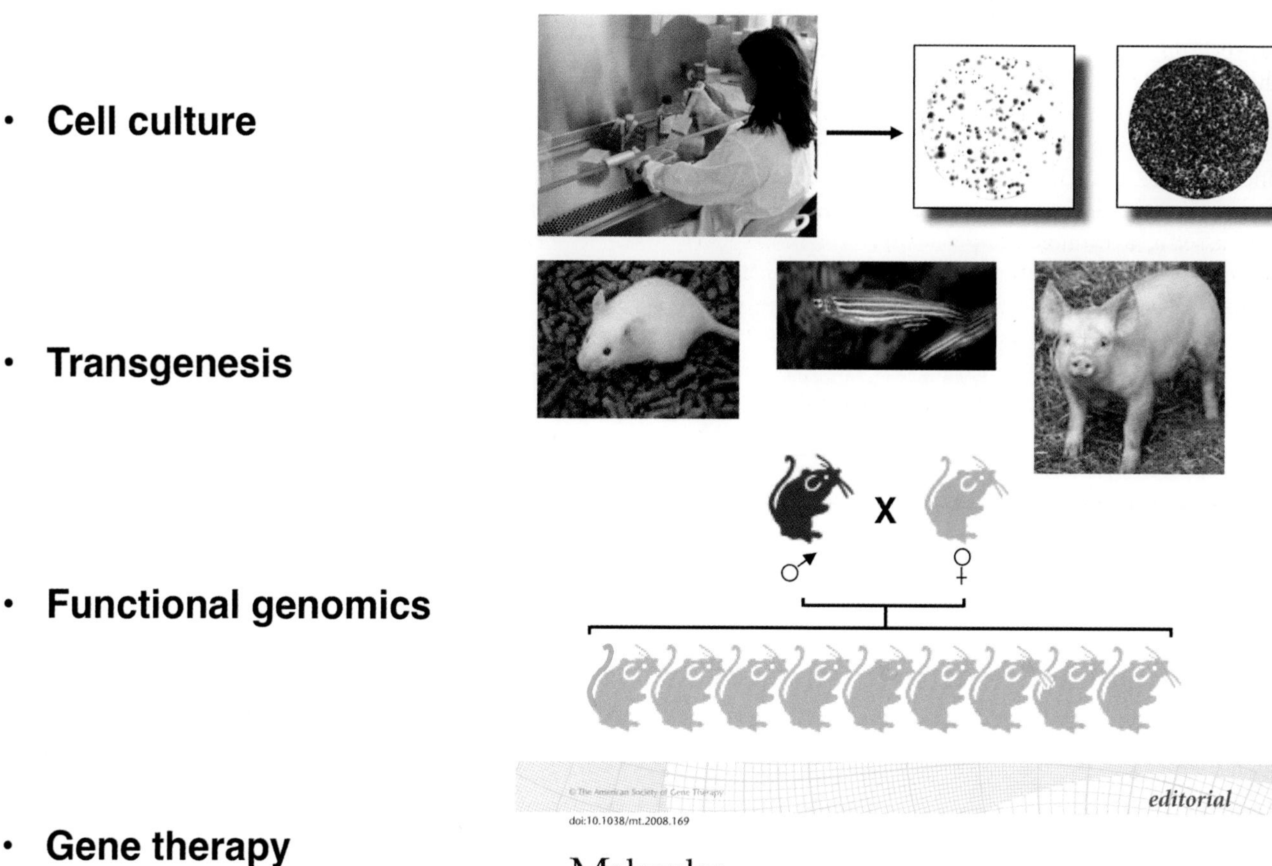

Figure 12 Broad applicability of *SB* transposon-based gene vectors in vertebrate genetics. Reprinted from *Mobile DNA* (168) with permission from the publisher. doi:10.1128/microbiolspec.MDNA3-0042-2014.f12

observed with the bacterial transposons IS*1* (153) and Tn*10* (154) and with *piggyBac* (56). Since larger transposons contain more potential target sites, they could be particularly attractive targets for autointegration. Indeed, increasing size was found to sensitize both *SB* and *piggyBac* transposition for autointegration (56). However, the competition between autointegration and productive transposition is unlikely to be the only factor responsible for sensitivity to size as transposon excision, a step prior to integration, is already affected by the size of the *SB* transposon (56).

A host-encoded protein, barrier-to-autointegration factor (BAF or BANF1) has been identified by its ability to protect retroviruses from autointegration (155, 156, 157, 158, 159). Intriguingly, BANF1 also inhibited transposon autointegration of *SB* and was detected in higher-order protein complexes containing the transposase in human cells (56). Thus, the *SB* transposon seems to be able to recruit phylogenetically conserved cellular factors such as BANF1 that protects against self-disruption in a new environment (a human cell is a naïve host for *SB*). In fact, BANF1 might be an ideal cellular factor for integrating elements in higher eukaryotes. As in retroviral integration (155, 156, 157, 158), BANF1 may compact the transposon genome to be a less accessible target for autointegration, thereby promoting the chromosomal integration step.

Sleeping Beauty AS A GENETIC TOOL

Transposons can be viewed as natural DNA transfer vehicles that, similar to integrating viruses, are capable of efficient genomic insertion. Transposition can be controlled by conditionally providing the transposase component of the transposition reaction. Thus, a DNA of interest (be it a fluorescent marker, an shRNA expression cassette, a mutagenic gene trap, or a therapeutic gene construct) cloned between the IR sequences of a transposon-based vector can be utilized for stable genomic insertion in a regulated and highly efficient manner. This methodological paradigm opened up a number of avenues for genome manipulations in vertebrates, including transgenesis for the generation of transgenic cells in tissue culture, the production of germline-transgenic animals for basic and applied research, forward genetic screens for functional gene annotation in model species, and therapy of genetic disorders in humans (Fig. 12). *SB* was the first transposon ever shown to be capable of gene transfer in vertebrate cells, and recent results confirm that *SB* supports a full spectrum of genetic engineering including transgenesis, insertional mutagenesis, and therapeutic somatic gene transfer, both *ex vivo* and *in vivo*. The first clinical application of the *SB* system will help to validate both the safety and efficacy of this approach. Applications of the *SB* system fall outside the scope of this chapter, and readers are referred to recent review articles (160, 161, 162, 163, 164, 165, 166, 167, 168, 169, 170, 171, 172).

Acknowledgments. *The authors thank past and present members of the "Transposition and Genome Engineering" group at the Paul Ehrlich Institute and the "Mobile DNA" group at the Max Delbrück Center for Molecular Medicine for their dedicated work and contribution. We are grateful to Irina Nesmelova for providing a color version of Fig. 3.*

Citation. Ivics Z, Izsvák Z, 2014. *Sleeping beauty* transposition. Microbiol Spectrum 3(2):MDNA3-0042-2014.

References

1. **Plasterk RH, Izsvak Z, Ivics Z.** 1999. Resident aliens: the Tc1/mariner superfamily of transposable elements. *Trends Genet* **15:**326–332.
2. **Ivics Z, Izsvák Z, Minter A, Hackett PB.** 1996. Identification of functional domains and evolution of Tc1-like transposable elements. *Proc Natl Acad Sci U S A* **93:** 5008–5013.
3. **Radice AD, Bugaj B, Fitch DH, Emmons SW.** 1994. Widespread occurrence of the Tc1 transposon family: Tc1-like transposons from teleost fish. *Mol Gen Genet* **244:**606–612.
4. **Goodier JL, Davidson WS.** 1994. Tc1 transposon-like sequences are widely distributed in salmonids. *J Mol Biol* **241:**26–34.
5. **Ivics Z, Hackett PB, Plasterk RH, Izsvak Z.** 1997. Molecular reconstruction of Sleeping Beauty, a Tc1-like transposon from fish, and its transposition in human cells. *Cell* **91:**501–510.
6. **Izsvák Z, Khare D, Behlke J, Heinemann U, Plasterk RH, Ivics Z.** 2002. Involvement of a bifunctional, paired-like DNA-binding domain and a transpositional enhancer in *Sleeping Beauty* transposition. *J Biol Chem* **277:**34581–34588.
7. **Pietrokovski S, Henikoff S.** 1997. A helix-turn-helix DNA-binding motif predicted for transposases of DNA transposons. *Mol Gen Genet* **254:**689–695.
8. **Vos JC, Plasterk RH.** 1994. Tc1 transposase of *Caenorhabditis elegans* is an endonuclease with a bipartite DNA binding domain. *EMBO J* **13:**6125–6132.
9. **Franz G, Loukeris TG, Dialektaki G, Thompson CR, Savakis C.** 1994. Mobile Minos elements from Drosophila hydei encode a two-exon transposase with similarity to the paired DNA-binding domain. *Proc Natl Acad Sci U S A* **91:**4746–4750.
10. **Czerny T, Schaffner G, Busslinger M.** 1993. DNA sequence recognition by Pax proteins: bipartite structure of the paired domain and its binding site. *Genes Dev* **7:** 2048–2061.
11. **Breitling R, Gerber JK.** 2000. Origin of the paired domain. *Dev Genes Evol* **210:**644–650.
12. **van Pouderoyen G, Ketting RF, Perrakis A, Plasterk RH, Sixma TK.** 1997. Crystal structure of the specific DNA-binding domain of Tc3 transposase of *C. elegans* in complex with transposon DNA. *EMBO J* **16:**6044–6054.
13. **Carpentier CE, Schreifels JM, Aronovich EL, Carlson DF, Hackett PB, Nesmelova IV.** 2014. NMR structural analysis of Sleeping Beauty transposase binding to DNA. *Protein Sci* **23:**23–33.
14. **Gehring WJ, Qian YQ, Billeter M, Furukubo-Tokunaga K, Schier AF, Resendez-Perez D, Affolter M, Otting G, Wuthrich K.** 1994. Homeodomain-DNA recognition. *Cell* **78:**211–223.
15. **Feng JA, Johnson RC, Dickerson RE.** 1994. Hin recombinase bound to DNA: the origin of specificity in major and minor groove interactions. *Science* **263:**348–355.
16. **Spanopoulou E, Zaitseva F, Wang FH, Santagata S, Baltimore D, Panayotou G.** 1996. The homeodomain region of Rag-1 reveals the parallel mechanisms of bacterial and V(D)J recombination. *Cell* **87:**263–276.
17. **Kulkosky J, Jones KS, Katz RA, Mack JP, Skalka AM.** 1992. Residues critical for retroviral integrative recombination in a region that is highly conserved among retroviral/retrotransposon integrases and bacterial insertion sequence transposases. *Mol Cell Biol* **12:**2331–2338.
18. **Doak TG, Doerder FP, Jahn CL, Herrick G.** 1994. A proposed superfamily of transposase genes: transposon-like elements in ciliated protozoa and a common "D35E" motif. *Proc Natl Acad Sci U S A* **91:**942–946.
19. **Kim DR, Dai Y, Mundy CL, Yang W, Oettinger MA.** 1999. Mutations of acidic residues in RAG1 define the active site of the V(D)J recombinase. *Genes Dev* **13:** 3070–3080.

20. **Landree MA, Wibbenmeyer JA, Roth DB.** 1999. Mutational analysis of RAG1 and RAG2 identifies three catalytic amino acids in RAG1 critical for both cleavage steps of V(D)J recombination. *Genes Dev* **13:**3059–3069.

21. **Haren L, Ton-Hoang B, Chandler M.** 1999. Integrating DNA: transposases and retroviral integrases. *Annu Rev Microbiol* **53:**245–281.

22. **Bujacz G, Alexandratos J, Wlodawer A, Merkel G, Andrake M, Katz RA, Skalka AM.** 1997. Binding of different divalent cations to the active site of avian sarcoma virus integrase and their effects on enzymatic activity. *J Biol Chem* **272:**18161–18168.

23. **Goldgur Y, Dyda F, Hickman AB, Jenkins TM, Craigie R, Davies DR.** 1998. Three new structures of the core domain of HIV-1 integrase: an active site that binds magnesium. *Proc Natl Acad Sci U S A* **95:**9150–9154.

24. **van Luenen HG, Colloms SD, Plasterk RH.** 1994. The mechanism of transposition of Tc3 in C. elegans. *Cell* **79:**293–301.

25. **Yant SR, Meuse L, Chiu W, Ivics Z, Izsvak Z, Kay MA.** 2000. Somatic integration and long-term transgene expression in normal and haemophilic mice using a DNA transposon system. *Nat Genet* **25:**35–41.

26. **Richardson JM, Colloms SD, Finnegan DJ, Walkinshaw MD.** 2009. Molecular architecture of the Mos1 paired-end complex: the structural basis of DNA transposition in a eukaryote. *Cell* **138:**1096–1108.

27. **Katayanagi K, Miyagawa M, Matsushima M, Ishikawa M, Kanaya S, Ikehara M, Matsuzaki T, Morikawa K.** 1990. Three-dimensional structure of ribonuclease H from *E. coli*. *Nature* **347:**306–309.

28. **Ariyoshi M, Vassylyev DG, Iwasaki H, Nakamura H, Shinagawa H, Morikawa K.** 1994. Atomic structure of the RuvC resolvase: a holliday junction-specific endonuclease from *E. coli*. *Cell* **78:**1063–1072.

29. **Lohe AR, De Aguiar D, Hartl DL.** 1997. Mutations in the mariner transposase: the D,D(35)E consensus sequence is nonfunctional. *Proc Natl Acad Sci U S A* **94:** 1293–1297.

30. **Mates L, Chuah MK, Belay E, Jerchow B, Manoj N, Acosta-Sanchez A, Grzela DP, Schmitt A, Becker K, Matrai J, Ma L, Samara-Kuko E, Gysemans C, Pryputniewicz D, Miskey C, Fletcher B, Vandendriessche T, Ivics Z, Izsvak Z.** 2009. Molecular evolution of a novel hyperactive Sleeping Beauty transposase enables robust stable gene transfer in vertebrates. *Nat Genet* **41:** 753–761.

31. **Bushman FD, Engelman A, Palmer I, Wingfield P, Craigie R.** 1993. Domains of the integrase protein of human immunodeficiency virus type 1 responsible for polynucleotidyl transfer and zinc binding. *Proc Natl Acad Sci U S A* **90:**3428–3432.

32. **Dyda F, Hickman AB, Jenkins TM, Engelman A, Craigie R, Davies DR.** 1994. Crystal structure of the catalytic domain of HIV-1 integrase: similarity to other polynucleotidyl transferases. *Science* **266:**1981–1986.

33. **Yant SR, Park J, Huang Y, Mikkelsen JG, Kay MA.** 2004. Mutational analysis of the N-terminal DNA-binding domain of sleeping beauty transposase: critical residues for DNA binding and hyperactivity in mammalian cells. *Mol Cell Biol* **24:**9239–9247.

34. **Geurts AM, Yang Y, Clark KJ, Liu G, Cui Z, Dupuy AJ, Bell JB, Largaespada DA, Hackett PB.** 2003. Gene transfer into genomes of human cells by the sleeping beauty transposon system. *Mol Ther* **8:**108–117.

35. **Zayed H, Izsvak Z, Walisko O, Ivics Z.** 2004. Development of hyperactive sleeping beauty transposon vectors by mutational analysis. *Mol Ther* **9:**292–304.

36. **Baus J, Liu L, Heggestad AD, Sanz S, Fletcher BS.** 2005. Hyperactive transposase mutants of the Sleeping Beauty transposon. *Mol Ther* **12:**1148–1156.

37. **Xue X, Huang X, Nodland SE, Mates L, Ma L, Izsvak Z, Ivics Z, LeBien TW, McIvor RS, Wagner JE, Zhou X.** 2009. Stable gene transfer and expression in cord blood-derived CD34+ hematopoietic stem and progenitor cells by a hyperactive Sleeping Beauty transposon system. *Blood* **114:**1319–1330.

38. **Belay E, Matrai J, Acosta-Sanchez A, Ma L, Quattrocelli M, Mates L, Sancho-Bru P, Geraerts M, Yan B, Vermeesch J, Rincon MY, Samara-Kuko E, Ivics Z, Verfaillie C, Sampaolesi M, Izsvak Z, Vandendriessche T, Chuah MK.** 2010. Novel hyperactive transposons for genetic modification of induced pluripotent and adult stem cells: a nonviral paradigm for coaxed differentiation. *Stem Cells* **28:**1760–1771.

39. **Jin Z, Maiti S, Huls H, Singh H, Olivares S, Mates L, Izsvak Z, Ivics Z, Lee DA, Champlin RE, Cooper LJ.** 2011. The hyperactive Sleeping Beauty transposase SB100X improves the genetic modification of T cells to express a chimeric antigen receptor. *Gene Ther* **18:** 849–856.

40. **Ivics Z, Garrels W, Mates L, Yau TY, Bashir S, Zidek V, Landa V, Geurts A, Pravenec M, Rulicke T, Kues WA, Izsvak Z.** 2014. Germline transgenesis in pigs by cytoplasmic microinjection of Sleeping Beauty transposons. *Nat Protoc* **9:**810–827.

41. **Ivics Z, Hiripi L, Hoffmann OI, Mates L, Yau TY, Bashir S, Zidek V, Landa V, Geurts A, Pravenec M, Rulicke T, Bosze Z, Izsvak Z.** 2014. Germline transgenesis in rabbits by pronuclear microinjection of Sleeping Beauty transposons. *Nat Protoc* **9:**794–809.

42. **Ivics Z, Mates L, Yau TY, Landa V, Zidek V, Bashir S, Hoffmann OI, Hiripi L, Garrels W, Kues WA, Bosze Z, Geurts A, Pravenec M, Rulicke T, Izsvak Z.** 2014. Germline transgenesis in rodents by pronuclear microinjection of Sleeping Beauty transposons. *Nat Protoc* **9:** 773–793.

43. **Izsvák Z, Ivics Z, Hackett PB.** 1995. Characterization of a Tc1-like transposable element in zebrafish (*Danio rerio*). *Mol Gen Genet* **247:**312–322.

44. **Franz G, Savakis C.** 1991. Minos, a new transposable element from Drosophila hydei, is a member of the Tc1-like family of transposons. *Nucleic Acids Res* **19:**6646.

45. **Merriman PJ, Grimes CD, Ambroziak J, Hackett DA, Skinner P, Simmons MJ.** 1995. *S* elements: a family of Tc1-like transposons in the genome of *Drosophila melanogaster*. *Genetics* **141:**1425–1438.

46. **Moschetti R, Caggese C, Barsanti P, Caizzi R.** 1998. Intra- and interspecies variation among *Bari*-1 elements of the *melanogaster* species group. *Genetics* **150:**239–250.
47. **Ke Z, Grossman GL, Cornel AJ, Collins FH.** 1996. *Quetzal*: A transposon of the Tc1 family in the mosquito *Anopheles albimanus*. *Genetica* **98:**141–147.
48. **Lam WL, Seo P, Robison K, Virk S, Gilbert W.** 1996. Discovery of amphibian Tc1-like transposon families. *J Mol Biol* **257:**359–366.
49. **Sinzelle L, Pollet N, Bigot Y, Mazabraud A.** 2005. Characterization of multiple lineages of Tc1-like elements within the genome of the amphibian Xenopus tropicalis. *Gene* **349:**187–196.
50. **Lampe DJ, Churchill ME, Robertson HM.** 1996. A purified *mariner* transposase is sufficient to mediate transposition *in vitro*. *EMBO J* **15:**5470–5479.
51. **Vos JC, De Baere I, Plasterk RH.** 1996. Transposase is the only nematode protein required for *in vitro* transposition of Tc1. *Genes Dev* **10:**755–761.
52. **Raz E, van Luenen HG, Schaerringer B, Plasterk RH, Driever W.** 1998. Transposition of the nematode Caenorhabditis elegans Tc3 element in the zebrafish Danio rerio. *Curr Biol* **8:**82–88.
53. **Zayed H, Izsvak Z, Khare D, Heinemann U, Ivics Z.** 2003. The DNA-bending protein HMGB1 is a cellular cofactor of *Sleeping Beauty* transposition. *Nucleic Acids Res* **31:**2313–2322.
54. **Walisko O, Schorn A, Rolfs F, Devaraj A, Miskey C, Izsvak Z, Ivics Z.** 2008. Transcriptional activities of the Sleeping Beauty transposon and shielding its genetic cargo with insulators. *Mol Ther* **16:**359–369.
55. **Izsvák Z, Stuwe EE, Fiedler D, Katzer A, Jeggo PA, Ivics Z.** 2004. Healing the wounds inflicted by sleeping beauty transposition by double-strand break repair in mammalian somatic cells. *Mol Cell* **13:**279–290.
56. **Wang Y, Wang J, Devaraj A, Singh M, Jimenez Orgaz A, Chen JX, Selbach M, Ivics Z, Izsvak Z.** 2014. Suicidal autointegration of sleeping beauty and piggyBac transposons in eukaryotic cells. *PLoS Genet* **10:**e1004103.
57. **Walisko O, Jursch T, Izsvak Z, Ivics Z.** 2008. Transposon-host cell interactions in the regulation of Sleeping Beauty transposition, p 109–132. *In* Volff J-N, Lankenau D-H (ed), *Transposons and the Dynamic Genome*, Springer, Berlin Heidelberg.
58. **Izsvák Z, Ivics Z, Plasterk RH.** 2000. *Sleeping Beauty*, a wide host-range transposon vector for genetic transformation in vertebrates. *J Mol Biol* **302:**93–102.
59. **Hozumi A, Mita K, Miskey C, Mates L, Izsvak Z, Ivics Z, Satake H, Sasakura Y.** 2013. Germline transgenesis of the chordate Ciona intestinalis with hyperactive variants of sleeping beauty transposable element. *Dev Dyn* **242:**30–43.
60. **Leaver MJ.** 2001. A family of Tc1-like transposons from the genomes of fishes and frogs: evidence for horizontal transmission. *Gene* **271:**203–214.
61. **Sijen T, Plasterk RH.** 2003. Transposon silencing in the Caenorhabditis elegans germ line by natural RNAi. *Nature* **426:**310–314.
62. **Moldt B, Yant SR, Andersen PR, Kay MA, Mikkelsen JG.** 2007. Cis-acting gene regulatory activities in the terminal regions of sleeping beauty DNA transposon-based vectors. *Hum Gene Ther* **18:**1193–1204.
63. **Rauschhuber C, Ehrhardt A.** 2012. RNA interference is responsible for reduction of transgene expression after Sleeping Beauty transposase mediated somatic integration. *PLoS One* **7:**e35389.
64. **Bewley CA, Gronenborn AM, Clore GM.** 1998. Minor groove-binding architectural proteins: structure, function, and DNA recognition. *Annu Rev Biophys Biomol Struct* **27:**105–131.
65. **Cui Z, Geurts AM, Liu G, Kaufman CD, Hackett PB.** 2002. Structure-function analysis of the inverted terminal repeats of the *Sleeping Beauty* transposon. *J Mol Biol* **318:**1221–1235.
66. **Leung PC, Teplow DB, Harshey RM.** 1989. Interaction of distinct domains in Mu transposase with Mu DNA ends and an internal transpositional enhancer. *Nature* **338:**656–658.
67. **van Gent DC, Hiom K, Paull TT, Gellert M.** 1997. Stimulation of V(D)J cleavage by high mobility group proteins. *EMBO J* **16:**2665–2670.
68. **Agrawal A, Schatz DG.** 1997. RAG1 and RAG2 form a stable postcleavage synaptic complex with DNA containing signal ends in V(D)J recombination. *Cell* **89:**43–53.
69. **Bustin M.** 1999. Regulation of DNA-dependent activities by the functional motifs of the high-mobility-group chromosomal proteins. *Mol Cell Biol* **19:**5237–5246.
70. **Claeys Bouuaert C, Liu D, Chalmers R.** 2011. A simple topological filter in a eukaryotic transposon as a mechanism to suppress genome instability. *Mol Cell Biol* **31:** 317–327.
71. **Yoder JA, Walsh CP, Bestor TH.** 1997. Cytosine methylation and the ecology of intragenomic parasites. *Trends Genet* **13:**335–340.
72. **Yusa K, Takeda J, Horie K.** 2004. Enhancement of Sleeping Beauty transposition by CpG methylation: possible role of heterochromatin formation. *Mol Cell Biol* **24:**4004–4018.
73. **Jursch T, Miskey C, Izsvak Z, Ivics Z.** 2013. Regulation of DNA transposition by CpG methylation and chromatin structure in human cells. *Mobile DNA* **4:**15.
74. **Mizuuchi K.** 1992. Polynucleotidyl transfer reactions in transpositional DNA recombination. *J Biol Chem* **267:** 21273–21276.
75. **Craig NL.** 1995. Unity in transposition reactions. *Science* **270:**253–254.
76. **Turlan C, Chandler M.** 2000. Playing second fiddle: second-strand processing and liberation of transposable elements from donor DNA. *Trends Microbiol* **8:**268–274.
77. **Bischerour J, Chalmers R.** 2009. Base flipping in tn10 transposition: an active flip and capture mechanism. *PLoS One* **4:**e6201.
78. **Bischerour J, Lu C, Roth DB, Chalmers R.** 2009. Base flipping in V(D)J recombination: insights into the mechanism of hairpin formation, the 12/23 rule, and the coordination of double-strand breaks. *Mol Cell Biol* **29:** 5889–5899.

79. **Mitra R, Fain-Thornton J, Craig NL.** 2008. piggyBac can bypass DNA synthesis during cut and paste transposition. *EMBO J* **27:**1097–1109.

80. **Zhou L, Mitra R, Atkinson PW, Hickman AB, Dyda F, Craig NL.** 2004. Transposition of hAT elements links transposable elements and V(D)J recombination. *Nature* **432:**995–1001.

81. **Hencken CG, Li X, Craig NL.** 2012. Functional characterization of an active Rag-like transposase. *Nat Struct Mol Biol* **19:**834–836.

82. **Gellert M.** 2002. V(D)J recombination: RAG proteins, repair factors, and regulation. *Annu Rev Biochem* **71:** 101–132.

83. **van Gent DC, Mizuuchi K, Gellert M.** 1996. Similarities between initiation of V(D)J recombination and retroviral integration. *Science* **271:**1592–1594.

84. **Bhasin A, Goryshin IY, Reznikoff WS.** 1999. Hairpin formation in Tn5 transposition. *J Biol Chem* **274:** 37021–37029.

85. **Kennedy AK, Guhathakurta A, Kleckner N, Haniford DB.** 1998. Tn10 transposition via a DNA hairpin intermediate. *Cell* **95:**125–134.

86. **Richardson JM, Dawson A, O'Hagan N, Taylor P, Finnegan DJ, Walkinshaw MD.** 2006. Mechanism of Mos1 transposition: insights from structural analysis. *EMBO J* **25:**1324–1334.

87. **Goryshin IY, Reznikoff WS.** 1998. Tn5 in vitro transposition. *J Biol Chem* **273:**7367–7374.

88. **Miskey C, Papp B, Mates L, Sinzelle L, Keller H, Izsvak Z, Ivics Z.** 2007. The Ancient Mariner Sails Again: Transposition of the Human Hsmar1 Element by a Reconstructed Transposase and Activities of the SETMAR Protein on Transposon Ends. *Mol Cell Biol* **27:**4589–4600.

89. **Engels WR, Johnson-Schlitz DM, Eggleston WB, Sved J.** 1990. High-frequency P element loss in *Drosophila* is homolog dependent. *Cell* **62:**515–525.

90. **Lohe AR, Timmons C, Beerman I, Lozovskaya ER, Hartl DL.** 2000. Self-inflicted wounds, template-directed gap repair and a recombination hotspot. Effects of the mariner transposase. *Genetics* **154:**647–656.

91. **Durocher D, Jackson SP.** 2001. DNA-PK, ATM and ATR as sensors of DNA damage: variations on a theme? *Curr Opin Cell Biol* **13:**225–231.

92. **Jackson SP, Jeggo PA.** 1995. DNA double-strand break repair and V(D)J recombination: involvement of DNA-PK. *Trends Biochem Sci* **20:**412–415.

93. **Luo G, Ivics Z, Izsvak Z, Bradley A.** 1998. Chromosomal transposition of a Tc1/mariner-like element in mouse embryonic stem cells. *Proc Natl Acad Sci U S A* **95:**10769–10773.

94. **Daniel R, Katz RA, Skalka AM.** 1999. A role for DNA-PK in retroviral DNA integration. *Science* **284:**644–647.

95. **Takata M, Sasaki MS, Sonoda E, Morrison C, Hashimoto M, Utsumi H, Yamaguchi-Iwai Y, Shinohara A, Takeda S.** 1998. Homologous recombination and non-homologous end-joining pathways of DNA double-strand break repair have overlapping roles in the maintenance of chromosomal integrity in vertebrate cells. *EMBO J* **17:**5497–5508.

96. **Richardson C, Jasin M.** 2000. Coupled homologous and nonhomologous repair of a double-strand break preserves genomic integrity in mammalian cells. *Mol Cell Biol* **20:**9068–9075.

97. **Plasterk RH.** 1991. The origin of footprints of the Tc1 transposon of *Caenorhabditis elegans*. *EMBO J* **10:** 1919–1925.

98. **Gloor GB, Moretti J, Mouyal J, Keeler KJ.** 2000. Distinct P-element excision products in somatic and germline cells of Drosophila melanogaster. *Genetics* **155:** 1821–1830.

99. **Nassif N, Penney J, Pal S, Engels WR, Gloor GB.** 1994. Efficient copying of nonhomologous sequences from ectopic sites via P-element-induced gap repair. *Mol Cell Biol* **14:**1613–1625.

100. **Downs JA, Jackson SP.** 1999. Involvement of DNA end-binding protein Ku in Ty element retrotransposition. *Mol Cell Biol* **19:**6260–6268.

101. **Sharon G, Burkett TJ, Garfinkel DJ.** 1994. Efficient homologous recombination of Ty1 element cDNA when integration is blocked. *Mol Cell Biol* **14:**6540–6551.

102. **Peukert K, Staller P, Schneider A, Carmichael G, Hanel F, Eilers M.** 1997. An alternative pathway for gene regulation by Myc. *EMBO J* **16:**5672–5686.

103. **Walisko O, Izsvak Z, Szabo K, Kaufman CD, Herold S, Ivics Z.** 2006. Sleeping Beauty transposase modulates cell-cycle progression through interaction with Miz-1. *Proc Natl Acad Sci U S A* **103:**4062–4067.

104. **Baldin V, Lukas J, Marcote MJ, Pagano M, Draetta G.** 1993. Cyclin D1 is a nuclear protein required for cell cycle progression in G1. *Genes Dev* **7:**812–821.

105. **Lee SE, Mitchell RA, Cheng A, Hendrickson EA.** 1997. Evidence for DNA-PK-dependent and -independent DNA double-strand break repair pathways in mammalian cells as a function of the cell cycle. *Mol Cell Biol* **17:**1425–1433.

106. **Lee J, Desiderio S.** 1999. Cyclin A/CDK2 regulates V(D)J recombination by coordinating RAG-2 accumulation and DNA repair. *Immunity* **11:**771–781.

107. **Yant SR, Kay MA.** 2003. Nonhomologous-end-joining factors regulate DNA repair fidelity during *Sleeping Beauty* element transposition in mammalian cells. *Mol Cell Biol* **23:**8505–8518.

108. **Emerman M.** 1996. HIV-1, Vpr and the cell cycle. *Curr Biol* **6:**1096–1103.

109. **Lomonte P, Everett RD.** 1999. Herpes Simplex Virus type 1 immediate-early protein Vmw110 inhibits progression of cells through mitosis and from G1 into S phase of the cell cycle. *J Virol* **73:**9456–9467.

110. **Lu M, Shenk T.** 1999. Human cytomegalovirus UL69 protein induces cells to accumulate in G1 phase of the cell cycle. *J Virol* **73:**676–683.

111. **Cayrol C, Flemington EK.** 1996. The Epstein-Barr virus bZIP transcription factor Zta causes G0/G1 cell cycle arrest through induction of cyclin-dependent kinase inhibitors. *EMBO J* **15:**2748–2759.

112. **Izumiya Y, Lin S-F, Ellison TJ, Levy AM, Mayeur GL, Izumiya C, Kung H-J.** 2003. Cell cycle regulation by Kaposi's sarcoma-associated herpesvirus K-bZIP: direct

interaction with Cyclin-CDK2 and induction of G1 growth arrest. *J Virol* 77:9652–9661.

113. **Chen CJ, Makino S.** 2004. Murine coronavirus replication induces cell cycle arrest in G0/G1 phase. *J Virol* **78:** 5658–5669.

114. **Moldt B, Miskey C, Staunstrup NH, Gogol-Doring A, Bak RO, Sharma N, Mates L, Izsvak Z, Chen W, Ivics Z, Mikkelsen JG.** 2011. Comparative genomic integration profiling of Sleeping Beauty transposons mobilized with high efficacy from integrase-defective lentiviral vectors in primary human cells. *Mol Ther* **19:**1499–1510.

115. **Li X, Ewis H, Hice RH, Malani N, Parker N, Zhou L, Feschotte C, Bushman FD, Atkinson PW, Craig NL.** 2013. A resurrected mammalian hAT transposable element and a closely related insect element are highly active in human cell culture. *Proc Natl Acad Sci U S A* **110:**E478–E487.

116. **Voigt K, Gogol-Doring A, Miskey C, Chen W, Cathomen T, Izsvak Z, Ivics Z.** 2012. Retargeting sleeping beauty transposon insertions by engineered zinc finger DNA-binding domains. *Mol Ther* **20:**1852–1862.

117. **Ammar I, Gogol-Doring A, Miskey C, Chen W, Cathomen T, Izsvak Z, Ivics Z.** 2012. Retargeting transposon insertions by the adeno-associated virus Rep protein. *Nucleic Acids Res* **40:**6693–6712.

118. **Liu G, Geurts AM, Yae K, Srinivasan AR, Fahrenkrug SC, Largaespada DA, Takeda J, Horie K, Olson WK, Hackett PB.** 2005. Target-site preferences of Sleeping Beauty transposons. *J Mol Biol* **346:**161–173.

119. **de Jong J, Akhtar W, Badhai J, Rust AG, Rad R, Hilkens J, Berns A, van Lohuizen M, Wessels LF, de Ridder J.** 2014. Chromatin landscapes of retroviral and transposon integration profiles. *PLoS Genet* **10:** e1004250.

120. **Junop MS, Haniford DB.** 1997. Factors responsible for target site selection in Tn10 transposition: a role for the DDE motif in target DNA capture. *EMBO J* **16:** 2646–2655.

121. **Katzman M, Sudol M.** 1995. Mapping domains of retroviral integrase responsible for viral DNA specificity and target site selection by analysis of chimeras between human immunodeficiency virus type 1 and visna virus integrases. *J Virol* **69:**5687–5696.

122. **Vigdal TJ, Kaufman CD, Izsvak Z, Voytas DF, Ivics Z.** 2002. Common physical properties of DNA affecting target site selection of sleeping beauty and other Tc1/mariner transposable elements. *J Mol Biol* **323:**441–452.

123. **Geurts AM, Hackett CS, Bell JB, Bergemann TL, Collier LS, Carlson CM, Largaespada DA, Hackett PB.** 2006. Structure-based prediction of insertion-site preferences of transposons into chromosomes. *Nucleic Acids Res* **34:**2803–2811.

124. **van Luenen HG, Plasterk RH.** 1994. Target site choice of the related transposable elements Tc1 and Tc3 of Caenorhabditis elegans. *Nucleic Acids Res* **22:**262–269.

125. **Yant SR, Wu X, Huang Y, Garrison B, Burgess SM, Kay MA.** 2005. High-resolution genome-wide mapping of transposon integration in mammals. *Mol Cell Biol* **25:**2085–2094.

126. **Grabundzija I, Irgang M, Mates L, Belay E, Matrai J, Gogol-Doring A, Kawakami K, Chen W, Ruiz P, Chuah MK, VandenDriessche T, Izsvak Z, Ivics Z.** 2010. Comparative analysis of transposable element vector systems in human cells. *Mol Ther* **18:**1200–1209.

127. **Huang X, Guo H, Tammana S, Jung YC, Mellgren E, Bassi P, Cao Q, Tu ZJ, Kim YC, Ekker SC, Wu X, Wang SM, Zhou X.** 2010. Gene transfer efficiency and genome-wide integration profiling of Sleeping Beauty, Tol2, and piggyBac transposons in human primary T cells. *Mol Ther* **18:**1803–1813.

128. **Li MA, Pettitt SJ, Eckert S, Ning Z, Rice S, Cadinanos J, Yusa K, Conte N, Bradley A.** 2013. The piggyBac transposon displays local and distant reintegration preferences and can cause mutations at noncanonical integration sites. *Mol Cell Biol* **33:**1317–1330.

129. **Zhang W, Muck-Hausl M, Wang J, Sun C, Gebbing M, Miskey C, Ivics Z, Izsvak Z, Ehrhardt A.** 2013. Integration profile and safety of an adenovirus hybrid-vector utilizing hyperactive sleeping beauty transposase for somatic integration. *PLoS One* **8:**e75344.

130. **de Silva S, Mastrangelo MA, Lotta LT Jr, Burris CA, Izsvak Z, Ivics Z, Bowers WJ.** 2010. Herpes simplex virus/Sleeping Beauty vector-based embryonic gene transfer using the HSB5 mutant: loss of apparent transposition hyperactivity in vivo. *Hum Gene Ther* **21:** 1603–1613.

131. **Zhang W, Solanki M, Muther N, Ebel M, Wang J, Sun C, Izsvak Z, Ehrhardt A.** 2013. Hybrid adeno-associated viral vectors utilizing transposase-mediated somatic integration for stable transgene expression in human cells. *PLoS One* **8:**e76771.

132. **Tower J, Karpen GH, Craig N, Spradling AC.** 1993. Preferential transposition of Drosophila P elements to nearby chromosomal sites. *Genetics* **133:**347–359.

133. **Fischer SE, Wienholds E, Plasterk RH.** 2001. Regulated transposition of a fish transposon in the mouse germ line. *Proc Natl Acad Sci U S A* **98:**6759–6764.

134. **Carlson CM, Dupuy AJ, Fritz S, Roberg-Perez KJ, Fletcher CF, Largaespada DA.** 2003. Transposon mutagenesis of the mouse germline. *Genetics* **165:**243–256.

135. **Horie K, Yusa K, Yae K, Odajima J, Fischer SE, Keng VW, Hayakawa T, Mizuno S, Kondoh G, Ijiri T, Matsuda Y, Plasterk RH, Takeda J.** 2003. Characterization of Sleeping Beauty transposition and its application to genetic screening in mice. *Mol Cell Biol* **23:**9189–9207.

136. **Liang Q, Kong J, Stalker J, Bradley A.** 2009. Chromosomal mobilization and reintegration of Sleeping Beauty and PiggyBac transposons. *Genesis* **47:**404–408.

137. **Kokubu C, Horie K, Abe K, Ikeda R, Mizuno S, Uno Y, Ogiwara S, Ohtsuka M, Isotani A, Okabe M, Imai K, Takeda J.** 2009. A transposon-based chromosomal engineering method to survey a large cis-regulatory landscape in mice. *Nat Genet* **41:**946–952.

138. **Dupuy AJ, Fritz S, Largaespada DA.** 2001. Transposition and gene disruption in the male germline of the mouse. *Genesis* **30:**82–88.

139. **Ruf S, Symmons O, Uslu VV, Dolle D, Hot C, Ettwiller L, Spitz F.** 2011. Large-scale analysis of the regulatory architecture of the mouse genome with a transposon-associated sensor. *Nat Genet* **43:**379–386.

140. **Keng VW, Yae K, Hayakawa T, Mizuno S, Uno Y, Yusa K, Kokubu C, Kinoshita T, Akagi K, Jenkins NA, Copeland NG, Horie K, Takeda J.** 2005. Region-specific saturation germline mutagenesis in mice using the Sleeping Beauty transposon system. *Nat Methods* **2:**763–769.

141. **Yergeau DA, Kelley CM, Kuliyev E, Zhu H, Johnson Hamlet MR, Sater AK, Wells DE, Mead PE.** 2011. Remobilization of Sleeping Beauty transposons in the germline of Xenopus tropicalis. *Mobile DNA* **2:**15.

142. **Belzile F, Yoder JI.** 1992. Pattern of somatic transposition in a high copy Ac tomato line. *Plant J* **2:**173–179.

143. **Osborne BI, Corr CA, Prince JP, Hehl R, Tanksley SD, McCormick S, Baker B.** 1991. Ac transposition from a T-DNA can generate linked and unlinked clusters of insertions in the tomato genome. *Genetics* **129:**833–844.

144. **Knapp S, Larondelle Y, Rossberg M, Furtek D, Theres K.** 1994. Transgenic tomato lines containing Ds elements at defined genomic positions as tools for targeted transposon tagging. *Mol Gen Genet* **243:**666–673.

145. **Kunze R, Weil CF.** 2002. The hAT and CACTA superfamilies of plant transposons, p 565–610. *In* Craig NL, Craigie R, Gellert M, Lambowitz AM (ed), *Mobile DNA II*, ASM Press, Washington, DC.

146. **Benjamin HW, Kleckner N.** 1989. Intramolecular transposition by Tn10. *Cell* **59:**373–383.

147. **Maxwell A, Craigie R, Mizuuchi K.** 1987. B protein of bacteriophage mu is an ATPase that preferentially stimulates intermolecular DNA strand transfer. *Proc Natl Acad Sci U S A* **84:**699–703.

148. **Garfinkel DJ, Stefanisko KM, Nyswaner KM, Moore SP, Oh J, Hughes SH.** 2006. Retrotransposon suicide: formation of Ty1 circles and autointegration via a central DNA flap. *J Virol* **80:**11920–11934.

149. **Shoemaker C, Hoffman J, Goff SP, Baltimore D.** 1981. Intramolecular integration within Moloney murine leukemia virus DNA. *J Virol* **40:**164–172.

150. **Fischer SE, van Luenen HG, Plasterk RH.** 1999. Cis requirements for transposition of Tc1-like transposons in C. elegans. *Mol Gen Genet* **262:**268–274.

151. **Lampe DJ, Grant TE, Robertson HM.** 1998. Factors affecting transposition of the *Himar*1 *mariner* transposon *in vitro*. *Genetics* **149:**179–187.

152. **Karsi A, Moav B, Hackett P, Liu Z.** 2001. Effects of insert size on transposition efficiency of the *Sleeping Beauty* transposon in mouse cells. *Mar Biotechnol* **3:**241–245.

153. **Chandler M, Clerget M, Galas DJ.** 1982. The transposition frequency of IS1-flanked transposons is a function of their size. *J Mol Biol* **154:**229–243.

154. **Morisato D, Way JC, Kim HJ, Kleckner N.** 1983. Tn10 transposase acts preferentially on nearby transposon ends in vivo. *Cell* **32:**799–807.

155. **Lee MS, Craigie R.** 1994. Protection of retroviral DNA from autointegration: involvement of a cellular factor. *Proc Natl Acad Sci U S A* **91:**9823–9827.

156. **Lee MS, Craigie R.** 1998. A previously unidentified host protein protects retroviral DNA from autointegration. *Proc Natl Acad Sci U S A* **95:**1528–1533.

157. **Suzuki Y, Craigie R.** 2002. Regulatory mechanisms by which barrier-to-autointegration factor blocks autointegration and stimulates intermolecular integration of Moloney murine leukemia virus preintegration complexes. *J Virol* **76:**12376–12380.

158. **Mansharamani M, Graham DR, Monie D, Lee KK, Hildreth JE, Siliciano RF, Wilson KL.** 2003. Barrier-to-autointegration factor BAF binds p55 Gag and matrix and is a host component of human immunodeficiency virus type 1 virions. *J Virol* **77:**13084–13092.

159. **Lin CW, Engelman A.** 2003. The barrier-to-autointegration factor is a component of functional human immunodeficiency virus type 1 preintegration complexes. *J Virol* **77:**5030–5036.

160. **Mates L, Izsvak Z, Ivics Z.** 2007. Technology transfer from worms and flies to vertebrates: transposition-based genome manipulations and their future perspectives. *Genome Biol* **8**(Suppl 1):S1.

161. **Ivics Z, Izsvak Z.** 2006. Transposons for gene therapy! *Curr Gene Ther* **6:**593–607.

162. **Ivics Z, Izsvák Z.** 2004. Transposable elements for transgenesis and insertional mutagenesis in vertebrates: a contemporary review of experimental strategies. *Methods Mol Biol* **260:**255–276.

163. **Ivics Z, Li MA, Mates L, Boeke JD, Nagy A, Bradley A, Izsvak Z.** 2009. Transposon-mediated genome manipulation in vertebrates. *Nat Methods* **6:**415–422.

164. **Izsvák Z, Ivics Z.** 2004. Sleeping beauty transposition: biology and applications for molecular therapy. *Mol Ther* **9:**147–156.

165. **Miskey C, Izsvak Z, Kawakami K, Ivics Z.** 2005. DNA transposons in vertebrate functional genomics. *Cell Mol Life Sci* **62:**629–641.

166. **Takeda J, Izsvak Z, Ivics Z.** 2008. Insertional mutagenesis of the mouse germline with Sleeping Beauty transposition. *Methods Mol Biol* **435:**109–125.

167. **VandenDriessche T, Ivics Z, Izsvak Z, Chuah MK.** 2009. Emerging potential of transposons for gene therapy and generation of induced pluripotent stem cells. *Blood* **114:**1461–1468.

168. **Ivics Z, Izsvak Z.** 2010. The expanding universe of transposon technologies for gene and cell engineering. *Mobile DNA* **1:**25.

169. **Ivics Z, Izsvak Z.** 2011. Nonviral gene delivery with the sleeping beauty transposon system. *Hum Gene Ther* **22:**1043–1051.

170. **Izsvák Z, Hackett PB, Cooper LJ, Ivics Z.** 2010. Translating Sleeping Beauty transposition into cellular therapies: victories and challenges. *BioEssays* **32:**756–767.

171. **Swierczek M, Izsvak Z, Ivics Z.** 2012. The Sleeping Beauty transposon system for clinical applications. *Expert Opin Biol Ther* **12:**139–153.

172. **Ammar I, Izsvak Z, Ivics Z.** 2012. The Sleeping Beauty transposon toolbox. *Methods Mol Biol* **859:**229–240.

Mobile DNA, 3rd Edition
Nancy L. Craig, Michael Chandler, Martin Gellert, Alan M. Lambowitz, Phoebe A. Rice and Suzanne Sandmeyer

doi:10.1128/microbiolspec.MDNA3-0028-2014

Kosuke Yusa[1]

piggyBac Transposon

39

INTRODUCTION

The *piggyBac* transposon superfamily is a relatively recently recognized transposon superfamily. The original *piggyBac* transposon was isolated from the genome of the cabbage looper moth, *Trichoplusia ni* in the 1980s. However, the second member of the *piggyBac*-like element superfamily was not identified until 2000. It was not described as a transposon superfamily in the previous edition of *Mobile DNA*. In the last decade or so, a number of sequenced genomes have revealed that *piggyBac*-like elements are actually widespread DNA transposons. Active copies of the transposon have also been identified from another moth species, from frogs, and for the first time, from a mammal. Moreover, because the *piggyBac* transposon has a broad host spectrum from yeast to mammals, this mobile element has been widely used for a variety of applications in a diverse range of organisms. In this chapter, we will describe the discovery and diversity of the *piggyBac* transposon, its mechanism of transposition, and its application as a genetic tool. We will also provide two examples of genetic screening that the *piggyBac* transposon has enabled.

DISCOVERY OF THE *piggyBac* TRANSPOSON

It was known in the late 1970s that when insect DNA viruses, namely *Galleria mellonella* or *Autographa californica* nuclear polyhedrosis viruses (species of Baculovirus), were serially passaged in *T. ni* cell line TN-368, mutant viruses that showed the Few Polyhedra (FP) plaque morphology phenotype appeared spontaneously but reproducibly (1, 2, 3). Analyses of these mutant viruses revealed that the host cell DNA had inserted into the viral genome, resulting in the loss of the 25 KDa viral protein (2). These mutant viruses could revert to the wild-type by serial passage in the TN-368 cell line. This was associated with the loss of the inserted DNAs in the revertant viral genome (2). The sizes of these inserted fragments varied between 0.5 and 2.8 kb (2). One of the most frequently inserted fragments was 2.5 kb in size and was present as multiple copies in the *T. ni* genome (2). These observations strongly indicated that the insertions in the viral genome are mobile elements that reside in the *T. ni* genome. Sequencing analyses of the inserted DNA revealed that the integration sites had typical DNA transposon signatures such as perfect duplication of the 4-bp viral DNA (TTAA) and terminal inverted repeats (TIRs) within the inserted fragments (4). In 1989, the entire sequence of the inserted DNA was reported, revealing the structure of the *piggyBac* transposon for the first time (5).

The *T. ni piggyBac* transposon is a 2,475-bp-long autonomous mobile element (GenBank Accession number. J04364.2; Fig. 1). It has 13-bp TIRs located at both ends and 19-bp subterminal inverted repeats located 3-bp and 31-bp 5′ and 3′ internally from the TIRs, respectively. Between the subterminal inverted repeats,

[1]Wellcome Trust Sanger Institute, Hinxton, Cambridge, UK.

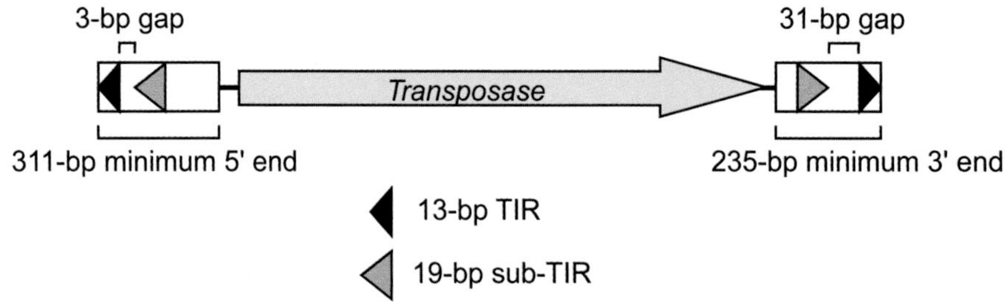

Figure 1 Structure of the *T. ni piggyBac* transposon (GenBank accession number J04364.2). TIR, terminal inverted repeat. The minimum TIR sequences are based on ref. (61). doi:10.1128/microbiolspec.MDNA3-0028-2014.f1

there is a single 1.8-kb open reading frame, which encodes the 594-amino acid *piggyBac* transposase with a molecular weight of 64 kDa. The functionality of the coded protein as a transposase was first confirmed in 1996 with the successful mobilization of a nonautonomous *piggyBac* element (6).

DIVERSITY AND DOMESTICATION OF THE *piggyBac* SUPERFAMILY TRANSPOSONS

Since its discovery in 1989, *T. ni piggyBac* had for a long time remained the only member of the currently-known *piggyBac* superfamily. However, this view has now changed completely owing to the discovery of a number of *piggyBac*-like elements in a variety of organisms. The *piggyBac* transposon superfamily is now recognized as a widespread DNA transposon superfamily. Surprisingly, a currently active copy of the *piggyBac* transposon was recently isolated from mammals. Furthermore, genome sequencing has revealed many genes that were derived from transposable elements. It has been shown that domesticated *piggyBac* transposases play an essential role in cellular functions in ciliates (7, 8). This section describes the diversity of the *piggyBac* transposon and the domestication of *piggyBac* transposases.

piggyBac Transposable Elements

In the 1980s and 1990s, several genomic DNA segments were observed to contain target site duplication of the tetranucleotide TTAA and an addition of 2-to-3 Cs. These included *Tx1* derived from *Xenopus* (9, 10) and TFP3 (4, 11) from *T ni.*, *Pokey* from *Daphnia*, and host DNA integration into the genome of *Autographa californica* nuclear polyhedrosis virus passaged in a *Spodoptera frugiperda* cell line (12, 13, 14). All these elements were nonautonomous and the transposases responsible for the mobilization of these nonautonomous elements could not be identified. It was thus not clear whether these TTAA-specific transposons shared molecular characteristics with *T. ni piggyBac* and belonged to the same transposon superfamily.

In the mid-1990s, transgenesis of nondrosophilid insects using *T. ni piggyBac* transposon-based vectors was successful, which popularized the approach (13, 14). Detection and genetic stability of the integrated transposon were often examined by Southern blot analysis. None of the early transgenesis work identified genetic elements that cross-hybridized with the *T. ni piggyBac* transposon in the test insects' genomes (15, 16), suggesting that *piggyBac* might be restricted in *T. ni*. However, surprising results were reported in 2000 in a paper describing the transgenesis of the Oriental fruit fly, *Bactrocera dorsalis* (17). The genome of the Oriental fruit fly contains sequences, some of which are apparently full-length elements, that cross-hybridized with *T. ni piggyBac*. Sequencing analysis of a transposase-coding region revealed that the cross-hybridizing elements are indeed *piggyBac*-like elements with 95% nucleotide identity to *T. ni piggyBac* (17). This remarkable similarity and the absence of the *piggyBac*-like elements in another bactrocerid species, *B. cucuribitae*, or Mediterranean fruit fly, *Ceratitis capitata*, suggested a very recent horizontal transmission between *T. ni* and *B. dorsalis*; this was probably via one or more intermediate species since these species are geographically distinct (17).

In the following several years, whole genome sequences were published for human (18), pufferfish *Fugu rubripes* (19), and the African malaria mosquito *Anopheles gambiae* (20). Analyses of these genome sequences revealed a number of *piggyBac*-like repetitive sequences. Furthermore, the full-length transposase-coding sequence of the *Pokey* element was isolated in 2002 and showed a clear similarity to *T. ni piggyBac* transposase (21). Thus far, *piggyBac*-like sequences have been found in the genomes of a wide range of organisms including fungi, plants, insects, crustaceans, urochordates, amphibians, fishes, and mammals (22). In addition, active or apparently intact full-length *piggyBac*-like transposons have been identified in another moth *Macdunnoughia crassisigna* (23), silkworm (24), ants (25), *Xenopus* (26), and the bat *Myotis lucifugus* (27).

piggyBat, The First Active DNA Transposon Isolated from Mammals

Around 35 to 50% of the mammalian genome is repetitive elements and a few percent typically comprise fossil DNA transposons derived from various transposon superfamilies (18, 28, 29, 30, 31, 32). A detailed analysis of the evolutionary history of DNA transposons found in the human genome revealed that DNA transposons were active during early primate evolution (approximately 80 to 40 million years ago); however, all these elements are now extinct and there is no evidence of transposition in the last 40 million years (33). A similar trend is also observed in the mouse genome (28). Like other organisms, the *M. lucifugus* genome contains numerous DNA transposons derived from almost every transposon superfamily (34). However, the DNA transposons in the *M. lucifugus* genome have clear differences from those found in the human genome. Firstly, the sequences of the bat DNA transposons are minimally diverged from their consensus sequence. Secondly, the bat genome contains potentially functional full-length *hAT* and *piggyBac* transposons (34). Finally, preintegration sites have also been identified in a related species *M. austroriparius* (34, 35), which separated from *M. lucifugus* approximately 6 million years ago. This strongly suggests that the bat genome was recently invaded by the DNA transposons. Among these transposons, *piggyBac*-like elements are markedly young. For example, a large fraction of the nonautonomous *piggyBac1_ML* element (*npiggy_156* and *npiggy_239*) is identical in sequence (34). Thirty-four full-length copies of the *piggyBac1_ML* element identified in the *M. lucifugus* genome show >99% nucleotide identity (27). Recently, the full-length *piggyBac1_ML* element was cloned and tested for its transposition activity. Similar to *T. ni piggyBac*, the bat-derived *piggyBac* transposon (named *piggyBat*) is able to transpose in yeast, bat, and human cells, albeit at a 2-to-10 times lower frequency than *T. ni piggyBac* (27). Cloning of the *piggyBat* element represented the first isolation of mammalian *piggyBac* derivative. Further analysis of this active mammalian transposon would give us deeper insight into how these parasitic genetic elements contribute to the diversity and evolution of the host species.

Domesticated *piggyBac* Transposases

Various organisms have occasionally utilized transposase genes to create new genes with beneficial cellular functions. This is called "domestication" of transposases, the process of which is well exemplified by RAG1 in V(D)J recombination (derived from *Transib* transposases [36, 37]) and the CENPB centromere protein (derived from *pogo* transposases [38]). The genome sequences of various species have also revealed a number of previously unrecognized genes that were derived from various transposable elements including *piggyBac* transposases (18, 22). For example, the human genome contains 5 genes that were apparently derived from *piggyBac* transposases, *PGBD1-5*. While *piggyBac*-derived sequences in *PGBD1-4* are found in one coding exon, those of *PGBD5* are separated by multiple introns. Furthermore, *PGBD5* can be found in all vertebrates including the lamprey *Petromyzon marinus* and also in the lancelet *Branchiostoma floridae*. Surprisingly, both amino acid sequences and synteny are highly conserved, indicating that *PGBD5* was domesticated more than 500 million years ago. Given that *PGBD5* is expressed mainly in the brain and central nervous system, this gene may have played an important role in the evolution of the central nervous system (39).

Most of the domesticated *piggyBac* transposases have substitutions in the catalytic core residues DDD (see below). They are thus unlikely to have transposase activity, suggesting that their DNA-binding capacity may be utilized by the host (22). Intriguing instances of domestication of *piggyBac* transposase can be found in the ciliates *Paramecium tetraurelia* (7) and *Tetrahymena thermophila* (8). Ciliates are unicellular eukaryotes and have two functionally distinct versions of nuclei in the cytoplasm, namely a micronucleus as the germ-line genome and a macronucleus that is derived from a micronucleus and responsible for somatic gene transcription. During macronucleus development, the genome undergoes massive DNA amplification and extensive genome rearrangement including elimination of internal eliminated sequences (IESs) and transposable elements. The precise excision of IESs is particularly essential as this is required for reconstruction of functional genes. Domesticated *piggyBac* transposases encoded by *piggyMac* in *P. tetraurelia* or *TPB2* in *T. Thermophila* play a central role in the genome rearrangement, as shown in RNA interference experiments (7, 8). Animals in which expression of the domesticated gene was knocked down by RNAi exhibited deficiency in initiating genome rearrangement in the developing macronucleus. Furthermore, both proteins carry a conserved DDD motif and the *TPB2* gene product has been shown to have an endonuclease activity *in vitro* (27). In addition to RAG1 recombinase, the domesticated *piggyBac* transposases in ciliates provide another example that the catalytic activity of transposases is utilized by the host organisms.

MECHANISM OF TRANSPOSITION

piggyBac Transposase

Many transposases and retroviral integrases contain a DDE/D domain, which includes two highly conserved aspartic acid (D) residues and either a glutamic acid (E) residue or a third D. This domain is also known to form an RNase H-like fold (40, 41). The essential amino acid triad coordinates divalent metal ions such as Mg^{2+}, which catalyses transposition/integration activities (40, 41). The disruption of any one of the triad completely abolishes transposase activity. An alignment of *piggyBac* transposases from *piggyBac*-like elements found in various species shows several highly conserved blocks of amino acids in the core region between positions 130 and 522 of the *T. ni piggyBac* transposase, which contains several conserved aspartic acid (D) and glutamic acid (E) residues (22, 42, 43). Although this region does not readily show similarity to the widespread DDE domains, a weak similarity to the IS4 family protein was identified (22), leading to predictions that D268 and D346 in the *T. ni piggyBac* transposase are the conserved aspartic acid in the DDE/D domain. Mutational analyses of D268, D346 as well as another highly conserved D447 of *T. ni piggyBac* transposase revealed that these residues are absolutely required for all steps of transposition including nicking, hairpin resolution, and target joining (see below). This allows us to conclude that the *piggyBac* transposase is a member of the DDE/D recombinase family (43).

The C-terminal region is variable but contains several conserved cysteine residues with regular spacing, forming a putative zinc-binding plant homeodomain (PHD) finger (43). A *T. ni piggyBac* transposase mutant lacking the C-terminal domain can show *in vitro* transposition activity at the level equivalent to the wild-type transposase. Given that PHD fingers bind to chromatin (44), the C-terminus of *piggyBac* transposases may facilitate binding to transposon DNA in a chromatin context.

Chemical Steps

The *piggyBac* transposon has a broad host range, suggesting that its mobilization could be host-factor independent. In 2008, Craig and colleagues purified bacterially expressed *piggyBac* transposase and reconstituted excision and integration reactions *in vitro* using the purified transposase and substrate DNA (43). Their detailed analyses revealed the unique chemical steps that *piggyBac* employs.

Figure 2 illustrates the chemical steps of the mechanism of *piggyBac* transposition. When the *piggyBac* transposase binds to the transposon end, it initiates the excision reaction by nicking the 3′ end of each strand of the transposon ends. The free 3′ OH then attacks the complementary strand of the 5′ end of the flanking TTAA, resulting in the formation of a hairpin on the transposon end and the release of the transposon from the host genome. The hairpin structures in the transposition intermediates are quickly resolved by the transposase and the exposed 3′ OH is then used for target joining. During the integration reaction, the exposed 3′ OH at the transposon end first attacks the 5′ end of target TTAA sites. This joins one strand of the transposon to the target site. Integration is completed when the ends of the transposon ligate to the complementary genomic strand.

At the excised genomic sites, 5′ overhangs of the tetranucletide TTAA are produced at both ends. These are simply paired and re-ligated by host factors, which restores the sequence to its original state. *piggyBac* thus shows precise excision and does not usually leave a footprint. However, there are occasional failures of donor site repair, which result in small insertions and/or deletions, at a frequency of around 1% of excision events during chromosomal transposition in mouse embryonic stem (ES) cells (45).

There are two unique features in *piggyBac* transposition. Firstly, it requires no DNA synthesis. Mobilization of many conventional DNA transposons is associated with DNA synthesis by host repair proteins, which results in target site duplication and a "footprint" (Fig. 3 right). However, *piggyBac* produces 5′ TTAA overhang attached to the transposition intermediates and uses it to base-pair with the 5′ TTAA single-strand gaps on the target DNA during integration (Fig. 3 left). Therefore, *piggyBac* does not require DNA synthesis. Secondly, the excised transposon ends form a hairpin structure. This mechanism was the first to be observed in eukaryotic transposons although it has been seen in bacterial NA transposons that belong to the IS4 family, namely Tn*5* and Tn*10*. In the Tn*5* and Tn*10* transposons, an essential step of hairpin processing is base flipping, which is often observed in reactions involving DNA modifying enzymes such as DNA methylases and DNA glycosylases. The cocrystal structure of Tn*5* synaptic complex revealed for the first time that base flipping is employed in transposition reactions. Bischerour and Chalmers further characterized base flipping at the molecular level and revealed the involvement of two W residues in Tn*5*

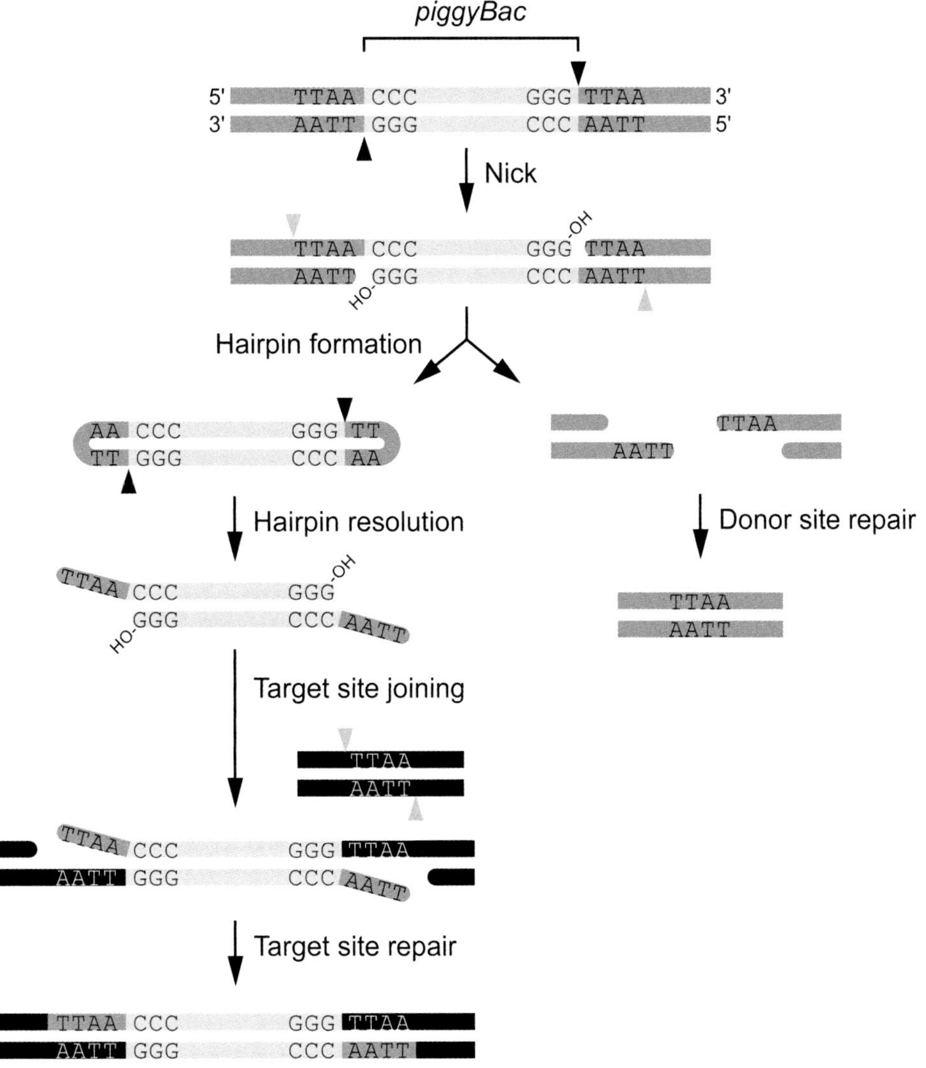

Figure 2 The chemical steps of *T. ni piggyBac* transposition. Black and grey arrowheads indicate positions of nicks or sites where 3′ OH groups attack, respectively. Modified from ref. (43). doi:10.1128/microbiolspec.MDNA3-0028-2014.f2

and one W and one E residue in Tn*10*. These residues are located at the equivalent position near the third E residue of the DDE triad (46, 47). Although these amino acids are not conserved in their corresponding positions in *T. ni piggyBac* transposase, there is one W residue (position 465 in *T. ni piggyBac* transposase, downstream of the third D residue of the DDD triad) that is highly conserved among *piggyBac* transposases and has been suggested to be involved in base flipping (48). However, W465A mutant showed much reduced nicking activity as well as deficiency in every subsequent step of transposition reaction: hairpin formation, hairpin resolution, and target joining (43). Therefore, W465 plays a central role in the transposase activity and cannot have a role only in DNA hairpin formation and resolution. Given that Tn*5*/Tn*10* transposons produce a blunt intermediate molecule (38, 39) whereas *piggyBac* intermediate carries 4-nucleotide 5′ overhangs (43), *piggyBac* may therefore use different amino acids or mechanism(s) to facilitate hairpin processing.

Integration Site Preference

The *piggyBac* target site, TTAA, is fairly frequent in the genome but the epigenetic status of the target sites may affect integration site preference of the *piggyBac* transposon. Understanding such a preference is important, especially when *piggyBac* is used as an insertional mutagen or a gene therapy vehicle.

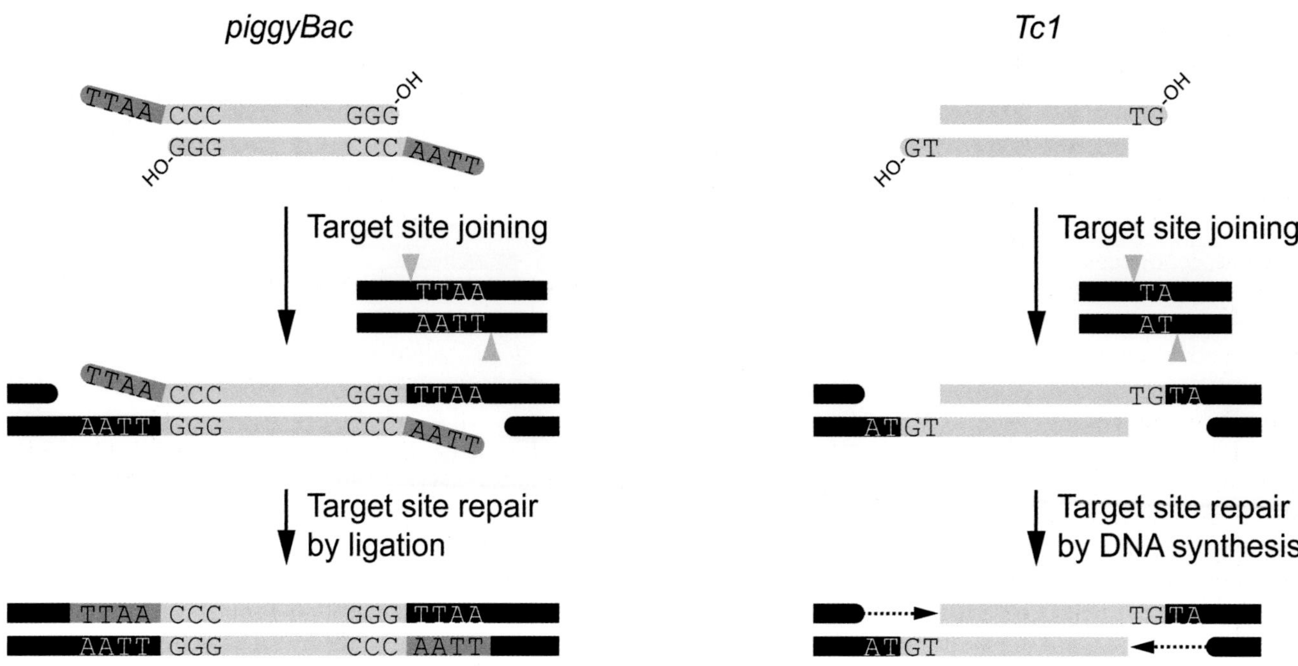

Figure 3 Comparison of target site joining and repair in *piggyBac* (left) and *Tc1* (right). Grey arrowheads indicate sites where 3′ OH groups attack. Modified from ref. (136). doi:10.1128/microbiolspec.MDNA3-0028-2014.f3

To analyze site preference with an adequate statistical power, a large number of integration sites are required. Ligation-mediated PCR is the most frequently used method to identify transposon integration sites (49). In this method, genomic DNA is typically digested with a 4-bp cutter restriction enzyme, ligated with an adaptor (known as a splinkerette) and amplified by PCR using transposon-specific and splinkerette-specific primers. The resulting PCR products, which contain junctions between the transposon end and the flanking genomic regions, are determined by capillary sequencing, thereby identifying the integration sites. An issue is that the throughput of this method is too low to collect statistically significant numbers of transposon integration sites. Recent advances in sequencing technologies, however, have transformed the way we identify transposon integration sites (50). It is now possible to identify tens of thousands of sites simultaneously. These can then be compared with an accumulating genomic dataset, such as those of the histone modification and DNase hypersensitive sites, allowing us to comprehensively analyze and identify the epigenetic elements that affect integration site preference.

Using the method described above, Li *et al.* identified more than 30 thousand *T. ni piggyBac* integration sites in mouse embryonic stem cells (51). When mobilized from chromosomal donor sites, the *piggyBac* transposon showed strong local hopping patterns; these are commonly observed in DNA transposon mobilization. Approximately 10 to 15% of integrations were found within 5 Mb from the donor sites. This also increased the frequency of integration into the chromosome that contains the donor sites, resulting in 25% of reintegrations in the original donor chromosome. When compared to the epigenetic features, the *piggyBac* integration sites are clearly associated with accessible chromatin structures. These sites include DNase hypersensitive sites, sites with trimethylated lysine 4 on histone H3 and sites with pol II binding. In sharp contrast, the *piggyBac* integration sites are negatively correlated with lamin-associated domains, which are heterochromatin domains. These trends are more obvious when the *piggyBac* transposons are mobilized from genomic donor sites rather than from transfected plasmid DNA. *piggyBac* preferentially integrates ingenic regions, especially regions that contain expressed genes.

These preferential integrations into the open chromatin structure were also observed in *piggyBat* transposon (the *M. lucifugus*-derived *piggyBac*-like element) in both human cells and bat fibroblasts (27). Interestingly, the *piggyBat* transposons are accumulated in genic regions of the bat genome, indicating that the *in vitro* studies using human or mouse cell lines truly reflect native transposition patterns.

piggyBac AS GENETIC TOOLS

DNA transposons have been used as versatile genetic tools in a wide range of organisms. For example, the *Tc*1 element was used in *Caenorhabditis elegans* mutagenesis (52) and the *P* element was used in *Drosophila melanogaster* (53, 54). Earlier studies revealed that *Drosophila*-derived transposons such as the *P* element and the *hobo* transposon were either not functional or limited in nondrosophilid insects (55). There was therefore a need to develop alternative transposon-based transgenesis systems for nondrosophilid insects. The Lepidopteran transposon, *piggyBac*, was one such candidate. An engineered *piggyBac*-based vector was first shown to be able to mobilize in a nonhost *S. frugiperda* cell line in 1995 (56, 57). Subsequently, it was shown that the *piggyBac* was able to mobilize in non-Lepidopteran insects (15). Today, *T. ni piggyBac* transposition has been confirmed in 5 orders of insects (Table 1) and in organisms including plants, yeasts, protozoa, and vertebrates (Table 2). The *piggyBac*-based transposon vector system is the most widely used transposon system for a variety of applications such as transgenesis and mutagenesis. Furthermore, *piggyBac* has potential as a gene therapy vehicle (58). The next section describes the *piggyBac* transposon in the context of genetic tools.

The Engineered Nonautonomous *piggyBac* Transposon System

Similar to the other transposon systems such as *P* element and *Tc*1 transposon, the autonomous *piggyBac* transposon can be separated into 2 components: a DNA segment flanked by *piggyBac* TIR and a transposase. The initial transgenesis experiments using the *piggyBac* transposon were conducted in a conservative vector system in which genetic marker genes were simply inserted into the middle of the transposase coding region (13, 14). Alternatively, the short fragment (0.75 kb) in the

Table 1 Studies in which *piggyBac* transposition has been confirmed in insect species

Order	Common name	Scientific name	Reference
Coleoptera	Ladybird beetle	*Harmonia axyridis*	(81)
Coleoptera	Red flour beetle	*Tribolium castaneum*	(82)
Diptera	La Crosse encephalitis vector mosquito	*Aedes triseriatus*	(83)
Diptera	Yellow fever mosquito	*Aedes aegypti*	(84)
Diptera	Dengue vector mosquito	*Aedes albopictus*	(83)
Diptera	Mosquito	*Aedes fluviatilis*	(85)
Diptera	Mexican fruit fly	*Anastrepha ludens*	(86)
Diptera	Malaria mosquito	*Anopheles albimanus*	(87)
Diptera	Malaria mosquito	*Anopheles gambiae*	(88)
Diptera	Mosquito	*Anopheles gambiae*	(89)
Diptera	Malaria mosquito	*Anopheles stephensi*	(90)
Diptera	Oriental fruit fly	*Bactrocera dorsalis*	(17)
Diptera	Queensland fruit fly	*Bactrocera tryoni*	(91)
Diptera	Mediterranean fruit fly	*Ceratitis capitata*	(15)
Diptera	Fruit fly	*Drosophila melanogaster*	(16, 84)
Diptera	Spotted wing drosophilid	*Drosophila suzukii*	(92)
Diptera	House fly	*Musca domestica*	(93)
Diptera	Stalk-eyed fly	*Teleopsis dalmanni*	(94)
Hymenoptera	Sawfly	*Athalia rosae*	(95)
Lepidoptera	Butterfly	*Bicyclus anynana*	(96)
Lepidoptera	Silkworm	*Bombyx mori*	(59)
Lepidoptera	Codling moth	*Cydia pomonella*	(97)
Lepidoptera	Tobacco budworm	*Heliothis virescens*	(98)
Lepidoptera	Cabbage moth	*Mamestra brassicae*	(99)
Lepidoptera	Asian corn borer	*Ostrinia furnacalis*	(100)
Lepidoptera	Pink bollworm	*Pectinophora gossypiella*	(101)
Lepidoptera	Potato tuber moth	*Phthorimaea operculella*	(102)
Lepidoptera	Diamondback moth	*Plutella xylostella*	(103)
Lepidoptera	Fall armyworm	*Spodoptera frugiperda*	(56)
Orthoptera	Two-spotted cricket	*Gryllus bimaculatus*	(104)

Table 2 Studies in which *piggyBac* transposition has been confirmed in noninsect species

Organism	Common name (cell type)	Scientific name	Reference
Yeast	Budding yeast	*Saccharomyces cerevisiae*	(43)
Yeast	Fission yeast	*Schizosaccharomyces pombe*	(105)
Protozoa	Malaria parasite	*Plasmodium falciparum*	(79)
Protozoa	Human blood fluke parasite	*Schistosoma mansoni*	(106)
Protozoa	Malaria parasite	*Plasmodium berghei*	(107)
Protozoa	Apicomplexan parasite	*Eimeria tenella*	(108)
Protozoa	Rat gastrointestinal parasite	*Strongyloides ratti*	(109)
Pranarian	Pranarian	*Girardia tigrina*	(110)
Fish	Zebrafish (Zygote)	*Danio rerio*	(111)
Bird	Chicken (PGC, spinal cord)	*Gallus gallus*	(112, 113, 114, 115)
Mammal	Mouse (ES cells, *in vivo*)	*Mus musculus*	(49, 62, 64)
Mammal	Rat (Zygote, *in vivo*)	*Rattus norvegicus*	(116, 117)
Mammal	Goat (Fetal fibroblasts)	*Capra aegagrus*	(118)
Mammal	Pig (Fetal fibroblasts)	*Sus scrofa*	(119)
Mammal	Horse (Fetal fibroblasts)	*Equus ferus*	(120)
Mammal	Human (Cancer cell lines, primary T lymphocyte, $CD34^+$ hematopoietic stem cells, ES cells, iPS cells)	*Homo sapiens*	(63, 70, 121, 122, 123)
Mammal	Macaque (ES cells)	*Macaca fascicularis* *M. mulatta*	(65)
Plant	Rice	*Oryza sativa*	(124)

transposase coding region was replaced with marker genes (59), retaining most of the original transposon sequences. To improve and increase the versatility of the vector system, the minimum terminal sequences needed to be identified. This was done by analysis of a series of internal deletion constructs. The originally identified minimum terminal sequences were 125 bp for the 5′ terminus and 162 bp for the 3′ terminus; such a vector had transposition activities that were comparable to the full-length transposon in *T. ni* embryos (60). However, this minimal *piggyBac* vector had a dramatic reduction in transformation efficiency in *Drosophila*. Further analysis revealed the necessity of internal domain sequences for efficient transposition and identified the optimal terminal length as 311 bp for the 5′ terminus and 235 bp for the 3′ terminus (61). This is the most commonly used vector configuration today.

Any DNA fragments can be inserted as a transposon cargo between these terminal repeats and mobilized by the transposase. Various elements have been used depending on the experimental purpose. To express a gene of interest exogenously, a transcription unit, containing a suitable promoter, a coding sequence, and a polyadenylation signal sequence, is inserted. For disrupting gene function, a gene trap element consisting of a splice acceptor site and a polyadenylation signal sequence can be used. The cargo capacity of the *piggyBac* transposon is fairly large; transposons with a cargo of up to 10 kb can be mobilized without losing transposition efficiency (62). Recently, it has been shown that *piggyBac* can transpose bacterial artificial chromosomes (BACs), which are 150 to 300 kb in length, in mouse and human pluripotent stem cells (50, 63) and in mouse zygotes (64). In the best result in the zygote injection, 45% of F0 mice carried a BAC transgene and most of them have transposon signature (64).

Transposase/Transposon Variants

Increasing transposition efficiency is the key to improving efficiency of transposon-mediated genetic manipulation. This has been done recently by a series of mutagenesis of the TIRs and the transposase, followed by screening for hyperactive variants.

(i) Transposon

Transposon TIR sequences were randomly mutagenized by error-prone PCR. These sequences were then screened for higher transposition activity. A mutant 5′ TIR carrying two substitutions, T53C and C136T, showed a 59% increase in overall transposition compared to the wild-type TIR (65). The mechanisms by which these substitutions increase transposition frequency remain elusive.

(ii) Transposase

Since *piggyBac* is derived from insects, the codon usage may not be suitable for expression in mammalian cells. One simple idea to increase the transposition efficiency is to optimize codon usage and increase the expression level of the transposase. Indeed, optimization of the transposase-coding sequences to the codon usage of mouse (66) or human (65) increased the transposition efficiency by several fold when compared to the original insect sequences. Another approach is random mutagenesis of the transposase by error-prone PCR followed by screening for hyperactive mutants (45). The yeast-based transposition assay system is particularly useful for large-scale screening. A total of 10,000 clones were screened and 17 hyperactive mutants were identified. The activity of each mutant was further tested in mouse ES cells and 5 mutants showed higher activity (I30V/G165S, S103P, M282V, S509G/N570S and N538K). All seven amino acid substitutions that were found in the five mutants were then combined into one coding sequence and the resulting transposase, named hyPBase, showed increases of approximately 20-fold in excision and 10-fold in integration in mouse ES cells (45). The hyPBase was also able to increase transposition efficiency in human cells (67). The mechanisms by which these mutations alter transposition activity remain elusive.

In addition to the developments described above, a variant called $Exc^{+}Int^{-}$ transposase that can excise a transposon but cannot integrate it back into the host genome has recently been generated (68). This variant was identified by site-directed mutagenesis of potentially DNA-interacting amino acids. Given that *piggyBac* can be excised seamlessly, this new variant transposase is useful for the removal of transgenes carried by *piggyBac*. The variant was found to have the R372A/K375A double mutation and was further mutagenized by error-prone PCR to screen for hyperactive variants. Two mutations, namely M194V and D450N, were found to increase the excision activity of the $Exc^{+}Int^{-}$ (R372A/K375A) transposase. When R372A/K375A/D450N mutations were introduced into the hyPBase background, the transposase showed a marked increase in its excision activity with no change in its integration activity.

In addition, transposase variants have also been generated by fusing functional protein domains. The ERT2 domain (a mutated version of the ligand-binding domain of the human estrogen receptor) allows temporal regulation of enzymatic activity of the fusion proteins by tamoxifen administration (69). The *piggyBac* transposase-ERT2 fusion protein can be activated upon induction, but remains inactive without tamoxifen (66). DNA-binding domains such as the Gal4 DNA-binding domain and custom-made zinc finger DNA-binding domains have also been used. These fusion transposes are not only active but also able to integrate transposons into sites that are in close proximity to their respective binding sites, thereby allowing site-directed transposition (65, 68, 70). *Sleeping Beauty* and *Tol2* transposases are not amenable to protein fusion (70). The flexibility of transposase modification by protein domain fusion is therefore another unique feature of the *piggyBac* transposon system.

piggyBac-MEDIATED GENETIC SCREENING

One major application of DNA transposons is insertional mutagenesis. The *piggyBac* transposon vectors have been used as an insertional mutagen in several organisms (Table 3). One of the best examples of mutagenesis using *piggyBac* is *in vivo* mutagenesis in mice for cancer gene discovery. Traditionally, murine leukemia virus and murine mammary tumor virus were used for oncogene discovery in hematopoietic cells (71, 72) and mammary tissues (73), respectively. However, such experiments were not possible in a large fraction of solid tumors due to the limited accessibility of these retroviruses. Analyses of such tumors therefore required the development of active transposon systems in mammalian cells. *Sleeping Beauty* is the first transposon system that has sufficient transposition efficiency for use in mutagenesis (74). Subsequently, *piggyBac* was also shown to be able to transpose efficiently in mice (62). The vector configuration that is typically used in mice for *in vivo* mutagenesis is shown in Fig. 4. It contains both gene inactivation and activation elements to identify tumor suppressors and oncogenes, respectively. Mobilization of this mutagenic transposon in mice could either significantly increase tu-

Table 3 Studies using the *piggyBac* transposon as an insertional mutagen

Organisms	Purpose	Species	Reference
Insect	Mutagenesis	*Tribolium castaneum*	(125, 126)
Insect	Mutagenesis	*Drosophila melanogaster*	(127, 128, 129)
Mouse	*In vivo* cancer gene discovery	*Mus musculus*	(76, 130, 131)
Mouse	*In vitro* screening in ES cells	*Mus musculus*	(132, 133, 134)
Yeast	Mutagenesis	*Fission yeast*	(105)
Malaria	Mutagenesis and phenotype-based screen	*Plasmodium berghei* *Plasmodium falciparum*	(80, 107, 135)

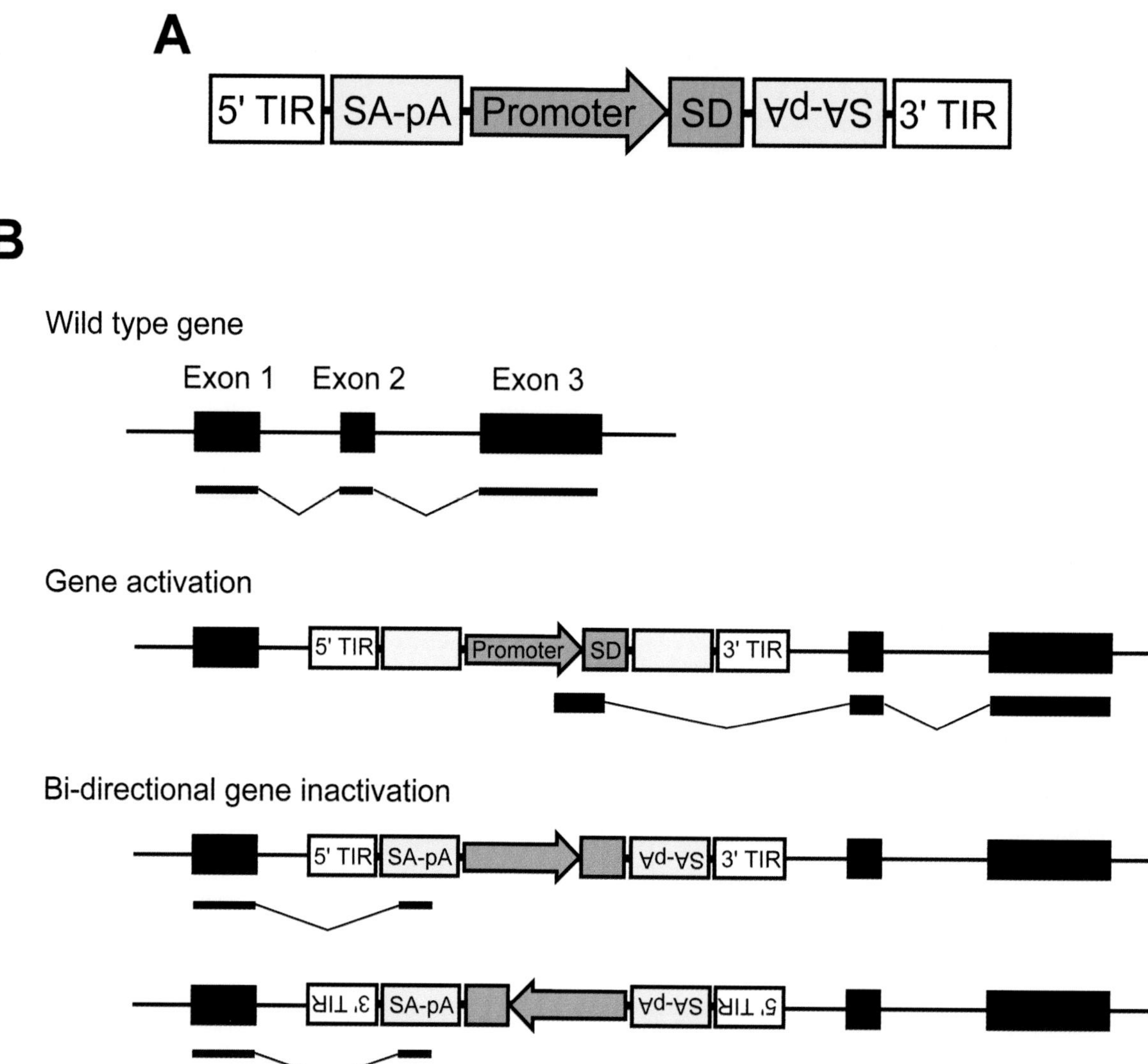

Figure 4 Transposon-mediated cancer gene discovery in mice. (A) Commonly used genetic elements. TIR, terminal inverted repeat; SA, splice acceptor site; pA, polyadenylation signal sequence; SD, splice donor site. (B) In gene activation, a strong constitutive promoter ectopically expresses or overexpresses a trapped gene. The transposon carries two splice acceptor sites in both directions; the trapped genes will be inactivated in spite of the transposon orientation relative to the gene. doi:10.1128/microbiolspec.MDNA3-0028-2014.f4

mor formation (75, 76) or promote tumorigenesis when used in conjunction with cancer-predisposing genetic backgrounds (77). These transposon-induced tumors typically hosted *piggyBac* integration events within the tumor suppressors and/or upstream of the oncogenes (76). Detailed analyses successfully identified novel oncogenes (76).

Another example of the use of *piggyBac* in mutagenesis is the malaria parasite, *Plasmodium falciparum*. Genetic manipulation is a fundamental experimental method for gene function studies in any organism. However, it is extremely difficult in *P. falciparum* for a number of reasons (78). The transfection efficiency is extremely low since DNA has to travel through multilayers of membranes to get into the nucleus of blood-stage parasites. Linear DNA is degraded before reaching the parasite nucleus, whereas circular DNA stays as episomes in the parasite nucleus. The *piggyBac* transposon can circumvent these issues and a *piggyBac*-based transformation system has recently been developed (79). By scaling up the experimental scale, one can generate a library of mutant parasites. A small library consisting of 189 parasites with *piggyBac* insertion was recently generated and, from this library, 29 parasites that were deficient in gametocytogenesis were successfully isolated (80).

CONCLUDING REMARKS

Recent advances in genome sequencing of various organisms have unveiled a number of previously unrecognized transposable elements and domesticated transposases, which has allowed us to greatly widen our understanding of these elements. The *piggyBac* transposon is one such good example, as exemplified by the discoveries of *piggyBat* and *piggyMac*. Genome sequence data will continue to accumulate and may provide even more surprising characteristics of these parasitic DNA elements. In terms of the molecular biology of *piggyBac*, a number of efforts have been made and the chemical steps of *piggyBac* transposition have been characterized. However, its uniqueness has left some unsolved questions; for instance, the molecular mechanisms of hairpin processing. Further biochemical characterization of *piggyBac* transposase and domesticated enzymes will reveal its unique DNA processing mechanisms. In particular, crystal structures of *piggyBac* transposases in a free or in a DNA-binding form will be of great interest to unveil the mechanisms of hairpin processing. In addition, the versatility of the *piggyBac* transposon system has allowed us to achieve previously difficult-to-perform transgenesis and mutagenesis. This will further expand our ability to address biological questions.

Citation. **Yusa K.** 2014. *piggyBac* transposon. Microbiol Spectrum 3(2):MDNA3-0028-2014.

References

1. **Potter KN, Faulkner P, MacKinnon EA.** 1976. Strain selection during serial passage of Trichoplusia in nuclear polyhedrosis virus. *J Virol* **18:**1040–1050.
2. **Fraser MJ, Smith GE, Summers MD.** 1983. Acquisition of Host Cell DNA Sequences by Baculoviruses: Relationship Between Host DNA Insertions and FP Mutants of Autographa californica and Galleria mellonella Nuclear Polyhedrosis Viruses. *J Virol* **47:**287–300.
3. **Fraser MJ, Hink WF.** 1982. The isolation and characterization of the MP and FP plaque variants of Galleria mellonella nuclear polyhedrosis virus. *Virology* **117:** 366–378.
4. **Fraser MJ, Brusca JS, Smith GE, Summers MD.** 1985. Transposon-mediated mutagenesis of a baculovirus. *Virology* **145:**356–361.
5. **Cary LC, Goebel M, Corsaro BG, Wang HG, Rosen E, Fraser MJ.** 1989. Transposon mutagenesis of baculoviruses: analysis of Trichoplusia ni transposon IFP2 insertions within the FP-locus of nuclear polyhedrosis viruses. *Virology* **172:**156–169.
6. **Elick TA, Bauser CA, Fraser MJ.** 1996. Excision of the piggyBac transposable element in vitro is a precise event that is enhanced by the expression of its encoded transposase. *Genetica* **98:**33–41.
7. **Baudry C, Malinsky S, Restituito M, Kapusta A, Rosa S, Meyer E, Betermier M.** 2009. PiggyMac, a domesticated piggyBac transposase involved in programmed genome rearrangements in the ciliate Paramecium tetraurelia. *Genes Dev* **23:**2478–2483. doi:10.1101/gad.547309.
8. **Cheng CY, Vogt A, Mochizuki K, Yao MC.** 2010. A domesticated piggyBac transposase plays key roles in heterochromatin dynamics and DNA cleavage during programmed DNA deletion in Tetrahymena thermophila. *Mol Biol Cell* **21:**1753–1762. doi:10.1091/mbc. E09-12-1079.
9. **Garrett JE, Carroll D.** 1986. Tx1: a transposable element from Xenopus laevis with some unusual properties. *Mol Cell Biol* **6:**933 941.
10. **Garrett JE, Knutzon DS, Carroll D.** 1989. Composite transposable elements in the Xenopus laevis genome. *Mol Cell Biol* **9:**3018–3027.
11. **Wang HH, Fraser MJ, Cary LC.** 1989. Transposon mutagenesis of baculoviruses: analysis of TFP3 lepidopteran transposon insertions at the FP locus of nuclear polyhedrosis viruses. *Gene* **81:**97–108.
12. **Schetter C, Oellig C, Doerfler W.** 1990. An insertion of insect cell DNA in the 81-map-unit segment of Autographa californica nuclear polyhedrosis virus DNA. *J Virol* **64:**1844–1850.
13. **Carstens EB.** 1987. Identification and nucleotide sequence of the regions of Autographa californica nuclear polyhedrosis virus genome carrying insertion elements derived from Spodoptera frugiperda. *Virology* **161:**8–17.
14. **Beames B, Summers MD.** 1990. Sequence comparison of cellular and viral copies of host cell DNA insertions found in Autographa californica nuclear polyhedrosis virus. *Virology* **174:**354–363.
15. **Handler AM, McCombs SD, Fraser MJ, Saul SH.** 1998. The lepidopteran transposon vector, piggyBac, mediates germ-line transformation in the Mediterranean fruit fly. *Proc Natl Acad Sci U S A* **95:**7520–7525.
16. **Handler AM, Harrell RA 2nd.** 1999. Germline transformation of Drosophila melanogaster with the piggyBac transposon vector. *Insect Mol Biol* **8:**449–457.
17. **Handler AM, McCombs SD.** 2000. The piggyBac transposon mediates germ-line transformation in the Oriental fruit fly and closely related elements exist in its genome. *Insect Mol Biol* **9:**605–612. doi:imb227.
18. **Lander ES, Linton LM, Birren B, Nusbaum C, Zody MC, Baldwin J, Devon K, Dewar K, Doyle M, FitzHugh W, Funke R, Gage D, Harris K, Heaford A, Howland J, Kann L, Lehoczky J, LeVine R, McEwan P, McKernan K, Meldrim J, Mesirov JP, Miranda C, Morris W, Naylor J, Raymond C, Rosetti M, Santos R, Sheridan A, Sougnez C, Stange-Thomann N, Stojanovic N, Subramanian A, Wyman D, Rogers J, Sulston J, Ainscough R, Beck S, Bentley D, Burton J, Clee C, Carter N, Coulson A, Deadman R, Deloukas P, Dunham A, Dunham I, Durbin R, French L, Grafham D, Gregory S, Hubbard T,

Humphray S, Hunt A, Jones M, Lloyd C, McMurray A, Matthews L, Mercer S, Milne S, Mullikin JC, Mungall A, Plumb R, Ross M, Shownkeen R, Sims S, Waterston RH, Wilson RK, Hillier LW, McPherson JD, Marra MA, Mardis ER, Fulton LA, Chinwalla AT, Pepin KH, Gish WR, Chissoe SL, Wendl MC, Delehaunty KD, Miner TL, Delehaunty A, Kramer JB, Cook LL, Fulton RS, Johnson DL, Minx PJ, Clifton SW, Hawkins T, Branscomb E, Predki P, Richardson P, Wenning S, Slezak T, Doggett N, Cheng JF, Olsen A, Lucas S, Elkin C, Uberbacher E, Frazier M, Gibbs RA, Muzny DM, Scherer SE, Bouck JB, Sodergren EJ, Worley KC, Rives CM, Gorrell JH, Metzker ML, Naylor SL, Kucherlapati RS, Nelson DL, Weinstock GM, Sakaki Y, Fujiyama A, Hattori M, Yada T, Toyoda A, Itoh T, Kawagoe C, Watanabe H, Totoki Y, Taylor T, Weissenbach J, Heilig R, Saurin W, Artiguenave F, Brottier P, Bruls T, Pelletier E, Robert C, Wincker P, Smith DR, Doucette-Stamm L, Rubenfield M, Weinstock K, Lee HM, Dubois J, Rosenthal A, Platzer M, Nyakatura G, Taudien S, Rump A, Yang H, Yu J, Wang J, Huang G, Gu J, Hood L, Rowen L, Madan A, Qin S, Davis RW, Federspiel NA, Abola AP, Proctor MJ, Myers RM, Schmutz J, Dickson M, Grimwood J, Cox DR, Olson MV, Kaul R, Shimizu N, Kawasaki K, Minoshima S, Evans GA, Athanasiou M, Schultz R, Roe BA, Chen F, Pan H, Ramser J, Lehrach H, Reinhardt R, McCombie WR, de la Bastide M, Dedhia N, Blocker H, Hornischer K, Nordsiek G, Agarwala R, Aravind L, Bailey JA, Bateman A, Batzoglou S, Birney E, Bork P, Brown DG, Burge CB, Cerutti L, Chen HC, Church D, Clamp M, Copley RR, Doerks T, Eddy SR, Eichler EE, Furey TS, Galagan J, Gilbert JG, Harmon C, Hayashizaki Y, Haussler D, Hermjakob H, Hokamp K, Jang W, Johnson LS, Jones TA, Kasif S, Kaspryzk A, Kennedy S, Kent WJ, Kitts P, Koonin EV, Korf I, Kulp D, Lancet D, Lowe TM, McLysaght A, Mikkelsen T, Moran JV, Mulder N, Pollara VJ, Ponting CP, Schuler G, Schultz J, Slater G, Smit AF, Stupka E, Szustakowski J, Thierry-Mieg D, Thierry-Mieg J, Wagner L, Wallis J, Wheeler R, Williams A, Wolf YI, Wolfe KH, Yang SP, Yeh RF, Collins F, Guyer MS, Peterson J, Felsenfeld A, Wetterstrand KA, Patrinos A, Morgan MJ, de Jong P, Catanese JJ, Osoegawa K, Shizuya H, Choi S, Chen YJ. 2001. Initial sequencing and analysis of the human genome. *Nature* **409**:860–921. doi:10.1038/35057062.

19. **Aparicio S, Chapman J, Stupka E, Putnam N, Chia JM, Dehal P, Christoffels A, Rash S, Hoon S, Smit A, Gelpke MD, Roach J, Oh T, Ho IY, Wong M, Detter C, Verhoef F, Predki P, Tay A, Lucas S, Richardson P, Smith SF, Clark MS, Edwards YJ, Doggett N, Zharkikh A, Tavtigian SV, Pruss D, Barnstead M, Evans C, Baden H, Powell J, Glusman G, Rowen L, Hood L, Tan YH, Elgar G, Hawkins T, Venkatesh B, Rokhsar D, Brenner S.** 2002. Whole-genome shotgun assembly and analysis of the genome of Fugu rubripes. *Science* **297**:1301–1310. doi:10.1126/science.1072104.

20. **Holt RA, Subramanian GM, Halpern A, Sutton GG, Charlab R, Nusskern DR, Wincker P, Clark AG, Ribeiro JM, Wides R, Salzberg SL, Loftus B, Yandell M, Majoros WH, Rusch DB, Lai Z, Kraft CL, Abril JF, Anthouard V, Arensburger P, Atkinson PW, Baden H, de Berardinis V, Baldwin D, Benes V, Biedler J, Blass C, Bolanos R, Boscus D, Barnstead M, Cai S, Center A, Chaturverdi K, Christophides GK, Chrystal MA, Clamp M, Cravchik A, Curwen V, Dana A, Delcher A, Dew I, Evans CA, Flanigan M, Grundschober-Freimoser A, Friedli L, Gu Z, Guan P, Guigo R, Hillenmeyer ME, Hladun SL, Hogan JR, Hong YS, Hoover J, Jaillon O, Ke Z, Kodira C, Kokoza E, Koutsos A, Letunic I, Levitsky A, Liang Y, Lin JJ, Lobo NF, Lopez JR, Malek JA, McIntosh TC, Meister S, Miller J, Mobarry C, Mongin E, Murphy SD, O'Brochta DA, Pfannkoch C, Qi R, Regier MA, Remington K, Shao H, Sharakhova MV, Sitter CD, Shetty J, Smith TJ, Strong R, Sun J, Thomasova D, Ton LQ, Topalis P, Tu Z, Unger MF, Walenz B, Wang A, Wang J, Wang M, Wang X, Woodford KJ, Wortman JR, Wu M, Yao A, Zdobnov EM, Zhang H, Zhao Q, Zhao S, Zhu SC, Zhimulev I, Coluzzi M, della Torre A, Roth CW, Louis C, Kalush F, Mural RJ, Myers EW, Adams MD, Smith HO, Broder S, Gardner MJ, Fraser CM, Birney E, Bork P, Brey PT, Venter JC, Weissenbach J, Kafatos FC, Collins FH, Hoffman SL.** 2002. The genome sequence of the malaria mosquito Anopheles gambiae. *Science* **298**:129–149. doi:10.1126/science.1076181.

21. **Penton EH, Sullender BW, Crease TJ.** 2002. Pokey, a new DNA transposon in Daphnia (cladocera: crustacea). *J Mol Evol* **55**:664–673. doi:10.1007/s00239-002-2362-9.

22. **Sarkar A, Sim C, Hong YS, Hogan JR, Fraser MJ, Robertson HM, Collins FH.** 2003. Molecular evolutionary analysis of the widespread piggyBac transposon family and related "domesticated" sequences. *Mol Genet Genomics* **270**:173–180. doi:10.1007/s00438-003-0909-0.

23. **Wu M, Sun Z-C, Hu C-L, Zhang G-F, Han Z-J.** 2008. An active piggyBac-like element in Macdunnoughia crassisigna. *Insect Sci* **15**:521–528.

24. **Xu HF, Xia QY, Liu C, Cheng TC, Zhao P, Duan J, Zha XF, Liu SP.** 2006. Identification and characterization of piggyBac-like elements in the genome of domesticated silkworm, Bombyx mori. *Mol Genet Genomics* **276**:31–40. doi:10.1007/s00438-006-0124-x.

25. **Bonasio R, Zhang G, Ye C, Mutti NS, Fang X, Qin N, Donahue G, Yang P, Li Q, Li C, Zhang P, Huang Z, Berger SL, Reinberg D, Wang J, Liebig J.** 2010. Genomic comparison of the ants Camponotus floridanus and Harpegnathos saltator. *Science* **329**:1068–1071. doi:10.1126/science.1192428.

26. **Hikosaka A, Kobayashi T, Saito Y, Kawahara A.** 2007. Evolution of the Xenopus piggyBac transposon family TxpB: domesticated and untamed strategies of transposon subfamilies. *Mol Biol Evol* **24**:2648–2656. doi:msm191.

27. **Mitra R, Li X, Kapusta A, Mayhew D, Mitra RD, Feschotte C, Craig NL.** 2013. Functional characterization of piggyBat from the bat Myotis lucifugus unveils an active mammalian DNA transposon. *Proc Natl Acad Sci U S A* **110**:234–239. doi:10.1073/pnas.1217548110.

28. **Waterston RH, Lindblad-Toh K, Birney E, Rogers J, Abril JF, Agarwal P, Agarwala R, Ainscough R, Alexandersson M, An P, Antonarakis SE, Attwood J, Baertsch R, Bailey J, Barlow K, Beck S, Berry E, Birren B, Bloom T, Bork P, Botcherby M, Bray N, Brent MR, Brown DG, Brown SD, Bult C, Burton J, Butler J, Campbell RD, Carninci P,**

Cawley S, Chiaromonte F, Chinwalla AT, Church DM, Clamp M, Clee C, Collins FS, Cook LL, Copley RR, Coulson A, Couronne O, Cuff J, Curwen V, Cutts T, Daly M, David R, Davies J, Delehaunty KD, Deri J, Dermitzakis ET, Dewey C, Dickens NJ, Diekhans M, Dodge S, Dubchak I, Dunn DM, Eddy SR, Elnitski L, Emes RD, Eswara P, Eyras E, Felsenfeld A, Fewell GA, Flicek P, Foley K, Frankel WN, Fulton LA, Fulton RS, Furey TS, Gage D, Gibbs RA, Glusman G, Gnerre S, Goldman N, Goodstadt L, Grafham D, Graves TA, Green ED, Gregory S, Guigo R, Guyer M, Hardison RC, Haussler D, Hayashizaki Y, Hillier LW, Hinrichs A, Hlavina W, Holzer T, Hsu F, Hua A, Hubbard T, Hunt A, Jackson I, Jaffe DB, Johnson LS, Jones M, Jones TA, Joy A, Kamal M, Karlsson EK, Karolchik D, Kasprzyk A, Kawai J, Keibler E, Kells C, Kent WJ, Kirby A, Kolbe DL, Korf I, Kucherlapati RS, Kulbokas EJ, Kulp D, Landers T, Leger JP, Leonard S, Letunic I, Levine R, Li J, Li M, Lloyd C, Lucas S, Ma B, Maglott DR, Mardis ER, Matthews L, Mauceli E, Mayer JH, McCarthy M, McCombie WR, McLaren S, McLay K, McPherson JD, Meldrim J, Meredith B, Mesirov JP, Miller W, Miner TL, Mongin E, Montgomery KT, Morgan M, Mott R, Mullikin JC, Muzny DM, Nash WE, Nelson JO, Nhan MN, Nicol R, Ning Z, Nusbaum C, O'Connor MJ, Okazaki Y, Oliver K, Overton-Larty E, Pachter L, Parra G, Pepin KH, Peterson J, Pevzner P, Plumb R, Pohl CS, Poliakov A, Ponce TC, Ponting CP, Potter S, Quail M, Reymond A, Roe BA, Roskin KM, Rubin EM, Rust AG, Santos R, Sapojnikov V, Schultz B, Schultz J, Schwartz MS, Schwartz S, Scott C, Seaman S, Searle S, Sharpe T, Sheridan A, Shownkeen R, Sims S, Singer JB, Slater G, Smit A, Smith DR, Spencer B, Stabenau A, Stange-Thomann N, Sugnet C, Suyama M, Tesler G, Thompson J, Torrents D, Trevaskis E, Tromp J, Ucla C, Ureta-Vidal A, Vinson JP, Von Niederhausern AC, Wade CM, Wall M, Weber RJ, Weiss RB, Wendl MC, West AP, Wetterstrand K, Wheeler R, Whelan S, Wierzbowski J, Willey D, Williams S, Wilson RK, Winter E, Worley KC, Wyman D, Yang S, Yang SP, Zdobnov EM, Zody MC, Lander ES. 2002. Initial sequencing and comparative analysis of the mouse genome. *Nature* 420:520–562. doi:10.1038/nature01262.

29. Gibbs RA, Weinstock GM, Metzker ML, Muzny DM, Sodergren EJ, Scherer S, Scott G, Steffen D, Worley KC, Burch PE, Okwuonu G, Hines S, Lewis L, DeRamo C, Delgado O, Dugan-Rocha S, Miner G, Morgan M, Hawes A, Gill R, Celera, Holt RA, Adams MD, Amanatides PG, Baden-Tillson H, Barnstead M, Chin S, Evans CA, Ferriera S, Fosler C, Glodek A, Gu Z, Jennings D, Kraft CL, Nguyen T, Pfannkoch CM, Sitter C, Sutton GG, Venter JC, Woodage T, Smith D, Lee HM, Gustafson E, Cahill P, Kana A, Doucette-Stamm L, Weinstock K, Fechtel K, Weiss RB, Dunn DM, Green ED, Blakesley RW, Bouffard GG, De Jong PJ, Osoegawa K, Zhu B, Marra M, Schein J, Bosdet I, Fjell C, Jones S, Krzywinski M, Mathewson C, Siddiqui A, Wye N, McPherson J, Zhao S, Fraser CM, Shetty J, Shatsman S, Geer K, Chen Y, Abramzon S, Nierman WC, Havlak PH, Chen R, Durbin KJ, Egan A, Ren Y, Song XZ, Li B, Liu Y, Qin X, Cawley S, Cooney AJ, D'Souza LM, Martin K, Wu JQ, Gonzalez-Garay ML, Jackson AR, Kalafus KJ, McLeod MP, Milosavljevic A, Virk D, Volkov A, Wheeler DA, Zhang Z, Bailey JA, Eichler EE, Tuzun E, Birney E, Mongin E, Ureta-Vidal A, Woodwark C, Zdobnov E, Bork P, Suyama M, Torrents D, Alexandersson M, Trask BJ, Young JM, Huang H, Wang H, Xing H, Daniels S, Gietzen D, Schmidt J, Stevens K, Vitt U, Wingrove J, Camara F, Mar Alba M, Abril JF, Guigo R, Smit A, Dubchak I, Rubin EM, Couronne O, Poliakov A, Hubner N, Ganten D, Goesele C, Hummel O, Kreitler T, Lee YA, Monti J, Schulz H, Zimdahl H, Himmelbauer H, Lehrach H, Jacob HJ, Bromberg S, Gullings-Handley J, Jensen-Seaman MI, Kwitek AE, Lazar J, Pasko D, Tonellato PJ, Twigger S, Ponting CP, Duarte JM, Rice S, Goodstadt L, Beatson SA, Emes RD, Winter EE, Webber C, Brandt P, Nyakatura G, Adetobi M, Chiaromonte F, Elnitski L, Eswara P, Hardison RC, Hou M, Kolbe D, Makova K, Miller W, Nekrutenko A, Riemer C, Schwartz S, Taylor J, Yang S, Zhang Y, Lindpaintner K, Andrews TD, Caccamo M, Clamp M, Clarke L, Curwen V, Durbin R, Eyras E, Searle SM, Cooper GM, Batzoglou S, Brudno M, Sidow A, Stone EA, Payseur BA, Bourque G, Lopez-Otin C, Puente XS, Chakrabarti K, Chatterji S, Dewey C, Pachter L, Bray N, Yap VB, Caspi A, Tesler G, Pevzner PA, Haussler D, Roskin KM, Baertsch R, Clawson H, Furey TS, Hinrichs AS, Karolchik D, Kent WJ, Rosenbloom KR, Trumbower H, Weirauch M, Cooper DN, Stenson PD, Ma B, Brent M, Arumugam M, Shteynberg D, Copley RR, Taylor MS, Riethman H, Mudunuri U, Peterson J, Guyer M, Felsenfeld A, Old S, Mockrin S, Collins F. 2004. Genome sequence of the Brown Norway rat yields insights into mammalian evolution. *Nature* 428: 493–521. doi:10.1038/nature02426.

30. Lindblad-Toh K, Wade CM, Mikkelsen TS, Karlsson EK, Jaffe DB, Kamal M, Clamp M, Chang JL, Kulbokas EJ 3rd, Zody MC, Mauceli E, Xie X, Breen M, Wayne RK, Ostrander EA, Ponting CP, Galibert F, Smith DR, DeJong PJ, Kirkness E, Alvarez P, Biagi T, Brockman W, Butler J, Chin CW, Cook A, Cuff J, Daly MJ, DeCaprio D, Gnerre S, Grabherr M, Kellis M, Kleber M, Bardeleben C, Goodstadt L, Heger A, Hitte C, Kim L, Koepfli KP, Parker HG, Pollinger JP, Searle SM, Sutter NB, Thomas R, Webber C, Baldwin J, Abebe A, Abouelleil A, Aftuck L, Ait-Zahra M, Aldredge T, Allen N, An P, Anderson S, Antoine C, Arachchi H, Aslam A, Ayotte L, Bachantsang P, Barry A, Bayul T, Benamara M, Berlin A, Bessette D, Blitshteyn B, Bloom T, Blye J, Boguslavskiy L, Bonnet C, Boukhgalter B, Brown A, Cahill P, Calixte N, Camarata J, Cheshatsang Y, Chu J, Citroen M, Collymore A, Cooke P, Dawoe T, Daza R, Decktor K, DeGray S, Dhargay N, Dooley K, Dorje P, Dorjee K, Dorris L, Duffey N, Dupes A, Egbiremolen O, Elong R, Falk J, Farina A, Faro S, Ferguson D, Ferreira P, Fisher S, FitzGerald M, Foley K, Foley C, Franke A, Friedrich D, Gage D, Garber M, Gearin G, Giannoukos G, Goode T, Goyette A, Graham J, Grandbois E, Gyaltsen K, Hafez N, Hagopian D, Hagos B, Hall J, Healy C, Hegarty R, Honan T, Horn A, Houde N, Hughes L, Hunnicutt L, Husby M, Jester B, Jones C, Kamat A, Kanga B, Kells C, Khazanovich D,

Kieu AC, Kisner P, Kumar M, Lance K, Landers T, Lara M, Lee W, Leger JP, Lennon N, Leuper L, LeVine S, Liu J, Liu X, Lokyitsang Y, Lokyitsang T, Lui A, Macdonald J, Major J, Marabella R, Maru K, Matthews C, McDonough S, Mehta T, Meldrim J, Melnikov A, Meneus L, Mihalev A, Mihova T, Miller K, Mittelman R, Mlenga V, Mulrain L, Munson G, Navidi A, Naylor J, Nguyen T, Nguyen N, Nguyen C, Nicol R, Norbu N, Norbu C, Novod N, Nyima T, Olandt P, O'Neill B, O'Neill K, Osman S, Oyono L, Patti C, Perrin D, Phunkhang P, Pierre F, Priest M, Rachupka A, Raghuraman S, Rameau R, Ray V, Raymond C, Rege F, Rise C, Rogers J, Rogov P, Sahalie J, Settipalli S, Sharpe T, Shea T, Sheehan M, Sherpa N, Shi J, Shih D, Sloan J, Smith C, Sparrow T, Stalker J, Stange-Thomann N, Stavropoulos S, Stone C, Stone S, Sykes S, Tchuinga P, Tenzing P, Tesfaye S, Thoulutsang D, Thoulutsang Y, Topham K, Topping I, Tsamla T, Vassiliev H, Venkataraman V, Vo A, Wangchuk T, Wangdi T, Weiand M, Wilkinson J, Wilson A, Yadav S, Yang S, Yang X, Young G, Yu Q, Zainoun J, Zembek L, Zimmer A, Lander ES. 2005. Genome sequence, comparative analysis and haplotype structure of the domestic dog. *Nature* 438:803–819. doi:10.1038/nature04338.

31. **Elsik CG, Tellam RL, Worley KC, Gibbs RA, Muzny DM, Weinstock GM, Adelson DL, Eichler EE, Elnitski L, Guigo R, Hamernik DL, Kappes SM, Lewin HA, Lynn DJ, Nicholas FW, Reymond A, Rijnkels M, Skow LC, Zdobnov EM, Schook L, Womack J, Alioto T, Antonarakis SE, Astashyn A, Chapple CE, Chen HC, Chrast J, Camara F, Ermolaeva O, Henrichsen CN, Hlavina W, Kapustin Y, Kiryutin B, Kitts P, Kokocinski F, Landrum M, Maglott D, Pruitt K, Sapojnikov V, Searle SM, Solovyev V, Souvorov A, Ucla C, Wyss C, Anzola JM, Gerlach D, Elhaik E, Graur D, Reese JT, Edgar RC, McEwan JC, Payne GM, Raison JM, Junier T, Kriventseva EV, Eyras E, Plass M, Donthu R, Larkin DM, Reecy J, Yang MQ, Chen L, Cheng Z, Chitko-McKown CG, Liu GE, Matukumalli LK, Song J, Zhu B, Bradley DG, Brinkman FS, Lau LP, Whiteside MD, Walker A, Wheeler TT, Casey T, German JB, Lemay DG, Maqbool NJ, Molenaar AJ, Seo S, Stothard P, Baldwin CL, Baxter R, Brinkmeyer-Langford CL, Brown WC, Childers CP, Connelley T, Ellis SA, Fritz K, Glass EJ, Herzig CT, Livanainen A, Lahmers KK, Bennett AK, Dickens CM, Gilbert JG, Hagen DE, Salih H, Aerts J, Caetano AR, Dalrymple B, Garcia JF, Gill CA, Hiendleder SG, Memili E, Spurlock D, Williams JL, Alexander L, Brownstein MJ, Guan L, Holt RA, Jones SJ, Marra MA, Moore R, Moore SS, Roberts A, Taniguchi M, Waterman RC, Chacko J, Chandrabose MM, Cree A, Dao MD, Dinh HH, Gabisi RA, Hines S, Hume J, Jhangiani SN, Joshi V, Kovar CL, Lewis LR, Liu YS, Lopez J, Morgan MB, Nguyen NB, Okwuonu GO, Ruiz SJ, Santibanez J, Wright RA, Buhay C, Ding Y, Dugan-Rocha S, Herdandez J, Holder M, Sabo A, Egan A, Goodell J, Wilczek-Boney K, Fowler GR, Hitchens ME, Lozado RJ, Moen C, Steffen D, Warren JT, Zhang J, Chiu R, Schein JE, Durbin KJ, Havlak P, Jiang H, Liu Y, Qin X, Ren Y, Shen Y, Song H, Bell SN, Davis C, Johnson AJ, Lee S, Nazareth LV, Patel BM, Pu LL, Vattathil S, Williams RL Jr, Curry S, Hamilton C, Sodergren E, Wheeler DA, Barris W, Bennett GL, Eggen A, Green RD, Harhay GP, Hobbs M, Jann O, Keele JW, Kent MP, Lien S, McKay SD, McWilliam S, Ratnakumar A, Schnabel RD, Smith T, Snelling WM, Sonstegard TS, Stone RT, Sugimoto Y, Takasuga A, Taylor JF, Van Tassell CP, Macneil MD, Abatepaulo AR, Abbey CA, Ahola V, Almeida IG, Amadio AF, Anatriello E, Bahadue SM, Biase FH, Boldt CR, Carroll JA, Carvalho WA, Cervelatti EP, Chacko E, Chapin JE, Cheng Y, Choi J, Colley AJ, de Campos TA, De Donato M, Santos IK, de Oliveira CJ, Deobald H, Devinoy E, Donohue KE, Dovc P, Eberlein A, Fitzsimmons CJ, Franzin AM, Garcia GR, Genini S, Gladney CJ, Grant JR, Greaser ML, Green JA, Hadsell DL, Hakimov HA, Halgren R, Harrow JL, Hart EA, Hastings N, Hernandez M, Hu ZL, Ingham A, Iso-Touru T, Jamis C, Jensen K, Kapetis D, Kerr T, Khalil SS, Khatib H, Kolbehdari D, Kumar CG, Kumar D, Leach R, Lee JC, Li C, Logan KM, Malinverni R, Marques E, Martin WF, Martins NF, Maruyama SR, Mazza R, McLean KL, Medrano JF, Moreno BT, More DD, Muntean CT, Nandakumar HP, Nogueira MF, Olsaker I, Pant SD, Panzitta F, Pastor RC, Poli MA, Poslusny N, Rachagani S, Ranganathan S, Razpet A, Riggs PK, Rincon G, Rodriguez-Osorio N, Rodriguez-Zas SL, Romero NE, Rosenwald A, Sando L, Schmutz SM, Shen L, Sherman L, Southey BR, Lutzow YS, Sweedler JV, Tammen I, Telugu BP, Urbanski JM, Utsunomiya YT, Verschoor CP, Waardenberg AJ, Wang Z, Ward R, Weikard R, Welsh TH Jr, White SN, Wilming LG, Wunderlich KR, Yang J, Zhao FQ.** 2009. The genome sequence of taurine cattle: a window to ruminant biology and evolution. *Science* **324:**522–528. doi:10.1126/science.1169588.

32. **Wade CM, Giulotto E, Sigurdsson S, Zoli M, Gnerre S, Imsland F, Lear TL, Adelson DL, Bailey E, Bellone RR, Blocker H, Distl O, Edgar RC, Garber M, Leeb T, Mauceli E, MacLeod JN, Penedo MC, Raison JM, Sharpe T, Vogel J, Andersson L, Antczak DF, Biagi T, Binns MM, Chowdhary BP, Coleman SJ, Della Valle G, Fryc S, Guerin G, Hasegawa T, Hill EW, Jurka J, Kiialainen A, Lindgren G, Liu J, Magnani E, Mickelson JR, Murray J, Nergadze SG, Onofrio R, Pedroni S, Piras MF, Raudsepp T, Rocchi M, Roed KH, Ryder OA, Searle S, Skow L, Swinburne JE, Syvanen AC, Tozaki T, Valberg SJ, Vaudin M, White JR, Zody MC, Lander ES, Lindblad-Toh K.** 2009. Genome sequence, comparative analysis, and population genetics of the domestic horse. *Science* **326:**865–867. doi:10.1126/science.1178158.

33. **Pace JK 2nd, Feschotte C.** 2007. The evolutionary history of human DNA transposons: evidence for intense activity in the primate lineage. *Genome Res* **17:**422–432. doi:10.1101/gr.5826307.

34. **Ray DA, Feschotte C, Pagan HJ, Smith JD, Pritham EJ, Arensburger P, Atkinson PW, Craig NL.** 2008. Multiple waves of recent DNA transposon activity in the bat, Myotis lucifugus. *Genome Res* **18:**717–728. doi:10.1101/gr.071886.107.

35. **Ray DA, Pagan HJ, Thompson ML, Stevens RD.** 2007. Bats with hATs: evidence for recent DNA transposon activity in genus Myotis. *Mol Biol Evol* **24:**632–639. doi:10.1093/molbev/msl192.

36. **Kapitonov VV, Jurka J.** 2005. RAG1 core and V(D)J recombination signal sequences were derived from Transib

transposons. *PLoS Biol* **3:**e181. doi:10.1371/journal.pbio.0030181.

37. **Hencken CG, Li X, Craig NL.** 2012. Functional characterization of an active Rag-like transposase. *Nat Struct Mol Biol* **19:**834–836. doi:10.1038/nsmb.2338.
38. **Casola C, Hucks D, Feschotte C.** 2008. Convergent domestication of pogo-like transposases into centromere-binding proteins in fission yeast and mammals. *Mol Biol Evol* **25:** 29–41. doi:10.1093/molbev/msm221.
39. **Pavelitz T, Gray LT, Padilla SL, Bailey AD, Weiner AM.** 2013. PGBD5: a neural-specific intron-containing piggyBac transposase domesticated over 500 million years ago and conserved from cephalochordates to humans. *Mobile DNA* **4:**23. doi:10.1186/1759-8753-4-23.
40. **Hickman AB, Chandler M, Dyda F.** 2010. Integrating prokaryotes and eukaryotes: DNA transposases in light of structure. *Crit Rev Biochem Mol Biol* **45:**50–69. doi.10.3109/10409230903505596.
41. **Yuan YW, Wessler SR.** 2011. The catalytic domain of all eukaryotic cut-and-paste transposase superfamilies. *Proc Natl Acad Sci U S A* **108:**7884–7889. doi:10.1073/pnas.1104208108.
42. **Keith JH, Schaeper CA, Fraser TS, Fraser MJ Jr.** 2008. Mutational analysis of highly conserved aspartate residues essential to the catalytic core of the piggyBac transposase. *BMC Mol Biol* **9:**73. doi:10.1186/1471-2199-9-73.
43. **Mitra R, Fain-Thornton J, Craig NL.** 2008. piggyBac can bypass DNA synthesis during cut and paste transposition. *EMBO J* **27:**1097–1109. doi:10.1038/emboj.2008.41.
44. **Bienz M.** 2006. The PHD finger, a nuclear protein-interaction domain. *Trends Biochem Sci* **31:**35–40. doi:10.1016/j.tibs.2005.11.001.
45. **Yusa K, Zhou L, Li MA, Bradley A, Craig NL.** 2011. A hyperactive piggyBac transposase for mammalian applications. *Proc Natl Acad Sci U S A* **108:**1531–1536. doi:10.1073/pnas.1008322108.
46. **Bischerour J, Chalmers R.** 2009. Base flipping in tn10 transposition: an active flip and capture mechanism. *PLoS One* **4:**e6201. doi:10.1371/journal.pone.0006201.
47. **Bischerour J, Chalmers R.** 2007. Base-flipping dynamics in a DNA hairpin processing reaction. *Nucleic Acids Res* **35:**2584–2595. doi:10.1093/nar/gkm186.
48. **Arkhipova IR, Meselson M.** 2005. Diverse DNA transposons in rotifers of the class Bdelloidea. *Proc Natl Acad Sci U S A* **102:**11781–11786. doi:10.1073/pnas.0505333102.
49. **Wang W, Lin C, Lu D, Ning Z, Cox T, Melvin D, Wang X, Bradley A, Liu P.** 2008. Chromosomal transposition of PiggyBac in mouse embryonic stem cells. *Proc Natl Acad Sci U S A* **105:**9290–9295. doi:10.1073/pnas.0801017105.
50. **Li MA, Turner DJ, Ning Z, Yusa K, Liang Q, Eckert S, Rad L, Fitzgerald TW, Craig NL, Bradley A.** 2011. Mobilization of giant piggyBac transposons in the mouse genome. *Nucleic Acids Res* **39:**e148. doi:10.1093/nar/gkr764.
51. **Li MA, Pettitt SJ, Eckert S, Ning Z, Rice S, Cadinanos J, Yusa K, Conte N, Bradley A.** 2013. The piggyBac transposon displays local and distant reintegration preferences and can cause mutations at noncanonical integration sites. *Mol Cell Biol* **33:**1317–1330. doi:10.1128/MCB.00670-12.
52. **Plasterk RH.** 1996. The Tc1/mariner transposon family. *Curr Top Microbiol Immunol* **204:**125–143.
53. **Spradling AC, Stern DM, Kiss I, Roote J, Laverty T, Rubin GM.** 1995. Gene disruptions using P transposable elements: an integral component of the Drosophila genome project. *Proc Natl Acad Sci U S A* **92:**10824–10830.
54. **Bellen HJ, O'Kane CJ, Wilson C, Grossniklaus U, Pearson RK, Gehring WJ.** 1989. P-element-mediated enhancer detection: a versatile method to study development in Drosophila. *Genes Dev* **3:**1288–1300.
55. **Handler AM.** 2002. Use of the piggyBac transposon for germ-line transformation of insects. *Insect Biochem Mol Biol* **32:**1211–1220. doi:S096517480200084X.
56. **Fraser MJ, Cary L, Boonvisudhi K, Wang HG.** 1995. Assay for movement of Lepidopteran transposon IFP2 in insect cells using a baculovirus genome as a target DNA. *Virology* **211:**397–407. doi:10.1006/viro.1995.1422
57. **Fraser MJ, Ciszczon T, Elick T, Bauser C.** 1996. Precise excision of TTAA-specific lepidopteran transposons piggyBac (IFP2) and tagalong (TFP3) from the baculovirus genome in cell lines from two species of Lepidoptera. *Insect Mol Biol* **5:**141–151.
58. **Feschotte C.** 2006. The piggyBac transposon holds promise for human gene therapy. *Proc Natl Acad Sci U S A* **103:**14981–14982. doi:0607282103.
59. **Tamura T, Thibert C, Royer C, Kanda T, Abraham E, Kamba M, Komoto N, Thomas JL, Mauchamp B, Chavancy G, Shirk P, Fraser M, Prudhomme JC, Couble P.** 2000. Germline transformation of the silkworm Bombyx mori L. using a piggyBac transposon-derived vector. *Nat Biotechnol* **18:**81–84. doi:10.1038/71978.
60. **Li X, Lobo N, Bauser CA, Fraser MJ Jr.** 2001. The minimum internal and external sequence requirements for transposition of the eukaryotic transformation vector piggyBac. *Mol Genet Genomics* **266:**190–198.
61. **Li X, Harrell RA, Handler AM, Beam T, Hennessy K, Fraser MJ Jr.** 2005. piggyBac internal sequences are necessary for efficient transformation of target genomes. *Insect Mol Biol* **14:**17–30. doi:IMB525.
62. **Ding S, Wu X, Li G, Han M, Zhuang Y, Xu T.** 2005. Efficient transposition of the piggyBac (PB) transposon in mammalian cells and mice. *Cell* **122:**473–483. doi:S0092-8674(05)00707-5.
63. **Rostovskaya M, Fu J, Obst M, Baer I, Weidlich S, Wang H, Smith AJ, Anastassiadis K, Stewart AF.** 2012. Transposon mediated BAC transgenesis in human ES cells. *Nucleic Acids Res* **40:**e150. doi:10.1093/nar/gks643.
64. **Rostovskaya M, Naumann R, Fu J, Obst M, Mueller D, Stewart AF, Anastassiadis K.** 2013. Transposon mediated BAC transgenesis via pronuclear injection of mouse zygotes. *Genesis* **51:**135–141. doi:10.1002/dvg.22362.

65. **Lacoste A, Berenshteyn F, Brivanlou AH.** 2009. An efficient and reversible transposable system for gene delivery and lineage-specific differentiation in human embryonic stem cells. *Cell Stem Cell* **5:**332–342. doi:10.1016/j.stem.2009.07.011.

66. **Cadinanos J, Bradley A.** 2007. Generation of an inducible and optimized piggyBac transposon system. *Nucleic Acids Res* **35:**e87. doi:gkm446.

67. **Doherty JE, Huye LE, Yusa K, Zhou L, Craig NL, Wilson MH.** 2012. Hyperactive piggyBac gene transfer in human cells and in vivo. *Hum Gene Ther* **23:**311–320. doi:10.1089/hum.2011.138.

68. **Li X, Burnight ER, Cooney AL, Malani N, Brady T, Sander JD, Staber J, Wheelan SJ, Joung JK, McCray PB Jr, Bushman FD, Sinn PL, Craig NL.** 2013. piggyBac transposase tools for genome engineering. *Proc Natl Acad Sci U S A* **110:**E2279–E2287. doi:10.1073/pnas.1305987110.

69. **Feil R, Wagner J, Metzger D, Chambon P.** 1997. Regulation of Cre recombinase activity by mutated estrogen receptor ligand-binding domains. *Biochem Biophys Res Commun* **237:**752–757. doi:10.1006/bbrc.1997.7124.

70. **Wu SC, Meir YJ, Coates CJ, Handler AM, Pelczar P, Moisyadi S, Kaminski JM.** 2006. piggyBac is a flexible and highly active transposon as compared to sleeping beauty, Tol2, and Mos1 in mammalian cells. *Proc Natl Acad Sci U S A* **103:**15008–15013. doi:0606979103.

71. **Cuypers HT, Selten G, Quint W, Zijlstra M, Maandag ER, Boelens W, van Wezenbeek P, Melief C, Berns A.** 1984. Murine leukemia virus-induced T-cell lymphomagenesis: integration of proviruses in a distinct chromosomal region. *Cell* **37:**141–150.

72. **Kool J, Uren AG, Martins CP, Sie D, de Ridder J, Turner G, van Uitert M, Matentzoglu K, Lagcher W, Krimpenfort P, Gadiot J, Pritchard C, Lenz J, Lund AH, Jonkers J, Rogers J, Adams DJ, Wessels L, Berns A, van Lohuizen M.** 2010. Insertional mutagenesis in mice deficient for p15Ink4b, p16Ink4a, p21Cip1, and p27Kip1 reveals cancer gene interactions and correlations with tumor phenotypes. *Cancer Res* **70:**520–531. doi:10.1158/0008-5472.CAN-09-2736.

73. **Peters G, Brookes S, Smith R, Dickson C.** 1983. Tumorigenesis by mouse mammary tumor virus: evidence for a common region for provirus integration in mammary tumors. *Cell* **33:**369–377.

74. **Ivics Z, Hackett PB, Plasterk RH, Izsvak Z.** 1997. Molecular reconstruction of Sleeping Beauty, a Tc1-like transposon from fish, and its transposition in human cells. *Cell* **91:**501–510.

75. **Dupuy AJ, Akagi K, Largaespada DA, Copeland NG, Jenkins NA.** 2005. Mammalian mutagenesis using a highly mobile somatic Sleeping Beauty transposon system. *Nature* **436:**221–226. doi:10.1038/nature03691.

76. **Rad R, Rad L, Wang W, Cadinanos J, Vassiliou G, Rice S, Campos LS, Yusa K, Banerjee R, Li MA, de la Rosa J, Strong A, Lu D, Ellis P, Conte N, Yang FT, Liu P, Bradley A.** 2010. PiggyBac transposon mutagenesis: a tool for cancer gene discovery in mice. *Science* **330:**1104–1107. doi:10.1126/science.1193004.

77. **Collier LS, Carlson CM, Ravimohan S, Dupuy AJ, Largaespada DA.** 2005. Cancer gene discovery in solid tumours using transposon-based somatic mutagenesis in the mouse. *Nature* **436:**272–276. doi:10.1038/nature03681.

78. **Balu B, Adams JH.** 2006. Functional genomics of Plasmodium falciparum through transposon-mediated mutagenesis. *Cell Microbiol* **8:**1529–1536. doi:10.1111/j.1462-5822.2006.00776.x.

79. **Balu B, Shoue DA, Fraser MJ Jr, Adams JH.** 2005. High-efficiency transformation of Plasmodium falciparum by the lepidopteran transposable element piggyBac. *Proc Natl Acad Sci U S A* **102:**16391–16396. doi:0504679102.

80. **Ikadai H, Shaw Saliba K, Kanzok SM, McLean KJ, Tanaka TQ, Cao J, Williamson KC, Jacobs-Lorena M.** 2013. Transposon mutagenesis identifies genes essential for Plasmodium falciparum gametocytogenesis. *Proc Natl Acad Sci U S A* **110:**E1676–E1684. doi:10.1073/pnas.1217712110.

81. **Kuwayama H, Yaginuma T, Yamashita O, Niimi T.** 2006. Germ-line transformation and RNAi of the ladybird beetle, Harmonia axyridis. *Insect Mol Biol* **15:**507–512. doi:10.1111/j.1365-2583.2006.00665.x.

82. **Lorenzen MD, Berghammer AJ, Brown SJ, Denell RE, Klingler M, Beeman RW.** 2003. piggyBac-mediated germline transformation in the beetle Tribolium castaneum. *Insect Mol Biol* **12:**433–440. doi:427.

83. **Lobo N, Li X, Hua-Van A, Fraser MJ Jr.** 2001. Mobility of the piggyBac transposon in embryos of the vectors of Dengue fever (Aedes albopictus) and La Crosse encephalitis (Ae. triseriatus). *Mol Genet Genomics* **265:**66–71.

84. **Lobo N, Li X, Fraser MJ Jr.** 1999. Transposition of the piggyBac element in embryos of Drosophila melanogaster, Aedes aegypti and Trichoplusia ni. *Mol Gen Genet* **261:**803–810.

85. **Rodrigues FG, Oliveira SB, Rocha BC, Moreira LA.** 2006. Germline transformation of Aedes fluviatilis (Diptera:Culicidae) with the piggyBac transposable element. *Mem Inst Oswaldo Cruz* **101:**755–757. doi:S0074-02762006000700008.

86. **Condon KC, Condon GC, Dafa'alla TH, Forrester OT, Phillips CE, Scaife S, Alphey L.** 2007. Germ-line transformation of the Mexican fruit fly. *Insect Mol Biol* **16:**573–580. doi:10.1111/j.1365-2583.2007.00752.x.

87. **Perera OP, Harrell IR, Handler AM.** 2002. Germ-line transformation of the South American malaria vector, Anopheles albimanus, with a piggyBac/EGFP transposon vector is routine and highly efficient. *Insect Mol Biol* **11:**291–297. doi:336.

88. **Grossman GL, Rafferty CS, Clayton JR, Stevens TK, Mukabayire O, Benedict MQ.** 2001. Germline transformation of the malaria vector, Anopheles gambiae, with the piggyBac transposable element. *Insect Mol Biol* **10:**597–604. doi:299.

89. **Grossman GL, Rafferty CS, Fraser MJ, Benedict MQ.** 2000. The piggyBac element is capable of precise excision and transposition in cells and embryos of the mosquito, Anopheles gambiae. *Insect Biochem Mol Biol* **30:**909–914. doi:S0965174800000928.

90. **Nolan T, Bower TM, Brown AE, Crisanti A, Catteruccia F.** 2002. piggyBac-mediated germline transformation of the malaria mosquito Anopheles stephensi using the red fluorescent protein dsRED as a selectable marker. *J Biol Chem* **277:**8759–8762. doi:10.1074/jbc.C100766200.

91. **Raphael KA, Shearman DC, Streamer K, Morrow JL, Handler AM, Frommer M.** 2011. Germ-line transformation of the Queensland fruit fly, Bactrocera tryoni, using a piggyBac vector in the presence of endogenous piggyBac elements. *Genetica* **139:**91–97. doi:10.1007/s10709-010-9500-x.

92. **Schetelig MF, Handler AM.** 2013. Germline transformation of the spotted wing drosophilid, Drosophila suzukii, with a piggyBac transposon vector. *Genetica* **141:**189–193. doi:10.1007/s10709-013-9717-6.

93. **Hediger M, Niessen M, Wimmer EA, Dubendorfer A, Bopp D.** 2001. Genetic transformation of the housefly Musca domestica with the lepidopteran derived transposon piggyBac. *Insect Mol Biol* **10:**113–119. doi:imb243.

94. **Warren IA, Fowler K, Smith H.** 2010. Germline transformation of the stalk-eyed fly, Teleopsis dalmanni. *BMC Mol Biol* **11:**86. doi:10.1186/1471-2199-11-86.

95. **Sumitani M, Yamamoto DS, Oishi K, Lee JM, Hatakeyama M.** 2003. Germline transformation of the sawfly, Athalia rosae (Hymenoptera: Symphyta), mediated by a piggyBac-derived vector. *Insect Biochem Mol Biol* **33:**449–458. doi:S0965174803000092.

96. **Marcus JM, Ramos DM, Monteiro A.** 2004. Germline transformation of the butterfly Bicyclus anynana. *Proc Biol Sci / The Royal Society* **271**(Suppl 5)**:**S263–S265. doi:10.1098/rsbl.2004.0175.

97. **Ferguson HJ, Neven LG, Thibault ST, Mohammed A, Fraser M.** 2011. Genetic transformation of the codling moth, Cydia pomonella L., with piggyBac EGFP. *Transgenic Res* **20:**201–214. doi:10.1007/s11248-010-9391-8.

98. **Ren X, Han Z, Miller TA.** 2006. Excision and transposition of piggyBac transposable element in tobacco budworm embryos. *Arch Insect Biochem Physiol* **63:**49–56. doi:10.1002/arch.20140.

99. **Mandrioli M, Wimmer EA.** 2003. Stable transformation of a Mamestra brassicae (lepidoptera) cell line with the lepidopteran-derived transposon piggyBac. *Insect Biochem Mol Biol* **33:**1–5. doi:S0965174802001893.

100. **Liu D, Yan S, Huang Y, Tan A, Stanley DW, Song Q.** 2012. Genetic transformation mediated by piggyBac in the Asian corn borer, Ostrinia furnacalis (Lepidoptera: Crambidae). *Arch Insect Biochem Physiol* **80:**140–150. doi:10.1002/arch.21035.

101. **Thibault ST, Luu HT, Vann N, Miller TA.** 1999. Precise excision and transposition of piggyBac in pink bollworm embryos. *Insect Mol Biol* **8:**119–123.

102. **Mohammed A, Coates CJ.** 2004. Promoter and piggyBac activities within embryos of the potato tuber moth, Phthorimaea operculella, Zeller (Lepidoptera: Gelechiidae). *Gene* **342:**293–301. doi:S0378-1119(04)00492-5.

103. **Martins S, Naish N, Walker AS, Morrison NI, Scaife S, Fu G, Dafa'alla T, Alphey L.** 2012. Germline transformation of the diamondback moth, Plutella xylostella L., using the piggyBac transposable element. *Insect Mol Biol* **21:**414–421. doi:10.1111/j.1365-2583.2012.01146.x.

104. **Shinmyo Y, Mito T, Matsushita T, Sarashina I, Miyawaki K, Ohuchi H, Noji S.** 2004. piggyBac-mediated somatic transformation of the two-spotted cricket, Gryllus bimaculatus. *Dev Growth Differ* **46:**343–349. doi:10.1111/j.1440-169x.2004.00751.x.

105. **Li J, Zhang JM, Li X, Suo F, Zhang MJ, Hou W, Han J, Du LL.** 2011. A piggyBac transposon-based mutagenesis system for the fission yeast Schizosaccharomyces pombe. *Nucleic Acids Res* **39:**e40. doi:10.1093/nar/gkq1358.

106. **Morales ME, Mann VH, Kines KJ, Gobert GN, Fraser MJ Jr, Kalinna BH, Correnti JM, Pearce EJ, Brindley PJ.** 2007. piggyBac transposon mediated transgenesis of the human blood fluke, Schistosoma mansoni. *FASEB J* **21:**3479–3489. doi:fj.07-8726com.

107. **Fonager J, Franke-Fayard BM, Adams JH, Ramesar J, Klop O, Khan SM, Janse CJ, Waters AP.** 2011. Development of the piggyBac transposable system for Plasmodium berghei and its application for random mutagenesis in malaria parasites. *BMC Genomics* **12:**155. doi:10.1186/1471-2164-12-155.

108. **Su H, Liu X, Yan W, Shi T, Zhao X, Blake DP, Tomley FM, Suo X.** 2012. piggyBac transposon-mediated transgenesis in the apicomplexan parasite Eimeria tenella. *PLoS One* **7:**e40075. doi:10.1371/journal.pone.0040075.

109. **Shao H, Li X, Nolan TJ, Massey HC Jr, Pearce EJ, Lok JB.** 2012. Transposon-mediated chromosomal integration of transgenes in the parasitic nematode Strongyloides ratti and establishment of stable transgenic lines. *PLoS Pathog* **8:**e1002871. doi:10.1371/journal.ppat.1002871.

110. **Gonzalez-Estevez C, Momose T, Gehring WJ, Salo E.** 2003. Transgenic planarian lines obtained by electroporation using transposon-derived vectors and an eye-specific GFP marker. *Proc Natl Acad Sci U S A* **100:**14046–14051. doi:10.1073/pnas.2335980100.

111. **Lobo NF, Fraser TS, Adams JA, Fraser MJ Jr.** 2006. Interplasmid transposition demonstrates piggyBac mobility in vertebrate species. *Genetica* **128:**347–357. doi:10.1007/s10709-006-7165-2.

112. **Lu Y, Lin C, Wang X.** 2009. PiggyBac transgenic strategies in the developing chicken spinal cord. *Nucleic Acids Res* **37:**e141. doi:10.1093/nar/gkp686.

113. **Park TS, Han JY.** 2012. piggyBac transposition into primordial germ cells is an efficient tool for transgenesis in chickens. *Proc Natl Acad Sci U S A* **109:**9337–9341. doi:10.1073/pnas.1203823109.

114. **Liu X, Li N, Hu X, Zhang R, Li Q, Cao D, Liu T, Zhang Y.** 2013. Efficient production of transgenic chickens based on piggyBac. *Transgenic Res* **22:**417–423. doi:10.1007/s11248-012-9642-y.

115. **Macdonald J, Taylor L, Sherman A, Kawakami K, Takahashi Y, Sang HM, McGrew MJ.** 2012. Efficient genetic modification and germ-line transmission of primordial germ cells using piggyBac and Tol2 transposons. *Proc Natl Acad Sci U S A* **109:**E1466–E1472. doi:10.1073/pnas.1118715109.

116. **Jang CW, Behringer RR.** 2007. Transposon-mediated transgenesis in rats. *CSH protocols* **2007**:pdb prot4866. doi:10.1101/pdb.prot4866.

117. **Chen F, LoTurco J.** 2012. A method for stable transgenesis of radial glia lineage in rat neocortex by piggyBac mediated transposition. *J Neurosci Methods* **207**:172–180. doi:10.1016/j.jneumeth.2012.03.016.

118. **Bai DP, Yang MM, Chen YL.** 2012. PiggyBac transposon-mediated gene transfer in Cashmere goat fetal fibroblast cells. *Biosci Biotechnol Biochem* **76**:933–937. doi:DN/JST.JSTAGE/bbb/110939.

119. **Wu Z, Xu Z, Zou X, Zeng F, Shi J, Liu D, Urschitz J, Moisyadi S, Li Z.** 2013. Pig transgenesis by piggyBac transposition in combination with somatic cell nuclear transfer. *Transgenic Res* **22**:1107–1118. doi:10.1007/s11248-013-9729-0.

120. **Nagy K, Sung HK, Zhang P, Laflamme S, Vincent P, Agha-Mohammadi S, Woltjen K, Monetti C, Michael IP, Smith LC, Nagy A.** 2011. Induced pluripotent stem cell lines derived from equine fibroblasts. *Stem Cell Rev* **7**:693–702. doi:10.1007/s12015-011-9239-5.

121. **Xue X, Huang X, Nodland SE, Mates L, Ma L, Izsvak Z, Ivics Z, LeBien TW, McIvor RS, Wagner JE, Zhou X.** 2009. Stable gene transfer and expression in cord blood-derived CD34+ hematopoietic stem and progenitor cells by a hyperactive Sleeping Beauty transposon system. *Blood* **114**:1319–1330. doi:10.1182/blood-2009-03-210005.

122. **Galvan DL, Nakazawa Y, Kaja A, Kettlun C, Cooper LJ, Rooney CM, Wilson MH.** 2009. Genome-wide mapping of PiggyBac transposon integrations in primary human T cells. *J Immunother* **32**:837–844. doi:10.1097/CJI.0b013e3181b2914c.

123. **Yusa K, Rashid ST, Strick-Marchand H, Varela I, Liu PQ, Paschon DE, Miranda E, Ordonez A, Hannan NR, Rouhani FJ, Darche S, Alexander G, Marciniak SJ, Fusaki N, Hasegawa M, Holmes MC, Di Santo JP, Lomas DA, Bradley A, Vallier L.** 2011. Targeted gene correction of alpha1-antitrypsin deficiency in induced pluripotent stem cells. *Nature* **478**:391–394. doi:10.1038/nature10424.

124. **Nishizawa-Yokoi A, Endo M, Osakabe K, Saika H, Toki S.** 2014. Precise marker excision system using an animal-derived piggyBac transposon in plants. *Plant J* **77**:454–463. doi:10.1111/tpj.12367.

125. **Trauner J, Schinko J, Lorenzen MD, Shippy TD, Wimmer EA, Beeman RW, Klingler M, Bucher G, Brown SJ.** 2009. Large-scale insertional mutagenesis of a coleopteran stored grain pest, the red flour beetle Tribolium castaneum, identifies embryonic lethal mutations and enhancer traps. *BMC Biol* **7**:73. doi:10.1186/1741-7007-7-73.

126. **Lorenzen MD, Kimzey T, Shippy TD, Brown SJ, Denell RE, Beeman RW.** 2007. piggyBac-based insertional mutagenesis in Tribolium castaneum using donor/helper hybrids. *Insect Mol Biol* **16**:265–275. doi:IMB727.

127. **Mathieu J, Sung HH, Pugieux C, Soetaert J, Rorth P.** 2007. A sensitized PiggyBac-based screen for regulators of border cell migration in Drosophila. *Genetics* **176**:1579–1590. doi:genetics.107.071282.

128. **Hacker U, Nystedt S, Barmchi MP, Horn C, Wimmer EA.** 2003. piggyBac-based insertional mutagenesis in the presence of stably integrated P elements in Drosophila. *Proc Natl Acad Sci U S A* **100**:7720–7725. doi:10.1073/pnas.1230526100.

129. **Bonin CP, Mann RS.** 2004. A piggyBac transposon gene trap for the analysis of gene expression and function in Drosophila. *Genetics* **167**:1801–1811. doi:10.1534/genetics.104.027557.

130. **Ni TK, Landrette SF, Bjornson RD, Bosenberg MW, Xu T.** 2013. Low-copy piggyBac transposon mutagenesis in mice identifies genes driving melanoma. *Proc Natl Acad Sci U S A* **110**:E3640–E3649. doi:10.1073/pnas.1314435110.

131. **Landrette SF, Cornett JC, Ni TK, Bosenberg MW, Xu T.** 2011. piggyBac transposon somatic mutagenesis with an activated reporter and tracker (PB-SMART) for genetic screens in mice. *PLoS One* **6**:e26650. doi:10.1371/journal.pone.0026650.

132. **Pettitt SJ, Rehman FL, Bajrami I, Brough R, Wallberg F, Kozarewa I, Fenwick K, Assiotis I, Chen L, Campbell J, Lord CJ, Ashworth A.** 2013. A genetic screen using the PiggyBac transposon in haploid cells identifies Parp1 as a mediator of olaparib toxicity. *PLoS One* **8**:e61520. doi:10.1371/journal.pone.0061520.

133. **Wang W, Hale C, Goulding D, Haslam SM, Tissot B, Lindsay C, Michell S, Titball R, Yu J, Toribio AL, Rossi R, Dell A, Bradley A, Dougan G.** 2011. Mannosidase 2, alpha 1 deficiency is associated with ricin resistance in embryonic stem (ES) cells. *PLoS One* **6**:e22993. doi:10.1371/journal.pone.0022993.

134. **Wang W, Bradley A, Huang Y.** 2009. A piggyBac transposon-based genome-wide library of insertionally mutated Blm-deficient murine ES cells. *Genome Res* **19**:667–673. doi:10.1101/gr.085621.108.

135. **Balu B, Chauhan C, Maher SP, Shoue DA, Kissinger JC, Fraser MJ Jr, Adams JH.** 2009. piggyBac is an effective tool for functional analysis of the Plasmodium falciparum genome. *BMC Microbiol* **9**:83. doi:10.1186/1471-2180-9-83.

136. **Vos JC, De Baere I, Plasterk RHA.** 1996. Transposase is the only nematode protein required for in vitro transposition of Tc1. *Gene Dev* **10**:755–761. doi: 10.1101/gad.10.6.755.

Mobile DNA, 3rd Edition
Nancy L. Craig, Michael Chandler, Martin Gellert, Alan M. Lambowitz, Phoebe A. Rice and Suzanne Sandmeyer

doi:10.1128/microbiolspec.MDNA3-0049-2014

Jainy Thomas[1]
Ellen J. Pritham[1]

40

Helitrons, the Eukaryotic Rolling-circle Transposable Elements

INTRODUCTION

Helitrons are one of three groups of eukaryotic class 2 transposable elements (TEs) so far described. Unique in structure and coding capacity, they are hypothesized to move by a rolling-circle-like replication mechanism via a single-stranded DNA intermediate (1, 2). The other two groups, the classic cut-and-paste and the *Maverick/Polinton* (3–5) both encode a transposase/integrase and are flanked by target site duplications (TSDs) (for review see reference 6). The repair resulting from the staggered double-stranded joining of the TE to the target DNA creates the TSD flanking the insertion (for reviews see references 7 and 8). *Helitrons* encode a putative protein called the Rep/Helicase (1), which is predicted to have both HUH endonuclease activity (for review see reference 9) and 5′ to 3′ helicase activity. The HUH endonuclease (the Rep of the Rep/Helicase) (Figure 1) would likely make a single-stranded nick in the host DNA, which is consistent with the lack of TSD observed flanking *Helitron* insertions. A related protein with an HUH endonuclease domain encoded by various bacterial Insertion Sequence families (IS*608*, IS*91*, and IS*CR1*) makes a single-stranded nick in the host DNA and the insertions are not flanked by TSDs (for review see reference 9).

Helitrons were the first group of TEs to be discovered by computational analysis of whole genome sequences (1). Homology of proteins encoded by *Helitrons* in concert with the conservation of critical residues in the Rep portion of the Rep/Helicase to that of entities that are proposed to undergo rolling-circle replication (e.g., IS*91*, single-stranded DNA viruses, and plasmids) led to the development of a transposition model (1, 2). The discovery of *Helitrons* marked a new era of computational-based TE discovery, which has gone hand in hand with a shift from studying single elements or families to the study of populations and/or entire genome complements. Computational studies have revolutionized what we know about the composition of genomes, and the distribution and life cycle of TEs. In addition, sequence analysis and comparative genomics have allowed substantial hypothesis generation. However, biochemical and *in vivo* studies of mechanism have not kept pace. Indeed, 13 years after the description of putatively autonomous *Helitrons* no mechanistic studies have been published. Hence, the

[1]Department of Human Genetics, University of Utah, Salt Lake City, UT 84112.

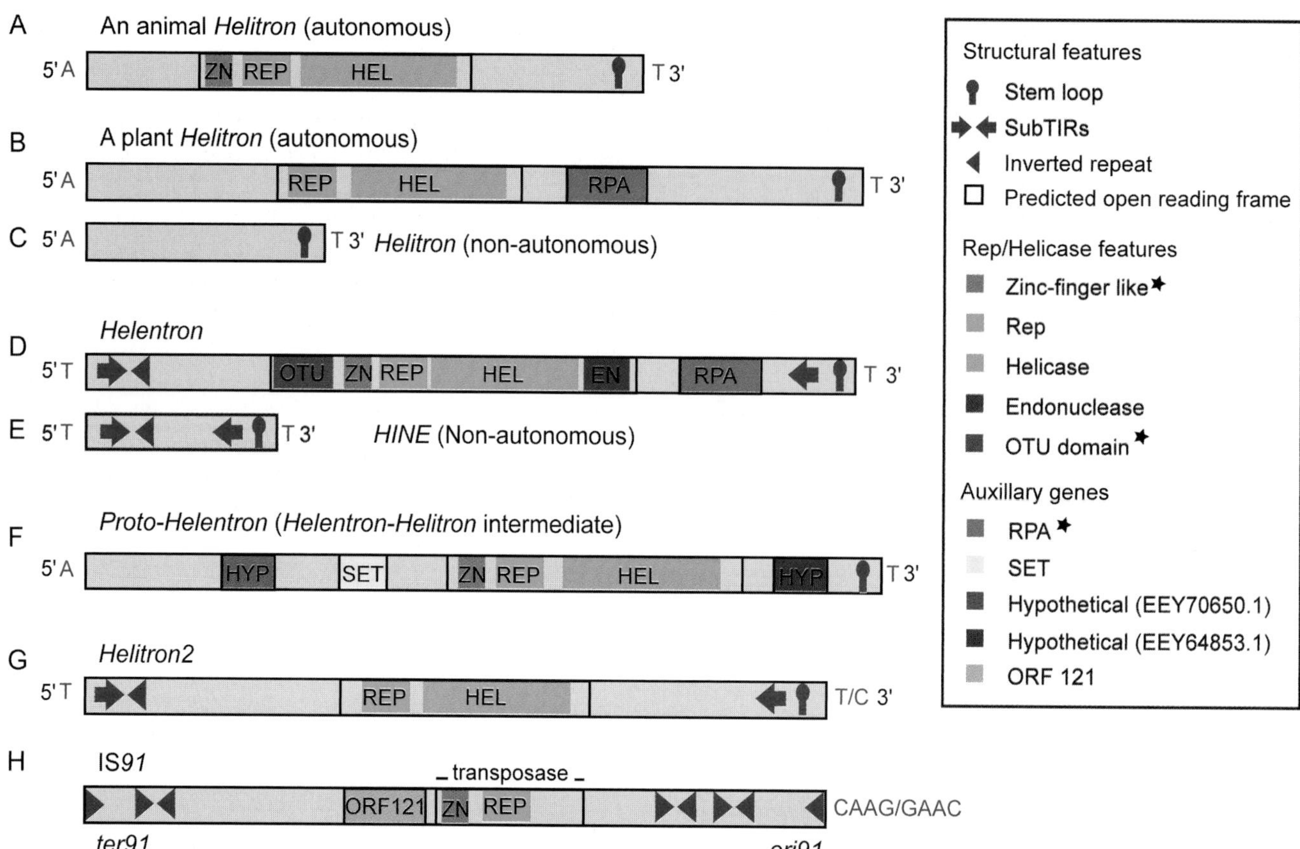

Figure 1 Structure and coding capacity of canonical animal and plant *Helitrons*, *Helentrons*, *Proto-Helentron*, *Helitron2* and IS*91*. (A) Structure of a typical animal *Helitron*. (B) A typical plant *Helitron* encoding Rep/Helicase and RPA proteins; one to three RPA genes can be found on either side of the Rep/Helicase gene. (C) Structure of a typical nonautonomous plant or animal *Helitron*; they do not encode the Rep/Helicase gene but share the common structural features. (D) Structure and coding capacity of a *Helentron*; *Helentrons* have sub terminal inverted repeats (subTIRs) (red), and a short palindrome at the 3′ end (stem loop). The subTIRs can either be palindromic or form a palindrome with the short inverted repeats (sideways triangle), near the subTIR, if present. (E) Structure of a *Helentron*-associated INterspersed Element (*HINE*); *HINE*s are nonautonomous but have the same structural features as that of the autonomous partner. (F) Structure and coding capacity of *Proto-Helentron*. (G) Structure and coding capacity of *Helitron2* (redrawn from reference 12). (H) Structure of a bacterial IS*91* element that is proposed to transpose by rolling-circle mechanism (redrawn from references 9, 46). The genes that are occasionally carried by *Helentrons* are indicated with a black asterisk (*) and are included only if they were found in multiple families or across species. Sequences flanking the elements are shown in red. doi:10.1128/microbiolspec.MDNA3-0049-2014.f1

rolling-circle mechanism of transposition remains a well-supported but not yet tested hypothesis. The lack of such studies has resulted in a large gap in our knowledge of *Helitron* biology.

In addition to canonical *Helitrons* (1), multiple structural and coding variants can be identified in eukaryotes. One variant called *Helentron* (10, 11) encodes an endonuclease containing the Rep/Helicase protein (detailed in later sections, Figure 1). Both *Helitrons* and *Helentrons* have given rise to thousands of nonautonomous elements, some of which carry host genes and/or fragments sometimes picked up from multiple places in the genome. Nonautonomous elements generally outnumber the protein-coding counterparts, allowing them to be more easily recognized in the genome. Other variants include the *Proto-Helentron* and *Helitron2* and do not encode the endonuclease (11). The *Proto-Helentron* shares the structure of canonical *Helitrons,* while

Helitron2 has the structure of a *Helentron*. Their Rep/Helicase proteins cluster in a well supported phylogenetic group with the *Helentron* proteins (11, 12) (detailed in the later sections, Figure 1). It is not clear if these families are true evolutionary intermediates, as they are known only from a limited group of organisms, or if they are families that simply lost coding potential for the endonuclease (11).

The purpose of this review is to summarize what we have learned about *Helitrons* in the decade since their discovery. First, we describe the history of autonomous *Helitrons*, *Helentrons*, and nonautonomous partner families. Second, we explain the common coding features and difference in structure of canonical *Helitrons* versus the endonuclease-encoding *Helentrons*. Third, we review how *Helitrons* and *Helentrons* are classified and discuss why the system used for other TE families is not applicable. We also touch upon how genome-wide identification and verification of candidate *Helitrons* is carried out. We then shift our focus to a model of transposition and the report of an excision event. We discuss the different proposed models for the mechanism of gene capture. Finally we will talk about where *Helitrons* are found, including discussions of vertical versus horizontal transfer and the propensity of *Helitrons* and *Helentrons* to capture and shuffle genes and how they impact the genome. We will end the review with a summary of open questions concerning the biology of this intriguing group of TEs.

HISTORY

The era of genome sequencing revolutionized the identification of TE families making genome-wide analysis a standard for discovery and classification (for review see reference 6). Nonautonomous *Helitrons* called Aie, AthE1, Atrep, and Basho (Table 1) were first described from the genome of *Arabidopsis thaliana* (13–16). These were the first *Helitrons* described; however, the classification was unknown until the discovery of protein coding-elements predicted to be the autonomous partners in 2001 (1). In this landmark paper, the structure and coding potential of canonical *Helitrons* were described (1). The rolling-circle mechanism of transposition was proposed as well as the possibility that some of the encoded genes captured from the host are now used for replication (1, 2). In addition, to describing the coding capacity of *Helitrons* in *A. thaliana*, *Oryza sativa*, and *Caenorhabditis elegans*, the connection to previously described nonautonomous families was forged (Table 1). These genome surveys revealed that *Helitron* activity could contribute to a significant fraction (~ 2%) of the plant and invertebrate genomes where they were found (1), but the extent of their distribution elsewhere was not clear.

In 2003, proteins related to the *Helitron* Rep/Helicase were described from *Danio rerio* and *Sphoeroides nephelus* (10) with an interesting twist. The Rep/Helicase proteins were predicted to be 500 to 700 amino acids longer because of a C-terminal fusion of a domain with homology to apurinic-apyrimidinic (AP) endonuclease. Phylogenetic analysis revealed that the AP endonuclease nested within the Chicken Repeat 1 (CR1) clade (17, 18) of non-long terminal repeat (non-LTR) retrotransposons (10). This relationship of the AP endonuclease to non-LTR retrotransposons suggested that it originated from a retrotransposon insertion either nearby or within a *Helitron* (10, 11). The authors were not able to identify the ends of the Rep/Helicase/Endonuclease unit, so the structure was not reported. They called these new *Helitron* relatives *Helentrons* (10).

The structure of *Helentrons* was a mystery until recently and is indeed completely different from that of the canonical *Helitrons* (11) (Figure 1). The description of the structure allowed a connection to be established between previously identified but unclassified nonautonomous families (11) (Figure 1) including *DINE-1*/*DNAREP1*, *INE-1*, *SGM*, *ISY*, *IS-amb*, *IS-gua*, *IS-sub*, *mini-me*, *GEM*, *PERI*, a part of S812, *ISBu*, *cDK27*, *MINE-1*,*MINE-2*, *Novel_NA*, *SGM*, *MgE*, *DTC84*, *Tsp*, and *pearl*, from *Drosophila*, other insects, and invertebrates (19–40) (Table 1). The nonautonomous elements associated with *Helentrons* are called *Helentron*-associated INterspersed Elements (*HINEs*) (11) (Figure 1).

STRUCTURE AND CODING CAPACITY

Helitrons and the endonuclease-encoding *Helentrons* are both structurally asymmetric and do not create TSD upon integration (Figure 1). The 5′ end of *Helitrons* and *Helentrons* was initially defined based on the orientation of the Rep/Helicase gene. However, the structures are different (Figure 1A–E). The canonical *Helitron* and *Proto-Helentron* typically begin with a 5′ T (C/T) and terminate with the nucleotides CTRR (most frequently CTAG, but occasionally variation has been noted) (Figure 1A–C and F). In addition, they frequently have a short palindromic sequence (16 to 20 nucleotides) about ~ 11 bp from the 3′ end. They integrate between an AT host dinucleotide (1, 2, 11).

Helentrons have inverted repeats (12 to 15 bp) subterminal to each end (subTIRs) (Figure 1D). The

Table 1 The classification of nonautonomous families as *Helitron* or *Helentron*

Name	Organism where first described	Autonomous (A)/ Non-autonomous (NA)	Structure described	*Helitron*/ *Helentron*	Citation
Aie	*Arabidopsis thaliana*	NA	partial	*Helitron*	13
AthE1	*A. thaliana*	NA	yes	*Helitron*	14
Atrep	*A. thaliana*	NA	–	*Helitron*	15
Basho	*A. thaliana*	NA	yes	*Helitron*	16
DINE-1/ DNAREP1/INE-1	*Drosophila melanogaster*	NA	yes	*Helentron*	19, 20, 38
Helitron	*A. thaliana, Oryza sativa, Caenorhabditis elegans*	A	yes	*Helitron*	1
Helentron	*Danio rerio, Sphoeroides neghelus*	A	no	*Helentron*	10
Helentron	*Metaseiulus occidentalis, Culex quinquefasciatus, D. ananassae, D. willistoni, D. yakuba*	A	yes	*Helentron*	11
Helitron2	*Clamydomonas reinhardtii*	A	yes	*Helentron* (Intermediate/ deletion derivative)	12
DINE-1	*D. willistoni, D. melanogaster, D. simulans, D. sechellia, D. grimshawi, D. yakuba, D. ananassae, D. mojavensis, D. erecta, D. persimilis, D. pseudoobscura, D. virilis*	NA	yes	*Helentron*	39
SGM	*D. subobscura, D. guanche, D. madeirensis*	NA	yes	*Helentron*	22
ISY	*D. miranda, D. pseudoobscura*	NA	partial	*Helentron*	23
IS-amb, IS-gua, IS-sub	*D. ambigua, D. guanche, D. subobscura*	NA	yes	*Helentron*	21
mini-me	*D. nigrodunni, D. dunni*	NA	yes	*Helentron*	24
A part of S812	*D. subobscura*	NA	partial	*Helentron*	40
GEM	*D. subobscura*	NA	partial	*Helentron*	25
PERI, pSsp400	*D. buzzatti, D. koepferae, D. serido, D. borborema, D. seriema, D. antonietae, D. gouveai*	NA	yes	*Helentron*	26
ISBu	*D. buzzatti*	NA	No	*Helentron*	27
cDk27	*D. buzzatti, D. koepferae, D. serido, D. borborema, D. starmeri, D. venezolana, D. uniseta, D. martensis*	NA	partial	*Helentron*	30
MINE-1	*Bombyx mori, Pectinophora gossypiella, Ostrinia nubilalis*	NA	yes	*Helentron*	32
MINE-2	*Bicyclus anynana, Helicoverpa armigera, Heliconius melpomene, Heliconius numata, Spodoptera frugiperda*	NA	yes	*Helentron*	31
NOVEL_NA	*Rhodnius prolixus, Bombyx mori, Danaus plexippus*	NA	yes	*Helentron*	33
MgE	*Mytilus galloprovincialis*	NA	yes	*Helentron*	34
Pearl	*Crassostrea virginica, Anadara trapezia*	NA	yes	*Helentron*	35
DTC84	*Donax trunculus*	NA	yes	*Helentron*	36
Tsp	*Strongylocentrotus purpuratus*	NA	yes	*Helentron*	37

5′ subTIR is palindromic with proximal sequence. In some families, a short inverted repeat occurs approximately 3 to 22 bp away from the 5′ subTIR (39) whereas in other families the subTIRs are themselves palindromic (11). Like *Helitrons*, *Helentrons* also have an additional palindromic sequence, which is located near the 3′ end of the element. They integrate between a TT host dinucleotide, frequently located in a stretch of T nucleotides. Often a different number of T nucleotides are found when the termini of related elements are compared (11). The *Helitron2* elements integrate between TT/C dinucleotides and have a structure similar to that of the *Helentrons* with short inverted repeats (12) (Figure 1G).

IS*91* and related elements (e.g., IS*CR*) that are also predicted to use a rolling-circle mechanism of replication (41–44) (for reviews see references 9 and 45) (Figure 1H) share some structural features with *Helitrons* and *Helentrons* (Figure 1D and E). IS*91* elements have short interrupted inverted repeats at the termini (for review see reference 46). Experimental evidence has pinpointed the origin of the RC replication to the right terminus of the IS*91* elements (*ori91*), and termination has been shown to occur at the left (*ter91*). The target sequence (C/GTTC/G) is adjacent to the *ori91* (for review see reference 46). The terminal AT-rich 23-bp sequences within the *ori91* are marked by the presence of a 9-bp inverted repeat. A second inverted repeat lies adjacent to it and is also functional and necessary for transposition (for review see reference 46). Within the *ter91*, there is a 6-bp inverted repeat within a 17-bp GC-rich sequence. These inverted repeats can form short palindromes (46). It is interesting to note that the structure of IS*91* elements (44) (for reviews see references 46 and 47) is more similar to that of *Helentrons* described from the *Metaseiulus occidentalis*, *Culex quinquefasciatus*, and *Drosophila* (one family) (Figure 1) (11). These *Helentrons* have palindromic subTIRs in addition to a palindrome at the 3′ end, and so have three sets of palindromic sequences (11) like IS*91* elements (for review see reference 46).

Some families of *Helitrons* and *Helentrons* carry tandem repeats, like microsatellites and minisatellites (11, 22, 24, 26, 32, 36, 39, 48–51). For some *HINE* families, the tandem repeats occupy ~ 50% of the total length of the element (11). Tandem repeat sequences are generally highly mutable (for review see reference 52) and even closely related copies tend to display variation.

The Rep/Helicase protein includes zinc finger motifs, the Rep domain (which has HUH endonuclease activity), and an eight-domain PiF1 family helicase (Super-Family1) (Figure 1) (1, 10, 11, 53, 54). The putative zinc-finger-like motif has been proposed to be involved in DNA binding (10, 54) (Figure 1). Members of the HUH family of endonucleases are involved in the breaking and joining of single-stranded DNA and are characterized by both the presence of the HUH motif (two histidine residues separated by a hydrophobic residue) and the Y motif (one or two tyrosine [Y1/Y2] residues that are separated by several amino acids) (for review see reference 9). *Helitrons*, *Helentrons*, and IS*91* encode a Y2 HUH endonuclease (11, for review

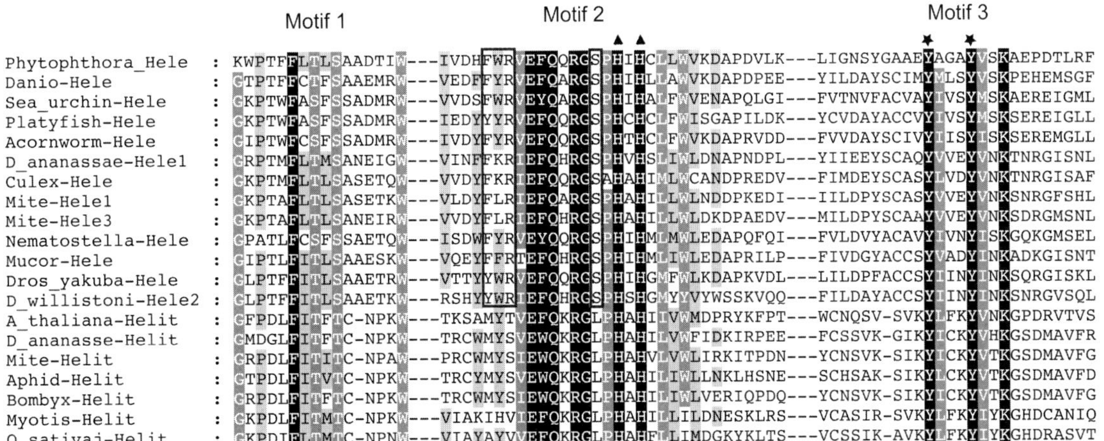

Figure 2 Amino acid alignment of the Rep motifs of select *Helitrons* and *Helentrons*. An alignment of the Rep motif of *Helentrons* from 12 species and *Helitrons* from seven species (redrawn from reference 11). Identical residues are shaded in black and conservative changes are shaded in gray. Amino acids that distinguish *Helentrons* from *Helitrons* are boxed in red. The black triangles and stars above the alignment denote the two histidine residues and the two tyrosines respectively, which are known to be critical for catalytic activity of the rolling-circle elements. Sequences representing *Helentrons* have "Hele" and *Helitrons* have "Helit" as suffix to the name of the organism. The accessions and coordinates of the sequences used in this alignment are available in reference 11.
doi:10.1128/microbiolspec.MDNA3-0049-2014.f2

see reference 9) (Figure 2). A model for how transposition occurs is presented in the *Putative mechanism of transposition* section.

The PiF1 family of helicases has 5′ to 3′ unwinding activity (55). For many rolling-circle entities this activity is host encoded (for review see reference 9). It is proposed that an ancestral *Helitron* (before the divergence of *Helitrons* and *Helentrons*) captured a helicase gene. The Helicase and Rep are apparently a single protein. The Helicase is proposed to aid transposition, so eliminating the need for the host-encoded protein. The plant and *C. elegans Helitron*-encoded genes have introns (1), but no introns have been identified in other animal *Helitrons* or *Helentrons* (11, 54, 56). The *Helentron* AP endonuclease has seven domains (10) and is proposed to be involved in the targeting and cleaving of the target sequence (11) as is the case of non-LTR encoded endonuclease (57–60).

Plant *Helitrons* and some *Helentrons* also encode an open reading frame with homology to single-stranded DNA-binding proteins (RPA) (Figure 1B and D) (1, 11) (for review see reference 56). The RPA gene is multicopy in some plant *Helitrons* (1, 61). The RPA gene of plant *Helitrons* and that encoded by *Helentrons* share little amino acid homology, suggesting an independent origin (JT and EJP, unpublished results). *Helentrons* also carry other highly conserved genes, suggesting that the encoded proteins probably function in the life cycle of these elements (11, 56). Some *Helentron* encoded Rep/Helicase/Endonuclease proteins have an additional two domains that share strong sequence identity with the Ovarian Tumor protein of *Drososphila* (OTU) (11, 54, 56, 62). The OTU proteins are a family of cysteine proteases (54, 63, 64). It is predicted that the cysteine protease proteins might play a role in modifying chromatin (63, 64) to favor transposition (11) or in cleaving the Rep, Helicase, and Endonuclease domains (56). The *Proto-Helentrons* identified from *Phytophthora* genomes carry a gene predicted to encode a SET domain protein (11). The SET domain proteins function as lysine methyltransferases that transfer a methyl group to an amino group of a lysine residue in a histone or other protein. This mark is a post-translational epigenetic modification (for review see reference 65). These proteins have diverse functions that depend on interaction partners as well as protein substrates (for review see reference 66). The *Proto-Helentrons* encode two more open reading frames of unknown function (11). The coding capacity of *Helentrons* differs between families and across genomes, which suggests that gene capture and loss may be common. It is predicted that gene capture must occur with an appreciable frequency to accumulate advantageous coding potential (2). Indeed, genes advantageous to the host (a variety of antibiotic-resistance genes) have been mobilized by IS*91* and IS*CR* families (41, 42) (for review see reference 47).

CLASSIFICATION OF *HELITRONS* AND *HELENTRONS*

A new classification system has been proposed for *Helitrons* (67, 68). A system that is different from that applied to other TEs is necessary because of their peculiar structure and propensity to switch or adopt new termini (69, 70). Because the sequence features are minimal and sequence heterogeneity among copies is high, the standard 80–80 rule (80% or more identity in at least 80% of the aligned nucleotide sequence) applied to other DNA TEs for family definition (71) is not valid. The new classification system based on genome-wide analysis in maize defines a family to include all elements that share > 80% sequence identity over the last 30 bp at the 3′ end (Figure 3A) (67, 68). Elements that also share > 80% sequence identity over the first 30 bp of the 5′ end belong to same subfamily (67) (Figure 3A). As *Helitrons* frequently capture and amplify host genome sequences, the internal sequence of elements that belong to a subfamily might be completely different. To represent the diversity of *Helitrons*, the term "exemplar" was introduced. Exemplars are defined as those elements that have a unique internal sequence that is at least 20% divergent compared with other exemplars (67) (Figure 3A).

Helentrons have different structural features compared with the canonical *Helitrons*, so they require yet another classification scheme (11). The subTIRS are nearly identical across different *Helentron*s that share < 70% identity at the nucleotide level within a species or across species. Therefore, a family is defined as a group of elements that display 100% identity over at least 11 bp subTIR (Figure 3B). The elements that also share > 80% sequence identity in the last 60 bp at the 3′ end belong to the same subfamily (11). As these elements frequently insert in a T-rich region and often display a variable number of Ts at the termini, the last 60 bp do not include the terminal T residues (Figure 3B). *Helentron* exemplars are defined using the same rules as for *Helitrons* (Figure 3B).

GENOME-WIDE IDENTIFICATION OF CANDIDATE *HELITRONS*

The atypical structure, lack of target site modification, and sequence heterogeneity of *Helitrons* have made

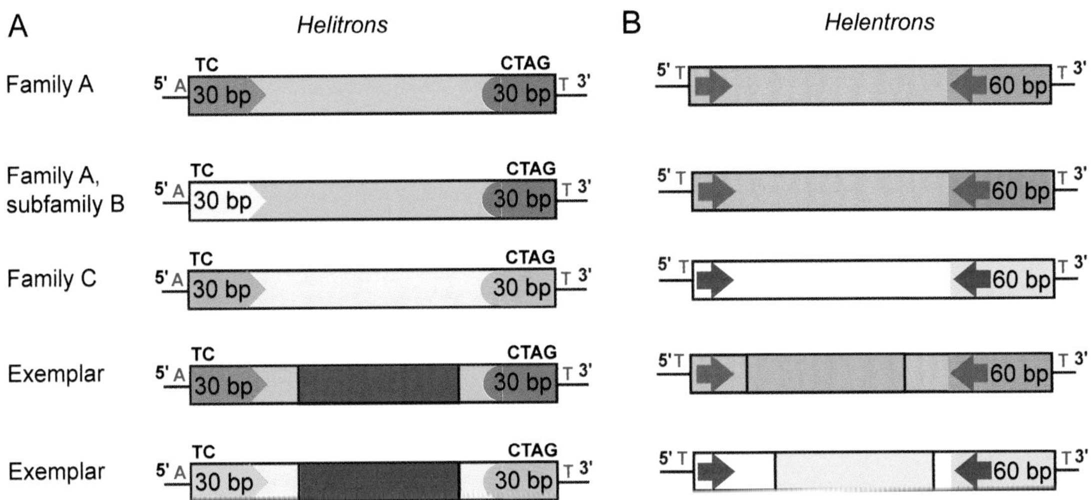

Figure 3 Criteria for classifying *Helitrons* and *Helentrons*. (A) Classification criteria for *Helitrons*; colors of the 5′ and 3′ ends denote common ancestry. *Helitrons* belong to the same family (Family A) when they share > 80% identity over the last 30 bp (denoted by an orange 3′ end). Subfamilies share 80% sequence identity in the 3′ end but have different 5′ ends (Family A, subfamily B). *Helitrons* belonging to family C have a different 3′ end. Exemplars have internal sequences that are > 20% diverged compared with any other exemplar. (B) Classification criteria for *Helentrons*. *Helentrons* belonging to the same family share 100% sequence identity across the 11-bp subterminal inverted repeats (subTIRs) (Family A). A subfamily has at least 80% identity over last 60 bp at the 3′ end (excluding variable Ts) (Family A, subfamily B). *Helentrons* belonging to Family C have a different subTIR. Exemplars have internal sequences that are > 20% diverged.
doi:10.1128/microbiolspec.MDNA3-0049-2014.f3

automated identification difficult. For genome-wide analysis there are two approaches that have been applied to find canonical *Helitrons* (Figure 4A). *De novo* repeat identification approaches like REPuter (72), RepeatFinder (73), RECON (74), RepeatScout (75), PILER (76), RepeatModeler (http://www.repeatmasker.org/RepeatModeler.html), and REPET (77) can be used to build consensus libraries of all repeated sequences (Figure 4A). Tools like REPCLASS (78), PILER (76), and PASTEC (79) can automate the classification of consensus repeat families. These approaches can be used to generate a library of candidate *Helitrons* (Figure 4A). *De novo* repeat finding approaches will identify *Helitrons* and *Helentrons* as long as they are present in multiple relatively homogeneous copies in the genome. Therefore the low copy and older *Helitrons* will tend to be fragmented and have poorly defined ends. These approaches are limited by the quality of the genome assembly and the homogeneity of the repeats.

Another set of approaches relies on the structural features of canonical *Helitrons* and includes programs like Helitronfinder (68, 80), HelSearch, (81), Helraizer (82), and HelitronScanner (83) (Figure 4A). The HelitronFinder and Helraizer programs were specifically designed to identify *Helitrons* in the maize genome (80, 82). As these programs are trained on known *Helitron* elements, they may not be efficient at identifying divergent families. Helsearch identifies sequences that have features of *Helitron* 3′ termini and occur in at least two copies in the genome. Helsearch was able to successfully identify *Helitrons* from other plant genomes like of *Arabidopsis*, *Medicago*, rice, and sorghum, and from the animal *C. elegans* (81) in addition to maize (67). A similar approach was used to identify *Helitrons* in many genomes including those of lepidopterans and brassicas (84, 85). However, this approach generates many false positives. The candidates generated by this approach should be vetted before the sequences are formally classified as *Helitrons*. This approach does not create consensus sequences of the candidate *Helitrons*, resulting in large data sets (68, 81). A recently published tool called HelitronScanner was able to identify novel *Helitrons* from many plant genomes (83). It predicts putative *Helitrons* by using a local combinational variable algorithm, which is used to extract patterns from sequences of diverse protein families (86). The copy number of the element is also used to assign scores based on the quality of prediction. The program has the

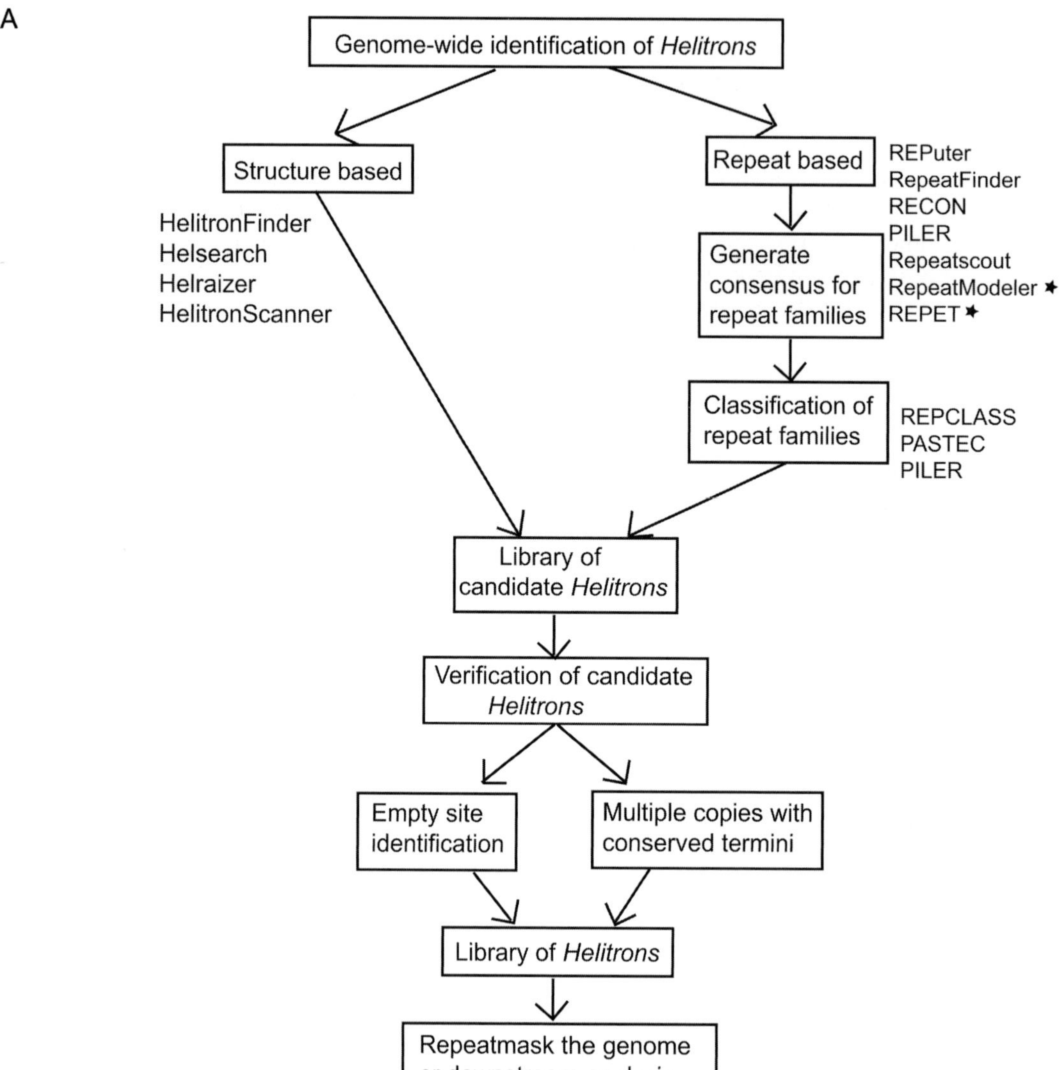

B

121946 *HelibatN542* 123135
AAPE02022450.1: AACTCTCAGTTACTTCCCTA | **TCCTATCTAA---GGGCCTCTAG** | TTATTCTATAACTCCTGAT
AAPE02058147.1: CTAAGCCGCCAGTTTATCTA | **TCCTATCTAA---GGGCCTCTAG** | TATATATATATATATAAAA
637 1837

C

73969 *HelibatN542* 75160
AAPE02002534.1: CATGATTATAAGATACAAGA | **TCCTATCTAA---GGGCCTCTAG** | TTAATATATAAAAATCCAGT
ALWT01094596.1: CATGATTATAAGATACAAGA------------------------ | TTAATATATAAAAATCCTCT
1850 1889

D

103093 *HINE-Da-41A* 101073
AAPP01018455.1: CAACAAAATCGAAGATATAT | **TTTGAAGTGA---ACCCTTTTTT** | TTTTTTTTTCAAAAGCTACA
AAPP01018365.1: ATGTTTTTTTTCAAAATATT | **TTTGAAGTGA---ACCCTTTT--** | ATTTATTTCTGTTGCATTTT
9393 11415

E

103093 *HINE-Da-41A* 101073
AAPP01018455.1: CAACAAAATCGAAGATATAT | **TTTGAAGTGA---ACCCTTTTTT** | TTTTTTTTTCAAAAGCTACA
AAPP01018623.1: TAACAAAATCGAAGATATAT--------------------------TTTTTTTTTCAAAAGCTACA
13245 13284

capability to test the *Helitrons* with medium to low score predictions by searching in randomized genomes (83). This approach might help to reduce the false positives.

As the structure of *Helentrons* has been described only recently (11), no automated structure-based tools have been specifically designed to identify them. The use of homology-based approaches to identify Rep helicase proteins, in conjunction with repeat-based and structure-based approaches, is a comprehensive strategy to identify all candidate *Helitrons and Helentrons* in a genome. Reiterated searches with verified elements also help to identify related families, subfamilies, and exemplars.

Candidate Verification

The candidate *Helitrons* created by both approaches need to be verified to avoid false positives. The lack of TSD requires that the boundaries of the elements be determined by aligning multiple independent copies (inserted at different genomic locations) (Figure 4B and D). The homology between the sequences drops at the boundary of the element (this procedure is complicated by the insertion of other TEs, poor genome assembly, low coverage, and degraded elements). Another approach to determine the proper termini of a candidate is the identification of an empty or insertion-free site in the genome (paralogous) or in another genome (orthologous) (Figure 4C and E). This will also help to identify whether the elements integrate between AT or TT dinucleotides. The sequence flanking the insertion should align completely with the empty site. This allows the termini of the insertion to be verified and helps to identify any modification to the target site caused by insertion. For single copy elements, identification of an orthologous or paralogous empty site is necessary to confirm the mobility and boundary. In addition, in certain cases (Figure 4B and D) only an empty site can reveal the precise boundary of the elements (e.g., to identify the exact number of Ts flanking the element or to identity the 5′ end TC for the *Helitron*). Empty sites can be identified through homology-based searches using the chimeric query created from upstream and downstream sequences (50 bp) of the candidate *Helitron* (Figure 4C and E). Hits with at least 80% identity over 85% of the query sequence are usually sufficient to serve this purpose.

Differentiating *Helitron* and *Helentron* Proteins

In some cases the structure of the element cannot be determined but whether the encoded protein belongs to a *Helitron* or *Helentron* may be quickly distinguished. Sequence alignments of the Rep/Helicase proteins revealed the presence of signature amino acids in the Rep motif that distinguish animal *Helentron* proteins from any *Helitron* protein (Figure 2). The diagnostic positions for animal *Helentron* proteins include three amino acids (F/Y w/l/y/k R) near the motif 2 "(V/I) ExQxRG(S/L)(P/L)HxH" (Figure 2). In addition, the three amino acids directly preceding the well-conserved histidine residues reveal another diagnostic site. The *Helentron* Rep includes the amino acid sequence "GSP" whereas the *Helitron* protein is "GLP" (Figure 2) (11). So far *Helentrons* are limited to animals and fungi. Because the structure of fungal *Helentrons* has yet to be reported, it is not known if these residues are sufficient to classify fungal elements. However, the Rep/Helicase proteins from animals can be easily classified as of *Helitron* or *Helentron* origin by the presence of this signature. In addition, the presence of these residues distinguishes *Helitron* proteins in other phylogenetic groups from *Helentron* proteins.

Figure 4 Genome-wide identification of *Helitrons*. (A) A pipeline for genome-wide identification of candidate *Helitrons* and their verification. Examples of structure and repeat-based tools that could be used for annotating candidate *Helitrons* are listed on the respective side. The black star denotes that they are pipelines that use a set of tools. (B) Alignment of two *Helitron* copies inserted at different locations to identify the boundary of the element. Homology drops at the boundary of the element and the sequences at the boundary have canonical *Helitron* features. (C) Empty site verification for a *Helitron*. The first line is host sequence with a *Helitron* insertion and the second line is the paralogous site without the *Helitron* insertion. (D) Alignment of two *Helentron*-associated INterspersed Element (*HINE*) copies at different locations to identify the boundary of the element. Homology drops at the boundary of the element. Since the insertion is in T-rich sequence, the precise boundary of the element can be unambiguously identified only through the identification of an empty site (E). Empty site verification of a *HINE* copy. The first line is the host sequence containing *HINE* insertion and second line is the paralogous site without *HINE* insertion. The accession numbers and coordinates are given in black.
doi:10.1128/microbiolspec.MDNA3-0049-2014.f4

PUTATIVE MECHANISM OF TRANSPOSITION

Helitrons and *Helentrons* are proposed to transpose by a mechanism similar to rolling-circle replication via a single-stranded DNA intermediate (1, 2). The specific details of the model are based on the proposed rolling-circle mechanism of bacterial transposons (Figure 5), e.g., IS*91* (2, 44, 87) (for reviews see references 46, 47 and 56). Two models for IS*91* transposition have been proposed, the concerted versus sequential (for review see reference 46). In the concerted model, the donor strand cleavage and ligation occurs simultaneously while in the sequential model they occur in a stepwise fashion. The concerted model does not require a circular intermediate although they could occur if a step fails or is bypassed during transposition. The sequential model differs in that a circular intermediate is a required step of transposition (for review see reference 46). Because circular intermediates are not known for *Helitrons* or *Helentrons*, the concerted model was adapted to explain transposition. It should be noted that although the presence of circular intermediates has not been formally demonstrated in any eukaryotic system, they have been described for IS*91* (88). (It is not known whether the ssDNA circles are functional intermediates or byproducts generated when there is a failure to create concerted cleavage at the target site (for review see reference 46)). Indeed, a couple of

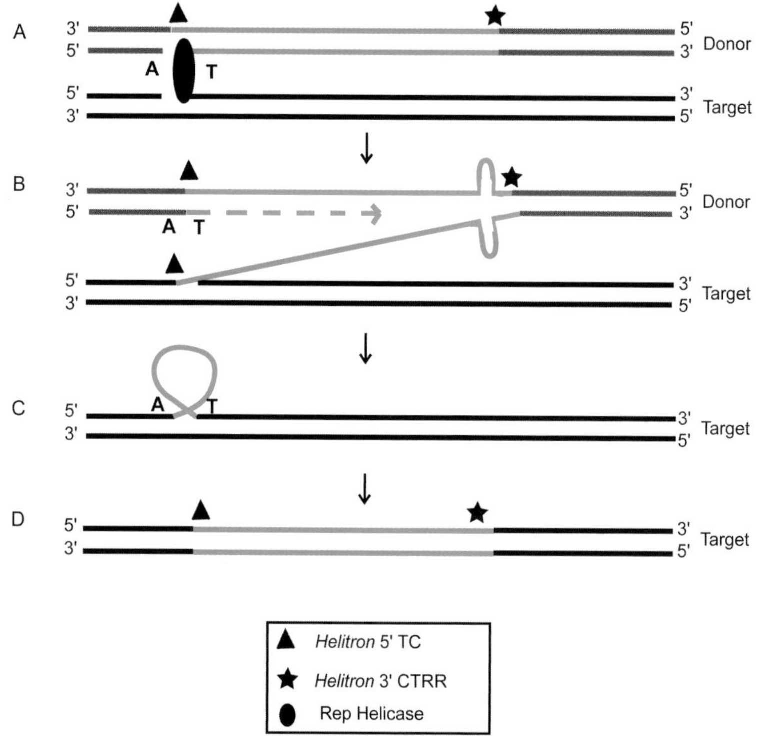

Figure 5 Proposed model for the transposition of *Helitrons*. The blue line indicates the *Helitron*, the small triangle indicates the 5′ end of the *Helitron* and the star indicates the 3′ end of the *Helitron*. (A) The first tyrosine (Y1) residue of the Rep protein (shown as black oval) cleaves at the 5′ end of the *Helitron* in the donor strand (shown as green lines) and the second tyrosine (Y2) residue cleaves the target DNA (shown as black lines). The tyrosine residues covalently join to the 5′ end of the respective strands. (B) The free 3′ hydroxyl in the target DNA attacks the DNA–Y1 bond and forms a covalent bond with the donor strand resulting in strand transfer. The free 3′ hydroxyl in the donor strand serves as a primer for DNA synthesis by host DNA polymerase. The strand is displaced by 5′ to 3′ activity of the Helicase protein and remains single-stranded (ss) with the help of ssDNA-binding protein. (C) At the termination site, the free Y1 residue cleave the 3′ end and becomes covalently linked to the 5′ end of the nicked strand and initiates the strand transfer when the 3′ hydroxyl of the cleaved *Helitron* attacks the Y2 at the 5′ end of the target DNA and forms a covalent bond. (D) The heteroduplex is passively resolved by DNA replication (redrawn from references 2, 46). doi:10.1128/microbiolspec.MDNA3-0049-2014.f5

independent lines of circumstantial evidence suggest that circular intermediates may exist at least occasionally in eukaryotes. This includes the description of tandem arrays of similar elements and the report of possible excision events (explained in more detail below).

Helitron replication is proposed to initiate when residues in the N-terminus of the Rep protein bind to the 5′ end of an element (2). The tyrosine residues (Figure 2) are proposed to cleave both the donor and target sites and initiate the strand transfer (for review see reference 46) (see Figure 5 for details). The strand transfer is complete when one tyrosine residue cleaves the donor strand at the termination signal (Figure 5). The two histidine residues (Figure 2) act as the ligands necessary for the divalent metal ion (Mn^{2+} or Mg^{2+}), which is required for catalyzing the cleavage reaction (1) (for review see reference 9) (Figure 2).

For IS*91* elements, only one end is necessary to initiate transposition and the other end is dispensable (44) (for review see reference 47). In the case of IS*91* elements, one-ended transposition in the absence of a precise termination signal occurs at high frequency in an experimental setting (for review see reference 47). One-ended transposition results in the generation of tandem copies of the donor plasmid (44) or in the capture of flanking sequence (for review see reference 47). In such cases, a variable end will terminate transposition and is defined as a target sequence that could be used as a termination site (for reviews see references 46 and 47). Similarly failure to recognize the termination signal for *Helitron* and *Helentron* transposition may result in the capture of flanking sequence (see *Models for gene capture* and *Where do they go in the genome?—Pattern of insertion* sections for further details). It is predicted that the variable number of T nucleotides at the termini of *Helentrons* is determined by which T nucleotide is cleaved during the termination of single-stranded DNA-mediated replication (11). However, further experimental validation is necessary to clarify the mechanism of transposition.

Example of Excision in Maize

Even though transposition is proposed to be a copy-and-paste mechanism via a single-stranded intermediate, a study conducted in maize inbred lines suggests that *Helitrons* are capable of excision like cut-and-paste DNA transposons (89). The evidence presented for excision is the presence/absence of a *Helitron* insertion at a location when genetically identical maize inbred progenies are compared (89). Maize inbred lines homozygous for every location should give rise to genetically identical offspring. Hence, a *Helitron* present in a parent should be present at that site in all the offspring. Li and Dooner observed empty sites in some offspring from the inbred lines that carry *Helitrons* in single copy regions (89). Cloning and sequencing of those regions further confirmed that the sites were empty in the offspring. In one case, the presence of unique single nucleotide polymorphisms confirmed that the empty sites that were compared were the same across the parents and offspring (89). In other instances, a variable number of TA repeats were observed instead of a *Helitron* in certain offspring of inbred lines. The *Helitron,* in this case, was identified in a TA-repeat-rich region. They also examined other likely *Helitron* excision events, which were not associated with repeats or footprints. They hypothesize that the generation of the TA repeats was associated with double-strand break repair (89). Excisions are described for the bacterial transposons belonging to IS*200*/*605* family (for reviews see references 9 and 45). They encode a Y1, HUH transposase and excise as ssDNA circles (90) during the host replication from the lagging strand (91). They also preferentially insert into the lagging strand (91), (for reviews see references 9 and 45). If *Helitron* excision occurred from only one strand during the S phase of meiosis 1, then it is formally possible that gametes could be produced that would not have the *Helitron* insertion. It may be that excision is more widespread than appreciated. If so this would suggest that *Helitrons* also transpose by a "peel-and-paste" mechanism, where a single-stranded transposon is peeled from the donor site (91), (for review see reference 45). It will be interesting to see if other examples of excision events can be identified or if excision can be reproduced in a cell-based system.

GENE CAPTURE OCCURS AT BOTH THE DNA AND RNA LEVEL

The presence of contiguous exons and introns within the host DNA carried by *Helitrons* suggested a DNA-based mechanism of acquisition (Figure 6A) (61, 67–69, 84, 92–104). *Helitrons* from maize (e.g., 61, 67, 68, 97–99, 101, 102), silk worm (*Bombyx mori*) (84), Pooideae grass (*Lolium perenne*) (95), rice (93), and vespertilionid bats including little brown bat (*Myotis lucifugus*), David's bat (*Myotis davidii*), and big brown bat (*Eptesicus fuscus*) (51) have been observed to carry fragments from multiple chromosomal locations. *Helitron* gene capture was proposed to occur in a stepwise or sequential manner, i.e., gene capture occurs during one transposition and capture of a second gene occurs during a subsequent transposition event (105). Stepwise

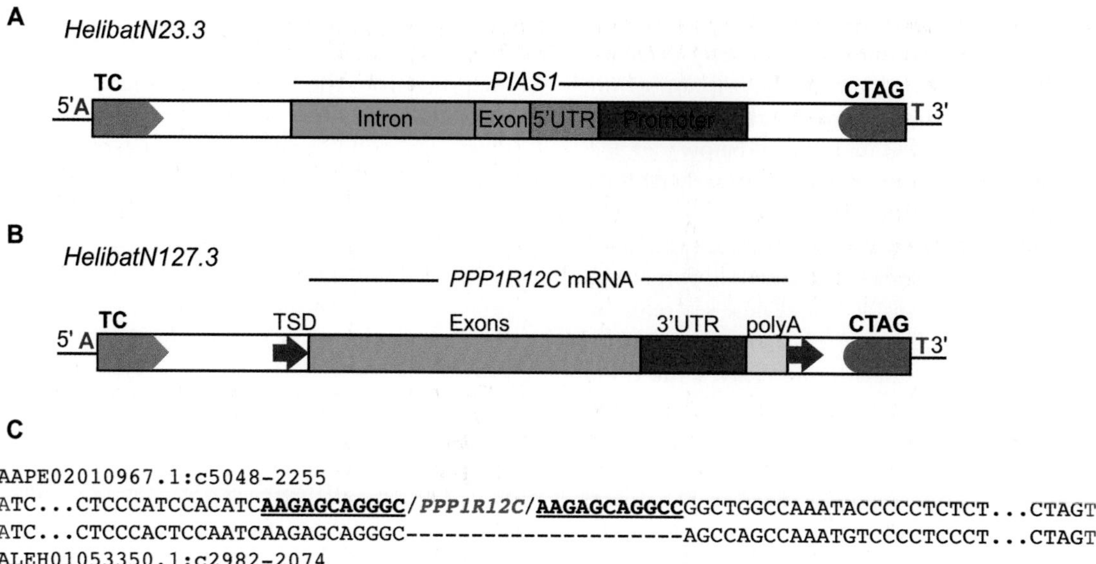

Figure 6 *Helitron*-containing gene fragment captured at the DNA level and RNA level. (A) The structure of (*HelibatN23.3*) exemplar that has captured the promoter, 5′ untranslated region (UTR), Exon1, and Intron1 of the *PIAS1* (protein inhibitor of activated signal transducer and activator of transcription 1 [STAT-1]) gene, which inhibits STAT1-mediated gene activation and the DNA-binding activity. (B) Structure of the *HelibatN127.3* containing the cDNA of protein phosphatase 1, regulatory (inhibitor) subunit 12C (*PPP1R12C*) gene; *Helitron* contains seven exons (blue box), 3′ UTR (pink box), polyAs (yellowish green box) and 11-bp target site duplication (TSD) (purple arrows). (C) Empty site for the retroposed mRNA; first line is the *Helitron* containing the *PPP1R12C* cDNA and second line is a paralogous site within another *Helitron* but without the retrogene. The black bold letters shows the TSD. The accession and coordinates of the *Helitrons* are given. The flanking AT dinucleotide of the *Helitron* insertion is shown in red.
doi:10.1128/microbiolspec.MDNA3-0049-2014.f6

capture would result in *Helitrons* that contain gene fragments from different locations (61, 67, 68, 93, 95). In maize, *Helitrons* that have high sequence identity (> 95%) in shared sequences but that differ due to additional captured regions were described, which supports the sequential capture model (61). Sequential gene captures that span speciation events were reported in the Vespertilionidae bat family (51). The sequential capture model may explain *Helitrons* carrying multiple gene fragments observed in other organisms (84, 93, 95).

A new study in *M. lucifugus* illustrates that host DNA can be captured at the RNA level as well as the DNA level, and that the two processes can happen in concert (51) (Figure 6B). It should be noted that the capture at the RNA level is not a direct outcome of *Helitron* transposition but rather mediated by retrotransposition. For example, the generation of a new *Helitron* exemplar (*HelibatN424*) resulted from the retroposition of a host mRNA (*RPLP0*) into another exemplar (*HelibatN211*). These secondary insertions display the hallmark of retrogenes: they lack introns, terminate by a stretch of adenines and are flanked by TSDs consistent with LINE-1-mediated retroposition events (106). The *HelibatN211* had previously captured the promoter, 5′ untranslated region (UTR), exon and intron of another gene (*SRPK1*) via a DNA-based mechanism. In the *M. lucifugus* genome, several examples of host DNA integrated by retrotransposition were described (SINEs (short interspersed repeated elements), in addition to retrogenes). The sequences were frequently further amplified to hundreds of copies through *Helitron* transposition (51). Retrotransposition events display telltale hallmarks (Figure 6B and C). The identification of such signatures was used to determine if a retrogene was captured at the DNA level or the result of a bona-fide retroposition event (Figure 6C). Because retrotransposition can occur in conjunction with DNA-based gene capture, *Helitrons* that carry multiple gene fragments can also originate by this process (51).

Models for Gene Capture

Several models have been proposed to explain the mechanism of gene capture at the DNA level (2, 56, 69, 107). To understand how much of the data can be explained by a model, it is helpful to consider whether the proposed mechanism is expected to be specific to *Helitrons*, TEs in general or a feature of all unconstrained DNA. The first group of models (end bypass and chimeric transposition) (2, 69) require active transposition to capture the host sequence so are *Helitron* specific (Figure 7). The filler DNA model (56) and site-specific recombination model (107) do not require active transposition to capture host sequences but depend on host factors (Figure 8). The different models are discussed below.

End Bypass Model

Based on the literature from bacterial rolling-circle transposons, Feschotte and Wessler (2) proposed an end bypass model based on that previously proposed for IS*91* (44, for review see reference 47) (Figure 7A). In this model, transposition initiates at the 5′ end and

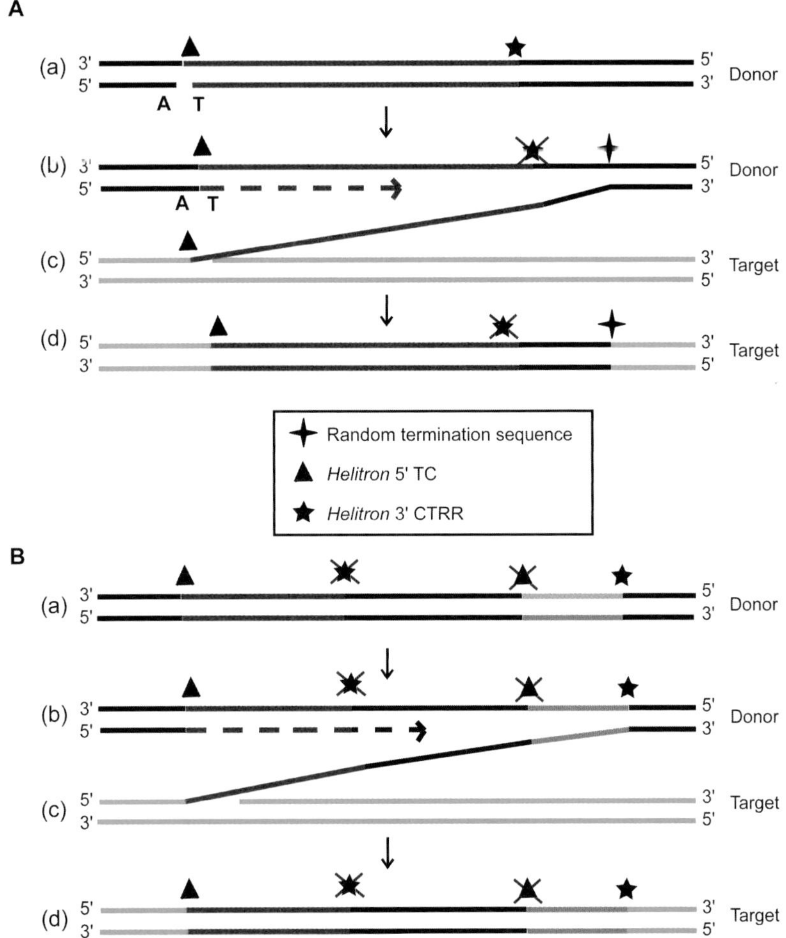

Figure 7 Active transposition-based models for gene capture. (A) End bypass model (2). (a) The Rep/Helicase protein cleaves the 5′ end of the *Helitron* (red line) and invades the target site (blue line) (shown in c) (see transposition model, Figure 5 for details). (b and c) Capture of flanking sequence (black line) occurs when the protein fails to recognize the termination signal (black star). Later the transposition is terminated by a cryptic random termination signal (four star) and the donor strand is cleaved and transferred to the target site (redrawn from reference 2). (B) Modified end bypass/chimeric transposition model (69). (a) The Rep/Helicase protein cleaves the 5′ end of the *Helitron* (red line) and invades the target site (blue line). (b and c) Capture of the flanking sequence occurs when the protein fails to recognize the termination signal or the 3′ end is truncated. The 3′ end of another *Helitron* (green line) in the proper orientation is recognized and is used a new 3′ end, thus creating a novel composite element. The protein cleaves the donor strand at the new termination signal and the donor strand is transferred to the target site.
doi:10.1128/microbiolspec.MDNA3-0049-2014.f7

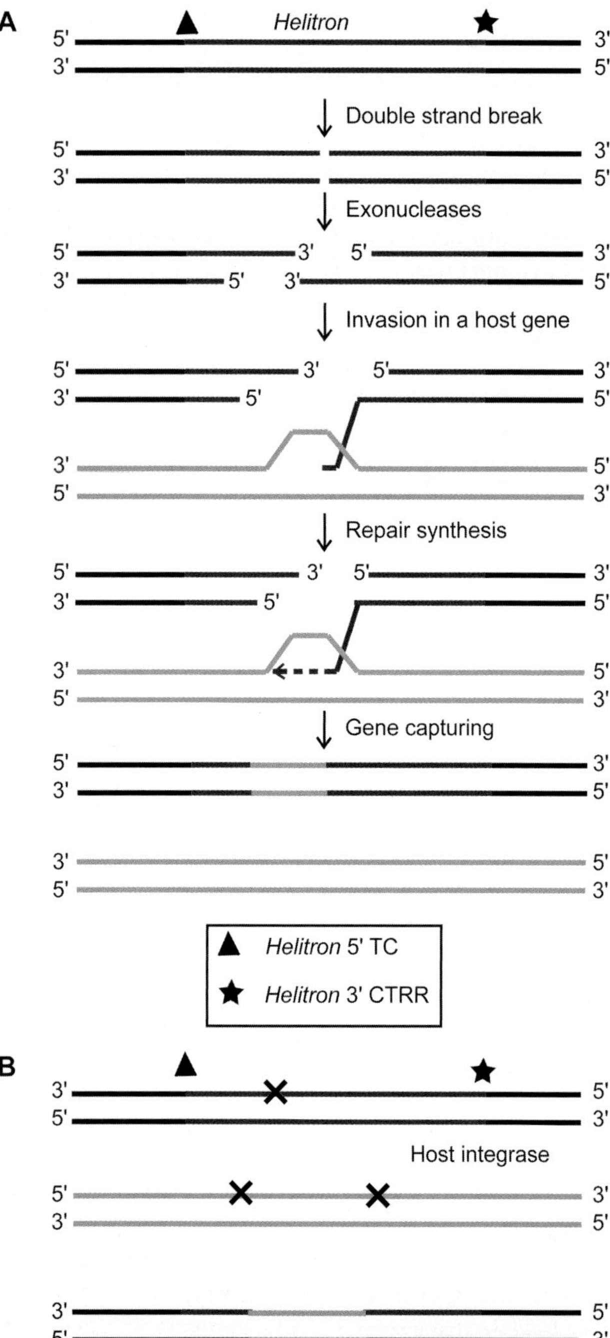

Figure 8 Filler DNA model and site-specific recombination model of gene captures. (A) Filler DNA model (for review see reference 56). When a double-strand break occurs in the acceptor DNA (*Helitron*) (red line), the host exonuclease creates 3′ single-stranded (ss) overhangs. The free 3′ ssDNA anneals to the donor DNA based on microhomology triggering the synthesis of new DNA, which then anneals back to the acceptor DNA. The new DNA acquired from the donor acts as the template for the other strand. Hence, the *Helitron* now contains host sequences from a random location, which could be of genic or nongenic origin depending on which region of the host was used for repair. (B) Site-specific recombination model (107). The sites of recombination are marked by crosses. The capture of the host sequence by the *Helitron* would require three recombinational events, one within *Helitron* and two flanking the host sequence. As *Helitrons* do not encode integrase a host integrase is used for the capture. doi:10.1128/microbiolspec.MDNA3-0049-2014.f8

gene capture occurs if the 3′ termination signal is missed. A random cryptic sequence located downstream would then act as the termination signal and all intervening sequence would be captured. Because a random sequence provides the novel termination signal, this model does not require a high density of *Helitrons* in the genome.

An Example for the End Bypass Model

The capture of a *xanA* gene fragment by a *Helitron* in the *Aspergillus nidulans* genome supports the end bypass model (2, 94, 108) (Figure 7A). The truncated 5′ end of a *Helitron* (98 bp) is identified upstream of the *xanA* parental gene (94). It is proposed that *Helitron* transposition began at the 5′ end but due to lack of the termination signal, the flanking gene promoter and the first exon and intron sequences were captured. The transposition was terminated by a serendipitous CTTG within exon 2 (94). The fate of the element with the novel end is not clear in this example, because only a single copy of the *Helitron* containing gene fragment is found in the genome.

Modified End Bypass/Chimeric Transposition Model

Tempel *et al.* reported many *Helitron* families from *Arabidopsis* that display alternate 5′ and 3′ ends. These subfamilies provided the basis for a slightly modified end bypass model called the chimeric transposition model (Figure 7B) (69). In this model, transposition is initiated at the 5′ end of an element and gene capture results if the 3′ end of that element is missing, so transposition is terminated at the next 3′ end of a *Helitron* in the correct orientation. The result is that all intervening sequence is captured (69). This model assumes the close proximity of *Helitron* insertions in the genome, which facilitates the use of another *Helitron* end for transposition.

An Example for the End Bypass/Chimeric Transposition Model

In the genome of the big brown bat, *E. fuscus*, a *Helitron* subfamily, *HelibatN539* has captured a piece of an intron from the *KCNQ5* gene (51). Examination of the intron in the *KCNQ5* gene revealed the presence of many *Helitrons* in close proximity to one another. One element had a nearly identical 5′ end to *HelibatN539* but a different 3′ end, whereas another element had a nearly identical 3′ end but different 5′ end. These elements are found in the same orientation and flank the intronic region captured by *HelibatN539*. Hence, it appears that a composite transposition event occurred using the 5′ end of one *Helitron* copy and the 3′ end of the other copy resulting in the creation of a compound transposon *HelibatN539*, which further amplified to 20 copies. The result is a novel *Helitron* exemplar containing the intervening host sequences (51). We cannot infer whether transposition initiated from the 5′ or 3′ end, but otherwise this example fits into the end bypass group of models.

Filler DNA Model

The Filler DNA model proposed by Kapitonov and Jurka (56) suggests that *Helitrons* acquire DNA from the host during the repair of double-strand breaks (DSBs) internal to the element. The capture is proposed to occur when repair of the DSBs occurs by synthesis-dependent strand annealing mechanism (Figure 8A) (109–111). The model predicts that short (2 to 8 bp) regions of microhomology exist between the regions that flank the DSB in the *Helitron* and that flank the original host sequence captured by the *Helitron*. When DSBs occur at multiple positions within a *Helitron*, sequences will be captured from multiple locations (Figure 8A). The DSBs are proposed to occur at fragile sites in the *Helitron* or by other phenomena, such as the transposition of cut-and-paste DNA transposons or external stress. This model does not require that the *Helitron* be in close proximity to the donor gene; however, if this were the mechanism at play then one would predict that all unconstrained DNA would be subject to this process and that it would not be a *Helitron*-specific phenomenon. To our knowledge no comparison of gene capture frequency across all forms of unconstrained DNA in a genome have been published. This kind of analysis would allow the role of this process in *Helitron* gene capture to be formally tested.

Site-Specific Recombination Model

Site-specific recombination was proposed as yet another potential mechanism of *Helitron* gene fragment capture in a manner akin to Integrons (Figure 8B) (107, 112). This model was proposed based on multiple common features shared by *Helitrons* and Integrons, for example, the lack of terminal inverted repeats (TIRs) and TSDs upon integration. Another requirement of this model is a site-specific recombinase not encoded by *Helitrons*. This model also proposes that a *Helitron* acquires multiple gene fragments in a stepwise manner. No examples have been described that support this model.

Transposable Element Capture

A recurrent phenomenon that is independent of TE class or superfamily is the integration of TEs via transposition into other TEs, also called TE nesting. When a TE integrates into a *Helitron*, the *Helitron* sometimes transposes with the nested TE (51). This phenomenon was frequently observed in the *M. lucifugus* genome (51) where there have been many waves of transposition that post-dated the major wave of *Helitron* amplification (113).

There is a lack of examples to limit the mechanism of gene capture to a single model. Examples for DNA-based and RNA-based mechanisms reveal that *Helitrons* capture gene fragments through multiple mechanisms. The frequency of gene captures varies between genomes (51, 67, 68, 84, 93, 96) and between families. In both maize and bat the most prevalent high-copy-number families also exhibit the highest frequency of gene capture. This might suggest that high density of elements facilitates end-bypass-mediated capture or that carrying host sequence might somehow allow *Helitrons* to slip under the radar of the host control. Further investigations will shed light on the molecular mechanisms behind gene capture and how it favors the amplification and survival of *Helitrons*.

DISTRIBUTION OF *HELITRONS* AND *HELENTRONS* ACROSS EUKARYOTES

Helitrons are widespread in eukaryotes with families found in at least one species from all major eukaryotic super groups except for Rhizaria—zero of three genomes in GenBank (Figure 9), which consists of unicellular "amoebae-like" organisms (114). Despite this, they often display a patchy distribution with some branches seemingly devoid of sequence related to protein-coding *Helitrons* despite representation in GenBank (Apicomplexa 0/30 genomes, Kinetoplastids 0/30 genomes). *Helitrons* occur in all three subgroups of Plantae (Streptophytes, Green algae, Red algae). They are also found in the two subgroups of Chromalveolates (Alveloates [e.g., Dinoflagellates and *Perkinsus*] and Stramenopiles [e.g., blight-causing microbes]) and in Unikonts (such as animals, fungi, and some amoebae). Although *Helitron* protein-coding genes but not the complete elements have been identified in the Unikont subgroup Ameobozoa (e.g., Entamoeba) and the Discicristate group of Excavates (e.g., *Trichomonas*) (Figure 9), among Ophistokonts, *Helitrons* are found in choanoflagellates, fungi, and multiple animals (56, 94, 115, 116). Among animals, *Helitrons* are also found in many vertebrate (e.g., lampreys, reptiles, fish, mammals) (56, 70, 117, 118) and invertebrate (e.g., insects, flatworms, nematodes, sea anemone, hydra, molluscs) (56, 70) genomes. *Helitrons* have been identified in the genomes of polydnaviruses associated with certain Hymenopteran wasps (70, 115). Mammals, among animals, are notable for the lack of *Helitrons*, with the exception of a few degenerated fragments in platypus and at least two (and maybe more) independent *Helitron* invasions over the last ~ 40 million years (my) in the Vespertilionidae family of bats (53, 70, 119). The abundance of *Helitrons* across genomes varies tremendously (10, 70, 83, 120) (Figure 10) and some genomes may harbor only degenerated remnants—e.g., *Tetrahymena thermophilla* (JT and EJP unpublished) and platypus (53). Most sequenced land plant and invertebrate genomes have experienced recent prolific *Helitron* activity (Figure 10) (56, 67, 70, 83, 84).

Helentrons seem to have a much narrower distribution compared with *Helitrons.* Land plants where *Helitrons* are most prevalent seem to be devoid of *Helentrons* despite good representation in GenBank (0/116 genomes). Genes encoding putative proteins related to the *Helentron* (Rep) have been identified in both green and red algae, and in the two subgroups of Chromalveolates (Alveolates and Stramenopiles) and in the subgroup Opisthokonts of Unikont. In most cases, the structural characteristics and full coding capacity of these elements are not known. Among Opisthokonts, *Helentrons* encoding the Rep/Helicase/Endonuclease gene have mostly been found in animal genomes and three fungal genomes (Mucorales: Zygomycetes) (Figure 5) (11). *Helentron* protein-coding genes are also found in other groups of fungi but the structure and complete coding capacity have not been described (11). Among animals, they are most prevalent in the genomes of insects and other invertebrates. In insects, they have been identified from Diptera, Lepidoptera, Hymenoptera, Coleoptera, Hemiptera, (11, 32, 56), Thysanoptera, Orthoptera, Megaloptera, Strepsitera, and Isoptera. In addition to insects, *Helentron* genes have been identified in the genome of the polydnavirus, *Cotesia sesamia* bracovirus associated with hymenopteran wasps (11). *Helentron* protein-coding genes are also found in other animal groups such as Arachnida, Nematoda, Mollusca, Annelida, Priapulida (11), Cnidaria (for review see reference 56), Echindodermata (54), Rotifera, and Hemichordata (11). *Helentrons* have been described from the Chordates including Cephalochordata (54), Tunicata (10, 54), and Coelacanth (11). Among vertebrates, *Helentrons* are found in Amphibia, numerous fish genomes—18 families of fish (10, 11, 54, 62), and lampreys (11). No *Helentrons* have been identified in any mammalian genome. *Proto-Helentrons* are found in *Phytophthora* genomes (Oomycetes; Stramenopiles Figure 9) (11) and *Helitron2* are reported from green algae (*Chlamydomonas reinhardtii*; Figure 9) (12). The known distribution of both types is very limited.

VERTICAL INHERITANCE OF *HELITRONS* AND *HELENTRONS*

Genome-wide analyses revealed that the bulk of *Helitrons* tend to be quite recent (67, 68, 84, 93, 96). The vast majority of plant *Helitron* families are less than 6 my old (67, 68, 93, 96). Indeed, many of the *Helitron* families from maize, rice, and sorghum amplified within the last 250 thousand years (67). The bulk of *B. moriHelitron* families amplified within the last 2 my, although some older (~ 10 my) families were reported (84). The young age of *Helitron* families is of course biased by the genomes that have been examined carefully, which are predominately plant and insect where the unconstrained DNA half-life (the average amount of time when half of DNA not conserved for function is lost) is quite rapid; 2.5 my for maize (121), ~ 14.3 my for insects (based on *Drosophila* comparisons) (122).

Vertical persistence and long-term activity of TEs are quite unusual because of the interplay between mutation rate, rate of unconstrained DNA loss, and effective population size. Vertical persistence has been documented for non-LTR retrotransposons in primates where activity has persisted throughout the diversification of anthropoid primates over tens of millions of years contributing to nearly one-third of the human genome (for review see reference 123). This is quite different from the usual rapid burst of animal cut-and-paste TE activity followed by silencing and eventual degradation (for reviews see references 124, 125 and 126). As an exception to DNA transposons, the *Helitrons* from some species have been reported to exhibit long-term activity perhaps related to the mechanism of transposition or the inability of the host to recognize as foreign either because of sequence heterogeneity or host gene capture. The *Helibat* family from vespertilionid bats and several *HINE* families from flies are reported to display patterns consistent with long-term activity (39, 51, 127).

It is estimated that the *Helibat* family entered the genome of a vespertilionid ancestor at least ~ 36 million years ago (mya) (53). This is based on the observation

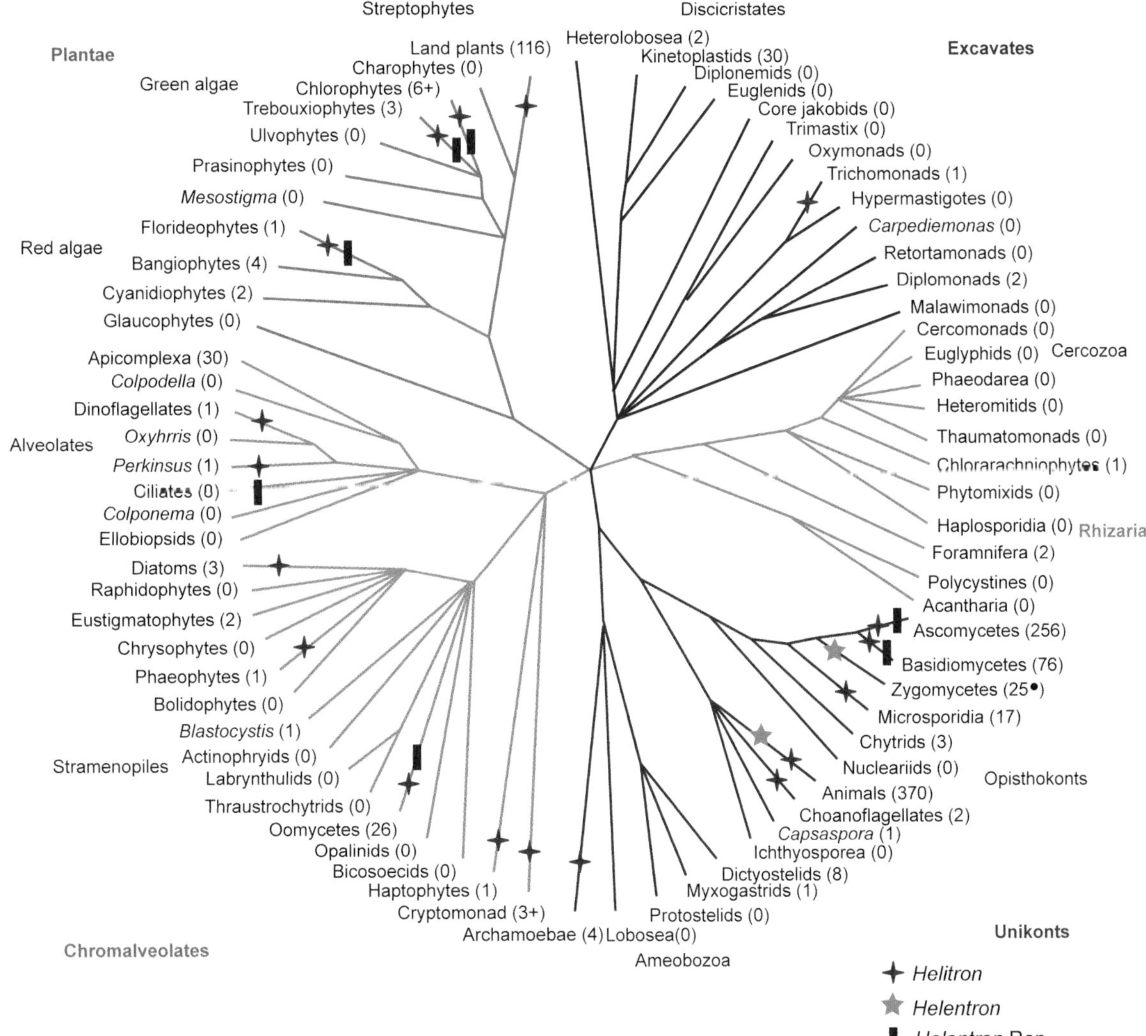

Figure 9 Distribution of *Helitrons* across the eukaryotic tree of life. The four-pointed star shows the presence of *Helitrons*, the five-pointed star represents the presence of *Helentrons*, and the rectangular bars represent the presence of the *Helentron* Rep protein. The tree of life was redrawn from reference 192. The numbers within parentheses represent the numbers of whole genome sequences available at the NCBI whole genome shotgun (wgs) database as per 18 June 2014 http://www.ncbi.nlm.nih.gov/. The plus sign within parentheses indicates that the respective element was identified from the transcriptome assembly deposited at the wgs database. The dot within parentheses indicates that *Helentrons* were identified from the Mucorales group of fungi traditionally classified as a Zygomycete but are not deposited as Zygomycetes at NCBI. TBLASTN searches (193) were employed against the wgs database to identify sequences homologous to the *Helitron* Rep protein query and the signature amino acids (see Figure 2) were used to differentiate *Helentron* from *Helitron* proteins (we do not know how this correlates with structure outside of animals and plants). Hits with very low copy number and of short contigs were not reported because of the possibility of contamination. doi:10.1128/microbiolspec.MDNA3-0049-2014.f9

that the majority of *Helibat* insertions are fixed and that *Helitrons* are only found in this family of bats (119). However, despite an apparent decrease in the level of transposition or fixation, the activity of the *Helibat* family has persisted throughout the diversification of this family (51). Insertions specific to the *M. lucifugus* genome were identified and are estimated to have amplified 1.8–6 mya. In fact, lineage-specific

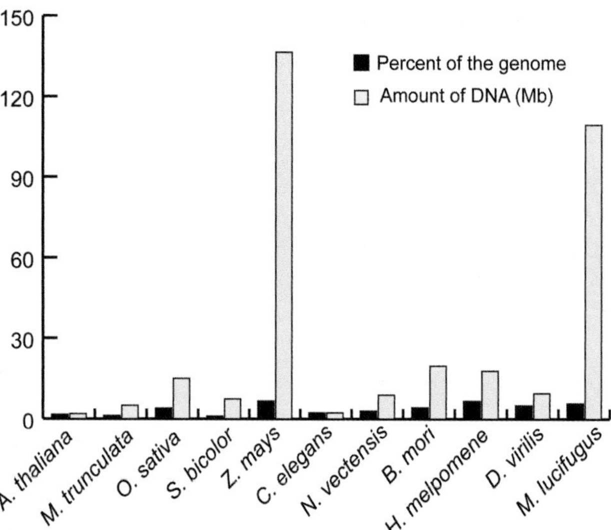

Figure 10 The abundance of *Helitron*-generated DNA in different organisms. The different organisms include *Arabidopsis thaliana* (1.6%; 1.9/120 Mb), *Medicago truncculata* (1.2%; 5/419 Mb), *Oryza sativa* spp. *japonica* (4%; 15/374 Mb), *Sorghum bicolor* (1%; 7.4/734 Mb), *Zea mays* (6.6%; 136.4/2066 Mb) (83), *Caenorhabditis elegans* (2.3%; 2.3/100 Mb) (81), *Nematostella vectensis* (3%; 8.9/297 Mb) (133), *Bombyx mori* (4.2%; 19.7/465.7 Mb) (84), *Heliconius melpomene* (6.62%; 17.1/260 Mb) (84), *Drosophila virilis* (5%; 9.5/189 Mb) (56), and *Myotis lucifugus* (5.8%; 109.5/1887 Mb) (51).
doi:10.1128/microbiolspec.MDNA3-0049-2014.f10

insertions were identified in every genome examined (51). There are two possible explanations for *Helibat* perdurance in mammals. In contrast, to the relatively faster unconstrained DNA half-life (2.5–14 my) of the plant and insect genomes analyzed, the mammalian rate is estimated to be much slower (884 my) (122). It may be that the minimal requirements of *Helitron* transposition and the slow rate of decay in mammals have shaped this pattern of vertical persistence. Although it cannot be ruled out that the *Helibat* family has invaded multiple times over the course of the diversification of the Vespertilionidae family, multiple invasions by the same TE family over the course of evolution is unprecedented and members of the *Helibat* family have not been reported outside bats.

Analysis of *HINEs* in *Drosophila* genomes revealed multiple independent lines of evidence suggestive of their potential to be active for long periods of time and to be vertically inherited (22, 39, 127). A phylogenetic analysis of the conserved regions of the *HINE* consensus sequences from the 12 *Drosophila* genomes revealed congruency between TE and host, which can be indicative of vertical inheritance (39). A study of dosage compensation binding sites revealed that they were coopted independently from different *HINE* insertions that were active at different times over the last ~ 1–15 my in *Drosophila miranda* (127) (see *Impact of Helitrons on genome architecture* section). The pattern of prolonged *HINE* activity has been previously reported in diverse species of the *D. obscura* group (21, 22). Further detailed analysis of *Helentrons* and *HINEs* in different *Drosophila* genomes is necessary to unequivocally confirm their long-term activity. Another possibility, as for the *Helibat* family in bats, is that there is recurrent horizontal invasion of the same family of TEs.

HORIZONTAL TRANSFER

Horizontal transfer (HT) is the movement of DNA between organisms in a nonvertical manner (i.e., not from parent to offspring). The role and importance of HT in the life cycle of TEs has long been discussed in a theoretical context. Indeed HT is proposed as a major mechanism by which TEs evade vertical extinction and loss. The recent avalanche of publicly available genome data has led to the discovery of many examples of TEs with sequence identity that can only be explained by the movement of genetic material between organisms in a nonvertical fashion (70, 128, 129). The HT of TEs is supported by the high sequence identity between elements present in divergent species, the patchy distribution of the elements among divergent taxa, and incongruence with the host phylogeny. The lack of any signatures of selection at the nucleotide level also supports the HT of elements when coupled with high sequence identity between elements, which is greater than would be expected based on divergence (for review see reference 125). These examples add empirical support to theoretical models, but leave many questions unanswered. For example, how does HT occur? Does it require a vector? What circumstances favor it?

Examples of Horizontal Transfer of *Helitrons* and *Helentrons*

Four different horizontally transferred families of *Helitrons* were reported in an unprecedented array of organisms, including mammals, reptiles, fish, invertebrates, and polydnaviruses (70) (Figure 11). One family *Heligloria* (~ 88% sequence identity) was described in the genomes of *M. lucifugus*, the blood sucking bug (*Rhodnius prolixus*), the green anole (*Anolis carolinensis*), and the sea lamprey (*Petromyzon marinus*), which diverged from a common ancestor > 750 mya (70) (Figure 11). Horizontally transferred *Helitrons* were identified in polydnaviruses (*Cotesia sesamia* Mombasa and Kitale bracoviruses, and *Cotesia plutellae* braco virus) and insects including *B. mori*, *R. prolixus*, *Drosophila willistoni*, *Drosophila ananassae* and *Acyrthosiphon pisum* (Figure 11) (70). Recently, a family of *Helitrons* (Lep1) (92) that were initially identified in lepidopterans were also identified in multiple non-lepidopteran species including polydnaviruses and the intracellular microsporidia parasite, *Nosema bombycis* (115) (Figure 11). The *R. prolixus and B. mori* genomes have also been reported to share *HINEs,* although no autonomous *Helentrons* have been described (11, 33).

No case of HT of *Helitrons* in plant genomes has been reported. However, maize contains some *Helitron* families that are shared between sorghum and rice (67), which diverged ~ 11 mya (130) and ~ 50 mya (131), respectively. Because of the highly diverged internal sequence it is difficult to make a clear case for common ancestry or recent horizontal transfer (67). It is possible that this could be the result of multiple independent horizontal transfers of the same *Helitron* family. Indeed, this pattern is similar to what was observed with the *Heligloria* family in animals (Figure 11) (70). Members of *Heligloria* family, (*HeligloriaAi, -Aii* and *-B*) share the 3′ end but have divergent internal sequences (70).

Proposed Mechanisms of Horizontal Transfer

Presence of multiple, horizontally transferred *Helitrons* revealed that some species were recurrent players in this process, although whether they are donors, receivers or both is not clear (see Figure 11). For example, the blood-sucking insect *R. prolixus* and the silkworm moth *B. mori* were found to harbor multiple families of horizontally transferred *Helitrons* and *HINEs* (Figure 11) (33, 70). The physiological or ecological factors favoring the high frequency of horizontal transfer is unknown. However, the role of a host–parasite relationship has been proposed recently as a major mechanism of horizontal DNA transfer (132, for review see reference 125). So far this model is supported by examples where a TE family is shared between a host and parasite. For example, *R. prolixus* is known to feed on bats. Other examples include the discovery of *Helitrons* and *Helentrons* in the genomes of viruses that are used to combat the immune systems lepidopteran prey during infection by parasitic wasps (11, 70, 115). Finally there has recently been a report of a *Helitron* family found in the genome of the microsporidian intracellular parasite *N. bombycis* and in the genome of *B. mori* (115). However, the presence of horizontally transferred elements in taxa (e.g., lizard and sea lamprey) (Figure 11) that do not have a host–parasite relationship illustrates that the process is much more complex that we understand (70, 129).

IMPACT OF *HELITRONS* ON GENOME ARCHITECTURE

The genomic impact of *Helitrons*, like all TEs, is intimately intertwined with its mechanism of transposition, its interactions within the host, competition with other TE families and population genetic features. Copy numbers are shaped by features such as transposition mechanism and whether a single- or double-stranded intermediate is mobilized. Transposition occurs in the confines of the host genome so host features like silencing and repair mechanisms, which may differ depending on the timing and pattern of tissue and cell type expression, are also at play. The rate of fixation or loss is dependent on effective population size, which is closely connected to environmental factors. Genome surveys provide a window into the complex interplay between the host, the TE landscape, and the environment. This snapshot has provided insight as to how *Helitron* activity has shaped genomes in the past and continues to impact them today.

Abundance

Both *Helitrons* and *Helentrons* are capable of reaching high copy numbers and therefore may compose a sizable fraction of the genomes where they are found. The total contribution varies between genera (81, 83, 84, 133) (Figure 10) and even between closely related species (134–136). To date, *Heliconius melpomene* and maize are reported to have the highest contributions at ~ 6.6% (17.1 Mb/260 Mb and 136 Mb/2 Gb, respectively) (83, 84). The second highest is the vespertilionid bat, *M. lucifugus,* where *Helitrons* have contributed ~ 110 Mb/1.89 Gb to the genome (51). Because the method of estimating the contribution of *Helitrons* to DNA in bat (51) and maize (83) differ, these estimates are not directly comparable. The ~ 110 Mb of DNA contributed by *Helitrons* is unique to vespertilionid bats among other mammals, which do not contain *Helitrons* (53, 119). *Helentrons/HINEs* are also estimated to occupy a significant portion of many genomes (1% to 10%) including many Drosophilids (Figure 10) (22, 137), the fly *Calliphora vicina* (138) and a mollusc genome (11, 36). Studies of the genomes of moss, some fungi, and vertebrates have revealed only a few low-copy-number families or individual insertions, which suggests that successful invasion does not necessarily translate into high copy number (10, 70, 83, 139).

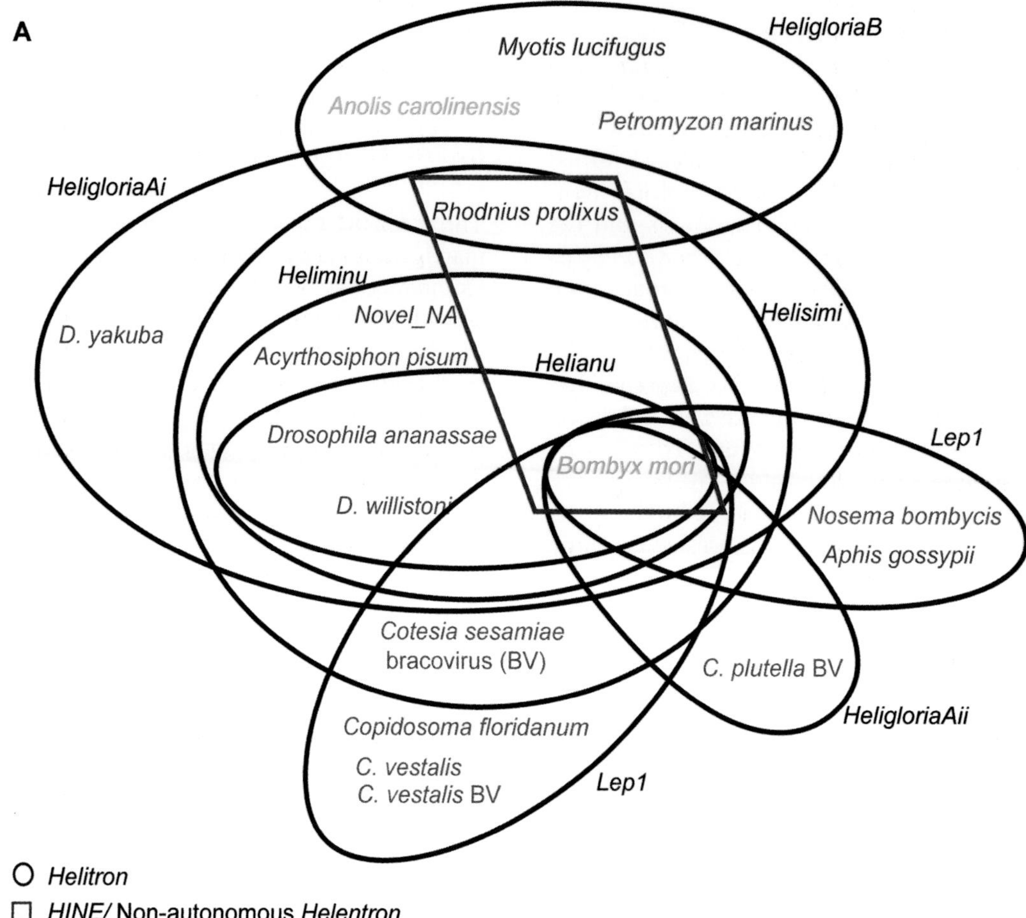

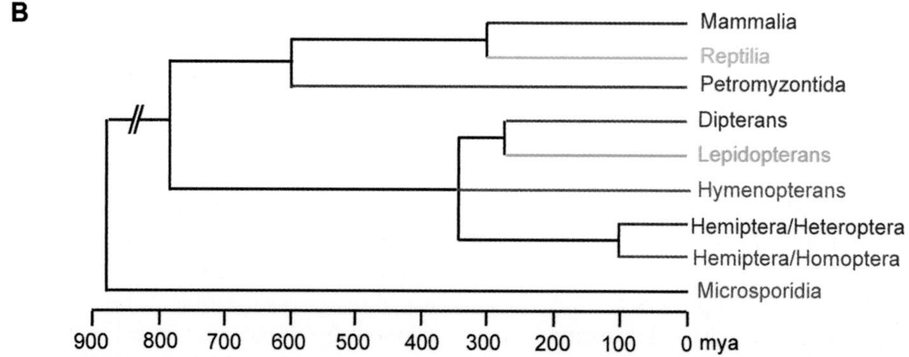

Figure 11 Distribution of horizontally transferred *Helitrons* and their phylogenetic relationship. (A) A venn diagram showing the distribution of different horizontally transferred *Helitron* families—*Heligloria, Helisimi, Heliminu,* and *Helianu* (70), Lep1 (115)—and non-autonomous *Helentrons*—*Helentron*-associated INterspersed Elements (*HINEs*) (11, 33). (B) Phylogenetic relationship between different organisms that carry horizontally transferred *Helitrons*. The exact divergence of Microsporidia from other groups is not known. The phylogenetic tree is redrawn from reference (70). The distribution of a *Helitron* family (black) is shown as organisms within the black ellipse. The colors of letters for each organism are related to the group in the phylogeny. The red rectangle shows the distribution of *HINEs* and red letters represent the *HINE* family.
doi:10.1128/microbiolspec.MDNA3-0049-2014.f11

Where Do They Go in the Genome?—Pattern of insertion

The insertion pattern of *Helitrons* and *Helentrons* is deterministic of their abundance and impact on the genome. In theory, insertions close to genes are more likely to impact gene expression and to be weeded out by natural selection (140). If the modified end bypass/chimeric transposition is the predominant mechanism for host gene capture at the DNA level, then increased *Helitron* density (especially in gene-rich regions) would also increase gene capture frequency. In maize, 6% of 1,930 intact *Helitrons* are inserted near genes (67). In the *Bombyx* genome, *Helitron* abundance within introns and within 0.1 kb of genes was higher than expected by chance (84). Insertion preference for genic region was observed for a random subset (~ 33% of 571 insertions) of *Helitrons* in the genome of *M. lucifugus* (51). In contrast, the *Helitrons* in the *Arabidopsis* and *C. elegans* genomes were enriched in the gene poor pericentromeric or terminal regions of each chromosome, respectively (67). However, this maybe the result of purifying selection, at least for *Arabidopsis*, as active families have been reported near genic regions (140, 141). *Helitrons* in the rice genome were distributed in both the pericentromeric as well as in the terminal regions of chromosomes (81). The abundance and insertion pattern observed is an outcome of multiple factors including insertion specificity, purifying selection on the insertions, and rate of DNA removal (for review see reference 142).

HINE insertions are most abundant in the centromeres and heterochromatic regions of drosophilid chromosomes (19, 22, 26, 49, 143), including the sex chromosomes (11, 26). However, recent studies suggest that the pattern of distribution is shaped by selection because young insertions are evenly distributed towards the chromosome arms and in the euchromatic regions (22, 39, 49). Indeed, *HINE* insertions are also found in genic regions in multiple drosophilids (39), lepidopterans (31), and molluscs (34, 35).

Helitrons in bat and a few other genomes have been found in tandem arrays (53, 70). Only the terminal elements in the tandem arrays display the typical sequence features. The internal copies are not found between an AT dinucleotide but rather the 3′ end abuts directly the 5′ end of another insertion sometimes of an identical exemplar (Figure 12). A similar phenomenon has been observed for bacterial IS*91* elements and is postulated to be linked to one-ended transposition via rolling-circle transposition (for review see reference 46). It should be noted that *Helitrons* can insert into each other creating this pattern but with different copies arrayed rather than sequences with high identity arrayed in this pattern (Figure 12). *Helitrons* are also found to insert near each other (51, 67). These insertions differ from the tandem arrays because they display the typical AT insertion site. Tandem arrays have also been identified for *HINE*s (Figure 12) (19, 23, 26, 36, 143, 144).

Gene Captures

One poorly understood process that likely has profound implications for genome evolution is the propensity of *Helitrons* to generate protogene families by picking up, shuffling and amplifying host genes. The formation of protogenes can occur through host capture of sequence at the DNA level, through retrotransposition, or via both processes in concert (see *Models for gene capture* section) (Figure 6). The protogenes are pieces of host genome merged into the *Helitron* forming a new exemplar. New exemplars are capable of transposition creating multi-protogene families. In some genomes this process has made a substantial impact on the total number of gene duplicates. For example, it is estimated that ~ 12,382 *Helitron*-generated protogenes carrying host gene fragment chimeras are present in the *M. lucifugus* genome (51). Because many of these protogene families are chimeric, they are *de novo* in the bat family Vespertilionidae. Indeed, many protogene families are lineage specific, suggesting that they might be the basis for genetic innovation in a species-specific fashion (51).

While most of the *de novo* protogenes are likely to degenerate into pseudogenes, some display patterns suggestive of purifying or positive selection. Indeed, in maize three gene fragments (4% of random subset of 85 gene fragments) exhibited significant evidence of purifying selection and three gene fragments are reported to be evolving under positive selection. Thomas *et al.* (51) also found that one retrogene amplified by a *Helitron* was evolving under purifying selection among the four copies with an intact open reading frame. These data suggest that the majority of the protogene duplicates are evolutionarily neutral (145) but occasionally a small fraction are transcribed and are translated, allowing functionalization (146).

The *Helitron*-mediated protogenes make up ~ 0.6% of the *M. lucifugus* genome (51), which is almost half of the predicted gene content of the human genome (147). The human genome is estimated to harbor ~ 20,000 pseudogenes produced by non-LTR retrotransposition, segmental duplication and unknown processes (but not *Helitron* activity) (148). All of these gene duplication processes are likely active in the *M. lucifugus* genome as well. In addition to these other

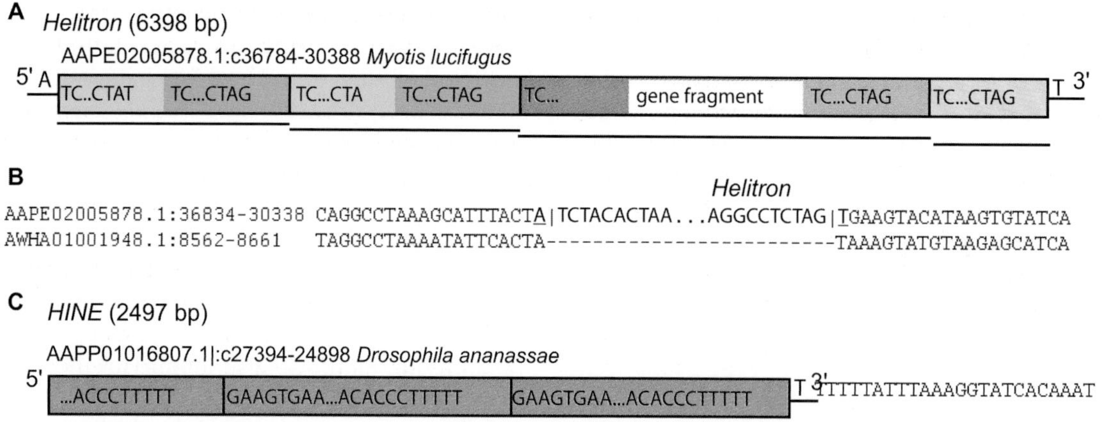

Figure 12 Tandem copies of *Helitrons* and *Helentron*-associated INterspersed Elements (*HINEs*). (A) Tandem copies of four *Helitrons* having 5′ TC and 3′ CTAG are counted as indicated by the horizontal black line underneath the box. Boxes with the same color indicate that they have > 85% sequence identity. Sequences homologous to multiple *Helitron* ends can be identified within a single *Helitron*. The sequences that are homologous to *Helitron 5′* ends are shown before the dots and sequences that are homologous to *Helitron 3′* ends are shown after the dots within the box. (B) Empty site of the tandem *Helitron* described above. The first line is host sequences with the *Helitron* insertion and second line is an orthologous site in another bat *Rhinolophus ferrumequinum*. (C) Three tandem copies of a *HINE* insertion in the *Drosophila ananassae* genome. One copy is truncated because it is at the end of the contig. The *HINE* copies are 99% identical to each other.
doi:10.1128/microbiolspec.MDNA3-0049-2014.f12

gene duplication processes, *Helitron* activity has profoundly impacted the repertoire of protogenes available in the genome. It is likely that relics of *Helitron* activity, such as *de novo* genes, are present in genomes where the causal *Helitron* families are no longer active.

Protogene formation has also occurred in other animal and plant genomes at different rates. In maize, Yang and Bennetzen (67) estimated that *Helitrons* have generated as many as ~ 24,000 protogenes. Most of the exemplars are low copy number, so contributing to low-copy-number protogene families. High-frequency gene capture has not been observed in other genome-wide analyses of plant genomes. The gene fragments found in *A. thaliana Helitrons* range from 30 to 350 bp and derive from only five genes (96). Similarly, the rice genome analysis did not reveal a high frequency of gene captures (93, 149, 150) (11 unique gene capture events in the 552 japonica rice *Helitrons*) (93). Cases of gene capture by *Helitrons* have also been reported in the fungus *Aspergillus nidulans*, (94) and from the plants *Lolium perenne* (95) and *Brachypodium distachyon* (150). Capture of gene fragments is reported in brassica and grape vine genomes (85, 151), but not enough evidence is provided to clearly show that the sequences reported are indeed part of *Helitrons*. In the two animal genomes *M. lucifugus* and *B. mori*, the gene captures by *Helitrons* occur at a higher frequency. In the silkworm genome, a total of 3,724 gene fragments from 268 genes were reported and > 18% (3,546/19,580) of the total intact elements carry one or more captured gene fragments (84). In all of these cases, the majority of the *Helitrons* had one gene fragment (51, 67, 68, 84, 93).

Some *Helitron*-amplified protogenes consist of nearly complete genes, not just fragments (100, 103). This is less likely to occur at the DNA level in genomes like mammals that harbor genes with big introns. *Helitron*-amplified copies of the cytidine deaminase (ZmCDA1) (100) and cytochrome P450 monooxygenase (P450) genes (103) were nearly complete. The transcriptional evidence for the three *Helitron*-amplified copies of the latter gene suggested that they could undergo selection for novel functions.

Helentrons are also found to capture and amplify gene fragments in many genomes (11). In the *Culex quinquefasciatus* genome, a *Helentron* that captured a histone gene fragment created a protogene family consisting of eight copies. In the Phytophthora genome, a *Proto-Helentron* that carries the gene for a putative transmembrane protein has been amplified to ~ 25 copies creating a large protogene family (11). Texim genes containing the SGNH hydrolase domain are also found to associate with some *Helentrons* in the platyfish genome; however, it is not clear whether they are mobilized by *Helentrons* or duplicated as part of segmental duplications (152). In *Drosophila*, *HINEs* are reported to have nonrandom associations with several gene duplicates, which are functional (144, 153–155). However it is not clear whether or not the duplications occur as a result of capture or by ectopic recombination (144, 153–155) because of the rapid sequence turnover and high deletion rates in the drosophilids (11, 156). *HINE* insertions were described near 11 gene duplicates from different *Drosophila* species (144, 153, 154, 157). To date, no genomes boast a frequency of protogene formation by *Helentrons* that rivals that observed with maize or bat *Helitrons* (e.g., 51, 81), but this may result from an ascertainment bias because the relationship of these elements to *Helitrons* was only recently unequivocally demonstrated (11). However, these findings suggest that *Helentron* amplification could also play an important role in the amplification and dispersal of genic fragments across the genome.

Protogene Family Generation via Capture at the RNA Level: The Amplification of Retrogenes

In the *M. lucifugus* genome, Thomas *et al.* (51) identified four distinct instances of mRNAs that had been reverse transcribed into *Helitrons* and further propagated (118 copies) (51) (Figure 6B). This is the first report of such retrogene propagation by *Helitrons* or, to our knowledge, any DNA transposon. The retrogenes are usually promoterless; therefore they most likely evolve as pseudogenes (for review see reference 158). One example in the *M. lucifugus* genome, the *RPLP0* mRNA was retrotransposed into the captured promoter region of *SRPK1* gene within a *Helitron* (exemplar HelibatN424). This illustrates one mechanism by which retrogenes could be brought into close proximity to prefabricated promoter regions, gaining the sequences necessary to support transcription (51). However, the transcription of *RPLP0* transcribed from this promoter remains to be investigated, as the *SRPK1* promoter typically drives expression in the direction opposite to the retrogene (159).

Generation of Structural Diversity and Genetic Variation

Helitrons have contributed to structural diversity and polymorphisms among maize inbred lines (for review see reference 160). Polymorphic *Helitron*, insertions carrying multiple gene fragments disrupt the genic collinearity between inbred lines (101, 104, 161). The whole genome comparison of gene content in allelic BAC contigs in the B73 and Mo17 maize inbred lines revealed that *Helitron* insertions introduced at least 10,000 genic polymorphisms (97, 105). Some of these insertions are present in multiple copies in the genomes (97, 98) and produced chimeric transcripts containing segments from different regions (97, 98, 101). Indeed, this pattern of generation of genetic variation is probably not limited to maize but may be at play in other plant genomes that harbor abundant *Helitrons*. Indeed, polymorphic insertions of *Helitrons* have been identified in rice, between indica and japonica cultivars (93), and among *A. thaliana* individuals (96) or haplotypes (162, 163). TEs including *Helitrons* even when no longer active are frequently subject to ectopic recombination events based on sequence homology. These events can lead to duplication, deletion, inversions, and translocations (for review see reference 6). Indeed, different TEs including *Helitrons* are identified at the break points of the large duplicated fragments in *Brachypodium*, rice, and sorghum (150). In addition, *Helitron* fragments flank nearly one-third of the 154 chromosomal inversions identified between *A. thaliana* and *Arabidopsis lyrata* genomes (164, for review see reference 142).

One mechanism of speciation is hypothesized to occur when recombination is inhibited due to chromosomal topology incongruencies like those described above (for reviews see references 165, 166, and 167). Indeed, He and Dooner reported that no recombination events could be detected in a 100-kb space containing multiple *Helitron* insertions and that recombination was inversely correlated with a pattern of dense *Helitron* methylation (168).

The contribution of *Helitrons* to the generation of structural diversity is not limited to plants. Novel lineage-specific *Helitron* insertions have been described that have resulted from the continued activity of the *Helibat* family during diversification of the Vespertilionidae family. Although it has not been specifically documented; it is likely that *Helentron* activity has contributed to genetic diversity among Drosophilid genomes (21, 22, 28, 39, 49, 127). Considering the widespread distribution, the potential for recent and persistent activity, *Helitrons* and *Helentrons* have likely contributed to genetic variation in other organisms.

SPONTANEOUS *HELITRON* MUTATIONS AND IMPACT ON GENE EXPRESSION

To understand the impact that *Helitron* activity has had on gene expression, regulation and the generation of novel genetic units, multiple transcription surveys have been undertaken in different plants, animals, and a fungus. The results are not directly comparable because some studies examined whole transcriptomes, some were RNA-seq based, while others used expressed sequence tag (EST) analysis or surveyed examples on a case-by-case basis. Even the global analysis suffered different limitations, for example whether the data sets include coding and noncoding transcripts and the relative contributions. However, some intriguing patterns emerge from these analyses implicating *Helitron* activity in the generation of variation in both coding and noncoding transcripts. In this section we will focus on known examples of how *Helitrons* shape the diversity and regulation of the transcriptome.

Contribution of *Helitrons* to Transcriptomes

One simple approach to assessing the impact is to measure the abundance of transcripts that contain a portion of a *Helitron*. The transcripts could be coding or noncoding. Masking the transcriptome from the bat salivary gland revealed that approximately 1.4% of the 29,493 transcripts contained a *Helitron* fragment (51). This data set comprised all transcripts including isoforms with a fragments per kilobase of exon per million reads mapped > 1 and longer than 200 bp regardless of coding potential. An analysis of multiple tissues from *B. mori* revealed that 123 of 8,654 of full-length cDNAs from multiple tissues contained a *Helitron* sequence (84). The full-length cDNAs were sampled from multiple tissues at multiple stages and represent a more thorough sampling than reported for the little brown bat. In maize, where *Helitrons* are much more recent, at least 9% of the identified *Helitrons* (~ 1800) are expressed in at least one tissue based on screening maize EST databases (67, 82). In addition, a reverse transcription-PCR analysis revealed that four randomly selected *Helitron*-containing gene fragments produced 24 alternatively spliced transcripts (six on average), most of which were not detected in the maize EST database (82).

How *Helitron* Insertions Have Altered Gene Expression

Many examples have been described that illustrate how *Helitron* activity impacts gene expression (Figure 13). *Helitrons*, like all TEs, are potent insertional mutagens. The *barren stalk-1* (102) and three *tasselseed4* (*ts4*) mutants (169) have suffered a *Helitron* insertion within the promoter region that results in the abolition of measurable transcripts and the observed phenotypes (102, 169) (Figure 13A). *Helitron* insertions can modulate the expression of nearby genes. Both the ETT/Auxin responsive Factor3 (*ARF3*) and *ARF4* genes in the *TEBICHI* gene mutant (*teb*) background in *Arabidopsis* display increased transcripts levels, which are linked to defects in the adaxial–abaxial polarity of leaves (170) (Figure 13B). Because these mutants have *Helitron* insertions in close proximity to the genes in the mutant background, the phenotypes have been linked to the insertions. In some cases it has been shown that a *Helitron* insertion has provided regulatory motifs necessary for transcription initiation. In one example, the CArG motif necessary for the transcription of the master regulator gene Leafy cotyledon 2 involved in seed development in *A. thaliana* is provided by a *Helitron* insertion (171). Similarly, a *HINE* insertion (in the ancestor of *Drosophila subobscura* and *Drosophila guache*) upstream of a domesticated P element "repressor-like" protein-coding gene drives the expression and provides *de novo* regulatory elements such as CAAT-box, GC-box, octamer motif, and TATA- box sites (22, 172, 173) (Figure 13C). In addition, multiple chimeric transcripts have been identified in maize and bat that initiate within *Helitrons* although phenotype associations have not been investigated or at least reported (51, 82, 97, 101, 103) (Figure 13D).

Helitrons alter the length and sequence of both 5′ UTRs and 3′ UTRs of the coding transcripts from *M. lucifugus* and *B. mori*, although the frequency of such transcripts differs between species. For example, transcripts containing *Helitrons* in the 5′ UTR are few in the *M. lucifugus* transcriptome (51) compared with *B. mori* (84) (Figure 13E). This discrepancy might be influenced by the age of the *Helitrons* in both genomes. The insertions in the *B. mori* genome are on average much younger and therefore purifying selection may not have shaped the distribution yet. Alternatively, transcripts containing *Helitrons* in the 3′ UTR are relatively abundant in both the *M. lucifugus* salivary gland transcriptome as well as in *B. mori*. In the tetraploid sour cherry (*Prunus cerasus*), a *Helitron* insertion in the 3′ UTR is proposed to disrupt polyadenylation, resulting in a loss of function of the SFB gene involved in gametophytic self-incompatibility (Figure 13F) (174). In *M. lucifugus*, *Helitron* insertions in the 3′ UTR have provided putative alternative polyadenylation signals and putative regulatory motifs including binding sites for conserved microRNA (Figure 13G and H) (51). Putative microRNA have been reported to arise from *Helitrons* that might regulate genes expressed in *E. fuscus* testes (175).

Because *Helitrons* frequently carry host regulatory sequences like promoters, it is of interest to understand

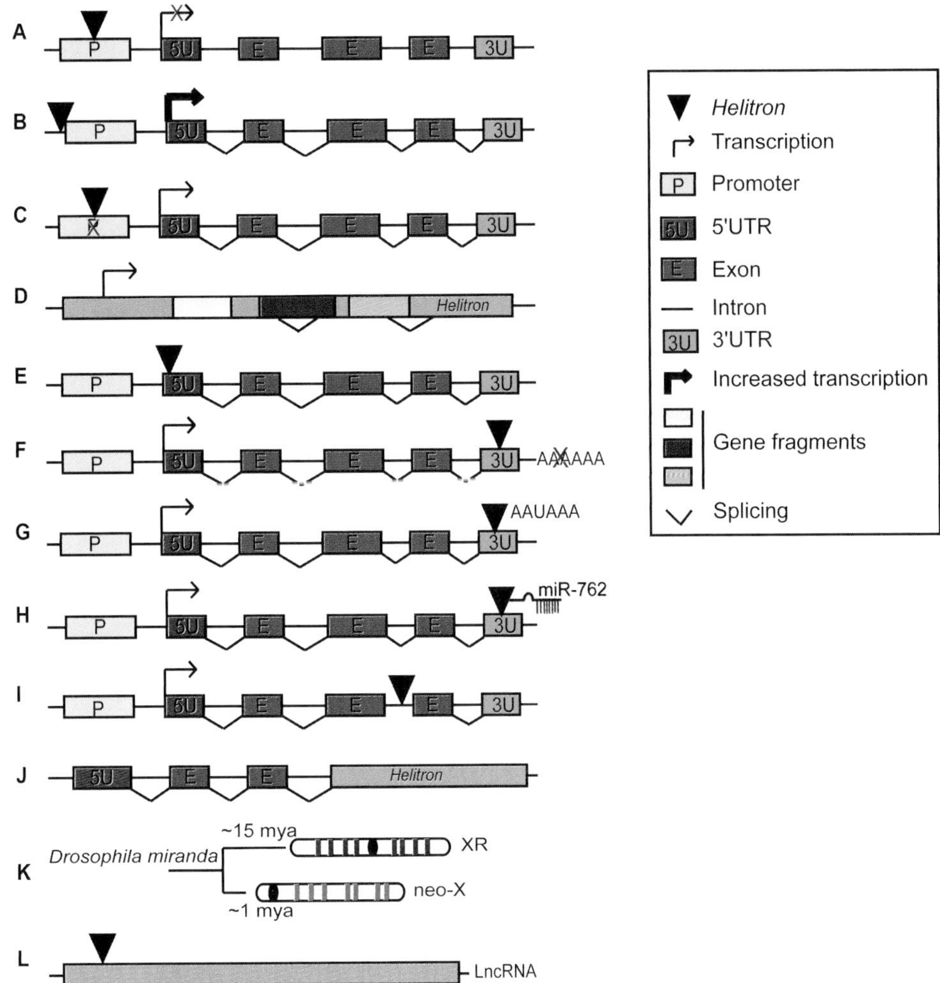

Figure 13 Examples of the impact of *Helitrons* on gene structure and expression. (A) Insertion of a *Helitron* in the promoter (Yellow box with letter P) disrupts the transcription of the gene (102, 169). (B) A *Helitron* insertion upstream of the promoter can increase the expression of the gene (170). (C) A *Helentron*-associated INterspersed Element (*HINE*) insertion in the promoter region of P element disrupts the promoter but provides a *de novo* promoter (172, 173). (D) Chimeric transcript generated from a *Helitron* containing multiple gene fragments (maroon, light orange and light pink boxes) (e.g., 82, 98, 101). (E) *Helitron* insertions in the 5′ UTR contributes to transcript diversity (51, 84). (F) Insertion in the 3′ UTR can disrupt the polyadenylation of the transcripts causing loss of function (174). (G) *Helitron* insertion provides novel alternative polyadenylation sites (51). (H) *Helitron* insertions in the 3′ UTR provides putative microRNA binding sites (51). (I) Insertion of a *Helitron* in an intron can disrupt or alter splicing increasing transcript diversity (51, 82) often causing loss of function (99, 177). (J) *Helitrons* provide cryptic splice sites and create novel fusion transcripts (51, 82). (K) Two different waves of *HINE* amplification provided binding sites for the protein involved in dosage compensation (shown in orange and red bars) on the X chromosomes that were generated ~ 15 and ~1 million years ago (127). (L) *Helitron* insertions contribute to long noncoding RNAs (51).
doi:10.1128/microbiolspec.MDNA3-0049-2014.f13

if these sequences can be repurposed to regulate protogenes. In the *M. lucifugus* salivary gland, no concrete evidence of transcription from *Helitron*-amplified host promoters could be found. Failure to detect such examples is likely influenced by the limitations of the data set, including low coverage, selection of transcripts with polyAs, and lack of information about strand specificity (51). No transcription was found from a *Helitron*-

amplified promoter of *xanA* gene in the fungus *Aspergillus nidulans* (94). However, *ex vivo* experiments have shown that the *Helitron*-amplified promoter is intact and capable of driving transcription (94). These studies have not produced any examples of repurposed host sequence but are limited by experimental design or scope; therefore this is still an open question that deserves to be addressed.

The Generation of Novel Splice Variants

Helitrons may contribute to novel splice variants by promoting alternative splicing and by providing cryptic splice sites (51, 82) (Figure 13I and J). A number of spontaneous mutations have been reported in plants that are caused by intronic *Helitron* insertions that result in the generation of chimeric transcript species. In maize, a *Helitron* insertion has been linked to the spontaneous mutant *shrunken2 (sh2)* phenotype. In this example, the insert ion into intron 11 generates splice variants, disrupting the synthesis of an enzyme key in the starch biosynthetic pathway (99, 176) (Figure 13I). The insertion of a putatively autonomous *Helitron* into intron 5 of the anthocyanin-producing gene in *Ipomoea tricolor* results in a white flower mutation (*pearly-s*) (177). As observed in the maize *sh2-7527* mutant (99) the *Helitron* insertion created a *de novo* chimeric transcript species with altered splicing pattern (177) (Figure 13I). In the yellow seed coat mutant in *Brassica rapa*, no transcripts were detected for the mutant allele with a *Helitron* insertion in intron 2 of the *Br TT8* gene (178). It is not clear why an insertion into an intron would disrupt gene expression. It is possible that a novel transcript with a premature stop is produced, which triggers the nonsense-mediated decay pathway (179). *Helitrons* can form fusion transcripts involving other genes by providing cryptic splice sites (51, 82) (Figure 13J). The downstream effects of these variants are not known, but such events have the potential for generation of fusion proteins with new function.

Rewiring of Regulatory Networks

Recently a report of *Helitron* involvement in the rewiring of regulatory networks illustrated that activity over an extended period of evolutionary time has led to innovation. In this example, *HINE*s were described that carry sites involved in the regulation of dosage compensation in *Drosophila* (110) (Figure 12K). The insertions were shown to provide binding sites for the male-specific lethal (MSL) complex which facilitates increased gene expression of genes on the male sex chromosome (127) (Figure 13K). An independent role for *HINEs* in gene expression has been proposed for insecticide-resistance genes (180). A nonrandom accumulation of *HINE* insertions was described upstream of the cytochrome P450 (*cyp*) gene cluster (180) that is proposed to play a role in the insecticide resistance (181–184). Interestingly, putative transcription factor binding sites were found to be carried by *HINEs* in the *cyp* genes associated with insecticide resistance both in *Drosophila simulans* and *Drosophila melanogaster* (180).

Host Gene Regulation as a Consequence of Gene Capture

Transcripts generated from *Helitrons* containing gene fragments have been hypothesized to regulate the expression of parental gene transcripts (185) possibly by generating small RNAs, which target both the *Helitron* as well as the parental genes (67, 186). Small RNAs homologous to *Helitron* sequences have been described from maize (67), *A. thaliana* (134), *Phytophthora* species, (187, 188) and *Physcosmitrella patens* (189). Generally it is surmised that one kind of small RNA (small interfering (siRNA)) regulate TE expression. If siRNAs are produced from host gene fragments captured by the *Helitron*, it is possible that host gene expression would therefore be mitigated as a side effect of TE control. This model sets up an interesting paradigm for the retention of genes captured that are involved in TE control. How the host controls *Helitrons* and the link to the capture and retention of gene fragments are not well understood and require further experimental studies.

Although these examples are limited, it is clear that *Helitron* insertions can impact gene expression and phenotype, at least in plants. In addition, *Helitrons* also form part of both conserved and novel long noncoding RNAs in *M. lucifugus* (51) (Figure 13). While TEs are known to play a major role in the evolution of lincRNA, the role of *Helitrons* has not been formally investigated, (190, 191). It is likely that *Helitrons*, will play an important role, especially in those genomes where these elements are abundant.

OPEN QUESTIONS

Genome-wide analysis has allowed the diversity of four structural and coding variants of *Helitrons* to be investigated. It has illustrated the role that *Helitron* and *Helentron* transposition has played in gene duplication and how activity has sculpted the genetic architecture but neither the various mechanisms by which this occurs or the frequency is well understood. Genetic analysis has been leveraged to identify recently active families, the polymorphism they have caused and possible excision events. Nonetheless, many questions remain. For example, why do some *Helitron* families spawn more gene duplicates than others? What features govern the apparent long-term activity that characterizes some *Helitrons* or *Helentrons*? Is this due to specific features of the element structure, abundance, repair mechanisms, or is it dependent on stochastic processes that influence the fixation of elements, is it entirely based on population genetic features? The development of a cell based transposition system would not only allow the dissection of the transposition mechanism and what features lead to excision but it would also allow the models of gene capture to be formally tested. Questions about the frequency and requirements of gene capture during transposition could be addressed for the structural variants. The purpose and necessity of the additional auxiliary genes could be delimited. Are there structural or coding variants of *Helitrons*? Beyond these fundamental questions about *Helitron* biology remain questions about impact on the host. For example, how does the host genome cope with the bulk of *Helitron*-amplified gene fragments? Are *Helitron* amplified promoter and regulatory regions active in the genome? If not how are they silenced? How is the host able to distinguish self and nonself when host DNA is captured and amplified? Are the genes and sequences carried by *Helitrons* domesticated in any genomes as previously observed for other cut-and-paste DNA transposons? Have *Helitrons* mediated the horizontal transfer of genes?

CONCLUSIONS

The eukaryotic rolling-circle transposons include canonical *Helitrons* and the structural and coding variants called *Helentron*, *Proto-Helentron*, and *Helitron2*. *Helitrons* and *Helentrons* have a patchy distribution among eukaryotes and are mainly characterized by the presence of nonautonomous elements. Many abundant but previously ambiguously classified nonautonomous repeats have been linked to *Helentrons*, for example *DINE-1* like elements. In contrast to other animal cut-and-paste DNA transposons, both *Helitrons* and *Helentrons* have the ability for long-term activity that has led to various genomic impacts with evolutionary significance. For example, the persistent activity of *Helentrons* has led to evolution of gene regulatory networks in *D. miranda* (127) whereas the activity of *Helitrons* over the last ~ 36 my has led to the generation of lineage-specific structural variation in the bat family Vespertilionidae (51). In addition, *Helitrons* and *Helentrons* can alter gene expression, in some cases with phenotypic consequences. Reports of excision of *Helitrons* suggest transposition may occur by a peel-and-paste mechanism (originally proposed for some bacterial transposons (e.g., IS*200*/*605*)) rather than by a copy-and-paste mechanism. Furthermore, the ability to transpose even when carrying diverse internal host sequence of different sizes suggests that *Helitrons* might make interesting tools for genome engineering.

Acknowledgments. *We would like to thank Rachel Leigh Cosby for the critical reviewing of the manuscript. This work was supported by startup funds from the university of Utah to EJP.*

Citation. Thomas J, Pritham EJ. 2014. *Helitrons*, the eukaryotic rolling-circle transposable elements. Microbiol Spectrum 3(1):MDNA3-0049-2014.

References

1. **Kapitonov VV, Jurka J.** 2001. Rolling-circle transposons in eukaryotes. *Proc Natl Acad Sci USA* **98:** 8714–8719.
2. **Feschotte C, Wessler SR.** 2001. Treasures in the attic: rolling circle transposons discovered in eukaryotic genomes. *Proc Natl Acad Sci USA* **98:**8923–8924.
3. **Feschotte C, Pritham EJ.** 2005. Non-mammalian c-integrases are encoded by giant transposable elements. *Trends Genet* **21:**551–552.
4. **Pritham EJ, Putliwala T, Feschotte C.** 2007. Mavericks, a novel class of giant transposable elements widespread in eukaryotes and related to DNA viruses. *Gene* **390:** 3–17.
5. **Kapitonov VV, Jurka J.** 2006. Self-synthesizing DNA transposons in eukaryotes. *Proc Natl Acad Sci U S A* **103:**4540–4545.
6. **Feschotte C, Pritham EJ.** 2007. DNA transposons and the evolution of eukaryotic genomes. *Annu Rev Genet* **41:**331–368.
7. **Mizuuchi K, Baker TA.** 2002. Chemical mechanisms for mobilizing DNA. *In* Craig NL, Craigie R, Gellert M, Lambowitz A (ed), *Mobile DNA II*, p 12–23. ASM Press, Washington D. C.
8. **Craig NL.** 2002. Tn7, p 423–454. *In* Craig NL, Craigie R, Gellert M, Lambowitz A (ed), *Mobile DNA II*. ASM Press, Washington D.C.
9. **Chandler M, de la Cruz F, Dyda F, Hickman AB, Moncalian G, Ton-Hoang B.** 2013. Breaking and joining single-stranded DNA: the HUH endonuclease superfamily. *Nat Rev Microbiol* **11:**525–538.
10. **Poulter RT, Goodwin TJ, Butler MI.** 2003. Vertebrate helentrons and other novel Helitrons. *Gene* **313:**201–212.
11. **Thomas J, Vadnagara K, Pritham EJ.** 2014. DINE-1, the highest copy number repeats in Drosophila melanogaster are non-autonomous endonuclease-encoding rolling-circle transposable elements (Helentrons). *Mob DNA* **5:**18.
12. **Bao W, Jurka J.** 2013. Homologues of bacterial TnpB_IS605 are widespread in diverse eukaryotic transposable elements. *Mob DNA* **4:**12.

13. **Doutriaux MP, Couteau F, Bergounioux C, White C.** 1998. Isolation and characterisation of the RAD51 and DMC1 homologs from *Arabidopsis thaliana. Mol Gen Genet* **257:**283–291.

14. **Surzycki SA, Belknap WR.** 1999. Characterization of repetitive DNA elements in *Arabidopsis. J Mol Evol* **48:** 684–691.

15. **Kapitonov VV, Jurka J.** 1999. Molecular paleontology of transposable elements from *Arabidopsis thaliana. Genetica* **107:**27–37.

16. **Le QH, Wright S, Yu Z, Bureau T.** 2000. Transposon diversity in *Arabidopsis thaliana. Proc Natl Acad Sci U S A* **97:**7376–7381.

17. **Silva R, Burch JB.** 1989. Evidence that chicken CR1 elements represent a novel family of retroposons. *Mol Cell Biol* **9:**3563–3566.

18. **Malik HS, Burke WD, Eickbush TH.** 1999. The age and evolution of non-LTR retrotransposable elements. *Mol Biol Evol* **16:**793–805.

19. **Locke J, Howard LT, Aippersbach N, Podemski L, Hodgetts RB.** 1999. The characterization of DINE-1, a short, interspersed repetitive element present on chromosome and in the centric heterochromatin of *Drosophila melanogaster. Chromosoma* **108:**356–366.

20. **Kapitonov VV, Jurka J.** 2003. Molecular paleontology of transposable elements in the *Drosophila melanogaster* genome. *Proc Natl Acad Sci U S A* **100:**6569–6574.

21. **Hagemann S, Miller WJ, Haring E, Pinsker W.** 1998. Nested insertions of short mobile sequences in *Drosophila* P elements. *Chromosoma* **107:**6–16.

22. **Miller WJ, Nagel A, Bachmann J, Bachmann L.** 2000. Evolutionary dynamics of the SGM transposon family in the *Drosophila obscura* species group. *Mol Biol Evol* **17:**1597–1609.

23. **Steinemann M, Steinemann S.** 1992. Degenerating Y chromosome of *Drosophila miranda*: a trap for retrotransposons. *Proc Natl Acad Sci U S A* **89:**7591–7595.

24. **Wilder J, Hollocher H.** 2001. Mobile elements and the genesis of microsatellites in dipterans. *Mol Biol Evol* **18:** 384–392.

25. **Vivas MV, García-Planells J, Ruiz C, Marfany G, Paricio N, Gonzàlez-Duarte R, de Frutos R.** 1999. GEM, a cluster of repetitive sequences in the *Drosophila subobscura* genome. *Gene* **229:**47–57.

26. **Kuhn GC, Heslop-Harrison JS.** 2011. Characterization and genomic organization of PERI, a repetitive DNA in the *Drosophila buzzatii* cluster related to DINE-1 transposable elements and highly abundant in the sex chromosomes. *Cytogenet Genome Res* **132:**79–88.

27. **Cáceres M, Ranz JM, Barbadilla A, Long M, Ruiz A.** 1999. Generation of a widespread *Drosophila* inversion by a transposable element. *Science* **285:**415–418.

28. **Negre B, Ranz JM, Casals F, Cáceres M, Ruiz A.** 2003. A new split of the Hox gene complex in *Drosophila*: relocation and evolution of the gene labial. *Mol Biol Evol* **20:**2042–2054.

29. **Marín I, Fontdevila A.** 1996. Evolutionary conservation and molecular characteristics of repetitive sequences of *Drosophila koepferae. Heredity (Edinb)* **76**(Pt 4)**:**355–366.

30. **Marin I, Labrador M, Fontdevila A.** 1992. The evolutionary history of *Drosophila buzzatii.* XXIII. High content of nonsatellite repetitive DNA in *D. buzzatii* and in its sibling *D. koepferae. Genome* **35:**967–974.

31. **Coates BS, Kroemer JA, Sumerford DV, Hellmich RL.** 2011. A novel class of miniature inverted repeat transposable elements (MITEs) that contain hitchhiking (GTCY)(n) microsatellites. *Insect Mol Biol* **20:**15–27.

32. **Coates BS, Sumerford DV, Hellmich RL, Lewis LC.** 2010. A helitron-like transposon superfamily from lepidoptera disrupts (GAAA)(n) microsatellites and is responsible for flanking sequence similarity within a microsatellite family. *J Mol Evol* **70:**275–288.

33. **Zhang HH, Xu HE, Shen YH, Han MJ, Zhang Z.** 2013. The origin and evolution of six miniature inverted-repeat transposable elements in *Bombyx mori* and *Rhodnius prolixus. Genome Biol Evol* **5:**2020–2031.

34. **Kourtidis A, Drosopoulou E, Pantzartzi CN, Chintiroglou CC, Scouras ZG.** 2006. Three new satellite sequences and a mobile element found inside HSP70 introns of the Mediterranean mussel (*Mytilus galloprovincialis*). *Genome* **49:**1451–1458.

35. **Gaffney PM, Pierce JC, Mackinley AG, Titchen DA, Glenn WK.** 2003. Pearl, a novel family of putative transposable elements in bivalve mollusks. *J Mol Evol* **56:** 308–316.

36. **Satovic E, Plohl M.** 2013. Tandem repeat-containing MITEs in the clam *Donax trunculus. Genome Biol Evol* **5:**2549–2559.

37. **Cohen JB, Liebermann D, Kedes L.** 1985. Tsp transposons: a heterogeneous family of mobile sequences in the genome of the sea urchin *Strongylocentrotus purpuratus. Mol Cell Biol* **5:**2814–2825.

38. **Wang J, Keightley PD, Halligan DL.** 2007. Effect of divergence time and recombination rate on molecular evolution of *Drosophila* INE-1 transposable elements and other candidates for neutrally evolving sites. *J Mol Evol* **65:**627–639.

39. **Yang HP, Barbash DA.** 2008. Abundant and species-specific DINE-1 transposable elements in 12 *Drosophila* genomes. *Genome Biol* **9:**R39.

40. **Marfany G, Gonzàlez-Duarte R.** 1992. Evidence for retrotranscription of protein-coding genes in the *Drosophila subobscura* genome. *J Mol Evol* **35:**492–501.

41. **Toleman MA, Bennett PM, Walsh TR.** 2006. ISCR elements: novel gene-capturing systems of the 21st century? *Microbiol Mol Biol Rev* **70:**296–316.

42. **Tavakoli N, Comanducci A, Dodd HM, Lett MC, Albiger B, Bennett P.** 2000. IS1294, a DNA element that transposes by RC transposition. *Plasmid* **44:** 66–84.

43. **Mendiola MV, Jubete Y, de la Cruz F.** 1992. DNA sequence of IS91 and identification of the transposase gene. *J Bacteriol* **174:**1345–1351.

44. **Mendiola MV, Bernales I, de la Cruz F.** 1994. Differential roles of the transposon termini in IS91 transposition. *Proc Natl Acad Sci U S A* **91:**1922–1926.

45. **Siguier P, Gourbeyre E, Chandler M.** 2014. Bacterial insertion sequences: their genomic impact and diversity. *FEMS Microbiol Rev* **38:**865–891.
46. **Garcillan-Barcia MP, Bernales I, Mendiola MV, de La Cruz F.** 2002. IS91 Rolling-circle transposition, p 891–904. *In* Craig NL, Craigie R, Gellert M, Lambowitz A (ed), *Mobile DNA II.* ASM Press, Washington D. C.
47. **Mahillon J, Chandler M.** 1998. Insertion sequences. *Microbiol Mol Biol Rev* **62:**725–774.
48. **Abdurashitov MA, Gonchar DA, Chernukhin VA, Tomilov VN, Tomilova JE, Schostak NG, Zatsepina OG, Zelentsova ES, Evgen'ev MB, Degtyarev SK.** 2013. Medium-sized tandem repeats represent an abundant component of the *Drosophila virilis* genome. *BMC Genomics* **14:**771.
49. **Yang HP, Hung TL, You TL, Yang TH.** 2006. Genomewide comparative analysis of the highly abundant transposable element DINE-1 suggests a recent transposition burst in *Drosophila yakuba. Genetics* **173:** 189–196.
50. **Coates BS, Sumerford DV, Hellmich RL, Lewis LC.** 2009. Repetitive genome elements in a European corn borer, *Ostrinia nubilalis*, bacterial artificial chromosome library were indicated by bacterial artificial chromosome end sequencing and development of sequence tag site markers: implications for lepidopteran genomic research. *Genome* **52:**57–67.
51. **Thomas J, Phillips CD, Baker RJ, Pritham EJ.** 2014. Rolling-circle transposons catalyze genomic innovation in a mammalian lineage. *Genome Biol Evol* **6:**2595–2610.
52. **Gemayel R, Vinces MD, Legendre M, Verstrepen KJ.** 2010. Variable tandem repeats accelerate evolution of coding and regulatory sequences. *Annu Rev Genet* **44:** 445–477.
53. **Pritham EJ, Feschotte C.** 2007. Massive amplification of rolling-circle transposons in the lineage of the bat *Myotis lucifugus. Proc Natl Acad Sci USA* **104:**1895–1900.
54. **Zhou Q, Froschauer A, Schultheis C, Schmidt C, Bienert GP, Wenning M, Dettai A, Volff JN.** 2006. Helitron transposons on the sex chromosomes of the platyfish *Xiphophorus maculatus* and their evolution in animal genomes. *Zebrafish* **3:**39–52.
55. **Bochman ML, Sabouri N, Zakian VA.** 2010. Unwinding the functions of the Pif1 family helicases. *DNA Repair (Amst)* **9:**237–249.
56. **Kapitonov VV, Jurka J.** 2007. Helitrons on a roll: eukaryotic rolling-circle transposons. *Trends Genet* **23:** 521–529.
57. **Feng Q, Moran JV, Kazazian HH, Boeke JD.** 1996. Human L1 retrotransposon encodes a conserved endonuclease required for retrotransposition. *Cell* **87:**905–916.
58. **Feng Q, Schumann G, Boeke JD.** 1998. Retrotransposon R1Bm endonuclease cleaves the target sequence. *Proc Natl Acad Sci U S A* **95:**2083–2088.
59. **Cost GJ, Boeke JD.** 1998. Targeting of human retrotransposon integration is directed by the specificity of the L1 endonuclease for regions of unusual DNA structure. *Biochemistry* **37:**18081–18093.
60. **Takahashi H, Fujiwara H.** 2002. Transplantation of target site specificity by swapping the endonuclease domains of two LINEs. *EMBO J* **21:**408–417.
61. **Dong Y, Lu X, Song W, Shi L, Zhang M, Zhao H, Jiao Y, Lai J.** 2011. Structural characterization of helitrons and their stepwise capturing of gene fragments in the maize genome. *BMC Genomics* **12:**609.
62. **Cocca E, De Iorio S, Capriglione T.** 2011. Identification of a novel helitron transposon in the genome of Antarctic fish. *Mol Phylogenet Evol* **58:**439–446.
63. **Balakirev MY, Tcherniuk SO, Jaquinod M, Chroboczek J.** 2003. Otubains: a new family of cysteine proteases in the ubiquitin pathway. *EMBO Rep* **4:**517–522.
64. **Makarova KS, Aravind L, Koonin EV.** 2000. A novel superfamily of predicted cysteine proteases from eukaryotes, viruses and *Chlamydia pneumoniae. Trends Biochem Sci* **25:**50–52.
65. **Dillon SC, Zhang X, Trievel RC, Cheng X.** 2005. The SET-domain protein superfamily: protein lysine methyltransferases. *Genome Biol* **6:**227.
66. **Herz HM, Garruss A, Shilatifard A.** 2013. SET for life: biochemical activities and biological functions of SET domain-containing proteins. *Trends Biochem Sci* **38:** 621–639.
67. **Yang LX, Bennetzen JL.** 2009. Distribution, diversity, evolution, and survival of Helitrons in the maize genome. *Proc Natl Acad Sci USA* **106:**19922–19927.
68. **Du C, Fefelova N, Caronna J, He LM, Dooner HK.** 2009. The polychromatic Helitron landscape of the maize genome. *Proc Natl Acad Sci USA* **106:**19916–19921.
69. **Tempel S, Nicolas J, El Amrani A, Couee I.** 2007. Model-based identification of Helitrons results in a new classification of their families in *Arabidopsis thaliana. Gene* **403:**18–28.
70. **Thomas J, Schaack S, Pritham EJ.** 2010. Pervasive horizontal transfer of rolling-circle transposons among animals. *Genome Biol Evol* **2:**656–664.
71. **Wicker T, Sabot F, Hua-Van A, Bennetzen JL, Capy P, Chalhoub B, Flavell A, Leroy P, Morgante M, Panaud O, Paux E, SanMiguel P, Schulman AH.** 2007. A unified classification system for eukaryotic transposable elements. *Nat Rev Genet* **8:**973–982.
72. **Kurtz S, Ohlebusch E, Schleiermacher C, Stoye J, Giegerich R.** 2000. Computation and visualization of degenerate repeats in complete genomes. *Proc Int Conf Intell Syst Mol Biol* **8:**228–238.
73. **Volfovsky N, Haas BJ, Salzberg SL.** 2001. A clustering method for repeat analysis in DNA sequences. *Genome Biol* **2:**RESEARCH0027.
74. **Bao Z, Eddy SR.** 2002. Automated de novo identification of repeat sequence families in sequenced genomes. *Genome Res* **12:**1269–1276.
75. **Price AL, Jones NC, Pevzner PA.** 2005. De novo identification of repeat families in large genomes. *Bioinformatics* **21:**I351–I358.
76. **Edgar RC, Myers EW.** 2005. PILER: identification and classification of genomic repeats. *Bioinformatics* **21** (Suppl 1)**:**i152–i158.

77. **Flutre T, Duprat E, Feuillet C, Quesneville H.** 2011. Considering transposable element diversification in *de novo* annotation approaches. *PLoS One* **6:**e16526.

78. **Feschotte C, Keswani U, Ranganathan N, Guibotsy ML, Levine D.** 2009. Exploring repetitive DNA landscapes using REPCLASS, a tool that automates the classification of transposable elements in eukaryotic genomes. *Genome Biol Evol* **1:**205–220.

79. **Hoede C, Arnoux S, Moisset M, Chaumier T, Inizan O, Jamilloux V, Quesneville H.** 2014. PASTEC: an automatic transposable element classification tool. *PLoS One* **9:**e91929.

80. **Du C, Caronna J, He L, Dooner HK.** 2008. Computational prediction and molecular confirmation of Helitron transposons in the maize genome. *BMC Genomics* **9:**51.

81. **Yang LX, Bennetzen JL.** 2009. Structure-based discovery and description of plant and animal Helitrons. *Proc Natl Acad Sci USA* **106:**12832–12837.

82. **Barbaglia AM, Klusman KM, Higgins J, Shaw JR, Hannah LC, Lal SK.** 2012. Gene capture by Helitron transposons reshuffles the transcriptome of maize. *Genetics* **190:**965–975.

83. **Xiong W, He L, Lai J, Dooner HK, Du C.** 2014. HelitronScanner uncovers a large overlooked cache of Helitron transposons in many plant genomes. *Proc Natl Acad Sci U S A* **111:**10263–10268.

84. **Han MJ, Shen YH, Xu MS, Liang HY, Zhang HH, Zhang Z.** 2013. Identification and evolution of the silkworm Helitrons and their contribution to transcripts. *DNA Res* **20:**471–484.

85. **Fu D, Wei L, Xiao M, Hayward A.** 2013. New insights into helitron transposable elements in the mesopolyploid species *Brassica rapa*. *Gene* **532:**236–245.

86. **Xiong W, Li T, Chen K, Tang K.** 2009. Local combinational variables: an approach used in DNA-binding helix-turn-helix motif prediction with sequence information. *Nucleic Acids Res* **37:**5632–5640.

87. **Galas DJ, Chandler M.** 1981. On the molecular mechanisms of transposition. *Proc Natl Acad Sci U S A* **78:** 4858–4862.

88. **Garcillán-Barcia MP, Bernales I, Mendiola MV, de la Cruz F.** 2001. Single-stranded DNA intermediates in IS91 rolling-circle transposition. *Mol Microbiol* **39:**494–501.

89. **Li Y, Dooner HK.** 2009. Excision of Helitron transposons in maize. *Genetics* **182:**399–402.

90. **Ton-Hoang B, Guynet C, Ronning DR, Cointin-Marty B, Dyda F, Chandler M.** 2005. Transposition of ISHp608, member of an unusual family of bacterial insertion sequences. *EMBO J* **24:**3325–3338.

91. **Ton-Hoang B, Pasternak C, Siguier P, Guynet C, Hickman AB, Dyda F, Sommer S, Chandler M.** 2010. Single-stranded DNA transposition is coupled to host replication. *Cell* **142:**398–408.

92. **Coates BS, Hellmich RL, Grant DM, Abel CA.** 2012. Mobilizing the genome of Lepidoptera through novel sequence gains and end creation by non-autonomous Lep1 Helitrons. *DNA Res* **19:**11–21.

93. **Sweredoski M, DeRose-Wilson L, Gaut BS.** 2008. A comparative computational analysis of nonautonomous Helitron elements between maize and rice. *Bmc Genomics* **9:**467.

94. **Cultrone A, Dominguez YR, Drevet C, Scazzocchio C, Fernandez-Martin R.** 2007. The tightly regulated promoter of the xanA gene of *Aspergillus nidulans* is included in a helitron. *Mol Microbiol* **63:**1577–1587.

95. **Langdon T, Thomas A, Huang L, Farrar K, King J, Armstead I.** 2009. Fragments of the key flowering gene GIGANTEA are associated with helitron-type sequences in the Pooideae grass *Lolium perenne*. *Bmc Plant Biol* **9:**70.

96. **Hollister JD, Gaut BS.** 2007. Population and evolutionary dynamics of Helitron transposable elements in *Arabidopsis thaliana*. *Mol Biol Evol* **24:**2515–2524.

97. **Morgante M, Brunner S, Pea G, Fengler K, Zuccolo A, Rafalski A.** 2005. Gene duplication and exon shuffling by helitron-like transposons generate intraspecies diversity in maize. *Nat Genet* **37:**997–1002.

98. **Lai JS, Li YB, Messing J, Dooner HK.** 2005. Gene movement by Helitron transposons contributes to the haplotype variability of maize. *Proc Natl Acad Sci USA* **102:**9068–9073.

99. **Lal SK, Giroux MJ, Brendel V, Vallejos CE, Hannah LC.** 2003. The maize genome contains a Helitron insertion. *Plant Cell* **15:**381–391.

100. **Xu JH, Messing J.** 2006. Maize haplotype with a helitron-amplified cytidine deaminase gene copy. *Bmc Genet* **7:**52.

101. **Brunner S, Pea G, Rafalski A.** 2005. Origins, genetic organization and transcription of a family of non-autonomous helitron elements in maize. *Plant J* **43:**799–810.

102. **Gupta S, Gallavotti A, Stryker GA, Schmidt RJ, Lal SK.** 2005. A novel class of Helitron-related transposable elements in maize contain portions of multiple pseudogenes. *Plant Mol Biol* **57:**115–127.

103. **Jameson N, Georgelis N, Fouladbash E, Martens S, Hannah LC, Lal S.** 2008. Helitron mediated amplification of cytochrome P450 monooxygenase gene in maize. *Plant Mol Biol* **67:**295–304.

104. **Wang Q, Dooner HK.** 2006. Remarkable variation in maize genome structure inferred from haplotype diversity at the bz locus. *Proc Natl Acad Sci U S A* **103:**17644–17649.

105. **Lal SK, Hannah LC.** 2005. Plant genomes – massive changes of the maize genome are caused by Helitrons. *Heredity* **95:**421–422.

106. **Esnault C, Maestre J, Heidmann T.** 2000. Human LINE retrotransposons generate processed pseudogenes. *Nat Genet* **24:**363–367.

107. **Lal S, Oetjens M, Hannah LC.** 2009. Helitrons: enigmatic abductors and mobilizers of host genome sequences. *Plant Sci* **176:**181–186.

108. **Clutterbuck AJ.** 2007. A tale of two dead ends: origin of a potential new gene and a potential new transposable element. *Mol Microbiol* **63:**1565–1567.

109. **Puchta H.** 2005. The repair of double-strand breaks in plants: mechanisms and consequences for genome evolution. *J Exp Bot* **56:**1–14.

110. **Gorbunova V, Levy AA.** 1997. Non-homologous DNA end joining in plant cells is associated with deletions and filler DNA insertions. *Nucleic Acids Res* **25:**4650–4657.

111. **Nassif N, Penney J, Pal S, Engels WR, Gloor GB.** 1994. Efficient copying of nonhomologous sequences from ectopic sites via P-element-induced gap repair. *Mol Cell Biol* **14:**1613–1625.

112. **Hall RM, Collis CM.** 1995. Mobile gene cassettes and integrons: capture and spread of genes by site-specific recombination. *Mol Microbiol* **15:**593–600.

113. **Ray DA, Feschotte C, Pagan HJT, Smith JD, Pritham EJ, Arensburger P, Atkinson PW, Craig NL.** 2008. Multiple waves of recent DNA transposon activity in the bat, *Myotis lucifugus*. *Genome Res* **18:**717–728.

114. **Brown MW, Kolisko M, Silberman JD, Roger AJ.** 2012. Aggregative multicellularity evolved independently in the eukaryotic supergroup *Rhizaria*. *Curr Biol* **22:**1123–1127.

115. **Guo X, Gao J, Li F, Wang J.** 2014. Evidence of horizontal transfer of non-autonomous Lep1 Helitrons facilitated by host–parasite interactions. *Sci Rep* **4:**5119.

116. **Hood ME.** 2005. Repetitive DNA in the automictic fungus *Microbotryum violaceum*. *Genetica* **124:**1–10.

117. **Sun C, Shepard DB, Chong RA, López Arriaza J, Hall K, Castoe TA, Feschotte C, Pollock DD, Mueller RL.** 2012. LTR retrotransposons contribute to genomic gigantism in plethodontid salamanders. *Genome Biol Evol* **4:**168–183.

118. **Novick PA, Smith JD, Floumanhaft M, Ray DA, Boissinot S.** 2011. The evolution and diversity of DNA transposons in the genome of the Lizard *Anolis carolinensis*. *Genome Biol Evol* **3:**1–14.

119. **Thomas J, Sorourian M, Ray D, Baker RJ, Pritham EJ.** 2011. The limited distribution of Helitrons to vesper bats supports horizontal transfer. *Gene* **474:**52–58.

120. **Rensing SA, Lang D, Zimmer AD, Terry A, Salamov A, Shapiro H, Nishiyama T, Perroud PF, Lindquist EA, Kamisugi Y, Tanahashi T, Sakakibara K, Fujita T, Oishi K, Shin-I T, Kuroki Y, Toyoda A, Suzuki Y, Hashimoto S, Yamaguchi K, Sugano S, Kohara Y, Fujiyama A, Anterola A, Aoki S, Ashton N, Barbazuk WB, Barker E, Bennetzen JL, Blankenship R, Cho SH, Dutcher SK, Estelle M, Fawcett JA, Gundlach H, Hanada K, Heyl A, Hicks KA, Hughes J, Lohr M, Mayer K, Melkozernov A, Murata T, Nelson DR, Pils B, Prigge M, Reiss B, Renner T, Rombauts S, Rushton PJ, Sanderfoot A, Schween G, Shiu SH, Stueber K, Theodoulou FL, Tu H, Van de Peer Y, Verrier PJ, Waters E, Wood A, Yang LX, Cove D, Cuming AC, Hasebe M, Lucas S, Mishler BD, Reski R, Grigoriev IV, Quatrano RS, Boore JL.** 2008. The *Physcomitrella* genome reveals evolutionary insights into the conquest of land by plants. *Science* **319:**64–69.

121. **Bennetzen JL.** 2009. Maize genome structure and evolution, p 179–200. *In* Bennetzen JL, Hake SC (ed), *Handbook of Maize: Genetics and Genomics*, **vol. II.** Springer, New York.

122. **Petrov DA, Hartl DL.** 1998. High rate of DNA loss in the *Drosophila melanogaster* and *Drosophila virilis* species groups. *Mol Biol Evol* **15:**293–302.

123. **Cordaux R, Batzer MA.** 2009. The impact of retrotransposons on human genome evolution. *Nat Rev Genet* **10:**691–703.

124. **Huang CR, Burns KH, Boeke JD.** 2012. Active transposition in genomes. *Annu Rev Genet* **46:**651–675.

125. **Schaack S, Gilbert C, Feschotte C.** 2010. Promiscuous DNA: horizontal transfer of transposable elements and why it matters for eukaryotic evolution. *Trends Ecol Evol* **25:**537–546.

126. **Robertson HM.** 2002. Evolution of DNA transposons in eukaryotes, p 1093–1110. *In* Craig NL, Craigie R, Geller M, Lambowitz AM (ed), *Mobile DNA II*. ASM Press, Washington, DC.

127. **Ellison CE, Bachtrog D.** 2013. Dosage compensation via transposable element mediated rewiring of a regulatory network. *Science* **342:**846–850.

128. **Pace JK, Gilbert C, Clark MS, Feschotte C.** 2008. Repeated horizontal transfer of a DNA transposon in mammals and other tetrapods. *Proc Natl Acad Sci USA* **105:**17023–17028.

129. **Loreto ELS, Carareto CMA, Capy P.** 2008. Revisiting horizontal transfer of transposable elements in *Drosophila*. *Heredity* **100:**545–554.

130. **Swigonová Z, Lai J, Ma J, Ramakrishna W, Llaca V, Bennetzen JL, Messing J.** 2004. Close split of sorghum and maize genome progenitors. *Genome Res* **14:**1916–1923.

131. **Wolfe KH, Gouy M, Yang YW, Sharp PM, Li WH.** 1989. Date of the monocot-dicot divergence estimated from chloroplast DNA sequence data. *Proc Natl Acad Sci U S A* **86:**6201–6205.

132. **Gilbert C, Schaack S, Pace JK, Brindley PJ, Feschotte C.** 2010. A role for host–parasite interactions in the horizontal transfer of transposons across phyla. *Nature* **464:** 1347–1350.

133. **Putnam NH, Srivastava M, Hellsten U, Dirks B, Chapman J, Salamov A, Terry A, Shapiro H, Lindquist E, Kapitonov VV, Jurka J, Genikhovich G, Grigoriev IV, Lucas SM, Steele RE, Finnerty JR, Technau U, Martindale MQ, Rokhsar DS.** 2007. Sea anemone genome reveals ancestral eumetazoan gene repertoire and genomic organization. *Science* **317:**86–94.

134. **Hollister JD, Smith LM, Guo YL, Ott F, Weigel D, Gaut BS.** 2011. Transposable elements and small RNAs contribute to gene expression divergence between *Arabidopsis thaliana* and *Arabidopsis lyrata*. *Proc Natl Acad Sci U S A* **108:**2322–2327.

135. **Gill N, SanMiguel P, Dhillon BDS, Abernathy B, Kim H, Stein L, Ware D, Jackson SA.** 2010. Dynamic *Oryza* genomes: repetitive DNA sequences as genome modeling agents. *Rice* **3:**251–269.

136. **de la Chaux N, Tsuchimatsu T, Shimizu KK, Wagner A.** 2012. The predominantly selfing plant *Arabidopsis thaliana* experienced a recent reduction in transposable element abundance compared to its outcrossing relative *Arabidopsis lyrata*. *Mob DNA* **3:**2.

137. **Kapitonov VV, Jurka J.** 2007. Helitrons in fruit flies. *Repbase Rep* **7:**129.

138. **Negre B, Simpson P.** 2013. Diversity of transposable elements and repeats in a 600 kb region of the fly *Calliphora vicina*. *Mob DNA* **4:**13.

139. **Rensing SA, Lang D, Zimmer AD, Terry A, Salamov A, Shapiro H, Nishiyama T, Perroud PF, Lindquist EA, Kamisugi Y, Tanahashi T, Sakakibara K, Fujita T, Oishi K, Shin-I T, Kuroki Y, Toyoda A, Suzuki Y, Hashimoto S, Yamaguchi K, Sugano S, Kohara Y, Fujiyama A, Anterola A, Aoki S, Ashton N, Barbazuk WB, Barker E, Bennetzen JL, Blankenship R, Cho SH, Dutcher SK, Estelle M, Fawcett JA, Gundlach H, Hanada K, Heyl A, Hicks KA, Hughes J, Lohr M, Mayer K, Melkozernov A, Murata T, Nelson DR, Pils B, Prigge M, Reiss B, Renner T, Rombauts S, Rushton PJ, Sanderfoot A, Schween G, Shiu SH, Stueber K, Theodoulou FL, Tu H, Van de Peer Y, Verrier PJ, Waters E, Wood A, Yang L, Cove D, Cuming AC, Hasebe M, Lucas S, Mishler BD, Reski R, Grigoriev IV, Quatrano RS, Boore JL.** 2008. The *Physcomitrella* genome reveals evolutionary insights into the conquest of land by plants. *Science* **319:**64–69.

140. **Hollister JD, Gaut BS.** 2009. Epigenetic silencing of transposable elements: a trade-off between reduced transposition and deleterious effects on neighboring gene expression. *Genome Res* **19:**1419–1428.

141. **Numa H, Kim JM, Matsui A, Kurihara Y, Morosawa T, Ishida J, Mochizuki Y, Kimura H, Shinozaki K, Toyoda T, Seki M, Yoshikawa M, Habu Y.** 2010. Transduction of RNA-directed DNA methylation signals to repressive histone marks in *Arabidopsis thaliana*. *EMBO J* **29:**352–362.

142. **Li Y, Dooner HK.** 2012. Helitron proliferation and gene-fragment capture, p 193–217. *In* Grandbastien MA, Casacuberta JM (ed), *Plant Transposable Elements, Topics in Current Genetics*, **vol. 24**. Springer-Verlag, Berlin Heidelberg.

143. **Slawson EE, Shaffer CD, Malone CD, Leung W, Kellmann E, Shevchek RB, Craig CA, Bloom SM, Bogenpohl J, Dee J, Morimoto ET, Myoung J, Nett AS, Ozsolak F, Tittiger ME, Zeug A, Pardue ML, Buhler J, Mardis ER, Elgin SC.** 2006. Comparison of dot chromosome sequences from *D. melanogaster* and *D. virilis* reveals an enrichment of DNA transposon sequences in heterochromatic domains. *Genome Biol* **7:**R15.

144. **Chen ST, Cheng HC, Barbash DA, Yang HP.** 2007. Evolution of hydra, a recently evolved testis-expressed gene with nine alternative first exons in *Drosophila melanogaster*. *PLoS Genet* **3:**e107.

145. **Juretic N, Hoen DR, Huynh ML, Harrison PM, Bureau TE.** 2005. The evolutionary fate of MULE-mediated duplications of host gene fragments in rice. *Genome Res* **15:**1292–1297.

146. **Hanada K, Vallejo V, Nobuta K, Slotkin RK, Lisch D, Meyers BC, Shiu SH, Jiang N.** 2009. The functional role of pack-MULEs in rice inferred from purifying selection and expression profile. *Plant Cell* **21:** 25–38.

147. **Lander ES, Linton LM, Birren B, Nusbaum C, Zody MC, Baldwin J, Devon K, Dewar K, Doyle M, FitzHugh W, Funke R, Gage D, Harris K, Heaford A, Howland J, Kann L, Lehoczky J, LeVine R, McEwan P, McKernan K, Meldrim J, Mesirov JP, Miranda C, Morris W, Naylor J, Raymond C, Rosetti M, Santos R, Sheridan A, Sougnez C, Stange-Thomann N, Stojanovic N, Subramanian A, Wyman D, Rogers J, Sulston J, Ainscough R, Beck S, Bentley D, Burton J, Clee C, Carter N, Coulson A, Deadman R, Deloukas P, Dunham A, Dunham I, Durbin R, French L, Grafham D, Gregory S, Hubbard T, Humphray S, Hunt A, Jones M, Lloyd C, McMurray A, Matthews L, Mercer S, Milne S, Mullikin JC, Mungall A, Plumb R, Ross M, Shownkeen R, Sims S, Waterston RH, Wilson RK, Hillier LW, McPherson JD, Marra MA, Mardis ER, Fulton LA, Chinwalla AT, Pepin KH, Gish WR, Chissoe SL, Wendl MC, Delehaunty KD, Miner TL, Delehaunty A, Kramer JB, Cook LL, Fulton RS, Johnson DL, Minx PJ, Clifton SW, Hawkins T, Branscomb E, Predki P, Richardson P, Wenning S, Slezak T, Doggett N, Cheng JF, Olsen A, Lucas S, Elkin C, Uberbacher E, Frazier M, Gibbs RA, Muzny DM, Scherer SE, Bouck JB, Sodergren EJ, Worley KC, Rives CM, Gorrell JH, Metzker ML, Naylor SL, Kucherlapati RS, Nelson DL, Weinstock GM, Sakaki Y, Fujiyama A, Hattori M, Yada T, Toyoda A, Itoh T, Kawagoe C, Watanabe H, Totoki Y, Taylor T, Weissenbach J, Heilig R, Saurin W, Artiguenave F, Brottier P, Bruls T, Pelletier E, Robert C, Wincker P, Rosenthal A, Platzer M, Nyakatura G, Taudien S, Rump A, Yang HM, Yu J, Wang J, Huang GY, Gu J, Hood L, Rowen L, Madan A, Qin SZ, Davis RW, Federspiel NA, Abola AP, Proctor MJ, Myers RM, Schmutz J, Dickson M, Grimwood J, Cox DR, Olson MV, Kaul R, Shimizu N, Kawasaki K, Minoshima S, Evans GA, Athanasiou M, Schultz R, Roe BA, Chen F, Pan HQ, Ramser J, Lehrach H, Reinhardt R, McCombie WR, de la Bastide M, Dedhia N, Blocker H, Hornischer K, Nordsiek G, Agarwala R, Aravind L, Bailey JA, Bateman A, Batzoglou S, Birney E, Bork P, Brown DG, Burge CB, Cerutti L, Chen HC, Church D, Clamp M, Copley RR, Doerks T, Eddy SR, Eichler EE, Furey TS, Galagan J, Gilbert JGR, Harmon C, Hayashizaki Y, Haussler D, Hermjakob H, Hokamp K, Jang WH, Johnson LS, Jones TA, Kasif S, Kaspryzk A, Kennedy S, Kent WJ, Kitts P, Koonin EV, Korf I, Kulp D, Lancet D, Lowe TM, McLysaght A, Mikkelsen T, Moran JV, Mulder N, Pollara VJ, Ponting CP, Schuler G, Schultz JR, Slater G, Smit AFA, Stupka E, Szustakowki J, Thierry-Mieg D, Thierry-Mieg J, Wagner L, Wallis J, Wheeler R, Williams A, Wolf YI, Wolfe KH, Yang SP, Yeh RF, Collins F, Guyer MS, Peterson J, Felsenfeld A, Wetterstrand KA, Patrinos A, Morgan MJ, Int Human Genome Sequencing C.** 2001. Initial sequencing and analysis of the human genome. *Nature* **409:**860–921.

148. **Torrents D, Suyama M, Zdobnov E, Bork P.** 2003. A genome-wide survey of human pseudogenes. *Genome Res* **13:**2559–2567.

149. **Fan C, Zhang Y, Yu Y, Rounsley S, Long M, Wing RA.** 2008. The subtelomere of *Oryza sativa* chromosome 3 short arm as a hot bed of new gene origination in rice. *Mol Plant* **1:**839–850.

150. **Wicker T, Buchmann JP, Keller B.** 2010. Patching gaps in plant genomes results in gene movement and erosion of colinearity. *Genome Res* **20:**1229–1237.

151. **Malacarne G, Perazzolli M, Cestaro A, Sterck L, Fontana P, Van de Peer Y, Viola R, Velasco R, Salamini F.** 2012. Deconstruction of the (paleo)polyploid grapevine genome based on the analysis of transposition events involving NBS resistance genes. *PLoS One* **7:**e29762.

152. **Tomaszkiewicz M, Chalopin D, Schartl M, Galiana D, Volff JN.** 2014. A multicopy Y-chromosomal SGNH hydrolase gene expressed in the testis of the platyfish has been captured and mobilized by a Helitron transposon. *BMC Genet* **15:**44.

153. **Yang S, Arguello JR, Li X, Ding Y, Zhou Q, Chen Y, Zhang Y, Zhao R, Brunet F, Peng L, Long M, Wang W.** 2008. Repetitive element-mediated recombination as a mechanism for new gene origination in Drosophila. *PLoS Genet* **4:**e3.

154. **Kogan GL, Usakin LA, Ryazansky SS, Gvozdev VA.** 2012. Expansion and evolution of the X-linked testis specific multigene families in the melanogaster species subgroup. *PLoS One* **7:**e37738.

155. **Ding Y, Zhao L, Yang S, Jiang Y, Chen Y, Zhao R, Zhang Y, Zhang G, Dong Y, Yu H, Zhou Q, Wang W.** 2010. A young *Drosophila* duplicate gene plays essential roles in spermatogenesis by regulating several Y-linked male fertility genes. *PLoS Genet* **6:**e1001255.

156. **Singh ND, Petrov DA.** 2004. Rapid sequence turnover at an intergenic locus in *Drosophila*. *Mol Biol Evol* **21:** 670–680.

157. **Steinemann M, Steinemann S.** 1993. A duplication including the Y allele of Lcp2 and the TRIM retrotransposon at the Lcp locus on the degenerating neo-Y chromosome of *Drosophila miranda*: molecular structure and mechanisms by which it may have arisen. *Genetics* **134:**497–505.

158. **Kaessmann H, Vinckenbosch N, Long M.** 2009. RNA-based gene duplication: mechanistic and evolutionary insights. *Nat Rev Genet* **10:**19–31.

159. **Amin EM, Oltean S, Hua J, Gammons MV, Hamdollah-Zadeh M, Welsh GI, Cheung MK, Ni L, Kase S, Rennel ES, Symonds KE, Nowak DG, Royer-Pokora B, Saleem MA, Hagiwara M, Schumacher VA, Harper SJ, Hinton DR, Bates DO, Ladomery MR.** 2011. WT1 mutants reveal SRPK1 to be a downstream angiogenesis target by altering VEGF splicing. *Cancer Cell* **20:**768–780.

160. **Buckler ES, Gaut BS, McMullen MD.** 2006. Molecular and functional diversity of maize. *Curr Opin Plant Biol* **9:**172–176.

161. **Lal SK, Hannah LC.** 2005. Helitrons contribute to the lack of gene colinearity observed in modern maize inbreds. *Proc Natl Acad Sci USA* **102:**9993–9994.

162. **Liu P, Sherman-Broyles S, Nasrallah ME, Nasrallah JB.** 2007. A cryptic modifier causing transient self-incompatibility in *Arabidopsis thaliana*. *Curr Biol* **17:** 734–740.

163. **Sherman-Broyles S, Boggs N, Farkas A, Liu P, Vrebalov J, Nasrallah ME, Nasrallah JB.** 2007. S locus genes and the evolution of self-fertility in *Arabidopsis thaliana*. *Plant Cell* **19:**94–106.

164. **Hu TT, Pattyn P, Bakker EG, Cao J, Cheng JF, Clark RM, Fahlgren N, Fawcett JA, Grimwood J, Gundlach H, Haberer G, Hollister JD, Ossowski S, Ottilar RP, Salamov AA, Schneeberger K, Spannagl M, Wang X, Yang L, Nasrallah ME, Bergelson J, Carrington JC, Gaut BS, Schmutz J, Mayer KF, Van de Peer Y, Grigoriev IV, Nordborg M, Weigel D, Guo YL.** 2011. The *Arabidopsis lyrata* genome sequence and the basis of rapid genome size change. *Nat Genet* **43:**476–481.

165. **Böhne A, Brunet F, Galiana-Arnoux D, Schultheis C, Volff JN.** 2008. Transposable elements as drivers of genomic and biological diversity in vertebrates. *Chromosome Res* **16:**203–215.

166. **Rebollo R, Horard B, Hubert B, Vieira C.** 2010. Jumping genes and epigenetics: Towards new species. *Gene* **454:**1–7.

167. **Oliver KR, Greene WK.** 2009. Transposable elements: powerful facilitators of evolution. *Bioessays* **31:**703–714.

168. **He L, Dooner HK.** 2009. Haplotype structure strongly affects recombination in a maize genetic interval polymorphic for Helitron and retrotransposon insertions. *Proc Natl Acad Sci U S A* **106:**8410–8416.

169. **Chuck G, Meeley R, Irish E, Sakai H, Hake S.** 2007. The maize tasselseed4 microRNA controls sex determination and meristem cell fate by targeting Tasselseed6/ indeterminate spikelet1. *Nat Genet* **39:**1517–1521.

170. **Inagaki S, Nakamura K, Morikami A.** 2009. A link among DNA replication, recombination, and gene expression revealed by genetic and genomic analysis of TEBICHI gene of *Arabidopsis thaliana*. *PLoS Genet* **5:** e1000613.

171. **Berger N, Dubreucq B, Roudier F, Dubos C, Lepiniec L.** 2011. Transcriptional regulation of *Arabidopsis* LEAFY COTYLEDON2 involves RLE, a cis-element that regulates trimethylation of histone H3 at lysine-27. *Plant Cell* **23:**4065–4078.

172. **Miller WJ, McDonald JF, Pinsker W.** 1997. Molecular domestication of mobile elements. *Genetica* **100:**261–270.

173. **Miller WJ, Paricio N, Hagemann S, Martínez-Sebastián MJ, Pinsker W, de Frutos R.** 1995. Structure and expression of clustered P element homologues in *Drosophila subobscura* and *Drosophila guanche*. *Gene* **156:**167–174.

174. **Tsukamoto T, Hauck NR, Tao R, Jiang N, Iezzoni AF.** 2010. Molecular and genetic analyses of four nonfunctional S haplotype variants derived from a common ancestral S haplotype identified in sour cherry (*Prunus cerasus* L.). *Genetics* **184:**411–427.

175. **Platt RN, Vandewege MW, Kern C, Schmidt CJ, Hoffmann FG, Ray DA.** 2014. Large numbers of novel miRNAs originate from DNA transposons and are coincident with a large species radiation in bats. *Mol Biol Evol* **31:**1536–1545.

176. **Eckardt NA.** 2003. A new twist on transposons: the maize genome harbors helitron insertion. *Plant Cell* **15:** 293–295.

177. **Choi JD, Hoshino A, Park KI, Park IS, Iida S.** 2007. Spontaneous mutations caused by a Helitron

transposon, Hel-It1, in morning glory, *Ipomoea tricolor*. *Plant J* **49:**924–934.

178. **Li X, Chen L, Hong M, Zhang Y, Zu F, Wen J, Yi B, Ma C, Shen J, Tu J, Fu T.** 2012. A large insertion in bHLH transcription factor BrTT8 resulting in yellow seed coat in *Brassica rapa*. *PLoS One* **7:**e44145.

179. **Chang YF, Imam JS, Wilkinson MF.** 2007. The nonsense-mediated decay RNA surveillance pathway. *Annu Rev Biochem* **76:**51–74.

180. **Carareto CM, Hernandez EH, Vieira C.** 2014. Genomic regions harboring insecticide resistance-associated Cyp genes are enriched by transposable element fragments carrying putative transcription factor binding sites in two sibling *Drosophila* species. *Gene* **537:**93–99.

181. **Catania F, Kauer MO, Daborn PJ, Yen JL, Ffrench-Constant RH, Schlotterer C.** 2004. World-wide survey of an Accord insertion and its association with DDT resistance in *Drosophila melanogaster*. *Mol Ecol* **13:** 2491–2504.

182. **Daborn P, Boundy S, Yen J, Pittendrigh B, ffrench-Constant R.** 2001. DDT resistance in *Drosophila* correlates with Cyp6g1 over-expression and confers cross-resistance to the neonicotinoid imidacloprid. *Mol Genet Genomics* **266:**556–563.

183. **Daborn PJ, Yen JL, Bogwitz MR, Le Goff G, Feil E, Jeffers S, Tijet N, Perry T, Heckel D, Batterham P, Feyereisen R, Wilson TG, ffrench-Constant RH.** 2002. A single p450 allele associated with insecticide resistance in *Drosophila*. *Science* **297:**2253–2256.

184. **Joussen N, Heckel DG, Haas M, Schuphan I, Schmidt B.** 2008. Metabolism of imidacloprid and DDT by P450 CYP6G1 expressed in cell cultures of *Nicotiana tabacum* suggests detoxification of these insecticides in Cyp6g1-overexpressing strains of *Drosophila melanogaster*, leading to resistance. *Pest Manag Sci* **64:**65–73.

185. **Li L, Petsch K, Shimizu R, Liu S, Xu WW, Ying K, Yu J, Scanlon MJ, Schnable PS, Timmermans MC, Springer NM, Muehlbauer GJ.** 2013. Mendelian and non-Mendelian regulation of gene expression in maize. *PLoS Genet* **9:**e1003202.

186. **Feschotte C, Pritham EJ.** 2009. A cornucopia of Helitrons shapes the maize genome. *Proc Natl Acad Sci USA* **106:**19747–19748.

187. **Fahlgren N, Bollmann SR, Kasschau KD, Cuperus JT, Press CM, Sullivan CM, Chapman EJ, Hoyer JS, Gilbert KB, Grünwald NJ, Carrington JC.** 2013. Phytophthora have distinct endogenous small RNA populations that include short interfering and microRNAs. *PLoS One* **8:** e77181.

188. **Vetukuri RR, Åsman AK, Tellgren-Roth C, Jahan SN, Reimegård J, Fogelqvist J, Savenkov E, Söderbom F, Avrova AO, Whisson SC, Dixelius C.** 2012. Evidence for small RNAs homologous to effector-encoding genes and transposable elements in the oomycete *Phytophthora infestans*. *PLoS One* **7:**e51399.

189. **Cho SH, Addo-Quaye C, Coruh C, Arif MA, Ma Z, Frank W, Axtell MJ.** 2008. *Physcomitrella patens* DCL3 is required for 22-24 nt siRNA accumulation, suppression of retrotransposon-derived transcripts, and normal development. *PLoS Genet* **4:**e1000314.

190. **Kelley D, Rinn J.** 2012. Transposable elements reveal a stem cell-specific class of long noncoding RNAs. *Genome Biol* **13:**R107.

191. **Kapusta A, Kronenberg Z, Lynch VJ, Zhuo X, Ramsay L, Bourque G, Yandell M, Feschotte C.** 2013. Transposable elements are major contributors to the origin, diversification, and regulation of vertebrate long noncoding RNAs. *PLoS Genet* **9:**e1003470.

192. **Keeling PJ, Burger G, Durnford DG, Lang BF, Lee RW, Pearlman RE, Roger AJ, Gray MW.** 2005. The tree of eukaryotes. *Trends Ecol Evol* **20:**670–676.

193. **Gertz EM, Yu YK, Agarwala R, Schäffer AA, Altschul SF.** 2006. Composition-based statistics and translated nucleotide searches: improving the TBLASTN module of BLAST. *BMC Biol* **4:**41.

Mobile DNA, 3rd Edition
Nancy L. Craig, Michael Chandler, Martin Gellert, Alan M. Lambowitz, Phoebe A. Rice and Suzanne Sandmeyer

doi:10.1128/microbiolspec.MDNA3-0053-2014

M. Joan Curcio[1]
Sheila Lutz[1]
Pascale Lesage[2]

41 The Ty1 LTR-Retrotransposon of Budding Yeast, *Saccharomyces cerevisiae*

WHY IS Ty1 A GREAT MODEL SYSTEM?

Ty1 Structure and Replication

Organization of the Ty1 Genome

Ty1 elements have a structure that is analogous to simple retroviruses, but they lack an envelope gene (Fig. 1) (1). The most highly characterized Ty1 element is Ty1-H3, which was isolated following its retrotransposition into plasmid DNA (2). Nucleotide coordinates provided in this review specifically refer to Ty1-H3, unless otherwise noted. Ty1 is 5918 base pairs (bp) in length with 334 bp direct repeats, or long-terminal repeats (LTRs), at each end. Ty1 LTRs, like that of most LTR-retrotransposons and retroviruses, have the dinucleotide inverted repeat, 5′-TG...CA-3′ at their termini, and are composed of three distinct domains-U3, R, and U5. These domains are defined by their position in the major sense-strand transcript expressed from Ty1 DNA. The 38-nucleotide U5 region and 240-nucleotide U3 region are unique to the 5′ and 3′ end of the Ty1 RNA, respectively, while the R region of 56 nucleotides is repeated at both ends of the processed transcript. Functional Ty1 elements encode two partially overlapping open reading frames: *GAG* (historically known as *TYA*) and *POL* (*TYB*). The last three nucleotides of the R region of the 5′ LTR encode the first codon of *GAG*. The *GAG* ORF encodes a single functional protein with capsid and nucleic acid chaperone functions. The *POL* ORF is in the +1 frame relative to *GAG* and overlaps the last 38 base pairs of *GAG*. *POL* encodes three proteins with catalytic activity: protease (PR), integrase (IN), and reverse transcriptase/RNase H (RT/RH).

The Ty1 Replication Cycle

The process of retrotransposition is replicative, resulting in the parental retrotransposon and a copy of the element in the genome. The major steps in Ty1 replication are analogous to those in retroviral replication, except

[1]Laboratory of Molecular Genetics, Wadsworth Center, and Department of Biomedical Sciences, University at Albany-SUNY; Center for Medical Sciences, 150 New Scotland Avenue, Albany, NY, 12208; [2]Université Paris Diderot, Sorbonne Paris Cité, Institut Universitaire d'Hématologie, Hôpital St. Louis, Institut National de la Santé et de la Recherche Médicale (INSERM) UMR 944, Centre National de la Recherche Scientifique (CNRS) UMR 7212, Paris cedex 10, France.

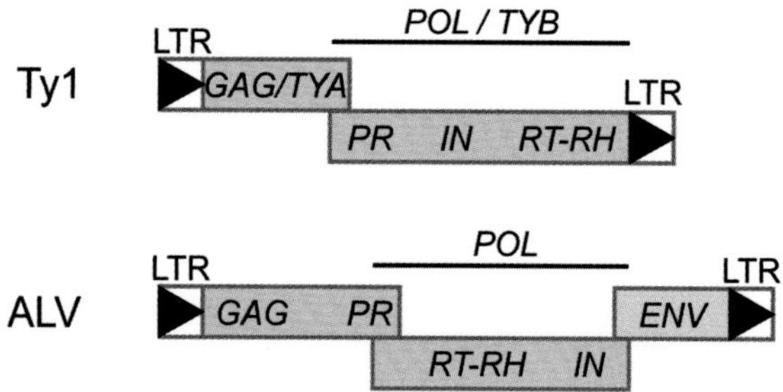

Figure 1 Structure of the Ty1 element relative to the simple retrovirus, avian leukemia virus (ALV). Ty1 consists of long terminal repeats (LTRs; boxed arrowheads) flanking a central coding region that contains two overlapping ORFs, *GAG* (*TYA*) and *POL* (*TYB*), which are analogous to retroviral *GAG* and *POL* genes, respectively. Separate functional domains of *POL* that are conserved in LTR-retrotransposons and retroviruses and are synthesized as part of the Gag-Pol polyprotein and cleaved into separate proteins posttranslationally include protease (PR), reverse transcriptase-RNase H (RT-RH), and integrase (IN). The retroviral envelope gene (*ENV*) is not present in Ty1.
doi:10.1128/microbiolspec.MDNA3-0053-2014.f1

that Ty1 replication is entirely intracellular (Fig. 2). Ty1 elements are transcribed by RNA polymerase II (Pol II), resulting in capped and polyadenylated transcripts that are exported to the cytoplasm. Translation of Ty1 RNA produces two primary gene products, p49-Gag and p199-Gag-Pol, the latter a product of a programmed translational frameshift from the *GAG* ORF to the *POL* ORF (3). Gag, Gag-Pol and Ty1 RNA assemble into nucleocapsids known as virus-like particles (VLPs). Within the VLP, PR is autocatalytically cleaved from p199-Gag-Pol and catalyzes all additional cleavages of the Gag and Gag-Pol precursors to yield p45-Gag, p20-PR, p71-IN and p63-RT/RH. Following maturation of Ty1 proteins, Ty1 RNA is reverse transcribed into a linear, double-stranded DNA. The resulting cDNA, presumably in association with p71-IN, is imported into the nucleus. IN interacts with host proteins to target Ty1 cDNA integration to specific regions of the host genome. The cDNA is integrated into chromosomal DNA by a nonhomologous strand transfer process (reviewed in [4]). Potentially, Ty1 cDNA is also an excellent substrate for gene conversion of Ty1 elements and nondegenerate solo LTRs in the genome. Nonetheless, Ty1 cDNA rarely enters the genome by gene conversion of endogenous Ty1 sequences (5, 6), unless integration is blocked by mutations in IN or cDNA terminal motifs that are bound by IN (7), or cells are grown at temperatures above 30°C (8). Together, the processes of IN-mediated retrotransposition of cDNA and insertion of cDNA by homologous recombination are known as retromobility.

The Toolbox for Studying Retromobility

Detection of RNA-Mediated Mobility Events

Repression of retromobility by the host cell is a nearly universal feature of retrotransposon biology. In laboratory strains, Ty1 retromobility occurs at a rate of 10^{-5} to 10^{-7} per element per generation (9), which is too infrequently to differentiate retromobility from DNA-based recombination events. Therefore, assays to detect retromobility have focused on increasing the frequency of retrotransposition or improving the detection of RNA-mediated events versus DNA-mediated events. Beginning with the innovation that led to the discovery of LTR-retrotransposon mobility through an RNA intermediate 30 years ago, an array of tools have been developed in *Saccharomyces cerevisiae* to facilitate characterization of retromobility processes. In 1985, Boeke et al. isolated the functional element, Ty1-H3 and expressed it from the galactose-inducible *GAL1* promoter on a high-copy plasmid (pGTy1). Expression of pGTy1 resulted in such a high level of retrotransposition that new genomic copies of Ty1 could be identified in cells chosen at random (2). To demonstrate that Ty1 was mobilized via an RNA intermediate, an intron-bearing gene fragment was inserted downstream of the *POL* ORF in pGTy1. Precise intron excision in transposed elements provided unambiguous evidence that a spliced Ty1 RNA was the template for synthesis of retrotransposed DNA. Next generation pGTy1 elements bearing selectable markers were used to detect newly integrated chromosomal Ty1 elements following segregation of the plasmid copy (10). Expression of Ty1 RNA from *GAL1* or other heterologous promoters also stimulates VLP formation, which facilitates their purification and functional and structural characterization (11, 12).

Precise splicing of an intron inserted into Ty1 provided the basis for development of retrotranscript indicator genes (RIGs). RIGs are synthetic genes constructed by inserting an intron into the ORF of a selectable marker gene in an antisense and thus unspliceable orientation (Fig. 3). When a RIG is inserted into a Ty1 element in the 3′ untranslated region such that Ty1 and the intron-bearing marker gene are in opposing transcriptional orientations, the intron is in a spliceable orientation in the Ty1 transcript. After the intron is spliced from the Ty1 transcript, reverse transcription of the spliced Ty1 transcript and integration of the resulting cDNA create a functional chromosomal copy of the marker gene lacking the intron. The recreated marker gene allows phenotypic detection of cells that sustain a transposition event. The RIG is activated only in newly mobilized Ty1 copies, and thus, can be used to detect

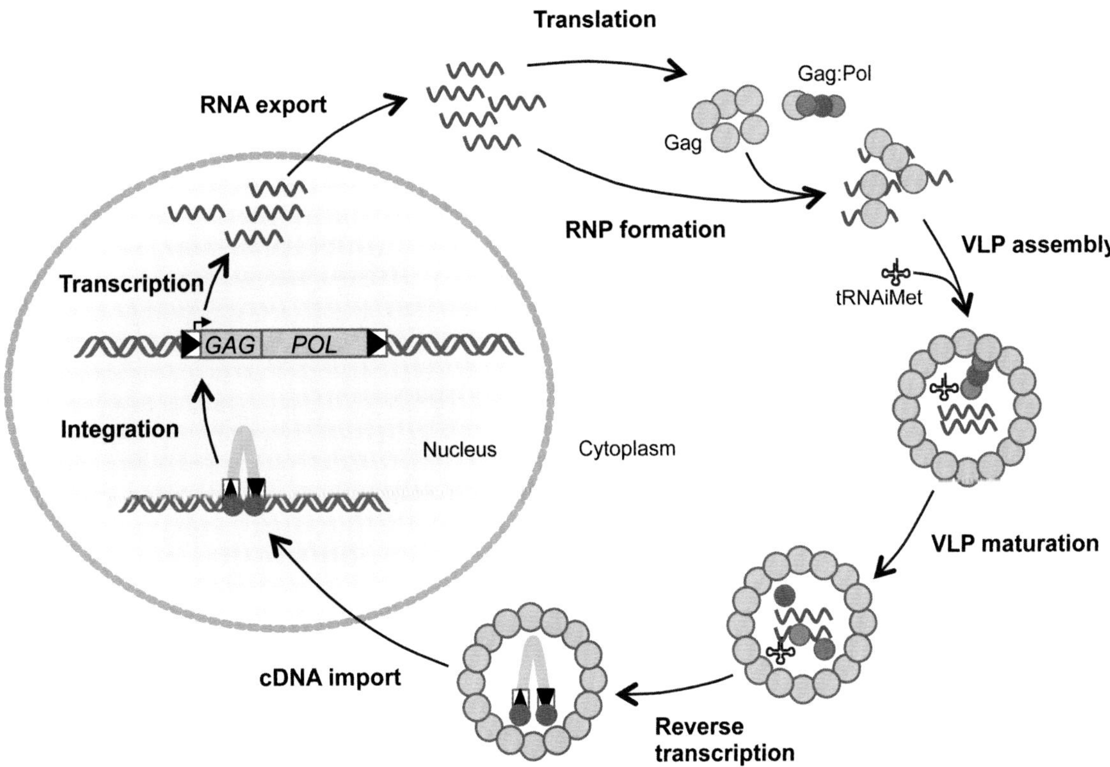

Figure 2 Ty1 replication cycle. The major steps in replication of Ty1, which results in the introduction of a new copy of Ty1 into the host genome, are illustrated. A Ty1 element in the host genome (blue double helix) is transcribed and the Ty1 RNA (wavy teal lines) is exported to the cytoplasm. The RNA is translated into Gag and Gag-Pol proteins and associates with these proteins to form Ty1 RNPs, also known as retrosomes. Ty1 RNPs give rise to virus-like particles (VLPs) that encapsidate a dimer of Ty1 RNA and $tRNA_i^{Met}$. Within the VLP, Gag and Pol proteins are cleaved by protease (PR) (maroon ball) to form mature Gag, PR, integrase (IN), and reverse transcriptase (RT) proteins. Following VLP maturation, Ty1 RNA is reverse transcribed into cDNA by RT (blue ball) using $tRNA_i^{Met}$ as a primer. The cDNA is bound by IN (purple ball) to form the preintegation complex, which is imported into the nucleus. IN integrates Ty1 cDNA into the yeast genome. doi:10.1128/microbiolspec.MDNA3-0053-2014.f2

retromobility of Ty1 in the presence of the marked parental element. Retromobility at a frequency as low as 1×10^{-9} has been detected in a simple Petri plate-based assay (Fig. 3) (13). First developed to detect retrotransposition of endogenous Ty1 elements, RIGs have been adapted to study a breadth of RNA-mediated mobility events such as group II intron retrohoming in bacteria and non-LTR-retrotransposon mobility in human cells and in mice (14, 15, 16, 17).

Separation of Ty1 RNA and Protein Function

LTR-retrotransposons encode an RNA that has binary functions as an mRNA and as the genomic RNA (gRNA) of VLPs. In instances in which LTR-retrotransposon proteins bind their encoding mRNA during or immediately after translation and promote its encapsidation into the VLP, the proteins are said to be *cis*-acting. A major consequence of preferential *cis* action is that only autonomous elements that encode functional proteins retrotranspose efficiently. In contrast, *trans*-acting retrotransposon proteins do not show a preference for encapsidating the RNA from which they are translated, and therefore, can encapsidate a gRNA from any element regardless of whether it serves as an mRNA. Ty1 proteins are *trans*-acting, resulting in the retrotransposition of both autonomous elements and nonautonomous elements that do not encode functional proteins (18). This feature allows the function of the Ty1 RNA as a protein-coding template to be separated from its function as a template for reverse transcription. The

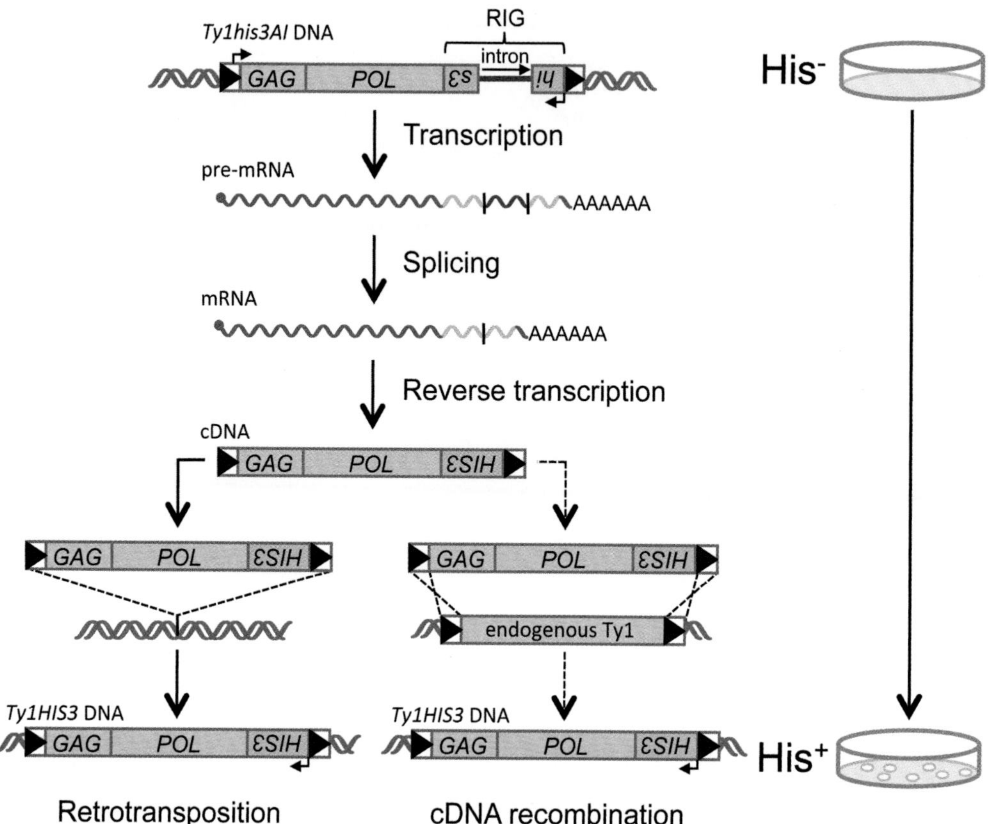

Figure 3 Assay for Ty1 retromobility using a retrotranscript indicator gene (RIG). A chromosomal Ty1 element tagged with the *his3AI* RIG gives rise to Ty1 cDNA bearing a functional *HIS3* gene. The cDNA can enter the host genome (represented by a blue double helix) by two retromobility pathways. Retrotransposition occurs when Ty1 integrase (IN) mediates the integration of cDNA into the genome, while cDNA recombination occurs when the cDNA recombines homologously with an endogenous Ty1 element. The dashed lines represent the low frequency of Ty1 cDNA recombination in wild-type cells. Cells that sustain a Ty1*HIS3* retromobility event give rise to His⁺ colonies. Other RIGs used in yeast include *kanMXAI*, which is selectable with G418 and *ade2AI*, which is selectable by adenine prototrophy (268, 269). doi:10.1128/microbiolspec.MDNA3-0053-2014.f3

trans-action of Ty1 proteins has been exploited to develop helper-donor assays. A retrotransposition-defective "helper" Ty1 element encodes functional Gag and Gag-Pol proteins required for retrotransposition but carries mutations that prevent its function as a template for reverse transcription (Fig. 4). "Donor" Ty1s are nonautonomous elements containing the minimal sequences necessary for the transcript to be recognized for encapsidation in VLPs, reverse transcription, and integration. In addition, the donor Ty1 is marked with a selectable marker gene or RIG to detect its retrotransposition into the host genome. The helper-donor assay has been employed to define the role of Ty1 RNA sequences and secondary structures that are required for packaging and reverse transcription (19, 20, 21) and to study Ty1 RNA dynamics (22).

An Element Poised for Action

Compared to the genomes of most other eukaryotic organisms, a relatively small proportion of the genomes of *S. cerevisiae* strains, ranging from about 1.3% to 3.4%, consists of transposon sequences (23). Five families of LTR-retrotransposons, Ty1 to Ty5 are the only transposable elements in the *S. cerevisiae* nuclear genome. In the laboratory reference strain S288C, Ty1 and Ty2 are the most and next most prevalent in the genome, respectively. Ty1 and Ty2 are closely related elements distinguished by a one-nucleotide deletion in the Ty2 LTR, a divergent *GAG* ORF and a ~300 nucleotide region of sequence divergence in *POL* (24). Ty2 elements were introduced into a recent ancestor of *S. cerevisiae* from a related species, *Saccharomyces mikitae*, as a result of horizontal transfer (23, 25).

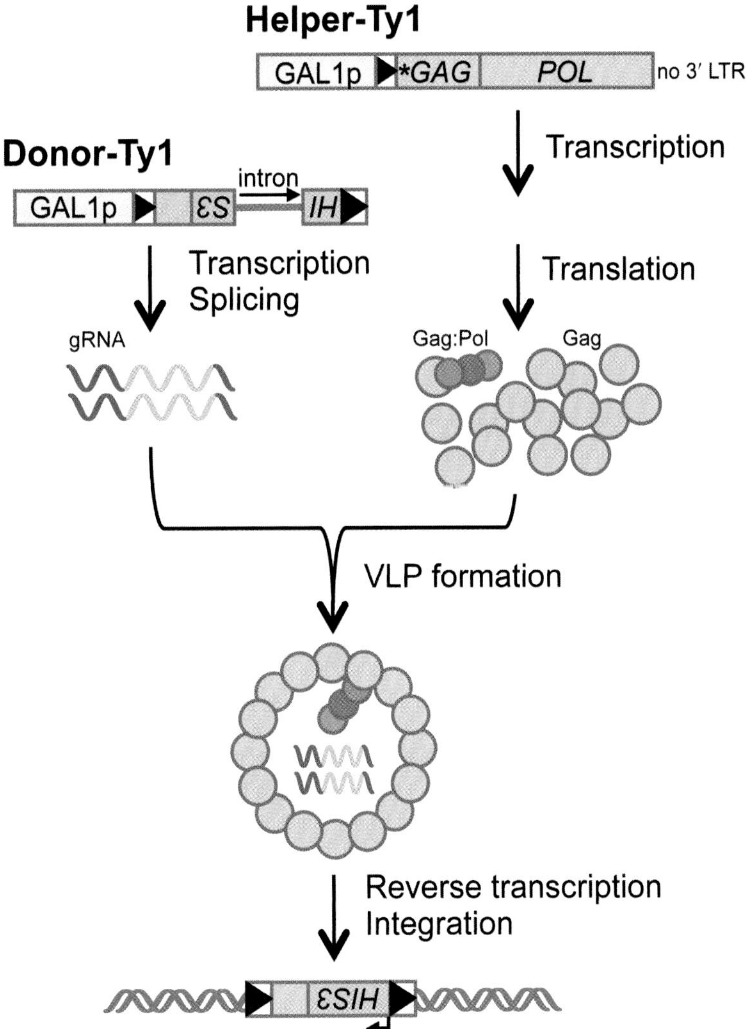

Figure 4 Helper-donor assay for separation of mRNA and gRNA functions. In this assay, expression of both Ty1 elements is driven from the *GAL1* promoter in a *spt3* strain, which lacks endogenous Ty1 expression. The helper-Ty1 encodes a functional mRNA with *GAG* and *POL* ORFs and silent mutations at the 5′ end of *GAG* (indicated by an asterisk) that disrupt *cis*-acting signals required for reverse transcription. The absence of a 3′ long terminal repeat (LTR) also precludes the use of the helper-Ty1 RNA as a template for reverse transcription. The donor-Ty1 RNA encodes a functional gRNA that lacks ORFs but contains *cis*-acting signals for dimerization, packaging, and reverse transcription. The *his3AI* RIG is also contained in this element to detect retromobility of mini-Ty1*HIS3* cDNA. The minimal donor element capable of retromobility is depicted.
doi:10.1128/microbiolspec.MDNA3-0053-2014.f4

There are 313 copies of Ty1 dispersed throughout the *S. cerevisiae* S288C genome, including 279 solo LTRs, two other truncated elements, and 32 full-length elements (23, 24). The Ty2 family consists of 46 elements, of which 13 are full-length elements, 31 are solo LTRs, and two are other truncations. Population genomic and phylogenetic analyses of S288C and other *S. cerevisiae* strains sequenced in the Saccharomyces Genome Resequencing Project (25) indicate that all full-length Ty1 and Ty2 elements, as well as all Ty2 solo LTRs in S288C, have recently transposed (23, 26). In contrast, the majority of Ty1 solo LTRs result from ancient transposition events and many are degenerate in sequence. Full-length Ty1 elements comprise three active lineages: Ty1, Ty1/Ty2 hybrids, and Ty1′, with Ty1 likely being the ancestral lineage. In Ty1/Ty2 hybrid elements, the U3 domain of both LTRs is derived from Ty2 elements, while the R-U5 domains and coding regions are derived from Ty1 (27). Ty1′ elements are distinguished by a high degree of sequence divergence in the *GAG* ORF (24, 27).

Consistent with the recent transposition of Ty1 and Ty2 elements, the coding regions of members of each family are very homogeneous in sequence, with 86% and 96% invariant amino acids in Ty1 and Ty2 ORFs, respectively (24, 26). Furthermore, Ty1 and Ty2 elements in laboratory strains S288C and GRF167 are predominantly autonomous elements that encode functional gRNA and proteins capable of promoting retrotransposition in the absence of other elements (18, 24). Remarkably, the coding regions of both element families appear to be evolving under purifying selection (26, 28). The predominance of autonomous elements is puzzling, given that Ty1 proteins are *trans*-acting and display no apparent preference for mobilizing autonomous elements (18). The mechanism of selection for autonomous Ty1 and Ty2 elements is not well understood but the robust conversion of full-length Ty elements to solo LTRs by homologous recombination between the 5′ LTR and 3′ LTR of individual elements is likely to be a significant factor in cleansing the genome of defective elements (23, 26, 29). In the diploid stage of the yeast life cycle, the predominant stage in the wild, Ty1 elements are removed with no footprint by loss of heterozygosity, a process in which the Ty1 containing allele undergoes gene conversion by the unoccupied allele on the homolog. In addition, a significant rate of ectopic gene conversion among nonallelic Ty1 elements is likely to homogenize Ty1 sequences and could be involved in eliminating defective copies (6). Finally, the low rate of retrotransposition in yeast limits the spread of defective copies (6, 30). The transpositional dormancy of Ty2 elements is, at least in part, a consequence of the low level of Ty2 RNA (31). Ty1 RNA is approximately twenty times more abundant than Ty2 RNA, but retromobility is repressed at a post-transcriptional level by a mechanism known as copy number control (32, 33). This mechanism, discussed in detail below, balances the rate of gain of Ty1 elements by retrotransposition with the rate of loss by solo LTR formation and loss of heterozygosity. Nonetheless, transpositional dormancy imposed by copy number control can be modulated by a variety of extrinsic and intrinsic parameters, including cell type, activation state of MAPK and DNA damage signaling pathways, and environmental signals, including temperature and nutrient availability (34, 35, 36, 37, 38, 39, 40, 41). Thus, the retrotransposon landscape is rife with sleeping elements, held in check but poised to mobilize when triggered by environmental or cellular cues.

Ty1 RNA AND PROTEIN EXPRESSION, LOCALIZATION, AND TURNOVER

Ty1 Expression

Ty1 Transcript Synthesis, Abundance, and Stability

The major Ty1 transcript starts precisely at position 241 of Ty1-H3, corresponding to the first nucleotide of the 5′ LTR-R region (42), and ends at the last nucleotide of the 3′-LTR-R region (43, 44), yielding a 5640 nt RNA with redundant termini (Fig. 5A). Two TATA boxes (T_1 and T_2) located upstream of the transcription start site contribute to transcription and their simultaneous mutation nearly abolishes transcription (45). The 5′ LTR has weak transcriptional activity, *per se* (46). The presence of transcription factor binding sites required for full expression define the Ty1 promoter as a 1 kb region, extending both upstream and downstream of the TATA boxes and including the 5′ LTR and a large portion of the *GAG* coding region (Fig. 5B). Transcription termination occurs by endonucleolytic cleavage and polyadenylation of Ty1 mRNA. Two sequences, TS_1 and TS_2, located in the U3 and R portions of the 3′ LTR, respectively, contribute equally to 3′ end formation (47).

Ty1 transcription is modulated by at least nine transcription factors that bind to the Ty1 promoter (i.e., Gcr1, Ste12, Tec1, Mcm1, Tea1/Ibf1, Rap1, Gcn4, Mot3, and Tye7) and three chromatin-remodeling complexes (Swi/Snf, SAGA and ISWI) (34, 48, 49, 50, 51, 52, 53, 54, 55, 56, 57, 58, 59). Deletion of *STE12*, *TEC1*, or genes encoding subunits of the SWI/SNF chromatin remodeling complex or the Spt/Ada/Gcn5 (SAGA) histone acetyltransferase complex, such as *SPT3*, leads to a severe decrease in Ty1 transcription (34, 54, 57, 60, 61). The other transcription factors have mild or synergistic effects on Ty1 transcription or act under specific growth conditions. For example, the Gcr1 transcriptional activator of genes involved in glycolysis activates Ty1 transcription in the presence of lactate and glycerol (62). Gcn4 is required for Ty1 transcription under amino acid starvation (56), and Tye7 activates Ty1 transcription under adenylic nucleotide stress (58).

Ty1 RNA accounts for 0.1% to 0.8% of total cellular RNA and 10% of polyadenylated mRNA (31, 63). This abundance is probably the consequence of the unusually long half-life of Ty1 RNA—higher than the

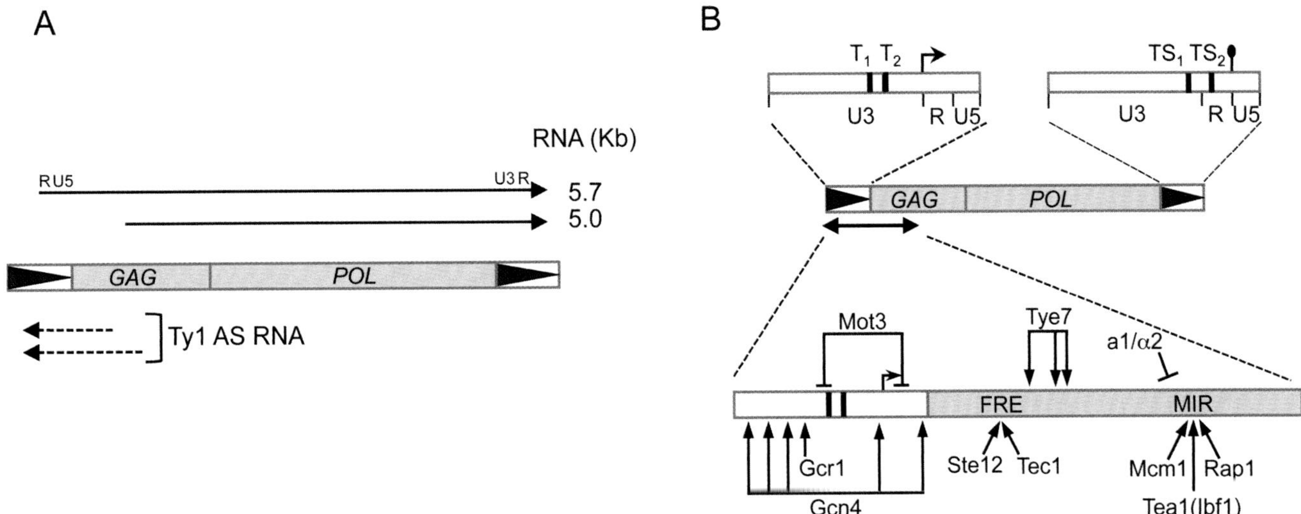

Figure 5 Ty1 transcription. (A) Sense and antisense RNAs transcribed from Ty1. The Ty1 sense-strand transcript starts at the U3/R junction of the 5′ LTR and ends at the R/U5 junction of the 3′ LTR. The 5 kb Ty1 RNA species is detected in *spt3* mutants. Ty1AS RNA 5′ and 3′ extremities have been mapped by RACE to positions 661 (68) and 760 (33) of Ty1-H3 and to positions 136 and 178 of Ty1-H3 (33), respectively. (B) Organization of the Ty1 promoter. Ty1 contains two TATA boxes, T_1 and T_2 (at positions 159 to 165 and 167 to 173, respectively) and two termination sequences TS_1 (5,776 to 5,781) and TS_2 (5,837 to 5,842). The arrow and lollipop indicate sites of transcription initiation and termination, respectively. The Ty1 promoter extends over 1 kb including the 5′ LTR and part of the *GAG* ORF. The positions of the Ty1 activator binding sites are: Gcn4 (five binding sites: 12 to 17, 79 to 84, 98 to 103, 155 to 160, and 318 to 323), Gcr1 (115 to 119), Ste12 (395 to 401), Tec1 (418 to 422), Tye7 (three binding sites: 463 to 468, 661 to 666, and 727 to 732), Mcm1 (833 to 848), Tea1/Ibf1 (884 to 899), and Rap1 (911 to 923). The filamentous response element (FRE) comprises Ste12 and Tec1 binding sites, while MIR comprises Mcm1, Tea1/Ibf1, and Rap1 binding sites. The positions of the Ty1 repressor binding sites are: Mot3 (several binding sites in the 5′ LTR; higher affinity site *in vitro* at positions 147 to 150) and a1/α2 (832 to 863). Positions are given relative to Ty1-H3 sequence (10). doi:10.1128/microbiolspec.MDNA3-0053-2014.f5

average half-life of yeast mRNAs and ~5 h when expressed from pGTy1 (20, 64)—rather than robust transcriptional activity (56). Only about 15% of total cellular Ty1 RNA is polyadenylated (65), suggesting that deadenylated Ty1 RNA, unlike most deadenylated mRNAs (66), is not rapidly degraded.

Expression of Individual Endogenous Ty1 Elements

The relative transcription level of endogenous Ty1 elements of S288C strain was resolved by tagging 31 of the 32 Ty1 elements with a *lacZ* reporter gene. Expression of individual Ty elements varied by up to 50-fold (Fig. 6) (56). Eleven highly expressed elements account for 75% of total Ty1 expression, and eight of these are Ty1/Ty2 hybrid elements. Ty1/Ty2 hybrids contain five potential binding sites for the Gcn4 transcription activator in their 5′ LTR, whereas weakly expressed elements generally have fewer Gcn4 binding sites (67). Overproduction of Gcn4 increases Ty1 mRNA levels and stimulates the transcription of Ty1/Ty2 hybrids by a Swi/Snf and SAGA-dependent mechanism. In contrast, the transcription of Ty1 elements lacking some or all Gcn4 binding sites is low and independent of Swi/Snf. Based on these observations, it was proposed that Gcn4 recruits Swi/Snf and SAGA, which in turn renders the chromatin accessible to transcriptional activators such as Tec1 and Ste12 (56).

Antisense Ty1 RNA Transcription

Two major species of antisense noncoding RNA, known as Ty1-RTL or Ty1AS RNA, are transcribed from Ty1 elements in *S. cerevisiae* strain S288C (Fig. 5A). These transcripts have heterogenous 5′ and 3′ termini with several transcription start sites in *GAG* and two termination positions in the 5′ LTR. Ty1AS RNA is

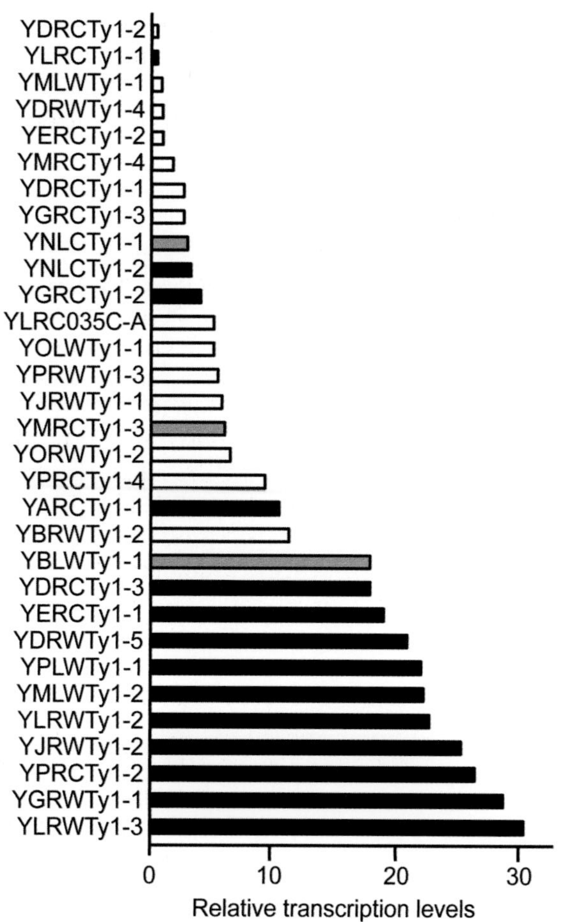

Figure 6 Expression of individual endogenous Ty1 elements. Relative transcriptional activities of 31 Ty1 elements present in the *Saccharomyces cerevisiae* S288C strain (adapted from Morillon et al. [56]). Ty1/Ty2 hybrid elements are indicated by black filled bars and Ty′ elements are indicated by gray filled bars. doi:10.1128/microbiolspec.MDNA3-0053-2014.f6

transcribed by Pol II, capped and polyadenylated (68, 69). The level of Ty1AS RNA is increased in cells lacking the SAGA subunit Spt3, indicating that transcription of sense-strand Ty1 RNA and Ty1AS RNA occurs by independent mechanisms (70, 71). Ty1AS RNA is a cryptic unstable transcript (CUT) whose stability is enhanced by deletion of genes required for 5′ to 3′ mRNA degradation, including *DCP1*, *DCP2*, and *XRN1* (68, 70, 71). Proposed roles for Ty1AS RNA in the regulation of Ty1 retrotransposition are discussed below.

Translation of Ty1 RNA

Little was known about the regulation of Ty1 RNA translation until recently. It was previously assumed that Ty1 RNA was translated inefficiently because of the apparent difference in abundance between Ty1 RNA and Ty1 proteins and VLPs (12, 31, 63, 72). Furthermore, a computational model of the 5′ end of Ty1 RNA, bolstered by mutational analyses, suggested that there is significant secondary structure in the 53-nucleotide 5′ UTR (19), which is unusual among yeast mRNAs and is potentially inhibitory to translation initiation (73). Recent secondary structure models of Ty1 RNA both *in vitro* and *in vivo*, which were developed using selective 2′-hydroxyl acylation of RNA analyzed by primer extension (SHAPE) data, indicate that nucleotides one to seven, 39 to 46, and 55 to 59 are involved in long-range pairing with downstream sequences, and nucleotides 15 to 29 form a stable stem-loop (Fig. 7) (20, 74). Moreover, most Ty1 RNA molecules migrate in sucrose gradients as very high molecular weight complexes that are stable when ribosomes are dissociated with EDTA. Thus, it was inferred that these Ty1 RNPs were nonpolysomal and translationally inactive (71). However, the view that Ty1 RNA is translationally inert is contradicted by a few recent findings that are consistent with active translation of Ty1 RNA. When a premature termination codon is introduced into the *GAG* ORF, Ty1 RNA is subject to nonsense-mediated decay, a translation-coupled process that targets defective mRNA for 5′ to 3′ degradation (22). In addition, Ty1 RNA copurifies with ribosome-nascent chain complexes (75, 76). Finally, the global translation efficiency of genomic Ty1 elements can be estimated using data from deep sequencing of ribosome-protected mRNA fragments in a ribosomal profiling experiment (77). Analysis of these data indicate that the translational efficiency of *GAG* [i.e., the ratio of the percentage of ribosome-protected mRNA-seq reads that align to any *GAG* ORF in the genome to the percentage of poly(A)$^+$ mRNA-seq reads that align to any *GAG* ORF] is substantially higher than the median translation efficiency of all yeast ORFs (R.J. Palumbo and M.J. Curcio, unpublished results). While this calculation does not consider the large pool of deadenylated Ty1 RNA in cells, the fact remains that at least a fraction of Ty1 RNA is actively translated.

Two nonessential ribosome biogenesis factors that promote 18S ribosomal RNA maturation and small ribosomal subunit formation, Bud21 and Bud22, are necessary for efficient translation of Ty1 RNA (76, 78, 79). In a *bud22Δ* mutant, the level of Gag, but not Ty1 RNA, is strongly reduced, and Ty1 RNA is shifted from polysomes containing six to eight ribosomes to polysomes of three to eight ribosomes (78). This reduction in the number of ribosomes per Ty1 RNA is consistent with initiation of Ty1 RNA translation being stalled by the limited availability of 40S rRNA subunits; however, this finding must be interpreted with caution, as the migration of Ty1 RNA in dense sucrose gradient fractions is not due solely to its translation on polysomes (71). In a *bud21*Δ mutant, newly synthesized Gag is reduced to about one quarter of the wild-type level (76).

The Role of Signal Recognition Particle in Gag Translation and Stability

A conserved pathway that targets membrane and secretory proteins to the endoplasmic reticulum (ER) was recently shown to be essential for the stability of Gag (76). Analysis of mutants in this pathway led to the unanticipated discovery that Gag is guided to the ER during translation by signal recognition particle (SRP), a universally conserved ribonucleoprotein complex that functions as a cotranslational chaperone (reviewed in [80]). SRP binding to ribosome-nascent chain complexes transiently pauses translation and targets translation complexes to the SRP receptor on the ER membrane, where the nascent peptide is translocated through the membrane to the ER lumen. Ty1 RNA is associated with affinity purified SRP-ribosome-nascent chain complexes, demonstrating that SRP recognizes ribosomes synthesizing Gag, but the sequences within nascent Gag that interact with SRP have not yet been identified (75, 76). Once Ty1 RNA-ribosome-nascent Gag complexes are targeted to the ER membrane, nascent Gag is translocated to the ER lumen. The half-life of newly synthesized Gag is markedly reduced when proteins required for ER translocation are depleted, including subunits of SRP, the SRP receptor, or the ER translocon. This finding suggests that Gag is rapidly degraded when it is synthesized in the cytoplasm (76).

Because translocation of Gag to the ER lumen is required for its stability, there must be a mechanism by which Gag is retrotranslocated to the cytoplasm to associate with Ty1 RNA to form VLPs. Virtually nothing is known about this mechanism, as Ty1 Gag is the first cytoplasmic yeast protein described that traverses the ER (76). In mammalian cells, several viral proteins are retrotranslocated from the ER lumen via the ER-associated degradation (ERAD) pathway, including the ORF2 protein of hepatitis E virus, the precore protein hepatitis B virus and REM protein of the betaretrovirus, MMTV (81, 82, 83). The ERAD pathway normally retrotranslocates resident ER and membrane proteins to the cytoplasm for degradation by the proteosome (84), but these cytoplasmic viral substrates escape degradation (81, 82). Much remains to be learned about retrotranslocation of cytosolic proteins from the ER and Ty1 can serve as a simple model system to explore this process. An important consequence of the cotranslational translocation of nascent Gag to the ER and retrotranslocation of Gag to the cytoplasm is that this trafficking sequesters nascent Gag from its encoding Ty1 RNA, and thereby enhances the interaction of Gag with Ty1 RNA in *trans* (76).

Expression of *POL* by Translational Frameshifting

The Ty1 *POL* ORF is expressed only as a Gag-Pol fusion protein that forms when translation "slips" forward one nucleotide from the *GAG* ORF to the *POL* ORF. Programmed ribosomal frameshifting occurs at a heptanucleotide sequence, CUU-AGG-C, in the 38-nucleotide overlap between the *GAG* and *POL* ORF. Both the CUU codon and the UUA codon in the +1 frame encode leucine and can be decoded by $tRNA^{UAG}$ (85). When $tRNA^{UAG}$ is in the P site, it can slip from the CUU codon to the +1 UUA codon (86). Slippage of $tRNA^{UAG}$ is enhanced by pausing of the ribosome with the CUU codon in the P site and the AGG codon in the A site. Pausing at this position is a consequence of the low abundance of the AGG-decoding $tRNA^{CCU}$, which is encoded by the single copy *HSX1* gene (87, 88).

Translational frameshifting is a common mechanism of optimizing the ratio of viral structural proteins to catalytic proteins for assembly of replication-competent viral particles (89). Indeed, expression of Ty1 Gag and Pol proteins at a specific ratio is critical for proteolytic processing of Ty1 proteins and retrotransposition (88, 90, 91). The amount of p199-Gag-Pol synthesized appears to be primarily a function of the efficiency of ribosomal frameshifting, which is 3% to 13%, depending on the assay method and experimental conditions used to measure Ty1 frameshifting efficiency (88, 92). The efficiency of Ty1 frameshifting genome-wide can be estimated from deep sequencing data generated from a ribosomal profiling experiment (77). Our compilation of mRNA-seq reads of ribosome-protected mRNA footprints that map to Ty1 *POL* versus *GAG* reveals an overall frameshifting efficiency of 7% for genomic Ty1 elements (Palumbo and Curcio, unpublished results). While ribosomal frameshifting is a potentially important control point for Ty1 retrotransposition, few host factors that influence frameshifting other than rRNA and the tRNAs that decode the heptanucleotide sequence have been characterized. Host factors required for biosynthesis of spermine and spermidine, Spe1 and Spe2, suppress frameshifting and are essential for efficient retrotransposition (93, 94). In addition, the 40S rRNA biogenesis factor Bud22 enhances Ty1 frameshifting but it is also necessary for efficient synthesis of Gag (78).

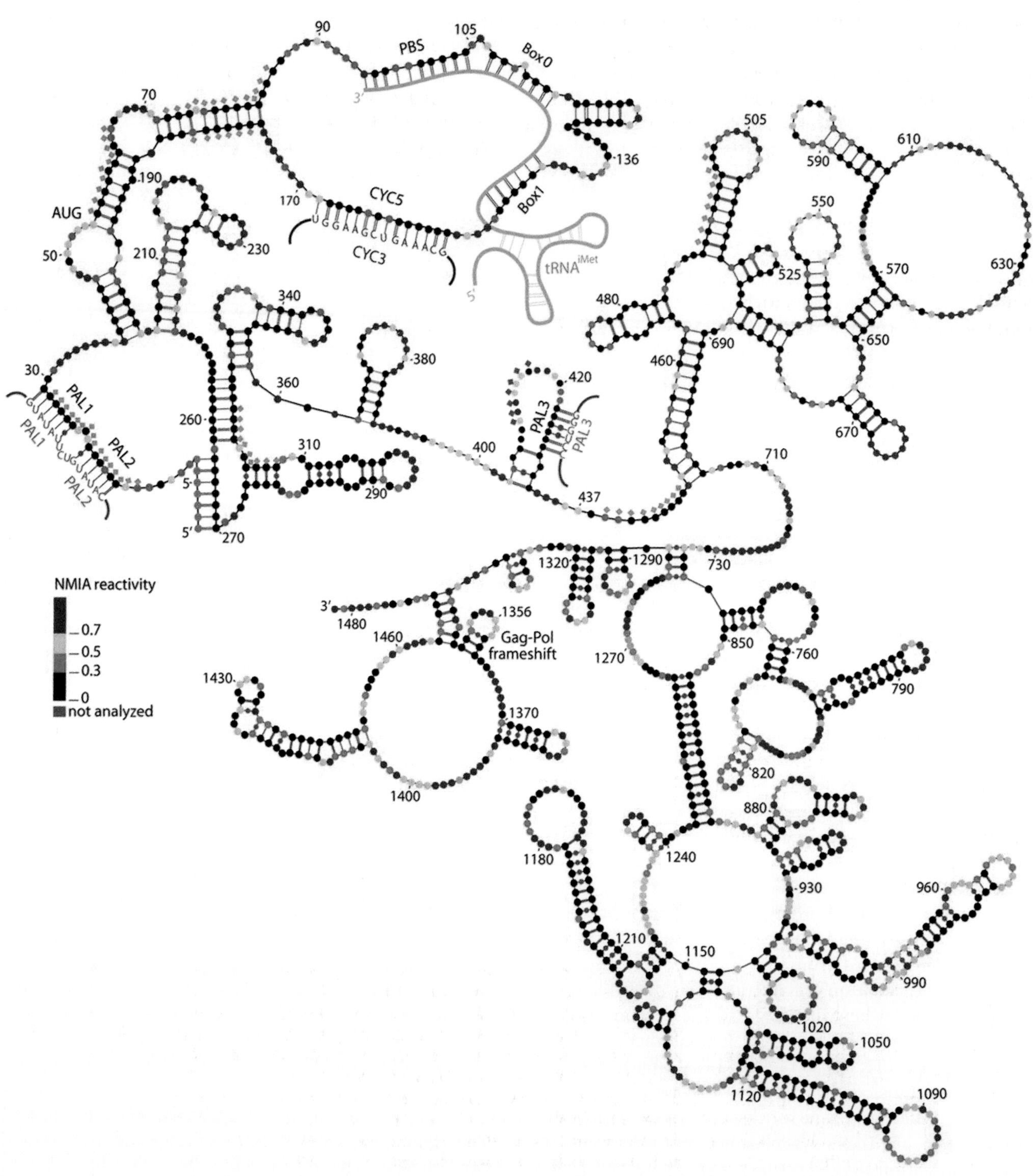
90
105
PBS
Box0
3'
70
136
505
610
590
550
190
170
CYC5
Box1
AUG
CYC3
tRNA^iMet
50
210
230
525
570
630
5'
480
340
690
650
460
380
30
PAL1
PAL1
PAL2
PAL2
360
420
PAL3
PAL3
260
400
670
310
710
290
437
5'
5'
270
1320
1290
730
NMIA reactivity
0.7
0.5
0.3
0
not analyzed
3'
1480
1356
Gag-Pol
frameshift
1460
1270
850
760
1430
790
1370
820
1400
880
1180
1240
930
960
1210
1150
990
1020
1050
1120
1090

Biogenesis of the VLP Assembly Site

Localization of Ty1 RNA and Gag Protein to Retrosomes

The components of retroviral particles are typically concentrated in subcellular domains known as nucleocapsid assembly sites or "virus factories." As the evolutionary progenitors of retroviruses, LTR-retrotransposons might also be expected to have cytoplasmic domains where gRNA and Gag and Gag-Pol proteins are concentrated to promote assembly of VLPs. However, little was known about the site of VLP assembly until the serendipitous discovery of microscopically distinct cytoplasmic foci, known as T-bodies or retrosomes, where Ty1 RNA and Gag proteins colocalize (65). In strain S288C, retrosomes are detected in 30% to 50% of cells at 20°C, a permissive temperature for retrotransposition (Fig. 8) (65, 70, 76). Notably, retrosomes are not detected in GRF167, the strain in which Ty1 retrotransposition was initially characterized (S. Lutz and M. J. Curcio, unpublished results). Ty1 retrosomes are not simply clusters of preformed VLPs, as VLP formation is very inefficient in S288C. Nonetheless, VLPs are thought to assemble within retrosomes, as clusters of VLPs are associated with Gag foci when VLP formation is induced by expressing a pGTy1 element (12, 70).

The trafficking of retroviral and LTR-retrotransposon RNAs and proteins to their sites of assembly has recently begun to be unraveled (95). HIV-1 and Rous sarcoma virus gRNAs are recruited to the presumptive assembly site by Gag, which interacts with gRNA in the cytoplasm or nucleus, respectively (96, 97). The scenario for nucleation of Ty1 assembly sites appears to be distinct, in that Ty1 RNA is targeted to the retrosome prior to or during translation (76). Translation of Ty1 RNA in retrosomes is supported by the finding that glucose deprivation, a global inhibitor of translational initiation, cause retrosomes to disperse (65, 70). Moreover, the abundance of retrosomes is increased by hypomorphic mutations in SRP subunits Srp54 and Srp72, which reduce the rate of cotranslational transport of nascent Gag to the ER. Suppression of these mild translocation defects by compensatory reductions in the rate of translational elongation rapidly reverses the increase in retrosomes (76). This modulation of retrosome abundance in response to small changes in the rate of translocation relative to translation indicates that the Ty1 retrosome is a dynamic ribonucleoprotein complex containing Ty1 RNA translation complexes awaiting translocation of nascent Gag to the ER.

SRP-associated Ty1 RNA translation complexes are bound by Gag, presumably via a direct interaction with Ty1 RNA (76). It is likely that multimerization of Gag bound to Ty1 RNA translation complexes is necessary for the stability of the retrosome, as retrosomes fail to appear when Gag levels are very low. For example, retrosomes are not visible in a *bud21* mutant, in which Gag is inefficiently synthesized, or in a *srp68-DAmP* mutant, in which newly synthesized Gag is rapidly degraded (76). Retrosomes are also not detected when Gag lacking the C-terminal region required for RNA binding and VLP formation is expressed in a strain lacking endogenous Ty1 expression (22, 98).

One study suggests that Gag facilitates export of Ty1 RNA from the nucleus to cytoplasmic retrosomes (22). When expressed in the absence of endogenous Ty1 RNA, a Ty1 RNA harboring a stop codon directly downstream of the *GAG* start codon accumulates in the nucleus. Expression of functional Ty1 RNA and Gag in *trans* restores the localization of this untranslatable Ty1 RNA to cytoplasmic retrosomes. At least a fraction of Gag is located at the nuclear membrane, the outer

Figure 7 Secondary structure model of the 5′ end of Ty1 RNA. (Reproduced from Purzycka et al. [74].) Full-length Ty1 RNA within virus-like particles (VLPs) (*in virio*) was analyzed to determine the SHAPE reactivities of nucleotides 1 to 1,482, which are individually represented by a color-coded ball. The color is indicative of the reactivity of that nucleotide to the SHAPE reagent, N-methyisatoic anhydride (NMIA). The secondary structure model in which regions predicted to be based-paired (illustrated as bars linking balls) and regions predicted to be single stranded are shown was obtained from the SHAPE reactivities using RNAstructure 4.6 software (270). Nucleotide positions at which SHAPE reactivities changed when the proteins were gently stripped away from VLP-associated Ty1 RNA (*ex virio*) are marked with blue diamonds (increased reactivity) or gray diamonds (decreased reactivity). Annotated regions include the following: PAL1, PAL2, and PAL3 motifs, at which the differences in nucleotide reactivity *in virio* versus *ex virio* suggest that the Ty1 RNA pairs with a second molecule of Ty1 RNA to form a dimer; the AUG codon of *GAG* (nucleotides 54 to 57); CYC5, a region that hybridizes to the CYC3 domain near the 3′ end of Ty1 RNA; primer binding sequence (PBS), Box 0, and Box 1, where the $tRNA_i^{Met}$ primer (shown in gray) hybridizes to Ty1 RNA to initiate reverse transcription; and the Gag-Pol frameshift site at nucleotides 1,356 to 1,362. doi:10.1128/microbiolspec.MDNA3-0053-2014.f7

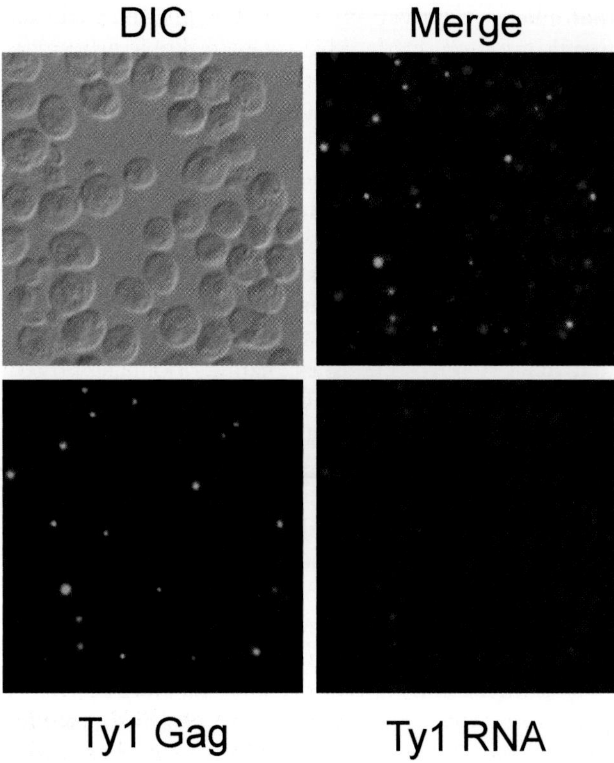

Figure 8 The Ty1 retrosome. Retrosomes are cytoplasmic foci in which Gag and retrotransposon RNA colocalize, as visualized by fluorescence microscopy. Ty1 retrosomes are detected in fixed cells after fluorescence *in situ* hybridization and immunofluorescence. Ty1 RNA is detected using a DNA oligomer end-labeled with Cy3, and Gag is detected using anti-VLP antibodies that are bound by a secondary antibody coupled to Alexa Fluor® 488 (Life Technologies, Grand Island, NY) doi:10.1128/microbiolspec.MDNA3-0053-2014.f8

membrane of which is continuous with the ER. Further studies are necessary to determine whether the retrosome is associated with the nuclear membrane, or whether Gag's transit through the ER plays a role in export of Ty1 RNA from the nucleus to the retrosome.

A Role for Ty1 RNA Subcellular Localization in Retrosome Formation

A cohort of host cofactors that are associated with mRNA processing bodies (P-bodies) is necessary for the localization of Ty1 RNA to retrosomes. Interestingly, these host-factors have little effect on the expression of sense-strand Ty1 RNA or the formation of VLPs. P-bodies are cytoplasmic RNA granules containing translationally repressed mRNAs and 5′ to 3′ mRNA decay factors (99). Lsm1 and Pat1 are associated with deadenylated RNA and activate decapping and decay of mRNAs (100, 101, 102, 103). Xrn1 (Kem1) is the cytoplasmic 5′ to 3′ exoribonuclease that degrades mRNA (104). In *lsm1*Δ and *pat1*Δ mutants, Ty1 RNA is dispersed throughout the cytoplasm, while in an *xrn1*Δ mutant, Ty1 RNA and Gag are in diffuse aggregates (70).

Based on the colocalization of Ty3 retrosomes with P-bodies (see Chapter 42 by S. Sandmeyer), it was initially thought that partial colocalization of Ty1 Gag with P-body proteins might have functional significance in the formation of the retrosome (71). However, unlike Ty3 retrosomes, Ty1 retrosomes form under different conditions than those that promote P-body formation, which reflects their divergent populations of actively translated and translationally repressed RNAs, respectively (65, 70, 71). Despite the fact that Ty1 Gag and RNA are not concentrated in P-bodies, P-body-associated deadenylation-dependent mRNA decay and nonsense-mediated decay factors are required for Ty1 cDNA accumulation and retrotransposition (70, 71). In *xrn1*Δ, *lsm1*Δ, and *pat1*Δ mutants, there are only minor reductions in Ty1 RNA and Gag protein, and proteolytic processing of Gag, carried out by Ty1 PR in the VLP, is efficient, indicating that these mRNA decay factors are not essential for VLP formation. However, VLPs tend to be dispersed throughout the cytoplasm in *xrn1*Δ and *lsm1*Δ mutants. (A deletion of *PAT1* did not have this effect.) The levels of processed p63-RT and p71-IN were also reduced in *xrn1*Δ and *lsm1*Δ mutants (70). Deletion of *XRN1* abolished Ty1 RNA packaging in VLPs, as measured by the lack of protection of Ty1 RNA from degradation by benzonase (71). Thus, the lack of retrosomes in these mutants is associated with formation of replication-incompetent VLPs with defects in Ty1 RNA packaging, stability of VLP-associated Pol proteins, and cDNA accumulation.

It remains to be determined whether Xrn1, Lsm1, and Pat1 interact directly with Ty1 components to promote retrotransposition, or whether they act indirectly via their general effects on RNA metabolism. No direct interaction between Ty1 Gag and Xrn1, Lsm1 or Pat1, or between Ty1 RNA and Lsm1 has been detected (71, 76). Several models have been proposed to explain the lack of Ty1 retrosomes and retrotransposition in 5′ to 3′ mRNA decay factor mutants. In the first model, accumulation of cellular mRNA in the absence of 5′ to 3′ mRNA decay may interfere with the specific recognition of Ty1 RNA by Gag and result in a high level of nonspecific RNA packaging (71). The second model, based on the observation that Ty1AS RNA is stabilized in *lsm1*Δ, *pat1*Δ and *xrn1*Δ mutants, posits that elevated levels of Ty1AS RNA suppress retrosome formation and retrotransposition (33, 70). A third possibility is that mRNA decay factors are required for the localization of Ty1 RNA to a specific subcellular domain of the cell prior to translation. The tethering of Ty1 RNA translation complexes to the ER by SRP and translocation of nascent Gag cannot, in and of itself, explain how Ty1 RNA and Gag are localized to a single address in the cell (76). Perhaps Ty1 RNA localization, like the localization of many mRNAs, requires translational repression and active transport (105), and P body factors Xrn1, Lsm1, and Pat1, may be regulators of this fate (70). Support for the idea that host factors traffic Ty1 RNA to the presumptive retrosome prior to translation comes from the analysis of a mutant lacking *RPL7A*, one of two paralogous genes that encode ribosomal protein L7. In a *rpl7a*Δ mutant, Ty1 RNA is translated in association with SRP, but Gag does not interact with Ty1 RNA translation complexes, and retrosomes are not formed (76). Therefore, Rpl7a and possibly Xrn1, Lsm1, and Pat1 may direct Ty1 RNA to a specific subcellular location prior to Gag synthesis in order for newly synthesized Gag that is exiting the ER to interact with Ty1 RNA to form the retrosome.

POSTTRANSLATIONAL STEPS IN RETROTRANSPOSITION

VLP Formation

Gag assembly and VLP structure

Ty1 VLPs are nucleocapsid particles formed by assembly of Gag and Gag-Pol precursors into particles that encapsidate Ty1 RNA. Gag is the only Ty1 protein domain necessary for the formation of VLPs, which assemble even when Gag is expressed in *Escherichia coli* (106, 107). The smallest fragments of the 440 amino acid-Gag protein that can assemble into particles span amino acids 41 to 346 or 31 to 363 (107, 108, 109). Within this minimal Gag fragment, several regions have been shown to be important in assembly, including amino acids 41 to 62, 114 to 147, 223 to 287, and 330 to 346 (reviewed in [110]). Remarkably, one or two amino acid substitutions can completely block assembly or increase the size of VLPs as much as 8-fold (109). The N-terminus of Gag is on the surface of the VLP, while the C-terminus of both mature and immature forms of Gag is buried in the core (108). This alignment ensures that the RNA binding domain at the C-terminus of Gag and the Pol domain of p199-Gag-Pol is within the capsid shell.

Ty1 VLPs are ~14 MDa in size and are composed of a spikey, symmetrical electron dense shell of ~30 to 80 nm in diameter, surrounding a luminal core. The shell is porous and allows globular proteins ~13 kDa, for example, RNaseA, but not larger ~30 kDa, for example, benzonase, to access the gRNA packaged within the VLP (111, 112). A remarkable feature of Ty1 particles is their variability in size (107, 112, 113). Expression of either unprocessed p49-Gag or mature p45-Gag, which is not subject to further proteolytic processing, results in particles of two to three different size classes. The p49-Gag precursor yields particles with icosahedral T numbers (subunits) of T=7 and T=9. C-terminal truncation of Gag results in polydisperse VLPs with a smaller average size than those formed by p49-Gag, but proteolytic processing of p49-Gag does not cause a reduction in the size of Ty1 VLPs; ergo, the heterogeneity in particle size is not a reflection of incomplete Gag processing, but rather manifests an extraordinarily flexible assembly process. The finding that VLPs formed in the presence and absence of Ty1 PR-mediated processing have the same size range provides direct experimental evidence that Gag is processed after VLPs assemble (111).

Formation of VLPs is a key step in the replication cycle of Ty1. The VLP protects Ty1 gRNA from attack by nucleases and concentrates factors that copy Ty1 gRNA into cDNA. In addition to Gag, *POL*-encoded PR, IN and RT/RH, Ty1 gRNA, the host-coded $tRNA_i^{Met}$ and deoxynucleotide triphosphates are necessary for the production of cDNA within the VLP. An exogenously expressed human antiretroviral protein, APOBEC3G, is the only non-Ty1-encoded protein that has been shown to cofractionate with Ty1 VLPs (114). APOBEC3G, a potent inhibitor of Ty1 retrotransposition, gains access to the VLP via its RNA-dependent interaction with Ty1 Gag (114, 115).

RNA Packaging

By analogy with retroviruses, Ty1 RNA is bound by the nucleocapsid domain of Gag and packaged into the VLP. Ty1 RNA sequences required for Gag binding and packaging have not been precisely defined. Xu and Boeke (21) demonstrated that an internally deleted mini-Ty1 RNA containing 380 nucleotides of the 5′ terminus and 357 nucleotides of the 3′ terminus of Ty1 RNA was mobilized when Gag and Pol were expressed in *trans* from a helper-Ty1 element (Fig. 4). Deletion of nucleotides 237 to 380 abolished retrotransposition and copurification of mini-Ty1 RNA with VLPs, suggesting that *cis*-acting sequences required for Ty1 RNA packaging reside in this domain.

A single nucleotide resolution secondary structure model of the 5′ terminus of Ty1 gRNA within VLPs was derived using SHAPE reactivities (Fig. 7). In this model, nucleotides one to 325 of Ty1 RNA extracted from VLPs (referred to as "*in virio*") form a long-range pseudoknot. The pseudoknot core consists of two 7 bp stems (S1 and S2) with a one-nucleotide interhelical connector, and long structured loops that bridge the stems. The essential packaging region between nucleotides 237 and 380 overlaps pseudoknot sequences in the S1 stem and the S2 stem and its structured loop (74). Surprisingly though, mutations that destabilize the S1 stem of the pseudoknot do not diminish Ty1 RNA packaging (20); *ergo*, pseudoknot formation may play a minor role in packaging of Ty1 gRNA.

Like the gRNA of retroviral particles, Ty1 gRNA is packaged in the VLP as a dimer (116). The dimer interaction is noncovalent. *In vitro*, Ty1 RNA dimerizes in the presence of $tRNA_i^{Met}$ and the RNA binding domain of Gag; however, single nucleotide resolution structural analyses of Ty1 gRNA indicated that the $tRNA_i^{Met}$ does not participate in dimerization of Ty1 RNA (74, 117). Retroviral RNA dimerization and packaging are tightly coupled processes, both facilitated by the nucleocapsid activity of Gag (118, 119). Purzycka et al. (74) identified three palindromic sequences (PAL1 to PAL3) in the 5′ terminus of Ty1 RNA that were less reactive in gRNA *in virio* than in gRNA *ex virio*. Based on analogy with retroviral dimerization sites, the authors proposed that PAL sequences are sites where the nucleic acid chaperone activity of Gag promotes a transition from intramolecular pairing to intermolecular pairing of Ty1 RNA (74, 120).

Ty1AS RNA copurifies with Ty1 VLPs, but there is no evidence that it interacts with Ty1 gRNA within the VLP (33, 74). Cellular mRNAs can be coincorporated into VLPs and subsequently copied into cDNA using Ty1 cDNA as a primer. Most mRNAs gain access to the VLP on the basis of their abundance; however, a small number of mRNAs, including the transcript of the subtelomeric repeat, Y′, which encodes a helicase of unknown function, are highly enriched in VLPs (121, 122, 123).

Protease-Dependent Gag and Gag-Pol Processing

Ty1 protease (PR) is a 20 kDa monomeric aspartyl protease encoded by the C-terminus of the *GAG* ORF and N-terminus of the *POL* ORF. Processed p20-PR is very difficult to detect, even when pGTy1 is expressed (124). Ty1 PR activity is required for all cleavages of Ty1 Gag and Gag-Pol proteins and for retrotransposition activity (124, 125, 126). The p199-Gag-Pol precursor has three distinct cleavage sites at the Gag/PR (RAH/NVS), PR/IN (TIN/NVH), and IN/RT (LIA/AVK) junctions, which were defined by sequencing of various termini of mature Ty1 proteins and mutagenesis (125, 127, 128). The 6-amino acid cleavage site, while necessary, is not sufficient to define the processing site. The Gag-Pol precursor must be cleaved at the Gag/PR site for subsequent cleavages of the Gag-Pol precursor or p49-Gag precursor to occur (125). Pulse-chase experiments of pGTy1 expression and mutagenesis of PR cleavage sites support semi-ordered processing of the p49-Gag and p199-Gag-Pol precursor proteins (124, 125). Mutations in Ty1 that abolish PR activity result in VLPs with significant reverse transcriptase activity when an exogenous primer and template are supplied, but reverse transcription of the endogenous template is undetectable, perhaps because VLPs produced from a PR-defective element contain significantly less Ty1 gRNA, and Ty1 RNA dimerization is reduced (116, 126).

cDNA Synthesis

Cis-Acting Sequences Required for Reverse Transcription

Cis-acting RNA motifs in the 5′ termini of Ty1 RNA participate in the initiation of reverse transcription. Critical sequences have been defined by introducing silent nucleotide substitutions into the 5′ end of a pGTy1 element or by using a nonautonomous Ty1 as an RT template donor (Fig. 4). Immediately adjacent to the 5′ LTR is the primer binding sequence (PBS; nucleotides 95 to 104), a 10-nucleotide sequence that is complementary to the 3′ end of $tRNA_i^{Met}$ (Fig. 7). This tRNA species is selectively packaged into the VLP, where it functions as the primer for initiation of reverse transcription (129, 130). Two adjacent seven- and six-nucleotide regions, known as Box 0 and Box 1, respectively, are complementary to sequences within the T or D hairpins of $tRNA_i^{Met}$ (131, 132). Analyses of mutations in both Ty1 RNA and $tRNA_i^{Met}$ have established a role for an extended interaction between $tRNA_i^{Met}$ and the PBS, Box 0 and Box 1 regions of Ty1 RNA in the initiation of reverse transcription (132, 133). Adjacent to Box 1 is a 14-nucleotide region known as CYC5, which is perfectly complementary to a sequence in the 3′ UTR known as CYC3. CYC5:CYC3 complementarity promotes efficient reverse transcription *in vitro* and retrotransposition *in vivo* (117, 134). In addition, intramolecular pairing of nucleotides one to seven to nucleotides 264 to 270 promotes efficient reverse transcription (19, 20). The secondary structure model of the 5′ terminus of Ty1 gRNA *in virio* supports many aspects of earlier structural models that were based on mutational analyses (19, 133), including pairing of the $tRNA_i^{Met}$ to the PBS, Box 0 and Box 1 regions and circularization of Ty1 RNA via the CYC5:CYC3 interaction. Moreover, the functionally defined pairing of nucleotides one to seven to nucleotides 264 to 270 forms the S1 stem of the Ty1 RNA pseudoknot. All of the RNA motifs that are known to be required for reverse transcription are in the multibranched loop that is formed by the S1 stem of the pseudoknot, suggesting that this subdomain may be functionally as well as structurally distinct (Fig. 7).

Reverse Transcriptase

The Ty1 reverse transcriptase is a 63 kDa protein encoded at the carboxy terminal end of the *POL* ORF. Like retroviral RTs, Ty1 RT has an amino-terminal domain with an RNA- or DNA-dependent DNA polymerase activity and a carboxy terminal RNase H domain whose activity specifically degrades the RNA strand of an RNA:DNA duplex. These dual activities allow a full-length, double-stranded Ty1 cDNA to be synthesized from the single-stranded Ty1 RNA.

Ty1 RT, like all RTs, is dependent on a divalent cation, preferably Mg^{2+}, for catalytic activity. The Ty1 reverse transcription reaction also occurs in the presence of Mn^{2+}, although to a much lesser extent (135, 136, 137). Ty1 RT contains a triad of aspartate residues in the polymerase active site that is conserved among RTs. A mutant bearing a substitution of asparagine for the aspartate at position 211 of Ty1 RT, the second aspartate in the canonical YXDD motif, retains some catalytic activity but has altered properties, including a preference for Mn^{2+} over Mg^{2+} and a marked defect in pyrophosphate binding and release, which affect the ability of polymerase to translocate on the template (136, 138). Interestingly, specific mutations in the RNase H domain increase the polymerase activity of Ty1 RT without affecting the RNase H activity (139).

Despite the fact that the RT and IN domains are arranged in an order opposite to that in the *POL* gene of retroviruses, Ty1 RT activity, like that of retroviruses, is dependent on an interaction between RT and IN. *In vitro*, the biochemical activity of recombinant Ty1 RT purified from *E. coli* requires the expression of the contiguous 115-amino acid C-terminal portion of IN fused to the N-terminus of the entire RT ORF (137). Fusion of a small acidic tail to the RT protein restores polymerase and RNase H activity *in vitro*, supporting the hypothesis that a highly acidic domain of IN, located between amino acids 521 and 607, interacts with basic residues at the N-terminus of RT to allow RT to adopt an active conformation (140). Consistent with this model, complementation assays based on the coexpression of two Ty1 elements show that IN acts exclusively in *cis* to activate RT *in vivo*. Furthermore, RT purified from Ty1 VLPs is not active unless it remains in close association with IN (141).

Reverse Transcription Reaction

Reverse transcription of the Ty1 RNA dimer within the VLP occurs by a series of ordered steps that are generally analogous to those of reverse transcription of retroviral RNA within virion cores (11, 12, 142) (Fig. 9). Reverse transcription is initiated from the 3′ OH of the $tRNA_i^{Met}$ primer that anneals to the complementary PBS sequence immediately downstream of the 5′ LTR (129). The initial DNA product, known as minus-strand strong-stop cDNA, contains the U5 and R regions of the LTR linked to the $tRNA_i^{Met}$ (143). Once RT reaches the 5′ end of the Ty1 RNA template, the Ty1 RNA portion of the RNA:DNA hybrid is degraded by the RNase H activity of RT. Degradation of the RNA facilitates transfer of minus-strand strong stop cDNA from the 5′ end of the Ty1 RNA template to the 3′ end , where it anneals to the R region. Minus-strand strong stop cDNA is thought to anneal to the full-length Ty1 RNA and not the Ty1 RNA whose 5′ end has been degraded by RNase H (144). Using minus-strand strong-stop cDNA as a primer, a minus-strand Ty1 cDNA that extends into the U5 and R regions present in the undegraded Ty1 RNA template is synthesized. The Ty1 RNA is then degraded by RNase H, except for two polypurine tracts, one directly upstream of U3 (PPT) and one near the center of Ty1 (cPPT) (145, 146). PPT serves as the primer for synthesis of plus-strand strong stop cDNA to the 5′ end of the minus-strand cDNA (144). This plus-strand cDNA fragment is transferred from the 5′ end to the 3′ end of the minus-strand cDNA, where it hybridizes to R and U5 sequences and serves as a primer for synthesis of the complete plus strand. Finally, the minus-strand cDNA is extended through the U3 region, using the plus-strand strong-stop cDNA as a template. These reactions synthesize linear double-stranded cDNA with complete 5′-U3-R-U5-3′ sequences at the 5′ and 3′ termini. Synthesis of plus-strand cDNA from cPPT leads to the formation of a flap in the plus strand of Ty1 cDNA, but is not essential for retrotransposition (145, 147, 148, 149).

The yeast RNA lariat debranching enzyme, Dbr1, initially identified by characterization of a Ty1 retrotransposition-defective host mutant (150), is thought to be required for Ty1 cDNA synthesis (151, 152, 153, 154). Notably, the human homolog of Dbr1 is required for HIV-1 cDNA synthesis at a stage subsequent to minus-strand strong-stop cDNA synthesis (155). A

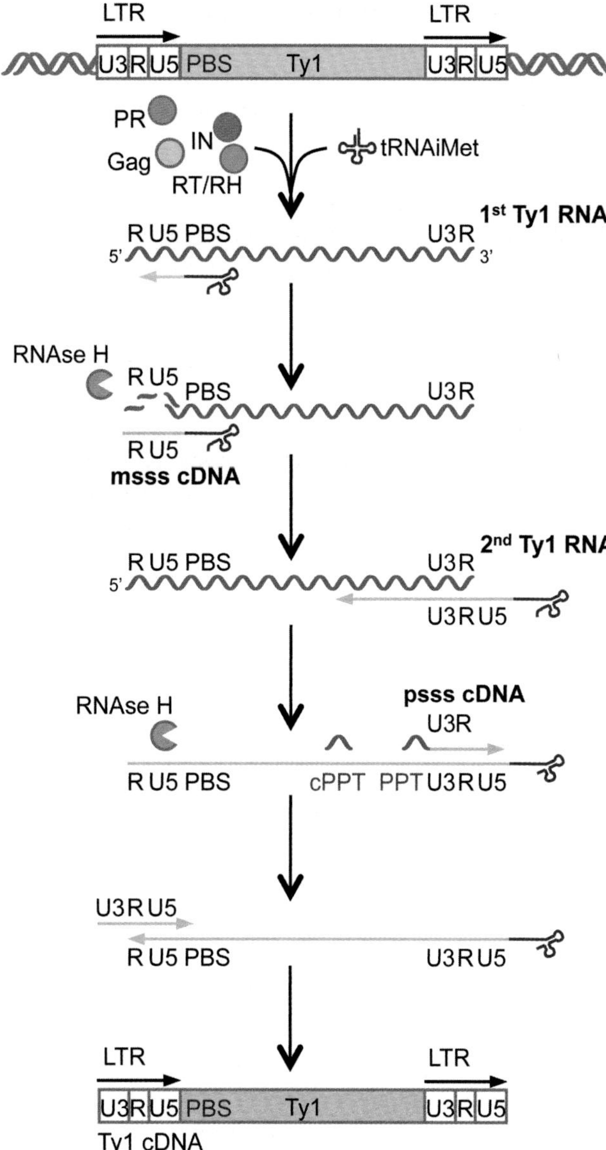

Figure 9 Steps in reverse transcription of Ty1 cDNA. Details of each step are provided in the text. Gag and protease (PR) have nucleic acid chaperone activities required for reverse transcription and integrase (IN) facilitates the reverse transcription reaction carried out by reverse transcriptase/RNase H (RT/RH). In this schematic, minus-strand strong-stop (msss) cDNA is shown being transferred to the second Ty1 RNA in the VLP for minus strand cDNA synthesis. However, it is formally possible that msss cDNA transfers to the 3′ end of the first Ty1 RNA, as long as this RNA, following RNAse H-mediated degradation, contains a remnant of R-U5 sequence to template minus-strand cDNA synthesis so that minus-strand cDNA can hybridize to plus-strand strong-stop (psss) cDNA in the final step of the RT reaction. Single strands of cDNA are represented by thin green lines; the presence of an arrowhead indicates strands being extended by reverse transcription. Blue wavy line, Ty1 RNA; short blue squiggles, polypurine tracts of Ty1 RNA, cPPT and PPT, remaining after RNase H endonucleolytic activity; blue pacman shape, RNase H activity of RT/RH; LTR, long terminal repeat; PBS, primer binding sequence.
doi:10.1128/microbiolspec.MDNA3-0053-2014.f9

hypothesis that Dbr1 was required for debranching of a 2′, 5′-branched form of Ty1 RNA that serves as a template for transfer of the minus-strand strong stop cDNA provided an alluring model to explain the role of Dbr1 in Ty1 cDNA synthesis (154). However, neither the existence of a lariat form of Ty1 RNA *in vivo* nor the proposition that Ty1 RT can read through a 2′, 5′ branched RNA were substantiated in further studies (156, 157).

Nucleocapsid Activities of Gag and PR

A nucleic acid chaperone activity that promotes annealing of the $tRNA_i^{Met}$ primer to the PBS of Ty1 RNA has been mapped to the C-terminus of p45-Gag. A peptide containing the C-terminal 103 amino acids of p45-Gag binds Ty1 RNA and promotes annealing of $tRNA_i^{Met}$ to the PBS and initiation of reverse transcription *in vitro*. Ty1 Gag lacks the canonical zinc finger motif that is found in the nucleocapsid domain of many retroviral and retrotransposon Gag proteins, instead it contains three stretches of basic amino acids in its nucleic acid chaperone domain (117). Remarkably, the Ty1 PR also has nucleic acid chaperone activity that is essential for Ty1 retrotransposition. The N-terminal 35 amino acids of PR, encoded within the overlapping *GAG* ORF upstream of the frameshifting site, is required for reverse transcription but is dispensable for the proteolytic activity of PR. Expression of a pGTy1 element lacking the N-terminal PR domain in the absence of endogenous Ty1 element expression results in defective minus-strand strong-stop cDNA synthesis, despite the presence of Ty1 RNA and $tRNA_i^{Met}$ within VLPs (158).

Ty1 cDNA

When the pGTy1 element is expressed, full-length, double-stranded Ty1 cDNA can be purified from cell fractions enriched for VLPs (12, 159). Unintegrated Ty1 cDNA can also be detected in total DNA from a wild-type yeast strain grown at 20°C, the permissive temperature for retrotransposition. Unintegrated Ty1 cDNA is differentiated from genomic Ty1 elements on the basis of its lack of flanking DNA (160). It is not known where the pool of unintegrated cDNA is in the cell, or what fraction can be utilized in retrotransposition. However, cDNA levels are correlated with rates of retrotransposition in a large number of mutants, suggesting that stable unintegrated Ty1 cDNA represents a functional retrotransposition intermediate (161, 162, 163, 164). Unintegrated Ty1 cDNA is present in less than one copy per haploid cell (160, 165, 166). Treatment of cells with phosphonoformate, an inhibitor of Ty1 RT, has been used to measure the rate of decay of Ty1 cDNA (167). The half-life of Ty1 cDNA in wild-type cells is between 93 and 252 min in different studies, and is similar to the doubling time of the cell (167, 168, 169), which suggests that degradation of Ty1 cDNA does not play a major role in regulating transposition.

cDNA Integration

Characteristics of the Integration Process

cDNA integration is performed by Ty1 IN, a 71 kDa protein of 635 amino acid residues produced by cleavage of the 140 kDa IN-RT processing intermediate (125). Although Ty1 integration probably relies on a variety of cellular cofactors *in vivo*, it is possible to faithfully reproduce this process *in vitro* using DNA substrates and VLPs as a source of IN (128, 159, 170, 171). The requirements for efficient Ty1 IN activity *in vitro* have been reviewed in detail (140, 172). *In vitro*, the integration machinery recognizes as little as 4 bp at Ty1 cDNA termini (170, 173). The conserved inverted dinucleotides 5′-TG...CA-3′ at the ends of the cDNA duplex are absolutely required for integration *in vivo*, although very few sequences alterations have been explored (7). During the integration process, Ty1 IN catalyzes a nucleophilic attack by the 3′ OH moieties at the blunt ends of the cDNA molecule on each strand of the target DNA at 5 bp staggered sites. These concerted transesterification reactions, called the strand transfer step (reviewed in [174]), result in a newly exposed 3′ OH and five-nucleotide gap on each strand of the target DNA at the cDNA:target DNA junctions. Repair of the gaps generates a 5 bp target site duplication (TSD) flanking the newly integrated Ty1 cDNA, which is a hallmark of Ty1 integration. Ty1 IN does not repair the gaps, suggesting that either Ty1 RT or an unidentified cellular DNA repair protein performs this function. Ty1 IN has been shown to dimerize *in vitro* and by analogy with retroviral integrases, is thought to assemble as a tetramer on cDNA to form the preintegration complex (PIC) (175).

In vivo, IN overexpression from the *GAL1* promoter allows the integration of blunt double-stranded DNA fragments, which have no similarity with Ty1 DNA, except for the conserved terminal dinucleotides (5′-TG-CA-3′) (176). Most insertion events present hallmarks of IN-mediated events, such as 5 bp TSDs and Ty1 target site preferences. IN-mediated insertion of non-Ty1 DNA fragments has also been observed in cells lacking the Ku70 protein, suggesting that a Ku70-dependent mechanism restricts endogenous IN activity to Ty1 cDNA (177).

IN Structure and Function

Ty1 IN has the same three-domain organization as that of other LTR-retrotransposon and retroviral integrases (Fig. 10A). The N-terminal domain contains a conserved $H(X_{3\text{-}7})H(X_{23\text{-}32})C(X_2)C$ zinc-binding motif (ZBD). Alanine substitution of each of the four amino acids ($H_{17}H_{22}C_{55}C_{58}$) of the Ty1 IN ZBD motif results in IN instability and inefficient proteolytic processing, leading to substantially decreased cDNA levels and retrotransposition. Alanine substitution of additional hydrophobic residues between H_{22} and C_{55}, which are conserved between LTR-retrotransposons and retroviruses, alters IN-IN interaction, *in vitro* integration and *in vivo* retrotransposition (175).

The catalytic core domain in the central region of Ty1 IN is characterized by a $DD(X_{35})E$ motif that is conserved among retroviral and retrotransposon integrases as well as some bacterial proteins (178) (Fig. 10A). This motif binds divalent metal ions (Mg^{2+} or Mn^{2+}) and is critical for the transesterification reaction. An in-frame linker insertion of four amino acids at position 187 in this motif prevents Ty1 retrotransposition *in vivo* and *in vitro* (159).

The C-terminal domain is less conserved and larger than that of other retroviral integrases (Fig. 10A). It begins approximately 35 amino acids downstream of the DDE motif and includes a GKGY motif unique to the *Pseudoviridae* family of LTR-retroelements (179). This domain contains a bipartite nuclear localization signal (NLS) in the last 74 C-terminal residues consisting of two identical basic clusters (KKR) separated by 29 amino acids (180, 181) (Fig. 10A). The NLS is recognized by the classical importin-α receptor and is absolutely required for nuclear localization of Ty1 IN and retrotransposition but not for VLP production, protein processing, reverse transcription, and *in vitro* IN activities (180, 181, 182). IN also interacts closely with RT during reverse transcription and contributes to the proper folding and active conformation of RT *in vivo* (141).

REGULATION OF Ty1 RETROTRANSPOSITION

A variety of mechanisms that limit the potentially deleterious effects of retrotransposition on the host genome

A

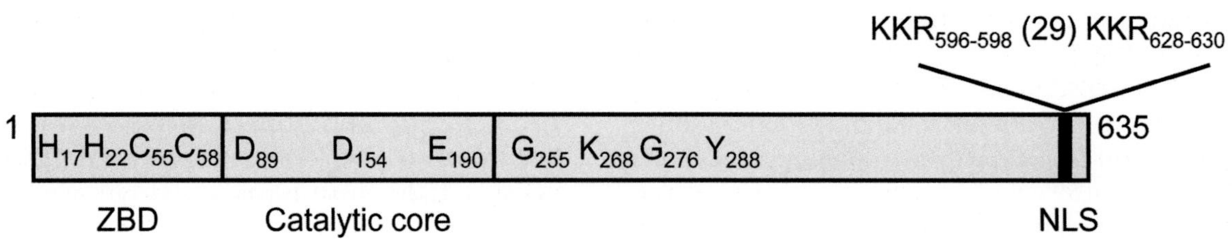

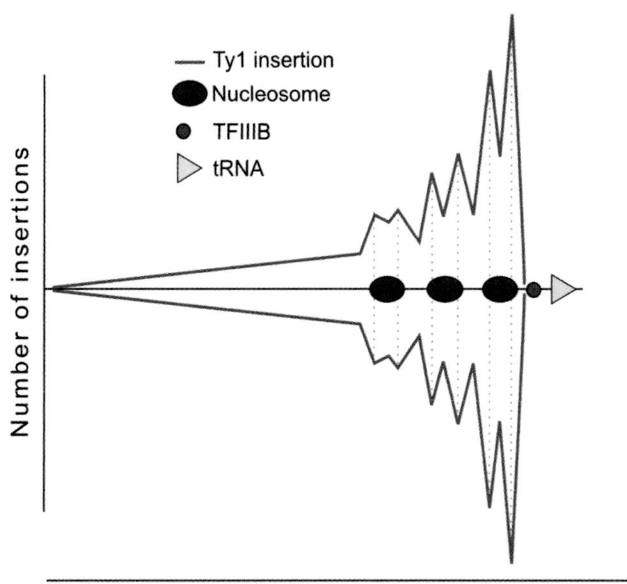

Distance from tRNA transcription start

Figure 10 Integrase and the integration target region. (A) Schematic representation of Ty1 integrase domains. Amino acid residues in the zinc binding motif (ZBD) and the catalytic core domain that are conserved in retroviral integrases, or residues in the C-terminal domain that are conserved in the *Pseudoviridiae* family of long terminal repeat-retrotransposons are indicated. Identical clusters of basic residues that define the bipartite nuclear localization signal (NLS) are also indicated. (B) Plot of Ty1 insertions upstream of tRNA genes. (Reproduced from Bridier-Nahmias and Lesage [271].) The blue curve above the midline represents Ty1 insertions in tandem with the tRNA genes and that below the midline represents elements inserted in inverted orientation.
doi:10.1128/microbiolspec.MDNA3-0053-2014.f10

have been described. These fall into two major categories: limiting the frequency of retrotransposition and confining integration events to specific regions of the genome that are gene-poor. A common theme that emerges from analysis of several types of host-mediated control of the frequency of Ty1 retrotransposition is the regulation of reverse transcriptase activity.

Control of Retromobility by Ty1 Copy Number

Transcriptional Cosuppression

All examined *S. cerevisiae* strains contain no more than 40 full-length Ty1 elements (23, 183). This copy number limit is not likely to be the consequence of a growth disadvantage conferred on cells harboring more Ty1 copies, as a 2- to 10-fold increase in the number of Ty1 elements, obtained experimentally, does not alter the growth rate, except when strains are challenged with DNA damaging agents (184). Two mechanisms that limit retrotransposition in strains with a high Ty1 copy number have been described. One type, called transcriptional cosuppression, was detected using Ty1 (*GAG:URA3*) reporter constructs and 5-fluorotic acid counterselection, and is characterized by transient and rapid switches between Ura$^+$ and Ura$^-$ states, in which all Ty1 elements are transcribed or silenced within a cell population, respectively (185). Cosuppression requires Ty1 elements in high copy number and actively transcribed from the native promoter. In contrast to this system that depends on counterselection to identify cells in which Ty1 RNA expression is suppressed, analysis of the expression of a Ty1(*GAG:GFP*) reporter construct in single cells by flow cytometry failed to detect a population of cells lacking expression of Gag-GFP (168). Thus, Ty1 transcriptional cosuppression may be strain or marker-specific.

Transcriptional cosuppression of Ty1 elements may be mediated by Ty1AS RNA (68). In different studies, stabilizing Ty1AS RNA by deleting *XRN1* decreased the level of Ty1 RNA to about 20% to 60% of that in a congenic wild-type strain (68, 70, 71). Notably, an internally initiated Ty1 sense-strand transcript identical in size to a 4.9 kb transcript previously reported in the *spt3*Δ mutant was detected in the *xrn1*Δ mutant (68, 186). Expression of Ty1AS RNA from the *GAL1* promoter silenced the expression of a Ty1(*GAG::URA3*) reporter in *trans*, which is consistent with a potential role for Ty1AS RNA in cosuppression. Interestingly, truncated Ty1AS RNA species lacking either U3-R, U5, or *GAG* sequences did not repress Ty1 transcription in *trans*, revealing that the integrity of Ty1AS RNA is critical for transcriptional repression. Although *S. cerevisiae* lack RNAi, introduction of Dicer and Argonaute proteins from *Saccharomyces castellii* results in strong repression of Ty1 sense-strand RNA expression and retrotransposition, suggesting that Ty1 RNA and Ty1AS RNA may form double-stranded RNAs that trigger RNAi-mediated repression (187).

Posttranscriptional Copy Number Control

Using a strain of the related species, *Saccharomyces paradoxus* that lacks full-length Ty1 elements, Garfinkel et al. (32) discovered a robust posttranscriptional form of Ty1 copy number control (CNC). This regulatory system results in repression of retrotransposition of a chromosomal RIG-marked Ty1 element when unmarked Ty1 elements are present in the same cell, with the extent of repression being roughly proportional to the copy number of Ty1 elements. The level of total Ty1 RNA increases with Ty1 copy number, demonstrating that repression occurs at a posttranscriptional step. A minimal segment of Ty1-H3 spanning nucleotides 238 to 1,702 is sufficient to confer posttranscriptional CNC, although inclusion of the 3′ LTR enhances CNC activity of this fragment. Deletion of nucleotides 238 to 281 or nucleotides 1,600 to 1,702 abolishes the repressing activity of this minimal construct (32, 33). A high copy number of Ty2, but not Ty3 or Ty5, confers CNC on Ty1, which is consistent with a homology-dependent mechanism. Ty1AS transcripts map to the minimal region required for CNC, and the strength of CNC is correlated with the abundance of Ty1AS RNA (33). Nevertheless, ectopic expression of two different Ty1AS RNA species from the *GAL1* promoter did not recapitulate the level of repression observed when Ty1AS RNA is expressed from its own undefined promoter. The failure of Ty1AS RNA to act in *trans* is not consistent with a causative role in CNC.

The mechanism of CNC-mediated repression of retrotransposition was investigated by constructing strains with an integrated *GAL1*-driven Ty1 element, which promotes the formation of VLPs, and either a high copy number of Ty1 elements whose sense-strand transcript was abolished by the *spt3*Δ mutation, and thus, high levels of Ty1AS RNA (CNC$^+$) or only one endogenous Ty1 element and low levels of Ty1AS RNA (CNC$^-$). VLP fractions from CNC$^+$ strains were enriched for Ty1AS RNA and had reduced levels of Gag and RT and barely detectable levels of IN (33). CNC$^+$ VLPs had reverse transcriptase activity on an exogenous primer-template added to VLPs, but no cDNA was synthesized from the endogenous primer and template within CNC$^+$ VLPs. Moreover, tRNA$_i^{Met}$ binding to Ty1 RNA in CNC$^+$ VLPs was robust, but no minus-strand strong-stop cDNA was detected.

SHAPE analysis of the overall structure of the minimal region required for CNC in Ty1 RNA extracted from VLPs (*in virio*) revealed similar secondary structures in the RNA from CNC$^+$ and CNC$^-$ samples, and provided no evidence for extensive annealing between Ty1 RNA and Ty1AS RNA *in virio* in CNC$^+$ strains (74). The major difference between CNC$^+$ and CNC$^-$ RNA samples was that the nucleotide sequence 5′-AUGAUGA-3′, present at nucleotide positions 321 to 327, 694 to 700, and 1,406 to 1,412, was more accessible to 2′-hydroxyl acylation in CNC$^+$ RNA. The same regions also became more accessible in CNC$^-$ RNA when the VLPs were gently stripped of protein before the 2′-hydroxyl acylation reactions were performed (*ex virio*). These findings suggest that Gag binds these regions in CNC^- RNA and that disruption of Gag binding in CNC^+ RNA blocks steps in the initiation of cDNA synthesis, perhaps by interfering with requisite chaperone activities of Gag (74). The mechanism of CNC-mediated repression of Ty1 cDNA synthesis, while correlated with high levels of Ty1AS RNA and specific alterations in the structure of Ty1 RNA *in virio*, remains to be elucidated.

CNC and Transpositional Dormancy

CNC is likely to be the major factor responsible for transpositional dormancy of Ty1 elements in laboratory strains. This view is consistent with the observation that a single RIG-marked Ty1 element mobilizes at a much higher rate in the absence of other elements than it does in the presence of multiple unmarked elements, even though the level of RIG-marked Ty1 RNA remains constant (32). Moreover, both CNC and transpositional dormancy are overridden at a posttranscriptional level by expression of a pGTy1 element (32, 72). Expression of a pGTy1 substantially increases the levels of processed IN and RT and cDNA, and retrotransposition is elevated as much as 100-fold per Ty1 transcript, which is consistent with the idea that CNC is negated (72). Deletion of many host repressors of Ty1 retrotransposition has been demonstrated to stimulate a posttranslational step in retrotransposition, and therefore, it is possible that the absence of these factors counteracts CNC as well. For example, deletion of *RTT101*, which causes activation of the DNA damage checkpoint, results in increased levels of p63-RT and p71-IN and cDNA, precisely the intermediates that are suppressed by CNC (162). Deletion of *FUS3*, which triggers the activation of the invasive-growth signaling pathway, causes similar, if not identical, alterations in Ty1 replication (166). This hypothesis implies that the activation state of the DNA damage checkpoint pathway or invasive growth pathway directly impacts the potency of CNC. Finally, a common mechanism of release from CNC-mediated repression in host mutants and in cells expressing pGTy1 would explain why deletion of many host repressors including Rtt101 and Fus3 only stimulates the retrotransposition of LTR-driven endogenous Ty1 elements and not that of pGTy1.

Host-Retrotransposon Interaction

S. cerevisiae is an unparalleled system for characterizing the role of the eukaryotic host in the mobility of LTR-retrotransposons and retroviruses (reviewed in [188]). The major reasons are the ease of connecting genotype and phenotype, the availability of simple genetic assays for retromobility, and the extensive variety of genomic tools available in yeast. A plethora of host factors required to maintain transpositional dormancy (known as restrictors of Ty1 transposition and encoded by *RTT* genes) and host factors that promote efficient Ty1 retromobility (known as retromobility host factors and encoded by *RHF* genes) have been identified. *RTT* genes were initially identified by screening for mutants with increased levels of retromobility of a chromosomal Ty1*his3AI* element (160, 164, 166) (Fig. 3). Because retromobility of a RIG-marked chromosomal Ty1 in a wild-type strain is detected in only about 1 in 10^7 cells, this strategy only identified mutants with elevated retromobility. Subsequent studies used various approaches to increase retrotransposition of genetically marked Ty1 elements so that hypotransposition mutants could be found. For example, Griffith et al. (151) introduced a plasmid-borne pGTy1*HIS3* element into each of the ~5,000 homozygous diploid nonessential ORF deletion strains, and screened for those with decreased retromobility of Ty1*HIS3* into the genome. 99 nonessential *RHF* genes were identified. Another screen employed an integrating plasmid-based Ty1*his3AI* element introduced into the haploid ORF deletion strain collection. While a single copy of this plasmid-based element is presumably integrated into the genome, retromobility of the integrating plasmid-based Ty1*his3AI* element is significantly higher than that of a chromosomal Ty1*his3AI* element for reasons that are not understood. This screen identified 168 nonessential *RHF* genes and 91 *RTT* genes (78, 161). Most recently, iterative synthetic genetic array screens in which two different *rtt* mutations were used to increase the mobility of a chromosomal Ty1*his3AI* element were performed. A chromosomal Ty1*his3AI* element and one of two *rtt* mutations was introduced into the haploid ORF deletion collection. Screens for ORF deletions that suppressed the hypermobility phenotype of either *rtt* mutation resulted in the identification of 275 nonessential *RHF* genes (163). The large number of host factors that influence retrotransposition is consistent with the fact that LTR-retrotransposons and retroviruses have a limited coding capacity and complex replication cycle (Fig. 2), and therefore, substantial host participation in replication is necessary.

More than 200 *RTT* and *RHF* genes have been verified by at least one of the following criteria: isolation in two independent screens; isolation in one screen and confirmation in a secondary screen for changes in the level of a retrotransposition intermediate, such as Gag or cDNA, that mirror the changes in retrotransposition; or characterization of effects on retromobility frequency and retrotransposition intermediates, irrespective of isolation in a genetic screen (Table 1). A few of these verified genes have been identified as *RTT* genes in one screen and *RHF* genes in another. The reason for this is not understood, but the observation underscores the variety of methodologies employed in individual screens and suggests that screen design influences the subset of host factors that are identified.

Regulation by Genome Integrity

Activation of Retrotransposition by DNA Lesions

Ty1 is an exquisitely sensitive sensor of the integrity of the genome. The majority of host restriction factors encoded by *RTT* genes are involved in different aspects of host genome maintenance (160, 161, 162, 164, 168, 169, 188, 189, 190, 191). These genome caretakers do not act directly to repress retrotransposition; rather, their absence results in DNA lesions that activate DNA damage signaling pathways that in turn induce Ty1 retrotransposition.

An example of this class of host cofactors is telomerase, an enzyme complex required for the lengthening of simple repeats at ends of chromosomes known as telomeres. The telomerase holoenzyme consists of a reverse transcriptase, Est2, associated with an RNA template, *TLC1* RNA. Telomerase adds telomeric DNA repeats to chromosome ends using *TLC1* RNA as template. In the absence of either telomerase component, telomeres become shorter with every generation because DNA polymerase cannot synthesize to the 5′ end of the DNA template. Telomere shortening has no major effect on cell growth or Ty1 retrotransposition until about 60 to 80 generations after either subunit of telomerase is rendered dysfunctional. When the telomere reaches a critically short length, it is recognized as a double-strand break, triggering damage signals that in turn result in progressive loss of growth, or senescence (192, 193). In the senescence phase, the levels of Ty1 cDNA and retrotransposition increase in parallel with the shortening of telomeres and gradual loss of cell viability (168). A fraction of yeast cells temporarily recover from senescence by invoking an alternative method of telomere lengthening. As survivors with elongated telomeres restore the viability of the culture, retrotransposition and cDNA are concomitantly reduced. The inverse correlation between retrotransposition levels and the length of telomeres and viability of the culture demonstrates that DNA lesions created by the absence of telomerase, rather than the absence of telomerase per se, activates Ty1 retrotransposition.

Spontaneous DNA lesions occurring in S-phase in the absence of genome caretakers stimulate Ty1 retrotransposition completely or partially through two S-phase specific checkpoint pathways (162). The components of these checkpoint pathways are not needed for the basal level of retromobility in wild-type strains, but are required to induce retromobility in genome caretaker mutants. In the absence of recombinational repair genes (e.g., *RAD50*, *MRE11*, *XRS2*, *RAD51*, *RAD52*, and *RAD55*), or the *RTT101* gene, which encodes a cullin involved in replication fork progression, activation of retrotransposition is Rad9-dependent, and thus, occurs primarily through the DNA damage checkpoint pathway. The absence of other genome caretakers, notably Elg1, a component of a replication factor C complex, results in Rad9-independent, Rad53-dependent hypermobility, indicating that the replication stress pathway induces hypermobility in these mutants. In a third class of mutants that includes *rad27*Δ, *rmm3*Δ and *rtt109*Δ, mobility is activated partially through one of these S-phase checkpoint pathways and partially by independent mechanisms, as discussed below.

Hypermobility of Ty1 in the absence of genome caretakers is usually accompanied by an increase in the level of Ty1 cDNA but not Ty1 RNA or Gag protein. The target of the S-phase checkpoint pathways has not been identified, but it is likely to be encoded by a domain of Ty1 that is divergent in Ty2, since Ty2 is not mobilized by these pathways. When the DNA damage checkpoint pathway was stimulated by deletion of *RTT101*, VLPs resulting from pGTy1 expression had elevated levels of p63-RT and p71-IN, and 20-fold higher levels of reverse transcriptase activity on an exogenous template relative to a *RTT101* strain (162). Notably, these increases were not observed when a pGTy1 lacking a domain encoding 30-amino acids of the N-terminus of p20-PR was expressed in the presence of the *rtt101*Δ mutation (162). This domain is divergent between Ty1 and Ty2 and has nucleocapsid function required for the initial steps in reverse transcription (158, 194). As such, increased pRT-63 and p71-IN and RT activity may be downstream effects of stimulating the nucleocapsid function of p20-PR, or another unknown function of this region within the Gag-Pol precursor protein.

Exposure of yeast cells to various DNA-damaging agents, such as UV light, γ-rays, 4-nitroquinoline-1-oxide (4-NQO), and methylmethane sulfonate (MMS), increases the level of Ty1 RNA and retrotransposition (38, 40, 41, 195, 196, 197). Treatment with γ-rays or

Table 1 List of verified host factors that regulate Ty1 retromobility

Cellular function	Host activators (*RHF* genes)	Host restrictors (*RTT* genes)
Chromatin/transcription	*ARP5* (78), *CTK1* (151), *NPL6* (78), *RPC53*[a] (151), *SPT2* (65, 151, 186)	*ASF1* (65, 161), *ELF1* (65, 161), *HHT1* (161), *RTT103* (65, 164), *RTT106* (65, 161, 164)
Pol II TFIIH subunits		*RAD3*[a] (167), *SSL2*[a] (160)
Transcription factors	*MCM1* (50, 262), *MIG3* (78, 163), *RAP1* (262), *STB5* (151), *STE12* (34, 48, 49, 56, 166), *TEC1* (54, 65), *TEA1* (52)	
Mediator complex	*SIN4* (65, 78, 151), *SRB8* (78, 151), *SSN2* (78, 151)	*CSE2* (161), *MED1* (65, 162, 164), *NUT2*[a] (164), *SOH1* (65, 161), *SRB5* (65, 161)
Paf1 complex		*CDC73* (161), *LEO1* (65, 161), *PAF1* (161), *RTF1*[b] (65, 151, 161)
SAGA complex	*SPT3* (65, 78, 163, 186), *SPT7* (186), *SPT8* (78, 163, 186)	
SPT4/5 complex	*SPT4* (65, 78, 151), *SPT5*[a,b] (78, 161, 163)	
SWI/SNF complex	*SNF2* (48, 56, 60), *SNF5* (48, 60, 163), *SNF6* (60, 163), *SWI3* (48, 60, 78, 163)	
THO/TREX complex	*THP2* (151, 163), *MFT1* (151)	
Histone acetylation/ deacetylation	*HDA3* (78, 163), *NAT4* (163)	*RTT109* (65, 161, 162, 164)
Rpd3S/L complex	*SAP30* (151)	*EAF3* (161), *RPD3* (161), *SIN3* (161)
SetC3 complex	*HOS2* (153), *SET3* (153), *SNT1* (163)	
Elongator complex	*ELP2* (163), *ELP4* (78, 151)	
Histone ubiquitylation		*BRE1* (161), *RAD6* (218, 263, 264)
Histone transcription	*SPT10* (65, 78, 151, 163)	*SPT21*[b] (65, 151, 161), *HIR1*[c] (265), *HIR2*[c] (65, 161, 265), *HIR3*[c] (161, 265)
Translation	*EFT2*[a] (266), *SPE1* (94), *SPE2* (94)	*ASC1* (161), *BUD27* (161)
Ribosome biogenesis factors	*BUD21* (76, 163), *BUD22* (78), *DBP7* (163), *HCR1* (163), *MRT4* (163), *RKM4* (163), *RPA49* (151), *HMO1* (78, 163), *NOP12* (78, 151)	
Ribosomal subunits	*RPL16B* (151, 163), *RPL19A* (163), *RPL19B* (151), *RPL20B* (151), *RPL27A* (78, 151), *RPL31A* (163), *RPL43A* (163), *RPL6A* (151), *RPL7A* (76, 163), *RPP0*[a] (266), *RPP1A* (151, 163), *RPS0B* (78), *RPS10A* (163), *RPS19B* (78, 163), *RPS25A* (78, 163), *RPS30A* (163)	
RNA metabolism	*DBR1* (65, 150, 151, 152, 153), *LSM1* (68, 70, 71, 78, 151, 163), *REF2* (78, 163), *SKI8* (78, 163), *TGS1* (78, 163)	
Cap binding	*CBC2* (151), *STO1* (78)	
RNA binding	*CTH1* (163), *LOC1* (78, 163, 252), *PUF6* (151, 163), *SCP160* (78, 151)	
RNA degradation	*CCR4* (68, 71, 163), *DCP1*[a] (68, 71), *DCP2*[a] (68, 71), *DHH1* (70, 71, 78, 163, 252), *NAM7* (68, 71, 78, 163), *NMD2* (71, 78), *PAT1* (68, 70, 71, 78, 151), *POP2* (78, 151), *PUB1* (76, 78), *UPF3* (68, 71, 78, 163), *XRN1* (68, 70, 71, 78)	
tRNA biogenesis	*HSX1* (88, 91, 252), *LOS1* (163), *NCL1* (163), *RIT1* (78, 151)	*RTT10* (161)

(Continued)

Table 1 *(Continued)*

Cellular function	Host activators (*RHF* genes)	Host restrictors (*RTT* genes)
Protein metabolism	*DFG10* (76, 163, 252), *DOA4* (78, 151), *MCK1* (78, 151)	*CKB2* (161)
Chaperones	*JJJ1*[a] (78, 151), *KAR2*[a] (76), *SRP68*[a] (76), *SRP101*[a] (76), *SRP102*[a] (76), *SSE1* (78, 163), *SSZ1* (78), *UMP1* (78, 163), *YKE2* (78)	
N-terminal acetyltransferase	*ARD1* (151), *NAT1* (151), *NAT3* (151)	
Peptidyl-prolyl *cis-trans* isomerase	*CPR7* (78, 151, 163), *ESS1*[a] (78)	*RRD2* (161, 162, 164)
Protein transport/sorting	*CLC1* (78), *LST7* (163), *SEC22* (151), *SIT4* (78), *SLA2* (78), *STP22* (78), *VPS9* (151), *VPS16* (78)	
Nuclear/cytoplasmic transport	*APQ12* (78, 163), *GSP1*[a] (182), *NTF2*[a] (182), *NUP133* (151, 163), *NUP170* (78, 163), *RSL1*[a] (182), *SRP1*[a] (182)	*KAP122* (164)
Cell polarity division	*AFR1* (163), *BEM1* (151), *BUD25* (78), *CDC50* (78, 163), *DBF20* (163), *HOF1* (151), *NUM1* (151)	*BEM4* (161), *BUD27* (161)
Dynactin complex	*JNM1* (151), *NIP100* (78, 163)	
Cell cycle progression		*CDC40* (161), *CLN2* (161), *SIC1* (161)
Cellular energetics/ion balance	*ATP17* (163), *PDE2* (78, 163), *PHO88* (78), *PMR1* (135), *QDR2* (78), *SPF1* (163), *TRK1* (78, 151, 163), *VPH1* (78, 163)	*AGP3* (161), *RNR1* (164)
Inositol metabolism	*KCS1* (151)	*ARG82* (161), *IPK1* (161), *KCS1* (161)
Cell signaling	*ASI1* (78), *KSS1* (166)	
HOG pathway		*HOG1* (36, 161), *PBS2* (161), *SSK2* (161), *SSK22* (161)
Mating response pathway	*RAM1* (78), *STE4* (78, 166), *STE5* (78, 166), *STE7* (49, 166, 267) *STE50* (78)	*STE11* (34, 36), *FUS3* (36, 65, 162, 166)
Genome stability		
DNA repair and replication	*APN1* (151)	*CDC9* (189), *CTF4* (162, 163), *ELG1* (65, 161, 162, 164), *MMS1* (65, 161, 162, 164), *MMS22* (161, 162, 163), *MRC1* (162), *RAD18* (161, 162), *RAD27* (161, 162, 169), *RRM3* (65, 161, 162, 164, 191), *RTT101* (65, 161, 162, 164), *RTT107* (65, 161, 162, 164), *SAE2* (161, 162, 164), *SGS1* (13, 161, 164), *WSS1* (162)
Telomere maintenance		*EST2* (164, 168), *TEL1* (162, 164)
Recombinational repair		*RAD51* (65, 161, 162, 189), *RAD52* (65, 161, 162, 189), *RAD54* (161, 189), *RAD55* (161, 162), *RAD57* (161, 162, 164, 189)
Mre11/MRX complex		*MRE11* (161, 162, 164), *RAD50* (161, 162, 164, 189), *XRS2* (162, 164)
Miscellaneous	*DGR2* (163), *GLO2* (78, 163), *HIT1* (163), *HSP31* (78), *KGD1* (163), *OCA4* (78, 163), *TPS2* (151), *YD124W* (163), *YIL102C* (78), *YOR292C* (151)	*RTT105* (164), *YOL159C* (151)

[a]Essential or nearly essential genes.
[b]Gene was identified as a *RHF* and as a *RTT* in different genetic screens.
[c]Phenotype of deleted gene requires a specific genetic background containing a *cac3* mutation.

4-NQO does not increase the level of Gag protein (38, 41), suggesting that DNA lesions that result from treatment with mutagens stimulate a posttranslational step in retrotransposition, like those generated in the absence of genome caretakers.

Repression of Multimeric cDNA Formation

Sgs1, Rrm3, and Rad27 are genome caretakers that suppress Ty1 retromobility by inhibiting the introduction of multiple Ty1 cDNAs as tandem arrays, or concatenates, into the host genome. Rad27, also known as flap-endonuclease 1 (FEN1), is a structure-specific nuclease that degrades a 5′ single-stranded flap that is generated during DNA replication of the lagging strand, as well as in base excision repair (198). Rad27/FEN1 plays critical roles in genome stability. Functional defects in Rad27 result in large increases in Ty1 retromobility that are only partially mediated via S-phase checkpoint pathways (162). In *rad27Δ* mutants, concatenates of cDNA are introduced into the genome at a high frequency (169). These concatenates presumably form by recombination between the 3′ LTR of one cDNA molecule and the 5′ LTR of another cDNA molecule, resulting in joints with a single LTR. Homologous recombination between cDNA and genomic Ty1 sequences is also substantially increased in the *rad27Δ* mutant (169), indicating that cDNA concatenates may enter the genome by integration or recombination. The phenotype of mutants lacking Sgs1, a nucleolar RecQ-family DNA helicase, or Rrm3, a Pif1-family helicase is distinct. In these mutants, increased incorporation of Ty1 cDNA into the genome is strongly dependent on the homologous recombination factor Rad52, but recombination between cDNA and chromosomal Ty1 elements remains at its normally low level (5, 190, 191). Instead, recombination generates concatenates of cDNAs that are inserted into the genome at typical Ty1 integration sites. Thus, the involvement of multiple cDNAs in each integration event explains the increased retromobility of Ty1 cDNA in *sgs1Δ* and *rrm3Δ* mutants. Multimeric Ty1 integration events are associated with aberrantly sized chromosomes in *rrm3Δ* mutants, suggesting that cDNA multimers are large enough to change the migration of chromosomes in pulse-field gels, or that their formation is associated with chromosomal rearrangements. Remarkably the formation of multimeric insertions in an *rrm3Δ* strain, but not a wild-type or *sgs1Δ* strain, is suppressed by overexpression of RNase H. This finding supports a model in which newly inserted Ty1 cDNA is broken within an RNA:DNA hybrid region, and the resulting break in Ty1 is repaired by homology-driven incorporation of multiple cDNAs (191). RNA:DNA hybrids may be present in the newly integrated Ty1 cDNA because of incomplete plus strand cDNA synthesis (143, 199).

Cell Cycle Restriction of Retromobility

There is some data to suggest that Ty1 retrotransposition is restricted to certain stages of the cell cycle. Treatment of cells with 25 to 100 mM hydroxyurea, which results in nucleotide depletion, replication fork pausing and delayed progression of S phase, dramatically increases Ty1 cDNA synthesis and retrotransposition (162). In contrast, transient cell-cycle arrest at G2/M by nocodazole has no effect on retrotransposition, and arrest of the cell cycle in G1 by mating pheromone strongly reduces retrotransposition of a pGTy1 element (200). Treatment of cells with mating pheromone results in a posttranslational block in Ty1 retrotransposition. VLPs prepared from mating pheromone treated cells have very low levels of reverse transcriptase activity both *in vivo* and *in vitro*. Together, these data are consistent with the model that retrotransposition is potentially restricted to S phase, or to the S and G2 phases of the cell cycle.

Three observations are consistent with the idea that a feature of the DNA replication fork may be recognized as a target for integration of Ty1 cDNA. First, Ty1 cDNA autointegration events are preferentially targeted to the region near the central PPT, where a DNA flap structure resembling a replication fork forms (148). Second, genes transcribed by RNA polymerase III (Pol III) are not only hotspots for integration of Ty1 cDNA but also barriers to replication fork progression (201, 202, 203). Third, the increase in multimeric Ty1 integration events in the absence of the Rrm3 helicase, which travels with the replication fork, suggests that integration might occur just prior to passage of the replication fork (191). It is tempting to speculate that replication pausing at Pol III-transcribed genes influences the efficiency of Ty1 targeting, although this hypothesis remains to be tested.

Integration Targeting

A wealth of evidence supports the idea that the host cell strictly controls not only Ty1 mobility but also target site selectivity with the consequence of limiting deleterious mutations and promoting repair of double-strand breaks. In the yeast genome, most Ty1 and Ty2 elements, and solo LTRs, reside upstream of genes transcribed by Pol III, in a window that generally begins ~80 bp upstream and extends over several hundred base pairs (24, 204, 205). Regions harboring tRNA genes, which comprise the major class of Pol III-transcribed genes, are generally devoid of protein coding genes; consequently, Ty1 integration into these regions is thought to minimize negative impacts on host cell growth and survival (206, 207). In addition, regions harboring tRNA genes are sites of replication pausing, chromosome fragility, and Ty1-mediated chromosomal rearrangement (201, 202, 203, 208, 209). The presence of Ty1 sequences at these fragile sites likely facilitates repair of deleterious double-strand breaks by recombination with nonallelic Ty1 elements, and therefore, is likely to promote host cell survival in conditions of replication stress (209).

Characterization of *de novo* integration events on chromosome III provided the first indication that the chromosomal position of Ty1 is due to a nonrandom integration pattern and not to selection (210). Ty1 integrates preferentially within a 700 bp window upstream of transcriptionally active RNA Pol III-transcribed genes (210, 211). An extensive analysis of several hundreds of *de novo* insertions events at $tRNA^{Gly}$ and $tRNA^{Thr}$ gene families uncovered a periodic insertion pattern upstream of tRNA genes, with approximately 80 bp between insertion peaks. Interestingly, the frequency of integration upstream of individual tRNA genes within the same family was highly variable, suggesting that local features near the tRNA genes also influence targeting (212). Two recent studies that employed deep-sequencing approaches to identify Ty1 integration sites showed that the upstream region of virtually all RNA Pol III-transcribed genes, including tRNA genes, *SNR6, RPR1, SNR52, SCR1* and the repeated *RDN5* loci, were targets for Ty1 integration (5, 213). Two loci, *RNA170* and *ZOD1*, which recruit incomplete Pol III transcription complexes (214, 215), were not targeted. These studies confirmed that there is significant variation in the frequency of insertion at RNA Pol-III transcribed genes, although genomic features specifically associated with hot or cold targets were not identified. Insertions into ORFs were significantly under-represented (5, 213). Importantly, the same insertion profiles were observed in diploid and haploid cells (5), demonstrating that avoidance of ORFs is not due to selection.

Deep sequencing further revealed that Ty1 integrates into nucleosomal DNA, with the first three nucleosomes upstream of Pol III genes being the hottest targets. Two integration hotspots per nucleosomal DNA separated by about 70 bp and located near the H2A/H2B interface were identified (Fig. 10B). The unexpected asymmetry in the Ty1 insertion sites relative to the nucleosome dyad axis suggests that a dynamic process of nucleosome remodeling exposes a specific H2A/H2B surface permissive to Ty1 integration. Mutation of the ATP-dependent chromatin remodeling factor Isw2 alters nucleosome positioning and Ty1 integration periodicity upstream of tRNA genes, suggesting that periodic integration is at least partially dependent on a specific chromatin structure imposed by the Isw2 complex (216). The Bdp1 subunit of the Pol III transcription factor TFIIIB is important to target Isw2 to tRNA genes and consequently influences the periodicity of Ty1 integration (217).

The mechanism underlying targeting of Ty1 integration to Pol III genes has been the subject of intense interest. Two histone deacetylases of the Set3C complex, Hos2 and Set3, enhance the efficiency of Ty1 integration upstream of Pol III-transcribed genes, but neither is sufficient to designate a Pol III-transcribed gene as a hotspot of integration (153). Ty1 integration into ORFs is increased in *rrm3*Δ, *rtt109*Δ and *rad6*Δ mutants (5, 161, 191, 218), but there is no change in the pattern of integration upstream of Pol III-transcribed genes (5). A physical interaction between Ty1 integrase and a subunit of Pol III was discovered recently, and several lines of evidence indicate that this interaction is essential for the targeting of Ty1 to the upstream region of Pol III-transcribed genes (A. Bridier-Nahmias, A. Tchalikian-Cosson, and P. Lesage, unpublished data). It remains to be determined how the interaction between Ty1 IN and Pol III facilitates integration into nucleosomal regions upstream of Pol III-transcribed genes.

IMPACT OF Ty1 ON THE YEAST GENOME

Regulation of Adjacent Gene Transcription by Stress

The yeast Ty1 element is arguably the best model to understand the activation of LTR-retrotransposons by environmental stress and the impact of this activation on adjacent gene expression (67, 219). In diploid cells, nitrogen starvation activates the Ty1 promoter through the invasive/filamentous signaling pathway by a mechanism involving the transcription factors Ste12 and Tec1 that bind Ty1 promoter sequences (34). In haploid cells, Ty1 transcription increases in adenine-deprived *bas1*Δ cells defective in *de novo*-AMP biosynthesis (a condition known as severe adenine starvation [35]). The low levels of ATP and ADP in these cells suggest that a decrease in adenylic nucleotides could be a signal for activation of Ty1 transcription (58). Although the mechanism of promoter activation has not yet been solved, it involves Snf2-dependent chromatin remodeling in the Ty1 promoter region and primarily affects weakly transcribed Ty1 elements (35). Transcriptional activation by severe adenine starvation also requires the 5′ LTR and a motif in the *GAG* ORF that is recognized by the Tye7 transcription factor (46, 58). Notably, severe adenine starvation represses transcription of the major Ty1AS RNA species in a Tye7-dependent manner (58).

Elevated Ty1 cDNA levels and a 15-fold increase in retrotransposition of Ty1 elements that are normally weakly expressed accompany the increase in Ty1 RNA in severe adenine starvation (35); however, the level of Gag protein is not altered (A.L. Todeschini and P. Lesage, unpublished data), suggesting that induction of Ty1 expression may not be the primary mechanism of stimulating retrotransposition. Alternatively, Ty1 retrotransposition may be stimulated by derepression of CNC or Ty1AS RNA-mediated control, as may also be the case with activation of retrotransposition via other environmental stress pathways (36, 162). Thus, the major significance of Ty1 promoter activation in severe adenine starvation could be its resulting effect on the transcription of adjacent genes. Although Ty1 insertions have modest effects, if any, on transcription of adjacent Pol III-transcribed genes (220), they can have substantial effects on expression of adjacent Pol II transcribed-genes, as reviewed in detail previously (35, 219). Briefly, integration of Ty1 upstream of an ORF and in the opposing orientation situates the Ty1 promoter near the 5′ end of the adjacent ORF and places expression of that ORF under the control of environmental signaling pathways that regulate Ty1 transcription. Indeed, transcription of ORFs from cryptic sites within the 5′ LTR of an adjacent Ty1 is stimulated in response to severe adenine deprivation (46). The transcriptional regulator, Tye7 is required for Ty1-dependent activation of adjacent target genes (58). Given that Tye7 also controls Ty1AS RNA transcription, it is possible that Ty1AS RNA directly influences the regulation of genes adjacent to Ty1 insertions.

Ty1 can exert its regulatory effect at a relatively long distance. For example, severe adenine starvation activates the *ESF1* gene, which is located 320 bp away from a full length Ty1 element in the yeast genome (58). A consequence of activating Ty1 transcription by stress could be the reprogramming of the yeast transcriptome by modulating the expression of coding or noncoding sequences that are adjacent to endogenous Ty1 sequences, as proposed by Servant et al. (46).

Contribution to Genome Plasticity

Ty1-Mediated Repair of Double-Strand Breaks (DSBs)

Ty1 elements in yeast, like repetitive elements in many organisms including humans, mediate many types of simple and complex chromosomal rearrangements, including deletions, segmental duplications, inversions and reciprocal and nonreciprocal translocations (221, 222, 223, 224, 225, 226, 227, 228, 229, 230, 231). Remarkably, the introduction of a single Ty1 element near the nonessential terminus of the left arm of chromosome V in haploid cells results in a 380-fold increase in gross chromosomal rearrangements (GCRs) with breakpoints near the introduced Ty1 element (221). The formation of GCRs at Ty1 elements is primarily a result of DSB repair within or near the Ty1 element, mainly by *RAD52*-mediated homologous recombination between nonallelic Ty1 elements (221, 232). In haploid cells, Ty1-mediated repair results in a remarkably high rate of cells that survive DSB formation (221, 232, 233). In diploid cells, where homologous recombination (HR) is highly favored over repair through nonhomologous end-joining (NHEJ), nearly all repair of DSBs leading to chromosomal aberrations results from recombination among nonallelic Ty1 sequences (230, 232). Notably, multiple Ty1 loci can be involved in rearrangement-associated recombination events, suggesting that both ends of the DSB may be able to find homology with different Ty1 loci (221). Alternatively, Ty1 cDNA or tandem cDNA arrays may be incorporated at sites of Ty1-mediated DSB repair (230). By activating DNA damage checkpoint pathways, DSBs and other DNA lesions can trigger the synthesis of Ty1 cDNA, which in turn can be captured at the sites of DNA damage (122, 162, 168, 191, 234). This capture can be initiated by retrotransposition or repair of DSBs in Ty1 elements or resected back to nearby Ty1 elements (122, 191, 234).

When homologous recombination is blocked, a small percentage of DSB repair events (1%) involves the capture of Ty cDNA fragments by NHEJ (235, 236, 237, 238). In some cases, the portion of Ty1 cDNA found at the breakpoint junction corresponds to minus-strand strong stop cDNA and few bases that are part of the tRNA primer, suggesting the possibility of formation of a RNA:DNA hybrid during the integration process. Capture of longer Ty1 fragments has also been described and requires RT but not IN (236). Ty1 cDNA capture may have occurred several times during *S. cerevisiae* evolution (235).

Ty1 as an Initiator of DSBs

In addition to being sites of ectopic homologous recombination, Ty1 elements can themselves be fragile sites where chromosomes tend to acquire DSBs. At least two types of fragility are associated with Ty1 sequences. First, two adjacent inverted Ty1 elements can induce

DSB formation, probably because of their capacity to form a long hairpin structure in the lagging strand (239). Fragility at inverted Ty1 elements is greatly enhanced in mutants with a low level of DNA polymerase, which retards replication fork progression. Fragile sites containing inverted Ty1 elements are naturally occurring in the yeast genome, and these sites are frequently at the breakpoints of GCRs (239, 240). A second mechanism is through the formation of R-loops, which are regions of the genome in which one strand of the DNA duplex is hybridized to RNA. Formation of RNA:DNA hybrids is associated with genome instability and replication stress across many species, likely because its presence results in DNA replication fork stalling (241, 242). Genome-wide profiling has shown that retrotransposon Ty1 and Ty2 sequences are greatly enriched in R loop regions (199). The same study found a strong correlation between regions of RNA:DNA hybrid formation and antisense transcription, suggesting that transcription of Ty1AS RNA contributes to R-loop formation in Ty1 sequences and may play a role in the formation of DSBs and spontaneous chromosomal aberrations at Ty1 elements.

Despite the well-established role of Ty1 in DSB repair by homologous recombination, Ty1 elements function as repressors of recombination at meiotic hotpots. This repression is a consequence of the compact chromatin structure of Ty1 and requires the first 4 kb of Ty1 sequence (243).

Consequences of Ty1 Activity for Genome Plasticity, Adaptation, and Longevity

Both in the wild and in laboratory evolution experiments, yeasts adapt readily to a variety of environmental stresses that limit their growth, including nutrient deprivation and exposure to toxic compounds (178, 222, 227, 234, 244, 245, 246, 247). Rapidly evolving adaptive phenotypes are often the result of duplicated chromosomal fragments that arise at high frequency and alter expression of multiple genes (222, 227, 234, 244, 246, 247, 248). The formation of segmental duplications by ectopic recombination between Ty1 elements is a common mechanism of creating mutations that answer selective pressures (222, 227, 234, 244, 249). Adaptive chromosomal duplications can have negative consequences for cells when they are grown in the absence of selective pressure, probably because of the expression changes in large numbers of genes (250, 251). Therefore, chromosomal duplications that can be rapidly reversed when the stress is relieved or when a more finely tuned mutation arises may be optimal for adaptation to rapidly changing environments (222, 244). Ty1-mediated chromosomal duplications are flanked on both sides by a Ty1 element or array, or bounded on one side by Ty1 elements, and on the other, by the telomere. They are highly dynamic and reversible by ectopic recombination between the Ty1 elements that flank the duplication (244). Thus, the presence of Ty1 elements within intergenic regions that are also fragile sites provides a dynamic mechanism by which cells can rapidly generate repeatable and reversible adaptive chromosomal duplications without significant insult to protein coding genes.

Not only is the presence of Ty1 in the genome, but also the mobility of Ty1 cDNA, likely to influence genome plasticity and adaptation. Retrotransposition events are commonly seen in strains that emerge from short-term evolution experiments. Ty1 insertion mutations, although often neutral or deleterious, can result in positive fitness effects by contributing to GCR formation or by modifying the expression of adjacent genes (183, 206, 234). The impact of Ty1 retrotransposition on genome plasticity has been observed during chronological aging in yeast. Ty1 retrotransposition is substantially increased in very old cells, and the resulting Ty1 insertions are frequently found on aberrant sized chromosomes. Moreover, aging-associated loss of heterozygosity and chromosome loss are consistently diminished by treatments or mutations that reduce Ty1 retrotransposition (252). These findings support the argument that Ty1 retrotransposition is a significant cause of the increased chromosomal instability that is characteristic of aging. Perhaps retrotransposition in aging cells is a response to stresses associated with chronological aging. In fact, the presence of multiple Ty1 elements in the genome can promote longevity of yeast (253). Under certain growth conditions, strains of *Saccharomyces paradoxus* containing approximately 20 *S. cerevisiae* Ty1 elements have a longer chronological life span than a congenic strain lacking Ty1 elements. Ty1 copy number-mediated longevity is correlated with reduced levels of reactive oxygen species, and accordingly, is phenocopied by treatment of the strain lacking Ty1 with antioxidants, which reduces reactive oxygen species. These findings suggest that the presence or expression of Ty1 retrotransposons can alter the level of reactive oxygen species. The relationship between Ty1 and reactive oxygen species is likely to be complex, as chemicals that increase reactive oxygen species are potent inducers of Ty1 retrotransposition (254). The longevity-promoting role of Ty1 elements, albeit context-dependent, is a unique example of a beneficial function associated with the presence of retrotransposons in a eukaryotic genome.

Retrosequence Formation

Occasionally, cellular mRNAs are used as templates to synthesize cDNAs that are introduced into the host genome, a process known as retrosequence or processed pseudogene formation. Retrosequence formation has been detected in yeast by expressing the *his3AI* RIG from the strong *GAL1* promoter or by embedding *his3AI* into the 3′ UTR of a cellular mRNA (121, 122, 255, 256). Reverse transcription-dependent duplication of the ATCase domain of the *URA2* gene has been detected by positive selection for an expressed *URA2* retrosequence (257, 258). Retrosequence formation is initiated by reverse transcription of the cellular mRNA within Ty1 VLPs, using Ty1 cDNA as a primer to initiate synthesis at the 3′ poly(A) tail of the mRNA (121, 122, 123, 258). Most often the poly(A) tail sequence is joined to the end of a complete LTR, suggesting that RT can use the 3′ OH end of a complete Ty1 cDNA as a primer, even when there is no homology with the mRNA template (121, 122). Overexpressing Ty1 RNA stimulates retrosequence formation, as does stimulating Ty1 RT activity posttranslationally by deleting either subunit of telomerase (121, 122, 203, 258). In telomerase-negative cells, cDNA of highly expressed genes joined to Ty1 LTR sequences, detected by polymerase chain reaction (PCR), is dramatically increased as telomeres erode (121, 122). The cDNA of cellular mRNAs can be introduced into the genome by three major mechanisms. First, gene conversion of mRNA-encoding genes by their homologous cDNAs replaces part of the original gene with an intronless copy (122, 123, 255, 256). These gene conversion events, which do not retain the poly(A) stretch or flanking Ty1 sequences, occur only in strains expressing Rad52 (122, 123, 255). Gene conversion of protein-coding genes by homologous cDNA has been proposed to explain the dearth of introns in *S. cerevisiae* (259). Second, chimeric retrosequence-Ty1 cDNA can undergo Rad52-dependent homologous recombination with genomic Ty1 elements at sites of Ty1-mediated DSB repair (188, 258). This mechanism results in the introduction of retrosequences that are flanked on both sides by Ty1 sequences. Ty1 cDNA sequences upstream of cellular retrosequences are usually at junctions with microhomology, suggesting that the upstream Ty1-retrosequence junction is created by template switching during plus-strand cDNA synthesis. Chimeric retrosequences, probably formed by RT switching between one cellular mRNA template and another, are frequently observed. Also commonly observed are large regions of Ty1 sequence, possibly Ty1 cDNA arrays, flanking the retrosequence on one or both sides (188). When Ty1 promoter sequences are present upstream of the retrosequence, they can contribute to its expression (258). Ty1-flanked retrosequences are introduced at the breakpoints of chromosomal rearrangements, suggesting that the retrosequences and flanking Ty1 cDNA are incorporated as "molecular band-aids" to facilitate DSB repair (122). Finally, retrosequences can enter the genome via rare *RAD52*-independent events such as nonhomologous end joining or Ty1 IN-mediated integration (122, 255).

The mRNA of the subtelomeric repeat element, Y′, is selectively packaged within Ty1 VLPs by an unknown mechanism. Thus, even though Y′ RNA is not particularly abundant, chimeric Y′-Ty1 cDNA can be detected by PCR and, like other retrosequence cDNAs, increases substantially as telomeres erode in telomerase-negative cells. Y′-Ty1 cDNA reaches levels as high as 10^{-2} to 10^{-4} copies per genome in the absence of telomerase (121), which are high enough for this cDNA to play a major role in extending telomeres. Y′ retrosequences are typically inserted at chromosome ends by homologous recombination between a subtelomeric Y′ element and Y′-Ty1 cDNA. This mechanism places Ty1 cDNA sequences at the unprotected chromosome end, and therefore, the ends frequently undergo ectopic recombination with internal Ty1 elements throughout the genome, resulting in a very high incidence of chromosomal rearrangements (121, 123).

PERSPECTIVES

From the time of their discovery as retrotransposons in 1985, Ty elements have been at the forefront of research that has elucidated fundamental mechanisms of retrotransposition and the impact of retrotransposons on their eukaryotic hosts. Since the publication of *Mobile DNA II* nearly 15 years ago, the trajectory of discoveries in Ty1 and retrotransposon biology has been shaped by the major advances of the postgenomic era, particularly DNA microarray analysis, high throughput sequencing, and functional genomics tools, especially in yeast (249). Studies have illuminated the role that Ty1 elements and the expression of Ty1 elements play in the rapid and reversible formation of chromosomal rearrangements in response to selective pressures. This type of abrupt large-scale change is similar to chromosomal aberrations that are often associated with human cancers (260, 261). As our understanding of Ty1-mediated mutations expands, we increasingly uncover examples of recurring mutations that are tailored to the stress by which they are evoked, such as the induction of Ty1 RT-mediated synthesis of chimeric cDNA that elongates telomeres when telomerase is inactivated (121).

These findings underscore the importance of retrotransposons in shaping the cell's response to stress. Looking ahead, we can expect the increasing use of Ty1 as a model to understand whether retrotransposons are drivers or passengers in the retrotransposon-related genomic alterations that accompany aging, neurodegenerative disorders, and the formation of tumors.

Other advances in the postgenomic era include the discovery of Ty1AS RNA and its correlation with copy number-mediated silencing of Ty1 elements (33, 68). This system has many parallels to RNA interference, which does not occur in *S. cerevisiae* (187). Future studies are likely to define the mechanism by which Ty1AS RNA mediates changes in the levels and functions of VLP-associated proteins. Addressing these questions will provide clues to the evolution of RNA interference and the roles that noncoding RNAs play in modulating gene expression. Recent studies have also begun to reveal the secondary structure of Ty1 RNA and its functional significance (20, 74). In the future, structure-function analysis will address a fundamental question in retrotransposon and retrovirus biology: how does the RNA function as both an mRNA and a gRNA?

A striking advance in retrotransposon biology over the last 15 years is the identification of hundreds of host factors that positively or negatively influence the rate of Ty1 retrotransposition. Evolutionarily conserved genes with functions in DNA replication and repair, histone modification, chromatin remodeling, transcription, and protein modifications were initially identified as RTT factors, illustrating the axiomatic contributions of Ty1 research to the discovery of fundamental host processes. Most retrotransposition host factors are conserved in eukaryotes and many, including mRNA silencing and decay proteins, have parallels in regulating the replication of mammalian retrotransposons and retroviruses. The sheer number of Ty1 host factors and the variety of cell processes in which they are involved illuminate the complex symbiosis between Ty1 and its host. Characterization of these host factors continues to reveal that the host carries out fundamental steps in Ty1 retrotransposition from transcription to protein trafficking and integration targeting. For example, recent work has shown that a universally conserved cotranslational chaperone is critical for the synthesis of Gag and the nucleation of VLP assembly sites (76). Future work will determine where VLP assembly takes place and how the transition of Ty1 RNA from translation to packaging is regulated. As the events that initiate assembly of retroviral nucleocapsids are only beginning to be described, studies on Ty1 are likely to provide an essential paradigm for understanding fundamental aspects of retroviral replication. Future studies will likely characterize additional essential genes that are required for Ty1 retrotransposition. The identification of essential genes holds the promise of revealing mechanisms by which the cell cycle controls retrotransposon and retroviral replication, which, in turn, will provide clues to the connection between retrotransposons and the development of cancer.

In summary, the golden age of Ty1 retrotransposons as a model is in its adolescence, with burgeoning opportunities for studies on Ty1 to advance our understanding of the tangled *tête-à-tête* between retroelements and their hosts. These studies hold promise for improving the treatment of a variety of human diseases by suggesting novel approaches to controlling retrovirus and retrotransposon-mediated genome rearrangements and cell pathologies.

Acknowledgments. *The authors thank David Garfinkel, Jessica Mitchell, and Hyo Won Ahn for their insightful comments on the manuscript.*

Citation. Curcio MJ, Lutz S, Lesage P. 2014. The Ty1 LTR-retrotransposon of budding yeast, Saccharomyces cerevisiae. Microbiol Spectrum 3(2):MDNA3-0053-2014

References

1. **Malik HS, Henikoff S, Eickbush TH.** 2000. Poised for contagion: evolutionary origins of the infectious abilities of invertebrate retroviruses. *Genome Res* **10:**1307–1318.
2. **Boeke JD, Garfinkel DJ, Styles CA, Fink GR.** 1985. Ty elements transpose through an RNA intermediate. *Cell* **40:**491–500.
3. **Clare JJ, Belcourt M, Farabaugh PJ.** 1988. Efficient translational frameshifting occurs within a conserved sequence of the overlap between the two genes of a yeast Ty1 transposon. *Proc Natl Acad Sci U S A* **85:**6816–6820.
4. **Curcio MJ, Derbyshire KM.** 2003. The outs and ins of transposition: from mu to kangaroo. *Nat Rev Mol Cell Biol* **4:**865–877.
5. **Baller JA, Gao J, Stamenova R, Curcio MJ, Voytas DF.** 2012. A nucleosomal surface defines an integration hotspot for the *Saccharomyces cerevisiae* Ty1 retrotransposon. *Genome Res* **22:**704–713.
6. **Garfinkel DJ, Nyswaner KM, Stefanisko KM, Chang C, Moore SP.** 2005. Ty1 copy number dynamics in Saccharomyces. *Genetics* **169:**1845–1857.
7. **Sharon G, Burkett TJ, Garfinkel DJ.** 1994. Efficient homologous recombination of Ty1 element cDNA when integration is blocked. *Mol Cell Biol* **14:**6540–6551.
8. **Radford SJ, Boyle ML, Sheely CJ, Graham J, Haeusser DP, Zimmerman L, Keeney JB.** 2004. Increase in Ty1 cDNA recombination in yeast sir4 mutant strains at high temperature. *Genetics* **168:**89–101.

9. **Curcio MJ, Garfinkel DJ.** 1991. Single-step selection for Ty1 element retrotransposition. *Proc Natl Acad Sci U S A* **88:**936–940.

10. **Boeke JD, Xu H, Fink GR.** 1988. A general method for the chromosomal amplification of genes in yeast. *Science* **239:**280–282.

11. **Mellor J, Malim MH, Gull K, Tuite MF, McCready S, Dibbayawan T, Kingsman SM, Kingsman AJ.** 1985. Reverse transcriptase activity and Ty RNA are associated with virus-like particles in yeast. *Nature* **318:**583–586.

12. **Garfinkel DJ, Boeke JD, Fink GR.** 1985. Ty element transposition: reverse transcriptase and virus-like particles. *Cell* **42:**507–517.

13. **Bryk M, Banerjee M, Murphy M, Knudsen KE, Garfinkel DJ, Curcio MJ.** 1997. Transcriptional silencing of Ty1 elements in the RDN1 locus of yeast. *Genes Dev* **11:**255–269.

14. **Cousineau B, Smith D, Lawrence-Cavanagh S, Mueller JE, Yang J, Mills D, Manias D, Dunny G, Lambowitz AM, Belfort M.** 1998. Retrohoming of a bacterial group II intron: mobility via complete reverse splicing, independent of homologous DNA recombination. *Cell* **94:**451–462.

15. **Moran JV, Holmes SE, Naas TP, DeBerardinis RJ, Boeke JD, Kazazian HH Jr.** 1996. High frequency retrotransposition in cultured mammalian cells. *Cell* **87:**917–927.

16. **An W, Han JS, Wheelan SJ, Davis ES, Coombes CE, Ye P, Triplett C, Boeke JD.** 2006. Active retrotransposition by a synthetic L1 element in mice. *Proc Natl Acad Sci U S A* **103:**18662–18667.

17. **Esnault C, Maestre J, Heidmann T.** 2000. Human LINE retrotransposons generate processed pseudogenes. *Nat Genet* **24:**363–367.

18. **Curcio MJ, Garfinkel DJ.** 1994. Heterogeneous functional Ty1 elements are abundant in the *Saccharomyces cerevisiae* genome. *Genetics* **136:**1245–1259.

19. **Bolton EC, Coombes C, Eby Y, Cardell M, Boeke JD.** 2005. Identification and characterization of critical *cis*-acting sequences within the yeast Ty1 retrotransposon. *RNA* **11:**308–322.

20. **Huang Q, Purzycka KJ, Lusvarghi S, Li D, Legrice SF, Boeke JD.** 2013. Retrotransposon Ty1 RNA contains a 5′-terminal long-range pseudoknot required for efficient reverse transcription. *RNA* **19:**320–332.

21. **Xu H, Boeke JD.** 1990. Localization of sequences required in *cis* for yeast Ty1 element transposition near the long terminal repeats: analysis of mini-Ty1 elements. *Mol Cell Biol* **10:**2695–2702.

22. **Checkley MA, Mitchell JA, Eizenstat LD, Lockett SJ, Garfinkel DJ.** 2013. Ty1 Gag enhances the stability and nuclear export of Ty1 mRNA. *Traffic* **14:**57–69.

23. **Carr M, Bensasson D, Bergman CM.** 2012. Evolutionary genomics of transposable elements in *Saccharomyces cerevisiae*. *PLoS One* **7:**e50978.

24. **Kim JM, Vanguri S, Boeke JD, Gabriel A, Voytas DF.** 1998. Transposable elements and genome organization: a comprehensive survey of retrotransposons revealed by the complete *Saccharomyces cerevisiae* genome sequence. *Genome Res* **8:**464–478.

25. **Liti G, Carter DM, Moses AM, Warringer J, Parts L, James SA, Davey RP, Roberts IN, Burt A, Koufopanou V, Tsai IJ, Bergman CM, Bensasson D, O'Kelly MJ, van Oudenaarden A, Barton DB, Bailes E, Nguyen AN, Jones M, Quail MA, Goodhead I, Sims S, Smith F, Blomberg A, Durbin R, Louis EJ.** 2009. Population genomics of domestic and wild yeasts. *Nature* **458:**337–341.

26. **Jordan IK, McDonald JF.** 1999. Tempo and mode of Ty element evolution in *Saccharomyces cerevisiae*. *Genetics* **151:**1341–1351.

27. **Jordan IK, McDonald JF.** 1998. Evidence for the role of recombination in the regulatory evolution of *Saccharomyces cerevisiae* Ty elements. *J Mol Evol* **47:**14–20.

28. **Jordan IK, McDonald JF.** 1999. Comparative genomics and evolutionary dynamics of *Saccharomyces cerevisiae* Ty elements. *Genetica* **107:**3–13.

29. **Moore SP, Liti G, Stefanisko KM, Nyswaner KM, Chang C, Louis EJ, Garfinkel DJ.** 2004. Analysis of a Ty1-less variant of Saccharomyces paradoxus: the gain and loss of Ty1 elements. *Yeast* **21:**649–660.

30. **Promislow DE, Jordan IK, McDonald JF.** 1999. Genomic demography: a life-history analysis of transposable element evolution. *Proc R Soc B* **266:**1555–1560.

31. **Curcio MJ, Hedge AM, Boeke JD, Garfinkel DJ.** 1990. Ty RNA levels determine the spectrum of retrotransposition events that activate gene expression in *Saccharomyces cerevisiae*. *Mol Gen Genet* **220:**213–221.

32. **Garfinkel DJ, Nyswaner K, Wang J, Cho JY.** 2003. Post-transcriptional cosuppression of Ty1 retrotransposition. *Genetics* **165:**83–99.

33. **Matsuda E, Garfinkel DJ.** 2009. Posttranslational interference of Ty1 retrotransposition by antisense RNAs. *Proc Natl Acad Sci U S A* **106:**15657–15662.

34. **Morillon A, Springer M, Lesage P.** 2000. Activation of the Kss1 invasive-filamentous growth pathway induces Ty1 transcription and retrotransposition in *Saccharomyces cerevisiae*. *Mol Cell Biol* **20:**5766–5776.

35. **Todeschini AL, Morillon A, Springer M, Lesage P.** 2005. Severe adenine starvation activates Ty1 transcription and retrotransposition in *Saccharomyces cerevisiae*. *Mol Cell Biol* **25:**7459–7472.

36. **Conte D Jr, Curcio MJ.** 2000. Fus3 controls Ty1 transpositional dormancy through the invasive growth MAPK pathway. *Mol Microbiol* **35:**415–427.

37. **Lawler JF Jr, Haeusser DP, Dull A, Boeke JD, Keeney JB.** 2002. Ty1 defect in proteolysis at high temperature. *J Virol* **76:**4233–4240.

38. **Staleva Staleva L, Venkov P.** 2001. Activation of Ty transposition by mutagens. *Mutat Res* **474:**93–103.

39. **Morawetz C.** 1987. Effect of irradiation and mutagenic chemicals on the generation of ADH2-constitutive mutants in yeast. Significance for the inducibility of Ty transposition. *Mutat Res* **177:**53–60.

40. **Bradshaw VA, McEntee K.** 1989. DNA damage activates transcription and transposition of yeast Ty retrotransposons. *Mol Gen Genet* **218:**465–474.

41. **Sacerdot C, Mercier G, Todeschini AL, Dutreix M, Springer M, Lesage P.** 2005. Impact of ionizing radiation

on the life cycle of *Saccharomyces cerevisiae* Ty1 retrotransposon. *Yeast* **22:**441–455.

42. **Mules EH, Uzun O, Gabriel A.** 1998. *In vivo* Ty1 reverse transcription can generate replication intermediates with untidy ends. *J Virol* **72:**6490–6503.

43. **Elder RT, Loh EY, Davis RW.** 1983. RNA from the yeast transposable element Ty1 has both ends in the direct repeats, a structure similar to retrovirus RNA. *Proc Natl Acad Sci U S A* **80:**2432–2436.

44. **Hou W, Russnak R, Platt T.** 1994. Poly(A) site selection in the yeast Ty retroelement requires an upstream region and sequence-specific titratable factor(s) *in vitro*. *EMBO J* **13:**446–452.

45. **Dudley AM, Gansheroff LJ, Winston F.** 1999. Specific components of the SAGA complex are required for Gcn4- and Gcr1-mediated activation of the his4-912delta promoter in *Saccharomyces cerevisiae*. *Genetics* **151:** 1365–1378.

46. **Servant G, Pennetier C, Lesage P.** 2008. Remodeling yeast gene transcription by activating the Ty1 long terminal repeat retrotransposon under severe adenine deficiency. *Mol Cell Biol* **28:**5543–5554.

47. **Yu K, Elder RT.** 1989. Some of the signals for 3′-end formation in transcription of the *Saccharomyces cerevisiae* Ty-D15 element are immediately downstream of the initiation site. *Mol Cell Biol* **9:**2431–2443.

48. **Ciriacy M, Freidel K, Lohning C.** 1991. Characterization of *trans*-acting mutations affecting Ty and Ty-mediated transcription in *Saccharomyces cerevisiae*. *Curr Genet* **20:**441–448.

49. **Company M, Errede B.** 1987. Cell-type-dependent gene activation by yeast transposon Ty1 involves multiple regulatory determinants. *Mol Cell Biol* **7:**3205–3211.

50. **Errede B.** 1993. MCM1 binds to a transcriptional control element in Ty1. *Mol Cell Biol* **13:**57–62.

51. **Grant PA, Duggan L, Cote J, Roberts SM, Brownell JE, Candau R, Ohba R, Owen-Hughes T, Allis CD, Winston F, Berger SL, Workman JL.** 1997. Yeast Gcn5 functions in two multisubunit complexes to acetylate nucleosomal histones: characterization of an Ada complex and the SAGA (Spt/Ada) complex. *Genes Dev* **11:** 1640–1650.

52. **Gray WM, Fassler JS.** 1996. Isolation and analysis of the yeast TEA1 gene, which encodes a zinc cluster Ty enhancer-binding protein. *Mol Cell Biol* **16:**347–358.

53. **Kent NA, Karabetsou N, Politis PK, Mellor J.** 2001. *In vivo* chromatin remodeling by yeast ISWI homologs Isw1p and Isw2p. *Genes Dev* **15:**619–626.

54. **Laloux I, Dubois E, Dewerchin M, Jacobs E.** 1990. TEC1, a gene involved in the activation of Ty1 and Ty1-mediated gene expression in *Saccharomyces cerevisiae*: cloning and molecular analysis. *Mol Cell Biol* **10:**3541–3550.

55. **Madison JM, Dudley AM, Winston F.** 1998. Identification and analysis of Mot3, a zinc finger protein that binds to the retrotransposon Ty long terminal repeat (delta) in *Saccharomyces cerevisiae*. *Mol Cell Biol* **18:**1879–1890.

56. **Morillon A, Benard L, Springer M, Lesage P.** 2002. Differential effects of chromatin and Gcn4 on the 50-fold range of expression among individual yeast Ty1 retrotransposons. *Mol Cell Biol* **22:**2078–2088.

57. **Pollard KJ, Peterson CL.** 1997. Role for ADA/GCN5 products in antagonizing chromatin-mediated transcriptional repression. *Mol Cell Biol* **17:**6212–6222.

58. **Servant G, Pinson B, Tchalikian-Cosson A, Coulpier F, Lemoine S, Pennetier C, Bridier-Nahmias A, Todeschini AL, Fayol H, Daignan-Fornier B, Lesage P.** 2012. Tye7 regulates yeast Ty1 retrotransposon sense and antisense transcription in response to adenylic nucleotides stress. *Nucleic Acids Res* **40:**5271–5282.

59. **Turkel S, Liao XB, Farabaugh PJ.** 1997. GCR1-dependent transcriptional activation of yeast retrotransposon Ty2-917. *Yeast* **13:**917–930.

60. **Happel AM, Swanson MS, Winston F.** 1991. The SNF2, SNF5 and SNF6 genes are required for Ty transcription in *Saccharomyces cerevisiae*. *Genetics* **128:**69–77.

61. **Winston F, Carlson M.** 1992. Yeast SNF/SWI transcriptional activators and the SPT/SIN chromatin connection. *Trends Genet* **8:**387–391.

62. **Lopez MC, Baker HV.** 2000. Understanding the growth phenotype of the yeast gcr1 mutant in terms of global genomic expression patterns. *J Bacteriol* **182:**4970–4978.

63. **Elder RT, St John TP, Stinchcomb DT, Davis RW, Scherer S.** 1981. Studies on the transposable element Ty1 of yeast. I. RNA homologous to Ty1. II. Recombination and expression of Ty1 and adjacent sequences. *Cold Spring Harb Symp Quant Biol* **45:**581–591.

64. **Munchel SE, Shultzaberger RK, Takizawa N, Weis K.** 2011. Dynamic profiling of mRNA turnover reveals gene-specific and system-wide regulation of mRNA decay. *Mol Biol Cell* **22:**2787–2795.

65. **Malagon F, Jensen TH.** 2008. The T body, a new cytoplasmic RNA granule in *Saccharomyces cerevisiae*. *Mol Cell Biol* **28:**6022–6032.

66. **Tucker M, Parker R.** 2000. Mechanisms and control of mRNA decapping in *Saccharomyces cerevisiae*. *Annu Rev Biochem* **69:**571–595.

67. **Lesage P, Todeschini AL.** 2005. Happy together: the life and times of Ty retrotransposons and their hosts. *Cytogenet Genome Res* **110:**70–90.

68. **Berretta J, Pinskaya M, Morillon A.** 2008. A cryptic unstable transcript mediates transcriptional *trans*-silencing of the Ty1 retrotransposon in *S. cerevisiae*. *Genes Dev* **22:**615–626.

69. **van Dijk EL, Chen CL, d'Aubenton-Carafa Y, Gourvennec S, Kwapisz M, Roche V, Bertrand C, Silvain M, Legoix-Ne P, Loeillet S, Nicolas A, Thermes C, Morillon A.** 2011. XUTs are a class of Xrn1-sensitive antisense regulatory non-coding RNA in yeast. *Nature* **475:**114–117.

70. **Checkley MA, Nagashima K, Lockett SJ, Nyswaner KM, Garfinkel DJ.** 2010. P-body components are required for Ty1 retrotransposition during assembly of retrotransposition-competent virus-like particles. *Mol Cell Biol* **30:**382–398.

71. **Dutko JA, Kenny AE, Gamache ER, Curcio MJ.** 2010. 5′ to 3′ mRNA decay factors colocalize with Ty1 gag

and human APOBEC3G and promote Ty1 retrotransposition. *J Virol* **84:**5052–5066.

72. **Curcio MJ, Garfinkel DJ.** 1992. Posttranslational control of Ty1 retrotransposition occurs at the level of protein processing. *Mol Cell Biol* **12:**2813–2825.

73. **Ringner M, Krogh M.** 2005. Folding free energies of 5′-UTRs impact post-transcriptional regulation on a genomic scale in yeast. *PLoS Comput Biol* **1:**e72.

74. **Purzycka KJ, Legiewicz M, Matsuda E, Eizentstat LD, Lusvarghi S, Saha A, Grice SF, Garfinkel DJ.** 2013. Exploring Ty1 retrotransposon RNA structure within virus-like particles. *Nucleic Acids Res* **41:**463–473.

75. **del Alamo M, Hogan DJ, Pechmann S, Albanese V, Brown PO, Frydman J.** 2011. Defining the specificity of cotranslationally acting chaperones by systematic analysis of mRNAs associated with ribosome-nascent chain complexes. *PLoS Biol* **9:**e1001100.

76. **Doh JH, Lutz S, Curcio MJ.** 2014. Co-translational Localization of an LTR-Retrotransposon RNA to the Endoplasmic Reticulum Nucleates Virus-Like Particle Assembly Sites. *PLoS Genet* **10:**e1004219.

77. **Ingolia NT, Ghaemmaghami S, Newman JR, Weissman JS.** 2009. Genome-wide analysis *in vivo* of translation with nucleotide resolution using ribosome profiling. *Science* **324:**218–223.

78. **Dakshinamurthy A, Nyswaner KM, Farabaugh PJ, Garfinkel DJ.** 2010. BUD22 affects Ty1 retrotransposition and ribosome biogenesis in *Saccharomyces cerevisiae*. *Genetics* **185:**1193–1205.

79. **Dragon F, Gallagher JE, Compagnone-Post PA, Mitchell BM, Porwancher KA, Wehner KA, Wormsley S, Settlage RE, Shabanowitz J, Osheim Y, Beyer AL, Hunt DF, Baserga SJ.** 2002. A large nucleolar U3 ribonucleoprotein required for 18S ribosomal RNA biogenesis. *Nature* **417:**967–970.

80. **Saraogi I, Shan SO.** 2011. Molecular mechanism of co-translational protein targeting by the signal recognition particle. *Traffic* **12:**535–542.

81. **Surjit M, Jameel S, Lal SK.** 2007. Cytoplasmic localization of the ORF2 protein of hepatitis E virus is dependent on its ability to undergo retrotranslocation from the endoplasmic reticulum. *J Virol* **81:**3339–3345.

82. **Duriez M, Rossignol JM, Sitterlin D.** 2008. The hepatitis B virus precore protein is retrotransported from endoplasmic reticulum (ER) to cytosol through the ER-associated degradation pathway. *J Biol Chem* **283:** 32352–32360.

83. **Byun H, Gou Y, Zook A, Lozano MM, Dudley JP.** 2014. ERAD and how viruses exploit it. *Front Microbiol* **5:**330.

84. **Brodsky JL, Skach WR.** 2011. Protein folding and quality control in the endoplasmic reticulum: Recent lessons from yeast and mammalian cell systems. *Curr Opin Cell Biol* **23:**464–475.

85. **Weissenbach J, Dirheimer G, Falcoff R, Sanceau J, Falcoff E.** 1977. Yeast tRNALeu (anticodon U–A–G) translates all six leucine codons in extracts from interferon treated cells. *FEBS Lett* **82:**71–76.

86. **Sundararajan A, Michaud WA, Qian Q, Stahl G, Farabaugh PJ.** 1999. Near-cognate peptidyl-tRNAs promote +1 programmed translational frameshifting in yeast. *Mol Cell* **4:**1005–1015.

87. **Belcourt MF, Farabaugh PJ.** 1990. Ribosomal frameshifting in the yeast retrotransposon Ty: tRNAs induce slippage on a 7 nucleotide minimal site. *Cell* **62:**339–352.

88. **Kawakami K, Pande S, Faiola B, Moore DP, Boeke JD, Farabaugh PJ, Strathern JN, Nakamura Y, Garfinkel DJ.** 1993. A rare tRNA-Arg(CCU) that regulates Ty1 element ribosomal frameshifting is essential for Ty1 retrotransposition in *Saccharomyces cerevisiae*. *Genetics* **135:**309–320.

89. **Dinman JD.** 2012. Mechanisms and implications of programmed translational frameshifting. *Wiley Interdiscip Rev RNA* **3:**661–673.

90. **Tumer NE, Parikh BA, Li P, Dinman JD.** 1998. The pokeweed antiviral protein specifically inhibits Ty1-directed +1 ribosomal frameshifting and retrotransposition in *Saccharomyces cerevisiae*. *J Virol* **72:** 1036–1042.

91. **Xu H, Boeke JD.** 1990. Host genes that influence transposition in yeast: the abundance of a rare tRNA regulates Ty1 transposition frequency. *Proc Natl Acad Sci U S A* **87:**8360–8364.

92. **Harger JW, Dinman JD.** 2003. An *in vivo* dual-luciferase assay system for studying translational recoding in the yeast *Saccharomyces cerevisiae*. *RNA* **9:** 1019–1024.

93. **Balasundaram D, Dinman JD, Tabor CW, Tabor H.** 1994. SPE1 and SPE2: two essential genes in the biosynthesis of polyamines that modulate +1 ribosomal frameshifting in *Saccharomyces cerevisiae*. *J Bacteriol* **176:**7126–7128.

94. **Balasundaram D, Dinman JD, Wickner RB, Tabor CW, Tabor H.** 1994. Spermidine deficiency increases +1 ribosomal frameshifting efficiency and inhibits Ty1 retrotransposition in *Saccharomyces cerevisiae*. *Proc Natl Acad Sci U S A* **91:**172–176.

95. **Jouvenet N, Laine S, Pessel-Vivares L, Mougel M.** 2011. Cell biology of retroviral RNA packaging. *RNA Biol* **8:** 572–580.

96. **Garbitt-Hirst R, Kenney SP, Parent LJ.** 2009. Genetic evidence for a connection between Rous sarcoma virus gag nuclear trafficking and genomic RNA packaging. *J Virol* **83:**6790–6797.

97. **Jouvenet N, Simon SM, Bieniasz PD.** 2009. Imaging the interaction of HIV-1 genomes and Gag during assembly of individual viral particles. *Proc Natl Acad Sci U S A* **106:**19114–19119.

98. **Malagon F, Jensen TH.** 2011. T-body formation precedes virus-like particle maturation in *S. cerevisiae*. *RNA Biol* **8:**184–189.

99. **Decker CJ, Parker R.** 2012. P-bodies and stress granules: possible roles in the control of translation and mRNA degradation. *Cold Spring Harb Perspect Biol* **4:**a012286.

100. **Coller J, Parker R.** 2005. General translational repression by activators of mRNA decapping. *Cell* **122:** 875–886.

101. **Bouveret E, Rigaut G, Shevchenko A, Wilm M, Seraphin B.** 2000. A Sm-like protein complex that participates in mRNA degradation. *EMBO J* **19:**1661–1671.
102. **Tharun S, He W, Mayes AE, Lennertz P, Beggs JD, Parker R.** 2000. Yeast Sm-like proteins function in mRNA decapping and decay. *Nature* **404:**515–518.
103. **Tharun S, Parker R.** 2001. Targeting an mRNA for decapping: displacement of translation factors and association of the Lsm1p-7p complex on deadenylated yeast mRNAs. *Mol Cell* **8:**1075–1083.
104. **Stevens A.** 1980. Purification and characterization of a *Saccharomyces cerevisiae* exoribonuclease which yields 5′-mononucleotides by a 5′ leads to 3′ mode of hydrolysis. *J Biol Chem* **255:**3080–3085.
105. **Buchan JR.** 2014. mRNP granules: assembly, function, and connections with disease. *RNA Biol* **11:**1019–1030.
106. **Luschnig C, Bachmair A.** 1997. RNA packaging of yeast retrotransposon Ty1 in the heterologous host, *Escherichia coli. Biol Chem* **378:**39–46.
107. **Luschnig C, Hess M, Pusch O, Brookman J, Bachmair A.** 1995. The gag homologue of retrotransposon Ty1 assembles into spherical particles in *Escherichia coli. Eur J Biochem* **228:**739–744.
108. **Brookman JL, Stott AJ, Cheeseman PJ, Adamson CS, Holmes D, Cole J, Burns NR.** 1995. Analysis of TYA protein regions necessary for formation of the Ty1 virus-like particle structure. *Virology* **212:**69–76.
109. **Martin-Rendon E, Marfany G, Wilson S, Ferguson DJ, Kingsman SM, Kingsman AJ.** 1996. Structural determinants within the subunit protein of Ty1 virus-like particles. *Mol Microbiol* **22:**667–679.
110. **Roth JF.** 2000. The yeast Ty virus-like particles. *Yeast* **16:**785–795.
111. **AL-Khayat HA, Bhella D, Kenney JM, Roth JF, Kingsman AJ, Martin-Rendon E, Saibil HR.** 1999. Yeast Ty retrotransposons assemble into virus-like particles whose T-numbers depend on the C-terminal length of the capsid protein. *J Mol Biol* **292:**65–73.
112. **Burns NR, Saibil HR, White NS, Pardon JF, Timmins PA, Richardson SM, Richards BM, Adams SE, Kingsman SM, Kingsman AJ.** 1992. Symmetry, flexibility and permeability in the structure of yeast retrotransposon virus-like particles. *EMBO J* **11:**1155–1164.
113. **Palmer KJ, Tichelaar W, Myers N, Burns NR, Butcher SJ, Kingsman AJ, Fuller SD, Saibil HR.** 1997. Cryo-electron microscopy structure of yeast Ty retrotransposon virus-like particles. *J Virol* **71:**6863–6868.
114. **Dutko JA, Schafer A, Kenny AE, Cullen BR, Curcio MJ.** 2005. Inhibition of a yeast LTR retrotransposon by human APOBEC3 cytidine deaminases. *Curr Biol* **15:**661–666.
115. **Schumacher AJ, Nissley DV, Harris RS.** 2005. APOBEC3G hypermutates genomic DNA and inhibits Ty1 retrotransposition in yeast. *Proc Natl Acad Sci U S A* **102:**9854–9859.
116. **Feng YX, Moore SP, Garfinkel DJ, Rein A.** 2000. The genomic RNA in Ty1 virus-like particles is dimeric. *J Virol* **74:**10819–10821.
117. **Cristofari G, Ficheux D, Darlix JL.** 2000. The GAG-like protein of the yeast Ty1 retrotransposon contains a nucleic acid chaperone domain analogous to retroviral nucleocapsid proteins. *J Biol Chem* **275:**19210–19217.
118. **Rein A, Datta SA, Jones CP, Musier-Forsyth K.** 2011. Diverse interactions of retroviral Gag proteins with RNAs. *Trends Biochem Sci* **36:**373–380.
119. **Lu K, Heng X, Summers MF.** 2011. Structural determinants and mechanism of HIV-1 genome packaging. *J Mol Biol* **410:**609–633.
120. **Purzycka KJ, Garfinkel DJ, Boeke JD, Le Grice SF.** 2013. Influence of RNA structural elements on Ty1 retrotransposition. *Mob Genet Elements* **3:**e25060.
121. **Maxwell PH, Coombes C, Kenny AE, Lawler JF, Boeke JD, Curcio MJ.** 2004. Ty1 mobilizes subtelomeric Y′ elements in telomerase-negative *Saccharomyces cerevisiae* survivors. *Mol Cell Biol* **24:**9887–9898.
122. **Maxwell PH, Curcio MJ.** 2007. Retrosequence formation restructures the yeast genome. *Genes Dev* **21:**3308–3318.
123. **Maxwell PH, Curcio MJ.** 2008. Incorporation of Y′-Ty1 cDNA destabilizes telomeres in *Saccharomyces cerevisiae* telomerase mutants. *Genetics* **179:**2313–2317.
124. **Garfinkel DJ, Hedge AM, Youngren SD, Copeland TD.** 1991. Proteolytic processing of pol-TYB proteins from the yeast retrotransposon Ty1. *J Virol* **65:**4573–4581.
125. **Merkulov GV, Lawler JF Jr, Eby Y, Boeke JD.** 2001. Ty1 proteolytic cleavage sites are required for transposition: all sites are not created equal. *J Virol* **75:**638–644.
126. **Youngren SD, Boeke JD, Sanders NJ, Garfinkel DJ.** 1988. Functional organization of the retrotransposon Ty from *Saccharomyces cerevisiae*: Ty protease is required for transposition. *Mol Cell Biol* **8:**1421–1431.
127. **Merkulov GV, Swiderek KM, Brachmann CB, Boeke JD.** 1996. A critical proteolytic cleavage site near the C terminus of the yeast retrotransposon Ty1 Gag protein. *J Virol* **70:**5548–5556.
128. **Moore SP, Garfinkel DJ.** 1994. Expression and partial purification of enzymatically active recombinant Ty1 integrase in *Saccharomyces cerevisiae. Proc Natl Acad Sci U S A* **91:**1843–1847.
129. **Chapman KB, Bystrom AS, Boeke JD.** 1992. Initiator methionine tRNA is essential for Ty1 transposition in yeast. *Proc Natl Acad Sci U S A* **89:**3236–3240.
130. **Keeney JB, Chapman KB, Lauermann V, Voytas DF, Astrom SU, von Pawel-Rammingen U, Bystrom A, Boeke JD.** 1995. Multiple molecular determinants for retrotransposition in a primer tRNA. *Mol Cell Biol* **15:**217–226.
131. **Wilhelm M, Wilhelm FX, Keith G, Agoutin B, Heyman T.** 1994. Yeast Ty1 retrotransposon: the minus-strand primer binding site and a *cis*-acting domain of the Ty1 RNA are both important for packaging of primer tRNA inside virus-like particles. *Nucleic Acids Res* **22:**4560–4565.
132. **Friant S, Heyman T, Wilhelm ML, Wilhelm FX.** 1996. Extended interactions between the primer tRNAi(Met) and genomic RNA of the yeast Ty1 retrotransposon. *Nucleic Acids Res* **24:**441–449.

133. **Friant S, Heyman T, Bystrom AS, Wilhelm M, Wilhelm FX.** 1998. Interactions between Ty1 retrotransposon RNA and the T and D regions of the tRNA(iMet) primer are required for initiation of reverse transcription *in vivo*. *Mol Cell Biol* **18:**799–806.

134. **Cristofari G, Bampi C, Wilhelm M, Wilhelm FX, Darlix JL.** 2002. A 5′-3′ long-range interaction in Ty1 RNA controls its reverse transcription and retrotransposition. *EMBO J* **21:**4368–4379.

135. **Bolton EC, Mildvan AS, Boeke JD.** 2002. Inhibition of reverse transcription *in vivo* by elevated manganese ion concentration. *Mol Cell* **9:**879–889.

136. **Pandey M, Patel S, Gabriel A.** 2004. Insights into the role of an active site aspartate in Ty1 reverse transcriptase polymerization. *J Biol Chem* **279:**47840–47848.

137. **Wilhelm M, Boutabout M, Wilhelm FX.** 2000. Expression of an active form of recombinant Ty1 reverse transcriptase in *Escherichia coli*: a fusion protein containing the C-terminal region of the Ty1 integrase linked to the reverse transcriptase-RNase H domain exhibits polymerase and RNase H activities. *Biochem J* **348:**337–342.

138. **Pandey M, Patel SS, Gabriel A.** 2008. Kinetic pathway of pyrophosphorolysis by a retrotransposon reverse transcriptase. *PLoS One* **3:**e1389.

139. **Yarrington RM, Chen J, Bolton EC, Boeke JD.** 2007. Mn2+ suppressor mutations and biochemical communication between Ty1 reverse transcriptase and RNase H domains. *J Virol* **81:**9004–9012.

140. **Wilhelm M, Wilhelm FX.** 2005. Role of integrase in reverse transcription of the *Saccharomyces cerevisiae* retrotransposon Ty1. *Eukaryot Cell* **4:**1057–1065.

141. **Wilhelm M, Wilhelm FX.** 2006. Cooperation between reverse transcriptase and integrase during reverse transcription and formation of the preintegrative complex of Ty1. *Eukaryot Cell* **5:**1760–1769.

142. **Le Grice SF.** 2003. "In the beginning": initiation of minus strand DNA synthesis in retroviruses and LTR-containing retrotransposons. *Biochemistry* **42:**14349–14355.

143. **Muller F, Laufer W, Pott U, Ciriacy M.** 1991. Characterization of products of TY1-mediated reverse transcription in *Saccharomyces cerevisiae*. *Mol Gen Genet* **226:**145–153.

144. **Lauermann V, Boeke JD.** 1997. Plus-strand strong-stop DNA transfer in yeast Ty retrotransposons. *EMBO J* **16:**6603–6612.

145. **Heyman T, Agoutin B, Friant S, Wilhelm FX, Wilhelm ML.** 1995. Plus-strand DNA synthesis of the yeast retrotransposon Ty1 is initiated at two sites, PPT1 next to the 3′ LTR and PPT2 within the pol gene. PPT1 is sufficient for Ty1 transposition. *J Mol Biol* **253:**291–303.

146. **Pochart P, Agoutin B, Rousset S, Chanet R, Doroszkiewicz V, Heyman T.** 1993. Biochemical and electron microscope analyses of the DNA reverse transcripts present in the virus-like particles of the yeast transposon Ty1. Identification of a second origin of Ty1DNA plus strand synthesis. *Nucleic Acids Res* **21:**3513–3520.

147. **Friant S, Heyman T, Wilhelm FX, Wilhelm M.** 1996. Role of RNA primers in initiation of minus-strand and plus-strand DNA synthesis of the yeast retrotransposon Ty1. *Biochimie* **78:**674–680.

148. **Garfinkel DJ, Stefanisko KM, Nyswaner KM, Moore SP, Oh J, Hughes SH.** 2006. Retrotransposon suicide: formation of Ty1 circles and autointegration via a central DNA flap. *J Virol* **80:**11920–11934.

149. **Heyman T, Wilhelm M, Wilhelm FX.** 2003. The central PPT of the yeast retrotransposon Ty1 is not essential for transposition. *J Mol Biol* **331:**315–320.

150. **Chapman KB, Boeke JD.** 1991. Isolation and characterization of the gene encoding yeast debranching enzyme. *Cell* **65:**483–492.

151. **Griffith JL, Coleman LE, Raymond AS, Goodson SG, Pittard WS, Tsui C, Devine SE.** 2003. Functional genomics reveals relationships between the retrovirus-like Ty1 element and its host *Saccharomyces cerevisiae*. *Genetics* **164:**867–879.

152. **Karst SM, Rutz ML, Menees TM.** 2000. The yeast retrotransposons Ty1 and Ty3 require the RNA Lariat debranching enzyme, Dbr1p, for efficient accumulation of reverse transcripts. *Biochem Biophys Res Commun* **268:**112–117.

153. **Mou Z, Kenny AE, Curcio MJ.** 2006. Hos2 and Set3 promote integration of Ty1 retrotransposons at tRNA genes in *Saccharomyces cerevisiae*. *Genetics* **172:**2157–2167.

154. **Cheng Z, Menees TM.** 2004. RNA branching and debranching in the yeast retrovirus-like element Ty1. *Science* **303:**240–243.

155. **Ye Y, De Leon J, Yokoyama N, Naidu Y, Camerini D.** 2005. DBR1 siRNA inhibition of HIV-1 replication. *Retrovirology* **2:**63.

156. **Coombes CE, Boeke JD.** 2005. An evaluation of detection methods for large lariat RNAs. *RNA* **11:**323–331.

157. **Pratico ED, Silverman SK.** 2007. Ty1 reverse transcriptase does not read through the proposed 2′,5′-branched retrotransposition intermediate *in vitro*. *RNA* **13:** 1528–1536.

158. **Lawler JF Jr, Merkulov GV, Boeke JD.** 2002. A nucleocapsid functionality contained within the amino terminus of the Ty1 protease that is distinct and separable from proteolytic activity. *J Virol* **76:**346–354.

159. **Eichinger DJ, Boeke JD.** 1988. The DNA intermediate in yeast Ty1 element transposition copurifies with virus-like particles: cell-free Ty1 transposition. *Cell* **54:**955–966.

160. **Lee BS, Lichtenstein CP, Faiola B, Rinckel LA, Wysock W, Curcio MJ, Garfinkel DJ.** 1998. Posttranslational inhibition of Ty1 retrotransposition by nucleotide excision repair/transcription factor TFIIH subunits Ssl2p and Rad3p. *Genetics* **148:**1743–1761.

161. **Nyswaner KM, Checkley MA, Yi M, Stephens RM, Garfinkel DJ.** 2008. Chromatin-associated genes protect the yeast genome from Ty1 insertional mutagenesis. *Genetics* **178:**197–214.

162. **Curcio MJ, Kenny AE, Moore S, Garfinkel DJ, Weintraub M, Gamache ER, Scholes DT.** 2007. S-phase checkpoint pathways stimulate the mobility of the

retrovirus-like transposon Ty1. *Mol Cell Biol* **27:**8874–8885.

163. **Risler JK, Kenny AE, Palumbo RJ, Gamache ER, Curcio MJ.** 2012. Host co-factors of the retrovirus-like transposon Ty1. *Mob DNA* **3:**12.

164. **Scholes DT, Banerjee M, Bowen B, Curcio MJ.** 2001. Multiple regulators of Ty1 transposition in *Saccharomyces cerevisiae* have conserved roles in genome maintenance. *Genetics* **159:**1449–1465.

165. **Curcio MJ, Garfinkel DJ.** 1999. New lines of host defense: inhibition of Ty1 retrotransposition by Fus3p and NER/TFIIH. *Trends Genet* **15:**43–45.

166. **Conte D Jr, Barber E, Banerjee M, Garfinkel DJ, Curcio MJ.** 1998. Posttranslational regulation of Ty1 retrotransposition by mitogen-activated protein kinase Fus3. *Mol Cell Biol* **18:**2502–2513.

167. **Lee BS, Bi L, Garfinkel DJ, Bailis AM.** 2000. Nucleotide excision repair/TFIIH helicases RAD3 and SSL2 inhibit short-sequence recombination and Ty1 retrotransposition by similar mechanisms. *Mol Cell Biol* **20:**2436–2445.

168. **Scholes DT, Kenny AE, Gamache ER, Mou Z, Curcio MJ.** 2003. Activation of a LTR-retrotransposon by telomere erosion. *Proc Natl Acad Sci U S A* **100:**15736–15741.

169. **Sundararajan A, Lee BS, Garfinkel DJ.** 2003. The Rad27 (Fen-1) nuclease inhibits Ty1 mobility in *Saccharomyces cerevisiae*. *Genetics* **163:**55–67.

170. **Braiterman LT, Boeke JD.** 1994. *In vitro* integration of retrotransposon Ty1: a direct physical assay. *Mol Cell Biol* **14:**5719–5730.

171. **Moore SP, Garfinkel DJ.** 2000. Correct integration of model substrates by Ty1 integrase. *J Virol* **74:**11522–11530.

172. **Voytas DF, Boeke JD.** 2002. Ty1 and Ty5 of *Saccharomyces cerevisiae*, p 631–662. *In* Craig N, Craigie R, Gellert M, Lambowitz A (ed), *Mobile DNA II*. ASM Press, Washington DC.

173. **Eichinger DJ, Boeke JD.** 1990. A specific terminal structure is required for Ty1 transposition. *Genes Dev* **4:**324–330.

174. **Wilhelm FX, Wilhelm M, Gabriel A.** 2005. Reverse transcriptase and integrase of the *Saccharomyces cerevisiae* Ty1 element. *Cytogenet Genome Res* **110:**269–287.

175. **Moore SP, Garfinkel DJ.** 2009. Functional analysis of N-terminal residues of Ty1 integrase. *J Virol* **83:**9502–9511.

176. **Friedl AA, Kiechle M, Maxeiner HG, Schiestl RH, Eckardt-Schupp F.** 2010. Ty1 integrase overexpression leads to integration of non-Ty1 DNA fragments into the genome of *Saccharomyces cerevisiae*. *Mol Genet Genom* **284:**231–242.

177. **Kiechle M, Friedl AA, Manivasakam P, Eckardt-Schupp F, Schiestl RH.** 2000. DNA integration by Ty integrase in yku70 mutant *Saccharomyces cerevisiae* cells. *Mol Cell Biol* **20:**8836–8844.

178. **Haren L, Ton-Hoang B, Chandler M.** 1999. Integrating DNA: transposases and retroviral integrases. *Annu Rev Microbiol* **53:**245–281.

179. **Peterson-Burch BD, Voytas DF.** 2002. Genes of the Pseudoviridae (Ty1/*copia* retrotransposons). *Mol Biol Evol* **19:**1832–1845.

180. **Kenna MA, Brachmann CB, Devine SE, Boeke JD.** 1998. Invading the yeast nucleus: a nuclear localization signal at the C terminus of Ty1 integrase is required for transposition *in vivo*. *Mol Cell Biol* **18:**1115–1124.

181. **Moore SP, Rinckel LA, Garfinkel DJ.** 1998. A Ty1 integrase nuclear localization signal required for retrotransposition. *Mol Cell Biol* **18:**1105–1114.

182. **McLane LM, Pulliam KF, Devine SE, Corbett AH.** 2008. The Ty1 integrase protein can exploit the classical nuclear protein import machinery for entry into the nucleus. *Nucleic Acids Res* **36:**4317–4326.

183. **Wilke CM, Adams J.** 1992. Fitness effects of Ty transposition in *Saccharomyces cerevisiae*. *Genetics* **131:**31–42.

184. **Scheifele LZ, Cost GJ, Zupancic ML, Caputo EM, Boeke JD.** 2009. Retrotransposon overdose and genome integrity. *Proc Natl Acad Sci U S A* **106:**13927–13932.

185. **Jiang YW.** 2002. Transcriptional cosuppression of yeast Ty1 retrotransposons. *Genes Dev* **16:**467–478.

186. **Winston F, Durbin KJ, Fink GR.** 1984. The SPT3 gene is required for normal transcription of Ty elements in *S. cerevisiae*. *Cell* **39:**675–682.

187. **Drinnenberg IA, Weinberg DE, Xie KT, Mower JP, Wolfe KH, Fink GR, Bartel DP.** 2009. RNAi in budding yeast. *Science* **326:**544–550.

188. **Maxwell PH, Curcio MJ.** 2007. Host factors that control long terminal repeat retrotransposons in *Saccharomyces cerevisiae*: implications for regulation of mammalian retroviruses. *Eukaryot Cell* **6:**1069–1080.

189. **Rattray AJ, Shafer BK, Garfinkel DJ.** 2000. The *Saccharomyces cerevisiae* DNA recombination and repair functions of the RAD52 epistasis group inhibit Ty1 transposition. *Genetics* **154:**543–556.

190. **Bryk M, Banerjee M, Conte D Jr, Curcio MJ.** 2001. The Sgs1 helicase of *Saccharomyces cerevisiae* inhibits retrotransposition of Ty1 multimeric arrays. *Mol Cell Biol* **21:**5374–5388.

191. **Stamenova R, Maxwell PH, Kenny AE, Curcio MJ.** 2009. Rrm3 protects the *Saccharomyces cerevisiae* genome from instability at nascent sites of retrotransposition. *Genetics* **182:**711–723.

192. **Lundblad V, Blackburn EH.** 1993. An alternative pathway for yeast telomere maintenance rescues est1- senescence. *Cell* **73:**347–360.

193. **Wellinger RJ, Zakian VA.** 2012. Everything you ever wanted to know about *Saccharomyces cerevisiae* telomeres: beginning to end. *Genetics* **191:**1073–1105.

194. **Lawler JF Jr, Merkulov GV, Boeke JD.** 2001. Frameshift signal transplantation and the unambiguous analysis of mutations in the yeast retrotransposon Ty1 Gag-Pol overlap region. *J Virol* **75:**6769–6775.

195. **McClanahan T, McEntee K.** 1984. Specific transcripts are elevated in *Saccharomyces cerevisiae* in response to DNA damage. *Mol Cell Biol* **4:**2356–2363.

196. **Morawetz C, Hagen U.** 1990. Effect of irradiation and mutagenic chemicals on the generation of ADH2- and

ADH4-constitutive mutants in yeast: the inducibility of Ty transposition by UV and ethyl methanesulfonate. *Mutat Res* **229:**69–77.

197. **Rolfe M, Spanos A, Banks G.** 1986. Induction of yeast Ty element transcription by ultraviolet light. *Nature* **319:**339–340.

198. **Liu Y, Kao HI, Bambara RA.** 2004. Flap endonuclease 1: a central component of DNA metabolism. *Annu Rev Biochem* **73:**589–615.

199. **Chan YA, Aristizabal MJ, Lu PY, Luo Z, Hamza A, Kobor MS, Stirling PC, Hieter P.** 2014. Genome-wide profiling of yeast DNA:RNA hybrid prone sites with DRIP-chip. *PLoS Genet* **10:**e1004288.

200. **Xu H, Boeke JD.** 1991. Inhibition of Ty1 transposition by mating pheromones in *Saccharomyces cerevisiae*. *Mol Cell Biol* **11:**2736–2743.

201. **Ivessa AS, Lenzmeier BA, Bessler JB, Goudsouzian LK, Schnakenberg SL, Zakian VA.** 2003. The *Saccharomyces cerevisiae* helicase Rrm3p facilitates replication past nonhistone protein-DNA complexes. *Mol Cell* **12:** 1525–1536.

202. **Sekedat MD, Fenyo D, Rogers RS, Tackett AJ, Aitchison JD, Chait BT.** 2010. GINS motion reveals replication fork progression is remarkably uniform throughout the yeast genome. *Mol Syst Biol* **6:**353.

203. **Deshpande AM, Newlon CS.** 1996. DNA replication fork pause sites dependent on transcription. *Science* **272:**1030–1033.

204. **Eigel A, Feldmann H.** 1982. Ty1 and delta elements occur adjacent to several tRNA genes in yeast. *EMBO J* **1:** 1245–1250.

205. **Hani J, Feldmann H.** 1998. tRNA genes and retroelements in the yeast genome. *Nucleic Acids Res* **26:** 689–696.

206. **Blanc VM, Adams J.** 2004. Ty1 insertions in intergenic regions of the genome of *Saccharomyces cerevisiae* transcribed by RNA polymerase III have no detectable selective effect. *FEMS Yeast Res* **4:**487–491.

207. **Boeke JD, Devine SE.** 1998. Yeast retrotransposons: finding a nice quiet neighborhood. *Cell* **93:**1087–1089.

208. **Admire A, Shanks L, Danzl N, Wang M, Weier U, Stevens W, Hunt E, Weinert T.** 2006. Cycles of chromosome instability are associated with a fragile site and are increased by defects in DNA replication and checkpoint controls in yeast. *Genes Dev* **20:**159–173.

209. **Cheng E, Vaisica JA, Ou J, Baryshnikova A, Lu Y, Roth FP, Brown GW.** 2012. Genome rearrangements caused by depletion of essential DNA replication proteins in *Saccharomyces cerevisiae*. *Genetics* **192:**147–160.

210. **Ji H, Moore DP, Blomberg MA, Braiterman LT, Voytas DF, Natsoulis G, Boeke JD.** 1993. Hotspots for unselected Ty1 transposition events on yeast chromosome III are near tRNA genes and LTR sequences. *Cell* **73:**1007–1018.

211. **Devine SE, Boeke JD.** 1996. Integration of the yeast retrotransposon Ty1 is targeted to regions upstream of genes transcribed by RNA polymerase III. *Genes Dev* **10:**620–633.

212. **Bachman N, Eby Y, Boeke JD.** 2004. Local definition of Ty1 target preference by long terminal repeats and clustered tRNA genes. *Genome Res* **14:**1232–1247.

213. **Mularoni L, Zhou Y, Bowen T, Gangadharan S, Wheelan SJ, Boeke JD.** 2012. Retrotransposon Ty1 integration targets specifically positioned asymmetric nucleosomal DNA segments in tRNA hotspots. *Genome Res* **22:**693–703.

214. **Moqtaderi Z, Struhl K.** 2004. Genome-wide occupancy profile of the RNA polymerase III machinery in *Saccharomyces cerevisiae* reveals loci with incomplete transcription complexes. *Mol Cell Biol* **24:**4118–4127.

215. **Soragni E, Kassavetis GA.** 2008. Absolute gene occupancies by RNA polymerase III, TFIIIB, and TFIIIC in *Saccharomyces cerevisiae*. *J Biol Chem* **283:**26568–26576.

216. **Gelbart ME, Bachman N, Delrow J, Boeke JD, Tsukiyama T.** 2005. Genome-wide identification of Isw2 chromatin-remodeling targets by localization of a catalytically inactive mutant. *Genes Dev* **19:**942–954.

217. **Bachman N, Gelbart ME, Tsukiyama T, Boeke JD.** 2005. TFIIIB subunit Bdp1p is required for periodic integration of the Ty1 retrotransposon and targeting of Isw2p to *S. cerevisiae* tDNAs. *Genes Dev* **19:**955–964.

218. **Picologlou S, Brown N, Liebman SW.** 1990. Mutations in RAD6, a yeast gene encoding a ubiquitin-conjugating enzyme, stimulate retrotransposition. *Mol Cell Biol* **10:** 1017–1022.

219. **Boeke J, Sandmeyer S.** 1991. Yeast transposable elements, p 193–261. *In* Broach J, Jones E, Pringle J (ed), *The Molecular and Cellular Biology of the Yeast Saccharomyces cerevisiae*. Cold Spring Harbor Laboratory Press, Cold Spring Harbor, NY.

220. **Bolton EC, Boeke JD.** 2003. Transcriptional interactions between yeast tRNA genes, flanking genes and Ty elements: a genomic point of view. *Genome Res* **13:**254–263.

221. **Chan JE, Kolodner RD.** 2011. A genetic and structural study of genome rearrangements mediated by high copy repeat Ty1 elements. *PLoS Genet* **7:**e1002089.

222. **Koszul R, Caburet S, Dujon B, Fischer G.** 2004. Eucaryotic genome evolution through the spontaneous duplication of large chromosomal segments. *EMBO J* **23:**234–243.

223. **Chaleff DT, Fink GR.** 1980. Genetic events associated with an insertion mutation in yeast. *Cell* **21:**227–237.

224. **Roeder GS, Fink GR.** 1980. DNA rearrangements associated with a transposable element in yeast. *Cell* **21:** 239–249.

225. **Roeder GS, Fink GR.** 1982. Movement of yeast transposable elements by gene conversion. *Proc Natl Acad Sci U S A* **79:**5621–5625.

226. **Downs KM, Brennan G, Liebman SW.** 1985. Deletions extending from a single Ty1 element in *Saccharomyces cerevisiae*. *Mol Cell Biol* **5:**3451–3457.

227. **Dunham MJ, Badrane H, Ferea T, Adams J, Brown PO, Rosenzweig F, Botstein D.** 2002. Characteristic genome rearrangements in experimental evolution of *Saccharomyces cerevisiae*. *Proc Natl Acad Sci U S A* **99:**16144–16149.

228. **Putnam CD, Pennaneach V, Kolodner RD.** 2005. *Saccharomyces cerevisiae* as a model system to define the chromosomal instability phenotype. *Mol Cell Biol* **25:** 7226–7238.

229. **Surosky RT, Tye BK.** 1985. Resolution of dicentric chromosomes by Ty-mediated recombination in yeast. *Genetics* **110:**397–419.

230. **Umezu K, Hiraoka M, Mori M, Maki H.** 2002. Structural analysis of aberrant chromosomes that occur spontaneously in diploid *Saccharomyces cerevisiae*: retrotransposon Ty1 plays a crucial role in chromosomal rearrangements. *Genetics* **160:**97–110.

231. **Zhang H, Zeidler AF, Song W, Puccia CM, Malc E, Greenwell PW, Mieczkowski PA, Petes TD, Argueso JL.** 2013. Gene copy-number variation in haploid and diploid strains of the yeast *Saccharomyces cerevisiae*. *Genetics* **193:**785–801.

232. **Argueso JL, Westmoreland J, Mieczkowski PA, Gawel M, Petes TD, Resnick MA.** 2008. Double-strand breaks associated with repetitive DNA can reshape the genome. *Proc Natl Acad Sci U S A* **105:**11845–11850.

233. **Scheifele LZ, Boeke JD.** 2008. From the shards of a shattered genome, diversity. *Proc Natl Acad Sci U S A* **105:**11593–11594.

234. **Gresham D, Desai MM, Tucker CM, Jenq HT, Pai DA, Ward A, DeSevo CG, Botstein D, Dunham MJ.** 2008. The repertoire and dynamics of evolutionary adaptations to controlled nutrient-limited environments in yeast. *PLoS Genet* **4:**e1000303.

235. **Moore JK, Haber JE.** 1996. Capture of retrotransposon DNA at the sites of chromosomal double-strand breaks. *Nature* **383:**644–646.

236. **Teng SC, Kim B, Gabriel A.** 1996. Retrotransposon reverse-transcriptase-mediated repair of chromosomal breaks. *Nature* **383:**641–644.

237. **Yu X, Gabriel A.** 1999. Patching broken chromosomes with extranuclear cellular DNA. *Mol Cell* **4:**873–881.

238. **Yu X, Gabriel A.** 2003. Ku-dependent and Ku-independent end-joining pathways lead to chromosomal rearrangements during double-strand break repair in *Saccharomyces cerevisiae*. *Genetics* **163:**843–856.

239. **Lemoine FJ, Degtyareva NP, Lobachev K, Petes TD.** 2005. Chromosomal translocations in yeast induced by low levels of DNA polymerase a model for chromosome fragile sites. *Cell* **120:**587–598.

240. **Casper AM, Greenwell PW, Tang W, Petes TD.** 2009. Chromosome aberrations resulting from double-strand DNA breaks at a naturally occurring yeast fragile site composed of inverted ty elements are independent of Mre11p and Sae2p. *Genetics* **183:**423–439.

241. **Aguilera A, Garcia-Muse T.** 2012. R loops: from transcription byproducts to threats to genome stability. *Mol Cell* **46:**115–124.

242. **Gan W, Guan Z, Liu J, Gui T, Shen K, Manley JL, Li X.** 2011. R-loop-mediated genomic instability is caused by impairment of replication fork progression. *Genes Dev* **25:**2041–2056.

243. **Ben-Aroya S, Mieczkowski PA, Petes TD, Kupiec M.** 2004. The compact chromatin structure of a Ty repeated sequence suppresses recombination hotspot activity in *Saccharomyces cerevisiae*. *Mol Cell* **15:**221–231.

244. **Chang SL, Lai HY, Tung SY, Leu JY.** 2013. Dynamic large-scale chromosomal rearrangements fuel rapid adaptation in yeast populations. *PLoS Genet* **9:**e1003232.

245. **Dhar R, Sagesser R, Weikert C, Yuan J, Wagner A.** 2011. Adaptation of *Saccharomyces cerevisiae* to saline stress through laboratory evolution. *J Evol Biol* **24:** 1135–1153.

246. **Kao KC, Sherlock G.** 2008. Molecular characterization of clonal interference during adaptive evolution in asexual populations of *Saccharomyces cerevisiae*. *Nat Genet* **40:**1499–1504.

247. **Selmecki AM, Dulmage K, Cowen LE, Anderson JB, Berman J.** 2009. Acquisition of aneuploidy provides increased fitness during the evolution of antifungal drug resistance. *PLoS Genet* **5:**e1000705.

248. **Lynch M.** 2008. The cellular, developmental and population-genetic determinants of mutation-rate evolution. *Genetics* **180:**933–943.

249. **Botstein D, Fink GR.** 2011. Yeast: an experimental organism for 21st Century biology. *Genetics* **189:**695–704.

250. **Torres EM, Sokolsky T, Tucker CM, Chan LY, Boselli M, Dunham MJ, Amon A.** 2007. Effects of aneuploidy on cellular physiology and cell division in haploid yeast. *Science* **317:**916–924.

251. **Pavelka N, Rancati G, Zhu J, Bradford WD, Saraf A, Florens L, Sanderson BW, Hattem GL, Li R.** 2010. Aneuploidy confers quantitative proteome changes and phenotypic variation in budding yeast. *Nature* **468:** 321–325.

252. **Maxwell PH, Burhans WC, Curcio MJ.** 2011. Retrotransposition is associated with genome instability during chronological aging. *Proc Natl Acad Sci U S A* **108:** 20376–20381.

253. **VanHoute D, Maxwell PH.** 2014. Extension of Saccharomyces paradoxus chronological lifespan by retrotransposons in certain media conditions is associated with changes in reactive oxygen species. *Genetics* **198:**531–545.

254. **Stoycheva T, Pesheva M, Venkov P.** 2010. The role of reactive oxygen species in the induction of Ty1 retrotransposition in *Saccharomyces cerevisiae*. *Yeast* **27:** 259–267.

255. **Derr LK, Strathern JN.** 1993. A role for reverse transcripts in gene conversion. *Nature* **361:**170–173.

256. **Derr LK, Strathern JN, Garfinkel DJ.** 1991. RNA-mediated recombination in *S. cerevisiae*. *Cell* **67:**355–364.

257. **Roelants F, Potier S, Souciet JL, de Montigny J.** 1997. Delta sequence of Ty1 transposon can initiate transcription of the distal part of the URA2 gene complex in *Saccharomyces cerevisiae*. *FEMS Microbiol Lett* **148:** 69–74.

258. **Schacherer J, Tourrette Y, Souciet JL, Potier S, De Montigny J.** 2004. Recovery of a function involving gene duplication by retroposition in *Saccharomyces cerevisiae*. *Genome Res* **14:**1291–1297.

259. **Fink GR.** 1987. Pseudogenes in yeast? *Cell* **49:**5–6.

260. **Darai-Ramqvist E, Sandlund A, Mueller S, Klein G, Imreh S, Kost-Alimova M.** 2008. Segmental duplications and evolutionary plasticity at tumor chromosome break-prone regions. *Genome Res* **18:**370–379.

261. **Ni X, Zhuo M, Su Z, Duan J, Gao Y, Wang Z, Zong C, Bai H, Chapman AR, Zhao J, Xu L, An T, Ma Q, Wang Y, Wu M, Sun Y, Wang S, Li Z, Yang X, Yong J, Su XD, Lu Y, Bai F, Xie XS, Wang J.** 2013. Reproducible copy number variation patterns among single circulating tumor cells of lung cancer patients. *Proc Natl Acad Sci U S A* **110:**21083–21088.

262. **Gray WM, Fassler JS.** 1993. Role of *Saccharomyces cerevisiae* Rap1 protein in Ty1 and Ty1-mediated transcription. *Gene Expr* **3:**237–251.

263. **Kang XL, Yadao F, Gietz RD, Kunz BA.** 1992. Elimination of the yeast RAD6 ubiquitin conjugase enhances base-pair transitions and G.C → T.A transversions as well as transposition of the Ty element: implications for the control of spontaneous mutation. *Genetics* **130:** 285–294.

264. **Liebman SW, Newnam G.** 1993. A ubiquitin-conjugating enzyme, RAD6, affects the distribution of Ty1 retrotransposon integration positions. *Genetics* **133:**499–508.

265. **Qian Z, Huang H, Hong JY, Burck CL, Johnston SD, Berman J, Carol A, Liebman SW.** 1998. Yeast Ty1 retrotransposition is stimulated by a synergistic interaction between mutations in chromatin assembly factor I and histone regulatory proteins. *Mol Cell Biol* **18:** 4783–4792.

266. **Harger JW, Meskauskas A, Nielsen J, Justice MC, Dinman JD.** 2001. Ty1 retrotransposition and programmed +1 ribosomal frameshifting require the integrity of the protein synthetic translocation step. *Virology* **286:**216–224.

267. **Dubois E, Jacobs E, Jauniaux JC.** 1982. Expression of the ROAM mutations in *Saccharomyces cerevisiae*: involvement of *trans*-acting regulatory elements and relation with the Ty1 transcription. *EMBO J* **1:**1133–1139.

268. **Bryk M, Briggs SD, Strahl BD, Curcio MJ, Allis CD, Winston F.** 2002. Evidence that Set1, a factor required for methylation of histone H3, regulates rDNA silencing in *S. cerevisiae* by a Sir2-independent mechanism. *Curr Biol* **12:**165–170.

269. **Dalgaard JZ, Banerjee M, Curcio MJ.** 1996. A novel Ty1-mediated fragmentation method for native and artificial yeast chromosomes reveals that the mouse steel gene is a hotspot for Ty1 integration. *Genetics* **143:**673–683.

270. **Wilkinson KA, Vasa SM, Deigan KE, Mortimer SA, Giddings MC, Weeks KM.** 2009. Influence of nucleotide identity on ribose 2′-hydroxyl reactivity in RNA. *RNA* **15:**1314–1321.

271. **Bridier-Nahmias A, Lesage P.** 2012. Two large-scale analyses of Ty1 LTR-retrotransposon de novo insertion events indicate that Ty1 targets nucleosomal DNA near the H2A/H2B interface. *Mob DNA* **3:**22.

Mobile DNA, 3rd Edition
Nancy L. Craig, Michael Chandler, Martin Gellert, Alan M. Lambowitz, Phoebe A. Rice and Suzanne Sandmeyer

doi:10.1128/microbiolspec.MDNA3-0057-2014

Suzanne Sandmeyer[1,2]
Kurt Patterson[1]
Virginia Bilanchone[1]

42

Ty3, a Position-specific Retrotransposon in Budding Yeast

INTRODUCTION

Long terminal repeat (LTR) retrotransposons occur throughout eukaryotic phyla, but vary greatly in both the types of elements and the representation among species. The majority of some genomes are composed of these elements. The LTR retrotransposons are taxonomically divided into *Pseudoviridae* (or Ty1/Copia) and *Metaviridae* (or Ty3/Gypsy) elements based on genome organization and relatedness of proteins encoded (reviewed in references 1–3) (Gypsy Database 2.0, gydb.org). The Ty3/Gypsy family has shared ancestry with retroviruses and some members encode envelope enabling intercellular transmission. The eponymous founding elements of these LTR retrotransposon families, Ty1/Copia and Ty3/Gypsy, exist in *Saccharomyces cerevisiae* (Ty1 and Ty3) and in *Drosophila* (Copia and Gypsy).

This review is focused on Ty3, a representative of the Ty3/Gypsy retrotransposon group, which has been extensively characterized at the molecular level. Ty3 LTRs flank two overlapping open reading frames (ORFs), *GAG3* and *POL3*, which encode structural and catalytic functions, respectively. Ty3 was discovered as a polymorphism associated with tRNA genes (tDNAs) and has been extensively studied as a model of targeted integration. A combination of approaches has implicated numerous host factors in control of Ty3 retrotransposition. Altogether, studies of Ty3 provide for a relatively comprehensive, albeit incomplete, picture of the ongoing remodeling of a eukaryotic genome by an LTR retrotransposon.

TY3 DIVERSITY

Ty3 is the sole representative of the Ty3/Gypsy family in *S. cerevisiae*. The reference *S. cerevisiae* strain BY4741 contains two closely related, full-length Ty3 elements one of which, YGRWTy3-1, is transpositionally active. In addition to the four LTRs associated with full-length elements, BY4741 contains 41 isolated LTRs (4–6) derived from recombination of the LTRs flanking full-length elements. These isolated LTRs were named

[1]Department of Biological Chemistry; [2]Department of Microbiology and Molecular Genetics, University of California, Irvine, CA.

sigma elements (7, 8) before the discovery of full-length Ty3 (9). Ty1/Copia elements in *S. cerevisiae* are represented by Ty1, -2, -4, and -5. Ty1 (see Curcio, Lutz, and Lesage, Chapter 41) is the most abundant of these, with 32 full-length elements, two truncated elements with single LTRs, and 185 isolated LTRs (4–6).

Phylogenetic analysis of full-length elements and isolated LTRs has provided insight into the time of appearance and retrotransposition activity relative to host phylogeny based on occurrence of integrants shared among diverging host species and the divergence of flanking LTRs of individual full-length copies, as these are regenerated from the same template during retrotransposition (10). Comparison of Ty3 elements and LTRs in related *Saccharomyces* species showed that LTRs flanking full-length elements in the reference strain were identical and LTRs overall had > 96% identity. In comparison, Ty1, Ty2, and Ty5 LTRs display 70 to 99% identity (5, 6). Although this suggests that Ty3 appeared recently relative to other elements, analysis of a large number of related *Saccharomyces* genomes identified ancient, shared Ty3 insertions pre-dating the appearance of *S. cerevisiae* (4). However, the presence of 24 highly similar, short-branch LTRs is consistent with recent retrotransposition of elements related to the full-length *S. cerevisiae* elements (4). Fifteen highly degenerate LTRs were identified in *S. cerevisiae* that were more similar to *Saccharomyces paradoxus* Ty3p than to *S. cerevisiae* Ty3 LTRs. Because common insertions preceding the divergence of *S. paradoxus* and *S. cerevisiae* were absent, these elements were concluded to have entered the *S. cerevisiae* genome via horizontal transmission from *S. paradoxus* or from some ancestral species shared between *S. paradoxus* and *S. cerevisiae* (4). Ty3 also has a greater proportion of solo LTRs relative to full-length elements than other Tys except Ty4. Although the reason for this is not clear, it is possible that this explains the absence of full-length copies of Ty3p from the modern *S. cerevisiae* genome (4).

TY3 OVERVIEW

Integrated, full-length copies of Ty3 are 5.4 kb in length, comprised of two LTRs of 340-bp each, flanking overlapping ORFs, *GAG3* and *POL3*. Ty3 is transcribed into a 5.2-kb genomic (g)RNA that begins and ends in the LTRs (Figure 1). Translation of this RNA yields polyprotein precursors, a 34-kDa Gag3 and a 178-kDa Gag3–Pol3 product of frameshifting between *GAG3* and *POL3*. Gag3 contains major structural protein domains CApsid (CA), SPacer (SP), and NucleoCapsid (NC). Gag3–Pol3, in addition to those structural domains, contains catalytic domains PRotease (PR), Reverse Transcriptase (RT), and INtegrase (IN), and a spacer "J" between the PR and RT domains. Gag3 and Gag3–Pol3 assemble together with gRNA into immature 44 to 53 nm diameter virus-like particles (VLPs). Assembly of these occurs in cytoplasmic foci referred to as retrosomes (11) in which Ty3 proteins and RNA are concentrated with host assembly factors. Retrosomes form over the first several hours of Ty3 expression and become generally perinuclear. Within the VLP, precursor polyproteins undergo proteolytic processing by Ty3 PR into mature species. These populations of VLPs are much more diverse in size and morphology and range from 25 to 52 nm with electron dense cores. Subsequently Ty3 RT reverse transcribes gRNA into cDNA and remodeling of the VLP accompanies nuclear entry of a preintegration complex (PIC) of unknown composition. Integration into the transcription start site (TSS) of genes transcribed by RNA polymerase III (RNAP III) completes the cycle.

The reference *S. cerevisiae* strain BY4741 contains two full-length elements, YGRWTy3-1 and YILWTy3-1 (9, 12). YGRWTy3-1 is transpositionally active. YILWTy3-1 is transcribed and translated, but is inactivated by a frameshift mutation in the IN-coding region. The active element, YGRWTy3-1, is comprised of 340-bp LTRs flanking an internal coding domain of 4671 bp for a total length of 5351 bp. Ty3 null strains have been derived by deletion of endogenous elements (13).

TY3 EXPRESSION

Mating Control of Ty3 Expression

Ty3 transcription initiates in the upstream LTR and terminates downstream of that point in the downstream LTR to yield the 5.2-kb RNA. This defines terminal RNA segments as U5 (unique to the 5′ end); R (repeated); and U3 (unique to the 3′ end) for a full-length element structure of U3-R-U5-internal domain-U3-R-U5 and RNA structure of R-U5-internal domain-U3-R (Figure 2A, B). This pattern of expression is characteristic of LTR retrotransposons and to retroviruses. The Ty3 *GAG3* ORF lies downstream of R-U5, but the *POL3* ORF overlaps U3.

Saccharomyces cerevisiae exists in haploid *MAT*a and *MAT*α mating types that secrete mating-type-specific pheromones that bind to receptors triggering mitogen-activated protein (MAP) kinase signaling. Ty3 transcription and retrotransposition are subject to mating type control (Figure 1A). In haploid cells, Ty3 is

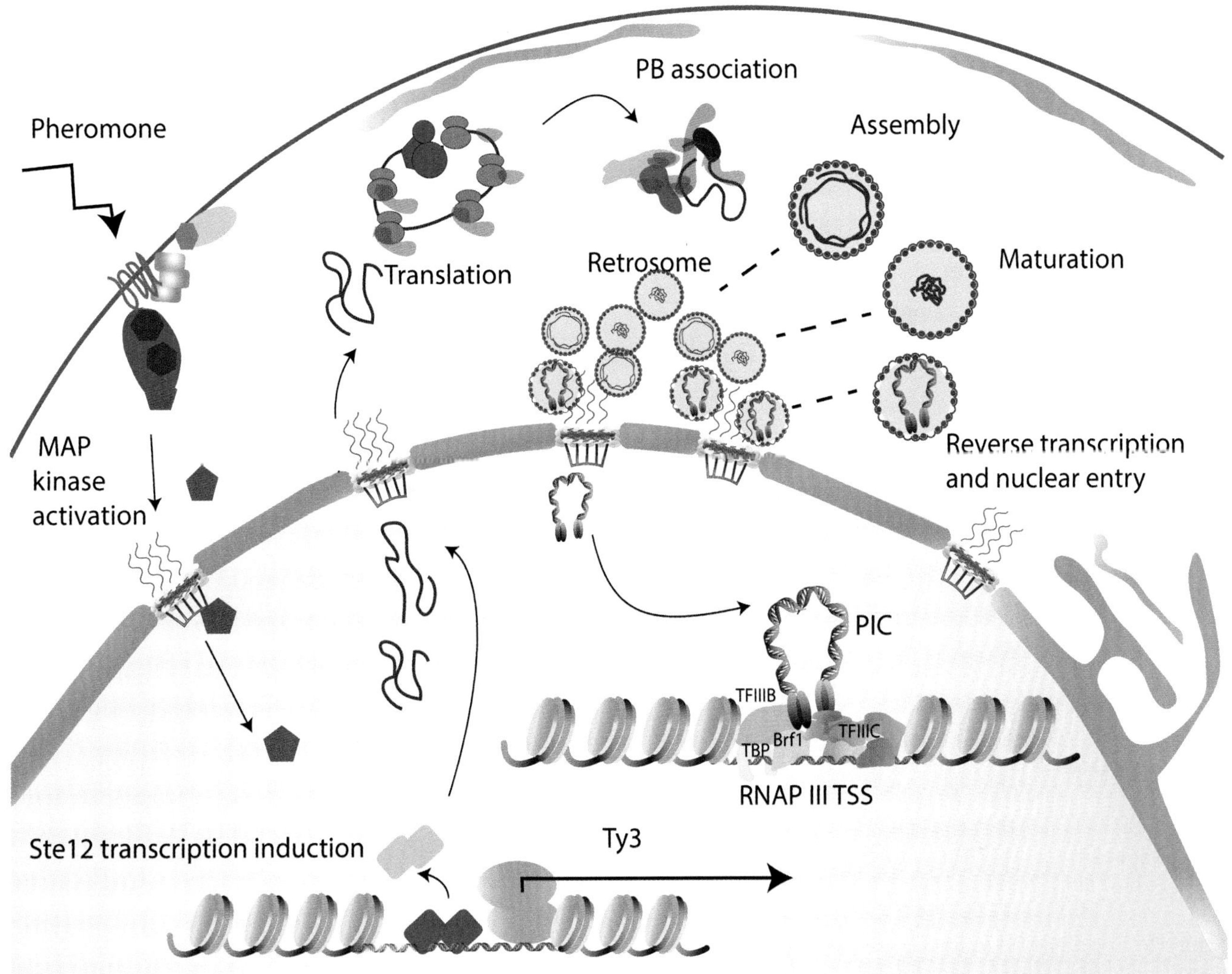

Figure 1 Ty3 retrotransposition. (A) Ty3 replication cycle. Pheromone binding to *MATa* or *MATα* pheromone receptors activates G protein-coupled mating signal transduction via mitogen-activated protein (MAP) kinase kinase kinase Ste7, MAP kinase kinase Ste11 and MAP kinase Fus3 (rose). Scaffold protein Ste5 (blue) supports specificity of their interaction preventing crosstalk with the filamentous growth pathway. Fus3 phosphorylates Dig1 and Dig2 negative regulators (gold) of Ste12 (dark blue), which then dissociate allowing Ste12 activation of RNA Pol II transcription of Ty3. Ty3 poly(A) RNA (maroon) is exported and translated into Gag3 and Gag3–Pol3 (tan), which then associate, together with the gRNA and RNA processing body (PB) factors, forming retrosomes within which Ty3 VLPs assemble. These foci become perinuclear over time. Assembly activates protease (PR) processing and maturation of the virus-like particles (VLPs). After cells mate (not shown) reverse transcription of the gRNA into cDNA occurs. Uncoating (dissociation of Gag3) presumably accompanies nuclear entry of the PIC. (A, B) Ty3 cDNA associates with RNAP III transcription initiation complexes composed of TFIIIB (yellow) and TFIIIC (green). *In vitro* TATA binding protein and Brf1 constitute the minimum target, but *in vivo* evidence suggests that TFIIIC can also be present. doi:10.1128/microbiolspec.MDNA3-0057-2014.f1

expressed at low levels. However, the full-length Ty3 elements (13, 14) and some LTRs are induced up to 80-fold in mating cells (15). This activates the MAP kinase Fus3 to relieve Dig1,2 repression of positive transcription factor Ste12 bound to pheromone response elements. These pheromone response elements occur in promoters of genes mediating physiological changes in preparation for mating including agglutinin

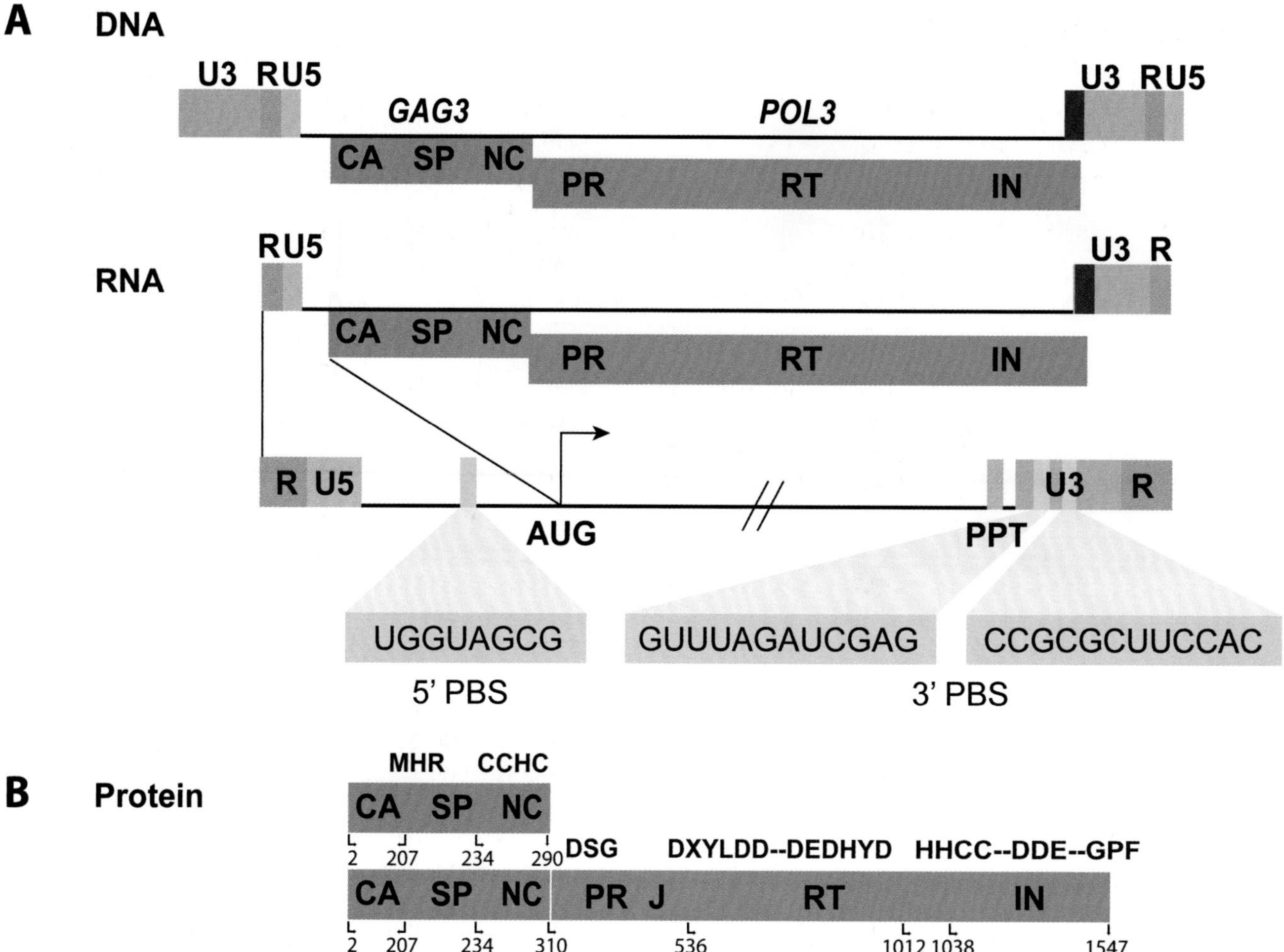

Figure 2 Ty3 DNA, RNA, and protein. (A) DNA and RNA. The 5.4-kbp Ty3 element is transcribed into a 5.2-kb RNA. The major 5′ end of Ty3 RNA maps to 118 nucleotides (nt) inside the upstream long terminal repeat (LTR), and the 3′ end to heterogeneous positions between 243 and 273 nt inside the downstream LTR, as well as beyond the downstream LTR (14). The overlap resulting from termination downstream of the position of initiation results in a sequence that is repeated ("R") and defines 5′ U5 and 3′ U3 sequences. The initiator AUG of the *GAG3* open reading frame (ORF) occurs at nucleotides 76 to 78 inside the Ty3 internal domain for a total 5′ untranslated region (UTR) of 193 nt; *POL3* extends into the downstream LTR to define a 3′ UTR of ~227 nt (12, 14). Candidate upstream TATA and downstream polyadenylation sites are identified, but not experimentally verified. The RNA contains a bipartite primer binding site (PBS), which anneals to initiator $tRNA^{Met}$ in the upstream untranslated region and the downstream LTR (gold boxes). The *GAG3* and *POL3* ORFs encode Gag3 and Gag3–Pol3. (B) Protein. Gag3 is 290 amino acids (aa) and contains domains that mature via Ty3 PR processing into 207-aa capsid (CA), 27-aa spacer (SP), and 57-aa nucleocapsid (NC). Gag3–Pol3 contains those and additionally, protease (PR); reverse transcriptase (RT) starting at amino acids 536; and two forms of integrase (IN) domains (starting at amino acids 1012 and 1038) produced via a programmed frameshift. The *POL3* ORF terminates within the downstream LTR so that the polypurine tract (PPT) plus strand primer is actually within the IN-coding region. (C) Reverse transcription of Ty3 genomic RNA. The tRNA primes synthesis of a minus-strand strong stop containing U5 and R segments, which then transfers to the 3′ end and primes extension of the minus strand. The plus-strand strong stop intermediate is initiated with cleavage by RNaseH at the downstream end of the PPT just outside the downstream LTR and is extended through U3, R, and U5 and likely copied into the 3′ end of the tRNA then transferred to the 5′ end of the RNA and extended to form the plus strand of the cDNA. Although as described in the text minus- and plus-strand strong-stop intermediates have been identified, the overall flow described is based on the retrovirus model. An additional possibility (not shown) is that the 5′ and 3′ ends are transiently joined in a lariat RNA (see text). Bottom, the full-length cDNA has two extra base pairs on each end derived from a 2-nt offset between the priming sites and the LTR ends of the integrated element. Integrase (IN) processes 2 nt from each 3′ end and mediates the nucleophilic attacks of the resulting hydroxyls at 5-bp staggered positions flanking the RNA polymerase III (RNAPIII) transcription initiation sites. The integration site is repaired, resulting in 5-bp direct repeats flanking the ends of the Ty3 element. doi:10.1128/microbiolspec.MDNA3-0057-2014.f2

Figure 2 *continues on next page*

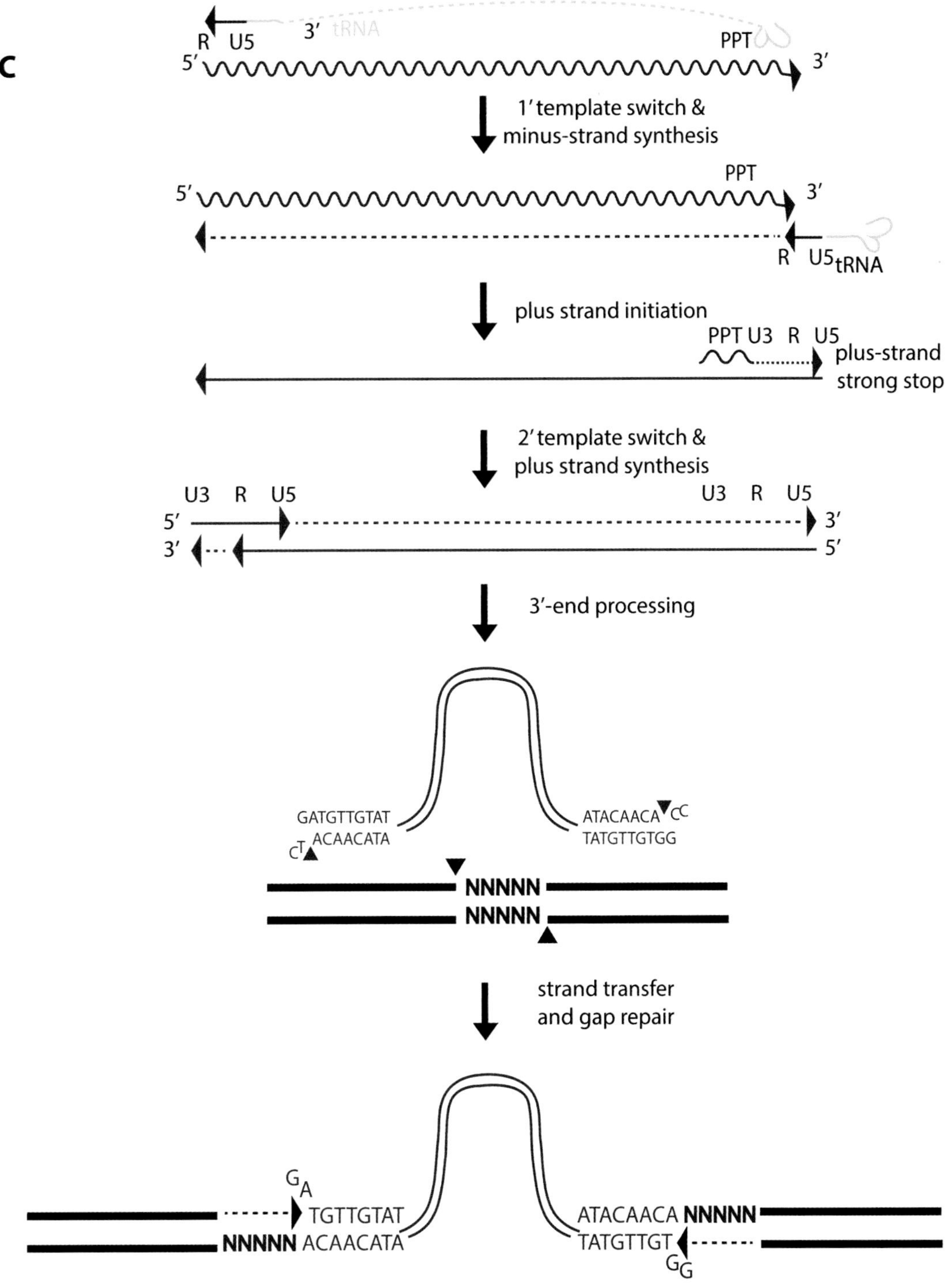

Figure 2 *continued*

production, enhanced vesicular trafficking, polarized cell growth (shmooing), and arrest in G1 of the cell cycle (reviewed in reference 16). Deletion analysis identified two adjacent PRE sequences within the Ty3 U3.

Pheromone induction of Ty3 transcription in G1-arrested cells is accompanied by translation of Ty3 proteins, and formation and maturation of VLPs. However, reverse transcription is delayed until cells resume cycling (17). Experiments in which transposing elements were activated in one mating type, and integrations of those elements were detected into plasmids carried in the opposite mating type demonstrated that this control of cDNA production promotes invasion of Ty3 into new host genomes (18). Once cells have mated, Ty3 is repressed by the action of the *MAT* a1/α2 heterodimer formed from proteins expressed from the *MAT* loci of the two haploid mating cell types. Deletion analysis identified *MAT* a1/α2 repressor binding sites in the U3 region (13).

Ty3 is also regulated in response to environmental conditions. Similar to Ty1, it is induced in *bas1Δ* cells during adenylic nucleotide deprivation. *BAS1* encodes a Myb-related factor that regulates transcription of genes in purine and histidine biosynthetic pathways and meiotic recombination at specific genes (19–21). Although the basis of induction has not been explored for Ty3, Ty1 induction is inversely correlated with anti-sense transcription and depends on both the absence of Bas1 and the presence of the bHLH transcription factor Tye7 (22).

Interaction of Ty3 RNA Polymerase II and Flanking tDNA RNA Polymerase III Promoters

The unique Ty3 specificity for integration at the RNAP III TSS provides a context in which to examine interactions of RNAP II and RNAP III promoters. Investigation of the interactions between the *SUP2* tDNA RNAP III promoter and the adjacent Ty3 LTR RNAP II promoter showed that mutations that inactivated the tDNA promoter greatly enhanced pheromone induction of the LTR promoter (23). The interference by RNAP III promoters with RNAP II repression was further defined and shown to be generalizable (24) and dependent upon localization of tDNA and flanking sequences to the nucleolus (25, 26). Hence, similar to expression of chromodomain retrotransposons that target heterochromatic regions, expression of Ty3, at least under noninducing conditions, might be attenuated by genomic context (18, 27).

As suggested in the foregoing discussion, repression of Ty3 by association with tDNAs is not absolute. In addition to Ty3 being inducible under certain conditions, integrations of genetically tagged Ty3 elements are readily selectable (28). A possible event which could relieve nucleolar repression of tDNA-associated LTR promoters is relocalization of transcriptionally active DNA to Nuclear Pore Complexes (NPC) referred to as gene gating (29). Although the nucleolus is located on the nuclear envelope in yeast, genes could transiently relocalize to other NPC. In addition to gene gating for RNAP II promoters such as the LTRs, transcriptionally active tDNAs have been shown to localize to NPC (30). Consistent with the latter possibility, chromatin conformation capture analysis of yeast chromosomal DNA indicated that a subpopulation of tDNAs are not associated with rDNA (31).

Ty1/Copia and Ty3/Gypsy Coexistence in Budding Yeast

Ty1 and Ty3 share a subset of host factors and are both induced by MAP kinase signaling in response to environmental stimuli. As described above, Ty3 is induced in haploid cells by pheromone signaling through the mating MAP kinase pathway, which activates transcription factor Ste12 (13, 18). Ty1 is induced under MAP kinase activation of Ste12/Tec1 in diploid cells, in response to nutritional deprivation (32). Despite common upstream components of these pathways, cross talk between them is controlled by mating MAP kinase Fus3-mediated degradation of Tec1 in haploid cells and mating type suppression of Ty3 in diploid cells (33). Ty1 constitutes a high fraction of RNA in haploid cells (34), but Ty1 proteins are actively degraded during the mating pheromone response (35–37). Hence, despite overlapping induction pathways and shared host factors, Ty1 and Ty3 occupy discrete retrotransposition niches.

TY3 RETROTRANSPOSITION ASSAYS

Because retrotransposition is a relatively rare event, occurring in less than 1% of induced cell cultures, the sensitivity of assays used in genetic screens affects the spectrum of Ty or host mutants recovered. Genetic screens require patching naive cells, replica plating patches onto one or more types of selective medium and screening for papillation frequency. Given the rarity of events, identification of mutants deficient in transposition becomes nontrivial. Identification of the step in Ty3 retrotransposition affected by mutations is further complicated because expression at the RNA, protein, and even cDNA level is relatively robust, but cells in which transposition occurs are infrequent. One possible explanation of this apparent disparity is that Ty3 VLPs assemble in cytoplasmic clusters. As clusters increase in size, the ability of individual VLPs to access nuclear pores may become limiting.

Ty3 can be induced under native or synthetic regulation. Induction under the native pheromone promoter allows evaluation of host factors in the physiologic context of retrotransposition, but transcription increases as cells accumulate in G1, and so is relatively asynchronous. Alternatively Ty3 has been engineered to be synchronously inducible in the presence of galactose by substitution of the *GAL1–10* upstream activating sequence for the upstream Ty3 U3 region (9, 13, 14). Galactose induces a synchronous response, but has the disadvantage that galactose metabolism is under catabolite repression, so that rapid induction requires pregrowth in a nonglucose carbon source in which the growth rate slows.

Ty3 targeting offers the possibility of monitoring retrotransposition by selecting for insertions into target traps. One such target is composed of a pair of divergent tDNAs, one of which is a suppressor tDNA (18). Expression from the target suppressor is blocked by mutual interference of the tDNA transcription complexes so that cells containing a suppressible allele in a nutritional marker are auxotrophic on selective medium. Ty3 integration between the tDNAs relieves interference and growth correlates with retrotransposition.

Genetic assays in which a marker is inserted into the retrotransposon in theory allow straightforward retrotransposition assays. However, in the case of Ty3, the *POL3* ORF, which encodes IN, overlaps important *cis*-acting sequences within and flanking the LTR so that there is no spacer into which a marker can be inserted without inactivating the element. This impediment was circumvented by introducing a short repeat of the downstream end of *POL3*, which supplies a second, noncoding, copy of the *cis*-acting polypurine tract (PPT) and LTR sequences required for replication and integration. A genetic marker, for example, *HIS3*, can be inserted between these repeats (38). This Ty3 can be expressed from a *URA3*-marked plasmid so that in a *his3 ura3* background, retrotransposed cells can be selected on medium lacking histidine and containing 5-fluoro-uracil, which selects against the *URA3*-marked plasmid (39) but for the transposed element. Incorporation of an anti-sense intron into *HIS3*, a strategy devised for Ty1 (40), further simplifies the assay. In this strategy, the marker, which is transcribed anti-sense to the Ty, is disrupted by an artificial intron oriented so that splicing signals are read in the Ty sense. Ty3 transcription, splicing, reverse transcription, and integration generate a genomic copy of the selectable intron-free marker (Figure 3, left panel). This assay has the limitation that it superimposes a requirement for splicing into genetic screens.

Physical retrotransposition assays have the major advantages of using native elements, not requiring long periods of selective cell growth, and the ability to examine the activity of specific targets. In the case of Ty3, PCR using primers annealing inside Ty3 and within or downstream of tDNAs allows rapid and specific measurement of retrotransposition (41). In a modification of this approach suitable for next generation sequencing, a short unique sequence was inserted into the test Ty3. After expression, genomic DNA fragmentation and ligation of adapters, PCR using primers annealing to the Ty3 tag and downstream adapter allowed next generation sequencing of transposon–genome junction sites (Tpn–Seq) (Figure 3, right panel) (38) (our unpublished results).

TY3 VIRUS-LIKE PARTICLES: COMPONENTS AND ASSEMBLY

Polyprotein Precursor Expression and Maturation

Ty3 mRNA is translated into a 290-amino-acid (aa) (34 kDa) Gag3 structural polyprotein precursor from the first ORF, *GAG3*, and a 1547-aa (178 kDa) Gag3–Pol3 fusion polyprotein precursor by a programmed plus-one frameshift between *GAG3* and *POL3* (28, 42). Gag3 contains structural domains CA, SP, and NC. The Gag3–Pol3 polyprotein precursor, in addition to Gag3 domains, contains critical catalytic domains PR, RT, and IN. These are encoded in that order, which is the same as in retroviruses. The N-termini of these proteins have been determined by protein sequencing and mass spectrometry (43, 44).

Precursor Gag3 and Gag3–Pol3 polyproteins multimerize via Gag3–Gag3 interactions into immature VLPs (44, 45). Within VLPs, PR is activated to process Gag3 and Gag3–Pol3 into protein domains of the mature VLPs (Figure 2B). Ty3 PR is a 15-kDa aspartyl protease with DSG active site residues, similar to retroviral PR, and required for polyprotein processing (43). Based on the retrovirus precedent, it is likely that dimerization of PR within the VLP promotes its activation. In addition to PR itself, Gag3–Pol3 processing by PR generates 55-kDa Ty3 RT and 61-kDa and 58-kDa IN proteins (46). The significance of the difference in N-terminal processing between these IN forms is

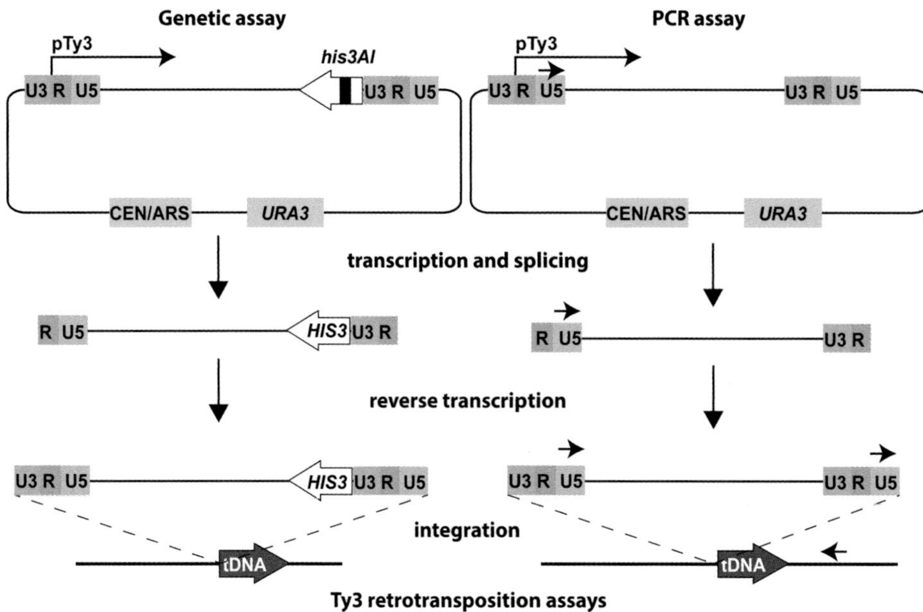

Figure 3 Retrotransposition assays. Ty3 retrotransposition can be assayed using the *his3AI* reporter embedded in the mobilized Ty3 (genetic) or by PCR. In the case of the Genetic Assay (left panel), Ty3 transcription is accompanied by splicing of a synthetic intron which is antisense to the *HIS3* marker, preventing *HIS3* from being productively expressed. After Ty3 transcription, splicing and reverse transcription, the intronless *HIS3* gene is expressed and cells in which transposition has occurred are selected on medium lacking histidine. Alternatively, retrotransposition of a tagged Ty3 element can be monitored by PCR assay (right panel) using one primer complementary to the Ty3 tag and one complementary to a sequence present in a tDNA family or in the unique sequence downstream of any tDNA target. doi:10.1128/microbiolspec.MDNA3-0057-2014.f3

unknown. However, the 61-kDa IN species expressed as a recombinant protein is capable of *in vitro* strand-transfer activity (47). Within cells expressing wild-type (WT) Ty3, VLPs are heterogeneous, representing a mixture of mature and immature forms (43).

In addition to catalytic domains with known functions encoded in *POL3*, a Junction (J) domain inferred to be 10 kDa is encoded between PR- and RT-coding regions (43, 48). This includes a 26-aa motif, which occurs twice in YGRWTy3-1 and three times in YILWTy3-1. Deletion of J within boundaries chosen to reconstruct the processing site decreased Ty3 cDNA levels dramatically, but did not abrogate retrotransposition (48). Chimeras constructed between the YGRWTy3-1 and YILWTy3-1 elements showed that the J domain does not account for the inactivity of the YILWTy3-1 (12).

Fluorescence microscopy provides a powerful tool for examining the appearance and subcellular localization of retrotransposon proteins and RNA. In cells undergoing pheromone induction, a monomeric cherry-tagged version of the Gag3 precursor polyprotein appears within about 10 min as a diffuse cytoplasmic signal and a green fluorescent protein (GFP) tagged version of the full-length Gag3–Pol3 precursor polyprotein appears within about 30 min (our unpublished results) (49). Immunoblotting shows that processed products, likely reflecting VLP formation, become more abundant than Gag3 between 2 to 4 hours (our unpublished results), and clusters of VLPs can also be detected during this time by transmission electron microscopy (our unpublished results). Anti-GFP immunoelectron microscopy shows that Ty3 proteins tagged with GFP localize to these clusters (49). Southern blotting detects cDNA within 6 hours.

Frameshifting

Greater amounts of structural than catalytic proteins are required to assemble VLPs (28). The ratio of these components is controlled by a programmed plus-one frameshift near the downstream end of *GAG3*. The relative abundance of Gag3 and the presence of this domain in Gag3–Pol3 not only drives assembly, but insures inclusion of catalytic proteins (28, 45, 50). The key frameshifting event is mediated by a 7-nucleotide (nt) sequence (GCG AGU U) read as Ala-Ser within the *GAG3* frame. Within this sequence, the Ser codon (AGU) is decoded by a rare tRNA, stimulating plus-one frameshifting into the GCG GUU frame. This sequence is translated as Ala-Val, resulting in formation of full-length Gag3–Pol3 (42). Shifting is further stimulated by a downstream enhancer structural element of 12 nt. The Ty3 recoding shift is essentially executed by the incoming amino-acyl tRNA. Hence, this programmed frameshift differs from both the retrovirus "slippery" minus-one frameshift and the plus-one Ty1 frameshift, which depend upon shifting of ribosome-bound tRNAs. The inefficiency of this process causes the Gag3–Pol3 precursor to be produced at 5% to 10% of the level of structural proteins (28, 42). Given this mechanism, it is not surprising that growth conditions affect the frequency of frameshifting and modulate retrotransposition. For example, frameshifting increases in cells grown in galactose-containing medium (51).

Virus-Like Particle Structure

Immature VLPs contain Gag3, Gag3–Pol3, dimeric Ty3 gRNA (52), tRNAs including Ty3 primer initiator $tRNA^{Met}$ ($tRNA^{Met}i$) (53) and probably other host RNAs and proteins (52, 54). Gag3 is the major structural protein of immature Ty3 VLPs, and its expression in *Escherichia coli* is sufficient to achieve formation of aberrant particles (55). Formation of VLPs correlates strongly with ability of mutants to undergo proteolytic processing and reverse transcription. The simplest interpretation is that the VLP shell not only triggers processing to activate protein domain functions, but also selectively sequesters, concentrates and protects enzymes and substrates. Ty3 PR and RT active site mutants, and WT VLP populations have been characterized by transmission electron microscopy and atomic force microscopy (44, 56–58). Immature VLPs generated from Ty3 PR active site mutants are relatively uniform, and appear roughly spherical with diameters measured by atomic force microscopy of 44 to 53 nm. Striking physical features of the VLPs, particularly evident in atomic force microscopy preparations of immature forms, are knob-like capsomeres with six-fold symmetry interspersed with knobs of five-fold symmetry suggesting a pseudo-icosahedral structure. Transmission electron microscopy of immature VLPs shows an outer layer that is presumably the Gag3 and associated RNA. Populations of VLPs in cells expressing WT Ty3 are more heterogeneous, with diameters ranging from 25 to 52 nm. Many of these have electron-dense centers, presumably reflecting NC and complexed RNA released from the shell. Smaller particles could be cores that have generated cDNA and have remodeled in some way.

Ty3 VLPs are extremely stable, and Gag3 spontaneously aggregates or forms particles (unpublished data). Ty1 particles have also been isolated and studied extensively, and must be similarly stable (59). In contrast, retroviral cores undergo maturation coincident with budding and so are enveloped in the mature form. These core particles can be stripped of envelope and readily dissociated (60). VLP stability may be a general feature of retrotransposons necessitated by intracellular maturation in the absence of envelope.

Capsid

The 290-aa Gag3 is processed into 207-aa CA (processed to remove the N-terminal Met), 26-aa SP and 57-aa NC species (43, 44). A CA-SP species is also prominent among Gag3 products of processing (56). CA is acetylated on the N-terminal Ser (unpublished data). Although retroviral Gag CA domains are not generally conserved in sequence, they are remarkably similar in structure. They share N-terminal domain (NTD) and C-terminal domain (CTD) bundles of alpha helices (61, 62) with an intervening linker. Modeling suggests that, similar to retroviral CA domains, the CA domain of Ty3 Gag3 forms alpha helical bundles with the NTD contributing the outer surface of the VLP (58). The N-terminal 20-aa of Ty3 Gag3 contains a late domain motif (YPXL). Late motifs are named after their function "late" in the retrovirus lifecycle when structural proteins interact with "Endosomal Sorting Complexes Required for Transport" (ESCRT) proteins to sequester maturing VLPs into plasma membrane budding structures (63). Intriguingly although Ty3 has not been observed budding into intracellular membranes, density equilibrium centrifugation suggests that it is transiently associated with membranes (our unpublished results), and ESCRT proteins are required for Ty3 retrotransposition (64). Although retroviral CA domains are mainly conserved at the structural level, a "Major Homology Region" (MHR) is found in Gag CA domains and also occurs in Ty3 CA from amino acids 86 to 100 ($QGX_2EX_5FX_3LX_3H$) (65). The Ty3 MHR occurs N-terminal in Ty3 Gag3 compared with the MHR in retroviral Gag, but mutations in the MHR cause similar defects in Ty3 and retroviral particle morphogenesis.

Analysis of Gag3 mutants generated by Ala scanning identified residues that affect assembly as evaluated by transmission electron microscopy visualization of VLPs, RNA packaging, and concentration of Ty3 protein into foci. Assembly was particularly sensitive to disruption of the N-terminal 100 aa (58). Only a few individual mutations completely abrogated particle formation and in one case this was accompanied by appearance of extensive cytoplasmic filaments. Visualization with RFP and GFP fusions showed that these filaments contained Gag3 and at least two components of processing bodies (PB) (58). Two-hybrid analysis showed that interactions occur between the NTD and NTD, as well as between the NTD and CTD and that introduction of mutations with effects on assembly disrupted these interactions (45, 58).

Nucleocapsid

The 57-aa Ty3 NC domain of Gag3 is critical for capture of the gRNA. The NC NTD contains 17 basic aa and is followed by one copy of the conserved zinc knuckle $CX_2CX_4HX_4C$ present in one or two copies in retroviral NC domains (11, 50). Mutations in conserved residues within and around the zinc knuckle eliminated concentration of Ty3 RNA into retrosome foci, and significantly reduced gRNA incorporation into VLPs (packaging) and retrotransposition (57). Unexpectedly, substitution of the first and second, but not the third and fourth, residues of the CCHC motif caused accumulation of Gag3 in nuclear, partially condensed particles. Because substitution of any one of the CCHC residues should disrupt zinc binding and gRNA packaging, such disparate phenotypes suggest that this subdomain has more than one role in assembly. In addition, the finding that Gag3 accumulates efficiently in the nucleus of these mutants suggests that, similar to the scenario proposed for Rous sarcoma virus gRNA packaging (66, 67), Ty3 Gag3 unbound to gRNA might at some stage invade the nucleus to scavenge for gRNA. Potentially related scenarios for Gag perinuclear capture of newly exported gRNA are proposed for human immunodeficiency virus 1 (HIV-1) (68, 69) and Ty1 (70).

Spacer

A SP domain separates Gag3 CA and NC domains, and a similar situation exists in a subset of retrovirus Gag proteins, including those of HIV-1 and Rous sarcoma virus (71, 72). These SP domains are not conserved, but are thought to participate in assembly, possibly substituting for functions performed by the CTD of CA for retroviruses lacking SP (73) (reviewed in reference 11). Ty3 SP is distinguished from retrovirus SP domains in that it is longer and more acidic. As is the case for retroviruses, removal of SP is important for morphogenesis. Mutations at the Ty3 SP–NC junction that disrupt PR processing also block processing of CA–SP, but the converse is not the case. This situation suggests that Gag3 processing is ordered (56), similar to what occurs for HIV-1 (74). Because failure to produce free NC does not block gRNA packaging but blocks appearance of cDNA, these observations argue that the NC domain of immature Gag3 functions in packaging, but mature NC must chaperone nucleic acid interactions during reverse transcription. Disruption of CA–SP processing also blocks retrotransposition but subsequent to cDNA production. Surprisingly, deletion of the SP domain by reconstructing a hybrid processed CA–NC mutant, allowed a low level of retrotransposition, indicating that SP is not absolutely essential. In contrast, conversion of acidic SP residues to Ala disrupted VLP formation and eliminated retrotransposition. Together these observations are consistent with a model in which the Ty3 acidic SP domain could interact with the basic NTD of NC in a nonessential condensation function that promotes assembly. However, if SP is present, it is essential that it be released from CA to allow progression through some step subsequent to reverse transcription, such as VLP uncoating.

Genomic RNA

The Ty3 gRNA contains *cis*-acting sequences, enabling Gag3-mediated packaging (*psi*), priming of reverse transcription, and integration. Ty3 gRNA is dimeric in VLPs (52) and by analogy with retroviruses has roles in packaging and reverse transcription. The primer for Ty3 minus-strand reverse transcription is $tRNA^{Met}i$ (53), which anneals to complementary primer binding site (PBS) split between just downstream of U5 and within U3 (75). The primer for plus-strand reverse transcription is a PPT just upstream of U3. *Cis*-acting sequences were collectively first delimited by determining that Ty3 nucleotides 429 to 4979 of the 5.2-kb RNA were dispensable for retrotransposition in the presence of a helper Ty3 to provide proteins (28). Subsequently, an assay in which packaged RNA was protected from exonuclease digestion further delimited the Ty3 packaging sequence (*psi*) (76). These experiments showed that an abbreviated RNA with the Ty3 5′ untranslated region (UTR) lacking both the upstream and downstream PBS and having a heterologous 3′ UTR is sufficient to mediate packaging of a short RNA encoding Gag3 (76) (Figure 2C). At least a subset of retroviral *psi* sequences, including that of murine leukemia virus, are upstream sequences containing hairpin structures. Within these hairpin loops, short inverted repeats seed gRNA dimer formation, and trigger annealing of the RNA sequences within the opposing gRNA hairpins. The resulting gRNA dimers expose binding sites for the NC domain of Gag (77). Although such features are not identified for Ty3, the 5′ end of the gRNA does contain the region required in *cis* for protection by Gag3.

To better understand the subcellular localization of VLP assembly, Ty3 RNA was fluorescently tagged. RNA tagged with tandem hairpin binding sites for RNA MS2 bacteriophage CApsid (MS2-CA) can be visualized if expressed in cells together with MS2-CA fused to a fluorescent protein reporter (78). Ty3 gRNA tagged with the MS2-CA binding site is active for retrotransposition and was visualized with MS2 CA fused to red fluorescent protein (RFP) or GFP (49, 57, 76). Examination of cells expressing Ty3 RNA visualized with MS2-CA-GFP showed that gRNA colocalizes in foci with Gag3 identified using anti-CA antibodies (49).

Retrosome Localization of Virus-Like Particle Assembly

One of the striking features of *Drosophila* (79) and *S. cerevisiae* (54) expressing Ty3/Gypsy and Ty1/Copia retrotransposons (80) is occurrence of VLP clusters or retrosomes. Several observations support the interpretation that, at least in the case of Ty3, retrosomes are associated with assembly of Ty3 protein and RNA into VLPs (Figure 1A). First, Ty3 RNA, proteins, and VLPs accumulate in clusters and individual VLPs are rarely observed outside clusters; second, mutations that disrupt assembly either abrogate formation of clusters or cause less dense clusters of poorly formed particles; third, deletion of genes encoding host proteins required for assembly also disrupts cluster formation (49, 56–58) (our unpublished results). Studies of Ty3 retrosome formation have used antibodies against individual Ty3 proteins as well as GFP, RFP, and monomeric cherry (mCherry) as *in vivo* reporters for Ty3 RNA and protein (49). Ty3 Gag3 and Gag3–Pol3 have been fused alternatively to GFP, RFP or mCherry. Within 1 hour of Ty3 induction, fluorescently tagged Ty3 gRNA and proteins progress from diffuse fluorescence to small foci, which aggregate over the next several hours into larger foci containing VLPs and representing retrosomes (49) (our unpublished results). Quantitation of the ability of Ty3 RNAs lacking various Ty3 gRNA attributes (e.g., frameshift, *POL3* ORF, PBS) or having UTRs from heterologous RNAs showed that in the presence of Gag3, RNAs localize to foci if they contain either the 5′ Ty3 UTR or a second long, poorly translated ORF (76). However, *GAG3–POL3* RNA in which the Ty3 UTR are substituted with heterologous UTRs is not efficiently packaged, consistent with the 5′ UTR containing *psi*. Hence, localization to foci does not insure assembly of Ty3 RNA into particles.

Retrosome Association with RNA Processing Body Components

Screens for Ty1 and Ty3 host factors identified conserved components of RNA PB as factors required for WT levels of retrotransposition (64, 81). PB are RNA granules containing concentrations of RNA deadenylation-dependent degradation components including deadenylation enzymes, translation suppressors, decapping activators and enzymes, helicases, and exonuclease. PB were initially identified as sites of mRNA degradation (82), but this is unlikely to be their exclusive function. A subset of RNAs in PB is competent to resume polysome-associated translation (83). In addition, at least some poly(A)-dependent degradation occurs cotranslationally (84). Indeed further investigation indicated that PB and stress granule (SG) ribonucleoprotein particles, which reversibly sequester nontranslating RNAs during stress (85), share a subset of components (86). A revised interpretation is that these granules represent different ends of a spectrum of ribonucleoprotein complexes (87). PB proteins colocalizing with Ty3 retrosomes (49) include translational repressors Pat1 and Dhh1 (88–90); decapping activator Lsm1 (91); decapping protein Dcp2 (92); and 5′ exonuclease Xrn1 (87, 93–95).

Galactose induction of Ty3 expression causes formation and/or enlargement of PB, suggesting that PB proteins interact directly with Ty3 proteins and/or RNA. Examination of galactose-induced Ty3 protein and RNA fluorescent reporters in cells expressing PB-GFP reporters showed that Ty3 RNA and protein and PB components colocalize (11, 49). PB formation is also observed under conditions independent of galactose induction of Ty3. These conditions include nutritional stress (86), stationary phase (96), and mating (97). Ty3 protein and PB factors also colocalize in mating cells (49).

On the surface, it is surprising that retroelement assembly is associated with factors engaged in RNA degradation. However, as discussed above, not all PB targeted RNAs are degraded and the natural sequestration of RNAs in PB away from translation could provide a mechanism for Ty3 RNA to transition from translation into assembly. The *POL3* ORF, downstream of the frameshift would have a lower density of ribosomes and could be subject to binding by PB translational repressors. Translational repression might enhance access of Gag3 to the 5′ packaging site, thereby protecting the 5′ end from decapping and degradation. Greater interaction with PB or trafficking factors would then promote assembly by concentrating VLP components (76). In the absence of such enhanced concentration, Ty3 proteins and RNA might fail to assemble. PB factors are also required for Ty1 assembly (98, 99). However, Ty1 retrosomes are not completely coincident with PB or SG foci (98, 99) (see Curcio, Lutz and Lesage, Chapter 41).

REVERSE TRANSCRIPTASE AND REVERSE TRANSCRIPTION

Reverse transcription is the defining feature of the retroelement replication. Non-LTR retrotransposon RT containing solely the polymerase domain (Pol) likely preceded LTR retrotransposon RTs that contain Pol and RNaseH domains (3). The Pol catalytic activity is responsible for cDNA production from both plus-strand RNA and minus-strand cDNA templates. The RNaseH domain is responsible for degradation of template RNA in DNA–RNA heteroduplexes as well as the cleavage at PPT to create the plus-strand primer. Pol and RNaseH domains, each have conserved carboxylate triad active sites competent to chelate two Mg^{2+} or Mn^{2+} ions. However, these reactions differ in both their structural context and in the reactions they mediate: in Pol, a hydroxyl attacks a dNTP; whereas in RNaseH, H_2O is activated to hydrolyze a polynucleotide chain (see Hughes, Chapter 46) (100, 101).

Relationship of Ty3 RT to Other RT Enzymes

Despite conservation of RT Pol and RNaseH domains among retroviral RT, there is diversity in quaternary structure: gamma-retroviruses, such as murine leukemia virus, have monomeric RT (Pol-RNaseH); alpha-retroviruses, such as Rous sarcoma virus, use an RT/RT-IN heterodimer; and lenti-retroviruses such as HIV-1, use a Pol-RNaseH/Pol (p61/p56) to highlight a few variations (see Hughes, Chapter 46; Skalka, Chapter 48). Retroviruses likely evolved from progenitor LTR retrotransposons related to Ty3/Gypsy. Ty3 RT has about 25% identity in the Pol domain with monomeric murine leukemia virus RT (9). Ty3/Gypsy retrotransposon RNaseH domains occur immediately after the Pol domain. Retroviral RNaseH domains are not only separated from the Pol domain by a RNaseH fold, that lacks the key carboxylate triad, but are actually more closely related to RNaseH domains of cellular enzymes than RNaseH domains of LTR retrotransposons. An attractive explanation is that retroviruses acquired a cellular RNaseH domain, but retained the progenitor retroelement RNaseH domain as a catalytically inactive connector between Pol and RNaseH (102).

Ty3 Reverse Transcriptase Structure and Active Sites

In vivo, Ty3 Gag3–Pol3 is processed into 55-kDa RT and 116-kDa RT-IN fusion species (46). *In vitro* recombinant 55-kDa RT displays an ability to extend synthetic primers on an RNA template comparable to that of recombinant retroviral RT (103, 104). Inspection shows Ty3 RT amino acids D151-X_{59}-Y211-L212-D213-D214 consistent with the conserved retroviral Pol active-site motif, D-Xn-Y-L-D-D. The Ty3 protein sequence D358-E401-D426-H427-Y459-D469 is consistent with the RNaseH carboxylate triad in other members of the Ty3/Gypsy family, D-E-DH-Y-D (104). The conserved D-E-D triad is required for metal binding and activity, as are the His and Tyr residues (105). However, consistent with different origins of the RNaseH domain, the retroviral RNaseH motif, D-E-D-HD, differs from the Ty3/Gypsy motif, in that Tyr is lacking and the His residue is moved away from the carboxylate triad (e.g. HIV-1 D443-E478-D498-H539-D549).

The crystal structure of the Ty3 55-kDa RT bound to a 16-bp RNA/DNA heteroduplex PPT was the first retrotransposon RT crystal structure. It revealed a novel, substrate-induced dimer structure (106). The bound nucleic acid is intermediate between the A and B forms, a polynucleotide duplex with little structural deformation. In the Ty3 RT, subunits "A" and "B" have similar conserved subdomain folds, but these are arranged asymmetrically. Subunit A displays the characteristic Pol right-handed topology with palm subdomain containing the active site carboxylates, fingers stabilizing the RNA primer and thumb interacting with template DNA. In contrast, subunit B thumb and RNaseH subdomains are rotated by approximately 90°. This arrangement obstructs interaction between the subunit A RNaseH active site and scissile phosphate, but leaves an unobstructed path for subunit B RNaseH. Altogether the structure suggests that catalytic activities are shared between the two subunits. In support of that interpretation, RT proteins mutated in the half-dimer interface or in the RNaseH active site lack *in vitro* activity, but complement when combined *in vitro*. Hence, Ty3 and HIV-1 RT have similar conserved subdomains, but the overall structures show that subunits play different roles. In the case of Ty3 RT, substrate binding induces RT dimerization, whereas in the case of HIV-1, RT forms a heterodimer in the absence of substrate. Ty3 RT enzymatic activity is split between Pol, contributed by subunit A, and RNaseH contributed by subunit B. In contrast, HIV-1 Pol and RNaseH activities are contained in p66 and the p51 subunit functions as a scaffold. Nonetheless, p51 of HIV RT and subunit B of Ty3 RT have similar subdomain arrangements, including the positions of the vestigial RNaseH p51 connector domain and active Ty3 subunit B RNaseH.

Mechanism of Reverse Transcription

The sites for initiation of reverse transcription of Ty3 minus-strand and plus-strand cDNA synthesis intermediates and full-length cDNA are identified. Overall, observations are consistent with the sequence of events proposed for replication of retrovirus gRNA (Figure 2C, see legend for detailed description) (see Skalka, Chapter 48; Hughes, Chapter 46). Ty3 minus-strand reverse transcription is primed from $tRNA^{Met}i$ (53) annealed to a bipartite PBS (75) and the plus-strand is primed from a downstream PPT. Overall, Ty3 RT performs an amazingly diverse set of activities: RNA- and DNA-dependent DNA synthesis, strand transfers, degradation of the RNA component of the RNA/DNA heteroduplex, and RNA endo-nucleolytic cleavages to create the plus-strand primer.

Ty3 cDNA Primers

Ty3 minus-strand synthesis is primed by $tRNA^{Met}i$. However, this tRNA rather than annealing to a 17-bp PBS as is the case for retrovirus tRNA primers, binds to a bipartite PBS, an 8-bp segment 2 nt downstream of U5 and two adjacent segments thousands of bases downstream within U3, for an additional 23 nt (75). Ty1 also uses a bipartite $tRNA^{Met}i$ PBS with a short upstream segment and two longer downstream segments. However, in the case of Ty1, the bipartite PBS is completely in the 5′ end (107). Ty retroelements differ from retroviruses, which typically have 17–18 bases of uninterrupted complementarity to the tRNA primer just downstream of U5 (108).

Ty3, Ty1 (109) and retrovirus plus-strand primers are PPT, but have distinguishing properties. The Ty3 PPT is GAGAGAGAGGAAGA. It differs from that of retroviral PPT in being slightly shorter and lacking homopolymeric tracts. Based on the sensitivity of RNaseH activities to mutation and the crystal structure, Ty3 RT thumb residues Q290-F292-G294-N297-Y298 (corresponding to HIV-1 RT Q258-L260-G262-N265-W266) interact with the DNA backbone and position the PPT into the RNaseH active site (106, 110). Nuclear magnetic resonance of unbound Ty3 PPT heteroduplex and the crystal structure of the RT-bound PPT heteroduplex agree on mostly base-paired and A-form duplex lacking major structural distortions. However, nuclear magnetic resonance revealed sugar puckers at position +1 (G) of the RNA strand (where *in vitro* cleavage is between −1 and +1) from C3′ endo to mixed C3′/C2′ endo (111). Perturbations using locked nucleic acid analogs to constrain sugar rings and abasic tetrahydrofuran linkages, which lack base pairing, indicated that the the 5′ and 3′ ends of the PPT are more sensitive than internal positions to substitution with novel constituents. This suggests that PPT ends contribute to RNaseH recognition (112, 113). Underscoring these apparent distinctions from retroviral PPT, Ty3 RNaseH fails to recognize at least one example of a retrovirus PPT, that of HIV (AAAAGAAAAGGGGGGA) (104, 114).

RNA–RNA Interactions Supporting Strand Transfer

Packaging into core particles and participation in strong-stop cDNA intermediates strand transfers during reverse transcription are two requirements of LTR retrotransposon and retrovirus gRNAs. Dimer formation by gRNA and direct interactions between the 5′ and 3′ ends of the gRNAs, respectively, have been proposed to participate in these activities. It is noteworthy that $tRNA^{Met}i$ is shared as the minus-strand primer by distantly related elements Ty3 (53), Ty1 (53) and Ty5 (115). In the case of Ty3, which has a bipartite PBS so that annealed $tRNA^{Met}i$ bridges the 5′ and 3′ gRNA ends, the primer is proposed to facilitate strand transfer (75). An alternative proposal is that a transient covalent structure connects 5′ and 3′ ends, thereby supporting transfer of intermediates. In the case of Ty1 and Ty3, the lariat debranching enzyme, Dbr1, is required for WT levels of cDNA production (116). Based on this requirement, and consistent with *in vitro* observations, it was proposed that a 5′ : 3′ lariat transiently links the Ty1 gRNA ends (117). Although the evidence for Ty1 is incomplete (118), experiments with HIV-1, which also requires Dbr1, appear supportive of an intermediate lariat in an alternative reverse transcription pathway (119).

Ty3 gRNA Dimerization

Ty3 VLPs contain gRNA dimers (52). In the case of retroviruses, the dimerization initiation sequence maps within the *psi* packaging sequence, and this duplex structure is proposed to be a feature that identifies gRNA for packaging (reviewed in references 77 and 120). Gabus *et al.* (75) suggested that palindromic 5′ ends of the two bridging $tDNA^{Met}i$ molecules form an interface between the two Ty3 gRNAs. In addition to linking 5′ and 3′ ends, this would mediate dimerization of the gRNA. In an *in vitro* system with mini-Ty3 element substrate, minus-strand strong-stop transfer and gRNA dimer formation were dependent upon the bipartite PBS and $tRNA^{Met}i$ primer (75, 103). However, the bipartite Ty3 PBS is not required for Ty3 gRNA packaging, suggesting either that dimer formation is not required as it is for retroviruses, or that $tRNA^{Met}i$ does not provide this interface (76). Ty1, which also uses $tRNA^{Met}i$ to prime reverse transcription, was similarly proposed to dimerize through the tRNA interface. However, recent analysis of Ty1 VLP RNA using Selective 2′ Hydroxyl Acylation Analyzed by Primer Extension (SHAPE) did not identify the proposed $tRNA^{Met}i$ interface, but rather a palindromic sequence within the 5′ end of the gRNA similar to the retroviral dimerization initiation sequence (121).

NUCLEAR ENTRY

Ty3 Accesses Integration Targets via the Nuclear Pore Complex

Ty3 VLPs assemble in the cytoplasm, but must access nuclear targets. In fungi the nuclear envelope does not break down during mitosis so that NPC are the gateway for multimolecular complexes. The yeast NPC is a conserved, eight-fold symmetric outer- and inner-ring and channel structure with a nuclear basket and a central meshwork of filamentous proteins rich in Phe-Gly (FG) repeats (122–124). NPC are dynamic and participate in multiple activities including gene gating, replication, and dsDNA break repair (29, 125). Because Ty3 VLPs are generally above the ~39-nm upper limit for complexes accommodated by the pore channel (126), it is likely that cDNA transit through the NPC is accompanied by significant remodeling also known as uncoating. This requirement is shared with fungal retroelements Ty1 (127, 128) and Tf1 (129). In nondividing animal cells, those retroviruses that are infectious, including HIV-1 and avian sarcoma leukosis virus), also access chromosomal DNA through the NPC (130, 131) (see Bushman and Craigie, Chapter 45). One unifying scenario is that the NPC plays a dynamic role in promoting entry by transforming both retrotransposon and VLPs and retrovirus cores into PIC.

Nuclear Pore Complex Components as Host Factors

Consistent with the idea that nuclear access of retroelements involves cooperation between retroelements and host factors, multiple factors that facilitate (cofactors) or antagonize (restriction factors) retrotransposition have been identified. Genomewide screens for Ty3 host factors identified members of the FG filaments (Nup159 and Nup100); and outer ring Nup84 complex (Nup120, Nup133, and Nup84) as Ty3 restriction factors. Gtr1, a negative regulator of Ran; importin Kap120; inner ring adapters Nup59 and Nup157; and FG filament Nup116 (combined with Nup100 constitutes hNup98) were identified as cofactors (64, 132, 133).

Deletions of genes encoding components of the Nup84 complex increased retrotransposition frequency. The NPC outer ring Nup84-Nup120-Nup133 (hNup107-hNup160-hNup107, respectively) subcomplex participates in NPC organogenesis, and yeast mutants lacking one of these components cluster pores on the nuclear envelope (134, 135). In *nup133Δ* and *nup120Δ* mutants, the Ty3 retrosome co-localizes with the clustered NPC. Physical colocalization between Ty3 VLP clusters and NPC is consistent with cumulative association of fluorescent Ty3 protein clusters on the periphery of DAPI-stained nuclei and with localization observed in some transmission electron microscopy images (133). Although it is attractive to think that mutations in the Nup84 complex enhance access of VLPs to NPC across a greater surface of the nuclear envelope, it is equally possible that distortion caused by loss of NPC outer ring coat proteins reduces the stringency of the NPC size filter.

Nups identified as Ty3 cofactors might mediate the VLP docking required for nuclear entry, or play a role in uncoating. Members of the GLFG subset (Nup110 and Nup116) of FG Nups were identified as Ty3 retrotransposition cofactors (64, 132). A complicating factor in examination of the role of FG Nups is the fact that individual proteins are essential and the different classes of FG repeats are redundant (122, 136). The roles of FG repeat families were dissected using mutants in which combinations of deletions of specific FG repeats have been achieved so that strains are viable, but lack certain types of FG repeats (136). These experiments highlighted a role for GLFG repeat Nups, Nup100 and Nup116 (133). *In vitro* investigation showed not only that assembly-competent immature and mature VLPs interact with GLFG repeat Nups, but that GLFG repeats specifically interact with recombinant Gag3 (133).

An attractive model is that FG hydrophobic filaments interact with the VLP surface and promote dissociation of CA. Although specific contacts for Gag3 subdomains in uncoating have not been identified, mutations in CA NTD and SP block retrotransposition subsequent to cDNA synthesis (56, 58). This phenotype is consistent with an uncoating defect.

Redundancy of Mechanisms of Nuclear Entry

A complicating factor in understanding nuclear entry of all retroelements is the multiplicity of potential VLP mediators of nuclear import. In the case of HIV-1, MA, Vpr, IN and CA have been implicated in nuclear entry (130, 137–139). In addition to Ty3 Gag (49) implication in nuclear entry, a bipartite basic nuclear localization sequence occurs in IN (amino acids 401–436 of IN) (140). Mutations in this motif block integration at a post-cDNA synthesis stage and also block nuclear localization of IN expressed as an independent protein. Hence, there could be redundant routes of PIC nuclear entry, or Gag3 and IN could cooperate in sequential NPC association and translocation to insure efficient and directional nuclear entry.

INTEGRASE AND INTEGRATION

Functions of Integrase

Mutations in Ty3 IN block retrotransposition at multiple steps: before or during cDNA synthesis, 3′ end processing, and strand transfer into the genome. The best understood of these steps is its role in targeted integration. The substrate for this activity is the ends of the Ty3 cDNA. As occurs for some retroelements including retroviruses, the Ty3 extra-chromosomal cDNA has two extra base pairs on each end and these are removed from the 3′ ends by IN before strand transfer (46). Integrated Ty3 has terminal 8-bp inverted repeats, the terminal dinucleotides of which are conserved among transposable elements (5′-*TG*TTGTAT/ATACAA*CA*-3′). The cDNA is joined to host DNA by IN with strand transfers staggered by 5 bp, which are filled in on each strand so that the integrated Ty3 ends are flanked by 5-bp direct repeats of host sequence (Figure 2C).

Comparison of retrotransposons and retroviruses suggests that the former benefit from integration that avoids disruption of host functions by avoiding ORFs or targeting heterochromatin or both, whereas the latter benefit from targeting expressed regions. Yeast Ty1–4 and *Dictyostelium* TRE5 and TRE3 are targeted to tDNAs (141–145); Ty5 integrates into heterochromatin (146); Maggy chromodomain element targets H3K9Me (147), and Tf1 targets intergenic regions (see Levin, Chapter 43). Of the characterized retroelements, Ty3 has the most specific pattern of integration.

Structural Features of Ty3 Integrase

Long terminal repeat retrotransposon and retrovirus integrases have related three-domain structures. Alignment of the Ty3 61-kDa IN with other retroelement IN proteins shows conservation of a catalytic core domain flanked by less-conserved NTD and CTD domains. The first crystal structure of a retrovirus IN, that of prototypic foamy virus IN complexed oligonucleotides representing cDNA ends and target DNA provided insight into this class of proteins (148, 149). Overall retroviral IN proteins function as a tetramer composed of dimers bound to the ends of the cDNA with the inner subunits participating in strand transfer. The NTD of each inner subunit functions in *trans* with the catalytic core domain of the other subunit. Outer subunits of each dimer are thought to have similar folds, but whether they have distinct functions in target association is not yet known (150). The crystal structure showed that prototypic foamy virus IN has an N-terminal extended domain of about 100 residues. This is not highly conserved among retroviral IN proteins but is thought to make contacts with the viral DNA backbone in the case of prototypic foamy virus IN. The IN NTD is typically defined by a conserved $HX_{3-7}H_{23-32}CX_2C$ motif near its N terminus. The position of the Ty3 zinc-binding motif, H93-X-H95-X_5-H101-X_{29}-C131-X_2-C134, indicates that Ty3 IN also has an N-terminal extended domain. The Ty3 catalytic core domain contains approximately a dozen residues highly conserved among Ty3/Gypsy and retrovirus IN proteins. These include Ty3 residues D164-X_{60}-D225-X_{35}-E261 representing the D-Xn-D-X_{35}-E Mg^{2+} binding motif common to polynucleotidyl esterases, including DNA cut-and-paste transposable elements (151, 152). Mutations in this motif do not affect assembly or reverse transcription, but abrogate 3′-end processing and integration (46). Similarity between Ty3 IN and retroviral IN proteins is sufficient to support *in vitro* strand-transfer activity of a small subset of 27 chimeras of Ty3, prototypic foamy virus, and HIV IN proteins (153).

Just C-terminal to the catalytic core domain, is the Ty3 GPF/Y motif, G398-P399-F400, which is found in a subset of retroelement IN proteins (154). In Tf1 this motif is located in a subdomain that participates in multimerization (155). Overall, the CTD of IN is poorly conserved among retroviruses and LTR retrotransposons (154).

Functions of Nonconserved Integrase Domains

Ala-scanning mutagenesis of the Ty3 IN NTD and CTD produced mutations in each domain that underwent processing, but lacked cDNA (156). This pattern of IN mutant phenotypes is similar to that of class II mutants of HIV-1 IN (157). These form immature cores. Retrovirus immature cores disassemble upon proteolytic processing and reassemble as mature cores inside the envelope (158). However, class II mutants fail to correctly re-assemble and fail to make cDNA (159). The similarities of the mutant phenotypes suggest that Ty3 IN NTD and CTD have a role in post-processing rearrangement of domains required for cDNA synthesis.

The Ty3 IN CTD contains a bipartite basic nuclear localization sequences (R401-RVVKKINDNAYELDLN SHKKKHRVINVQFLKKFVYR-436) sufficient to mediate nuclear import of ectopic IN (140). As indicated above, the Ty3 nuclear localization sequence is required at some post-cDNA synthesis step, consistent with a role in nuclear import (140).

A large fraction of Ty3/Gypsy elements have IN proteins with chromodomains in the CTD and these have been taxonomically classified as *Chromoviruses*. The chromodomain is ~50 aa and mediates interactions with dimethyl and trimethyl H3K9, a signature of heterochromatin (147). This domain occurs in Ty3/Gypsy Maggy and Tf1 IN (147), but has been replaced by an alternative domain in Ty3 (154). A plausible explanation for the replacement is that the targeted dimethyl and trimethyl H3K9 modification is lacking in *S. cerevisiae* (160).

Mechanism of Ty3 Targeting

Motivation for studies of Ty3 stemmed in large part from its remarkable integration specificity for the TSS of genomic RNAP III-transcribed genes including tDNAs (8, 141, 161), *SNR6* (U6 RNA), and *RDN5* genes (5S RNA) (162) (Figure 1B). The RNAP III TSS occurs in a favored context rather than an absolute consensus, and is flanked by transcription factors. This suggested early on that transcription factors might similarly position Ty3 integration. *In vivo S. cerevisiae* RNAP III transcription requires general factors TFIIIC and TFIIIB. TFIIIC associates with promoter elements boxA and boxB of about 10 bp each located 20 bp and ~20 to ~100 bp downstream of the TSS, respectively. TFIIIC directs sequence independent binding of TFIIIB upstream of the TSS (163, 164). TFIIIB is comprised of subunits TATA binding protein (TBP), Brf1 and Bdp1. It is considered the initiation factor because once bound, it is sufficient *in vitro* to mediate multiple rounds of RNAP III transcription (165, 166). *SNR6* and a subset of tDNAs, have an upstream TATA element that contributes to transcription and TSS selection (167, 168). *In vitro*, in a purified system with a TATA-containing template, TBP mediates binding of Brf1, and the two are sufficient for transcription (169). Genomewide chromatin immunoprecipitation–DNA microarray experiments showed that in growing cultures, yeast RNAP III genes are mostly occupied by TFIIIB, TFIIIC, and RNAP III (170–172).

The target requirements of Ty3 for integration are similar to requirements for transcription. Mutagenesis of targets showed that promoter elements are required for Ty3 integration *in vivo*, and underscored roles for the TATA box and boxA in copositioning the TSS and integration site (162, 173). *In vitro* Ty3 integration can be detected by PCR using primers binding within Ty3 and downstream of the target. Early experiments used VLPs as a source of cDNA and IN protein. Experiments showed that TFIIIC and TFIIIB are required to target integration to TATA-less tDNAs (174), but that TFIIIC is dispensable at genes where the upstream TATA element directly mediates binding of TBP and Brf1 (175, 176). *In vitro* at least, RNAP III is not required and competes with the PIC for access to targets (177). A refined system uses recombinant IN and a fusion of Brf1 and TBP constructed as a structural mimetic ($Brf1_{NTD}$-TBP-$Brf1_{CTD}$) (178). In this reaction, in the presence of Mg^{2+}, IN targets strand transfer of the 3′ end of a DNA duplex representing the terminal 20 bp of Ty3 cDNA into the *SNR6* TSS (47). Surprisingly, in the presence of Mn^{2+}, strand transfer independent of Brf1-TBP-Brf1 occurs within the plasmid target proximal to sequences resembling the Ty3 LTR ends, rather than the RNAP III TSS (47).

Genomewide Identification of Ty3 Targets and RNAP III Factor Binding Sites

Next generation sequencing enables genomewide assessment of retroelement targeting and use of Ty3 as a probe for RNAP III-transcribed genes, as well as an entrée to additional features of RNAP III activation (38). Next generation sequencing analysis of genomic DNA from cells representing ~10,000 independent Ty3-*HIS3* retrotransposition events identified ~300 significant insertion sites. These included 275 tDNAs, and the eight additional types of RNAP III-transcribed genes (*iYGR033C*, *RNA170*, *RPR1*, *SCR1*, *SNR52*, *SNR6*, *ZOD1*, and *RDN5*) and 18 integrations associated with LTRs. Because of the frequent occurrence of boxA and boxB sequences in the *S. cerevisiae* genome, it was possible that integrations would associate with boxA–boxB pairs resembling RNAP III promoters. Five insertion sites were identified that were not known RNAP III-transcribed genes. However, insertions were flanked by boxA- and boxB-like sequences. Two of the five sites were subcloned and shown to act as plasmid-borne targets. Targeting was sensitive to mutations in conserved positions of the boxB promoter element. There are boxB sites in the *S. cerevisiae* genome which are bound by "*E*xtra *T*FIIIC" (ETC), but little or no TFIIIB (170–172). These are not targeted by Ty3, further underscoring the requirement for TFIIIB of Ty3 integration.

One of the striking observations enabled by next generation sequencing of Ty3 insertion sites was that tDNA families, within which sequences are similar or identical, generated widely different numbers of sequencing reads (38). Ty3 could use identical genes at widely differing frequencies because of different chromatin contexts. RNAP III genes undergo specific types of epigenetic modifications and bind chromatin remodelers and condensins (179–190). In addition to possible differences in chromatin context, tDNAs also act as replication fork barriers (191), and subpopulations localize to the nucleolus (31, 192) and nuclear pores (30). Overall, it seems that we are just beginning to learn about the dynamic activities of RNAP III-transcribed genes. Ty3 offers a potentially exquisitely sensitive, indelible, *in vivo* probe for transient interactions between these genes and their chromatin caretakers.

Integration Host Factors

Although host RNAP III machinery plays a clear role in targeting, most genes encoding subunits of key complexes are essential and were not identified in host factor screens (64, 132). Exceptions were the La protein, a small-RNA chaperone (41) and Tfc1, a subunit of TFIIIC (193). Despite the association of La with RNAP III-transcribed precursor RNAs (194), the *lhp1Δ* mutant had reduced Ty3 cDNA, so that it seems likely that the role of La precedes the integration step. A mutagenized strain expressing a C-terminal truncation of Tfc1 was identified in a screen for genes which affected recovery of integrations that activated expression of the suppressor tDNA in the divergent tDNA target described above. Although Ty3 cDNA integrates into this target in both orientations, Ty3 integration occurred in only one orientation in this mutant. This result indicates that the Ty3 PIC itself is asymmetric. A combination of genetic experiments, mutagenesis and *in vitro* pull-down assays indicated that the N-terminal extended domain of IN interacts directly with Tfc1 (193).

The example of fungal retrotransposons and availability of RNAi and high throughput sequencing spurred productive investigation into host factors involved in targeting retrovirus integration. Although there are no known examples of the highly specific targeting or targeting of RNAP III genes, as observed for a subset of retrotransposons, retroviral integration is also nonrandom. Retrotransposons generally avoid coding regions and a subset target inactive chromatin via interactions between IN chromodomains and H3K9Me. In contrast, retroviruses integrate into chromatin associated with epigenetic marks of transcriptionally active chromatin (195). HIV-1 associates with chromatin by interaction between the IN catalytic core domain and LEDGF/P75. This factor interacts with an epigenetic mark of active chromatin, H3K36Me3, via the LEDGF PWWP domain (reviewed in reference 196) (see Bushman and Craigie, Chapter 45). The murine leukemia virus IN associates with targets via interaction between the IN CTD and Bromo-domain and ExtraTerminal domains proteins, which bind acetylated H3 and H4 tails (197) (see Bushman and Craigie, Chapter 45).

TY3–HOST INTERACTIONS

Genetic screens and cytological analyses have identified large numbers of candidate Ty host factor genes. Importantly, many of these cluster by function (64, 81, 132, 198–200). Overall there was about 15% overlap between the genes identified in Ty3 and Ty1 screens despite the relatively different assays used. Genes identified in Ty3 genomewide screens were grouped according to the gene ontology process (Figure 4). Cofactors assisting retrotransposition included those involved in catabolism, PB components, nucleoporins, checkpoint control, and vesicular trafficking proteins including ESCRT factors. The most striking class of restriction factors antagonizing retrotransposition were those related to DNA replication, particularly helicases, and certain RNA transporters, including Npl3 and Sac3. A small, but significant, percentage of cofactors and restriction factors that overlapped with those identified in Ty1 screens included replication factors (64, 199, 201), PB proteins (81, 198), and several RTT genes of various functions including chromatin remodelers and histone modification factors (64, 199).

Genes involved in replication and responses to DNA damage were the major class of restriction factors. Deletions of genes encoding replication factors Csm3, Rrm3, Mrc1, Rad6, Bre1, Sgs1 and Hpr5 enhanced Ty3 retrotransposition. Work in the Curcio laboratory on Ty1 showed that replication stress or DNA damage caused by the absence of a subset of DNA maintenance factors triggers increased Ty1 cDNA synthesis. This was traced to activation of checkpoint proteins Rad24 and Rad9, as the effect was significantly reduced in backgrounds lacking one of the checkpoint genes (201). Although epistasis analysis was not conducted

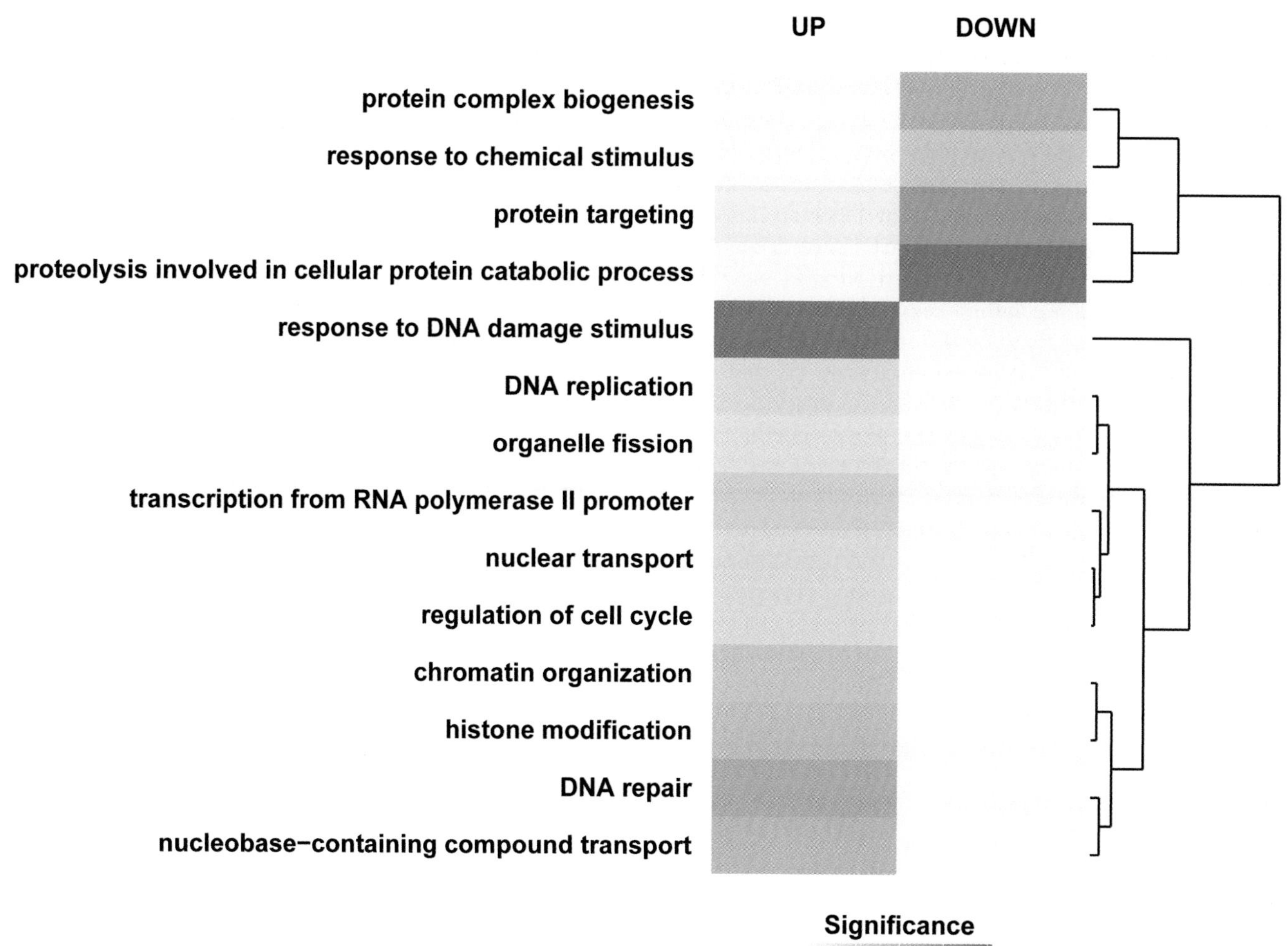

Figure 4 Hierarchical clustering of Ty3 cofactors and restriction factors by gene ontology groups. Gene ontology (GO) analysis was performed using the GO SLIM Biological Process mapping tool available through *Saccharomyces* Genome Database (http://www.yeastgenome.org). Knockout mutants identified as having either increased "Up" or decreased "Down" Ty3 retrotransposition phenotypes were analyzed for GO: Biological Process terms. Enriched categories were determined using chi-squared test. GO categories were considered enriched if two criteria were met: (i) the *P*-value was <0.05 and (ii) the number of genes in the enriched category exceeded 10% of the total number of genes in the Up or Down list. Enriched categories were converted to a heat map with hierarchical clustering using R; values represent the [–Log(*P*-value)] scaled from 0 (no significance) to 1, blue coloring reflects the intensity of significance. doi:10.1128/microbiolspec.MDNA3-0057-2014.f4

for Ty3, retrotransposition was decreased in the *rad24Δ* background (64) consistent with checkpoint control promoting Ty3 retrotransposition. However, tDNA targeting might also be affected directly by mutations in replication factors. Transfer RNA genes block replication forks coming from downstream (191). If incoming replication forks disrupt targeting, fork stalling could increase integration. Alternatively, fork breakage caused by loss of key helicases could trigger localization to NPC (125), enhancing accessibility of tDNA targets to incoming integrases.

As discussed above, Ty3 retrotransposition increased in mutants with deletions of genes encoding a subset of Nup84 complex components, *nup84Δ*, *nup120Δ*, and *nup133Δ* (64). We speculate that clustering of mutant pores and retrosomes could enhance nuclear access. Alternatively, these mutations could simply make the pore more permissive for PIC translocation (133).

Among the Ty3 cofactor genes were vesicular trafficking proteins. In budding yeast, ESCRT proteins were identified as essential for the late stages of trafficking of ubiquitinated surface proteins from the plasma membrane into multivesicular bodies (202). Subsequently, these factors were shown to be critical for retrovirus budding, which is the topological equivalent of multivesicular body formation (62). ESCRT proteins identified as Ty3 cofactors include ESCRTI (Vps28), ESCRTII, (Vps25, Vps22, Vps36), and ESCRTIII (Vps20, Snf7), as well as ESCRT-related proteins Vps4 and Bro1 (64) and Vps27. Although there is no evidence of Ty3 budding, it could associate with membranes in the course of morphogenesis. An interesting alternative possibility is raised by the recent implication of ESCRT components in NPC maintenance (203).

Retrovirus genomewide host screens, with HIV-1 being a major focus, have identified large numbers of host factors, among them, many nuclear proteins (see Bushman and Craigie, Chapter 45). Metadata analysis of nine such screens concluded that ~5% of all human genes encode factors related to HIV-1 and showed low, but statistically significant, overlap (204). As in the case of the Ty elements, overlap is limited by differences among screening strategies. Ty3 host factors (64, 132) were compared with factors identified in a subset of retrovirus siRNA host factor screens (205–208) using the Overlapper tool (http://hivsystemsbiology.org/GeneListOverlapper/app). These results are shown in Table 1. Functional overlaps include overlaps involved in vesicular trafficking, nuclear access (porins and translocation factors), DNA repair, ubiquitin signaling, mediator complex/transcription, RNA binding, and DNA and RNA unwinding (reviewed in references 62, 209–211).

Nucleoporins and NPC translocation factors were identified as host cofactors for Ty3 and for HIV-1, which infects nondividing cells. Prominent among these are the hydrophobic Nups enriched in FG repeats (122, 123). Human HsNup98, an FG Nup (212, 213), interacts with HIV-1 Gag and is required for HIV-1 nuclear entry. Budding yeast homologs of HsNup98, *Sc*Nup116, and *Sc*Nup100 are required for WT levels of Ty3 nuclear entry and recombinant *Sc*Nup116 interacts with Ty3 VLPs and recombinant Gag3 *in vitro*. A role in for this class of Nups in moving the uncoating VLP through the pore was proposed (133). *Schizosaccharomyces pombe* basket NUP, *Sp*Nup124, was isolated as a factor required for nuclear entry of Ty3/Gypsy element Tf1 Gag cores (129). The *S. cerevisiae* homolog, *Sc*Nup1, is essential, and so was underrepresented in Ty screens. However, HsNup153, the human homolog, is required for HIV-1 nuclear entry and mutations shift the integration bias, suggesting that basket Nups could affect NPC-proximal integration site selection (214). These observations suggest that interactions between Gag and FG Nups are a common feature of retroelement nuclear pore translocation and possibly uncoating.

Ribonucleoprotein granule components are common themes in retrotransposon and retrovirus assembly, but in seemingly distinct roles. During assembly, RNA and protein from LTR retrotransposon Tys and non-LTR retrotransposon LI LINEs, interact productively with PB and SG components (133, 215, 216). Mouse mammary tumor virus assembles intracellularly in association with YB-1-enriched ribonucleoprotein granules (217). YB-1 is a translational regulatory protein which accumulates in SG during stress, but is also found in PB. Expression of mouse mammary tumor virus Gag causes granules to increase in size, and these are distinct from SG and PB. HIV-1, which assembles into cores during extrusion into budding plasma membrane, actively antagonizes formation of SG, but associates with Staufen in cytoplasmic ribonucleoprotein particles (218). Host HIV-1 restriction factors, including apolipoprotein B mRNA editing enzyme catalytic polypeptide-like (APOBEC) cytidine deaminase family members, overexpressed Mov10 helicase (219–221), and RNAi components, associate with PB (210, 222). The PB factors also restrict endogenous retroviruses (223). However, studies disagree about the positive contributions of PB factors to retrovirus infection. DEAD box helicase DDX6 (*Sc*Dhh1) is ascribed a positive role in the intracellular packaging of prototypic foamy virus RNA, but studies differ as to roles in HIV-1 core assembly (224, 225). Overall these studies suggest that individual components of ribonucleoprotein granules, including PB and SG, interact with assembling retrovirus RNA and proteins, but assembly is not associated with these granules *per se*. Although the examples are limited, they are consistent with ribonucleoprotein granules sequestering proteins to protect host RNAs from inappropriate degradation or translational suppression, while providing a source of soluble proteins to interact with specific components of viruses. These associate with specialized granule components intracellularly, or complete assembly during budding on the plasma membrane.

The presence of a mixture of components in RNA granules that variously restrict and promote retroelement assembly might extend to the retroelement expression in germ cell-related lineages. In these cells, cycles of genome demethylation are accompanied by activation of retrotransposon transcription and post-transcriptional

Table 1 Host factors in common between Ty3 and HIV-1

ORF	Gene name	Full name	Ty3 ΔTPN[1]	Human homolog	Retrovirus host factor[2]
YBR284W	YBR284W		up, down	AMPD1, AMPD2, AMPD3	
YAR003W	SWD1	Set1c, WD40 repeat protein	up, down	RBBP5, WDR72	
YPR119W	CLB2	CycLin B	up	CCNA1, CCNA2, CCNB1, CCNB2, CCNB3, CCNE1, CCNE2, CNTD2	
YPL256C	CLN2	CycLiN	up	CCNB1, CCNB2, CNTD2	
YPL024W	RMI1	RecQ Mediated genome Instability	up		
YOR144C	ELG1	Enhanced Level of Genomic instability	up		
YOR039W	CKB2	Casein Kinase Beta' subunit	up	CSNK2B	
YNR052C	POP2	PGK promoter directed OverProduction	up	CNOT7, CNOT8	
YNL242W	ATG2	AuTophaGy related	up	ATG2A, ATG2B	
YNL138W	SRV2	Suppressor of RasVal19	up	CAP1, CAP2	
YMR190C	SGS1	Slow Growth Suppressor	up	BLM, RECQL, RECQL5, WRN	
YMR048W	CSM3	Chromosome Segregation in Meiosis	up		
YML051W	GAL80	GALactose metabolism	up	GFOD1, GFOD2	
YLR399C	BDF1	BromoDomain Factor	up	BRD2, BRD3, BRD4, BRDT, CREBBP, EP300	(207)
YLR268W	SEC22	SECretory	up	SEC22A, SEC22C	
YLR235C	YLR235C		up		
YLR015W	BRE2	BREfeldin A sensitivity	up	ASH2L	
YLL044W	YLL044W		up		
YLL030C	RRT7	Regulator of rDNA Transcription	up		
YLL002W	RTT109	Regulator of Ty1 Transposition	up		
YKR082W	NUP133	NUclear Pore	up	NUP133	(206)
YKL135C	APL2	clathrin Adaptor Protein complex Large chain	up	AP1B1, AP2B1, AP4B1	
YKL068W	NUP100	NUclear Pore	up	NUP98, NUPL1	(207)
YKL057C	NUP120	NUclear Pore	up		
YKL041W	VPS24	Vacuolar Protein Sorting	up	CHMP3	
YJL092W	SRS2	Suppressor of Rad Six	up		
YJL047C	RTT101	Regulator of Ty1 Transposition	up	CUL1	
YIR021W	MRS1	Mitochondrial RNA Splicing	up		
YIL090W	ICE2	Inheritance of Cortical ER	up		
YIL069C	RPS24B	Ribosomal Protein of the Small subunit	up	RPS24	
YIL015W	BAR1	BARrier to the alpha factor response	up	CTSD, CTSE, NAPSA, PGA3, PGA4, PGA5, PGC, REN	
YIL011W	TIR3	TIp1-Related	up		
YIL009C-A	EST3	Ever Shorter Telomeres	up		
YHR177W	YHR177W		up		
YHR031C	RRM3	rDNA Recombination Mutation	up	PIF1	
YHL030W	ECM29	ExtraCellular Mutant	up	KIAA0368	
YHL003C	LAG1	Longevity Assurance Gene	up	CERS1, CERS2, CERS3, CERS4, CERS5, CERS6, TRAM1, TRAM1L1, TRAM2	
YGR270W	YTA7	Yeast Tat-binding Analog	up	ATAD2, ATAD2B	
YGR056W	RSC1	Remodel the Structure of Chromatin	up	PBRM1	
YGL174W	BUD13	BUD site selection	up	BUD13	
YGL066W	SGF73	SaGa associated Factor, 73 kDa	up	ATXN7, ATXN7L1, ATXN7L2, ATXN7L3, ATXN7L3B	
YGL058W	RAD6	RADiation sensitive	up	UBE2A, UBE2B, UBE2S, UBE2W	(207)

(Continued)

Table 1 *(Continued)*

ORF	Gene name	Full name	Ty3 ΔTPN[1]	Human homolog	Retrovirus host factor[2]
YER155C	BEM2	Bud EMergence	up	ARHGAP21, ARHGAP23, ARHGAP35, CHN1, GMIP, HMHA1, RACGAP1, SYDE1, SYDE2	
YER060W-A	FCY22	FluoroCYtosine resistance	up		
YER057C	HMF1	Homologous Mmf1p Factor	up	HRSP12	
YEL013W	VAC8	VACuole related	up	ANKAR, ARMC3, ARMC4, CTNNB1, JUP	
YEL004W	YEA4		up	SLC35B4	
YDR448W	ADA2	transcriptional ADAptor	up	TADA2A, TADA2B	
YDR432W	NPL3	Nuclear Protein Localization	up	CIRBP, DMKN, HNRNPA1, HNRNPA1L2, HNRNPA2B1, HNRNPA3, LOR, RBMX, RBMXL1, RBMXL2, RBMY1A1, RBMY1B, RBMY1D, RBMY1E, RBMY1F, RBMY1J, SRSF1, SRSF9	
YDR290W	YDR290W		up		
YDR289C	RTT103	Regulator of Ty1 Transposition	up	RPRD1A, RPRD1B, RPRD2	
YDR176W	NGG1		up	TADA3	
YDR159W	SAC3	Suppressor of ACtin	up	MCM3AP, SAC3D1	
YDR074W	TPS2	Trehalose-6-phosphate PhoSphatase	up		
YDL116W	NUP84	NUclear Pore	up	NUP107	(206)
YDL074C	BRE1	BREfeldin A sensitivity	up	RNF20, RNF40	
YDL056W	MBP1	MluI-box Binding Protein	up		
YCR020W-B	HTL1	High-Temperature Lethal	up		
YCL061C	MRC1	Mediator of the Replication Checkpoint	up		
YCL037C	SRO9	Suppressor of rho3	up		
YBR277C	YBR277C		up		
YBR269C	FMP21	Found in Mitochondrial Proteome	up	C6orf57	
YBL094C	YBL094C		up		
YBL006C	LDB7	Low Dye Binding	up		
YPL084W	BRO1	BCK1-like Resistance to Osmotic shock	down	PDCD6IP, PTPN23, RHPN1, RHPN2	
YNL236W	SIN4	Switch INdependent	down		
YMR154C	RIM13	Regulator of IME2	down		
YPR173C	VPS4	Vacuolar Protein Sorting	down	KATNA1, KATNAL1, KATNAL2, VPS4A, VPS4B	(209)
YPL225W	YPL225W		down	PBDC1	
YPL150W	YPL150W		down	MELK	
YPL125W	KAP120	KAryoPherin	down	IPO11	
YPL065W	VPS28	Vacuolar Protein Sorting	down	VPS28	
YPL042C	SSN3	Suppressor of SNf1	down	CDK19, CDK8	
YPL002C	SNF8	Sucrose NonFermenting	down	SNF8	
YOR275C	RIM20	Regulator of IME2	down	PDCD6IP, PTPN23	
YOR270C	VPH1	Vacuolar pH	down	ATP6V0A1, ATP6V0A2, ATP6V0A4, TCIRG1	(206)
YOR055W	YOR055W		down		
YOR054C	VHS3	Viable in a Hal3 Sit4 background	down	PPCDC	
YOL083W	ATG34	AuTophaGy related	down		
YNR059W	MNT4	MaNnosylTransferase	down		

(Continued)

Table 1 Host factors in common between Ty3 and HIV-1 *(Continued)*

ORF	Gene name	Full name	Ty3 ΔTPN[1]	Human homolog	Retrovirus host factor[2]
YNR006W	VPS27	Vacuolar Protein Sorting	down	GGA1, GGA2, GGA3, HGS, TOM1, TOM1L1, TOM1L2, WDFY1, WDFY2, ZFYVE21	(209) (206)
YNL224C	SQS1	SQuelch of Splicing suppression	down	SON	
YNL136W	EAF7	Esa1-Associated Factor	down	MRGBP	
YNL059C	ARP5	Actin-Related Protein	down	ACTR5	
YMR158W-B	YMR158W-B		down		
YMR123W	PKR1	Pichia farinosa Killer toxin Resistance	down		
YMR116C	ASC1	Absence of growth Suppressor of Cyp1	down	GNB2L1, WDR31	
YMR077C	VPS20	Vacuolar Protein Sorting	down	CHMP6, CHMP7	
YMR058W	FET3	FErrous Transport	down		
YML121W	GTR1	GTp binding protein Resemblance	down	RRAGA, RRAGB	(206)
YML097C	VPS9	Vacuolar Protein Sorting	down	GAPVD1, RABGEF1, RIN1, RIN2, RIN3, RINL, VPS9D1	(206)
YML081C-A	ATP18	ATP synthase	down		
YML073C	RPL6A	Ribosomal Protein of the Large subunit	down	RPL6	
YLR423C	ATG17	AuTophaGy related	down		
YLR417W	VPS36	Vacuolar Protein Sorting	down	VPS36	
YLR242C	ARV1	ARE2 Required for Viability	down	ARV1	
YLR025W	SNF7	Sucrose NonFermenting	down	CHMP4A, CHMP4B, CHMP4C	
YKR099W	BAS1	BASal	down	MYB, MYBL1, MYBL2, SNAPC4	
YKR020W	VPS51	Vacuolar Protein Sorting	down		
YKL149C	DBR1	DeBRanching	down	DBR1	
YJR137C	MET5	METhionine requiring	down		
YJR121W	ATP2	ATP synthase	down	ATP5A1, ATP5B	(209)
YJR120W	YJR120W		down		
YJR102C	VPS25	Vacuolar Protein Sorting	down	VPS25	
YJR032W	CPR7	Cyclosporin-sensitive Proline Rotamase	down	NKTR, PPID, PPIG	
YJL006C	CTK2	Carboxy-Terminal domain Kinase	down	CCNT1, CCNT2	(209) (206)
YIL134W	FLX1	FLavin eXchange	down	SLC25A32	
YIL115C	NUP159	NUclear Pore	down		
YIL097W	FYV10	Function required for Yeast Viability	down	MAEA	
YIL073C	SPO22	SPOrulation	down		
YHR082C	KSP1		down	CHEK2	
YHR026W	VMA16	Vacuolar Membrane Atpase	down	ATP6V0B	(208)
YHR008C	SOD2	SuperOxide Dismutase	down	SOD2	
YGR246C	BRF1	B-Related Factor	down	BRF1	
YGR188C	BUB1	Budding Uninhibited by Benzimidazole	down	BUB1, BUB1B	
YGR167W	CLC1	Clathrin Light Chain	down	CLTA, CLTB	(206)
YGL212W	VAM7	VAcuolar Morphogenesis	down	STX8	
YGL173C	XRN1	eXoRiboNuclease	down	XRN1	
YGL156W	AMS1	Alpha-MannoSidase	down	MAN2C1	
YGL078C	DBP3	Dead Box Protein	down	DDX21, DDX50	(208)
YGL057C	GEP7	GEnetic interactors of Prohibitins	down		
YFR019W	FAB1	Forms Aploid and Binucleate cells	down	PIKFYVE, ZFYVE21, ZFYVE28	
YFR012W	DCV1	Demands Cdc28 kinase activity for Viability	down		

(Continued)

Table 1 *(Continued)*

ORF	Gene name	Full name	Ty3 ΔTPN[1]	Human homolog	Retrovirus host factor[2]
YFR009W	GCN20	General Control Nonderepressible	down	ABCF1, ABCF3	
YFL007W	BLM10	BLeoMycin resistance	down	PSME4	
YFL001W	DEG1	DEpressed Growth rate	down	PUS3	
YER173W	RAD24	RADiation sensitive	down	RAD17	
YER148W	SPT15	SuPpressor of Ty	down	TBP, TBPL1, TBPL2	(209)
YER145C	FTR1	Fe TRansporter	down		
YER105C	NUP157	NUclear Pore	down	NUP155	(206)
YEL044W	IES6	Ino Eighty Subunit	down	INO80C	
YDR493W	MZM1	Mitochondrial Zinc Maintenance	down		
YDR323C	PEP7	carboxyPEPtidase Y-deficient	down	ZFYVE20	
YDR139C	RUB1	Related to UBiquitin	down	NEDD8	
YDR067C	OCA6	Oxidant-induced Cell-cycle Arrest	down		
YDL160C	DHH1	DEAD box Helicase Homolog	down	DDX6	
YDL159W	STE7	STErile	down	MAP2K1, MAP2K2, MAP2K3, SBK2, SLK, STK10, TAOK1, TAOK2, TAOK3	(206)
YDL118W	YDL118W		down		
YDL088C	ASM4		down	NUP35	
YBR293W	VBA2	Vacuolar Basic Amino acid transporter	down	SLC18A1, SLC18A2, SLC18A3, SLC18B1	
YBR263W	SHM1	Serine HydroxyMethyltransferase	down	SHMT1, SHMT2	
YBR262C	AIM5	Altered Inheritance rate of Mitochondria	down		
YBR103W	SIF2	Sir4p-Interacting Factor	down	TBL1X, TBL1XR1, TBL1Y, THOC3, WDR17	
YBR044C	TCM62	TriCarboxylic acid cycle Mutant	down	HSPD1	
YBR021W	FUR4	5-FlUoRouridine sensitivity	down		
YBL024W	NCL1	NuCLear protein	down	NSUN2	
YBL007C	SLA1	Synthetic Lethal with ABP1	down	EPS15, EPS15L1, GRAP, GRAP2, GRB2, ITSN1, ITSN2, REPS1, REPS2, SLA, SLA2	
YAL027W	SAW1	Single-strand Annealing Weakened	down		
YAL019W	FUN30	Function Unknown Now	down	SMARCAD1	
YAL013W	DEP1	Disability in regulation of Expression of genes involved in Phospholipid biosynthesis	down	BFSP2, DES, GFAP, IFLTD1, INA, KRT1, KRT10, KRT12, KRT13, KRT14, KRT15, KRT16, KRT17, KRT18, KRT19, KRT2, KRT20, KRT23, KRT24, KRT25, KRT26, KRT27, KRT28, KRT3, KRT31, KRT32, KRT33A, KRT33B, KRT34, KRT35, KRT36, KRT37, KRT38, KRT39, KRT4, KRT40, KRT5, KRT6A, KRT6B, KRT6C, KRT7, KRT71, KRT72, KRT73, KRT74, KRT75, KRT76, KRT77, KRT78, KRT79, KRT8, KRT80, KRT81, KRT82, KRT83, KRT84, KRT85, KRT9, LMNA, LMNB1, LMNB2, NEFM, NES, SYNC, SYNM, VIM	(208)

[1]Genes identified in genomewide screens for Ty3 host factors (64, 133).
[2]Retrovirus host genes identified in Ty3 screens and in retrovirus host screens (206–209).

control in perinuclear granules containing PB components. Mouse, fly, nematode, and zebra fish germ cell-related lineages accumulate in these granules. When RNAi is suppressed, these cells undergo retrotransposition (95, 226–228). Based on the role of components of these granules in budding yeast retrotransposition, where RNAi is naturally absent, we speculate that in retrotransposing germ cell systems, at least a subset of these components interact with assembling retrotransposon proteins and RNAs.

PERSPECTIVES

The dawn of genomewide analysis enabled by next generation sequencing shows host genes embedded in a sea of retroelements. Together with other retrotransposons, LTR retrotransposons account for ongoing *expansion*, *adaptation*, and *remodeling* of eukaryotic genomes. Many consequences, noncoding RNAs and RNAi among them, were unexpected and raise new questions.

In the rush to apply these findings, retrotransposition itself remains largely unexplored. In animal cells, despite RNAi, LTR element retrotransposition is ongoing in germ cells, early-stage embryos, the brain, and some cancers. We know little of the mechanics of retrotransposition in these cells, and less about real-time effects of retrotransposition on those and descendant cells.

Ty3/Gypsy elements represent a major class of LTR retrotransposons, and dominate many plant and some animal genomes. Ty3, the sole representative in budding yeast, is one of the more completely understood of these elements at the molecular level. However, it is virtually a total anomaly. Classified as a Chromovirus, Ty3 lacks the defining IN chromodomain, and stands out for the specificity of its targeting. Most retrotransposons are expressed and retrotranspose in single cells, but Ty3 in the natural context of mating, retrotransposes only after diploids are formed. LTR retrotransposons are suppressed by RNAi, but due to the evolutionary loss of Dicer and Argonaute functions, this system is absent in budding yeast. Perhaps most surprisingly, Ty3 VLP assembly requires components of the RNA processing system that supposedly suppresses and degrades LTR retrotransposon RNAs in metazoans. What can we expect to learn from such an exceptional element?

Work in the Ty3 system has provided considerable information about the biochemistry of retrotransposition, much of which is proving generalizable. Three areas especially illustrate this point. First, Ty3 targeting provided the first clear evidence for interaction with host chromatin/transcription factors, and still offers the most precise tool for understanding the interaction between IN and target protein complexes and DNA. Targets in this system can be precisely predicted and manipulated to understand requirements of strand transfer. Second, Ty3/Tf1 Gypsy element retrotransposons show clear interactions with nuclear pores, and combined with targeting, can be used to explore the links between entry and integration. Known and essentially identical targets of Ty3 used at widely different frequencies, offer potential insights into unknown areas of nuclear dynamics. Third, because Ty3 is not regulated by post transcriptional RNA interference, it offers a rare opportunity to study retrotransposition itself. It cannot be a coincidence that Ty3 retrotransposes in the yeast analog of germ cells during cell fusion, and animal retrotransposons are activated in germ cells. We know quite a bit about how retrotransposition is suppressed. Now we should think about what happens when it occurs.

At the center of all this, we have little understanding of how retrotransposition affects cells in real time. In part this is because retrotransposition occurs in a few cells per population. The recent advent of single cell biology will surely produce a wave of insights into those rare cells that are undergoing retrotransposition. Are there clusters of retrotransposition in those cells? Does retrotransposition target broken replication forks as suggested by mutant screens? One DNA break induces checkpoint arrest; do retrotransposon cDNAs have the same effect? What determines the nuclear winner between competing processes of integration and destruction of invading cDNA? Does retrotransposition encrypt an individual cell record, as theorized for neuronal tissues? Is retrotransposition a trigger in the cancers with which it is associated? We know many mechanisms through which retrotransposons are tolerated by hosts. Is there also a kind of retrotransposon race that determines how retrotransposon composition differs so dramatically among eukaryotic species? Answers to these and other questions will lead us to a better understanding of our genomes.

Acknowledgments. *Ty3 research from the Sandmeyer laboratory was supported by PHS grant GM33281. We thank Ivan Chang and Parth Sitlani (UCI) for helpful discussions and other past and present members of the Sandmeyer laboratory who made this work possible. We thank Stuart Le Grice, NCI Frederick, MD, for helpful discussions, and Mercy Chang for artistic illustrations.*

Citation. Sandmeyer S, Patterson K, Bilanchone V. 2015. Ty3, a position-specific retrotransposon in budding yeast. Microbiol Spectrum 3(2):MDNA3-0057-2014.

References

1. **Llorens C, Fares MA, Moya A.** 2008. Relationships of gag–pol diversity between Ty3/Gypsy and Retroviridae LTR retroelements and the three kings hypothesis. *BMC Evol Biol* **8:**276.
2. **Havecker ER, Gao X, Voytas DF.** 2004. The diversity of LTR retrotransposons. *Genome Biol* **5:**225.
3. **Eickbush TH, Jamburuthugoda VK.** 2009. The diversity of retrotransposons and the properties of their reverse transcriptases. *Virus Res* **134:**221–234.
4. **Carr M, Bensasson D, Bergman CM.** 2012. Evolutionary genomics of transposable elements in *Saccharomyces cerevisiae*. *PLoS One* **7:**e50978.
5. **Goffeau A, Barrell BG, Bussey H, Davis RW, Dujon B, Feldmann H, Galibert F, Hoheisel JD, Jacq C, Johnston M, Louis EJ, Mewes HW, Murakami Y, Philippsen P, Tettelin H, Oliver SG.** 1996. Life with 6000 genes. *Science* **274:**546, 563–547.
6. **Kim JM, Vanguri S, Boeke JD, Gabriel A, Voytas DF.** 1998. Transposable elements and genome organization: a comprehensive survey of retrotransposons revealed by the complete *Saccharomyces cerevisiae* genome sequence. *Genome Res* **8:**464–478.
7. **del Rey FJ, Donahue TF, Fink GR.** 1982. sigma, a repetitive element found adjacent to tRNA genes of yeast. *Proc Natl Acad Sci U S A* **79:**4138–4142.
8. **Sandmeyer SB, Olson MV.** 1982. Insertion of a repetitive element at the same position in the 5′-flanking regions of two dissimilar yeast tRNA genes. *Proc Natl Acad Sci U S A* **79:**7674–7678.
9. **Hansen LJ, Chalker DL, Sandmeyer SB.** 1988. Ty3, a yeast retrotransposon associated with tRNA genes, has homology to animal retroviruses. *Mol Cell Biol* **8:**5245–5256.
10. **Jordan IK, McDonald JF.** 1999. Tempo and mode of Ty element evolution in Saccharomyces cerevisiae. *Genetics* **151:**1341–1351.
11. **Sandmeyer SB, Clemens KA.** 2010. Function of a retrotransposon nucleocapsid protein. *RNA Biol* **7:**642–654.
12. **Hansen LJ, Sandmeyer SB.** 1990. Characterization of a transpositionally active Ty3 element and identification of the Ty3 integrase protein. *J Virol* **64:**2599–2607.
13. **Bilanchone VW, Claypool JA, Kinsey PT, Sandmeyer SB.** 1993. Positive and negative regulatory elements control expression of the yeast retrotransposon Ty3. *Genetics* **134:**685–700.
14. **Clark DJ, Bilanchone VW, Haywood LJ, Dildine SL, Sandmeyer SB.** 1988. A yeast sigma composite element, TY3, has properties of a retrotransposon. *J Biol Chem* **263:**1413–1423.
15. **Van Arsdell SW, Stetler GL, Thorner J.** 1987. The yeast repeated element sigma contains a hormone-inducible promoter. *Mol Cell Biol* **7:**749–759.
16. **Bardwell L.** 2005. A walk-through of the yeast mating pheromone response pathway. *Peptides* **26:**339–350.
17. **Menees TM, Sandmeyer SB.** 1994. Transposition of the yeast retroviruslike element Ty3 is dependent on the cell cycle. *Mol Cell Biol* **14:**8229–8240.
18. **Kinsey PT, Sandmeyer SB.** 1995. Ty3 transposes in mating populations of yeast: a novel transposition assay for Ty3. *Genetics* **139:**81–94.
19. **Mieczkowski PA, Dominska M, Buck MJ, Gerton JL, Lieb JD, Petes TD.** 2006. Global analysis of the relationship between the binding of the Bas1p transcription factor and meiosis-specific double-strand DNA breaks in *Saccharomyces cerevisiae*. *Mol Cell Biol* **26:**1014–1027.
20. **Daignan-Fornier B, Fink GR.** 1992. Coregulation of purine and histidine biosynthesis by the transcriptional activators BAS1 and BAS2. *Proc Natl Acad Sci U S A* **89:**6746–6750.
21. **Arndt KT, Styles C, Fink GR.** 1987. Multiple global regulators control HIS4 transcription in yeast. *Science* **237:**874–880.
22. **Servant G, Pinson B, Tchalikian-Cosson A, Coulpier F, Lemoine S, Pennetier C, Bridier-Nahmias A, Todeschini AL, Fayol H, Daignan-Fornier B, Lesage P.** 2012. Tye7 regulates yeast Ty1 retrotransposon sense and antisense transcription in response to adenylic nucleotides stress. *Nucleic Acids Res* **40:**5271–5282.
23. **Kinsey PT, Sandmeyer SB.** 1991. Adjacent pol II and pol III promoters: transcription of the yeast retrotransposon Ty3 and a target tRNA gene. *Nucleic Acids Res* **19:**1317–1324.
24. **Hull MW, Erickson J, Johnston M, Engelke DR.** 1994. tRNA genes as transcriptional repressor elements. *Mol Cell Biol* **14:**1266–1277.
25. **Wang L, Haeusler RA, Good PD, Thompson M, Nagar S, Engelke DR.** 2005. Silencing near tRNA genes requires nucleolar localization. *J Biol Chem* **280:**8637–8639.
26. **Haeusler RA, Pratt-Hyatt M, Good PD, Gipson TA, Engelke DR.** 2008. Clustering of yeast tRNA genes is mediated by specific association of condensin with tRNA gene transcription complexes. *Genes Dev* **22:**2204–2214.
27. **Bolton EC, Boeke JD.** 2003. Transcriptional interactions between yeast tRNA genes, flanking genes and Ty elements: a genomic point of view. *Genome Res* **13:**254–263.
28. **Kirchner J, Sandmeyer SB, Forrest DB.** 1992. Transposition of a Ty3 GAG3–POL3 fusion mutant is limited by availability of capsid protein. *J Virol* **66:**6081–6092.
29. **Casolari JM, Brown CR, Komili S, West J, Hieronymus H, Silver PA.** 2004. Genome-wide localization of the nuclear transport machinery couples transcriptional status and nuclear organization. *Cell* **117:**427–439.
30. **Chen M, Gartenberg MR.** 2014. Coordination of tRNA transcription with export at nuclear pore complexes in budding yeast. *Genes Dev* **28:**959–970.
31. **Duan Z, Andronescu M, Schutz K, McIlwain S, Kim YJ, Lee C, Shendure J, Fields S, Blau CA, Noble WS.** 2010. A three-dimensional model of the yeast genome. *Nature* **465:**363–367.
32. **Morillon A, Springer M, Lesage P.** 2000. Activation of the Kss1 invasive-filamentous growth pathway induces Ty1 transcription and retrotransposition in *Saccharomyces cerevisiae*. *Mol Cell Biol* **20:**5766–5776.

33. **Saito H.** 2010. Regulation of cross-talk in yeast MAPK signaling pathways. *Curr Opin Microbiol* **13:**677–683.
34. **Elder RT, St John TP, Stinchcomb DT, Davis RW, Scherer S.** 1981. Studies on the transposable element Ty1 of yeast. I. RNA homologous to Ty1. II. Recombination and expression of Ty1 and adjacent sequences. *Cold Spring Harb Symp Quant Biol* **45**(Pt 2):581–591.
35. **Xu H, Boeke JD.** 1991. Inhibition of Ty1 transposition by mating pheromones in *Saccharomyces cerevisiae*. *Mol Cell Biol* **11:**2736–2743.
36. **Conte D Jr, Barber E, Banerjee M, Garfinkel DJ, Curcio MJ.** 1998. Posttranslational regulation of Ty1 retrotransposition by mitogen-activated protein kinase Fus3. *Mol Cell Biol* **18:**2502–2513.
37. **Conte D Jr, Curcio MJ.** 2000. Fus3 controls Ty1 transpositional dormancy through the invasive growth MAPK pathway. *Mol Microbiol* **35:**415–427.
38. **Qi X, Daily K, Nguyen K, Wang H, Mayhew D, Rigor P, Forouzan S, Johnston M, Mitra RD, Baldi P, Sandmeyer S.** 2012. Retrotransposon profiling of RNA polymerase III initiation sites. *Genome Res* **22:**681–692.
39. **Boeke JD, Trueheart J, Natsoulis G, Fink GR.** 1987. 5-Fluoroorotic acid as a selective agent in yeast molecular genetics. *Methods Enzymol* **154:**164–175.
40. **Curcio MJ, Garfinkel DJ.** 1991. Single-step selection for Ty1 element retrotransposition. *Proc Natl Acad Sci U S A* **88:**936–940.
41. **Aye M, Sandmeyer SB.** 2003. Ty3 requires yeast La homologous protein for wild-type frequencies of transposition. *Mol Microbiol* **49:**501–515.
42. **Farabaugh PJ, Zhao H, Vimaladithan A.** 1993. A novel programed frameshift expresses the POL3 gene of retrotransposon Ty3 of yeast: frameshifting without tRNA slippage. *Cell* **74:**93–103.
43. **Kirchner J, Sandmeyer S.** 1993. Proteolytic processing of Ty3 proteins is required for transposition. *J Virol* **67:** 19–28.
44. **Kuznetsov YG, Zhang M, Menees TM, McPherson A, Sandmeyer S.** 2005. Investigation by atomic force microscopy of the structure of Ty3 retrotransposon particles. *J Virol* **79:**8032–8045.
45. **Zhang M, Larsen LS, Irwin B, Bilanchone V, Sandmeyer S.** 2010. Two-hybrid analysis of Ty3 capsid subdomain interactions. *Mob DNA* **1:**14.
46. **Kirchner J, Sandmeyer SB.** 1996. Ty3 integrase mutants defective in reverse transcription or 3′-end processing of extrachromosomal Ty3 DNA. *J Virol* **70:**4737–4747.
47. **Qi X, Sandmeyer SB.** 2012. In vitro targeting of strand transfer by the Ty3 retroelement integrase. *J Biol Chem* **287:**18589–18595.
48. **Claypool JA, Malik HS, Eickbush TH, Sandmeyer SB.** 2001. Ten-kilodalton domain in Ty3 Gag3–Pol3p between PR and RT is dispensable for Ty3 transposition. *J Virol* **75:**1557–1560.
49. **Beliakova-Bethell N, Beckham C, Giddings TH Jr, Winey M, Parker R, Sandmeyer S.** 2006. Virus-like particles of the Ty3 retrotransposon assemble in association with P-body components. *RNA* **12:**94–101.
50. **Orlinsky KJ, Sandmeyer SB.** 1994. The Cys-His motif of Ty3 NC can be contributed by Gag3 or Gag3–Pol3 polyproteins. *J Virol* **68:**4152–4166.
51. **Turkel S, Kaplan G, Farabaugh PJ.** 2011. Glucose signalling pathway controls the programmed ribosomal frameshift efficiency in retroviral-like element Ty3 in *Saccharomyces cerevisiae*. *Yeast* **28:**799–808.
52. **Nymark-McMahon MH, Beliakova-Bethell NS, Darlix JL, Le Grice SF, Sandmeyer SB.** 2002. Ty3 integrase is required for initiation of reverse transcription. *J Virol* **76:**2804–2816.
53. **Keeney JB, Chapman KB, Lauermann V, Voytas DF, Astrom SU, von Pawel-Rammingen U, Bystrom A, Boeke JD.** 1995. Multiple molecular determinants for retrotransposition in a primer tRNA. *Mol Cell Biol* **15:** 217–226.
54. **Hansen LJ, Chalker DL, Orlinsky KJ, Sandmeyer SB.** 1992. Ty3 GAG3 and POL3 genes encode the components of intracellular particles. *J Virol* **66:**1414–1424.
55. **Larsen LS, Kuznetsov Y, McPherson A, Hatfield GW, Sandmeyer S.** 2008. TY3 GAG3 protein forms ordered particles in *Escherichia coli*. *Virology* **370:**223–227.
56. **Clemens K, Larsen L, Zhang M, Kuznetsov Y, Bilanchone V, Randall A, Harned A, Dasilva R, Nagashima K, McPherson A, Baldi P, Sandmeyer S.** 2011. The Ty3 Gag3 spacer controls intracellular condensation and uncoating. *J Virol* **85:**3055–3066.
57. **Larsen LS, Beliakova-Bethell N, Bilanchone V, Zhang M, Lamsa A, Dasilva R, Hatfield GW, Nagashima K, Sandmeyer S.** 2008. Ty3 nucleocapsid controls localization of particle assembly. *J Virol* **82:**2501–2514.
58. **Larsen LS, Zhang M, Beliakova-Bethell N, Bilanchone V, Lamsa A, Nagashima K, Najdi R, Kosaka K, Kovacevic V, Cheng J, Baldi P, Hatfield GW, Sandmeyer S.** 2007. Ty3 capsid mutations reveal early and late functions of the amino-terminal domain. *J Virol* **81:**6957–6972.
59. **Martin-Rendon E, Marfany G, Wilson S, Ferguson DJ, Kingsman SM, Kingsman AJ.** 1996. Structural determinants within the subunit protein of Ty1 virus-like particles. *Mol Microbiol* **22:**667–679.
60. **Forshey BM, von Schwedler U, Sundquist WI, Aiken C.** 2002. Formation of a human immunodeficiency virus type 1 core of optimal stability is crucial for viral replication. *J Virol* **76:**5667–5677.
61. **Briggs JA, Krausslich HG.** 2011. The molecular architecture of HIV. *J Mol Biol* **410:**491–500.
62. **Ganser-Pornillos BK, Yeager M, Sundquist WI.** 2008. The structural biology of HIV assembly. *Curr Opin Struct Biol* **18:**203–217.
63. **Sundquist WI, Hill CP.** 2007. How to assemble a capsid. *Cell* **131:**17–19.
64. **Irwin B, Aye M, Baldi P, Beliakova-Bethell N, Cheng H, Dou Y, Liou W, Sandmeyer S.** 2005. Retroviruses and yeast retrotransposons use overlapping sets of host genes. *Genome Res* **15:**641–654.
65. **Orlinsky KJ, Gu J, Hoyt M, Sandmeyer S, Menees TM.** 1996. Mutations in the Ty3 major homology region

affect multiple steps in Ty3 retrotransposition. *J Virol* 70:3440–3448.

66. **Garbitt-Hirst R, Kenney SP, Parent LJ.** 2009. Genetic evidence for a connection between Rous sarcoma virus gag nuclear trafficking and genomic RNA packaging. *J Virol* **83:**6790–6797.
67. **Parent LJ.** 2011. New insights into the nuclear localization of retroviral Gag proteins. *Nucleus* **2:**92–97.
68. **Grigorov B, Decimo D, Smagulova F, Pechoux C, Mougel M, Muriaux D, Darlix JL.** 2007. Intracellular HIV-1 Gag localization is impaired by mutations in the nucleocapsid zinc fingers. *Retrovirology* **4:**54.
69. **Levesque K, Halvorsen M, Abrahamyan L, Chatel-Chaix L, Poupon V, Gordon H, DesGroseillers L, Gatignol A, Mouland AJ.** 2006. Trafficking of HIV-1 RNA is mediated by heterogeneous nuclear ribonucleoprotein A2 expression and impacts on viral assembly. *Traffic* **7:**1177–1193.
70. **Checkley MA, Mitchell JA, Eizenstat LD, Lockett SJ, Garfinkel DJ.** 2013. Ty1 gag enhances the stability and nuclear export of Ty1 mRNA. *Traffic* **14:**57–69.
71. **Keller PW, Johnson MC, Vogt VM.** 2008. Mutations in the spacer peptide and adjoining sequences in Rous sarcoma virus Gag lead to tubular budding. *J Virol* **82:** 6788–6797.
72. **Yeager M.** 2011. Design of *in vitro* symmetric complexes and analysis by hybrid methods reveal mechanisms of HIV capsid assembly. *J Mol Biol* **410:**534–552.
73. **Qualley DF, Stewart-Maynard KM, Wang F, Mitra M, Gorelick RJ, Rouzina I, Williams MC, Musier-Forsyth K.** 2010. C-terminal domain modulates the nucleic acid chaperone activity of human T-cell leukemia virus type 1 nucleocapsid protein via an electrostatic mechanism. *J Biol Chem* **285:**295–307.
74. **Pettit SC, Moody MD, Wehbie RS, Kaplan AH, Nantermet PV, Klein CA, Swanstrom R.** 1994. The p2 domain of human immunodeficiency virus type 1 Gag regulates sequential proteolytic processing and is required to produce fully infectious virions. *J Virol* **68:** 8017–8027.
75. **Gabus C, Ficheux D, Rau M, Keith G, Sandmeyer S, Darlix JL.** 1998. The yeast Ty3 retrotransposon contains a 5′–3′ bipartite primer-binding site and encodes nucleocapsid protein NCp9 functionally homologous to HIV-1 NCp7. *EMBO J* **17:**4873–4880.
76. **Clemens K, Bilanchone V, Beliakova-Bethell N, Larsen LS, Nguyen K, Sandmeyer S.** 2013. Sequence requirements for localization and packaging of Ty3 retroelement RNA. *Virus Res* **171:**319–331.
77. **D'Souza V, Summers MF.** 2005. How retroviruses select their genomes. *Nat Rev Microbiol* **3:**643–655.
78. **Bertrand E, Chartrand P, Schaefer M, Shenoy SM, Singer RH, Long RM.** 1998. Localization of ASH1 mRNA particles in living yeast. *Mol Cell* **2:**437–445.
79. **Lecher P, Bucheton A, Pelisson A.** 1997. Expression of the *Drosophila* retrovirus gypsy as ultrastructurally detectable particles in the ovaries of flies carrying a permissive flamenco allele. *J Gen Virol* **78**(Pt 9)**:**2379–2388.
80. **Garfinkel DJ, Boeke JD, Fink GR.** 1985. Ty element transposition: reverse transcriptase and virus-like particles. *Cell* **42:**507–517.
81. **Griffith JL, Coleman LE, Raymond AS, Goodson SG, Pittard WS, Tsui C, Devine SE.** 2003. Functional genomics reveals relationships between the retrovirus-like Ty1 element and its host *Saccharomyces cerevisiae. Genetics* **164:**867–879.
82. **Sheth U, Parker R.** 2003. Decapping and decay of messenger RNA occur in cytoplasmic processing bodies. *Science* **300:**805–808.
83. **Brengues M, Teixeira D, Parker R.** 2005. Movement of eukaryotic mRNAs between polysomes and cytoplasmic processing bodies. *Science* **310:**486–489.
84. **Hu W, Sweet TJ, Chamnongpol S, Baker KE, Coller J.** 2009. Co-translational mRNA decay in *Saccharomyces cerevisiae. Nature* **461:**225–229.
85. **Thomas MG, Loschi M, Desbats MA, Boccaccio GL.** 2011. RNA granules: the good, the bad and the ugly. *Cell Signal* **23:**324–334.
86. **Buchan JR, Nissan T, Parker R.** 2010. Analyzing P-bodies and stress granules in *Saccharomyces cerevisiae. Methods Enzymol* **470:**619–640.
87. **Mitchell SF, Jain S, She M, Parker R.** 2013. Global analysis of yeast mRNPs. *Nat Struct Mol Biol* **20:**127–133.
88. **Nissan T, Rajyaguru P, She M, Song H, Parker R.** 2010. Decapping activators in *Saccharomyces cerevisiae* act by multiple mechanisms. *Mol Cell* **39:**773–783.
89. **Coller J, Parker R.** 2005. General translational repression by activators of mRNA decapping. *Cell* **122:** 875–886.
90. **Rajyaguru P, Parker R.** 2009. CGH-1 and the control of maternal mRNAs. *Trends Cell Biol* **19:**24–28.
91. **Tharun S, Parker R.** 2001. Targeting an mRNA for decapping: displacement of translation factors and association of the Lsm1p–7p complex on deadenylated yeast mRNAs. *Mol Cell* **8:**1075–1083.
92. **Fillman C, Lykke-Andersen J.** 2005. RNA decapping inside and outside of processing bodies. *Curr Opin Cell Biol* **17:**326–331.
93. **Nagarajan VK, Jones CI, Newbury SF, Green PJ.** 2013. XRN 5′—>3′ exoribonucleases: structure, mechanisms and functions. *Biochim Biophys Acta* **1829:**590–603.
94. **Parker R.** 2012. RNA degradation in *Saccharomyces cerevisae. Genetics* **191:**671–702.
95. **Anderson P, Kedersha N.** 2009. RNA granules: post-transcriptional and epigenetic modulators of gene expression. *Nat Rev Mol Cell Biol* **10:**430–436.
96. **Jain S, Parker R.** 2013. The discovery and analysis of P Bodies. *Adv Exp Med Biol* **768:**23–43.
97. **Ka M, Park YU, Kim J.** 2008. The DEAD-box RNA helicase, Dhh1, functions in mating by regulating Ste12 translation in *Saccharomyces cerevisiae. Biochem Biophys Res Commun* **367:**680–686.
98. **Dutko JA, Kenny AE, Gamache ER, Curcio MJ.** 2010. 5′ to 3′ mRNA decay factors colocalize with Ty1 gag and human APOBEC3G and promote Ty1 retrotransposition. *J Virol* **84:**5052–5066.

99. **Checkley MA, Nagashima K, Lockett SJ, Nyswaner KM, Garfinkel DJ.** 2010. P-body components are required for Ty1 retrotransposition during assembly of retrotransposition-competent virus-like particles. *Mol Cell Biol* **30:**382–398.

100. **Le Grice SFJ, Nowotny M.** 2014. Nucleic acid polymerases: reverse transcriptases. *Nucleic Acids Mol Biol* **30:** 189–214.

101. **Rausch JW, Le Grice SF.** 2004. 'Binding, bending and bonding': polypurine tract-primed initiation of plus-strand DNA synthesis in human immunodeficiency virus. *Int J Biochem Cell Biol* **36:**1752–1766.

102. **Malik HS, Eickbush TH.** 2001. Phylogenetic analysis of ribonuclease H domains suggests a late, chimeric origin of LTR retrotransposable elements and retroviruses. *Genome Res* **11:**1187–1197.

103. **Cristofari G, Gabus C, Ficheux D, Bona M, Le Grice SF, Darlix JL.** 1999. Characterization of active reverse transcriptase and nucleoprotein complexes of the yeast retrotransposon Ty3 *in vitro*. *J Biol Chem* **274:**36643–36648.

104. **Rausch JW, Grice MK, Henrietta M, Nymark M, Miller JT, Le Grice SF.** 2000. Interaction of p55 reverse transcriptase from the *Saccharomyces cerevisiae* retrotransposon Ty3 with conformationally distinct nucleic acid duplexes. *J Biol Chem* **275:**13879–13887.

105. **Lener D, Budihas SR, Le Grice SF.** 2002. Mutating conserved residues in the ribonuclease H domain of Ty3 reverse transcriptase affects specialized cleavage events. *J Biol Chem* **277:**26486–26495.

106. **Nowak E, Miller JT, Bona MK, Studnicka J, Szczepanowski RH, Jurkowski J, Le Grice SF, Nowotny M.** 2014. Ty3 reverse transcriptase complexed with an RNA–DNA hybrid shows structural and functional asymmetry. *Nat Struct Mol Biol* **21:**389–396.

107. **Friant S, Heyman T, Wilhelm ML, Wilhelm FX.** 1996. Extended interactions between the primer tRNAi(Met) and genomic RNA of the yeast Ty1 retrotransposon. *Nucleic Acids Res* **24:**441–449.

108. **Le Grice SF.** 2003. "In the beginning": initiation of minus strand DNA synthesis in retroviruses and LTR-containing retrotransposons. *Biochemistry* **42:**14349–14355.

109. **Heyman T, Agoutin B, Friant S, Wilhelm FX, Wilhelm ML.** 1995. Plus-strand DNA synthesis of the yeast retrotransposon Ty1 is initiated at two sites, PPT1 next to the 3′ LTR and PPT2 within the pol gene. PPT1 is sufficient for Ty1 transposition. *J Mol Biol* **253:**291–303.

110. **Bibillo A, Lener D, Tewari A, Le Grice SF.** 2005. Interaction of the Ty3 reverse transcriptase thumb subdomain with template-primer. *J Biol Chem* **280:**30282–30290.

111. **Yi-Brunozzi HY, Brabazon DM, Lener D, Le Grice SF, Marino JP.** 2005. A ribose sugar conformational switch in the LTR-retrotransposon Ty3 polypurine tract-containing RNA/DNA hybrid. *J Am Chem Soc* **127:**16344–16345.

112. **Dash C, Marino JP, Le Grice SF.** 2006. Examining Ty3 polypurine tract structure and function by nucleoside analog interference. *J Biol Chem* **281:**2773–2783.

113. **Lener D, Kvaratskhelia M, Le Grice SF.** 2003. Nonpolar thymine isosteres in the Ty3 polypurine tract DNA template modulate processing and provide a model for its recognition by Ty3 reverse transcriptase. *J Biol Chem* **278:**26526–26532.

114. **Nair GR, Dash C, Le Grice SF, DeStefano JJ.** 2012. Viral reverse transcriptases show selective high affinity binding to DNA–DNA primer-templates that resemble the polypurine tract. *PLoS One* **7:**e41712.

115. **Ke N, Gao X, Keeney JB, Boeke JD, Voytas DF.** 1999. The yeast retrotransposon Ty5 uses the anticodon stem–loop of the initiator methionine tRNA as a primer for reverse transcription. *RNA* **5:**929–938.

116. **Chapman KB, Boeke JD.** 1991. Isolation and characterization of the gene encoding yeast debranching enzyme. *Cell* **65:**483–492.

117. **Cheng Z, Menees TM.** 2004. RNA branching and debranching in the yeast retrovirus-like element Ty1. *Science* **303:**240–243.

118. **Coombes CE, Boeke JD.** 2005. An evaluation of detection methods for large lariat RNAs. *RNA* **11:**323–331.

119. **Galvis AE, Fisher HE, Nitta T, Fan H, Camerini D.** 2014. Impairment of HIV-1 cDNA Synthesis by DBR1 Knockdown. *J Virol* **88:**7054–7069.

120. **Moore MD, Hu WS.** 2009. HIV-1 RNA dimerization: It takes two to tango. *AIDS Rev* **11:**91–102.

121. **Purzycka KJ, Legiewicz M, Matsuda E, Eizentstat LD, Lusvarghi S, Saha A, Le Grice SF, Garfinkel DJ.** 2013. Exploring Ty1 retrotransposon RNA structure within virus-like particles. *Nucleic Acids Res* **41:**463–473.

122. **Wente SR, Rout MP.** 2010. The nuclear pore complex and nuclear transport. *Cold Spring Harbor Persp Biol* **2:**a000562.

123. **Hoelz A, Debler EW, Blobel G.** 2011. The structure of the nuclear pore complex. *Annu Rev Biochem* **80:** 613–643.

124. **Taddei A, Gasser SM.** 2012. Structure and function in the budding yeast nucleus. *Genetics* **192:**107–129.

125. **Nagai S, Dubrana K, Tsai-Pflugfelder M, Davidson MB, Roberts TM, Brown GW, Varela E, Hediger F, Gasser SM, Krogan NJ.** 2008. Functional targeting of DNA damage to a nuclear pore-associated SUMO-dependent ubiquitin ligase. *Science* **322:**597–602.

126. **Pante N, Kann M.** 2002. Nuclear pore complex is able to transport macromolecules with diameters of about 39 nm. *Mol Biol Cell* **13:**425–434.

127. **Kenna MA, Brachmann CB, Devine SE, Boeke JD.** 1998. Invading the yeast nucleus: a nuclear localization signal at the C terminus of Ty1 integrase is required for transposition *in vivo*. *Mol Cell Biol* **18:**1115–1124.

128. **Moore SP, Rinckel LA, Garfinkel DJ.** 1998. A Ty1 integrase nuclear localization signal required for retrotransposition. *Mol Cell Biol* **18:**1105–1114.

129. **Dang VD, Levin HL.** 2000. Nuclear import of the retrotransposon Tf1 is governed by a nuclear localization signal that possesses a unique requirement for the FXFG nuclear pore factor Nup124p. *Mol Cell Biol* **20:** 7798–7812.

130. **Levin A, Loyter A, Bukrinsky M.** 2011. Strategies to inhibit viral protein nuclear import: HIV-1 as a target. *Biochim Biophys Acta* **1813:**1646–1653.

131. **Katz RA, Greger JG, Darby K, Boimel P, Rall GF, Skalka AM.** 2002. Transduction of interphase cells by avian sarcoma virus. *J Virol* **76:**5422–5434.

132. **Aye M, Irwin B, Beliakova-Bethell N, Chen E, Garrus J, Sandmeyer S.** 2004. Host factors that affect Ty3 retrotransposition in *Saccharomyces cerevisiae*. *Genetics* **168:**1159–1176.

133. **Beliakova-Bethell N, Terry LJ, Bilanchone V, DaSilva R, Nagashima K, Wente SR, Sandmeyer S.** 2009. Ty3 nuclear entry is initiated by viruslike particle docking on GLFG nucleoporins. *J Virol* **83:**11914–11925.

134. **Heath CV, Copeland CS, Amberg DC, Del Priore V, Snyder M, Cole CN.** 1995. Nuclear pore complex clustering and nuclear accumulation of poly(A)+ RNA associated with mutation of the *Saccharomyces cerevisiae* RAT2/NUP120 gene. *J Cell Biol* **131:**1677–1697.

135. **Li O, Heath CV, Amberg DC, Dockendorff TC, Copeland CS, Snyder M, Cole CN.** 1995. Mutation or deletion of the *Saccharomyces cerevisiae* RAT3/NUP133 gene causes temperature-dependent nuclear accumulation of poly(A)+ RNA and constitutive clustering of nuclear pore complexes. *Mol Biol Cell* **6:**401–417.

136. **Strawn LA, Shen T, Shulga N, Goldfarb DS, Wente SR.** 2004. Minimal nuclear pore complexes define FG repeat domains essential for transport. *Nat Cell Biol* **6:** 197–206.

137. **Fassati A.** 2012. Multiple roles of the capsid protein in the early steps of HIV-1 infection. *Virus Res* **170:**15–24.

138. **Matreyek KA, Engelman A.** 2011. The requirement for nucleoporin NUP153 during human immunodeficiency virus type 1 infection is determined by the viral capsid. *J Virol* **85:**7818–7827.

139. **Krishnan L, Matreyek KA, Oztop I, Lee K, Tipper CH, Li X, Dar MJ, Kewalramani VN, Engelman A.** 2010. The requirement for cellular transportin 3 (TNPO3 or TRN-SR2) during infection maps to human immunodeficiency virus type 1 capsid and not integrase. *J Virol* **84:**397–406.

140. **Lin SS, Nymark-McMahon MH, Yieh L, Sandmeyer SB.** 2001. Integrase mediates nuclear localization of Ty3. *Mol Cell Biol* **21:**7826–7838.

141. **Chalker DL, Sandmeyer SB.** 1990. Transfer RNA genes are genomic targets for *de novo* transposition of the yeast retrotransposon Ty3. *Genetics* **126:**837–850.

142. **Sandmeyer SB, Hansen LJ, Chalker DL.** 1990. Integration specificity of retrotransposons and retroviruses. *Annu Rev Genet* **24:**491–518.

143. **Hani J, Feldmann H.** 1998. tRNA genes and retroelements in the yeast genome. *Nucleic Acids Res* **26:** 689–696.

144. **Natsoulis G, Thomas W, Roghmann MC, Winston F, Boeke JD.** 1989. Ty1 transposition in *Saccharomyces cerevisiae* is nonrandom. *Genetics***123:**269–279.

145. **Hofmann J, Schumann G, Borschet G, Gosseringer R, Bach M, Bertling WM, Marschalek R, Dingermann T.** 1991. Transfer RNA genes from *Dictyostelium discoideum* are frequently associated with repetitive elements and contain consensus boxes in their 5′ and 3′-flanking regions. *J Mol Biol* **222:**537–552.

146. **Zou S, Voytas DF.** 1997. Silent chromatin determines target preference of the *Saccharomyces* retrotransposon Ty5. *Proc Natl Acad Sci U S A* **94:**7412–7416.

147. **Gao X, Hou Y, Ebina H, Levin HL, Voytas DF.** 2008. Chromodomains direct integration of retrotransposons to heterochromatin. *Genome Res* **18:**359–369.

148. **Hare S, Gupta SS, Valkov E, Engelman A, Cherepanov P.** 2010. Retroviral intasome assembly and inhibition of DNA strand transfer. *Nature* **464:**232–236.

149. **Maertens GN, Hare S, Cherepanov P.** 2010. The mechanism of retroviral integration from X-ray structures of its key intermediates. *Nature* **468:**326–329.

150. **Li X, Krishnan L, Cherepanov P, Engelman A.** 2011. Structural biology of retroviral DNA integration. *Virology* **411:**194–205.

151. **Rice P, Craigie R, Davies DR.** 1996. Retroviral integrases and their cousins. *Curr Opin Struct Biol* **6:** 76–83.

152. **Kulkosky J, Jones KS, Katz RA, Mack JP, Skalka AM.** 1992. Residues critical for retroviral integrative recombination in a region that is highly conserved among retroviral/retrotransposon integrases and bacterial insertion sequence transposases. *Mol Cell Biol* **12:**2331–2338.

153. **Qi X, Vargas E, Larsen L, Knapp W, Hatfield GW, Lathrop R, Sandmeyer S.** 2013. Directed DNA shuffling of retrovirus and retrotransposon integrase protein domains. *PLoS One* **8:**e63957.

154. **Malik HS, Eickbush TH.** 1999. Modular evolution of the integrase domain in the Ty3/Gypsy class of LTR retrotransposons. *J Virol* **73:**5186–5190.

155. **Ebina H, Chatterjee AG, Judson RL, Levin HL.** 2008. The GP(Y/F) domain of TF1 integrase multimerizes when present in a fragment, and substitutions in this domain reduce enzymatic activity of the full-length protein. *J Biol Chem* **283:**15965–15974.

156. **Nymark-McMahon MH, Sandmeyer SB.** 1999. Mutations in nonconserved domains of Ty3 integrase affect multiple stages of the Ty3 life cycle. *J Virol* **73:**453–465.

157. **Engelman A, Englund G, Orenstein JM, Martin MA, Craigie R.** 1995. Multiple effects of mutations in human immunodeficiency virus type 1 integrase on viral replication. *J Virol* **69:**2729–2736.

158. **Briggs JA, Grunewald K, Glass B, Forster F, Krausslich HG, Fuller SD.** 2006. The mechanism of HIV-1 core assembly: insights from three-dimensional reconstructions of authentic virions. *Structure* **14:**15–20.

159. **Jurado KA, Wang H, Slaughter A, Feng L, Kessl JJ, Koh Y, Wang W, Ballandras-Colas A, Patel PA, Fuchs JR, Kvaratskhelia M, Engelman A.** 2013. Allosteric integrase inhibitor potency is determined through the inhibition of HIV-1 particle maturation. *Proc Natl Acad Sci U S A* **110:**8690–8695.

160. **Millar CB, Grunstein M.** 2006. Genome-wide patterns of histone modifications in yeast. *Nat Rev Mol Cell Biol* **7:**657–666.

161. **Brodeur GM, Sandmeyer SB, Olson MV.** 1983. Consistent association between sigma elements and tRNA genes in yeast. *Proc Natl Acad Sci U S A* **80:**3292–3296.

162. **Chalker DL, Sandmeyer SB.** 1992. Ty3 integrates within the region of RNA polymerase III transcription initiation. *Genes Dev* **6:**117–128.

163. **Dieci G, Fiorino G, Castelnuovo M, Teichmann M, Pagano A.** 2007. The expanding RNA polymerase III transcriptome. *Trends Genet* **23:**614–622.

164. **Geiduschek EP, Kassavetis GA.** 2001. The RNA polymerase III transcription apparatus. *J Mol Biol* **310:**1–26.

165. **Kassavetis GA, Geiduschek EP.** 2006. Transcription factor TFIIIB and transcription by RNA polymerase III. *Biochem Soc Trans* **34:**1082–1087.

166. **Kassavetis GA, Letts GA, Geiduschek EP.** 2001. The RNA polymerase III transcription initiation factor TFIIIB participates in two steps of promoter opening. *EMBO J* **20:**2823–2834.

167. **Giuliodori S, Percudani R, Braglia P, Ferrari R, Guffanti E, Ottonello S, Dieci G.** 2003. A composite upstream sequence motif potentiates tRNA gene transcription in yeast. *J Mol Biol* **333:**1–20.

168. **Eschenlauer JB, Kaiser MW, Gerlach VL, Brow DA.** 1993. Architecture of a yeast U6 RNA gene promoter. *Mol Cell Biol* **13:**3015–3026.

169. **Kassavetis GA, Braun BR, Nguyen LH, Geiduschek EP.** 1990. S. cerevisiae TFIIIB is the transcription initiation factor proper of RNA polymerase III, while TFIIIA and TFIIIC are assembly factors. *Cell* **60:**235–245.

170. **Roberts DN, Stewart AJ, Huff JT, Cairns BR.** 2003. The RNA polymerase III transcriptome revealed by genome-wide localization and activity-occupancy relationships. *Proc Natl Acad Sci U S A* **100:**14695–14700.

171. **Moqtaderi Z, Struhl K.** 2004. Genome-wide occupancy profile of the RNA polymerase III machinery in *Saccharomyces cerevisiae* reveals loci with incomplete transcription complexes. *Mol Cell Biol* **24:**4118–4127.

172. **Harismendy O, Gendrel CG, Soularue P, Gidrol X, Sentenac A, Werner M, Lefebvre O.** 2003. Genome-wide location of yeast RNA polymerase III transcription machinery. *EMBO J* **22:**4738–4747.

173. **Chalker DL, Sandmeyer SB.** 1993. Sites of RNA polymerase III transcription initiation and Ty3 integration at the U6 gene are positioned by the TATA box. *Proc Natl Acad Sci U S A* **90:**4927–4931.

174. **Kirchner J, Connolly CM, Sandmeyer SB.** 1995. Requirement of RNA polymerase III transcription factors for *in vitro* position-specific integration of a retrovirus-like element. *Science* **267:**1488–1491.

175. **Yieh L, Hatzis H, Kassavetis G, Sandmeyer SB.** 2002. Mutational analysis of the transcription factor IIIB-DNA target of Ty3 retroelement integration. *J Biol Chem* **277:**25920–25928.

176. **Yieh L, Kassavetis G, Geiduschek EP, Sandmeyer SB.** 2000. The Brf and TATA-binding protein subunits of the RNA polymerase III transcription factor IIIB mediate position-specific integration of the gypsy-like element, Ty3. *J Biol Chem* **275:**29800–29807.

177. **Connolly CM, Sandmeyer SB.** 1997. RNA polymerase III interferes with Ty3 integration. *FEBS Lett* **405:**305–311.

178. **Kassavetis GA, Soragni E, Driscoll R, Geiduschek EP.** 2005. Reconfiguring the connectivity of a multiprotein complex: fusions of yeast TATA-binding protein with Brf1, and the function of transcription factor IIIB. *Proc Natl Acad Sci U S A* **102:**15406–15411.

179. **Parnell TJ, Huff JT, Cairns BR.** 2008. RSC regulates nucleosome positioning at Pol II genes and density at Pol III genes. *EMBO J* **27:**100–110.

180. **Orioli A, Pascali C, Pagano A, Teichmann M, Dieci G.** 2012. RNA polymerase III transcription control elements: themes and variations. *Gene* **493:**185–194.

181. **Good PD, Kendall A, Ignatz-Hoover J, Miller EL, Pai DA, Rivera SR, Carrick B, Engelke DR.** 2013. Silencing near tRNA genes is nucleosome-mediated and distinct from boundary element function. *Gene* **526:**7–15.

182. **Kirkland JG, Raab JR, Kamakaka RT.** 2013. TFIIIC bound DNA elements in nuclear organization and insulation. *Biochim Biophys Acta* **1829:**418–424.

183. **Acker J, Conesa C, Lefebvre O.** 2013. Yeast RNA polymerase III transcription factors and effectors. *Biochim Biophys Acta* **1829:**283–295.

184. **Mahapatra S, Dewari PS, Bhardwaj A, Bhargava P.** 2011. Yeast H2A.Z, FACT complex and RSC regulate transcription of tRNA gene through differential dynamics of flanking nucleosomes. *Nucleic Acids Res* **39:**4023–4034.

185. **Nagarajavel V, Iben JR, Howard BH, Maraia RJ, Clark DJ.** 2013. Global 'bootprinting' reveals the elastic architecture of the yeast TFIIIB-TFIIIC transcription complex *in vivo*. *Nucleic Acids Res* **41:**8135–8143.

186. **Van Bortle K, Corces VG.** 2012. tDNA insulators and the emerging role of TFIIIC in genome organization. *Transcription* **3:**277–284.

187. **Kumar Y, Bhargava P.** 2013. A unique nucleosome arrangement, maintained actively by chromatin remodelers facilitates transcription of yeast tRNA genes. *BMC Genomics* **14:**402.

188. **Mou Z, Kenny AE, Curcio MJ.** 2006. Hos2 and Set3 promote integration of Ty1 retrotransposons at tRNA genes in *Saccharomyces cerevisiae*. *Genetics* **172:**2157–2167.

189. **Morse RH, Roth SY, Simpson RT.** 1992. A transcriptionally active tRNA gene interferes with nucleosome positioning *in vivo*. *Mol Cell Biol* **12:**4015–4025.

190. **D'Ambrosio C, Schmidt CK, Katou Y, Kelly G, Itoh T, Shirahige K, Uhlmann F.** 2008. Identification of cis-acting sites for condensin loading onto budding yeast chromosomes. *Genes Dev* **22:**2215–2227.

191. **Deshpande AM, Newlon CS.** 1996. DNA replication fork pause sites dependent on transcription. *Science* **272:**1030–1033.

192. **Thompson M, Haeusler RA, Good PD, Engelke DR.** 2003. Nucleolar clustering of dispersed tRNA genes. *Science* **302:**1399–1401.

193. **Aye M, Dildine SL, Claypool JA, Jourdain S, Sandmeyer SB.** 2001. A truncation mutant of the 95-kilodalton

subunit of transcription factor IIIC reveals asymmetry in Ty3 integration. *Mol Cell Biol* **21**:7839–7851.

194. **Rinke J, Steitz JA.** 1982. Precursor molecules of both human 5S ribosomal RNA and transfer RNAs are bound by a cellular protein reactive with anti-La lupus antibodies. *Cell* **29**:149–159.

195. **Wang GP, Ciuffi A, Leipzig J, Berry CC, Bushman FD.** 2007. HIV integration site selection: analysis by massively parallel pyrosequencing reveals association with epigenetic modifications. *Genome Res* **17**:1186–1194.

196. **Eidahl JO, Crowe BL, North JA, McKee CJ, Shkriabai N, Feng L, Plumb M, Graham RL, Gorelick RJ, Hess S, Poirier MG, Foster MP, Kvaratskhelia M.** 2013. Structural basis for high-affinity binding of LEDGF PWWP to mononucleosomes. *Nucleic Acids Res* **41**:3924–3936.

197. **Sharma A, Larue RC, Plumb MR, Malani N, Male F, Slaughter A, Kessl JJ, Shkriabai N, Coward E, Aiyer SS, Green PL, Wu L, Roth MJ, Bushman FD, Kvaratskhelia M.** 2013. BET proteins promote efficient murine leukemia virus integration at transcription start sites. *Proc Natl Acad Sci U S A* **110**:12036–12041.

198. **Risler JK, Kenny AE, Palumbo RJ, Gamache ER, Curcio MJ.** 2012. Host co-factors of the retrovirus-like transposon Ty1. *Mob DNA* **3**:12.

199. **Scholes DT, Banerjee M, Bowen B, Curcio MJ.** 2001. Multiple regulators of Ty1 transposition in *Saccharomyces cerevisiae* have conserved roles in genome maintenance. *Genetics* **159**:1449–1465.

200. **Nyswaner KM, Checkley MA, Yi M, Stephens RM, Garfinkel DJ.** 2008. Chromatin-associated genes protect the yeast genome from Ty1 insertional mutagenesis. *Genetics* **178**:197–214.

201. **Curcio MJ, Kenny AE, Moore S, Garfinkel DJ, Weintraub M, Gamache ER, Scholes DT.** 2007. S-phase checkpoint pathways stimulate the mobility of the retrovirus-like transposon Ty1. *Mol Cell Biol* **27**:8874–8885.

202. **Saksena S, Sun J, Chu T, Emr SD.** 2007. ESCRTing proteins in the endocytic pathway. *Trends Biochem Sci* **32**:561–573.

203. **Webster BM, Colombi P, Jager J, Lusk CP.** 2014. Surveillance of nuclear pore complex assembly by ESCRT-III/Vps4. *Cell* **159**:388–401.

204. **Bushman FD, Malani N, Fernandes J, D'Orso I, Cagney G, Diamond TL, Zhou H, Hazuda DJ, Espeseth AS, Konig R, Bandyopadhyay S, Ideker T, Goff SP, Krogan NJ, Frankel AD, Young JA, Chanda SK.** 2009. Host cell factors in HIV replication: meta-analysis of genome-wide studies. *PLoS Pathog* **5**:e1000437.

205. **Brass AL, Dykxhoorn DM, Benita Y, Yan N, Engelman A, Xavier RJ, Lieberman J, Elledge SJ.** 2008. Identification of host proteins required for HIV infection through a functional genomic screen. *Science* **319**:921–926.

206. **Konig R, Zhou Y, Elleder D, Diamond TL, Bonamy GM, Irelan JT, Chiang CY, Tu BP, De Jesus PD, Lilley CE, Seidel S, Opaluch AM, Caldwell JS, Weitzman MD, Kuhen KL, Bandyopadhyay S, Ideker T, Orth AP, Miraglia LJ, Bushman FD, Young JA, Chanda SK.** 2008. Global analysis of host–pathogen interactions that regulate early-stage HIV-1 replication. *Cell* **135**: 49–60.

207. **Yeung ML, Houzet L, Yedavalli VS, Jeang KT.** 2009. A genome-wide short hairpin RNA screening of jurkat T-cells for human proteins contributing to productive HIV-1 replication. *J Biol Chem* **284**:19463–19473.

208. **Zhou H, Xu M, Huang Q, Gates AT, Zhang XD, Castle JC, Stec E, Ferrer M, Strulovici B, Hazuda DJ, Espeseth AS.** 2008. Genome-scale RNAi screen for host factors required for HIV replication. *Cell Host Microbe* **4**:495–504.

209. **Wolf D, Goff SP.** 2008. Host restriction factors blocking retroviral replication. *Annu Rev Genet* **42**:143–163.

210. **Goff SP.** 2007. Host factors exploited by retroviruses. *Nat Rev Microbiol* **5**:253–263.

211. **Freed EO, Mouland AJ.** 2006. The cell biology of HIV-1 and other retroviruses. *Retrovirology* **3**:77.

212. **Ebina H, Aoki J, Hatta S, Yoshida T, Koyanagi Y.** 2004. Role of Nup98 in nuclear entry of human immunodeficiency virus type 1 cDNA. *Microbes Infect / Inst Pasteur* **6**:715–724.

213. **Di Nunzio F, Fricke T, Miccio A, Valle-Casuso JC, Perez P, Souque P, Rizzi E, Severgnini M, Mavilio F, Charneau P, Diaz-Griffero F.** 2013. Nup153 and Nup98 bind the HIV-1 core and contribute to the early steps of HIV-1 replication. *Virology* **440**:8–18.

214. **Koh Y, Wu X, Ferris AL, Matreyek KA, Smith SJ, Lee K, KewalRamani VN, Hughes SH, Engelman A.** 2013. Differential effects of human immunodeficiency virus type 1 capsid and cellular factors nucleoporin 153 and LEDGF/p75 on the efficiency and specificity of viral DNA integration. *J Virol* **87**:648–658.

215. **Goodier JL, Mandal PK, Zhang L, Kazazian HH Jr.** 2010. Discrete subcellular partitioning of human retrotransposon RNAs despite a common mechanism of genome insertion. *Hum Mol Genet* **19**:1712–1725.

216. **Doucet AJ, Hulme AE, Sahinovic E, Kulpa DA, Moldovan JB, Kopera HC, Athanikar JN, Hasnaoui M, Bucheton A, Moran JV, Gilbert N.** 2010. Characterization of LINE-1 ribonucleoprotein particles. *PLoS Genet* **6**(10):pii: e1001150.

217. **Bann DV, Beyer AR, Parent LJ.** 2014. A murine retrovirus co-Opts YB-1, a translational regulator and stress granule-associated protein, to facilitate virus assembly. *J Virol* **88**:4434–4450.

218. **Abrahamyan LG, Chatel-Chaix L, Ajamian L, Milev MP, Monette A, Clement JF, Song R, Lehmann M, DesGroseillers L, Laughrea M, Boccaccio G, Mouland AJ.** 2010. Novel Staufen1 ribonucleoproteins prevent formation of stress granules but favour encapsidation of HIV-1 genomic RNA. *J Cell Sci* **123**:369–383.

219. **Goodier JL, Cheung LE, Kazazian HH Jr.** 2012. MOV10 RNA helicase is a potent inhibitor of retrotransposition in cells. *PLoS Genet* **8**:e1002941.

220. **Furtak V, Mulky A, Rawlings SA, Kozhaya L, Lee K, Kewalramani VN, Unutmaz D.** 2010. Perturbation of

the P-body component Mov10 inhibits HIV-1 infectivity. *PLoS One* 5:e9081.

221. **Arjan-Odedra S, Swanson CM, Sherer NM, Wolinsky SM, Malim MH.** 2012. Endogenous MOV10 inhibits the retrotransposition of endogenous retroelements but not the replication of exogenous retroviruses. *Retrovirology* **9:**53.

222. **Malim MH, Bieniasz PD.** 2012. HIV Restriction Factors and Mechanisms of Evasion. *Cold Spring Harbor Persp Med* **2:**a006940.

223. **Lu C, Contreras X, Peterlin BM.** 2011. P bodies inhibit retrotransposition of endogenous intracisternal a particles. *J Virol* **85:**6244–6251.

224. **Phalora PK, Sherer NM, Wolinsky SM, Swanson CM, Malim MH.** 2012. HIV-1 replication and APOBEC3 antiviral activity are not regulated by P bodies. *J Virol* **86:**11712–11724.

225. **Reed JC, Molter B, Geary CD, McNevin J, McElrath J, Giri S, Klein KC, Lingappa JR.** 2012. HIV-1 Gag co-opts a cellular complex containing DDX6, a helicase that facilitates capsid assembly. *J Cell Biol* **198:**439–456.

226. **Saito K, Siomi MC.** 2010. Small RNA-mediated quiescence of transposable elements in animals. *Dev Cell* **19:** 687–697.

227. **Aravin AA, Hannon GJ, Brennecke J.** 2007. The Piwi-piRNA pathway provides an adaptive defense in the transposon arms race. *Science* **318:**761–764.

228. **Voronina E, Seydoux G, Sassone-Corsi P, Nagamori I.** 2011. RNA granules in germ cells. *Cold Spring Harbor Perspectives in Biology* **3:**a002774.

Mobile DNA, 3rd Edition
Nancy L. Craig, Michael Chandler, Martin Gellert, Alan M. Lambowitz, Phoebe A. Rice and Suzanne Sandmeyer

doi:10.1128/microbiolspec.MDNA3-0040-2014

Caroline Esnault[1]
Henry L. Levin[1]

43 The Long Terminal Repeat Retrotransposons Tf1 and Tf2 of *Schizosaccharomyces pombe*

INTRODUCTION

The fission yeast *Schizosaccharomyces pombe*, discovered in the late 1800s in East Africa and genetically characterized in the 1950s by Urs Leupold (97), has became a central model for studies of cell cycle, gene expression, and the complex relationship between transposable elements (TEs) and their host. *Schizosaccharomyces pombe*, also known as fission yeast, can be studied with a sophisticated toolbox of molecular and genetic techniques. The haploid genome is 12.57 Mbp and encodes 5,052 genes distributed among three chromosomes (1). The complete genome sequence of the Leupold isolate revealed that TEs constitute 1.1% of the genome (2) (Table 1). All TE-related sequences in *S. pombe* derive from long terminal repeat (LTR) retrotransposons. The intact elements present in the Leupold strain are 13 full-length copies of the Tf2 element (3). Recombination that occurs between the LTRs of a full-length element results in solo LTRs that serve as a fossil record of TEs that are no longer present. The Leupold strain contains 249 solo LTRs or LTR fragments that are derived from nine clades of LTR retrotransposons. The youngest clades are the 35 LTR sequences from Tf2 and the 28 LTRs from Tf1, an element related to Tf2 but that is no longer present in the Leupold strain (2). Full-length Tf1 elements are present in wild isolates of *S. pombe* collected from different geographic regions (3). The transposition activity of Tf1 and the function of its proteins is measured by expressing a plasmid-encoded copy of Tf1 that contains *neo* (4, 5). Levels of Tf1 transposition correspond to amounts of G418 resistance. *wtfs* are another form of repeat identified in the Leupold strain (6). They are present in 25 copies and are generally 250 bp downstream of an LTR (2). Their function is unknown, but they appear to encode protein and their transcription is strongly induced during meiosis (7, 8).

THE CYCLE OF Tf1 TRANSPOSITION

The full-length Tf1 and Tf2 elements are flanked by two LTRs and encode Gag, protease (PR), reverse transcriptase (RT), and integrase (IN) (Fig. 1, Table 2). The LTR sequences of Tf1 and Tf2 share stretches of sequence

[1]Section on Eukaryotic Transposable Elements, Program in Cellular Regulation and Metabolism, Eunice Kennedy Shriver National Institute of Child Health and Human Development, National Institutes of Health, Bethesda, MD 20892.

Table 1 Features and content in transposable elements of the *Schizosaccharomyces pombe* genome

Feature	Number
Genome size (sequenced portion, excluding rDNA)[a]	12.57 Mb
Chromosome number[a]	3
Chromosome size range[a]	2.45–5.58 Mb
Overall GC content[a]	36.06%
Gene count (protein coding genes on Chr I, II and III)[a]	5,123
Gene count excluding dubious genes (protein coding genes on Chr I, II and III)[a]	5,052
Gene density (genes per Mb)[a]	554.41
Transposon related sequences in strains derived from 972	
Full-length Tf2 elements[b]	13
Full-length Tf1 elements[b]	0
Solo Tf2-related LTRs[b,d]	35
Solo Tf1-related LTRs[b,d]	28
Solo LTRs non-Tf1 nor Tf2[b,d]	111
Fragments related to LTR[b,e]	75
wtf elements[c]	25

[a]Ref. (1).
[b]Ref. (2).
[c]Ref. (95).
[d]> 200 bp in length.
[e]< 200 bp in length.

identity and although the Gags of these two elements are divergent, the coding sequences for RT and IN are virtually identical (3, 9). The structure of these TEs and the function of their proteins are closely related to those of retroviruses. Indeed, the order of RT before IN in the open reading frame (ORF) and phylogenetic analysis of the coding sequences place Tf1 and Tf2 firmly within the Gypsy/Ty3 family of LTR retrotransposons, a clade that includes elements with envelope-like proteins that appear to mediate infection (10–14).

The transposition cycle of Tf1 begins with transcription of the full-length mRNA, which is then exported to the cytoplasm where it is translated (Fig. 1). Most retroviruses and LTR retrotransposons encode two separate ORFs for the *pol* and *gag* proteins, and PR cleaves the polyproteins into their mature size. Mechanisms such as frame shifting that result in low levels of the Pol proteins RT and IN relative to Gag are thought to be important to allow efficient particle assembly. However, Tf1 encodes one long polyprotein. The processing activity of PR cleaves the polyprotein into the mature sized Gag, PR, RT, and IN (15). An unknown activity degrades RT and IN to produce the appropriate low levels relative to Gag that allow for particle assembly (16). Gag forms the coat of the virus-like particle, a 50-nm structure that contains at least two copies of Tf1 mRNA, RT, and IN (15, 17, 18). RT copies the mRNA into double-stranded cDNA, which is transported into the nucleus where IN mediates the integration of the cDNA fragment at a new genomic position and generates a 5-bp target site duplication (5).

As is true for the Gag proteins of retroviruses, Tf1 Gag is sufficient for the formation of virus-like particles and is required for the packaging of Tf1 mRNA and reverse transcription (18). Deletion analyses of Gag identified an N-terminal nuclear localization signal and adjacent residues that modulate the mode of import (19, 20). The import of Tf1 Gag is governed by a unique requirement for the FXFG nuclear pore factor Nup124p and mutations that compromise virus-like particle structure bypass the requirement for Nup124p (19–21). These data, together with the finding that Gag interacts directly with Nup124p, led to the model that Nup124p is required for Tf1 import because it accommodates the particle-like structure of Tf1 Gag (21). Interestingly, the role of Nup124p in particle import may be conserved in humans. Nup153, the component of the human nuclear pore complex that is most closely related to Nup124p, contains a domain that can mediate the import of Tf1 Gag in *S. pombe* that lack Nup124p (22, 23). Similar to the role of Nup124p in the import of Tf1 Gag, Nup153p directly engages the capsid protein of HIV-1 and is required for infectivity (24–27).

Another host factor that mediates the import of Tf1 Gag into the nucleus is the histone acetyltransferase interacting factor Pst1p and experiments indicate that acetyltransferase activity is necessary for transport of Gag into the nucleus (28).

The reverse transcription of Tf1 mRNA into full-length cDNA relies on the same intermediates and strand transfers of these intermediates as other LTR retroelements (see Mobile DNA III chapter 46, *Reverse Transcription of Retroviruses and LTR Retrotransposons* by S. Hughes). However, unlike typical LTR retrotransposons and retroviruses, Tf1 does not require a tRNA to initiate reverse transcription. Instead, Tf1 possesses a unique mechanism of self-priming that relies on the priming activity of the first 11 bp of the Tf1 mRNA (4, 29, 30). This 11-bp segment of Tf1 RNA anneals to the primer binding site in the mRNA and the RNaseH domain of RT cleaves the mRNA between bases 11 and 12 (Fig. 2A) (30–32). The self-primer is then used by RT to initiate minus-strand cDNA synthesis.

Self-priming of reverse transcription requires the formation of a complex RNA structure that includes duplexes of several segments of the Tf1 mRNA (Fig. 2B) (33). One element of the RNA structure termed the U5-IR stem loop includes a downstream segment of the LTR

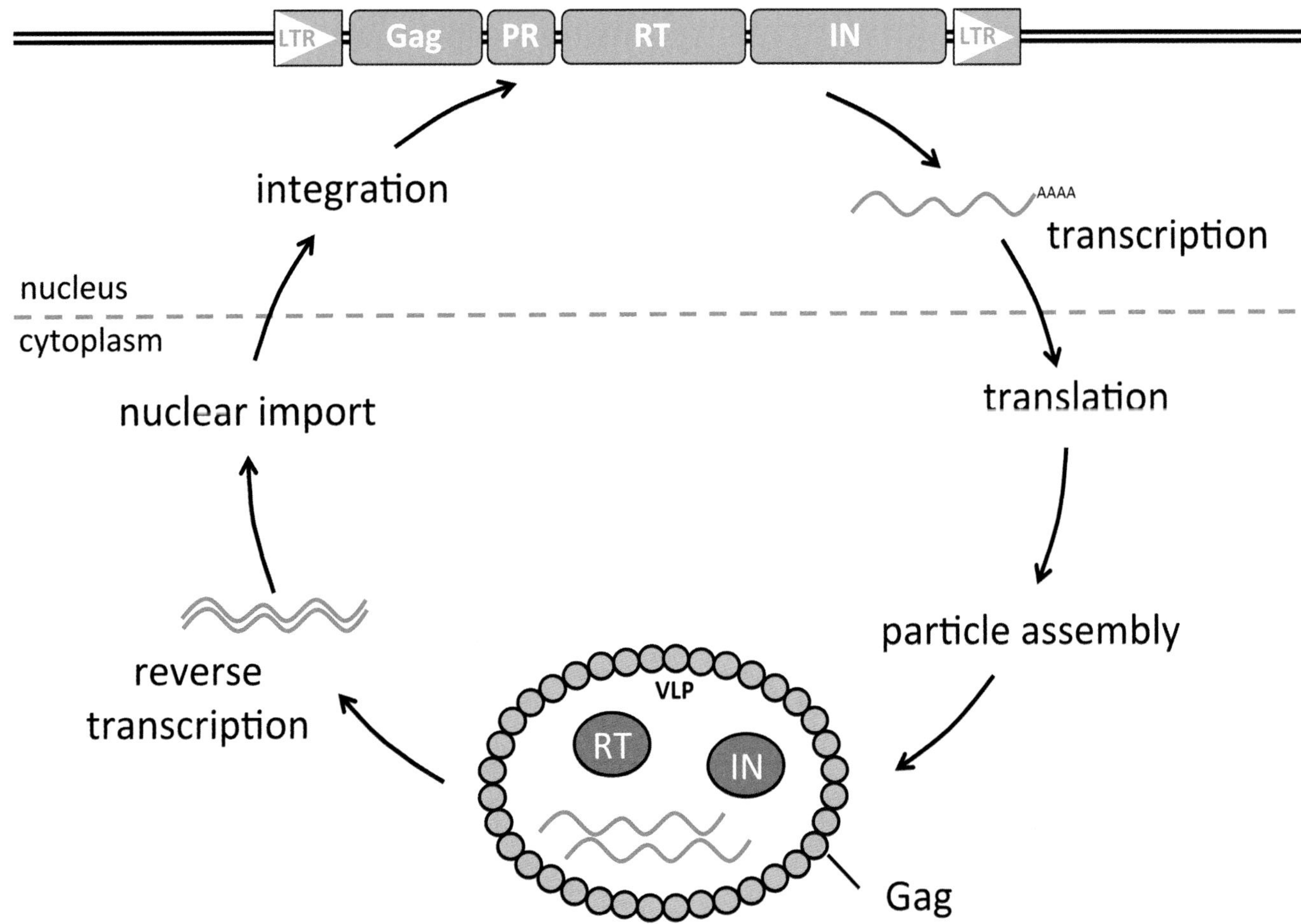

Figure 1 Tf1 transposition in *Schizosaccharomyces pombe*. Full-length Tf1 (blue rectangles) is transcribed and translated. The polyprotein assembles into particle precursors and PR processes the protein into Gag (pink), protease (PR), reverse transcriptase (RT) (orange), and integrase (IN) (orange). The mature virus-like particles (VLP) contain the processed proteins and two copies of mRNA (blue wavy lines). The RT reverse-transcribes the mRNA into double-stranded cDNA that associates with IN. Once transported into the nucleus (above dotted line), the IN integrates the cDNA at a new position in the genome. The long terminal repeats are symbolized by triangles. The protein coding sequences are represented by rectangles. doi:10.1128/microbiolspec.MDNA3-0040-2014.f1

(U5) and is required for self-priming. This structure is surprisingly similar to a feature in the mRNA of Rous sarcoma virus that is also necessary for reverse transcription (34). This structural similarity indicates that there is strong conservation in the way RT recognizes the minus-strand primer. Nevertheless, the lack of a potential tRNA primer and the formation of a duplex between the primer binding site and the 5′ end of the mRNA are hallmarks of Tf1 that are central to the self-priming mechanism. Scrutiny of many other LTR retrotransposons reveals that Tf1 and Tf2 are the founding members of a large class of elements that also use self-priming to initiate reverse transcription (35, 36). While virtually all of the self-priming elements belong to the Ty3-gypsy family of retrotransposons, one element from the Ty1-copia family also has the potential to self-prime (37).

In another twist of Tf1 reverse transcription the self-primer is not removed during reverse transcription, which results in 11 additional bases at the 3′ end of the plus strand cDNA that are complementary to the self-primer (38). Biochemical experiments with recombinant protein show that IN has an unusual processing activity

Table 2 Sizes of transposable elements and the molecular weights of their components in *Schizosaccharomyces pombe*

Feature	Size
Tf1	4,941 bp[a]
Tf1 ORF	1,330 aa[a]
Tf1 LTR	358 bp[a]
Tf2	4,916 bp[b]
Tf2 ORF	1,333 aa[b]
Tf2 LTR	349 bp[b]
Tf1 Gag	27 kDa[c]
Tf1 RT	60 kDa[d]
Tf1 IN	56 kDa[c]

[a]Ref. (3).
[b]Ref. (9).
[c]Ref. (15).
[d]Ref. (54).
Abbreviations: IN, integrase; LTR, long terminal repeat; ORF, open reading frame; RT, reverse transcriptase.

that removes these 11 bases from the 3′ end of the plus strand cDNA and allows integration to occur (38, 39). The plus-strand primer of reverse transcription or polypurine tract (PPT) contains 13 ribonucleotides positioned immediately 5′ of the downstream LTR. Analysis with single nucleotide resolution of the 5′ plus-strand and 3′ minus-strand ends of the cDNA indicate the PPT consists of UUGGGGAGGGCAA (31). *In vivo* and biochemical experiments show that the PPT is removed by RT before reverse transcription is complete and that purified RT can remove the self-primer from the 5′ end of the minus-strand (31, 40). Interestingly, the PPT removal activity of RT is greatly stimulated by IN (40). Other biochemical assays demonstrate that Tf1 RT possesses several unusual activities such as template-independent DNA synthesis and the ability to clamp very short primer terminal sequences onto the 3′ termini of template strands (41–44).

The IN of Tf1 has many similarities with INs of retroviruses and other LTR retrotransposons including the 3′ processing activity that exposes the 3′ CA of the cDNA and the strand transfer activity that simultaneously cleaves the phosphodiester bonds in the target DNA and catalyzes formation of the covalent bond between the 3′ end of the LTR and the 5′ end of the target DNA (see Mobile DNA III chapter 45, *Host Factors in Retroviral Integration and the Selection of Integration Target Sites* by F. Bushman and R. Craigie) (39). The structure of retrovirus and LTR retrotransposon INs is divided into three domains that contain conserved motifs. The N-terminal domain binds zinc and contains an

A

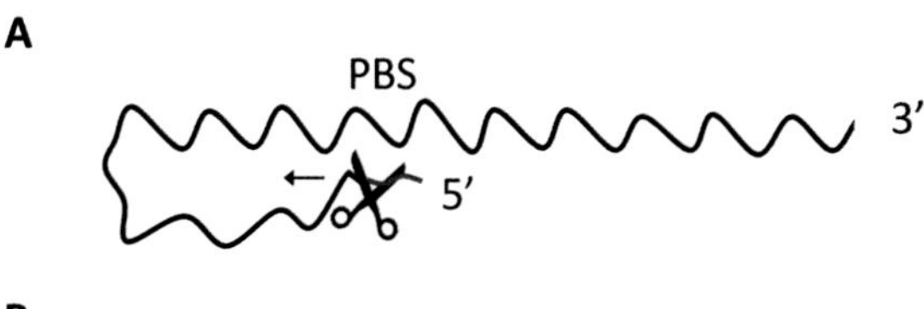

B

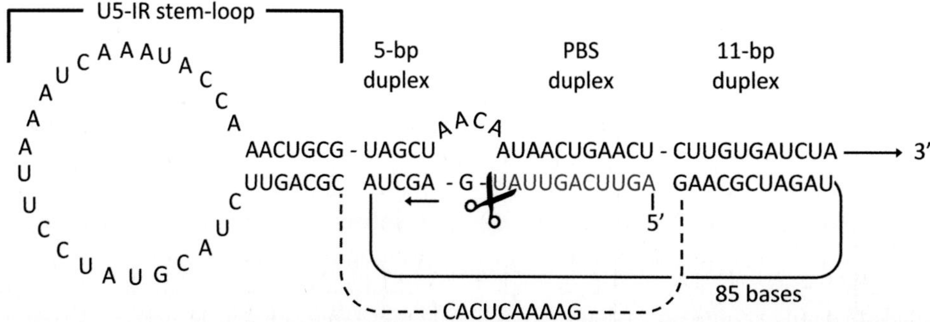

Figure 2 Reverse transcription of Tf1 is initiated by a self-priming mechanism. (A) The 5′ end of the Tf1 mRNA anneals to the primer binding site (PBS). The first 11 nucleotides (red) are cleaved from the Tf1 mRNA (black) and prime reverse transcription towards the 5′ end of the mRNA. The scissors indicate the position of cleavage. The arrow indicates the direction of reverse transcription. (B) The 5′ end of the Tf1 mRNA folds into a complex duplex structure. The scissors indicate the position of the cleavage that liberates the self-primer (red). The position of the U5-IR stem-loop is indicated.
doi:10.1128/microbiolspec.MDNA3-0040-2014.f2

HHCC motif; the center domain has the catalytic residues and contains the D-D-E motif; and the C-terminal domain has DNA binding activity. Some INs in their C-terminal domain have subdomains, a GP(Y/F) domain, a chromodomain, or both (13). Tf1 possesses both the GP (Y/F) domain and the chromodomain (45). Both of these features are necessary for integration *in vivo* (45, 46). *In vitro* experiments demonstrate that the GP(Y/F) domain is necessary for strand transfer activity, binds DNA, and mediates IN multimerization (45). The chromodomains of INs are similar in structure to the chromodomains of chromatin factors and in cases of plant retrotransposons they direct binding to heterochromatin (47). The chromodomain of the fungal element Maggy directs integration to heterochromatin by interacting with histone H3 methylated at lysine 9 (47). Although no interaction has been detected between the chromodomain of Tf1 IN and histones, it does play a role in binding IN to cDNA *in vivo* (48).

SITE SELECTION OF Tf1 INTEGRATION

Integration of TEs can damage the genetic content of the host. LTR retrotransposons feature strategies to avoid disrupting coding sequences during integration. In *Saccharomyces cerevisiae* the IN of Ty3 interacts with components of TFIIIB to position integration one to four nucleotides upstream of RNA pol III-transcribed genes (see Mobile DNA III chapter 42, *Ty3, A Position-Specific Retrotransposon in Budding Yeast* by S. Sandmeyer and K. Patterson) (49). The IN of Ty5 directs integration to heterochromatin by binding Sir4, a component of heterochromatin (50). In *S. pombe* Tf1 avoids integrating in coding sequences by a different mechanism that is not fully understood. The choice of integration sites is thought to depend on features of Tf1 IN as well as on features of the host factors and sequence.

Existing TE sequences provide evidence of past integration events and can provide information about the integration behavior of elements. The genome sequence of *S. pombe* contains 13 full-length Tf2 elements and 249 solo LTRs or LTR fragments (2) (Table 1). Phylogenetic analysis of solo LTRs demonstrates that Tf1 and Tf2 are the most recently active transposable elements in *S. pombe*. Tf1 and Tf2 sequences are in promoter regions of RNA polymerase II-transcribed genes, in a window located 100–400 bp upstream of ORFs (2). The elements are absent from tRNA gene clusters and are greatly underrepresented in intergenic sequences between convergent genes. Although it is possible that these integrated copies persist because of selective pressure, it is now known that the positions of the existing copies result from integration directed to promoter sequences.

Studies of Tf1 integration within the *S. pombe* genome rely on heterologous promoters to induce the expression of Tf1 with a drug-resistance gene. Selection of cells with *de novo* insertions allowed a total of 27 sites to be isolated and sequenced (51). Schemes that lack selection of insertions in *S. pombe* are also used to generate and sequence insertion sites (52, 53). In two studies not using genetic selection, 51 and 9 insertions were sequenced. Regardless of whether selection is used, the insertions distribute throughout the genome and cluster in promoter regions upstream of ORFs (52, 53).

Unlike Tf1, Tf2 does not depend on IN for the majority of its activity, and mobilizes with a frequency 10- to 20-fold lower than Tf1 (54). When Tf2 marked with *neo* is expressed from a heterologous promoter 70% of events that introduce Tf2-*neo* into the genome result from homologous recombination between Tf2 cDNA and pre-existing copies of Tf2. This mechanism is a different strategy to avoid disrupting the host coding sequences, which has been viewed as a recycling of the integration sites. In addition to integration site recycling, the remaining 30% of mobilization events do occur through true integration events (54). Although the amino acid sequences of Tf1 and Tf2 INs are 98% identical it was possible that the low levels of Tf2 integration were due to the small number of amino acid differences. The results of domain swapping experiments between Tf1 and Tf2 coding sequences revealed that Tf2 IN is fully active. Instead, the low levels of Tf2 integration relative to Tf1 were the result of sequence differences in capsid and PR (55).

To understand what features of promoter sequences are recognized by Tf1, and to facilitate the isolation of independent insertions, integration events are isolated in target plasmids. These plasmids are targets of integration when introduced into cells that express Tf1. Transposition assays with plasmids containing a collection of genes confirm that insertions occur in promoter sequences and reveal that each promoter possesses a unique and reproducible pattern of integration sites (46). By generating a series of deletions in target plasmids, minimal sequences as short as 160 bp and 70 bp are found to function as efficient targets (46, 56). In specific cases the minimal sequences contain binding sites for transcription factors and in the case of insertions in the *fbp1* promoter, binding of the transcription activator Atf1p to the promoter mediates integration (46). The minimal target sequences from the promoters do not support transcription and artificial promoters that do drive transcription are not targets of integration

(46, 56). These results indicate that specific transcription factors, not general transcription machinery, may mediate integration. Although Atf1p mediates integration in the *fpb1* promoter it does not contribute to integration at the promoter of *ade6* and the overall frequency of genome-wide transposition is not reduced in the absence of Atf1p.

The target plasmid assay can also be used to test domains of IN for a role in positioning integration. Tf1 IN lacking the chromodomain can be expressed from a version of Tf1 that has a frameshift mutation in IN. Although this substantially reduces the frequency of transposition, integration events can be isolated in the target plasmid (48). Tf1 IN lacking the chromodomain greatly modifies the insertion sites in target plasmids containing a variety of target promoters. These data suggest that the chromodomain plays an important role in positioning integration. However, the pattern of Tf1 integration genome-wide is not altered when the chromodomain is absent (57). These data indicate that with natural genome targets, the chromodomain does not contribute to the selection of insertion sites.

The development of deep-sequencing technologies has radically increased the number of TE insertions that can be mapped genome-wide. Cells with insertions are generated by overexpressing Tf1 containing *neo*, extracting DNA from G418-resistant cells, and using ligation-mediated polymerase chain reaction to sequence libraries of insertions. A saturated map of Tf1 integration with reproducible numbers of events in each intergenic region can be obtained using 454 pyrosequencing (58). Insertion profiles generated in haploid and diploid strains are indistinguishable. Over 95% of 73,125 independent integration events target intergenic regions, showing again a clear preference for promoters of RNA polymerase II-transcribed genes. An improved method for sequencing Tf1 integration (described below) mapped 800,723 independent insertions and these showed the same preference for promoter regions as observed earlier (Fig. 3). Just 3.8% of the integrations occur in ORFs. The integration pattern relative to ORFs as shown in Fig. 3 is highly similar to the one of existing integration events, indicating that the distribution of pre-existing Tf elements results from integration mechanisms, not selection pressure (2).

The dense profiles of integration produced with deep sequencing show that a specific set of promoters receive the bulk of the insertion events (58). Approximately 20% of the intergenic sequences have higher levels of integration than would occur by random selection. Comparisons with genome-wide profiles of transcription do not identify any correlation between amounts of integration in promoters and the levels of transcription (58). Importantly, the intergenic sequences with enhanced integration are enriched with promoters of stress-induced genes (57, 58). Promoters of genes that are induced by cadmium, hydrogen peroxide, heat, methylmethane sulfonate, and sorbitol display high levels of Tf1 integration. This pattern supports the model that specific transcription factors mediate the pattern of integration. The bias favoring integration in stress-induced promoters is also seen in diploid strains,

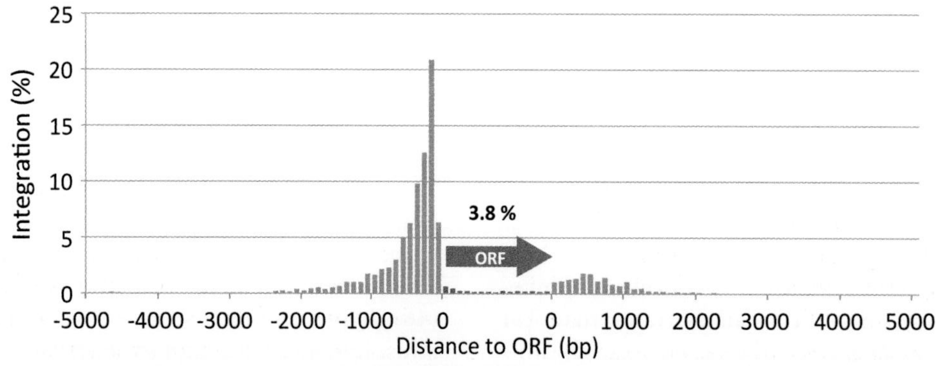

Figure 3 Distribution of the distance from Tf1 integration to the nearest open reading frame (ORF). A collection of 800,723 independent integration events was positioned according to the distance to the nearest ORF (red) (57). The *x*-axis represents the distance from the 5′ end and 3′ end of ORFs in bins that are 100 bp wide; distances within ORFs are divided into 15 bins of equal proportion. The *y*-axis represents the percentage of independent integration events positioned within the bins. The majority of integration events are located near the 5′ end of ORFs, 3.8% of all integration events targeted coding sequences. Figure reproduced with permission from ref. (57).
doi:10.1128/microbiolspec.MDNA3-0040-2014.f3

indicating that this insertion preference is the result of integration bias, not selection occurring after integration (57, 58). In contrast to the integration in promoter sequences, the 3.5% of insertion events that occur in ORFs exhibit no gene-specific bias and levels of integration in individual ORFs vary between experiments (58). This indicates that Tf1 integration may occur by two discrete mechanisms, one that directs the bulk of integration to specific promoters, and one that distributes low levels of integration randomly.

Another type of factor that may play a role in positioning integration is the DNA binding protein Sap1. This essential factor binds to DNA with sequence specificity, blocks replication forks in rDNA repeats, and is required for the imprinting and recombination mechanisms that cause mating type switching (59–64). Interestingly, Sap1 binding sites show a strong correlation with positions of Tf1 integration (65). This link with Sap1 binding sites suggests that Sap1 plays a role in integration.

Recent efforts to characterize patterns of Tf1 integration include a new method of sequencing that maps significantly more insertion events than previous techniques. This method, termed the serial number system, provides more integration data by distinguishing events that occur at the same nucleotide position as independent (57). Tf1 expression plasmids are induced in cultures of *S. pombe* to generate large numbers of insertions. But in contrast to previous strategies, diverse libraries of expression plasmids are used that contain 8-bp serial numbers consisting of random nucleotides embedded near the downstream end of the LTR. Insertions that occur at the same nucleotide position and orientation have different serial numbers and can be identified as independent events because the sequence reads that reveal the position of integration also include the serial number sequence. With this method as many as 455,000 insertion events can be sequenced from a single culture of *S. pombe* (57). The integration data from the serial number system have the added dimension of tabulating numbers of insertions at each nucleotide coordinate, creating profiles of peaks, spikes, and flat terrains. Comparison of integration profiles between independent experiments shows very high reproducibility. The tabulation of the number of integration events at individual positions reveals that 75% of all the integration occurs at just 35% of the insertion positions. The integration positions with the highest number of repeated insertions have strong nucleotide preferences, traces of which are observed at the 35% of the positions where 75% of the events occur (57).

HOST DEFENSES AGAINST Tf ELEMENTS IN *S. POMBE*

Unrestricted transposition is detrimental to the host. Even when integration is directed to gene-free regions the over-accumulation of TEs is a burden on the DNA replication of the host and reduces genome stability. As a result, host organisms such as *S. pombe* have evolved a variety of defense mechanisms that inhibit TE activity (Table 3) (66, 67). RNA interference (RNAi) is a highly conserved system that inhibits the expression of TEs either by degrading their mRNA or by assembling TE sequences into heterochromatin (68, 69). In *S. pombe*, RNAi is mediated by the Argonaute family protein Ago1, the RNA cleaving factor Dcr1, and the RNA-dependent RNA polymerase Rdr1 (70, 71). In vertebrates, insects, nematodes, and plants, RNAi strongly inhibits the expression of TEs (72). Surprisingly, deletion of *ago1*, *dcr1*, or *rdr1* resulted only in slight increases in Tf2 expression and did not affect the expression of genes adjacent to LTRs, suggesting that RNAi has a relatively minor role in inhibiting the transcription of TEs in *S. pombe* (73). The chromatin features that determine heterochromatin-mediated silencing, histone H3 methylated on lysine 9 and the HP1 homolog swi6, are not associated with Tf sequences (74). Furthermore, deep sequencing of small interfering RNAs (siRNAs) does not detect Tf-derived sequences (74–76). Interestingly, another yeast from the *Schizosaccharomyces* genus, *Schizosaccharomyces japonicus*, possesses an extensive population of LTR retrotransposons and mRNAs of these elements are processed into siRNAs (77).

Although the RNAi machinery of *S. pombe* does not appear to limit TE activity under optimal conditions, cells with defects in nuclear exosome activity possess a strong RNAi response that processes Tf2 mRNA into siRNAs and assembles Tf2 sequences into heterochromatin (Table 3) (78). The targeting of Tf2s by RNAi relies on the recognition of a cryptic intron by Nrl1, a protein associated with splicing factors (79). Splicing of the cryptic intron is predicted to remove 20 amino acids from RT near the N terminus of the protein. The incorporation of Tf2 into heterochromatin when cells lack the exosome demonstrates that there is competition between the exosome and RNAi for Tf2 mRNA. Under conditions of nutrient deprivation, such as low glucose, the competition is altered such that Tf2 mRNAs are processed into siRNAs and Tf2 sequences assemble into heterochromatin even when cells possess functional exosomes (78).

The transcription apparatus of the host inhibits the expression of genes that are not necessary during

Table 3 Host factors that restrict transposable element activity

Host defense machinery	Host factors / Function	Regulatory impact on Tf elements	Refs
Exosome	Pla1: canonical polyA polymerase	Elimination of transcripts via the exosome	(78)
	Red1: RNA surveillance factor		(96)
	Pab2: polyA binding protein		
	Rrp6: exosome subunit		
	Mtl1: component of Red1 associated MTREC complex		(79)
RNAi	Nrl1: splicing factor binding protein	Activation of RNAi response through recognition of cryptic introns	(79)
Under exosome deficiency or nutrient starvation	Clr4: histone methyltransferase	Activation of RNAi response through histone methylation	(78)
	Ago1: argonaute, RNA binding Dcr1: dicer, ribonuclease Rdr1: RNA-directed RNA polymerase	Processing of mRNA into siRNAs and the formation of heterochromatin	(78)
	Swi6: HP1 chromodomain protein	Heterochromatinization	(74)
Nucleosome deposition by histone chaperone complex HIRA	Hip1: hira histone chaperone Hip3: HIRA interacting protein Slm9: HIRA protein	Inhibition of transcription through nucleosome deposition	(80,81)
CENP-B proteins and associated factors	Abp1/Cbp1: CENP-B homolog Cbh1: CENP-B homolog Cbh2: CENP-B homolog	Inhibition of transcription Restriction of homologous recombination Recruitment of HDACs Clr3 and Clr6	(86,87)
	Clr3: histone deacetylase Clr6: histone deacetylase	Inhibition of transcription through histone hypoacetylation	(73,86,87)
	Set1: histone methyltransferase	Inhibition of sense and antisense transcription	(87)
	Ku70: Ku domain protein, DNA binding Ku80: Ku domain protein, DNA binding	Clustering into Tf-bodies	(88)

vegetative growth in ideal conditions. The histone chaperone complex Histone Regulator A (HIRA) deposits nucleosomes independent of DNA replication and this prevents the transcription of many genes when cells grow in ideal conditions (80). Interestingly, HIRA also plays a key role in inhibiting TE expression in *S. pombe*. Cells lacking either of the two HIRA factors, Hip1 or Slm9, express high levels of Tf2 mRNA and noncoding transcripts that initiate at solo LTRs (80, 81).

Another class of proteins that inhibit the expression of TEs in *S. pombe* is the centromere protein B (CENP-B) proteins. CENP-B factors are a conserved set of proteins that bind DNA repeats in many species from *S. pombe* to humans (82). Interestingly, CENP-B proteins are thought to have evolved from an ancient pogo-like DNA transposon (83, 84). In human cells CENP-B binds to a 17-bp sequence present in pericentromeric α-satellite repeats (85). The three CENP-B homologs in *S. pombe* are Abp1/Cbp1, Cbh1 and Cbh2. Together, these proteins bind to specific sequences in the LTRs throughout the genome (86). The CENP-B proteins, in particular, Abp1, greatly inhibit the transcription of Tf1, Tf2 and genes adjacent to LTRs (Table 3) (86). The reduced transcription can result in inhibition of Tf1 integration events. Abp1 inhibits transcription of Tf2 in part by recruiting the histone deacetylases Clr3 and Clr6. In addition, the histone H3K4 methyltransferase Set1 cooperates with Abp1 to regulate the sense and antisense transcription of Tf2 (87). Importantly, Abp1-mediated inhibition of Tf2 transcription is independent of the repressive function of the HIRA complex (86). If CENP-B proteins are derived from ancient DNA transposons it is interesting to speculate that the DNA-binding activity of the transposase evolved to recognize sequences within LTR retrotransposons and this resulted in an inhibitory activity that was co-opted by the host to control TEs.

The mobility of LTR retrotransposons can include efficient homologous recombination between transposon cDNA and existing copies of the TE in the host genome. This recycling of insertion sites is the primary mode of action for Tf2 and it is thought to repair damaged copies of the TE (54). The CENP-B proteins not only defend *S. pombe* by inhibiting Tf2 transcription, they also restrict the homologous recombination between cDNA and existing Tf2s (86).

An intriguing feature of the CENP-B proteins is that they play a role in organizing the Tf2 sequences into one to three subnuclear foci called Tf bodies (86). The clustering of Tf bodies also requires the Ku heterodimer complex of Ku70 and Ku80 and the condensin complex (88). In a chain of interactions, Abp1 recruits the Ku complex to Tf2 and LTR sequences and in turn Ku recruits condensin. Tf bodies are dynamic and disassociate during S-phase and after DNA damage when histone H3 Lys56 acetylation inhibits Ku recruitment (88). Although it is clear that Tf2 elements and presumably solo LTRs organize into clusters in the nucleus, and that this must have significant impact on the three-dimensional structure of the genome, the function of Tf bodies remains to be determined. Although the Ku complex and the dimerization domain of Abp1 are required for the formation of Tf bodies they are not required for restricting Tf2 transcription (86).

High copy number is a feature of TEs that inflict harm on the host by reducing genome stability by way of chromosome rearrangements including duplications and deletions. The LTRs in *S. pombe* impair DNA replication by blocking fork progression, a property that results from the binding of Sap1 to LTRs (65). Although the function of this replication block is not clear, it also occurs in rDNA repeats and has the intriguing feature of occurring in only one orientation. In the absence of the three CENP-B factors the stalled replication forks collapse and this leads to a high frequency of double-stranded chromosomal breaks that results in lethality (65, 89).

THE BIOLOGICAL IMPACT OF TRANSPOSITION IN *S. POMBE*

The relationship between TEs and the host is often described as one of genetic conflict where each evolves machinery to combat the other. The examples of host mechanisms that reduce the expression of Tf2 described above clearly show that *S. pombe* has methods to combat TE activity. However, the relation is far more complex, in part because the propagation of TEs is entirely dependent on the host. In addition, TEs have been proposed to benefit the host greatly by providing regulatory sequences and by reorganizing the genome in response to environmental stress (90). Examples of TE insertions that alter the expression of adjacent genes are observed in many organisms and include *Saccharomyces cerevisiae* in which copies of Ty1 or one representative copy of Ty3 increase expression of tRNA genes (91, 92). Genome-wide studies of RNA in *S. pombe* reveal examples of Tf2 LTRs that regulate adjacent genes. The LTRs function as transcription start sites for mRNAs that read through neighboring genes. When cells are grown in low oxygen conditions, the transcription activator Sre1 binds to Tf2 LTRs at the sequence 5′-ATCGTACCAT-3′, and induces transcription of Tf2 (93). Importantly, solo Tf2 LTRs are also bound by Sre1 and this generates transcripts that initiate in the LTR and read through into downstream genes (Fig. 4A). However, the Sre1 binding sequence does not exist in Tf1 LTRs and Sre1 does not bind or regulate Tf1 LTRs. To determine whether Tf1 insertion can alter the expression of adjacent genes a target plasmid was studied that contains a Tf1 insertion in the

A

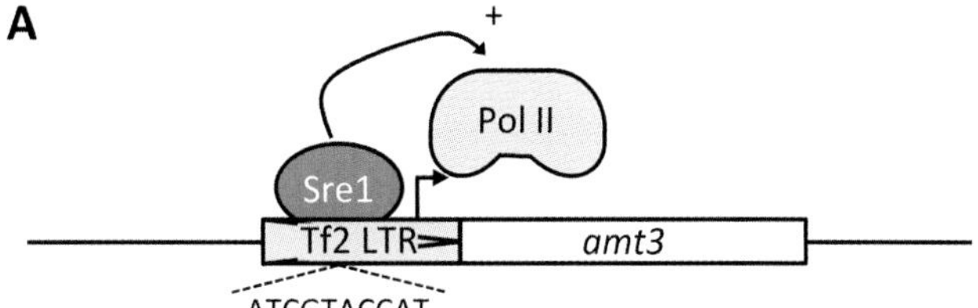

B

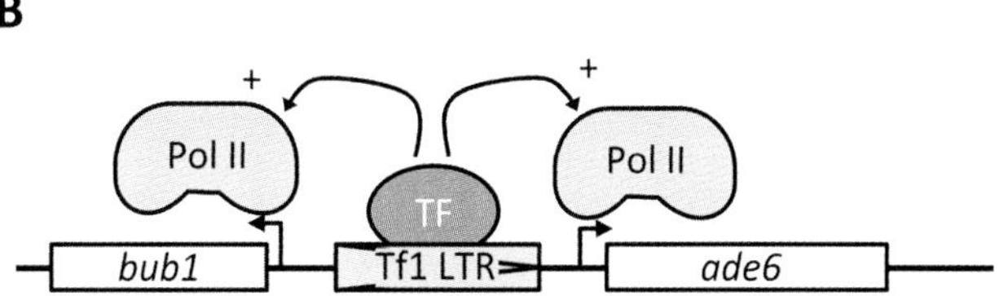

C

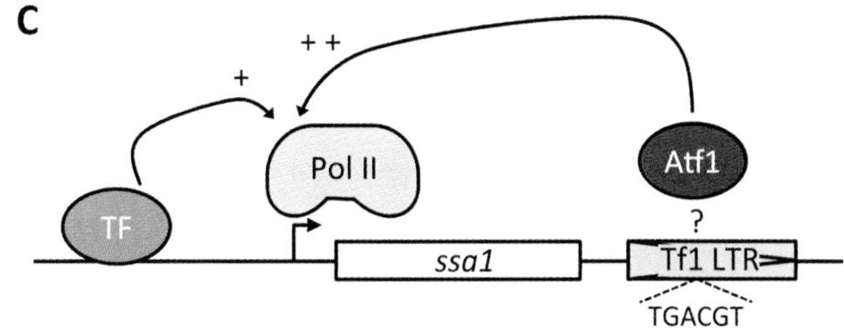

Figure 4 Impact of Tf elements on gene expression. Tf element integration can activate the expression of adjacent genes. (A) Tf2 long terminal repeat (LTR) (blue) possesses a motif ATCGTACCAT bound by the transcription factor Sre1 (green), which activates the transcription of oxygen-dependent genes such as *amt3*. Pol II (pink); RNA polymerase II. (B) Sequences within the Tf1 LTR substitute for the elements in the *ade6* promoter that are disrupted by Tf1 integration. Tf1 integration results in a 2.1- to 3.6-fold increase in *ade6* expression. TF (grey): transcription factor. (C) Tf1 integration near the heat shock gene *ssa1* acts as an enhancer of transcription, through a mechanism that may involve the transcription activation factor Atf1 (red). A motif TGACGT similar to the sequence bound by Atf1 is within the Tf1 LTR (blue). doi:10.1128/microbiolspec.MDNA3-0040-2014.f4

intergenic sequence between the genes *bub1* and *ade6*. Northern blots show that the Tf1 insertion increases the expression of both *bub1* and *ade6* (Fig. 4B) (46).

To understand the biological impact of TE insertion it is important to know what fraction of transposition events alter the expression of neighboring genes. In general, this is not known because insertions studied are initially identified using phenotypes that result from altered expression. To gauge what fraction of Tf1 insertions alter gene expression of adjacent genes an unbiased collection of integration events can be studied. A systematic analysis of 14 strains with single Tf1 insertions reveals that 40% of the insertions increase expression of adjacent genes (94). Surprisingly, none of the insertions reduce the expression of adjacent genes. The insertions that increase expression of adjacent genes do so by introducing a transcriptional enhancer. This results in higher amounts of transcripts that initiate from the original start site (Fig. 4C). The genes that are induced by the Tf1 enhancer turn out to be the genes that, in the absence of Tf1, are induced by heat shock and oxidative stress (94). The Tf1 promoter itself is activated by heat shock and oxidative stress, suggesting that the Tf1 enhancer synergizes with stress response promoters to increase the transcription of genes adjacent to Tf1 (94).

As more information accumulates about the properties of Tf1 it becomes clear that this transposable element is wired into a network of stress response mechanisms. First, Tf1 transcription increases substantially when cells are exposed to heat and oxidative stress (94). This suggests a mechanism that activates transposition in response to stress. Second, the profile of Tf1 integration shows a pattern of insertion sites that clearly favors stress response promoters (57, 58). Third, the expression of stress response genes is increased by adjacent copies of Tf1 (94). Together, these features suggest the intriguing possibility that Tf1 integration is a mechanism that is induced by stress and efficiently generates mutations that have the potential to improve survival of individual cells. While this scenario may sound compelling, much must be done to test it. Does environmental stress actually induce Tf1 integration? Can Tf1 integration provide a competitive benefit to cells exposed to environmental stress? If Tf1 integration does improve survival of individual cells, is the frequency of cells with this outcome any greater than the spontaneous rate of resistance mutations that are selected during stress. Once these questions are answered it will be clear whether Tf1 possesses specific mechanisms that provide direct benefit to cells exposed to stress.

CONCLUSIONS

The strains of *S. pombe* studied in laboratories around the world all originated from an individual isolate that contained 13 Tf2 LTR retrotransposons together with about 250 single LTRs. Although these 13 retrotransposons represent the only TEs present in the laboratory strains, it is important to remember that these sequences are a single snapshot of the dynamic and polymorphic mobile collection that exist in wild populations of *S. pombe*. The 250 single LTRs in the laboratory strain are evidence of many different families of LTR retrotransposons including Tf1, the highly active TE that was isolated from a wild strain. The static nature of Tf2 elements in laboratory strains could be a result of the inhibitory functions of CENP-Bs, the exosome, and the heterochromatization of Tf2 that occurs under nutrient deprivation. However, conditions of environmental stress overcome the mechanisms that restrict these TEs and result in induction of Tf1 and Tf2 mRNA. Once Tf1 integrates into new positions it can overcome host mechanisms that inhibit its own expression and increase transcription of adjacent genes. Because integration often occurs next to stress response genes we propose that the ability of Tf1 to induce the expression of adjacent genes may be beneficial to the host when exposed to environmental stress. This idea is consistent with the observation that environmental stress induces Tf1 and Tf2 transcription, a process that may lead to *de novo* insertions. The mechanisms that restrict TE activity and the ability of Tf1 and Tf2 to escape the restrictions are vivid examples of the conflict that exists between the TEs and *S. pombe*. The possibility that Tf1 integration may be beneficial to the host may explain the balance that has been struck between restriction and mobility.

FUNDING

This research was supported by the Intramural Research Program of the NIH from the Eunice Kennedy Shriver National Institute of Child Health and Human Development.

Citation. Esnault C, Levin HL. 2014. The long terminal repeat retrotransposons Tf1 and Tf2 of Schizosaccharomyces pombe. Microbiol Spectrum 3(2):MDNA3-0040-2014.

References

1. **Wood V, Harris MA, McDowall MD, Rutherford K, Vaughan BW, Staines DM, Aslett M, Lock A, Bahler J, Kersey PJ, Oliver SG.** 2011. PomBase: a comprehensive online resource for fission yeast. *Nucleic Acids Res* **40:** D695–D699.

2. **Bowen NJ, Jordan I, Epstein J, Wood V, Levin HL.** 2003. Retrotransposons and their recognition of pol ii promoters: a comprehensive survey of the transposable elements derived from the complete genome sequence of *Schizosaccharomyces pombe*. *Genome Res* **13:**1984–1997.
3. **Levin HL, Weaver DC, Boeke JD.** 1990. Two related families of retrotransposons from *Schizosaccharomyces pombe* [published erratum appears in *Mol Cell Biol* 1991;**11**(4):2334]. *Mol Cell Biol* **10:**6791–6798.
4. **Levin HL.** 1995. A novel mechanism of self-primed reverse transcription defines a new family of retroelements. *Mol Cell Biol* **15:**3310–3317.
5. **Levin HL, Boeke JD.** 1992. Demonstration of retrotransposition of the Tf1 element in fission yeast. *EMBO J* **11:**1145–1153.
6. **Lespinet O, Wolf YI, Koonin EV, Aravind L.** 2002. The role of lineage-specific gene family expansion in the evolution of eukaryotes. *Genome Res* **12:**1048–1059.
7. **Mata J, Lyne R, Burns G, Bahler J.** 2002. The transcriptional program of meiosis and sporulation in fission yeast. *Nat Genet* **32:**143–147.
8. **Watanabe T, Miyashita K, Saito TT, Yoneki T, Kakihara Y, Nabeshima K, Kishi YA, Shimoda C, Nojima H.** 2001. Comprehensive isolation of meiosis-specific genes identifies novel proteins and unusual non-coding transcripts in *Schizosaccharomyces pombe*. *Nucleic Acids Res* **29:**2327–2337.
9. **Weaver DC, Shpakovski GV, Caputo E, Levin HL, Boeke JD.** 1993. Sequence analysis of closely related retrotransposon families from fission yeast. *Gene* **131:**135–139.
10. **Song SU, Kurkulos M, Boeke JD, Corces VG.** 1997. Infection of the germ line by retroviral particles produced in the follicle cells: a possible mechanism for the mobilization of the gypsy retroelement of *Drosophila*. *Development* **124:**2789–2798.
11. **Song SU, Gerasimova T, Kurkulos M, Boeke JD, Corces VG.** 1994. An env-like protein encoded by a *Drosophila* retroelement: evidence that gypsy is an infectious retrovirus. *Genes Dev* **8:**2046–2057.
12. **Teysset L, Burns JC, Shike H, Sullivan BL, Bucheton A, Terzian C.** 1998. A moloney murine leukemia virus-based retroviral vector pseudotyped by the insect retroviral gypsy envelope can infect *Drosophila* cells. *J Virol* **72:** 853–856.
13. **Malik HS, Eickbush TH.** 1999. Modular evolution of the integrase domain in the Ty3/Gypsy class of LTR retrotransposons. *J Virol* **73:**5186–5190.
14. **Malik HS, Henikoff S, Eickbush TH.** 2000. Poised for contagion: evolutionary origins of the infectious abilities of invertebrate retroviruses. *Genome Res* **10:**1307–1318.
15. **Levin HL, Weaver DC, Boeke JD.** 1993. Novel gene expression mechanism in a fission yeast retroelement: Tf1 proteins are derived from a single primary translation product [published erratum appears in *EMBO J* 1994;**13:** 1494]. *EMBO J* **12:**4885–4895.
16. **Atwood A, Lin JH, Levin HL.** 1996. The retrotransposon Tf1 assembles virus-like particles that contain excess Gag relative to integrase because of a regulated degradation process. *Mol Cell Biol* **16:**338–346.
17. **Haag AL, Lin JH, Levin HL.** 2000. Evidence for the packaging of multiple copies of Tfl mRNA into particles and the trans priming of reverse transcription. *J Virol* **74:** 7164–7170.
18. **Teysset L, Dang VD, Kim MK, Levin HL.** 2003. A long terminal repeat-containing retrotransposon of Schizosaccharomyces pombe expresses a Gag-like protein that assembles into virus-like particles which mediate reverse transcription. *J Virol* **77:**5451–5463.
19. **Dang VD, Levin HL.** 2000. Nuclear import of the retrotransposon Tf1 is governed by a nuclear localization signal that possesses a unique requirement for the FXFG nuclear pore factor Nup124p. *Mol Cell Biol* **20:**7798–7812.
20. **Kim MK, Claiborn KC, Levin HL.** 2005. The long terminal repeat-containing retrotransposon Tf1 possesses amino acids in Gag that regulate nuclear localization and particle formation. *J Virol* **79:**9540–9555.
21. **Balasundaram D, Benedik MJ, Morphew M, Dang VD, Levin HL.** 1999. Nup124p is a nuclear pore factor of *Schizosaccharomyces pombe* that is important for nuclear import and activity of retrotransposon Tf1. *Mol Cell Biol* **19:**5768–5784.
22. **Sistla S, Pang JV, Wang CX, Balasundaram D.** 2007. Multiple conserved domains of the nucleoporin Nup124p and its orthologs Nup1p and Nup153 are critical for nuclear import and activity of the fission yeast Tf1 retrotransposon. *Mol Biol Cell* **18:**3692–3708.
23. **Varadarajan P, Mahalingam S, Liu P, Ng SB, Gandotra S, Dorairajoo DS, Balasundaram D.** 2005. The functionally conserved nucleoporins Nup124p from fission yeast and the human Nup153 mediate nuclear import and activity of the Tf1 retrotransposon and HIV-1 Vpr. *Mol Biol Cell* **16:**1823–1838.
24. **Matreyek KA, Yucel SS, Li X, Engelman A.** 2013. Nucleoporin NUP153 phenylalanine-glycine motifs engage a common binding pocket within the HIV-1 capsid protein to mediate lentiviral infectivity. *PLoS Pathog* **9:** e1003693.
25. **Matreyek KA, Engelman A.** 2013. Viral and cellular requirements for the nuclear entry of retroviral preintegration nucleoprotein complexes. *Viruses* **5:**2483–2511.
26. **Koh Y, Wu X, Ferris AL, Matreyek KA, Smith SJ, Lee K, KewalRamani VN, Hughes SH, Engelman A.** 2013. Differential effects of human immunodeficiency virus type 1 capsid and cellular factors nucleoporin 153 and LEDGF/p75 on the efficiency and specificity of viral DNA integration. *J Virol* **87:**648–658.
27. **Matreyek KA, Engelman A.** 2011. The requirement for nucleoporin NUP153 during human immunodeficiency virus type 1 infection is determined by the viral capsid. *J Virol* **85:**7818–7827.
28. **Dang VD, Benedik MJ, Ekwall K, Choi J, Allshire RC, Levin HL.** 1999. A new member of the sin3 family of corepressors is essential for cell viability and required for

retroelement propagation in fission yeast. *Mol Cell Biol* **19:**2351–2365.

29. **Levin HL.** 1997. It's prime time for reverse transcriptase. *Cell* **88:**5–8.
30. **Levin HL.** 1996. An unusual mechanism of self-primed reverse transcription requires the RNase H domain of reverse transcriptase to cleave an RNA duplex. *Mol Cell Biol* **16:**5645–5654.
31. **Atwood-Moore A, Ejebe K, Levin HL.** 2005. Specific recognition and cleavage of the plus-strand primer by reverse transcriptase. *J Virol* **79:**14863–14875.
32. **Hizi A.** 2008. The reverse transcriptase of the Tf1 retrotransposon has a specific novel activity for generating the RNA self-primer that is functional in cDNA synthesis. *J Virol* **82:**10906–10910.
33. **Lin JH, Levin HL.** 1997. A complex structure in the mRNA of Tf1 is recognized and cleaved to generate the primer of reverse transcription. *Genes Dev* **11:**270–285.
34. **Lin JH, Levin HL.** 1998. Reverse transcription of a self-primed retrotransposon requires an RNA structure similar to the U5-IR stem-loop of retroviruses. *Mol Cell Biol* **18:**6859–6869.
35. **Lin JH, Levin HL.** 1997. Self-primed reverse transcription is a mechanism shared by several LTR-containing retrotransposons [letter]. *RNA* **3:**952–953.
36. **Butler M, Goodwin T, Simpson M, Singh M, Poulter R.** 2001. Vertebrate LTR retrotransposons of the Tf1/Sushi group. *J Mol Evol* **52:**260–274.
37. **SanMiguel P, Tikhonov A, Jin YK, Motchoulskaia N, Zakharov D, Melake-Berhan A, Springer PS, Edwards KJ, Lee M, Avramova Z, Bennetzen JL.** 1996. Nested retrotransposons in the intergenic regions of the maize genome [see comments]. *Science* **274:**765–768.
38. **Atwood-Moore A, Yan K, Judson RL, Levin HL.** 2006. The self primer of the long terminal repeat retrotransposon Tf1 is not removed during reverse transcription. *J Virol* **80:**8267–8270.
39. **Hizi A, Levin HL.** 2005. The integrase of the long terminal repeat-retrotransposon tf1 has a chromodomain that modulates integrase activities. *J Biol Chem* **280:**39086.
40. **Herzig E, Voronin N, Hizi A.** 2012. The removal of RNA primers from DNA synthesized by the reverse transcriptase of the retrotransposon Tf1 is stimulated by Tf1 integrase. *J Virol* **86:**6222–6230.
41. **Oz-Gleenberg I, Herzig E, Voronin N, Hizi A.** 2012. Substrate variations that affect the nucleic acid clamp activity of reverse transcriptases. *FEBS J* **279:**1894–1903.
42. **Oz-Gleenberg I, Herschhorn A, Hizi A.** 2011. Reverse transcriptases can clamp together nucleic acids strands with two complementary bases at their 3′-termini for initiating DNA synthesis. *Nucleic Acids Res* **39:**1042–1053.
43. **Oz-Gleenberg I, Herzig E, Hizi A.** 2012. Template-independent DNA synthesis activity associated with the reverse transcriptase of the long terminal repeat retrotransposon Tf1. *FEBS J* **279:**142–153.
44. **Kirshenboim N, Hayouka Z, Friedler A, Hizi A.** 2007. Expression and characterization of a novel reverse transcriptase of the LTR retrotransposon Tf1. *Virology* **366:** 263–276.
45. **Ebina H, Chatterjee AG, Judson RL, Levin HL.** 2008. The GP(Y/F) domain of TF1 integrase multimerizes when present in a fragment, and substitutions in this domain reduce enzymatic activity of the full-length protein. *J Biol Chem* **283:**15965–15974.
46. **Leem YE, Ripmaster TL, Kelly FD, Ebina H, Heincelman ME, Zhang K, Grewal SIS, Hoffman CS, Levin HL.** 2008. Retrotransposon Tf1 is targeted to pol II promoters by transcription activators. *Mol Cell* **30:**98–107.
47. **Gao X, Hou Y, Ebina H, Levin HL, Voytas DF.** 2008. Chromodomains direct integration of retrotransposons to heterochromatin. *Genome Res* **18:**359–369.
48. **Chatterjee AG, Leem YE, Kelly FD, Levin HL.** 2009. The chromodomain of Tf1 integrase promotes binding to cDNA and mediates target site selection. *J Virol* **83:**2675–2685.
49. **Qi X, Sandmeyer S.** 2012. In vitro targeting of strand transfer by the Ty3 retroelement integrase. *J Biol Chem* **287:**18589–18595.
50. **Dai J, Xie W, Brady TL, Gao J, Voytas DF.** 2007. Phosphorylation regulates integration of the yeast Ty5 retrotransposon into heterochromatin. *Molecular Cell* **27:**289–299.
51. **Behrens R, Hayles J, Nurse P.** 2000. Fission yeast retrotransposon Tf1 integration is targeted to 5′ ends of open reading frames. *Nucleic Acids Res* **28:**4709–4716.
52. **Singleton TL, Levin HL.** 2002. A long terminal repeat retrotransposon of fission yeast has strong preferences for specific sites of insertion. *Eukaryotic Cell* **1:**44–55.
53. **Cherry KE, Hearn WE, Seshie OY, Singleton TL.** 2014. Identification of Tf1 integration events in S. pombe under nonselective conditions. *Gene* **542:**221–231.
54. **Hoff EF, Levin HL, Boeke JD.** 1998. *Schizosaccharomyces pombe* retrotransposon Tf2 mobilizes primarily through homologous cDNA recombination. *Mol Cell Biol* **18:**6839–6852.
55. **Hoff EKF.** 1997. *Dissertation*, Johns Hopkins University, Baltimore, MD.
56. **Majumdar A, Chatterjee AG, Ripmaster TL, Levin HL.** 2011. The determinants that specify the integration pattern of retrotransposon Tf1 in the fbp1 promoter of *Schizosaccharomyces pombe*. *J Virol* **85:**519–529.
57. **Chatterjee AG, Esnault C, Guo Y, Hung S, McQueen PG, Levin HL.** 2014. Serial number tagging reveals a prominent sequence preference of retrotransposon integration. *Nucleic Acids Res* **42:**8449–8460.
58. **Guo Y, Levin HL.** 2010. High-throughput sequencing of retrotransposon integration provides a saturated profile of target activity in *Schizosaccharomyces pombe*. *Genome Res* **20:**239–248.
59. **Arcangioli B, Copeland TD, Klar AJ.** 1994. Sap1, a protein that binds to sequences required for mating-type switching, is essential for viability in *Schizosaccharomyces pombe*. *Mol Cell Biol* **14:**2058–2065.

60. **Arcangioli B, Klar AJ.** 1991. A novel switch-activating site (SAS1) and its cognate binding factor (SAP1) required for efficient mat1 switching in *Schizosaccharomyces pombe*. *EMBO J* **10:**3025–3032.

61. **de Lahondes R, Ribes V, Arcangioli B.** 2003. Fission yeast Sap1 protein is essential for chromosome stability. *Eukaryot Cell* **2:**910–921.

62. **Mejia-Ramirez E, Sanchez-Gorostiaga A, Krimer DB, Schvartzman JB, Hernandez P.** 2005. The mating type switch-activating protein Sap1 Is required for replication fork arrest at the rRNA genes of fission yeast. *Mol Cell Biol* **25:**8755–8761.

63. **Krings G, Bastia D.** 2005. Sap1p binds to Ter1 at the ribosomal DNA of *Schizosaccharomyces pombe* and causes polar replication fork arrest. *J Biol Chem* **280:** 39135–39142.

64. **Noguchi C, Noguchi E.** 2007. Sap1 promotes the association of the replication fork protection complex with chromatin and is involved in the replication checkpoint in *Schizosaccharomyces pombe*. *Genetics* **175:**553–566.

65. **Zaratiegui M, Vaughn MW, Irvine DV, Goto D, Watt S, Bahler J, Arcangioli B, Martienssen RA.** 2011. CENP-B preserves genome integrity at replication forks paused by retrotransposon LTR. *Nature* **469:**112–115.

66. **Fedoroff NV.** 2012. Presidential address. Transposable elements, epigenetics, and genome evolution. *Science* **338:**758–767.

67. **Levin HL, Moran JV.** 2011. Dynamic interactions between transposable elements and their hosts. *Nat Rev Genet* **12:**615–627.

68. **Wilson RC, Doudna JA.** 2013. Molecular mechanisms of RNA interference. *Annu Rev Biophys* **42:**217–239.

69. **Guzzardo PM, Muerdter F, Hannon GJ.** 2013. The piRNA pathway in flies: highlights and future directions. *Curr Opin Genet Dev* **23:**44–52.

70. **Lejeune E, Allshire RC.** 2011. Common ground: small RNA programming and chromatin modifications. *Curr Opin Cell Biol* **23:**258–265.

71. **Reyes-Turcu FE, Grewal SI.** 2012. Different means, same end-heterochromatin formation by RNAi and RNAi-independent RNA processing factors in fission yeast. *Curr Opin Genet Dev* **22:**156–163.

72. **Slotkin RK, Martienssen R.** 2007. Transposable elements and the epigenetic regulation of the genome. *Nat Rev Genet* **8:**272–285.

73. **Hansen KR, Burns G, Mata J, Volpe TA, Martienssen RA, Bahler J, Thon G.** 2005. Global effects on gene expression in fission yeast by silencing and RNA interference machineries. *Mol Cell Biol* **25:**590–601.

74. **Cam HP, Sugiyama T, Chen ES, Chen X, FitzGerald PC, Grewal SI.** 2005. Comprehensive analysis of heterochromatin- and RNAi-mediated epigenetic control of the fission yeast genome. *Nat Genet* **37:**809–819.

75. **Halic M, Moazed D.** 2010. Dicer-independent primal RNAs trigger RNAi and heterochromatin formation. *Cell* **140:**504–516.

76. **Zaratiegui M, Castel SE, Irvine DV, Kloc A, Ren J, Li F, de Castro E, Marin L, Chang AY, Goto D, Cande WZ, Antequera F, Arcangioli B, Martienssen RA.** 2011. RNAi promotes heterochromatic silencing through replication-coupled release of RNA Pol II. *Nature* **479:**135–138.

77. **Rhind N, Chen Z, Yassour M, Thompson DA, Haas BJ, Habib N, Wapinski I, Roy S, Lin MF, Heiman DI, Young SK, Furuya K, Guo Y, Pidoux A, Chen HM, Robbertse B, Goldberg JM, Aoki K, Bayne EH, Berlin AM, Desjardins CA, Dobbs E, Dukaj L, Fan L, FitzGerald MG, French C, Gujja S, Hansen K, Keifenheim D, Levin JZ, Mosher RA, Müller CA, Pfiffner J, Priest M, Russ C, Smialowska A, Swoboda P, Sykes SM, Vaughn M, Vengrova S, Yoder R, Zeng Q, Allshire R, Baulcombe D, Birren BW, Brown W, Ekwall K, Kellis M, Leatherwood J, Levin H, Margalit H, Martienssen R, Nieduszynski CA, Spatafora JW, Friedman N, Dalgaard JZ, Baumann P, Niki H, Regev A, Nusbaum C.** 2011. Comparative functional genomics of the fission yeasts. *Science* **332:** 930–936.

78. **Yamanaka S, Mehta S, Reyes-Turcu FE, Zhuang F, Fuchs RT, Rong Y, Robb GB, Grewal SI.** 2013. RNAi triggered by specialized machinery silences developmental genes and retrotransposons. *Nature* **493:**557–560.

79. **Lee NN, Chalamcharla VR, Reyes-Turcu F, Mehta S, Zofall M, Balachandran V, Dhakshnamoorthy J, Taneja N, Yamanaka S, Zhou M, Grewal SI.** 2013. Mtr4-like protein coordinates nuclear RNA processing for heterochromatin assembly and for telomere maintenance. *Cell* **155:**1061–1074.

80. **Anderson HE, Wardle J, Korkut SV, Murton HE, Lopez-Maury L, Bahler J, Whitehall SK.** 2009. The fission yeast HIRA histone chaperone is required for promoter silencing and the suppression of cryptic antisense transcripts. *Mol Cell Biol* **29:**5158–5167.

81. **Greenall A, Williams ES, Martin KA, Palmer JM, Gray J, Liu C, Whitehall SK.** 2006. Hip3 interacts with the HIRA proteins Hip1 and Slm9 and is required for transcriptional silencing and accurate chromosome segregation. *J Biol Chem* **281:**8732–8739.

82. **Casola C, Hucks D, Feschotte C.** 2008. Convergent domestication of pogo-like transposases into centromere-binding proteins in fission yeast and mammals. *Mol Biol Evol* **25:**29–41.

83. **Smit AF.** 1996. The origin of interspersed repeats in the human genome. *Curr Opin Genet Dev* **6:**743–748.

84. **Tudor M, Lobocka M, Goodell M, Pettitt J, O'Hare K.** 1992. The pogo transposable element family of *Drosophila melanogaster*. *Mol Gen Genet* **232:**126–134.

85. **Masumoto H, Nakano M, Ohzeki J.** 2004. The role of CENP-B and alpha-satellite DNA: de novo assembly and epigenetic maintenance of human centromeres. *Chromosome Res* **12:**543–556.

86. **Cam HP, Noma K, Ebina H, Levin HL, Grewal SIS.** 2008. Host genome surveillance for retrotransposons by transposon-derived proteins. *Nature* **451:**U431–U432.

87. **Lorenz DR, Mikheyeva IV, Johansen P, Meyer L, Berg A, Grewal SI, Cam HP.** 2012. CENP-B cooperates with Set1

in bidirectional transcriptional silencing and genome organization of retrotransposons. *Mol Cell Biol* **32:**4215–4225.

88. **Tanaka A, Tanizawa H, Sriswasdi S, Iwasaki O, Chatterjee AG, Speicher DW, Levin HL, Noguchi E, Noma K.** 2012. Epigenetic regulation of condensin-mediated genome organization during the cell cycle and upon DNA damage through histone H3 lysine 56 acetylation. *Mol Cell* **48:**532–546.
89. **Baum M, Clarke L.** 2000. Fission yeast homologs of human CENP-B have redundant functions affecting cell growth and chromosome segregation. *Mol Cell Biol* **20:** 2852–2864.
90. **McClintock B.** 1984. The significance of responses of the genome to challenge. *Science* **226:**792–801.
91. **Bolton EC, Boeke JD.** 2003. Transcriptional interactions between yeast tRNA genes, flanking genes and Ty elements: a genomic point of view. *Genome Res* **13:**254–263.
92. **Kinsey PT, Sandmeyer SB.** 1991. Adjacent pol II and pol III promoters: transcription of the yeast retrotransposon Ty3 and a target tRNA gene. *Nucleic Acids Res* **19:** 1317–1324.
93. **Sehgal A, Lee CY, Espenshade PJ.** 2007. SREBP controls oxygen-dependent mobilization of retrotransposons in fission yeast. *PLoS Genet* **3:**e131.
94. **Feng G, Leem YE, Levin HL.** 2013. Transposon integration enhances expression of stress response genes. *Nucleic Acids Res* **41:**775–789.
95. **Wood V, Gwilliam R, Rajandream MA, Lyne M, Lyne R, Stewart A, Sgouros J, Peat N, Hayles J, Baker S, Basham D, Bowman S, Brooks K, Brown D, Brown S, Chillingworth T, Churcher C, Collins M, Connor R, Cronin A, Davis P, Feltwell T, Fraser A, Gentles S, Goble A, Hamlin N, Harris D, Hidalgo J, Hodgson G, Holroyd S, Hornsby T, Howarth S, Huckle EJ, Hunt S, Jagels K, James K, Jones L, Jones M, Leather S, McDonald S, McLean J, Mooney P, Moule S, Mungall K, Murphy L, Niblett D, Odell C, Oliver K, O'Neil S, Pearson D, Quail MA, Rabbinowitsch E, Rutherford K, Rutter S, Saunders D, Seeger K, Sharp S, Skelton J, Simmonds M, Squares R.** 2002. The genome sequence of *Schizosaccharomyces pombe*. *Nature* **415:**871–880.
96. **Sugiyama T, Sugioka-Sugiyama R.** 2011. Red1 promotes the elimination of meiosis-specific mRNAs in vegetatively growing fission yeast. *EMBO J* **30:**1027–1039.
97. **Leupold U.** 1993. The origins of *Schizosaccharomyces pombe* genetics, p 125–128. *In* Hall MN, Linder P (ed), *The Early Days of Yeast Genetics*. Cold Spring Harbor Laboratory Press, New York.

Mobile DNA, 3rd Edition
Nancy L. Craig, Michael Chandler, Martin Gellert, Alan M. Lambowitz, Phoebe A. Rice and Suzanne Sandmeyer

doi:10.1128/microbiolspec.MDNA3-0024-2014

Alan Engelman[1]
Peter Cherepanov[2]

44 Retroviral Integrase Structure and DNA Recombination Mechanism

INTRODUCTION

Retroviruses are the only animal viruses that require the stable integration of genetic information into the genome of the host cell as an obligate step in replication. All members of the virus family Retroviridae accordingly carry with them integrase, which is a specialized DNA recombination enzyme. Integration is required for efficient expression of retroviral genes by the host transcriptional machinery and hence productive virus replication. The integrase encoded by human immunodeficiency virus type 1 (HIV-1) is thus an important antiviral target in the fight against HIV/AIDS (1). Integration additionally ensures replication and segregation of viral genes to daughter cells during cell division. Stable association of HIV-1 with cellular DNA underlies the notoriously incurable nature of AIDS despite highly active antiretroviral therapy (HAART) (2).

Retroviridae is composed of two virus subfamilies, Orthoretrovirinae and Spumaretrovirinae. While six viral genera, including alpha through epsilon and lenti, belong to Orthoretrovirinae, the spumaviruses solely comprise Spumaretrovirinae. Spumavirus biology accordingly differs somewhat from the other retroviruses. For brevity, this chapter describes generalities that apply to most, if not all, retroviruses; specific viruses or genera are discussed as applicable. Viruses pathogenic to humans include the lentiviruses HIV-1 and HIV-2 and the deltaretroviruses human T-lymphotropic virus type 1 (HTLV-1) and HTLV-2.

Retroviruses carry two copies of plus-sense genomic RNA. Integrase is encoded at the 3′ end of the *pol* gene (3, 4, 5), which also encodes for the RNA-dependent DNA polymerase reverse transcriptase (RT) enzyme that converts the genomic RNA into linear, double-stranded DNA. Integrase and RT are translated as part of a Gag-Pol precursor polyprotein in virus producer cells, which is cleaved by the viral protease as particles bud from the plasma membrane and mature into infectious virions. Integrase and RT, together with viral RNA in association with the viral nucleocapsid protein, are situated within the viral capsid core during particle maturation.

Viral DNA synthesis occurs within the context of the reverse transcription complex, which is a large subviral

[1]Department of Cancer Immunology and AIDS, Dana-Farber Cancer Institute, 450 Brookline Avenue, CLS-1010, Boston, MA 02215; [2]Cancer Research UK, London Research Institute, Clare Hall Laboratories, Blanche Lane, Potters Bar, EN6 3LD, United Kingdom.

complex derived from the virus core (6, 7). Retroviral RNAs harbor a terminal repeat (R) element that adjoins unique sequences at the 5′ and 3′ ends (U5 and U3, respectively). U3 and U5 become duplicated during reverse transcription such that the DNA contains a copy of 5′-U3RU5-3′, dubbed the long terminal repeat (LTR), at both ends. Integrase engages approximately 16 to 20 bp of the LTR termini to integrate the viral DNA substrate into host DNA (8, 9, 10, 11, 12) (Fig. 1A).

INTEGRASE ACTIVITIES

Two integrase activities, 3′ processing and strand transfer, are required for productive virus replication. The associated DNA cutting and joining steps were initially deciphered using extracts of acutely infected cells. All retroviruses harbor an invariant CA dinucleotide in the immediate proximity to each 3′ end of the unintegrated viral DNA (Fig. 1A, underline), and downstream sequences (up to three nucleotides depending on the viral species) are removed during 3′ processing (13, 14, 15,

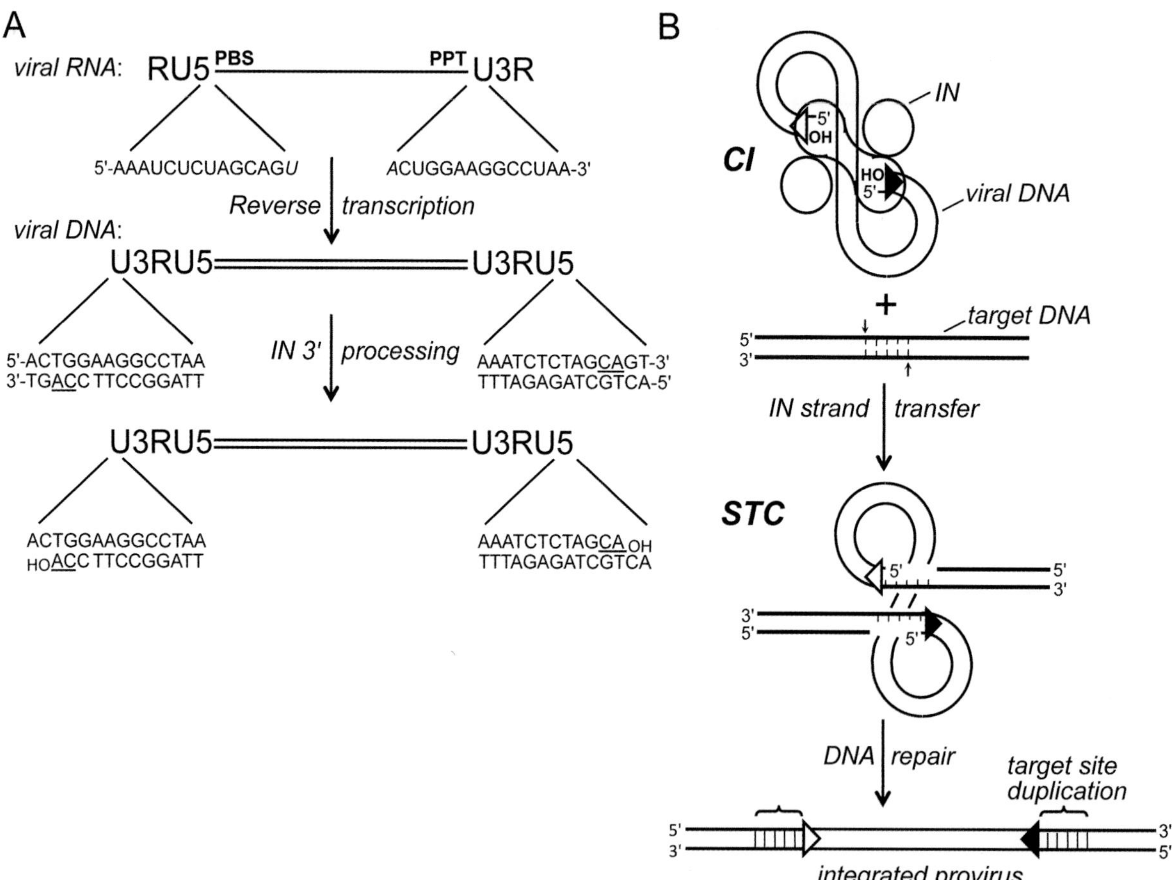

Figure 1 Integration substrate and integrase activities. (A) Reverse transcription yields linear, double-stranded DNA with U3RU5 long terminal repeats. The 14 terminal bp of HIV-1 DNA are compared with their originating positions in viral RNA (italicized bases form part of the primer binding site (PBS) and polypurine tract (PPT) that are important for DNA synthesis). The positions of the invariant CA dinucleotides (underlined in the DNA) relative to the 3′ ends of the PBS and PPT determines whether 3′ processing is required to yield a CA_{OH} 3′ terminus; 3′ processing by HIV-1 integrase liberates a $pGpT_{OH}$ dinucleotide from each viral DNA end. (B) Two monomers (oblong shape) of an HIV-1 integrase tetramer within the cleaved intasome (CI) use the viral DNA 3′-hydroxyl groups to cut the target DNA (thick bold lines) with a 5 bp stagger, which concomitantly joins the LTR ends to target DNA 5′ phosphates. Repair of the strand transfer complex (STC) yields a 5 bp duplication of target DNA flanking the integrated HIV-1 provirus. Open and filled triangles, U3 and U5 termini, respectively. Integrase was omitted from the drawing of the STC for simplicity. IN, integrase. doi:10.1128/microbiolspec.MDNA3-0024-2014.f1

16, 17). A key breakthrough came with the demonstration that the DNA-containing nucleoprotein complex, termed the preintegration complex (PIC), can integrate the endogenous reverse transcript into heterologous target DNA *in vitro* (18). The processed viral DNA CA-3′ ends are joined to target DNA during strand transfer, while the unprocessed viral DNA 5′ ends remain unjoined (14, 15, 17) (Fig. 1B). Cellular enzymes remove the two unpaired nucleotides from the viral DNA 5′ ends and repair the single-stranded gaps in the recombination intermediate, which yields a short (4 to 6 bp, depending on the virus) duplication of target DNA flanking the integrated provirus (Fig. 1B). PIC activity requires divalent metal ions such as Mn^{2+} or Mg^{2+}, but does not require a high-energy cofactor (14, 18, 19, 20, 21).

Purified recombinant integrase proteins display 3′ processing and strand transfer activities *in vitro* (22, 23, 24, 25, 26, 27, 28, 29, 30). Simplified assay systems utilized relatively short (~17 to 22 bp) double-stranded oligonucleotides that model the ends of the LTRs for 3′ processing reaction substrates (22, 24) and also for acceptor target DNA during strand transfer (23, 26, 27, 29, 30). Because pre-processed substrates that lacked sequences normally removed during 3′ processing supported strand transfer activity, the integration of retroviral DNA ends is mechanistically separable from dinucleotide cleavage (23, 26, 27, 29, 30). Concordantly, 3′ processing of HIV-1 DNA ends can occur soon after the LTRs are synthesized (31, 32), which precedes integration into chromosomal DNA minimally by several hours (33, 34).

Purified integrase proteins display two additional endonucleolytic activities *in vitro*, disintegration (35) and alcoholysis (36, 37). Though it seems unlikely that either of these activities occurs in the context of virus infection, their study has nevertheless yielded valuable information on integrase domain organization and reaction mechanism.

3′ Processing and Strand Transfer Reaction Mechanisms

Oligonucleotide-based biochemical assays enabled relatively rapid assessment of DNA substrate and integrase protein requirements for 3′ processing and strand transfer activities, as well as the mechanism of DNA recombination. Because integrase activity does not require a high-energy cofactor, the energy to drive the formation of the new viral-target DNA bond during strand transfer must come from the pre-existing target DNA phosphodiester bond. Two different mechanisms can be considered: the viral DNA could attack the target DNA directly, which through isoenergetic transesterification would concomitantly drive the formation of the viral-target DNA phosphodiester bond, or the bond energy could be temporarily stored in the form of an integrase-DNA covalent intermediate prior to viral-target DNA bond formation (38). Essentially all enzyme-mediated phosphoryl transfer reactions proceed by bimolecular nucleophilic substitution (S_N2) displacement (39); as the hallmark of S_N2 chemistry is inversion of chirality, monitoring the stereochemical course of the strand transfer reaction afforded a means by which to address its mechanism. The phosphodiester group can be made chiral by substituting one of its non-bridging oxygen atoms (Pro-*S*p or Pro-*R*p) with sulfur. Finding the retention of phosphorothioate chirality in strand transfer reaction products would be consistent with a protein-DNA covalent intermediate reaction mechanism, as formation of the integrase-DNA covalent bond would invert chirality, and its subsequent resolution would invert chirality a second time to overall retention (40). Because phosphorothioate chirality was inverted in HIV-1 DNA reaction products, strand transfer proceeds via an odd number of transesterification reactions (41). Though convoluted models that consider 3 or 5 separate chemical reactions could be entertained, it was evident that strand transfer proceeds via a single transesterification reaction (41).

HIV-1 integrase uses a water molecule under Mg^{2+}-dependent conditions to hydrolyze CA\GT (backslash denotes scissile phosphodiester bond throughout the chapter) during 3′ processing (42). Reaction specificity is loosened somewhat in the presence of Mn^{2+}, such that two or three carbon-containing diols, or the 3′ end of the viral DNA itself, can additionally serve as the nucleophile (36, 41). Determining phosphorothioate chirality in the cyclic dinucleotide reaction product formed by DNA end-mediated processing provided a means to monitor the stereochemical course of the reaction. Because chirality was inverted, the HIV-1 integrase 3′ processing reaction, like strand transfer, proceeds via a single S_N2 transesterification (41).

Similarities with other DNA Recombination Systems

Retroviral and LTR retrotransposon integrase proteins are evolutionarily related to a large variety of metal ion-dependent polynucleotidyl transferase enzymes that include bacterial transposases, RuvC resolvase, RNase H (43), the Argonaute component of RISC (RNA-induced silencing complex) (44), the RAG1 component of the RAG1/2 recombinase that catalyzes V(D)J recombination (45, 46), and RT (47). The key similarities include the enzyme active sites and reaction mechanisms.

The bacteriophage MuA transposase protein catalyzes DNA cutting and joining reactions that are analogous to retroviral integrase 3′ processing and strand transfer activities (38). Monitoring the stereochemical course of the phage Mu strand transfer reaction revealed inversion of phosphorothioate chirality (40). The utilization of $H_2{}^{18}O$ afforded the monitoring of the stereochemical course of Mu DNA end hydrolysis. Similar to the results obtained for DNA-mediated 3′ processing of HIV-1 DNA, MuA transposase-mediated hydrolysis yielded inversion of phosphorothioate chirality (48).

The RAG1/2 recombinase yields a DNA hairpin product during the first step of V(D)J recombination (see the chapter by David Roth in this monograph). Monitoring the stereochemical course of the reaction revealed the inversion of phosphorothioate chirality, which is consistent with a single step transesterification (49). The phosphorothioate stereoselectivity of the various reactions catalyzed by the Tn10 transposase protein were moreover similar to those catalyzed by MuA transposase, HIV-1 integrase, and RAG1/2 recombinase (50, 51). Thus, similar mechanisms underlie the DNA cutting and joining reactions that are catalyzed by elements as seemingly disparate as HIV-1 integrase, MuA transposase, and RAG1/2 recombinase.

Challenges of Working with Recombinant Integrase Proteins

Integration proceeds through the pairwise or concerted integration of both ends of linear viral DNA into chromosomal DNA (Fig. 1B). However, purified integrase proteins vary greatly in their ability to catalyze the concerted integration of substrate DNA *in vitro*. Heterologous circular target DNA is used to distinguish strand transfer reaction products that result from the integration of single ends of oligonucleotide duplexes from those that result from the concerted integration of two independent duplex oligonucleotide ends: single end integration yields a nicked circular DNA product, while concerted integration yields a linearized product after deproteinization (23). Many integrase proteins yield mixtures of the two types of reaction products (23, 52, 53, 54, 55) whereas others, such as those derived from the spumavirus prototype foamy virus (PFV) (56) and lentivirus equine infectious leukemia virus (54), predominantly display concerted integration activity. HIV-1 integrase by contrast predominantly catalyzes the integration of single oligonucleotide DNA ends *in vitro* (26). The reason behind the relatively poor behavior of HIV-1 integrase in concerted integration reactions remains unclear. Modifications that included relatively long viral DNA substrates (57, 58, 59) and/or the addition of viral nucleocapsid (60) or host lens epithelium-derived growth factor (LEDGF)/p75 protein (61) have enhanced concerted integration activity. Purification under conditions that disfavor protein aggregation has also been reported to enhance the ability of the HIV-1 enzyme to integrate oligonucleotide substrate DNA in concerted fashion (62).

Retroviral integrase proteins display a range of solubility properties. Epsilonretrovirus integrase was reportedly insoluble following its expression in *Escherichia coli* under a variety of conditions (55). HIV-1 integrase purified following its expression in bacteria can attain concentrations of 1 mg/ml or greater, however this strictly depends on non-physiological concentrations of salt (e.g., 1 M NaCl). PFV integrase is by contrast highly soluble, with concentrations in excess of 10 mg/ml achieved in buffer containing 200 mM NaCl (56, 63, 64). The favorable solubility of PFV integrase is at least partially responsible for its utility as a crystallography substrate with DNA (see below).

Various modifications, most notably the use of solubilizing mutations, can improve the solubility of HIV-1 integrase (42, 65, 66). The lentiviral integration cofactor LEDGF/p75 possesses favorable solubility, and binding to HIV-1 integrase yields a protein complex that displays generally favorable solubility properties (67, 68). Such observations have sparked interest in the structural biology of lentiviral integrase-LEDGF/p75 complexes (61, 68, 69, 70).

INTEGRASE PROTEIN STRUCTURES

Domain Organization of Integrase Proteins

Retroviral integrase proteins harbor three common domains, the N-terminal domain (NTD), catalytic core domain (CCD), and C-terminal domain (CTD) (71, 72, 73, 74, 75, 76) (Fig. 2A). All evidence suggests that the three-dimensional structures of the individual domains are preserved across the different integrase proteins (77). At the level of amino acid sequence, the CCD and CTD display the greatest and least extents of conservation, respectively.

The CCD harbors the enzyme active site, at the heart of which are the invariant amino acid residues of the DDE catalytic triad (73, 78, 79, 80) that coordinate a pair of magnesium ions during catalysis (see below).

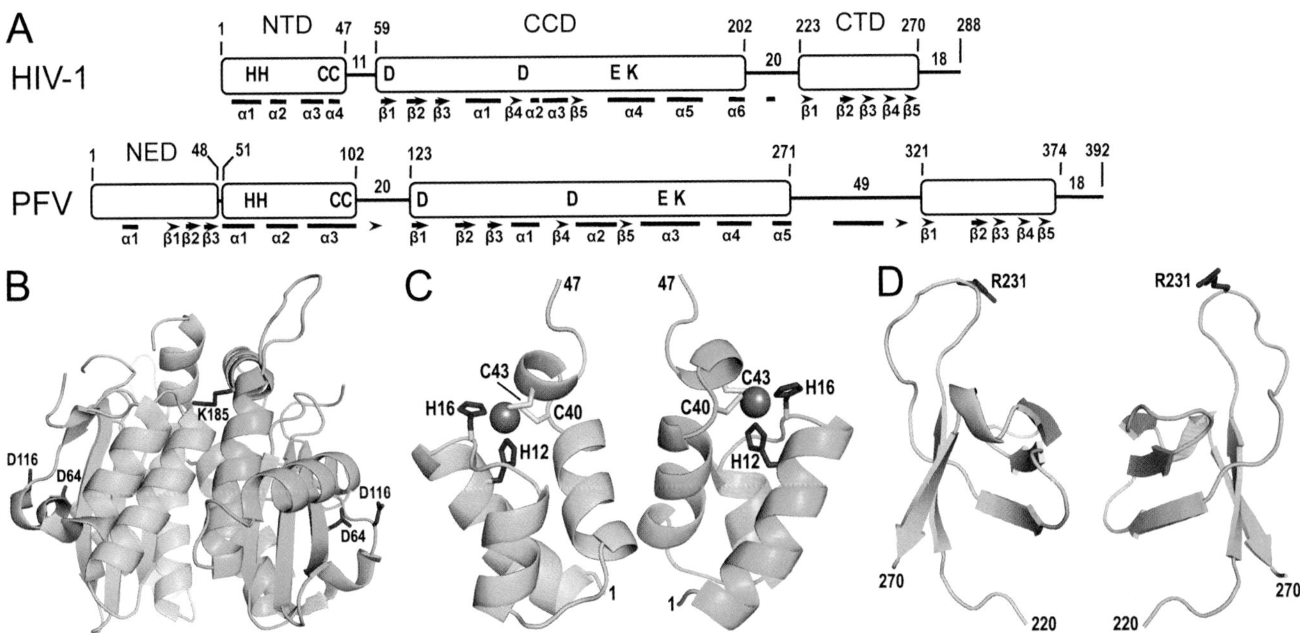

Figure 2 Retroviral integrase domain organization and HIV-1 integrase domain structures. (A) The N-terminal domain (NTD), catalytic core domain (CCD), and C-terminal domain (CTD) are common among all retroviral integrase proteins, whereas sequence analysis indicates that gammaretrovirus and epsilonretrovirus in addition to spumavirus proteins harbor an N-terminal extension domain (NED) (55, 94). HIV-1 and PFV integrases were aligned by NTD N-termini, with positions of domain boundaries and lengths of interdomain linker and C-terminal tail regions indicated. Residues conserved across all retroviral integrase proteins are shown in single letter code. Bars and arrows indicate alpha helix and beta strand secondary elements as determined by X-ray crystallography for PFV integrase (94) and by a combination of X-ray crystallography (66, 149) and molecular modeling (105) for HIV-1 integrase. (B) The X-ray crystal structure of the HIV-1 integrase CCD [protein database (pdb) code 1ITG] (43) highlights in red sticks the aspartate residues of the DDE catalytic triad (the glutamic acid was not visualized in this structure) and in blue sticks the Lys185 substitution that enhanced protein solubility and enabled protein crystallization (the Lys residue at the rear face of the dimer is barely visible in this projection). (C) The NMR structure of the integrase NTD (pdb code 1WJC) highlights the His (blue sticks) and Cys (yellow sticks) residues that coordinate a single zinc atom (grey sphere) (99). (D) The structure of the HIV-1 integrase CTD as determined by NMR (pdb code 1IHV) (142) highlights Arg231, which has been implicated in target DNA binding (108, 166).
doi:10.1128/microbiolspec.MDNA3-0024-2014.f2

Initially recognized as a $DX_{39-58}DX_{35}E$ motif conserved among retroviral and retrotransposon integrases and bacterial IS3 insertion sequences (74, 78, 81), the advent of genomic sequencing has since expanded the DD(E/D) superfamily of polynucleotidyl transferases to include the transposase proteins of numerous prokaryotic and eukaryotic transposable elements (47, 82, 83, 84). To date, the roles of the active site residues during integrative recombination are most thoroughly understood from X-ray crystallographic analyses of active PFV integrase-DNA complexes (85).

The CCD harbors additional conserved residues that contact viral DNA (86, 87, 88, 89, 90, 91, 92, 93, 94) and target DNA (90, 95, 96, 97, 98) during integration. The lysine residue that engages the phosphate backbone of the invariant CA dinucleotide in viral DNA (86, 94) is conserved across retroviral integrase proteins and some bacterial insertion elements (Fig. 2A), although not among the retrotransposon integrase proteins, despite the conservation of the terminal CA sequence (78, 86).

Retroviral and retrotransposon integrase NTDs harbor a conserved HHCC motif that coordinates the binding of a single zinc atom (72, 74, 99, 100). Metal binding stabilizes the native fold of the NTD and concordantly stimulates HIV-1 integrase catalytic activity

(61, 101, 102). The NTD is involved in integrase multimerization and interacts with viral DNA during integration (94, 102, 103, 104, 105).

Although the extent of amino acid sequence conservation among integrase CTDs is less prevalent than for the other common domains, certain conserved amino acid motifs are recognizable across subsets of retroviral integrase proteins (106). The CTD binds DNA in a sequence non-specific fashion (75, 107, 108, 109) and contributes to the functional multimerization of full-length integrase proteins (42, 108, 110).

The NTD and CTD play important roles in integrase 3′ processing and DNA strand transfer activities *in vitro* (74, 76, 111, 112, 113, 114, 115, 116). The isolated CCD can catalyze disintegration activity (74, 113, 114), which was consistent with the notion that it housed the enzyme active site (73, 74). The integrase proteins from three of the viral genera, including the gammaretroviruses, epsilonretroviruses, and spumaviruses, carry a fourth domain, termed the N-terminal extension domain (NED), based on sequence and X-ray crystallographic analyses (55, 94) (Fig. 2A).

Integrase Functions as a Multimer

Kinetic measures of integrase activities (117, 118) and chemical cross-linking (58, 111, 119, 120, 121) provided initial evidence that integrase functions as a multimer. The discovery that mixtures of defective integrase deletion mutant proteins restored as much as 50% of wild type activity provided a means to probe the functional organization of the different integrase domains within the multimer (111, 115, 116, 122, 123). For example, the function of the NTD was required in *trans* to the catalytically active protomer (111, 116, 122).

Mutagenesis experiments revealed that the invariant CA/TG bp is the most critical sequence for oligonucleotide-based integrase activity *in vitro* (29, 124, 125, 126). Titers of viruses that contained one mutated LTR end were modestly reduced, whereas double end-mutant viruses were essentially dead (11, 12, 127). Sequence analysis revealed large insertions or deletions of cellular DNA associated with the integration of mutant LTR ends (128, 129). Though virus replication normally proceeds through the pairwise insertion of both LTRs (Fig. 1B), integrase-mediated insertion of one end can apparently template the integration of the second, mutated end via cell-mediated DNA recombination.

Retroviral integration and DNA transposition proceed through a series of stable nucleoprotein complexes (85) (for reviews, see 130 and 131). In the case of integration, the two ends of viral DNA are bridged together by a tetramer of integrase in a complex that is called the intasome or stable synaptic complex (SSC) (31, 58, 94, 120, 121, 132, 133, 134, 135). Processing of the viral DNA ends converts the SSC to the cleaved intasome (CI) (85) (Fig. 1B), which is referred to as the cleaved donor complex or type 1 transpososome in the transposition literature (130). The CI subsequently morphs into the target capture complex (TCC) upon target DNA binding. Covalent joining of the viral DNA ends to target DNA yields the strand transfer complex (STC) (85, 98).

Whereas an integrase tetramer catalyzes strand transfer activity (58, 94, 118, 120), the form of the enzyme that catalyzes 3′ processing activity during virus infection has been debated. Although an integrase dimer appears to suffice to process single LTR ends *in vitro* (58, 136), a variety of information needs to be taken into account when considering whether 3′ processing is mediated by an integrase dimer bound to a single viral DNA end or by the integrase tetramer during infection. DNA end processing prior to SSC formation could account for the different apparent rates of U5 versus U3 cleavage during HIV-1 infection (31) and is consistent with integrase-mediated integration of the sole wild type ends of single LTR-end mutant viruses (128, 129). However, 3′ processing and strand transfer of single viral DNA ends can occur in the context of the SSC *in vitro* (120). Crystallized PFV intasomes are functional (see below), and the integrase tetramer catalyzed 3′ processing activity in crystallo (85). The PFV integrase NTD moreover functioned in *trans* with the CCD during 3′ processing (85). Therefore, if an integrase dimer were active for 3′ processing, it would need to assume a cis–trans conformation that is different from the active form observed in the crystals. While not impossible, we favor the straightforward interpretation that 3′ processing activity is catalyzed by the integrase tetramer in the context of the SSC during viral infection. Of note, results of small angle X-ray scattering (SAXS) have yielded novel "reaching dimer" solution conformations for full-length HIV-1 and avian-sarcoma leukosis virus (ASLV) integrase in the absence of DNA (137, 138). Although extensive CTD–CTD interactions preclude these structures from functionally engaging DNA, they nevertheless could represent intermediates on the pathway to SSC formation (see the chapter by Anna Marie Skalka).

Integrase Domain Structures

A high-resolution structure of a full-length retroviral integrase protein in the absence of DNA has yet to be reported, which is likely due to the inherent flexibility of integrase interdomain linkers (Fig. 2A) (64, 104). Over the years, structures of isolated integrase domains and 2-domain constructs have been elucidated. Because the integrase CCD could catalyze disintegration activity (74, 113, 114), it was earmarked early on as a target of interest. The HIV-1 integrase CCD was however insoluble following its expression in *E. coli* (74, 139). An important advance came from the identification of amino acid substitutions, including F185K, which significantly increased the solubility of HIV-1 integrase CCD protein (65) and afforded its crystallization and structural determination by X-ray diffraction (43). The CCD harbors an RNase H fold, which situates the active site residues of the DDE catalytic triad in close proximity to one another (Fig. 2B). Retroviral integrase CCDs in the vast majority of cases crystallize as a homodimer with a large dimeric interface (43, 140) (reviewed in 77), and the Lys185 side-chain composed part of the HIV-1 integrase CCD/F185K interface (Fig. 2B). The ~35 Å distance between the active sites within the homodimer was incompatible with pairwise insertion of two ends of HIV-1 DNA across a major groove (43), which is ~17 Å in canonical B-form DNA. More recent studies of PFV intasomes have helped to clarify the role of the CCD dimer during integration (see below).

The F185K change also enhanced the solubility of full-length HIV-1 integrase, and the mutant enzyme retained integrase 3′ processing and strand transfer activities *in vitro* (42). The mutation however rendered HIV-1 replication-defective, which was attributed to pleiotropic defects at the steps of virus particle assembly and reverse transcription (42). The CCD from the alpharetrovirus ASLV integrase was sufficiently soluble to permit its crystallization in the absence of solubility-enhancing mutations (140).

The NTD adopts a helix-turn-helix fold around a single zinc atom, which is chelated by the side chains of the conserved residues of the HHCC motif (99, 100, 141) (Fig. 2C). The CTD folds into a 5-stranded beta barrel with homology to Src homology 3 (SH3) domains (142, 143, 144) (Fig. 2D). Though SH3 domains in general interact with Pro-rich regions of proteins (145), some, such as Sso7d from *Sulfolobus solfatarius*, also mediate DNA binding (146). Each integrase domain notably engages DNA in the context of the active PFV intasome (94).

Integrase Two-Domain Structures

Structures of 2-domain integrase constructs provided initial insight into the organization of the different protein domains during integration. The HIV-1 integrase CCD-CTD fragment containing five solubilizing mutations (including F185K) crystallized as a homodimer, with the CTDs positioned at the ends of alpha helical extensions that emanated from the common CCD dimer (66) (Fig. 3A). The structures of two other integrase CCD-CTD proteins that were solved at around the same time, including those from simian immunodeficiency virus (147) and ASLV (148), revealed different positions of these CTDs relative to the integrase CCD dimer (reviewed in 77). Protein contacts made during crystallization therefore likely influenced the positions of the CTDs relative to the CCD dimer in each of the 2-domain CCD-CTD structures (77). Consistent with this interpretation, the CCD-CTD linker region of PFV integrase adopts an extended conformation that is largely devoid of secondary structure in the presence of viral DNA (see below).

Perhaps the most interesting of the partial integrase structures was that of the HIV-1 NTD-CCD construct, where the asymmetric unit harbored four protein molecules (Fig. 3B) (149). Although each CCD participated in canonical dimer interface formation (yellow-cyan and green-yellow dimers in the figure), a novel interface was observed between two "inner" molecules of the tetramer (Fig. 3B, green and cyan). The NTDs of the cyan and green integrase protomers were moreover seemingly positioned to work in trans with the apposing CCD, though incomplete electron density maps precluded assignments of interdomain NTD-CCD connectors in this structure. The active sites of the two inner monomers additionally seemed too far apart to catalyze pairwise insertion of HIV-1 DNA ends (149).

LEDGF/p75-Integrase Structures

LEDGF/p75 is a lentiviral-specific integrase-binding protein that helps to guide integration to active genes (reviewed in 150 and 151; also see the chapter by Craigie and Bushman in this monograph). LEDGF/p75 harbors a conserved domain, called the integrase-binding domain (IBD), which is necessary and sufficient for binding to integrase (152). On the integrase side, the CCD is minimally required for LEDGF/p75 binding, with the NTD contributing to high affinity binding (61, 153).

The LEDGF/p75 IBD is a compact alpha helical domain, with the hairpins from two helix-hairpin-helix folds situated on one end of the elongated structure (154). A crystal structure of the HIV-1 integrase CCD-IBD complex revealed that the host factor predominantly utilizes hairpin residues, most notably Ile365 and Asp366 at the tip of the first hairpin, to nestle into the CCD-CCD dimer interface (69) (Fig. 4A). A series of quinoline-based antiviral compounds has been

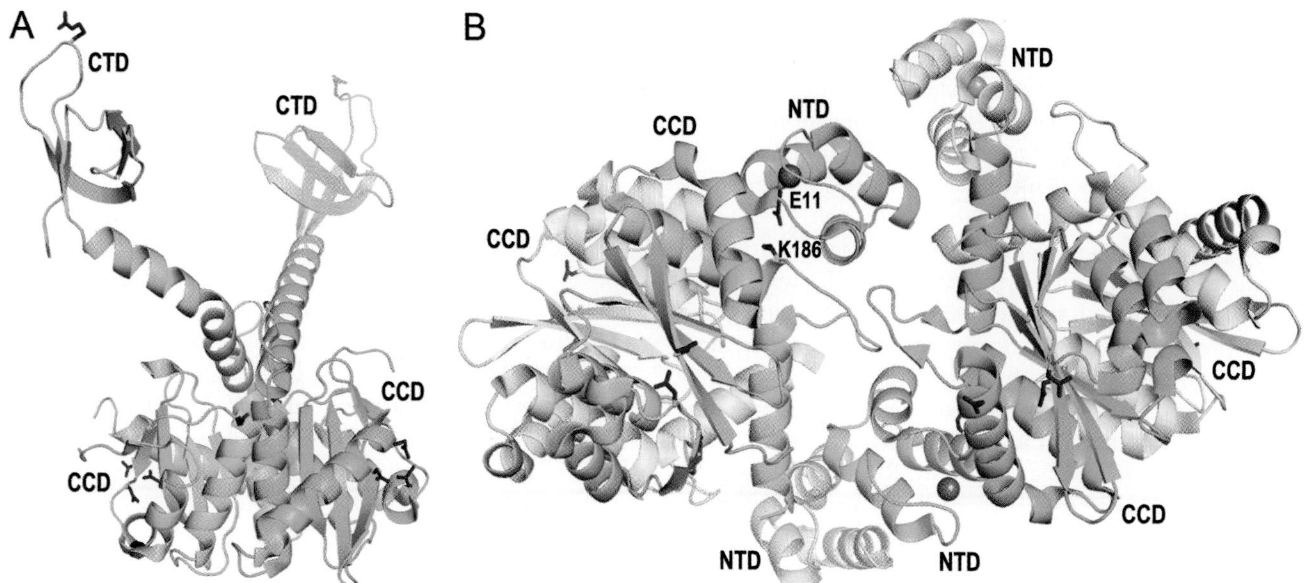

Figure 3 Structures of 2-domain HIV-1 integrase constructs. (A) The X-ray crystal structure of the HIV-1 integrase CCD-CTD dimer (pdb code 1EX4) (66), highlighting the CCD and CTD side chains that were shown in Fig. 2. (B) The crystal structure of the NTD-CCD asymmetric unit (pdb code 1K6Y) (149) highlights NTD residue Glu11 and CCD residue Lys186 of the green and cyan molecules, respectively, which play important roles in integrase concerted integration activity and HIV-1 infection (70). The other pair of interacting residues (Glu11 from the cyan NTD and Lys186 from the green CCD) is not visible in this projection. The side chains of the DDE catalytic triad (red sticks) and NTD-coordinated zinc atoms (grey spheres) are also shown. doi:10.1128/microbiolspec.MDNA3-0024-2014.f3

developed that mimics the amino acid contacts of the IBD-CCD interaction, and the small molecules accordingly compete for the binding of LEDGF/p75 to HIV-1 integrase *in vitro* (155) (reviewed in 156). It is however currently unclear to what extent this inhibition contributes to the antiviral activities of the compounds (see Craigie and Bushman chapter).

Structures of lentiviral integrase NTD-CCD 2-domain constructs with the LEDGF/p75 IBD elucidated the structural basis of the NTD-IBD interaction (61) and clarified the organization of the NTD and CCD within the active tetramer (70). Owing to the favorable solubility of LEDGF/p75 protein, crystallization in these cases proceeded in the absence of solubility-enhancing mutations. Electronegative side chains on one face of the HIV-2 integrase NTD alpha 1 helix apposed electropositive side chains from the second helix-hairpin-helix repeat of the IBD (Fig. 4B). The electrostatic interaction is mutagenetically reversible, as HIV-1 integrase containing Lys residues in place of Glu10 and Glu13 was active only in the presence of reverse-charge LEDGF/p75 mutant protein (61). Such mutant co-dependent function can in theory guide the customized integration of defective integrase mutant viruses in cells that express complementary mutant LEDGF/p75 protein (61, 157).

Similar to the HIV-1 integrase 2-domain construct (149), the maedi-visna virus integrase NTD-CCD fragment crystallized as a dimer of dimers in the presence of the LEDGF/p75 IBD (70). Because an NTD-CCD interdomain linker was resolved in one of two crystal forms, all four NTD-CCD linkers could be unambiguously assigned in this structure. Within the inner dimer, CCD residue Lys188 formed an intermolecular salt bridge with NTD residue Glu11. Mutating the analogous HIV-1 integrase residues (Lys186 and Glu11; Fig. 3B) revealed that the salt bridge played an important role in integrase concerted integration activity and the establishment of HIV-1 infection (70). Thus, although there was limited information on the positioning of the CTDs or viral DNA strands within the intasome, these studies established the basic geometry of the CCD and NTD within a functional integrase tetramer.

FOAMY VIRUS INTASOME STRUCTURES

A key breakthrough in the field of retroviral integrase structural biology came with the crystallization and structural determination of functional PFV integrase-DNA complexes. As alluded to above, PFV integrase is

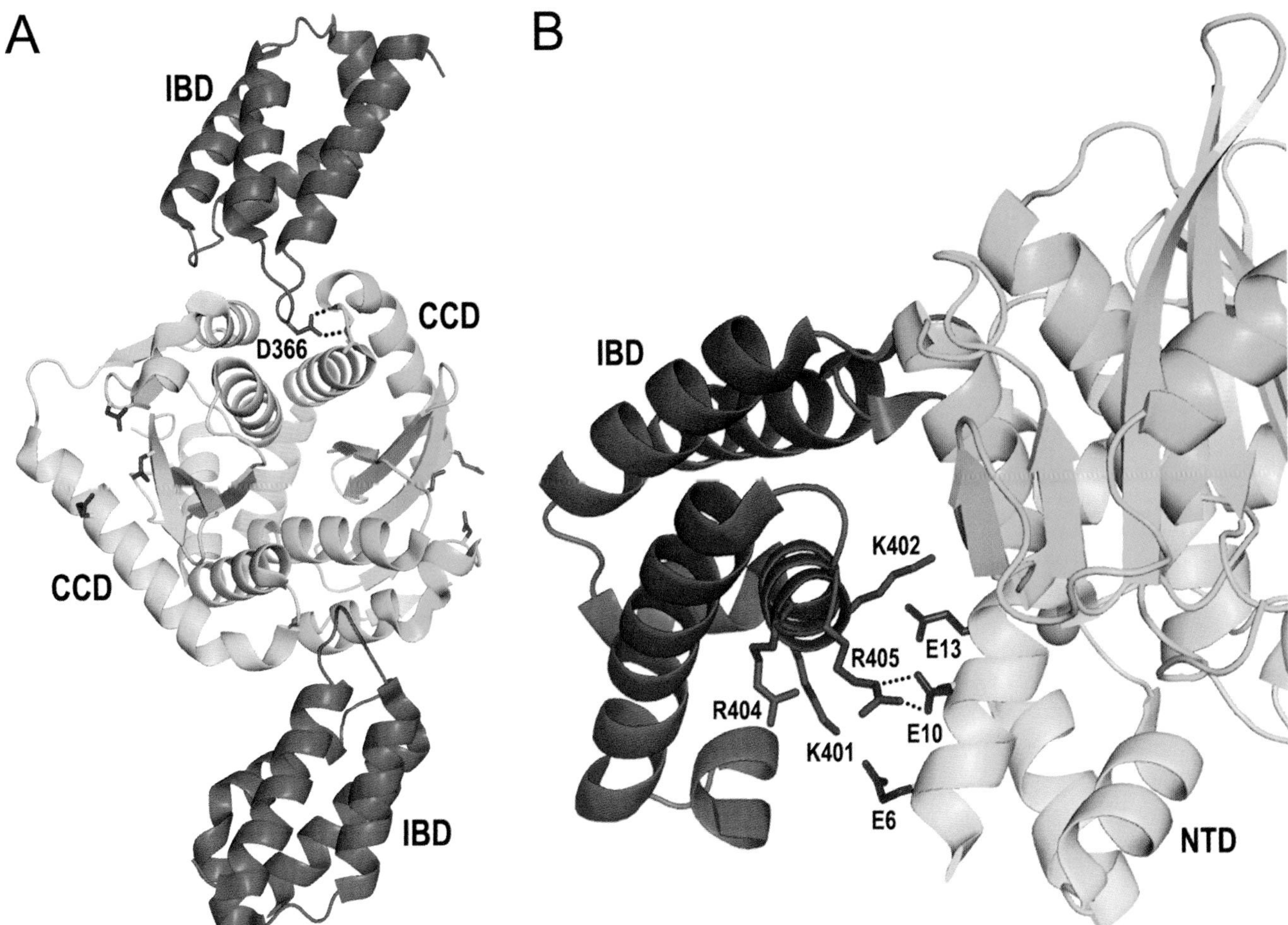

Figure 4 Lentiviral integrase-LEDGF/p75 IBD structures. (A) X-ray crystal structure of the IBD (magenta cartoon)-HIV-1 integrase CCD/F185K (cyan/green dimer) complex (pdb code 2B4J) (69). Highlighted is the Asp366 side-chain from the upper IBD molecule (red stick) hydrogen bonding (dashed lines) to backbone amides of integrase residues within the linker between CCD α helices 4 and 5. The extent to which integrase CCD α4/5 connector regions contribute to forming analogous CCD-CCD dimer interface pockets at least in part accounts for the lentiviral specificity of the LEDGF/p75-integrase interaction (54, 69). The Asp and Glu side chains of the catalytic DDE triad are also painted red. (B) The crystal structure of the HIV-2 integrase NTD-CCD–IBD complex highlights the electrostatic interaction between the integrase NTD and the second helix-hairpin-helix unit of the IBD (pdb code 3F9K, chains A, B, and C) (61). Salt bridges between IBD residue Arg405 and integrase residue Glu10 are indicated by dashed lines; IBD residue Asp366 is behind the green CCD, hidden from view. doi:10.1128/microbiolspec.MDNA3-0024-2014.f4

well behaved in solution, and the intasome complex moreover retained its structural and functional integrity upon challenge with relatively high concentrations of salt (94).

Dimer-of-Dimers Architecture

As hinted at from prior NTD-CCD structures that were solved in the absence of DNA (70, 149), the PFV intasome contains a tetramer of integrase with the dimer-of-dimers architecture (Fig. 5A) (94). Each dimer harbors the canonical CCD dimer interface (green-yellow and cyan-yellow dimers in the figure). The inner integrase protomers (green and cyan) within the tetramer make all the contacts with the viral DNA ends and contribute the active sites that process and integrate the viral DNA (85, 94, 98). The NED and NTD of each inner monomer span the structure to interact with the apposing DNA duplex, and, in the case of the

NTD, the apposing CCD (Fig. 5A). The CTD from each inner monomer positions in between each NTD and CCD, poised to interact with target DNA (94) (Fig. 5).

TCC and STC Structures

Sequence analysis of retroviral integration sites revealed weakly conserved palindromes that center on the staggered cut in target DNA, indicating that an integrase multimer with dyad symmetry prefers particular nucleotides at the sites of viral DNA joining (158, 159, 160, 161, 162). PFV preferentially integrates at sites that on average harbor the sequence $_{-3}$KWK\VYRBMWM$_6$ (written using International Union of Biochemistry base codes; the backslash indicates the position of U5 DNA joining, which occurs at the −1\0 position; the target DNA sequence that is duplicated after integration is underlined) (56). PFV integrase catalyzed the integration of pre-cleaved U5 DNA into a 30 bp symmetric target DNA duplex during intasome crystallogenesis, which allowed the structural determination of the STC (98). TCC structures were determined by omitting divalent metal ion, or by using viral DNA that terminated in dideoxy adenylate and therefore lacked reactive CA_{OH}-3′ ends (98) (Fig. 5B).

The overall geometry of the PFV integrase-viral DNA complex does not change during target DNA binding and integration (98) (Fig. 5). Approximately 26.5 Å separates the two active sites of the inner monomers; the intasome accordingly accommodates target DNA in a severely bent conformation to enable integration across the expanded major groove. The deformation of the target DNA duplex is localized at the central base pair step, which incurs a negative roll of approximately 55° (98) (Fig. 5B). The severe kink impressively occurs in the absence of direct protein–DNA stacking interactions.

As anticipated from the relatively weak nature of target DNA palindrome sequence conservation, the majority of PFV integrase-target DNA interactions are mediated through the phosphodiester backbone (98). Two amino acids, Ala188 in the integrase CCD and Arg329 in the CTD, by contrast make base-specific contacts: Ala188 interacts with cytosine at position 6 through van der Waals interaction whereas Arg329 hydrogen bonds with guanosine 3, guanosine −1, and thymine −2 (see Fig. 5B for target DNA sequence). Thus, within the tetramer, Ala188 and Arg329 interact with all $_{-3}$KWK \VYRBMWM$_6$ consensus bases aside from the central dinucleotide (YR). The sequencing of *in vitro* concerted integration products accordingly revealed that R329S, R329E, and A188S mutant integrases displayed novel target DNA nucleotide preferences (98). The Arg329 mutants additionally preferred sites that harbored relatively flexible dinucleotides at the center.

Depending on their composition, DNA dinucleotides differ in their base-stacking propensity, and hence vary in their ability to accommodate a distortion of a DNA duplex. Pyrimidine-purine (YR) and RY are the most and least flexible, respectively, with intermediate flexibilities for YY and RR dinucleotides (163). Thus, the consensus target DNA palindrome for PFV integrase underscores features of DNA bendability, including bases that register with interacting amino acids Arg329 and Ala188 surrounding a central, flexible YR step (98). Bendability may moreover underlie a general property of retroviral integration and DNA transposition sites (164, 165). The consensus HIV-1 integrase-target DNA sequence $_{-3}$TDG\(G/V)TWA(C/B)CHA$_7$ harbors the central nucleotide signature $_0$RYXRY$_4$ (166). Though seemingly enriched for rigid RY dinucleotides, the pattern actually ensures for relatively flexible sequences at the center of the integration site: Y at the center X position yields YY and YR at nucleotide positions 1 and 2 and at positions 2 and 3, respectively, whereas R at the center X yields YR and RR at these respective positions. Due to the lack of HIV-1 intasome structures, less is known about HIV-1 integrase-target DNA interactions than is known for PFV integrase. Nevertheless, mutagenesis experiments suggest that Ser119 of HIV-1 integrase, a residue that is structurally equivalent to Ala188 in PFV integrase, also interacts with bases that lie three positions upstream from the positions of viral DNA joining (166). Altered patterns of strand transfer reaction products on sequencing gels notably first highlighted a role for Ser119 in HIV-1 integrase and the analogous residue Ser124 in ASLV integrase in target DNA binding (96, 97, 167).

Structural Basis for Integrase Enzymatic Activities

The chemistry of transesterification starts with deprotonation of an attacking nucleophile and concludes with the protonation of a leaving group (168, 169, 170). The key role of the integrase active site residues in this process is to coordinate the binding of divalent metal ions; the metal ions in turn orchestrate the chemistry. One metal ion accordingly positions and deprotonates the attacking nucleophile, which is water for 3′ processing and the 3′-OH of cleaved viral DNA for strand transfer, whereas the other metal ion helps to destabilize the scissile phosphodiester bond and promote the formation of the pentavalent phosphorane reaction intermediate (168, 169, 171).

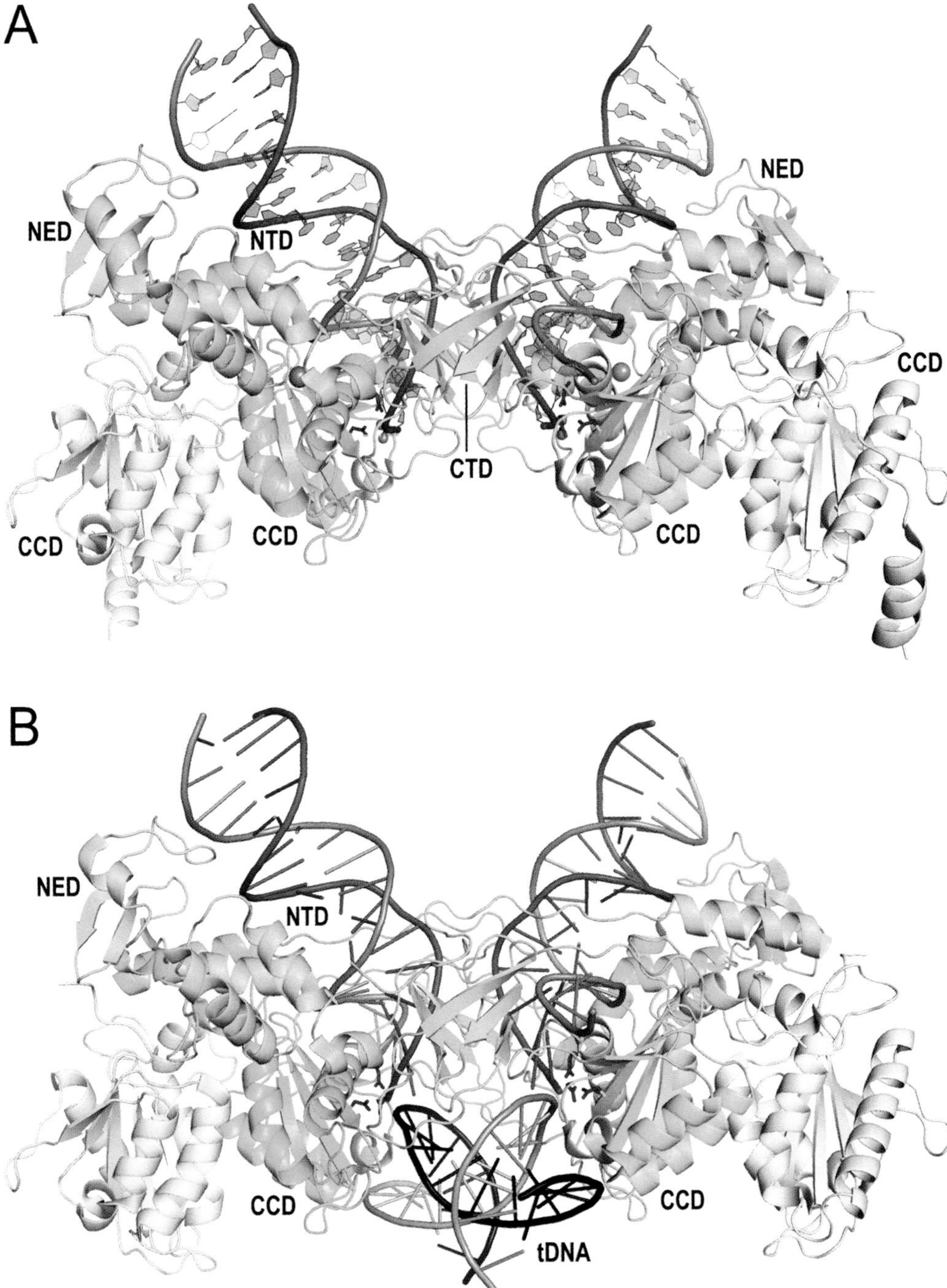

Figure 5 PFV intasome structures. (A) Structure containing 19 bp pre-cleaved U5 DNA end (94, 207) (pdb code 3OY9). The inner integrase monomers of the tetramer, which contact the viral DNA, are painted cyan and green; the outer integrase molecules are yellow. The transferred DNA strands with CA_{OH} 3′ ends are painted magenta whereas the non-transferred strands are orange. The large grey spheres are NTD-coordinated zinc; small grey spheres are Mn atoms coordinated by the DDE active site residues (red sticks) and viral DNA end. (B) Structure of the TCC (98) (pdb code 3OS1). Although a 30 bp target DNA (tDNA) was utilized during crystallization, only 18 bp (grey plus strand sequence $_{-7}$GCACGTG\CTAGCACGTGC$_{10}$) was resolved in the electron density maps.
doi:10.1128/microbiolspec.MDNA3-0024-2014.f5

The co-crystallization of PFV integrase with uncleaved U5 DNA led to the structural determination of the SSC (85). Soaking the SSC crystals with divalent metal ions (Mg^{2+} or Mn^{2+}) supported integrase 3′ processing activity in crystallo (85). Brief exposure of the crystals to manganese chloride allowed freezing out the active form of the SSC just prior to viral DNA cleavage and elucidating the roles of the divalent metal ions in 3′ processing activity. Metal ion A is in near perfect octahedral coordination though its engagement of active site residues Asp128 and Asp185, the nonbridging Pro-*Sp* oxygen of the scissile CA\AT phosphodiester bond, and three water molecules, including the nucleophilic water (Fig. 6A). The distance between the attacking water and scissile phosphodiester bond, 3.3 Å (red dashed line in the figure), was notably identical to that observed in a metal-bound structure of the RNase H enzyme (172). Metal ion B, coordinated through active site residues Asp128 and Glu221, a water molecule, and a bridging oxygen atom in addition to Pro-*Sp*, is in a less ideal environment, which may aid scissile phosphodiester bond destabilization (Fig. 6A) (85). The structure is consistent with the stereoselectivity of the HIV-1 integrase 3′ processing reaction, as the substitution of the Pro-*Rp* oxygen with sulfur was tolerated by integrase to a much greater extent than was the substitution of the Pro-*Sp* position (50).

The crystallized PFV TCC was also catalytically proficient. Due to the relatively rapid kinetics of the strand transfer reaction, numerous crystals were surveyed at early times post-metal ion exposure to determine manganese-bound TCC and STC structures (85). The overlay of both structures clarified the roles of the metal ions during strand transfer as well as the irreversible nature of the reaction (Fig. 6B). The roles of the metal ions in 3′ processing activity reverse during strand transfer, even though the majority of coordination contacts, including those with active site residues and Pro-*Sp* oxygen, remain. Thus, reaction specificity is in large part determined by the positions of attacking nucleophile and scissile bond relative to the metals. Metal ion B accordingly activates the 3′-OH of viral DNA for nucleophilic attack, whereas metal ion A helps to destabilize the scissile bond through its contact with a bridging oxygen atom (Fig. 6B). The structure is again consistent with the stereoselectivity of the HIV-1 integrase enzyme, as the substitution of sulfur for the Pro-*Sp* as compared to Pro-*Rp* oxygen preferentially inhibited strand transfer activity (50).

The phosphodiester bond that is formed between the viral and target DNA during strand transfer is displaced from the active site relative to the position

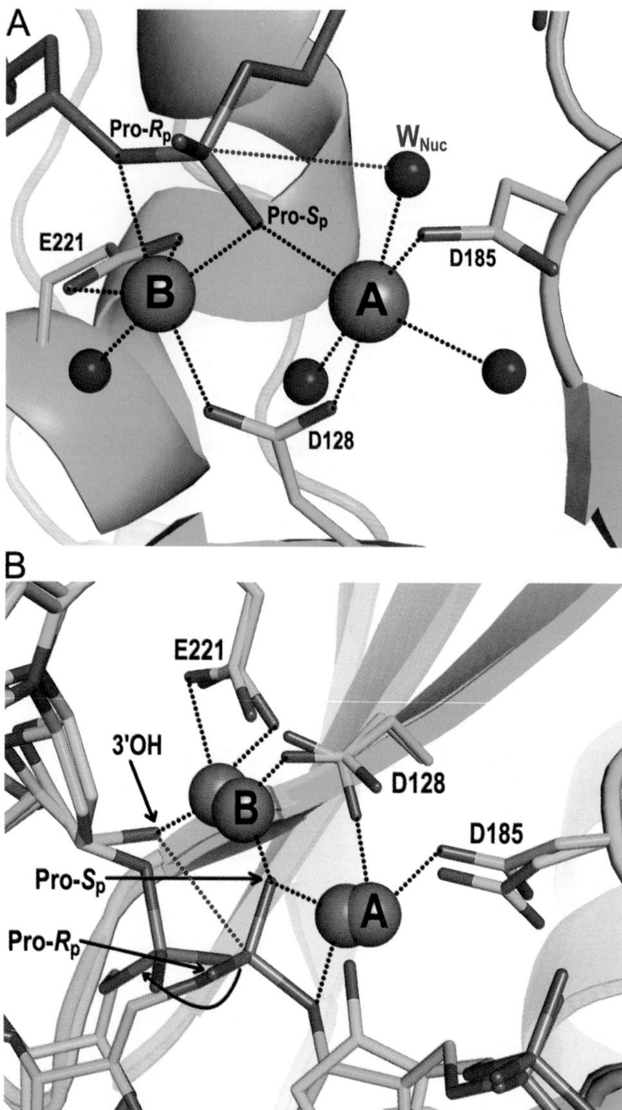

Figure 6 Structural basis of integrase 3′ processing and strand transfer activities. (A) Structure of the manganese-bound SSC (pdb code 4E7I). The DNA and integrase backbones are colored magenta and green, respectively; red and orange sticks are oxygen and phosphorus atoms, respectively. Gray and red spheres are manganese ions and water molecules, respectively, with the nucleophilic water labeled W_{Nuc}. Black dashed lines indicate metal ion interactions; the red dashed line connects the nucleophile and scissile phosphodiester bond. (B) Overlay of metal ion-bound TCC (pdb code 4E7K; DNA and protein in green) and STC (pdb code 4E7L; elements painted in cyan) structures. Both sets of metal ions are shown; the 3.8 Å spacing between ions in the TCC contracts to 3.2 Å in the STC (85). The curved black line indicates the displacement of the viral-target DNA phosphodiester bond after strand transfer relative to the scissile bond in target DNA. Other labeling is the same as in panel A.
doi:10.1128/microbiolspec.MDNA3-0024-2014.f6

of the scissile bond in target DNA prior to catalysis (Fig. 6B, curved arrow). Because isoenergetic reactions like strand transfer are in theory reversible, the displacement ensures that virus integration into chromosomal DNA is a largely irreversible process. The torsional stress that is applied to the target DNA by integrase (Fig. 5B) is the likely driving force behind the displacement (85, 98).

Future Directions in Integrase Structural Biology

The NED, NTD, and CTD of the outer integrase protomers were not resolved in the PFV intasome X-ray crystal structures (Fig. 5) (85, 94, 98). The interaction of these domains with the inner integrase dimer and/or target DNA therefore seems dispensable for integrase 3′ processing and strand transfer activities. We accordingly infer that the main role of the outer two molecules in PFV integrase catalysis *in vitro* is architectural, to frame the critical inner integrase dimer and viral DNA ends together.

Retroviral integration prefers nucleosomal DNA both *in vitro* (173, 174, 175, 176) and during virus infection (177, 178, 179). One potential role for the outer in tegrase protomers may be to orchestrate interactions with nucleoprotein targets as compared to the naked target DNA utilized during intasome crystallogenesis (85, 98). SAXS and small angle neutron scattering studies have yielded models for the outer integrase NED, NTD, and CTD in the PFV intasome (64), and it could accordingly prove useful to conduct similar experiments with nucleoprotein targets such as reconstituted nucleosomes.

Four integrase molecules comprise the intasome, yet retrovirus particles harbor ~100 to 200 copies of integrase (180). It is unclear what fraction of this population may associate with the viral DNA as the PIC forms and traffics through the cell. Retroviral integrase proteins have been reported to bind to numerous cellular proteins (181, 182, 183) and some of these, like LEDGF/p75, have been confirmed to play an important role in integration. Other potential roles for integrase-binding partners include the facilitation of reverse transcription (184) and PIC nuclear import (185, 186, 187, 188, 189), though the biological relevance of some of these interactions, for example that between HIV-1 integrase and the beta-karyopherin transportin 3, has been brought into question (190, 191). Complexes of integrase and cellular binding proteins that are verified to play important roles in virus replication are obvious candidates for structural biology studies moving forward.

Despite the advances afforded from recent success with PFV integrase structural biology, the spumaviruses are but one of seven retroviral genera. The PFV integrase interdomain linkers are relatively long (Fig. 2A), bringing into question as to whether integrase proteins that harbor shorter linker regions, like those derived from HIV-1 (Fig. 2A) or ASLV (77), will support the formation of the dimer-of-dimers architecture observed with the PFV intasome. Numerous groups have assembled HIV-1 models based on the PFV structure (105, 135, 192, 193), indicating reasonable potential for concordance (64). With ASLV integrase NTD-CCD and CCD-CTD linkers as seemingly short as 13 and 8 amino acid residues, respectively (77), the structure of the ASLV intasome should shed light on the potential universality of the PFV intasome architecture as a virus family-wide model.

THE INTEGRASE ACTIVE SITE AS A TARGET FOR ANTI-HIV DRUGS

The critical requirement of integrase for productive HIV-1 replication highlighted the enzyme as a target for drug development (194, 195). Initial attempts however yielded compounds with limited antiviral specificity that predominantly inhibited the assembly of active integrase-DNA complexes, often via electrostatic interactions (1). Screening assays staged with integrase pre-bound to immobilized viral DNA turned out to be key for the discovery and the development of the first clinically useful HIV-1 integrase inhibitors (1, 196, 197).

Integrase Strand Transfer Inhibitors (INSTIs)

Whereas 3′ processing of HIV-1 LTR ends occurs soon after reverse transcription (31, 32), strand transfer is delayed until the HIV-1 TCC is formed in the nucleus. Depending on the activation state of the infected cell, retroviral PICs are accordingly vulnerable to small molecule inhibitors of integrase strand transfer activity for a period of several hours to days (33, 34, 198). The initial compounds in this drug class, which contained a common diketo acid moiety (199), chelated divalent metal ions in the active site and competed with target DNA for binding to the intasome *in vitro* (200, 201).

There are currently three INSTIs approved for the treatment of HIV/AIDS, raltegravir (202), elvitegravir (203), and dolutegravir (204) (Fig. 7). Commonalities among the molecules define aspects of the pharmacophore critical for antiviral activity. Each compound harbors co-planar oxygen atoms (highlighted in red in the figure) that resemble the diketo acid moiety of progenitor INSTIs. The compounds additionally harbor a

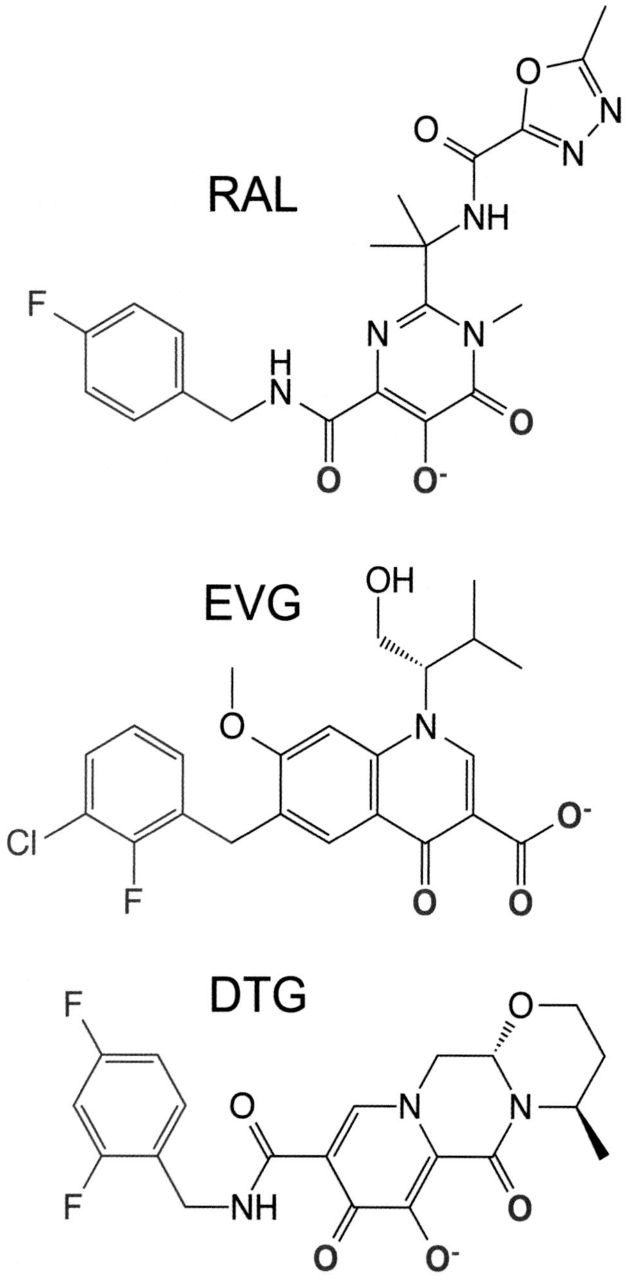

Figure 7 Chemical structures of INSTIs raltegravir (RAL), elvitegravir (EVG), and dolutegravir (DTG).
doi:10.1128/microbiolspec.MDNA3-0024-2014.f7

halogenated benzyl group on a flexible linker (Fig. 7, blue). INSTIs display generally broad-spectrum antiretroviral activity (56, 205, 206), and co-crystallization with the PFV CI accordingly revealed critical aspects of the mechanism of INSTI action (94, 207, 208). As predicted from prior solution-based measures (201), the co-planar oxygen atoms engage the divalent metal ions at the integrase active site. Through stacking interactions with the cytosine of the invariant CA dinucleotide and its guanosine partner on the non-transferred DNA strand, the INSTI halobenzyl group supplants the invariant adenosine base and accordingly ejects the 3′ deoxyadenylate from the active site (compare Fig. 8B with 8A) (94, 207, 208). Ejection of the 3′-OH strand transfer nucleophile from the active site underscores the mechanism of INSTI action. The structure of the uncleaved dinucleotide in the active site of the integrase SSC accounts for the mechanistic selectivity of INSTIs, as the binding sites of the DNA and compounds coincide (Fig. 8B and 8C). Inhibition of 3′ processing activity would accordingly require the ejection of a trinucleotide from the integrase active site (85).

Overlaying drug-bound structures with those of the metal-bound SSC and TCC revealed additional insight into the function of INSTI co-planar oxygen atoms. The position of the raltegravir oxygen atom that is distal from the halobenzyl group mimics the positions of the nucleophilic water molecule for 3′ processing activity and a bridging oxygen atom of the scissile phosphodiester bond in target DNA. Vice versa, the position of the raltegravir oxygen atom proximal to the halobenzyl group mimics those of the 3′-OH strand transfer nucleophile and a bridging oxygen atom of the viral DNA scissile phosphodiester bond (Fig. 8D). Therefore, the INSTIs are in fact substrate mimics of the 3′ processing and strand transfer reactions (85).

Mechanisms of Viral Resistance to INSTIs

The PFV intasome model has additionally proved useful in probing the structural basis of INSTI drug resistance. Raltegravir, elvitegravir, and dolutegravir, as of this writing, have been in the clinic for roughly 7, 2, and 1 years, respectively, and the mechanism of raltegravir resistance has accordingly been investigated most thoroughly. Three resistance pathways have been described, including those involving changes of HIV-1 integrase residues Gln148, Asn155, and Tyr143 (209); PFV integrase harbors Ser217, Asn224, and Tyr212 at these respective positions. Tyr212 stacked against a unique 5-member oxadiazole ring in raltegravir (Fig. 8B), so loss of a direct drug binding contact likely accounts for the Tyr143 resistance pathway (94). Perturbation of local active site structure as compared to loss of direct drug contact is potentially the basis for Gln148 and Asn155 resistance pathways. In particular, INSTI binding to the PFV integrase mutant S217H, which mimicked the HIV-1 integrase resistance mutation Q148H, correlated with an approximate 1-Å shift in the conformation of the active site. INSTI binding requires the mutant

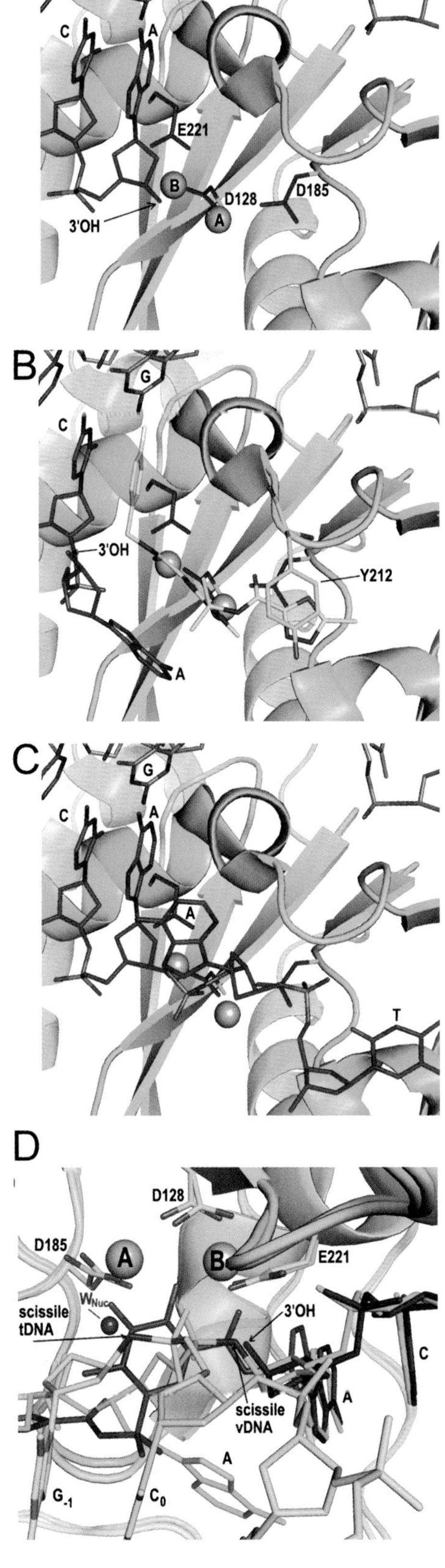

Figure 8 Structural basis of INSTI mechanism of action. (A) The active site from pdb code 3OY9 highlights PFV integrase DDE residues, Mn^{2+} ions A and B, and the 3′-OH of the terminal deoxyadenylate. (B) Structure of raltegravir (cyan)-bound integrase active site (pdb code 3OYA) highlighting positions of supplanted deoxyadenylate 3′-OH, magnesium ions (grey), and integrase residue Tyr212. (C) The integrase active site in the context of the PFV SSC highlights the position of the unprocessed AT dinucleotide. Additional panel A-C coloring: green, integrase; magenta, transferred DNA strand; orange, non-transferred strand. (D) Overlaid structures of the PFV SSC (pdb code 4E7I; integrase and viral DNA in green), TCC (pdb code 4E7K; integrase and target DNA in cyan and viral DNA in blue), and raltegravir-bound CI (pdb code 3OYA; raltegravir in magenta) highlights the common positioning of raltegravir oxygen atoms with strand transfer and 3′ processing attacking and leaving groups. Subscript numbers denote target DNA bases. tDNA, target DNA; vDNA, viral DNA.
doi:10.1128/microbiolspec.MDNA3-0024-2014.f8

active site to adopt a wild type-like conformation, and the associated energy cost should reduce drug binding affinity (207).

The capacity of a small molecule to trap the HIV-1 PIC in a long-lived inhibited state correlates with its antiviral activity, and extended dissociative half-time ($td_{1/2}$) emerged as an important benchmark in INSTI development (210, 211). The second-generation INSTIs dolutegravir and MK-2048 (an experimental compound), which possess $td_{1/2}$s of roughly 71 and 32 hours, respectively, dissociate much more slowly from wild type-integrase DNA complexes *in vitro* than either raltegravir (with a $td_{1/2}$ of 8.8 hours) or elvitegravir (2.7 hours) (210, 211) and are much more potent against viruses with Q148H or N155H resistance mutations (212, 213). Plausibly, the tighter binding INSTIs are less affected by the energetic penalty associated with the need for the active site to reconfigure to a wild type-like conformation (207). In addition, the higher flexibility of dolutegravir was suggested to play a role in its reduced susceptibility to canonical INSTI resistance mutations (208).

Future Directions in Integrase Active Site Drug Development

Though INSTIs are ineffective 3′ processing inhibitors, designs that side step the requirement for nucleotide ejection could be entertained. For example, small molecules that engage the viral DNA-integrase interface in the SSC could insert a chemically inert moiety at the position for the nucleophilic water molecule. Molecular dynamic simulations of the PFV integrase strand transfer reaction could moreover suggest improved positions for INSTI metal chelating groups, which in turn could improve INSTI potency (85). Toward these ends, we suspect that the pharmaceutical industry will employ the PFV model system to help develop novel inhibitors of the integrase active site and HIV-1 DNA integration.

Acknowledgments. *We are grateful to Robert Craigie and Frederic Bushman for critically reading the manuscript. This work was supported by U.S. National Institutes of Health grants AI039394 and AI070042 (to A.E.) and by Medical Research Council UK grants G0900116 and G1000917 (to P.C.).*

Citation. Engelman A. 2014. Retroviral integrase structure and DNA recombination mechanism. Microbiol Spectrum 2(6):MDNA3-0024-2014.

References

1. **Métifiot M, Marchand C, Pommier Y.** 2013. HIV integrase inhibitors: 20-year landmark and challenges. *Adv Pharmacol* **67:**75–105.
2. **Ruelas DS, Greene WC.** 2013. An integrated overview of HIV-1 latency. *Cell* **155:**519–529.
3. **Donehower LA, Varmus HE.** 1984. A mutant murine leukemia virus with a single missense codon in pol is defective in a function affecting integration. *Proc Natl Acad Sci USA* **81:**6461–6465.
4. **Panganiban AT, Temin HM.** 1984. The retrovirus pol gene encodes a product required for DNA integration: Identification of a retrovirus int locus. *Proc Natl Acad Sci USA* **81:**7885–7889.
5. **Schwartzberg P, Colicelli J, Goff SP.** 1984. Construction and analysis of deletion mutations in the pol gene of Moloney murine leukemia virus: A new viral function required for productive infection. *Cell* **37:**1043–1052.
6. **Fassati A, Goff SP.** 1999. Characterization of intracellular reverse transcription complexes of Moloney murine leukemia virus. *J Virol* **73:**8919–8925.
7. **Fassati A, Goff SP.** 2001. Characterization of intracellular reverse transcription complexes of human immunodeficiency virus type 1. *J Virol* **75:**3626–3635.
8. **Panganiban AT, Temin HM.** 1983. The terminal nucleotides of retrovirus DNA are required for integration but not virus production. *Nature* **306:**155–160.
9. **Colicelli J, Goff SP.** 1985. Mutants and pseudorevertants of Moloney murine leukemia virus with alterations at the integration site. *Cell* **42:**573–580.
10. **Colicelli J, Goff SP.** 1988. Sequence and spacing requirements of a retrovirus integration site. *J Mol Biol* **199:**47–59.
11. **Masuda T, Kuroda MJ, Harada S.** 1998. Specific and independent recognition of U3 and U5 att sites by human immunodeficiency virus type 1 integrase in vivo. *J Virol* **72:**8396–8402.
12. **Brown HEV, Chen H, Engelman A.** 1999. Structure-based mutagenesis of the human immunodeficiency virus type 1 DNA attachment site: effects on integration and cDNA synthesis. *J Virol* **73:**9011–9020.
13. **Roth MJ, Schwartzberg PL, Goff SP.** 1989. Structure of the termini of DNA intermediates in the integration of retroviral DNA: Dependence on IN function and terminal DNA sequence. *Cell* **58:**47–54.
14. **Fujiwara T, Mizuuchi K.** 1988. Retroviral DNA integration: structure of an integration intermediate. *Cell* **54:**497–504.
15. **Brown PO, Bowerman B, Varmus HE, Bishop JM.** 1989. Retroviral integration: structure of the initial covalent product and its precursor, and a role for the viral IN protein. *Proc Natl Acad Sci USA* **86:**2525–2529.
16. **Pauza CD.** 1990. Two bases are deleted from the termini of HIV-1 linear DNA during integrative recombination. *Virology* **179:**886–889.
17. **Lee YM, Coffin JM.** 1991. Relationship of avian retrovirus DNA synthesis to integration in vitro. *Mol Cell Biol* **11:**1419–1430.
18. **Brown PO, Bowerman B, Varmus HE, Bishop JM.** 1987. Correct integration of retroviral DNA in vitro. *Cell* **49:**347–356.
19. **Lee YM, Coffin JM.** 1990. Efficient autointegration of avian retrovirus DNA in vitro. *J Virol* **64:**5958–5965.
20. **Ellison V, Abrams H, Roe T, Lifson J, Brown P.** 1990. Human immunodeficiency virus integration in a cell-free system. *J Virol* **64:**2711–2715.
21. **Farnet CM, Haseltine WA.** 1990. Integration of human immunodeficiency virus type 1 DNA in vitro. *Proc Natl Acad Sci USA* **87:**4164–4168.
22. **Katzman M, Katz RA, Skalka AM, Leis J.** 1989. The avian retroviral integration protein cleaves the terminal sequences of linear viral DNA at the in vivo sites of integration. *J Virol* **63:**5319–5327.
23. **Craigie R, Fujiwara T, Bushman F.** 1990. The IN protein of Moloney murine leukemia virus processes the viral DNA ends and accomplishes their integration in vitro. *Cell* **62:**829–837.
24. **Sherman PA, Fyfe JA.** 1990. Human immunodeficiency virus integration protein expressed in Escherichia coli possesses selective DNA cleaving activity. *Proc Natl Acad Sci USA* **87:**5119–5123.
25. **Bushman FD, Fujiwara T, Craigie R.** 1990. Retroviral DNA integration directed by HIV integration protein in vitro. *Science* **249:**1555–1558.
26. **Bushman FD, Craigie R.** 1991. Activities of human immunodeficiency virus (HIV) integration protein in vitro: Specific cleavage and integration of HIV DNA. *Proc Natl Acad Sci USA* **88:**1339–1343.
27. **Katz RA, Merkel G, Kulkosky J, Leis J, Skalka AM.** 1990. The avian retroviral IN protein is both necessary and sufficient for integrative recombination in vitro. *Cell* **63:**87–95.
28. **Vora AC, Fitzgerald ML, Grandgenett DP.** 1990. Removal of 3′-OH-terminal nucleotides from blunt-ended long terminal repeat termini by the avian retrovirus integration protein. *J Virol* **64:**5656–5659.
29. **LaFemina RL, Callahan PL, Cordingley MG.** 1991. Substrate specificity of recombinant human immunodeficiency virus integrase protein. *J Virol* **65:**5624–5630.

30. **Pahl A, Flugel RM.** 1993. Endonucleolytic cleavages and DNA–joining activities of the integration protein of human foamy virus. *J Virol* **67:**5426–5434.
31. **Miller M, Farnet C, Bushman F.** 1997. Human immunodeficiency virus type 1 preintegration complexes: studies of organization and composition. *J Virol* **71:** 5382–5390.
32. **Munir S, Thierry S, Subra F, Deprez E, Delelis O.** 2013. Quantitative analysis of the time–course of viral DNA forms during the HIV-1 life cycle. *Retrovirology* **10:**87.
33. **Butler SL, Hansen MS, Bushman FD.** 2001. A quantitative assay for HIV DNA integration in vivo. *Nat Med* **7:** 631–634.
34. **Vandegraaff N, Kumar R, Burrell CJ, Li P.** 2001. Kinetics of human immunodeficiency virus type 1 (HIV) DNA integration in acutely infected cells as determined using a novel assay for detection of integrated HIV DNA. *J Virol* **75:**11253–11260.
35. **Chow SA, Vincent KA, Ellison V, Brown PO.** 1992. Reversal of integration and DNA splicing mediated by integrase of human immunodeficiency virus. *Science* **255:**723–726.
36. **Vink C, Yeheskiely E, van der Marel GA, Van Boom JH, Plasterk RHA.** 1991. Site-specific hydrolysis and alcoholysis of human immunodeficiency virus DNA termini mediated by the viral integrase protein. *Nucleic Acids Res* **19:**6691–6698.
37. **Katzman M, Sudol M.** 1995. Mapping domains of retroviral integrase responsible for viral DNA specificity and target site selection by analysis of chimeras between human immunodeficiency virus type 1 and visna virus integrases. *J Virol* **69:**5687–5696.
38. **Mizuuchi K.** 1992. Transpositional recombination: Mechanistic insights from studies of Mu and other elements. *Annu Rev Biochem* **61:**1011–1051.
39. **Knowles JR.** 1980. Enzyme-catalyzed phosphoryl transfer reactions. *Annu Rev Biochem* **49:**877–919.
40. **Mizuuchi K, Adzuma K.** 1991. Inversion of the phosphate chirality at the target site of Mu DNA strand transfer: Evidence for a one-step transesterification mechanism. *Cell* **66:**129–140.
41. **Engelman A, Mizuuchi K, Craigie R.** 1991. HIV-1 DNA integration: Mechanism of viral DNA cleavage and DNA strand transfer. *Cell* **67:**1211–1221.
42. **Jenkins TM, Engelman A, Ghirlando R, Craigie R.** 1996. A soluble active mutant of HIV-1 integrase: Involvement of both the core and the C-terminal domains in multimerization. *J Biol Chem* **271:**7712–7718.
43. **Dyda F, Hickman AB, Jenkins TM, Engelman A, Craigie R, Davies DR.** 1994. Crystal structure of the catalytic domain of HIV-1 integrase: similarity to other polynucleotidyl transferases. *Science* **266:**1981–1986.
44. **Song JJ, Smith SK, Hannon GJ, Joshua–Tor L.** 2004. Crystal structure of Argonaute and its implications for RISC slicer activity. *Science* **305:**1434–1437.
45. **Landree MA, Wibbenmeyer JA, Roth DB.** 1999. Mutational analysis of RAG1 and RAG2 identifies three catalytic amino acids in RAG1 critical for both cleavage steps of V(D)J recombination. *Genes Dev* **13:**3059–3069.
46. **Kim DR, Dai Y, Mundy CL, Yang W, Oettinger MA.** 1999. Mutations of acidic residues in RAG1 define the active site of the V(D)J recombinase. *Genes Dev* **13:** 3070–3080.
47. **Nowotny M.** 2009. Retroviral integrase superfamily: The structural perspective. *EMBO Rep* **10:**144–151.
48. **Mizuuchi K, Nobbs TJ, Halford SE, Adzuma K, Qin J.** 1999. A new method for determining the stereochemistry of DNA cleavage reactions: Application to the SfiI and HpaII restriction endonucleases and to the MuA transposase. *Biochemistry* **38:**4640–4648.
49. **van Gent DC, Mizuuchi K, Gellert M.** 1996. Similarities between initiation of V(D)J recombination and retroviral integration. *Science* **271:**1592–1594.
50. **Gerton JL, Herschlag D, Brown PO.** 1999. Stereospecificity of reactions catalyzed by HIV–1 integrase. *J Biol Chem* **274:**33480–33487.
51. **Kennedy AK, Haniford DB, Mizuuchi K.** 2000. Single active site catalysis of the successive phosphoryl transfer steps by DNA transposases: insights from phosphorothioate stereoselectivity. *Cell* **101:**295–305.
52. **Vora AC, McCord M, Fitzgerald ML, Inman RB, Grandgenett DP.** 1994. Efficient concerted integration of retrovirus–like DNA in vitro by avian myeloblastosis virus integrase. *Nucleic Acids Res* **22:**4454–4461.
53. **Yang F, Roth MJ.** 2001. Assembly and catalysis of concerted two–end integration events by Moloney murine leukemia virus integrase. *J Virol* **75:**9561–9670.
54. **Cherepanov P.** 2007. LEDGF/p75 interacts with divergent lentiviral integrases and modulates their enzymatic activity in vitro. *Nucleic Acids Res* **35:**113–124.
55. **Ballandras-Colas A, Naraharisetty H, Li X, Serrao E, Engelman A.** 2013. Biochemical characterization of novel retroviral integrase proteins. *PLoS One* **8:**e76638.
56. **Valkov E, Gupta SS, Hare S, Helander A, Roversi P, McClure M, Cherepanov P.** 2009. Functional and structural characterization of the integrase from the prototype foamy virus. *Nucleic Acids Res* **37:**243–255.
57. **Sinha S, Pursley MH, Grandgenett DP.** 2002. Efficient concerted integration by recombinant human immunodeficiency virus type 1 integrase without cellular or viral cofactors. *J Virol* **76:**3105–3113.
58. **Faure A, Calmels C, Desjobert C, Castroviejo M, Caumont–Sarcos A, Tarrago–Litvak L, Litvak S, Parissi V.** 2005. HIV-1 integrase crosslinked oligomers are active in vitro. *Nucleic Acids Res* **33:**977–986.
59. **Li M, Craigie R.** 2005. Processing of viral DNA ends channels the HIV-1 integration reaction to concerted integration. *J Biol Chem* **280:**29334–29339.
60. **Carteau S, Gorelick RJ, Bushman FD.** 1999. Coupled integration of human immunodeficiency virus type 1 cDNA ends by purified integrase in vitro: Stimulation by the viral nucleocapsid protein. *J Virol* **73:**6670–6679.
61. **Hare S, Shun MC, Gupta SS, Valkov E, Engelman A, Cherepanov P.** 2009. A novel co-crystal structure affords the design of gain-of-function lentiviral integrase mutants in the presence of modified PSIP1/LEDGF/p75. *PLoS Pathog* **5:**e1000259.

62. **Pandey KK, Bera S, Grandgenett DP.** 2011. The HIV-1 integrase monomer induces a specific interaction with LTR DNA for concerted integration. *Biochemistry* **50:** 9788–9796.
63. **Delelis O, Carayon K, Guiot E, Leh H, Tauc P, Brochon JC, Mouscadet JF, Deprez E.** 2008. Insight into the integrase-DNA recognition mechanism. A specific DNA-binding mode revealed by an enzymatically labeled integrase. *J Biol Chem* **283:**27838–27849.
64. **Gupta K, Curtis JE, Krueger S, Hwang Y, Cherepanov P, Bushman FD, Van Duyne GD.** 2012. Solution conformations of prototype foamy virus integrase and its stable synaptic complex with U5 viral DNA. *Structure* **20:** 1918–1928.
65. **Jenkins T, Hickman A, Dyda F, Ghirlando R, Davies D, Craigie R.** 1995. Catalytic domain of human immunodeficiency virus type 1 integrase: Identification of a soluble mutant by systematic replacement of hydrophobic residues. *Proc Natl Acad Sci USA* **92:**6057–6061.
66. **Chen JC-H, Krucinski J, Miercke LJW, Finer-Moore JS, Tang AH, Leavitt AD, Stroud RM.** 2000. Crystal structure of the HIV-1 integrase catalytic core and C-terminal domains: A model for viral DNA binding. *Proc Natl Acad Sci USA* **97:**8233–8238.
67. **Busschots K, Vercammen J, Emiliani S, Benarous R, Engelborghs Y, Christ F, Debyser Z.** 2005. The interaction of LEDGF/p75 with integrase is lentivirus-specific and promotes DNA binding. *J Biol Chem* **280:** 17841–17847.
68. **Michel F, Crucifix C, Granger F, Eiler S, Mouscadet JF, Korolev S, Agapkina J, Ziganshin R, Gottikh M, Nazabal A, Emiliani S, Benarous R, Moras D, Schultz P, Ruff M.** 2009. Structural basis for HIV-1 DNA integration in the human genome, role of the LEDGF/P75 cofactor. *EMBO J* **28:**980–991.
69. **Cherepanov P, Ambrosio ALB, Rahman S, Ellenberger T, Engelman A.** 2005. From the Cover: Structural basis for the recognition between HIV–1 integrase and transcriptional coactivator p75. *Proc Natl Acad Sci USA* **102:**17308–17313.
70. **Hare S, Di Nunzio F, Labeja A, Wang J, Engelman A, Cherepanov P.** 2009. Structural basis for functional tetramerization of lentiviral integrase. *PLoS Pathog* **5:** e1000515.
71. **Khan E, Mack JPG, Katz RA, Kulkosky J, Skalka AM.** 1991. Retroviral integrase domains: DNA binding and the recognition of LTR sequences. *Nucleic Acids Res* **19:**851–860.
72. **Burke CJ, Sanyal G, Bruner MW, Ryan JA, LaFemina RL, Robbins HL, Zeft AS, Middaugh CR, Cordingley MG.** 1992. Structural implications of spectroscopic characterization of a putative zinc finger peptide from HIV-1 integrase. *J Biol Chem* **267:**9639–9644.
73. **Engelman A, Craigie R.** 1992. Identification of conserved amino acid residues critical for human immunodeficiency virus type 1 integrase function in vitro. *J Virol* **66:**6361–6369.
74. **Bushman FD, Engelman A, Palmer I, Wingfield P, Craigie R.** 1993. Domains of the integrase protein of human immunodeficiency virus type 1 responsible for polynucleotidyl transfer and zinc binding. *Proc Natl Acad Sci USA* **90:**3428–3432.
75. **Woerner AM, Klutch M, Levin JG, Marcus-Sekura CJ.** 1992. Localization of DNA binding activity of HIV-1 integrase to the C-terminal half of the protein. *AIDS Res Hum Retroviruses* **8:**297–304.
76. **Vink C, Oude Groeneger AM, Plasterk RHA.** 1993. Identification of the catalytic and DNA-binding region of the human immunodeficiency virus type I integrase protein. *Nucleic Acids Res* **21:**1419–1425.
77. **Li X, Krishnan L, Cherepanov P, Engelman A.** 2011. Structural biology of retroviral DNA integration. *Virology* **411:**194–205.
78. **Kulkosky J, Jones KS, Katz RA, Mack JP, Skalka AM.** 1992. Residues critical for retroviral integrative recombination in a region that is highly conserved among retroviral/retrotransposon integrases and bacterial insertion sequence transposases. *Mol Cell Biol* **12:**2331–2338.
79. **Drelich M, Wilhelm R, Mous J.** 1992. Identification of amino acid residues critical for endonuclease and integration activities of HIV-1 IN protein *in vitro*. *Virology* **188:**459–468.
80. **Leavitt AD, Shiue L, Varmus HE.** 1993. Site-directed mutagenesis of HIV–1 integrase demonstrates differential effects on integrase functions in vitro. *J Biol Chem* **268:**2113–2119.
81. **Fayet O, Ramond P, Polard P, Prère MF, Chandler M.** 1990. Functional similarities between retroviruses and the IS3 family of bacterial insertion sequences? *Mol Microbiol* **4:**1771–1777.
82. **Yuan Y-W, Wessler SR.** 2011. The catalytic domain of all eukaryotic cut–and–paste transposase superfamilies. *Proc Natl Acad Sci USA* **108:**7884–7889.
83. **Kojima KK, Jurka J.** 2013. A superfamily of DNA transposons targeting multicopy small RNA genes. *PLoS One* **8:**e68260.
84. **Guérillot R, Siguier P, Gourbeyre E, Chandler M, Glaser P.** 2014. The diversity of prokaryotic DDE transposases of the mutator superfamily, insertion specificity, and association with conjugation machineries. *Genome Biol Evol* **6:**260–272.
85. **Hare S, Maertens GN, Cherepanov P.** 2012. 3′–Processing and strand transfer catalysed by retroviral integrase in crystallo. *EMBO J* **31:**3020–3028.
86. **Jenkins TM, Esposito D, Engelman A, Craigie R.** 1997. Critical contacts between HIV-1 integrase and viral DNA identified by structure-based analysis and photo-crosslinking. *EMBO J* **16:**6849–6859.
87. **Esposito D, Craigie R.** 1998. Sequence specificity of viral end DNA binding by HIV-1 integrase reveals critical regions for protein–DNA interaction. *EMBO J* **17:** 5832–5843.
88. **Gerton JL, Brown PO.** 1997. The core domain of HIV-1 integrase recognizes key features of its DNA substrates. *J Biol Chem* **272:**25809–25815.
89. **Gerton JL, Ohgi S, Olsen M, DeRisi J, Brown PO.** 1998. Effects of mutations in residues near the active

site of human immunodeficiency virus type 1 integrase on specific enzyme-substrate interactions. *J Virol* **72:** 5046–5055.

90. **Heuer TS, Brown PO.** 1997. Mapping features of HIV–1 integrase near selected sites on viral and target DNA molecules in an active enzyme–DNA complex by photo–cross–linking. *Biochemistry* **36:**10655–10665.
91. **Chen A, Weber IT, Harrison RW, Leis J.** 2006. Identification of amino acids in HIV–1 and avian sarcoma virus integrase subsites required for specific recognition of the long terminal repeat ends. *J Biol Chem* **281:** 4173–4182.
92. **Johnson AA, Santos W, Pais GCG, Marchand C, Amin R, Burke TR Jr, Verdine G, Pommier Y.** 2006. Integration requires a specific interaction of the donor DNA terminal 5′–cytosine with glutamine 148 of the HIV–1 integrase flexible loop. *J Biol Chem* **281:**461–467.
93. **Hobaika Z, Zargarian L, Boulard Y, Maroun RG, Mauffret O, Fermandjian S.** 2009. Specificity of LTR DNA recognition by a peptide mimicking the HIV–1 integrase α4 helix. *Nucleic Acids Res* **37:**7691–7700.
94. **Hare S, Gupta SS, Valkov E, Engelman A, Cherepanov P.** 2010. Retroviral intasome assembly and inhibition of DNA strand transfer. *Nature* **464:**232–236.
95. **Appa RS, Shin C-G, Lee P, Chow SA.** 2001. Role of the nonspecific DNA-binding region and alpha helices within the core domain of retroviral integrase in selecting target DNA sites for integration. *J Biol Chem* **276:**45848–45855.
96. **Harper AL, Skinner LM, Sudol M, Katzman M.** 2001. Use of patient-derived human immunodeficiency virus type 1 integrases to identify a protein residue that affects target site selection. *J Virol* **75:**7756–7762.
97. **Harper AL, Sudol M, Katzman M.** 2003. An amino acid in the central catalytic domain of three retroviral integrases that affects target site selection in nonviral DNA. *J Virol* **77:**3838–3845.
98. **Maertens GN, Hare S, Cherepanov P.** 2010. The mechanism of retroviral integration through X-ray structures of its key intermediates. *Nature* **468:**326–329.
99. **Cai M, Zheng R, Caffrey M, Craigie R, Clore GM, Gronenborn AM.** 1997. Solution structure of the N-terminal zinc binding domain of HIV-1 integrase. *Nat Struct Biol* **4:**567–577.
100. **Eijkelenboom AP, van den Ent FM, Vos A, Doreleijers JF, Hard K, Tullius TD, Plasterk RH, Kaptein R, Boelens R.** 1997. The solution structure of the amino–terminal HHCC domain of HIV-2 integrase: a three-helix bundle stabilized by zinc. *Curr Biol* **7:**739–746.
101. **Lee SP, Han MK.** 1996. Zinc stimulates Mg2+–dependent 3′–processing activity of human immunodeficiency virus type 1 integrase in vitro. *Biochemistry* **35:**3837–3844.
102. **Zheng R, Jenkins TM, Craigie R.** 1996. Zinc folds the N-terminal domain of HIV-1 integrase, promotes multimerization, and enhances catalytic activity. *Proc Natl Acad Sci USA* **93:**13659–13664.
103. **van den Ent FMI, Vos A, Plasterk RHA.** 1999. Dissecting the role of the N-terminal domain of human immunodeficiency virus integrase by trans–complementation analysis. *J Virol* **73:**3176–3183.
104. **Zhao Z, McKee CJ, Kessl JJ, Santos WL, Daigle JE, Engelman A, Verdine G, Kvaratskhelia M.** 2008. Subunit-specific protein footprinting reveals significant structural rearrangements and a role for N-terminal Lys-14 of HIV-1 integrase during viral DNA binding. *J Biol Chem* **283:**5632–5641.
105. **Krishnan L, Li X, Naraharisetty HL, Hare S, Cherepanov P, Engelman A.** 2010. Structure–based modeling of the functional HIV-1 intasome and its inhibition. *Proc Natl Acad Sci USA* **107:**15010–15915.
106. **Cannon PM, Byles ED, Kingsman SM, Kingsman AJ.** 1996. Conserved sequences in the carboxyl terminus of integrase that are essential for human immunodeficiency virus type 1 replication. *J Virol* **70:**651–657.
107. **Engelman A, Hickman AB, Craigie R.** 1994. The core and carboxyl–terminal domains of the integrase protein of human immunodeficiency virus type 1 each contribute to nonspecific DNA binding. *J Virol* **68:**5911–5917.
108. **Puras Lutzke RA, Plasterk RHA.** 1998. Structure-based mutational analysis of the C–terminal DNA-binding domain of human immunodeficiency virus type 1 integrase: Critical residues for protein oligomerization and DNA binding. *J Virol* **72:**4841–4848.
109. **Puras Lutzke RA, Vink C, Plasterk RHA.** 1994. Characterization of the minimal DNA-binding domain of the HIV integrase protein. *Nucleic Acids Res* **22:**4125–4131.
110. **Andrake MD, Skalka AM.** 1995. Multimerization determinants reside in both the catalytic core and C terminus of avian sarcoma virus integrase. *J Biol Chem* **270:** 29299–29306.
111. **Engelman A, Bushman FD, Craigie R.** 1993. Identification of discrete functional domains of HIV-1 integrase and their organization within an active multimeric complex. *EMBO J* **12:**3269–3275.
112. **van Gent DC, Oude Groeneger AAM, Plasterk RHA.** 1993. Identification of amino acids in HIV-2 integrase involved in site–specific hydrolysis and alcoholysis of viral DNA termini. *Nucleic Acids Res* **21:**3373–3377.
113. **Bushman FD, Wang B.** 1994. Rous sarcoma virus integrase protein: Mapping functions for catalysis and substrate binding. *J Virol* **68:**2215–2223.
114. **Kulkosky J, Katz RA, Merkel G, Skalka AM.** 1995. Activities and substrate specificity of the evolutionarily conserved central domain of retroviral integrase. *Virology* **206:**448–456.
115. **Pahl A, Flügel RM.** 1995. Characterization of the human spuma retrovirus integrase by site–directed mutagenesis, by complementation analysis, and by swapping the zinc finger domain of HIV-1. *J Biol Chem* **270:**2957–2966.
116. **Jonsson C, Donzella G, Gaucan E, Smith C, Roth M.** 1996. Functional domains of Moloney murine leukemia virus integrase defined by mutation and complementation analysis. *J Virol* **70:**4585–4597.
117. **Jones KS, Coleman J, Merkel GW, Laue TM, Skalka AM.** 1992. Retroviral integrase functions as a multimer and can turn over catalytically. *J Biol Chem* **267:** 16037–16040.

118. **Bao KK, Wang H, Miller JK, Erie DA, Skalka AM, Wong I.** 2003. Functional oligomeric state of avian sarcoma virus integrase. *J Biol Chem* **278:**1323–1327.

119. **Cherepanov P, Maertens G, Proost P, Devreese B, Van Beeumen J, Engelborghs Y, De Clercq E, Debyser Z.** 2003. HIV-1 integrase forms stable tetramers and associates with LEDGF/p75 protein in human cells. *J Biol Chem* **278:**372–381.

120. **Li M, Mizuuchi M, Burke TRJ, Craigie R.** 2006. Retroviral DNA integration: Reaction pathway and critical intermediates. *EMBO J* **25:**1295–1304.

121. **Bera S, Pandey KK, Vora AC, Grandgenett DP.** 2009. Molecular interactions between HIV–1 integrase and the two viral DNA ends within the synaptic complex that mediates concerted integration. *J Mol Biol* **389:** 183–198.

122. **van Gent DC, Vink C, Groeneger AAMO, Plasterk RHA.** 1993. Complementation between HIV integrase proteins mutated in different domains. *EMBO J* **12:** 3261–3267.

123. **Ellison V, Gerton J, Vincent KA, Brown PO.** 1995. An essential interaction between distinct domains of HIV-1 integrase mediates assembly of the active multimer. *J Biol Chem* **270:**3320–3326.

124. **Vink C, van Gent DC, Elgersma Y, Plasterk RH.** 1991. Human immunodeficiency virus integrase protein requires a subterminal position of its viral DNA recognition sequence for efficient cleavage. *J Virol* **65:**4636–4644.

125. **Leavitt AD, Rose RB, Varmus HE.** 1992. Both substrate and target oligonucleotide sequences affect in vitro integration mediated by human immunodeficiency virus type 1 integrase protein produced in Saccharomyces cerevisiae. *J Virol* **66:**2359–2368.

126. **Sherman PA, Dickson ML, Fyfe JA.** 1992. Human immunodeficiency virus type 1 integration protein: DNA sequence requirements for cleaving and joining reactions. *J Virol* **66:**3593–3601.

127. **Masuda T, Planelles V, Krogstad P, Chen IS.** 1995. Genetic analysis of human immunodeficiency virus type 1 integrase and the U3 att site: unusual phenotype of mutants in the zinc finger–like domain. *J Virol* **69:**6687–6696.

128. **Oh J, Chang KW, Alvord WG, Hughes SH.** 2006. Alternate polypurine tracts affect Rous sarcoma virus integration in vivo. *J Virol* **80:**10281–10284.

129. **Oh J, Chang KW, Hughes SH.** 2006. Mutations in the U5 sequences adjacent to the primer binding site do not affect tRNA cleavage by Rous sarcoma virus RNase H but do cause aberrant integrations in vivo. *J Virol* **80:** 451–459.

130. **Chaconas G.** 1999. Studies on a "jumping gene machine": Higher–order nucleoprotein complexes in Mu DNA transposition. *Biochem Cell Biol* **77:**487–491.

131. **Krishnan L, Engelman A.** 2012. Retroviral integrase proteins and HIV–1 DNA integration. *J Biol Chem* **287:** 40858–40866.

132. **Wei S–Q, Mizuuchi K, Craigie R.** 1997. A large nucleoprotein assembly at the ends of the viral DNA mediates retroviral DNA integration. *EMBO J* **16:**7511–7520.

133. **Chen H, Wei S–Q, Engelman A.** 1999. Multiple integrase functions are required to form the native structure of the human immunodeficiency virus type I intasome. *J Biol Chem* **274:**17358–17364.

134. **Kotova S, Li M, Dimitriadis EK, Craigie R.** 2010. Nucleoprotein intermediates in HIV–1 DNA integration visualized by atomic force microscopy. *J Mol Biol* **399:** 491–500.

135. **Kessl JJ, Li M, Ignatov M, Shkriabai N, Eidahl JO, Feng L, Musier–Forsyth K, Craigie R, Kvaratskhelia M.** 2011. FRET analysis reveals distinct conformations of IN tetramers in the presence of viral DNA or LEDGF/p75. *Nucleic Acids Res* **39:**9009–9022.

136. **Guiot E, Carayon K, Delelis O, Simon F, Tauc P, Zubin E, Gottikh M, Mouscadet J–F, Brochon J–C, Deprez E.** 2006. Relationship between the oligomeric status of HIV-1 integrase on DNA and enzymatic activity. *J Biol Chem* **281:**22707–22719.

137. **Bojja RS, Andrake MD, Weigand S, Merkel G, Yarychkivska O, Henderson A, Kummerling M, Skalka AM.** 2011. Architecture of a full-length retroviral integrase monomer and dimer, revealed by small angle X-ray scattering and chemical cross-linking. *J Biol Chem* **286:**17047–17059.

138. **Bojja RS, Andrake MD, Merkel G, Weigand S, Dunbrack RLJ, Skalka AM.** 2013. Architecture and assembly of HIV integrase multimers in the absence of DNA substrates. *J Biol Chem* **288:**7373–7386.

139. **Hickman AB, Palmer I, Engelman A, Craigie R, Wingfield P.** 1994. Biophysical and enzymatic properties of the catalytic domain of HIV-1 integrase. *J Biol Chem* **269:**29279–29287.

140. **Bujacz G, Jaskolski M, Alexandratos J, Wlodawer A, Merkel G, Katz RA, Skalka AM.** 1995. High-resolution structure of the catalytic domain of avian sarcoma virus integrase. *J Mol Biol* **253:**333–346.

141. **Eijkelenboom AP, van den Ent FM, Wechselberger R, Plasterk RH, Kaptein R, Boelens R.** 2000. Refined solution structure of the dimeric N-terminal HHCC domain of HIV-2 integrase. *J Biomol NMR* **18:**119–128.

142. **Lodi PJ, Ernst JA, Kuszewski J, Hickman AB, Engelman A, Craigie R, Clore GM, Gronenborn AM.** 1995. Solution structure of the DNA binding domain of HIV-1 integrase. *Biochemistry* **34:**9826–9833.

143. **Eijkelenboom APAM, Puras Lutzke RA, Boelens R, Plasterk RHA, Kaptein R, Hård K.** 1995. The DNA-binding domain of HIV–1 integrase has an SH3-like fold. *Nat Struct Biol* **2:**807–810.

144. **Eijkelenboom APAM, Sprangers R, Hård K, Puras Lutzke RA, Plasterk RHA, Boelens R, Kaptein R.** 1999. Refined solution structure of the C-terminal DNA-binding domain of human immunovirus-1 integrase. *Proteins* **36:**556–564.

145. **Kaneko T, Li L, Li SS.** 2008. The SH3 domain—a family of versatile peptide- and protein-recognition module. *Front Biosci* **13:**4938–4952.

146. **Baumann H, Knapp S, Lundbäck T, Ladenstein R, Härd T.** 1994. Solution structure and DNA-binding properties of a thermostable protein from the archaeon Sulfolobus solfataricus. *Nat Struct Biol* **1:**808–819.

147. **Chen Z, Yan Y, Munshi S, Li Y, Zugay-Murphy J, Xu B, Witmer M, Felock P, Wolfe A, Sardana V.** 2000. X-ray structure of simian immunodeficiency virus integrase containing the core and C-terminal domain (residues 50–293) – an initial glance of the viral DNA binding platform. *J Mol Biol* **296:**521–533.

148. **Yang Z-N, Mueser TC, Bushman FD, Hyde CC.** 2000. Crystal structure of an active two-domain derivative of Rous sarcoma virus integrase. *J Mol Biol* **296:**535–548.

149. **Wang J-Y, Ling H, Yang W, Craigie R.** 2001. Structure of a two-domain fragment of HIV-1 integrase: implications for domain organization in the intact protein. *EMBO J* **20:**7333–7343.

150. **Engelman A, Cherepanov P.** 2008. The lentiviral integrase binding protein LEDGF/p75 and HIV-1 replication. *PLoS Pathog* **4:**e1000046.

151. **Poeschla EM.** 2008. Integrase, LEDGF/p75 and HIV replication. *Cell Mol Life Sci* **65:**1403–1424.

152. **Cherepanov P, Devroe E, Silver PA, Engelman A.** 2004. Identification of an evolutionarily-conserved domain in LEDGF/p75 that binds HIV-1 integrase. *J Biol Chem* **279:**48883–48892.

153. **Maertens G, Cherepanov P, Pluymers W, Busschots K, De Clercq E, Debyser Z, Engelborghs Y.** 2003. LEDGF/p75 is essential for nuclear and chromosomal targeting of HIV-1 integrase in human cells. *J Biol Chem* **278:** 33528–33539.

154. **Cherepanov P, Sun Z-YJ, Rahman S, Maertens G, Wagner G, Engelman A.** 2005. Solution structure of the HIV-1 integrase-binding domain in LEDGF/p75. *Nat Struct Mol Biol* **12:**526–532.

155. **Christ F, Voet A, Marchand A, Nicolet S, Desimmie BA, Marchand D, Bardiot D, Van der Veken NJ, Van Remoortel B, Strelkov SV, De Maeyer M, Chaltin P, Debyser Z.** 2010. Rational design of small-molecule inhibitors of the LEDGF/p75-integrase interaction and HIV replication. *Nat Chem Biol* **6:**442–448.

156. **Jurado KA, Engelman A.** 2013. Multimodal mechanism of action of allosteric HIV-1 integrase inhibitors. *Expert Rev Mol Med* **15:**e14.

157. **Wang H, Shun MC, Li X, Di Nunzio F, Hare S, Cherepanov P, Engelman A.** 2014. Efficient transduction of LEDGF/p75 mutant cells by complementary gain-of-function HIV-1 integrase mutant viruses. *Mol Ther Methods Clin Dev* **1:**2.

158. **Stevens SW, Griffith JD.** 1996. Sequence analysis of the human DNA flanking sites of human immunodeficiency virus type 1 integration. *J Virol* **70:**6459–6462.

159. **Carteau S, Hoffmann C, Bushman F.** 1998. Chromosome structure and human immunodeficiency virus type 1 cDNA integration: Centromeric alphoid repeats are a disfavored target. *J Virol* **72:**4005–4014.

160. **Holman AG, Coffin JM.** 2005. Symmetrical base preferences surrounding HIV-1, avian sarcoma/leukosis virus, and murine leukemia virus integration sites. *Proc Natl Acad Sci USA* **102:**6103–6107.

161. **Wu X, Li Y, Crise B, Burgess SM, Munroe DJ.** 2005. Weak palindromic consensus sequences are a common feature found at the integration target sites of many retroviruses. *J Virol* **79:**5211–5214.

162. **Berry C, Hannenhalli S, Leipzig J, Bushman FD.** 2006. Selection of target sites for mobile DNA integration in the human genome. *PLoS Comput Biol* **2:**e157.

163. **Johnson RC, Stella S, Heiss JK.** 2008. Bending and compaction of DNA by proteins, p 176–220. *In* Rice PA, Correll CC (ed), *Protein-Nucleic Acid Interactions: Structural Biology*. RSC Publishing, London.

164. **Liao G-c, Rehm EJ, Rubin GM.** 2000. Insertion site preferences of the P transposable element in Drosophila melanogaster. *Proc Natl Acad Sci USA* **97:**3347–3351.

165. **Haapa-Paananen S, Rita H, Savilahti H.** 2002. DNA transposition of bacteriophage Mu: A quantitative analysis of target site selection in vitro. *J Biol Chem* **277:** 2843–2851.

166. **Serrao E, Krishnan L, Shun M-C, Li X, Cherepanov P, Engelman A, Maertens GN.** 2014. Integrase residues that determine nucleotide preferences at sites of HIV-1 integration: implications for the mechanism of target DNA binding. *Nucleic Acids Res* **42:**5146–5176.

167. **Nowak MG, Sudol M, Lee NE, Konsavage WMJ, Katzman M.** 2009. Identifying amino acid residues that contribute to the cellular-DNA binding site on retroviral integrase. *Virology* **389:**141–148.

168. **Mizuuchi K.** 1992. Polynucleotidyl transfer reactions in transpositional DNA recombination. *J Biol Chem* **267:** 21273–21276.

169. **Yang W, Lee JY, Nowotny M.** 2006. Making and breaking nucleic acids: two-Mg2+-ion catalysis and substrate specificity. *Mol Cell* **22:**5–13.

170. **Rosta E, Nowotny M, Yang W, Hummer G.** 2011. Catalytic mechanism of RNA backbone cleavage by ribonuclease H from quantum mechanics/molecular mechanics simulations. *J Am Chem Soc* **133:**8934–8941.

171. **Nowotny M, Yang W.** 2006. Stepwise analyses of metal ions in RNase H catalysis from substrate destabilization to product release. *EMBO J* **25:**1924–1933.

172. **Nowotny M, Gaidamakov SA, Crouch RJ, Yang W.** 2005. Crystal structures of RNase H bound to an RNA/DNA hybrid: substrate specificity and metal-dependent catalysis. *Cell* **121:**1005–1016.

173. **Pryciak PM, Varmus HE.** 1992. Nucleosomes, DNA-binding proteins, and DNA sequence modulate retroviral integration target site selection. *Cell* **69:**769–780.

174. **Pryciak PM, Sil A, Varmus HE.** 1992. Retroviral integration into minichromosomes in vitro. *EMBO J* **11:** 291–303.

175. **Pruss D, Bushman F, Wolffe A.** 1994. Human immunodeficiency virus integrase directs integration to sites of severe DNA distortion within the nucleosome core. *Proc Natl Acad Sci USA* **91:**5913–5917.

176. **Pruss D, Reeves R, Bushman FD, Wolffe AP.** 1994. The influence of DNA and nucleosome structure on integration events directed by HIV integrase. *J Biol Chem* **269:** 25031–25041.

177. **Pryciak PM, Müller HP, Varmus HE.** 1992. Simian virus 40 minichromosomes as targets for retroviral integration in vivo. *Proc Natl Acad Sci USA* **89:**9237–9241.

178. **Wang GP, Ciuffi A, Leipzig J, Berry CC, Bushman FD.** 2007. HIV integration site selection: Analysis by massively parallel pyrosequencing reveals association with epigenetic modifications. *Genome Res* **17:**1186–1194.

179. **Roth SL, Malani N, Bushman FD.** 2011. Gammaretroviral integration into nucleosomal target DNA in vivo. *J Virol* **85:**7393–7401.

180. **Coffin JM, Hughes SH, Varmus HE.** 1997. *Retroviruses.* Cold Spring Harbor Laboratory Press, Plainville.

181. **Turlure F, Devroe E, Silver PA, Engelman A.** 2004. Human cell proteins and human immunodeficiency virus DNA integration. *Front Biosci* **9:**3187–3208.

182. **Van Maele B, Busschots K, Vandekerckhove L, Christ F, Debyser Z.** 2006. Cellular co-factors of HIV-1 integration. *Trends Biochem Sci* **31:**98–105.

183. **Engelman A.** 2007. Host cell factors and HIV-1 integration. *Future HIV Ther* **1:**415–426.

184. **Nishitsuji H, Hayashi T, Takahashi T, Miyano M, Kannagi M, Masuda T.** 2009. Augmentation of reverse transcription by integrase through an interaction with host factor, SIP1/Gemin2 is critical for HIV-1 infection. *PLoS One* **4:**e7825.

185. **Gallay P, Hope T, Chin D, Trono D.** 1997. HIV-1 infection of nondividing cells through the recognition of integrase by the importin/karyopherin pathway. *Proc Natl Acad Sci USA* **94:**9825–9830.

186. **Fassati A, Gorlich D, Harrison I, Zaytseva L, Mingot JM.** 2003. Nuclear import of HIV-1 intracellular reverse transcription complexes is mediated by importin 7. *EMBO J* **22:**3675–3685.

187. **Ao Z, Danappa Jayappa K, Wang B, Zheng Y, Kung S, Rassart E, Depping R, Kohler M, Cohen EA, Yao X.** 2010. Importin alpha3 interacts with HIV–1 integrase and contributes to HIV–1 nuclear import and replication. *J Virol* **84:**8650–8663.

188. **Ao Z, Huang G, Yao H, Xu Z, Labine M, Cochrane AW, Yao X.** 2007. Interaction of human immunodeficiency virus type 1 integrase with cellular nuclear import receptor importin 7 and its impact on viral replication. *J Biol Chem* **282:**13456–13467.

189. **Christ F, Thys W, De Rijck J, Gijsbers R, Albanese A, Arosio D, Emiliani S, Rain JC, Benarous R, Cereseto A, Debyser Z.** 2008. Transportin-SR2 imports HIV into the nucleus. *Curr Biol* **18:**1192–1202.

190. **Krishnan L, Matreyek KA, Oztop I, Lee K, Tipper CH, Li X, Dar MJ, Kewalramani VN, Engelman A.** 2010. The requirement for cellular transportin 3 (TNPO3 or TRN-SR2) during infection maps to human immunodeficiency virus type 1 capsid and not integrase. *J Virol* **84:**397–406.

191. **Maertens GN, Cook NJ, Wang W, Hare S, Gupta SS, Öztop I, Lee K, Pye VE, Cosnefroy O, Snijders AP, KewalRamani VN, Fassati A, Engelman A, Cherepanov P.** 2014. Structural basis for nuclear import of splicing factors by human Transportin 3. *Proc Natl Acad Sci USA* **111:**2728–2733.

192. **Quashie PK, Mesplède T, Han Y-S, Oliveira M, Singhroy DN, Fujiwara T, Underwood MR, Wainberg MA.** 2012. Characterization of the R263K mutation in HIV-1 integrase that confers low-level resistance to the second-generation integrase strand transfer inhibitor dolutegravir. *J Virol* **86:**2696–2705.

193. **Johnson BC, Métifiot M, Ferris A, Pommier Y, Hughes SH.** 2013. A homology model of HIV-1 integrase and analysis of mutations designed to test the model. *J Mol Biol* **425:**2133–2146.

194. **LaFemina R, Schneider C, Robbins H, Callahan P, LeGrow K, Roth E, Schleif W, Emini E.** 1992. Requirement of active human immunodeficiency virus type 1 integrase enzyme for productive infection of human T-lymphoid cells. *J Virol* **66:**7414–7419.

195. **Sakai H, Kawamura M, Sakuragi J, Sakuragi S, Shibata R, Ishimoto A, Ono N, Ueda S, Adachi A.** 1993. Integration is essential for efficient gene expression of human immunodeficiency virus type 1. *J Virol* **67:**1169–1174.

196. **Wolfe AL, Felock PJ, Hastings JC, Blau CU, Hazuda DJ.** 1996. The role of manganese in promoting multimerization and assembly of human immunodeficiency virus type 1 integrase as a catalytically active complex on immobilized long terminal repeat substrates. *J Virol* **70:**1424–1432.

197. **Hazuda D, Felock PJ, Hastings JC, Pramanik B, Wolfe AL.** 1997. Discovery and analysis of inhibitors of the human immunodeficiency integrase. *Drug Des Discov* **15:**17–24.

198. **Pierson TC, Zhou Y, Kieffer TL, Ruff CT, Buck C, Siliciano RF.** 2002. Molecular characterization of preintegration latency in human immunodeficiency virus type 1 infection. *J Virol* **76:**8518–8531.

199. **Hazuda DJ, Felock P, Witmer M, Wolfe A, Stillmock K, Grobler JA, Espeseth A, Gabryelski L, Schleif W, Blau C, Miller MD.** 2000. Inhibitors of strand transfer that prevent integration and inhibit HIV-1 replication in cells. *Science* **287:**646–650.

200. **Espeseth AS, Felock P, Wolfe A, Witmer M, Grobler J, Anthony N, Egbertson M, Melamed JY, Young S, Hamill T, Cole JL, Hazuda DJ.** 2000. HIV-1 integrase inhibitors that compete with the target DNA substrate define a unique strand transfer conformation for integrase. *Proc Natl Acad Sci USA* **97:**11244–11249.

201. **Grobler JA, Stillmock K, Hu B, Witmer M, Felock P, Espeseth AS, Wolfe A, Egbertson M, Bourgeois M, Melamed J, Wai JS, Young S, Vacca J, Hazuda DJ.** 2002. Diketo acid inhibitor mechanism and HIV-1 integrase: Implications for metal binding in the active site of phosphotransferase enzymes. *Proc Natl Acad Sci USA* **99:**6661–6666.

202. **Summa V, Petrocchi A, Bonelli F, Crescenzi B, Donghi M, Ferrara M, Fiore F, Gardelli C, Gonzalez Paz O, Hazuda DJ, Jones P, Kinzel O, Laufer R, Monteagudo E, Muraglia E, Nizi E, Orvieto F, Pace P, Pescatore G, Scarpelli R, Stillmock K, Witmer MV, Rowley M.** 2008. Discovery of raltegravir, a potent, selective orally bioavailable HIV-integrase inhibitor for the treatment of HIV-AIDS infection. *J Med Chem* **51:**5843–5855.

203. **Sato M, Motomura T, Aramaki H, Matsuda T, Yamashita M, Ito Y, Kawakami H, Matsuzaki Y, Watanabe W, Yamataka K, Ikeda S, Kodama E,**

Matsuoka M, Shinkai H. 2006. Novel HIV-1 integrase inhibitors derived from quinolone antibiotics. *J Med Chem* **49:**1506–1508.

204. **Johns BA, Kawasuji T, Weatherhead JG, Taishi T, Temelkoff DP, Yoshida H, Akiyama T, Taoda Y, Murai H, Kiyama R, Fuji M, Tanimoto N, Jeffrey J, Foster SA, Yoshinaga T, Seki T, Kobayashi M, Sato A, Johnson MN, Garvey EP, Fujiwara T.** 2013. Carbamoyl pyridone HIV-1 integrase inhibitors 3. A diastereomeric approach to chiral nonracemic tricyclic ring systems and the discovery of dolutegravir (S/GSK1349572) and (S/GSK1265744). *J Med Chem* **56:**5901–5916.

205. **Shimura K, Kodama E, Sakagami Y, Matsuzaki Y, Watanabe W, Yamataka K, Watanabe Y, Ohata Y, Doi S, Sato M, Kano M, Ikeda S, Matsuoka M.** 2008. Broad antiretroviral activity and resistance profile of the novel human immunodeficiency virus integrase inhibitor elvitegravir (JTK–303/GS–9137). *J Virol* **82:**764–774.

206. **Koh Y, Matreyek KA, Engelman A.** 2011. Differential sensitivities of retroviruses to integrase strand transfer inhibitors. *J Virol* **85:**3677–3682.

207. **Hare S, Vos AM, Clayton RF, Thuring JW, Cummings MD, Cherepanov P.** 2010. Molecular mechanisms of retroviral integrase inhibition and the evolution of viral resistance. *Proc Natl Acad Sci USA* **107:**20057–20062.

208. **Hare S, Smith SJ, Métifiot M, Jaxa–Chamiec A, Pommier Y, Hughes SH, Cherepanov P.** 2011. Structural and functional analyses of the second-generation integrase strand transfer inhibitor dolutegravir (S/GSK1349572). *Mol Pharmacol* **80:**565–572.

209. **Cooper DA, Steigbigel RT, Gatell JM, Rockstroh JK, Katlama C, Yeni P, Lazzarin A, Clotet B, Kumar PN, Eron JE, Schechter M, Markowitz M, Loutfy MR, Lennox JL, Zhao J, Chen J, Ryan DM, Rhodes RR, Killar JA, Gilde LR, Strohmaier KM, Meibohm AR, Miller MD, Hazuda DJ, Nessly ML, DiNubile MJ, Isaacs RD, Teppler H, Nguyen B–Y, the BENCHMRK Study Teams.** 2008. Subgroup and resistance analyses of raltegravir for resistant HIV–1 infection. *N Engl J Med* **359:**355–365.

210. **Grobler JA, McKenna PM, Ly S, Stillmock KA, Bahnck CM, Danovich RM, Dornadula G, Hazuda DJ, Miller MD.** 2009. HIV integrase inhibitor dissociation rates correlate with efficacy in vitro. *Antiviral Ther* **14**(Suppl 1)**:** A27.

211. **Hightower KE, Wang R, Deanda F, Johns BA, Weaver K, Shen Y, Tomberlin GH, Carter HLr, Broderick T, Sigethy S, Seki T, Kobayashi M, Underwood MR.** 2011. Dolutegravir (S/GSK1349572) exhibits significantly slower dissociation than raltegravir and elvitegravir from wild-type and integrase inhibitor-resistant HIV-1 integrase-DNA complexes. *Antimicrob Agents Chemother* **55:** 4552–4559.

212. **Goethals O, Vos A, Van Ginderen M, Geluykens P, Smits V, Schols D, Hertogs K, Clayton R.** 2010. Primary mutations selected in vitro with raltegravir confer large fold changes in susceptibility to first-generation integrase inhibitors, but minor fold changes to inhibitors with second-generation resistance profiles. *Virology* **402:**338–346.

213. **Canducci F, Ceresola ER, Boeri E, Spagnuolo V, Cossarini F, Castagna A, Lazzarin A, Clementi M.** 2011. Cross-resistance profile of the novel integrase inhibitor dolutegravir (S/GSK1349572) using clonal viral variants selected in patients failing raltegravir. *J Infect Dis* **204:**1811–1815.

Mobile DNA, 3rd Edition
Nancy L. Craig, Michael Chandler, Martin Gellert, Alan M. Lambowitz, Phoebe A. Rice and Suzanne Sandmeyer

doi:10.1128/microbiolspec.MDNA3-0026-2014

Robert Craigie[1]
Frederic D. Bushman[2]

45 Host Factors in Retroviral Integration and the Selection of Integration Target Sites

INTRODUCTION

Retroviruses integrate a DNA copy of the viral genome into cellular DNA as an obligatory step in the viral replication cycle. Once integrated, the viral DNA is stably replicated with cellular DNA through cycles of DNA replication and cell division. The first clues regarding the mechanism of integration came from genetic experiments (1, 2, 3). Mutations at two locations within the viral genome resulted in a phenotype in which reverse transcription occurred normally but the viral DNA failed to integrate. These mutations mapped to regions which we now know encode the viral integrase (IN) protein and the ends of the viral DNA sequence recognized by IN. The finding that viral DNA within extracts of infected cells efficiently integrated into exogenously added target DNA *in vitro* (4, 5, 6) facilitated biochemical studies of integration. This *in vitro* integration system enabled the DNA breaking and joining events to be unambiguously determined (6, 7). It also established that the viral DNA forms part of a large nucleoprotein complex termed the preintegration complex (PIC) (8). Later biochemical experiments showed that viral IN protein is necessary and sufficient to carry out the DNA cutting and joining steps of integration in the presence of divalent metal ions (9, 10, 11, 12, 13). Subsequent studies established reaction conditions that facilitated efficient concerted integration of both viral DNA ends into the target DNA molecule *in vitro* (14, 15, 16, 17, 18, 19). This chapter focuses on mechanisms of targeting integration and the contributions of viral and cellular proteins. For structural information on nucleoprotein complexes involved in retroviral DNA integration see the chapter by Engelman and Cherepanov. For detailed discussions of the mechanisms of DNA transposition of related elements see other chapters in *Mobile DNA III*.

COMPOSITION OF PREINTEGRATION COMPLEXES

The composition and architecture of PICs remains poorly understood, largely due to their low abundance in extracts of infected cells; typically only a few PICs are present per cell. Knowledge of the protein composi-

[1]Laboratory of Molecular Biology, National Institute of Diabetes and Digestive and Kidney Diseases, National Institutes of Health, Bethesda, Maryland 20892-0560; [2]Department of Microbiology, Perelman School of Medicine at the University of Pennsylvania, Philadelphia, Pennsylvania.

tion is largely limited to immunoprecipitation studies which can identify protein components, but not their abundance, organization, or functional roles. Even then, some identified components border on the limits of detection. Viral DNA within PICs is much easier to monitor because of highly sensitive assays such as Southern blotting and PCR. PICs have been valuable for studying the fate of the viral DNA *in vitro* and defining the DNA cutting and joining steps (6, 7).

A near-full complement of viral proteins have been detected within HIV-1 PICs, including IN, RT, CA, NC, Vpr, and PR (20, 21, 22, 23, 24, 25, 26, 27). Whereas IN, MA, RT, and Vpr are reported to be present in most studies, CA, NC, and PR are only detected in a few studies. These inconsistencies are likely due to differences in methods of preparation and sensitivity of the assays. Cellular proteins implied to be associated with PICs by biochemical studies include LEDGF/p75 (28), barrier-to-autointegration factor (BAF) (29), and HMGA1 (30). Mass spectroscopy screens have identified many additional cellular proteins potentially associated with HIV-1 PICs (31, 32). The functional role, if any, of the majority of these proteins is unknown.

The key viral protein within PICs is IN, which is responsible for the initial steps of covalent joining of viral DNA into host DNA. CA within PICs has been implicated in nuclear import (see below). Reverse transcriptase is also present in PICs but it is unclear if it plays any functional role once reverse transcription is complete. Functional studies of HIV-1 and murine leukemia virus (MLV) PICs demonstrate a tight association of IN to the viral DNA—PICs remain active for *in vitro* DNA integration even after treatment with greater than 500 mM NaCl. A tetramer of HIV-1 IN forms a tight complex with viral DNA ends *in vitro* (33) and this likely bridges the viral DNA ends in the PIC. Footprinting of the viral DNA within MLV and HIV PICs (34, 35, 36) revealed extensive protection extending up to approximately 1 kb from the viral DNA ends. This protection is dependent on IN and is not observed in PICs derived from viruses that lack IN. However, it is unclear if the protection is mediated by IN itself or by recruitment of other factors by IN.

The viral DNA within PICs is sensitive to endonucleases, suggesting that it adopts a relatively open structure (27). In contrast, it is not sensitive to digestion with exonucleases (27), as expected if the DNA ends are stably bridged by IN. A cellular factor termed the barrier-to-autointegration factor protects MLV PICs from autointegration in *in vitro* PIC integration assays (37). BAF is an essential cellular protein involved in DNA management within the nucleus (38, 39) which makes it difficult to study the effects of BAF depletion on integration in the cell. BAF is a DNA bridging protein that compacts DNA (40). Stripping BAF from MLV PICs by high-salt treatment leads to slower sedimentation in sucrose gradients (41), consistent with decompaction of the viral DNA. Addition of BAF to salt-stripped PICs restores the normal sedimentation behavior in parallel with restoring the preference for intermolecular integration and avoidance of autointegration. Consistent with these findings, treatment of MLV PICs with VRK1, the major kinase responsible for phosphorylation of BAF, promotes autointegration in a similar manner to salt-stripping (42). Since phosphorylation of BAF abrogates DNA binding (40, 43), it is expected that VRK1 would promote loss of BAF from the viral DNA within the PIC. A similar role of BAF in preventing autointegration of HIV-1 PICs has not been unambiguously demonstrated.

NUCLEAR IMPORT OF PICS

The large size of PICs presents a formidable challenge to nuclear entry. In the case of gammaretroviruses, the chromosomes are accessed during mitosis when the nuclear envelope is disassembled. However, HIV-1 and other lentiviruses can integrate in non-dividing cells, so entry into the nucleus must be an active process involving the nuclear pores. HIV-1 PICs have comparable sedimentation properties to ribosomes (44) and an estimated Stokes radius of about 28 nM based on size exclusion chromatography (27). The mechanism has been the subject of a controversy that is still not completely resolved. Many of the protein components of the PIC possess nuclear localization signals that, when fused to normally cytoplasmic proteins, result in their localization to the nucleus. MA (45), IN (23, 46), and Vpr (24, 47, 48) have all been proposed to mediate nuclear localization of HIV-1 PICs, as has a DNA structure near the center of the viral DNA called the central flap (49). However, other studies have shown that each of these can be dispensible for nuclear entry under some circumstances. It is possible that the interpretation of the data may be complicated by redundancy of nuclear import mechanisms. More recent data strongly suggests an important role for CA in mediating nuclear entry (reviewed in 50).

The involvement of CA in nuclear entry was suggested by the finding that infection by chimeric HIV-1 harboring an MLV CA is cell cycle-dependent (51) and this conclusion has been further substantiated by subsequent studies. Certain point mutations in HIV-1 CA also confer a cell cycle-dependent phenotype. Genome-wide screens for cellular co-factors identified a large number of proteins known to be involved in nuclear import and related functions (52, 53, 54, 55). Of these, TNPO3, NUP358, and NUP153 are of particular interest. Knockdown of these proteins diminishes HIV-1 replication, but does not affect replication of MLV, suggesting the mechanism is specific to HIV-1 rather than an indirect effect on cellular function (53, 56, 57, 58, 59). Experiments with HIV-1/MLV chimeric viruses map the TNPO3 and NUP153 dependency to CA (56, 58).

HOST FACTORS AFFECTING INTEGRATION TARGETING

Intranuclear Trafficking

Following import into the nucleus, the PIC must access the chromosomes of an infected cell. Though not fully clarified, several studies suggest that this may not be purely diffusion but rather a controlled process (60, 61). Imaging studies have suggested that integration may commonly be localized to certain areas of the nucleus, particularly at the nuclear periphery (62). Consistent with this idea, data suggest that active genes may be enriched near the nuclear periphery in yeast (63); however, other data from studies in animal cells suggest that in fact active genes may be enriched more in the center of the nucleus (64, 65, 66).

A study of nuclear localization and integration targeting suggested a possible model in which retroviral integration complexes engaged a "railroad track" in the cytoplasm that transfers PICs to specific regions of the nucleus rich in favored sites of integration (57, 67). The CA mutant N74D, which alters the interaction with the cellular splicing factor CPSF6, also altered integration targeting, suggesting involvement of this system. Modulating the level or activity of the cyclophillin protein, another protein that binds CA, also affected targeting. These findings point to greater involvement of CA in PIC functioning and targeting than was previously appreciated (57). Several additional nuclear-localized cell factors have been reported to influence integration (68, 69, 70) and could act at this step. More work is needed to clarify the mechanisms underlying the possible relationship between nuclear positioning of chromosomes and frequency of retroviral integration.

Favored Target Sites for Retroviral DNA Integration in Cellular Chromosomes

Studies mapping the distribution of integration sites in cellular chromosomes have led to a detailed picture of favored and disfavored target sites (for some previous reviews in this area see 33, 36, 71). The first genome-wide mapping study, carried out shortly after the completion of the first draft human genome sequence, showed that HIV favored integration particularly strongly within transcription units of the SupT1 T-cell line (72). Comparison to microarray data quantifying transcriptional activity in the same cells showed that integration was particularly strongly favored in active transcription units. Subsequent studies have shown that this pattern is maintained in most cell lines studied (73, 74, 75, 76, 77, 78, 79, 80), with the exception of a few experimentally manipulated examples described below (Fig. 1). A variety of features on the human genome correlate with active transcription units (81, 82) such as sites of histone modification, including H3K4 mono and dimethylation; H3, K9, and K27 monomethylation; H3K36 trimethylation; DNAseI hypersensitive sites; and sites of histone acetylation generally (67, 82, 83). In addition, gene-rich regions are associated with a syndrome of associated features, such as high G/C content, high Alu element content, low LINE element content, and Geimsa dark banding patterns, all of which are positively associated with integration (72).

The above collection of favorable features characterizes all the lentiviruses studied to date, but other retroviral genera show different patterns. The gammaretroviruses, which include the MLVs and XMRV, show favored integration close to transcription start sites rather than throughout transcription units as is seen with lentiviruses (74, 84). The alpha-retroviruses, which include RSV and the allied ALVs, show near-random distributions (76, 85). A study of the viral determinants of integration targeting implicated IN as a major determinant. Swapping the MLV (gammaretroviral) IN for that of HIV, in a background of an otherwise wild-type HIV provirus, resulted in chimeric viruses that showed the integration pattern expected for MLV (86). Although these viruses showed reduced ability to replicate, these studies suggested that IN was a major determinant of integration target site specificity. As is described below, targeting by the lentiviruses and gammaretroviruses can be understood as engagement of different cellular tethering factors that are distinctive for each group.

LEDGF/p75—A Cellular Tethering Factor for HIV Integration

The LEDGF/p75 protein (a product of the PSIP1 gene) was first implicated in HIV replication when it was identified as a tight binder to HIV IN protein (87, 88, 89) These studies involved artificial systems, such as overexpression of IN in animal cells or yeast two-hybrid assays, leaving the significance of the interaction uncertain initially. However, subsequent experiments eventually documented that reduction of LEDGF/p75 function

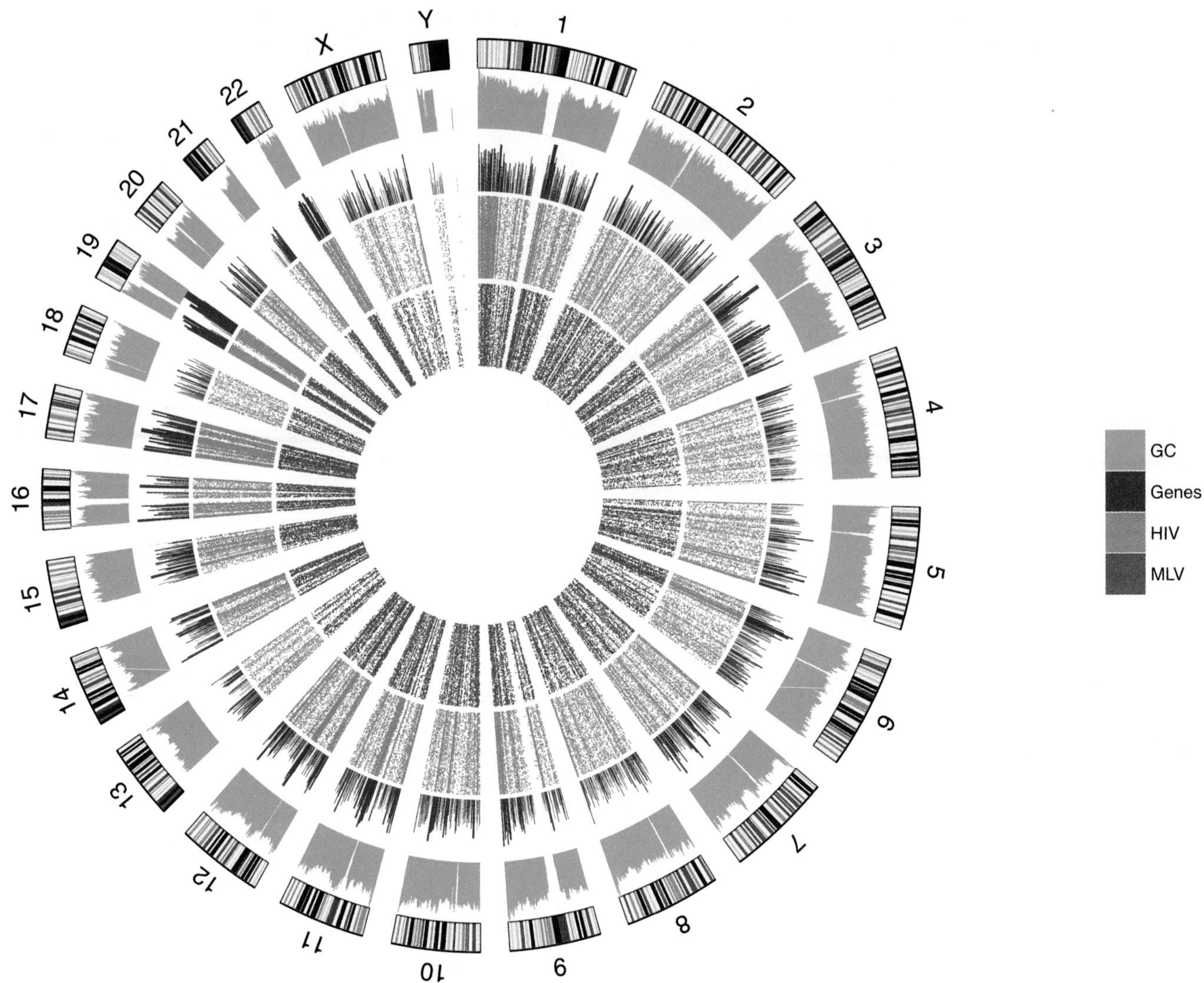

Figure 1 Sites of HIV DNA integration in chromosomes. The human chromosomes are shown in the outermost ring. The green histograms indicated relative G/C content; red indicates gene density, orange indicates density of HIV integration sites in T-cells (78), and purple indicates MLV integration site density in T-cells (115).
doi:10.1128/microbiolspec.MDNA3-0026-2014.f1

did have strong negative effects on viral replication (79, 90, 91), establishing its importance.

Mapping studies showed that LEDGF/p75 is composed of several protein domains. Multiple N-terminal domains mediate chromatin-binding, potentially in part to the H3 histone trimethylated at the K36 position (92, 93). The C-terminal domain binds IN ("integrase binding domain" or IBD). X-ray crystallography reveals that the LEDGF-IBD engages the catalytic and N-terminal domains of integrase. The most extensive interactions involve a pocket formed at the dimer interface of two monomers of the integrase catalytic domain and the tip of the IBD helical bundle structure (94, 95).

Given that LEDGF/p75 was a cellular protein implicated in both transcriptional regulation and HIV integration, it was of interest to test whether integration in active transcription units required LEDGF/p75. Knockdown studies showed that depleting LEDGF/p75 in fact caused a significant shift in integration targeting toward a more random pattern, consistent with the idea that the LEDGF/p75-IN interaction mediated targeting to transcription units (77, 79, 96, 97).

LEDGF/p75 is a member of a protein family, and the related protein HRP2 has also been shown to act as an integration tether. However, effects of HRP2 on HIV integration efficiency and targeting are only prominent in cells lacking LEDGF/p75, and the influence of HRP2 even in this setting is modest, supporting the idea that LEDGF/p75 is the more significant cofactor (98, 99, 100, 101).

Evidence supporting the idea that LEDGF/p75 acts as a simple tether was provided by results with LEDGF/p75 chimeras (102, 103, 104). Several groups substituted new chromatin-binding domains for the LEDGF/p75 N-terminal chromatin-binding domain. These chimeras, which retained the LEDGF/p75 IN binding C-terminal domain but grafted to new domains conferring new chromatin-binding specificities, were introduced into cells depleted for wild-type LEDGF/p75, and analysis of HIV integration showed new favored locations. A variety of such novel tethers have now been reported, collectively supporting the idea that LEDGF/p75 acts as a tethering protein for PICs. Studies of the yeast Ty elements Ty1-5, described elsewhere in this book, have disclosed additional examples of targeting by tethering of integrase to cellular chromatin-binding proteins.

BET Proteins—Cellular Factors Directing Gammaretroviral Integration

The observation that HIV targeting is directed by tethering to LEDGF/p75 raised the question of whether other retroviral genera with different targeting preferences might dock with their own group-specific tethering factors. The gammaretroviruses favor integration near transcription start sites, and within the past year cellular BET proteins (Brd2, 3, and 4) have been identified as targeting factors (BET stands for "Bromodomain and extra terminal" and Brd stands for "bromodomain containing"; 105, 106, 107, 108). Brd2–4 are known to bind to sites of histone acetylation, explaining their association with transcription start regions. Initially these proteins were identified as binders to MLV IN in yeast two-hybrid screens (109), but the importance was uncertain. Later studies found that BET proteins also bound MLV IN in affinity capture-mass spec protocols, and that the addition of BET proteins could affect the activity of MLV IN *in vitro* (105, 106, 107, 108).

Definitive data on the effects of BET proteins came with experiments modulating the activity of BET proteins in cells (105, 106, 107, 108). Reducing BET protein activity was difficult with siRNA because three proteins were involved, and full elimination of BET activity was lethal to cells. A key breakthrough came with studies based on the anticancer drug JQ1, which binds to BET proteins and competes with binding of actetylated histone tails. This allowed activity of all BET proteins to be abrogated acutely, allowing experimental analysis of cells before the onset of lethality. Using JQ1, it was possible to show that the frequency of MLV integration was reduced in the region around transcription start sites. Effects could also be detected with siRNA knockdowns of the three genes, though with lower magnitude. In follow-up experiments, hybrid tethering proteins were synthesized that contained the BET IN binding region and novel chromatin-binding domains. In cells depleted for normal BET proteins but containing the chimeric tethers, integration sites were notably redistributed, again strongly supporting the tethering model.

Retroviral Integration and Nucleosomal Association

Several studies have explored the association between nucleosome packaging and integration bias. Early studies suggesting a loose association of retroviral integration sites with chromosomal sites sensitive to DNAseI digestion (reviewed in 16, 19, 20, 34), which led to the idea that packaging DNA in nucleosomes would inhibit integration. However, once assays were available that recapitulated integration *in vitro*, nucleosomal wrapping was found instead to favor integration (110, 111, 112, 113, 114). Subsequent studies *in vivo* have suggested that DNA predicted to be facing outward when wrapped on nucleosomal particles is favored for integration (82, 115). However, complicating the analysis, not all methods for calling nucleosome positions *in vivo* agree well, and experimental mapping of nucleosome positions is imperfect and fraught with lab-specific noise. Additional studies have investigated possible roles of cellular factors in modulating interactions with nucleosomes (116, 117). Different integrating elements, described elsewhere in this book, show a wide range of responses to nucleosome binding to integration target DNA.

Role of DNA Repair Enzymes in Completing Provirus Formation

The initial steps in the integration reaction are catalyzed by the viral-encoded IN enzymes, but this only takes the reaction to the point of joining one DNA strand at each end of the provirus to host DNA. The integration intermediate produced by IN has DNA gaps at each host-virus DNA junction, requiring intervention of additional enzymes to generate the fully integrated provirus. Cocktails of host-encoded DNA repair enzymes have been shown to be able to carry out the needed polymerization, branch excision, and ligation steps *in vitro* on substrate DNAs modeling the expected gaps (118). These enzymes are ubiquitous in cell nuclei and quite active, so their involvement seems likely. However, another possibility is that the virus-encoded RT and IN enzymes actually complete gap repair. Some *in vitro* evidence supports this (119), although the observed enzymatic activities may not be adequately robust. So far no studies have succeeded in creating mutant combinations that result in production of stalled gapped intermediates inside cells, leaving the activities involved incompletely clarified. In addition, several further cellular DNA repair enzymes have been proposed to act on retroviral DNA and influence integration or replication (120, 121, 123, 124, 125, 126, 127).

INHIBITORS OF HIV INTEGRATION

Inhibitors Acting at the Active Site

Antiviral agents are now available that target many steps in the HIV replication cycle, greatly improving the quality of life of HIV-positive individuals, and IN is among the targets (128). Initial work improving assays for integration allowed high-throughput screening under relatively authentic conditions, yielding starting points for drug development (129, 130, 131). All IN inhibitors targeting the active site contain chemical groups that can bind both catalytic metal atoms together with pendant groups that contact other parts of the IN-DNA complex. Several chemical families were explored before the pyrimidinone group was identified. Extensive development eventually yielded raltegravir, which received FDA approval in 2007 (132).

Mechanistic studies show that raltegravir acts by blocking binding of the integration target DNA to the PIC (133). In addition, structural studies of raltegravir bound to the PFV IN/DNA complex show that the drug displaces the LTR 3' end from its normal position in the active site, thereby obstructing the chemical steps of DNA transesterification (134).

As with all antiviral agents, resistance mutations have arisen in viruses in subjects treated with raltegravir. The effects of these amino acid substitutions can also be understood from the PFV structure as altering the drug binding site (see chapter by Engelman and Cherepanov). The observation of escape mutations has motivated a search for second generation IN inhibitors, resulting in promising new compounds such as dolutegravir, which shows high potency and favorable pharmacokinetic properties (135, 136).

Inhibitors Acting at the LEDGF/p75 Binding Site on IN

Another site of HIV IN, the LEDGF/p75 binding site, was recently shown to be a promising target. The observation that removal of LEDGF/p75 from cells or overexpression of isolated LEDGF/p75 IBD could inhibit HIV-1 replication, motivated searches for compounds that obstructed the LEDGF/p75 interaction (79, 90, 91, 137). Further motivating this search was a report of a small molecule inhibitor of HIV integration bound to a site at the dimer interface of the HIV IN catalytic domain (138) that was later shown to also be the LEDGF/p75 IBD binding site. Several groups carried out large-scale screens and identified small molecules that bound at this site and blocked binding of the LEDGF/p75 IBD to HIV-1 IN (137, 139, 140, 141, 142, 143). Some of these molecules were found to have low nanomolar affinities, acceptable toxicity to cells, and quite potent inhibition of virus infection. Structural studies using X-ray crystallography showed the drugs to be bound as expected in a pocket at the catalytic domain dimer interface (Fig. 2). Growth of HIV in the presence of the drugs elicited escape mutations that mapped to regions encoding segments of IN near the drug binding site, genetically specifying IN as the drug target during viral replication.

Mechanistic studies of the steps affected in the viral replication cycle by these drugs yielded a surprise.

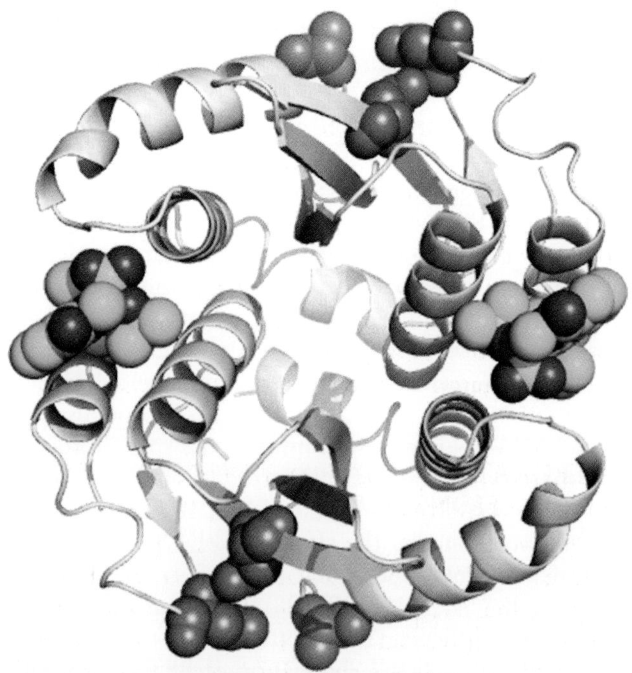

Figure 2 The inhibitor GSK1264 bound to the LEDGF-binding site on the HIV IN catalytic domain dimer. The alpha carbon backbone of the IN catalytic domain dimer is shown in gold. The GSK1264 compound is show in cyan (carbons) and red (oxygens). Active site residues are shown in orange. Details on the structure and function of GSK1264 are reported in (137).
doi:10.1128/microbiolspec.MDNA3-0026-2014.f2

Unexpectedly, the drugs were not particularly potent during the early phase of the viral replication cycle (entry through integration). In some cases the small molecules reduced targeting to transcription units, but this has been variable across the drug series. Compounds inhibited potently during the late steps (gene expression through assembly and release). It turned out that drugs of this class were promoting abnormal assembly of HIV polyprotein precursors and/or IN, so that viral particles had a disrupted structure and were fatally impaired during early steps of the next infectious cycle. LEDGF does not seem to be involved in normal assembly, because assembly proceeds normally, and the inhibitors are active in the absence of LEDGF. At the time of writing, large-scale efforts are under way by several pharmaceutical companies to complete development and bring this class of inhibitor to patients.

Controlling Retroviral Integration Site Selection with Small Molecules

As is described above, it is now possible to control the distributions of retroviral integration target sites by adding small molecules to cells during infections. For both lentiviruses and gammaretroviruses, addition of the drugs—some LEDGF site blockers (139, 140, 141, 142, 143) or JQ1 (105, 106, 107, 108)—reduces targeting biases, so that integration site patterns more closely resemble random distributions. These molecules may have applications in human gene therapy, where targeting integration in or near transcription units is undesirable because of the risk of genotoxicity. If the efficiency of gene expression after integration is acceptable at these new locations, and if the small molecules are not too toxic to primary cells, these small molecules may be useful additives during cell transduction prior to transplantation of modified cells into patients.

INTEGRATION AND HIV LATENCY: IMPLICATIONS FOR CURE RESEARCH

Recent work has focused attention on the possible influence on HIV latency of the chromosomal environment of the integrated proviruses. The formation of latent (transcriptionally silent) proviruses represents a major block to eradicating HIV from infected cells (144, 145). Studies show that HIV+ subjects who have viral replication suppressed by anti-retroviral therapy to below the limit of detection nevertheless harbor rare cells containing integrated proviruses that are transcriptionally silent. Some of these are known to be resting memory CD4+ T-cells which, upon exposure to antigen, can resume active transcription that allows proviral induction. As a result, renewed viral replication can restart active infection.

The observation that several individuals may have been cured of HIV has resulted in intense interest in developing generalizable means of eradicating HIV from infected people (146, 147, 148). One idea holds that cells harboring latent viruses might be selectively killed if latent proviruses could be coaxed to resume transcription, provided that the host was equipped to destroy these cells efficiently once the latent cells were disclosed to the immune system. A recent clinical trial tested this by treating well-suppressed HIV+ subjects with the histone deacetylase inhibitor SAHA (vorinostat) with the goal of altering the chromatin environment in latent cells to promote HIV transcription (149). Studies to date have reported an increase in cell-associated HIV RNA following SAHA treatment, though no increase in circulating virus, providing an early suggestion that the shock-and-kill strategy might be feasible.

The idea that SAHA would boost transcription presupposes that a lack of histone acetylation is the limiting factor for transcription of the HIV provirus. This assumption can be addressed experimentally using cell-based models of latency and proviral induction (150). In such studies, cells are infected with HIV or an HIV-based vector. Cells are then fractionated based on whether or not they express the HIV provirus. The nonexpressing cells can then be induced to express HIV with an agent that activates T-cells. Those cells that respond and transcribe the proviral genome are operationally defined as the inducibly latent population.

Several studies have explored integration site distributions in these cell populations (150, 151, 152). In general, inducible proviruses are found in locations similar to typical HIV favored integration sites. A recent meta-analysis showed that no single feature characterized latent cells studied in five different cell models (150). Significant trends were seen within each model, but the chromosomal features near favored sites were mostly dissimilar between models. The most prominent shared feature of latent integration sites was proximity to cellular alphoid repeats, which was observed to be positively associated with latency in four of five models. Alphoid repeats are enriched near centromeres, so the silencing detected may resemble position effect variagation seen in early studies in Drosophila, where the apposition of an active gene to centromeric heterochromatin by inversion resulted in reduction in expression (153). However, latent sites near alphoid repeats comprised only a minority of all inducible integration sites. SAHA shock-and-kill protocols predicted a negative association between latent sites and histone acetylation, but this association was not consistently observed. Latency in these models was not associated with a single integration site pattern, indicating that transcriptional latency in HIV may have many molecular mechanisms, a conclusion also reached with other experimental approaches (154). Unfortunately these findings complicate efforts to expose and remove cells harboring latent HIV.

Other studies suggest additional possible roles for integration targeting in HIV latency. Ikeda et al. reported in 2007 that rare cells from HIV+ patients on long-term successful drug therapy showed clustering of integration sites (155), and recent reports have developed this picture further (156, 157, 158). Some of these integration sites were in genes related to cell growth or persistence, such as the transcription factor BACH2, which is a lymphoid cancer-related gene. These results raise the possibility that gene disruption by HIV integration may promote growth or persistence of latently infected cells. An example of lentiviral integration correlated with outgrowth of a cell clone has been reported in a gene therapy trial. In this case, the vector integrated in the expanded cells was located in the proto-oncogene HMGA2, providing a possible analogy for clonal outgrowth during latency (159).

GENOME-WIDE STUDIES TO IDENTIFY CANDIDATE HOST FACTORS

Numerous genome-wide screens have been carried out to identify host factors important for HIV replication (collected in 161, 162, 163). A wide variety of methods have been applied. In a few cases proteins have been identified that are candidates for affecting the integration step. Loss-of-function screens have been carried out using siRNA knockdowns, in which siRNAs targeting most human genes are assayed individually for their ability to either reduce or increase HIV infection (14, 53, 54). Gain of function studies have been carried out by overexpressing cDNAs and assessing whether they increased or decreased HIV infection (163). Focused screens have queried interferon-inducible genes, a particularly rich source of antiviral factors (164, 165).

Another approach involves genome-wide studies to identify host proteins that bind to viral proteins (32). In these protocols, HIV proteins are expressed in cells, then captured using appropriate tagging systems. Bound proteins are then eluted and analyzed by mass spectrometry approaches. Recent comprehensive studies have identified many more cellular proteins binding to the HIV proteins, in some cases with strong statistical support for specificity. Another approach involves yeast two-hybrid assays, in which candidate interactors are expressed together as fusions in yeast and interactions scored by effects on yeast transcription. For both LEDGF/p75 and BET proteins, early studies used yeast two-hybrid assays to document binding.

Given the gigantic size of the genome-wide data sets, the development of useful summaries and convenient web browsers becomes as important as generating the data to begin with. Table 1 summarizes some of the web-enabled sites which may be used to investigate these data. The GuavaH web site allows one to query HIV and human genotypes and their relationship to phenotypes (166). The GeneOverlapper site can be used to query lists of genes in groups to find genes in common (162). More than 40 lists are included, which can be queried in over 200 trillion combinations, so a blog site is included for readers to post comments on their findings, allowing "crowd sourcing" of discovery. The NCBI keeps a list of HIV proteins and their interactions. So far, with respect to integration biology, these large data sets have helped identify tethering factors and a variety of cellular factors implicated in

Table 1 Some resources for working with genome-wide data on HIV research

Name of resource	Link	Comments
GuavaH	http://www.GuavaH.org	Relates HIV and human genotype to phenotype
NCBI	http://www.ncbi.nlm.nih.gov/	A national repository of biomedial data and literature
Los Alamos HIV Databases	www.hiv.lanl.gov/	Repository of HIV sequence and epitope information
UNAIDS	http://www.unaids.org/en/	Information and policy materials from the UN
Stanford University HIV Drug Resistance Database	http://hivdb.stanford.edu/index.html	Database of mutations in HIV conferring resistance to antiviral agents
GPS-prot	http://www.gpsprot.org	A site for exploring interactions between, HIV and cellular proteins, with rich tools for followup
HIV Replication Cycle	http://www.hivsystemsbiology.org/wiki/index.php/Introduction	A web-based account of the HIV replication cycle, emphasizing integration
Gene Overlapper	http://www.hivsystemsbiology.org/GeneListOverlapper/	Data from genome-wide surveys together with tools for exploring overlaps and blogging about results

PIC trafficking and nuclear import. A rich array of additional cellular proteins have been implicated, and some bind IN, offering many new starting points for investigating the roles of cellular proteins in integration.

OTHER FATES OF THE VIRAL DNA

The normal fate of retroviral DNA is integration into the host genome. However, forms of viral DNA are also found that are thought to be abortive products not on the integration pathway (167, 168). Some viral DNA is found in the form of 2-LTR circles that likely arise by nonhomologous end joining of the linear viral DNA after preintegration complexes have fallen apart (125). 1-LTR circles appear to be the product of homologous recombination between the two LTRs in the linear viral DNA. Other types of aberrant product, including full length circular viral DNA with a rearrangement and smaller circles with deletion, are the result of autointegration.

Several models have suggested that the unrepaired integration of intermediate and/or unintegrated forms of viral DNA may trigger pathways leading to cell death. Retroviral infection was reported by some to induce cell death in NHEJ-deficient cell lines (120, 121, 169). Cell death in these cell lines was reported to be dependent on the presence of functional IN, suggesting that the integration intermediate is recognized as DNA damage. Consistent with this idea, inhibition of DNA-PK by wortmannin makes normal cells sensitive to killing by retroviral vectors with an active IN, but not by vectors with an inactive IN. Results with these models have varied (125, 170). In another study, infection of activated primary $CD4^+$ lymphocytes by HIV-1 also induced cell killing (171). Killing was not observed in the absence of active IN or the presence of IN inhibitors, suggesting a requirement for integration. However, in a recent study from another group, unintegrated viral DNA was implicated as the signal in killing of $CD4^+$ lymphocytes (172). Knockdown of the interferon–γ-inducible protein 16 (IFI16), a DNA sensor, abrogated cell killing in this model, suggesting that the integrated viral DNA triggers an innate immune response that leads to caspase 1 activation and pyroptosis. Killing was still observed with virus with an active site mutation in IN, but not in the presence of a nonnucleoside reverse transcriptase inhibitor, demonstrating a requirement for DNA synthesis but not integration. Further work is required to reconcile and advance understanding in this area.

CONCLUSIONS

The issue of retroviral host factors and their roles in integration targeting has been a very active area of study. Genome-wide screens have yielded many new starting points for investigating HIV biology, including HIV integration. The discoveries of tethering factors for HIV and MLV provide a simple mechanism for targeting. In contrast, the finding that drugs targeting the LEDGF/p75 binding site in fact disrupt assembly in a fashion that is mostly independent of LEDGF/p75 was not at all expected, and focuses attention on the apparent role of IN in late replication steps. The role of integration in latency, and the role of replication intermediates in cell toxicity and immunodepletion, are further areas where additional investigation would be valuable.

Acknowledgments. *We are grateful to Karen Ocwieja, Scott Sherrill-Mix, Alan Engelman, and Peter Cherepanov for comments on the manuscript. This work was supported by the Intramural Program of NIDDK, NIH, by the AIDS Targeted Antiviral Program of the Office of the Director of the NIH RC), AI 052845 and AI 090935 (FDB), and by the HINT collaboratory.*

Citation. Craigie R, Bushman FD. 2014. Host factors in retroviral integration and the selection of integration target sites. Microbiol Spectrum **2**(6):MDNA3-0026-2014.

References

1. **Donehower LA, Varmus HE.** 1984. A mutant murine leukemia virus with a single missense codon in pol is defective in a function affecting integration. *Proc Natl Acad Sci USA* **81:**6461–6465.
2. **Panganiban AT, Temin HM.** 1984. The retrovirus pol gene encodes a product required for DNA integration: Identification of a retrovirus int locus. *Proc Natl Acad Sci USA* **81:**7885–7889.
3. **Schwartzberg P, Colecilli J, Goff SP.** 1984. Construction and analysis of deletion mutations in the pol gene of Moloney murine leukemia virus: A new viral function required for productive infection. *Cell* **37:**1043–1052.
4. **Brown PO, Bowerman B, Varmus HE, Bishop JM.** 1987. Correct integration of retroviral DNA *in vitro*. *Cell* **49:**347–356.
5. **Farnet CM, Haseltine WA.** 1990. Integration of human immunodeficiency virus type 1 DNA *in vitro*. *Proc Natl Acad Sci USA* **87:**4164–4168.
6. **Fujiwara T, Mizuuchi K.** 1988. Retroviral DNA integration: Structure of an integration intermediate. *Cell* **54:**497–504.
7. **Brown PO, Bowerman B, Varmus HE, Bishop JM.** 1989. Retroviral integration: Structure of the initial covalent complex and its precursor, and a role for the viral IN protein. *Proc Natl Acad Sci USA* **86:**2525–2529.
8. **Bowerman B, Brown PO, Bishop JM, Varmus HE.** 1989. A nucleoprotein complex mediates the integration of retroviral DNA. *Genes and Development* **3:**469–478.
9. **Bushman FD, Craigie R.** 1990. Sequence requirements for integration of Moloney murine leukemia virus DNA *in vitro*. *J Virol* **64:**5645–5648.

10. **Katz RA, Merkel G, Kulkosky J, Leis J, Skalka AM.** 1990. The avian retroviral IN protein is both necessary and sufficient for integrative recombination *in vitro*. *Cell* **63:**87–95.
11. **Bushman FD, Fujiwara T, Craigie R.** 1990. Retroviral DNA integration directed by HIV integration protein *in vitro*. *Science* **249:**1555–1558.
12. **Sherman PA, Fyfe JA.** 1990. Human immunodeficiency virus integration protein expressed in Escherichia coli possesses selective DNA cleaving activity. *Proc Natl Acad Sci USA* **87:**5119–5123.
13. **Katzman M, Katz RA, Skalka AM, Leis J.** 1989. The avian retroviral integration protein cleaves the terminal sequences of linear viral DNA at the *in vivo* sites of integration. *J Virol* **63:**5319–5327.
14. **Hindmarsh P, Ridky T, Reeves R, Andrake M, Skalka AM, Leis J.** 1999. HMG protein family members stimulate human immunodeficiency virus type 1 avain sarcoma virus concerted DNA integration *in vitro*. *J Virol* **73:** 2994–3003.
15. **Li M, Craigie R.** 2005. Processing of viral DNA ends channels the HIV-1 integration reaction to concerted integration. *J Biol Chem* **280:**29334–29339.
16. **Sinha S, Grandgenett DP.** 2005. Recombinant human immunodeficiency virus type 1 integrase exhibits a capacity for full-site integration *in vitro* that is comparable to that of purified preintegration complexes from virus-infected cells. *J Virol* **79:**8208–8216.
17. **Sinha S, Pursley MH, Grandgenett DP.** 2002. Efficient concerted integration by recombinant human immunodeficiency virus type 1 integrase without cellular or viral cofactors. *J Virol* **76:**3105–3113.
18. **Valkov E, Gupta SS, Hare S, Helander A, Roversi P, McClure M, Cherepanov P.** 2009. Functional and structural characterization of the integrase from the prototype foamy virus. *Nucleic Acids Res* **37:**243–255.
19. **Carteau S, Gorelick R, Bushman FD.** 1999. Coupled integration of human immunodeficiency virus cDNA ends by purified integrase *in vitro*: stimulation by the viral nucleocapsid protein. *J Virol* **73:**6670–6679.
20. **Iordanskiy S, Berro R, Altieri M, Kashanchi F, Bukrinsky M.** 2006. Intracytoplasmic maturation of the human immunodeficiency virus type 1 reverse transcription complexes determines their capacity to integrate into chromatin. *Retrovirology* **3:**4.
21. **Bukrinsky MI, Sharova N, McDonald TL, Pushkarskaya T, Tarpley GW, Stevenson M.** 1993. Association of integrase, matrix, and reverse transcriptase antigens of human immunodeficiency virus type 1 with viral nucleic acids following acute infection. *Proc Natl Acad Sci USA* **90:**6125–6129.
22. **Gallay P, Swingler S, Song J, Bushman F, Trono D.** 1995. HIV nuclear import is governed by the phosphotyrosine-mediated binding of matrix to the core domain of integrase. *Cell* **17:**569–576.
23. **Gallay P, Hope T, Chin D, Trono D.** 1997. HIV-1 infection of nondividing cells through the recognition of integrase by the importin/karyopherin pathway. *Proc Natl Acad Sci USA* **94:**9825–9830.
24. **Heinzinger NK, Bukrinsky MI, Haggerty SA, Ragland AM, V K, Lee MA, Gendelman HE, Ratner L, Stevenson M, Emerman M.** 1994. The Vpr protein of human immunodeficiency virus type 1 influences nuclear localization of viral nucleic acids in nondividing host cells. *Proc Natl Acad Sci USA* **91:**7311–7315.
25. **Karageorgos L, Li P, Burrell C.** 1993. Characterization of HIV replication complexes early after cell-to-cell infection. *AIDS Res Human Retrovir* **9:**817–823.
26. **Farnet CM, Haseltine WA.** 1991. Determination of viral proteins present in the human immunodeficiency virus type 1 preintegration complex. *J Virol* **65:**1910–1915.
27. **Miller MD, Farnet CM, Bushman FD.** 1997. Human Immunodeficiency Virus Type 1 preintegration complexes: Studies of organization and composition. *J Virol* **71:**5382–5390.
28. **Llano M, Vanegas M, Fregoso O, Saenz D, Chung S, Peretz M, Poeschla EM.** 2004. LEDGF/p75 determines cellular trafficking of diverse lentiviral but not murine oncoretroviral integrase proteins and is a component of functional lentiviral preintegration complexes. *J Virol* **78:**9524–9537.
29. **Chen H, Engelman A.** 1998. The barrier-to-autointegration protein is a host factor for HIV type 1 integration. *Proc Natl Acad Sci USA* **95:**15270–15274.
30. **Farnet C, Bushman FD.** 1997. HIV-1 cDNA integration: Requirement of HMG i(y) protein for function of preintegration complexes *in vitro*. *Cell* **88:**1–20.
31. **Raghavendra NK, Shkriabai N, Graham R, Hess S, Kvaratskhelia M, Wu L.** 2010. Identification of host proteins associated with HIV-1 preintegration complexes isolated from infected CD4+ cells. *Retrovirology* **7:**66.
32. **Jager S, Cimermancic P, Gulbahce N, Johnson JR, McGovern KE, Clarke SC, Shales M, Mercenne G, Pache L, Li K, Hernandez H, Jang GM, Roth SL, Akiva E, Marlett J, Stephens M, D'Orso I, Fernandes J, Fahey M, Mahon C, O'Donoghue AJ, Todorovic A, Morris JH, Maltby DA, Alber T, Cagney G, Bushman FD, Young JA, Chanda SK, Sundquist WI, Kortemme T, Hernandez RD, Craik CS, Burlingame A, Sali A, Frankel AD, Krogan NJ.** 2012. Global landscape of HIV-human protein complexes. *Nature* **481:**365–370.
33. **Li M, Mizuuchi M, Burke TR Jr, Craigie R.** 2006. Retroviral DNA integration: Reaction pathway and critical intermeidates. *Embo J* **25:**1295–1304.
34. **Wei SQ, Mizuuchi K, Craigie R.** 1997. A large nucleoprotein assembly at the ends of the viral DNA mediates retroviral DNA integration. *Embo J* **16:**7511–7520.
35. **Wei SQ, Mizuuchi K, Craigie R.** 1998. Footprints of the viral DNA ends in Moloney murine leukemia virus preintegration complexes reflect a specific association with integrase. *Proc Natl Acad Sci USA* **95:**10535–10540.
36. **Chen H, Wei SQ, Engelman A.** 1999. Multiple integrase functions are required to form the native structure of the human immunodeficiency virus type I intasome. *J Biol Chem* **274:**17358–17364.
37. **Lee MS, Craigie R.** 1998. A previously unidentified host protein protects retroviral DNA from autointegration. *Proc Natl Acad Sci USA* **95:**1528–1533.

38. **Zheng R, Ghirlando R, Lee MS, Mizuuchi K, Krause M, Craigie R.** 2000. Barrier-to-autointegration factor (BAF) bridges DNA in a discrete, higher-order nucleoprotein complex. *Proc Natl Acad Sci USA* **97:**8997–9002.

39. **Margalit A, Segura-Totten M, Gruenbaum Y, Wilson KL.** 2005. Barrier-to-autointegration factor is required to segregate and enclose chromosomes within the nuclear envelope and assemble the nuclear lamina. *Proc Natl Acad Sci USA* **102:**3290–3295.

40. **Skoko D, Li M, Huang Y, Mizuuchi M, Cai M, Bradley CM, Pease PJ, Xiao B, Marko JF, Craigie R, Mizuuchi K.** 2009. Barrier-to-autointegration factor (BAF) condenses DNA by looping. *Proc Natl Acad Sci USA* **106:**16610–16615.

41. **Suzuki Y, Craigie R.** 2002. Regulatory mechanisms by which barrier-to-autointegration factor blocks autointegration and stimulates intermolecular integration of Moloney murine leukemia virus preintegration complexes. *J Virol* **76:**12376–12380.

42. **Suzuki Y, Ogawa K, Koyanagi Y, Suzuki Y.** 2010. Functional disruption of the moloney murine leukemia virus preintegration complex by vaccinia-related kinases. *J Biol Chem* **285:**24032–24043.

43. **Nichols RJ, Wiebe MS, Traktman P.** 2006. The vaccinia-related kinases phosphorylate the N′ terminus of BAF, regulating its interaction with DNA and its retention in the nucleus. *Molecular Biology of the Cell* **17:**2451–2464.

44. **Bushman FD.** 1995. Targeting retroviral integration. *Science* **267:**1443–1444.

45. **Bukrinsky MI, Haggerty S, Dempsey MP, Sharova N, Adzhubel A, Spitz L, Lewis P, Goldfarb D, Emerman M, Stevenson M.** 1993. A nuclear localization signal within HIV-1 matrix protein that governs infection of non-dividing cells. *Nature* **365:**666–669.

46. **Bouyac-Bertoia M, Dvorin JD, Fouchier RA, Jenkins Y, Meyer BE, Wu LI, Emerman M, Malim MH.** 2001. HIV-1 infection requires a functional integrase NLS. *Molecular Cell* **7:**1025–1035.

47. **Fouchier RA, Meyer BE, Simon JH, Fischer U, Albright AV, Gonzalez-Scarano F, Malim MH.** 1998. Interaction of the human immunodeficiency virus type 1 Vpr protein with the nuclear pore complex. *J Virol* **72:**6004–6013.

48. **Vodicka MA, Koepp DM, Silver PA, Emerman M.** 1998. HIV-1 Vpr interacts with the nuclear transport pathway to promote macrophage infection. *Genes & Development* **12:**175–185.

49. **Zennou V, Petit C, Guetard D, Nerhbass U, Montagnier L, Charneau P.** 2000. HIV-1 genome nuclear import is mediated by a central DNA flap. *Cell* **101:**173–185.

50. **Matreyek KA, Engelman A.** 2013. Viral and cellular requirements for the nuclear entry of retroviral preintegration nucleoprotein complexes. *Viruses* **5:**2483–2511.

51. **Yamashita M, Emerman M.** 2004. Capsid is a dominant determinant of retrovirus infectivity in nondividing cells. *J Virol* **78:**5670–5678.

52. **Brass AL, Dykxhoorn DM, Benita Y, Yan N, Engelman A, Xavier RJ, Lieberman J, Elledge SJ.** 2008. Identification of host proteins required for HIV infection through a functional genomic screen. *Science* **319:**921–926.

53. **Konig R, Zhou Y, Elleder D, Diamond TL, Bonamy GM, Irelan JT, Chiang CY, Tu BP, De Jesus PD, Lilley CE, Seidel S, Opaluch AM, Caldwell JS, Weitzman MD, Kuhen KL, Bandyopadhyay S, Ideker T, Orth AP, Miraglia LJ, Bushman FD, Young JA, Chanda SK.** 2008. Global analysis of host-pathogen interactions that regulate early-stage HIV-1 replication. *Cell* **135:**49–60.

54. **Zhou H, Xu M, Huang Q, Gates AT, Zhang XD, Castle JC, Stec E, Ferrer M, Strulovici B, Hazuda DJ, Espeseth AS.** 2008. Genome-Scale RNAi Screen for Host Factors Required for HIV Replication. *Cell Host & Microbe* **4:**495–504.

55. **Yeung ML, Houzet L, Yedavalli VS, Jeang KT.** 2009. A genome-wide short hairpin RNA screening of jurkat T-cells for human proteins contributing to productive HIV-1 replication. *J Biol Chem* **284:**19463–19473.

56. **Krishnan L, Matreyek KA, Oztop I, Lee K, Tipper CH, Li X, Dar MJ, Kewalramani VN, Engelman A.** 2010. The requirement for cellular transportin 3 (TNPO3 or TRN-SR2) during infection maps to human immunodeficiency virus type 1 capsid and not integrase. *J Virol* **84:**397–406.

57. **Schaller T, Ocwieja KE, Rasaiyaah J, Price AJ, Brady TL, Roth SL, Hue S, Fletcher AJ, Lee K, KewalRamani VN, Noursadeghi M, Jenner RG, James LC, Bushman FD, Towers GJ.** 2011. HIV-1 capsid-cyclophilin interactions determine nuclear import pathway, integration targeting and replication efficiency. *PLoS Pathogens* **7:**e1002439.

58. **Matreyek KA, Engelman A.** 2011. The requirement for nucleoporin NUP153 during human immunodeficiency virus type 1 infection is determined by the viral capsid. *J Virol* **85:**7818–7827.

59. **Thys W, De Houwer S, Demeulemeester J, Taltynov O, Vancraenenbroeck R, Gerard M, De Rijck J, Gijsbers R, Christ F, Debyser Z.** 2011. Interplay between HIV entry and transportin-SR2 dependency. *Retrovirology* **8:**7.

60. **Greene WC, Peterlin BM.** 2002. Charting HIV's remarkable voyage through the cell: Basic science as a passport to future therapy. *Nature Medicine* **8:**673–680.

61. **Campbell EM, Hope TJ.** 2008. Live cell imaging of the HIV-1 life cycle. *Trends in Microbiology* **16:**580–587.

62. **Di Primio C, Quercioli V, Allouch A, Gijsbers R, Christ F, Debyser Z, Arosio D, Cereseto A.** 2013. Single-cell imaging of HIV-1 provirus (SCIP). *Proc Natl Acad Sci USA* **110:**5636–5641.

63. **Casolari JM, Brown CR, Komili S, West J, Hieronymus H, Silver PA.** 2004. Genome-wide localization of the nuclear transport machinery couples transcriptional status and nuclear organization. *Cell* **117:**427–439.

64. **Boyle S, Gilchrist S, Bridger JM, Mahy NL, Ellis JA, Bickmore WA.** 2001. The spatial organization of human chromosomes within the nuclei of normal and emerin-mutant cells. *Hum Mol Genet* **10:**211–219.

65. **Chubb JR, Bickmore WA.** 2003. Considering nuclear compartmentalization in light of nuclear dynamics. *Cell* **112:**403–406.

66. **Wei Z, Huang D, Gao F, Chang WH, An W, Coetzee GA, Wang K, Lu W.** 2013. Biological implications and regulatory mechanisms of long-range chromosomal interactions. *J Biol Chem* **288:**22369–22377.

67. **Ocwieja KE, Brady TL, Ronen K, Huegel A, Roth SL, Schaller T, James LC, Towers GJ, Young JA, Chanda SK, Konig R, Malani N, Berry CC, Bushman FD.** 2011. HIV integration targeting: A pathway involving Transportin-3 and the nuclear pore protein RanBP2. *PLoS pathogens* **7:**e1001313.

68. **Cereseto A, Manganaro L, Gutierrez MI, Terreni M, Fittipaldi A, Lusic M, Marcello A, Giacca M.** 2005. Acetylation of HIV-1 integrase by p300 regulates viral integration. *Embo J* **24:**3070–3081.

69. **Allouch A, Di Primio C, Alpi E, Lusic M, Arosio D, Giacca M, Cereseto A.** 2011. The TRIM family protein KAP1 inhibits HIV-1 integration. *Cell Host Microbe* **9:** 484–495.

70. **Kalpana GV, Goff SP.** 1993. Genetic analysis of homomeric interactions of human immunodeficiency virus type 1 integrase using the yeast two-hybrid system. *Proc Natl Acad Sci USA* **90:**10593–10597.

71. **Lewinski M, Bushman FD.** 2005. Retroviral DNA integration—mechanism and consequences. *Adv Genet* **55:** 147–181.

72. **Schroder ARW, Shinn P, Chen HM, Berry C, Ecker JR, Bushman F.** 2002. HIV-1 integration in the human genome favors active genes and local hotspots. *Cell* **110:** 521–529.

73. **Mitchell R, Chiang C, Berry C, Bushman FD.** 2003. Global effects on cellular transcription following infection with an HIV-based vector. *Mol Ther* **8:**674–687.

74. **Wu X, Li Y, Crise B, Burgess SM.** 2003. Transcription start regions in the human genome are favored targets for MLV integration. *Science* **300:**1749–1751.

75. **Barr SD, Ciuffi A, Leipzig J, Shinn P, Ecker JR, Bushman FD.** 2006. HIV integration site selection: Targeting in macrophages and the effects of different routes of viral entry. *Mol Ther* **14:**218–225.

76. **Barr SD, Leipzig J, Shinn P, Ecker JR, Bushman FD.** 2005. Integration targeting by avian sarcoma-leukosis virus and human immunodeficiency virus in the chicken genome. *J Virol* **79:**12035–12044.

77. **Marshall HM, Ronen K, Berry C, Llano M, Sutherland H, Saenz D, Bickmore W, Poeschla E, Bushman FD.** 2007. Role of PSIP1/LEDGF/p75 in lentiviral infectivity and integration targeting. *PLoS One* **2:**e1340.

78. **Berry CC, Ocwieja KE, Malani N, Bushman FD.** 2014. Comparing DNA integration site clusters with scan statistics. *Bioinformatics* **30:**1493–1500.

79. **Shun MC, Raghavendra NK, Vandegraaff N, Daigle JE, Hughes S, Kellam P, Cherepanov P, Engelman A.** 2007. LEDGF/p75 functions downstream from preintegration complex formation to effect gene-specific HIV-1 integration. *Genes & Development* **21:**1767–1778.

80. **Han Y, Lassen K, Monie D, Sedaghat AR, Shimoji S, Liu S, Pierson TC, Margolick JB, Siliciano RF, Siliciano JD.** 2004. Resting CD4+ T cells from human immunodeficiency virus type 1 (HIV-1)-infected inndividuals carry integrated HIV-1 genomes within actively transcribed host genes. *J Virol* **78:**6122–6133.

81. **Berry C, Hannenhalli S, Leipzig J, Bushman FD.** 2006. Selection of target sites for mobile DNA integration in the human genome. *PLoS Computational Biology* **2:**e157.

82. **Wang GP, Ciuffi A, Leipzig J, Berry CC, Bushman FD.** 2007. HIV integration site selection: Analysis by massively parallel pyrosequencing reveals association with epigenetic modifications. *Genome Research* **17:** 1186–1194.

83. **Ciuffi A, Ronen K, Brady T, Malani N, Wang G, Berry CC, Bushman FD.** 2009. Methods for integration site distribution analyses in animal cell genomes. *Methods* **47:**261–268.

84. **Mitchell RS, Beitzel BF, Schroder ARW, Shinn P, Chen HM, Berry CC, Ecker JR, Bushman FD.** 2004. Retroviral DNA integration: ASLV, HIV, and MLV show distinct target site preferences. *PLoS Biol* **2:**1127–1137.

85. **Narezkina A, Taganov KD, Litwin S, Stoyanova R, Hayashi J, Seeger C, Skalka AM, Katz RA.** 2004. Genome-wide analyses of avain sarcoma virus integration sites. *J Virol* **78:**11656–11663.

86. **Lewinski MK, Yamashita M, Emerman M, Ciuffi A, Marshall H, Crawford G, Collins F, Shinn P, Leipzig J, Hannenhalli S, Berry CC, Ecker JR, Bushman FD.** 2006. Retroviral DNA integration: Viral and cellular determinants of target-site selection. *PLoS Pathogens* **2:**e60.

87. **Cherepanov P, Maertens G, Proost P, Devreese B, Van Beeumen J, Engelborghs Y, De Clercq E, Debyser Z.** 2003. HIV-1 Integrase Forms Stable Tetramers and Associates with LEDGF/p75 Protein in Human Cells. *J Biol Chem* **278:**372–381.

88. **Turlure F, Devroe E, Silver PA, Engelman A.** 2004. Human cell proteins and human immunodeficiency virus DNA integration. *Front Biosci* **9:**3187–3208.

89. **Emiliani S, Mousnier A, Busschots K, Maroun M, Van Maele B, Tempe D, Vandekerckhove L, Moisant F, Ben-Slama L, Witvrouw M, Christ F, Rain JC, Dargemont C, Debyser Z, Benarous R.** 2005. Integrase mutants defective for interaction with LEDGF/p75 are impaired in chromosome tethering and HIV-1 replication. *J Biol Chem* **280:**25517–25523.

90. **Llano M, Saenz DT, Meehan A, Wongthida P, Peretz M, Walker WH, Teo WL, Poeschla EM.** 2006. An essential role for LEDGF/p75 in HIV integration. *Science* **314:**461–464.

91. **De Rijck J, Vandekerckhove L, Gijsbers R, Hombrouck A, Hendrix J, Vercammen J, Engelborghs Y, Christ F, Debyser Z.** 2006. Overexpression of the lens epithelium-derived growth factor/p75 integrase binding domain inhibits human immunodeficiency virus replication. *J Virol* **80:**11498–11509.

92. **Eidahl JO, Crowe BL, North JA, McKee CJ, Shkriabai N, Feng L, Plumb M, Graham RL, Gorelick RJ, Hess S, Poirier MG, Foster MP, Kvaratskhelia M.** 2013. Struc-

tural basis for high-affinity binding of LEDGF PWWP to mononucleosomes. *Nucleic Acids Res* **41**:3924–3936.

93. **van Nuland R, van Schaik FM, Simonis M, van Heesch S, Cuppen E, Boelens R, Timmers HM, van Ingen H.** 2013. Nucleosomal DNA binding drives the recognition of H3K36-methylated nucleosomes by the PSIP1-PWWP domain. *Epigenetics & Chromatin* **6**:12.

94. **Cherepanov P, Ambrosio AL, Rahman S, Ellenberger T, Engelman A.** 2005. Structural basis for the recognition between HIV-1 integrase and transcriptional coactivator p75. *Proc Natl Acad Sci USA* **102**:17308–17313.

95. **Hare S, Shun MC, Gupta SS, Valkov E, Engelman A, Cherepanov P.** 2009. A novel co-crystal structure affords the design of gain-of-function lentiviral integrase mutants in the presence of modified PSIP1/LEDGF/p75. *PLoS Pathogens* **5**:e1000259.

96. **Ciuffi A, Llano M, Poeschla E, Hoffmann C, Leipzig J, Shinn P, Ecker JR, Bushman F.** 2005. A role for LEDGF/p75 in targeting HIV DNA integration. *Nat Med* **11**:1287–1289.

97. **Ciuffi A, Diamond T, Hwang Y, Marshall H, Bushman FD.** 2006. Fusions of LEDGF/p75 to lambda repressor promote HIV DNA integration near lambda operators *in vitro*. *Hum Gene Ther* **17**:960–967.

98. **Vandegraaff N, Devroe E, Turlure F, Silver PA, Engelman A.** 2006. Biochemical and genetic analyses of integrase-interacting protein lens epithelium-derived growth factor (LEDGF)/p75 and hepatoma-derived growth factor related protein 2 (HRP2) in preintegration complex function and HIV-1 replication. *Virology* **346**:415–426.

99. **Schrijvers R, Vets S, De Rijck J, Malani N, Bushman FD, Debyser Z, Gijsbers R.** 2012. HRP-2 determines HIV-1 integration site selection in LEDGF/p75 depleted cells. *Retrovirology* **9**:84.

100. **Vandegraaff N, Devroe E, Turlure F, Silver PA, Engelman A.** 2005. Biochemical and genetic analysis of integrase-interacting protein lens epithelium-derived growth factor (LEDGF)/p75 and hepatoma-derived growth factor related protein 2 (HRP2) in preintegration complex function and HIV-1 replication. *Virology* Epub.

101. **Schrijvers R, De Rijck J, Demeulemeester J, Adachi N, Vets S, Ronen K, Christ F, Bushman FD, Debyser Z, Gijsbers R.** 2012. LEDGF/p75-independent HIV-1 replication demonstrates a role for HRP-2 and remains sensitive to inhibition by LEDGINs. *PLoS Pathogens* **8**: e1002558.

102. **Silvers RM, Smith JA, Schowalter M, Litwin S, Liang Z, Geary K, Daniel R.** 2010. Modification of integration site preferences of an HIV-1-based vector by expression of a novel synthetic protein. *Hum Gene Ther* **21**:337–349.

103. **Gijsbers R, Ronen K, Vets S, Malani N, De Rijck J, McNeely M, Bushman FD, Debyser Z.** 2010. LEDGF hybrids efficiently retarget lentiviral integration into heterochromatin. *Mol Ther* **18**:552–560.

104. **Ferris AL, Wu X, Hughes CM, Stewart C, Smith SJ, Milne TA, Wang GG, Shun MC, Allis CD, Engelman A, Hughes SH.** 2010. Lens epithelium-derived growth factor fusion proteins redirect HIV-1 DNA integration. *Proc Natl Acad Sci USA* **107**:3135–3140.

105. **De Rijck J, de Kogel C, Demeulemeester J, Vets S, El Ashkar S, Malani N, Bushman FD, Landuyt B, Husson SJ, Busschots K, Gijsbers R, Debyser Z.** 2013. The BET family of proteins targets Moloney murine leukemia virus integration near transcription start sites. *Cell Reports* **5**:886–894.

106. **Sharma A, Larue RC, Plumb MR, Malani N, Male F, Slaughter A, Kessl JJ, Shkriabai N, Coward E, Aiyer SS, Green PL, Wu L, Roth MJ, Bushman FD, Kvaratskhelia M.** 2013. BET proteins promote efficient murine leukemia virus integration at transcription start sites. *Proc Natl Acad Sci USA* **110**:12036–12041.

107. **Larue RC, Plumb MR, Crowe BL, Shkriabai N, Sharma A, Difiore J, Malani N, Aiyer SS, Roth MJ, Bushman FD, Foster MP, Kvaratskhelia M.** 2014. Bimodal high-affinity association of Brd4 with murine leukemia virus integrase and mononucleosomes. *Nucleic Acids Res* **42**:4868–4881.

108. **Gupta SS, Maetzig T, Maertens GN, Sharif A, Rothe M, Weidner-Glunde M, Galla M, Schambach A, Cherepanov P, Schulz TF.** 2013. Bromo- and extra-terminal domain chromatin regulators serve as cofactors for murine leukemia virus integration. *J Virol* **87**:12721–12736.

109. **Studamire B, Goff SP.** 2008. Host proteins interacting with the Moloney murine leukemia virus integrase: Multiple transcriptional regulators and chromatin binding factors. *Retrovirology* **5**:48.

110. **Pryciak PM, Varmus HE.** 1992. Nucleosomes, DNA-binding proteins, and DNA sequence modulate retroviral integration target site selection. *Cell* **69**:769–780.

111. **Pryciak PM, Sil A, Varmus HE.** 1992. Retroviral integration into minichromosomes *in vitro*. *Embo J* **11**: 291–303.

112. **Pryciak P, Muller HP, Varmus HE.** 1992. Simian Virus 40 minichromosomes as targets for retroviral integration *in vivo*. *Proc Natl Acad Sci USA* **89**:9237–9241.

113. **Pruss D, Bushman FD, Wolffe AP.** 1994. Human immunodeficiency virus integrase directs integration to sites of severe DNA distortion within the nucleosome core. *Proc Natl Acad Sci USA* **91**:5913–5917.

114. **Pruss D, Reeves R, Bushman FD, Wolffe AP.** 1994. The influence of DNA and nucleosome structure on integration events directed by HIV integrase. *J Biol Chem* **269**: 25031–25041.

115. **Roth SL, Malani N, Bushman FD.** 2011. Gammaretroviral integration into nucleosomal target DNA *in vivo*. *J Virol* **85**:7393–7401.

116. **Lesbats P, Botbol Y, Chevereau G, Vaillant C, Calmels C, Arneodo A, Andreola ML, Lavigne M, Parissi V.** 2011. Functional coupling between HIV-1 integrase and the SWI/SNF chromatin remodeling complex for efficient *in vitro* integration into stable nucleosomes. *PLoS pathogens* **7**:e1001280.

117. **Botbol Y, Raghavendra NK, Rahman S, Engelman A, Lavigne M.** 2008. Chromatinized templates reveal the requirement for the LEDGF/p75 PWWP domain during HIV-1 integration *in vitro*. *Nucleic Acids Res* **36**: 1237–1246.

118. **Yoder K, Bushman FD.** 2000. Repair of gaps in retroviral DNA integration intermediates. *J Virol* **74:**11191–11200.

119. **Chow SA, Vincent KA, Ellison V, Brown PO.** 1992. Reversal of integration and DNA splicing mediated by integrase of human immunodeficiency virus. *Science* 255:723–726.

120. **Daniel R, Katz RA, Skalka AM.** 1999. A role for DNA-PK in retroviral DNA integration. *Science* **284:**644–647.

121. **Daniel R, Katz RA, Merkel G, Hittle JC, Yen TJ, Skalka AM.** 2001. Wortmannin potentiates integrase-mediated killing of lymphocytes and reduces the efficency of stable transduction by retroviruses. *Mol Cell Biol* **21:**1164–1172.

122. **Lau A, Kanaar R, Jackson SP, O'Connor MJ.** 2004. Suppression of retroviral infection by the RAD52 DNA repair protein. *Embo J* **23:**3421–3429.

123. **Daniel R, Kao G, Taganov K, Greger JG, Favorova O, Merkel G, Yen TJ, Katz RA, Skalka AM.** 2003. Evidence that the retroviral DNA integration process triggers an ATR-dependent DNA damage response. *Proc Natl Acad Sci USA* **100:**4778–4783.

124. **Yoder K, Sarasin A, Kraemer K, McIlhatton M, Bushman F, Fishel R.** 2006. The DNA repair genes XPB and XPD defend cells from retroviral infection. *Proc Natl Acad Sci USA* **103:**4622–4627.

125. **Li L, Olvera JM, Yoder K, Mitchell RS, Butler SL, Lieber MR, Martin SL, Bushman FD.** 2001. Role of the non-homologous DNA end joining pathway in retroviral infection. *Embo J* **20:**3272–3281.

126. **Espeseth AS, Fishel R, Hazuda D, Huang Q, Xu M, Yoder K, Zhou H.** 2011. siRNA screening of a targeted library of DNA repair factors in HIV infection reveals a role for base excision repair in HIV integration. *PLoS One* **6:**e17612.

127. **Yoder KE, Espeseth A, Wang XH, Fang Q, Russo MT, Lloyd RS, Hazuda D, Sobol RW, Fishel R.** 2011. The base excision repair pathway is required for efficient lentivirus integration. *PLoS One* **6:**e17862.

128. **Arts EJ, Hazuda DJ.** 2012. HIV-1 antiretroviral drug therapy. *Cold Spring Harbor Perspectives in Medicine* **2:**a007161.

129. **Hazuda DJ, Hastings JC, Wolfe AL, Emini EA.** 1994. A novel assay for the DNA strand-transfer reaction of HIV-1 integrase. *Nuc Acids Res* **22:**1121–1122.

130. **Hazuda DJ, Felock P, Witmer M, Wolfe A, Stillmock K, Grobler JA, Espeseth A, Gabryelski L, Schleif W, Blau C, Miller MD.** 2000. Inhibitors of strand transfer that prevent integration and inhibit HIV-1 replication in cells. *Science* **287:**646–650.

131. **Hazuda DJ, Anthony NJ, Gomez RP, Jolly SM, Wai JS, Zhuang L, Fisher TE, Embrey M, P GJ Jr, Egbertson MS, Vacca JP, Huff JR, Felock PJ, Witmer MV, Stillmock KA, Danovich R, Grobler J, Miller MD, Espeseth AS, Jin L, Chen IW, Lin JH, Kassahun K, Ellis JD, Wong BK, Xu W, Pearson PG, Schleif WA, Cortese R, Emini E, V S, Holloway MK, Young SD.** 2004. A naphthyridine carboxamide provides evidence for discordant resistance between mechanistically identical inhibitors of HIV-1 integrase. *Proc Natl Acad Sci USA* **101:**11233–11238.

132. **Metifiot M, Marchand C, Pommier Y.** 2013. HIV integrase inhibitors: 20-year landmark and challenges. *Advances in Pharmacology* **67:**75–105.

133. **Espeseth AS, Felock P, Wolfe A, Witmer M, Grobler J, Anthony N, Egbertson M, Melamed JY, Young S, Hamill T, Cole JL, Hazuda DJ.** 2000. HIV-1 integrase inhibitors that compete with the target DNA substrate define a unique strand transfer conformation for integrase. *Proc Natl Acad Sci USA* **97:**11244–11249.

134. **Krishnan L, Li X, Naraharisetty HL, Hare S, Cherepanov P, Engelman A.** 2010. Structure-based modeling of the functional HIV-1 intasome and its inhibition. *Proc Natl Acad Sci USA* **107:**15910–15915.

135. **Min S, Sloan L, DeJesus E, Hawkins T, McCurdy L, Song I, Stroder R, Chen S, Underwood M, Fujiwara T, Piscitelli S, Lalezari J.** 2011. Antiviral activity, safety, and pharmacokinetics/pharmacodynamics of dolutegravir as 10-day monotherapy in HIV-1-infected adults. *Aids* **25:**1737–1745.

136. **Walmsley SL, Antela A, Clumeck N, Duiculescu D, Eberhard A, Gutierrez F, Hocqueloux L, Maggiolo F, Sandkovsky U, Granier C, Pappa K, Wynne B, Min S, Nichols G, Investigators S.** 2013. Dolutegravir plus abacavir-lamivudine for the treatment of HIV-1 infection. *New England Journal of Medicine* **369:**1807–1818.

137. **Gupta K, Brady T, Dyer BM, Malani N, Hwang Y, Male F, Nolte RT, Wang L, Velthuisen E, Jeffrey J, Van Duyne GD, Bushman FD.** 2014. Allosteric inhibition of human immunodeficiency virus integrase: late block during viral replication and abnormal multimerization involving specific protein domains. *J Biol Chem* **289:**20477–20488.

138. **Molteni V, Greenwald J, Rhodes D, Hwang Y, Kwiatkowski W, Bushman FD, Siegel JS, Choe S.** 2001. Identification of a small molecule binding site at the dimer interface of the HIV integrase catalytic domain. *Acta Crystallogr D Biol Crystallogr* **57:**536–544.

139. **Le Rouzic E, Bonnard D, Chasset S, Bruneau JM, Chevreuil F, Le Strat F, Nguyen J, Beauvoir R, Amadori C, Brias J, Vomscheid S, Eiler S, Levy N, Delelis O, Deprez E, Saib A, Zamborlini A, Emiliani S, Ruff M, Ledoussal B, Moreau F, Benarous R.** 2013. Dual inhibition of HIV-1 replication by integrase-LEDGF allosteric inhibitors is predominant at the post-integration stage. *Retrovirology* **10:**144.

140. **Jurado KA, Wang H, Slaughter A, Feng L, Kessl JJ, Koh Y, Wang W, Ballandras-Colas A, Patel PA, Fuchs JR, Kvaratskhelia M, Engelman A.** 2013. Allosteric integrase inhibitor potency is determined through the inhibition of HIV-1 particle maturation. *Proc Natl Acad Sci USA* **110:**8690–8695.

141. **Feng L, Sharma A, Slaughter A, Jena N, Koh Y, Shkriabai N, Larue RC, Patel PA, Mitsuya H, Kessl JJ, Engelman A, Fuchs JR, Kvaratskhelia M.** 2013. The A128T resistance mutation reveals aberrant protein multimerization as the primary mechanism of action of allosteric HIV-1 integrase inhibitors. *J Biol Chem* **288:**15813–15820.

142. **Engelman A, Kessl JJ, Kvaratskhelia M.** 2013. Allosteric inhibition of HIV-1 integrase activity. Current opinion in chemical biology. *Curr Opin Chem* **17:**339–345.

143. **Desimmie BA, Schrijvers R, Demeulemeester J, Borrenberghs D, Weydert C, Thys W, Vets S, Van Remoortel B, Hofkens J, De Rijck J, Hendrix J, Bannert N, Gijsbers R, Christ F, Debyser Z.** 2013. LEDGINs inhibit late stage HIV-1 replication by modulating integrase multimerization in the virions. *Retrovirology* **10:**57.

144. **Finzi D, Hermankova M, Pierson T, Carruth LM, Buck C, Chaisson RE, Quinn TC, Chadwick K, Margolick J, Brookmeyer R, Gallant J, Markowitz M, Ho DD, Richman DD, Siliciano RF.** 1997. Identification of a reservoir for HIV-1 in patients on highly active antiretroviral therapy. *Science* **278:**1295–1300.

145. **Siliciano RF, Greene WC.** 2011. HIV latency. *Cold Spring Harbor Perspectives in Medicine* **1:**a007096.

146. **Hutter G, Nowak D, Mossner M, Ganepola S, Mussig A, Allers K, Schneider T, Hofmann J, Kucherer C, Blau O, Blau IW, Hofmann WK, Thiel E.** 2009. Long-term control of HIV by CCR5 Delta32/Delta32 stem-cell transplantation. *New England Journal of Medicine* **360:**692–698.

147. **Xing S, Siliciano RF.** 2013. Targeting HIV latency: Pharmacologic strategies toward eradication. *Drug Discovery Today* **18:**541–551.

148. **Saez-Cirion A, Bacchus C, Hocqueloux L, Avettand-Fenoel V, Girault I, Lecuroux C, Potard V, Versmisse P, Melard A, Prazuck T, Descours B, Guergnon J, Viard JP, Boufassa F, Lambotte O, Goujard C, Meyer L, Costagliola D, Venet A, Pancino G, Autran B, Rouzioux C, Group AVS.** 2013. Post-treatment HIV-1 controllers with a long-term virological remission after the interruption of early initiated antiretroviral therapy ANRS VISCONTI Study. *PLoS Pathogens* **9:**e1003211.

149. **Archin NM, Liberty AL, Kashuba AD, Choudhary SK, Kuruc JD, Crooks AM, Parker DC, Anderson EM, Kearney MF, Strain MC, Richman DD, Hudgens MG, Bosch RJ, Coffin JM, Eron JJ, Hazuda DJ, Margolis DM.** 2012. Administration of vorinostat disrupts HIV-1 latency in patients on antiretroviral therapy. *Nature* **487:**482–485.

150. **Sherrill-Mix S, Lewinski MK, Famiglietti M, Bosque A, Malani N, Ocwieja KE, Berry CC, Looney D, Shan L, Agosto LM, Pace MJ, Siliciano RF, O'Doherty U, Guatelli J, Planelles V, Bushman FD.** 2013. HIV latency and integration site placement in five cell-based models. *Retrovirology* **10:**90.

151. **Lewinski MK, Bisgrove D, Shinn P, Chen H, Hoffmann C, Hannenhalli S, Verdin E, Berry CC, Ecker JR, Bushman FD.** 2005. Genome-wide analysis of chromosomal features repressing human immunodeficiency virus transcription. *J Virol* **79:**6610–6619.

152. **Han Y, Lin YB, An W, Xu J, Yang HC, O'Connell K, Dordai D, Boeke JD, Siliciano JD, Siliciano RF.** 2008. Orientation-dependent regulation of integrated HIV-1 expression by host gene transcriptional readthrough. *Cell Host & Microbe* **4:**134–146.

153. **Elgin SC, Reuter G.** 2013. Position-effect variegation, heterochromatin formation, and gene silencing in Drosophila. *Cold Spring Harbor Perspectives in Biology* **5:**a017780.

154. **Spina CA, Anderson J, Archin NM, Bosque A, Chan J, Famiglietti M, Greene WC, Kashuba A, Lewin SR, Margolis DM, Mau M, Ruelas D, Saleh S, Shirakawa K, Siliciano RF, Singhania A, Soto PC, Terry VH, Verdin E, Woelk C, Wooden S, Xing S, Planelles V.** 2013. An in-depth comparison of latent HIV-1 reactivation in multiple cell model systems and resting $CD4^+$ T cells from aviremic patients. *PLoS Pathogens* **9:**e1003834.

155. **Ikeda T, Shibata J, Yoshimura K, Koito A, Matsushita S.** 2007. Recurrent HIV-1 integration at the BACH2 locus in resting $CD4^+$ T cell populations during effective highly active antiretroviral therapy. *J Infect Dis* **195:**716–725.

156. **Maldarelli F, Wu X, Su L, Simonetti FR, Shao W, Hill S, Spindler J, Ferris AL, Mellors JW, Kearney MF, Coffin JM, Hughes SH.** 2014. HIV latency. Specific HIV integration sites are linked to clonal expansion and persistence of infected cells. *Science* **345:**179–183.

157. **Wagner TA, McLaughlin S, Garg K, Cheung CY, Larsen BB, Styrchak S, Huang HC, Edlefsen PT, Mullins JI, Frenkel LM.** 2014. HIV latency. Proliferation of cells with HIV integrated into cancer genes contributes to persistent infection. *Science* **345:**570–573.

158. **Margolis D, Bushman F.** 2014. HIV/AIDS. Persistence by proliferation? *Science* **345:**143–144.

159. **Cavazzana-Calvo M, Payen E, Negre O, Wang G, Hehir K, Fusil F, Down J, Denaro M, Brady T, Westerman K, Cavallesco R, Gillet-Legrand B, Caccavelli L, Sgarra R, Maouche-Chretien L, Bernaudin F, Girot R, Dorazio R, Mulder GJ, Polack A, Bank A, Soulier J, Larghero J, Kabbara N, Dalle B, Gourmel B, Socie G, Chretien S, Cartier N, Aubourg P, Fischer A, Cornetta K, Galacteros F, Beuzard Y, Gluckman E, Bushman F, Hacein-Bey-Abina S, Leboulch P.** 2010. Transfusion independence and HMGA2 activation after gene therapy of human beta-thalassaemia. *Nature* **467:**318–322.

160. **Goff SP.** 2008. Knockdown screens to knockout HIV-1. *Cell* **135:**417–420.

161. **Bushman FD, Malani N, Fernandes J, D'Orso I, Cagney G, Diamond TL, Zhou H, Hazuda DJ, Espeseth AS, Konig R, Bandyopadhyay S, Ideker T, Goff SP, Krogan NJ, Frankel AD, Young JA, Chanda SK.** 2009. Host cell factors in HIV replication: meta-analysis of genome-wide studies. *PLoS Pathogens* **5:**e1000437.

162. **Bushman FD, Barton S, Bailey A, Greig C, Malani N, Bandyopadhyay S, Young J, Chanda S, Krogan N.** 2013. Bringing it all together: Big data and HIV research. *Aids* **27:**835–838.

163. **Nguyen DG, Yin H, Zhou Y, Wolff KC, Kuhen KL, Caldwell JS.** 2007. Identification of novel therapeutic targets for HIV infection through functional genomic cDNA screening. *Virology* **362:**16–25.

164. **Kane M, Yadav SS, Bitzegeio J, Kutluay SB, Zang T, Wilson SJ, Schoggins JW, Rice CM, Yamashita M, Hatziioannou T, Bieniasz PD.** 2013. MX2 is an interferon-induced inhibitor of HIV-1 infection. *Nature* **502:**563–566.

165. **Goujon C, Moncorge O, Bauby H, Doyle T, Ward CC, Schaller T, Hue S, Barclay WS, Schulz R, Malim MH.**

2013. Human MX2 is an interferon-induced post-entry inhibitor of HIV-1 infection. *Nature* **502:**559–562.

166. **Bartha I, McLaren PJ, Ciuffi A, Fellay J, Telenti A.** 2014. GuavaH: A compendium of host genomic data in HIV biology and disease. *Retrovirology* **11:**6.

167. **Coffin JM, Hughes SH, Varmus HE.** 1997. *Retroviruses.* Cold Spring Harbor Laboratory Press, Cold Spring Harbor, NY.

168. **Farnet CM, Haseltine WA.** 1991. Circularization of human immunodeficiency virus type 1 DNA in vitro. *J Virol* **65:**6942–6952.

169. **Daniel R, Greger JG, Katz RA, Taganov KD, Wu X, Kappes JC, Skalka AM.** 2004. Evidence that stable retroviral transduction and cell survival following DNA integration depend on components of the nonhomologous end joining repair pathway. *J Virol* **78:**8573–8581.

170. **Ariumi Y, Turelli P, Masutani M, Trono D.** 2005. DNA damage sensors ATM, ATR, DNA-PKcs, and PARP-1 are dispensable for human immunodeficiency virus type 1 integration. *J Virol* **79:**2973–2978.

171. **Cooper A, Garcia M, Petrovas C, Yamamoto T, Koup RA, Nabel GJ.** 2013. HIV-1 causes CD4 cell death through DNA-dependent protein kinase during viral integration. *Nature* **498:**376–379.

172. **Monroe KM, Yang Z, Johnson JR, Geng X, Doitsh G, Krogan NJ, Greene WC.** 2014. IFI16 DNA sensor is required for death of lymphoid CD4 T cells abortively infected with HIV. *Science* **343:**428–432.

Mobile DNA, 3rd Edition
Nancy L. Craig, Michael Chandler, Martin Gellert, Alan M. Lambowitz, Phoebe A. Rice and Suzanne Sandmeyer
© 2014 American Society for Microbiology, Washington, DC
doi:10.1128/microbiolspec.MDNA3-0027-2014

Stephen H. Hughes[1]

46 Reverse Transcription of Retroviruses and LTR Retrotransposons

INTRODUCTION

The conversion, well over a billion years ago, of the RNA world into the modern configuration, in which genetic information is maintained primarily in DNA, required reverse transcriptases (RTs), enzymes that were able to copy genetic information from RNA into DNA, a process called reverse transcription. With minor (but important) exceptions, for example telomerases, normal cellular processes no longer require reverse transcription, which is now primarily employed in the replication of hepadnaviruses, retroviruses, and retrotransposons. This chapter will cover the process of reverse transcription, and the RTs that are involved in the replication of retroviruses and the related long terminal repeat (LTR) retrotransposons, which have lifestyles that are similar to a retrovirus that has either lost, or never acquired, the ability to be transmitted horizontally from one cell to another. The RTs of, and reverse transcription by, non-LTR retrotransposons will be considered in the chapters that describe these elements (49–55). A substantial fraction of the work that has been done on reverse transcription and RT has focused on human immunodeficiency virus type 1 (HIV-1); this is entirely appropriate given the extent of the HIV epidemic and the fact that HIV-1 RT is the target of two important classes of anti-HIV drugs. Thus, a substantial portion of this review will describe data and insights obtained in experiments that were done with HIV-1 and HIV-1 RT. However, there are some important differences in the RTs, and the process of reverse transcription, among the different retroviruses and LTR retrotransposons; these differences will also be considered, at least briefly. The literature on RT and reverse transcription is both vast and complex. Any review, including this one, can present no more than a superficial overview of what is known. Much that is important has been omitted, some intentionally, some inadvertently; for these omissions, the author apologizes. For those who are interested, a number of helpful reviews have already been published, most of which are focused on retroviral RTs (1, 2, 3, 4).

THE PROCESS OF REVERSE TRANSCRIPTION

In retroviruses and LTR retrotransposons, reverse transcription is the conversion of a single-stranded RNA (ssRNA) copy of the genome into a double-stranded

[1]HIV Drug Resistance Program, Center for Cancer Research, National Cancer Institute at Frederick, 1050 Boyles St., Building 539 Rm. 130A, Frederick, MD 21702.

DNA (dsDNA). To avoid the loss of genetic information, the dsDNA copy is longer, on both ends, than the ssRNA from which it is derived (Fig. 1). Although genomic RNA has both a 5′ cap and a poly(A) tail, the ends of the genomic RNA that are transcribed from DNA have short duplications, called R. After reverse transcription, the DNA copy is flanked by longer duplications called the long terminal repeats, or LTRs (5, 6, 7). The LTRs have in them the sequences (enhancers, a promoter, poly-adenylation signals, etc.) that are needed for the host machinery to generate the genomic RNA from the reverse-transcribed DNA copy. The basic scheme by which retroviruses and LTR retrotransposons synthesize the DNA copies of their genomes is the same. It is theoretically possible to produce the dsDNA version of the genome from a single copy of the ssRNA genome (Fig. 1); in practice, retroviruses and LTR retrotransposons contain two copies of the RNA genome (8). It is likely that the primary reason for the presence of two copies of the RNA genome is that the second copy makes it possible to produce a complete dsDNA copy even if the RNA genome is nicked; RT can use the second RNA strand as an alternate template to bypass the nick. This switching of DNA synthesis between the two RNAs also means, if the two RNAs differ in their sequences, that the resulting dsDNA will be a recombinant; this will be discussed in more detail in a later section. Although retrotransposons and retroviruses encode other proteins that increase the efficiency of reverse transcription, RT has both of the enzymatic activities (DNA polymerase and RNase H) that are needed to carry out all of the steps in the process. The DNA polymerase of RT can copy either an RNA or a DNA template; RNase H cleaves RNA if, and only if, it is part of an RNA/DNA hybrid. Like many other DNA polymerases, the polymerase of RT requires both a template and a primer. Many retroviruses and LTR retrotransposons encapsidate a host tRNA that RT uses as a primer to initiate the synthesis of the first DNA strand; however, there are alternative strategies, including the use of a half tRNA primer (9) and self-priming using a processed form of the RNA template (10). Because the RNA genome is the same sense as the mRNAs, it is a plus strand; the first DNA strand to be synthesized by RT is a minus strand. Retroviruses and retrotransposons use a number of different host tRNAs as minus-strand primers; in such cases, the RNA genome carries, near its 5′ end, the appropriate complementary sequence, called the primer-binding site (PBS) (Fig. 1). This means that only a short piece of DNA (sometimes called the minus-strand strong stop) can be synthesized before the growing minus strand reaches the 5′ end of the RNA template. It would appear, at least for HIV-1 RT, that the initiation of reverse transcription is a particularly difficult step, probably at least in part because HIV RT cannot use an RNA/RNA template/primer efficiently (11, 12). This issue is discussed in more detail in the section on the structure of RT.

Synthesis of minus-strand DNA creates an RNA/DNA hybrid that is a substrate for RNase H. There is some controversy about whether there is significant RNase H cleavage by the RT that carries out minus-strand viral DNA synthesis. Cleavage by the polymerizing RT is not required; complete infectious copies of the dsDNA can be made (albeit inefficiently) by a mixture of RTs, some of which have only polymerase activity and some of which have only RNase H activity (13, 14). Moreover, although there is some disagreement in the literature, the bulk of the evidence suggests that there is little, if any, RNase H cleavage by an actively polymerizing RT in *in vitro* assays (15, 16). This issue is discussed in the section on the structure of RT. The short duplicated sequences at the ends of genomic RNA sequences (R) allow the complementary portion of minus-strand DNA to bind to the R sequence at the 3′ end of the viral RNA after the RNA template has been degraded by RNase H. It appears that minus-strand DNA is equally able to bind to (transfer to) the 3′ end of either of the two viral RNAs; thus, retroviral minus-strand DNA synthesis can either continue on the same RNA template, or switch to the other RNA (17, 18, 19); a similar result was obtained in experiments done with the retrotransposon Ty1 (20). After the minus strand is bound to the 3′ end of one of the two the RNAs (this step is called the first or minus-strand transfer), minus-strand DNA synthesis can continue.

The RNA genomes of retroviruses and retrotransposons all have at least one, and some have two, copies of purine-rich sequences that are relatively resistant to cleavage by RNase H. The essential copy of these polypurine tracts (PPTs) is found near the 3′ end of the genome; the second copy, if one is present, is usually found near the middle of the genome. After minus-strand DNA synthesis has copied the RNA past the PPT, RNase H degrades most of the RNA that has been copied but spares the PPT. The PPT can then act as the primer for plus-strand DNA synthesis (see Fig. 1). The PPT primer is subsequently removed from the plus-strand DNA by RNase H. If a central PPT is present (as it is in HIV-1), it can act as a second site for the initiation of plus-strand DNA synthesis. Based on experiments that were done primarily with HIV vectors, the second PPT does not appear to be essential for virus replication, although it does appear to help the efficien-

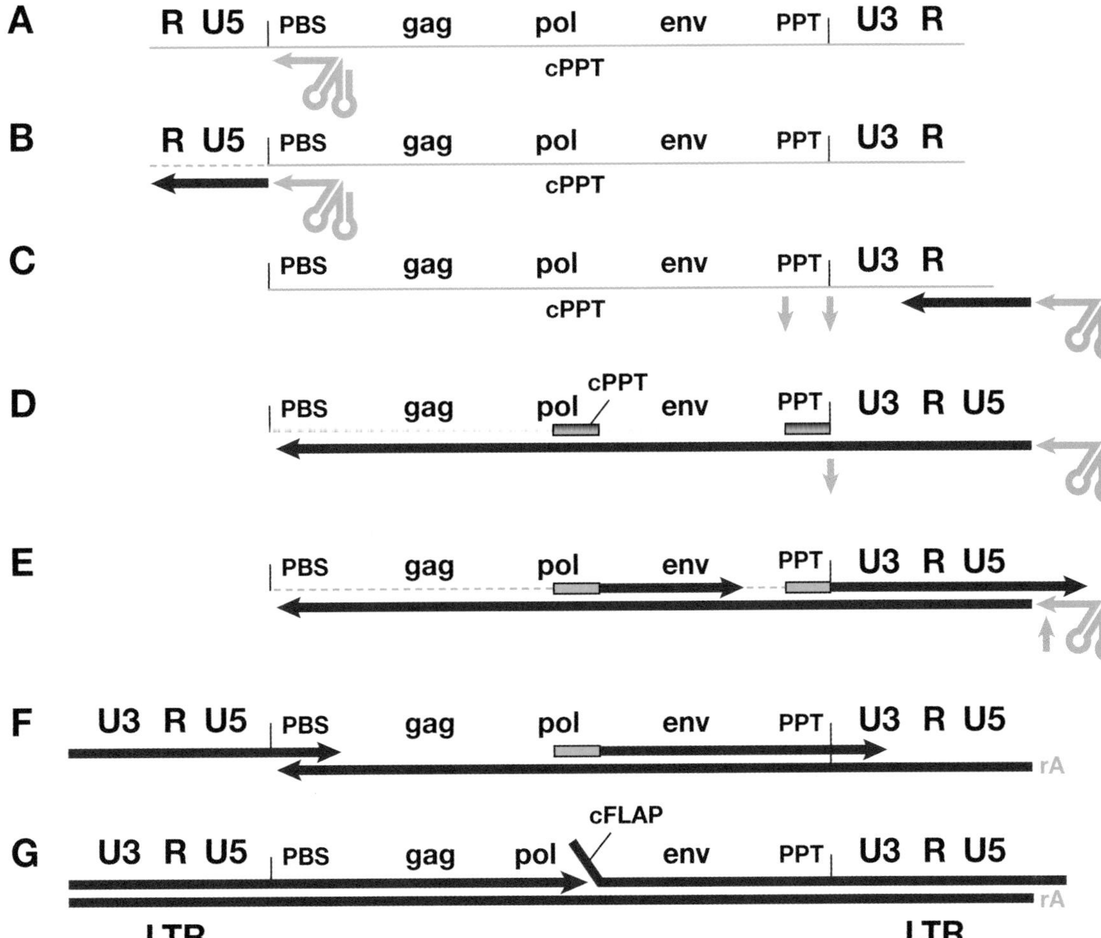

Figure 1 Reverse transcription of the genome of HIV-1. This figure shows, in cartoon form, the steps that are involved in the conversion of the ssRNA genome found in virions into dsDNA. In the figure, RNA is shown in green and DNA in purple. For simplicity, the 5′ cap and the poly(A) tail, which are present on the viral genomic RNA, have been omitted. The various sequence elements in the viral genome, including the genes, are not drawn to scale. (**A**) The tRNA primer (green arrow on the left) is base paired to the primer-binding site (PBS). (**B**) RT has initiated minus-strand DNA synthesis from the tRNA primer, and has copied the U5 and R sequences at the 5′ end of the genome. This creates an RNA/DNA duplex, which allows RNase H to degrade the RNA that has been copied (dotted green line). (**C**) The minus-strand DNA (see text) has been transferred, using the R sequence found at both ends of the viral RNA, to the 3′ end of the viral RNA, and minus-strand DNA synthesis can resume. The HIV-1 genome has two purine-rich sequences (polypurine tracts, or PPTs, one immediately adjacent to U3; the other, the central PPT [cPPT] is in the *pol* gene). (**D–G**) The two PPTs are relatively resistant to RNase H (**D**) and they serve as primers for plus-strand DNA synthesis (**E**). Once plus-strand DNA synthesis has been initiated, RNase H removes the PPT from the plus-strand DNAs (**F, G**). The plus-strand that is initiated in U3 is extended until the first 18 bases to the tRNA primer are copied (see text); RNase H then removes the tRNA primer (**E**). It appears that the entire PPT primers are removed by RNase H; however, RNase H leaves a single riboA on the 5′ end of the minus-strand (**E, F**). Removing the tRNA primer sets the stage for the second-strand transfer (**F**). Both the plus- and minus-strand DNAs are then elongated. The plus-strand that was initiated at the U3 junction (on the left in the figure) displaces a segment of the plus-strand that was initiated from the cPPT, creating a small flap, called the central flap (cFLAP).
doi:10.1128/microbiolspec.MDNA3-0027-2014.f1

cy of plus-strand DNA synthesis (21, 22). Extension of the plus-strand DNA along the minus-strand continues past the end of minus-strand DNA. This second transfer creates the LTRs (U3-R-U5) that will be found at the ends of the DNA genome when it is completed. Once plus-strand DNA synthesis copies the end of U5, synthesis continues until the portion of the tRNA primer that binds the PBS is copied. Experiments done with avian sarcoma leukosis virus (ASLV) showed that the portion of the tRNA that is copied extends to a modified A, which cannot properly base pair with a T; the segment that is copied into plus-strand DNA corresponds exactly to the PBS (23). The minus-strand DNA is also extended and copies the PBS from the RNA genome. It is likely that, in many cases, the RNA 5′ of the PBS has already been copied by RT and degraded by RNase H; in such cases, the 3′ end of the growing minus strand will stop at the 5′ end of the PBS. Copying the PBS and/or the tRNA primer requires RT to displace the tRNA from the RNA genome; RT is capable of this kind of displacement activity in *in vitro* reactions. After RT copies the end of the tRNA primer, the primer is cleaved from the minus-strand DNA by RNase H, setting the stage for the second (or plus-) strand transfer reaction. In contrast to the first strand transfer, the second strand transfer occurs *in cis*, presumably because there is only one long minus-strand DNA that can participate in the plus-strand transfer reaction (24). Full extension of the minus and plus strands leads to the synthesis of the dsDNA form of the genome, which is flanked on each end by the LTRs (Fig. 1).

Although the general outline of the process of reverse transcription for retroviruses and LTR retrotransposons has been known for about 35 years, there are some additional points that should be considered. There is the question of whether, at least in some cases, plus-strand DNA synthesis is initiated at positions other than the PPT, making use of residual fragments of the RNA genome that were not completely degraded by RNase H as plus-strand primers. There is quite good evidence that, in ASLVs, plus-strand DNA synthesis is initiated at many sites (25, 26). Conversely, murine leukemia viruses (MLVs) appear to have only one site at which plus-strand DNA synthesis is initiated, the PPT (5). HIV is somewhat controversial; however, there are published data that suggest that, like the ASLVs, HIV plus-strand DNA synthesis has multiple initiation sites (27, 28, 29). However, this raises an interesting question. ASLVs do not have a second PPT (nor, for that matter, do MLVs). If HIV can initiate plus-strand DNA synthesis at multiple points, why does it need a second central PPT?

It also appears, at least in the case of HIV, that plus-strand viral DNA synthesis displaces a portion of 5′ end of the plus-strand viral DNA that was initiated from the central PPT, creating a structure called the central flap. Various functions have been attributed to this flap in the literature. In particular, there are claims from several laboratories that the central flap is involved in nuclear import; however, recent experiments have called this idea into question (30). Although vectors that lack the central PPT (and, by extension, the central flap) replicate reasonably well, several groups have reported that the presence of the central PPT increases the titer of the virus (22, 31, 32). It does appear that the presence of the central PPT allows the virus to complete second-strand DNA synthesis more quickly, making it less susceptible to the host restriction by members of the APOBEC gene family (33, 34) (the APOBECs and their effects are discussed, briefly, in a later section).

In addition, it should be noted that the ends of the viral DNA, which need to be precise because they are the substrates for the integrase (IN) that inserts the dsDNA into the host genome, are defined by the specificity of RNase H cleavages. The U5 end of the genome is defined by the cleavage that removes the tRNA primer (Fig. 1); the U3 end is defined both by the cleavage that creates the 3′ end of the PPT and also by the subsequent cleavage that removes the PPT primer. What is remarkable is that these cleavages are usually made with single-nucleotide precision by an enzyme that has no obvious ability, based on both extensive structural and biochemical analysis, to bind to or otherwise recognize specific sequences. At first glance, it would appear that removing the tRNA primer would be a simple task; all RNase H needs to do is to find the RNA/DNA junction at the 3′ end of the tRNA and remove the entire primer. In most retroviruses, that is exactly what happens; however, HIV-1 is an exception. The RNase H of HIV-1 actually cleaves its tRNA primer one nucleotide from the 3′ end, and the last nucleotide at the 5′ end of the minus-strand of the completed linear viral DNA is a riboA (Fig. 1) (35, 36, 37). This riboA is lost during integration, but its presence, at the very 5′ end of the minus strand in the linear form of HIV-1 viral DNA, shows that RNase H, in removing an RNA primer, does not always simply look for the junction between RNA and DNA. As far as we know (this is a problem that is difficult to investigate directly), the entire PPT primer is always removed; here the problem is how RNase H manages to cleave the RNA genome with single-nucleotide precision. As has already been mentioned, the single-nucleotide precision of these cleavages does not appear to involve sequence recognition in any direct sense; rather the specificity appears to be the result of the structure of the RNA/DNA duplex and its interactions with RT. Altering the sequence of either the PPT (which would alter its structure) or of the portion of the RNase H that contacts the nucleic acid near the PPT can cause a loss of cleavage specificity. Thus, it appears to be the interactions between the RNA/DNA duplex and the polymerase domain that determines the specificity of RNase H cleavage (38, 39, 40, 41, 42); this issue will be discussed in more detail in the next section.

REVERSE TRANSCRIPTASES OF RETROVIRUSES AND LTR RETROTRANSPOSONS

The RTs of retroviruses and LTR retrotransposons are synthesized as polyproteins that are processed by the corresponding viral/retrotransposon proteases (43). This processing takes place in the context of an assembled particle and, depending on which retrovirus/retrotransposon is being considered, can happen either within the producer cell or outside of it. The details of assembly of the polyproteins, their processing, and subsequent rearrangement to form mature particles are fascinating, but are issues that are beyond the scope of this review. It is important, for the reverse transcription reaction, that RT and its nucleic acid substrates are contained within a particle; why this is important will be considered later. Replication-competent retroviruses/retrotransposons encode three enzymes, RT, the protease needed to process the polyproteins, and the IN that inserts the dsDNA form of the viral genome into the host genome. These three proteins are most commonly expressed as part of a Gag-Pro-Pol polyprotein, with *gag* being the gene that encodes the structural components of the particle. Gag is encoded 5′ of Pro and Pol; in most but not all cases, the genomes of retroviruses and retrotransposons have special features that allow protein synthesis to occasionally bypass the stop codon at the end of *gag*, leading to the synthesis of Gag-Pro-Pol. Gag self-assembles, and Gag-Pro-Pol co-assembles with Gag. The protease processes both Gag and Gag-Pro-Pol to produce the mature forms of the structural proteins and the enzymes. There are exceptions; for example, the foamy viruses, despite being retroviruses, synthesize Pro-Pol separately from Gag, and Pro-Pol co-assembles with Gag by some other mechanism (44, 45). The mature form of RT also differs among various retroviruses and retrotransposons. In its simplest form, RT is a monomer (at least in solution) that contains both the polymerase and RNase H (46). There are also more complex forms of RT; the RT of prototype foamy virus, which also appears to be a monomer, is composed of both protease and RT (47). There are forms of RT that appear to be homodimers of a complete copy of RT (polymerase+RNase H); however, there is recent structural evidence to suggest, based on the structure of Ty3 RT, that, in at least some homodimeric RTs, the two subunits assume different structures, and it has been suggested that the polymerase activity and the RNaseH activities of Ty3 RT are contributed by different subunits (48). There are also heterodimeric forms of RT. ASLV RT appears to exist as a heterodimer that is composed of a larger subunit (RT+IN) and a smaller subunit (RT alone), and as two types of homodimer composed of the larger and smaller subunits of the heterodimer. All three forms are active in *in vitro* assays (49, 50, 51); whether all three forms are also active in infected cells is not clear. The RT of HIV-1, which is by far the best characterized, is a heterodimer in which the larger subunit is a complete copy of RT and the smaller subunit is a protease cleavage product that has lost almost all of the RNase H domain (52, 53, 54).

The polymerase domain of viral and retrotransposon RTs are related, both at the structural and sequence levels, to DNA polymerases, including *Escherichia coli* DNA polymerase I (55). Although the polymerase of RT and DNA polymerase I (Pol I) are only distantly related, the similarity in both structure and function suggests that they had a common ancestor. The RNase H domain of RT has considerable similarity to a cellular RNase H; however, it has been suggested, based in part on comparing RNase H sequences, that retroviruses and LTR retrotransposons were derived from a non-LTR retrotransposon, perhaps by fusion with a DNA transposon (56). This idea is supported by the observation that in many RTs there is a subdomain (the connection; see Fig. 2) that links polymerase and RNase H. It would appear that the connection subdomain was derived from RNase H. Although the two domains of the RTs that are now found in retroviruses and retrotransposons can be linked to cellular enzymes, whether they are related in any direct way to the original RTs that participated in the conversion of the RNA world into a DNA world is not clear. If retroviruses and LTR retrotransposons were derived, at least in part, from non-LTR retrotransposons, then the path back would have to go through the RTs of non-LTR retrotransposons.

HIV-1 RT

As already noted, the RT of HIV-1 is by far the best-studied RT, and it has been extensively analyzed

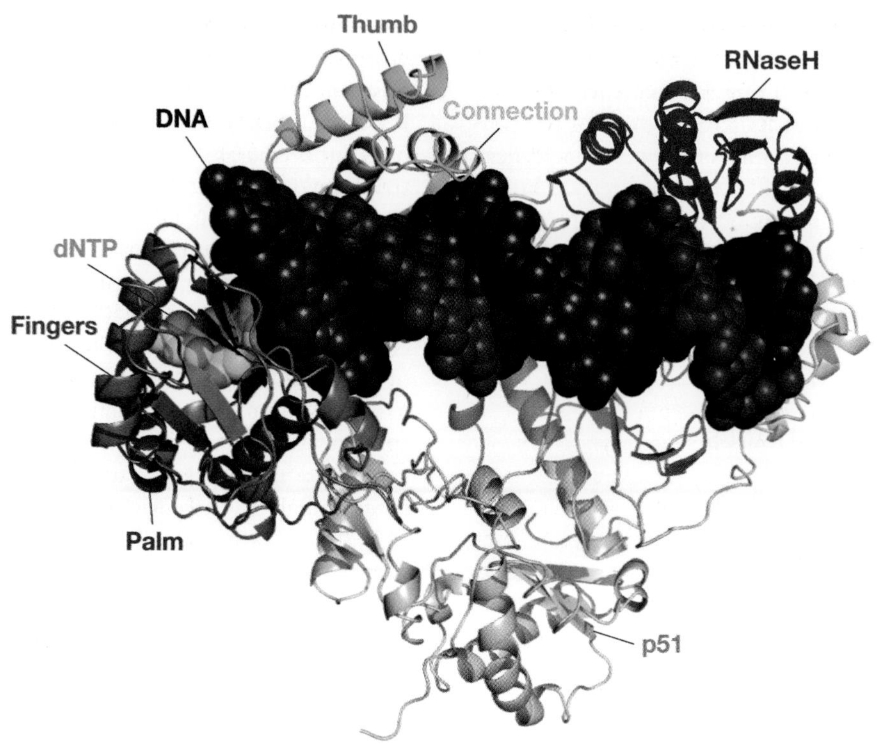

Figure 2 HIV-1 RT in a complex with dsDNA and an incoming dNTP. RT is shown as a ribbon diagram; the DNA and the incoming dNTP are shown as space filling models. The p51 subunit (at the bottom) is shown in gray. The RNase H domain is shown in pink, and the four subdomains to the polymerase domain are shown in different colors: fingers, blue; thumb, green; palm, red; and connection, yellow. The DNA template strand (the strand that is being copied) is dark red and the primer strand (the strand that is being extended) is purple. The incoming dNTP is light blue. doi:10.1128/microbiolspec.MDNA3-0027-2014.f2

biochemically, genetically, and structurally. It will not be possible, in this review, to do justice to the wealth of biochemical and genetic data that is available, and the description of the structural data will, of necessity, be superficial. The two subunits of HIV-1 RT are, respectively, 560 and 440 amino acids (52). The larger subunit (often called p66) is composed of two domains, the polymerase and RNase H. The polymerase domain, which includes approximately the first 430 amino acids of the p66 subunit, is folded into four subdomains: the fingers, palm, thumb, and connection (53, 54). The smaller subunit (p51) corresponds approximately, but not exactly, to the polymerase domain of the larger subunit and is folded into the same four subdomains. However, the relationship of the four subdomains to each other is quite different in the two subunits (Fig. 2). The active site of the polymerase is composed of a triad of aspartic acids (D110, D185, and D186) in the p66 subunit. The subdomains of p51 are arranged quite differently from their relative positions in p66; one consequence of this difference is that D110, D185, and D186 in the smaller subunit are not involved in any enzymatic function. In the p66 subunit, these three aspartates chelate the two Mg^{2+} ions that carry out the chemical step of adding nucleotides to the growing DNA chain. There are a number of other amino acids near the catalytic triad that play important roles in the polymerase reaction. For example Y115, and the F at the corresponding position in MLV RT, are situated immediately beneath the 2′ position of incoming dNTPs. These aromatic amino acids act as a steric gate that allows the incorporation of dNTPs, but, because of a clash with a 2′-OH, greatly reduce the ability of the polymerases of these RTs to incorporate NTPs (57, 58).

The RNase H domain is present only in the larger subunit of HIV-1 RT; the RNase H active site is composed of four amino acids that chelate the two Mg^{2+} ions (D443, E478, D498, and D549). Thus, both of the enzymatic functions of RT are in p66; p51 appears to play a structural role (53, 54). We are fortunate not only to have structures of HIV-1 RT as an unliganded protein (59, 60, 61), but also structures in which RT has bound dsDNA (54, 62) or RNA/DNA hybrids (63, 64). We also have structures in which RT has bound dsDNA and an incoming dNTP or a dNTP analog (65, 66, 67, 68), and numerous structures of RT bound to non-nucleoside RT inhibitors (NNRTIs), only some of which are cited here (53, 69, 70, 71, 72, 73, 74, 75). Although there are still some modest holes in the overall structural picture (it would, for example, be nice to have a structure of HIV-1 RT bound to an RNA/RNA duplex, particularly one that mimicked the initiation of minus-strand DNA synthesis), there are enough different structures available that we have a good idea of the structural changes that take place during the binding of a nucleic acid substrate and the incorporation of a dNTP.

One of the important themes of reverse transcription is that some of the structural elements of HIV-1 RT undergo considerable movement during the polymerase reaction. These movements are an integral part of polymerization: RT really is a small molecular machine. In the unliganded state, the thumb of p66 is closed and reaches over to touch the fingers (Fig. 3). Binding of a template/primer is accompanied by a major movement of the thumb; this allows the double-stranded nucleic acid access to a long groove in the protein where it binds. The double-stranded nucleic acid to which RT binds is relatively long, 17 to 18 bp (54, 63, 65). Binding bends the double-stranded nucleic acid substrate by approximately 40°. This bend involves a segment of the dsDNA several base pairs long and occurs near the base of the thumb of p66. In contrast to most DNA polymerases, which only have to copy DNA, RTs must be able to copy both a DNA and an RNA template. dsDNA can exist as either an A-form or a B-form duplex, although the B form is usually the preferred structure in solution. Because of the 2′-OH, dsRNA cannot exist as a B-form duplex. RNA/DNA duplexes often form a sort of hybrid structure, intermediate between A and B, called H form. RT appears to have resolved the problem of dealing with more than one type of double-stranded nucleic acid substrate by constraining DNA/DNA duplexes in an A-form configuration near the polymerase active site. The transition from primarily A form to B form occurs at the bend in the DNA near the base of the p66 thumb.

HIV-1 RT prefers to bind a nucleic acid substrate with a 5′ extension that can serve as a template for the addition of dNTPs; this type of substrate is preferentially bound in a fashion that places the 3′ end of the strand that will be extended (the primer stand) immediately adjacent to the polymerase active site, in what is called the P or priming site. In the absence of an incoming dNTP, the fingers of p66 are open (54). Binding an incoming dNTP at the active site (also called the N or nucleoside triphosphate-binding site) induces the fingers to close; this closing of the fingers upon dNTP binding is a feature that HIV-1 RT shares with other DNA polymerases (65, 76). The closing of the fingers sets the stage for the chemical step in which the dNMP portion of the incoming dNTP is incorporated into DNA and pyrophosphate is released. This is essentially an energy-neutral reaction; the total number of high-energy phosphate bonds is unchanged, and the polymerization reaction can be run backwards *in vitro* in the presence of an appropriate nucleic acid substrate and sufficient pyrophosphate. It is the release and degradation of the pyrophosphate that drives the polymerization reaction. Once pyrophosphate is released, the fingers of p66 open, and the end of the primer, now temporarily in the N site, can move back to the P site, in a process called translocation. This permits the binding and incorporation of the next incoming dNTP, allowing the polymerase of HIV-1 RT to carry out multiple rounds of dNTP incorporation without releasing the nucleic acid substrate. However, *in vitro*, HIV-1 RT is only modestly processive, usually copying at most a few hundred nucleotides before it falls off the nucleic acid substrate.

It is important to remember that the reactions catalyzed by DNA polymerases depend directly on the sequence of the template that is being copied. As will be discussed in more detail later, RTs carry out this process with considerable precision. Thus, the active site, in the sense that it is the binding site for an incoming dNTP, is composed of both protein and nucleic acid. For the copying reaction to be faithful, the nucleic acid must be held both tightly and precisely by RT. As might be expected, both the polymerase domain and the RNase H domain make extensive contacts with their nucleic acid substrates. Elements have been identified in the polymerase domain (for example, the primer grip and the template grip) that play important roles in positioning the nucleic acid substrate precisely and appropriately (54). However, it is also important to point out that elements of both the RNase H domain and the polymerase domain that are involved in binding the nucleic acid substrates help to position these substrates accurately at the polymerase active site. Conversely, as will be discussed briefly in the next section, the nucleic acid-binding elements in both domains are needed to position the nucleic acid substrate appropriately for proper RNase H cleavage (39, 40, 77, 78, 79, 80, 81, 82, 83).

The structural basis of RNase H activity is somewhat less clear, in part because we have only two published structures of HIV-1 RT in a complex with an

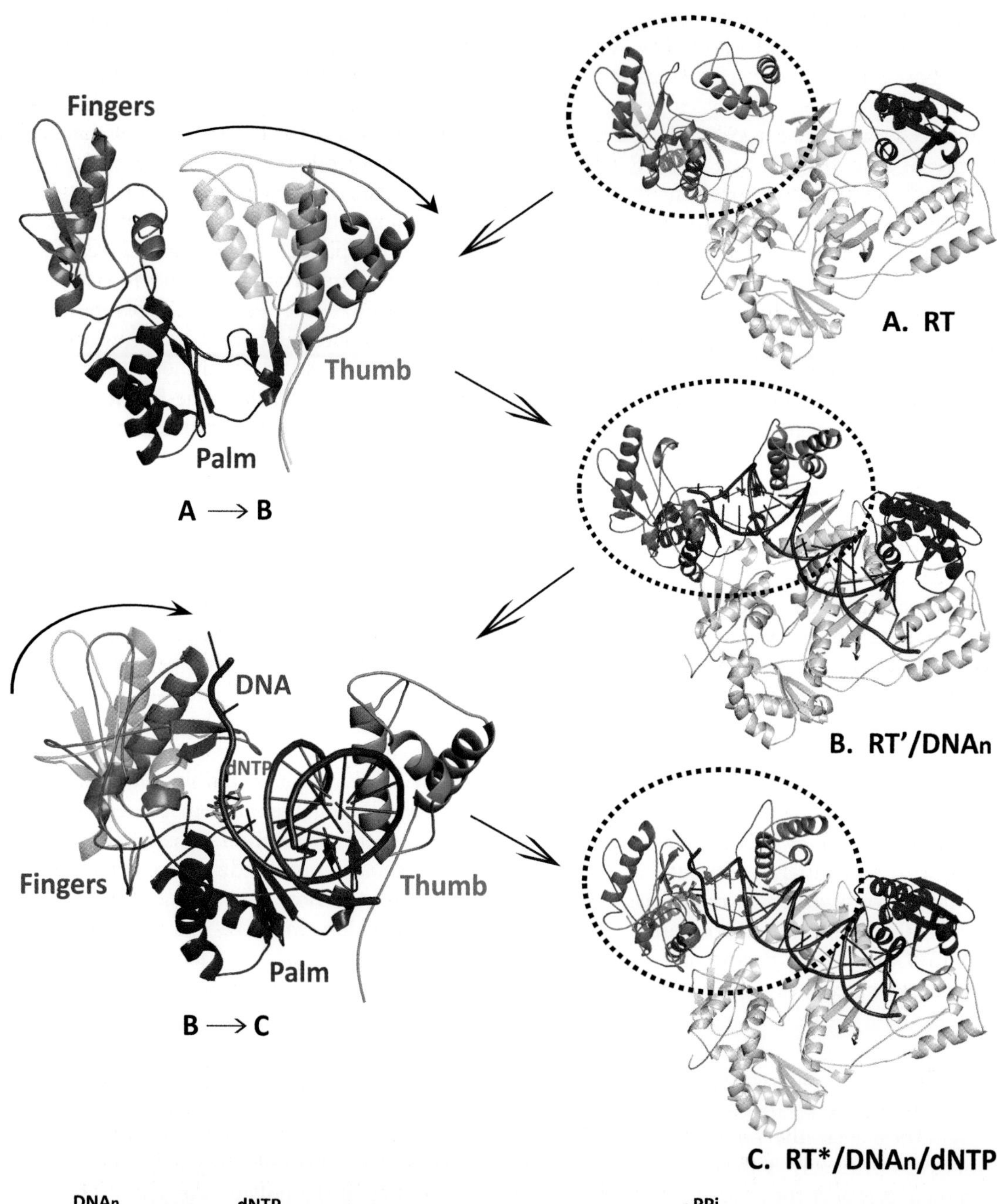

RT $\underset{}{\overset{DNA_n}{\rightleftarrows}}$ RT'/DNA$_n$ $\overset{dNTP}{\rightleftarrows}$ RT'/DNA$_n$/dNTP $\rightleftarrows$ RT*/DNA$_n$/dNTP $\overset{-\,PPi}{\rightleftarrows}$ RT'/DNA$_{n+1}$ $\rightleftarrows$ RT + DNA

RT	RT'/DNAn	RT'/DNAn/dNTP	RT*/DNAn/dNTP	RT'/DNAn+1
Thumb Closed	Thumb Open	Fingers Open	Fingers Closed	Catalysis, Translocation

RNA/DNA duplex (63, 64). When RT binds a dsDNA duplex, although the RNase H active site is near the template strand (Fig. 2), the active site is not close enough to the template strand for cleavage to take place, even if the template strand was RNA. This separation of the DNA template from the RNase H active site may be important because, strictly speaking, RNase H is not an RNase but a nuclease (84, 85), and it could conceivably cleave a DNA template strand if it contacted the RNase H active site. However, in neither of the structures in which RT is bound to RNA/DNA is there close contact between the active site of RNase H and the RNA strand. In the older structure, the RNA sequence in the nucleic acid duplex was derived from the PPT, and the alignment of the sequence with the enzyme places the RNase H active site over a portion of the PPT where RNase H does not cleave (63). The overall trajectory of the RNA/DNA duplex as it passes through RT is remarkably similar to the overall trajectory of a bound dsDNA whose sequence is unrelated. In particular, there is a similar bend in both nucleic acid duplexes near where they pass the thumb of p66. There is an unusual feature of the PPT RNA/DNA RT structure: in the middle of the nucleic acid duplex there is an unpaired base, two mispaired bases, and then a second unpaired base that brings the duplex back into proper register. There has been considerable speculation over whether this unusual mispaired feature is related to the sequence of the PPT, whether (and to what extent) the mispairing is induced by binding to RT, and whether the mispairing is related to the fact that RNase H cannot cleave within the PPT. We do not have clear answers to any of these questions; however, there are experiments that suggest that portions of the PPT are only weakly base paired when it is not bound to RT, so that binding to RT could easily create (or enhance) the tendency of this segment to be mispaired (41, 42).

As mentioned earlier, it is still unclear exactly how the RNase H of RT is able to make the cleavages that generate and remove the PPT primer (or, for that matter, remove the tRNA primer) with single-nucleotide precision. Although the cleavage site that is chosen is sequence dependent in the sense that, if the sequence is changed, the cleavage site(s) also changes, it appears that this is due to the structure of the PPT when it is bound to RT rather than to the sequence *per se*. Similarly, sites have been identified in RT, particularly a set of amino acids called the RNase H primer grip, that interact with the DNA strand of an RNA/DNA duplex and are needed to position the RNA strand properly for cleavage (39, 40). As has already been discussed, in thinking about the proper positioning of the RNA/DNA duplex, contacts in both the RNase H domain and the polymerase domain are important. As might be expected, it would appear that the most important contacts are in the RNase H domain. However, it is worth pointing out that if the RNase H domain of HIV-1 RT is expressed separately from the polymerase domain, it is unable to cleave an RNA/DNA heteroduplex, despite the fact that it is properly folded, because, by itself, the RNase H domain is not able to bind an RNA/DNA heteroduplex. In this sense, RT acts as an integrated whole: two domains work together to properly position the various nucleic acid substrates relative to the two active sites.

Although the more recent structure of RT bound to an RNA/DNA duplex did not involve a duplex whose sequence is based on the PPT, the RNA strand of this structure still does not make close contact with the RNase H active site (64). There are some significant differences in the structures of the two RT/RNA/DNA complexes, for example, there is no evidence of mispaired bases in the newer structure. However, the RNA strand is nicked in this structure and questions have

Figure 3 Movement of HIV-1 RT during polymerization. The colors used for the various subunits, domains, and subdomains of RT, and for the two DNA strands, are as in Fig. 2. The p51 subunit is gray. The RNase H domain is shown in pink, and the four subdomains to the polymerase domain are as follows: fingers, blue; thumb, green; palm, red; and connection, yellow. The DNA template is dark red and the primer strand is purple. The incoming dNTP is light blue. The structural changes in RT can be correlated with specific steps in the binding of the substrates and the incorporation of the incoming dNTP. In unliganded RT, the fingers and thumb are closed (A). Before the dsDNA (or other nucleic acid substrate) can be bound, the thumb must move (A → B); this allows the dsDNA to be bound (B). This sets the stage for the binding of the incoming dNTP. When the incoming dNTP binds, the fingers close, which allows the dNTP to be incorporated, with the release of pyrophosphate (PPi). The incorporation of the dNTP temporarily leaves the end of the primer stand in the N or nucleoside triphosphate-binding site (see text). Translocation moves the nucleic acid by 1 bp, which allows the next dNTP to be bound and incorporated.
doi:10.1128/microbiolspec.MDNA3-0027-2014.f3

been raised about whether the nick and/or crystal contacts may have affected the structure of this RT/RNA/DNA complex. Thinking about the fact that there are no structures in which a nucleic acid makes close contact with the active site of RNase H brings back the question, raised earlier, of whether an RT that is actively polymerizing can simultaneously carry out RNase H cleavage. As was mentioned previously, there is controversy in the literature, and the available data are based entirely on *in vitro* assays; however, the bulk of the data suggests that there is little, if any, cleavage of an RNA/DNA duplex in an *in vitro* reaction if RT is actively polymerizing (15, 16). If this really is the correct interpretation, and if RT must pause before RNase H can cleave, it is possible that there needs to be some repositioning of an RNA/DNA duplex to bring the RNA strand into close contact with the RNase H active site and, conversely, the nucleic acid, when it is properly positioned for RNase H cleavage, might not be properly positioned for polymerization.

There is a large and complicated volume of literature describing the effects of mutations in RT and in critical portions of the RNA genome (such as the PBS and the PPT) on both the polymerase and RNase H activities of RT *in vitro* and on the process of reverse transcription in infected cells. Most, but not all, of this literature is based on experiments done with purified recombinant HIV-1 RT and with HIV-1-based vectors. All of the obvious features of the enzyme have been mutated (for example, the amino acids in both of the active sites, amino acids that are involved in holding and positioning the nucleic acid in both the polymerase domain and the RNase H domain, and amino acids at the interface between p66 and p51). The good news is that, in general, the genetic and related biochemical data nicely match the structural data. Thus, we can conclude that the interpretations that have been made about what various parts of RT are doing based on where they are in the structure and whether they contact either the nucleic acid or the dNTP substrates are largely correct. There is, in general, relatively good agreement between the effects of mutations on the behavior of purified RT *in vitro* and their impact on reverse transcription in infected cells. However, as might be expected, there are also some mutations that appear to differentially affect the behavior of purified RT and virus replication. In many of the cases where there appears to be a disagreement, the mutations have a more profound effect on virus replication than on the activity of purified RT.

One simple example involves mutations that affect the ability of RT to be degraded by the protease in virions (86, 87, 88, 89, 90, 91). Many of these RT mutants are temperature sensitive, both in the sense that the extent to which RT is degraded is much greater at the temperature at which the virus normally grows, 37.5°C, than at 32°C, and in the sense that, if the mutant RTs are prepared as purified recombinant proteins, they melt at a lower temperature than wild-type RT. Basically, these temperature-sensitive mutations allow the RT to partially unfold at 37.5°C, and the partially unfolded RTs are susceptible to cleavage by the viral protease. Mutations that confer this phenotype are not difficult to generate; however, as expected, they are, for the most part, mutations that are not usually found in viruses isolated from patients. There is one interesting exception: a mutation, G190E, that confers resistance to NNRTIs. In a standard genetic background, the G190E mutation leads to extensive degradation of RT (88). However, G190E mutants have been found in a small number of patients; in these viruses, the G190E mutation is accompanied by several additional mutations that allow the mutant RT to escape degradation. Taken together, the data suggest that the evolutionary space that is available to RT is constrained not only by the need for the enzyme to be able to carry out its appropriate enzymatic functions efficiently but also by a requirement that RT needs to survive in virions in the presence of the viral protease.

REVERSE TRANSCRIPTION IN CELLS

By the standards of host DNA polymerases, the polymerase of HIV RT is relatively slow. Completing the minus-strand DNA in an infected cell, which involves RT extending a single 3′ end, takes several hours, and the rate of nucleotide addition during minus-strand synthesis in cells has been estimated to be approximately 70 nucleotides per minute (92). There is evidence that it takes less time to complete plus-strand DNA synthesis; the most obvious explanation is that the plus-strand DNA is made in segments. If there are multiple 3′ ends, several RTs can collaborate to complete the task of plus-strand DNA synthesis and the job can be completed more quickly.

Although it is convenient to study reverse transcription *in vitro*, using purified recombinant RT, this approach is a considerable simplification relative to what happens in cells. Reverse transcription is initiated in the cytoplasm, within a complex structure, usually called the reverse transcription complex (RTC), which is comprised of RT and other viral proteins (93, 94, 95, 96). In addition, there are host proteins that can affect the process of reverse transcription and the outcome. At least initially, the RTC has an outer shell, called the capsid; it is composed of a single protein, also called capsid, or CA, that was derived from Gag by protease cleavage. In the literature, there is much speculation about what happens to the capsid shell during reverse transcription, and there is evidence that the shell is modified during reverse transcription (94, 97, 98). Much of the discussion of what happens to the RTC during reverse transcription involves a process that is called "uncoating." This name is misleading; there is growing evidence that the capsid shell is largely intact for most or all of the process of reverse transcription, although it appears to be remodeled, either during reverse transcription or perhaps by the process of reverse transcription.

Both the foamy retroviruses and the hepadnaviruses carry out reverse transcription in producer cells, rather than in newly infected cells (45, 99, 100). Thus, for these viruses, it is a dsDNA copy of the viral genome that is inside the virion and inside the capsid shell. This shows that the capsids of these viruses can accommodate a complete, or nearly complete, dsDNA copy of their viral genomes inside their capsid shells, and that, for these viruses reverse transcription is either entirely, or almost entirely, cytoplasmic. The total amount of DNA (number of nucleotides) that is present following reverse transcription is only slightly larger than the total amount of ssRNA in the two strands of genomic RNA that were initially packaged. If, when the plus-strand DNA is synthesized, it is segmented, then the dsDNA could be folded into a relatively compact structure. It has recently been shown that the capsid of some retroviruses protects their viral DNA from host sensors that are part of the innate immune system (101). These sensors are designed to detect the presence of DNA in the cytoplasm; the fact that the capsid protects (or at least partially protects) the viral DNA from these sensors suggests that the capsid remains intact until relatively late in the reverse transcription process. Finally, and probably most importantly, the capsid shell serves to keep the proteins that are needed to carry out both the reverse transcription and integration together with their nucleic acid substrates. Structural data suggests that the mature capsids of retroviruses have holes in them that are large enough to permit dNTPs to enter and pyrophosphate to leave but small enough to prevent the loss of proteins (102, 103). Not surprisingly, there are a number of mutations in the capsid protein that completely block reverse transcription. In particular, mutations in the capsid protein that disrupt the structure of the mature capsid shell block reverse transcription, even if the virions contain a full complement of the other proteins (including active RT) needed to carry out the process (94, 104). Capsid mutants that do not form a normal capsid shell that properly encapsidates the genomic RNA, nucleocapsid (NC), RT, and integrase (IN) fail to initiate reverse transcription, presumably because the necessary components fail to remain associated in the cytoplasm of the newly infected cell. In thinking about the role of the capsid shell, it should be remembered that the RTCs of most retroviruses and retrotransposons contain a fairly large number of copies of RT and IN (estimated to be 50 to 100 for HIV-1). In the case of HIV, reducing the number of enzymatically active copies of RT more than 2- to 3-fold dramatically reduces the ability of the virus to complete the process of reverse transcription (14). This result illustrates the need to retain most or all of the RTs (and, by extension, the other protein components) that were present inside the capsid shell when the particle was formed.

There are other capsid mutants that are able to form what appears to be a reasonably normal capsid shell but a shell that appears to be either too unstable or too stable; both types of mutant can affect the process of reverse transcription (94). Either during the process of reverse transcription or concomitant with the completion of reverse transcription, the RTC is converted into the preintegration complex (PIC) (94, 97, 98, 101, 105). The exact composition of the PIC is not well understood, and whether the PICs of different retroviruses and retrotransposons are or are not different is also unclear. Retroviruses in which reverse transcription is carried out in the infected cell can complete the process in the cytoplasm; fully competent PICs can be isolated from the cytoplasm of infected cells (106, 107). However, it is not clear that reverse transcription must always be completed in the cytoplasm. In the end, the type and the state of the infected cell could play a critical role in determining whether any reverse transcription takes place in the nucleus. In addition, for those retroviruses/retrotransposons that synthesize their plus-strand DNA in multiple segments (25, 26, 108), it is likely that the final filling in and ligation of a complete, intact plus-strand takes place in the nucleus, and is carried out by host DNA-repair enzymes.

In addition to RT and capsid, the other viral protein that is known to play a major role in reverse transcription is the NC. Like CA, the NC is synthesized as part of the Gag polyprotein; in most cases,the processing of Gag by protease separates the CA and NC. However, the processing of the Gag protein of foamy viruses, while essential for virus replication, removes only a small piece from the C terminus, and, in foamy viruses, the NC equivalent is part of the capsid (109, 110). NC plays the role of a nucleic acid chaperone, facilitating the formation of low-energy RNA and DNA structures. Because NC (in the context of Gag) has an important role in the packaging of genomic RNA, it has not been a simple matter to determine exactly what NC does during reverse transcription in cells. *In vitro*, NC can, under the right circumstances, help RT get though regions of secondary structure, facilitate transfers between templates, and help prevent "turnaround" DNA synthesis in which the 3′ end of a ssDNA folds back on itself, leading to the synthesis of a DNA hairpin (111, 112, 113, 114, 115, 116, 117, 118).

There are several other viral proteins that have been proposed to have a role in reverse transcription; however, for these other proteins, the data are less compelling. There is no question that a significant fraction of the IN mutants that have been analyzed have defects in reverse transcription, and there are IN mutants that have a profound negative impact on reverse transcription in both HIV and retrotransposons (119, 120, 121, 122, 123, 124, 125). There are claims that an association between RT and IN is needed for reverse transcription. There are, however, no clear data showing that IN helps the process of reverse transcription or the enzymatic activities of RT *in vitro*. It is likely that the IN mutants have an indirect effect on reverse transcription. At least in the case of HIV-1, the IN mutations that block reverse transcription also affect the structure of the capsid shell. It is unclear why or how IN is involved in forming a proper capsid shell; however, it does appear that the disruption of the capsid shell is the root cause of the failure of these IN mutants to carry out reverse transcription. The mutant virions contain normal amounts of properly processed viral proteins, including RT, and if the virions are broken open *in vitro*, the RT is fully active. However, as judged by electron microscopy images of the virions that carry these IN mutants, the capsid shell does not enclose the RNA. It is likely that this leads to a dissociation of RT and the RNA early in infection; the IN mutants that have defective capsid shells do not initiate reverse transcription and thus resemble, in their phenotypic behavior, the CA mutants that fail to form a proper capsid shell. This interpretation supports the idea that an intact capsid shell, which surrounds the RNA, helps the RTC retain RT and NC, making it essential for reverse transcription. There have also been claims that other viral proteins, for example Vpr and Tat, can affect the activity of RT *in vitro*; however, it is not clear that these proteins play an important role in reverse transcription in infected cells (126, 127, 128, 129, 130, 131). These factors are present in only a small subset of retroviruses, and it seems, for that reason, unlikely that they play a fundamental role in reverse transcription.

There are host factors that have been reported to help reverse transcription. The best example is dUTPase. Some retroviruses encode their own dUTPases (132, 133, 134); others appear to use Vpr to interact with host dUTPases (135, 136, 137, 138). However, the use of a dUTPase of either host or viral origin is not universal, and it is unclear why some retroviruses make (or steal) them and others do not. However, the fact that some retroviruses make use of host or viral dUTPases does suggest that these viruses are using these enzymes to increase the fidelity of their replication, and proposals that the replication of the viruses is deliberately sloppy are probably incorrect. This issue will be considered later, in the section devoted to fidelity. It has also been claimed that, for HIV, the host lysyl aminoacyl-tRNA synthetase plays a role in the selective packaging of the correct tRNA primer (139, 140); however, HIV can replicate reasonably efficiently using other tRNA primers (141, 142, 143). It has also been proposed that an RNA debranching enzyme is involved in the first-strand transfer during reverse transcription (144, 145); this claim should also be viewed with some caution, at least until more and better evidence has been presented.

There are host proteins that can negatively affect reverse transcription. The well-characterized negative host factors appear to be involved in host defense: the host would like to protect itself both from retroviral infections and from uncontrolled replication of retrotransposons. Three of the best-known host defense factors, Trim5a, TrimCyp, and FV-1, appear to affect the stability of the capsid shell in the cytoplasm (146, 147, 148, 149, 150, 151, 152, 153, 154, 155, 156). This reprises the theme that has been touched on several times already, that the capsid shell needs to pass the Goldilocks test: not too unstable or it will fall apart and the proteins, including RT, that are needed to complete reverse transcription (and subsequently, integration) will be lost; but not too stable or there will be problems, less well defined but possibly related to the completion of reverse transcription and/or the conversion of the RTC into a PIC. Another well-characterized host factor known to have a significant negative impact on the replication of some retrotransposons and retroviruses that acts at the stage of reverse transcription is SAMHD1. SAMHD1, which is found, in an active form, in some non-dividing cells, catalyzes the conversion of dNTPs to their corresponding nucleosides and free triphosphate, and it can reduce the levels of dNTPs to a point where the reverse transcription of retroviruses is blocked (157, 158, 159, 160, 161, 162). Although SAMHD1 was discovered because it can block the replication of HIV-1 in some cultured cells and because HIV-2 makes a protein that counteracts it, based on the phenotype of humans who fail to make SAMHD1, its primary role may be to help control the level of replication of retrotransposons and retroposons (163, 164).

The other well-defined group of host proteins that negatively affects reverse transcription is the APOBECs. These are cytosine deaminases, and the members of the family that affect reverse transcription act on dC residues in a ssDNA template, converting dC to dU (165, 166, 167, 168, 169, 170). There are a number of credible reports that some APOBECs can affect reverse transcription (and apparently integration), even in the absence of an ability to deaminate dC (171, 172, 173, 74). Exactly how this is accomplished is unclear, but the APOBECs bind single-stranded nucleic acids, and it is possible that binding alone is sufficient, at least in some cases, to interfere with reverse transcription and/or integration. However, it is also clear that the conversion of dC to dU is an important part of the antiviral/anti-retrotransposon effects of at least some of the APOBECs. To be active they (usually) need to be packaged into the virion in the producer cell. The antiviral/antiretrotransposon APOBECs are DNA specific: they act primarily on minus-strand DNA after the RNA template has been removed by cleavage by RNase H but before the plus-strand DNA has been synthesized. As was mentioned briefly in a previous section, this could help to explain the role of the central PPT in HIV and perhaps in other retroviruses and retrotransposons. The presence of the central PPT reduces the time that the first DNA strand is single stranded, thus reducing the susceptibility of the virus to APOBECs (33, 34).

The primary signature of the APOBECs is the conversion, in the newly synthesized plus-strand DNA, and subsequently in the RNA, of multiple G residues to As. The various APOBECs have slightly different preferences for the sequences adjacent to the dC residues they deaminate; the preference of the best studied of the antiviral APOBECs, APOBEC3G, causes it to have a tendency to generate termination codons. The fact that, in most cases, APOBECs are only able to modify the genomes of retroviruses if they are expressed in the producer cells and packaged into virions suggests that the capsid shell protects the viral DNA not only from cytoplasmic host-cell DNA sensors in a newly infected cell but also from any APOBEC proteins that might be present. This provides one more piece of evidence that the capsid shell remains largely intact, at least until the RNA genome has been converted into dsDNA. There are a modest number of exceptions to this general rule, and there are some circumstances (or perhaps more accurately, there are some cells) in which APOBEC proteins that are present in a newly infected cell are able to modify the viral genome. As intriguing as these examples are, they are exceptions, and it is not clear how they fit into the broad picture in which most of reverse transcription takes place inside the capsid shell.

FIDELITY

A number of factors contribute to the complex array of mutations that are seen in populations of retroviruses and retrotransposons. Some endogenous elements spend much of their existence hiding in the host genome and, if they are expressed at a high level, they can be deleterious to the host (163, 164). For endogenous viruses and retrotransposons that are primarily transmitted vertically, the position at which the element is integrated can be a major factor in determining whether it will be selected against; in such cases, mutations in the genome can be fixed because the element occupies a fortuitously favorable integration site, rather than because the mutations the element carries help it replicate efficiently. Conversely, for viruses that are transmitted horizontally and replicate rapidly, like HIV, the replicative fitness of the virus is quite important. HIV-1 infections are often initiated by a single virion (175); however, in an untreated patient, within a few years, the virus population has diverged to the point where no two viral RNAs isolated from the blood have the same sequence. This rapid diversification is due to the fact that virus replicates rapidly, and to the relatively large population of replicating virus (176). The viruses that are seen in the blood are those in which mutations have had, in general, a minimal impact on the overall replicative fitness of the virus. In addition, there is selection, in patients, for variants that reduce the immunological susceptibility of the virus. However, all of the variation that is seen in these virus populations derives from errors that are made during replication. Three enzymes participate in the replication of the genomes of retroviruses and retrotransposons: the host RNA Pol II, RT, and the host's replicative DNA polymerase. The degree to which the host DNA polymerase makes a significant contribution depends on the life-style of the element in question. For actively replicating viruses, like HIV-1, the contribution of the host DNA polymerase, which, in combination with its ancillary proofreading machinery, has a very low error rate, is negligible. However, for some endogenous retroviruses and retrotransposons that are vertically transmitted and are passively replicated for many generations, the host DNA polymerase is the major source of errors. Because the process of reverse transcription creates two LTRs that have identical sequences when they are first synthesized, comparing the sequences of the two LTRs of ancient proviruses/retrotransposons is an important tool in molecular paleontology.

Although HIV-1 RT can inefficiently excise the nucleotide analogs that are used to treat HIV infections after they have been incorporated into viral DNA (discussed briefly later in this chapter), and there are RT mutants that are significantly better at excising these analogs than wild-type RT (177, 178), these mutations do not appear to improve the fidelity of HIV replication, and it appears that RT has no editing function that improves its fidelity (179). In contrast, RNA Pol II is able to back up and excise a misincorporated nucleotide (180). This, coupled with the knowledge that HIV-1 RT (and most other retroviral RTs) are relatively error prone *in vitro*, having an error rate of approximately 10^{-4}, has led to widespread claims that the errors made during virus replication can be attributed to RT. There are several potential problems with this simple and convenient picture. First, the actual error rate of RNA Pol II has not been well defined, and its contribution to the overall errors made during the active replication of retroviruses and retrotransposons is not known (181). Secondly, the overall error rate of retrovirus replication is about 10^{-5}, a much lower error rate than has been measured for HIV-1 RT *in vitro* (182, 183, 184, 185). This suggests that, in the environment of the RTC and perhaps because of the participation of other viral and/or host proteins, reverse transcription is a higher-fidelity process in an infected cell than it is *in vitro*. If the overall error rate is about 10^{-5}, then the fidelity of both RNA Pol II and RT cannot be less than 10^{-5}; however, we still do not know which of the two enzymes makes the majority of the errors, or if they make roughly equal contributions. In thinking about the problem, it is important to remember that the flow of information is from RNA Pol II to RT. For this reason, RT cannot be selected to have a higher fidelity, in terms of copying information, than RNA Pol II. If we assume that most retroviruses and retrotransposons are trying to copy their genetic information faithfully, an idea that is supported by the fact that some viruses encode dUTPases, and others make use of host-cell dUTPases, and that retroviruses carry proteins that counteract the APOBEC proteins, then the fidelity of RT would be limited by the fidelity of RNA Pol II, and the fidelity of the two polymerases would be similar.

There are, however, additional layers of complexity. Although it is convenient to think about fidelity in terms of an overall error rate, this does not provide a clear picture of the errors that arise during one round of replication. In the first place, not all substitutions are equally likely. As would be expected, at least for HIV-1, the best-studied system, missense mutations are more frequent than frameshift mutations, and the majority of the missense mutations are transitions rather than transversions (184, 185). In the case of HIV, the most common transition (in the plus-strand) is G to A. If this preference for making G to A mutations is largely due to errors made by RT, which is not clear, it probably reflects a tendency of RT to insert a T opposite a G during first-strand DNA synthesis. (These experiments were done in cells that do not express significant levels of the APOBEC proteins that affect HIV replication, and essentially all of the G-to-A mutations were made in the wrong sequence context to be APOBEC mutations.) However, the mutations were not uniformly distributed but preferentially occurred at specific sites or hotspots. In some cases, exactly the same target sequences were used in the *in vitro* assays done with purified RT and the virus replication assays; however, the hotspots that were reported *in vitro* do not match those seen during virus replication (186, 187, 188, 189). To make matters worse, the hotspots seen in experiments done by different laboratories using purified RT to copy the same template do not match. This result, taken together with the fact, already mentioned, that the fidelity of RT, as measured *in vitro*, is at least 10-fold too low, should serve as a warning to those who are trying to understand fidelity as it applies to the replication of retroviruses/retrotransposons using *in vitro* fidelity data that was obtained with purified RT. Not surprisingly, all of the problems that have just been discussed apply with equal or greater force to the very large volume of literature that describes the impact of mutations in RT on fidelity.

As might be expected, mutations in RT affect its fidelity, measured in *in vitro* assays, and the fidelity of the replication of the parental virus. Here too, most of the experiments have been done with HIV-1, and HIV-1 RT, although there are some data for some other retroviruses and their respective RTs. This is a large and complicated literature, and it is complicated, at least in part, because there have been a large number of assays that have been used to try to measure fidelity *in vitro*. As was the case with the experiments done with wild-type RT and a vector that replicates using wild-type RT, there is relatively little agreement between the *in vitro* data obtained with purified RT and the corresponding data, obtained with the same RT mutants, using a viral vector (179, 185, 186, 187, 188, 189, 190, 191, 192, 193, 194, 195, 196, 197, 198, 199, 200, 202). There are several simple conclusions. Although only a small number of RT mutants have been tested in the context of virus replication, mutations that have been reported to increase the fidelity of RT *in vitro* caused a small to moderate increase in the overall error rate in virus replication assays (185). Some of the mutations in RT affected the relative proportion of missense and frameshift errors. In addition, mutations in RT that did not cause a substantial increase in the overall error rate still caused substantial increases in the errors made at specific positions in a target gene. This result demonstrates that the overall error rate is not, in many cases, a good measure of the impact of a mutation in RT on the errors made during virus replication. In particular, changes in RT that have little effect on the overall error rate can still increase the chances that mutations will arise that can allow the virus to escape from immunological surveillance and/or develop resistance to particular drugs.

RECOMBINATION

As has already been discussed briefly, there are, during minus-strand DNA synthesis, frequent transfers between the two RNA templates (203, 204, 205, 206, 207, 208, 209, 210, 211, 212). Most or all of these transfers are probably caused by nicks in the genomic RNAs (213). By switching DNA synthesis to the other RNA strand, RT can generate an intact DNA copy of the genome from two nicked RNA templates, so long as both of the RNA strands are not nicked at the same place. The strand transfers that lead to recombination are similar to the first-strand transfer reaction, which has been described in an earlier section of this review. If, as is often the case, the two RNA copies are identical, these multiple strand transfers have no genetic consequences. However, if there are, in the same host cell, DNAs that encode two variants of the same retrovirus or retrotransposon, particles can be produced that contain two related RNA genomes. In such cases, strand transfers that occur during first-strand DNA synthesis will produce a recombinant genome (203). This can have important consequences, if, for example, recombination involves RNAs from two HIV-1s that carry different drug-resistance mutations. Because recombination can only occur if the two parental RNAs are copackaged into a single particle, recombination between different but fairly closely related retroviruses or retrotransposons is rare; however, it does happen. For example, HIVcpz, the chimpanzee retrovirus that gave rise in humans to HIV-1, is a recombinant between two different simian immunodeficiency viruses, one of which is normally found in red-capped mangabeys, and the other in a *Cercopithecus* monkey (214).

It appears that there are at least three factors that can affect the generation of recombinant retroviruses and retrotransposons: (i) the copackaging of the RNAs; (ii) the similarity of the genomic sequences, which is required for the transfer of DNA synthesis between the templates; and (iii) the viability of any recombinant virus that emerges. The packaging of the retroviral RNA genome is a complex topic, the details of which lie outside of the scope of this review. Briefly, there is a structured segment of the RNA near the 5′ end of the genome, called Ψ, which is important for packaging. The RNA dimer forms before it is packaged, and there is a secondary condensation step that happens after the RNA dimer is packaged (215). In the case of HIV, a kissing loop, called the dimer initiation signal (DIS), is a critical component that has a significant role in the formation of the RNA dimer (216, 217, 218). Even within HIV strains, there is variation in the sequence of the DIS (219, 220), and the compatibility of the DIS (the ability to form a kissing-loop dimer) affects RNA dimerization and, by extension, recombination. However, sequences outside the DIS also affect the ability of two RNA genomes to dimerize and be copackaged (221). Although the compatibility of the DIS affects the ability of two genomic RNAs to be copackaged, the similarity of the sequences of the two RNA genomes will affect the efficiency of the strand transfer. Perhaps even more important, in terms of the ability of any recombinant viruses to grow out, is whether the proteins encoded by the recombinant viruses are fully functional and are able to work together. Studies with artificial HIV/MLV recombinants show that even carefully constructed recombinants of such distantly related viruses replicate very poorly. Even recombinants between different strains of HIV-1 often replicate much less efficiently than the parental viruses.

Not all transfers between templates are precise. Imprecise transfers can create either an insertion or a deletion. It is likely that most of these deletions and duplications are created by nonhomologous strand transfers during first-strand DNA synthesis. In many but not all cases, there are small regions of homology at the site of duplications/deletions. In experiments done with retroviral vectors that carried sizable duplications, the duplications were rapidly lost if the vector was allowed to replicate, and the rate at which duplications were lost was related to both the length of the duplicated sequences and how far they were separated in the genome of the vector (222, 223). There are two RNA targets at which a nonhomologous strand transfer would create a deletion; in contrast, there is only one RNA target at which a strand transfer could create a duplication. In addition, it is likely that many of the erroneous strand transfers are the result of having a break in both RNA strands at the same, or almost exactly the same, site. In such cases, homologous strand transfers are not possible. Thus, any strand transfer that can occur will create either a deletion or a duplication. However, if the breaks are at the same site in the two RNAs, a strand transfer that creates a duplication will not get the growing DNA strand past the breaks in the two RNAs. Conversely, a strand transfer that bypasses the breaks in both RNA strands will, of necessity, create a deletion. These constraints, taken together, nicely account for the large excess of deletions that are seen relative to duplications.

RT DRUGS AND DRUG RESISTANCE

AZT, which inhibits reverse transcription, was the first successful anti-HIV drug; now, some 25 years later, drugs that block reverse transcription are still fundamentally important and widely used components of successful combination anti-HIV therapy. There are, broadly speaking, two major classes of drugs that block reverse transcription, nucleoside analogs (NRTIs) and nonnucleoside RT inhibitors (NNRTIs). NRTIs are, as their name implies, analogs of the normal nucleosides that are used to synthesize DNA; however, all of the NRTIs that have been approved for human use lack a 3′-OH, and, when incorporated into viral DNA by RT, act as chain terminators, blocking viral DNA synthesis. In the strictest sense, NRTIs are not RT inhibitors; their true target is the viral DNA. In contrast, NNRTIs are RT inhibitors; they do not interfere with the ability of RT to bind its substrates but block the chemical step of DNA synthesis (224, 225). With one important exception, NRTIs are given to patients as free nucleosides, and are converted to the triphosphate form in cells in the patient. Tenofovir disoproxil fumarate is the exception; it is given as a monophosphate prodrug. Tenofovir disoproxil fumarate can be taken up by cells because, in the prodrug form, the phosphate is masked by two hydrophobic modifications. In general, NRTIs tend to have toxic effects, particularly after long-term use; a significant part of their toxicity comes from the fact that the triphosphate forms of the NRTIs can, to a greater or lesser extent, be used by host DNA polymerases; the host gamma DNA polymerase found in mitochondria appears to be particularly vulnerable (226, 227).

The other class of approved RT inhibitors, NNRTIs, is a diverse group of hydrophobic compounds that bind in a pocket in the p66 palm of HIV-1 RT, about 10 Å away from the polymerase active site (53, 59, 71, 73). The NNRTI-binding pocket does not exist in unliganded HIV-1 RT, and the binding of an NNRTI alters the structure of RT (60, 61). A bound NNRTI pushes up the palm in a region of the polymerase domain that underlies the cleft where the nucleic acid binds. This, in turn, pushes the end of the nucleic acid substrate upward, moving it away from the polymerase active site (228). In contrast to NRTIs, NNRTIs are quite selective; not only do they not block the activity of host DNA polymerases, they do not in general bind to or block polymerization by other retroviral RTs, including HIV-2 RT. This is both good, in the sense that, for the most part, NNRTIs are relatively non-toxic; and bad, because the first-generation NNRTIs were very susceptible to mutations that made the virus drug resistant. To cite a specific example, attempting to block mother-to-child transmission of HIV at birth by giving the mother a single dose of the first-generation NNRTI nevirapine led to the development of resistance in the majority of treated women (229, 230, 231). Fortunately, it has been possible to develop more advanced NNRTIs that are less susceptible to the development of HIV resistance (232, 233).

Resistance is a major problem in HIV drug therapy. All of the available anti-HIV drugs, including all of the NRTIs and NNRTIs, can, under the proper circumstances, select for resistance mutations that substantially reduce the efficacy of the drugs. Resistance to NRTIs implies that the mutant form of RT, which must still be able to synthesize viral DNA, has an increased ability to discriminate between the NRTI and the corresponding normal nucleoside compared with the wild-type RT. There are, broadly speaking, two mechanisms of NRTI resistance. The first, exclusion, is typified by the M184I/V mutations that reduce the susceptibility of the virus to 2′,3′-dideoxy-3′-thiacytidine (3TC or lamivudine) and 2′,3′-dideoxy-5-fluoro-3′-thiacytidine (FTC or emtricitabine) (234, 235). As the name implies, exclusion mutations reduce the ability of the mutant RT to bind and/or incorporate the triphosphate form of the affected NRTI. 3TC and FTC differ from a normal nucleoside both because their pseudosugar ring contains sulfur, and also because the ring is in the L rather than in the normal D configuration. These differences mean that when the triphosphates of these analogs bind at the polymerase active site (N site), a portion of the pseudosugar ring is closer to position 184 than is the case for a normal dNTP. There is room for the analogs to bind and be incorporated if the amino acid at position 184 is M, but there is a steric clash, caused by the β-branch on the side chain of the amino acid, if either I or V is present at this position. Thus, the M184I/V mutations allow RT to selectively discriminate against the incorporation of 3TC triphosphate and FTC triphosphate.

The second mechanism is more interesting, because it depends on the mutant form of HIV-1 RT developing a new ATP-binding site that allows it to excise the AZT monophosphate (and, if appropriate additional mutations are present, other NRTIs) that is blocking the end of the viral DNA after the analog has been incorporated (67, 177, 178, 236). This frees the blocked end of the viral DNA, allowing a complete copy of the genome to be synthesized. Basically, the excision reaction is related to pyrophosphorolysis, which is polymerization run backwards. As described earlier, the incorporation of a dNMP, with the concomitant release of pyrophosphate, is essentially energy neutral. Thus, the polymerization reaction can be run backwards *in vitro* in the presence of sufficient pyrophosphate. However, it is not possible, because of microscopic reversibility, to speed up an excision reaction that uses pyrophosphate without causing an equivalent increase in the incorporation of the AZT triphosphate. The AZT resistance mutations avoid this problem by creating a novel ATP-binding site: the excision reaction differs from the incorporation reaction because it uses ATP as the pyrophosphate donor. ATP-mediated excision creates a dinucleoside tetraphosphate, AZTppppA. The two primary AZT resistance mutations, T215F/Y and K70R, help create the new ATP-binding site: the F or Y at 215 makes π/π stacking interactions with the adenine ring of ATP, while the R at position 70 interacts with both the ribose ring and the α-phosphate of ATP (67). For this reason, these two mutations (and the other associated AZT resistance mutations) selectively enhance the binding of the ATP that participates in the excision reaction but have little or no effect on the binding or release of the pyrophosphate that is produced during normal polymerization. This means that these mutations selectively enhance the excision of AZT monophosphate without causing a concomitant increase in the incorporation of AZT triphosphate, and thus the mutations cause resistance.

In contrast to NRTI resistance, in which RT must be able to retain the ability to bind and incorporate normal dNTPs, developing resistance to NNRTIs is a simpler problem: almost any mutation that will interfere with NNRTI binding and not distort or disrupt the overall structure and function of RT will be acceptable. As might be expected, the polymerase active site is much more conserved and constrained than the NNRTI-binding site, and the spectrum of mutations that are acceptable to the virus is considerably greater at the NNRTI-binding site. Most of the common NNRTI-resistance mutations affect the binding of NNRTIs, either directly or indirectly. For example, mutations that replace the tyrosine at position 181 or 188 with a non-aromatic amino acid reduce the hydrophobic interactions with a number of NNRTIs (53, 237, 238, 239). There is an interesting exception: the K103N mutation creates a new hydrogen bond (with Y188). This bond helps to keep the entrance to the NNRTI-binding pocket closed; the K103N mutation makes it more difficult for NNRTIs to enter the pocket, and the K103N mutation confers resistance to a number of different NNRTIs (240). The good news is that considerable progress has been made, both in the development of drugs that can effectively block reverse transcription (and other steps in the viral life cycle) and in the development of combination therapies that, if the patients are compliant, can completely suppress the replication of virus and the emergence of resistance. Progress is also being made in the development of new drugs that are effective against many of the extant drug-resistant mutants. There is also now the hope, even in the absence of an effective anti-HIV vaccine that can prevent transmission, that anti-HIV drugs, including RT inhibitors, can help prevent new infections, both because they reduce the viral load in the donor and because, if taken prophylactically, they can block infection in the recipient.

SUMMARY

We have, over the approximately 45 years since RT was discovered, learned a great deal about these fascinating enzymes, their structures, and how they are able to convert a ssRNA genome into a dsDNA copy from which new copies of the genomic RNA can be produced. RTs were essential tools that played a key role in the origins of molecular biology and in the creation of the biotechnology industry, and RTs are widely used in both research laboratories and clinical laboratories today. Although the fundamentals of how reverse transcription works and structures of HIV-1 RT have been available for some time, there are still unsolved puzzles, particularly in relation to understanding how the capsid shell assists the process of reverse transcription in cells and how the shell is remodeled in the process. What is the exact composition and structure of the

RTC? How does it help RT carry out DNA synthesis efficiently and effectively? How does the RTC transition into the PIC? Moreover, we are just now beginning to understand how the host's innate immunity targets the reverse transcription process.

There has been good progress in slowing the spread of HIV not only in the developed world, but also in the developing world; drugs that block reverse transcription have been an essential part of this progress. However, there are still problems with both resistance and toxicity. We will need to supplement our armamentarium of anti-HIV drugs so that we do not fall behind in the battle, particularly in the developing world. We also need to be able to provide less-toxic drugs to those who, at least for the foreseeable future, will need to stay on anti-HIV drug therapy throughout their lives. If the past is any guide, at least some of these new drugs will be RT inhibitors; the more we know about RT and how it works, the more likely we will be to succeed.

Acknowledgments. *The author would like to thank Teresa Burdette for help with the manuscript, and Karen Kirby and Allen Kane for help with the figures.*

Citation. Hughes SH. 2014. Reverse transcription of retroviruses and LTR retrotransposons. Microbiol Spectrum 3(**2**): MDNA3-0027-2014

References

1. **Telesnitsky A, Goff GP.** 1997. Reverse transcriptase and the generation of retroviral DNA, p 121–160. *In* Coffin JM, Hughes SH, Varmus HE (ed), *Retroviruses*. Cold Spring Harbor Laboratory Press, Cold Spring Harbor, NY.
2. **Sarafianos SG, Marchand B, Das K, Himmel DM, Parniak MA, Hughes SH, Arnold E.** 2009. Structure and function of HIV-1 reverse transcriptase: molecular mechanisms of polymerization and inhibition. *J Mol Biol* **385:**693–713.
3. **Hu W, Hughes SH.** 2012. HIV-1 Reverse Transcription, p 37–58. *In* Bushman FD, Nabel GJ, Swanstrom R (ed), *HIV from Biology to Prevention and Treatment*. Cold Spring Harbor Laboratory Press, Cold Spring Harbor, NY.
4. **Skalka AM, Goff SP.** 1993. *Reverse Transcriptase*. Cold Spring Harbor Laboratory Press, Cold Spring Harbor, NY.
5. **Gilboa E, Mitra SW, Goff S, Baltimore D.** 1979. A detailed model of reverse transcription and tests of crucial aspects. *Cell* **18:**93–100.
6. **Shank PR, Hughes SH, Kung HJ, Majors JE, Quintrell N, Guntaka RV, Bishop JM, Varmus HE.** 1978. Mapping unintegrated avian sarcoma virus DNA: termini of linear DNA bear 300 nucleotides present once or twice in two species of circular DNA. *Cell* **15:**1383–1395.
7. **Hughes SH, Shank PR, Spector DH, Kung HJ, Bishop JM, Varmus HE, Vogt PK, Breitman ML.** 1978. Proviruses of avian sarcoma virus are terminally redundant, co-extensive with unintegrated linear DNA and integrated at many sites. *Cell* **15:**1397–1410.
8. **Nikolaitchik OA, Dilley KA, Fu W, Gorelick RJ, Tai SH, Soheilian F, Ptak RG, Nagashima K, Pathak VK, Hu WS.** 2013. Dimeric RNA recognition regulates HIV-1 genome packaging. *PLoS Pathog* **9:**e1003249.
9. **Ke N, Gao X, Keeney JB, Boeke JD, Voytas DF.** 1999. The yeast retrotransposon Ty5 uses the anticodon stem-loop of the initiator methionine tRNA as a primer for reverse transcription. *RNA* **5:**929–938.
10. **Lin JH, Levin HL.** 1998. Reverse transcription of a self-primed retrotransposon requires an RNA structure similar to the U5–IR stem–loop of retroviruses. *Mol Cell Biol* **18:**6859–6869.
11. **Isel C, Lanchy JM, Le Grice SF, Ehresmann C, Ehresmann B, Marquet R.** 1996. Specific initiation and switch to elongation of human immunodeficiency virus type 1 reverse transcription require the post-transcriptional modifications of primer tRNA3Lys. *EMBO J* **15:** 917–924.
12. **Lanchy JM, Ehresmann C, Le Grice SF, Ehresmann B, Marquet R.** 1996. Binding and kinetic properties of HIV-1 reverse transcriptase markedly differ during initiation and elongation of reverse transcription. *EMBO J* **15:**7178–7187.
13. **Telesnitsky A, Goff SP.** 1993. Two defective forms of reverse transcriptase can complement to restore retroviral infectivity. *EMBO J* **12:**4433–4438.
14. **Julias JG, Ferris AL, Boyer PL, Hughes SH.** 2001. Replication of phenotypically mixed human immunodeficiency virus type 1 virions containing catalytically active and catalytically inactive reverse transcriptase. *J Virol* **75:**6537–6546.
15. **Driscoll MD, Golinelli MP, Hughes SH.** 2001. In vitro analysis of human immunodeficiency virus type 1 minus-strand strong-stop DNA synthesis and genomic RNA processing. *J Virol* **75:**672–686.
16. **Purohit V, Roques BP, Kim B, Bambara RA.** 2007. Mechanisms that prevent template inactivation by HIV-1 reverse transcriptase RNase H cleavages. *J Biol Chem* **282:**12598–12609.
17. **Panganiban AT, Fiore D.** 1988. Ordered interstrand and intrastrand DNA transfer during reverse transcription. *Science* **241:**1064–1069.
18. **Hu WS, Temin HM.** 1990. Retroviral recombination and reverse transcription. *Science* **250:**1227–1233.
19. **van Wamel JL, Berkhout B.** 1998. The first strand transfer during HIV-1 reverse transcription can occur either intramolecularly or intermolecularly. *Virology* **244:** 245–251.
20. **Wilhelm M, Boutabout M, Heyman T, Wilhelm FX.** 1999. Reverse transcription of the yeast Ty1 retrotransposon: the mode of first strand transfer is either intermolecular or intramolecular. *J Mol Biol* **288:** 505–510.
21. **Hungnes O, Tjotta E, Grinde B.** 1992. Mutations in the central polypurine tract of HIV-1 result in delayed replication. *Virology* **190:**440–442.

22. **Charneau P, Alizon M, Clavel F.** 1992. A second origin of DNA plus-strand synthesis is required for optimal human immunodeficiency virus replication. *J Virol* **66:** 2814–2820.

23. **Swanstrom R, Varmus HE, Bishop JM.** 1981. The terminal redundancy of the retrovirus genome facilitates chain elongation by reverse transcriptase. *J Biol Chem* **256:**1115–1121.

24. **Yu H, Jetzt AE, Ron Y, Preston BD, Dougherty JP.** 1998. The nature of human immunodeficiency virus type 1 strand transfers. *J Biol Chem* **273:**28384–28391.

25. **Kung HJ, Fung YK, Majors JE, Bishop JM, Varmus HE.** 1981. Synthesis of plus strands of retroviral DNA in cells infected with avian sarcoma virus and mouse mammary tumor virus. *J Virol* **37:**127–138.

26. **Hsu TW, Taylor JM.** 1982. Single-stranded regions on unintegrated avian retrovirus DNA. *J Virol* **44:**47–53.

27. **Miller MD, Wang B, Bushman FD.** 1995. Human immunodeficiency virus type 1 preintegration complexes containing discontinuous plus strands are competent to integrate in vitro. *J Virol* **69:**3938–3944.

28. **Klarmann GJ, Yu H, Chen X, Dougherty JP, Preston BD.** 1997. Discontinuous plus-strand DNA synthesis in human immunodeficiency virus type 1-infected cells and in a partially reconstituted cell-free system. *J Virol* **71:** 9259–9269.

29. **Thomas JA, Ott DE, Gorelick RJ.** 2007. Efficiency of human immunodeficiency virus type 1 postentry infection processes: evidence against disproportionate numbers of defective virions. *J Virol* **81:**4367–4370.

30. **Burdick RC, Hu WS, Pathak VK.** 2013. Nuclear import of APOBEC3F-labeled HIV-1 preintegration complexes. *Proc Natl Acad Sci U S A* **110:**E4780–E4789.

31. **Ao Z, Yao X, Cohen EA.** 2004. Assessment of the role of the central DNA flap in human immunodeficiency virus type 1 replication by using a single-cycle replication system. *J Virol* **78:**3170–3177.

32. **Van Maele B, De Rijck J, De Clercq E, Debyser Z.** 2003. Impact of the central polypurine tract on the kinetics of human immunodeficiency virus type 1 vector transduction. *J Virol* **77:**4685–4694.

33. **Hu C, Saenz DT, Fadel HJ, Walker W, Peretz M, Poeschla EM.** 2010. The HIV-1 central polypurine tract functions as a second line of defense against APOBEC3G/F. *J Virol* **84:**11981–11993.

34. **Wurtzer S, Goubard A, Mammano F, Saragosti S, Lecossier D, Hance AJ, Clavel F.** 2006. Functional central polypurine tract provides downstream protection of the human immunodeficiency virus type 1 genome from editing by APOBEC3G and APOBEC3B. *J Virol* **80:** 3679–3683.

35. **Whitcomb JM, Kumar R, Hughes SH.** 1990. Sequence of the circle junction of human immunodeficiency virus type 1: implications for reverse transcription and integration. *J Virol* **64:**4903–4906.

36. **Pullen KA, Ishimoto LK, Champoux JJ.** 1992. Incomplete removal of the RNA primer for minus-strand DNA synthesis by human immunodeficiency virus type 1 reverse transcriptase. *J Virol* **66:**367–373.

37. **Smith JS, Roth MJ.** 1992. Specificity of human immunodeficiency virus-1 reverse transcriptase-associated ribonuclease H in removal of the minus-strand primer, tRNA(Lys3). *J Biol Chem* **267:**15071–15079.

38. **Pullen KA, Rattray AJ, Champoux JJ.** 1993. The sequence features important for plus strand priming by human immunodeficiency virus type 1 reverse transcriptase. *J Biol Chem* **268:**6221–6227.

39. **Julias JG, McWilliams MJ, Sarafianos SG, Arnold E, Hughes SH.** 2002. Mutations in the RNase H domain of HIV-1 reverse transcriptase affect the initiation of DNA synthesis and the specificity of RNase H cleavage in vivo. *Proc Natl Acad Sci U S A* **99:**9515–9520.

40. **Rausch JW, Lener D, Miller JT, Julias JG, Hughes SH, Le Grice SF.** 2002. Altering the RNase H primer grip of human immunodeficiency virus reverse transcriptase modifies cleavage specificity. *Biochemistry* **41:** 4856–4865.

41. **Dash C, Rausch JW, Le Grice SF.** 2004. Using pyrrolo-deoxycytosine to probe RNA/DNA hybrids containing the human immunodeficiency virus type-1 3′ polypurine tract. *Nucleic Acids Res* **32:**1539–1547.

42. **Yi-Brunozzi HY, Le Grice SF.** 2005. Investigating HIV-1 polypurine tract geometry via targeted insertion of abasic lesions in the (–)-DNA template and (+)-RNA primer. *J Biol Chem* **280:**20154–20162.

43. **Swanstrom R, Willis JW.** 1997. Synthesis, assembly, and processing of viral proteins, p 263–334. *In* Coffin JM, Hughes SH, Varmus HE (ed), *Retroviruses*. Cold Spring Harbor Laboratory Press, Cold Spring Harbor, NY.

44. **Linial ML.** 1999. Foamy viruses are unconventional retroviruses. *J Virol* **73:**1747–1755.

45. **Yu SF, Baldwin DN, Gwynn SR, Yendapalli S, Linial ML.** 1996. Human foamy virus replication: a pathway distinct from that of retroviruses and hepadnaviruses. *Science* **271:**1579–1582.

46. **Roth MJ, Tanese N, Goff SP.** 1985. Purification and characterization of murine retroviral reverse transcriptase expressed in *Escherichia coli*. *J Biol Chem* **260:** 9326–9335.

47. **Boyer PL, Stenbak CR, Clark PK, Linial ML, Hughes SH.** 2004. Characterization of the polymerase and RNase H activities of human foamy virus reverse transcriptase. *J Virol* **78:**6112–6121.

48. **Nowak E, Miller JT, Bona MK, Studnicka J, Szczepanowski RH, Jurkowski J, Le Grice SF, Nowotny M.** 2014. Ty3 reverse transcriptase complexed with an RNA–DNA hybrid shows structural and functional asymmetry. *Nat Struct Mol Biol* **21:**389–396.

49. **Hizi A, Gazit A, Guthmann D, Yaniv A.** 1982. DNA-processing activities associated with the purified α, β2, and αβ molecular forms of avian sarcoma virus RNA-dependent DNA polymerase. *J Virol* **41:**974–981.

50. **Hizi A, Leis JP, Joklik WK.** 1977. RNA-dependent DNA polymerase of avian sarcoma virus B77. II. Comparison of the catalytic properties of the α, β2, and αβ enzyme forms. *J Biol Chem* **252:**2290–2295.

51. **Hizi A, Leis JP, Joklik WK.** 1977. The RNA-dependent DNA polymerase of avian sarcoma virus B77. Binding

of viral and nonviral ribonucleic acids to the α, β2, and αβ forms of the enzyme. *J Biol Chem* **252:**6878–6884.

52. **Lightfoote MM, Coligan JE, Folks TM, Fauci AS, Martin MA, Venkatesan S.** 1986. Structural characterization of reverse transcriptase and endonuclease polypeptides of the acquired immunodeficiency syndrome retrovirus. *J Virol* **60:**771–775.

53. **Kohlstaedt LA, Wang J, Friedman JM, Rice PA, Steitz TA.** 1992. Crystal structure at 3.5 Å resolution of HIV-1 reverse transcriptase complexed with an inhibitor. *Science* **256:**1783–1790.

54. **Jacobo-Molina A, Ding J, Nanni RG, Clark AD Jr, Lu X, Tantillo C, Williams RL, Kamer G, Ferris AL, Clark P, Hizi A, Hughes SH, Arnold E.** 1993. Crystal structure of human immunodeficiency virus type 1 reverse transcriptase complexed with double-stranded DNA at 3.0 Å resolution shows bent DNA. *Proc Natl Acad Sci U S A* **90:**6320–6324.

55. **Poch O, Sauvaget I, Delarue M, Tordo N.** 1989. Identification of four conserved motifs among the RNA--dependent polymerase encoding elements. *EMBO J* **8:** 3867–3874.

56. **Malik HS, Eickbush TH.** 2001. Phylogenetic analysis of ribonuclease H domains suggests a late, chimeric origin of LTR retrotransposable elements and retroviruses. *Genome Res* **11:**1187–1197.

57. **Gao G, Goff SP.** 1998. Replication defect of moloney murine leukemia virus with a mutant reverse transcriptase that can incorporate ribonucleotides and deoxyribonucleotides. *J Virol* **72:**5905–5911.

58. **Boyer PL, Sarafianos SG, Arnold E, Hughes SH.** 2000. Analysis of mutations at positions 115 and 116 in the dNTP binding site of HIV-1 reverse transcriptase. *Proc Natl Acad Sci U S A* **97:**3056–3061.

59. **Esnouf R, Ren J, Ross C, Jones Y, Stammers D, Stuart D.** 1995. Mechanism of inhibition of HIV-1 reverse transcriptase by non-nucleoside inhibitors. *Nat Struct Biol* **2:**303–308.

60. **Rodgers DW, Gamblin SJ, Harris BA, Ray S, Culp JS, Hellmig B, Woolf DJ, Debouck C, Harrison SC.** 1995. The structure of unliganded reverse transcriptase from the human immunodeficiency virus type 1. *Proc Natl Acad Sci U S A* **92:**1222–1226.

61. **Hsiou Y, Ding J, Das K, Clark AD Jr, Hughes SH, Arnold E.** 1996. Structure of unliganded HIV-1 reverse transcriptase at 2.7 Å resolution: implications of conformational changes for polymerization and inhibition mechanisms. *Structure* **4:**853–860.

62. **Sarafianos SG, Clark AD Jr, Das K, Tuske S, Birktoft JJ, Ilankumaran P, Ramesha AR, Sayer JM, Jerina DM, Boyer PL, Hughes SH, Arnold E.** 2002. Structures of HIV-1 reverse transcriptase with pre- and post-translocation AZTMP-terminated DNA. *EMBO J* **21:** 6614–6624.

63. **Sarafianos SG, Das K, Tantillo C, Clark AD Jr, Ding J, Whitcomb JM, Boyer PL, Hughes SH, Arnold E.** 2001. Crystal structure of HIV-1 reverse transcriptase in complex with a polypurine tract RNA:DNA. *EMBO J* **20:** 1449–1461.

64. **Lapkouski M, Tian L, Miller JT, Le Grice SF, Yang W.** 2013. Complexes of HIV-1 RT, NNRTI and RNA/DNA hybrid reveal a structure compatible with RNA degradation. *Nat Struct Mol Biol* **20:**230–236.

65. **Huang H, Chopra R, Verdine GL, Harrison SC.** 1998. Structure of a covalently trapped catalytic complex of HIV-1 reverse transcriptase: implications for drug resistance. *Science* **282:**1669–1675.

66. **Lansdon EB, Samuel D, Lagpacan L, Brendza KM, White KL, Hung M, Liu X, Boojamra CG, Mackman RL, Cihlar T, Ray AS, McGrath ME, Swaminathan S.** 2010. Visualizing the molecular interactions of a nucleotide analog, GS-9148, with HIV-1 reverse transcriptase–DNA complex. *J Mol Biol* **397:**967–978.

67. **Tu X, Das K, Han Q, Bauman JD, Clark AD Jr, Hou X, Frenkel YV, Gaffney BL, Jones RA, Boyer PL, Hughes SH, Sarafianos SG, Arnold E.** 2010. Structural basis of HIV-1 resistance to AZT by excision. *Nat Struct Mol Biol* **17:**1202–1209.

68. **Tuske S, Sarafianos SG, Clark AD Jr, Ding J, Naeger LK, White KL, Miller MD, Gibbs CS, Boyer PL, Clark P, Wang G, Gaffney BL, Jones RA, Jerina DM, Hughes SH, Arnold E.** 2004. Structures of HIV-1 RT-DNA complexes before and after incorporation of the anti-AIDS drug tenofovir. *Nat Struct Mol Biol* **11:**469–474.

69. **Lansdon EB, Brendza KM, Hung M, Wang R, Mukund S, Jin D, Birkus G, Kutty N, Liu X.** 2010. Crystal structures of HIV-1 reverse transcriptase with etravirine (TMC125) and rilpivirine (TMC278): implications for drug design. *J Med Chem* **53:**4295–4299.

70. **Das K, Bauman JD, Clark AD Jr, Frenkel YV, Lewi PJ, Shatkin AJ, Hughes SH, Arnold E.** 2008. High-resolution structures of HIV-1 reverse transcriptase/TMC278 complexes: strategic flexibility explains potency against resistance mutations. *Proc Natl Acad Sci U S A* **105:**1466–1471.

71. **Das K, Ding J, Hsiou Y, Clark AD Jr, Moereels H, Koymans L, Andries K, Pauwels R, Janssen PA, Boyer PL, Clark P, Smith RH Jr, Kroeger Smith MB, Michejda CJ, Hughes SH, Arnold E.** 1996. Crystal structures of 8-Cl and 9-Cl TIBO complexed with wild-type HIV-1 RT and 8-Cl TIBO complexed with the Tyr181Cys HIV-1 RT drug-resistant mutant. *J Mol Biol* **264:** 1085–1100.

72. **Esnouf RM, Ren J, Hopkins AL, Ross CK, Jones EY, Stammers DK, Stuart DI.** 1997. Unique features in the structure of the complex between HIV-1 reverse transcriptase and the bis(heteroaryl)piperazine (BHAP) U-90152 explain resistance mutations for this non-nucleoside inhibitor. *Proc Natl Acad Sci U S A* **94:** 3984–3989.

73. **Ren J, Esnouf R, Hopkins A, Ross C, Jones Y, Stammers D, Stuart D.** 1995. The structure of HIV-1 reverse transcriptase complexed with 9-chloro-TIBO: lessons for inhibitor design. *Structure* **3:**915–926.

74. **Ren J, Nichols C, Bird LE, Fujiwara T, Sugimoto H, Stuart DI, Stammers DK.** 2000. Binding of the second generation non-nucleoside inhibitor S-1153 to HIV-1 reverse transcriptase involves extensive main chain hydrogen bonding. *J Biol Chem* **275:**14316–14320.

75. **Ren J, Esnouf RM, Hopkins AL, Stuart DI, Stammers DK.** 1999. Crystallographic analysis of the binding modes of thiazoloisoindolinone non-nucleoside inhibitors to HIV-1 reverse transcriptase and comparison with modeling studies. *J Med Chem* **42:**3845–3851.

76. **Doublie S, Tabor S, Long AM, Richardson CC, Ellenberger T.** 1998. Crystal structure of a bacteriophage T7 DNA replication complex at 2.2 Å resolution. *Nature* **391:**251–258.

77. **Palaniappan C, Wisniewski M, Jacques PS, Le Grice SF, Fay PJ, Bambara RA.** 1997. Mutations within the primer grip region of HIV-1 reverse transcriptase result in loss of RNase H function. *J Biol Chem* **272:**11157–11164.

78. **Gao HQ, Boyer PL, Arnold E, Hughes SH.** 1998. Effects of mutations in the polymerase domain on the polymerase, RNase H and strand transfer activities of human immunodeficiency virus type 1 reverse transcriptase. *J Mol Biol* **277:**559–572.

79. **Powell MD, Ghosh M, Jacques PS, Howard KJ, Le Grice SF, Levin JG.** 1997. Alanine-scanning mutations in the "primer grip" of p66 HIV-1 reverse transcriptase result in selective loss of RNA priming activity. *J Biol Chem* **272:**13262–13269.

80. **Sevilya Z, Loya S, Adir N, Hizi A.** 2003. The ribonuclease H activity of the reverse transcriptases of human immunodeficiency viruses type 1 and type 2 is modulated by residue 294 of the small subunit. *Nucleic Acids Res* **31:**1481–1487.

81. **Sevilya Z, Loya S, Hughes SH, Hizi A.** 2001. The ribonuclease H activity of the reverse transcriptases of human immunodeficiency viruses type 1 and type 2 is affected by the thumb subdomain of the small protein subunits. *J Mol Biol* **311:**957–971.

82. **Ghosh M, Jacques PS, Rodgers DW, Ottman M, Darlix JL, Le Grice SF.** 1996. Alterations to the primer grip of p66 HIV-1 reverse transcriptase and their consequences for template-primer utilization. *Biochemistry* **35:**8553–8562.

83. **Powell MD, Beard WA, Bebenek K, Howard KJ, Le Grice SF, Darden TA, Kunkel TA, Wilson SH, Levin JG.** 1999. Residues in the αH and αI helices of the HIV-1 reverse transcriptase thumb subdomain required for the specificity of RNase H-catalyzed removal of the polypurine tract primer. *J Biol Chem* **274:**19885–19893.

84. **Yang W, Steitz TA.** 1995. Recombining the structures of HIV integrase, RuvC and RNase H. *Structure* **3:** 131–134.

85. **Nowotny M, Gaidamakov SA, Crouch RJ, Yang W.** 2005. Crystal structures of RNase H bound to an RNA/DNA hybrid: substrate specificity and metal-dependent catalysis. *Cell* **121:**1005–1016.

86. **Dunn LL, McWilliams MJ, Das K, Arnold E, Hughes SH.** 2009. Mutations in the thumb allow human immunodeficiency virus type 1 reverse transcriptase to be cleaved by protease in virions. *J Virol* **83:**12336–12344.

87. **Dunn LL, Boyer PL, Clark PK, Hughes SH.** 2013. Mutations in HIV-1 reverse transcriptase cause misfolding and miscleavage by the viral protease. *Virology* **444:** 241–249.

88. **Huang W, Gamarnik A, Limoli K, Petropoulos CJ, Whitcomb JM.** 2003. Amino acid substitutions at position 190 of human immunodeficiency virus type 1 reverse transcriptase increase susceptibility to delavirdine and impair virus replication. *J Virol* **77:**1512–1523.

89. **Takehisa J, Kraus MH, Decker JM, Li Y, Keele BF, Bibollet-Ruche F, Zammit KP, Weng Z, Santiago ML, Kamenya S, Wilson ML, Pusey AE, Bailes E, Sharp PM, Shaw GM, Hahn BH.** 2007. Generation of infectious molecular clones of simian immunodeficiency virus from fecal consensus sequences of wild chimpanzees. *J Virol* **81:**7463–7475.

90. **Huang M, Zensen R, Cho M, Martin MA.** 1998. Construction and characterization of a temperature-sensitive human immunodeficiency virus type 1 reverse transcriptase mutant. *J Virol* **72:**2047–2054.

91. **Wang J, Bambara RA, Demeter LM, Dykes C.** 2010. Reduced fitness in cell culture of HIV-1 with nonnucleoside reverse transcriptase inhibitor-resistant mutations correlates with relative levels of reverse transcriptase content and RNase H activity in virions. *J Virol* **84:** 9377–9389.

92. **Thomas DC, Voronin YA, Nikolenko GN, Chen J, Hu WS, Pathak VK.** 2007. Determination of the ex vivo rates of human immunodeficiency virus type 1 reverse transcription by using novel strand-specific amplification analysis. *J Virol* **81:**4798–4807.

93. **Fassati A, Goff SP.** 2001. Characterization of intracellular reverse transcription complexes of human immunodeficiency virus type 1. *J Virol* **75:**3626–3635.

94. **Forshey BM, von Schwedler U, Sundquist WI, Aiken C.** 2002. Formation of a human immunodeficiency virus type 1 core of optimal stability is crucial for viral replication. *J Virol* **76:**5667–5677.

95. **Nermut MV, Fassati A.** 2003. Structural analyses of purified human immunodeficiency virus type 1 intracellular reverse transcription complexes. *J Virol* **77:** 8196–8206.

96. **Iordanskiy S, Berro R, Altieri M, Kashanchi F, Bukrinsky M.** 2006. Intracytoplasmic maturation of the human immunodeficiency virus type 1 reverse transcription complexes determines their capacity to integrate into chromatin. *Retrovirology* **3:**4.

97. **Dismuke DJ, Aiken C.** 2006. Evidence for a functional link between uncoating of the human immunodeficiency virus type 1 core and nuclear import of the viral preintegration complex. *J Virol* **80:**3712–3720.

98. **Hulme AE, Perez O, Hope TJ.** 2011. Complementary assays reveal a relationship between HIV-1 uncoating and reverse transcription. *Proc Natl Acad Sci U S A* **108:**9975–9980.

99. **Summers J, Mason WS.** 1982. Replication of the genome of a hepatitis B-like virus by reverse transcription of an RNA intermediate. *Cell* **29:**403–415.

100. **Yu SF, Sullivan MD, Linial ML.** 1999. Evidence that the human foamy virus genome is DNA. *J Virol* **73:** 1565–1572.

101. **Rasaiyaah J, Tan CP, Fletcher AJ, Price AJ, Blondeau C, Hilditch L, Jacques DA, Selwood DL, James LC,**

Noursadeghi M, Towers GJ. 2013. HIV-1 evades innate immune recognition through specific cofactor recruitment. *Nature* **503:**402–405.

102. **Pornillos O, Ganser-Pornillos BK, Yeager M.** 2011. Atomic-level modelling of the HIV capsid. *Nature* **469:** 424–427.

103. **Li S, Hill CP, Sundquist WI, Finch JT.** 2000. Image reconstructions of helical assemblies of the HIV-1 CA protein. *Nature* **407:**409–413.

104. **Tang S, Murakami T, Agresta BE, Campbell S, Freed EO, Levin JG.** 2001. Human immunodeficiency virus type 1 N-terminal capsid mutants that exhibit aberrant core morphology and are blocked in initiation of reverse transcription in infected cells. *J Virol* **75:** 9357–9366.

105. **Arhel NJ, Souquere-Besse S, Munier S, Souque P, Guadagnini S, Rutherford S, Prevost MC, Allen TD, Charneau P.** 2007. HIV-1 DNA Flap formation promotes uncoating of the pre-integration complex at the nuclear pore. *EMBO J* **26:**3025–3037.

106. **Brown PO, Bowerman B, Varmus HE, Bishop JM.** 1987. Correct integration of retroviral DNA in vitro. *Cell* **49:**347–356.

107. **Brown PO, Bowerman B, Varmus HE, Bishop JM.** 1989. Retroviral integration: structure of the initial covalent product and its precursor, and a role for the viral IN protein. *Proc Natl Acad Sci U S A* **86:**2525–2529.

108. **Taylor JM, Cywinski A, Smith JK.** 1983. Discontinuities in the DNA synthesized by an avian retrovirus. *J Virol* **48:**654–659.

109. **Lochelt M, Flugel RM.** 1996. The human foamy virus pol gene is expressed as a Pro-Pol polyprotein and not as a Gag-Pol fusion protein. *J Virol* **70:**1033–1040.

110. **Konvalinka J, Lochelt M, Zentgraf H, Flugel RM, Krausslich HG.** 1995. Active foamy virus proteinase is essential for virus infectivity but not for formation of a Pol polyprotein. *J Virol* **69:**7264–7268.

111. **Feng YX, Copeland TD, Henderson LE, Gorelick RJ, Bosche WJ, Levin JG, Rein A.** 1996. HIV-1 nucleocapsid protein induces "maturation" of dimeric retroviral RNA in vitro. *Proc Natl Acad Sci U S A* **93:**7577–7581.

112. **Zhang WH, Hwang CK, Hu WS, Gorelick RJ, Pathak VK.** 2002. Zinc finger domain of murine leukemia virus nucleocapsid protein enhances the rate of viral DNA synthesis in vivo. *J Virol* **76:**7473–7484.

113. **Driscoll MD, Hughes SH.** 2000. Human immunodeficiency virus type 1 nucleocapsid protein can prevent self-priming of minus-strand strong stop DNA by promoting the annealing of short oligonucleotides to hairpin sequences. *J Virol* **74:**8785–8792.

114. **Buckman JS, Bosche WJ, Gorelick RJ.** 2003. Human immunodeficiency virus type 1 nucleocapsid Zn^{2+} fingers are required for efficient reverse transcription, initial integration processes, and protection of newly synthesized viral DNA. *J Virol* **77:**1469–1480.

115. **Thomas JA, Gagliardi TD, Alvord WG, Lubomirski M, Bosche WJ, Gorelick RJ.** 2006. Human immunodeficiency virus type 1 nucleocapsid zinc-finger mutations cause defects in reverse transcription and integration. *Virology* **353:**41–51.

116. **Thomas JA, Bosche WJ, Shatzer TL, Johnson DG, Gorelick RJ.** 2008. Mutations in human immunodeficiency virus type 1 nucleocapsid protein zinc fingers cause premature reverse transcription. *J Virol* **82:** 9318–9328.

117. **Thomas JA, Gorelick RJ.** 2008. Nucleocapsid protein function in early infection processes. *Virus Res* **134:** 39–63.

118. **Guo J, Wu T, Anderson J, Kane BF, Johnson DG, Gorelick RJ, Henderson LE, Levin JG.** 2000. Zinc finger structures in the human immunodeficiency virus type 1 nucleocapsid protein facilitate efficient minus- and plus-strand transfer. *J Virol* **74:**8980–8988.

119. **Engelman A, Englund G, Orenstein JM, Martin MA, Craigie R.** 1995. Multiple effects of mutations in human immunodeficiency virus type 1 integrase on viral replication. *J Virol* **69:**2729–2736.

120. **Masuda T, Planelles V, Krogstad P, Chen IS.** 1995. Genetic analysis of human immunodeficiency virus type 1 integrase and the U3 att site: unusual phenotype of mutants in the zinc finger-like domain. *J Virol* **69:** 6687–6696.

121. **Leavitt AD, Robles G, Alesandro N, Varmus HE.** 1996. Human immunodeficiency virus type 1 integrase mutants retain in vitro integrase activity yet fail to integrate viral DNA efficiently during infection. *J Virol* **70:** 721–728.

122. **Wu X, Liu H, Xiao H, Conway JA, Hehl E, Kalpana GV, Prasad V, Kappes JC.** 1999. Human immunodeficiency virus type 1 integrase protein promotes reverse transcription through specific interactions with the nucleoprotein reverse transcription complex. *J Virol* **73:** 2126–2135.

123. **Kirchner J, Sandmeyer SB.** 1996. Ty3 integrase mutants defective in reverse transcription or 3′-end processing of extrachromosomal Ty3 DNA. *J Virol* **70:**4737–4747.

124. **Nymark-McMahon MH, Beliakova-Bethell NS, Darlix JL, Le Grice SF, Sandmeyer SB.** 2002. Ty3 integrase is required for initiation of reverse transcription. *J Virol* **76:**2804–2816.

125. **Nymark-McMahon MH, Sandmeyer SB.** 1999. Mutations in nonconserved domains of Ty3 integrase affect multiple stages of the Ty3 life cycle. *J Virol* **73:**453–465.

126. **Harrich D, Ulich C, Garcia-Martinez LF, Gaynor RB.** 1997. Tat is required for efficient HIV-1 reverse transcription. *EMBO J* **16:**1224–1235.

127. **Kameoka M, Morgan M, Binette M, Russell RS, Rong L, Guo X, Mouland A, Kleiman L, Liang C, Wainberg MA.** 2002. The Tat protein of human immunodeficiency virus type 1 (HIV-1) can promote placement of tRNA primer onto viral RNA and suppress later DNA polymerization in HIV-1 reverse transcription. *J Virol* **76:**3637–3645.

128. **Liang C, Wainberg MA.** 2002. The role of Tat in HIV-1 replication: an activator and/or a suppressor? *AIDS Rev* **4:**41–49.

129. **Apolloni A, Meredith LW, Suhrbier A, Kiernan R, Harrich D.** 2007. The HIV-1 Tat protein stimulates reverse transcription in vitro. *Curr HIV Res* **5:**473–483.

130. **Henriet S, Sinck L, Bec G, Gorelick RJ, Marquet R, Paillart JC.** 2007. Vif is a RNA chaperone that could temporally regulate RNA dimerization and the early steps of HIV-1 reverse transcription. *Nucleic Acids Res* **35:**5141–5153.

131. **Carr JM, Coolen C, Davis AJ, Burrell CJ, Li P.** 2008. Human immunodeficiency virus 1 (HIV-1) virion infectivity factor (Vif) is part of reverse transcription complexes and acts as an accessory factor for reverse transcription. *Virology* **372:**147–156.

132. **Elder JH, Lerner DL, Hasselkus-Light CS, Fontenot DJ, Hunter E, Luciw PA, Montelaro RC, Phillips TR.** 1992. Distinct subsets of retroviruses encode dUTPase. *J Virol* **66:**1791–1794.

133. **Wagaman PC, Hasselkus-Light CS, Henson M, Lerner DL, Phillips TR, Elder JH.** 1993. Molecular cloning and characterization of deoxyuridine triphosphatase from feline immunodeficiency virus (FIV). *Virology* **196:** 451–457.

134. **Lerner DL, Wagaman PC, Phillips TR, Prospero-Garcia O, Henriksen SJ, Fox HS, Bloom FE, Elder JH.** 1995. Increased mutation frequency of feline immunodeficiency virus lacking functional deoxyuridine-triphosphatase. *Proc Natl Acad Sci U S A* **92:**7480–7484.

135. **Selig L, Benichou S, Rogel ME, Wu LI, Vodicka MA, Sire J, Benarous R, Emerman M.** 1997. Uracil DNA glycosylase specifically interacts with Vpr of both human immunodeficiency virus type 1 and simian immunodeficiency virus of sooty mangabeys, but binding does not correlate with cell cycle arrest. *J Virol* **71:** 4842–4846.

136. **Mansky LM, Preveral S, Selig L, Benarous R, Benichou S.** 2000. The interaction of vpr with uracil DNA glycosylase modulates the human immunodeficiency virus type 1 In vivo mutation rate. *J Virol* **74:**7039–7047.

137. **Chen R, Le Rouzic E, Kearney JA, Mansky LM, Benichou S.** 2004. Vpr-mediated incorporation of UNG2 into HIV-1 particles is required to modulate the virus mutation rate and for replication in macrophages. *J Biol Chem* **279:**28419–28425.

138. **Schrofelbauer B, Yu Q, Zeitlin SG, Landau NR.** 2005. Human immunodeficiency virus type 1 Vpr induces the degradation of the UNG and SMUG uracil-DNA glycosylases. *J Virol* **79:**10978–10987.

139. **Cen S, Javanbakht H, Kim S, Shiba K, Craven R, Rein A, Ewalt K, Schimmel P, Musier-Forsyth K, Kleiman L.** 2002. Retrovirus-specific packaging of aminoacyl-tRNA synthetases with cognate primer tRNAs. *J Virol* **76:**13111–13115.

140. **Javanbakht H, Cen S, Musier-Forsyth K, Kleiman L.** 2002. Correlation between tRNALys3 aminoacylation and its incorporation into HIV-1. *J Biol Chem* **277:** 17389–17396.

141. **Kelly MC, Kosloff BR, Morrow CD.** 2007. Forced selection of tRNA(Glu) reveals the importance of two adenosine-rich RNA loops within the U5-PBS for SIV (smmPBj) replication. *Virology* **366:**330–339.

142. **Djekic UV, Morrow CD.** 2007. Analysis of the replication of HIV-1 forced to use tRNAMet(i) supports a link between primer selection, translation and encapsidation. *Retrovirology* **4:**10.

143. **Li M, Eipers PG, Ni N, Morrow CD.** 2006. HIV-1 designed to use different tRNAGln isoacceptors prefers to select tRNAThr for replication. *Virol J* **3:**80.

144. **Galvis AE, Fisher HE, Nitta T, Fan H, Camerini D.** 2014. Impairment of HIV-1 cDNA synthesis by DBR1 knockdown. *J Virol* **88:**7054–7069.

145. **Ye Y, De Leon J, Yokoyama N, Naidu Y, Camerini D.** 2005. DBR1 siRNA inhibition of HIV-1 replication. *Retrovirology* **2:**63.

146. **Frankel WN, Stoye JP, Taylor BA, Coffin JM.** 1989. Genetic analysis of endogenous xenotropic murine leukemia viruses: association with two common mouse mutations and the viral restriction locus Fv-1. *J Virol* **63:**1763–1774.

147. **Best S, Le Tissier P, Towers G, Stoye JP.** 1996. Positional cloning of the mouse retrovirus restriction gene Fv1. *Nature* **382:**826–829.

148. **Nisole S, Lynch C, Stoye JP, Yap MW.** 2004. A Trim5–cyclophilin A fusion protein found in owl monkey kidney cells can restrict HIV-1. *Proc Natl Acad Sci U S A* **101:**13324–13328.

149. **Sayah DM, Sokolskaja E, Berthoux L, Luban J.** 2004. Cyclophilin A retrotransposition into TRIM5 explains owl monkey resistance to HIV-1. *Nature* **430:**569–573.

150. **Stremlau M, Owens CM, Perron MJ, Kiessling M, Autissier P, Sodroski J.** 2004. The cytoplasmic body component TRIM5α restricts HIV-1 infection in Old World monkeys. *Nature* **427:**848–853.

151. **Stremlau M, Perron M, Lee M, Li Y, Song B, Javanbakht H, Diaz-Griffero F, Anderson DJ, Sundquist WI, Sodroski J.** 2006. Specific recognition and accelerated uncoating of retroviral capsids by the TRIM5α restriction factor. *Proc Natl Acad Sci U S A* **103:** 5514–5519.

152. **Wu X, Anderson JL, Campbell EM, Joseph AM, Hope TJ.** 2006. Proteasome inhibitors uncouple rhesus TRIM5α restriction of HIV-1 reverse transcription and infection. *Proc Natl Acad Sci U S A* **103:**7465–7470.

153. **Brennan TP, Woods JO, Sedaghat AR, Siliciano JD, Siliciano RF, Wilke CO.** 2009. Analysis of human immunodeficiency virus type 1 viremia and provirus in resting CD4$^+$ T cells reveals a novel source of residual viremia in patients on antiretroviral therapy. *J Virol* **83:** 8470–8481.

154. **Newman RM, Hall L, Kirmaier A, Pozzi LA, Pery E, Farzan M, O'Neil SP, Johnson W.** 2008. Evolution of a TRIM5–CypA splice isoform in old world monkeys. *PLoS Pathog* **4:**e1000003.

155. **Virgen CA, Kratovac Z, Bieniasz PD, Hatziioannou T.** 2008. Independent genesis of chimeric TRIM5–cyclophilin proteins in two primate species. *Proc Natl Acad Sci U S A* **105:**3563–3568.

156. **Wilson SJ, Webb BL, Ylinen LM, Verschoor E, Heeney JL, Towers GJ.** 2008. Independent evolution of an antiviral TRIMCyp in rhesus macaques. *Proc Natl Acad Sci U S A* **105:**3557–3562.

157. **Ahn J, Hao C, Yan J, DeLucia M, Mehrens J, Wang C, Gronenborn AM, Skowronski J.** 2012. HIV/simian immunodeficiency virus (SIV) accessory virulence factor Vpx loads the host cell restriction factor SAMHD1 onto the E3 ubiquitin ligase complex CRL4DCAF1. *J Biol Chem* **287:**12550–12558.

158. **Hrecka K, Hao C, Gierszewska M, Swanson SK, Kesik-Brodacka M, Srivastava S, Florens L, Washburn MP, Skowronski J.** 2011. Vpx relieves inhibition of HIV-1 infection of macrophages mediated by the SAMHD1 protein. *Nature* **474:**658–661.

159. **St Gelais C, Wu L.** 2011. SAMHD1: a new insight into HIV-1 restriction in myeloid cells. *Retrovirology* **8:**55.

160. **St Gelais C, de Silva S, Amie SM, Coleman CM, Hoy H, Hollenbaugh JA, Kim B, Wu L.** 2012. SAMHD1 restricts HIV-1 infection in dendritic cells (DCs) by dNTP depletion, but its expression in DCs and primary CD4+ T-lymphocytes cannot be upregulated by interferons. *Retrovirology* **9:**105.

161. **Lahouassa H, Daddacha W, Hofmann H, Ayinde D, Logue EC, Dragin L, Bloch N, Maudet C, Bertrand M, Gramberg T, Pancino G, Priet S, Canard B, Laguette N, Benkirane M, Transy C, Landau NR, Kim B, Margottin-Goguet F.** 2012. SAMHD1 restricts the replication of human immunodeficiency virus type 1 by depleting the intracellular pool of deoxynucleoside triphosphates. *Nat Immunol* **13:**223–228.

162. **Laguette N, Sobhian B, Casartelli N, Ringeard M, Chable-Bessia C, Segeral E, Yatim A, Emiliani S, Schwartz O, Benkirane M.** 2011. SAMHD1 is the dendritic- and myeloid-cell-specific HIV-1 restriction factor counteracted by Vpx. *Nature* **474:**654–657.

163. **Rice GI, Bond J, Asipu A, Brunette RL, Manfield IW, Carr IM, Fuller JC, Jackson RM, Lamb T, Briggs TA, Ali M, Gornall H, Couthard LR, Aeby A, Attard-Montalto SP, Bertini E, Bodemer C, Brockmann K, Brueton LA, Corry PC, Desguerre I, Fazzi E, Cazorla AG, Gener B, Hamel BC, Heiberg A, Hunter M, van der Knaap MS, Kumar R, Lagae L, Landrieu PG, Lourenco CM, Marom D, McDermott MF, van der Merwe W, Orcesi S, Prendiville JS, Rasmussen M, Shalev SA, Soler DM, Shinawi M, Spiegel R, Tan TY, Vanderver A, Wakeling EL, Wassmer E, Whittaker E, Lebon P, Stetson DB, Bonthron DT, Crow YJ.** 2009. Mutations involved in Aicardi–Goutieres syndrome implicate SAMHD1 as regulator of the innate immune response. *Nat Genet* **41:**829–832.

164. **Rice GI, Forte GM, Szynkiewicz M, Chase DS, Aeby A, Abdel-Hamid MS, Ackroyd S, Allcock R, Bailey KM, Balottin U, Barnerias C, Bernard G, Bodemer C, Botella MP, Cereda C, Chandler KE, Dabydeen L, Dale RC, De Laet C, De Goede CG, Del Toro M, Effat L, Enamorado NN, Fazzi E, Gener B, Haldre M, Lin JP, Livingston JH, Lourenco CM, Marques W Jr, Oades P, Peterson P, Rasmussen M, Roubertie A, Schmidt JL, Shalev SA, Simon R, Spiegel R, Swoboda KJ, Temtamy SA, Vassallo G, Vilain CN, Vogt J, Wermenbol V, Whitehouse WP, Soler D, Olivieri I, Orcesi S, Aglan MS, Zaki MS, Abdel-Salam GM, Vanderver A, Kisand K, Rozenberg F, Lebon P, Crow YJ.** 2013. Assessment of interferon-related biomarkers in Aicardi–Goutieres syndrome associated with mutations in TREX1, RNASEH2A, RNASEH2B, RNASEH2C, SAMHD1, and ADAR: a case–control study. *Lancet Neurol* **12:**1159–1169.

165. **Bishop KN, Holmes RK, Sheehy AM, Davidson NO, Cho SJ, Malim MH.** 2004. Cytidine deamination of retroviral DNA by diverse APOBEC proteins. *Curr Biol* **14:**1392–1396.

166. **Bishop KN, Holmes RK, Sheehy AM, Malim MH.** 2004. APOBEC-mediated editing of viral RNA. *Science* **305:**645.

167. **Zheng YH, Irwin D, Kurosu T, Tokunaga K, Sata T, Peterlin BM.** 2004. Human APOBEC3F is another host factor that blocks human immunodeficiency virus type 1 replication. *J Virol* **78:**6073–6076.

168. **Harris RS, Bishop KN, Sheehy AM, Craig HM, Petersen-Mahrt SK, Watt IN, Neuberger MS, Malim MH.** 2003. DNA deamination mediates innate immunity to retroviral infection. *Cell* **113:**803–809.

169. **Harris RS, Sheehy AM, Craig HM, Malim MH, Neuberger MS.** 2003. DNA deamination: not just a trigger for antibody diversification but also a mechanism for defense against retroviruses. *Nat Immunol* **4:**641–643.

170. **Mangeat B, Turelli P, Caron G, Friedli M, Perrin L, Trono D.** 2003. Broad antiretroviral defence by human APOBEC3G through lethal editing of nascent reverse transcripts. *Nature* **424:**99–103.

171. **Bishop KN, Holmes RK, Malim MH.** 2006. Antiviral potency of APOBEC proteins does not correlate with cytidine deamination. *J Virol* **80:**8450–8458.

172. **Bishop KN, Verma M, Kim EY, Wolinsky SM, Malim MH.** 2008. APOBEC3G inhibits elongation of HIV-1 reverse transcripts. *PLoS Pathog* **4:**e1000231.

173. **Holmes RK, Malim MH, Bishop KN.** 2007. APOBEC-mediated viral restriction: not simply editing? *Trends Biochem Sci* **32:**118–128.

174. **Mbisa JL, Barr R, Thomas JA, Vandegraaff N, Dorweiler IJ, Svarovskaia ES, Brown WL, Mansky LM, Gorelick RJ, Harris RS, Engelman A, Pathak VK.** 2007. Human immunodeficiency virus type 1 cDNAs produced in the presence of APOBEC3G exhibit defects in plus-strand DNA transfer and integration. *J Virol* **81:**7099–7110.

175. **Keele BF, Giorgi EE, Salazar-Gonzalez JF, Decker JM, Pham KT, Salazar MG, Sun C, Grayson T, Wang S, Li H, Wei X, Jiang C, Kirchherr JL, Gao F, Anderson JA, Ping LH, Swanstrom R, Tomaras GD, Blattner WA, Goepfert PA, Kilby JM, Saag MS, Delwart EL, Busch MP, Cohen MS, Montefiori DC, Haynes BF, Gaschen B, Athreya GS, Lee HY, Wood N, Seoighe C, Perelson AS, Bhattacharya T, Korber BT, Hahn BH, Shaw GM.** 2008. Identification and characterization of transmitted and early founder virus envelopes in primary HIV-1 infection. *Proc Natl Acad Sci U S A* **105:**7552–7557.

176. **Coffin JM.** 1995. HIV population dynamics in vivo: implications for genetic variation, pathogenesis, and therapy. *Science* **267:**483–489.

177. **Meyer PR, Matsuura SE, Mian AM, So AG, Scott WA.** 1999. A mechanism of AZT resistance: an increase in nucleotide-dependent primer unblocking by mutant HIV-1 reverse transcriptase. *Mol Cell* **4:**35–43.

178. **Boyer PL, Sarafianos SG, Arnold E, Hughes SH.** 2001. Selective excision of AZTMP by drug-resistant human immunodeficiency virus reverse transcriptase. *J Virol* **75:**4832–4842.

179. **Mansky LM, Bernard LC.** 2000. 3′-Azido-3′- deoxythymidine (AZT) and AZT-resistant reverse transcriptase can increase the in vivo mutation rate of human immunodeficiency virus type 1. *J Virol* **74:**9532–9539.

180. **Koyama H, Ito T, Nakanishi T, Sekimizu K.** 2007. Stimulation of RNA polymerase II transcript cleavage activity contributes to maintain transcriptional fidelity in yeast. *Genes Cells* **12:**547–559.

181. **O'Neil PK, Sun G, Yu H, Ron Y, Dougherty JP, Preston BD.** 2002. Mutational analysis of HIV-1 long terminal repeats to explore the relative contribution of reverse transcriptase and RNA polymerase II to viral mutagenesis. *J Biol Chem* **277:**38053–38061.

182. **Mansky LM, Temin HM.** 1995. Lower in vivo mutation rate of human immunodeficiency virus type 1 than that predicted from the fidelity of purified reverse transcriptase. *J Virol* **69:**5087–5094.

183. **Mansky LM.** 1996. Forward mutation rate of human immunodeficiency virus type 1 in a T lymphoid cell line. *AIDS Res Hum Retroviruses* **12:**307–314.

184. **Abram ME, Ferris AL, Shao W, Alvord WG, Hughes SH.** 2010. The nature, position and frequency of mutations made in a single-cycle of HIV-1 replication. *J Virol* **84:**9864–9878.

185. **Abram ME, Ferris AL, Das K, Quinones O, Shao W, Tuske S, Alvord WG, Arnold E, Hughes SH.** 2014. Mutations in HIV-1 RT affect the errors made in a single cycle of viral replication. *J Virol* **88:**7589–7601.

186. **Roberts JD, Bebenek K, Kunkel TA.** 1988. The accuracy of reverse transcriptase from HIV-1. *Science* **242:** 1171–1173.

187. **Bebenek K, Abbotts J, Roberts JD, Wilson SH, Kunkel TA.** 1989. Specificity and mechanism of error-prone replication by human immunodeficiency virus-1 reverse transcriptase. *J Biol Chem* **264:**16948–16956.

188. **Rezende LF, Curr K, Ueno T, Mitsuya H, Prasad VR.** 1998. The impact of multidideoxynucleoside resistance-conferring mutations in human immunodeficiency virus type 1 reverse transcriptase on polymerase fidelity and error specificity. *J Virol* **72:**2890–2895.

189. **Boyer PL, Hughes SH.** 2000. Effects of amino acid substitutions at position 115 on the fidelity of human immunodeficiency virus type 1 reverse transcriptase. *J Virol* **74:**6494–6500.

190. **Mansky LM, Pearl DK, Gajary LC.** 2002. Combination of drugs and drug-resistant reverse transcriptase results in a multiplicative increase of human immunodeficiency virus type 1 mutant frequencies. *J Virol* **76:**9253–9259.

191. **Martin-Hernandez AM, Gutierrez-Rivas M, Domingo E, Menendez-Arias L.** 1997. Mispair extension fidelity of human immunodeficiency virus type 1 reverse transcriptases with amino acid substitutions affecting Tyr115. *Nucleic Acids Res* **25:**1383–1389.

192. **Jonckheere H, De Clercq E, Anne J.** 2000. Fidelity analysis of HIV-1 reverse transcriptase mutants with an altered amino-acid sequence at residues Leu74, Glu89, Tyr115, Tyr183 and Met184. *Eur J Biochem* **267:** 2658–2665.

193. **Wainberg MA, Drosopoulos WC, Salomon H, Hsu M, Borkow G, Parniak M, Gu Z, Song Q, Manne J, Islam S, Castriota G, Prasad VR.** 1996. Enhanced fidelity of 3TC-selected mutant HIV-1 reverse transcriptase. *Science* **271:**1282–1285.

194. **Pandey VN, Kaushik N, Rege N, Sarafianos SG, Yadav PN, Modak MJ.** 1996. Role of methionine 184 of human immunodeficiency virus type-1 reverse transcriptase in the polymerase function and fidelity of DNA synthesis. *Biochemistry* **35:**2168–2179.

195. **Feng JY, Anderson KS.** 1999. Mechanistic studies examining the efficiency and fidelity of DNA synthesis by the 3TC-resistant mutant (184V) of HIV-1 reverse transcriptase. *Biochemistry* **38:**9440–9448.

196. **Rezende LF, Drosopoulos WC, Prasad VR.** 1998. The influence of 3TC resistance mutation M184I on the fidelity and error specificity of human immunodeficiency virus type 1 reverse transcriptase. *Nucleic Acids Res* **26:** 3066–3072.

197. **Oude Essink BB, Berkhout B.** 1999. The fidelity of reverse transcription differs in reactions primed with RNA versus DNA primers. *J Biomed Sci* **6:**121–132.

198. **Hsu M, Inouye P, Rezende L, Richard N, Li Z, Prasad VR, Wainberg MA.** 1997. Higher fidelity of RNA-dependent DNA mispair extension by M184V drug-resistant than wild-type reverse transcriptase of human immunodeficiency virus type 1. *Nucleic Acids Res* **25:** 4532–4536.

199. **Drosopoulos WC, Prasad VR.** 1998. Increased misincorporation fidelity observed for nucleoside analog resistance mutations M184V and E89G in human immunodeficiency virus type 1 reverse transcriptase does not correlate with the overall error rate measured in vitro. *J Virol* **72:**4224–4230.

200. **Ji J, Loeb LA.** 1994. Fidelity of HIV-1 reverse transcriptase copying a hypervariable region of the HIV-1 *env* gene. *Virology* **199:**323–330.

201. **Ji JP, Loeb LA.** 1992. Fidelity of HIV-1 reverse transcriptase copying RNA in vitro. *Biochemistry* **31:** 954–958.

202. **Stuke AW, Ahmad-Omar O, Hoefer K, Hunsmann G, Jentsch KD.** 1997. Mutations in the SIV env and the M13 lacZa gene generated in vitro by reverse transcriptases and DNA polymerases. *Arch Virol* **142:** 1139–1154.

203. **Hu WS, Temin HM.** 1990. Genetic consequences of packaging two RNA genomes in one retroviral particle: pseudodiploidy and high rate of genetic recombination. *Proc Natl Acad Sci U S A* **87:**1556–1560.

204. **Robertson DL, Sharp PM, McCutchan FE, Hahn BH.** 1995. Recombination in HIV-1. *Nature* **374:**124–126.

205. **Anderson JA, Bowman EH, Hu WS.** 1998. Retroviral recombination rates do not increase linearly with marker distance and are limited by the size of the recombining subpopulation. *J Virol* **72:**1195–1202.

206. **Jetzt AE, Yu H, Klarmann GJ, Ron Y, Preston BD, Dougherty JP.** 2000. High rate of recombination throughout the human immunodeficiency virus type 1 genome. *J Virol* **74:**1234–1240.

207. **Zhuang J, Jetzt AE, Sun G, Yu H, Klarmann G, Ron Y, Preston BD, Dougherty JP.** 2002. Human immunodeficiency virus type 1 recombination: rate, fidelity, and putative hot spots. *J Virol* **76:**11273–11282.

208. **Dykes C, Balakrishnan M, Planelles V, Zhu Y, Bambara RA, Demeter LM.** 2004. Identification of a preferred region for recombination and mutation in HIV-1 gag. *Virology* **326:**262–279.

209. **Levy DN, Aldrovandi GM, Kutsch O, Shaw GM.** 2004. Dynamics of HIV-1 recombination in its natural target cells. *Proc Natl Acad Sci U S A* **101:**4204–4209.

210. **Rhodes T, Wargo H, Hu WS.** 2003. High rates of human immunodeficiency virus type 1 recombination: near-random segregation of markers one kilobase apart in one round of viral replication. *J Virol* **77:** 11193–11200.

211. **Rhodes TD, Nikolaitchik O, Chen J, Powell D, Hu WS.** 2005. Genetic recombination of human immunodeficiency virus type 1 in one round of viral replication: effects of genetic distance, target cells, accessory genes, and lack of high negative interference in crossover events. *J Virol* **79:**1666–1677.

212. **Galli A, Kearney M, Nikolaitchik OA, Yu S, Chin MP, Maldarelli F, Coffin JM, Pathak VK, Hu WS.** 2010. Patterns of Human Immunodeficiency Virus type 1 recombination ex vivo provide evidence for coadaptation of distant sites, resulting in purifying selection for intersubtype recombinants during replication. *J Virol* **84:**7651–7661.

213. **Coffin JM.** 1979. Structure, replication, and recombination of retrovirus genomes: some unifying hypotheses. *J Gen Virol* **42:**1–26.

214. **Etienne L, Hahn BH, Sharp PM, Matsen FA, Emerman M.** 2013. Gene loss and adaptation to hominids underlie the ancient origin of HIV-1. *Cell Host Microbe* **14:** 85–92.

215. **Fu W, Rein A.** 1993. Maturation of dimeric viral RNA of Moloney murine leukemia virus. *J Virol* **67:**5443–5449.

216. **Chin MP, Rhodes TD, Chen J, Fu W, Hu WS.** 2005. Identification of a major restriction in HIV-1 intersubtype recombination. *Proc Natl Acad Sci U S A* **102:** 9002–9007.

217. **Chen Y, Wu B, Musier-Forsyth K, Mansky LM, Mueller JD.** 2009. Fluorescence fluctuation spectroscopy on viral-like particles reveals variable gag stoichiometry. *Biophys J* **96:**1961–1969.

218. **Moore MD, Fu W, Nikolaitchik O, Chen J, Ptak RG, Hu WS.** 2007. Dimer initiation signal of human immunodeficiency virus type 1: its role in partner selection during RNA copackaging and its effects on recombination. *J Virol* **81:**4002–4011.

219. **Hussein IT, Ni N, Galli A, Chen J, Moore MD, Hu WS.** 2010. Delineation of the preferences and requirements of the human immunodeficiency virus type 1 dimerization initiation signal by using an in vivo cell-based selection approach. *J Virol* **84:**6866–6875.

220. **St Louis DC, Gotte D, Sanders-Buell E, Ritchey DW, Salminen MO, Carr JK, McCutchan FE.** 1998. Infectious molecular clones with the nonhomologous dimer initiation sequences found in different subtypes of human immunodeficiency virus type 1 can recombine and initiate a spreading infection in vitro. *J Virol* **72:** 3991–3998.

221. **Chin MP, Chen J, Nikolaitchik OA, Hu WS.** 2007. Molecular determinants of HIV-1 intersubtype recombination potential. *Virology* **363:**437–446.

222. **Delviks KA, Pathak VK.** 1999. Effect of distance between homologous sequences and 3′ homology on the frequency of retroviral reverse transcriptase template switching. *J Virol* **73:**7923–7932.

223. **Delviks KA, Pathak VK.** 1999. Development of murine leukemia virus-based self-activating vectors that efficiently delete the selectable drug resistance gene during reverse transcription. *J Virol* **73:**8837–8842.

224. **Rittinger K, Divita G, Goody RS.** 1995. Human immunodeficiency virus reverse transcriptase substrate-induced conformational changes and the mechanism of inhibition by nonnucleoside inhibitors. *Proc Natl Acad Sci U S A* **92:**8046–8049.

225. **Spence RA, Kati WM, Anderson KS, Johnson KA.** 1995. Mechanism of inhibition of HIV-1 reverse transcriptase by nonnucleoside inhibitors. *Science* **267:**988–993.

226. **Brown JA, Pack LR, Fowler JD, Suo Z.** 2011. Pre-steady-state kinetic analysis of the incorporation of anti-HIV nucleotide analogs catalyzed by human X- and Y-family DNA polymerases. *Antimicrob Agents Chemother* **55:**276–283.

227. **Lim SE, Ponamarev MV, Longley MJ, Copeland WC.** 2003. Structural determinants in human DNA polymerase gamma account for mitochondrial toxicity from nucleoside analogs. *J Mol Biol* **329:**45–57.

228. **Das K, Martinez SE, Bauman JD, Arnold E.** 2012. HIV-1 reverse transcriptase complex with DNA and nevirapine reveals non-nucleoside inhibition mechanism. *Nat Struct Mol Biol* **19:**253–259.

229. **Richman DD, Havlir D, Corbeil J, Looney D, Ignacio C, Spector SA, Sullivan J, Cheeseman S, Barringer K, Pauletti D, Shih C-K, Myers M, Griffin J.** 1994. Nevirapine resistance mutations of human immunodeficiency virus type 1 selected during therapy. *J Virol* **68:** 1660–1666.

230. **Johnson JA, Li JF, Morris L, Martinson N, Gray G, McIntyre J, Heneine W.** 2005. Emergence of drug-resistant HIV-1 after intrapartum administration of single-dose nevirapine is substantially underestimated. *J Infect Dis* **192:**16–23.

231. **Palmer S, Boltz V, Martinson N, Maldarelli F, Gray G, McIntyre J, Mellors J, Morris L, Coffin J.** 2006. Persis-

tence of nevirapine-resistant HIV-1 in women after single-dose nevirapine therapy for prevention of maternal-to-fetal HIV-1 transmission. *Proc Natl Acad Sci U S A* **103:**7094–7099.

232. **Das K, Lewi PJ, Hughes SH, Arnold E.** 2005. Crystallography and the design of anti-AIDS drugs: conformational flexibility and positional adaptability are important in the design of non-nucleoside HIV-1 reverse transcriptase inhibitors. *Prog Biophys Mol Biol* **88:** 209–231.

233. **Das K, Clark AD Jr, Lewi PJ, Heeres J, De Jonge MR, Koymans LM, Vinkers HM, Daeyaert F, Ludovici DW, Kukla MJ, De Corte B, Kavash RW, Ho CY, Ye H, Lichtenstein MA, Andries K, Pauwels R, De Bethune MP, Boyer PL, Clark P, Hughes SH, Janssen PA, Arnold E.** 2004. Roles of conformational and positional adaptability in structure-based design of TMC125-R165335 (etravirine) and related non-nucleoside reverse transcriptase inhibitors that are highly potent and effective against wild-type and drug-resistant HIV-1 variants. *J Med Chem* **47:**2550–2560.

234. **Sarafianos SG, Das K, Clark AD Jr, Ding J, Boyer PL, Hughes SH, Arnold E.** 1999. Lamivudine (3TC) resistance in HIV-1 reverse transcriptase involves steric hindrance with β-branched amino acids. *Proc Natl Acad Sci U S A* **96:**10027–10032.

235. **Gao HQ, Boyer PL, Sarafianos SG, Arnold E, Hughes SH.** 2000. The role of steric hindrance in 3TC resistance of human immunodeficiency virus type-1 reverse transcriptase. *J Mol Biol* **300:**403–418.

236. **Meyer PR, Matsuura SE, Tolun AA, Pfeifer I, So AG, Mellors JW, Scott WA.** 2002. Effects of specific zidovudine resistance mutations and substrate structure on nucleotide-dependent primer unblocking by human immunodeficiency virus type 1 reverse transcriptase. *Antimicrob Agents Chemother* **46:**1540–1545.

237. **Ren J, Nichols C, Bird L, Chamberlain P, Weaver K, Short S, Stuart DI, Stammers DK.** 2001. Structural mechanisms of drug resistance for mutations at codons 181 and 188 in HIV-1 reverse transcriptase and the improved resilience of second generation non-nucleoside inhibitors. *J Mol Biol* **312:**795–805.

238. **Das K, Sarafianos SG, Clark AD Jr, Boyer PL, Hughes SH, Arnold E.** 2007. Crystal structures of clinically relevant Lys103Asn/Tyr181Cys double mutant HIV-1 reverse transcriptase in complexes with ATP and non-nucleoside inhibitor HBY 097. *J Mol Biol* **365:** 77–89.

239. **Hsiou Y, Das K, Ding J, Clark AD Jr, Kleim JP, Rosner M, Winkler I, Riess G, Hughes SH, Arnold E.** 1998. Structures of Tyr188Leu mutant and wild-type HIV-1 reverse transcriptase complexed with the non-nucleoside inhibitor HBY 097: inhibitor flexibility is a useful design feature for reducing drug resistance. *J Mol Biol* **284:**313–323.

240. **Hsiou Y, Ding J, Das K, Clark AD Jr, Boyer PL, Lewi P, Janssen PA, Kleim JP, Rosner M, Hughes SH, Arnold E.** 2001. The Lys103Asn mutation of HIV-1 RT: a novel mechanism of drug resistance. *J Mol Biol* **309:**437–445.

Mobile DNA, 3rd Edition
Nancy L. Craig, Michael Chandler, Martin Gellert, Alan M. Lambowitz, Phoebe A. Rice and Suzanne Sandmeyer

doi:10.1128/microbiolspec.MDNA3-0009-2014

Dixie L. Mager[1]
Jonathan P. Stoye[2]

47 Mammalian Endogenous Retroviruses

INTRODUCTION

Mammalian genomes have accumulated millions of retrotransposed sequences during evolution. This material can be divided into long terminal repeat (LTR) retrotransposons that include the endogenous retroviruses (ERVs), as well as long and short retrotransposons lacking LTRs, known as LINEs and SINEs, respectively. ERVs are defined as inherited genetic elements closely resembling the proviruses formed following exogenous retrovirus infection. In this chapter we describe the discovery, classification, and origins of ERVs in mammals, consider cellular mechanisms that have evolved to control their expression, and discuss the biological consequences, both positive and negative from the host's standpoint, of ERV inheritance.

DISCOVERY OF ERVs

In the mid-1960s, prompted in large part by the economic impact of a leukosis-causing retrovirus in chicken flocks, serological assays for antibodies to viral Gag proteins were developed to identify birds exposed to avian leukosis virus (ALV) infection. Strangely, a number of birds scored positive in the absence of overt infection and apparent exposure followed Mendelian inheritance. Simultaneously, further virological studies revealed the expression of functional Env proteins in certain birds; expression of viral Gag and Env cosegregated in genetic crosses. Together these findings suggested that the viral proteins were encoded in the chicken genome (1). In retrospect, with our current knowledge of the retroviral life cycle, this suggestion would be regarded as uncontroversial. However, at that time retroviruses were still thought of purely as RNA viruses and the suggestions of proviruses in DNA form, despite the knowledge that phage could lysogenize, were met with considerable skepticism. It was only with the discovery of an enzyme capable of converting RNA into DNA, reverse transcriptase, that the idea of an ERV, inherited in the germ line as DNA, became palatable (Fig. 1A).

At about the same time, similar experiments were being conducted in the murine system, with evidence accumulating that retroviruses could be induced by radiation or chemicals from apparently virus-free cells (2) and that retrovirus-related sequences could be detected by liquid hybridization in genomic DNA (3). Two classes of murine ERVs were detected in this way, mouse

[1]Terry Fox Laboratory, British Columbia Cancer Agency and Dept. of Medical Genetics, University of British Columbia, Vancouver, BC, Canada; [2]MRC National Institute for Medical Research, The Ridgeway, Mill Hill, London, UK.

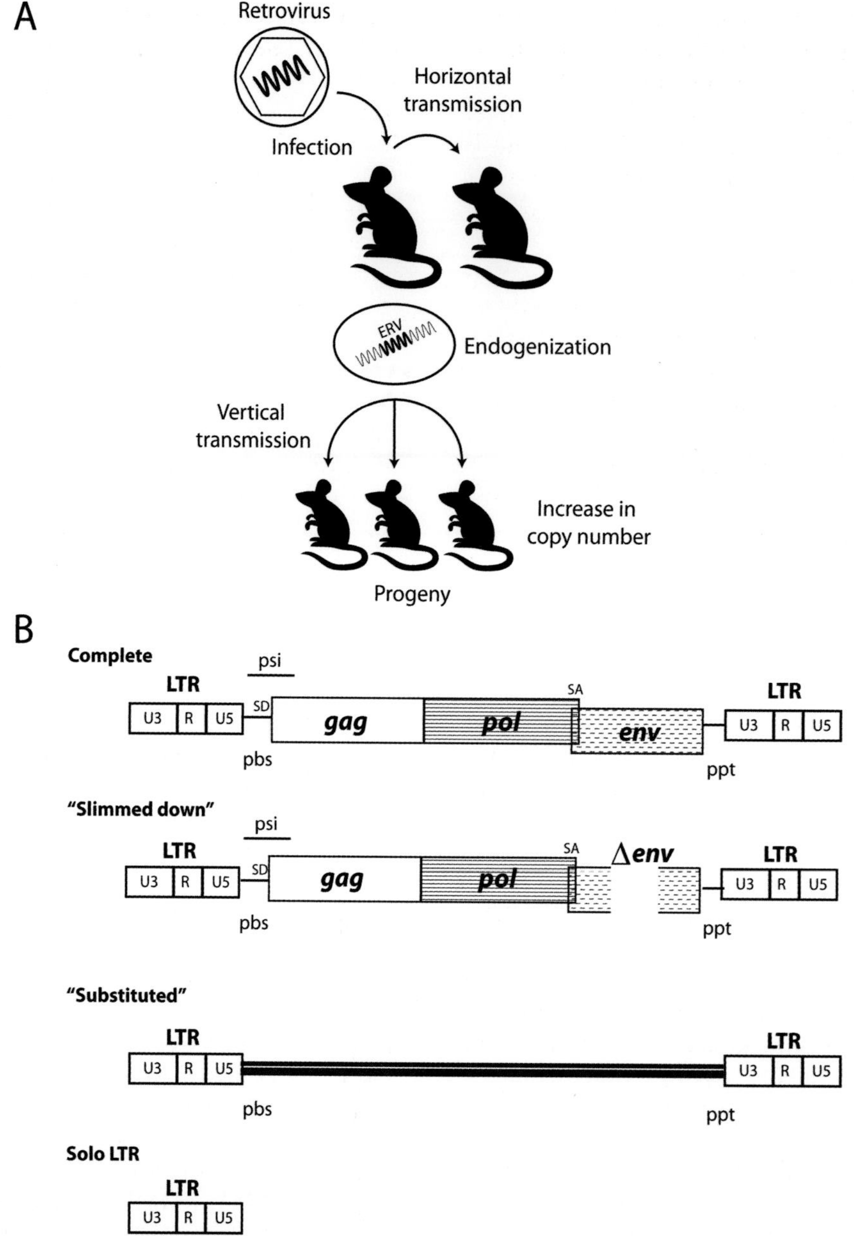

Figure 1 (A) Formation of ERVs. Exogenous retroviruses typically infect their host and spread to other individuals via horizontal transmission. When retroviruses infect and integrate into the genome of germ line cells, the provirus can be vertically transmitted and become endogenous to the host. ERVs can amplify in the host genome through reinfection or through intracellular retrotransposition (see text). (B) Basic structures of ERVs. Complete ERVs are essentially identical to the integrated proviruses of simple exogenous retroviruses; they contain two LTRs made up of unique 3′ (U3), repeat (R), and unique 5′ (U5) regions, a primer binding site (pbs) and polypurine tract (ppt), as well as a full complement of coding sequences (*gag*, *pol*, and *env*), splice donor (SD) and acceptor (SA) sites, and an RNA packaging signal (psi). "Slimmed down" ERVs are elements lacking coding sequences compared to a complete ERV—here illustrated with a deletion in *env*. "Substituted" ERVs are elements in which the ERV coding sequences have been replaced with nonviral sequences. "Solo LTRs" are single LTRs generated by homologous recombination between the two LTRs of a complete element. doi:10.1128/microbiolspec.MDNA3-0009-2014.f1

mammary tumor virus (MMTV) and murine leukemia virus (MLV). As their names suggest, ALV, MMTV, and MLV can cause cancer or, if they do not, are related to cancer-causing viruses. This raised the question, still unresolved (see "Pathogenic effects of ERVs and related sequences on the host" below), of ERV involvement in human cancer, and prompted an ongoing series of searches for ERVs in other species, particularly in humans.

Two different approaches have been utilized in these experiments (4). The first, which might be labeled biological, is based on the functional properties of an active retrovirus and involves the search for a replicating virus. The second, which might be considered structural, looks for nucleic acid sequences related to known retroviruses or having the sequence organization of a retrovirus. An example of the first approach involves treatment of cells with inducing agents, agents that have already been shown to activate ERVs in other species (such as the demethylating agent 5-azacytidine), and coculture of these cells or the cell-free supernatant with an appropriate indicator cell to monitor for the appearance of viral markers such as reverse transcriptase activity (5). In this way, infectious ERVs have been recovered from multiple species but not from humans. Alternately, cells from a tumor might be introduced into animals unable to mount an immune response and the virus production monitored. This approach has mainly been employed to try to find agents involved in tumor causation; ironically, viruses isolated in this fashion tend to represent ERVs from the transplanted animal rather than from the tumor itself (6). A number of false associations between retroviruses and cancer have been reported as a result of such studies; these are frequently referred to as "rumor viruses" (7).

Structural approaches are exemplified by early experiments to clone human ERVs using low-stringency hybridization with cloned ERV probes obtained from other species (8, 9). In this way sequences related to the murine viruses MLV and MMTV were cloned and characterized. Later, ERVs from chickens lacking endogenous ALVs were cloned in a similar fashion (10). Subsequently, PCR-based approaches using primers complementary to conserved regions of *pol* became more widely used (11, 12, 13). These experiments began to reveal the number and diversity of ERVs present in the DNA of different species.

Ultimately, it was the analysis of sequence data (14, 15), particularly when whole genome sequences became available (16, 17, 18, 19), that revealed the full contribution of ERVs to the compositions of vertebrate genomes. Indeed, the initial analyses of the human genome sequence showed that a larger percentage of our DNA comprises ERV-related material (8%) than encodes proteins (1 to 2%). Since the completion of the first human sequence, DNA-sequencing technology has improved dramatically with a corresponding decrease in costs. This has led to the completion of multiple genome-sequencing projects. Simultaneously, a variety of *in silico* techniques for identifying and annotating potential ERVs have been developed (20). These entail searches for conserved sequences and/or sequence motifs organized in the same manner as exogenous retroviruses (21). Using such mining techniques, comprehensive ERV collections have been obtained from ever-increasing numbers of species (22, 23). Thus, one recent paper refers to 87,750 defined ERVs identified in an analysis of 60 vertebrate genomes (24), and these numbers can only increase with time. Studies of their relationship to one another and their biological impact therefore present significant challenges.

STRUCTURE AND CLASSIFICATION

Examination of the genetic structures of different LTR-retrotransposons identified in genomic DNA reveals four fundamental forms (Fig. 1B). First are structures essentially indistinguishable from the integrated proviruses of exogenous retroviruses and will be referred to as "complete" ERVs. They contain two LTRs ranging in size from 300 to 1000 bp with signals to regulate transcription separated by around 6 to 9kb encoding the viral Gag, Pol, and Env proteins. Signals for RNA polyadenylation, splicing, and packaging in virions are also present, as well as the tRNA primer-binding site (PBS) and polypurine tract that define boundaries of the LTRs. The ERV sequences are flanked by a short (4 to 6 bp) duplication of host sequences. The endogenous MLVs of mice (25) are well-characterized members of this form of ERVs. With a relatively few exceptions, notably HERV-K (HML-2) in primate genomes (26), genes encoding viral accessory proteins are not present in ERVs (27).

A second group, which might be thought of as "slimmed down" retroviruses that lack one or more coding sequence (usually the *env* gene) essential for autonomous extracellular replication, but not ruling out an intracellular replication cycle resembling the Ty elements of yeast. In contrast to the retrotransposable elements of yeast, it seems likely that these elements derive from complete proviruses by the loss of specific genes and they may carry an altered assembly/budding signal in *gag*. Examples include the intracisternal A type particle (IAP) (28) and MusD (29) ERVs of mice.

A third group contains elements with two LTRs and an appropriately positioned PBS and polypurine tract, but no other recognizable homology with retroviral proteins. Perhaps such elements arose by mechanisms of read-through transcription and illegitimate recombination similar to those thought to be associated with oncogene transduction by retroviruses (30). Provided they contain appropriate RNA packaging sequences, such elements might be replicated in a manner similar to the genomes of retroviral vectors. Examples include the early transposons (ETns) of mice (31), which are retrotransposed by the related MusD ERVs (29), and THE1 mammalian apparent LTR-retrotransposon (MaLR) elements in humans (14). Overall, these non-autonomous elements have received relatively little attention, although ETn elements are still retrotransposing in mice (32).

The final group of sequences, present in an approximately 10-fold excess in the genome over the other groups, are the so-called "solo LTRs." Although they contain a number of short motifs essential for integration into chromosomal DNA and the control of RNA transcription, they are difficult to identify from scratch (33) and are usually found by their sequence similarity to the LTRs of complete viruses described above. These are thought to arise by homologous recombination between the LTRs of complete elements (34). In one case, an MLV associated with the dilute coat-color mutation of mice, in which recombination between the LTRs causes a readily identifiable phenotypic effect, the rate of proviral loss has been estimated at 4.5×10^{-6} excisions per meiosis (35).

Despite the abundance of ERV elements revealed by genome sequencing only a few replicating ERVs have been isolated by induction/culture techniques. Examination of the predicted ERV coding sequences provides a ready explanation for this observation. Very few elements have complete open reading frames for Gag, Pol, and Env (36, 37). Most elements contain numerous mutations, consistent with neutral evolution for significant periods of time and lengthy periods of residence in the germ line, although some proviruses reveal evidence for inactivating mutations that occurred prior to integration (38). In the human genome, even the most recent ERV insertions, those of the HERV-K (HML-2) group, require multiple changes to allow the rescue of infectious viruses (39, 40).

Classifying and naming ERVs, particularly down to the level of individual proviral loci, remains a considerable problem (41, 42), so far without final resolution. ERV sequences with internal homology to retroviruses can be placed into one of three classes based on phylogenetic analyses of the conserved regions of the *pol* gene. Class I elements are most closely related to the exogenous gamma- and epsilonretroviruses, Class II to the alpha- and betaretroviruses, and Class III to the spumaviruses (see Table 1 for ERV representatives of each class). The large category of MaLR LTR elements, although they lack a detectable *pol*-related sequence, is sometimes also considered "Class III" due to the slight homology of some members to Class III *gag* sequences. These classes can be further subdivided into groups (not families, a term reserved for higher order taxa of viruses) each apparently derived from individual germ

TABLE 1 Representative ERVs

Class	Related exogenous class	Group	Species[a]
I	gamma	MLV (murine leukemia virus)	Mouse
	gamma	VL30 (virus like 30)	Mouse
	gamma	KoRV (koala retrovirus)	Koala
	gamma	HERV-H (human ERV with H-tRNA PBS)	Human
	gamma	HERV-W (human ERV with W-tRNA PBS)	Human
	gamma	HERV-E (human ERV with E-tRNA PBS)	Human
II	beta	MMTV (mouse mammary tumor virus)	Mouse
	beta	IAP (intracisternal A-type particle)	Mouse
	beta	MusD (Mus type-D related retrovirus)	Mouse
	beta	HERV-K (human ERV with K-tRNA PBS)	Human
	beta	enJSRV (Jaagsiekte sheep retrovirus)	Sheep
	alpha	ALV (avian leukosis virus)	Chicken
III	spuma	MERV-L (mouse ERV with L-tRNA PBS)	Mouse
	spuma	HERV-L (human ERV with L-tRNA PBS)	Human
	spuma	MaLR (mammalian apparent LTR retrotransposon)	Human/mouse

[a]All ERVs in human are also found in other primates.

line infections (43). However, a number of discrepancies arise when ERVs are analyzed on the basis of their *env* genes, a finding that illustrates the confounding effects of recombination on phylogenetic analysis (44).

The nomenclature system for ERVs has developed in a manner reflecting the history of their discovery. It first became common to name endogenous proviruses after the most closely related exogenous retrovirus, such as MLV, as well as subgroups, like xenotropic MLV. It is now standard to add one or two letters before the designation ERV to indicate the species in which they were initially identified; thus, HERV indicates an ERV first seen in human DNA and MERV or MuERV implies one originally found in mice. HERVs are further classified on the basis of the tRNA that binds to the viral PBS to prime reverse transcription. Hence, HERV-K implies a provirus or groups of proviruses that utilize a lysine tRNA, no matter their relationship to one another. In some cases the PBS sequence was not available when novel elements were first discovered, leading to the names based on neighboring genes or the probe used for cloning. A further problem is to distinguish different members of a phylogenetic group that differ by integration site rather than coding sequence and possibly are present in hundreds of copies. Nomenclature became even more complicated with the realization that the majority of ERVs entered the germ line before modern species were established (see below). If a provirus integrated at a given position in the genome is present in both chimpanzee and human, should it be called a CERV or a HERV? Further, how do you indicate that these orthologous proviruses in different species are more closely related to one another than to other proviruses of the same viral group within one species? Another complicated aspect of ERV nomenclature is that Repbase, the widely used and valuable database of genomic repeats (45), by necessity designates groups of ERVs and solitary LTRs with numerical names that can be confusing. For example, mouse LTRs related to MLV are designated RLTR4 in Repbase and LTRs related to the primate HERV9 group are designated as LTR12. Helpful tables listing various naming schemes for human ERVs and associated LTRs have been published (46, 47). However, considerable confusion remains; the development of a rational scheme of classification and nomenclature for all ERVs would be of great value.

Studies of the distribution of different ERVs in the genomes of multiple species, using *in silico* methods such as RepeatMasker and RetroTektor, are now providing a series of valuable insights into the evolution of retrovirus–host interactions. Class I, II, and III ERVs, along with the ancient MaLR elements, are present in all mammalian species tested, but their relative proportions and abundance differ from species to species (48, 49), reflecting different evolutionary dynamics within the host (Fig. 2). One notable example of different ERV distributions among mammals is the significant expansion of Class II ERVs in the mouse. The number of different groups of sequences belonging to each Class ranges from one to 20 (43) and each group can vary in size from one to in excess of a thousand. It is possible to examine their integration sites relative to other features of the genome to investigate the genetic impact of inherited elements (see below). One can define novel elements and seek to define the origins of ERVs (49). Transmissions between species and, more rarely, between different evolutionary orders can be demonstrated (24) and hypotheses regarding their possible pathogenic potential developed (50). One interesting set of data reveals that ERVs lacking an *env* gene have been particularly successful in amplifying their copy number (23), thus constituting genomic superspreaders.

ACQUISITION AND AMPLIFICATION

Two methods have been used to estimate the age of specific ERV elements, both based on the properties of retroviral replication, thought to be the mechanism underlying ERV acquisition and amplification (see below). The first relies on the observation that the choice of retroviral integration site is essentially random. It follows that proviruses mapped at exactly the same position in two different species are likely to be descendants from the same integration event. The presence of such orthologous ERVs, which can readily be detected by PCR using primers directed to conserved flanking sequences, implies integration before divergence and provides a minimum estimate for the time of integration. This method, although it requires integration site data from multiple species, has found wide use in dating primate proviruses (51, 52). The second method takes advantage of the fact that the process of reverse transcription generates two LTRs with identical sequences and thus no differences are present in the LTRs of a newly integrated provirus. Differences will subsequently accumulate at a rate proportional to the host organism's neutral mutation rate. Thus, the number of differences between the two LTRs of a given provirus will provide an estimate of the time that element was integrated (43). This method can be applied to single genomic sequences but can be compromised by reverse transcriptase errors and, particularly for older ERVs, by

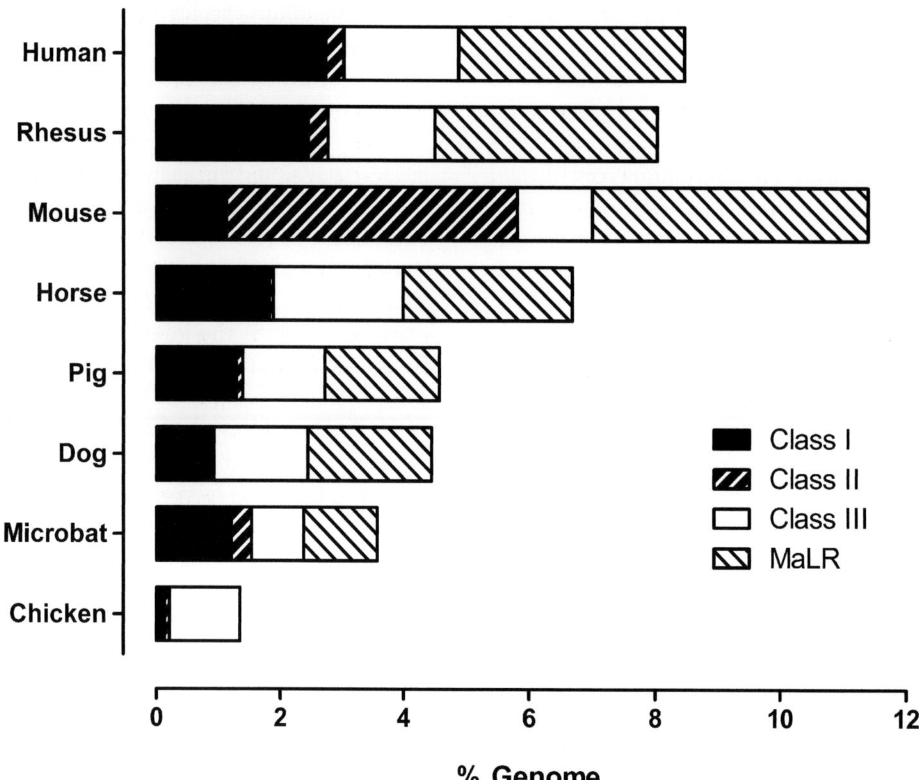

Figure 2 ERV/LTR content in different species. Overall genomic fractions occupied by ERVs and solo LTRs in various species. Data were obtained from the RepeatMasker web site (http://www.repeatmasker.org/genomicDatasets/RMGenomicDatasets.html) and are updated from the original genome publications—for example, the human and mouse genome papers (18, 19) due to a more-sensitive repeat detection.
doi:10.1128/microbiolspec.MDNA3-0009-2014.f2

the insertion of other elements, by back mutation, and by gene conversion (53).

Using these methods a number of conclusions may be drawn about the history of ERV and their hosts. These include: (i) the formation of novel ERVs has been occurring continuously during the course of mammalian evolution (54); (ii) in their modern form retroviruses essentially have existed for tens, if not hundreds, of millions of years (27, 55); (iii) amplification of a given ERV group can continue gradually for tens of millions of years, sometimes interspersed with bursts of amplification, either through intercell infections or retrotransposition, but it seems to come to a stop eventually (52, 56); (iv) in some species, including humans, germ line colonization by novel ERVs has virtually ceased, whereas in others, like mice, it continues apparently unabated (57, 58). It is unclear why mammalian species differ in their level of present-day ERV activity but such differences could result from population bottlenecks, host restriction genes, and/or random chance.

Most original endogenizations took place in the distant past, but a series of experiments, mainly carried out in mice in the 1970s and 1980s, allows us to deduce the likely events leading to the current repertoire of ERV elements. First came experiments showing that the *in vitro* infection of preimplantation embryos with Moloney MLV followed by reimplantation leads to the development of novel Moloney MLV-derived ERVs in up to half the exposed embryos (59). Analogous experiments were later performed with murine embryonic stem (ES) cells and fertilized chicken eggs. Next came experiments to examine the spontaneous germ line amplification of MLV proviruses that had been observed to occur in strains of mice naturally expressing high levels of endogenous MLV in somatic tissues. It rapidly became apparent that germ line amplification only occurred in the progeny of viremic females but not in viremic males, implying a requirement for cells or tissues of the female germ line in amplification (60). This idea was tested directly in transplant experiments (61). Ovaries from nonviremic females were implanted into the ovarian bursas of viremic animals. Novel ERVs were shown to be present in the progeny of the virus-negative transplanted oocytes, demonstrating infection after transplantation. Further transplantation experiments of virus-negative oocytes into females that were viremic as a result of infection rather than genomic inheritance also revealed the generation of novel ERVs. It therefore seems reasonable to conclude that virus infection, possibly from a different species, could result in viremia and the infection of oocytes. Following fertilization and the birth of progeny, the novel endogenous elements, provided that they do not prove harmful, could then be fixed by random breeding. Thus, a new group of ERVs could colonize a novel species. Once established, amplification could then occur by extracellular reinfection of oocytes recapitulating the original infection.

Different ERV groups are present in very different copy numbers, indicating that amplification can follow very different trajectories. The factors responsible for these different fates are unclear, but might involve the rapid mutational loss of genes required for amplification or the development of host mechanisms to suppress replication (54, 56). Alternately, the ERV might undergo a more dramatic genetic change. The evolutionary success of the superspreaders, such as IAP elements in mice, suggests the development of an intracellular replication strategy resembling the retrovirus life cycle but lacking budding from and reentering the cell, might be advantageous (23). Perhaps replication in this manner is intrinsically more efficient; alternately, it avoids host responses to infection (see below). A further evolution of ERVs might involve loss of the sequences encoding replicative functions, such as Gag and Pol. If these can be supplied in *trans* in a manner resembling the preparation of a retroviral vector, and are replaced by an RNA that is efficiently packaged and reversed transcribed, evolution of a nonautonomous element might take place.

Until recently, the opportunity for following the process of endogenization in a natural setting had not arisen. However, this situation has now changed with the observation of a spreading wave of germ line colonization in the koala by a virus called KoRV-A. This virus is present in variable copy number in genomic DNA of all koalas present in the north of Australia but appears absent in a fraction of the animals found in the south (62, 63). High levels of expression are seen in newborn joeys, but it remains unclear whether this results from the endogenous virus or the concomitant exogenous virus spreading. KoRV expression is associated with lymphomas in captive koalas (64). However, here the relative contributions of endogenous and exogenous virus are also unclear. Initial estimates suggested that KoRV first became endogenous around 100 years ago and was derived from an ERV present in rodent species found in Southeast Asia that also gave rise to a virus that caused disease in gibbon apes (63). Although neither inference can be considered conclusive (65, 24), the current epidemic offers a number of opportunities for studying the process of cross-species infection leading to endogenization. For example, it should present opportunities to identify the source of the virus in koalas, assess how it reached the geographically isolated target population, how copy number is controlled, and whether KoRV shows a tendency to become less pathogenic when establishing itself as a neo-ERV. Adding urgency to these experiments is the possibility that KoRV might contribute to the extinction of an iconic species.

HOST DEFENSES AGAINST ERVs

Given their potential for harm, genomes have evolved multiple lines of defense against ERVs involving both epigenetic modifications that curtail ERV transcription as well as host protein restriction factors that target other phases of the ERV replication cycle (54, 66, 67, 68, 69, 70).

Epigenetic Mechanisms

Epigenetics of ERV Suppression

In the mouse it has been shown that, depending on tissue type, ERV transcription is suppressed by DNA methylation and/or repressive histone modifications (for reviews, see references 68, 69, 71). DNA methylation in mammals is generally associated with transcriptional silencing and nearly always involves the addition of a methyl group to the fifth carbon of cytosine at CpG sites, catalyzed by three DNA methyltransferases (Dnmts), the *de novo* Dnmt3a and Dnmt3b, and the maintenance Dnmt1 (72). DNA methylation plays an important role in suppressing mouse ERV transcription in somatic cells (73, 74, 75), as well as in the later stages of germ line development (76). During epigenetic reprogramming in preimplantation mouse embryos, loss of DNA methylation occurs and numerous ERVs, including MERV-L and ETn/MusD elements, are derepressed and expressed at very high levels (32, 77, 78). In *Dnmt1–/–* mouse embryos, both MLV and IAP ERVs become demethylated, and IAP transcripts are expressed up to 100-fold higher relative to wild-type embryos (75, 79).

In humans, although ERV transcripts are detectable in many cell types (80), several investigations have shown that most ERVs are heavily methylated in somatic cells with demethylation associated with transcription or enhancer function (81, 82, 83, 84, 85, 86). The CpG substitution rate in primate genomes is significantly higher in repeats such as L1 retroelements and ERVs compared to nonrepetitive DNA, suggesting a homology-dependent methylation mechanism (87). Indeed, a specialized small RNA (piRNA) pathway promotes DNA methylation of mouse ERVs in the developing the male germ line (88). There is evidence that small RNA pathways promote DNA methylation and suppression of the activity of human LINE1 non-LTR retroelements (89), but their role in human ERV suppression is unclear.

Although DNA methylation is important for the silencing of ERVs in somatic tissues, histone modifications also play major roles, particularly in undifferentiated and/or stem cells. H3K9me2 or H3K9me3, both associated with transcriptional repression (90), have major roles in mouse ERV suppression. In mouse ES cells, H3K9me3 is associated with Class I and II ERVs (but not Class III ERVs) (91) and is required for the silencing of these ERVs, including IAP elements (92, 93). Many ERV groups are derepressed in H3K9me3-deficient but not Dnmt triple knockout (*Dnmt1–/–*, *Dnmt3a–/–*, *Dnmt3b–/–*) ES cells. This derepression leads to upregulation of several genes as a result of ERV LTRs acting as alternative promoters (94). H3K9me3 is deposited on mouse Class I and II ERVs by the SETDB1/KAP1 complex (92, 93). As KAP1 interacts with Krüppel-associated box zinc finger proteins (KRAB-ZFPs), of which there are several hundred in mammals (95), it has been proposed that the SETDB1/KAP1 complex is recruited to distinct ERVs by a variety of different KRAB-ZFPs (69, 71, 96). Indeed, a particular KRAB-ZFP (ZNF809) is required for the silencing of MLV in ES cells through binding to the MLV primer binding site (PBS) (96). Moreover, genomic comparative studies suggest that the rapid evolution and expansion of KRAB-ZFPs in mammalian genomes has been an "arms-race" response to invasions of retroviruses in the past (95).

H3K9me3 is not the only repressive mark associated with ERV silencing. H3K9me2 is required for the silencing of a distinct set of ERVs, predominantly Class III MERV-L elements, in mouse ES cells, and, when this mark is depleted, some genes are upregulated via upstream MERV-L LTRs (97). Aberrant expression of MERV-L-driven chimeric gene transcripts also occurs in ES cells deficient for the H3K4 demethylase Lsd1 (98). Upregulation of MLV and IAP ERVs, associated with loss of the repressive mark H3K27me3, was also observed in mouse ES cells deficient for H3K27 methyltransferases (99). Thus, both DNA methylation and various repressive histone modifications are involved in ERV transcriptional silencing.

Epigenetics of Active ERVs

Although most studies have focused on the epigenetic suppression of ERVs, those elements that do show promoter or enhancer activity appear to be associated with active epigenetic marks, as would be expected (100). For example, tissue-specific hypomethylated LTRs and other retroelements gain H3K4me1 (an enhancer mark) and exhibit enhancer activity in reporter assays (86). Mouse ES cells depleted of Kap1 lose the repressive mark H3K9me3 at IAP ERVs and gain enhancer marks H3K27ac and H3K4me1, which correlates with their upregulation and the induction of a subset of nearby genes (101). Mouse LTRs close to actively transcribing genes can acquire the active promoter mark H3K4me3, are unmethylated, and can act as alternative promoters for the neighboring gene (102). In mice harboring a large portion of human chromosome 21, many normally silent LTRs on this chromosome are transcribed and are associated with a lower DNA methylation and active epigenetic marks (103), suggesting that the lack of species-specific repressive mechanisms can lead to widespread transcriptional activation of ERV sequences.

Restriction Factors

A number of cellular proteins interfering with retrovirus replication have been described. Several were first identified as factors blocking HIV-1, but more detailed studies revealed activity against multiple kinds of virus including ERVs (54, 104, 105). Foremost on this list are members of the APOBEC3 family of cytidine deaminases. APOBEC3G was first shown to inhibit HIV-1 replication (106) with plausible mechanisms, including lethal mutation of viral genomes by cytosine deamination and interference with reverse transcription (107). There are 11 members of this protein family, several with antiretrovirus or antiretrotransposon activity. TRIM5α (108) belongs to an even larger family of cellular proteins, many of which have roles in the innate immune system (109). TRIM5α binds to the capsid protein of retroviruses shortly after the virus enters target cells and acts to block reverse transcriptase and/or nuclear entry (110). Another restriction factor is tetherin (111), a dimeric protein with membrane anchors at its N- and C-terminal ends that tethers newly budded enveloped virions from cells. SAMHD1 seems to block virus replication in certain cell types by reducing the levels of deoxyribonucleoside 5′-triphosphates below those needed by reverse transcriptase (112).

These factors can act to limit the replication of both exogenous and endogenous viruses; their primary function thus remains open to question (54). Nevertheless, it is clear that they can restrict ERVs that are expressed despite epigenetic controls. Thus, APOBEC3B can act as a brake on LINE amplification (113) and a protective allele of Fv1, a mouse protein functionally related to TRIM5α, can prevent ERV-mediated "spontaneous" leukemia in mice (114). One attractive idea is that recently acquired ERVs might act to "educate" their hosts, their presence in the germ line driving the evolution of restriction factors with specificity for currently circulating external threats.

An alternative mechanism for host restriction of the virus life cycle is the mutation of host proteins required for viral replication. One common theme is involvement of the cellular receptor for the virus. Mutations affecting virus binding to the receptor without affecting its normal function arise with significant frequency (115). This provides a convenient means for the host to prevent ERV amplification and provides an explanation for the common phenomenon of "xenotropism," whereby ERVs are unable to reinfect the cells of a host in which they have established a home. For example, MLVs are classified as ecotropic (able to infect mouse), xenotropic (unable to infect mouse), or polytropic (able to infect both mouse and other species), based on the ability of their env glycoproteins to recognize the host-encoded receptor. This phenomenon provides a simple explanation for the frequent isolation of ERVs following the passage of human tumors in immunodeficient animals (6) and suggests a selection pressure for the evolution of ERVs with amplification cycles lacking an extracellular phase.

PATHOGENIC EFFECTS OF ERVs AND RELATED SEQUENCES ON THE HOST

Cancer

It is well established that exogenous retroviruses can cause diseases such as AIDS in humans and cancer in animals. The carcinogenic effects of retroviruses are usually the result of either introducing a transduced oncogene or via insertional activation of a host protooncogene by providing ectopic enhancers or promoters (Fig. 3A) (116). Indeed, retroviruses have been widely used as a tool to detect new oncogenes via mapping of common insertion sites (117). Replication-competent ERVs closely related to exogenous counterparts, such as MMTV and MLV, can cause malignancy via reinfection and insertional activation of protooncogenes (116). In AKR mice, specifically selected for a high incidence of cancer, the retroviruses that cause malignancy are the product of recombination between different types of endogenous MLV (118). Mobilization of IAP ERVs in mice, which have adopted an intracellular life cycle as a retrotransposon (28), can also activate protooncogenes, inducing lymphomas in mice with reduced levels of the maintenance DNA methyltransferase DNMT1 (73) and myeloid leukemia after irradiation in certain inbred strains (119).

Human ERVs are often transcriptionally upregulated in malignancies, likely due, at least in part, to genome-wide DNA hypomethylation (120), but a causative role for such sequences in cancer remains largely speculative (26, 121). HERV-K (HML-2), the youngest family of ERVs in the human genome, has been extensively investigated for potential pathogenicity and some evidence suggests that the HERV-K-encoded accessory proteins Rec and Np9 have a carcinogenic role, particularly in germ cell tumors, but also in some other cancers (26, 122, 123). As ongoing retrotranspositional activity of ERVs is low or nonexistent in modern humans (46), it is not surprising that no instances of activation of oncogenes via new ERV insertions have been found in human malignancy. However, it is possible that existing ERVs or solo LTRs could become transcriptionally active and drive ectopic protooncogene expression, as shown in Fig. 3A. Indeed, this scenario has been reported in Hodgkin lymphoma. In this case, the growth-factor receptor CSF1R is ectopically expressed from a hypomethylated solo LTR of the MaLR THE group specifically in Hodgkin lymphoma cells (124). Moreover, these cells are dependent on CSF1R for survival (124). Unlike cases of evolutionary "exaptation" of LTRs where they participate in normal gene regulation (see "ERV/LTR mediated gene regulation" below), this particular THE LTR is transcriptionally silent in normal cells. While intriguing, it remains to be determined how prevalent or significant this phenomenon is with respect to human cancer.

Germ Line Mutations

As insertional mutagens, ERVs have obvious potential to disrupt genes and cause mutations. In the human, no

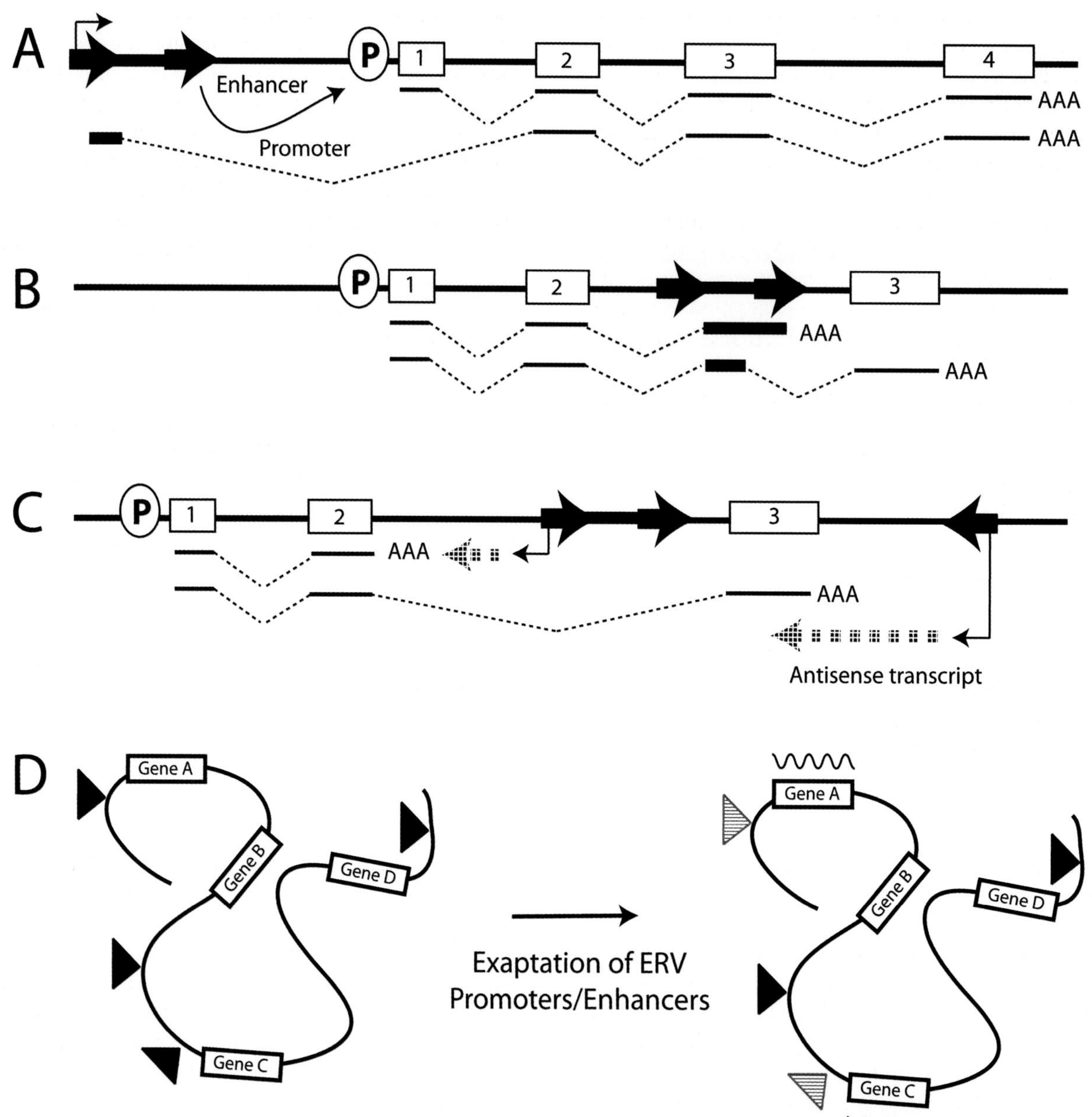

Figure 3 Common ways in which retroviral insertions can affect genes. (A) Proviral elements are shown as thick lines, with arrows for the LTRs, and gene exons are numbered boxes with P showing the gene promoter. Retroviral elements inserted near genes can donate their LTR enhancers or promoters to affect gene expression. This mechanism occurs when new somatic retroviral insertions activate oncogenes. This mechanism also occurs in normal cells as an evolutionary adaptation and in mouse cells defective for LTR epigenetic silencing. This can occur even if only a solitary LTR remains. (B) When inserted in an intron, splice sites and polyA signals within ERVs can perturb splicing. Most of the documented germ line-detrimental ERV insertions in mice are due to this mechanism. (C) ERVs/LTRs located downstream of a gene or within an intron have the potential to promote antisense transcripts, possibly regulating the gene or causing premature polyadenylation. (D) Closely related LTRs/ERVs distributed across the genome (black triangles) that bind the same transcription factors can become exapted (shown as gray triangles) and regulate expression of sets of genes (gene A and C in this example), establishing a regulatory network.
doi:10.1128/microbiolspec.MDNA3-0009-2014.f3

disease-causing mutations have been linked to new ERV insertions, but it is a different story in inbred mice, where at least 10% of all documented germ line mutations with an observable phenotype are due to new ERV insertions (19, 32). These mutations can be either loss of function via ablation or reduction in normal gene expression, or gain of function by causing ectopic gene expression. Straightforward loss of normal protein expression can occur due to ERV insertions into coding exons. However, exonic insertions are relatively rare, with the vast majority of documented loss-of-function mutations being due to intronic ERV insertions. Such insertions can disrupt normal gene transcript processing by providing splice sites and/or polyadenylation signals, the use of which creates aberrant transcript forms (Fig. 3B) (for review, see reference 32). Intronic ERVs that significantly disrupt transcript processing are most often oriented in the same transcriptional direction as the enclosing gene, whereas intronic ERVs fixed in a species (and therefore presumably neutral) are heavily biased to be in the opposite or antisense direction, indicating stronger selection against sense-oriented insertions (125, 126, 127). These findings are expected because the canonical transcript processing signals in a provirus that may be aberrantly recognized during gene transcription typically only operate in sense. One exception is an antisense promoter that exists in IAP LTRs, transcription from which can cause premature gene transcript termination (128, 129) as shown in Fig. 3C. LTRs located downstream of a gene can also promote transcripts antisense to that gene (130), which could have regulatory effects (Fig. 3C).

In addition to gene disruption mutations, germ line insertions of IAP elements can cause ectopic gene expression leading to a measurable phenotype with nine such cases reported, four affecting the *agouti* gene (32, 131, 132). Most of these cases, including the four impacting *agouti*, involve the antisense promoter of the IAP 5′ LTR, which drives abnormal gene expression and which shows variable expressivity in genetically identical mice. Such alleles are therefore termed metastable epialleles (133) and are due to the stochastic establishment of variable methylation levels of the LTR. However, studies on the epigenetic inheritance of the LTR epigenetic state, i.e., transgenerational inheritance, suggest that DNA methylation is not the mark that is directly inherited (134).

Auto-Immune Disorders

As exogenous retroviruses typically elicit an immune response, ERVs have the potential to be involved in autoimmunity and, indeed, have been implicated in the autoimmune disease systemic lupus erythematosus in mice (135). In addition, in immunodeficient mice lacking antibodies, recombinant and infectious ecotropic MLVs can emerge and induce lymphomas (136). The lack of functional toll-like receptors, involved in innate immunity (137), can also lead to the generation of infectious MLV and eventual leukemia (138). These two studies in mice indicate an important role for the innate and adaptive immune systems in controlling ERVs. As human ERVs are much older than MLV sequences in mice, it might be assumed that the host immune system would have evolved in concert and be unable to recognize human ERV nucleic acids or proteins as foreign. Nonetheless, some human ERV-encoded proteins have been shown to be immunogenic, with antibodies and cytotoxic T cells against ERV proteins being detected in some cancers (139). Although a true causative role for ERVs in autoimmune diseases is difficult to demonstrate, there is suggestive evidence (140, 141, 142). For example, the HERV-W-encoded Env protein syncytin1 is upregulated in multiple sclerosis (MS) lesions and its expression in astrocytes induces the production of proinflammatory cytokines and oxidative damage (140). Such a process leads to oligodendrocyte death, which is a key feature of MS.

ERV PROTEIN DOMESTICATIONS

The widespread distribution and stable genetic inheritance of ERVs has prompted speculation that the provision of "useful" protein functions might provide a partial explanation for their long-term survival. Properties one might expect for ERV gene products playing an essential physiological role would be: (i) conservation of an open reading frame for the protein product in all members of a species or group of species; (ii) a plausible biochemical function: and (iii) expression in the appropriate cell or tissue.

One tissue that has attracted continued interest is the placenta. Placenta formation involves extensive cell fusion to form the syncytial trophoblast layer and it is easy to see how the fusogenic properties of the Env protein might be put to use in this regard. Moreover, the transmembrane (TM) protein of Env contains an immunosuppressive peptide that might also play a role in protecting the developing fetus from maternal immune responses. The first candidate for such a role was the Env protein encoded by the ERV-3 provirus (143). This provirus is conserved in most ape and Old World monkey families and shows specific expression in the placenta. However, sequencing studies revealed that around 1% of the human population have a premature stop codon in the ERV-3 with no evidence for counterselection, thereby indicating that ERV-3 cannot play an essential physiological role (144). Subsequently, two other Env proteins, encoded by conserved proviruses from the HERV-W (145, 146) and HERV-FRD (147) groups of ERVs, were put forwards as putative placental factors. They are expressed in placental cells and show fusogenic properties. Hence, these ENV proteins were dubbed syncytin-1 and syncytin-2. Based on their species distribution, it seems likely that the proviruses encoding syncytin-1 and -2 have been present in the germ line for approximately 30 million and 45 million years, respectively (147). The fact that the *env* genes of these two proviruses have maintained their open reading frames, in contrast to their *gag* and *pol* genes, implies selection for the retention of functional proteins. In addition, small interfering RNA-mediated suppression of syncytin expression in spontaneously differentiating trophoblasts interferes with cell–cell fusion, suggesting role(s) in placenta formation (148).

Direct experimental evidence for the role(s) of ERV-encoded proteins in trophoblast formation comes from studies of mice. Although they do not have ERVs at chromosomal locations of syncytin-1 or syncytin-2, genomic sequence analyses of the mouse genome identified two distinct fusogenic *env* genes with placenta-specific expression, called syncytin-A and -B, that have been conserved for more than 25 million years. Knocking out syncytin-A results in a failure to form organized placentae and midgestational death of homozygotes (149). Deletion of syncytin-B results in a less dramatic phenotype, although trophoblast fusion is also affected (150). Apparently, a mechanism involving gap junction formation compensates for fusion loss.

Further examples of ERV capture and utilization in placenta formation have been reported in a variety of mammalian orders, including another rodent species (squirrels), ruminants, carnivores, and lagomorphs (Fig. 4). All are independent and appear to represent different ERV captures. It is tempting to suggest that the variety

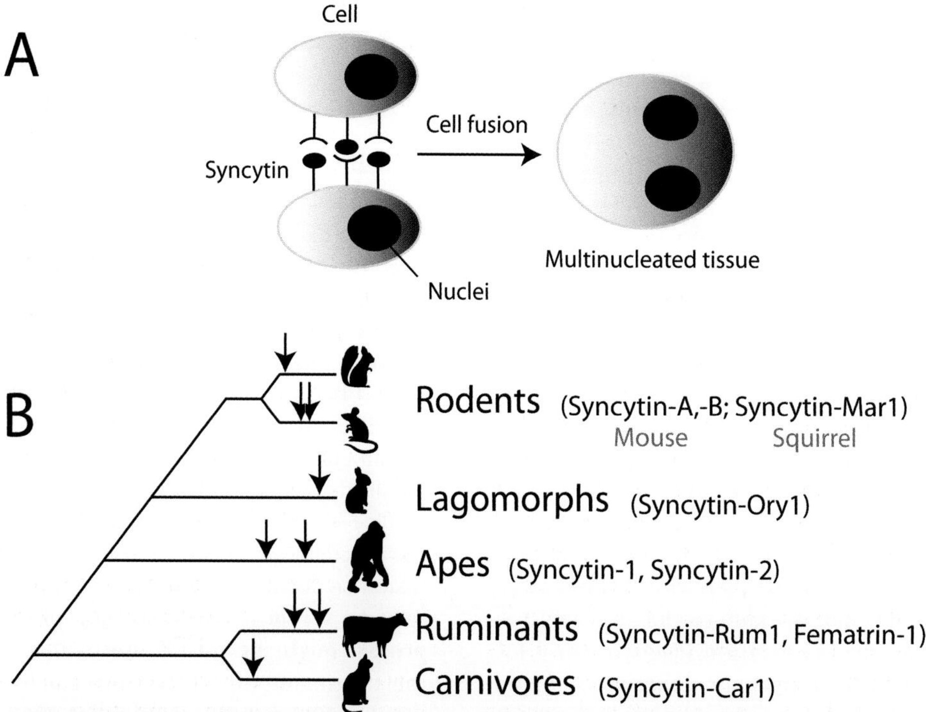

Figure 4 Independent instances of ERV Env domestication involved in placenta formation. (A) Domestication of Env proteins (which have natural fusogenic properties) can promote the formation of multinucleated tissue, as occurs in the developing placenta. Env syncytin-like molecules are shown as balls binding to receptors. (B) Schematic evolutionary tree showing different instances of ERV Env domestication (shown as arrows): Syncytin-A and Syncytin-B (197), Syncytin-Ory1 (198), Syncytin-1 (146), Syncytin-2 (147), Syncytin-Rum1 (199), Fematrin-1 (200), Syncytin-Car1 (201), and Syncytin-Mar1 (202).
doi:10.1128/microbiolspec.MDNA3-0009-2014.f4

of structures of placenta seen in different orders of mammals can be explained by the expression, receptor, and fusion properties of the specific ERVs put to use in each case. This would imply that ERV capture has played a major role in the evolutionary history of placental mammals (151).

The other area in which ERV gene products have been set to work by the host involves resistance to exogenous infection. The first, and perhaps best-known, example is the murine Fv1 (Friend virus susceptibility-1) gene (152). This gene shows many functional similarities to the capsid-binding restriction factor TRIM5α, blocking exogenous virus infection at a postentry but preintegration stage in the viral life cycle (110). It appears to be derived from an ERV related to the MERV-L group, with an open reading frame of about 1.4 kb showing 40% sequence identity to the *gag* gene of MERV-L (153). Interestingly, the remaining viral sequences are missing; both promoter and polyadenylation sequences are provided by the host (154). How Fv1 selects it target and restricts replication are still open questions. Fv1 restriction of infecting viruses involves binding to the retroviral capsid protein (155); however, the detailed mechanisms of target recognition and restriction remain to be elucidated.

A second group of resistance genes correspond to the *env* gene of certain ERVs. The property of superinfection resistance, whereby cells infected with one retrovirus are resistant to reinfection by a second virus using the same receptor, has been well-described for many years. This phenomenon results from newly synthesized Env binding to a cellular receptor resulting either in downregulation of the receptor on the cell surface or the saturation of virus binding. In either event, fresh virus binding is not possible. Expression of the Env protein from an ERV can mimic this effect, providing protection against infection. This strategy has been used on multiple occasions in a number of different species, including mice, chickens, cats, and sheep, with a variety of ERVs (156, 157, 158). Under certain circumstances, endogenous Env protein expression may also act to modulate immune responses to infection by acting as a selfantigen (159). Interestingly, as well as blocking infection of its exogenous counterpart by receptor interference, at least two loci of endogenous Jaagsiekte sheep retrovirus (enJSRV) encode defective Gag proteins that act in a dominant-negative way to block late stages of virus replication (160).

A different mechanism leading to resistance is associated with expression of the superantigen (*sag*) gene of endogenous MMTVs. Sag is a type II TM glycoprotein encoded by the MMTV LTR (161). Expression is needed for exogenous MMTVs to stimulate cell proliferation of target lymphocytes that allow efficient infection as they transfer from gut to mammary gland after infection with a milk-borne virus. When expressed from an ERV and presented with the correct major histocompatibility complex molecule, it can lead to depletion of specific T-cell subsets, sometimes corresponding to 30% of the mature T-cells of the mouse by a mechanism involving stimulation and apoptosis (162). The absence of target T-cells has the effect of rendering such mice resistant to infection by an exogenous MMTV carrying the same *sag* gene, but does not appear to compromise immunity to other pathogens (163).

ERV/LTR MEDIATED GENE REGULATION

Although such protein domestications are remarkable, a wider and more common role for existing ERV sequences is in gene regulation. As discussed above, pathogenic consequences of ERV-mediated gene regulation can occur when they integrate near potential oncogenes and/or ectopically activate such genes via their LTR promoters and/or enhancers (Fig. 3A). Very similar phenomena can occur in normal cells as an evolutionary adaptation. Notably, although some ERVs retain the general structure of an integrated provirus with retroviral genes and two LTRs, the vast majority of older elements exist currently as solo LTRs due to recombination over time between the 5′ and 3′ LTRs of the provirus (34). As encoded retroviral LTRs naturally contain promoters and enhancers/transcription factor binding sites, these abundant LTR sequences have inherent potential to regulate genes. A listing of such cases is available (164) and some examples are mentioned here. Among the first well-documented cases of ERV-derived genic enhancers are the human amylase loci, where a HERV-E LTR acts as a parotid-specific enhancer (165), and the sex-limited protein (*slp*) gene in mouse, which is hormone responsive due to an upstream androgen-response element within an LTR (166, 167). Another well-studied case occurs in the human beta-globin gene cluster. Strong evidence suggests that an ERV-9 (LTR12) LTR upstream of the locus control region in the cluster recruits transcription factors to downstream globin gene promoters via long range chromatin interactions and expression of a long noncoding (lnc) RNA (168). Indeed, as discussed further below, ERV-mediated involvement in gene regulation via the production of lncRNAs is an intriguing emerging topic.

Although LTR enhancer effects are likely more common due to less positional constraint, it is technically more straightforward to detect LTR promoter exaptations because the LTR sequence forms the 5′ end of the resulting mRNA. By direct screening for LTR-gene chimeric mRNAs or through other methods, numerous individual examples of LTR "promoter donation" in normal cells have been documented (164, 169, 170, 171, 172). In cases where an LTR acts as a gene promoter, it is most often an alternative promoter with variable contributions to total gene expression and which occasionally creates new or altered tissue specificity of expression (170). Interesting examples of novel tissue-specific gene expression enabled by an LTR promoter include testis-specific expression of the p63 gene in humans and other primates driven by an ERV-9 LTR (169) and LTR-promoted placental expression of several human genes (for a review, see reference 170). In rarer cases, LTRs can also apparently serve as the primary or sole promoter for a gene, presumably replacing the original or "native" genic promoter due to a favorable integration location and compatible or more advantageous transcription factor binding sites. Examples of such genes include rat *Ocm2* (173) and the rodent *Naip* multigene family, which has adopted an ancient LTR as its primary promoter (174). LTRs have also been shown to provide polyadenylation sites for mRNAs of cellular genes (175).

In addition to the many examples of ERVs controlling single genes, a growing number of findings implicate ERVs in whole regulatory networks (Fig. 3D) (for recent reviews, see references 176, 177, 178), supporting the visionary ideas of McClintock (179) and Britten and Davidson (180), who postulated major roles for transposable elements (TEs) in gene and genome regulation. For example, in 2004 it was reported that Class III mouse ERVs (MT and MERV-L elements) function as alternative promoters for many genes in full-grown oocytes and two-cell embryos, suggesting that such elements may participate in reprogramming of the embryonic genome (78). The use of computational predictions (181) and genomic approaches such as ChIP-seq (182, 183) has revealed that binding sites for some classes of transcription factors are enriched in ERV LTRs, a finding which is perhaps expected given that LTRs of functioning retroviruses naturally contain transcriptional regulatory modules. One of the first such ChIP-seq studies showed that certain types of human ERV LTRs are enriched in p53 binding sites (183). Another provocative study showed that binding sites for the pluripotency factors OCT4 and NANOG are largely not conserved between human and mouse ES cells with a large fraction of such nonconserved sites occurring within species-specific ERVs (182). The presence of such ERV-associated sites is correlated with species-specific gene expression patterns, promoting the suggestion that ERVs have "rewired" the regulatory network in ES cells (182). A study in mouse has found that RLTR13D5 LTRs are enriched for species-specific enhancer chromatin marks and bind core-transcription factors involved in the regulatory program of trophoblast stem cells (184). Genome-wide analyses of DNase I hypersensitive sites indicative of open chromatin (185), as well as patterns of DNA hypomethylation and chromatin marks associated with enhancers (86), have found significant enrichment in certain classes of human ERVs, suggesting that ERVs are enriched for tissue-specific enhancers.

ERVs can also regulate genes via the production of noncoding RNAs. ERVs and other TEs can promote expression of RNAs antisense to coding genes (Fig. 3C) (130) and a number of more-recent studies indicate that ERV LTRs are overrepresented as a source of transcriptional promoters for lncRNAs (186, 187, 188). Although many such RNAs could simply be the result of "transcriptional noise" and have no function, regulatory roles for some specific cases are being revealed. As mentioned above, an ERV-9 (LTR12) LTR-promoted lncRNA appears to have a role in the regulation of the beta-globin gene cluster (168). HERV-H (LTR7)-promoted transcripts/lncRNAs are particularly abundant in undifferentiated human ES cells (187, 189), correlating with their association with DNAse I hypersensitive sites, hypomethylation, active chromatin marks, and the binding of pluripotency transcription factors (182, 185, 189, 190). Indeed, there is growing evidence that HERV-H-derived RNAs play an important, regulatory role in human stem cell identity and maintenance of pluripotency (191, 192, 193, 194). In mouse, the nonautonomous VL30 ERVs have been shown to produce enhancer-associated noncoding RNAs controlled by Trim24 and involved in gene regulation in hepatocytes (195). As the field of noncoding RNA continues to grow, it is likely that many cellular roles for ERV-derived RNAs will be uncovered.

CONCLUDING REMARKS

ERVs and ERV remnants are vestiges of the continual bombardment of their host genome by exogenous retroviruses during evolution. It is worth emphasizing that the ERVs in present-day mammalian genomes [e.g., ~400,000 loci in humans (18)] represent only a tiny fraction of the retroviral insertions that have occurred in the germ line, have survived natural selection, and have gone to fixation. Therefore, it is interesting to contemplate the level of activity by retroviruses and other viruses (and their associated diseases) that must have shaped the population history of a species and driven the rapid evolution of host genes to counter new viral threats (196). Although the activity of ERVs appears to have died out in some species, notably human, several ERV groups are active in mouse, the organism most widely used to model human disease. Here we have reviewed the discovery of ERVs in mammals, mechanisms that curtail their expression, and their biological effects. ERVs have an enigmatic relationship with their host species. On the one hand, the negative effects of active ERVs as genomic mutagens and cancer-causing agents are well established. On the other hand, examples of ERV-protein domestications to serve host functions, as in placental development, are also clear. Moreover, there is growing evidence that the gene regulatory potential of ERV LTRs has been exploited multiple times during evolution to regulate genes and gene networks. Thus, although recently endogenized retroviral elements are often pathogenic, those that survive the forces of negative selection or loss by genetic drift become neutral components of the host genome or can be harnessed to serve beneficial roles.

Acknowledgments. *We thank R. Rebollo for help with the figures and for reading the manuscript and A. Babaian for generating Fig. 2. We regret that, due to space limitations, we could not cite all the relevant work on ERVs. Work in the Mager laboratory is supported by the Canadian Institutes of Health Research and the Natural Sciences and Engineering Research Council of Canada. Work in the Stoye laboratory is funded by the UK Medical Research Council.*

Citation. Mager DL, Stoye JP. 2014. Mammalian endogenous retroviruses. Microbiol Spectrum 3(1):MDNA3-0009-2014.

References

1. **Weiss RA.** 2006. The discovery of endogenous retroviruses. *Retrovirology* **3:**67.
2. **Lowy DR, Rowe WP, Teich N, Hartley JW.** 1971. Murine leukemia virus: high frequency activation in vitro by 5-iododeoxyuridine and 5-bromodeoxyuridine. *Science* **174:**155–156.
3. **Benveniste RE, Todaro GJ.** 1974. Multiple divergent copies of endogenous type C virogenes in mammalian cells. *Nature* **252:**170–173.
4. **Boeke JD, Stoye JP.** 1997. Retrotransposons, endogenous retroviruses and the evolution of retroelements, p 343–435. *In* Coffin JM, Hughes SH, Varmus HE (ed), *Retroviruses.* Cold Spring Harbor Laboratory, Cold Spring Harbor, NY.
5. **Khan AS, Ma W, Ma Y, Kumar A, Williams DK, Muller J, Ma H, Galvin TA.** 2009. Proposed algorithm to investigate latent and occult viruses in vaccine cell substrates by chemical induction. *Biologicals* **37:**196–201.
6. **Paprotka T, Delviks-Frankenberry KA, Cingoz O, Martinez A, Kung HJ, Tepper CG, Hu WS, Fivash MJ Jr, Coffin JM, Pathak VK.** 2011. Recombinant origin of the retrovirus XMRV. *Science* **333:**97–101.
7. **Voisset C, Weiss RA, Griffiths DJ.** 2008. Human RNA "rumor" viruses: the search for novel human retroviruses in chronic disease. *Microbiol Mol Biol Rev* **72:** 157–196.
8. **Callahan R, Drohan W, Tronick S, Schlom J.** 1982. Detection and cloning of human DNA sequences related to the mouse mammary tumor virus genome. *Proc Natl Acad Sci U S A* **79:**5503–5507.
9. **Martin MA, Bryan T, Rasheed S, Khan AS.** 1981. Identification and cloning of endogenous retroviral sequences present in human DNA. *Proc Natl Acad Sci U S A* **78:**4892–4896.
10. **Dunwiddie CT, Resnick R, Boyce-Jacino M, Alegre JN, Faras AJ.** 1986. Molecular cloning and characterization of *gag-*, *pol-*, and *env*-related gene sequences in the *ev*$^{-}$ chicken. *J Virol* **59:**669–675.
11. **Medstrand P, Blomberg J.** 1993. Characterization of novel reverse transcriptase encoding human endogenous retroviral sequences similar to type A and type B retroviruses: differential transcription in normal human tissues. *J Virol* **67:**6778–6787.
12. **Shih A, Misra R, Rush MG.** 1989. Detection of multiple, novel reverse transcriptase coding sequences in human nucleic acids: relation to primate retroviruses. *J Virol* **63:**64–75.
13. **Tristem M, Kabat P, Lieberman L, Linde S, Karpas A, Hill F.** 1996. Characterization of a novel murine leukemia virus-related subgroup within mammals. *J Virol* **70:** 8241–8246.
14. **Smit AFA.** 1993. Identification of a new, abundant superfamily of mammalian LTR-retrotransposons. *Nucl Acids Res* **21:**1863–1872.
15. **Smit AFA.** 1996. The origin of interspersed repeats in the human genome. *Curr Opin Genet Dev* **6:**743–748.
16. **Chimpanzee Sequencing and Analysis Consortium.** 2005. Initial sequence of the chimpanzee genome and comparison with the human genome. *Nature* **437:** 69–87.
17. **International Chicken Genome Sequencing Consortium.** 2004. Sequence and comparative analysis of the chicken genome provide unique perspectives on vertebrate evolution. *Nature* **432:**695–716.
18. **International Human Genome Sequencing Consortium.** 2001. Initial sequencing and analysis of the human genome. *Nature* **409:**860–921.
19. **International Mouse Genome Sequencing Consortium.** 2002. Initial sequencing and comparative analysis of the mouse genome. *Nature* **420:**520–562.
20. **Lerat E.** 2010. Identifying repeats and transposable elements in sequenced genomes: how to find your way through the dense forest of programs. *Heredity* **104:** 520–522.
21. **Sperber GO, Airola T, Jern P, Blomberg J.** 2007. Automated recognition of retroviral sequences in genomic data—RectoTector. *Nucl Acids Res* **35:**4964–4976.

22. **Katzourakis A, Gifford RJ.** 2010. Endogenous viral elements in animal genomes. *PLoS Genet* **6:**e1001191.
23. **Magiorkinis G, Gifford RJ, Katzourakis A, De Ranter J, Belshaw R.** 2012. Env-less endogenous retroviruses are genomic superspreaders. *Proc Natl Acad Sci U S A* **109:** 7385–7390.
24. **Hayward A, Grabherr M, Jern P.** 2013. Broad-scale phylogenomics provides insights into retrovirus-host evolution. *Proc Natl Acad Sci U S A* **110:**20146–20151.
25. **Stoye JP, Coffin JM.** 1987. The four classes of endogenous murine leukemia viruses: structural relationships and potential for recombination. *J Virol* **61:**2659–2669.
26. **Hohn O, Hanke K, Bannert N.** 2013. HERV-K (HML-2), the best preserved family of HERVs: endogenisation, expression and implications in health and disease. *Front Oncol* **3:**1–12.
27. **Katzourakis A, Tristem M, Pybus OG, Gifford RJ.** 2007. Discovery and analysis of the first endogenous lentivirus. *Proc Natl Acad Sci U S A* **104:**6261–6265.
28. **Ribet D, Harper F, Dupressoir A, Dewannieux M, Pierron G, Heidmann T.** 2008. An infectious progenitor for the murine IAP retrotransposon: emergence of an intracellular genetic parasite from an ancient retrovirus. *Genome Res* **18:**597–609.
29. **Ribet D, Harper F, Dewannieux M, Pierron G, Heidmann T.** 2007. Murine MusD retrotransposon: structure and molecular evolution of an "intracellularized" retrovirus. *J Virol* **81:**1888–1898.
30. **Swain A, Coffin JM.** 1992. Mechanism of transduction by retroviruses. *Science* **255:**841–845.
31. **Sonigo P, Wain-Hobson S, Bougueleret L, Tiollais P, Jacob F, Brulet P.** 1987. Nucleotide sequence and evolution of ETn elements. *Proc Natl Acad Sci U S A* **84:**3768–3771.
32. **Maksakova IA, Romanish MT, Gagnier L, Dunn CA, van de Lagemaat LN, Mager DL.** 2006. Retroviral elements and their hosts: insertional mutagenesis in the mouse germ line. *PLoS Genet* **2:**e2.
33. **Benachenhou F, Jern P, Oja M, Sperber G, Blikstad V, Somervuo P, Kaski S, Blomberg J.** 2009. Evolutionary conservation of orthoretroviral long terminal repeats (LTRs) and *ab initio* detection of single LTRs in genomic data. *PLoS One* **4:**e5179.
34. **Belshaw R, Watson J, Katzourakis A, Howe A, Woolven-Allen J, Burt A, Tristem M.** 2007. Rate of recombinational deletion among human endogenous retroviruses. *J Virol* **81:**9437–9442.
35. **Seperack PK, Strobel MC, Corrow DJ, Jenkins NA, Copeland NG.** 1988. Somatic and germ-line reverse mutation rates of the retrovirus-induced dilute coat-color mutation of DBA mice. *Proc Natl Acad Sci U S A* **85:** 189–192.
36. **de Parseval N, Lazr V, Casella JF, Benit L, Heidmann T.** 2003. Survey of human genes of retroviral origin: identification and transcriptome of the genes with coding capacity for complete envelope proteins. *J Virol* **77:** 10414–10422.
37. **Villeson P, Aagaard L, Wiuf C, Pedersen FS.** 2004. Identification of endogenous retroviral reading frames in the human genome. *Retrovirology* **1:**32.
38. **Jern P, Stoye JP, Coffin JM.** 2007. Role of APOBEC3 in genetic diversity among endogenous murine leukemia viruses. *PLoS Genet* **3:**e183.
39. **Dewannieux M, Harper F, Richaud A, Letzelter C, Ribet D, Pierron G, Heidmann T.** 2006. Identification of an infectious progenitor for the multiple-copy HERV-K human endogenous retroelements. *Genome Res* **16:**1548–1556.
40. **Lee YN, Bieniasz PD.** 2007. Reconstitution of an infectious human endogenous retrovirus. *PLoS Pathog* **3:**e10.
41. **Blomberg J, Benachenhou F, Blikstad V, Sperber G, Meyer J.** 2009. Classification and nomenclature of endogenous retroviral sequences (ERVs): problems and recommendations. *Gene* **448:**115–123.
42. **Meyer J, Blomberg J, Seal RL.** 2011. A revised nomenclature for transcribed endogenous retroviral loci. *Mob DNA* **2:**7.
43. **Tristem M.** 2000. Identification and characterization of novel human endogenous retrovirus families by phylogenetic screening of the human genome mapping project database. *J Virol* **74:**3715–3730.
44. **Henzy JE, Johnson WE.** 2013. Pushing the endogenous envelope. *Phil Trans R Soc Lond B Biol Sci* **368:** 20120506.
45. **Jurka J, Kapitonov VV, Pavlicek A, Klonowski P, Kohany O, Walichiewicz J.** 2005. Repbase Update, a database of eukaryotic repetitive elements. *Cytogenet Genome Res* **110:**462–467.
46. **Bannert N, Kurth R.** 2006. The evolutionary dynamics of human endogenous retroviral families. *Annu Rev. Genomics Hum Genet* **7:**149–173.
47. **de Parseval N, Heidmann T.** 2005. Human endogenous retroviruses: from infectious elements to human genes. *Cytogenet Genome Res* **110:**318–332.
48. **Gifford R, Tristem M.** 2003. The evolution, distribution and diversity of endogenous retroviruses. *Virus Genes* **26:**291–315.
49. **Zhuo X, Rho M, Feschotte C.** 2013. Genome-wide characterization of endogenous retroviruses in the bat *Myotis lucifugus* reveals recent and diverse infections. *J Virol* **87:**8493–8501.
50. **Magiorkinis G, Belshaw R, Katzourakis A.** 2013. "There and back again": revisiting the pathophysiological roles of human endogenous retroviruses in the post-genomic era. *Phil Trans R Soc Lond B Biol Sci* **368:**20120504.
51. **Gilbert C, Feschotte C.** 2010. Genomic fossils calibrate the long term evolution of hepadnaviruses. *PLoS Biol* **8:** e1000495.
52. **Goodchild NL, Wilkinson DA, Mager DL.** 1993. Recent evolutionary expansion of a subfamily of RTVL-H human endogenous retrovirus-like elements. *Virology* **196:**778–788.
53. **Hughes JF, Coffin JM.** 2005. Human endogenous retroviral elements as indicators of ectopic recombination events in the primate genome. *Genetics* **171:**1183–1194.
54. **Stoye JP.** 2012. Studies of endogenous retroviruses reveal a continuing evolutionary saga. *Nat Rev Microbiol* **10:**395–406.

55. **Lee A, Nolan A, Watson J, Tristem M.** 2013. Identification of an ancient endogenous retrovirus, predating the divergence of placental animals. *Phil Trans R Soc Lond B Biol Sci* **368:**20120503.

56. **Katzourakis A, Rambaut A, Pybus OG.** 2005. The evolutionary dynamics of endogenous retroviruses. *Trends Microbiol* **13:**463–468.

57. **Stocking C, Kozak CA.** 2008. Murine endogenous retroviruses. *Cell Mol Life Sci* **65:**3383–3398.

58. **Subramanian RP, Wildschutte JH, C R, Coffin JM.** 2011. Identification, characterization, and comparative genomic distribution of the HERV-K(HML-2) group of human endogenous retroviruses. *Retrovirology* **8:**90.

59. **Soriano P, Jaenisch R.** 1986. Retroviruses as probes for mammalian development: allocation of cells to the somatic and germ cell linages. *Cell* **46:**19–29.

60. **Rowe WP, Kozak CA.** 1980. Germ-line reinsertions of AKR murine leukemia virus genomes in AKV-1 congenic mice. *Proc Natl Acad Sci U S A* **77:**4871–4874.

61. **Lock LF, Keshet E, Gilbert DJ, Jenkins NA, Copeland NG.** 1988. Studies on the mechanism of spontaneous germline ecotropic provirus acquisition in mice. *EMBO J* **7:**4169–4177.

62. **Simmons GS, Young PR, Hanger JJ, Jones K, Clarke D, McKee JJ, Meers J.** 2012. Prevalence of koala retrovirus in geographically diverse populations in Australia. *Aust Vet J* **90:**404–409.

63. **Tarlinton RE, Meers J, Young PR.** 2006. Retroviral invasion of the koala genome. *Nature* **442:**79–81.

64. **Xu W, Stadler CK, Gorman K, Jensen N, Kim DH, Zheng H, Tang S, Switzer WM, Pye GW, Eiden MV.** 2013. An exogenous retrovirus isolated from koalas with malignant neoplasias in a US zoo. *Proc Natl Acad Sci U S A* **119:**11547–11552.

65. **Avila-Arcos M, Ho SYW, Ishida Y, Nikolaidis N, Tsangaras K, Hönig K, Medina R, Rasmussen M, Fordyce SL, Calvignac-Spencer S, Willersley E, Gilbert MTP, Helgen KM, Roca AL, Greenwood AD.** 2013. One hundred twenty years of koala retrovirus evolution determined from museum skins. *Mol Biol Evol* **30:**299–304.

66. **Arias JF, Koyama T, Kinomoto M, Tokunaga K.** 2012. Retroelements versus apobec3 family members: no great escape from the magnificent seven. *Front Microbiol* **3:**275.

67. **Crichton JH, Dunican DS, Maclennan M, Meehan RR, Adams IR.** 2014. Defending the genome from the enemy within: mechanisms of retrotransposon suppression in the mouse germline. *Cell Mol Life Sci* **71:**1581–1605.

68. **Maksakova I, Mager D, Reiss D.** 2008. Keeping active endogenous retroviral-like sequences in check: the epigenetic perspective. *Cell Mol Life Sci* **65:**3329–3347.

69. **Rowe HM, Trono D.** 2011. Dynamic control of endogenous retroviruses during development. *Virology* **411:**273–287.

70. **Wolf D, Goff SP.** 2008. Host restriction factors blocking retroviral replication. *Annu Rev Gen* **42:**143–163.

71. **Leung DC, Lorincz MC.** 2012. Silencing of endogenous retroviruses: when and why do histone marks predominate? *Trends Biochem Sci* **37:**127–133.

72. **Law JA, Jacobsen SE.** 2010. Establishing, maintaining and modifying DNA methylation patterns in plants and animals. *Nat Rev Genet* **11:**204–220.

73. **Howard G, Eiges R, Gaudet F, Jaenisch R, Eden A.** 2008. Activation and transposition of endogenous retroviral elements in hypomethylation induced tumors in mice. *Oncogene* **27:**404–408.

74. **Jaenisch R, Schnieke A, Harbers K.** 1985. Treatment of mice with 5-azacytidine efficiently activates silent retroviral genomes in different tissues. *Proc Natl Acad Sci* **82:**1451–1445.

75. **Walsh CP, Chaillet JR, Bestor TH.** 1998. Transcription of IAP endogenous retroviruses is constrained by cytosine methylation. *Nat Genet* **20:**116–117.

76. **Bourc'his D, Bestor TH.** 2004. Meiotic catastrophe and retrotransposon reactivation in male germ cells lacking Dnmt3L. *Nature* **431:**96–99.

77. **Brûlet P, Condamine H, Jacob F.** 1985. Spatial distribution of transcripts of the long repeated ETn sequence during early mouse embryogenesis. *Proc Natl Acad Sci U S A* **82:**2054–2058.

78. **Peaston AE, Evsikov AV, Graber JH, de Vries WN, Holbrook AE, Solter D, Knowles BB.** 2004. Retrotransposons regulate host genes in mouse oocytes and preimplantation embryos. *Dev Cell* **7:**597–606.

79. **Li E, Bestor TH, Jaenisch R.** 1992. Targeted mutation of the DNA methyltransferase gene results in embryonic lethality. *Cell* **69:**915–926.

80. **Seifarth W, Frank O, Zeilfelder U, Spiess B, Greenwood AD, Hehlmann R, Leib-Mösch C.** 2005. Comprehensive analysis of human endogenous retrovirus transcriptional activity in human tissues with a retrovirus-specific microarray. *J Virol* **79:**341–352.

81. **Lavie L, Kitova M, Maldener E, Meese E, Mayer J.** 2005. CpG methylation directly regulates transcriptional activity of the human endogenous retrovirus family HERV-K(HML-2). *J Virol* **79:**876–883.

82. **Reiss D, Zhang Y, Mager DL.** 2007. Widely variable endogenous retroviral methylation levels in human placenta. *Nucleic Acids Res* **35:**4743–4754.

83. **Schulz WA, Steinhoff C, Florl AR.** 2006. Methylation of endogenous human retroelements in health and disease. *Curr Top Microbiol Immunol* **310:**211–250.

84. **Stengel S, Fiebig U, Kurth R, Denner J.** 2010. Regulation of human endogenous retrovirus-K expression in melanomas by CpG methylation. *Genes Chromosomes Cancer* **49:**401–411.

85. **Szpakowski S, Sun X, Lage JM, Dyer A, Rubinstein J, Kowalski D, Sasaki C, Costa J, Lizardi PM.** 2009. Loss of epigenetic silencing in tumors preferentially affects primate-specific retroelements. *Gene* **448:**151–167.

86. **Xie M, Hong C, Zhang B, Lowdon RF, Xing X, Li D, Zhou X, Lee HJ, Maire CL, Ligon KL, Gascard P, Sigaroudinia M, Tlsty TD, Kadlecek T, Weiss A, O'Geen H, Farnham PJ, Madden PAF, Mungall AJ, Tam A, Kamoh B, Cho S, Moore R, Hirst M, Marra MA, Costello JF, Wang T.** 2013. DNA hypomethylation within specific transposable element families associates with tissue-specific enhancer landscape. *Nat Genet* **45:**836–841.

87. **Meunier J, Khelifi A, Navratil V, Duret L.** 2005. Homology-dependent methylation in primate repetitive DNA. *Proc Natl Sci Adad U S A* **102:**5471–5476.

88. **Aravin AA, Bourc'his D.** 2008. Small RNA guides for *de novo* DNA methylation in mammalian germ cells. *Genes Dev* **22:**970–975.

89. **Yang N, Kazazian HH.** 2006. L1 retrotransposition is suppressed by endogenously encoded small interfering RNAs in human cultured cells. *Nat Struct Mol Biol* **13:** 763–771.

90. **Shilatifard A.** 2006. Chromatin modifications by methylation and ubiquitination: implications in the regulation of gene expression. *Annu Rev Biochem* **75:**243–269.

91. **Mikkelsen TS, Ku M, Jaffe DB, Issac B, Lieberman E, Giannoukos G, Alvarez P, Brockman W, Kim T-K, Koche RP, Lee W, Mendenhall E, O/'Donovan A, Presser A, Russ C, Xie X, Meissner A, Wernig M, Jaenisch R, Nusbaum C, Lander ES, Bernstein BE.** 2007. Genome-wide maps of chromatin state in pluripotent and lineage-committed cells. *Nature* **448:**553–560.

92. **Matsui T, Leung D, Miyashita H, Maksakova I, Miyachi H, Kimura H, Tachibana M, Lorincz MC, Shinkai Y.** 2010. Proviral silencing in embryonic stem cells requires the histone methyltransferase ESET. *Nature* **464:**927–931.

93. **Rowe H, Jakobsson J, Mesnard D, Rougemont J, Reynard S, Aktas T, Maillard P, Layard-Liesching H, Verp S, Marquis J, Spitz F, Constam D, Trono D.** 2010. KAP1 controls endogenous retroviruses in embryonic stem cells. *Nature* **463:**237–240.

94. **Karimi MM, Goyal P, Maksakova IA, Bilenky M, Leung D, Tang JX, Shinkai Y, Mager DL, Jones S, Hirst M, Lorincz MC.** 2011. DNA methylation and SETDB1/H3K9me3 regulate predominantly distinct sets of genes, retroelements, and chimeric transcripts in mESCs. *Cell Stem Cell* **8:**676–687.

95. **Thomas JH, Schneider S.** 2011. Coevolution of retroelements and tandem zinc finger genes. *Genome Res* **21:** 1800–1812.

96. **Wolf D, Goff SP.** 2009. Embryonic stem cells use ZFP809 to silence retroviral DNAs. *Nature* **458:** 1201–1204.

97. **Maksakova IA, Thompson PJ, Goyal P, Jones SJ, Singh PB, Karimi MM, Lorincz MC.** 2013. Distinct roles of KAP1, HP1 and G9a/GLP in silencing of the two-cell-specific retrotransposon MERVL in mouse ES cells. *Epigenetics Chromatin* **6:**15.

98. **Macfarlan TS, Gifford WD, Agarwal S, Driscoll S, Lettieri K, Wang J, Andrews SE, Franco L, Rosenfeld MG, Ren B, Pfaff SL.** 2011. Endogenous retroviruses and neighboring genes are coordinately repressed by LSD1/KDM1A. *Genes Dev* **25:**594–607.

99. **Leeb M, Pasini D, Novatchkova M, Jaritz M, Helin K, Wutz A.** 2010. Polycomb complexes act redundantly to repress genomic repeats and genes. *Genes Dev* **24:** 265–276.

100. **Huda A, Bowen NJ, Conley AB, Jordan IK.** 2011. Epigenetic regulation of transposable element derived human gene promoters. *Gene* **475:**39–48.

101. **Rowe HM, Kapopoulou A, Corsinotti A, Fasching L, Macfarlan TS, Tarabay Y, Viville SP, Jakobsson J, Pfaff SL, Trono D.** 2013. TRIM28 repression of retrotransposon-based enhancers is necessary to preserve transcriptional dynamics in embryonic stem cells. *Genome Res* **23:**452–461.

102. **Rebollo R, Miceli-Royer K, Zhang Y, Farivar S, Gagnier L, Mager DL.** 2012. Epigenetic interplay between mouse endogenous retroviruses and host genes. *Genome Biol* **13:**R89.

103. **Ward MC, Wilson MD, Barbosa-Morais NL, Schmidt D, Stark R, Pan Q, Schwalie PC, Menon S, Lukk M, Watt S, Thybert D, Kutter C, Kirschner K, Flicek P, Blencowe BJ, Odom DT.** 2013. Latent regulatory potential of human-specific repetitive elements. *Mol Cell* **49:**262–272.

104. **Bieniasz PD.** 2004. Intrinsic immunity: a front-line defense against viral attack. *Nat Immunol* **5:**1109–1115.

105. **Malim MH, Bieniasz PD.** 2012. HIV restriction factors and mechanisms of evasion. *Cold Spring Harb Perspect Med* **2:**a006940.

106. **Sheehy AM, Gaddis NC, Choi JD, Malim MH.** 2002. Isolation of a human gene that inhibits HIV-1 infection and is suppressed by the viral Vif protein. *Nature* **418:** 646–650.

107. **Malim MH.** 2009. APOBEC proteins and intrinsic resistance to HIV-1 infection. *Phil Trans R Soc Lond B Biol Sci* **364:**675–687.

108. **Stremlau M, Owens CM, Perron MJ, Kiessling M, Autissler P, Sodroski J.** 2004. The cytoplasmic body component TRIM5a restricts HIV-1 infection in Old World monkeys. *Nature* **427:**848–853.

109. **Nisole S, Stoye JP, Saïb A.** 2005. Trim family proteins: retroviral restriction and antiviral defence. *Nat Rev Microbiol* **3:**799–808.

110. **Sanz-Ramos M, Stoye JP.** 2013. Capsid-binding retrovirus restriction factors: discovery, restriction specificity and implications for the development of novel therapeutics. *J Gen Virol* **94:**2587–2898.

111. **Neil SJD, Zang T, Bieniasz PD.** 2008. Tetherin inhibits retrovirus release and is antagonized by HIV-1 Vpu. *Nature* **451:**425–430.

112. **Goldstone DC, Ennis-Adeniran V, Hedden JJ, Groom HC, Rice GI, Christodoulou E, Walker PA, Kelly G, Haire LF, Yap MW, de Carvalho LP, Stoye JP, Crow YJ, Taylor IA, Webb M.** 2011. HIV-1 restriction factor SAMHD1 is a deoxynucleoside triphosphate triphosphohydrolase. *Nature* **480:**379–382.

113. **Koito A, Ikeda T.** 2011. Intrinsic restriction activity of AID/APOBEC family of enzymes against the mobility of retroelements. *Mob Genet Elements* **1:**197–202.

114. **Haran-Ghera N, Peled A, Brightman BK, Fan H.** 1993. Lymphomagenesis in AKR.$Fv\text{-}1^b$ congenic mice. *Cancer Res* **53:**3433–3438.

115. **Kozak CA.** 2013. Evolution of different antiviral strategies in wild mouse populations exposed to different gammaretroviruses. *Curr Opin Virol* **3:**657–663.

116. **Rosenberg N, Jolicoeur P.** 1997. Retroviral pathogenesis, p 475–586. *In* Coffin JM, Hughes SH, Varmus H

(ed), *Retroviruses*. Cold Spring Harbor Laboratory, Cold Spring Harbor, NY.

117. **Kool J, Berns A.** 2009. High-throughput insertional mutagenesis screens in mice to identify oncogenic networks. *Nat Rev Cancer* **9:**389–399.

118. **Stoye JP, Moroni C, Coffin JM.** 1991. Virological events leading to spontaneous AKR thymomas. *J Virol* **65:**1273–1285.

119. **Ishihara H, Tanaka I, Wan H, Nojima K, Yoshida K.** 2004. Retrotransposition of limited deletion type of intracisternal A-particle elements in the myeloid leukemia cells of C3H/He mice. *J Radiat Res (Tokyo)* **45:** 25–32.

120. **De Smet C, Loriot A.** 2010. DNA hypomethylation in cancer: Epigenetic scars of a neoplastic journey. *Epigenetics* **5:**206–213.

121. **Romanish MT, Cohen CJ, Mager DL.** 2010. Potential mechanisms of endogenous retroviral-mediated genomic instability in human cancer. *Semin Cancer Biol* **20:** 246–253.

122. **Chen T, Meng Z, Gan Y, Wang X, Xu F, Gu Y, Xu X, Tang J, Zhou H, Zhang X, Gan X, Van Ness C, Xu G, Huang L, Zhang X, Fang Y, Wu J, Zheng S, Jin J, Huang W, Xu R.** 2013. The viral oncogene Np9 acts as a critical molecular switch for co-activating beta-catenin, ERK, Akt and Notch1 and promoting the growth of human leukemia stem/progenitor cells. *Leukemia* **27:**1469–1478.

123. **Galli UM, Sauter M, Lecher B, Maurer S, Herbst H, Roemer K, Mueller-Lantzsch N.** 2005. Human endogenous retrovirus rec interferes with germ cell development in mice and may cause carcinoma in situ, the predecessor lesion of germ cell tumors. *Oncogene* **24:** 3223–3228.

124. **Lamprecht B, Walter K, Kreher S, Kumar R, Hummel M, Lenze D, Kochert K, Bouhlel MA, Richter J, Soler E, Stadhouders R, Johrens K, Wurster KD, Callen DF, Harte MF, Giefing M, Barlow R, Stein H, Anagnostopoulos I, Janz M, Cockerill PN, Siebert R, Dorken B, Bonifer C, Mathas S.** 2010. Derepression of an endogenous long terminal repeat activates the CSF1R proto-oncogene in human lymphoma. *Nat Med* **16:**571–579.

125. **Medstrand P, van de Lagemaat LN, Mager DL.** 2002. Retroelement distributions in the human genome: variations associated with age and proximity to genes. *Genome Res* **12:**1483–1495.

126. **Smit AF.** 1999. Interspersed repeats and other mementos of transposable elements in mammalian genomes. *Curr Opin Genet Dev* **9:**657–663.

127. **Zhang Y, Romanish MT, Mager DL.** 2011. Distributions of transposable elements reveal hazardous zones in mammalian introns. *PLoS Comput Biol* **7:**e1002046.

128. **Druker R, Bruxner TJ, Lehrbach NJ, Whitelaw E.** 2004. Complex patterns of transcription at the insertion site of a retrotransposon in the mouse. *Nucl Acids Res* **32:**5800–5808.

129. **Li J, Akagi K, Hu Y, Trivett AL, Hlynialuk CJ, Swing DA, Volfovsky N, Morgan TC, Golubeva Y, Stephens RM, Smith DE, Symer DE.** 2012. Mouse endogenous retroviruses can trigger premature transcriptional termination at a distance. *Genome Res* **22:**870–884.

130. **Conley AB, Miller WJ, Jordan IK.** 2008. Human *cis* natural antisense transcripts initiated by transposable elements. *Trends Genet* **24:**53–56.

131. **Morgan HD, Sutherland HGE, Martin DIK, Whitelaw E.** 1999. Epigenetic inheritance at the agouti locus in the mouse. *Nat Genet* **23:**314–318.

132. **Rakyan VK, Chong S, Champ ME, Cuthbert PC, Morgan HD, Luu KV, Whitelaw E.** 2003. Transgenerational inheritance of epigenetic states at the murine Axin(Fu) allele occurs after maternal and paternal transmission. *Proc Natl Acad Sci U S A* **100:**2538–2543.

133. **Rakyan VK, Blewitt ME, Druker R, Preis JI, Whitelaw E.** 2002. Metastable epialleles in mammals. *Trends Genet* **18:**348–351.

134. **Blewitt ME, Vickaryous NK, Paldi A, Koseki H, Whitelaw E.** 2006. Dynamic reprogramming of DNA methylation at an epigenetically sensitive allele in mice. *PLoS Genet* **2:**e49.

135. **Baudino L, Yoshinobu K, Morito N, Santiago-Raber M-L, Izui S.** 2010. Role of endogenous retroviruses in murine SLE. *Autoimmun Reviews* **10:**27–34.

136. **Young GR, Eksmond U, Salcedo R, Alexopoulou L, Stoye JP, Kassiotis G.** 2012. Resurrection of endogenous retroviruses in antibody-deficient mice. *Nature* **491:**774–778.

137. **Sasai M, Yamamoto M.** 2013. Pathogen recognition receptors: ligands and signaling pathways by toll-like receptors. *Int Rev Immunol* **32:**116–133.

138. **Yu P, Lubben W, Slomka H, Gebler J, Konert M, Cai C, Neubrandt L, Prazeres da Costa O, Paul S, Dehnert S, Dohne K, Thanisch M, Storsberg S, Wiegand L, Kaufmann A, Nain M, Quintanilla-Martinez L, Bettio S, Schnierle B, Kolesnikova L, Becker S, Schnare M, Bauer S.** 2012. Nucleic acid-sensing Toll-like receptors are essential for the control of endogenous retrovirus viremia and ERV-induced tumors. *Immunity* **37:**867–879.

139. **Cherkasova E, Weisman Q, Childs RW.** 2012. Endogenous retroviruses as targets for antitumor immunity in renal cell cancer and other tumors. *Front Oncol* **3:**243.

140. **Antony JM, DesLauriers AM, Bhat RK, Ellestad KK, Power C.** 2011. Human endogenous retroviruses and multiple sclerosis: Innocent bystanders or disease determinants? *Biochim Biophys Acta* **1812:**162–176.

141. **Balada E, Vilardell-Tarrés M, Ordi-Ros J.** 2010. Implication of human endogenous retroviruses in the development of autoimmune diseases. *Int Rev Immunol* **29:** 351–370.

142. **Perl A, Fernandez D, Telarico T, Phillips PE.** 2010. Endogenous retroviral pathogenesis in lupus. *Curr Opin Rheumatol* **22:**483–492.

143. **Venables PJW, Brookes SM, Griffiths D, Weiss RA, Boyd MT.** 1995. Abundance of an endogenous retroviral envelope protein in placental trophoblasts suggests a biological function. *Virology* **211:**589–592.

144. **de Parseval N, Heidmann T.** 1998. Physiological knockout of the envelope gene of the single-copy ERV-3

human endogenous retrovirus in a fraction of the Caucasian population. *J Virol* 72:3442–3445.

145. **Blond JL, Lavillette D, Cheynet V, Bouton O, Oriol G, Chapel-Fernandes S, Mandrand B, Mallet F, Cosser F-L.** 2000. An envelope glycoprotein of the human endogenous retrovirus HERV-W is expressed in the human placenta and fuses cells expressing the type D mammalian retrovirus receptor. *J Virol* **74:**3321–3329.

146. **Mi S, Lee X, Li X, Veldman GM, Finnerty H, Racie L, LaVallie E, Tang XY, Edouard P, Howes S, Keith JC Jr, McCoy JM.** 2000. Syncytin is a captive retroviral envelope protein involved in human placental morphogenesis. *Nature* **403:**785–789.

147. **Blaise S, de Parseval N, Bénit L, Heidmann T.** 2003. Genomewide screening for fusogenic human endogenous retrovirus envelopes identifies syncytin 2, a gene conserved on primate evolution. *Proc Natl Acad Sci U S A* **100:**13013–13018.

148. **Vargas A, Moreau J, Landry S, LeBellego F, Toutaily C, Rassart E, Lafond J, Barbeau B.** 2009. Sybcytin-2 plays an important role in the fusion of human trophoblast cells. *J Mol Biol* **392:**301–308.

149. **Dupressoir A, Vernochet C, Bawa O, Harper F, Pierron G, Opolon P, Heidmann T.** 2009. Syncytin-A knockout mice demonstrate the critical role in placentation of a fusogenic, endogenous retrovirus-derived, envelope gene. *Proc Natl Acad Sci U S A* **106:**12127–12132.

150. **Dupressoir A, Vernochet C, Harper F, Guegan J, Dessen P, Pierron G, Heidmann T.** 2011. A pair of co-opted retroviral envelope syncytin genes is required for formation of the two-layered murine placental syncytiotrophoblast. *Proc Natl Acad Sci U S A* **108:** E1164–E1173.

151. **Lavialle C, Cornelis G, Dupressoir A, Esnault C, Heidmann O, Vernochet C, Heidmann T.** 2013. Paleovirology of 'syncytins', retroviral *env* genes exapted for a role in placentation. *Phil Trans R Soc Lond B Biol Sci* **368:**20120507.

152. **Lilly F.** 1970. Fv-2: Identification and location of a second gene governing the spleen focus response to Friend leukemia virus in mice. *J Natl Cancer Inst* **45:**163–169.

153. **Best S, Le Tissier P, Towers G, Stoye JP.** 1996. Positional cloning of the mouse retrovirus restriction gene *Fv1*. *Nature* **382:**826–829.

154. **Best S, Le Tissier PR, Stoye JP.** 1997. Endogenous retroviruses and the evolution of resistance to retroviral infection. *Trends Microbiol* **4:**313–318.

155. **Dodding MP, Bock M, Yap MW, Stoye JP.** 2005. Capsid processing requirements for abrogation of Fv1 and Ref1 restriction. *J Virol* **79:**10571–10577.

156. **McDougall AS, Terry A, Tzavaras T, Cheney C, Rojko J, Neil JC.** 1994. Defective endogenous proviruses are expressed in feline lymphoid cells: evidence for a role in natural resistance to subgroup B feline leukemia virus. *J Virol* **68:**2151–2160.

157. **Odaka T, Ikeda H, Akatsuka T.** 1980. Restricted expression of endogenous N-tropic XC-positive leukemia virus in hybrids between G and AKR mice: an effect of the Fv-4r gene. *Int J Cancer* **25:**757–762.

158. **Wu T, Yan Y, Kozac CA.** 2005. *Rmcf2*, a xenotropic provirus in the Asian mouse species *Mus castaneus*, blocks infection by mouse gammaretroviruses. *J Virol* 79:9677–9684.

159. **Young GR, Ploquin MJ, Eksmond U, Wadwa M, Stoye JP, Kassiotis G.** 2012. Negative selection by an endogenous retrovirus promotes a higher-avidity CD4+ T cell response to retroviral antigen. *PLoS Pathog* **8:** e1002709.

160. **Arnaud F, Murcia PR, Palmarini M.** 2007. Mechanisms of late restriction induced by an endogenous retrovirus. *J Virol* **81:**11441–11451.

161. **Acha-Orbea H, Held W, Wanders GA, Shakhov AN, Scarpellino L, Lees RK, MacDonald HR.** 1993. Exogenous and endogenous mouse mammary tumor virus superantigens. *Immunol Rev* **131:**5–25.

162. **Czarneski J, Rassa JC, Ross SR.** 2003. Mouse mammary tumor virus and the immune system. *Immunol Res* 27:469–480.

163. **Frankel WN, Rudy C, Coffin JM, Huber BT.** 1991. Linkage of Mls genes to endogenous mammary tumour viruses of inbred mice. *Nature* **349:**526–528.

164. **Rebollo R, Farivar S, Mager DL.** 2012. C-GATE—catalogue of genes affected by transposable elements. *Mob DNA* **3:**9.

165. **Samuelson LC, Phillips RS, Swanberg LJ.** 1996. Amylase gene structures in primates: retroposon insertions and promoter evolution. *Mol Biol Evol* **13:**767–779.

166. **Ramakrishnan C, Robins DM.** 1997. Steroid hormone responsiveness of a family of closely related mouse proviral elements. *Mamm Genome* **8:**811–817.

167. **Stavenhagen JB, Robins DM.** 1988. An ancient provirus has imposed androgen regulation on the adjacent mouse sex-limited protein gene. *Cell* **55:**247–254.

168. **Pi W, Zhu X, Wu M, Wang Y, Fulzele S, Eroglu A, Ling J, Tuan D.** 2010. Long-range function of an intergenic retrotransposon. *Proc Natl Acad Sci U S A* **107:** 12992–12997.

169. **Beyer U, Moll-Rocek J, Moll UM, Dobbelstein M.** 2011. Endogenous retrovirus drives hitherto unknown proapoptotic p63 isoforms in the male germ line of humans and great apes. *Proc Natl Acad Sci U S A* **108:** 3624–3629.

170. **Cohen CJ, Lock WM, Mager DL.** 2009. Endogenous retroviral LTRs as promoters for human genes: a critical assessment. *Gene* **448:**105–114.

171. **Conley AB, Piriyapongsa J, Jordan IK.** 2008. Retroviral promoters in the human genome. *Bioinformatics* **24:** 1563–1567.

172. **van de Lagemaat LN, Landry J-R, Mager DL, Medstrand P.** 2003. Transposable elements in mammals promote regulatory variation and diversification of genes with specialized functions. *Trends Genet* **19:** 530–536.

173. **Banville D, Boie Y.** 1989. Retroviral long terminal repeat is the promoter of the gene encoding the tumor-associated calcium-binding protein oncomodulin in the rat. *Journal of Molecular Biology* **207:**481–490.

174. **Romanish MT, Lock WM, van de Lagemaat LN, Dunn CA, Mager DL.** 2007. Repeated recruitment of LTR retrotransposons as promoters by the anti-apoptotic locus NAIP during mammalian evolution. *PLoS Genetics* **3:**e10.

175. **Conley A, Jordan I.** 2012. Cell type-specific termination of transcription by transposable element sequences. *Mob DNA* **3:**15.

176. **Bourque G.** 2009. Transposable elements in gene regulation and in the evolution of vertebrate genomes. *Curr Opin Genet Dev* **19:**607–612.

177. **Feschotte C.** 2008. Transposable elements and the evolution of regulatory networks. *Nat Rev Genet* **9:**397–405.

178. **Rebollo R, Romanish MT, Mager DL.** 2012. Transposable Elements: An Abundant and Natural Source of Regulatory Sequences for Host Genes. *Annu Rev Genet* **46:**21–42.

179. **McClintock B.** 1956. Controlling elements and the gene. *Cold Spring Harb Symp Quant Biol* **21:**197–216.

180. **Britten RJ, Davidson EH.** 1969. Gene regulation for higher cells: a theory. *Science* **165:**349–357.

181. **Thornburg BG, Gotea V, Makalowski W.** 2006. Transposable elements as a significant source of transcription regulating signals. *Gene* **365:**104–110.

182. **Kunarso G, Chia NY, Jeyakani J, Hwang C, Lu X, Chan YS, Ng HH, Bourque G.** 2010. Transposable elements have rewired the core regulatory network of human embryonic stem cells. *Nat Genet* **42:**631–634.

183. **Wang T, Zeng J, Lowe CB, Sellers RG, Salama SR, Yang M, Burgess SM, Brachmann RK, Haussler D.** 2007. Species-specific endogenous retroviruses shape the transcriptional network of the human tumor suppressor protein p53. *Proc Natl Acad Sci U S A* **104:** 18613–18618.

184. **Chuong EB, Rumi MA, Soares MJ, Baker JC.** 2013. Endogenous retroviruses function as species-specific enhancer elements in the placenta. *Nat Genet* **45:**325–329.

185. **Jacques P-E, Jeyakani J, Bourque G.** 2013. The majority of primate-specific regulatory sequences are derived from transposable elements. *PLoS Genet* **9:**e1003504.

186. **Kapusta A, Kronenberg Z, Lynch VJ, Zhuo X, Ramsay L, Bourque G, Yandell M, Feschotte C.** 2013. Transposable elements are major contributors to the origin, diversification, and regulation of vertebrate long noncoding RNAs. *PLoS Genet* **9:**e1003470.

187. **Kelley D, Rinn J.** 2013. Transposable elements reveal a stem cell specific class of long noncoding RNAs. *Genome Biol* **13:**R107.

188. **St Laurent G, Shtokalo D, Dong B, Tackett M, Fan X, Lazorthes S, Nicolas E, Sang N, Triche T, McCaffrey T, Xiao W, Kapranov P.** 2013. VlincRNAs controlled by retroviral elements are a hallmark of pluripotency and cancer. *Genome Biol* **14:**R73.

189. **Santoni F, Guerra J, Luban J.** 2012. HERV-H RNA is abundant in human embryonic stem cells and a precise marker for pluripotency. *Retrovirology* **9:**111.

190. **Xie W, Schultz MD, Lister R, Hou Z, Rajagopal N, Ray P, Whitaker JW, Tian S, Hawkins RD, Leung D, Yang H, Wang T, Lee AY, Swanson SA, Zhang J, Zhu Y, Kim A, Nery JR, Urich MA, Kuan S, Yen C-a, Klugman S, Yu P, Suknuntha K, Propson NE, Chen H, Edsall LE, Wagner U, Li Y, Ye Z, Kulkarni A, Xuan Z, Chung W-Y, Chi NC, Antosiewicz-Bourget JE, Slukvin I, Stewart R, Zhang MQ, Wang W, Thomson JA, Ecker JR, Ren B.** 2013. Epigenomic analysis of multilineage differentiation of human embryonic stem cells. *Cell* **153:**1134–11348.

191. **Fort A, Hashimoto K, Yamada D, Salimullah M, Keya CA, Saxena A, Bonetti A, Voineagu I, Bertin N, Kratz A, Noro Y, Wong C-H, de Hoon M, Andersson R, Sandelin A, Suzuki H, Wei C-L, Koseki H, The FC, Hasegawa Y, Forrest ARR, Carninci P.** 2014. Deep transcriptome profiling of mammalian stem cells supports a regulatory role for retrotransposons in pluripotency maintenance. *Nat Genet* **46:**558–566.

192. **Loewer S, Cabili MN, Guttman M, Loh YH, Thomas K, Park IH, Garber M, Curran M, Onder T, Agarwal S, Manos PD, Datta S, Lander ES, Schlaeger TM, Daley GQ, Rinn JL.** 2010. Large intergenic non--coding RNA-RoR modulates reprogramming of human induced pluripotent stem cells. *Nat Genet* **42:** 1113–1117.

193. **Lu X, Sachs F, Ramsay L, Jacques PE, Goke J, Bourque G, Ng HH.** 2014. The retrovirus HERVH is a long noncoding RNA required for human embryonic stem cell identity. *Nat Struct Mol Biol* **21:**423–425.

194. **Wang Y, Xu Z, Jiang J, Xu C, Kang J, Xiao L, Wu M, Xiong J, Guo X, Liu H.** 2013. Endogenous miRNA sponge lincRNA-RoR regulates Oct4, Nanog, and Sox2 in human embryonic stem cell self-renewal. *Dev Cell* **25:**69–80.

195. **Herquel B, Ouararhni K, Martianov I, Le Gras S, Ye T, Keime C, Lerouge T, Jost B, Cammas F, Losson R, Davidson I.** 2013. Trim24-repressed VL30 retrotransposons regulate gene expression by producing noncoding RNA. *Nat Struct Mol Biol* **20:**339–346.

196. **Emerman M, Malik HS.** 2010. Paleovirology–modern consequences of ancient viruses. *PLoS Biol* **8:**e1000301.

197. **Dupressoir A, Marceau G, Vernochet C, Benit L, Kanellopoulos C, Sapin V, Heidmann T.** 2005. Syncytin-A and syncytin-B, two fusogenic placenta-specific murine envelope genes of retroviral origin conserved in *Muridae*. *Proc Natl Sci Adad U S A* **102:** 725–730.

198. **Heidmann O, Vernochet C, Dupressoir A, Heidmann T.** 2009. Identification of an endogenous retroviral envelope gene with fusogenic activity and placenta-specific expression in the rabbit: a new "syncytin" in a third order of mammals. *Retrovirology* **6:**107.

199. **Cornelis G, Heidmann O, Degrelle SA, Vernochet C, Lavialle C, Letzelter C, Bernard-Stoecklin S, Hassanin A, Mulot B, Guillomot M, Hue I, Heidmann T, Dupressoir A.** 2013. Captured retroviral envelope syncytin gene associated with the unique placental structure of higher ruminants. *Proc Natl Sci Adad U S A* **110:** E828–E837.

200. **Nakaya Y, Koshi K, Nakagawa S, Hashizume K, Miyazawa T.** 2013. Fematrin-1 is involved in

fetomaternal cell-to-cell fusion in *Bovinae* placenta and has contributed to diversity of ruminant placentation. *J Virol* 87:10563–10572.

201. **Cornelis G, Heidmann O, Bernard-Stoecklin S, Reynaud K, Veron G, Mulot B, Dupressoir A, Heidmann T.** 2012. Ancestral capture of syncytin-Car1, a fusogenic endogenous retroviral envelope gene involved in placentation and conserved in *Carnivora*. *Proc Natl Sci Adad U S A* **109:**E432–E441.

202. **Redelsperger F, Cornelis G, Vernochet C, Tennant BC, Catzeflis F, Mulot B, Heidmann O, Heidmann T, Dupressoir A.** 2014. Capture of syncytin-Mar1, a fusogenic endogenous retroviral envelope gene involved in placentation in the rodentia squirrel-related clade. *J Virol* **88:**7915–7928.

Mobile DNA, 3rd Edition
Nancy L. Craig, Michael Chandler, Martin Gellert, Alan M. Lambowitz, Phoebe A. Rice and Suzanne Sandmeyer

doi:10.1128/microbiolspec.MDNA3-0005-2014

Anna Marie Skalka[1]

48 Retroviral DNA Transposition: Themes and Variations

PROVIRAL GENE ORGANIZATION AND EXPRESSION

Two unique features of retroviruses are their genome organization and strategies for gene expression. While the gene order of retroviruses is conserved, the synthesis of retroviral proteins can be controlled by different mechanisms. Variations include the ways in which RNA splicing is used to produce mRNAs from a single long transcript, a portion of which must remain unspliced to serve as the viral genome. With these general principles in mind, an overview of the retrovirus family follows, along with a description of how the gene organization of its members relates to gene expression and the viral entry process. Detailed descriptions of the molecular aspects of these processes can be found in the comprehensive *Retroviruses* (2) an overview chapter in *Fields Virology* (3), or a number of recent reviews that focus mainly on the early steps in the reproductive cycle of human immunodeficiency virus type-1 (HIV-1) (4–11).

The family *Retroviridae* (Fig. 1) includes two subfamilies. The subfamily Oncoretrovirinae comprises six genera, five of which are designated alpha- to epsilon-retroviruses, and the sixth comprises the lentiviruses. The subfamily Spumavirinae includes the "unconventional" spumaviruses (12). The organization of all retroviral proviruses follows the general pattern illustrated at the top of Figure 2. Open reading frames are flanked by long terminal repeats (LTRs) of varying lengths, which include sequences that direct transcription by the host cell RNA polymerase II. The gene order starts with those encoding structural proteins (*gag*) and enzymes (*pro, pol*), followed by a gene (*env*) for an envelope protein that is inserted into the membrane that surrounds the viral capsid, and which determines host cell receptor specificity. Retroviruses with "simple" genomes (i.e. alpharetroviruses and gammaretroviruses) have only these essential genes. The genomes of the other retroviruses include additional sequences that specify auxiliary proteins, most encoded in alternative reading frames that begin on either side of *env*.

As illustrated for a representative alpharetrovirus, gammaretrovirus, and spumavirus (Fig. 2), one long transcript, which includes the R and U5 regions from the upstream LTR and the U3 and R regions from the downstream LTR, serves as *gag* mRNA for all retroviruses. Env protein is always translated from a spliced mRNA, with sub-optimal splicing signals regulating the amounts of *gag* and *env* mRNAs (and the corresponding major structural and envelope polyproteins) that are appropriate for particle production. For all but the spumaviruses, *pol* gene expression is regulated at the level of translation. Gag-Pol polyproteins are

[1]Fox Chase Cancer Center, 333 Cottman Avenue, Philadelphia, PA 19111, United States.

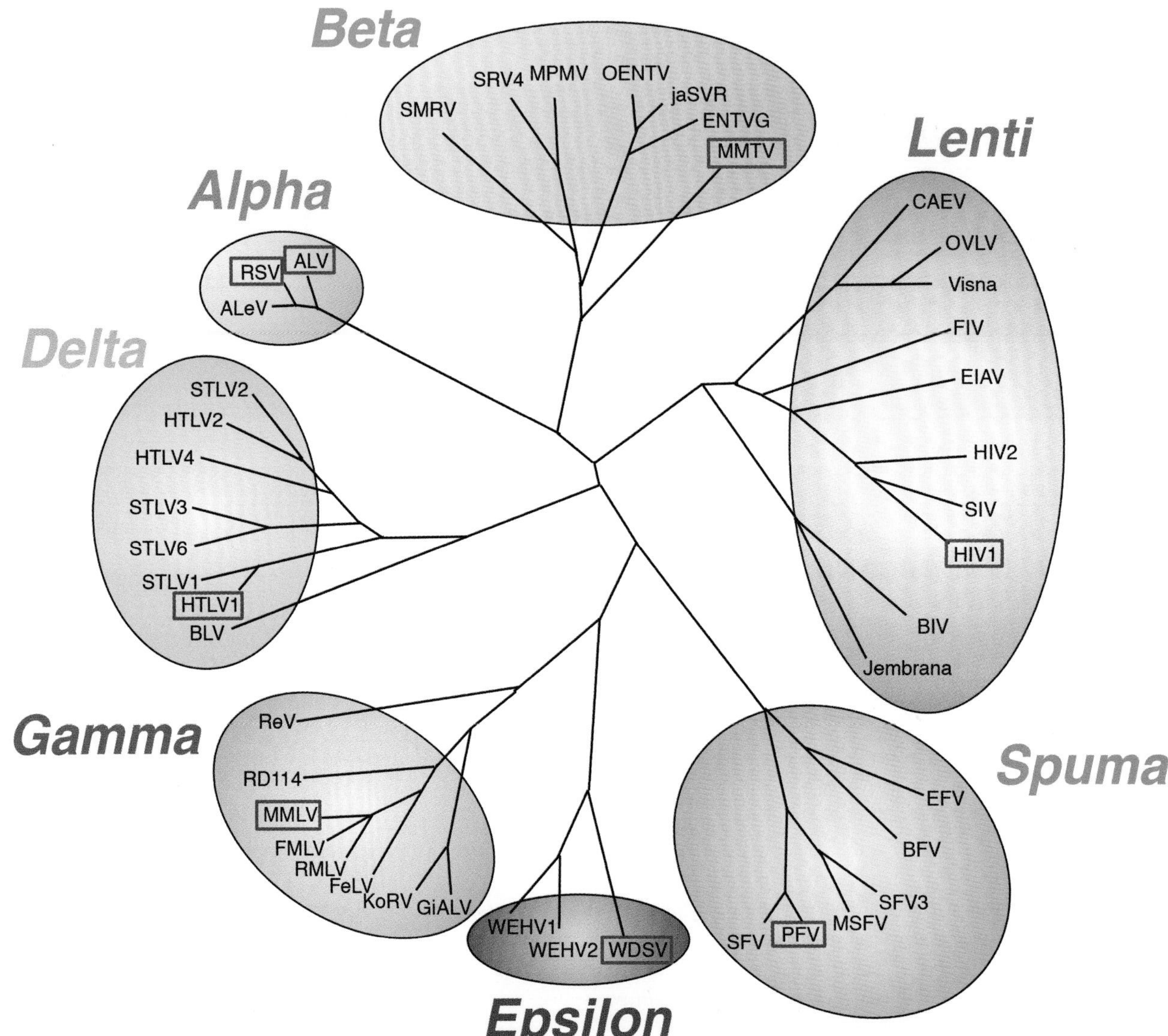

Figure 1 A phylogenetic tree based on the collection of RT-IN sequences for all retro-transcribing viruses in the NCBI taxonomy and NCBI RefSeq databases. Full-length Pol or Gag-Pol sequences for all viruses were downloaded and truncated based on their alignment with the RT-IN region of HIV-1. ClustalX algorithm with neighbor joining clustering was then used for tree reconstruction. The tree in the illustration is an artistic representation, based on the results. Red boxes identify viral species discussed in this overview: for alpharetroviruses, RSV is Rous sarcoma virus and ALV avian leukosis virus; betaretrovirus MMTV is mouse mammary tumor virus; gammaretrovirus MMLV is Moloney murine leukemia virus; deltaretrovirus HTLV-1 is human T-lymphotropic virus 1; epsilonretrovirus WDSV is walleye dermal sarcoma virus; lentivirus HIV-1 is human immunodeficiency virus type 1; and spumaretrovirus PFV is the prototype foamy virus. The analysis and Figure were kindly provided by Dr. Vladimir Belyi, Rutgers Cancer Institute of New Jersey.
doi:10.1128/microbiolspec.MDNA3-0005-2014.f1

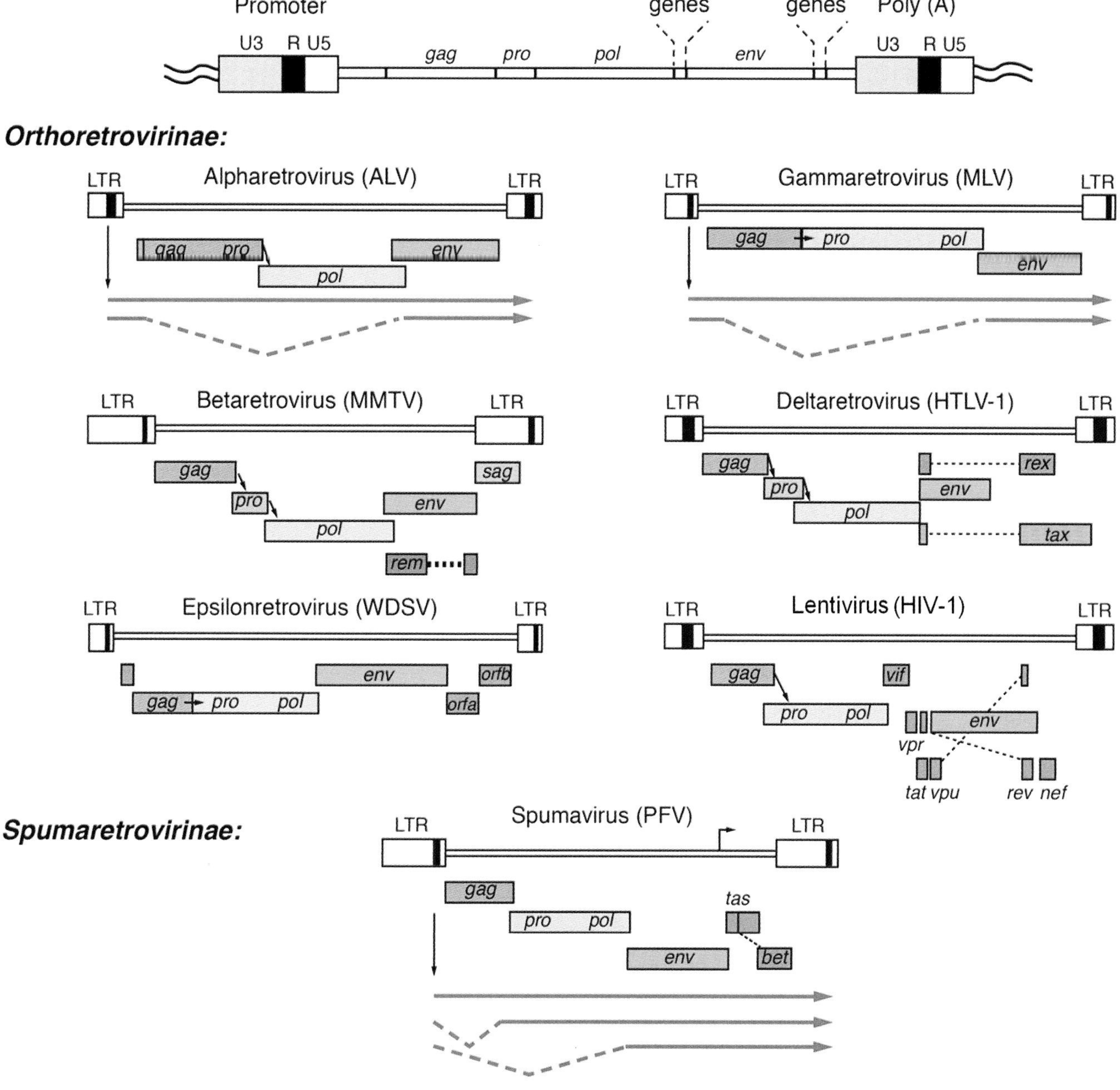

Figure 2 The organization of proviruses in each of the retroviral genera. A generic proviral map is included at the top. Representative genomes have been aligned to allow comparisons, and are not to scale. Viral species are identified in the Figure 1 legend. Origins of the major transcripts of ALV, MLV, and PFV are represented by green arrows below the maps. Translational frameshifts are indicated by descending arrows, and read-throughs by vertical arrows in the gene coding regions. doi:10.1128/microbiolspec.MDNA3-0005-2014.f2

produced from the full-length *gag* transcripts by occasional ribosomal frameshifting or codon read-through at the *gag-pro* or *pro-pol* borders. This mechanism limits the amount of Gag-Pol to approximately 5 to 10% of Gag alone, a ratio that is optimal for packaging into progeny particles. More recent studies have revealed an exception to this paradigm: the Pol protein of spumaviruses is translated from a spliced (*pro, pol*) mRNA (Fig. 2), so that the Gag to Pol ratio is regulated by splicing rather than by ribosomal frameshifting. Multiply spliced mRNAs, characteristic of retroviruses with complex genomes, typically encode the auxiliary proteins that are important for control of viral gene expression, replication, and pathogenesis of these viruses.

The retroviral mRNAs include *cis*-acting signals that facilitate nuclear exit and translation. These mRNAs are translated either by cytoplasmic ribosomes (*gag and gag-pro-pol*) or those attached to the rough endoplasmic reticulum (*env*) (Fig. 3, Late Phase). For the assembly of each virus particle, these viral polyproteins and two copies of full-length viral RNA accumulate at the plasma membrane, as illustrated for the alpharetrovirus, avian leukosis virus (ALV), or in the cytoplasm, for betaretroviruses and spumaviruses. Progeny virus particles are released by budding at sites where the processed Env proteins (surface, SU, and trans membrane, TM, components) have accumulated, either at the plasma membrane or into intracellular vesicles, via co-option of the host cell's intracellular vesicle trafficking machinery. With all but the spumaviruses, mature, infectious virions are generated, during or following budding, by cleavage of the Gag and Gag-Pol polyprotein precursors by the viral protease, which produces the mature viral structural proteins (matrix, MA, capsid, CA, nucleocapsid, NC, and protease, PR) and active viral enzymes (reverse transcriptase, RT, and integrase, IN). This proteolytic processing reaction is associated with condensation of the capsid structures.

The assembly and maturation pathway of all retroviruses positions the processed Pol products, RT and IN, within the viral capsids in preparation for the early steps in infection. Exit from a particle-producing cell and subsequent entry into the cytoplasm of a naïve cell is a major distinction between the retroviruses and the intra-cellular cycling, LTR retrotransposons. With the latter, particles accumulate in the cytoplasm of the producing cell and DNA copies are inserted at new sites when a preintegration complex enters the nucleus of the same cell (see Chapter 42 Sandmeyer). Retroviral particles attach to specific receptors on host cells via interaction with their envelope proteins (Fig. 3, Early Phase). Capsid entry into the cytoplasm is mediated via membrane fusion, which is triggered by major rearrangements in the viral envelope protein. Such fusion takes place either at the cell surface or following particle uptake into endosomes, depending on the virus and, in some cases, the target cell type.

SYNTHESIS OF RETROVIRAL DNA

Although the biochemistry of the RT enzymes is fairly well understood (see Chapter 46 Hughes), and routinely exploited in biotechnology, details of the retroviral reverse transcription process that takes place within infected cells have not been fully elucidated. The current state of the field is a dichotomy: seminal and atomic level studies of isolated RT enzymes, versus the continuing exploration of RT complexes in infected cells, which may include viral and host proteins that participate in the reaction. The latter studies have been driven forward by the use of new, post genomics-era technologies.

The Reverse Transcriptase Complex

The steps following virus capsid entry (Fig. 3, Early Phase) are poorly understood, even for the most intensely studied retrovirus, the lentivirus HIV-1. The capsids of HIV-1 and the gammaretrovirus murine leukemia virus (MLV) are known to be incompletely disassembled following entry and to remain associated with the sub-viral, **"reverse transcription complex" (RTC)**. The RTC contains two copies of viral RNA coated with the viral nucleocapsid (NC) protein, a specific tRNA primer, ~50–100 molecules each of RT and IN, and other viral and host proteins. Live cell imaging studies with fluorescently labeled viral components have shown that the HIV-1 RTC moves through the cytoplasm by interaction with the host cell cytoskeletal fibers (13) as DNA is synthesized and an integration-competent nucleoprotein complex, called a **pre-integration complex (PIC)**, is formed. Genetic and biochemical studies have shown that two host proteins that bind to the HIV-1 capsid protein (CA), a cleavage and polyadenylation factor (CPSF6) and the peptidyl-prolyl isomerase cyclophilin A (CypA), provide structural stability to the RTC and PIC, helping to suppress premature capsid disassembly ("uncoating") by host restriction factors, such as TRIM5α and TRIMCypA, as well as the detection of viral nucleic acids by intrinsic immune sensors in the host cell (14–16). HIV-1 mutants with capsids that are either fragile or abnormally stable, are replication defective (17). The HIV-1 capsid also interacts with nuclear pore proteins; CA binding to the cyclophilin domain of one such protein, Nup358/Ranbp2, is proposed to facilitate transfer of a PIC to

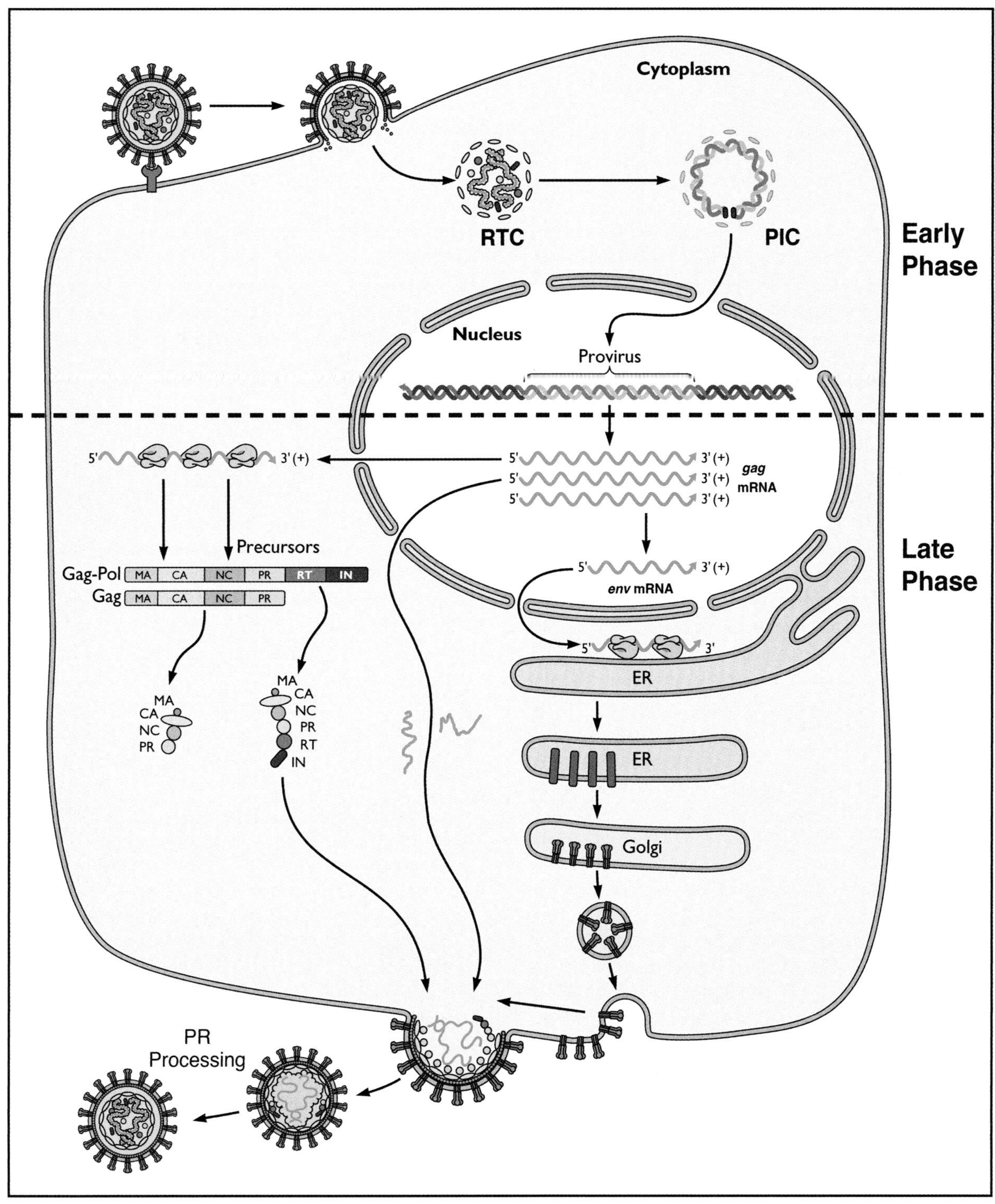

Figure 3 The single cell reproductive cycle of the alpharetrovirus, ALV. The virus life cycle is divided into an early phase that includes steps from virus infection to establishment of the provirus, and a late phase that includes expression of the provirus and formation of progeny virions. Adapted from: Principles of Virology, 3rd edition Vol. I. 2009. S.J Flint, L.W. Enquist V.R Racaniello, and A.M. Skalka ASM Press Washington DC, Appendix A, Figure 20. doi:10.1128/microbiolspec.MDNA3-0005-2014.f3

the nuclear compartment in non-dividing cells, such as mature macrophages (18; see also Chapter 45 Craigie & Bushman). Control of capsid disassembly is also important for the PIC of retroviruses like MLV, which requires nuclear membrane breakdown during mitosis to gain access to host DNA. The host restriction factor Fv1, itself the product of a co-opted endogenous *gag* gene, confers resistance to MLV in certain mouse strains by interacting with capsid protein lattices in the PIC, and suppressing uncoating (19). The question of how host proteins might affect this process with other retroviral family members has received little attention to date, but such investigations seem likely to reveal additional variations in mechanism.

Viral DNA in Infected Cells

Almost nothing is known about the molecular organization of components within retroviral RTCs and PICs. Analysis of these sub-viral particles is difficult because they are minor components in infected cell extracts, and their existence is transient. Furthermore, interpretation of results is confounded by the knowledge that most infecting particles may be non-functional, as only a minority give rise to proviruses. *In vitro* studies with retrovirus particles purified from infected cell supernatants, and permeabilized by treatment with mild detergents or the chaotropic protein melittin, suggest that reverse transcription is poised to begin immediately upon infection. Upon addition of deoxyribonucleotide triphosphates to such permeabilized virus particles, an endogenous reaction in which both RNA- and DNA-templated viral DNA synthesis occurs (20–22). Studies with cultured cells infected with HIV-1 or the alpharetrovirus, avian sarcoma and leucosis virus (ASLV) have demonstrated that synthesis of viral DNAs can be detected within 2–3 hours following virus entry into the cytoplasm (23, 24). However, the structures of the DNA products isolated from such cells vary among retroviruses, indicating that virus-specific differences in the details or timing of the process exist. While RNA-templated minus-strand DNA of the beta- and gammaretroviruses, mouse mammary tumor virus (MMTV) and MLV, are mostly full length, both have large gaps in their plus-strand DNAs. Plus-strands of ASLV are also discontinuous, but include many overlapping "flaps", which are produced by RT-catalyzed "strand displacement synthesis" as the 5′ end of a previously synthesized plus-strand is removed from the minus-strand template by 3′ the leading edge of a newly synthesized strand. The plus-strand of HIV-1 contains a central flap produced by strand displacement synthesis through a central PPT, which encodes as an additional primer for plus-strand synthesis in this virus. Studies with PICs isolated from cytoplasmic extracts of HIV-1 infected cells have suggested that discontinuities in DNA may persist even after integration (25).

RT-Mediated Recombination

The high rate of genetic recombination, a hallmark of retrovirus replication, is facilitated by the presence of two RNA templates within the RTC. Although only one viral DNA molecule is normally produced by each infecting virion, recombination can occur if RT switches from one RNA template to a homologous sequence on the second RNA template, either by chance or "forced" by a break in the first template strand. While this "copy choice" mechanism is probably the most frequent (26), recombination may also occur among DNA products copied from both RNA templates. In the latter model, the 5′-end of a DNA plus-strand that is displaced from the minus-strand DNA synthesized from one RNA template (a DNA "flap"), binds to a complementary sequence on the minus-strand DNA synthesized from the second RNA template. A recombinant is formed when the invading plus-strand is joined to a plus-strand of the second template (27). The ability to form a complete provirus from two damaged RNA templates was probably strongly selected during evolution. If the two RNA templates in an RTC are identical, recombination events are invisible. However, the incorporation of two distinct RNA templates in a single virus particle can lead to the production of new combinations of sequences. Combinations of viral and cellular sequences may also arise in RTCs from proviral transcripts that include both viral and downstream host sequences that are acquired via abnormal RNA splicing, transcription read-through, or following deletion of downstream proviral sequences. Such transcripts will be incorporated into progeny viral particles by virtue of their viral components. Although rare, such events can give rise to oncogene transducing proviruses that are almost always replication defective, but can have profound effects on host cell biology and evolution. Exactly how the multiple, highly-ordered reactions necessary for producing a single double stranded viral DNA copy from two RNA templates are catalyzed in infected cells is an area of ongoing research.

RT Composition and Organization

While similar reverse transcription reactions are catalyzed by all retroviruses, the structures of the responsible RTs can vary considerably. These findings raise intriguing evolutionary and biochemical questions. For example, retrovirus-specific differences in proteolytic processing of the Gag-Pol polyprotein precursors can lead to the inclusion of additional RT sequences or domains (Fig. 4). Furthermore, although the RTs of most retroviral genera function as monomers, the enzymes of ASLV, and HIV-1, function as dimers (28). Both αβ and αα dimers of RT are detected in ASLV particles. Although the α and β subunits have identical polymerase and RNase H domain sequences, the β subunit also includes the IN domain. Genetic studies indicate that the smaller, α subunit is the catalytically active component of the αβ heterodimer (29). Nevertheless, the integrase (IN) domain at the C-terminal end of the β subunit likely contributes importantly to reverse transcription as the αα homodimer exhibits reduced processivity and RNase H activity when compared to the heterodimer *in vitro*. Consequently the αβ heterodimer, which is most abundant in virus particles, is considered to be the physiologically relevant isoform. In the HIV-1 RT heterodimer, comprising the polymerase and RNase H domains (p66) and the polymerase alone (p51), the larger subunit is the catalytically active component and the smaller subunit provides stability to the complex. Although the HIV-1 p66 and p51 subunits have the same amino acid sequences in all but the RNase H domain, their components are organized differently. It would be of interest to know if dichotomy also exists for the ASLV RT heterodimer.

The *pol* genes of several non-primate lentiviruses such as equine infectious anemia virus (EIAV) and feline immunodeficiency viruses (FIV) encode a dUTPase between RT and IN. The genome of the betaretrovirus MMTV also encodes a UTPase, but in this case the sequences are in a reading frame upstream of RT, between the ends of *gag* and *pro*. Programmed alternative frameshifting yields different Gag fusion products to be incorporated into the particles of different retroviruses. In the case of beta- and deltaretroviruses, frameshifting yields Gag-Pro and Gag-Pro-Pol fusions that differ in amino acid sequence downstream of the shift site such that distinct unshifted C-terminal PR and shifted N-terminal RT sequences are encoded from the same nucleotide sequence. The first 27 amino acids of the MMTV RT are derived via frameshift from sequences near the end of the *pro* gene, that encodes PR, and a 26 amino acid N-terminal stretch in the RT of the deltaretrovirus, bovine leukemis virus (BLV) is acquired by a similar frameshift in *pro*. The most extreme of these fusions is

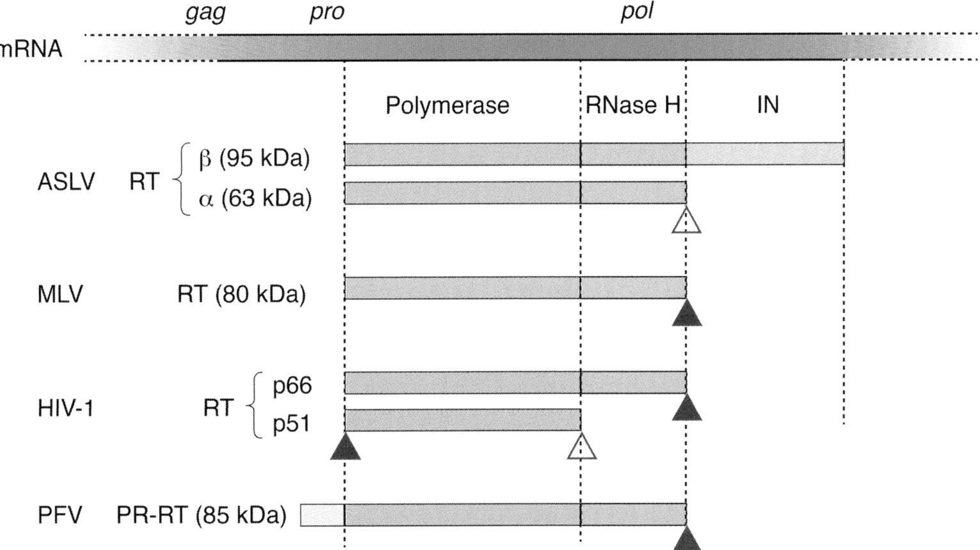

Figure 4 Domain and subunit relationships of the RTs of different retroviruses. The organization of open reading frames in the mRNAs of all but the spumavirus PFV, is indicated at the top. PFV RT is expressed from a spliced *pro-pol* mRNA (Fig. 2). Protein products (not to scale) are shown below, with arrows pointing to the sites of proteolytic processing that produce the diversity of RT subunit composition. Open red arrows indicate partial (asymmetric) processing, and the solid arrows indicate complete processing. ASLV, avian sarcoma/leucosis viruses; others are identified in Figure 1. Adapted from: Principles of Virology, 3rd edition Vol. I. 2009. S.J Flint, L.W. Enquist V.R Racaniello, and A.M. Skalka, ASM Press Washington DC, Figure 7.9. doi:10.1128/microbiolspec.MDNA3-0005-2014.f4

found in the prototype foamy virus (PFV) where the entire protease region (PR) is attached to the N-terminus of RT. This latter arrangement is somewhat of a puzzle, as PR functions as a dimer but RT as a monomer, and it has been proposed that the PFV PR domain may be dimerized transiently by binding to viral RNA (30). Biochemical studies indicate that the C-terminal 42 amino acids of the PR domain are important for PFV RT activity and stability (31).

Biochemistry of Reverse Transcription

A detailed, biochemical analysis of the reverse transcription is provided by Hughes in Chapter 46. This section provides a general description of the process, noting the similarities and differences observed among the various retroviral genera. An outline of the distinct stages in the multi-step process of reverse transcription (Fig. 4) was deduced mainly from study of the structure and function of purified, bacterially produced enzymes. DNA synthesis is primed by a tRNA that is annealed to the primer binding site (pbs) near the 5′-end of the viral RNA template and extends to the 5′ end of the RNA including a sequence (R), which is repeated at the 3′ end of the RNA. Following RNase H digestion, an exchange of templates occurs via annealing of the newly synthesized DNA (called minus-strand strong-stop DNA) to the repeated sequence, R, at the 3′-end of the viral RNA (Fig. 5A and B). This first template exchange allows continued synthesis of minus-strand DNA and digestion of the viral RNA template by RNase H. Synthesis of plus-strand DNA is primed by a purine rich, RNase H-resistant fragment of RNA (ppt), followed by a second strand exchange at the primer binding site in minus-strand DNA (PBS) (Fig. 5C). The ends of the completed linear DNA product are produced by strand displacement synthesis in which sequences corresponding to the LTRs are copied at each end (Fig. 5D). A single RT molecule contains all of the enzymatic activities required to catalyze each of these steps, but the number of molecules in the RTC that actually take part in viral DNA synthesis *in vivo* is unknown. Genetic and structural studies suggest that the polymerase and RNase H catalytic sites do not act simultaneously; conformational changes are required to optimize each function (32). *In vitro* studies of purified HIV-1 RT have revealed the amazing dynamic capabilities of the enzyme, which is able to bind to the primer/substrate in a position poised for catalysis in one direction and to flip 180 degrees to initiate synthesis from the ppt primer. RT flipping and sliding can occur without disengaging from the template (33, 34). Such large-scale molecular contortions may be required to accomplish the various steps in reverse transcription.

Comparison of the biochemical activities of RTs from the various retroviral families has revealed similarities but also notable differences, even among those of closely related viruses, such as the lentiviruses HIV-1 and HIV-2 (28, 35). In some cases, the differences in catalytic activity, or sensitivity to inhibitors, can be explained by particular variations in specific amino acid side chains. The fingers-palm-thumb domains characteristic of all nucleic acid polymerases are also present in all retrovirus RT structures, but amino acid differences in the palm modules can account for the fact that non-nucleoside inhibitors (NNRTIs) that bind to this module and inactivate HIV-1 RT are ineffective against HIV-2 RT.

Despite their structural and biochemical differences, recent X-ray crystallographic analyses of enzymes bound to cognate DNA-RNA hybrids reveal a strikingly similar topological arrangement of the modular structure among RTs as distinct as the heterodimer of HIV-1, the monomer of MLV-related XMRV, and the homodimer of the yeast retrotransposon Ty3 (Fig. 6). The fingers-palm-thumb modules, and the positions of the nucleic acids, are almost superimposable in the catalytic subunits of these enzymes, even though the dimer architectures of HIV-1 and Ty3 RTs are distinct (36). The RNase H topologies are also similar but in the case of Ty3 the catalytic polymerase and RNase H domains are derived from separate monomers. Equally striking is the similarity in overall topology of the non-catalytic subunits of HIV-1 and Ty3 RTs, in which the RNase H domain of the catalytic polymerase subunit of Ty3 occupies the same position as the connection domain in p51 of HIV-1. The conserved architectures are most evident in features that are critical to the catalytic reactions common to these enzymes. The differences in sequence and quaternary architectures that exist among specific RTs must therefore relate to other functions or interactions that are important in the replication of these viruses and retrotransposons, most of which remain to be discovered.

FROM REVERSE TRANSCRIPTION TO A PRE-INTEGRATION COMPLEX

The nucleoprotein dynamics that lead to the production of viral DNA poised for integration are incompletely understood. Among the key questions are: How do RT and IN function coordinately? How do IN-DNA complexes move into the nucleus? Which host proteins participate in, or block these steps? Our current understanding represents an amalgam of historical and recent findings.

A. Initiation of (−) strand DNA synthesis

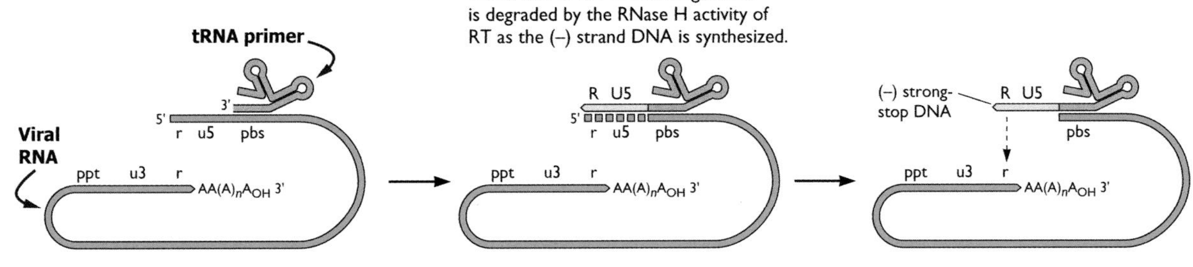

B. First template exchange

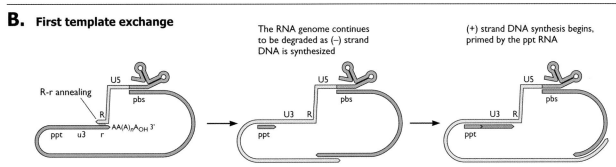

The RNA genome continues
to be degraded as (–) strand
DNA is synthesized

(+) strand DNA synthesis begins,
primed by the ppt RNA

R-r annealing
U5
pbs
R
AA(A)$_n$A$_{OH}$ 3'
ppt u3 r
U5
pbs
U3 R
ppt
U5
pbs
U3 R
ppt

C. (+) strand DNA synthesis

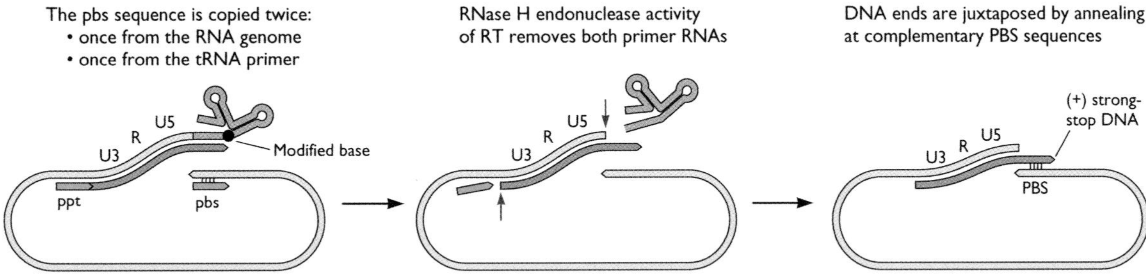

D. Strand Displacement Synthesis
Linear Product

(+) strand synthesis continues
from (+) strong-stop DNA

Sequences bordered by
the (–) and (+) strand
start sites (pbs and ppt)
are copied twice

pbs
ppt U3 R U5 pbs
ppt U3 R U5
Strand displacement
synthesis by RT extends
the (–) strand DNA
pbs
ppt U3 R U5
U3 R U5 pbs

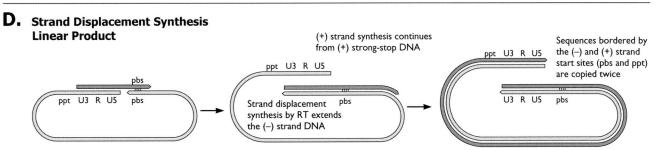

Figure 5 Major steps in retroviral reverse transcription. For simplicity, reverse transcription from a single RNA template is shown, and potential (+) strand synthesis from viral RNA fragments other than the ppt are omitted. RNA is represented by green lines, with key regions identified in lower case: pbs, tRNA primer binding site; u5, unique 5′-end sequence; r, repeated sequence; u3, unique 3′-end sequence, ppt, polypurine tract. Light blue lines represent (-) strand DNA, and dark blue, (+) strand DNA: key regions are identified in uppercase. A modified base in the tRNA primer (C.) blocks further reverse transcription of the tRNA. Adapted from: Principles of Virology, 3rd edition Vol. I. 2009. S.J Flint, L.W. Enquist V.R Racaniello, and A.M. Skalka, ASM Press Washington DC, Figures 7.3–7.6.
doi:10.1128/microbiolspec.MDNA3-0005-2014.f5

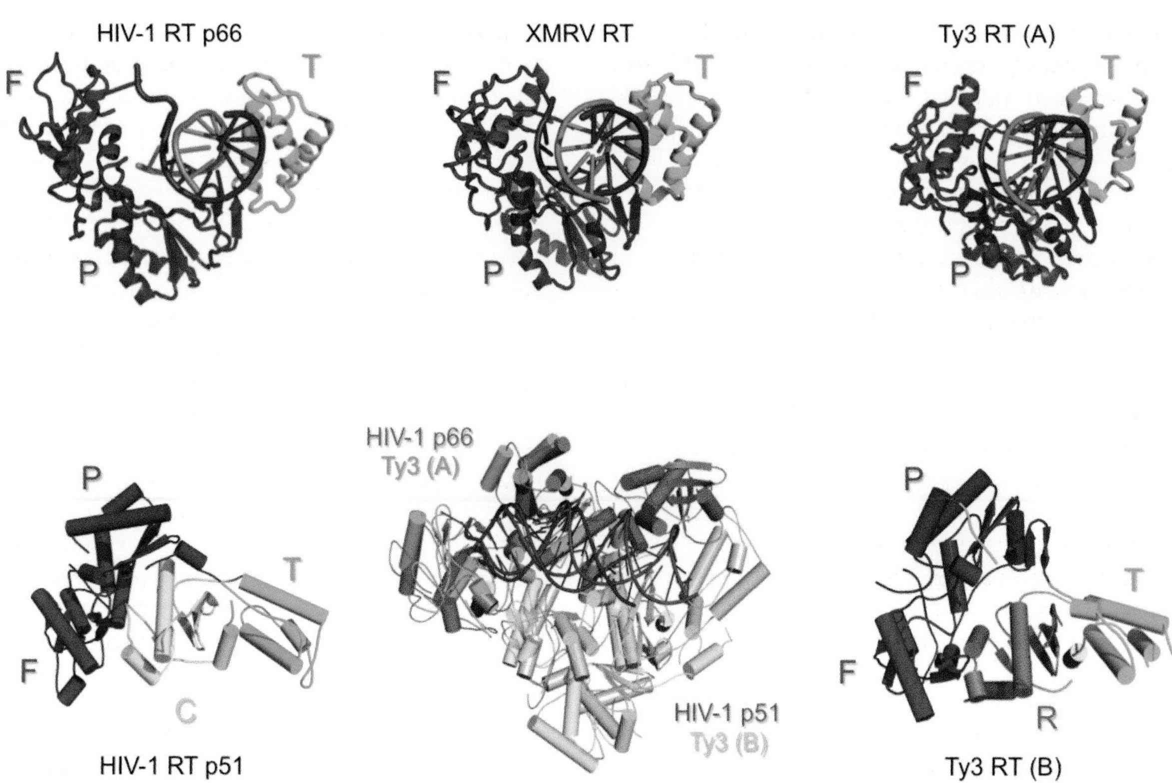

Figure 6 Comparison of the structures of three RTs. Top row: The DNA polymerase domains of lentiviral (HIV-1 p66, *left*), gammaretroviral (XMRV, *center*) and LTR-retrotransposon (Ty3 subunit A, *right*) RTs. Fingers, palm and thumb subdomains are designated F, P and T, respectively. Positions of the RNA (magenta) and DNA strands (teal) of the bound RNA/DNA hybrids are shown. Bottom row: Architectures of the non-catalytic subunits of the dimeric RTs: HIV-1 p51 (*left*) and Ty3 subunit B (*right*). Both subunits contain F, P and T subdomains in analogous positions; the p51 connection and Ty3 (subunit B) RNase H domain, denoted C and R respectively, are also in analogous positions. Superposition of the asymmetric p66/p51 HIV-1 RT heterodimer and the symmetric Ty3 (A)/(B) homodimer is shown in the center. HIV RT subunits are in orange and grey, and Ty3 subunits in green and yellow. The illustration was prepared by Drs. Jason Rausch, and Stuart Le Grice, NCI-Frederick, and Dr. Marcin Nowotny, International Institute of Molecular and Cell Biology, Poland. Structure details for HIV-1 RT are in (32, 121), for XMRV RT in (122, 123), and for Ty3 RT in (36). doi:10.1128/microbiolspec.MDNA3-0005-2014.f6

It seems likely that structural changes accompany the transition from an RTC to a PIC in an infected cell. Differences in the representation of proteins associated with these complexes have been observed (37), but the details of molecular arrangements within them remain obscure. As the DNA product of RT is the substrate for IN, functional coupling between the two proteins at early stages in the retrovirus life cycle may be expected, and several lines of evidence support this idea.

RT-Integrase (IN) Interactions

The earliest biochemical studies of retroviral RTs focused on proteins purified from alpharetroviruses, because large quantities of particles could be prepared from the plasma of chickens infected with avian myoblastosis virus or from cultured cells infected with the Rous ASLV. The presence of IN at the C-terminus of the larger, β subunit of RTs of these ASLVs (Fig. 4), prompted initial speculation that RT might be able to catalyze both viral DNA synthesis and its integration. Indeed, the first site-selective nicking of viral DNA end sequences was actually detected with this RT (38). Although it was later shown that the isolated ASLV and MLV IN proteins could catalyze the integration reaction *in vitro* (39, 40), subsequent genetic and biochemical studies have suggested that retroviral RT and IN proteins may, indeed, interact in functionally relevant ways. For example, physical association between the

RT and IN proteins of MLV and HIV-1 has been detected by a variety of assays *in vitro* (28, 41–46). Investigations with purified HIV-1 RT indicate that IN can stimulate both the initiation and elongation activities of the RT by enhancing its processivity *in vitro* (47). Site-directed mutagenesis studies have identified residues in the C-terminal domain (CTD) of HIV-1 IN that are critical for RT binding (48). Experiments in yeast cells indicate that the IN protein of Ty3 is required for the initiation of reverse transcription during retrotransposition (49). However, because RT and IN interactions may occur in the context of polyproteins as well as in the processed forms, the biological relevance of particular contacts can be difficult to determine.

Genetic analyses have shown that HIV-1 IN mutants are pleiotropic: some mutations (called Class I mutants) block integration specifically, for example by affecting residues in the active site or residues important for substrate DNA binding. But others (Class II mutants), which can be mapped throughout the IN coding region, affect additional stages of the life cycle including reverse transcription, and virus particle assembly and release (50). A significant, as yet unexplained, effect on capsid morphogenesis is suggested by the fact that viral RNA is excluded from the capsid in particles formed by some HIV-1 IN mutants, and in particles that lack IN protein entirely (51). Some Ty3 IN mutants appear to be similar to these HIV class II mutants, as processing defects that might be linked to assembly were observed (39). Despite the potential complexities, complementation assays in which the reverse transcription defects of certain HIV-1 Class II IN mutants were corrected by inclusion of a Class I mutant as a Vpr-IN fusion *in trans*, have confirmed the importance of IN-RT interactions at early stages in HIV-1 replication (43, 52, 53). The impact of such interactions for other retroviruses remains obscure.

First Steps of IN Catalysis

Retroviral IN proteins catalyze two sequential reactions, which are separable both biochemically and temporally (Fig. 7, left). In the first reaction, IN removes nucleotides (usually two) that follow the conserved dinucleotide CA at the 3′-ends of the double-stranded viral DNA termini produced by RT. Analysis of cells infected with MLV, showed that “processed,” recessed viral DNA ends can be detected in the cytoplasm shortly after blunt ends are formed at viral DNA termini (54). The second step catalyzed by IN, concerted joining of the two processed CA_{OH}-3′-ends of viral DNA to host DNA, can be detected by providing a target DNA (e.g. from bacteriophage lambda) to PICs isolated from the cytoplasm of both MLV and HIV-1 infected cells. Because this reaction was reported to be inefficient with ASLV PICs, unless deoxyribonucleotide triphosphates are added, it is possible that for some retroviruses synthesis of blunt viral DNA ends, and processing by IN, may not be completed until the PIC enters the nucleus (55).

The activity and function of retroviral IN proteins may also be affected by post-translational modifications (e.g. ubiquitination, SUMOylation, acetylation, phosphorylation) that can occur in the cytoplasm or nuclei of infected cells. The pathways for the placement and removal of such modifications in cellular proteins are complex. The interplay among them, and the net downstream effects, regulate critical events in cell biology including gene expression, DNA repair, and the cell cycle. While a variety of potential of post-translational modification sites can be identified at various locations in retroviral IN proteins, studies of the potential effects of such modifications have focused primarily on HIV-1 IN. This protein is reported to be modifiable by ubiquitination of residues in the CTD (Fig. 8) and the preceding linker (K211, 215, 219, 273), acetylation in the CTD (K258, 264, 266, 273), SUMOylation in all three domains (K46, 136, 244), and by phosphorylation of a serine in the central, catalytic core domain (S57) (reviewed in 56). Such modifications could affect not only IN stability or catalytic activity, but also interactions with other proteins at various stages of the retroviral life cycle. HIV-1 IN is subject to degradation by the ubiquitin-proteosome pathway, and it has been suggested that proteasome degradation following interaction of IN with the von Hippel-Lindau binding protein (VBP1) and ubiquitination by the Cul2/VHL ligase, may be required for cellular components of post-integration repair and gene transcription to gain access to the provirus after its integration (57). Two histone acetyl transferases, p300 and GCN5, have been shown to bind to HIV-1 IN, *in vitro* and in infected cells, and acetylation of lysines in the CTD is reported to enhance integration by increasing DNA binding affinity *in vitro* (58, 59). In another study, HIV-1 SUMOylation site mutants were shown to be defective in a step after reverse transcription, but before integration (60), and it was suggested that this modification may enhance IN binding to cellular proteins that contain SUMO interaction motifs (SIMs), and are required for efficient viral replication. However, unraveling the functions of any of these modifications in the HIV-1 life cycle will require further study.

Nuclear Entry

In non-dividing cells, the nuclear membrane separates the retroviral PIC from its host DNA target. For MLV,

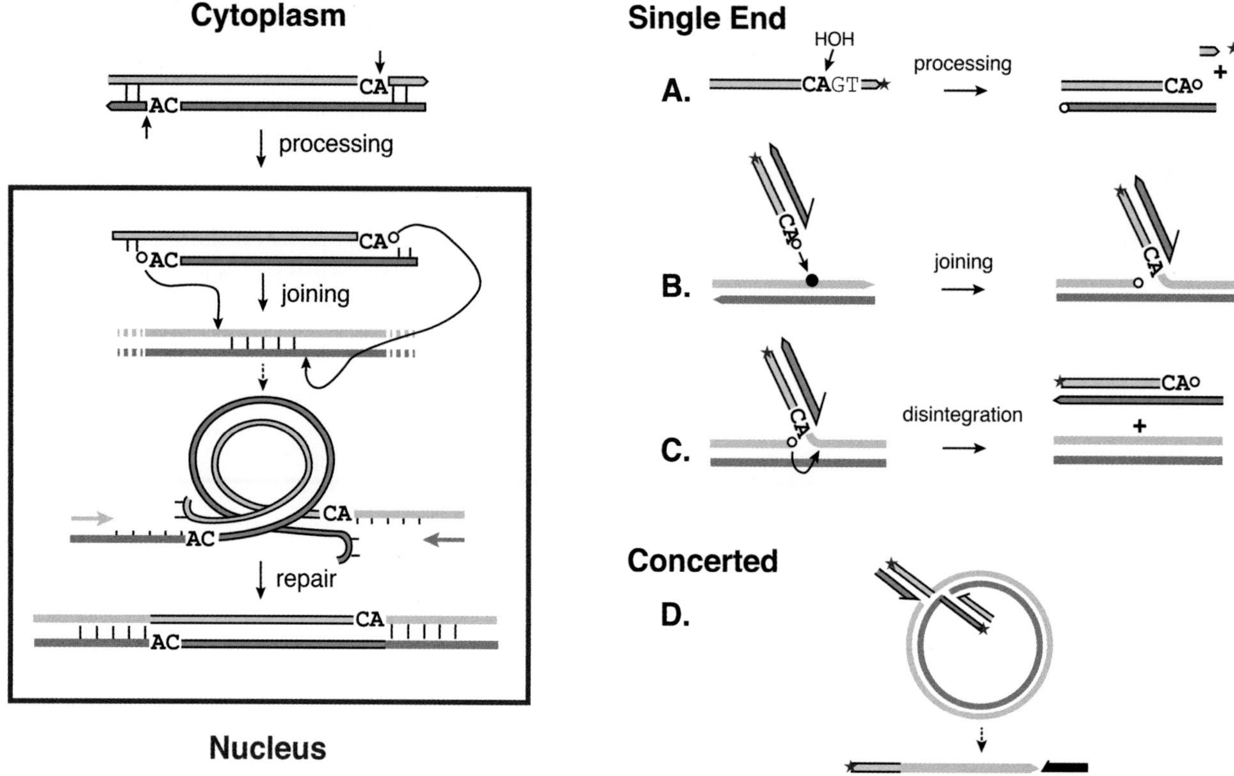

Figure 7 Reactions catalyzed by retroviral integrases. Left: Reactions at the retroviral DNA ends produced by RT in infected cells. The processing reaction takes place in the cytoplasm as soon as DNA synthesis is completed at the termini. Following nuclear entry the two processed viral DNA ends are joined to host DNA in concerted cleavage and ligation reactions at staggered positions in the target site. Repair of the resulting gaps is catalyzed by host enzymes. The integrated provirus is flanked by short repeats (indicated by vertical lines) of the host DNA, with length determine by the distance between the staggered cuts made by IN. Right: Reactions as assayed *in vitro* using duplex oligonucleotides containing viral DNA end sequences and target DNAs. Red stars indicate radioactive or fluorescent labels that can be used for distinguishing reactions (A) and identifying recombinant molecules (B–D) following electrophoresis in denaturing or non-denaturing (D) gels.
doi:10.1128/microbiolspec.MDNA3-0005-2014.f7

this is a nearly impenetrable barrier. Integration of MLV DNA is only efficient in cycling cells, in which the nuclear membrane breaks down during mitosis and reassembles before G_1. Tethering to mitotic chromosomes by the MLV *gag*-encoded protein p12 in the PIC, allows its selective accumulation in newly formed nuclei (61–63). In contrast, HIV-1 and ASLV integration can occur in non-cycling cells, although the process is more efficient with the former (64–66). Indeed for HIV-1, nuclear transport entry is likely to be the major mechanism of PIC access to target host DNA, even in dividing cells. Considerable effort has been expended in numerous laboratories to identify the viral and cellular components that are critical for nuclear entry of the HIV-1 PIC. Results have been contradictory and sometimes controversial, perhaps because there are compensating modes of transport into the nucleus for this virus. Nevertheless, the current consensus is that interaction of HIV-1 CA, and perhaps other viral components, with nuclear pore proteins help to position the PIC for transport (reviewed in 6); (see also Chapter 45, Craigie & Bushman). As the cytoplasmic HIV-1 PIC exceeds the size of the nuclear pore, one attractive model is that, as with herpesviruses and adenoviruses (67), attachment to the pore facilitates (partial) disassembly of the complex allowing transport of a streamlined PIC into the nuclear compartment. A transferable nuclear localization signal has been identified in ASLV IN, and studies with permeabilized cells indicate that nuclear import of this protein relies on one or more of

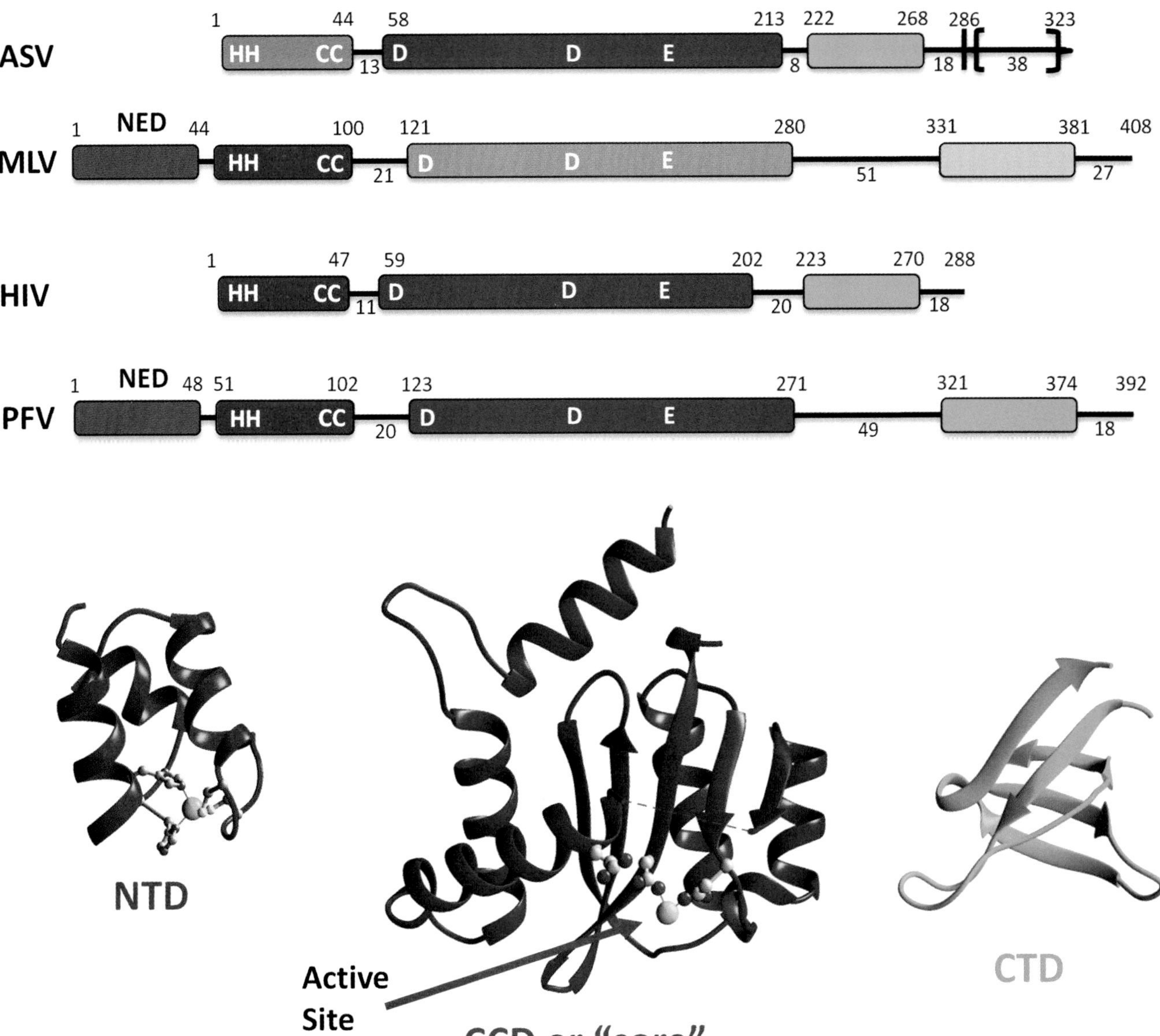

Figure 8 Domain organization of IN proteins from different retroviruses. Maps for the organization of IN proteins are shown with amino acid numbers that delineate the start and end of each domain. The lengths of linkers that connect the domains are also indicated below the lines between domains. The domain models below are from the crystal structures of the HIV-1 N-terminal domain (NTD), catalytic core domain (CCD), and C-terminal domain (CTD), PDB codes 1K6Y, 1BIU, 1EX4, respectively. The Zn^{2+} ion in the NTD is shown as a blue sphere, and the Mg^{2+} ion in the active site of the CCD, as a green sphere. Domains in proteins for which there is no experimentally determined structure from crystallography are shown in muted colors in the maps above. The domain pictures were generated using Chimera software (UCSF) and the figure kindly provided by Dr. M.D. Andrake, Fox Chase Cancer Center, Philadelphia, PA. doi:10.1128/microbiolspec.MDNA3-0005-2014.f8

the cellular components that mediate transport of the linker histone H1 (68). Whether IN is the major viral determinant of nuclear transport for the ASLV PIC remains to be resolved. Nevertheless, while it is clear that ASLV DNA can be integrated in non-dividing cells, it is often misstated in the literature that the lentiviruses are "unique" in this property. In fact, very little is known about the manner in which PICs other than those of ASLV, MLV, and HIV-1, gain access to their cellular DNA targets. Moreover, as early steps in retrovirus replication can be restricted in non-dividing cells by factors other than nuclear access (reviewed in 69), cell-cycle dependence cannot be used as a surrogate criterion for transit through the nuclear pore.

THE MECHANICS OF INTEGRATION

Molecular cloning and characterization of MLV and ASLV proviruses in the 1980s identified the essential hallmarks of the integration reaction (Fig. 7), and revealed their similarity with DNA transposable elements of bacteria and eukaryotic cells (70–73). Like these transposable elements, proviral DNAs all contain the dinucleotide the CA at their 3′-ends. The finding that these dinucleotides are sub-terminal in unintegrated viral DNA indicated that two base pairs must be lost from either end upon integration. As with DNA transposable elements, the ends of the provirus are flanked by direct repeats of host DNA that were formed during integration. These duplications are characteristic of the virus (i.e. 4-6 base pairs). The genesis of these features became clear in subsequent analyses of the viral DNAs formed in infected cells, and of the activities of purified IN proteins.

Target Site Selection

Retroviral DNA arrives in the nuclear compartment with processed ends tightly bound to IN and poised for joining to a target DNA. This IN-viral DNA complex has been called an "intasome." While integration can occur at many sites in host cell DNA, it is now clear that site selection is not random. LTR retrotransposons are known to be integrated at very specific target sites. For example, the Ty1 and Ty3 retrotransposons of the budding yeast *Saccharomyces cerevisiae* are inserted preferentially upstream of genes transcribed by RNA polymerase III (e.g. tRNA and 5S ribosomal genes), and the Ty5 retrotransposon into regions of heterochromatin at telomeres and the silent mating-type loci. The availability of sequence data for human, mouse, and avian genomes, and use of next generation sequencing methods have now made it possible to determine integration site preferences for representatives of most retroviral genera, which also vary in characteristic ways (74). The gammaretroviruses and spumaretroviruses tested all show a marked preference for integration in, or near, transcription start sites and CpG islands, which are enriched in the promoter regions of highly-expressed genes. The lentiviruses show strong preference for integration within active transcription units (but not start sites or CpG islands), and the alpharetroviruses and deltaretroviruses exhibit only weak preference for transcription units and CpG islands (but not transcription start sites). Finally, the betaretroviruses tested show no significant deviation from random insertion in integration site preference. These differences may be explained, in part, by dissimilar requirements for chromatin structure or accessibility, and/or by IN-specific cellular protein attachments that tether the intasome to chromatin. For HIV-1, interaction of the chromatin-targeting, lens epithelium-derived growth factor (LEDGF/p75) with the a catalytic core dimer interface of IN is clearly a major determinant of integration site selection (for review see 75). Recent studies with gammaretroviruses MLV and FeLV have identified bromodomain and extraterminal domain (BET) proteins (Brd 2, −3, −4) as virus-specific cellular tethers. In these cases, the conserved ET domains bind to a region of IN that includes portions of the catalytic core and C-terminal domains, and direct the intasomes to transcription start sites (76, 77); (see Chapter 45 Craigie & Bushman). This interaction may also affect IN structure in some beneficial way, as the binding of an isolated ET domain was shown to increase the catalytic activity of MLV IN *in vitro*, and its over-expression in cultured cells stimulated MLV integration.

Comparison of the integration sites of MLV and two other gammaretrovirus provirus in various cell types showed strong associations with binding sites for a particular transcription factor (STAT1) and specific covalent modifications of histone H3, consistent with the idea that chromosomal features *per se*, may also affect integration site selection (78). Recent *in vitro* studies with chromatinized target DNA indicate that nucleosome density or stability can influence target selection differently for ASLV, MLV, HIV-1, and PFV intasomes in the absence of tethering (79). DNA sequence can also influence integration site selection. Weak, but characteristic palindromic consensus sequences have been identified at the integration target sites for members of the different retroviral families (for review see 74). Moreover, the preference for a specific consensus sequence appears to be distinct from IN tethering, as HIV-1 infection of cells that lack LEDGF show the same palindromic preference as cells that posses the targeting protein, despite the fact that that transcription units are no longer preferred and integration efficiency is reduced substantially in LEDGF deficient cells (80). These findings suggest that nucleosome context and sequence preferences are dictated by distinct features of the intasomes.

IN Domain Composition and Multimerization

Retroviral IN proteins contain three characteristic structural domains (Fig. 8): an N-terminal domain (NTD) that includes a helix-turn-helix fold, which is characterized by conserved HHCC motif that binds a single Zn ion; a larger catalytic core domain (CCD or core) that includes a conserved D, D(35)E motif, which chelates the two divalent metal ions (Mg^{+2} or Mn^{+2}) required for catalysis; and a C-terminal domain (CTD) with an SH3-like fold. However, as with RT, some IN proteins (e.g. MLV and the spumavirus prototype foamy virus, PFV) include additional domains. Furthermore, linkers of different length and sequence composition separate the three major IN domains in proteins from different retroviruses. These distinct features and variations in the non-conserved residues contribute to the properties that distinguish IN proteins from different retroviruses. For example, the bacterially-produced IN proteins from particular retroviruses vary in their specific activities and solubility (81). The proteins can also differ in multimerization properties (82, 83). Under similar conditions at concentration of ~2 mg/ml (~ 60 µM) ASLV IN is a dimer, HIV-1 IN is a tetramer, while PFV IN is a monomer even at twice this concentration (84). Assuming that no other macromolecules are present, a conservative calculation for the intracapsid concentration of ASLV IN in a virus particle is ~150 µM (82), a concentration at which a dimer should predominate. However, until more is known about the internal organization of retroviral capsids, RTCs, and PICs, the biological significance of differences in the solution properties of retroviral IN proteins is impossible to gauge.

Biochemistry of Integration

Despite uncertainties regarding PIC composition and organization, much crucial information has been garnered from *in vitro* studies with reconstructed systems comprising purified IN proteins, cognate viral DNA substrates, and model targets. An early, major breakthrough in these efforts was the development of simple assays that used short duplex oligonucleotides to monitor both the processing of viral DNA end sequences and their joining to a target (85) (Fig. 7, right A-C). Biochemical, genetic, and complementation studies subsequently revealed that a single active site catalyzes both steps, and that IN proteins of ASLV and HIV-1 function as multimers. Later investigations, and the development of simple concerted integration assays (Fig. 7 right D), identified a tetramer as the minimal functional unit for ASLV and HIV-1 IN (86, 87). The use of model target DNAs assembled in nucleosome arrays revealed differences in the tolerance of the ASLV and HIV-1 integration reactions to chromatin compaction by histone H1 (88). More recent use of a similar target has suggested a role for cellular chromatin remodeling proteins (SWI/SNF) in HIV-1 integration (89).

Determination of the crystal structures of the isolated CCDs of ASV IN and HIV-1 IN (90, 91) revealed superimposable architectures, and their relationship to other polynucleotidyl transferases (Fig. 8). Subsequent solution of the isolated flanking terminal domain structures, and of two-domain IN fragments, provided important insights into IN function. However, the more recent solution of crystal structures of a tetramer of PFV IN in complex with viral DNA end sequences, and with a target DNA fragment, were major advances in the field (92–94) (see also Chapter 44 Engelman & Cherapanov). The PFV structures not only provided valuable templates from which to model IN-DNA complexes of other retroviruses (95), but also revealed the molecular details of how active site inhibitors block integration. In addition to being extraordinarily illuminating, the PFV structures also raised some interesting new questions. For example, although an IN tetramer is required for integration, only two monomers, comprising an "inner" dimer (Fig. 9A), appear to participate in catalysis. As predicted from earlier studies (96) there is a reciprocal (*trans*) arrangement in the inner dimer of the crystal structure: Each viral DNA end is bound by the two terminal domains of one monomer (with tips frayed (97, 98) via interaction of the 5′-ends with the CTD), but processed in the CCD of the second monomer. Removal of the two nucleotides at the 3′-ends in the processing reaction allows binding of a bent target DNA, and the concerted joining of the 3′-ends of viral DNA to both strands of the target, five base pairs apart. Two other monomers are bound to either end of the inner dimer by CCD-CCD interactions, which define the "outer" dimer interfaces, but only the CCDs of these outer monomers were resolved in the crystals and these subunits do not appear to contribute to the reaction in any way. Subsequent studies verified that the unexpected arrangement of the PFV IN tetramer in the crystal structure is not due to crystal packing constraints. A combination of biophysical approaches, including small angle X-ray and neutron scattering (SAXS/SANS), revealed a similar structure for the PFV IN-DNA complex in solution (99). In the solution structure, the NTDs and CTDs domains of the outer monomers are extended outwards on either side of the complex (Fig. 9A). It has been speculated that these outer subunits may contribute to tetramer stability, or

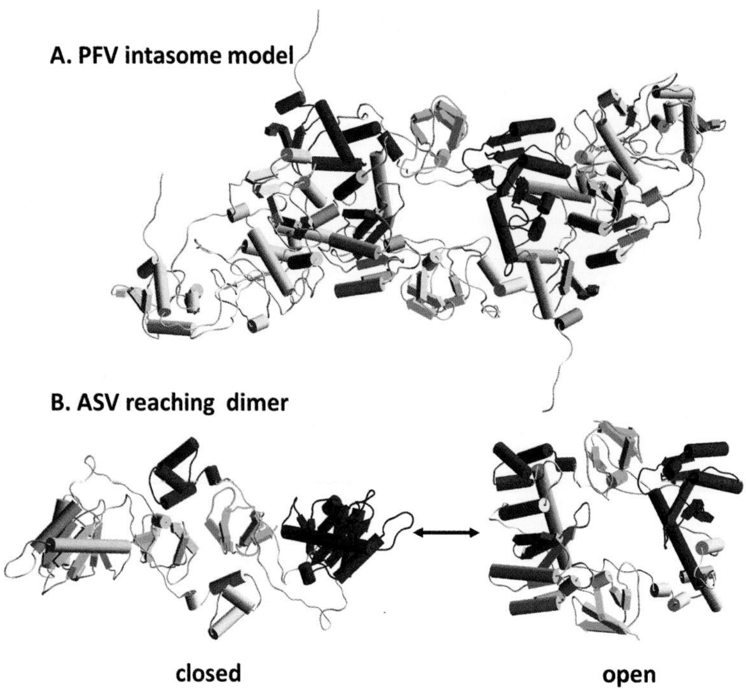

Figure 9 Solution models for the PFV IN tetramer in an intasome and the ASV IN reaching dimer. A. The PFV IN tetramer in an intasome, with DNA omitted to draw attention to IN subunit organization of the inner dimer. Color codes for IN domains in the inner dimer are as in Figure 8, with one subunit in dim pastels for ease of distinction. All three domains of the outer monomers are in yellow. Coordinates for the PFV intasome solution structure were kindly provided by Dr. Kushol Gupta. B. The left side shows the solution structure of the ASV IN reaching dimer structure in the absence of DNA, color coded as in the PFV IN inner dimer. In this structure the CTDs interact with each other, in a "closed" configurations. The right side shows a hypothetical "open" configuration of the reaching dimer formed by rotation of the domain linkers. This open configuration resembles that of the PFV inner IN dimer and could bind viral DNA ends in a similar manner. Pictures were generated using Chimera software (UCSF) and the figure kindly provided by Drs. R. Bojja and M.D. Andrake, Fox Chase Cancer Center, Philadelphia, PA. Structural details are found in (99, 100). doi:10.1128/microbiolspec.MDNA3-0005-2014.f9

that they may be required for appropriate assembly of the tetramer in the presence of DNA. Further study will be required to answer these interesting questions for PFV and other retroviral IN proteins.

Biophysical studies of the ASLV IN dimer in the absence of DNA, which employed a combination of SAXS, chemical crosslinking and mass spectrometry, revealed another unexpected structure, named a "reaching dimer" (Fig. 9B). As with the inner dimer of the PFV IN-DNA complex, the CCDs of the ASLV IN reaching dimer lie at opposite poles, and the structure is stabilized by interactions of the NTD of one monomer with CCD and CTD of the second monomer (100). However, in contrast to the PFV IN inner dimer, the ASLV IN reaching dimer is also stabilized by CTD-CTD interactions, which must disassociate to bind viral DNA ends. A similar reaching dimer was identified for HIV-1 IN, but with this protein, dimers that are stabilized by CCD-CCD interactions were also detected (Fig. 10) (84). The latter finding is consistent with preliminary calculations for free energy of formation, which predict similar stabilities for the HIV-1 IN reaching dimer and CCD-CCD interfaces. Based on SAXS data, a model was proposed for the HIV-1 IN tetramer in the absence of DNA, in which the CCD-CCD interfaces lie at opposite poles (Fig. 10). As with the reaching dimers there is no way to accommodate viral DNA in this tetramer, and major conformational changes would be required to form an intasome. Biochemical studies indicate that HIV-1 IN multimer dissociation is required for function (101), and that the HIV-1-DNA complex may also be formed by dissociated monomers (102), as is the case for the PFV intasome. However, given the wide variation in self-association

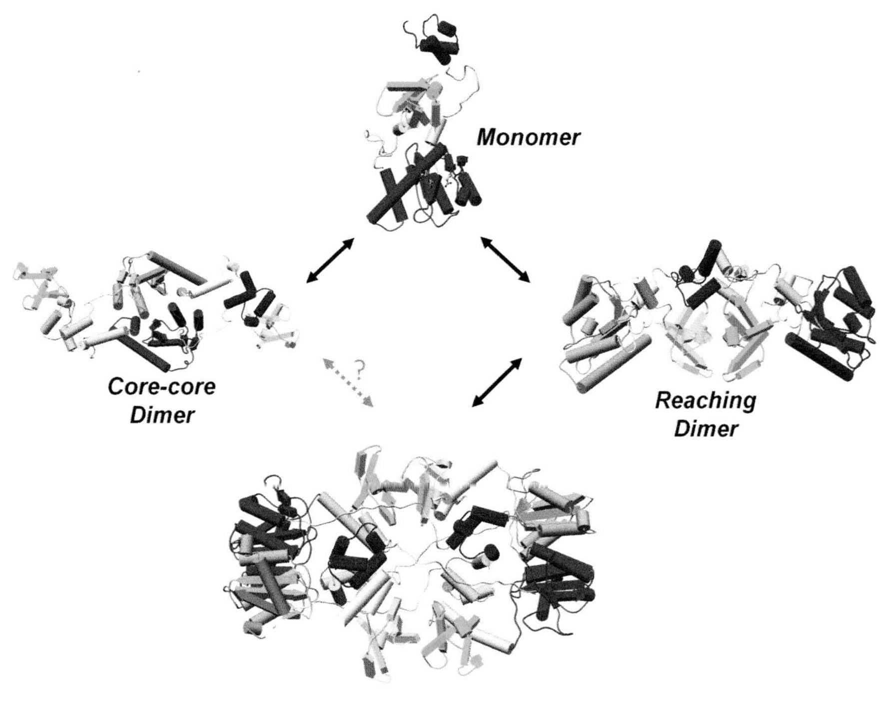

Figure 10 Models for architectures of full length HIV-1 IN protein. Structures for HIV-1 IN protein in the absence of DNA substrates were derived by HADDOCK data-driven modeling of the HIV-1 IN monomer, dimer, and tetramer in solution, based on Small Angle X-Ray Scattering and protein cross-linking data (84). It is not yet known which of these multimeric forms are competent for viral DNA binding in the formation of an HIV-1 intasome. Figures were generated using Chimera software (UCSF) and kindly provided by Drs. R. Bojja and M.D. Andrake, Fox Chase Cancer Center, Philadelphia, PA. doi:10.1128/microbiolspec.MDNA3-0005-2014.f10

properties among IN proteins, and the possible contributions of other components in RTCs and PICs, it is not yet possible to delineate the mode of intasome assembly for either the HIV-1 or ASLV IN.

Post-Integration Events

Retroviral DNA integration produces a double strand break in the genome of its target cell, in which host DNA ends are held together by single strand attachments to the viral DNA (Fig 7). Interruption of existing chromatin conformation and composition by the insertion of a large stretch of newly synthesized viral DNA comprises a major assault on the genomic integrity of the cell. While the 3′-end of viral DNA are linked covalently to host DNA, the 5′-ends are free and adjacent to a short stretches of single-stranded host DNA. In MLV-infected cells, repair of this gap and covalent joining of the 5′-ends of the integrated viral DNA to host DNA could be detected within an hour after 3′-end attachment (103). Subsequent studies with MLV, ASLV, and HIV-1 infected cells have indicated that such post-integration repair requires signaling through the two major DNA damage sensing kinases ATM, ATR, and the function of components in the non-homologous-end joining (NHEJ) pathway (24, 104–106). As it has been reported that the cellular Ku protein, a component of the NHEJ complex, is bound to the HIV-1 PIC (107), and that the repair protein Rad 18 interacts with HIV-1 IN (108), it is possible that binding of such proteins to PIC components may facilitate post-integration repair for some viruses. This idea is supported by suppressive effects of the Rad 52 protein, a component of the homologous recombination repair pathway, on HIV-1 DNA integration (109). Decrease in the amount of Rad 52 protein enhanced HIV-1 DNA integration, and its overexpression reduced integration, but other components of this repair pathway had no effect. From these and other observations, it was proposed that the RAD52 protein may bind to viral DNA ends in the PIC in a way that blocks loading or recruitment of Ku or other proteins that are required for efficient reverse transcription or integration.

A selected screen with siRNAs for DNA repair pathway genes, revealed that knockdown of several enzymes in the short patch, base excision repair (BER) pathway reduces the efficiency of HIV-1 infection (110). Knockdown of other DNA repair proteins or NHEJ components had no effect in these assays, but a negative result is inconclusive because small amounts of proteins that remain after knockdown may be sufficient for function. Subsequent studies with murine cells carrying deletions of specific genes in the BER pathway (Ogg1, Myh, Neil1 and Polß) showed that these proteins are required for efficient integration of the lentiviruses HIV-1 and feline immunodeficiency virus, FIV, but not the gammaretrovirus MLV (111). Exactly how these proteins affect integration efficiency, and the reason for the observed differences among retroviruses, remain to be elucidated.

Finally, the expression of proviral genes depends on integration in numerous ways. The local chromatin environment may have an influence by promoting (euchromatin) or silencing (heterochromatin) proviral gene expression, and it is generally believed that features of neighboring chromatin can "spread" into the integrated retroviral DNA along with the acquisition of histones. Alternatively, or in addition, autonomous chromatin domains may be established on the provirus, either rapidly or over many cell divisions (112, 113). One example of an autonomous mechanism is silencing of the MLV provirus in mouse stem cells. In this case the proviral PBS acts as a *cis*-acting sequence to promote silencing by binding a stem cell-specific complex, comprising a DNA-binding Zn-finger protein (ZFP809), the co-repressor TRIM 28, and the ErbB3-binding protein 1 (EBP1) (114, 115). Another example is the human Daxx protein, identified as an interactor with ASLV IN. Daxx binds to the ASLV IN in the PIC and recruits repressive histone deacetylase (HDAC1) and DNA methyl transferases to proviruses in mammalian cells (116). A similar role for Daxx has been suggested for HIV-1 (117). Deposition of histone H3, and repressive histone marks, can be detected on the ASLV provirus as early as 12 hours post-infection, suggesting that its chromatization occurs quite rapidly. Epigenetic repression is also prominent in the establishment of HIV-1 latency, and evidence suggests that transcription factor binding and DNA methylation may cooperate to maintain HIV latency (118, 119). As the establishment and maintenance of a silent provirus may be advantageous under some conditions, mechanisms that promote this response are likely to be widespread in the different retroviral genera.

POSTSCRIPT

Early studies of retroviral DNA transposition centered on the avian and murine retroviruses viruses following infection of chicken and mouse cells. These were valuable experimental systems that established some of the main features of the retroviral life cycle. Furthermore, as these viruses were known to cause cancer in their host species, biochemical and genetic studies with these viruses did much to move that field along. The focus of retrovirus research changed dramatically in the 1980's with the identification of retroviral species that infect humans, particularly the AIDS virus HIV-1. Because of the critical need to understand as much as possible about its replication and pathogenesis, a substantial increase in financial resources, and the continuing pressure to develop anti-viral drugs and a vaccine, the vast majority of retrovirologists today are HIV-1 virologists. Yet a world of potential variety and nuance is suggested by consideration of the relatively little we know about some of the other, non-lentiviral genera, and by consideration of the striking contrasts in lifestyle among others: from the epsilonretrovirus WDSV, which must promote oncogenesis to propagate (120), to the unconventional spumaviruses, in which reverse-transcription begins before virus particle release (12). Further elucidation of the similarities and variations by which different retroviruses accomplish transposition will surely enhance our knowledge of both the viruses and the cells that they infect.

Acknowledgments. *It is a pleasure to acknowlege the expertise and generosity of Drs. Vladimir Belyi, Stuart Le Grice, and Fox Chase colleagues Dr. Mark Andrake, and RaviShankar Bojja, in compiling and analyzing information, and preparing several figures. Ms. Karen Trush, Fox Chase Cancer Center Special Services, also provided valuable assistance with several illustrations. Drs. Le Grice, Andrake, and Richard Katz read early drafts of the text and offered many helpful suggestions. The assistance of Ms. Marie Estes in preparation of the manuscript and bibliography is also most gratefully acknowledged. This work was supported by National Institutes of Health grants AI40385, CA71515, CA006927, and the WW Smith Charitable Trust.*

Citation. Skalka AM. 2014. Retroviral DNA transposition: themes and variations. Microbiol Spectrum **2**(5):MDNA3-0005-2014.

References

1. **Eickbush TH, Malik HS.** 2002. Origin and evolution of retrotransposons, p 1111–1144. *In* Craig NL, Craigie R, Gellert M, Lambowitz AM (ed), *Mobile DNA II*. ASM Press, Washington, DC.
2. **Coffin JM, Hughes SH, Varmus HE, (eds).** 1997. *Retroviruses*. Cold Spring Harbor Laboratory Press.

3. **Goff SP.** 2013. Retroviridae, p 1424–1473. Knipe DM, Howley PM (ed), *Fields Virology*, Chapter 47, 6th edition, **vol. II**, Wolters Kluwer, Lippincott Williams & Wilkens.

4. **Nisole S, Saib A.** 2004. Early steps of retrovirus replicative cycle. *Retrovirology* **1:**9.

5. **Fassati A.** 2012. Multiple roles of the capsid protein in the early steps of HIV-1 infection. *Virus Res* **170:**15–24.

6. **Matreyek KA, Engelman A.** 2013. Viral and cellular requirements for the nuclear entry of retroviral preintegration nucleoprotein complexes. *Viruses* **5:**2483–2511.

7. **Le Grice SF.** 2012. Human immunodeficiency virus reverse transcriptase: 25 years of research, drug discovery, and promise. *J Biol Chem* **287:**40850–40857.

8. **Hu WS, Hughes SH.** 2012. HIV-1 reverse transcription. *Cold Spring Harb Perspect Med* **2.**

9. **Levin JG, Mitra M, Mascarenhas A, Musier-Forsyth K.** 2010. Role of HIV-1 nucleocapsid protein in HIV-1 reverse transcription. *RNA Biol* **7:**754–774.

10. **Krishnan L, Engelman A.** 2012. Retroviral integrase proteins and HIV-1 DNA integration. *J Biol Chem* **287:**40858–40866.

11. **Craigie R, Bushman FD.** 2012. HIV DNA integration. *Cold Spring Harb Perspect Med* **2:**a006890.

12. **Linial ML.** 1999. Foamy viruses are unconventional retroviruses. *J Virol* **73:**1747–1755.

13. **McDonald D, Vodicka MA, Lucero G, Svitkina TM, Borisy GG, Emerman M, Hope TJ.** 2002. Visualization of the intracellular behavior of HIV in living cells. *J Cell Biol* **159:**441–452.

14. **Rasaiyaah J, Tan CP, Fletcher AJ, Price AJ, Blondeau C, Hilditch L, Jacques DA, Selwood DL, James LC, Noursadeghi M, Towers GJ.** 2013. HIV-1 evades innate immune recognition through specific cofactor recruitment. *Nature* **503:**402–405.

15. **Hatziioannou T, Perez-Caballero D, Cowan S, Bieniasz PD.** 2005. Cyclophilin interactions with incoming human immunodeficiency virus type 1 capsids with opposing effects on infectivity in human cells. *J Virol* **79:**176–183.

16. **De Iaco A, Santoni F, Vannier A, Guipponi M, Antonarakis S, Luban J.** 2013. TNPO3 protects HIV-1 replication from CPSF6-mediated capsid stabilization in the host cell cytoplasm. *Retrovirology* **10:**20.

17. **Forshey BM, von Schwedler U, Sundquist WI, Aiken C.** 2002. Formation of a human immunodeficiency virus type 1 core of optimal stability is crucial for viral replication. *J Virol* **76:**5667–5677.

18. **Schaller T, Ocwieja KE, Rasaiyaah J, Price AJ, Brady TL, Roth SL, Hue S, Fletcher AJ, Lee K, KewalRamani VN, Noursadeghi M, Jenner RG, James LC, Bushman FD, Towers GJ.** 2011. HIV-1 capsid-cyclophilin interactions determine nuclear import pathway, integration targeting and replication efficiency. *PLoS Pathog* **7:**e1002439.

19. **Sanz-Ramos M, Stoye JP.** 2013. Capsid-binding retrovirus restriction factors: discovery, restriction specificity and implications for the development of novel therapeutics. *J Gen Virol* **94:**2587–2598.

20. **Boone LR, Skalka AM.** 1981. Viral DNA synthesized *in vitro* by avian retrovirus particles permeabilized with melittin. I. Kinetics of synthesis and size of minus- and plus-strand transcripts. *J Virol* **37:**109–116.

21. **Boone LR, Skalka AM.** 1981. Viral DNA synthesized in vitro by avian retrovirus particles permeabilized with melittin. II. Evidence for a strand displacement mechanism in plus-strand synthesis. *J Virol* **37:**117–126.

22. **Yong WH, Wyman S, Levy JA.** 1990. Optimal conditions for synthesizing complementary DNA in the HIV-1 endogenous reverse transcriptase reaction. *AIDS* **4:**199–206.

23. **Vandegraaff N, Kumar R, Burrell CJ, Li P.** 2001. Kinetics of human immunodeficiency virus type 1 (HIV) DNA integration in acutely infected cells as determined using a novel assay for detection of integrated HIV DNA. *J Virol* **75:**11253–11260.

24. **Daniel R, Ramcharan J, Rogakou E, Taganov KD, Greger JG, Bonner W, Nussenzweig A, Katz RA, Skalka AM.** 2004. Histone H2AX is phosphorylated at sites of retroviral DNA integration but is dispensable for postintegration repair. *J Biol Chem* **279:**45810–45814.

25. **Miller MD, Wang B, Bushman FD.** 1995. Human immunodeficiency virus type 1 preintegration complexes containing discontinuous plus strands are competent to integrate *in vitro*. *J Virol* **69:**3938–3944.

26. **Zhang J, Tang LY, Li T, Ma Y, Sapp CM.** 2000. Most retroviral recombinations occur during minus-strand DNA synthesis. *J Virol* **74:**2313–2322.

27. **Junghans RP, Boone LR, Skalka AM.** 1982. Retroviral DNA H structures: displacement-assimilation model of recombination. *Cell* **30:**53–62.

28. **Herschhorn A, Hizi A.** 2010. Retroviral reverse transcriptases. *Cell Mol Life Sci* **67:**2717–2747.

29. **Werner S, Wohrl BM.** 2000. Asymmetric subunit organization of heterodimeric Rous sarcoma virus reverse transcriptase alphabeta: localization of the polymerase and RNase H active sites in the alpha subunit. *J Virol* **74:**3245–3252.

30. **Hartl MJ, Schweimer K, Reger MH, Schwarzinger S, Bodem J, Rosch P, Wohrl BM.** 2010. Formation of transient dimers by a retroviral protease. *Biochem J* **427:**197–203.

31. **Schneider A, Peter D, Schmitt J, Leo B, Richter F, Rosch P, Wohrl BM, Hartl MJ.** 2013. Structural requirements for enzymatic activities of foamy virus protease-reverse transcriptase. *Proteins* **82:**375–385.

32. **Lapkouski M, Tian L, Miller JT, Le Grice SF, Yang W.** 2013. Complexes of HIV-1 RT, NNRTI and RNA/DNA hybrid reveal a structure compatible with RNA degradation. *Nat Struct Mol Biol* **20:**230–236.

33. **Abbondanzieri EA, Bokinsky G, Rausch JW, Zhang JX, Le Grice SF, Zhuang X.** 2008. Dynamic binding orientations direct activity of HIV reverse transcriptase. *Nature* **453:**184–189.

34. **Liu S, Abbondanzieri EA, Rausch JW, Le Grice SF, Zhuang X.** 2008. Slide into action: dynamic shuttling

of HIV reverse transcriptase on nucleic acid substrates. *Science* **322:**1092–1097.

35. **Hizi A, Herschhorn A.** 2008. Retroviral reverse transcriptases (other than those of HIV-1 and murine leukemia virus): a comparison of their molecular and biochemical properties. *Virus Res* **134:**203–220.

36. **Nowak E, Miller JT, Bona M, Studnicka J, Szczepanowski RH, Jurkowski J, Le Grice SFJ, Nowotny M.** 2014. Ty3 reverse transcriptase complexed with an RNA-DNA hybrid shows structural and functional asymmetry. *Nature Struct Mol Biol* **21:** 389–396.

37. **Schweitzer CJ, Jagadish T, Haverland N, Ciborowski P, Belshan M.** 2013. Proteomic analysis of early HIV-1 nucleoprotein complexes. *J Proteome Res* **12:**559–572.

38. **Duyk G, Leis J, Longiaru M, Skalka AM.** 1983. Selective cleavage in the avian retroviral long terminal repeat sequence by the endonuclease associated with the alpha beta form of avian reverse transcriptase. *Proc Natl Acad Sci USA* **80:**6745–6749.

39. **Craigie R, Fujiwara T, Bushman F.** 1990. The IN protein of Moloney murine leukemia virus processes the viral DNA ends and accomplishes their integration *in vitro*. *Cell* **62:**829–837.

40. **Katz RA, Merkel G, Kulkosky J, Leis J, Skalka AM.** 1990. The avian retroviral IN protein is both necessary and sufficient for integrative recombination *in vitro*. *Cell* **63:**87–95.

41. **Herschhorn A, Oz-Gleenberg I, Hizi A.** 2008. Quantitative analysis of the interactions between HIV-1 integrase and retroviral reverse transcriptases. *Biochem J* **412:** 163–170.

42. **Lai L, Liu H, Wu X, Kappes JC.** 2001. Moloney murine leukemia virus integrase protein augments viral DNA synthesis in infected cells. *J Virol* **75:**11365–11372.

43. **Wu X, Liu H, Xiao H, Conway JA, Hehl E, Kalpana GV, Prasad V, Kappes JC.** 1999. Human immunodeficiency virus type 1 integrase protein promotes reverse transcription through specific interactions with the nucleoprotein reverse transcription complex. *J Virol* **73:** 2126–2135.

44. **Engelman A.** 1999. *In vivo* analysis of retroviral integrase structure and function. *Adv Virus Res* **52:** 411–426.

45. **Zhu K, Dobard C, Chow SA.** 2004. Requirement for integrase during reverse transcription of human immunodeficiency virus type 1 and the effect of cysteine mutations of integrase on its interactions with reverse transcriptase. *J Virol* **78:**5045–5055.

46. **Hehl EA, Joshi P, Kalpana GV, Prasad VR.** 2004. Interaction between human immunodeficiency virus type 1 reverse transcriptase and integrase proteins. *J Virol* **78:** 5056–5067.

47. **Dobard CW, Briones MS, Chow SA.** 2007. Molecular mechanisms by which human immunodeficiency virus type 1 integrase stimulates the early steps of reverse transcription. *J Virol* **81:**10037–10046.

48. **Wilkinson TA, Januszyk K, Phillips ML, Tekeste SS, Zhang M, Miller JT, Le Grice SF, Clubb RT, Chow SA.** 2009. Identifying and characterizing a functional HIV-1 reverse transcriptase-binding site on integrase. *J Biol Chem* **284:**7931–7939.

49. **Nymark-McMahon MH, Beliakova-Bethell NS, Darlix JL, Le Grice SF, Sandmeyer SB.** 2002. Ty3 integrase is required for initiation of reverse transcription. *J Virol* **76:**2804–2816.

50. **Engelman A.** 2011. Pleiotropic nature of HIV-1 integrase mutations, p 67–81. *In* Neamati N (ed), *HIV-1 Integrase: Mechanism and Inhibitor Design*. John Wiley & Sons, Inc., Hoboken, NJ.

51. **Jurado KA, Wang H, Slaughter A, Feng L, Kessl JJ, Koh Y, Wang W, Ballandras-Colas A, Patel PA, Fuchs JR, Kvaratskhelia M, Engelman A.** 2013. Allosteric integrase inhibitor potency is determined through the inhibition of HIV-1 particle maturation. *Proc Natl Acad Sci USA* **110:**8690–8695.

52. **Fletcher TM 3rd, Soares MA, McPhearson S, Hui H, Wiskerchen M, Muesing MA, Shaw GM, Leavitt AD, Boeke JD, Hahn BH.** 1997. Complementation of integrase function in HIV-1 virions. *EMBO J* **16:** 5123–5138.

53. **Lu R, Limon A, Devroe E, Silver PA, Cherepanov P, Engelman A.** 2004. Class II integrase mutants with changes in putative nuclear localization signals are primarily blocked at a postnuclear entry step of human immunodeficiency virus type 1 replication. *J Virol* **78:** 12735–12746.

54. **Roth MJ, Schwartzberg PL, Goff SP.** 1989. Structure of the termini of DNA intermediates in the integration of retroviral DNA: dependence on IN function and terminal DNA sequence. *Cell* **58:**47–54.

55. **Lee YM, Coffin JM.** 1991. Relationship of avian retrovirus DNA synthesis to integration in vitro. *Mol Cell Biol* **11:**1419–1430.

56. **Zheng Y, Yao X.** 2013. Posttranslational modifications of HIV-1 integrase by various cellular proteins during viral replication. *Viruses* **5:**1787–1801.

57. **Mousnier A, Kubat N, Massias-Simon A, Segeral E, Rain JC, Benarous R, Emiliani S, Dargemont C.** 2007. von Hippel Lindau binding protein 1-mediated degradation of integrase affects HIV-1 gene expression at a postintegration step. *Proc Natl Acad Sci USA* **104:** 13615–13620.

58. **Cereseto A, Manganaro L, Gutierrez MI, Terreni M, Fittipaldi A, Lusic M, Marcello A, Giacca M.** 2005. Acetylation of HIV-1 integrase by p300 regulates viral integration. *EMBO J* **24:**3070–3081.

59. **Terreni M, Valentini P, Liverani V, Gutierrez MI, Di Primio C, Di Fenza A, Tozzini V, Allouch A, Albanese A, Giacca M, Cereseto A.** 2010. GCN5-dependent acetylation of HIV-1 integrase enhances viral integration. *Retrovirology* **7:**18.

60. **Zamborlini A, Coiffic A, Beauclair G, Delelis O, Paris J, Koh Y, Magne F, Giron ML, Tobaly-Tapiero J, Deprez E, Emiliani S, Engelman A, de The H, Saib A.** 2011. Impairment of human immunodeficiency virus type-1 integrase SUMOylation correlates with an early replication defect. *J Biol Chem* **286:**21013–21022.

61. **Elis E, Ehrlich M, Prizan-Ravid A, Laham-Karam N, Bacharach E.** 2012. p12 tethers the murine leukemia virus pre-integration complex to mitotic chromosomes. *PLoS Pathog* **8:**e1003103.
62. **Prizan-Ravid A, Elis E, Laham-Karam N, Selig S, Ehrlich M, Bacharach E.** 2010. The Gag cleavage product, p12, is a functional constituent of the murine leukemia virus pre-integration complex. *PLoS Pathog* **6:** e1001183.
63. **Schneider WM, Brzezinski JD, Aiyer S, Malani N, Gyuricza M, Bushman FD, Roth MJ.** 2013. Viral DNA tethering domains complement replication-defective mutations in the p12 protein of MuLV Gag. *Proc Natl Acad Sci USA* **110:**9487–9492.
64. **Hatziioannou T, Goff SP.** 2001. Infection of nondividing cells by Rous sarcoma virus. *J Virol* **75:**9526–9531.
65. **Katz RA, Greger JG, Boimel P, Skalka AM.** 2003. Human immunodeficiency virus type 1 DNA nuclear import and integration are mitosis independent in cycling cells. *J Virol* **77:**13412–13417.
66. **Katz RA, Greger JG, Darby K, Boimel P, Rall GF, Skalka AM.** 2002. Transduction of interphase cells by avian sarcoma virus. *J Virol* **76:**5422–5234.
67. **Greber UF, Fornerod M.** 2004. Nuclear import in viral infections. *Curr Top Microbiol Immunol* **285:**109–138.
68. **Andrake MD, Sauter MM, Boland K, Goldstein AD, Hussein M, Skalka AM.** 2008. Nuclear import of Avian Sarcoma Virus integrase is facilitated by host cell factors. *Retrovirology* **5:**73.
69. **Katz RA, Greger JG, Skalka AM.** 2005. Effects of cell cycle status on early events in retroviral replication. *J Cell Biochem* **94:**880–889.
70. **Dhar R, McClements WL, Enquist LW, Vande Woude GF.** 1980. Nucleotide sequences of integrated Moloney sarcoma provirus long terminal repeats and their host and viral junctions. *Proc Natl Acad Sci USA* **77:** 3937–3941.
71. **Hishinuma F, DeBona PJ, Astrin S, Skalka AM.** 1981. Nucleotide sequence of acceptor site and termini of integrated avian endogenous provirus ev1: integration creates a 6 bp repeat of host DNA. *Cell* **23:**155–164.
72. **Hughes SH, Shank PR, Spector DH, Kung HJ, Bishop JM, Varmus HE, Vogt PK, Breitman ML.** 1978. Proviruses of avian sarcoma virus are terminally redundant, co-extensive with unintegrated linear DNA and integrated at many sites. *Cell* **15:**1397–1410.
73. **Ju G, Skalka AM.** 1980. Nucleotide sequence analysis of the long terminal repeat (LTR) of avian retroviruses: structural similarities with transposable elements. *Cell* **22:**379–386.
74. **Desfarges S, Ciuffi A.** 2010. Retroviral integration site selection. *Viruses* **2:**111–130.
75. **Engelman A, Cherepanov P.** 2008. The lentiviral integrase binding protein LEDGF/p75 and HIV-1 replication. *PLoS Pathog* **4:**e1000046.
76. **Gupta SS, Maetzig T, Maertens GN, Sharif A, Rothe M, Weidner-Glunde M, Galla M, Schambach A, Cherepanov P, Schulz TF.** 2013. Bromo- and extraterminal domain chromatin regulators serve as cofactors for murine leukemia virus integration. *J Virol* **87:** 12721–12736.
77. **Sharma A, Larue RC, Plumb MR, Malani N, Male F, Slaughter A, Kessl JJ, Shkriabai N, Coward E, Aiyer SS, Green PL, Wu L, Roth MJ, Bushman FD, Kvaratskhelia M.** 2013. BET proteins promote efficient murine leukemia virus integration at transcription start sites. *Proc Natl Acad Sci USA* **110:**12036–12041.
78. **Santoni FA, Hartley O, Luban J.** 2010. Deciphering the code for retroviral integration target site selection. *PLoS Comput Biol* **6:**e1001008.
79. **Benleulmi MS, Matysiak J, Lesbats P, Calmels C, Henriquez D, Naughtin M, Leon O, Vaillant C, Skalka AM, Ruff M, Lavigne M, Andreola ML, Parissi V.** 2014. Intasome architectures and chromatin density modulate retroviral integration into nucleosomes. Submitted.
80. **Shun MC, Raghavendra NK, Vandegraaff N, Daigle JE, Hughes S, Kellam P, Cherepanov P, Engelman A.** 2007. LEDGF/p75 functions downstream from preintegration complex formation to effect gene-specific HIV-1 integration. *Genes Dev* **21:**1767–1778.
81. **Ballandras-Colas A, Naraharisetty H, Li X, Serrao E, Engelman A.** 2013. Biochemical characterization of novel retroviral integrase proteins. *PLoS One* **8:**e76638.
82. **Jones KS, Coleman J, Merkel GW, Laue TM, Skalka AM.** 1992. Retroviral integrase functions as a multimer and can turn over catalytically. *J Biol Chem* **267:**16037–16040.
83. **Jenkins TM, Engelman A, Ghirlando R, Craigie R.** 1996. A soluble active mutant of HIV-1 integrase: involvement of both the core and carboxyl-terminal domains in multimerization. *J Biol Chem* **271:**7712–7718.
84. **Bojja RS, Andrake MD, Merkel G, Weigand S, Dunbrack RL Jr, Skalka AM.** 2013. Architecture and assembly of HIV integrase multimers in the absence of DNA substrates. *J Biol Chem* **288:**7373–7386.
85. **Katzman M, Katz RA, Skalka AM, Leis J.** 1989. The avian retroviral integration protein cleaves the terminal sequences of linear viral DNA at the *in vivo* sites of integration. *J Virol* **63:**5319–5327.
86. **Bao KK, Wang H, Miller JK, Erie DA, Skalka AM, Wong I.** 2003. Functional oligomeric state of avian sarcoma virus integrase. *J Biol Chem* **278:**1323–13277.
87. **Li M, Mizuuchi M, Burke TR Jr, Craigie R.** 2006. Retroviral DNA integration: reaction pathway and critical intermediates. *EMBO J* **25:**1295–1304.
88. **Taganov KD, Cuesta I, Daniel R, Cirillo LA, Katz RA, Zaret KS, Skalka AM.** 2004. Integrase-specific enhancement and suppression of retroviral DNA integration by compacted chromatin structure in vitro. *J Virol* **78:** 5848–5855.
89. **Lesbats P, Botbol Y, Chevereau G, Vaillant C, Calmels C, Arneodo A, Andreola ML, Lavigne M, Parissi V.** 2011. Functional coupling between HIV-1 integrase and the SWI/SNF chromatin remodeling complex for efficient *in vitro* integration into stable nucleosomes. *PLoS Pathog* **7:**e1001280.

90. **Bujacz G, Jaskolski M, Alexandratos J, Wlodawer A, Merkel G, Katz RA, Skalka AM.** 1995. High-resolution structure of the catalytic domain of avian sarcoma virus integrase. *J Mol Biol* **253:**333–346.

91. **Dyda F, Hickman AB, Jenkins TM, Engelman A, Craigie R, Davies DR.** 1994. Crystal structure of the catalytic domain of HIV-1 integrase: similarity to other polynucleotidyl transferases. *Science* **266:**1981–1986.

92. **Hare S, Gupta SS, Valkov E, Engelman A, Cherepanov P.** 2010. Retroviral intasome assembly and inhibition of DNA strand transfer. *Nature* **464:**232–236.

93. **Maertens GN, Hare S, Cherepanov P.** 2010. The mechanism of retroviral integration from X-ray structures of its key intermediates. *Nature* **468:**326–329.

94. **Hare S, Maertens GN, Cherepanov P.** 2012. 3′-processing and strand transfer catalysed by retroviral integrase in crystallo. *EMBO J* **31:**3020–3028.

95. **Krishnan L, Li X, Naraharisetty HL, Hare S, Cherepanov P, Engelman A.** 2010. Structure-based modeling of the functional HIV-1 intasome and its inhibition. *Proc Natl Acad Sci USA* **107:**15910–15915.

96. **van Gent DC, Vink C, Groeneger AA, Plasterk RH.** 1993. Complementation between HIV integrase proteins mutated in different domains. *EMBO J* **12:**3261–3267.

97. **Katz RA, DiCandeloro P, Kukolj G, Skalka AM.** 2001. Role of DNA end distortion in catalysis by avian sarcoma virus integrase. *J Biol Chem* **276:**34213–34220.

98. **Katz RA, Merkel G, Andrake MD, Roder H, Skalka AM.** 2011. Retroviral integrases promote fraying of viral DNA ends. *J Biol Chem* **286:**25710–25718.

99. **Gupta K, Curtis JE, Krueger S, Hwang Y, Cherepanov P, Bushman FD, Van Duyne GD.** 2012. Solution conformations of prototype foamy virus integrase and its stable synaptic complex with U5 viral DNA. *Structure* **20:**1918–1928.

100. **Bojja RS, Andrake MD, Weigand S, Merkel G, Yarychkivska O, Henderson A, Kummerling M, Skalka AM.** 2011. Architecture of a full-length retroviral integrase monomer and dimer, revealed by small angle X-ray scattering and chemical cross-linking. *J Biol Chem* **286:**17047–17059.

101. **Kessl JJ, Li M, Ignatov M, Shkriabai N, Eidahl JO, Feng L, Musier-Forsyth K, Craigie R, Kvaratskhelia M.** 2011. FRET analysis reveals distinct conformations of IN tetramers in the presence of viral DNA or LEDGF/p75. *Nucleic Acids Res* **39:**9009–9022.

102. **Pandey KK, Bera S, Grandgenett DP.** 2011. The HIV-1 integrase monomer induces a specific interaction with LTR DNA for concerted integration. *Biochemistry* **50:**9788–9796.

103. **Roe T, Chow SA, Brown PO.** 1997. 3′-end processing and kinetics of 5′-end joining during retroviral integration in vivo. *J Virol* **71:**1334–1340.

104. **Daniel R, Kao G, Taganov K, Greger JG, Favorova O, Merkel G, Yen TJ, Katz RA, Skalka AM.** 2003. Evidence that the retroviral DNA integration process triggers an ATR-dependent DNA damage response. *Proc Natl Acad Sci USA* **100:**4778–4783.

105. **Sakurai Y, Komatsu K, Agematsu K, Matsuoka M.** 2009. DNA double strand break repair enzymes function at multiple steps in retroviral infection. *Retrovirology* **6:**114.

106. **Skalka AM, Katz RA.** 2005. Retroviral DNA integration and the DNA damage response. *Cell Death Differ* **12:**971–978.

107. **Li L, Olvera JM, Yoder KE, Mitchell RS, Butler SL, Lieber M, Martin SL, Bushman FD.** 2001. Role of the non-homologous DNA end joining pathway in the early steps of retroviral infection. *EMBO J* **20:**3272–3281.

108. **Mulder LC, Chakrabarti LA, Muesing MA.** 2002. Interaction of HIV-1 integrase with DNA repair protein hRad18. *J Biol Chem* **277:**27489–27493.

109. **Lau A, Kanaar R, Jackson SP, O'Connor MJ.** 2004. Suppression of retroviral infection by the RAD52 DNA repair protein. *EMBO J* **23:**3421–3429.

110. **Espeseth AS, Fishel R, Hazuda D, Huang Q, Xu M, Yoder K, Zhou H.** 2011. siRNA screening of a targeted library of DNA repair factors in HIV infection reveals a role for base excision repair in HIV integration. *PLoS One* **6:**e17612.

111. **Yoder KE, Espeseth A, Wang XH, Fang Q, Russo MT, Lloyd RS, Hazuda D, Sobol RW, Fishel R.** 2011. The base excision repair pathway is required for efficient lentivirus integration. *PLoS One* **6:**e17862.

112. **Ellis J.** 2005. Silencing and variegation of gammaretrovirus and lentivirus vectors. *Hum Gene Ther* **16:**1241–1246.

113. **Mok HP, Lever AM.** 2007. Chromatin, gene silencing and HIV latency. *Genome Biol* **8:**228.

114. **Wang GZ, Wolf D, Goff SP.** 2014. EBP1, a novel host factor involved in primer binding site-dependent restriction of Moloney Murine Leukemia Virus in embryonic cells. *J Virol* **88:**1825–1829.

115. **Wolf D, Goff SP.** 2009. Embryonic stem cells use ZFP809 to silence retroviral DNAs. *Nature* **458:**1201–1204.

116. **Shalginskikh N, Poleshko A, Skalka AM, Katz RA.** 2013. Retroviral DNA methylation and epigenetic repression are mediated by the antiviral host protein Daxx. *J Virol* **87:**2137–2150.

117. **Huang L, Xu GL, Zhang JQ, Tian L, Xue JL, Chen JZ, Jia W.** 2008. Daxx interacts with HIV-1 integrase and inhibits lentiviral gene expression. *Biochem Biophys Res Commun* **373:**241–245.

118. **Blazkova J, Trejbalova K, Gondois-Rey F, Halfon P, Philibert P, Guiguen A, Verdin E, Olive D, Van Lint C, Hejnar J, Hirsch I.** 2009. CpG methylation controls reactivation of HIV from latency. *PLoS Pathog* **5:**e1000554.

119. **Kauder SE, Bosque A, Lindqvist A, Planelles V, Verdin E.** 2009. Epigenetic regulation of HIV-1 latency by cytosine methylation. *PLoS Pathog* **5:**e1000495.

120. **Rovnak J, Quackenbush SL.** 2010. Walleye dermal sarcoma virus: molecular biology and oncogenesis. *Viruses* **2:**1984–1999.

121. **Lapkouski M, Tian L, Miller JT, Le Grice SF, Yang W.** 2013. Reply to "Structural requirements for RNA degradation by HIV-1 reverse transcriptase". *Nat Struct Mol Biol* **20:**1342–1343.

122. **Nowak E, Potrzebowski W, Konarev PV, Rausch JW, Bona MK, Svergun DI, Bujnicki JM, Le Grice SF, Nowotny M.** 2013. Structural analysis of monomeric retroviral reverse transcriptase in complex with an RNA/DNA hybrid. *Nucleic Acids Res* **41:**3874–3887.

123. **Zhou D, Chung S, Miller M, Grice SF, Wlodawer A.** 2012. Crystal structures of the reverse transcriptase-associated ribonuclease H domain of xenotropic murine leukemia-virus related virus. *J Struct Biol* **177:** 638–645.

Non-LTR Retrotransposons

VI

Mobile DNA, 3rd Edition
Nancy L. Craig, Michael Chandler, Martin Gellert, Alan M. Lambowitz, Phoebe A. Rice and Suzanne Sandmeyer

doi:10.1128/microbiolspec.MDNA3-0011-2014

Thomas H. Eickbush[1]
Danna G. Eickbush[1]

49 Integration, Regulation, and Long-Term Stability of R2 Retrotransposons

INTRODUCTION

R2 elements exclusively insert into 28S rRNA genes (Figure 1). As a result of this specificity, R2 is one of the more tractable mobile elements to study and, thus, is now among the best understood elements both in terms of its mechanism and its population dynamics. The R2 element was first identified in the rDNA loci of *Drosophila melanogaster* in the early 1980's (1, 2), when little was known of the structure or abundance of mobile elements in eukaryotes. In fact, the exclusive residence of the element at a specific site in the 28S gene initially suggested that it might be an intron. However, the findings that only a fraction of the genes contained the insertion, that 28S genes containing the insertion did not appear to be transcribed, and that many of the insertions had a sizeable deletion at the 5′ end all argued against its role as an intron. Insertions were soon identified at the same position of the 28S rRNA gene in many other species of insects (3, 4, 5). The complete sequence of the insertions in both *D. melanogaster* and *Bombyx mori* revealed a large open reading frame (ORF) encoding a reverse transcriptase that had greatest sequence similarity to that of non-LTR retrotransposons (6, 7). R2 differed from most non-LTR retrotransposons, however, in that it only contained a single ORF. Furthermore, rather than an encoded apurinic endonuclease (APE) located amino-terminal to the reverse transcriptase (8), R2 encoded carboxyl terminal to the reverse transcriptase an endonuclease with an active site more similar to that of certain restriction enzymes (9).

The search for R2 in additional species was simple because the 28S gene sequences to either side of the R2 insertion site have undergone almost no substitutions in the entire evolution of eukaryotes. Thus it was straightforward to determine whether a species contained R2 insertions by direct cloning of 28S genes, PCR amplification of the insertion region, or computer searches of whole genome shotgun sequences. Such analyses have revealed R2 elements in most lineages of insects and arthropods (10, 11) and in many other taxa of animals including nematodes, tunicates, and birds (12, 13, 14; unpublished data, DE Stage); however, there have been no reports of R2 elements in plants, fungi, or protozoans. The presence of R2 elements within a group can be spotty, for example only 4 out of 7 fish species

[1]Department of Biology, University of Rochester, Rochester, NY 14627.

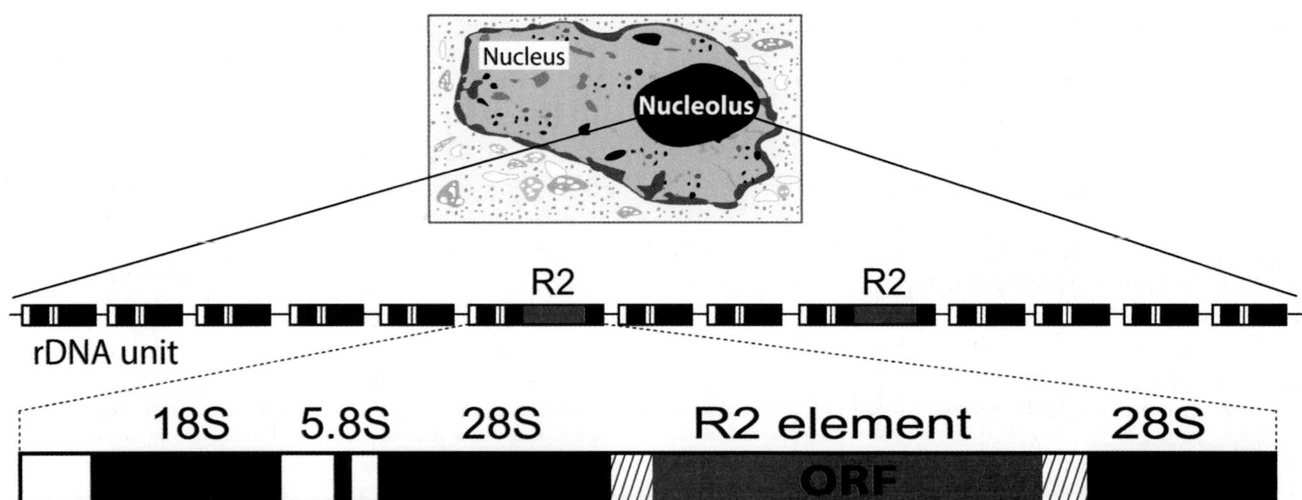

Figure 1 R2 elements insert within the 28S rRNA genes. The nucleolus, the site of rRNA transcription and processing, is organized around the hundreds of tandem units (rDNA units) that comprise the rDNA locus. Each rDNA unit is composed of a single transcription unit containing the 18S, 5.8S, and 28S genes (black boxes) and external and internal transcribed spacers (white boxes). The transcription units are separated by intergenic spacers (thin lines). A subset of the 28S genes in many animals contain R2 insertions near the middle of the gene (red box). R2 elements encode a single open reading frame (ORF). doi:10.1128/microbiolspec.MDNA3-0011-2014.f1

examined have R2. Thus the apparent absence of R2 from some animal taxa may simply reflect the small numbers of species whose genomes have been tested. The large number of mammalian species examined without detecting R2 insertions does suggest with some confidence, however, that R2 is not present in this group.

The 3′ junctions of the 28S gene with the R2 insertions in all but two species are identical suggesting that the R2 endonuclease is highly specific and that it has rarely changed the specificity of the initial DNA cleavage since its origin. The two exceptions are the R2 elements of hydra, named R8, which insert into a specific sequence of the 18S rRNA gene (13), and the R2 elements of rotifer, named R9, which insert into a different site in the 28S rRNA gene (15). The ORF of all R2 elements is also very similar in coding capacity; the only significant difference is the number of zinc-finger motifs associated with DNA binding at the amino-terminal end of the protein (11, 13, 16). As described by Fujiwara in this volume (17), many other lineages of non-LTR retrotransposons have evolved sequence specificity for the rRNA genes or for other repeated sequences in the genomes of eukaryotes (18–22). Some of these site-specific elements are like R2 and encode a carboxyl-terminal restriction-like endonuclease, while others contain an amino-terminal APE domain. Among the latter, R1 elements insert in the 28S rRNA gene 74 bp downstream of the R2 insertion site. R1 elements were first identified along with the R2 elements of *D. melanogaster* (1, 2) and subsequently in most lineages of arthropods (10). The turnover and evolution of R1 elements in the rDNA loci of Drosophila species is similar in most respects to that of the R2 elements (23–25).

Reconstructing the evolutionary history of R2 elements based on the sequence of their ORF, first in the genus *Drosophila* (26, 27), then in all of arthropods (28), and finally in all animals (12, 13, 14), has suggested the R2 elements have evolved entirely by vertical descent. The absence of horizontal jumps between species has enabled the divergence of R2 elements to be used as a molecular clock to time the age of its various lineages as well as a guide to estimate the time of divergence of other retrotransposons (29). Unfortunately, because the R2 protein sequence eventually reaches maximal divergence, this dating can only be done with confidence for a time frame of less than 200–300 million years. Remarkably, multiple lineages of R2 have propagated in some animal lineages for the entire 200–300 million year time estimate (10, 12, 30, 31). Why some animals are able to maintain multiple R2 lineages and other animals only one R2 lineage is unknown.

The long history and wide distribution of R2 is remarkable for a mobile element. Its success has been interpreted by some to indicate that R2 provides a useful function to the host. One possibility is that the R2 endonuclease initiates the recombinations that give rise to the concerted evolution of the locus (32). However, in instances where R2 elements are known to be active, the DNA cleavages generated by the endonuclease appear to lead to large deletions of the rDNA locus which are detrimental to the host (33). A second premise is that the inactivation of rDNA units by insertions and the subsequent reduction in 28S rRNA synthesis could influence the rate of development (34). However, the original findings leading to this conclusion have been challenged (35). The many species containing either multiple lineages of R2 or multiple classes of non-LTR retrotransposons inserted into the rDNA locus (10, 12, 28, 31) are more consistent with the propagation of selfish genetic elements than a means used by animals to regulate gene expression. Finally, the frequently suggested argument that mobile elements provide useful genetic diversity seems unlikely for R2; comparisons of species with or without R2 elements reveal no sequence differences in 28S rRNA genes near the insertion site and no detectable difference in the mechanism of rRNA regulation. Thus R2 remains one of the best examples of an element that endures because it simply has the ability to make copies of itself (i.e. a selfish genetic element). R2 has likely been present since the origin of the metazoan radiation, est. 500–800 million years ago, making it one of the oldest known mobile genetic elements. Perhaps most remarkable is that throughout this long period R2 has undergone essentially no changes in its insertion site or in its mechanism of insertion. Clearly, R2 has found a niche in which it can hold its own in the genomic battle between element and host.

MECHANISM OF R2 INTEGRATION

Studies of the R2 integration mechanism have been conducted with the ORF of the R2 element from *Bombyx mori* expressed in and purified from *E. coli* (36, 37). The purified protein (120 kilodaltons) was found to have all the RNA and DNA binding properties as well as enzymatic activities needed to complete a retrotransposition reaction. The critical first step in this reaction is the ability of the protein to nick one strand of the DNA target site and use the 3′ end of the DNA exposed by this cleavage to prime reverse transcription of the R2 RNA. The integration reaction was termed target DNA-primed reverse transcription, or TPRT (37), to distinguish it from the retrotransposition mechanism that had been previously discovered for retroviruses and LTR retrotransposons (reviewed in 38, 39). The TPRT mechanism can explain three unusual properties revealed by the initial sequencing of non-LTR retrotransposons. First, the absence of the integrase domain usually seen in mobile elements can be explained because the new copies of the element are synthesized onto the target site rather than inserted into the target site. Second, the absence of the integrase also explains why many non-LTR retrotransposons generate variable in length or no target site duplications. Third, the uniform presence of a complete 3′ end but frequent presence of truncations at the 5′ end of non-LTR retrotransposons can be explained by the reverse transcriptase falling off before reaching the 5′ end of the RNA or reaching the end of a degraded RNA. Characterizations of the initial steps in the R2 TPRT reaction were previously summarized in Mobile DNA II (40). Here, experiments conducted since that publication are the primary focus.

Properties of the R2 Reverse Transcriptase

R2 elements from all lineages of animals encode a protein with a domain structure similar to that shown in Figure 2. The central domain of the R2 protein corresponds to the reverse transcriptase (RT). R2 RT has a number of properties that differentiate it from the RTs encoded by LTR retrotransposons and retroviruses. Clearly the most distinctive property is the ability of R2 RT to use the 3′ end of DNA to prime reverse transcription (41, 42). This reaction is most efficient at the 28S gene insertion site with RNA containing the 3′ UTR of the R2 element as template. In the absence of these components and at a lower efficiency, R2 RT can prime the reverse transcription of any RNA using the 3′ end of any other RNA or single-stranded DNA as the primer. This priming also occurs in the absence of complementarity between the template and primer (43).

A second interesting property of R2-RT is its higher processivity than that of most RTs. Processivity refers to the product length that can be catalyzed by the enzyme before it dissociates from the template. In single cycle reactions on RNA and DNA templates, R2 RT synthesizes cDNA that is 2 to 5 times the length of that synthesized by retroviral RTs (44, 45). This difference in processivity is likely a result of the different demands placed on these enzymes. DNA synthesis by retroviral or LTR RTs occurs within virus-like particles within the cytoplasm. In these particles, the RT is able to undergo rounds of dissociation from and reassociation to the template before giving rise to full-length products (38, 39). In contrast, DNA synthesis by R2 RT occurs in the nucleus at the DNA target site. If R2 RT dissociates from the RNA template then reassociation could be difficult, and the result is a truncated (dead) copy. Consistent with this model, R2 RT has been shown to initiate only poorly at the 3′ end of long DNA primers

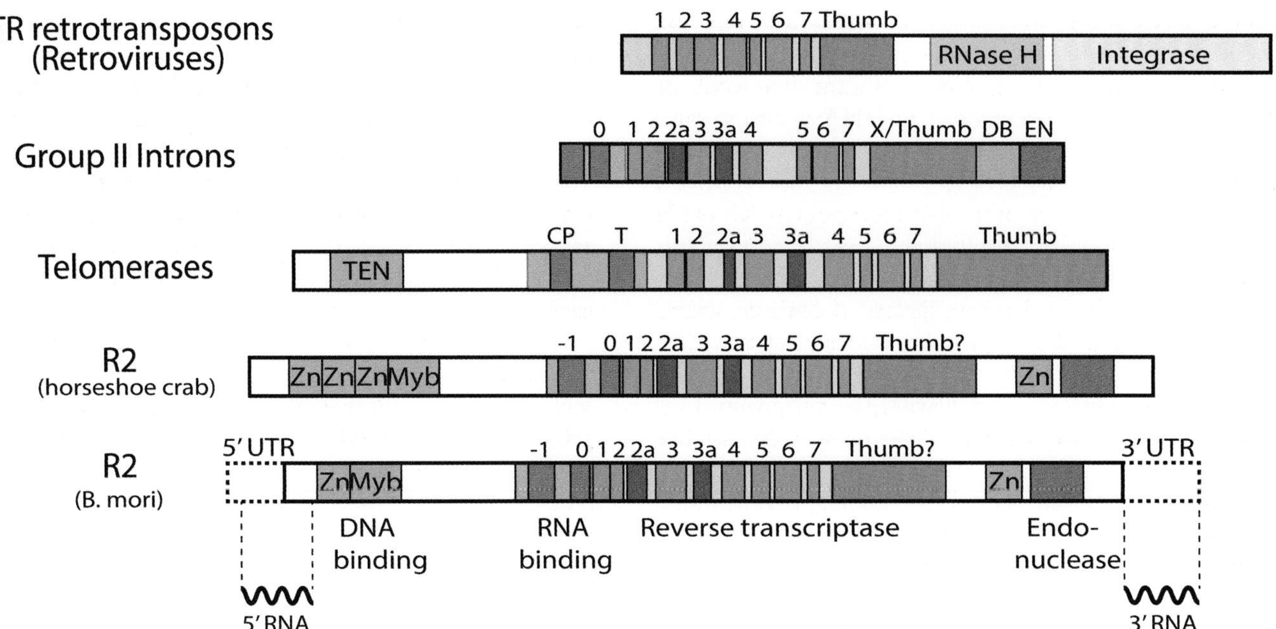

Figure 2 Domain structure of the R2 protein and its similarity to other elements. At the bottom is the R2 element from *B. mori* with the 5′ and 3′ untranslated regions indicated by dotted lines. The central region of the encoded protein contains the reverse transcriptase domain. The various conserved motifs within the fingers and palm regions (motifs 1–7) and the predicted thumb are indicated. An RNA binding domain is immediately N-terminal to the reverse transcriptase and conserved motifs within this domain are labeled 0 and −1. The N-terminal region of the protein contains zinc finger (Zn) and c-myb (Myb) motifs, while the C-terminal region encodes a putative zinc-binding domain and the R2 endonuclease. Shown below the R2 diagram are the 5′ and 3′ regions of the R2 RNA that are bound by the R2 protein during a retrotransposition reaction (see Figure 4). The major difference among R2 elements from different species is the presence of one, two, or three zinc finger domains at the N-terminal end. The R2 element from horseshoe crab is an example of the latter. Comparison of the R2 protein with the *pol* gene of LTR retrotransposons (and retroviruses) reveals little in common except for 7 out of the 9 motifs in the reverse transcriptase domain. Most LTR retrotransposon *pol* genes also encode an RNase H and integrase not found in R2. The R2 protein has greater similarity to the proteins encoded by group II introns and telomerases. These three groups share all nine motifs of the reverse transcriptase. In the case of telomerase, these motifs are frequently termed 1, 2, 3, A, IFD, B, C, D, and E (from left to right) (69). Group II introns, telomerases, and R2 also share an RNA binding domain upstream of the reverse transcriptase (purple segment). Group II introns and R2 both encode an endonuclease domain at the 3′ end, while R2 and some telomerases have DNA binding domains (TEN) at the N-terminal end. doi:10.1128/microbiolspec.MDNA3-0011-2014.f2

annealed to RNA templates (43). Thus there appears to have been selective pressure on R2 and likely other non-LTR retrotransposons to evolve RTs with high processivity.

Another unusual ability of R2 RT is that it can jump from the 5′ end of one RNA template to the 3′ end of another RNA template. These "end-to-end template jumps" do not require sequence identity between the templates (43, 46). Instead, R2 RT adds up to five non-templated nucleotides to the cDNA when it reaches the end of a template. Microhomologies between these overhanging nucleotides and sequences near the 3′ end of the acceptor template enable the polymerase to jump between templates (46). R2 end-to-end template jumps are similar to the template jumps observed for viral RNA directed RNA polymerases (47) as well as for the Mauriceville retroplasmid and group II intron RTs (48, 49) but differ from the template switching reaction associated with retroviral DNA synthesis, which does require sequence identity between the donor and acceptor RNA templates (50).

The R2 protein has no identifiable RNase H domain (9, 51) and no RNase H activity has been detected *in vitro* (37, 45) suggesting the template for second strand DNA synthesis in a retrotransposition reaction is an RNA:DNA duplex. Consistent with this model R2 RT has the ability to displace an annealed RNA or DNA strand as it uses an RNA or DNA strand as template (43, 45). Remarkably, the processivity of R2 RT is not reduced by the presence of an annealed strand. Retroviral RTs, on the other hand, do possess RNase H activity and show only limited ability to displace RNA annealed to DNA (52, 53). Because most non-LTR retrotransposons do not have an RNase H domain, strand displacement may be a common property of non-LTR RTs. However, the acquisition of an RNase H domain by some elements (51) and the finding that group II intron RTs have a strong strand displacement activity but depend on the host RNase H for mobility (54) leave the question open as to the extent non-LTR retrotransposons may rely on host RNase H activity.

Recently, it was shown that R2 RT has a level of nucleotide misincorporation (mutation rate) similar to that of HIV-1 RT (55, 56). Like HIV-1 the low fidelity of R2 RT is a result of its ability to extend a mismatch if the wrong nucleotide is incorporated into the product. For HIV-1 the high error rate has been suggested to enable the virus to escape the immune system of the host (57). In the case of R2 RT, the high error rate could be a consequence of the unusual flexibility needed at the active site to enable priming in the absence of sequence complementarity. However, because R2 retrotransposition is infrequent relative to the many germ line replications that occur each generation, the long-term nucleotide substitution rate for R2 is not significantly above that associated with typical genes (55).

DNA and RNA Binding Domains of the R2 protein

The C-terminal end of the R2 protein contains the endonuclease domain (Figure 2). A similar domain appears to be present in all non-LTR retrotransposons that do not encode an APE-like endonuclease at the N-terminal end of their protein (29). The active site of the R2 endonuclease was found to be similar to that of type IIs restriction enzymes (9). The catalytic and DNA binding domains of type IIs restriction enzymes are separate, thus, these enzymes bind the DNA a short distance from the cleavage site. The separation of cleavage from binding is also suggested for R2 as protein footprint analyses indicted that most DNA contacts by the R2 protein are located upstream and downstream of the insertion site (58, 59). The C-terminal end of the R2 protein also encodes a potential zinc-binding domain that could play a role in DNA binding (59, 60). Whether this motif is involved in nucleic acid binding or simply involved in protein folding remains unresolved as mutations in this motif of R2 gave rise to an unstable protein, precluding further analysis (unpublished data, SM Christensen). A potential zinc-binding domain downstream of the RT domain is found in many non-LTR retrotransposons. *In vivo* integration assays have revealed that mutations in this putative zinc-binding domain eliminate retrotransposition activity in L1 elements of mammals (61).

At the N-terminal end of the R2 protein (Figure 2) are classic C_2-H_2 zinc-finger and Myb-like nucleic acid binding motifs (9, 62). DNA-binding and DNase footprint analyses of wild type and mutant polypeptides spanning the 140 amino acid N-terminal end revealed that the Zn-finger motif binds the DNA target from 1 to 3 bases upstream of the cleavage site while the Myb-motif binds DNA sequences from 10 to 15 base pair downstream of the insertion site (62). Because the complete R2 protein also protects a region of DNA from 10–40 bp upstream of the cleavage site, the C-terminal domain of the protein was postulated to be responsible for this upstream binding (62). Recently, however, the N-terminal domain of the R2 protein from the horseshoe crab was shown to bind this upstream region (16). The horseshoe crab R2 protein differs from the *B. mori R2* in having three zinc-finger domains instead of one (Figure 2). It is possible that both the N-terminal and C-terminal domains of all R2 proteins contribute to protein binding 10–40 bp upstream of the insertion site, but the isolated N-terminal domain of the *B. mori* protein binds too weakly to be detected *in vitro*. Analysis of the DNA-binding motifs that are N-terminal to the RT domain in other non-LTR retrotransposons show considerable flexibility in their binding to the target site (63).

An RNA binding domain of the R2 protein was recently identified immediately N-terminal to the RT domain (64). This domain comprises two conserved sequence motifs that have been termed 0 and −1 because they are encoded before motifs 1 through 7 of the RT domain (Figure 2). Mutations within either motif affect all properties of the R2 protein that require RNA binding. These include the ability of the protein to conduct the TPRT reaction, the ability of the protein to bind RNA in gel mobility shift assays, and the ability of the protein to cleave the second strand of the target site. Sequence similarity to the 0 motif can be found in all lineages of non-LTR retrotransposons (29, 65, 66), suggesting a similar RNA binding domain. As diagramed in Figure 2, an RNA binding domain of telomerase and of the group-II introns is also located upstream of the

RT domain (67–69). Indeed, sequence similarity exists between the 0 motifs of R2 and group-II introns (11, 70, 71). The similar location of these RNA binding domains provides support to the model, originally based on the sequences of the reverse transcriptase domain, for a close evolutionary relationship among these three groups of genetic elements (70, 72, 73). This common origin and structural similarity suggests that future studies of non-LTR retrotransposition mechanisms should use telomerase and group II introns as guides.

Nature of the RNA Template: The R2 Ribozyme

An understanding of the R2 integration reaction requires an understanding of the RNA template that is used for reverse transcription. The exact 3′ end of the RNA did not appear critical to the TPRT mechanism, as *in vitro* experiments showed that similar TPRT integrated products were formed whether the RNA templates ended at the precise 3′ end of the R2 element or contained downstream 28S sequences (41, 42). On the other hand, the exact 5′ end of the RNA did appear important, as *in vivo* integration results (74, 75) indicated the products differed depending on whether 28S sequences were present or absent at this end of the transcript.

An analysis of *Drosophila simulans* stocks that supported frequent R2 retrotransposition events (described in greater detail below) revealed R2 transcripts of the approximate length of a full-length R2 element (76). Several lines of evidence suggested these transcripts were derived by co-transcription with the 28S rRNA gene (Figure 3A). First, previous studies could not identify a promoter at the 5′ end of full-length *D. melanogaster* elements using transient transcription assays in tissue culture cells (77). Second, the level of full-length R2 transcripts usually correlated with the level of transcription of extensively 5′ truncated elements (78). Third, in stocks with significant levels of R2 transcripts low levels of 28S/R2 co-transcripts could also be observed (76). Detailed RT-PCR analysis of these stocks revealed that the co-transcripts were derived from inserted rDNA units containing R2 5′ junctions with small deletions. Thus not only was processing of the R2 transcript from a 28S co-transcript suggested, but processing at the 5′ end was dependent upon the sequence at the 28S/R2 5′ junction (79).

To examine processing, RNA templates comprising different lengths of the consensus *D. simulans* 28S/R2 5′ junction were generated *in vitro* with T7 RNA polymerase (79). In addition to the expected RNA product, these simple RNA synthesis reactions also produced RNA fragments resulting from cleavage at the exact 5′ junction of the R2 element with the upstream 28S rRNA sequences. The autocatalysis was efficient with up to 98% of the RNA cleaved. RNA self-cleavage required the first 184 nucleotides of the R2 element but was not dependent upon the upstream 28S sequences. As shown in Figure 3B, the first 184 nucleotides of the R2 RNA could be folded into a secondary structure containing a double pseudoknot and five base paired regions that was similar to the self-cleaving ribozyme previously characterized from the hepatitis delta virus (HDV) (80, 81). Remarkably, 21 of the 27 nucleotide positions of the HDV ribozyme that had been shown to generate the active site of the HDV ribozyme were identical to the R2 sequence (80–82). Using an entirely structure-based bioinformatics approach, the R2 ribozyme was independently identified by Ruminski and co-workers (83).

An analysis of R2 from many other animals spanning its entire host range revealed that the 5′ end of all R2 elements could be folded into an HDV-like ribozyme structure (84). Comparison of these different R2 ribozymes revealed considerable flexibility in some aspects of the structure and in the sequence. Surprisingly, however, several distantly related R2 lineages had sequences comprising the active site that were highly similar to those observed for the *D. simulans* R2 and HDV ribozymes. Presumably the limited parameter space afforded by using only four nucleotides to make the ribozyme has resulted in the R2 ribozyme converging upon the same sequences on multiple occasions. HDV-like ribozymes have also been found in other non-LTR retrotransposon lineages including several that are not site-specific (83–86).

The vast majority of the 28S/R2 5′ junction RNAs that were directly tested for activity *in vitro* showed detectable levels of self-cleavage. The major difference in the activity of the R2 ribozyme from diverse animals to that observed in Drosophila was the location of RNA self-cleavage. Unlike the *D. simulans* R2 ribozyme, which cleaves at the precise 28S/R2 5′junction, many R2 ribozymes cleaved in GC-rich regions of the 28S rRNA either 13 or 28 nucleotides upstream of the R2 insertion site (Figure 3B). As described below in the discussion of the R2 integration mechanism, self-cleavage by the ribozyme in the upstream 28S sequences in some species gives rise to insertions with uniform 5′ junctions while self-cleavage at the 5′ end of R2 in other species gives rise to variable junctions (11, 84, 87).

The discovery of the R2 ribozyme also helped to explain the presence of several non-autonomous parasites of R2 in the rDNA locus of some Drosophila species (88). For example, a short (530 bp) element propagates at the precise R2 insertion site of the 28S rRNA genes in *D. willistoni*. Divergent but clearly derived from the typical R2 elements found in this species, this short element had lost the entire ORF but had retained the 5′ ribozyme to enable processing from the 28S co-transcript and the structure of the 3′ UTR to enable binding by the R2 integration machinery. The elements were termed SIDEs for Short Internally Deleted Elements. Surprisingly, a more often encountered type of SIDE contained an R2 ribozyme at the 5′ end and the 3′ UTR from an R1 element (88). These R2/R1 hybrid SIDEs were located in the typical R1 insertion site suggesting that while their RNA was processed by the R2 ribozyme it was the R1 integration machinery generating the new insertions. In many respects these SIDEs are similar to the non-autonomous SINE elements that parasitize other non-LTR retrotransposons (LINEs) (89, 90). The presence of R2s, R1s, and their SIDEs suggest that in some species there can be intense competition between selfish elements for the limited number of 28S rRNA insertion sites.

Current Model of R2 Integration

The initiation of the R2 integration reaction, i.e. using the initial nick at the target site to prime reverse transcription, was deduced soon after the R2 protein of *B. mori* was purified (37, 41, 42). The key to understanding the second half of the integration reaction was revealed over 10 years later when it was discovered that in addition to binding RNA from the 3′ end of the R2 transcript, R2 protein could also bind a segment of RNA from near the 5′ end of the R2 element (60). This 300 nt segment of RNA starts within the 5′ UTR and ends just before the sequences encoding the N-terminal zinc-finger (Figure 2). A distinctive structure of this RNA is a pseudoknot that is conserved across silk moths (91, 92). The segment of R2 RNA that is bound by the protein determines its function in the integration reaction. In the presence of the 3′ RNA, the R2 protein binds the 28S gene upstream of the insertion site. In the presence of the 5′ RNA the R2 protein binds the 28S gene downstream of the insertion site.

In the complete model for R2 retrotransposition, these two protein/RNA complexes are proposed to perform symmetric reactions as diagramed in Figure 4. The subunit bound upstream initiates the retrotransposition reaction by both cleaving the bottom (first) strand of the DNA target and polymerizing the first DNA strand onto the released 3′ OH. The R2 subunit bound downstream of the insertion site appears to remain inactive until after the bound 5′ RNA is removed (60). Presumably this RNA is 'pulled' from the subunit as the RNA is used for first strand DNA synthesis. The downstream subunit then initiates the second half of the reaction by cleaving the top (second) DNA strand. The protein again utilizes the released 3′ end of the DNA as a primer and polymerizes the second DNA strand. Second strand synthesis involves displacement of the annealed RNA strand by the R2 protein (43, 45). Gel shift experiments suggest single R2 protein subunits separately bind the 3′ and 5′ ends of the RNA (59, 93). However the stoichiometry of a complete integration reaction is not known.

The initiation of second strand DNA synthesis has not yet been documented in the *in vitro* retrotransposition reaction (37, 45, 46, 59). Because this step is so inefficient *in vitro*, we cannot exclude the possibility that second strand synthesis is accomplished *in vivo* by a host polymerase as appears to be the case for group II intron retrohoming (54). Based on the variation observed at the 5′ junction of endogenous R2 insertions in different species, it also appears to be a step that has evolved. In *B. mori* and many other animals, the R2 5′ junctions within a species show little variation except for the occasional direct duplication of 28S gene sequences of a characteristic length for each species (11, 84). However, in species of Drosophila and other animals, most 5′ junctions contain variable deletions of the 28S target site and variable additions of non-templated nucleotides (87, 94). As described above in the discussion of the R2 ribozyme, this difference is correlated with the location of the R2 self-cleavage.

A model to explain the difference in the uniformity of the 5′ junction in different animals is diagramed in Figure 5. If the RNA template contains upstream 28S sequences, then the cDNA that is generated by reverse transcription can anneal to the upper strand of the DNA target. This annealing allows efficient and precise priming of second-strand synthesis and uniform 5′ junctions. However, if the 5′ end of the transcript contains only R2 sequences, then the cDNA generated by reverse transcription has no sequence identity with the upper strand of the DNA target. Second strand DNA synthesis is then postulated to initiate at a region of microhomology between the DNA strand upstream of the insertion site and the extra nucleotides added to the cDNA strand as the reverse transcriptase runs off the RNA template (46). Because these extra nucleotides are random, the microhomology used in each integration event can vary giving rise to the different length deletions of 28S sequences and/or non-templated nucleotide additions observed in some species.

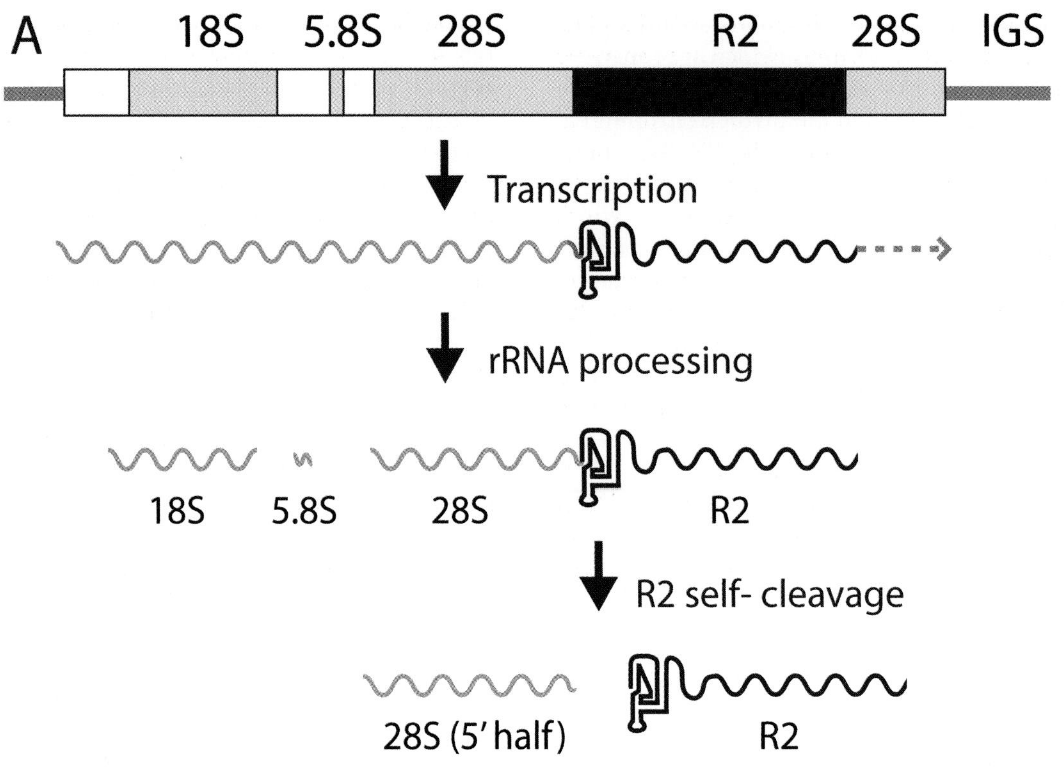
A
18S
5.8S
28S
R2
28S
IGS
Transcription
rRNA processing
18S
5.8S
28S
R2
R2 self- cleavage
28S (5′ half)
R2

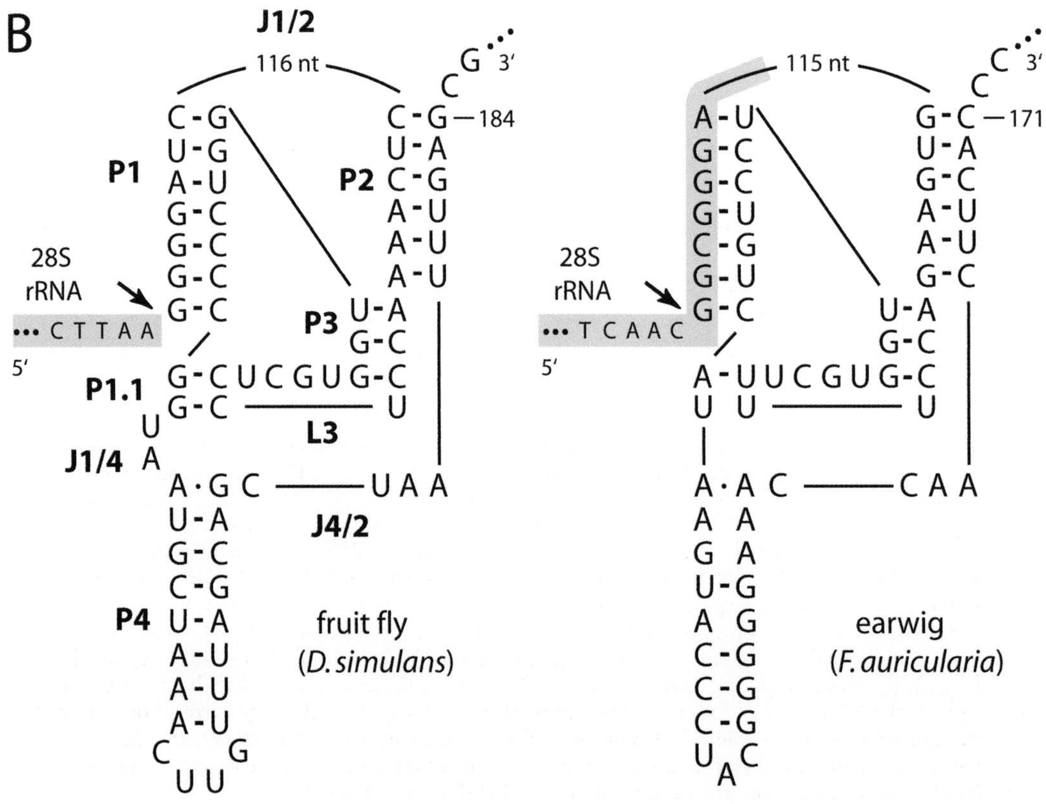
B
J1/2
116 nt
184
P1
P2
28S
rRNA
••• C T T A A
5′
P3
P1.1
L3
J1/4
J4/2
P4
fruit fly
(D. simulans)
115 nt
171
28S
rRNA
••• T C A A C
5′
earwig
(F. auricularia)

Consistent with this model, in all animals the junctions of 5′ truncated R2 insertions, which are generated when the reverse transcriptase doesn't reach the 5′ end of the RNA template, are found to involve microhomologies and non-templated additions (84, 87, 94). Direct support for this model has also been provided by *in vivo* R2 injection experiments in Drosophila, which yielded integrations with precise 5′ junctions when the injected R2 RNA contained upstream 28S sequences and highly variable junctions when the injected R2 RNA did not contain upstream 28S sequences (74). Similarly a tissue culture R2 integration system based on a baculovirus vector required the R2 transcript to contain upstream 28S sequence to give rise to precise R2 integrations into the 28S rRNA gene (75). Models for the formation of a heteroduplex between the cDNA strand and the DNA template have also been hypothesized during initiation of second strand synthesis for other non-LTR retrotransposons (87, 95).

EXPRESSION OF R2 ELEMENTS

R2 Copies are Rapidly Gained and Lost from rDNA Loci

Theoretical studies and computer simulations have suggested that repeated crossovers over time are responsible for the nearly identical sequence of the rRNA genes within a species, but allow these rRNA sequences to evolve in unison over time (concerted evolution) (reviewed in 96). These crossovers result in a wide range in the number of rDNA units per loci for the individuals in a population. This concerted evolution of the rDNA locus would also affect the sequence and number of R2 elements residing within the rDNA units. Thus, while copies of mobile elements that insert throughout a genome can remain for long periods at locations where they do no harm, all R2 elements would be predicted to be like the rDNA units themselves and undergo turnover. Evidence for this turnover was found in the low levels of nucleotide sequence divergence (range 0.0–0.6 %) between R2 copies within a species (26, 27, 87).

It would appear technically difficult to monitor the turnover of R2 elements because all elements insert into the same site within the 28S genes and all R2 copies and 28S rRNA genes are nearly identical in sequence (87, 97). Fortunately in species of Drosophila, individual R2 insertions can be monitored because the RNA transcript does not contain upstream 28S sequences (79), and thus as diagramed in Figure 5, have highly variable 5′ junctions. These variants include small deletions of the 28S gene and/or additional non-templated sequences as well as R2 5′ truncations ranging from less than 100 bp to more than 3 kb.

A PCR based assay was developed to generate profiles of the distinctive "lengths" associated with the 5′ end of the R2 elements in individual flies (24). The assay utilizes a forward primer that anneals to the 28S gene about 75 bp upstream of the R2 insertion site as well as a series of reverse primers that anneal to R2 sequences spaced at 200 to 400 bp intervals throughout the R2 element. The PCR amplification products are separated on 5% acrylamide gels to monitor the R2 5′ truncations or on high resolution sequencing gels to obtain the single base pair resolution needed to differentiate between the full-length R2 inserts. Such assays revealed most stocks of *D. melanogaster* and *D. simulans* contained 15 to 25 different 5′ truncated R2 copies and 20 to 30 full-length R2 copies of which 10–15 copies contained junctions that differed in length (23, 24, 98). The "5′ profiles" for flies from different populations or even from individuals within the same population could completely differ suggesting old copies were lost and new copies gained; that is, the R2 elements in these species were turning over (24, 98).

Figure 3 The R2 ribozyme. (A) An rDNA transcription unit is diagramed with 18S, 5.8S, and 28.S genes (gray boxes), transcribed spacers (white boxes), and R2 insertion (black box). All three rRNAs are normally processed from the single primary transcript. When a unit contains an R2 insertion, a self-cleaving ribozyme encoded at the 5′ end of the element releases the 5′ end of the R2 transcript from the upstream 28S rRNA sequence. It is not known if transcription ends at the 3′ end of the R2 element, or if this end is processed from downstream 28S gene sequences. (B) On the left is the *D. simulans* R2 ribozyme folded in a structure similar to that of the hepatitis delta virus (HDV) ribozyme (80, 81). The various components of the ribozyme are labeled as in the HDV ribozyme: P, base-paired region; L, loop; J, nucleotides joining paired regions. 28S gene sequences are shaded with gray. On the right is the R2 ribozyme from *Forficula auricularia* (earwig). Self-cleavage (arrow) occurs at the precise junction of the R2 element with the 28S gene in the case of the *D. simulans* element and upstream of the junction in the 28S gene sequences in the case of the R2 element from *F. auricularia*. doi:10.1128/microbiolspec.MDNA3-0011-2014.f3

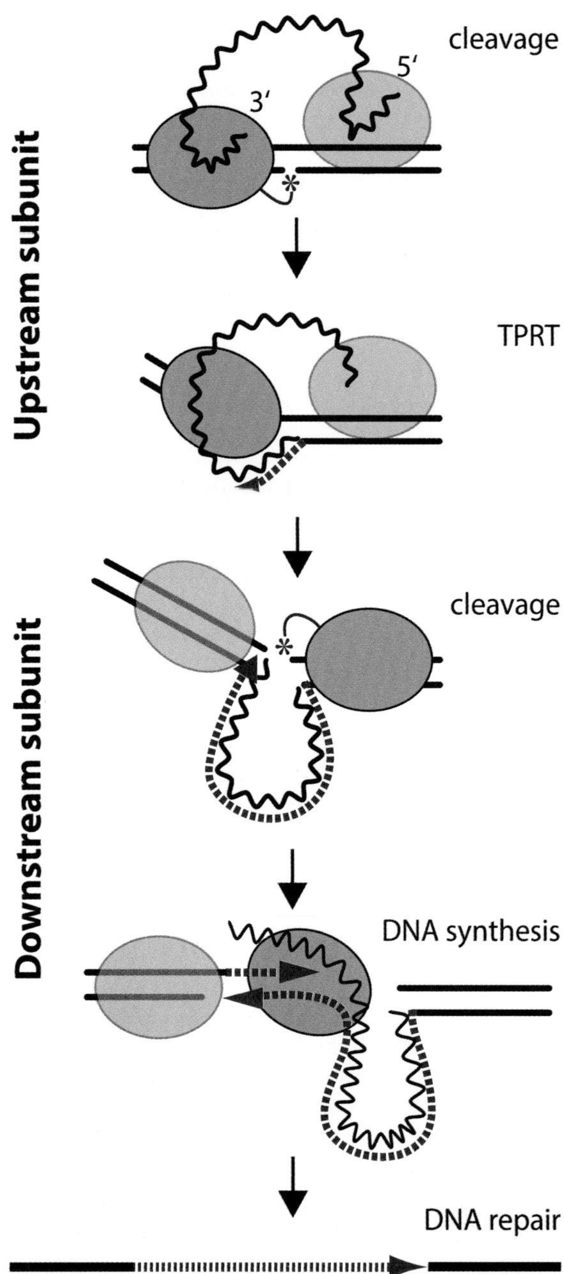

Figure 4 The R2 retrotransposition model. An R2 integration reaction is proposed to involve symmetric cleavage/DNA synthesis steps by R2 proteins bound upstream and downstream of the insertion site. From top to bottom, protein bound upstream of the insertion site is associated with the 3′ end of the R2 transcript. This protein both cleaves the bottom stand of DNA and catalyzes the reverse transcription of the R2 RNA using the cleaved DNA target as primer, target primed reverse transcription (TPRT). R2 protein bound downstream of the insertion site is associated with the 5′ end of the R2 transcript. When the reverse transcription reaction catalyzed by the upstream protein dislodges the 5′ RNA, the downstream protein cleaves the top DNA strand and again uses the cleaved DNA to prime second strand DNA synthesis. Second strand synthesis requires the polymerase to displace the R2 RNA. Because in the absence of bound RNA the downstream protein does not bind tightly to the DNA target, it is shown dissociated from the target site during polymerization. The integration reaction is completed by the host repair machinery which fills in the single stranded gaps at the target site. Blue oval, protein subunit (dark, active; light, inactive); wavy black line, R2 RNA; dashed red lines, synthesized DNA; solid black lines; target DNA.
doi:10.1128/microbiolspec.MDNA3-0011-2014.f4

A similar analysis of 5′ truncated elements in the tadpole shrimp likewise revealed R2 copies were being gained and lost (99). PCR assays were also conducted to monitor the R1 non-LTR retrotransposons that are also present in the rDNA of the two Drosophila species. While most full-length R1 copies are of identical length, and thus cannot be monitored, the profiles of 5′ truncated copies of R1 from individual flies suggested a level of turnover similar to that of R2 (24).

To directly monitor the rate of turnover, the R2 and R1 elements were examined in the Harwich mutation accumulation lines of *D. melanogaster* generated by Mackay and co-workers (100). Those flies had first undergone 45 generations of inbreeding to remove any variation before multiple sublines were started (generation 0) and maintained by mass mating a small number of flies at each generation. At generation 350, the R2 and R1 5′ profiles were determined in 19 sublines (25, 101). Over this ~17 year period, each subline was found to have gained 0 to 2 new R2 5′ variants (mean 0.8 insertions/line) and to have lost 0 to 8 of the 34 ancestral R2 variants (mean 2.9 deletions/line) indicating the R2 elements were slowly being lost from these lines. In contrast, a mean of 9.5 new insertions and 2.1 deletions of ancestral R1 elements were detected in the sublines indicating that R1 elements were increasing in number. Equally important, the sizes of the rDNA loci in the sublines after 350 generations varied from 140 to 310 units, indicating that many crossover events must have occurred. However, almost all of the R2 and R1 length variants, whether new or ancestral, were present at one copy per locus indicating that these crossovers had seldom duplicated individual insertions (102). The variation in rDNA number among sublines was almost exclusively associated with the number of uninserted rDNA units. Thus copies of R1 and R2 were lost from the rDNA locus, presumably by recombination, but copies of these insertions were seldom duplicated by recombination.

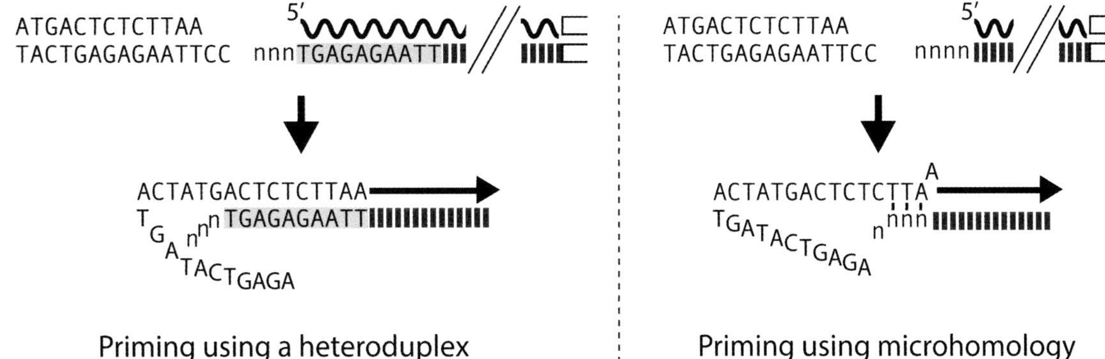

Figure 5 Variation in the priming of second-strand DNA synthesis. R2 elements differ in whether the 5′ end of the RNA template used in the integration reaction ends at the boundary between R2 and the 28S gene or extends a short distance upstream in the 28S rRNA sequence. This difference is dependent upon the location of the self-cleavage site by the R2 ribozyme (see text). Left panel. If self-cleavage by the R2 ribozyme is upstream in the 28S gene sequences, the resulting cDNA strand can form a heteroduplex with the upstream target DNA. This heteroduplex can stabilize the integration intermediate resulting in precise initiation of second strand synthesis (arrow) and uniform 5′ ends for different R2 copies. Right panel. If self-cleavage is at the 28S/R2 junction, there are no 28S sequences on the DNA strand (cDNA) generated by reverse transcription. As a consequence, the R2 protein must use regions of microhomology to initiate second strand synthesis (arrow). Priming frequently involves the 3–5 non-templated nucleotides added to the cDNA strand as the enzyme ran off the RNA template (lower case n's). This use of chance microhomologies to prime second strand DNA synthesis gives rise to sequence variation at the 5′ junctions of different integrated copies of R2. Wavy black line, RNA with 5′ end denoted; red dashed line, first strand DNA composed of R2 sequences; gray box, first strand DNA sequences complementary to upstream DNA target sequences.
doi:10.1128/microbiolspec.MDNA3-0011-2014.f5

Active R2 Retrotransposition and Its Developmental Timing

The question thus became: are new R2 insertions generated at a low rate in possibly all individuals, or can R2 elements undergo higher rates of retrotransposition in individuals of a specific genetic composition or under certain physiological conditions? The search for flies with R2 activity was easier to conduct in *D. simulans* than in *D. melanogaster*. Unlike *D. melanogaster* where rDNA loci are on both the X and Y chromosomes, *D. simulans* encodes an rDNA locus only on the X chromosome. Therefore, *D. simulans* males will have a single rDNA locus that can be scanned for new insertions. A survey of laboratory stocks originally derived from a single population in California revealed that the males in most of the *D. simulans* stocks showed only minimal differences in their R2 5′ profiles, consistent with little or no retrotransposition and slow rates of deletion (98). However, the males in a few stocks showed extensive variation among their R2 5′ profiles. Indeed, in two stocks virtually every male had a 5′ profile that differed from every other tested male suggesting very high rates of R2 turnover.

To directly monitor the rates of new insertions in these lines, the R2 5′ profiles were determined for the male progeny of single pair crosses (33). New R2 insertions, typically only 1 or 2 per locus, were observed in about 10% of the sons from each cross. Two lines examined in detail showed R2 insertion rates of 0.12 and 0.15 insertions per locus per generation. Surprisingly, large numbers of parental R2 copies were also being deleted in these crosses with an average loss of over four R2 copies (range 1 to 15 copies) per deletion event. Thus the rates of R2 deletions in these R2 active lines based on the single generation experiments (0.22 and 0.44 deletions/locus/generation) were actually higher than the rates of new insertions. High R2 deletion rates were never seen in lines in which new insertions were not detected suggesting that R2 activity itself was causing the deletions. Presumably, the presence of multiple R2 endonucleases attempting to initiate integration in multiple rDNA units results in the deletion of large segments of the rDNA locus.

To determine if the high rates of R2 activity seen in these stocks could be maintained over multiple generations, the progeny of individual pairs were monitored after 30 generations (33). The rates of new insertions per generation in these 30-generation experiments were similar to those found after the one-generation experiment, however, the rates of R2 deletions determined from these long-term experiments was significantly less. In fact, the ~0.10 deletions per locus per generation found in these long term experiments was now similar to the rate of new insertions. Thus many of the large deletions of the rDNA locus that were seen in the one generation experiment presumably gave rise to less viable flies which were then lost from the small populations being maintained.

The ability to detect new R2 insertions in a single generation made it possible to ask when and where R2 retrotranspositions occurred. The above studies detected retrotranspositions in the female germ line since sons were compared to their mother. To look at the timing of the retrotransposition events in the development of the germ cells over 200 male progeny from a single female were screened for new R2 insertions (33). Thirty-one of the 32 new R2 insertions that were scored were only found in one or two males indicating that most retrotransposition events occurred late in the development of the egg, i.e. during oogenesis rather than during germ cell propagation. To look for activity in the male germ line, males from each of the R2 active stocks were crossed to attached-X (XXY) females. The only surviving male progeny from this cross inherit their Y chromosome from their mother and their X chromosome from their father. New R2 insertions were also detected through the male germ line but at 3–4 fold lower levels than observed through the female germ line.

Finally, R2 insertions were also assayed in somatic tissues (103). Somatic insertions were defined as an R2 variant not inherited from either the father or mother and present in only a subset of the tissues tested from an animal. The tissues examined were individual imaginal discs dissected from third instar larvae as well as various adult tissues (e.g. leg, wing, antenna, proboscis). Somatic R2 insertions were detected in about one quarter of the animals tested from the *D. simulans* stocks that contained highly active R2 elements. Remarkably in one third of these somatic events, the same new insertion was detected in more than one body segment. Because determination of body segments occurs at the blastoderm stage in 2–3 hour embryos, this implied that the R2 retrotransposition events were occurring in the first two hours of embryo development. During this time, embryonic nuclei are entering the pole plasma to become the germ line. Therefore, an R2 insertion event that occurred at this time could give rise to germ line mosaics, which at the next generation would be scored as a germ line event. Thus the low number of R2 retrotransposition events that were originally scored in the male germ line as well as a fraction of the retrotranspositions scored in the female germ line (33) may actually represent somatic events rather than true germ line events.

Control Over R2 Activity is at the Level of Transcription

Since the first discovery of R2 insertions in *D. melanogaster* a question frequently asked has been when are they transcribed (78, 104, 105). Even though they reside in a significant fraction of the genes in arguably the most actively transcribed loci, the levels of R2 transcripts were found to be extremely low to nonexistent. Nuclear run-on experiments (106) as well as the direct microscopic observation of transcribing rDNA loci (107) indicated that the low level of detectable R2 transcripts was due to transcription repression and not the rapid degradation of RNA transcripts. The chromatin associated with R2 insertions contained the typical epigenetic marks linked to heterochromatin (106). These studies revealed that most of the uninserted rDNA units also appeared to be packaged into heterochromatin. These findings are consistent with numerous studies conducted in many eukaryotic taxa that found only a fraction of the total number of rDNA units are actively transcribed (108–110). The fraction of transcriptionally active rDNA units in *D. melanogaster* has been estimated at only 30–40 units of the several hundred units typically present in the rDNA loci (32, 106, 111). Thus in most flies the rDNA units containing an R2 insertion are not transcribed.

The transcriptional state of the R2-inserted units could be changed, however, by manipulating the composition of the rDNA locus (78). R2 protein/RNA complexes of *B. mori* (similar to those used in the *in vitro* studies of the TPRT reaction) were injected into *D. melanogaster* embryos to give rise to flies with *B. mori* R2 insertions. Several of these flies also contained large deletions of the rDNA loci which was presumably the result of the injected R2 endonuclease cleaving multiple rDNA units within a locus and the subsequent loss of large sections. In these stocks, the level of transcript from the *B. mori* R2 insertion was inversely correlated with the total number of uninserted units in the locus. For example, a stock that contained about 20 uninserted units on the X chromosome rDNA locus had a level of *B. mori* R2 transcription in females that was about 100-fold higher than stocks that had greater than

80 uninserted units (78). Interestingly, transcription of specific endogenous *D. melanogaster* R2 insertions was also inversely correlated with the number of uninserted units.

The recovery of naturally occurring stable stocks of *D. simulans* that exhibited frequent R2 retrotranspositions (98) provided a means to directly study control over R2 activity. Northern blots of RNA from stocks with frequent R2 retrotranspositions (R2 active) and stocks with infrequent/no retrotranspositions (R2 inactive) revealed a clear correlation between the level of full-length R2 transcripts and the level of R2 retrotransposition events (76). Consistent with the studies of when and where R2 retrotranspositions were occurring (33, 103) R2 transcripts were detected in both males and females and in both germ line and somatic tissues of larvae and adults. Nuclear run-on experiments again suggested that control was at the level of transcription rather than at a post-transcription step (76).

To assess the genetic control over R2 transcription, crosses were conducted between the R2 active and R2 inactive stocks (76). In both the F1 and F2 progeny of these crosses, the pattern of high versus low R2 transcript level suggested that control over R2 transcription mapped to the single rDNA locus on the X chromosome and was otherwise little influenced by the genetic background of the fly. For example, monitoring individual male progeny from crosses between animals with R2 active loci ($X^{R2\text{-}A}$) and animals with R2 inactive loci ($X^{R2\text{-}I}$) revealed that even after many generations of random mating all males with the $X^{R2\text{-}A}$ locus had a high R2 transcript level, and all males with the $X^{R2\text{-}I}$ locus had low levels of R2 transcript. Surprisingly, females that contained both an $X^{R2\text{-}I}$ and an $X^{R2\text{-}A}$ locus showed nucleolar dominance (112). These heterozygous females supported only low R2 transcript levels suggesting the $X^{R2\text{-}I}$ locus showed dominance over the $X^{R2\text{-}A}$ locus. Microscopic analysis of the secondary constrictions in these heterozygous females revealed a transcriptionally active rDNA locus on only one X chromosome, directly demonstrating nucleolar dominance. The dominance of an $X^{R2\text{-}I}$ locus over an $X^{R2\text{-}A}$ locus only occurred when the latter had very high levels of R2 transcription. Females with two rDNA loci that each supported more intermediate levels of R2 transcription in males typically had a transcriptionally active rDNA locus on both X chromosomes and R2 transcript levels near the average of the two loci (76).

While the extremely high levels of R2 transcription seen in some *D. simulans* stocks have to date not been observed in *D. melanogaster*, crosses between *D. melanogaster* stocks with moderate and extremely low levels of R2 transcription again revealed intermediate levels of transcription in the female progeny (113). However, in *D. melanogaster* an rDNA locus is also present on the Y chromosome. Surprisingly, the level of R2 transcripts detected in the male progeny was found to be dependent upon the level of R2 transcription in the father and independent of the level of R2 transcription in the mother. This suggested that the Y chromosome rDNA locus showed nucleolar dominance over the X chromosome rDNA locus (113, 114). Microscopic analysis confirmed that only the rDNA locus on the Y chromosome was transcriptionally active. While the key to R2 activity, it is interesting to note that control over rDNA locus expression also appears to have an influence on the regulation of genes throughout the genome. The size of the rDNA locus and thus the number of genes that must be turned-off has been shown to affect the level of expression of many genes, in particular those that are influenced by position-effect variegation (113, 115, 116).

A Model for the Regulation of R2 Activity

Little is known about the factors affecting the long-term stability of a retrotransposable element. Do the defensive mechanisms used by the host to control the element occasionally breakdown, thus, the element experiences windows of opportunity to replenish its number? Or, are the host's defensive mechanisms not completely effective, thus, there are continuous but low levels of element activity? In the case of R2, the former appears more likely as most laboratory stocks exhibited no R2 activity, while a few stocks demonstrated high levels of R2 activity over many generations. Because higher numbers of R2 elements were found associated with the rDNA loci of these active stocks and because host control over R2 activity mapped to the rDNA locus itself, one plausible model was that the many R2 elements were simply overwhelming the cellular defensive machinery (76, 98). However the alternative model in which the activation of R2 elements gave rise to the higher numbers of copies was also a viable possibility. To directly address this issue, it was necessary to obtain a snapshot of the dynamic nature of the rDNA locus in natural populations.

To this end, ~100 *D. simulans* lines were generated that contained individual rDNA loci (iso-rDNA lines) from each of two populations (Atlanta, San Diego) (117). For each population, about one half of the iso-rDNA lines were found to have no detectable level of R2 transcription. The remaining lines had levels of R2 transcription that varied over a factor of 100. Indeed, each population gave rise to one or two lines with levels of R2 transcripts that equaled or even exceeded that seen in our most active laboratory stocks. A subset of the lines encompassing the full range of R2 transcript levels was tested for their ability to generate new R2 insertions. As found in the laboratory stocks, the R2 transcript level correlated with the frequency of new retrotranspositions events, again suggesting that regulation was at the level of transcription.

To determine the differences among these lines that could be responsible for R2 activity, 100 lines were selected for a more detailed analysis of the physical properties of their rDNA loci (118). The mean number of R2 elements in the lines was 50 (range 25 to 80), and the mean locus size was 250 units (range 150 to 400). A small but significant trend towards higher R2 transcription was associated with higher numbers of R2 copies and with a smaller rDNA locus size (i.e. fewer rDNA units), however, there was a wide range of R2 transcript levels associated with all values of locus size and R2 copy number. Thus, a simple model in which R2 number or the fraction of inserted rDNA units in the locus would determine whether a stock contained active R2 elements did not appear valid.

The property of the rDNA loci that could be directly linked to R2 activity was revealed by genomic blots which monitored the distribution of R2 elements in the rDNA locus. A *Not*I restriction enzyme site is located in the R2 element but not in either the rDNA unit or its spacer region. Therefore, a pulsed-field gel of *Not*I-digested genomic DNA probed with 18S sequences reveals the spacing between R2-inserted rDNA units in the locus, with the largest *Not*I fragment representing the most extensive region of the rDNA locus that is free of R2 insertions. Those lines without detectable R2 transcription, both iso-rDNA lines from natural populations as well as laboratory stocks, had a region greater than 40-units (>400 kb) of their rDNA loci free of R2 insertions. For those stocks with measurable levels of R2 transcription, the largest R2-free region was less (frequently much less) than 40 units in size (76, 117). Thus the critical property predicting R2 activity was whether the R2-inserted rDNA units were distributed uniformly throughout the rDNA locus or were clustered, the latter producing at least one large region free of R2 insertions.

These findings gave rise to the "transcription domain" model as shown in Figure 6A. Fundamental to this model were the previous findings that only 30 to 40 rDNA units within the rDNA loci of Drosophila are transcribed (106, 111) and the electron microscopic observations that actively transcribed rDNA units are contiguous (107, 111). The transcription domain model suggests that the host activates for transcription a block of ~40 rDNA units that contains the fewest R2 insertions. The remaining rDNA units are packaged into transcriptionally inactive heterochromatin. If the area activated for transcription is entirely free of R2 insertions, then R2 transcripts are not produced. If a large region without R2 insertions is unavailable in the locus, the cell is required to include R2-inserted units in the transcription domain. Loci with transcribed R2-inserted units give rise to the generation of new R2 copies by retrotransposition. How the host is able to differentiate inserted and uninserted units is not known. However, small RNA silencing pathways (119, 120) which induce heterochromatin formation of the R2-inserted units, potentially spreading into flanking units, would seem a likely mechanism. The finding that an $X^{R2\text{-}I}$ locus can be dominant over an $X^{R2\text{-}A}$ locus suggests that in extreme instances heterochromatin formation can spread through the whole ribosomal array.

Interestingly, variation in the ability of an organism to transcribe a region of the rDNA locus free of R2 insertions can explain why species have very different percentages of their rDNA units inserted with R2 (10, 27, 31, 121, 122). If an organism is less capable at differentiating between inserted and uninserted rDNA units, than all R2 inserts are potentially harmful and will be selected against. This selection against all copies means that the level of R2 elements in the locus will remain low. However, if the organism is adept at differentiating between inserted and uninserted units, than larger numbers of R2 elements can accumulate in the locus with minimal effects on the host.

Simulating rDNA Loci and a Stable Population of R2 Elements

Many studies have suggested that the driving force in the concerted evolution of the rRNA genes is random crossovers between rDNA loci (123–125). The presence of R2 elements, each inactivating a rDNA unit, were incorporated into computer simulations (126). By varying the crossover rate, the rate of R2 retrotransposition, and the number of uninserted units needed for full viability of the host, it was possible to simulate stable populations that contained from low to high mean levels of R2 elements. In all simulations, with or without the R2 element, the mean number of functional rDNA units per locus for the population far exceeded the minimum number required for full viability. The presence of R2-inserted units simply added to the total number of rDNA units present in each locus. The large excess in the number of uninserted rDNA units

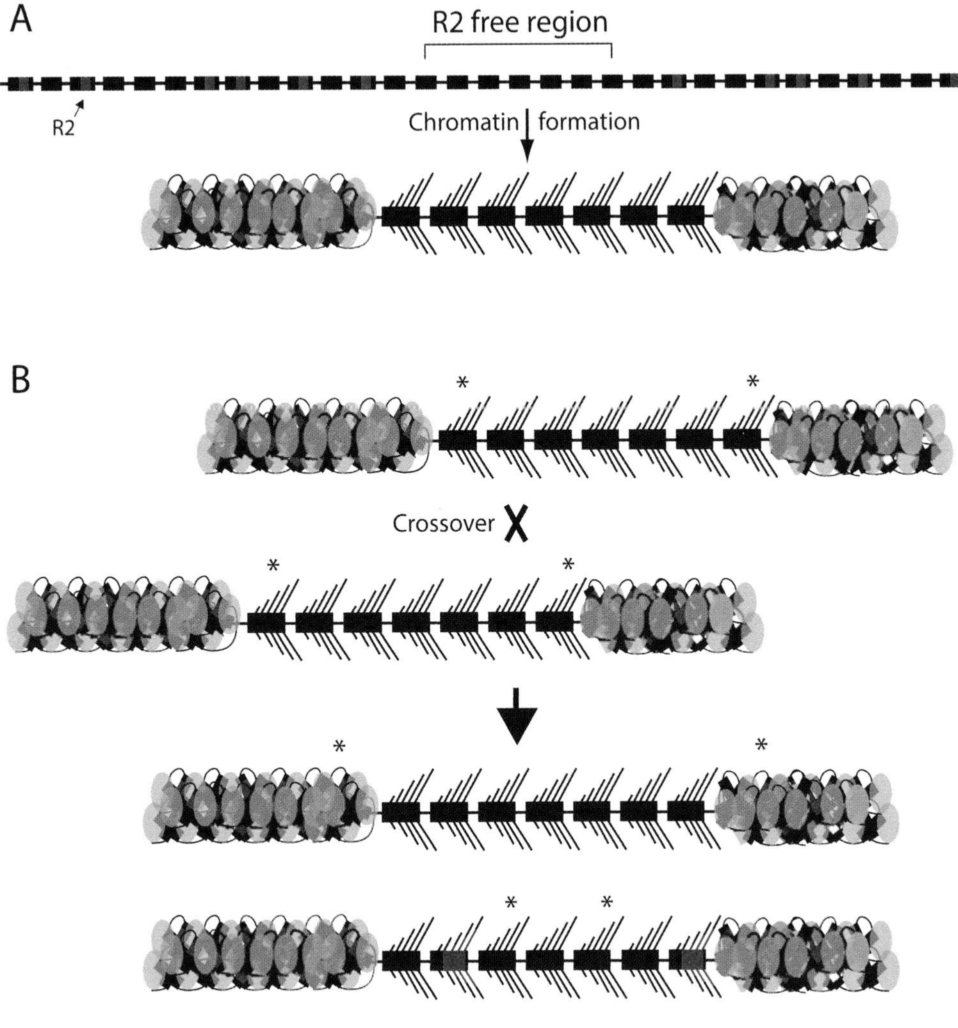

Figure 6 The transcription domain model of the rDNA locus and the long-term stability of R2 elements. (A) Uninserted (black boxes) and R2-inserted (black boxes with red insert) rDNA units are interspersed throughout the tandem array of rRNA genes. In Drosophila, a contiguous region of the rDNA locus with the lowest level of R2 insertions is selected for transcription. For simplicity this region is drawn as only seven units, but in *D. simulans* it is believed to be about 40 units. The remainder of the locus is packaged into heterochromatin (the compacted DNA plus protein flanking the active region). If the region selected as the transcription domain is free of R2-inserted units, then there is no R2 transcription and no R2 retrotransposition. (B) The driving force in the concerted evolution of the rDNA locus is crossovers between chromosomes. Most of these crossovers occur within the transcription domain (see text). The diagramed crossover produces one chromosome with an expanded R2 free region. Because the same number of rDNA units is still activated for transcription, some of the units that were transcribed before the crossover are packaged into heterochromatin after the crossover. Asterisks marking the original boundary of the transcription domain show this shift. The other chromosome product of the recombination contains an rDNA locus with a smaller R2-free region. In this case, rDNA units originally flanking the transcription domain are now activated for transcription. These flanking units contain R2 inserted units and thus copies of the R2 element are transcribed and retrotranspositions result. doi:10.1128/microbiolspec.MDNA3-0011-2014.f6

observed in organisms (106, 108–110) was explained by the wide range in rDNA locus size that is generated by the crossovers. In any population, these crossovers generate a small number of individuals (the extreme low end of the range) that have insufficient numbers of rDNA units for peak fitness and are therefore lost from the population. The loss of those individuals with the smallest loci results in a continual increase in the mean number of rDNA units per locus over generations. This gradual increase in mean locus size was previously noted by Lyckegaard and Clark (125). To avoid this increase, all simulations of the rDNA locus assume either selection against very large locus size or a low rate of chromosomal loop-deletion (i.e. recombination between rDNA units on the same chromosome).

More sophisticated simulations were next attempted that could reproduce the structure of the rDNA loci and the dynamics of the R2 elements observed for the natural populations of *D. simulans* (118). The characterized properties of the rDNA loci that had to be reproduced included: a) the frequency range of R2 elements in the loci (118), b) the frequency range of all rDNA units (118), c) the infrequent duplication of individual copies of R2 by crossover events (102, 117), d) the more rapid change in the number of uninserted units in a locus than the number of inserted units (102), e) the finding that about 45% of the individuals in a population have some level of R2 transcription (117), f) the linkage of R2 activity to the distribution of the R2 inserted units in the rDNA locus (i.e. the transcription domain model) (76, 117), and g) the size of the transcription domain at about 40 units (76, 111, 117). Simulated populations that duplicated the natural populations could be generated and most importantly required R2 retrotransposition rates only slightly higher than that which had been estimated from direct observations (33, 117).

The simulations suggested two other properties of the system. First, the only means found to minimize the number of duplicated R2 copies in the simulations was to localize the crossover events to the transcription domain. Clustering crossovers within the transcription domain is consistent with studies that suggested transcription increased crossover rates while heterochromatin typically inhibited recombination (127, 128) and with predictions that regions free of insertions would allow multiple contiguous rDNA units to align and thus more likely to be involved in chromosome pairing. Second, in order to match the natural populations the simulations suggested that selection against individuals with active R2 elements had to be quite low (118). Even a 1% reduction in fitness for individuals that were transcribing a single R2-inserted unit was sufficient to rapidly drive R2 elements out of the simulated populations. Such a low effect on fitness would suggest that those individuals in natural populations forced to transcribe R2-inserted units could readjust the size of the transcription domain and compensate for any potential imbalance in 28S, 18S and 5.8S rRNA levels.

As shown in Figure 6B, the key to the long-term survival of R2 in the rDNA loci of many organisms is the ability of fully functional R2 copies to reside in rDNA units outside the transcription domain for many generations with essentially no effect on the host. Invariably, however, crossovers within the transcription domain increase and decrease the area free of R2 insertions and generate rDNA loci in which the organism is forced to place R2-inserted units within the transcription domain. The resultant production of R2 transcripts and subsequent R2 retrotransposition events can then repopulate the locus. Thus it is the stochastic nature of crossover within the rDNA locus that enables R2 elements to survive in many lineages.

Ultimately, it has been the exclusive niche of R2 elements for 28S rRNA genes that has made possible the wide range of studies described here. These long-term hitchhikers of eukaryotic genomes have provided valuable insights into the origin of non-LTR retrotransposons, their mechanism of integration, and their mechanism of expression. In studying R2 activity and control over that activity, R2 elements have also served as a useful tool to better understand both the evolution and transcriptional control of the rDNA locus itself.

Acknowledgments. *The authors thank William Burke for his comments and assistance on this manuscript. We also want to thank Alan Lambowitz for his comments. We especially thank all the past and present members of the Eickbush lab whose invaluable contributions over the years have furthered our understanding of the R2 element. This work was made possible by the many years of support through National Institutes of Health Grant Number R01 GM42790.*

Citation. Eickbush TH, Eickbush DG. 2014. Integration, regulation, and long-term stability of R2 retrotransposons. Microbiol Spectrum 3(2):MDNA3-0011-2014

References

1. **Dawid IB, Rebbert ML.** 1981. Nucleotide sequence at the boundaries between gene and insertion regions in the rDNA of *D. melanogaster. Nucleic Acids Res* **9:** 5011–5020.
2. **Roiha H, Miller JR, Woods LC, Glover DM.** 1981. Arrangements and rearrangements of sequence flanking the two types of rDNA insertion in *D. melanogaster. Nature* **290:**749–753.

3. **Smith VL, Beckingham K.** 1984. The intron boundaries and flanking rRNA coding sequences of *Calliphora erythrocephala* rDNA. *Nucleic Acids Res* **12:**1707–1724.

4. **Fujiwara H, Orgura T, Takada N, Miyajima N, Ishikawa H, Maekawa H.** 1984. Introns and their flanking sequences of *B. mori* rDNA. *Nucleic Acids Res* **12:**6861–6869.

5. **Eickbush TH, Robins B.** 1985. *B. mori* 28S genes contain insertion elements similar to the type I and type II elements of *D. melanogaster*. *EMBO J* **4:**2281–2285.

6. **Burke WD, Calalang CC, Eickbush TH.** 1987. The site-specific ribosomal insertion element type II of *Bombyx mori* (R2Bm) contains the coding sequence for a reverse transcriptase-like enzyme. *Mol Cell Biol* **7:**2221–2230.

7. **Jakubczak JL, Xiong Y, Eickbush TH.** 1990. Type I (R1) and Type II (R2) ribosomal DNA insertions of *Drosophila melanogaster* are retrotransposable elements closely related to those of *Bombyx mori*. *J Mol Biol* **212:**37–52.

8. **Feng Q, Moran JV, Kazazian HH, Boeke JD.** 1996. Human L1 retrotransposon encodes a conserved endonuclease required for retrotransposition. *Cell* **87:**905–916.

9. **Yang J, Malik HS, Eickbush TH.** 1999. Identification of the endonuclease domain encoded by R2 and other site-specific, non-long terminal repeat retrotransposable elements. *Proc Natl Acad Sci USA* **96:**7847–7852.

10. **Jakubczak JL, Burke WD, Eickbush TH.** 1991. Retrotransposable elements R1 and R2 interrupt the rRNA genes of most insects. *Proc Natl Acad Sci USA* **88:**3295–3299.

11. **Burke WD, Malik HS, Jones JP, Eickbush TH.** 1999. The domain structure and retrotransposition mechanism of R2 elements are conserved throughout arthropods. *Mol Biol Evol* **16:**502–511.

12. **Kojima KK, Fujiwara H.** 2004. Cross-genome screening of novel sequence-specific non-LTR retrotransposons: various multicopy RNA genes and microsatellites are selected as targets. *Mol Biol Evol* **21:**207–217.

13. **Kojima KK, Kuma K, Toh H, Fujiwara H.** 2006. Identification of rDNA-specific non-LTR retrotransposons in Cnidaria. *Mol Biol Evol* **23:**1984–1993.

14. **Luchetti A, Mantovani B.** 2013. Non-LTR R2 element evolutionary patterns: phylogenetic incongruences, rapid radiation and the maintenance of multiple lineages. *PLoS ONE* **8:**e57076.

15. **Gladyshev EA, Arkhipova IR.** 2009. Rotifer rDNA-specific R9 retrotransposable elements generate an exceptionally long target site duplication upon insertion. *Gene* **448:**145–150.

16. **Thompson BK, Christensen SM.** 2011. Independently derived targeting of 28S rDNA by A- and D-clade R2 retrotransposons. *Mobile Genetic Elements* **1:**29–37.

17. **Fujiwara H.** 2014. *Mobile DNA III*. [to be completed]

18. **Burke WD, Müller F, Eickbush TH.** 1995. R4, a non-LTR retrotransposon specific to the large subunit rRNA gene of nematodes. *Nucleic Acids Res* **23:**4628–4634.

19. **Malik HS, Eickbush TH.** 2000. NeSL-1, an ancient lineage of site-specific non-LTR retrotransposons from *Caenorhabditis elegans*. *Genetics* **154:**193–203.

20. **Burke WD, Singh D, Eickbush TH.** 2003. R5 retrotransposons insert into a family of infrequently transcribed 28S rRNA genes of Planaria. *Mol Biol Evol* **20:** 1260–1270.

21. **Burke WD, Malik HS, Rich SM, Eickbush TH.** 2002. Ancient lineages of non-LTR retrotransposons in the primitive eukaryote, *Giardia lamblia*. *Mol Biol Evol* **19:** 619–630.

22. **Kojima KK, Fujiwara H.** 2003. Evolution of target specificity in R1 clade non-LTR retrotransposons. *Mol Biol Evol* **20:**351–361.

23. **Jakubczak JL, Zenni MK, Woodruff RC, Eickbush TH.** 1992. Turnover of R1 (Type I) and R2 (Type II) retrotransposable elements in the ribosomal DNA of *Drosophila melanogaster*. *Genetics* **131:**129–142.

24. **Pérez-González CE, Eickbush TH.** 2001. Dynamics of R1 and R2 Elements in the rDNA locus of *Drosophila simulans*. *Genetics* **158:**1557–1567.

25. **Pérez-González CE, Eickbush TH.** 2002. Rates of R1 and R2 retrotransposition and elimination from the rDNA locus of *Drosophila melanogaster*. *Genetics* **162:** 799–811.

26. **Eickbush DG, Eickbush TH.** 1995. Vertical transmission of the retrotransposable elements R1 and R2 during the evolution of the *Drosophila melanogaster* species subgroup. *Genetics* **139:**671–684.

27. **Lathe WC III, Eickbush TH.** 1997. A single lineage of R2 retrotransposable elements is an active, evolutionarily stable component of the *Drosophila* rDNA locus. *Mol Biol Evol* **14:**1232–1241.

28. **Burke WD, Malik HS, Lathe WC, Eickbush TH.** 1998. Are retrotransposons long term hitchhikers? *Nature* **239:**141–142.

29. **Malik HS, Burke WD, Eickbush TH.** 1999. The age and evolution of non-LTR retrotransposable elements. *Mol Biol Evol* **16:**793–805.

30. **Burke WD, Eickbush DG, Xiong Y, Jakubczak JL, Eickbush TH.** 1993. Sequence relationship of retrotransposable elements R1 and R2 within and between divergent insect species. *Mol Biol Evol* **10:**163–185.

31. **Stage DE, Eickbush TE.** 2010. Maintenance of multiple lineages of R1 and R2 retrotransposable elements in the ribosomal RNA gene loci of *Nasonia*. *Insect Mol Biol* **19**(Suppl. 1)**:**37–48.

32. **Hawley RS, Marcus CH.** 1989. Recombinational controls of rDNA redundancy in *Drosophila*. *Annu Rev Genet* **23:**87–120.

33. **Zhang X, Zhou J, Eickbush TH.** 2008. Rapid R2 retrotransposition leads to the loss of previously inserted copies via large deletions of the rDNA locus. *Mol Biol Evol* **25:**229–237.

34. **Hollocher H, Templeton AR.** 1994. The molecular through ecological genetics of abnormal abdomen in *Drosophila mercatorum*. VI. The non-neutrality of the Y chromosome rDNA polymorphism. *Genetics* **136:**1373–1384.

35. **Malik HS, Eickbush TH.** 1999. Retrotransposable elements R1 and R2 in the rDNA units of *Drosophila mercatorum*: *abnormal abdomen* revisited. *Genetics* **151:**653–665.

36. **Xiong Y, Eickbush TH.** 1988. Functional expression of a sequence-specific endonuclease encoded by the retrotransposon R2Bm. *Cell* **55:**235–246.
37. **Luan DD, Korman MH, Jakubczak JL, Eickbush TH.** 1993. Reverse transcription of R2Bm RNA is primed by a nick at the chromosomal target site: a mechanism for non-LTR retrotransposition. *Cell* **72:**595–605.
38. **Craigie R.** 2002. Retroviral DNA integration, p 613–630. *In* Craig NL, Craigc R, Gellert M, Lambowitz AM (ed), *Mobile DNA 11.* ASM Press, Washington, DC.
39. **Voytas DF, Boeke JD.** 2002. Ty1 and Ty5 of *Saccharomyces cerevisiae*, p 631–662. *In* Craig NL, Craige R, Gellert M, Lambowitz AM (ed), *Mobile DNA 11.* ASM Press, Washington, DC.
40. **Eickbush TH.** 2002. R2 and related site-specific non-long terminal repeat retrotransposons, p 813–835. *In* Craig NL, Craige R, Gellert M, Lambowitz AM (ed), *Mobile DNA 11.* ASM Press, Washington, DC.
41. **Luan DD, Eickbush TH.** 1995. RNA template requirements for target DNA-primed reverse transcription by the R2 retrotransposable element. *Mol Cell Biol* **15:** 3882–3891.
42. **Luan DD, Eickbush TH.** 1996. Downstream 28S gene sequences on the RNA template affect the choice of primer and the accuracy of initiation by the R2 reverse transcriptase. *Mol Cell Biol* **16:**4726–4734.
43. **Bibillo A, Eickbush TH.** 2002. The reverse transcriptase of the R2 non-LTR retrotransposon: continuous synthesis of cDNA on non-continuous RNA templates. *J Mol Biol* **316:**459–473.
44. **Bibillo A, Eickbush TH.** 2002. High processivity of the reverse transcriptase from a non-long terminal repeat retrotransposon. *J Biol Chem* **277:**34836–34845.
45. **Kurzynska-Kokorniak A, Jamburuthugoda VK, Bibillo A, Eickbush TH.** 2007. DNA-directed DNA polymerase and strand displacement activity of the reverse transcriptase encoded by the R2 retrotransposon. *J Mol Biol* **374:**322–333.
46. **Bibillo A, Eickbush TH.** 2004. End-to-end template jumping by the reverse transcriptase encoded by the R2 retrotransposon. *J Biol Chem* **279:**14945–14953.
47. **Arnold JJ, Cameron CE.** 1999. Poliovirus RNA-dependent RNA polymerase (3Dpol) is sufficient for template switching in vitro. *J Biol Chem* **274:**2706–2716.
48. **Chen B, Lambowitz AM.** 1997. *De novo* and DNA primer-mediated initiation of cDNA synthesis by the Mauriceville retroplasmid reverse transcriptase involve recognition of a 3′ CCA sequence. *J Mol Biol* **271:** 311–332.
49. **Mohr S, Ghanem E, Smith W, Sheeter D, Qin Y, King O, Polioudakis D, Iyer VR, Hunicke-Smith S, Swamy S, Kuersten S, Lambowitz AM.** 2013. Thermostable group II intron reverse transcriptase fusion proteins and their use in cDNA synthesis and next-generation RNA sequencing. *RNA* **19:**958–970.
50. **Peliska JA, Benkovic SJ.** 1992. Mechanism of DNA strand transfer reactions catalyzed by HIV-1 reverse transcriptase. *Science* **258:**1112–1118.
51. **Malik HS, Eickbush TH.** 2001. Phylogenetic analysis of Ribonuclease H domains suggests a late, chimeric origin of LTR retrotransposable elements and retroviruses. *Genome Res* **11:**1187–1197.
52. **Kelleher CD, Champoux JJ.** 1998. Characterization of RNA strand displacement synthesis by Moloney Murine Leukemia virus reverse transcriptase. *J Biol Chem* **273:** 9976–9986.
53. **Lanciault C, Champoux JJ.** 2004. Single unpaired nucleotides facilitate HIV-1 reverse transcriptase displacement synthesis through duplex RNA. *J Biol Chem* **279:** 32252–32261.
54. **Yao J, Truong DM, Lambowitz AM.** 2013. Genetic and biochemical assays reveal a key role for replication restart proteins in group II intron retrohoming. *PLOS Genetics* **9:**e1003469.
55. **Jamburuthugoda VK, Eickbush TH.** 2011. The reverse transcriptase encoded by the non-LTR retrotransposon R2 is as error-prone as that encoded by HIV-1. *J Mol Biol* **407:**661–672.
56. **Kim B, Ayran JC, Sagar SG, Adman ET, Fuller SM, Tran NH, Horrigan J.** 1999. New human immunodeficiency virus, type 1 reverse transcriptase (HIV-1 RT) mutants with increased fidelity of DNA synthesis. Accuracy, template binding, and processivity. *J Biol Chem* **274:**27666–27673.
57. **Preston BD, Poiesz BJ, Loeb LA.** 1988. Fidelity of HIV-1 reverse transcriptase. *Science* **242:**1168–1171.
58. **Christensen S, Eickbush TH.** 2004. Footprint of the R2Bm protein on its target site before and after cleavage in the presence and absence of RNA. *J Mol Biol* **336:** 1035–1045.
59. **Christensen SM, Eickbush TH.** 2005. R2 target primed reverse transcription: ordered cleavage and polymerization steps by protein subunits asymmetrically bound to the target DNA. *Mol Cell Biol* **25:**6617–6628.
60. **Christensen SM, Ye J, Eickbush TH.** 2006. RNA from the 5′ end of the R2 retrotransposon controls R2 protein binding to and cleavage of its DNA target site. *Proc Natl Acad Sci USA* **104:**17602–17607.
61. **Moran JV, Holmes SE, Naas TP, DeBerardinis RJ, Boeke JD, Kazazian HH.** 1996. High frequency retrotransposition in cultured mammalian cells. *Cell* **87:**917–927.
62. **Christensen SM, Bibillo A, Eickbush TH.** 2005. Role of the R2 element amino-terminal domain in the target-primed reverse transcription reaction. *Nucleic Acids Res* **33:**6461–6468.
63. **Shivram H, Cawley D, Christensen SM.** 2011. Targeting novel sites: the N-terminal DNA binding domain of non-LTR retrotransposons is an adaptable module that is implicated in changing site specificities. *Mob Genet Elements* **1:**169–178.
64. **Jamburuthugoda VK, Eickbush TH.** 2014. Identification of RNA binding motifs in the R2 retrotransposon-encoded reverse transcriptase (*Nuc Acids Res*, in press).
65. **Clements AP, Singer MF.** 1998. The human LINE-1 reverse transcriptase: effects of deletions outside the common reverse transcriptase domain. *Nucleic Acids Res* **26:**3528–3535.

66. **Moran JV, Gilbert N.** 2002. Mammalian LINE-1 retrotransposons and related elements, p 836–869. *In* Craig NL, Craige R, Gellert M, Lambowitz AM (ed), *Mobile DNA II*. ASM Press, Washington, DC.

67. **Gu SQ, Cui X, Mou S, Mohr S, Yao J, Lambowitz AM.** 2010. Genetic identification of potential RNA-binding regions in a group II intron-encoded reverse transcriptase. *RNA* **16:**732–747.

68. **Rouda S, Skordalakes E.** 2007. Structure of the RNA binding domain of telomerase: implications for RNA recognition and binding. *Structure* **15:**1403–1412.

69. **Mitchell M, Gillis A, Futahashi M, Fujiwara H, Skordalakes E.** 2010. Structural basis for telomerase catalytic subunit TERT binding to RNA template and telomeric DNA. *Nat Struct Mol Biol* **17:**513–518.

70. **Xiong Y, Eickbush TH.** 1990. Origin and evolution of retroelements based upon their reverse transcriptase sequences. *EMBO J* **9:**3353–3362.

71. **Blocker FJ, Mohr G, Conlan LH, Qi L, Belfort M, Lambowitz AM.** 2005. Domain structure and three-dimensional model of a group II intron-encoded reverse transcriptase. *RNA* **11:**14–28.

72. **Eickbush TH.** 1997. Telomerase and retrotransposons: which came first? *Science* **277:**911–912.

73. **Arkhipova IA, Pyatkov KI, Meselson M, Evgenev MB.** 2003. Retroelements containing introns in diverse invertebrate taxa. *Nature Genetics* **33:**123–124.

74. **Eickbush DG, Luan DD, Eickbush TH.** 2000. Integration of *Bombyx mori* R2 sequences into the 28S ribosomal DNA loci of *D. melanogaster*. *Mol Cell Biol* **20:** 213–223.

75. **Fujimoto H, Hirukawa Y, Tani H, Matsuura Y, Hashido K, Tsuchida K, Takada N, Kobayashi M, Maekawa H.** 2004. Integration of the 5′ end of the retrotransposon, R2Bm, can be complemented by homologous recombination. *Nucleic Acids Res* **32:**1555–1565.

76. **Eickbush DG, Ye J, Zhang X, Burke WD, Eickbush TH.** 2008. Epigenetic regulation of retrotransposons within the nucleolus of Drosophila. *Mol Cell Biol* **28:** 6452–6461.

77. **George JA, Eickbush TH.** 1999. Conserved features at the 5′ end of Drosophila R2 retrotransposable elements: implications for transcription and translation. *Insect Mol Biol* **8:**3–10.

78. **Eickbush DG, Eickbush TH.** 2003. Transcription of endogenous and exogenous R2 elements in the rDNA gene locus of *Drosophila melanogaster*. *Mol Cell Biol* **23:** 3825–3836.

79. **Eickbush DG, Eickbush TH.** 2010. R2 retrotransposons encode a self-cleaving ribozyme for processing from an rRNA co-transcript. *Mol Cell Biol* **30:**3142–3150.

80. **Been MD, Wickham GS.** 1997. Self-cleaving ribozymes of hepatitis delta virus RNA. *Eur J Biochem* **247:** 741–753.

81. **Ferré-D'Amaré AR, Zhou K, Doudna JA.** 1998. Crystal structure of a hepatitis delta virus ribozyme. *Nature* **395:**567–574.

82. **Nehdi A, Perreault J-P.** 2006. Unbiased in vitro selection reveals the unique character of the self-cleaving antigenomic HDV RNA sequence. *Nucleic Acids Res* **34:**584–592.

83. **Ruminski DJ, Webb C-HT, Riccitelli NJ, Lupták A.** 2012. Processing and translation initiation of non-long terminal repeat retrotransposons by hepatitis delta virus (HDV)-like self-cleaving ribozymes. *J Biol Chem* **286:** 41286–41295.

84. **Eickbush DG, Burke WD, Eickbush TH.** 2013. Evolution of the R2 retrotransposon ribozyme and its self-cleavage site. *PLoS One* **8:**e66441.

85. **Webb C-H, Riccitelli NJ, Ruminski DJ, Lupták A.** 2009. Widespread occurrence of self-cleaving ribozymes. *Science* **326:**953.

86. **Sánchez-Luque FJ, López MC, Macias F, Alonso C, Thomas MC.** 2011. Identification of an hepatitis delta virus-like ribozyme at the mRNA 5′ end of the L1Tc retrotransposon from *Trypanosoma cruzi*. *Nucleic Acids Res* **39:**8065–8077.

87. **Stage DE, Eickbush TH.** 2009. Origin of nascent lineages and the mechanisms used to prime second-strand DNA synthesis in the R1 and R2 retrotransposons of *Drosophila*. *Genome Biology* **10:**R49.

88. **Eickbush DG, Eickbush TH.** 2012. R2 and R1/R1 hybrid non-autonomous retrotransposons derived by internal deletions of full-length elements. *Mobile DNA* **3:**10.

89. **Ohshima K, Okada N.** 2005. SINEs and LINEs: symbionts of eukaryotic genomes with a common tail. *Cytogenet Genome Res* **110:**475–490.

90. **Belancio VP, Hedges DJ, Deininger P.** 2008. Mammalian non-LTR retrotransposons: for better or worse, in sickness and in health. *Genome Res* **18:**343–358.

91. **Kierzek E, Kierzek R, Moss WN, Christensen SM, Eickbush TH, Turner DH.** 2008. Isoenergetic penta- and hexanucleotide microarray probing and chemical mapping provide a secondary structure model for an RNA element orchestrating R2 retrotransposon protein function. *Nucleic Acids Res* **36:**1770–1782.

92. **Kierzek E, Christensen SM, Eickbush TE, Kierzek R, Turner DH, Moss WN.** 2009. Secondary structures for 5′ regions of R2 retrotransposon RNAs reveal a novel conserved pseudoknot and regions that evolve under different constraints. *J Mol Biol* **374:**322–333.

93. **Yang J, Eickbush TH.** 1998. RNA-induced changes in the activity of the endonuclease encoded by the R2 retrotransposable element. *Mol Cell Biol* **18:**3455–3465.

94. **George JA, Burke WD, Eickbush TH.** 1996. Analysis of the 5′ junctions of R2 insertions with the 28S gene: implications for non-LTR retrotransposition. *Genetics* **142:**853–863.

95. **Ostertag EM, Kazazian HH.** 2001. Biology of mammalian L1 retrotransposons. *Annu Rev Genet* **35:**501–538.

96. **Eickbush TH, Eickbush DG.** 2007. Finely orchestrated movements: evolution of the ribosomal RNA genes. *Genetics* **175:**477–485.

97. **Stage DE, Eickbush TH.** 2007. Sequence variation within the rRNA gene loci of 12 *Drosophila* species. *Genome Res* **17:**1888–1897.

98. **Zhang X, Eickbush TH.** 2005. Characterization of active R2 retrotransposition in the rDNA locus of *Drosophila simulans*. *Genetics* **170:**195–205.

99. **Mingazzini V, Luchetti A, Mantovani B.** 2011. R2 dynamics in *Triops cancriformis* (Bosc, 1801) (Crustacea, Branchiopoda, Notostraca): turnover rate and 28S concerted evolution. *Heredity* **106:**567–575.

100. **Mackay TFC, Lyman RF, Jackson MS, Terzian C, Hill WG.** 1992. Polygenic mutation in *Drosophila melanogaster*: estimates from divergence among inbred strains. *Evolution* **46:**300–316.

101. **Pérez-González CE, Burke WD, Eickbush TH.** 2003. R1 and R2 retrotransposition and deletion in the rDNA loci on the X and Y chromosomes of *Drosophila melanogaster*. *Genetics* **165:**675–685.

102. **Averbeck KT, Eickbush TH.** 2005. Monitoring the mode and tempo of concerted evolution in the *Drosophila melanogaster* rDNA locus. *Genetics* **171:**1837–1846.

103. **Eickbush MT, Eickbush TH.** 2011. Retrotransposition of R2 elements in somatic nuclei during the early development of Drosophila. *Mobile DNA* **2:**11.

104. **Long EO, Dawid IB.** 1979. Expression of ribosomal DNA insertions in *Drosophila melanogaster*. *Cell* **18:** 1185–1196.

105. **Kidd SJ, Glover DM.** 1981. *D. melanogaster* ribosomal DNA containing type II insertions is variably transcribed in different strains and tissues. *J Mol Biol* **151:** 645–662.

106. **Ye J, Eickbush TH.** 2006. Chromatin structure and transcription of the R1- and R2-inserted rRNA genes of *Drosophila melanogaster*. *Mol Cell Biol* **23:**8781–8790.

107. **Jamrich M, Miller OL.** 1984. The rare transcripts of interrupted rDNA genes in *Drosophila melanogaster* are processed or degraded during synthesis. *EMBO J* **3:** 1541–1545.

108. **Conconi A, Widmer RM, Koller T, Sogo JM.** 1989. Two different chromatin structures coexist in ribosomal RNA genes throughout the cell cycle. *Cell* **57:**753–761.

109. **Conconi A, Sogo JM, Ryan CA.** 1992. Ribosomal gene clusters are uniquely proportioned between open and closed chromatin structures in both tomato leaf cells and exponentially growing suspension cultures. *Proc Natl Acad Sci USA* **89:**5256–5260.

110. **Dammann R, Lucchini R, Koller T, Sogo JM.** 1993. Chromatin structures and transcription of rDNA in yeast *Saccharomyces cerevisiae*. *Nucleic Acids Res* **21:** 2331–2338.

111. **McKnight SL, Miller OL.** 1976. Ultrastructural patterns of RNA synthesis during early embryogenesis of *Drosophila melanogaster*. *Cell* **8:**305–319.

112. **Tucker S, Vitins A, Pikaard CS.** 2010. Nucleolar dominance and ribosomal RNA gene silencing. *Curr Opin Cell Biol* **22:**351–356.

113. **Zhou J, Sackton TB, Martinsen L, Lemos B, Eickbush TH, Hartl DL.** 2012. Y chromosome mediates ribosomal DNA silencing and modulates the chromatin state in Drosophila. *Proc Natl Acad Sci USA* **109:**9941–9946.

114. **Greil F, Ahmad K.** 2012. Nucleolar dominance of the Y chromosome in *Drosophila melanogaster*. *Genetics* **191:**1119–1128.

115. **Paredes S, Branco AT, Hartl DL, Maggert KA, Lemos B.** 2011. Ribosomal DNA deletions modulate genome-wide gene expression: "rDNA-sensitive" genes and natural variation. *PLoS Genetics* **7:**e1001376.

116. **Paredes S, Maggert KA.** 2009. Expression of *I-Cre*I endonuclease generates deletions within the rDNA of Drosophila. *Genetics* **181:**1661–1671.

117. **Zhou J, Eickbush TH.** 2009. The pattern of R2 retrotransposon activity in natural populations of *Drosophila simulans* reflects the dynamic nature of the rDNA locus. *PLoS Genetics* **5:**e1000386.

118. **Zhou J, Eickbush MT, Eickbush TH.** 2013. A population genetic model for the maintenance of R2 retrotransposons in rRNA gene loci. *PLoS Genetics* **8:**e1003179.

119. **Girard A, Hannon GJ.** 2008. Conserved themes in small-RNA-mediated transposon control. *Trends Cell Biol* **18:**136–148.

120. **Senti K-A, Brennecke J.** 2010. The piRNA pathway: a fly's perspective on the guardian of the genome. *Trends in Genetics* **26:**499–509.

121. **Ghesini S, Luchetti A, Marini M, Mantovani B.** 2011. The non-LTR retrotransposon R2 in termites (Insecta, Isoptera): characterization and dynamics. *J Mol Evol* **72:**296–305.

122. **Montiel EE, Cabrero J, Ruiz-Estévez M, Burke WD, Eickbush TH, Camacho JPM, López-León MD.** 2014. Preferential occupancy of R2 retroelements on the B chromosome of the grasshopper *Eyprepocnemis plorans*.

123. **Ohta T.** 1980. *Evolution and variation of multigene families*. Springer-Verlag, Berlin/Heidelberg, Germany/ New York.

124. **Ohta T, Dover GA.** 1983. Population genetics of multigene families that are dispersed into two or more chromosomes. *Proc Natl Acad Sci USA* **80:**4079–4083.

125. **Lyckegaard EMS, Clark AG.** 1991. Evolution of ribosomal RNA gene copy number on the sex chromosomes of *Drosophila melanogaster*. *Mol Biol Evol* **8:**458–474.

126. **Zhang X, Eickbush MT, Eickbush TH.** 2008. Role of recombination in the long-term retention of transposable elements in rRNA gene loci. *Genetics* **180:**1617–1626.

127. **Aguilera A, Gómez-González B.** 2008. Genome instability: a mechanistic view of its causes and consequences. *Nature Rev Genetics* **9:**204–217.

128. **Voelket-Meiman K, Keil RL, Roeder GS.** 1987. Recombination-stimulating sequences in yeast ribosomal DNA correspond to sequences regulating transcription by RNA polymerase I. *Cell* **48:**1071–1079.

Mobile DNA, 3rd Edition
Nancy L. Craig, Michael Chandler, Martin Gellert, Alan M. Lambowitz, Phoebe A. Rice and Suzanne Sandmeyer

doi:10.1128/microbiolspec.MDNA3-0001-2014

Haruhiko Fujiwara[1]

50

Site-specific non-LTR retrotransposons

OVERVIEW OF NON-LTR RETROTRANSPOSONS

DNA transposons are the mobile elements that move by a "cut and paste" mechanism (1, 2). In contrast, retrotransposons encode reverse transcriptase, and move by a "copy and paste" mechanism. The process of retrotransposon insertion into genomic locations involves an RNA intermediate. Retrotransposons can be classified into long terminal repeat (LTR) and non-LTR retrotransposons. LTR retrotransposons have LTRs at both ends and resemble retroviruses in both structure and integration mechanisms. Non-LTR retrotransposons comprise two subtypes, long interspersed nuclear elements (LINEs) and short interspersed nuclear elements (SINEs). Non-LTR retrotransposons are in general 4 to 7 kb long and do not carry LTRs, and their retrotransposition mechanism is different from that of LTR retrotransposons. SINEs are nonautonomous retrotransposons of 100 to 500 bp that do not encode proteins. It has been proposed that the proteins encoded by LINEs are the source of the enzymatic retrotransposition machinery of SINEs (3, 4, 5).

The process of retrotransposition of non-LTR elements is initiated by an encoded endonuclease domain that nicks one strand of DNA at the target site and creates a 3′-hydroxyl end, which is used as a primer for reverse transcription of the retrotransposon mRNA onto the DNA target (6, 7). This unique process, target primed reverse transcription (TPRT), is peculiar to non-LTR retrotransposons (6). These retrotransposons have been classified into two large groups on the basis of their structural and phylogenetic features (8, 9). One of the groups encodes a restriction enzyme-like endonuclease (RLE) in the C-terminal region of the single open-reading frame (ORF) (10). The RLE-encoding non-LTR elements are a phylogenetically ancient class and are further categorized into five clades. The other group usually encodes an endonuclease with homology to apurinic/apyrimidinic endonuclease (APE) in one of its two ORFs (11). This group is thought to be younger and is categorized into at least 22 clades (8, 9). One clade (Dualen/RandI), which encodes both APE and RLE, is positioned phylogenetically at the midpoint between the early branched and recently branched groups (9, 12).

Retrotransposons are widespread in metazoan genomes. Recent and large-scale genome projects have revealed that the proportion of the genome that contains retrotransposons often exceeds that containing DNA transposons, particularly in higher vertebrates such as humans (13). Non-LTR retrotransposon L1 integrates essentially throughout the whole genome and

[1]Department of Integrated Biosciences, Graduate School of Frontier Sciences, University of Tokyo, Kashiwa 277-8562, Japan.

occupies more than 16% of the human genome (14). The integration of L1 has been shown to be involved in genetic diseases and cancers and affects genome reconstitution and gene evolution (14, 15). Most retrotransposons integrate into random sites of the host genome, but some have a sequence preference. In particular, a few non-LTR subclades integrate into the genome in a highly sequence-specific manner.

In this chapter, I summarize site-specific non-LTR retrotransposons, focusing on the mechanisms of their integration.

OVERVIEW OF SITE-SPECIFIC NON-LTR RETROTRANSPOSONS

Some non-LTR retrotransposons have very restricted integration targets within the genome. Targets of site-specific non-LTR retrotransposons are usually located within multicopy RNA genes, such as clusters of rRNA gene (rDNA) or repetitive genomic sequences (16). Targeted integration of site-specific elements into repetitive sequences is considered to be a symbiotic strategy, which allows spreading through the genome with very little damage to essential genes of the host (Fig. 1). In contrast to this strategy, most transposable elements integrate in a parasitic-like, random manner (Fig. 1). The copy number of target sequences is considered to be a primary factor in restricting various sequence specificity of non-LTR elements (17).

How are site-specific non-LTR elements found in the genome? We could find both the 5′ and 3′ ends of genomic copies of non-LTR retrotransposons because they usually end with a poly (A) stretch and have target-site duplication (TSD) at both ends. Thus, it is possible to search for the site-specific non-LTR elements in DNA databases by investigating the boundary sequences flanking multiple genomic copies of the non-LTR elements. If all the junction sequences of the multiple copies are the same, they are accepted as "site-specific non-LTR retrotransposons." However, the 5′ portion of non-LTR elements is often deleted by incomplete reverse transcription (5′ truncation), which means that the same genomic repetitive sequences flank copies of target-specific non-LTR elements 5′ truncated at different positions. Many new examples of site-specific non-LTR elements have been found from sequenced genomes (16, 17).

There are five clades of RLE-encoding elements based on their RT sequence similarity. Most of three clades (NeSL, R2, and R4) are target specific and two clades (HERO and CRE) have some site-specific elements (Fig. 2 and Table 1). CRE1/CRE2/SLACS/CZAR

A. Randomly retrotransposed non-LTR elements

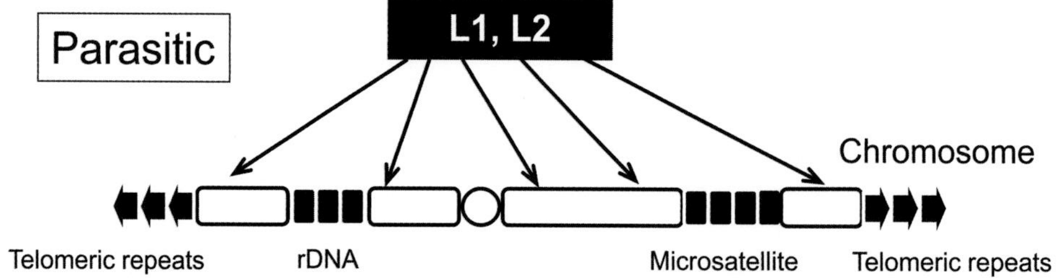

B. Site-specific non-LTR elements

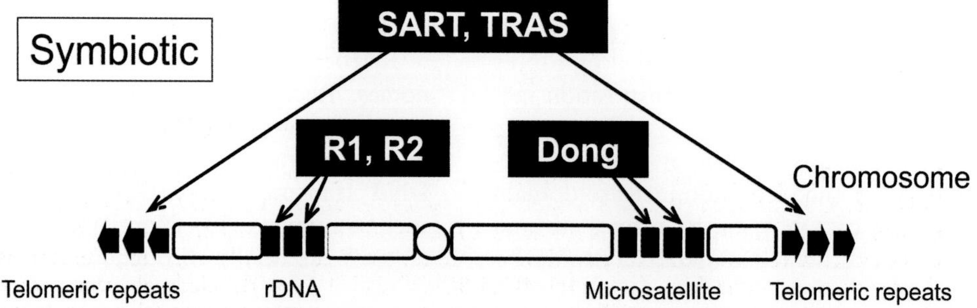

Figure 1 Two types of genomic insertion of non-LTR elements.
doi:10.1128/microbiolspec.MDNA3-0001-2014.f1

in trypanosomes (CRE clade) and NeSL-1 in nematodes (NeSL clade) were found in the spliced leader exons (18, 19, 20). R2 (R2 clade) was found at a specific sequence in the 28S rDNA of many invertebrate and vertebrate species (21, 22, 23) (Fig. 3A). R4 in *Ascaris* and Dong (R4 clade) were found at another site of the 28S rDNA (Fig. 3A) and microsatellite TAA repeats, respectively (24, 25) (Table 1). Apart from these already characterized site-specific elements, some novel site-specific RLE-encoding elements have been found *in silico* using BLAST searches of databases or degenerate PCR cloning (17). HERO-1_HR (HERO clade) was recently found in a microsatellite, $(ATT)_n$ repeats (26) (Table 1). Table 1 lists site-specific non-LTR elements from various clades and their targets.

In contrast to RLE-encoding elements, most of the APE-encoding non-LTR elements do not insert themselves in a sequence-specific manner, but do have weak target-site specificity, e.g., human L1 for TAAA repeats (30, 31, 32). APE is considered to be closely related to class II of the AP endonuclease family of highly conserved multifunctional DNA repair enzymes (33, 34). Among over 20 clades of APE-encoding non-LTR elements, members of only two clades, Tx1 and R1, are known to be sequence-specific (Fig. 2 and Table 1). The Tx1 clade, which has many types of site-specific non-LTR elements, targets rRNA genes (Mutsu), tRNA genes (Dewa), snRNA genes (Keno), telomeric repeats (Tx1-1_ACar) (Fig. 3B), and microsatellites (Kibi and Koshi) (17, 35) (Table 1). Tx1L and Tx2L insert into other transposons, Tx1D and Tx2D, respectively (36). The R1 clade also contains several site-specific elements. Extensive screening of novel R1 clade elements found different targets for sequence-specific integration: different locations of the rRNA genes (e.g., R1) (Fig. 3A), telomeric repeats (e.g., TRAS) (Fig. 3B), and microsatellite (e.g., Waldo) (16). To the best of our knowledge, the sequence-specific R1 clade elements target only three genomic locations: rDNA, telomeric repeats, and ACAY (or AC) repeats (Fig. 3 and Table 1). Phylogenetic trees for R1 clade elements reveal that the target specificity has been independently altered

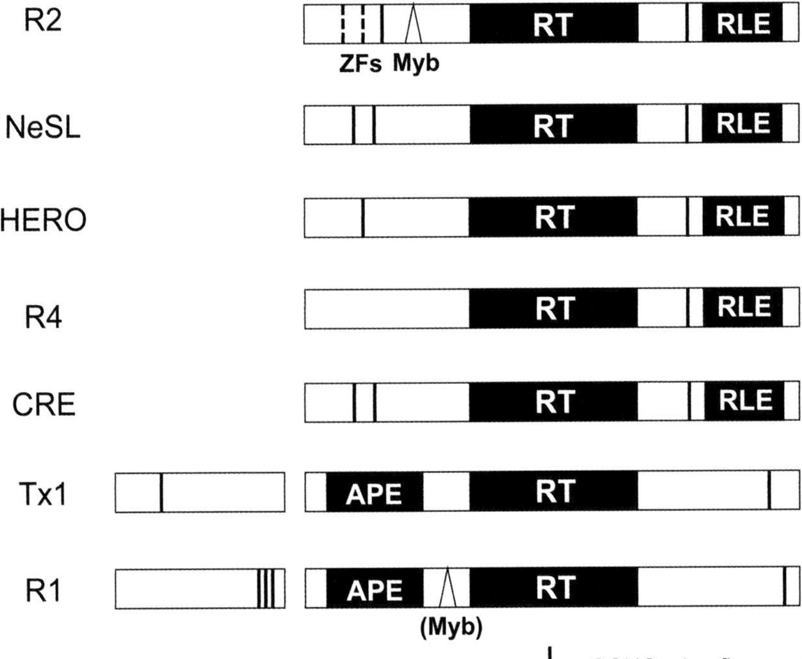

Figure 2 Schematic structure of each clade of site-specific non-LTR retrotransposons. The ORF1 in APE-encoding elements is shown as a short rectangle. An ORF2 in APE-encoding elements or ORF in RLE-encoding elements is shown as a long rectangle. A ZF-like structure is shown as a vertical bold line. A ZF-like structure in ORF1 of APE-encoding elements is shown as a zinc knuckle: Cx2Cx4Hx4Cx5-8Cx2Cx3Hx4C. In R2 clade elements, some elements have an additional ZF-like structure shown as a dotted vertical line. Myb-like (Myb) domains are found in R2 clade-elements and in TRAS families in the R1-clade. doi:10.1128/microbiolspec.MDNA3-0001-2014.f2

TABLE 1 Targets of site-specific non-LTR retrotransposons

Endonuclease	APE		RLE				
Target \ Clade	R1	Tx1	R2	NeSL	HERO	R4	CRE
rRNA gene	R1/R6/R7/RT	(Mutsu)**	R2/R8/R9	R5		R4	
tRNA gene		Dewa				R4-2_Sra*	
snRNA gene		Keno					
Spliced leader				NeSL			CRE
Telomeric repeat	TRAS/SART	Tx1-1_Acar					MoTeR
Microsatellite	Waldo	Kibi			HERO-1_HR	Dong	CRE-1_NV*
Transposon		Tx1		R5-2_SM*			

Only representative elements are shown.
APE: endonuclease with homology to apurinic/apyrimidinic-endonuclease, RLE: restriction enzyme-like endonuclease
* : References are shown in the reference list 27–29. **: Mutsu targets 5S rRNA genes.

several times during evolution (16) (Fig. 4). We note that some elements (Hida, Noto, HOPE, Kaga, and Hal) of this "site-specific clade" have lost their specificity of integration.

Dr. Eickbush's chapter describes in detail the retrotransposition mechanisms of R2 and the related site-specific RLE-encoding non-LTR elements. In this chapter, I describe the retrotransposition mechanism of mainly site-specific APE-encoding retrotransposons.

TELOMERE-SPECIFIC NON-LTR ELEMENTS

Telomeres are the regions at the ends of chromosomes that are essential for the complete replication, meiotic paring, and stability of chromosomes (37). It has been hypothesized that the terminal sequences of chromosome ends shorten with each cell division, and a reduction in telomere length causes cellular senescence associated with the cessation of cell division. The telomeres of most eukaryotic cells are composed of simple repeated sequences called telomeric repeats. One of the strands is G-rich and is synthesized by a specialized enzyme, telomerase (37). In many organisms, the addition of telomeric repeats by telomerase is essential to compensate for the critical telomere shortening that occurs with aging (37, 38).

Drosophila Telomere-Specific Elements

However, alternative telomere maintenance has been suggested for some insects (39, 40). All dipteran insects have lost the telomeric repeats; it has been suggested that they use a telomerase-independent telomere maintenance system (41, 42, 43, 44, 45). In *D. melanogaster*, telomeres are composed of three specialized non-LTR retrotransposons, TART, HeT-A, and TAHRE and maintained mainly by retrotransposition of these elements, which have been extensively studied to date (41, 42, 43, 46, 47, 48, 49, 50, 51, 52) (Fig. 3B). Phylogenetically, the three telomere-specific elements belong to the APE-encoding Jockey clade, but HeT-A has no RT (pol) domain. It is noteworthy that the retrotransposition of these elements does not depend on a specific DNA at the target site. Although the mechanism as to how they target chromosome ends is not understood clearly, a recent report showed that the ORF1p of HeT-A localizes in the nucleus and forms spherical structures at telomeres, which may support targeting to the chromosome ends (52). In *Drosophila*, repeated but random retrotransposition of the three elements with their 5′ ends toward the chromosome ends form long arrays of elements. Distantly related *Drosophila* species, *D. yakuba*, *D. virilis*, and *D. melanogaster*, all have this robust mechanism of telomere maintenance, suggesting establishment before the divergence of the *Drosophila* genus (48).

Telomere-Specific Non-LTR Elements in *Bombyx* and *Tribolium*

In contrast to *D. melanogaster*, the silkworm *B. mori* retains the telomeric $(TTAGG)_n$ repeats at the ends of chromosomes (39, 53). Immediately proximal to a 6 to 8-kb stretch of TTAGG repeats, there are more than 1,000 copies of non-LTR retrotransposons, which are designated the TRAS and SART families of the R1 clade (54, 55, 56, 57) (Fig. 3B and 4). They insert into both strands of telomeric repeats so these elements are inserting in different orientations (i.e., CCTAA is the reverse complement of TTAGG) (Fig. 4). TRAS1, a predominant element in the TRAS family, is 7.8 kb in length and is inserted specifically into the site between the C and T of the CCTAA telomeric strand (58, 59). A major element in the silkworm SART family, SART1

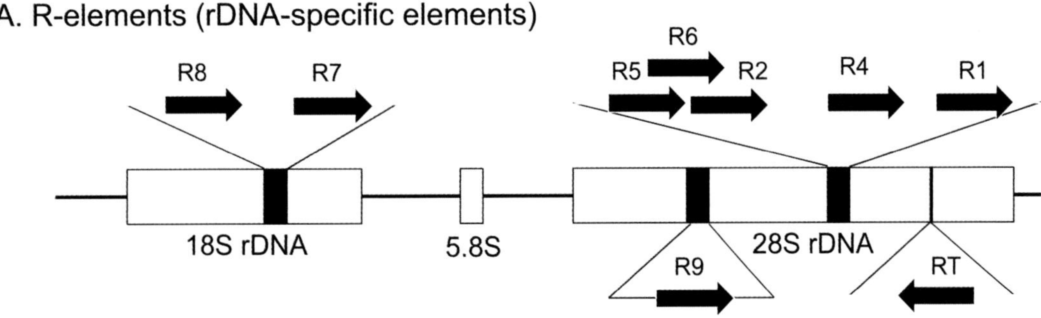

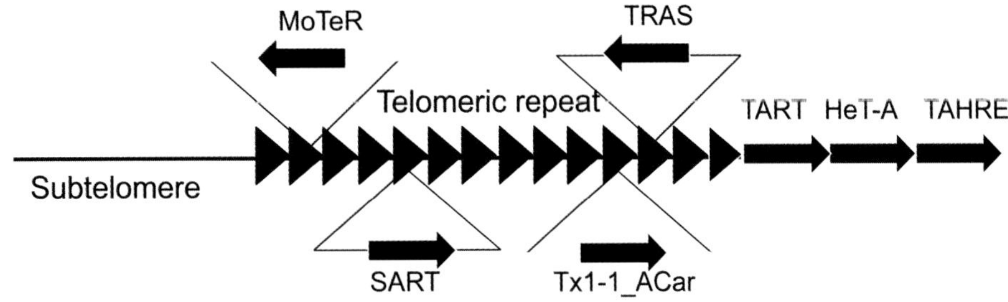

Figure 3 rDNA-specific-elements (R-elements) and telomere-specific-elements. Locations and insertion sites of rDNA-specific non-LTR elements (A) and telomere-specific non-LTR elements (B) are shown schematically. Arrows indicate the 5′ to 3′ orientation of non-LTR element insertion. Three telomere-specific elements, TART, HeT-A, and TAHRE, in *Drosophila* are located at extreme ends of chromosome (their locations are in random order). doi:10.1128/microbiolspec.MDNA3-0001-2014.f3

(SART1Bm), is 6.7 kb in length and is inserted specifically between the T and A of the TTAGG strand (60, 61). The TRAS and SART families occupy 3% of the silkworm genome and more than 300 kb of each of the chromosome ends (39). These telomeric repeat-specific non-LTR families are also found in other insects (40, 61, 62) and in arthropods (K. K. Kojima, personal communication). The TRAS and SART elements are transcribed abundantly in comparison with other non-LTR elements (63). The translational regulation of bicistronic (ORF1 and ORF2) mRNA of SART1 is analogous to translational coupling in prokaryotes and viruses (64). Although most insects retain the TTAGG telomeric repeats (except for *Diptera*), some have different pentanucleotide telomeric repeats, such as TCAGG in the flour beetle *T. castaneum* (65) (Fig. 5). Recently, seven types of non-LTR retrotransposons of the R1 clade were found in the TCAGG telomeric repeats of *T. castaneum* (61). Based on their amino acid sequences, these elements constitute a monophyletic group. They are phylogenetically closer to SART1Bm than to TRAS1Bm, and named SARTTc1 to SARTTc7 (SARTTc family). All of these TCAGG-repeat-specific non-LTR elements are inserted between C and A of TCAGG telomeric repeats and the orientation of the insertion is the same as that observed for the SART1Bm family (61).

To investigate the functional roles of these telomere-specific non-LTR elements, we measured the telomerase activity in various insects using a modified TRAP (telomeric repeat amplification protocol) method (66). Insects retaining standard telomeric repeats undisturbed by retrotransposons, such as cockroaches or crickets, showed significant telomerase activities in several tissues. However, the telomerase activity could not be detected in two cultured cell types of *D. melanogaster,* or in three cell lines and tissues from *B. mori*. Recently, we also found by TRAP analysis that *Tribolium* shows only weak telomerase activity in the elongation step (67; M. Osanai-Futahashi and H. Fujiwara, unpublished data). These results suggest that telomere-specific non-LTR elements rescue the telomere crisis in *D. melanogaster*, and provide a backup for the telomere function in *B. mori* and *T. castaneum*, which means that *B. mori* does not need as much telomerase activity.

We analyzed the gene structure of *B. mori* (*Lepidoptera*) and *T. castaneum* (*Coleoptera*) TERT (telomerase reverse transcriptase) and found some unusual shared characteristics compared to the TERTs of other insects

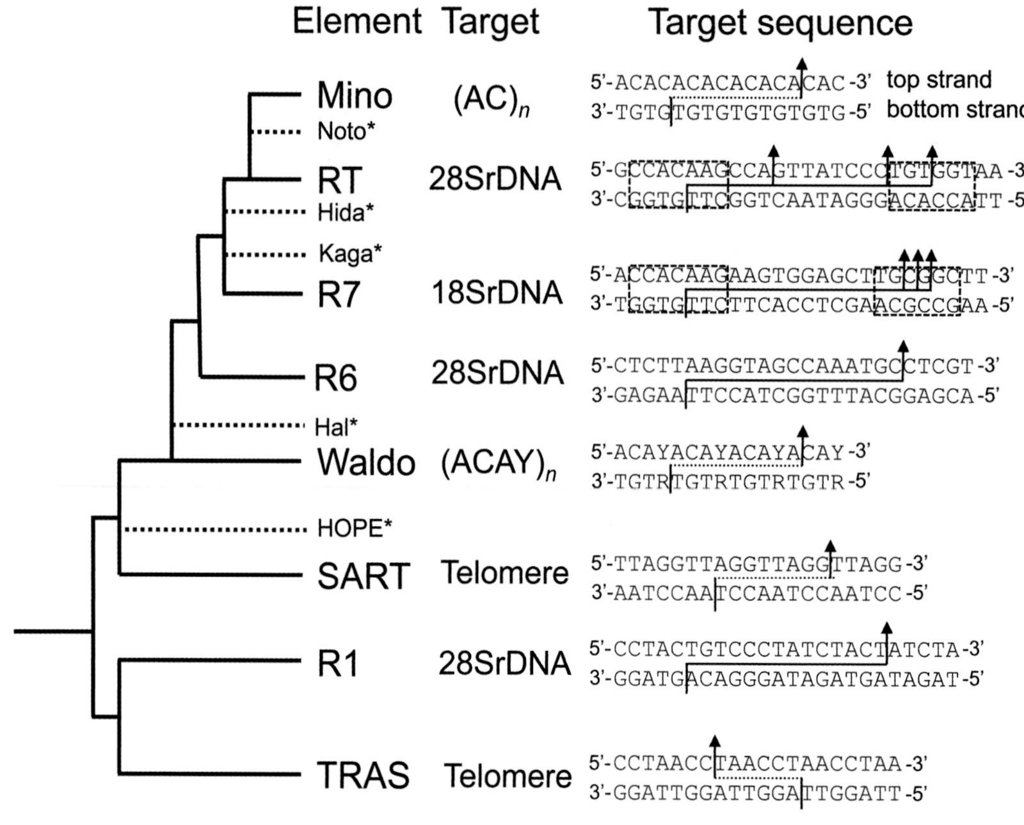

Figure 4 Evolution of target sequences in R1-clade elements. The phylogeny is constructed on the basis of data in K. K. Kojima and H. Fujiwara (16). Nonsequence-specific elements are shown as dotted lines and asterisks. Target and flanking sequences of each site-specific element are shown on the right. In the TPRT model, the bottom strand is first nicked and then the top strand is cleaved. The SART and TRAS elements target the same (TTAGG/CCATT)$_n$ telomeric repeats. However, TRAS first cut the TTAGG bottom strand, whereas SART first cut the CCTAA strand. The bottom-to-top strand cleavage in each target sequence is represented by bent lines. Arrows indicate variation of top-strand cleavage. Broken lines between the top and bottom strands, in TRAS, SART, Waldo, and Mino, indicate unidentified exact cleavage sites because they target the tandem repeats TTAGG and ACAY (or AC). Broken boxes indicate homologous sequences near cleavage sites among RT and R7. doi:10.1128/microbiolspec.MDNA3-0001-2014.f4

without TRAS and SART elements: (i) an intronless gene (Fig. 5), (ii) 3′-poly A tails in the genomic copy (Fig. 6B), (iii) upstream ATG codons in the 5′ UTR (five ATGs for *BmoTERT* and four ATGs for *TcTERT*) (Fig. 6A), (iv) loss of N-terminal GQ motifs (Fig. 5 and 6), and (v) repressed transcription (40, 65). It is reported that multiple upstream ATGs (or upper AUG) reduce the translation of some genes (68). In addition, N-terminal GQ motifs are shown to be involved in the telomeric repeat processivity in yeast and human TERT (69, 70). Thus, the above features may cause inefficient elongation of their telomeric repeats. Hymenoptera (such as the honey bee *Apis melifera*), an ancestral group among higher insects (71, 72), have an usual TERT structure with many introns and long telomeric repeats undisturbed by retrotransposons (Fig. 5) (73). This suggests that TERT mRNA has been reverse-transcribed in a common ancestor of *Coleoptera* and *Lepidoptera* after branching from Hymenoptera, and integrated into the genome as a processed gene. Owing to 5′ truncation in the process of reverse transcription, the N-terminal including the GQ motif was deleted, and this unusual gene replaced the normal TERT gene (Fig. 6C). The original TERT genes with introns could not be found in the silkworm and *Tribolium* genome databases, and we do not yet know how these genes

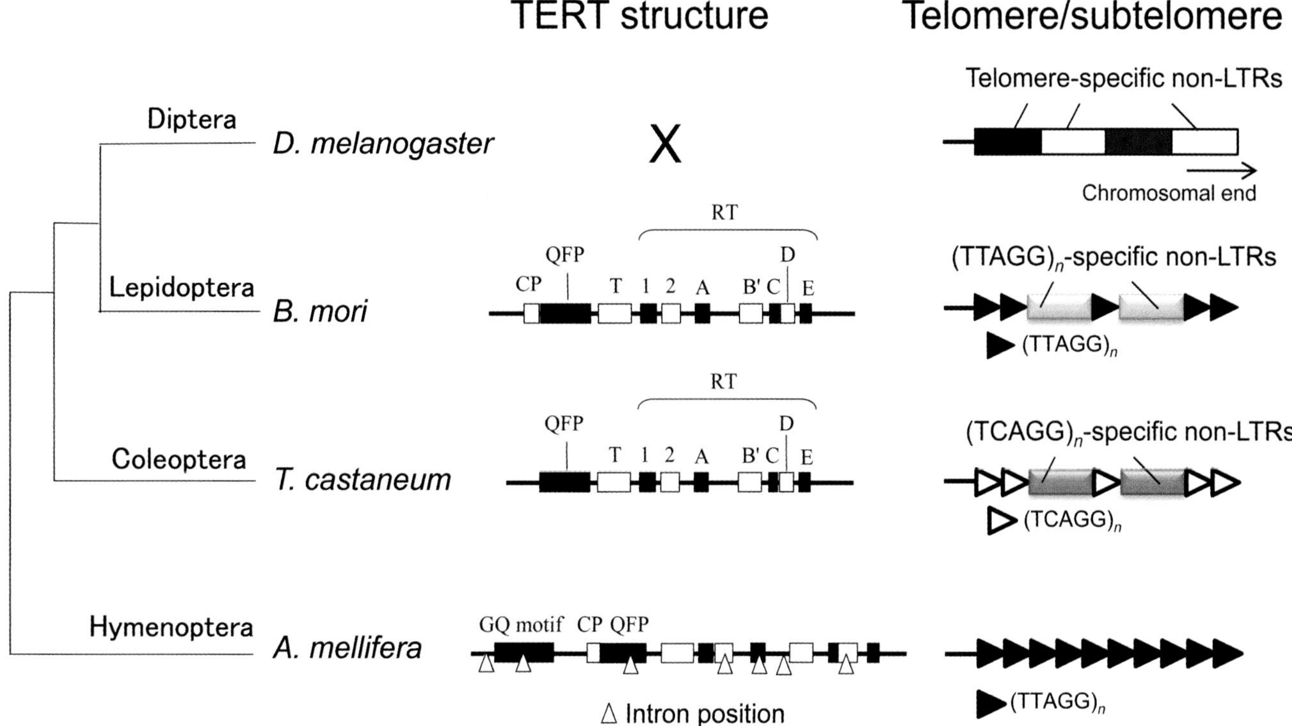

Figure 5 Evolution of telomere and TERT structure in higher insects. X in *D. melanogaster* indicates the loss of a TERT gene from the genome. In the silkworm (*B. mori*) and the flour beetle (*T. castaneum*), introns and GQ domains are lost from TERT genes (see text). The phylogenetic tree was constructed based on recent reports (71, 72). doi:10.1128/microbiolspec.MDNA3-0001-2014.f5

were lost. The telomeric repeat-specific non-LTR retrotransposons may have evolved to backup the weak activity of telomerase in *Lepidoptera* and *Coleoptera*, which helped them to survive weak telomerase activity during insect evolution (40).

Telomere-Associated Non-LTR Elements in Other Organisms

Telomere-associated non-LTR retrotransposons have also been found in organisms other than insects (47) (Fig. 3B and Table 1). Zepp is an L1-clade non-LTR element found in the telomeres of all 16 chromosomes in *Chlorella vulgaris;* this element exhibits telomere region specificity but not sequence specificity (74). In the genome of *Giardia lamblia*, two RLE-encoding non-LTR elements capped by TAGGG repeats, GilM and GilT, are located at the ends of the chromosome (75). Recently, two related RLE-encoding non-LTR retrotransposons, MoTeR elements (clade CRE), were found in the telomeric repeat $(TTAGGG)_n$ of the fungus *Magnaporthe oryzae* (76). In the genome of the green anole lizard, an APE-encoding element Tx1-1_ACar (clade Tx1) inserted in the telomeric repeats $(TTAGGG)_n$ has been found recently (Table 1) (35).

rDNA-SPECIFIC NON-LTR ELEMENTS (R-ELEMENTS)

As mentioned above, many sequence-specific non-LTR elements accumulate in and target the ribosomal RNA (28S, 18S, and 5.8S rRNAs) gene array (rDNA). These elements are called R-elements. Thus far, nine different R elements inserted into various sites of rDNA have been reported (Fig. 3A and Table 1). R7 in R1 clade (16) and R8 in R2 clade (77) insert into 18S rDNA, while the remaining seven R elements, R1 (78, 79), R2 (7, 78), R4 (24), R5 (80), R6 (16), R9 (81), and RT (82), are inserted into the 28S rDNA. It is of interest that all R-element target sites within 28S and 18S rDNA are highly conserved among organisms.

The APE-encoding elements R1, R6, R7, and RT, which are structurally related and classified in the R1 clade, are believed to share a common sequence-specific ancestor (Fig. 4) (16). In phylogenetic trees based on

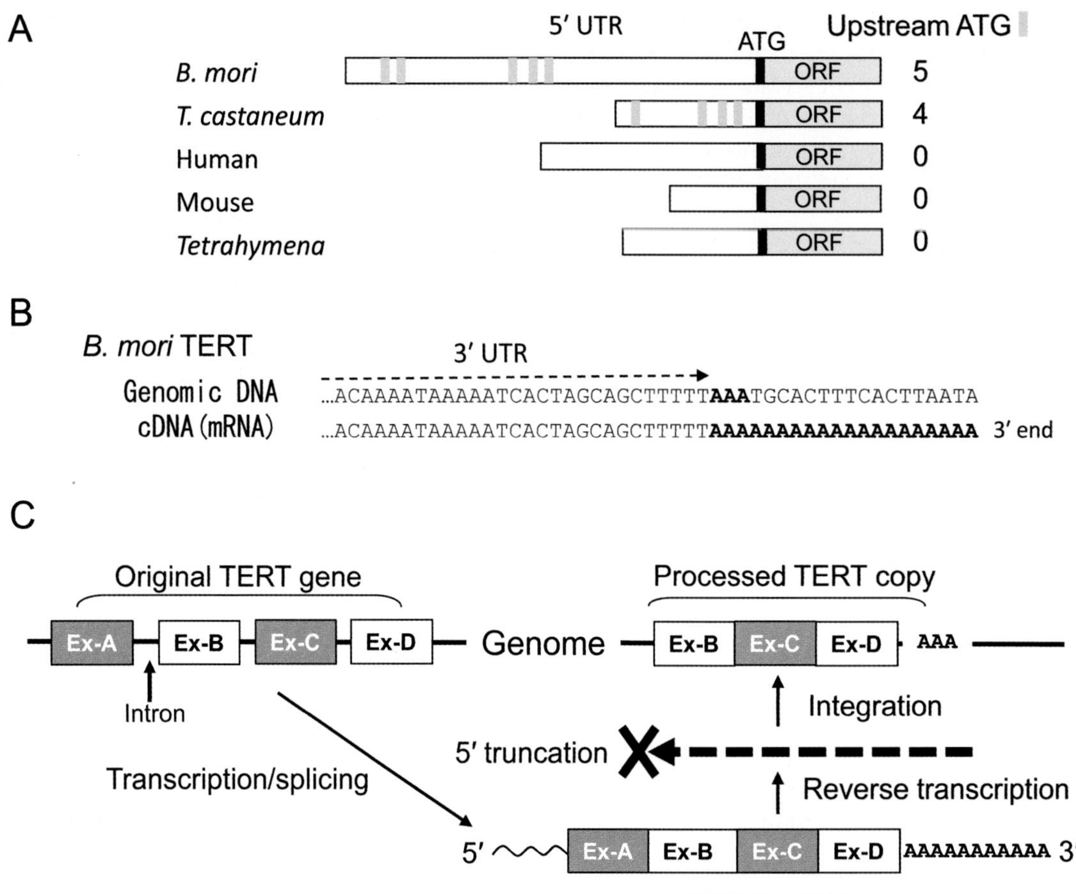

Figure 6 Structures of the silkworm TERT gene. (A) The position of upstream ATG in several TERT genes. The numbers of upstream ATG in each 5′ UTR of TERT are shown on the right. (B) Polyadenylation site of *B. mori* TERT. Poly (A) sequences at the 3′ ends of the TERT gene in genomic DNA and cDNA (corresponding to the mRNA sequence) are shown in bold. (C) The schematic model for the generation of a processed TERT gene from the original TERT with introns by reverse transcription. Four exons are shown as Ex-A to Ex-D. doi:10.1128/microbiolspec.MDNA3-0001-2014.f6

amino acid sequences of the RT and endonuclease domains, these elements and telomeric repeat-specific TRAS and SART are closely related (Fig. 4) (16, 61). The tree also shows that R1, TRAS, SART, and Waldo (ACAY-specific element) are ancient retrotransposons that branched at a very early stage. In fact, these elements exist in a relatively wide variety of insect and arthropod species. The 28S rDNA-specific RT and 18S rDNA-specific R7 are closely related and their target sequences are similar (Fig. 4).

R2, R8, and R9 are the R2 clade elements containing the RLE domain. R2 constitutes a large group that includes many retrotransposons with the same sequence specificity for 28S rDNA and is distributed over at least six animal phyla: *Arthropoda*, *Nematoda*, *Chordata*, *Echinodermata*, *Platyhelminthes*, and *Cnidaria* (17, 21, 22, 77, 83). R2 elements have been found recently in many fishes and in some birds and reptiles, but not in mammal genomes (22, 23, 84). The R2 clade can be further classified into four subclades, A, B, C, and D. The R2-A, -B, -C, and –D subclades have three, two, two, and one zinc-finger motif(s), respectively, in the N-terminal region (Fig. 2) (22, 23). R8 identified in *Hydra magnipapillata* in the *Cnidaria* phylum is categorized as a member of the R2-A subclade and is likely to have been derived from R2, changing its target specificity from 28S rDNA to 18S rDNA (Fig. 3A) (77). In contrast to the target changes between RT and R7 described above, we found no similarities between the sequences around the target sites of R2 and R8. The target recognition of longer sequence may explain a lack of target similarity between R2 and R8 (77). Recent reports by Christensen and coworkers suggest that the members of R2-A and R2-D subclades use different targeting mechanisms (85, 86).

R9 has been recently identified from bdelloid rotifers (81). R9 is a member of R2-A subclade and inserts within the newly characterized 28S rDNA target (Fig. 3A and Table 1). The long TSD (126 bp) found at the insertion site of R9 is unique; a short (up to 20 bp) TSD or a small deletion (R2Bm) are usually observed at the target sites. R4 is closely related to some elements that insert in TAA repeats and to some elements with no evident specificity (clade R4). Another rDNA-specific non-LTR-element, R5, is a member of NeSL clade.

MOLECULAR MECHANISM OF SEQUENCE-SPECIFIC RETROTRANSPOSITION

APE-Encoding Non-LTR Retrotransposons

APE Domain

The process of TPRT, which is unique to non-LTR retrotransposons, is initiated by the targeting endonuclease that cleaves one (bottom) strand of the target site and creates a free 3′-hydroxyl end, which is used as a primer for reverse transcription. Thus, target-site selection of sequence-specific non-LTR retrotransposons is primarily or mainly determined by the endonuclease domain itself (87).

L1Tc from *Trypanosoma cruzi* encodes a domain that has a true AP endonuclease activity; it has been shown to cleave apurinic/apyrimidinic (AP) sites, indicating a possible role of L1Tc in DNA repair (88). However, other APE-encoding non-LTR elements appear to have lost this intrinsic enzymatic activity. Detailed analysis of the APE domain of the human nonsequence-specific retrotransposon L1 has shown that it preferentially cleaves the T–A junction similar to *in vivo* target sequences (5′-TTTT/AA-3′) (11, 89). APEs purified from R1Bm and Tx1L APEs have been also shown to cleave the 28SrDNA and Tx1D, respectively, target sequences *in vitro* (90, 91). Purified APE of TRAS1 could generate nicks in a highly specific manner on both strands of the telomeric repeats, between T and A on the $(TTAGG)_n$ bottom strand and between C and T on the $(CCTAA)_n$ top strand (Fig. 4) (58). These sites are consistent with insertion sites expected from the genomic structure of boundary regions of TRAS1. Time-course studies of the nicking activity of TRAS1 APE showed that the cleavages of the bottom strand preceded those on the top strand (58). TRAS1 APE cleaves not only the bottom strand but also the top strand in a sequence-specific manner *in vitro*, while the mechanism of the top strand cleavage *in vivo* is not clarified. The cleavage of TRAS1 APE is not affected by the flanking sequence, suggesting that the target-site specificity of TRAS1 is mainly determined by its endonuclease domain. In R1Bm, Tx1L, and TRAS1, the bottom-strand cleavage is faster than the top-strand cleavage. This behavior is consistent with the TPRT model of Luan et al. (6), in which the bottom-strand cleavage defines the initial target site for reverse transcription.

Crystal structures of APEs from some non-LTR elements have revealed folding patterns similar to those of AP endonucleases (34, 59, 92). However, the structures have an extra beta-hairpin at the DNA-binding surface, which is a common feature among APEs of non-LTR elements (34, 59, 92). The crystal structure of APE of human L1 suggests that the prominent betaB6–betaB5 hairpin loop may insert into the DNA minor groove at the T–A junction, which is important for recognizing the DNA target (34). Further mutational approaches have revealed that variations in the loop sequence result in altered DNA-nicking profiles, including novel sites (34). Similarly, the extra beta hairpin (beta10–beta11) of APE of TRAS1 fits into the minor groove at the telomeric repeats, and Asp-130 in the hairpin is suggested to be involved in specific recognition of telomeric repeats (59). Structural analysis of APE of R1Bm has shown that mutations on the DNA binding surface decrease the cleavage activity but do not affect the sequence recognition in most residues. However, amino acids changes at Tyr-98 and Asn-180 had altered cleavage patterns of targets, suggesting important roles of these residues for the sequence recognition (92).

A novel retrotransposition assay system using baculovirus has demonstrated that APE-encoding elements, TRAS1 (60), SART1 (60), R1Bm (93), and R7 (94), are capable of *in vivo* retrotransposition into their specific targets. Using this system, replacing the APE domain of SART1 with the APE domain from TRAS1 changed the SART1 specificity to that of TRAS1, suggesting that the primary determinant of *in vivo* target selection is the APE domain (60). More recently, similar swapping experiments have shown that the APE domains of SARTTc1 and SARTBm1 are involved in recognition of the target site for $(TCAGG)_n$ and $(TTAGG)_n$ telomeric repeats, respectively (61). It has been also shown that protein-engineered APE of TRAS1 combined with the telomere-binding proteins can cleave the human telomeric repeat $(TTAGGG)_n$ strands in a sequence-specific manner (95).

mRNA/Target DNA Interaction

It is noteworthy that sequence-specific non-LTR elements usually target the same sequence in different repetitive units in the genome, in contrast to randomly integrated non-LTR elements. This means that a read-through mRNA product of sequence-specific elements can easily bind the same DNA sequence of a different repetitive target (Fig. 7A). Importantly, it has been suggested that some rDNA-specific non-LTR elements are transcribed by RNA polymerase I, not by RNA polymerase II (96). R2 of *D. simulans* can be co-transcribed with 28S rRNA and processed at or near the 5′ end of the retrotransposon unit by a self-cleavage mechanism (97, 98) (see Fig. 7B, which shows similar steps in R2Ol). The self-cleavage site in most R2 elements from different species was shown not at the 5′ end of the element but at 28S sequences up to 36 nucleotides upstream of the 5′ junction (98). It has been postulated that the co-transcribed 28S sequences of the first cDNA strand at the 5′ junction may anneal to the target site and prime the synthesis of the second DNA strand (98).

In R1Bm, the read-through transcript of 28S rRNA target sequence at the 3′ junction is suggested to be base-paired with the DNA target nicked by endonuclease of R1Bm and exposed as a "primer" of reverse transcription (93). During TPRT, this annealing of

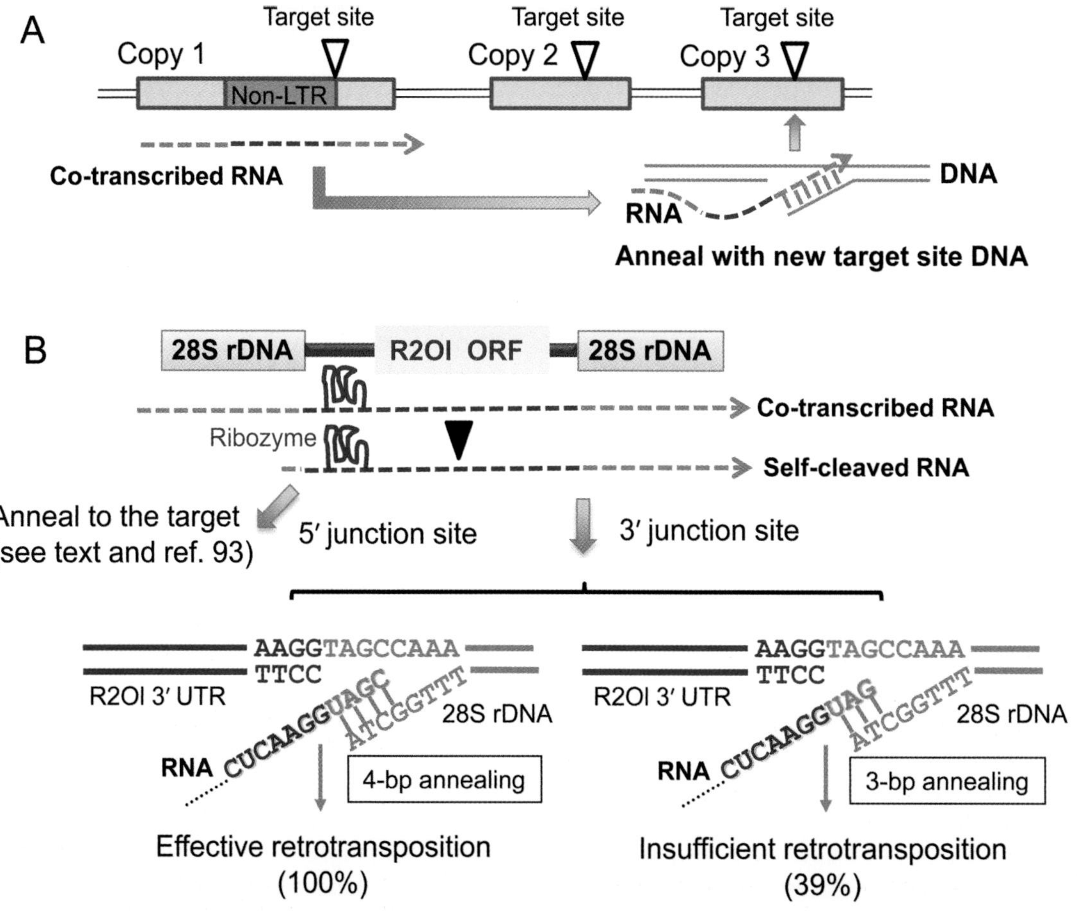

Figure 7 Schematic model for interaction between mRNA of a site-specific non-LTR element and target DNA at the 3′ junction. (A) Interaction between the read-through transcript of a non-LTR element (from copy1) and the target site DNA at the 3′ junction of copy 3. (B) Annealing of R2Ol mRNA with the target DNA (28S rDNA) at the 3′ junction. *In vitro*-transcribed R2Ol with 4 bp of the 28S target sequence at its 3′ end showed the accurate and efficient retrotransposition in the zebrafish embryo. However, the R2Ol mRNA with 3 bp of the 28S target sequence showed insufficient retrotransposition (39%) (H. Mizuno and H. Fujiwara, unpublished data). These observations indicate that annealing of the read-though product of R2Ol to the target DNA at the 3′ junction seems important for its efficient and accurate retrotransposition. doi:10.1128/microbiolspec.MDNA3-0001-2014.f7

mRNA to the exposed end of nicked DNA target places the RNA template at the accurate position for site-specific retrotransposition. Reverse transcription of R1Bm starts from the position next to the read-through UGU sequence, the 5′ end sequence of TSD (28S sequence adjacent to the 3′ end of R1Bm) (93). It has been also shown that R2Ol is co-transcribed with 28S rRNA, and its transcription appears to continue past the 3′ end of the retrotransposon unit in several tissues of the fish (H. Mizuno and H. Fujiwara, unpublished data) (Fig. 7B). A similar interaction between mRNA and target DNA and an accurate reverse transcription were observed at R2Ol, and at least a 4-bp interaction at the 3′ junction is necessary for its efficient retrotransposition (Fig. 7B).

It has been also suggested that base-pairing between the 3′-end sequence of the RNA and the DNA target supports the formation of the primer–template complex (90). The 3′-end sequence of CR1 RNA could be involved in target-site selection by hybridization to homologous sequences at nicked chromosomal sites (99, 100). In telomeric repeat-specific SART1Bm, short telomeric repeat-like GGUU sequences in the 3′ UTR of mRNA may anneal the bottom strand of $(TTAGG)_n$ repeats, which helps the element to access the target DNA (101).

Access to the Target Site

Most of ORF1p appears indispensable for the activity of APE-type elements (60, 102). ORF1 proteins of the mammalian L1 and the silkworm SART1 have been shown to form a ribonucleoprotein (RNP) complex with their mRNA (103, 104, 105), and contain nucleic acid chaperone activity (106, 107). In some site-specific non-LTR elements, evidence has been obtained that suggests ORF1p could be involved in gaining access to the target genomic site. ORF1 of telomere-specific non-LTR retrotransposons of *Drosophila*, HeT-A, TART, and TAHRE encodes a Gag-like protein and transports it back into the nucleus (108). The Gag-like protein from the HeT-A element localizes to chromosome ends in interphase nuclei (109). It has been suggested that the HeT-A ORF1p specifically localizes to telomeric ends in interphase nuclei and that the TART ORF1p moves to chromosome ends if assisted by the HeT-A ORF1p (109). ORF1p of TAHRE also requires the help of HeT-A ORF1p to localize to chromosome ends (110). A recent report has shown that ORF1p of HeT-A is expressed in the early S phase and forms spherical structures at telomeres undergoing replication (52). In the telomeric repeat-specific SART1 of the silkworm, ORF1p includes telomeric repeat interaction domains *in vitro* and may be found in the telomere dot structures in the nucleus (111).

Figure 8 summarizes each process of the retrotransposition of APE-encoding elements, especially focusing on three factors involved in the sequence specific integration.

Other Factors

Zinc knuckle-like motifs in ORF1

ORF1 proteins with one to three cysteine-rich motifs (CCHC motifs) encoded at carboxy terminal are found in many clades of APE-encoding non-LTR elements (Fig. 2 and Fig. 8). In retroviruses, the similar CCHC motifs in Gag proteins, called zinc knuckle motifs, participate in interaction between retroviral RNA and Gag proteins (112, 113), and thus the CCHC motifs in ORF1p of APE-encoding elements are also called zinc knuckle motifs (zinc knuckle-like motifs in this manuscript). It is intriguing that the zinc knuckle-like motif is conserved in both Tx1 and R1 clades, which include sequence-specific non-LTR elements (9). Detailed mutation analyses have shown that three CCHC motifs in ORF1p from SART1 are involved in the interaction with its mRNA in a sequence-specific manner; this suggests that the motifs may have played important roles in the evolution of closely related site-specific elements (105). Matsumoto et al. have also shown that the SART1 ORF1p includes domains for the ORF1p–ORF1p and ORF1p–ORF2p interactions (105). These data suggest that the formation of SART1 RNP complex composed of ORF1p, ORF2p, and mRNA is mediated by the ORF1p.

Zinc finger-like motifs in ORF2

The ORF2 proteins of most APE-type retrotransposons encode at least one CCHC motif at their C-terminal regions, but the structure is different from that of the CCHC motifs in ORF1 (Fig. 2 and Fig. 8) (105, 114). This motif in ORF2 is also conserved in both Tx1 and R1 clades. Point mutations in this motif in human L1 and *Bombyx* TRAS1 cause the loss of the retrotransposition activity (60, 102). However, the elements from the CR1 group, which has no CCHC in OFR2, retain the retroposition activity (114). It is generally assumed that the zinc finger (ZF)-like motif is involved in the interaction with the RNA template, DNA target, or proteins, but the function of CCHC in ORF2 has not been elucidated yet.

Myb-like domain

A Myb-like domain is found between endonuclease and RT domains in ORF2 of telomere-specific TRAS

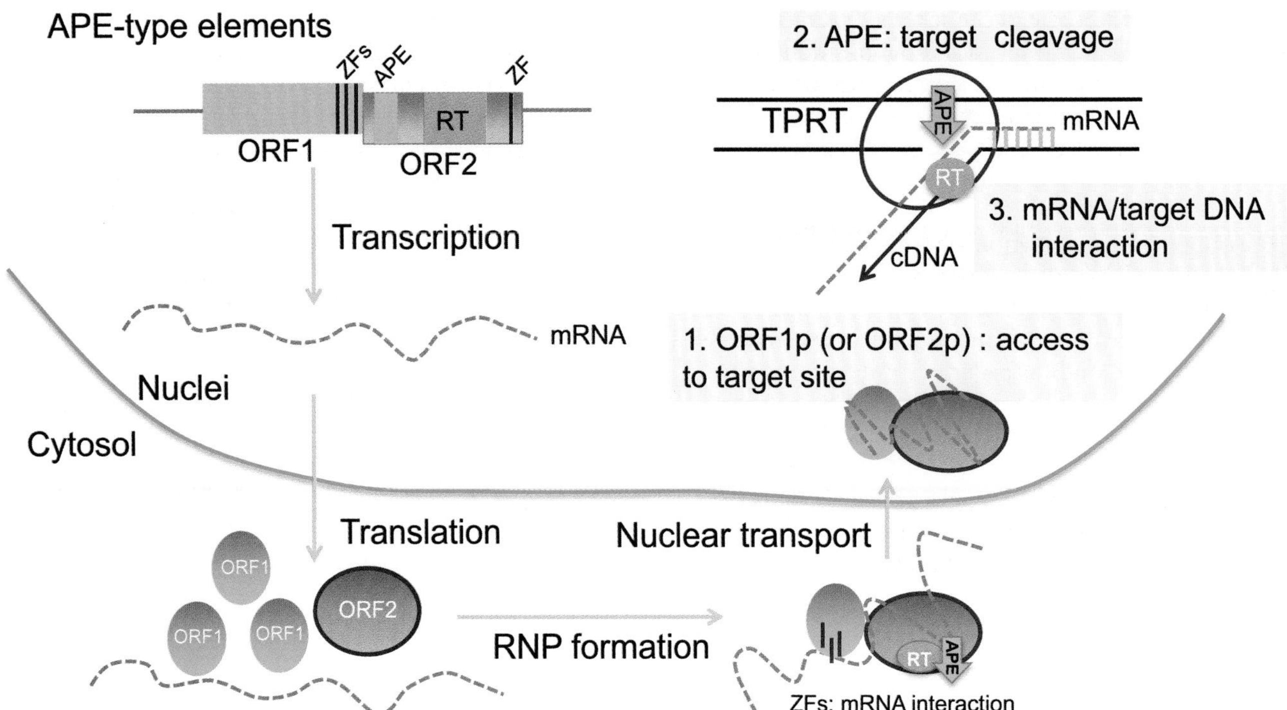

Figure 8 Main factors involved in the target-site selection of site-specific APE-encoding elements. Four main steps of the retrotransposition of non-LTR elements in the cells (transcription, translation, RNP formation, and nuclear import) are shown schematically. Three factors involved in the target-site selection are indicated.
doi:10.1128/microbiolspec.MDNA3-0001-2014.f8

elements (Fig. 2) (56). Many telomere-binding proteins, like RAP1, TAZ1, TRF1, and TRF2 share the Myb-like three-helix motif (115). Myb-like motifs in TRAS1 may interact with the telomere structure, although further study is needed to confirm this assumption.

RLE-Encoding Non-LTR Retrotransposons

Many of the RLE-encoding non-LTR retrotransposons are site-specific. All clades of RLE-encoding non-LTR elements (R2, R4, NeSL, CRE, and HERO) encode some DNA-binding motifs, one to three ZFs at the N-terminal end and one CCHC (cysteine–histidine) motif C-terminal to the RT domain within a multifunctional single ORF. R2-clade elements also encode a Myb-like domain (Fig. 2). In contrast to APE-retrotransposons, the functional roles of RLE-type endonuclease itself in sequence specificity is not known clearly to date (86, 116, 117), and the target-site recognition is achieved primarily through the DNA-binding motifs mentioned above.

R2 elements, particularly R2Dm and R2Bm, have been used as model systems to investigate the mechanism of target-site selection of RLE retrotransposons. R2Bm (classified as subclade R2-D) encodes a single ZF motif, while some other R2 elements, such as R2Lp, R2Ol, R8, and R9 (classified as subclade R2-A), encode three ZF motifs. A study of R2Bm has shown that two subunits of the ORF protein are bound to the R2Bm mRNA (86, 118). One protein subunit at the 3′ UTR of the RNA is bound upstream of the insertion site of the target DNA through the function of an undetermined domain. It has been reported that the RLE of the upstream protein cleaves the bottom strand of the 28S target DNA and the RT domain of this protein catalyzes first-strand cDNA synthesis using the nicked DNA as a primer (118, 119). The other protein subunit bound to the 5′ UTR of the RNA is bound downstream from the insertion site through the N-terminal ZF and Myb motifs and may be involved in the second DNA-strand synthesis (86, 120). Myb is a major contributor to the specificity and ZF provides some DNA contacts in R2, while other clades of RLE elements have no Myb motif (85, 86, 120). Recent studies suggest that R2-A clade elements, such as R2Lp and R9, may use Myb and ZF (mainly the third ZF from the N-terminal) motifs to bind to upstream sequences of the insertion site (86). The different binding modes of R2-A and R2-D elements suggest a plasticity of integration mechanisms of R2-clade elements. The authors of the study also extend the targeting mechanism of R2 elements to the members of another RLE-clade NeSL element (86).

APPLICATION OF SEQUENCE SPECIFIC RETROTRANSPOSONS AS A GENE DELIVERY TOOL

Precise transgene integration to a specific target site is necessary to avoid unpredictable side effects, particularly in therapeutic applications for human cells. Recently developed novel methods such as TALEN (transcription activator-like effector nuclease) or CRISPR (clustered regularly interspaced short palindromic repeats)/Cas9 have been used to cleave and edit the genome regions in a sequence-specific manner, but the transgene integration achieved using these methods is still ongoing (121, 122). Also, transposons combined with ZF protein or adeno-associated virus proteins have been used as target-specific gene delivery tools (123).

Site-specific non-LTR retrotransposons may be good candidates for the sequence-specific gene delivery (94). The site-specific non-LTR elements TRAS, SART, R1, R2, R7, and R8, when recombined in the AcNPV vector, could integrate into their respective specific targets very effectively. Many non-LTR elements recognize the 3′ UTR of their mRNA during the initial step of the TPRT process. A sequence such as an EGFP reporter with an added 3′ UTR sequence can be retrotransposed in a target-specific manner with a helper non-LTR element by trans-complementation, a useful feature for a gene-delivery tool (61, 101). Baculovirus AcNPV has a wide host range and can infect many species, including human cells (124). It has been shown that R1 and SART1 recombined in AcNPV can integrate at specific target sites in the genome of various tissues when injected into the larva of *B. mori* (125); this gene-transfer system is also useful in the honey bee (126).

In vitro-transcribed R2Ol retrotransposes into the 28S rDNA target sequence of the zebrafish embryo with a high frequency and accuracy and is transmitted to the next generation (A. Kuroki and H. Fujiwara, unpublished data). It has been shown that fluorescent signals are seen in the G2 generation of transgenic fish whose 28S gene has been integrated with R2Ol carrying the reporter EGFP gene in its 3′ UTR. Moreover, the adenovirus-mediated and AcNPV-mediated R2Ol retrotransposition results in an accurate integration into the 28S target in several types of cultured human cells. These results demonstrate that site-specific non-LTR retrotransposons, such as R2Ol, could become a novel type of sequence-specific gene-delivery vectors in gene-therapy application.

Acknowledgments. *I gratefully appreciate Dr. K. K. Kojima for a critical reading of the manuscript and for helpful comments. My research was supported by grants from the Ministry of Education, Culture, Sports and Technology of Japan.*

Citation. Fujiwara H. 2014. Site-specific non-LTR retrotransposons. Microbiol Spectrum 3(2):MDNA3-0001-2014.

References

1. **Craig NL.** 2002. Tn7, p 423–456 *In* Craig NL, Cragie R, Gellert M, Lambowitz AM (ed), *Mobile DNA II*. ASM Press, Washington, DC.
2. **Rio DC.** 2002. P transposable elements in *Drosophila melanogaster*, p 484–518. *In* Craig NL, Cragie R, Gellert M, Lambowitz AM (ed), *Mobile DNA II*. ASM Press, Washington, DC.
3. **Kajikawa M, Okada N.** 2002. LINEs mobilize SINEs in the eel through a shared 3′ sequence. *Cell* **111:**433–444.
4. **Dewannieux M, Heidmann T.** 2005. LINEs, SINEs and processed pseudogenes: parasitic strategies for genome modeling. *Cytogenet Genome Res* **110:**35–48.
5. **Roy-Engel AM.** 2012. A tale of an A-tail: the lifeline of SINE. *Mob Genet Elements* **2:**282–286.
6. **Luan DD, Korman MH, Jakubczak JL, Eickbush TH.** 1993. Reverse transcription of R2Bm RNA is primed by a nick at the chromosomal target site: a mechanism for non-LTR retrotransposition. *Cell* **72:**595–605.
7. **Eickbush TM.** 2002. R2 and related site-specific non-long terminal repeat retrotransposons, p 813–835. *In* Craig NL, Cragie R, Gellert M, Lambowitz AM (ed), *Mobile DNA II*. ASM Press, Washington, DC.
8. **Malik HS, Burke WD, Eickbush TH.** 1999. The age and evolution of non-LTR retrotransposable elements. *Mol Biol Evol* **16:**793–805.
9. **Kapitonov VV, Tempel S, Jurka J.** 2009. Simple and fast classification of non-LTR retrotransposons based on phylogeny of their RT domain protein sequences. *Gene* **448:**207–213.
10. **Yang J, Malik HS, Eickbush TH.** 1999. Identification of the endonuclease domain encoded by R2 and other site-specific, non-long terminal repeat retrotransposable elements. *Proc Natl Acad Sci U S A* **96:**7847–7852.
11. **Feng Q, Moran JV, Kazazian HH Jr, Boeke JD.** 1996. Human L1 retrotransposon encodes a conserved endonuclease required for retrotransposition. *Cell* **87:**905–916.
12. **Kojima KK, Fujiwara H.** 2005. An extraordinary retrotransposon family encoding dual endonucleases. *Genome Res* **15:**1106–1117.
13. **Huang CR, Burns KH, Boeke JD.** 2012. Active transposition in genomes. *Annu Rev Genet* **46:**651–675.
14. **Cordaux R, Batzer MA.** 2009. The impact of retrotransposons on human genome evolution. *Nat Rev Genet* **10:**691–703.

15. **Rodić N, Burns KH.** 2013. Long interspersed element-1 (LINE-1): passenger or driver in human neoplasms? *PLoS Genet* **9:**e1003402.
16. **Kojima KK, Fujiwara H.** 2003. Evolution of target specificity in R1 clade non-LTR retrotransposons. *Mol Biol Evol* **20:**351–361.
17. **Kojima KK, Fujiwara H.** 2004. Cross-genome screening of novel sequence-specific non-LTR retrotransposons: various multicopy RNA genes and microsatellites are selected as targets. *Mol Biol Evol* **21:** 207–217.
18. **Aksoy S, Williams S, Chang S, Richards FF.** 1990. SLACS retrotransposon from *Trypanosoma brucei gambiense* is similar to mammalian LINEs. *Nucleic Acids Res* **18:**785–792.
19. **Gabriel A, Yen TJ, Schwartz DC, Smith CL, Boeke JD, Sollner-Webb B, Cleveland DW.** 1990. A rapidly rearranging retrotransposon within the miniexon gene locus of *Crithidia fasciculata*. *Mol Cell Biol* **10:**615–624.
20. **Malik HS, Eickbush TH.** 2000. NeSL-1, an ancient lineage of site-specific non-LTR retrotransposons from *Caenorhabditis elegans*. *Genetics* **154:**193–203.
21. **Jakubczak JL, Burke WD, Eickbush TH.** 1991. Retrotransposable elements R1 and R2 interrupt the rRNA genes of most insects. *Proc Natl Acad Sci U S A* **88:** 3295–3299.
22. **Kojima KK, Fujiwara H.** 2005. Long-term inheritance of the 28S rDNA-specific retrotransposon R2. *Mol Biol Evol* **22:**2157–2165.
23. **Luchetti A, Mantovani B.** 2013. Non-LTR R2 element evolutionary patterns: phylogenetic incongruences, rapid radiation and the maintenance of multiple lineages. *PLoS One* **8:**e57076.
24. **Burke WD, Müller F, Eickbush TH.** 1995. R4, a non-LTR retrotransposon specific to the large subunit rRNA genes of nematodes. *Nucleic Acids Res* **23:**4628–4634.
25. **Xiong Y, Eickbush TH.** 1993. Dong, a non-long terminal repeat (non-LTR) retrotransposable element from *Bombyx mori*. *Nucleic Acids Res* **21:**1318.
26. **Kapitonov VV, Jurka J.** 2014. A family of HERO non-LTR retrotransposons from the Californian leech genome. *Repbase Reports* **14:**311.
27. **Jurka J.** 1997. Sequence patterns indicate an enzymatic involvement in integration of mammalian retroposons. *Proc Natl Acad Sci USA* **94:**1872–1877.
28. **Kapitonov VV, Jurka J.** 2009. *R5-2_SM*, http://www.girinst.org/protected/repbase_extract.php?access=R5-2_SM, Repbase Update (04-Aug-2009)
29. **Kojima KK, Jurka J.** 2013. R4, a 28S ribosomal RNA gene-specific non-LTR retrotransposon family from nematodes. *Repbase Reports* **13:**821–834.
30. **Kapitonov VV, Jurka J.** 2009. First examples of CRE non-LTR retrotransposons in animals. *Repbase Reports* **9:**2157–2160.
31. **Gilbert N, Lutz-Prigge S, Moran JV.** 2002. Genomic deletions created upon LINE-1 retrotransposition. *Cell* **110:**315–325.
32. **Symer DE, Connelly C, Szak ST, Caputo EM, Cost GJ, Parmigiani G, Boeke JD.** 2002. Human l1 retrotransposition is associated with genetic instability *in vivo*. *Cell* **110:**327–338.
33. **Martín F, Marañón C, Olivares M, Alonso C, López MC.** 1995. Characterization of a non-long terminal repeat retrotransposon cDNA (L1Tc) from *Trypanosoma cruzi*: homology of the first ORF with the ape family of DNA repair enzymes. *J Mol Biol* **247:**49–59.
34. **Weichenrieder O, Repanas K, Perrakis A.** 2004. Crystal structure of the targeting endonuclease of the human LINE-1 retrotransposon. *Structure* **12:**975–986.
35. **Kojima KK, Jurka J.** 2013. Telomere-specific Tx1 non-LTR retrotransposons from green anole. *Repbase Reports* **13:**843.
36. **Garrett JE, Knutzon DS, Carroll D.** 1989. Composite transposable elements in the *Xenopus laevis* genome. *Mol Cell Biol* **9:**3018–3027.
37. **Blackburn EH.** 1991. Structure and function of telomeres. *Nature* **350:**569–573.
38. **Lundblad V, Wright WE.** 1996. Telomeres and telomerase: a simple picture becomes complex. *Cell* **87:** 369–375.
39. **Fujiwara H, Osanai M, Matsumoto T, Kojima KK.** 2005. Telomere-specific non-LTR retrotransposons and telomere maintenance in the silkworm, *Bombyx mori*. *Chromosome Res* **13:**455–467.
40. **Fujiwara H.** 2014. Accumulation of telomeric-repeat-specific retrotransposons in subtelomere of *Bombyx mori* and *Tribolium castaneum*, p 227–241. *In* Louis EJ, Becker MM (ed), *Subtelomeres*. Springer, Berlin.
41. **Biessmann H, Valgeirsdottir K, Lofsky A, Chin C, Ginther B, Levis RW, Pardue ML.** 1992. HeT-A, a transposable element specifically involved in "healing" broken chromosome ends in *Drosophila melanogaster*. *Mol Cell Biol* **12:**3910–3918.
42. **Levis RW, Ganesan R, Houtchens K, Tolar LA, Sheen FM.** 1993. Transposons in place of telomeric repeats at a *Drosophila* telomere. *Cell* **75:**1083–1093.
43. **Pardue ML, DeBaryshe PG.** 2003. Retrotransposons provide an evolutionarily robust non-telomerase mechanism to maintain telomeres. *Annu Rev Genet* **37:** 485–511.
44. **Cohn M, Edström JE.** 1992. Telomere-associated repeats in *Chironomus* form discrete subfamilies generated by gene conversion. *J Mol Evol* **35:**114–122.
45. **Roth CW, Kobeski F, Walter MF, Biessmann H.** 1997. Chromosome end elongation by recombination in the mosquito *Anopheles gambiae*. *Mol Cell Biol* **17:** 5176–5183.
46. **Abad JP, De Pablos B, Osoegawa K, De Jong PJ, Martín-Gallardo A, Villasante A.** 2004. TAHRE, a novel telomeric retrotransposon from *Drosophila melanogaster*, reveals the origin of *Drosophila* telomeres. *Mol Biol Evol* **21:**1620–1624.
47. **Pardue ML, DeBaryshe PG.** 2002. Telomeres and transposable elements, p 870–887. *In* Craig NL, Cragie R, Gellert M, Lambowitz AM (ed), *Mobile DNA II*, ASM Press, Washington, DC.

48. **Pardue ML, DeBaryshe PG.** 2012. Retrotransposons that maintain chromosome ends. *Proc Natl Acad Sci U S A* **108:**20317–20324.

49. **Silva-Sousa R, Casacuberta E.** 2013. The JIL-1 kinase affects telomere expression in the different telomere domains of *Drosophila*. *PLoS One* **8:**e81543.

50. **Capkova Frydrychova R, Biessmann H, Mason JM.** 2008. Regulation of telomere length in *Drosophila*. *Cytogenet Genome Res* **122:**356–364.

51. **Raffa GD, Cenci G, Ciapponi L, Gatti M.** 2013. Organization and evolution of *Drosophila* terminin: similarities and differences between *Drosophila* and human telomeres. *Front Oncol* **3:**e112.

52. **Zhang L, Beaucher M, Cheng Y, Rong YS.** 2014. Coordination of transposon expression with DNA replication in the targeting of telomeric retrotransposons in *Drosophila*. *EMBO J* **33:**1148–1158.

53. **Okazaki S, Tsuchida K, Maekawa H, Ishikawa H, Fujiwara H.** 1993. Identification of a pentanucleotide telomeric sequence, $(TTAGG)_n$, in the silkworm, *Bombyx mori* and in other insects. *Mol Cell Biol* **13:** 1424–1432.

54. **Okazaki S, Ishikawa H, Fujiwara H.** 1995. Structural analysis of TRAS1, a novel family of telomeric repeat associated retrotransposons in the silkworm, *Bombyx mori*. *Mol Cell Biol* **15:**4545–4552.

55. **Takahashi H, Okazaki S, Fujiwara H.** 1997. A new family of site-specific retrotransposons, SART1, is inserted into telomeric repeats of the silkworm, *Bombyx mori*. *Nucleic Acids Res* **25:**1578–1584.

56. **Kubo Y, Okazaki S, Anzai T, Fujiwara H.** 2001. Structural and phylogenetic analysis of TRAS, telomeric repeat-specific non-LTR retrotransposon families in lepidopteran insects. *Mol Biol Evol* **18:**848–857.

57. **Osanai-Futahashi M, Suetsugu Y, Mita K, Fujiwara H.** 2008. Genome-wide screening and characterization of transposable elements and their distribution analysis in the silkworm, *Bombyx mori*. *Insect Biochem Mol Biol* **38:**1046–1146.

58. **Anzai T, Takahashi H, Fujiwara H.** 2001. Sequence-specific recognition and cleavage of telomeric repeat $(TTAGG)_n$ by endonuclease of non-LTR retrotransposon, TRAS1. *Mol Cell Biol* **21:**100–108.

59. **Maita N, Anzai T, Aoyagi H, Mizuno H, Fujiwara H.** 2004. Crystal structure of the endonuclease domain encoded by the telomere-specific LINE, TRAS1. *J Biol Chem* **279:**41067–41076.

60. **Takahashi H, Fujiwara H.** 2002. Transplantation of target site specificity by swapping endonuclease domains of two LINEs. *EMBO J* **21:**408–417.

61. **Osanai-Futahashi M, Fujiwara H.** 2011. Coevolution of telomeric repeats and telomeric-repeat-specific non-LTR retrotransposons in insects. *Mol Biol Evol* **28:**2983–2986.

62. **Monti V, Serafini C, Manicardi GC, Mandrioli M.** 2013. Characterization of non-LTR retrotransposable TRAS elements in the aphids *Acyrthosiphon pisum* and *Myzus persicae* (Aphididae, Hemiptera). *J Hered* **104:** 547–553.

63. **Takahashi H, Fujiwara H.** 1999. Transcription analysis of the telomeric repeat-specific retrotransposons TRAS1 and SART1 of the silkworm *Bombyx mori*. *Nucleic Acids Res* **27:**2015–2021.

64. **Kojima KK, Matsumoto T, Fujiwara H.** 2005. Eukaryotic translational coupling in UAAUG stop-start codons for the bicistronic RNA translation of the non-long terminal repeat retrotransposon SART1. *Mol Cell Biol* **25:** 7675–7686.

65. **Osanai M, Kojima KK, Futahashi R, Yaguchi S, Fujiwara H.** 2006. Identification and characterization of the telomerase reverse transcriptase of *Bombyx mori* (silkworm) and *Tribolium castaneum* (flour beetle). *Gene* **376:**281–289.

66. **Sasaki T, Fujiwara H.** 2000. Detection and distribution patterns of telomerase activity in insects. *Eur J Biochem* **267:**3025–3031.

67. **Mitchell M, Gillis A, Futahashi M, Fujiwara H, Skordalakes E.** 2010. Structural basis for telomerase catalytic subunit TERT binding to RNA template and telomeric DNA. *Nat Struct Mol Biol* **17:**513–518.

68. **Jin X, Turcott E, Englehardt S, Mize GJ, Morris DR.** 2003. The two upstream open reading frames of oncogene mdm2 have different translational regulatory properties. *J Biol Chem* **278:**25716–25721.

69. **Moriarty TJ, Marie-Egyptienne DT, Autexier C.** 2004. Functional organization of repeat addition processivity and DNA synthesis determinants in the human telomerase multimer. *Mol Cell Biol* **24:**3720–3733.

70. **Lue NF.** 2005. A physical and functional constituent of telomerase anchor site. *J Biol Chem* **280:**26586–26591.

71. **Savard J, Tautz D, Richards S, Weinstock GM, Gibbs RA, Werren JH, Tettelin H, Lercher MJ.** 2006. Phylogenomic analysis reveals bees and wasps (Hymenoptera) at the base of the radiation of holometabolous insects. *Genome Res* **16:**1334–1338.

72. **Tribolium Genome Sequencing Consortium.** 2008. The genome of the model beetle and pest *Tribolium castaneum*. *Nature* **452:**949–955.

73. **Robertson HM, Gordon KH.** 2006. Canonical TTAGG-repeat telomeres and telomerase in the honey bee, *Apis mellifera*. *Genome Res* **16:**1345–1351.

74. **Higashiyama T, Noutoshi Y, Fujie M, Yamada T.** 1997. Zepp, a LINE-like retrotransposon accumulated in the *Chlorella* telomeric region. *EMBO J* **16:**3715–3723.

75. **Arkhipova IR, Morrison HG.** 2001. Three retrotransposon families in the genome of *Giardia lamblia*: two telomeric, one dead. *Proc Natl Acad Sci U S A* **98:** 14497–14502.

76. **Starnes JH, Thornbury DW, Novikova OS, Rehmeyer CJ, Farman ML.** 2012. Telomere-targeted retrotransposons in the rice blast fungus *Magnaporthe oryzae*: agents of telomere instability. *Genetics* **191:**389–406.

77. **Kojima KK, Kuma K, Toh H, Fujiwara H.** 2006. Identification of rDNA-specific non-LTR retrotransposons in *Cnidaria*. *Mol Biol Evol* **23:**1984–1994.

78. **Fujiwara H, Ogura T, Takada N, Miyajima N, Ishikawa H, Maekawa H.** 1984. Introns and their

flanking sequences of *Bombyx mori* rDNA. *Nucleic Acids Res* **12:**6861–6869.

79. **Xiong Y, Eickbush TH.** 1988. The site-specific ribosomal DNA insertion element R1Bm belongs to a class of non-long-terminal-repeat retrotransposons. *Mol Cell Biol* **8:**114–123.
80. **Burke WD, Singh D, Eickbush TH.** 2003. R5 retrotransposons insert into a family of infrequently transcribed 28S rRNA genes of *Planaria*. *Mol Biol Evol* **20:** 1260–1270.
81. **Gladyshev EA, Arkhipova IR.** 2009. Rotifer rDNA-specific R9 retrotransposable elements generate an exceptionally long target site duplication upon insertion. *Gene* **448:**145–150.
82. **Besansky NJ, Paskewitz SM, Hamm DM, Collins FH.** 1992. Distinct families of site-specific retrotransposons occupy identical positions in the rRNA genes of *Anopheles gambiae*. *Mol Cell Biol* **12:**5102–5110.
83. **Kapitonov VV, Jurka J.** 2009a. A family of R2 non-LTR retrotransposons in the non-segmented roundworm genome. *Repbase Reports* **9:**1150.
84. **Kapitonov VV, Jurka J.** 2009b. R2 non-LTR retrotransposons in the bird genome. *Repbase Reports* **9:**1329.
85. **Thompson BK, Christensen SM.** 2011. Independently derived targeting of 28S rDNA by A- and D-clade R2 retrotransposons: plasticity of integration mechanism. *Mob Genet Elements* **1:**29–37.
86. **Shivram H, Cawley D, Christensen SM.** 2011. Targeting novel sites: the N-terminal DNA binding domain of non-LTR retrotransposons is an adaptable module that is implicated in changing site specificities. *Mob Genet Elements* **1:**169–178.
87. **Zingler N, Weichenrieder O, Schumann GG.** 2005. APE-type non-LTR retrotransposons: determinants involved in target site recognition. *Cytogenet Genome Res* **110:**250–268.
88. **Olivares M, Thomas MC, Alonso C, López MC.** 1999. The L1Tc, long interspersed nucleotide element from *Trypanosoma cruzi*, encodes a protein with 3′-phosphatase and 3′-phosphodiesterase enzymatic activities. *J Biol Chem* **274:**23883–22886.
89. **Cost GJ, Boeke JD.** 1998. Targeting of human retrotransposon integration is directed by the specificity of the L1 endonuclease for regions of unusual DNA structure. *Biochemistry* **37:**18081–18093.
90. **Feng Q, Schumann G, Boeke JD.** 1998. Retrotransposon R1Bm endonuclease cleaves the target sequence. *Proc Natl Acad Sci U S A* **95:**2083–2088.
91. **Christensen S, Pont-Kingdon G, Carroll D.** 2000. Target specificity of the endonuclease from the *Xenopus laevis* non-long terminal repeat retrotransposon, Tx1L. *Mol Cell Biol* **20:**1219–1226.
92. **Maita N, Aoyagi H, Osanai M, Shirakawa M, Fujiwara H.** 2007. Characterization of the sequence specificity of the R1Bm endonuclease domain by structural and biochemical studies. *Nucleic Acids Res* **35:**3918–3927.
93. **Anzai T, Osanai M, Hamada M, Fujiwara H.** 2005. Functional roles of read-through 28S rRNA sequence in *in vivo* retrotransposition of non-LTR retrotransposon, R1Bm. *Nucleic Acids Res* **33:**1993–2002.
94. **Schumann GG, Elena V, Gogvadze EV, Osanai-Futahashi M, Kuroki A, Münk C, Fujiwara H, Ivics Z, Buzdin A.** 2010. Unique functions of repetitive transcriptome. *Int Rev Cell Mol Biol* **285:**115–188.
95. **Yoshitake K, Aoyagi H, Fujiwara H.** 2010. Creation of a novel telomere-cutting endonuclease based on the EN domain of telomere-specific non-LTR retrotransposon, TRAS1. *Mobile DNA* **1:**e13.
96. **Ye J, Eickbush TH.** 2006. Chromatin structure and transcription of the R1- and R2-inserted rRNA genes of *Drosophila melanogaster*. *Mol Cell Biol* **26:**8781–8790.
97. **Eickbush DG, Eickbush TH.** 2010. R2 retrotransposons encode a self-cleaving ribozyme for processing from an rRNA cotranscript. *Mol Cell Biol* **30:**3142–3150.
98. **Eickbush DG, Burke WD, Eickbush TH.** 2013. Evolution of the R2 retrotransposon ribozyme and its self-cleavage site. *PLoS One* **8:**e66441.
99. **Burch JB, Davis DL, Haas NB.** 1993. Chicken repeat 1 elements contain a pol-like open reading frame and belong to the non-long terminal repeat class of retrotransposons. *Proc Natl Acad Sci U S A* **90:**8199–8203.
100. **Volff JN, Körting C, Sweeney K, Schartl M.** 1999. The non-LTR retrotransposon Rex3 from the fish *Xiphophorus* is widespread among teleosts. *Mol Biol Evol* **16:**1427–1438.
101. **Osanai M, Takahashi H, Kojima KK, Hamada M, Fujiwara H.** 2004. Novel motifs in 3′-untranslated region required for precise reverse transcription start of telomere specific LINE, SART1. *Mol Cell Biol* **24:** 7902–7913.
102. **Moran JV, Holmes SE, Naas TP, DeBerardinis RJ, Boeke JD, Kazazian HH Jr.** 1996. High frequency retrotransposition in cultured mammalian cells. *Cell* **87:** 917–927.
103. **Hohjoh H, Singer MF.** 1997. Sequence-specific single-strand RNA binding protein encoded by the human LINE-1 retrotransposon. *EMBO J* **16:**6034–6043.
104. **Doucet AJ, Hulme AE, Sahinovic E, Kulpa DA, Moldovan JB, Kopera HC, Athanikar JN, Hasnaoui M, Bucheton A, Moran JV, Gilbert N.** 2010. Characterization of LINE-1 ribonucleoprotein particles. *PLoS Genet* **6:**e1001150.
105. **Matsumoto T, Hamada M, Osanai M, Fujiwara H.** 2006. Essential domains for ribonucleoprotein complex formation required for retrotransposition of a telomere specific non-LTR retrotransposon SART1. *Mol Cell Biol* **26:**5168–5179.
106. **Martin SL, Bushman D, Wang F, Li PW, Walker A, Cummiskey J, Branciforte D, Williams MC.** 2008. A single amino acid substitution in ORF1 dramatically decreases L1 retrotransposition and provides insight into nucleic acid chaperone activity. *Nucleic Acids Res* **36:**5845–5854.
107. **Martin SL.** 2010. Nucleic acid chaperone properties of ORF1p from the non-LTR retrotransposon, LINE-1. *RNA Biol* **7:**706–711.

108. **Rashkova S, Karam SE, Pardue ML.** 2002. Element-specific localization of *Drosophila* retrotransposon Gag proteins occurs in both nucleus and cytoplasm. *Proc Natl Acad Sci U S A* **99:**3621–3626.

109. **Rashkova S, Karam SE, Kellum R, Pardue ML.** 2002. Gag proteins of the two *Drosophila* telomeric retrotransposons are targeted to chromosome ends. *J Cell Biol* **159:**397–402.

110. **Fuller AM, Cook EG, Kelley KJ, Pardue ML.** 2010. Gag proteins of *Drosophila* telomeric retrotransposons: collaborative targeting to chromosome ends. *Genetics* **184:** 629–636.

111. **Matsumoto T, Takahashi H, Fujiwara H.** 2004. Targeted nuclear import of ORF1 protein is required for *in vivo* retrotransposition of telomere-specific non-long terminal repeat retrotransposon, SART1. *Mol Cell Biol* **24:**105–122.

112. **Williams MC, Gorelick RJ, Musier-Forsyth K.** 2002. Specific zinc-finger architecture required for HIV-1 nucleocapsid protein's nucleic acid chaperone function. *Proc Natl Acad Sci U S A* **99:**8614–8619.

113. **D'Souza V, Summers MF.** 2004. Structural basis for packaging the dimeric genome of Moloney murine leukaemia virus. *Nature* **431:**586–590.

114. **Kajikawa M, Ohshima K, Okada N.** 1997. Determination of the entire sequence of turtle CR1: the first open reading frame of the turtle CR1 element encodes a protein with a novel zinc finger motif. *Mol Biol Evol* **14:** 1206–1217.

115. **König P, Fairall L, Rhodes D.** 1998. Sequence-specific DNA recognition by the Myb-like domain of the human telomere binding protein TRF1: a model for the protein–DNA complex. *Nucleic Acids Res* **26:**1731–1740.

116. **Mandal PK, Bagchi A, Bhattacharya A, Bhattacharya S.** 2004. An *Entamoeba histolytica* LINE/SINE pair inserts at common target sites cleaved by the restriction enzyme-like LINE-encoded endonuclease. *Eukaryot Cell* **3:**170–179.

117. **Volff JN, Körting C, Froschauer A, Sweeney K, Schartl M.** 2001. Non-LTR retrotransposons encoding a restriction enzyme-like endonuclease in vertebrates. *J Mol Evol* **52:**351–360.

118. **Christensen SM, Ye J, Eickbush TH.** 2006. RNA from the 5′ end of the R2 retrotransposon controls R2 protein binding to and cleavage of its DNA target site. *Proc Natl Acad Sci U S A* **103:**17602–17607.

119. **Christensen SM, Eickbush TH.** 2005. R2 target-primed reverse transcription: ordered cleavage and polymerization steps by protein subunits asymmetrically bound to the target DNA. *Mol Cell Biol* **25:**6617–6628.

120. **Christensen SM, Bibillo A, Eickbush TH.** 2005. Role of the *Bombyx mori* R2 element N-terminal domain in the target-primed reverse transcription (TPRT) reaction. *Nucleic Acids Res* **33:**6461–6468.

121. **Sun N, Abil Z, Zhao H.** 2012. Recent advances in targeted genome engineering in mammalian systems. *Biotechnol J* **7:**1074–1087.

122. **Gaj T, Gersbach CA, Barbas CF III.** 2013. ZFN, TALEN, and CRISPR/Cas-based methods for genome engineering. *Trends Biotechnol* **31:**397–405.

123. **Ammar I, Gogol-Döring A, Miskey C, Chen W, Cathomen T, Izsvák Z, Ivics Z.** 2012. Retargeting transposon insertions by the adeno-associated virus Rep protein. *Nucleic Acids Res* **40:**6693–6712.

124. **Airenne KJ, Hu YC, Kost TA, Smith RH, Kotin RM, Ono C, Matsuura Y, Wang S, Ylä-Herttuala S.** 2013. Baculovirus: an insect-derived vector for diverse gene transfer applications. *Mol Ther* **21:**739–749.

125. **Kawashima T, Osanai M, Futahashi R, Kojima T, Fujiwara H.** 2007. A novel target-specific gene delivery system combining baculovirus and sequence-specific LINEs. *Virus Res* **127:**49–60.

126. **Ando T, Fujiyuki T, Kawashima T, Morioka M, Kubo T, Fujiwara H.** 2007. *In vivo* gene transfer into the honeybee using a nucleopolyhedrovirus vector. *Biochem Biophys Res Commun* **352:**335–340.

Mobile DNA, 3rd Edition
Nancy L. Craig, Michael Chandler, Martin Gellert, Alan M. Lambowitz, Phoebe A. Rice and Suzanne Sandmeyer

doi:10.1128/microbiolspec.MDNA3-0061-2014

Sandra R. Richardson,[1] Aurélien J. Doucet,[1] Huira C. Kopera,[1]
John B. Moldovan,[2] José Luis Garcia-Perez,[3] and John V. Moran[1,2,4,5]

51

The Influence of LINE-1 and SINE Retrotransposons on Mammalian Genomes

INTRODUCTION

Transposable elements (TEs) or "jumping genes" historically have been disparaged as a class of "junk DNA" in mammalian genomes (1, 2). The advent of whole genome DNA sequencing, in conjunction with molecular genetic, biochemical, and modern genomic and functional studies, is revealing that TEs are biologically important components of mammalian genomes. TEs are classified by whether they mobilize via a DNA or an RNA intermediate (detailed in reference 3). Classical DNA transposons, such as the maize Activator/Dissociation elements originally discovered by Barbara McClintock, move via a DNA intermediate (4, 5). Their mobility (*i.e.*, transposition) can impact organism phenotypes such as corn kernel variegation. Retrotransposons, the predominant class of TEs in most mammalian genomes, mobilize via an RNA intermediate by a process termed retrotransposition (6).

The completion of the human genome reference sequence (HGR) (7, 8) confirmed the results of DNA hybridization-based re-annealing studies (9, 10) and revealed that retrotransposons have been a major force in shaping the structure and function of mammalian genomes. The mobility of non-long terminal repeat (non-LTR) retrotransposons, namely autonomously active Long INterspersed Element-1 sequences (LINE-1s, also known as L1s) and non-autonomous Short INterspersed Elements (SINEs, such as Alu and SINE-R/VNTR/Alu-like retrotransposons [SVA elements]) continues to generate both intra-individual and inter-individual genetic variation in the human population (reviewed in reference 11). Remarkably, LINE-1s and SINEs constitute at least one-third (~1 billion bp) of human genomic DNA (7).

Since the publication of *Mobile DNA II* in 2002, much has been learned about the mechanism of LINE-1 and SINE retrotransposition, the impact of these elements on the human genome, and how the cell defends itself from unabated retrotransposition. A number of outstanding reviews have been published that discuss advances in each of the above areas (for example see refs. [11–19]). Here, we provide a brief background

[1]Department of Human Genetics; [2]Cellular and Molecular Biology Graduate Program, University of Michigan Medical School, Ann Arbor, MI 48109-5618; [3]Department of Human DNA Variability, GENYO (Pfizer-University of Granada & Andalusian Regional Government Genomics & Oncology Center), PTS Granada, 18016 Granada, Spain; [4]Department of Internal Medicine; [5]Howard Hughes Medical Institute, University of Michigan Medical School, Ann Arbor, MI 48109-5618.

about the types of TEs in the human genome. Our discussions then focus on major advances in LINE-1 and SINE biology that have occurred in the past 13 years.

TRANSPOSABLE ELEMENTS IN MAMMALIAN GENOMES

Overview

Genomes are not simply static catalogues of genes. Instead, they are ever changing, genetically dynamic entities. While a typical cellular gene resides at a discrete chromosomal locus, TEs are present in multiple copies that reside at numerous genomic locations. TEs can invade new chromosomal locations, often increasing their copy number in the genome. The evolutionary impact of TEs is readily apparent by examining recently completed mammalian whole genome DNA sequences. For example, at least 46% of human DNA, 31% of canine DNA, and 37% of mouse DNA are derived from TEs (7, 20, 21). Computer algorithms that allow the detection of ancient, highly mutated TEs suggest that TEs could possibly account for as much as 70% of human genomic DNA (22).

Most individual TE-derived sequences in mammalian genomes cannot mobilize because they have been riddled by mutations over the course of evolution; hence, they can be considered as molecular fossils (7, 20, 21). It is now well established that most mammalian genomes contain active LINE-1 and SINE retrotransposons. Other chapters in *Mobile DNA III* discuss the impact of DNA transposons and LTR retrotransposons on mammalian genomes. Here, we briefly discuss the structure and abundance of DNA transposons and LTR retrotransposons in the human genome. We focus the remainder of this chapter on how non-LTR retrotransposons mobilize and how the resultant insertions contribute to disease, genetic variation, and genomic evolution.

DNA Transposons: Abundance and Structure in the Human Genome

DNA transposons can move (*i.e.*, transpose) to new genomic locations via a DNA intermediate (3) and they comprise approximately 3% of the human genome (7). In their simplest form, DNA transposons contain inverted terminal repeat sequences that surround an open reading frame that encodes a protein (*i.e.*, transposase) (3). During a round of transposition, transcription of a DNA transposon is initiated from sequences within the transposable element. The resultant mRNA is transported to the cytoplasm where it undergoes translation, leading to the synthesis of a transposase protein. A generic transposase protein, such as that encoded by the eukaryotic Sleeping Beauty DNA transposon (23), contains a nuclear localization signal as well as DNA binding and integrase activities. After translation, transposase is imported into the nucleus, where it binds either within or near the transposon inverted terminal repeat sequences to promote transposition by a "cut and paste" (or in some cases, a replicative "copy and paste") mechanism. A given transposase protein can mobilize both protein coding (*i.e.*, autonomous) and non-protein coding (*i.e.*, non-autonomous) DNA transposons to new genomic locations (3). As a consequence of transposase activity, the newly inserted DNA transposon is generally flanked by short target-site duplications of a defined size for a given class of element (3, 24).

DNA transposons are active in numerous organisms, (*e.g.*, insertion sequences, or IS elements, in bacteria [reviewed in reference 25], P-elements in *Drosophila* [reviewed in reference 26], Activator/Dissociation elements in maize [reviewed in reference 27], and PiggyBac elements from the cabbage looper moth [28–30]). Due to their mutagenic potential, DNA transposons have been exploited as genetic tools for various molecular biological applications (reviewed in references 31–33). Although it was assumed that DNA transposons were no longer active in mammalian genomes, studies have identified potentially active DNA transposons (*e.g.*, *hobo-Ac*-TAM [hAT] and helitron elements) in the genomes of certain bat species, suggesting that their transposition may continue to generate genetic diversity in these species (34–36). In contrast, virtually all human DNA transposons are mutated, incapable of transposition, and can be considered as molecular fossils (7, 37). Despite this fact, it is clear that DNA transposons have made enduring contributions to mammalian genomes over evolutionary time and that the recurrent domestication of transposase-derived genes has led to genomic innovations (7, 38). For example, the recombination activating gene-1 (*RAG-1*) and *RAG-2*, which are critical for V(D)J recombination and immune system development, may have been co-opted from an ancient Transib DNA transposon (39). Moreover, recent studies also have revealed that the human *THAP9* gene encodes a protein that is capable of mobilizing *Drosophila* P-elements to new genomic locations in both *Drosophila* and human cells (40).

LTR Retrotransposons in the Human and Mouse Genomes

Long terminal repeat retrotransposons (also known as endogenous retroviruses or ERVs) and their nonautonomous derivatives are present at >450,000 copies in the human genome and comprise approximately 8% of human DNA (7). ERVs resemble simple retroviruses in structure. They retrotranspose via a "copy and paste" mechanism, but generally contain a nonfunctional envelope gene or lack the gene completely, which relegates them to an intracellular fate (reviewed in reference 41).

Virtually all ERVs in the human genome have been rendered inactive by mutations and cannot undergo autonomous retrotransposition (7). Certain human-specific ERVs (HERVs) from the HERV-K subfamily (where the K denotes the lysine tRNA needed to prime (–) strand cDNA synthesis from an ERV RNA template) are polymorphic with respect to presence in the human population (42–46). This fact, coupled with the identification of polymorphic ERVs in both the chimpanzee and gorilla genomes (47), suggests that ERVs have been active since the divergence of humans and chimpanzees. Despite concentrated efforts, no one has reported the identification of an active HERV. However, recent studies have demonstrated that HERV-K proviruses that have been reanimated using recombinant DNA technology are infectious in cultured human cells (48, 49). Since retrovirus-encoded proteins can work efficiently *in trans*, it is formally possible that *trans*-complementation might allow the assembly of functional virus-like particles from partially defective HERVs, allowing the generation of new retrotransposition events. Advances in DNA sequencing technologies may reveal rare, active HERV-K elements or *de novo* germline or somatic HERV-K retrotransposition events in individual human genomes.

LTR-retrotransposons are present at >600,000 copies in mouse DNA and comprise approximately 10% of the genome (21). In contrast to the human genome, the mouse genome contains multiple, active ERV subfamilies (50, 51). These include autonomously active MusD and intracisternal A particle (IAP) elements, as well as non-autonomous early transposons (ETns) and mammalian apparent LTR retrotransposons (MaLRs). It is estimated that ERV insertions are responsible for approximately 10% of spontaneously arising mouse mutations (reviewed in reference 51) (discussed in greater detail in other chapters of *Mobile DNA III*).

LINE-1 Retrotransposons: Abundance and Structure

A Brief Overview of Human LINE-1 Evolution and Nomenclature

LINE-1 retrotransposons have been amplifying in mammalian genomes for more than 160 million years (52–54). In humans, the vast majority of LINE-1 sequences have amplified since the divergence of the ancestral mouse and human lineages approximately 65 to 75 million years ago (7). As a consequence, LINE-1-derived sequences now account for approximately 17% of human genomic DNA (7) (Figure 1).

Sequence comparisons between individual genomic LINE-1 sequences and a consensus sequence derived from modern, active LINE-1s can be used to estimate the age of genomic LINE-1s. These analyses uncovered 16 LINE-1 primate-specific subfamilies (termed PA1 to PA16) (53, 55). These subfamilies have a monophyletic origin, suggesting that older LINE-1 subfamilies are replaced over evolutionary time by the emergence of new LINE-1 subfamilies—a phenomenon known as subfamily succession. Indeed, recent studies suggest that host proteins that restrict LINE-1 expression may drive subfamily succession (56, 57) (see below).

Functional analyses have revealed that only certain human-specific LINE-1s (termed L1Hs elements) from the PA1 subfamily remain retrotransposition-competent (58–61). The majority of active L1Hs elements belong to a small population of elements termed the transcribed-active subset (Ta-subset) (62, 63). Ta-subset LINE-1s contain a diagnostic 5′-ACA-3′ trinucleotide sequence and G nucleotide in their 3′ untranslated regions (UTRs) (at positions 5930–5932 and 6015, respectively, based on the sequence of L1.2, an active LINE-1: GenBank accession number M80343 [64]). The diagnostic ACA and G nucleotides allow the discrimination of Ta-subset LINE-1s from older, retrotransposition-incompetent LINE-1s, which generally contain a GAG trinucleotide at the same position of their respective 3′ UTRs. The exploitation of the ACA trinucleotide present in Ta-subset LINE-1s has been instrumental in allowing the identification of polymorphic, human-specific LINE-1 insertion alleles in both the HGR and individual human genomic sequences (7, 63, 65–72).

The Ta-subset of LINE-1s can be further grouped into finer subdivisions (*e.g.*, Ta-1, Ta-0, and pre-Ta) based on more subtle sequence distinctions (63). Of these, the Ta-1 subset contains the greatest number of active LINE-1s, followed by the Ta-0 and pre-Ta subfamilies (58, 59, 61, 63). Notably, LINE-1s from the Ta-subset are responsible for most of the retrotransposition activity in modern human genomes (reviewed in reference 11).

Active Human LINE-1s: Abundance

The vast majority of LINE-1-derived sequences in the HGR predate the emergence of the human lineage, are "fixed" with respect to presence in the human population, and have been rendered inactive by 5′ truncations,

Non-LTR retrotransposons in the human genome

	size	copy number	% of genome	active subfamilies
LINE-1 (5'UTR, ORF1: CC RRM CTD, ORF2: EN RT C, A_n)	~6 kb	516,000	17%	L1 PA-1 Ta-1, Ta-0, and pre-Ta
Alu (A, B, AAA, AAAA, TTTT; Left Monomer, Right Monomer)	~300 bp	1,090,000	11%	Alu Y (Ya5, Yb8, and Yd8) (Alu Sx ?)
SVA ($(CCCTCT)_n$, Alu-like, VNTR, SINE-R: ENV LTR (HERV-K10), CPSF, A_n)	~2 kb	2,700	0.2%	SVA-E, SVA-F1 (SVA-D, SVA-F ?)
Processed pseudogene (cellular mRNA, A_n)	variable	8,000 to 17,000		

Non-LTR retrotransposons in the mouse genome

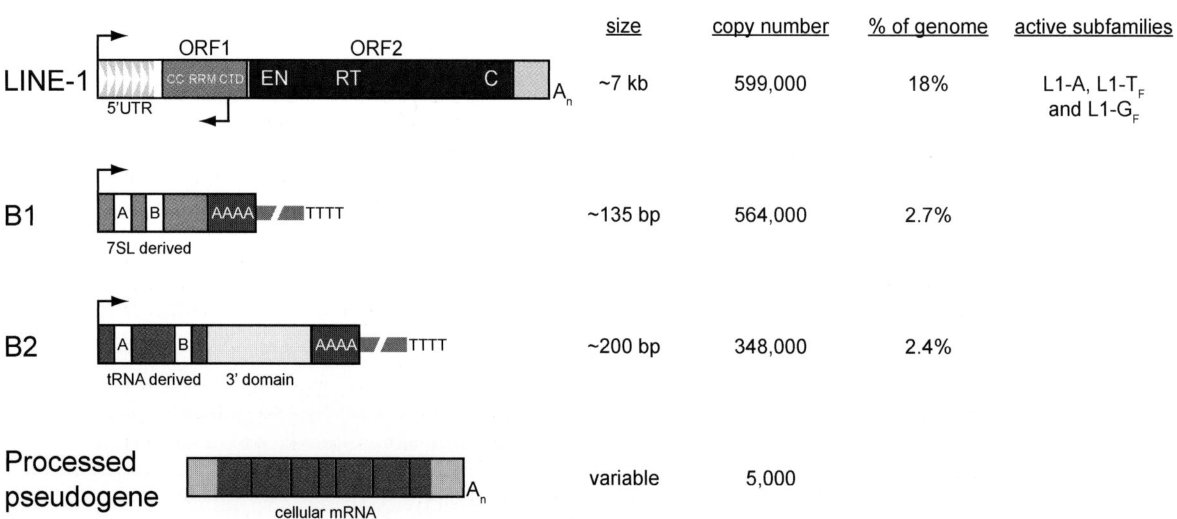

internal rearrangements (such as inversion/deletion events), and point mutations that prevent the production of active forms of the LINE-1 encoded proteins (ORF1p and ORF2p) (7, 73). The completion of the HGR, in combination with functional assays to assess retrotransposition potential in cultured human HeLa cells, has revealed that the average human genome contains approximately 80 to 100 LINE-1s that remain retrotransposition-competent (59, 61). Intriguingly, only a handful of Ta-subset LINE-1s (termed "hot L1s") were found to account for the bulk of retrotransposition activity in a given genome (59). Subsequent genomic and computationally based studies have revealed that Ta-subset LINE-1s, which are present at low allele frequencies in the human population, are highly enriched with active LINE-1s (58). When considering that there are more than 7 billion people on the planet, these studies raise the possibility that there are perhaps millions of rare, active LINE-1 alleles in the human population (58). When combined with the number of active non-autonomous non-LTR retrotransposons, it is undeniable that there are many more active retrotransposons in the human population than previously thought.

Active Human LINE-1s: Structure

A retrotransposition-competent human LINE-1 (RC-L1) is ~6 kb in length (64, 74) (Figure 1). An RC-L1 encodes a 5′ UTR containing an internal RNA polymerase II sense strand promoter (75), as well as an antisense promoter of unknown function (76). The 5′ UTR is followed by two open reading frames (ORF1 and ORF2) and a 3′ UTR that terminates in a poly (A) tract (64). Human ORF1 encodes a 40 kDa RNA binding protein (ORF1p) that has nucleic acid chaperone activity (77–80). Human ORF2 encodes a 150 kDa protein (ORF2p) (81, 82) with demonstrated endonuclease (L1 EN) (83) and reverse transcriptase (L1 RT) activities (84). The ORF2p C-terminus contains a cysteine-rich domain of unknown function (60, 85). Functional studies have revealed that both ORF1p and ORF2p are required for retrotransposition (60, 83).

Figure 1 Non-long terminal repeat (LTR) retrotransposons of the human and mouse genomes. The top and bottom panels represent non-LTR retrotransposons in the human and mouse genomes, respectively. Each non-LTR retrotransposon is listed with its name, structure, average size, copy number, percentage of the genome reference sequence occupied by the element, and, if applicable, the active subfamilies (question marks [?] denote uncertainty in whether Alu Sx and SVA-D and F elements are active *in vivo*). Details of the structure and abbreviations for human and mouse Long INterspersed Element-1 retrotransposons (LINE-1s): Untranslated regions (UTRs) (gray boxes); sense and antisense internal promoters (black arrows); monomeric repeats (white triangles) are followed by an untranslated linker sequence (white box) just upstream of open reading frame 1 (ORF1) in the mouse 5′ UTR; ORF1 (yellow box for human LINE-1; brown box for mouse LINE-1) includes a coiled-coil domain (CC), an RNA recognition motif (RRM), and a C-terminal domain (CTD); inter-ORF spacer (gray box between ORF1 and ORF2); ORF2 (blue boxes) includes endonuclease (EN), reverse transcriptase (RT), and cysteine-rich domains (C); poly (A) tract (A_n downstream of 3′ UTR). For human Alu: 7SL-derived monomers (orange boxes); RNA polymerase III transcription start site (black arrow) and conserved *cis*-acting sequences required for transcription (A and B white boxes in left 7SL-derived monomer); adenosine-rich fragment (AAA gray box between left and right 7SL-derived monomers); terminal poly (A) tract (AAAA gray box); variable sized flanking genomic DNA (interrupted small gray box) followed by the RNA pol III termination signal (TTTT). For human SVA: hexameric CCCTCT repeat ($(CCCTCT)_n$ light green box); inverted Alu-like repeat (green box with backward arrows); GC-rich VNTR (striped green box); SINE-R sequence sharing homology with HERV-K10, (envelope [ENV] and LTR); cleavage polyadenylation specific factor (CPSF) binding site; terminal poly (A) tract (A_n). For human and mouse processed pseudogenes: spliced cellular mRNA with UTR (gray boxes) and coding ORF (red boxes for human and purple boxes for mouse, boxes are interrupted by exon–exon junctions [vertical black lines]). For mouse B1 and B2: 7SL-derived monomer (light orange boxes) or tRNA derived sequence (dark orange boxes); RNA pol III transcription start site (black arrow) and conserved *cis*-acting sequences required for transcription (A and B white boxes); terminal poly (A) tract (AAAA dark gray box); variable sized flanking genomic DNA (interrupted gray box) followed by the RNA polymerase III termination signal (TTTT). The 3′ end of B2 also contains a non-tRNA derived sequence (3′ domain light gray box). Mouse ID and B4 elements are not represented in the figure. References are provided in the text.
doi:10.1128/microbiolspec.MDNA3-0061-2014.f1

Mouse LINE-1s: Abundance and Structure

The mouse genome is replete with LINE-1-derived sequences and they comprise approximately 18% of mouse DNA (21) (Figure 1). As in the HGR, the vast majority of LINE-1s in the *Mus musculus* reference genome have been rendered inactive by mutation (21). Functional studies in cultured cells have revealed that approximately 3,000 mouse LINE-1s remain retrotransposition-competent (86).

Mouse LINE-1s are structurally similar to human LINE-1s; however, mouse LINE-1s contain different RNA polymerase II promoter structures at their respective 5′ ends (Figure 1). The mouse LINE-1 promoter consists of a series of repeats and is followed by an untranslated linker sequence located immediately upstream of ORF1 (reviewed in reference 87). As in humans, mouse LINE-1 sequences can be stratified into subfamilies: at least five structurally different subfamilies (V, F, A, T_F, and G_F) of mouse LINE-1s exist, and they differ in the DNA sequences of the monomeric repeats present in their respective 5′ UTRs (88). LINE-1s from the V and F subfamilies are thought to be inactive (89), whereas LINE-1s from the A, T_F, and G_F subfamilies remain retrotransposition-competent (86, 90, 91).

Non-Autonomous Retrotransposons: Prevalence and Structure

The proteins encoded by LINE-1s (ORF1p and/or ORF2p) can act *in trans* to mobilize non-autonomous retrotransposons (*e.g.*, human Alu and SVA elements and mouse B1 and B2 elements) (92–96) and cellular mRNAs to new genomic locations, with the latter giving rise to processed pseudogenes (97, 98) (Figure 1). As expected, each of the above elements contains structural hallmarks that are consistent with being mobilized by the LINE-1 encoded proteins (reviewed in reference 11). They generally: (i) terminate with a poly (A) or adenosine-rich sequence; (ii) are flanked by target-site duplications that vary both in their size and sequence; and (iii) integrate at a LINE-1 endonuclease consensus site in genomic DNA (99). A brief description of each of the major types of non-autonomous retrotransposons present in the human and mouse genomes is provided below.

Human Alu Elements

Alu elements represent the most abundant class of human non-autonomous retrotransposons (7). They are present at more than 1 million copies in the HGR and comprise approximately 11% of genomic DNA (7) (Figure 1). Alu elements are approximately 300 bp (100) in length and are derived from the 7SL RNA component of the signal recognition particle (101). A full-length Alu element exhibits a dimeric structure that consists of highly similar left and right monomers that are separated by an adenosine-rich linker sequence. The left Alu monomer contains conserved A and B box sequences that are required for RNA polymerase III dependent transcription (102). The right monomer lacks conserved A and B boxes and ends in a poly (A) tract. Genomic DNA flanking the 3′ ends of active Alu elements contains an RNA polymerase III terminator sequence (*i.e.*, a stretch of four to six consecutive thymidines) (102, 103). Both the size of an Alu poly (A) tract and genomic sequences that reside downstream, and in some cases upstream, of an individual Alu element can influence its expression and retrotransposition potential (104–109).

Like LINE-1s, Alu elements can be stratified into three major subfamilies: Alu J, S, and Y (110–114); these subfamilies can be further grouped into finer subdivisions (103, 115). Alu J represents the oldest Alu lineage; its retrotransposition activity peaked approximately 65 million years ago (reviewed in reference 103). Alu S represents the second oldest lineage; its retrotransposition activity peaked approximately 30 million years ago (reviewed in reference 103). Alu Y elements represent the youngest Alu lineage (reviewed in reference 103). The analysis of individual human DNA sequences, in conjunction with molecular genetics and genomics-based approaches, has revealed that certain Alu Y subdivisions (*e.g.*, Alu Ya5, Alu Yb8, and Alu Yd8) are polymorphic with respect to presence in the human population (116, 117). Some of these Alus may represent "source elements" or "master genes" that presently are amplifying in the human population. Indeed, the identical-by-descent mode of transmission, homoplasy-free nature, and directionality of integration (*i.e.*, the absence of the element represents the ancestral state) have allowed the use of polymorphic Alu elements and LINE-1s as genetic markers in population-based and phylogenetic studies (118–123) (reviewed in reference 124).

Although it is clear that the Alu Y subfamily is responsible for the bulk of Alu retrotransposition in the human genome, studies have demonstrated that an average human genome may contain thousands of active Alu "core elements" derived from both the Alu Y and Alu S subfamilies (125) and that interactions between Alu RNA and the signal recognition proteins (SRP9 and SRP14) are required to undergo efficient retrotransposition (125, 126).

Human SVA Elements

SVA elements comprise an evolutionarily young, non-autonomous retrotransposon family that arose in primate lineages approximately 25 million years ago (127–129). SVA elements are present at approximately 2,700 copies in the HGR, comprising approximately 0.2% of human genomic DNA (Figure 1). A typical SVA element is approximately 2,000 bp in length and has a composite structure that consists of: (i) a hexameric CCCTCT repeat; (ii) an inverted Alu-like element repeat; (iii) a set of GC-rich variable nucleotide tandem repeats (VNTRs); (iv) a SINE-R sequence that shares homology with HERV-K10, an inactive LTR retrotransposon; and (v) a canonical cleavage polyadenylation specificity factor binding site that is followed by a poly (A) tract (reviewed in reference 129) (Figure 1). This latter feature suggests that SVA elements are transcribed by RNA polymerase II; whether SVA elements contain an internal RNA polymerase II promoter is not yet known. SVA elements also can be stratified into subfamilies based on sequence similarities. For example, approximately 40% of SVA elements from the youngest SVA subfamilies (SVA-D, SVA-E, SVA-F, and SVA-F1) are polymorphic with respect to presence in the human population, suggesting that members of these subfamilies may retain the ability to retrotranspose and contribute to human genetic diversity (129). Interestingly, recent reports suggest that certain SVA elements can influence reporter gene expression in *in vitro* assays (56, 130); so, in principle, SVA elements might alter the expression of neighboring genes.

Gibbons and certain other nonhuman primates contain composite retrotransposons called LINE-Alu-VNTR-Alu (LAVA) elements (131–133). LAVA elements are similar in structure to SVA elements, suggesting that they are mobilized *in trans* by the LINE-1 encoded proteins. In contrast to SVA elements, LAVA elements lack a HERV-K sequence and instead contain a sequence motif that consists of unique DNA, as well as sequences derived from ancient Alu S and LINE-1 elements (131–133). Interestingly, LAVA elements have undergone a massive expansion in gibbon genomes. Recent reports demonstrate that LAVA elements are concentrated at centromeres and may contribute to gibbon genomic plasticity (134).

Mouse B1 and B2 Elements and Other SINEs

The completion of the *Mus musculus* reference genome revealed the presence of various abundant classes of non-autonomous SINEs (21). The first, B1, is an ~135 bp, 7SL-derived monomeric SINE (135, 136) (Figure 1). B1 elements have amplified to more than 500,000 copies and comprise approximately 2.7% of mouse DNA (21). Like Alu elements, an active B1 element contains conserved A and B box sequences that are required for RNA polymerase III dependent transcription and end in a poly (A) tract that is flanked by genomic DNA sequences containing an RNA polymerase III terminator. The second, B2, is an ~200 bp, tRNA-derived SINE (137) (Figure 1). B2 elements have amplified to more than 300,000 copies and comprise approximately 2.4% of mouse DNA (21). The tRNA-derived region of B2 contains conserved A and B box sequences that are required for RNA polymerase III dependent transcription. The 3′ end of B2 ends in a poly (A) tract and is flanked by genomic DNA sequences containing an RNA polymerase III terminator (137). A third class of mouse SINE, the ID elements, appears to be derived from a tRNA; an ID-element-derived RNA (BC1 RNA) is expressed abundantly in the nervous system (137, 138). A fourth class, known as B4 elements, appears to represent an ancient hybrid SINE derived from a fusion between B1 and ID elements and does not seem to be active in the mouse genome (139, 140). Although cell culture-based assays have demonstrated the existence of active B1 and B2 SINEs in the mouse genome (93), the number of active elements remains unknown.

Processed Pseudogenes

LINE-1 ORF1p and ORF2p occasionally can act *in trans* to mobilize matured cellular mRNAs to new genomic locations, leading to the formation of processed pseudogenes that bear LINE-1 structural hallmarks (97, 98, 141, 142) (Figure 1). Recent studies have revealed that there may be at least 8,000 to 17,000 processed pseudogenes in the HGR (143–145). The majority of processed pseudogenes appear to be derived from housekeeping genes or ribosomal protein-encoding genes. Whole genome DNA sequence studies have further revealed the presence of segregating processed pseudogene insertions in individual genomes, indicating that processed pseudogene formation continues to contribute to inter-individual genomic diversity (146–150). Finally, several studies have demonstrated how processed pseudogenes can evolve to acquire a cellular function (reviewed in reference 151). In extreme cases, this process can lead to the evolution of new genes that participate in host defense against exogenous viruses. For example, independent LINE-1-mediated retrotransposition events of cyclophilin mRNA into the TRIM5 locus have led to the production of a TRIM5/cyclophilin fusion protein that can restrict the mobility of HIV in owl monkeys (152, 153).

Other Cellular RNAs Mobilized by the LINE-1-Encoded Proteins

The LINE-1-encoded proteins have been implicated in the mobility of other cellular RNAs, such as uracil-rich small nuclear RNAs (snRNAs) (*i.e.*, U6, U6atac, and to a lesser extent U1, U2, U4, and U4atac) and small nucleolar RNAs (snoRNAs) (*i.e.*, U3 snoRNA) (154–158). In contrast to the mechanism of processed pseudogene formation, the structures of chimeric U6/L1 pseudogenes suggest that U6 snRNA was reverse transcribed onto a 5′ truncated LINE-1 cDNA during the process of LINE-1 integration (155) (see below). Chimeric U6/LINE-1 pseudogenes have been identified in many primate genomes (159). Moreover, the recapitulation of U6/LINE-1 chimeric pseudogene formation in cultured human cells strongly supports the hypothesis that U6/LINE-1 pseudogene formation is ongoing in the human population (156, 157).

LINE-1-Mediated Retrotransposition Events and Human Disease

A Brief Historical Perspective

Historically, it was assumed that LINE-1s could be dismissed as a class of repetitive "junk DNA". This view changed radically in 1988, when the Kazazian laboratory identified two independent, mutagenic LINE-1 insertions into the *Factor VIII* genes of unrelated boys afflicted with hemophilia A (160). Although those LINE-1 insertions were predicted to be inactive (*i.e.*, one was 5′ truncated, while the other was both 5′ truncated and internally rearranged), DNA sequencing revealed that each of the disease-producing insertions likely was derived from an active full-length progenitor LINE-1 that contained intact open reading frames. The derivation of an oligonucleotide probe that specifically recognized the disease-producing LINE-1 insertion and related elements subsequently led to the identification of a cohort of full-length LINE-1s with intact open reading frames (61, 64, 161). One of these LINE-1s (L1.2B), isolated from genomic DNA derived from the mother of one of the hemophiliac patients, was identical (over the ~3.8 kb length of insertion) to the disease-producing LINE-1 insertion in her son (64). The demonstration that an allele of L1.2 (L1.2A) could retrotranspose in cultured human cells subsequently confirmed that LINE-1s were active in the human genome (60, 162).

Disease-Producing LINE-1-Mediated Insertions

Mutagenic LINE-1 retrotransposition events have been implicated in at least 25 cases of human disease, including Duchenne muscular dystrophy, hemophilia B, chronic granulomatous disease, X-linked retinitis pigmentosa, and β-thalassemia trait (reviewed in reference 14). In general, these LINE-1 retrotransposition events either disrupt coding exons or occur into introns, which can result in mis-splicing or exon skipping, and lead to the generation of null or hypomorphic expression alleles (reviewed in references 11, 151). As predicted by Maxine Singer during her studies of LINE-1 expression in human teratocarcinoma cell lines (62, 163), all but two of the 25 disease-producing LINE-1 insertions were derived from the LINE-1 Ta-subset. The remaining insertions were derived from the slightly older pre-Ta subset (which contains an ACG at nucleotide positions 5930–5932 as opposed to the ACA in L1.2 [64]). These observations, as well as functional studies (reviewed in reference 11), provide evidence that the LINE-1 Ta-subfamily comprises the bulk of active LINE-1s in the human genome and that some pre-Ta LINE-1s remain active.

LINE-1-mediated non-autonomous retrotransposition events also have caused sporadic cases of human disease. For example, deleterious Alu retrotransposition events are responsible for over 60 disease-producing mutations in man (14, 164). Almost all of the disease-producing insertions are derived from members of the Alu Y subfamily, and the majority of these belong to the Alu Ya5 and Alu Yb8 subdivisions. Similarly, deleterious SVA retrotransposition events (derived from the SVA-E and SVA-F1 subfamilies) are responsible for at least 10 disease-producing mutations in man (14, 165–167). Recent studies have identified a processed pseudogene insertion into the *CYBB* locus that is responsible for a sporadic case of chronic granulomatous disease (168). *De novo* processed pseudogene insertions also have been identified in cancer genomes (169). Finally, there are four examples where human diseases have been caused by the insertion of poly (A) tracts, which likely are derived from severely truncated LINE-1-mediated retrotransposition events (reviewed in reference 14).

The above studies have revealed that LINE-1-mediated retrotransposition events are responsible for approximately 100 cases of sporadic human disease and provoke the following question: how often does LINE-1-mediated retrotransposition lead to human disease? A union of the above data, in conjunction with recent efforts to catalog the spectrum of mutations of the *NF1* gene causing the autosomal dominant disease neurofibromatosis, suggests that LINE-1-mediated retrotransposition events are responsible for approximately 1 in 250 disease-producing mutations in man (170).

LINE-1-Mediated Retrotransposition Events as Mutagens in Other Mammals

Deleterious LINE-1-mediated retrotransposition events are implicated as disease-producing mutations in nonhuman mammals. For example, LINE-1 retrotransposition events into genes are linked to mutagenic phenotypes in at least seven different mouse strains (171–177). Two of these mutagenic insertions, one into the gene encoding the glycine receptor β-subunit (*i.e.*, $L1_{spa}$) and another into the *reeler* gene (*i.e.*, $L1_{Orl}$), represent full-length LINE-1 insertions (171–173). Subsequent studies revealed that both $L1_{spa}$ and $L1_{Orl}$ could retrotranspose in cultured human and mouse cells, and led to the discovery of active LINE-1s from the T_F subfamily (90). Likewise, a mutagenic B1 insertion into the mouse *Atcay* locus is responsible for the jittery mouse, further demonstrating that B1 element retrotransposition continues to impact the mouse genome (178, 179).

LINE-1 retrotransposition events have been implicated in various dog phenotypes. For example, a LINE-1 retrotransposition event into the *c-myc* gene has been identified in a canine transmissible venereal tumor (180, 181), although it remains uncertain if the insertion event is involved in tumorigenesis. Likewise, LINE-1 retrotransposition events into the *Factor IX*, *dystrophin*, and *DLX6* genes are implicated in a mild case of hemophilia B in German Wirehaired Pointers, Duchenne-like muscular dystrophy in the Pembroke Welsh Corgi, and cleft palate and mandibular abnormalities in the Nova Scotia Duck Tolling Retriever (182–184).

The retrotransposition of a canine tRNA-derived SINE (*i.e.*, SINEC_Cf) is responsible for various phenotypes in dog breeds. These include narcolepsy in Doberman Pinschers, centronuclear myopathy in Labrador Retrievers, "merle" coat color pigmentation in Shetland Sheepdogs, and progressive retinal atrophy in Tibetan Spaniels and Tibetan Terriers (185–188). Finally, the expression of an *FGF4* pseudogene is responsible for the "short-legged" phenotypes of at least 19 dog breeds (189). It will be interesting to determine the extent to which dog breeders have inadvertently selected for mutagenic LINE-1-mediated retrotransposition events that lead to "desirable" phenotypes within dog breeds. Indeed, selective breeding leads to genetic bottlenecks, which may allow the ready identification of retrotransposon insertions that dramatically affect phenotypic traits in mammals and other organisms.

MECHANISTIC STUDIES OF LINE-1 RETROTRANSPOSITION

An Assay to Study LINE-1 Retrotransposition

Almost 20 years ago, a functional assay was developed to assess the retrotransposition potential of LINE-1s in cultured mammalian cells (60). The assay builds upon a rationale developed by Boeke and colleagues to demonstrate that the yeast Ty1 retrotransposon mobilizes via an RNA intermediate (6). Subsequent enhancements of the assay by the Heidmann and Curcio laboratories then led to the development of retrotransposition indicator cassettes that could only become activated for expression upon a successful round of retrotransposition (190, 191).

Briefly, the 3′ UTR sequences of candidate full-length LINE-1s are tagged with a retrotransposition indicator cassette that consists of a backward copy of a neomycin phosphotransferase reporter gene equipped with its own promoter and polyadenylation signals (*i.e.*, the *mneoI* cassette; Figure 2). Importantly, the reporter gene is disrupted by an intron that resides in the same transcriptional orientation as the LINE-1 (60, 192, 193). This arrangement ensures that the expression of the neomycin phosphotransferase gene only occurs upon a successful round of LINE-1 retrotransposition, which ultimately leads to the generation of clonal foci that grow in the presence of the neomycin analog G418. Hence, the assay allows a simple, yet powerful way to monitor LINE-1 retrotransposition efficiency by counting G418-resistant foci (60).

Since the inception of the cultured cell retrotransposition assay, a battery of retrotransposition indicator cassettes has been developed to assess LINE-1 retrotransposition by either exploiting drug selection (*i.e.*, a blasticidin resistance cassette) or screening for reporter gene activation (*i.e.*, green fluorescent protein and luciferase cassettes) (194–197) (Figure 2). Moreover, engineered LINE-1 elements, that contain epitope tags on the C-termini of ORF1p and ORF2p and an MS2 binding site in the LINE-1 mRNA, have allowed the direct detection of the LINE-1-encoded proteins and mRNA in cultured cells using both biochemical approaches and fluorescence microscopy (82, 198–201) (Figure 2). Finally, retrotransposition indicator cassettes have been developed that allow the direct recovery of engineered LINE-1 retrotransposition events as autonomously replicating plasmids in *Escherichia coli* (156, 202, 203) (Figure 2).

In summary, the cultured cell retrotransposition assay, in conjunction with complementary molecular genetic and biochemical studies, has: (i) allowed the

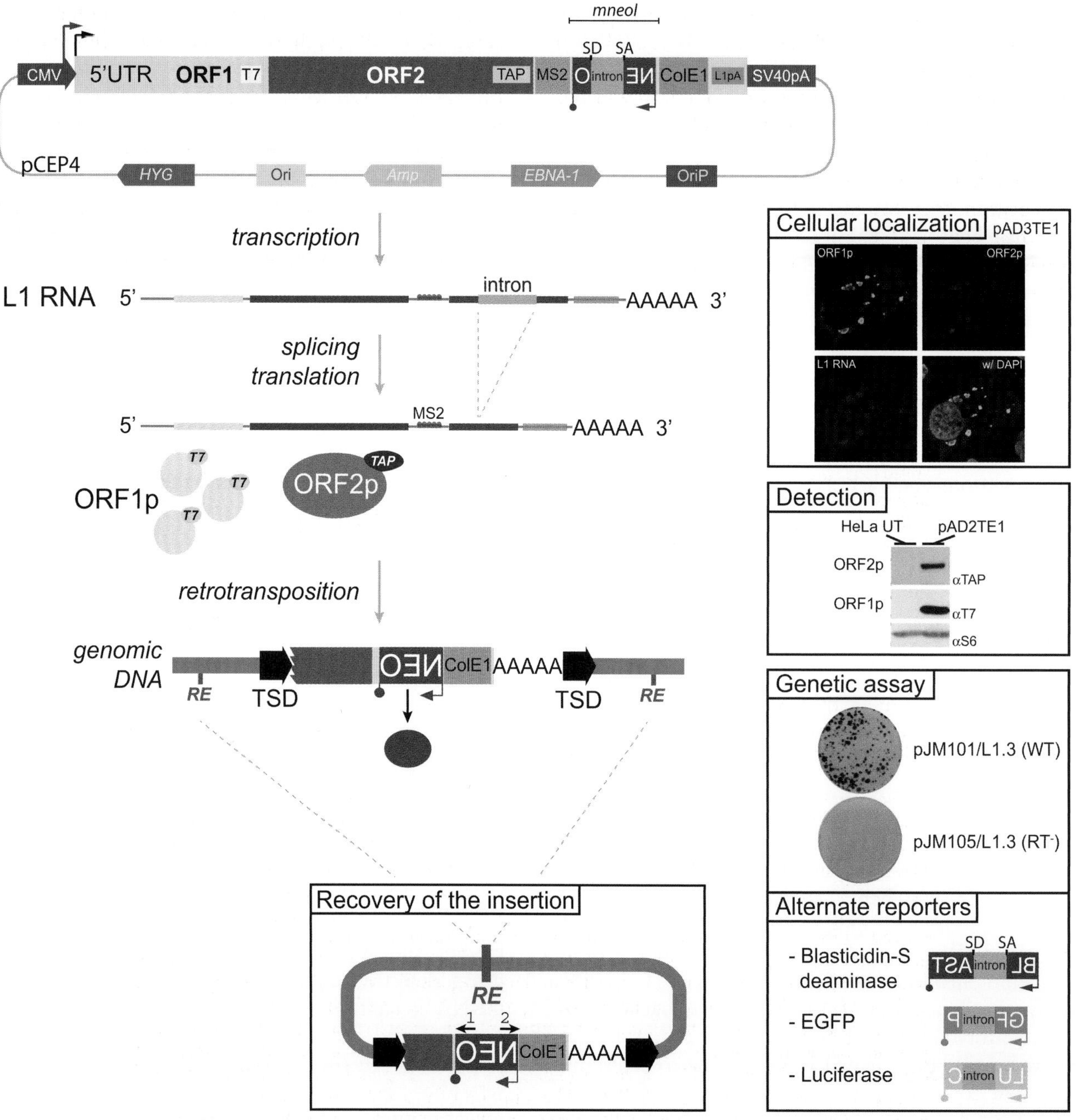
mneol
SD SA
CMV
5'UTR
ORF1
T7
ORF2
TAP
MS2
intron
ColE1
L1pA
SV40pA
pCEP4
HYG
Ori
Amp
EBNA-1
OriP
transcription
L1 RNA 5'
intron
AAAAA 3'
splicing
translation
5'
MS2
AAAAA 3'
ORF1p
ORF2p
retrotransposition
genomic DNA
RE
TSD
ColE1 AAAAA
TSD
RE
Recovery of the insertion
RE
1 2
ColE1 AAAA
Cellular localization
pAD3TE1
ORF1p
ORF2p
L1 RNA
w/ DAPI
Detection
HeLa UT
pAD2TE1
ORF2p
αTAP
ORF1p
αT7
αS6
Genetic assay
pJM101/L1.3 (WT)
pJM105/L1.3 (RT-)
Alternate reporters
- Blasticidin-S deaminase
SD SA
intron
- EGFP
intron
- Luciferase
intron

identification of active LINE elements from mammalian and vertebrate genomes (58–61, 86, 90, 91, 204, 205); (ii) shown that allelic heterogeneity affects LINE-1 retrotransposition (162, 206); (iii) facilitated experimental illumination of the LINE-1 retrotransposition mechanism (reviewed in reference 11); (iv) demonstrated that LINE-1 retrotransposition generates genomic structural variation (156, 202, 203); (v) revealed that the LINE-1-encoded proteins (ORF1p and/or ORF2p) could act *in trans* to mediate SINE retrotransposition and processed pseudogene formation (92–98); and (vi) allowed the identification of host factors that may restrict and/or promote retrotransposition (see below). Clearly, the cultured cell assay has been and continues to be instrumental in allowing a deeper mechanistic understanding of LINE-1 biology.

Functional Studies of LINE-1 Retrotransposition

The LINE-1 5′ UTR

Although the promoter structures of human and mouse LINE-1s differ (Figure 1), it is clear that the acquisition of an internal RNA polymerase II promoter has ensured that full-length retrotransposed LINE-1s retain the potential to undergo subsequent amplification in the genome. The human LINE-1 5′ UTR is approximately 910 bp in length and contains an internal RNA polymerase II promoter that directs transcription of LINE-1 mRNA at or near the first nucleotide of the element (75, 207). Experimental studies have revealed that a YY1-binding site at the 5′ end of the 5′ UTR is critical for accurate transcriptional initiation and that most LINE-1 mRNAs contain a 7-methyl guanosine

Figure 2 Engineered Long INterspersed Element-1 (LINE-1) structure and cell based strategies to study retrotransposition. The LINE-1 expression vector consists of a retrotransposition-competent LINE-1 subcloned into pCEP4 (flanked by a CMV promoter and an SV40 polyadenylation signal). The pCEP4 vector is an episomal plasmid that has protein encoding (EBNA-1) and *cis*-acting (OriP) sequences necessary for replication in mammalian cells; it also has a hygromycin resistance gene (HYG) that allows for the selection of mammalian cells containing the vector, as well as a bacterial origin of replication (Ori) and ampicillin selection marker (Amp) for plasmid amplification in bacteria. The *mneoI* reporter cassette, located in the LINE-1 3′ UTR, contains the neomycin phosphotransferase gene (NEO, purple box, with its own promoter and polyadenylation signals, purple arrow and lollipop, respectively) in the opposite transcriptional orientation of LINE-1 transcription. The reporter gene is interrupted by an intron (light purple box) with splice donor (SD) and splice acceptor (SA) sites in the same transcriptional orientation as the LINE-1. This arrangement of the reporter cassette ensures that the reporter gene will only be expressed after a successful round of retrotransposition. *De novo* retrotransposition of the *mneoI* reporter cassette will result in G418-resistant colonies that can be quantified—genetic assay panel with pJM101/L1.3 (wild-type [WT]) and pJM105/L1.3 (RT mutant [RT-]) LINE-1 constructs. Alternative reporters can be used instead of *mneoI* to allow different drug-resistance, fluorescent, or luminescent read-outs (alternative reporters panel, with blasticidin-S deaminase [BLAST], enhanced green fluorescent protein [EGFP] or luciferase [LUC]) retrotransposition indicator cassettes. The addition of the ColE1 bacterial origin of replication (recovery of the insertion panel, green box) to a modified version of the *mneoI* reporter cassette allows the recovery from cultured cell genomic DNA of engineered LINE-1 retrotransposition events as autonomously replicating plasmids in *Escherichia coli*. The insertions also can be characterized by inverse polymerase chain reaction using divergent oligonucleotide primers (recovery of the insertion panel, black arrows: 1 and 2) that anneal to the reporter gene. RE indicates restriction enzyme cleavage sites in flanking genomic DNA (gray lines). The use of epitope tags (T7-tag in C-terminus of ORF1, yellow box, and TAP-tag in C-terminus of ORF2, blue box) allow the immunoprecipitation (not shown) and detection of LINE-1 proteins by western blot and immunofluorescence (IF) (detection panel, with western blot data obtained with pAD2TE1, a vector expressing ORF1-T7p and ORF2-TAPp, compared to untransfected [UT] HeLa cells [82]). The addition of the RNA-stem loops that bind the bacteriophage MS2 coat protein (409) (orange box) in the 3′ UTR of LINE-1 can be used to detect the cellular localization of LINE-1 RNA by fluorescent *in situ* hybridization (FISH). Both IF and FISH strategies can be combined to detect the subcellular localization of ORF1p, ORF2p, and LINE-1 RNA (cellular localization panel, with pAD3TE1 vector containing ORF1-T7p, ORF2-TAPp, and LINE-1 RNA-MS2 [82]). The images shown in the cellular localization and the detection boxes originally were published in (82). Additional references are provided in the text. doi:10.1128/microbiolspec.MDNA3-0061-2014.f2

cap structure, which facilitates their translation (208, 209). The 5′ UTR harbors *cis*-acting binding sites for the following transcription factors: Runx3, Sp1, and SRY-related (Sox) proteins (75, 207, 210, 211). Studies in cultured cells have revealed that mutations in these *cis*-acting sequences reduce LINE-1 transcription and retrotransposition. In addition, it is likely that other host factors bind the 5′ UTR and regulate LINE-1 expression (see below).

In addition to containing a sense strand promoter, the human LINE-1 5′ UTR contains a conserved RNA polymerase II antisense promoter (76). Transcription from the LINE-1 antisense promoter can lead to the generation of chimeric transcripts comprising LINE-1 sequences conjoined to sequences derived from the 5′ genomic flank of a given LINE-1 locus (76, 212). These chimeric transcripts have been used as a proxy to identify transcriptionally active LINE-1 elements in human embryonic stem cells (213). Although mammalian bidirectional promoters have been identified to be the source of some noncoding RNAs (214), the function of the human LINE-1 antisense chimeric transcripts, if any, requires elucidation.

The mouse LINE-1 5′ UTR consists of a series of up to seven and two-thirds monomeric repeats that are followed by an untranslated linker sequence immediately upstream of ORF1 (87, 90). Reporter gene assays have revealed that 5′ UTRs from the A, T_F, and G_F LINE-1 subfamilies remain transcriptionally active (90, 215, 216), whereas 5′ UTRs from the V and F LINE-1 subfamilies generally lack transcription activity (89). Cell culture-based and biochemical assays have revealed that mRNAs derived from the A, T_F, and G_F subfamilies are enriched in ribonucleoprotein particles (RNPs) and that select LINE-1 elements from these remain retrotransposition-competent (86, 90, 91). Hence, it appears that the ability of mouse LINE-1s to capture new promoter sequences has, in part, led to their evolutionary success.

Mouse LINE-1s contain a transcriptionally active antisense RNA polymerase promoter (217, 218). Unlike human LINE-1s, the antisense promoter is located within ORF1. Transcription from the mouse antisense promoter leads to the generation of chimeric transcripts containing LINE-1 sequences conjoined to genomic sequences that flank the 5′ end of the LINE-1 locus. Moreover, the overexpression of LINE-1 antisense mRNA could lead to a reduction of retrotransposition in cultured cells (218). Hence, it is intriguing to speculate that LINE-1 antisense mRNA may play a role in regulating mouse LINE-1 retrotransposition *in vivo*.

Recent studies indicate that a large fraction of mammalian long noncoding RNAs contain retrotransposon-derived sequences and that some are transcribed from LTR-retrotransposon-derived promoters (219, 220), (reviewed in reference 221). Intriguingly, select long noncoding RNAs are involved in maintaining the pluripotency of embryonic stem cells by yet unidentified mechanisms (222). It will be interesting to determine if the LINE-1 antisense promoter contributes to the transcriptional regulatory network regulating stem cell identity (reviewed in reference 223).

ORF1p

ORF1p is an ~40 kDa protein (also known as p40 for human LINE-1s) that is translated from LINE-1 mRNA by a traditional cap-dependent mechanism (77, 209, 224). Early biochemical studies demonstrated that mouse and human ORF1p resides in cytoplasmic RNPs and binds single-strand RNA in a sequence-independent manner (78, 225–228). Biochemical and genetic analyses clearly demonstrate that ORF1p is required for LINE-1 RNP formation, and that LINE-1 RNP formation is a necessary step in the retrotransposition process (198).

Structural studies have revealed that the N-terminus of LINE-1 ORF1p contains a coiled-coil domain, which facilitates trimerization of ORF1p molecules (79, 229–232). The central region of ORF1p contains a noncanonical RNA recognition motif that, with assistance of its C-terminal domain, is required for ORF1p RNA binding (79, 231, 233). Notably, missense mutations in highly conserved amino acid residues in both the RNA recognition motif and C-terminal domain either abolish or adversely affect LINE-1 retrotransposition in cultured cells (60, 231).

Human and mouse ORF1p contain nucleic acid chaperone activities that can facilitate the re-annealing of single-strand DNAs *in vitro* (80, 234–236). Notably, several studies have shown that, despite the lack of sequence homology, proteins encoded by non-mammalian LINEs also contain nucleic acid chaperone activity (237, 238). For example, ORF1p from a zebrafish LINE (ZfL2-1) has nucleic acid chaperone activity (239). It is hypothesized that this nucleic acid chaperone activity facilitates the initial steps of LINE-1 integration *in vivo*. Somewhat unexpectedly, the deletion of ORF1 does not abolish ZfL2-1 retrotransposition activity in cultured human cells (240). These data, coupled with the fact that Alu retrotransposition only requires the protein encoded by LINE-1 ORF2 (92), raise questions regarding how ORF1p nucleic acid chaperone activity participates in LINE-1 retrotransposition.

The development of cell-free systems to monitor LINE-1 retrotransposition would allow a more rigorous examination of the ORF1p functions required for retrotransposition.

ORF2p

ORF2p is an ~150 kDa protein (81, 82, 200, 241) that contains endonuclease (L1 EN) (83) and reverse transcriptase (L1 RT) (84) activities that are critical for retrotransposition (60, 83). The L1 EN domain resides near the N-terminus of the protein, and bears similarity to apurinic/apyrimidinic endonucleases (APEs) (242–244). *In vitro* and bioinformatics analyses (71, 83, 99, 196, 245) suggest that L1 EN makes a single-strand endonucleolytic nick at a loosely defined consensus sequence in genomic DNA (5′-TTTT/A-3′; where the slash indicates the scissile phosphate), exposing a 5′ phosphate and 3′ hydroxyl group (83). Crystallographic studies suggest that L1 EN recognizes an extra helical "flipped" adenine residue 3′ of the scissile bond to mediate cleavage using a mechanism similar to that employed by other APE proteins (244). In addition, it is likely that epigenetic modifications of target DNA (*e.g.*, nucleosome accessibility) might affect ORF2p accessibility and L1 EN cleavage activity (246).

The L1 RT domain is located downstream of the EN domain in ORF2p, and shares sequence similarity to the RT domains encoded by telomerase, Penelope-like retrotransposons, group II introns, other non-LTR retrotransposons, LTR retrotransposons, and retroviruses (247–249). Biochemical and genetic assays originally were used to demonstrate that Ty1/LINE-1 ORF2p fusion proteins possess reverse transcriptase activity *in vitro* (84, 250). The subsequent purification of recombinant ORF2p produced in a baculovirus expression system revealed that full-length ORF2p could efficiently generate reverse transcripts from poly rA/oligo dT_{12} primer template complexes, that L1 RT activity exhibited a preference for Mg^{2+} over Mn^{2+}, and that L1 RT exhibited both RNA-dependent and DNA-dependent polymerase activities (251). Additional studies revealed that, like the RT encoded by the R2Bm retrotransposon (252), L1 RT is highly processive (when compared to Moloney murine leukemia virus RT) and lacks detectable RNase H activity (253).

L1 RT activity has been detected in LINE-1 RNP preparations derived from cells transfected with engineered LINE-1 expression vectors (199). Importantly, this work confirmed that ORF2p preferentially reverse transcribes its own mRNA template (*i.e.*, it exhibits *cis*-preference for its encoding RNA) and that point mutations in ORF1 and the L1 EN domain, which adversely affect LINE-1 retrotransposition, retain reverse transcriptase activity (199). Finally, these and subsequent studies confirmed previous inferences (156) that L1 RT can extend terminally mismatched primer–template complexes (199, 254). The latter property distinguishes L1 RT from Moloney murine leukemia virus and other retroviral reverse transcriptase enzymes.

Although LINE-1 ORF1p and L1 RT activity were readily detectable in RNP preparations, the detection of the LINE-1 ORF2p had been notoriously difficult. Epitope-tagging strategies have allowed the detection of ORF2p in whole cell extract and RNP preparations derived from cells transfected with engineered LINE-1 expression vectors (82, 200, 241). Using a similar strategy, immunofluorescence microscopy studies revealed that engineered ORF2p co-localizes in cytoplasmic foci with both ORF1p and LINE-1 mRNA (82, 200, 241) (Figure 2). Despite progress in detecting ORF2p in cultured cells, debate continues regarding the stoichiometry of ORF1p and ORF2p bound to LINE-1 mRNA (82, 200, 241). It appears that ORF1p is much more abundant in LINE-1 RNPs than ORF2p. Additionally, the composition of a functional LINE-1 RNP and how the LINE-1 mRNA transitions to a retrotransposition intermediate require elucidation. Clearly, the development of reconstituted *in vitro* target-site primed reverse transcription reactions would greatly advance the understanding of the detailed molecular mechanism of LINE-1 retrotransposition.

LINE-1 ORF2p contains an ill-defined cysteine-rich domain (C-domain) at its C-terminus, which has been suggested to function as a zinc-knuckle domain (85). Consistent with its biological importance, cysteine to serine mutations in the C-domain interfere with LINE-1 RNP formation and strongly inhibit LINE-1 retrotransposition in cultured cells (60, 82). Recent studies indicate that a recombinant protein containing the last 180 amino acids of ORF2p exhibits nonsequence-specific RNA binding *in vitro*, and that cysteine to serine mutations in the C-domain do not adversely affect RNA binding (255). Hence, future studies are required to elucidate the exact function of the C-domain in LINE-1 retrotransposition.

Additional functional domains are likely to exist within LINE-1 ORF2p. Indeed, PCNA, which is the sliding clamp protein essential for DNA replication, recently was found to directly interact with ORF2p through a conserved sequence known as a PCNA interaction protein domain (PIP box), which is located between the L1 EN and L1 RT domains (200). Mutating the PIP box abolished LINE-1 retrotransposition (200);

however, how PCNA functions in LINE-1 retrotransposition requires further elucidation.

How ORF2p is translated from bicistronic LINE-1 mRNA remains an active area of study and recent reports suggest that human and mouse LINE-1 ORF2p may be translated by distinct mechanisms (256, 257). In human LINE-1s, a 63-nucleotide spacer that contains two in-frame stop codons separates ORF1 and ORF2. Genetic studies in cultured cells suggest that ORF2p translation occurs by an unconventional termination/re-initiation mechanism where a translating ribosome must be able to scan from the stop codon of ORF1 to the start codon of ORF2 (256). Remarkably, studies show that human ORF2 can be translated in an AUG-independent manner (256). By comparison, evidence from luciferase reporter assays suggests that the presence of an internal ribosome entry site, which is located near the 3′ end of mouse ORF1, is used to facilitate translation of mouse ORF2 (257). In addition, cell culture assays have revealed that mouse ORF2 may be translated in an AUG-independent manner (256, 257). Notably, it is unlikely that the 3′ end of human ORF1 has an internal ribosome entry site (256). Indeed, it has been demonstrated that the sequence of the ORFs encoded by human and mouse LINE-1s can be subjected to substantial sequence changes by codon optimization without affecting retrotransposition in cultured cells (258, 259), suggesting that strict *cis*-acting sequences are not required for ORF2 translation.

It is unlikely that LINE-1s have evolved a novel mechanism to mediate ORF2 translation. Instead, we hypothesize that LINE-1s have evolved to exploit translation mechanisms inherent to their hosts to mediate ORF2 translation (256). Interestingly, recent ribosomal profiling studies have uncovered an increasing number of unannotated reading frames that reside 5′ of annotated ORFs (260); some short ORFs also may be translated via an AUG-independent mechanism (261). Clearly, additional studies are warranted to elucidate the ORF2 translation mechanism and to determine if ORF2 translation differs among mammalian LINE-1s.

The LINE-1 3′ UTR

The human LINE-1 3′ UTR is ~206 bp in length and contains a conserved polypurine tract that is predicted to form a G-quadruplex structure (262). Intriguingly, the polypurine tract is not required for LINE-1 retrotransposition in cultured cells (60); but the polypurine tract can inhibit LINE-1 RT activity in *in vitro* biochemical assays (253). Despite its evolutionary conservation, how the polypurine tract functions in LINE-1 biology remains unknown.

LINE-1 3′ UTRs contain a functional RNA polymerase II polyadenylation signal near their 3′ ends. Experiments in cultured cells have revealed that the LINE-1 poly (A) signal is relatively weak, is often bypassed by RNA polymerase II, and that RNA polymerase II frequently uses canonical polyadenylation sites fortuitously present in 3′ flanking genomic DNA sequences (263). The use of these genomic polyadenylation sequences can lead to the generation of chimeric LINE-1 transcripts containing genomic DNA sequences at their 3′ end (see below). Finally, recent data suggest that the human and mouse 3′ UTRs have promoter activity that leads to the generation of alternative LINE-1 transcripts in various tissues (264). The field awaits a better definition of this promoter activity and the role of the resultant transcripts in LINE-1 biology.

An Overview of the LINE-1 Replication Pathway

LINE-1 retrotransposition occurs via a "copy and paste" process termed target-site primed reverse transcription (TPRT; Figure 3), a mechanism originally described by the Eickbush laboratory for the related site-specific non-LTR retrotransposon, R2Bm, from the silkworm *Bombyx mori* genome (265). After transcription from a chromosomal locus, a full-length bicistronic LINE-1 mRNA is exported to the cytoplasm. Upon translation, ORF1p and ORF2p exhibit a strong *cis*-preference (97, 98) and bind to their respective encoding mRNA, forming an RNP (78, 82, 198, 227, 228). The LINE-1 RNP minimally consists of LINE-1 mRNA, multiple ORF1p trimers, and as few as one molecule of ORF2p (82, 256), but also likely contains numerous cellular proteins and RNAs (200, 201, 266) (Figure 3).

Intriguingly, subsequent studies revealed that ORF1p, ORF2p, and LINE-1 RNA accumulate in dense cytoplasmic foci, which are closely associated with stress granule proteins (82, 197). In yeast, proteins encoded by Ty1 and Ty3 retrotransposons are associated with cytoplasmic foci called processing bodies (P-bodies) and experiments suggest that P-body localization is important for RNP assembly and may represent a host mechanism that regulates retrotransposition (267–269).

How the LINE-1 RNP enters the nucleus is not fully understood. Experiments using modified second-generation adenoviral expression vectors containing an active human LINE-1 have demonstrated that LINE-1 retrotransposition can occur in G1/S arrested cells (270). Similarly, LINE-like sequences from *Candida albicans* (271, 272) and *Neurospora crassa* (273), which undergo closed mitosis, can retrotranspose independently of nuclear envelope breakdown. So, it does not appear that cell division is a requisite for LINE-1 retrotransposition. Some reports suggest that cell division augments the retrotransposition of engineered LINE-1s in cultured cells (274, 275). Notably, the cultured cell retrotransposition assay generally requires the detection of retrotransposition events as a function of reporter gene expression. Hence, as cells divide they may produce more of the reporter gene product, leading to an apparent increase in LINE-1 retrotransposition potential, thereby explaining the apparent discrepancies among the above studies.

Once in the nucleus, L1 EN makes a single-strand endonucleolytic nick in genomic DNA at a degenerate consensus sequence (5′-TTTT/A: where the "/" indicates the scissile phosphate), exposing a 3′ hydroxyl group that serves as a primer for the reverse transcription of the LINE-1 mRNA by the L1 RT activity encoded in ORF2p (83, 276) (Figure 3). Whether the LINE-1 mRNA simply acts as a template for retrotransposition or whether it plays additional roles during TPRT requires more study. It is notable that codon-optimized synthetic mouse and human LINE-1s, in which ~25% of the nucleotide sequence has been replaced to increase the G-C content of LINE-1 RNA while retaining the amino acid sequence of the LINE-1 encoded proteins, can readily retrotranspose in cultured human cells (258, 259).

Studies in cultured human cells have revealed that the LINE-1 RT has a misincorporation error rate of ~1 in 6,500 bases (156). By using the binomial distribution, it has been estimated that ~40% of full-length LINE-1 retrotransposition events represent faithful copies of the progenitor LINE-1 element (~37% contain one mutation, and ~16% contain two mutations) (156). Subsequent steps in the retrotransposition process, including second-strand target-site DNA cleavage and second-strand LINE-1 cDNA synthesis, require additional investigation. By analogy to the evolutionarily related R2 retrotransposon of *Bombyx mori*, ORF2p may play a role in each of the above processes (277). The net result of TPRT is the integration of a new LINE-1 copy at a new chromosomal location (Figure 3).

Genomic Rearrangements Generated During LINE-1 Retrotransposition

LINE-1-mediated retrotransposition events are sometimes accompanied by intra-LINE-1 rearrangements (*e.g.*, 5′ truncations and 5′ truncations associated with inversion/deletion events) or genomic structural rearrangements (Figure 4). The features of these events suggest that host processes, such as DNA repair and/or DNA replication, may ultimately impact the structure of newly retrotransposed LINE-1s. Below we discuss some of these rearrangements.

Intra-LINE-1 Alterations

The examination of the HGR reveals that ~30% to 35% of human-specific Ta-subset LINE-1 insertions are full-length, ~40% to 45% are truncated at their 5′ ends, and ~25% contain internal rearrangements known as inversion/deletions (7, 63, 71, 278). The characterization of engineered LINE-1 retrotransposition events from cultured cells has led to the proposition that two pathways of LINE-1 retrotransposition exist: conventional and abortive retrotransposition (156). Conventional retrotransposition accounts for the generation of full-length LINE-1 insertions and can lead to the formation of new "master genes" that can serve as a source of retrotransposition events in subsequent generations (Figure 4A). In general, full-length LINE-1 insertions are characterized by typical LINE-1 structural hallmarks (*i.e.*, they terminate with a poly (A) tract; are flanked by variable size target-site duplications; and integrate at a LINE-1 endonuclease consensus site) (reviewed in reference 11).

The generation of 5′ truncated LINE-1 elements is proposed to occur via abortive retrotransposition. Here, the L1 RT becomes dissociated from the (–) strand LINE-1 cDNA during TPRT. Annealing of the LINE-1 cDNA to top-strand genomic DNA then may specify the placement of top-strand (also referred to as second-strand) genomic DNA cleavage, generating a 3′ hydroxyl group needed for DNA-dependent (+) strand LINE-1 cDNA synthesis (156, 279). How the L1 RT may become dissociated from the LINE-1 cDNA requires clarification; however, it is intriguing to speculate that the process of LINE-1 integration represents a battleground between LINE-1 and the host, and that the Y-branch intermediate generated during (–) strand LINE-1 cDNA synthesis may elicit a DNA repair response(s) by the host (156). Recent studies suggest that the ataxia telangiectasia mutated (ATM) and excision repair cross-complementation group 1 (ERCC1) proteins modulate LINE-1 retrotransposition (280–282); hence, it is reasonable to speculate that DNA repair pathways might influence the generation of 5′ truncated LINE-1s.

The formation of inversion/deletion structures represents an alternative form of conventional retrotransposition termed "twin-priming" (278). Here, the LINE-1 mRNA anneals to single-strand DNA exposed at both the cleaved bottom- and top-strand genomic

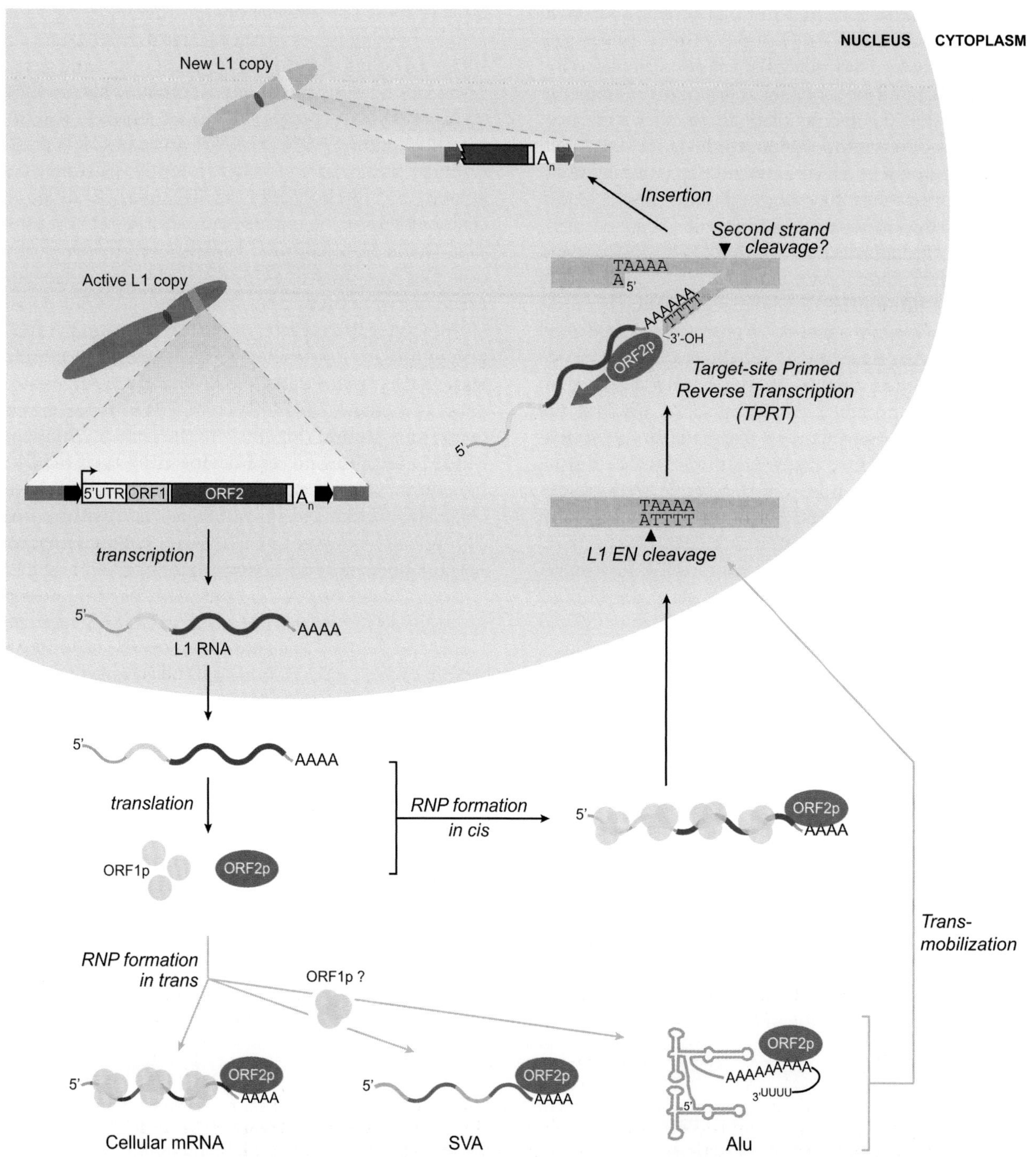
NUCLEUS
CYTOPLASM
New L1 copy
A
n
Insertion
Second strand
cleavage?
TAAAA
A 5'
AAAAAA
TTTT
3'-OH
ORF2p
Target-site Primed
Reverse Transcription
(TPRT)
5'
Active L1 copy
5'UTR
ORF1
ORF2
A
n
TAAAA
ATTTT
L1 EN cleavage
transcription
5'
AAAA
L1 RNA
5'
AAAA
translation
RNP formation
in cis
5'
ORF2p
AAAA
ORF1p
ORF2p
Trans-
mobilization
RNP formation
in trans
ORF1p ?
5'
ORF2p
AAAA
5'
ORF2p
AAAA
ORF2p
AAAAAAAAA
3'UUUU
5'
Cellular mRNA
SVA
Alu

DNA sequences (Figure 4A). Template switching of the L1 RT during RNA-dependent (–) strand LINE-1 cDNA synthesis, or perhaps a second molecule of ORF2p, then allows the use of the 3′ hydroxyl group generated at the top-strand LINE-1 mRNA/genomic DNA duplex to serve as a primer for convergent RNA-dependent (–) strand LINE-1 cDNA synthesis. Microhomology-mediated annealing of the resultant cDNAs followed by the completion of cDNA synthesis (by either a LINE-1 RT DNA-dependent DNA polymerase activity or a host-encoded DNA polymerase) then can lead to the formation of inversion/deletion structures (Figure 4A). Notably, virtually all of the predictions of the twin-priming model have been confirmed by examining engineered LINE-1 integration events from cultured cells (156). How microhomology-mediated annealing occurs needs further study, although one can hypothesize that it is carried out by an alternative, microcomplementarity-mediated non-homologous end joining pathway of DNA repair (283).

The incorporation of untemplated nucleotides, presumably added after the completion of (–) strand LINE-1 cDNA synthesis by the LINE-1 RT, can result in short stretches of nontemplated sequence at the 5′ genomic DNA/LINE-1 junction (156, 203, 208, 284), which may facilitate annealing of the LINE-1 cDNA to single-strand DNA exposed at the top-strand genomic DNA target-site (279, 285, 286). If so, we reason that the resultant LINE-1 cDNA/genomic DNA hybrid then may specify the placement of top-strand genomic DNA cleavage, generating the 3′ hydroxyl group needed for DNA-dependent (+) strand cDNA synthesis by either the LINE-1 reverse transcriptase or a host-encoded DNA polymerase.

LINE-1-Mediated Transduction Events

Active LINE-1s mobilize sequences that are derived from their 5′ and 3′ flanking genomic DNA by a process termed LINE-1-mediated transduction (Figure 4A). LINE-1s containing 5′ transduction events occur when a cellular promoter, which resides upstream of an active genomic full-length LINE-1 copy, is used to initiate LINE-1 transcription. Retrotransposition of the chimeric 5′ genomic/LINE-1 mRNA transcript then leads to the transduction of the 5′ derived genomic DNA sequence to a new chromosomal location. If a conventional RNA polymerase II promoter in genomic DNA is used to initiate LINE-1 transcription, the resultant 5′ transduced LINE-1 will lack the genomic promoter and generally will be transcribed using the internal promoter present in the LINE-1 5′ UTR in successive rounds of retrotransposition. Full-length LINE-1s containing 5′ transduced genomic DNA sequences originally were detected in the HGR (7). A 5′ transduction event is relatively rare and can only be identified by examining the sequences of full-length LINE-1s. Notably, the Nathans laboratory demonstrated that a full-length mouse LINE-1 insertion carrying a 28 bp 5′ transduction led to the mis-splicing of the *Nr2e3* gene in a retinal degeneration 7-mouse model (287).

Figure 3 Long INterspersed Element-1 (LINE-1) retrotransposition cycle. An active copy of LINE-1 is present at one chromosomal locus (light blue box in dark gray chromosome) and consists of a 5′ untranslated region (UTR) (light gray box) with an internal promoter (thin black arrow), two open reading frames (ORF1, yellow box, and ORF2, blue box), a 3′ UTR (light gray box) followed by a poly (A) tract (A_n) and is flanked by target-site duplications (thick black arrows). Transcription of LINE-1 occurs in the nucleus and produces a bicistronic RNA (wavy line). Upon translation in the cytoplasm, ORF1p and ORF2p (yellow circles and blue oval, respectively) bind back to their encoding RNA (*cis*-preference) to form a ribonucleoprotein particle (RNP) complex. ORF1p and/or ORF2p also can retrotranspose cellular RNAs (mRNA, SVA, and Alu, in red, green, and orange wavy lines, respectively). The retrotransposition of Alu RNA only requires ORF2p (92). There is some debate as to whether ORF1p augments Alu retrotransposition (410), and if SVA retrotransposition requires both ORF1p and ORF2p (94, 95). The LINE-1 RNP enters the nucleus where *de novo* insertion occurs by a mechanism termed target-site primed reverse transcription (TPRT). The ORF2p endonuclease activity makes a single-strand endonucleolytic nick at the genomic DNA target (L1 EN cleavage), at a loosely defined consensus site (5′-TTTT/A-3′, with "/" indicating the scissile phosphate). The ORF2p RT activity then uses the exposed 3′-OH group to initiate first-strand LINE-1 cDNA synthesis using the bound RNA as a template. The final steps of TPRT (*i.e.*, top-strand cleavage, second-strand LINE-1 cDNA synthesis, and repair of the DNA ends) lead to the insertion of a *de novo* LINE-1 copy at a new chromosomal locus (light yellow box in light gray chromosome). The new LINE-1 copy is often 5′ truncated, contains a variable-sized poly (A) tract (A_n), and generally is flanked by target-site duplications (thick gray arrows). Additional references are provided in the text.
doi:10.1128/microbiolspec.MDNA3-0061-2014.f3

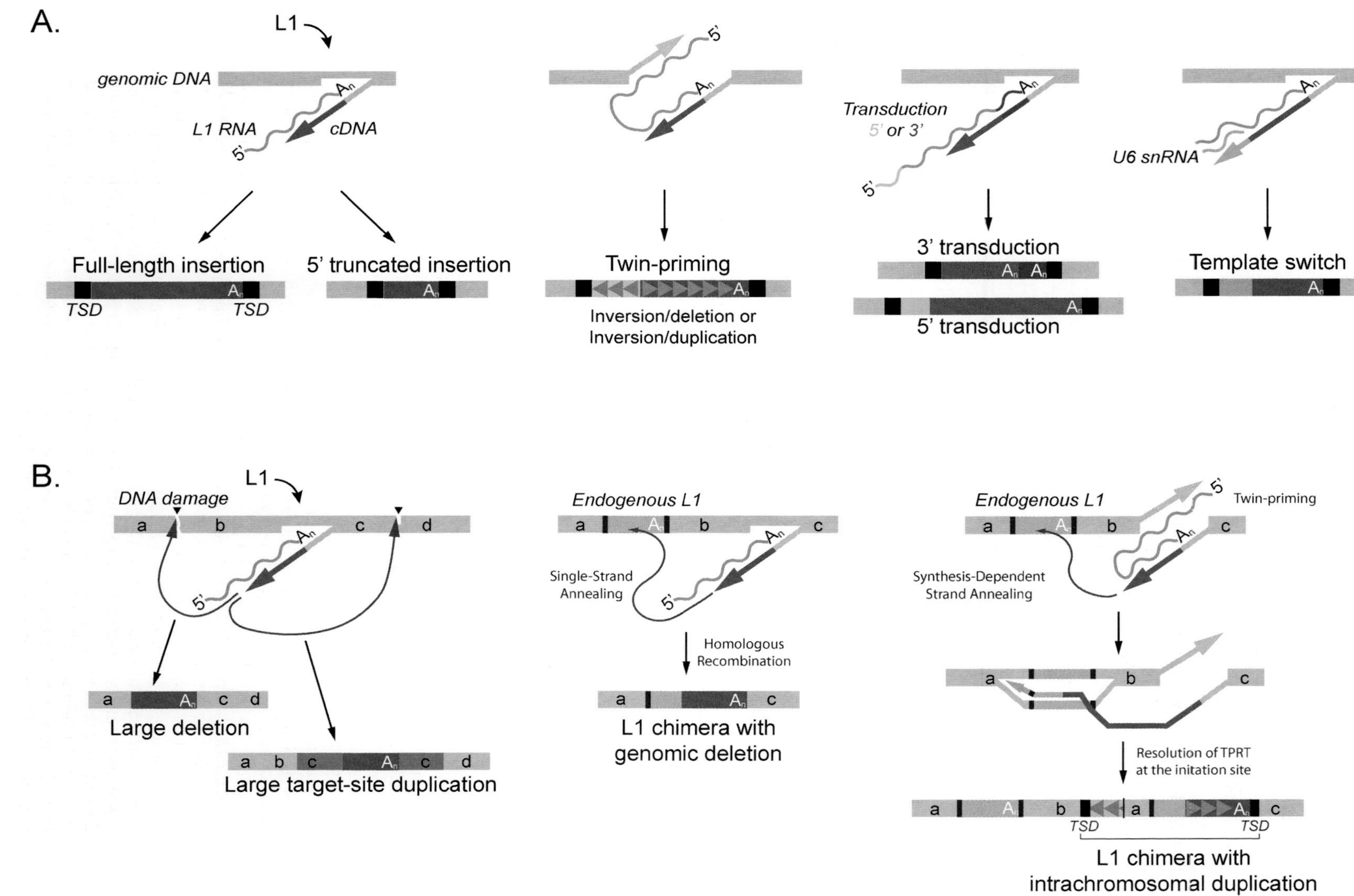
A.
L1
genomic DNA
L1 RNA
cDNA
5'
Full-length insertion
TSD
TSD
5' truncated insertion
Twin-priming
Inversion/deletion or
Inversion/duplication
Transduction
5' or 3'
3' transduction
5' transduction
U6 snRNA
Template switch
B.
DNA damage
L1
Large deletion
Large target-site duplication
Endogenous L1
Single-Strand Annealing
Homologous Recombination
L1 chimera with genomic deletion
Endogenous L1
Twin-priming
Synthesis-Dependent Strand Annealing
Resolution of TPRT at the initation site
TSD
TSD
L1 chimera with intrachromosomal duplication

Due to the presence of inherently weak polyadenylation signals in their 3′ UTRs, LINE-1s also can mobilize sequences that are derived from their 3′ flanks, including exons, that range in size from tens of base pairs to at least 1.6 kb in length by 3′ transduction (263, 288–290) (Figure 4A). This class of insertion is generated when RNA polymerase II bypasses the weak polyadenylation signal present at the 3′ end of a full-length LINE-1 and instead uses a fortuitous polyadenylation signal in the 3′ flanking genomic DNA. Retrotransposition of the resultant LINE-1/genomic hybrid mRNA leads to the insertion of the 3′ flanking genomic DNA downstream of the new LINE-1 copy at a new chromosomal location. Due to the structure of most mammalian genes, which contain long introns with significant numbers of LINE-1 and SINE insertions, we speculate that LINE-1s have evolved to contain a weak polyadenylation signal to minimize premature polyadenylation of intron-containing genes (263) (reviewed in references 291–293).

The fact that LINE-1s can retrotranspose sequences derived from their 3′ genomic flanks to new genomic locations was first appreciated while characterizing a mutagenic LINE-1 insertion into the dystrophin gene (290). The 3′ transduction "genomic tag" in the mutagenic insertion then was used to isolate the likely progenitor LINE-1, named LRE2 (290). Experiments in cultured cells subsequently showed the LINE-1 3′ transduction events occur frequently and could, in principle, mobilize exons and promoters to new genomic locations, providing a possible mechanism for exon shuffling (60, 263). Since that time, it has become apparent that ~20% to 25% of LINE-1 retrotransposition events are accompanied by 3′ transduced sequences (288, 289). The presence of 3′ transductions also have been used as "genomic tags" to identify progeny/offspring

Figure 4 Alterations generated upon Long INterspersed Element-1 (LINE-1) retrotransposition. (A) LINE-1 retrotransposition events can alter target-site genomic DNA. *De novo* insertion of LINE-1 occurs at a genomic DNA target (thick gray line). LINE-1 RNA (blue wavy line) is followed by a poly (A) tail (A_n); LINE-1 cDNA (blue arrow); and a new LINE-1 copy (blue box including a poly (A) tract (A_n)). Insertions can occur by either conventional (full-length, left) or abortive (5′ truncated, right) retrotransposition and generally result in the formation of variable-length target-site duplications (TSD, black boxes). "Twin-priming" generates LINE-1 inversion/deletions or inversion/duplications (represented by opposing arrows in the new LINE-1 copy). The priming of LINE-1 cDNA synthesis from the cleaved top-strand genomic DNA is represented by the light blue arrow. The transduction of genomic DNA sequences can occur when either 5′ or 3′ flanking genomic sequences are incorporated into LINE-1 RNAs and are mobilized by retrotransposition. The 5′ and 3′ transductions are depicted in both LINE-1 RNA (green or pink wavy lines) and the new LINE-1 copy (green or pink boxes). The 3′ transduction events contain two poly (A) sequences (A_n). The LINE-1 enzymatic machinery also can mobilize small nuclear RNAs (snRNAs) such as U6 snRNA to new genomic locations. The proposed model involves an L1 RT template switch from LINE-1 RNA to the U6 snRNA (orange wavy line) to generate U6 cDNA (orange arrow) during target-site primed reverse transcription (TPRT). (B) LINE-1 retrotransposition events associated with genomic structural variation. LINE-1 RNA, cDNA, and a *de novo* LINE-1 insertion are depicted as in panel A. Lower case letters (a, b, c, or d) in genomic DNA (gray boxes) are used to depict deletions or duplications (by alteration of the alphabetical order). The resolution of TPRT at the site of DNA damage (left panel, black arrowhead upstream of the integration site) is hypothesized to result in a large genomic deletion (the loss of segment "b"), whereas the resolution of TPRT at a single-strand endonucleolytic nick downstream from the LINE-1 integration site (left panel, black arrowhead) is hypothesized to lead to a large target-site duplication (the duplication of segment "c"). The resolution of TPRT by single-strand annealing (middle) can lead to the generation of a chimeric LINE-1, where an endogenous LINE-1 (light purple box) is fused to a new LINE-1 (dark blue box); the formation of the chimera results in the loss of segment "b". Similarly, the resolution of "twin-priming" intermediates by synthesis-dependent strand annealing (right) can lead to the generation of an L1 chimera with an intrachromosomal duplication (the duplication of both segment "a" and the endogenous L1 sequence). The entire insertion is flanked by target-site duplications (black boxes). Notably, synthesis-dependent strand annealing can occasionally repair LINE-1 insertions generated in cultured cells by "twin-priming" (156). Details on how chimeric LINE-1 integration events are formed can be found elsewhere (156, 202, 203). Additional references are provided in the text. doi:10.1128/microbiolspec.MDNA3-0061-2014.f4

relationships among LINE-1 elements, to stratify full-length LINE-1 elements into subdivisions, and to identify clusters of LINE-1s that are actively mobilizing in the human population (58, 294).

The severe 5′ truncation of the LINE-1/genomic mRNA upon TPRT can lead to the generation of "orphan transductions" that lack LINE-1 sequence (263). A mutagenic "orphan transduction" recently was identified as a cause of Duchene muscular dystrophy (295). Notably, and consistent with cultured cell studies, a recent study reported that ~24% of somatic LINE-1 retrotransposition events in tumors derived from 244 patients were accompanied by 3′ transduction events, and that many of these events represented "orphan transductions" (296). Finally, it is noteworthy that transduction is not peculiar to LINE-1s. Both 5′ and 3′ transduction events have been observed with SVA retrotransposons (127, 297–299). Indeed, SVA retrotransposons provided a vehicle to shuffle the acyl-malonyl condensing enzyme-1 (*AMAC*) gene to three different locations in primate genomes (298).

LINE-1 Target-Site Alterations: Local Alterations at the Integration Site

The mechanism of top-strand target-site cleavage at the genomic LINE-1 integration site requires elucidation. It is clear that the placement of second-strand DNA cleavage can influence the structure of the resultant LINE-1 integration events. After characterizing 100 engineered LINE-1 insertions in cultured cells, Gilbert and colleagues proposed a model that accounts for a number of observed target-site alterations (156, 202). In this model, top-strand DNA cleavage upstream from the bottom-strand endonucleolytic nick can lead to small deletions of genomic DNA at the LINE-1 integration site. Likewise, top-strand DNA cleavage directly opposite to bottom-strand cleavage can lead to LINE-1 integration events that lack target-site alterations, whereas top-strand cleavage downstream of the initial endonucleolytic nick can lead to the generation of target-site duplications (Figure 4A). Interestingly, ~10% of LINE-1 insertions in cultured cells were accompanied by large (greater than 50 bp) target-site duplications (TSDs); these large TSDs rarely are observed flanking LINE-1s in the human genome reference sequence (156, 202) (Figure 4B). Otherwise, the cultured cell retrotransposition assay largely recapitulates the spectrum of structural outcomes observed among endogenous germline retrotransposition events. However, the cellular milieu in which retrotransposition takes place, perhaps defined by the presence and activity of DNA repair machinery and other host factors, likely influences the range of possible LINE-1-mediated target-site alterations. Notably, the mechanisms described above also are likely to account for local target-site alterations accompanying SINE retrotransposition and processed pseudogene formation (reviewed in reference 11).

LINE-1-Mediated Retrotransposition Target-Site Alterations: The Generation of Structural Variants

In addition to minor target-site alterations, LINE-1 retrotransposition can lead to more substantial target-site genomic DNA modifications. The examination of LINE-1 integration events in cultured human cells has revealed that approximately 10% of retrotransposition events are accompanied by rearrangements of target-site DNA, creating genomic structural variation (156, 202, 203) (Figure 4B). Comparisons of pre-integration and post-integration sites in genomic DNA have revealed that the resolution of TPRT intermediates can lead to the generation of chimeric LINE-1 sequences (156, 202, 203). Single nucleotide polymorphism analyses revealed that the resultant integration events contain an endogenous genomic LINE-1 fused to the engineered LINE-1 and concomitant genomic alterations (156, 202, 203) (Figure 4B).

The formation of chimeric LINE-1 elements can occur by various mechanisms. For example, the resolution of TPRT intermediates by single-strand annealing or synthesis-dependent strand annealing can lead to the formation of LINE-1 retrotransposition-mediated deletions or duplications, respectively (156, 202, 203) (Figure 4B). Importantly, large-scale genomic alterations observed in cultured cells are reflective of events occurring in humans. For example, a LINE-1 retrotransposition event was responsible for a 46 kb deletion in the *PDHX* gene of a human patient, resulting in pyruvate dehydrogenase deficiency (300).

Genomic alterations also can accompany the retrotransposition of both Alu and SVA elements. For example, an Alu retrotransposition event that occurred ~2.7 million years ago led to the deletion of an internal 92 bp exon within the CMP-Neu5Ac hydroxylase gene. As a result, humans are genetically deficient for *N*-glyconeuraminic acid (301). Similarly, an SVA retrotransposition event led to an ~14 kb deletion that resulted in the loss of the *HLA-A* gene in a cohort of Japanese families afflicted with leukemia (302). Finally, recent studies showed that two independent post-zygotic SVA retrotransposition events into the *NF1* gene were associated with large deletions of ~1 Mb and 867 kb, respectively (167).

On a larger scale, comparative genomics approaches between the human and chimpanzee reference sequences led to the identification of 50 LINE-1 retrotransposition events responsible for the deletion of ~18 kb from the human genome and ~15 kb from the chimpanzee genome (303). Similar approaches uncovered 33 Alu retrotransposition events that eliminated approximately 9 kb of human DNA (304). Hence, although relatively rare when compared with conventional retrotransposition events, LINE-1 retrotransposition-mediated deletion events continue to sculpt the landscape of the human genome.

Post-Integration Recombination Events Between Genomic Retrotransposons

The sheer mass of LINE-1 and Alu sequences in the genome also can provide substrates for post-integration recombination, generating structural variation in the human genome. For example, non-allelic homologous recombination, non-homologous DNA end joining, and other types of recombination events between genomic LINE-1 or Alu elements can lead to genomic alterations that result in human disease (reviewed in references 11, 151, and 305–307). Clearly, these examples will continue to grow as individual whole genome DNA sequencing continues during the coming years.

Endonuclease-Independent LINE-1 Retrotransposition

LINE-1 retrotransposition by TPRT is usually initiated by the cleavage of genomic DNA by L1 EN. The examination of LINE-1 retrotransposition in Chinese hamster ovary (CHO) cells deficient in components of the non-homologous end joining pathway of DNA double-stranded break repair led to the discovery of an alternative integration pathway termed endonuclease-independent (ENi) LINE-1 retrotransposition (196). The ENi pathway of LINE-1 retrotransposition is reminiscent of a type of RNA-mediated DNA repair in which LINE-1 elements that lack L1 EN function presumably can use genomic lesions to initiate TPRT (196, 308). ENi retrotransposition events bear structural hallmarks distinct from canonical TPRT-mediated LINE-1 insertions in that they are frequently both 5′ and 3′ truncated, do not occur at typical LINE-1 endonuclease sites in genomic DNA, generally lack target-site duplications, and often are accompanied by the deletion of genomic DNA at the integration site (196). ENi LINE-1 retrotransposition events also occasionally are accompanied by the insertion of short cDNA fragments at both their 5′ and 3′ LINE-1/genomic DNA junctions, which appear to be derived from the reverse transcription of cellular mRNAs (196, 309).

Subsequent studies in DNA protein kinase catalytic subunit-deficient CHO cells revealed that ENi retrotransposition events could occur at dysfunctional telomeres, highlighting similarities between ENi retrotransposition and telomerase activity (309–312). Indeed, these results parallel the situation in *Drosophila* and bdelloid rotifer genomes, where "domesticated" retrotransposons function to maintain telomere length in place of a conventional telomerase activity (311, 313–315).

As with other phenomena discovered using engineered LINE-1 retrotransposons in the cultured cell retrotransposition assay, putative ENi retrotransposition events also have been identified in the human and mouse genomes (174, 316). Indeed, a likely ENi retrotransposition event into the *EYA1* gene was accompanied by an ~17 kb deletion, leading to a sporadic case of human oto-renal syndrome (317). It will be interesting to learn whether deficiencies in other DNA repair pathways lead to increased ENi LINE-1 retrotransposition.

Notably, some group II introns lack an EN domain and can use 3′ hydroxyl groups at nascent DNA strands present at DNA replication forks to initiate retrotransposition (318). Moreover, it has been proposed that the L1 EN domain was acquired after the L1 RT domain during LINE-1 evolution (248). Together, these data indicate that the ENi pathway of LINE-1 retrotransposition may represent an ancient mechanism of LINE-1 retrotransposition before the acquisition of an APE-like endonuclease domain.

THE IDENTIFICATION OF HOST FACTORS THAT REGULATE LINE-1 RETROTRANSPOSITION

In the face of the deleterious consequences of LINE-1-mediated retrotransposition, the host cell appears to have evolved a variety of mechanisms to restrict LINE-1 activity (reviewed in reference 13). Several of these processes are briefly described below.

LINE-1 DNA Methylation

DNA Methyltransferases

The methylation of CpG sequences in the LINE-1 5′ UTR is associated with the suppression of LINE-1 expression in a variety of cell types (319). DNA methylation is established in primordial germ cells, is maintained throughout the life of an organism, and is thought to control the expression of LINE-1 in somatic tissues (320). *De novo* DNA methylation is catalyzed by the DNA methyltransferases Dnmt3a and Dnmt3b, but targeting of methylation to retrotransposon sequences requires the non-catalytic paralog DNA methyltransferase 3-like (Dnmt3L) (reviewed in reference 321). Notably, Dnmt3L-deficient male mice exhibit meiotic catastrophe, concomitant with an aberrant overexpression of LINE-1 and other transposon sequences (322). Similar phenotypes also are observed in Maelstrom-deficient male mice (323). However, the mechanistic link between the loss of transposon methylation, transcriptional reactivation, and meiotic catastrophe in germ cells requires further investigation. Finally, recent reports suggest that 5-hydroxymethylation of cytidine, which is a mark enriched on the LINE-1 5′ UTR in pluripotent cells, might further regulate LINE-1 transcription (reviewed in reference 324).

piRNAs, PIWI Proteins, and Small RNAs Derived from LINE-1

Other factors also are proposed to regulate *de novo* methylation of LINE-1 sequences in the developing germline (13). P-element induced wimpy testes (PIWI) proteins, representing a subclade of the Argonaute family of small RNA binding proteins, interact with 26 to 31 nucleotide RNAs termed piwi-interacting RNAs (piRNAs) (reviewed in references 325 and 326). In the *Drosophila* germline, PIWI proteins and piRNAs comprise an effective transposon defense system, wherein transposon-derived piRNAs direct PIWI proteins to cleave active transposon transcripts (327, 328). The cleavage and subsequent processing of expressed transposon RNAs give rise to additional piRNAs that fuel additional rounds of this robust and adaptive transposon defense cycle, which is aptly named the "ping-pong" amplification cycle (329, 330) (reviewed in references 13, 326, 331, and 332).

Small piRNAs actively defend the mammalian germline from transposons (333) (reviewed in references 326 and 334). In male mice, a deficiency in either of two murine PIWI clade proteins, murine piwi (MIWI2) or miwi-like (MILI), results in a similar phenotype to Dnmt3L deficiency marked by meiotic catastrophe, aberrant retrotransposon transcriptional upregulation, and failure to establish methylation on retroelements in the genome (335, 336). A similar phenotype also is observed in MILI-interacting Tudor domain-containing protein-1 deficient mice (337). MIWI2 and MILI are proposed to participate in a "ping-pong-like" amplification cycle in the male primordial germline to generate piRNAs against actively expressed retrotransposons. In addition, MIWI2, directed by piRNA, is proposed to participate in the targeting of *de novo* methylation to retrotransposon sequences (333, 335).

Several RNA-based restriction mechanisms also have been implicated in LINE-1 control. For example, the antisense promoter within the LINE-1 5′ UTR can give rise to an antisense RNA transcript that may form a complex with the LINE-1 mRNA to create a substrate for siRNA biogenesis, resulting in siRNAs that target LINE-1 transcripts (338). In addition, the miRNA biogenesis factor Microprocessor/Drosha-DGCR8 has recently been demonstrated to specifically bind LINE-1, Alu, and SVA transcripts and cleave LINE-1 RNA *in vitro* (339). Accordingly, cultured cell experiments reveal that Microprocessor can restrict LINE-1 and Alu retrotransposition (339). In mouse embryonic stem cells, Dicer-dependent and Ago2-dependent RNAi mechanisms may possibly participate in LINE-1 regulation by limiting LINE-1 transcript accumulation (340, 341).

LINE-1 RNA Transcription and Splicing

KRAB zinc-finger proteins

Recent studies have revealed that members of the Krüppel-associated box (KRAB) zinc-finger (KZNF) protein family can recruit KRAB-associated protein-1 (KAP1) and its associated repressive complex to LINE-1 and SVA retrotransposons, which, in turn, inhibits their expression in embryonic stem cells (56, 57). The human KZNF protein, ZNF91, requires the SVA VNTR to bind and potently inhibit SVA expression (56). By comparison, ZNF93 binds to a sequence within the 5′ UTR of older LINE-1 subfamilies (*i.e.*, L1PA3, L1PA4, and older elements) to inhibit their expression (56). Interestingly, the ZNF93 binding site has been deleted from the 5′ UTR of active L1Hs elements, suggesting that L1Hs elements evaded the repressive effects of ZNF93 (56). Consistent with this idea, ZNF93 overexpression potently inhibited the retrotransposition of an engineered L1Hs element containing a reconstituted ZNF93 binding site in its 5′ UTR (56). Indeed, these data suggest that select KZNF proteins may be locked in an evolutionary arms race with LINE-1 and SVA retrotransposons and that the ability of LINE-1 to evade ZNF93 binding may represent a mechanism to drive LINE-1 subfamily succession (56, 57).

LINE-1 RNA Splicing and Premature Polyadenylation

TPRT often leads to the retrotransposition of a full-length human LINE-1 mRNA. Hence, it was somewhat surprising to find that full-length human LINE-1 RNAs contain a conserved splice donor sequence at position +98 of the 5′ UTR (342). Recent studies have revealed that the splice donor site is functional and that its use can lead to the generation of shorter LINE-1 mRNAs, which may be compromised for retrotransposition (342, 343). Indeed, LINE-1 mRNA splicing may serve as a regulatory mechanism to restrict LINE-1 expression and/or retrotransposition in a tissue-specific manner.

A full-length LINE-1 also contains a number of potential polyadenylation signals, and their use can lead to the generation of shorter LINE-1 mRNAs that are compromised for retrotransposition (344). Indeed, it also is proposed that the adenosine-rich nature of the LINE-1 sense strand transcript, as well as the presence of the abovementioned polyadenylation signals, may act as "molecular rheostats" to fine-tune the expression of genes containing full-length LINE-1s (345).

Cellular RNA Binding Proteins

Recent evidence suggests that RNA binding proteins may play a direct role in modulating LINE-1 retrotransposition. For example, heterogeneous nuclear ribonucleoprotein L (hnRNPL), a protein that facilitates alternative splicing, was shown to associate with both mouse LINE-1 RNA (346) and human LINE-1 RNPs (201). The knockdown of hnRNPL led to an increase in the levels of mouse LINE-1 RNA and ORF1p and an increase in the retrotransposition of engineered LINE-1 in cultured cells (346). In addition, the poly (A) binding protein C1 (PABPC1) interacts with the LINE-1 RNA and is required for LINE-1 RNP formation and efficient retrotransposition (347). Finally, cell culture assays have shown that human LINE-1 and mouse IAP mobility are restricted by the RNase activity of human RNase L, a member of the 2′,5′-oligoadenylate (2-5A) synthetase (OAS)-RNase L system, initially described for restricting viral infections during the interferon antiviral response (348). Clearly, it will be interesting to determine how these RNA binding proteins affect LINE-1 retrotransposition at the mechanistic level.

Other Proteins that Restrict LINE-1 Retrotransposition

APOBEC3 cytidine deaminases

The human *APOBEC3* gene family encodes seven proteins that can catalyze the deamination of cytidine to uridine residues in single-strand DNA substrates (reviewed in reference 349). Landmark findings by Malim and colleagues revealed that APOBEC3G potently restricts *vif*-deficient HIV infectivity (350). Subsequent studies demonstrated that several members of the APOBEC3 family, most potently APOBEC3A (A3A) and APOBEC3B (A3B), robustly inhibit LINE-1 and Alu retrotransposition in cultured cells (351–356). How A3B restricts LINE-1 retrotransposition remains unknown, although it may do so by both cytidine deamination-dependent and -independent pathways (351). Similarly, APOBEC3C (A3C) modestly inhibits LINE-1 retrotransposition in cultured cells by a cytidine deaminase-independent pathway (357). In contrast, a recent study demonstrated that A3A-mediated LINE-1 inhibition occurs by a deaminase-dependent pathway and that A3A can deaminate single-strand DNAs that are exposed transiently during LINE-1 TPRT (358). This mechanism is consistent with recent reports of APOBEC3-mediated deamination of single-strand genomic DNA in several types of cancer (359, 360).

Trex1/SAMHD1

Aicardi–Goutières syndrome (AGS) is a rare childhood inflammatory disorder that can lead to neurodevelopmental deficiencies (reviewed in references 361 and 362). AGS can be caused by mutations in several genes, including *Trex1*, *SAMHD1*, and *RNaseH2* (363–365). Intriguingly, the overexpression of Trex1, a 3′ to 5′ DNA exonuclease, can restrict both LINE-1 and IAP retrotransposition in cultured cells (366). Similarly, the over-expression of SAMHD1, a triphosphohydrolase that can reduce intracellular dNTP pools, can inhibit LINE-1 retrotransposition in cultured cells (367). Detailed studies are needed to determine the mechanism by which Trex1 and SAMHD1 inhibit LINE-1 retrotransposition. It also will be interesting to determine whether LINE-1 retrotransposition is elevated in tissues (*e.g.*, brain regions [see below]) of AGS patients and if mutations in any of the three subunits of human RNaseH2 or other genes leading to AGS (*i.e.*, *ADAR1* and *IFIH1*) regulate LINE-1 retrotransposition (368–370).

MOV10

Moloney leukemia virus 10 (MOV10) is an RNA helicase that inhibits the activity of several retroviruses including HIV-1 (reviewed in reference 371). MOV10 also restricts the retrotransposition of human LINE-1, Alu, SVA, and mouse IAP elements in cultured cells (372–374). The mechanism by which MOV10 inhibits LINE-1 retrotransposition requires elucidation. However, recent experiments have shown that MOV10 colocalizes with LINE-1 ORF1p in RNPs, directly binds LINE-1 RNA, and that mutations in the conserved helicase domains of MOV10 inhibit its ability to restrict LINE-1 retrotransposition (373–375). Notably, the MOV10-like-1 (MOV10L1) putative helicase also has been reported to play a role in piRNA-directed retrotransposon silencing in male germ cells (376, 377).

THE IMPACT OF LINE-1-MEDIATED RETROTRANSPOSITION EVENTS ON THE GENOME

Developmental Timing of LINE-1-Mediated Retrotransposition Events

Despite the myriad of host defense mechanisms in place to restrict LINE-1 retrotransposition, *de novo* heritable LINE-1 insertions continue to occur and contribute to inter-individual human variation (reviewed in references 11 and 13). Although it remains unclear how frequently new, heritable LINE-1-mediated retrotransposition events arise in the human population, estimates suggest that ~1/100 humans harbors a *de novo* LINE-1 insertion and ~1/20 humans harbors a new Alu insertion (reviewed in reference 378). Moreover, the developmental timing of heritable retrotransposition events remains an area of active investigation. Retrotransposition in the germline lineage (*i.e.*, gametes or precursor cells giving rise to gametes) presents an attractive and logical hypothesis for the generation of heritable insertions and is supported by evidence for full-length LINE-1 mRNA expression in the male germline and LINE-1 ORF1p expression in male and female germ cells (379–381). On the other hand, recent studies employing transgenic animal models, human patient samples, and cell culture models have provided substantial evidence that pluripotent cells of the early embryo (*i.e.*, embryonic stem cells) express LINE-1 mRNA and ORF1p, and provide a permissive milieu for accumulating potentially heritable *de novo* retrotransposition events (see below) (Figure 5).

Transgenic Animal Models

The first study to experimentally recapitulate LINE-1 retrotransposition *in vivo* used transgenic mice harboring an engineered human LINE-1 tagged with an *EGFP* retrotransposition indicator cassette (195). A pPolII promoter augmented the expression of the LINE-1, whereas a sperm-specific acrosin promoter drove the expression of the *EGFP* gene. The enhanced green fluorescent protein (EGFP) was equipped with a signal peptide, which would allow EGFP to localize to the acrosome of developing spermatids (382). The examination of 135 offspring derived from crosses of transgenic males to wild-type females allowed the identification of two *de novo* LINE-1 retrotransposition events. Interestingly, one of these animals contained a *de novo* LINE-1 retrotransposition event but lacked the transgene, suggesting that LINE-1 retrotransposition could occur in the male germline before the onset of meiosis II (382, 383).

Subsequent work, employing transgenic mouse and rat lines in which LINE-1 expression was driven by the native LINE-1 promoter, once again led to the identification of *de novo* LINE-1 insertions (384). As above, a number of offspring contained *de novo* LINE-1 retrotransposition events, but lacked the transgene. Unexpectedly, these animals were unable to pass the *de novo* retrotransposition event on to their progeny, suggesting that animals were mosaic—that is, they contained the *de novo* retrotransposition event in somatic cells, but not in their germline cells (384). How might such a scenario happen? The authors suggested that LINE-1 RNPs that are formed in either the male or female germline can be carried over to the zygote and subsequently can undergo retrotransposition in the early embryo. Clearly, these results require verification; however, this study provided tantalizing evidence that a substantial number of LINE-1 retrotransposition events may occur postzygotically.

It remains difficult to study LINE-1 retrotransposition in female oocytes. However, recent data suggest that LINE-1 expression and/or retrotransposition may lead to fetal oocyte attrition in mice (385). Hence, additional research in this area might provide mechanistic information about the timing of heritable LINE-1 retrotransposition events.

Human patients

Although approximately 100 cases of human genetic disease have been attributed to LINE-1-mediated retrotransposition events (reviewed in reference 14), it remains difficult to identify the developmental timing of a retrotransposition event. It was long assumed that LINE-1 actively retrotransposes in the male and female germline. Recent reports suggest that endogenous LINE-1s are expressed in human female oocytes and that engineered LINE-1 constructs readily retrotranspose in oocytes (381). In contrast, despite concerted efforts, it has been difficult to find evidence for endogenous LINE-1 retrotransposition in human sperm (386).

The identification of disease-producing LINE-1-mediated retrotransposition events has provided additional evidence for postzygotic, somatic insertions. In one instance, a LINE-1 retrotransposition event into the *CHM* gene led to X-linked choroideremia in a male patient (387). The examination of DNA from his family members revealed that his two sisters shared the same maternally inherited haplotype in the *CHM* region, but only one sister contained the retrotransposition event (387). Surprisingly, the mother of the patient was a somatic mosaic with respect to the LINE-1 retrotransposition event. These data, combined with the failure to pass the LINE-1 retrotransposition event to one of her

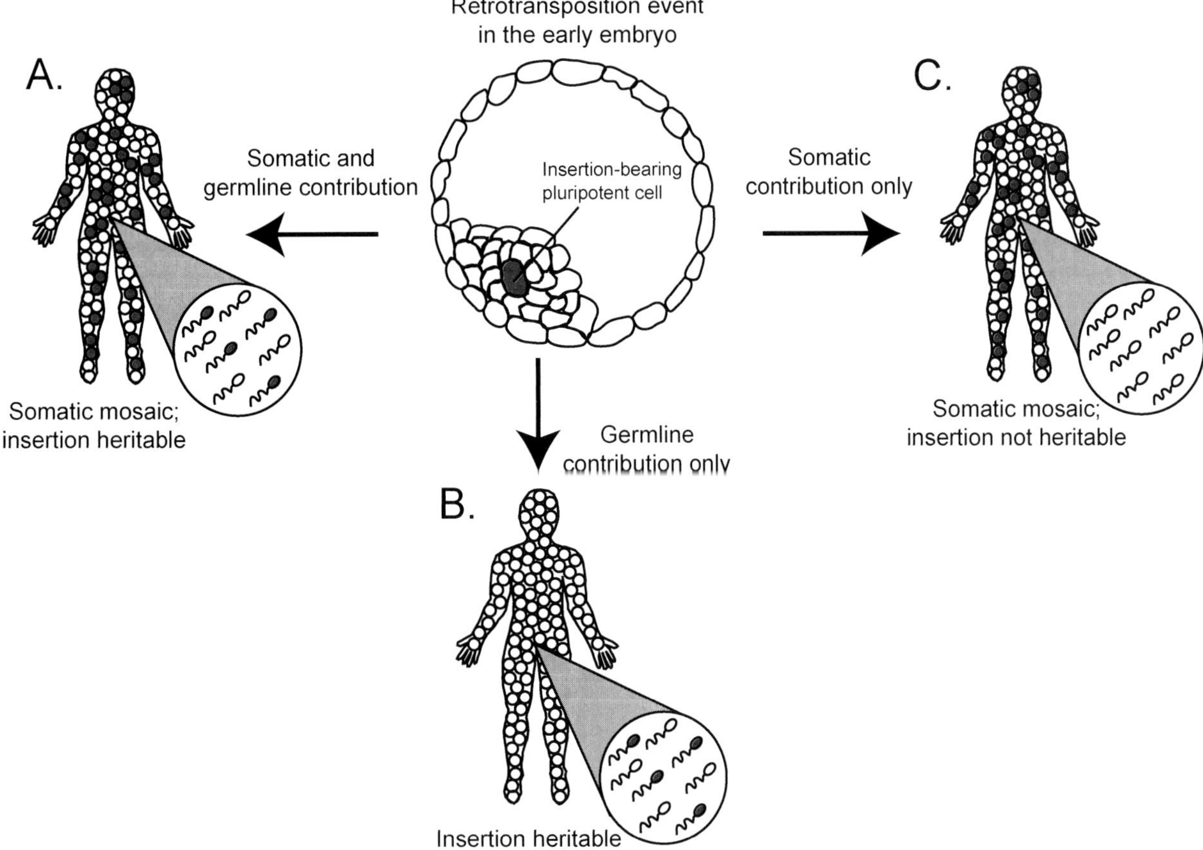

Figure 5 Hypothetical consequences of retrotransposition in pluripotent cells of the early embryo. (A) Cells harboring a *de novo* retrotransposition event could contribute both to the soma and germline, resulting in an individual with somatic as well as germline mosaicism and a heritable insertion. (B) Conceivably, cells harboring the retrotransposon insertion could contribute solely to the germline, giving rise to germline mosaicism, thereby rendering the insertion heritable. (C) Retrotransposon insertion-bearing cells could contribute to the somatic lineage but not to the germline, resulting in somatic mosaicism. Such an event would not be transmissible to the next generation. Red and white shaded circles in the human figures and sperm represent retrotransposon insertion-bearing and non-insertion-bearing cells in the soma and germline, respectively. (This figure was reproduced from Sandra Richardson's doctoral thesis [408]). Additional references are provided in the text.
doi:10.1128/microbiolspec.MDNA3-0061-2014.f5

daughters (see above), provided unequivocal evidence that LINE-1 retrotransposition can occur in the early embryo to generate both somatic and germline mosaicism and that the resultant retrotransposition event can be passed on to subsequent generations (387) (Figure 5). Indeed, recent data suggest that SVA and processed pseudogene insertions also occur postzygotically (167, 168).

Experiments in Developmentally Relevant Cultured Cell Lines

The notion that LINE-1-mediated retrotransposition events can occur early in development is supported by studies in cultured cells. For example, human embryonic stem cells express endogenous LINE-1 mRNA and ORF1p and can accommodate the retrotransposition of engineered human and mouse LINE-1s (213, 388). Similar results also have been obtained from studies conducted with human embryonic-carcinoma-derived cell lines (389) and induced pluripotent stem cells (390). Together, the above studies suggest that cells of the early embryo may represent an important developmental stage for heritable retrotransposition events in mammals.

Somatic LINE-1 Retrotransposition

LINE-1 Retrotransposition in the Brain

In order to ensure its continued existence, it was widely believed that LINE-1 retrotransposition must occur in cells (*e.g.*, germ cells) that have the potential to contribute to subsequent generations. This view was radically overturned by the unexpected observations that: (i) the differentiation of adult rat hippocampal neural stem cells into neuronal precursor cells and neurons leads to an increase in LINE-1 transcript abundance, and (ii) engineered LINE-1 retrotransposons could retrotranspose in cultured rat neuronal precursor cells and in the brain of transgenic mice (391) (Figure 6A). Subsequent studies revealed that engineered LINE-1s could retrotranspose in both fetal and human embryonic stem cell-derived neuronal progenitor cells (392), that sensitive quantitative polymerase chain reaction-based experiments could detect an increase in LINE-1 copy number in several human brain regions (*e.g.*, hippocampus) when compared with matched liver or heart samples (392), and that engineered human LINE-1 retrotransposons exhibit enhanced somatic retrotransposition in the mouse models lacking either the methyl-CpG-binding protein 2 (*MeCP2*) or *ATM* genes (282, 393) (Figure 6).

The use of next-generation sequencing is now providing additional insights into LINE-1 retrotransposition in the brain. The development of retrotransposon-capture sequencing (RC-seq), a technique in which custom oligonucleotide probes are used to enrich sequencing libraries for fragments containing LINE-1/genome junctions, demonstrated that endogenous LINE-1 insertions contribute to somatic mosaicism in the human brain (394) (Figure 6). The examination of DNAs derived from the hippocampus and caudate nucleus of three advanced-age post-mortem brain samples revealed ~7,700 potential somatic LINE-1 insertions and thousands of potential Alu and SVA somatic insertions (394). Sanger sequencing verified a handful of these putative insertions and showed that several insertions exhibited structural LINE-1 hallmarks.

One of the limitations of RC-seq and other methods that analyze bulk tissue samples is the inability to accurately quantify the extent of somatic LINE-1 mosaicism in the brain. Single-cell genomic analysis, a technically challenging undertaking, has the potential to provide key insights regarding this facet of neuronal retrotransposition. Indeed, a recent study using whole cell amplified genomic DNAs from 300 single neurons from the cerebral cortex and caudate nucleus led to the identification of somatic retrotransposition events in the brain (395). However, the level of LINE-1 retrotransposition, which was estimated to be ~0.6 insertions per cell (395), was lower than the level detected in previous studies (392, 394). Methodological differences, as well as differences in the brain regions that were analyzed (*e.g.*, the hippocampus, caudate nucleus, or cortex) may account for the different number of retrotransposition events detected in the above studies.

Clearly, more studies are needed to determine if the rate of LINE-1 retrotransposition varies in different brain regions or cell types, why neuronal precursor cells apparently are permissive for LINE-1 retrotransposition, and whether LINE-1 retrotransposition in the brain has biological consequences. However, a union of the above results indicates that LINE-1 can retrotranspose in the human brain. It also is notable that the mobility of TEs may not be restricted to the mammalian brain, as recent reports suggest that TEs also are active in the *Drosophila* brain (396, 397).

LINE-1 Retrotransposition in Cancer

The discovery of a mutagenic LINE-1 retrotransposition event into the adenomatous polyposis coli (*APC*) gene led to the realization that LINE-1 retrotransposition may have a role in tumorigenesis or tumor progression (398). In recent years, the creation of polymerase chain reaction amplicon libraries enriched for L1Hs elements and their associated genomic flanking DNA sequences, in conjunction with next-generation DNA sequencing, has led to the identification of *de novo* LINE-1-mediated retrotransposition events in a variety of tumor types, including lung, liver, ovarian, colorectal, and prostate cancers (69, 296, 399–402). Whether these insertions represent "driver" or "passenger" mutations in tumors requires elucidation (403) (Figure 6B); however, it is clear that advances in DNA sequencing technology should allow an answer to this question in coming years.

CONCLUDING REMARKS

The discovery of TEs in the mid-twentieth century by Barbara McClintock elegantly demonstrated that TEs were arbiters of genetic diversity and that their mobility could influence gene expression and genome structure. It is now clear that TEs are a dynamic component of mammalian genomes that have a profound effect on genome evolution. LINE-1s and SINEs generate both intra-individual and inter-individual genetic variation and continue to sculpt the human genome. Advances made during the previous 13 years have elucidated mechanistic features of LINE-1-mediated retrotransposition, the impact of LINE-1-mediated retrotransposition on the human genome, and how the host defends itself from the onslaught of unabated LINE-1-mediated retrotransposition events. Future research should elucidate how, when, and where LINE-1-mediated retrotransposition events

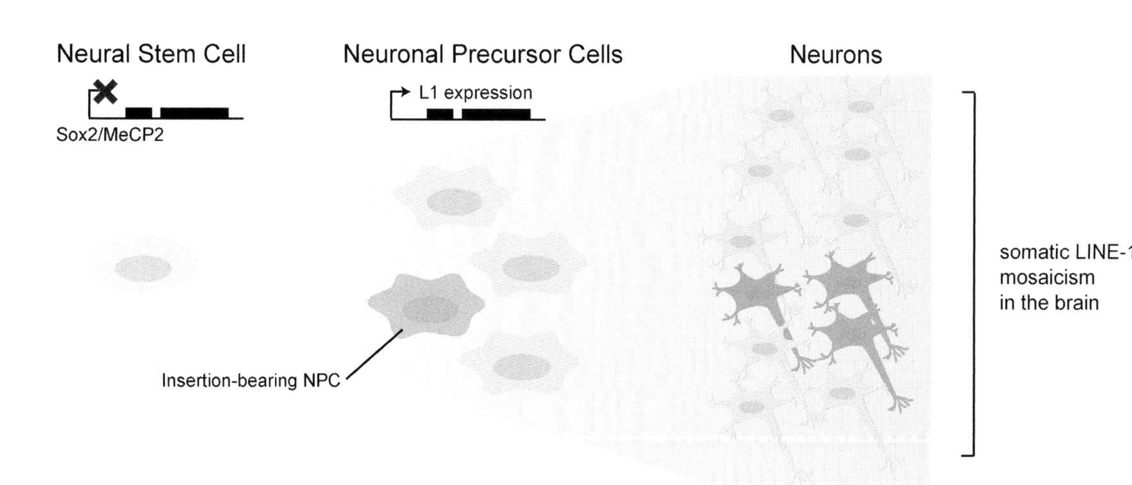

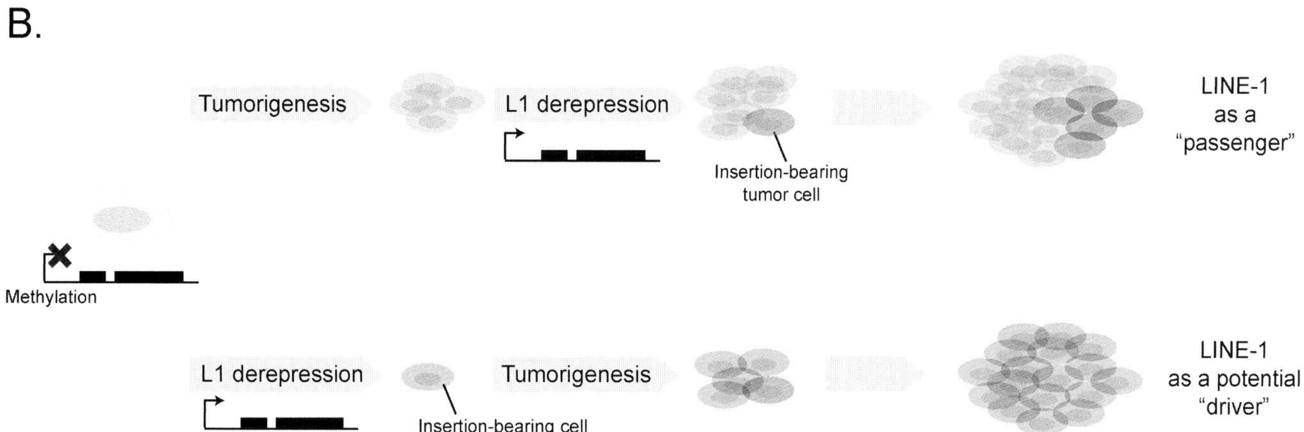

Figure 6 Long INterspersed Element-1 (LINE-1) retrotransposition in the brain and in cancer. (A) Model for how LINE-1 generates somatic mosaicism in the brain. Sox2, MeCP2, and promoter methylation (red X over the LINE-1 5′ untranslated region [UTR]) are hypothesized to repress LINE-1 expression in neural stem cells (yellow cell). The differentiation of neural stem cells into neuronal precursor cells (NPCs) correlates with a reduction in LINE-1 promoter methylation and a derepression of LINE-1 expression, allowing a permissive milieu for retrotransposition (insertion-bearing NPC [blue cell]). Subsequent differentiation of NPCs into neurons leads to somatic LINE-1 mosaicism in the brain (insertion-bearing neurons [blue cells]). It is unknown whether LINE-1 retrotransposition occurs in postmitotic neurons. (B) Model for how LINE-1 may act as a "driver" or "passenger" mutation during cancer progression. In a somatic cell (yellow cell), LINE-1 expression generally is repressed by promoter methylation (red X over the LINE-1 5′ UTR). After oncogenic transformation (top panel), the derepression of LINE-1 expression in some tumor cells (green cells), allows *de novo* LINE-1 retrotransposition events that act as "passenger" mutations (insertion-bearing tumor cell in red), leading to somatic mosaicism in the resultant tumor. Alternatively, tumorigenesis can be triggered by a *de novo* LINE-1 retrotransposition event that acts as a potential "driver" mutation (bottom panel), leading to the clonal amplification of the insertion-bearing cell (red cell). Additional references are provided in the text. doi:10.1128/microbiolspec.MDNA3-0061-2014.f6

occur in humans, how epigenetic mechanisms act to regulate LINE-1 expression (for example see refs. [389, 404]), and how LINE-1s alter the epigenetic landscape of the genome (*e.g.*, perhaps by playing a role in X-inactivation [405–407]). Indeed, we look forward to gaining a greater understanding about how LINE-1-mediated retrotransposition events contribute to diseases, such as cancer, and if they play a role in human neuronal plasticity. Although once disparaged as a class of "junk DNA", it is now clear that a deep understanding of LINE-1 biology is paramount to our understanding of the evolutionary forces that have shaped human genomes.

Acknowledgments. *We thank Nancy Leff for outstanding editorial assistance during the preparation of this chapter. We thank Drs. Haig Kazazian, Mark Batzer, and Tomo-ichiro Miyoshi, and members of the Moran laboratory for helpful discussions and for help with editing this Chapter. S.R.R. was supported in part by a National Institutes of Health Training Grant (T32-GM07544). A.J.D. was supported in part by a Fondation pour la Recherche Médicale (FRM) fellowship. J.B.M. was supported in part by a National Institutes of Health Training Grant (T32 GM007315). The laboratory of J.L.G.-P. is supported by CICE-FEDER-P09-CTS-4980, FIS-FEDER-PI11/01489, the European Research Council (ERC-Consolidator ERC-STG-2012-233764), and by an International Early Career Scientist grant from the Howard Hughes Medical Institute (IECS-55007420). J.V.M. is supported by National Institutes of Health Grant GM060518 and is an Investigator in the Howard Hughes Medical Institute. Portions of this chapter originally appeared in draft form in Dr. Sandra Richardson's PhD thesis dissertation (408); they have been rewritten and adapted for this chapter. The funders had no role in the decision to submit this work for publication.Conflicts of interest: J.V.M. is an inventor on the following patent: "Kazazian, H.H., Boeke, J.D., Moran, J.V., and Dombroski, B.A. Compositions and methods of use of mammalian retrotransposons. Application No. 60/006,831; Patent number 6,150,160; Issued November 21, 2000." He has not made any money from this patent and discloses this information voluntarily.*

Citation. Richardson SR, Doucet AJ, Kopera HC, Moldovan JB, Garcia-Perez JL, Moran JV. 2014. The influence of LINE-1 and SINE retrotransposons on mammalian genomes. Microbiol Spectrum 3(2):MDNA3-0061-2014.

References

1. **Orgel LE, Crick FH, Sapienza C.** 1980. Selfish DNA. *Nature* **288:**645–646.
2. **Doolittle WF, Sapienza C.** 1980. Selfish genes, the phenotype paradigm and genome evolution. *Nature* **284:** 601–603.
3. **Craig NL, Craigie R, Gellert M, Lambowitz AM.** 2002. p 1–1204. *Mobile DNA II.* ASM Press, Washington, D.C.
4. **McClintock B.** 1950. The origin and behavior of mutable loci in maize. *Proc Natl Acad Sci U S A* **36:**344–355.
5. **Fedoroff N, Wessler S, Shure M.** 1983. Isolation of the transposable maize controlling elements Ac and Ds. *Cell* 35:235–242.
6. **Boeke JD, Garfinkel DJ, Styles CA, Fink GR.** 1985. Ty elements transpose through an RNA intermediate. *Cell* 40:491–500.
7. **Lander ES, Linton LM, Birren B, Nusbaum C, Zody MC, Baldwin J, Devon K, Dewar K, Doyle M, FitzHugh W, Funke R, Gage D, Harris K, Heaford A, Howland J, Kann L, Lehoczky J, LeVine R, McEwan P, McKernan K, Meldrim J, Mesirov JP, Miranda C, Morris W, Naylor J, Raymond C, Rosetti M, Santos R, Sheridan A, Sougnez C, Stange-Thomann N, Stojanovic N, Subramanian A, Wyman D, Rogers J, Sulston J, Ainscough R, Beck S, Bentley D, Burton J, Clee C, Carter N, Coulson A, Deadman R, Deloukas P, Dunham A, Dunham I, Durbin R, French L, Grafham D, Gregory S, Hubbard T, Humphray S, Hunt A, Jones M, Lloyd C, McMurray A, Matthews L, Mercer S, Milne S, Mullikin JC, Mungall A, Plumb R, Ross M, Shownkeen R, Sims S, Waterston RH, Wilson RK, Hillier LW, McPherson JD, Marra MA, Mardis ER, Fulton LA, Chinwalla AT, Pepin KH, Gish WR, Chissoe SL, Wendl MC, Delehaunty KD, Miner TL, Delehaunty A, Kramer JB, Cook LL, Fulton RS, Johnson DL, Minx PJ, Clifton SW, Hawkins T, Branscomb E, Predki P, Richardson P, Wenning S, Slezak T, Doggett N, Cheng JF, Olsen A, Lucas S, Elkin C, Uberbacher E, Frazier M, Gibbs RA, Muzny DM, Scherer SE, Bouck JB, Sodergren EJ, Worley KC, Rives CM, Gorrell JH, Metzker ML, Naylor SL, Kucherlapati RS, Nelson DL, Weinstock GM, Sakaki Y, Fujiyama A, Hattori M, Yada T, Toyoda A, Itoh T, Kawagoe C, Watanabe H, Totoki Y, Taylor T, Weissenbach J, Heilig R, Saurin W, Artiguenave F, Brottier P, Bruls T, Pelletier E, Robert C, Wincker P, Smith DR, Doucette-Stamm L, Rubenfield M, Weinstock K, Lee HM, Dubois J, Rosenthal A, Platzer M, Nyakatura G, Taudien S, Rump A, Yang H, Yu J, Wang J, Huang G, Gu J, Hood L, Rowen L, Madan A, Qin S, Davis RW, Federspiel NA, Abola AP, Proctor MJ, Myers RM, Schmutz J, Dickson M, Grimwood J, Cox DR, Olson MV, Kaul R, Raymond C, Shimizu N, Kawasaki K, Minoshima S, Evans GA, Athanasiou M, Schultz R, Roe BA, Chen F, Pan H, Ramser J, Lehrach H, Reinhardt R, McCombie WR, de la Bastide M, Dedhia N, Blocker H, Hornischer K, Nordsiek G, Agarwala R, Aravind L, Bailey JA, Bateman A, Batzoglou S, Birney E, Bork P, Brown DG, Burge CB, Cerutti L, Chen HC, Church D, Clamp M, Copley RR, Doerks T, Eddy SR, Eichler EE, Furey TS, Galagan J, Gilbert JG, Harmon C, Hayashizaki Y, Haussler D, Hermjakob H, Hokamp K, Jang W, Johnson LS, Jones TA, Kasif S, Kaspryzk A, Kennedy S, Kent WJ, Kitts P, Koonin EV, Korf I, Kulp D, Lancet D, Lowe TM, McLysaght A, Mikkelsen T, Moran JV, Mulder N, Pollara VJ, Ponting CP, Schuler G, Schultz J, Slater G, Smit AF, Stupka E, Szustakowski J, Thierry-Mieg D, Thierry-Mieg J, Wagner L, Wallis J, Wheeler R, Williams A, Wolf YI, Wolfe KH, Yang SP, Yeh RF, Collins F, Guyer MS, Peterson J, Felsenfeld A, Wetterstrand KA, Patrinos A, Morgan MJ, de Jong P, Catanese JJ,**

Osoegawa K, Shizuya H, Choi S, Chen YJ. 2001. Initial sequencing and analysis of the human genome. *Nature* **409:**860–921.

8. **Venter JC, Adams MD, Myers EW, Li PW, Mural RJ, Sutton GG, Smith HO, Yandell M, Evans CA, Holt RA, Gocayne JD, Amanatides P, Ballew RM, Huson DH, Wortman JR, Zhang Q, Kodira CD, Zheng XH, Chen L, Skupski M, Subramanian G, Thomas PD, Zhang J, Gabor Miklos GL, Nelson C, Broder S, Clark AG, Nadeau J, McKusick VA, Zinder N, Levine AJ, Roberts RJ, Simon M, Slayman C, Hunkapiller M, Bolanos R, Delcher A, Dew I, Fasulo D, Flanigan M, Florea L, Halpern A, Hannenhalli S, Kravitz S, Levy S, Mobarry C, Reinert K, Remington K, Abu-Threideh J, Beasley E, Biddick K, Bonazzi V, Brandon R, Cargill M, Chandramouliswaran I, Charlab R, Chaturvedi K, Deng Z, Di Francesco V, Dunn P, Eilbeck K, Evangelista C, Gabrielian AE, Gan W, Ge W, Gong F, Gu Z, Guan P, Heiman TJ, Higgins ME, Ji RR, Ke Z, Ketchum KA, Lai Z, Lei Y, Li Z, Li J, Liang Y, Lin X, Lu F, Merkulov GV, Milshina N, Moore HM, Naik AK, Narayan VA, Neelam B, Nusskern D, Rusch DB, Salzberg S, Shao W, Shue B, Sun J, Wang Z, Wang A, Wang X, Wang J, Wei M, Wides R, Xiao C, Yan C, Yao A, Ye J, Zhan M, Zhang W, Zhang H, Zhao Q, Zheng L, Zhong F, Zhong W, Zhu S, Zhao S, Gilbert D, Baumhueter S, Spier G, Carter C, Cravchik A, Woodage T, Ali F, An H, Awe A, Baldwin D, Baden H, Barnstead M, Barrow I, Beeson K, Busam D, Carver A, Center A, Cheng ML, Curry L, Danaher S, Davenport L, Desilets R, Dietz S, Dodson K, Doup L, Ferriera S, Garg N, Gluecksmann A, Hart B, Haynes J, Haynes C, Heiner C, Hladun S, Hostin D, Houck J, Howland T, Ibegwam C, Johnson J, Kalush F, Kline L, Koduru S, Love A, Mann F, May D, McCawley S, McIntosh T, McMullen I, Moy M, Moy L, Murphy B, Nelson K, Pfannkoch C, Pratts E, Puri V, Qureshi H, Reardon M, Rodriguez R, Rogers YH, Romblad D, Ruhfel B, Scott R, Sitter C, Smallwood M, Stewart E, Strong R, Suh E, Thomas R, Tint NN, Tse S, Vech C, Wang G, Wetter J, Williams S, Williams M, Windsor S, Winn-Deen E, Wolfe K, Zaveri J, Zaveri K, Abril JF, Guigo R, Campbell MJ, Sjolander KV, Karlak B, Kejariwal A, Mi H, Lazareva B, Hatton T, Narechania A, Diemer K, Muruganujan A, Guo N, Sato S, Bafna V, Istrail S, Lippert R, Schwartz R, Walenz B, Yooseph S, Allen D, Basu A, Baxendale J, Blick L, Caminha M, Carnes-Stine J, Caulk P, Chiang YH, Coyne M, Dahlke C, Mays A, Dombroski M, Donnelly M, Ely D, Esparham S, Fosler C, Gire H, Glanowski S, Glasser K, Glodek A, Gorokhov M, Graham K, Gropman B, Harris M, Heil J, Henderson S, Hoover J, Jennings D, Jordan C, Jordan J, Kasha J, Kagan L, Kraft C, Levitsky A, Lewis M, Liu X, Lopez J, Ma D, Majoros W, McDaniel J, Murphy S, Newman M, Nguyen T, Nguyen N, Nodell M, Pan S, Peck J, Peterson M, Rowe W, Sanders R, Scott J, Simpson M, Smith T, Sprague A, Stockwell T, Turner R, Venter E, Wang M, Wen M, Wu D, Wu M, Xia A, Zandieh A, Zhu X.** 2001. The sequence of the human genome. *Science* **291:** 1304–1351.

9. **Waring M, Britten RJ.** 1966. Nucleotide sequence repetition: a rapidly reassociating fraction of mouse DNA. *Science* **154:**791–794.

10. **Britten RJ, Kohne DE.** 1968. Repeated sequences in DNA. Hundreds of thousands of copies of DNA sequences have been incorporated into the genomes of higher organisms. *Science* **161:**529–540.

11. **Beck CR, Garcia-Perez JL, Badge RM, Moran JV.** 2011. LINE-1 elements in structural variation and disease. *Annu Rev Genomics Hum Genet* **12:**187–215.

12. **Belancio VP, Roy-Engel AM, Deininger PL.** 2010. All y'all need to know 'bout retroelements in cancer. *Semin Cancer Biol* **20:**200–210.

13. **Levin HL, Moran JV.** 2011. Dynamic interactions between transposable elements and their hosts. *Nat Rev Genet* **12:**615–627.

14. **Hancks DC, Kazazian HH Jr.** 2012. Active human retrotransposons: variation and disease. *Curr Opin Genet Dev* **22:**191–203.

15. **Richardson SR, Morell S, Faulkner GJ.** 2014. L1 Retrotransposons and somatic mosaicism in the brain. *Annu Rev Genet* **48:**1–27.

16. **Goodier JL, Kazazian HH Jr.** 2008. Retrotransposons revisited: the restraint and rehabilitation of parasites. *Cell* **135:**23–35.

17. **Feschotte C.** 2008. Transposable elements and the evolution of regulatory networks. *Nat Rev Genet* **9:**397–405.

18. **Babushok DV, Kazazian HH Jr.** 2007. Progress in understanding the biology of the human mutagen LINE-1. *Hum Mutat* **28:**527–539.

19. **Burns KH, Boeke JD.** 2012. Human transposon tectonics. *Cell* **149:**740–752.

20. **Lindblad-Toh K, Wade CM, Mikkelsen TS, Karlsson EK, Jaffe DB, Kamal M, Clamp M, Chang JL, Kulbokas EJ 3rd, Zody MC, Mauceli E, Xie X, Breen M, Wayne RK, Ostrander EA, Ponting CP, Galibert F, Smith DR, DeJong PJ, Kirkness E, Alvarez P, Biagi T, Brockman W, Butler J, Chin CW, Cook A, Cuff J, Daly MJ, DeCaprio D, Gnerre S, Grabherr M, Kellis M, Kleber M, Bardeleben C, Goodstadt L, Heger A, Hitte C, Kim L, Koepfli KP, Parker HG, Pollinger JP, Searle SM, Sutter NB, Thomas R, Webber C, Baldwin J, Abebe A, Abouelleil A, Aftuck L, Ait-Zahra M, Aldredge T, Allen N, An P, Anderson S, Antoine C, Arachchi H, Aslam A, Ayotte L, Bachantsang P, Barry A, Bayul T, Benamara M, Berlin A, Bessette D, Blitshteyn B, Bloom T, Blye J, Boguslavskiy L, Bonnet C, Boukhgalter B, Brown A, Cahill P, Calixte N, Camarata J, Cheshatsang Y, Chu J, Citroen M, Collymore A, Cooke P, Dawoe T, Daza R, Decktor K, DeGray S, Dhargay N, Dooley K, Dorje P, Dorjee K, Dorris L, Duffey N, Dupes A, Egbiremolen O, Elong R, Falk J, Farina A, Faro S, Ferguson D, Ferreira P, Fisher S, FitzGerald M, Foley K, Foley C, Franke A, Friedrich D, Gage D, Garber M, Gearin G, Giannoukos G, Goode T, Goyette A, Graham J, Grandbois E, Gyaltsen K, Hafez N, Hagopian D, Hagos B, Hall J, Healy C, Hegarty R, Honan T, Horn A, Houde N, Hughes L, Hunnicutt L,**

Husby M, Jester B, Jones C, Kamat A, Kanga B, Kells C, Khazanovich D, Kieu AC, Kisner P, Kumar M, Lance K, Landers T, Lara M, Lee W, Leger JP, Lennon N, Leuper L, LeVine S, Liu J, Liu X, Lokyitsang Y, Lokyitsang T, Lui A, Macdonald J, Major J, Marabella R, Maru K, Matthews C, McDonough S, Mehta T, Meldrim J, Melnikov A, Meneus L, Mihalev A, Mihova T, Miller K, Mittelman R, Mlenga V, Mulrain L, Munson G, Navidi A, Naylor J, Nguyen T, Nguyen N, Nguyen C, Nicol R, Norbu N, Norbu C, Novod N, Nyima T, Olandt P, O'Neill B, O'Neill K, Osman S, Oyono L, Patti C, Perrin D, Phunkhang P, Pierre F, Priest M, Rachupka A, Raghuraman S, Rameau R, Ray V, Raymond C, Rege F, Rise C, Rogers J, Rogov P, Sahalie J, Settipalli S, Sharpe T, Shea T, Sheehan M, Sherpa N, Shi J, Shih D, Sloan J, Smith C, Sparrow T, Stalker J, Stange-Thomann N, Stavropoulos S, Stone C, Stone S, Sykes S, Tchuinga P, Tenzing P, Tesfaye S, Thoulutsang D, Thoulutsang Y, Topham K, Topping I, Tsamla T, Vassiliev H, Venkataraman V, Vo A, Wangchuk T, Wangdi T, Weiand M, Wilkinson J, Wilson A, Yadav S, Yang S, Yang X, Young G, Yu Q, Zainoun J, Zembek L, Zimmer A, Lander ES. 2005. Genome sequence, comparative analysis and haplotype structure of the domestic dog. *Nature* 438:803–819.

21. **Waterston RH, Lindblad-Toh K, Birney E, Rogers J, Abril JF, Agarwal P, Agarwala R, Ainscough R, Alexandersson M, An P, Antonarakis SE, Attwood J, Baertsch R, Bailey J, Barlow K, Beck S, Berry E, Birren B, Bloom T, Bork P, Botcherby M, Bray N, Brent MR, Brown DG, Brown SD, Bult C, Burton J, Butler J, Campbell RD, Carninci P, Cawley S, Chiaromonte F, Chinwalla AT, Church DM, Clamp M, Clee C, Collins FS, Cook LL, Copley RR, Coulson A, Couronne O, Cuff J, Curwen V, Cutts T, Daly M, David R, Davies J, Delehaunty KD, Deri J, Dermitzakis ET, Dewey C, Dickens NJ, Diekhans M, Dodge S, Dubchak I, Dunn DM, Eddy SR, Elnitski L, Emes RD, Eswara P, Eyras E, Felsenfeld A, Fewell GA, Flicek P, Foley K, Frankel WN, Fulton LA, Fulton RS, Furey TS, Gage D, Gibbs RA, Glusman G, Gnerre S, Goldman N, Goodstadt L, Grafham D, Graves TA, Green ED, Gregory S, Guigo R, Guyer M, Hardison RC, Haussler D, Hayashizaki Y, Hillier LW, Hinrichs A, Hlavina W, Holzer T, Hsu F, Hua A, Hubbard T, Hunt A, Jackson I, Jaffe DB, Johnson LS, Jones M, Jones TA, Joy A, Kamal M, Karlsson EK, Karolchik D, Kasprzyk A, Kawai J, Keibler E, Kells C, Kent WJ, Kirby A, Kolbe DL, Korf I, Kucherlapati RS, Kulbokas EJ, Kulp D, Landers T, Leger JP, Leonard S, Letunic I, Levine R, Li J, Li M, Lloyd C, Lucas S, Ma B, Maglott DR, Mardis ER, Matthews L, Mauceli E, Mayer JH, McCarthy M, McCombie WR, McLaren S, McLay K, McPherson JD, Meldrim J, Meredith B, Mesirov JP, Miller W, Miner TL, Mongin E, Montgomery KT, Morgan M, Mott R, Mullikin JC, Muzny DM, Nash WE, Nelson JO, Nhan MN, Nicol R, Ning Z, Nusbaum C, O'Connor MJ, Okazaki Y, Oliver K, Overton-Larty E, Pachter L, Parra G, Pepin KH, Peterson J, Pevzner P, Plumb R, Pohl CS, Poliakov A, Ponce TC, Ponting CP, Potter S, Quail M, Reymond A, Roe BA, Roskin KM, Rubin EM, Rust AG, Santos R, Sapojnikov V, Schultz B, Schultz J, Schwartz MS, Schwartz S, Scott C, Seaman S, Searle S, Sharpe T, Sheridan A, Shownkeen R, Sims S, Singer JB, Slater G, Smit A, Smith DR, Spencer B, Stabenau A, Stange-Thomann N, Sugnet C, Suyama M, Tesler G, Thompson J, Torrents D, Trevaskis E, Tromp J, Ucla C, Ureta-Vidal A, Vinson JP, Von Niederhausern AC, Wade CM, Wall M, Weber RJ, Weiss RB, Wendl MC, West AP, Wetterstrand K, Wheeler R, Whelan S, Wierzbowski J, Willey D, Williams S, Wilson RK, Winter E, Worley KC, Wyman D, Yang S, Yang SP, Zdobnov EM, Zody MC, Lander ES.** 2002. Initial sequencing and comparative analysis of the mouse genome. *Nature* **420:**520–562.

22. **de Koning AP, Gu W, Castoe TA, Batzer MA, Pollock DD.** 2011. Repetitive elements may comprise over two-thirds of the human genome. *PLoS Genet* **7:** e1002384.

23. **Ivics Z, Hackett PB, Plasterk RH, Izsvak Z.** 1997. Molecular reconstruction of Sleeping Beauty, a Tc1-like transposon from fish, and its transposition in human cells. *Cell* **91:**501–510.

24. **Shapiro JA.** 1979. Molecular model for the transposition and replication of bacteriophage Mu and other transposable elements. *Proc Natl Acad Sci U S A* **76:** 1933–1937.

25. **Kleckner N.** 1990. Regulation of transposition in bacteria. *Annu Rev Cell Biol* **6:**297–327.

26. **Rio DC.** 2002. P transposable elements in *Drosophila melanogaster*, p 484–518. *In* Craig NL, Craigie R, Gellert M, Lambowitz AM (ed), *Mobile DNA II*. ASM Press, Washington, D.C.

27. **Lazarow K, Doll ML, Kunze R.** 2013. Molecular biology of maize Ac/Ds elements: an overview. *Methods Mol Biol* **1057:**59–82.

28. **Cary LC, Goebel M, Corsaro BG, Wang HG, Rosen E, Fraser MJ.** 1989. Transposon mutagenesis of baculoviruses: analysis of Trichoplusia ni transposon IFP2 insertions within the FP-locus of nuclear polyhedrosis viruses. *Virology* **172:**156–169.

29. **Fraser MJ, Cary L, Boonvisudhi K, Wang HG.** 1995. Assay for movement of Lepidopteran transposon IFP2 in insect cells using a baculovirus genome as a target DNA. *Virology* **211:**397–407.

30. **Ding S, Wu X, Li G, Han M, Zhuang Y, Xu T.** 2005. Efficient transposition of the piggyBac (PB) transposon in mammalian cells and mice. *Cell* **122:**473–483.

31. **Rad R, Rad L, Wang W, Cadinanos J, Vassiliou G, Rice S, Campos LS, Yusa K, Banerjee R, Li MA, de la Rosa J, Strong A, Lu D, Ellis P, Conte N, Yang FT, Liu P, Bradley A.** 2010. PiggyBac transposon mutagenesis: a tool for cancer gene discovery in mice. *Science* **330:** 1104–1107.

32. **Ivics Z, Li MA, Mates L, Boeke JD, Nagy A, Bradley A, Izsvak Z.** 2009. Transposon-mediated genome manipulation in vertebrates. *Nat Methods* **6:**415–422.

33. **Largaespada DA.** 2009. Transposon mutagenesis in mice. *Methods Mol Biol* **530:**379–390.

34. **Ray DA, Feschotte C, Pagan HJ, Smith JD, Pritham EJ, Arensburger P, Atkinson PW, Craig NL.** 2008. Multiple

waves of recent DNA transposon activity in the bat, *Myotis lucifugus*. *Genome Res* **18:**717–728.

35. **Ray DA, Pagan HJ, Thompson ML, Stevens RD.** 2007. Bats with hATs: evidence for recent DNA transposon activity in genus Myotis. *Mol Biol Evol* **24:**632–639.

36. **Mitra R, Li X, Kapusta A, Mayhew D, Mitra RD, Feschotte C, Craig NL.** 2013. Functional characterization of piggyBat from the bat *Myotis lucifugus* unveils an active mammalian DNA transposon. *Proc Natl Acad Sci U S A* **110:**234–239.

37. **Pace JK 2nd, Feschotte C.** 2007. The evolutionary history of human DNA transposons: evidence for intense activity in the primate lineage. *Genome Res* **17:**422–432.

38. **Volff JN.** 2006. Turning junk into gold: domestication of transposable elements and the creation of new genes in eukaryotes. *Bioessays* **28:**913–922.

39. **Kapitonov VV, Jurka J.** 2005. RAG1 core and V(D)J recombination signal sequences were derived from Transib transposons. *PLoS Biol* **3:**e181.

40. **Majumdar S, Singh A, Rio DC.** 2013. The human THAP9 gene encodes an active P-element DNA transposase. *Science* **339:**446–448.

41. **Bannert N, Kurth R.** 2006. The evolutionary dynamics of human endogenous retroviral families. *Annu Rev Genomics Hum Genet* **7:**149–173.

42. **Belshaw R, Dawson AL, Woolven-Allen J, Redding J, Burt A, Tristem M.** 2005. Genomewide screening reveals high levels of insertional polymorphism in the human endogenous retrovirus family HERV-K(HML2): implications for present-day activity. *J Virol* **79:**12507–12514.

43. **Macfarlane C, Simmonds P.** 2004. Allelic variation of HERV-K(HML-2) endogenous retroviral elements in human populations. *J Mol Evol* **59:**642–656.

44. **Moyes D, Griffiths DJ, Venables PJ.** 2007. Insertional polymorphisms: a new lease of life for endogenous retroviruses in human disease. *Trends Genet* **23:**326–333.

45. **Shin W, Lee J, Son SY, Ahn K, Kim HS, Han K.** 2013. Human-specific HERV-K insertion causes genomic variations in the human genome. *PLoS One* **8:**e60605.

46. **Hughes JF, Coffin JM.** 2004. Human endogenous retrovirus K solo-LTR formation and insertional polymorphisms: implications for human and viral evolution. *Proc Natl Acad Sci U S A* **101:**1668–1672.

47. **Yohn CT, Jiang Z, McGrath SD, Hayden KE, Khaitovich P, Johnson ME, Eichler MY, McPherson JD, Zhao S, Paabo S, Eichler EE.** 2005. Lineage-specific expansions of retroviral insertions within the genomes of African great apes but not humans and orangutans. *PLoS Biol* **3:**e110.

48. **Dewannieux M, Harper F, Richaud A, Letzelter C, Ribet D, Pierron G, Heidmann T.** 2006. Identification of an infectious progenitor for the multiple-copy HERV-K human endogenous retroelements. *Genome Res* **16:**1548–1556.

49. **Lee YN, Bieniasz PD.** 2007. Reconstitution of an infectious human endogenous retrovirus. *PLoS Pathog* **3:**e10.

50. **Stocking C, Kozak CA.** 2008. Murine endogenous retroviruses. *Cell Mol Life Sci* **65:**3383–3398.

51. **Maksakova IA, Romanish MT, Gagnier L, Dunn CA, van de Lagemaat LN, Mager DL.** 2006. Retroviral elements and their hosts: insertional mutagenesis in the mouse germ line. *PLoS Genet* **2:**e2.

52. **Burton FH, Loeb DD, Voliva CF, Martin SL, Edgell MH, Hutchison CA 3rd.** 1986. Conservation throughout mammalia and extensive protein-encoding capacity of the highly repeated DNA long interspersed sequence one. *J Mol Biol* **187:**291–304.

53. **Smit AF, Toth G, Riggs AD, Jurka J.** 1995. Ancestral, mammalian-wide subfamilies of LINE-1 repetitive sequences. *J Mol Biol* **246:**401–417.

54. **Yang L, Brunsfeld J, Scott L, Wichman H.** 2014. Reviving the dead: history and reactivation of an extinct L1. *PLoS Genet* **10:**e1004395.

55. **Khan H, Smit A, Boissinot S.** 2006. Molecular evolution and tempo of amplification of human LINE-1 retrotransposons since the origin of primates. *Genome Res* **16:**78–87.

56. **Jacobs FM, Greenberg D, Nguyen N, Haeussler M, Ewing AD, Katzman S, Paten B, Salama SR, Haussler D.** 2014. An evolutionary arms race between KRAB zinc-finger genes ZNF91/93 and SVA/L1 retrotransposons. *Nature* **516:**242–245.

57. **Castro-Diaz N, Ecco G, Coluccio A, Kapopoulou A, Yazdanpanah B, Friedli M, Duc J, Jang SM, Turelli P, Trono D.** 2014. Evolutionally dynamic L1 regulation in embryonic stem cells. *Genes Dev* **28:**1397–1409.

58. **Beck CR, Collier P, Macfarlane C, Malig M, Kidd JM, Eichler EE, Badge RM, Moran JV.** 2010. LINE-1 retrotransposition activity in human genomes. *Cell* **141:**1159–1170.

59. **Brouha B, Schustak J, Badge RM, Lutz-Prigge S, Farley AH, Moran JV, Kazazian HH Jr.** 2003. Hot L1s account for the bulk of retrotransposition in the human population. *Proc Natl Acad Sci U S A* **100:**5280–5285.

60. **Moran JV, Holmes SE, Naas TP, DeBerardinis RJ, Boeke JD, Kazazian HH Jr.** 1996. High frequency retrotransposition in cultured mammalian cells. *Cell* **87:**917–927.

61. **Sassaman DM, Dombroski BA, Moran JV, Kimberland ML, Naas TP, DeBerardinis RJ, Gabriel A, Swergold GD, Kazazian HH Jr.** 1997. Many human L1 elements are capable of retrotransposition. *Nat Genet* **16:**37–43.

62. **Skowronski J, Fanning TG, Singer MF.** 1988. Unit-length Line-1 transcripts in human teratocarcinoma cells. *Mol Cell Biol* **8:**1385–1397.

63. **Boissinot S, Chevret P, Furano AV.** 2000. L1 (LINE-1) retrotransposon evolution and amplification in recent human history. *Mol Biol Evol* **17:**915–928.

64. **Dombroski BA, Mathias SL, Nanthakumar E, Scott AF, Kazazian HH Jr.** 1991. Isolation of an active human transposable element. *Science* **254:**1805–1808.

65. **Ovchinnikov I, Troxel AB, Swergold GD.** 2001. Genomic characterization of recent human LINE-1 insertions: evidence supporting random insertion. *Genome Res* **11:**2050–2058.

66. **Sheen FM, Sherry ST, Risch GM, Robichaux M, Nasidze I, Stoneking M, Batzer MA, Swergold GD.** 2000. Reading between the LINEs: human genomic variation induced by LINE-1 retrotransposition. *Genome Res* **10:**1496–1508.

67. **Badge RM, Alisch RS, Moran JV.** 2003. ATLAS: a system to selectively identify human-specific L1 insertions. *Am J Hum Genet* **72:**823–838.

68. **Ewing AD, Kazazian HH Jr.** 2010. High-throughput sequencing reveals extensive variation in human-specific L1 content in individual human genomes. *Genome Res* **20:**1262–1270.

69. **Iskow RC, McCabe MT, Mills RE, Torene S, Pittard WS, Neuwald AF, Van Meir EG, Vertino PM, Devine SE.** 2010. Natural mutagenesis of human genomes by endogenous retrotransposons. *Cell* **141:** 1253–1261.

70. **Boissinot S, Entezam A, Young L, Munson PJ, Furano AV.** 2004. The insertional history of an active family of L1 retrotransposons in humans. *Genome Res* **14:** 1221–1231.

71. **Myers JS, Vincent BJ, Udall H, Watkins WS, Morrish TA, Kilroy GE, Swergold GD, Henke J, Henke L, Moran JV, Jorde LB, Batzer MA.** 2002. A comprehensive analysis of recently integrated human Ta L1 elements. *Am J Hum Genet* **71:**312–326.

72. **Durbin RM, Abecasis GR, Altshuler DL, Auton A, Brooks LD, Gibbs RA, Hurles ME, McVean GA.** 2010. A map of human genome variation from population-scale sequencing. *Nature* **467:**1061–1073.

73. **Grimaldi G, Skowronski J, Singer MF.** 1984. Defining the beginning and end of KpnI family segments. *EMBO J* **3:**1753–1759.

74. **Scott AF, Schmeckpeper BJ, Abdelrazik M, Comey CT, O'Hara B, Rossiter JP, Cooley T, Heath P, Smith KD, Margolet L.** 1987. Origin of the human L1 elements: proposed progenitor genes deduced from a consensus DNA sequence. *Genomics* **1:**113–125.

75. **Swergold GD.** 1990. Identification, characterization, and cell specificity of a human LINE-1 promoter. *Mol Cell Biol* **10:**6718–6729.

76. **Speek M.** 2001. Antisense promoter of human L1 retrotransposon drives transcription of adjacent cellular genes. *Mol Cell Biol* **21:**1973–1985.

77. **Holmes SE, Singer MF, Swergold GD.** 1992. Studies on p40, the leucine zipper motif-containing protein encoded by the first open reading frame of an active human LINE-1 transposable element. *J Biol Chem* **267:** 19765–19768.

78. **Hohjoh H, Singer MF.** 1996. Cytoplasmic ribonucleoprotein complexes containing human LINE-1 protein and RNA. *Embo J* **15:**630–639.

79. **Khazina E, Truffault V, Buttner R, Schmidt S, Coles M, Weichenrieder O.** 2011. Trimeric structure and flexibility of the L1ORF1 protein in human L1 retrotransposition. *Nat Struct Mol Biol* **18:**1006–1014.

80. **Martin SL, Bushman FD.** 2001. Nucleic acid chaperone activity of the ORF1 protein from the mouse LINE-1 retrotransposon. *Mol Cell Biol* **21:**467–475.

81. **Ergun S, Buschmann C, Heukeshoven J, Dammann K, Schnieders F, Lauke H, Chalajour F, Kilic N, Stratling WH, Schumann GG.** 2004. Cell type-specific expression of LINE-1 open reading frames 1 and 2 in fetal and adult human tissues. *J Biol Chem* **279:**27753–27763.

82. **Doucet AJ, Hulme AE, Sahinovic E, Kulpa DA, Moldovan JB, Kopera HC, Athanikar JN, Hasnaoui M, Bucheton A, Moran JV, Gilbert N.** 2010. Characterization of LINE-1 ribonucleoprotein particles. *PLoS Genet* **6:**e1001150.

83. **Feng Q, Moran J, Kazazian H, Boeke JD.** 1996. Human L1 retrotransposon encodes a conserved endonuclease required for retrotransposition. *Cell* **87:**905–916.

84. **Mathias SL, Scott AF, Kazazian HH Jr, Boeke JD, Gabriel A.** 1991. Reverse transcriptase encoded by a human transposable element. *Science* **254:**1808–1810.

85. **Fanning T, Singer M.** 1987. The LINE-1 DNA sequences in four mammalian orders predict proteins that conserve homologies to retrovirus proteins. *Nucleic Acids Res* **15:**2251–2260.

86. **Goodier JL, Ostertag EM, Du K, Kazazian HH Jr.** 2001. A novel active L1 retrotransposon subfamily in the mouse. *Genome Res* **11:**1677–1685.

87. **Furano AV.** 2000. The biological properties and evolutionary dynamics of mammalian LINE-1 retrotransposons. *Prog Nucleic Acid Res Mol Biol* **64:**255–294.

88. **Sookdeo A, Hepp CM, McClure MA, Boissinot S.** 2013. Revisiting the evolution of mouse LINE-1 in the genomic era. *Mob DNA* **4:**3.

89. **Adey NB, Tollefsbol TO, Sparks AB, Edgell MH, Hutchison CA 3rd.** 1994. Molecular resurrection of an extinct ancestral promoter for mouse L1. *Proc Natl Acad Sci U S A* **91:**1569–1573.

90. **Naas TP, DeBerardinis RJ, Moran JV, Ostertag EM, Kingsmore SF, Seldin MF, Hayashizaki Y, Martin SL, Kazazian HH.** 1998. An actively retrotransposing, novel subfamily of mouse L1 elements. *EMBO J* **17:**590–597.

91. **DeBerardinis RJ, Goodier JL, Ostertag EM, Kazazian HH Jr.** 1998. Rapid amplification of a retrotransposon subfamily is evolving the mouse genome. *Nat Genet* **20:** 288–290.

92. **Dewannieux M, Esnault C, Heidmann T.** 2003. LINE-mediated retrotransposition of marked Alu sequences. *Nat Genet* **35:**41–48.

93. **Dewannieux M, Heidmann T.** 2005. L1-mediated retrotransposition of murine B1 and B2 SINEs recapitulated in cultured cells. *J Mol Biol* **349:**241–247.

94. **Raiz J, Damert A, Chira S, Held U, Klawitter S, Hamdorf M, Lower J, Stratling WH, Lower R, Schumann GG.** 2012. The non-autonomous retrotransposon SVA is trans-mobilized by the human LINE-1 protein machinery. *Nucleic Acids Res* **40:**1666–1683.

95. **Hancks DC, Goodier JL, Mandal PK, Cheung LE, Kazazian HH Jr.** 2011. Retrotransposition of marked SVA elements by human L1s in cultured cells. *Hum Mol Genet* **20:**3386–3400.

96. **Hancks DC, Mandal PK, Cheung LE, Kazazian HH Jr.** 2012. The minimal active human SVA retrotransposon

requires only the 5′-hexamer and Alu-like domains. *Mol Cell Biol* 32:4718–4726.

97. **Esnault C, Maestre J, Heidmann T.** 2000. Human LINE retrotransposons generate processed pseudogenes. *Nat Genet* **24**:363–367.
98. **Wei W, Gilbert N, Ooi SL, Lawler JF, Ostertag EM, Kazazian HH, Boeke JD, Moran JV.** 2001. Human L1 retrotransposition: cis preference versus trans complementation. *Mol Cell Biol* **21**:1429–1439.
99. **Jurka J.** 1997. Sequence patterns indicate an enzymatic involvement in integration of mammalian retroposons. *Proc Natl Acad Sci U S A* **94**:1872–1877.
100. **Rubin CM, Houck CM, Deininger PL, Friedmann T, Schmid CW.** 1980. Partial nucleotide sequence of the 300-nucleotide interspersed repeated human DNA sequences. *Nature* **284**:372–374.
101. **Ullu E, Tschudi C.** 1984. Alu sequences are processed 7SL RNA genes. *Nature* **312**:171–172.
102. **Chu WM, Liu WM, Schmid CW.** 1995. RNA polymerase III promoter and terminator elements affect Alu RNA expression. *Nucleic Acids Res* **23**:1750–1757.
103. **Batzer MA, Deininger PL.** 2002. Alu repeats and human genomic diversity. *Nat Rev Genet* **3**:370–379.
104. **Chesnokov I, Schmid CW.** 1996. Flanking sequences of an Alu source stimulate transcription in vitro by interacting with sequence-specific transcription factors. *J Mol Evol* **42**:30–36.
105. **Comeaux MS, Roy-Engel AM, Hedges DJ, Deininger PL.** 2009. Diverse cis factors controlling Alu retrotransposition: what causes Alu elements to die? *Genome Res* **19**:545–555.
106. **Dewannieux M, Heidmann T.** 2005. Role of poly(A) tail length in Alu retrotransposition. *Genomics* **86**:378–381.
107. **Goodier JL, Maraia RJ.** 1998. Terminator-specific recycling of a B1-Alu transcription complex by RNA polymerase III is mediated by the RNA terminus-binding protein La. *J Biol Chem* **273**:26110–26116.
108. **Liu WM, Schmid CW.** 1993. Proposed roles for DNA methylation in Alu transcriptional repression and mutational inactivation. *Nucleic Acids Res* **21**:1351–1359.
109. **Ullu E, Weiner AM.** 1985. Upstream sequences modulate the internal promoter of the human 7SL RNA gene. *Nature* **318**:371–374.
110. **Jurka J, Smith T.** 1988. A fundamental division in the Alu family of repeated sequences. *Proc Natl Acad Sci U S A* **85**:4775–4778.
111. **Slagel V, Flemington E, Traina-Dorge V, Bradshaw H, Deininger P.** 1987. Clustering and subfamily relationships of the Alu family in the human genome. *Mol Biol Evol* **4**:19–29.
112. **Britten RJ, Baron WF, Stout DB, Davidson EH.** 1988. Sources and evolution of human Alu repeated sequences. *Proc Natl Acad Sci U S A* **85**:4770–4774.
113. **Willard C, Nguyen HT, Schmid CW.** 1987. Existence of at least three distinct Alu subfamilies. *J Mol Evol* **26**: 180–186.
114. **Batzer MA, Deininger PL, Hellmann-Blumberg U, Jurka J, Labuda D, Rubin CM, Schmid CW, Zietkiewicz E, Zuckerkandl E.** 1996. Standardized nomenclature for Alu repeats. *J Mol Evol* **42**:3–6.
115. **Batzer MA, Schmid CW, Deininger PL.** 1993. Evolutionary analyses of repetitive DNA sequences. *Methods Enzymol* **224**:213–232.
116. **Carroll ML, Roy-Engel AM, Nguyen SV, Salem AH, Vogel E, Vincent B, Myers J, Ahmad Z, Nguyen L, Sammarco M, Watkins WS, Henke J, Makalowski W, Jorde LB, Deininger PL, Batzer MA.** 2001. Large-scale analysis of the Alu Ya5 and Yb8 subfamilies and their contribution to human genomic diversity. *J Mol Biol* **311**:17–40.
117. **Witherspoon DJ, Zhang Y, Xing J, Watkins WS, Ha H, Batzer MA, Jorde LB.** 2013. Mobile element scanning (ME-Scan) identifies thousands of novel Alu insertions in diverse human populations. *Genome Res* **23**:1170–1181.
118. **Minghetti PP, Dugaiczyk A.** 1993. The emergence of new DNA repeats and the divergence of primates. *Proc Natl Acad Sci U S A* **90**:1872–1876.
119. **Batzer MA, Stoneking M, Alegria-Hartman M, Bazan H, Kass DH, Shaikh TH, Novick GE, Ioannou PA, Scheer WD, Herrera RJ, et al.** 1994. African origin of human-specific polymorphic Alu insertions. *Proc Natl Acad Sci U S A* **91**:12288–12292.
120. **Nikaido M, Rooney AP, Okada N.** 1999. Phylogenetic relationships among cetartiodactyls based on insertions of short and long interpersed elements: hippopotamuses are the closest extant relatives of whales. *Proc Natl Acad Sci U S A* **96**:10261–10266.
121. **Okada N.** 1991. SINEs. *Curr Opin Genet Dev* **1**: 498–504.
122. **Ray DA, Xing J, Salem AH, Batzer MA.** 2006. SINEs of a nearly perfect character. *Syst Biol* **55**:928–935.
123. **Witherspoon DJ, Marchani EE, Watkins WS, Ostler CT, Wooding SP, Anders BA, Fowlkes JD, Boissinot S, Furano AV, Ray DA, Rogers AR, Batzer MA, Jorde LB.** 2006. Human population genetic structure and diversity inferred from polymorphic L1(LINE-1) and Alu insertions. *Hum Hered* **62**:30–46.
124. **Konkel MK, Walker JA, Batzer MA.** 2010. LINEs and SINEs of primate evolution. *Evol Anthropol* **19**: 236–249.
125. **Bennett EA, Keller H, Mills RE, Schmidt S, Moran JV, Weichenrieder O, Devine SE.** 2008. Active Alu retrotransposons in the human genome. *Genome Res* **18**: 1875–1883.
126. **Sarrowa J, Chang DY, Maraia RJ.** 1997. The decline in human Alu retroposition was accompanied by an asymmetric decrease in SRP9/14 binding to dimeric Alu RNA and increased expression of small cytoplasmic Alu RNA. *Mol Cell Biol* **17**:1144–1151.
127. **Ostertag EM, Goodier JL, Zhang Y, Kazazian HH Jr.** 2003. SVA elements are nonautonomous retrotransposons that cause disease in humans. *Am J Hum Genet* **73**:1444–1451.
128. **Wang H, Xing J, Grover D, Hedges DJ, Han K, Walker JA, Batzer MA.** 2005. SVA elements: a hominid-specific retroposon family. *J Mol Biol* **354**:994–1007.

129. **Hancks DC, Kazazian HH Jr.** 2010. SVA retrotransposons: Evolution and genetic instability. *Semin Cancer Biol* **20:**234–245.

130. **Savage AL, Bubb VJ, Breen G, Quinn JP.** 2013. Characterisation of the potential function of SVA retrotransposons to modulate gene expression patterns. *BMC Evol Biol* **13:**101.

131. **Hara T, Hirai Y, Baicharoen S, Hayakawa T, Hirai H, Koga A.** 2012. A novel composite retrotransposon derived from or generated independently of the SVA (SINE/VNTR/Alu) transposon has undergone proliferation in gibbon genomes. *Genes Genet Syst* **87:**181–190.

132. **Ianc B, Ochis C, Persch R, Popescu O, Damert A.** 2014. Hominoid Composite Non-LTR Retrotransposons-Variety, Assembly, Evolution, and Structural Determinants of Mobilization. *Mol Biol Evol* **31:**2847–2864.

133. **Carbone L, Harris RA, Mootnick AR, Milosavljevic A, Martin DI, Rocchi M, Capozzi O, Archidiacono N, Konkel MK, Walker JA, Batzer MA, de Jong PJ.** 2012. Centromere remodeling in *Hoolock leuconedys* (Hylobatidae) by a new transposable element unique to the gibbons. *Genome Biol Evol* **4:**648–658.

134. **Carbone L, Harris RA, Gnerre S, Veeramah KR, Lorente-Galdos B, Huddleston J, Meyer TJ, Herrero J, Roos C, Aken B, Anaclerio F, Archidiacono N, Baker C, Barrell D, Batzer MA, Beal K, Blancher A, Bohrson CL, Brameier M, Campbell MS, Capozzi O, Casola C, Chiatante G, Cree A, Damert A, de Jong PJ, Dumas L, Fernandez-Callejo M, Flicek P, Fuchs NV, Gut I, Gut M, Hahn MW, Hernandez-Rodriguez J, Hillier LW, Hubley R, Ianc B, Izsvak Z, Jablonski NG, Johnstone LM, Karimpour-Fard A, Konkel MK, Kostka D, Lazar NH, Lee SL, Lewis LR, Liu Y, Locke DP, Mallick S, Mendez FL, Muffato M, Nazareth LV, Nevonen KA, O'Bleness M, Ochis C, Odom DT, Pollard KS, Quilez J, Reich D, Rocchi M, Schumann GG, Searle S, Sikela JM, Skollar G, Smit A, Sonmez K, ten Hallers B, Terhune E, Thomas GW, Ullmer B, Ventura M, Walker JA, Wall JD, Walter L, Ward MC, Wheelan SJ, Whelan CW, White S, Wilhelm LJ, Woerner AE, Yandell M, Zhu B, Hammer MF, Marques-Bonet T, Eichler EE, Fulton L, Fronick C, Muzny DM, Warren WC, Worley KC, Rogers J, Wilson RK, Gibbs RA.** 2014. Gibbon genome and the fast karyotype evolution of small apes. *Nature* **513:**195–201.

135. **Krayev AS, Kramerov DA, Skryabin KG, Ryskov AP, Bayev AA, Georgiev GP.** 1980. The nucleotide sequence of the ubiquitous repetitive DNA sequence B1 complementary to the most abundant class of mouse fold-back RNA. *Nucleic Acids Res* **8:**1201–1215.

136. **Quentin Y.** 1994. A master sequence related to a free left Alu monomer (FLAM) at the origin of the B1 family in rodent genomes. *Nucleic Acids Res* **22:**2222–2227.

137. **Daniels GR, Deininger PL.** 1985. Repeat sequence families derived from mammalian tRNA genes. *Nature* **317:** 819–822.

138. **DeChiara TM, Brosius J.** 1987. Neural BC1 RNA: cDNA clones reveal nonrepetitive sequence content. *Proc Natl Acad Sci U S A* **84:**2624–2628.

139. **Lee IY, Westaway D, Smit AF, Wang K, Seto J, Chen L, Acharya C, Ankener M, Baskin D, Cooper C, Yao H, Prusiner SB, Hood LE.** 1998. Complete genomic sequence and analysis of the prion protein gene region from three mammalian species. *Genome Res* **8:**1022–1037.

140. **Serdobova IM, Kramerov DA.** 1998. Short retroposons of the B2 superfamily: evolution and application for the study of rodent phylogeny. *J Mol Evol* **46:**202–214.

141. **Weiner AM, Deininger PL, Efstratiadis A.** 1986. Nonviral retroposons: genes, pseudogenes, and transposable elements generated by the reverse flow of genetic information. *Annu Rev Biochem* **55:**631–661.

142. **Vanin EF.** 1985. Processed pseudogenes: characteristics and evolution. *Annu Rev Genet* **19:**253–272.

143. **Torrents D, Suyama M, Zdobnov E, Bork P.** 2003. A genome-wide survey of human pseudogenes. *Genome Res* **13:**2559–2567.

144. **Zhang Z, Harrison P, Gerstein M.** 2002. Identification and analysis of over 2000 ribosomal protein pseudogenes in the human genome. *Genome Res* **12:**1466–1482.

145. **Zhang Z, Harrison PM, Liu Y, Gerstein M.** 2003. Millions of years of evolution preserved: a comprehensive catalog of the processed pseudogenes in the human genome. *Genome Res* **13:**2541–2558.

146. **Ewing AD, Ballinger TJ, Earl D, Harris CC, Ding L, Wilson RK, Haussler D.** 2013. Retrotransposition of gene transcripts leads to structural variation in mammalian genomes. *Genome Biol* **14:**R22.

147. **Schrider DR, Navarro FC, Galante PA, Parmigiani RB, Camargo AA, Hahn MW, de Souza SJ.** 2013. Gene copy-number polymorphism caused by retrotransposition in humans. *PLoS Genet* **9:**e1003242.

148. **Abyzov A, Iskow R, Gokcumen O, Radke DW, Balasubramanian S, Pei B, Habegger L, Lee C, Gerstein M.** 2013. Analysis of variable retroduplications in human populations suggests coupling of retrotransposition to cell division. *Genome Res* **23:**2042–2052.

149. **Richardson SR, Salvador-Palomeque C, Faulkner GJ.** 2014. Diversity through duplication: whole-genome sequencing reveals novel gene retrocopies in the human population. *Bioessays* **36:**475–481.

150. **Kazazian HH Jr.** 2014. Processed pseudogene insertions in somatic cells. *Mob DNA* **5:**20.

151. **Hulme AE, Kulpa DA, Garcia-Perez JL, Moran JV.** 2006. p 35–72. *In* Lupski JR, Stankiewicz P (ed), *The impact of LINE-1 retrotransposition on the human genome.* Humana Press, Totowa, New Jersey.

152. **Sayah DM, Sokolskaja E, Berthoux L, Luban J.** 2004. Cyclophilin A retrotransposition into TRIM5 explains owl monkey resistance to HIV-1. *Nature* **430:**569–573.

153. **Malfavon-Borja R, Wu LI, Emerman M, Malik HS.** 2013. Birth, decay, and reconstruction of an ancient TRIMCyp gene fusion in primate genomes. *Proc Natl Acad Sci U S A* **110:**E583–E592.

154. **Buzdin A, Gogvadze E, Kovalskaya E, Volchkov P, Ustyugova S, Illarionova A, Fushan A, Vinogradova T, Sverdlov E.** 2003. The human genome contains many

types of chimeric retrogenes generated through in vivo RNA recombination. *Nucleic Acids Res* **31**:4385–4390.

155. **Buzdin A, Ustyugova S, Gogvadze E, Vinogradova T, Lebedev Y, Sverdlov E.** 2002. A new family of chimeric retrotranscripts formed by a full copy of U6 small nuclear RNA fused to the 3′ terminus of l1. *Genomics* **80**: 402–406.

156. **Gilbert N, Lutz S, Morrish TA, Moran JV.** 2005. Multiple fates of L1 retrotransposition intermediates in cultured human cells. *Mol Cell Biol* **25**:7780–7795.

157. **Garcia-Perez JL, Doucet AJ, Bucheton A, Moran JV, Gilbert N.** 2007. Distinct mechanisms for trans-mediated mobilization of cellular RNAs by the LINE-1 reverse transcriptase. *Genome Res* **17**:602–611.

158. **Weber MJ.** 2006. Mammalian small nucleolar RNAs are mobile genetic elements. *PLoS Genet* **2**:e205.

159. **Hasnaoui M, Doucet AJ, Meziane O, Gilbert N.** 2009. Ancient repeat sequence derived from U6 snRNA in primate genomes. *Gene* **448**:139–144.

160. **Kazazian HH Jr, Wong C, Youssoufian H, Scott AF, Phillips DG, Antonarakis SE.** 1988. Haemophilia A resulting from de novo insertion of L1 sequences represents a novel mechanism for mutation in man. *Nature* **332**:164–166.

161. **Dombroski BA, Scott AF, Kazazian HH Jr.** 1993. Two additional potential retrotransposons isolated from a human L1 subfamily that contains an active retrotransposable element. *Proc Natl Acad Sci U S A* **90**:6513–6517.

162. **Lutz SM, Vincent BJ, Kazazian HH Jr, Batzer MA, Moran JV.** 2003. Allelic heterogeneity in LINE-1 retrotransposition activity. *Am J Hum Genet* **73**:1431–1437.

163. **Skowronski J, Singer MF.** 1985. Expression of a cytoplasmic LINE-1 transcript is regulated in a human teratocarcinoma cell line. *Proc Natl Acad Sci U S A* **82**: 6050–6054.

164. **Wallace MR, Andersen LB, Saulino AM, Gregory PE, Glover TW, Collins FS.** 1991. A de novo Alu insertion results in neurofibromatosis type 1. *Nature* **353**:864–866.

165. **Hassoun H, Coetzer TL, Vassiliadis JN, Sahr KE, Maalouf GJ, Saad ST, Catanzariti L, Palek J.** 1994. A novel mobile element inserted in the alpha spectrin gene: spectrin dayton. A truncated alpha spectrin associated with hereditary elliptocytosis. *J Clin Invest* **94**: 643–648.

166. **van der Klift HM, Tops CM, Hes FJ, Devilee P, Wijnen JT.** 2012. Insertion of an SVA element, a nonautonomous retrotransposon, in PMS2 intron 7 as a novel cause of Lynch syndrome. *Hum Mutat* **33**:1051–1055.

167. **Vogt J, Bengesser K, Claes KB, Wimmer K, Mautner VF, van Minkelen R, Legius E, Brems H, Upadhyaya M, Hogel J, Lazaro C, Rosenbaum T, Bammert S, Messiaen L, Cooper DN, Kehrer-Sawatzki H.** 2014. SVA retrotransposon insertion-associated deletion rep resents a novel mutational mechanism underlying large genomic copy number changes with non-recurrent breakpoints. *Genome Biol* **15**:R80.

168. **de Boer M, van Leeuwen K, Geissler J, Weemaes CM, van den Berg TK, Kuijpers TW, Warris A, Roos D.** 2014. Primary immunodeficiency caused by an exonized retroposed gene copy inserted in the CYBB gene. *Hum Mutat* **35**:486–496.

169. **Cooke SL, Shlien A, Marshall J, Pipinikas CP, Martincorena I, Tubio JM, Li Y, Menzies A, Mudie L, Ramakrishna M, Yates L, Davies H, Bolli N, Bignell GR, Tarpey PS, Behjati S, Nik-Zainal S, Papaemmanuil E, Teixeira VH, Raine K, O'Meara S, Dodoran MS, Teague JW, Butler AP, Iacobuzio-Donahue C, Santarius T, Grundy RG, Malkin D, Greaves M, Munshi N, Flanagan AM, Bowtell D, Martin S, Larsimont D, Reis-Filho JS, Boussioutas A, Taylor JA, Hayes ND, Janes SM, Futreal PA, Stratton MR, McDermott U, Campbell PJ.** 2014. Processed pseudogenes acquired somatically during cancer development. *Nat Commun* **5**:3644.

170. **Wimmer K, Callens T, Wernstedt A, Messiaen L.** 2011. The NF1 gene contains hotspots for L1 endonuclease-dependent de novo insertion. *PLoS Genet* **7**:e1002371.

171. **Kingsmore SF, Giros B, Suh D, Bieniarz M, Caron MG, Seldin MF.** 1994. Glycine receptor beta-subunit gene mutation in spastic mouse associated with LINE-1 element insertion. *Nat Genet* **7**:136–141.

172. **Mulhardt C, Fischer M, Gass P, Simon-Chazottes D, Guenet JL, Kuhse J, Betz H, Becker CM.** 1994. The spastic mouse: aberrant splicing of glycine receptor beta subunit mRNA caused by intronic insertion of L1 element. *Neuron* **13**:1003–1015.

173. **Takahara T, Ohsumi T, Kuromitsu J, Shibata K, Sasaki N, Okazaki Y, Shibata H, Sato S, Yoshiki A, Kusakabe M, Muramatsu M, Ueki M, Okuda K, Hayashizaki Y.** 1996. Dysfunction of the Orleans reeler gene arising from exon skipping due to transposition of a full-length copy of an active L1 sequence into the skipped exon. *Hum Mol Genet* **5**:989–993.

174. **Kojima T, Nakajima K, Mikoshiba K.** 2000. The disabled 1 gene is disrupted by a replacement with L1 fragment in yotari mice. *Brain Res Mol Brain Res* **75**:121–127.

175. **Perou CM, Pryor RJ, Naas TP, Kaplan J.** 1997. The bg allele mutation is due to a LINE1 element retrotransposition. *Genomics* **42**:366–368.

176. **Yajima I, Sato S, Kimura T, Yasumoto K, Shibahara S, Goding CR, Yamamoto H.** 1999. An L1 element intronic insertion in the black-eyed white (Mitf[mi-bw]) gene: the loss of a single Mitf isoform responsible for the pigmentary defect and inner ear deafness. *Hum Mol Genet* **8**:1431–1441.

177. **Kohrman DC, Harris JB, Meisler MH.** 1996. Mutation detection in the med and medJ alleles of the sodium channel Scn8a. Unusual splicing due to a minor class AT-AC intron. *J Biol Chem* **271**:17576–17581.

178. **Bomar JM, Benke PJ, Slattery EL, Puttagunta R, Taylor LP, Seong E, Nystuen A, Chen W, Albin RL, Patel PD, Kittles RA, Sheffield VC, Burmeister M.** 2003. Mutations in a novel gene encoding a CRAL-TRIO domain cause human Cayman ataxia and ataxia/dystonia in the jittery mouse. *Nat Genet* **35**:264–269.

179. **Gilbert N, Bomar JM, Burmeister M, Moran JV.** 2004. Characterization of a mutagenic B1 retrotransposon insertion in the jittery mouse. *Hum Mutat* **24**:9–13.

180. **Katzir N, Rechavi G, Cohen JB, Unger T, Simoni F, Segal S, Cohen D, Givol D.** 1985. "Retroposon" insertion into the cellular oncogene c-myc in canine transmissible venereal tumor. *Proc Natl Acad Sci U S A* **82:** 1054–1058.

181. **Choi Y, Ishiguro N, Shinagawa M, Kim CJ, Okamoto Y, Minami S, Ogihara K.** 1999. Molecular structure of canine LINE-1 elements in canine transmissible venereal tumor. *Anim Genet* **30:**51–53.

182. **Brooks MB, Gu W, Barnas JL, Ray J, Ray K.** 2003. A Line 1 insertion in the Factor IX gene segregates with mild hemophilia B in dogs. *Mamm Genome* **14:**788–795.

183. **Smith BF, Yue Y, Woods PR, Kornegay JN, Shin JH, Williams RR, Duan D.** 2011. An intronic LINE-1 element insertion in the dystrophin gene aborts dystrophin expression and results in Duchenne-like muscular dystrophy in the corgi breed. *Lab Invest* **91:**216–231.

184. **Wolf ZT, Leslie EJ, Arzi B, Jayashankar K, Karmi N, Jia Z, Rowland DJ, Young A, Safra N, Sliskovic S, Murray JC, Wade CM, Bannasch DL.** 2014. A LINE-1 insertion in DLX6 is responsible for cleft palate and mandibular abnormalities in a canine model of Pierre Robin sequence. *PLoS Genet* **10:**e1004257.

185. **Downs LM, Mellersh CS.** 2014. An Intronic SINE insertion in FAM161A that causes exon-skipping is associated with progressive retinal atrophy in Tibetan Spaniels and Tibetan Terriers. *PLoS One* **9:**e93990.

186. **Clark LA, Wahl JM, Rees CA, Murphy KE.** 2006. Retrotransposon insertion in SILV is responsible for merle patterning of the domestic dog. *Proc Natl Acad Sci U S A* **103:**1376–1381.

187. **Lin L, Faraco J, Li R, Kadotani H, Rogers W, Lin X, Qiu X, de Jong PJ, Nishino S, Mignot E.** 1999. The sleep disorder canine narcolepsy is caused by a mutation in the hypocretin (orexin) receptor 2 gene. *Cell* **98:** 365–376.

188. **Pele M, Tiret L, Kessler JL, Blot S, Panthier JJ.** 2005. SINE exonic insertion in the PTPLA gene leads to multiple splicing defects and segregates with the autosomal recessive centronuclear myopathy in dogs. *Hum Mol Genet* **14:**1417–1427.

189. **Parker HG, VonHoldt BM, Quignon P, Margulies EH, Shao S, Mosher DS, Spady TC, Elkahloun A, Cargill M, Jones PG, Maslen CL, Acland GM, Sutter NB, Kuroki K, Bustamante CD, Wayne RK, Ostrander EA.** 2009. An expressed fgf4 retrogene is associated with breed-defining chondrodysplasia in domestic dogs. *Science* **325:**995–998.

190. **Heidmann T, Heidmann O, Nicolas JF.** 1988. An indicator gene to demonstrate intracellular transposition of defective retroviruses. *Proc Natl Acad Sci U S A* **85:** 2219–2223.

191. **Curcio MJ, Garfinkel DJ.** 1991. Single-step selection for Ty1 element retrotransposition. *Proc Natl Acad Sci USA* **88:**936–940.

192. **Wei W, Morrish TA, Alisch RS, Moran JV.** 2000. A transient assay reveals that cultured human cells can accommodate multiple LINE-1 retrotransposition events. *Anal Biochem* **284:**435–438.

193. **Freeman JD, Goodchild NL, Mager DL.** 1994. A modified indicator gene for selection of retrotransposition events in mammalian cells. *Biotechniques* **17:**46, 8–9, 52.

194. **Xie Y, Rosser JM, Thompson TL, Boeke JD, An W.** 2011. Characterization of L1 retrotransposition with high-throughput dual-luciferase assays. *Nucleic Acids Res* **39:**e16.

195. **Ostertag EM, Prak ET, DeBerardinis RJ, Moran JV, Kazazian HH Jr.** 2000. Determination of L1 retrotransposition kinetics in cultured cells. *Nucleic Acids Res* **28:**1418–1423.

196. **Morrish TA, Gilbert N, Myers JS, Vincent BJ, Stamato TD, Taccioli GE, Batzer MA, Moran JV.** 2002. DNA repair mediated by endonuclease-independent LINE-1 retrotransposition. *Nat Genet* **31:**159–165.

197. **Goodier JL, Zhang L, Vetter MR, Kazazian HH Jr.** 2007. LINE-1 ORF1 protein localizes in stress granules with other RNA-binding proteins, including components of RNA interference RNA-induced silencing complex. *Mol Cell Biol* **27:**6469–6483.

198. **Kulpa DA, Moran JV.** 2005. Ribonucleoprotein particle formation is necessary but not sufficient for LINE-1 retrotransposition. *Hum Mol Genet* **14:**3237–3248.

199. **Kulpa DA, Moran JV.** 2006. Cis-preferential LINE-1 reverse transcriptase activity in ribonucleoprotein particles. *Nat Struct Mol Biol* **13:**655–660.

200. **Taylor MS, Lacava J, Mita P, Molloy KR, Huang CR, Li D, Adney EM, Jiang H, Burns KH, Chait BT, Rout MP, Boeke JD, Dai L.** 2013. Affinity proteomics reveals human host factors implicated in discrete stages of LINE-1 retrotransposition. *Cell* **155:**1034–1048.

201. **Goodier JL, Cheung LE, Kazazian HH Jr.** 2013. Mapping the LINE1 ORF1 protein interactome reveals associated inhibitors of human retrotransposition. *Nucleic Acids Res* **41:**7401–7419.

202. **Gilbert N, Lutz-Prigge S, Moran JV.** 2002. Genomic deletions created upon LINE-1 retrotransposition. *Cell* **110:**315–325.

203. **Symer DE, Connelly C, Szak ST, Caputo EM, Cost GJ, Parmigiani G, Boeke JD.** 2002. Human L1 retrotransposition is associated with genetic instability in vivo. *Cell* **110:**327–338.

204. **Kajikawa M, Okada N.** 2002. LINEs mobilize SINEs in the eel through a shared 3′ sequence. *Cell* **111:**433–444.

205. **Sugano T, Kajikawa M, Okada N.** 2006. Isolation and characterization of retrotransposition-competent LINEs from zebrafish. *Gene* **365:**74–82.

206. **Seleme MC, Vetter MR, Cordaux R, Bastone L, Batzer MA, Kazazian HH Jr.** 2006. Extensive individual variation in L1 retrotransposition capability contributes to human genetic diversity. *Proc Natl Acad Sci U S A* **103:** 6611–6616.

207. **Becker KG, Swergold GD, Ozato K, Thayer RE.** 1993. Binding of the ubiquitous nuclear transcription factor YY1 to a cis regulatory sequence in the human LINE-1 transposable element. *Hum Mol Genet* **2:**1697–1702.

208. **Athanikar JN, Badge RM, Moran JV.** 2004. A YY1-binding site is required for accurate human LINE-1

transcription initiation. *Nucleic Acids Res* 32:3846–3855.

209. **Dmitriev SE, Andreev DE, Terenin IM, Olovnikov IA, Prassolov VS, Merrick WC, Shatsky IN.** 2007. Efficient translation initiation directed by the 900-nucleotide-long and GC-rich 5′ untranslated region of the human retrotransposon LINE-1 mRNA is strictly cap dependent rather than internal ribosome entry site mediated. *Mol Cell Biol* 27:4685–4697.

210. **Tchenio T, Casella JF, Heidmann T.** 2000. Members of the SRY family regulate the human LINE retrotransposons. *Nucleic Acids Res* 28:411–415.

211. **Yang N, Zhang L, Zhang Y, Kazazian HH Jr.** 2003. An important role for RUNX3 in human L1 transcription and retrotransposition. *Nucleic Acids Res* 31:4929–4940.

212. **Nigumann P, Redik K, Matlik K, Speek M.** 2002. Many human genes are transcribed from the antisense promoter of L1 retrotransposon. *Genomics* 79:628–634.

213. **Macia A, Munoz-Lopez M, Cortes JL, Hastings RK, Morell S, Lucena-Aguilar G, Marchal JA, Badge RM, Garcia-Perez JL.** 2011. Epigenetic control of retrotransposon expression in human embryonic stem cells. *Mol Cell Biol* 31:300–316.

214. **Uesaka M, Nishimura O, Go Y, Nakashima K, Agata K, Imamura T.** 2014. Bidirectional promoters are the major source of gene activation-associated non-coding RNAs in mammals. *BMC Genomics* 15:35.

215. **DeBerardinis RJ, Kazazian HH Jr.** 1999. Analysis of the promoter from an expanding mouse retrotransposon subfamily. *Genomics* 56:317–323.

216. **Severynse DM, Hutchison CA 3rd, Edgell MH.** 1992. Identification of transcriptional regulatory activity within the 5′ A-type monomer sequence of the mouse LINE-1 retroposon. *Mamm Genome* 2:41–50.

217. **Zemojtel T, Penzkofer T, Schultz J, Dandekar T, Badge R, Vingron M.** 2007. Exonization of active mouse L1s: a driver of transcriptome evolution? *BMC Genomics* 8:392.

218. **Li J, Kannan M, Trivett AL, Liao H, Wu X, Akagi K, Symer DE.** 2014. An antisense promoter in mouse L1 retrotransposon open reading frame-1 initiates expression of diverse fusion transcripts and limits retrotransposition. *Nucleic Acids Res* 42:4546–4562.

219. **Fort A, Hashimoto K, Yamada D, Salimullah M, Keya CA, Saxena A, Bonetti A, Voineagu I, Bertin N, Kratz A, Noro Y, Wong CH, de Hoon M, Andersson R, Sandelin A, Suzuki H, Wei CL, Koseki H, Hasegawa Y, Forrest AR, Carninci P.** 2014. Deep transcriptome profiling of mammalian stem cells supports a regulatory role for retrotransposons in pluripotency maintenance. *Nat Genet* 46:558–566.

220. **Kapusta A, Kronenberg Z, Lynch VJ, Zhuo X, Ramsay L, Bourque G, Yandell M, Feschotte C.** 2013. Transposable elements are major contributors to the origin, diversification, and regulation of vertebrate long noncoding RNAs. *PLoS Genet* 9:e1003470.

221. **Kapusta A, Feschotte C.** 2014. Volatile evolution of long noncoding RNA repertoires: mechanisms and biological implications. *Trends Genet* 30:439–452.

222. **Guttman M, Donaghey J, Carey BW, Garber M, Grenier JK, Munson G, Young G, Lucas AB, Ach R, Bruhn L, Yang X, Amit I, Meissner A, Regev A, Rinn JL, Root DE, Lander ES.** 2011. lincRNAs act in the circuitry controlling pluripotency and differentiation. *Nature* 477:295–300.

223. **Macia A, Blanco-Jimenez E, Garcia-Perez JL.** 2014. Retrotransposons in pluripotent cells: Impact and new roles in cellular plasticity. *Biochim Biophys Acta* [Epub ahead of print].

224. **Leibold DM, Swergold GD, Singer MF, Thayer RE, Dombroski BA, Fanning TG.** 1990. Translation of LINE-1 DNA elements in vitro and in human cells. *Proc Natl Acad Sci U S A* 87:6990–6994.

225. **Kolosha VO, Martin SL.** 2003. High-affinity, non-sequence-specific RNA binding by the open reading frame 1 (ORF1) protein from long interspersed nuclear element 1 (LINE-1). *J Biol Chem* 278:8112–8117.

226. **Kolosha VO, Martin SL.** 1997. In vitro properties of the first ORF protein from mouse LINE-1 support its role in ribonucleoprotein particle formation during retrotransposition. *Proc Natl Acad Sci U S A* 94:10155–10160.

227. **Martin SL.** 1991. Ribonucleoprotein particles with LINE-1 RNA in mouse embryonal carcinoma cells. *Mol Cell Biol* 11:4804–4807.

228. **Hohjoh H, Singer MF.** 1997. Ribonuclease and high salt sensitivity of the ribonucleoprotein complex formed by the human LINE-1 retrotransposon. *J Mol Biol* 271:7–12.

229. **Martin SL, Branciforte D, Keller D, Bain DL.** 2003. Trimeric structure for an essential protein in L1 retrotransposition. *Proc Natl Acad Sci U S A* 100:13815–13820.

230. **Basame S, Wai-lun Li P, Howard G, Branciforte D, Keller D, Martin SL.** 2006. Spatial assembly and RNA binding stoichiometry of a LINE-1 protein essential for retrotransposition. *J Mol Biol* 357:351–357.

231. **Khazina E, Weichenrieder O.** 2009. Non-LTR retrotransposons encode noncanonical RRM domains in their first open reading frame. *Proc Natl Acad Sci U S A* 106:731–736.

232. **Callahan KE, Hickman AB, Jones CE, Ghirlando R, Furano AV.** 2012. Polymerization and nucleic acid-binding properties of human L1 ORF1 protein. *Nucleic Acids Res* 40:813–827.

233. **Januszyk K, Li PW, Villareal V, Branciforte D, Wu H, Xie Y, Feigon J, Loo JA, Martin SL, Clubb RT.** 2007. Identification and solution structure of a highly conserved C-terminal domain within ORF1p required for retrotransposition of long interspersed nuclear element-1. *J Biol Chem* 282:24893–24904.

234. **Martin SL.** 2010. Nucleic acid chaperone properties of ORF1p from the non-LTR retrotransposon, LINE-1. *RNA Biol* 7:706–711.

235. **Martin SL, Bushman D, Wang F, Li PW, Walker A, Cummiskey J, Branciforte D, Williams MC.** 2008. A single amino acid substitution in ORF1 dramatically decreases L1 retrotransposition and provides insight into nucleic acid chaperone activity. *Nucleic Acids Res* 36:5845–5854.

236. **Evans JD, Peddigari S, Chaurasiya KR, Williams MC, Martin SL.** 2011. Paired mutations abolish and restore the balanced annealing and melting activities of ORF1p that are required for LINE-1 retrotransposition. *Nucleic Acids Res* **39:**5611–5621.

237. **Heras SR, Lopez MC, Garcia-Perez JL, Martin SL, Thomas MC.** 2005. The L1Tc C-terminal domain from Trypanosoma cruzi non-long terminal repeat retrotransposon codes for a protein that bears two C2H2 zinc finger motifs and is endowed with nucleic acid chaperone activity. *Mol Cell Biol* **25:**9209–9220.

238. **Dawson A, Hartswood E, Paterson T, Finnegan DJ.** 1997. A LINE-like transposable element in *Drosophila*, the I factor, encodes a protein with properties similar to those of retroviral nucleocapsids. *EMBO J* **16:**4448–44455.

239. **Nakamura M, Okada N, Kajikawa M.** 2012. Self-interaction, nucleic acid binding, and nucleic acid chaperone activities are unexpectedly retained in the unique ORF1p of zebrafish LINE. *Mol Cell Biol* **32:** 458–469.

240. **Kajikawa M, Sugano T, Sakurai R, Okada N.** 2012. Low dependency of retrotransposition on the ORF1 protein of the zebrafish LINE, ZfL2-1. *Gene* **499:**41–47.

241. **Goodier JL, Mandal PK, Zhang L, Kazazian HH Jr.** 2010. Discrete subcellular partitioning of human retrotransposon RNAs despite a common mechanism of genome insertion. *Hum Mol Genet* **19:**1712–1725.

242. **Martin F, Maranon C, Olivares M, Alonso C, Lopez MC.** 1995. Characterization of a non-long terminal repeat retrotransposon cDNA (L1Tc) from *Trypanosoma cruzi*: homology of the first ORF with the ape family of DNA repair enzymes. *J Mol Biol* **247:**49–59.

243. **Mol CD, Kuo CF, Thayer MM, Cunningham RP, Tainer JA.** 1995. Structure and function of the multifunctional DNA-repair enzyme exonuclease III. *Nature* **374:**381–386.

244. **Weichenrieder O, Repanas K, Perrakis A.** 2004. Crystal structure of the targeting endonuclease of the human LINE-1 retrotransposon. *Structure (Camb)* **12:**975–986.

245. **Cost GJ, Boeke JD.** 1998. Targeting of human retrotransposon integration is directed by the specificity of the L1 endonuclease for regions of unusual DNA structure. *Biochemistry* **37:**18081–18093.

246. **Cost GJ, Golding A, Schlissel MS, Boeke JD.** 2001. Target DNA chromatinization modulates nicking by L1 endonuclease. *Nucleic Acids Res* **29:**573–577.

247. **Xiong Y, Eickbush TH.** 1990. Origin and evolution of retroelements based upon their reverse transcriptase sequences. *EMBO J* **9:**3353–3362.

248. **Malik HS, Burke WD, Eickbush TH.** 1999. The age and evolution of non-LTR retrotransposable elements. *Mol Biol Evol* **16:**793–805.

249. **Evgen'ev MB, Arkhipova IR.** 2005. Penelope-like elements—a new class of retroelements: distribution, function and possible evolutionary significance. *Cytogenet Genome Res* **110:**510–521.

250. **Dombroski BA, Feng Q, Mathias SL, Sassaman DM, Scott AF, Kazazian HH Jr, Boeke JD.** 1994. An in vivo assay for the reverse transcriptase of human retrotransposon L1 in *Saccharomyces cerevisiae*. *Mol Cell Biol* **14:**4485–4492.

251. **Piskareva O, Denmukhametova S, Schmatchenko V.** 2003. Functional reverse transcriptase encoded by the human LINE-1 from baculovirus-infected insect cells. *Protein Expr Purif* **28:**125–130.

252. **Bibillo A, Eickbush TH.** 2002. High processivity of the reverse transcriptase from a non-long terminal repeat retrotransposon. *J Biol Chem* **277:**34836–34845.

253. **Piskareva O, Schmatchenko V.** 2006. DNA polymerization by the reverse transcriptase of the human L1 retrotransposon on its own template in vitro. *FEBS Lett* **580:**661–668.

254. **Monot C, Kuciak M, Viollet S, Mir AA, Gabus C, Darlix JL, Cristofari G.** 2013. The specificity and flexibility of l1 reverse transcription priming at imperfect T-tracts. *PLoS Genet* **9:**e1003499.

255. **Piskareva O, Ernst C, Higgins N, Schmatchenko V.** 2013. The carboxy-terminal segment of the human LINE-1 ORF2 protein is involved in RNA binding. *FEBS Open Bio* **3:**433–437.

256. **Alisch RS, Garcia-Perez JL, Muotri AR, Gage FH, Moran JV.** 2006. Unconventional translation of mammalian LINE-1 retrotransposons. *Genes Dev* **20:**210–224.

257. **Li PW, Li J, Timmerman SL, Krushel LA, Martin SL.** 2006. The dicistronic RNA from the mouse LINE-1 retrotransposon contains an internal ribosome entry site upstream of each ORF: implications for retrotransposition. *Nucleic Acids Res* **34:**853–864.

258. **An W, Dai L, Niewiadomska AM, Yetil A, O'Donnell KA, Han JS, Boeke JD.** 2011. Characterization of a synthetic human LINE-1 retrotransposon ORFeus-Hs. *Mob DNA* **2:**2.

259. **Han JS, Boeke JD.** 2004. A highly active synthetic mammalian retrotransposon. *Nature* **429:**314–318.

260. **Ingolia NT, Ghaemmaghami S, Newman JR, Weissman JS.** 2009. Genome-wide analysis in vivo of translation with nucleotide resolution using ribosome profiling. *Science* **324:**218–223.

261. **Slavoff SA, Mitchell AJ, Schwaid AG, Cabili MN, Ma J, Levin JZ, Karger AD, Budnik BA, Rinn JL, Saghatelian A.** 2013. Peptidomic discovery of short open reading frame-encoded peptides in human cells. *Nat Chem Biol* **9:**59–64.

262. **Usdin K, Furano AV.** 1989. The structure of the guanine-rich polypurine:polypyrimidine sequence at the right end of the rat L1 (LINE) element. *J Biol Chem* **264:**15681–15687.

263. **Moran JV, DeBerardinis RJ, Kazazian HH Jr.** 1999. Exon shuffling by L1 retrotransposition. *Science* **283:** 1530–1534.

264. **Faulkner GJ, Kimura Y, Daub CO, Wani S, Plessy C, Irvine KM, Schroder K, Cloonan N, Steptoe AL, Lassmann T, Waki K, Hornig N, Arakawa T, Takahashi H, Kawai J, Forrest AR, Suzuki H, Hayashizaki Y, Hume DA, Orlando V, Grimmond SM, Carninci P.** 2009. The regulated retrotransposon transcriptome of mammalian cells. *Nat Genet* **41:**563–571.

265. **Luan DD, Korman MH, Jakubczak JL, Eickbush TH.** 1993. Reverse transcription of R2Bm RNA is primed by a nick at the chromosomal target site: a mechanism for non-LTR retrotransposition. *Cell* **72:**595–605.

266. **Mandal PK, Ewing AD, Hancks DC, Kazazian HH Jr.** 2013. Enrichment of processed pseudogene transcripts in L1-ribonucleoprotein particles. *Hum Mol Genet* **22:** 3730–3748.

267. **Dutko JA, Kenny AE, Gamache ER, Curcio MJ.** 2010. 5′ to 3′ mRNA decay factors colocalize with Ty1 gag and human APOBEC3G and promote Ty1 retrotransposition. *J Virol* **84:**5052–5066.

268. **Larsen LS, Beliakova-Bethell N, Bilanchone V, Zhang M, Lamsa A, Dasilva R, Hatfield GW, Nagashima K, Sandmeyer S.** 2008. Ty3 nucleocapsid controls localization of particle assembly. *J Virol* **82:**2501–2514.

269. **Larsen LS, Zhang M, Beliakova-Bethell N, Bilanchone V, Lamsa A, Nagashima K, Najdi R, Kosaka K, Kovacevic V, Cheng J, Baldi P, Hatfield GW, Sandmeyer S.** 2007. Ty3 capsid mutations reveal early and late functions of the amino-terminal domain. *J Virol* **81:**6957–6972.

270. **Kubo S, Seleme MC, Soifer HS, Perez JL, Moran JV, Kazazian HH Jr, Kasahara N.** 2006. L1 retrotransposition in nondividing and primary human somatic cells. *Proc Natl Acad Sci U S A* **103:**8036–8041.

271. **Goodwin TJ, Ormandy JE, Poulter RT.** 2001. L1-like non-LTR retrotransposons in the yeast *Candida albicans*. *Curr Genet* **39:**83–91.

272. **Dong C, Poulter RT, Han JS.** 2009. LINE-like retrotransposition in *Saccharomyces cerevisiae*. *Genetics* **181:**301–311.

273. **Kinsey JA.** 1993. Transnuclear retrotransposition of the Tad element of Neurospora. *Proc Natl Acad Sci U S A* **90:**9384–9387.

274. **Xie Y, Mates L, Ivics Z, Izsvak Z, Martin SL, An W.** 2013. Cell division promotes efficient retrotransposition in a stable L1 reporter cell line. *Mob DNA* **4:**10.

275. **Shi X, Seluanov A, Gorbunova V.** 2007. Cell divisions are required for L1 retrotransposition. *Mol Cell Biol* **27:**1264–1270.

276. **Cost GJ, Feng Q, Jacquier A, Boeke JD.** 2002. Human L1 element target-primed reverse transcription in vitro. *EMBO J* **21:**5899–5910.

277. **Christensen SM, Eickbush TH.** 2005. R2 target-primed reverse transcription: ordered cleavage and polymerization steps by protein subunits asymmetrically bound to the target DNA. *Mol Cell Biol* **25:**6617–6628.

278. **Ostertag EM, Kazazian HH Jr.** 2001. Twin priming: a proposed mechanism for the creation of inversions in L1 retrotransposition. *Genome Res* **11:**2059–2065.

279. **Zingler N, Willhoeft U, Brose HP, Schoder V, Jahns T, Hanschmann KM, Morrish TA, Lower J, Schumann GG.** 2005. Analysis of 5′ junctions of human LINE-1 and Alu retrotransposons suggests an alternative model for 5′-end attachment requiring microhomology-mediated end-joining. *Genome Res* **15:**780–789.

280. **Gasior SL, Wakeman TP, Xu B, Deininger PL.** 2006. The human LINE-1 retrotransposon creates DNA double-strand breaks. *J Mol Biol* **357:**1383–1393.

281. **Gasior SL, Roy-Engel AM, Deininger PL.** 2008. ERCC1/XPF limits L1 retrotransposition. *DNA Repair (Amst)* **7:**983–989.

282. **Coufal NG, Garcia-Perez JL, Peng GE, Marchetto MC, Muotri AR, Mu Y, Carson CT, Macia A, Moran JV, Gage FH.** 2011. Ataxia telangiectasia mutated (ATM) modulates long interspersed element-1 (L1) retrotransposition in human neural stem cells. *Proc Natl Acad Sci U S A* **108:**20382–20387.

283. **Boboila C, Alt FW, Schwer B.** 2012. Classical and alternative end-joining pathways for repair of lymphocyte-specific and general DNA double-strand breaks. *Adv Immunol* **116:**1–49.

284. **Lavie L, Maldener E, Brouha B, Meese EU, Mayer J.** 2004. The human L1 promoter: variable transcription initiation sites and a major impact of upstream flanking sequence on promoter activity. *Genome Res* **14:**2253–2260.

285. **Babushok DV, Ostertag EM, Courtney CE, Choi JM, Kazazian HH Jr.** 2006. L1 integration in a transgenic mouse model. *Genome Res* **16:**240–250.

286. **Ichiyanagi K, Nakajima R, Kajikawa M, Okada N.** 2007. Novel retrotransposon analysis reveals multiple mobility pathways dictated by hosts. *Genome Res* **17:** 33–41.

287. **Chen J, Rattner A, Nathans J.** 2006. Effects of L1 retrotransposon insertion on transcript processing, localization and accumulation: lessons from the retinal degeneration 7 mouse and implications for the genomic ecology of L1 elements. *Hum Mol Genet* **15:**2146–2156.

288. **Goodier JL, Ostertag EM, Kazazian HH Jr.** 2000. Transduction of 3′-flanking sequences is common in L1 retrotransposition. *Hum Mol Genet* **9:**653–657.

289. **Pickeral OK, Makalowski W, Boguski MS, Boeke JD.** 2000. Frequent human genomic DNA transduction driven by LINE-1 retrotransposition. *Genome Res* **10:** 411–415.

290. **Holmes SE, Dombroski BA, Krebs CM, Boehm CD, Kazazian HH Jr.** 1994. A new retrotransposable human L1 element from the LRE2 locus on chromosome 1q produces a chimaeric insertion. *Nat Genet* **7:**143–148.

291. **Moran JV.** 1999. Human L1 retrotransposition: insights and peculiarities learned from a cultured cell retrotransposition assay. *Genetica* **107:**39–51.

292. **Moran JV, Gilbert N.** 2002. Mammalian LINE-1 retrotransposons and related elements, p 836–869. *In* Craig NL, Craigie R, Gellert M, Lambowitz AM (ed), *Mobile DNA II*. ASM Press, Washington, D.C.

293. **Eickbush T.** 1999. Exon shuffling in retrospect. *Science* **283:**1465–1467.

294. **Macfarlane CM, Collier P, Rahbari R, Beck CR, Wagstaff JF, Igoe S, Moran JV, Badge RM.** 2013. Transduction-Specific ATLAS Reveals a Cohort of Highly Active L1 Retrotransposons in Human Populations. *Hum Mutat* **34:**974–985.

295. **Solyom S, Ewing AD, Hancks DC, Takeshima Y, Awano H, Matsuo M, Kazazian HH Jr.** 2012. Pathogenic orphan transduction created by a nonreference LINE-1 retrotransposon. *Hum Mutat* **33:**369–371.

296. **Tubio JM, Li Y, Ju YS, Martincorena I, Cooke SL, Tojo M, Gundem G, Pipinikas CP, Zamora J, Raine K, Menzies A, Roman-Garcia P, Fullam A, Gerstung M, Shlien A, Tarpey PS, Papaemmanuil E, Knappskog S, Van Loo P, Ramakrishna M, Davies HR, Marshall J, Wedge DC, Teague JW, Butler AP, Nik-Zainal S, Alexandrov L, Behjati S, Yates LR, Bolli N, Mudie L, Hardy C, Martin S, McLaren S, O'Meara S, Anderson E, Maddison M, Gamble S, Foster C, Warren AY, Whitaker H, Brewer D, Eeles R, Cooper C, Neal D, Lynch AG, Visakorpi T, Isaacs WB, van't Veer L, Caldas C, Desmedt C, Sotiriou C, Aparicio S, Foekens JA, Eyfjord JE, Lakhani SR, Thomas G, Myklebost O, Span PN, Borresen-Dale AL, Richardson AL, Van de Vijver M, Vincent-Salomon A, Van den Eynden GG, Flanagan AM, Futreal PA, Janes SM, Bova GS, Stratton MR, McDermott U, Campbell PJ.** 2014. Mobile DNA in cancer. Extensive transduction of nonrepetitive DNA mediated by L1 retrotransposition in cancer genomes. *Science* **345:**1251343.

297. **Hancks DC, Ewing AD, Chen JE, Tokunaga K, Kazazian HH Jr.** 2009. Exon-trapping mediated by the human retrotransposon SVA. *Genome Res* **19:**1983–1991.

298. **Xing J, Wang H, Belancio VP, Cordaux R, Deininger PL, Batzer MA.** 2006. Emergence of primate genes by retrotransposon-mediated sequence transduction. *Proc Natl Acad Sci U S A* **103:**17608–17613.

299. **Damert A, Raiz J, Horn AV, Lower J, Wang H, Xing J, Batzer MA, Lower R, Schumann GG.** 2009. 5′-Transducing SVA retrotransposon groups spread efficiently throughout the human genome. *Genome Res* **19:** 1992–2008.

300. **Mine M, Chen JM, Brivet M, Desguerre I, Marchant D, de Lonlay P, Bernard A, Ferec C, Abitbol M, Ricquier D, Marsac C.** 2007. A large genomic deletion in the PDHX gene caused by the retrotranspositional insertion of a full-length LINE-1 element. *Hum Mutat* **28:** 137–142.

301. **Chou HH, Hayakawa T, Diaz S, Krings M, Indriati E, Leakey M, Paabo S, Satta Y, Takahata N, Varki A.** 2002. Inactivation of CMP-*N*-acetylneuraminic acid hydroxylase occurred prior to brain expansion during human evolution. *Proc Natl Acad Sci U S A* **99:**11736–11741.

302. **Takasu M, Hayashi R, Maruya E, Ota M, Imura K, Kougo K, Kobayashi C, Saji H, Ishikawa Y, Asai T, Tokunaga K.** 2007. Deletion of entire HLA-A gene accompanied by an insertion of a retrotransposon. *Tissue Antigens* **70:**144–150.

303. **Han K, Sen SK, Wang J, Callinan PA, Lee J, Cordaux R, Liang P, Batzer MA.** 2005. Genomic rearrangements by LINE-1 insertion-mediated deletion in the human and chimpanzee lineages. *Nucleic Acids Res* **33:**4040–4052.

304. **Callinan PA, Wang J, Herke SW, Garber RK, Liang P, Batzer MA.** 2005. Alu retrotransposition-mediated deletion. *J Mol Biol* **348:**791–800.

305. **Chen JM, Cooper DN, Ferec C, Kehrer-Sawatzki H, Patrinos GP.** 2010. Genomic rearrangements in inherited disease and cancer. *Semin Cancer Biol* **20:**222–233.

306. **Ade C, Roy-Engel AM, Deininger PL.** 2013. Alu elements: an intrinsic source of human genome instability. *Curr Opin Virol* **3:**639–645.

307. **Deininger PL, Moran JV, Batzer MA, Kazazian HH Jr.** 2003. Mobile elements and mammalian genome evolution. *Curr Opin Genet Dev* **13:**651–658.

308. **Eickbush TH.** 2002. Repair by retrotransposition. *Nat Genet* **31:**126–127.

309. **Morrish TA, Garcia-Perez JL, Stamato TD, Taccioli GE, Sekiguchi J, Moran JV.** 2007. Endonuclease-independent LINE-1 retrotransposition at mammalian telomeres. *Nature* **446:**208–212.

310. **Kopera HC, Moldovan JB, Morrish TA, Garcia-Perez JL, Moran JV.** 2011. Similarities between long interspersed element-1 (LINE-1) reverse transcriptase and telomerase. *Proc Natl Acad Sci U S A* **108:**20345–20350.

311. **Belfort M, Curcio MJ, Lue NF.** 2011. Telomerase and retrotransposons: reverse transcriptases that shaped genomes. *Proc Natl Acad Sci U S A* **108:**20304–20310.

312. **Eickbush TH.** 1997. Telomerase and retrotransposons: which came first? *Science* **277:**911–912.

313. **Levis RW, Ganesan R, Houtchens K, Tolar LA, Sheen FM.** 1993. Transposons in place of telomeric repeats at a *Drosophila* telomere. *Cell* **75:**1083–1093.

314. **Biessmann H, Valgeirsdottir K, Lofsky A, Chin C, Ginther B, Levis RW, Pardue ML.** 1992. HeT-A, a transposable element specifically involved in "healing" broken chromosome ends in *Drosophila melanogaster*. *Mol Cell Biol* **12:**3910–3918.

315. **Gladyshev EA, Arkhipova IR.** 2007. Telomere-associated endonuclease-deficient Penelope-like retroelements in diverse eukaryotes. *Proc Natl Acad Sci U S A* **104:** 9352–9357.

316. **Sen SK, Huang CT, Han K, Batzer MA.** 2007. Endonuclease-independent insertion provides an alternative pathway for L1 retrotransposition in the human genome. *Nucleic Acids Res* **35:**3741–3751.

317. **Morisada N, Rendtorff ND, Nozu K, Morishita T, Miyakawa T, Matsumoto T, Hisano S, Iijima K, Tranebjaerg L, Shirahata A, Matsuo M, Kusuhara K.** 2010. Branchio-oto-renal syndrome caused by partial EYA1 deletion due to LINE-1 insertion. *Pediatr Nephrol* **25:**1343–1348.

318. **Zhong J, Lambowitz AM.** 2003. Group II intron mobility using nascent strands at DNA replication forks to prime reverse transcription. *EMBO J* **22:**4555–4565.

319. **Yoder JA, Walsh CP, Bestor TH.** 1997. Cytosine methylation and the ecology of intragenomic parasites. *Trends Genet* **13:**335–340.

320. **Bestor TH, Bourc'his D.** 2004. Transposon silencing and imprint establishment in mammalian germ cells. *Cold Spring Harb Symp Quant Biol* **69:**381–387.

321. **Ooi SK, O'Donnell AH, Bestor TH.** 2009. Mammalian cytosine methylation at a glance. *J Cell Sci* **122:** 2787–2791.

322. **Bourc'his D, Bestor TH.** 2004. Meiotic catastrophe and retrotransposon reactivation in male germ cells lacking Dnmt3L. *Nature* **431:**96–99.

323. **Soper SF, van der Heijden GW, Hardiman TC, Goodheart M, Martin SL, de Boer P, Bortvin A.** 2008. Mouse maelstrom, a component of nuage, is essential for spermatogenesis and transposon repression in meiosis. *Dev Cell* **15:**285–297.

324. **Branco MR, Ficz G, Reik W.** 2012. Uncovering the role of 5-hydroxymethylcytosine in the epigenome. *Nat Rev Genet* **13:**7–13.

325. **Siomi MC, Sato K, Pezic D, Aravin AA.** 2011. PIWI-interacting small RNAs: the vanguard of genome defence. *Nat Rev Mol Cell Biol* **12:**246–258.

326. **Aravin AA, Hannon GJ, Brennecke J.** 2007. The Piwi-piRNA pathway provides an adaptive defense in the transposon arms race. *Science* **318:**761–764.

327. **Malone CD, Brennecke J, Dus M, Stark A, McCombie WR, Sachidanandam R, Hannon GJ.** 2009. Specialized piRNA pathways act in germline and somatic tissues of the *Drosophila* ovary. *Cell* **137:**522–535.

328. **Brennecke J, Malone CD, Aravin AA, Sachidanandam R, Stark A, Hannon GJ.** 2008. An epigenetic role for maternally inherited piRNAs in transposon silencing. *Science* **322:**1387–1392.

329. **Brennecke J, Aravin AA, Stark A, Dus M, Kellis M, Sachidanandam R, Hannon GJ.** 2007. Discrete small RNA-generating loci as master regulators of transposon activity in *Drosophila*. *Cell* **128:**1089–1103.

330. **Gunawardane LS, Saito K, Nishida KM, Miyoshi K, Kawamura Y, Nagami T, Siomi H, Siomi MC.** 2007. A slicer-mediated mechanism for repeat-associated siRNA 5′ end formation in *Drosophila*. *Science* **315:**1587–1590.

331. **Ghildiyal M, Zamore PD.** 2009. Small silencing RNAs: an expanding universe. *Nat Rev Genet* **10:**94–108.

332. **Malone CD, Hannon GJ.** 2009. Small RNAs as guardians of the genome. *Cell* **136:**656–668.

333. **Aravin AA, Sachidanandam R, Bourc'his D, Schaefer C, Pezic D, Toth KF, Bestor T, Hannon GJ.** 2008. A piRNA pathway primed by individual transposons is linked to de novo DNA methylation in mice. *Mol Cell* **31:**785–799.

334. **Aravin AA, Bourc'his D.** 2008. Small RNA guides for de novo DNA methylation in mammalian germ cells. *Genes Dev* **22:**970–975.

335. **Carmell MA, Girard A, van de Kant HJ, Bourc'his D, Bestor TH, de Rooij DG, Hannon GJ.** 2007. MIWI2 is essential for spermatogenesis and repression of transposons in the mouse male germline. *Dev Cell* **12:**503–514.

336. **Kuramochi-Miyagawa S, Watanabe T, Gotoh K, Totoki Y, Toyoda A, Ikawa M, Asada N, Kojima K, Yamaguchi Y, Ijiri TW, Hata K, Li E, Matsuda Y, Kimura T, Okabe M, Sakaki Y, Sasaki H, Nakano T.** 2008. DNA methylation of retrotransposon genes is regulated by Piwi family members MILI and MIWI2 in murine fetal testes. *Genes Dev* **22:**908–917.

337. **Reuter M, Chuma S, Tanaka T, Franz T, Stark A, Pillai RS.** 2009. Loss of the Mili-interacting Tudor domain-containing protein-1 activates transposons and alters the Mili-associated small RNA profile. *Nat Struct Mol Biol* **16:**639–646.

338. **Yang N, Kazazian HH Jr.** 2006. L1 retrotransposition is suppressed by endogenously encoded small interfering RNAs in human cultured cells. *Nat Struct Mol Biol* **13:**763–771.

339. **Heras SR, Macias S, Plass M, Fernandez N, Cano D, Eyras E, Garcia-Perez JL, Caceres JF.** 2013. The Microprocessor controls the activity of mammalian retrotransposons. *Nat Struct Mol Biol* **20:**1173–1181.

340. **Ciaudo C, Jay F, Okamoto I, Chen CJ, Sarazin A, Servant N, Barillot E, Heard E, Voinnet O.** 2013. RNAi-dependent and independent control of LINE1 accumulation and mobility in mouse embryonic stem cells. *PLoS Genet* **9:**e1003791.

341. **Faulkner GJ.** 2013. Retrotransposon silencing during embryogenesis: dicer cuts in LINE. *PLoS Genet* **9:**e1003944.

342. **Belancio VP, Hedges DJ, Deininger P.** 2006. LINE-1 RNA splicing and influences on mammalian gene expression. *Nucleic Acids Res* **34:**1512–1521.

343. **Belancio VP, Roy-Engel AM, Deininger P.** 2008. The impact of multiple splice sites in human L1 elements. *Gene* **411:**38–45.

344. **Perepelitsa-Belancio V, Deininger P.** 2003. RNA truncation by premature polyadenylation attenuates human mobile element activity. *Nat Genet* **35:**363–366.

345. **Han JS, Szak ST, Boeke JD.** 2004. Transcriptional disruption by the L1 retrotransposon and implications for mammalian transcriptomes. *Nature* **429:**268–274.

346. **Peddigari S, Li PW, Rabe JL, Martin SL.** 2013. hnRNPL and nucleolin bind LINE-1 RNA and function as host factors to modulate retrotransposition. *Nucleic Acids Res* **41:**575–585.

347. **Dai L, Taylor MS, O'Donnell KA, Boeke JD.** 2012. Poly(A) binding protein C1 is essential for efficient L1 retrotransposition and affects L1 RNP formation. *Mol Cell Biol* **32:**4323–4336.

348. **Zhang A, Dong B, Doucet AJ, Moldovan JB, Moran JV, Silverman RH.** 2014. RNase L restricts the mobility of engineered retrotransposons in cultured human cells. *Nucleic Acids Res* **42:**3803–3820.

349. **Chiu YL, Greene WC.** 2008. The APOBEC3 cytidine deaminases: an innate defensive network opposing exogenous retroviruses and endogenous retroelements. *Annu Rev Immunol* **26:**317–353.

350. **Sheehy AM, Gaddis NC, Choi JD, Malim MH.** 2002. Isolation of a human gene that inhibits HIV-1 infection and is suppressed by the viral Vif protein. *Nature* **418:**646–650.

351. **Bogerd HP, Wiegand HL, Hulme AE, Garcia-Perez JL, O'Shea KS, Moran JV, Cullen BR.** 2006. Cellular inhibitors of long interspersed element 1 and Alu retrotransposition. *Proc Natl Acad Sci U S A* **103:**8780–8785.

352. **Chen H, Lilley CE, Yu Q, Lee DV, Chou J, Narvaiza I, Landau NR, Weitzman MD.** 2006. APOBEC3A is a potent inhibitor of adeno-associated virus and retrotransposons. *Curr Biol* **16:**480–485.

353. **Hulme AE, Bogerd HP, Cullen BR, Moran JV.** 2007. Selective inhibition of Alu retrotransposition by APOBEC3G. *Gene* **390:**199–205.

354. **Muckenfuss H, Hamdorf M, Held U, Perkovic M, Lower J, Cichutek K, Flory E, Schumann GG, Munk C.** 2006. APOBEC3 proteins inhibit human LINE-1 retrotransposition. *J Biol Chem* **281:**22161–22172.

355. **Schumann GG.** 2007. APOBEC3 proteins: major players in intracellular defence against LINE-1-mediated retrotransposition. *Biochem Soc Trans* **35:**637–642.

356. **Wissing S, Montano M, Garcia-Perez JL, Moran JV, Greene WC.** 2011. Endogenous APOBEC3B restricts LINE-1 retrotransposition in transformed cells and human embryonic stem cells. *J Biol Chem* **286:**36427–36437.

357. **Horn AV, Klawitter S, Held U, Berger A, Vasudevan AA, Bock A, Hofmann H, Hanschmann KM, Trosemeier JH, Flory E, Jabulowsky RA, Han JS, Lower J, Lower R, Munk C, Schumann GG.** 2014. Human LINE-1 restriction by APOBEC3C is deaminase independent and mediated by an ORF1p interaction that affects LINE reverse transcriptase activity. *Nucleic Acids Res* **42:**396–416.

358. **Richardson SR, Narvaiza I, Planegger RA, Weitzman MD, Moran JV.** 2014. APOBEC3A deaminates transiently exposed single-strand DNA during LINE-1 retrotransposition. *Elife* **3:**e02008.

359. **Nik-Zainal S, Wedge DC, Alexandrov LB, Petljak M, Butler AP, Bolli N, Davies HR, Knappskog S, Martin S, Papaemmanuil E, Ramakrishna M, Shlien A, Simonic I, Xue Y, Tyler-Smith C, Campbell PJ, Stratton MR.** 2014. Association of a germline copy number polymorphism of APOBEC3A and APOBEC3B with burden of putative APOBEC-dependent mutations in breast cancer. *Nat Genet* **46:**487–491.

360. **Helleday T, Eshtad S, Nik-Zainal S.** 2014. Mechanisms underlying mutational signatures in human cancers. *Nat Rev Genet* **15:**585–598.

361. **Crow YJ.** 2013. Aicardi–Goutières syndrome. *Handb Clin Neurol* **113:**1629–1635.

362. **Crow YJ, Rehwinkel J.** 2009. Aicardi–Goutières syndrome and related phenotypes: linking nucleic acid metabolism with autoimmunity. *Hum Mol Genet* **18:** R130–R136.

363. **Crow YJ, Leitch A, Hayward BE, Garner A, Parmar R, Griffith E, Ali M, Semple C, Aicardi J, Babul-Hirji R, Baumann C, Baxter P, Bertini E, Chandler KE, Chitayat D, Cau D, Dery C, Fazzi E, Goizet C, King MD, Klepper J, Lacombe D, Lanzi G, Lyall H, Martinez-Frias ML, Mathieu M, McKeown C, Monier A, Oade Y, Quarrell OW, Rittey CD, Rogers RC, Sanchis A, Stephenson JB, Tacke U, Till M, Tolmie JL, Tomlin P, Voit T, Weschke B, Woods CG, Lebon P, Bonthron DT, Ponting CP, Jackson AP.** 2006. Mutations in genes encoding ribonuclease H2 subunits cause Aicardi–Goutières syndrome and mimic congenital viral brain infection. *Nat Genet* **38:**910–916.

364. **Crow YJ, Hayward BE, Parmar R, Robins P, Leitch A, Ali M, Black DN, van Bokhoven H, Brunner HG, Hamel BC, Corry PC, Cowan FM, Frints SG, Klepper J, Livingston JH, Lynch SA, Massey RF, Meritet JF, Michaud JL, Ponsot G, Voit T, Lebon P, Bonthron DT, Jackson AP, Barnes DE, Lindahl T.** 2006. Mutations in the gene encoding the 3′-5′ DNA exonuclease TREX1 cause Aicardi–Goutières syndrome at the AGS1 locus. *Nat Genet* **38:**917–920.

365. **Rice GI, Bond J, Asipu A, Brunette RL, Manfield IW, Carr IM, Fuller JC, Jackson RM, Lamb T, Briggs TA, Ali M, Gornall H, Couthard LR, Aeby A, Attard-Montalto SP, Bertini E, Bodemer C, Brockmann K, Brueton LA, Corry PC, Desguerre I, Fazzi E, Cazorla AG, Gener B, Hamel BC, Heiberg A, Hunter M, van der Knaap MS, Kumar R, Lagae L, Landrieu PG, Lourenco CM, Marom D, McDermott MF, van der Merwe W, Orcesi S, Prendiville JS, Rasmussen M, Shalev SA, Soler DM, Shinawi M, Spiegel R, Tan TY, Vanderver A, Wakeling EL, Wassmer E, Whittaker E, Lebon P, Stetson DB, Bonthron DT, Crow YJ.** 2009. Mutations involved in Aicardi–Goutières syndrome implicate SAMHD1 as regulator of the innate immune response. *Nat Genet* **41:**829–832.

366. **Stetson DB, Ko JS, Heidmann T, Medzhitov R.** 2008. Trex1 prevents cell-intrinsic initiation of autoimmunity. *Cell* **134:**587–598.

367. **Zhao K, Du J, Han X, Goodier JL, Li P, Zhou X, Wei W, Evans SL, Li L, Zhang W, Cheung LE, Wang G, Kazazian HH Jr, Yu XF.** 2013. Modulation of LINE-1 and Alu/SVA retrotransposition by Aicardi–Goutières syndrome-related SAMHD1. *Cell Rep* **4:**1108–1115.

368. **Rice GI, Kasher PR, Forte GM, Mannion NM, Greenwood SM, Szynkiewicz M, Dickerson JE, Bhaskar SS, Zampini M, Briggs TA, Jenkinson EM, Bacino CA, Battini R, Bertini E, Brogan PA, Brueton LA, Carpanelli M, De Laet C, de Lonlay P, del Toro M, Desguerre I, Fazzi E, Garcia-Cazorla A, Heiberg A, Kawaguchi M, Kumar R, Lin JP, Lourenco CM, Male AM, Marques W Jr, Mignot C, Olivieri I, Orcesi S, Prabhakar P, Rasmussen M, Robinson RA, Rozenberg F, Schmidt JL, Steindl K, Tan TY, van der Merwe WG, Vanderver A, Vassallo G, Wakeling EL, Wassmer E, Whittaker E, Livingston JH, Lebon P, Suzuki T, McLaughlin PJ, Keegan LP, O'Connell MA, Lovell SC, Crow YJ.** 2012. Mutations in ADAR1 cause Aicardi–Goutières syndrome associated with a type I interferon signature. *Nat Genet* **44:**1243–1248.

369. **Oda H, Nakagawa K, Abe J, Awaya T, Funabiki M, Hijikata A, Nishikomori R, Funatsuka M, Ohshima Y, Sugawara Y, Yasumi T, Kato H, Shirai T, Ohara O, Fujita T, Heike T.** 2014. Aicardi–Goutières syndrome is caused by IFIH1 mutations. *Am J Hum Genet* **95:**121–125.

370. **Rice GI, del Toro Duany Y, Jenkinson EM, Forte GM, Anderson BH, Ariaudo G, Bader-Meunier B, Baildam EM, Battini R, Beresford MW, Casarano M, Chouchane M, Cimaz R, Collins AE, Cordeiro NJ, Dale RC, Davidson JE, De Waele L, Desguerre I, Faivre L, Fazzi E, Isidor B, Lagae L, Latchman AR, Lebon P, Li C, Livingston JH, Lourenco CM, Mancardi MM, Masurel-Paulet A, McInnes IB, Menezes MP, Mignot C, O'Sullivan J, Orcesi S, Picco PP, Riva E, Robinson RA, Rodriguez D, Salvatici E, Scott C, Szybowska M, Tolmie JL, Vanderver A, Vanhulle C, Vieira JP, Webb K, Whitney RN, Williams SG, Wolfe LA, Zuberi SM, Hur S, Crow YJ.** 2014. Gain-of-function mutations in

IFIH1 cause a spectrum of human disease phenotypes associated with upregulated type I interferon signaling. *Nat Genet* **46:**503–509.

371. **Zheng YH, Jeang KT, Tokunaga K.** 2012. Host restriction factors in retroviral infection: promises in virus–host interaction. *Retrovirology* **9:**112.

372. **Arjan-Odedra S, Swanson CM, Sherer NM, Wolinsky SM, Malim MH.** 2012. Endogenous MOV10 inhibits the retrotransposition of endogenous retroelements but not the replication of exogenous retroviruses. *Retrovirology* **9:**53.

373. **Goodier JL, Cheung LE, Kazazian HH Jr.** 2012. MOV10 RNA helicase is a potent inhibitor of retrotransposition in cells. *PLoS Genet* **8:**e1002941.

374. **Li X, Zhang J, Jia R, Cheng V, Xu X, Qiao W, Guo F, Liang C, Cen S.** 2013. The MOV10 helicase inhibits LINE-1 mobility. *J Biol Chem* **288:**21148–21160.

375. **Gregersen LH, Schueler M, Munschauer M, Mastrobuoni G, Chen W, Kempa S, Dieterich C, Landthaler M.** 2014. MOV10 Is a 5′ to 3′ RNA helicase contributing to UPF1 mRNA target degradation by translocation along 3′ UTRs. *Mol Cell* **54:**573–585.

376. **Frost RJ, Hamra FK, Richardson JA, Qi X, Bassel-Duby R, Olson EN.** 2010. MOV10L1 is necessary for protection of spermatocytes against retrotransposons by Piwi-interacting RNAs. *Proc Natl Acad Sci U S A* **107:**11847–11852.

377. **Zheng K, Xiol J, Reuter M, Eckardt S, Leu NA, McLaughlin KJ, Stark A, Sachidanandam R, Pillai RS, Wang PJ.** 2010. Mouse MOV10L1 associates with Piwi proteins and is an essential component of the Piwi-interacting RNA (piRNA) pathway. *Proc Natl Acad Sci U S A* **107:**11841–11846.

378. **Cordaux R, Batzer MA.** 2009. The impact of retrotransposons on human genome evolution. *Nat Rev Genet* **10:**691–703.

379. **Branciforte D, Martin SL.** 1994. Developmental and cell type specificity of LINE-1 expression in mouse testis: implications for transposition. *Mol Cell Biol* **14:**2584–2592.

380. **Trelogan SA, Martin SL.** 1995. Tightly regulated, developmentally specific expression of the first open reading frame from LINE-1 during mouse embryogenesis. *Proc Natl Acad Sci U S A* **92:**1520–1524.

381. **Georgiou I, Noutsopoulos D, Dimitriadou E, Markopoulos G, Apergi A, Lazaros L, Vaxevanoglou T, Pantos K, Syrrou M, Tzavaras T.** 2009. Retrotransposon RNA expression and evidence for retrotransposition events in human oocytes. *Hum Mol Genet* **18:**1221–1228.

382. **Ostertag EM, DeBerardinis RJ, Goodier JL, Zhang Y, Yang N, Gerton GL, Kazazian HH Jr.** 2002. A mouse model of human L1 retrotransposition. *Nat Genet* **32:**655–660.

383. **Athanikar JN, Morrish TA, Moran JV.** 2002. Of man in mice. *Nat Genet* **32:**562–563.

384. **Kano H, Godoy I, Courtney C, Vetter MR, Gerton GL, Ostertag EM, Kazazian HH Jr.** 2009. L1 retrotransposition occurs mainly in embryogenesis and creates somatic mosaicism. *Genes Dev* **23:**1303–1312.

385. **Malki S, van der Heijden GW, O'Donnell KA, Martin SL, Bortvin A.** 2014. A role for retrotransposon LINE-1 in fetal oocyte attrition in mice. *Dev Cell* **29:**521–533.

386. **Freeman P, Macfarlane C, Collier P, Jeffreys AJ, Badge RM.** 2011. L1 hybridization enrichment: a method for directly accessing de novo L1 insertions in the human germline. *Hum Mutat* **32:**978–988.

387. **van den Hurk JA, Meij IC, Seleme MC, Kano H, Nikopoulos K, Hoefsloot LH, Sistermans EA, de Wijs IJ, Mukhopadhyay A, Plomp AS, de Jong PT, Kazazian HH, Cremers FP.** 2007. L1 retrotransposition can occur early in human embryonic development. *Hum Mol Genet* **16:**1587–1592.

388. **Garcia-Perez JL, Marchetto MC, Muotri AR, Coufal NG, Gage FH, O'Shea KS, Moran JV.** 2007. LINE-1 retrotransposition in human embryonic stem cells. *Hum Mol Genet* **16:**1569–1577.

389. **Garcia-Perez JL, Morell M, Scheys JO, Kulpa DA, Morell S, Carter CC, Hammer GD, Collins KL, O'Shea KS, Menendez P, Moran JV.** 2010. Epigenetic silencing of engineered L1 retrotransposition events in human embryonic carcinoma cells. *Nature* **466:**769–773.

390. **Wissing S, Munoz-Lopez M, Macia A, Yang Z, Montano M, Collins W, Garcia-Perez JL, Moran JV, Greene WC.** 2012. Reprogramming somatic cells into iPS cells activates LINE-1 retroelement mobility. *Hum Mol Genet* **21:**208–218.

391. **Muotri AR, Chu VT, Marchetto MC, Deng W, Moran JV, Gage FH.** 2005. Somatic mosaicism in neuronal precursor cells mediated by L1 retrotransposition. *Nature* **435:**903–910.

392. **Coufal NG, Garcia-Perez JL, Peng GE, Yeo GW, Mu Y, Lovci MT, Morell M, O'Shea KS, Moran JV, Gage FH.** 2009. L1 retrotransposition in human neural progenitor cells. *Nature* **460:**1127–1131.

393. **Muotri AR, Marchetto MC, Coufal NG, Oefner R, Yeo G, Nakashima K, Gage FH.** 2010. L1 retrotransposition in neurons is modulated by MeCP2. *Nature* **468:**443–446.

394. **Baillie JK, Barnett MW, Upton KR, Gerhardt DJ, Richmond TA, De Sapio F, Brennan PM, Rizzu P, Smith S, Fell M, Talbot RT, Gustincich S, Freeman TC, Mattick JS, Hume DA, Heutink P, Carninci P, Jeddeloh JA, Faulkner GJ.** 2011. Somatic retrotransposition alters the genetic landscape of the human brain. *Nature* **479:**534–537.

395. **Evrony GD, Cai X, Lee E, Hills LB, Elhosary PC, Lehmann HS, Parker JJ, Atabay KD, Gilmore EC, Poduri A, Park PJ, Walsh CA.** 2012. Single-neuron sequencing analysis of L1 retrotransposition and somatic mutation in the human brain. *Cell* **151:**483–496.

396. **Li W, Prazak L, Chatterjee N, Gruninger S, Krug L, Theodorou D, Dubnau J.** 2013. Activation of transposable elements during aging and neuronal decline in *Drosophila*. *Nat Neurosci* **16:**529–531.

397. **Perrat PN, DasGupta S, Wang J, Theurkauf W, Weng Z, Rosbash M, Waddell S.** 2013. Transposition-driven genomic heterogeneity in the Drosophila brain. *Science* **340:**91–95.

398. **Miki Y, Nishisho I, Horii A, Miyoshi Y, Utsunomiya J, Kinzler KW, Vogelstein B, Nakamura Y.** 1992. Disruption of the APC gene by a retrotransposal insertion of L1 sequence in a colon cancer. *Cancer Res* **52:**643–645.

399. **Solyom S, Ewing AD, Rahrmann EP, Doucet T, Nelson HH, Burns MB, Harris RS, Sigmon DF, Casella A, Erlanger B, Wheelan S, Upton KR, Shukla R, Faulkner GJ, Largaespada DA, Kazazian HH Jr.** 2012. Extensive somatic L1 retrotransposition in colorectal tumors. *Genome Res* **22:**2328–2338.

400. **Lee E, Iskow R, Yang L, Gokcumen O, Haseley P, Luquette LJ 3rd, Lohr JG, Harris CC, Ding L, Wilson RK, Wheeler DA, Gibbs RA, Kucherlapati R, Lee C, Kharchenko PV, Park PJ.** 2012. Landscape of somatic retrotransposition in human cancers. *Science* **337:**967–971.

401. **Helman E, Lawrence MS, Stewart C, Sougnez C, Getz G, Meyerson M.** 2014. Somatic retrotransposition in human cancer revealed by whole-genome and exome sequencing. *Genome Res* **24:**1053–1063.

402. **Shukla R, Upton KR, Munoz-Lopez M, Gerhardt DJ, Fisher ME, Nguyen T, Brennan PM, Baillie JK, Collino A, Ghisletti S, Sinha S, Iannelli F, Radaelli E, Dos Santos A, Rapoud D, Guettier C, Samuel D, Natoli G, Carninci P, Ciccarelli FD, Garcia-Perez JL, Faivre J, Faulkner GJ.** 2013. Endogenous retrotransposition activates oncogenic pathways in hepatocellular carcinoma. *Cell* **153:**101–111.

403. **Rodic N, Burns KH.** 2013. Long interspersed element-1 (LINE-1): passenger or driver in human neoplasms? *PLoS Genet* **9:**e1003402.

404. **Bulut-Karslioglu A, De La Rosa-Velazquez IA, Ramirez F, Barenboim M, Onishi-Seebacher M, Arand J, Galan C, Winter GE, Engist B, Gerle B, O'Sullivan RJ, Martens JH, Walter J, Manke T, Lachner M, Jenuwein T.** 2014. Suv39h-dependent H3K9me3 marks intact retrotransposons and silences LINE elements in mouse embryonic stem cells. *Mol Cell* **55:** 277–290.

405. **Chow JC, Ciaudo C, Fazzari MJ, Mise N, Servant N, Glass JL, Attreed M, Avner P, Wutz A, Barillot E, Greally JM, Voinnet O, Heard E.** 2010. LINE-1 activity in facultative heterochromatin formation during X chromosome inactivation. *Cell* **141:**956–969.

406. **Bailey JA, Carrel L, Chakravarti A, Eichler EE.** 2000. Molecular evidence for a relationship between LINE-1 elements and X chromosome inactivation: the Lyon repeat hypothesis. *Proc Natl Acad Sci U S A* **97:** 6634–6639.

407. **Lyon MF.** 1998. X-chromosome inactivation: a repeat hypothesis. *Cytogenet Cell Genet* **80:**133–137.

408. **Richardson SR.** 2013. *A mechanistic examination of APOBEC3-mediated LINE-1 inhibition, Doctoral Dissertation Thesis*, University of Michigan, Ann Arbor.

409. **Bertrand E, Chartrand P, Schaefer M, Shenoy SM, Singer RH, Long RM.** 1998. Localization of ASH1 mRNA particles in living yeast. *Mol Cell* **2:** 437–445.

410. **Wallace N, Wagstaff BJ, Deininger PL, Roy-Engel AM.** 2008. LINE-1 ORF1 protein enhances Alu SINE retrotransposition. *Gene* **419:**1–6.

Mobile DNA, 3rd Edition
Nancy L. Craig, Michael Chandler, Martin Gellert, Alan M. Lambowitz, Phoebe A. Rice and Suzanne Sandmeyer

doi:10.1128/microbiolspec.MDNA3-0050-2014

Alan M. Lambowitz[1]
Marlene Belfort[2]

52 Mobile Bacterial Group II Introns at the Crux of Eukaryotic Evolution

INTRODUCTION

Group II introns are remarkable mobile retroelements that use the combined activities of an autocatalytic RNA and an intron-encoded reverse transcriptase (RT) to propagate efficiently within genomes. But perhaps their most noteworthy feature is the pivotal role they are thought to have played in eukaryotic evolution. Mobile group II introns are ancestrally related to nuclear spliceosomal introns, retrotransposons and telomerase, which collectively comprise more than half of the human genome. Additionally, group II introns are postulated to have been a major driving force in the evolution of eukaryotes themselves, including for the emergence of the nuclear envelope to separate transcription from translation.

In this review, we focus on recent developments in our understanding of group II intron function, the relationships of these introns to retrotransposons and spliceosomes, and how their common features inform our thinking about bacterial group II introns at the crux of eukaryotic evolution. We rely on previous reviews for more detailed coverage of history, structure, mechanism and biotechnological applications of group II introns (1, 2, 3, 4, 5, 6).

BACKGROUND

Group II introns are found predominantly in bacteria and in the mitochondrial (mt) and chloroplast (cp) genomes of some eukaryotes, particularly fungi and plants, but are rare in archaea and absent from eukaryotic nuclear genomes (4). Mobile group II introns consist of a catalytically active intron RNA (a ribozyme) and an intron-encoded protein (IEP), which is a multifunctional RT. The IEP functions in intron mobility by synthesizing a cDNA copy of the intron RNA and as a "maturase" that promotes folding of the intron RNA into a catalytically active ribozyme structure required for both RNA splicing and mobility reactions. Some IEPs also have a DNA endonuclease (En) activity that plays a role in intron mobility.

Group II Intron Splicing

The splicing pathway, which is assisted by the IEP, involves two reversible transesterifications catalyzed by the intron RNA (7). In the first transesterification, the 2′-OH of a "branch-point" adenosine near the 3′ end of the intron attacks the 5′-splice site (Fig. 1A).

[1]Institute for Cellular Molecular Biology and Department of Molecular Biosciences, University of Texas at Austin, Austin, TX 78712;
[2]Department of Biological Sciences and RNA Institute, University at Albany, State University of New York, Albany, NY 12222.

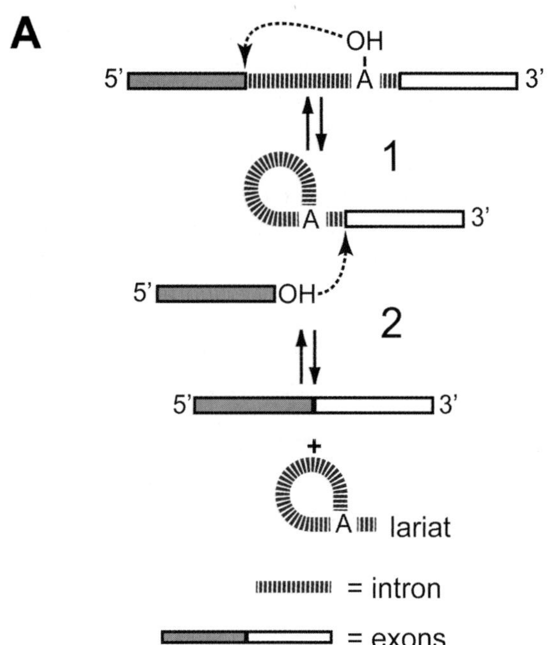
A
OH
5'
A
3'
1
A
3'
5'
OH
2
5'
3'
+
A
lariat
= intron
= exons

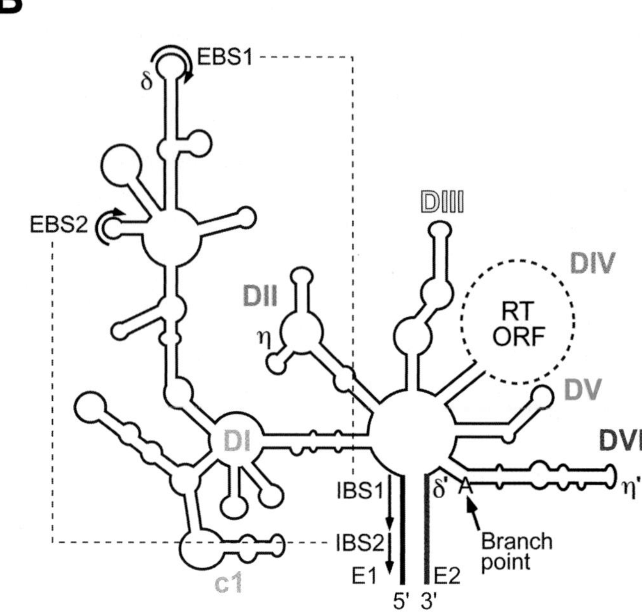
B
EBS1
δ
EBS2
DIII
DIV
DII
RT
ORF
η
DV
DI
DVI
IBS1
δ'
A
η'
IBS2
Branch
point
E1
E2
5'
3'
c1

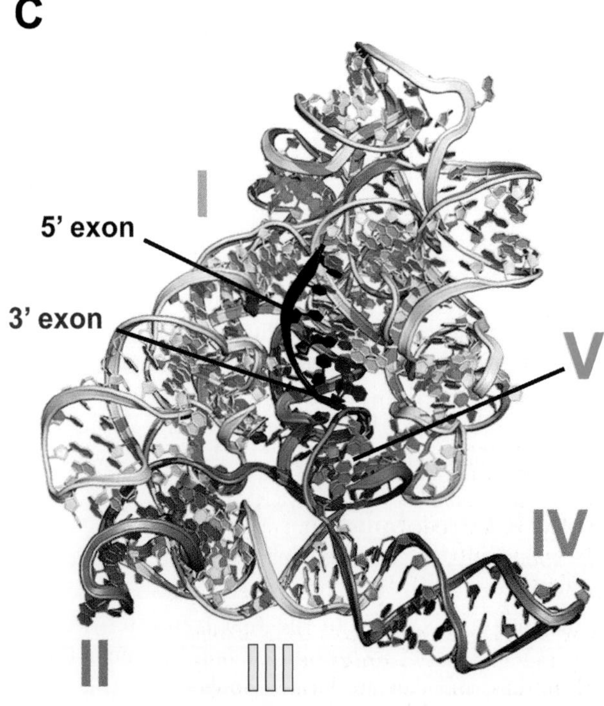
C
I
5' exon
3' exon
V
IV
II
III

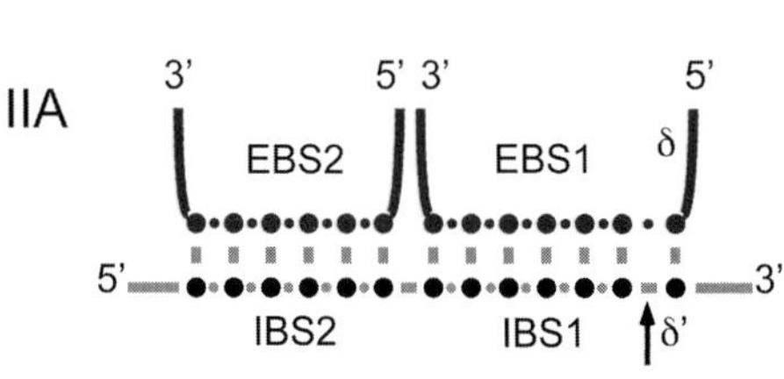
D
IIA
3'
5'
3'
5'
EBS2
EBS1
δ
5'
3'
IBS2
IBS1
δ'

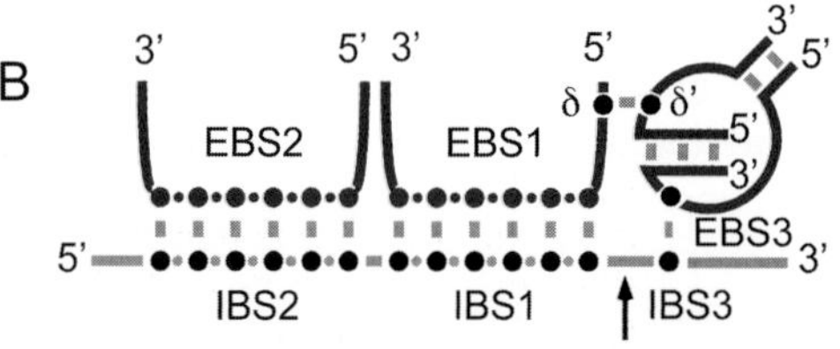
IIB
3'
5'
3'
5'
3'
5'
δ
δ'
EBS2
EBS1
5'
3'
EBS3
5'
3'
IBS2
IBS1
IBS3

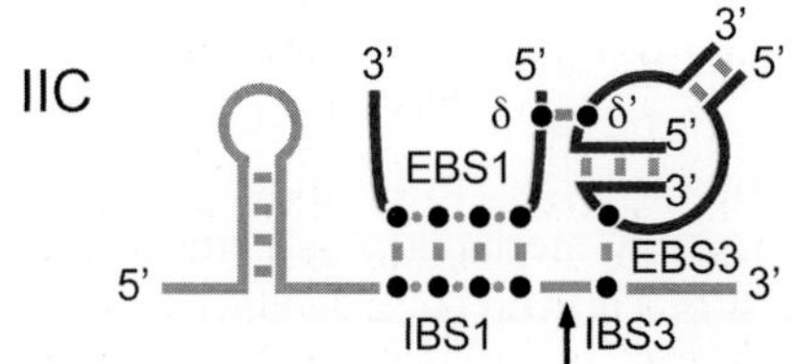
IIC
3'
5'
3'
5'
δ
δ'
EBS1
5'
3'
EBS3
5'
3'
IBS1
IBS3

This reaction releases the 5′ exon and produces a branched intermediate in which the attacking adenosine is linked to the 5′ intron residue by a 2′-5′ phosphodiester bond. In the second transesterification, the newly released 3′-OH of the 5′-exon attacks the 3′ splice site, resulting in ligation of the 5′ and 3′ exons and excision of the intron lariat. A linear intron can result from hydrolysis rather than transesterification at the 5′-splice site, or by a lariat reopening reaction (8, 9). Circular introns can also form (10). The reversibility of the transesterifications (Fig. 1A) enables "reverse splicing" of the excised intron into RNA or DNA containing the ligated-exon sequence, and may also provide a proof-reading mechanism for 5′-splice site selection (11). Reverse splicing into DNA plays a key role in intron mobility.

Intron Architecture

Group II intron RNAs have conserved 5′- and 3′-end sequences (GUGYG and AY, respectively), which resemble those of spliceosomal introns (GU and AG, respectively), and fold into a conserved three-dimensional structure consisting of six interacting secondary structure domains (DI to DVI) (Fig. 1B, C). This folded RNA forms the ribozyme active site, which uses specifically bound Mg^{2+} ions to catalyze RNA splicing and reverse splicing reactions.

Biochemical studies and X-ray crystal structures of a group II intron from the halophile *Oceanobacillus iheyensis* have provided insight into the function of group II introns domains (Fig. 1C) (5, 10, 12, 13). DI, the largest domain, is a scaffold, which contains sequence motifs that base pair with exon sequences to align them at the active site. These exon-recognition motifs differ for group II intron subgroups (IIA, IIB and IIC; see below) and are denoted exon-binding sites (EBSs) and δ with the complementary exon motifs denoted intron-binding sites (IBSs) and δ' (Fig. 1D). DV is the active site helix that binds metal ions at the catalytic center of the intron. DVI contains the branch-point A residue and undergoes a conformational change to reposition the branch-point A between the two steps of splicing (14, 15, 16). DII and DIII engage in stabilizing interactions with other domains, with DIII functioning as an effector to increase the rate of catalysis. The ORF encoding the RT protrudes from DIV, which projects away from the catalytic core.

Intron-Encoded Reverse Transcriptase

Fig. 2 compares two well-studied group II intron RTs (the LtrA protein encoded by the *Lactococcus lactis* Ll.LtrB intron and the *Sinorhizobium meliloti* RmInt1 RT) with two non-LTR-retrotransposon RTs (R2Bm and LINE-1), telomerase, and a retroviral (HIV-1) RT. The Ll.LtrB RT contains four conserved domains, RT, X/thumb, DNA binding (D), and DNA endonuclease (En), whereas the RmInt1 RT belongs to a subset of group II intron RTs that lacks the En domain. The RT domain contains conserved amino acid sequence blocks 1 to 7 that are present in the fingers and palm regions of retroviral RTs, while the X/thumb domain has a predicted structure similar to the thumb domain of retroviral RTs (17). Although homologous to retroviral RTs, group II intron RTs have an N-terminal extension containing an additional conserved sequence block (RT-0), as well as insertions between other conserved

Figure 1 Group II intron RNA splicing mechanism and structure. (A) Splicing and reverse splicing. Step 1. The 2′-OH of the branch-point adenosine acts as nucleophile to attack the 5′ splice site. Step 2. The 3′-OH of the upstream exon is the nucleophile that attacks the 3′ splice site to generate ligated exons and an excised intron lariat. Both reactions are reversible. (B) Group II intron secondary structure. The *L. lactis* Ll.LtrB group IIA intron is shown, with the six domains DI-DVI. Exons are represented by thicker lines with the EBS-IBS pairings shown by dashed black lines. The IEP ORF is looped out of DIV (not drawn to scale). (C) Group II intron crystal structure. The representation is of DI-DV of the *O. iheyensis* group IIC intron (PDB:4E8K) bound to ligated exons, before the spliced exon reopening reaction, provided by Marcia and Pyle (13). Colors are coded to the domain labels in panel B, although these are different introns that belong to different structural subgroups. The 5′ exon is black and the 3′ exon is dark blue. (D) Base-pairing interactions of group IIA, IIB, and IIC introns with flanking exons. Group IIA and IIB recognize 5′ exons by similar IBS1-EBS1 and IBS2-EBS2 interactions, but use different interactions to recognize 3′ exons (δ-δ' in IIA introns and EBS3/IBS3 in IIB introns (214). Group IIC ribozymes are only ~400 nt long, considerably smaller than IIA and IIB introns, and they are located downstream of inverted repeats, such as transcription terminators, which contribute to exon recognition along with short EBS1/IBS1 and EBS3/IBS3 interactions similar to those of IIB introns (215, 216). Panel D is adapted from reference 4, with permission of the publisher (© Cold Spring Harbor Laboratory Press). doi:10.1128/microbiolspec.MDNA3-0050-2014.f1

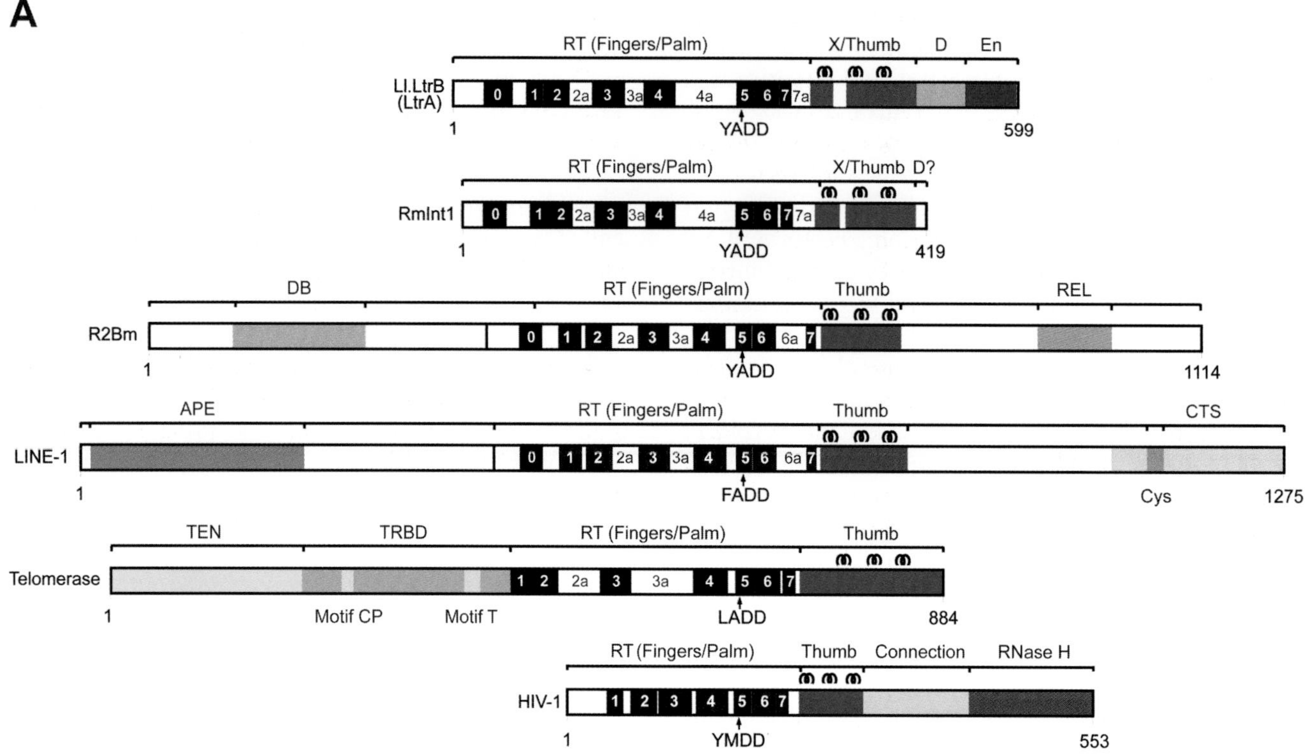
A
RT (Fingers/Palm)
X/Thumb
D
En
Ll.LtrB (LtrA)
0 1 2 2a 3 3a 4 4a 5 6 7 7a
1
YADD
599
RT (Fingers/Palm)
X/Thumb
D?
RmInt1
0 1 2 2a 3 3a 4 4a 5 6 7 7a
1
YADD
419
DB
RT (Fingers/Palm)
Thumb
REL
R2Bm
0 1 2 2a 3 3a 4 5 6 6a 7
1
YADD
1114
APE
RT (Fingers/Palm)
Thumb
CTS
LINE-1
0 1 2 2a 3 3a 4 5 6 6a 7
1
FADD
Cys
1275
TEN
TRBD
RT (Fingers/Palm)
Thumb
Telomerase
1 2 2a 3 3a 4 5 6 7
1
Motif CP
Motif T
LADD
884
RT (Fingers/Palm)
Thumb
Connection
RNase H
HIV-1
1 2 3 4 5 6 7
1
YMDD
553

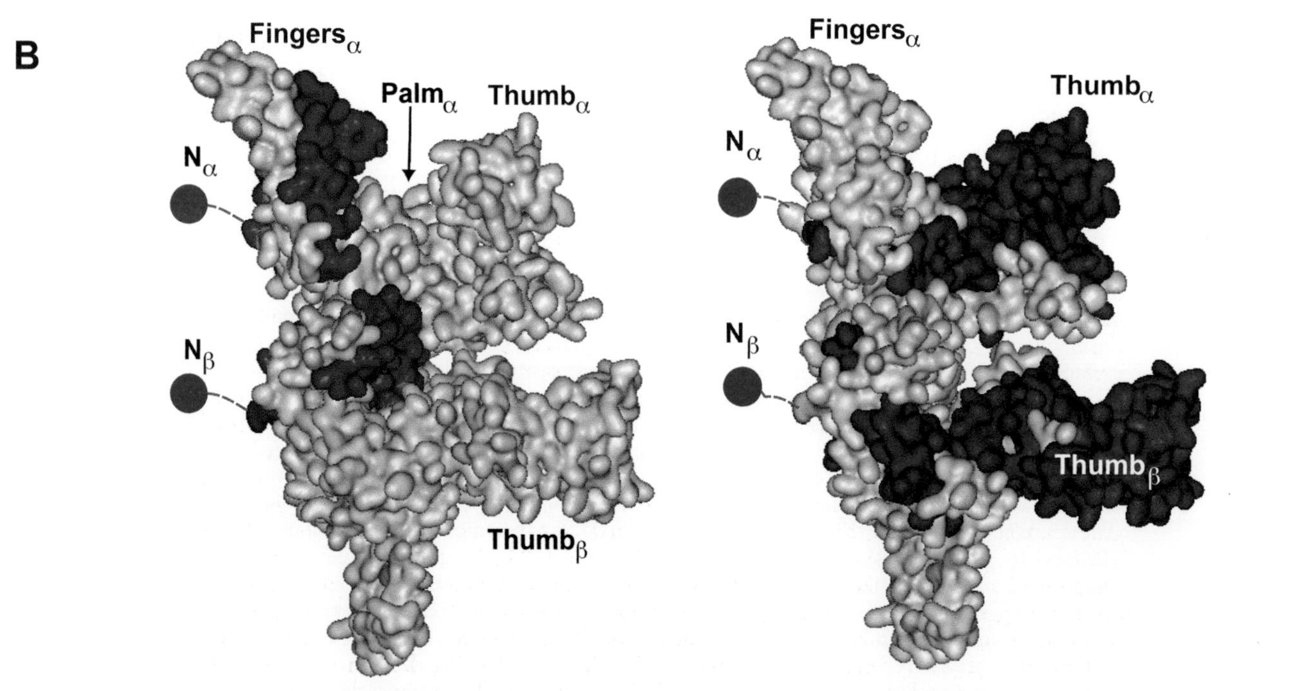
B
Fingers$_\alpha$
Palm$_\alpha$
Thumb$_\alpha$
N$_\alpha$
N$_\beta$
Thumb$_\beta$
Fingers$_\alpha$
Thumb$_\alpha$
N$_\alpha$
N$_\beta$
Thumb$_\beta$

sequence blocks (RT-2a, -3a, -4a, and -7a), some of which are conserved in non-LTR-retrotransposons and telomerase RTs (17, 18). A three-dimensional model of a group II intron RT predicts a right hand-like structure similar to that of retroviral RTs with the N-terminal extension comprising part of a larger fingers region and the other insertions lying outside the RT active site (Fig. 2B) (17).

In addition to reverse transcription, the RT and X/thumb domains of group II intron RTs bind the intron RNA for splicing. Mutational and high-throughput unigenic evolution analyses suggest an extended RNA-binding surface that includes distal parts of the fingers, regions in and around the template-primer binding tract, and patches on the back of the hand (Fig. 2B, red left and dark blue right highlight regions that potentially interact with different parts of the intron RNA; see legend and below) (17, 19). The RT active site is not required for splicing, with mutations in the conserved YADD metal binding motif having little effect on splicing activity (19, 20). The overlap between regions of the IEP required for splicing with those that bind RNA templates for reverse transcription suggests how an RT that became associated with a catalytic RNA could evolve a secondary function in RNA splicing (19, 21).

The D domain of group II IEPs functions in DNA target site binding during intron mobility. Studies with the Ll.LtrB IEP identified two regions of this domain that are functionally important and conserved in other group II IEPs, an upstream cluster of basic amino acids and a downstream predicted α-helix (22). Mutations in these regions affect both the efficiency and target specificity of retrohoming, consistent with their involvement in DNA target site recognition (22, 23). Distinctive variations of the D domain have been described for RmInt1 and related group II intron lineages whose IEPs lack an En domain (22, 24, 25).

The En domain belongs to the H-N-H family of DNA endonucleases, which includes bacterial colicins, phage T4 endonuclease VII, and some group I intron homing endonucleases (22, 26, 27, 28, 29). In group II intron RTs, residues of the H-N-H motif coordinate a catalytic Mg^{2+} ion and are interspersed with two pairs of conserved cysteine residues, which may coordinate another metal ion to stabilize the protein fold (22). These cysteine residues are important for the endonuclease activity of the Ll.LtrB RT, but have diverged in some group II intron IEP lineages (22).

Group II Intron RNPs

Group II intron RNAs and IEPs function together in a ribonucleoprotein (RNP) complex, which forms when the IEP binds to the intron in unspliced precursor RNA to promote RNA-catalyzed splicing (30, 31). Group II intron IEPs typically function as intron-specific splicing factors, discriminating even among closely related group II introns (32, 33, 34). Ll.LtrB RNPs contain two molecules of IEP per intron RNA, suggesting that the IEP functions as a dimer, similar to HIV-1 RT (30, 35, 36).

The IEP has a high-affinity binding site in intron subdomain DIVa, a variable stem loop structure that

Figure 2 Group II intron and related reverse transcriptase (RTs). (A) Schematics of RTs. Two group II intron RTs, *L. lactis* Ll.LtrB (denoted LtrA protein; GenBank: AAB06503) and *Sinorhizobium meliloti* RmInt (NCBI Reference Sequence: NP_438012) are compared with two non-LTR-retrotransposon RTs, *Bombyx mori* R2Bm (GenBank: AAB59214) and human LINE-1 (UniProtKB/Swiss-Prot: O00370) RTs; yeast telomerase RT (GenBank: AAB64520); and retrovirus HIV-1 RT (PDB:2HMI). Conserved sequence blocks in the RT domain are numbered, and the sequence motif containing two of the conserved aspartic acid residues at the RT active site is shown below for each protein. (B) Three-dimensional model of the Ll.LtrB RT. Regions identified by unigenic evolution analysis as being required for binding the high-affinity binding site DIVa and catalytic core regions of group II intron RNAs are highlighted in red and dark blue in the left and right panels, respectively. Pink in the left panel indicates a region of the protein that may contribute to DIVa binding by stabilizing the structure of neighboring regions (46). The model was constructed by threading the amino acid sequence of the Ll.LtrB RT onto a HIV-1 RT crystal structure, with one subunit (denoted α; gray) modeled based on the catalytic p66 subunit of HIV-1 RT and the other subunit (denoted β; cyan) modeled based on the p51 subunit of HIV-1 RT (17). The N-terminal 36 amino acid residues of the Ll.LtrB RT could not be modeled based on the HIV-1 RT and are represented as spheres. APE, apurinic endonuclease domain; CTS, conserved carboxy-terminal segment found to bind RNA nonspecifically in human LINE-1 RT (217); Cys, cysteine-rich sequence conserved in LINE-1 RT; DB, DNA-binding domain in R2Bm RT; REL, restriction endonuclease-like domain; TEN, telomerase N-terminal domain; TRBD, telomerase RNA-binding domain including motifs CP and T, which contact telomerase RNA (139, 218). doi:10.1128/microbiolspec.MDNA3-0050-2014.f2

lies outside the intron's catalytic core near the beginning of DIV and is a feature that contributes to intron specificity (37, 38, 39). The IEP also makes weaker secondary contacts with conserved core regions, including in DI, DII, and DVI, that stabilize the active ribozyme structure for RNA splicing and reverse splicing (31, 40). This mode of interaction in which the IEP is anchored by binding to DIVa and binds more weakly to the catalytic core may enable the IEP to accommodate conformational changes within the intron RNA during RNA splicing and different steps in intron mobility (31, 36, 41, 42, 43).

When mapped onto a three-dimensional model of the Ll.LtrB intron RNA, putative IEP contact sites identified by RNA footprinting, site-specific cross-linking and fluorescence quenching revealed a broad IEP binding surface that extends from DIVa across the interface of DI, DII, and DVI (31, 40). The location of the putative IEP contact sites and analysis of conformational changes that occur upon IEP binding suggest that the IEP stabilizes interactions between DI, II, and VI, as well as the folded structure of DI (31, 41). Although much group II intron RNA tertiary structure can form at physiological Mg^{2+} concentrations in the absence of the IEP (31, 41), continued binding of the Ll.LtrB IEP is required to stabilize the active ribozyme structure even for lariat RNA, which rapidly reverts to an inactive structure if the IEP is removed (44, 45).

Group II Intron Classification

Group II intron IEPs have a strong *cis*-preference for splicing and mobilizing the intron RNA in which they are encoded, likely reflecting that the nascent IEP binds to the intron during or just after translation and then remains bound to the excised intron during RNA splicing and intron mobility (19, 46). As a result of this *cis*-preference, group II intron RNAs and IEPs have coevolved, leading to divergence of mobile group II introns into distinct evolutionary lineages (47, 48). Group II intron RNAs are classified into three major subgroups (IIA, IIB, and IIC) distinguished by their size, secondary structure and mode of exon recognition by DI (Fig. 1D) (discussed in detail in reference 4). Whereas EBS1-IBS1 interactions are common to all group II introns, EBS2-IBS2 base pairings between the intron and 5′ exon are confined to IIA and IIB introns. The 3′ exon is recognized by the δ-δ' interaction in IIA introns and by the EBS3-IBS3 interaction involving another region of the intron RNA in IIB and IIC introns (Fig. 1D). The IEPs have diverged into at least nine subclasses [A, B, C, D, E, F, chloroplast-like 1 and 2 (CL1 and CL2, respectively) and mitochondria-like (ML)], each associated with a particular RNA structure (25, 49, 50, 51, 52). The En domain, which enables an efficient mode of intron mobility (see below), is present only in IEPs of subclasses B, ML, and CL, suggesting that it may have been acquired by a common ancestor of these subclasses (25).

GROUP II INTRON RETROHOMING MECHANISMS

Group II intron retromobility occurs by a target DNA-primed reverse transcription (TPRT) mechanism in which the excised intron RNA reverse splices into one strand of a DNA target site and is then reverse transcribed by the IEP to produce an intron cDNA that is integrated into the genome (53, 54, 55, 56, 57). This mechanism is used by group II introns both for retrohoming to specific DNA target sites and for retrotransposition (also referred to as ectopic retrohoming) to sites that resemble the normal homing site (Fig. 3A, B, C, D). In contrast to retrohoming, which occurs at frequencies that can approach 100% of recipients acquiring the intron, retrotransposition occurs at lower frequency, typically $<10^{-4}$/recipient (58, 59). The ability of the intron to insert into different genes and then remove itself by RNA splicing minimizes damage to the host. Variations of the retrohoming mechanism discussed below illustrate how group II introns can evolve different modes of action and adapt to different hosts.

IEP Expression

Building upon earlier genetic studies (reviewed in reference 1), the major features of group II intron retromobility mechanisms were elucidated for the *Saccharomyces cerevisiae* mtDNA introns aI1 and aI2, in the *COX1* gene encoding cytochrome oxidase subunit I (53, 54, 55, 56), and in bacteria, for the Ll.LtrB intron, in a relaxase gene in a conjugative element (57), and the *Sinorhizobium meliloti* RmInt1 intron, in an insertion sequence (IS) element (60, 61, 62). The aI1, aI2 and Ll.LtrB introns are group IIA, whose IEPs contain an En domain, whereas RmInt1 is group IIB and does not. For all these introns, the IEP is translated from the intron-containing precursor RNA, binds to the unspliced intron, and promotes formation of the active ribozyme structure for RNA splicing. Cryo-EM and small-angle X-ray scattering show that precursor RNP forms a loosely packed structure that undergoes a dramatic conformational change upon splicing, resulting in a compact excised intron RNP particle (36, 43).

The nascent IEP is thought to bind first to its primary high-affinity binding site in DIVa, using regions near its N-terminus, including the N-terminal extension upstream of RT-0 (46) (Fig. 2B left, regions highlighted in red). In the Ll.LtrB intron, DIVa contains the Shine-Dalgarno sequence and initiation codon of the intron ORF, as it does in many other bacterial group II introns, and the binding of the IEP to DIVa down-regulates its own translation (38, 63). This binding to DIVa halts ribosome entry into the intron, enabling it to fold into the active ribozyme structure for RNA splicing, and also prevents accumulation of excess free IEP, which would likely be deleterious to the host cell. In the yeast aI1 and aI2 introns and some other fungal mtDNA introns, the intron ORF is continuous with that of the upstream exon and translated as part of a preprotein that is proteolytically processed to yield the active IEP (38, 64). DIVa nevertheless remains a high-affinity binding site for the IEP, reflecting a critical role of this interaction in nucleating RNP assembly and positioning the IEP for initiation of TPRT (65). After splicing, the IEP remains tightly bound to the intron lariat in a stable RNP that promotes retrohoming (30, 42).

Endonuclease-Dependent Retrohoming

Group II intron RNPs initiate retrohoming by recognizing a DNA target sequence using both the IEP and motifs within the intron RNA that base pair with the DNA target site. The intron RNA then reverse splices into the DNA strand to which it is base paired (Fig. 3A, B, step 1). In the En-dependent retrohoming pathway, the IEP uses its En domain to nick the opposite strand (Fig. 3A, step 1′) and then uses the 3′ DNA end generated at the nick as a primer for TPRT of the inserted intron RNA (Fig. 3A, step 2). The resulting intron cDNA is integrated into the genome by host enzymes, and after intron degradation and second-strand cDNA synthesis (Fig. 3A, step 3), the nicks are ligated to complete the reaction (Fig. 3A, step 4). Unlike retroviruses, whose RTs have low fidelity and processivity that introduce and propagate mutations for evasion of host defenses (66), group II intron retrohoming requires an RT with high processivity and fidelity (error rate of $\sim 10^{-5}$/nt for the Ll.LtrB RT) in order to preserve the functional integrity of the intron (57, 67, 68).

Group II introns can use different mechanisms for cDNA integration, depending upon both the intron and DNA repair pathways that are available in different hosts. For the *S. cerevisiae* mtDNA introns, most retrohoming events are completed via a recombination mechanism in which a nascent intron cDNA strand invades an intron-containing allele for the completion of intron DNA synthesis and a second cross-over occurs in the homologous upstream exon (56, 69). This process results in co-conversion of the intron with sequence polymorphisms in the upstream exon (20, 70). A smaller proportion of retrohoming events occurs without co-conversion of exon sequences, possibly via synthesis of a full-length intron cDNA that is integrated by DNA repair (69).

By contrast to the *S. cerevisiae* introns, which rely heavily on homologous recombination for cDNA integration, the Ll.LtrB intron and other bacterial group II introns use RecA-independent DNA repair (57, 60, 71). The preference of bacterial introns for this mechanism may reflect less proficient homologous recombination than in fungal mitochondria.

Genetic screens and biochemical assays in which purified group II intron RNPs were combined with *Escherichia coli* extracts to reconstitute the complete retrohoming reaction *in vitro* suggested a model for host contributions to Ll.LtrB retrohoming shown in Fig. 3A (72, 73). According to this model, enzymes including nucleases directed against RNA and DNA, replicative and repair DNA polymerases and components of the replisome act in their various roles to resect DNA, degrade the template RNA, traverse RNA-DNA junctions, carry out second-strand DNA synthesis, and ligate DNA (Fig. 3A and Table 1). A key feature is that after TPRT, the intron cDNA is extended into the upstream exon, either by the group II intron RT or by a switch to a host DNA repair polymerase, resulting in a branched intermediate that resembles a stalled DNA replication fork (Fig. 3A, step 2). The latter is recognized by the replication-restart proteins, PriA or PriC, which act on branched structures having different length gaps between the branch and the nascent DNA strand. PriA and PriC are thought to function in retrohoming much as they would during replication restart at a stalled or collapsed replication fork by initiating a replisome loading cascade leading to recruitment of DNA polymerase PolI and the host replicative polymerase, PolIII (Fig. 3A, step 3).

Because group II introns encode RTs lacking an RNase H domain (17), they must rely on a host RNase H to degrade the intron RNA template. For the Ll.LtrB intron in *E. coli*, this is achieved by RNase H1, with a likely contribution from the 5′ to 3′ exonuclease activity of PolI, which removes RNA primers during DNA replication (Fig. 3A, step 3) (72, 73). DNA overhangs are trimmed by host 5′-3′ exonucleases, such as RecJ, and after completion of second-strand synthesis by PolIII, nicks are sealed by host DNA ligase (Fig. 3A, step 4). The requirement for host DNA polymerases for second-strand synthesis is again dictated by the properties of

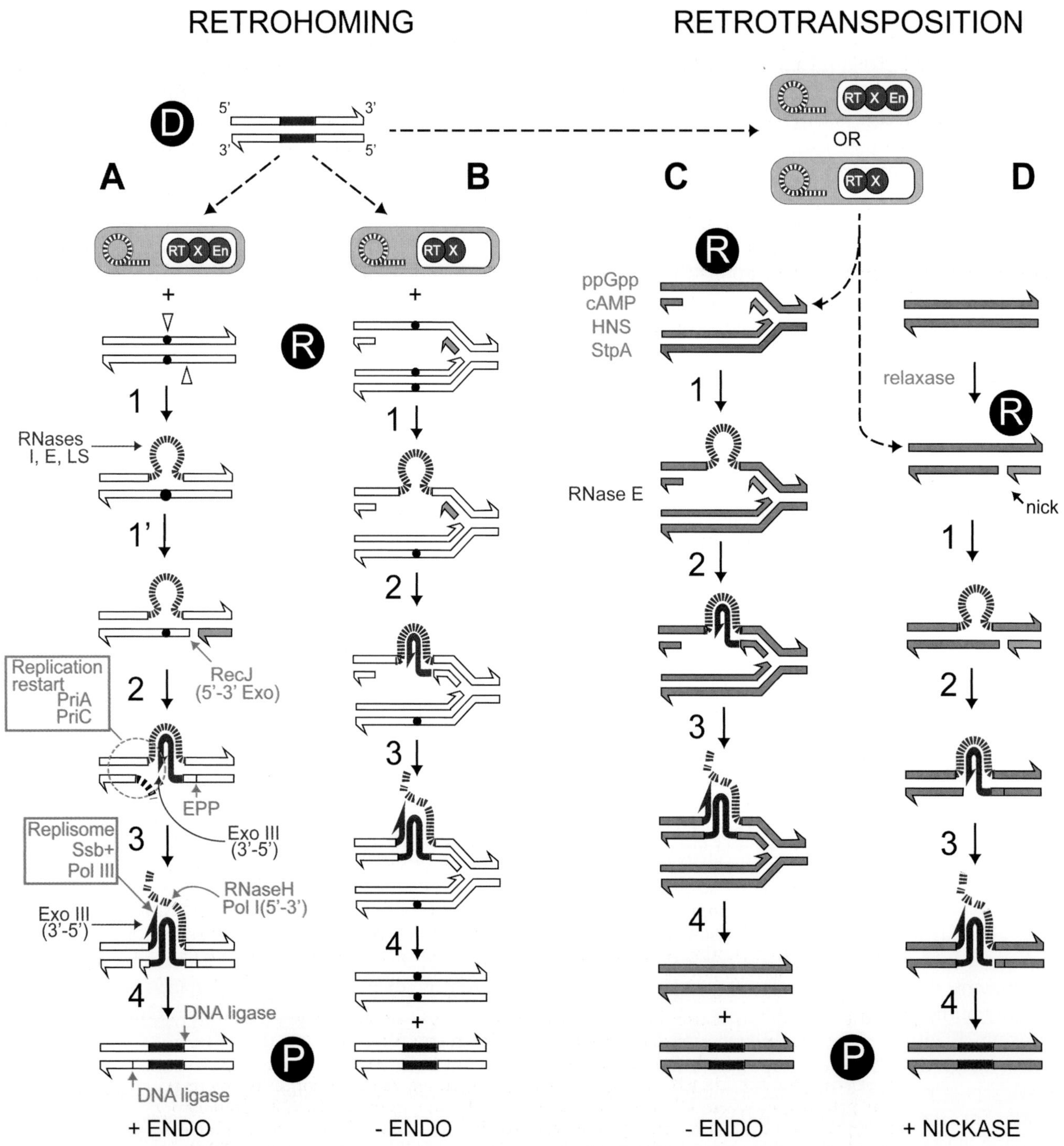

RETROHOMING

RETROTRANSPOSITION

D

5' 3'

3' 5'

OR

A

B

C

D

RT X En

RT X

R

ppGpp
cAMP
HNS
StpA

relaxase

RNases
I, E, LS

RNase E

nick

RecJ
(5'-3' Exo)

Replication
restart
PriA
PriC

EPP

Replisome
Ssb+
Pol III

Exo III
(3'-5')

RNaseH
Pol I(5'-3')

Exo III
(3'-5')

DNA ligase

DNA ligase

P

P

+ ENDO

- ENDO

- ENDO

+ NICKASE

group II intron RTs, which do not initiate efficiently from an annealed primer on DNA templates (72).

Whereas hosts provide housekeeping functions, such as the single-strand binding protein Ssb and nucleoid components H-NS and StpA, that promote group II intron retromobility (73, 74), they also mount counterattacks, to inhibit intron proliferation (75). Both RNases (RNase E, RNase I and RNase LS) and DNases (ExoIII) are degradative enzymes that keep retromobility in check (45, 72, 73, 76). The RNases act by degrading the intron RNA, whereas ExoIII is thought to resect newly synthesized cDNA.

Endonuclease-Independent Retrohoming

Group II introns whose IEPs lack DNA endonuclease activity retrohome by a mechanism in which a nascent leading or lagging strand at a DNA replication fork rather than a cleaved DNA strand is used to prime reverse transcription. The RmInt1 group IIB intron and group IIC introns, which encode IEPs lacking an En domain, preferentially use lagging DNA strand primers (Okazaki fragments), likely reflecting that these introns reverse splice into transiently ssDNA at DNA replication forks (62) (Fig. 3B). The relatively high efficiency with which RmInt1 retrohomes by this mechanism (20 to 45% per recipient target site) raises the possibility of a yet-to-be-demonstrated interaction between the intron RNP and DNA replication machinery (77). In the case of Ll.LtrB, which reverse splices efficiently into dsDNA, En-deficient mutants show the opposite bias for using the nascent leading strand as a primer for reverse transcription of the intron RNA (78). This bias likely reflects that after reverse splicing into dsDNA, an intron RNP integrated into the leading template strand is positioned to directly use a nascent leading strand from an approaching replication fork to prime reverse transcription; by contrast, an intron RNP integrated into the lagging template strand could not be reverse transcribed until after the potentially disruptive passage of the replication fork (78).

Retrohoming of Linear Group II Intron RNAs

Experiments in which Ll.LtrB intron RNPs were microinjected into *Xenopus laevis* oocyte nuclei or *Drosophila melanogaster* embryos showed that linear group II intron RNAs can retrohome by a mechanism

Figure 3 Representative retrohoming and retrotransposition pathways. (A, B) Retrohoming into cognate sites is represented by the two pathways on the left. (C, D) Retrotransposition to ectopic sites is represented by the two pathways on the right. For all panels, the intron is red (DNA solid lines, RNA hatched lines); the intron RNP is represented by a grey rectangle with a red intron lariat and an IEP with RT and maturase (X) domains and either containing or lacking an En domain. Exons of the donor (encircled D) are white, and those of the recipient (encircled R) are either white (retrohoming) or grey (retrotransposition). Each pathway ends with a product (encircled P). Green exon fragment represents primer for reverse transcription. The retrohoming pathways (A, B) differ by the presence of the En domain, and whether dsDNA (A) or ssDNA, such as at a replication fork (B), is the target. The black dot in the recipient strand represents the intron-insertion site. In pathway A, the IEP contains an En domain and after reverse splicing into the top strand, cleavage of the bottom strand occurs 9 or 10 nt downstream (step 1′), as for the Ll.LtrB and yeast aI2 introns, respectively (54, 101). In pathway B, the IEP lacks an En domain and integrates into DNA at a replication fork (step 1) as demonstrated for RmInt1 intron (62) and En-deficient mutants of the Ll.LtrB intron (78), which preferentially use lagging or leading strand primers, respectively. Use of a leading strand primer is not shown in the figure (see reference 78). In both pathways, cDNA synthesis proceeds with the intron as template, using the 3′-OH of either the cleaved bottom strand (A) or an Okazaki fragment (B) to prime reverse transcription (step 2). Intron degradation and second-strand cDNA synthesis (step 3) is followed by DNA repair to generate the retrohoming products for both pathways (step 4). Host factors that participate in the process are indicated on pathway A, as established for the Ll.LtrB intron in *E. coli*, with those that silence the pathway shown in red, and those that promote retrohoming indicated in green (Table 1) (72, 73). Retrotransposition (C, D) occurs when the intron integrates into ectopic sites with reduced specificity. This occurs for the Ll.LtrB intron into the lagging strand template for DNA synthesis, as in pathway B, where Okazaki fragments prime cDNA synthesis (C) (58). Stimulatory and repressive host factors are again represented in green and red, respectively (Table 1) (45, 74, 97). Alternatively, primers can be provided by nicks introduced into dsDNA by relaxase, the product of the *ltrB* gene that hosts the intron (pathway D) (93). Steps 1 to 4 in pathway D are as for retrohoming. EPP, error-prone polymerase.
doi:10.1128/microbiolspec.MDNA3-0050-2014.f3

Table 1 Cellular factors that affect group II intron retromobility[a]

Factor[b]	E[c]	Identified Function	Putative effect on group II intron	Reference
cAMP	S*	Global small-molecular regulator	Promotes retromobility	(97)
DnaB	S	Helicase	Primosome assembly	(73)
DnaC	S	Loading factor for DnaB	Primosome assembly	(73)
DnaG	S	Primase	Second-strand DNA synthesis; possible role in initiation of TPRT	(73)
DnaT	S	Loading factor with DnaC-DnaB	Primosome assembly	(73)
Exo III	I	3′-5′ exonuclease	Degrade nascent cDNA	(72)
H-NS	S*	Nucleoid component, transcription regulator	Promotes retromobility – global	(74)
Ligase	S	DNA ligase	Sealing of DNA nicks	(72, 73)
LtrB	S	Relaxase from conjugative plasmid	Introduces spurious nicks into DNA	(93)
MutD	S	3′-5′ exonuclease subunit of Pol III (*dnaQ*)	Repair second-strand DNA synthesis	(72)
Pol I	S	5′-3′ exonuclease; removal of RNA primer from Okazaki fragments	Remove intron RNA template	(72, 73)
Pol II	S	Repair polymerase (*polB*)	Repair polymerization across DNA-RNA junctions	(72)
Pol III	S	Replicative polymerase	Second-strand DNA synthesis	(72, 73)
Pol IV	S	Repair polymerase (*dinB*)	Repair polymerization	(72)
Pol V	S	Repair polymerase (*umuDC*)	Repair polymerization	(72)
poly(P)	S*	Global regulator	Alters IEP localization and intron integration bias	(96)
PriA	S	Replication restart	Recognizes branched DNA with short gap and initiates replisome loading	(73)
PriC	S	Replication restart	Recognizes branched DNA with long gap and initiates replisome loading	(73)
ppGpp	S*	Global small-molecule regulator	Can promote retromobility	(97)
RbfA	S	Ribosome 30S and 50S association	Protects intron from RNase E degradation	(45)
RecJ	S	5′-3′ exonuclease	5′-3′ resection of DNA	(72, 73)
RNase E	I	Ribonuclease; part of RNA degradosome	Reduces half-life of intron RNA	(45, 72, 76)
RNase H1	S	Ribonuclease; cleaves RNA strand in RNA/DNA hybrid	Removes intron RNA template	(72, 73)
RNase I	I	Ribonuclease	Reduces half-life of intron RNA	(72)
RNase LS	I	Ribonuclease	Reduces half-life of intron RNA	(73)
Ssb	S	Single-stranded binding protein	Stabilizes ssDNA and may interact with PriA to recruit DnaB	(73)
StpA	S*	Nucleoid component, RNA chaperone	Promotes retromobility – global	(74)

[a]Revised and updated from Beauregard et al. (75).
[b]All factors shown to function in *E. coli* except the LtrB relaxase, which was shown to function in L. lactis.
[c]E, effect; S, stimulates retromobility; I, inhibits retromobility; S*, stimulates retromobility into the chromosome only.

in which group II intron RNPs catalyze the first step of reverse splicing, resulting in attachment of the linear intron RNA to the downstream exon of the DNA target site. TPRT then yields an intron cDNA whose unattached 3′ end is ligated to the upstream exon DNA by nonhomologous end-joining (NHEJ), yielding a mixture of precise and aberrant 5′ junctions (79). In *D. melanogaster*, linear intron RNA retrohoming occurs by both major Lig4-dependent and minor Lig4-independent mechanisms, which appear to be related to classical and alternate NHEJ, respectively (80). The DNA repair polymerase θ plays a crucial role in both pathways, presumably by adding extra nucleotides to the 3′ end of the intron cDNA to generate microhomologies that enable annealing of the cDNA end to the upstream exon. However, linear intron RNA retrohoming can occur independently of Ku70, which functions in capping chromosome ends during classical NHEJ.

RT-Independent Homing

A subset of filamentous fungal mt group IIB introns and a recently identified giant sulfur bacterial group IIC intron encode a LAGLIDADG DNA homing endonuclease instead of an RT (81, 82, 83, 84). The LAGLIDADG ORF in the bacterial intron is located in DIV, whereas those in the fungal introns are located in DIII, indicating two separate insertions. Biochemical assays show that both the ribozyme and endonuclease of the fungal introns are active (82). *S. cerevisiae* mtDNA introns with mutations in the RT active site have been shown to home efficiently by a conventional double-strand break repair mechanism, similar to group I introns (56, 69), and group II introns encoding LAGLIDADG proteins are assumed to use the same mechanism (85). Phylogenetic analysis indicates horizontal transfer of LAGLIDADG introns, suggesting that they are actively mobile in fungal populations (83).

MECHANISM AND REGULATION OF GROUP II INTRON RETROTRANSPOSITION

Whereas retrohoming is important for maintaining introns at a fixed site in genomes, retrotransposition to ectopic sites plays a major role in intron dissemination to novel locations, thereby increasing the diversity of intron-containing host genes on an evolutionary timeframe (58, 86, 87, 88, 89). Retrotransposition is distinguished by relaxed sequence requirements for intron integration. Retrotransposition mechanisms of the Ll.LtrB intron have been analyzed in both *L. lactis* and *E. coli* by using powerful selection systems based on inserting a retrotransposition indicator gene (RIG) into intron DIV (58, 59). RIGs, which were first developed for other retroelements (90, 91), allow direct selection of a drug resistance marker after insertion into DNA via an RNA intermediate. It was thereby shown that retrotransposition of the Ll.LtrB intron in its native *L. lactis* host is biased toward the template for lagging-strand DNA synthesis, suggesting the replication folk as a source of ssDNA and Okazaki fragments as primers for cDNA synthesis (Fig. 3C, step 1). Many bacterial group II introns reside on the lagging-strand template, implicating a role for DNA replication in intron spread in nature. Also consistent with a role of the replication fork is the preponderance of events that occurs into plasmid targets, where plasmids have a higher number of forks per unit length of DNA than does the chromosome (92).

The bacterial host in which retrotransposition occurs not only influences the mobility pathway, but also the chromosomal location of intron-insertion sites. In contrast to *L. lactis*, retrotranspositon of the Ll.LtrB intron in *E. coli* is predominantly En-dependent occurring frequently into dsDNA. Furthermore, whereas events are scattered in *L. lactis*, with some preference for the replication terminus (*ter* domain), their localization in *E. coli* is bipolar, to the replication origin (*ori*) and *ter* domains (59, 74, 93). A polar pattern in *E. coli*, with preference for the *ori* domain, was also found for retrohoming of an Ll.LtrB intron with randomized EBS and δ sequences (94) and attributed to localization of the IEP to the cellular poles (95), an explanation that may also apply to the bipolar pattern of retrotransposition events (74). These studies also gave the first hint that culture conditions can regulate retrotransposition, as slow growth eliminated all events except those around *ori* and *ter* and confined them to an endonuclease-dependent pathway into dsDNA (59). Additionally, a mutagenesis screen showed that Ll.LtrB IEP localization is related to its interaction with negatively charged intracellular polyphosphates, which are normally pole-localized but delocalize away from the poles in response to cellular stress, leading to a more uniform distribution of intron-insertion sites (96). These studies portended growth phase and cellular stress as regulatory forces in retrotransposition.

Importantly, nutritional stress stimulates retrotransposition. Both amino acid and glucose starvation promote retrotransposition via the small molecule alarmones ppGpp and cAMP, respectively (76, 97). Their stimulatory effects are indirect. Rather than promote RNP production, they appear to act on the target DNA, facilitating intron access. Precisely how the stress is transduced to the retroelement is an unanswered question. The same mutant screen that indicated a role for these alarmones and for RNase E showed that retromobility is stimulated by mutations in cytochrome oxidase, raising the possibility that oxidative stress promotes intron movement (76).

Both retrohoming and retrotransposition are suppressed by RNase E, which degrades the intron RNA (45, 72, 76). Because RNase E is part of the degradosome, which is tuned to the metabolic status of the cell, there is a potential for regulation of retromobility in response to changes in the physiological state of the cell. In *L. lactis*, the Ll.LtrB intron associates with the ribosome, particularly the 30S subunit, which affords protection to the degradative silencing imposed by RNase E (45).

Finally, retrotransposition of the Ll.LtrB intron, which resides within the relaxase gene of a conjugative plasmid, pRS01, is boosted in the presence of the plasmid (93, 98). The relaxase initiates conjugation by nicking DNA at the transfer origin and is a stimulatory factor for retrotransposition. By developing a genomic retrotransposition detection system called RIG-Seq, in which events are profiled by high-throughput sequencing, it was shown that relaxase, expressed either from pRS01 or from a plasmid vector in the absence of pRS01, increases both the frequency of retrotransposition and the diversity of DNA target sites. Specific point mutations and relaxase inhibitors support the hypothesis that relaxase elevates retrotransposition by inducing spurious nicks in the host DNA (93). Thus, the plasmid and mobile group II intron interact, with intron splicing required for relaxase expression and conjugation, and relaxase stimulating retrotransposition and intron spread via horizontal transfer.

DNA TARGET SITE RECOGNITION

For retrohoming, group II intron RNPs recognize DNA target sequences by using both the IEP and base pairing of motifs within the intron RNA (56, 99, 100, 101). However, there are differences in detail for different intron subgroups and lineages. DNA target site interactions for representative group IIA, IIB, and IIC introns are diagrammed in Fig. 4. In each case, the intron RNA base pairs to a central region of the DNA target site that encompasses the intron-insertion site, while the IEP recognizes exon sequences flanking the region recognized by the intron. For each subgroup, the intron RNA motifs that base pair to the DNA target site are the same as those that base pair to flanking 5′- and 3′-exon sequences for RNA splicing (Fig. 1D), favoring intron insertion at sites that allow RNA splicing with minimal impairment to gene expression.

In group IIA and IIB introns, the intron RNA base-pairing interactions typically span 12 to 16 nts, while the IEP recognizes a small number of nucleotide residues in the distal 5′ and 3′ exons, which differ even for closely related introns (99, 102, 103, 104). The stringency of IEP recognition varies for different nucleotides, and typically, none is required for retrohoming. For the *S. cerevisiae* aI2 and Ll.LtrB group IIA introns, IEP recognition of the sequences in the distal 5′-exon region likely involves the D domain and is needed for efficient reverse splicing into dsDNA but not ssDNA, indicating a contribution to local DNA melting (78, 99, 105). IEP recognition of the 3′ exon is required only for bottom-strand cleavage (99, 101). Group II introns that reverse splice preferentially into transiently ssDNA, such as RmInt1, are less dependent upon IEP interactions with the distal 5′ exon for DNA melting, and correspondingly their IEPs recognize fewer distal 5′-exon nucleotide residues (77, 106). In group IIC introns, intron RNA base pairing is limited to short EBS1/IBS1 and EBS3/IBS3 interactions, and the region upstream of IBS1 contains a hairpin structure corresponding to a transcription terminator or integron attachment (*att*C) site that is recognized by an as-yet unknown mechanism (107). The formation of this hairpin structure is presumably favored in transiently ssDNA at replication forks, thereby directing group II introns to this location where they can use nascent DNA strands to prime reverse transcription (see above).

For the Ll.LtrB intron, DNA target site interactions of RNPs have been investigated biochemically suggesting a model in which the RNPs bind DNA nonspecifically and scan for target sites by facilitated diffusion along the DNA (105, 108). The IEP is thought to first recognize a small number of bases, including T-23, G-21, and A-20, in the distal 5′ exon via major-groove interactions. These base contacts bolstered by phosphate-backbone interactions along one face of the helix promote local DNA melting, enabling the intron RNA to base pair to the target DNA (105). The small number of 5′-exon nucleotides recognized initially by the IEP suggests that the intron RNA may contribute to initial target site recognition, either by triplex formation or by processively base pairing to the IBS and δ' sequences. Bottom-strand cleavage between positions +9 and +10 of the DNA target site occurs after a lag and requires additional interactions of the IEP with the 3′ exon, most critically recognition of T+5. Atomic force microscopy showed that the binding of RNPs bends the target DNA into two progressively sharper bend angles, the first correlated with 3′-exon interactions that position the scissile phosphate at the En active site for bottom-strand cleavage, and the second with repositioning of the 3′ end of the cleaved strand from the En to the RT active site for initiation of cDNA synthesis (42). Notably, even with this DNA bending, the distance between DNA target site residues T-23 and T+5 recognized by the IEP is longer than can be spanned readily by a single IEP molecule. This suggests either that T-23 and T+5 may be recognized sequentially by a single IEP or simultaneously by different subunits of an RT homodimer (17).

Retrotransposition to ectopic sites has relaxed sequence requirements. For all three classes of group II introns, the relatively limited contribution of IEP recognition facilitates dispersal to new target sites compatible with base-pairing interactions of the intron RNA. The Ll.LtrB intron in its native *L. lactis* host retrotransposes into ssDNA, and thus, the residues recognized by the IEP for either DNA unwinding in the upstream exon (T-23, G-21, A-20) or for second-strand DNA cleavage in the downstream exon (T+5) are not required (58, 59, 92, 93). Furthermore, although there is a preference for native IBS1 and δ' sequences that base pair with sequences in the intron, there appears to be minimal dependence on an IBS2-EBS2 interaction. There is, however, a strong bias toward -6C in IBS1, a residue also required for reverse splicing into dsDNA (59, 92, 93, 101).

GROUP II INTRON PROLIFERATION TO HIGH COPY NUMBER

The distribution of group II introns in bacteria suggests waves of intron invasion, proliferation, and extinction, with the latter resulting from a combination of mutational inactivation, limiting intron spread, and removal of genomes harboring deleterious group II introns by purifying selection (109). As a result, group II intron distribution in bacteria is highly variable, even among different strains of the same species, with most bacteria containing only one or a small number of group II introns (109, 110, 111, 112). In several organisms, however, group II introns have escaped host defenses and proliferated to high copy number in the genome. These systems are of interest because the evolution of spliceosomal introns is hypothesized to have involved proliferation of invasive group II introns to high copy number in the genomes of early eukaryotes.

Group II introns have been successful mobile elements not only because of their self-contained mobility apparatus, but also because they have evolved mechanisms for minimizing host damage, including high DNA target specificity, ability to remove themselves by RNA splicing and the sequestration of potentially deleterious IEPs by tight binding to the intron RNA. Factors that act to prevent group II intron proliferation include saturation of potential target sites (113), host defense mechanisms, including bacterial ribonucleases (see above), and negative selection due to deleterious effects on the host. Such deleterious effects might include inefficient splicing from some genomic sites, negative effects of high expression of group II intron RNPs, such as promiscuous breaks or integrations in chromosomal DNA, recombination between multiple dispersed copies of the introns leading to genomic rearrangements (86, 87, 114, 115), and replicative disadvantage due to larger genome size.

The balance of factors that favor mobility and act against proliferation varies for different introns and different hosts. For example, group IIC introns have lower DNA target specificity than do IIA or IIB introns and preferentially insert downstream of transcription terminators, enabling them to proliferate in some bacteria to moderately high copy number (116). Some strains of *Wolbachia*, a bacterial endosymbiont found in arthropods and insects, resemble organelles in containing relatively high numbers of group II introns of different families (115). This accumulation has been attributed to multiple intron invasions combined with inefficient purifying selection due to the limited population size of intracellular symbionts (115).

A dramatic example of group II intron proliferation to high copy number is provided by *Euglena gracilis* cpDNA, which contains more than 150 group II introns that interrupt almost every protein-coding gene (117, 118). Most of the introns are degenerate lacking elements of the conserved group II intron RNA structure, including entire RNA domains, and only two contain intact ORFs that could potentially encode functional proteins (49, 118). The smallest of these introns, referred to as group III introns, consist only of DI and DVI and lack DV, which is essential for ribozyme activity (117, 119). Comparisons of cpDNA sequences of different euglenoid species show that early branching species contain fewer introns and suggest at least two bursts of intron proliferation (120, 121). These proliferations may have been enabled by an efficient common splicing apparatus that provides key proteins and possibly RNA domains in *trans* (117). Other contributing factors include inefficient purifying selection against organellar genomes harboring group II introns and the ability of some group II introns to avoid host genes by inserting into other group II introns, forming "twintrons" (117). The euglenoid cpDNA introns most likely proliferated by a promiscuous reverse splicing-based retrotransposition mechanism that enables dispersal to large numbers of new sites, with bursts potentially reflecting acquisition of new actively mobile introns by horizontal transfer (122).

The thermophilic cyanobacterium *Thermosynechococcus elongatus* contains 28 closely related copies of a group IIB intron constituting ~1.3% of the genome (123). A combination of bioinformatics and mobility assays at different temperatures identified four mechanisms that contributed to the proliferation of *T. elongatus* group II introns (114). First, the *T. elongatus* introns have diverged into six families with different EBS sequences that target the introns to different sites. This divergence suggests waves of retrohoming following intron mutations that enable insertion into different target sites, thereby circumventing the limitation of target-site saturation. Second, some of the *T. elongatus* IEPs have evolved relaxed intron specificity and can efficiently splice and mobilize other introns that have lost their own ORFs, forming a rudimentary common splicing apparatus for multiple dispersed introns. Further, deletion of the intron ORF favors proliferation because the smaller, more compact ORF-less introns splice more efficiently and are less susceptible to nuclease degradation. More efficient splicing in turn enables intron insertion into housekeeping genes at lower fitness cost to the host. Third, some *T. elongatus* introns have evolved to insert at sites that are not deleterious to the host, such as a conserved sequence in another group II intron, leading to twintrons, or into an IS element.

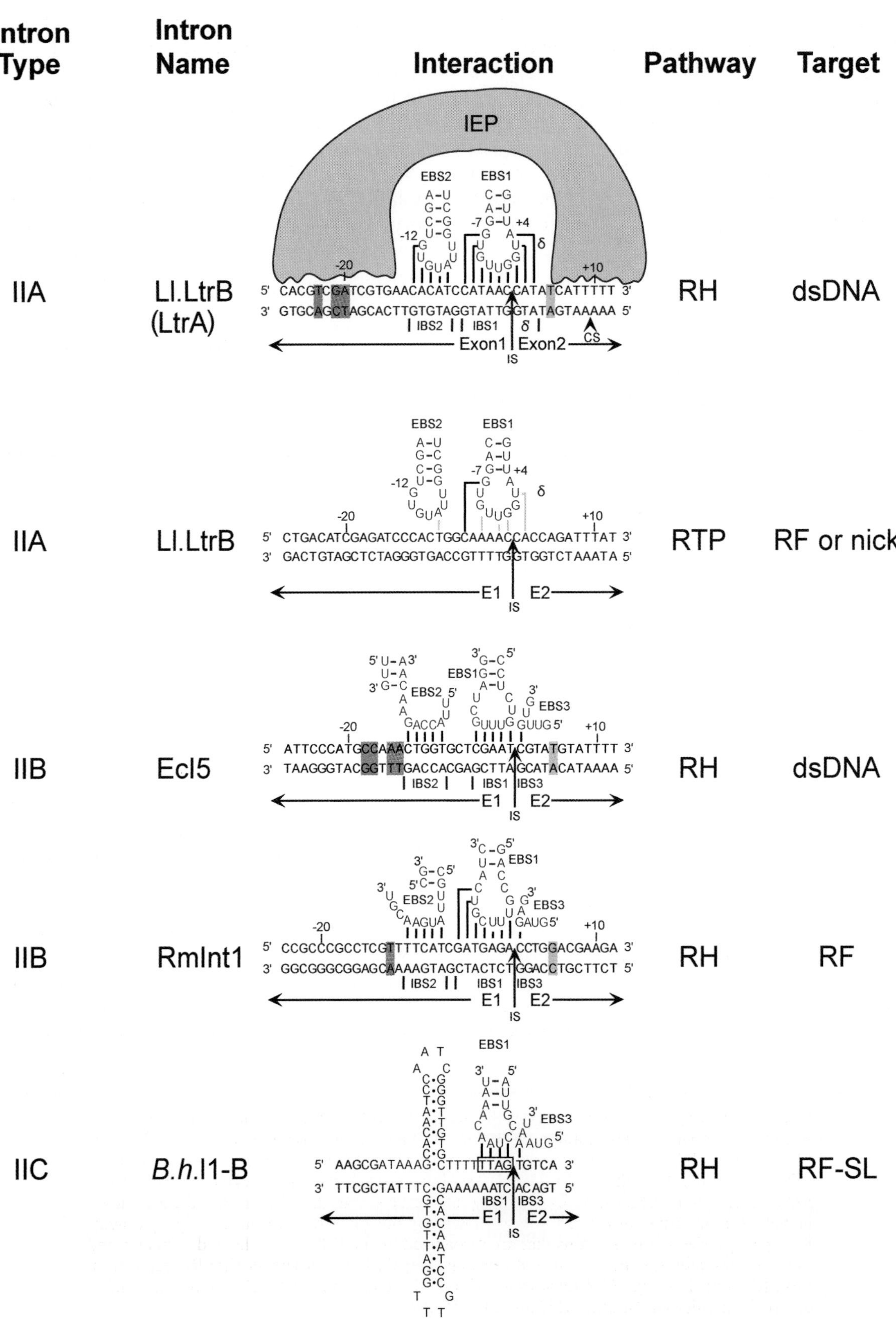
Intron Type
Intron Name
Interaction
Pathway
Target
IEP
EBS2
EBS1
δ
IIA
Ll.LtrB (LtrA)
5' CACGTCGATCGTGAACACATCCATAACCATATCATTTTT 3'
3' GTGCAGCTAGCACTTGTGTAGGTATTGGTATAGTAAAAA 5'
IBS2
IBS1
Exon1
Exon2
CS
IS
RH
dsDNA
IIA
Ll.LtrB
5' CTGACATCGAGATCCCACTGGCAAAACCACCAGATTTAT 3'
3' GACTGTAGCTCTAGGGTGACCGTTTTGGTGGTCTAAATA 5'
E1
E2
RTP
RF or nick
IIB
Ecl5
5' ATTCCCATGCCAAACTGGTGCTCGAATCGTATGTATTTT 3'
3' TAAGGGTACGGTTTGACCACGAGCTTAGCATACATAAAA 5'
EBS3
IBS3
RH
dsDNA
IIB
RmInt1
5' CCGCCCGCCTCGTTTTCATCGATGAGACCTGGACGAAGA 3'
3' GGCGGGCGGAGCAAAAGTAGCTACTCTGGACCTGCTTCT 5'
RH
RF
IIC
B.h.I1-B
5' AAGCGATAAAG•CTTTTTTAGTGTCA 3'
3' TTCGCTATTTC•GAAAAAATCACAGT 5'
RH
RF-SL

Finally, unlike mesophilic group II introns, the thermophilic *T. elongatus* introns rely on elevated temperatures to promote DNA strand separation, enabling access to a larger number of DNA target sites by base pairing of the intron RNA, with minimal requirement for recognition by the IEP to promote DNA melting. Thus, higher temperatures, which are thought to have prevailed on Earth during the emergence of eukaryotes, favor intron proliferation by increasing the accessibility of DNA target sites.

GROUP II INTRON BEHAVIOR IN EUKARYOTES

Although mobile group II introns are thought to have proliferated in the nuclear genomes of early eukaryotes before evolving into spliceosomal introns (124, 125) and integration of organellar DNA fragments containing group II introns into the nuclear genome is an on-going process (126, 127), functional group II introns have not been found in eukaryotic nuclear genomes. In order to survive a potentially overwhelming group II intron invasion early in their evolution, eukaryotes had to evolve mechanisms that restrain group II intron proliferation and expression. One such restraint may have been lower free Mg^{2+} concentrations, which limit group II intron ribozyme activity (128, 129, 130). Intracellular free Mg^{2+} concentrations are lower in eurkayotes (0.2 to 1 mM) than in bacteria (1 to 4 mM) and may be particularly low in eukaryotic nuclei, where Mg^{2+} is chelated to chromosomal DNA (130, 131). Efficient group II intron retrohoming in *X. laevis* oocyte nuclei or *D. melanogaster* or zebrafish embryos occurs only after injection of $MgCl_2$ (129). Spliceosomal introns evolved to function at the lower Mg^{2+} concentrations in eukaryotes, possibly by their disintegration into snRNAs, and by their increased reliance on protein cofactors, which can substitute for Mg^{2+} to promote RNA folding (130).

Studies in which the Ll.LtrB group IIA intron and its IEP were inserted within the *S. cerevisiae* nuclear genes *CUP1* or *URA3* showed that expression of these genes is silenced. The pre-mRNAs harboring the intron are not spliced in the nucleus but are spliced accurately in the cytoplasm (132), possibly reflecting higher free Mg^{2+} concentrations in that compartment. Interestingly, the mRNA from which Ll.LtrB is spliced in the cytoplasm is subjected to nonsense-mediated decay, translational repression of mRNA, and targeting of the RNA to cytoplasmic foci (132, 133). By contrast, a spliceosomal intron assayed in parallel evaded these surveillance mechanisms (132). Insertion of the brown algal P1.LSU/2 group IIB intron into a yeast nuclear gene results in similar silencing of spliced mRNAs (134). Tenacious intermolecular interactions between the mRNA and the pre-mRNA and/or excised intron, based on IBS-EBS base pairings, contribute to translational silencing of the spliced mRNAs (133). Whereas disruption of these pairings by mutation promoted gene expression, compensatory mutations to restore the pairings again resulted in silencing. Thus, impaired expression of host genes containing inserted group II introns could have selected against ancestral eukaryotes carrying intact group II introns in their nuclear genomes and favored the evolution of splicesomal introns.

MOBILE GROUP II INTRON-EUKARYOTIC RETROTRANSPOSON RELATIONSHIPS

Mobile group II intron RTs are closely related to non-LTR-retrotransposon and telomerase RTs of eukaryotes (Fig. 2A). All three types of RT use analogous reverse transcription mechanisms, in which the site of initiation of cDNA synthesis is dictated by specific binding of the RT to the RNA template and the 3′ end of a target DNA is used as the primer for reverse transcription (17, 135, 136, 137, 138). Recent studies extend these similarities by showing that all three RTs employ regions upstream of the RT1 motif that are not present in retroviral RTs for the specific binding of RNA templates (46, 139, 140) (Fig. 2A). In the case of group II introns and non-LTR-retrotransposons, the

Figure 4 DNA target site recognition by group IIA, IIB, and IIC introns. Target site interactions are shown for retrohoming (100, 101, 103) and retrotransposition (59, 92, 93) of the Ll.LtrB group IIA intron, and for retrohoming of group IIB introns EcI5 (104) and RmInt1 (77), and group IIC intron *B.h.*I1-B (107). Intron RNA regions involved in EBS1-IBS1, EBS2-IBS2, δ-δ' or EBS3-IBS3 base-pairing interactions with the DNA target site are shown in red. A representative ectopic site is shown for the Ll.LtrB retrotransposition pathway. Base-pairs in the 5′ and 3′ exons that are recognized by the IEP are highlighted in mauve and blue, respectively. CS, bottom-strand cleavage site; IS, intron-insertion site; RF, replication fork; RH, retrohoming; RTP, retrotransposition; SL, stem-loop. CS for EcI5 is not known.
doi:10.1128/microbiolspec.MDNA3-0050-2014.f4

primer for TPRT is typically generated by cleavage of the target DNA by an En-domain appended to the RT [H-N-H for group II introns, restriction enzyme-like endonuclease (REL) for R2, and apurinic endonuclease (APE) for LINEs; Fig. 2A]. The identity of the En is variable, with some nuclear non-LTR-retrotransposons encoding both APE and REL domains, and others, the *Penelope* retrotransposons, encoding a GIY-YIG endonucålease domain related to group I intron homing endonucleases (141, 142, 143). Both group II introns and non-LTR-retrotransponsons can also use random nicks in DNA as primers for reverse transcription (93, 144, 145). In contrast, telomerase uses the 3′ end of a chromosomal DNA as the primer, an ability shared by non-LTR-retrotransposons that lack En activity (142, 146, 147), but not yet demonstrated for mobile group II introns.

Like group II intron RTs (19), LINE-1 elements and other non-LTR-retrotransposon RTs show a *cis*-preference for mobilizing the element in which they are encoded, presumably reflecting that their RTs bind rapidly and preferentially to the RNA from which they are translated (148, 149, 150). The assembled RNP is then transported into the nucleus or nucleolus (for elements that insert into rDNA), which are sites of TPRT. Also like group II intron RTs (72, 73), mammalian LINE-1 elements and most other non-LTR-retrotransposon RTs lack an RNase H domain and may rely at least in part on a host RNase H for degradation of the RNA template strand in addition to strand displacement by the RT (136, 151). Additionally, the TPRT mechanism used by non-LTR-retrotransposons results in a cDNA whose 3′ end may be ligated to upstream sequences by NHEJ enzymes (152), analogous to the retrohoming mechanism used by linear group II introns (see above), or linked to upstream sequences via template switching, a proficient activity of both group II intron and non-LTR-retrotransposon RTs (68, 136, 153, 154). Finally, the mechanism used for second-strand DNA synthesis by non-LTR-retrotransposons is not known, but given their nuclear localization could involve a host DNA polymerase, similar to mobile group II introns (73), rather than the DNA-dependent polymerase activity of the RT.

STRUCTURAL AND MECHANISTIC SIMILARITIES BETWEEN GROUP II AND SPLICEOSOMAL INTRONS

Since the discovery of self-splicing RNAs and the realization that group II introns splice via a lariat pathway, group II introns have been suspected of being the ancestors of spliceosomal introns (7, 155, 156, 157). In the ensuing years, evidence for this relationship has strengthened, based on shared splicing pathways and chemistries, together with sequence and structural similarities.

Structural Similarities Between Group II And Spliceosomal Introns

The spliceosome comprises a plethora of proteins and five small nuclear RNAs (snRNAs), U1, U2, U4, U5, and U6 (158, 159, 160). During spliceosome assembly, there is a flux of both proteins and snRNAs with U2, U5, and U6 being the only snRNAs remaining in the catalytic particle (Fig. 5A). All three of these snRNAs have counterparts in the group II intron ribozyme, with U6 base-paired to U2 similar to the active site helix DV, U2 in complex with the branch-point region of the intron resembling DVI, and U5 base pairing to 5′ and 3′ exons paralleling the DI-exon interactions of group II intron RNAs (Fig. 5A) (4, 161, 162, 163). Indeed, several group II intron domains (DIc, DIII, DI-DIV, DV) have been shown to be modular and promote splicing of group II introns lacking them when provided in *trans* (164, 165, 166) or to be interchangeable with analogous spliceosomal snRNAs: DV for U6 and U5 for the DId stem-loop containing EBS1 (162, 167). Additionally, there are many examples of naturally occurring fragmented group II introns that consist of two or three unlinked segments that reassemble for splicing *in vivo* (reviewed in references 4, 157, and 168). Systematic studies with the Ll.LtrB intron identified numerous functional breakpoints for bi- and tripartite group II introns (169, 170, 171).

Besides these structural similarities, group II introns, like spliceosomal introns, can utilize alternative splicing with potential to generate different protein isoforms (172). Additionally, both group II and spliceosomal introns can splice via 5′-splice site hydrolysis when regions required for branching (DVI and U2 snRNA) are absent or compromised (173, 174).

Mechanistic Similarities Between Group II And Spliceosomal Intron RNAs

The parallels between the group II autocatalytic pathway and spliceosome-mediated splicing (Fig. 1A), and the high degree of snRNA conservation over evolutionary time resulted in the proposal that the spliceosome is a ribozyme (175, 176). Details of the reaction in spliceosomes and group II introns are identical down to stereochemistry of the oxygen atoms that bind the two metals and the role of these divalent metals in stabilizing the leaving group in both steps of splicing

(Fig. 5B, C) (12, 13, 175, 177). U6 base-paired with U2 catalyzes both steps of splicing by positioning divalent metals in precisely the same way as DV, with an internal stem-loop (ISL) of U6 resembling the DV helix in both structure and sequence (Fig. 5C) (178). In group II introns, the catalytic Mg^{2+} ions bind to DV via a CGC triad and bulge, whereas U6 contains an AGC triad and bulge that position the catalytic Mg^{2+} (Fig. 5B, C) (12, 163, 179). Finally, U6, like DV, uses base-triple interactions to present the catalytic metals to the scissile phosphates for the two steps of splicing (Fig. 5C, D, E) (163, 180). The Mg^{2+}-positioning triple helix in group II introns is formed between the DV triad and two nucleotides in the J2/3 linker (between DII and DIII), and one nucleotide in the DV bulge. In precise analogy, the U6 AGC triad, which base-pairs with U2, forms two base triples with the highly conserved 5′-ACAGAGA-3′ box in U6, and one nucleotide in the U6 ISL bulge (Fig. 5C, D, E). These parallels leave little doubt as to an evolutionary relationship between these two intron types.

Prp8, A Conserved Protein At The Active Site Of The Spliceosome, Has Similarities To Group II Intron RTs

Beyond the structural and mechanistic similarities of group II intron and spliceosomal RNAs, recent studies have shown a relationship between group II intron RTs and Prp8, a highly conserved, 280 kDa protein at the heart of the spliceosome (181, 182, 183). Prp8 interacts with the three snRNAs, U2, U5, and U6, which form the catalytic core of the spliceosome (159, 184). The structure of a large fragment of *S. cerevisiae* Prp8 in complex with the U5 snRNP assembly factor Aar2 showed that the central conserved region contains RT and thumb/X domains related to those of group II intron RTs, separated by a relatively long linker from a type II restriction enzyme-like (REL) En domain (Fig. 6). These domains are followed by an RNase H-like domain and a domain with a Jab1/MPN fold found in de-ubiquitinating enzymes. None of these domains appear to be enzymatically active, with the active site of the RT-like domain containing only one of three aspartates involved in Mg^{2+} coordination in active RTs.

The crystal structure of Prp8 shows that the RT, thumb/X, linker, En and RNase H-like domains form a large internal cavity that is the site of mutations that suppress splicing defects caused by mutations in the 5′- and 3′-splice sites and branch-point nucleotide, as well as the site of cross-links with the 5′- and 3′-splice sites, the branch point, and U5 and U6 snRNA (185, 186). Moreover, this internal cavity can accommodate the catalytic core of a group II intron RNA, which is thought to have evolved into that of the spliceosome (182). Based on these findings, it was suggested that Prp8 may have been derived from an ancestral group II intron RT that evolved to function as an assembly platform for snRNAs and pre-mRNA (182). The similarities of the RT and X/thumb domain suggest an evolutionary relationship to group II IEPs, with the lack of conservation of conserved aspartates at the RT active site consistent with the finding that mutations in the group II intron RT active site have minimal effect on splicing activity (19).

Plant nMat proteins provide previous examples of group II intron RTs that were subsumed into a nuclear genome without their associated intron RNAs and continue to function as splicing factors, albeit for group II introns within organelles (187, 188, 189, 190). nMat2 proteins have a conventional H-N-H En domain, whereas in nMat1 proteins the En domain has been replaced by or diverged into a novel C-terminal domain that is conserved in different plant species and may contribute to the RNA splicing function (Fig. 6).

Despite these tantalizing similarities, a more complex situation is suggested by the finding that the En domain of Prp8 is a REL domain rather than an H-N-H endonuclease as in group II intron RTs and that group II intron RTs lack an associated RNase H domain (Fig. 6). As both REL and RNase H domains are found in some subgroups of nuclear non-LTR-retrotransposons (135, 191), these findings raise the possibility that Prp8 may be derived from another non-LTR-retroelement that is evolutionarily related to and possibly a missing link between group II introns and the spliceosome (192).

GROUP II INTRON EVOLUTION

Group II Intron Origins And Other Bacterial RT-Containing Elements

Although it is often conjectured that group II introns and other ribozymes arose in a primordial "RNA world," wherein catalysis was performed by RNA, the relationship of group II introns to the RNA world remains uncertain (193). The order of events for evolving mobile group II introns is also obscure, and appears to have occurred independently in at least two ways: through acquisition of an RT that imparted RNA-based mobility needed for intron dispersal in bacteria, and through acquisition of a LAGLIDADG DNA endonuclease that likely conferred the ability to mobilize

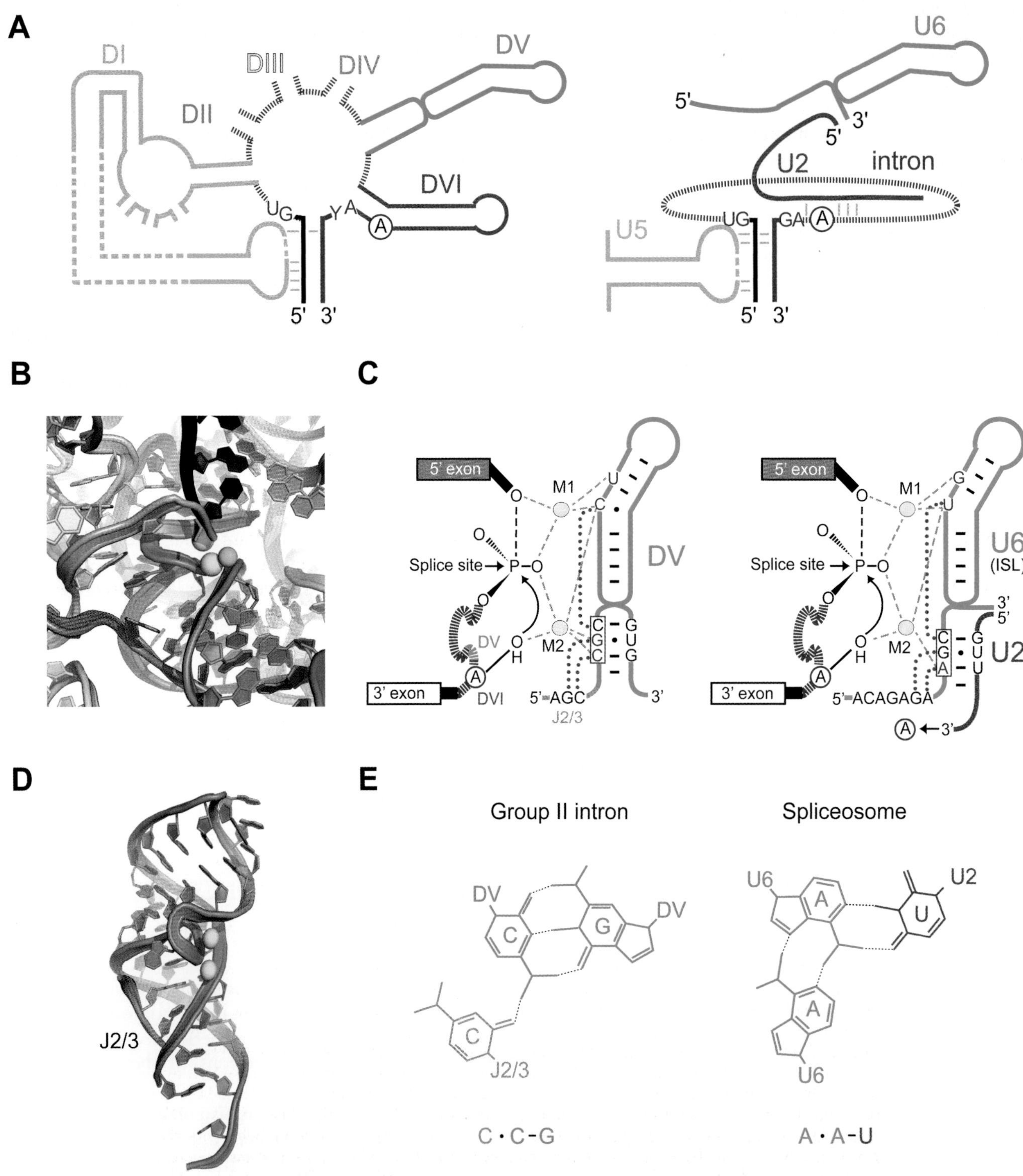
A
DI
DII
DIII
DIV
DV
DVI
UG
YA
A
5'
3'
U6
5'
3'
5'
U2
intron
U5
UG
GA
A
5'
3'
B
C
5' exon
M1
U
C
O
Splice site
P-O
DV
M2
C
G
C
G
U
G
DV
O
H
A
3' exon
DVI
5'-AGC
J2/3
3'
5' exon
M1
G
U
U6
(ISL)
Splice site
3'
5'
M2
C
G
A
G
U
U
U2
3' exon
5'-ACAGAGA
A
3'
D
J2/3
E
Group II intron
Spliceosome
DV
C
G
DV
C
J2/3
C • C-G
U6
A
U
U2
A
U6
A • A-U

the intron via DNA-based recombination in both bacteria and fungi. In both cases, the DNA encoding the catalytic RNA splicing apparatus is assumed to have preexisted an invasion event by different genes coding for mobilizing activities. But for RT-driven mobility, a coevolutionary scenario is also possible, where self-splicing ability developed from a retroelement under selective pressure to minimize transposon damage to the genome (194).

In addition to group II introns, bacteria contain a variety of other retroelements and RTs, which may have evolved from or into group II intron RTs (195, 196). These include retrons (197), abortive phage infection (Abi) retroelements (196), diversity generating retroelement RTs (DGRs) (198, 199), and RTs associated with some CRISPR-Cas systems (195, 196). Abi retroelments are host-encoded elements that defend against phage infection (200), while DGRs are bacterial and phage-encoded elements that diversify protein-encoding DNA sequences, such as those involved in cell surface display or phage tropism switching, via an error prone retrohoming mechanism (198, 201, 202). The RTs associated with CRISPR-Cas systems may function in cellular immunity by synthesizing cDNAs of RNA phages or transcripts of invading DNA elements that are then incorporated into the CRISPR repeats (195, 196). Thus, bacterial RTs are engaged in the phage-host arms race, to evade cellular defenses on one hand and to develop immunity to invasive DNA on the other.

Role Of Group II Introns In The Origin Of Eukaryotes

A favored scenario for the origin of a eukaryotic cell is endosymbiosis, based on an archaeon hosting a bacterial invader that carried group II introns and evolved into mitochondria (124, 125, 203, 204, 205). Although this scenario remains a matter of debate particularly among those who favor bacterial-archaeal fusion scenarios for the origin of eukaryotes (206, 207, 208), the bacterial species thought to have evolved into mitochondria and chloroplasts, α-proteobacteria and cyanobacteria, respectively, both harbor copious numbers of group II introns (209). Bacteria contain all group II intron lineages whereas organelles contain only the IIA/ML and IIB/CL lineages, possibly reflecting that group II introns of these lineages were present in the bacteria that gave rise to mitochondria and chloroplasts (49). The early occurrence of spliceosomal introns in eukaryotes is supported by studies inferring the existence of at least a primitive spliceosome in the last common ancestor of extant eukaryotes (210).

Martin and Koonin (211) proposed that intron proliferation in a primitive eukaryote stimulated the evolution of the nuclear envelope in order to separate

Figure 5 Similar RNA active sites of group II introns and the spliceosome. (A) Schematic of group II intron (left) and spliceosome (right). Group II intron elements are colored as in Fig. 1B and C, with corresponding spliceosomal elements in the same color (D1 and U5 blue; DV and U6 green; DVI and U2, purple). Conserved residues at the splice sites are in red and the branch-point adenosine (A) is circled. Elements are not drawn to scale. (B) Crystal structure of *O. iheyensis* group II intron in the precatalytic state (from PDB file 4FFAQ). The 5′ exon (black) is shown before 5′ splice site hydrolysis. Color-coding is as in Fig. 1B and C and Fig. 5A, with the two catalytic Mg^{2+} ions shown as yellow spheres bound to DV (green) (13, 219). The putative water nucleophile is a cyan sphere. Images in panel B and D provided by Marcia and Pyle. (C) Secondary structures and Mg^{2+} interactions in *O. iheyensis* intron DV and spliceosomal U6 snRNA. DV (left, shown also in panel B) corresponds to the internal stem loop (ISL) of U6 (right). M1 and M2 are catalytic Mg^{2+} ions coordinated by phosphate oxygens of the nucleotides shown in red. The circled A in the intron is the adenosine nucleophile with the 2′-OH corresponding to the water molecule in B, which attacks the 5′ splice site (arrow). The introns are depicted in red hatch marks, with a small segment in green representing DV of the group II intron detailed to the right. U2 3′ to the U2-U6 pairing extends to interact with the branch-point region of the intron (arrow to circled A). The three boxed nucleotides in each case comprise the conserved catalytic triad (163, 178, 183, 220). Dotted brown lines join residues involved in base triples, which are formed by two pairings between the catalytic triad and ether J2/3 or the 5′-ACAGAGA-3′ box and the third pairing to the bulge in each structure (163, 180). (D) Base triples in DV of the *O. iheyensis* intron. The J2/3 nucleotides (orange ribbon) form a triple helix with the major groove at the base of DV (from reference 10). Two metal ions (yellow spheres) are bound near the twisted bulge loop. (E) Example of base triples in DV and U2-U6. The color-coded base triples are shown for the lower-most base-pair in the catalytic triad with a nucleotide in J2/3 in the group II intron or the 5′-ACAGAGA-3′ box in U6 of the spliceosome, as diagramed in panel C. doi:10.1128/microbiolspec.MDNA3-0050-2014.f5

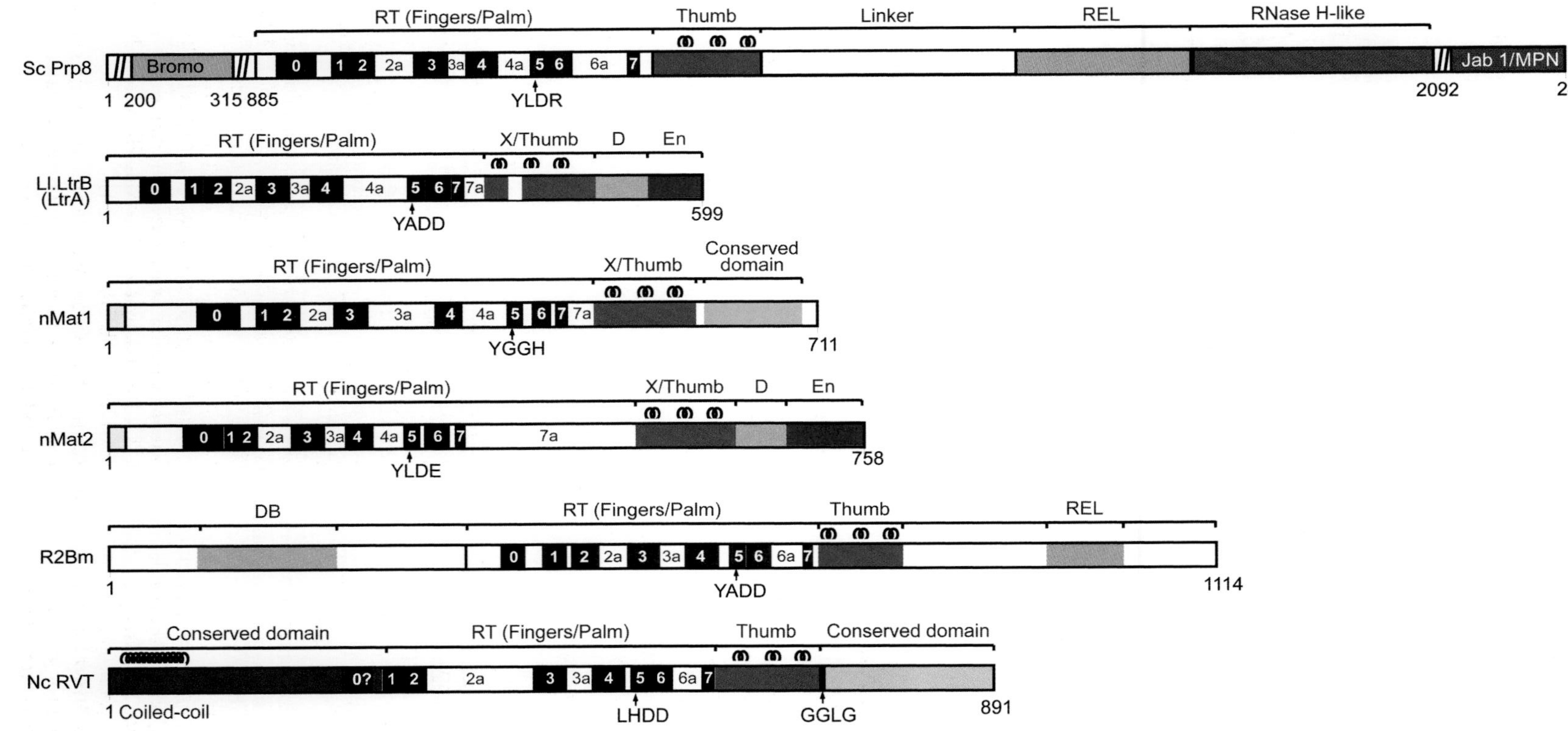

Figure 6 Comparison of Prp8 to group II intron and related RTs. Schematic comparing *S. cerevisiae* Prp8 (PDB:4I43) with the Ll.LtrB group II intron RT (LtrA protein; GenBank:AAB06503), *Arabidopis thaliana* nMat1 and nMat2 proteins (NCBI Reference Sequence: NP_174294 and NP_177575, respectively), non-LTR-retrotranspson R2Bm RT (GenBank: AAB59214), and *Neurospora crassa* RVT RT (GenBank:CAE76174). The plant nMat1 and nMat2 proteins are previous examples of group II intron RTs that were subsumed into nuclear genomes and retained RNA splicing function, with the nMat1 proteins acquiring a novel conserved domain (green) in place of the En domain. The Prp8 configuration of the thumb, long linker, and REL domain is similar to that in the R2Bm RT. Although group II intron RTs and the R2Bm RT lack an RNase H domain, RNase H domains are known to be acquired sporadically in different non-LTR-retrotransposon lineages (135, 191). RVT RTs, another potential candidate for an ancestor of Prp8 (181), lack En and integrase domains and are further distinguished from group II intron and non-LTR-retrotransposon RTs by a large acidic insertion within RT-2a and by conserved N- and C-terminal domains that are not found in other proteins (212). Conserved sequence blocks in the RT domain are numbered, and the sequence motif containing two of the conserved aspartic acid residues at the RT active site is shown below for each protein. The locations of conserved RT sequence blocks in Prp8 are from structure-based sequence alignments by G Mohr (UT Austin). DB, DNA binding domain; REL, restriction endonuclease-like domain. doi:10.1128/microbiolspec.MDNA3-0050-2014.f6

splicing from translation and prevent ribosomes from traversing the intron, resulting in mistranslated proteins, and to counter further intron invasions. At the same time, the separation of transcription and splicing in the nucleus from translation of spliced mRNAs in the cytosol necessitated the evolution of machinery that could function in *trans* to splice multiple dispersed introns. The presence of large numbers of introns in protein-coding genes also necessitated mechanisms like nonsense-mediated decay, which helps ensure that only accurately spliced mRNAs are translated.

Next, group II introns are hypothesized to have fragmented and reassembled into a spliceosome, which has been augmented by continued evolution into a megamachine. Consistent with this hypothesis is the absence of functional group II introns from extant nuclear genes, where they appear unable to function and where silencing of gene expression and nuclear compartmentalization may have added impetus to the evolution of the spliceosome through fragmentation and reassembly (130, 132, 133).

Evolution Of Intron-Related Retroelements And Their Role In Sculpting Our Genomes

The discovery that group II intron retrohoming occurs by TPRT raised the possibility that group II introns are the ancestors of eukaryotic non-LTR retrotransposons and telomerase (54). This hypothesis is supported by the phylogenetic distribution of group II introns, which suggests that they evolved in eubacteria and entered eukaryotes with bacterial endosymbionts that evolved into mitochondria and chloroplasts. However, bacteria contain diverse retroelements and RTs (see above), as well as single-copy *rvt* genes. RVTs are of unknown function and found sporadically in all eukaryotic kingdoms (195, 196, 212, 213) (Fig. 6). Consistent with their single-copy nature, they are not associated with known endonucleases or integrases (212). These considerations leave one pondering whether cellular RT genes originated from retrotransposons or vice versa. Regardless, retroelements related to group II introns have sculpted our genomes, together with the spliceosomal introns to which these RNA catalysts almost certainly gave rise.

Acknowledgments. *We thank Anne Pyle and Marco Marcia for structure representations in Figures 1 and 5, Matt Stanger and Georg Mohr for rendering the figures, and Rebecca McCarthy and Susan Snyder for help with the manuscript. We thank Georg Mohr (UT Austin), John Moran (U. Michigan), Anna Pyle and Marco Marcia (Yale), Steven Zimmerly (U. Calgary), and Carol Lyn Piazza and Dorie Smith (U. Albany) for comments on the manuscript. Work in our labs is supported by NIH grants GM37949 and GM37951 and Welch Foundation Grant F-1607 to AML and NIH grants GM39422 and GM44844 to MB.*

Citation. Lambowitz AM, Belfort M. 2014. Mobile bacterial group II introns at the crux of eukaryotic evolution. Microbiol Spectrum 3(1):MDNA3-0050-2014.

References

1. **Lambowitz AM, Zimmerly S.** 2004. Mobile group II introns. *Annu Rev Genet* **38:**1–35.
2. **Pyle AM, Lambowitz AM.** 2006. Group II introns: ribozymes that splice RNA and invade DNA, p 469–505. *The RNA World*, 3rd ed. Cold Spring Harbor Laboratory Press, Plainview, NY.
3. **Toro N, Jiménez-Zurdo JI, García-Rodríguez FM.** 2007. Bacterial group II introns: not just splicing. *FEMS Microbiol Rev* **31:**342–358.
4. **Lambowitz AM, Zimmerly S.** 2011. Group II introns: mobile ribozymes that invade DNA. *Cold Spring Harb Perspect Biol* **3:**a003616.
5. **Marcia M, Somarowthu S, Pyle AM.** 2013. Now on display: a gallery of group II intron structures at different stages of catalysis. *Mob DNA* **4:**14.
6. **Enyeart PJ, Mohr G, Ellington AD, Lambowitz AM.** 2014. Biotechnological applications of mobile group II introns and their reverse transcriptases: gene targeting, RNA-seq, and non-coding RNA analysis. *Mob DNA* **5:**2.
7. **Peebles CL, Perlman PS, Mecklenburg KL, Petrillo ML, Tabor JH, Jarrell KA, Cheng H-L.** 1986. A self-splicing RNA excises an intron lariat. *Cell* **44:**213–223.
8. **Jarrell KA, Peebles CL, Dietrich RC, Romiti SL, Perlman PS.** 1988. Group II intron self-splicing. Alternative reaction conditions yield novel products. *J Biol Chem* **263:**3432–3439.
9. **Podar M, Chu VT, Pyle AM, Perlman PS.** 1998. Group II intron splicing *in vivo* by first-step hydrolysis. *Nature* **391:**915–918.
10. **Pyle AM.** 2010. The tertiary structure of group II introns: implications for biological function and evolution. *Crit Rev Biochem Mol Biol* **45:**215–232.
11. **Chin K, Pyle AM.** 1995. Branch-point attack in group II introns is a highly reversible transesterification, providing a potential proofreading mechanism for 5′-splice site selection. *RNA* **1:**391–406.
12. **Toor N, Keating KS, Taylor SD, Pyle AM.** 2008. Crystal structure of a self-spliced group II intron. *Science* **320:**77–82.
13. **Marcia M, Pyle AM.** 2012. Visualizing group II intron catalysis through the stages of splicing. *Cell* **151:**497–507.
14. **Li CF, Costa M, Michel F.** 2011. Linking the branchpoint helix to a newly found receptor allows lariat formation by a group II intron. *EMBO J* **30:**3040–3051.
15. **Somarowthu S, Legiewicz M, Keating KS, Pyle AM.** 2014. Visualizing the ai5γ group IIB intron. *Nucl Acids Res* **42:**1947–1958.
16. **Robart AR, Chan RT, Peters JK, Rajashankar KR, Toor N.** 2014. Crystal structure of a eukaryotic group II intron lariat. *Nature* **514:**193–197.

17. **Blocker FH, Mohr G, Conlan LH, Qi L, Belfort M, Lambowitz AM.** 2005. Domain structure and three-dimensional model of a group II intron-encoded reverse transcriptase. *RNA* **11:**14–28.
18. **Malik HS, Burke WD, Eickbush TH.** 1999. The age and evolution of non-LTR retrotransposable elements. *Mol Biol Evol* **16:**793–805.
19. **Cui X, Matsuura M, Wang Q, Ma H, Lambowitz AM.** 2004. A group II intron-encoded maturase functions preferentially *in cis* and requires both the reverse transcriptase and X domains to promote RNA splicing. *J Mol Biol* **340:**211–231.
20. **Moran JV, Zimmerly S, Eskes R, Kennell JC, Lambowitz AM, Butow RA, Perlman PS.** 1995. Mobile group II introns of yeast mitochondrial DNA are novel site-specific retroelements. *Mol Cell Biol* **15:**2828–2838.
21. **Kennell JC, Moran JV, Perlman PS, Butow RA, Lambowitz AM.** 1993. Reverse transcriptase activity associated with maturase-encoding group II introns in yeast mitochondria. *Cell* **73:**133–146.
22. **San Filippo J, Lambowitz AM.** 2002. Characterization of the C-terminal DNA-binding/DNA endonuclease region of a group II intron-encoded protein. *J Mol Biol* **324:**933–951.
23. **San Filippo J.** 2003. *The DNA-binding and DNA endonuclease domains of a group II intron-encoded protein: characterization and application to the engineering of gene targeting vectors. PhD Thesis*, University of Texas at Austin, Austin.
24. **Molina-Sánchez M, Martínez-Abarca F, Toro N.** 2010. Structural features in the C-terminal region of the *Sinorhizobium meliloti* RmInt1 group II intron-encoded protein contribute to its maturase and intron DNA-insertion function. *FEBS J* **277:**244–254.
25. **Toro N, Martínez-Abarca F.** 2013. Comprehensive phylogenetic analysis of bacterial group II intron-encoded ORFs lacking the DNA endonuclease domain reveals new varieties. *PLoS One* **8:**e55102.
26. **Gorbalenya AE.** 1994. Self-splicing group I and group II introns encode homologous (putative) DNA endonucleases of a new family. *Protein Sci* **3:**1117–1120.
27. **Belfort M, Roberts RJ.** 1997. Homing endonucleases: keeping the house in order. *Nucl Acids Res* **25:**3379–3388.
28. **Belfort M, Bonocora RP.** 2014. Homing endonucleases: from genetic anomalies to programmable genomic clippers. *Methods Mol Biol* **1123:**1–26.
29. **Stoddard BL.** 2014. Homing endonucleases from mobile group I introns: discovery to genome engineering. *Mobile DNA* **5:**7.
30. **Saldanha R, Chen B, Wank H, Matsuura M, Edwards J, Lambowitz AM.** 1999. RNA and protein catalysis in group II intron splicing and mobility reactions using purified components. *Biochemistry* **38:**9069–9083.
31. **Matsuura M, Noah JW, Lambowitz AM.** 2001. Mechanism of maturase-promoted group II intron splicing. *EMBO J* **20:**7259–7270.
32. **Anziano PQ, Hanson DK, Mahler HR, Perlman PS.** 1982. Functional domains in introns: trans-acting and cis-acting regions of intron 4 of the *cob* gene. *Cell* **30:** 925–932.
33. **Carignani G, Groudinsky O, Frezza D, Schiavon E, Bergantino E, Slonimski PP.** 1983. An mRNA maturase is encoded by the first intron of the mitochondrial gene for the subunit I of cytochrome oxidase in *S. cerevisiae*. *Cell* **35:**733–742.
34. **Moran JV, Mecklenburg KL, Sass P, Belcher SM, Mahnke D, Lewin A, Perlman P.** 1994. Splicing defective mutants of the *COX1* gene of yeast mitochondrial DNA: initial definition of the maturase domain of the group II intron aI2. *Nucl Acids Res* **22:**2057–2064.
35. **Rambo RP, Doudna JA.** 2004. Assembly of an active group II intron-maturase complex by protein dimerization. *Biochemistry* **43:**6486–6497.
36. **Gupta K, Contreras LM, Smith D, Qu G, Huang T, Spruce LA, Seeholzer SH, Belfort M, Van Duyne GD.** 2014. Quaternary arrangement of an active, native group II intron ribonucleoprotein complex revealed by small-angle X-ray scattering. *Nucl Acids Res* **42:** 5347–5360.
37. **Wank H, SanFilippo J, Singh RN, Matsuura M, Lambowitz AM.** 1999. A reverse-transcriptase/maturase promotes splicing by binding at its own coding segment in a group II intron RNA. *Mol Cell* **4:**239–250.
38. **Singh RN, Saldanha RJ, D'Souza LM, Lambowitz AM.** 2002. Binding of a group II intron-encoded reverse transcriptase/maturase to its high affinity intron RNA binding site involves sequence-specific recognition and autoregulates translation. *J Mol Biol* **318:**287–303.
39. **Watanabe K, Lambowitz AM.** 2004. High-affinity binding site for a group II intron-encoded reverse transcriptase/maturase within a stem-loop structure in the intron RNA. *RNA* **10:**1433–1443.
40. **Dai L, Chai D, Gu SQ, Gabel J, Noskov SY, Blocker FJ, Lambowitz AM, Zimmerly S.** 2008. A three-dimensional model of a group II intron RNA and its interaction with the intron-encoded reverse transcriptase. *Mol Cell* **30:** 472–485.
41. **Noah JW, Lambowitz AM.** 2003. Effects of maturase binding and Mg^{2+} concentration on group II intron RNA folding investigated by UV cross-linking. *Biochemistry* **42:**12466–12480.
42. **Noah JW, Park S, Whitt JT, Perutka J, Frey W, Lambowitz AM.** 2006. Atomic force microscopy reveals DNA bending during group II intron ribonucleoprotein particle integration into double-stranded DNA. *Biochemistry* **45:**12424–12435.
43. **Huang T, Shaikh TR, Gupta K, Contreras-Martin LM, Grassucci RA, Van Duyne GD, Frank J, Belfort M.** 2011. The group II intron ribonucleoprotein precursor is a large, loosely packed structure. *Nucl Acids Res* **39:** 2845–2854.
44. **Mohr S, Matsuura M, Perlman PS, Lambowitz AM.** 2006. A DEAD-box protein alone promotes group II intron splicing and reverse splicing by acting as an RNA chaperone. *Proc Natl Acad Sci U S A* **103:**3569–3574.
45. **Contreras LM, Huang T, Piazza CL, Smith D, Qu G, Gelderman G, Potratz J, Russell R, Belfort M.** 2013. Group II intron-ribosome association protects intron RNA from degradation. *RNA* **19:**1497–1509.

46. **Gu SQ, Cui X, Mou S, Mohr S, Yao J, Lambowitz AM.** 2010. Genetic identification of potential RNA-binding regions in a group II intron-encoded reverse transcriptase. *RNA* **16:**732–747.

47. **Fontaine JM, Goux D, Kloareg B, Loiseaux-de Goër S.** 1997. The reverse-transcriptase-like proteins encoded by group II introns in the mitochondrial genome of the brown alga *Pylaiella littoralis* belong to two different lineages which apparently coevolved with the group II ribosyme lineages. *J Mol Evol* **44:**33–42.

48. **Toor N, Hausner G, Zimmerly S.** 2001. Coevolution of group II intron RNA structures with their intron-encoded reverse transcriptase. *RNA* **7:**1142–1152.

49. **Zimmerly S, Hausner G, Wu X.** 2001. Phylogenetic relationships among group II intron ORFs. *Nucl Acids Res* **29:**1238–1250.

50. **Simon DM, Clarke NA, McNeil BA, Johnson I, Pantuso D, Dai L, Chai D, Zimmerly S.** 2008. Group II introns in eubacteria and archaea: ORF-less introns and new varieties. *RNA* **14:**1704–1713.

51. **Simon DM, Kelchner SA, Zimmerly S.** 2009. A broad-scale phylogenetic analysis of group II intron RNAs and intron-encoded reverse transcriptases. *Mol Biol Evol* **26:** 2795–2808.

52. **Nagy V, Pirakitikulr N, Zhou KI, Chillón I, Luo J, Pyle AM.** 2013. Predicted group II intron lineages E and F comprise catalytically active ribozymes. *RNA* **19:** 1266–1278.

53. **Zimmerly S, Guo H, Eskes R, Yang J, Perlman PS, Lambowitz AM.** 1995. A group II intron RNA is a catalytic component of a DNA endonuclease involved in intron mobility. *Cell* **83:**529–538.

54. **Zimmerly S, Guo H, Perlman PS, Lambowitz AM.** 1995. Group II intron mobility occurs by target DNA-primed reverse transcription. *Cell* **82:**545–554.

55. **Yang J, Zimmerly S, Perlman PS, Lambowitz AM.** 1996. Efficient integration of an intron RNA into double-stranded DNA by reverse splicing. *Nature* **381:** 332–335.

56. **Eskes R, Yang J, Lambowitz AM, Perlman PS.** 1997. Mobility of yeast mitochondrial group II introns: engineering a new site specificity and retrohoming via full reverse splicing. *Cell* **88:**865–874.

57. **Cousineau B, Smith D, Lawrence-Cavanagh S, Mueller JE, Yang J, Mills D, Manias D, Dunny G, Lambowitz AM, Belfort M.** 1998. Retrohoming of a bacterial group II intron: mobility via complete reverse splicing, independent of homologous DNA recombination. *Cell* **94:**451–462.

58. **Ichiyanagi K, Beauregard A, Lawrence S, Smith D, Cousineau B, Belfort M.** 2002. Retrotransposition of the Ll.LtrB group II intron proceeds predominantly via reverse splicing into DNA targets. *Mol Microbiol* **46:** 1259–1272.

59. **Coros CJ, Landthaler M, Piazza CL, Beauregard A, Esposito D, Perutka J, Lambowitz AM, Belfort M.** 2005. Retrotransposition strategies of the *Lactococcus lactis* Ll.LtrB group II intron are dictated by host identity and cellular environment. *Mol Microbiol* **56:** 509–524.

60. **Martínez-Abarca F, García-Rodríguez FM, Toro N.** 2000. Homing of a bacterial group II intron with an intron-encoded protein lacking a recognizable endonuclease domain. *Mol Microbiol* **35:**1405–1412.

61. **Muñoz-Adelantado E, San Filippo J, Martínez-Abarca F, García-Rodriguez FM, Lambowitz AM, Toro N.** 2003. Mobility of the *Sinorhizobium meliloti* group II intron RmInt1 occurs by reverse splicing into DNA, but requires an unknown reverse transcriptase priming mechanism. *J Mol Biol* **327:**931–943.

62. **Martínez-Abarca F, Barrientos-Durán A, Fernández-López M, Toro N.** 2004. The RmInt1 group II intron has two different retrohoming pathways for mobility using predominantly the nascent lagging strand at DNA replication forks for priming. *Nucl Acids Res* **32:**2880–2888.

63. **Candales MA, Duong A, Hood KS, Li T, Neufeld RA, Sun R, McNeil BA, Wu L, Jarding AM, Zimmerly S.** 2012. Database for bacterial group II introns. *Nucl Acids Res* **40:**D187–D190.

64. **Zimmerly S, Moran JV, Perlman PS, Lambowitz AM.** 1999. Group II intron reverse transcriptase in yeast mitochondria. Stabilization and regulation of reverse transcriptase activity by the intron RNA. *J Mol Biol* **289:**473–490.

65. **Huang HR, Chao MY, Armstrong B, Wang Y, Lambowitz AM, Perlman PS.** 2003. The DIVa maturase binding site in the yeast group II intron aI2 is essential for intron homing but not for *in vivo* splicing. *Mol Cell Biol* **23:**8809–8819.

66. **Hu WS, Hughes SH.** 2012. HIV-1 reverse transcription. *Cold Spring Harb Perspect Med* **2:**a006882.

67. **Conlan LH, Stanger MJ, Ichiyanagi K, Belfort M.** 2005. Localization, mobility and fidelity of retrotransposed group II introns in rRNA genes. *Nucl Acids Res* **33:** 5262–5270.

68. **Mohr S, Ghanem E, Smith W, Sheeter D, Qin Y, King O, Polioudakis D, Iyer VR, Hunicke-Smith S, Swamy S, Kuersten S, Lambowitz AM.** 2013. Thermostable group II intron reverse transcriptase fusion proteins and their use in cDNA synthesis and next-generation RNA sequencing. *RNA* **19:**958–970.

69. **Eskes R, Liu L, Ma H, Chao MY, Dickson L, Lambowitz AM, Perlman PS.** 2000. Multiple homing pathways used by yeast mitochondrial group II introns. *Mol Cell Biol* **20:**8432–8446.

70. **Lazowska J, Meunier B, Macadre C.** 1994. Homing of a group II intron in yeast mitochondrial DNA is accompanied by unidirectional co-conversion of upstream-located markers. *EMBO J* **13:**4963–4972.

71. **Mills DA, Manias DA, McKay LL, Dunny GM.** 1997. Homing of a group II intron from *Lactococcus lactis* subsp. *lactis* ML3. *J Bacteriol* **179:**6107–6111.

72. **Smith D, Zhong J, Matsuura M, Lambowitz AM, Belfort M.** 2005. Recruitment of host functions suggests a repair pathway for late steps in group II intron retrohoming. *Genes Dev* **19:**2477–2487.

73. **Yao J, Truong DM, Lambowitz AM.** 2013. Genetic and biochemical assays reveal a key role for replication restart proteins in group II intron retrohoming. *PLoS Genet* **9:**e1003469.

74. **Beauregard A, Chalamcharla VR, Piazza CL, Belfort M, Coros CJ.** 2006. Bipolar localization of the group II intron Ll.LtrB is maintained in *Escherichia coli* deficient in nucleoid condensation, chromosome partitioning and DNA replication. *Mol Microbiol* **62:**709–722.

75. **Beauregard A, Curcio MJ, Belfort M.** 2008. The take and give between retrotransposable elements and their hosts. *Annu Rev Genet* **42:**587–617.

76. **Coros CJ, Piazza CL, Chalamcharla VR, Belfort M.** 2008. A mutant screen reveals RNase E as a silencer of group II intron retromobility in *Escherichia coli. RNA* **14:**2634–2644.

77. **Jiménez-Zurdo JI, García-Rodríguez FM, Barrientos-Durán A, Toro N.** 2003. DNA target site requirements for homing *in vivo* of a bacterial group II intron encoding a protein lacking the DNA endonuclease domain. *J Mol Biol* **326:**413–423.

78. **Zhong J, Lambowitz AM.** 2003. Group II intron mobility using nascent strands at DNA replication forks to prime reverse transcription. *EMBO J* **22:**4555–4565.

79. **Zhuang F, Mastroianni M, White TB, Lambowitz AM.** 2009. Linear group II intron RNAs can retrohome in eukaryotes and may use nonhomologous end-joining for cDNA ligation. *Proc Natl Acad Sci U S A* **106:** 18189–18194.

80. **White TB, Lambowitz AM.** 2012. The retrohoming of linear group II intron RNAs in *Drosophila melanogaster* occurs by both DNA ligase 4-dependent and -independent mechanisms. *PLoS Genet* **8:**e1002534.

81. **Toor N, Zimmerly S.** 2002. Identification of a family of group II introns encoding LAGLIDADG ORFs typical of group I introns. *RNA* **8:**1373–1377.

82. **Mullineux ST, Costa M, Bassi GS, Michel F, Hausner G.** 2010. A group II intron encodes a functional LAGLIDADG homing endonuclease and self-splices under moderate temperature and ionic conditions. *RNA* **16:**1818–1831.

83. **Mullineux ST, Willows K, Hausner G.** 2011. Evolutionary dynamics of the mS952 intron: a novel mitochondrial group II intron encoding a LAGLIDADG homing endonuclease gene. *J Mol Evol* **72:**433–449.

84. **Salman V, Amann R, Shub DA, Schulz-Vogt HN.** 2012. Multiple self-splicing introns in the 16S rRNA genes of giant sulfur bacteria. *Proc Natl Acad Sci U S A* **109:** 4203–4208.

85. **Lambowitz AM, Belfort M.** 1993. Introns as mobile genetic elements. *Annu Rev Biochem* **62:**587–622.

86. **Mueller MW, Allmaier M, Eskes R, Schweyen RJ.** 1993. Transposition of group II intron *aI1* in yeast and invasion of mitochondrial genes at new locations. *Nature* **366:**174–176.

87. **Sellem CH, Lecellier G, Belcour L.** 1993. Transposition of a group II intron. *Nature* **366:**176–178.

88. **Martínez-Abarca F, Toro N.** 2000. RecA-independent ectopic transposition *in vivo* of a bacterial group II intron. *Nucl Acids Res* **28:**4397–4402.

89. **Dickson L, Huang HR, Liu L, Matsuura M, Lambowitz AM, Perlman PS.** 2001. Retrotransposition of a yeast group II intron occurs by reverse splicing directly into ectopic DNA sites. *Proc Natl Acad Sci U S A* **98:** 13207–13212.

90. **Curcio MJ, Garfinkel DJ.** 1991. Single-step selection for Ty1 element retrotransposition. *Proc Natl Acad Sci U S A* **88:**936–940.

91. **Heidmann T, Heidmann O, Nicolas JF.** 1988. An indicator gene to demonstrate intracellular transposition of defective retroviruses. *Proc Natl Acad Sci U S A* **85:** 2219–2223.

92. **Ichiyanagi K, Beauregard A, Belfort M.** 2003. A bacterial group II intron favors retrotransposition into plasmid targets. *Proc Natl Acad Sci U S A* **100:** 15742–15747.

93. **Novikova O, Smith D, Hahn I, Beauregard A, Belfort M.** 2014. Interaction between conjugative and retrotransposable elements in horizontal gene transfer. *PLoS Genet* **10:**e1004853.

94. **Zhong J, Karberg M, Lambowitz AM.** 2003. Targeted and random bacterial gene disruption using a group II intron (targetron) vector containing a retrotransposition-activated selectable marker. *Nucl Acids Res* **31:**1656–1664.

95. **Zhao J, Lambowitz AM.** 2005. A bacterial group II intron-encoded reverse transcriptase localizes to cellular poles. *Proc Natl Acad Sci U S A* **102:**16133–16140.

96. **Zhao J, Niu W, Yao J, Mohr S, Marcotte EM, Lambowitz AM.** 2008. Group II intron protein localization and insertion sites are affected by polyphosphate. *PLoS Biology* **6:**e150.

97. **Coros CJ, Piazza CL, Chalamcharla VR, Smith D, Belfort M.** 2009. Global regulators orchestrate group II intron retromobility. *Mol Cell* **34:**250–256.

98. **Belhocine K, Plante I, Cousineau B.** 2004. Conjugation mediates transfer of the Ll.LtrB group II intron between different bacterial species. *Mol Microbiol* **51:** 1459–1469.

99. **Guo H, Zimmerly S, Perlman PS, Lambowitz AM.** 1997. Group II intron endonucleases use both RNA and protein subunits for recognition of specific sequences in double-stranded DNA. *EMBO J* **16:**6835–6848.

100. **Guo H, Karberg M, Long M, Jones JP III, Sullenger B, Lambowitz AM.** 2000. Group II introns designed to insert into therapeutically relevant DNA target sites in human cells. *Science* **289:**452–457.

101. **Mohr G, Smith D, Belfort M, Lambowitz AM.** 2000. Rules for DNA target-site recognition by a lactococcal group II intron enable retargeting of the intron to specific DNA sequences. *Genes Dev* **14:**559–573.

102. **Yang J, Mohr G, Perlman PS, Lambowitz AM.** 1998. Group II intron mobility in yeast mitochondria: target DNA-primed reverse transcription activity in aI1 and reverse splicing into DNA transposition sites *in vitro. J Mol Biol* **282:**505–523.

103. **Perutka J, Wang W, Goerlitz D, Lambowitz AM.** 2004. Use of computer-designed group II introns to disrupt *Escherichia coli* DExH/D-box protein and DNA helicase genes. *J Mol Biol* **336:**421–439.

104. **Zhuang F, Karberg M, Perutka J, Lambowitz AM.** 2009. EcI5, a group IIB intron with high retrohoming

frequency: DNA target site recognition and use in gene targeting. *RNA* **15**:432–449.

105. **Singh NN, Lambowitz AM.** 2001. Interaction of a group II intron ribonucleoprotein endonuclease with its DNA target site investigated by DNA footprinting and modification interference. *J Mol Biol* **309**:361–386.

106. **Barrientos-Durán A, Chillón I, Martínez-Abarca F, Toro N.** 2011. Exon sequence requirements for excision *in vivo* of the bacterial group II intron RmInt1. *BMC Mol Biol* **12**:12–24.

107. **Robart AR, Seo W, Zimmerly S.** 2007. Insertion of group II intron retroelements after intrinsic transcriptional terminators. *Proc Natl Acad Sci U S A* **104**: 6620–6625.

108. **Aizawa Y, Ziang Q, Lambowitz AM, Pyle AM.** 2003. The pathway for DNA recognition and RNA integration by a group II intron retrotransposon. *Mol Cell* **11**: 795–805.

109. **Leclercq S, Cordaux R.** 2012. Selection-driven extinction dynamics for group II introns in *Enterobacteriales*. *PLoS One* **7**:e52268.

110. **Dai L, Zimmerly S.** 2002. Compilation and analysis of group II intron insertions in bacterial genomes: evidence for retroelement behavior. *Nucl Acids Res* **30**: 1091–1102.

111. **Fernández-López M, Muñoz-Adelantado E, Gillis M, Willems A, Toro N.** 2005. Dispersal and evolution of the *Sinorhizobium meliloti* group II RmInt1 intron in bacteria that interact with plants. *Mol Biol Evol* **22**: 1518–1528.

112. **Tourasse NJ, Kolstø AB.** 2008. Survey of group I and group IIX introns in 29 sequenced genomes of the *Bacillus cereus* group: insights into their spread and evolution. *Nucl Acids Res* **36**:4529–4548.

113. **Goddard MR, Burt A.** 1999. Recurrent invasion and extinction of a selfish gene. *Proc Natl Acad Sci U S A* **96**:13880–13885.

114. **Mohr G, Ghanem E, Lambowitz AM.** 2010. Mechanisms used for genomic proliferation by thermophilic group II introns. *PLoS Biol* **8**:e1000391.

115. **Leclercq S, Giraud I, Cordaux R.** 2011. Remarkable abundance and evolution of mobile group II introns in *Wolbachia* bacterial endosymbionts. *Mol Biol Evol* **28**: 685–697.

116. **Ueda K, Yamashita A, Ishikawa J, Shimada M, Watsuji TO, Morimura K, Ikeda H, Hattori M, Beppu T.** 2004. Genome sequence of *Symbiobacterium thermophilum*, an uncultivable bacterium that depends on microbial commensalism. *Nucl Acids Res* **32**:4937–4944.

117. **Copertino DW, Hallick RB.** 1993. Group II and group III introns of twintrons: potential relationships to nuclear pre-mRNA introns. *Trends Biochem Sci* **18**: 467–471.

118. **Hallick RB, Hong L, Drager RG, Favreau MR, Monfort A, Orsat B, Spielmann A, Stutz E.** 1993. Complete sequence of *Euglena gracilis* chloroplast DNA. *Nucl Acids Res* **21**:3537–3544.

119. **Christopher DA, Hallick RB.** 1989. *Euglena gracilis* chloroplast ribosomal protein operon: a new chloroplast gene for ribosomal protein L5 and description of a novel organelle intron category designated group III. *Nucl Acids Res* **17**:7591–7608.

120. **Pombert JF, James ER, Janoukovec J, Keeling PJ.** 2012. Evidence for transitional stages in the evolution of euglenid group II introns and twintrons in the *Monomorphina aenigmatica* plastid genome. *PLoS One* **7**:e53433.

121. **Wiegert KE, Bennett MS, Triemer RE.** 2013. Tracing patterns of chloroplast evolution in euglenoids: contributions from *Colacium vesiculosum* and *Strombomonas acuminata* (Euglenophyta). *J Euk Micro* **60**:214–221.

122. **Sheveleva EV, Hallick RB.** 2004. Recent horizontal intron transfer to a chloroplast genome. *Nucl Acids Res* **32**:803–810.

123. **Nakamura Y, Kaneko T, Sato S, Ikeuchi M, Katoh H, Sasamoto S, Watanabe A, Iriguchi M, Kawashima K, Kimura T, Kishida Y, Kiyokawa C, Kohara M, Matsumoto M, Matsuno A, Nakazaki N, Shimpo S, Sugimoto M, Takeuchi C, Yamada M, Tabata S.** 2002. Complete genome structure of the thermophilic cyanobacterium *Thermosynechococcus elongatus* BP-1. *DNA Res* **9**:123–130.

124. **Cavalier-Smith T.** 1991. Intron phylogeny: a new hypothesis. *Trends Genet* **7**:145–148.

125. **Palmer JD, Logsdon JM Jr.** 1991. The recent origins of introns. *Curr Opin Genet Devel* **1**:470–477.

126. **Knoop V, Brennicke A.** 1994. Promiscuous mitochondrial group II intron sequences in plant nuclear genomes. *J Mol Evol* **39**:144–150.

127. **Lin X, Kaul S, Rounsley S, Shea TP, Benito MI, Town CD, Fujii CY, Mason T, Bowman CL, Barnstead M, Feldblyum TV, Buell CR, Ketchum KA, Lee J, Ronning CM, Koo HL, Moffat KS, Cronin LA, Shen M, Pai G, Van Aken S, Umayam L, Tallon LJ, Gill JE, Adams MD, Carrera AJ, Creasy TH, Goodman HM, Somerville CR, Copenhaver GP, Preuss D, Nierman WC, White O, Eisen JA, Salzberg SL, Fraser CM, Venter JC.** 1999. Sequence and analysis of chromosome 2 of the plant *Arabidopsis thaliana*. *Nature* **402**:761–768.

128. **Gregan J, Kolisek M, Schweyen RJ.** 2001. Mitochondrial Mg^{2+} homeostasis is critical for group II intron splicing *in vivo*. *Genes Dev* **15**:2229–2237.

129. **Mastroianni M, Watanabe K, White TB, Zhuang F, Vernon J, Matsuura M, Wallingford J, Lambowitz AM.** 2008. Group II intron-based gene targeting reactions in eukaryotes. *PLoS One* **3**:e3121.

130. **Truong DM, Sidote DJ, Russell R, Lambowitz AM.** 2013. Enhanced group II intron retrohoming in magnesium-deficient *Escherichia coli* via selection of mutations in the ribozyme core. *Proc Natl Acad Sci U S A* **110**:E3800–E3809.

131. **Günther T.** 2006. Concentration, compartmentation and metabolic function of intracellular free Mg^{2+}. *Magnes Res* **19**:225–236.

132. **Chalamcharla VR, Curcio MJ, Belfort M.** 2010. Nuclear expression of a group II intron is consistent with spliceosomal intron ancestry. *Genes Dev* **24**:827–836.

133. **Qu G, Dong X, Piazza CL, Chalamcharla V, Lutz S, Curcio MJ, Belfort M.** 2014. RNA-RNA interactions and pre-RNA mislocalization as the drivers of group II intron loss from nuclear genomes. *Proc Natl Acad Sci U S A* **111:**6612–6617.

134. **Zerbato M, Holic N, Moniot-Frin S, Ingrao D, Galy A, Perea J.** 2013. The brown algae Pl.LSU/2 group II intron-encoded protein has functional reverse transcriptase and maturase activities. *PLoS One* **8:**e58263.

135. **Eickbush T, Malik H.** 2002. Origins and evolution of retrotransposons, p 1111–1146. *In* Craig N, Craigie R, Gellert M, Lambowitz A (ed), *Mobile DNA II*. ASM Press, Washington DC.

136. **Eickbush TH, Eickbush DG.** 2014. Integration, regulation, and long-term stability of R2 retrotransposons. *Mobile DNA III*. ASM Press, Washington DC.

137. **Fujiwara H.** 2014. Site-specific non-LTR elements. *Mobile DNA III*. ASM Press, Washington DC.

138. **Richardson SR, Doucet AJ, Kopera HC, Moldovan JB, Garcia-Perez JL, Moran JV.** 2014. The influence of LINE-1 and SINE retrotransposons on mammalian genomes. *Mobile DNA III*. ASM Press, Washington DC.

139. **Mitchell M, Gillis A, Futahashi M, Fujiwara H, Skordalakes E.** 2010. Structural basis for telomerase catalytic subunit TERT binding to RNA template and telomeric DNA. *Nat Struct Mol Biol* **17:**513–518.

140. **Jamburuthugoda VK, Eickbush TH.** 2014. Identification of RNA binding motifs in the R2 retrotransposon-encoded reverse transcriptase. *Nucl Acids Res* **42:**8405–8415.

141. **Kojima KK, Fujiwara H.** 2005. An extraordinary retrotransposon family encoding dual endonucleases. *Genome Res* **15:**1106–1117.

142. **Gladyshev EA, Arkhipova IR.** 2007. Telomere-associated endonuclease-deficient *Penelope*-like retroelements in diverse eukaryotes. *Proc Natl Acad Sci U S A* **104:** 9352–9357.

143. **Arkhipova IR, Yushenova IA, Rodriguez F.** 2013. Endonuclease-containing *Penelope* retrotransposons in the bdelloid rotifer *Adineta vaga* exhibit unusual structural features and play a role in expansion of host gene families. *Mobile DNA* **4:**19.

144. **Morrish TA, Gilbert N, Myers JS, Vincent BJ, Stamato TD, Taccioli GE, Batzer MA, Moran JV.** 2002. DNA repair mediated by endonuclease-independent LINE-1 retrotransposision. *Nat Genet* **31:**159–165.

145. **Onozawa M, Zhang Z, Kim YJ, Goldberg L, Varga T, Bergsagel PL, Kuehl WM, Aplan PD.** 2014. Repair of DNA double-strand breaks by templated nucleotide sequence insertions derived from distant regions of the genome. *Proc Natl Acad Sci U S A* **111:**7729–7734.

146. **Morrish TA, Garcia-Perez JL, Stamato TD, Taccioli GE, Sekiguchi J, Moran JV.** 2007. Endonuclease-independent LINE-1 retrotransposition at mammalian telomeres. *Nature* **446:**208–212.

147. **Curcio MJ, Belfort M.** 2007. The beginning of the end: Links between ancient retroelements and modern telomerases. *Proc Nat Acad Sci U S A* **104:**9107–9108.

148. **Esnault C, Maestre J, Heidmann T.** 2000. Human LINE retrotransposons generate processed pseudogenes. *Nat Genet* **24:**363–367.

149. **Wei W, Gilbert N, Ooi SL, Lawler JF, Ostertag EM, Kazazian HH, Boeke JD, Moran JV.** 2001. Human L1 retrotransposition: *cis* preference versus *trans* complementation. *Mol Cell Biol* **21:**1429–1439.

150. **Kulpa DA, Moran JV.** 2006. *Cis*-preferential LINE-1 reverse transcriptase activity in ribonucleoprotein particles. *Nat Struct Mol Biol* **13:**655–660.

151. **Richardson SR, Narvaiza I, Planegger RA, Weitzman MD, Moran JV.** 2014. APOBEC3A deaminates transiently exposed single-strand DNA during LINE-1 retrotransposition. *Elife* **3:**e02008.

152. **Suzuki J, Yamaguchi K, Kajikawa M, Ichiyanagi K, Adachi N, Koyama H, Takeda S, Okada N.** 2009. Genetic evidence that the non-homologous end-joining repair pathway is involved in LINE retrotransposition. *PLoS Genet* **5:**e1000461.

153. **Bibillo A, Eickbush TH.** 2002. The reverse transcriptase of the R2 non-LTR retrotransposon: continuous synthesis of cDNA on non-continuous RNA templates. *J Mol Biol* **316:**459–473.

154. **Bibillo A, Eickbush TH.** 2004. End-to-end template jumping by the reverse transcriptase encoded by the R2 retrotransposon. *J Biol Chem* **279:**14945–14953.

155. **Sharp PA.** 1985. On the origin of RNA splicing and introns. *Cell* **42:**397–400.

156. **Cech TR.** 1986. The generality of self-splicing RNA: relationship to nuclear mRNA splicing. *Cell* **44:**207–210.

157. **Sharp PA.** 1991. Five easy pieces. *Science* **254:**663.

158. **Valadkhan S.** 2010. Role of the snRNAs in spliceosomal active site. *RNA Biol* **7:**345–353.

159. **Will CL, Lührmann R.** 2011. Spliceosome structure and function. *Cold Spring Harb Perspect Biol* **3:**a003707.

160. **Valadkhan S.** 2013. The role of snRNAs in spliceosomal catalysis. *Prog Mol Biol Transl Sci* **120:**195–228.

161. **Madhani HD, Guthrie C.** 1992. A novel base-pairing interaction between U2 and U6 snRNAs suggests a mechanism for the catalytic activation of the spliceosome. *Cell* **71:**803–817.

162. **Shukla GC, Padgett RA.** 2002. A catalytically active group II intron domain 5 can function in the U12-dependent spliceosome. *Mol Cell* **9:**1145–1150.

163. **Keating KS, Toor N, Perlman PS, Pyle AM.** 2010. A structural analysis of the group II intron active site and implications for the spliceosome. *RNA* **16:**1–9.

164. **Jarrell KA, Dietrich RC, Perlman PS.** 1988. Group II intron domain 5 facilitates a *trans*-splicing eeaction. *Mol Cell Biol* **8:**2361–2366.

165. **Goldschmidt-Clermont M, Choquet Y, Girard-Bascou J, Michel F, Schirmer-Rahire M, Rochaix J-D.** 1991. A small chloroplast RNA may be required for *trans*-splicing in Chlamydomonas reinhardtii. *Cell* **65:**135–143.

166. **Suchy M, Schmelzer C.** 1991. Restoration of the self-splicing activity of a defective group II intron by a small *trans*-acting RNA. *J Mol Biol* **222:**179–187.

167. **Hetzer M, Wurzer G, Schweyen RJ, Mueller MW.** 1997. *Trans*-activation of group II intron splicing by nuclear U5 snRNA. *Nature* **386:**417–420.

168. **Glanz S, Kück U.** 2009. *Trans*-splicing of organelle introns–a detour to continuous RNAs. *Bioessays* **31:**921–934.

169. **Belhocine K, Mak AB, Cousineau B.** 2008. *Trans*-splicing versatility of the Ll.LtrB group II intron. *RNA* **14:**1782–1790.
170. **Quiroga C, Kronstad L, Ritlop C, Filion A, Cousineau B.** 2011. Contribution of base-pairing interactions between group II intron fragments during *trans*-splicing *in vivo*. *RNA* **17:**2212–2221.
171. **Ritlop C, Monat C, Cousineau B.** 2012. Isolation and characterization of functional tripartite group II introns using a Tn5-based genetic screen. *PLoS One* **7:**e41589.
172. **McNeil BA, Simon DM, Zimmerly S.** 2014. Alternative splicing of a group II intron in a surface layer protein gene in *Clostridium tetani*. *Nucl Acids Res* **42:**1959–1969.
173. **Chu VT, Liu Q, Podar M, Perlman PS, Pyle AM.** 1998. More than one way to splice an RNA: branching without a bulge and splicing without branching in group II introns. *RNA* **4:**1186–1202.
174. **Tseng CK, Cheng SC.** 2008. Both catalytic steps of nuclear pre-mRNA splicing are reversible. *Science* **320:**1782–1784.
175. **Gordon PM, Sontheimer EJ, Piccirilli JA.** 2000. Metal ion catalysis during the exon-ligation step of nuclear pre-mRNA splicing: extending the parallels between the spliceosome and group II introns. *RNA* **6:**199–205.
176. **Collins CA, Guthrie C.** 2000. The question remains: is the spliceosome a ribozyme? *Nat Struct Biol* **7:**850–854.
177. **Sontheimer EJ, Sun S, Piccirilli JA.** 1997. Metal ion catalysis during splicing of premessenger RNA. *Nature* **388:**801–805.
178. **Fica SM, Tuttle N, Novak T, Li NS, Lu J, Koodathingal P, Dai Q, Staley JP, Piccirilli JA.** 2013. RNA catalyses nuclear pre-mRNA splicing. *Nature* **503:**229–234.
179. **Gordon PM, Piccirilli JA.** 2001. Metal ion coordination by the AGC triad in domain 5 contributes to group II intron catalysis. *Nat Struct Biol* **8:**893–898.
180. **Fica SM, Mefford MA, Piccirilli JA, Staley JP.** 2014. Evidence for a group II intron-like catalytic triplex in the spliceosome. *Nat Struct Mol Biol* **21:**464–471.
181. **Dlakic M, Mushegian A.** 2011. Prp8, the pivotal protein of the spliceosomal catalytic center, evolved from a retroelement-encoded reverse transcriptase. *RNA* **17:**799–808.
182. **Galej WP, Oubridge C, Newman AJ, Nagai K.** 2013. Crystal structure of Prp8 reveals active site cavity of the spliceosome. *Nature* **493:**638–643.
183. **Galej WP, Nguyen TH, Newman AJ, Nagai K.** 2014. Structural studies of the spliceosome: zooming into the heart of the machine. *Curr Opin Struct Biol* **25C:**57–66.
184. **Anokhina M, Bessonov S, Miao Z, Westhof E, Hartmuth K, Lührmann R.** 2013. RNA structure analysis of human spliceosomes reveals a compact 3D arrangement of snRNAs at the catalytic core. *EMBO J* **32:**2804–2818.
185. **Reyes JL, Gustafson EH, Luo HR, Moore MJ, Konarska MM.** 1999. The C-terminal region of hPrp8 interacts with the conserved GU dinucleotide at the 5′ splice site. *RNA* **5:**167–179.
186. **Turner IA, Norman CM, Churcher MJ, Newman AJ.** 2006. Dissection of Prp8 protein defines multiple interactions with crucial RNA sequences in the catalytic core of the spliceosome. *RNA* **12:**375–386.
187. **Mohr G, Lambowitz AM.** 2003. Putative proteins related to group II intron reverse transcriptase/maturases are encoded by nuclear genes in higher plants. *Nucl Acids Res* **31:**647–652.
188. **Nakagawa N, Sakurai N.** 2006. A mutation in At-nMat1a, which encodes a nuclear gene having high similarity to group II intron maturase, causes impaired splicing of mitochondrial NAD4 transcript and altered carbon metabolism in *Arabidopsis thaliana*. *Plant Cell Physiol* **47:**772–783.
189. **Keren I, Bezawork-Geleta A, Kolton M, Maayan I, Belausov E, Levy M, Mett A, Gidoni D, Shaya F, Ostersetzer-Biran O.** 2009. AtnMat2, a nuclear-encoded maturase required for splicing of group-II introns in *Arabidopsis* mitochondria. *RNA* **15:**2299–2311.
190. **Brown GG, Colas des Francs-Small C, Ostersetzer-Biran O.** 2014. Group II intron splicing factors in plant mitochondria. *Front Plant Sci* **5:**35.
191. **Smyshlyaev G, Voigt F, Blinov A, Barabas O, Novikova O.** 2013. Acquisition of an Archaea-like ribonuclease H domain by plant L1 retrotransposons supports modular evolution. *Proc Nat Acad Sci U S A* **110:**20140–20145.
192. **Query CC, Konarska MM.** 2013. Spliceosome's core exposed. *Nature* **493:**615–616.
193. **Doolittle WF.** 2013. The spliceosomal catalytic core arose in the RNA world...or did it? *Genome Biol* **14:**141.
194. **Curcio MJ, Belfort M.** 1996. Retrohoming: cDNA-mediated mobility of group II introns requires a catalytic. *RNA Cell* **84:**9–12.
195. **Kojima KK, Kanehisa M.** 2008. Systematic survey for novel types of prokaryotic retroelements based on gene neighborhood and protein architecture. *Mol Biol Evol* **25:**1395–1404.
196. **Simon DM, Zimmerly S.** 2008. A diversity of uncharacterized reverse transcriptases in bacteria. *Nucl Acids Res* **36:**7219–7229.
197. **Lampson BC, Inouye M, Inouye S.** 2005. Retrons, msDNA, and the bacterial genome. *Cytogenet Genome Res* **110:**491–499.
198. **Doulatov S, Hodes A, Dai L, Mandhana N, Liu M, Deora R, Simons RW, Zimmerly S, Miller JF.** 2004. Tropism switching in *Bordetella* bacteriophage defines a family of diversity-generating retroelements. *Nature* **431:**476–481.
199. **Guo H, Arambula D, Ghosh P, Miller JF.** 2014. Diversity-generating retroelements in phage and bacterial genomes. *Mobile DNA III*. ASM Press, Washington DC.
200. **Fortier LC, Bouchard JD, Moineau S.** 2005. Expression and site-directed mutagenesis of the lactococcal abortive phage infection protein AbiK. *J Bacteriol* **187:**3721–3730.
201. **Medhekar B, Miller JF.** 2007. Diversity-generating retroelements. *Curr Opin Microbiol* **10:**388–395.
202. **Minot S, Grunberg S, Wu GD, Lewis JD, Bushman FD.** 2012. Hypervariable loci in the human gut virome. *Proc Natl Acad Sci U S A* **109:**3962–3966.

203. **Martin W, Koonin EV.** 2006. Introns and the origin of nucleus-cytosol compartmentalization. *Nature* **440:**41–45.
204. **Williams TA, Foster PG, Cox CJ, Embley TM.** 2013. An archaeal origin of eukaryotes supports only two primary domains of life. *Nature* **504:**231–236.
205. **Irimia M, Roy SW.** 2014. Origin of spliceosomal introns and alternative splicing. *Cold Spring Harb Perspect Biol* **6:**a01607.
206. **Poole AM.** 2006. Did group II intron proliferation in an endosymbiont-bearing archaeon create eukaryotes? *Biol Direct* **1:**36.
207. **Koonin EV.** 2009. Intron-dominated genomes of early ancestors of eukaryotes. *J Hered* **100:**618–623.
208. **Doolittle WF.** 2014. The trouble with (group II) introns. *Proc Natl Acad Sci U S A* **111:**6536–6537.
209. **Robart AR, Zimmerly S.** 2005. Group II intron retroelements: function and diversity. *Cytogenet Genome Res* **110:**589–597.
210. **Collins L, Penny D.** 2005. Complex spliceosomal organization ancestral to extant eukaryotes. *Mol Biol Evol* **22:**1053–1066.
211. **Martin W, Koonin EV.** 2006. A positive definition of prokaryotes. *Nature* **442:**868.
212. **Gladyshev EA, Arkhipova IR.** 2011. A widespread class of reverse transcriptase-related cellular genes. *Proc Natl Acad Sci U S A* **108:**20311–20316.
213. **Zimmerly S, Wu L.** 2014. An unexplored diversity of reverse transcriptases in bacteria. *Mobile DNA III*. ASM Press, Washington DC.
214. **Costa M, Michel F, Westhof E.** 2000. A three-dimensional perspective on exon binding by a group II self-splicing intron. *EMBO J* **19:**5007–5018.
215. **Toor N, Robart AR, Christianson J, Zimmerly S.** 2006. Self-splicing of a group IIC intron: 5′ exon recognition and alternative 5′ splicing events implicate the stem-loop motif of a transcriptional terminator. *Nucl Acids Res* **34:**6461–6471.
216. **Michel F, Costa M, Doucet AJ, Ferat JL.** 2007. Specialized lineages of bacterial group II introns. *Biochimie* **89:** 542–553.
217. **Piskareva O, Ernst C, Higgins N, Schmatchenko V.** 2013. The carboxy-terminal segment of the human LINE-1 ORF2 protein is involved in RNA binding. *FEBS Open Bio* **3:**433–437.
218. **Rouda S, Skordalakes E.** 2007. Structure of the RNA-binding domain of telomerase: implications for RNA recognition and binding. *Structure* **15:**1403–1412.
219. **Marcia M, Pyle AM.** 2014. Principles of ion recognition in RNA: insights from the group II intron structures. *RNA* **20:**516–527.
220. **Strobel SA.** 2013. Biochemistry: metal ghosts in the splicing machine. *Nature* **503:**201–202.

Mobile DNA, 3rd Edition
Nancy L. Craig, Michael Chandler, Martin Gellert, Alan M. Lambowitz, Phoebe A. Rice and Suzanne Sandmeyer

doi:10.1128/microbiolspec.MDNA3-0029-2014

Huatao Guo,[1] Diego Arambula,[2] Partho Ghosh,[3] and Jeff F. Miller[2,4]

53 Diversity-generating Retroelements in Phage and Bacterial Genomes

INTRODUCTION

Mobile genetic elements have repeatedly been called to duty in life-and-death struggles between hosts and their pathogens (1, 2, 3, 4). One of their greatest utilities is the capacity to create DNA sequence diversity in protein-encoding genes, thereby generating protective shields to defend against enemies, or to create arsenals of weapons to exploit potential hosts. After decades of research, considerable evidence now suggests that the V(D)J recombination system, which is essential for generating adaptive immunity in vertebrates, has evolved from an ancestral DNA transposon (2, 3, 4). The site-specific recombinases responsible for V(D)J recombination, RAG1 and RAG2, are able to catalyze DNA transposition in a manner analogous to DNA transposons (2), and the RAG1 core and V(D)J recombination signals are likely derived from the transposase and terminal repeats of an ancient DNA transposon similar to *Transib* (3, 4). Ironically, pathogens also exploit mobile genetic elements to generate protein diversity, altering their antigenic characteristics to evade host immunity (1). This process of antigenic variation is employed by *Borrelia* species, *Neisseria gonorrhoeae*, and other pathogens. Bacterial antigenic variation often involves a single, highly expressed gene encoding an abundant surface protein and dozens of archived ones that are homologous but different from each other. Replacing all or part of the expressed copy by DNA transposition leads to antigenic variation on the surface of the pathogen.

Diversity-generating retroelements (DGRs) are a recently discovered class of beneficial mobile element that diversify DNA sequences and the proteins they encode (5, 6). DGRs function through a template-dependent, reverse transcriptase (RT)-mediated mechanism that introduces nucleotide substitutions at defined locations in specific genes (5, 6, 7). DGRs were initially discovered during studies of pathogenesis by *Bordetella* species, which cause respiratory diseases in humans and other animals (5). The cell surfaces of these bacteria are highly dynamic due to changes in gene expression that accompany their infectious cycles (8). In a search for transducing vectors, a group of temperate bacteriophages was discovered that possess a remarkable ability

[1]Department of Molecular Microbiology and Immunology, University of Missouri School of Medicine Columbia, MO 65212; [2]Department of Microbiology, Immunology and Molecular Genetics, David Geffen School of Medicine at UCLA, Los Angeles, CA 90095; [3]Department of Chemistry and Biochemistry, University of California at San Diego, La Jolla, CA 92093; [4]The California NanoSystems Institute, University of California at Los Angeles, Los Angeles, CA 90095.

to generate tropic variants that use different cell-surface molecules for infection (5, 9). Subsequent genetic and genomic studies with the prototype phage, BPP-1, showed that tropism switching is mediated by a phage-encoded DGR. This DGR introduces nucleotide substitutions in a gene that specifies a host-cell-binding protein, which is positioned at the distal tips of phage tail fibers (Fig. 1) (5, 9). As a result, BPP-1 can adapt to dynamic changes on the surfaces of *Bordetella* species. Guided by the sequences of phage DGR components, homologous elements have been identified in numerous bacterial, plasmid, and phage genomes (6, 10, 11, 12, 13). Most DGRs are bacterial chromosomal elements and they are distributed throughout the bacterial domain, with representatives in all phyla that have significant sequence coverage. Although variations in architectures and associated components appear to mediate adaptations to particular needs, all DGRs are predicted to function in a fundamentally similar way. The BPP-1 phage serves as a model for this entire family of retroelements, and our discussion begins with a brief description of its features.

TROPISM-SWITCHING *BORDETELLA* PHAGE

Bordetella species are aerobic, Gram-negative bacterial pathogens that colonize ciliated respiratory epithelial surfaces. *Bordetella pertussis* and *Bordetella parapertussis* are human-restricted and cause whooping cough (pertussis), while *Bordetella bronchiseptica* infects a broad range of wild and domesticated mammals (8). The infectious cycle of these closely related species is regulated by a conserved, environmentally responsive phosphorelay system composed of the BvgS sensor protein and the BvgA response regulator, which control expression of an extensive array of cell-surface and secreted molecules. In the so-called Bvg^+ phase, BvgAS is active and induces expression of adhesins, toxins, a type III secretion system, and numerous additional factors involved in colonization of respiratory surfaces. In the Bvg^- phase, the BvgAS phosphorelay is suppressed, virulence genes are quiescent, and a distinct set of loci are induced. In *B. bronchiseptica*, Bvg^--phase genes are responsible for motility, chemotaxis, and survival under conditions of nutrient deprivation, presumably representing an *ex vivo* phase of the infectious cycle. Dynamic changes in surface molecule expression are critical for the lifestyles of *Bordetella* species, and they likely pose a challenge to infecting phage.

BPP-1 (Bvg-plus trophic phage 1), isolated from a *B. bronchiseptica* strain cultured from the upper respiratory tract of a rabbit, preferentially forms plaques on Bvg^+-phase as opposed to Bvg^--phase *Bordetella* (5, 9). Using a combination of deletion, complementation, and cell-binding assays, Liu *et al.* identified the receptor for BPP-1 as pertactin (Prn), an outer-membrane autotransporter protein that is included as a protective antigen in acellular pertussis vaccines (5). Expression of *prn* is activated by BvgAS, thereby explaining BPP-1 tropism. At a frequency of $\sim10^{-5}$, however, BPP-1 formed normal-sized plaques on Bvg^--phase *Bordetella* (5). Since plaque formation requires repeated rounds of infection, replication, and reinfection, this simple observation suggested that a tropism switch had occurred (5). Indeed, tropic variants fell into two classes. The first, designated BMP (Bvg-minus tropic phage), preferentially infected Bvg^--phase cells, while the second, BIP (Bvg indiscriminant phage), infected both Bvg^+ and Bvg^- cells with similar efficiencies. Although these phenotypes are inheritable, tropic variants are continuously generated at characteristic frequencies by BPP, BMP, and BIP.

THE BPP-1 DGR

To identify genetic changes responsible for tropism switching, whole-genome sequences were obtained for a collection of BPP, BIP, and BMP variants (5, 9). In each case, tropism switching was accompanied by nucleotide substitutions within a 134-bp sequence, designated the variable repeat (VR), located at the 3′ end of the *mtd* (major tropism determinant) gene (Fig. 1). Mtd is a trimeric tail fiber protein responsible for phage binding to *Bordetella* surfaces, and changes in its coding sequence confer new ligand specificities (14). Comparison of phage variants showed that variable nucleotides were located at a subset of positions in the VR and most often at the first two positions of codons, thereby maximizing amino acid diversity (5, 14). There are 23 variable positions at which any of the four nucleotides can be found, giving a theoretical DNA diversity of 10^{14} sequences and a resulting repertoire of ~10 trillion polypeptides. This rivals the estimated diversity that can be generated by mammalian antibody and T-cell-receptor genes (15, 16).

An imperfect repeat of the VR, designated the template repeat (TR), is found downstream of *mtd* (Fig. 1) (5). Liu *et al.* made the simple but seminal observation that variable residues in the VRs corresponded to adenine residues in the TR, which remained unchanged during phage tropism switching (5). This prompted the hypothesis that tropism switching is associated with a mutagenic mechanism that is adenine specific, and that

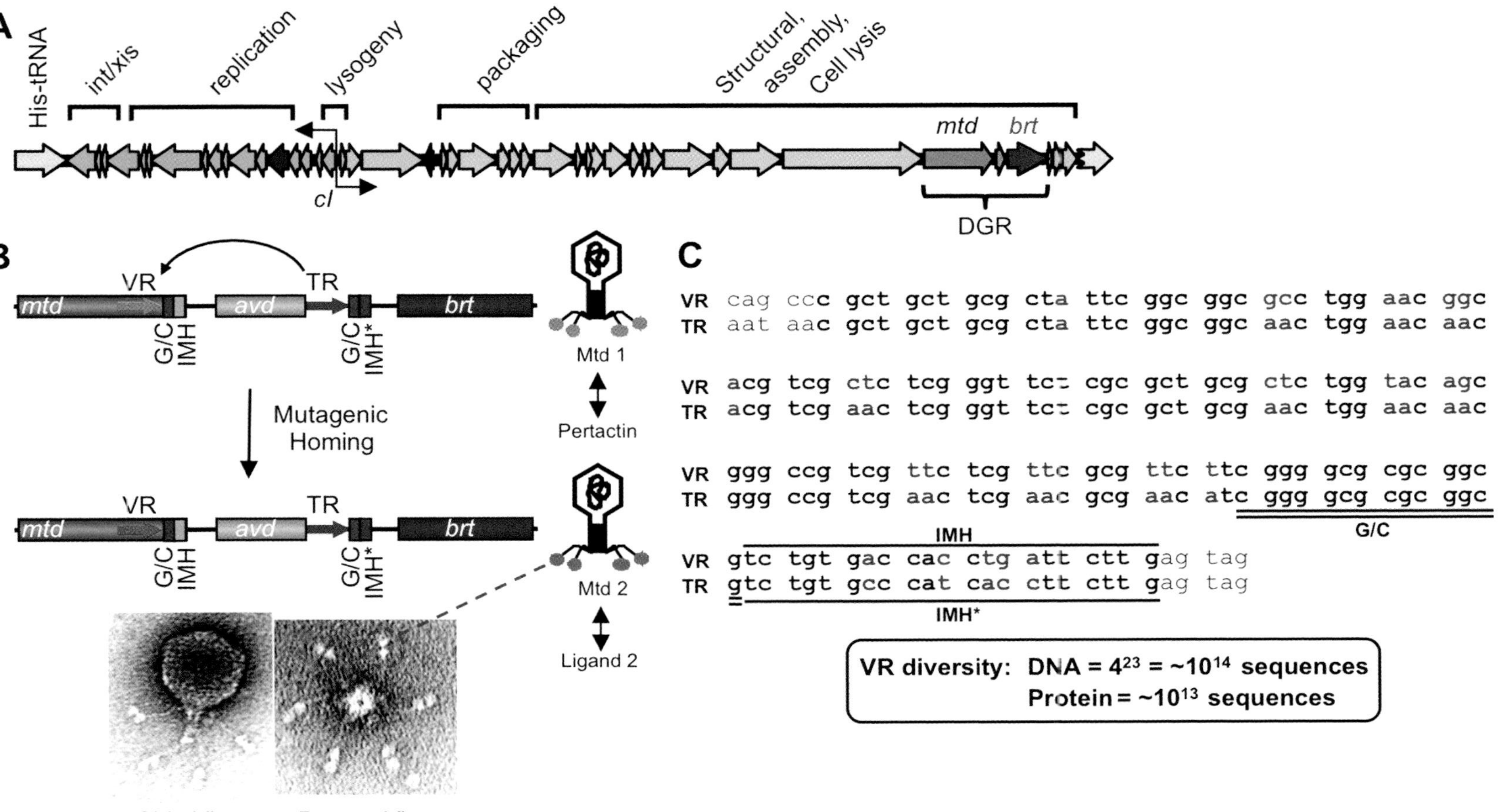

Figure 1 BPP-1 phage and its diversity-generating retroelement (DGR). (**A**) The BPP-1 genome is represented in the prophage form flanked by a duplication of the His-tRNA gene formed during integration. Functional assignments for most gene clusters are indicated, along with the *cI*-like repressor and the DGR cassette. (**B**) Schematic representation of the DGR cassette and its function in phage tropism switching. The cassette contains three genes (*mtd*, *avd*, and *brt*) and two 134-bp repeats (template and variable repeats, or TR and VR, respectively). The VR is located at the 3′ end of the *mtd* gene, which encodes the distal tail fiber protein responsible for receptor recognition. Located at the 3′ ends of the VR and TR are IMH (initiation of mutagenic homing) and IMH* elements, respectively, in addition to a GC-rich element. Phage tropism switching occurs through DGR-mediated mutagenic homing, in which TR sequence information is transferred to the VR with adenine residues in the TR appearing as random nucleotides in the VR. Shown on the bottom are electron micrographs of the BPP-1 phage; globular structures at the distal ends of tail fibers are Mtd trimers (two per fiber). (**C**) Comparison of the BPP-1 TR and VR. The TR and VR sequences are shown in bold. The VR variable positions and the corresponding adenine residues in the TR are shown in red. IMH, IMH*, and GC-rich elements are also indicated. There are 23 adenines in the TR, which can theoretically generate ~10^{14} different DNA sequences, or ~10^{13} different peptides. (Adapted from 7 and 9.) doi:10.1128/microbiolspec.MDNA3-0029-2014.f1

the TR somehow provides a template for this process. Indeed, precise deletion of the BPP-1 TR resulted in fully infectious phage particles for Bvg[+]*Bordetella* but abolished the ability to switch tropism. Furthermore, single-nucleotide substitutions introduced into the TR, which corresponded to silent mutations in the VR, appeared in the VR during tropism switching and were accompanied by mutations at adenines (5). These results demonstrated that DNA sequence information was "transferred" from the TR to the VR during tropism switching and was accompanied by adenine-specific mutagenesis.

An additional piece of the puzzle came from the unexpected finding of an RT gene, *brt* (*Bordetella* reverse transcriptase), in the double-stranded (ds)DNA BPP-1 genome (Fig. 1) (5). Its position adjacent to the VR/TR prompted the hypothesis that reverse transcription could be involved in tropism switching. To test this, an in-frame deletion was introduced into *brt* and the resulting phenotype was identical to the ΔTR deletion: infectious phage particles were produced but tropism switching was abolished. This suggested an intriguing link between reverse transcription and adenine mutagenesis. The *brt* gene encodes a 38-kDa protein with similarity to RTs found in other retroelements, including a conserved YMDD box found in RT catalytic centers. Substitution of the YMDD box with SMAA resulted in the loss of phage tropism switching *in vivo* and RT activity *in vitro*. It was proposed that tropism switching occurs through an RT-dependent transfer of sequence information from an invariant template (the TR) to a region of variability (the VR), and is accompanied by adenine mutagenesis in a process called "mutagenic homing." The RT dependency predicted the involvement of an RNA intermediate, leading Liu *et al.* to designate the tropism-switching cassette shown in Fig. 1 as a DGR (5). An additional gene in the BPP-1 DGR, designated *avd* (accessory variability determinant), encodes a small polypeptide (Avd) that plays an essential role in DGR function as described below (6, 9).

DGRs ARE WIDESPREAD IN BACTERIAL AND PHAGE GENOMES

The ability of the BPP-1 DGR to accelerate the evolution of novel ligand-binding specificities suggested that similar elements would be found elsewhere in nature. Not surprisingly, using Brt as a template, Doulatov *et al.* identified additional DGRs in phage and bacterial genomes (6). These included a *Vibrio harveyi* phage and the chromosomes of human commensals (*Bifidobacterium longum* and *Bacteroides thetaiotaomicron*), a human oral pathogen (*Treponema denticola*), and cyanobacteria. Some cyanobacterial species contained multiple DGRs, or multiple target open reading frames that are diversified *in trans* by a single DGR. DGR RTs were found to be closely related to RTs of mobile group II introns, and in all cases repeated elements corresponding to VR/TR cognate pairs could be found in nearby loci. Predicted target proteins and VR sequences were diverse, but the VRs differed from cognate TRs almost exclusively at sites corresponding to TR adenines, suggesting that DGRs function through a conserved mechanism.

More recently, Minot *et al.* performed high-throughput DNA sequencing on the gut virome of healthy humans (12). Metagenomic analysis of phage populations from stool samples identified 36 unique TR/VR pairs, 29 of which were adjacent to a DGR-type RT gene. In total, DGRs were identified in 11 out of 12 subjects studied. As with other DGRs, TR/VR sequences differed almost exclusively at sites corresponding to TR adenines. This discovery demonstrated that DGRs are common in the genomes of bacteriophages found in the lower gastrointestinal tract. Using a custom-made script called DiGReF, Schillinger *et al.* conducted a large-scale search of sequence databases and identified 155 DGRs in phage and bacteria, both Gram positive and Gram negative (13). DGRs are particularly abundant in certain phyla, with bacteroides (27.7%), firmicutes (31.0%), and proteobacteria (25.2%) containing the greatest numbers of unique elements. DGRs were also found in actinobacteria, cyanobacteria, *Deinococcus-Thermus*, nitrospirae, spirochaetes, chlamydiae/verrumicrobia, and other bacterial groups. These observations indicated that DGR host organisms are highly diverse and occupy varied environmental niches. In this large dataset, nearly all the VRs were located at the 3′ end of protein-coding genes and they ranged from 50 to 150 bp. The relatively short length of VRs might reflect constraints on DNA sequence hypervariation in protein-encoding genes, as highly dense nucleotide substitutions over a long stretch would be more likely to result in a loss of function. This study confirmed the earlier observation by Doulatov *et al.* that many bacterial genomes contain multiple DGRs, or encode single DGRs that diversify multiple target genes (6).

Although DGRs are often considered in the context of *Bordetella* phage, this is an artifact of their discovery and the choice of BPP-1 as a prototype for mechanistic studies. In reality, DGRs are widely distributed in bacterial chromosomes, in addition to phage and plasmid genomes.

DGR TARGET PROTEINS

The cellular localization and physiological functions of the vast majority of DGR-diversified target proteins are uncharacterized, but common themes are beginning to emerge. Arambula *et al.* recently analyzed a DGR found on the chromosome of *Legionella pneumophila*, an opportunistic pathogen (Fig. 2) (17). The *L. pneumophila* DGR is located on a conjugative transposable element and was found to be capable of mutagenic homing with characteristic adenine mutagenesis. Remarkably, its TR contains 43 adenine residues and is theoretically capable of generating 10^{26} unique DNA sequences, creating a repertoire of 10^{19} distinct proteins. The DGR-encoded target protein LdtA contains both a TAT (twin arginine transport) motif and a lipobox at its N terminus. The TAT pathway is an alternative secretion system that can translocate folded proteins or protein complexes across bacterial cytoplasmic membranes (18), and lipobox motifs mediate signal peptide cleavage, lipid modification, and anchoring to inner or outer membranes (19). Genetic and biochemical analysis showed that LdtA is indeed a TAT-secreted protein that is lipid modified and localized to the outer leaflet of the *Legionella* outer membrane, with C-terminal VR sequences exposed to the extracellular milieu (Fig. 2) (17). Bioinformatic analysis predicts that target proteins of many bacterial DGRs, including those of *T. denticola*, *Bacteroides* species, *Vibrio angustum*, and *Shewanella baltica*, also contain N-terminal lipobox motifs preceded by either TAT or Sec secretion signals (17). This suggests that lipid modification and surface display on bacterial outer membranes will be a common feature of proteins diversified by bacterial DGRs.

From both structural and functional perspectives, the BPP-1 Mtd protein is by far the most extensively characterized DGR target protein (14, 20, 21). To understand structure–function relationships, McMahon *et al.* determined the atomic structures of five different Mtd tropic variants (Fig. 3) (14). The Mtd variants formed intertwined, pyramid-shaped homotrimers with nearly identical secondary and tertiary structures, indicating that VR diversification did not lead to gross conformational changes. Each monomer was organized into three discrete domains: a β-prism, a β-sandwich, and a C-type lectin (CLec), arranged from the N to the C terminus. The β-prism domains converge to form a vertex on the top, and the CLec domains interact with each other to form the bottom part of the pyramid-shaped trimer. The VR-encoded variable amino acids are presented by CLec folds on the bottom surface of the Mtd trimer, with their side chains solvent exposed and accessible to ligand interactions. The most

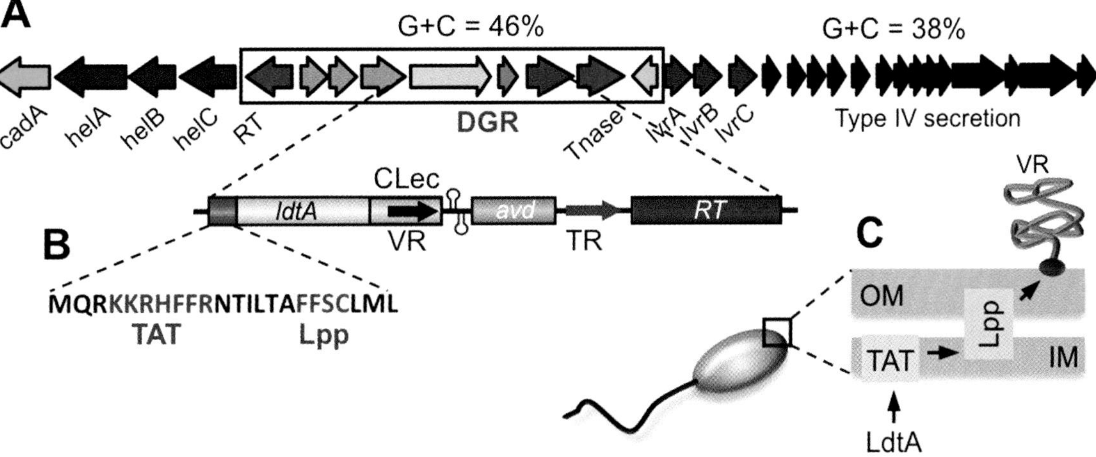

Figure 2 Diversification of a surface-displayed lipoprotein by a *Legionella* DGR. (**A**) The *L. pneumophila* strain Corby DGR is encoded on a genomic island that differs in G+C content from the rest of the genome. The VR sequences at the 3′ end of the diversified locus, *ldtA*, are flanked by tandem hairpin/cruciform structures that are essential for efficient homing. The TR contains 43 adenine residues, which can create a potential repertoire of 10^{26} different VR DNA sequences. (**B**) LdtA contains atypical TAT (twin arginine transport) and Lpp (lipobox, lipid modification) signals at the N terminus. (**C**) Cellular localization studies demonstrated that LdtA is exported through the inner membrane via the TAT pathway, lipid modified, and anchored on the outer surface of the outer membrane via an Lpp-like lipoprotein processing pathway. VR-encoded residues are surface displayed by a C-terminal CLec fold. (Adapted from 17.) doi:10.1128/microbiolspec.MDNA3-0029-2014.f2

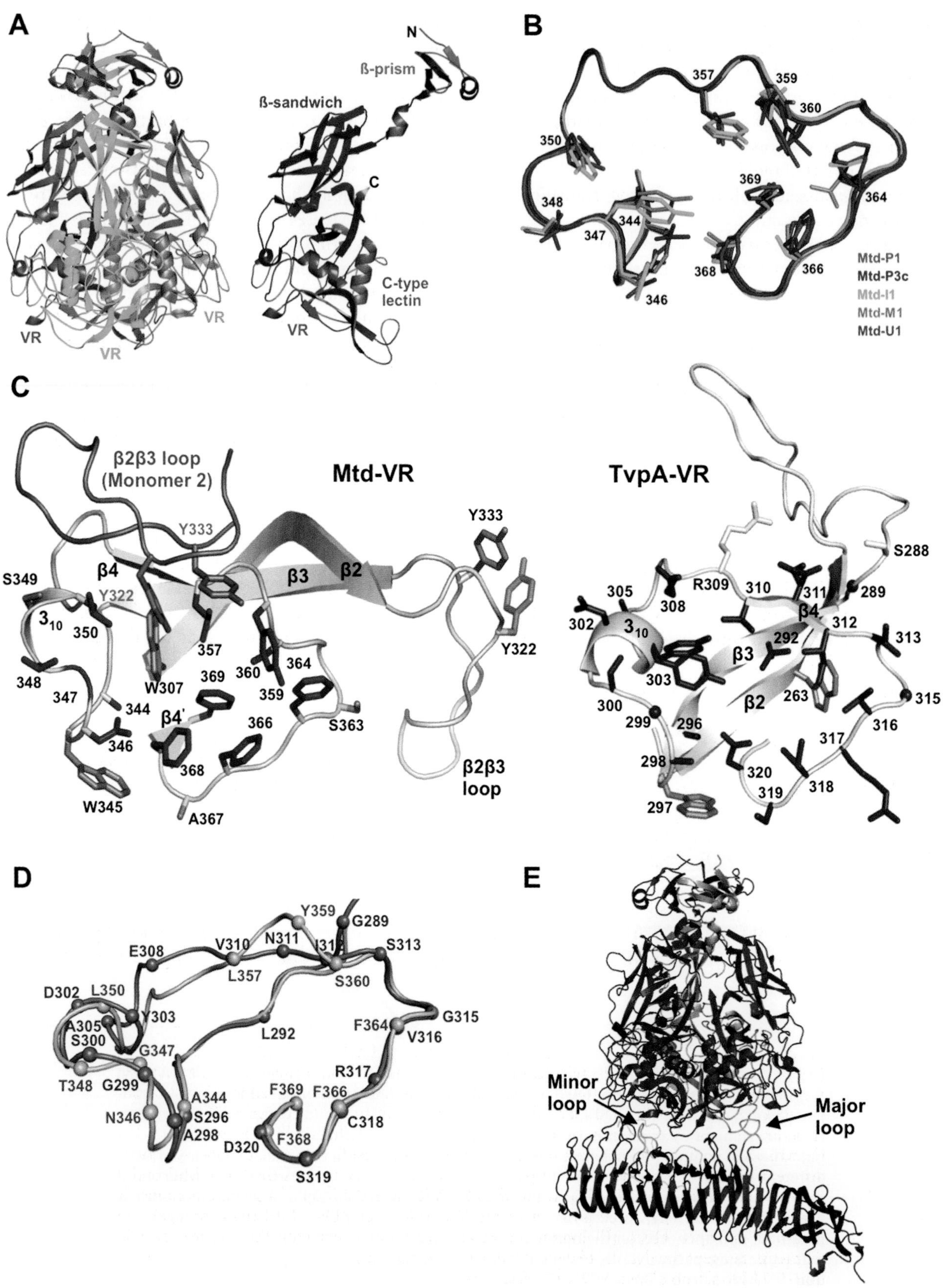
A
N
ß-prism
ß-sandwich
C
C-type lectin
VR
VR
VR
VR
B
357
359
360
350
369
364
348
344
347
368
366
346
Mtd-P1
Mtd-P3c
Mtd-I1
Mtd-M1
Mtd-U1
C
β2β3 loop (Monomer 2)
Mtd-VR
TvpA-VR
Y333
S349
β4
Y322
β3
β2
3₁₀
350
357
360
364
348
347
W307
369
359
344
β4'
366
S363
346
368
W345
A367
Y322
β2β3 loop
S288
305
R309
310
311
289
308
β4
302
312
313
292
β3
303
263
β2
315
300
299
296
316
317
298
320
318
319
297
D
Y359
G289
E308
V310
N311
I312
S313
L357
S360
D302
L350
Y303
G315
A305
F364
L292
V316
S300
G347
R317
T348
G299
F369
F366
A344
N346
S296
C318
A298
F368
D320
S319
E
Minor loop
Major loop

remarkable feature of the comparative analysis of Mtd structures is that the five distinct tropic variants showed virtually no conformational variation in the VR-encoded regions. As shown in Fig. 3, the main chain conformations of these tropic variants are nearly superimposable. This is in striking contrast to antigen-binding regions of immunoglobulin (Ig) molecules, where conformational flexibility is associated with the ability to recognize diverse antigens.

The CLec fold appears to be a conserved feature of many DGR-diversified proteins. Although DGR target proteins generally share little sequence similarity, structure-based sequence alignments predicted that VR-encoded variable residues are often presented in the context of C-terminal CLec domains (14). This was recently confirmed by X-ray crystallography using the *T. denticola* TvpA protein (Fig. 3C, D), which is predicted to be a DGR-diversified lipoprotein localized to the spirochaetal surface (22). TvpA contains a CLec fold that is highly homologous to Mtd, and VR residues are positioned in a remarkably similar manner. Instead of forming homotrimers like BPP-1 Mtd, however, TvpA exists as a monomer and does not contain a β-prism or β-sandwich domain as found in Mtd.

Interestingly, a subset of phage DGRs were predicted to use Ig folds, similar to those found in Igs, instead of CLec folds to display variable residues (12). These Ig folds were usually located in the middle, as opposed to the 3′ ends of VR-containing open reading frames. Although Igs and these predicted DGR variable proteins both use Ig folds, they appear to have evolved different means for displaying variable residues (12). In the former, diversified residues are located in flexible loop regions positioned between β-sheets. In contrast, structural modeling of Ig-type DGR target proteins indicated that diversified residues are displayed on one face of a β-sheet and the linker region connecting it to the adjacent domain. These observations suggest that DGR target proteins and antigen receptors may have evolved different solutions to accommodate sequence diversity in the context of Ig folds.

To understand better the basis of ligand recognition by DGR variable proteins, Miller *et al.* co-crystalized BPP-1 Mtd and its outer-membrane ligand, pertactin (Fig. 3E) (20). Structural analysis showed that each Mtd trimer bound to one molecule of pertactin, whose extracellular domain has an extended β-helix structure. An asymmetric mode of interaction was observed: two identical VR regions from two of the Mtd monomers in the trimer each bound a different loop from pertactin, with these loops having no sequence similarity to one another.

The fundamental basis of the evolvability of Mtd–ligand interactions was revealed, at least in part, through an analysis of binding constants (20). The K_d value for the interaction between purified Mtd trimers and the ectodomain of pertactin was 3.5 μM as determined by surface plasmon resonance. In contrast, the K_d for the intact phage was ~6.9 pM, reflecting a nearly 10^6-fold increase in binding strength. The BPP-1 phage contains six tail fibers with two Mtd trimers at their tips (21). With 12 Mtd trimers on each phage particle, and the ectodomain of pertactin displayed at high density on the *Bordetella* surface, ligand–receptor binding is multivalent and driven by avidity, which results in the exponential amplification of individual binding strengths (20). This amplification relaxes the demand for optimal complementarity between partners, greatly expanding the scope of molecules that can be recognized by DGR-diversified proteins. Avidity-driven interactions are inherently evolvable, and are predicted to characterize most, if not all, DGR variable protein–ligand interactions.

DIRECTIONALITY OF DGR MUTAGENIC HOMING

Mutagenic homing is an RT-mediated, adenine-specific, error-prone process that unidirectionally transfers sequence information from the TR to the VR (5, 6). In

Figure 3 The CLec fold as a scaffold for display of DGR-generated protein diversity. (**A**) Left: The BPP-1 Mtd protein forms a pyramid-shaped homotrimer with VR-encoded residues exposed on the bottom surface. Right: An Mtd monomer containing β-prism, β-sandwich, and VR-encoded CLec domains, from the N to the C terminus. (**B**) Backbone structures of the VR regions of five Mtd variants with different ligand specificities are shown. Despite side-chain variations in diversified VR residues, the backbone structures are nearly superimposable. (**C**) Comparison of the CLec VR regions of BPP-1 Mtd and a *T. denticola* variable protein, TvpA. For the Mtd VR, the β2β3 loop of a second monomer is also shown (blue). (**D**) Superposition of the VR regions of BPP-1 Mtd (light orange) and *T. denticola* TvpA (blue). (**E**) Interaction of an Mtd homotrimer with the receptor protein pertactin. See text for details. (Adapted from 14, 20, and 22.)
doi:10.1128/microbiolspec.MDNA3-0029-2014.f3

BPP-1, substitution of adenine residues in the TR with nonadenine nucleotides eliminated VR mutagenesis at cognate positions, while replacing nonadenine residues with adenines resulted in novel sites of mutagenesis in the VR, proving that sequence diversification is intrinsic to TR adenines.

A prominent feature of mutagenic homing is that diversity is specifically targeted to the VR while the TR sequences remain unchanged (5, 6). What determines this directionality? A comparison of the two repeats in BPP-1 revealed that, in addition to differences corresponding to adenine residues in the TR, they also differ by five base pairs at their 3′ ends (6). These polymorphisms are located within a 21 bp segment downstream of a 14-bp GC-rich element common to both the TR and VR (Figs 1 and 4). During mutagenic homing, the polymorphisms are never co-converted. Swapping experiments by Doulatov *et al.* revealed that they are required for the unidirectional transfer of sequence information (6). When the 21-bp element at the 3′ end of the VR was swapped with analogous sequences from the TR, the VR was no longer diversified. Conversely, replacing the 21-bp TR element with the corresponding VR sequences resulted in TR diversification at adenines, albeit at a low level. The 21-bp element in the VR was named the initiation of mutagenic homing (or IMH) element, while the corresponding sequence in TR was called IMH*. Further studies showed that similar polymorphisms could be found at the 3′ ends of cognate TRs and VRs in many DGRs (6). As detailed below, DNA structural determinants that follow IMH are also required for efficient homing.

DGR HOMING OCCURS THROUGH AN RNA INTERMEDIATE

The presence of conserved RTs in DGRs and the required role of Brt in tropism switching suggested that mutagenic homing occurs through an RNA intermediate. To test this hypothesis, the BPP-1 TR was "tagged" with a self-splicing group I intron from bacteriophage T4 (the *td* intron) (7). Intron tagging was first used by Boeke *et al.* (23) to probe retrotransposition by the *Saccharomyces cerevisiae* transposon Ty1, and has subsequently been used to prove that other retroelements, including human long interspersed nuclear elements (LINEs) and mobile group II introns, function through RNA intermediates (24, 25, 26). As intron excision takes place only at the RNA level, detection of ligated exons in DNA homing products provides genetic proof that sequence information flows from DNA to RNA to DNA.

Polymerase chain reaction (PCR)-based assays were used to detect homing products (i.e., VR sequences) derived from *td* intron-tagged TRs supplied *in trans* on a replicating plasmid (7). As predicted, homing products consisted of VR sequences that contained ligated *td* exons, and their detection required functional Brt and a splicing-competent *td* intron. Sequence analysis verified precise intron excision and adenine mutagenesis in progeny VRs. These observations conclusively showed that DGR homing had occurred through a TR RNA intermediate. Regions of the TR that are important for function were analyzed by similar PCR-based assays in which donor TRs were sequence tagged to allow detection and characterization of homing products (7). Mutational analysis demonstrated that sequences internal to the TR are largely dispensable, while sequences at either terminus are essential for the function of the RNA intermediate (7). By expressing wild-type *avd* on a compatible plasmid *in trans* to the TR and *brt*, a 300-bp portion of the *avd*-coding sequence that extends upstream of the limit of TR/VR homology was found to be required for optimal TR RNA function (27). Shorter sequences upstream of TR, of as little as 48 bp, provided partially functional TR RNA species that supported DGR homing at lower efficiencies (27). Sequences downstream of TR that are required for optimal function extend for at least 110 bp (7). Although the TR-containing RNA species that serves as an intermediate in the retrotransposition reaction has yet to be characterized, the observation that sequences that lie beyond the limits of TR/VR homology influence activity suggests that RNA stability and/or structural determinants are important for DGR-mediated homing.

cDNA SYNTHESIS AND INTEGRATION

cDNA synthesis could occur through one of several mechanisms. First, it could be initiated through a target DNA-primed RT (TPRT) reaction similar to those used by non-LTR (long terminal repeat) retroelements and mobile group II introns, which are closely related to DGRs (28, 29, 30, 31). Group II introns are site-specific retrotransposons found in bacterial, fungal, and plant organelle genomes that insert at cognate intronless alleles in a process called intron homing (32). Intron mobility is catalyzed by a ribonucleoprotein complex consisting of both the spliced intron RNA and the intron-encoded protein with both RT and endonuclease activities. The intron RNA cleaves the DNA sense strand in an RNA-catalyzed reverse splicing reaction, while the intron-encoded protein cleaves the DNA antisense strand shortly downstream of the exon junction

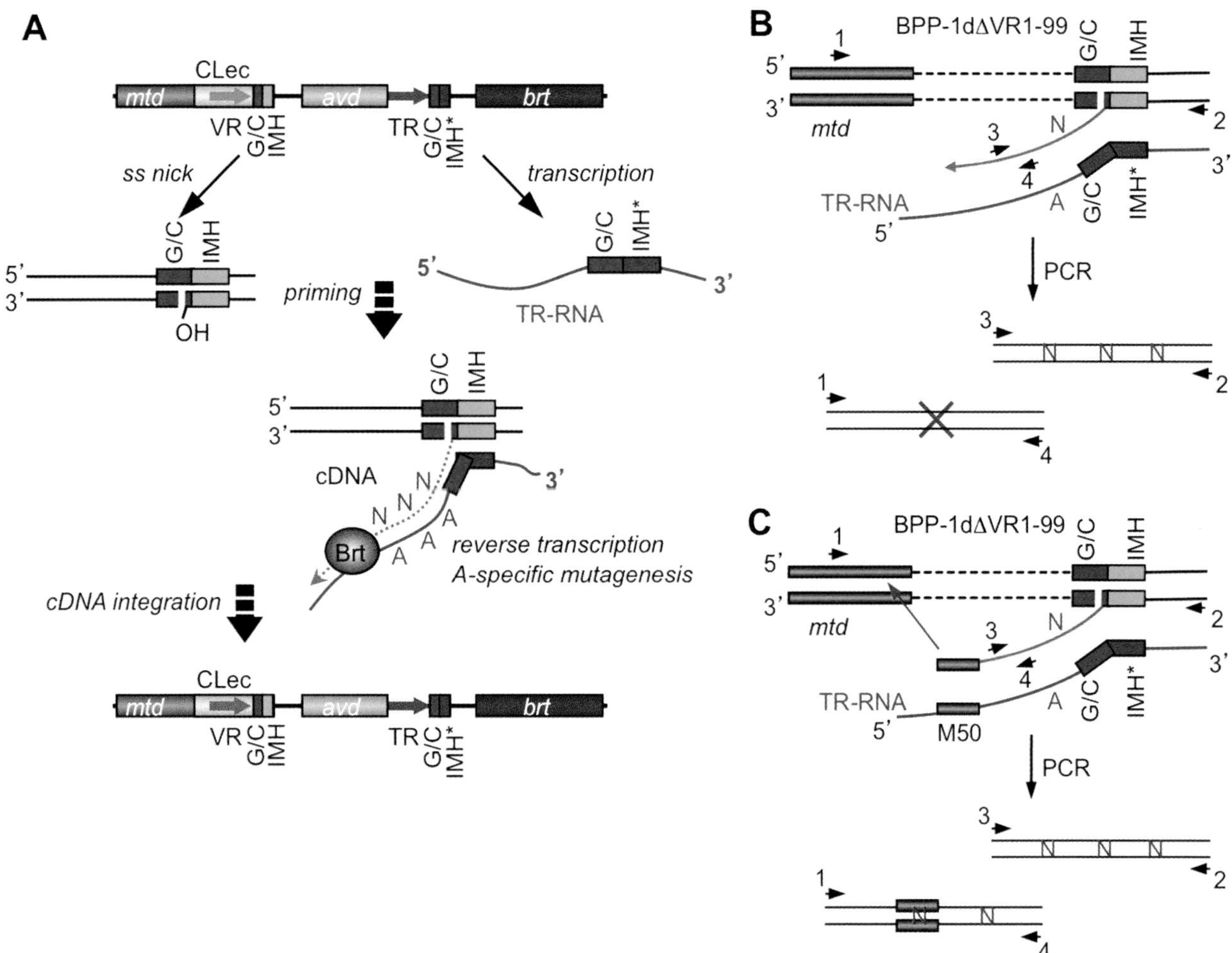

Figure 4 The TPRT model of BPP-1 DGR-mediated mutagenic homing. (A) Mutagenic homing occurs through a TR RNA intermediate and is RecA independent, similar to group II intron homing. A marker co-conversion assay mapped the cDNA transfer boundary to a narrow region within the GC-rich element at the 3′ end, which may represent 3′ cDNA integration site(s). The marker transfer boundary at the 5′ end was more heterogeneous. A target DNA-primed reverse transcription model, similar to that of group II intron homing, has been proposed to explain these observations. The DNA target site was hypothesized to be nicked within the GC-rich element, with the exposed 3′-hydroxyl group serving as a primer for adenine-specific error-prone reverse transcription of the TR RNA. Integration of cDNA products at the 5′ end requires short stretches of homology between the VR and the cDNA and may occur through strand displacement or template switching followed by break repair. Subsequent DNA replication would then create progeny genomes with mutagenized variable regions. (B) Deletion of the VR sequence upstream of GC and IMH elements appeared to block 5′ cDNA integration but not 3′ cDNA integration, as analyzed by PCR with primer sets 1/4 and 2/3, respectively. Sequence analysis showed adenine mutagenesis in PCR products generated with primers 2&3. (C) 5′ cDNA integration in ΔVR1-99 was restored by inserting a 50 bp *mtd* sequence, which is homologous to the region upstream of the deletion junction, in the TR. (Adapted from reference 7)
doi:10.1128/microbiolspec.MDNA3-0029-2014.f4

and uses it as a primer to reverse transcribe the RNA, leading to intron mobility. Alternative mechanisms for cDNA initiation during DGR homing include priming by an RNA moiety, a nontarget DNA, or even a protein capable of donating a free hydroxyl group. The close evolutionary relationship between RTs from DGRs and group II introns, the site-specificity of mutagenic homing, and other observations led to the hypothesis

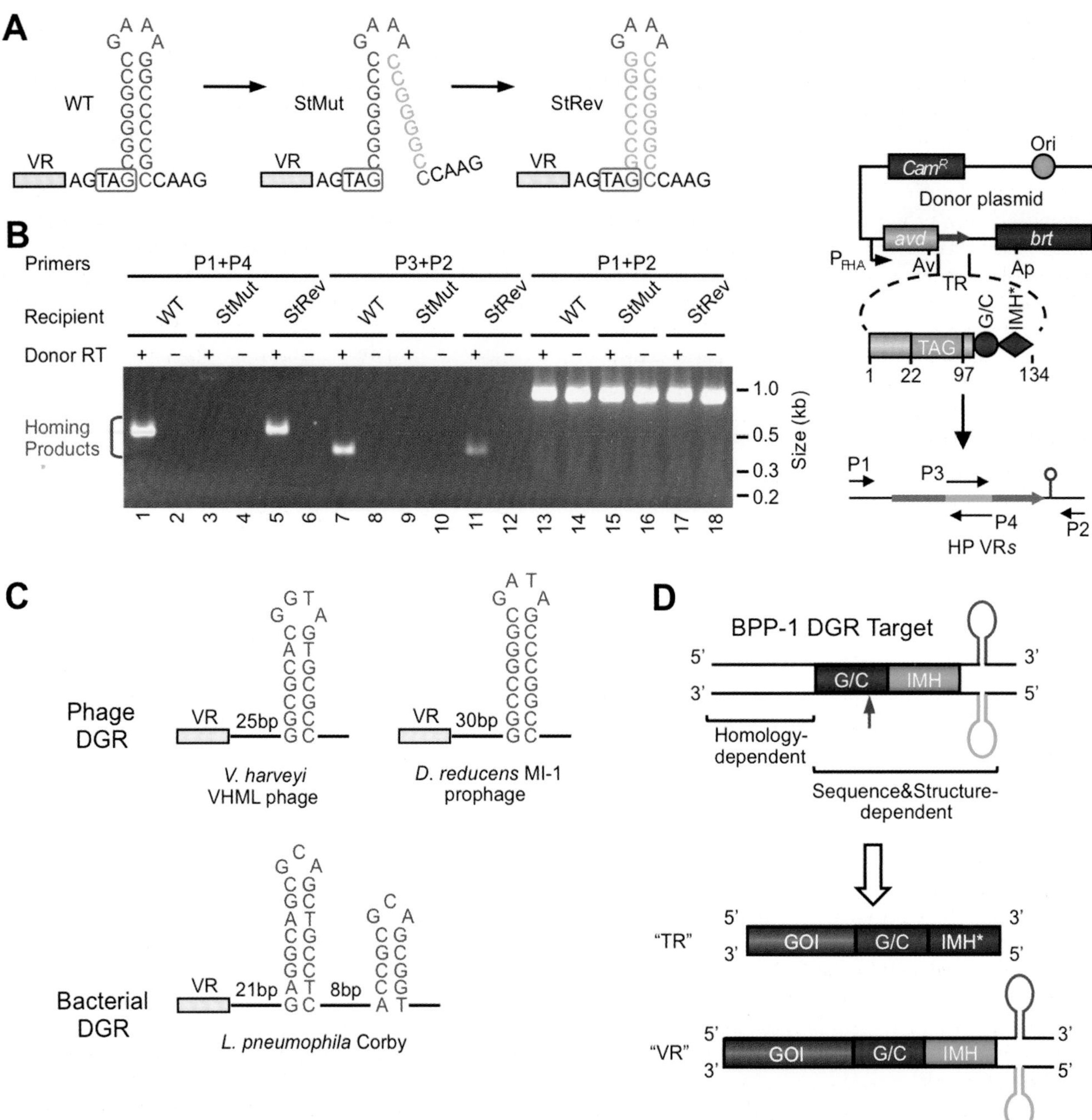

Figure 5 Role of a DNA secondary structure in DGR target recognition. **(A)** A DNA hairpin/cruciform structure downstream of the VR is required for BPP-1 DGR target recognition. The wild-type (WT) structure contains an 8-bp GC-rich stem and a 4-nt GAAA loop and is located 4 bp downstream of the VR. Mutating the 3′ half of the stem (StMut) dramatically reduced DGR mutagenic homing **(B)** and phage tropism switching (not shown), while complementary changes to the 5′ half of the stem (StRev) restored DGR activity in both assays. **(B)** PCR-based DGR homing assays with sequence-tagged TRs and VRs flanked by WT or mutant stem sequences. Shown on the right is a diagram of the PCR assay. Green represents the tag sequence transferred from the TR to the VR. P1 to P4 are primers annealing to the tag or flanking regions. **(C)** Similar DNA structures are found at analogous positions in a number of other phage (two shown) and bacterial (one shown) DGRs. The phage stems are GC rich and range from 7 to 10 bp, and loops have a conserved 4-nt sequence, G(A/G)NA. The *L. pneumophila* Corby DGR has a more complex tandem structure that is required for homing. **(D)** BPP-1 DGR target recognition at the 3′ end is both sequence and structure dependent, requiring GC, IMH, and a hairpin/cruciform structure. Target recognition at the 5′ end is homology mediated. By inserting a gene of interest (GOI) upstream of GC, IMH, and the DNA structure, the heterologous gene can be diversified by the BPP-1 DGR through appropriate engineering of the TR. (Adapted from 41.) doi:10.1128/microbiolspec.MDNA3-0029-2014.f5

that DGRs also function through a TPRT-type mechanism (7).

To determine the site at which TR-derived cDNA integrates at the 3′ end of the VR, a marker co-conversion assay was developed (7). Single-nucleotide substitutions were introduced into sequence-tagged TRs, and DGR-mediated transfer to the VR was determined by sequencing homing products amplified by PCR. A clear boundary for marker transfer was detected and localized to a 5-bp sequence within the GC element at the 3′ end of the VR (Fig. 4). This could represent the site at which cDNA synthesis initiates in a TPRT reaction, or possibly the site of integration of cDNA that was primed by a non-TPRT mechanism. Marker co-conversion at the 5′ end of the VR was more heterogeneous, suggesting that 5′ cDNA integration can occur at different VR locations, possibly mediated by homology between VR and cDNA sequences.

The TPRT model predicts that homing occurs through sequential steps, with 3′ cDNA integration taking place first (7). Indeed, when the first 99 bp of the BPP-1 VR were deleted (ΔVR1–99) to prevent 5′ cDNA integration, cDNA products linked to VR sequences at their 3′ end with free, unlinked 5′ ends were detected (Fig. 4B). These products were IMH and Brt dependent and contained adenine-specific mutations, suggesting that adenine mutagenesis occurs during reverse transcription and is an intrinsic property of the DGR RT. In addition, these observations suggested that cDNA integration at the 3′ end of the VR can occur independently of 5′ integration, consistent with the hypothesis that DGR homing occurs through a TPRT reaction. It is important to note, however, that neither single-strand nicks nor double-strand breaks have been detected within the GC-rich element where marker co-conversion begins, and no endonuclease gene, domain, or activity has been identified in a DGR cassette, raising the possibility that DGR homing could occur through an alternative mechanism.

Is 5′ cDNA integration homology dependent? To test this, Guo *et al.* inserted *mtd* sequences into the TR that were homologous to a 50-bp region immediately upstream of the deletion junction in the ΔVR1–99 recipient (Fig. 4C) (7). This restored 5′ cDNA integration and DGR homing as confirmed by PCR-based assays and sequence analysis. cDNA integration into cryptic sites was observed at low frequency, and the sites of integration mapped to short stretches of identity (4 to 12 bp) between TR–*mtd* and sequences upstream of the ΔVR1–99 deletion. These results showed that 5′ integration can be mediated by short stretches of homology between TR-derived cDNA and target DNA. One model that could account for this is template switching during reverse transcription. It has been observed previously that RTs encoded by the group II intron Ll.LtrB and other non-LTR retroelements, including the Mauriceville retroplasmid and the silkworm *Bombyx mori* R2 element, are capable of template switching during reverse transcription (33, 34, 35, 36, 37, 38, 39, 40). Interestingly, these reactions appear to occur primarily from the 5′ end of the first template to the 3′ end of the second template (DNA or RNA), and involve little or no base-pairing interactions between the latter and the 3′ end of the cDNA. Whatever the mechanism of 5′ cDNA integration might be, it appears to be RecA independent, as DGR homing and phage tropism switching occur at similar efficiencies in wild-type and RecA-deficient *Bordetella* (7).

DGR TARGET RECOGNITION

In addition to the GC-rich sequence and IMH, deletion analysis revealed a third element located downstream of the BPP-1 VR that is required for efficient homing (Fig. 5) (41). This region contains two 8-bp GC-rich inverted repeats capable of forming a DNA hairpin or cruciform structure with a 4-nt loop. Mutations that disrupt stem formation greatly decreased DGR homing and phage tropism switching, while restoration of base pairing with heterologous complementary sequences restored DGR activity. This suggested that the DNA stem–loop structure, rather than its primary sequence, is important for target recognition. Further characterization of the BPP-1 structure showed that the length and G+C content of the stem modulate the efficiency of target recognition. The sequence of the 4-nt loop, however, appeared to be critical, as changes in either sequence or size dramatically reduced homing. Structure-specific nuclease digestion assays confirmed DNA structure formation *in vitro* in dsDNA, which required negative supercoiling. The position of the hairpin/cruciform, which is normally located 4 bp downstream of the VR, is also important. Moving it 4 bp in the 5′ direction or 15 bp in the 3′ direction significantly reduced DGR homing, although shorter insertions were tolerated. Marker transfer studies showed that extending the length between VR and the structured element did not affect the 3′ marker co-conversion boundary, demonstrating that the cruciform influences the efficiency of homing, but not the site of cDNA integration. Although the exact function of the stem–loop structure is unknown, it is predicted to represent a binding site for host and/or DGR-encoded factors involved in cDNA priming and integration. Comparisons with other phage DGRs, as well as DGRs in bacterial chromosomes, suggested that similar structures are a conserved feature of target sequences (Fig. 5C). This was proved to be correct by Arambula *et al.* who demonstrated the required roles of analogous stem–loop sequences for homing by the *L. pneumophila* DGR (Figs 2 and 5C) (17).

As shown in Fig. 4, mutagenic homing is a "copy-and-replace" process that replaces parental VRs with TR-derived cDNAs that are mutagenized at specific sites (7). Target recognition at the 3′ end is both sequence and structure dependent, while target recognition at the 5′ end is dependent on short stretches of homology (Fig. 5D) (7, 41). A feature of the mechanism that seems key to its beneficial nature is that all *cis*-acting sequences and *trans*-acting factors required for additional rounds of diversification are precisely reconstituted during mutagenic homing. This allows repeated rounds of diversification and, presumably, the optimization of beneficial traits.

Guided by the TPRT model and the VR recognition rules described above (Fig. 5D), the BPP-1 DGR was engineered to target a kanamycin resistance gene (*aph3′ Ia*) (41). A defective *allele* with a 3′ truncation was placed upstream of the GC, IMH, and cruciform structure to form a recipient VR. A donor TR was then engineered to contain the missing *aph3′ Ia* sequences along with a short stretch of upstream homology. As predicted, the donor TR was able to repair the defective gene, conferring kanamycin resistance in an RT-dependent reaction. Thus, DGRs can be designed to target genes of interest, with potentially broad applications for protein engineering.

STRUCTURE AND FUNCTION OF THE Avd ACCESSORY PROTEIN

The BPP-1 DGR includes the *avd* locus, which encodes a positively charged 15-kDa protein that is essential for homing and is conserved in numerous bacterial and phage DGRs (6, 9, 10, 27). An X-ray crystal structure of BPP-1 Avd was determined by Alayyoubi *et al.* (Fig. 6A) to provide a guide for functional analysis (27). Avd forms a barrel-shaped homopentamer, which is positively charged on all surfaces, including an hourglass-shaped pore that runs through the center of the barrel and constricts to an ~8 Å diameter. Each Avd monomer forms a four-helix bundle with the α-helices running up and down in anti-parallel fashion.

A number of amino acid residues located on the surface or near the pore of the BPP-1 Avd pentamer were subjected to mutagenesis (27). Mutations at two conserved residues on the side of the pentamer, R79A and R83A, eliminated detectable homing, and mutations of two residues on the top of the pentamer, R36A and K37A, resulted in a partial loss of DGR activity. Defects in homing were not due to a loss of protein structure or stability, as identical CD spectra were obtained with wild-type and homing-defective Avd mutant proteins. Other mutations, including R19A (side), P35A (top), E43A (near the pore), and Q64A (bottom), had little effect on activity. These results suggest that positively charged residues on the side and top of the Avd pentamer have important functional roles.

As predicted by its positive charge, BPP-1 Avd was found to associate *in vitro* with single-stranded (ss) RNA, ssDNA, DNA:RNA complexes, and dsDNA, although the interactions were nonspecific (27). In contrast, Avd pentamers associated with purified Brt, which is also predicted to be positively charged, in a manner that was abolished by the R79A and R83A mutations (i.e., the ones that eliminated DGR homing), but not by other point mutations. These results suggested a correlation between Avd–Brt binding and DGR function. From these and other results, Alayyoubi *et al.* proposed that the pentameric nature of Avd may be involved in organizing a multivalent assembly consisting of Brt and nucleic acid components to somehow coordinate the multiple events required for homing (27).

DGR RTs AND ADENINE MUTAGENESIS

Adenine mutagenesis is a unique hallmark of DGRs, but its mechanism remains enigmatic. In the *td* intron-tagging experiment described above, no adenine-specific changes were observed in spliced RNA, suggesting that adenine mutagenesis does not involve RNA editing (7). However, adenine mutagenesis was observed in cDNA products when VR1–99 was deleted to "trap" cDNA intermediates between the 3′ and 5′ integration steps, suggesting that adenine mutagenesis occurs during cDNA synthesis. These observations, coupled with the fact that DGR RTs are both highly conserved and unique, support the hypothesis that adenine mutagenesis is a property inherent to DGR RTs.

DGR RTs range from 260 to 527 amino acids, with an average length of ~380 residues (Fig. 6B) (13). They have divergent sequences at their N and C termini, with a conserved central core that includes common structural motifs found in most other RTs. They do not contain domains encoding RNase H activity, as found in retroviral RTs, or a DNA endonuclease activity as found in RTs of group II introns and LINEs. Within

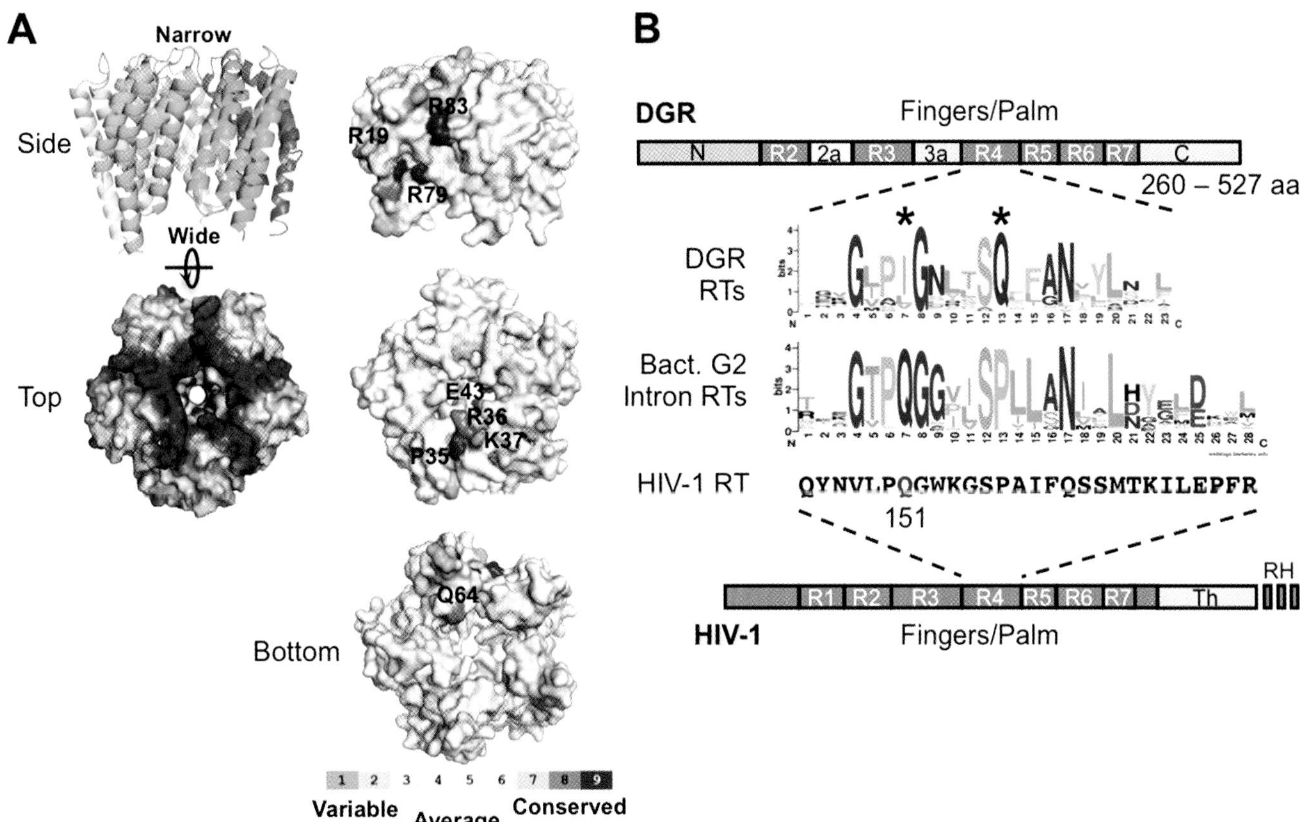

Figure 6 Avd and Brt. (**A**) Left: The BPP-1 Avd protein forms a homopentameric structure, with each monomer containing four helices running up and down (side view). The pentamer is highly positively charged (top view; blue, positively charged; red, negatively charged). Right: Amino acid residues on the side, top, and bottom of the Avd pentamer that were tested for Avd–Brt binding and/or DGR homing (27). (**B**) DGR RT domains and the sequence logo of its highly conserved domain R4. R1 to R7 are conserved sequence blocks found in the finger and palm domains of retroviral RTs, such as human immunodeficiency virus type 1 (HIV-1) RT (bottom). DGR RTs contain sequence insertions between R2 and R3 (R2a), and between R3 and R4 (R3a), as well as divergent N and C termini. They do not contain the thumb (Th) and RNase H (RH) domains that are found in HIV-1 RT. The domain R4 sequence logo of 155 DGR RTs was generated by Schillinger *et al.* using WebLogo (13, 42, 43). Comparison with the domain R4 sequence logo that we generated from 93 bacterial group II intron RTs (group II intron database: http://webapps2.ucalgary.ca/~groupii/orf/orfalignment.html; [44, 45, 46]), which are most closely related to DGR RTs, showed several characteristic differences, including the two highly conserved positions labeled with an asterisk (*). Also included for comparison is the corresponding amino acid sequence block of HIV-1 RT (strain BRU; GenBank accession no. K02013). The glutamine residue at position 151, which plays a role in nucleotide and template preference during reverse transcription, is highlighted in blue. (Adapted from 13.)
doi:10.1128/microbiolspec.MDNA3-0029-2014.f6

their shared sequences, DGR RTs have some intriguing features. The most prominent is located within the finger 4 region, which differs from those of group II intron RTs and human immunodeficiency virus type 1(HIV-1) RT at several highly conserved positions (13). The finger 4 region of HIV-1 RT forms part of the nucleotide-binding pocket that positions incoming dNTPs and influences specificity and error-prone polymerization (13, 47). Mutations at Q151 (highlighted in blue in Fig. 6B) in HIV-1 RT, which corresponds to conserved isoleucine/leucine/valine residues in DGR RTs, change nucleotide and template preferences and confer resistance to inhibitory nucleoside analogs like azidothymidine (AZT) (48, 49, 50). It seems likely that unique features of this motif found in DGR RTs play a role in adenine mutagenesis.

At least two potential mechanisms could account for adenine mutagenesis. In the first, DGR RTs are hypothesized to carry out a variation of error-prone reverse transcription that inserts random dNTPs opposite adenines but rarely other residues, while synthesizing cDNA from TR-containing RNA templates. An alternative mechanism could be imagined in which DGR RTs have an increased propensity to incorporate dUTP, as opposed to dTTP when copying adenines. dUTP residues in cDNA products would then be recognized by host-encoded uracil DNA glycosylases and excised, leaving abasic sites. If these cDNA products are subsequently used as templates for second-strand cDNA sysnthesis, which could be catalyzed by a DGR RT or a host-encoded DNA polymerase, random nucleotides would be incorporated opposite these abasic sites.

Understanding the mechanism of adenine mutagenesis may have broad implications. During mutagenic homing by the BPP-1 DGR, about 30% of the TR adenines are converted to other nucleotides in progeny VRs (7). This error rate is far greater (~10,000 times higher) than that of HIV-1 RT, which is $\sim1.4\times10^{-5}$ to 3.0×10^{-5} *in vivo* (51, 52). The low fidelity of HIV-1 RT enables the virus to generate vast numbers of sequence variants, enabling escape from immune surveillance and drug resistance. If DGR RTs are truly responsible for adenine mutagenesis, they are likely to be the most error-prone DNA-polymerizing enzymes yet discovered.

SUMMARY

DGRs are beneficial retroelements present in diverse bacteria and phage (6, 10, 11, 12, 13). Their apparent function is to accelerate the evolution of ligand-binding interactions. Structural, bioinformatic, and biochemical studies indicate that CLec folds are a conserved solution to display variable residues of DGR-diversified proteins (14, 20, 22). Unexpectedly, this structural fold is highly static in DGR-diversified proteins studied to date. Despite major differences in amino acid side chains, the backbones of BPP-1 Mtd variants are nearly identical to each other and their CLec folds superimpose with the diversified domain in *T. denticola* TvpA. This limited structural variability suggests that most DGR variable proteins will have relatively weak interactions with their respective ligands. For the BPP-1 phage, successful host recognition relies on multivalent interactions (i.e. avidity) between Mtd trimers on phage particles and arrays of receptors on the bacterial surface, which results in immense amplification of an otherwise weak monovalent ligand–receptor interaction. Avidity-driven binding is likely a conserved feature that contributes to the evolvability of DGR variable proteins (20).

A significant amount has been learned regarding the mechanism of DGR-mediated mutagenic homing (5, 6, 7, 17, 27, 41). Sequence diversification is mediated by a unique class of RTs associated with the conversion of adenine residues in an RNA intermediate into random nucleotides in a cDNA that ultimately replaces the variable region of a target gene with a diversified derivative. This process is independent of the host RecA-mediated homologous recombination machinery, similar to the mobility mechanism of group II introns (7). Current evidence is consistent with the hypothesis that mutagenic homing initiates through a TPRT mechanism, although other possibilities exist. BPP-1 DGR target recognition is both sequence and structure dependent at the 3′ end, requiring GC and IMH sequence elements and a DNA cruciform structure, while target recognition at the 5′ end is mediated by short stretches of homology (41). Based on these and other observations, the BPP-1 DGR was successfully engineered to target a heterologous gene.

Despite these advances, many questions remain unanswered. Hundreds of DGRs have been identified in diverse bacteria and phage, but their biological functions remain largely unknown (6, 10, 13, 17, 53). Understanding the functions of diverse DGRs is a major challenge. On the mechanistic side, the TPRT model is valuable for guiding experiments, but it is only one of several models that could explain DGR homing, and the evidence supporting it is incomplete. In addition, it is unclear what specific role Avd plays in the homing reaction. The mechanism of adenine mutagenesis is of great interest as it represents a hallmark of DGR function. It has been hypothesized to be an intrinsic property of DGR-encoded RTs and, if true, understanding how it occurs may help us understand RT fidelity in general. Mutagenic homing is clearly a complex process that requires the participation of host factors, although none has been identified to date. Finally, the utility of DGRs for protein engineering has yet to be exploited. In addition to providing prodigious levels of diversity, mutagenic homing is a regenerative process that can operate through unlimited rounds to optimize protein functions. This may be particularly advantageous for directed protein evolution since desired traits can be selected and continuously evolved in iterative cycles, without the need for library construction or other interventions, and through a process that takes place entirely within bacterial cells.

Acknowledgments. *This work was supported by NIH grant (R01 AI069838) to J.F.M. and P.G. and University of Missouri startup funds to H.G. J.F.M. is a co-founder of AvidBiotics Corp. and a member of its Board of Directors and Scientific Advisory Board.*

Citation. Guo H, Arambula D, Ghosh P, Miller JF. 2014. Diversity generating retroelements in phage and bacterial genomes. Microbiol Spectrum **2**(6):MDNA3-0029-2014.

References

1. **Vink C, Rudenko G, Seifert HS.** 2012. Microbial antigenic variation mediated by homologous DNA recombination. *FEMS Microbiol Rev* **36:**917–948.
2. **Agrawal A, Eastman QM, Schatz DG.** 1998. Transposition mediated by RAG1 and RAG2 and its implications for the evolution of the immune system. *Nature* **394:** 744–751.
3. **Kapitonov VV, Jurka J.** 2005. RAG1 core and V(D)J recombination signal sequences were derived from Transib transposons. *PLoS Biol* **3:**e181.
4. **Hencken CG, Li X, Craig NL.** 2012. Functional characterization of an active Rag-like transposase. *Nat Struct Mol Biol* **19:**834–836.
5. **Liu M, Deora R, Doulatov SR, Gingery M, Eiserling FA, Preston A, Maskell DJ, Simons RW, Cotter PA, Parkhill J, Miller JF.** 2002. Reverse transcriptase-mediated tropism switching in *Bordetella* bacteriophage. *Science* **295:** 2091–2094.
6. **Doulatov S, Hodes A, Dai L, Mandhana N, Liu M, Deora R, Simons RW, Zimmerly S, Miller JF.** 2004. Tropism switching in *Bordetella* bacteriophage defines a family of diversity-generating retroelements. *Nature* **431:** 476–481.
7. **Guo H, Tse LV, Barbalat R, Sivaamnuaiphorn S, Xu M, Doulatov S, Miller JF.** 2008. Diversity-generating retroelement homing regenerates target sequences for repeated rounds of codon rewriting and protein diversification. *Mol Cell* **31:**813–823.
8. **Melvin JA, Scheller EV, Miller JF, Cotter PA.** 2014. Bordetella pertussis pathogenesis: current and future challenges. *Nat Rev Microbiol* **12:**274–288.
9. **Liu M, Gingery M, Doulatov SR, Liu Y, Hodes A, Baker S, Davis P, Simmonds M, Churcher C, Mungall K, Quail MA, Preston A, Harvill ET, Maskell DJ, Eiserling FA, Parkhill J, Miller JF.** 2004. Genomic and genetic analysis of Bordetella bacteriophages encoding reverse transcriptase-mediated tropism-switching cassettes. *J Bacteriol* **186:** 1503–1517.
10. **Medhekar B, Miller JF.** 2007. Diversity-generating retroelements. *Curr Opin Microbiol* **10:**388–395.
11. **Simon DM, Zimmerly S.** 2008. A diversity of uncharacterized reverse transcriptases in bacteria. *Nucleic Acids Res* **36:**7219–7229.
12. **Minot S, Grunberg S, Wu GD, Lewis JD, Bushman FD.** 2012. Hypervariable loci in the human gut virome. *Proc Natl Acad Sci U S A* **109:**3962–3966.
13. **Schillinger T, Lisfi M, Chi J, Cullum J, Zingler N.** 2012. Analysis of a comprehensive dataset of diversity generating retroelements generated by the program DiGReF. *BMC Genomics* **13:**430.
14. **McMahon SA, Miller JL, Lawton JA, Kerkow DE, Hodes A, Marti-Renom MA, Doulatov S, Narayanan E, Sali A, Miller JF, Ghosh P.** 2005. The C-type lectin fold as an evolutionary solution for massive sequence variation. *Nat Struct Mol Biol* **12:**886–892.
15. **Murphy K.** 2012. The Generation of Lymphocyte Antigen Receptors, p 171. *In Janeway's Immunobiology*, 8th ed. Garland Science, Taylor & Francis Group, LLC, London and New York.
16. **Abbas AK, Lichtman AH, Pillai S.** 2012. Lymphocyte Development and Antigen Receptor Gene Rearrangement, p 186. *In Cellular and Molecular Immunology*, 7th ed. Elsevier Saunders, Philadelphia, PA.
17. **Arambula D, Wong W, Medhekar BA, Guo H, Gingery M, Czornyj E, Liu M, Dey S, Ghosh P, Miller JF.** 2013. Surface display of a massively variable lipoprotein by a *Legionella* diversity-generating retroelement. *Proc Natl Acad Sci U S A* **110:**8212–8217.
18. **Stanley NR, Palmer T, Berks BC.** 2000. The twin arginine consensus motif of Tat signal peptides is involved in Sec-independent protein targeting in *Escherichia coli. J Biol Chem* **275:**11591–11596.
19. **Narita S, Tokuda H.** 2010. Sorting of bacterial lipoproteins to the outer membrane by the Lol system. *Methods Mol Biol* **619:**117–129.
20. **Miller JL, Le Coq J, Hodes A, Barbalat R, Miller JF, Ghosh P.** 2008. Selective ligand recognition by a diversity-generating retroelement variable protein. *PLoS Biol* **6:**e131.
21. **Dai W, Hodes A, Hui WH, Gingery M, Miller JF, Zhou ZH.** 2010. Three-dimensional structure of tropism-switching *Bordetella* bacteriophage. *Proc Natl Acad Sci U S A* **107:**4347–4352.
22. **Le Coq J, Ghosh P.** 2011. Conservation of the C-type lectin fold for massive sequence variation in a *Treponema* diversity-generating retroelement. *Proc Natl Acad Sci U S A* **108:**14649–14653.
23. **Boeke JD, Garfinkel DJ, Styles CA, Fink GR.** 1985. Ty elements transpose through an RNA intermediate. *Cell* **40:**491–500.
24. **Moran JV, Holmes SE, Naas TP, DeBerardinis RJ, Boeke JD, Kazazian HH Jr.** 1996. High frequency retrotransposition in cultured mammalian cells. *Cell* **87:**917–927.
25. **Cousineau B, Smith D, Lawrence-Cavanagh S, Mueller JE, Yang J, Mills D, Manias D, Dunny G, Lambowitz AM, Belfort M.** 1998. Retrohoming of a bacterial group II intron: mobility via complete reverse splicing, independent of homologous DNA recombination. *Cell* **94:**451–462.
26. **Guo H, Karberg M, Long M, Jones JP 3rd, Sullenger B, Lambowitz AM.** 2000. Group II introns designed to insert into therapeutically relevant DNA target sites in human cells. *Science* **289:**452–457.
27. **Alayyoubi M, Guo H, Dey S, Golnazarian T, Brooks GA, Rong A, Miller JF, Ghosh P.** 2013. Structure of the essential diversity-generating retroelement protein bAvd and its functionally important interaction with reverse transcriptase. *Structure* **21:**266–276.

28. **Luan DD, Korman MH, Jakubczak JL, Eickbush TH.** 1993. Reverse transcription of R2Bm RNA is primed by a nick at the chromosomal target site: a mechanism for non-LTR retrotransposition. *Cell* **72:**595–605.

29. **Cost GJ, Feng Q, Jacquier A, Boeke JD.** 2002. Human L1 element target-primed reverse transcription in vitro. *EMBO J* **21:**5899–5910.

30. **Zimmerly S, Guo H, Perlman PS, Lambowitz AM.** 1995. Group II intron mobility occurs by target DNA-primed reverse transcription. *Cell* **82:**545–554.

31. **Zimmerly S, Guo H, Eskes R, Yang J, Perlman PS, Lambowitz AM.** 1995. A group II intron RNA is a catalytic component of a DNA endonuclease involved in intron mobility. *Cell* **83:**529–538.

32. **Lambowitz AM, Zimmerly S.** 2011. Group II introns: mobile ribozymes that invade DNA. *Cold Spring Harb Perspect Biol* **3:**a003616.

33. **Kennell JC, Wang H, Lambowitz AM.** 1994. The Mauriceville plasmid of *Neurospora* spp. uses novel mechanisms for initiating reverse transcription in vivo. *Mol Cell Biol* **14:**3094–3107.

34. **Chen B, Lambowitz AM.** 1997. De novo and DNA primer-mediated initiation of cDNA synthesis by the Mauriceville retroplasmid reverse transcriptase involve recognition of a 3′ CCA sequence. *J Mol Biol* **271:**311–332.

35. **George JA, Burke WD, Eickbush TH.** 1996. Analysis of the 5′ junctions of R2 insertions with the 28S gene: implications for non-LTR retrotransposition. *Genetics* **142:** 853–863.

36. **Bibillo A, Eickbush TH.** 2002. The reverse transcriptase of the R2 non-LTR retrotransposon: continuous synthesis of cDNA on non-continuous RNA templates. *J Mol Biol* **316:**459–473.

37. **Bibillo A, Eickbush TH.** 2004. End-to-end template jumping by the reverse transcriptase encoded by the R2 retrotransposon. *J Biol Chem* **279:**14945–14953.

38. **Stage DE, Eickbush TH.** 2009. Origin of nascent lineages and the mechanisms used to prime second-strand DNA synthesis in the R1 and R2 retrotransposons of *Drosophila. Genome Biol* **10:**R49.

39. **Zhuang F, Mastroianni M, White TB, Lambowitz AM.** 2009. Linear group II intron RNAs can retrohome in eukaryotes and may use nonhomologous end-joining for cDNA ligation. *Proc Natl Acad Sci U S A* **106:**18189–18194.

40. **White TB, Lambowitz AM.** 2012. The retrohoming of linear group II intron RNAs in *Drosophila melanogaster* occurs by both DNA ligase 4-dependent and -independent mechanisms. *PLoS Genet* **8:**e1002534.

41. **Guo H, Tse LV, Nieh AW, Czornyj E, Williams S, Oukil S, Liu VB, Miller JF.** 2011. Target site recognition by a diversity-generating retroelement. *PLoS Genet* **7:** e1002414.

42. **Crooks GE, Hon G, Chandonia JM, Brenner SE.** 2004. WebLogo: a sequence logo generator. *Genome Res* **14:** 1188–1190.

43. **Schneider TD, Stephens RM.** 1990. Sequence logos: a new way to display consensus sequences. *Nucleic Acids Res* **18:**6097–6100.

44. **Dai L, Toor N, Olson R, Keeping A, Zimmerly S.** 2003. Database for mobile group II introns. *Nucleic Acids Res* **31:**424–426.

45. **Simon DM, Clarke NA, McNeil BA, Johnson I, Pantuso D, Dai L, Chai D, Zimmerly S.** 2008. Group II introns in eubacteria and archaea: ORF-less introns and new varieties. *RNA* **14:**1704–1713.

46. **Candales MA, Duong A, Hood KS, Li T, Neufeld RA, Sun R, McNeil BA, Wu L, Jarding AM, Zimmerly S.** 2012. Database for bacterial group II introns. *Nucleic Acids Res* **40:**D187–D190.

47. **Huang H, Chopra R, Verdine GL, Harrison SC.** 1998. Structure of a covalently trapped catalytic complex of HIV-1 reverse transcriptase: implications for drug resistance. *Science* **282:**1669–1675.

48. **Kaushik N, Talele TT, Pandey PK, Harris D, Yadav PN, Pandey VN.** 2000. Role of glutamine 151 of human immunodeficiency virus type-1 reverse transcriptase in substrate selection as assessed by site-directed mutagenesis. *Biochemistry* **39:**2912–2920.

49. **Sarafianos SG, Hughes SH, Arnold E.** 2004. Designing anti-AIDS drugs targeting the major mechanism of HIV-1 RT resistance to nucleoside analog drugs. *Int J Biochem Cell Biol* **36:**1706–1715.

50. **Boyer PL, Sarafianos SG, Clark PK, Arnold E, Hughes SH.** 2006. Why do HIV-1 and HIV-2 use different pathways to develop AZT resistance? *PLoS Pathog* **2:**e10.

51. **Mansky LM, Temin HM.** 1995. Lower in vivo mutation rate of human immunodeficiency virus type 1 than that predicted from the fidelity of purified reverse transcriptase. *J Virol* **69:**5087–5094.

52. **Abram ME, Ferris AL, Shao W, Alvord WG, Hughes SH.** 2010. Nature, position, and frequency of mutations made in a single cycle of HIV-1 replication. *J Virol* **84:** 9864–9878.

53. **Schillinger T, Zingler N.** 2012. The low incidence of diversity-generating retroelements in sequenced genomes. *Mob Genet Elements* **2:**287–291.

Mobile DNA, 3rd Edition
Nancy L. Craig, Michael Chandler, Martin Gellert, Alan M. Lambowitz, Phoebe A. Rice and Suzanne Sandmeyer

doi:10.1128/microbiolspec.MDNA3-0058-2014

Steven Zimmerly[1]
Li Wu[1]

54 An Unexplored Diversity of Reverse Transcriptases in Bacteria

INTRODUCTION

Reverse transcriptase (RT) is generally considered a eukaryotic enzyme because it is prevalent in eukaryotes and was first characterized from eukaryotic sources. Discovered in 1970 in the Rous Sarcoma and murine leukemia viruses (1, 2), RT has since been studied for its central role in the replication of many eukaryotic genetic elements including retroviruses (e.g., HIV-1), pararetroviruses, hepadnaviruses, long terminal repeat (LTR), and non-LTR retroelements, Penelope-like elements, and telomerase (3, 4, 5, 6, 7, 8, 9, 10). Over the years, the accumulated studies of RT have painted a picture in which the enzyme functions primarily as the replicative enzyme of selfish DNAs (viruses, retrotransposons), while occasionally becoming domesticated to perform useful cellular functions. These functions include the maintenance of chromosomal ends (telomerase, *Drosophila* Het-A elements) (10, 11) and contributions to genomic change (both beneficial and deleterious) through pseudogene formation or other retroprocessing events (12, 13, 14, 15).

RT was discovered in bacteria in 1989 as a component of an element named a retron (16, 17, 18). Retrons produce small nucleic acid products in the cell called multicopy single-stranded DNAs (msDNAs), whose function remains elusive to this day. In 1993, a second class of bacterial RTs was discovered, group II introns, which had previously been found in the mitochondria and chloroplasts of fungi, algae, and plants (19). Group II introns were subsequently found to have features typical of mobile DNAs, in that they amplify themselves and spread in genomes without providing obvious benefits to their hosts (20, 21, 22). About a decade later, the third type of bacterial retroelement was discovered, the diversity-generating retroelement (DGR) (23). Unlike most retroelements, DGRs are not actively mobile but carry out a directed-mutagenesis reaction that is advantageous to its host genome.

More recently, it has become evident that bacteria, in fact, harbor a plethora of overlooked RTs and RT-related sequences. Most of these putative RTs do not show signs of being mobile DNAs because they do not accumulate to multiple copies in genomes. While the functions of these elements remain unclear, they do not appear to fit into the paradigm of eukaryotic selfish retroelements. The aim of this article is to summarize

[1]Department of Biological Sciences, University of Calgary, Calgary, Alberta T2N 1N4, Canada.

what is known about this enigmatic universe of retroelements in bacteria and to consider their relationships to characterized retroelements.

GROUP II INTRONS: THE PROTOTYPICAL BACTERIAL RETROELEMENT

Group II introns are the prototypical retroelements in bacteria because they are the most abundant and best understood (reviewed in references 20, 21, 22, 24, and 25). They consist of a ~500 to 800 bp sequence corresponding to a catalytic, self-splicing intron RNA (ribozyme) and an internally encoded ~1.0 to 1.5 kb ORF that is translated into an intron-encoded protein (IEP) (Fig. 1A). The IEP is an RT, containing seven conserved sequence blocks that are alignable across RT types, as well as an X domain that is structurally analogous to the polymerase's thumb domain but is not conserved in sequence. Downstream of the X domain is a DNA-binding domain (D), and sometimes, an endonuclease (En) domain (Fig. 1A). The RT and X domains together promote splicing of the intron RNA, while all four IEP domains participate in the mobility reaction (26, 27, 28, 29, 30, 31).

In bacterial cells, the intron is initially transcribed along with its exons as part of a precursor mRNA transcript, after which the intron RNA folds into a three-dimensional structure that catalyzes splicing (32, 33, 34, 35, 36, 37). Self-splicing of the intron occurs *in vitro* under conditions of elevated magnesium and salt; however, to efficiently splice *in vivo*, the IEP must first be translated from the unspliced mRNA and bind to the intron RNA to help it achieve its catalytic conformation (27, 28, 30) (Fig. 1B). After splicing, the protein

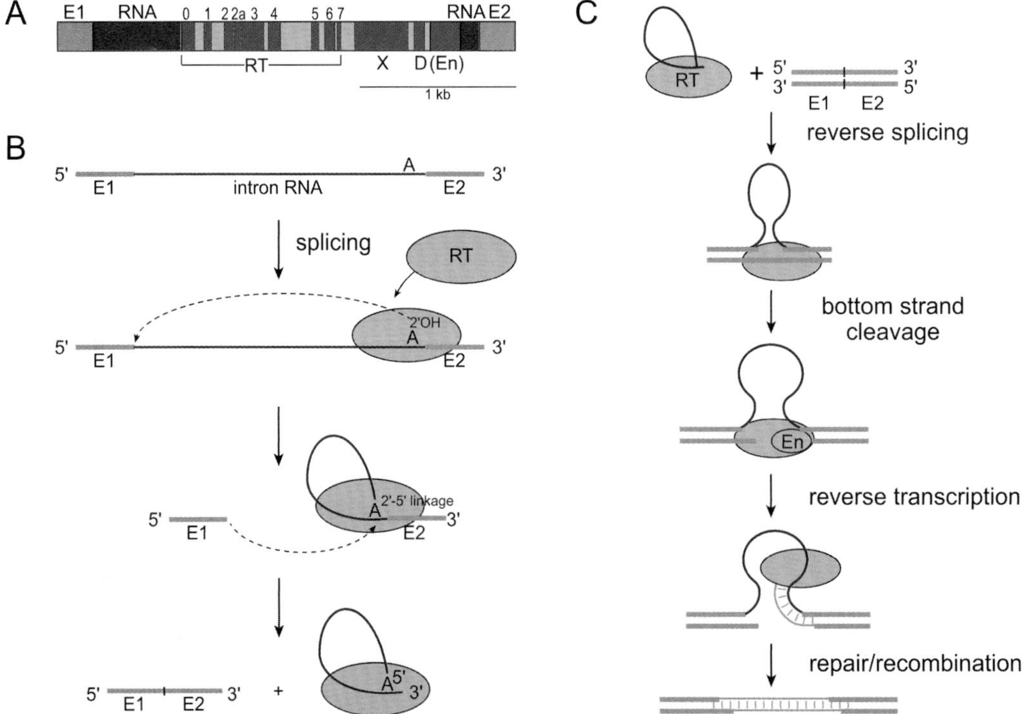

Figure 1 Group II introns. (A) The genomic structure of a group II intron consists of sequence for an RNA structure (~500 to 800 bp; red boxes) and an ORF for an intron-encoded protein (green). The protein contains a reverse transcriptase (RT) domain with motifs 0 to 7, an X/thumb domain, a DNA-binding domain (D), and sometimes, an endonuclease domain (En). The intron is flanked by exons E1 and E2 (blue). The structure is drawn to scale for the Ll.LtrB intron of *Lactococcus lactis*. (B) After transcription of the intron, the intron-encoded protein is translated from unspliced transcript and binds to the RNA structure to facilitate a two-step splicing reaction, yielding spliced exons and an RNP consisting of the RT and intron lariat RNA. (C) The RNP inserts intron sequence into new genomic targets. To do this, the RNP binds to the double-stranded DNA target, the intron lariat reverse splices into the top strand, and the En domain cleaves the bottom strand to produce a primer that is reverse transcribed by the RT. Cellular repair activities convert the insertion product to dsDNA. doi:10.1128/microbiolspec.MDNA3-0058-2014.f1

remains bound to the intron lariat, forming a stable ribonucleoprotein (RNP) particle.

Mobility of group II introns has been well studied. The overall process is called retrohoming, and it occurs through the mechanism of target-primed reverse transcription (TPRT) (20, 38, 39, 40). The TPRT mechanism is initiated by reverse splicing of the lariat RNA into the top strand of a double-stranded DNA target, followed by cleavage of the bottom strand by the En domain of the IEP to form a primer, and finally reverse transcription of the inserted intron (Fig. 1C). The final steps of the integration, which vary across organisms, are carried out by cellular repair mechanisms to generate double-stranded DNA (41, 42, 43). Some IEPs lack the En domain, and require an alternative primer for bottom strand synthesis (44). This has been shown to be a nascent DNA strand provided by a replication fork (45). Other variations of mobility have been described as well (46, 47, 48, 49, 50, 51).

An important characteristic of group II introns in bacteria is that they behave primarily as retroelements rather than introns. Their selfish nature is evident in several ways. First, the introns are generally excluded from housekeeping or conserved genes, suggesting that they inhibit gene expression in some way (52, 53). Second, over half of intron copies in bacteria are truncated or nonfunctional, with the introns often located among or within other mobile DNAs, suggesting a migratory rather than stable lifestyle (52, 54, 55). These features contrast with group II introns in mitochondrial and chloroplast genomes, where the introns are located in housekeeping genes, and many are nonmobile splicing units (56). Third, the distribution of group II introns across bacterial species and strains is sporadic (44, 54, 57, 58). For example, related strains of *Escherichia coli* can harbor from one to 15 copies of group II introns (59), and some genomes contain over 20 identical copies of an intron (60).

There are limited exceptions to the introns' selfish character. Some bacterial introns are immobile (54, 61) or are located in housekeeping genes such as RecA, DNA polymerase III, helicases, and DNA repair enzymes, where they presumably do not impair gene expression (53). Still, the overall pattern of group II introns in bacterial genomes indicates that the introns survive by constant movement, spreading faster than they are lost, as opposed to taking up residence in conserved genes, as occurs in organellar genomes where splicing can potentially help to regulate gene expression. Notably, group II introns are by far the most numerous of RT types in bacteria (below), consistent with their robust retromobility.

DGRs: RETROELEMENTS THAT EVOLVED A USEFUL FUNCTION

DGRs do not spread selfishly but they have a clearly useful function in creating sequence diversity in a target gene (62). DGRs are comprised of multiple components that make up a functional cassette of genes: an RT gene, a 100 to 150 bp TR (template repeat) gene, a target gene ending in a VR (variable region) sequence that is ~90% identical to the TR sequence, and usually, the accessory gene *avd* (accessory variability determinant) (Fig. 2). During the so-called mutagenic retrohoming reaction, the RT reverse transcribes the TR transcript, during which every A in the TR template is mutagenized to any nucleotide in the resulting cDNA. The cDNA is integrated into the target gene's DNA, replacing the VR sequence and creating randomization

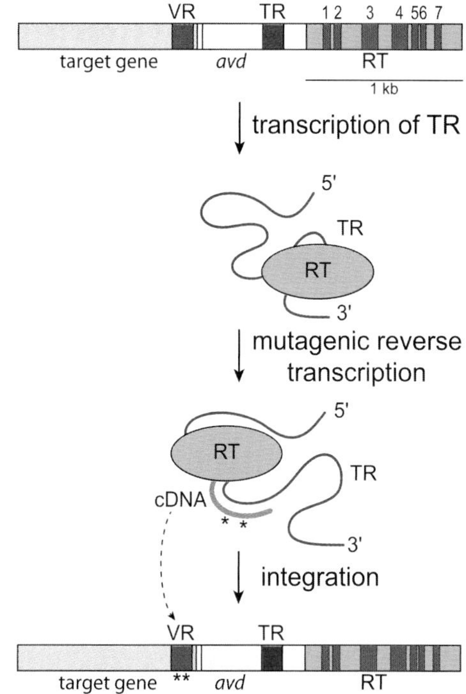

Figure 2 Diversity-generating retroelements (DGRs). A DGR consists of a reverse transcriptase (RT) gene with seven motifs, a target gene with a C-terminal variable region (VR), a template repeat (TR), and usually, an accessory variability determinant gene (*avd*) (drawn to scale for the *Bordetella* phage DGR [23]). The RT's thumb motif is not defined in sequence but presumably would be present downstream of motif 7. For the mutagenic homing reaction, the RT reverse transcribes the TR transcript and the resulting cDNA is integrated into the target gene to replace the previous VR sequence. During this process, each A in the TR sequence is mutagenized to any nucleotide, producing directed randomization of the VR sequence in the target gene.
doi:10.1128/microbiolspec.MDNA3-0058-2014.f2

in the amino acid sequence at the C-terminus of the target protein (23, 63, 64, 65, 66). In the case of the *Bordetalla* phage, the target protein is the phage's tail protein, and specifically the region that adheres to the cell surface of *Bordetella* during infection (67, 68). This is beneficial to the phage because *Bordetella* has two growth phases, virulent and avirulent, each bearing a characteristic cellular surface (69). The DGR allows the phage to adapt its tropism to the changing surface of its host bacterium by creating sequence diversity in the region of the tail protein that binds to the bacterial surface. Other DGR examples are known that are encoded on a bacterial chromosome rather than by a phage. Among these is the chromosomally encoded DGR of *Legionella pneumophila*, which creates sequence diversity in the C-terminus of a variable cell surface lipoprotein (70). Thus, DGRs represent a general mechanism of adaptation in bacteria that can benefit either phages or bacteria.

The genomic components of DGRs can vary considerably. Some DGRs lack the *avd* gene or have another accessory gene belonging to the helicase and RNase D C-terminal (HRDC) family. DGRs can have either one or two target genes, and there is variation as well in the order and orientations of the RT, TR, *avd*, and target genes (62). Relevant to this article is the fact that even without experimental validation, DGRs can be identified bioinformatically in genomic sequences through the detection of the two repeat sequences, TR and VR (~90% identity), which are adjacent to an RT gene, with the VR sequence bearing A-to-N differences relative to the TR sequence (71). This illustrates how a retroelement and its potential properties can be identified and/or classified based on sequence alone, a pertinent point when considering the many uncharacterized RT and RT-like sequences (below).

RETRONS: RETROELEMENTS IN SEARCH OF A FUNCTION

Retrons consist of an RT gene named *ret* and an adjacent inverted repeat sequence corresponding to the overlapping genes for the RNA and DNA segments of an msDNA (Fig. 3) (18, 72, 73, 74, 75). Both the RT and inverted repeat sequences are transcribed together as a single operon. The translated RT binds to the RNA's inverted repeat sequence. A specific G residue within the structure provides its 2′OH as a primer for reverse transcription (76, 77). The RNA is partially reverse transcribed, with most of the RNA template removed by a cellular RNase H activity. The resulting chimeric RNA-DNA molecule is covalently linked via the 2′OH priming bond and is also base paired at the 3′ ends of the RNA and DNA where the RNA is not digested by RNase H. The RT remains bound to this structure after its formation.

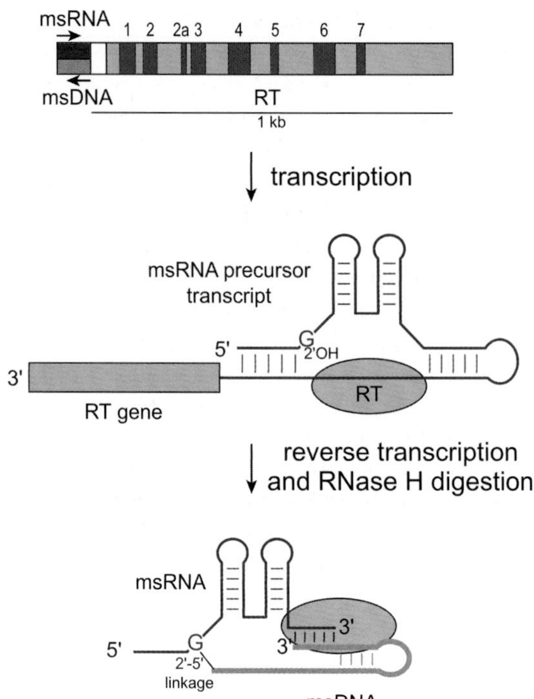

Figure 3 Retrons. A retron consists of an inverted repeat sequence corresponding to msRNA and msDNA genes, and a reverse transcriptase (RT) with seven conserved motifs (drawn to scale for retron Ec86 [77]). The thumb domain is presumably located directly downstream of motif 7. All three genes are transcribed in a single transcript and the RT binds to the RNA structure formed by the inverted repeat sequence. A specific G residue presents a 2′OH that acts as the primer for reverse transcription. After removal of the RNA template by cellular RNase H, the final msDNA consists of one RNA and one DNA linked by a 2′OH bond, and base paired at the 3′ ends of both.
doi:10.1128/microbiolspec.MDNA3-0058-2014.f3

Retromobility of retrons was suspected from the start, but this has not been demonstrated. The only experimental evidence for mobility is for the retron Ec73, which resides within the cryptic prophage R73 in *E. coli* (78). When R73 was mobilized by coinfection with a helper phage, the retron spread to another strain of *E. coli* and formed msDNA in the new host. This observation is consistent with retrons not being independently mobile, but being dispersed passively among strains or species. In another mobility-related observation, a comparison of different *E. coli* genome sequences led to the conclusion that the Ec107 retron replaced a 34 bp palindromic sequence through a *de novo* insertion; however, the mechanism was not elucidated (79). Similarly, two *Vibrio* species contain distinct retrons substituted at the same chromosomal locus, but again the mechanism of integration, loss, and/or replacement is unclear (80).

While the function of retrons remains unknown, there are two distribution patterns that are relevant. The first is vertical inheritance, which is found in soil-dwelling myxobacteria. The retron, Mx162, was found in all *Myxococcus xanthus* strains tested and in nearly all other myxobacterial species, while Mx65 is additionally found in some strains (the numbers refer to the length of the DNA component of the msDNA) (81, 82). The conclusion of vertical inheritance is supported by similar codon usage for the RTs and the host genomes (75). Vertical inheritance of the retrons implies that they carry out an essential—or at least a useful—function in their bacterial hosts.

The second distribution pattern is exemplified by retrons in *E. coli*, where presence of retrons is patchy. Most strains do not carry retrons, but there are seven distinct retrons among isolates. As a rough measure of frequency among *E. coli* strains, only 11 strains of the 72-strain ECOR collection contain a retron (83). The *E. coli* retron sequences are not closely related to each other, and they appear to have spread via horizontal transfers, a hypothesis supported by their codon usage, which differs from that of cellular genes (75). Because they are not universally present in the strains, the *E. coli* retrons are unlikely to carry out an essential function, although they may still carry out a useful function. The dynamics of their horizontal spread may be analogous to the propagation of antibiotic resistance genes among bacteria.

Two useful functions have been proposed for retrons, neither of which provides a satisfying general explanation for their existence. First, some retrons in *E. coli* have mutagenic effects on their host strains. The retrons, Ec83 and Ec86, have stem loops with mismatched base pairs that sequester the mismatch-repairing MutS protein, thereby leading to a higher mutation frequency. In certain contexts, an increased mutation rate might aid in bacterial adaptation (84, 85). Opposing this rationale is the fact that not all retrons contain mismatched bases in their stem-loops (86). A second potential function is based on the anecdotal observation that all 12 pathogenic *Vibrio* strains out of 21 isolates tested contain the retron Vp96, providing an apparent correlation between the retron and pathogenicity in this bacterium (86). However, the correlation does not hold for pathogenic and nonpathogenic strains of *E. coli*. Together, it would appear that retrons have an, as yet, unidentified biological function. Supporting this prediction is the presence of a putative protease domain at the C-terminus of some retron RTs (reference 87 and below), which suggests an undetected biochemical reaction of retrons.

A UNIVERSE OF UNCHARACTERIZED RTs AND RT-RELATED ENZYMES IN BACTERIA

Over the last decade, mounting sequence data have led to the identification of many new RTs and RT-like sequences in bacterial genomes. Two studies in 2008 reported numerous examples of novel bacterial RTs and retroelements (87, 88). An update of the compilation presented in reference 88 is summarized in Fig. 4, which is based on a set of 3,044 bacterial and archaebacterial RT sequences collected from GenBank (L. Wu and S. Zimmerly, unpublished). Despite their considerable sequence diversity, all putative RTs were identified as belonging to the superfamily of RTs (as opposed to other polymerases) by the NCBI Conserved Domain Database (CDD) (89). The classifications depicted in Fig. 4 are based on a combination of criteria, including the alignment of RT motif sequences, the presence of domain motifs or sequence extensions appended to the RT domain, and published data about characterized RTs (see reference 88 for a more detailed explanation).

As observed previously, approximately 90% of RT copies in bacterial genomes belong to characterized types of retroelements: group II introns (75%), retrons (12%), and DGRs (3%). Twenty-five additional groupings are now defined, which include five classes designated group II-like (G2L), which are related to group II introns but do not have an associated ribozyme structure. Due to the lack of experimental characterization, the classifications are not meant to imply unique functions for each class, but rather are an attempt to organize the diversity of uncharacterized RT-like sequences.

Five RTs remain unclassified, meaning that they have no close relatives among the set of 3044 RTs (E values to closest relatives range from e-6 to e-26 in BLASTP searches [Wu and Zimmerly, unpublished]). The five orphan sequences presumably represent classes that will become better defined when more DNA sequences are available. The existence of orphans indicates that sequence databases have not saturated coverage of the RTs, and we do not yet know all varieties of bacterial RTs and RT-like proteins that exist in nature. The sequences and alignments of all classified and unclassified RTs can be accessed in the Supplementary Data file.

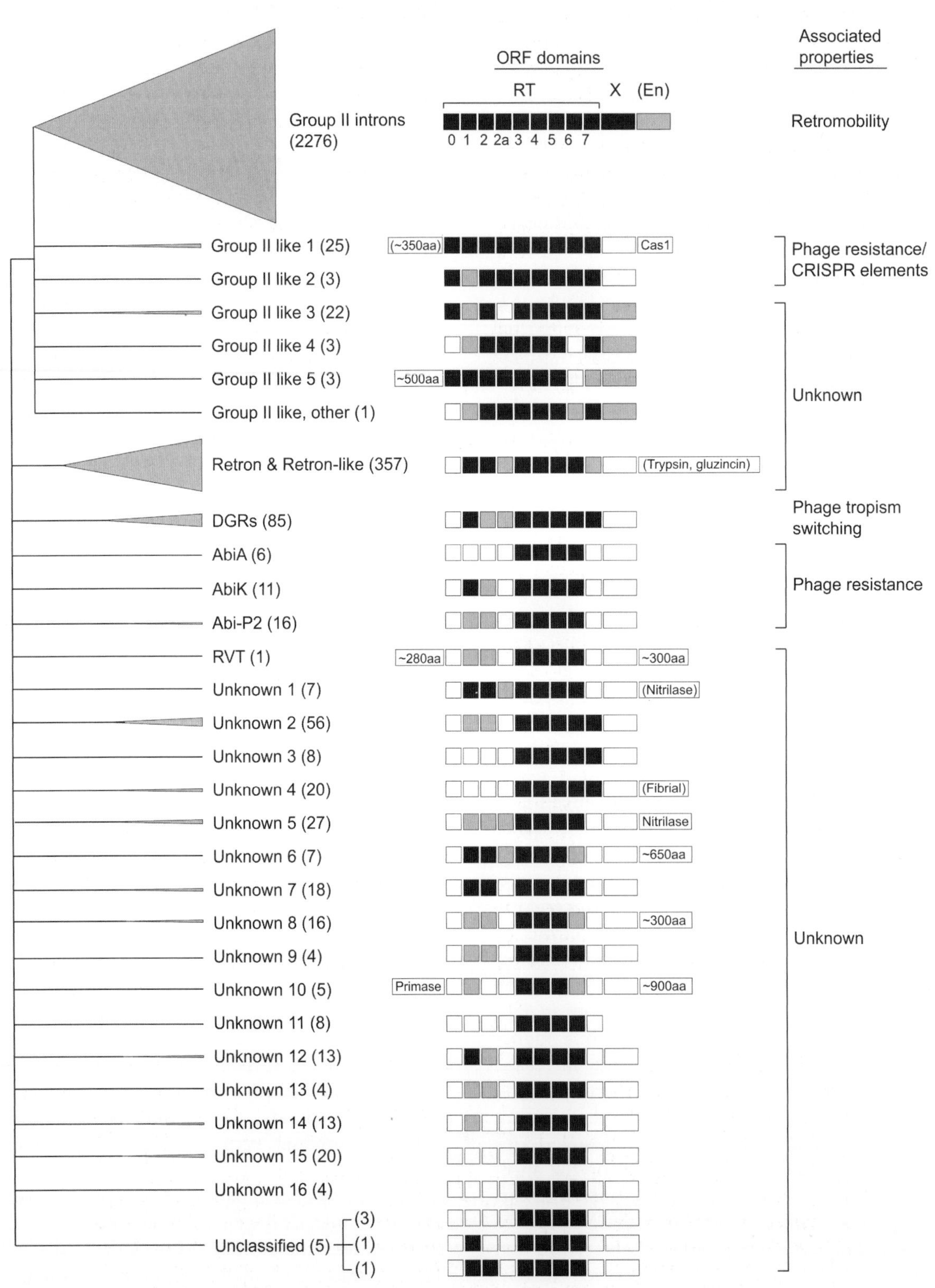

ORF domains
Associated properties
RT
X (En)
Group II introns (2276)
0 1 2 2a 3 4 5 6 7
Retromobility
Group II like 1 (25)
(~350aa)
Cas1
Group II like 2 (3)
Phage resistance/ CRISPR elements
Group II like 3 (22)
Group II like 4 (3)
Group II like 5 (3)
~500aa
Group II like, other (1)
Unknown
Retron & Retron-like (357)
(Trypsin, gluzincin)
DGRs (85)
Phage tropism switching
AbiA (6)
AbiK (11)
Abi-P2 (16)
Phage resistance
RVT (1)
~280aa
~300aa
Unknown 1 (7)
(Nitrilase)
Unknown 2 (56)
Unknown 3 (8)
Unknown 4 (20)
(Fibrial)
Unknown 5 (27)
Nitrilase
Unknown 6 (7)
~650aa
Unknown 7 (18)
Unknown 8 (16)
~300aa
Unknown 9 (4)
Unknown 10 (5)
Primase
~900aa
Unknown 11 (8)
Unknown 12 (13)
Unknown 13 (4)
Unknown 14 (13)
Unknown 15 (20)
Unknown 16 (4)
Unclassified (5)
(3)
(1)
(1)
Unknown

All 3,044 putative RTs align clearly with other RTs across RT motifs 3, 4, and 5, which correspond to the palm and finger domains of the polymerase, and include the three aspartate residues that coordinate the two divalent metals at the active site (Fig. 5). In addition, most RTs have a recognizable motif 6, whose conserved lysine is also an active site residue that acts as a general acid to facilitate the pyrophosphate leaving group (90). The remaining motifs (0, 1, 2, 2a, and 7) are less conserved, with motifs 0 and 2a being the least conserved (Fig. 4 and Fig. 5). Apart from the core polymerase structure (motifs 3 to 6), the putative RTs can be predicted to have a ~50 to 150 aa thumb domain located directly downstream of the RT domain. Thumb domains have little or no sequence conservation across RT types, but their location in sequence and three-dimensional structure is generally conserved across polymerases (91). An exception is unknown class 11, which does not have sufficient C-terminal sequence to encode a thumb domain. Overall, all RT classes except for unknown class 11 are expected to have very similar three-dimensional RT core structures with a thumb domain, consistent with their common functions as polymerases (92).

About two-thirds of uncharacterized RT classes have no motifs or extensions in addition to the predicted RT structure (Fig. 4). These RTs range in size from ~200 aa (unknown class 11) to ~650 aa, with the variability in size due mainly to indels within motifs 1 to 7 and X. None of the indel sequences match domains defined in either the CDD database of NCBI or the Pfam database (89, 93; Wu and Zimmerly, unpublished). About a third of the RT classes have sizeable extensions, containing either conserved domain motifs identified by CDD or Pfam, or minimally conserved sequence extensions of >300 aa found among members of the class. In all there are 11 domain architectures of putative RT proteins, with six conserved domains appended to RT domains and five less conserved extensions of >300 amino acids (Fig. 4).

RETROELEMENTS ASSOCIATED WITH PHAGE IMMUNITY

While most RT classes in Fig. 4 are wholly uncharacterized, there is limited data for several proteins that give indications of their properties. Three bacterial RT-related proteins are implicated in phage resistance: AbiA, AbiK, and Abi-P2. AbiA and AbiK are encoded by *Lactococcus lactis* plasmids and provide phage immunity through abortive infection (Abi), a process in which phage DNA enters a bacterium but phage multiplication is subsequently blocked (Fig. 6A). A third RT-related protein, found in the P2 prophage of some *E. coli* strains and here called Abi-P2, operates through phage exclusion, which is a process that prevents new phage infections of cells that contain the protective mechanism. For AbiK, it has been demonstrated that the RT protein alone confers phage resistance. This issue is not clarified for AbiA or Abi-P2, but the genetic loci conferring resistance are small (<3 kb).

AbiK

The *abiK* gene is encoded by the natural plasmid pSRQ800, which was discovered in a *L. lactis* strain isolated from raw milk (94). The single copy, constitutively transcribed gene, confers resistance against the three major classes of lactococcal phages (936, c2, and P335), typically at a level of six orders of magnitude over sensitive cells (95). The conserved RT motifs are in the N-terminal half of the 599 aa AbiK protein, while a C-terminal region of ~275 aa contains the presumed thumb domain of the polymerase as well as additional sequence with no identified domain motifs (Fig. 6B). Nevertheless, the C-terminal region has an essential function because a 42 amino acid C-terminal deletion eliminates abortive infection (94). Point mutations in the RT motifs, including the predicted catalytic aspartate residues, disrupt or abolish abortive infection, consistent with AbiK being a polymerase (95, 96).

AbiK expressed in *E. coli* and purified as a GST fusion protein showed polymerase activity *in vitro*, but

Figure 4 Classes of reverse transcriptases (RTs) and RT-like sequences in bacterial genomes. The figure is an update of Fig. 1 in reference 88. A set of 3,044 RTs were collected from GenBank and classified according to alignability of RT motif sequences, phylogenetic analyses, and the presence of additional domains. The number of members in each class is indicated in parentheses and by the area of the gray triangles. RT motifs are denoted by boxes that are either black (clearly alignable with group II introns), gray (ambiguously alignable), or white (not alignable, although an analogous structure is expected to be present). Sizeable extensions to the RTs are indicated by amino acid sizes, and are in parentheses when the motif is present in fewer than half of the examples. Protein motifs identified by either CDD of NCBI or Pfam are Cas1, trypsin, gluzincin, nitrilase, fimbrial, and primase. Biological properties associated with the different classes are indicated to the right. doi:10.1128/microbiolspec.MDNA3-0058-2014.f4

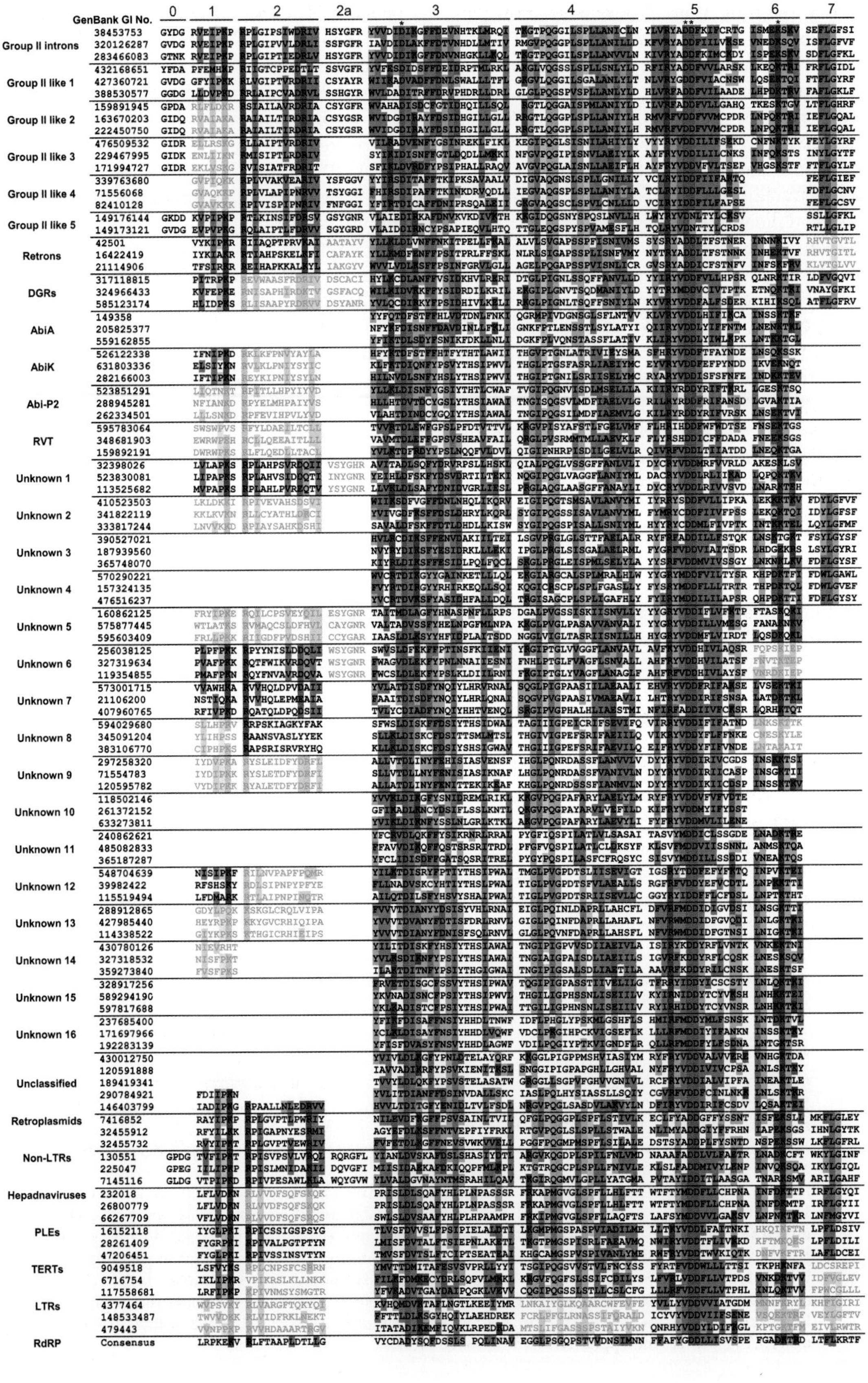

	GenBank GI No.	0	1	2	2a	3	4	5	6	7
						*		**	*	
Group II introns	38453753	GYDG	RVEIPKP	RPLGIPSIWDRIV	HSYGFR	YVVDIDIKGFFDEVNHTKLMRQI	TKGTPQGGILSPLLANICLN	YLVRYADDFKIFCRTG	ISMEKSKV	SEFLGFSI
	320126287	GVDG	RVEIPKP	RPLGIPVVLDRLI	SSFGFR	IAVDIDLAKFFDTVNHDLLMTIV	RMGVPQGGPLSPLLANILLD	KFVRYADDFIILVKSE	VNEDKSQV	ISFLGFVF
	283466083	GTNK	RVEIPKP	RPLGIPTIEDRLV	HSYGFR	YVVDIDIKGFFDNVNHGKLLKQL	TKGTPQGGILSPLLANVVLN	KIVRYADDFKIMCKDY	ISPEKSKV	SDFLGFKL
Group II like 1	432168651	YFDA	PFEMHKP	RIIGTCPPEDTLT	SSVGFR	YVFESDIDSFFDEIDRPTMLRKL	AKGLVQGSSLSPLLSNLYLD	RFIRYADDFVVLARSK	LKEQKTRI	FRFLGIDL
	427360721	GVDG	GFYIPKK	RLVGIPTVRDRII	CSYAYR	WIIKADVADFFDNLSWALLLTFL	GKGVLQGGILSGALANLYLT	NLVRYGDDFVIACNSW	LQSEKTQI	FTFLGYRF
	388530577	GGDG	LLDVPKD	RRLAIPCVADRVL	SSHGYR	WVLDADITRFFDRVPHDRLLDRL	GLGLPQGSPVSPLLANLYLD	HLVRFADDFVILAADE	LHPDKTRV	FAFLGKLF
Group II like 2	159891945	GPDA	RLFLDKR	RSIAILAVRDRIA	CSYGFR	WVAHADISDCFGTIDHQILLSQL	RGTLQGGAISPMLANIYLD	ILVRFADDFVLLGAHQ	TKESKTGV	LTFLGHRF
	163670203	GIDQ	RVAIAKA	RAIAILTIRDRIA	CSYGSR	WVIDGDIRAYFDSIDHGILLGLL	RRGTLQGGPLSPLLANIYLH	RMVRFVDDFVVMCPDR	LNPQKTRI	IEFLGQAL
	222450750	GIDQ	RVAIAKA	RAIAILTIRDRIA	CSYGSR	WVIDGDIRAYFDSIDHGILLGLL	RRGTLQGGPLSPLLANIYLH	RMVRFVDDFVVMCPDR	LNPQKTRI	IEFLGQAL
Group II like 3	476509532	GIDR	ELLRSKG	RLLAIPTVRDRIV		VILRADVENFYGSINREKLFIKL	KEGIPQGLSISNILAHIYLY	VYYRYVDDILIFSEKD	CNFNKTYK	FEYLGYRF
	229467995	GIDK	ENLIIKN	RMISIPTLRDRIV		SYIKIDISNFFGTLDQDLLMKKI	NFGVPQGIPISNVLAEIYLK	AYFRYVDDILILCNKS	INFQKSTS	INYLGYTF
	171994727	GIDR	EKLVSKG	RVISIATFRDRIT		SFIKLDVRDFYPSIPHALLRAQV	AVGVPQGLAISNLLAEIFLH	AYFRYVDDVFVLTSEP	VHGSKSTF	FTFLGYLF
Group II like 4	339763680		GVPIQKK	RPLVVAKVEARIV	YSFGGV	YYIRSDITAFFTKIPKSAVAALV	DIGVAQGNSLSPLLGNILLY	VCLRYIDDFIIFARTQ		FEFLGIEF
	71556068		GVAQKRP	RPLVLAPIPNRVV	TSYGGI	FHIRSDIPAFFTKINKDRVQDLL	IEGVAQGSPLSPLLANIYLA	TCLRYIDDFLLLGESL		FDFLGCNV
	82410128		GVAVKKK	RPIVISPIPNRIV	FNFGGI	YFIRTDICAFFDNIPRSQALEII	GKGVAQGSCLSPVLCNLLLD	VCIRYIDDFILFAPSE		FEFLGCSV
Group II like 5	149176144	GKDD	KVPIPKP	RTLKINSIFDRSV	GSYGNR	VLAIEDIRKAFDNVKVKDIVKTH	KKGIDQGSNYSPQSLNVLLH	LWYRYVDNLTYLCKSV		SSLLGFKL
	149173121	GVDG	EVPVPKG	RQLAIPTLFDRVV	SGYGRD	VLAIDDIRNCYPSAPIEQVLHTQ	TTGLEQGSPYSPVAMESFLH	TQLRYVDNTTYLCRDS		RTLLGLIP
Retrons	42501		VYKIPKR	RIIAQPTPRVKAI	AATAYV	YLLKLDLVNFFNKITPELLFKAL	ALVLSVGAPSSPFISNIVMS	SYSRYADDLTFSTNER	INNNKIVY	RHVTGVTL
	16422419		IYKIAKR	RTIAHPSKELKFI	CAFAYK	YLLKMDFENFFPSITPRLFFSKL	NLRLSIGAPSSPLISNFVMY	NYTRYADDLTFSTNNK	INHEKTVF	RHVTGITL
	21114906		TFSIRKR	REIHAPKKALKYL	IAKGYV	WVVLVDLKSFFPSINFGRVLGLL	AGELPQGAPSSPVISNLICR	GVSRYADDICFSTNFV	INFSKFRV	KLVTGLVV
DGRs	317118815		PITRPKP	REVWAASFRDRIV	DSCACI	HYLKCDLANFFVSIDKHVLRERI	DTGLPIGNLSSQFFANVLLD	YYIRYVDDFVLLHPSR	QLNRKTIR	LDFVGQVI
	324966433		KVFEPKE	RNISAPHIRDKTV	GSFACQ	WILKIDVKKFFYSIDRDILKRIL	EKGIPLGNVTSQDMANIYLD	YYTRYMDDVCIVTPTK	ETNQKTKI	VNAYGFKI
	585123174		HLIDPKS	RLISAAPYRDRVV	DSYANR	YVLQCDIRKYFPSIDHIVLKELI	RKGLPIGNLTSQFFSNIYLN	KYVRYVDDFALFSDER	KIHIKSQL	ATFLGFRV
AbiA	149358					YYFQTDFSTFFHLVDTDNLFNKI	QGRMPIVDGNSGLSFLNTVV	KLVRYVDDLHIFIKCA	INSSKTKF	
	205825377					NFYKFDISNFFDAVDINLLFKLI	GNKFPTLENSSTLSYLATYI	QIIRYVDDLYIFFNTM	LNENKTKL	
	559162855					YFIKTDLSDYFSNIKFDKLLNLI	DGKFPLVQNSTASSFLATIV	KLIRYVDDLYIWLKPK	LNTKKTGL	
AbiK	526122338		IFNIPKD	RKLKFPNVYAYLA		HFYKTDFSTFFHTFYTHTLAWII	THGVPTGNLATRIVIEYSMA	SFHRYVDDFTFAYNDE	LNSQKSSK	
	631803336		ELSIYKN	RVLKLPNIYSYIC		KLFKTDIQNFYPSVYTHSIPWVL	THGLPTGSFASRLIAEIYMC	EYVRYVDDFEFPYNDD	IKVERNQT	
	282166003		IFTIPKN	REYKIPNIYSYLN		HILNVDLSNFYHSLYTHSIPWVI	THGIPTGNILSRIISELYMC	RYARYVDDISFSFNFE	INDKKTEV	
Abi-P2	523851291		LIQTNKT	RPITLLHPYIYVD		YLLKLDISNFYGSIYTHTLCWAF	TVGIPQGNVISDLMSELLLA	KIIRYRDDYRIFTKRL	LGESKTSQ	
	288945281		NFIANKD	RPYELMHPAIYVS		HLLHTDVTDCYGSLYTHSIAWAI	TNGISQGSVLMDFIAELVLG	RILKYRDDFRIFANSD	LGVAKTIA	
	262334501		LLLSNKD	RPFEVIHPVLYVD		VLAHTDINDCYGQIYTHSIAWAL	TNGIPQGSILMDFIAEMVLG	KILRYRDDYRIFVRSK	LNSEKTVI	
RVT	595783064		SWSWPVS	RFYLDAEILTCLL		TVVRTDLEWFGPSLPFDTVTTVL	KRGVPISYAFSTLFGELVMF	FLHRIHDDFWFWDTSE	FNSEKTGS	
	348681903		EWRWPEH	RCLLQEEAITLLL		VAVMTDLEFFGPSVSHEAVFAIL	QRGLPVSRMMTMLLAEVKLF	FLYRSHDDICFFDADA	FNEEKSGS	
	159892191		DWRWPKS	RLFLQEDLLTACL		YVLKTDFRDYYPSLNQQFVLDVL	QIGIPNHRPISDILGELVLR	QIVRFVDDLTIIATDD	LNEQKTGA	
Unknown 1	32398026		LVLAPKS	RPLAHPSVRDQII	VSYGHR	AVITADLSQFYDRVRPSLLHSKL	QIALPQGLVSSGFFANLVLI	DYCRYVDDMRFVVRLD	AKESKLSV	
	523830081		LIPAPKS	RPLAHVSIDDQTI	INYGNR	YEIHLDFSKFYDSVDRTILTEKI	NQGIPQGLVAGGFLANIYML	DACRYVDDLRLIIKAD	LQPQKTKV	
	113525682		MVPAPKS	RPLAHLPVREQTV	YSYGNR	LLVKLDLSAFYDNIDVGRLIESL	PRGLAQGLAASGFFANAYLI	DYCRYVDDLRIVVSVD	LNARKTEH	
Unknown 2	410523503		LKLDKKI	RPIVEVAHSDSVI		WIIKSDFVGFFDNLNHQLIKQRV	EIGIPQGTSMSAVLANVYMI	IYRRYSDDFVLLIPKA	LEKKKTKV	FDYLGFVF
	341822119		KKLKVKN	RLLCYATHLDRCI		YVIVGDFKSFFDSLDHRYLKQRL	SYGIPQGSSISAVLANVYML	FYMRYCDDFIIVFPSS	LEKQKTQI	IDYLGFSF
	333817244		LNVVKKD	RPIAYSAHKDSHI		SAVALDFSKFFDTLDHDLLKISW	SYGIPQGSPISALLSNIYML	HYYRYCDDMLFIVPTK	INTKKTEL	LQYLGFMF
Unknown 3	390527021					HVLRCDIKSFFENVDAKIILTEI	LSGVPRGLGLSTTFAELALR	RYFRFADDILLFSTQK	LNSKTGKT	FSYLGYSF
	187939560					NVYRYDIKSFYESIDRKLLLEKI	IPGLPRGLSISGALAELRML	FYGRFVDDVIAITSDR	LHDGEKRS	LSYLGYRI
	365748070					KIYRLDIKSFFESIDLPQLFQCL	SKGLPRGLEISPMLSELYLA	YYSRFVDDMVIVSSGY	LNKNKLKV	FDFLGYSF
Unknown 4	570290221					WVCRTDIKGYYGAINKETLLLQL	ERGIARGCALSPLMRALHLW	YYGRYMDDFVILTYSR	KHPDKTFI	FDWLGAWL
	157324135					FVYRTDIRGYYRHIRKEQLLSQI	KQGICRSCPLSPLFGASLLY	FYSRYMDDFLLLTRTR	THPDKTQL	FDWLGVEF
	476516237					YVCRTDVKSFYASIDHFALLDQL	TRGISAGCPLSPLIGAFHLY	FYIRYMDDILILAPSR	QHPDKTTI	FDFLGYSY
Unknown 5	160862125		FRYIPKE	RQILCPSVEYQIL	ESYGNR	TAITMDLAGFYHNASPNFLLRPS	DGALPVGSSISKIISNVLLY	YYGRYVDDIFLVFKTP	FTASKQKI	
	575877445		WTLATKS	RVMAQCSLDFHVL	CAYGNR	VALTADVSSFYHELNPGFMLNPA	KKGLPVGLPASAVVANVALI	YYGRYVDDILLVMESG	FANAKNKV	
	595603409		FRLLPKK	RIIGDFPVDSHII	CCYGAR	IAASLDLKSYYHFIDPLAITSDD	NGGLVIGLTASRIISNILLH	HYGRYVDDMFLVIRDT	LQSDKQKL	
Unknown 6	256038125		PLPFPKK	RPYYNISLDDQLI	WSYGNR	SWVSLDFEKFFPTINSFKIIENI	YRGIPTGLVVGGFLANVAVL	AVFKYVDDHIVLAQSR	FQPSKIEP	
	327319634		PVAFPKK	RQTFWIKVRDQVT	WSYGNR	FWAGVDLEKFYPNLNNAIIESNI	FNHLPTGLFVAGFLSNVALL	AHFRYVDDHVILATSF	FNVTKTEP	
	119354855		PMAFPKN	RQYFNVAVRDQVA	WSYGNR	FWCSLDLEKFYPSLKLDIILRNI	FKGIPTGLYVAGFLANAGLF	AHFRFVDDHIVLAYSF	VNRDKIEP	
Unknown 7	573001715		VVAWHKA	RVVHQLDPVDAII		YVLATDISDFYNQIYLHRVRNAL	SQGLPIGPAASIILAEAALI	EHVRYVDDFRIFAKSE	LVSEKTKI	
	21106200		NSTIQKA	RVVHQLEPMEAIA		FAAITDISDFYNQIYLHRLQNAI	SQGVPVGPAASIVMAEAVLI	LHTRYVDDIRIFSNSA	LATDKTKL	
	407960765		RFIVPKD	RQATQLDPQDSII		TVLYCDIADFYNQIYHHTVENQL	SRGIPVGPHALHLIAESTMI	NFIRFADDIIVFCKSR	LQRHKTQT	
Unknown 8	594029680		SLLHPRV	RRPSKIAGKYFAK		SFWSLDISKFFDSIYTHSIDWAL	TAGIIIGPEICRIFSEVIFQ	VIRRYVDDIFIFATND	LNKSRTTK	
	345091204		YLIHPSS	RAANSVASLYYEK		SLLKLDISKCFDSITTSMLMTSL	THGIVIGPEFSRIFAEIILQ	VIKRYVDDYFLFFNKE	CNESKYLE	
	383106770		CIPHPKS	RAPSRISRVRYHQ		KLLKLDISKCFDSIYSHSIGWAV	THGIIIGPEFSRIFAEVILQ	EIFRYVDDYFIFVNDE	LNTAKAIT	
Unknown 9	297258320		IYDVPKA	RYSLEIDFYDRFL		ALLVTDLINYFEHISIASVENSF	IHGLPQNRDASSFLANVVLV	DYYRYVDDIRIVCGDS	INSKKTSI	
	71554783		IYDIPKK	RYSLETDFYDRFI		SLLVTDLLNYFENISIASIKNAF	LHGLPQNRDASSFVANIVLN	DYYRYVDDIRIICASP	INSGKTII	
	120595782		VYDIPKK	RYALETDFYDRFI		ALLATDLINYFENITTEKIKEAF	KHGLPQNRDSSSFIANMVLN	DYYRYVDDIKIICDSP	INSSKTKV	
Unknown 10	118502146					YVVKLDIKGFYSNIDREMLRIKL	KRGVPQGPAFARYLAELYLM	RYFRYVDDVFVFVDTE		
	261372152					GFIKADLKNCYDSIFLSSLKNTL	QKGVPQGPAYARVLVEFILD	KIFRYVDDMYIFYDST		
	633273811					YVIKLDIKNFYSSLNLGRLKTKL	AKGVPQGPAFARYLAEVYLI	FYFRYVDDMVLILENE		
Unknown 11	240862621					YFCRVDLQKFFYSIKRNRLRRAL	PYGFIQSPILATLVLSASAI	TASVYMDDICLSSGDE	LNADKTRE	
	485082833					FFAVVDIKQFFQSTSRSRITRDL	PFGFVQSPILATLCLDKSYF	KLSVFMDDVIISSNNL	ANMSKTQA	
	365187287					YFCLIDISDFFGATSQSRITREL	PYGYPQSPILASFCFRQSYC	SISVYMDDILLSSDDI	VNEAKTQS	
Unknown 12	548704639		NISIPKF	RILNVPAPFPQMR		YILKTDISRYFPTIYTHSIPWAL	TMGLPVGPDTSLIISEVIGT	IGSRYTDDFEFYFKTQ	INPVKTEI	
	39982422		RFSHSKY	RDLSIPNPYPFYE		FLLNADVSKCYHTIYTHSLPWAL	TIGLPVGPDTSFVLAEALLS	RGFRFVDDYEFVCDTL	LNPKKTTI	
	115519494		LFDMARK	RTLAIPNPINQTR		AILQTDILSFYHSVYSHAIPWAI	TIGLPVGPDTSRIISEVLLC	GGYRYIDDFFLCFDSL	LNPTKTHT	
Unknown 13	288912865		GDYLPQK	RSKGLCRQLVIPA		FVVVTDIANYYDSISYVHLRNAI	EIGLPQINLDAPRLLAHCFL	DFVRFMDDIDIGVDSI	LNSGKTTI	
	427985440		HEYRPKP	RKYGVCRHIQIPA		YVVVTDVANYFDTISFDRLRNVL	GIGLPQINFDAPRLLAHAFL	NFVRWMDDIDFGVQDI	LNLGKTKI	
	114338522		GIYKPKS	RTHGICRHIEIPS		YVVVTDIANYFDNISFSQLRNVL	GLGLPQVNFDAPRLLAHSFL	NFVRWMDDIDFGTNSI	LNIGKTKI	
Unknown 14	430780126		NIEVRHT			YILITDISKFYHSIYTHSIAWAL	TNGIPIGPVVSDIIAEIVLA	ISIRYKDDYRFLVNTK	VNKEKTNI	
	327318532		NISFPKT			YVLKSDIKNFYPSIYTHSIAWAI	TNGIAIGPAISDLIAEIILS	IGVRFKDDYRFLCQSK	LNESKSQV	
	359273840		FVSFPKS			ILAKTDITNFYPSIYTHGIGWAI	TNGIPIGSALSDLIAETILA	AAVRFKDDYRILCNSK	LNESKTSF	
Unknown 15	328917256					FRVETDISGCFSSVYTHSIPWAV	TQGIPIGPASSTIIVELILG	TFRRYIDDYICSCSTY	LNLQKTKI	
	589294190					YKVNADISNCFPSIYTHSIPWVL	THGILIGPHSSNLISEIILV	KYIRNIDDYTCYVKSH	LNHKKTEI	
	597817688					YKLKADISTCFPSIYTHSIPWAI	TTGILIGPHNSNLISEIILV	RYIRHIDDYTCYVNSR	LNHKKTKI	
Unknown 16	237685400					YFIKFDISAFFNSIYHHDLTNWF	IDFLPHGLYPSKMLGSHFLS	HMIRFMDDYMLFSNSK	LNTDKTVL	
	171697966					YCLKLDISAYFNSIYHHDLVQWF	VDCLPKGIHPCKVIGSEFLK	LLLRFMDDIYIFANKN	INSSKTKY	
	192283139					YFISFDVASYFNSVYHHDLAGWF	VDILPQGIYPTKVIGNDFLR	QLLRFMDDFYLFSDNA	LNTGKTSR	
Unclassified	430012750					YVIVLDLKGFYPNLDTELAYQRF	KKGGLPIGPPMSHVIASIYM	RYFRYVDDVALVVERE	VNHGKTDA	
	120591888					IAVVADIKRFYPSVKIENITKSL	SNGGIPIGPAPGHLLGHVAL	NYFRYVDDIVIVCPSA	LNLSKTRY	
	189419341					TVVYLDLQKFYPSVSTELASATW	GRGGLLSGPVFGHVVGNIVL	RCFRYVDDIALVIPFA	INEAKTLE	
	290784921		FDIIPKN			YVLITDIANFFDSINVDALLSKC	IASLPQLHYSIASSLLSQIY	CGVRFVDDFCINLNKR	LNLSKTKR	
	146403799		IADIPKG	RPAALLNLEDRVV		HVVLTDITGFYENIDLTVLFSDL	NRGVPQGLSASDVLAKVYLN	DFIRYVDDIRIFCSDV	LQSAKTEI	
Retroplasmids	7416852		RAYIPKP	RPLGVPTLPWRIY		NILEVDFKGFFPSVSAINLTVIL	QFGLPQGGPLSPFLSTIVLK	ECLFYADDGFFYSSNT	ISFEKSLL	MKFLGLEY
	32455912		RFYILKK	RPIGAPNYESRMI		AGYEFDLKSFFNTVEPFIYFRKL	RTGVPQGLSLSPLLSTWALE	NLIMYADDGIYFFRHN	IAPEKSGS	IHNLGYTK
	32455732		RVYIPKT	RPLGVPTVEWRIV		FVFETDLKGFFNEVSVWKVVELL	PGGFPQGMPMSPFLSILALE	DSTSYADDPLFYSNTD	NSPEKSSW	LKFLGFRL
Non-LTRs	130551	GPDG	TVFIPKT	RPISVPSVLVRQL	RQRGFL	YIANLDVSKAFDSLSHASIYDTL	ARGVKQGDPLSPILFNLVMD	NAAAFADDLVLFAETR	LNADKCFT	WKYLGINF
	225047	GPEG	IILIPKP	RPISLMNIDAKIL	DQVGFI	MIISIDAEKAFDKIQQPFMLKPL	KTGTRQGCPLSPLLFNIVLE	KLSLFADDMIVYLENP	INVQKSQA	IKYLGIQL
	7145116	GLDG	VTPIPKD	RPIVPESAWLKLA	WQYGVW	YLVALDGVNAYNTMSRAHILQAV	TKGIRQGMVLGPLLYATGMA	PVTAYIDDITLAASGA	TNARKSMV	ARILGAHF
Hepadnaviruses	232018		LFLVDKN	RLVVDFSQFSKQK		PRISLDLSQAFYHLPLNPASSSR	FRKAPMGVGLSPFLLHLFTT	WTFTYMDDFLLCHPNA	INFDKTTP	IRFLGYQI
	26800779		LFLVDKN	RLVVDFSQFSKQK		PRISLDLSQAFYHLPLNPASSSR	FRKAPMGVGLSPFLLHLFTT	WTFTYMDDFLLCHPNA	INFDKMTP	IRFLGYII
	66267709		VFLVDKN	RLVVDFSQFSRQS		SWLSLDVSAAFYHLPLHPAAMPH	FRKIPMGVGLSPFLLAQFTS	LAFSYMDDVVLGAKSV	LNPNKTKR	LNFMGYVI
PLEs	16152118		IYGLPKI	RPICSSIGSPSYG		TLVSFDVVSLFPSIPIELALDTI	LKGMPMGSPASPVIADILME	LLTKYVDDLFAITNKI	HKQIKFTN	LPFLDSIV
	28261409		FYGRPKI	RPIVALPGTPTYN		LMISFDVTALFTSIEPNLAKETL	TKGTPMGSPISRLFAEAVMQ	NWIRYVDDTFLIVKKD	KFTMKQES	LPFLDILI
	47206451		FYGLPKI	RPIVSSINSVTYN		TMVSFDVTSLFTCIPTSEATEAI	KHGCAMGSPVSPIVANLYME	RWFRYVDDTWVKIQTK	DNFVKFTR	LAFLDCEI
TERTs	9049518		LSFVYKS	RPLCNPSFCSRRN		YMVTTDMITAFESVSVPRLLYYI	TSGIPQGSVVSTVLFNCYSS	FYRTFVDDWLLLTTSI	TKPHKNFA	LDCSREPI
	6716754		IKLIPKR	VPIKRSLKLLNKK		FILKFDMKECYDRLSQPVLMKKL	KRGVFQGFSLSSIFCDILYS	LFVRLVDDFLLVTPDS	VNKDKTVV	IDFVGLEV
	117558681		LRFIPKP	RPIVNMSYSMGTR		YFVKADVTGAYDAIPQGKLVEVV	CQGIPQGSSLSTLLCSLCFG	LLLRFVDDFLLVTPHL	INLQKTVV	FPWCGLLL
LTRs	4377464		WVPSVKY	RLVARGFTQKYQI		KVHQMDVKTAFLNGTLKEEIYMR	LNKAIYGLKQAARCWFEVFE	YVLLYVDDVVIATGDM	MNNFKRYL	KHFIGIRI
	148533487		TWVVDKK	RLVIDFRKLNEKT		FFTTLDLKSGYHQIYLAEHDREK	FCRLPFGLRNASSIFQRALD	ICYVYVDDVIIFSENE	VSQEKTNF	VEYLGFTV
	479443		VVNPPKP	RVVHDAAARTRGV		ITATADIKEMFIQVKLRPEDRDA	MTSLIFGASSSPSTAIYVKN	QNRHYVDDYLDIFKGL	KPTGKTFM	EIVLRWTR
RdRP	Consensus		LRPKEKV	RLFTAAPLDTLLG		VYCDADYSQFDSSLS PQLINAV	EGGLPSGQPSTVVDNSIMNN	FEAFYGDDLLISVSPE	FGADKTRD	LTFLKRTF

surprisingly, not the properties of a reverse transcriptase. Even without an RNA or DNA template, long DNAs were polymerized by AbiK having a "random" sequence, making AbiK's activity analogous to a terminal transferase (96). The newly synthesized DNA was covalently attached to the AbiK protein, because an amino acid of AbiK was the primer for polymerization. A similar self-priming reaction has been observed for the hepadnavirus RT (97), which uses the OH of a tyrosine residue as a primer. For the hepadnavirus enzyme, a priming domain is located in an N-terminal region upstream of the RT domains. By comparison, the priming site on the AbiK protein may lie in the 250 aa C-terminal domain, because the domain has an essential but undefined function.

Interestingly, phage mutants can be readily isolated in the laboratory that acquire resistance to the AbiK antiphage mechanism, giving clues as to how AbiK interacts with the phage replication process. For four different phages, AbiK-resistant mutations mapped to the *sak* genes (sensitivity to AbiK), *sak1*, *sak2*, *sak3*, and *sak4* (98). The four *sak* genes are not closely related to each other, but *sak1* and *sak2* have sequence similarity to single-strand annealing proteins (SSAPs) of the RAD52 and Erf families, respectively, while *sak4* is related to RAD51 recombinases (99, 100). Sak3 protein, having no known motifs, was shown experimentally to have SSAP activity, as was Sak1 (100, 101, 102). Taken together, the Sak proteins appear to participate in a step of phage genome replication.

The interaction between AbiK and Sak that blocks phage multiplication is predicted to be a functional interaction rather than direct binding. This conclusion is based on the diverse sequences of the four *sak* genes and the fact that the resistance mutations occur throughout the *sak* sequences (98). However, the precise nature of the interaction remains unclear, as does the overall mechanism by which AbiK mediates immunity to phages. Also unresolved is whether AbiK exhibits true RT activity (i.e., RNA-dependent DNA polymerization) under conditions not tested or whether AbiK is a highly derived enzyme that has lost its RT activity and evolved a new polymerase reaction.

AbiA and *Abi-P2*

AbiA shares many similarities with AbiK, although its sequence is distantly related. It is encoded by the lactococcal plasmid pTR2030, and at 628 amino acids, AbiA is comparable in size to AbiK, with a C-terminal extension lacking protein motifs. AbiA protects against the major classes of lactococcal phages with a sensitivity profile that is similar but not identical to the set of AbiK-sensitive phages (103, 104). AbiA is similarly inferred to act at the stage of phage DNA replication, because DNA products do not accumulate in infected bacteria (104, 105).

Phage mutations lending resistance to the AbiA mechanism mapped to a gene (ORF245) having motifs for RecA-like NTPases and phage P loops (nucleotide binding), which resemble features of the Sak proteins. A second group of resistance mutations mapped to an intergenic inverted repeat sequence lying 500 bp upstream of phage ORF245 (105), for which there is no reported parallel for AbiK.

The *abi-P2* gene is not plasmid-encoded but is found in two P2 prophages in *E. coli* strains ECOR30 and ECOR58 of the ECOR collection. The two *abi-P2* genes share ~75% identity. When ORF570 of ECOR30 was expressed on a plasmid in *E. coli* strain BL21 (DE3), the strain was rendered resistant at levels of 10^{-7} relative to infection by T5 phage, but there was no effect on infection by lambda, T2, T4, or T6 phages (106). The expressed ORF570 was reported to have RT activity in a biochemical assay, but the assays were with crude extracts under conditions where artifactual incorporation can occur (96).

Figure 5 Amino acid alignment of reverse transcriptase (RT) motifs 0 to 7 for different classes of RTs in bacteria and eukaryotes. Three example sequences are presented for each class for motifs 0 to 7. Sequences in black lettering and bold color shading are clearly alignable with group II introns, while sequences in gray and light color shading are ambiguously alignable. Sequences not shown indicate unalignability with group II RT sequence motifs, although similar structures are likely present in the proteins. Positions with >30% identity across the entire alignment are back-shaded in colors to highlight the most conserved residues across RT classes. For comparison, the sequences of major classes of eukaryotic RTs are listed, as well as a consensus sequence for the Pfam group, RNA-dependent RNA polymerase (RdRP) 1, which among RdRPs has the greatest alignability to group II RTs. Asterisks above the alignment mark the three catalytic aspartate residues in motifs 3 and 5 and the active site lysine in motif 6. DGRs, diversity-generating retroelements; LTR, long terminal repeat; PLEs, Penelope-like elements; TERT, telomerase reverse transcriptase. doi:10.1128/microbiolspec.MDNA3-0058-2014.f5

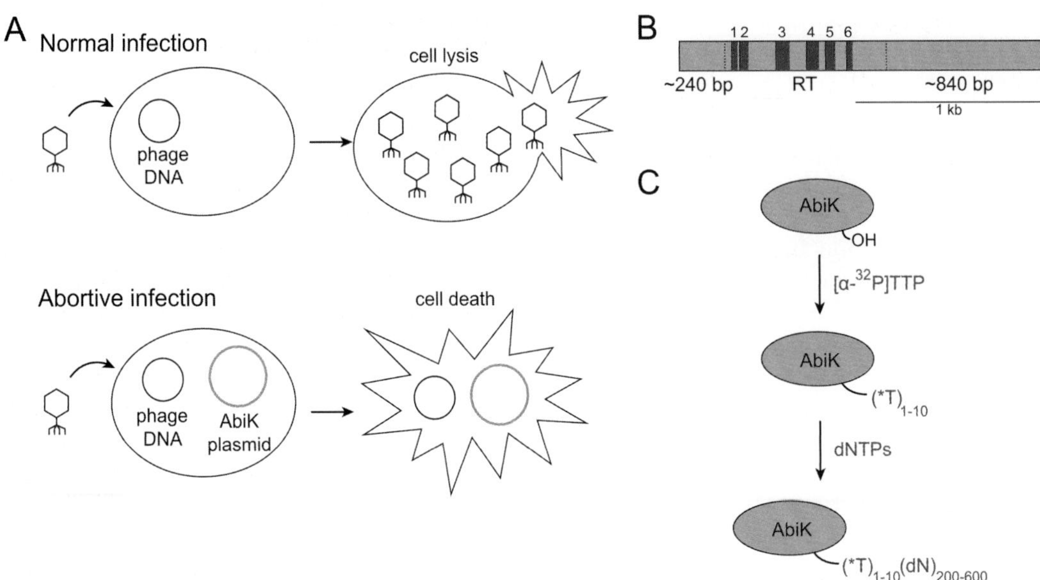

Figure 6 AbiK and abortive infection. (A) During abortive infection by AbiK, the phage injects its DNA into a *Lactococcus lactis* cell, but the multiplication cycle is blocked through an undefined mechanism by the AbiK protein. Although not necessarily a suicide mechanism, most cells still die and infective phages are not released. (B) The AbiK protein contains reverse transcriptase (RT) motifs 1 to 6 and 7 is essentially unalignable with group II introns. Estimates for the boundaries of the RT domain and thumb domain of the polymerase are indicated with dotted lines. In addition, the proteins contain a short N-terminal extension and a ~840 bp C-terminal extension (drawn to scale for AbiK of *Lactococcus lactis* [96]). (C) Purified AbiK protein has a terminal transferase activity, with the synthesized DNA becoming covalently linked to the AbiK protein. In the "label" reaction with low concentrations of [α-^{32}P]TTP, AbiK produce a short poly T DNA that is covalently linked to the AbiK protein. In the "chase" reaction, high concentrations of dNTPs cause polymerization of hundreds of nucleotides of heterogeneous sequence.
doi:10.1128/microbiolspec.MDNA3-0058-2014.f6

Together, the AbiK, AbiA, and Abi-P2 all give similar properties of resistance against phage infections. It is tempting to suspect that they use similar mechanisms; however, this cannot be automatically assumed because their RT sequences are quite distantly related (Fig. 5). For example, only RT motifs 4, 5, and 6 align convincingly between AbiA and AbiK in a pairwise alignment (Wu and Zimmerly, unpublished). Interestingly, homologs of AbiK, AbiA, and Abi-P2 are present in a wide range of species across multiple eubacterial phyla, including in clinical isolates (107; Wu and Zimmerly, unpublished). At the very least, AbiA, AbiK, and Abi-P2 point to a family of mechanisms of phage resistance conferred by RT-related proteins, which are not yet understood.

rvt RETROELEMENTS

Discovered only in 2011, *rvt* elements are found primarily in eukaryotes, but there are a few examples in bacteria (108). Their distribution is sporadic, with most examples found in diverse fungi. Homologs are identified in animals (rotifers), plants (moss), stramenophiles (water molds), and a bacterium (*Herpetosiphon aurantiacus*). The RVT proteins are 800 to 1,000 aa long, and include a ~300 aa N-terminal extension and ~200 aa C-terminal extension, neither of which contains a discernible protein motif (Fig. 7A). In most cases, *rvt* genes are found as single copies in genomes, although sometimes there are two nonidentical copies (108).

The only full-length bacterial *rvt* gene is found in *H. aurantiacus*, a predatory, filamentous gliding bacterium of the phylum Chloroflexi. Additionally, two partial sequences are reported from environmental samples deemed to be bacterial. The closest relatives of the *H. aurantiacus rvt* gene are not bacterial RTs but eukaryotic RTs, suggesting that the *rvt* gene was transferred horizontally from a eukaryote to a bacterium rather than vice versa.

The *rvt* gene of *Neurospora crassa* has been studied genetically and biochemically. Interestingly, an *rvt*

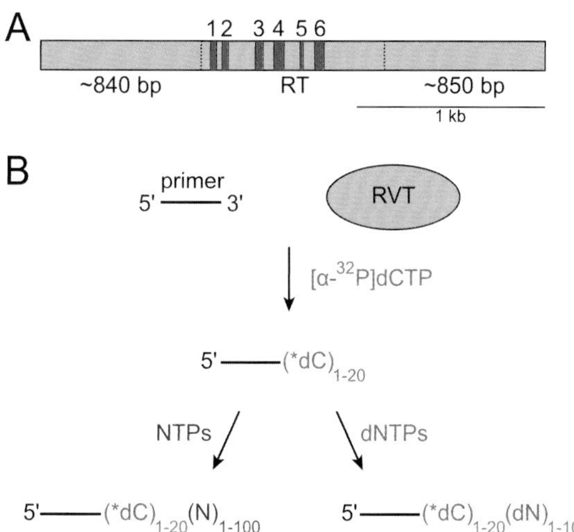

Figure 7 The *rvt* element. (A) The RVT ORF contains reverse transcriptase (RT) motifs 1 to 6, while motif 7 and thumb domains are unalignable with group II RTs but are presumably present in the polymerase structure. Estimates for the boundaries of the RT domain and thumb domain of the polymerase are noted with dotted lines. The large N-terminal and C-terminal extensions have no detectable protein motifs (drawn to scale for the *N. crassa* RVT [108]). (B) Purified RVT protein has terminal transferase activity that requires an RNA or DNA primer and has a preference for nucleoside triphosphates (NTPs) over deoxynucleoside triphosphates (dNTPs). When purified RVT protein is incubated with [α-^{32}P]dCTP, a short sequence is synthesized, which is extended by either NTPs or dNTPs in a chase reaction.
doi:10.1128/microbiolspec.MDNA3-0058-2014.f7

knockout in *N. crassa* exhibited no phenotype, revealing a nonessential role. Under normal laboratory growth conditions, the gene was transcribed at low levels, but transcription was elevated 50-fold when *N. crassa* was grown in the presence of drugs that disrupted translation or blocked histidine biosynthesis, implying that the element is responsive to stress conditions (108).

Like AbiK, purified *N. crassa* RVT showed terminal transferase activity and inability of template-dependent polymerization (Fig. 7B). Contrasting with AbiK's polymerization activity, RVT showed a strong preference for nucleoside triphosphates (NTPs) over deoxynucleoside triphosphates (dNTPs). A primer of either RNA or DNA was inferred to be used in the *in vitro* reactions but was not defined. Similar to AbiK, it remains possible that RVT possesses RNA-dependent DNA polymerization activity under conditions not tested. Also left unresolved is the biological function of the element. Paralleling retrons, the *rvt* elements presumably benefit their hosts through an unknown activity in order to account for their maintenance as single-copy genes over evolutionary time in the absence of retromobility.

RTs ASSOCIATED WITH CRISPR-Cas SYSTEMS

Although not experimentally investigated, a subset of bacterial RTs can be concluded to contribute to cellular defense against phages and foreign DNA through CRISPR/Cas systems, which are a large family of adaptive immunity systems in bacteria (109, 110, 111). The RTs classified as G2L1 and G2L2 are associated with *cas1* genes of CRISPR/Cas loci. Most G2L1 RTs are fused to the ORF of *cas1*, whereas G2L2 RTs are freestanding genes located adjacent to *cas1* genes (88). The *cas1* gene is universally present in all subtypes of CRISPR/Cas systems and encodes a metal-dependent nuclease that forms a complex with the Cas2 protein. The resulting nuclease activity is essential for integrating new spacer sequences into CRISPR arrays (112), giving the bacterium immunity against phages or plasmids containing those sequences. It seems likely that the CRISPR-associated RTs are involved in the insertion of new spacer sequences into CRISPR arrays using a mechanism similar to, and derived from, group II intron retromobility.

FUNCTIONS OF THE OTHER UNCHARACTERIZED RETROELEMENTS IN BACTERIA

Apart from group II introns, there is no evidence for active retromobility of bacterial RTs. None are present in genomes in more than one (identical) copy, nor are there easily discernible repeats flanking the RT genes that could be remnants of insertion events (Wu and Zimmerly, unpublished). If the RTs are, in fact, part of mobile DNAs, the mobility levels would have to be much lower than for group II introns or most other mobile DNAs. It is tempting to predict that most, if not all, of the putative RTs carry out useful functions for their host organisms and are retained and spread horizontally for the benefits they provide, rather than their retromobility.

A number of uncharacterized RTs have appended protein motifs that give hints as to their biochemical reactions. Peptidase motifs are found at the C-terminus of two subsets of retron RTs—either a trypsin-2 motif or a gluzincin (zinc-dependent) proteinase motif (87) (Fig. 4). Unknown classes 1 and 5 have an appended C-terminal nitrilase (C-N hydrolase) motif (87). Proteins with nitrilase domains break C-N bonds in nonpeptide cleavage reactions and they are implicated in processes of small molecule metabolism, detoxification, signaling, and posttranslational modification (89). The fimbrial domain found in unknown class 4 RTs is a pilus-related protein that serves structural roles in many processes including conjugation, intracellular trafficking, adhesion, and secretion (88). The primase domain in unknown class 10 suggests a concerted priming and reverse transcription reaction by the two polymerase domains of the protein.

A second source of information about the function of genomic elements can come from flanking genes or gene neighborhoods (87). Interestingly, RTs of unknown class 3 and unknown class 8 are found adjacent to each other in genomic sequences (87), suggesting that they operate together, perhaps forming a heterodimer and/or performing two distinct polymerization reactions in a process. More recent attempts to identify homologous flanking genes within the unknown classes did not identify additional conserved flanking genes as candidate cofactors for RT functions, although it is possible that such genes were overlooked (Wu and Zimmerly, unpublished).

Overall, despite the clues provided by the motifs appended to RT domains, there remains an exceedingly wide range of possibilities for the reactions and functions of the RTs. Given the range of properties uncovered for some of the example RTs, often involving phage resistance (Fig. 4), it seems highly likely that they are involved in novel biochemical reactions. We can look forward to interesting biochemistry and biological phenomena from these elements.

EVOLUTIONARY CONSIDERATIONS

The origin and evolution of RT has been the subject of considerable speculation (8, 113, 114, 115, 116, 117). The enzyme has often been postulated to date back to the transition from the RNA to the DNA world (118, 119, 120). According to this scenario, the first polymerase would have consisted of RNA molecule (ribozyme), which exhibited RNA-dependent RNA polymerase (RdRP) activity (121). The RNA-only enzyme would have evolved into an RNP enzyme and it would have eventually been replaced by a protein-only polymerase. One hypothesis holds that the catalytic region of contemporary polymerases is a remnant of that ancient, substrate-binding polypeptide, which then gained catalytic activity and supplanted the ribozyme polymerase (118). RT polymerases would have derived from this ancient protein RdRP, and helped to drive the transition from ancient RNA-based life forms to modern organisms with DNA as the dominant genetic material.

It is worth keeping in mind that all right hand polymerases had a common ancestor. These include, in addition to RTs, the A, B, and Y families of replicative DNA polymerases (e.g., *E. coli* pol I, *E. coli* pol II, and human pol I, respectively), single-subunit DNA-dependent RNA polymerases (e.g., T7 RNA polymerase), and RNA-dependent RNA polymerases (RdRPs) (91, 92). A recent study superimposed representative polymerase crystal structures and revealed that all six families had a common structural core of 57 superimposable amino acids, corresponding to the heart of the palm domain and including the aspartate residues that coordinate the catalytic metal ions (92). It can be reasonably assumed that the ancestral RT contained this 57 aa structural core. On the other hand, it may be grist for debate whether the first protein polymerase is best represented by RdRPs or RTs, or was a distinct enzyme that was less specialized than all contemporary polymerases.

Bacterial RTs are usually considered to be older than eukaryotic RTs. Specifically, bacterial group II intron RTs are thought to have given rise to eukaryotic non-LTR RTs, because of similarities in their sequences and mechanisms (8, 116, 117). In their sequences, group II and non-LTR RTs share motifs 0 and 2a (in addition to the conserved RT motifs 1 to 7), making them sister families of RTs (122) (Fig. 5). Mechanistically, group II introns and non-LTR elements share the retromobility mechanism of TPRT, in which the DNA target is cleaved by the RT's endonuclease domain to form a primer for reverse transcription (38, 123). Given that group II introns are widely considered to be the ancestors of nuclear spliceosomal introns (124), it is quite plausible that their invasion of the ancestral nuclear genome gave rise to nuclear non-LTR retroelements as well as spliceosomal introns (38). Other classes of eukaryotic retroelements (e.g., LTR elements, hepadnaviruses), which have more elaborate mechanisms and machineries, might have derived from non-LTR retroelements (8, 116, 125).

In any case, the many new types of RTs and RT-related proteins in bacteria raise new possibilities for the evolution of reverse transcriptase. The number of bacterial classes raises the possibility that RT might have migrated multiple times from bacteria to eukaryotes (in addition to the group II/non-LTR migration) because eukaryotic RTs themselves are not closely related (Fig. 5). The low sequence similarity of eukaryotic RTs might reflect different ancestors rather than a sequential pathway of eukaryotic diversification and evolution from a non-LTR retroelement. Another provocative possibility is suggested by the fact that among bacterial RTs, only group II introns appear to be actively mobile. This suggests that the oldest RTs may have been nonreplicative polymerases that performed useful functions in their bacterial hosts. From these RTs emerged a selfish DNA (group II intron) that was capable of multiplying through retrotransposition, which eventually gave rise to eukaryotic retroelements. Of course, the precise history of RT may be ultimately unknowable. Still, these bacterial RT-related enzymes expand the biochemical and biological properties possible for an ancestral reverse transcriptase, and are fodder for consideration of the origin and evolution of RTs and polymerases.

Citation. Zimmerly S, Wu L. 2014. An unexplored diversity of reverse transcriptases in bacteria. Microbiol Spectrum 3(1): MDNA3-0058-2014

References

1. **Baltimore D.** 1970. RNA-dependent DNA polymerase in virions of RNA tumour viruses. *Nature* **226:**1209–1211.
2. **Temin HM, Mizutani S.** 1970. RNA-dependent DNA polymerase in virions of Rous sarcoma virus. *Nature* **226:**1211–1213.
3. **Le Grice SF.** 2012. Human immunodeficiency virus reverse transcriptase: 25 years of research, drug discovery, and promise. *J Biol Chem* **287:**40850–40857.
4. **Hohn T, Rothnie H.** 2013. Plant pararetroviruses: replication and expression. *Curr Opin Virol* **3:**621–628.
5. **Glebe D, Bremer CM.** 2013. The molecular virology of hepatitis B virus. *Semin Liver Dis* **33:**103–112.
6. **Roy-Engel AM.** 2012. LINEs, SINEs and other retroelements: do birds of a feather flock together? *Front Biosci (Landmark Ed)* **17:**1345–1361.
7. **Eickbush TH, Jamburuthugoda VK.** 2008. The diversity of retrotransposons and the properties of their reverse transcriptases. *Virus Res* **134:**221–234.
8. **Eickbush TH.** 1994. Origin and evolutionary relationships of retroelements, p 121–157. *In* Morse SS (ed), *The Evolutionary Biology of Viruses*. Raven Press, New York, NY.
9. **Evgen'ev MB, Arkhipova IR.** 2005. Penelope-like elements–a new class of retroelements: distribution, function and possible evolutionary significance. *Cytogenet Genome Res* **110:**510–521.
10. **Blackburn EH, Collins K.** 2011. Telomerase: an RNP enzyme synthesizes DNA. *Cold Spring Harb Perspect Biol* **3:**a003558.
11. **Pardue ML, DeBaryshe PG.** 2003. Retrotransposons provide an evolutionarily robust non-telomerase mechanism to maintain telomeres. *Annu Rev Genet* **37:**485–511.
12. **Cordaux R, Batzer MA.** 2009. The impact of retrotransposons on human genome evolution. *Nat Rev Genet* **10:**691–703.
13. **Hancks DC, Kazazian HH Jr.** 2012. Active human retrotransposons: variation and disease. *Curr Opin Genet Dev* **22:**191–203.
14. **Konkel MK, Batzer MA.** 2010. A mobile threat to genome stability: the impact of non-LTR retrotransposons upon the human genome. *Semin Cancer Biol* **20:**211–221.
15. **Belfort M, Curcio MJ, Lue NF.** 2011. Telomerase and retrotransposons: reverse transcriptases that shaped genomes. *Proc Natl Acad Sci U S A* **108:**20304–20310.
16. **Lampson BC, Sun J, Hsu MY, Vallejo-Ramirez J, Inouye S, Inouye M.** 1989. Reverse transcriptase in a clinical strain of *Escherichia coli*: production of branched RNA-linked msDNA. *Science* **243:**1033–1038.
17. **Lim D, Maas WK.** 1989. Reverse transcriptase-dependent synthesis of a covalently linked, branched DNA-RNA compound in *E. coli* B. *Cell* **56:**891–904.
18. **Lampson BC.** 2007. Prokaryotic reverse transcriptases, p 403–420. *In* Polaina J, MacCabe AP (ed), *Industrial Enzymes: Structure, Function and Applications*. Springer, The Netherlands.
19. **Ferat JL, Michel F.** 1993. Group II self-splicing introns in bacteria. *Nature* **364:**358–361.
20. **Lambowitz AM, Zimmerly S.** 2011. Group II introns: mobile ribozymes that invade DNA. *Cold Spring Harb Perspect Biol* **3:**a003616.
21. **Lambowitz AM, Zimmerly S.** 2004. Mobile group II introns. *Annu Rev Genet* **38:**1–35.
22. **Belfort M, Derbyshire V, Parker MM, Cousineau B, Lambowitz AM.** 2002. Mobile introns: pathways and proteins, p 761–783. *In* Craig NL, Craigie R, Gellert M, Lambowitz AM (ed), *Mobile DNA II*. ASM Press, Washington DC.
23. **Liu M, Deora R, Doulatov SR, Gingery M, Eiserling FA, Preston A, Maskell DJ, Simons RW, Cotter PA, Parkhill J, Miller JF.** 2002. Reverse transcriptase-mediated tropism switching in *Bordetella* bacteriophage. *Science* **295:**2091–2094.
24. **Pyle AM, Lambowitz AM.** 2006. Group II introns: ribozymes that splice RNA and invade DNA, p 469–506. *In* Gesteland RF, Cech TR, Atkins JF (ed), *The RNA World*, 3rd ed. Cold Spring Harbor Laboratory Press, Cold Spring Harbor, New York, NY.
25. **Lehmann K, Schmidt U.** 2003. Group II introns: structural and catalytic versatility of large natural ribozymes. *Crit Rev Biochem Mol Biol* **38:**249–303.
26. **Mohr G, Perlman PS, Lambowitz AM.** 1993. Evolutionary relationships among group II intron-encoded proteins and identification of a conserved domain that may be related to maturase function. *Nucleic Acids Res* **21:**4991–4997.
27. **Cui X, Matsuura M, Wang Q, Ma H, Lambowitz AM.** 2004. A group II intron-encoded maturase functions preferentially in *cis* and requires both the reverse transcriptase and X domains to promote RNA splicing. *J Mol Biol* **340:**211–231.
28. **Saldanha R, Chen B, Wank H, Matsuura M, Edwards J, Lambowitz AM.** 1999. RNA and protein catalysis in group II intron splicing and mobility reactions using purified components. *Biochemistry* **38:**9069–9083.
29. **Wank H, San Filippo J, Singh RN, Matsuura M, Lambowitz AM.** 1999. A reverse transcriptase/maturase promotes splicing by binding at its own coding segment in a group II intron RNA. *Mol Cell* **4:**239–250.
30. **Matsuura M, Noah JW, Lambowitz AM.** 2001. Mechanism of maturase-promoted group II intron splicing. *EMBO J* **20:**7259–7270.

31. **Singh RN, Saldanha RJ, D'Souza LM, Lambowitz AM.** 2002. Binding of a group II intron-encoded reverse transcriptase/maturase to its high affinity intron RNA binding site involves sequence-specific recognition and autoregulates translation. *J Mol Biol* **318:**287–303.

32. **Michel F, Ferat JL.** 1995. Structure and activities of group II introns. *Annu Rev Biochem* **64:**435–461.

33. **Pyle AM.** 2010. The tertiary structure of group II introns: implications for biological function and evolution. *Crit Rev Biochem Mol Biol* **45:**215–232.

34. **Fedorova O, Zingler N.** 2007. Group II introns: structure, folding and splicing mechanism. *Biol Chem* **388:** 665–678.

35. **Marcia M, Pyle AM.** 2012. Visualizing group II intron catalysis through the stages of splicing. *Cell* **151:**497–507.

36. **Marcia M, Somarowthu S, Pyle AM.** 2013. Now on display: a gallery of group II intron structures at different stages of catalysis. *Mob DNA* **4:**14.

37. **Robart AR, Chan RT, Peters JK, Rajashankar KR, Toor N.** 2014. Crystal structure of a eukaryotic group II intron lariat. *Nature* **514:**193–197.

38. **Zimmerly S, Guo H, Perlman PS, Lambowitz AM.** 1995. Group II intron mobility occurs by target DNA-primed reverse transcription. *Cell* **82:**545–554.

39. **Zimmerly S, Guo H, Eskes R, Yang J, Perlman PS, Lambowitz AM.** 1995. A group II intron RNA is a catalytic component of a DNA endonuclease involved in intron mobility. *Cell* **83:**529–538.

40. **Cousineau B, Smith D, Lawrence-Cavanagh S, Mueller JE, Yang J, Mills D, Manias D, Dunny G, Lambowitz AM, Belfort M.** 1998. Retrohoming of a bacterial group II intron: mobility via complete reverse splicing, independent of homologous DNA recombination. *Cell* **94:**451–462.

41. **Smith D, Zhong J, Matsuura M, Lambowitz AM, Belfort M.** 2005. Recruitment of host functions suggests a repair pathway for late steps in group II intron retrohoming. *Genes Dev* **19:**2477–2487.

42. **Coros CJ, Landthaler M, Piazza CL, Beauregard A, Esposito D, Perutka J, Lambowitz AM, Belfort M.** 2005. Retrotransposition strategies of the *Lactococcus lactis* Ll.LtrB group II intron are dictated by host identity and cellular environment. *Mol Microbiol* **56:** 509–524.

43. **Yao J, Truong DM, Lambowitz AM.** 2013. Genetic and biochemical assays reveal a key role for replication restart proteins in group II intron retrohoming. *PLoS Genet* **9:**e1003469.

44. **Toro N, Martinez-Abarca F.** 2013. Comprehensive phylogenetic analysis of bacterial group II intron-encoded ORFs lacking the DNA endonuclease domain reveals new varieties. *PLoS ONE* **8:**e55102.

45. **Zhong J, Lambowitz AM.** 2003. Group II intron mobility using nascent strands at DNA replication forks to prime reverse transcription. *EMBO J* **22:**4555–4565.

46. **Cousineau B, Lawrence S, Smith D, Belfort M.** 2000. Retrotransposition of a bacterial group II intron. *Nature* **404:**1018–1021.

47. **Robart AR, Seo W, Zimmerly S.** 2007. Insertion of group II intron retroelements after intrinsic transcriptional terminators. *Proc Natl Acad Sci U S A* **104:**6620–6625.

48. **Eskes R, Liu L, Ma H, Chao MY, Dickson L, Lambowitz AM, Perlman PS.** 2000. Multiple homing pathways used by yeast mitochondrial group II introns. *Mol Cell Biol* **20:**8432–8446.

49. **Mastroianni M, Watanabe K, White TB, Zhuang F, Vernon J, Matsuura M, Wallingford J, Lambowitz AM.** 2008. Group II intron-based gene targeting reactions in eukaryotes. *PLoS ONE* **3:**e3121.

50. **White TB, Lambowitz AM.** 2012. The retrohoming of linear group II intron RNAs in *Drosophila melanogaster* occurs by both DNA ligase 4-dependent and -independent mechanisms. *PLoS Genet* **8:**e1002534.

51. **Muñoz-Adelantado E, San Filippo J, Martínez-Abarca F, García-Rodríguez FM, Lambowitz AM, Toro N.** 2003. Mobility of the *Sinorhizobium meliloti* group II intron RmInt1 occurs by reverse splicing into DNA, but requires an unknown reverse transcriptase priming mechanism. *J Mol Biol* **327:**931–943.

52. **Dai L, Zimmerly S.** 2002. Compilation and analysis of group II intron insertions in bacterial genomes: evidence for retroelement behavior. *Nucleic Acids Res* **30:**1091–1102.

53. **Candales MA, Duong A, Hood KS, Li T, Neufeld RA, Sun R, McNeil BA, Wu L, Jarding AM, Zimmerly S.** 2012. Database for bacterial group II introns. *Nucleic Acids Res* **40:**D187–D190.

54. **Simon DM, Clarke NA, McNeil BA, Johnson I, Pantuso D, Dai L, Chai D, Zimmerly S.** 2008. Group II introns in Eubacteria and Archaea: ORF-less introns and new varieties. *RNA* **14:**1704–1713.

55. **Toro N, Martinez-Rodriguez L, Martinez-Abarca F.** 2014. Insights into the history of a bacterial group II intron remnant from the genomes of the nitrogen-fixing symbionts *Sinorhizobium meliloti* and *Sinorhizobium medicae*. *Heredity* **113:**306–315.

56. **Michel F, Umesono K, Ozeki H.** 1989. Comparative and functional anatomy of group II catalytic introns–a review. *Gene* **82:**5–30.

57. **Toro N, Jiménez-Zurdo JI, García-Rodríguez FM.** 2007. Bacterial group II introns: not just splicing. *FEMS Microbiol Rev* **31:**342–358.

58. **Toro N, Martinez-Abarca F, Fernandez-Lopez M, Munoz-Adelantado E.** 2003. Diversity of group II introns in the genome of *Sinorhizobium meliloti* strain 1021: splicing and mobility of RmInt1. *Mol Genet Genomics* **268:**628–636.

59. **Dai L, Zimmerly S.** 2002. The dispersal of five group II introns among natural populations of *Escherichia coli*. *RNA* **8:**1294–1307.

60. **Ueda K, Yamashita A, Ishikawa J, Shimada M, Watsuji TO, Morimura K, Ikeda H, Hattori M, Beppu T.** 2004. Genome sequence of *Symbiobacterium thermophilum*, an uncultivable bacterium that depends on microbial commensalism. *Nucleic Acids Res* **32:**4937–4944.

61. **McNeil BA, Simon DM, Zimmerly S.** 2014. Alternative splicing of a group II intron in a surface layer protein

gene in *Clostridium tetani*. *Nucleic Acids Res* **42:** 1959–1969.

62. **Medhekar B, Miller JF.** 2007. Diversity-generating retroelements. *Curr Opin Microbiol* **10:**388–395.
63. **Doulatov S, Hodes A, Dai L, Mandhana N, Liu M, Deora R, Simons RW, Zimmerly S, Miller JF.** 2004. Tropism switching in *Bordetella* bacteriophage defines a family of diversity-generating retroelements. *Nature* **431:**476–481.
64. **Alayyoubi M, Guo H, Dey S, Golnazarian T, Brooks GA, Rong A, Miller JF, Ghosh P.** 2013. Structure of the essential diversity-generating retroelement protein bAvd and its functionally important interaction with reverse transcriptase. *Structure* **21:**266–276.
65. **Guo H, Tse LV, Barbalat R, Sivaamnuaiphorn S, Xu M, Doulatov S, Miller JF.** 2008. Diversity-generating retroelement homing regenerates target sequences for repeated rounds of codon rewriting and protein diversification. *Mol Cell* **31:**813–823.
66. **Guo H, Tse LV, Nieh AW, Czornyj E, Williams S, Oukil S, Liu VB, Miller JF.** 2011. Target site recognition by a diversity-generating retroelement. *PLoS Genet* **7:** e1002414.
67. **McMahon SA, Miller JL, Lawton JA, Kerkow DE, Hodes A, Marti-Renom MA, Doulatov S, Narayanan E, Sali A, Miller JF, Ghosh P.** 2005. The C-type lectin fold as an evolutionary solution for massive sequence variation. *Nat Struct Mol Biol* **12:**886–892.
68. **Miller JL, Le Coq J, Hodes A, Barbalat R, Miller JF, Ghosh P.** 2008. Selective ligand recognition by a diversity-generating retroelement variable protein. *PLoS Biol* **6:** e131.
69. **Cummings CA, Bootsma HJ, Relman DA, Miller JF.** 2006. Species- and strain-specific control of a complex, flexible regulon by *Bordetella* BvgAS. *J Bacteriol* **188:** 1775–1785.
70. **Arambula D, Wong W, Medhekar BA, Guo H, Gingery M, Czornyj E, Liu M, Dey S, Ghosh P, Miller JF.** 2013. Surface display of a massively variable lipoprotein by a *Legionella* diversity-generating retroelement. *Proc Natl Acad Sci U S A* **110:**8212–8217.
71. **Schillinger T, Lisfi M, Chi J, Cullum J, Zingler N.** 2012. Analysis of a comprehensive dataset of diversity generating retroelements generated by the program DiGReF. *BMC Genomics* **13:**430.
72. **Lampson BC, Inouye M, Inouye S.** 2005. Retrons, msDNA, and the bacterial genome. *Cytogenet Genome Res* **110:**491–499.
73. **Lampson B, Inouye M, Inouye S.** 2001. The msDNAs of bacteria. *Prog Nucleic Acid Res Mol Biol* **67:**65–91.
74. **Inouye S, Inouye M.** 1993. The retron: a bacterial retroelement required for the synthesis of msDNA. *Curr Opin Genet Dev* **3:**713–718.
75. **Inouye M, Inouye S.** 1991. msDNA and bacterial reverse transcriptase. *Annu Rev Microbiol* **45:**163–186.
76. **Inouye M, Ke H, Yashio A, Yamanaka K, Nariya H, Shimamoto T, Inouye S.** 2004. Complex formation between a putative 66-residue thumb domain of bacterial reverse transcriptase RT-Ec86 and the primer recognition RNA. *J Biol Chem* **279:**50735–50742.
77. **Inouye S, Hsu MY, Xu A, Inouye M.** 1999. Highly specific recognition of primer RNA structures for 2′-OH priming reaction by bacterial reverse transcriptases. *J Biol Chem* **274:**31236–31244.
78. **Inouye S, Sunshine MG, Six EW, Inouye M.** 1991. Retronphage phi R73: an *E. coli* phage that contains a retroelement and integrates into a tRNA gene. *Science* **252:**969–971.
79. **Herzer PJ, Inouye S, Inouye M.** 1992. Retron-Ec107 is inserted into the *Escherichia coli* genome by replacing a palindromic 34bp intergenic sequence. *Mol Microbiol* **6:**345–354.
80. **Shimamoto T, Ahmed AM, Shimamoto T.** 2013. A novel retron of *Vibrio parahaemolyticus* is closely related to retron-Vc95 of *Vibrio cholerae*. *J Microbiol* **51:**323–328.
81. **Lampson BC, Inouye M, Inouye S.** 1991. Survey of multicopy single-stranded DNAs and reverse transcriptase genes among natural isolates of *Myxococcus-xanthus*. *J Bacteriol* **173:**5363–5370.
82. **Rice SA, Lampson BC.** 1995. Phylogenetic comparison of retron elements among the myxobacteria: evidence for vertical inheritance. *J Bacteriol* **177:**37–45.
83. **Herzer PJ, Inouye S, Inouye M, Whittam TS.** 1990. Phylogenetic distribution of branched RNA-linked multicopy single-stranded DNA among natural isolates of *Escherichia coli*. *J Bacteriol* **172:**6175–6181.
84. **Maas WK, Wang C, Lima T, Hach A, Lim D.** 1996. Multicopy single-stranded DNA of *Escherichia coli* enhances mutation and recombination frequencies by titrating MutS protein. *Mol Microbiol* **19:**505–509.
85. **Maas WK, Wang C, Lima T, Zubay G, Lim D.** 1994. Multicopy single-stranded DNAs with mismatched base pairs are mutagenic in *Escherichia coli*. *Mol Microbiol* **14:**437–441.
86. **Yamanaka K, Shimamoto T, Inouye S, Inouye M.** 2002. Retrons, p 784–795. *In* Craig NL, Craigie R, Gellert M, Lambowitz AM (ed), *Mobile DNA II*. ASM Press, Washington DC.
87. **Kojima KK, Kanehisa M.** 2008. Systematic survey for novel types of prokaryotic retroelements based on gene neighborhood and protein architecture. *Mol Biol Evol* **25:**1395–1404.
88. **Simon DM, Zimmerly S.** 2008. A diversity of uncharacterized reverse transcriptases in bacteria. *Nucleic Acids Res* **36:**7219–7229.
89. **Marchler-Bauer A, Zheng C, Chitsaz F, Derbyshire MK, Geer LY, Geer RC, Gonzales NR, Gwadz M, Hurwitz DI, Lanczycki CJ, Lu F, Lu S, Marchler GH, Song JS, Thanki N, Yamashita RA, Zhang D, Bryant SH.** 2013. CDD: conserved domains and protein three-dimensional structure. *Nucleic Acids Res* **41:**D348–D352.
90. **Castro C, Smidansky ED, Arnold JJ, Maksimchuk KR, Moustafa I, Uchida A, Gotte M, Konigsberg W, Cameron CE.** 2009. Nucleic acid polymerases use a general acid for nucleotidyl transfer. *Nat Struct Mol Biol* **16:**212–218.

91. **Steitz TA.** 1999. DNA polymerases: structural diversity and common mechanisms. *J Biol Chem* **274:**17395–17398.

92. **Mönttinen HA, Ravantti JJ, Stuart DI, Poranen MM.** 2014. Automated structural comparisons clarify the phylogeny of the right-hand-shaped polymerases. *Mol Biol Evol* **31:**2741–2752.

93. **Finn RD, Bateman A, Clements J, Coggill P, Eberhardt RY, Eddy SR, Heger A, Hetherington K, Holm L, Mistry J, Sonnhammer EL, Tate J, Punta M.** 2014. Pfam: the protein families database. *Nucleic Acids Res* **42:**D222–D230.

94. **Emond E, Holler BJ, Boucher I, Vandenbergh PA, Vedamuthu ER, Kondo JK, Moineau S.** 1997. Phenotypic and genetic characterization of the bacteriophage abortive infection mechanism AbiK from *Lactococcus lactis*. *Appl Environ Microbiol* **63:**1274–1283.

95. **Fortier LC, Bouchard JD, Moineau S.** 2005. Expression and site-directed mutagenesis of the lactococcal abortive phage infection protein AbiK. *J Bacteriol* **187:**3721–3730.

96. **Wang C, Villion M, Semper C, Coros C, Moineau S, Zimmerly S.** 2011. A reverse transcriptase-related protein mediates phage resistance and polymerizes untemplated DNA *in vitro*. *Nucleic Acids Res* **39:**7620–7629.

97. **Wang GH, Seeger C.** 1992. The reverse transcriptase of hepatitis B virus acts as a protein primer for viral DNA synthesis. *Cell* **71:**663–670.

98. **Bouchard JD, Moineau S.** 2004. Lactococcal phage genes involved in sensitivity to AbiK and their relation to single-strand annealing proteins. *J Bacteriol* **186:**3649–3652.

99. **Lopes A, Amarir-Bouhram J, Faure G, Petit MA, Guerois R.** 2010. Detection of novel recombinases in bacteriophage genomes unveils Rad52, Rad51 and Gp2.5 remote homologs. *Nucleic Acids Res* **38:**3952–3962.

100. **Ploquin M, Bransi A, Paquet ER, Stasiak AZ, Stasiak A, Yu X, Cieslinska AM, Egelman EH, Moineau S, Masson JY.** 2008. Functional and structural basis for a bacteriophage homolog of human RAD52. *Curr Biol* **18:**1142–1146.

101. **Scaltriti E, Moineau S, Launay H, Masson JY, Rivetti C, Ramoni R, Campanacci V, Tegoni M, Cambillau C.** 2010. Deciphering the function of lactococcal phage ul36 Sak domains. *J Struct Biol* **170:**462–469.

102. **Scaltriti E, Launay H, Genois MM, Bron P, Rivetti C, Grolli S, Ploquin M, Campanacci V, Tegoni M, Cambillau C, Moineau S, Masson JY.** 2011. Lactococcal phage p2 ORF35-Sak3 is an ATPase involved in DNA recombination and AbiK mechanism. *Mol Microbiol* **80:**102–116.

103. **Hill C, Miller LA, Klaenhammer TR.** 1990. Nucleotide sequence and distribution of the pTR2030 resistance determinant (hsp) which aborts bacteriophage infection in lactococci. *Appl Environ Microbiol* **56:**2255–2258.

104. **Tangney M, Fitzgerald GF.** 2002. Effectiveness of the lactococcal abortive infection systems AbiA, AbiE, AbiF and AbiG against P335 type phages. *FEMS Microbiol Lett* **210:**67–72.

105. **Dinsmore PK, Klaenhammer TR.** 1997. Molecular characterization of a genomic region in a *Lactococcus* bacteriophage that is involved in its sensitivity to the phage defense mechanism AbiA. *J Bacteriol* **179:**2949–2957.

106. **Odegrip R, Nilsson AS, Haggård-Ljungquist E.** 2006. Identification of a gene encoding a functional reverse transcriptase within a highly variable locus in the P2-like coliphages. *J Bacteriol* **188:**1643–1647.

107. **Wattam AR, Abraham D, Dalay O, Disz TL, Driscoll T, Gabbard JL, Gillespie JJ, Gough R, Hix D, Kenyon R, Machi D, Mao C, Nordberg EK, Olson R, Overbeek R, Pusch GD, Shukla M, Schulman J, Stevens RL, Sullivan DE, Vonstein V, Warren A, Will R, Wilson MJ, Yoo HS, Zhang C, Zhang Y, Sobral BW.** 2014. PATRIC, the bacterial bioinformatics database and analysis resource. *Nucleic Acids Res* **42:**D581–D591.

108. **Gladyshev EA, Arkhipova IR.** 2011. A widespread class of reverse transcriptase-related cellular genes. *Proc Natl Acad Sci U S A* **108:**20311–20316.

109. **Barrangou R, Marraffini LA.** 2014. CRISPR-Cas systems: prokaryotes upgrade to adaptive immunity. *Mol Cell* **54:**234–244.

110. **Chylinski K, Makarova KS, Charpentier E, Koonin EV.** 2014. Classification and evolution of type II CRISPR-Cas systems. *Nucleic Acids Res* **42:**6091–6105.

111. **van der Oost J, Westra ER, Jackson RN, Wiedenheft B.** 2014. Unravelling the structural and mechanistic basis of CRISPR-Cas systems. *Nat Rev Microbiol* **12:**479–492.

112. **Nuñez JK, Kranzusch PJ, Noeske J, Wright AV, Davies CW, Doudna JA.** 2014. Cas1-Cas2 complex formation mediates spacer acquisition during CRISPR-Cas adaptive immunity. *Nat Struct Mol Biol* **21:**528–534.

113. **Curcio MJ, Belfort M.** 2007. The beginning of the end: links between ancient retroelements and modern telomerases. *Proc Natl Acad Sci U S A* **104:**9107–9108.

114. **Inouye S, Inouye M.** 1995. Structure, function, and evolution of bacterial reverse transcriptase. *Virus Genes* **11:**81–94.

115. **Nakamura TM, Cech TR.** 1998. Reversing time: origin of telomerase. *Cell* **92:**587–590.

116. **Eickbush TH, Malik HS.** 2002. Origins and evolution of retrotransposons, p 1111–1144. *In* Craig NL, Craigie R, Gellert M, Lambowitz AM (ed), *Mobile DNA II*. ASM Press, Washington DC.

117. **Eickbush TH.** 1997. Telomerase and retrotransposons: which came first? *Science* **277:**911–912.

118. **Iyer LM, Koonin EV, Aravind L.** 2003. Evolutionary connection between the catalytic subunits of DNA-dependent RNA polymerases and eukaryotic RNA-dependent RNA polymerases and the origin of RNA polymerases. *BMC Struct Biol* **3:**1.

119. **Darnell JE, Doolittle WF.** 1986. Speculations on the early course of evolution. *Proc Natl Acad Sci U S A* **83:**1271–1275.

120. **Cech TR, Golden BL.** 1999. Building a catalytic active site using only RNA, p 321–349. *In* Gesteland RF, Cech TR, Atkins JF (ed), *The RNA World*, 2nd ed.

Cold Spring Harbor Laboratory Press, Cold Spring Harbor, NY.

121. **Johnston WK, Unrau PJ, Lawrence MS, Glasner ME, Bartel DP.** 2001. RNA-catalyzed RNA polymerization: accurate and general RNA-templated primer extension. *Science* **292:**1319–1325.

122. **Malik HS, Burke WD, Eickbush TH.** 1999. The age and evolution of non-LTR retrotransposable elements. *Mol Biol Evol* **16:**793–805.

123. **Luan DD, Korman MH, Jakubczak JL, Eickbush TH.** 1993. Reverse transcription of R2Bm RNA is primed by a nick at the chromosomal target site: a mechanism for non-LTR retrotransposition. *Cell* **72:**595–605.

124. **Martin W, Koonin EV.** 2006. Introns and the origin of nucleus-cytosol compartmentalization. *Nature* **440:** 41–45.

125. **Eickbush TH.** 1999. Mobile introns: retrohoming by complete reverse splicing. *Curr Biol* **9:**R11–R14.

Mobile DNA, 3rd Edition
Nancy L. Craig, Michael Chandler, Martin Gellert, Alan M. Lambowitz, Phoebe A. Rice and Suzanne Sandmeyer

doi:10.1128/microbiolspec.MDNA3-0036-2014

Russell T. M. Poulter[1]
Margi I. Butler[1]

55 Tyrosine Recombinase Retrotransposons and Transposons

INTRODUCTION

Eukaryote retrotransposons have been organized into four major groups on the basis of their mechanistic features, open reading frame organization and reverse transcriptase (RT) phylogeny: long terminal repeat (LTR) retrotransposons, tyrosine recombinase (YR) encoding elements, Penelope-like elements (PLEs) and long interspersed nuclear elements (LINEs) (1). The major feature distinguishing the tyrosine recombinase-encoding elements from other retrotransposons is that the YR elements encode a tyrosine recombinase (2, 3) that performs the role of integration. Other retroelements employ integrases (LTR retrotransposons) or endonucleases (LINEs and PLEs). Tyrosine recombinases (YRs) are widespread in prokaryotes, typically involved in site-specific recombination between similar or identical DNA sequences (4). Representative examples include the Cre recombinase of bacteriophage P1, the FLP recombinase of yeast 2-micron circle plasmids, and the XerC and XerD recombinases of *Escherichia coli*.

The YR-encoding retroelements can be classified into three groups: DIRS, Ngaro and PAT. Together, the DIRS, Ngaro and PAT groups of tyrosine recombinase-encoding retrotransposons contain elements with a large diversity of structures and a variety of unusual features, such as extensive overlapping open reading frames (ORFs), novel protein-coding domains and spliceosomal introns.

The first tyrosine recombinase retrotransposon to be described was DIRS from the slime mold *Dictyostelium discoideum* (5). The presence of the tyrosine recombinase (YR) was recognized by Goodwin and Poulter (3). Related retrotransposons have been described from diverse organisms, including PAT from the nematode *Panagrellus redivivus* (6), *Prt1* from the zygomycete fungus *Phycomyces blakesleeanus* (7) and more recently from all major eukaryote lineages, including a few chlorophytes (3, 8, 9). A group of DNA transposons, cryptons, also carry a YR ORF which is related to the YRs in retrotransposons (10). Here we will describe these novel elements and discuss recent advances in our understanding of YR-encoding elements.

The following terminology will be used in this review:

YR elements: All those retroelements that have a tyrosine recombinase.

[1]Department of Biochemistry, University of Otago, Dunedin 9054, New Zealand.

DIRS1: The DIRS1 retrotransposon from *Dictyostelium.*

DIRS-like: A sub-group of YR elements phylogenetically close to DIRS1.

Ngaro-like: A sub-group of YR elements phylogenetically close to DrNgaro1 from *Danio rerio.*

PAT-like: A sub-group of YR elements phylogenetically close to PAT from *Panagrellus.*

DIRS1: STRUCTURE AND REPLICATION

The *D. discoideum* transposon DIRS1 is a well described YR retrotransposon (5). There are about 40 intact copies in the genome and 200 to 300 fragments. DIRS1 is about 4,813 base pairs (bp) in length (there is some variation) and has terminal repeats that are inverted (ITRs) and non-identical.

Between the ITRs DIRS1 contains three long ORFs (Fig. 1). These are ORF1 (putative *gag*-like), ORF2 (tyrosine recombinase or YR ORF) and ORF3 (reverse transcriptase/RNAaseH/N6 deoxy-adenosine methylase or RT/RH/DAM ORF). The ORF1 is in an appropriate position and of an appropriate size to correspond to the *gag* ORFs of other retrotransposons although they may not be homologous. The second ORF (ORF2 or YR ORF) overlaps the 3′ end of the putative *gag* ORF and extends to near the 3′ end of the element. The ORF3 is entirely overlapped by ORF2, but in a different reading frame and encodes an RT/RH/DAM (2, 8). The DAM methylase motif of ORF3 resembles the prokaryote N6-adenine methyltransferase (pfam 05689). The YR recombinase is encoded by that part of ORF2 that does not overlap ORF3. DIRS1 does not encode an aspartic protease or a DDE-type integrase.

The unusual structure of the DIRS1 retrotransposon is the basis of a proposed distinct replication process (5). The left ITR of DIRS1 is approximately 320 bp and the right ITR is approximately 350 bp. The 4,158-bp internal region includes (Fig. 1), at its 3′ end, a non-coding 88-bp sequence known as the internal complementary region (ICR). The left 33 bp of the ICR are complementary to the beginning of the left ITR. The right 55 bp of the ICR are complementary to the end of the right ITR. Cappello et al. (5) demonstrated experimentally that transcription begins near the (internal) end of the 316-bp left ITR (position 303) and terminates near the (external) end of the right ITR to give a 4.5-kb mRNA. The mRNA would therefore have very little of the left ITR sequence and most of the right ITR sequence. Cappello et al. proposed a model by which the sub-genomic 4.5-kb DIRS1 mRNA could be used to generate a complete 4,813-kb genomic DNA copy with non-identical inverted terminal repeats (5). In this model the mRNA was copied into a single-stranded DNA by reverse transcriptase. The primer for this has not been determined but there is a potential internal hairpin primer at the 3′ end of the mRNA (positions 479 to 4805). Once copied into the complementary single-stranded DNA, this sub-genomic DNA could have most of the left ITR restored by using the right ITR sequence as a template. However, this would not restore the very beginning and end of the DIRS1 sequence. These sequences could be restored by the incomplete left and right ITRs annealing with an ICR sequence and replicating the missing sequence using the ICR as a template. Following this replication a ligation would occur to produce a single-stranded circular full-length DNA, which could subsequently be converted to a double strand circle. The proposal that the product of replication is a complete circular DNA fits well with one further feature that distinguishes DIRS1 from LTR retrotransposons. Most LTR retrotransposons create short (4- to 6-bp) duplications of their genomic integration site. In contrast, DIRS1 does not generate a target-site duplication. The recombination of a circular DNA into the genome using a site-specific recombinase would not generate a target site duplication. This model of replication, based on the work of Cappello et al. (5), has already been described in detail (11).

IDENTIFICATION OF ELEMENTS RELATED TO DIRS1

The whole genome sequences of many eukaryotes are now available in public databases and this has facilitated the description of DIRS elements in these genomes.

DIRS1 Related Elements in Fish: DrDirs1, TnDirs1

The first vertebrate DIRS1-like elements were recognized in the whole genome sequence data of *Danio rerio* (zebrafish) and the freshwater pufferfish *Tetraodon nigroviridis* (3). These elements were found by searching the whole genome sequences using the DIRS1 RT/RH/DAM and YR sequences as queries. High-quality matches to the motifs were found to a sequence within the odorant receptor gene cluster of the zebrafish *Danio rerio* (bases 54803 to 56384 of accession AF112374) (12). This element was named DrDirs1 (*Danio rerio*DIRS1-like element no. 1). A reciprocal search performed using the predicted DrDirs1 RT/RH

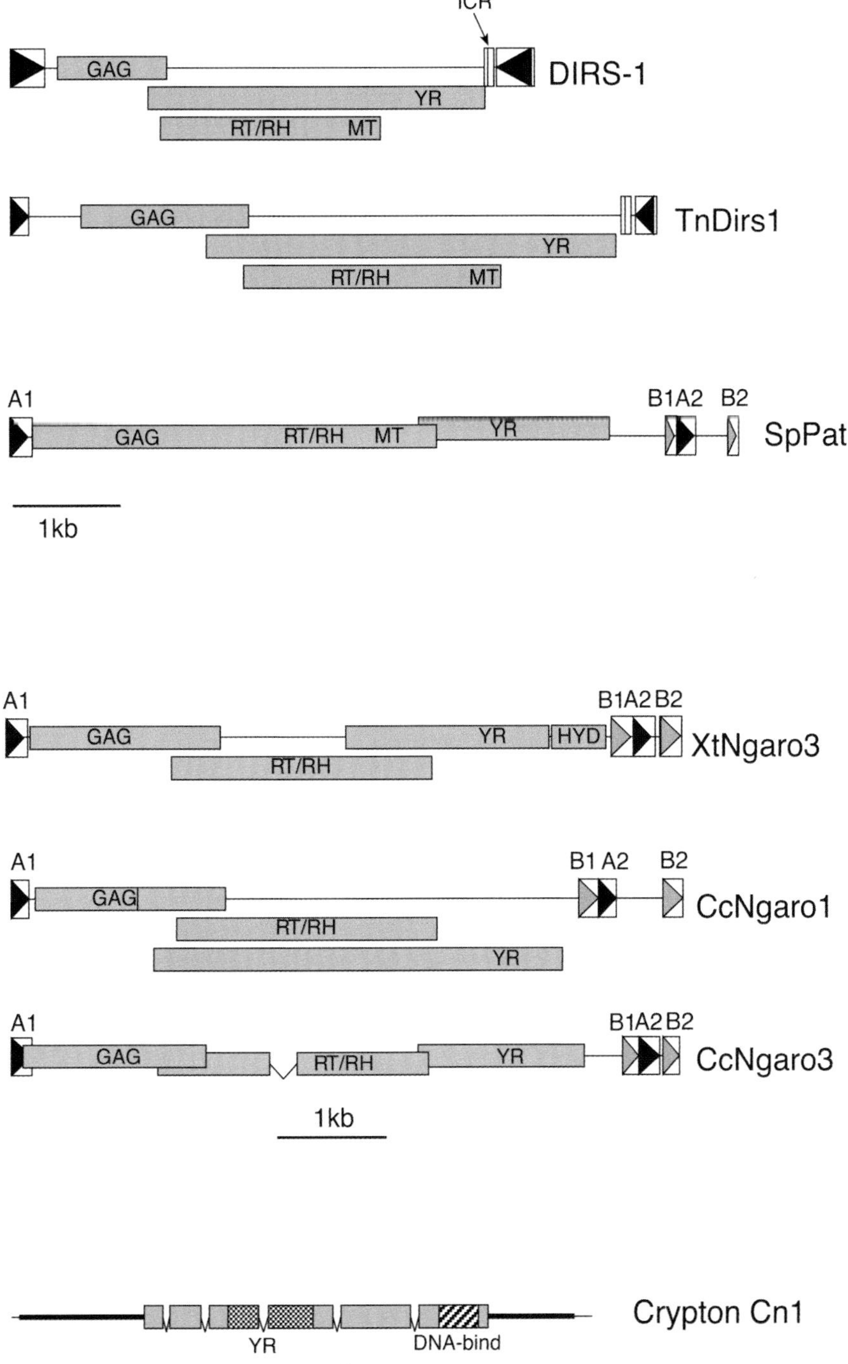

Figure 1 Structures of YR-encoding elements discussed in this chapter. These include: members of the DIRS-like group with ITRs and an ICR, a PAT-like element with 'split' direct repeats (SpPat1); Ngaro elements with 'split' direct repeats; a YR-encoding DNA transposon from *Cryptococcus* (Crypton_Cn1). Repeat sequences are represented by boxed triangles. Shaded boxes represent ORFs. V-shaped lines represent introns. In the crypton, the stippled box represents the YR-encoding region, while the hatched box represents a putative DNA-binding domain. doi:10.1128/microbiolspec.MDNA3-0036-2014.f1

sequence of this zebrafish element as a query in a TBLASTN search of the whole database detected DIRS1 as the top hits. These findings, including especially the reciprocal result, strongly suggested that this zebrafish sequence was a DIRS1-related retrotransposon. The repeat structures of this element also resemble those of DIRS1. BLAST searches indicated that there are over a thousand different DrDirs1 elements in the *Danio* genome. DrDirs1 has an overall structure remarkably similar to that of DIRS1 itself, although at ~6.1 kb in length it is somewhat longer than DIRS1 (4.8 kb). Like DIRS1, it is bordered by inverted repeats. These repeats are slightly different in sequence. The right ITR is 27 bp longer (211 bp) having additional bases at its 3′ end which are not present in the left copy. This difference in length resembles the situation in DIRS1. The zebrafish DrDirs1 element also has a 97-bp ICR in the 3′ end of its internal region. This ICR contains adjacent sequences exactly complementary to the outer edge of each ITR, similar to the 88-bp ICR of DIRS1.

The internal region of DrDirs1 contains several long ORFs. The first ORF is termed *gag* as it is of an appropriate size (480 codons) and is located in a similar position to the *gag* ORFs of other retrotransposons, although no sequence similarity to previously identified *gag* ORFs has been found. The second ORF encodes a putative protein bearing all the expected highly conserved residues of RT/RH and an N6-adenine methyltransferase. The YR ORF overlaps the majority of the RT/RH/DAM ORF, but is in a different reading frame, similar to the YR ORF of DIRS1. The recombinase is in the –1 phase with respect to the RT as in DIRS1. The putative tyrosine recombinase encoded by this ORF is similar to the predicted product of the YR ORF of DIRS1. The overall structures of the other copies of DrDirs1 are generally similar to that of DrDirs1. It is also of interest to note that there are several EST (expressed sequence tag) sequences apparently derived from copies of DrDirs1, suggesting that some copies of DrDirs1 are transcriptionally active.

There are multiple high-quality matches to DIRS1 related RT/RH/DAM and YR proteins in the sequences of the freshwater pufferfish *T. nigroviridis* present in the public databases. Most of the *T. nigroviridis* sequences belong to elements of one, apparently fairly abundant, family (TnDirs1). The overall structure of TnDirs1 is similar to the structures of DIRS1 and DrDirs1 (Fig. 1). The element is bordered by ITRs that are slightly different from each other, and the internal region has a 91-bp ICR containing sequences complementary to each end of the element.

Like DIRS1 and DrDirs1, the *T. nigroviridis* TnDirs1 element contains three long ORFs. The predicted product of the first ORF is similar in sequence to that of the putative *gag* ORF of the zebrafish DrDirs1 element. Both these two proteins both have a region containing cysteine and histidine residues with similar spacing (Cx3Cx9Hx2Cx2Cx4Hx9-10CxHC) near their N-termini that may form zinc fingers; these sequences are commonly found in retrotransposon Gag proteins. The putative DIRS1 Gag protein, in contrast, contains no obvious potential zinc fingers. The second ORF of TnDirs1 encodes a protein with the conserved motifs of the RT/RH/DAM domains. The YR ORF of the *T. nigroviridis* element overlaps the RT/RH/DAM ORF for much of its length. The predicted product of this ORF is a tyrosine recombinase similar to the predicted products of the YR ORF of DIRS1 and DrDirs1 and expressed as a –1 frame shift relative to the RT/RH ORF (3).

More recently, DIRS-like elements have been described from a diversity of fish, including the agnathan sea lamprey (*Petromyzon marinus*), the Australian lungfish *Neoceratodus forsteri* (13), the coelocanth *Latimeria* (14) and various Actinopterygii, including Notothenioid fish species (15) (Table 1). There are also DIRS1 elements in the elasmobranch *Leucoraja erinacea* (little skate).

DIRS1-like Elements from Amphibian and Reptiles

DIRS1-like RT/RH/DAM and YR ORFs are found in the genomes of *Xenopus*, including complete elements from *Xenopus tropicalis* (AC144974). Most of the sequences have overlapping RT/RH and YR ORFs as in DrDirs1. The *Xenopus laevis* sequences differ from each other quite substantially suggesting that *X. laevis* contains several distinct families of DIRS1-related elements. Many *Xenopus* sequences are found in the EST database, suggesting that they are transcriptionally active elements. According to Hellsten et al. (16), the genome of *X. tropicalis* contains over a hundred highly diverse families of DIRS-like elements (where the sequences of family members are >90% similar). They suggest that >50 of these families contain active members, since the sequence divergence is low and all contain three uncorrupted ORFs.

DIRS1-like elements are present in the plethodontid salamanders, where they make a major contribution (2.0% to 5.7% of the genome) to these highly expanded genomes (17). For example, in the 40 Gb genome of *Aneides flavipunctatus*, there are several hundred

Table 1 DIRS-like elements in fish. All those in this table are from the Actinopterygii except *Latimeria chalumnae* (Sarcopterygii) and the little skate, *Leucoraja* (Chondrichthyes)

Species	Name	Accession	Reference
DIRS-like			
Danio rerio	Zebrafish	AF112374	3
Tetraodon nigroviridis	Green spotted pufferfish	AF442732	8
Poecilia formosa	Amazon molly	AYCK01014057	
Gadus morhua	Atlantic cod	CAEA01505858	
Neolamprologus brichardi	Cichlid	AFNY01032878	
Gasterosteus aculeatus	Three-spined stickleback	AANH01008719	
Oreochromis niloticus	Nile tilapia	AERX01068420	
Xiphophorus maculatus	Southern platyfish	AGAJ0104998	
Haplochromis burtoni	Burton's mouthbreeder	GBDH01091653	
Pundamilia nyereri	Cichlid	AFNX01021957	
Pimephales promelas	Fathead minnow	JNCD01001357	
Stegastes partitus	Bicolour damselfish	JMKM01002005	
Mchenga conophorus	Cichlid	ABPJ01025120	
Maylandia zebra	Zebra mbuna Cichlid	AGTA02023338	
Dissostichus mawsoni	Antarctic toothfish (notothenoid)	HQ447060	
Nothobranchius furzeri	Turquoise killifish	GAIB01104168	
Latimeria chalumnae	West Indian Ocean coelocanth	BAHO01326816	
Leucoraja erinacea	Little skate	AESE010643923	
Neoceratodus forsteri	Australian lungfish		13
Sinocyclocheilus angustiporus	Cavefish	GAHO01055858	
Ngaro-like			
Danio rerio	Zebrafish		8
Astyanax mexicanus	Mexican tetra	APWO01060904	
Thunnus orientalis	Pacific Bluefin tuna	BADN01064445	
Lepisosteus oculatus	Spotted gar	AHAT01041850	
Oryzias latipes	Medaka	BAAF04075296	
Sebastes rubrivinctus	Flag rockfish	AUPQ01010767	
Sinocyclocheilus angustiporus	Cavefish	GAHO01122442	
Neoceratodus forsteri	Australian lungfish		13
Latimeria chalumnae	Indian Ocean coelocanth	BAHO01173054	

thousand DIRS1-like elements (17). The DIRS elements show the expected inverted terminal repeats (ITRs).

In the reptiles, DIRS1 elements are widespread and abundant. For example, they are found in the genomes of the king cobra *Ophiophagus hannah* and the Burmese python (*Python molurus bivittatus*) (18), in the green anole lizard *Anolis carolinensis* and in turtles such as *Chrysemys picta bellii* and *Pelodiscus sinensis*. Many of these elements appear uncorrupted and are presumably functional. Matches to retroelement protein sequences in ~1.5 Mb of genome data suggest that there are DIRS-like elements in the tuatara (*Sphenodon punctatus*, from the order *Rhyncocephalia*) (19).

DIRS1-like Elements in Deuterostomes

Apart from the vertebrate elements mentioned above, DIRS1-like elements have been detected in urochordates (*Ciona* and *Oikopleura*), cephalochordates (*Branchiostoma*) (9), hemichordates (*Saccoglossus*) (9) and echinoderms (*Strongylocentrotus* and *Lytechinus*). Many of these elements are apparently uncorrupted and presumably functional. For example, some of the DIRS1-like elements from *Strongylocentrotus purpuratus* are essentially identical in structure to the DIRS1 element, with ITRs and the RT/RH ORF entirely overlapped by the YR ORF. Some of the SpDIRS elements (SpDIRS1, 3 and 4) differ slightly in structure in that their RT/RH ORF is not overlapped by a long 5′ extension of the YR ORF (11). These elements may have developed an alternative method for expressing the YR to that likely to operate in the other members of the DIRS group.

DIRS1-like Elements in Protostomes

An insect DIRS1-like element, TcDirs1, was detected in the genome of the red flour beetle, *Tribolium castaneum*

(20). TcDirs1 is broadly similar in structure to DIRS1 itself, with terminal inverted repeats and an internal complementary region. The internal region contains three long ORFs: the first encodes a putative Gag protein; the second contains domains specifying RT/RH and a putative methyltransferase, and the third encodes the expected tyrosine recombinase. TcDirs1 is widespread in *T. castaneum* and exhibits several features suggestive of recent activity, such as a high level of sequence similarity between different elements, sequence identity between the repeat sequences, and inter-strain polymorphisms in insertion sites. Apparently uncorrupted DIRS1-like elements are widespread in insects other than the *Coleoptera* (beetles; *Tribolium castaneum*). These include *Lepidoptera* (butterflies: *Manduca sexta*), *Diptera* (flies: *Phlebotomus papatasi*), *Hemiptera* (aphids: *Diaphorina citri*) and *Hymenoptera* (wasps and ants: *Glyptapanteles indiensis*).

Other arthropods whose genomes include DIRS1-like elements are arachnids (*Metaseiulus occidentalis*, a mite; *Ixodes scapularis*, a tick), myriapods (*Stigamia maritima*), the Atlantic horseshoe crab, *Limulus polyphemus*, and crustaceans (*Daphnia* and many decapods). In a large, PCR-based study, Piednoël et al. (21) detected DIRS elements in 15 diverse decapod species (shrimps, lobsters, crabs and galatheid crabs), especially species from hydrothermal vents. Phylogenetic analysis of the RT/RH sequences of these decapod DIRS elements indicates that they form a group distinct from other previously described DIRS elements (21). In *Daphnia pulex*, 19 intact DIRS elements were detected; there was also some evidence of transcriptional activity in two of these elements, suggesting they may be active retrotransposons (22).

Other protostome groups whose genomes carry DIRS-like elements include nematodes such as *Caenorhabditis* and *Ancylostoma*, bivalve molluscs such as *Crassostrea gigas* (23) and *Pinctada fucata* (24) as well as annelids such as *Capitella teleta* (9). Other animal groups that contain DIRS1-like elements include *Porifera* such as *Amphimedon queenslandica* and *Cnidaria* (the coral, *Acropora digitifera*, and the sea anemone, *Nematostella vectensis*).

DIRS1-like Elements in the Fungi

In the fungi, DIRS1-like elements occur in three divisions, the *Blastocladiomycota* (*Allomyces macrogynus, Catenaria anguillulae*), *Kickxellomycotina* (*Coemansia reversa*) and *Mucormycotina* (*Rhizopus* species, *Phycomyces blakesleeanus, Mucor circinelloides, Mortierella* species, *Backusella circina*). DIRS elements are also present in the mycorrhizal fungus *Rhizophagus irregularis* (*Glomus intraradices*). Remarkably, there are no DIRS1-like elements in the *Ascomycota, Basidiomycota* or *Chytridiomycota* (25).

The first fungal DIRS-like element to be detected was *Prt1* from *Phycomyces blakesleeanus* in the order *Mucorales* (7). This element was originally described as containing two somewhat corrupt ORFs, the first of unknown function, and the second encoding DIRS-like RT/RH domains. Later, the remnants of a third ORF encoding a tyrosine recombinase were also identified (3). Like the other DIRS-like elements, *Prt1* also contains the methyltransferase-like domain C-terminal to the RH domain. The *Prt1* element, as originally described, contains only short inverted repeats at the termini (i.e., it lacks the long ITRs and the ICR of other DIRS1 elements). These unusual features may not accurately reflect the structure of a functional *Prt1* element but may result from mutations subsequent to its last transposition.

The structure of *DIRS*-like elements in fungi was clarified by genome sequence data from *Rhizopus oryzae*, also from the *Mucorales* (3). The *Rhizopus* elements have an overall structure very similar to that of DIRS1, containing ITRs and an ICR at the 3′ end of the unique internal region. These elements contain all the expected coding sequences, arranged in a similar fashion to those of DIRS1 itself.

DIRS1-like Elements in Other Eukaryotes

There are apparently few or no DIRS1 elements in the *Stramenopile* phylum; for example they are absent from the oomycete *Phytophthora infestans*. Finally, DIRS itself is present in the *Amoebozoa*, including the phylum *Mycetozoa* (*Dictyostelium*). Piednoël et al. provide a valuable summary of the presence of DIRS-like elements among the *Eukarya* (9).

PAT-LIKE ELEMENTS

While the overall complement of coding regions is similar in the DIRS and PAT-like elements, their arrangements of repeat sequences are different (Fig. 1). Instead of the ITRs found in DIRS1, PAT has a series of direct repeats in which the termini of the element are repeated adjacent to each other (*A1, B1, A2, B2*). The ORFs of the sequenced PAT from *Panagrellus* are slightly corrupted by frameshifts and nonsense mutations, but a putative *gag* ORF, an RT/RH/DAM ORF and an ORF encoding a tyrosine recombinase are apparent. Several copies of a PAT-related element (CbPat1) are present in the nematode *Caenorhabditis briggsae* (3). The element is somewhat shorter than PAT itself but otherwise has a very similar overall structure. Like PAT it contains 'split' direct repeats. The element also has three recognizable ORFs. The predicted product of the *gag* ORF contains a putative Cx2Cx4Hx4C zinc finger, similar to those often found in Gag proteins. A similar zinc finger is also encoded by the putative *gag* ORF of PAT itself (6). A second ORF encodes an RT/RH/DAM while the third ORF encodes a tyrosine recombinase. While PAT and CbPat1 are very similar in structure, the two elements are not highly similar in their actual sequences. For instance, the predicted products of their RT/RH/DAM ORFs are ~29% identical over a 580 amino acid range and their repeat sequences bear little resemblance to each other. This suggests that their 'split' direct repeat structures have been maintained over a considerable period of time and are believed to be important for the replication of these elements (8).

PAT-like Elements in the Echinoderms: SpPAT

A full-length element, SpPAT1, was identified in the *Strongylocentrotus purpuratus* sequence. The RT/RH of this element is most similar to PAT (rather than SpDirs or SpNgaro). Like PAT it has 'split' direct repeats, but the YR ORF is located between the A1 and B1 repeats as in the Ngaro elements. SpPAT encodes a methyltransferase like other DIRS-like and PAT elements (Fig. 1).

Other animals with PAT-like elements in their genomes include the poriferan *Amphimedon queenslandica* and the hemichordate *Saccoglossus kowalevskii*. Both the *Saccoglossus* (SkowPat) and *Aphimedon* (AquePat) PAT-like elements have 'split' direct repeats of about 200 bp and ~120 bp. The ORFs of AquePat are somewhat corrupt, while those of SkowPat are likely to be intact.

PAT-like elements are present in *Naegleria gruberi*, a member of the *Heterolobosea*, a group most closely related to *Euglenozoa* (Kingdom *Excavata*). These elements encode RT/RH, a methylase and YR and contain direct terminal repeats. The RT/RH/DAM ORF is the most divergent of those found in PAT-like elements (see PHYLOGENETIC ANALYSES EMPLOYING RT).

PAT-like Elements in Plants

YR-encoding retroelements have been described from plants; these are kangaroo, TOC1 and TOC3 in the algae *Volvox carteri* and *Chlamydomonas reinhardtii*. Duncan et al. (2) described a YR-encoding element, Kangaroo, in the genome of the green alga *Volvox carteri*. The element is active: it was isolated as a result of its insertion into a nitrate-reductase-encoding gene, and it has long, uninterrupted ORFs, containing all the expected coding domains, including the methyltransferase. The overall structure of kangaroo is similar to that of the DIRS-like element, PAT, from the nematode *P. redivivus*. Kangaroo has nested direct repeats, with the recombinase ORF located between repeats *A2* and *B2*. Uniquely, the recombinase ORF of kangaroo is located on the opposite strand to the remaining ORFs, and therefore must be translated from a distinct transcript. Kangaroo, like some of the basidiomycete Ngaro elements described below, is one of the rare examples of intron-containing retrotransposons. As part of their analysis of kangaroo, Duncan et al. also detected by PCR a fragment of DNA containing juxtaposed termini of the element. This DNA molecule is identical in structure to the proposed extrachromosomal circular intermediate in the replication of DIRS-like elements. This work thus provides experimental evidence in support of the model.

Several YR retroelements can be found in the *Chlamydomonas reinhardtii* genome. One of these elements, TOC3, resembles a PAT-like element. Immediately upstream of the *B1* repeat in TOC3 is a recombinase gene similar in sequence to those of PAT-like elements. TOC1 is a non-autonomous version of TOC3 (26, 8). There are also abundant elements related to kangaroo in the *Chlamydomonas reinhardtii* genome (termed Pioneer 2). Pioneer 2 are full-length PAT-like retroelements.

THE NGARO ELEMENTS: A DISTINCT LINEAGE OF YR RETROTRANSPOSONS

Analysis of the *Danio* sequence data detected distinct families of retrotransposons that, while clearly encoding a tyrosine recombinase, had very distinct sequences and structure (8). A representative example is DrNgaro1 (Accession AY152729). Ngaro is the New Zealand Maori word for fly. There are numerous DrNgaro elements in the Danio genome, although they are not as abundant as DrDirs1.

DrNgaro1 contains three long ORFs. The first encodes a putative *gag*, the predicted product of the ORF contains three zinc finger-like motifs: Cx2Cx4Hx4C, Cx7Cx2HxC, and Cx2Cx3Hx4C. In DrNgaro1 the second ORF encodes RT/RH, but lacks the methyltransferase domain characteristic of DIRS1-like elements. In most DrNgaro elements the *gag* and the RT/RH form a single ORF. The start of the YR ORF overlaps a little with the end of the RT/RH ORF in the –1 reading frame. It is probable that translation of the YR ORF is achieved by programmed ribosomal frameshifts in the region of overlap. The YR ORF encodes a putative tyrosine recombinase bearing the conserved tetrad of catalytic residues (RHRY, 4) characteristic of these proteins.

The termini of DrNgaro1 and related elements were defined by comparison between the copies of these elements at different genomic loci. In addition comparisons were made between the various elements and the sequences of allelic 'empty' genomic sites. The termini of DrNgaro elements are 'split' direct repeats (Fig. 1). The repeat sequences are designated *A1*, *B1*, *A2* and *B2*, where *A1* and *A2* form a pair of direct repeats, as do *B1* and *B2*. In the case of DrNgaro1 the *A1* and *A2* repeats are 157 bp long and are 100% identical. The *B1* and *B2* repeats are 158 bp long and are also 100% identical. The *A* and *B* repeats share no significant sequence similarity with each other. The structure of DrNgaro1 is quite distinct from that of any previously known YR elements.

Interestingly, some elements have similar repeat structures to that of DrNgaro1, and even have sequence similarity with the repeats of DrNgaro1, yet do not appear to have any protein-coding capabilities. An example of such an element is DrNgaro3 (AL591180). It is possible that these elements are non-autonomous, but can still transpose when supplied with the appropriate proteins in *trans*.

Ngaro Elements from Other Fish

In addition to the elements detected in *Danio*, full-length, uncorrupted Ngaro elements have also been found in bony fish (*Osteicthys*) such as *Oryzias latipes*, *Lepisosteus oculatus* (the spotted gar), and the cavefish *Sinocyclocheilus angustiporus* (Table 1). Among the sarcopterygian fish, both the coelacanth *Latimeria* and the Australian lungfish (*Neoceratodus*) have Ngaro elements, but those of *Latimeria* are mostly corrupted. As yet there are no Ngaro elements described from elasmobranchs or the agnathans. It would seem, therefore, that the Ngaro elements are not as widely encountered in fish as the DIRS elements.

Ngaro-like Elements from Amphibian and Reptiles

Ngaro-like elements are abundant in *X. laevis*; these are referred to as XlNgaro (*Xenopus laevis Ngaro1*-like). Ngaro elements are also present in *X. tropicalis* (Fig. 1) and in other frogs such as *Rana clamitans*. Ngaro elements are abundant in the genome of the Chinese salamander, *Hynobius chinensis*. Full-length Ngaro elements have also been found in the expanded genomes of several plethodontid salamanders (17). These Ngaro elements show direct repeat termini.

The amphibian Ngaro elements each contain an additional ORF compared to Ngaro elements from fish) between the end of the YR gene and the start of the repeat sequences (Fig. 1). The product of this ORF is highly conserved among the different Ngaro elements. The sequence is related to SGNH hydrolases (also referred to as GDSL hydrolases) from a diverse family of lipases and esterases (8).

In *Anolis carolinensis* there are abundant, uncorrupted Ngaro elements which all carry the additional hydrolase ORF. There do not seem to be Ngaro elements in the snakes *Ophiophagus hannah* (king cobra) or *Python bivittatus* (Burmese python).

Ngaro-like Elements from the Echinoderms: SpNgaro1 and LvNgaro1

A Ngaro-like element was found in a sequence from the sea urchin *Lytechinus variegatus* (AC131494). This element, LvNgaro1, has a similar overall structure to the zebrafish Ngaro elements but resembles the XlNgaro elements in having an ORF between the recombinase gene and the *B1* repeat which encodes a putative hydrolase protein related to the ORF1 proteins of several vertebrate non-LTR retrotransposons. There are several Ngaro-like elements in the genome of the echinoderm, *Strongylocentrotus purpuratus*, some of which are full-length.

Ngaro Elements in Deuterostomes

The cephalochordate, *Branchiostoma floridus* also has several examples of Ngaro elements (27); these encode generally uncorrupted RT/RH and YR sequences that are quite divergent from each other (that is, there are several different families of Ngaro elements in this species). The *Branchiostoma* (BfloNgaro) elements do not encode hydrolases. The genome of the hemichordate *Saccoglossus* (acorn worm) also contains uncorrupted Ngaro-like elements, but like those in *Branchiostoma*, they do not encode hydrolases. No Ngaro elements are apparent in the genomes of the urochordates for which there are sequences in the public databases.

Ngaro Elements in the Protostomes

There are Ngaro elements in the crustaceans; these are abundant and expressed in *Calanus* and *Pontastacus* (28, 29). These elements lack the hydrolase ORF. In contrast to their abundance in crustacea, there are no Ngaro elements in any insects. In the molluscs there are abundant Ngaro elements in the bivalve *Crassostrea* and a smaller numbers in other molluscs, such as the gastropod *Lottia gigantea*. All the mollusc

Ngaro elements encode a hydrolase ORF. In the annelids, Ngaro elements without a hydrolase ORF are present in the polychaete *Capitella teleta*. There are apparently no Ngaro elements in the nematodes. Other animal groups that contain Ngaro elements (without the hydrolase ORF) include Porifera such as *Amphimedon queenslandica* and Cnidaria (the coral *Acropora digitifera* and the sea anemone *Nematostella vectensis*).

Ngaro Elements in *Stramenopiles*

Ngaro elements (without hydrolases) occur in the *Stramenopiles*. They are abundant in several species of the oomycete *Phytophthora* and in *Pythium*.

Ngaro-like Elements from Fungi

Ngaro-like elements in fungi are confined to the basidiomycetes. Full-length and potentially intact Ngaro-like retrotransposons are present in the draft genome sequences of many basidiomycete fungi; for example, *Phanerochaete chrysosporium*, *Coprinopsis cinerea* and *Laccaria bicolor*. The elements are often abundant, for example there are 50 to 100 elements in *Laccaria*. These basidiomycete elements have overall structures similar to those of the other Ngaro elements, with nested direct repeats, and a similar complement of coding domains. The fungal Ngaro elements do not carry the hydrolase ORF encountered in some metazoan Ngaro elements.

Ngaro elements have recently been detected in many basidiomycete genomes (25). Phylogenetic analyses based on a concatenation of the RT/RH and YR protein sequences revealed that the Ngaro elements in basidiomycetes form four major groups (25). Each group contained elements with potentially active RT, RH and YR domains. Each group also contains elements in which the YR ORF is in the same frame as the RT/RH ORF as well as elements where there is a frame shift between these ORFs (25).

In contrast to the metazoan Ngaro elements, some of the fungal Ngaro elements have their RT/RH ORF entirely overlapped by the recombinase ORF. This arrangement resembles that in the DIRS-like elements. Despite sharing a similar arrangement of ORFs with DIRS-like elements, these fungal elements are clearly most closely related to Ngaro elements as they group with them on phylogenetic trees based on RT/RH and YR sequences. They also lack the putative methyltransferase domain characteristic of DIRS-like elements. It seems likely that the arrangement of overlapping ORFs has arisen twice, independently. In all DIRS-like elements the recombinase is in the –1 phase with respect to the RT whereas in all the Ngaro elements showing extensive overlap of RT/RH and YR ORFs the recombinase is in the +1 phase with respect to the RT.

A second unusual feature of many of the basidiomycete Ngaro elements is the unexpected presence (for a retrotransposon) of spliceosomal introns within the RT-RH gene. Some of the *Laccaria* Ngaro elements have no intron but a large group have an intron at a precise site in the RT, while a further group have an intron at an alternative site in the RT. Introns have been previously detected in the retrotransposon Penelope from *Drosophila virilis* and a Penelope-related element from a rotifer (30). The presence of non-coding spliceosomal introns in basidiomycete Ngaro elements is the first report of introns in an YR retrotransposon. The presence of introns was first suggested when comparisons of various related elements detected the presence of poorly conserved regions, often of variable length and containing frameshifts and stop codons, in otherwise well conserved ORFs (8). Comparisons of the putative protein products of these elements revealed that these poorly conserved regions appear as inserts relative to the predicted proteins encoded by other elements. More in depth analysis revealed that the inserts are flanked by the canonical 5′-GT....AG-3′ nucleotides of spliceosomal introns, and the identification of these sequences as introns has been essentially confirmed by the identification of intron-less copies of some elements, in which the predicted intron has been precisely removed. The presence of spliceosomal introns raises questions about the replication cycle of these retrotransposons. The disruption they cause to the ORF indicates that splicing must occur prior to translation. However, their very presence, and furthermore their presence in identical sites in considerably diverged elements from different species, indicates that the spliced form is generally not used as a substrate for reverse transcription. One possible explanation is that following transcription the full-length RNA can follow one of two pathways. The transcript could be spliced, exported from the nucleus and translated. Alternatively the transcript might not be spliced but instead be retained in the nucleus to become associated with the proteins specified by a spliced mRNA followed by entry into the replication cycle. This proposal clearly requires experimental testing, but if confirmed it might represent an interesting and significant departure from the replication cycle of elements such as Ty1 and vertebrate retroviruses where reverse transcriptase acts in the cytoplasm. If correct, it would mean the RT acted within the nucleus (as in LINE elements).

DIRS1- AND NGARO-RELATED ELEMENTS ENCODE TYROSINE RECOMBINASES

Clearly the YR ORF is central to our understanding of the relationships, evolution and functioning of YR retroelements (31). Determining if newly described DIRS- or Ngaro-related retroelements carry a YR ORF and if these ORFs are uncorrupted is made difficult by the very diverse nature of tyrosine recombinases. Only four highly conserved residues have been identified that are present in all, or nearly all, of the members of this family (4). These residues are together known as the RHRY tetrad and are believed to contribute to the active sites of the enzymes. The tyrosine recombinase family (31) includes tyrosine recombinase from bacteriophage, the integrases or resolvases of some bacterial plasmids and transposons and the XerC and XerD recombinases of *Escherichia coli*, which promote the stable inheritance of the *E. coli* chromosome (4). The yeast FLP-recombinases, the *Tec* transposons from *Euplotes crassus* (32) and cryptons (10) are the only other members of the tyrosine recombinase family identified in eukaryotes.

The uncorrupted YR ORFs of all the full-length DIRS-like, PAT-like and Ngaro-like retroelements encode proteins bearing highly conserved RHRY tetrads similar to those of tyrosine recombinases (Fig. 2). The highly conserved RHRY residues in the DIRS1- and Ngaro-related elements are spaced similarly to those of known tyrosine recombinases and sequence similarities between the DIRS1 elements, Ngaro elements and members of the tyrosine recombinase family are also evident in the regions surrounding the four conserved residues. Experimental evidence of site-specific recombination generated by examples of these YR proteins is yet to be shown, however. It seems probable that the YR of retrotransposons have a broad site specificity, allowing integration at multiple sites in a genome.

A METHYLTRANSFERASE IN DIRS-LIKE ELEMENTS

All DIRS-like and PAT-related elements have a conserved domain downstream from the RT/RH domain that is completely absent from the Ngaro elements. This domain is similar in sequence to the DNA N6 adenine methyltransferases encoded by various bacteriophages, and can be detected using PF05869 from the Pfam database. Alignment of the two sets of proteins indicates that the DIRS element proteins share a number of conserved residues with the phage methyltransferases. Included among these is a motif similar in sequence to the conserved phage (D/N)PP(Y/F) motif known to be important in binding the S-adenosylmethionine substrate (33). This level of sequence similarity suggests that the two sets of proteins are homologous and may have related activities. The role of these putative methyltransferase domains in the replication of DIRS elements is, as yet, completely unknown.

NGARO ELEMENTS ENCODE A PUTATIVE HYDROLASE

Ngaro elements encode a Gag-like protein, an RT/RH and a YR. The Ngaro elements from some echinoderms and all the amphibian and reptile Ngaro elements, encode an additional ORF (8). The sequences of the products of these ORFs are highly conserved among the different elements and are related to SGNH hydrolases from a diverse family of lipases and esterases (8). Sequence comparisons suggest that these ORFs encode proteins most closely related to the C-terminal halves of the ORF1 proteins of several vertebrate members of the CR1 clade of non-LTR retrotransposons, such as CR1 from the chicken (34) and Maui from the pufferfish *Fugu rubripes* (35). As has been noted by Kapitonov and Jurka (36), the ORF1 proteins of these non-LTR retrotransposons are related to a variety of cellular enzymes, including TEP-I (thioesterase) from *E. coli* (37), brain platelet-activating factor acetyl hydrolase (PAF-AH) from the cow (38), and the isoamyl acetate-hydrolyzing esterase (IAH1) from *Saccharomyces cerevisiae* (39). Several of these related enzymes, including brain PAF-AH (38) and TEP-I (37), have been experimentally characterized and found to possess catalytic triads containing conserved serine, aspartate, and histidine residues. The proteins encoded by the Ngaro retrotransposons contain the same conserved residues, and these appear within similar contexts and with similar spacing (8). These proteins may therefore have some related enzymatic function.

REPLICATION OF THE PAT-LIKE AND NGARO 'SPLIT' DIRECT REPEAT ELEMENTS

Whereas DIRS-like elements have ITRs, the PAT-like and Ngaro elements have 'split' direct repeat structures. The replication of DIRS-like elements must differ from the replication of Ngaro and PAT-like elements. The replication model of Cappello et al. (5) described above explains many features of the DIRS1 structure. An alternative replication model applicable to the elements with 'split' direct repeats has been proposed (8). As

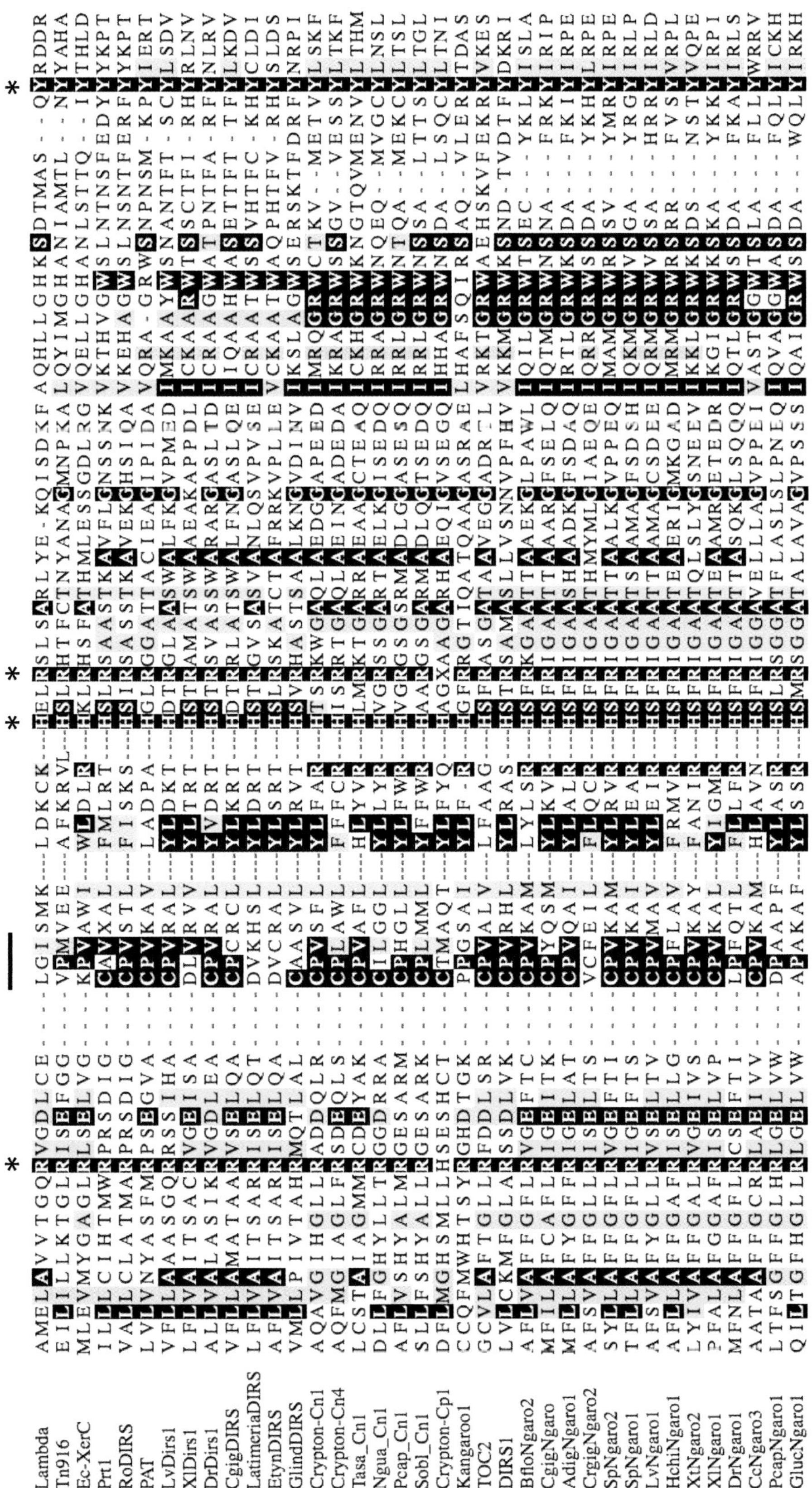

Figure 2 Alignment of tyrosine recombinase conserved domains. A comparison of aligned tyrosine recombinase sequences from retrotransposons and DNA transposons with those from prokaryotes. Four regions of the recombinases are illustrated; the dashed lines common to all elements represent intervening regions of variable length. The conserved RHRY tetrad is denoted by *; the conserved CPV motif is overlined. doi:10.1128/microbiolspec.MDNA3-0036-2014.f2

with the replication of other retrotransposons, the first step in this model is the production of a sub-genomic transcript, beginning within the *A1* repeat and terminating within the *B2* repeat. This RNA molecule is then copied into minus-strand DNA by reverse transcriptase. The primer for this reaction is not known, but presumably acts at the 3′ end of the RNA. Subsequently, the RNA of the RNA/DNA hybrid is partially degraded by the element's RH, leaving a sub-genomic minus-strand DNA molecule. The ends of this minus-strand DNA molecule correspond to incomplete copies of the *A1* and *B2* repeats. These repeats could anneal to the internal *B1/A2* region of a sub-genomic mRNA (or mRNA fragment). It is proposed that the replication of the 'split' repeat elements requires two molecules (8). Binding of the ends of the minus-strand DNA to the *B1/A2* mRNA would form a largely single-stranded, gapped, circular molecule. The gap in the minus-strand DNA could be closed using the *B1/A2* RNA as a template and plus-strand DNA synthesis could be primed using the *B1/A2* RNA as a primer. The final result would be a full-length, double-stranded, circular molecule that could be inserted into the host genome by recombination using the encoded tyrosine recombinase. This model thus (1) allows for the synthesis of a full-length DNA molecule from a sub-genomic RNA, (2) results in the production of a circular, double-stranded DNA, (3) provides a primer for plus-strand DNA synthesis, and (4) utilizes all the identified repeat structures within the Ngaro and PAT-like elements. The model shares some features with the model for the replication of DIRS1, such as the use of internal copies of the element's termini to regenerate the ends and circularize the resulting DNA. However, it differs in other respects, such as using a separate RNA molecule as the template for regenerating the ends, rather than an ICR within the same DNA.

To summarize, consideration of DIRS1, PAT and the Ngaro YR elements establishes that there are two basic genomic patterns: DIRS1 type (inverted repeats with an internal ICR) and Ngaro/PAT ('split' direct repeats). These two forms must have somewhat different replication systems.

INTEGRATION OF DIRS1-RELATED ELEMENTS

Given that DIRS1-related elements encode proteins related to site-specific recombinases it is likely that the recombinase mediates the insertion of the putative extrachromosomal intermediates into the host genome. The insertion sites of DIRS1-related elements were examined to see if this would provide information relevant to the integration of these elements. The availability of multiple sequences allowed us to analyze the insertion sites of the zebrafish element DrDirs1, the pufferfish element TnDirs1, and the DIRS1 element itself. All the sequenced copies of DrDirs1 are bordered at both their 5′ and 3′ ends by GTT sequences and there is little sequence similarity in the broader flanking regions. Empty allelic sites that have not undergone the insertions of the DrDirs1 elements have the sequence GTT at the point of insertion (3).

A preference for insertion at GTT sequences in turn suggests a possible mechanism for integration of DrDirs1 elements. If, as deduced by Cappello et al. (5), the final extrachromosomal replication intermediate of a DIRS1 is a circular molecule, then the junction between the ends of the DrDirs1 element would be likely to consist of the sequence GTT. The sequence similarity between the circular junction of the ends of the element and the sequence of the insertion site could then be used by the recombinase to perform a recombination reaction and thus insert the element into the host genome. Such a process would result in the full-length element being flanked by GTT sequences. One of these repeats, or more likely the proximal part of one and the distal part of the other, would be derived from the target site, while the remaining sequences would be derived from the DIRS1-related element.

In support of this proposed mechanism of integration for DrDirs1, we found that the TnDirs1 elements are also flanked by GTT trinucleotides and appear to preferentially integrate at GTT sequences. Likewise, from an alignment of ten unoccupied DIRS1 target sites we detected a preference for DIRS1 insertions at sequences of the 3-bp form (A or T)TT, similar to the termini of DIRS1 elements. The two sequenced copies of the PAT element (6) are both flanked by AAC sequences and the one sequenced copy of an unoccupied PAT target site also contains the sequence AAC at the insertion site. PAT may therefore insert by recombination between the AAC trinucleotide at the circle junction and an AAC sequence in the target site.

In summary, it appears from several systems that the retrotransposon YR is recognizing and integrating the circular dsDNA retroelement into a trinucleotide target that is specific for the particular YR. It is theoretically possible that the YR could catalyse the reverse process, the excision of a closed dsDNA circular element, but there is no evidence to support such a pathway. It is possible that the YR retroelements that retain an intron are replicating via this DNA pathway, rather than via an RNA intermediate.

PHYLOGENETIC ANALYSES EMPLOYING RT

The widespread distribution of DIRS-like and Ngaro-like elements among the eukaryotes, and their distinct structures, makes it interesting to explore the evolutionary relationships between the two types of YR retroelement and between these YR elements and the LTR retrotransposons. The phylogenetic relationships among a large number of retroelements were examined using an alignment of their RT and RH encoding ORFs (8, 11). The analysis indicates that the major groups of elements Ty1/copia, Ty3/gypsy, BEL, VIPER, Ngaro, PAT-like and DIRS-like elements were all resolved into monophyletic groups with high levels of bootstrap support (8, 11). With respect to the tyrosine recombinase retrotransposons, the most important aspect of these RT/RH phylogenetic trees was that they clearly distinguished the Ngaro, DIRS and PAT groups.

Recent analyses have contributed many new YR retrotransposons. A phylogenetic tree, based on an alignment of RT and RH sequences from DIRS-like, PAT-like and Ngaro elements from a wide representation of species (Table 2) is illustrated in Fig. 3. The analysis was restricted to retrotransposons with apparently uncorrupted ORFs (or only lightly corrupted). Only elements clearly associated with YR ORFs are included. Three lineages can be distinguished that correspond to DIRS, PAT-like and Ngaro. The Ngaro elements fall into three main groups, held together with significant bootstrap support (95%). One group consists of Ngaro elements in animal genomes, one group includes only Ngaro from basidiomycete fungi, while the third contains Ngaro elements from oomycetes. The branch lengths within each group indicate that there is significant sequence divergence among the Ngaro elements. Within the animal group of Ngaro elements, there are examples of species with multiple families of Ngaro elements (for example, CgiNgaro1 and CgiNgaro2 from the oyster *Crassostrea gigas*). The PAT-like elements group together with moderate bootstrap support. These elements are from a phylogenetically very diverse group of organisms, including fungi, metazoans and plants (all described plant YR retrotransposons fall into the PAT-like group). The DIRS-like elements are quite diverse. The DIRS elements fall into three broad groups. One group includes all vertebrate elements; within this group, the bony fish fall into a well-supported sub-group. LchaDirs, from the coelacanth/lobe finned fish and the amphibian and reptile elements fall outside this sub-group. The non-vertebrate metazoan DIRS-like elements form a second, quite diverse group. The fungal DIRS elements form a third, distinct group. A number of DIRS-like elements (including DIRS1 itself) fall outside these three major groups.

YR-ENCODING DNA TRANSPOSONS: CRYPTONS

Cryptons from the Basidiomycete *Cryptococcus Neoformans*

In addition to the retrotransposons described above, a number of eukaryotic tyrosine recombinase-encoding DNA transposons have been described. For example, several distinct YR genes were found in *C. neoformans* (10). The matches to YR sequences were frequently found in different reading frames, and often with stop codons in the intervening regions. Comparison of the genomic and cDNA sequences revealed that there are introns in these YR genes. These appear to be typical spliceosomal introns, similar to those found in other fungi (40). Conceptual removal of these introns from the genomic sequence creates an uninterrupted ORF with the potential to code for a 691-amino-acid-residue protein.

The *C. neoformans* tyrosine recombinase genes form part of transposable elements referred to as cryptons. cryptons are present in all *C. neoformans* strains that have been sequenced, typically in less than five copies. None of these cryptons has been found in association with a reverse transcriptase gene, suggesting that these are DNA transposons rather than retrotransposons. Cryptons also do not appear to be associated with any extensive repeat sequences. Indeed, the only repeat sequences consistently found in association with cryptons are short (4- to 6-bp) direct repeats at each extremity. A single copy of this repeat appears at the uninserted target site. This is consistent with insertion of cryptons occurring via a recombination reaction between short matching sequences in the element and the host genome, mediated by the encoded tyrosine recombinase. This resembles the predicted activity of the YRs described from retrotransposons.

Cryptons share significant sequence similarities with members of the tyrosine recombinase family. In particular, most of them contain the four very highly conserved residues characteristic of these enzymes: the RHRY tetrad (4). The C-terminal most region of each crypton gene encodes a conserved domain that is similar in sequence to several fungal transcription factors. These related proteins share a highly conserved set of residues, GRIER (10). The matching proteins include *Saccharomyces cerevisiae* Gcr1p (transcriptional

TABLE 2 Tyrosine recombinase-encoding (YR) elements used for phylogenetic analyses in this study. The sources of the sequence data are held in the Genbank accessions, references or database sources shown. Element names are generated from the initial of the genus and either 2 or 3 initial letters from the species name of the organism in which the element is found

Element name	Species	Accession	Element name	Species	Accession
AmDirs	*Apis mellifera*	AADG02016821	CcNgaro1	*Coprinopsis cinerea*	BK001716
GinDirs	*Glyptapanteles indiensis*	CE75468	CcNgaro3	*Coprinopsis cinerea*	BK001748
TcDirs	*Tribolium castaneum*	AY531876, 7	LbicNgaro	*Laccaria bicolor*	ABFE01003074
SpDirs1	*Strongylocentrotus purpuratus*	Retrobase[a]	PcarNgaro	*Phanerochaete carnosa*	AEHB01000886
SpDirs2	*S. purpuratus*	Retrobase	GlucNgaro	*Ganoderma lucidum*	AHGX01000842
SpDirs3	*S. purpuratus*	Retrobase	TveNgaro	*Trametes versicolor*	AEJI01000671
SpDirs4	*S. purpuratus*	Retrobase	PinfNgaro	*Phytophthora infestans*	AATU01012611
AdigDirs	*Acropora digitifera*	BACK01048443	PrubNgaro	*Phytophthora rubi*	JMRJ01001886
AceyDirs	*Ancylostoma ceylanicum*	EYC23896	PvexNgaro	*Pythium vexans*	AKYC02008204
SmaDirs	*Strigamia maritima (centipede)*	AFFK01020370	PramNgaro	*Phytophthora ramorum*	AAQX01006816
BfloDirs	*Branchiostoma floridae*	ABEP02014390	PcapNgaro	*Phytophthora capsici*	ADVJ01000933
CtelDirs	*Capitella teletus*	AMQN01027463	SsueNgaro	*Sistrotremastrum suecicum*	Sissu1 scaffold_215, JGI[b]
DpuDirs	*Daphnia pulex*	ACB38666	AqueNgaro	*Amphimedon queenslandica*	ACUQ01007268
RirrDirs	*Rhizophagus irregularis*	JEMT01009395	DrNgaro1	*Danio rerio*	AY152729
AmacDirs	*Allomyces macrogynus*	ACDU01003872	SkowNgaro1	*Saccoglossus kowalevskii*	ACQM0103978
Prt1	*Phycomyces blakesleeanus*	Z54337	BfloNgaro1	*Branchiostoma floridae*	ABEF02018380
RoDirs1	*Rhizopus oryzae*	JNDY01003873 retrobase	BfloNgaro2	*Branchiostoma floridae*	ABEF02002342
MraDirs	*Mucor ramosissimus*	JNEF01002880	CgiNgaro1	*Crassostrea gigas*	AFTI01004305
DciDirs	*Diaphorina citri*	AWGM01004308	CgiNgaro2	*C. gigas*	AFTI01000321
CgiDirs	*C. gigas*	AFTI01001587	PlepNgaro	*Pontastacus leptodactylus*	GAFS01025312
OhanDirs	*Ophiophagus hannah*	AZIM01001342	AdigNgaro	*Acropora digitifera*	BACK01006747
XtDirs1	*Xenopus tropicalis*	AC144974	XtNgaro1	*X. tropicalis*	AAMC02024023
XtDirs2	*X. tropicalis*	AC145807	XtNgaro2	*X. tropicalis*	AC175582

AcarDirs1-1	*Anolis carolinensis*	AAWZ02041142	XtNgaro3	*X. tropicalis*	AAMC02000484
AcarDirs1-2	*A. carolinensis*	AAWZ02027059	XtNgaro4	*X. tropicalis*	AC151347
EtynDirs	*Eurycea tynerensis*	(17)	SpNgaro1	*S. purpuratus*	Retrobase
LchaDirs	*Latimeria chalumnae*	BAHO01326816	SpNgaro2	*S. purpuratus*	Retrobase
TnDirs1	*Tetraodon nigriviridis*	Retrobase	SpNgaro3	*S. purpuratus*	AAGJ04006304
PforDIRS	*Poecilia formosa*	AYCK01014057	LvNgaro1	*Lytechinus variegatus*	BK001253
DmawDirs	*Dissostichus mawsoni*	HQ447060	LvNgaro2	*Lytechinus variegatus*	AGCV01398517
DrDirs1	*D. rerio*	AL590134	HchiNgaro	*Hynobius chinensis*	GAQK01079872
DrDirs2	*D. rerio*	BK001257	AnolisNgaro1-1	*A. carolinensis*	AAWZ02012912
DrDirs3	*D. rerio*	BK001259	AnolisNgaro1-2	*A. carolinensis*	AAWZ02027192
DIRS1	*Dictyostelium discoideum*	M11339	PcNgaro1	*Phanerochaete chrysosporium*	AADS01000526
DfaDirs	*Dictyostelium fasciculatum*	ADHC01000030	PcNgaro2	*P. chrysosporium*	AADS01000761
EvenDirs	*Echinogammarus veneris*	GARO01000003	PcNgaro3	*P. chrysosporium*	AADS01000332
PAT	*Panagrellus redivivus*	X60774	PcNgaro4	*P. chrysosporium*	AADS01000801
CbPat	*Caenorhabditis briggsae*	AC090521	crypton_Cn1	*Cryptococcus neoformans*	AY248893
TOC3	*Chlamydomonas reinhardtii*	Scaffold 2543, JGI	crypton_Cn2	*C. neoformans*	(10)
AquePAT	*Amphimedon queenslandica*	ACUQ01003527	crypton_Cn3	*C. neoformans*	(10)
SpPat1	*S. purpuratus*	Retrobase	crypton_Cn4	*C. neoformans*	(10)
SkowPat	*Saccoglossus kowalevskii*	ACQM01123180	crypton_Tas	*Trichosporon asahii*	ALBS01000010
NgruPat	*Naegleria gruberi*	ACER01000154	crypton_Sob	*Saksenae oblongispora*	JNEV01000931
Kangaroo	*Volvox carteri*	AY137241	crypton_Cgl	*Chaetomium globosum*	XM_001226232
Catenaria YR	*Catenaria anguillulae*	Catan1 scaffold_494:1-5927, JGI	crypton_Nga	*Neotyphodium gansuense*	AFRE01000827
			crypton_Tst	*Talaromyces stipitatus*	XM_002483890
			crypton_Cpo	*Coccidioides posadasii*	Goodwin et al., 2003
			crypton_Cim	*Coccidioides immitis*	XM_001239641

[a]Retrobase: http://biocadmin.otago.ac.nz/fmi/xsl/retrobase/home.xsl
[b]JGI, Joint Genome Institute: http://genome.jgi-psf.org

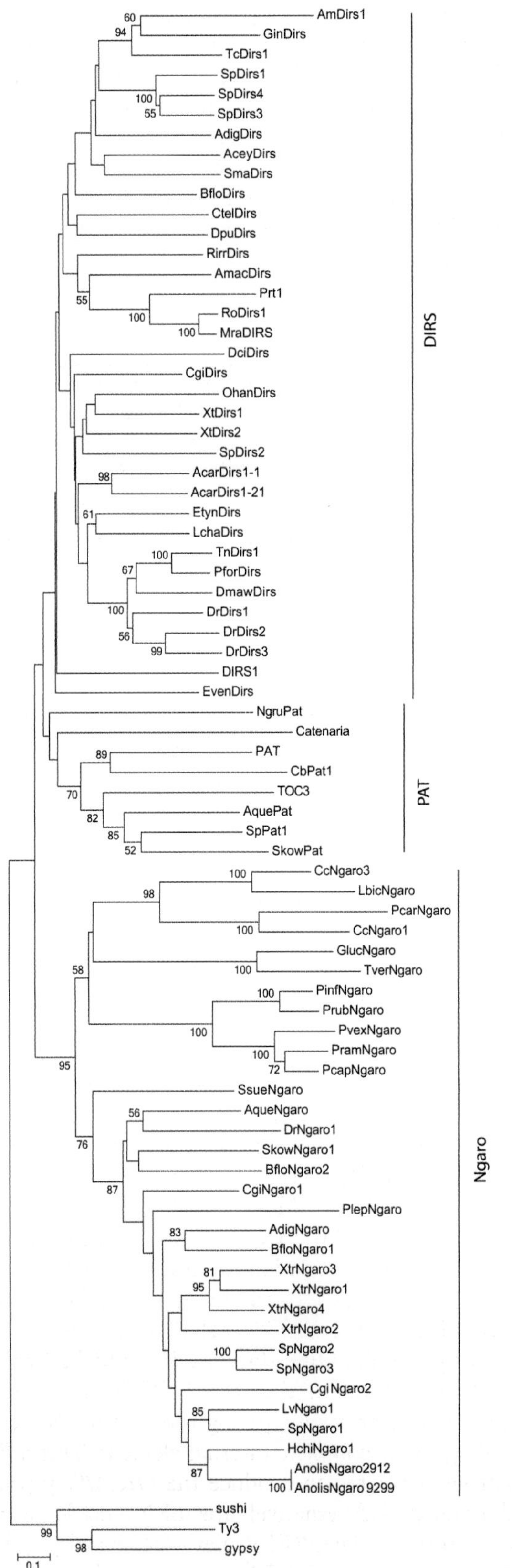

regulator of glycolytic genes) which has been shown to bind DNA with high affinity (41, 42). The observed sequence similarity suggests that the C-terminal domains of the crypton proteins are likely to be involved in binding to DNA. None of the other genes commonly found in TEs, such as DDE-type transposases or integrases were identified in cryptons.

Cryptons that are similar in sequence to those from *C. neoformans* are present in other species of *Cryptococcus*. For example, expressed sequences of cryptons have been detected in *Cryptococcus gattii* (43). *Tremella mesenterica*, a basidiomycete moderately closely related to *Cryptococcus*, also contains cryptons.

Cryptons in Other Fungi

Several cryptons were detected in the genome sequences of the ascomycetes *Coccidioides posadasii*, *Coccidioides immitis* and *Histoplasma capsulatum* (10). The elements from *C. posadasii* all appear to be members of the one family, crypton Cp1, typically sharing 85% to 95% identity at the DNA level. The elements appear at various locations in the *C. posadasii* genome, have distinct termini, and at the extremities of these termini are short direct repeats. In several cases related empty sites have been detected suggesting recent mobility. The elements are very similar in structure to their counterparts in *C. neoformans* and encode similar tyrosine recombinases and putative DNA-binding domains (but without introns).

Cryptons appear to be fairly abundant in *H. capsulatum*. Each strain has a unique set of insertion sites suggesting that the elements have been very active since these strains diverged from a common ancestor. *H. capsulatum* cryptons often have introns.

Other ascomycete fungi with cryptons include *Chaetomium globosum*, *Neotyphodium gansuense*, *Talaromyces stipitatus* and *Fusarium oxysporum* (44).

Cryptons are present in several *Mucormycotina*, such as species of *Cunninghamella*, *Rhizopus* and *Saksenae*. For example, the genome of *Saksenae oblongispora* contains 10 to 15 copies of a crypton that has the conserved RHRY tetrad of the tyrosine recombinase

Figure 3 Relationships among YR-encoding retroelements. This phylogenetic tree is based on an alignment of the conserved RT and RH protein sequences. Sequences from three LTR retrotransposons have been used as an outgroup: sushi (AF030881), Ty3 (M23367) and gypsy (AF033821). The tree was constructed by the Neighbour-joining method using MEGA5 (46). Bootstrap support from 1050 replicates is indicated for branches with >50% support. Element descriptions can be found in Table 2.
doi:10.1128/microbiolspec.MDNA3-0036-2014.f3

domain and the conserved features of the C-terminal DNA-binding domain, GRIER. These cryptons include multiple introns.

Cryptons in *Stramenopiles*

Cryptons are present in the *Stramenopiles*, including many oomycete species such as *Phytophthora, Pythium, Albugo laibachii, Saprolegnia parasitica* and *Hyalanospora arabidopsis* (44) and in the genome of the diatom *Nitzschia*. For example, the genome of *Phytophthorainfestans* contains four copies of a crypton that are uncorrupted and have the conserved RHRY tetrad of the tyrosine recombinase domain and the conserved features of the C-terminal DNA-binding domain, GRIER. The *Phytophthora* cryptons do not include introns. There are sequence data for multiple species and strains of *Phytophthora* available. Intriguingly, the integration sites of cryptons in these strains are apparently conserved. This may suggest that these cryptons have a very restricted range of target sites for integration, a feature of many tyrosine recombinases, such as XerC of *E. coli*.

PHYLOGENETIC ANALYSES EMPLOYING YR

The relationships between the tyrosine recombinases from retrotransposons and cryptons were studied by phylogenetic analysis (10). Trees were constructed based on an alignment of the regions encompassing the RHRY tetrads of a large number of bacterial, archaeal, and phage recombinases, and representatives of the diversity of tyrosine recombinases found in eukaryotes. These latter sequences include those from cryptons, the DIRS-like, PAT-like and Ngaro-like groups of retrotransposons, yeast 2-micron circle plasmids, and the recently described Tec transposons of *Euplotes crassus*, as well as some tyrosine recombinase-like sequences from baculoviruses and mitochondria. Phylogenetic analyses of tyrosine recombinases are hindered by the high level of sequence diversity within the group. This usually means that the nature of the more distant relationships within the group cannot be resolved with certainty. Nevertheless, on trees constructed by a variety of methods, the crypton recombinases consistently formed a well-supported monophyletic group within a larger clade that contains the recombinases of the DIRS-like, PAT-like and Ngaro-like retrotransposons, the Cre recombinase of bacteriophage P1 and several other prokaryotic sequences. The crypton and retrotransposon recombinases are not closely related to the other eukaryotic recombinases, including those of the Tec DNA transposons from *E. crassus*.

A tree based on an alignment of the YR sequences from recently described cryptons and retroelements is shown in Fig. 4. This tree (in agreement with the previous analysis) separates the crypton, Ngaro-like, PAT-like and DIRS-like elements into monophyletic groups. It is clear from this analysis that the YR elements found in eukaryote retrotransposons are related to the YR ORF present in the eukaryote crypton DNA transposons.

ORIGIN OF TYROSINE RECOMBINASE RETROTRANSPOSONS

The similarities in sequence, and the putative mechanistic similarities, between cryptons and YR retrotransposons suggest that these elements are related. We propose that the YR retrotransposons arose from the combination of a crypton-like element and the RT/RH gene of a pre-existing Ty3/gypsy retrotransposon. This is analogous to the evolution of a DDE transposon into an LTR retrotransposon by acquisition of a reverse transcriptase gene. The Ngaro, DIRS and PAT retrotransposon groups are characterized by distinct internal structures, believed to be critical for the reverse transcriptase conversion of the genomic copy of the retrotransposon into a free, circular DNA. The DIRS and PAT elements share the presence of a dam methylase motif as part of the RT/RH/DAM ORF. These methylases are related to each other (8) and located at the same position in the RT/RH/DAM ORF. This shared feature may suggest a shared origin. The Ngaro elements lack this methylase. Many Ngaro elements have the distinctive characteristic of a hydrolase encoded by a distinct ORF downstream of the YR ORF. These Ngaro elements, therefore, can have four ORFs. The very distinct structural features of the three types of YR retroelements suggests that the transition of a eukaryote DNA transposon (crypton) into a eukaryote retrotransposon may have occurred several times, giving rise to the Ngaro, DIRS and PAT-like elements.

There is as yet little experimental evidence relating to the control of translation in these retroelements. It is most unusual to have overlapping ORFs: for example, in DIRS1 itself the RT/RH ORF is completely overlapped by the YR ORF (Fig. 1). It could be hypothesized that all three of the DIRS1 ORFs are expressed from the same promoter, giving rise to a mRNA corresponding to the complete internal element. This mRNA might be translated to produce the ORF1/Gag protein and terminated. Alternatively the mRNA could be translated to produce the ORF1 protein, followed by a phaseshift to produce the RT/RH protein (and terminated).

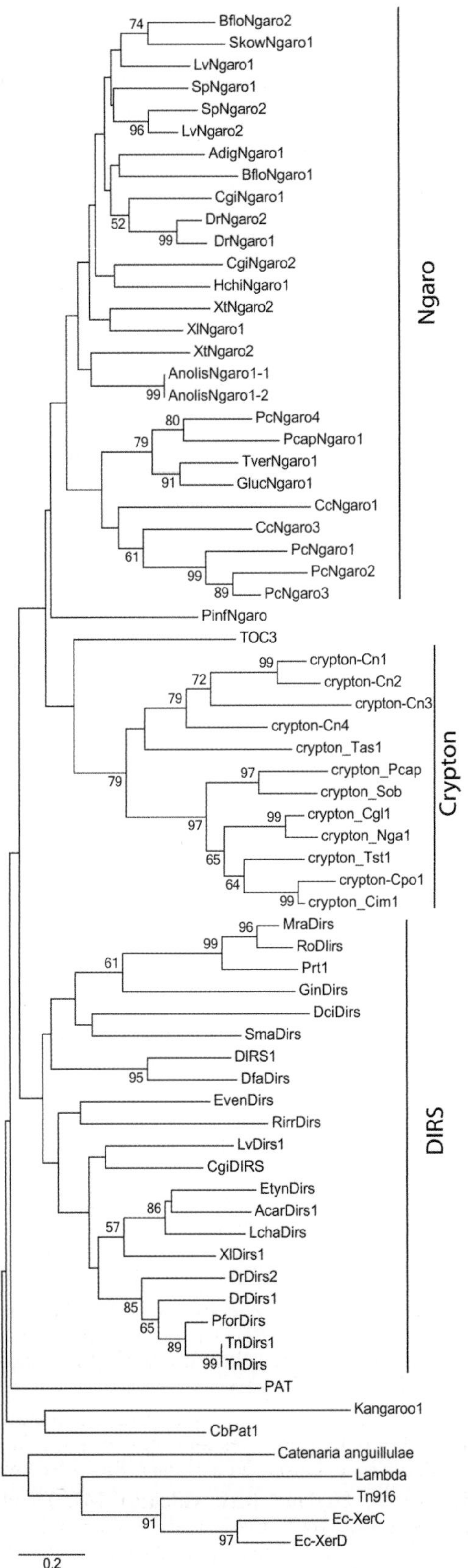

Figure 4 Relationships among YR-encoding retrotransposons and DNA transposons. This phylogenetic tree is based on an alignment of the YR protein sequences. Sequences from prokaryote tyrosine recombinases have been used as an outgroup: Lambda recombinase (KDT52537), Tn916 from *Enterococcus faecalis* (U09422) and two *E. coli* tyrosine recombinases (XerC, CDL28161; XerD, CDL49882). The tree was constructed by the Neighbour-joining method using MEGA5 (46). Bootstrap support from 1050 replicates is indicated for branches with >50% support. Element descriptions can be found in Table 2.
doi:10.1128/microbiolspec.MDNA3-0036-2014.f4

The third possibility would be that the mRNA could be translated to produce the ORF1 protein, followed by an alternative phase-shift to produce the YR. This arrangement may be necessary to express the RT/RH and YR ORFs independently.

Another unresolved question is the fourth ORF, encoding a hydrolase, found in some Ngaros. The gap between the YR and the hydrolase ORFs is typically very short and it may be that translation can continue across this gap. The hydrolases are most closely related to the C-terminal halves of the ORF1 proteins of several vertebrate members of the CR1 clade of non-LTR retrotransposons, such as CR1 from the chicken (34) and Maui from the pufferfish *F. rubripes* (35) Many non-LTR retrotransposons have a small gap separating ORFs, for example L1 LINE elements and Zorro (45). The function of the hydrolase, like the function of the methyltransferase, is unknown.

CONCLUDING REMARKS

The study of YR retrotransposons is attracting increasing interest. There are at least three reasons why this is so. Firstly, retrotransposons can be distinguished by the mechanism by which new copies are in integrated into the host genome: non-LTR retrotransposons integrate as they replicate, via target-primed retrotransposition; LTR retrotransposons integrate using a DDE-type integrase, while YR retrotransposons integrate using a tyrosine recombinase with relaxed specificity. The mechanism by which PLEs (the fourth major retrotransposon group) integrate remains uncertain. These retrotransposons have distinct methods of producing full-length DNA copies of an RNA replicative intermediate. Non-LTR retrotransposons replicate a full-length RNA transcript, the mechanism by which the complete double-stranded DNA copy is generated remains unclear (but frequent 5′ truncation occurs). LTR retrotransposons replicate using a tRNA primer and a complex series of reactions to replicate the initial, incomplete transcript. The YR retrotransposons have apparently developed two distinct methods of replicating an incomplete RNA transcript, involving either ITRs (DIRS) or direct repeats. YR elements, therefore, represent a very distinct type of mobile element.

The second reason for the increased interest in YR elements is their widespread distribution across eukaryotes. In particular, their presence in some vertebrates but not others is intriguing. For example, DIRS and Ngaro elements are present in fish, amphibian and reptiles, but are completely absent from birds and mammals. This distribution mimics that of LTR retrotransposons in the vertebrates and remains puzzling. LTR retroviruses are, of course, found widely in vertebrates, including both birds and mammals. They are, however, infectious particles that can easily be transferred between individuals of a species or related species. It is apparent that the phylogeny of the YR retrotransposons broadly reiterates the phylogeny of their hosts. This suggests that horizontal transfer is infrequent. It is probable that YR retrotransposons arose (possibly more than once) in a basal eukaryote group, probably as a result of a crypton DNA transposon acquiring a retrotransposition capability. The YR retrotransposons have been conserved across many phylogenetic groups; the DIRS-like elements in particular are persistent in in many phylogenies, whereas the distribution of Ngaro elements is more sporadic. In some organisms the YR organisms are very abundant and can consist of a number of diverse families.

The third area of interest is the biochemistry of the elements; this includes the unusual YR element itself, which has a permissive recombinase (probably responding to a 3-bp target). The way in which the overlapping ORFs are expressed is unresolved. The additional capabilities found in some YR retrotransposons, for example the methylase and hydrolase, play an undefined role in retrotransposition and may have influences on the host genome.

In summary, we still have much to learn about these intriguing elements and their significance for eukaryote evolution.

Acknowledgments. *We would like to acknowledge the contribution to the field made by Dr Timothy Goodwin, whose creative insight recognized the presence of tyrosine recombinases in eukaryote transposons and retrotransposons.*

Citation. Poulter RTM, Butler MI. 2014. Tyrosine recombinase retrotransposons and transposons. Microbiol Spectrum 3(2):MDNA3-0036-2014

References

1. **Wicker T, Sabot F, Hua-Van A, Bennetzen JL, Capy P, Chalhoub B, Flavell A, Leroy P, Morgante M, Panaud O, Paux E, SanMiguel P, Schulman AH.** 2007. A unified classification system for eukaryotic transposable elements. *Nat Rev Genet* 8(12):973–982.
2. **Duncan L, Bouckaert K, Yeh F, Kirk DL.** 2002. *kangaroo* a mobile element from *Volvox carteri* is a member of a newly recognized third class of retrotransposons. *Genetics* **162:**1617–1630.
3. **Goodwin TJ, Poulter RT.** 2001. The DIRS1 group of retrotransposons. *Mol Biol Evol* **18:**2067–2082.
4. **Nunes-Düby SE, Kwon HJ, Tirumalai RS, Ellenberger T, Landy A.** 1998. Similarities and differences among 105 members of the Int family of site-specific recombinases. *Nucleic Acids Res* **26:**391–406.
5. **Cappello J, Handelsman K, Lodish HF.** 1985. Sequence of *Dictyostelium* DIRS-1: an apparent retrotransposon with inverted terminal repeats and an internal circle junction sequence. *Cell* **43:**105–115.
6. **de Chastonay Y, Felder H, Link C, Aeby P, Tobler H, Muller F.** 1992. Unusual features of the retroid element PAT from the nematode *Panagrellus redivivus. Nucleic Acids Res* **20:**1623–1628.
7. **Ruiz-Perez VL, Murillo FJ, Torres-Martinez S.** 1996. *Prt1* an unusual retrotransposon-like sequence in the fungus *Phycomyces blakesleeanus. Mol Gen Genet* **253:**324–333.
8. **Goodwin TJ, Poulter RT.** 2004. A new group of tyrosine recombinase-encoding retrotransposons. *Mol Biol Evol* **21:**746–759.
9. **Piednoël M, Gonçalves IR, Higuet D, Bonnivard E.** 2011. Eukaryote DIRS1-like retrotransposons: an overview. *BMC Genomics* **12:**621.
10. **Goodwin TJD, Butler MI, Poulter RTM.** 2003. *Cryptons*: a group of tyrosine recombinase-encoding DNA transposons from pathogenic fungi. *Microbiology* **149:**3099–3109.
11. **Poulter RT, Goodwin TJ.** 2005. DIRS-1 and the other tyrosine recombinase retrotransposons. *Cytogenet Genome Res* **110:**575–588.
12. **Dugas JC, Ngai J.** 2001. Analysis and characterization of an odorant receptor gene cluster in the zebrafish genome. *Genomics* **71:**53–65.
13. **Metcalfe CJ, Filée J, Germon I, Joss J, Casane D.** 2012. Evolution of the Australian lungfish (*Neoceratodus forsteri*) genome: a major role for CR1 and L2 LINE elements. *Mol Biol Evol* **29:**3529–3539.
14. **Chalopin D, Fan S, Simakov O, Meyer A, Schartl M, Volff JN.** 2013. Evolutionary active transposable elements in the genome of the coelacanth. *J Exp Zool, Part B* **322:**322–333. doi:10.1002/jez.b.22521.
15. **Detrich HW, Amemiya CT.** 2010. Antarctic Notothenioid Fishes: Genomic Resources and Strategies for Analyzing an Adaptive Radiation. *Integr Comp Biol* **50:**1009–1017.
16. **Hellsten U, Harland RM, Gilchrist MJ, Hendrix D, Jurka J, Kapitonov V, Ovcharenko I, Putnam NH, Shu S, Taher L, Blitz IL, Blumberg B, Dichmann DS, Dubchak I, Amaya E, Detter JC, Fletcher R, Gerhard DS,**

Goodstein D, Graves T, Grigoriev IV, Grimwood J, Kawashima T, Lindquist E, Lucas SM, Mead PE, Mitros T, Ogino H, Ohta Y, Poliakov AV, Pollet N, Robert J, Salamov A, Sater AK, Schmutz J, Terry A, Vize PD, Warren WC, Wells D, Wills A, Wilson RK, Zimmerman LB, Zorn AM, Grainger R, Grammer T, Khokha MK, Richardson PM, Rokhsar DS.** 2010. The genome of the Western clawed frog *Xenopus tropicalis*. *Science* **328** (5978):633–636.

17. **Sun C, Shepard DB, Chong RA, López Arriaza J, Hall K, Castoe TA, Feschotte C, Pollock DD, Mueller RL.** 2012. LTR retrotransposons contribute to genomic gigantism in plethodontid salamanders. *Genome Biol Evol* **4**(2): 168–183.

18. **Castoe TA, Hall KT, Guibotsy Mboulas ML, Gu W, de Koning AP, Fox SE, Poole AW, Vemulapalli V, Daza JM, Mockler T, Smith EN, Feschotte C, Pollock DD.** 2011. Discovery of highly divergent repeat landscapes in snake genomes using high-throughput sequencing. *Genome Biol Evol* **3**:641–653.

19. **Shedlock AM.** 2006. Phylogenomic investigation of CR1 LINE diversity in reptiles. *Syst Biol* **55**(6):902–911.

20. **Goodwin TJ, Poulter RT, Lorenzen MD, Beeman RW.** 2004. DIRS retroelements in arthropods: identification of the recently active TcDirs1 element in the red flour beetle *Tribolium castaneum*. *Mol Genet Genomics* **272**:47–56.

21. **Piednoël M, Bonnivard E.** 2009. DIRS1-like retrotransposons are widely distributed among Decapoda and are particularly present in hydrothermal vent organisms. *BMC Evol Biol* **9**:86.

22. **Rho M, Schaack S, Gao X, Kim S, Lynch M, Tang H.** 2010. LTR retroelements in the genome of *Daphnia pulex*. *BMC Genomics* **11**:425.

23. **Zhang G, Fang X, Guo X, Li L, Luo R, Xu F, Yang P, Zhang L, Wang X, Qi H, Xiong Z, Que H, Xie Y, Holland PW, Paps J, Zhu Y, Wu F, Chen Y, Wang J, Peng C, Meng J, Yang L, Liu J, Wen B, Zhang N, Huang Z, Zhu Q, Feng Y, Mount A, Hedgecock D, Xu Z, Liu Y, Domazet-Lošo T, Du Y, Sun X, Zhang S, Liu B, Cheng P, Jiang X, Li J, Fan D, Wang W, Fu W, Wang T, Wang B, Zhang J, Peng Z, Li Y, Li N, Wang J, Chen M, He Y, Tan F, Song X, Zheng Q, Huang R, Yang H, Du X, Chen L, Yang M, Gaffney PM, Wang S, Luo L, She Z, Ming Y, Huang W, Zhang S, Huang B, Zhang Y, Qu T, Ni P, Miao G, Wang J, Wang Q, Steinberg CE, Wang H, Li N, Qian L, Zhang G, Li Y, Yang H, Liu X, Wang J, Yin Y, Wang J.** 2012. The oyster genome reveals stress adaptation and complexity of shell formation. *Nature* **490**(7418):49–54.

24. **Takeuchi T, Kawashima T, Koyanagi R, Gyoja F, Tanaka M, Ikuta T, Shoguchi E, Fujiwara M, Shinzato C, Hisata K, Fujie M, Usami T, Nagai K, Maeyama K, Okamoto K, Aoki H, Ishikawa T, Masaoka T, Fujiwara A, Endo K, Endo H, Nagasawa H, Kinoshita S, Asakawa S, Watabe S, Satoh N.** 2013. Draft genome of the pearl oyster *Pinctada fucata*: a platform for understanding bivalve biology. *DNA Res* **19**(2):117–130.

25. **Muszewska A, Steczkiewicz K, Ginalski K.** 2013. DIRS and Ngaro Retrotransposons in Fungi. *PLoS One* **8**(9): e76319.

26. **Day A, Rochaix JD.** 1991. A transposon with an unusual LTR arrangement from *Chlamydomonas reinhardtii* contains an internal tandem array of 76 bp repeats. *Nucleic Acids Res* **19**:1259–1266.

27. **Dishaw LJ, Mueller MG, Gwatney N, Cannon JP, Haire RN, Litman RT, Amemiya CT, Ota T, Rowen L, Glusman G, Litman GW.** 2008. Genomic complexity of the variable region containing chitin-binding proteins in amphioxus. *BMC Genet* **9**:78.

28. **Manfrin C, Tom M, De Moro G, Gerdol M, Guarnaccia C, Mosco A, Pallavicini A, Giulianini PG.** 2013. Application of D-Crustacean Hyperglycemic Hormone Induces Peptidases Transcription and Suppresses Glycolysis-Related Transcripts in the Hepatopancreas of the Crayfish *Pontastacus leptodactylus* - Results of a Transcriptomic Study. *PLoS One* **8**(6):e65176.

29. **Lenz PH, Roncalli V, Hassett RP, Wu LS, Cieslak MC, Hartline DK, Christie AE.** 2014. De Novo Assembly of a Transcriptome for *Calanus finmarchicus* (Crustacea, Copepoda), The Dominant Zooplankter of the North Atlantic Ocean. *PLoS One* **9**(2):e88589.

30. **Arkhipova IR, Pyatkov KI, Meselson M, Evgen'ev MB.** 2003. Retroelements containing introns in diverse invertebrate taxa. *Nat Genet* **33**:123–124.

31. **Curcio MJ, Derbyshire KM.** 2003. The outs and ins of transposition: from Mu to kangaroo. *Nat Rev Mol Cell Biol* **4**:865–877.

32. **Doak TG, Witherspoon DJ, Jahn CL, Herrick G.** 2003. Selection on the genes of *Euplotes crassus* Tec1 and Tec2 transposons: evolutionary appearance of a programmed frameshift in a Tec2 gene encoding a tyrosine family site-specific recombinase. *Eukaryotic Cell* **2**:95–102.

33. **Kossykh VG, Schlagman SL, Hattman S.** 1993. Conserved sequence motif DPPY in region IV of the phage T4 DAM DNA-[N6-adenine]-methyltransferase is important for S-adenosyl-L-methionine binding. *Nucleic Acids Res* **21**:4659–4662.

34. **Haas NB, Grabowski JM, Sivitz AB, Burch JBE.** 1997. Chicken repeat 1 (CR1) elements which define an ancient family of vertebrate non-LTR retrotransposons. *Microbiol Rev* **21**:157–178.

35. **Poulter R, Butler M, Ormandy J.** 1999. A LINE element from the pufferfish (fugu) *Fugu rubripes* that shows similarity to the CR1 family of non-LTR retrotransposons. *Gene* **227**:169–179.

36. **Kapitonov VV, Jurka J.** 2003. The esterase and PHD domains in CR1-like non-LTR retrotransposons. *Mol Biol Evol* **20**:38–46.

37. **Huang Y-T, Liaw Y-C, Gorbatyuk VY, Huang TH.** 2001. Backbone dynamics of *Escherichia coli* Thioesterase/Protease I: evidence of a flexible active-site environment for a serine protease. *J Mol Biol* **307**:1075–1090.

38. **Ho YS, Swenson L, Derewenda L, Serre L, Wei Y, Dauter Z, Hattori M, Adachi T, Aoki J, Arai H, Inoue K, Derewenda ZS.** 1997. Brain acetylhydrolase that inactivates platelet-activating factor is a G-protein-like trimer. *Nature* **385**:89–93.

39. **Fukuda K, Kiyokawa Y, Yanagiuchi T, Wakai Y, Kitamoto K, Inoue Y, Kimura A.** 2000. Purification and

characterization of isoamyl acetate-hydrolyzing esterase encoded by the IAH1 gene of Saccharomyces cerevisiae from a recombinant *Escherichia coli*. *Appl Microbiol Biotechnol* 53:596–600.

40. **Bon E, Casaregola S, Blandin G, Llorente B, Neuvéglise C, Munsterkotter M, Guldener U, Mewes HW, Van Helden J, Dujon B, Gaillardin C.** 2003. Molecular evolution of eukaryotic genomes: hemiascomycetous yeast spliceosomal introns. *Nucleic Acids Res* **31**(4):1121–1135.

41. **Huie MA, Baker HV.** 1996. DNA-binding properties of the yeast transcriptional activator, Gcr1p. *Yeast* **12:** 307–317.

42. **Huie MA, Scott EW, Drazinic CM, Lopez MC, Hornstra IK, Yang TP, Baker HV.** 1992. Characterization of the DNA-binding activity of GCR1: *in vivo* evidence for two GCR1- binding sites in the upstream activating sequence of TP1 of *Saccharomyces cerevisiae*. *Mol Cell Biol* **12:** 2690–2700.

43. **D'Souza CA, Kronstad JW, Taylor G, Warren R, Yuen M, Hu G, Jung WH, Sham A, Kidd SE, Tangen K, Lee N, Zeilmaker T, Sawkins J, McVicker G, Shah S, Gnerre S, Griggs A, Zeng Q, Bartlett K, Li W, Wang X, Heitman J, Stajich JE, Fraser JA, Meyer W, Carter D, Schein J, Krzywinski M, Kwon-Chung KJ, Varma A, Wang J, Brunham R, Fyfe M, Ouellette BF, Siddiqui A, Marra M, Jones S, Holt R, Birren BW, Galagan JE, Cuomo CA.** 2011. Genome variation in *Cryptococcus gattii*, an emerging pathogen of immunocompetent hosts. *mBio* **2** (1):e00342-10.

44. **Kojima KK, Jurka J.** 2011. Crypton transposons: identification of new diverse families and ancient domestication events. *Mobile DNA* **2**(1):12.

45. **Goodwin TJD, Ormandy JE, Poulter RT.** 2001. L1-like non-LTR retrotransposons in the yeast *Candida albicans*. *Curr Genet* **39**(2):83–91.

46. **Tamura K, Peterson D, Peterson N, Stecher G, Nei M, Kumar S.** 2011. MEGA5: Molecular Evolutionary Genetics Analysis using Maximum Likelihood, Evolutionary Distance, and Maximum Parsimony Methods. *Mol Biol Evol* **28:**2731–2739.

Index

A

B

D

H

J

K

L

M

N

R

S

T